Symbols, Weights, and Measures

Units of Length

m	meter
km	kilometer (10^3 m)
cm	centimeter (10^{-2} m)
mm	millimeter (10^{-3} m)
μm	micrometer (10^{-6} m)
nm	nanometer (10^{-9} m)

Units of Mass and Weight

amu	atomic mass unit
MW	molecular weight
mole	MW in grams
g	gram
kg	kilograms (10^3 g)
mg	milligrams (10^{-3} g)

Units of Pressure

atm	atmospheres (1 atm = 760 mmHg)
mmHg	millimeters of mercury
P_x	partial pressure of gas x (as in P_{O_2})

Conversion Factors

1 in.	= 2.54 cm	1 cm	= 0.394 in.
1 fl oz	= 29.6 mL	1 mL	= 0.034 fl oz
1 qt	= 0.946 L	1 L	= 1.057 qt
1 g	= 0.0035 oz	1 oz	= 28.38 g
1 lb	= 0.45 kg	1 kg	= 2.2 lb
°C	= (5/9)(°F – 32)	°F	= (9/5)(°C) + 32

Units of Volume

L	liter
dL	deciliter (= 100 mL) (10^{-1} L)
mL	milliliter (10^{-3} L)
μL	microliter (= 1 mm^3) (10^{-6} L)

Units of Concentration

mEq/L	milliequivalents per liter
Osm/L	osmoles per liter
mOsm/L	milliosmoles per liter
M	molar
mM	millimolar
pH	negative log of H^+ molarity

Units of Heat

cal	"small" calories
kcal	kilocalories (Calories; 1 kcal = 1,000 cal)
Cal	"large" (dietary) calories (1 Cal = 1,000 cal)

Greek Letters

α	alpha
β	beta
γ	gamma
Δ	delta (uppercase)
δ	delta (lowercase)
η	eta
θ	theta
μ	mu

Anatomy & Physiology

Second Edition

The Unity of Form and Function

Kenneth S. Saladin
Georgia College and State University

with

Donna Van Wynsberghe
Clinical Consultant
University of Wisconsin–Milwaukee

Boston Burr Ridge, IL Dubuque, IA Madison, WI New York San Francisco St. Louis
Bangkok Bogotá Caracas Lisbon London Madrid
Mexico City Milan New Delhi Seoul Singapore Sydney Taipei Toronto

McGraw-Hill Higher Education

A Division of The **McGraw-Hill** Companies

ANATOMY & PHYSIOLOGY: THE UNITY OF FORM AND FUNCTION
SECOND EDITION

Published by McGraw-Hill, an imprint of The McGraw-Hill Companies, Inc., 1221 Avenue of the Americas, New York, NY 10020.

Some ancillaries, including electronic and print components, may not be available to customers outside the United States.

This book is printed on acid-free paper.

2 3 4 5 6 7 8 9 0 VNH/VNH 0 9 8 7 6 5 4 3 2 1 0

ISBN 0–07–290786–X
ISBN 0–07–118086–9 (ISE)

Vice president and editor-in-chief: *Kevin T. Kane*
Publisher: *Michael D. Lange*
Senior development manager: *Kristine Tibbetts*
Developmental editors: *Connie Balius-Haakinson/Patricia Hesse*
Marketing managers: *Michelle Watnick/Heather K. Wagner*
Project manager: *Mary E. Powers*
Senior production supervisor: *Sandra Hahn*
Designer: *K. Wayne Harms*
Interior designer: *Kaye Farmer*
Cover illustration: *Imagineering*
Cover photo: *Chris Golly/SharpShooters*
Photo research coordinator: *John C. Leland*
Photo research: *Connie Mueller*
Senior supplement coordinator: *Audrey A. Reiter*
Compositor: *Carlisle Communications, Ltd.*
Typeface: *10/12 Melior*
Printer: *Von Hoffmann Press, Inc.*

The credits section for this book begins on page C–1 and is considered an extension of the copyright page.

Library of Congress Cataloging-in-Publication Data

Saladin, Kenneth S.
Anatomy & physiology : the unity of form and function / Kenneth S. Saladin. — 2nd ed.
p. cm.
Includes index.
ISBN 0–07–290786–X
1. Human physiology. 2. Human anatomy. 3. Paramedical education. I. Title: Anatomy and physiology. II. Title.

QP34.5 .S23 2001
612—dc21

00–024602
CIP

INTERNATIONAL EDITION ISBN 0–07–118086–9

www.mhhe.com

About the Author

Ken Saladin is professor of biology at Georgia College and State University in Milledgeville, Georgia, where he has taught since 1977. He earned his B.S. degree in zoology from Michigan State University and his Ph.D. in parasitology from Florida State University. In addition to human anatomy and physiology, he teaches histology, neuroanatomy, biomedical etymology, animal behavior, and sociobiology. Ken is a five-time recipient of the Phi Kappa Phi Honor Professor award for outstanding mentor to his undergraduate students. He also received a university Excellence in Research and Publication Award for the first edition of this book. Ken is an active member of the Human Anatomy and Physiology Society. He served as a developmental reviewer and wrote instructors' supplements for several other McGraw-Hill anatomy and physiology textbooks for a number of years before beginning his own textbook. Ken is married to Diane Saladin, a registered nurse; they have two children.

This book is dedicated to my wife

Diane
in loving appreciation
for her dedication to the author.

Brief Contents

Contents

Chapter 28

Chapter 29

Student Preface

My job as an author is not merely to provide you with information, but also to make it as easy and enjoyable to learn as possible. I try to achieve this through a clear writing style; beautiful illustrations; the way I structure each chapter around certain learning aids; interesting side notes on the history, personalities, and clinical relevance of the subject; helpful supplements to the book; and personal e-mail correspondence with students who are using the book.

The human body is fascinating and very complex. It's not easy to learn what you need to know for a career in the health and wellness professions; it takes real effort and mental discipline. But if I am effective as a writer and you are diligent as a student, you will acquire a knowledge that is immensely satisfying. Following are some tools that McGraw-Hill Higher Education and I have designed to make your goals attainable.

Chapter Structure and Learning Aids. It is a good idea in any course to get acquainted with the layout and features of a book before you begin to study the content. On the following pages, you will find a "Visual Guide" to this book that shows the typical structure of a chapter and calls your attention to features designed to aid your learning. I urge you to examine the guide so you will develop a sense of how to plan your study sessions around a chapter's structure and how to evaluate your understanding before you take your instructor's exams.

Art Program. We have gone to great care and expense to produce a beautiful and thorough collection of drawings and photographs for this book. Don't dismiss the illustrations as mere decoration or "eye candy." Every time I introduced a major idea, I tried to conceive of a good way to illustrate it. If you study the illustrations carefully and review them before exams, you are likely to have a good grasp of most of the important points I wanted to make in a chapter. Many students are "visual learners" who find that on exams, they can picture a drawing or photo and get information from that mental image more easily than they can remember words of text. Careful attention to the illustrations thus pays off at exam time.

The Essential Study Partner CD-ROM. In addition to the features built into the book itself, we offer a great variety of supplements. One of these is an interactive CD-ROM called the Essential Study Partner (ESP), included with every new copy of the book. On the opening page of each chapter, you will find a box that refers you to relevant sections of the ESP. The ESP makes anatomy and physiology come alive. Here you will find animations and activities that help to make concepts in the chapter more understandable and memorable.

Website and Online Learning Center. The website for this book is at www. mhhe.com/saladin/. Among other features of this site, you will find an Online Learning Center (OLC) that has a wealth of activities to supplement each chapter and test your understanding. The OLC provides supplemental readings, links to related websites, histological and other photographs, a downloadable *Student Study Guide,* anatomical art that you can label for practice, other self-testing exercises, and a complete practice exam for each chapter. This is a tremendous resource that should greatly enhance your learning of A&P and your enjoyment of the total learning package centered around this book. Our website also has a Message Center where you can discuss A&P issues with other students, post messages, and receive personal replies from me and the customer service representatives at McGraw-Hill Higher Education.

Survival Skills for A&P Students. In 23 years of teaching A&P, I have counseled countless students on how to study A&P more efficiently and earn higher grades on exams. I am happy to share that advice with you. I have consolidated all this into a "Survival Skills" section of the *Student Study Guide* at the Online Learning Center. Many of my own students dramatically improved their performance and felt much happier about the course when they read and applied the suggestions given there.

Supplements. This book is just part of a large multimedia package of learning tools. You can order additional study aids by telephone (800-338-3987)—including anatomy atlases, flashcards, coloring guides, and additional CD-ROMs. These are described on page xx as Learning Supplements for Students.

E-mail me! In a way, I consider everyone studying from this book to be my student, and not just those I see in my classroom each week. I have greatly enjoyed and benefited from e-mail correspondence with students around the world. Not only would I like to know what you find to be the best features of our book and total learning package, but I would also like to know your suggestions for improvement. In addition, I am available to help you with difficult concepts in the book. Some of the explanations in this edition developed out of my correspondence with student users. Who knows?—perhaps your questions will result in something that is new or better explained in the third edition.

With best wishes for your success,

Kenneth S. Saladin

Kenneth S. Saladin
Georgia College & State University
Milledgeville, Georgia
ksaladin@mail.gcsu.edu

A Visual Guide to

Anatomy & Physiology

The Unity of Form and Function

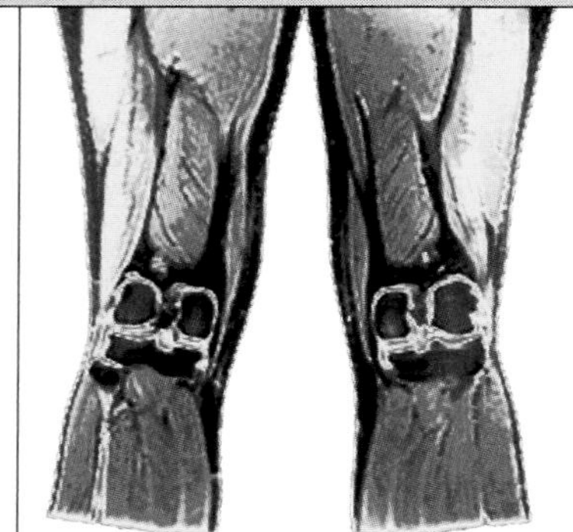

Muscles of the thigh to upper calf (MRI)

CHAPTER 11

The Muscular System

Chapter Outline

Insights

Study with Your ESP CD-ROM

Refer to your ESP Tutorial CD-ROM for helpful animations, activities, and quizzes related to the material covered in this chapter. The following are suggested areas to go to on the CD and are presented according to "Unit/Topic/":
- Support and Movement/Muscular System

Brushing Up

To understand this chapter, it is important that you understand or brush up on the following concepts:
- Gross anatomy of the skeleton (chapter 9)
- Movements of synovial joints (pp. 316–321)

Before You Begin

a chapter, it is important to have a broad overview of what it covers. This is provided by an outline on the first page of each chapter, page-referenced to facilitate your later review and study.

Icons At the Beginning

of each chapter encourage you to use the interactive CD-ROM that accompanies your book. The *Essential Study Partner* is packed with hundreds of animations and learning activities. The quizzes will help you to reinforce difficult topics. The animations bring to life those concepts that are difficult to envision.

Knowledge Doesn't Come

in little compartments that you can forget as soon as an exam is over. In the health professions, you must have a comprehensive understanding of the entire body, and as you begin each new book chapter, you need to remember concepts covered in earlier ones.

Brushing Up boxes list major concepts (with page references) that you should be familiar with before continuing. This feature demonstrates the interrelationship of concepts and helps give you the flexibility to cover topics in a different order than the book does.

Each Chapter Has

three to five *Insights* intended as diversionary readings for enjoyment. These essays illuminate the clinical relevance of a concept, give short historical sketches of the personalities and events behind the facts and findings in anatomy and physiology, and explore evolutionary insights into the body's structure and function.

Clinical Insight 11.3

Hernias

A hernia is any condition in which the viscera protrude through a weak point in the muscular wall of the abdominopelvic cavity. The most common type to require treatment is an *inguinal hernia.* In the male fetus, each testis descends from the pelvic cavity into the scrotum by way of a passage called the *inguinal canal* through the muscles of the groin. This canal remains a weak point in the pelvic floor, especially in infants and children. When pressure rises in the abdominal cavity, it can force part of the intestine or bladder into this canal or even into the scrotum. This also sometimes occurs in men who hold their breath while lifting heavy weights. When the diaphragm and abdominal muscles contract, pressure in the abdominal cavity can soar to 1,500 pounds per square inch—more than 100 times the normal pressure and quite sufficient to produce an inguinal hernia, or "rupture." Inguinal hernias rarely occur in women.

Two other sites of hernia are the diaphragm and navel. A *hiatal hernia* is a condition in which part of the stomach protrudes through the diaphragm into the thoracic cavity. This is most common in overweight people over 40. It may cause heartburn due to the regurgitation of stomach acid into the esophagus, but most cases go undetected. In an *umbilical hernia,* abdominal viscera protrude through the navel.

Before You Go On

Answer the following questions to test your understanding of the preceding section:

12. Which muscles are used more often, the external intercostals or internal intercostals? Explain.
13. Explain how pulmonary ventilation affects abdominal pressure and vice versa.
14. Name a major superficial muscle and two major deep muscles of the back.
15. Define *perineum, urogenital triangle,* and *anal triangle.*
16. Name one muscle in the superficial perineal space, one in the urogenital diaphragm, and one in the pelvic diaphragm. State the function of each.

Muscles Acting on the Shoulder and Upper Limb

Objectives

When you have completed this section, you should be able to

- Name and locate the muscles that act on the pectoral girdle, shoulder, elbow, wrist, and hand;
- Relate the actions of these muscles to the joint movements described in chapter 10; and
- Describe the origin, insertion, and innervation of each muscle.

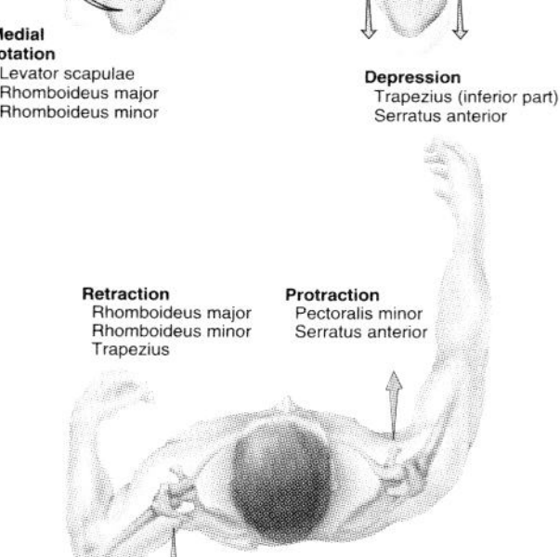

Figure 11.21 Actions of Some Thoracic Muscles on the Scapula. Note that an individual muscle can contribute to multiple actions, depending on which fibers contract and what synergists act with it.

Each Section Ends

with a few self-testing questions called *Before You Go On.* These questions help you to test your progress at frequent intervals. Use the questions as a test of your memory and ability to explain an idea.

Table 11.11 Actions of the Shoulder (Humeroscapular) Joint

Boldface indicates prime movers; others are synergists. Parentheses indicate only a slight effect.

Flexion	Extension
Anterior deltoid **Pectoralis major** Coracobrachialis Biceps brachii	**Posterior deltoid** **Latissimus dorsi** Teres major
Abduction	**Adduction**
Lateral deltoid Supraspinatus	**Pectoralis major** **Latissimus dorsi** Coracobrachialis Triceps brachii Teres major (Teres minor)
Medial Rotation	**Lateral Rotation**
Subscapularis Teres major Latissimus dorsi Deltoid Pectoralis major	**Infraspinatus** **Teres minor** Deltoid

extends it as in pointing at something behind you—thus, they are antagonists.

The other seven muscles of the shoulder are called *scapular muscles* because they originate on the scapula. Among these, the prime mover is the **deltoid**—the thick muscle that caps the shoulder. It acts like three different muscles. Its anterior fibers flex the shoulder, its posterior fibers extend it, and its lateral fibers abduct it. Abduction by the deltoid is antagonized by the combined action of the pectoralis major and latissimus dorsi. The **teres major** assists extension of the shoulder and the **coracobrachialis** assists flexion and adduction.

Tendons of the other four scapular muscles form the **rotator cuff**—the **infraspinatus, supraspinatus, subscapularis,** and **teres minor** (fig. 11.24). The subscapularis fills most of the subscapular fossa on the anterior surface of the scapula. The other three originate on the posterior surface. The supraspinatus and infraspinatus occupy the corresponding fossae above and below the scapular spine, and the teres minor lies inferior to the infraspinatus. The tendons of these muscles merge with the joint capsule of the shoulder as they pass it en route to the humerus. They insert on the proximal end of the humerus, forming a partial sleeve around it. The rotator cuff reinforces the joint capsule and holds the head of the humerus in the glenoid cavity. These muscles act as synergists in shoulder movements. The rotator cuff, especially the tendon of the supraspinatus, is easily damaged by strenuous circumduction.

Since the humeroscapular joint is capable of such a wide range of movements and is acted upon by so many muscles, its actions are summarized in table 11.11.

Think About It

Since a muscle can only pull, and not push, antagonistic muscles are needed to produce opposite actions at a joint. Reconcile this fact with the observation that the deltoid muscle both flexes and extends the shoulder.

There's no escaping the fact

that human anatomy and physiology encompasses a lot of information, but you don't have to swallow it all in one bite. Each chapter is divided into five or six digestible sections. Each section begins with a short list of objectives. This makes it easier for you to break down the subject matter and plan your goals for a study session.

Success in the Health Professions

requires far more than memorization. More important is your insight and ability to apply what you remember to new cases and problems. *Think About It* questions, which can be found strategically distributed throughout each chapter, encourage you to stop and think more deeply about the meaning or broader significance of what you have just read.

The Human Organ Systems

do not exist in isolation from each other. Diseases of the circulatory system can lead to failure of the urinary system, and aging of the skin can lead to weakening of the skeleton. For each organ system, a page called *Connective Issues* shows how it affects other systems of the body and is affected by them. These pages help you get the big picture and appreciate the body as an integrated whole.

340 Part Two Support and Movement

The **muscular system** consists of about 600 skeletal muscles—striated muscles that are usually attached to bone. (The term does not include smooth or cardiac muscle.) The form and function of the muscular system occupy a place of central importance in several fields of health care and fitness. Physical and occupational therapists must be well acquainted with the muscular system to design and carry out rehabilitation programs. Nurses and other health-care providers often move patients who are physically incapacitated, and to do this safely and effectively requires an understanding of joints and muscles. Even to give intramuscular injections safely requires a knowledge of the muscles and the nerves and blood vessels associated with them. Coaching, movement science, sports medicine, and dance benefit from a knowledge of skeletomuscular anatomy and mechanics.

Myology,[1] the study of muscles, is closely related to what we have covered in the preceding chapters. It relates muscle attachments to the bone structures described in chapter 9 and muscle function to the joint movements described in chapter 10. In this chapter, we consider the gross anatomy of the muscular system and how it relates to joint movements. In the following chapter, we examine the mechanisms of muscle contraction at the cellular and molecular levels.

- **Movement.** Most obviously, the muscles enable us to move from place to place and to move individual body parts. Muscular contractions also move body contents in the course of respiration, circulation, digestion, defecation, urination, and childbirth.
- **Stability.** Muscles maintain posture by resisting the pull of gravity and preventing unwanted movements. They hold some articulating bones in place by maintaining tension on the tendons.
- **Communication.** Muscles are used for facial expression, other body language, writing, and speech.
- **Control of body openings and passages.** Ringlike *sphincter muscles* around the eyelids, pupils, and mouth control the admission of light, food, and drink into the body; others that encircle the urethral and anal orifices control elimination of waste; and other sphincters control the movement of food, bile, and other materials through the body.
- **Heat production.** The skeletal muscles produce as much as 85% of our body heat, which is vital to the functioning of enzymes and therefore to all of our metabolism.

Some of these functions are shared by skeletal, cardiac, and smooth muscle. The remainder of this chapter, however, is concerned only with skeletal muscles.

Connective Tissues of a Muscle

A skeletal muscle is composed of both muscular tissue and connective tissue (fig. 11.1). A skeletal muscle cell (muscle fiber) is about 10 to 100 μm in diameter and up to 30 cm long. It is surrounded by a sparse layer of areolar connective tissue called the **endomysium**[2] (EN-doe-MIZ-ee-um), which allows room for blood capillaries and nerve fibers to reach each muscle fiber. Muscle fibers are grouped in bundles called **fascicles**[3] (FASS-ih-culs), which are visible to the naked eye as parallel strands. These are the "grain" in a cut of meat; tender meat is easily pulled apart along its fascicles. Each fascicle is separated from neighboring ones by a connective tissue sheath called the **perimysium**,[4] usually somewhat thicker than the endomysium. The muscle as a whole is surrounded by still another connective tissue layer, the **epimysium**.[5] The epimysium grades imperceptibly into connective tissue

2. *endo* = within + *mys* = muscle
3. *fasc* = bundle + *icle* = little
4. *peri* = around
5. *epi* = upon, above

Connective Issues

Interactions Between the MUSCULAR SYSTEM and Other Organ Systems

← Indicates ways in which this system affects other systems
→ Indicates ways in which other systems affect this one

Integumentary System
← Facial muscles pull on skin to provide facial expressions
→ Covers and protects superficial muscles; initiates synthesis of calcitriol, which promotes absorption of calcium needed for muscle contraction; dissipates heat generated by muscles

Skeletal System
← Muscles move and stabilize joints and produce stress that affects ossification, bone remodeling, and shapes of mature bones
→ Provides levers that enable muscles to act; stores calcium needed for muscle contraction

Nervous System
← Muscles give expression to thoughts, emotions, and motor commands that arise in the central nervous system
→ Stimulates muscle contraction; monitors and adjusts muscle tension; adjusts cardiopulmonary functions to meet needs of muscles during exercise

Endocrine System
← Exercise stimulates secretion of stress hormones; skeletal muscles protect some endocrine organs
→ Hormones stimulate growth and development of muscles and regulate levels of glucose and electrolytes important for muscle contraction

Circulatory System
← Muscle contractions help to move blood through veins; exercise stimulates growth of new blood vessels
→ Delivers O_2 and nutrients; carries away wastes and heat generated by muscles; cardiovascular efficiency, RBC count, hemoglobin level, and density of blood capillaries in muscle greatly affect muscular endurance

Lymphatic and Immune Systems
← Muscle contractions promote lymph flow; exercise elevates level of immune cells and antibodies; excess exercise inhibits immune responses
→ Lymphatic system drains fluid from muscles; immune system protects muscles from pathogens and promotes tissue repair

Respiratory System
← Muscle contractions ventilate lungs; muscles of larynx and pharynx regulate air flow; CO_2 generated by exercise stimulates respiratory rate and depth
→ Provides O_2 and removes CO_2; respiratory efficiency greatly affects muscular endurance

Urinary System
← Muscles control voluntary urination; muscles of pelvic floor support bladder
→ Eliminates wastes generated by muscles; regulates levels of electrolytes important for muscle contraction

Digestive System
← Muscles enable chewing and swallowing; control voluntary defecation; abdominal and lumbar muscles protect lower digestive organs
→ Absorbs nutrients needed by muscles; liver regulates blood glucose levels and metabolizes lactic acid generated by anaerobic muscle metabolism

Reproductive System
← Muscles contribute to erection and ejaculation; abdominal and pelvic muscles aid childbirth
→ Gonadal steroids affect muscular growth and development

401

Key Terms Are Set Off

in boldface when they first appear in the narrative and are compiled and page-referenced at the end of each chapter. Frequently, when a new term is introduced, it is followed by a pronunciation guide and a footnote is provided that gives you the derivation of the word. You will soon recognize that such elements as *hypo-*, *natri*, *-cyte*, and *-itis* occur over and over in the book. As their meanings become more familiar, you will approach pronunciation, spelling, and comprehension with more confidence and proficiency.

The Web Address

located within the *Chapter Review* is a reminder to you that additional study questions and links to anatomy and physiology–related topics appear on the web site.

Each Chapter Ends With

a chapter review. Part of this is a *Study Outline* that can be used as a basis for organizing your thoughts.

The Boldface Terms

in a chapter are listed in a page-referenced vocabulary checklist. Words are listed in the same order as the chapter discusses them so that related concepts appear together.

Testing Your Recall

contains 10 multiple choice and 10 short answer questions you can use to check your knowledge.

Each Chapter Has Five

Testing Your Comprehension questions that go beyond memorization to require a deeper level of analysis and application of your knowledge to clinical situations.

---

402 Part Two Support and Movement

Chapter Review

Online Learning Center

Study resources for this chapter are available at www.mhhe.com/saladin/

Study Outline

The Structural and Functional Organization of Muscles (p. 340)

1. Functions of muscles
   a. Movement
   b. Stability
   c. Communication
   d. Control of body openings and passages
   e. Heat production
2. Connective tissues of a muscle
   a. Endomysium
   b. Perimysium
   c. Epimysium
   d. Deep fascia
3. Muscle-bone attachments
   a. Direct (fleshy)
   b. Indirect (via tendon)
   c. Aponeuroses
   d. Retinacula
4. General anatomy of muscles
   a. Origin, insertion, and belly
   b. Fascicle orientation
5. Coordinated action of muscles
   a. Prime movers
   b. Synergists
   c. Antagonists
   d. Fixators
6. Intrinsic and extrinsic muscles
7. Muscle innervation
8. How muscles are named
   a. By size
   b. By shape
   c. By location
   d. By number of heads
   e. By orientation
   f. By action

Muscles of the Head and Neck (p. 348)

1. Muscles of facial expression (table 11.2)
2. Muscles of chewing and swallowing (table 11.3)
3. Muscles acting on the head (table 11.4)

Muscles of the Trunk (p. 359)

1. Muscles of respiration (table 11.5)
2. Muscles of the abdomen (table 11.6)
3. Muscles of the back (table 11.7)
4. Muscles of the pelvic floor (table 11.8)

Muscles Acting on the Shoulder and Upper Limb (p. 370)

1. Muscles acting on the scapula (table 11.9)
2. Muscles acting on the humerus (table 11.10)
3. Muscles acting on the forearm (table 11.12)
4. Muscles acting on the wrist and hand (table 11.14)

Muscles Acting on the Hip and Lower Limb (p. 386)

1. Muscles acting on the hip and femur (table 11.17)
2. Muscles acting on the knee (table 11.18)
3. Muscles acting on the foot (table 11.19)

Selected Vocabulary

To review the names of individual muscles, see the tables cited in the Study Outline.

muscular system 340
myology 340
endomysium 340
fascicle 340
perimysium 340
epimysium 340
deep fascia 341
superficial fascia 341
direct attachment 341
indirect attachment 341
tendon 341
aponeurosis 342
retinaculum 342
origin 342
head of a muscle 342
insertion 342
belly 342
fusiform muscle 342
parallel muscle 342
convergent muscle 342
pennate muscle 342
circular muscle 342
sphincter 342
action of a muscle 343
prime mover 343
synergist 343
antagonist 343
antagonistic pair 343
fixator 343
intrinsic muscle 344
extrinsic muscle 344
innervation 344
galea aponeurotica 350
suprahyoid group 352
infrahyoid group 352
anterior triangle 357
posterior triangle 357
central tendon of diaphragm 359
tendinous intersections of rectus abdominis 361
rectus sheath 361
linea alba 361
inguinal ligament 363
serratus posterior 365
thoracolumbar fascia 365
perineum 367
urogenital triangle 367
anal triangle 367
superficial perineal space 367
central tendon of perineum 367
urogenital diaphragm 367
pelvic diaphragm 367
rotator cuff 375
flexor retinaculum 378
extensor retinaculum 378
carpal tunnel 378
thenar group 383
hypothenar group 383
midpalmar group 383

---

lateral rotators 389
fascia lata 389
iliotibial band 389

Testing Your Recall

1. Which of the following muscles is the prime mover in spitting out a mouthful of liquid?
   a. platysma
   b. buccinator
   c. risorius
   d. masseter
   e. palatoglossus
2. Each muscle fiber has a sleeve of areolar connective tissue around it called
   a. the deep fascia.
   b. the superficial fascia.
   c. the perimysium.
   d. the epimysium.
   e. the endomysium.
3. Which of these is *not* a suprahyoid muscle?
   a. genioglossus
   b. geniohyoid
   c. stylohyoid
   d. mylohyoid
   e. digastric
4. Which of these muscles is an extensor of the neck?
   a. external oblique
   b. sternocleidomastoid
   c. splenius capitis
   d. iliocostalis
   e. latissimus dorsi
5. Which of these muscles of the pelvic floor is the deepest?
   a. superficial transverse perineus
   b. bulbospongiosus
   c. ischiocavernosus
   d. deep transverse perineus
   e. levator ani
6. Which of these actions is *not* performed by the trapezius?
   a. extension of the neck
   b. depression of the scapula
   c. elevation of the scapula
   d. rotation of the scapula
   e. adduction of the humerus
7. Both the hands and feet are acted upon by a muscle or muscles called
   a. the extensor digitorum.
   b. the abductor digiti minimi.
   c. the flexor digitorum profundus.
   d. the abductor hallucis.
   e. the flexor digitorum longus.
8. Which of the following muscles does *not* extend the hip joint?
   a. quadriceps femoris
   b. gluteus maximus
   c. biceps femoris
   d. semitendinosus
   e. semimembranosus
9. Both the gastrocnemius and ____ muscles insert on the heel by way of the calcaneal tendon.
   a. semimembranosus
   b. tibialis posterior
   c. tibialis anterior
   d. soleus
   e. plantaris
10. Which of the following muscles raises the upper lip?
    a. levator palpebrae superioris
    b. orbicularis oris
    c. zygomaticus minor
    d. masseter
    e. mentalis
11. The ________ of a muscle is the point where it attaches to a relatively stationary bone.
12. A bundle of muscle fibers surrounded by perimysium is called a/an ________.
13. The ________ is the muscle primarily responsible for a given movement at a joint.
14. The three large muscles on the posterior side of the thigh are commonly known as the ________ muscles.
15. Connective tissue bands called ________ prevent flexor tendons from rising like bowstrings.
16. The anterior half of the perineum is a region called the ________.
17. The abdominal aponeuroses converge on a midsagittal fibrous band on the abdomen called the ________.
18. A muscle that works with another to produce the same or similar movement is called a/an ________.
19. A muscle somewhat like a feather, with fibers obliquely approaching its tendon from both sides, is called a/an ________ muscle.
20. A circular muscle that closes a body opening is called a/an ________.

Answers in appendix B

Testing Your Comprehension

1. Radical mastectomy, once a common treatment for breast cancer, involved removal of the pectoralis major along with the breast. What functional impairments would result from this? What synergists could a physical therapist train a patient to use to recover some lost function?
2. Removal of cancerous lymph nodes from the neck sometimes requires removal of the sternocleidomastoid on that side. How would this affect a patient's range of head movement?
3. In a disease called tick paralysis, the saliva from a tick bite paralyzes skeletal muscles beginning with the lower limbs and progressing superiorly. What would be the most urgent threat to the life of a tick paralysis patient?
4. Women who habitually wear high heels may suffer painful "high heel syndrome" when they go barefoot or wear flat shoes. What muscle(s) and tendon(s) are involved? Explain.
5. A student moving out of a dormitory kneels down, in correct fashion, to lift a heavy box of books. What prime movers are involved as he straightens his legs to lift the box?

Answers at the Online Learning Center

Instructor Preface

In my 23 years in the classroom, nothing has been more satisfying than writing and teaching from *Anatomy & Physiology: The Unity of Form and Function.* The generous reception given to the first edition now gives me the opportunity and pleasure of fine-tuning the approach. This edition reflects the suggestions of many students in my own classroom and on other campuses; the knowledge and advice of more than 400 A&P instructors who have evaluated the two editions; and a commitment to excellence on the part of McGraw-Hill Higher Education.

This preface describes the book's intended audience, how we determined what students and instructors want in the ideal A&P textbook, what has changed in this edition to accommodate those needs, how this book differs from others, and what supplements are available to round out the total teaching package.

Audience

This book is for students who plan to pursue such careers as nursing, therapy, sports medicine, health education, and other health professions. It is designed for a two-semester combined anatomy and physiology course and assumes little or no scientific background on the student's part. Many A&P students have never had a college chemistry or biology course, and others are returning after long interruptions to raise families or pursue other careers. With this in mind, I provide all of the background in chemistry and cell biology (chapters 2–5) that one needs to understand the rest of the book. Many A&P students also are still developing the intellectual skills and study habits necessary for success in the health science curriculum. Therefore, I endeavor to write in a style that is clear, concise, and enjoyable to read and to enliven the facts of science with analogies, historical notes, clinical remarks, biographical vignettes, and other "seasoning" that will make the book enjoyable to students and instructors alike. Each chapter is built around certain pedagogic strategies that will make the subject attainable for a wide variety of students and instill study and thinking skills conducive to success in more advanced courses.

How We Evaluated Your Needs

The changes in this edition reflect an extensive effort to evaluate the needs of A&P instructors and students, to learn how well the first edition was meeting those needs, and to determine how it could even better do so.

We began our formal market research by commissioning 150 instructors, both users and nonusers of the first edition, to review the book in detail and submit suggestions for improvement. Another 140 instructors returned a questionnaire on the first-edition content, organization, and supplements. These reviews and questionnaires constituted a major basis for my revision plan.

We also contracted the textbook development firm of Burrston House for a wide-ranging research and marketing project. Burrston House organized focus groups in eight U.S. cities in 1999 and 2000, most of them composed of instructors using A&P textbooks other than my own, and one consisting of students from eight colleges. Participants reviewed the first edition in comparison to the books they were using and then met with Burrston House and me for several hours to discuss ideas for improving the second edition. This input was an important element in my second round of manuscript and art revisions.

Finally, Burrston House commissioned 50 of these instructors to review the entire second-edition manuscript during its development and 12 of them as an art panel to specifically critique the revised illustrations. These reviewers generated more than 2,000 pages of additional commentary that became the dominant influence in the final revision. Considering the two editions collectively, this book now embodies the collective knowledge, insights, and high teaching standards of more than 400 A&P instructors. Their influence is seen in everything from the size, color, and labeling of the illustrations to the accuracy and currency of the book's anatomical descriptions and physiological explanations.

What We Found

A consensus emerging from this research was that instructors choose a textbook primarily by four criteria: writing style, clarity in explaining difficult topics, illustrations, and supplements.

Writing Style

I steer a middle course in my writing, avoiding two extremes: one of them a rigid formality that disdains, for example, any mention of "we" or "you," and the other a chattiness that some find condescending to the student. The students in our focus group, none of whom were in courses using my book, commonly said they wished the textbooks they were using had a more conversational tone. I feel I have succeeded when colleagues describe the tone as suitably colloquial or students say that it makes them feel as if the author were addressing them personally:

> "I find him interesting to read. . . . His style is straightforward and conversational. In my opinion he writes on a level that the typical A&P student would find entertaining as well as informative."
>
> —C. Sternberg, Riverside Community College

> "This is where Saladin's book shines. His writing is clear, interesting, and in some cases entertaining. His writing is concise, insightful and consistent, and always with an enthusiasm for the topic."
>
> —H. Burnette, McLennan Community College

> "Professor Saladin has an exceptionally clear and fluid style. Most of what he says is clear and avoids confusing side issues. [The textbook we previously used] often lost the students this way. Student frustration has been much lower since we have started using Saladin's text."
>
> —D. Evans, Pennsylvania College of Technology

In this edition I have striven for even greater clarity through simpler language, more effective use of transitions and sentence and paragraph breaks, and more extensive use of bulleted and numbered lists to present related concepts and sequential processes in a way that is easy for students to grasp initially and review later.

If you are using or considering another textbook, the best way to judge which of them will better serve your students' needs is to choose some topics that they often find hard to understand—perhaps muscle contraction, action potentials, blood clotting, aerobic respiration, or the renal countercurrent multiplier—and compare the books side by side. Several instructors have had students themselves compare this book to others, and students have often written that they would rather be studying from this one.

Explanations of Difficult Concepts

The concepts that reviewers most often cited as especially difficult for their students were the muscular, nervous, endocrine, and urinary systems and metabolism. The key to getting students through especially difficult areas is especially effective writing and art. In response to reviews of these chapters, I developed several new illustrations, modified others, and fine-tuned the writing to clarify a number of points. Our final round of reviews indicated that the finished product succeeds at this task:

> "Saladin's explanations of the topics are excellent. He is very clear and insightful in his writing."
>
> —M. Gottlieb, La Guardia Community College

> "Clearly, the major strength of the manuscript is the writing. I am greatly impressed with the clarity of the prose. . . . The sections on synapses, neurotransmitters and synaptic transmission are excellent; those on neural integration and neural coding are very good."
>
> —W. Jones, Loyola University

Illustrations

The visual appeal and artistic clarity of the first edition was the second most significant factor in its success. I developed many original illustrative concepts, and the professional artists then digitally executed these, as well as the classic themes of human A&P, in a captivating way that set an entirely new standard for A&P textbook art. Yet here too, our reviews and focus groups identified room for improvement: larger illustrations and brighter coloration, especially. Reviewers also widely reported that they feel the illustrations in A&P textbooks in general are getting too cluttered with labels, to the extent that it is difficult for students to pick out the key points. I deleted many labels from this edition so that students can more quickly grasp the essential points of an illustration.

The illustration program is more than line art. I replaced most of the histological photomicrographs and added scale bars when I had sufficient information to do so, and I replaced most of the cadaver photographs with a new set of very clear and skillful dissections commissioned specifically for this book.

> "One of the major strengths of the Saladin text, one that prompted me to adopt the text, was the quality and quantity of the illustrations. In my view, this text is a hands down winner in this area."
>
> —R. Symmons, California State University at Hayward

Supplements

The most highly rated supplement to the first edition was our set of 500 overhead transparencies. The response to this was so enthusiastic, and the appeal for more was so compelling, that we have now expanded the set to 950 transparencies. These include every item of line art in the book, some of the tables, and unlabeled duplicates of many of the drawings, especially anatomical figures, so that you can use them for testing purposes or label them with the degree of simplicity or detail, or the terminology, that you prefer.

In addition, students who purchase a new copy of the book now get a new interactive *Essential Study Partner* CD-ROM. A great variety of additional learning aids are now offered in the Online Learning Center at www.mhhe.com/saladin/.

What Else Is New?

The transition from the first to the second edition is the most critical stage in a textbook's history, when a book not only demonstrates its staying power in a highly competitive market but also typically undergoes more changes than it will at any later edition. True to pattern, this second edition has undergone hundreds of changes, far beyond what I can list in this preface. I can only give examples here, but a more complete list of changes is provided at the aforementioned website.

Insights

This edition has three types of sidebars: 9 "Evolutionary Insights" that show the relevance of the biological history of the body to its present form and function; 17 "Historical Insights" that enliven and humanize the science of A&P through engrossing stories of the men and women behind the facts; and 104 "Clinical Insights" that make the science of A&P relevant to the student's career aspirations and to disorders they may have heard about or experienced. None of these topics—evolution, history, or pathology—are limited to these insight boxes. There are many others, scattered throughout the text in roughly these proportions, ranging from brief mention to extended discussions such as those on diabetes mellitus and immune deficiency diseases.

Expanded Clinical Content

This edition discusses more than 600 clinical topics, not just in the Clinical Insight boxes but also in the main body of discussion and a new series of tables of common dysfunctions of each organ system. The Clinical Applications list is found in the Instructor's Manual as well as the book website. Dr. Donna Van Wynsberghe, University of Wisconsin–Milwaukee, joined the project with this edition and reviewed the book from cover to cover for clinical accuracy. The new *Clinical Applications Manual* available for this book expands upon some of the topics briefly described in the book and introduces several new topics and case studies for each organ system.

Issues of Terminology

I have changed a number of terms for closer, but not dogmatic, conformity to the Nomina Anatomica. For example, I replaced *spinal column* with the more official term *vertebral column,* but I did not replace *Schwann cell* with the more official but obscure *neurolemmocyte.* When there are common, synonymous terms for something, I have been careful to use one term consistently throughout the book—not to use *ventral horn* in one place and *anterior horn* in another, for example. Synonyms are introduced parenthetically at the first use of a word. I have also taken added pains to be sure a word is explicitly defined when it is introduced. It is all too easy to name something such as the Golgi complex and immediately begin to describe what it looks like and what it does without ever saying exactly what it *is.*

Eponyms present a sticky problem for scientific writers. I have avoided most of them, substituting *tactile disc* for *Merkel disc,* for example, but some are unavoidable, such as *Golgi complex* and *Alzheimer disease.* Stylists have grappled for years with possessive eponyms and the contrary tastes of those who would insist on an apostrophe-*s* in *Cushing's syndrome,* yet take offense at the parallel expression, *Down's syndrome.* The American Medical Association's latest *Manual of Style* (9th edition, 1998) urges writers to delete the possessive in almost all cases, and I have followed that advice. Thus, you will find in this edition a usage that may seem unfamiliar at first (*Bell palsy, Peyer patches*) but that is more internally consistent and is now recommended by the AMA. I make exception, as does the AMA, when the nonpossessive form is overly awkward (as *Broca area* would be), and I retain the possessive form for eponymous laws (*Boyle's law, Charles' law*).

Content Changes

Aside from these global changes in the book, users of the first edition may notice many specific topics that have been relocated, expanded, condensed, added, clarified, or otherwise modified in this edition. I can give but a few examples here. The numbers in parentheses, except for illustration references, are chapter numbers.

Relocated: Second-messenger systems (4), osmolarity and tonicity (4), and edema (20) have been moved up to earlier chapters because students need an earlier introduction to these topics in order to better understand subsequent chapters. On the other hand, some topics have been moved back to later chapters where they fit better, including the overview of the skeleton and surface markings of bones (9), the ionic basis of membrane potentials (13), subdivisions of the peripheral nervous system (15), auditory ossicles (16), comparison of the nervous and endocrine systems (17), and obesity (26).

Many other topics have been shifted to different places within their original chapters or otherwise regrouped for a more integrated presentation: ionizing radiation (2); endoplasmic reticulum and membrane transport processes (4); rib-vertebra articulations (9); limb muscles (11); neurons, glia, and myelination (13); the ventricles and CSF (14); hypothalamo-pituitary relationships, diabetes mellitus, the synthesis, transport, and action of each class of hormones, and stress physiology (17); the body's three lines of defense, MHC proteins, passive and active immunity, and AIDS (21); intestinal physiology (25); and the ovarian and menstrual cycles (28).

Chapter 1 has been extensively reorganized for better flow. Chapter 6 (Histology) has been restructured to put the specific characteristics of each tissue type into tabular format while retaining a narrative style for more general points about each tissue class, and to present the tissues in the order most instructors say they present them: epithelial, connective, nervous, and muscular.

Added: Topic additions include pericardial, pleural, and peritoneal fluids (atlas A); wound healing (6); stress fractures and pathologic fractures (8); titin and dystrophin (12); disseminated intravascular coagulation (18); cardiac tamponade (19); vaccination (21); a reference table of the gas laws (22); a tabular summary of fluid imbalances (24); the effect of exercise on cholesterol level (25); behavioral thermoregulation (25); gestational diabetes (28); and immunological adaptation of the neonate (29).

New illustrations include coenzyme function (3.19), transmembrane proteins (4.7), the cAMP system (4.9), secondary active transport (4.27), Pap smears (6.13 and in clinical insight 28.1), wound healing (6.35), intramembranous ossification (8.8), feedback loops in calcium homeostasis (8.15), eccentric muscle contraction (12.16*c*), isometric and isotonic phases of muscle contraction (12.17), modes of ATP synthesis during exercise (12.18), temporal and spatial summation (13.22), the duplicity theory of vision (16.36), Cushing syndrome (17.14), the molecular basis of ABO blood types (18.12), and the physiology of Viagra (in clinical insight 27.4). Many other figures were redrawn for clarity, accuracy, or simplicity.

Expanded: Homeostasis (1); free radicals and antioxidants (2); secondary active transport, symport, and antiport systems (4); the regulation of cell division (5); atrophy, apoptosis, necrosis, and tissue repair (6); calcium homeostasis (8); knee injuries (10); ion gate inactivation in action potentials (13); the brain barrier system (14); spinal gating of pain (16); thyroid hormone, somatomedins, cytoplasmic hormone receptors, stress physiology, and the history of insulin (17); the Rh blood group (18); the infant larynx and respiratory diseases (22); the renal countercurrent exchanger (23); control of gastric function (25); and essential fatty acids and ascites (26).

Updated: Some new topics from the recent literature include sunscreens and skin cancer (7); the mechanoreceptive function of osteocytes, the action of parathyroid hormone, and treatment of osteoporosis (8); the genetics of Alzheimer disease (13); the origins of cerebrospinal fluid (14); the umami taste modality (16); melatonin, SAD, and PMS (17); psychoneuroimmunology (21); the genetics and endocrinology of obesity (25); the physiology of Viagra, environmental endocrine disruptors, and sexual transmission of hepatitis B and C (27); and the telomere theory of senescence (29).

Condensed: Some topics were deleted or condensed because reviewers said the first edition had gone into unnecessary detail, and some because it was more important to make room for larger illustrations and new information. Topics now treated more concisely include biomedical history (1), molarity (2), enzymology (3), the sliding filament mechanism (12), cellular mechanisms of memory (13), the evolution and physiology of sleep (14), cranial nerve routes (15), leukopoiesis (18), and infectious diseases of the respiratory system (22).

What Sets This Book Apart?

With all the A&P textbooks that were already available, I had to think more than twice about my publisher's original invitation, in the late 1980s, to write another one. When I accepted in 1992, it was with the conviction that I could contribute something to the curriculum that other authors were not. The features described here, taken collectively, offer a perspective and pedagogy unlike that of any other A&P textbook.

Historical Themes

I found long ago that students especially enjoyed lectures in which I introduced brief stories of the personal dramas that enliven the history of biomedical science. Thus, I incorporated that approach into my writing as well. Reviews of the first edition have shown that students elsewhere, like my own, especially like these stories. Some said that these were their favorite parts of a chapter and that they read the sidebars first. The stories in this book, selected for their poignant and inspiring quality, give students a more humanistic perspective on the field they've chosen to study and, I hope, cultivate an appropriately thoughtful attitude toward the discipline.

Profiles of Marie Curie, Rosalind Franklin, and Charles Drew, while telling of the struggles and ironies of their scientific careers, may also inspire students in their own career aspirations and ideals. Some of my favorite stories are of William Beaumont's digestive experiments on Alexis St. Martin; Crawford Long's first surgical use of ether, known in his time mainly as a party drug; the radical alteration of Phineas Gage's personality by a construction accident; and newly added to this edition, the feud between Nobel Prize corecipients Frederick Banting and J. J. R. MacLeod over the discovery of insulin. I am happy to find that instructors elsewhere regard this as a welcome addition to A&P teaching.

> "I favor the addition of historical and biographical information. The history tells an interesting story and it makes the text more enjoyable reading for both students

and instructors. This information gives an overall picture of how the scientific process works and how we have arrived at our current state of knowledge and theory."

—G. Brady, Spokane Falls Community College

"I like Saladin's presentation because I feel an understanding of how medicine and science have changed throughout history is part of becoming a 'well educated,' not just a 'well trained' student."

—R. Pope, Miami-Dade Community College

Evolutionary Themes

I feel that human form and function can never be *fully* appreciated without a sense of how and why it came to be as it is. Several books of "darwinian (evolutionary) medicine" appeared in the mid-1990s, and the medical journals since then have continued to show considerable interest in the evolutionary perspective on human structure, function, and disease. Yet other A&P textbooks are so far largely disregarding this development and point of view. I give a brief introduction to natural selection as applied to humans in chapter 1, nine Evolutionary Insight sidebars in other chapters, and occasional evolutionary remarks elsewhere. Students will thus find in these pages a novel and intriguing way of looking at such topics as mitochondria, body odors, bipedalism, hair, erythrocytes, the nephron loop, our taste for sweets, lactose intolerance, menopause, and senescence. Users of the book have often written that this perspective has been long overdue in A&P.

"I am particularly impressed with Saladin's inclusion of evolution and history of medicine. I often try to interject this information into my lectures, and find most modern texts lacking in these areas. Kudos to you!"

—M. Dawson, Kingsboro Community College

Clinical Applications

It is our primary task in A&P textbooks to teach the basic biology of the human body, not pathology. Yet students want to know the relevance of this biology—how it relates to their career aims. Furthermore, disease often gives us our most revealing windows on the importance of normal structure and function. What could serve better than cystic fibrosis, for example, to drive home the importance of membrane ion pumps? The clinical remarks and insight boxes in this book therefore substantially outnumber the historical and evolutionary ones.

"There are many tidbits of clinical information that are in this book, but not in others that I have seen. I think that's great! I have learned a thing or two. I also think that the author has tried to choose clinical examples that are commonly dealt with and therefore most useful to the student."

—L. Steele, Ivy Tech State College

Organization

Principles of Heredity

This book differs from others in its sequence of certain topics. The principles of heredity are presented early, for example. Students are ill-prepared to understand color blindness, blood types, hemophilia, sickle-cell disease, sex determination, and many other topics if they do not already understand dominant and recessive alleles, codominance, sex chromosomes, and sex linkage. Most textbooks postpone these concepts until the last chapter. I present them in chapter 5.

"One of the major strengths of chapter 5 is that it contains current information about all genetics—from molecular to Mendelian. We have found that students can better grasp phenotypic variation when they understand what a chromosome is and how it replicates. It is refreshing that Saladin has 'broken out of the mold.' "

—H. Mobley, Kilgore College

"I really like the idea of getting mendelian genetics started earlier than . . . it is in some texts because we can then talk about genetic diseases all through the course."

—M. Lockwood, University of New Hampshire

Muscle Anatomy and Physiology

Other textbooks treat muscle physiology before anatomy. I reverse this order so that the anatomy of the muscles (chapter 11) is closely tied to the joints on which they act (chapter 10), whereas muscle physiology (chapter 12)—which is where membrane potentials, synapses, and neurotransmitters are first introduced—is closely tied to the physiology of neurons (chapter 13).

"I think this is an excellent idea to better connect and integrate muscle and skeletal anatomy together and also muscle physiology and neurophysiology together. I have always felt this is a more logical and natural approach."

—B. Martin, La Sierra University

"I really like the arrangement of the joint, skeleton, and muscle anatomy chapters (contiguous) and the muscle physiology and nervous system chapters (contiguous). This is a better approach than the traditional one."

—M. Tenneson, Evangel University

Nervous System

Many instructors told us that nervous system coverage has gotten out of hand in A&P textbooks. This is often cited as the most difficult system for students to understand, yet most textbooks overwhelm the students with six chapters of it. This system typically concludes the first semester of A&P, and it is simply unmanageable to cover so much material at the end of the term. I have condensed the nervous system to four chapters—50 pages less than other leading books—yet retained its human interest through biographical and historical notes and discussions of such topics as speech defects, emotion, sensory deprivation, human pheromones, and other subjects missing from most textbooks.

> "I never have enough time to cover all the information included in a standard textbook on the nervous system. I prefer Saladin's concise approach to the nervous system."
>
> —L. Ritt, Burlington County College

> "I feel that in general [chapter 15] is the finest treatment of the [peripheral nervous system] that I have ever read at this level. . . . As usual the greatest strength of the chapter is the lucidity of the exposition."
>
> —W. Jones, Loyola University

Urinary System

My last major departure from the sequence found in other A&P textbooks is placement of the urinary system. Other textbooks place this near the end because of its anatomical association with the reproductive system. I feel that its intimate physiological ties with the circulatory and respiratory systems are far more important than this anatomical issue. I place the respiratory and urinary systems and the chapter on water, electrolyte, and acid-base balance as close as possible to the circulatory system (only a chapter on the lymphatic and immune systems intervenes). There are several compelling reasons for doing this: the respiratory and urinary systems collaborate to regulate the pH of the body fluids; the urinary system has more influence than any other on blood volume and blood pressure; and understanding glomerular filtration requires a knowledge of the forces of capillary fluid exchange explained in the chapter on the blood vessels. This sequence gives students a much better-integrated view of these system interactions than textbooks that relegate the urinary system to nearly the end.

> "Again, you must have read my mind. If I showed you my syllabus, that is my method. Urinary comes right with respiratory, as they are hand in hand in acid-base [balance], which is a concept that students always find difficult to grasp."
>
> —M. Van Cura, Brunswick Community College

System Interrelationships

Most instructors probably agree on the need to emphasize the interrelationships among organ systems and to discourage the idea that a system can be put out of one's mind after an exam is over. This book helps to underscore the connections in three ways. (1) Beginning with chapter 3, each has a Brushing Up box that lists concepts from earlier chapters that a student should understand before beginning to study the new one. This can also be useful for students returning to college after a break in their education and to instructors teaching the systems in a different order than the book does, and it helps reinforce the continuity between A&P I and II. (2) For each organ system, there is a Connective Issues page that summarizes many ways in which that system affects others and is affected by them. (3) In chapter 29, the section on Senescence of the Organ Systems compellingly shows how the age-related degenerative changes of each organ system contribute to the senescence of other systems. Senescence is an increasingly important topic for health-care providers as our population increases in average age. This section should help sensitize students not only to issues of gerontology but also to measures they can take at a young age to promote a better quality of life in their old age. These descriptions can be used throughout the course by those who prefer to teach aging as they teach each organ system, or used as a fine capstone topic at the end of A&P II to review and relate ideas that have been studied throughout the two semesters.

Learning Objectives

Some textbooks present all of the learning objectives at the beginning of a chapter and all of the review questions at the end. I divide each chapter into four to eight short, digestible segments, typically about five pages long, with a list of objectives at the beginning and a list of review questions at the end of each segment. This enables students to set concrete, attainable goals for a study session and to test their comprehension before going on to the next section.

Self-Testing Questions

Each chapter of *Anatomy & Physiology* has 50 to 65 self-testing questions at three levels of difficulty: recall, description, and application. The *ability to recall* terms and facts is tested by 10 multiple choice questions and 10 sentence completion questions in the chapter review. The *ability to describe* concepts is tested by 20 to 30 "Before You Go On" questions at the ends of the chapter subdivisions. The *ability to apply* ideas to new clinical situations and relate concepts in different chapters to each other is tested by 5 "Testing Your Comprehension" questions at the end of each chapter and an average of 5 "Think About It" questions dispersed through the chapter.

Etymological Aids

A&P students must assimilate a large working vocabulary. This is made far easier and more meaningful if they can pronounce words correctly and if they understand the roots that compose biomedical words. Appendix C is a two-page synopsis of my biological etymology course, enough to give students an appreciation of the value of dissecting words into their roots. The inside back cover has a lexicon of the 400 roots and affixes used most frequently in the book. The roots and affixes of new terms are given throughout the book in footnotes. Pronunciation guides are given parenthetically when new words are first introduced, using a "pro-NUN-see-AY-shun" format that is easy for students to interpret.

Learning Supplements for Students

There is much more to *Anatomy & Physiology* than this book. Numerous study aids for the student round out the learning package. The first two of the following items are packaged with every new copy of the book. The others can be ordered from the McGraw-Hill Customer Service Department at 800-338-3987.

- The **Essential Study Partner** prepared by Susanne Kalup, Westmoreland County Community College, is a two-disc CD-ROM study tool packaged with every new copy of the book. It contains 120 animations and more than 800 learning activities to help your students grasp complex concepts. Throughout this book, you will find a CD-ROM icon beside topics that are animated on the ESP, and at the beginning of each chapter we provide a list of related topics that can be found on the ESP. Interactive diagrams and quizzes will make learning stimulating and fun for your students. When a question is answered incorrectly, the CD-ROM provides a link to a tutorial section for further review.
- The **Clinical Applications Manual** is a collaborative work by Ken Saladin & Colleen Nolan of St. Mary's University. It follows *Anatomy & Physiology* chapter by chapter, expanding on its clinical themes, introducing new clinical topics of its own, and providing test questions and case studies to develop the student's ability to apply his or her knowledge to realistic situations.
- The **Online Learning Center** at www.mhhe.com/saladin offers an extensive array of learning and teaching tools. The site includes quizzes for each chapter, links to websites related to each chapter, supplemental reading lists, clinical applications, interactive activities, art labeling exercises, and case studies. Students can click on a diagram of the human body and get case studies related to the regions they select. Instructor resources at the site include lecture outlines, supplemental reading lists, technology resources, clinical applications, and case studies.

The *Student Study Guide* that I wrote for the first edition has been updated and adapted by Robert Beck, University of Illinois–Chicago, for web access at the Online Learning Center. Students can also download a copy from this site. The Study Guide includes an extensive discussion of academic survival skills and study tips for A&P students, along with review exercises and practice tests for each chapter of the book. I feel that testbanks are more effective when written by the author of the main textbook, and so I wrote all test questions for both the Study Guide and the Instructor's Manual.

- The **Anatomy and Physiology Laboratory Manual** by Eric Wise, Santa Barbara City College, correlates human and cat anatomy using the same line art as this book and supplementing it with photographs of the dissected cat. It is accompanied by a separate Instructor's Manual that includes a guide for grading laboratory reports. (0-07-038896-2)
- The **Anatomy and Physiology Laboratory Manual—Fetal Pig** by Terry R. Martin, Kishwaukee College, provides excellent full-color photos of the dissected fetal pig with corresponding labeled art. It includes World Wide Web activities for many chapters. (0-07-231199-1)
- The **Web-Based Cat Dissection Review for Human Anatomy and Physiology** by John Waters, Pennsylvania State University, is an online multimedia program containing vivid, high-quality labeled cat dissection photographs. The program helps students easily identify and review the corresponding structures and functions between the cat and the human body. (0-07-232157-1)
- **Dynamic Human Version 2.0** is a set of two interactive CD-ROMs that cover each body system and demonstrate clinical concepts, histology, and physiology with animated three-dimensional and other images. (0-697-38935-9)
- The **Interactive Histology CD-ROM** by Bruce Wingerd and Paul Paolini, San Diego State University, includes 135 full-color, high-resolution LM images and 35 SEM images of selected tissue sections typically studied in A&P. Each image has labels that can be clicked on or off, has full explanatory legends, offers views at two magnifications, and has links to study questions. The CD also has a glossary with pronunciation guides. (0-07-237308-3)
- An **Atlas to Human Anatomy** by Dennis Strete, McLennan Community College and Christopher H. Creek takes a systems approach with references to regional anatomy, thereby making it a great complement to your regular course structure, as well as to your laboratory. (0-697-38793-3)

- The **Atlas of the Skeletal Muscles,** third edition by Robert and Judith Stone, Suffolk County Community College is a guide to the structure and function of human skeletal muscles. The illustrations help students locate muscles and understand their actions. (0-07-290332-5)
- **Human Anatomy and Physiology Study Cards** by Kent Van De Graaff, R. Ward Rhees, and Christopher Creek is a set of 300 3″ x 5″ cards with terminology, pronunciation guides, word origins, diagrams, and concise descriptions of anatomical and physiological concepts. (0-07-290818-1)
- The **Coloring Guide to Anatomy and Physiology** by Robert and Judith Stone, Suffolk County Community College emphasizes learning through the process of color association. The *Coloring Guide* provides a thorough review of anatomical and physiological concepts. (0-697-17109-4)
- **E-Mail Support from the Author.** Many students took advantage of my offer in the first edition to e-mail me for personal discussion, clarification of physiological concepts, and suggestions for future editions. I welcome students and instructors to continue contacting me personally at ksaladin@mail.gcsu.edu.

Teaching Supplements for Instructors

The *Anatomy & Physiology* teaching package includes the following supplements for instructors, available from your McGraw-Hill sales representative or by calling 800-338-3987.

- We provide 950 **Overhead Transparencies,** including fully labeled copies of every item of line art in the book, unlabeled duplicates of many of them for testing purposes or custom labeling, and some of the tables. (0-07-290791-6)
- The **Instructor's Manual and Test Item File** by David Evans, Pennsylvania College of Technology includes supplemental topics and demonstration ideas for your lectures, suggested readings, critical thinking questions, teaching strategies, and lists of media sources and supply houses. For each chapter, I wrote a testbank of 75 questions in 4 formats. Although few textbook authors write their own testbanks, I felt that a test bank written by the author could better reflect the textbook than one contracted out to an independent writer. (0-07-290788-6)
- **Microtest** is a computerized test generator free upon request to qualified adopters of this book. The test generator contains the complete test item file on computer disk for Windows and Macintosh formats. (Windows 0-07-290790-8; Macintosh 0-07-290789-4)
- **Course Solutions** is an integrated answer to your specific teaching needs. A content expert in A&P will correlate your syllabus with this book and our supplemental media. You will be provided with a printed *Course Integration Guide* written by Mark Taylor, Baylor University, that enables you to easily locate the book pages and supplemental materials tied to the order of presentation in your course. Course Solutions also offers an online Student Tutorial Service and a Consultant Service for Instructors at www.mhhe.com/saladin. (0-07-235892-0)
- **PageOut** is McGraw-Hill's exclusive tool for creating your own website for your A & P course. It requires no knowledge of coding. Simply type your course information into the templates provided. PageOut is hosted by McGraw-Hill.
- The **Visual Resource Library** is a CD-ROM that contains labeled and unlabeled versions of all line art in the book. You can quickly preview images and incorporate them into PowerPoint or other presentation programs to create your own multimedia presentations. You can also remove and replace labels to suit your own preferences in terminology or level of detail. A prepared PowerPoint lecture outline by Sharon Simpson, Broward Community College, is available on the CD-ROM. (0-07-290792-4)
- The **Life Science Animation CD-ROM** contains 125 animations of major biological concepts and processes such as the sliding filament mechanism, active transport, genetic transcription and translation, and other topics that may be difficult for students to visualize. (0-07-234296-X)
- The **Life Science Animations 3D Videotape** contains 42 key biological processes that are narrated and animated in vibrant full color with dynamic three-dimensional graphics. (0-07-290652-9)
- The **Life Science Animations (LSA)** videotape series contains 53 animations on five VHS videocassettes; Chemistry, The Cell, and Energetics (0-697-25068-7); Cell Division, Heredity, Genetics, Reproduction, and Development (0-697-25069-5); Animal Biology No. 1 (0-697-25070-9); Animal Biology No. 2 (0-697-25071-7); and Plant Biology, Evolution, and Ecology (0-697-26600-1). Another available videotape is Physiological Concepts of Life Science (0-697-21512-1).

Essential Study Partner CD-ROM

A free study partner that engages, investigates, and reinforces what you are learning in your book. You'll find the **Essential Study Partner** for *Anatomy & Physiology: The Unity of Form and Function* to be a complete, interactive student study tool packed with hundreds of animations and learning activities. From quizzes to interactive diagrams, you'll find that there has never been a better study partner to ensure that you learn core concepts. Best of all, it's **FREE** with your new book purchase.

The topic menu contains an interactive list of the available topics. Clicking on any of the listings within this menu opens your selection and will show the specific concepts presented within this topic. Clicking on any of the concepts moves you to your selection. You can use the UP and DOWN arrow keys to move through the topics.

The unit pop-up menu is accessible at any time within the program. Clicking on the current unit brings up a menu of other units available in the program.

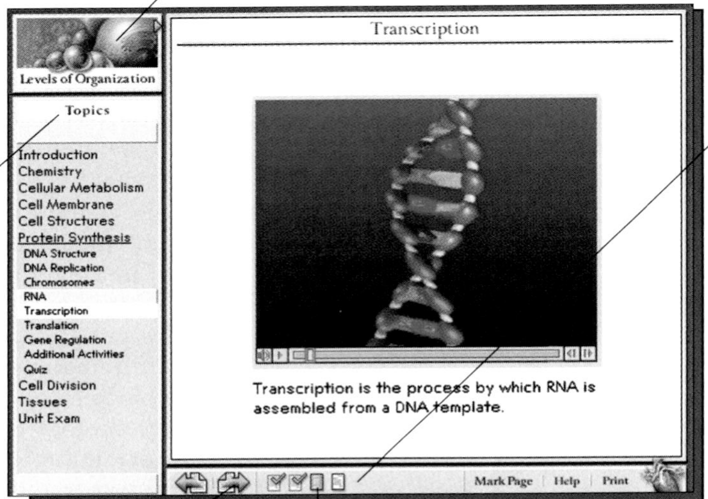

To the right of the arrows is a row of icons that represent the number of screens in a concept. There are three different icons, each representing different functions that a screen in that section will serve. The screen that is currently displayed will highlight yellow and visited ones will be checked.

The film icon represents an animation screen.

Along the bottom of the screen are various navigational aids. At the left are arrows that allow you to page forward and backward through text screens or interactive exercise screens. You can also use the LEFT and RIGHT arrows on your keyboard to perform the same function.

The activity icon represents an interactive learning activity.

The page icon represents a page of informational text.

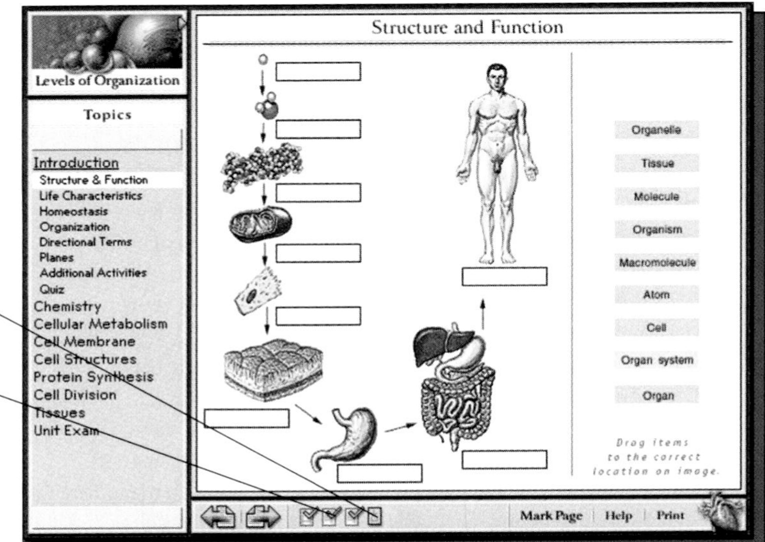

Contact your McGraw-Hill sales representative for more information or visit *www.mhhe.com*.

Visual Resource Library CD-ROMs

These CD-ROMs are electronic libraries of educational presentation resources that instructors can use to enhance their lectures. View, sort, search, and print catalog images, play chapter-specific slideshows using PowerPoint, or create customized presentations when you:

- Find and sort thumbnail image records by name, type, location, and user-defined keywords
- Search using keywords or terms
- View images at the same time with the Small Gallery View
- Select and view images at full size
- Display all the important file information for easy file identification
- Drag and place or copy and paste into virtually any graphics, desktop publishing, presentation, or multimedia application

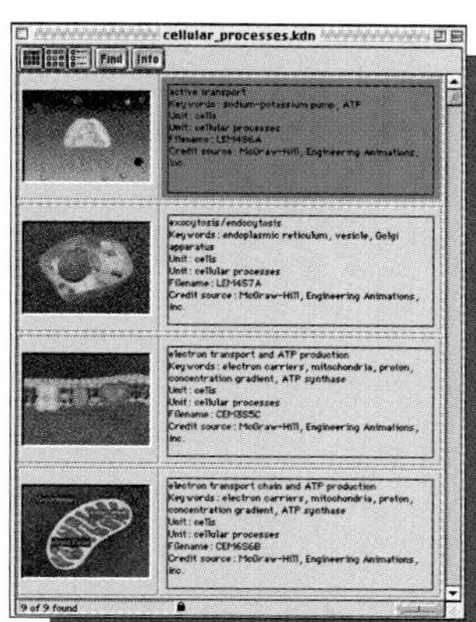

Life Science Animations Visual Resource Library CD-ROM

This instructor's tool, containing more than 125 animations of important biological concepts and processes—found in the *Essential Study Partner* and *Dynamic Human CD-ROMs*—is perfect to support your lecture. The animations contained in this library are not limited to subjects covered in the text, but include an expansion of general life science topics.

Anatomy & Physiology: The Unity of Form and Function Visual Resource Library CD-ROM

This helpful CD-ROM contains ALL illustrations from the text. You'll be able to create interesting multimedia presentations with the use of these images, and students will have the ability to easily access the same images in their texts to later review the content covered in class.

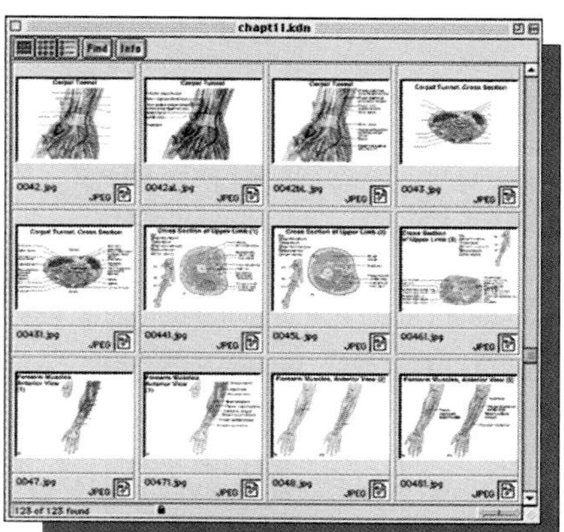

Contact your McGraw-Hill sales representative for more information or visit *www.mhhe.com*.

The Online Learning Center

Your Password to Success

www.mhhe.com/saladin

This text-specific website allows students and instructors from all over the world to communicate. Instructors can create a more interactive course with the integration of this site, and students will find tools that help them improve their grades and learn that biology can be fun.

Student Resources

Study questions
Quizzing with immediate feedback
Links to chapter-related websites
Case studies
Interactive art labeling exercises
Critical thinking exercises
Additional Readings

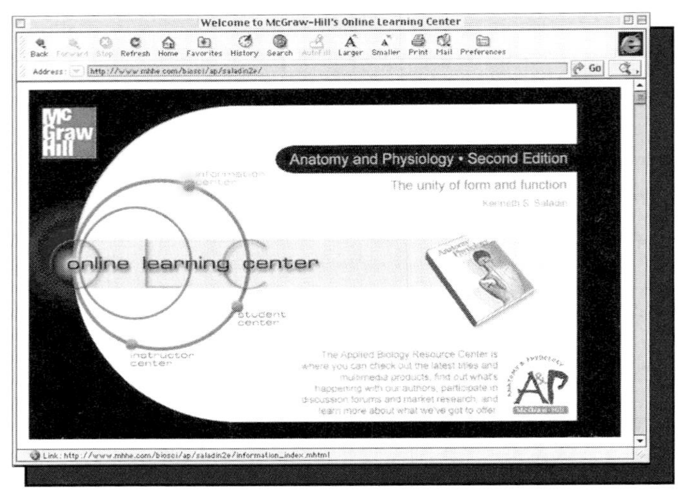

Instructor Resources

PowerPoint Lecture Outlines
Instructor's Manual
Activities that can be assigned as coursework
Links to related websites to expand on particular topics
Classroom activities
Lecture outlines
Case studies
Message Board

Imagine the advantages of having so many learning and teaching tools all in one place—all at your fingertips—FREE.

Contact your McGraw-Hill sales representative for more information or visit *www.mhhe.com*.

Course Solutions

Simplify your life.

Course Solutions 2000 is the answer to your teaching needs. With a full range of multimedia products and special services married to our market-respected, time-tested textbooks, the Course Solutions 2000 program is designed to make your life easier. Everything you need for effective, interactive teaching is at your fingertips.

BENEFITS TO USING COURSE SOLUTIONS:

- **McGraw-Hill Learning Architecture** Each *McGraw-Hill Online Learning Center* is ready to be imported into our *McGraw-Hill Learning Architecture*—a full course management software system for *Local Area Networks* and *Distance Learning Classes*. Developed in conjunction with Top Class software, *McGraw-Hill Learning Architecture* is a powerful course management system available upon special request.
- **Course Consultant Service** In addition to the *Course Integration Guide*, instructors using Course Solutions textbooks can access a special curriculum-based Course Consultant Service via a web-based threaded discussion list within each *Online Learning Center*.
- **Instructor Syllabus Service** For *new* adopters of Course Solutions textbooks, McGraw-Hill will help correlate all text, ancillary, and appropriate materials and services to your course syllabus. Simply call your McGraw-Hill representative for assistance.
- **PageOut** Use this software to create your own course website.
- **Other Delivery Options** *Online Learning Centers* are also compatible with a number of full-service online course delivery systems or outside educational service providers.
- **Student Tutorial Service** Within each *Online Learning Center* resides a free student tutorial service.
- **e-BOOKS** Each Course Solutions title will feature its own e-book—an onscreen version of the actual textbook with numerous in-text links to other valuable digital resources. E-books—which are extremely valuable and very affordable—are available in one of three formats, depending on the textbook:
    - Interactive e-Source CD-ROM
    - e-Text CD-ROM
    - Online Learning Center
- **Web CT Linkage** Specially prepared *Online Learning Centers*, available with Course Solutions textbooks, load easily into any *Web CT Course Management* system.
- **Improved New Media Integration Guide** Each Course Solutions title features a valuable Integration Guide for instructors, significantly improved over past *Integration Guides*. Each Guide indicates where and how to use available media resources with the text. Each now includes detailed descriptions of the content of each relevant New Media module or exercise.
- **Online Animations** Selected Course Solutions titles now include high-quality animations online in their *Online Learning Centers*.

Contact your local McGraw-Hill sales representative for more information or visit *www.mhhe.com*.

Acknowledgments

Until taking on a project of this magnitude, one can scarcely imagine the scale of the enterprise and the number and diverse talents of the people involved. Textbook writing is no simple relationship between an author, an editor, and a publisher. To be competitive in today's market requires a well-coordinated team effort to bring all the pieces together, as well as a staggering investment in product development.

I am deeply indebted to those at McGraw-Hill who have shown such faith in this book by marshaling the necessary resources to ensure its success—particularly to my Publishers Colin Wheatley and Michael Lange, and my Sponsoring Editor Kristine Tibbetts.

Connie Balius-Haakinson, my developmental editor, deserves praise for her keen attention to detail and undaunted dedication to accuracy and quality. If not for e-mail, our correspondence would have consumed a sizeable forest. A project of this size can be overwhelming at times, and I owe a great deal to Connie for her unfailing faith and encouragement and her uncompromising commitment to quality. I know from my correspondence with other authors that a good developmental editor can be a rare treasure, and throughout this project, I have felt very grateful to have Connie on my team.

So many people work on so many aspects of a project of such complexity that it is no small task to keep everyone coordinated, monitor every phase of the project, and keep it moving toward a timely release. For their admirable and indispensible efficiency in this, I thank Kimberly Meriwether David, director of editorial, design, and production; Mary Powers, project manager; Sandy Hahn, production supervisor; Pat Hesse, developmental editor; and Darlene Schueller, our archangel of communication and travel.

Eagle-eyed proofreaders and copyeditors are essential to catch and eliminate errors and inconsistencies before they get into print. Proofreader Sue Dillon and copyeditor Laura Beaudoin kept a sharp eye out for everything from errant punctuation marks to consistency of synonyms within the text and between the text and art, not only sparing me from embarrassing errors but making me a better writer in the process.

Visual appeal and clarity are indispensible to the survival of any A&P textbook, and I am immeasurably indebted to the people who produced and acquired illustrations for this book. The line art was drawn by the artists of Imagineering Scientific and Technical Artworks, Inc., of Toronto. I am especially grateful to Jack Haley for production management and Dustin Holmes for the cover art. Illustrations carried over from the first edition were produced by the Observatory Group, Inc., in Cincinnati. Reviewers asked for a brighter color and design in this edition, and it was designer Wayne Harms who so effectively gave the book a fresh design while preserving the features that made the first edition so successful.

For the photographs in this edition, my thanks go to Connie Mueller, photo researcher; and John Leland, photo research coordinator. For the elegant cadaver photographs new to this edition, I thank anatomists Don Kincaid and Rebecca Gray of the Ohio State University Department of Anatomy and Medical Education Morgue, who dissected specimens to my specifications and produced the photographs for this book.

It takes more than one content expert to produce a successful A&P textbook. Donna Van Wynsberghe performed a valuable service in reviewing the entire book for clinical accuracy, while going beyond that call of duty to also make many other good suggestions on presentation and style.

This book is but the flagship of a whole fleet of educational aids developed by multiple authors. For developing such an attractive package of supplements, I am thankful to David Evans *(Instructor's Manual),* Eric Wise *(Laboratory Manual),* Susanne Kalup *(The Essential Study Partner CD-ROM),* and Colleen Nolan *(Clinical Applications Manual).*

It's one thing to produce a good book and quite another to get the world to sit up and take notice. I have already mentioned the enormous undertaking by the esteemed textbook consulting firm of Burrston House to research the needs of A&P instructors and students, collect the feedback that became the basis for my final-draft manuscript, and increase the book's visibility among my colleagues. A toast to Glenn and Meg Turner, especially, for enthusiastically investing so much of themselves in the success of this edition. My thanks also to my Marketing Co-managers Heather Wagner and Michelle Watnick for tirelessly promoting the book from coast to coast and extolling its virtues to our sales force and prospective users.

This book also gained a great deal from informal dialog and feedback from students and colleagues. Many of the descriptions, explanations, and thought exercises in this edition stem from discussions with my students and e-mail exchanges with students on other campuses. Also of enormous value was the lively exchange of ideas among A&P instructors on HAPP-L, the e-mail list of the Human Anatomy and Physiology Society, conceived and maintained by Jim Pendley. Dr. Pendley has performed an inestimable service to us all and, perhaps without realizing it, contributed significantly to the quality of this book. I thank all of the colleagues on HAPP-L who have made the list such a lively, stimulating, informative venue for discussion and lent their expertise to fill in the gaps in my own.

Anyone who has written a book of this size and scope knows the importance of support at home. I could hardly have found the energy and perseverence to do this without the love and encouragement of my wife, Diane, and our children, Emory and Nicole. With our son on the threshold of his career, our daughter embarking on college as this edition is released, and Diane dreaming of the adventures we can now share, I hope and trust that this book will turn their dreams into reality. That will be the greatest reward of all.

Kenneth S. Saladin
Georgia College & State University
Milledgeville, Georgia
ksaladin@mail.gcsu.edu
www.mhhe.com/saladin/

Reviewers

No words could ever fully express my indebtedness and gratitude to the hundreds of first- and second-edition reviewers who provided such a wealth of scientific information, suggestions for effective presentation, corrections, and encouragement throughout this project. I have learned a great deal from these colleagues, not only about human structure and function itself, but also about how to express this knowledge in a way that will best inform and engage a broad range of students. For making the book beautiful, I am thankful to the team named earlier. For making it *right,* I am thankful to the colleagues listed below.

Reviewers

Mohammed Abbas
Schoolcraft College

Janice Asel
Mitchell Community College

E. B. Ashbrook
Patrick Henry Community College

Timothy A. Ballard
University of North Carolina–Wilmington

Debra J. Barnes
Contra Costa College

Steven G. Bassett
Seton Hill College

Stephen T. Bishoff
University of South Carolina–Sumter

Leonard Borack
Rutgers–The State University of New Jersey

Denise Bowen
Aurora College

Jimmy L. Boyd
Hinds Community College–Vicksburg/Warren County Branch

Gary Brady
Spokane Falls Community College

James Bridger
Prince George's Community College

James M. Britton
Arkansas State University–Beebe

Richard D. Brown
Brunswick Community College

Hoyt Burnette
McLennan Community College

Jennifer Carr Burtwistle
Northeast Community College

John R. Capeheart
University of Houston–Downtown

Miriam Chávez
University of New Mexico–Valencia Campus

Sai Chidambaram
Canisius College

William H. Chrouser
Warner Southern College

Homer L. Cissell
Lincoln College at Normal

Redding I. Corbett III
Midlands Technical College

Juville Dario-Becker
Central Virginia Community College

Donna O. Daugherty
Floyd College

John W. Davis
Benedictine College

Judith K. Davis
Florida Community College–Jacksonville

Mary E. Dawson
Kingsborough Community College

Nicholas G. Despo
Thiel College

Gurmeet K. Dhaliwal
Wilbur Wright College

Charles J. Dick
Pasco-Hernando Community College

John G. Dziak
Community College of Allegheny County, Allegheny Campus

Eugene J. Fenster
Longview Community College

Margaret D. Folsom
Methodist College

Leslie Foster
Sault College

Pamela B. Fouché
Walters State Community College

Marc Franco
South Seattle Community College

Dalia Giedrimiene
Saint Joseph College

Mary Anne S. Gummere
Lyndon State College

Margery K. Herrington
Adams State College

William Hightower
Southside Virginia Community College

Leland N. Holland, Jr.
Pasco-Hernando Community College

Timothy T. Horan
Xavier University

S. Beth Howard
Rutgers University

Randall G. Howell
Southern Union State Community College

James Junker
Campbell University School of Pharmacy

June A. Kinsley
Harcum College

Beverly P. Kirk
Northeast Mississippi Community College

J. Krontiris-Litowitz
Youngstown State University

Stephen G. Lebsack
Linn-Benton Community College

Florence Levin
Edison Community College

Jerri K. Lindsey
Tarrant County College–NE Campus

Mary Katherine Lockwood
University of New Hampshire

D.J. Lowrie, Jr.
University of Cincinnati College of Medicine/Xavier University

Jennifer Lundmark
California State University–Sacramento

Bradford D. Martin
La Sierra University

Grace D. Matzen
Molloy College

Harry S. McDonald
Stephen F. Austin State University

Michael L. McMahan
Union University

SuEarl McReynolds
San Jacinto College–Central Campus

Julie A. Medlin
Northwestern Michigan College

Sharon Michael
Pierce College

Nathaniel Mills
Texas Woman's University

Robert Moldenhauer
St. Clair County Community College

David P.S. Mork
St. Cloud State University

C. Aubrey Morris
Pensacola Junior College

Richard A. Mostardi
University of Akron

Alfredo Muñoz
University of Texas at Brownsville and Texas Southmost College

Margaret Nordlie
University of Mary

Weston Opitz
Kansas Wesleyan University

Charles M. Page
El Camino College

Betsy Peitz
California State University–Los Angeles

Stephen W. Perry
Liberty University

Jon Pigage
University of Colorado–Colorado Springs

Dan Porter
Amarillo College

Laura H. Ritt
Burlington County College

Delia Roberts
Selkirk College

Donald W. Rodd
University of Evansville

Geraldine Y. Ross
Highline Community College

Thomas J. Russell
Delgado Community College

Laura Jones Rutter
Lindenwood University

Jessica Sand
Union County College

Thomas J. Sarro
Mount Saint Mary College

Miriam Satern
Western Illinois University

Joseph R. Schiller
Austin Peay State University

Fred H. Schindler
Indian Hills Community College

Marilyn M. Shannon
Indiana University–Purdue University, Fort Wayne

Marilyn L. Shaver
University of Kentucky/Ashland Community College

Carl J. Shuster
Amarillo College

Michael E. Smith
Valdosta State University

Dianne Snyder
Augusta State University

Keith Snyder
Southern Adventist University

William G. Sproat, Jr.
Northern Virginia Community College–Manassas

Laura Steele
Ivy Tech State College

Dennis M. Sullivan
Cedarville College

Mark Taylor
Baylor University

Michael Tenneson
Evangel University

Kenneth Thomas
Hillsborough Community College

Kent R. Thomas
Wichita State University

Diane G. Tice
SUNY–Morrisville College of Agriculture & Technology

Mary Truex
Indiana University South Bend

David E. Turnbull
Lake Land College

Mark X. Van Cura
Brunswick Community College/ Cape Fear Community College

Michael A. Vitale
Daytona Beach Community College

Phil Walcott
Grand Valley State University

William G. Walther
Lake Erie College

Terry P. Wheeler
Halifax Community College

Callie A. Vanderbilt White
San Juan College

Philip Clason Whitford
Capital University

Peter J. Wilkin
Purdue University North Central

Esther K. Wilson
University of Wisconsin–Parkside

Linda A. Winkler
University of Pittsburgh

Hamilton Wise
Hinds Community College–Rankin Campus

David M. Wolfrom
Paducah Community College

Ferne Zabezensky
Chandler-Gilbert Community College

Jeffrey A. Zuiderveen
Columbus State University

Allen N. Hunt
Elizabethtown Community College

Warren R. Jones
Loyola University Chicago

Susanne J. Kalup
Westmoreland County Community College

Mary Katherine Lockwood
University of New Hampshire

Jane R. Marone
University of Illinois–Chicago

Harry S. McDonald
Stephen F. Austin State University

Ben Meyer
Grand Rapids Community College

Melissa Ann Mills
Anoka-Ramsey Community College

Helena T. Mobley
Kilgore College

James O'Leary
South Suburban College

Justicia Opoku
University of Maryland–College Park

Charles M. Page
El Camino College

Beverly J. Perry
Houston Community College

Robert Pope
Miami-Dade Community College

Gregory K. Reeder
Broward Community College-DTC

Jackie Reynolds
Richland College

Mark A. Shoop
Tennessee Wesleyan College

Sharon Simpson
Broward Community College

Charles F. Sternburg
Riverside Community College

Richard A. Symmons
California State University–Hayward

Patricia J. Turner
Howard Community College

Althea Tyndall
Prince George's Community College

Mark Weber
Grove City College

Reviewers of First-Draft Manuscript

Debra Barnes
Contra Costa College

Robert J. Beck
University of Illinois–Chicago

Sonya Conway
Northern Illinois University

Diana L. Curley
Community College of Baltimore County–Catonsville Campus

Mary E. Dawson
Kingsborough Community College

Teresa DiFiori
Pasadena City College

M. Gore Ervin
Middle Tennessee State University

David L. Evans
Pennsylvania College of Technology/Pennsylvania State University

Susan Finazzo
Broward Community College

Michael D. Gottlieb
LaGuardia Community College

Survey Participants

Constance L. Allen
Edison Community College

Joe Alls
Northwest-Shoals Community College

Temma al-Mukhtar
San Diego Mesa College

L. Amini-Kormi
Worcester State College

Mark Aronson
Scott Community College

Timothy A. Ballard
University of North Carolina–Wilmington

Maryam Bamshad
CUNY–Lehman College

Bruce D. Barnhart
California University of Pennsylvania

Anne Marie Basso
St. John's University, School of Pharmacy & Allied Health

Regina Betette
Midlands Technical College

Wendy D. Bircher
San Juan College

Richard M. Blaney
Brevard Community College

Franklyn F. Bolander, Jr.
University of South Carolina

Matthew M. Brady
Luzerne County Community College

Ann Brown
CUNY–Medgar Evers College

Alease S. Bruce
University of Massachusetts–Lowell

Edmund Burke
University of Nevada School of Medicine

Jamie D. Campbell
Truckee Meadows Community College

Leslie R. Carlson
Iowa State University

Frances M. Cashner
University of New Orleans

William M. Clark
Kingwood College

John Patrick Click
Indiana University Southeast

William H. Cliff
Niagara University

Shirley A. Colvin
Gadsden State Community College

Sam M. Coppoletti
Southwest Baptist University

Redding I. Corbett III
Midlands Technical College

Dave Dapkus
Winona State University

Brent DeMars
Lakeland Community College

Anne G. Denner
University of Southern Indiana

James L. Dolan
Lock Haven University

Charles Duggins, Jr.
University of South Carolina

William E. Dunscombe
Union County College

David M. Enock
Community College of Allegheny County

M. Gore Ervin
Middle Tennessee State University

David L. Evans
Pennsylvania College of Technology/Pennsylvania State University

Henriette K. Evans
Pennsylvania College of Technology/Pennsylvania State University

Ann M. Findley
Northeast Louisiana University

Esther M. Fleischmann
University of Maryland Baltimore County

Helen M. Foster
Santa Fe Community College

Kim T. Fredricks
Viterbo College

Purti P. Gadkari
Wharton County Junior College

Louis A. Giacinti
Milwaukee Area Technical College

John S. Green
Texas A&M University

Misty D. Gregg
Northeast State Community College

Jennifer Grunzweig
University of Delaware

Deke Gundersen
University of Southern Indiana

R. Christopher Harvey
Brevard Community College–Palm Bay Campus

Maurice Heller
El Paso Community College

Kristine Herbert
Westark College

Bob Herrington
Georgia Southwestern State University

Donald L. Hilscher
Lincoln College

Scott H. Huber
Truckee Meadows Community College

Yvette Ingram
Lock Haven University

Mary M. Januschka
Anoka-Ramsey Community College

Amy E. Jetton
Middle Tennessee State University

Carol J. Kafer
Pennsylvania College of Technology

Susanne J. Kalup
Westmoreland County Community College

D. T. Kidwell
Southeast Community College

Christine Kisiel
Mt. Wachusett Community College

Tom Knoedler
The Ohio State University–Lima

Karin E. Krieger
Iowa State University

Marian G. Langer
St. Francis College

Philip S. LaPolt
California State University–Los Angeles

Arthur Lavigne
Middlesex Community College

Carolyn J. Lebsack
Linn-Benton Community College

Steve Mackie
Pima Community College

Katherine Manley-Buser
Palmer College of Chiropractic

Gary M. Mawe
University of Vermont

Francis M. Maxin
Community College of Allegheny County

Karen McCort
El Paso Community College

Harry S. McDonald
Stephen F. Austin State University

Iain McGaw
University of Nevada–Las Vegas

Gail O. McKenzie
Samford University

Wilma Jo McNamara
Milwaukee Area Technical College

Melissa Meador
Arkansas State University–Beebe Branch

John A. Mecham
Catawba College

Anne M. Miller
Middlesex Community College

Steven Miller
Rochester Community and Technical College

Alice Mills
Middle Tennessee State University

Melissa Ann Mills
Anoka-Ramsey Community College

Javanika Mody
Anne Arundel Community College

Robert Moldenhauer
St. Clair County Community College

Paul F. Nicoletto
Lamar University

Margaret Nordlie
University of Mary

Francis Nussbaum
Kent State University–Tuscarawas

Saeb Omar
Southeast Community College

Mary Theresa Ortiz
CUNY–Kingsborough Community College

Amy G. Ouchley
University of Louisiana at Monroe

Brian K. Paulson
California University of Pennsylvania

Steven J. Person
Lake Superior State University

Polly K. Phillips
Miami-Dade Community College

Michael Porter
Lock Haven University

Frances Ragsdale
Winona State University

Sharon Raines
Northwest-Shoals Community College

Patricia Pierce Ramsey
Norfolk State University

Kim Raun
Wharton County Junior College

Laura H. Ritt
Burlington County College

Joanne Roehrs
Owens Community College

Mattie Roig
Broward Community College

Allan P. Russell
Mt. Wachusett Community College

Eugene P. Rutheny
Westchester Community College

Soma Sanyal
Pennsylvania State University–Altoona

Douglas P. Schelhaas
University of Mary

Norman A. Scherzer
Essex County College

Melvin Schmidt
McNeese State University

Timothy P. Scott
Texas A&M University

Harold E. Secor
Wharton County Junior College

Aida Shehata
Prince George's Community College

Janet Anne Sherman
Pennsylvania College of Technology

Stan Smith
Bowling Green State University

Jeff Spencer
The University of Akron

Charles F. Sternburg
Riverside Community College

Rick Stewart
Fresno City College

Hazel Taylor
Delgado Community College

Todd Templeton
Metro Community College

Eric Thomas
University of the Pacific

Ed W. Thompson
Winona State University

Janis G. Thompson
Lorain County Community College

Jane Toft
Rochester Community and Technical College

Kim Marie Tolson
The University of Louisana at Monroe

Kimberly T. Turk
Mitchell Community College

Robert Vick
Elon College

F. R. Voorhees
Central Missouri State University

Douglas Walker
Wharton County Junior College

O. Eugene Walton
Tallahassee Community College

Mary Jane Wheeler
Cuyahoga Community College

Eddie L. Whitson
Gadsden State Community College

Louis Wigginton
St. Clair County Community College

Avery A. Williams
Louisiana State University–Eunice

Mark Wygoda
McNeese State University

Focus Group Participants

James E. Aldridge
Palm Beach Community College

George R. Aliaga
Tarrant County College–NE Campus

Benja Allen
El Centro College

Gail Astle
Rider University

David Bastedo
San Bernardino Valley College

Robert J. Beck
University of Illinois–Chicago

Lillian Blaschke
Fullerton College

Eileen Boltz
Widener University

Spencer R. Bowers
Oakton Community College

Ann Brown
CUNY–Medgar Evers College

Darla H. Culmer
Broward Community College

Dianne R. S. Daniel
Burlington County College

Russell E. DiFiori
Pasadena City College

Teresa DiFiori
Pasadena City College

William E. Dunscombe
Union County College

Natalie Esser
Temple University

Wilton J. Evans
Malcolm X College

Richard L. Faircloth
Anne Arundel Community College

Pamela Favenesi
College of DuPage

Michael S. Glasgow
Anne Arundel Community College

Michael D. Gottlieb
LaGuardia Community College

Brian M. Guzzetti
Rowan University

Terry Hamilton
Broward Community College

Jean Helgeson
Collin County Community College

Timothy L. Henry
University of Texas–Arlington

Doug Hogan
Mount San Jacinto College

James Horwitz
Palm Beach Community College

Dale P. Layman
Joliet Junior College

Claire Lee
Villanova University

Leigh Levitt
Union County College

Jerri K. Lindsey
Tarrant County College–NE Campus

Reolee Yan Lor
Villanova University

Jane R. Marone
University of Illinois–Chicago

Kenneth A. Michaelis
San Bernardino Valley College

Rashied Moody
Temple University

Ebere Nduka
CUNY–Medgar Evers College

William A. Olexik
Montgomery College

Justicia Opoku
University of Maryland–College Park

Susan M. Caley Opsal
Illinois Valley Community College

Andrew Overman
Columbia University

Charles M. Page
El Camino College

Mary Lou Percy
Navarro College

Robert Pope
Miami-Dade Community College

Lee Porch
Richard J. Daley College

Charles B. Pumpuni
Northern Virginia Community College

Ronald Rak
Moraine Valley Community College

Cecil Ramsundar
University of District of Columbia

Jackie Reynolds
Richland College

Robyn V. Reynolds
Temple University

Cecilia F. Romaldini
Temple University

Nidia Stella Romer
Miami-Dade Community College

Jasbir K. Saluja
City Colleges of Chicago–Malcolm X Campus

Henning Schneider
William Paterson University of New Jersey

Kelly J. Sexton
North Lake College

Margaret Rose Silva
Mountain View College

Sharon Simpson
Broward Community College

Marc Skyler
New York City Technical College/CUNY

Jeff M. Slepski
Mt. San Jacinto College

Charles F. Sternburg
Riverside Community College

Patricia J. Turner
Howard Community College

Althea P. Tyndall
Prince George's Community College

Anthony Udeogalanya
CUNY–Medgar Evers College

Joe H. Wheeler III
North Lake College

Clarence C. Wolfe
Northern Virginia Community College

Art Consultants

Michael D. Gottlieb
LaGuardia Community College

Kurt Olson
College of St. Catherine

Brian K. Paulson
California University of Pennsylvania

Jon Pigage
University of Colorado–Colorado Springs

Jackie Reynolds
Richland College

Donald L. Rubbelke
Lakeland Community College

Sharon Simpson
Broward Community College

Marc Skyler
New York City Technical College/CUNY

Jane Toft
Rochester Community and Technical College

Anthony Udeogalanya
CUNY–Medgar Evers College

Joseph P. Yavornitzky
Baldwin-Wallace College

Online Learning Center Contributors

Robert J. Beck
University of Illinois–Chicago

Jane R. Marone
University of Illinois–Chicago

Sharon Simpson
Broward Community College

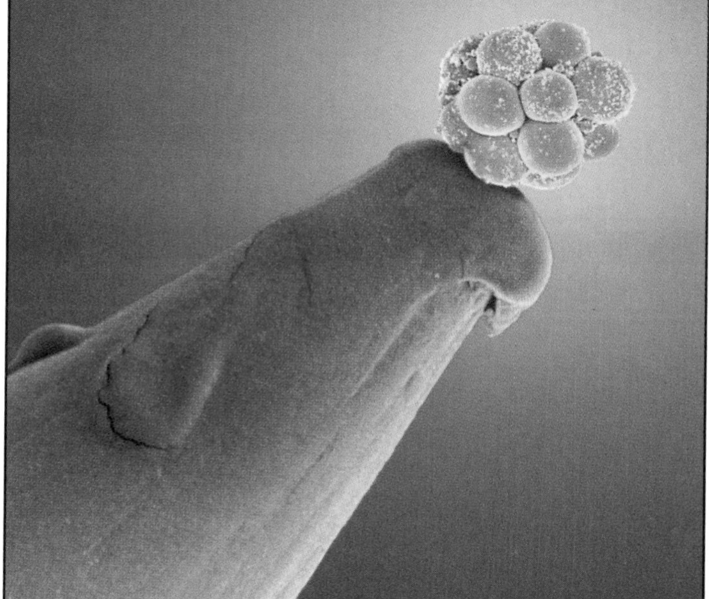

A new life begins—a human embryo on the point of a pin

CHAPTER 1

Major Themes of Anatomy and Physiology

Chapter Outline

Insights

 Study with Your ESP CD-ROM

Refer to your ESP Tutorial CD-ROM for helpful animations, activities, and quizzes related to the material covered in this chapter. The following are suggested areas to go to on the CD and are presented according to "Unit/Topic":
- Levels of Organization/Introduction

No branch of science hits as close to home as the science of our own bodies. We gratefully marvel at the dependability of our hearts; as we watch the Olympic Games, we are awed by the capabilities of muscles and joints; and we ponder with philosophers the ancient mysteries of mind and emotion. We want to know how our body works, and when it functions less than perfectly, we want to know what is happening and what we can do about it. To discuss all this, we've invented a rich vocabulary for our body parts and processes. Even in the most ancient writings of civilization, we find medical documents that attest to humanity's drive to know itself. Thus, you are embarking on a subject that is as old as civilization, yet one that grows by thousands of scientific publications every week.

This book is an introduction to human structure and function. It is meant primarily to give you a foundation for advanced study in health care, exercise physiology, pathology, and other fields related to health and fitness. Beyond that purpose, however, it can also provide you with a satisfying sense of self-understanding.

As rewarding and fascinating as this subject can be, the human body is highly complex and a knowledge of it requires us to comprehend a great deal of detail. The details become more manageable when we can relate them to a few broad, unifying concepts. The aim of this chapter, therefore, is to introduce such concepts and put the rest of the book into perspective. We consider the historical development of anatomy and physiology, the thought processes that led to the knowledge in this book, the meaning of human life, and a central concept of physiology called *homeostasis.* The Preface to Students describes some ways in which this book and its companion materials can be used to learn this subject most effectively. If you have not already read it, I urge you to do so before continuing.

The Scope of Anatomy and Physiology

The human body is the domain of the twin sciences of **anatomy,** the study of structure, and **physiology,** the study of function. These two approaches are complementary and are never entirely separable. When we study a structure, we naturally want to know, What does it do? Anatomy is not only the basis of physiology, but also its effect—it results from cells building tissues and organs. Conversely, anatomy is what makes all physiology possible. This *unity of form and function* is an important point to bear in mind as you study the body. Many examples of it will be apparent throughout the book—some of them pointed out for you, and others that you will notice for yourself.

Anatomy—The Study of Form

The structure of the human body is studied in several ways. The simplest is the observation of surface structure, for example in performing a physical examination or making a clinical diagnosis from surface appearance. But a deeper understanding of the body depends on **dissection**—the careful cutting and separation of tissues to reveal their relationships. Both *anatomy*[1] and *dissection*[2] literally mean "cutting apart"; until recently, dissecting was called "anatomizing." Many insights into human structure are obtained from **comparative anatomy**—the study of more than one species in order to learn generalizations and evolutionary trends. Students of anatomy often begin by dissecting other animals with which we share a common ancestry and many structural similarities. The dissection of a dead human body, or **cadaver,**[3] is an essential part of the training of many health science students today (fig. 1.1).

Figure 1.1 Early Medical Students in the Gross Anatomy Laboratory with Three Cadavers. Students of the health sciences have long begun their professional training by dissecting cadavers.

Dissection, of course, is not the method of choice when studying a living person! Physical examinations involve not only looking at the body for signs of normalcy or disease but also touching and listening to it. **Palpation**[4] is the process of feeling structure with the fingertips, such as palpating a swollen lymph node or taking a pulse. **Auscultation**[5] (AWS-cul-TAY-shun) is listening to the natural sounds made by the body, such as heart and lung sounds. In **percussion,** the examiner taps on the body and listens to the echo for signs of abnormalities such as pockets of fluid or air.

Structure that can be seen with the naked eye, whether by surface observation or dissection, is called **gross anatomy.** Ultimately, though, the functions of the body result from its individual cells. To see those, we usually take tissue specimens, thinly slice and stain them, and

1. *ana* = apart + *tom* = cut
2. *dis* = apart + *sect* = cut
3. *cadere* = to fall or die
4. *palp* = touch, feel
5. *auscult* = listen

observe them under the microscope. This approach is called **microscopic anatomy,** or **histology.**[6] **Histopathology** is the microscopic examination of tissues for signs of disease. **Ultrastructure** refers to fine details, down to the molecular level, revealed by the electron microscope.

Physiology—The Study of Function

Physiology[7] uses the methods of experimental science discussed later. It has many subdisciplines such as *neurophysiology* (physiology of the nervous system), *endocrinology* (physiology of hormones), and *pathophysiology* (mechanisms of disease). Partly because of limitations on experimentation with humans, much of what we know about bodily function has been gained through **comparative physiology,** the study of how different species have solved problems of life such as water balance, respiration, and reproduction. Comparative physiology is also the basis for the development of new drugs and medical procedures. For example, a cardiac surgeon cannot practice on humans without first learning animal surgery, and a vaccine cannot be used on human subjects until it has been demonstrated through animal research that it confers significant benefits without unacceptable risks.

The Origins of Biomedical Science

Objectives

When you have completed this section, you should be able to

- give examples of how modern biomedical science emerged from an era of superstition and authoritarianism; and
- describe the contributions of some key people who helped to bring about this transformation.

Health science has progressed far more in the last 25 years than in the 2,500 years before that, but the field did not spring up overnight. It is built upon centuries of thought and controversy, triumph and defeat. We cannot fully understand its present state without understanding its past—people who had the curiosity to try new things, the vision to look at human form and function in new ways, and the courage to question authority.

The Beginnings of Medicine

Physicians in Mesopotamia and Egypt treated patients with herbal drugs, salts, physical therapy, and faith healing as early as 3,000 years ago. The "father of medicine," however, is usually considered to be the Greek physician **Hippocrates** (c. 460–c. 375 B.C.E.). He and his followers established a code of ethics for physicians, the Hippocratic Oath, that is still recited in modern form by many graduating medical students. Hippocrates urged physicians to stop attributing disease to the activities of gods and demons and to seek their natural causes, which could afford the only rational basis for therapy. **Aristotle** (384–322 B.C.E.) believed that diseases and other natural events could have either supernatural causes, which he called *theologi,* or natural ones, which he called *physici* or *physiologi.* We derive such terms as *physician* and *physiology* from the latter. Until the nineteenth century, physicians were called "doctors of physic." In his anatomy book, *Of the Parts of Animals,* Aristotle tried to identify unifying themes in nature. He argued that complex structures are built from a smaller variety of simple components—a point of view that we will find useful later in this chapter.

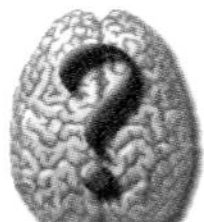

Think About It

When you have completed this chapter, discuss the relevance of Aristotle's philosophy to our current thinking about human structure.

Claudius Galen (129–c. 199), physician to the Roman gladiators, wrote the most noteworthy medical textbook of the ancient era—a book that was worshiped to excess by medical professors for centuries to follow. Aside from what he could learn by treating the gladiators' wounds, Galen was limited to learning anatomy by dissecting pigs, monkeys, and other animals. Cadaver dissection was banned in his time because of some horrid excesses that preceded him, including public dissection of living slaves and prisoners for demonstration and the satisfaction of the anatomist's curiosity. Galen saw science as a method of discovery, not as a body of fact to be taken on faith. He advised his followers to trust their own observations more than they trusted any book and warned that even his own books could be wrong. Unfortunately, his advice was not heeded. For nearly 1,500 years, medical professors dogmatically taught what they read in Aristotle and Galen, and few dared to question the authority of these "ancient masters."

The Birth of Modern Medicine

Medical science advanced very little during the Middle Ages. Even though many of the most famous medical schools of Europe were founded during this era, the professors taught medicine primarily as a dogmatic commentary on Galen and Aristotle, not as a field of original research. Medieval medical illustrations were crude representations of the body that served more to decorate a page than to give any meaningful instruction to the reader (fig. 1.2). Often, they were astrological charts that showed

6. *histo* = tissue + *logy* = study of
7. *physio* = nature + *logy* = study of

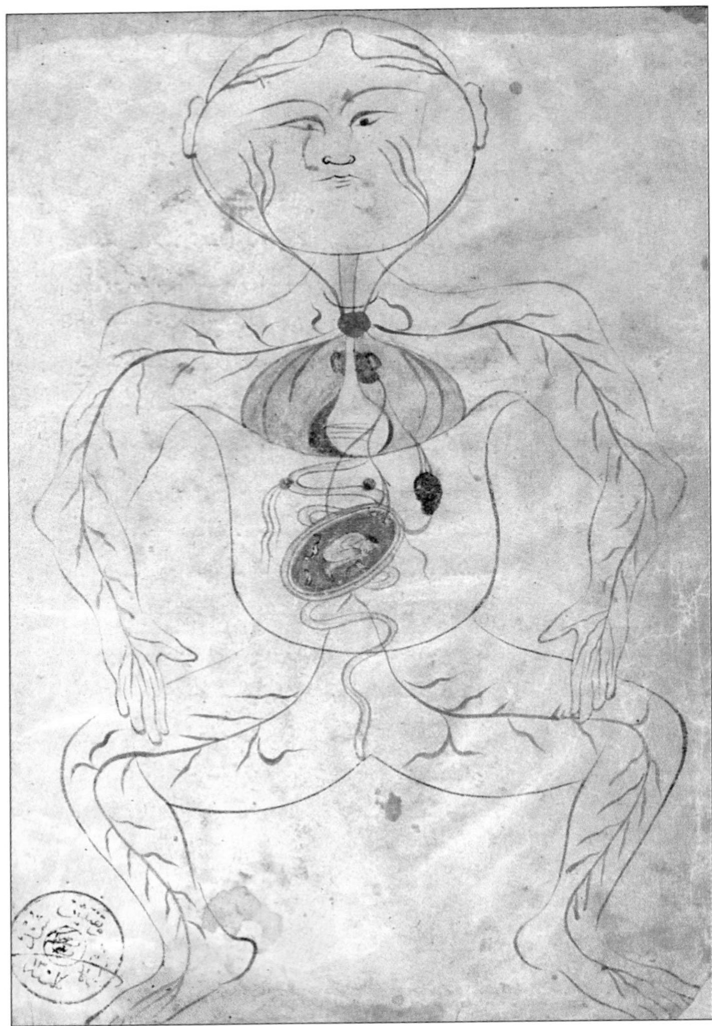

Figure 1.2 **Medieval Medical Illustration.** Anatomical illustrations of the Middle Ages were usually crude, not only because of a sketchy knowledge of anatomy, but also because they were intended more to decorate a page than to depict the body realistically. This figure depicts a pregnant woman with a fetus in the uterus and with the heart, lungs, arteries, and digestive tract.

which sign of the zodiac was thought to influence each organ of the body. From such pseudoscience came the word *influenza,* Italian for *influence.*

Free inquiry was less inhibited in the Muslim world. Ibn Sina (980–1037), known in the West as **Avicenna** or "the Galen of Islam," studied Galen and Aristotle, combined their findings with original discoveries, and questioned authority when the evidence demanded it. Muslim medicine soon became superior to Western medicine, and Avicenna's textbook, *The Canon of Medicine,* became the leading authority in European medical schools until the sixteenth century.

Modern medicine is often said to have begun around the sixteenth century in the innovative minds of such men as the anatomist Andreas Vesalius and the physiologist William Harvey. **Andreas Vesalius** (1514–64) was a Flemish physician who taught anatomy in Italy. In his time, cadaver dissection had resumed for legal purposes (autopsies) and gradually found its way into the training of medical students throughout Europe. Dissection was an unpleasant business, however, and most professors considered it beneath their dignity. In these days before refrigeration or embalming, the odor from the decaying cadaver was unbearable. Dissections were conducted outdoors in a nonstop 4-day race against decay. Bleary medical students had to fight the urge to vomit, afraid that they would incur the wrath of an overbearing professor. Professors typically sat in an elevated chair, the cathedra, reading dryly from Galen or Aristotle while a lower-ranking *barber-surgeon* removed putrefying organs from the cadaver and held them up for the students to see. Barbering and surgery were considered to be "kindred arts of the knife"; today's barber poles date from this era, their red and white stripes symbolizing blood and bandages.

Vesalius broke with tradition by coming down from the cathedra and doing the dissections himself. He was quick to point out that much of the anatomy in Galen's books was wrong, and he was the first to publish accurate illustrations for teaching anatomy (fig. 1.3). When others began to plagiarize his illustrations, Vesalius published the first atlas of anatomy, *De Humani Corporis Fabrica (On the Structure of the Human Body),* in 1543. This book began a rich tradition of medical illustration that has been handed down to you through such milestones as *Gray's Anatomy* (1856) and the vividly illustrated atlases and textbooks of today.

Anatomy preceded physiology and was a necessary foundation for the latter. What Vesalius was to anatomy, the Englishman **William Harvey** (1578–1657) was to physiology. Harvey is remembered especially for a little book he published in 1628, *On the Movement of the Heart and Blood in Animals.* Authorities before him believed that digested food traveled to the liver, turned into blood, and then traveled through the veins to the rest of the body. Harvey measured cardiac output in snakes and other animals, however, and concluded that the amount of food eaten could not possibly account for so much blood. Thus, he inferred that blood must be recycled—pumped out of the heart by way of arteries and returned to the heart by way of veins. Capillaries, the connections between arteries and veins, had not been discovered yet, but Harvey predicted their existence.

Modern medicine also owes an enormous debt to two inventors from this era. **Antony van Leeuwenhoek** (an-TOE-nee vahn LAY-wen-hook) (1632–1723), a Dutch textile merchant, was the first to invent a microscope capable of visualizing single cells. In order to examine the weave

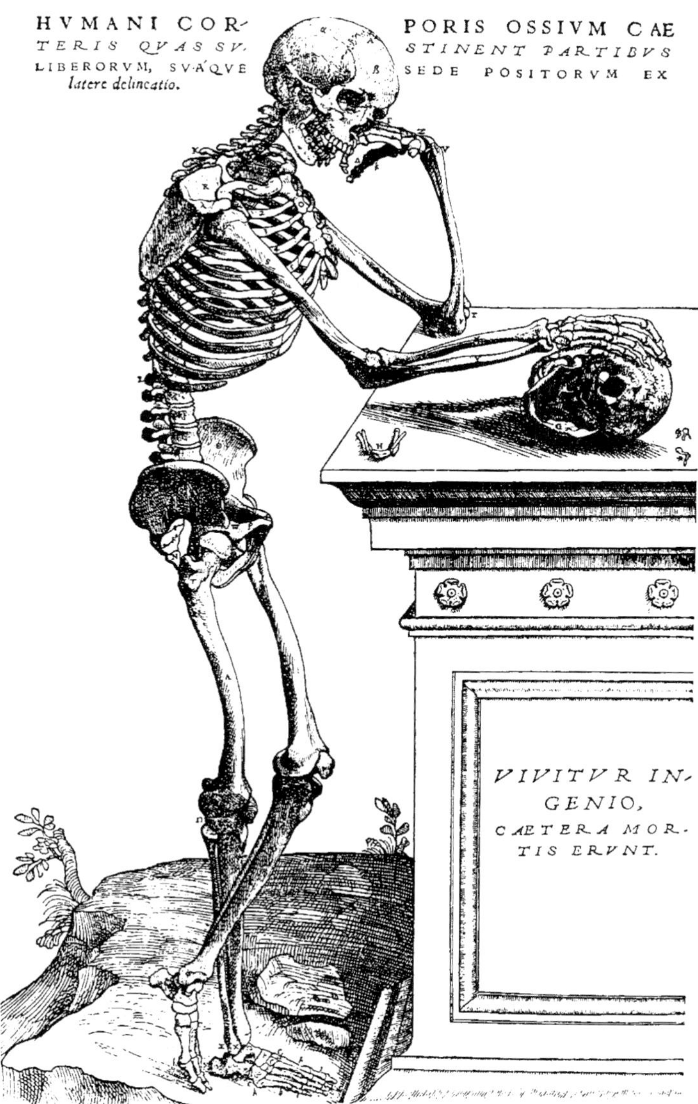

Figure 1.3 **The Art of Vesalius.** Andreas Vesalius revolutionized medical illustration with the comparatively realistic art prepared for his 1543 book, *De Humani Corporis Fabrica*.

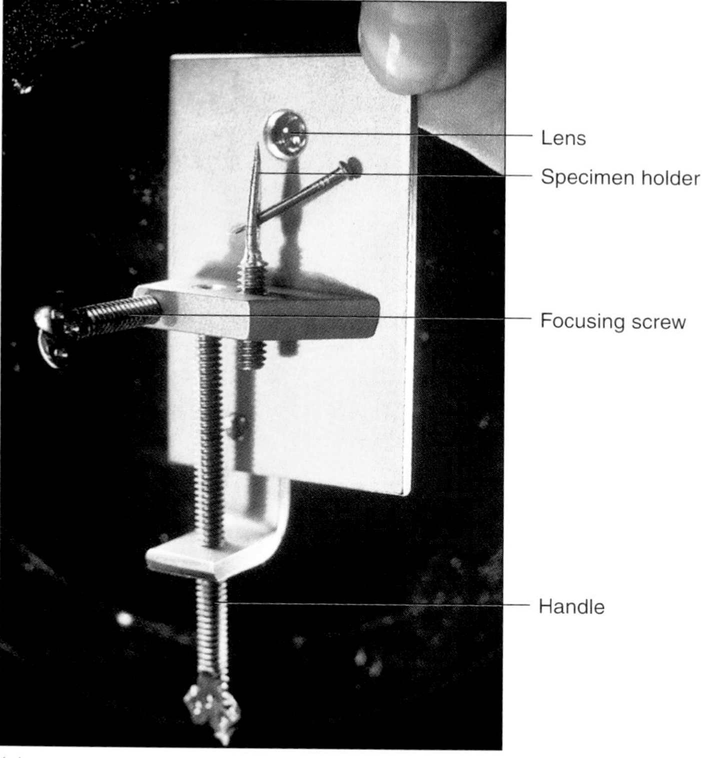

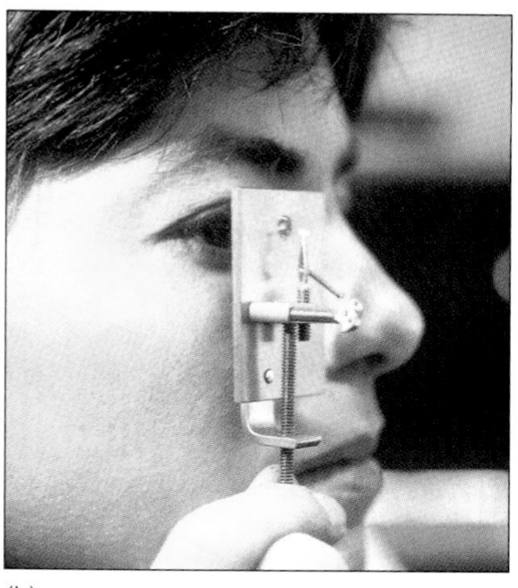

Figure 1.4 **Leeuwenhoek's Simple Microscope.** (*a*) Replica made by Dr. William Walter of Montana State University. (*b*) Viewing a specimen with a Leeuwenhoek microscope.

of fabrics more closely, he ground a beadlike lens and mounted it in a metal plate equipped with a movable specimen clip (fig. 1.4). This **simple** (single-lens) **microscope** magnified objects 200 to 300 times. Out of curiosity, Leeuwenhoek examined a drop of lake water and was astonished to find a variety of microorganisms—"little animalcules," he called them, "very prettily a-swimming." He went on to observe practically everything he could get his hands on, including blood cells, blood capillaries, sperm, and muscular tissue. Probably no one in history had looked at nature in such a revolutionary way. Leeuwenhoek opened the door to an entirely new understanding of human structure and the causes of disease.

He was praised at first, and reports of his observations were eagerly received by scientific societies, but this public enthusiasm did not last. By the end of the seventeenth century, the microscope was treated as a mere toy for the upper classes, as amusing and meaningless as a kaleidoscope. Leeuwenhoek had even become the brunt of satire.

Leeuwenhoek's most faithful admirer was the Englishman **Robert Hooke** (1635–1703), who developed the first practical **compound microscope**—a tube with a lens at each end, with the second lens used to further magnify the image produced by the first (fig. 1.5). Galileo had already made several compound microscopes by 1612, but Hooke invented many of the features found in microscopes used by students today: a stage to hold the specimen, an illuminator, and coarse and fine focus controls. His microscopes produced poor images with blurry edges *(spherical aberration)* and rainbow-colored distortions *(chromatic aberration)*, but poor images were better than none. Although Leeuwenhoek was the first to see cells, Hooke named them. In 1663, he observed thin shavings of cork with his microscope and observed that they "consisted of a great many little boxes," which he called *cells* after the cubicles of a monastery (fig. 1.5*b*). He published these observations in his book, *Micrographia,* in 1665.

In the early 1800s, **Carl Zeiss** (1818–88) of Germany greatly improved the compound microscope to reduce distortions of the image. Biologists then began eagerly examining a wider variety of specimens. Botanist **Matthias Schleiden** (1804–81) and zoologist **Theodor Schwann** (1810–82) concluded by 1839 that all organisms were composed of cells. This was the first tenet of the **cell theory,** added to by later biologists and summarized in chapter 4. The cell theory was perhaps the most important breakthrough in biomedical history, because all functions of the body are now interpreted as the effects of cellular function.

Figure 1.5 **Hooke's Compound Microscope.** (*a*) The compound microscope had a lens at each end of a tubular body. (*b*) Hooke's drawing of cork cells, showing the thick cell walls characteristic of plants.

Although the philosophical foundation for modern medicine was largely established by the time of Leeuwenhoek, Hooke, and Harvey, clinical practice was still in a dismal state. Only a few doctors attended medical school or received any formal education in basic science or human anatomy. Physicians tended to be ignorant, ineffective, and pompous. Their practice was heavily based on expelling imaginary toxins from the body by bleeding their patients or inducing vomiting, sweating, or diarrhea. They performed operations with dirty hands and instruments, spreading lethal infections from one patient to another. Fractured limbs often became gangrenous and had to be amputated, and there was no anesthesia to lessen the pain. Disease was still widely attributed to demons and witches, and many people felt they would be interfering with God's will if they tried to cure it.

Living in a Revolution

This short history brings us only to the threshold of modern biomedical science; it stops short of such momentous discoveries as the germ theory of disease, the mechanisms of heredity, and the structure of DNA. In the twentieth century, basic biology and biochemistry have given us a much deeper understanding of how the body works, and technological advances such as noninvasive medical imaging (see clinical insight 1.3, p. 22) have enhanced our diagnostic ability and life-support strategies. We have witnessed monumental developments in chemotherapy, immunization, anesthesia, surgery, organ transplants, and human genetics. We are on the threshold of having a library of the molecular structure of every human gene, and gene therapy is being used to treat children born with diseases recently considered incurable. As historians look back on the twentieth century, they may exult about the Genetic Revolution in which you lived.

Several discoveries of the nineteenth and twentieth centuries, and the men and women behind them, are covered in short historical sketches in later chapters. Yet, the stories told in this chapter are different in a significant way. The people discussed here were pioneers in establishing the scientific way of thinking. They helped to replace superstition with an appreciation of natural law. They bridged the chasm between mystery and medication. Without this intellectual revolution, those who followed could not have conceived of the right questions to ask, much less a method for answering them.

Before You Go On

Answer the following questions to test your understanding of the preceding section:

1. In what way did the followers of Galen disregard his own advice? How does Galen's advice apply to you?
2. Describe two ways in which Vesalius improved medical education and set standards that remain relevant today.
3. How is our concept of human form and function today affected by the inventions of Leeuwenhoek and Hooke?

Scientific Method

Objectives

When you have completed this section, you should be able to

- describe the inductive and hypothetico-deductive methods of obtaining scientific knowledge;
- describe some aspects of experimental design that help to ensure objective and reliable results; and
- explain what is meant by *hypothesis, fact, law,* and *theory* in science.

Prior to the seventeenth century, science was done in a haphazard way by a small number of isolated individuals. The philosophers **Francis Bacon** (1561–1626) in England and **René Descartes** (1596–1650) in France envisioned science as a far greater, systematic enterprise with enormous possibilities for human health and welfare. They detested those who endlessly debated ancient philosophy without creating anything new. Bacon argued for discarding biases in people's way of thinking and for giving science a more objective basis. He outlined a systematic way of seeking similarities, differences, and trends in nature and drawing useful generalizations from observable facts. You will see echoes of Bacon's philosophy in the discussion of scientific method that follows.

Though the followers of Bacon and Descartes argued bitterly with each other, both men wanted science to become a public, cooperative enterprise, supported by governments and conducted by an international community of scholars rather than a few isolated amateurs. Inspired by their vision, the French and English governments established academies of science that still flourish today. Bacon and Descartes are credited with putting science on the path to modernity, not by discovering anything new in nature or inventing any techniques—for neither man was a scientist—but by inventing new habits of scientific thought.

When we say "scientific," we mean that such thinking is based on assumptions and methods that yield reliable, objective, testable information about nature. The assumptions of science are ideas that have proven fruitful in the past—for example, the idea that natural phenomena have natural causes and nature is therefore predictable and understandable. The methods of science are highly variable. **Scientific method** refers less to observational procedures than to certain habits of disciplined creativity, careful observation, logical thinking, and honest analysis of one's observations and conclusions. It is important in health science to understand these habits. This field is more littered with fads and frauds than any other. We are called upon constantly to judge which claims are trustworthy and which are bogus. To make such judgments depends on an appreciation of how scientists think, how they set standards for truth, and why their claims are more reliable than others.

The Inductive Method

The **inductive method,** first prescribed by Bacon, is a process of making numerous observations until one feels confident in drawing generalizations and predictions from them. What we know of anatomy is a product of the inductive method. We describe the normal structure of the body based on observations of many bodies.

This raises the issue of what is considered proof in science. We can never prove a claim beyond all possible refutation. We can, however, consider a statement as proven *beyond reasonable doubt* if it was arrived at by reliable methods of observation, tested and confirmed repeatedly, and not falsified by any credible observation. In science, all truth is tentative; there is no room for dogma. We must always be prepared to abandon yesterday's truth if tomorrow's facts disprove it.

The Hypothetico-Deductive Method

Most physiological knowledge was obtained by the **hypothetico-deductive method.** An investigator begins by asking a question and formulating a **hypothesis**—an educated speculation or possible answer to the question. A good hypothesis must be (1) consistent with what is already known and (2) capable of being tested and possibly falsified by evidence. **Falsifiability** means that if we claim something is scientifically true, we must be able to specify what evidence it would take to prove it wrong. If nothing could possibly prove it wrong, then it is not scientific.

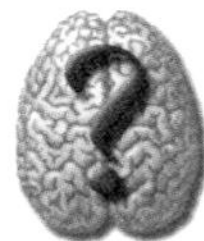

Think About It

The ancients thought that gods or invisible demons caused epilepsy. Today, epileptic seizures are attributed to bursts of abnormal electrical activity in nerve cells of the brain. Explain why one of these claims is falsifiable (and thus scientific), while the other claim is not.

The purpose of a hypothesis is to suggest a method for answering a question. From the hypothesis, a researcher makes a **deduction,** typically in the form of an "if-then" prediction: *If* my hypothesis is correct and I record the brain waves of patients during epileptic seizures, *then* I should observe abnormal bursts of activity. A properly conducted experiment yields observations that either support a hypothesis or require the scientist to modify or abandon it, formulate a better hypothesis, and test that one. Hypothesis testing operates in cycles of conjecture and disproof until one is found that is supported by the evidence.

Experimental Design

Doing an experiment properly involves several important considerations. What shall I measure and how can I measure it? What effects should I watch for and which ones should I ignore? How can I be sure that my results are due to the factors (*variables*) that I manipulate and not due to something else? When working on human subjects, how can I prevent the subject's expectations or state of mind from influencing the results? Most importantly, how can I eliminate my own biases and be sure that even the most skeptical critics will have as much confidence in my conclusions as I do? Several elements of experimental design address these issues:

- **Sample size.** The number of subjects (animals or people) used in a study is the sample size. An adequate sample size controls for chance events and individual variations in response and thus enables us to place more confidence in the outcome. For example, would you rather trust your health to a drug that was tested on 5 people or one tested on 5,000?
- **Controls.** Biomedical experiments require comparison between treated and untreated individuals so that we can judge whether the treatment has any effect. A **control group** consists of subjects that are as much like the **treatment group** as possible except with respect to the variable being tested. For example, there is evidence that garlic lowers blood cholesterol levels. In one study, a group of people with high cholesterol were given 800 mg of garlic powder daily for 4 months and exhibited an average 12% reduction in cholesterol. Is this a significant reduction, and is it due to the garlic? It is impossible to say without comparison to a control group of similar people who received no treatment. In this study, the control group averaged only a 3% reduction in cholesterol, so garlic *seems* to have made a difference.
- **Psychosomatic effects.** Psychosomatic effects (effects of the subject's state of mind on his or her physiology) can have an undesirable impact on experimental results if we do not control for them. In drug research, it is therefore customary to give the control group a **placebo** (pla-SEE-bo)—a substance with no significant physiological effect on the body. If we were testing a drug, for example, we could give the treatment group the drug and the control group identical-looking cornstarch tablets. Neither group must know which tablets it is receiving. If the two groups showed significantly different effects, we could feel confident that it did not result from a knowledge of what they were taking.
- **Experimenter bias.** In the competitive, high-stakes world of medical research, experimenters may want certain results so much that their biases, even subconscious ones, can affect their interpretation of the data. One way to control for this is the **double-blind method.** In this procedure, neither the subject to whom a treatment is given nor the person giving it and recording the results knows whether that subject is receiving the experimental treatment or placebo. A researcher might prepare identical-looking tablets, some with the drug and some with placebo, label them with code numbers, and distribute them to participating physicians. The physicians themselves do not know whether they are administering drug or placebo, so they cannot give the subjects even accidental hints of which substance they are taking. When the data are collected, the researcher can correlate them with the composition of the tablets and determine whether the drug had more effect than the placebo.
- **Statistical testing.** If you tossed a coin 100 times, you would expect it to come up about 50 heads and 50 tails. If it actually came up 48:52, you would probably attribute this to random error rather than bias in the coin. But what if it came up 40:60? At what point would you begin to suspect bias? This type of problem is faced routinely in research—how great a difference must there be between a control and experimental group before we feel confident that the treatment really had an effect? What if a treatment group exhibited a 12% reduction in cholesterol level and the placebo group a 10% reduction? Would this be enough to conclude that the treatment was effective? Scientists are well grounded in **statistical tests** that can be applied to the data. Perhaps you have heard of the chi-square test, the *t* test, or analysis of variance, for example. A typical outcome of a statistical test might be expressed, "We can be 99.5% sure that the difference between group A and group B was due to the experimental treatment and not to random variation."

Peer Review

When a scientist applies for funds to support a research project or submits results for publication, the application or manuscript is submitted to **peer review**—a critical evaluation by other experts in that field. Even after a report is published, if the results are important or unconventional, other scientists may attempt to reproduce them to see if the author was correct. At every stage from planning to post-publication, scientists are therefore subject to intense scrutiny by their colleagues. Peer review is one mechanism for ensuring honesty, objectivity, and quality in science.

Facts, Laws, and Theories

The most important product of scientific research is understanding how nature works—whether it be the nature of a pond to an ecologist or the nature of a liver cell to a physiologist. We express our understanding as *facts, laws,* and *theories* of nature. It is important to appreciate the differences between these.

A scientific **fact** is information that can be independently verified by any trained person—for example, the fact that an iron deficiency leads to anemia. A **law of nature** is a generalization about the predictable ways in which matter and energy behave. It is the result of inductive reasoning based on repeated, confirmed observations. Some laws are expressed as concise verbal statements, such as the *first law of thermodynamics:* Energy can be converted from one form to another but cannot be created or destroyed. Others are expressed as mathematical formulae, such as the *law of Laplace:* $F = 2T/r$, where F is a force that tends to cause a microscopic air sac of the lung to collapse, T is the surface tension of the fluid lining the sac, and r is the sac's radius.

A **theory** is an explanatory statement, or set of statements, derived from facts, laws, and confirmed hypotheses. Some theories have names, such as the *cell theory,* the *fluid-mosaic theory* of cell membranes, and the *sliding filament theory* of muscle contraction. Most, however, remain unnamed. The purpose of a theory is not only to concisely summarize what we already know but, moreover, to suggest directions for further study and to help predict what the findings should be if the theory is correct.

Law and *theory* mean something different in science than they do to most people. In common usage, a law is a rule created and enforced by people; we must obey it or risk a penalty. A law of nature, however, is a description; laws do not *govern* the universe, they *describe* it. Laypeople tend to use the word *theory* for what a scientist would call a hypothesis—for example, "I have a theory why my car won't start." The difference in meaning causes significant confusion when it leads people to think that a scientific theory (such as the theory of evolution) is merely a guess or conjecture, instead of recognizing it as a summary of conclusions drawn from a large body of observed facts. The concepts of gravity and electrons are theories, too, but this does not mean they are merely speculations.

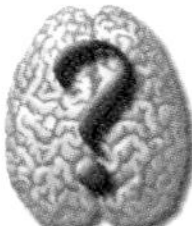

Think About It

Was the cell theory proposed by Schleiden and Schwann more a product of the hypothetico-deductive method or of the inductive method? Explain your answer.

Before You Go On

Answer the following questions to test your understanding of the preceding section:

4. Describe the general process involved in the inductive method.
5. Describe some sources of potential bias in biomedical research. What are some ways of minimizing such bias?
6. Is there more information in an individual scientific fact or in a theory? Explain.

Human Evolution

Objectives

When you have completed this section, you should be able to

- define *evolution* and *natural selection;*
- describe some human characteristics that can be attributed to the tree-dwelling habits of earlier primates;
- describe some human characteristics that evolved later in connection with upright walking; and
- explain why evolution is relevant to understanding human form and function.

If any two theories have the broadest implications for understanding the human body, they are probably the cell theory and the theory of natural selection. *Natural selection,* an explanation of how species originate and change through time, was the brainchild of **Charles Darwin** (1809–82) (fig. 1.6)—probably the most influential biologist who ever lived. His book, *On the Origin of Species by Means of Natural Selection* (1859), has been called "the book that shook the world." In presenting the first well-supported theory of evolution, *On the Origin of Species* not only caused the restructuring of all of biology but also profoundly changed the prevailing view of our origin, nature, and place in the universe.

While *On the Origin of Species* scarcely touched upon human biology, its unmistakable implications for humans created an intense storm of controversy that continues even today. In *The Descent of Man* (1871), Darwin directly addressed the issue of human evolution and emphasized features of anatomy and behavior that reveal our relationship to other animals. No understanding of human form and function is complete without an understanding of our evolutionary history.

Figure 1.6 Charles Darwin at the Age of 60.

Evolution, Selection, and Adaptation

Evolution simply means change in the genetic composition of a population of organisms. Examples include the evolution of bacterial resistance to antibiotics, the appearance of new strains of the AIDS virus, and the emergence of new species of organisms. The theory of **natural selection** is essentially this: Some individuals within a species have hereditary advantages over their competitors—for example, better camouflage, disease resistance, or ability to attract mates—that enable them to produce more offspring. They pass these advantages on to their offspring, and such characteristics therefore become more and more common in successive generations. This brings about the genetic change in a population that constitutes evolution.

Natural forces that promote the reproductive success of some individuals more than others are called **selection pressures.** They include such things as climate, predators, disease, competition, and the availability of food. **Adaptations** are features of an organism's anatomy, physiology, and behavior that have evolved in response to these selection pressures and enable the organism to cope with the challenges of its environment. We will consider shortly some selection pressures and adaptations that were important to human evolution.

Darwin could scarcely have predicted the overwhelming mass of genetic, molecular, fossil, and other evidence of human evolution that would accumulate in the twentieth century and further substantiate his theory. A technique called DNA hybridization, for example, suggests a difference of only 1.6% in DNA structure between humans and chimpanzees. Chimpanzees and gorillas differ by 2.3%. DNA structure suggests that a chimpanzee's closest living relative is not the gorilla or any other ape—it is us.

Several aspects of our anatomy make little sense without an awareness that the human body has a history (see evolutionary insight 1.1). Our evolutionary relationship to other species is also important in choosing animals for biomedical research. If there were no issues of cost, availability, or ethics, we might test drugs on our nearest living relatives, the chimpanzees, before approving them for human use. Their genetics, anatomy, and physiology are most similar to ours, and their reactions to drugs therefore afford the best prediction of how the human body would react. On the other hand, if we had no kinship with any other species, the selection of a test species would be arbitrary; we might as well use frogs or snails. In reality, we compromise. Rats and mice are used extensively for research because they are fellow mammals with a physiology similar to ours, but they present fewer of the aforementioned issues than chimpanzees or other mammals do. An animal species or strain selected for research on a particular problem is called a **model**—for example, a mouse model for leukemia.

Life in the Trees

We belong to an order of mammals called the Primates, which also includes the monkeys and apes. The story of primate origins begins 55 to 60 million years ago, when certain squirrel-sized, insect-eating mammals (insectivores) took up life in the trees of Africa. This **arboreal**[8] (treetop) habitat probably afforded greater safety from predators, less competition, and a rich food supply of leaves, fruit, insects, and lizards. But the forest canopy is a challenging world, with dim and dappled sunlight, swaying branches, and prey darting about in the dense foliage. Any new feature that enabled arboreal animals to move about more easily in the treetops would have been strongly favored by natural selection. Thus, the shoulder became more mobile and enabled primates to reach out in any direction (even overhead, which few other mammals can do). The thumbs became **opposable**—they could cross the palm to touch the fingertips—and enabled primates to hold small objects and manipulate them more precisely than other mammals can. Opposable thumbs made the hands **prehensile**[9]—able to grasp branches by encircling

8. *arbor* = tree + *eal* = pertaining to
9. *prehens* = to seize

Evolutionary Insight 1.1

Vestiges of Human Evolution

One of the classic lines of evidence for evolution, debated even before Darwin was born, is *vestigial organs.* These structures are the reduced remnants of organs that apparently were more functional in the ancestors of a species. They now serve little or no purpose or, in some cases, have been converted to new functions.

Our bodies, for example, are covered with millions of hairs, each equipped with a useless little muscle called an *arrector pili.* In other mammals, these muscles fluff the hair and conserve heat. In humans, they merely produce goosebumps. Above each ear, we have three *auricularis muscles.* In other mammals, they move the ears to receive sounds better, but most people cannot contract them at all. As Darwin said, it makes no sense that humans would have such structures were it not for the fact that we came from ancestors in which they were functional.

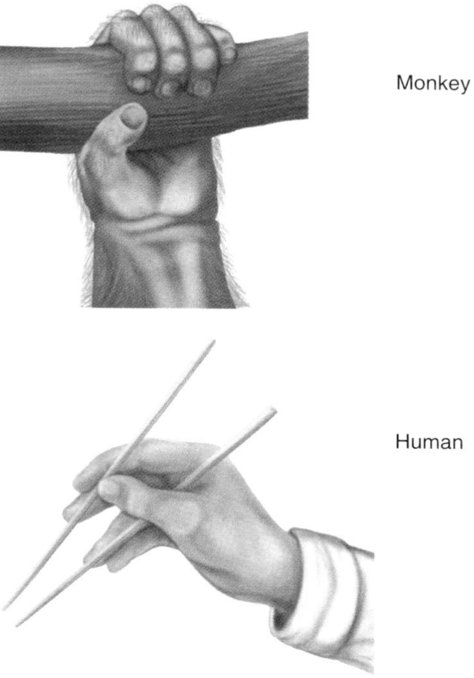

Figure 1.7 Primate Hands. The opposable thumb makes the primate hand prehensile, able to encircle and grasp objects.

Figure 1.8 Primitive Tool Use in a Primate. Chimpanzees exhibit the prehensile hands and forward-facing eyes typical of primates. Such traits endow primates with stereoscopic vision (depth perception) and good hand-eye coordination, two supremely important factors in human evolution.

them with the thumb and fingers (fig. 1.7). The thumb is so important that it receives highest priority in the repair of hand injuries. If the thumb can be saved, the hand can be reasonably functional; if it is lost, hand functions are severely diminished.

The eyes of primates moved to a more forward-facing position (fig. 1.8), which allowed for **stereoscopic**[10] **vision** (depth perception). This adaptation provided better hand-eye coordination in catching and manipulating prey, with the added advantage of making it easier to judge distances accurately in leaping from tree to tree. Color vision, rare among mammals, is also a primate hallmark. Primates eat mainly fruit and leaves. The ability to distinguish subtle

10. *stereo* = solid + *scop* = vision

shades of orange and red enables them to distinguish ripe, sugary fruits from unripe ones. Distinguishing subtle shades of green helps them to differentiate between tender young leaves and tough, more toxic older foliage.

Various fruits ripen at different times and in widely separated places in the tropical forest. This requires a good memory of what will be available when and how to get there. Larger brains may have evolved in response to the challenge of efficient food finding and, in turn, laid the foundation for more sophisticated social organization.

None of this is meant to imply that humans evolved from monkeys or apes—a common misconception about evolution that no biologist believes. Observations of monkeys and apes, however, provide insight into how primates adapt to the arboreal habitat and how certain human adaptations may have originated.

Walking Upright

About 4 to 5 million years ago, Africa became hotter and drier and much of the forest was replaced by savanna (grassland). Some primates adapted to living on the savanna, but this was a dangerous place with more predators and less protection. Just as squirrels and monkeys stand briefly on their hind legs to look around for danger, so would these early ground-dwellers. Being able to stand up not only helps an animal stay alert but also frees the forelimbs for purposes other than walking. Chimpanzees sometimes walk upright to carry food or weapons (sticks and rocks), and it is reasonable to suppose that our early ancestors did so too. They could also carry their infants.

These advantages are so great that they favored skeletal modifications that made **bipedalism**[11]—standing and walking on two legs—easier. The anatomy of the human pelvis, femur, knee, great toe, foot arches, vertebral column, skull, arms, and many muscles became adapted for bipedal locomotion, as did many aspects of human family life and society. As the skeleton and muscles became adapted for bipedalism, brain volume increased dramatically (table 1.1). It must have become increasingly difficult for a fully developed, large-brained infant to pass through the mother's pelvic outlet at birth. This may explain why humans are born in a relatively immature, helpless state compared to other mammals, before their nervous systems have matured and the bones of the skull have fused.

The oldest bipedal primates (family Hominidae) are classified in the genus *Australopithecus* (aus-TRAL-oh-PITH-eh-cus). About 2.5 million years ago, *Australopithecus* gave rise to *Homo habilis,* the earliest member of our own genus. *Homo habilis* differed from *Australopithecus* in height, brain volume, some details of skull anatomy, and tool-making ability. It was probably the first primate able to speak. *Homo habilis* gave rise to *Homo erectus* about 1.1 million years ago, which in turn led to our own species, *Homo sapiens,* about 300,000 years ago (fig. 1.9). *Homo sapiens* includes the extinct Neanderthal and Cro-Magnon people as well as modern humans.

This brief account barely begins to explain how human anatomy, physiology, and behavior have been shaped by ancient selection pressures. Later chapters further demonstrate that the evolutionary perspective provides a meaningful understanding of why humans are the way we are. Evolution is the basis for comparative anatomy and physiology, which have been so fruitful for the understanding of human biology. If we were not related to any other species, those sciences would be pointless. The emerging science of **darwinian medicine** traces some of our diseases and imperfections to our evolutionary past.

11. *bi* = two + *ped* = foot

Table 1.1 Brain Volumes of the Hominidae

Genus or Species	Time of Origin (millions of years ago)	Brain Volume (milliliters)
Australopithecus	3–4	400
Homo habilis	2.5	650
Homo erectus	1.1	1,100
Homo sapiens	0.3	1,350

Before You Go On

Answer the following questions to test your understanding of the preceding section:

7. Define *adaptation* and *selection pressure.* Why are these concepts important in understanding human anatomy and physiology?
8. Select any two human characteristics and explain how they may have originated in primate adaptations to an arboreal habitat.
9. Select two other human characteristics and explain how they may have resulted from adaptation to a grassland habitat.

The Nature of Human Life

Objectives

When you have completed this section, you should be able to

- state the characteristics that distinguish living organisms from nonliving objects;
- outline the classification of humans within the animal kingdom;

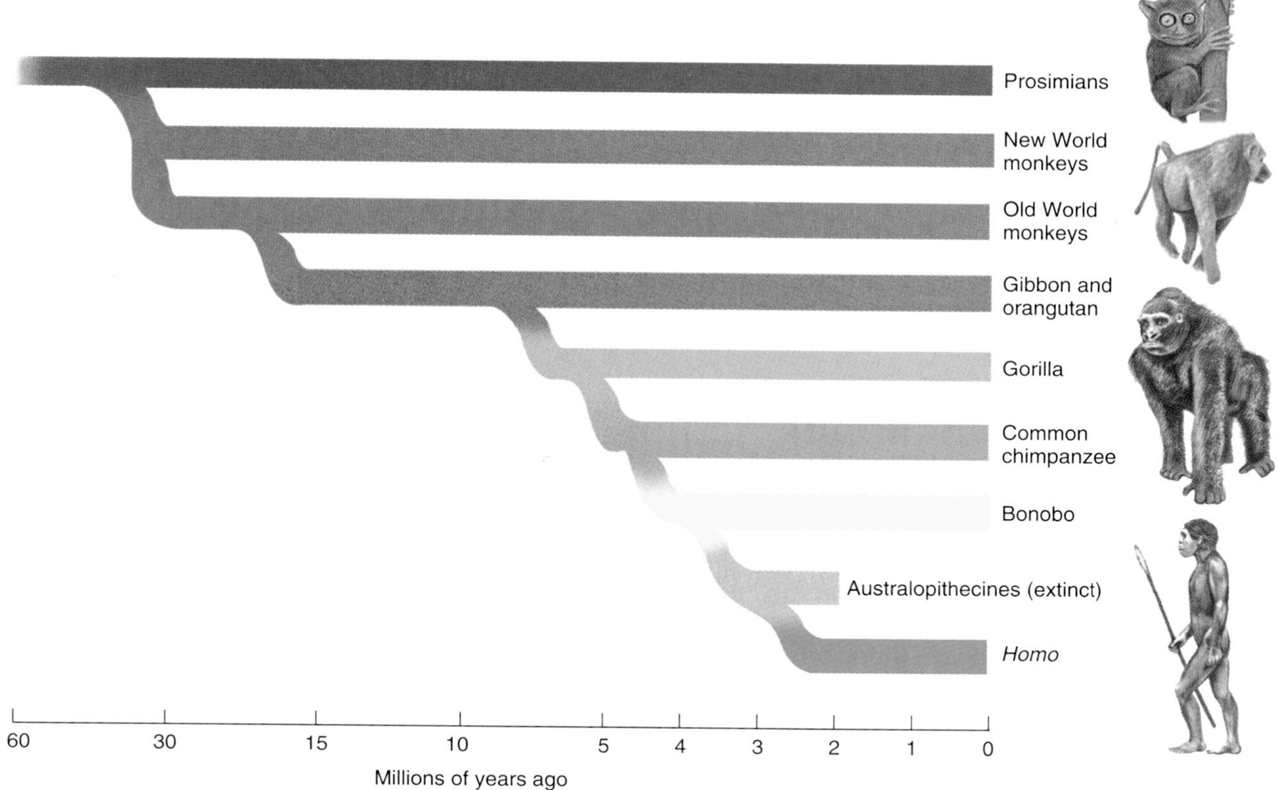

Figure 1.9 Primate Phylogeny. Figures at the *right* show representative primates alive today. The branch points in this "family tree" show the approximate times that different lines diverged from a common ancestor. Note that the time scale is not uniform; recent events are expanded for clarity.

- describe the anatomical features that define "human" and distinguish humans from other animals;
- list the levels of human structure from the simplest to the most complex; and
- discuss the value of both reductionistic and holistic viewpoints to understanding human form and function.

This book is a study of human life, so before we go much further, we should consider what we mean by the expression "human life." This is really two questions: What is life? and What is a human?

What Is Life?

Why do we consider a growing child to be alive but not a growing crystal? Is abortion the taking of a human life? If so, what about a contraceptive foam that kills only sperm? As a patient is dying, at what point does it become ethical to disconnect life-support equipment and remove organs for donation? If these organs are alive, as they must be to serve someone else, then why isn't the donor considered alive? Such questions have no easy answers, but they demand a concept of what life is—a concept that may differ with one's biological, medical, or legal perspective.

From a biological viewpoint, life is not a single property. It is a collection of properties that help to distinguish living from nonliving things:

- **Organization.** Living things exhibit a far higher level of organization than the nonliving world around them. They expend a great deal of energy to maintain order, and a breakdown in this order is accompanied by disease and often death.
- **Cellular composition.** Living matter is always compartmentalized into one or more cells.
- **Biochemical unity.** All living things have a universal chemical composition that includes DNA, proteins, lipids, and carbohydrates. Such compounds are not found in anything of nonbiological origin.

- **Metabolism.** Living things take in molecules from the environment and chemically change them into molecules that form their own structures, control their physiology, or provide them with energy. **Metabolism**[12] is the sum of all this internal chemical change. It consists of two classes of reactions: *anabolism,*[13] in which relatively complex molecules are synthesized from simpler ones (for example, protein synthesis), and *catabolism,*[14] in which relatively complex molecules are broken down into simpler ones (for example, protein digestion). There is a constant turnover of molecules in the body; few of the molecules now in your body have been there for more than a year. It is food for thought that although you sense a continuity of personality and experience from your childhood to the present, nearly all of your body has been replaced within the past year.
- **Responsiveness.** The ability of organisms to sense and react to **stimuli** (changes in their environment) is called *responsiveness, irritability,* or *excitability.* It occurs at all levels from the single cell to the entire body, and it characterizes all living things from bacteria to yourself. Responsiveness is especially obvious in animals because of nerve and muscle cells that exhibit high sensitivity to environmental stimuli, rapid transmission of information, and quick reactions.
- **Homeostasis.** While the environment around an organism changes, the organism maintains relatively stable internal conditions. This ability to maintain internal stability, called *homeostasis,* is explored in more depth later in this chapter.
- **Development.** Development is any change in form or function over the lifetime of the organism. In most organisms, it involves two major processes: (1) **growth,** an increase in size, and (2) **differentiation,** the transformation of cells with no specialized function to cells that are committed to a particular task.
- **Growth.** Some nonliving things grow, but not in the way your body does. If you let a saturated sugar solution evaporate, crystals will grow from it, but not through a change in the composition of the sugar. They merely add more sugar molecules from the solution to the crystal surface. The growth of the body, by contrast, occurs through chemical change; for the most part, your body is not composed of the molecules you ate but of molecules made by chemically altering your food.
- **Reproduction.** All living organisms can produce copies of themselves, thus passing their genes on to new, younger containers—their offspring.
- **Evolution.** All living species exhibit genetic change from generation to generation and therefore evolve. This occurs because *mutations* (changes in DNA structure) are inevitable and because environmental selection pressures endow some individuals with greater reproductive success than others. Unlike the other characteristics of life, evolution is a characteristic seen only in the population as a whole. No single individual evolves over the course of its life.

Clinical and legal criteria of life differ from these biological criteria. A person who has shown no brain waves for 24 hours, and has no reflexes, respiration, or heartbeat other than what is provided by artificial life support, can be declared legally dead. At such time, however, most of the body is still biologically alive and its organs may be useful for transplant.

What Is a Human?

Our second question was, What is a human? We belong to the animal kingdom—as opposed to plants, fungi, protists, or bacteria—but what distinguishes us from other animals? To answer this, it helps to begin with an outline of our classification within the kingdom Animalia. We belong to each of the following progressively smaller groups (fig. 1.10):

Phylum Chordata
 Subphylum Vertebrata
  Class Mammalia
   Order Primates
    Family Hominidae
     Genus *Homo*
      Species *Homo sapiens*

Our Chordate Characteristics

The animal kingdom is divided into 30 to 35 major categories called *phyla* (singular, *phylum*). Humans belong to the phylum **Chordata** (cor-DAH-tuh), which is far from the largest phylum but contains the most complex species. The following characteristics distinguish chordates from other animals:

- A **notochord,** a flexible rod present on the upper (dorsal) side of the body at some stage of the animal's development. It is usually replaced by the vertebral column as the embryo develops.
- A **dorsal hollow nerve cord,** a column of nervous tissue that passes along the animal's dorsal side and

12. *metabol* = change + *ism* = process
13. *ana* = up
14. *cata* = down

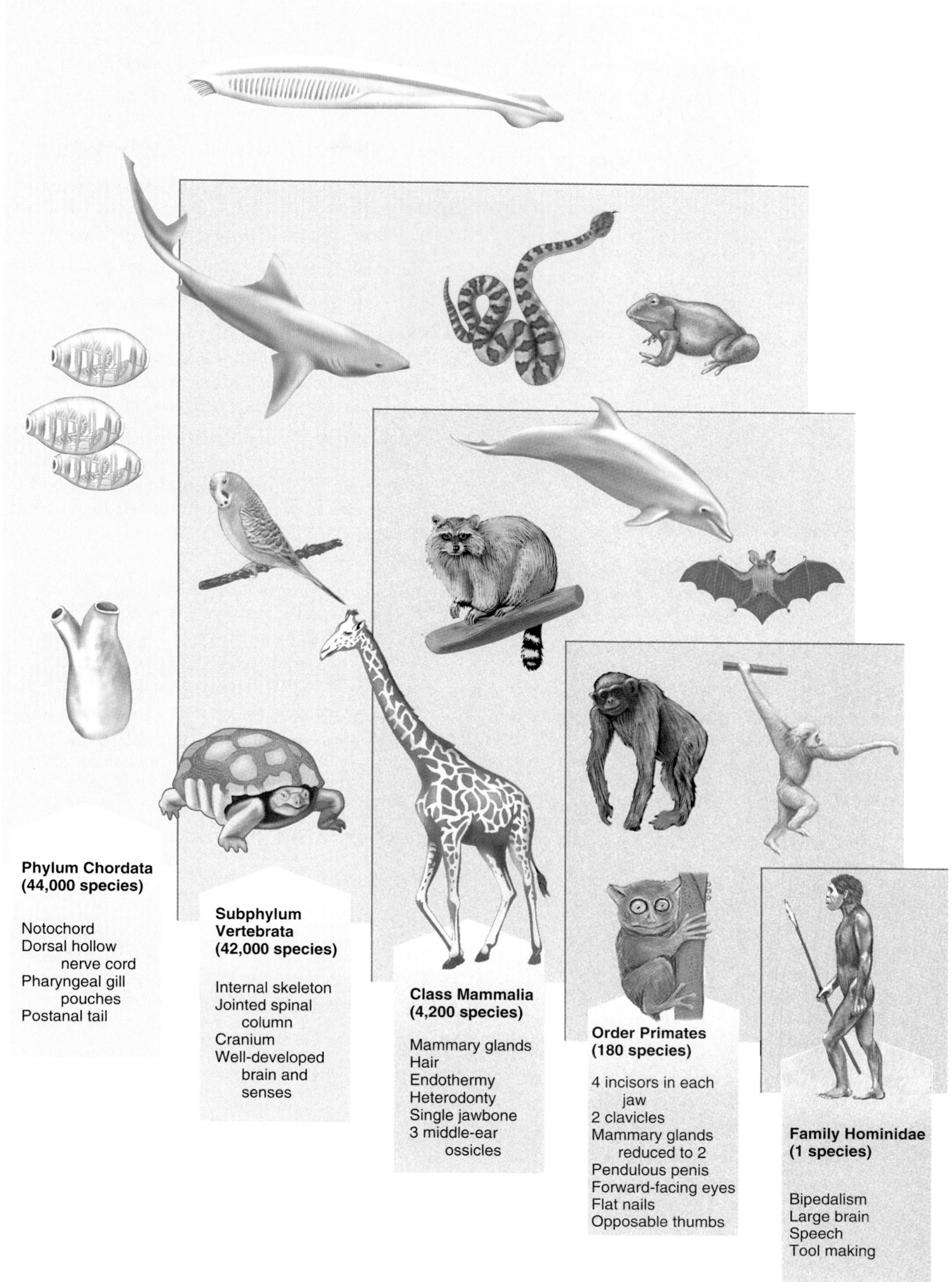

Figure 1.10 **Human Classification.** Each smaller box is a subdivision of the larger category above it. The characteristics that distinguish each group from other members of the larger group are listed across the bottom. Humans have all the characteristics listed there, but it is those at the far right that distinguish them from all other chordates.

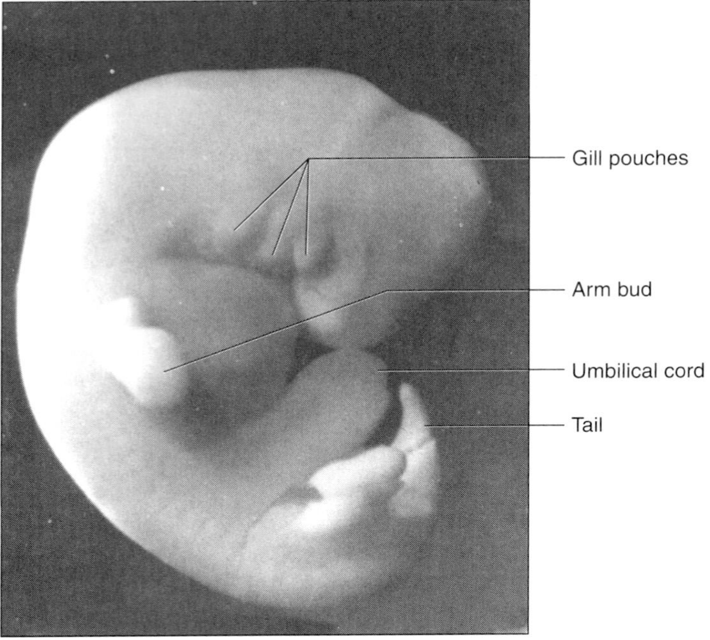

Figure 1.11 Primitive Chordate Characteristics in Humans. This human embryo at 38 days shows the tail and gill pouches that characterize all members of the Chordata at some stage of their development.

has a central canal filled with fluid. Nonchordates either lack nerve cords or have solid threadlike cords, usually on the ventral (lower) side of the body.

- **Gill pouches,** a series of bulges that develop in the pharyngeal (throat) region of all chordate embryos. They open and form gill slits in fish and amphibians, but human gill pouches (fig. 1.11) seldom break through to form slits.
- A **postanal tail,** which extends beyond the anus and usually aids in locomotion or balance. Many nonchordates have tails, but their digestive tracts generally extend to the end of the tail and the anus is located there. In humans, the tail is present only in the embryo (fig. 1.11), but the small bones of the coccyx ("tailbone") are a remnant of it.

Our Vertebrate Characteristics

Except for a few marine species, all chordates belong to the subphylum **Vertebrata.**[15] Vertebrates, including humans, are characterized by the following:

- An internal skeleton.
- A jointed vertebral (spinal) column.
- A well-developed brain and sense organs.
- A *cranium,* a protective bony enclosure for the brain.

15. *vertebr* = backbone + *ata* = possessing

Our Mammalian Characteristics

The Vertebrata are divided into fish, amphibians, reptiles, birds, and mammals. We belong to the class **Mammalia,**[16] of which most members have the following characteristics:

- Mammary glands for nourishment of the young.
- Hair, which serves to retain body heat.
- **Endothermy,**[17] the ability to generate most of their body heat by metabolic means. Fish, amphibians, and reptiles, by contrast, are *ectothermic*—they must obtain heat from the environment.
- **Heterodonty,**[18] the possession of varied types of teeth. Having incisors, canines, premolars, and molars specialized to puncture, cut, and grind food makes the food easier to digest. Rapid digestion is essential if mammals are to maintain the high metabolic rate needed for endothermy.
- A single lower jawbone. In reptiles, the jaw joint has three bones and is not very strong. As mammals evolved from reptiles, the lower jaw (mandible) retained only one of these bones and became better adapted for chewing.
- Three middle-ear bones (commonly called the hammer, anvil, and stirrup). Reptiles have only one. The other two bones in mammals arose from the two bones lost from the mandible.

Our Primate Characteristics

Mammals are divided into 19 orders. We belong to the order **Primates,** which also contains the apes, monkeys, lemurs, and a few other species. Most primates have the following characteristics:

- Four upper and four lower incisors, the front cutting teeth.
- A pair of clavicles (collarbones).
- Only two mammary glands.
- A pendulous penis, attached only at the base.
- Forward-facing eyes with stereoscopic vision.
- Flat nails in place of claws.
- Opposable thumbs.

16. *mamma* = breast + *alia* = possessing
17. *endo* = within + *therm* = heat
18. *hetero* = different, varied + *odont* = teeth

Our Hominid Characteristics

Humans are in the family **Hominidae** (ho-MIN-ih-dee), which may be defined as the bipedal primates. That is, in addition to the previously listed characteristics, hominids habitually walk on two legs. Only *Australopithecus* and *Homo* are classified in this family, and *Homo sapiens* is the only surviving species. There is no unanimously accepted definition of *human.* Some authorities treat all Hominidae as humans; some restrict the word *human* to the genus *Homo;* and some go even further and limit it to *Homo sapiens.*

Human Structure—A Hierarchy of Complexity

Consider for the moment an analogy to human structure: All words in the English language are composed of various combinations of just 26 letters. We organize these letters into syllables, words, sentences, and paragraphs. We could say that the language consists of a hierarchy of complexity, with letters, syllables, words, and so forth being successive levels of the hierarchy. From the simplicity of the alphabet comes the complexity and variety of all English literature. Humans have an analogous hierarchy of complexity (fig. 1.12), as follows:

Subatomic particles compose atoms,
atoms compose molecules,
molecules compose organelles,
organelles compose cells,
cells compose tissues,
tissues compose organs,
organs compose organ systems, and
organ systems compose the organism.

The **subatomic particles**—protons, neutrons, and electrons—are assembled into **atoms** and atoms into **molecules.** The largest molecules, such as proteins, fats, and DNA, are called *macromolecules.* Human structure and function can fully be understood only by considering this molecular level of organization, which is introduced in chapters 2 and 3 but forms a major theme of this entire book.

Organelles[19] are microscopic structures in a cell that carry out its individual functions. Examples include mitochondria, centrioles, and lysosomes. **Cells** are the smallest units of an organism that carry out all the basic functions of life; nothing simpler than a cell is considered alive. A cell is enclosed in a *plasma membrane* composed of lipids and protein. Most cells have one nucleus, an organelle that contains its DNA. Cytology, the study of cells and organelles, is the subject of chapters 4 and 5.

19. *elle* = little

A **tissue** is a mass of similar cells and cell products that forms a discrete region of an organ and performs a specific function. The body is composed of only four primary classes of tissue—epithelial, connective, nervous, and muscular tissues. Histology, the study of tissues, is the subject of chapter 6.

An **organ** is a structure composed of two or more tissue types that work together to carry out a particular function. Organs have definite anatomical boundaries and are visibly distinguishable from adjacent structures. Most organs and higher levels of structure are within the domain of gross anatomy. However, there are organs within organs—the large organs visible to the naked eye often contain smaller organs visible only with the microscope. The skin, for example, is the body's largest organ. Included within it are thousands of smaller organs: each hair, nail, gland, nerve, and blood vessel of the skin is an organ in itself.

An **organ system** is a group of organs with a unique collective function, such as circulation, respiration, or digestion. The human body has 11 organ systems, illustrated in atlas A immediately following this chapter: the integumentary, skeletal, muscular, nervous, endocrine, circulatory, lymphatic, respiratory, urinary, digestive, and reproductive systems. Usually, the organs of one system are physically interconnected, such as the kidneys, ureters, urinary bladder, and urethra, which compose the urinary system. The endocrine system, however, is a group of hormone-secreting glands that, for the most part, have no physical connection to each other. Beginning with chapter 7, this book is organized around the organ systems.

The **organism,** finally, is a single, complete individual.

The theory that a large, complex system such as the human body can be understood by studying its simpler components is called **reductionism.** First espoused by Aristotle, this has proven to be a highly productive approach; indeed, it is essential to scientific thinking. Yet the reductionistic view is not the last word in understanding human life. Just as it would be very difficult to predict the workings of an automobile transmission merely by looking at a pile of its disassembled gears and levers, one could never predict the human personality from a complete knowledge of the circuitry of the brain or the genetic sequence of DNA. **Holism**[20] is the complementary theory that there are "emergent properties" of the whole organism that cannot be predicted from the properties of its separate parts—human beings are more than the sum of their parts. To be most effective, a health-care provider does not treat merely a disease or an organ system but a whole person. A patient's perceptions, emotional responses to life, and confidence in the nurse, therapist, or physician profoundly affect the outcome of treatment. In fact, these psychological factors often play a greater role in a patient's recovery than the physical treatments administered.

20. *holo* = whole, entire

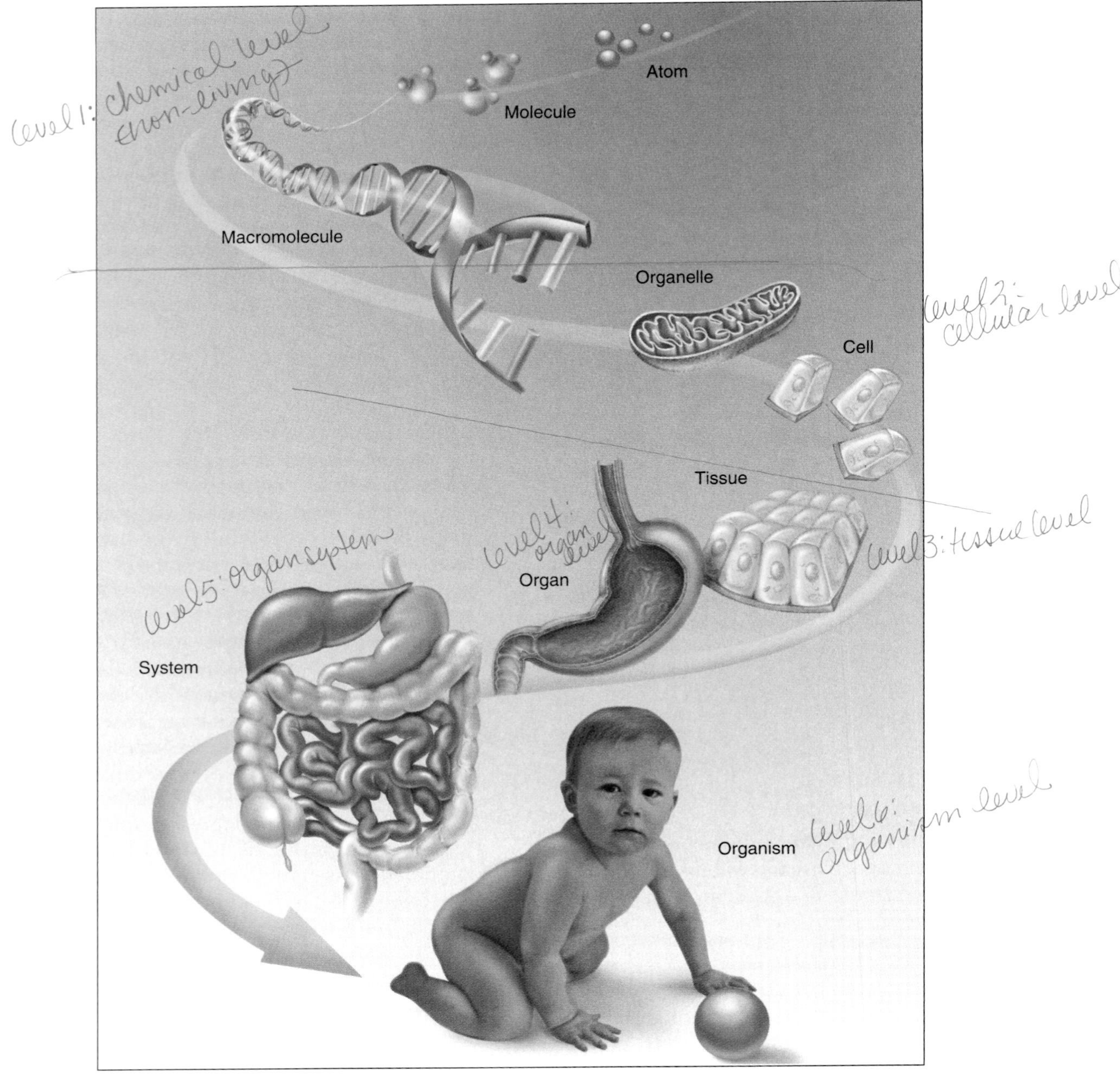

Figure 1.12 The Body's Structural Hierarchy.

Before You Go On

Answer the following questions to test your understanding of the preceding section:

10. List four biological criteria of life and one clinical criterion. Explain how a person could be clinically dead but biologically alive.
11. Explain why humans are classified as animals, why they are classified as mammals, and why they are classified as primates.
12. Do you consider *Australopithecus* to be human? Why or why not?
13. List the levels of human organization from atom to organism.
14. Explain why both reductionism and holism are important to human biology and therapy.

Homeostasis and Feedback

Objectives

When you have completed this section, you should be able to

- define *homeostasis* and explain why this concept is central to physiology;
- define *negative feedback*, give an example of it, and explain its importance to homeostasis; and
- define *positive feedback* and give examples of its beneficial and harmful effects.

The human body has a remarkable capacity for self-restoration. Hippocrates commented that it usually returns to a state of equilibrium by itself, and people recover from most illnesses even without the help of a physician. This tendency results from the ability to detect change and activate mechanisms that oppose it. This is the last major theory we explore in this chapter, but it will repeat itself throughout this book as we explore mechanisms of health and disease.

Homeostasis

French physiologist **Claude Bernard** (1813–78) observed that the internal conditions of the body remain fairly stable even when external conditions vary greatly. For example, whether it is freezing cold or swelteringly hot outdoors, the internal temperature of your body stays within a range of about 36° to 37°C (97°–99°F). American physiologist **Walter Cannon** (1871–1945) coined the term **homeostasis**[21] (ho-me-oh-STAY-sis) for this tendency to maintain internal stability. Homeostasis has been one of the most enlightening concepts in physiology. Physiology is largely a group of mechanisms for maintaining homeostasis, and the loss of homeostatic control tends to cause illness or death. Pathophysiology is essentially the study of unstable conditions that result when our homeostatic controls go awry.

Do not, however, overestimate the degree of internal stability. Internal conditions are not absolutely constant but fluctuate within a limited range, such as the range of body temperatures noted earlier. The internal state of the body is best described as a **dynamic equilibrium** (balanced change), in which there is a certain **set point** or average value for a given variable (such as 37°C for body temperature) and conditions fluctuate slightly around this point.

Negative Feedback and Stability

The fundamental mechanism that keeps a variable close to its set point is **negative feedback**—a process in which the body senses a change and activates mechanisms that negate or reverse it. By maintaining stability, negative feedback is the key mechanism for maintaining health.

These principles can be understood by comparison to a home heating system (fig. 1.13). Suppose it is a cold winter day and you have set your thermostat for 20°C (68°F)—the set point. If the room becomes too cold, a temperature-sensitive switch in the thermostat turns on the furnace. The temperature rises until it is slightly above the set point, and then the switch breaks the circuit and turns off the furnace. This is a negative feedback process that reverses the falling

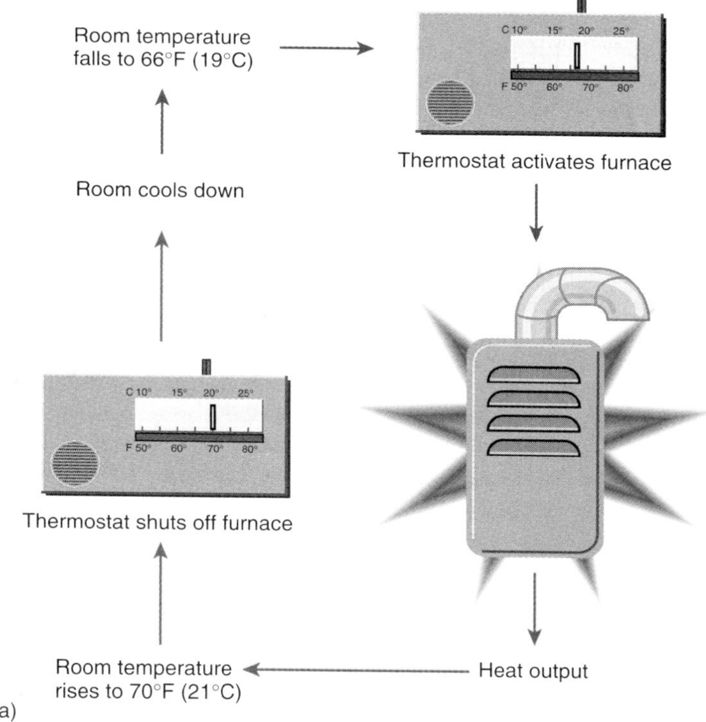

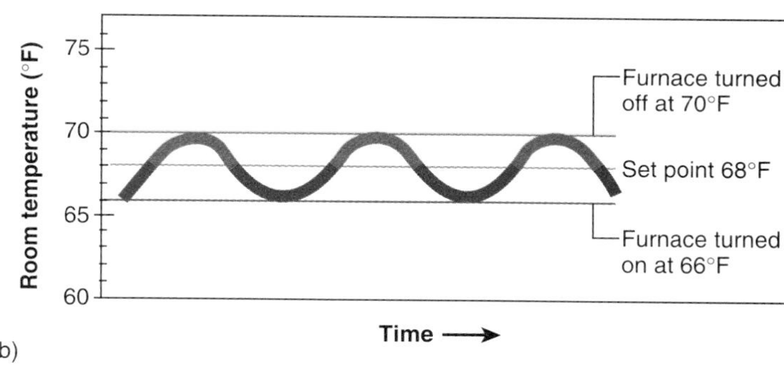

Figure 1.13 Negative Feedback in a Home Heating System. (*a*) The thermostat serves as the receptor and integrating center in this system, and the furnace as the effector. (*b*) Negative feedback keeps the room temperature within a few degrees of the thermostat's set point.

21. *homeo* = the same + *stas* = to place, stand, stay

Historical Insight 1.2

Men in the Oven

English physician Charles Blagden (1748–1820) staged a rather theatrical demonstration of homeostasis long before Cannon coined the word. In 1775, Blagden, some associates, a dog, and a beefsteak spent 45 minutes in a chamber heated to 127°C (260°F). Being alive and capable of evaporative cooling, the dog panted and the men sweated. The beefsteak, being dead and unable to maintain homeostatis, was cooked.

temperature and restores it to something close to the set point. When the furnace turns off, the temperature slowly drops again until the switch is reactivated—thus, the furnace cycles on and off all day. The room temperature does not stay at exactly 20°C but *fluctuates* a few degrees either way—the system maintains a state of dynamic equilibrium in which the temperature averages 20°C and deviates from the set point by only a few degrees. Because feedback mechanisms alter the original changes that triggered them (temperature, for example), they are often called **feedback loops.**

Body temperature is also regulated by a "thermostat"—a group of nerve cells in the base of the brain that monitors the temperature of the blood. If you become overheated, the thermostat triggers heat-losing mechanisms (fig. 1.14). One of these is **vasodilation** (VAY-zo-dy-LAY-shun), the widening of blood vessels. When blood vessels of the skin dilate, warm blood flows closer to the body surface and radiates heat into the surrounding air. If this is not enough to return your temperature to normal, sweating occurs; the evaporation of water from the skin has a powerful cooling effect (see historical insight 1.2). Conversely, if it is cold outside and your body temperature drops much below 37°C, these nerve cells activate heat-conserving mechanisms. The first to be activated is **vasoconstriction,** a narrowing of the blood vessels in the skin, which serves to retain warm blood deeper in your body and reduce heat loss. If this is not enough, the brain activates shivering—muscle tremors that generate heat.

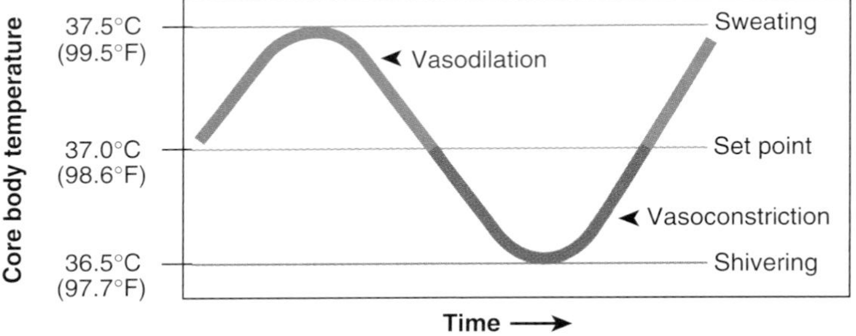

Figure 1.14 Negative Feedback in Human Thermoregulation. Negative feedback keeps the human body temperature homeostatically regulated within about 0.5°C of a 37°C set point. Sweating and cutaneous vasodilation lower the body temperature; shivering and cutaneous vasoconstriction raise it.

To take another example, a rise in blood pressure is sensed by stretch receptors in the wall of the heart and the major arteries above it. These receptors send nerve signals to a *cardiac center* in the brainstem. The cardiac center integrates this input with other information and sends nerve signals back to the heart to slow it and lower the blood pressure. Thus we can see that homeostasis is maintained by self-correcting negative feedback loops. Many more examples are found throughout this book.

It is common, although not universal, for feedback loops to include three components: a receptor, an integrator, and an effector. The **receptor** is a structure that senses a change in the body, such as the stretch receptors that monitor blood pressure. The **integrating (control) center,** such as the cardiac center of the brain, is a mechanism that processes this information, relates it to other available information (for example, comparing what the blood pressure is with what it should be), and "makes a decision" about what the appropriate response should be. The **effector,** in this case the heart, is the structure that carries out the response that restores homeostatis. The response, such as a lowering of the blood pressure, is then sensed by the receptor, and the feedback loop is complete.

Positive Feedback and Rapid Change

Positive Feedback is a self-amplifying cycle in which a physiological change leads to even greater change in the same direction, rather than producing the corrective effects of negative feedback. Positive feedback is often a normal way of producing rapid change. When a woman is giving birth, for example, the head of the baby pushes against her cervix (the neck of the uterus) and stimulates its nerve endings (fig. 1.15). Nerve signals travel to the brain, which, in turn, stimulates the pituitary gland to secrete the hormone oxytocin. Oxytocin travels in the blood and stimulates the uterus to contract. This pushes the baby downward, stimulating the cervix still more and causing the positive feedback loop to be repeated. Labor contractions therefore become more and more intense until the baby is expelled. Other cases of beneficial positive feedback are

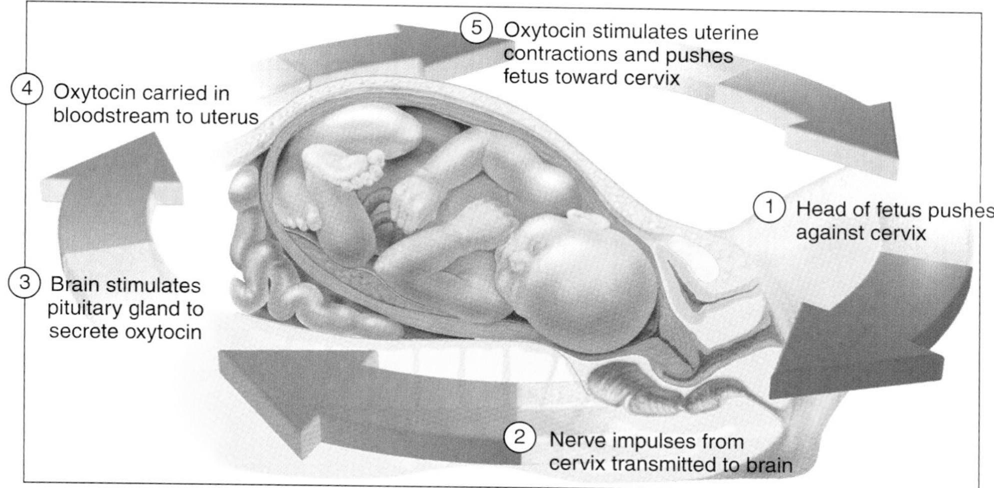

Figure 1.15 Positive Feedback in Childbirth. This is one of several cases in which positive feedback produces beneficial rapid change.

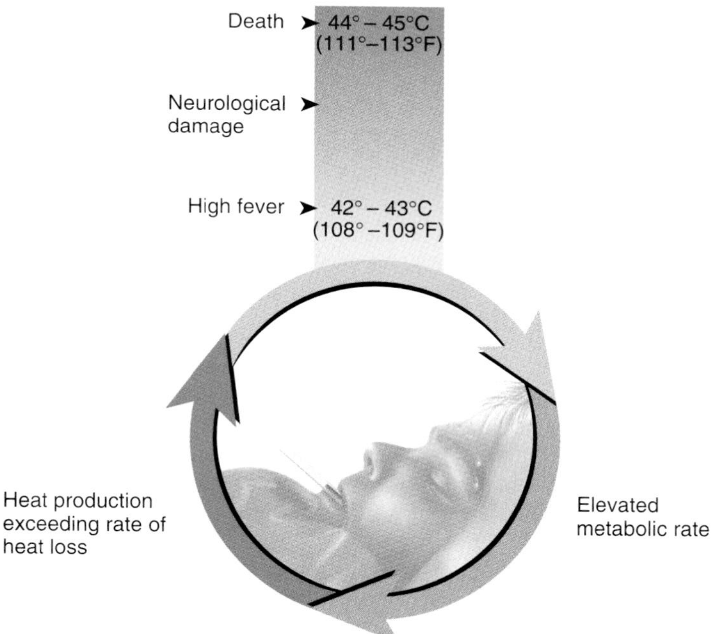

Figure 1.16 Positive Feedback in Fever. In such cases as this, positive feedback can produce a life-threatening loss of homeostatic control.

seen later in the book; for example, in blood clotting, protein digestion, and the generation of nerve signals.

Frequently, however, a positive feedback is a harmful or even life-threatening process. This is because its self-amplifying nature can quickly change the internal state of the body to something far from its homeostatic set point. Consider a high fever, for example. A fever triggered by infection is beneficial up to a point, but if the body temperature rises much above 42°C (108°F), it may create a dangerous positive feedback loop (fig. 1.16). This high temperature raises the metabolic rate, which makes the body produce heat faster than it can get rid of it. Thus, temperature rises still further, increasing the metabolic rate and heat production still more. This "vicious circle" becomes fatal at approximately 45°C (113°F). Thus, positive feedback loops often create dangerously out-of-control situations that require emergency medical treatment.

Before You Go On

Answer the following questions to test your understanding of the preceding section:

15. What is meant by dynamic equilibrium? Why would it be wrong to say homeostasis prevents internal change?
16. Explain why stabilizing mechanisms are called negative feedback.
17. Explain why positive feedback is more likely than negative feedback to disturb homeostasis.

Review of Major Themes

To close this chapter, let's distill a few major points from it. These themes can provide you with a sense of perspective that will make the rest of the book more meaningful and not just a collection of disconnected facts. These are some key unifying principles behind all study of human anatomy and physiology:

- **Cell theory.** *All structure and function result from the activity of cells.* Every physiological concept in this book ultimately must be understood from the standpoint of how cells function. Even anatomy is a result of cellular function. If cells are damaged or destroyed, we see the results in disease symptoms of the whole person.

- **Homeostasis.** *The purpose of most normal physiology is to maintain stable conditions within the body.* Human physiology is essentially a group of mechanisms that produce stable internal conditions favorable to cellular function. Any serious departure from these conditions can be harmful or fatal to cells.
- **Evolution.** *The human body is a product of evolution.* Like every other living species, we have been molded by millions of years of natural selection to function in a changing environment. Many aspects of human anatomy and physiology reflect our ancestors' adaptations to their environment. Human form and function cannot be fully understood except in light of our evolutionary history.
- **Hierarchy of structure.** *Human structure can be viewed as a series of levels of complexity.* Each level is composed of a smaller number of simpler subunits than the level above it. These subunits are arranged in different ways to form diverse structures of higher complexity. For example, all the body's organs are made of just four primary classes of tissue, and the thousands of proteins are made of various combinations of just 20 amino acids. Understanding these simpler components is the key to understanding higher levels of structure.
- **Unity of form and function.** *Form and function complement each other; physiology cannot be divorced from anatomy.* This unity holds true even down to the molecular level. Our very molecules, such as DNA and proteins, are structured in ways that enable them to carry out their functions. Slight changes in molecular structure can destroy their activity and threaten life.

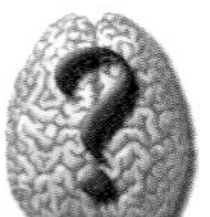

Think About It

Architect Louis Henri Sullivan coined the phrase, "Form ever follows function." What do you think he meant by this? Discuss how this idea could be applied to the human body and cite a specific example of human anatomy to support it.

Clinical Insight 1.3

Noninvasive Medical Imaging

The development of techniques for looking into the body without having to do exploratory surgery has greatly accelerated progress in medicine. A few of these techniques are described here.

Radiography

X rays, a form of high-energy radiation, were discovered by William Roentgen in 1885. X rays can penetrate soft tissues of the body and darken photographic film on the other side. They are absorbed, however, by dense tissues such as bone, teeth, tumors, and tuberculosis nodules, which leave the film lighter in these areas (fig. 1*a*). The process of examining the body with X rays is called *radiography.* The term *X ray* also applies to a photograph *(radiograph)* made by this method. Radiography is commonly used in dentistry, mammography, diagnosis of fractures, and examination of the chest. Hollow organs can be visualized by filling them with a *radiopaque* substance that absorbs X rays. Barium sulfate is given orally for examination of the esophagus, stomach, and small intestine or by enema for examination of the large intestine. Other substances are given by injection for *angiography,* the examination of blood vessels (fig. 1*b*). Some disadvantages of radiography are that images of overlapping organs can be confusing and slight differences in tissue density are not easily detected. Nevertheless, radiography still accounts for over 50% of all clinical imaging. Until the 1960s, it was the only method widely available.

Sonography

Sonography[22] is the second oldest and second most widely used method of imaging. It is an outgrowth of sonar technology developed in World War II. A handheld device held firmly to the skin produces high-frequency ultrasound waves and receives the signals that echo back from internal organs. Although sonography was first used medically in the 1950s, images of significant clinical value had to wait until computer technology had developed enough to analyze differences in the way tissues reflect ultrasound. Sonography is not very useful for examining bones or lungs, but it is the method of choice in obstetrics, where the image *(sonogram)* can be used to locate the placenta and evaluate fetal age, position, and development. Sonography avoids the harmful effects of X rays, and the equipment is inexpensive and portable. Its primary disadvantage is that it does not produce a very sharp image (fig. 2).

Computed Tomography (CT)

The *CT scan,* formerly called a computerized axial tomographic[23] (CAT) scan, is a more sophisticated application of X rays developed in 1972. The patient is moved through a ring-shaped machine that emits low-intensity X rays on one side and receives them with a detector on the opposite side. A computer analyzes signals from the detector and produces an image of a "slice" of the body about as thin as a coin (fig. 3). The computer can "stack" a series of these images to

22. *sono* = sound + *graphy* = recording process
23. *tomo* = section, cut, slice + *graphic* = pertaining to a recording

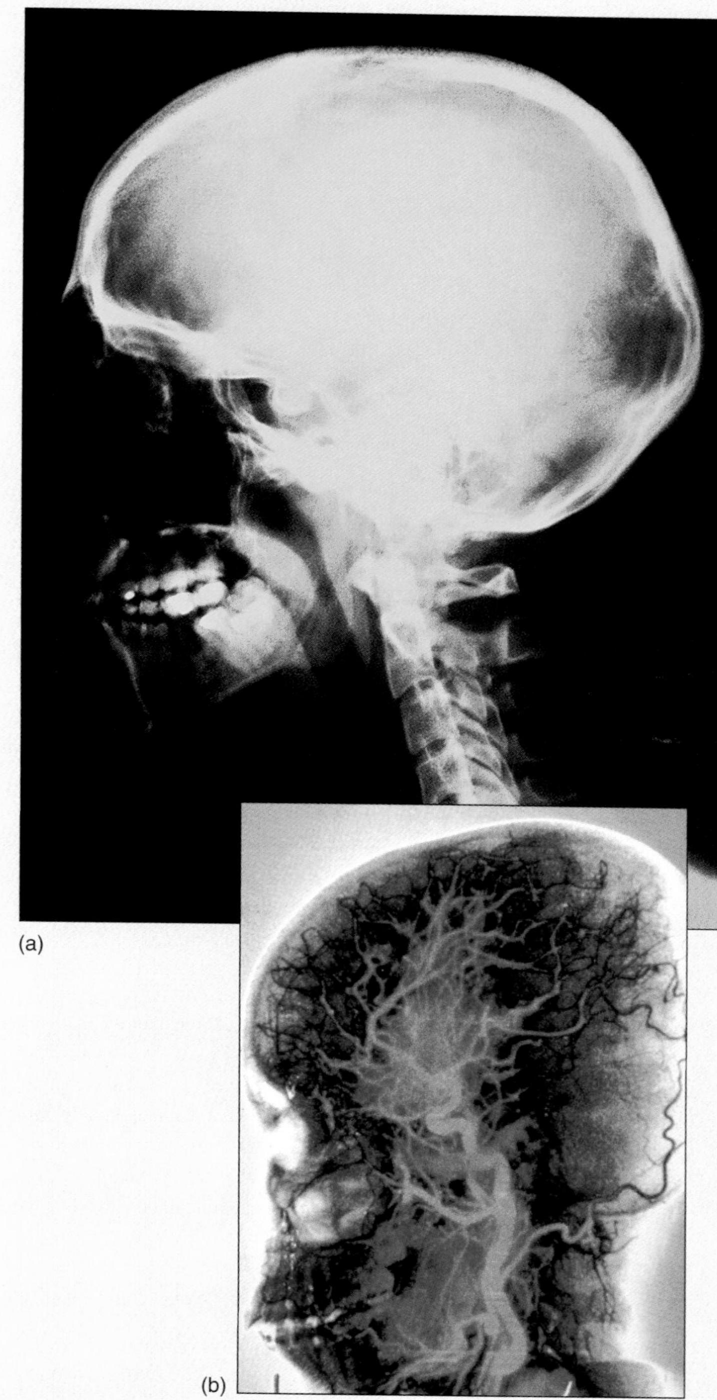

Figure 1 Radiography. (*a*) An X ray (radiograph) of the head and neck. (*b*) A cerebral angiogram, made by injecting a substance opaque to X rays into the circulation and then taking an X ray of the head to visualize the blood vessels. The arteries are enhanced with false color in this photograph.

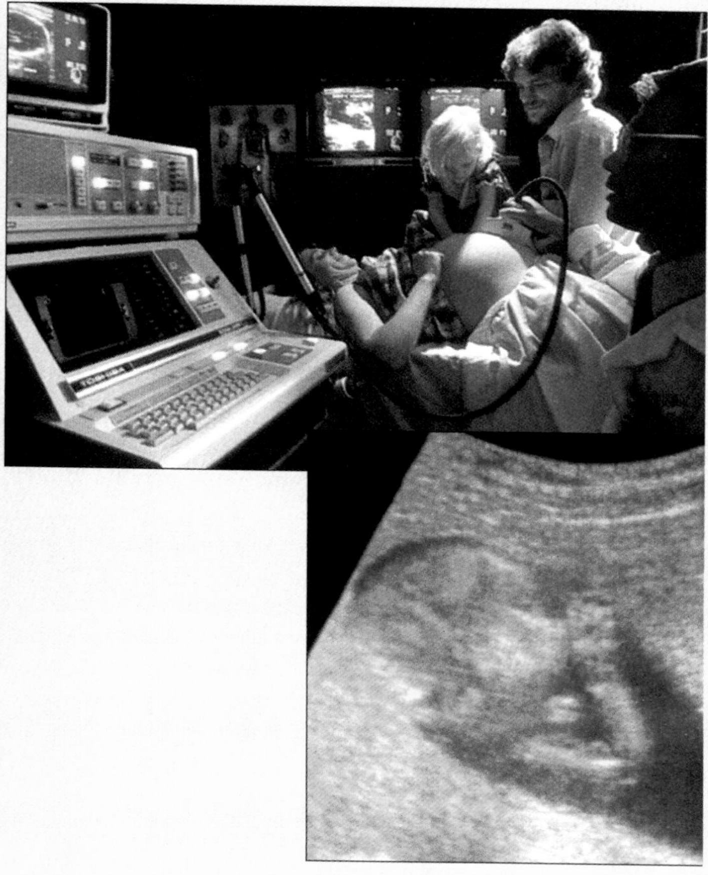

Figure 2 Fetal Sonogram. Shows the head and right arm of a 28-week-old fetus sucking its thumb.

construct a three-dimensional image of the body. CT scanning has the advantage of imaging thin sections of the body, so there is little overlap of organs and the image is much sharper than a conventional X ray. CT scanning is useful for identifying tumors, aneurysms, cerebral hemorrhages, kidney stones, and other abnormalities. It has virtually eliminated exploratory surgery.

Magnetic Resonance Imaging (MRI)

MRI, until recently known as *nuclear magnetic resonance (NMR) imaging,* was developed in the 1970s as a technique superior to CT scanning for visualizing soft tissues. The patient lies within a cylindrical chamber surrounded by a large electromagnet that creates a magnetic field 3,000 to 60,000 times as strong as the earth's. Hydrogen atoms in the tissues align themselves with the magnetic field. The patient is then bombarded with radio waves, which cause the hydrogen atoms to absorb additional energy and align in a different direction. When the radio waves are turned off, the hydrogen atoms abruptly realign themselves to the magnetic field, giving off their excess energy at different rates that depend on the type of tissue. A computer analyzes

Continued

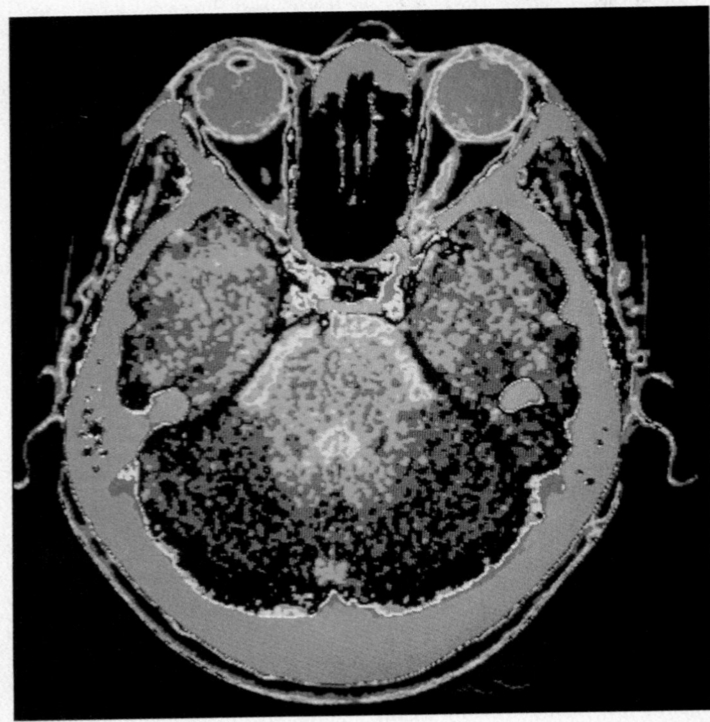

Figure 3 Computed Tomographic (CT) Scan of the Head at the Level of the Eyes. The eyes and skin are shown in blue, bone in red, and the brain in green.

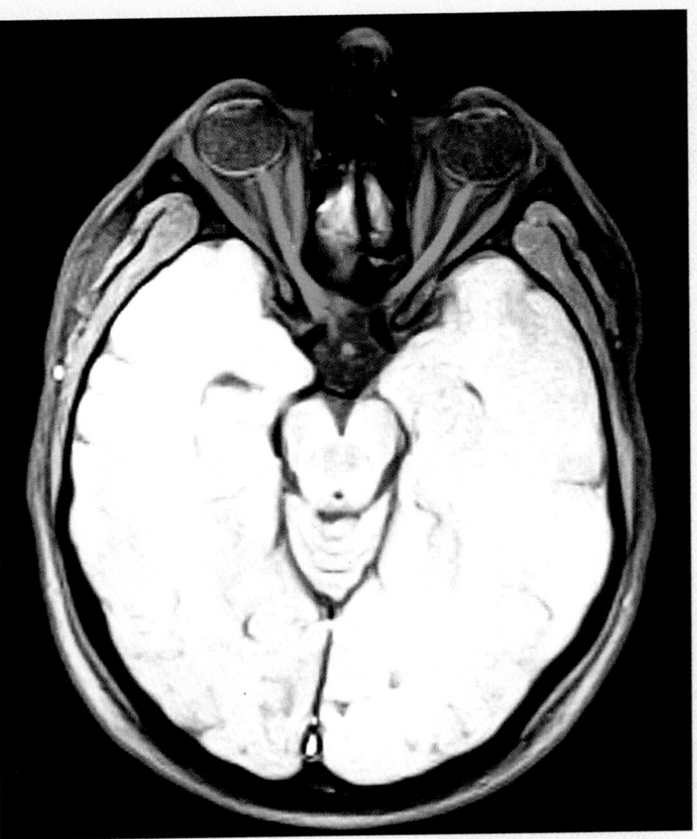

Figure 4 Magnetic Resonance Image (MRI) of the Head at the Level of the Eyes. The optic nerves appear in red and the muscles that move the eyes appear in green.

the emitted energy to produce an image of the body. MRI can "see" clearly through the skull and vertebral column to produce images of the nervous tissue. Moreover, it is better than CT for distinguishing between soft tissues such as the white and gray matter of the nervous system (fig. 4).

Positron Emission Tomography (PET)

The *PET scan,* developed in the 1970s, is used to assess the metabolic state of a tissue and to distinguish which tissues are most active at a given moment (fig. 5). The procedure begins with an injection of radioactively labeled glucose, which emits positrons (electron-like particles with a positive charge). When a positron and electron meet, they annihilate each other and give off a pair of gamma rays that can be detected by sensors and analyzed by computer. The result is a color image that shows which tissues were using the most glucose at the moment. In cardiology, PET scans can show the extent of damaged heart tissue. Since it consumes little or no glucose, the damaged tissue appears dark. Some applications to neuroscience are described in chapter 14. The PET scan is an example of *nuclear medicine*—the use of radioactive isotopes to treat disease or to form diagnostic images of the body (see clinical insight 2.4, p. 72).

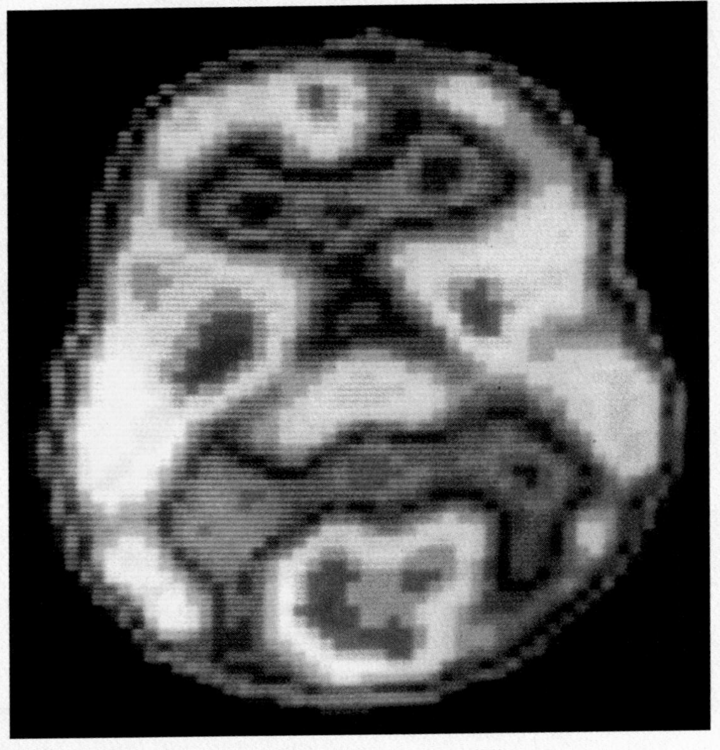

Figure 5 Positron Emission Tomographic (PET) Scan of the Brain of an Unmedicated Schizophrenic Patient. Red areas indicate high glucose consumption (high metabolism). In this patient, the visual center of the brain (rear of head, *bottom of photo*) was especially active when the scan was made.

Chapter Review

Online Learning Center

Study resources for this chapter are available at www.mhhe.com/saladin/

Study Outline

The Scope of Anatomy and Physiology (p. 2)

1. Anatomy: methods and subdisciplines
2. Physiology: methods and subdisciplines

The Origins of Biomedical Science (p. 3)

1. The beginnings of medicine
   a. Hippocrates
      - Medical ethics
      - Search for natural causes

   b. Aristotle
      - Natural versus supernatural causes
      - Unifying themes of nature

   c. Claudius Galen
      - Influential textbook
      - Anatomical studies
      - Science as a method of discovery
      - Encouragement of skepticism
2. The birth of modern medicine
   a. The Middle Ages
   b. Muslim medicine: Avicenna
   c. Birth of modern anatomy: Vesalius
   d. Birth of experimental physiology: Harvey
   e. The microscope: Leeuwenhoek, Hooke, and Zeiss
   f. Cell theory: Schleiden and Schwann
3. Living in a revolution: What's happening now?

Scientific Method (p. 7)

1. Bacon and Descartes
   a. Science as a public enterprise
   b. Quest for objectivity and method
2. The inductive method
   a. Making generalizations
   b. Meaning of proof
   c. Tentativeness of truth
3. The hypothetico-deductive method
   a. Criteria for a good hypothesis
      - Consistency with prior knowledge
      - Testability
      - Falsifiability

   b. Making a deduction
4. Experimental design
   a. Sample size
   b. Controls
   c. Psychosomatic effects
   d. Experimenter bias
   e. Statistical testing
5. Peer review
6. Facts, laws, and theories

Human Evolution (p. 9)

1. Darwin's impact
2. Evolution, selection, and adaptation
   a. Evolution: meaning and examples
   b. Natural selection
      - Selection pressures
      - Adaptations

   c. Relevance to medical research
3. Human evolution
   a. Adaptations to arboreal habitat
      - Prehensile hands
      - Stereoscopic vision
      - Color vision
      - Larger brain

   b. Adaptations for bipedalism
      - Skeletal changes
      - Larger brain
      - Earlier birth

   c. *Australopithecus* and *Homo*
   d. Darwinian medicine

The Nature of Human Life (p. 12)

1. What is life?
   a. Organization
   b. Cellular composition
   c. Biochemical unity
   d. Metabolism
   e. Responsiveness
   f. Homeostasis
   g. Development
   h. Growth
   i. Reproduction
   j. Evolution
2. What is a human?
   a. Our chordate characteristics
   b. Our vertebrate characteristics
   c. Our mammalian characteristics
   d. Our primate characteristics
   e. Our hominid characteristics
3. Structural hierarchy: from subatomic particles to organism
4. Reductionism and holism

Homeostasis and Feedback (p. 19)

1. Homeostasis
   a. Bernard and Cannon
   b. Set point and dynamic equilibrium
2. Negative feedback and stability
3. Positive feedback and rapid change
   a. Beneficial aspects
   b. Pathological aspects

Review of Major Themes (p. 21)

1. Cell theory
2. Homeostasis
3. Evolution
4. Hierarchy of structure
5. Unity of form and function

Selected Vocabulary

anatomy 2
physiology 2
dissection 2
comparative anatomy 2
cadaver 2
palpation 2
auscultation 2
percussion 2
gross anatomy 2
microscopic anatomy 3
histology 3
histopathology 3
ultrastructure 3
comparative physiology 3
Hippocrates 3
Aristotle 3
Claudius Galen 3
Avicenna 4
Andreas Vesalius 4
William Harvey 4
Antony van Leeuwenhoek 4
simple microscope 5
Robert Hooke 6
compound miscroscope 6
Carl Zeiss 6
Mathias Schleiden 6
Theodor Schwann 6
cell theory 6
Francis Bacon 7
René Descartes 7
scientific method 7
inductive method 7
hypothetico-deductive method 7
hypothesis 7
falsifiability 7
deduction 7
control group 8
treatment group 8
psychosomatic effect 8
placebo 8
double-blind method 8
statistical test 8
peer review 8
fact 9
law of nature 9
theory 9
Charles Darwin 9
evolution 10
natural selection 10
selection pressure 10
adaptation 10
model 10
arboreal 10
opposable 10
prehensile 10
stereoscopic vision 11
bipedalism 11
darwinian medicine 11
metabolism 14
stimuli 14
growth 14
differentiation 14
Chordata 14
notochord 14
gill pouch 16
Vertebrata 16
Mammalia 16
endothermy 16
heterodonty 16
Primates 16
Hominidae 17
subatomic particle 17
atom 17
molecule 17
organelle 17
cell 17
tissue 17
organ 17
organ system 17
organism 17
reductionism 17
holism 17
Claude Bernard 19
Walter Cannon 19
homeostasis 19
dynamic equilibrium 19
set point 19
negative feedback 19
feedback loop 20
vasodilation 20
vasoconstriction 20
receptor 20
integrating center 20
effector 20
positive feedback 20

Testing Your Recall

1. Structure that can be observed with the naked eye is called
   a. gross anatomy.
   b. ultrastructure.
   c. microscopic anatomy.
   d. macroscopic anatomy.
   e. cytology.

2. The prefix *homeo-* means
   a. tissue.
   b. metabolism.
   c. change.
   d. human.
   e. same.

3. The simplest structures considered to be alive are
   a. organisms.
   b. organs.
   c. tissues.
   d. cells.
   e. organelles.

4. Which of the following people revolutionized the teaching of gross anatomy?
   a. Vesalius
   b. Aristotle
   c. Hippocrates
   d. Leeuwenhoek
   e. Cannon

5. Which of the following embodies the greatest amount of scientific information?
   a. a fact
   b. a law of nature
   c. a theory
   d. a deduction
   e. a hypothesis

6. An informed, uncertain, but testable conjecture is
   a. a natural law.
   b. a scientific theory.
   c. a hypothesis.
   d. a deduction.
   e. a scientific fact.

7. A self-amplifying chain of physiological events is called
   a. positive feedback.
   b. negative feedback.
   c. dynamic constancy.
   d. homeostasis.
   e. metabolism.

8. Which of the following is *not* a human organ system?
   a. integumentary
   b. muscular
   c. epithelial
   d. nervous
   e. endocrine

9. Which of the following characteristics do humans *not* share with all other chordates?
   a. pharyngeal gill pouches
   b. a dorsal hollow nerve cord
   c. a tail extending beyond the anus
   d. a notochord
   e. a vertebral column

10. The prefix *hetero-* means
    a. same.
    b. different.
    c. both.
    d. solid.
    e. below.

11. Cutting and separating tissues to reveal structural relationships is called __________ .

12. ___________ invented many components of the compound microscope and named the cell.
13. By the process of ___________ , a scientist predicts what the result of a certain experiment will be if his or her hypothesis is correct.
14. Physiological effects of a person's mental state are called ___________ effects.
15. The tendency of the body to maintain stable internal conditions is called ___________ .
16. Blood pH averages 7.4 but fluctuates from 7.35 to 7.45. A pH of 7.4 can therefore be considered the ___________ for this system.
17. Self-corrective mechanisms in physiology are called ___________ loops.
18. A/an ___________ is the simplest body structure to be composed of two or more types of tissue.
19. Depth perception, or the ability to form three-dimensional images, is called ___________ vision.
20. Our hands are said to be ___________ because they can encircle an object such as a branch or tool. The presence of an ___________ thumb is important to this ability.

Answers in appendix B

Testing Your Comprehension

1. What aspect of William Harvey's view of blood circulation could be considered a scientific hypothesis? What would you predict from that hypothesis? What observation could you carry out today to test this hypothesis?
2. Which of the characteristics of living things are possessed by an automobile? What bearing does this have on our definition of life?
3. About 1 out of every 120 live-born infants has a structural defect in the heart such as a hole between two heart chambers. Such infants often suffer pulmonary congestion and heart failure, and about one-third of them die as a result. Which of the major themes in this chapter does this illustrate? Explain your answer.
4. Why is a monkey not classified as human? Why is a horse not classified as a primate? (Hint: What characteristics must an animal have to be a human or a primate, and which of these are monkeys or horses lacking?)
5. Suppose you have been doing heavy yard work on a hot day and sweating profusely. You become very thirsty, so you drink a tall glass of lemonade. Explain how your thirst relates to the concept of homeostasis. Which type of feedback—positive or negative—does this illustrate?

Answers at the Online Learning Center

ATLAS

General Orientation to Human Anatomy

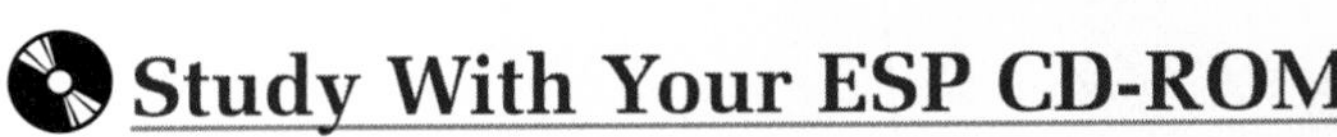

Study With Your ESP CD-ROM

Refer to your ESP Tutorial CD-ROM for helpful animations, activities, and quizzes related to the material covered in this chapter. The following are suggested areas to go to on the CD and are presented according to "Unit/Topic":
- Levels of Organization/Introduction

Anatomical Position

Anatomical position is a stance in which a person stands erect with the feet flat on the floor, arms at the sides, and the palms, face, and eyes facing forward (fig. A.1). This position provides a precise and standard frame of reference for anatomical description and dissection. Without such a frame of reference, to say that a structure such as the sternum, thymus, or aorta is "above the heart" would be vague, since it would depend on whether the subject was standing, lying face down, or lying face up. From the perspective of anatomical position, however, we can describe the thymus as *superior* to the heart, the sternum as *anterior* or *ventral* to it, and the aorta as *posterior* or *dorsal* to it. These descriptions remain valid regardless of the subject's position.

Unless stated otherwise, assume that all anatomical descriptions refer to anatomical position. Bear in mind that if a subject is facing you in anatomical position, the subject's left will be on your right and vice versa. In most anatomical illustrations, for example, the left atrium of the heart appears toward the right side of the page, and while the appendix is located in the right lower quadrant of the abdomen, it appears on the left side of most illustrations.

The forearm is said to be **supine** when the palms face forward or upward and **prone** when they face rearward or downward (fig. A.2). The difference is particularly important to descriptions of anatomy of this region. In the supine position, the two forearm bones (radius and ulna) are parallel and the radius is lateral to the ulna. In the prone position, the radius and ulna cross; the radius is lateral to the ulna at the elbow but medial to it at the wrist. Descriptions of nerves, muscles, blood vessels, and other structures of the arm assume that the arm is supine.

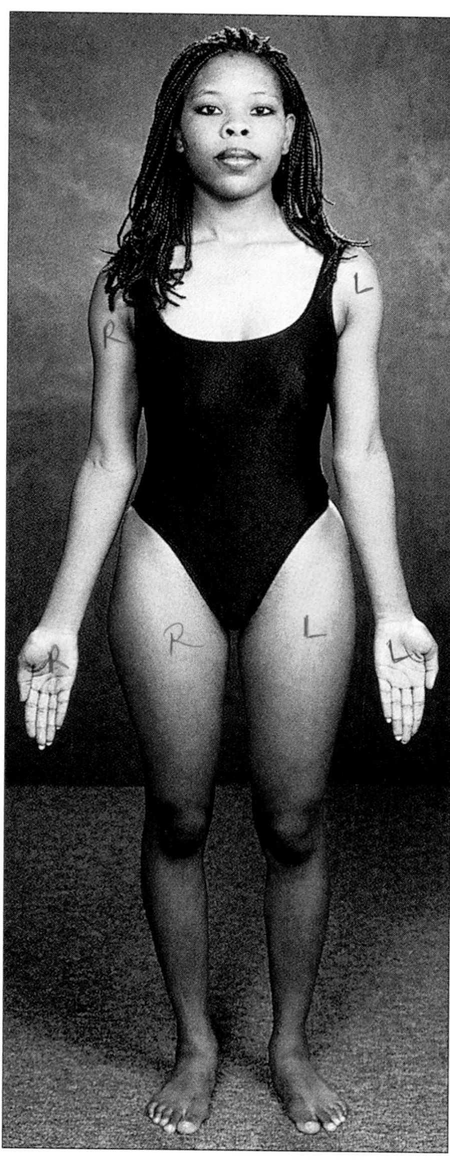

Figure A.1 Anatomical Position. The feet are flat on the floor, the arms are held downward and supine, and the face and eyes are directed forward.

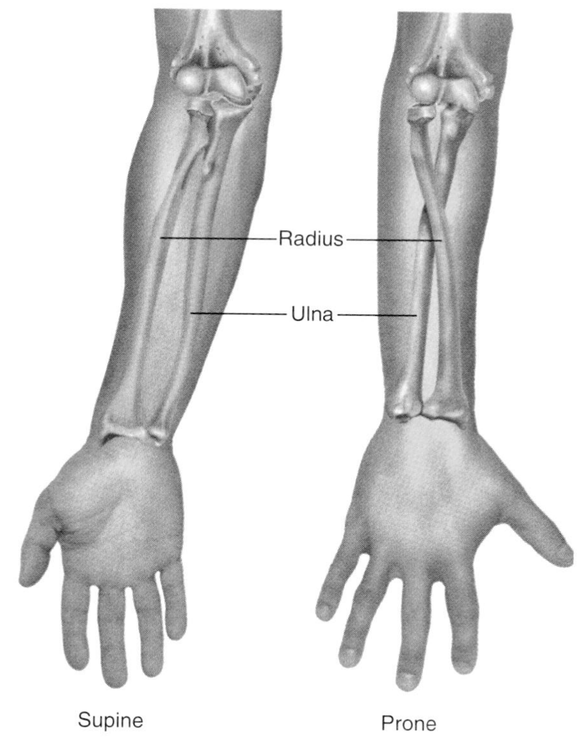

Figure A.2 Positions of the Forearm. When the forearm is supine, the palm faces forward; when prone, the palm faces to the rear. Note the difference in the relationship of the radius to the ulna.

Anatomical Planes

Many views of the body are based on real or imaginary "slices" called *sections* or *planes*. "Section" implies an actual cut or slice to reveal internal anatomy, whereas "plane" implies an imaginary flat surface passing through the body. The three major anatomical planes are *sagittal, frontal,* and *transverse* (fig. A.3)

A **sagittal**[1] (SADJ-ih-tul) **plane** extends vertically and divides the body or an organ into right and left portions. A **midsagittal plane** passes through the midline of the body and divides it into *equal* right and left halves. A **parasagittal**[2] **plane** lies to the right or left of the midsagittal plane and divides the body into *unequal* right and left portions (fig. A.4). While there is only one true midsagittal plane, there are an infinite number of parasagittal planes. The head and pelvic organs are commonly illustrated in midsagittal views (fig. A.5*a*).

A **frontal (coronal) plane** also extends vertically, but it is perpendicular to the sagittal plane and divides the body into anterior (front) and posterior (back) por-

1. *sagitta* = arrow
2. *para* = next to

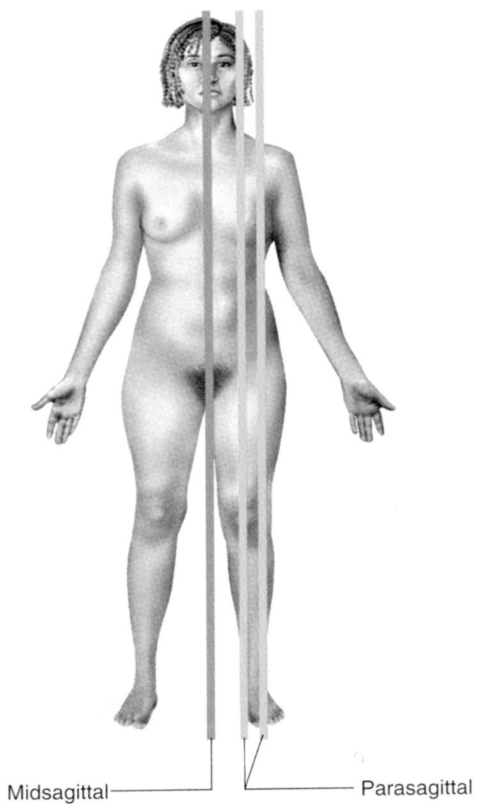

Figure A.4 **Sagittal Planes.** The midsagittal plane divides the body into equal halves. Parasagittal planes lie off center but parallel to the midsagittal plane.

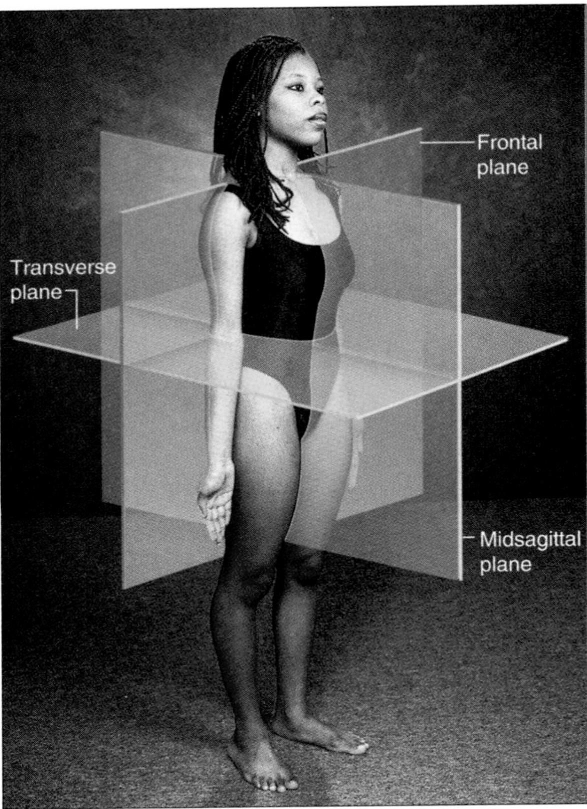

Figure A.3 Anatomical Planes of Reference.

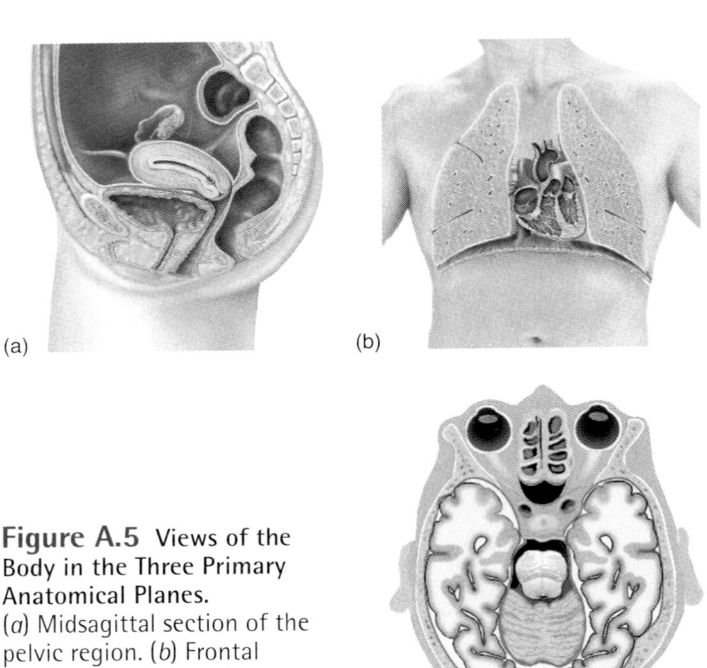

Figure A.5 **Views of the Body in the Three Primary Anatomical Planes.** (*a*) Midsagittal section of the pelvic region. (*b*) Frontal section of the thoracic region. (*c*) Transverse section of the head at the level of the eyes.

tions. A frontal section of the head, for example, would divide it into one portion bearing the face and another bearing the back of the head. Contents of the thoracic and abdominal cavities are most commonly shown in frontal section (fig. A.5*b*).

A **transverse plane**—also called a **horizontal** or **cross-sectional** plane—passes across the body or an organ perpendicular to its long axis (fig. A.5*c*); therefore, it divides the body or organ into superior (upper) and inferior (lower) portions. CT scans are typically transverse sections (see fig. 3, page 24).

Directional Terms

Table A.1 summarizes frequently used terms that describe the position of one structure relative to another. Intermediate directions are often indicated by combinations of these terms. For example, one structure may be described as *dorsolateral* to another (toward the back and side).

Because of the bipedal, upright stance of humans, some directional terms have different meanings for humans than they do for other animals. *Anterior,* for example, denotes the region of the body that leads the way in normal locomotion. For a four-legged animal such as a cat, this is the head end of the body; for a human, however, it is the area of the chest and abdomen. Thus, *anterior* has the same meaning as *ventral* for a human but not for a cat. *Posterior* denotes the region of the body that comes last in normal locomotion—the tail end of a cat but the dorsal side (back) of a human. These differences must be kept in mind when dissecting other animals for comparison to human anatomy.

Surface Anatomy

Knowledge of the external anatomy and landmarks of the body is important in performing a physical examination and many other clinical procedures. For purposes of study, the body is divided into two major regions called the *axial* and *appendicular regions.* Smaller areas within the major regions are described in the following paragraphs and illustrated in figure A.6.

Axial Region

The **axial region** consists of the **head, neck** (*cervical*[3] *region*), and **trunk.** The trunk is further divided into the **thoracic region** above the diaphragm and the **abdominal region** below it.

One way of referring to the locations of abdominal structures is to divide the region into quadrants. Two perpendicular lines intersecting at the umbilicus (navel) divide the abdomen into a **right upper quadrant (RUQ), right lower quadrant (RLQ), left upper quadrant (LUQ),** and **left lower quadrant (LLQ)** (fig. A.7*a, b*). The quadrant scheme is often used to describe the site of an abdominal pain or abnormality.

The abdomen also can be divided into nine regions defined by four lines that intersect like a tic-tac-toe grid (fig. A.7*c, d*). Each vertical line is called a **midclavicular line** because it passes through the midpoint of the clavicle (collarbone). The superior horizontal line is called the **subcostal**[4] **line** because it connects the inferior borders of

3. *cervic* = neck
4. *sub* = below + *cost* = rib

Table A.1 Directional Terms in Human Anatomy

Term	Meaning	Examples of Usage
Ventral	Toward the front* or belly	The aorta is *ventral* to the vertebral column.
Dorsal	Toward the back or spine	The vertebral column is *dorsal* to the aorta.
Anterior	Toward the ventral side*	The sternum is *anterior* to the heart.
Posterior	Toward the dorsal side*	The esophagus is *posterior* to the trachea.
Superior	Above	The heart is *superior* to the diaphragm.
Inferior	Below	The liver is *inferior* to the diaphragm.
Medial	Toward the midsagittal plane	The heart is *medial* to the lungs.
Lateral	Away from the midsagittal plane	The eyes are *lateral* to the nose.
Proximal	Closer to the point of attachment or origin	The elbow is *proximal* to the wrist.
Distal	Farther from the point of attachment or origin	The fingernails are at the *distal* ends of the fingers.
Superficial	Closer to the body surface	The skin is *superficial* to the muscles.
Deep	Farther from the body surface	The bones are *deep* to the muscles.

*In humans only; definition differs for other animals.

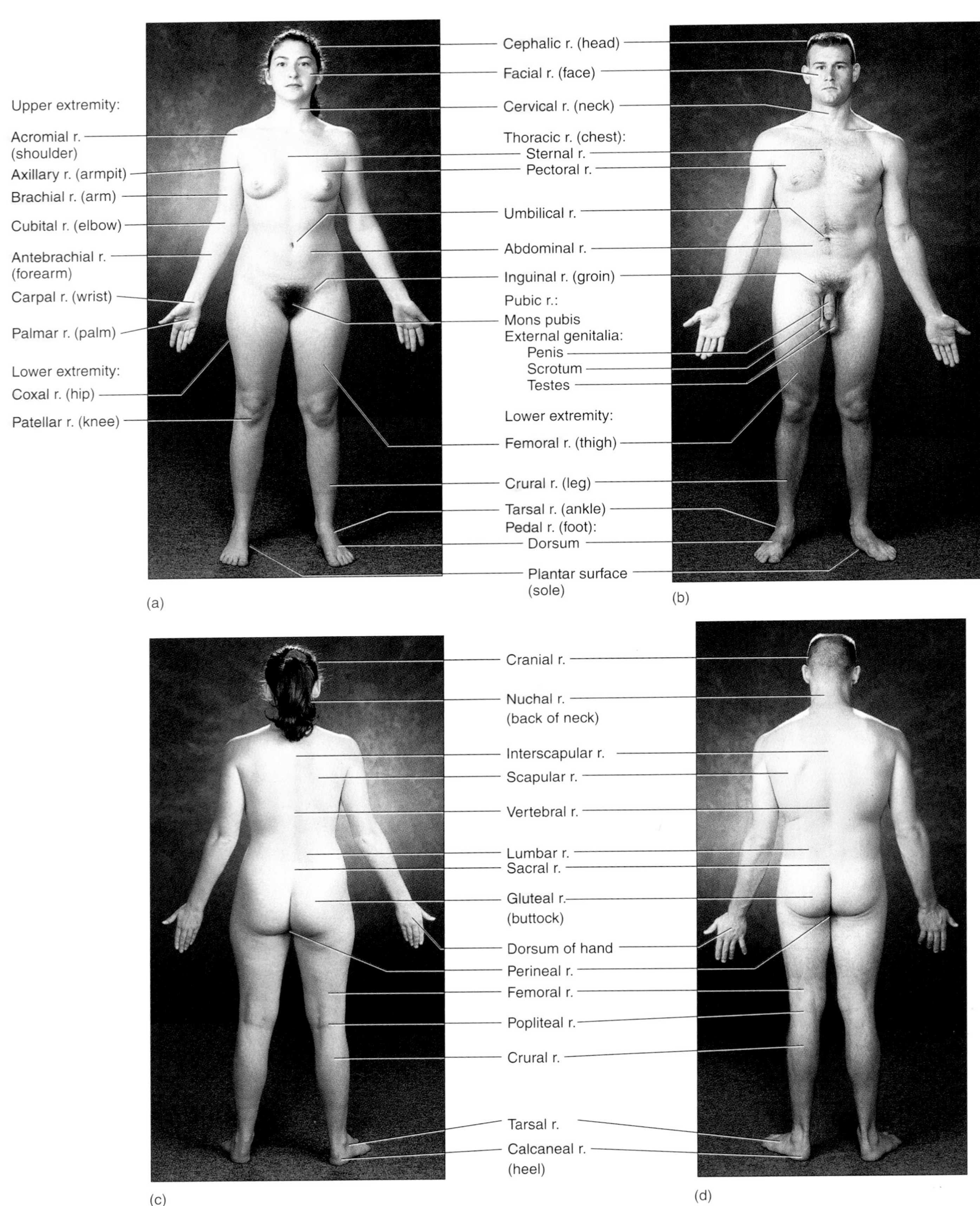

Figure A.6 The Adult Female and Male Bodies. (*a* and *b*) Ventral aspect. (*c* and *d*) Dorsal aspect. (*r.* = region.)

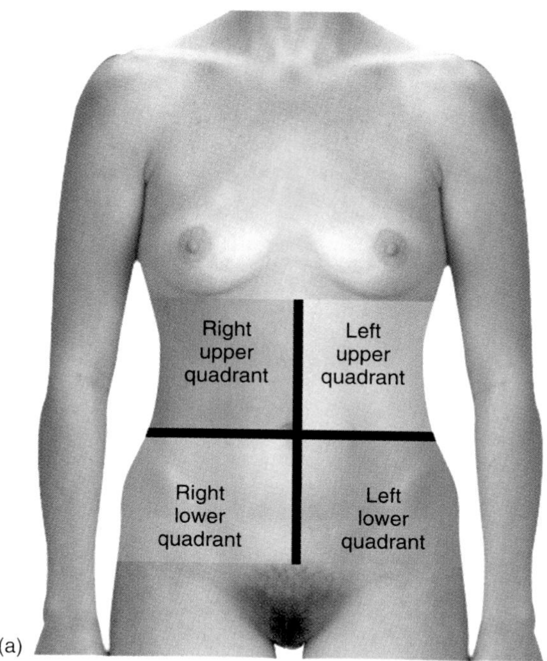

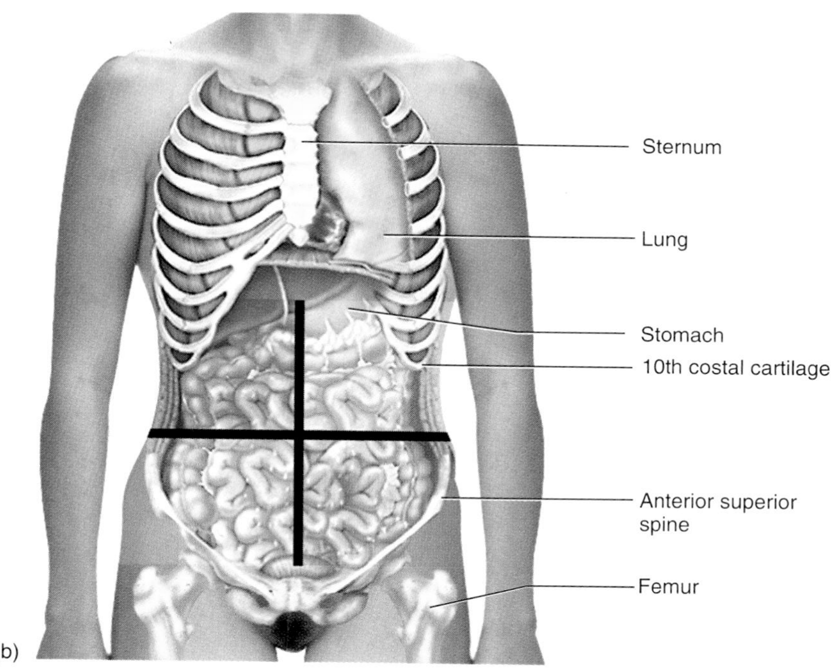

Figure A.7 Division of the Abdomen into Quadrants and Regions. (*a*) External division into four abdominal quadrants. (*b*) Internal anatomy correlated with the four quadrants. *(continued)*

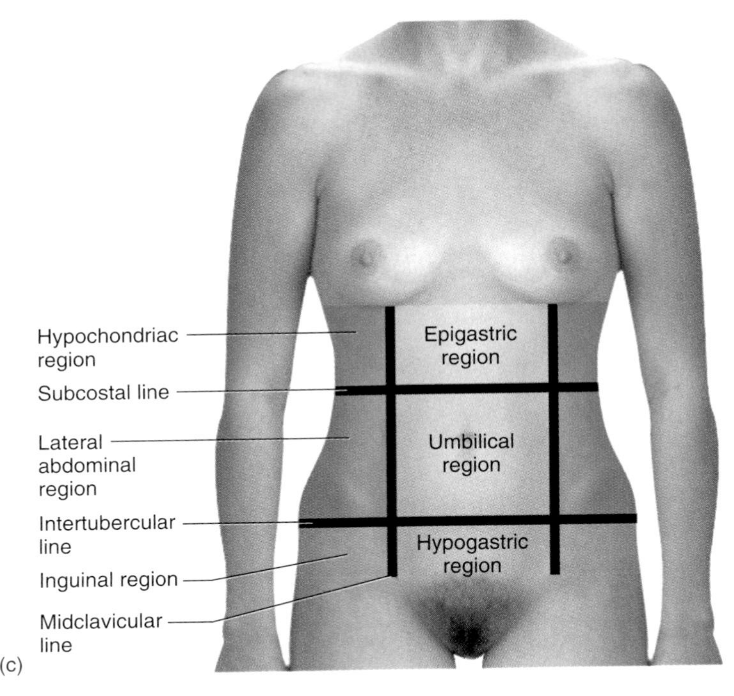

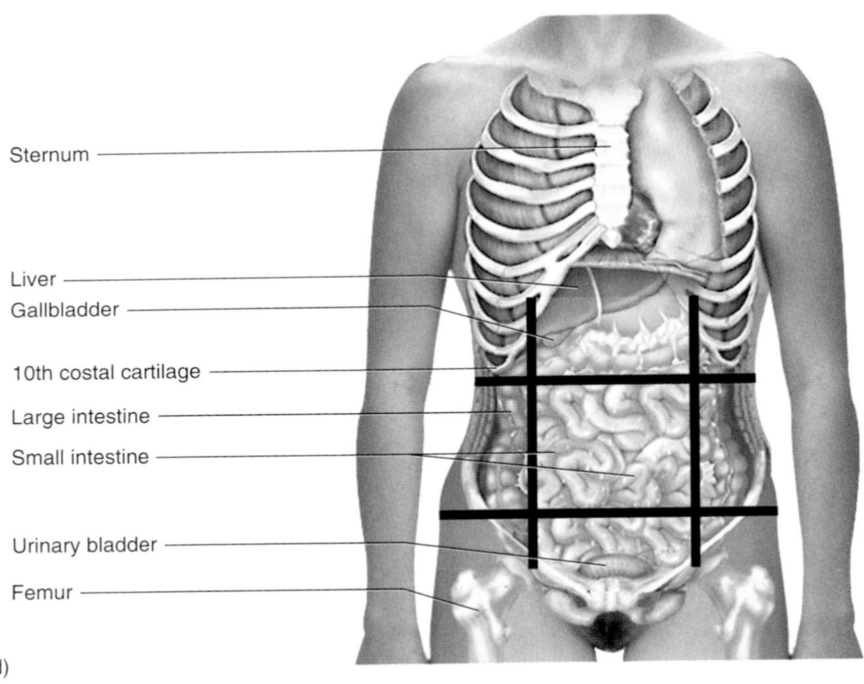

Figure A.7 Division of the Abdomen into Quadrants and Regions (continued). (*c*) External division into nine abdominal regions. (*d*) Internal anatomy correlated with the nine regions.

the lowest costal cartilages (cartilage connecting the tenth rib on each side to the inferior end of the sternum). The inferior horizontal line is called the **intertubercular**[5] **line** because it passes from left to right between the tubercles *(anterior superior spines)* of the pelvis—two points of bone located about where the front pockets open on most pants. The three lateral regions of this grid, from upper to lower, are the **hypochondriac,**[6] **lateral abdominal (lumbar),** and **inguinal**[7] **(iliac) regions.** The three medial regions from upper to lower are the **epigastric,**[8] **umbilical,** and **hypogastric (pubic)** regions.

Appendicular Region

The **appendicular** (AP-en-DIC-you-lur) **region** of the body consists of the appendages (also called *limbs* or *extremities*): the **upper limbs** and the **lower limbs.** The upper limb includes the **brachium** (BRAY-kee-um) (arm), **antebrachium**[9] (AN-teh-BRAY-kee-um) (forearm), **carpus** (wrist), **manus** (hand), and **digits** (fingers). The lower limb includes the **thigh, crus** (leg), **tarsus** (ankle), **pes** (foot), and **digits** (toes).

In strict anatomical terms, *arm* refers only to that part of the upper limb between the shoulder and elbow. *Leg* refers only to that part of the lower limb between the knee and ankle.

5. *inter* = between + *tubercul* = little swelling
6. *hypo* = below + *chondr* = cartilage
7. *inguin* = groin
8. *epi* = above, over + *gastr* = stomach
9. *ante* = fore, before + *brachi* = arm

Body Cavities and Membranes

The body is internally divided into three major **body cavities** (fig. A.8). The organs within them are called the **viscera** (VISS-er-uh) (singular, *viscus*[10]). Various membranes line the cavities, cover the viscera, and hold the viscera in place (table A.2).

Dorsal Body Cavity

The **dorsal body cavity** has two subdivisions: (1) the **cranial** (CRAY-nee-ul) **cavity,** which is enclosed by the cranium (braincase) and contains the brain, and (2) the **vertebral canal,** which is enclosed by the vertebral column (backbone) and contains the spinal cord. The dorsal body cavity is lined by three membrane layers called the **meninges** (meh-NIN-jeez). Among other functions, the meninges protect the delicate nervous tissue from the hard protective bone that encloses it.

Ventral Body Cavity

During embryonic development, a space called the **coelom** (SEE-loam) forms within the trunk and eventually gives rise to the **ventral body cavity.** This cavity later becomes partitioned by a muscular sheet, the **diaphragm,** into a superior **thoracic cavity** and an inferior **abdominopelvic cavity.** The thoracic and abdominopelvic cavities are lined

10. *viscus* = body organ

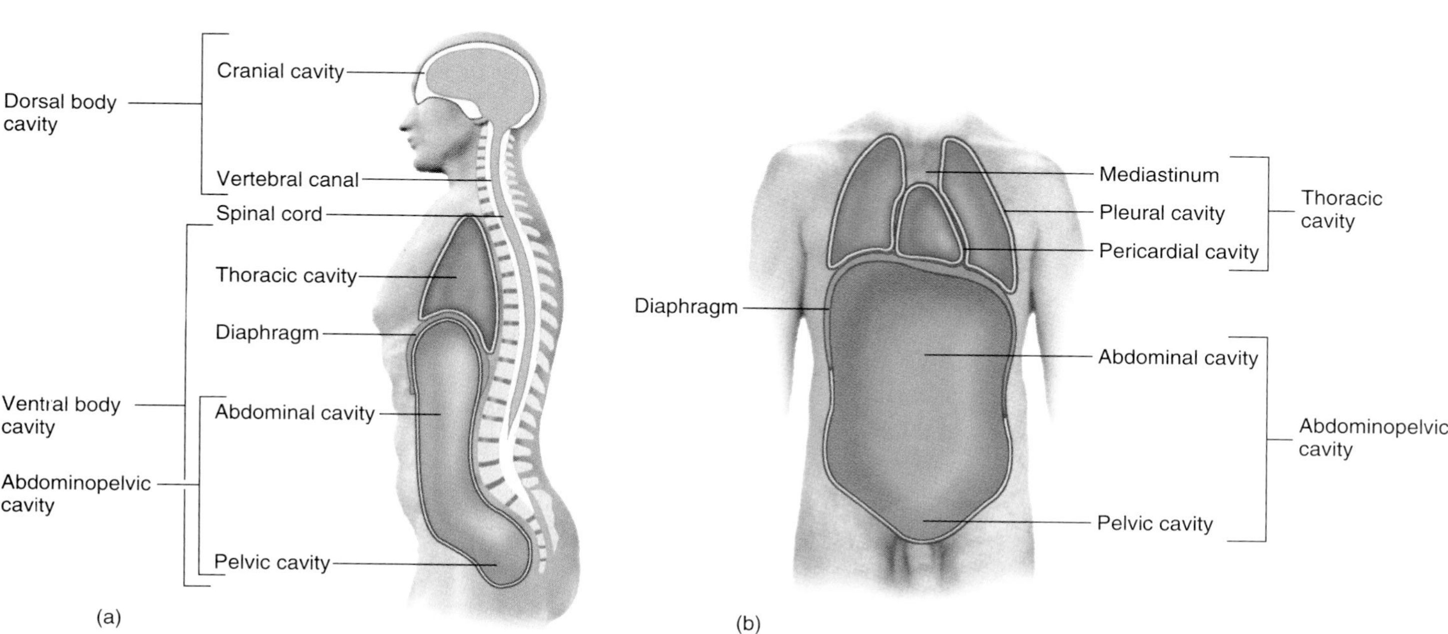

Figure A.8 **The Major Body Cavities.** (*a*) Left lateral view. (*b*) Anterior view.

Table A.2 Body Cavities and Membranes

Name of Cavity	Principal Contents	Membranous Lining
Dorsal Body Cavity		
Cranial cavity	Brain	Meninges
Vertebral canal	Spinal cord	Meninges
Ventral Body Cavity		
Thoracic Cavity		
Pleural cavities (2)	Lungs	Pleurae
Pericardial cavity	Heart	Pericardium
Abdominopelvic Cavity		
Abdominal cavity	Digestive organs, spleen, kidneys	Peritoneum
Pelvic cavity	Bladder, rectum, reproductive organs	Peritoneum

with thin **serous membranes.** These membranes secrete a lubricating film of moisture similar to blood serum (hence the name *serous*).

Thoracic Cavity

The thoracic cavity is divided into right, left, and medial portions by a partition called the **mediastinum**[11] (ME-dee-ass-TY-num) (fig. A.8). The right and left sides contain the lungs and are lined by a two-layered membrane called the **pleura**[12] (PLOOR-uh) (fig. A.9*a*). The outer layer, or **parietal**[13] (pa-RY-eh-tul) **pleura,** lies against the inside of the rib cage; the inner layer, or **visceral** (VISS-er-ul) **pleura,** forms the external surface of the lung. The narrow, moist space between the visceral and parietal pleurae is called the **pleural cavity** (see fig. A.20). It is lubricated by a slippery **pleural fluid.**

The medial portion, or mediastinum, is occupied by the esophagus and trachea, a gland called the thymus, and the heart and major blood vessels connected to it. The heart is enclosed by a two-layered membrane called the **pericardium.**[14] The **visceral pericardium** forms the heart surface while the **parietal pericardium** is separated from it by a space called the **pericardial cavity** (fig. A.9*b*). This space is lubricated by **pericardial fluid.**

Abdominopelvic Cavity

The abdominopelvic cavity consists of the **abdominal cavity** above the brim of the pelvis and the **pelvic cavity** below the brim (see fig. A.17). The abdominal cavity contains most of the digestive organs as well as the kidneys and ureters. The pelvic cavity is markedly narrower and its lower end tilts posteriorly (see fig. A.8*a*). It contains the distal part of the large intestine, the urinary bladder and urethra, and the reproductive organs.

The abdominopelvic cavity contains a moist serous membrane called the **peritoneum**[15] (PERR-ih-toe-NEE-

11. *mediastinum* = in the middle

12. *pleur* = rib, side
13. *pariet* = wall
14. *peri* = around + *cardi* = heart
15. *peri* = around + *tone* = stretched

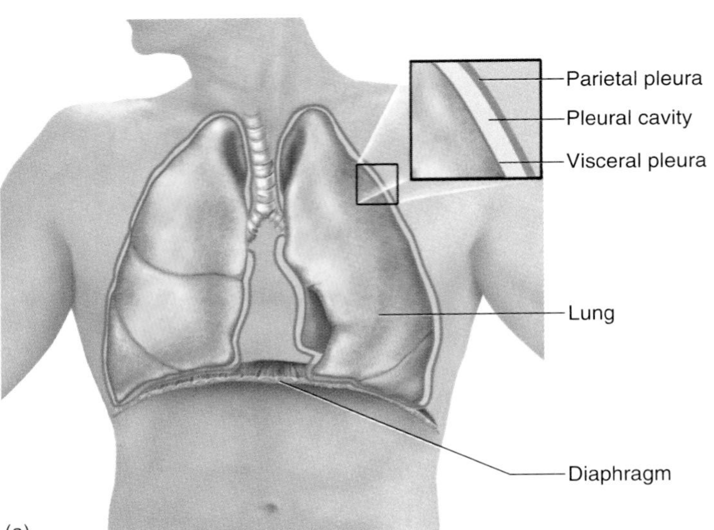

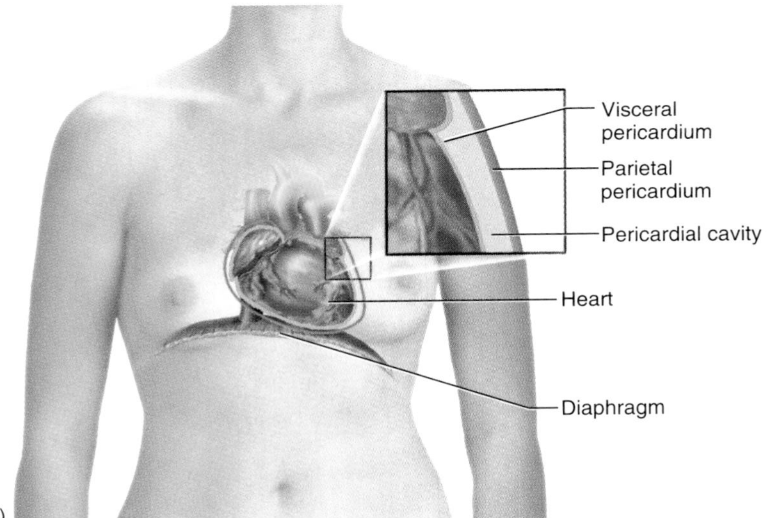

Figure A.9 Parietal and Visceral Layers of Double-Walled Membranes. (*a*) The pleura. (*b*) The pericardium.

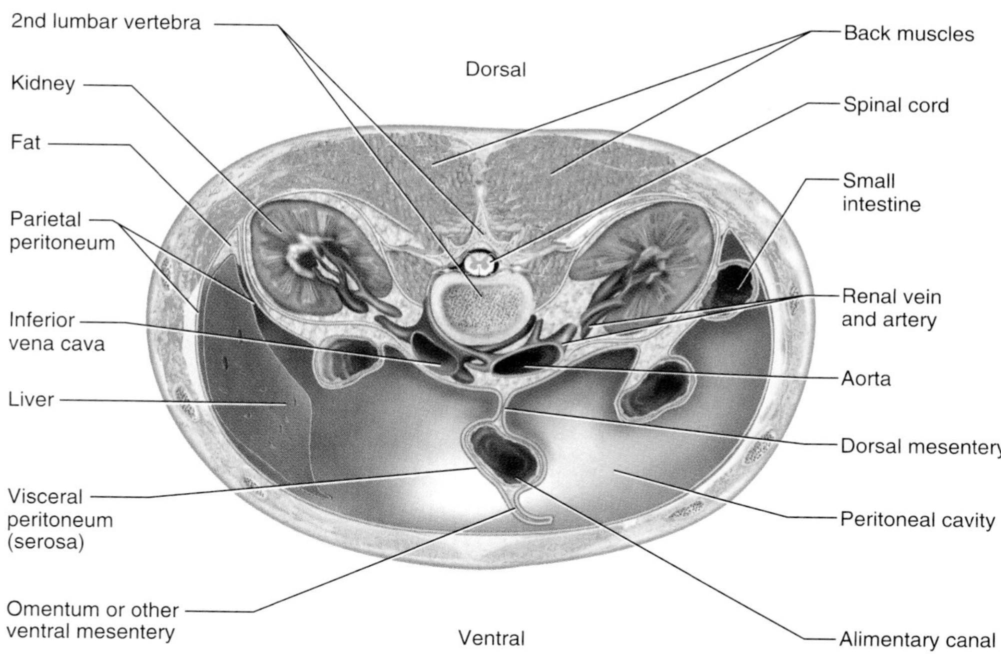

Figure A.10 **Transverse Section Through the Abdominal Cavity.** Shows the peritoneum, peritoneal cavity (with most viscera omitted), and some retroperitoneal organs.

um). The **parietal peritoneum** lines the walls of the cavity, while the **visceral peritoneum** covers the external surfaces of most digestive organs. The **peritoneal cavity** is the space between the parietal and visceral layers. It is lubricated by **peritoneal fluid.**

Some organs of the abdominal cavity lie between the peritoneum and dorsal body wall (outside of the peritoneal cavity), so they are said to have a **retroperitoneal**[16] position (fig. A.10). These include the kidneys, ureters, adrenal glands, most of the pancreas, and abdominal portions of two major blood vessels—the aorta and inferior vena cava (see fig. A.16).

The intestines are suspended from the dorsal abdominal wall by a translucent membrane called the **mesentery**[17] (MESS-en-tare-ee). This is a continuation of the peritoneum that wraps around the intestines, forming a moist membrane called the **serosa** (seer-OH-sa) on their outer surfaces (fig. A.11). The mesentery of the large intestine is called the **mesocolon.** The visceral peritoneum consists of the mesenteries and serosae.

A fatty membrane called the **greater omentum**[18] hangs like an apron from the inferolateral margin of the stomach and overlies the intestines (figs. A.11 and A.14). It is unattached at its inferior border and can be lifted to reveal the intestines. A smaller **lesser omentum** extends from the superomedial border of the stomach to the liver.

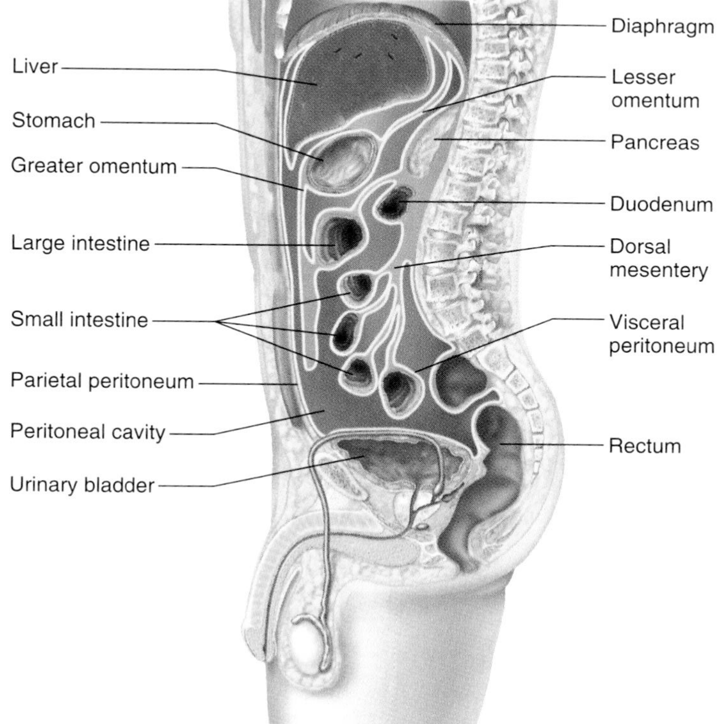

Figure A.11 **Serous Membranes of the Abdominal Cavity.** Midsagittal section, left lateral view.

16. *retro* = behind
17. *mes* = in the middle + *enter* = intestine
18. *omentum* = covering

Organ Systems

The human body has 11 **organ systems** (fig. A.12) and an immune system, which is better described as a population of cells than as an organ system. These systems are classified in the following list by their principal functions, but this is an unavoidably flawed classification. Some organs belong to two or more systems—for example, the male urethra is part of both the urinary and reproductive systems; the pharynx is part of the respiratory and digestive systems; and the mammary glands can be considered part of the integumentary and female reproductive systems.

Protection, Support, and Movement
Integumentary system
Skeletal system
Muscular system

Internal Communication and Integration
Nervous system
Endocrine system

Fluid Transport
Circulatory system
Lymphatic system

Defense
Immune system

Input and Output
Respiratory system
Urinary system
Digestive system

Reproduction
Reproductive system

A Visual Survey of the Body

Figures A.13 through A.17 provide an overview of the anatomy of the trunk and internal organs of the thoracic and abdominopelvic cavities. Figures A.18 through A.24 are photographs of the cadaver showing the major organs of the dorsal and ventral body cavities.

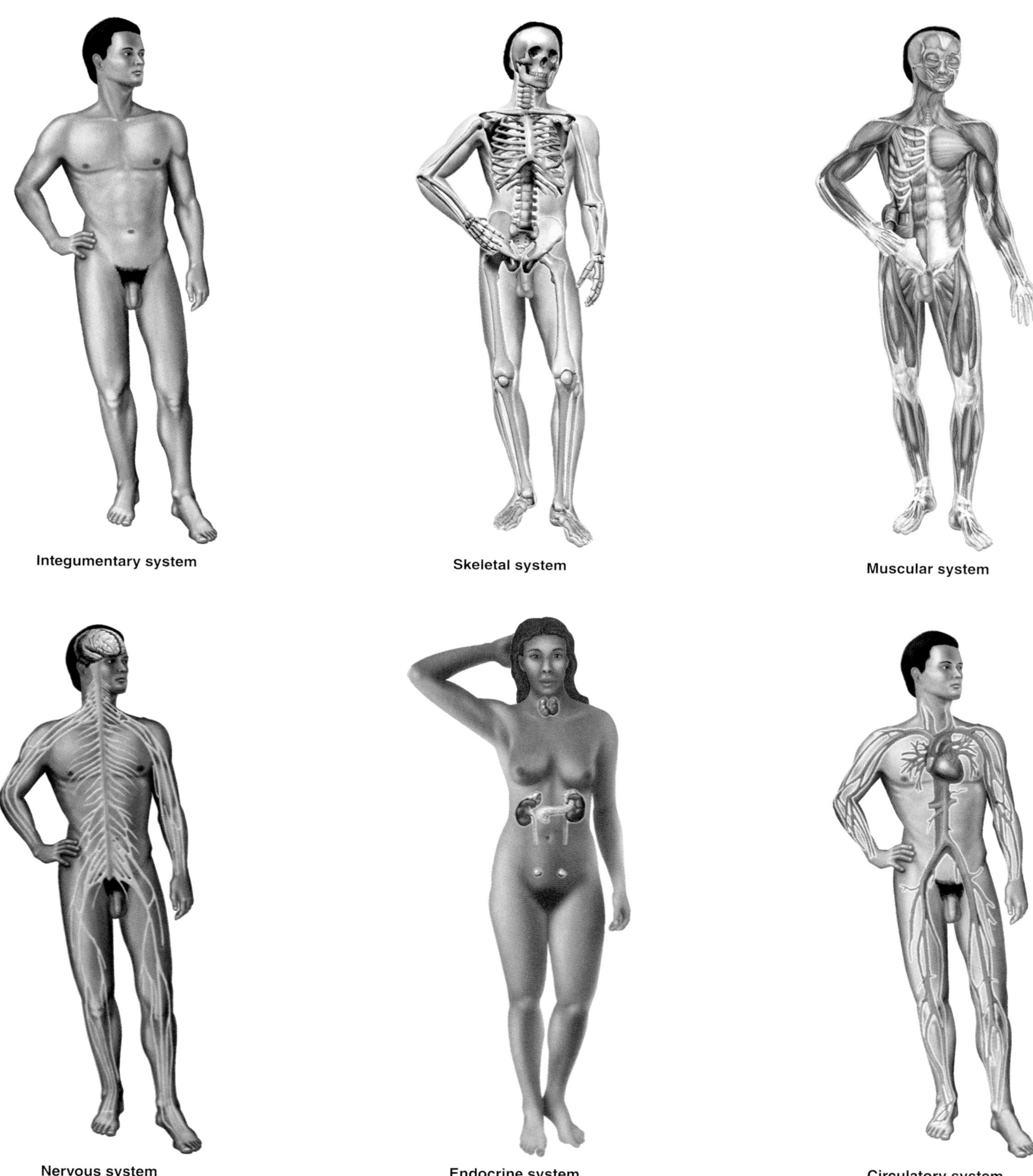

Figure A.12 The 11 Organ Systems.

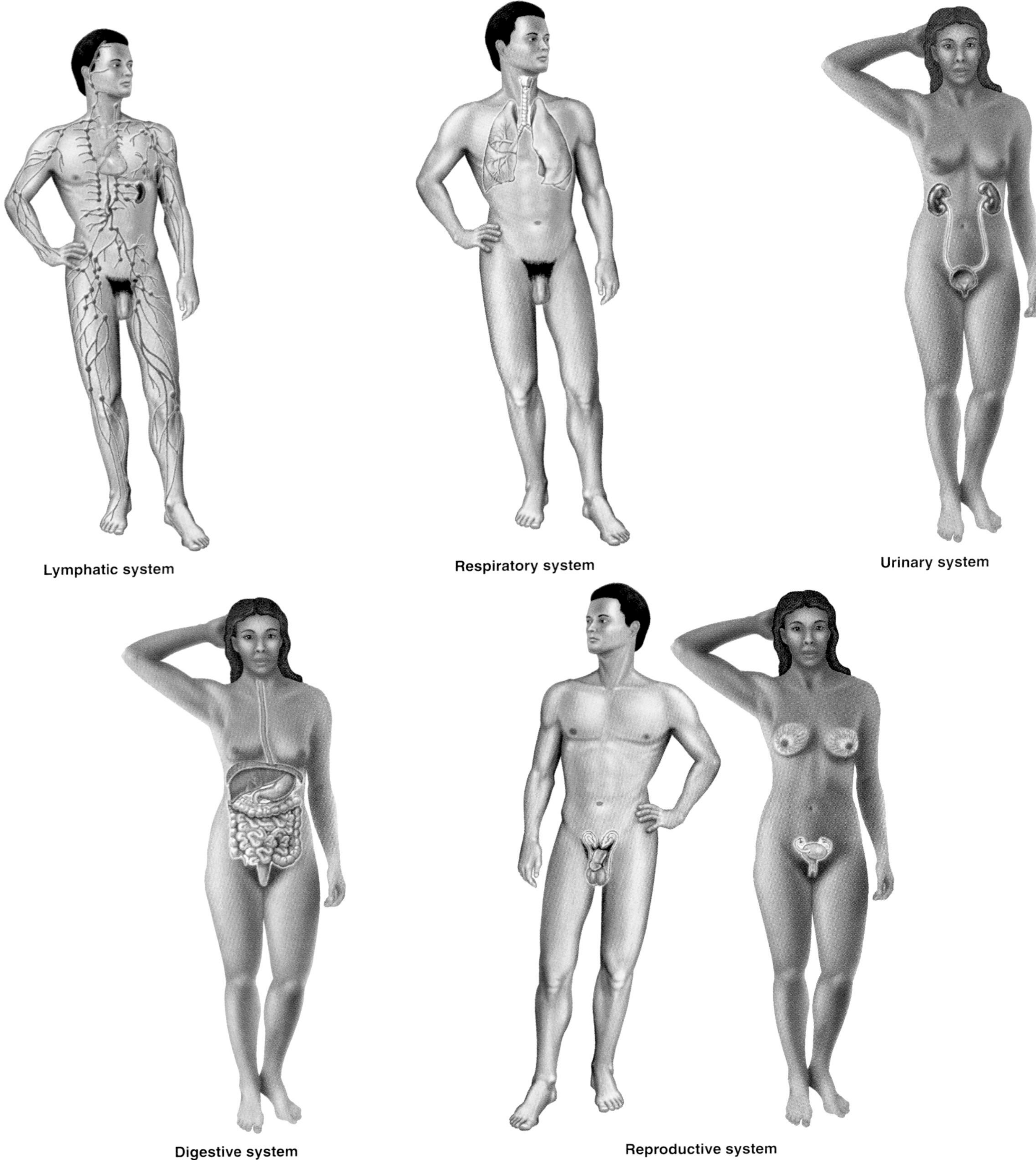

Figure A.12 The 11 Organ Systems (continued).

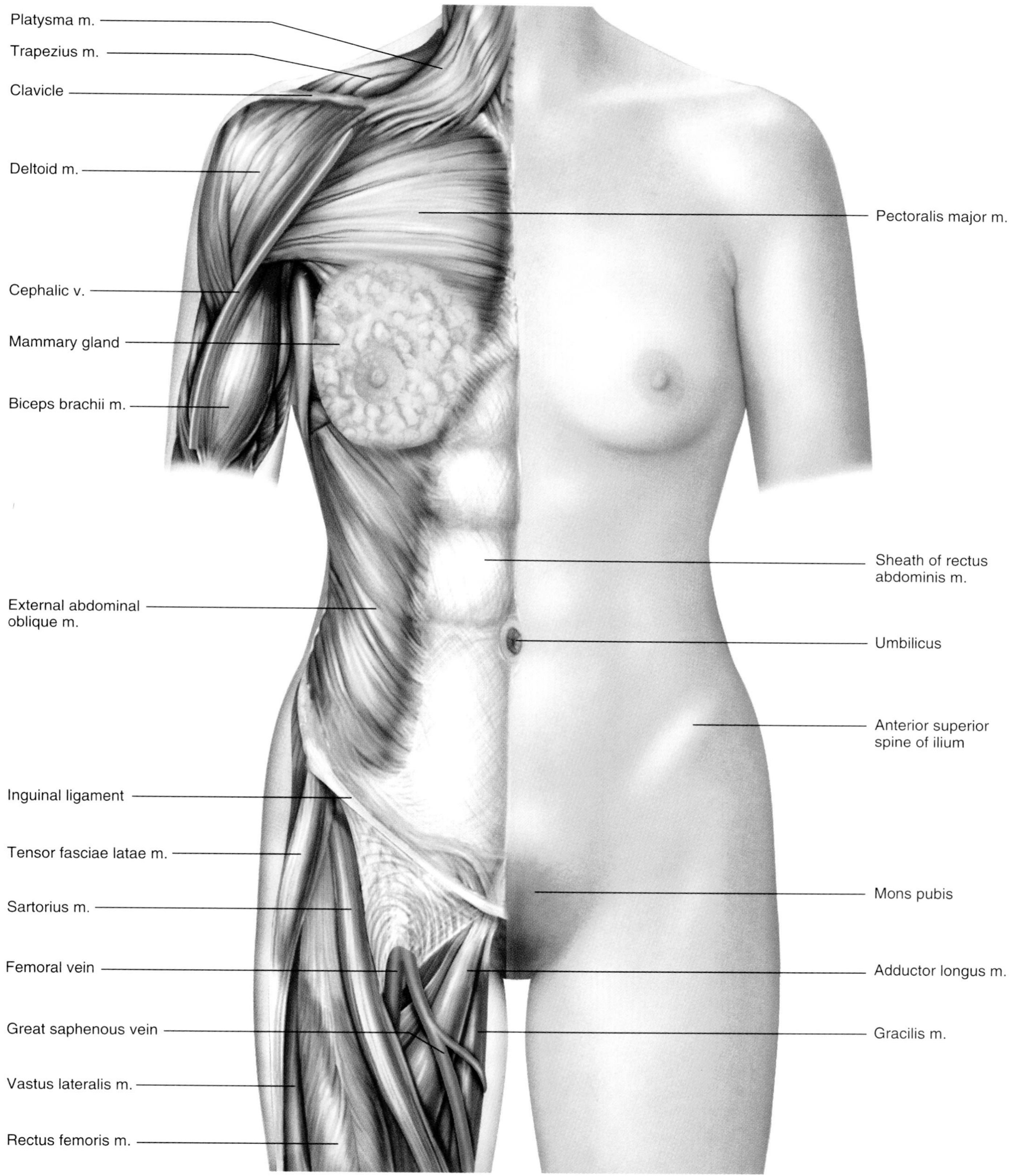

Figure A.13 **Superficial Anatomy of the Trunk (female).** Surface anatomy is shown on the anatomical left, and structures immediately deep to the skin on the right (*m.* = muscle; *v.* = vein).

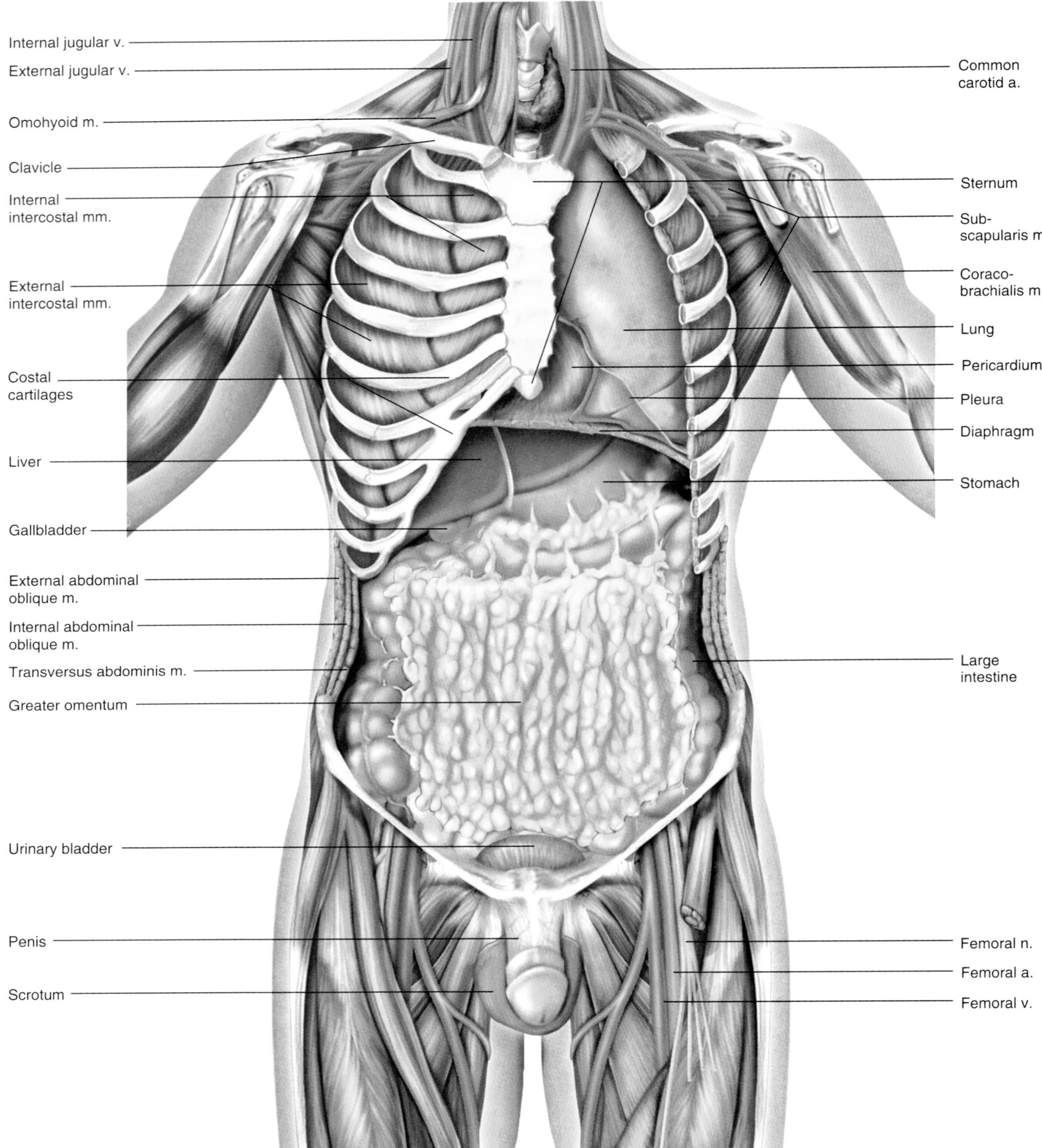

Figure A.14 Anatomy at the Level of the Rib Cage and Greater Omentum (male). The anterior body wall is removed, and the ribs, intercostal muscles, and pleura are removed from the anatomical left. (*a.* = artery; *v.* = vein; *m.* = muscle; *n.* = nerve.)

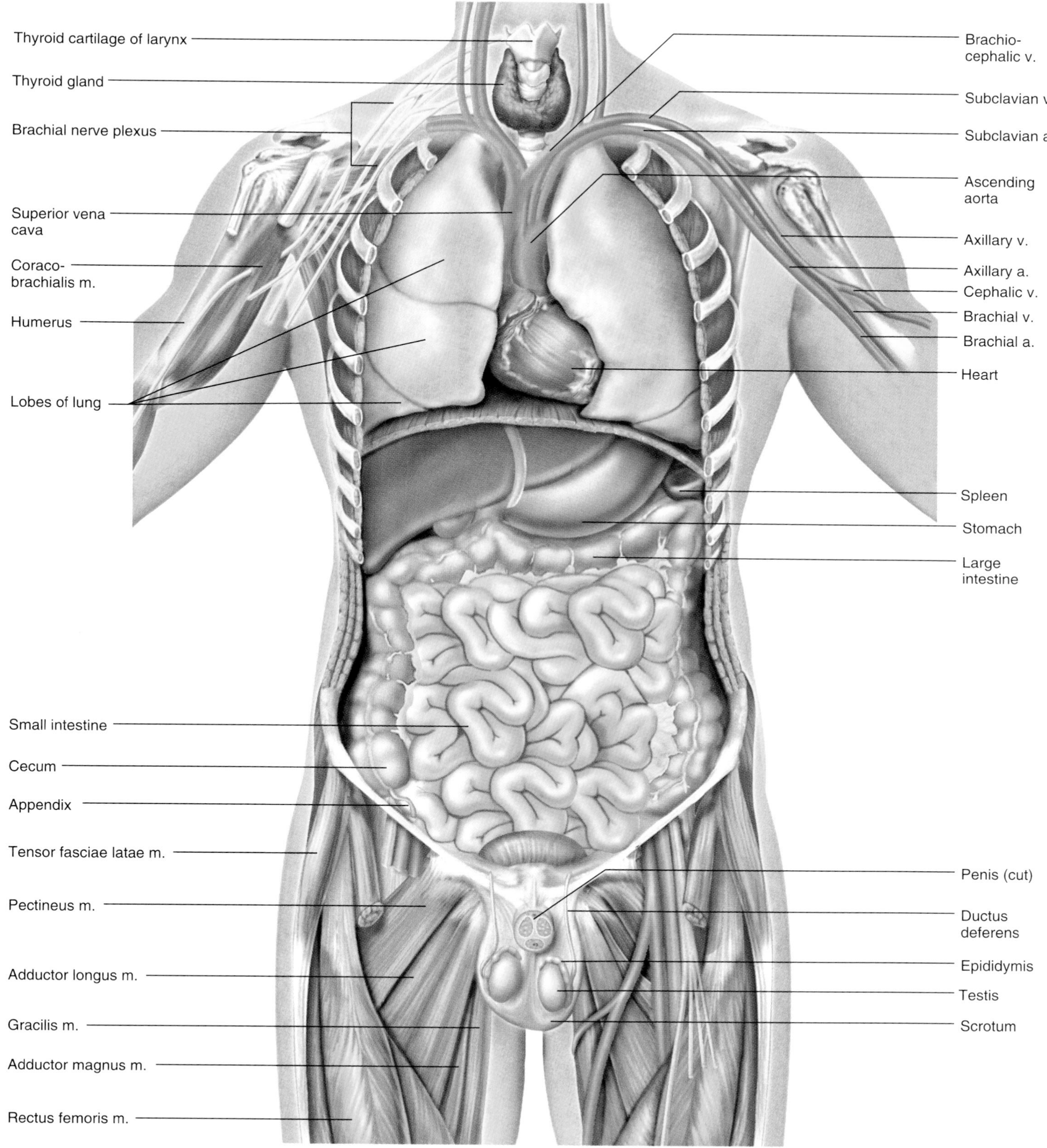

Figure A.15 Anatomy at the Level of the Lungs and Intestines (male). The sternum, ribs, and greater omentum are removed. (*a.* = artery; *v.* = vein; *m.* = muscle.)

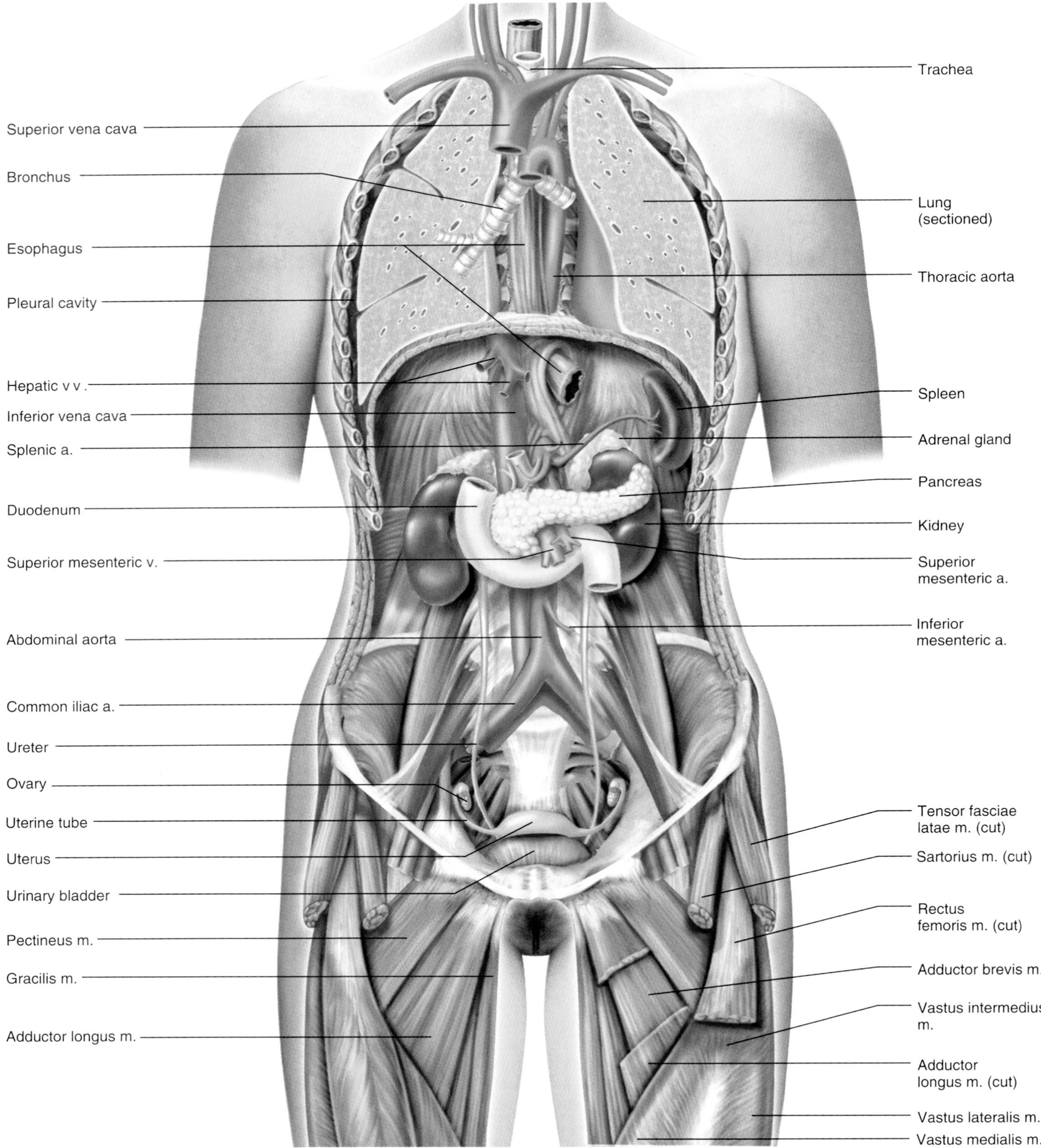

Figure A.16 **Anatomy at the Level of the Retroperitoneal Organs (female).** The heart is removed, the lungs are frontally sectioned, and the viscera of the peritoneal cavity and the peritoneum itself are removed. (*a.* = artery; *v.* = vein; *vv.* = veins; *m.* = muscle.)

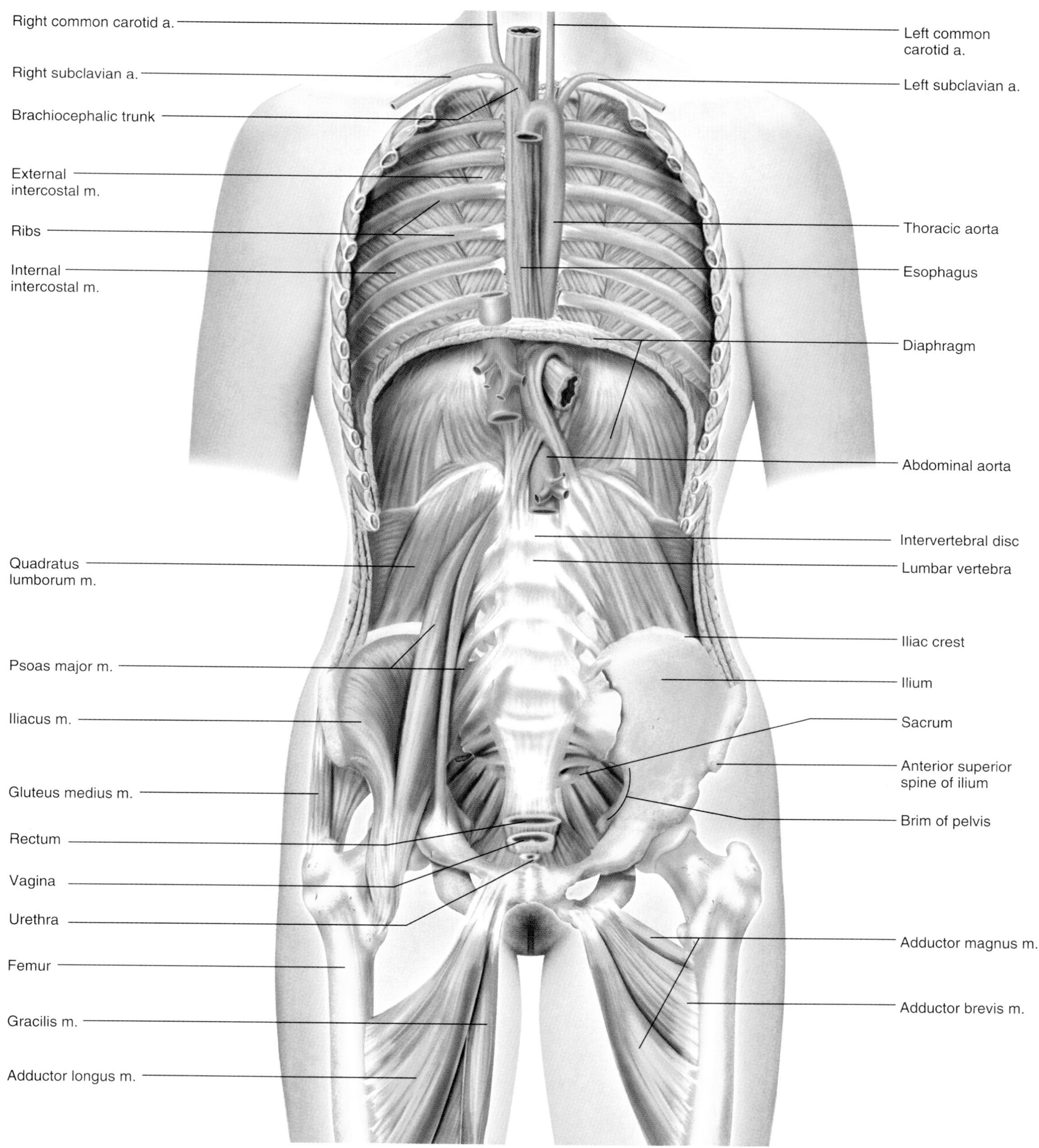

Figure A.17 **Anatomy at the Level of the Dorsal Body Wall (female).** The lungs and retroperitoneal viscera are removed. (*a.* = artery; *m.* = muscle.)

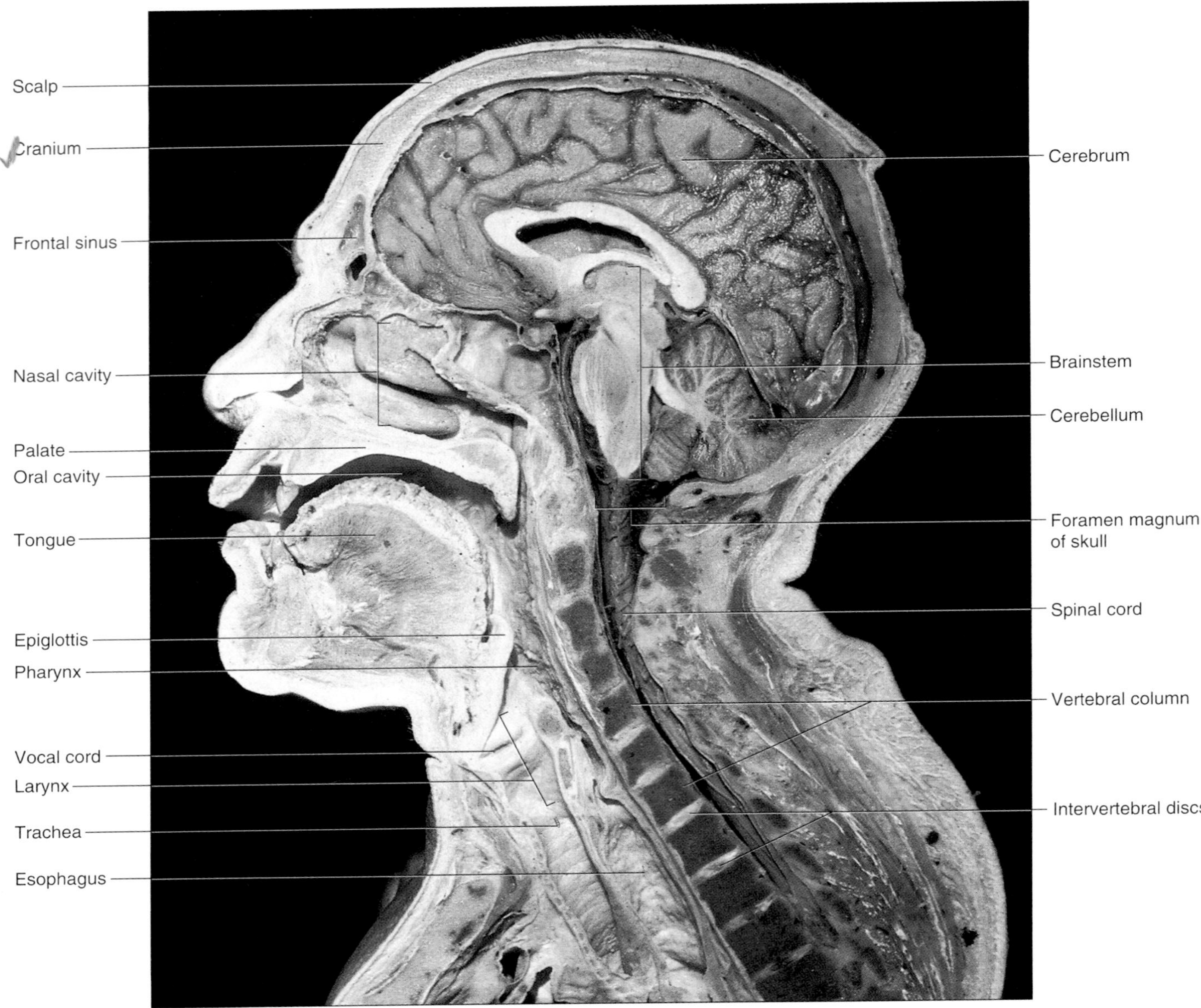

Figure A.18 **Midsagittal Section of the Head.** Shows contents of the cranial, nasal, and buccal cavities.

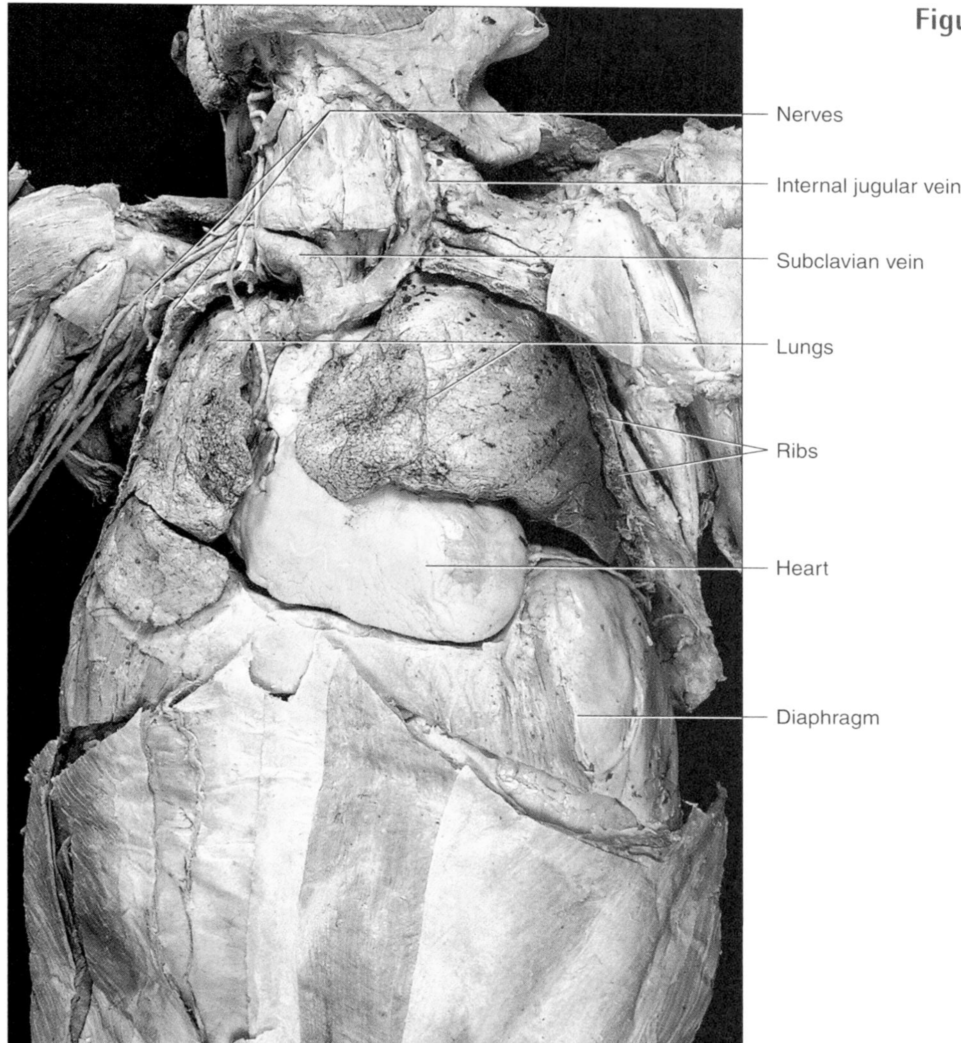

Figure A.19 Frontal View of the Thoracic Cavity.

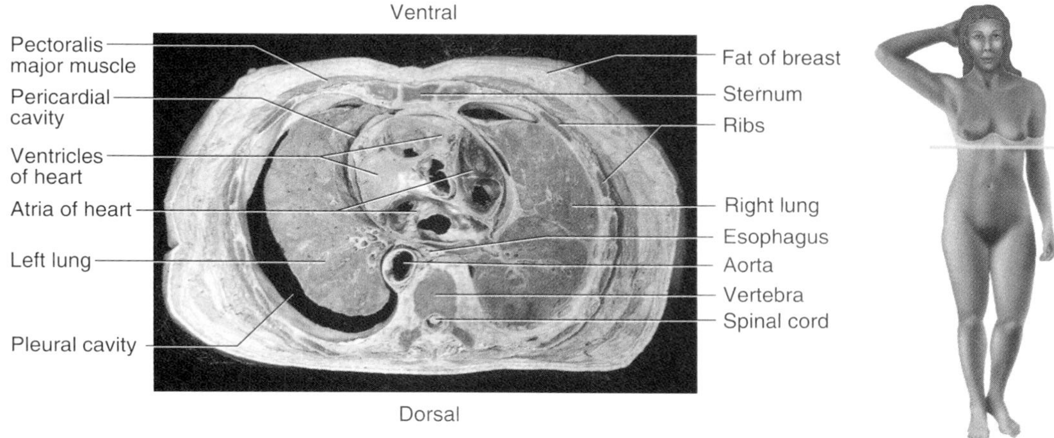

Figure A.20 Transverse Section of the Thoracic Cavity. Section taken at the level shown by the inset and oriented the same as the reader's body.

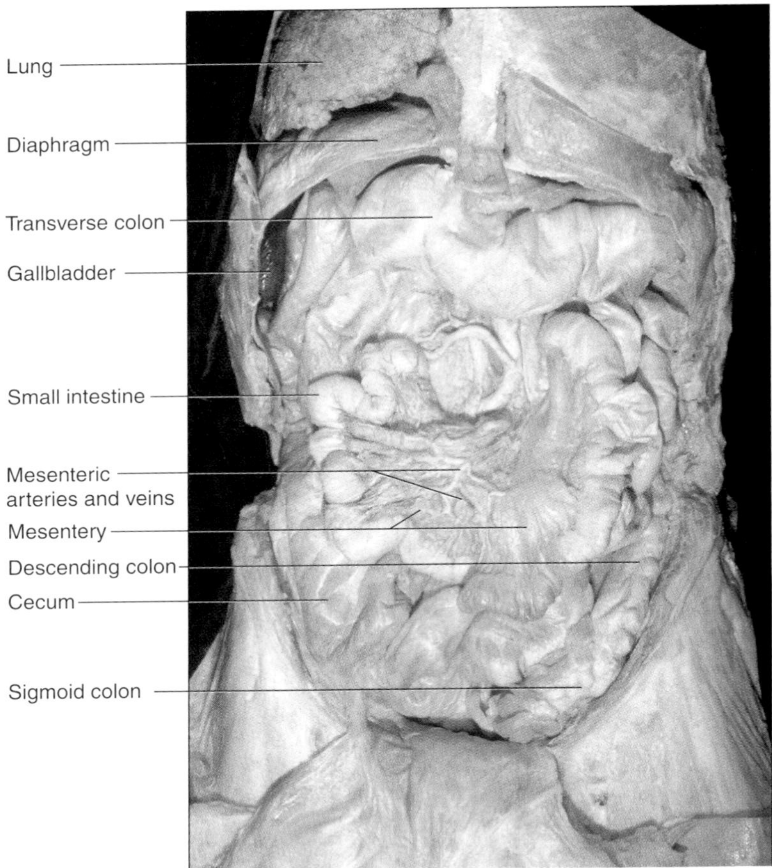

Figure A.21 Frontal View of the Abdominal Cavity.

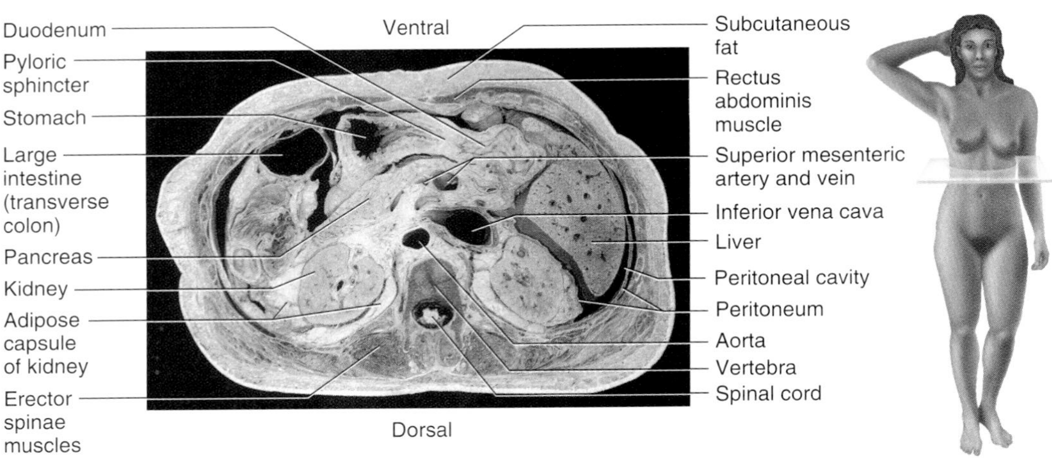

Figure A.22 **Transverse Section of the Abdominal Cavity.** Section taken at the level shown by the inset and oriented the same as the reader's body.

Urinary bladder
Pubic symphysis
Seminal vesicle
Prostate
Penis
Root
Bulb
Shaft
Corpus cavernosum
Corpus spongiosum
Glans
Sigmoid colon
Rectum
Anal canal
Anus
Epididymis
Scrotum
Testis

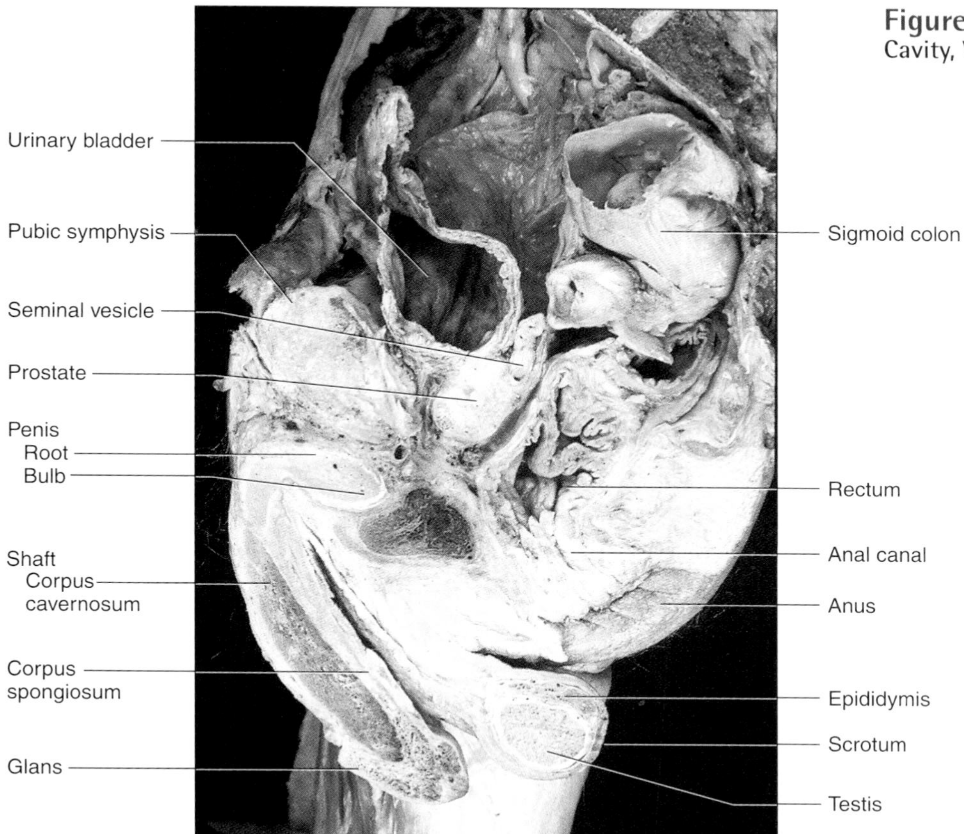

Figure A.23 Midsagittal Section of the Male Pelvic Cavity, Viewed from the Left.

Intervertebral disc
Vertebra
Red bone marrow
Mesentery
Small intestine
Sacrum
Sigmoid colon
Uterus
Cervix
Urinary bladder
Pubic symphysis
Urethra
Vagina
Rectum
Labium minus
Prepuce
Labium majus
Anal canal
Anus

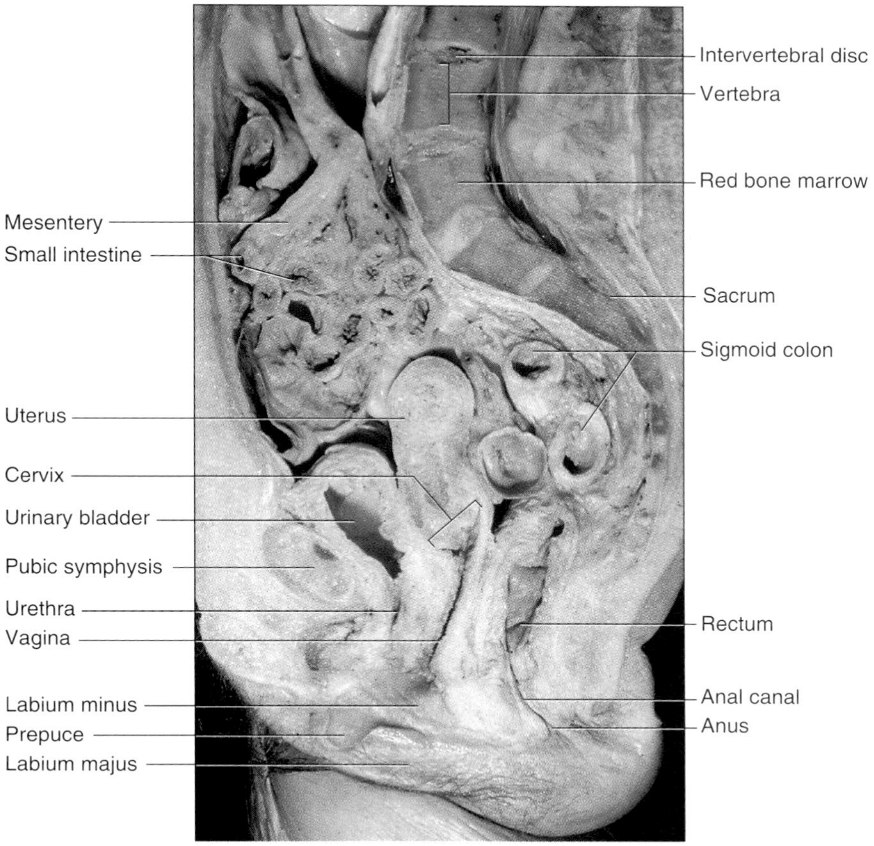

Figure A.24 Midsagittal Section of the Female Pelvic Cavity, Viewed from the Left.

Testing Your Recall

1. Which of the following is *not* an essential part of anatomical position?
   a. eyes facing forward
   b. feet flat on the floor
   c. forearms supine
   d. mouth closed
   e. arms down to the sides

2. A ring-shaped section of the small intestine would be a _____ section.
   a. parasagittal
   b. midsagittal
   c. transverse
   d. frontal
   e. medial

3. The tarsal region is _____ to the popliteal region.
   a. medial
   b. superficial
   c. superior
   d. dorsal
   e. distal

4. The greater omentum is _____ to the small intestine.
   a. posterior
   b. parietal
   c. deep
   d. superficial
   e. proximal

5. A _____ line passes through the sternum, umbilicus, and mons pubis.
   a. central
   b. proximal
   c. midclavicular
   d. midsagittal
   e. intertubercular

6. The _____ region is immediately medial to the coxal region.
   a. inguinal
   b. hypochondriac
   c. umbilical
   d. popliteal
   e. antecubital

7. Which of the following regions is *not* part of the upper limb?
   a. plantar
   b. carpal
   c. antecubital
   d. brachial
   e. palmar

8. Which of these organs is within the peritoneal cavity?
   a. urinary bladder
   b. kidneys
   c. heart
   d. small intestine
   e. brain

9. In which area do you think pain from the gallbladder would be felt?
   a. umbilical region
   b. right upper quadrant
   c. hypogastric region
   d. left hypochondriac region
   e. left lower quadrant

10. Which of the following is *not* an organ system?
    a. muscular system
    b. integumentary system
    c. endocrine system
    d. lymphatic system
    e. immune system

11. The forearm is said to be __________ when the palms are facing forward.

12. The more superficial layer of the pleura is called the __________ pleura.

13. The right and left pleural cavities are separated by a thick wall called the __________ .

14. The back of the neck is called the __________ region.

15. The manus is more commonly known as the __________ and the pes is more commonly known as the __________ .

16. The dorsal body cavity is lined by membranes called the __________ .

17. Organs that lie within the abdominal cavity but not within the peritoneal cavity are said to have a __________ position.

18. The sternal region is __________ to the pectoral region.

19. The pelvic cavity can be described as __________ to the abdominal cavity in position.

20. The anterior pit of the elbow is the __________ region, and the corresponding (but posterior) pit of the knee is the __________ fossa.

Answers in appendix B

Testing Your Comprehension

1. Identify which anatomical plane—sagittal, frontal, or transverse—is the only one that could *not* show (a) both the brain and tongue, (b) both eyes, (c) both the hypogastric and gluteal regions, (d) both kidneys, (e) both the sternum and vertebral column, and (f) both the heart and uterus.

2. Laypeople often misunderstand anatomical terminology. What do you think people really mean when they say they have "planter's warts"?

3. Name one structure or anatomical feature that could be found in each of the following locations relative to the ribs: medial, lateral, superior, inferior, deep, superficial, posterior, and anterior. Try not to use the same example twice.

4. Based on the illustrations in this atlas, identify an internal organ that is (a) in the upper left quadrant and retroperitoneal, (b) in the lower right quadrant of the peritoneal cavity, (c) in the hypogastric region, (d) in the right hypochondriac region, and (e) in the pectoral region.

5. Why do you think people with imaginary illnesses came to be called hypochondriacs?

Answers at Online Learning Center

Sodium hydroxide crystals seen through a polarizing microscope

CHAPTER 2

Matter and Energy

Chapter Outline

Insights

 Study with Your ESP CD-ROM

Refer to your ESP Tutorial CD-ROM for helpful animations, activities, and quizzes related to the material covered in this chapter. The following are suggested areas to go to on the CD and are presented according to "Unit/Topic":
- Levels of Organization/Chemistry
- Levels of Organization/Cellular Metabolism

Why is too much sodium or cholesterol harmful? Why does an iron deficiency cause anemia and an iodine deficiency cause a goiter? Why does a pH imbalance make some drugs less effective? Why do some pregnant women suffer convulsions after several days of vomiting? How can radiation cause cancer as well as cure it?

None of these questions can be answered, nor would the rest of this book be intelligible, without understanding the chemistry of life. A little knowledge of chemistry can help you choose a healthy diet, use medications more wisely, avoid worthless health fads and frauds, and explain treatments and procedures to your patients or clients. Thus, we begin our study of the human body with basic chemistry, the simplest level of the body's structural organization.

Chemical Elements and Atomic Structure

Objectives

When you have completed this section, you should be able to

- list the major chemical elements of the human body;
- describe the structure of atoms; and
- define *ion*, *electrolyte*, and *free radical.*

Matter and the Elements

Matter is anything that has mass and occupies space—the substance of which all things are composed. The simplest form of matter to have unique chemical properties is an **element.** Water, for example, has unique properties, but it can be broken down into two elements, hydrogen and oxygen, that have unique chemical properties of their own. If we carry this process further, however, we find that hydrogen and oxygen are made of protons, neutrons, and electrons—and none of these are unique. A proton of gold is identical to a proton of oxygen. Hydrogen and oxygen are the simplest chemically unique components of water and are thus elements.

Each element is identified by an **atomic number,** which is the number of protons in its nucleus. The atomic number of carbon is 6, and that of oxygen is 8, for example. The periodic table of the elements (see appendix A) arranges the elements in order by their atomic numbers. As the table also shows, the elements are represented by one- or two-letter symbols, usually based on their English names: C for carbon, Mg for magnesium, Cl for chlorine, and so forth. A few symbols are based on Latin names, such as K for potassium (*kalium*), Na for sodium (*natrium*), and Fe for iron (*ferrum*).

There are 91 naturally occurring elements on earth, 24 of which play normal physiological roles in humans. Table 2.1 groups these 24 according to their abundance in the body. Six of them account for 98.5% of the body's weight: oxygen, carbon, hydrogen, nitrogen, calcium, and phosphorus. Hydrogen, the lightest of the elements, is the most abundant in terms of number of atoms. It constitutes only 10% of the body's weight but 63% of its atoms (fig. 2.1).

Another 6 elements make up the next 0.8% of body weight: sulfur, potassium, sodium, chlorine, magnesium, and iron. The remaining 12 account for 0.7% of body weight, and no one of them accounts for more than 0.02%; thus they are known as **trace elements.** Despite their minute quantities, trace elements play vital roles in physiology. Other elements without natural physiological roles can contaminate the body and severely disrupt its functions (see clinical insight 2.1).

As figure 2.1 shows, the body is not simply composed of whatever is most readily available in the environment. Except for oxygen, what is most abundant on earth bears little resemblance to what is most abundant in the body. Silicon and aluminum are among earth's most abundant elements, for example, but the body normally contains no aluminum and only a tiny amount of silicon. Carbon constitutes about 18% of the body weight but only 0.09% of the earth's crust. Such differences show that the body is selective in its use of elements to form its own molecules.

Table 2.1 Elements of the Human Body

Name	Symbol	Percentage of Body Weight
Major Elements (total 98.5%)		
Oxygen	O	65.0
Carbon	C	18.0
Hydrogen	H	10.0
Nitrogen	N	3.0
Calcium	Ca	1.5
Phosphorus	P	1.0
Lesser Elements (total 0.8%)		
Sulfur	S	0.25
Potassium	K	0.20
Sodium	Na	0.15
Chlorine	Cl	0.15
Magnesium	Mg	0.05
Iron	Fe	0.006

Trace Elements (Total 0.7%)			
Chromium	Cr	Molybdenum	Mo
Cobalt	Co	Selenium	Se
Copper	Cu	Silicon	Si
Fluorine	F	Tin	Sn
Iodine	I	Vanadium	V
Manganese	Mn	Zinc	Zn

Clinical Insight 2.1

Lead Poisoning

Until recently, there was a high incidence of lead poisoning in the United States, much of it resulting from exposure to leaded gasoline, lead-based paints, and water from lead pipes. Indeed, the word *plumbing* is derived from the Latin word for lead, *plumbum*. Although the gasoline and paint sold today are lead-free and water lines are now made of other materials, many young children still become seriously ill from eating bits of old, peeling paint found in houses built before 1950. (Compulsive chewing on nonnutritive matter such as paint chips is called *pica*.) Lead interferes with a variety of metabolic processes. It causes degenerative changes in the brain, resulting in mental retardation, depression, delirium, or convulsions. It also causes anemia and sometimes irreversible damage to the liver and kidneys.

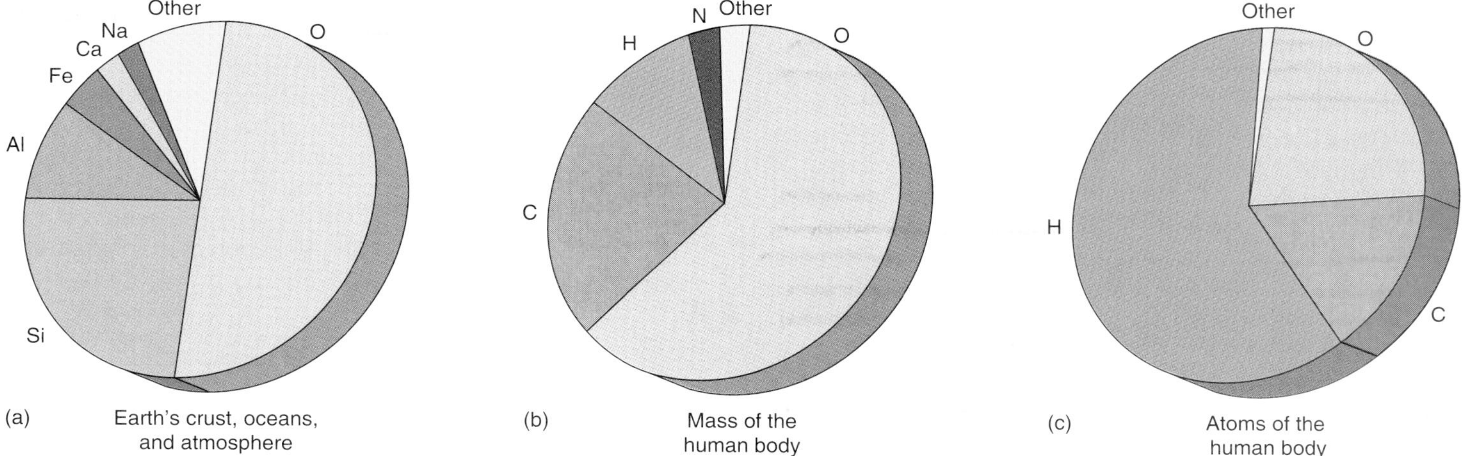

Figure 2.1 **Relative Abundance of Elements.** (*a*) At the surface of the earth, by mass; (*b*) in the human body, by mass; and (*c*) in the human body, by number of atoms.

Atomic Structure

In the fifth century B.C.E., the Greek philosopher Democritus reasoned that we can cut matter such as a gold nugget into smaller and smaller pieces, but there must ultimately be particles so small that nothing could cut them. He called these imaginary particles atoms[1] ("indivisible"). The concept of atoms remained a mere philosophical abstraction until 1803, when English chemist John Dalton began to develop an atomic theory based on experimental evidence. In 1913, Danish physicist Niels Bohr proposed a model of atomic structure similar to planets orbiting the sun (fig. 2.2). Although this *planetary model* is too simple to account for many of the properties of atoms, it remains useful for elementary purposes.

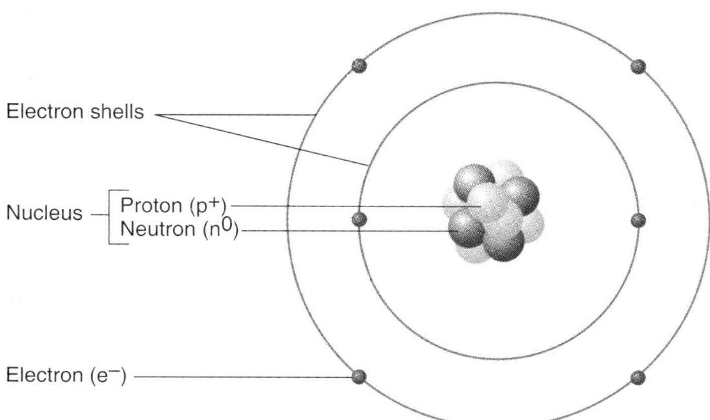

Figure 2.2 **Bohr Planetary Model of a Carbon Atom.**

1. *a* = not + *tom* = cut

Table 2.2 The Primary Subatomic Particles

Particle	Symbol	Charge	Mass	Location
Proton	p^+	+1	1.0073 amu	Nucleus
Neutron	n^0	0	1.0087 amu	Nucleus
Electron	e^-	−1	0.0005 amu	Energy shells around nucleus

At the center of an atom is the **nucleus,** composed of protons and neutrons. **Protons (p^+)** have a single positive charge and **neutrons (n^0)** have no charge. Each proton or neutron weighs approximately 1 **atomic mass unit (amu),** defined as one-twelfth the mass of an atom of carbon-12 (table 2.2). The **atomic mass** of an element is approximately equal to its total number of protons and neutrons.

Around the nucleus are one or more concentric clouds of **electrons (e^-),** tiny particles with a single negative charge and very low mass. It takes 1,836 electrons to equal 1 amu, so for most purposes we can disregard their mass. A person who weighs 64 kg (140 lb) contains less than 24 g (1 oz) of electrons. This hardly means that we can ignore electrons, however. They determine the chemical properties of an atom, thereby governing what molecules can exist and what chemical reactions can occur. The number of electrons equals the number of protons, so their charges cancel each other and an atom is electrically neutral.

Electrons swarm about the nucleus in concentric regions called **electron shells (energy levels).** The more energy an electron has, the farther away from the nucleus its orbit lies. Each shell holds a limited number of electrons (fig. 2.3). The one closest to the nucleus holds a maximum of 2 electrons, the second one holds a maximum of 8, and the third holds a maximum of 18. The outermost shell never holds more than 8 electrons, but a shell can acquire more electrons after another one, farther out, begins to fill. Thus, the third shell will hold 18 electrons only in atoms with four or more shells. The elements known to date have up to seven electron shells, but those ordinarily involved in human physiology do not exceed four.

The electrons of the outermost shell, called **valence electrons,** determine the chemical bonding properties of an atom. An atom tends to bond with other atoms that will fill its outer shell and produce a stable number of valence electrons. A hydrogen atom, with only one electron shell and one electron, tends to react with other atoms that provide another electron and fill this shell with a stable number of two electrons. All other atoms react in ways that produce eight electrons in the valence shell. This tendency is called the **octet rule,** or **rule of eights.**

Atoms of biological interest range from about 0.07 to 0.35 nm in diameter (1 nm = 10^{-9} m). It would take about 500,000 atoms side by side to be barely visible to the naked eye. (The limit of our vision is about 0.1 mm.)

Isotopes and Atomic Weight

Dalton believed that every atom of an element was identical, but we now know this to be untrue. All elements have varieties called **isotopes**[2] that differ from each other only in number of neutrons and therefore in atomic mass. Most hydrogen atoms, for example, have only one proton, but an isotope called *deuterium* has one proton and one neutron, and another called *tritium* has one proton and two neutrons (fig. 2.4). These isotopes are symbolized ^1H, ^2H, and ^3H to indicate their atomic masses. Over 99% of carbon atoms have an atomic mass of 12 ($6p^+$, $6n^0$) and are called carbon-12 (^{12}C), but a small percentage of carbon atoms are ^{13}C, with seven neutrons, and ^{14}C, with eight. All isotopes of a given element behave the same chemically. Deuterium (^2H), for example, reacts with oxygen the same way ^1H does to produce water.

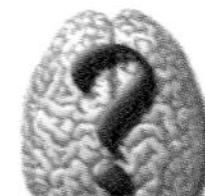

Think About It

Why is there no difference in the chemical behavior of different isotopes of the same element?

The **atomic weight** of an element accounts for the fact that an element is a mixture of isotopes. If all carbon were ^{12}C, the atomic weight of carbon would be the same as its atomic mass, 12.000. But since a sample of carbon also contains small amounts of the heavier isotopes ^{13}C and ^{14}C, the atomic weight is slightly higher, 12.011. These concepts are summarized in table 2.3.

Ions

Ions are charged particles with unequal numbers of protons and electrons. Elements with one to three valence electrons tend to give them up, and those with four to seven electrons tend to gain more. If an atom of the first kind is exposed to an atom of the second, electrons may transfer from one to the other and turn both of them into ions. This process is called **ionization.** The particle that gains electrons acquires a negative charge and is called an **anion** (AN-eye-on). The one that loses electrons acquires a positive charge (because it then has a surplus of protons) and is called a **cation** (CAT-eye-on).

2. *iso* = same + *top* = place (same position in the periodic table)

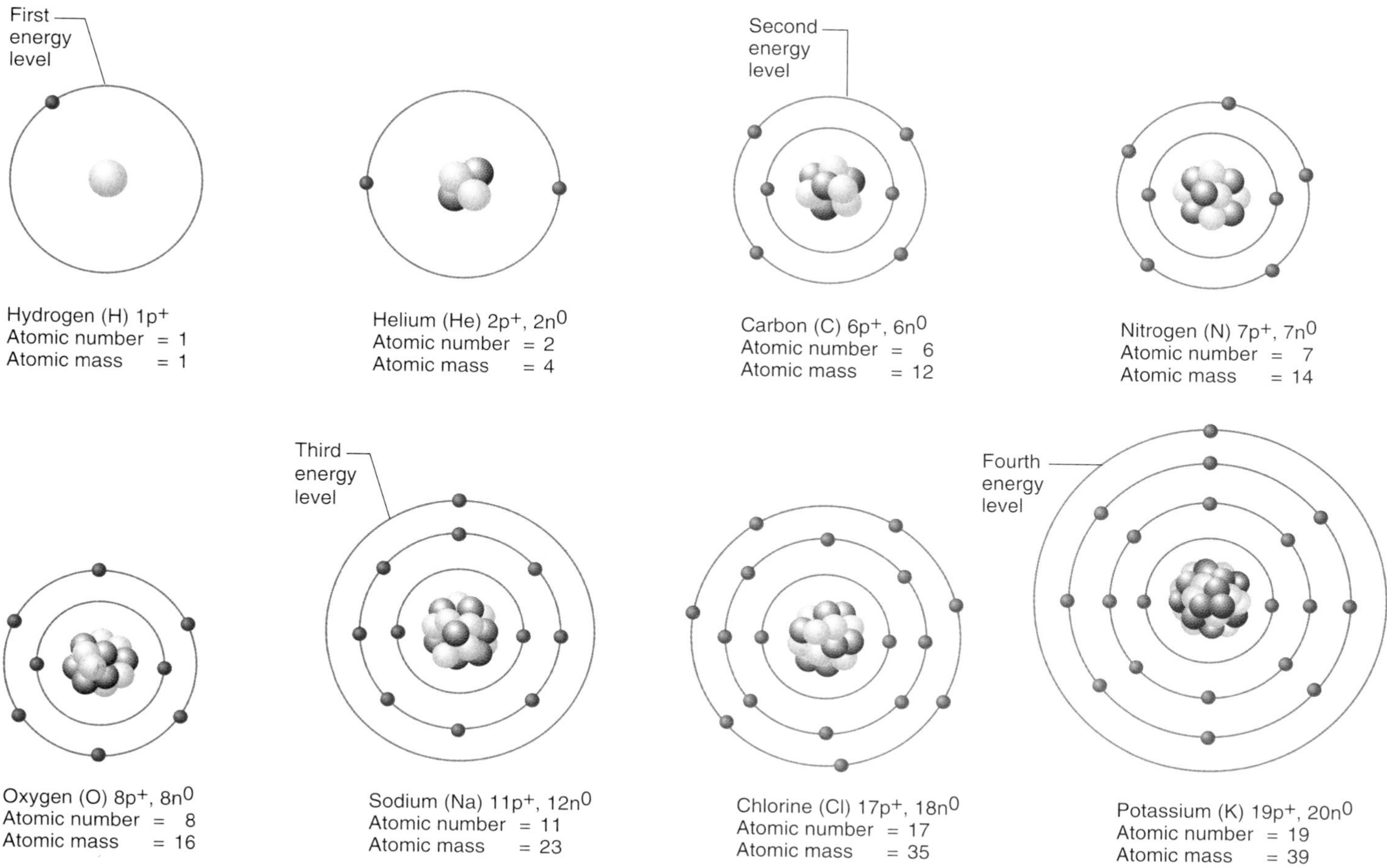

Figure 2.3 **Planetary Models of Eight Selected Elements.** Note the filling of successive electron shells as atomic number increases. Protons are represented as p^+ and neutrons are represented as n^0.

Consider, for example, what happens when sodium and chlorine meet. Sodium has three electron shells with a total of 11 electrons: 2 in the first shell, 8 in the second, and 1 in the third (fig. 2.5). If it gives up the electron in the third shell, its second shell becomes the valence shell and has the stable configuration of 8 electrons described by the octet rule. Chlorine, by contrast, has 17 electrons: 2 in the first shell, 8 in the second, and 7 in the third. If it can gain one more electron, it can fill the third shell with 8 electrons and become stable. Sodium and chlorine seem "made for each other"—one needs to lose an electron and the other needs to gain one. This is just what they do. When they interact, an electron transfers from sodium to chlorine. Now, sodium has 11 protons in its nucleus but only 10 electrons. This imbalance gives it a positive charge, so we symbolize the sodium ion Na^+. Chlorine has been changed to the chloride ion with a surplus negative charge, symbolized Cl^-.

Some elements can exist in two or more ionized forms. Iron, for example, has ferrous (Fe^{2+}) and ferric (Fe^{3+}) ions. Note that some ions have a single positive or negative charge, while others have charges of ±2 or ±3 because they gain or lose more than one electron. The charge on an ion is called its **valence.** Ions are not always single atoms that have become charged; some are groups of atoms—phosphate (PO_4^{3-}) and bicarbonate (HCO_3^-) ions, for example.

Ions with opposite charges are attracted to each other and tend to follow each other through the body. Thus, when Na^+ is excreted in the urine, Cl^- tends to follow it. The attraction of cations and anions to each other is important in maintaining the excitability of nerve and muscle cells, as we shall see in chapters 12 and 13.

Electrolytes

Electrolytes are molecules that ionize in water and form solutions capable of conducting electricity. The most abundant electrolytes in the body are salts of sodium (Na^+), potassium (K^+), calcium (Ca^{2+}), magnesium (Mg^{2+}), chloride (Cl^-), phosphate (PO_4^{3-}), and bicarbonate (HCO_3^-).

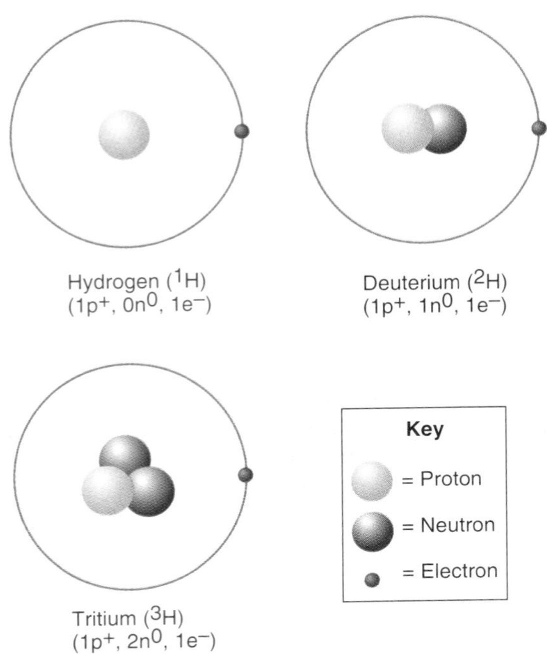

Figure 2.4 Isotopes of Hydrogen. The three isotopes differ only in the number of neutrons present.

We can detect electrical activity of the muscles, heart, and brain with electrodes on the skin because electrolytes in the body fluids conduct currents from these organs to the skin surface. Electrolytes are essential to nerve and muscle function, and electrolyte balance is one of the most important considerations in patient care. Electrolyte imbalances have effects ranging from muscle cramps and brittle bones to coma and cardiac arrest.

Free Radicals

A **free radical** is a particle with an odd number of electrons. For example, oxygen normally exists as a stable molecule composed of two oxygen atoms, O_2; but if an additional electron is added, it becomes a free radical called the *superoxide anion*, $O_2^{-}\cdot$. Free radicals are represented with a dot to symbolize the odd electron.

Free radicals are produced by some normal metabolic reactions of the body (such as a reaction that some white blood cells use to kill bacteria), by radiation (sunlight, ultraviolet radiation, and X rays), and by chemicals (such as carbon tetrachloride, a cleaning solvent, and nitrites, present as preservatives in some wine, meat, and other foods). Free radicals are short-lived and combine quickly with molecules such as fats, proteins, and DNA, thus converting these into free radicals and triggering chain reactions that destroy still more molecules. Among the damages caused by free radicals are some forms of cancer and myocardial infarction,

Table 2.3 Terms Related to Atomic Weight

Term	Definition
Atomic Number	Number of protons in an element, equal to its number of electrons.
Atomic Mass	Total number of protons and neutrons in an element (approximately).
Atomic Mass Unit (amu)	One-twelfth the mass of a ^{12}C atom, or approximately the mass of one proton or neutron.
Atomic Weight	Average atomic mass of the atoms in a sample of an element. Considers the atomic masses of all isotopes in the sample and relative abundance of each isotope.

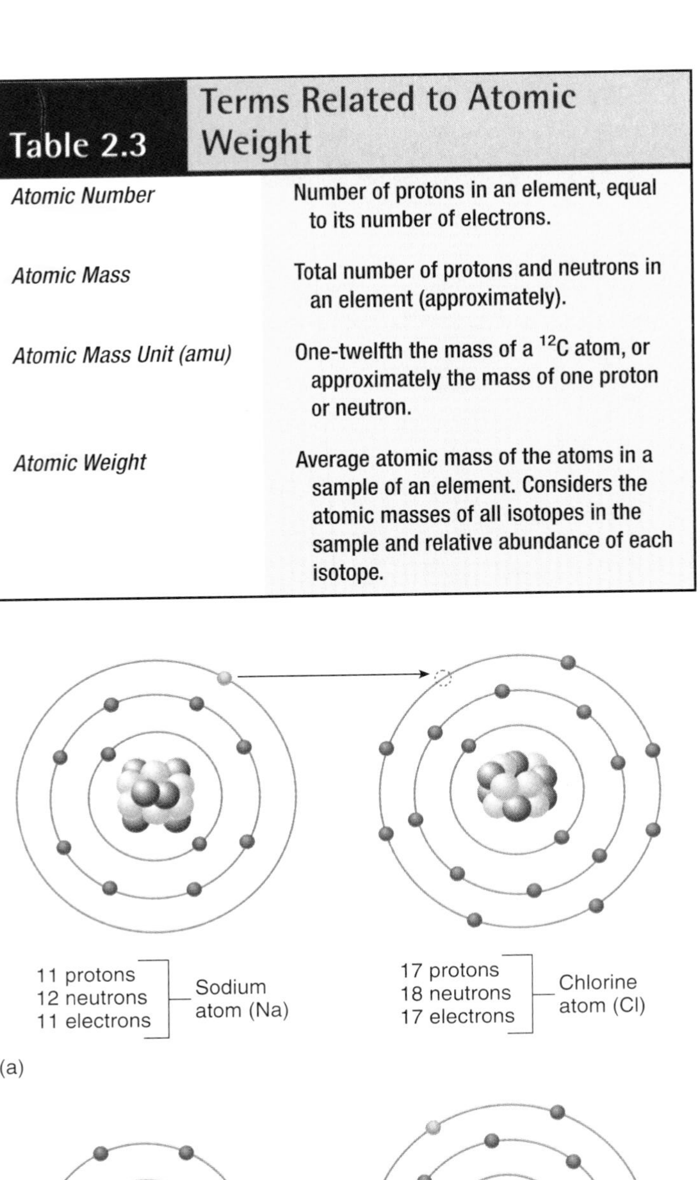

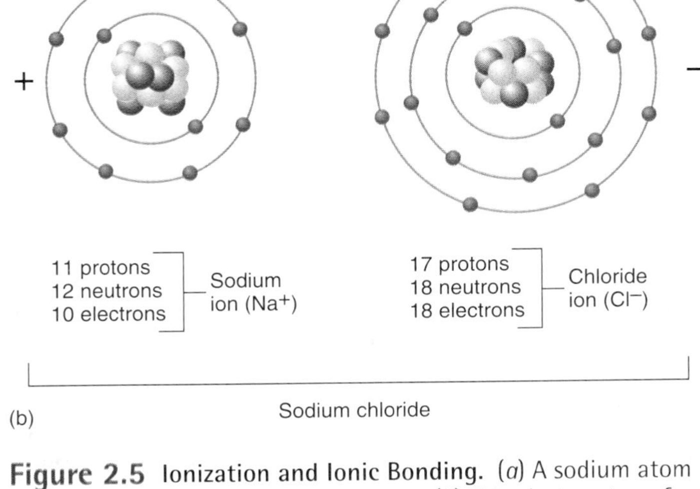

Figure 2.5 Ionization and Ionic Bonding. (*a*) A sodium atom donates an electron to a chlorine atom. (*b*) This electron transfer converts the atoms to a positive sodium ion (Na^+) and a negative chloride ion (Cl^-), which are mutually attracted by their opposite charges and form an ionic bond.

the death of heart tissue. One theory of aging is that it results in part from lifelong cellular damage by free radicals.

Because free radicals are so common and destructive, there are a number of mechanisms for neutralizing them. An **antioxidant** is a substance that neutralizes free radicals. *Superoxide dismutase (SOD)* is an enzyme that converts superoxide into oxygen and hydrogen peroxide. Selenium, vitamin E (α-tocopherol), vitamin C (ascorbic acid), and carotenoids (such as β-carotene) are some antioxidants obtained from the diet. These antioxidants can work together in a chain reaction. Vitamin E reacts with a free radical and converts it to a less harmful form, but this converts vitamin E itself into a potentially harmful radical. Carotenoids then inactivate the vitamin E radical and, in turn, vitamin C neutralizes the carotenoid radical. The resulting vitamin C radical is water-soluble and quickly eliminated in the urine.

In some parts of the country, the soil is deficient in selenium and people who grow much of their own food have a higher incidence of heart attacks as a result. The food distribution system in the United States ensures that food on the public market is pooled from a variety of locations so that food sold in a given locality is not consistently deficient in selenium or other nutrients. Animals deprived of vitamin E exhibit sterility, muscular dystrophy, and other signs of free radical damage. Fortunately, few people suffer a vitamin E deficiency because it is so abundant in vegetables, eggs, and meat.

Before You Go On

Answer the following questions to test your understanding of the preceding section:

1. What is an element? What six elements are most abundant in the human body? What is a trace element?
2. What is an electron shell?
3. Define *atomic number, atomic mass,* and *atomic weight.*
4. How do isotopes of the same element differ from each other?
5. How is an ion different from an atom?
6. Distinguish between an electrolyte and a free radical.
7. Explain why free radicals are harmful and how the body disposes of them.

Molecules, Compounds, and Chemical Bonds

Objectives

When you have completed this section, you should be able to

- show how compounds are represented by chemical formulae;
- define *isomer* and discuss the significance of isomers;
- calculate the molecular weight of compounds; and
- define the types of chemical bonds and explain how they form.

Molecules and Chemical Formulae

Atoms are most biologically interesting when they chemically bond with each other to form **molecules.** Some molecules consist of two or more atoms of the same element, such as oxygen (O_2) and nitrogen (N_2). Molecules composed of two or more different elements are called **compounds,** such as carbon dioxide (CO_2) and glucose ($C_6H_{12}O_6$).

Molecules are represented by **chemical formulae** that indicate the types and numbers of atoms present. A *molecular formula,* such as those just listed, itemizes each element only once, with subscripts indicating how many atoms of each are present. Molecular formulae conveniently fit a single line of print and are the shortest way to represent compounds, but they are sometimes not informative enough. For example, both ethanol (grain alcohol) and ethyl ether have the molecular formula C_2H_6O, but no wine connoisseur would consider them interchangeable! They have the same numbers of the same atoms, but their chemical properties are different because their atoms are arranged differently (fig. 2.6).

Molecules with identical atomic compositions but different arrangements of their atoms are called **structural isomers**[3] of each other. To show the difference between them, we use *structural formulae,* like those in figure 2.6, which show the location of each atom. We can also show their differences with a *condensed structural formula,* which can be written on one line. Ethanol is CH_3CH_2OH and ether is CH_3OCH_3. Note how these condensed structural

3. iso = same + *mers* = parts

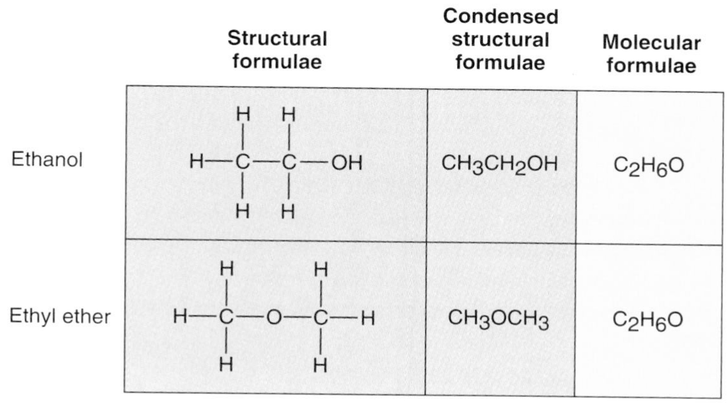

	Structural formulae	Condensed structural formulae	Molecular formulae
Ethanol	H—C(H)(H)—C(H)(H)—OH	CH_3CH_2OH	C_2H_6O
Ethyl ether	H—C(H)(H)—O—C(H)(H)—H	CH_3OCH_3	C_2H_6O

Figure 2.6 Structural Isomers, Ethanol and Ethyl Ether. The molecular formulae are identical, but the structures and chemical properties are different.

formulae relate to the structural formulae in figure 2.6. With a little practice, you can look at a formula of either type and visualize the other one.

Molecular Weight

The **molecular weight (MW)** of a compound is the sum of the atomic weights of its atoms. Rounding the atomic mass units (amu) to whole numbers, we can calculate the approximate MW of glucose ($C_6H_{12}O_6$), for example, as

6 C atoms × 12 amu each	=	72 amu
12 H atoms × 1 amu each	=	12 amu
6 O atoms × 16 amu each	=	96 amu
Molecular weight (MW)	=	180 amu

Molecular weight is needed to compute some measures of concentration, as we shall see later.

Chemical Bonds

A **chemical bond** is a force that attracts one atom to another. The three types of bonds of greatest physiological interest are *ionic bonds, covalent bonds,* and *hydrogen bonds.*

Ionic Bonds

In the earlier discussion of ionization, we saw that sodium readily gives up an electron to chlorine. The transfer of an electron creates two stable ions, the sodium ion (Na^+) and the chloride ion (Cl^-). Because these ions have opposite charges, they tend to cling together and form the compound sodium chloride (NaCl), common table salt. Such an attraction of oppositely charged ions to each other is called an **ionic bond.** Ionic compounds tend to form geometric arrays called *crystals* (fig. 2.7).

Some compounds form by the ionic bonding of three or more ions at once. For example, calcium has two valence electrons. It can become stable by donating one electron to one chlorine atom, and the other electron to another chlorine, thus producing a calcium ion, Ca^{2+}, and two chloride ions, Cl^-. The result is calcium chloride, $CaCl_2$.

Ionic bonds are weak and therefore tend to break up, or *dissociate,* in the presence of something more attractive, such as water. The ionic bonds of NaCl break down easily as salt dissolves in water, because both Na^+ and Cl^- are more attracted to water molecules than they are to each other.

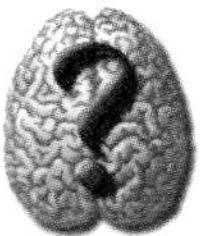

Think About It

Do you think ionic bonds are common in the human body? Explain your answer.

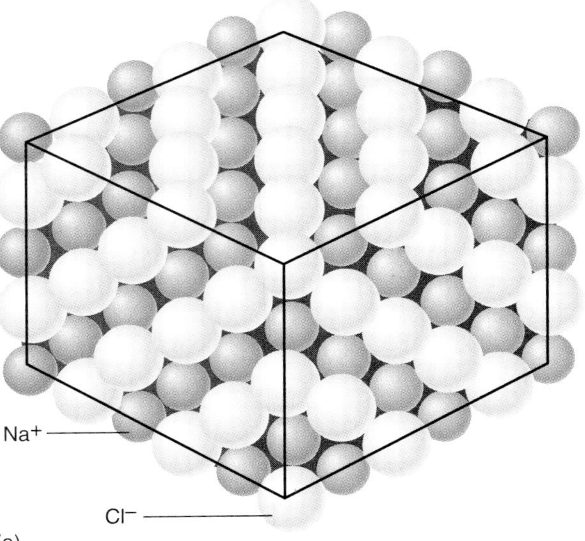

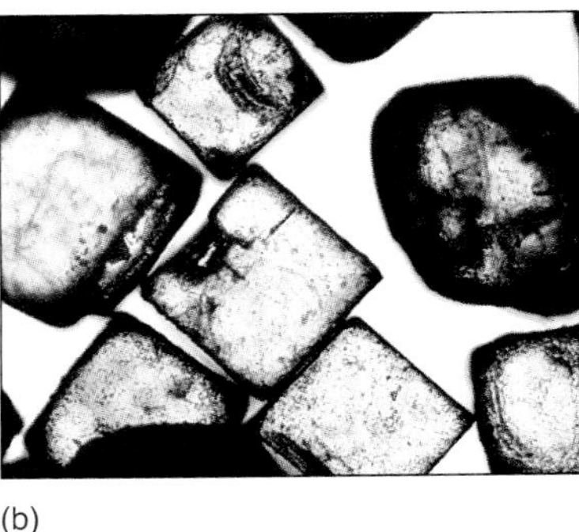

Figure 2.7 Sodium Chloride, an Ionic Compound. (*a*) The lattice of sodium and chloride ions in the NaCl crystal. (*b*) Sodium chloride (table salt) crystals (x400).

Covalent Bonds

Another way for atoms to form stable valence shells is to share electrons. For example, two hydrogen atoms can share valence electrons to form a hydrogen molecule, H_2 (fig. 2.8). The two electrons, one donated by each atom, swarm around both nuclei in a dumbbell-shaped cloud. Bonds formed by the sharing of electrons are called **covalent bonds.** A *single covalent bond* is the sharing of a single pair of electrons. It is symbolized by a single line between atomic symbols, for example H—H. A *double covalent bond* is the sharing of two pairs of electrons. In carbon dioxide, for example, a central carbon atom shares two electron pairs with each oxygen atom. Such bonds are symbolized by two lines—for example, O=C=O.

When shared electrons spend approximately equal time around each nucleus, they form a *nonpolar covalent bond* (fig. 2.9), the strongest of all chemical bonds. Carbon

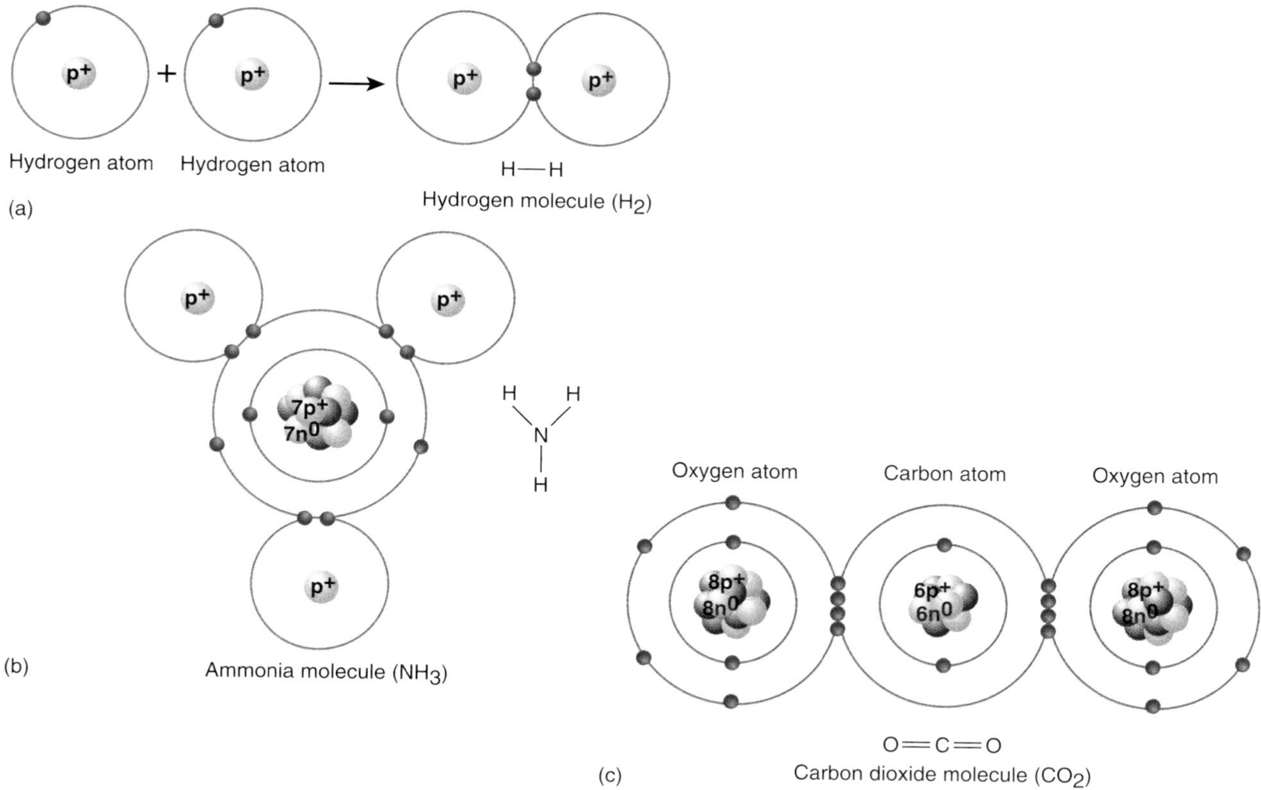

Figure 2.8 Covalent Bonding. (*a*) Two hydrogen atoms share a single pair of electrons to form a hydrogen molecule. (*b*) An ammonia molecule, in which a nitrogen atom shares a pair of valence electrons with each of three hydrogen atoms. Hydrogen and ammonia molecules exhibit single covalent bonds. (*c*) A carbon dioxide molecule, in which a carbon atom shares two pairs of electrons with each oxygen atom, forming double covalent bonds.

atoms bond to each other with nonpolar covalent bonds. If shared electrons spend significantly more time orbiting one nucleus than they do the other, they lend their negative charge to the region where they spend the most time, and they form a *polar covalent bond.* When hydrogen bonds with oxygen, for example, the electrons are more attracted to the oxygen nucleus and orbit it more than they do the hydrogen. This makes the oxygen region of the molecule slightly negative and the hydrogen regions slightly positive. The Greek letter delta (δ) is used to symbolize a charge less than that of one electron or proton. A slightly negative region of a molecule is represented $\delta-$ and a slightly positive region is represented $\delta+$.

Hydrogen Bonds

A weak attraction called a **hydrogen bond** often forms between a slightly positive hydrogen atom in one molecule and a slightly negative oxygen or nitrogen atom in another. Water molecules, for example, are weakly attracted to each other by hydrogen bonds (fig. 2.10). This attraction is responsible for many of the biologically important properties of water to be discussed in the next chapter. Hydrogen bonds can also form between different regions of the same molecule, especially in very large molecules such as proteins and DNA. Hydrogen bonding causes such molecules to fold or coil into precise three-dimensional shapes. Hydrogen bonds are represented by dotted or broken lines between atoms: –C=O···H–N–. Hydrogen bonds are the weakest of all the bond types we have considered, but they are enormously important to physiology.

Chemical bonds are summarized in table 2.4.

Before You Go On

Answer the following questions to test your understanding of the preceding section:

8. How do isomers differ from each other? What is the difference between an isomer and an isotope?
9. What does chemical bonding have to do with full electron shells?
10. How is an ionic bond formed?
11. What is a covalent bond? How do single and double covalent bonds differ? How do nonpolar and polar covalent bonds differ?
12. What is a hydrogen bond? Why do hydrogen bonds depend on the existence of polar covalent bonds?

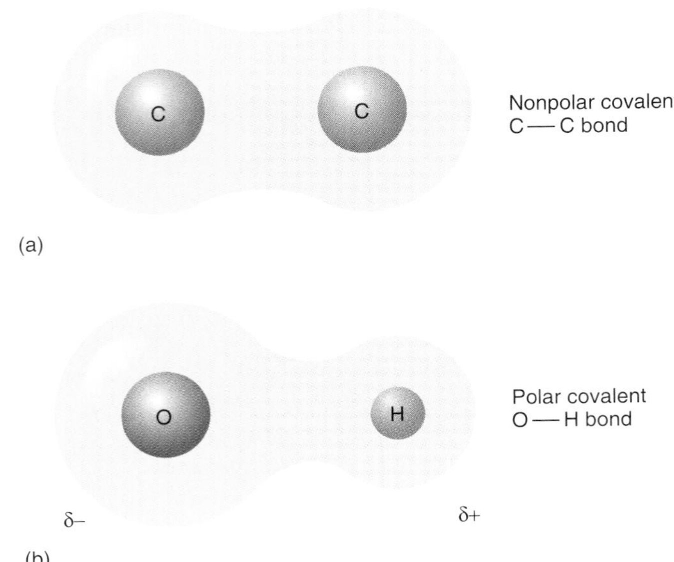

Figure 2.9 Nonpolar and Polar Covalent Bonds. (*a*) A nonpolar covalent bond between two carbon atoms, formed by electrons that spend an equal amount of time around each nucleus, as represented by the symmetric blue cloud. (*b*) A polar covalent bond, in which electrons orbit one nucleus significantly more than they do the other, as represented by the asymmetric cloud. This results in a slight negative charge (δ–) in the region where the electrons spend most of their time and a slight positive charge (δ+) at the other pole.

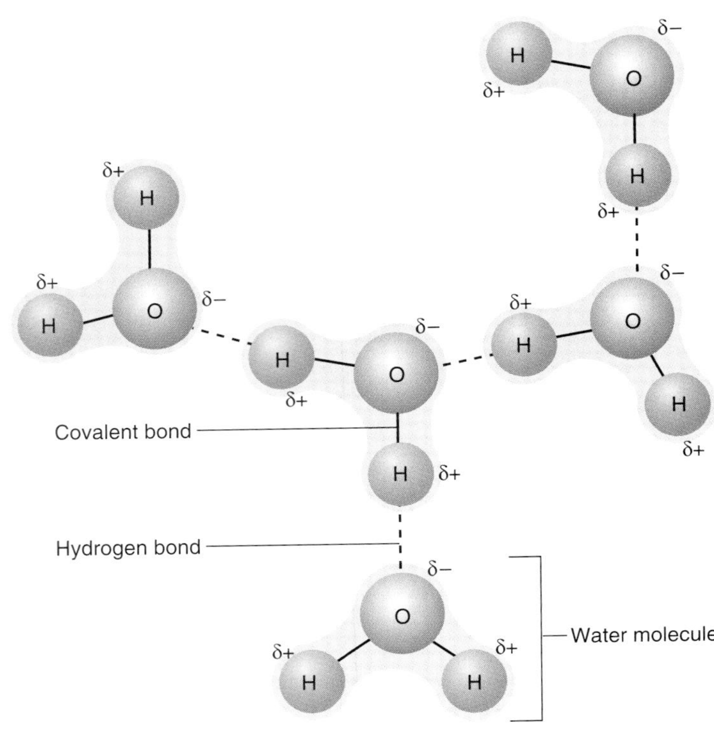

Figure 2.10 Hydrogen Bonding in Water. The polar covalent bonds of water molecules enable each oxygen atom to form a hydrogen bond with a hydrogen atom of a neighboring molecule. Thus, the water molecules are weakly attracted to each other.

Table 2.4 Types of Chemical Bonds

Bond Type	Definition and Remarks
Ionic bond	Relatively weak attraction between an anion and a cation. Easily disrupted in water, as when salt dissolves.
Covalent bond	Sharing of one or more pairs of electrons between nuclei. Strongest type of chemical bond.
Single covalent	Sharing of one electron pair.
Double covalent	Sharing of two electron pairs. Often occurs between carbon atoms, between carbon and oxygen, and between carbon and nitrogen.
Nonpolar covalent	Covalent bond in which electrons are equally attracted to both nuclei.
Polar covalent	Covalent bond in which electrons are more attracted to one nucleus than to the other, resulting in slightly positive and negative regions in one molecule.
Hydrogen bond	Weak attraction between polarized molecules or between polarized regions of the same molecule. Important in the three-dimensional folding and coiling of large molecules. Weakest of all bonds; easily disrupted by temperature and pH changes.

Mixtures

Objectives

When you have completed this section, you should be able to

- define *mixture* and distinguish between mixtures and compounds;
- show how three kinds of mixtures differ from each other; and
- discuss some ways in which the concentration of a mixture can be expressed, and explain why different expressions of concentration are used for different purposes.

A **mixture** consists of substances that are physically blended but not chemically combined. Each substance retains its own chemical properties. To contrast a mixture with a compound, consider sodium chloride again. Sodium is a lightweight metal that bursts into flame if exposed to water, and chlorine is a yellow-green poisonous gas that was used for chemical warfare in World War I. When these elements chemically react, they form common table salt—sodium chloride. Clearly, the compound has properties much different from the properties of its elements. But if you were to put a little salt on your watermelon, the watermelon would taste salty and sweet because the sugar of the melon and the salt you added would merely form a mixture in which each compound retained its individual properties.

Solutions, Colloids, and Suspensions

In human physiology, we are mostly concerned with mixtures of various substances in water. These mixtures can be classified as *solutions, colloids,* and *suspensions.*

Solutions

A **solution** consists of particles of matter called the **solute** mixed with a more abundant substance (usually water) called the **solvent.** The solute can be a gas, solid, or liquid—as in a solution of oxygen, sodium chloride, or alcohol in water, respectively. Solutions are defined by the following properties:

- The solute particles are under 1 nanometer (nm) in size. They are too small to scatter light noticeably, so solutions are usually transparent (fig. 2.11*a*).
- The solute and solvent cannot be visually distinguished from each other, even with a microscope.
- The solute does not separate from the solvent when the solution is allowed to stand.
- The solute particles will pass through most artificial selectively permeable membranes (as discussed in chapter 4).

Colloids

A **colloid**[4] is defined by the following properties:

- The colloidal particles range from 1 to 100 nm in size. This is large enough to scatter light, so colloids are usually cloudy (fig. 2.11*b*).
- The particles are still small enough to remain permanently mixed with the solvent when the mixture stands.
- The particles are too large to pass through most selectively permeable membranes.

Many colloids can change from liquid to gel states—gelatin desserts, agar culture media, and the fluids between and within our cells, for example. Proteins are the most common colloidal particles in the body.

Suspensions

A **suspension** is defined by the following properties:

- The suspended particles exceed 100 nm in size; therefore, suspensions are cloudy or opaque.
- The particles are too heavy to remain permanently suspended, so suspensions separate on standing.

Blood is a suspension of cells in plasma. If allowed to stand, blood cells settle to the bottom of a tube (fig. 2.11*c, d*). An **emulsion** is a suspension of one liquid in another, such as oil and vinegar salad dressing. Many medications, including Kaopectate and milk of magnesia, are emulsions.

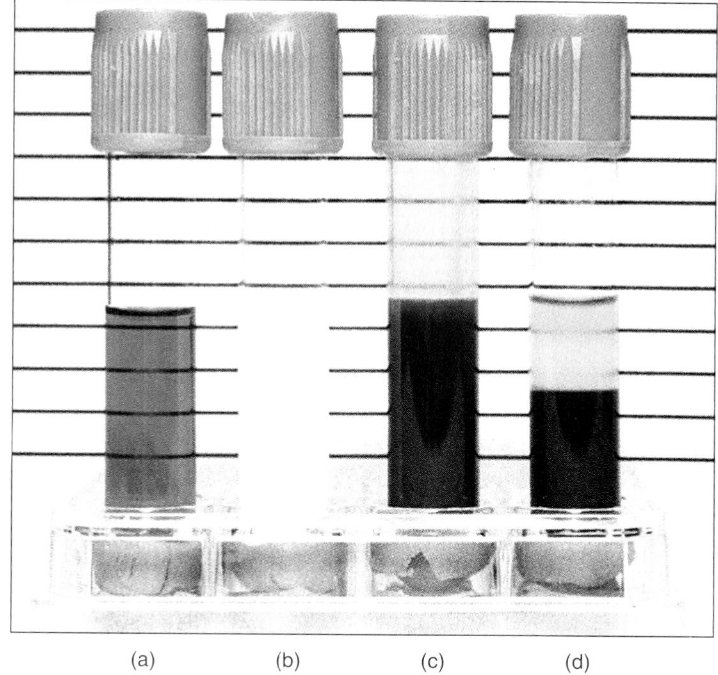

Figure 2.11 A Solution, a Colloid, and a Suspension. (*a*) In this copper sulfate solution, the solute particles are so small that they remain permanently mixed and the solution is transparent. (*b*) In colloids such as this milk, the particles are still small enough to remain permanently mixed, but they are large enough to scatter light, so we cannot see through the colloid. (*c*) In suspensions such as this freshly mixed blood, the particles also scatter light and make the mixture opaque. Furthermore (*d*), they are too large to remain permanently mixed, thus they spontaneously settle out of the mixture, as seen in this blood sample. This sample stood overnight, allowing the cells to settle out of the translucent yellowish plasma.

4. *collo* = glue + *oid* = like, resembling

Table 2.5 Types of Mixtures

	Solution	Colloid	Suspension
Particle Size	< 1 nm	1–100 nm	> 100 nm
Appearance	Clear	Often cloudy	Cloudy-opaque
Will particles settle out?	No	No	Yes
Will particles pass through a selectively permeable membrane?	Yes	No	No
Examples	Glucose in blood	Proteins in blood	Blood cells
	O_2 in water	Intracellular fluid	Cornstarch in water
	Saline solutions	Milk protein	Fats in blood
	Sugar in coffee	Gelatin	Kaopectate

A single mixture can fit into more than one of these categories. Blood is a perfect example—it is a solution of sodium chloride, a colloid of protein, and a suspension of cells. Another example, milk, is a solution of calcium, a colloid of protein, and an emulsion of butterfat. Table 2.5 summarizes the types of mixtures and provides additional examples.

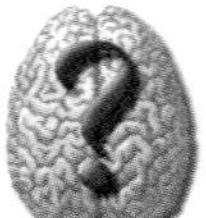

Think About It

Air is a mixture of nitrogen, oxygen, carbon dioxide, and other gases. Would you classify it as a solution, colloid, or suspension? Why?

Measures of Concentration

Solutions are often described in terms of their concentration—the amount of solute in a given volume of solution. Concentration is expressed in different ways for different purposes, some of which are explained here. You may find the table of symbols and measures inside the front cover to be helpful as you study this section.

Weight per Volume

A simple way to express concentration is the weight of solute in a given volume of solution. For example intravenous (I.V.) saline typically contains 8.5 grams of NaCl per liter of solution (8.5 g/L). For many biological purposes, however, we deal with smaller quantities such as milligrams per deciliter (mg/dL; 1 dL = 100 mL). For example, a typical serum cholesterol concentration may be 200 mg/dL, also expressed 200 mg/100 mL or 200 milligram-percent (mg-%).

Percentages

Percentage concentrations are also simple to compute, but it is necessary to specify whether the percentage refers to the weight or the volume of solute in a given volume of solution. For example, if we begin with 5 g of dextrose (an isomer of glucose) and add enough water to make 100 mL of solution, the resulting concentration will be 5% weight per volume (w/v). A common intravenous fluid is D5W, which stands for 5% w/v dextrose in distilled water. If the solute is a liquid, such as ethanol, percentages refer to volume of solute per volume of solution. Thus, 70 mL of ethanol diluted with water to 100 mL of solution produces 70% volume per volume (v/v) ethanol.

Molarity

Percent concentrations are easy to prepare, but that unit of measurement is inadequate for many purposes. The physiological effect of a chemical depends on how many molecules of it are present in a given volume, not on the weight of the chemical. A solution of 5% glucose, for example, contains almost twice as many glucose molecules as the same volume of 5% sucrose (fig. 2.12*a*). Each solution contains 50 g of sugar per liter, but glucose has a MW of 180 and sucrose has a MW of 342. Since each molecule of glucose is lighter, 50 g of glucose contains more molecules than 50 g of sucrose.

To produce solutions with a known number of molecules per volume, we must take molecular weight into consideration. If we know the MW, and we weigh out that many grams of the substance, we have a quantity known as its gram molecular weight, or 1 **mole.** One mole of glucose is 180 g and 1 mole of sucrose is 342 g. Each quantity contains the same number of molecules of the respective sugar—a number known as Avogadro's[5] number, 6.023×10^{23}. Such a large number is hard to imagine. If each molecule were the size of a pea, 6.023×10^{23} molecules would cover 60 earth-sized planets 3 m (10 ft) deep!

Molarity (M) is the number of moles of solute per liter of solution. A *one-molar* (1.0 M) solution of glucose contains 180 g/L, and a 1.0 M solution of sucrose contains

5. Amedeo Avogadro (1776–1856), Italian chemist

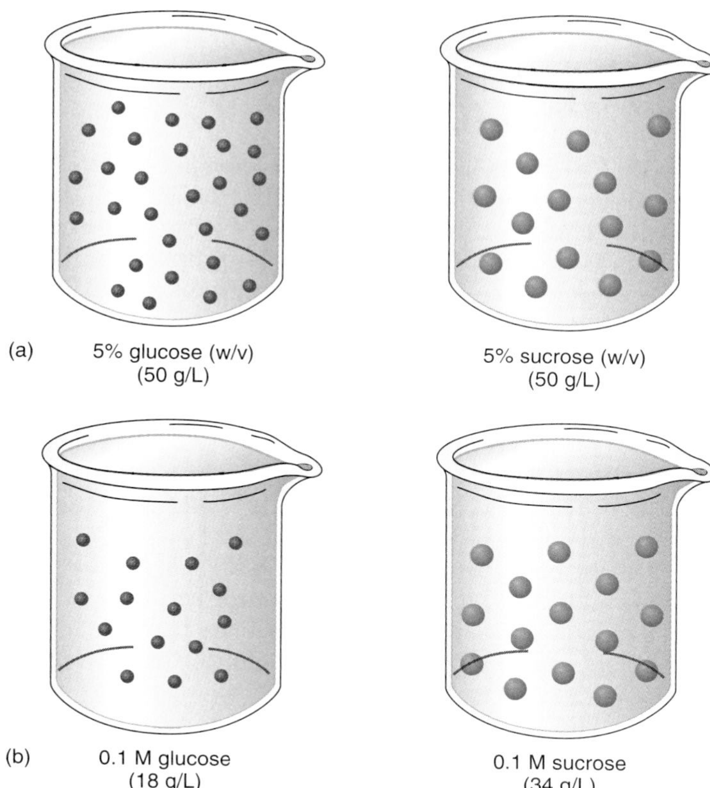

Figure 2.12 Comparison of Percentage and Molar Concentrations. (*a*) Solutions with the same percentage concentrations can differ greatly in the number of molecules per volume because of differences in molecular weights of the solutes. Fifty grams of sucrose has about half as many molecules as 50 g of glucose, for example. (*b*) Solutions with the same molarity have the same number of molecules per volume because molarity takes differences in molecular weight into account.

342 g/L. Both have the same number of solute molecules in a given volume. Body fluids and laboratory solutions usually are less concentrated than 1 M, so biologists and clinicians more often work with millimolar (mM) and micromolar (μM) concentrations. These are, respectively, one-thousandth and one-millionth molar. (A 1.0 mM solution of glucose is 0.18 g/L.)

Figure 2.12 contrasts solutions of equal molarity with solutions of equal percentage concentrations. Note that solutions of equal percentage have the same weight of solute in a given volume, but more importantly, solutions of equal molarity have the same number of molecules in a given volume.

Electrolyte Concentrations

Electrolytes are important for both their chemical and electrical effects on the body. Their electrical effects, which determine such things as nerve, heart, and muscle actions, depend not only on their concentration but also on their electrical charge. A calcium ion (Ca^{2+}) has twice the electrical effect of a sodium ion (Na^+), for example, because it carries twice the charge. When we measure electrolyte concentrations, we must therefore take the charges into account.

One **equivalent (Eq)** of an electrolyte is the amount that would electrically neutralize 1 mole of hydrogen ions (H^+) or hydroxyl ions (OH^-). For example, 1 mole (58.4 g) of NaCl yields 1 mole, or 1 Eq, of Na^+ in solution. Thus, an NaCl solution of 58.4 g/L contains 1 equivalent of Na^+ per liter (1 Eq/L). One mole (98 g) of sulfuric acid (H_2SO_4) yields 2 moles of positive charges (H^+). Thus, 98 g of sulfuric acid per liter would be a solution of 2 Eq/L.

The electrolytes in our body fluids have concentrations less than 1 Eq/L, so we more often express their concentrations in **milliequivalents per liter (mEq/L).** If you know the millimolar concentration of an electrolyte, you can easily convert this to mEq/L by multiplying it by the valence of the ion:

$$1 \text{ mM } Na^+ = 1 \text{ mEq/L}$$
$$1 \text{ mM } Ca^{2+} = 2 \text{ mEq/L}$$
$$1 \text{ mM } Fe^{3+} = 3 \text{ mEq/L}$$

Before You Go On

Answer the following questions to test your understanding of the preceding section:

13. Define and distinguish between *mixture, solution, colloid,* and *suspension.* Give examples of the last three.
14. Explain why it may be inappropriate for some purposes to express chemical concentrations as percentages.
15. What is a mole? How many particles are in 1 mole?
16. Explain how molarity is calculated.
17. What is meant by milliequivalents per liter, and what substances are measured in this way?

Acids, Bases, and pH

Objectives

When you have completed this section, you should be able to

- define *acid, base,* and *pH;*
- describe the pH scale; and
- discuss the role of buffers in the body's acid-base balance.

Acids and Bases

Most people have some sense of what acids and bases are. Advertisements are full of references to excess stomach acid and pH-balanced shampoo. We know that drain cleaner (a strong base) and battery acid can cause serious chemical burns. But what exactly do "acidic" and "basic" mean, and how can they be quantified?

An **acid** is any *proton donor,* a molecule that releases a proton (H^+) in water. A **base** is a proton acceptor. Since hydroxyl ions (OH^-) accept H^+, many bases are substances that release hydroxyl ions—sodium hydroxide (NaOH), for example. A base does not have to be a hydroxyl donor, however. Ammonia (NH_3) is also a base. It does not release hydroxyl ions, but it readily accepts hydrogen ions to become the ammonium ion (NH_4^+).

In pure water, 1 out of 10 million molecules spontaneously ionizes to form hydrogen and hydroxyl ions: $H_2O \leftrightarrow H^+ + OH^-$. Strictly speaking, hydrogen ions do not exist in this free state; rather, they combine with other water molecules to form hydronium ions (H_3O^+). Thus, the preceding equation could be rewritten: $2\ H_2O \leftrightarrow H_3O^+ + OH^-$. For convenience, however, we will use H^+ in chemical equations throughout this book.

pH

Acidity is expressed in terms of **pH,** a measure derived from the molarity of H^+. Molarity is represented by square brackets, so the molarity of H^+ is symbolized $[H^+]$. pH is defined as the negative logarithm of hydrogen ion molarity—that is, $pH = -\log [H^+]$.

Pure water has a neutral pH because it contains equal amounts of hydrogen (or hydronium) and hydroxyl ions. Since 1 in 10 million water molecules ionize, the molarity of H^+ and the pH of water are therefore

$$[H^+] = 0.0000001 \text{ molar} = 10^{-7}\ M$$

$$\log [H^+] = -7$$

$$pH = -\log [H^+] = 7$$

The pH scale (table 2.6) was invented in 1909 by Danish biochemist and brewer Sören Sörensen to measure the acidity of beer. The scale extends from 0.0 to 14.0. A solution with a pH of 7.0 is **neutral;** solutions with pH below 7 are **acidic;** and solutions with pH above 7 are **basic,** or **alkaline.** The lower the pH value, the more hydrogen ions a solution has and the more acidic it is. Since the pH scale is logarithmic, a change of one whole number on the scale represents a 10-fold change in H^+ concentration. In other words, a solution with a pH of 4 is 10 times as acidic as one with a pH of 5 and 100 times as acidic as one with a pH of 6. The pH values of some body fluids and common household substances are listed in table 2.7.

Slight disturbances of pH can seriously disrupt normal physiological functions and alter drug actions (see clinical insight 2.2), so it is important that the body be able to carefully control pH. Blood, for example, normally has a pH ranging from 7.35 to 7.45. Deviations from this range cause tremors, fainting, paralysis, or even death. Chemical solutions that resist changes in pH are called **buffers.** Buffers and pH regulation are considered in detail in chapter 24.

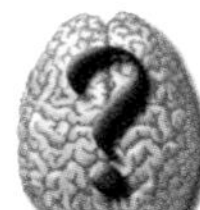

Think About It

A pH of 7.20 is slightly alkaline, yet a blood pH of 7.20 is called *acidosis.* Why do you think it is called this?

Table 2.6 The pH Scale

Molarity of OH^-		Molarity of H^+		Log $[H^+]$	$pH = -\log [H^+]$
0.00000000000001	$= 10^{-14}$	1.0	$= 10^{0}$	0	0
0.0000000000001	$= 10^{-13}$	0.1	$= 10^{-1}$	−1	1
0.000000000001	$= 10^{-12}$	0.01	$= 10^{-2}$	−2	2
0.00000000001	$= 10^{-11}$	0.001	$= 10^{-3}$	−3	3
0.0000000001	$= 10^{-10}$	0.0001	$= 10^{-4}$	−4	4
0.000000001	$= 10^{-9}$	0.00001	$= 10^{-5}$	−5	5
0.00000001	$= 10^{-8}$	0.000001	$= 10^{-6}$	−6	6
0.0000001	$= 10^{-7}$	0.0000001	$= 10^{-7}$	−7	7
0.000001	$= 10^{-6}$	0.00000001	$= 10^{-8}$	−8	8
0.00001	$= 10^{-5}$	0.000000001	$= 10^{-9}$	−9	9
0.0001	$= 10^{-4}$	0.0000000001	$= 10^{-10}$	−10	10
0.001	$= 10^{-3}$	0.00000000001	$= 10^{-11}$	−11	11
0.01	$= 10^{-2}$	0.000000000001	$= 10^{-12}$	−12	12
0.1	$= 10^{-1}$	0.0000000000001	$= 10^{-13}$	−13	13
1.0	$= 10^{0}$	0.00000000000001	$= 10^{-14}$	−14	14

Clinical Insight 2.2

pH and Drug Action

The pH of our body fluids has a direct bearing on how we react to drugs. Depending on pH, drugs such as aspirin, phenobarbital, and penicillin can exist in charged (ionized) or uncharged forms. Whether a drug is charged or not can determine whether it will pass through cell membranes. When aspirin is in the acidic environment of the stomach, for example, it is uncharged and passes easily through the stomach lining into the bloodstream. Here it encounters a basic pH, whereupon it ionizes. In this state, it is unable to pass back through the membrane, so it accumulates in the blood. This effect, called *ion trapping* or *pH partitioning,* can be controlled to help clear poisons from the body. The pH of the urine, for example, can be manipulated so that poisons become trapped there and thus rapidly excreted from the body.

Table 2.7 pH Values of Selected Body Fluids and Other Substances at 25°C

Substance	pH
1 M hydrochloric acid (HCl)	0.0
Gastric juice	0.8–3.0
Carbonated soft drinks, citrus juices	3.0–3.5
Coffee (black)	5.0
Urine	4.6–8.0
Cow's milk	6.5–6.6
Saliva	6.3–7.3
Pure water	7.0
Blood	7.35–7.45
Egg white	7.6–8.0
Pancreatic juice	7.1–8.2
Bile	7.6–8.6
Milk of magnesia	10.0–11.0
Household ammonia	10.5–11.8
Oven cleaner, drain cleaner, lye	13.5
1 M sodium hydroxide (NaOH)	14.0

Before You Go On

Answer the following questions to test your understanding of the preceding section:

18. Define *acid* and *base.*
19. Explain why each digit of the pH scale represents a 10-fold difference in hydrogen ion concentration.
20. What is a buffer? Why are buffers important?

Chemical Reactions

Objectives

When you have completed this section, you should be able to:

- define *chemical reaction;*
- recognize the reactants and products in a chemical equation;
- list and define the fundamental types of chemical reactions; and
- describe the factors that govern the speed and direction of a reaction.

Chemical Equations

A **chemical reaction** is a process in which a chemical bond is formed or broken. **Reactants** are substances that enter into a reaction and are consumed by it; **products** are substances produced by the reaction. A **chemical equation** is a symbolic representation of the course of a reaction. It typically shows the reactants on the left, the products on the right, and an arrow pointing from the reactants to the products. But the left-to-right rule is not always followed; in complex biochemical equations, reaction chains are often written vertically or even in circles.

As an example of a chemical reaction, consider this common occurrence: If you open a bottle of wine and let it stand for several days, you may be disappointed to find that it tastes sour. Wine "turns to vinegar" because oxygen gets into the bottle and reacts with ethanol to produce acetic acid and water. Acetic acid gives the tart flavor to vinegar and spoiled wine. The equation for this reaction is

$$\underset{\text{Ethanol}}{CH_3CH_2OH} + \underset{\text{Oxygen}}{O_2} \rightarrow \underset{\text{Acetic acid}}{CH_3COOH} + \underset{\text{Water}}{H_2O}$$

Ethanol and oxygen are the reactants, and acetic acid and water are the products. In any *balanced* chemical equation, it is essential that every atom on one side of the arrow be accounted for on the other side. Atoms never disappear, or appear from nothing, in the course of a reaction.

Classes of Reactions

Chemical reactions can be classified as *decomposition, synthesis,* or *exchange reactions.* In **decomposition reactions,** a large molecule breaks down into two or more smaller ones (fig. 2.13*a*); symbolically, $AB \rightarrow A + B$. When you eat a potato, for example, digestive enzymes decompose its starch into thousands of glucose molecules, and most cells further decompose glucose to water and carbon dioxide. Starch, a very large molecule, ultimately yields about 36,000 molecules of H_2O and CO_2.

Synthesis reactions are just the opposite of decomposition reactions. Two or more small molecules combine to form a larger one; symbolically, $A + B \rightarrow AB$ (fig. 2.13*b*). When the body synthesizes proteins, for example, it combines several hundred molecules called amino acids into one protein molecule.

In **exchange reactions,** two molecules exchange atoms or groups of atoms; $AB + CD \rightarrow AC + BD$ (fig. 2.14).

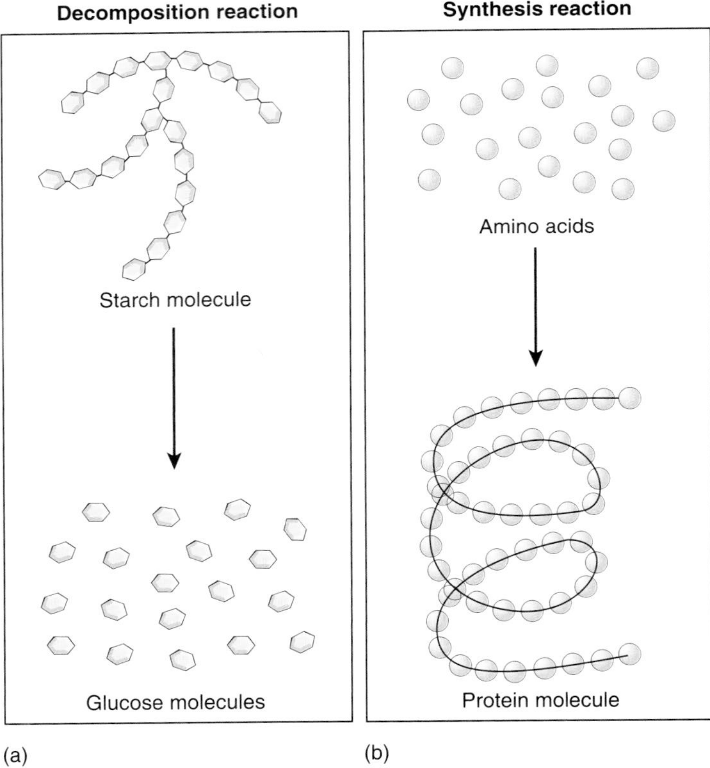

Figure 2.13 Decomposition and Synthesis Reactions. (*a*) In a decomposition reaction, large molecules are broken down into simpler ones. (*b*) In a synthesis reaction, smaller molecules are joined to form larger ones.

For example, when stomach acid (HCl) enters the small intestine, the pancreas secretes sodium bicarbonate ($NaHCO_3$) to neutralize it. The reaction between the two is $NaHCO_3 + HCl \rightarrow NaCl + H_2CO_3$. We could say the sodium atom has exchanged its bicarbonate group ($-HCO_3$) for a chlorine atom.

Reversible Reactions

Reversible reactions can go in either direction under different circumstances and are represented with double-headed arrows. For example, carbon dioxide combines with water to produce carbonic acid, which in turn decomposes into bicarbonate ions and hydrogen ions:

CO_2	+	H_2O	$\leftrightarrow$	H_2CO_3	$\leftrightarrow$	HCO_3^-	+	H^+
Carbon dioxide		Water		Carbonic acid		Bicarbonate ion		Hydrogen ion

This reaction appears in this book more often than any other, especially as we discuss respiratory, urinary, and digestive physiology.

The direction in which a reversible reaction goes is determined by the relative abundance of substances on each side of the equation. If there is a surplus of CO_2, this reaction proceeds to the right and produces bicarbonate and hydrogen ions. If bicarbonate and hydrogen ions are present in excess, the reaction proceeds to the left and generates CO_2 and H_2O. Reversible reactions follow the **law of mass action:** they proceed from the side with the greater quantity of reactants to the side with the lesser quantity. This law will help to explain processes discussed in later chapters, such as why hemoglobin binds oxygen in the lungs yet releases it to muscle tissue.

In the absence of upsetting influences, reversible reactions exist in a state of **equilibrium,** in which the ratio of products to reactants is stable. The carbonic acid reaction, for example, normally maintains a 20:1 ratio of bicarbonate ions to carbonic acid molecules. This equilibrium can be upset, however, by a surplus of hydrogen ions, which drive the

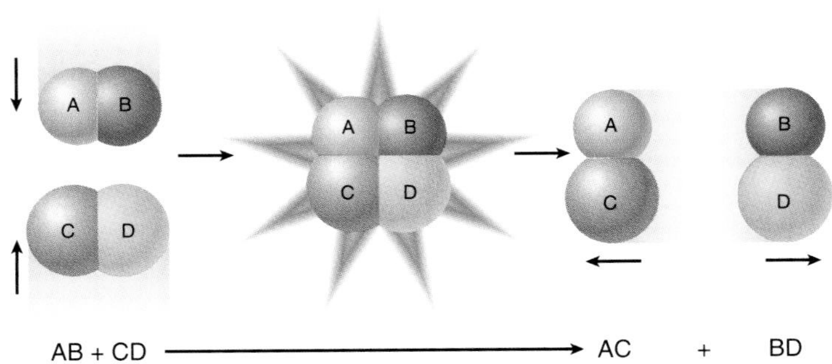

Figure 2.14 Exchange Reaction. Such reactions occur when two molecules collide and exchange atoms.

reaction to the left, or adding carbon dioxide and driving it to the right. Another equilibrium reaction we have already touched upon is the ionization of water, $H_2O \leftrightarrow H^+ + OH^-$.

Reaction Rates

The basis for chemical reactions is molecular motion and collisions. All molecules are in constant motion, and reactions occur when mutually reactive molecules collide with each other (fig. 2.14). The rate of a reaction depends on the nature of the reactants and on the frequency and force of these collisions. Some factors that affect reaction rates are:

- **Concentration.** Reaction rates increase when the reactants are more concentrated. This is because the molecules are more crowded and collide with each other more frequently.
- **Temperature.** Reaction rate increases as the temperature rises. This is because heat causes molecules to move more rapidly and collide with greater force and frequency.
- **Catalysts** (CAT-uh-lists). These are substances that temporarily bind to reactants and hold them in a favorable position to react with each other. By reducing the element of chance in molecular collisions, a catalyst speeds up a reaction. It then releases the products and is available to repeat the process with more reactants. The catalyst itself is not permanently consumed or changed by the reaction. The most important biological catalysts are *enzymes,* discussed in the next chapter.

Before You Go On

Answer the following questions to test your understanding of the preceding section:

21. What are the substances that enter into a chemical reaction called? What is the term for new substances produced by a reaction?
22. What are the differences between decomposition, synthesis, and exchange reactions? Write a generalized equation for each type using the letters A to D.
23. List three factors that affect the rate of a chemical reaction. Explain how each factor does so.

Energy and Metabolism

Objectives

When you have completed this section, you should be able to

- define *work* and *energy;*
- list some biologically important forms of energy;
- distinguish between exergonic and endergonic reactions and relate these to oxidation and reduction; and
- define *metabolism* and its two divisions, *anabolism* and *catabolism.*

Work and Energy

To do **work** means to move something, whether it is a muscle or a molecule. Some examples of physiological work are breaking chemical bonds, building molecules, pumping blood, and contracting skeletal muscles. All of the body's activities are forms of work.

Energy is the capacity to do work. It is broadly classified as *kinetic* or *potential energy.* **Kinetic energy** is the energy of motion. It is observed in skeletomuscular movements, the flow of ions into a cell, and vibration of the eardrum, for example. **Potential energy** is energy contained in an object because of its position or internal state but that is not doing work at the time. Stretch a rubber band and hold it, and you increase its potential energy by altering the internal state of its molecules. Release it and let it fly across the room, and it displays kinetic energy. The water behind a dam has potential energy because of its position. Open a spillway and let the water flow through, and it exhibits kinetic energy that can be tapped for generating electricity. Like water behind a dam, ions concentrated on one side of a cell membrane have potential energy that can be released by opening gates in the membrane. As the ions flow through the gates, their kinetic energy can be tapped to create a nerve signal or make the heart beat.

Within the two broad categories of potential and kinetic energy, there are several forms of energy relevant to human physiology. **Chemical energy** is potential energy stored in the bonds of molecules. Chemical reactions release this energy and make it available for physiological work. **Heat** is the kinetic energy of molecular motion. The temperature of a substance is a measure of rate of this motion, and adding heat to a substance increases this rate. **Electromagnetic energy** is the kinetic energy of moving "packets" of radiation called *photons.* The most familiar form of electromagnetic energy is light, but other, invisible forms—infrared and ultraviolet rays, X rays, and gamma rays—also have important physiological effects. **Electrical energy** has both potential and kinetic forms. It is potential energy when charged particles have accumulated at a point such as a battery terminal or on one side of a cell membrane; it becomes kinetic energy when these particles begin to move and create an electrical current—for example, when electrons move through your household wiring or sodium ions moving through a cell membrane.

Thermodynamics and Free Energy

The rest of this section concerns chemical energy, particularly its transfer from one molecule to another in chemical reactions. The science of energy transfer is called **thermodynamics.**

The **first law of thermodynamics** states that *energy can be converted from one form to another but cannot be created or destroyed.* This is not strictly true of nuclear reactions, where small amounts of matter are converted into enormous amounts of energy, but it is valid for the purposes of chemistry and physiology. It means, for example, that if you use a certain amount of glucose to jog a few laps around the track, the energy that was once in the chemical bonds of glucose is converted to other forms, such as the movement of your body and the heat you feel; however, none of it simply disappears.

The **second law of thermodynamics** states that *in every energy transfer, some energy becomes heat and can no longer do useful work.* For example, when we oxidize glucose we transfer much of its energy to a molecule called ATP. The ATP is then available to do other work. It might be nice if we could transfer *all* the energy of our dietary glucose to ATP, but the second law says this is impossible—some always escapes. Indeed, we conserve only 40% of it and lose the rest as body heat; thus, we say this process has a 40% efficiency. This may seem poor, but it is much better than the efficiency of most machines. Your automobile, for example, probably has about a 17% efficiency. For every $100 you spend on gasoline, $17 moves your car down the road and $83 results merely in air pollution and engine heat. This heat may provide winter comfort, however, so it is not a total waste. Similarly, our bodies require a certain amount of heat for our metabolic processes to work normally. We sometimes burn fat or other fuel just for the heat it provides.

Free energy is the energy available in a system to do work. The second law says that unless a system exchanges energy with its surroundings, it loses free energy with time. The less free energy a system has, the more disordered it becomes—or to put it another way, the greater its **entropy** becomes. We see this principle at work all around us. If we do not maintain our house, for example, it falls apart. The human body, too, is a highly ordered state of matter, and if we do not maintain it, it falls apart. We forestall this by putting energy (food) into the system, but entropy inevitably wins in the end, and we are forced to reconcile ourselves to the inevitability of death.

Metabolism

Metabolism refers to all the chemical reactions of the body. It has two divisions—*catabolism* and *anabolism.* **Catabolism**[6] (ca-TAB-oh-lizm) consists of energy-releasing decomposition reactions. Such reactions break covalent bonds, produce smaller molecules from larger ones, and release energy that can be used for other physiological work. Energy-releasing reactions are called **exergonic**[7] reactions. If you hold a beaker of water in your hand and pour sulfuric acid into it, for example, the beaker will get so hot you may have to put it down. If you break down energy-storage molecules to run a race, you too will get hot. In both cases, the heat signifies that exergonic reactions are occurring.

Anabolism[8] (ah-NAB-oh-lizm) consists of energy-consuming synthesis reactions, such as the production of protein or fat molecules. Reactions that require an energy input, such as these, are called **endergonic**[9] reactions. Anabolism is driven by the energy that catabolism releases, so endergonic and exergonic processes, anabolism and catabolism, are inseparably linked.

Oxidation and Reduction

Oxidation means any chemical reaction in which a molecule gives up electrons and releases free energy. A molecule is *oxidized* by this process, and whatever molecule takes the electrons from it is an **oxidizing agent,** or **electron acceptor.** The term *oxidation* stems from the fact that oxygen is often involved as the oxidizing agent. Thus, we can sometimes recognize an oxidation reaction from the fact that oxygen has been added to a molecule. The rusting of iron, for example, is a slow oxidation process in which oxygen is added to iron to form iron oxide (Fe_2O_3). Many oxidation reactions, however, do not involve oxygen at all. For example, when yeast ferments glucose to alcohol, no oxygen is required; indeed, the alcohol *contains less oxygen* than the sugar originally did, but it is *more oxidized* than the sugar:

$$\underset{\text{Glucose}}{C_6H_{12}O_6} \rightarrow \underset{\text{Ethanol}}{2CH_3CH_2OH} + \underset{\text{Carbon dioxide}}{2CO_2}$$

Reduction is a chemical reaction in which a molecule gains electrons and free energy. When a molecule accepts electrons, it is said to be *reduced;* a molecule that donates electrons to another is therefore called a **reducing agent,** or **electron donor.** The oxidation of one molecule is always accompanied by the reduction of another, so these electron transfers are known as **oxidation-reduction (redox) reactions.**

It is not necessary that *only* electrons be transferred in a redox reaction. Often, the electrons are transferred in the form of hydrogen atoms. The fact that a proton (the hydrogen nucleus) is also transferred is immaterial to whether we consider a reaction oxidation or reduction.

6. *cata* = down, to break down

7. *ex, exo* = out + *erg* = work
8. *ana* = up, to build up
9. *end* = in

Table 2.8 Energy-Transfer Reactions in the Human Body

Exergonic Reactions	Reactions in which there is a net release of energy. The products have less total free energy than the reactants did.
Oxidation	An exergonic reaction in which electrons are removed from a reactant. Electrons may be removed one or two at a time and may be removed in the form of hydrogen atoms (H or H_2). The product is then said to be oxidized.
Decomposition	A reaction such as digestion and cell respiration, in which larger molecules are broken down into smaller ones.
Catabolism	The sum of all decomposition reactions in the body.
Endergonic Reactions	Reactions in which there is a net input of energy. The products have more total free energy than the reactants did.
Reduction	An endergonic reaction in which electrons are donated to a reactant. The product is then said to be reduced.
Synthesis	A reaction such as protein and glycogen synthesis, in which two or more smaller molecules are combined into a larger one.
Anabolism	The sum of all synthesis reactions in the body.

Table 2.8 summarizes these energy-transfer reactions. We can symbolize oxidation and reduction as follows, letting H_2 represent a pair of electrons (hydrogen atoms):

AH_2	+	B	→	A	+	BH_2
High-energy reduced state		Low-energy oxidized state		Low-energy oxidized state		High-energy reduced state

AH_2 is a reducing agent because it reduces B, and B is an oxidizing agent because it oxidizes AH_2.

Before You Go On

Answer the following questions to test your understanding of the preceding section:

24. Define *energy.* Distinguish potential energy from kinetic energy.
25. Define *metabolism, catabolism,* and *anabolism.*
26. What does *oxidation* mean? What does *reduction* mean? Which of them is endergonic and which is exergonic?
27. When sodium chloride is formed, which element—sodium or chlorine—is oxidized? Which one is reduced?

Ionizing Radiation

Objectives

When you have completed this section, you should be able to

- explain what radioisotopes are;
- list and define the types of ionizing radiation produced by radioisotopes;
- discuss the relevance of the physical and biological half-lives of radioisotopes; and
- list and discuss environmental sources of radiation exposure.

Radioisotopes and Radioactivity

We saw earlier that different isotopes of an element exhibit identical chemical behavior. This is because chemical behavior is determined by the number of electrons in an atom, whereas isotopes differ only in number of neutrons. Isotopes differ, however, in their physical behavior. Many of them are unstable and **decay** (break down) to more stable isotopes by giving off radiation. Unstable isotopes are therefore called **radioisotopes,** and the process of decay is called **radioactivity** (see historical insight 2.3). Every element has at least one radioisotope. Oxygen, for example, has five radioisotopes and three stable isotopes. All of us contain radioisotopes such as ^{14}C and ^{40}K—that is, we are all mildly radioactive!

There are many forms of radiation, including light and radio waves, that have low energy and are harmless. High-energy radiation, however, ejects electrons from atoms and is consequently called **ionizing radiation.** It destroys molecules and produces dangerous free radicals and ions in human tissues. Examples of ionizing radiation include ultraviolet rays, X rays, and three kinds of radiation produced by nuclear decay: *alpha* (α) *particles, beta* (β) *particles,* and *gamma* (γ) *rays.*

An **α particle** is composed of two protons and two neutrons (equivalent to a helium nucleus), and a **β particle** is a free electron. Alpha particles are too large to penetrate the skin, and β particles can penetrate only a few millimeters. Thus they are relatively harmless when emitted by sources outside the body. They are very dangerous, however, when emitted by radioisotopes that have gotten into the body. Strontium-90 (^{90}Sr), for example, has been released by atmospheric testing of nuclear weapons and by accidents such as the 1986 explosion and fire at the Chernobyl nuclear power plant in the Ukraine. Strontium-90 settles onto pastures, gets into cow's milk, and behaves chemically like calcium. It becomes incorporated into the bones and emits β particles for many years. Uranium and plutonium emit electromagnetic **γ rays,** which have high energy and penetrating power. Gamma rays are very dangerous even when emitted by sources outside the body.

Historical Insight 2.3

Radiation and Madame Curie

In 1896, French scientist Henri Becquerel (1852–1908) discovered that uranium darkened photographic plates through several thick layers of paper. Marie Curie (1867–1934) and her husband, Pierre (1859–1906), discovered that polonium and radium did likewise. Marie Curie coined the term *radioactivity* for the emission of energy by these elements. The Curies and Becquerel shared a Nobel Prize in 1903 for this discovery. Marie Curie was not only the first woman in the world to receive a Nobel Prize but also the first woman in France even to receive a Ph.D. She received a second Nobel Prize in 1911 for inventing radiation therapy for breast and uterine cancer. Curie crusaded to train women for careers in science, and in World War I, she and her daughter, Irène Joliot-Curie (1897–1956), trained physicians in the use of X-ray machines.

In the wake of such discoveries, radium was regarded as a wonder drug. Slow to realize its danger, people drank radium tonics and flocked to health spas to bathe in radium-enriched waters. Marie Curie herself, oblivious to the danger, suffered extensive damage to her hands from handling radioactive minerals. She died at age 67 from radiation poisoning. The year after her mother died, Irène and her husband, Frédéric Joliot (1900–1958) were awarded a Nobel Prize for work in artificial radioactivity and synthetic radioisotopes. Perhaps also a martyr to her science, Irène died of leukemia, possibly induced by radiation exposure.

Marie Curie (1867–1934). This portrait was made in 1911, when Curie received her second Nobel Prize.

Each radioisotope has a characteristic **physical half-life,** the time required for 50% of the atoms to decay to a more stable state. One gram of ^{90}Sr, for example, would be half gone in 28 years. In 56 years, there would still be 0.25 g left, in 84 years 0.125 g, and so forth. Many radioisotopes are much longer-lived. The half-life of ^{40}K, for example, is 1.3 billion years. Nuclear power plants produce hundreds of radioisotopes that will be intensely radioactive for at least 10,000 years—longer than the life of any disposal container yet conceived. The **biological half-life** of a radioisotope is the time required for half of it to disappear from the body. This is a function of both physical decay and physiological clearance from the body. Cesium-137, for example, has a physical half-life of 30 years but a biological half-life of only 17 days. Chemically, it behaves like potassium; it is quite mobile and rapidly excreted by the kidneys.

Radiation Exposure

There are several ways to measure the intensity of ionizing radiation, the amount absorbed by the body, and its biological effects. To understand the units of measurement requires a grounding in physics beyond the scope of this book, but we can appreciate some relative values in millirems[10] (mrem). The average American absorbs about 295 mrem of naturally occurring **background radiation** annually. About two-thirds of this comes from *radon,* a gas produced by the radioactive decay of uranium in granite. Radon enters buildings through cracks and well water. It emits α particles and can cause lung cancer if inhaled in sufficient quantity. In recent years, there has been a move to make buildings more airtight to conserve energy, but

10. *rem* = Roentgen Equivalent, man. 1 rem = 1,000 mrem.

this has also resulted in the indoor trapping of more radon. The best protection against radon is well-sealed foundations and adequate ventilation.

Some lesser sources of background radiation include cosmic rays from outer space and other radioisotopes in the earth, air, and our tissues. We are also exposed to radiation from medical X rays, radiation therapy, and consumer products such as color televisions, smoke detectors, and luminous watch dials. Such voluntary exposure must be considered from the standpoint of their risk-to-benefit ratio. The benefits of a smoke detector or mammogram far outweigh the risk from the low levels of radiation involved. Radiation therapists and radiologists face a greater risk than their patients, however, and astronauts and airline flight crews receive more than average exposure. U.S. federal standards set a limit of 5 rem/year as acceptable occupational exposure to ionizing radiation.

Before You Go On

Answer the following questions to test your understanding of the preceding section:

28. Name three forms of ionizing radiation and explain the distinctions among them.
29. Distinguish between the physical half-life and biological half-life of a radioisotope. Explain why the latter is shorter than the former.
30. What is the greatest source of background radiation for the average U.S. resident? How can this exposure be reduced?

Chapter Review

Online Learning Center

Study resources for this chapter are available at www.mhhe.com/saladin/

Study Outline

Chemical Elements and Atomic Structure (p. 52)

1. Matter and the elements
   a. Meaning of matter
   b. Chemical elements
      - Atomic numbers and symbols
      - Abundance in body
2. Atomic structure
   a. Planetary model
   b. Electron shells
   c. Valence electrons
   d. Octet rule
3. Isotopes and atomic weight
4. Ions
   a. Ionization process
   b. Anions and cations
   c. Valence of an ion
   d. Mutual attraction of cations and anions
5. Electrolytes
   a. Examples
   b. Physiological effects
6. Free radicals
   a. Formation
   b. Damaging effects
   c. Neutralization by antioxidants

Molecules, Compounds, and Chemical Bonds (p. 57)

1. Molecules and chemical formulae
   a. Molecules and compounds
   b. Chemical formulae
      - Molecular formulae
      - Structural isomers
      - Structural formulae
      - Condensed structural formulae
2. Molecular weight
3. Chemical bonds
   a. Ionic bonds
   b. Covalent bonds
      - Single and double
      - Nonpolar and polar
   c. Hydrogen bonds

Mixtures (p. 60)

1. Distinguished from compounds
2. Types and properties of mixtures
   a. Solutions
   b. Colloids
   c. Suspensions
   d. Emulsions
3. Measures of concentration
   a. Weight per volume
   b. Percentages
   c. Molarity
      - Meaning of a mole
      - Molar solutions
   d. Milliequivalents per liter
      - Used for electrolytes
      - Method of computation

Acids, Bases, and pH (p. 63)

1. Definitions of acid and base
2. Definition of pH
3. The pH scale
4. Buffers

Chemical Reactions (p. 65)

1. Meaning of a chemical reaction
2. Reactants and products
3. Chemical equations
4. Classes of reactions
   a. Decomposition reactions
   b. Synthesis reactions
   c. Exchange reactions
5. Reversible reactions
   a. Law of mass action
   b. Equilibrium states
6. Factors that affect reaction rates
   a. Reactant concentrations

Clinical Insight 2.4

Radioisotopes in Sickness and Health

Radioactivity is a double-edged sword that can cause and cure disease. At high doses, it can kill immediately, while at lower doses it can produce delayed effects such as leukemia, lung cancer, and birth defects. Yet the field of *nuclear medicine* uses controlled doses of radioisotopes for medical diagnosis and treatment.

High, and usually accidental, doses of ionizing radiation cause *acute radiation sickness.* Whole-body doses up to 100 rem produce nausea and vomiting but are not usually fatal. Doses of 100 to 200 rem are typically followed by cancer. Doses of 200 to 1,000 rem produce *hemopoietic*[11] (HE-mo-poy-ET-ic) *syndrome*—death of blood-forming lymphatic tissue and bone marrow, resulting in a deficiency of infection-fighting white blood cells. At this dosage, victims typically die within 8 to 50 days from overwhelming infection or hemorrhage, especially around the body orifices. Doses of 600 to 3,000 rem produce *GI syndrome,* in which the intestinal lining is destroyed, massive amounts of blood plasma seep into the intestines, the body becomes swollen, and there is copious vomiting and diarrhea. So much fluid leaves the bloodstream for the intestines and tissues that the circulatory system fails, and death usually occurs within 3 to 10 days. A whole-body dose of 3,000 rem or more causes *cerebral syndrome,* characterized by vomiting and drowsiness, followed in a few hours by convulsions and death. The *median lethal dose* (*LD_{50}*) of radiation—about 400 to 500 rem—is the dose that kills 50% of exposed individuals.

Although radioisotopes can be dangerous, they also have beneficial uses in medicine and biology. In research, isotopes are used as *tracers.* Suppose, for example, we wish to find out which cells of the uterus are stimulated by estrogen and where in the cell the estrogen becomes concentrated. Estrogen molecules can be synthesized with a radioisotope such as tritium (^3H) in place of some of the hydrogen (^1H). This "tagged" hormone is injected into animals such as rats, and tissue sections (thin slices) of the uterus are made later. The sections are mounted on slides coated with photographic emulsion. Radiation emitted by the tritium makes tiny black dots in the emulsion wherever the hormone molecules are concentrated. The slides are then developed like a photograph and examined under the microscope. Black dots on the emulsion reveal which cells took up the hormone and where the hormone became localized. This procedure, called *autoradiography,* has yielded a great deal of knowledge of hormone action and other physiology.

Radioactive tracers are also used in clinical diagnosis. Thallium-201 is used to determine the location and extent of cardiac damage in heart attack patients, and cobalt-58 is used to evaluate intestinal absorption of vitamin B_{12}. A procedure called a *thyroid scan* uses radioactive iodine (^{131}I) to determine the size and activity of the thyroid gland and detect thyroid cancer. The patient swallows a solution of ^{131}I, which becomes concentrated in the thyroid—the only organ that uses iodine. Twenty-four hours later, the patient is scanned with a machine called a *scintillation counter.* A computer integrates signals from the counter and produces an image for interpretation by the radiologist (fig. 1). Since ^{131}I has a physical half-life of only 8.1 days, it presents little long-term hazard to the patient. The slight risk from radiation exposure is overridden by the value of the diagnosis.

Radiotherapy, the field pioneered by Marie Curie, is the use of ionizing radiation to treat cancer. Large doses of ^{131}I are used to treat thyroid cancer. This isotope is selectively absorbed by the gland and by any cancerous thyroid cells that have spread to other parts of the body. The radiation kills those cells with little risk to tissues that do not absorb iodine. Radium can be introduced into a tumor to destroy it. The α particles emitted by radium are highly destructive to cells in the immediate vicinity but penetrate so poorly through human tissues that they present little risk to adjacent normal tissues. Cobalt-60, which emits β particles and γ rays, is also used to destroy tumor cells.

Continued

11. *hemo* = blood + *poietic* = forming

- b. Temperature
- c. Catalysts

Energy and Metabolism (p. 67)

1. Work
2. Kinetic and potential energy
3. Thermodynamics and free energy
    - a. First and second laws of thermodynamics
    - b. Free energy and entropy
4. Metabolism
    - a. Catabolism and exergonic reactions
    - b. Anabolism and energonic reactions
5. Oxidation and reduction
    - a. Oxidizing agents (electron acceptors)
    - b. Reducing agents (electron donors)

Ionizing Radiation (p. 69)

1. Radioisotopes and radioactivity
2. Ionizing radiation
3. Products of nuclear decay
    - a. α particle: two protons and two neutrons
    - b. β particle: electron
    - c. γ ray: high-energy photon
4. Physical and biological half-life
5. Radiation exposure
    - a. Background radiation
        - Radon
        - Other natural sources
    - b. Radiation from consumer products and medical uses

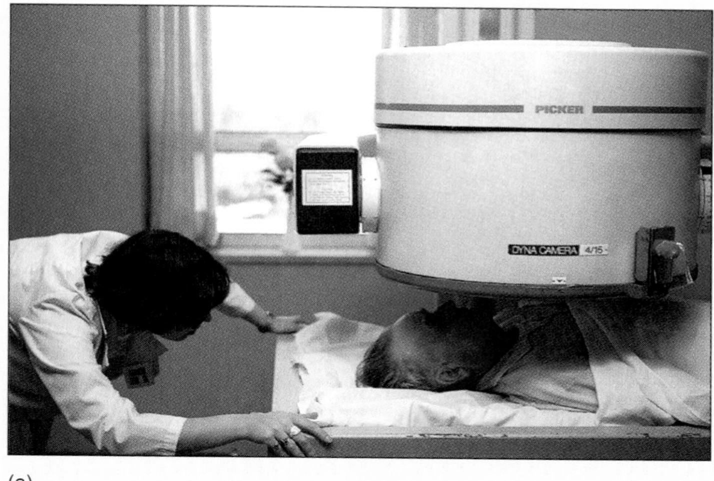

(a)

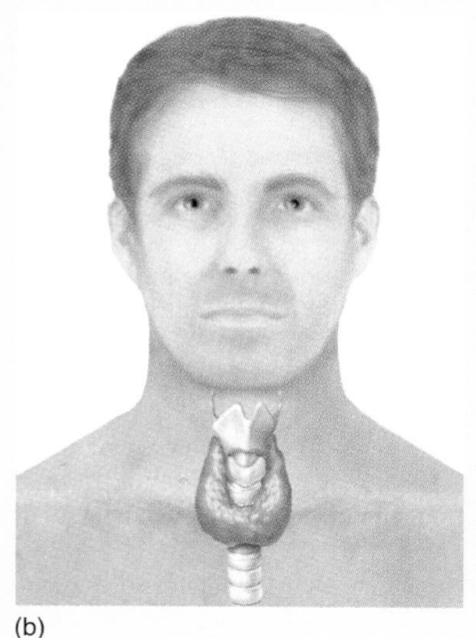

(b)

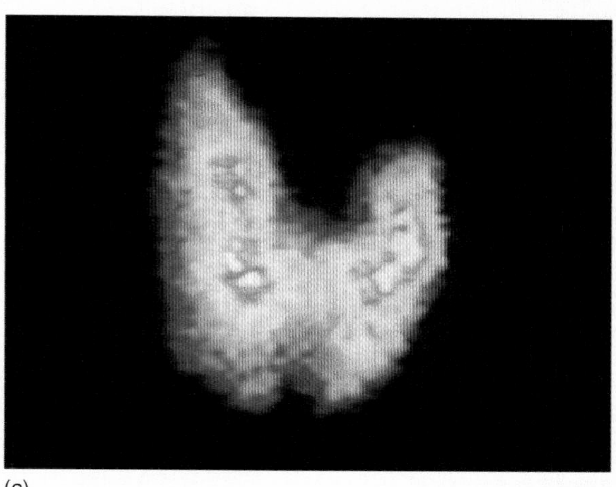

(c)

Figure 1 **Use of a Scintillation Counter.** (*a*) Scanning a patient's radioactively labeled thyroid gland. (*b*) Location of the thyroid gland. (*c*) Image of the thyroid gland produced by the scintillation counter.

Selected Vocabulary

matter 52
element 52
atomic number 52
trace element 52
nucleus 54
proton 54
neutron 54
atomic mass unit 54
atomic mass 54
electron 54
electron shell 54
valence electron 54
octet rule 54
isotope 54
atomic weight 54
ion 54
ionization 54
anion 54
cation 54
valence 55
electrolyte 55
free radical 56
antioxidant 57
molecule 57
compound 57
chemical formula 57
structural isomer 57
molecular weight 58
chemical bond 58
ionic bond 58
covalent bond 58
hydrogen bond 59
mixture 60
solution 61
solute 61
solvent 61
colloid 61
suspension 61
emulsion 61
mole 62

Testing Your Recall

1. The simplest substance with unique chemical properties is
   a. an isotope.
   b. an element.
   c. a subatomic particle.
   d. a molecule.
   e. a compound.
2. Which of the following substances does *not* play a role in normal human physiology?
   a. fluorine
   b. sulfur
   c. silver
   d. phosphorus
   e. iron
3. Which of these particles determine the chemical properties of an atom?
   a. electrons
   b. protons
   c. neutrons
   d. electrons and protons only
   e. protons and neutrons only
4. How many electrons must carbon gain from other atoms to have a stable valence shell?
   a. one
   b. two
   c. four
   d. six
   e. eight
5. Which of the following is the least penetrating type of ionizing radiation?
   a. α particles
   b. β particles
   c. γ rays
   d. π particles
   e. λ rays
6. A substance that _____ is considered to be a chemical compound.
   a. contains at least two different elements
   b. contains at least two atoms
   c. has a chemical bond
   d. has a stable valence shell
   e. has covalent bonds
7. An ionic bond is formed when
   a. two anions meet.
   b. two cations meet.
   c. electrons are shared equally between two nuclei.
   d. electrons are unequally shared between nuclei.
   e. electrons transfer completely from one atom to another.
8. The molecular weight of ethanol (CH_3CH_2OH) is
   a. 6.
   b. 9.
   c. 29.
   d. 46.
   e. 58.
9. The ionization of a sodium atom to produce Na^+ is an example of
   a. oxidation.
   b. reduction.
   c. catabolism.
   d. anabolism.
   e. a decomposition reaction.
10. A solution that resists changes in pH is
    a. a catalyst.
    b. an oxidizing agent.
    c. a reducing agent.
    d. an enzyme.
    e. a buffer.
11. The electrons of an atom's outermost shell are called its _________ electrons.
12. When an atom gives up an electron and acquires a positive charge, it is called a/an ___________.
13. If a pair of shared electrons are more attracted to one nucleus than to the other, they form a/an ___________ bond.
14. Two molecules with the same atoms arranged in different order are called ___________.
15. Dietary antioxidants are important because they neutralize ___________.
16. Any substance that increases the rate of a reaction without being consumed by it is a/an ___________.
17. A/an ___________ is a mixture of two liquids that separate from each other on standing.
18. Avogadro's number is the number of molecules in 1 ___________ of a substance.
19. A solution with pH 5 has a hydroxyl ion molarity of ___________.
20. All the synthesis reactions in the body form a division of metabolism called ___________.

Answers in appendix B

Testing Your Comprehension

1. Suppose a pregnant woman with severe morning sickness has been vomiting steadily for several days. How will her loss of stomach acid affect the pH of her body fluids? Explain.
2. Suppose a person with a severe anxiety attack hyperventilates and exhales CO_2 faster than his body produces it. Consider the carbonic acid reaction on page 66 and explain what effect this hyperventilation will have on his blood pH. (Hint: Remember the law of mass action.)
3. Glycine has a molecular weight of 75. How much glycine would be needed to make 500 mL of a 100 mM solution?
4. In one form of nuclear decay, a neutron breaks down into a proton and electron and emits a γ ray. Is this an endergonic or exergonic reaction, or neither? Is it an anabolic or catabolic reaction, or neither? Explain both answers.
5. How would the body's metabolic rate be affected if there were no such thing as enzymes? Explain.

Answers at the Online Learning Center

3 CHAPTER

The Molecules of Life

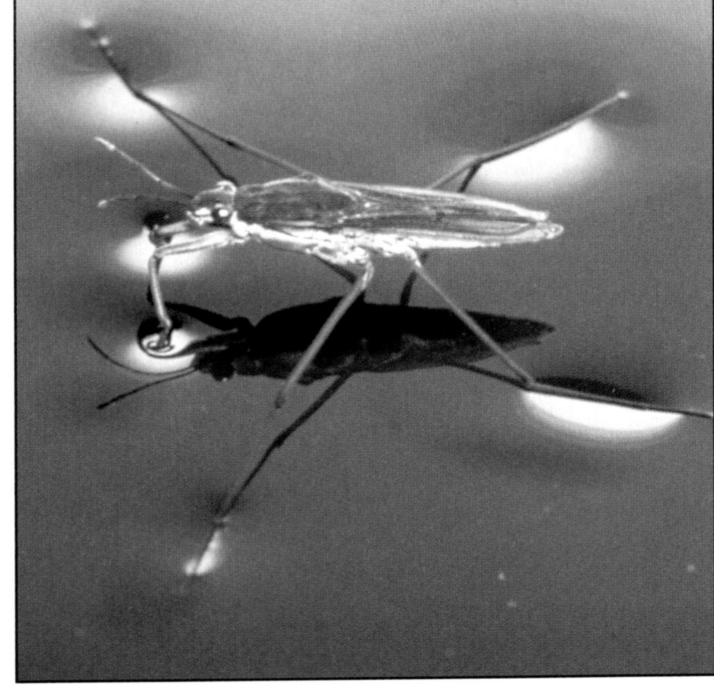

A water strider supported by the elastic surface film of water

Chapter Outline

Insights

Study with Your ESP CD-ROM

Refer to your ESP Tutorial CD-ROM for helpful animations, activities, and quizzes related to the material covered in this chapter. The following are suggested areas to go to on the CD and are presented according to "Unit/Topic":
- Levels of Organization/Chemistry
- Levels of Organization/Cellular Metabolism

Brushing Up

To understand this chapter, it is important that you understand or brush up on the following concepts:
- Polar and nonpolar covalent bonds (pp. 58–59)
- Hydrogen bonds (p. 59)
- Decomposition and synthesis reactions (p. 66)
- Catalysts (p. 67)
- Oxidation and reduction (p. 68)

It is common knowledge that we need proteins, fats, carbohydrates, vitamins, and minerals in our diet. We know, too, that we should avoid consuming too much saturated fat and cholesterol. Many people have only a vague concept of what these molecules are, much less how they function in the body—yet such knowledge is very helpful in matters of personal fitness and patient education, and is essential to the comprehension of the rest of this book.

Study of the molecules that compose living organisms is called **biochemistry.** Biochemistry overlaps with **organic chemistry,** the study of carbon-containing (organic) compounds, as well as with **inorganic chemistry,** the chemistry of compounds such as water that contain no carbon but are important to biological processes. This chapter surveys some concepts of biochemistry needed for a further understanding of the body's structure and function.

Inorganic Molecules

Objectives

When you have completed this section, you should be able to

- discuss the properties of water, why they are important to human life, and how they result from its molecular structure; and
- explain the biological roles of minerals and gases.

Water

Water is a deceptively simple substance with deep biological implications in its structure and behavior. It makes up 50% to 75% of the body weight, depending on a person's age and sex. Fluid intake and output, along with related variables such as blood pressure, are critically important to our health.

The unique properties of water stem from two aspects of its structure: (1) its atoms are joined by polar covalent bonds, and (2) the molecule is V-shaped, with a 105° bond angle (fig. 3.1). This makes the molecule as a whole polar, with a slight negative charge (δ−) on the oxygen atom and a slight positive charge (δ+) on each hydrogen atom. Like little magnets, water molecules are attracted to each other by hydrogen bonds (see fig. 2.10, p. 60). This gives water a set of properties that account for its ability to support life, namely *solvency, cohesion, adhesion, thermal stability,* and *chemical reactivity* (table 3.1).

Solvency

Solvency is the ability to dissolve matter. Water is sometimes called the *universal solvent* because it dissolves a broader range of substances than any other liquid does. Substances that dissolve in water, such as sugar, are said to be **hydrophilic**[1] (HY-dro-FILL-ic); the relatively few substances that do not, such as fats, are **hydrophobic**[2] (HY-dro-FOE-bic).

1. *hydro* = water + *philic* = loving, attracted to
2. *phobic* = fearing, avoiding

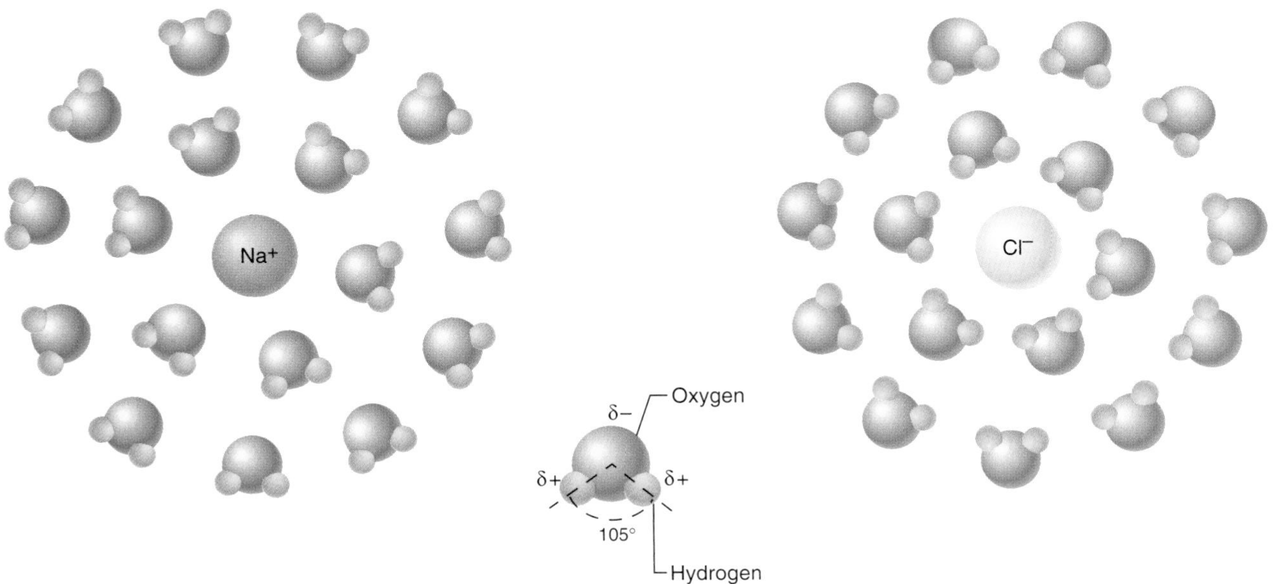

Figure 3.1 Hydration Spheres. Water molecules aggregate around a sodium ion with their negatively charged oxygen poles facing the Na^+ and aggregate around a chloride ion with their positively charged hydrogen poles facing the Cl^-. These hydration spheres keep the sodium and chloride ions separated and dissolved. The inset (*center*) shows the polarity and bond angle of the water molecule.

Table 3.1	Summary of the Properties of Water
Polarity	Polar covalent bonds and a 105° bond angle make the water molecule polar; this is the basis for its other properties.
Solvency	Water is called the *universal solvent* because it dissolves a wide range of biological molecules. As a solvent, water provides an effective medium for transport of other chemicals and for a wide variety of metabolic reactions.
Cohesion	Water molecules are attracted to each other by hydrogen bonds; this is the basis for the surface tension and thermal stability of water.
Adhesion	Water adheres to tissues and forms a lubricating film on body membranes.
Surface Tension	Hydrogen bonding produces surface tension, which creates an elastic surface film on water.
Thermal Stability	The high heat capacity of water stabilizes body temperature and makes water an effective coolant.
Chemical Reactivity	Water ionizes many molecules, including itself. It thus promotes chemical reactions and participates in many of them as a reactant or product.

To be soluble in water, a molecule must be charged so that its charges can interact with those of the water molecule. When NaCl is dropped into water, for example, the ionic bonds between Na^+ and Cl^- are overpowered by the attraction of each ion to water molecules. Water molecules form a cluster, or **hydration sphere,** around each sodium ion with the $O^{\delta-}$ pole of each water molecule facing the sodium ion. They also form a hydration sphere around each chloride ion, with the $H^{\delta+}$ poles facing it. This isolates the sodium ions from the chloride ions and keeps them dissolved (fig. 3.1).

Virtually all metabolic reactions depend on the solvency of water. Biological molecules must be dissolved in water to move freely, come together, and react. The solvency of water also makes it the body's primary means of transporting substances from place to place.

Adhesion and Cohesion

Adhesion is the tendency of one substance to cling to another, and **cohesion** is the tendency of molecules of the same substance to cling to each other. Water adheres to the body's tissues and forms a lubricating film on membranes such as the pleura and pericardium. This helps reduce friction as the lungs and heart contract, expand, and rub against these membranes. Water also is a very cohesive liquid because of the hydrogen bonds that attract water molecules to each other. Because of this property, when you spill water on the floor, it forms a puddle and evaporates slowly. By contrast, if you spill a nonpolar substance such as liquid nitrogen on the floor, it dances about and evaporates in seconds, like a drop of water in a hot dry skillet. This is because nitrogen molecules have no attraction for each other, so the little bit of heat provided by the floor is enough to disperse them into the air.

The cohesion of water is especially evident at its surface, where it forms an elastic layer called the *surface film* held together by a force called **surface tension** (see p. 76). Surface tension results from water molecules below the surface pulling on those at the surface. This force causes water to hang in drops from a leaky faucet and travel in rivulets down a window.

Thermal Stability

Water helps to stabilize the internal temperature of the body. This ability stems from its high **heat capacity**—the amount of heat required to raise the temperature of 1 g of a substance by 1°C. Heat is measured in **calories**[3] (cal)—1 cal is the amount of heat that raises the temperature of 1 g of water 1°C, whereas the same amount of heat would raise the temperature of a nonpolar substance such as nitrogen by about four times as much. The difference stems from the presence or absence of hydrogen bonding. To increase in temperature, the molecules of a substance must move around more actively. The hydrogen bonds of water molecules inhibit their movement, so water can absorb a given amount of heat without changing temperature (molecular motion) as much.

The high heat capacity of water also makes it a very effective coolant. When it changes from a liquid to a vapor, water carries a large amount of heat with it. Just 1 mL of perspiration evaporating from the skin removes about 500 calories of heat from the body. This effect is very apparent when you are sweaty and stand in front of a fan.

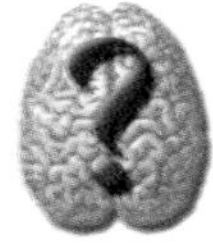

Think About It

Why are heat and temperature not the same thing?

3. *calor* = heat

Clinical Insight 3.1

Communicating with Gas

Nitric oxide (NO) was formerly regarded only as an air pollutant, destroyer of ozone, and possibly a cause of cancer. The discovery that it functions as an intercellular messenger has greatly expanded our view of cellular communication and led to the development of new drugs such as Viagra, used to treat erectile dysfunction.

Nitric oxide is a free radical synthesized from the amino acid arginine by an enzyme called *nitric oxide synthetase*. It stimulates the muscular walls of blood vessels to relax and can thereby lower blood pressure or increase blood flow. Some bacterial infections stimulate overproduction of NO, causing blood pressure to fall dangerously low. In such cases, NO inhibitors are often used to restore normal blood pressure. Nitroglycerin relieves heart pain (angina) by releasing NO, which dilates the coronary arteries and enhances blood flow to the cardiac muscle. During sexual arousal, nitric oxide dilates the arteries of the male and female genitalia and causes the increased blood flow needed for erection. NO inhibitors prevent erection, while Viagra prolongs and intensifies erection by inhibiting the breakdown of NO.

The functions of nitric oxide are not limited to blood circulation. Some forms of learning and memory apparently involve nerve cells communicating with each other by means of NO.

Chemical Reactivity

Reactivity is the ability to participate in chemical reactions. Not only does water ionize many other chemicals such as acids and salts, but water itself ionizes into H^+ and OH^- ions. Thus it facilitates chemical reactions, and water itself participates as a reactant or product in many of them, especially the important metabolic reactions, *hydrolysis* and *condensation*, which we study later in this chapter.

Minerals

Minerals are inorganic elements that are extracted from the soil by plants and passed up the food chain to humans and other organisms. They constitute about 4% of the human body by weight. Nearly three-quarters of this is calcium and phosphorus; the rest is mainly chlorine, magnesium, potassium, sodium, and sulfur. Others minerals present in lesser quantities (trace elements) were listed in table 2.1 (p. 52).

Minerals contribute significantly to body structure. The bones and teeth consist partly of crystals of calcium, phosphorus, magnesium, fluoride, and sulfate ions. Sulfur is a component of many proteins, and phosphorus is a major component of nucleic acids, ATP, and cell membranes.

Minerals also enable enzymes and other organic molecules to function. Iodine is a component of thyroid hormone; iron is a component of hemoglobin; and some enzymes function only when manganese, zinc, copper, or other minerals are bound to them. The electrolytes needed for nerve and muscle function are mineral salts. The biological roles of minerals are discussed in more detail in chapters 24 and 26.

Gases

The gases of greatest physiological importance are oxygen and carbon dioxide. Oxygen is carried mainly by the red blood cells, but there is some dissolved oxygen in all body fluids. It serves as the final electron acceptor in aerobic respiration (see chapter 4) and some other oxidation reactions. Carbon dioxide is a product of aerobic respiration and must be constantly expelled from the body to prevent it from upsetting the body's pH balance.

Biochemists have long studied organic chemicals that cells use to communicate with each other. More recently, they have discovered that some cells communicate with nitric oxide (NO), an inorganic gas (not to be confused with nitrous oxide, or "laughing gas," N_2O) (see clinical insight 3.1).

Before You Go On

Answer the following questions to test your understanding of the preceding section:

1. Explain how the bond angle and polar covalent bonds of water account for its biologically important properties.
2. Define *calorie*.
3. Describe the major roles of minerals in the body.
4. Name three physiologically important gases and state what roles they play in the body.

Carbon and Organic Molecules

Objectives

When you have completed this section, you should be able to

- explain why carbon is better suited than any other element to serve as the structural foundation of most organic molecules;
- name and diagram some common functional groups of organic molecules; and
- explain what monomers and polymers are and how they are related to each other through dehydration synthesis and hydrolysis.

Carbon

Compounds of carbon are called *organic* because it was once thought that only living organisms made them. With minor exceptions, this is true. Carbon is an especially versatile atom that serves as the basis of a wide variety of structures. It has four valence electrons, so it bonds with other atoms that can provide it with four more to complete its valence shell. Carbon atoms readily bond with each other and can form long chains, branched molecules, and rings—an enormous variety of "backbones" for organic molecules. It also forms covalent bonds with hydrogen, oxygen, nitrogen, sulfur, and other elements. The only other element with such versatility is silicon (Si), but it would be a poor substitute for carbon because Si–Si bonds are unstable in the presence of oxygen.

Functional Groups

Carbon backbones also are able to carry a variety of **functional groups**—small clusters of atoms that determine many of the properties of an organic molecule. For example, organic acids bear a **carboxyl** (car-BOC-sil) **group,** and adenosine triphosphate (ATP) is named for its three **phosphate groups.** Other common functional groups include **hydroxyl, methyl,** and **amino groups** (fig. 3.2).

Monomers and Polymers

Since carbon can form long chains, some organic molecules are gigantic **macromolecules** with molecular weights that range from the thousands (as in starch and proteins) to the millions (as in DNA). Most macromolecules are **polymers**[4]—molecules made of a repetitive series of identical or similar subunits called **monomers** (MON-oh-murs). A starch molecule, for example, is a polymer of about 3,000 glucose monomers. In some polymers, such as starch, the monomers are identical, while in other cases they have a basic structural similarity but differ in detail. DNA, for example, is made of 4 different kinds of monomers (nucleotides), and proteins are made of 20 kinds (amino acids).

The joining of monomers to form a polymer is called **polymerization.** Living cells achieve this by means of a reaction called **dehydration synthesis** (fig. 3.3*a*). A hydroxyl (–OH) group is removed from one monomer and a hydrogen (–H) from another, leaving each monomer in an unstable, reactive state. The two monomers become joined by a covalent bond, forming a *dimer*.[5] The –OH and –H unite to form water as a by-product. This reaction is repeated for each monomer added to the growing end of the chain, potentially leading to a chain long enough to be considered a polymer.

Name and Symbol	Structure	Occurs in
Hydroxyl (—OH)	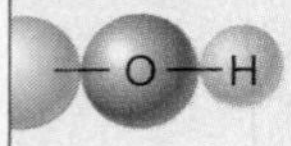	Sugars, alcohols
Methyl (—CH_3)	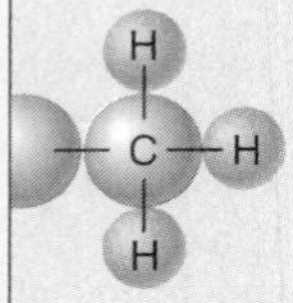	Fats, oils, steroids, amino acids
Carboxyl (—COOH)	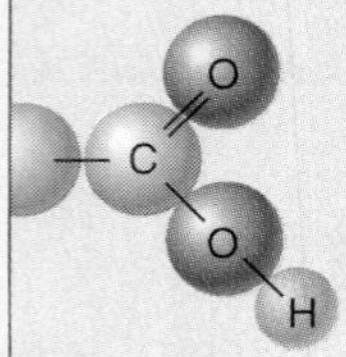	Amino acids, sugars, proteins
Amino (—NH_2)	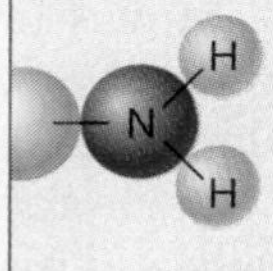	Amino acids, proteins
Phosphate (–H_2PO_4)	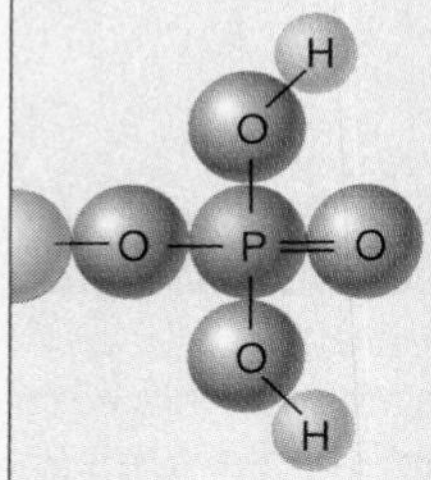	Nucleic acids, ATP

Figure 3.2 Functional Groups of Organic Molecules.

The opposite of dehydration synthesis is **hydrolysis**[6] (fig. 3.3*b*). In a hydrolysis reaction, a water molecule ionizes into OH^- and H^+. A covalent bond linking one monomer to another is then broken, the OH^- is added to one monomer, and the H^+ is added to the other one. All digestion consists of hydrolysis reactions.

4. *poly* = many + *mer* = part
5. *di* = two + *mer* = part
6. *hydro* = water + *lysis* = splitting apart

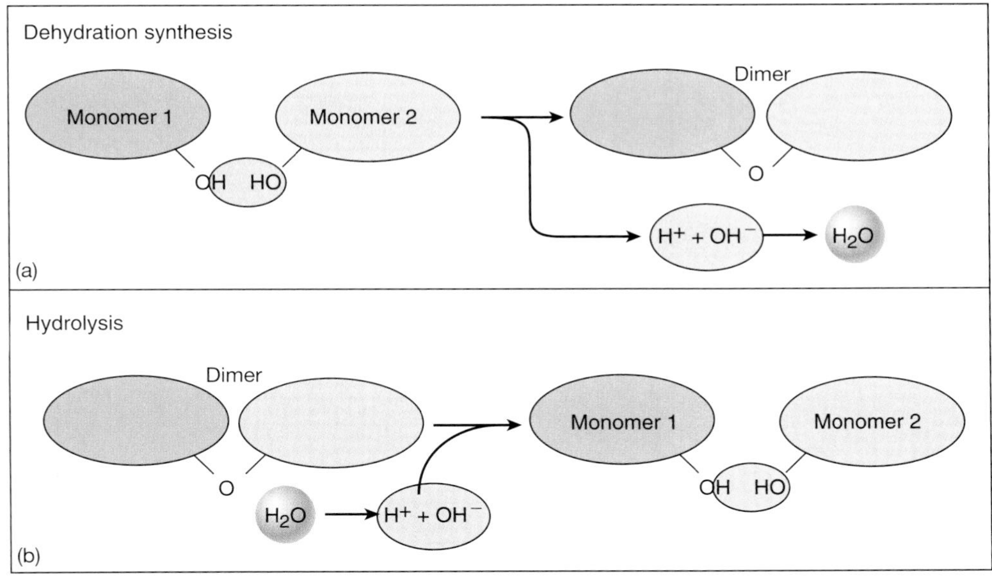

Figure 3.3 Synthesis and Hydrolysis Reactions. (*a*) In dehydration synthesis, a hydrogen atom is removed from one monomer and a hydroxyl group is removed from another. These combine to form water as a by-product. The monomers become joined by a covalent bond to form a dimer. (*b*) In hydrolysis, a covalent bond between two monomers is broken. Water donates a hydrogen atom to one monomer and a hydroxyl group to the other.

Before You Go On

Answer the following questions to test your understanding of the preceding section:

5. Explain why carbon serves better than any other element could do as a backbone for organic molecules.
6. Define *monomer* and *polymer.*
7. To what branch of metabolism do you think polymerization belongs—oxidation or reduction? Synthesis or decomposition? Explain each answer.

Carbohydrates

Objectives

When you have completed this section, you should be able to

- explain how carbohydrates differ from other organic molecules;
- list the major monosaccharides, disaccharides, and polysaccharides; and
- list the physiological functions of carbohydrates.

By 1900, biochemists had classified the biological macromolecules into four primary categories: *carbohydrates, lipids, proteins,* and *nucleic acids.* We examine the first three of these in this chapter but describe the details of nucleic acids, which are concerned with genetics, in chapter 5.

A **carbohydrate**[7] is a hydrophilic organic molecule with the general formula $(CH_2O)_n$, where *n* represents the number of carbon atoms. In the sugar glucose, for example, n = 6 and the formula is therefore $C_6H_{12}O_6$. As the generic formula shows, carbohydrates have a 2:1 ratio of hydrogen to oxygen. The names of individual carbohydrates are often built on the word root *sacchar-* or the suffix *-ose,* both of which mean "sugar" or "sweet." The most familiar carbohydrates are the sugars and starches.

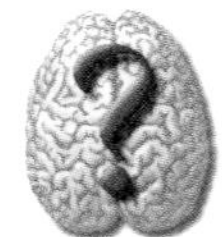

Think About It

Why is *carbohydrate* an appropriate name for this class of compounds? Relate this name to the general formula of carbohydrates.

Monosaccharides

The simplest carbohydrates are called simple sugars or **monosaccharides**[8] (MON-oh-SAC-uh-rides). The three of primary importance are **glucose, fructose,** and **galactose,** all of which have the molecular formula $C_6H_{12}O_6$. They differ only in the arrangement of these 24 atoms and are therefore

7. *carbo* = carbon + *hydr* = water
8. *mono* = one + *sacchar* = sugar

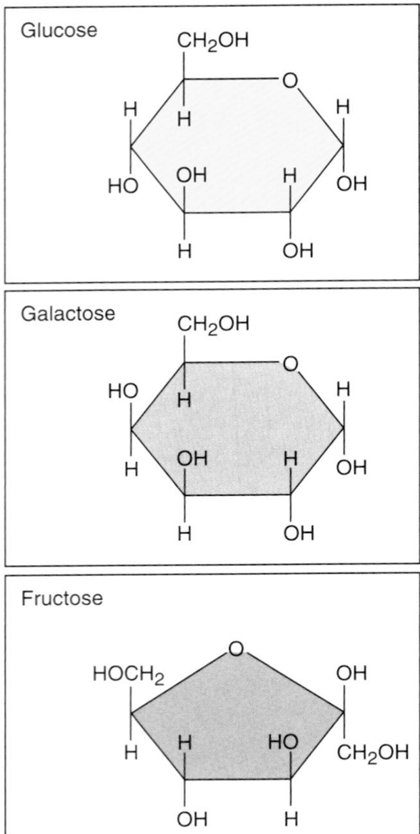

Figure 3.4 The Three Major Monosaccharides. All three have the molecular formula $C_6H_{12}O_6$. Each angle in the six-sided ring represents a carbon atom except the one where oxygen is shown. This is a conventional way of representing carbon in the structural formulae of organic compounds.

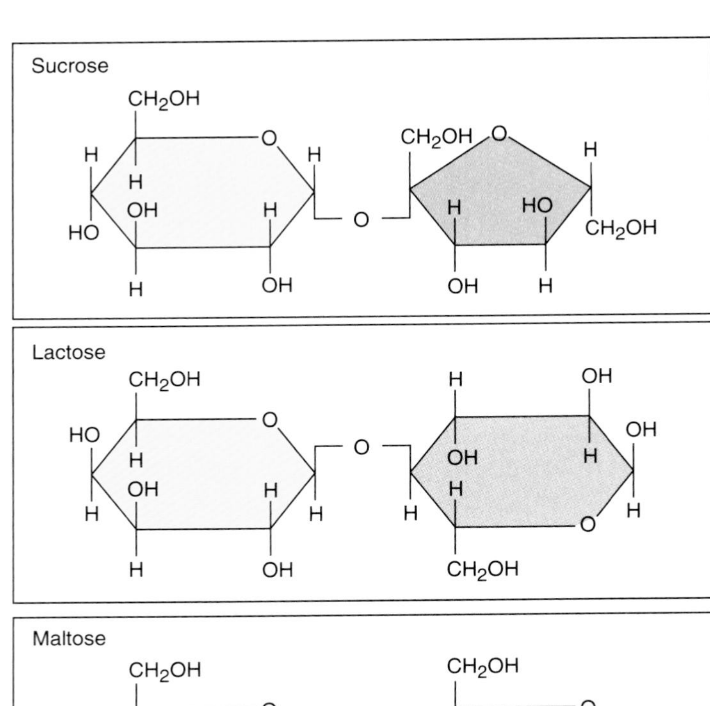

Figure 3.5 The Three Major Disaccharides. All three have the molecular formula $C_{12}H_{22}O_{11}$.

structural isomers of each other (fig. 3.4). We obtain these sugars mainly by the digestion of more complex carbohydrates. This book refers most often to glucose, which is the "blood sugar" that provides energy to most of our cells.

Disaccharides

Monosaccharides can be combined in pairs to form sugars called **disaccharides.** The three of greatest importance are **sucrose** (made of glucose + fructose), **lactose** (glucose + galactose), and **maltose** (glucose + glucose) (fig. 3.5). Sucrose is produced by sugarcane and sugar beets and used as common table sugar. Lactose is milk sugar. Maltose is a product of starch digestion and is present in a few foods such as germinating wheat.

The C–O–C bonds that hold disaccharides together are called **glycosidic bonds.** Figure 3.6*a* shows how a glycosidic bond is formed by dehydration synthesis, and figure 3.6*b* shows how these bonds are broken by hydrolysis.

Polysaccharides

Glucose can also form long chains called **polysaccharides** (POL-ee-SAC-uh-rides), some of which have molecular weights of 500,000 or more. Three polysaccharides of interest to human physiology are starch, cellulose, and glycogen. Plants make starch and cellulose while animals, including ourselves, make glycogen.

Starch functions as an energy-storage polysaccharide that plants store when sunlight and nutrients are available and draw from when photosynthesis is not possible (for example, at night and in winter, when a plant has shed its leaves). There are two forms of starch: a straight-chain molecule called *amylose* (fig. 3.7) and a branched form called *amylopectin.* Both forms are digested by an enzyme called *amylase,* present in our saliva and pancreatic juice.

Cellulose is a *structural polysaccharide* that gives strength to the cell walls of plants. It is the principal component of wood, cotton, and paper. A cellulose molecule consists of a few thousand glucose monomers joined together, with every other one "upside down" relative to the

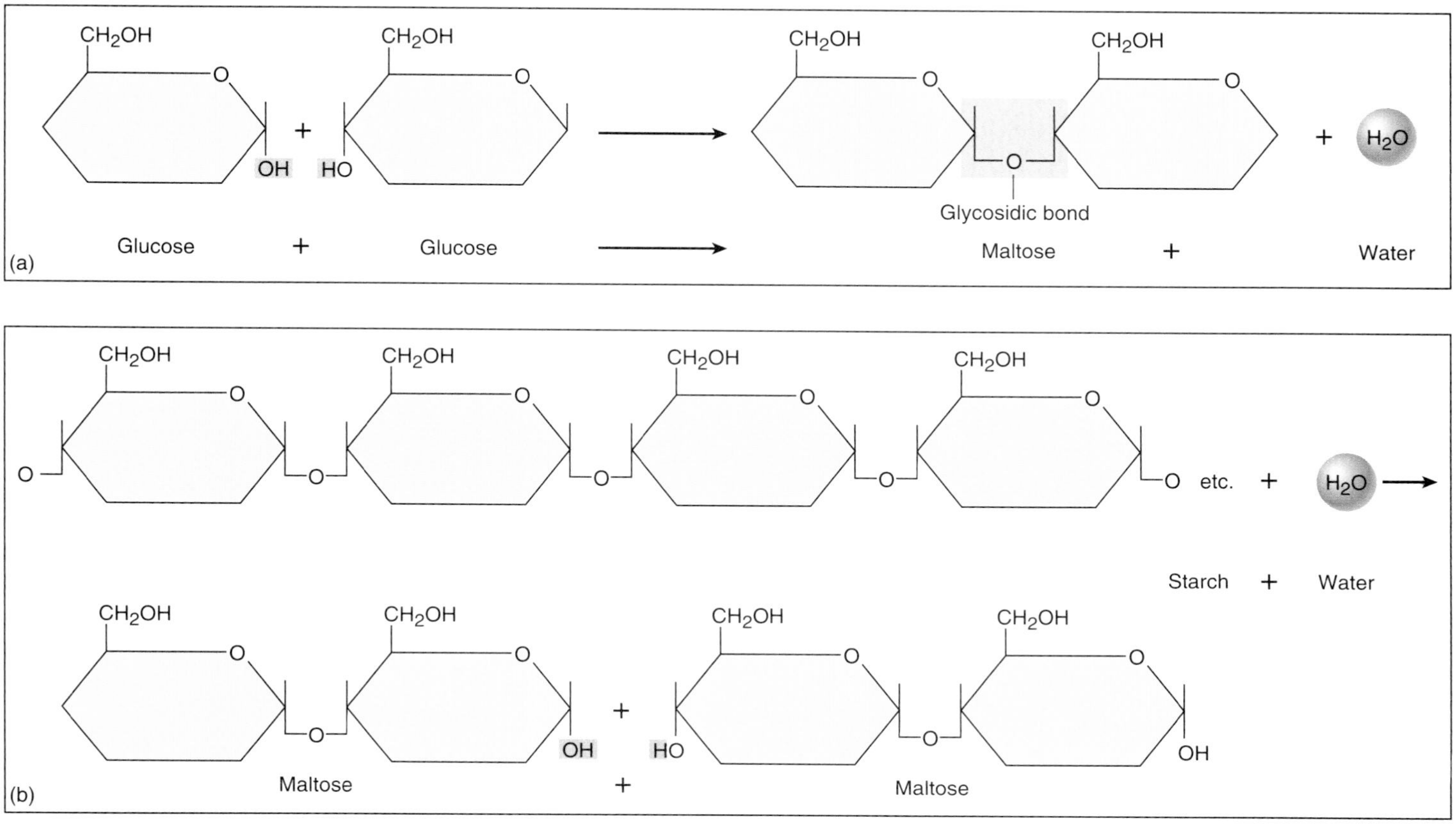

Figure 3.6 Maltose Synthesis and Starch Hydrolysis. (*a*) Maltose is formed from two glucose molecules by dehydration synthesis. (*b*) Starch is hydrolyzed to yield maltose molecules.

next. Cellulose is the most abundant organic compound on earth and it is a common component of the diets of humans and other animals—yet very few animals have the enzyme needed to digest it, and humans derive no energy or nutrition from it. In spite of being indigestible, it is important to the diet as "fiber," "bulk," or "roughage." It swells with water in the digestive tract and helps move other materials through the intestine that might otherwise stagnate, ferment, and generate toxins. A diet poor in cellulose increases the risk of colon cancer because it lacks the bulk to eliminate potentially harmful substances from the colon.

Glycogen[9] is an energy-storage polysaccharide made by animals. In humans, it is synthesized by cells of the liver, muscles, uterus, and vagina. Glycogen is similar to amylopectin but more highly branched. Liver cells produce glycogen after a meal, when the blood glucose level is high, and then break it down between meals to maintain blood glucose levels when there is no food intake. Muscle stores glycogen for its own energy needs, and the uterus uses it in pregnancy to nourish the embryo.

Carbohydrate Functions

Carbohydrates are, above all, a source of energy that can be quickly mobilized. All digested carbohydrate is ultimately converted to glucose, and glucose is oxidized to make ATP, a high-energy compound discussed at the end of this chapter. But carbohydrates have other functions as well (table 3.2). They are often **conjugated**[10] with (covalently bound to) proteins and lipids. Many of the lipid and protein molecules at the external surface of the cell membrane have chains of up to 12 sugars attached to them, thus forming **glycolipids** and **glycoproteins**, respectively. Among other functions, glycoproteins are a major component of mucus, which traps particles in the respiratory system, resists infection, and protects the digestive tract from its own acid and enzymes.

Proteoglycans (once called mucopolysaccharides) are macromolecules in which the carbohydrate component is dominant and a peptide or protein forms a smaller component. Proteoglycans form gels that help hold cells and tissues together, form a gelatinous filler in

9. *glyco* = sugar + *gen* = producing

10. *con* = together + *jug* = join

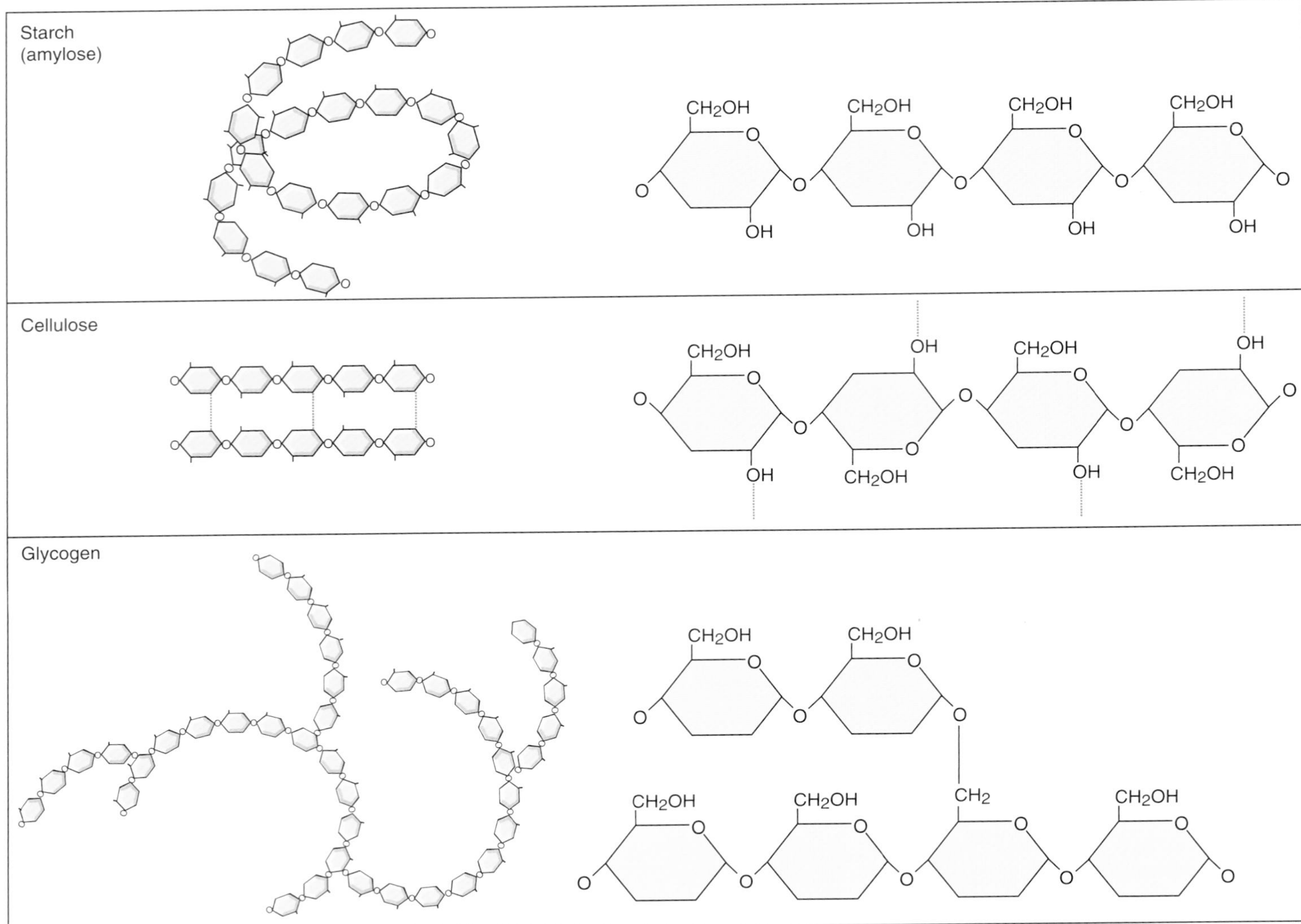

Figure 3.7 **The Three Major Polysaccharides.** *Dotted red lines* indicate hydrogen bonds between adjacent cellulose chains.

the umbilical cord and eye, lubricate the joints of the skeletal system, and account for the tough rubbery texture of cartilage. Their structure and functions are further considered in chapter 6.

When discussing conjugated macromolecules it is convenient to refer to each chemically different component as a **moiety**[11] (MOY-eh-tee). Proteoglycans have a protein moiety and a carbohydrate moiety, for example.

11. *moiet* = half

Before You Go On

Answer the following questions to test your understanding of the preceding section:

8. What is a carbohydrate?
9. Use the concepts of dehydration synthesis and hydrolysis to explain how monosaccharides are related to polysaccharides.
10. List the three major disaccharides of physiological importance, and name the monosaccharides that each is made of.
11. State five functions of carbohydrates.

Table 3.2 Carbohydrate Functions

Type	Function
Monosaccharides	
Glucose	Blood sugar—energy source for most cells
Galactose	Converted to glucose and metabolized
Fructose	Fruit sugar—converted to glucose and metabolized
Disaccharides	
Sucrose	Cane sugar—digested to glucose and fructose
Lactose	Milk sugar—digested to glucose and galactose; important in infant nutrition
Maltose	Malt sugar—product of starch digestion, further digested to glucose
Polysaccharides	
Cellulose	Structural polysaccharide of plants; dietary fiber
Starch	Energy storage in plant cells
Glycogen	Energy storage in animal cells (liver, muscle, uterus, vagina)
Conjugated Carbohydrates	
Glycoproteins	Component of the cell surface coat and mucus, among other roles
Glycolipids	Component of the cell surface coat
Proteoglycans	Cell adhesion; lubrication; supportive filler of some tissues and organs

Lipids

Objectives

When you have completed this section, you should be able to

- explain how lipids differ from other organic molecules;
- list the major classes of lipids and describe their differences; and
- state the major functions of lipids.

A **lipid** is a hydrophobic organic molecule, usually composed only of carbon, hydrogen, and oxygen, with a high ratio of hydrogen to oxygen. A fat called *tristearin* (tri-STEE-uh-rin), for example, has the molecular formula $C_{57}H_{110}O_6$—more than 18 hydrogens for every oxygen. Lipids are less oxidized than carbohydrates and thus have more calories per gram. Beyond these criteria, it is difficult to generalize about lipids; they are much more variable in structure than the other macromolecules we are considering. We consider the five primary types of lipids in humans—*fatty acids, triglycerides, phospholipids, eicosanoids,* and *steroids* (table 3.3).

Table 3.3 Some Lipid Functions

Type	Function
Fatty acids	Precursor of triglycerides; source of energy
Triglycerides	Energy storage; thermal insulation; filling space; binding organs together; cushioning organs
Phospholipids	Major component of cell membranes; aid in fat digestion
Eicosanoids	Chemical messengers between cells
Fat-soluble vitamins	Involved in a variety of functions including blood clotting, wound healing, vision, and calcium absorption
Cholesterol	Component of cell membranes; precursor of other steroids
Steroid hormones	Chemical messengers between cells
Bile acids	Steroids that aid in fat digestion and nutrient absorption

Fatty Acids

A **fatty acid** is a chain of usually 4 to 24 carbon atoms with a carboxyl group at one end and a methyl group at the other. Fatty acids are polymers of two-carbon *acetyl groups,* which gives each fatty acid an even number of carbons. Fatty acids, and the fats made from them, are classified as *saturated* or *unsaturated.* A **saturated fatty acid,** such as palmitic acid, has as much hydrogen as it can possibly carry. No more could be added without exceeding four covalent bonds per carbon atom; thus it is "saturated" with hydrogen. In **unsaturated fatty acids** such as linoleic acid, however, some carbon atoms are joined by double covalent bonds (fig. 3.8). Each of these could potentially share one pair of electrons with another hydrogen atom instead of with the adjacent carbon, so hydrogen could be added to this molecule.

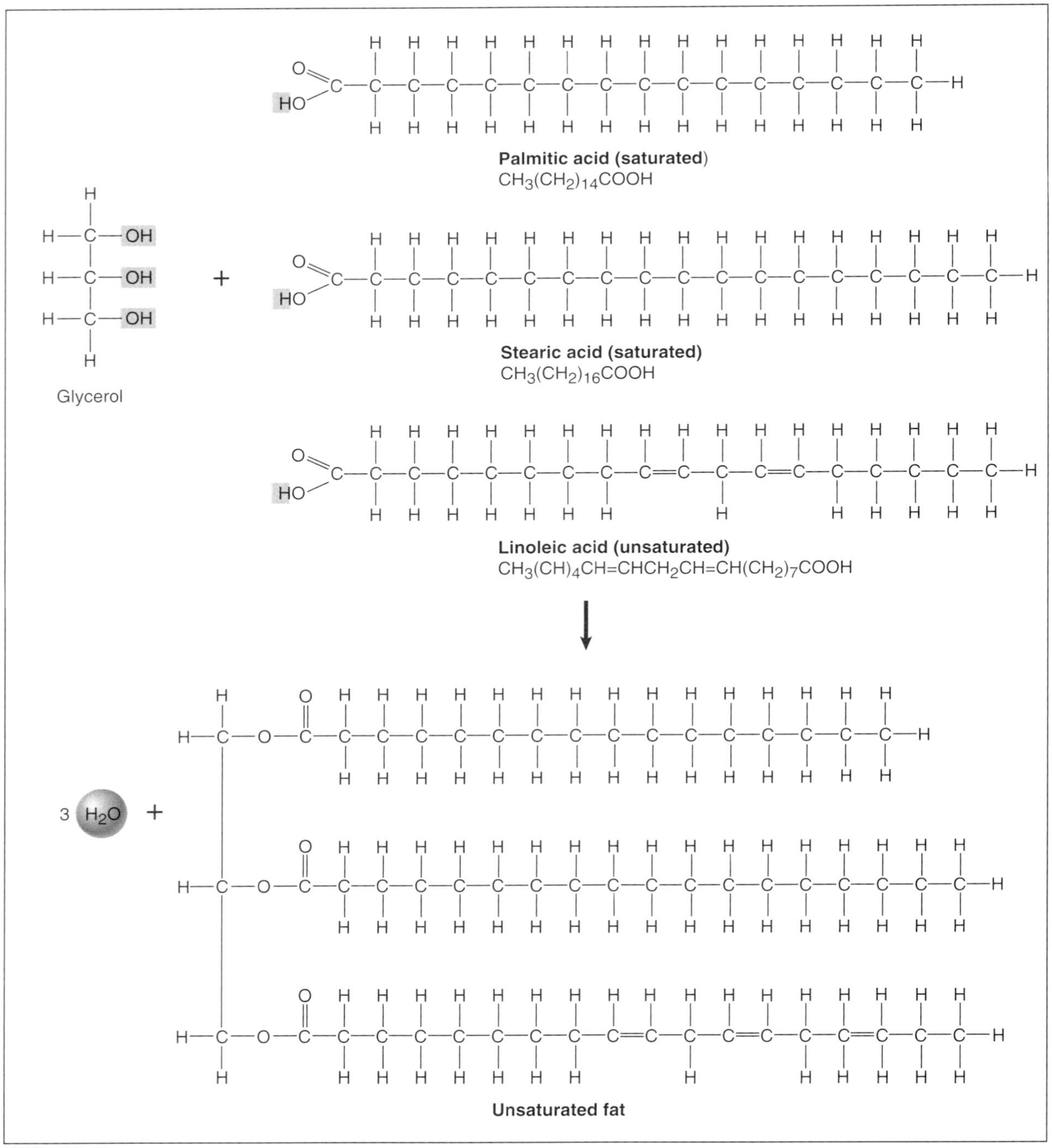

Palmitic acid (saturated)
$CH_3(CH_2)_{14}COOH$

Glycerol

Stearic acid (saturated)
$CH_3(CH_2)_{16}COOH$

Linoleic acid (unsaturated)
$CH_3(CH)_4CH{=}CHCH_2CH{=}CH(CH_2)_7COOH$

3 H_2O +

Unsaturated fat

Figure 3.8 Triglyceride (fat) Synthesis. Note the difference between saturated and unsaturated fatty acids and the production of 3 H_2O as a by-product of this dehydration synthesis reaction.

Polyunsaturated fatty acids are those with many C=C bonds. Most fatty acids can be synthesized by the human body, but a few, called **essential fatty acids,** must be obtained from the diet because we cannot synthesize them (see chapter 26).

Triglycerides

A **triglyceride** (try-GLISS-ur-ide) is a molecule consisting of three fatty acids covalently bonded to a three-carbon alcohol called **glycerol;** triglycerides are more correctly, although less widely, also known as *triacylglycerols.* Each bond between a fatty acid and glycerol is formed by dehydration synthesis (fig. 3.8). Once joined to glycerol, a fatty acid can no longer donate a proton to solution and is therefore no longer an acid. For this reason, triglycerides are also called **neutral fats.** Triglycerides are broken down by hydrolysis reactions, which split each of these bonds apart by the addition of water. Enzymes that hydrolyze fats are called **lipases.**

Triglycerides that are liquid at room temperature are also called *oils,* but the difference between a fat and oil is fairly arbitrary. Coconut oil, for example, is solid at room temperature. Animal fats are usually made of saturated fatty acids, so they are called *saturated fats.* They are solid at room or body temperature. Most plant triglycerides are *polyunsaturated fats,* which generally remain liquid at room temperature. Examples include peanut, olive, corn, and linseed oils. Saturated fats contribute more to cardiovascular disease than unsaturated fats, and for this reason it is healthier to cook with vegetable oils than with lard or bacon fat.

The primary function of fat is energy storage, but when concentrated in *adipose tissue,* it also provides thermal insulation and acts as a shock-absorbing cushion for vital organs (see chapter 6).

Phospholipids

Phospholipids are similar to neutral fats except that, in place of one fatty acid, they have a phosphate group which, in turn, is linked to other functional groups. Lecithin is a common phospholipid in which the phosphate is bonded to a nitrogenous group called *choline* (CO-leen) (fig. 3.9). Phospholipids have a dual nature. The two fatty acid "tails" of the molecule are hydrophobic, but the phosphate "head" is hydrophilic. Thus, phospholipids are said to be **amphiphilic**[12] (AM-fih-FIL-ic). The most important function of phospholipids is to serve as the structural foundation of cell membranes; in chapter 4, you will see how this role depends on their amphiphilic character.

Eicosanoids

Eicosanoids[13] (eye-CO-sah-noyds) are 20-carbon compounds derived from a fatty acid called *arachidonic* (ah-RACK-ih-DON-ic) *acid.* They function primarily as hormonelike chemical signals between cells. The most functionally diverse eicosanoids are the **prostaglandins,** in which five of the carbon atoms are arranged in a ring (fig. 3.10). They were originally found in the secretions of bovine prostate glands, hence their name, but they are now known to be produced in almost all tissues. They play a variety of signaling roles in inflammation, blood clotting, hormone action, labor contractions, control of blood vessel diameter, and other processes. These and other eicosanoids are further discussed in chapter 17.

Steroids

A **steroid** is a lipid with 17 of its carbon atoms arranged in four rings (fig. 3.11). Steroids differ from each other in the location of C=C bonds within the rings and in the functional groups attached to the rings. **Cholesterol** is the "parent" steroid from which the other steroids are synthesized. The others include cortisol, progesterone, estrogens, testosterone, and bile acids. The average adult contains over 200 g (half a pound) of cholesterol. Cholesterol is synthesized only by animals (especially in liver cells) and is not present in vegetable oils or other plant products.

Cholesterol has a bad reputation as a factor in cardiovascular disease (see clinical insight 3.2), and it is true that hereditary and dietary factors can elevate blood cholesterol to dangerously high levels. Nevertheless, cholesterol is a natural product of the body. Only about 15% of the body's cholesterol comes from the diet; the other 85% is internally synthesized. In addition to being the precursor of the other steroids, cholesterol is an important component of cell membranes and is required for proper nervous system function.

The primary lipids and their functions are summarized in table 3.3.

Before You Go On

Answer the following questions to test your understanding of the preceding section:

12. What are lipids? What do they have in common with carbohydrates, and how do they differ from them?
13. How do prostaglandins differ from other fatty acids?
14. What is the difference between saturated and unsaturated fats?
15. How do phospholipids differ from triglycerides?
16. State a function played by each of the following molecules: triglycerides, phospholipids, prostaglandins, and steroids.

Proteins

Objectives

When you have completed this section, you should be able to

- diagram the general structure of amino acids and show how bonds form between them;
- define the four levels of protein structure;
- describe some nonprotein components often bound to proteins; and
- list the physiological functions of proteins.

12. *amphi* = both + *philic* = loving
13. *eicosa* = 20

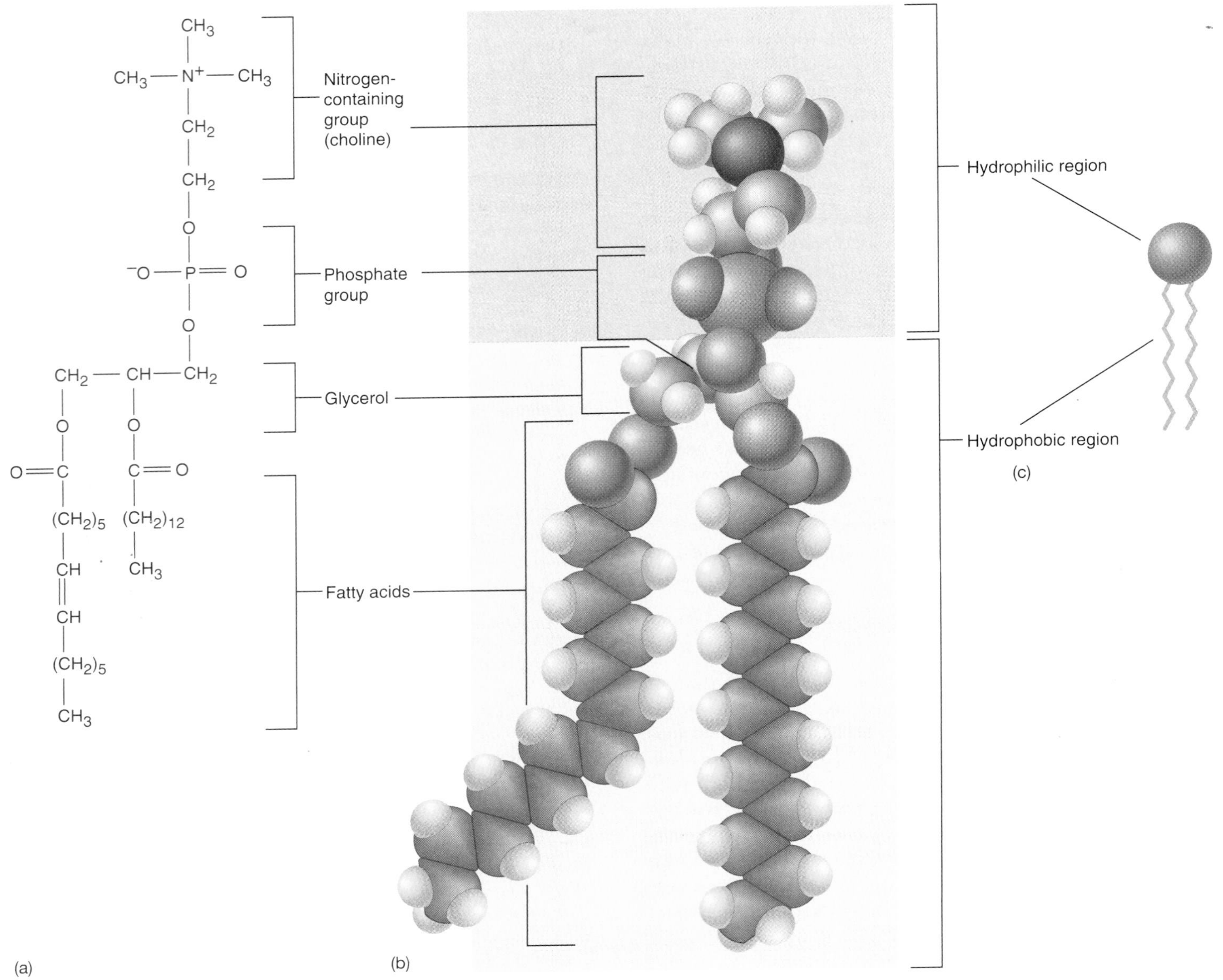

CH3
CH3—N+—CH3
CH2
CH2
O
−O—P=O
O
CH2—CH—CH2
O O
O=C C=O
(CH2)5 (CH2)12
CH CH3
CH
(CH2)5
CH3
Nitrogen-containing group (choline)
Phosphate group
Glycerol
Fatty acids
Hydrophilic region
Hydrophobic region
(a)
(b)
(c)

Figure 3.9 **Lecithin, a Representative Phospholipid.** (*a*) Structural formula, (*b*) a space-filling model that gives some idea of the actual shape of the molecule, and (*c*) a simplified representation of the phospholipid molecule used in diagrams of cell membranes.

O
COOH
OH
OH

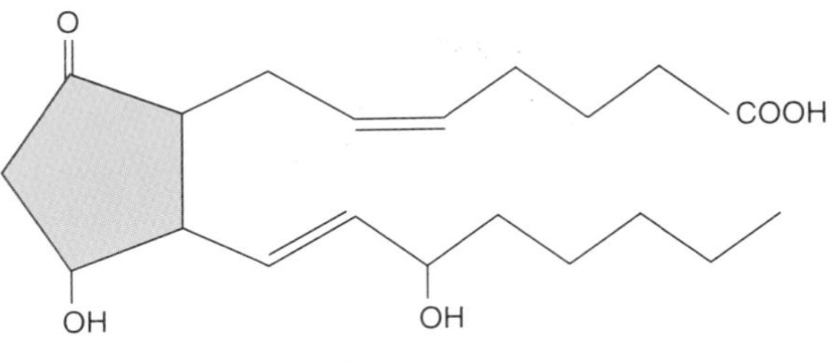

Figure 3.10 **A Prostaglandin.** This is a modified fatty acid with five of its carbon atoms arranged in a ring.

The word *protein* is derived from the Greek word *proteios,* meaning "of first importance." Proteins are the most versatile molecules in the body, and many discussions in this book draw on your understanding of protein structure and behavior.

Amino Acids

A protein is a polymer of **amino acids.** An amino acid has a central carbon atom with an amino ($-NH_2$) and a carboxyl ($-COOH$) group bound to it. The 20 amino acids

Clinical Insight 3.2

"Good" and "Bad" Cholesterol

There is only one kind of cholesterol, and it does far more good than harm. When the popular press refers to "good" and "bad" cholesterol, it is actually referring to droplets in the blood called *lipoproteins,* which are a complex of cholesterol, fat, phospholipids, and protein. So-called bad cholesterol refers to low-density lipoprotein (LDL), which has a high ratio of lipid to protein and contributes to cardiovascular disease. So-called good cholesterol refers to high-density lipoprotein (HDL), which has a lower ratio of lipid to protein and may help to prevent cardiovascular disease. Even when food products are advertised as cholesterol-free, they may be high in saturated fat, which stimulates the body to produce more cholesterol. Palmitic acid seems to be the greatest culprit in stimulating elevated cholesterol levels, while linoleic acid has a cholesterol-lowering effect. Both are shown in figure 3.8. Cardiovascular disease is further discussed at the end of chapter 19, and LDLs and HDLs are more fully explained in chapter 26.

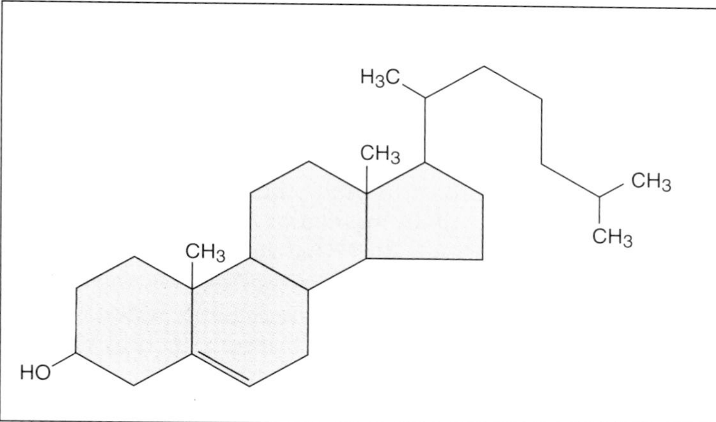

Figure 3.11 Cholesterol. All steroids have this basic four-ringed structure, with variations in the functional groups and locations of double bonds within the rings.

used to make proteins are identical except for a third functional group called the *radical,* or *R group,* attached to that central carbon. In the simplest amino acid, glycine, R is merely a hydrogen atom, while in the largest amino acids it includes rings of carbon. Some R groups are hydrophilic and some are hydrophobic. Being composed of many amino acids, proteins as a whole are therefore often amphiphilic. A few representative amino acids are illustrated in figure 3.12, and all 20 of those involved in proteins are listed in table 3.4 along with their abbreviations.

Peptides

A **peptide** is any molecule composed of two or more amino acids joined by **peptide bonds.** A peptide bond, formed by dehydration synthesis, joins the amino group of one amino acid to the carboxyl group of the next (fig. 3.13). Peptides are named for the number of amino acids they have—for example, dipeptides have two and tripeptides

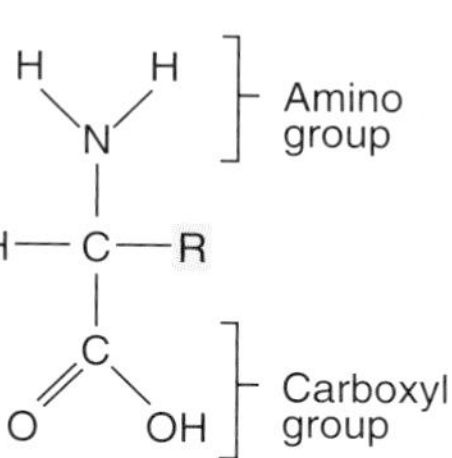

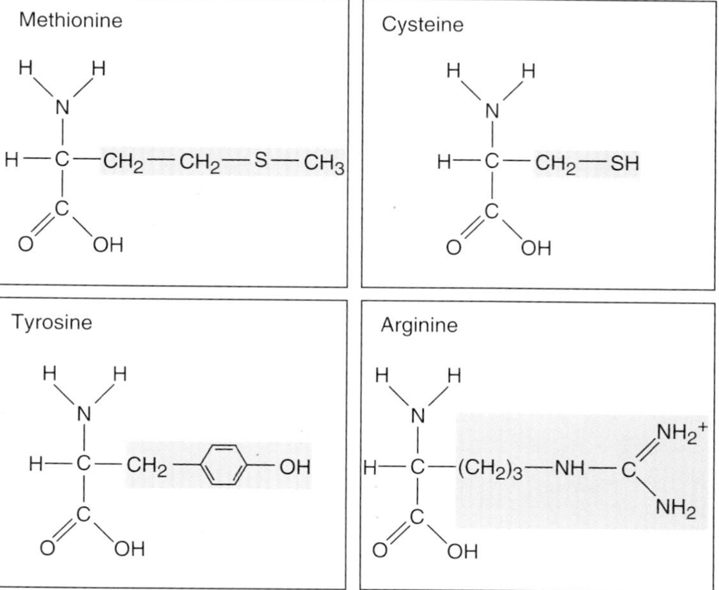

Figure 3.12 Amino Acid Structure. *Top:* General structure of an amino acid. Only the *R* group varies from one amino acid to another. *Below:* Four representative amino acids.

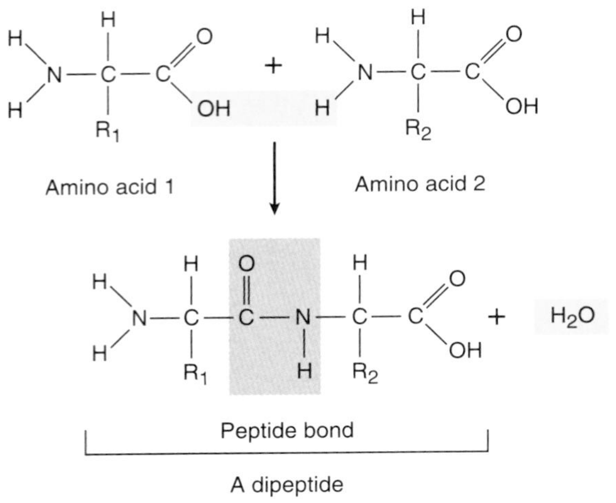

Amino acid 1

Amino acid 2

Peptide bond

A dipeptide

Figure 3.13 Dipeptide Synthesis. Dehydration synthesis removes water and creates a peptide bond linking two amino acids together. This is repeated to make long amino acid chains such as proteins.

Table 3.4 The 20 Amino Acids That Compose Most Proteins

Alanine	Ala	Leucine	Leu
Arginine	Arg	Lysine	Lys
Asparagine	Asn	Methionine	Met
Aspartic acid	Asp	Phenylalanine	Phe
Cysteine	Cys	Proline	Pro
Glutamine	Gln	Serine	Ser
Glutamic acid	Glu	Threonine	Thr
Glycine	Gly	Tryptophan	Trp
Histidine	His	Tyrosine	Tyr
Isoleucine	Ile	Valine	Val

have three. Chains of fewer than 10 or 15 amino acids are called **oligopeptides,**[14] and chains larger than that are called **polypeptides.** An example of an oligopeptide is the childbirth hormone oxytocin, composed of 9 amino acids. A representative polypeptide is adrenocorticotropic hormone (ACTH), which is 39 amino acids long. A **protein** is a polypeptide of 100 amino acids or more. A typical amino acid has a molecular weight of about 80 amu, and the molecular weights of the smallest proteins are around 4,000 to 8,000 amu. The average protein weighs in at about 30,000 amu, and some of them have molecular weights in the hundreds of thousands.

14. *oligo* = a few

Levels of Protein Structure

Proteins have complex coiled and folded structures that are critically important to the roles they play. Even slight changes in three-dimensional shape can render a protein unable to function. Protein molecules have three to four levels of complexity, from primary through quaternary structure (fig. 3.14).

Primary structure is the order in which the amino acids occur—a sequence that is encoded in the genes. In chapter 5 we examine how the genetic code directs the assembly of a protein.

Secondary structure is a coiled or folded shape held together by hydrogen bonds between the slightly negative C=O group of one peptide bond and the slightly positive N–H group of another peptide bond some distance away. The most common secondary structures are a springlike shape called the **α helix** and a pleated, ribbonlike shape, the **β sheet** (or **β-pleated sheet**). Many proteins have multiple α-helical and β-pleated regions joined by short segments with a less orderly geometry. A single protein molecule may fold back on itself and have two or more β-pleated regions linked to each other by hydrogen bonds. Separate, parallel protein molecules also may be linked to each other through their β-pleated regions.

Tertiary[15] (TUR-she-air-ee) **structure** is formed by the further bending and folding of proteins into various globular and fibrous shapes. It results from interaction of the R groups with each other and with the surrounding water. Hydrophobic R groups avoid water and tend to face the inside of the protein, while hydrophilic R groups are attracted to water and tend to face the outside. **Globular proteins,** somewhat resembling a wadded ball of yarn, have a compact tertiary structure well suited for proteins embedded in cell membranes and proteins that must move around freely in the body fluids, such as enzymes and antibodies. **Fibrous proteins** such as myosin, keratin, and collagen are slender filaments better suited for such roles as muscle contraction and providing strength to skin, hair, and tendons.

The amino acid cysteine (Cys), whose R group is $-CH_2-SH$ (see fig. 3.12), often stabilizes a protein's tertiary structure by forming covalent **disulfide bridges.** When two cysteines align with each other, each can release a hydrogen atom, leaving the sulfur atoms to form a disulfide (–S–S–) bridge:

$$\text{Cys}-CH_2-SH + HS-CH_2-\text{Cys} \rightarrow$$
$$\text{Cys}-CH_2-S-S-CH_2-\text{Cys} + 2\text{ H}$$

Disulfide bridges hold separate polypeptide chains together in such molecules as antibodies and the hormone insulin (fig. 3.15).

15. *tert* = third

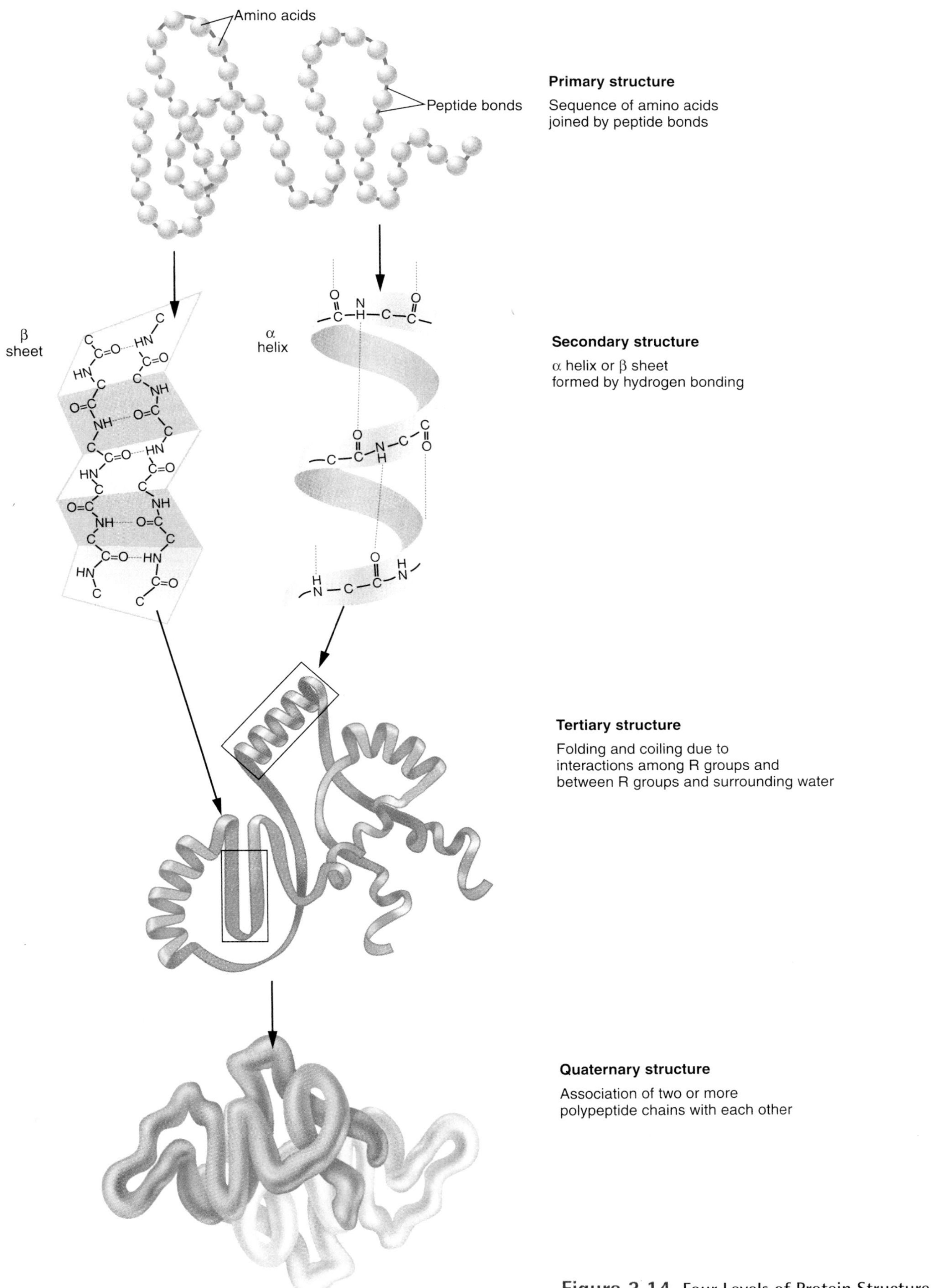

Figure 3.14 Four Levels of Protein Structure.

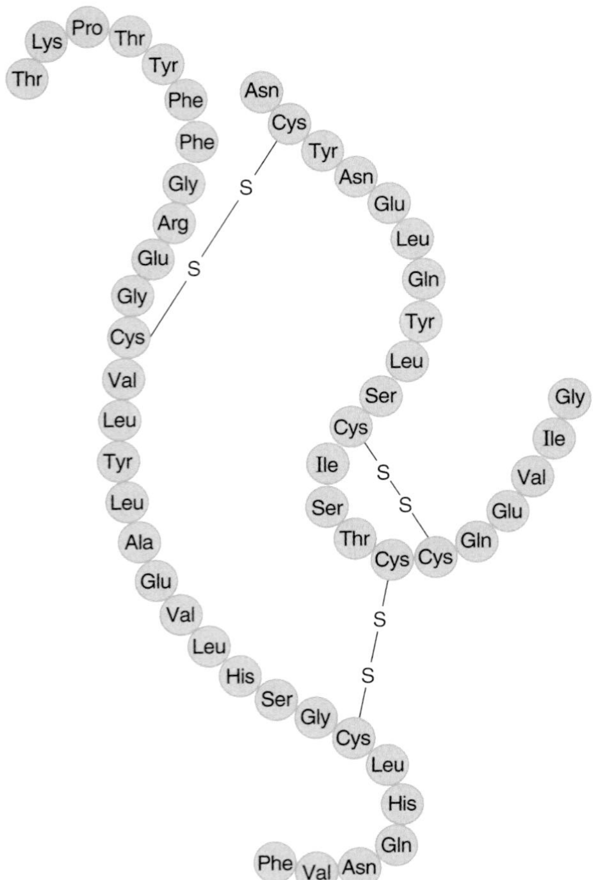

Figure 3.15 Primary Structure of Insulin. Insulin is composed of two polypeptide chains joined by disulfide bridges.

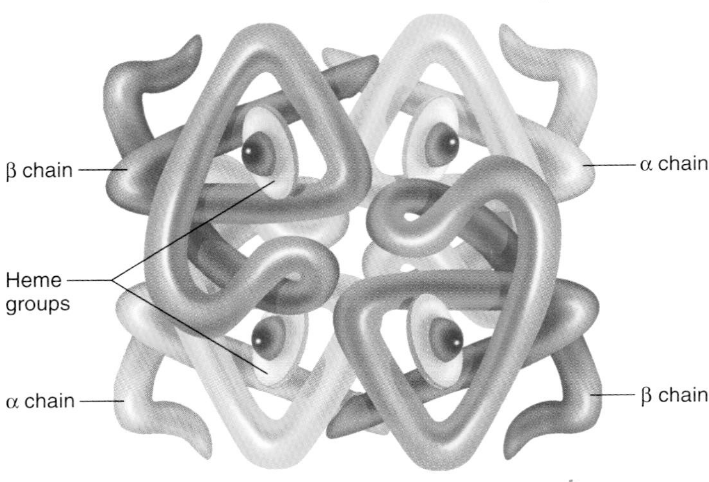

Figure 3.16 Hemoglobin. This is a conjugated protein with a nonprotein heme moiety in each polypeptide chain. Hemoglobin also shows a quaternary structure (multiple polypeptide chains joined together).

Quaternary[16] (QUA-tur-nare-ee) **structure** is the association of two or more polypeptide chains with each other by noncovalent forces such as ionic bonds and hydrophilic-hydrophobic interactions. It occurs in only some proteins. Hemoglobin, for example, consists of four polypeptides—two identical α chains and two identical, slightly longer β chains (fig. 3.16).

Protein Conformation and Denaturation

The **conformation** of a macromolecule, its overall three-dimensional shape, is highly important to its ability to function. In proteins, conformation refers especially to tertiary structure. One of the most important properties of proteins is their ability to change conformation. This can be triggered by such influences as voltage changes on a cell membrane during the action of nerve cells, the binding of a hormone to a protein, or the dissociation (detachment) of a molecule from a protein. Subtle conformational changes are important in processes such as enzyme function, muscle contraction, and the opening and closing of pores in cell membranes. **Denaturation** is a more drastic conformational change in response to conditions such as extreme heat or pH. It is seen, for example, when you cook an egg and the egg white protein (albumin) turns from clear to opaque. Denaturation is sometimes reversible when normal conditions are restored, but often, it permanently destroys the function of a protein.

Conjugated Proteins

Conjugated proteins have a non–amino acid moiety called a **prosthetic**[17] **group** covalently bound to them. Hemoglobin, for example, not only has the four polypeptide chains described earlier, but each chain also has a complex iron-containing ring called a *heme* moiety attached to it (fig. 3.16). Hemoglobin cannot transport oxygen unless this group is present. In glycoproteins, as described earlier, the carbohydrate moiety is a prosthetic group.

Protein Functions

Proteins have more diverse functions than other macromolecules. These include:

- **Structure.** *Keratin*, a tough structural protein, gives strength to the nails, hair, and skin surface. Deeper layers of the skin, as well as bones, cartilage, and teeth, contain an abundance of the durable protein *collagen.*
- **Communication.** Some hormones and other cell-to-cell signals are proteins, as are the receptors to which the signal molecules bind in the receiving cell. A

16. *quater* = fourth
17. *prosthe* = appendage, addition

Historical Insight 3.3

The Discovery of Enzymes

Yeast has been used since ancient times to ferment sugar to alcohol, but even as late as the nineteenth century, no one had figured out how to duplicate this or other biochemical reactions in the absence of living cells. Many biochemists fell back on a *vitalist* philosophy—the belief that organisms or cells contain a mysterious "vital force" beyond the realm of science and the physical world. The German biochemist Eduard Buchner (1860–1917), however, was less content with easy answers. He eventually extracted a juice from yeast that fermented glucose. Not knowing what chemicals in the extract did this, he named the hypothetical substances *enzymes*, from the Greek *en* = within + *zym* = yeast.

hormone or other molecule that reversibly binds to a protein is called a **ligand**[18] (LIG-and).

- **Membrane transport.** Some proteins form channels in cell membranes that govern what passes through the membranes and when. Other proteins act as carriers that briefly bind to solute particles and transport them to the other side of the membrane. Among their other roles, such proteins turn nerve and muscle activity on and off.
- **Catalysis.** Most metabolic pathways of the body are controlled by enzymes, which are globular proteins that function as catalysts.
- **Recognition and protection.** The role of glycoproteins in immune recognition was mentioned earlier. Antibodies and other proteins attack and neutralize organisms that invade the body. Clotting proteins protect the body against blood loss.
- **Movement.** Muscle contraction, the beating of cilia in the respiratory tract, and other kinds of movement are due to special contractile proteins.
- **Cell adhesion.** Proteins bind cells to each other, which enables sperm to fertilize eggs, enables immune cells to bind to enemy cancer cells, and keeps tissues from falling apart.

Before You Go On

Answer the following questions to test your understanding of the preceding section:

17. Draw two amino acids side by side and show how they would be joined by a peptide bond.
18. Describe the four levels of protein structure and the types of chemical bonds involved in each.
19. What level of protein structure could be disrupted by a chemical that breaks disulfide bonds?
20. Discuss how the tertiary structure of proteins exemplifies the unity of form and function.

18. *lig* = to bind

Enzymes and Metabolism

Objectives

When you have completed this section, you should be able to

- define *enzyme* and explain how enzymes are named;
- explain how enzymes accelerate metabolic reactions;
- describe the structure of an enzyme and relate this to its function; and
- explain how enzymes regulate metabolic pathways.

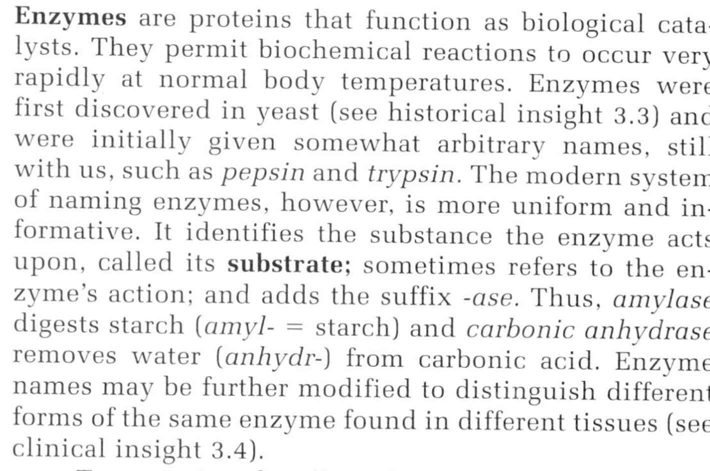

Enzymes are proteins that function as biological catalysts. They permit biochemical reactions to occur very rapidly at normal body temperatures. Enzymes were first discovered in yeast (see historical insight 3.3) and were initially given somewhat arbitrary names, still with us, such as *pepsin* and *trypsin*. The modern system of naming enzymes, however, is more uniform and informative. It identifies the substance the enzyme acts upon, called its **substrate;** sometimes refers to the enzyme's action; and adds the suffix *-ase*. Thus, *amylase* digests starch (*amyl-* = starch) and *carbonic anhydrase* removes water (*anhydr-*) from carbonic acid. Enzyme names may be further modified to distinguish different forms of the same enzyme found in different tissues (see clinical insight 3.4).

To appreciate the effect of an enzyme, think of what happens when paper burns. Paper is composed mainly of glucose (in the form of its polymer, cellulose). The burning of glucose can be represented by the equation

$$C_6H_{12}O_6 + 6\ O_2 \rightarrow 6\ CO_2 + 6\ H_2O$$

Paper does not spontaneously burst into flame because few of its molecules have enough kinetic energy to react. Lighting the paper with a match, however, raises the kinetic energy enough to initiate combustion (rapid oxidation). The energy needed to get the reaction started, supplied by the match, is called **activation energy** (fig. 3.17*a*).

In the body, we carry out the same reaction and oxidize glucose to water and carbon dioxide to extract its energy. We could not tolerate the heat of combustion in our

Clinical Insight 3.4

The Diagnostic Use of Isoenzymes

A given enzyme may exist in slightly different forms, called *isoenzymes,* in different cells. Isoenzymes catalyze the same chemical reactions but have enough structural differences that they can be distinguished by standard laboratory techniques. This is useful in the diagnosis of disease. When organs are diseased, some of their cells break down and release specific isoenzymes that can be detected in the blood. Normally, these isoenzymes would not be present in the blood or would have very low concentrations. If their blood levels are elevated, it can help pinpoint what cells in the body have been damaged.

For example, *creatine kinase (CK)* occurs in different forms in different cells. An elevated serum level of CK-1 indicates a breakdown of skeletal muscle and is one of the signs of muscular dystrophy. An elevated CK-2 level indicates heart disease, because this isoenzyme comes only from cardiac muscle. There are five isoenzymes of *lactate dehydrogenase (LDH).* High serum levels of LDH-1 may indicate a tumor of the ovaries or testes, while LDH-5 may indicate liver disease or muscular dystrophy. *Phosphatase* occurs in two forms—acid phosphatase, named for the fact that it functions best at pH 5.5, and alkaline phosphatase, which has an optimal pH of 9.0. Acid phosphatase is found in the prostate gland and alkaline phosphatase in bone, so elevated levels of one or the other may indicate prostate or bone disease.

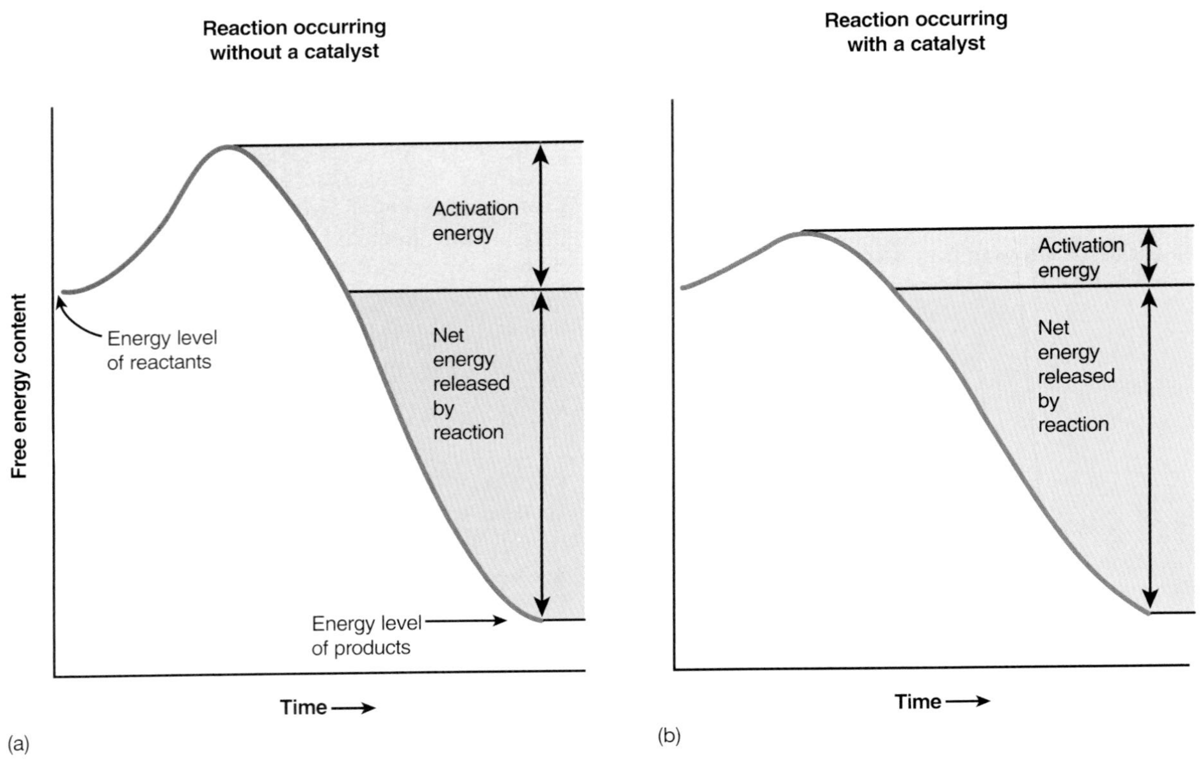

Figure 3.17 Effect of an Enzyme on Activation Energy. (*a*) Without catalysts, some chemical reactions proceed slowly because of the high activation energy needed to get molecules to react. (b) A catalyst facilitates molecular interaction, thus lowering the activation energy and making the reaction proceed more rapidly.

bodies, however, so we must oxidize glucose in a more controlled way at a biologically feasible and safe temperature. Enzymes enable this to happen by lowering the activation energy—that is, by reducing the barrier to glucose oxidation (fig. 3.17*b*)—and by releasing the energy in small steps rather than a single burst of heat.

Enzyme Structure and Action

Substrates bind to pockets or grooves called **active sites** in the enzyme surface and create a temporary **enzyme-substrate complex.** The enzyme may break covalent bonds and convert the substrate to a reaction product, or it may hold two or more substrates close together in ad-

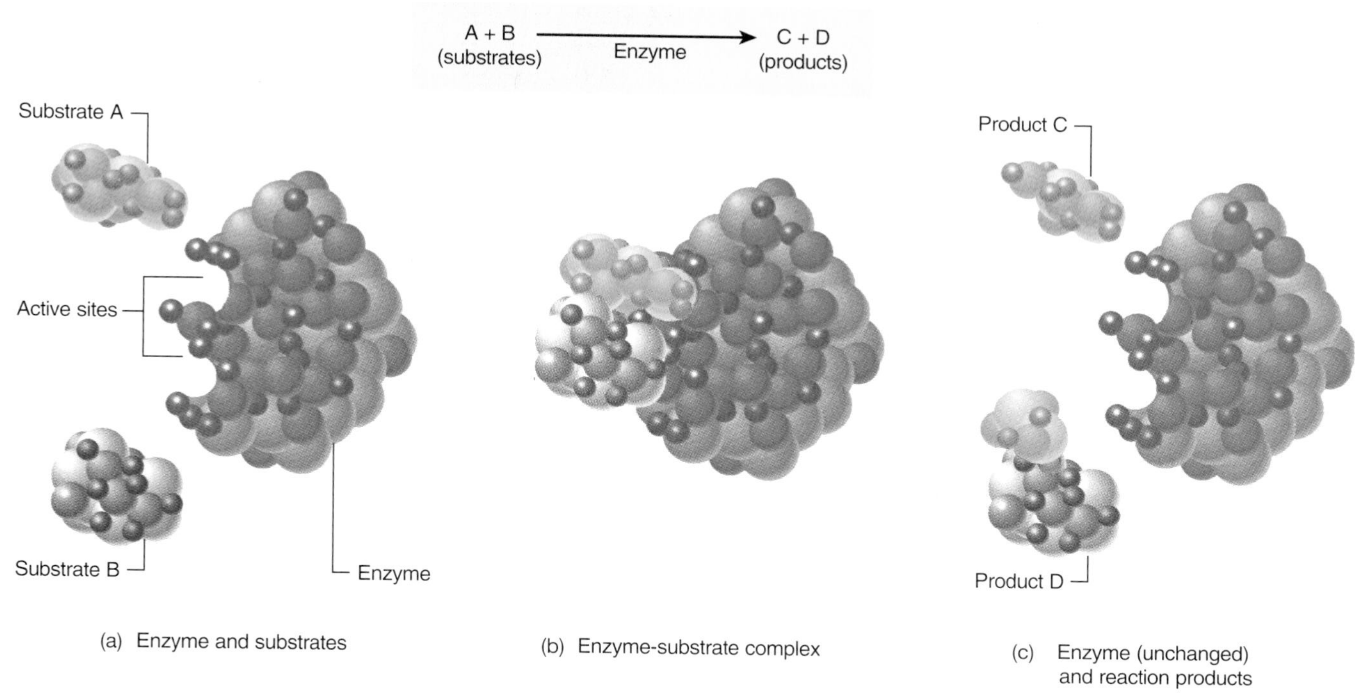

Figure 3.18 The Three Steps of an Enzymatic Reaction. (*a*) One or more substrate molecules bind to the enzyme's active sites. (*b*) The substrates and enzyme form a temporary enzyme-substrate complex and the substrates react chemically with each other. (*c*) The enzyme releases the reaction products and is available to catalyze the same reaction again.

jacent active sites, thus enabling the substrates to react with each other (fig. 3.18). The enzyme then releases the reaction products and is free to begin the process again. Since enzymes are not consumed by the reactions they catalyze, one enzyme molecule can convert millions of substrate molecules. Enzymes also work at astonishing speeds. A single molecule of carbonic anhydrase, for example, breaks carbonic acid (H_2CO_3) down to H_2O and CO_2 at a rate of 36 million molecules per minute.

A substrate fits an enzyme somewhat like a key fits a lock. A given enzyme is very selective—that is, it exhibits **enzyme-substrate specificity.** An enzyme that oxidizes glucose, for example, will not act on the similar sugar galactose, which does not fit its active site.

Factors that change the shape of an enzyme—notably temperature and pH extremes—tend to alter or destroy the ability of the enzyme to bind its substrate. They disrupt the hydrogen bonds and other weak forces that hold the enzyme in its proper conformation, essentially changing the shape of the "lock" so that the "key" no longer fits. Enzymes vary in optimum pH according to where in the body they normally function. Thus salivary amylase, which digests starch in the mouth, functions best at pH 7 and is inactivated when it is exposed to stomach acid; pepsin, which works in the acidic environment of the stomach, functions best around pH 2; and trypsin, a digestive enzyme that works in the alkaline environment of the small intestine, has an optimum pH of 9.5. Our internal body temperature is nearly the same everywhere, however, and all human enzymes have a temperature optimum (that is, they produce their fastest reaction rates) around 37°C.

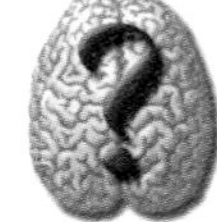

Think About It

Why is homeostasis important for enzyme function?

Cofactors

Many enzymes cannot function without nonprotein partners called **cofactors**—for example, iron, copper, zinc, magnesium, or calcium ions. By binding to an enzyme, a cofactor may stimulate it to fold into a shape that activates its active site. **Coenzymes** are organic cofactors usually derived from niacin, riboflavin, and other water-soluble vitamins. They accept electrons from an enzyme in one metabolic pathway and transfer them to an enzyme in another pathway. For example, cells partially oxidize glucose through a pathway called *glycolysis.* A coenzyme called NAD^+,[19] derived from niacin, shuttles electrons from this pathway to another one called *aerobic respiration,* which uses energy from the electrons to make ATP (fig. 3.19). If NAD^+ is unavailable, the glycolysis pathway shuts down.

19. nicotinamide adenine dinucleotide

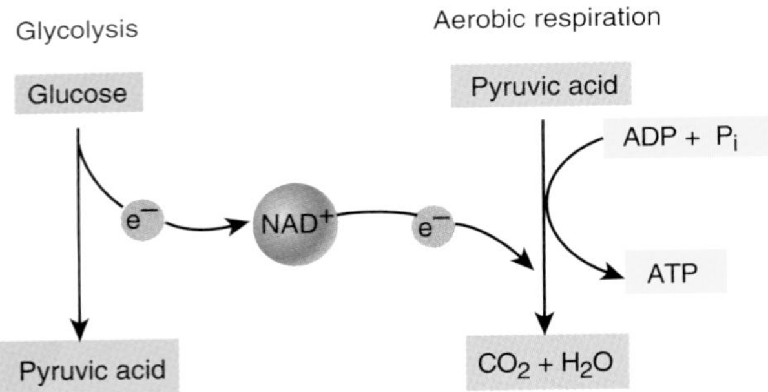

Figure 3.19 The Action of a Coenzyme. A coenzyme such as NAD^+ acts as a shuttle that picks up electrons from one metabolic pathway (in this case, glycolysis) and delivers them to another pathway (in this case, aerobic respiration).

Metabolic Pathways

A **metabolic pathway** is a chain of reactions with each step usually catalyzed by a different enzyme. A simple metabolic pathway can be symbolized

$$A \xrightarrow{\alpha} B \xrightarrow{\beta} C \xrightarrow{\gamma} D$$

where *A* is the initial *reactant*, *B* and *C* are *intermediates*, and *D* is the *end product*. The Greek letters above the reaction arrows represent enzymes that catalyze each step of the reaction. *A* is the substrate for enzyme α, *B* is the substrate for enzyme β, and *C* for enzyme γ. Such a pathway can be turned on or off by altering the conformation of any of these enzymes, thereby activating or deactivating them. This can be done by such means as the binding or dissociation of a cofactor or by an end product of the pathway binding to an enzyme at an earlier step (product *D* binding to enzyme α and shutting off the reaction chain at that step, for example). In these and other ways, cells are able to turn on metabolic pathways when their end products are needed and shut them down when the end products are not needed.

Table 3.5 summarizes some functional properties of enzymes.

Before You Go On

Answer the following questions to test your understanding of the preceding section:

21. How is enzyme function related to activation energy?
22. Explain how tertiary protein structure is related to enzyme-substrate specificity.
23. Explain why denaturation destroys enzyme activity.
24. How does the function of a coenzyme differ from that of an inorganic cofactor?

Table 3.5 Functional Properties of Enzymes

Enzymes lower activation energy.
Enzymes enable molecules with relatively low kinetic energy to react more easily, speeding up the rate of a reaction.
Enzymes are not consumed by the reactions they catalyze.
An enzyme molecule is temporarily changed during a reaction but is restored to its original conformation and can be reused. Thus, very small quantities of enzyme can process large amounts of substrate.
Enzymes are substrate-specific.
The binding site of an enzyme will only fit a particular substrate or a few similar substrates.
Enzymes are temperature- and pH-sensitive.
Abnormal temperature or pH can denature an enzyme and make it incapable of carrying out its catalytic function.
Enzymes can be reversibly inhibited.
Some molecules and ions can bind to an enzyme, or dissociate from it, causing a reversible change in conformation that temporarily activates or deactivates its enzymatic activity. Metabolic pathways can be turned on or off by means of reversible inhibition and activation.

Nucleotides and Nucleic Acids

Objectives

When you have completed this section, you should be able to

- describe ATP, a universal energy-carrying molecule in the nucleotide family;
- discuss the metabolic importance of ATP;
- explain how ATP releases energy;
- distinguish between aerobic and anaerobic pathways of ATP synthesis;
- identify some nucleotides other than ATP; and
- identify the major nucleic acids and explain how they are related to nucleotides.

Adenosine Triphosphate

Many of the body's metabolic reactions involve a universal energy-carrying molecule called **adenosine triphosphate (ATP).** ATP briefly stores energy gained from exergonic reactions such as glucose oxidation and releases it for physiological work such as polymerization reactions, muscle contraction, and pumping ions through cell membranes.

ATP is in a class of molecules called **nucleotides.** A nucleotide has three principal components—a single or double carbon-nitrogen ring called a *nitrogenous base,* a sugar (monosaccharide), and one or more phosphate groups. In ATP (fig. 3.20), the nitrogenous base is a double ring called *adenine,* the sugar is *ribose,* and there are three

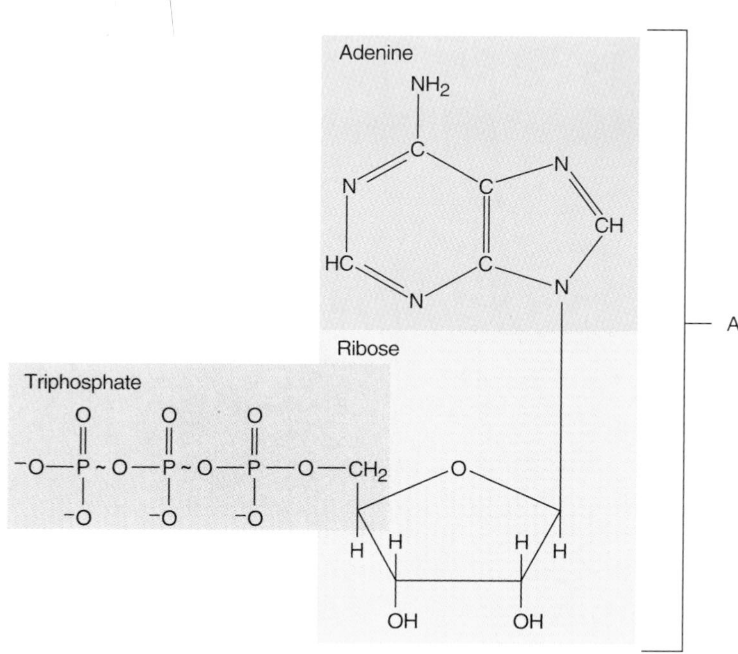

Figure 3.20 **Adenosine Triphosphate.** The two P~O bonds indicated by wavy lines are high-energy bonds.

phosphate groups. We examine some other nucleotides and their roles at the end of this section.

ATP As an Energy Source

The second and third phosphate groups of ATP are attached to the rest of the molecule by high-energy covalent bonds traditionally indicated by a wavy line in the molecular formula. Since phosphate groups are negatively charged, they repel each other. It requires a high-energy bond to overcome that repulsive force and hold them together—especially to add the third phosphate group to a chain that already has two negatively charged phosphates. Most energy transfers to and from ATP involve adding or removing that third phosphate.

Enzymes called **adenosine triphosphatases (ATPases)** are specialized to hydrolyze the third high-energy phosphate bond to produce adenosine diphosphate (ADP) and an inorganic phosphate group (P_i). This reaction releases 7.3 kilocalories of energy for every mole (505 g) of ATP. Most of this energy escapes as heat, but we live on the portion of it that does useful work. We can summarize this as follows:

$$\text{ATP} + H_2O \rightarrow \text{ADP} + P_i + \text{Energy} \begin{cases} \nearrow \text{Heat} \\ \searrow \text{Work} \end{cases}$$

The free phosphate groups released by ATP hydrolysis are often added to enzymes or other molecules to activate them. This addition of P_i, called **phosphorylation**, is carried out by enzymes called **kinases (phosphokinases).** The phosphorylation of an enzyme is sometimes the "switch" that turns on a metabolic pathway.

ATP is a short-lived molecule, usually consumed within 60 seconds of its formation. The entire amount in the body would support life for less than 1 minute if it were not continually replenished. At a moderate rate of physical activity, a full day's supply of ATP would weigh twice as much as you do. Even if you never got out of bed, you would need about 45 kg (99 lb) of ATP to stay alive for a day. The reason cyanide is so lethal is that it halts ATP synthesis.

ATP Synthesis

Most of the energy for ATP synthesis comes from glucose oxidation (fig. 3.21). ATP synthesis is explained in detail in chapter 26, but you will find it necessary to understand the general idea of it before you reach that chapter—especially in understanding muscle physiology (see chapter 12).

The first stage in glucose oxidation (fig. 3.22) is the reaction pathway known as **glycolysis** (gly-COLL-ih-sis). This literally means "sugar splitting," and indeed its major effect is to split the six-carbon glucose molecule into two three-carbon molecules of *pyruvic acid.* A little ATP is produced in this stage (a net yield of two ATPs per glucose), but most of the chemical energy of the glucose is still in the pyruvic acid.

What happens to pyruvic acid depends on whether or not oxygen is available. If it is not, pyruvic acid is converted to lactic acid by a pathway called **anaerobic**[20] (AN-err-OH-bic) **fermentation.** This pathway has two

20. *an* = without + *aer* = air + *obic* = pertaining to life

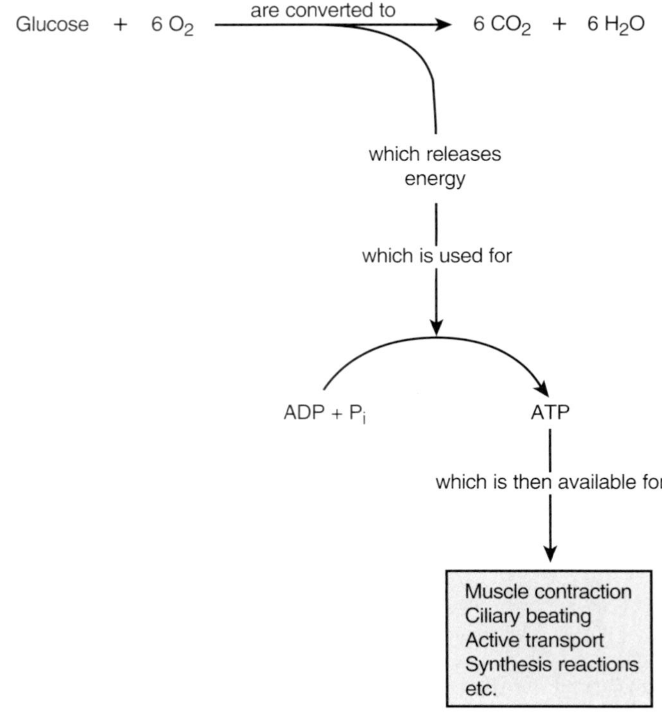

Figure 3.21 The Production and Uses of ATP.

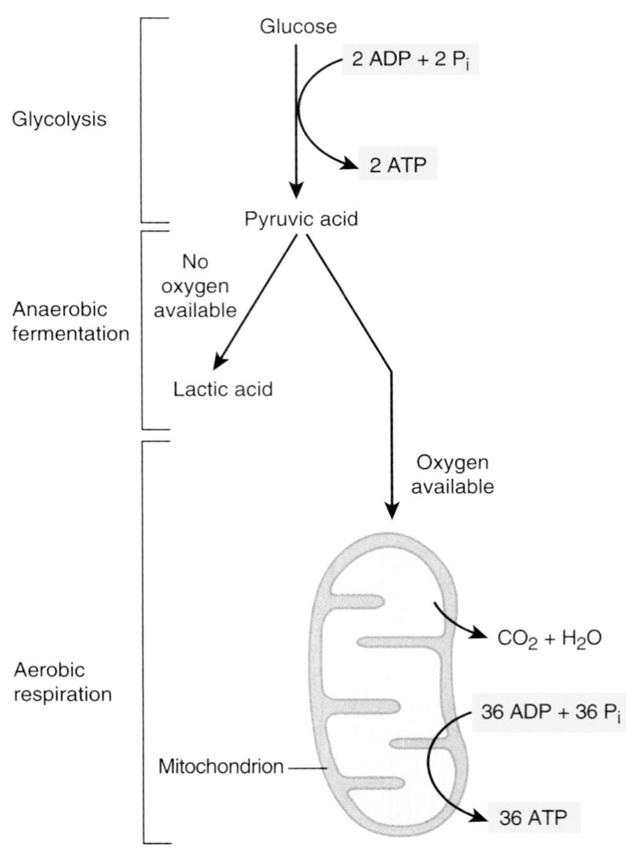

Figure 3.22 **ATP Production.** Glycolysis produces pyruvic acid and a net gain of two ATPs. In the absence of oxygen, anaerobic fermentation is necessary to keep glycolysis running and producing a small amount of ATP. In the presence of oxygen, aerobic respiration occurs in the mitochondria and produces a much greater amount of ATP.

noteworthy disadvantages: first, it does not extract any more energy from pyruvic acid; second, the lactic acid it produces is toxic, so most cells can use anaerobic fermentation only as a temporary measure. The only advantage to this pathway is that it enables glycolysis to continue (for reasons explained in chapter 26) and thus enables a cell to continue producing a small amount of ATP.

If oxygen is available, a much more efficient pathway called **aerobic respiration** occurs. This breaks pyruvic acid down to carbon dioxide and water and generates up to 36 more molecules of ATP for each of the original glucose molecules. The reactions of aerobic respiration are carried out in the cell's *mitochondria* (described in the next chapter), so mitochondria are regarded as a cell's principal "ATP factories."

Other Nucleotides

Guanosine (GWAH-no-seen) **triphosphate (GTP)** is another nucleotide involved in energy transfers. In some reactions, it donates phosphate groups to other molecules. In some pathways, it donates its third phosphate group to ADP to regenerate ATP.

Cyclic adenosine monophosphate (cAMP) is a molecule formed by the removal of both the second and third phosphate groups from ATP. In some cases, when a hormone or other chemical signal ("first messenger") binds to a cell surface, it triggers an internal reaction that converts ATP to cAMP. The cAMP then acts as a "second messenger" to activate metabolic effects within the cell.

Nucleic Acids

Nucleic (new-CLAY-ic) **acids** are polymers of nucleotides. The largest of them, **deoxyribonucleic acid (DNA),** is typically 100 million to 1 billion nucleotides long. It constitutes our genes, transfers hereditary information from cell to cell when cells divide and from generation to generation when organisms reproduce, and gives instructions for synthesizing all of the body's proteins. Three forms of **ribonucleic acid (RNA),** which range from 70 to 10,000 nucleotides long, carry out those instructions and synthesize the proteins, assembling amino acids in the right order to produce each protein "described" by the DNA. The detailed structure of DNA and RNA and the mechanisms of protein synthesis and heredity are described in chapter 5.

Clinical Insight 3.5

Anabolic Steroids

The sex hormone testosterone stimulates muscular growth and aggressive behavior, especially in males. In Nazi Germany, testosterone was given to SS troops in an effort to make them more aggressive, but with no proven success. In the 1950s, pharmaceutical companies developed compounds similar to testosterone, called *anabolic steroids,* to treat anemia, breast cancer, osteoporosis, and some muscle diseases and to prevent the shrinkage of muscles in immobilized patients. By the early 1960s, anabolic steroids were being used illegitimately by athletes to stimulate muscle growth, accelerate the repair of tissues damaged in training or competition, and stimulate the aggressive behavior needed to excel in some contact sports such as football and hockey. Such usage is officially condemned by the American Medical Association and American College of Sports Medicine, and it is banned by the National Football League, National Collegiate Athletic Association, and International Olympic Committee.

Aside from the fact that muscular mass or strength gained from steroid use is "ill-gotten gain" compared to the benefits of honest exercise and training, there are compelling medical reasons for the widespread condemnation of steroid use in sports. The doses used to stimulate muscle development are 10 to 1,000 times higher than the doses prescribed for medical purposes. They cause a rise in cholesterol levels, which promotes fatty degeneration of the arteries *(atherosclerosis).* This often leads in turn to coronary disease, heart and kidney failure, and stroke. Deteriorating blood circulation may result in gangrene and require amputation of the extremities. As the liver attempts to dispose of the high concentration of steroids, liver cancer and other liver diseases may ensue. In addition, steroids suppress the immune system, so the user is more subject to infection and cancer. They cause a premature end to bone elongation, so adolescents who abuse steroids may be permanently stunted.

Anabolic steroids have the same effect on the brain as natural testosterone. Thus, when steroid levels are high, the brain and pituitary gland stop producing the hormones that stimulate sperm production and testosterone secretion. In men, this leads to atrophy of the testes, impotence (inability to achieve or maintain an erection), low sperm count, and infertility. Ironically, anabolic steroids have feminizing effects on men and masculinizing effects on women. Men may develop enlarged breasts *(gynecomastia),* while in women the breasts and uterus atrophy, the clitoris may enlarge, ovulation is inhibited, and menstrual periods become irregular. Women develop excessive facial and body hair and a deeper voice, and both sexes show an increased tendency toward baldness.

Especially in men, anabolic steroid use can be linked to severe emotional disorders. The steroids themselves stimulate heightened aggressiveness and unpredictable mood swings, so the abuser may vacillate between depression and violence. It surely doesn't help matters that impotence, shrinkage of the testes, infertility, and enlargement of the breasts are so incongruous with the "macho" self-image of a male athlete who abuses steroids.

One would think that the dangers of steroids would be enough to deter abuse, yet some athletes still swear by them, and steroids remain available through unscrupulous coaches, physicians, and other sources. By some estimates, as many as 80% of weight lifters, 30% of college and professional athletes, and 20% of male high school athletes now use anabolic steroids.

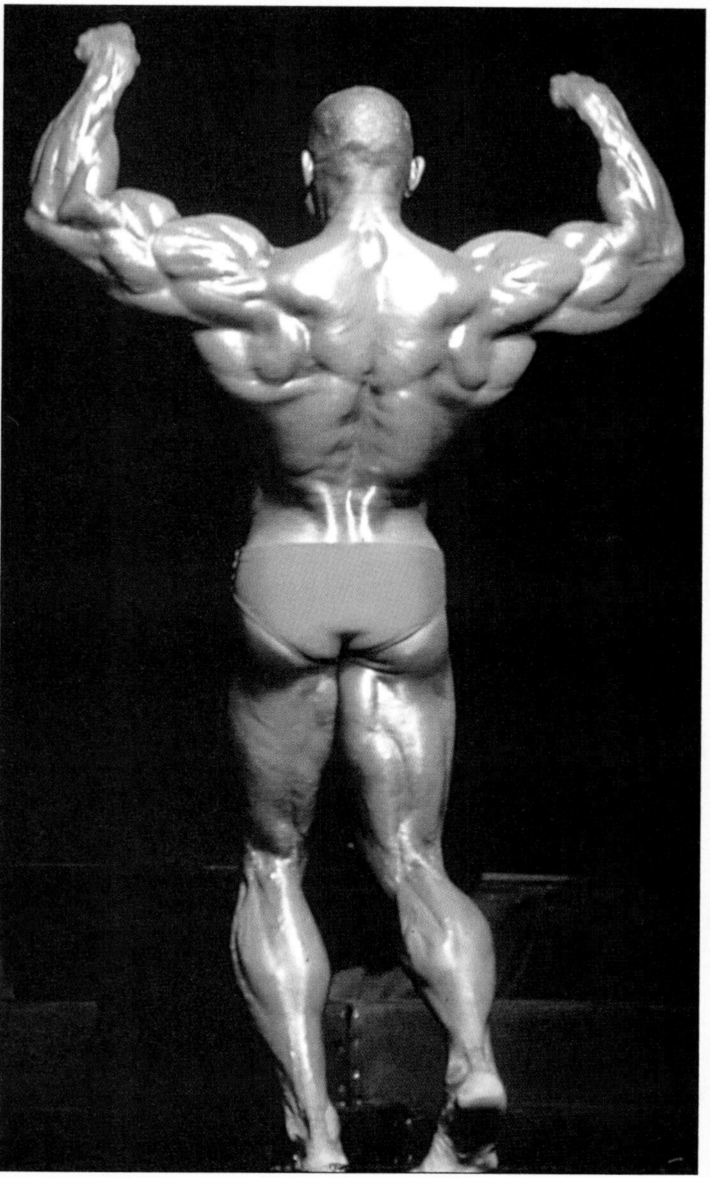

Before You Go On

Answer the following questions to test your understanding of the preceding section:

25. What are the three main structural components of an ATP molecule?
26. How does ATP change structure in the process of releasing energy?
27. Name two metabolic pathways in which ATP is produced. Compare the amount of ATP produced by each one.
28. How are GMP and cAMP related to ATP?
29. How are DNA and RNA related to nucleotides?

Chapter Review

Online Learning Center

Study resources for this chapter are available at www.mhhe.com/saladin/

Study Outline

Inorganic Molecules (p. 77)

1. Water
   a. Molecular structure
   b. Solvency
      - Hydrophilic and hydrophobic substances
      - Hydration spheres
   c. Adhesion, cohesion, and surface tension
   d. Thermal stability
      - Heat capacity
      - Calories
   e. Chemical reactivity
2. Minerals
   a. As structural components
   b. As activators of organic compounds
   c. As electrolytes
3. Gases
   a. Oxygen and carbon dioxide
   b. Nitric oxide

Carbon and Organic Molecules (p. 79)

1. Special properties of carbon
2. Functional groups of organic compounds
3. Monomers, polymers, and macromolecules
   a. Dehydration synthesis
   b. Hydrolysis

Carbohydrates (p. 81)

1. Definition and general formula
2. Monosaccharides: glucose, fructose, and galactose
3. Disaccharides
   a. Sucrose, lactose, and maltose
   b. Synthesis and hydrolysis
4. Polysaccharides: starch, cellulose, and glycogen
5. Carbohydrate functions
   a. Energy sources
   b. Conjugated carbohydrates
      - Glycolipids
      - Glycoproteins
      - Proteoglycans
6. Moieties of macromolecules

Lipids (p. 85)

1. Definition and general properties
2. Fatty acids
   a. General properties
   b. Saturated and unsaturated forms
3. Triglycerides (neutral fats)
   a. Synthesis and hydrolysis
   b. Fats and oils
   c. Functions
4. Phospholipids
   a. Structure
   b. Amphiphilic character
5. Eicosanoids
6. Steroids

Proteins (p. 87)

1. Amino acids
   a. General structure
   b. Variety of functional R groups
2. Peptides
   a. Peptide bonds
   b. Oligopeptides
   c. Polypeptides
3. Levels of protein structure
   a. Primary structure: amino acid sequence
   b. Secondary structure
      - α helix
      - β sheet
   c. Tertiary structure
      - Globular and fibrous proteins
      - Role of disulfide bridges
   d. Quaternary structure
4. Protein conformation and denaturation
5. Conjugated proteins
6. Protein functions
   a. As structural components
   b. In cellular communication
   c. In membrane transport
   d. As catalysts (enzymes)
   e. In recognition and protection
   f. In movement
   g. In cell adhesion

Enzymes and Metabolism (p. 93)

1. Function of enzymes
   a. Definition of an enzyme
   b. Enzyme substrates
   c. Effect on activation energy
2. Enzyme structure and action
   a. Active sites
   b. Enzyme-substrate complex
   c. Reusability of enzymes
   d. Enzyme-substrate specificity
   e. Effects of temperature and pH
3. Cofactors and coenzymes
4. Enzymes and metabolic pathways
   a. Reactants, intermediates, and end products
   b. Regulation of metabolic pathways

Nucleotides and Nucleic Acids (p. 96)

1. Adenosine triphosphate (ATP)
   a. Energy-transfer function
   b. Structure
      - Adenine (nitrogenous base)
      - Ribose (sugar)
      - Three phosphate groups
   c. High-energy bonds
   d. ATPases and hydrolysis
   e. Kinases and phosphorylation
   f. ATP synthesis
      - Glycolysis
      - Anaerobic fermentation
      - Aerobic respiration
2. Other nucleotides
   a. Guanosine triphosphate (GTP)
   b. Cyclic adenosine monophosphate (cAMP)
3. Nucleic acids
   a. DNA
   b. RNA

Selected Vocabulary

biochemistry 77
organic chemistry 77
inorganic chemistry 77
solvency 77
hydrophilic 77
hydrophobic 77
hydration sphere 78
adhesion 78
cohesion 78
surface tension 78
heat capacity 78
calorie 78
reactivity 79
mineral 79
functional group 80
carboxyl group 80
phosphate group 80
hydroxyl group 80
methyl group 80
amino group 80
macromolecule 80
polymer 80
monomer 80
polymerization 80
dehydration synthesis 80
hydrolysis 80
carbohydrate 81
monosaccharide 81
glucose 81
fructose 81
galactose 81
disaccharide 82
sucrose 82
lactose 82
maltose 82
glycosidic bond 82
polysaccharide 82
starch 82
cellulose 82
glycogen 83
conjugated 83
glycolipid 83
glycoprotein 83
proteoglycan 83
moiety 84
lipid 85
fatty acid 85
saturated fatty acid 85
unsaturated fatty acid 85
polyunsaturated fatty acid 86
essential fatty acid 86
triglyceride 86
glycerol 86
neutral fat 87
lipase 87
phospholipid 87
amphiphilic 87
eicosanoid 87
prostaglandin 87
steroid 87
cholesterol 87
amino acid 88
peptide 89
peptide bond 89
oligopeptide 90
polypeptide 90
protein 90
primary structure 90
secondary structure 90
α helix 90
β sheet 90
tertiary structure 90
globular protein 90
fibrous protein 90
disulfide bridge 90
quaternary structure 92
conformation 92
denaturation 92
prosthetic group 92
ligand 93
enzyme 93
substrate 93
activation energy 93
active site 94
enzyme-substrate complex 94
enzyme-substrate specificity 95
cofactor 95
coenzyme 95
metabolic pathway 96
adenosine triphosphate (ATP) 96
nucleotide 96
adenosine triphosphatase 97
phosphorylation 97
kinase 97
glycolysis 97
anaerobic fermentation 97
aerobic respiration 98
guanosine triphosphate (GTP) 98
cyclic adenosine monophosphate (cAMP) 98
nucleic acid 98
deoxyribonucleic acid (DNA) 98
ribonucleic acid (RNA) 98

Testing Your Recall

1. A substance capable of dissolving freely in water is
   a. hydrophilic.
   b. hydrophobic.
   c. hydrolyzed.
   d. hydrated.
   e. amphiphilic.
2. The major effect of glycolysis is
   a. glycogen hydrolysis.
   b. glycogen synthesis.
   c. glucose synthesis.
   d. breakdown of glucose into CO_2.
   e. splitting of glucose into pyruvic acid.
3. A carboxyl group is symbolized
   a. $-OH$.
   b. $-NH_2$.
   c. $-CH_3$.
   d. $-CH_2OH$.
   e. $-COOH$.

4. Which of the following is produced only by plants?
   a. starch
   b. glycogen
   c. cholesterol
   d. triglycerides
   e. polypeptides
5. The arrangement of a polypeptide into a fibrous or globular shape is called its
   a. primary structure.
   b. secondary structure.
   c. tertiary structure.
   d. quaternary structure.
   e. conjugated structure.
6. Which of the following functions is more characteristic of carbohydrates than of proteins?
   a. contraction
   b. energy storage
   c. catalyzing reactions
   d. immune defense
   e. intercellular communication
7. A substance acted upon and changed by an enzyme is called the enzyme's
   a. coenzyme.
   b. cofactor.
   c. substrate.
   d. reactant.
   e. intermediate.
8. The reason liquid nitrogen evaporates so easily is that it lacks
   a. polarity.
   b. cohesion.
   c. hydrogen bonds.
   d. surface tension.
   e. all of these.
9. Glycogen is most structurally similar to
   a. glucose.
   b. starch.
   c. phospholipid.
   d. cellulose.
   e. maltose.
10. The feature that most distinguishes a lipid from a carbohydrate is that a lipid has
    a. more phosphate.
    b. more sulfur.
    c. a lower ratio of carbon to oxygen.
    d. a lower ratio of oxygen to hydrogen.
    e. a greater molecular weight.
11. Molecules dissolve in water if the water can form zones called ________ around them.
12. A chemical reaction that produces water as a by-product is called ________.
13. Organic compounds get their unique chemical characteristics less from the carbon backbone than from their ________.
14. The suffix ________ denotes a sugar, while the suffix ________ denotes an enzyme.
15. The monomers of a polysaccharide are called ________ and those of a protein are called ________.
16. A saturated fat is one to which no more ________ atoms can be added.
17. Amino acids are joined by a unique type of bond called a/an ________ bond.
18. The amphiphilic lipids of cell membranes are called ________.
19. A chemical named ________ is derived from ATP and widely employed as a "second messenger" in cellular signaling.
20. When oxygen is unavailable, cells employ a metabolic pathway called _____ to produce ATP.

Answers in appendix B

Testing Your Comprehension

1. Explain why water would have none of its biologically important properties if its three atoms were in a straight line rather than a 105° angle.
2. Although nitric oxide is now known to be a chemical messenger between cells, it is very doubtful that carbon dioxide, another gas, will be found to play a similar role. Why?
3. Examine the structures of cellulose and starch. Based on your observation, why do you think the starch-digesting enzyme amylase cannot digest cellulose? In light of this conclusion, what do you think is the functional group on starch to which amylase binds?
4. Some metabolic conditions such as diabetes mellitus cause disturbances in the acid-base balance of the body, which gives the body fluids an abnormally low pH. Explain how this could affect enzyme-substrate reactions and metabolic pathways in the body.
5. What are the similarities and differences between a glycoprotein and a proteoglycan?

Answers at the Online Learning Center

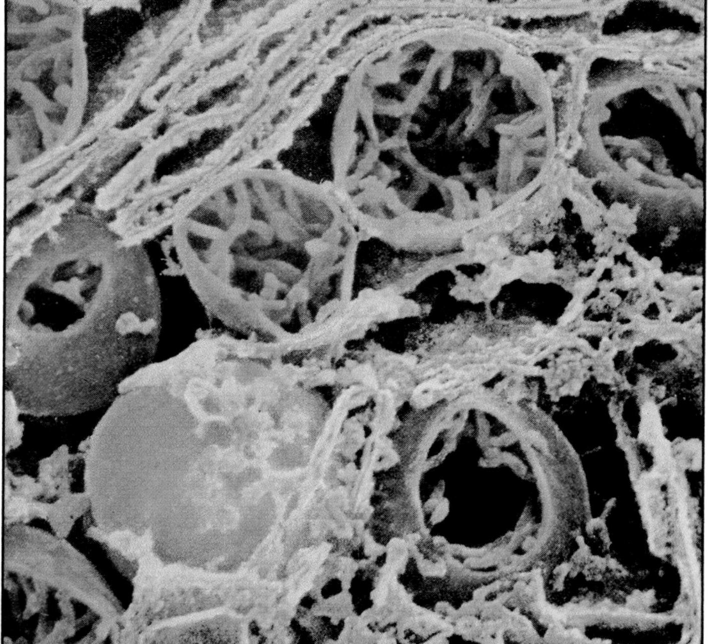

Mitochondria *(pink)* and rough endoplasmic reticulum *(green)* in a liver cell (SEM)

CHAPTER 4

Cellular Form and Function

Chapter Outline

Insights

 Study with Your ESP CD-ROM

Refer to your ESP Tutorial CD-ROM for helpful animations, activities, and quizzes related to the material covered in this chapter. The following are suggested areas to go to on the CD and are presented according to "Unit/Topic":
- Levels of Organization/Cell Membrane
- Levels of Organization/Cell Structures

Brushing Up

To understand this chapter, it is important that you understand or brush up on the following concepts:
- Glycolipids and glycoproteins (p. 83)
- Phospholipids and their amphiphilic character (p. 87)
- Protein conformation in relation to enzyme function (p. 92)
- Functions of proteins (p. 92)

All organisms, from the simplest to the most complex, are composed of cells—whether the single cell of a bacterium or the trillions of cells that constitute the human body. These cells are responsible for all structural and functional properties of a living organism. **Cytology,**[1] the study of cell structure and function, is therefore indispensable to any true understanding of the workings of the human body, the mechanisms of disease, and the rationale of therapy. Thus, this chapter and the next one introduce the basic cell biology of the human body, and subsequent chapters expand upon this information as we examine the specialized cellular functions of specific organs.

Concepts of Cellular Structure

Objectives

When you have completed this section, you should be able to

- discuss the development and modern tenets of the cell theory;
- describe cell shapes from their descriptive terms;
- state how big human cells are and discuss factors that limit cell size; and
- discuss the way that developments in microscopy have changed our view of cell structure.

Development of the Cell Theory

As you may recall from chapter 1, Robert Hooke had observed only the empty cell walls of cork when he first named the cell in 1663. Later, he studied thin slices of fresh wood and saw cells "filled with juices"—a fluid later named *protoplasm.*[2] Two centuries later, Theodor Schwann studied a wide range of animal tissues and concluded that all animals are made of cells. Schwann and other biologists originally believed that cells came from nonliving body fluid that somehow congealed and acquired a membrane and nucleus. The idea of *spontaneous generation*—that living things arise from nonliving matter—was rooted in the scientific thought of the times. For centuries, it had been considered simple common sense that decaying meat turns into maggots, stored grain into rodents, and mud into frogs. Schwann and his contemporaries merely extended this idea to cells. But common sense often proves wrong—for example, it would have us believe that the sun orbits around the earth—and only careful scientific observation is able to prove otherwise. So it was with the question of cell origins.

The idea of spontaneous generation was still controversial in 1859, when the French Academy of Sciences sponsored a competition to "try by well-performed experiment to throw new light on the question of spontaneous generation." French microbiologist Louis Pasteur (1822–95) accepted the challenge. His experiments proved that sterile meat broth gave rise to bacteria only if their airborne spores were able to get into it. By the end of the nineteenth century, it had been established beyond all reasonable doubt that cells do indeed arise only from other cells. The development of biochemistry from the late nineteenth to the twentieth century made it further apparent that all physiological processes of the body are based on cellular activity and that the cells of all species exhibit remarkable biochemical unity. Thus emerged the modern cell theory (table 4.1).

Table 4.1 Tenets of the Modern Cell Theory

1. All organisms are composed of cells and cell products.
2. The cell is the simplest structural and functional unit of life. There are no smaller subdivisions of a cell or organism that, in themselves, are alive.
3. An organism's structure and all of its functions are ultimately due to the activities of its cells.
4. Cells come only from preexisting cells, not from nonliving matter. All life, therefore, traces its ancestry to the same original cells.
5. Because of this common ancestry, the cells of all species have many fundamental similarities in their chemical composition and metabolic mechanisms.

Cell Shapes

The shapes of cells vary widely (fig. 4.1). **Squamous**[3] (SQUAY-mus) cells are thin, flat, and often have angular contours when viewed from above. Such cells line the esophagus and cover the skin. **Polygonal**[4] (pa-LIG-ah-nul) cells have irregularly angular shapes with four, five, or more sides. Some nerve cells have multiple extensions that give them a starlike, or **stellate,**[5] shape. **Cuboidal**[6] (cue-BOY-dul) cells are squarish and approximately as tall as they are wide; liver cells are a good example. **Columnar** cells, such as those lining the intestines, are markedly taller than wide. Egg cells and fat cells are **spheroid** to **ovoid** (round to oval). Red blood cells are **discoid** (disc-shaped). Smooth muscle cells are **fusiform**[7] (FEW-zih-form)—thick in the middle and tapered toward the ends. Skeletal muscle cells are described as **fibrous** because of their long, slender shape.

Epithelial cells line organ surfaces. An epithelial cell rests on a lower **basal surface,** the sides of the cell are

1. *cyto* = cell + *logy* = study of
2. *proto* = first + *plasm* = formed

3. *squam* = scale + *ous* = characterized by
4. *poly* = many + *gon* = angles
5. *stell* = star + *ate* = characterized by
6. *cub* = cube + *oidal* = like, resembling
7. *fusi* = spindle + *form* = shape

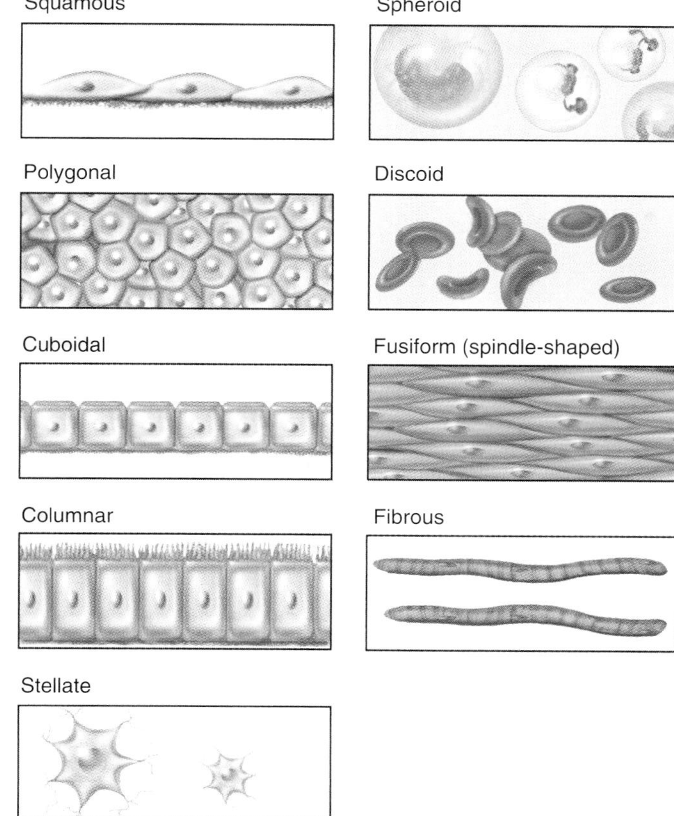

Figure 4.1 Common Cell Shapes.

its **lateral surfaces,** and its exposed upper surface is the **apical surface.** These three surfaces can be quite different from each other in membrane composition and function.

Cell Size

The size of a cell is usually expressed in micrometers (μm) and sizes of the smaller structures within a cell in nanometers (nm). (See the inside front book cover.) Most human cells range from 10 to 15 μm in diameter. The human egg cell, an exceptionally large 100 μm in diameter, is barely visible to the naked eye. The longest human cells are nerve cells (sometimes over a meter long) and muscle cells (up to 30 cm long), but both are too slender to be seen with the naked eye.

There is a limit to how large a cell can be, partly due to the relationship between its volume and surface area. The surface area of a cell is proportional to the square of its diameter, while volume is proportional to the cube of its diameter. Thus, for a given increase in diameter, cell volume increases much faster than surface area. Picture a cuboidal cell 10 μm on each side (fig. 4.2). It would have a surface area of 600 μm^2 (10 μm $\times$ 10 μm $\times$ 6 sides) and a volume of 1,000 μm^3 (10 $\times$ 10 $\times$ 10 μm). Now, suppose it grew by another 10 μm on each side. Its new surface area would be 20 μm $\times$ 20 μm $\times$ 6 = 2,400 μm^2, and its volume would be 20 $\times$ 20 $\times$ 20 μm = 8,000 μm^3. The 20 μm cell has eight times as much protoplasm needing nourishment and waste removal, but only four times as much membrane surface through which wastes and nutrients can be exchanged. A cell that is too big cannot support itself.

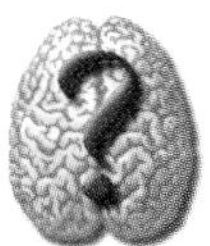

Think About It

Can you conceive of some other reasons for an organ to consist of many small cells rather than fewer larger ones?

An Evolving Perspective on Cells

In Schwann's time, little was known about cytology except that cells were enclosed in a membrane and contained a nucleus. The fluid between the nucleus and surface membrane, called **cytoplasm,** was thought to be little more than a gelatinous mixture of chemicals. This concept changed little until the **transmission electron microscope (TEM)** was invented in the mid-twentieth century. Using a beam of electrons in place of light, the TEM enabled biologists to see a cell's *ultrastructure* (fig. 4.3), a fine degree of detail extending even to the molecular level. The cytoplasm is now understood to be crowded with fibers, passageways, and compartments, all embedded in a clear gel called the **cytosol,** or **intracellular fluid (ICF)** (fig. 4.4). The fluid outside the cell is **extracellular fluid (ECF).**

(*Cytoplasm* means all the material between the nucleus and cell surface; *cytosol* means the clear gelatinous component of the cytoplasm; and *protoplasm* means all of the cellular material including its surface membrane, cytosol, nucleus, and other contents.)

Schwann and his contemporaries were unaware of the details of cell structure simply because most of the structures are too small to be seen with the ordinary light microscope (LM). The most important thing about a good microscope is not magnification but **resolution**—the ability to reveal detail. Any image can be photographed and enlarged as much as we wish, but if enlargement fails to reveal any more useful detail, it is *empty magnification.* A big fuzzy image is not nearly as informative as one that is small and sharp. The TEM reveals far more detail than the LM, even at the same magnification (fig. 4.5). The TEM allows for much greater magnifications only because of its superior resolution.

The minute structures within a cell are called **organelles** (literally "little organs") because they are to the cell what organs are to the body—structures that play individual roles in the survival of the whole. A cell may have 10 billion protein molecules, some of which are potent enzymes with the potential to destroy the cell if they

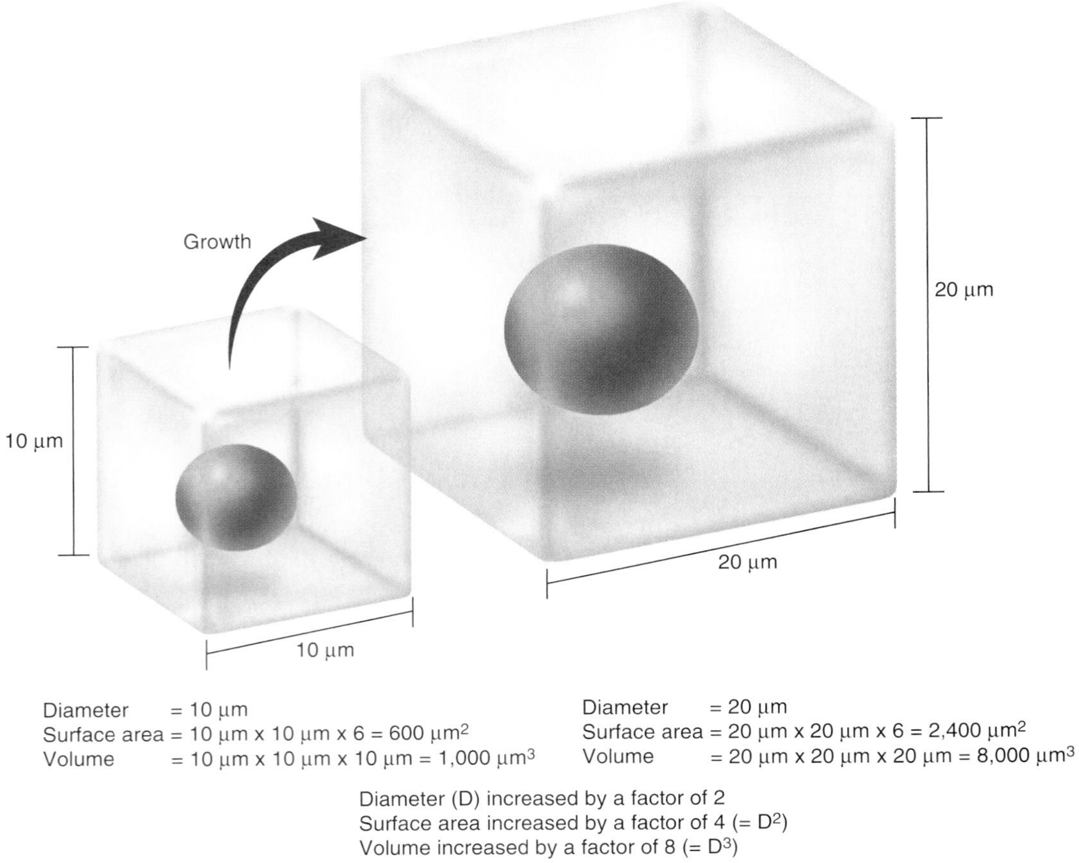

Figure 4.2 The Relationship Between Cell Surface Area and Volume. As a cell doubles in diameter, its volume increases eightfold, but its surface area increases only fourfold. A cell that is too large may have too little plasma membrane to serve the metabolic needs of its increased volume of cytoplasm.

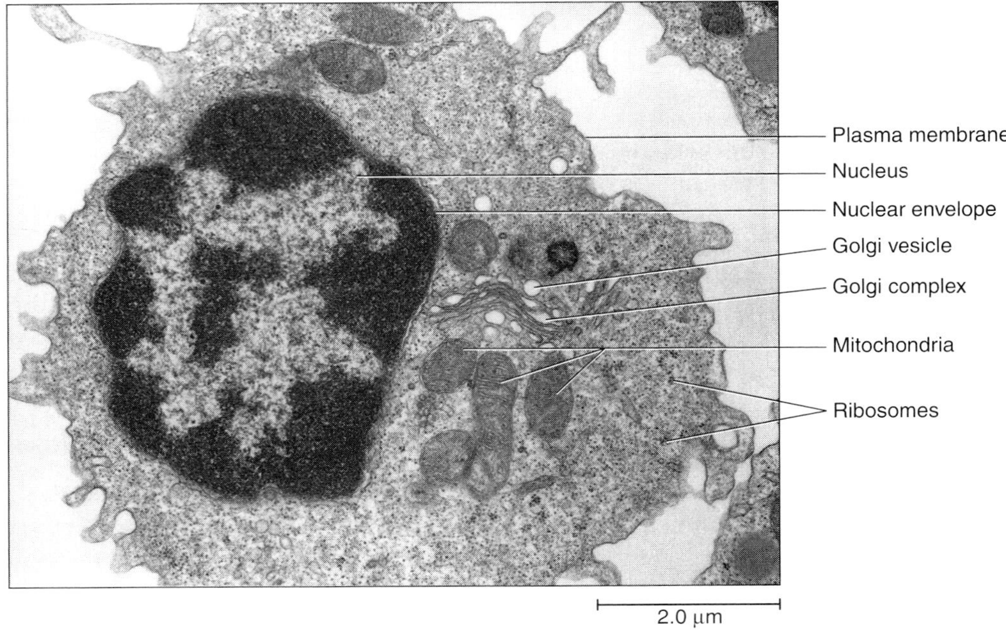

Figure 4.3 Ultrastructure of a White Blood Cell as Seen by TEM.

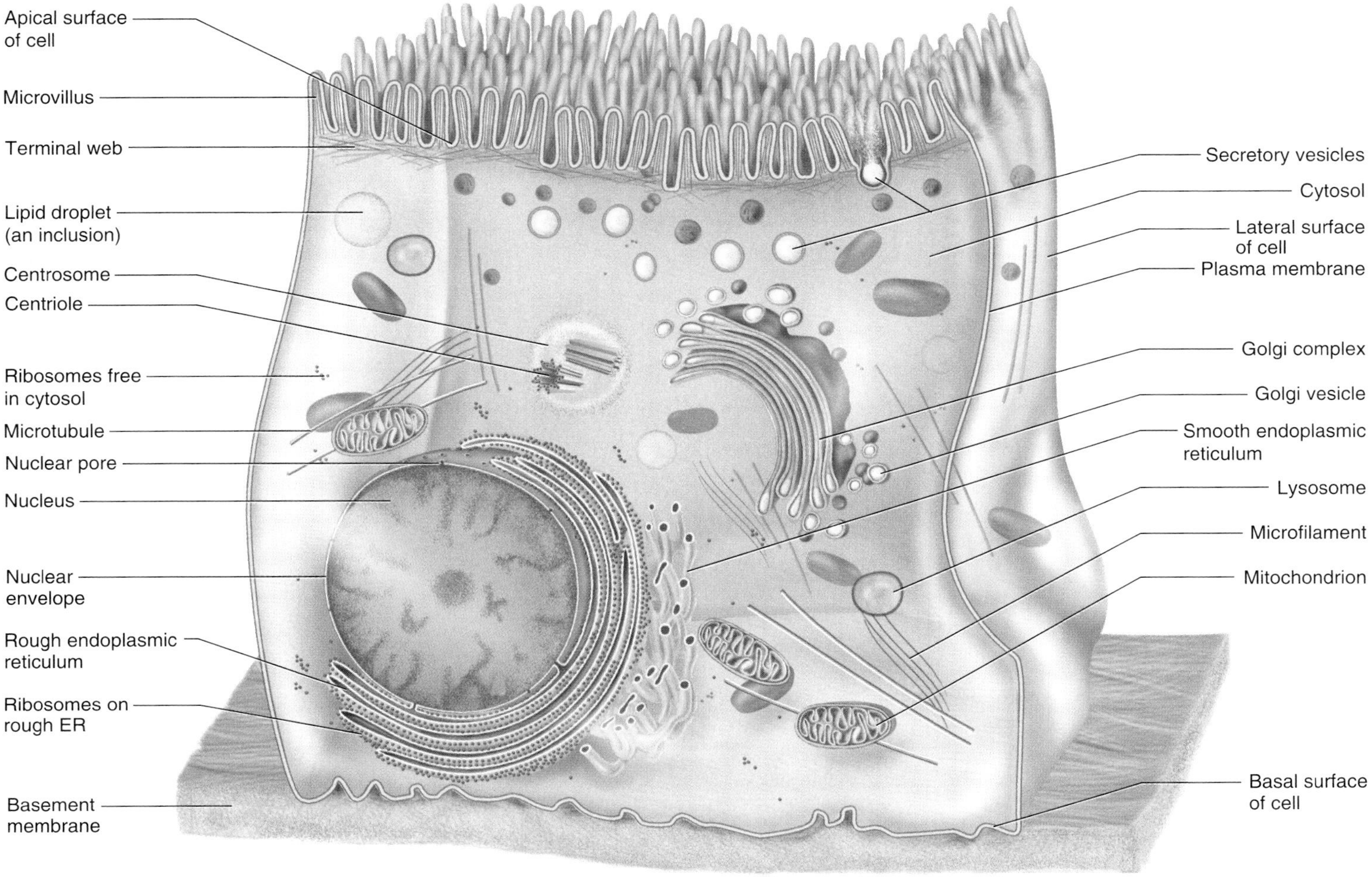

Apical surface of cell
Microvillus
Terminal web
Lipid droplet (an inclusion)
Centrosome
Centriole
Ribosomes free in cytosol
Microtubule
Nuclear pore
Nucleus
Nuclear envelope
Rough endoplasmic reticulum
Ribosomes on rough ER
Basement membrane
Secretory vesicles
Cytosol
Lateral surface of cell
Plasma membrane
Golgi complex
Golgi vesicle
Smooth endoplasmic reticulum
Lysosome
Microfilament
Mitochondrion
Basal surface of cell

Figure 4.4 Structure of a Representative Cell.

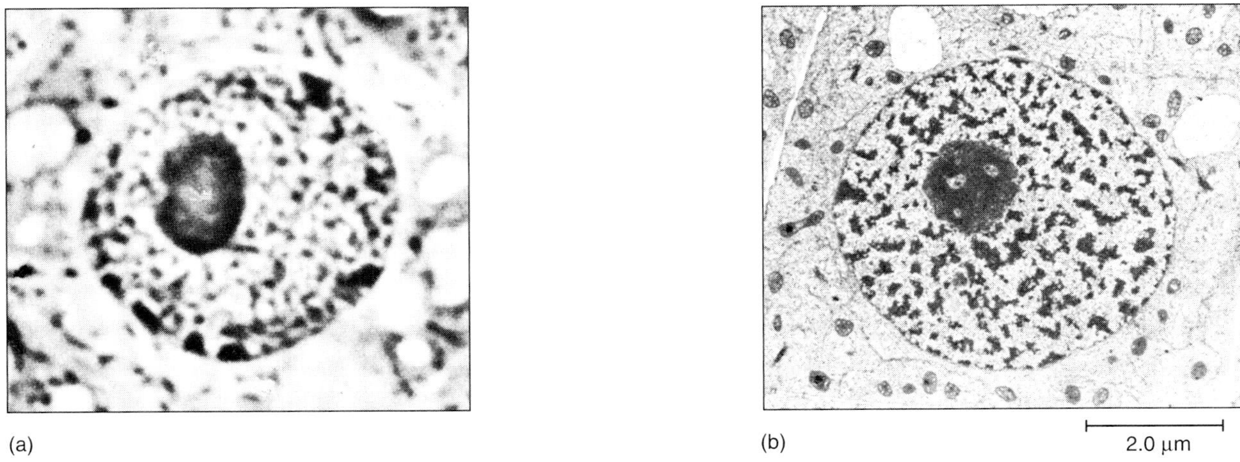

Figure 4.5 Magnification Versus Resolution. These cells were photographed at the same magnification (about ×750) through (*a*) a light microscope and (*b*) a transmission electron microscope.

Table 4.2 Sizes of Some Biological Structures in Relation to the Resolving Power of the Human Eye, Light Microscope (LM), and Transmission Electron Microscope (TEM)

Object	Size	Eye	LM	TEM
Human egg, diameter	100 μm			
Resolution of the unaided eye	70–100 μm			
Most human cells, diameter	10–15 μm			
Cilia, length	7–10 μm			
Mitochondria, width × length	0.2 × 4 μm			
Bacteria (*E. coli*), length	1–3 μm			
Microvilli, length	1–2 μm			
Lysosomes, diameter	0.5 μm = 500 nm			
Resolution of the light microscope	200 nm			
Nuclear pores, diameter	30–100 nm			
Centriole, diameter × length	20 × 50 nm			
Polio virus, diameter	30 nm			
Ribosomes, diameter	15 nm			
Globular proteins, diameter	5–10 nm			
Plasma membrane, thickness	7.5 nm			
DNA molecule, diameter	2.0 nm			
Plasma membrane channels, diameter	0.8 nm			
Resolution of the TEM	0.5 nm			
Carbon atom, diameter	0.15 nm			
Hydrogen atom, diameter	0.07 nm			

are not contained and isolated from other cellular components. You can imagine the enormous problem of keeping track of all this material, directing molecules to the correct destinations, and maintaining order against the incessant tug of disorder. Cells maintain order partly by compartmentalizing their contents in organelles.

Table 4.2 gives the sizes of some cells and subcellular objects relative to the resolution of the naked eye, light microscope, and electron microscope. You can see why the very existence of cells was unsuspected until the light microscope was invented and why little was known about the organelles until the TEM became available. Another instrument, the **scanning electron microscope (SEM),** produces dramatic three-dimensional images at high magnification and resolution (see historical insight 4.5, p. 136).

Before You Go On

Answer the following questions to test your understanding of the preceding section:

1. When was the idea of spontaneous generation of cells laid to rest? Who disproved the idea?
2. What are the tenets of the cell theory?
3. Define *cytoplasm, cytosol,* and *organelle.*
4. What is the main advantage of an electron microscope over a light microscope?

The Cell Surface

Objectives

When you have completed this section, you should be able to

- describe the structure of plasma membranes;
- explain the functions of the lipid, protein, and carbohydrate components of the plasma membrane;
- describe a second messenger system and discuss its importance in human physiology; and
- describe the structure and functions of microvilli, cilia, and flagella.

The Plasma Membrane

The electron microscope reveals that the cell and many of the organelles within it are bordered by a **unit membrane,** which appears as a pair of dark parallel lines with a total thickness of about 7.5 nm (fig. 4.6*a*). The unit membrane around the cell as a whole is called the **plasma (cell) membrane.** It defines the boundaries of the cell, governs its interactions with other cells, and controls the passage of materials into and out of the cell. The side that faces the cytoplasm is the *intracellular face* of the membrane, and the side that faces outward is the *extracellular face.*

By the 1930s, it was known that the plasma membrane consisted mainly of phospholipids and proteins. Multiple theories were proposed as to how these molecules were arranged, but the one that best accounted for the evidence was the **fluid-mosaic theory,** put forth by S. J. Singer and G. Nicolson in 1972. It describes the plasma membrane (and other unit membranes) as an array of mobile globular proteins embedded in an oily film of phospholipids. Further refinements of this model have given rise to the concept illustrated in figure 4.6*b*.

Membrane Lipids

Lipids constitute 90% to 99% of the molecules in the plasma membrane. About 75% of the lipids are phospholipids.

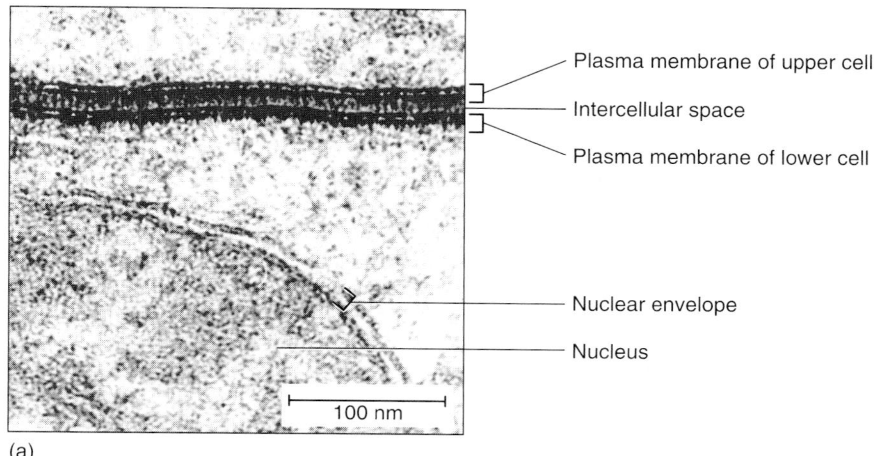

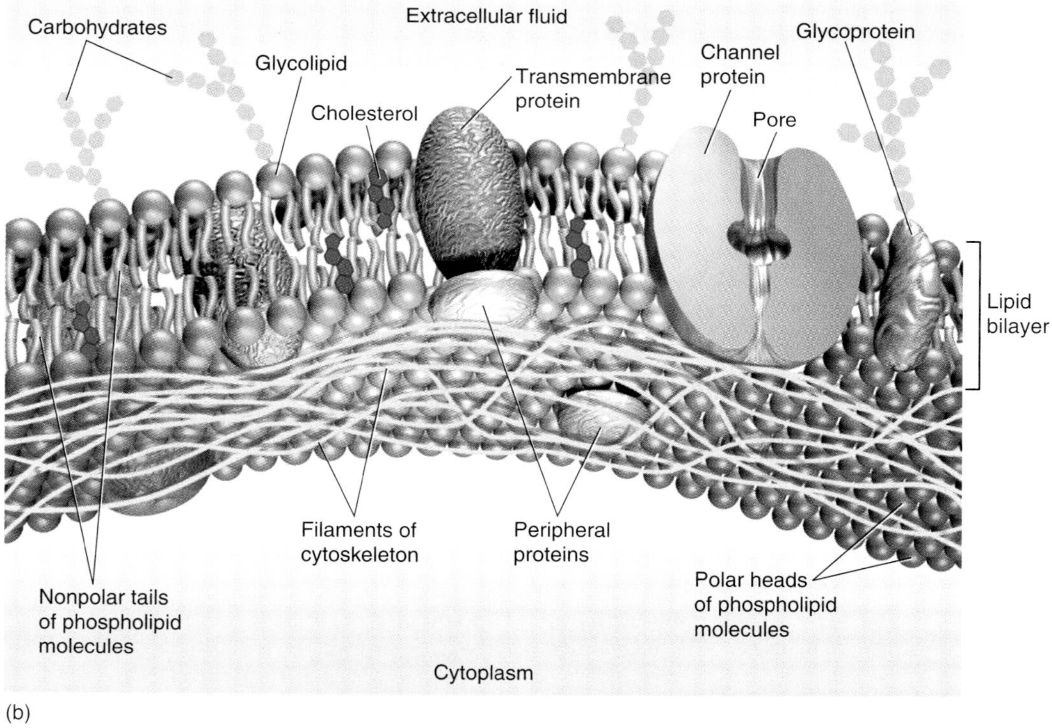

Figure 4.6 The Plasma Membrane. (*a*) Plasma membranes of two adjacent cells (electron micrograph). (*b*) Molecular structure of the plasma membrane.

These amphiphilic molecules arrange themselves into a bilayer, with their hydrophilic phosphate-containing heads facing the water on each side of the membrane and their hydrophobic tails directed toward the center of the membrane, avoiding the water. The phospholipids drift laterally from place to place, they spin on their axes, and their tails flex. These movements keep the membrane fluid.

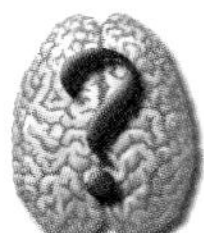

Think About It

What would happen if the plasma membrane were made primarily of a hydrophilic substance such as carbohydrate? Which of the major themes at the end of chapter 1 does this point exemplify?

Cholesterol molecules, found amid the fatty acid tails, constitute about 20% of the membrane lipids. By interacting with the phospholipids and "holding them still," cholesterol can stiffen the membrane (make it less fluid) in spots. Higher concentrations of cholesterol, however, can increase membrane fluidity by preventing the phospholipids from becoming packed too closely together.

The remaining 5% of the membrane lipids are glycolipids—phospholipids with short carbohydrate chains covalently bound to them. These are found only on the extracellular face of the membrane. Here, they contribute to the *glycocalyx,* a carbohydrate coating on the cell surface with multiple functions described shortly.

An important quality of the plasma membrane is its capacity for self-repair. When a physiologist inserts a probe into a cell, it does not pop the cell like a balloon. The probe slips through the oily film and the membrane seals itself around it. When cells take in matter by endocytosis (described later in this chapter), they pinch off bits of their own membrane, which form bubblelike vesicles in the cytoplasm. As these vesicles pull away from the membrane, they do not leave gaping holes; the lipids immediately flow together to seal the break.

Membrane Proteins

Although proteins constitute only 1% to 10% of the molecules of the plasma membrane, they are larger than lipids and constitute about half the membrane weight. Some of them, called **integral (transmembrane) proteins,** pass through the membrane. They have hydrophilic regions in contact with the cytoplasm and extracellular fluid and hydrophobic regions that pass back and forth through the lipid of the membrane (fig. 4.7). Most integral proteins are glycoproteins, which are conjugated with oligosaccharides on the extracellular side of the membrane. Some of them are anchored to the *cytoskeleton*—an intracellular system of tubules and filaments (discussed later). **Peripheral proteins** are those that do not protrude into the phospholipid layer but adhere to the intracellular face of the membrane. A peripheral protein is typically associated with an integral protein and tethered to the cytoskeleton.

The functions of membrane proteins fall within the general scope of protein functions described in the previous chapter:

- **Receptors** (fig. 4.8*a*). Cells communicate with each other by means of chemical signals such as hormones and neurotransmitters. Some of these messengers (epinephrine, for example) cannot enter target cells, but must bind to a membrane protein called a **receptor** on the cell surface. Receptors are usually specific for

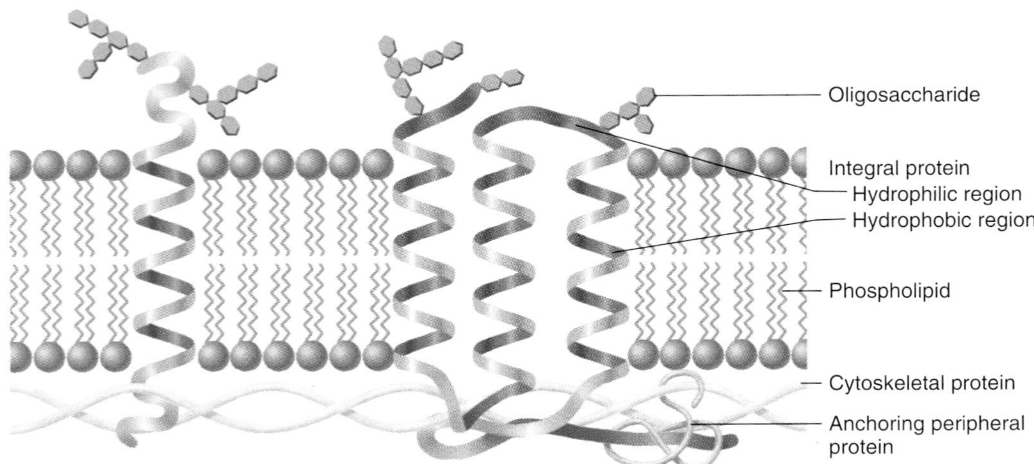

Figure 4.7 Transmembrane Proteins. A transmembrane protein has hydrophobic regions embedded in the phospholipid bilayer and hydrophilic regions projecting into the intracellular and extracellular fluids. The protein may cross the membrane once (*left*) or multiple times (*right*). The intracellular regions are often anchored to the cytoskeleton by peripheral proteins.

one particular messenger, much like an enzyme that is specific for one substrate.

- **Second messenger systems.** When a messenger binds to a surface receptor, it may trigger changes within the cell that produce a *second messenger* in the cytoplasm. This process involves both transmembrane proteins (the receptor) and peripheral proteins. Second messenger systems are discussed shortly in more detail.
- **Enzymes** (fig. 4.8*b*). Enzymes embedded in the plasma membranes of cells carry out the final stages of starch and protein digestion in the small intestine, play a part in producing second messengers, and break down hormones and neurotransmitters once their job is done, thus stopping them from excessively stimulating a cell.
- **Channel proteins** (fig. 4.8*c*). **Channel proteins** are integral proteins with pores that allow passage of water and hydrophilic solutes through the membrane. Some channels are always open, while others are **gates** that open and close under different circumstances, thus determining when solutes can pass through (fig. 4.8*d*). These gates open or close in response to three types of stimuli: **ligand-regulated gates** respond when chemical messengers bind to them; **voltage-regulated gates** respond to changes in electrical potential (voltage) across the plasma membrane; and **mechanically regulated gates** respond to physical stress on a cell, such as stretch and pressure. By controlling the movement of electrolytes through the plasma membrane, gated channels play an important role in the timing of nerve signals and muscle contraction (see clinical insight 4.1).
- **Carriers** (see figs. 4.25 and 4.26). Carriers are integral proteins that bind to glucose, electrolytes, and other solutes and transfer them to the other side of the membrane. Some carriers, called **pumps,** consume ATP in the process. Carriers, pumps, and transport processes are explained in detail later in this chapter.
- **Motor molecules** (fig. 4.8*e*). **Motor molecules** are proteins that produce movement by changing shape and pulling on other molecules. They move materials within a cell, as in transporting molecules and organelles to their destinations; they enable some cells, such as white blood cells, to crawl around in the body's tissues; and they make cells change shape, as when a cell surrounds and engulfs foreign particles or when it divides in two. Such processes depend on the action of fibrous proteins, especially *actin* and *myosin,* that pull on the integral proteins of the plasma membrane.

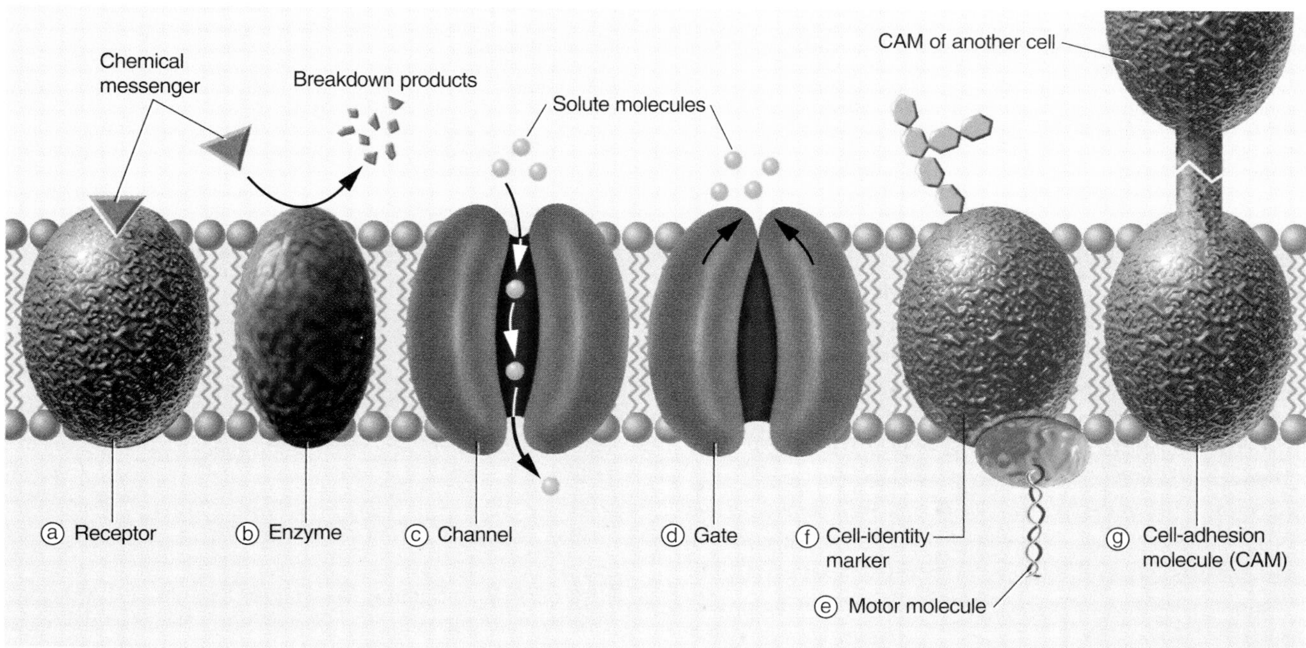

Figure 4.8 Some Functions of Membrane Proteins. (*a*) A receptor that binds to chemical messengers such as hormones sent by other cells. (*b*) An enzyme that breaks down a chemical messenger and terminates its effect on the target cell. (*c*) A channel protein that is constantly open and allows solutes to pass in and out of the cell. (*d*) A gated channel that opens and closes to allow solutes through only at certain times. (*e*) A motor molecule, a filamentous protein arising deeper in the cytoplasm that pulls on membrane proteins and causes cell movement. (*f*) A glycoprotein serving as a cell-identity marker. (*g*) A cell-adhesion molecule (CAM) that binds one cell to another.

Clinical Insight 4.1

Calcium Channel Blockers

The walls of our arteries contain smooth muscle that contracts or relaxes to change the diameter of the artery. These changes modify the blood flow through an artery and strongly influence blood pressure. Blood pressure rises when the arteries constrict and falls when the arteries relax and dilate. Excessive, widespread vasoconstriction can cause hypertension (high blood pressure), and vasoconstriction in the coronary blood vessels of the heart can cause pain (angina) due to inadequate blood flow to the cardiac muscle. In order to contract, a smooth muscle cell must open calcium channels in its plasma membrane and allow calcium to enter from the extracellular fluid. Drugs called *calcium channel blockers* prevent calcium channels from opening. Thus they help to relax the arteries, relieve angina, and lower blood pressure.

- **Cell-identity markers** (fig. 4.8*f*). Glycoproteins contribute to the *glycocalyx,* a carbohydrate surface coating discussed shortly. Among other functions, this acts like an "identification tag" that enables our bodies to tell which cells belong to it and which are foreign invaders.
- **Cell-adhesion molecules** (fig. 4.8*g*). Cells adhere to one another and to extracellular material through certain membrane proteins called **cell-adhesion molecules (CAMs).** With few exceptions (such as blood cells and metastasizing cancer cells), cells do not grow or survive normally unless they are mechanically linked to the extracellular material. Special events such as sperm-egg binding and the binding of an immune cell to a cancer cell also require CAMs.

Second Messengers

Second messengers are of such importance that they require a closer look. You will find this information essential for your later understanding of hormone and neurotransmitter action. Let's consider how the hormone epinephrine stimulates a cell. Epinephrine, the "first messenger," cannot pass through plasma membranes, so it binds to a surface receptor. The receptor is linked on the intracellular side to a peripheral protein called a **G protein** (fig. 4.9). G proteins are named for the ATP-like chemical, guanosine triphosphate (GTP), from which they get their energy. When activated by the receptor, a G protein relays the signal to another membrane protein, **adenylate cyclase** (ah-DEN-ih-late SY-clase). Adenylate cyclase removes two phosphate groups from ATP and converts it to cyclic AMP (cAMP), the second messenger. Cyclic AMP then activates enzymes called *kinases* (KY-nace-es) in the cytosol. Kinases add phosphate groups to other cellular enzymes, thereby activating those and triggering a great variety of physiological changes within the cell.

G proteins play such an enormous range of roles in physiology and disease that Martin Rodbell and Alfred Gilman received a 1994 Nobel Prize for discovering them. Up to 60% of currently used drugs function by altering the activity of G proteins.

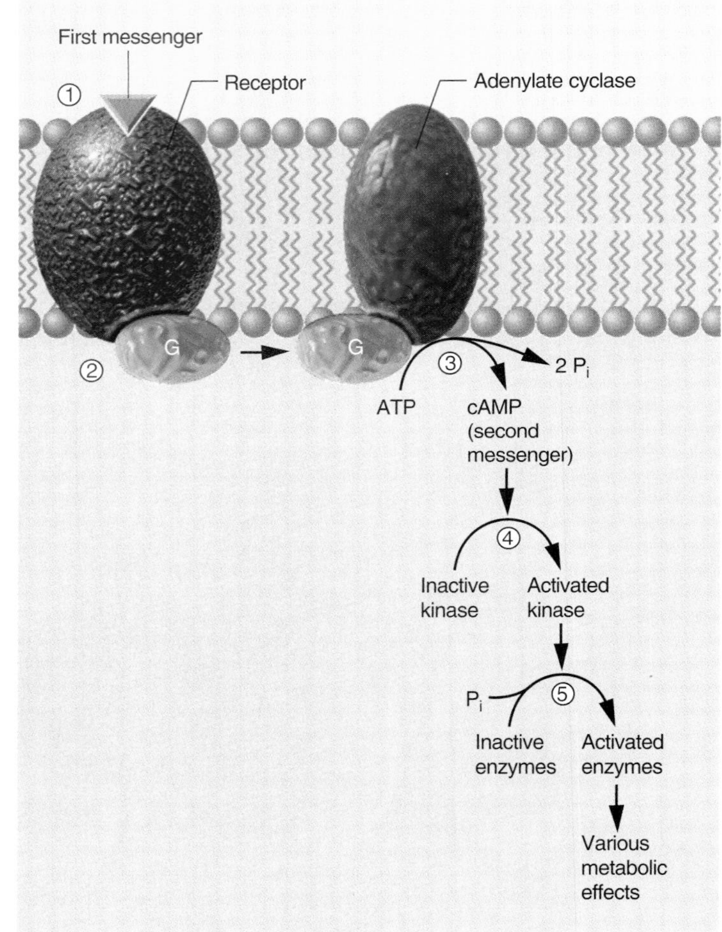

Figure 4.9 A Second Messenger System. (*1*) A messenger such as epinephrine (*red triangle*) binds to a receptor in the plasma membrane. (*2*) The receptor releases a G protein, which then travels freely in the cytoplasm and can have various effects in the cell. (*3*) The G protein binds to an enzyme, adenylate cyclase, in the plasma membrane. Adenylate cyclase converts ATP to cyclic AMP (cAMP), the second messenger. (*4*) cAMP activates a cytoplasmic enzyme called a kinase. (*5*) Kinases add phosphate groups (P_i) to other cytoplasmic enzymes. This activates some enzymes and deactivates others, leading to varied metabolic effects within the cell.

The Glycocalyx

External to the plasma membrane, all animal cells have a fuzzy coat called the **glycocalyx**[8] (GLY-co-CAY-licks) (fig. 4.10), which consists of the carbohydrate moieties of the membrane glycolipids and glycoproteins. It is chemically unique in everyone but identical twins, and acts like a personal identification tag that enables the body to distinguish its own healthy cells from transplanted tissues, invading organisms, and diseased cells. Human blood types and transfusion compatibility are determined by glycoproteins. The glycocalyx includes the cell-adhesion molecules that enable cells to adhere to each other and guide the movement of cells in embryonic development. Because the glycocalyx is hydrophilic, it retains a layer of water at the cell surface and enables the cell to absorb water and dissolved solutes from the extracellular fluid. The functions of the glycocalyx are summarized in table 4.3.

Table 4.3	Functions of the Glycocalyx
Immunity to Infection	Enables the immune system to recognize and selectively attack foreign organisms
Defense Against Cancer	Changes in the glycocalyx of cancerous cells enable the immune system to recognize and destroy them
Transplant Compatibility	Forms the basis for compatibility of blood transfusions, tissue grafts, and organ transplants
Cellular Uptake	Retains a film of water at the cell surface, enabling cells to absorb water and the solutes dissolved in it
Cell Adhesion	Binds cells together so that tissues do not fall apart
Fertilization	Enables sperm cells to recognize and bind to egg cells
Embryonic Development	Guides embryonic cells to their destinations in the body

Microvilli, Cilia, and Flagella

Many cells have surface extensions called *microvilli, cilia,* and *flagella.* These aid in absorption, movement, and sensory processes.

8. *glyco* = sugar + *calyx* = cup, vessel

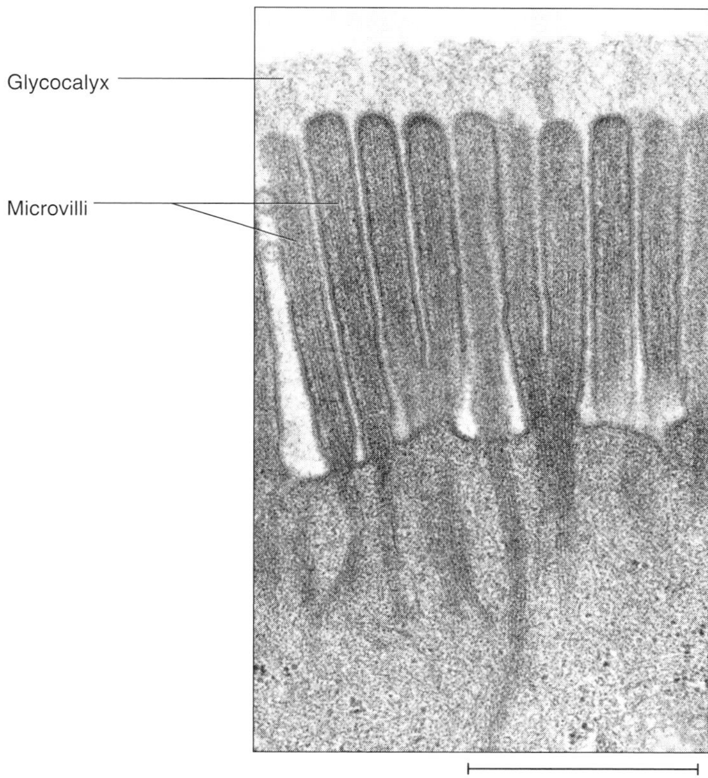

Figure 4.10 Microvilli and the Glycocalyx. This is the apical surface of an absorptive cell of the small intestine.

Microvilli

Microvilli[9] (MY-cro-VIL-eye; singular, *microvillus*) are extensions of the plasma membrane that serve primarily to increase a cell's surface area (figs. 4.10 and 4.11*a, b*). Microvilli are best developed in cells specialized for absorption, such as the epithelial cells of the intestines and kidney tubules. They give such cells 15 to 40 times as much absorptive surface area as they would have if their apical surfaces were flat. On many cells, microvilli are little more than tiny bumps on the plasma membrane. On cells of the taste buds and inner ear, they are well developed but serve sensory rather than absorptive functions.

Individual microvilli cannot be distinguished very well with the light microscope because they are only 1 to 2 μm long. On some cells, they are very dense and appear as a fringe called the **brush border** at the apical cell surface. With the scanning electron microscope, they resemble a deep-pile carpet. With the transmission electron microscope, microvilli typically look like finger-shaped projections of the cell surface. They show little internal structure, but often have a bundle of stiff filaments of a protein called *actin.* Actin filaments attach to the inside of the plasma membrane at the tip of the microvillus, and at its base they extend a little way into the cell and anchor the microvillus to a protein mesh called the **terminal web.** When tugged by another protein in the cytoplasm, actin can shorten a microvillus to "milk" its absorbed contents downward into the cell.

9. *micro* = small + *villi* = hairs

Figure 4.11 Microvilli and Cilia. (*a* and *b*) Electron micrograph and diagram of microvilli in cross section. (*c* and *d*) Electron micrograph and diagram of cilia in cross section. (*e*) Three-dimensional structure of a cilium at the point where it meets the cell surface.

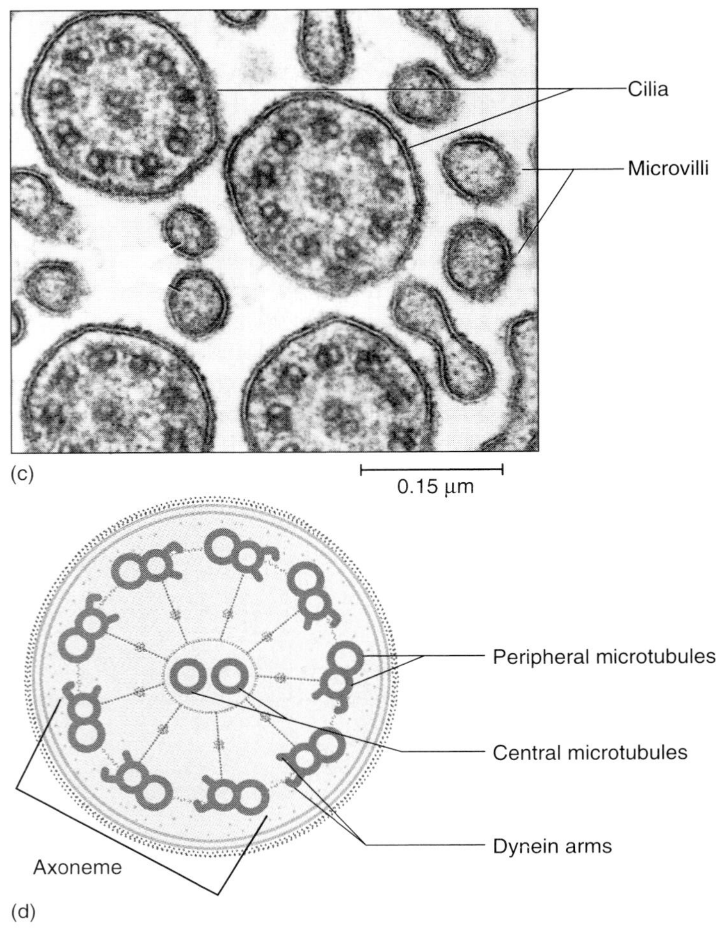

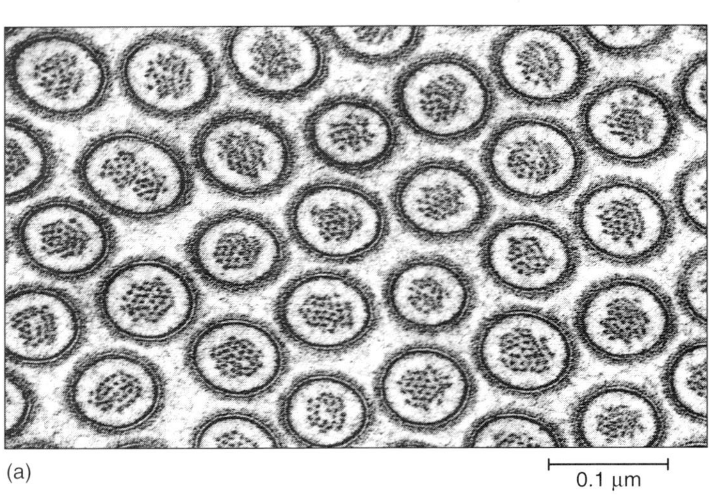

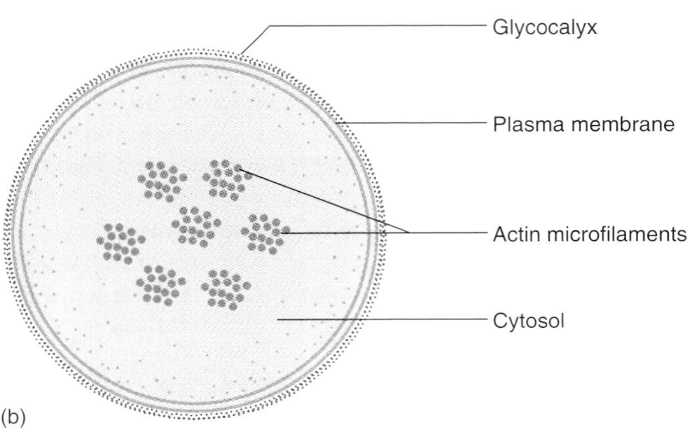

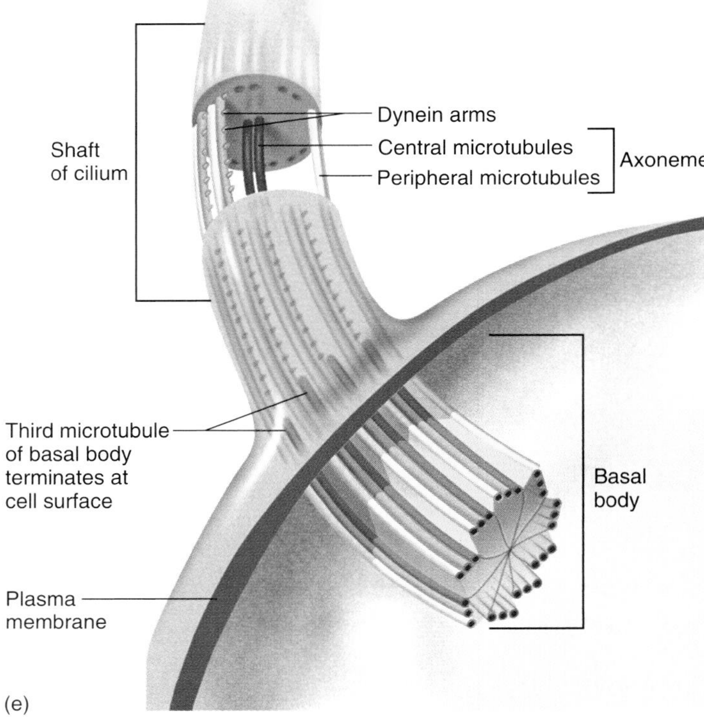

Cilia

Cilia (SIL-ee-uh; singular, *cilium*[10]) (figs. 4.11*c–e* and 4.12) are hairlike processes about 7 to 10 μm long. There may be 50 to 200 cilia on the surface of one cell. Cilia are usually motile and serve primarily to move mucus through the respiratory tract and egg cells through the uterine (fallopian) tubes.

Cilia beat in waves that sweep across the surface of an epithelium, always in the same direction (fig. 4.13). They bend stiffly forward and produce a **power stroke** that pushes along the mucus or other matter riding on their tips. Shortly after a cilium begins its power stroke, the one just ahead of it begins, and the next and the next—collectively producing a wavelike motion. After a cilium completes its power stroke, it is pulled limply back by a **recovery stroke** that restores it to the upright position, ready to flex again.

Think About It

How would the movement of mucus in the respiratory tract be affected if cilia were equally stiff on both their power and recovery strokes?

Cilia could not beat freely if they were embedded in sticky mucus (see clinical insight 4.2). Instead, they beat within a saline (saltwater) layer at the cell surface. *Chloride pumps* in the apical plasma membrane produce this layer by pumping Cl^- into the extracellular fluid. Sodium ions follow because of the attraction of oppositely charged particles to each other, and water follows the Na^+ by osmosis. Mucus essentially floats on the surface of this layer and is pushed along by the tips of the cilia.

The structural basis for ciliary contraction is a core called the **axoneme**[11] (ACK-so-neem), which consists of an array of thin protein cylinders called *microtubules*. There are two central microtubules surrounded by a pinwheel-like ring of nine microtubule pairs—an arrangement called the **9 + 2 structure** (see fig. 4.11*c*). The central microtubules stop at the cell surface, but the peripheral microtubules

10. *cilium* = eyelash
11. *axo* = axis + *neme* = thread

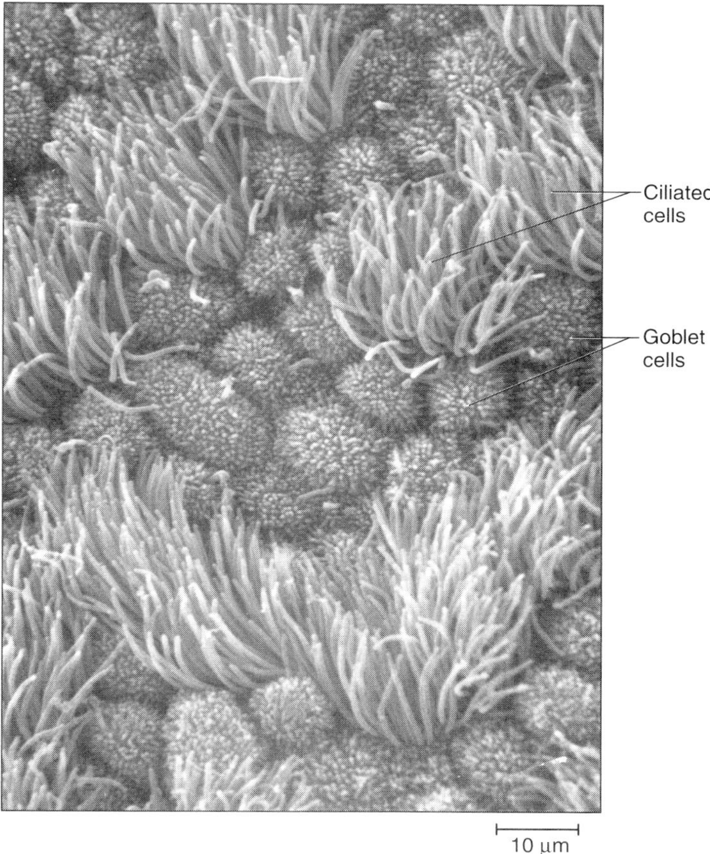

Figure 4.12 Cilia of the Trachea. Several nonciliated, mucus-secreting goblet cells are visible among the ciliated cells. The goblet cells have short microvilli on their surface.

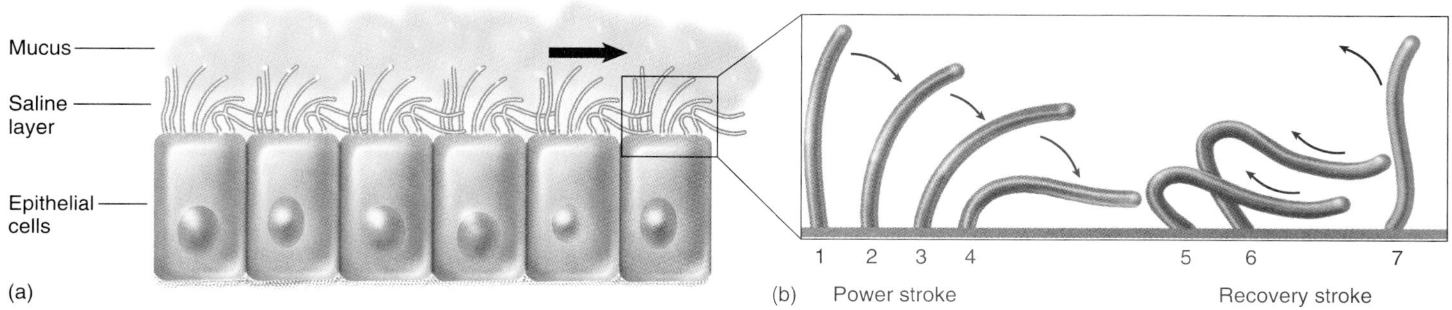

Figure 4.13 **Ciliary action.** (*a*) Cilia of an epithelium moving mucus along a surface layer of saline. (*b*) Power and recovery strokes of a cilium.

Clinical Insight 4.2

Cystic Fibrosis

The significance of chloride pumps becomes especially evident in *cystic fibrosis (CF)*, a hereditary disease affecting especially white children of European descent. CF is usually caused by a defect in which cells make chloride pumps but fail to install them in the plasma membrane. Consequently, there is an inadequate saline layer on the cell surface and the mucus is dehydrated and overly sticky. This thick mucus plugs the ducts of the pancreas and prevents it from secreting digestive enzymes into the small intestine, so digestion and nutrition are inadequate. In the respiratory tract, the mucus clogs the cilia and prevents them from beating freely. The respiratory tract becomes congested with thick mucus, often leading to chronic infection and pulmonary collapse. The mean life expectancy of people with CF is now about 30 years.

continue a short distance into the cell as part of a **basal body** that anchors the cilium to the cell. In each pair of peripheral microtubules, one tubule has two little **dynein**[12] (DINE-een) **arms.** Dynein, a motor protein, uses energy from ATP to "crawl" up the adjacent pair of microtubules. When microtubules on the front of the cilium crawl up the microtubules behind them, the cilium bends toward the front.

In some organs, cilia have lost their motility and assumed sensory roles. The outer half of each sensory cell in the retina of the eye, for example, is a modified cilium specialized for absorbing light. Some cilia in the nasal cavity bind to odor molecules and initiate the sense of smell. In the inner ear, cilia function along with microvilli in the senses of hearing and balance.

Flagella

A **flagellum**[13] (fla-JEL-um) is a long whiplike structure with an axoneme identical to that of a cilium. The only functional flagellum in humans is the tail of a sperm cell, but vestigial (stunted, nonfunctional) flagella are found on the epithelial cells of several organs, including the uterus, testis, kidney, pancreas, and pituitary gland.

Before You Go On

Answer the following questions to test your understanding of the preceding section:

5. How does the structure of a plasma membrane depend on the amphiphilic nature of phospholipids?
6. Distinguish between integral and peripheral proteins.
7. Identify several reasons why the glycocalyx is important to human survival.
8. How do microvilli and cilia differ in structure and function?

12. *dyn* = power, energy + *in* = protein
13. *flagellum* = whip

The Cytoplasm

Objectives

When you have completed this section, you should be able to

- list the main organelles of a cell and explain their functions;
- describe the cytoskeleton and its functions; and
- give some examples of cell inclusions and explain how inclusions differ from organelles.

We now probe more deeply into the cell to study the structures in the cytoplasm. These are classified into three groups—*organelles, cytoskeleton,* and *inclusions*—all embedded in the clear, gelatinous cytosol.

Organelles

Organelles are internal structures of a cell that carry out specialized metabolic tasks. Some are surrounded by one or two layers of unit membrane and are therefore referred to as *membranous organelles.* These are the nucleus, mitochondria, lysosomes, peroxisomes, rough endoplasmic reticulum, smooth endoplasmic reticulum, and Golgi complex. Organelles that are not surrounded by membranes include the ribosomes, centrosome, centrioles, and basal bodies.

The Nucleus

The **nucleus** is the largest organelle and usually the only one visible with the light microscope. With the TEM, it can be distinguished by the *two* unit membranes surrounding it, which together form the **nuclear envelope** (see fig. 4.4). The nucleus houses the DNA, a genetic program for the cell's structure and function. Nuclear structure and function are described in detail in the next chapter.

Endoplasmic Reticulum

Endoplasmic reticulum (ER) literally means "little network within the cytoplasm." It is a system of intercon-

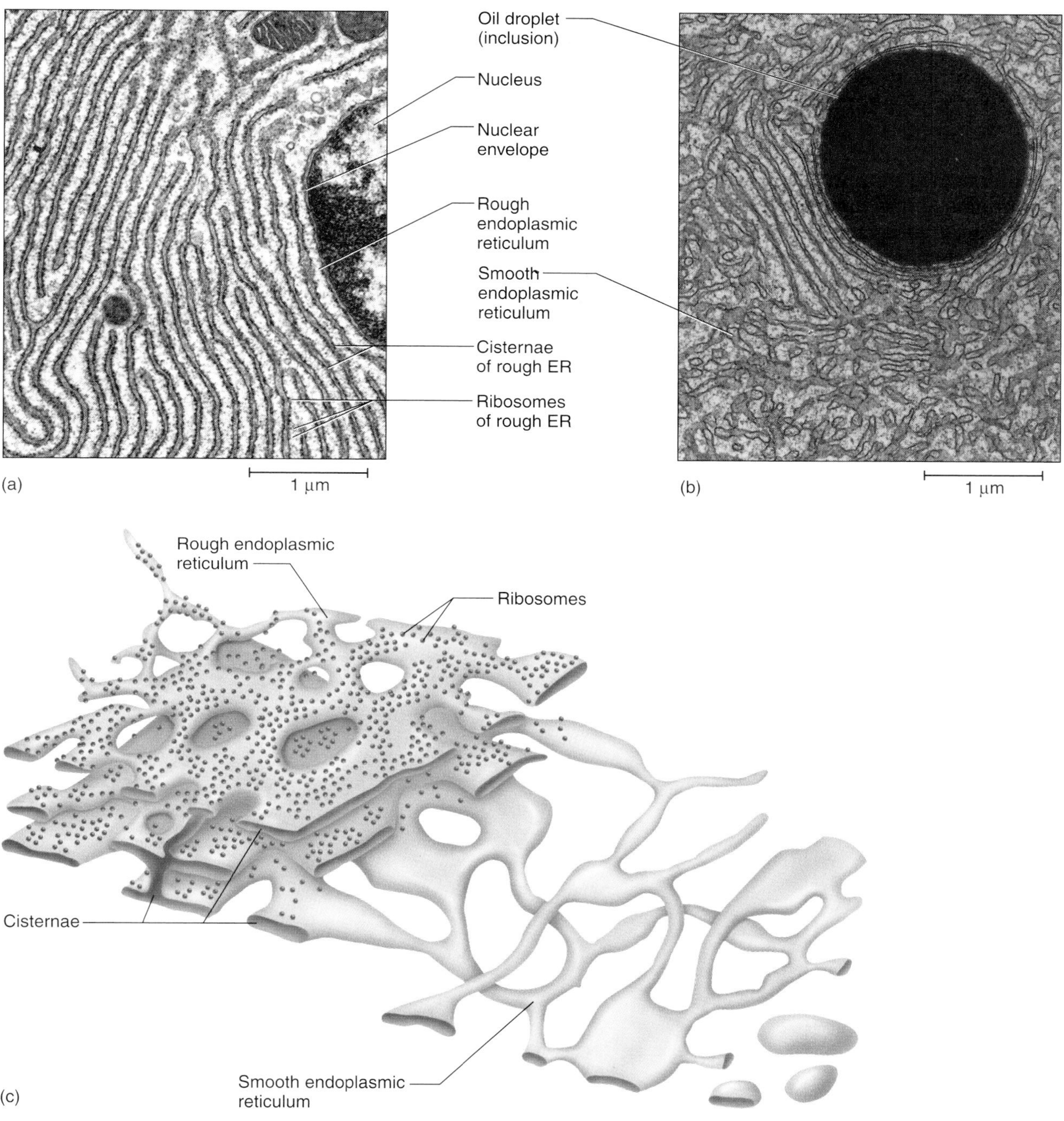

Figure 4.14 Endoplasmic Reticulum (ER). (*a*) Rough ER. (*b*) Smooth ER. (*c*) Structure of the endoplasmic reticulum, with rough and smooth regions.

nected channels called **cisternae**[14] (sis-TUR-nee) enclosed by a unit membrane (fig. 4.14). In areas called **rough endoplasmic reticulum,** the network is composed of parallel, flattened sacs covered with granules called *ribosomes.* The rough ER is continuous with the outer membrane of the nuclear envelope, and adjacent cisternae are often connected by perpendicular bridges. In areas called **smooth endoplasmic reticulum,** the membrane lacks ribosomes, the cisternae are more tubular in shape, and they branch more extensively. The cisternae of the smooth ER are thought to be continuous with those of the rough ER, so the two are functionally different parts of the same cytoplasmic network.

14. *cistern* = reservoir

The endoplasmic reticulum synthesizes steroids and other lipids, detoxifies alcohol and other drugs, and manufactures all of the membranes of the cell. The rough ER produces the phospholipids and proteins of the plasma membrane, and synthesizes proteins that are either packaged in other organelles such as lysosomes or secreted from the cell. Rough ER is most abundant in cells that synthesize large amounts of protein, such as antibody-producing cells and cells of the digestive glands. This role is discussed in detail in the next chapter.

Most cells have only a scanty smooth ER. It is relatively abundant, however, in cells that engage extensively in detoxification, such as liver and kidney cells. Long-term abuse of alcohol, barbiturates, and other drugs leads to tolerance partly because the smooth ER proliferates and detoxifies the drugs more quickly. Smooth ER is also abundant in cells of the testes and ovaries that synthesize steroid hormones. Skeletal muscle and cardiac muscle contain extensive networks of smooth ER that release calcium to trigger muscle contraction and store the calcium between contractions.

Ribosomes

Ribosomes are small granules of protein and RNA found in the cytosol and on the outer surfaces of the rough ER and nuclear envelope. They "read" coded genetic messages (messenger RNA) from the nucleus and assemble amino acids into proteins specified by the code. This process is detailed in the next chapter.

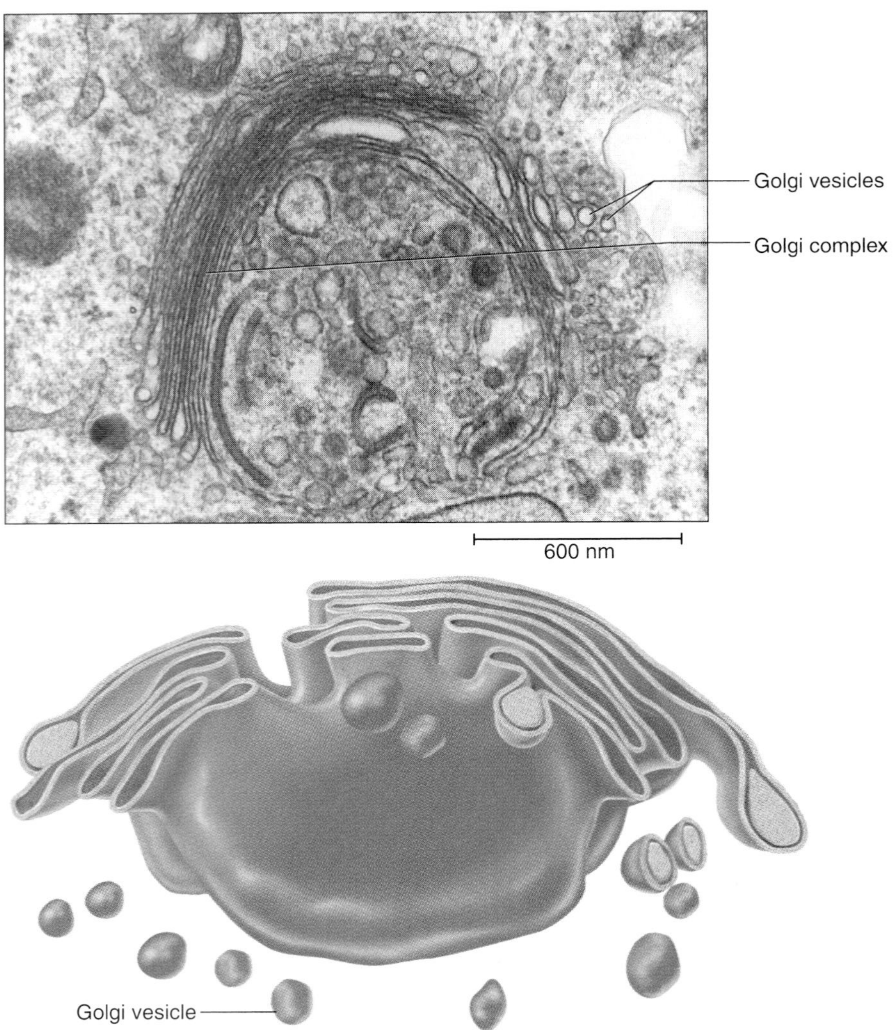

Figure 4.15 The Golgi Complex.

Golgi Complex

The **Golgi**[15] (GOAL-jee) **complex** is a small system of cisternae that synthesize carbohydrates and put the finishing touches on protein and glycoprotein synthesis. The complex resembles a stack of pita bread. Typically, the complex consists of about six cisternae, slightly separated from each other; each is a flattened, slightly curved sac with swollen edges (fig. 4.15). The Golgi complex receives the newly synthesized proteins from the rough ER. It sorts them, cuts and splices some of them, adds carbohydrate moieties to some of them, and finally packages the proteins in membrane-bounded **Golgi vesicles.** These vesicles bud off the swollen rim of a cisterna and are seen in abundance in the neighborhood of the Golgi complex. Some of them become *lysosomes,* the organelle discussed next; some migrate to the plasma membrane and fuse with it, contributing fresh protein and phospholipid to the membrane; and some become **secretory vesicles** that store a cell product, such as breast milk or digestive enzymes, for later release. The role of the Golgi complex in protein synthesis and secretion is detailed in the next chapter.

Lysosomes

A **lysosome**[16] (LY-so-some) (fig. 4.16*a*) is a package of enzymes contained in a single unit membrane. Although often round or oval, lysosomes are extremely variable in shape. When viewed with the TEM, they often exhibit dark gray contents devoid of structure, but sometimes show crystals or parallel layers of protein. At least 50 lysosomal enzymes have been identified. They hydrolyze proteins, nucleic acids, complex carbohydrates, phospholipids, and other substrates. In the liver, lysosomes break down stored glycogen to release glucose into the bloodstream. White blood cells called *neutrophils* engulf bacteria and then use their lysosomes to digest them. Lysosomes also digest and dispose of worn-out mitochondria and other organelles; this process is called **autophagy**[17] (aw-TOFF-uh-jee). Some cells are meant to do a certain job and then die. The uterus, for example, weighs about 900 g at full-term pregnancy and shrinks to 60 g within 5 or 6 weeks after birth. This shrinkage is due to **autolysis,**[18] the digestion of surplus cells by their own lysosomal enzymes. Such *programmed cell death* is further discussed in chapter 6.

Peroxisomes

Peroxisomes (fig. 4.16*b*) resemble lysosomes but contain different enzymes and are not produced by the Golgi complex. They are especially abundant in liver and kidney

15. Camillo Golgi (1843–1926), Italian histologist
16. *lyso* = loosen, dissolve + *some* = body
17. *auto* = self + *phagy* = eating
18. *auto* = self + *lysis* = dissolving

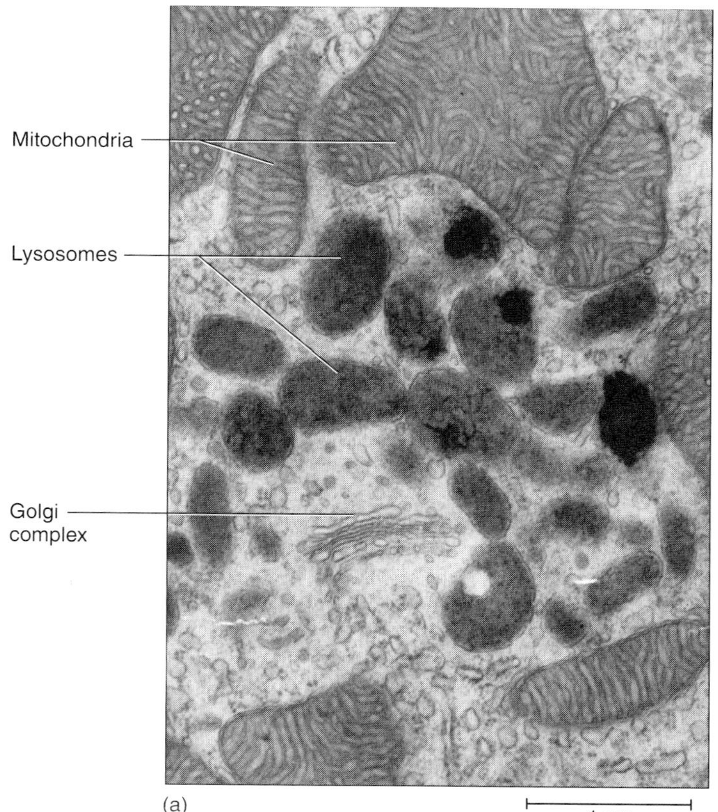

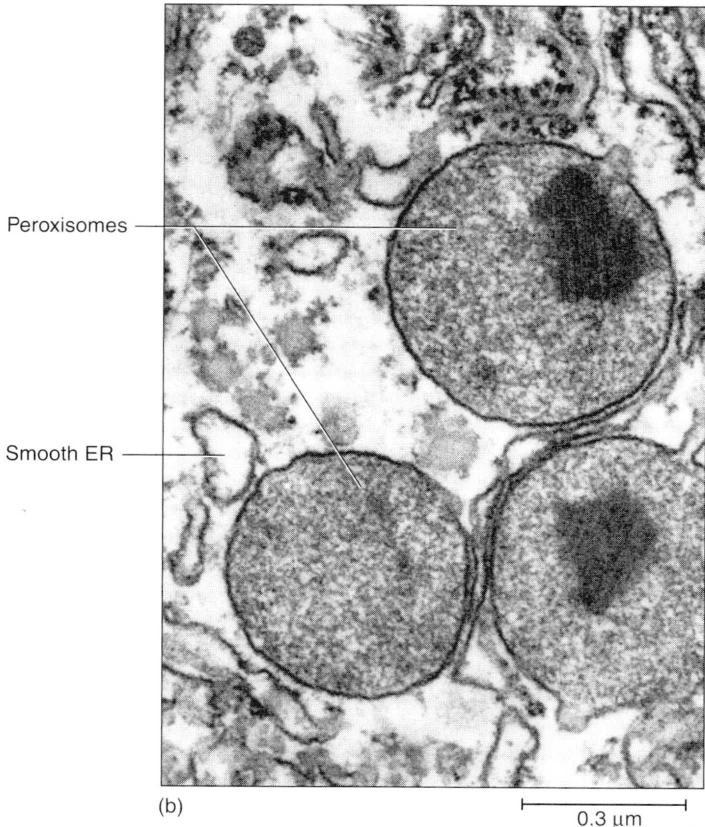

Figure 4.16 Lysosomes and Peroxisomes. (*a*) Lysosomes. (*b*) Peroxisomes.

Evolutionary Insight 4.3

The Origin of Mitochondria

There is evidence that mitochondria evolved from bacteria that either invaded or were engulfed by other ancient cells, then evolved a mutually beneficial metabolic relationship with them. In size and physiology, mitochondria resemble certain bacteria that live within other cells in a state of symbiosis. Mitochondrial DNA (mtDNA) resembles bacterial DNA more than it does nuclear DNA, and it is replicated independently of nuclear DNA.

While nuclear DNA is reshuffled in each generation by the process of sexual reproduction, mtDNA remains unchanged from generation to generation except by the slow pace of random mutation. Biologists and anthropologists have used mtDNA as a "molecular clock" to trace evolutionary lineages in humans and other species. They have gained some evidence, although still controversial, that all humans today are the descendants of a single female who lived in Africa about 200,000 years ago.

cells, where they neutralize free radicals and detoxify alcohol and other drugs. They are named for the hydrogen peroxide (H_2O_2) they produce in the course of detoxifying alcohol and killing bacteria. They break down excess H_2O_2 with an enzyme called *catalase.* Peroxisomes also break down fatty acids into two-carbon acetyl groups, which the mitochondria then use as an energy source for ATP synthesis.

Mitochondria

Mitochondria[19] (MY-toe-CON-dree-uh) (fig. 4.17) are organelles specialized for synthesizing ATP. They have a variety of shapes: spheroid, rod-shaped, bean-shaped, or threadlike. Like the nucleus, a mitochondrion is surrounded by a double unit membrane. The inner membrane usually has folds called **cristae**[20] (CRIS-tee), which project like shelves across the organelle. Mitochondria are the "powerhouses" of the cell. Energy is not *made* here, but it is extracted from organic compounds and transferred to ATP, primarily by enzymes located on the cristae. The role of mitochondria in ATP synthesis is explained in detail in chapter 26.

The space between the cristae is called the **matrix.** It contains ribosomes and a small, circular DNA molecule called *mitochondrial DNA (mtDNA),* which is genetically different from the DNA in the cell's nucleus. Mutations in mtDNA are responsible for some muscle, heart, and eye diseases. Mitochondria reproduce independently of their host cell (see evolutionary insight 4.3) and live for about 10 days.

Centrioles

A **centriole** (SEN-tree-ole) is a short cylindrical assembly of microtubules, arranged in nine groups of three microtubules each (fig. 4.18). Two centrioles lie perpendicular

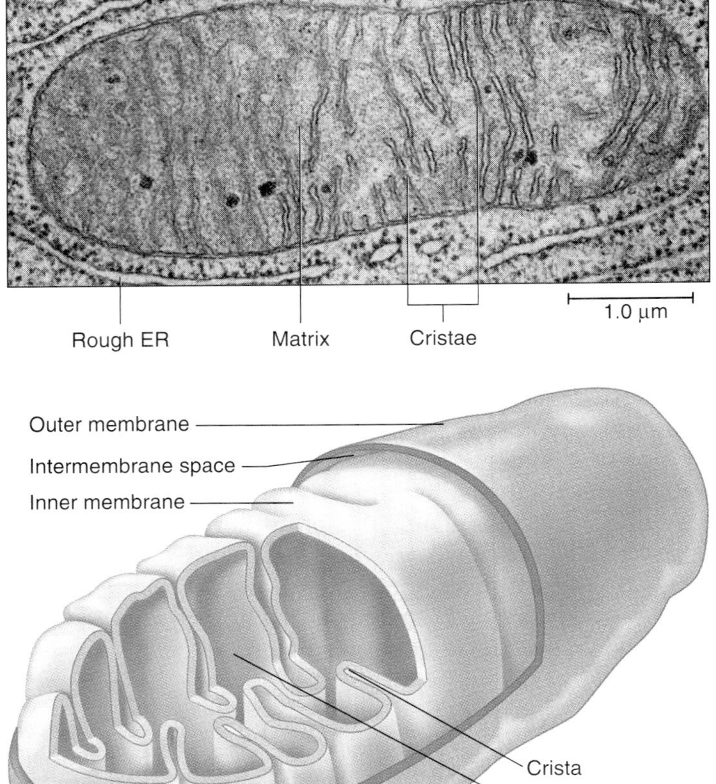

Figure 4.17 A Mitochondrion.

19. *mito* = thread + *chondr* = grain
20. *crista* = crest

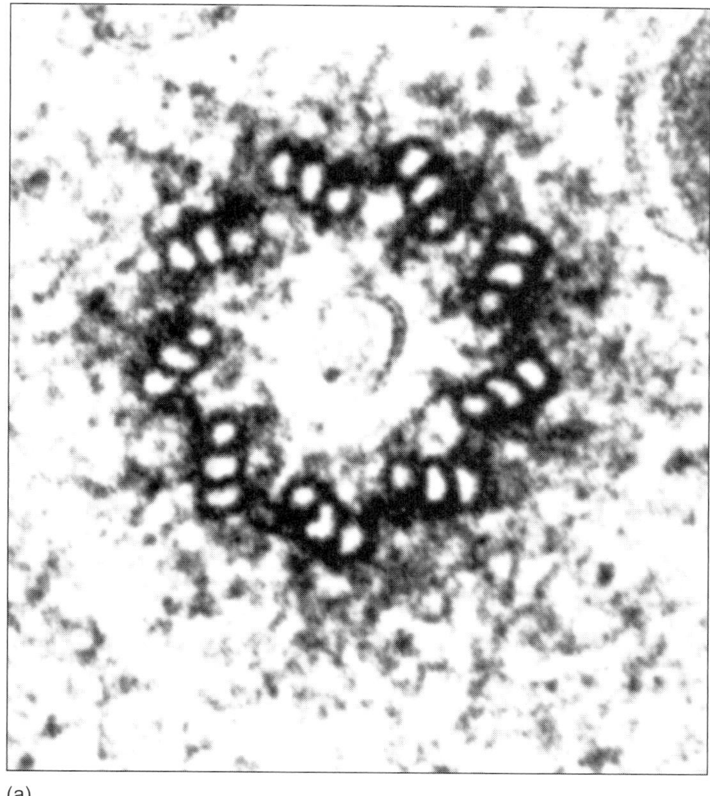

(a)

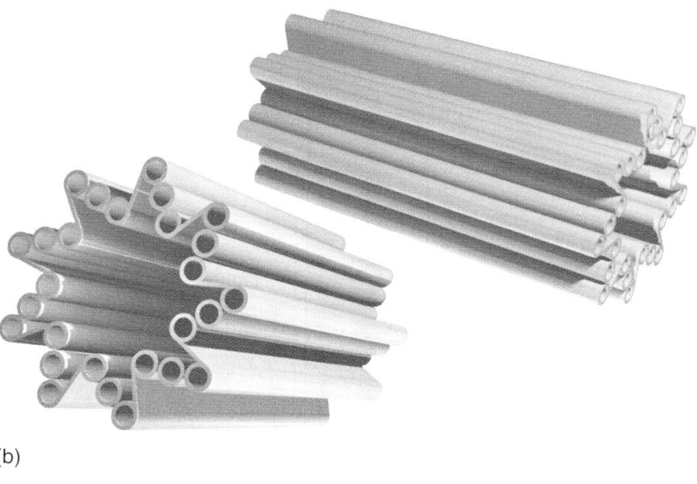

(b)

Figure 4.18 Centrioles. (*a*) Electron micrograph of a centriole as seen in cross section. (*b*) A pair of perpendicular centrioles.

to each other within a small clear area of cytoplasm called the **centrosome**[21] (see fig. 4.4). They play a role in cell division described in the next chapter. Each basal body of a flagellum or cilium is a single centriole oriented perpendicular to the plasma membrane. Basal bodies originate in a *centriolar organizing center* and migrate to the plasma membrane. Two microtubules of each triplet then elongate to form the nine pairs of peripheral microtubules of the axoneme. A cilium can grow to its full length in less than an hour.

The Cytoskeleton

The **cytoskeleton** is a collection of protein filaments and cylinders that determine the shape of a cell, lend it structural support, organize its contents, move substances through the cell, and contribute to movements of the cell as a whole. It can form a very dense supportive scaffold in the cytoplasm (fig. 4.19). It is connected to integral proteins of the plasma membrane, and they in turn are connected to protein fibers external to the cell, so there is a strong structural continuity from extracellular material to the cytoplasm. Cytoskeletal elements may even connect to chromosomes in the nucleus, enabling physical tension on a cell to move nuclear contents and mechanically stimulate genetic function.

The cytoskeleton is composed of *microfilaments, intermediate filaments,* and *microtubules.* **Microfilaments** are about 6 nm thick and are made of the protein *actin.* They form a network on the cytoplasmic side of the plasma membrane called the **membrane skeleton.** The phospholipids of the plasma membrane spread out over the membrane skeleton like butter on a slice of bread. It is thought that the phospholipids would break up into little droplets without this support. The roles of actin in supporting microvilli and producing cell movements were discussed earlier. In conjunction with another protein, *myosin,* microfilaments are also responsible for muscle contraction.

Intermediate filaments (8–10 nm in diameter) are thicker and stiffer than microfilaments. They participate in junctions that hold epithelial cells together and resist stresses placed on a cell. In epidermal cells, they are made of the tough protein *keratin* and occupy most of the cytoplasm.

21. *centro* = central + *some* = body

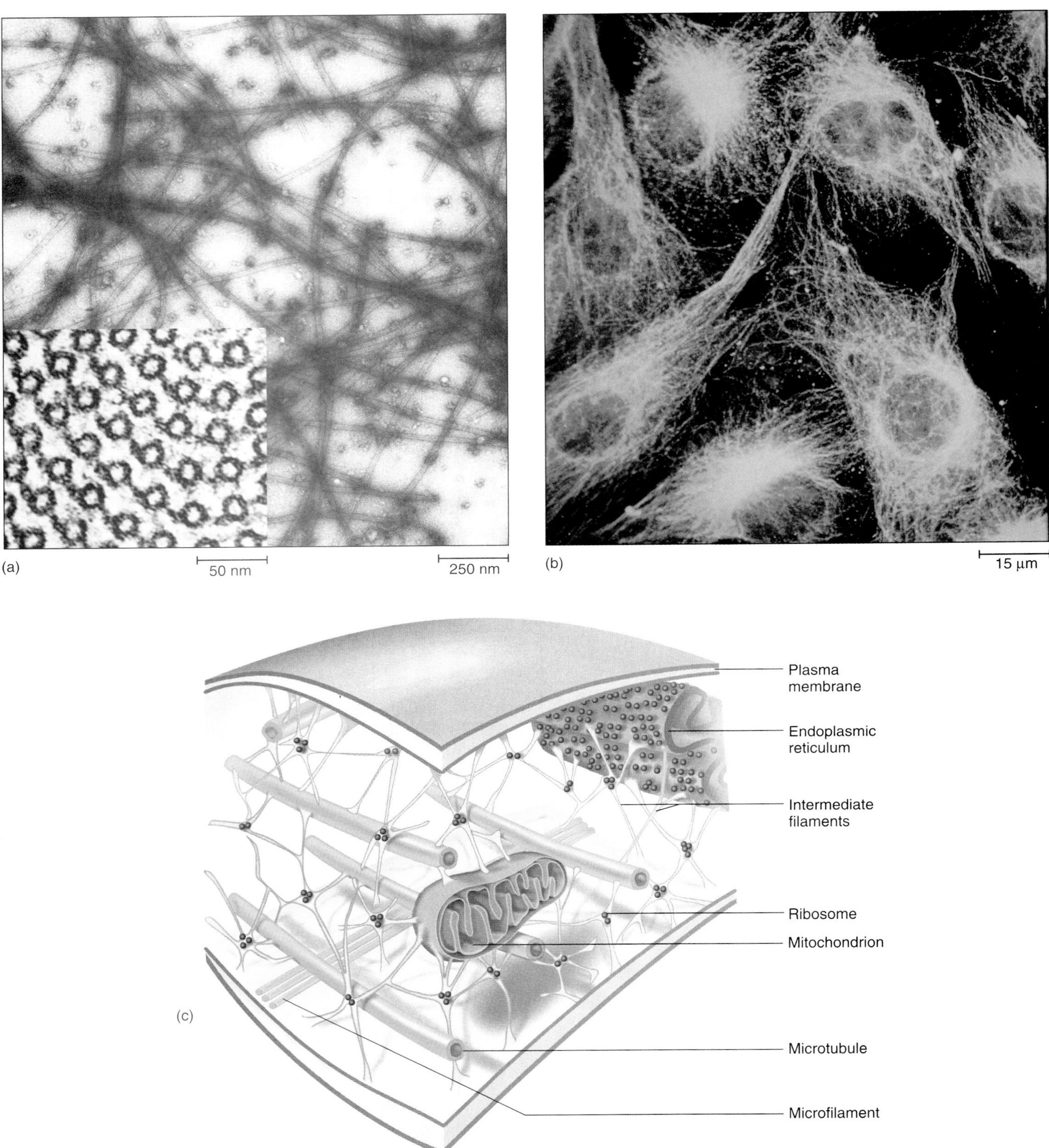

Figure 4.19 The Cytoskeleton. (*a*) Electron micrograph of a cell of the testis showing numerous microtubules in longitudinal section and cross section. (*b*) Fluorescent antibodies are used here to label the fibrous cytoskeleton of a cell and make it visible through a fluorescence microscope. (*c*) Diagram of the cytoskeleton.

A **microtubule** (25 nm in diameter) is a cylinder made of 13 parallel strands called *protofilaments.* Each protofilament is a long chain of globular proteins called *tubulin* (fig. 4.20). Microtubules radiate from the centrosome and hold organelles in place, form bundles that maintain cell shape and rigidity, and act somewhat like railroad tracks to guide organelles and molecules to specific destinations in a cell. They form the axonemes of cilia and flagella and are responsible for their beating movements. They also form the mitotic spindle that guides chromosome movement during cell division. Microtubules are not permanent structures. They come and go moment by moment as tubulin molecules assemble into a tubule and then suddenly break apart again to be used somewhere else in the cell. The double and triple sets of microtubules in cilia, flagella, basal bodies, and centrioles, however, are more stable.

Inclusions

Inclusions are of two kinds: stored cellular products such as glycogen granules, pigments, and fat droplets (see fig. 4.14*b*), and foreign bodies such as dust particles, viruses, and intracellular bacteria. Inclusions are never enclosed in a unit membrane, and unlike the organelles and cytoskeleton, they are not essential to cell survival.

The major features of a cell are summarized in table 4.4.

Before You Go On

Answer the following questions to test your understanding of the preceding section:

9. Distinguish between organelles and inclusions. State two examples of each.
10. Briefly state how each of the following cell components can be recognized in electron micrographs: the nucleus, a mitochondrion, a lysosome, and a centriole. What is the primary function of each?
11. What three organelles are involved in protein synthesis?
12. In what ways do rough and smooth endoplasmic reticulum differ?
13. Define *centriole, microtubule, cytoskeleton,* and *axoneme.* How are these structures related to each other?

Membrane Transport I: Filtration and Simple Diffusion

Objectives

When you have completed this section, you should be able to

- explain what is meant by a selectively permeable membrane;
- describe some processes of membrane transport that do not consume ATP; and
- define *osmolarity* and *tonicity* and explain their importance.

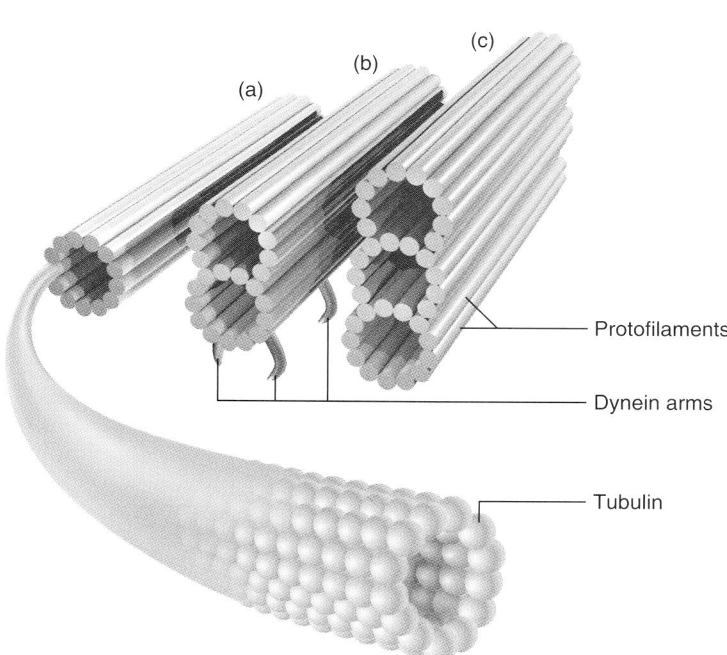

Figure 4.20 Microtubules. (*a*) A microtubule is composed of 13 protofilaments. Each protofilament is a helical chain of globular proteins called tubulin. (*b*) One of the nine microtubule pairs that form the axonemes of cilia and flagella. (*c*) One of the nine microtubule triplets that form a centriole.

Table 4.4 Summary of Organelles and Other Cellular Structures

Structure	Appearance to TEM	Function
Plasma membrane (figs. 4.3 and 4.6)	Two dark lines at cell surface, separated by narrow light space	Prevents escape of cell contents; regulates exchange of materials between cytoplasm and extracellular fluid; involved in intercellular communication
Microvilli (figs. 4.10 and 4.11)	Short, densely spaced, hairlike processes or scattered bumps on cell surface; interior featureless or with bundle of microfilaments	Increase absorptive surface area; some sensory roles (hearing, equilibrium, taste)
Cilia (figs. 4.11 and 4.12)	Long hairlike projections of apical cell surface; axoneme with 9 + 2 array of microtubules	Move substances along cell surface; some sensory roles (hearing, equilibrium, smell, vision)
Flagellum	Long, single, whiplike process with axoneme	Sperm motility
Nucleus (figs. 4.3 and 4.5)	Largest organelle in most cells, surrounded by double unit membrane with nuclear pores	Genetic control center of cell; directs protein synthesis
Rough ER (fig. 4.14*a*)	Extensive sheets of parallel unit membranes with cisterna between them and ribosomes on outer surface	Protein synthesis and manufacture of cellular membranes
Smooth ER (fig. 4.14*b*)	Branching network of tubules with smooth surface (no ribosomes); usually broken into numerous small segments in TEM photos	Lipid synthesis, detoxification, calcium storage
Ribosomes (fig. 4.14*a*)	Small dark granules free in cytosol or on surface of rough ER	Interpret the genetic code and synthesize polypeptides
Golgi complex (fig. 4.15)	Several closely spaced, parallel cisternae with thick edges, usually near nucleus, often with many Golgi vesicles nearby	Receives and modifies newly synthesized polypeptides, synthesizes carbohydrates, adds carbohydrates to glycoproteins; packages cell products into Golgi vesicles
Golgi vesicles (fig. 4.15)	Round to irregular sacs near Golgi complex, usually with light, featureless contents	Become secretory vesicles and carry cell products to apical surface for exocytosis, or become lysosomes
Lysosomes (fig. 4.16*a*)	Round to oval sacs with single unit membrane, often a dark featureless interior but sometimes with protein layers or crystals	Contain enzymes for intracellular digestion, autophagy, programmed cell death, and glucose mobilization
Peroxisomes (fig. 4.16*b*)	Similar to lysosomes; often lighter in color	Contain enzymes for detoxification of free radicals, alcohol, and other drugs; oxidize fatty acids
Mitochondria (figs. 4.16*a* and 4.17)	Round, rod-shaped, bean-shaped, or threadlike structures with double unit membrane and shelflike infoldings called cristae	ATP synthesis
Centrioles (fig. 4.18)	Short cylindrical bodies, each composed of a circle of nine triplets of microtubules	Form mitotic spindle during cell division; unpaired centrioles form basal bodies of cilia and flagella
Centrosome (fig. 4.4)	Clear area near nucleus containing a pair of centrioles	Organizing center for formation of microtubules of cytoskeleton and mitotic spindle
Basal body (fig. 4.11*e*)	Unpaired centriole at the base of a cilium or flagellum	Point of origin, growth, and anchorage of a cilium or flagellum; produces axoneme
Microfilaments (figs. 4.11*a* and 4.19)	Thin protein filaments (6 nm diameter), often in parallel bundles or dense networks in cytoplasm	Support microvilli; involved in muscle contraction and other cell motility, endocytosis, and cell division
Intermediate filaments (fig. 4.19)	Thicker protein filaments (8–10 nm diameter) extending throughout cytoplasm or concentrated at cell-to-cell junctions	Give shape and physical support to cell; anchor cells to each other and to extracellular material; compartmentalize cell contents
Microtubules (figs. 4.19 and 4.20)	Hollow protein cylinders (25 nm diameter)	Form axonemes of cilia and flagella, centrioles, basal bodies, and mitotic spindles; enable motility of cell parts; direct organelles and macromolecules to their destinations within a cell
Inclusions (fig. 4.14*b*)	Highly variable—fat droplets, glycogen granules, protein crystals, dust, bacteria, viruses; never enclosed in unit membranes	Storage products or other products of cellular metabolism, or foreign matter retained in cytoplasm

The plasma membrane is both a barrier and gateway between the cytoplasm and extracellular fluid (ECF). It is **selectively permeable**—it allows some things through, such as nutrients and wastes, but prevents other things, such as proteins and phosphates, from escaping the cell or getting into it.

The methods of moving substances into or out of a cell can be classified in two overlapping ways: as *passive* or *active* mechanisms and as *carrier-mediated* or not. Passive mechanisms require no energy (ATP) expenditure by the cell. In most cases, the random molecular motion of the particles themselves provides the energy. Passive mechanisms include filtration and simple diffusion (including a special case of simple diffusion, osmosis). Active mechanisms, however, require the cell to consume ATP. These include active transport and bulk transport. Carrier-mediated mechanisms use a membrane protein to transport substances from one side of the membrane to the other. The mechanisms discussed in this section—filtration and simple diffusion—are not carrier-mediated. The carrier-mediated mechanisms—facilitated diffusion and active transport—are described in the section following this.

Filtration

Filtration is a process in which particles are driven through a selectively permeable membrane by **hydrostatic pressure,** the force exerted on a membrane by water. A coffee filter provides an everyday example. The weight of the water on the filter is a force that drives water and dissolved matter through it, while the filter holds back larger particles (the coffee grounds). In physiology, the most important instance of filtration is in the blood capillaries, where blood pressure forces fluid through gaps in the capillary wall. This is how water, salts, nutrients, and other solutes are transferred from the bloodstream to the tissue fluid and how the kidneys filter wastes from the blood. Capillaries hold back larger particles such as blood cells and proteins.

Simple Diffusion

Simple diffusion is the movement of particles from place to place as a result of their constant, spontaneous motion. It can be observed by dropping a dye crystal in a dish of still water. As the crystal dissolves, it forms a colored zone in the water that gets larger and larger with time (fig. 4.21*a*). This occurs because the dye molecules exhibit *net* diffusion (more molecules diffusing in one direction than in another) from the point of origin, where their concentration is high, toward the edges of the dish, where their concentration is low. When the concentration of a substance differs from one point to another, we say that it exhibits a **concentration gradient.** Particle movement from a region of high concentration toward a region of lower concentration is said to go *down,* or *with,* the gradient, and movement in the other direction is said to go *up,* or *against,* the gradient.

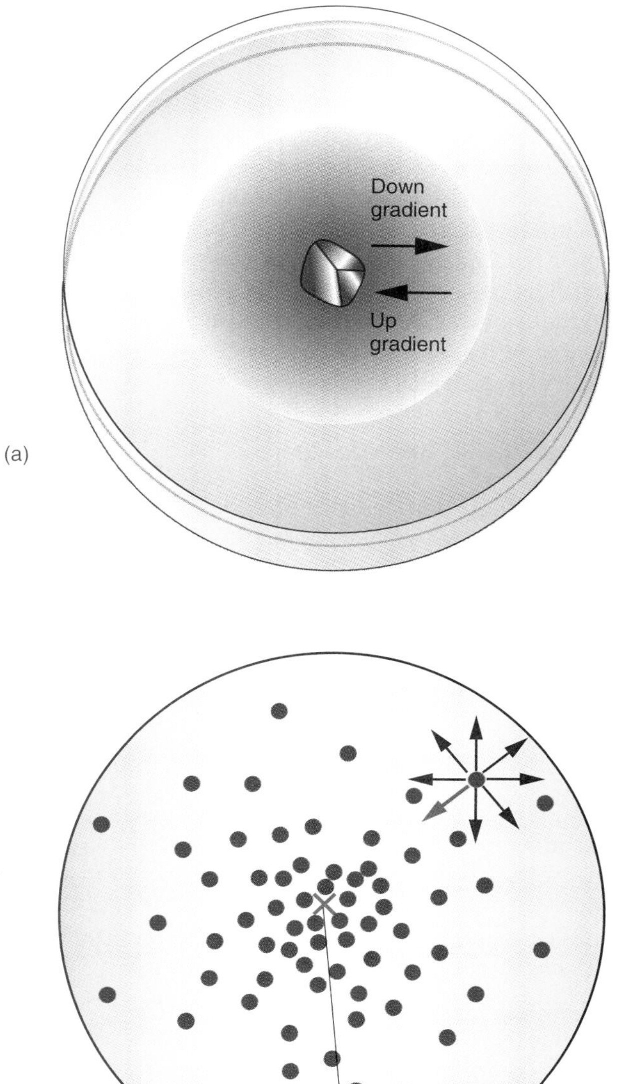

Figure 4.21 Diffusion and Concentration Gradients. (*a*) Dye molecules diffusing away from a crystal dissolving in water. The direction from high concentration (near the crystal) to low concentration is described as "down the concentration gradient"; the opposite direction is described as "up the concentration gradient." (*b*) A molecule at any point in space, like the one at the *upper right,* can move randomly in innumerable directions, but only one of these (*red arrow*) would take it closer to the point of origin. Thus, any molecule is more likely to diffuse down a concentration gradient than up it.

The reason net diffusion proceeds down a concentration gradient, and not up, can be explained in terms of probability. Each particle is in a state of constant random motion and changes directions every time it collides with another particle. Of all the possible directions in which a particle may move, there is only one that would take it toward the place where it originated—toward the crystal, for example. There are innumerable directions, however, that would take it farther away from its point of origin (fig. 4.21*b*). Thus, there is a much greater likelihood that a particle will move away from its origin than back to it.

Diffusion occurs in air, water, and even (slowly) in solids; it has no need of a membrane. However, if there is a selectively permeable membrane in the path of the diffusing molecules, and if it is permeable to that substance, the molecules will pass from one side of the membrane to the other. This is how oxygen leaves the bloodstream and enters cells of the surrounding tissue. Dialysis treatment for kidney disease patients is based on diffusion of solutes through artificial *dialysis membranes.*

Diffusion rates are very important to cell survival because they determine how quickly a cell can acquire nutrients or rid itself of wastes. Some factors that affect the rate of diffusion through a membrane are as follows:

- **Temperature.** Diffusion is driven by the kinetic energy of the particles, and temperature is a measure of that kinetic energy. The warmer a substance is, the more rapidly its particles diffuse. This is why sugar diffuses more quickly through hot tea than through iced tea.
- **Molecular weight.** Heavy molecules such as proteins move more sluggishly and diffuse more slowly than lighter particles such as electrolytes and gases. Smaller molecules also pass through membrane pores more easily than larger ones.
- **"Steepness" of the concentration gradient.** The steepness of a gradient refers to how great a concentration difference there is between two points. Just as a ball rolls down a steep hill faster than it rolls down a gentle slope, particles diffuse more rapidly if there is a greater concentration difference between two points. For example, we can increase the rate of oxygen diffusion into a patient's blood by using an oxygen mask, thus increasing the difference in oxygen concentration between the air and blood.
- **Membrane surface area.** As noted earlier, the apical surface of cells specialized for absorption (for example, in the small intestine) is often extensively folded into microvilli. This makes more membrane available for particles to diffuse through.
- **Membrane permeability.** Diffusion through a membrane depends on how permeable it is to the particles. For example, potassium ions diffuse more rapidly than sodium ions through a plasma membrane because the membrane is more permeable to potassium. Nonpolar, hydrophobic, lipid-soluble substances such as oxygen, nitric oxide, alcohol, and steroids diffuse through the phospholipid regions of a plasma membrane. Significant amounts of water, however, also "sneak" through the phospholipid regions by passing through tiny spaces created by the perpetual movement of the fatty acid tails. Charged, hydrophilic particles such as electrolytes do not mix with lipids, but diffuse through channel proteins in the membrane. Cells can adjust their permeability to a substance by adding channel proteins to the plasma membrane or taking them away. Kidney tubules, for example, do this as a way of controlling the amount of water eliminated from the body.

Osmosis

Osmosis[22] (oz-MO-sis) is the net diffusion of water through a selectively permeable membrane, from the side where it is more concentrated to the side where it is less so. Cells exchange a tremendous amount of water by osmosis. For example, red blood cells pass 100 times their own volume in water through the plasma membrane *every second.*

It is important to note that the higher the concentration of solutes is, the lower the concentration of water will be, since the solutes take up some of the space that would otherwise be occupied by water molecules. Therefore, the direction of osmosis will be from a more dilute solution (where there is more water) to a more concentrated one (where there is less water). In figure 4.22*a*, for example, we see a chamber divided by a selectively permeable membrane. Side A contains a mixture of large particles that cannot pass through the membrane pores—a *nonpermeating* solute such as albumin (egg white protein). Side B contains distilled water. Since albumin takes up some of the space on side A, water is more concentrated in B than in A, and it diffuses down its concentration gradient from B to A (fig. 4.22*b*). This is because more water molecules strike the membrane per second on side B than they do on side A, where water is less abundant.

Under these conditions, the water level on side B would fall and the level on side A would rise. It might seem as if this would go on indefinitely until side B dried up. This would not happen, however, because as water accumulated on side A, it would become heavier and heavier and exert more hydrostatic pressure on that side of the membrane. This would cause some filtration of water from side A back to B. At some point, the rate of filtration would equal the rate of osmosis, water would pass through the membrane equally in both directions, and net osmosis would slow down and stop. At this point, an equilibrium (balance between opposing forces) would exist. The hy-

22. *osm* = push, thrust + *osis* = condition, process

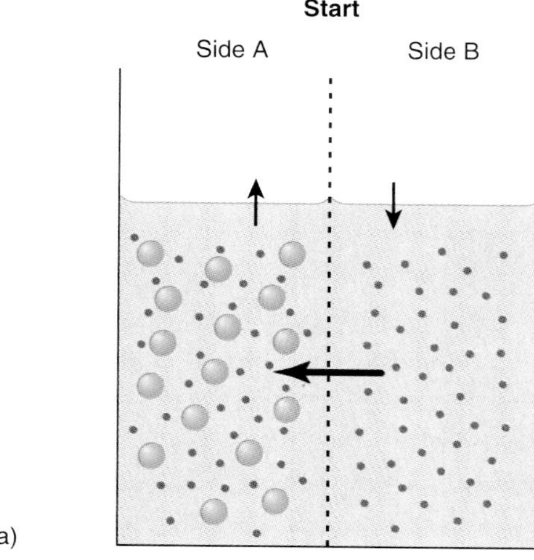

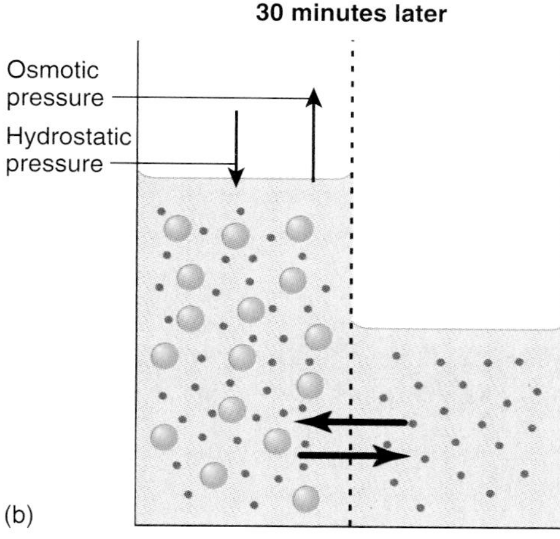

Figure 4.22 Osmosis. The *dashed line* represents a selectively permeable membrane dividing the chamber in half. The large particles on side A represent any solute, such as albumin, too large to pass through the membrane. The small particles are water molecules. (*a*) Water diffuses from side B, where it is relatively concentrated, to side A, where it is less concentrated. Fluid level rises in side A and falls in side B. (*b*) Net diffusion stops when the weight (hydrostatic pressure) of the fluid in side A balances the osmotic pressure. At this point, water passes at equal rates from A to B by filtration and from B to A by osmosis. The two processes are then in equilibrium.

drostatic pressure on side A that would stop osmosis is called **osmotic pressure.** The more solute there is on side A, the greater its osmotic pressure will be.

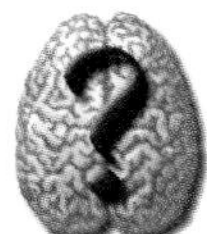

Think About It

If the albumin concentration on side A were half what it was in the original experiment, would the fluid on that side reach a higher or lower level than before? Explain.

The equilibrium between osmosis and filtration will be an important consideration as we study fluid exchange through blood capillaries in chapter 20. Blood plasma also contains albumin. In the preceding discussion, side A is analogous to the bloodstream and side B is analogous to the tissue fluid surrounding the capillaries (although tissue fluid is not distilled water). Water leaves the capillaries by filtration, but this is approximately balanced by water moving back into the capillaries by osmosis.

Osmolarity and Tonicity

The osmotic concentration of the body fluids has such a great effect on cellular function that it is important to understand the units in which it is measured. One **osmole** is 1 mole of dissolved particles. If a solute does not ionize in water, then 1 mole of the solute yields 1 osmole (osm) of dissolved particles. A solution of 1 molar (1 M) glucose, for example, is also 1 osm/L. If a solute does ionize, it yields two or more dissolved particles in solution. A 1 M solution of NaCl, for example, contains 1 mole of sodium ions and 1 mole of chloride ions per liter. Both ions affect osmosis and must be separately counted in a measure of osmotic concentration. Thus, 1 M NaCl = 2 osm/L. Calcium chloride ($CaCl_2$) would yield three ions if it dissociated completely (one Ca^{2+} and two Cl^-), so 1 M $CaCl_2$ = 3 osm/L.

Osmolality is the number of osmoles of solute *per kilogram of water,* and **osmolarity** is the number of osmoles *per liter of solution.* Most clinical calculations are based on osmolarity, since it is easier to measure the volume of a solution than the weight of water it contains. At the concentrations of human body fluids, there is less than 1% difference between osmolality and osmolarity, and the two terms are nearly interchangeable. All body fluids and many clinical solutions are mixtures of many chemicals. The osmolarity of such a solution is the total osmotic concentration of all of its dissolved particles.

A concentration of 1 osm/L is substantially higher than we find in most body fluids, so physiological concentrations are usually expressed in terms of **milliosmoles per liter** (mOsm/L) (1 mOsm/L = 10^{-3} osm/L). Blood plasma, tissue fluid, and intracellular fluid measure about 300 mOsm/L.

Tonicity is the ability of a solution to affect the fluid volume and pressure within a cell. If a solute cannot pass through a plasma membrane, but remains more concentrated on one side of the membrane than on the other, it will cause osmosis. A **hypotonic**[23] solution has a lower concentration of nonpermeating solutes than the intracellular fluid (ICF). Cells in a hypotonic solution absorb water, swell, and may burst open (*lyse*) (fig. 4.23*a*). Distilled water is the extreme example; given to a person intravenously, it would

23. *hypo* = less + *ton* = tension

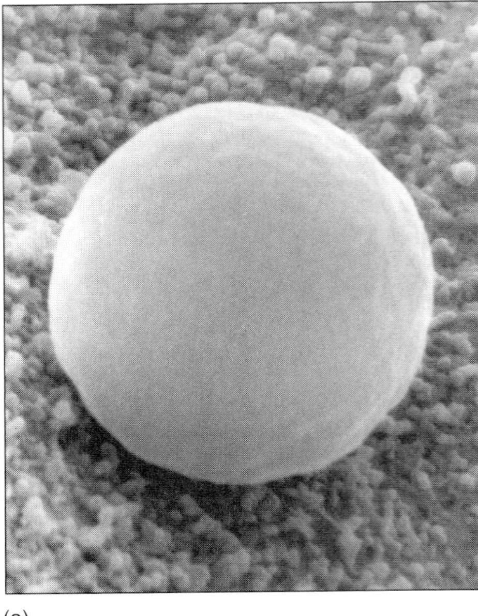
(a)

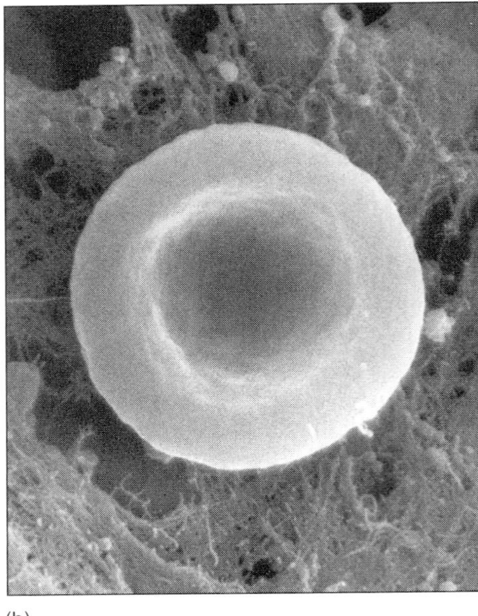
(b)

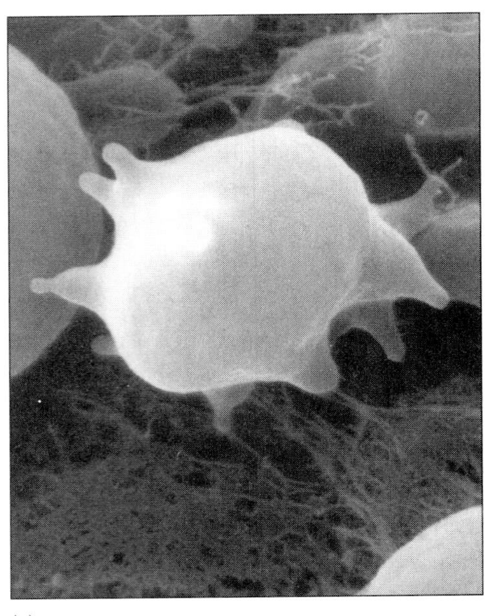
(c)

Figure 4.23 **Effects of Tonicity on Red Blood Cells (RBCs).** (*a*) In a hypotonic medium such as distilled water, RBCs absorb water, swell, and may burst. (*b*) In an isotonic medium such as 0.9% NaCl, RBCs gain and lose water at equal rates and maintain their normal, concave disc shape. (*c*) In a hypertonic medium such as 2% NaCl, RBCs lose more water than they gain and become shrunken and spiky (crenated).

cause lysis of the blood cells. A **hypertonic**[24] solution is one with a higher concentration of nonpermeating solutes than the ICF. It causes cells to lose water and shrivel (*crenate*) (fig. 4.23*c*). Their membranes may tear, and the cells may die. In **isotonic**[25] solutions, the total concentration of nonpermeating solutes is the same as in the ICF—hence, isotonic solutions cause no change in cell volume or shape (fig. 4.23*b*).

It is essential for cells to be in a state of osmotic equilibrium with the fluid around them, and this requires that the ECF have the same concentration of nonpermeating solutes as the ICF. Intravenous fluids given to patients are usually isotonic solutions, but hypertonic or hypotonic fluids are sometimes given for special purposes. A 0.9% solution of NaCl, called *normal saline*, is isotonic to human blood cells.

It is important to note that osmolarity and tonicity are not the same. Urea, for example, is a small organic molecule that easily penetrates plasma membranes. If cells are placed in 300 mOsm/L urea, urea diffuses into them (down its concentration gradient), water follows by osmosis, and the cells swell and burst. Thus, 300 mOsm/L urea is not isotonic to the cells. Sodium chloride, by contrast, penetrates plasma membranes poorly. In 300 mOsm/L NaCl, there is little change in cell volume; this solution is isotonic to cells.

24. *hyper* = more + *ton* = tension
25. *iso* = equal + *ton* = tension

Before You Go On

Answer the following questions to test your understanding of the preceding section:

14. What is the importance of filtration to human physiology?
15. How is net diffusion different from simple diffusion? How are they similar?
16. How does osmosis help to maintain blood volume?
17. Define *osmolarity* and *tonicity*, and explain the difference between them.
18. Define *hypotonic*, *isotonic*, and *hypertonic*, and explain why these concepts are important in clinical practice.

Membrane Transport II: Carrier-Mediated and Bulk Transport

Objectives

When you have completed this section, you should be able to

- describe the functional properties of carrier proteins in the plasma membrane;
- describe some membrane transport processes that use carrier proteins; and
- describe the forms of bulk transport through plasma membranes.

The processes of membrane transport described up to this point do not necessarily require a cell membrane; they can occur just as well through artificial membranes. We now, however, come to processes for which a cell membrane is essential, because they employ proteins in the membrane to carry solutes through it. Thus, the next two processes are cases of **carrier-mediated transport.** The carrier proteins act like enzymes in some ways: the solute is a ligand that binds to a specific receptor site on the carrier, like a substrate binding to the active site of an enzyme. The carrier exhibits **specificity** for a certain ligand, just as an enzyme does for its substrate. A glucose carrier, for example, cannot transport fructose. Carriers also exhibit **saturation;** as the solute concentration rises, its rate of transport through a membrane increases, but only up to a point. When every carrier is occupied, adding more solute cannot make the process go any faster. The carriers are saturated, no more are available to handle the increased demand, and the rate of transport levels off at a level called **transport maximum (T_m)** (fig. 4.24). As we will see later in the book, the transport maximum explains why glucose appears in the urine of people with diabetes mellitus. An important difference between a membrane carrier and an enzyme is that carriers do not chemically change their ligands; they simply pick them up on one side of the membrane and release them, unchanged, on the other. The two types of carrier-mediated transport are *facilitated diffusion* and *active transport.*

Facilitated Diffusion

Facilitated[26] diffusion is the carrier-mediated transport of a solute through a membrane *down its concentration gradient.* It is a passive transport process; that is, it does not consume ATP. It transports solutes such as glucose that cannot pass through the membrane unaided. The solute attaches to a binding site on the carrier, then the carrier changes confirmation and releases the solute on the other side of the membrane (fig. 4.25).

Active Transport

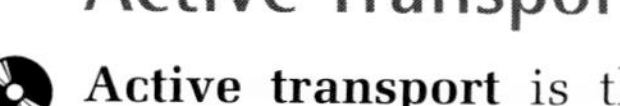

Active transport is the carrier-mediated transport of a solute through a cellular membrane *up its concentration gradient,* using energy provided by ATP. All cells continually pump out calcium ions by active transport, for example, even though Ca^{2+} is already more concentrated in the extracellular fluid than within the cell. Active transport also enables cells to absorb amino acids that are already more concentrated in the cytoplasm than in the ECF.

Active transport is a four-step process: (1) the carrier binds to a ligand; (2) the carrier is phosphorylated with a phosphate (P_i) group donated by ATP; (3) the carrier changes conformation; and (4) the carrier releases the

26. *facil* = easy

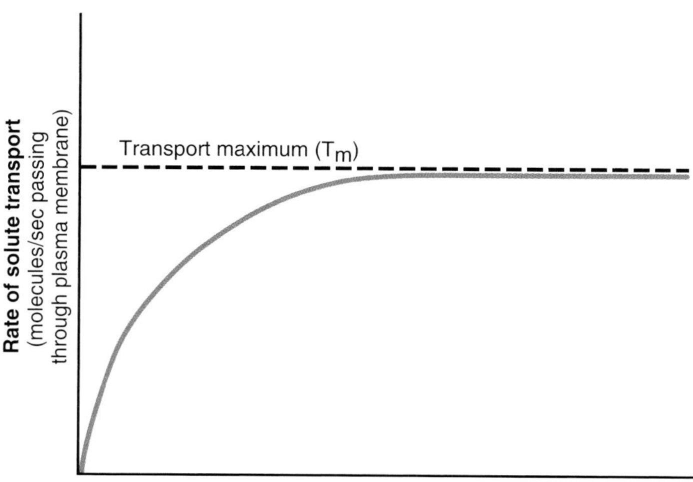

Figure 4.24 Saturation of a Membrane Carrier. Up to a point, increasing the solute concentration increases the rate of transport through a membrane. At the transport maximum (T_m), however, all carrier proteins are busy and cannot transport the solute any faster, even if more solute is added.

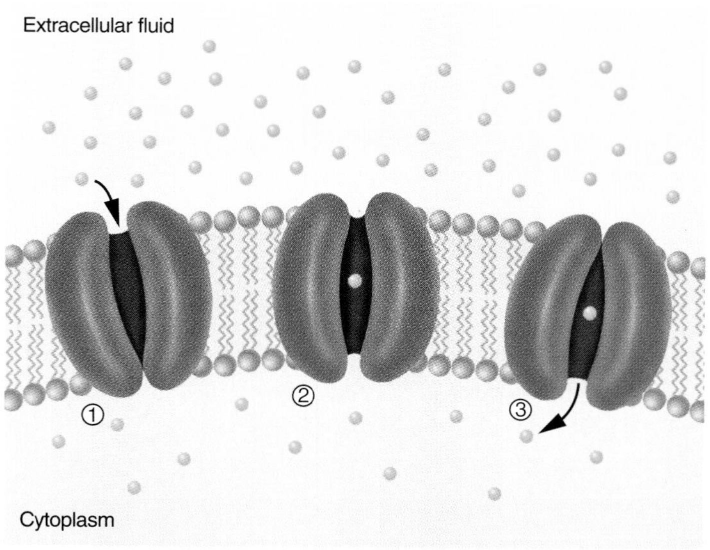

Figure 4.25 Facilitated Diffusion. (*1*) A solute particle enters the channel of a membrane protein (carrier). (*2*) The solute binds to a receptor site on the carrier and the carrier changes conformation. (*3*) The carrier releases the solute on the other side of the membrane. Note that the solute moves down its concentration gradient.

ligand to the other side of the membrane, releases the P_i group, and returns to its original conformation, ready to repeat the process.

A prominent example of active transport is the **sodium-potassium (Na^+-K^+) pump,** also known as **Na^+-K^+ ATPase** because of its consumption of ATP. The Na^+-K^+ pump follows the basic steps just described, but with some unusual characteristics (fig. 4.26). It binds three sodium ions simultaneously on the cytoplasmic side of the membrane, releases these to the extracellular fluid (ECF), binds two potassium ions simultaneously from the ECF, and releases these into the cell. Each cycle of the pump consumes one ATP and exchanges three Na^+ for two K^+. This keeps the K^+ concentration higher and the Na^+ concentration lower within the cell than in the ECF. These ions continually leak through the membrane, and the Na^+-K^+ pump compensates like bailing out a leaky boat.

Lest you question the importance of the Na^+-K^+ pump, about half of the calories you burn each day are used just to carry out this process. Beyond compensating for a leaky plasma membrane, the Na^+-K^+ pump has at least four functions:

1. **Regulation of cell volume.** Certain anions are "fixed" within the cell and cannot penetrate the plasma membrane. These anions, such as proteins and phosphates, attract and retain cations. If there were nothing to correct for it, the retention of these ions would cause the cell to swell by osmosis and eventually rupture. Cellular swelling, however, stimulates the Na^+-K^+ pumps. Since each cycle of the pump removes one ion more than it brings in, the pumps are part of a negative feedback loop that reduces ion concentration, osmolarity, and cellular swelling.
2. **Secondary active transport.** The Na^+-K^+ pump maintains a steep concentration gradient of Na^+ and K^+ between one side of the membrane and the other. Like water behind a dam that can be tapped to generate electricity, this gradient has a high potential energy that can drive other processes. Since Na^+ has a high concentration outside the cell, it tends to diffuse back in. Some cells exploit this to move other solutes into the cell. In kidney tubules, for example, the cells have Na^+-K^+ pumps in the basal membrane that remove Na^+ from the

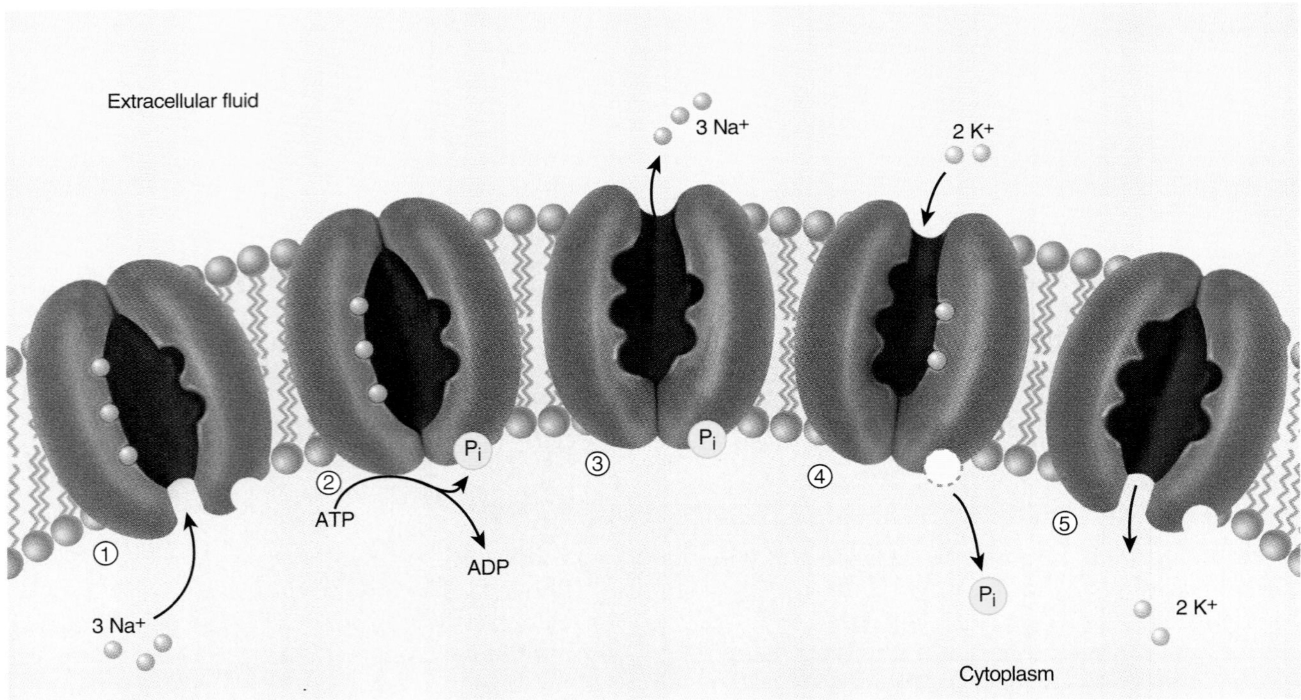

Figure 4.26 The Sodium-Potassium Pump. This membrane protein moves Na^+ and K^+ in opposite directions through the plasma membrane by active transport. (1) Three sodium ions from the cytoplasm bind to receptor sites on the carrier. (2) The carrier hydrolyzes ATP and becomes phosphorylated. (3) The carrier changes conformation and releases the Na^+ into the extracellular fluid. (4) Two potassium ions bind to receptor sites on the carrier, and the carrier releases the phosphate group. (5) The carrier returns to its original conformation and releases K^+ into the cytoplasm.

cytoplasm and maintain a low intracellular Na^+ concentration. In the apical membrane, the cells have carriers that simultaneously bind Na^+ and glucose and carry both into the cell at once by facilitated diffusion (fig. 4.27). These carriers are called *sodium-glucose transport* proteins *(SGLTs).* By exploiting the tendency of Na^+ to diffuse down its concentration gradient into these cells, the cells are also able to absorb glucose and prevent it from being wasted in the urine. The SGLT in itself does not consume ATP, but it does depend on the ATP-consuming Na^+-K^+ pumps at the base of the cell. We can say that glucose is absorbed by secondary active transport, as opposed to the primary active transport carried out by the Na^+-K^+ pump.

3. **Heat production.** When the weather turns chilly, we not only turn up the furnace in our home but also the "furnace" in our body. Thyroid hormone stimulates cells to produce more Na^+-K^+ pumps. As these pumps consume ATP, they release heat, compensating for the body heat we lose to the cold air around us.
4. **Maintenance of a membrane potential.** All living cells have an electrical charge difference called the *resting membrane potential* across the plasma membrane. Like the two poles of a battery, the inside of the membrane is negatively charged and the outside is positively charged. This difference stems from the unequal distribution of ions on the two sides of the membrane, maintained by the Na^+-K^+ pump. The membrane potential is essential to the function of nerve and muscle cells, as we will study in later chapters.

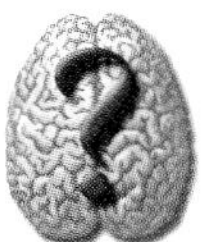

Think About It

An important characteristic of proteins is their ability to change configuration in response to the binding or dissociation of a ligand (see chapter 3). Explain how this characteristic is essential to carrier-mediated transport.

Cotransport

Cotransport[27] is a process in which a membrane carrier transports two solutes through a membrane simultaneously (or at least within one cycle of action). The Na^+-K^+ pump and sodium-glucose transport protein are examples of cotransport systems. When both solutes move in the same direction, as sodium and glucose do in the SGLT, the

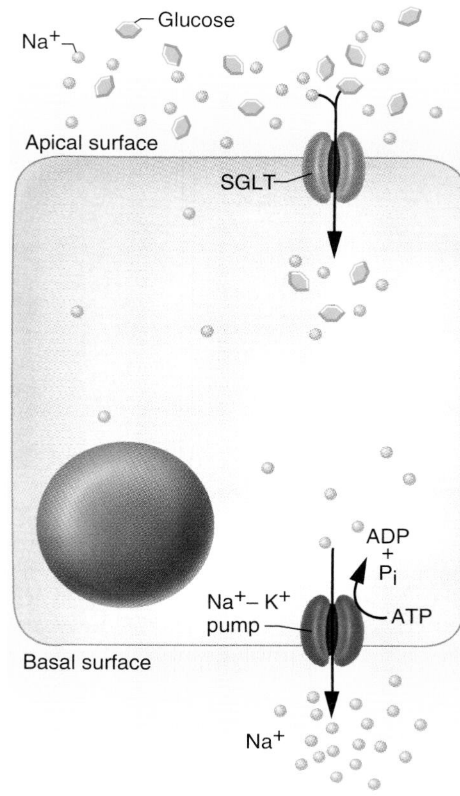

Figure 4.27 **Secondary Active Transport.** At the basal surface of the cell, an Na^+-K^+ pump removes sodium ions from the cytoplasm, maintaining a low sodium concentration within the cell. At the apical surface, sodium enters the cell by facilitated diffusion, following its concentration gradient. It can gain entry to the cell only by binding to a carrier, the sodium-dependent glucose transporter (SGLT), which simultaneously binds and transports glucose. The SGLT does not consume ATP, but does depend on the ATP-consuming pump at the base of the cell.

carrier is called a **symport.**[28] When the solutes move in opposite directions, as in the Na^+-K^+ pump, the carrier is called an **antiport.**[29]

Bulk Transport

So far, we have considered processes that move from one to a few particles through the plasma membrane at a time. We conclude this chapter with processes called **bulk transport** that move large particles, droplets of fluid, or numerous molecules at once through the membrane. Processes that bring matter into a cell are called **endocytosis**[30]

27. *co* = together + *trans* = across + *port* = carry

28. *sym* = together + *port* = carry
29. *anti* = opposite + *port* = carry
30. *endo* = into + *cyt* = cell + *osis* = process

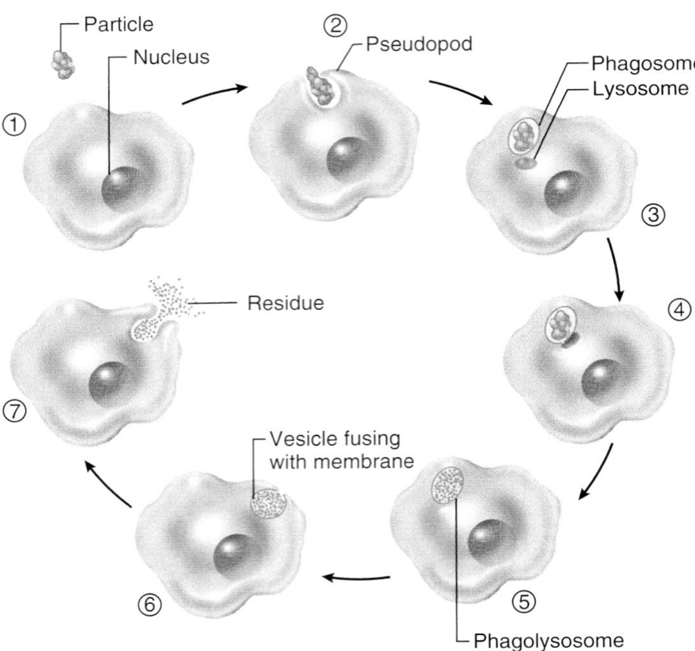

Figure 4.28 Phagocytosis, Intracellular Digestion, and Exocytosis. (*1*) A phagocytic cell encounters a particle of foreign matter. (*2*) The cell surrounds the particle with its pseudopods. (*3*) The particle is phagocytized and becomes incorporated into a phagosome. (*4*) A lysosome fuses with the phagosome and forms a phagolysosome. (*5*) Enzymes from the lysosome digest the foreign matter. (*6*) The phagolysosome fuses with the plasma membrane. (*7*) The indigestible residue is voided by exocytosis.

(EN-doe-sy-TOE-sis), and those that release material from a cell are called **exocytosis**[31] (EC-so-sy-TOE-sis).

There are three forms of endocytosis: *phagocytosis, fluid-phase pinocytosis,* and *receptor-mediated endocytosis.* **Phagocytosis**[32] (FAG-oh-sy-TOE-sis), or "cell eating," is the process of engulfing particles such as bacteria, dust, and cellular debris. Neutrophils (a class of white blood cells), for example, protect the body from infection by phagocytizing and killing bacteria. A neutrophil spends most of its life crawling about in the connective tissues by means of blunt footlike extensions called **pseudopods**[33] (SOO-doe-pods). When a neutrophil encounters a bacterium, it surrounds it with its pseudopods and traps it in a *phagosome*[34] (*vacuole*)—a "bubble" in the cytoplasm surrounded by a unit membrane (fig. 4.28). A lysosome subsequently merges with the vacuole, converting it to a *phagolysosome,* and contributes enzymes that destroy the invader. Several other kinds of phagocytic cells are described in chapter 21. In general, phagocytosis is a way of keeping the tissues free of debris and infectious microorganisms. Some cells called *macrophages* (literally, "big eaters") phagocytize the equivalent of 25% of their own volume per hour.

Fluid-phase pinocytosis[35] (PIN-oh-sy-TOE-sis), or "cell drinking," is the process of taking in droplets of extracellular fluid (ECF) containing molecules of some use to the cell. While phagocytosis occurs in only a few specialized cells, pinocytosis occurs in all human cells. The process begins as the plasma membrane becomes dimpled, or caved in, at points. These pits soon separate from the surface membrane and form small membrane-bounded **pinocytotic vesicles** in the cytoplasm. The vesicles contain droplets of the ECF with whatever molecules happened to be there.

Receptor-mediated endocytosis[36] (fig. 4.29) is more selective. It enables a cell to take in specific molecules from the ECF with a minimum of unnecessary fluid. Molecules in the ECF bind to specific receptors on the plasma membrane. The receptors then cluster together and the membrane sinks in at this point, creating a pit coated with a peripheral membrane protein called *clathrin.*[37] The pit soon pinches off to form a *clathrin-coated vesicle* in the cytoplasm. Clathrin may serve as an "address label" on the coated vesicle that directs it to an appropriate destination in the cell, or it may inform other structures in the cell what to do with the vesicle.

One example of receptor-mediated endocytosis is the uptake of *low-density lipoproteins (LDLs)*—protein-coated droplets of cholesterol and other lipids in the blood (see chapter 26). The thin endothelial cells that line our blood vessels have LDL receptors on their surfaces and absorb LDLs in clathrin-coated vesicles. Inside the cell, the LDL is freed from the vesicle and metabolized, and the membrane with its receptors is recycled to the cell surface. Most of what we know about receptor-mediated endocytosis comes from studies of a hereditary disease called *familial hypercholesterolemia,* which dramatically illustrates the significance of this process to our cardiovascular health (see clinical insight 4.4).

Endothelial cells also imbibe insulin by receptor-mediated endocytosis. Insulin is too large a molecule to pass through channels in the plasma membrane, yet it must somehow get out of the blood and reach the surrounding cells if it is to have any effect. Endothelial cells take up insulin by receptor-mediated endocytosis, transport the vesicles across the cell, and release the insulin on the other side, where tissue cells await it. Such transport of a substance across a cell is called **transcytosis**[38] (fig. 4.30).

31. *exo* = out of + *cyt* = cell + *osis* = process
32. *phago* = eating + *cyt* = cell + *osis* = process
33. *pseudo* = false + *pod* = foot
34. *phago* = eating + *some* = body
35. *pino* = drinking + *cyt* = cell + *osis* = process
36. *endo* = into + *cyt* = cell + *osis* = process
37. *clathr* = lattice + *in* = protein
38. *trans* = across + *cyt* = cell + *osis* = process

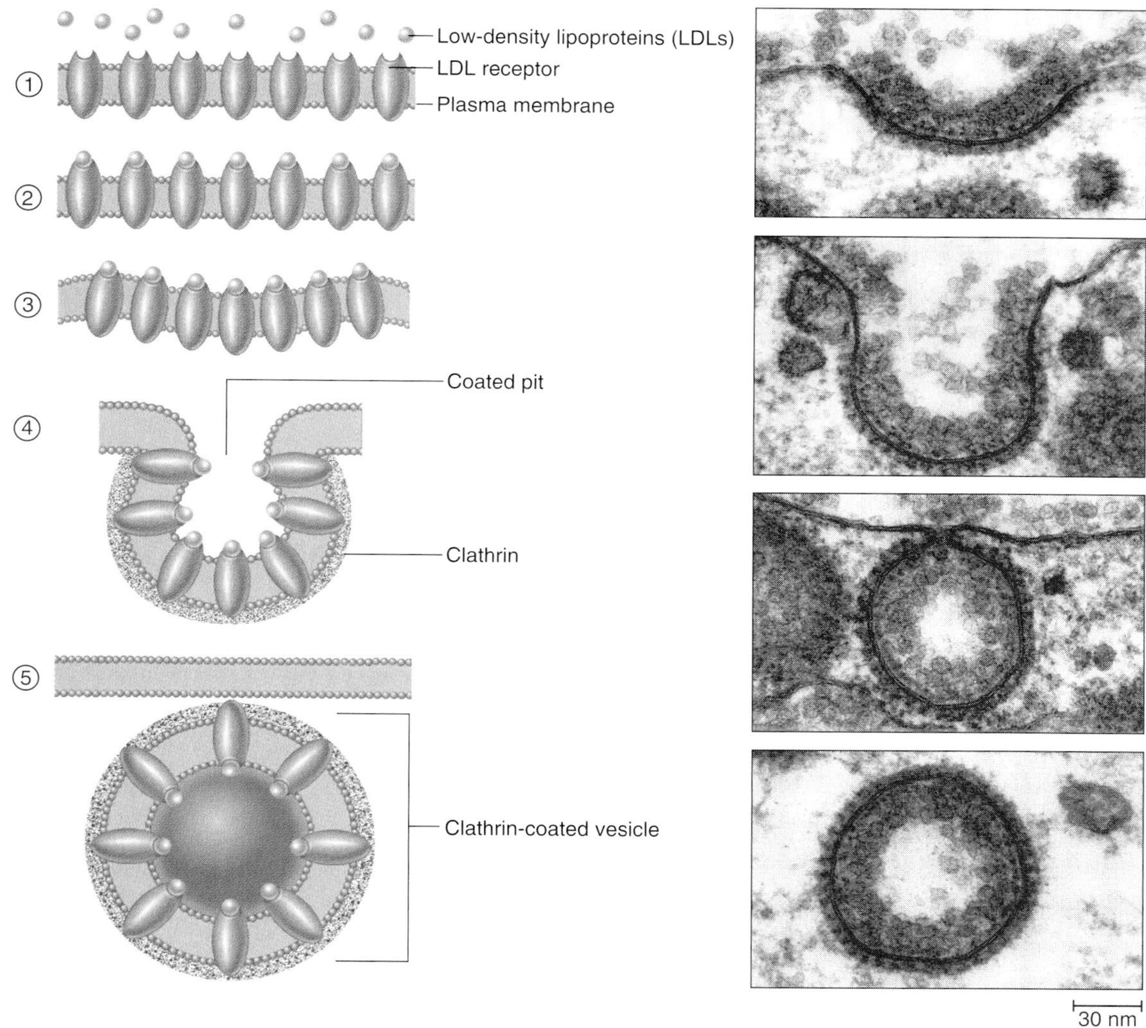

Figure 4.29 **Receptor-Mediated Endocytosis.** (*1*) Low-density lipoproteins (LDLs) are suspended in the blood. The endothelial cells of blood vessels have LDL receptor proteins in their plasma membranes. (*2*) LDLs bind to their receptors. (*3*) LDL-bearing receptors become clustered together. (*4*) The plasma membrane sinks inward at that point and forms a clathrin-coated pit. (*5*) The pit separates from the membrane and becomes a clathrin-coated vesicle containing concentrated LDLs. The electron micrographs show stages 3 to 5 of this process.

Receptor-mediated endocytosis is not always to our benefit; hepatitis, polio, and AIDS viruses "trick" our cells into taking them in by receptor-mediated endocytosis.

Exocytosis (fig. 4.31) is the process of discharging material from a cell. It occurs, for example, when endothelial cells release insulin to the tissue fluid, breast cells secrete milk, gland cells release hormones, and sperm cells release enzymes for penetrating an egg. It bears a superficial resemblance to endocytosis in reverse. A secretory vesicle in the cell migrates to the surface and "docks" on peripheral proteins of the plasma membrane. These proteins pull the membrane inward and create a dimple that eventually fuses with the vesicle and allows it to release its contents.

The question might occur to you, If endocytosis continually takes away bits of plasma membrane to form intracellular vesicles and vacuoles, why doesn't the membrane grow smaller and smaller? Another purpose of exocytosis, however, is to replace plasma membrane that has been removed by endocytosis or become damaged or worn out. Plasma membrane is continually recycled from the cell surface into the cytoplasm and back to the surface.

Table 4.5 summarizes the mechanisms of transport we have discussed.

Clinical Insight 4.4

Familial Hypercholesterolemia

The significance of LDL receptors and receptor-mediated endocytosis is dramatically illustrated by a hereditary disease called *familial hypercholesterolemia.*[39] People with this disease have an abnormally low number of LDL receptors. Their cells therefore absorb less cholesterol than normal, and the cholesterol remains in the blood. Their blood cholesterol levels may be as high as 1,200 mg/dL, compared to a normal level of about 200 mg/dL. People who inherit the gene from both parents typically have heart attacks before the age of 20 (sometimes even in infancy) and seldom survive beyond the age of 30.

39. *familial* = running in the family; *hyper* = above normal + cholesterol + *emia* = blood condition

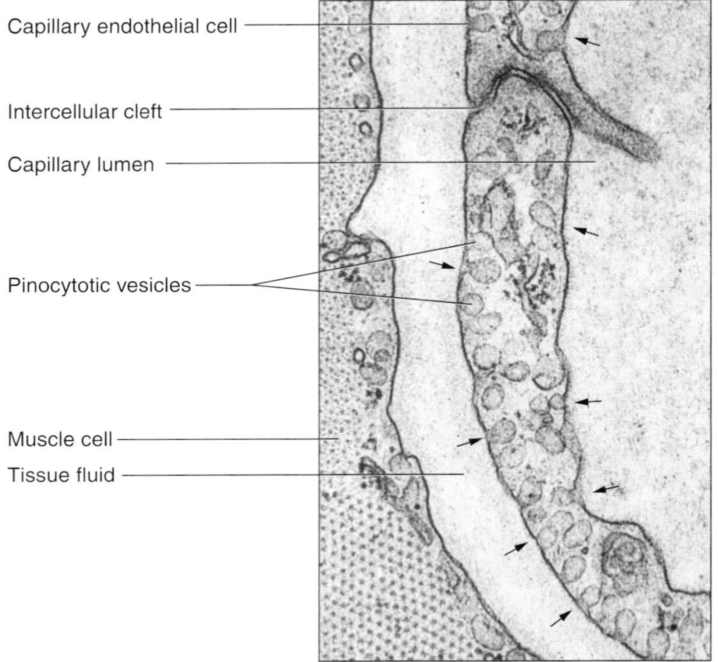

Figure 4.30 **Transcytosis.** An endothelial cell of a capillary imbibes droplets of blood plasma at sites indicated by *arrows* along the *right.* This forms pinocytotic vesicles, which the cell transports to the other side. Here, it releases the contents by exocytosis at sites indicated by *arrows* along the *left* side of the cell. This process is especially active in muscle capillaries and transfers a significant amount of blood albumin into the tissue fluid.

Before You Go On

Answer the following questions to test your understanding of the preceding section:

19. Compare and contrast facilitated diffusion and active transport.
20. How does the Na^+-K^+ pump exchange sodium ions for potassium ions across the plasma membrane? What are some purposes served by this pump?
21. How does phagocytosis differ from pinocytosis?
22. How does fluid-phase pinocytosis differ from receptor-mediated endocytosis?
23. Describe the process of exocytosis. What are some of its purposes?

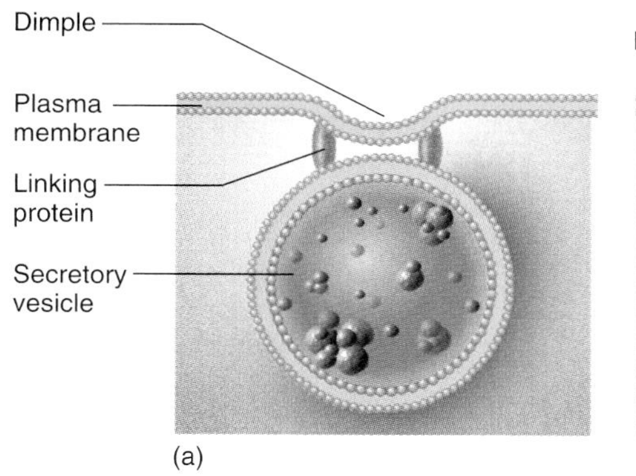

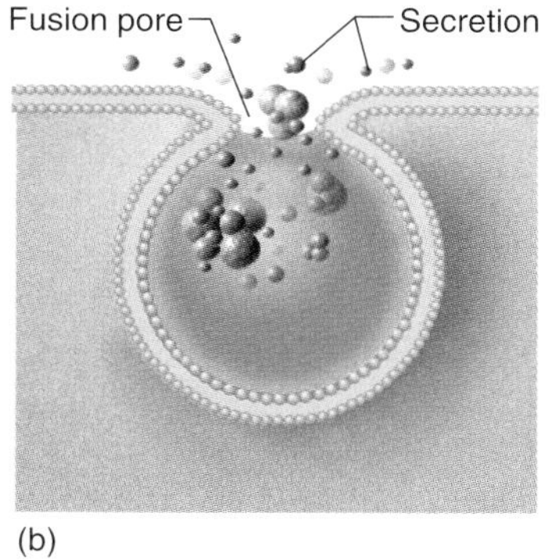

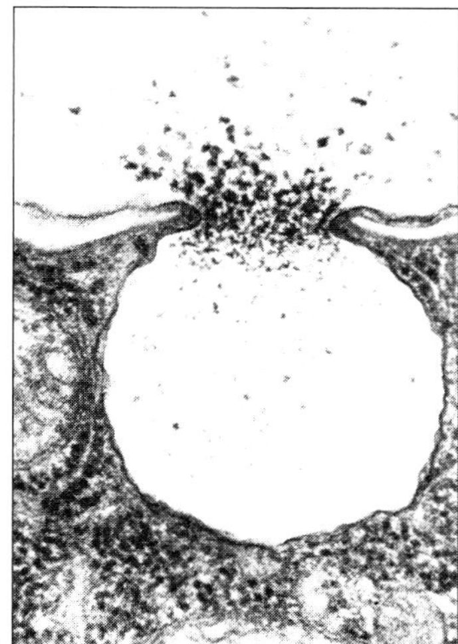

Figure 4.31 Exocytosis. (*a*) A secretory vesicle approaches the plasma membrane and docks on it by means of linking proteins. The plasma membrane caves in at that point to meet the vesicle. (*b*) The plasma membrane and vesicle unite to form a fusion pore through which the vesicle contents are released. (*c*) Electron micrograph of exocytosis.

Table 4.5 Methods of Membrane Transport

Filtration	Movement of particles through a selectively permeable membrane as a result of hydrostatic pressure
Passive Transport	Movement of particles down a concentration gradient as a result of their own kinetic energy; involves no expenditure of ATP
Simple Diffusion	Diffusion of particles through water or air or through a living or artificial membrane, down their concentration gradient, without the aid of membrane carriers
Osmosis	Simple diffusion of water through a selectively permeable membrane
Facilitated Diffusion	Diffusion of particles through a selectively permeable membrane, down their concentration gradient, with the aid of membrane carriers
Active Transport	Transport of particles through a selectively permeable membrane, up their concentration gradient, with the aid of membrane carriers (pumps) and with the expenditure of ATP
Primary Active Transport	Direct transport of solute particles by an ATP-using membrane pump
Secondary Active Transport	Transport of solute particles by a carrier that does not in itself use ATP but depends on concentration gradients produced by active transport pumps
Cotransport	Any carrier-mediated transport mechanism that moves two different solutes simultaneously, or at least within one cycle of transport
Symport	A cotransport that moves different solutes in the same direction through a plasma membrane
Antiport	A cotransport that moves different solutes in opposite directions through a plasma membrane
Bulk Transport	Movement of fluid and particles through a plasma membrane by way of vacuoles or vesicles of plasma membrane; consumes ATP
Endocytosis	Bulk transport of particles or fluid into a cell
Phagocytosis	Process of engulfing large particles by means of pseudopods; "cell eating"
Fluid-Phase Pinocytosis	Process of imbibing droplets of extracellular fluid in which the plasma membrane sinks in and pinches off small vesicles containing droplets of fluid
Receptor-Mediated Endocytosis	Process in which specific solute molecules bind to receptors on the plasma membrane and are then taken into the cell in clathrin-coated vesicles with a minimal amount of fluid
Exocytosis	Process of eliminating material from a cell by means of a vesicle approaching the cell surface, fusing with the plasma membrane, and expelling its contents; used to release cell secretions, replace worn-out plasma membrane, and replace membrane that has been internalized by endocytosis

Historical Insight 4.5

Evolution of the Microscope

The simple microscope of Antony van Leeuwenhoek magnified specimens about 200 times, and modern electron microscopes magnify up to 600,000 times. Yet the most significant aspect of the history of microscopy has been improvements not in magnification, but in *resolution*, the ability to see detail.

Light Microscope (LM)

The modern light microscope was developed in the early nineteenth century, chiefly by Carl Zeiss. By using many lenses in combination, he corrected for the deficiencies of each, including the spherical and chromatic aberrations of Robert Hooke's seventeenth-century microscopes (see chapter 1). The light microscope is still the most routinely used type. It illuminates the specimen with visible light, which has wavelengths from about 400 nm at the violet end of the spectrum to 750 nm at the red end. This range of wavelengths limits the resolution of the light microscope to about 200 nm (0.2 μm) and its useful magnification to about 1,200 times. There are many varieties of light microscope, including the fluorescence microscope used to photograph figure 4.19*b*.

Transmission Electron Microscope (TEM)

One of the laws of optics is that resolution improves when objects are viewed with radiation of shorter wavelength. This is the reason electron microscopes use an electron beam instead of visible light. In the transmission electron microscope, a beam of electrons, focused with electromagnets, passes through a specimen and strikes a fluorescent screen or photographic film. The electrons have a wavelength of 0.005 nm (compare visible light), and the TEM can resolve biological objects as small as 0.5 nm. This makes it possible to obtain meaningful detail at magnifications up to 600,000 times, which is sufficient to see proteins, nucleic acids, and other large molecules.

In order to absorb electrons, specimens for the TEM must be sliced ultrathin with diamond knives and stained with heavy metals such as osmium. All images seen by the TEM are really images of where these metal ions have collected in the specimen, not images of the specimen itself. One disadvantage of the TEM is that living cells usually cannot be studied. Since the electrons would be scattered by air, the specimen must be placed in a vacuum chamber. Recent advances have somewhat overcome this limitation with TEMs that operate at ultrahigh voltages (1–2 million volts) and use wet specimen chambers. The TEM produces two-dimensional black-and-white images, but TEM photomicrographs are often colorized for instructional purposes.

Scanning Electron Microscope (SEM)

The scanning electron microscope uses a specimen coated with vaporized metal ions (usually gold). An electron beam strikes the specimen and discharges secondary electrons from the metal coating. These electrons then produce an image on a fluorescent screen or photographic film. The SEM yields less resolution than the TEM and is used at lower magnification. It can be used only to view the surface of an object; it does not see through an object like the LM and TEM. Cell interiors can be viewed, however, by a freeze-fracture method in which a cell is frozen, cracked open, and then coated with gold vapor. An advantage of the SEM is that it produces dramatic three-dimensional images that are sometimes more informative than the flat images of the LM and TEM (see, for example, the contrasting views of nuclear pores in figure 5.1 in the next chapter).

Micrographs produced by the LM, TEM, and SEM are used throughout this book. Figure 1 shows red blood cells photographed by all three methods. See if you can identify at least two other TEM photomicrographs and two other SEM photomicrographs in this chapter.

Red blood cells

(a) 10.0 µm

Red blood cells

Blood vessel

(b) 10.0 µm

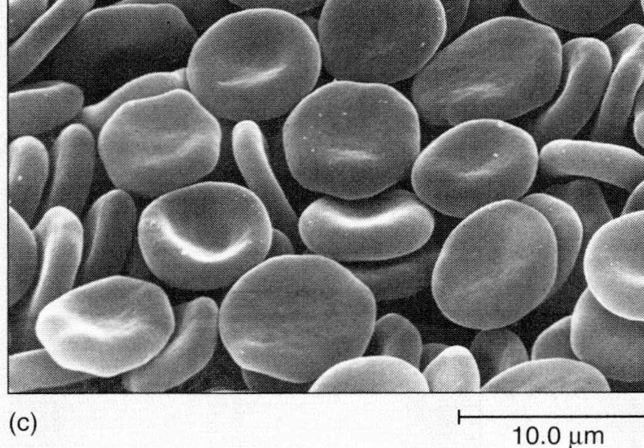

(c) 10.0 µm

Figure 1 Images of Red Blood Cells Produced by Three Kinds of Microscopes. (*a*) The light microscope (LM), (*b*) the transmission electron microscope (TEM), and (*c*) the scanning electron microscope (SEM).

Chapter Review

Online Learning Center

Study resources for this chapter are available at www.mhhe.com/saladin/

Study Outline

Concepts of Cellular Structure (p. 104)

1. Development of the cell theory
   a. Studies of Hooke and Schwann
   b. The spontaneous generation issue
   c. Tenets of the cell theory
2. Cell shapes
   a. Terms for cell shapes
   b. Surfaces of epithelial cells
3. Cell size
   a. Sizes of human cells
   b. Limitations on size
4. An evolving perspective on cells
   a. Microscopes and resolution
   b. Internal organization of a cell

The Cell Surface (p. 108)

1. The plasma membrane
   a. Membrane lipids
      - Phospholipid bilayer
      - Cholesterol
      - Glycolipids
   b. Membrane proteins
      - Integral proteins
      - Peripheral proteins
   c. Functions of membrane proteins
      - Receptors
      - Second messenger systems
      - Enzymes
      - Channel proteins
         •• Ligand-regulated gates
         •• Voltage-regulated gates
         •• Mechanically regulated gates
      - Carriers and pumps
      - Motor molecules
      - Cell-identity markers
      - Cell-adhesion molecules
   d. Second messengers
      - Role of G protein
      - Role of adenylate cyclase and cAMP
      - Role of kinases
2. The glycocalyx
   a. Composition
   b. Functions
3. Microvilli, cilia, and flagella
   a. Microvilli
      - Functions
      - Core of actin filaments
   b. Cilia
      - Mode of movement
      - Importance of saline layer
      - Structure of axoneme
      - Basal body
      - Sensory cilia
   c. Flagella

The Cytoplasm (p. 116)

1. Organelles
   a. The nucleus
   b. Endoplasmic reticulum
      - Rough ER
      - Smooth ER
   c. Ribosomes
   d. Golgi complex
      - Golgi vesicles
      - Secretory vesicles
   e. Lysosomes
   f. Peroxisomes
   g. Mitochondria
   h. Centrioles
2. The cytoskeleton
   a. Microfilaments
   b. Intermediate filaments
   c. Microtubules
3. Inclusions

Membrane Transport I: Filtration and Simple Diffusion (p. 123)

1. Selective permeability
2. Filtration
3. Simple diffusion
   a. Concentration gradients
   b. Determinants of diffusion rate
4. Osmosis
   a. Mechanism
   b. Equilibrium between osmosis and filtration
   c. Osmotic pressure
5. Osmolarity and tonicity

Membrane Transport II: Carrier-Mediated and Bulk Transport (p. 128)

1. Carrier-mediated transport
   a. Definition
   b. Properties of membrane carriers
      - Specificity
      - Saturation and the transport maximum
2. Facilitated diffusion
   a. Solutes move down gradient
   b. No ATP consumed
3. Active transport
   a. Criteria
      - Solutes move up gradient
      - ATP is consumed
   b. Steps in process
      - Carrier binds ligand
      - ATP phosphorylates carrier
      - Carrier changes conformation
      - Carrier releases ligand
   c. The sodium-potassium pump
      - Process
      - Functions
         •• Regulation of cell volume
         •• Secondary active transport
         •• Heat production
         •• Maintenance of membrane potential
4. Cotransport
   a. Symports
   b. Antiports
5. Bulk transport
   a. Endocytosis
      - Phagocytosis
      - Fluid-phase pinocytosis
      - Receptor-mediated endocytosis
      - Transcytosis
   b. Exocytosis

Selected Vocabulary

cytology 104
squamous cell 104
polygonal cell 104
stellate cell 104
cuboidal cell 104
columnar cell 104
spheroid cell 104
ovoid cell 104
discoid cell 104
fusiform cell 104
fibrous cell 104
basal cell surface 104
lateral cell surface 105
apical cell surface 105
cytoplasm 105
transmission electron microscope 105
cytosol 105
intracellular fluid 105
extracellular fluid 105
resolution 105
organelle 105
scanning electron microscope 108
unit membrane 108
plasma membrane 109
fluid-mosaic theory 109
integral protein 110
peripheral protein 110
receptor 110
channel protein 111
gate 111
ligand-regulated gate 111
voltage-regulated gate 111
mechanically regulated gate 111
pump 111
motor molecule 111
cell-identity marker 112
cell-adhesion molecule 112
second messenger 112
G protein 112
adenylate cyclase 112
glycocalyx 113
microvillus 113
brush border 113
terminal web 113
cilium 115
power stroke 115
recovery stroke 115
axoneme 115
9 + 2 structure 115
basal body 116
dynein arm 116
flagellum 116
nucleus 116
nuclear envelope 116
cisterna 117
rough endoplasmic reticulum 117
smooth endoplasmic reticulum 117
ribosome 118
Golgi complex 119
Golgi vesicle 119
secretory vesicle 119
lysosome 119
autophagy 119
autolysis 119
peroxisome 119
mitochondrion 120
crista 120
mitochondrial matrix 120
centriole 120
centrosome 121
cytoskeleton 121
microfilament 121
membrane skeleton 121
intermediate filament 121
microtubule 123
inclusion 123
selectively permeable membrane 125
filtration 125
hydrostatic pressure 125
simple diffusion 125
concentration gradient 125
osmosis 126
osmotic pressure 127
osmole 127
osmolality 127
osmolarity 127
milliosmoles per liter 127
tonicity 127
hypotonic 127
hypertonic 128
isotonic 128
carrier-mediated transport 129
specificity 129
saturation 129
transport maximum (T_m) 129
facilitated diffusion 129
active transport 129
sodium-potassium pump 130
Na^+-K^+ ATPase 130
secondary active transport 130
cotransport 131
symport 131
antiport 131
bulk transport 131
endocytosis 131
exocytosis 132
phagocytosis 132
pseudopod 132
fluid-phase pinocytosis 132
pinocytotic vesicle 132
receptor-mediated endocytosis 132
transcytosis 132

Testing Your Recall

1. The clear, structureless gel in a cell is its
   a. nucleoplasm.
   b. protoplasm.
   c. cytoplasm.
   d. neoplasm.
   e. cytosol.

2. The Na^+-K^+ pump is
   a. a peripheral protein.
   b. an integral protein.
   c. a G protein.
   d. a glycolipid.
   e. a phospholipid.

3. Which of the following processes could occur only in the plasma membrane of a living cell?
   a. facilitated diffusion
   b. simple diffusion
   c. filtration
   d. active transport
   e. osmosis

4. Cells specialized for absorption of matter from the ECF are likely to show an abundance of
   a. lysosomes.
   b. microvilli.
   c. mitochondria.
   d. secretory vesicles.
   e. ribosomes.

5. Osmosis is a special case of
   a. pinocytosis.
   b. carrier-mediated transport.
   c. active transport.
   d. facilitated diffusion.
   e. simple diffusion.

6. Membrane carriers resemble enzymes except for the fact that carriers
   a. are not proteins.
   b. do not have binding sites.
   c. are not selective for particular ligands.
   d. change conformation when they bind a ligand.
   e. do not chemically change their ligands.

7. The cotransport of glucose derives energy from
   a. a Na^+ concentration gradient.
   b. the glucose being transported.
   c. a Ca^{2+} gradient.
   d. the membrane voltage.
   e. body heat.

8. The function of cAMP in a cell is
   a. to activate a G protein.
   b. to remove phosphate groups from ATP.
   c. to activate kinases.
   d. to bind to the first messenger.
   e. to add phosphate groups to enzymes.

9. Most cellular membranes are made by
   a. the nucleus.
   b. the cytoskeleton.
   c. enzymes in the peroxisomes.
   d. the endoplasmic reticulum.
   e. replication of existing membranes.
10. Matter can leave a cell by any of the following means *except*
    a. active transport.
    b. fluid-phase pinocytosis.
    c. an antiport.
    d. simple diffusion.
    e. exocytosis.
11. Most human cells are 10 to 15 ___________ in diameter.
12. When a hormone cannot enter a cell, it activates the formation of a/an ___________ inside the cell.
13. ___________ channels in the plasma membrane open or close in response to changes in the electrical charge difference across the membrane.
14. The force exerted on a membrane by water is called ___________.
15. A concentrated solution that causes a cell to shrink is ___________ to the cell.
16. Fusion of a secretory vesicle with the plasma membrane, and release of the vesicle's contents, is called ___________.
17. Two organelles that are surrounded by a double unit membrane are the ___________ and the ___________.
18. Liver cells can detoxify alcohol with two organelles, the ___________ and ___________.
19. An ion gate in the plasma membrane that opens or closes when a chemical binds to it is called a/an ___________.
20. The space enclosed by the unit membrane of the Golgi complex and endoplasmic reticulum is called the ___________.

Answers in appendix B

Testing Your Comprehension

1. If someone bought a saltwater fish in a pet shop and put it in a freshwater aquarium at home, what would happen to the fish's cells? What would happen if someone put a freshwater fish in a saltwater aquarium? Explain.
2. A farmer's hand and forearm are badly crushed in a hay bailer. Upon hospital examination, his blood potassium level is found to be highly abnormal. Would you expect it to be higher or lower than normal? Explain.
3. Many children worldwide suffer from a severe deficiency of dietary protein. As a result, they have very low levels of blood albumin. How do you think this affects the water content and volume of their blood? Explain.
4. It is often said that mitochondria make energy for a cell. Why is this statement false? What law of thermodynamics bears on this question?
5. Kartagener syndrome is a hereditary disease in which dynein arms are lacking from the axonemes of cilia and flagella. Predict the effect of Kartagener syndrome on a man's ability to father a child. Predict its effect on his respiratory health. Explain both answers.

Answers at the Online Learning Center

5 CHAPTER

Genetics and Cellular Function

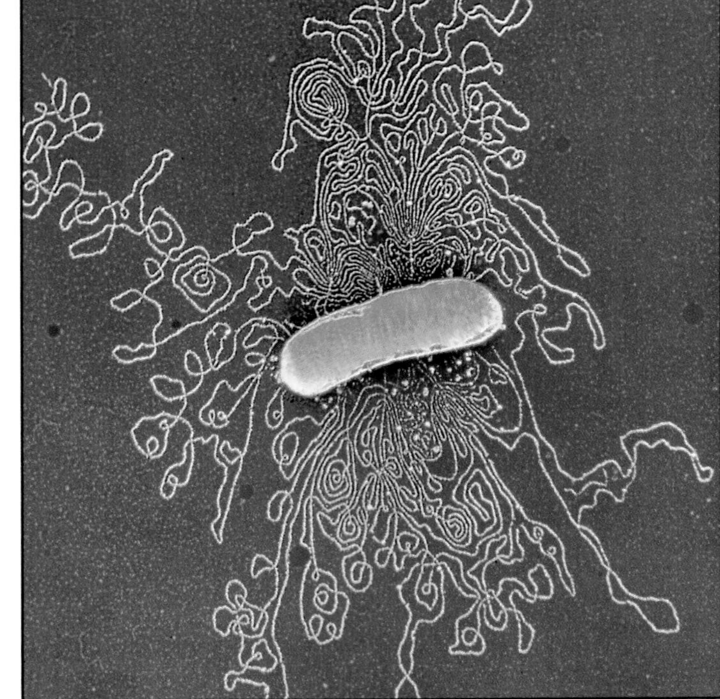

A single DNA molecule spilling from a ruptured bacterial cell (TEM)

Chapter Outline

Insights

Study with Your ESP CD-ROM

Refer to your ESP Tutorial CD-ROM for helpful animations, activities, and quizzes related to the material covered in this chapter. The following are suggested areas to go to on the CD and are presented according to "Unit/Topic":

- Levels of Organization/Protein Synthesis
- Levels of Organization/Cell Division

Brushing Up

To understand this chapter, it is important that you understand or brush up on the following concepts:

- Levels of protein structure (p. 90)
- Functions of proteins (p. 92)
- Ribosomes, rough endoplasmic reticulum, and Golgi complex (pp. 116–119)
- Centrioles and microtubules (pp. 120, 123)
- Exocytosis (p. 133)

Some of the basic ideas of heredity have been known since antiquity, but a scientific understanding of how traits are passed from parent to offspring began with the Austrian monk Gregor Mendel (1822–84) and his famous experiments on garden peas. In the early twentieth century, the importance of Mendel's work was realized and chromosomes were first seen with the microscope. **Cytogenetics** now uses techniques of cytology and microscopy to study chromosomes and their relationship to hereditary traits. **Molecular genetics** uses the techniques of biochemistry to study the structure and replication of DNA and its control of cellular function. In this chapter, we bring together some of the findings of molecular genetics, cytogenetics, and Mendelian heredity to explore what the genes are, how they regulate cellular function, and how they are passed on when cells divide and people reproduce. Some basic concepts of heredity are also introduced to provide a foundation for understanding concepts ranging from color blindness to blood types in the chapters that follow.

The Nucleus and Nucleic Acids

Objectives

When you have completed this section, you should be able to

- describe the structure of the cell nucleus;
- describe how DNA is organized in the nucleus; and
- compare the structures and functions of DNA and RNA.

With improvements in the microscope, nineteenth-century cytologists saw that the nucleus divides in preparation for cell division, and they came to regard the nucleus as the most likely center of heredity. We now recognize that it contains deoxyribonucleic acid (DNA), which directly or indirectly regulates all cellular form and function.

Nuclear Structure

Most cells have a single nucleus, but there are exceptions. Mature red blood cells have none; they are **enucleate.** A few cell types are **multinucleate**—having 2 to 50 nuclei—including some liver cells, skeletal muscle cells, and certain bone-dissolving and platelet-producing cells. The nucleus is usually spheroid to elliptical in shape and averages about 5 μm in diameter. It is surrounded by a **nuclear envelope** consisting of two parallel unit membranes (fig. 5.1). The envelope is perforated with **nuclear pores,** about 30 to 100 nm in diameter, formed by a ring-shaped complex of proteins. These proteins regulate molecular traffic through the envelope and act like a rivet to hold the two unit membranes together. Hundreds of molecules—especially ribonucleic acid (RNA) and proteins—pass through the nuclear pores every minute. Coming into the nucleus are raw materials for DNA and RNA synthesis, enzymes that are made in the cytoplasm but function in the nucleus, and hormones that activate certain genes. Going the other way, RNA is made in the nucleus but leaves to perform its job in the cytoplasm.

The material within the nucleus is called **nucleoplasm.** The DNA and its associated proteins constitute the

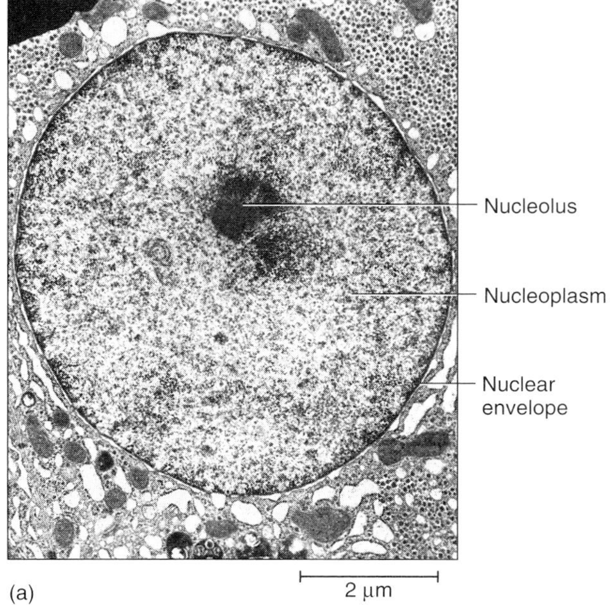

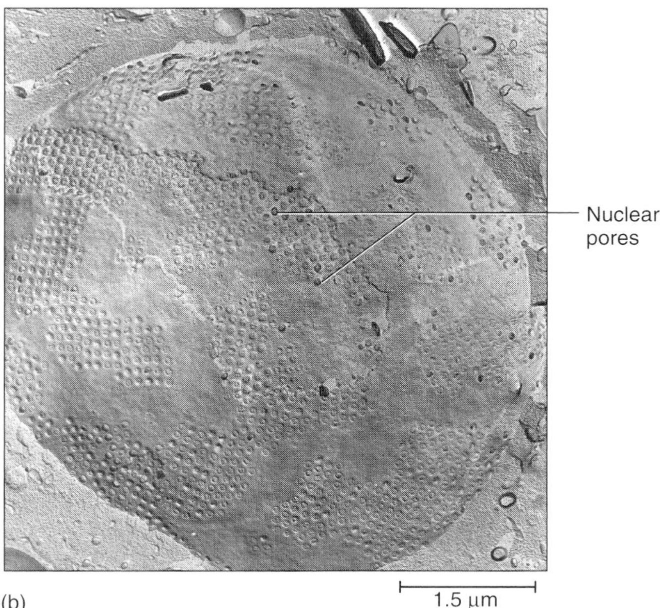

Figure 5.1 The Nucleus. (*a*) Transmission electron micrograph showing the nuclear envelope and internal structure. (*b*) Scanning electron micrograph of the nuclear envelope showing the high density of nuclear pores.

Historical Insight 5.1

Miescher and the Discovery of DNA

Swiss biochemist Johann Friedrich Miescher (1844–95) was one of the first scientists intent on identifying the hereditary material in nuclei. In order to isolate nuclei with minimal contamination, Miescher chose to work with cells that have large nuclei and very little cytoplasm. At first he chose white blood cells extracted from the pus in used bandages from a hospital; later, he used the sperm of salmon (probably more agreeable to work with than used bandages!). Miescher isolated an acidic substance rich in phosphorus, which he named *nuclein.* His student, Richard Altmann, later called it *nucleic acid*—a term we now use to include RNA. Miescher correctly guessed that "nuclein" (DNA) was the hereditary matter of the cell, but he was unable to provide strong evidence for this conjecture, and his work was harshly criticized. He died of tuberculosis at the age of 51.

chromatin[1] (CRO-muh-tin)—fine threadlike matter dispersed throughout the nucleoplasm. The nucleus usually contains one or more dark-staining masses called **nucleoli** (singular, *nucleolus*). Subunits of the cytoplasmic ribosomes are produced here and then transported into the cytoplasm.

Organization of the Chromatin

A human cell usually has 46 molecules of DNA with a total length of slightly over 2 m. Each DNA molecule is 2 nm in diameter and an average of 44 mm long. To put this in

1. *chromat* = color

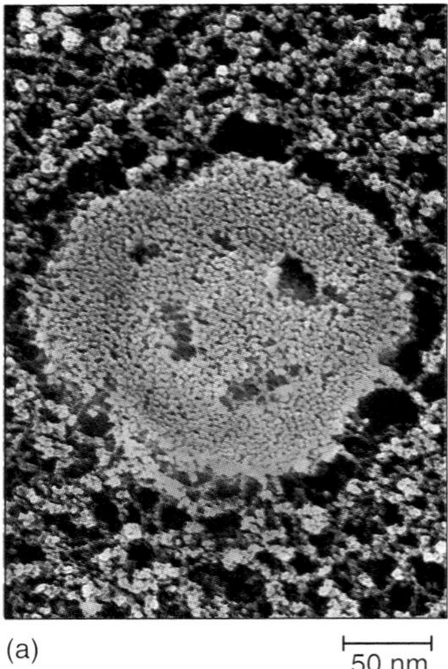

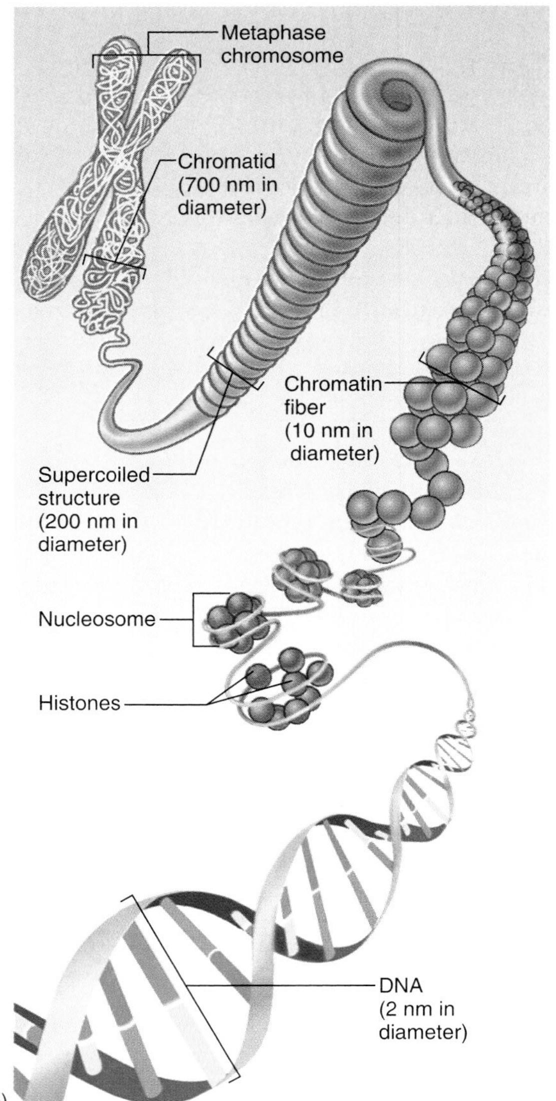

Figure 5.2 Chromatin Structure. (*a*) Electron micrograph of the chromatin. Each beadlike swelling is a nucleosome, and the thin strands connecting them are DNA. (*b*) The coiling of chromatin and its relationship to the histones. Supercoiling beyond the 10 nm level occurs only during mitosis.

perspective, if a DNA molecule were the thickness of a telephone pole (20 cm, or 8 in), it would reach about 4,400 km (2,700 mi) into space—far higher than the orbits of satellites and space shuttles. Imagine trying to make a pole 20 cm thick and 4,400 km long without breaking it! The problem for a cell is even greater. It has 46 DNA molecules packed together in a single nucleus, and it has to make an exact copy of every one of them and distribute these equally to its two daughter cells when the cell divides. Keeping the DNA organized and intact is a tremendous feat.

Molecular biology and high-resolution electron microscopy have provided some insight into how this task is accomplished. Chromatin looks like a granular thread (fig. 5.2*a*). The granules, called **nucleosomes,** consist of a cluster of eight proteins called **histones,** with the DNA molecule wound around the cluster. Histones serve as spools that protect and organize the DNA. Other nuclear proteins called **nonhistones** seem to provide structural support for the chromatin and regulate gene activity.

Winding DNA around the nucleosomes makes the chromatin shorter and more compact, but chromatin also has higher orders of structure. The "granular thread," about 10 nm wide, further coils and thickens to about 30 nm. When a cell prepares to undergo division, the chromatin supercoils into a fiber about 200 nm in diameter (fig. 5.2*b*). Thus, the 2 m of DNA in each cell becomes shortened and compacted in an orderly way that prevents tangling and breakage without interfering with genetic function.

DNA Structure and Function

Nucleic acids are polymers of **nucleotides** (NEW-clee-oh-tides). A nucleotide consists of a sugar, a phosphate group, and a single- or double-ringed **nitrogenous** (ny-TRODJ-eh-nus) **base.** Three bases—**cytosine (C), thymine (T),** and **uracil (U)**—have a single carbon-nitrogen ring and are classified as *pyrimidines* (py-RIM-ih-deens). The other two bases—**adenine (A)** and **guanine (G)**—have double rings and are classified as *purines* (fig. 5.3). DNA contains the bases C, T, A, and G, whereas RNA contains C, U, A, and G.

The structure of DNA resembles a ladder (fig. 5.4*a*). Each sidepiece is a backbone composed of phosphate groups and the sugar *deoxyribose* alternating with each other. The steplike connections between the backbones are pairs of nitrogenous bases. Imagine this as a soft rubber ladder that you can twist, so that the two backbones become entwined to resemble a spiral staircase. This is analogous to the shape of the DNA molecule, described as a *double helix.*

The nitrogenous bases face the inside of the helix and hold the two backbones together with hydrogen bonds. Across from a purine on one backbone, there is a pyrimidine on the other. A given purine cannot arbitrarily bind

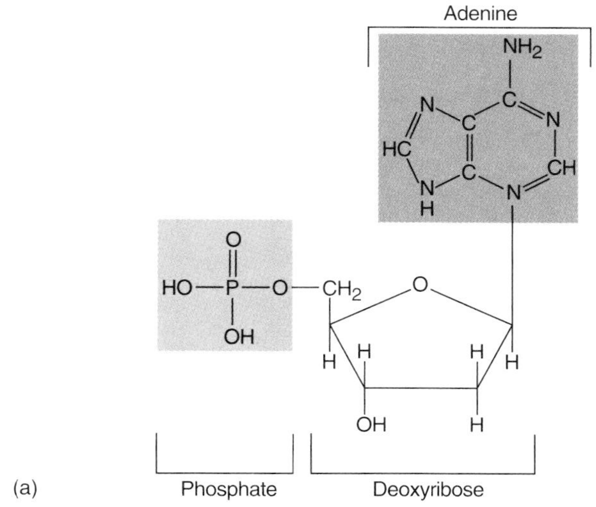

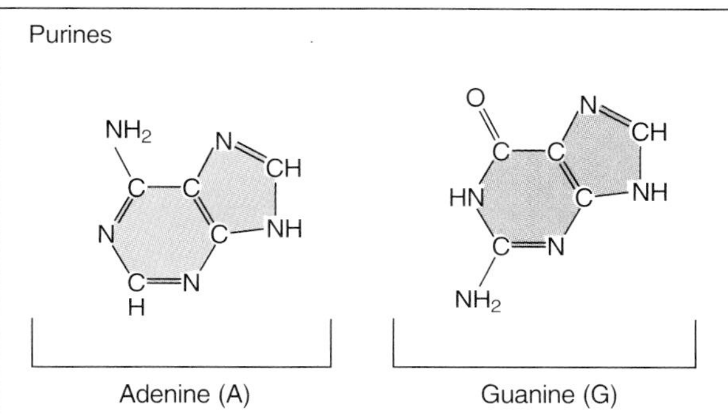

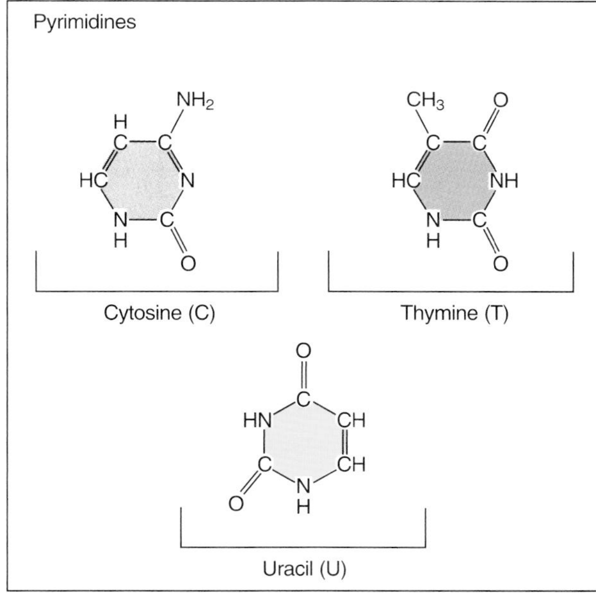

Figure 5.3 Nucleotides and Nitrogenous Bases. (*a*) The structure of a nucleotide, one of the monomers of DNA and RNA. In RNA, the sugar is ribose. (*b*) The five nitrogenous bases found in DNA and RNA nucleotides.

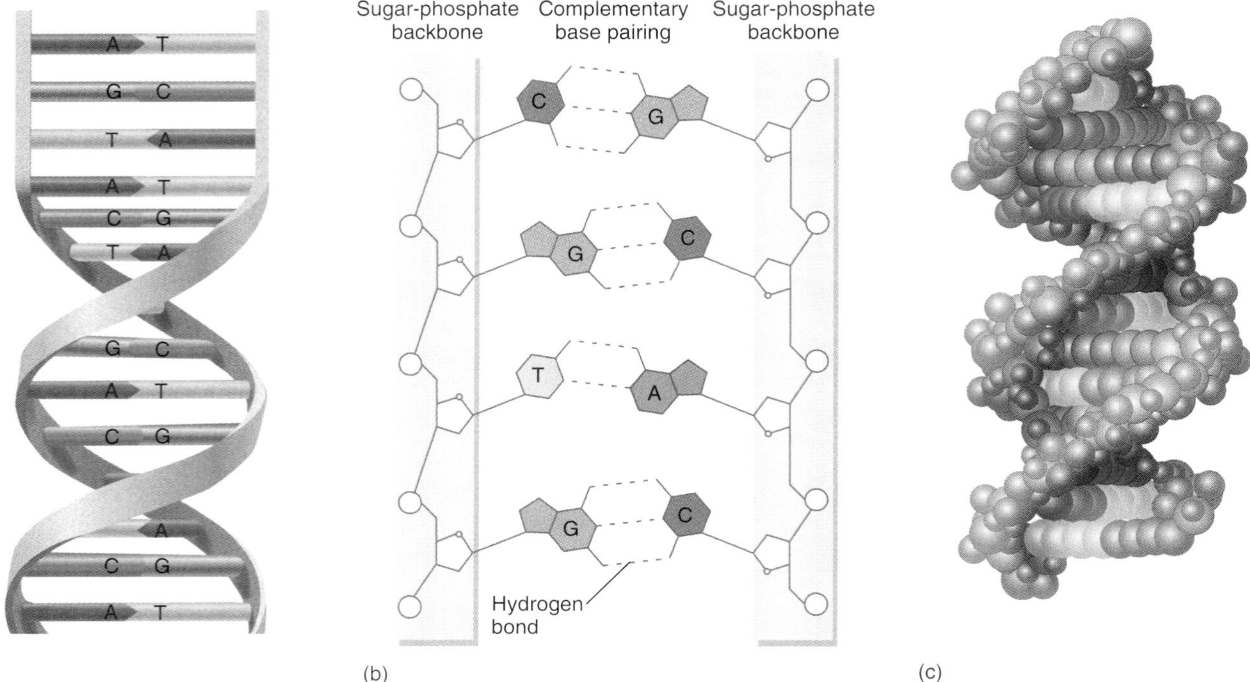

Figure 5.4 DNA Structure. (*a*) The "twisted ladder" structure. The two sugar-phosphate backbones twine around each other while complementary bases *(colored bars)* face each other on the inside of the double helix. (*b*) A small segment of DNA showing the composition of the backbone and complementary pairing of the nitrogenous bases. (*c*) A molecular space-filling model of DNA giving some impression of its actual geometry.

to just any pyrimidine. Adenine and thymine form two hydrogen bonds with each other, and guanine and cytosine form three, as shown in figure 5.4*b*. Therefore, wherever there is an A on one backbone, there is a T across from it, and every C is paired with a G. A–T and C–G are called the **base pairs.** The fact that one strand governs the base sequence of the other is called the **law of complementary base pairing.** It enables us to predict the base sequence of one strand if we know the sequence of the complementary strand. The pairing of each small, single-ringed pyrimidine with a large, double-ringed purine gives the DNA molecule its uniform 2 nm width.

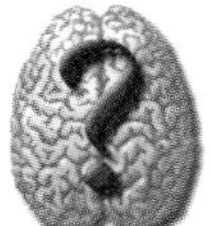

Think About It

What would be the base sequence of the DNA strand across from ATTGACTCG? If a DNA molecule were known to be 20% adenine, predict its percentage of cytosine and explain your answer.

The essential function of DNA is to serve as a code for the structure of proteins synthesized by a cell. A **gene** is a sequence of DNA nucleotides that codes for one polypeptide. The next section of this chapter explains in detail how the genes direct polypeptide synthesis. All the genes of one person are called the **genome** (JEE-nome); geneticists estimate that we each have 60,000 to 150,000 genes. These account for only 3% of our DNA, however; the other 97% does not code for anything. Some of this serves important organizing roles in the chromatin, and much of it is useless "junk DNA" that has accumulated over the course of human evolution.

RNA Structure and Function

DNA directs the synthesis of proteins by means of its smaller cousins, the ribonucleic acids (RNAs). There are three types of RNA: *messenger RNA (mRNA), ribosomal RNA (rRNA),* and *transfer RNA (tRNA).* Their individual roles will be described shortly. For now we consider what they have in common and how they differ from DNA (table 5.1).

The most significant difference is that while DNA is a double helix, RNA consists of only one nucleotide chain, not held together by complementary base pairs except in certain regions of tRNA where the molecule folds back on itself. RNA molecules are also much smaller, ranging from about 70 to 90 bases in tRNA to slightly over 10,000 bases in the largest mRNA. DNA, by contrast, may be over a billion base pairs long. The sugar in RNA

Historical Insight 5.2

Discovery of the Double Helix

The components of DNA were known by 1900—the sugar, phosphate, and bases—but the technology did not exist then to determine how they were put together. The credit for that discovery went mainly to James Watson and Francis Crick in 1953. The events surrounding their discovery of the double helix represent one of the most dramatic stories of modern science—the subject of many books and a movie. When Watson and Crick came to share a laboratory at Cambridge University in 1951, both had barely begun their careers. Watson, age 23, had just completed his Ph.D. in the United States, and Crick, 11 years older, was a doctoral candidate. Yet the two were about to become the most famous molecular biologists of the twentieth century, and the discovery that won them such acclaim came without a single laboratory experiment of their own.

Others were fervently at work on DNA, including Rosalind Franklin and Maurice Wilkins at King's College in London. Using a technique called X-ray diffraction, Franklin had determined that DNA had a repetitious helical structure with sugar and phosphate on the outside of the helix. Without her permission, Wilkins showed one of Franklin's best X-ray photographs to Watson. Watson said, "The instant I saw the picture my mouth fell open and my pulse began to race." It provided a flash of insight that allowed the Watson and Crick team to beat Franklin to the goal. They were quickly able to piece together a scale model from cardboard and sheet metal that fully accounted for the known geometry of DNA. They rushed a paper into print in 1953 describing the double helix, barely mentioning the importance of Franklin's two years of painstaking X-ray diffraction work in unlocking the mystery of life's most important molecule.

For this discovery, Watson, Crick, and Wilkins shared the Nobel Prize in 1962. Nobel Prizes are awarded only to the living, and in the final irony of her career, Rosalind Franklin had died in 1958, at the age of 37, of a cancer possibly induced by the X rays that were her window on DNA architecture.

(a)

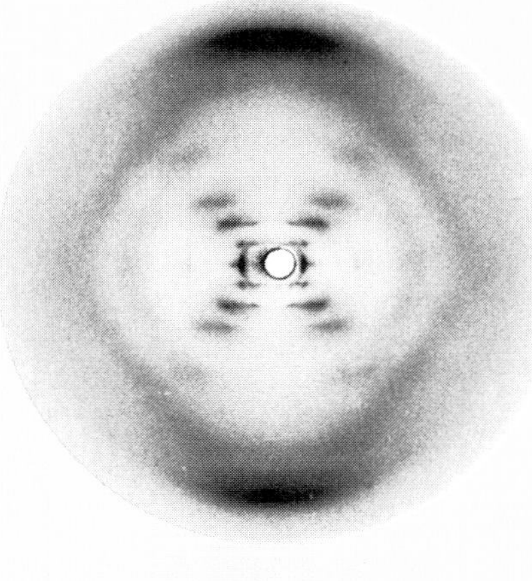
(b)

(c)

Figure 1 **Discoverers of the Double Helix.** (*a*) Rosalind Franklin (1920–58), whose painstaking X-ray diffraction photographs revealed important information about the basic geometry of DNA. (*b*) One of Franklin's X-ray photographs. (*c*) James Watson (1928–) *(left)* and Francis Crick (1916–) *(right)* with their model of the double helix.

is ribose instead of deoxyribose, and one of the pyrimidines of DNA, thymine, is replaced by uracil (U) in RNA (see fig. 5.3).

The essential function of RNA is to interpret the code in DNA and direct the synthesis of proteins. RNA works in the cytoplasm, while DNA remains safely behind in the nucleus, "giving orders" from there. This process is described in the next section of this chapter.

Before You Go On

Answer the following questions to test your understanding of the preceding section:

1. What is the difference between DNA and chromatin?
2. What are the three components of a nucleotide? Which component varies from one nucleotide to another in DNA?
3. What two factors govern the pattern of base pairing in DNA?
4. Summarize the differences between DNA and RNA.

Table 5.1 Comparison of DNA and RNA

Feature	DNA	RNA
Sugar	Deoxyribose	Ribose
Nitrogenous bases	A, T, C, G	A, U, C, G
Number of nucleotide chains	Two (double helix)	One
Number of nitrogenous bases	10^8 to 10^9 base pairs	70 to 10^4 unpaired bases
Site of action	Functions in nucleus; cannot leave	Leaves nucleus; functions in cytoplasm
Function	Codes for synthesis of RNA and protein	Carries out the instructions in DNA; assembles proteins

Protein Synthesis and Secretion

Objectives

When you have completed this section, you should be able to

- define *genetic code* and describe how DNA codes for protein structure;
- describe the process of assembling amino acids to form a protein;
- explain what happens to a protein after its amino acid sequence has been synthesized; and
- explain how DNA indirectly regulates the synthesis of nonprotein molecules.

Everything a cell synthesizes ultimately results from the action of enzymes, and DNA directs the synthesis of all enzymes as well as other proteins. Cells, of course, synthesize many other substances as well—glycogen, fat, phospholipids, steroids, pigments, and so on. There are no genes for these cell products, but their synthesis depends on enzymes that are coded for by the genes. For example, even though a cell of the testis has no genes for testosterone, testosterone synthesis is indirectly under genetic control (fig. 5.5). Since testosterone strongly influences such behaviors as sexual drive (in both sexes), we can see that genes also make a significant contribution to behavior. In this section, we examine how protein synthesis results from the instructions given in the genes.

Preview

Before studying the details of protein synthesis, it is helpful to consider the big picture. In brief, DNA contains a genetic code that specifies which proteins a cell can make. All the body's cells except the sex cells contain identical genes, but different genes are activated in different cells; for example, the genes for blood-clotting proteins are not active in muscle cells. When a gene is activated, a molecule of **messenger RNA (mRNA),** a sort of mirror-image copy of the gene, is made. Messenger RNA migrates from the nucleus to the cytoplasm, where its code is "read" by a ribosome. Ribosomes are composed of **ribosomal RNA (rRNA)** and enzymes. **Transfer RNA (tRNA)** delivers amino acids to the ribosome, and the ribosome chooses from among these to assemble amino acids in the order directed by the mRNA. In summary, you can think of the process of protein synthesis as DNA → mRNA → protein, with each arrow reading as "codes for the production of." After the protein is made, it may be modified in various ways by the rough ER and Golgi complex, then go to any of several destinations discussed later.

The Genetic Code

The human genome codes for about 100,000 proteins. All of these are made of the same 20 amino acids, and all genes are made of the same four nucleotides (A, T, C, G)—a striking illustration of how a great variety of complex structures can be made from a small variety of simpler components. The **genetic code** is a system that enables four nucleotides to code for the amino acid sequences of all proteins. It is not unusual for simple codes to represent complex information. The complete text of an encyclopedia could be transmitted in Morse code, if we wished, with only the two symbols · and –. Computers store and transmit information, including pictures and sounds, in a binary code with only the symbols 1 and 0. It is not surprising, then, that a mere 20 amino acids can be represented by a code of four nucleotides; all that is required is various combinations of these symbols. A combination of two nucleotides for each amino acid would not suffice, because A, U, C, and G can combine in only 16 different ways (AA, AU, AC, AG, UA, UU, etc.). The minimum code to symbolize 20 amino acids would require at least three symbols, and indeed this is the case (table 5.2). A sequence of three DNA nucleotides that stands for one amino acid is called a **base triplet.** The "mirror-image" sequence in mRNA is called a **codon.** The genetic code is expressed in terms of codons.

A few representative triplets and codons are listed in table 5.2, along with the amino acids they represent. You can see from this listing that two or more codons sometimes represent the same amino acid. The reason for this is easy to explain mathematically. Four symbols (N) taken three at a time (x) can be combined in N^x different ways; that is, there are $4^3 = 64$ possible codons available to represent 20 amino acids. Only 61 of these 64 codons code for amino acids. The other three—UAG, UGA, and UAA—are called **stop codons;** they signal "end of message," like the period at the end of a sentence. A stop codon enables the cell's protein-synthesizing machinery to sense that it has

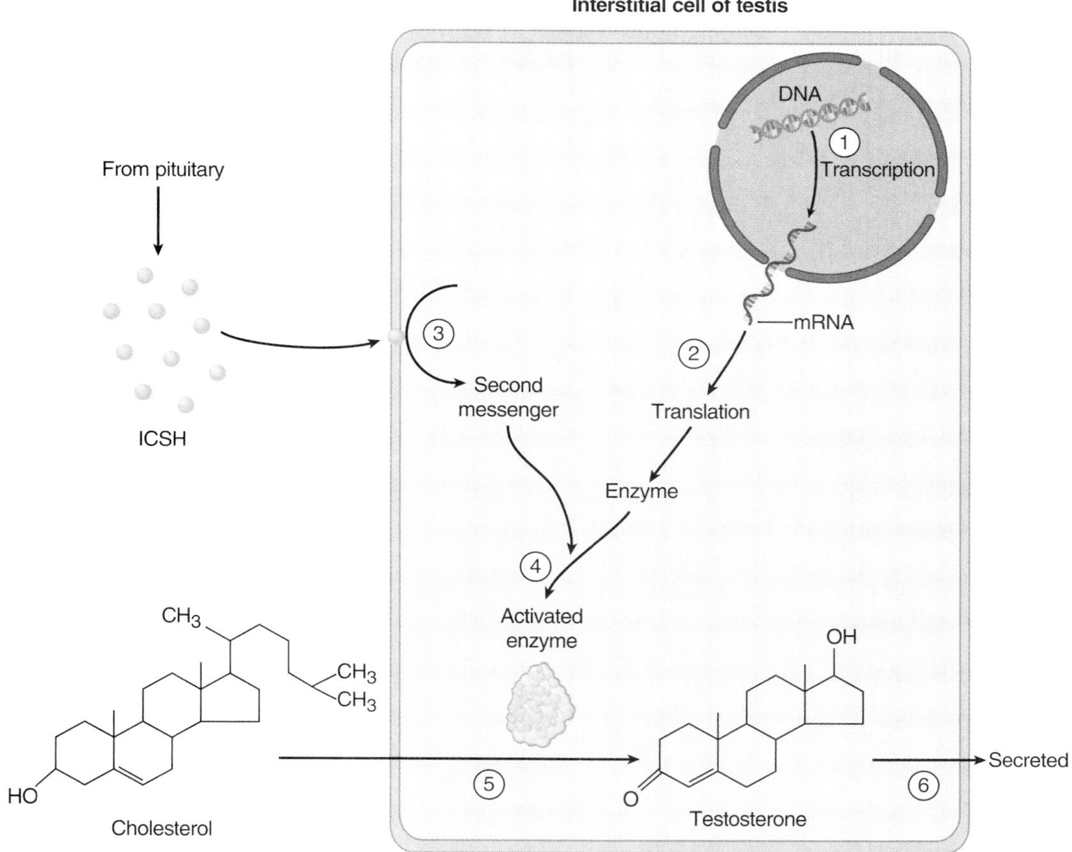

Figure 5.5 Indirect Control of Synthesis of Nonproteins by DNA. There is no gene for testosterone, but DNA regulates its synthesis through the enzymes for which it does code. (*1*) DNA codes for mRNA. (*2*) In the cytoplasm, mRNA directs the synthesis of an enzyme. (*3*) When testosterone is needed, interstitial cell–stimulating hormone (ICSH) stimulates production of a second messenger within cells of the testis. (*4*) The second messenger system activates the enzyme encoded by the mRNA. (*5*) The enzyme converts cholesterol to testosterone. (*6*) Testosterone is secreted from the cell and exerts various anatomical, physiological, and behavioral effects.

Table 5.2 Examples of the Genetic Code

Base Triplet of DNA	Codon of mRNA	Name of Amino Acid	Abbreviation for Amino Acid
CCT	GGA	Glycine	Gly
CCA	GGU	Glycine	Gly
CCC	GGG	Glycine	Gly
CTC	GAG	Glutamic acid	Glu
CGC	GCG	Alanine	Ala
CGT	GCA	Alanine	Ala
TGG	ACC	Threonine	Thr
TGC	ACG	Threonine	Thr
GTA	CAU	Valine	Val
TAC	AUG	Methionine	Met

reached the end of the gene for a particular protein. The codon AUG plays two roles—it serves as a code for methionine and as a **start codon.** This dual function is explained shortly.

Transcription

Protein synthesis occurs in the cytoplasm, but DNA is too large a molecule to leave the nucleus. It is necessary, therefore, to make a small RNA copy that can migrate through a nuclear pore into the cytoplasm. Just as we might transcribe (copy) a document, **transcription** in genetics means the process of copying genetic instructions from DNA to RNA. It is triggered by chemical messengers from the cytoplasm that enter the nucleus and bind to the chromatin at the site of the relevant gene. An enzyme called **RNA polymerase** (po-LIM-ur-ase) then binds to the DNA at this point and begins making RNA. Certain base sequences (often TATATA or TATAAA) inform the polymerase where to begin transcription.

RNA polymerase opens up the DNA helix about 17 base pairs at a time. It transcribes the bases from one strand of the DNA and makes a corresponding RNA. Where it finds a C on the DNA, it adds a G to the RNA; where it finds an A, it adds a U; and so forth. The enzyme then rewinds the DNA helix behind it. Another RNA polymerase may follow closely behind the first one; thus, a gene may be transcribed by several polymerase molecules at once, and numerous copies of the same RNA are made. At the end of the gene is a base sequence that serves as a terminator. It signals the polymerase to release the RNA molecule and separate from the DNA.

The RNA produced by transcription is not quite ready to direct protein synthesis. It is in a form called **pre-mRNA,** which contains "sense" portions called **exons** that will be translated into an amino acid sequence, and "nonsense" portions called **introns** that must be removed. There are 2 to 50 introns in a single gene, and each intron is a sequence of 100 to 1,000 bases. In a process called **posttranscriptional modification,** enzymes remove these introns and splice the exons together (fig. 5.6). The result is a functional mRNA molecule that leaves the nucleus and carries its genetic message into the cytoplasm.

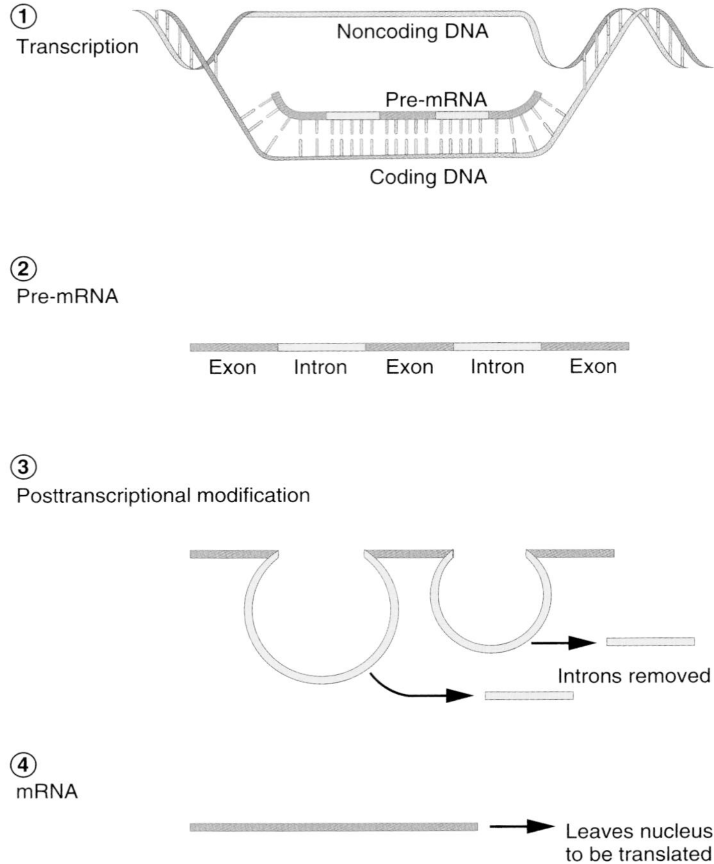

Figure 5.6 Transcription and Posttranscriptional Modification. (*1*) DNA uncoils and its coding strand is transcribed to make pre-mRNA. (*2*) Pre-mRNA has segments called exons that code for protein synthesis, and noncoding segments called introns that must be removed. (*3*) In posttranscriptional modification, enzymes remove the introns and splice together the exons. (*4*) The result is mRNA, which leaves the nucleus to be translated in the cytoplasm.

Translation

Just as we might translate a work from Spanish into English, **genetic translation** translates the language of nucleotides into the language of amino acids (fig. 5.7). This job is done by ribosomes, which are found in the cytosol and on the rough ER and nuclear envelope. A ribosome consists of two granular subunits, large and small, each made of several rRNA and enzyme molecules.

The mRNA molecule begins with a **leader sequence** of bases that are not translated to protein but serve as a binding site for the ribosome. The small ribosomal subunit binds to it, the large subunit joins the complex, and the ribosome begins pulling the mRNA through it like a ticker-tape, reading bases as it goes. When it reaches the start codon, AUG, it begins making protein. Since AUG codes for methionine, all proteins begin with methionine when first synthesized, although this may be removed later.

Translation requires the participation of 61 types of transfer RNA (tRNA), one for each codon (except stop codons). Transfer RNA is a small RNA molecule that turns back and coils on itself to form a cloverleaf shape, which is then twisted into an angular L shape (fig. 5.8). One end of the L includes three nucleotides called an **anticodon,** and the other end has a binding site specific for one amino acid. Each tRNA picks up an amino acid from a pool of free amino acids in the cytosol. One ATP molecule is used to bind the amino acid to this site and provide the energy that is used later to join that amino acid to the growing protein. Thus, protein synthesis consumes one ATP for each peptide bond formed.

When the small ribosomal subunit reads a codon such as CGC, it must find an activated tRNA with the corresponding anticodon; in this case, GCG. This particular tRNA would have the amino acid alanine at its other end. The ribosome binds and holds this tRNA and then reads the next codon—say GGU. Here, it would bind a tRNA with anticodon CCA, which carries glycine.

The large ribosomal subunit contains an enzyme that forms peptide bonds, and now that alanine and glycine are side by side, it links them together. The first tRNA is no longer needed, so it is released from the ribosome. The second tRNA is used, temporarily, to anchor the growing peptide to the ribosome. Now, the ribosome reads the third

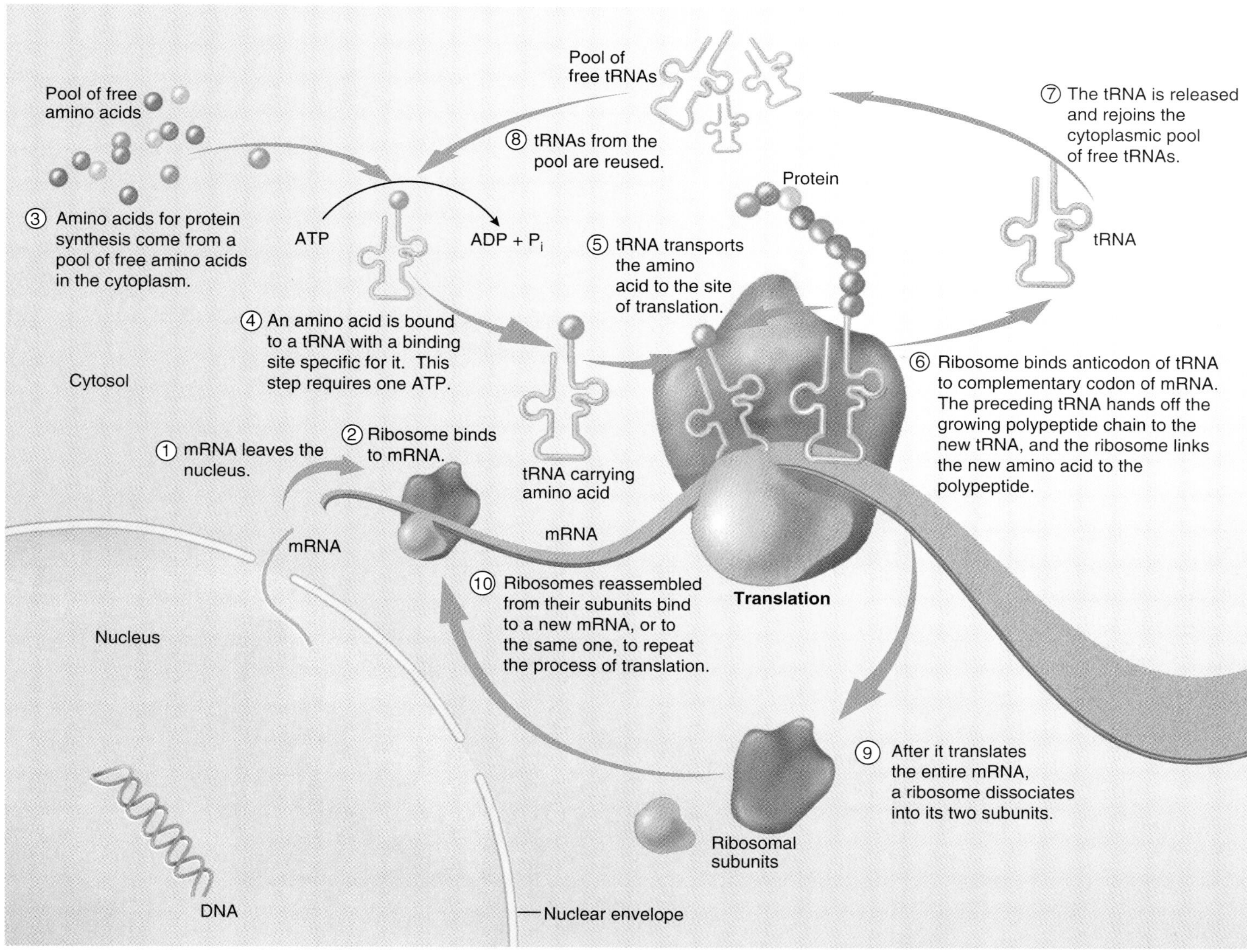

Figure 5.7 Translation of mRNA.

codon—say GUA. It finds the tRNA with the anticodon CAU, which carries the amino acid valine. The large subunit adds valine to the growing chain, now three amino acids long. By repetition of this process, the entire protein is assembled. Eventually, the ribosome reaches a stop codon and is finished translating this mRNA. The polypeptide is turned loose, and the ribosome dissociates into its two subunits.

One ribosome can assemble a protein of 400 amino acids in about 20 seconds, but it does not work at the task alone. After the mRNA leader sequence passes through one ribosome, a neighboring ribosome takes it up and begins translating the mRNA before the first ribosome has finished. One mRNA often holds 10 or 20 ribosomes together in a cluster called a **polyribosome** (fig. 5.9). Not only is each mRNA translated by all these ribosomes at once, but a cell may have 300,000 identical mRNA molecules undergoing simultaneous translation. Thus, a cell may produce over 150,000 protein molecules per second—a remarkably productive protein factory! In liver cells, which are highly active in protein synthesis, ribosomes make up as much as 25% of the dry weight.

Figure 5.10 summarizes transcription and translation and shows how a nucleotide sequence translates to a hypothetical peptide of 6 amino acids. A protein 500 amino acids long would have to be represented, at a minimum, by a sequence of 1,503 nucleotides (3 for each amino acid, plus a stop codon). The average gene is probably around 1,200 nucleotides long; a few may be as much as 10 times this long.

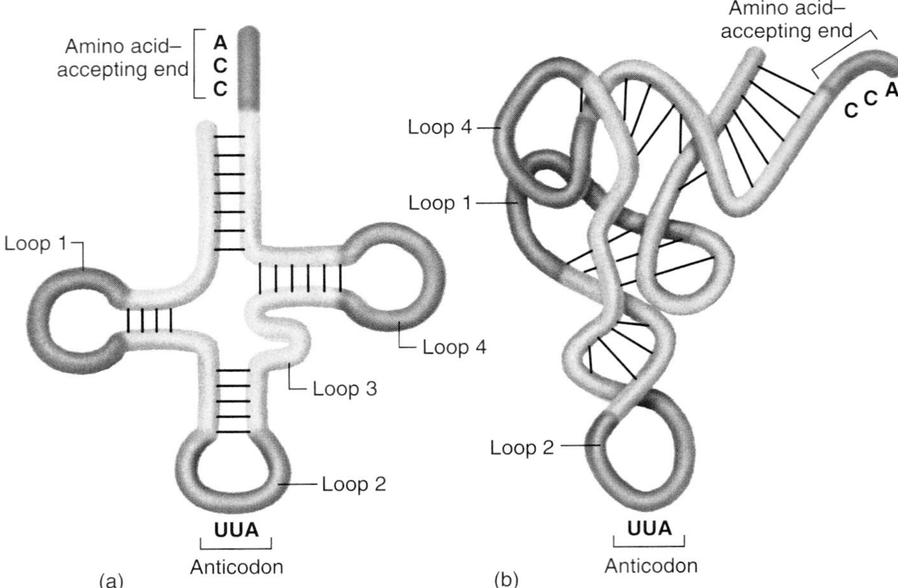

Figure 5.8 **Transfer RNA (tRNA).** (*a*) tRNA has an amino acid–accepting end that binds to one specific amino acid, and an anticodon that binds to a complementary codon of mRNA. (*b*) The three-dimensional shape of a tRNA molecule.

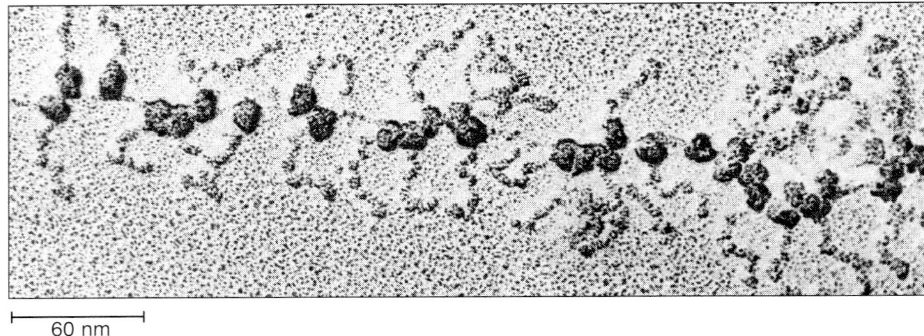

Figure 5.9 **Several Ribosomes Attached to a Single mRNA Molecule, Forming a Polyribosome.** The fine horizontal filament is mRNA; the large granules attached to it are ribosomes; and the beadlike chains projecting from each ribosome are newly formed proteins.

Chaperones and Protein Structure

The amino acid sequence of a protein (primary structure) is only the beginning; the end of translation is not the end of protein synthesis. The protein now coils or folds into its secondary and tertiary structures and, in some cases, associates with other polypeptide chains (quaternary structure) or conjugates with a nonprotein moiety, such as a vitamin or carbohydrate. It is essential that these processes not begin prematurely as the amino acid sequence is being assembled, since the correct final shape may depend on amino acids that have not been added yet. Therefore, as proteins are assembled by ribosomes, they are sometimes picked up by other proteins called **chaperones.** A chaperone prevents a new protein from folding prematurely and assists in its proper folding once the amino acid sequence has been completed. The chaperone may also escort a newly synthesized protein to the correct destination in a cell, such as the plasma membrane, and help to prevent improper associations between different proteins. As in the colloquial sense of the word, a chaperone is an older protein that escorts and regulates the behavior of the "youngsters." Some chaperones are also called *stress proteins* or *heat-shock proteins* because they are produced in response to heat or other stress on a cell, and because they

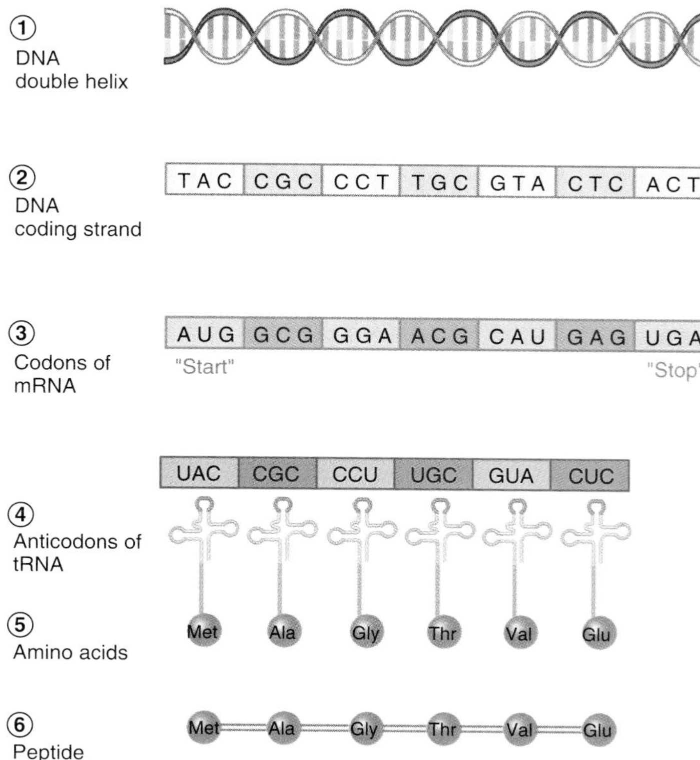

Figure 5.10 Relationship of a DNA Base Sequence to Protein Structure. (*1*) DNA. (*2*) A series of base triplets in the coding strand of DNA. (*3*) The corresponding codons that would be in an mRNA molecule transcribed from this DNA sequence. (*4*) Binding of mRNA to the complementary anticodons of six tRNA molecules. (*5*) The amino acids bound to these tRNAs. (*6*) Linkage of the amino acids into the peptide that was encoded in the DNA.

help stressed proteins fold back into their correct functional shapes.

Posttranslational Modification

If a protein is going to be used in the cytosol (for example, the enzymes of glycolysis), it is likely to be made by free ribosomes in the cytosol. If it is going to be packaged into a lysosome or secreted from the cell, however, the protein begins with a special amino acid sequence (*signal peptide*) that causes the entire polyribosome to migrate to the rough ER and dock on its surface. Protein synthesis is then completed on the rough ER and in the Golgi complex. Thus, we turn to the functions of these organelles in the modification, packaging, and secretion of a protein (fig. 5.11).

When a protein is produced on the rough ER, its signal peptide threads itself through a pore in the ER membrane and drags the rest of the growing protein into the cisterna. Enzymes in the rough ER then remove the signal peptide and modify the new protein in a variety of ways—removing some amino acids segments, folding the protein and stabilizing it with disulfide bridges, adding carbohydrate moieties, and so forth. Such changes are called **posttranslational modification.** Insulin, for example, is first synthesized as a polypeptide of 86 amino acids. In posttranslational modification, the chain folds back on itself, three disulfide bridges are formed, and 35 amino acids are removed. The final insulin molecule is therefore made of two chains of 21 and 30 amino acids held together by disulfide bridges (see fig. 17.22).

When the rough ER is finished modifying and transporting a protein, it pinches off clathrin-coated **transport vesicles.** Like the address on a letter, clathrin may direct the vesicle to its destination, the Golgi complex. The Golgi complex removes the clathrin, fuses with the vesicle, and takes the protein into its cisterna. Here, it may further modify the protein, for example by adding carbohydrate to it. Such modifications begin in the cisterna closest to the rough ER. Each cisterna forms transport vesicles that carry the protein to the next cisterna, where different enzymes may further modify the new protein.

Packaging and Secretion

When the protein is processed by the last Golgi cisterna, farthest from the rough ER, that cisterna pinches off spherical, membrane-bounded **Golgi vesicles** containing the finished product. Some Golgi vesicles become **secretory vesicles,** which migrate to the plasma membrane and release the product by exocytosis. This is how a cell of the salivary gland, for example, secretes mucus and digestive enzymes. The destinations of these and some other newly synthesized proteins are summarized in table 5.3.

Before You Go On

Answer the following questions to test your understanding of the preceding section:

5. Define *genetic code, codon,* and *genome.*
6. Describe the genetic role of RNA polymerase.
7. Describe the genetic role of ribosomes and tRNA.
8. Why are chaperones important in ensuring correct tertiary protein structure?
9. What roles do the rough ER and Golgi complex play in protein production?

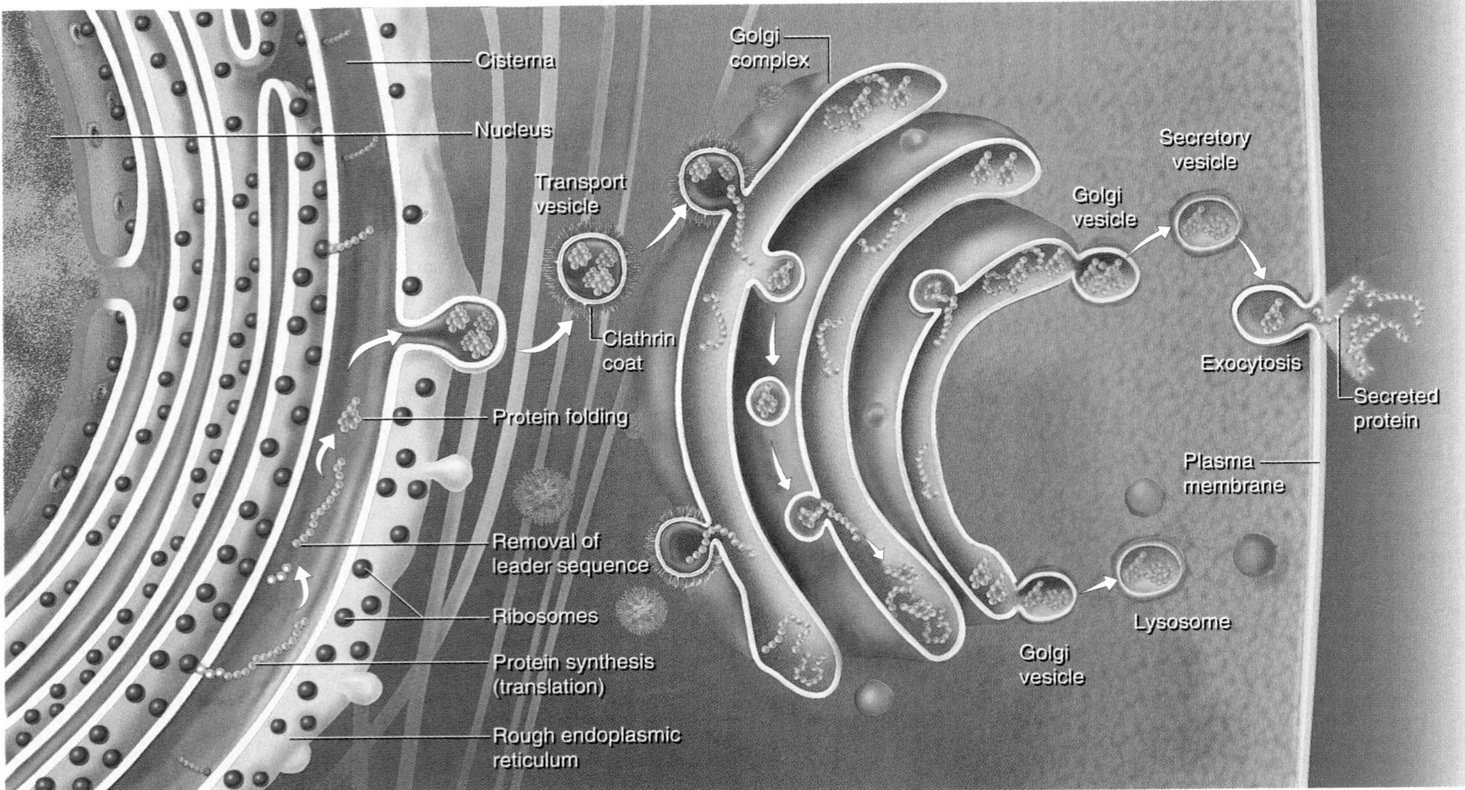

Figure 5.11 Protein Packaging and Secretion. Proteins are synthesized by ribosomes on the rough ER and carried in transport vesicles to the nearest cisterna of the Golgi complex. The Golgi complex modifies the structure of the protein, transferring it from one cisterna to the next, and finally packages it in Golgi vesicles. Some Golgi vesicles may remain within the cell and become lysosomes, while others may migrate to the plasma membrane and release the cell product by exocytosis.

Table 5.3 Some Destinations and Functions of Newly Synthesized Proteins

Destination or Function	Proteins
Deposited as a structural protein within cells	Actin of cytoskeleton Keratin of epidermis
Used in the cytosol as a metabolic enzyme	ATPase Carbonic anhydrase
Returned to the nucleus for use in nuclear metabolism	Histones of chromatin RNA polymerase
Packaged in lysosomes for use in autophagy and programmed cell death	Numerous lysosomal enzymes
Delivered to other organelles for intracellular use	Catalase of peroxisomes Mitochondrial enzymes
Delivered to plasma membrane to serve transport and other functions	Hormone receptors Sodium-potassium pumps
Secreted by exocytosis for extracellular functions	Digestive enzymes Casein of breast milk

DNA Replication and the Cell Cycle

Objectives

When you have completed this section, you should be able to

- describe how DNA is replicated;
- discuss the consequences of replication errors;
- describe the life history of a cell, including the events of mitosis;
- explain how the timing of cell division is regulated;
- describe some genetic and environmental causes of cancer; and
- describe some mechanisms behind the fatal effects of cancer.

When a cell divides, it distributes half of its DNA to each daughter cell. Following division, these daughter cells must therefore replace the missing half to restore the original amount and be able to divide again. Since DNA controls all cellular function, this replication process must be very exact. We now examine how it is accomplished and consider the consequences of mistakes.

DNA Replication

The law of complementary base pairing shows that we can predict the base sequence of one DNA strand if we know the sequence of the other. More importantly, it enables a cell to reproduce one strand based on information in the other. This immediately occurred to Watson and Crick when they discovered the structure of DNA. Watson was hesitant to make such a grandiose claim in their first publication, but Crick implored, "Well, we've got to say *something!* Otherwise people will think these two unknown chaps are so dumb they don't even realize the implications of their own work!" Thus, the last sentence of their first paper modestly states, "It has not escaped our notice that the specific pairing we have postulated . . . immediately suggests a possible copying mechanism for the genetic material." Five weeks later they published a second paper pressing this point more vigorously.

The basic idea of DNA replication is evident from its base pairing, but the way in which DNA is organized in the chromatin introduces some complications that were not apparent when Watson and Crick first wrote. The fundamental steps of the replication process are as follows:

1. The double helix unwinds from the histones.
2. Like a zipper, an enzyme called **DNA helicase** opens up a short segment of the helix, exposing its nitrogenous bases. The point where one strand of DNA is "unzipped" and separates from its complementary strand is called a *replication fork* (fig. 5.12*a*).
3. Enzymes called **DNA polymerase** move along the opened strands, read the exposed bases, and like matchmakers, arrange "marriages" with complementary free nucleotides in the nucleoplasm. If the polymerase finds the sequence TCG, for example, it assembles AGC across from it. One polymerase molecule moves away from the replication fork replicating one strand of the opened DNA, and another polymerase molecule moves in the opposite direction, replicating the other strand. Thus, from the old DNA molecule, two new ones are made. Each new DNA consists of one new helix synthesized from free nucleotides and one helix conserved from the parent DNA (fig. 5.12*b*). The process is therefore called **semiconservative replication.**
4. While DNA is synthesized in the nucleus, new histones are synthesized in the cytoplasm. Millions of histones are transported into the nucleus within a few minutes after DNA replication, and each new DNA helix wraps around them to make new nucleosomes.

Despite the complexity of this process, each DNA polymerase works at an impressive rate of about 100 base pairs per second. Even at this rate, however, it would take weeks for one polymerase molecule to replicate even one chromosome. All 46 chromosomes are replicated in a mere 6 to 8 hours, however, because thousands of polymerase molecules work simultaneously on each DNA molecule.

Errors and Mutations

DNA polymerase is fast and accurate, but it makes mistakes. For example, it might read A and place a C across from it where it should have placed a T. In *Escherichia coli,* a bacterial species in which DNA replication has been most thoroughly studied, about three errors occur for every 100,000 bases copied. At this rate of error, every generation of cells would have about 1,000 faulty proteins, coded for by DNA that had been miscopied. To help prevent such catastrophic damage to the organism, the DNA is continuously scanned for errors. After DNA polymerase has replicated a strand, a smaller polymerase comes along, "proofreads" it, and makes corrections where needed—for example, removing C and replacing it with T. This improves the accuracy of replication to one error for every billion bases copied—only one faulty protein for every 10 cell divisions (in *E. coli*).

Changes in DNA structure, called **mutations,**[2] can result from replication errors or environmental factors. Uncorrected mutations can be passed on to the descendents of that cell, but some of them have no adverse effect. One reason is that a new base sequence sometimes codes for the

2. *muta* = change

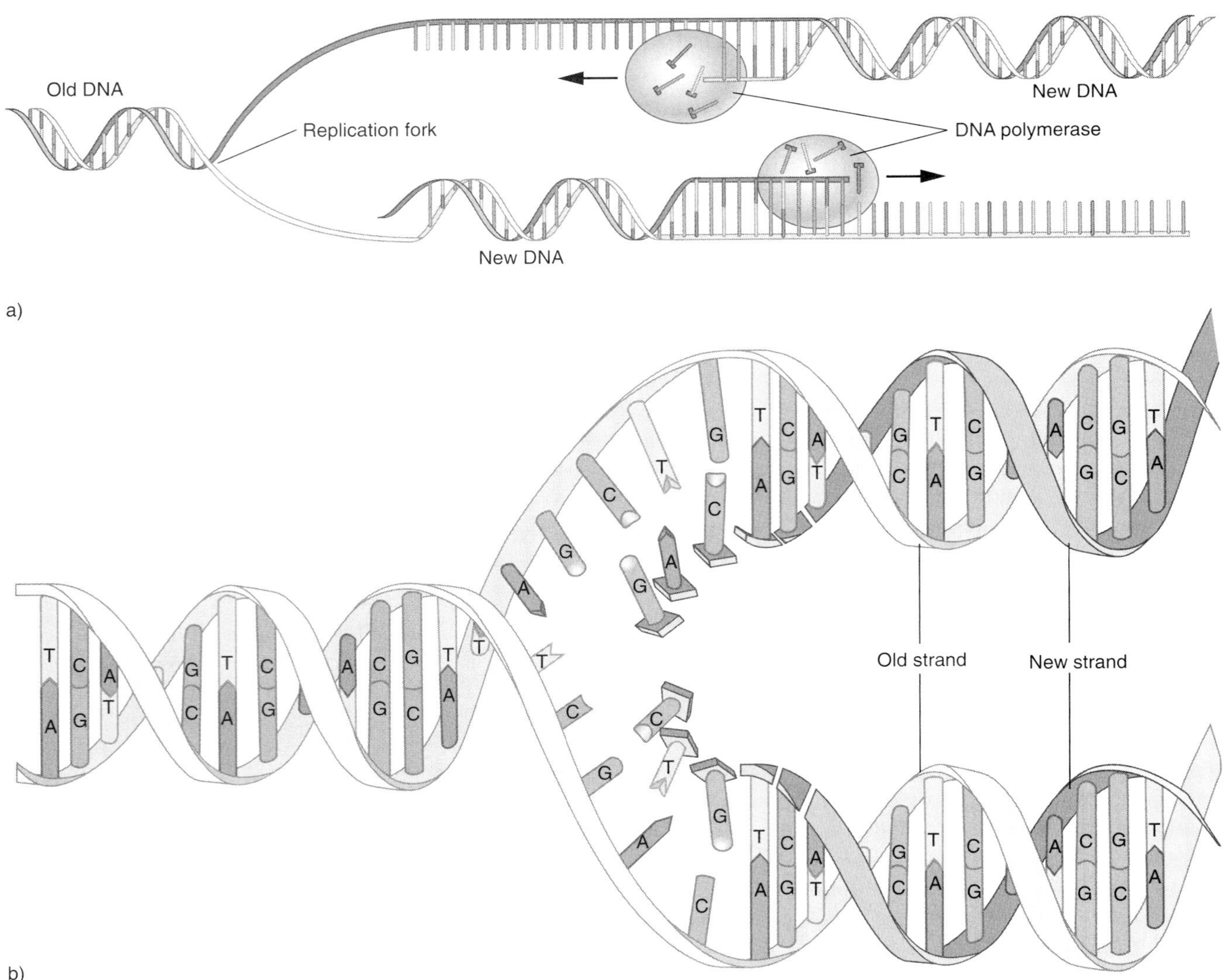

Figure 5.12 Semiconservative DNA Replication. (*a*) At the replication fork, DNA helicase (not shown) unwinds the double helix and exposes the bases. DNA polymerases begin assembling new bases across from the existing ones, moving away from the replication fork on one strand and toward it on the other strand. (*b*) The result is two DNA double helices, each composed of one strand of the original DNA and one newly synthesized strand.

same thing as the old one. For example, ACC and ACG both code for threonine (see table 5.2), so a mutation from C to G in the third place would cause no change in protein structure. Another reason is that a change in protein structure is not always critical to its function. For example, humans and horses differ in 25 of the 146 amino acids that make up their β hemoglobin, yet the hemoglobin is fully functional in both species. Some mutations, however, may kill a cell, turn it cancerous, or cause genetic defects in future generations. When a mutation changes the sixth amino acid of β hemoglobin from glutamic acid to valine, for example, the result is a crippling disorder called sickle-cell disease. Clearly some amino acid substitutions are more critical than others, and this affects the severity of a mutation.

The Cell Cycle

Most cells periodically divide into two daughter cells, so a cell has a life cycle extending from one division to the next. This **cell cycle** (fig. 5.13) is divided into four main phases: G_1, S, G_2, and M.

G_1 is the **first gap phase,** an interval between cell division and DNA replication. During this time, a cell synthesizes proteins, grows, and carries out its preordained tasks for the body. Almost all of the discussion in this book relates to what cells do in the G_1 phase. Cells in G_1 also begin to replicate their centrioles in preparation for the next mitosis and they accumulate the materials needed to replicate their DNA in the next phase. In cultured cells called

Clinical Insight 5.3

Can We Replace Brain Cells?

Until recently, neurons (nerve cells) of the brain were thought to be irreplaceable; when they died, we thought, they were gone forever. We believed, indeed, that there was good reason for this. Motor skills and memories are encoded in intricate neural circuits, and the growth of new neurons might disrupt those circuits. Now we are not so sure.

A chemical called BrDU (bromodeoxyuridine) can be used to trace the birth of new cells, because it becomes incorporated into their DNA. BrDU is too toxic, however, to use ordinarily in human research. However, in cancer patients, BrDU is sometimes used to monitor the growth of tumors. Peter Eriksson, at Göteborg University in Sweden, obtained permission from the families of cancer patients to examine the brain tissue of BrDU-treated patients who had died. In the hippocampus, a region of the brain concerned with memory, he and coworker Fred Gage found as many as 200 new neurons per cubic millimeter of tissue and estimated that up to 1,000 new neurons may be born per day even in people in their 50s to 70s. These new neurons apparently arise not by mitosis of mature neurons (which are believed to be incapable of mitosis), but from a reserve pool of embryonic stem cells. It remains unknown whether new neurons are produced late in life in other regions of the brain.

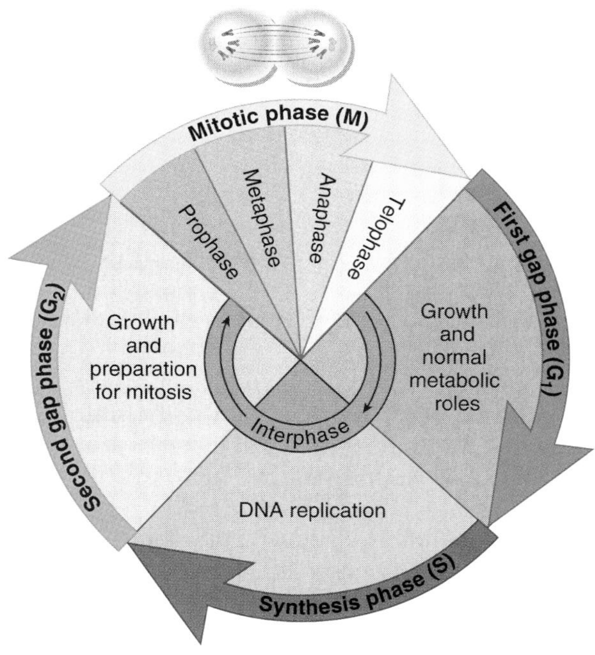

Figure 5.13 The Cell Cycle.

fibroblasts, which divide every 18 to 24 hours, G_1 lasts 8 to 10 hours.

S is the **synthesis phase,** in which a cell carries out DNA replication. This produces two identical sets of DNA molecules, which are then available, like the centrioles, to be divided up between daughter cells at the next cell division. This phase takes 6 to 8 hours in cultured fibroblasts.

G_2, the **second gap phase,** is a relatively brief interval (4 to 6 hours) between DNA replication and cell division. In G_2, a cell finishes replicating its centrioles and synthesizes enzymes that control cell division.

M is the **mitotic phase,** in which a cell replicates its nucleus and then pinches in two to form two new daughter cells. In cultured fibroblasts, the M phase takes 1 to 2 hours. The details of this phase are considered in the next section. Phases G_1, S, and G_2 are collectively called **interphase**—the time between M phases.

The length of the cell cycle varies greatly from one cell type to another. Stomach and skin cells divide rapidly, bone and cartilage cells slowly, and skeletal muscle cells and nerve cells not at all. Some cells leave the cell cycle for a "rest" and cease to divide for days, years, or the rest of one's life. Such cells are said to be in the **G_0 (G-zero) phase.** The balance between cells that are actively cycling and those standing by in G_0 is an important factor in determining the number of cells in the body. An inability to stop cycling and enter G_0 is characteristic of cancer cells.

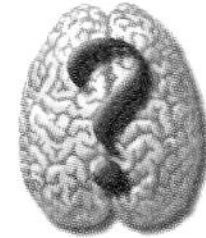

Think About It

What is the maximum number of DNA molecules ever contained in a cell over the course of its life cycle? (Assume the cell has only one nucleus.)

Mitosis

Mitosis (my-TOE-sis), in the sense used here, is the process by which a cell divides into two daughter cells with identical copies of its DNA. (Some define it as division of the nucleus only and do not include the subsequent cell division.) Mitosis has four main functions:

1. formation of a multicellular embryo from a fertilized egg;
2. tissue growth;
3. replacement of old and dead cells; and
4. repair of injured tissues.

Egg and sperm cells are produced by a combination of mitosis and another form of cell division, *meiosis,* described in chapter 27. Otherwise, all cells of the body are produced entirely by mitosis. Four phases of mitosis are recognizable—*prophase, metaphase, anaphase,* and *telophase* (fig. 5.14).

In **prophase,**[3] at the outset of mitosis, the chromatin becomes supercoiled into short, dense **chromosomes** (fig. 5.15), which are easier to distribute to daughter cells than the long, delicate chromatin. A chromosome consists of two *genetically identical* bodies called **sister chromatids,** joined together at a pinched spot called the **centromere.** At prophase, there are 46 chromosomes, two chromatids per chromosome, and one molecule of DNA in each chromatid. The nuclear envelope disintegrates during prophase and releases the chromosomes into the cytosol. The centrioles begin to sprout elongated microtubules, which push the centrioles apart as they grow. Eventually, a pair of centrioles come to lie at each pole of the cell.

In **metaphase,**[4] the chromosomes line up at random along the midline of the cell. Microtubules grow from each centriole toward the chromosomes, and some attach to the centromeres. This forms a football-shaped array called the **mitotic spindle.** Shorter microtubules also radiate from each centriole pair to form a star-shaped array called an *aster.*[5] These microtubules anchor the centrioles to the nearby plasma membrane.

In **anaphase,**[6] each centromere divides in two and the chromosome separates into two chromatids. One chromatid migrates to each pole of the cell, with its centromere leading the way and the arms trailing behind. There is some evidence that the spindle fiber acts a little like a railroad track, and a protein complex in the centromere called the **kinetochore**[7] (kih-NEE-toe-core) acts as a molecular motor that propels the chromatid along the track. One of the kinetochore proteins is dynein, the same motor molecule that causes movement of cilia and flagella (see chapter 4). Since sister chromatids are genetically identical, and since each daughter cell receives one chromatid from each chromosome, you can see why the daughter cells of mitosis are genetically identical.

In **telophase,**[8] the chromatids cluster on each side of the cell. The rough ER produces a new nuclear envelope around each cluster, and the chromatids begin to uncoil and return to the thinly dispersed chromatin form. The mitotic spindle breaks up and vanishes. Each new nucleus forms nucleoli, indicating it has already begun making RNA and preparing for protein synthesis.

Telophase is the end of nuclear division but overlaps with **cytokinesis**[9] (SY-toe-kih-NEE-sis), division of the cytoplasm. Cytokinesis is achieved by the motor protein myosin pulling on microfilaments of actin in the membrane skeleton. This creates a crease called the *cleavage furrow* around the equator of the cell, and the cell eventually pinches in two. Interphase has now begun for these new cells.

Timing of Cell Division

One of the most important questions in biology is what signals cells when to divide and when to stop. The activation and inhibition of cell division are subjects of intense research for obvious reasons such as management of cancer and tissue repair. Cells divide when (1) they grow large enough to have enough cytoplasm to distribute to their two daughter cells; (2) they have replicated their DNA, so they can give each daughter cell a duplicate set of genes; (3) they receive an adequate supply of nutrients; (4) they are stimulated by **growth factors,** chemical signals secreted by blood platelets, kidney cells, and other sources; and (5) neighboring cells die, opening up space in a tissue to be occupied by new cells. Cells stop dividing when nutrients or growth factors are withdrawn or when they snugly contact neighboring cells. The cessation of cell division in response to contact with other cells is called **contact inhibition.**

Cancer

Proper tissue development depends on a balance between cell division and cell death. When cells multiply faster than they die, they sometimes produce abnormal growths called tumors, or **neoplasms.**[10] The study of tumors is called **oncology.**[11]

Benign[12] (beh-NINE) tumors are surrounded by a connective tissue capsule, grow slowly, and do not spread to other organs. They are still potentially lethal—even slow-growing tumors can kill by compressing brain tissue, nerves, blood vessels, or airways. The term **cancer** refers to **malignant**[13] (muh-LIG-nent) tumors, which are unencapsulated, fast-growing, and spread easily to other organs by way of the blood or lymph. The word *cancer* dates to Hippocrates, who compared the distended veins in some breast tumors to the outstretched legs of a crab.[14] Malignant cells exhibit no contact inhibition or respect for tissue boundaries; they readily grow into other tissues and replace

3. *pro* = first
4. *meta* = next in a series
5. *aster* = star
6. *ana* = apart
7. *kineto* = motion + *chore* = place
8. *telo* = end, final
9. *cyto* = cell + *kinesis* = action, motion
10. *neo* = new + *plasm* = growth, formation
11. *onco* = tumor = *logy* = study of
12. *benign* = mild, gentle
13. *mal* = bad
14. *cancer* = crab

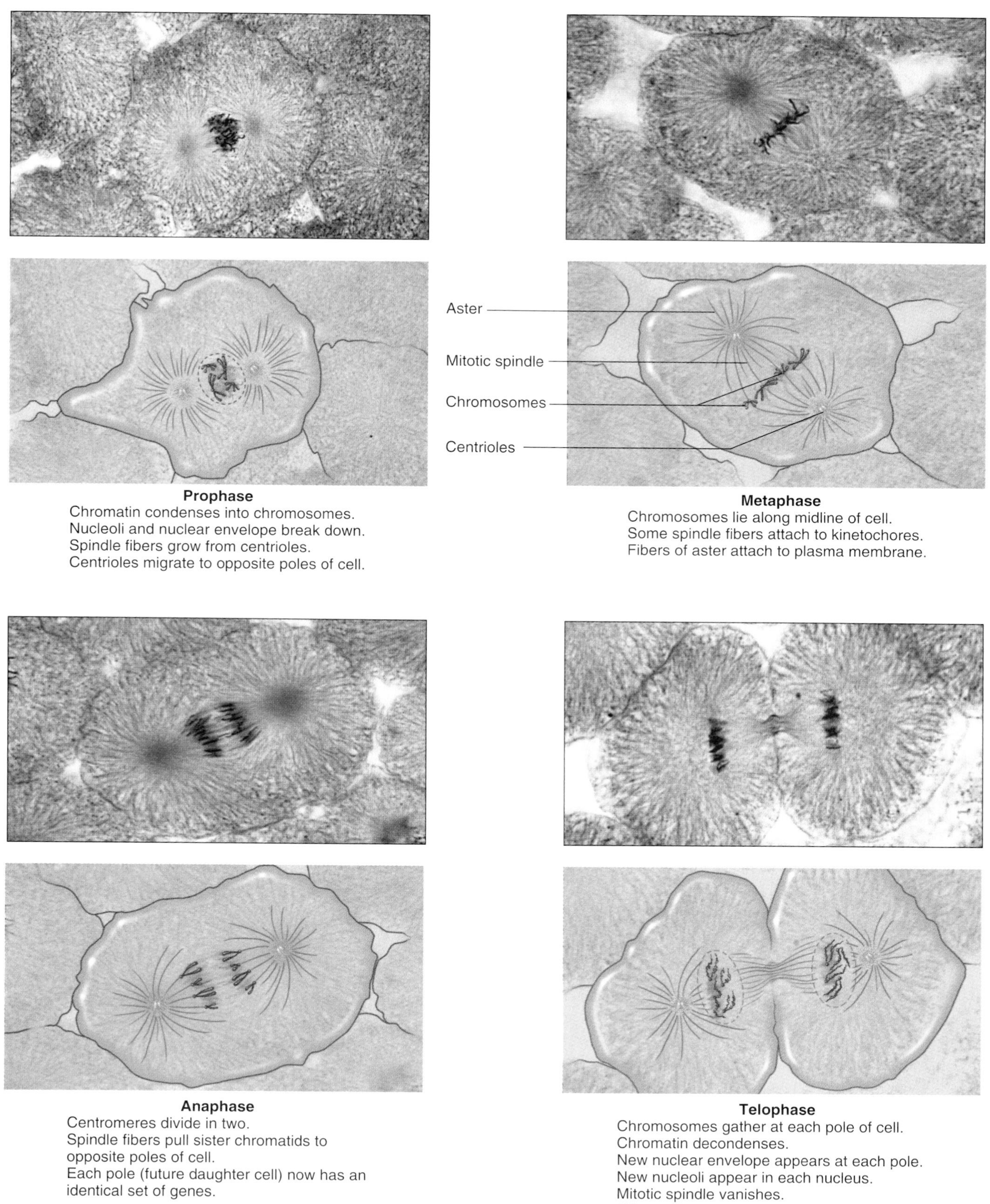

Figure 5.14 Mitosis. The photographs show mitosis in whitefish eggs, where chromosomes are relatively easy to observe. The drawings show a hypothetical cell with only two chromosome pairs; in humans, there are 23 pairs.

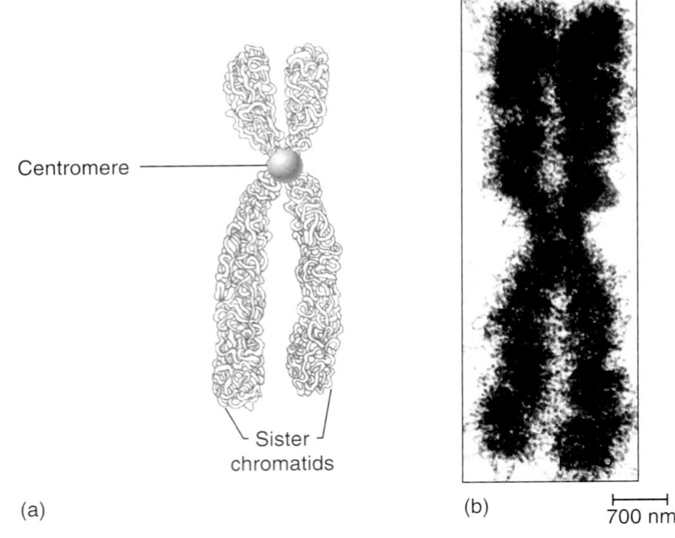

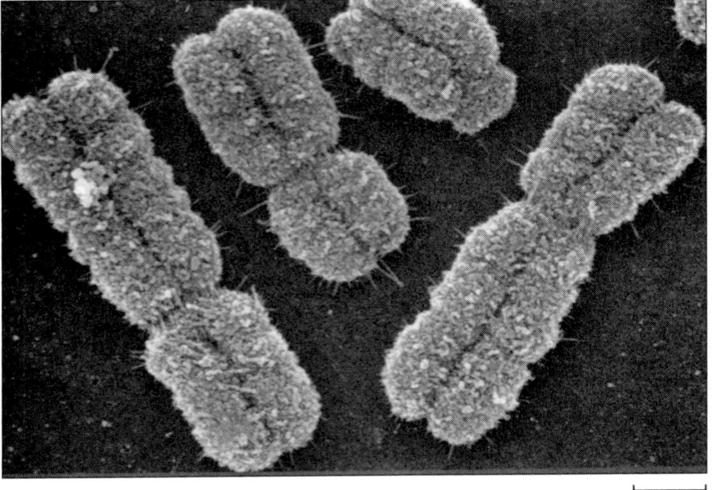

Figure 5.15 Chromosome Structure. (*a*) A metaphase chromosome. (*b*) Transmission electron micrograph. (*c*) Scanning electron micrograph.

healthy cells. About 90% of cancer deaths result from this spreading, called **metastasis**[15] (meh-TASS-tuh-sis), rather than from the primary (original) tumor. The cells involved in some common types of cancer are listed in table 5.4.

Causes of Cancer

The World Health Organization estimates that 60% to 70% of cancer is caused by environmental agents called **carcinogens**[16] (car-SIN-oh-jens). These fall into three categories:

Table 5.4 Classes of Cancer

Name	Cells, Tissues, and Organs Involved
Carcinoma	Epithelial cells
Melanoma	Pigment-producing skin cells (melanocytes)
Sarcoma	Bone, other connective tissue, or muscle
Leukemia	Blood-forming cells
Lymphoma	Lymph nodes

1. **Chemical carcinogens.** The most notorious chemical carcinogen is cigarette tar, responsible for 30% of cancer deaths in the United States. Others include food preservatives, such as nitrites, and asbestos fibers from old insulation and other sources.
2. **Radiation.** Cancer can be caused by ultraviolet radiation, γ rays, α particles, and β particles (see chapter 2).
3. **Viruses.** Some cases of uterine cancer have been attributed to the type 2 herpes simplex virus, and some liver cancer to the hepatitis B virus. The human immunodeficiency virus (HIV) promotes cancer in many AIDS patients by destroying immune cells that would otherwise destroy malignant cells.

Carcinogens are **mutagens**[17] (MEW-tuh-jens)—they trigger gene mutations. We have several defenses against mutagens: (1) scavenger cells may remove them before they cause genetic damage; (2) peroxisomes neutralize nitrites, free radicals, and other carcinogenic oxidizing agents; and (3) nuclear enzymes detect and repair damaged DNA. If these mechanisms fail, or if they are overworked by heavy exposure to mutagens, a cell may die of genetic damage, it may be recognized and destroyed by the immune system before it can multiply, or it may multiply and produce a tumor. Even then, tumors may be destroyed by a substance called *tumor necrosis factor (TNF),* secreted by macrophages and certain white blood cells.

Current cancer research has identified two families of genes that are especially important in the development of malignant tumors—*oncogenes* and *tumor suppressor genes*—and has linked many forms of cancer to abnormal growth factors or growth factor receptors. Most cells cannot divide unless a growth factor binds to a receptor on their surface. The receptor is a transmembrane protein of the plasma membrane. When a growth factor binds to its extracellular side, the receptor activates cell division enzymes on the intracellular side. This stimulates a cell to leave the G_0 phase, undergo mitosis, and develop *(differentiate)* into various kinds of mature, functional cells.

15. *meta* = beyond + *stasis* = being stationary
16. *carcino* = cancer + *gen* = producing

17. *muta* = change + *gen* = producing

Oncogenes are mutated, "misbehaving" forms of normal genes called **proto-oncogenes.** Healthy proto-oncogenes code for growth factors or growth factor receptors, whereas mutated oncogenes cause malfunctions in the growth factor mechanism. An oncogene called *sis,* for example, causes excessive production of growth factors that stimulate blood vessels to grow into a tumor and supply it with the rich blood supply that it requires. An oncogene known as *ras,* responsible for about one-quarter of human cancers, codes for abnormal growth factor receptors. These receptors act like a switch stuck in the on position, sending constant cell division signals even when there is no growth factor bound to them. Many cases of breast and ovarian cancer are caused by an oncogene called *erbB2.*

Tumor suppressor (TS) genes inhibit the development of cancer. They may act by opposing the action of oncogenes, promoting DNA repair, or controlling the normal histological organization of tissues, which is notably lacking in malignancies. A TS gene called *p16* acts by inhibiting one of the cyclin-dependent kinases of the cell cycle. Thus, if oncogenes are thought of as the "accelerator" of the cell cycle, then TS genes are like the "brake." Like other genes, TS genes occur in *homologous pairs* (explained in the following section), and even one normal TS gene in a pair is sometimes sufficient to suppress cancer. Damage to both members of a pair, however, removes normal controls over cell division and tends to trigger cancer. The first TS gene discovered was the *Rb* gene, which causes *retinoblastoma,* a cancer of the eyes of infants. Retinoblastoma occurs only if both copies of the *Rb* gene are damaged. In the case of a TS gene called *p53,* however, cancer occurs when only one copy is damaged. Gene *p53* is a large gene vulnerable to many cancer-causing mutations; it is involved in leukemia and in colon, lung, breast, liver, brain, and esophageal tumors.

Cancer often occurs only when several mutations have accumulated at different gene sites. Colon cancer, for example, is caused by damage to at least three TS genes on chromosomes 5, 17, and 18 and activation of an oncogene on chromosome 12. It takes time for so many mutations to accumulate, and this is one reason why cancer afflicts elderly people more than the young. In addition, the longer we live, the more carcinogens we are exposed to, the less efficient our DNA and tissue repair mechanisms become, and the less effective our immune system is at recognizing and destroying malignant cells.

Effects of Cancer

Untreated cancer is almost always fatal. Malignant tumors have several lethal effects:

- Cancer displaces normal tissue, so the function of the affected organ deteriorates. Lung cancer, for example, can destroy so much tissue that oxygenation of the blood becomes inadequate to support life. You might think that an increased number of cells in an organ would enable it to function better. In malignant tumors, however, the cells are in an immature state and unable to carry out the functions of mature cells of the same organ. Also, a tumor often consists of cells that have metastasized from elsewhere and are not typical of the host organ anyway. A colon cancer whose malignant cells have metastasized to the liver, for example, destroys liver tissue but cannot perform any of the liver's essential functions.
- Tumors can block vital passageways. The growth of a tumor can put pressure on a bronchus of the lung, obstructing air flow and causing pulmonary collapse, or it can compress a major blood vessel, reducing the delivery of blood to a vital organ or its return to the heart.
- Tumors have a high metabolic rate and compete with healthy tissues for nutrition. Other organs of the body may even break down their own proteins to nourish the tumor. This leads to general weakness, fatigue, emaciation, and susceptibility to infections.

Before You Go On

Answer the following questions to test your understanding of the preceding section:

10. Describe the genetic roles of DNA helicase and DNA polymerase. Contrast the function of DNA polymerase with that of RNA polymerase.
11. Explain why DNA replication is called *semiconservative.*
12. Define *mutation.* Explain why some mutations are harmless and others can be lethal.
13. List the stages of the cell cycle and summarize what occurs in each one.
14. Describe the structure of a chromosome at metaphase.
15. Contrast the actions of oncogenes and tumor suppressor genes. Explain the role of each one in cancer.

Chromosomes and Heredity

Objectives

When you have completed this section, you should be able to

- describe the structure and paired arrangement of human chromosomes;
- define *allele* and discuss how alleles affect the traits of an individual; and
- discuss the interaction of heredity and environment in producing individual traits.

Heredity is the transmission of genetic characteristics from parent to offspring. Several traits and diseases discussed in the forthcoming chapters are hereditary: baldness, blood types, color blindness, and hemophilia, for example. Thus it is appropriate at this point to lay the groundwork for these discussions by introducing a few basic principles of normal heredity. Hereditary defects are described in chapter 29.

The Karyotype

A **karyotype** (fig. 5.16) is a chart of the chromosomes isolated from a cell at metaphase, arranged in order by size and structure. It reveals that most human cells, with the exception of germ cells (described shortly), contain 23 pairs of similar-looking chromosomes (except for X and Y chromosomes). The two chromosomes in each pair are called **homologous**[18] (ho-MOLL-uh-gus) **chromosomes.** One is inherited from the mother and one from the father. Two chromosomes, designated X and Y, are called **sex chromosomes** and the other 22 pairs are called **autosomes** (AW-toe-somes). A female normally has a homologous pair of X chromosomes, whereas a male has one X chromosome and a much smaller Y chromosome.

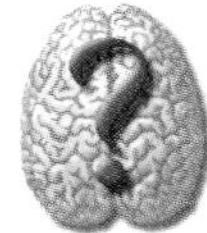

Think About It

Why would a cell in metaphase be more useful than a cell in interphase for producing a karyotype?

The paired state of the homologous chromosomes results from the fact that a sperm cell bearing 23 chromosomes fertilizes an egg, which also has 23. Sperm and egg cells, and cells on their way to becoming sperm and eggs, are called **germ cells.** All other cells of the body are called **somatic cells.** Somatic cells are described as **diploid**[19]

18. *homo* = same + *log* = relation

19. *diplo* = double

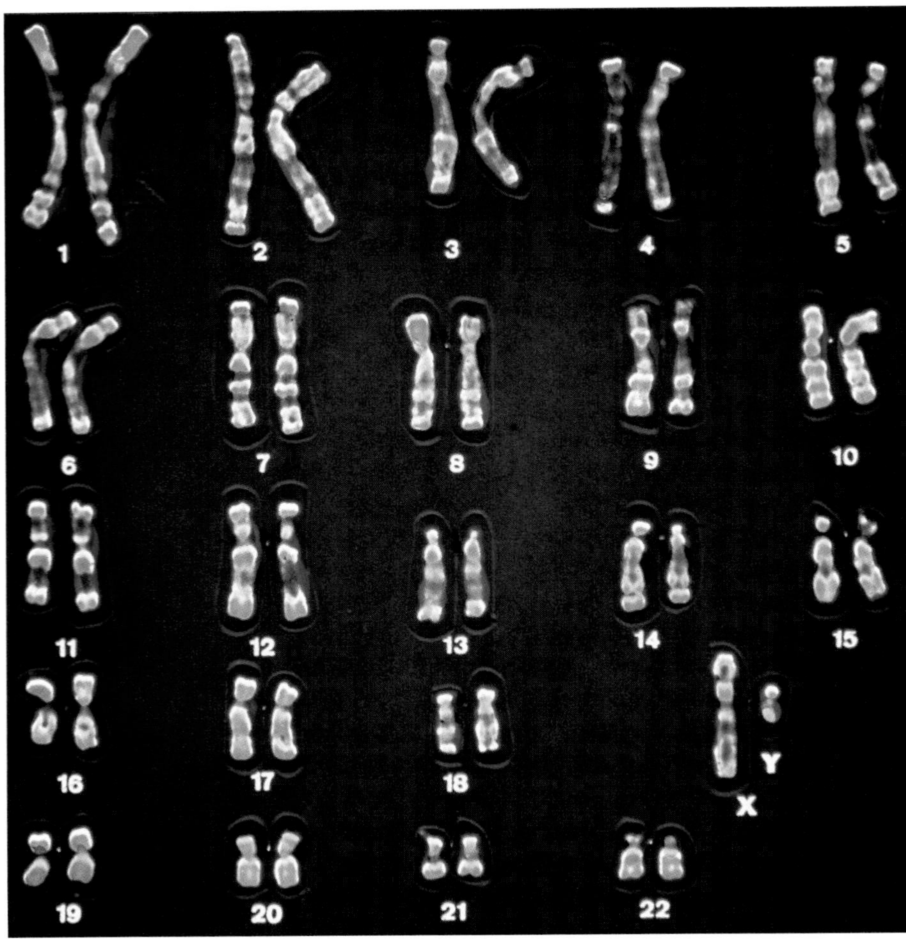

Figure 5.16 Karyotype of a Normal Human Male. This is a false-color micrograph of chromosomes stained to accentuate their banding patterns. The two chromosomes of each homologous pair exhibit similar size, shape, and banding. The only difference in a female karyotype would be the presence of a pair of X chromosomes in place of the XY pair of the male.

Historical Insight 5.4

The Human Genome Project

The Human Genome Project is a massive effort to map the locations and determine the nucleotide sequences of all genes of the genome. The project was proposed in 1985 and began operation on October 1, 1990. Now involving numerous laboratories in North America, Europe, and Japan, it is expected to be completed by 2003. In essence, the outcome will be an "instruction book" for human development and maintenance. Understanding the instructions, however, is likely to take much longer than obtaining them. At a cost of $3 billion, this is the first biological research project of such magnitude, and it was not born without controversy. One thing is certain, however—we cannot advance very far in the treatment of genetic diseases until we know where the responsible genes lie. The Human Genome Project is providing much of the information we need on the locations of these genes, as well as genes that underlie many normal aspects of human form and function.

because their chromosomes are in pairs, whereas germ cells beyond a certain stage of development are **haploid,**[20] which means they contain half as many chromosomes as the somatic cells. In meiosis (described in detail in chapter 27), homologous chromosomes become **segregated** from each other into separate daughter cells leading to the haploid sex cells. At fertilization, one set of *paternal* (sperm) chromosomes unites with one set of *maternal* (egg) chromosomes to restore the diploid number to the fertilized egg and the somatic cells that arise from it. Although the two chromosomes of a homologous pair appear to be identical, they come from different parents and therefore are not *genetically* identical.

Genes and Alleles

The location of a particular gene on a chromosome is called its **locus.** Two homologous chromosomes have the same gene at the same locus, although they may carry different forms of that gene, called **alleles**[21] (ah-LEELS), which produce alternative forms of a particular trait. Frequently, one allele will be **dominant** and the other one **recessive.** If at least one of the chromosomes carries the dominant allele, the body produces the normal, functional protein encoded by that gene. If both homologous chromosomes carry the recessive allele, the body either fails to produce that protein or produces an abnormal, often nonfunctional protein. The recessive allele then has a detectable influence on the person's characteristics.

For example, when the ears are developing in a fetus, a "death signal" is often activated in cells that attach the earlobe to the side of the head. These cells die, causing the earlobe to separate from the head. A person will then have "detached earlobes." This occurs in people who have either one or two copies of a dominant allele which we will denote *D.* If both homologous chromosomes have the recessive version of this gene, *d,* the cell suicide program is not activated, and the earlobes remain attached (fig. 5.17*a*). (It is customary to represent a dominant allele with a capital letter and a recessive allele with its lowercase equivalent.)

Individuals with two identical alleles, such as *DD* or *dd,* are said to be **homozygous**[22] (HO-mo-ZY-gus) for that trait. If the two homologous chromosomes have different alleles for that gene *(Dd),* the individual is **heterozygous**[23] (HET-er-oh-ZY-gus). The alleles that an individual possesses for a particular trait constitute the **genotype** (JEE-no-type). A detectable trait resulting from the genotype, such as attached or detached earlobes, is called the **phenotype**[24] (FEE-no-type).

We say that an allele is *expressed* if it shows in the phenotype of an individual. Earlobe allele *d* is expressed only when it is present in a homozygous state *(dd);* allele *D* is expressed whether it is homozygous *(DD)* or heterozygous *(Dd).* The only way most recessive alleles can be expressed is for an individual to inherit them from both parents.

Recessive traits can "skip" one or more generations. A diagram called a *Punnett square* (fig. 5.17*b*) shows how two heterozygous parents with detached earlobes can produce a child with attached lobes. Across the top are the two genetically possible types of eggs the mother could produce, and on the left side are the possible types of sperm cells from the father. The four cells of the square show the genotypes and phenotypes that would result from each possible combination of sperm and egg. You can see that three of the possible combinations would produce a child with detached lobes (genotypes *DD* and *Dd*), but one combination *(dd)* would produce a child with attached lobes. Therefore, the attached-lobe trait skipped

20. *haplo* = half
21. *allo* = different
22. *homo* = same + *zyg* = union, joined
23. *hetero* = different
24. *pheno* = showing, evident

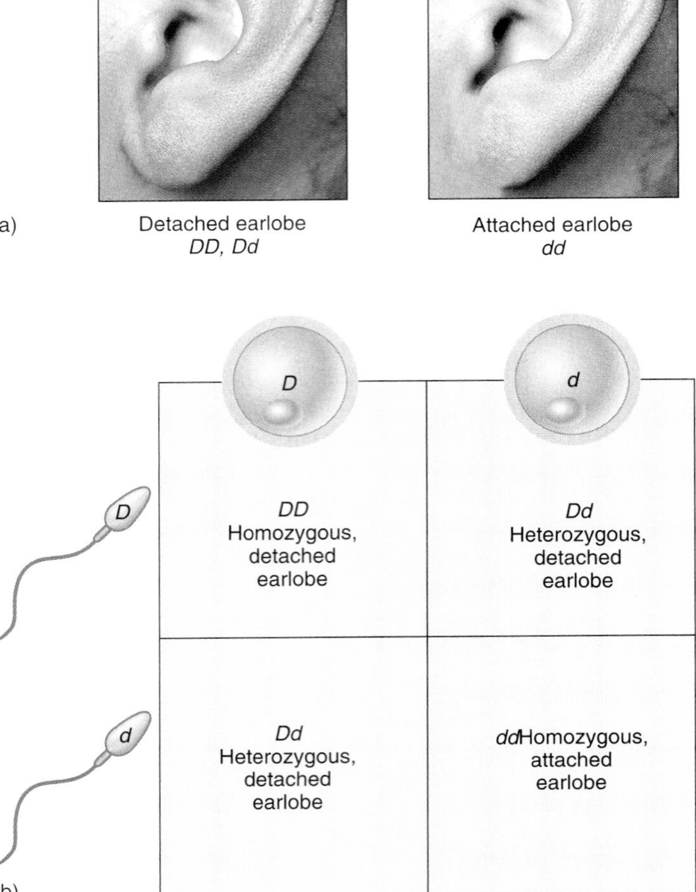

Figure 5.17 Genetics of Attached and Detached Earlobes. (*a*) Detached earlobes occur if even one allele of the pair is dominant *(D)*. Detached earlobes occur only when both alleles are recessive *(d)*. (*b*) A Punnett square shows why such a trait can "skip a generation." Both parents in this case have heterozygous genotypes *(Dd)* and detached earlobes, but there is a one in four chance that their offspring could have attached earlobes. Each parent is a carrier for attached earlobes.

the parental generation in this case but could be expressed in their child.

This phenomenon becomes more significant when parents are heterozygous **carriers** of hereditary diseases such as cystic fibrosis—individuals who carry a recessive allele and may pass it on, but do not phenotypically express it in themselves. For some hereditary diseases, tests are available to detect carriers and allow couples to weigh their risk for having children with genetic disorders. *Genetic counselors* perform genetic testing or refer clients for tests, advise couples on the probability of transmitting genetic diseases, and assist people in coping with genetic disease. *Gene therapy,* although still in its infancy, may hold increasing promise for treating people with hereditary diseases (see clinical insight 5.5, p. 166).

Think About It

Would it be possible for a woman with attached earlobes to have children with detached lobes? Use a Punnett square and one or more hypothetical genotypes for the father to demonstrate your point.

Multiple Alleles, Codominance, and Incomplete Dominance

Some genes exist in more than two allelic forms—that is, there are **multiple alleles** within the collective genetic makeup, or **gene pool,** of the population as a whole. For example, there are over 100 alleles responsible for cystic fibrosis, and there are 3 alleles for ABO blood types. Two of the ABO blood type alleles are dominant and symbolized with a capital *I* (for immunoglobulin) and a superscript: I^A and I^B. There is one recessive allele, symbolized with a lowercase *i*. Which two alleles one inherits determines the blood type, as follows:

Genotype	Phenotype
I^AI^A	Type A
I^Ai	Type A
I^BI^B	Type B
I^Bi	Type B
I^AI^B	Type AB
ii	Type O

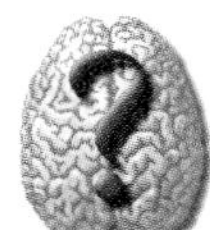

Think About It

Why can't one person have all three of the ABO alleles?

Some alleles are equally dominant, or **codominant.** When both of them are present, both are phenotypically expressed. For example, a person who inherits allele I^A from one parent and I^B from the other has blood type AB. These alleles code for enzymes that produce the surface glycoproteins of red blood cells. Type AB means that both A and B glycoproteins are present, and type O means that neither of them is present.

Other alleles exhibit **incomplete dominance.** When two different alleles are present, the phenotype is intermediate between the traits that each allele would produce alone. *Familial hypercholesterolemia,* the disease discussed in clinical insight 4.4 (p. 134), is a good example. Individuals with two abnormal alleles die of heart attacks in childhood, those with only one abnormal allele typically die as young adults, and those with two normal alleles have normal life expectancies. Thus, the heterozygous individuals suffer an effect between the two extremes.

Polygenic Inheritance and Pleiotropy

Polygenic (multiple-gene) inheritance (fig. 5.18*a*) is a phenomenon in which genes at two or more loci, or even on different chromosomes, contribute to a single phenotypic trait. Human eye and skin colors are normal polygenic traits, for example. They result from the combined expression of all the genes for each trait. Several diseases are also thought to stem from polygenic inheritance, including some forms of alcoholism, mental illness, cancer, and heart disease.

Pleiotropy (ply-OT-roe-pee) (fig. 5.18*b*) is a phenomenon in which one gene produces multiple phenotypic effects. Sickle-cell disease, for example, is caused by a recessive allele that changes one amino acid in hemoglobin. It causes red blood cells (RBCs) to assume an abnormally elongated, pointed shape when oxygen levels are low, and it makes them sticky and fragile. As RBCs rupture, a person becomes anemic and the spleen becomes enlarged as it attempts to dispose of the ruptured cells. Because of the deficiency of RBCs, the blood carries insufficient oxygen to the tissues, resulting in multiple, far-reaching effects on different parts of the body (see chapter 18).

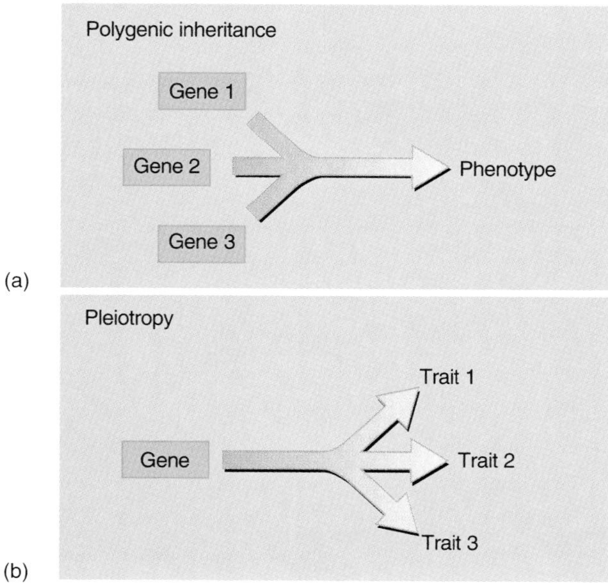

Figure 5.18 Polygenic Inheritance and Pleiotropy. (*a*) In polygenic inheritance, two or more genes combine their effects to produce a single phenotypic trait, such as skin color. (*b*) In pleiotropy, a single gene causes multiple phenotypic traits, as in sickle-cell disease.

Sex Linkage

Sex-linked traits are carried on the X or Y chromosome and therefore tend to be inherited by one sex more than the other. Red-green color blindness and hemophilia, for example, are more common in men than in women because the allele for each is recessive and located on the X chromosome *(X-linked)*. Women have two X chromosomes. If a woman inherits the recessive hemophilia allele *(h)* on one of her X chromosomes, there is still a good chance that her other X chromosome will carry a dominant allele *(H)*, which codes for normal blood-clotting proteins, and her blood will clot normally. Men, on the other hand, have only one X chromosome and normally express any recessive allele found there (fig. 5.19). Ironically, even though this hemophilia is far more common among men than women, a man can inherit it only from his mother. Why? Because only his mother contributes an X chromosome to him. If he inherits *h* on his mother's X chromosome, he will have hemophilia. He has no "second chance" to inherit a normal allele on a second X chromosome. A woman, however, gets an X chromosome from both parents. Even if one parent transmits the recessive allele to her, the chances are high that she will inherit a normal allele from her other parent. She would have to have the extraordinarily bad luck to inherit it from both parents in order for her to have a trait such as hemophilia or red-green color blindness.

The X chromosome is thought to carry about 260 genes, most of which have nothing to do with determining

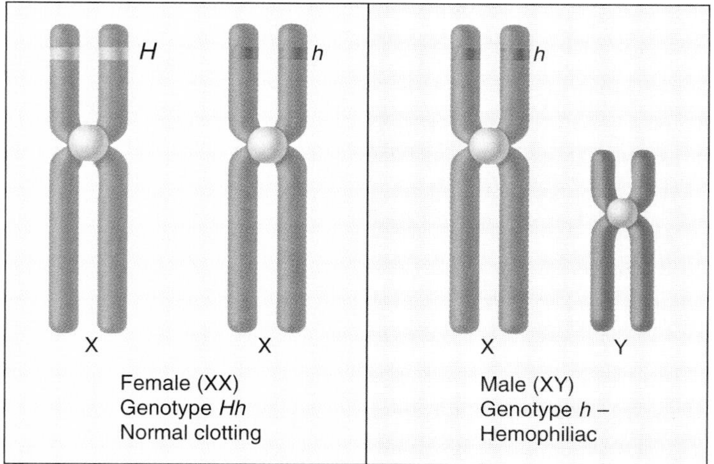

Figure 5.19 Sex-Linked Inheritance of Hemophilia. *Left:* A female who inherits a recessive allele *(h)* for hemophilia from one parent may not exhibit the trait, because she is likely to inherit the dominant allele *(H)* for a normal blood-clotting protein from her other parent. *Right:* A male who inherits *h* from his mother will exhibit hemophilia, because the Y chromosome inherited from his father does not have a gene locus for the clotting protein and therefore has no ability to mask the effect of *h*.

an individual's sex. There are so few functional genes on the Y chromosome—concerned mainly with development of the testes—that virtually all sex-linked traits are associated with the X chromosome.

Penetrance and Environmental Effects

People do not inevitably exhibit the phenotypes that would be predicted from their genotypes. For example, there is a dominant allele that causes *polydactyly,*[25] the presence of extra fingers or toes. We might predict that since it is dominant, anyone who inherited the allele would exhibit this trait. Most do, but others known to have the allele have the normal number of digits. **Penetrance** is the percentage of a population with a given genotype that actually exhibits the predicted phenotype. If 80% of people with the polydactyly allele actually exhibit extra digits, the allele has 80% penetrance.

Another reason the connection between genotype and phenotype is not inevitable is that environmental factors play an important role in the expression of all genes. At the very least, all gene expression depends on nutrition (fig. 5.20). Children born with the hereditary disease *phenylketonuria* (FEN-il-KEE-toe-NEW-ree-uh) *(PKU),* for example, become retarded if they eat a normal diet. However, if PKU is detected early, a child can be placed on a diet low in phenylalanine (an amino acid) and enjoy normal mental development.

No gene can produce a phenotypic effect without nutritional and other environmental input, and no nutrients can produce a body or specific phenotype without genetic instructions that tell cells what to do with them. In short, heredity + environment → phenotype.

25. *poly* = many + *dactyl* = fingers, toes

Dominant and Recessive Alleles at the Population Level

It is a common misconception that dominant alleles must be more common in the gene pool than recessive alleles. The truth is that dominance and recessiveness have little to do with how common an allele is. For example, type O is the most common ABO blood type in North America, but it is caused by the recessive allele *i*. Blood type AB, caused by the two dominant ABO alleles, is the rarest. Polydactyly, caused by a dominant allele, also is rare in the population.

Definitions of some basic genetic terms are summarized in table 5.5.

Before You Go On

Answer the following questions to test your understanding of the preceding section:

16. Why must the carrier of a genetic disease be heterozygous?
17. State at least three reasons why a person's phenotype can't always be determined from the genotype.
18. A man can only inherit color blindness from his mother, whereas a woman must inherit it from both her father and mother to show the trait. Explain this apparent paradox.
19. Cover the left side of table 5.5 with a blank strip of paper, look at the definitions, and fill in the term to which each definition refers. Check your spelling.

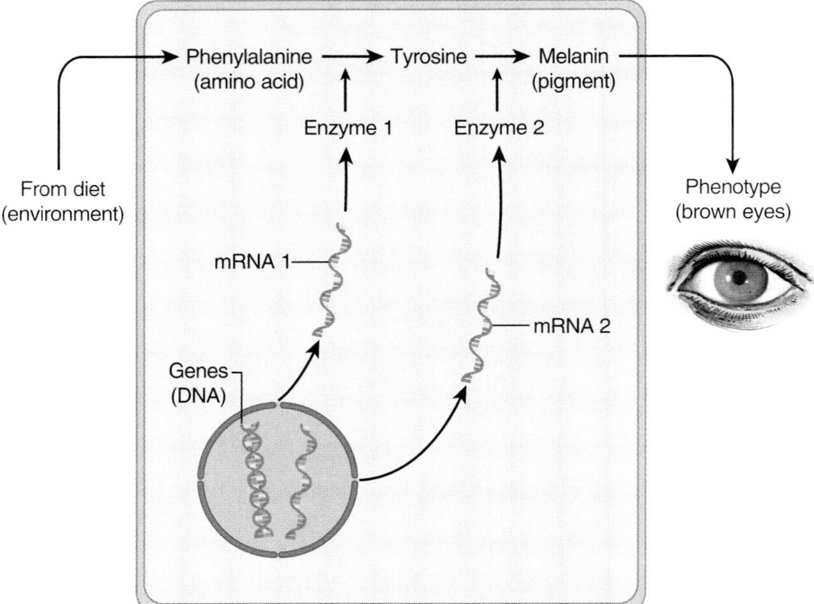

Figure 5.20 The Roles of Environment and Heredity in Producing a Phenotype. Brown eye color requires phenylalanine from the diet (environment) and two genetically coded (hereditary) enzymes to convert phenylalanine to melanin, the eye pigment.

Table 5.5 Basic Terminology of Genetics

Term	Definition
Gene	A segment of DNA that codes for a polypeptide
Genome	All genes possessed by one individual
Gene pool	All genes present in a population
Homologous chromosomes	Two physically identical chromosomes with the same gene loci but not necessarily the same alleles; one is of maternal origin and the other paternal
Sex chromosomes	Two chromosomes (X and Y) that determine a person's sex
Autosomes	All chromosomes except the sex chromosomes; occur in 22 homologous pairs
Locus	The site on a chromosome where a particular gene is located
Allele	Any of the alternative forms that a particular gene can take
Genotype	The alleles that a person possesses for a particular trait
Phenotype	The detectable trait resulting from a particular genotype, such as eye color or blood type
Recessive allele	An allele that is not phenotypically expressed in the presence of a dominant allele; represented with a lowercase letter
Dominant allele	An allele that is phenotypically expressed even in the presence of any other allele; represented with a capital letter
Homozygous	Having identical alleles for a given gene
Heterozygous	Having two different alleles for a given gene
Carrier	A person who carries a recessive allele but does not phenotypically express it
Codominance	A condition in which two alleles are both fully expressed when present in the same individual
Incomplete dominance	A condition in which two alleles are both expressed when present in the same individual, and the phenotype is intermediate between that which either allele would produce alone
Polygenic inheritance	A condition in which a single phenotype results from the combined action of genes at two or more different loci, as in eye color
Pleiotropy	A condition in which a single gene produces multiple phenotypic effects, as in sickle-cell disease
Sex linkage	Inheritance of a gene on the X or Y chromosome, so that the associated phenotype is expressed more in one sex than in the other
Penetrance	The percentage of individuals with a given genotype who actually exhibit the phenotype predicted from it

Clinical Insight 5.5

Genetic Engineering

Hemophilia, cystic fibrosis, muscular dystrophy, and sickle-cell disease are a few of the genetic diseases discussed at various places in this book. Historically, such diseases have been regarded as incurable, and treatment has centered mainly on controlling their symptoms. This is still largely the case, but *genetic engineering* has lately opened the door to new therapeutic approaches.

Recombinant DNA Technology

The most successful form of genetic engineering so far is *recombinant DNA technology. Recombinant DNA (rDNA)* is a DNA molecule produced by splicing together the DNA of two different species. One benefit that has resulted from this is the commercial production of blood-clotting enzyme called *factor VIII*, which is lacking in people with the most common form of hemophilia. Genes for human factor VIII have been engineered into bacterial cells, and industrial cultures of bacteria now synthesize human factor VIII in abundance.

Figure 1 shows the procedure for doing this. First, factor VIII mRNA is extracted from human liver cells. An enzyme called *reverse transcriptase* is employed to copy this mRNA in the form of short DNA molecules called *complementary DNA (cDNA)*. These cDNA molecules are copies of the original gene for factor VIII, except that they lack the introns, or "junk DNA," of the original gene.

The next thing needed is a vehicle, or *gene vector*, that can get this cDNA into bacterial cells. For this purpose, genetic engineers use *plasmids*—small circular DNA molecules isolated from bacteria. The plasmids are snipped open with *restriction enzymes*, and *DNA ligase* is used to join human cDNA to the bacterial plasmid and close the loop. This plasmid—now composed partly of bacterial DNA and partly of human DNA—is the rDNA molecule. When bacteria are exposed to recombinant plasmids, a few of them take up the rDNA. Those that do can be isolated and cultured in industrial vats. The bacteria reproduce every 20 to 30 minutes, and with each cell division, they replicate not only their own DNA but also the human factor VIII gene. Thus, in a short time, millions of copies of the factor VIII gene are produced; hence this procedure is known as *gene amplification*, or *gene cloning*. These cultured bacteria produce human factor VIII, which can be isolated from the culture and purified for pharmaceutical use.

The procedure is simple to understand, but in practice it can be tedious and frustrating. It often fails, but the many failures are more than compensated for by the success stories. Besides factor VIII, some valuable products produced in this way include insulin, growth hormone, interferon (a virus-fighting agent), erythropoietin (a hormone that stimulates red blood cell production), and a number of vaccines.

Gene Therapy

Another exciting dimension of genetic engineering is *gene therapy*—the introduction of new genes into a patient's body to take over the functions of a defective gene. This, too, sounds simple in principle: Remove specific cells from the body such as bone marrow stem cells, replace the defective gene, reintroduce the cells into the patient's body, and let them multiply and produce a gene product that was formerly lacking—perhaps normal hemoglobin or factor VIII. In reality, there are formidable obstacles at each of these steps, and gene therapy has produced only modest results to date.

First of all, isolating target cells from the patient is difficult. Bone marrow stem cells, for example, are scarce and hard to obtain in quantity. Then, even if the target cells can be isolated and the desired gene cloned, getting the gene into them presents a problem. Unlike bacteria, human cells do not take up plasmids. Something else must serve as a vector. Since viruses insert their genes into other cells as part of their normal reproductive cycle, they have been the most promising candidates: get a functional human gene into a virus, and let the virus bind to the target cells and inject the gene into them. However, it is hard to find viruses that not only are capable of this but also can be purified and concentrated and still retain their activity. Also, viruses are too small to carry many of the genes that require replacement therapy—the gene for Duchenne muscular dystrophy, for example, is gigantic. Furthermore, viruses bind only to certain target cells that have receptors for them, and we don't know enough yet about viral receptors on human cells to design viruses that will deliver DNA to the right cells. It is also hard to control what happens to the DNA when it enters a host cell; for every DNA that inserts at the right place on the right chromosome, a thousand times as many trials result in DNA insertion at the wrong place.

There are still other problems to overcome. For example, some of the viruses with the greatest vector potential may cause cancer if they multiply in the host body, so it is necessary to engineer viruses that will deliver the DNA and then perish without reproducing. In addition, the incorporation of foreign DNA into the host cell's DNA requires that the host cells divide, so gene replacement therapy may not be feasible for cells that divide slowly or not at all, such as nerve and muscle cells. These are just a few of the technical problems confronting molecular biologists in this field. The social and ethical implications of manipulating human genes are yet another consideration.

Despite all the obstacles, gene substitution therapy has scored a success in the treatment of a disease called *severe combined immunodeficiency disease (SCID)*. A child with SCID is born without the ability to ward off infection (see chapter 21). About 25% of cases of SCID are caused by a defective gene for the enzyme *adenosine deaminase (ADA)*. Fortunately, this gene is small and relatively manageable. The first federally approved clinical trials of gene substitution therapy for ADA deficiency began in September 1990. White blood cells called *T cells* were isolated from children with SCID, and viruses were used to introduce normal ADA genes into the cells. The engineered T cells were then reintroduced into the children, where they produced ADA and allowed them to live normal lives. This is not a lifetime cure, however. These children require periodic treatment with more T cells as the engineered cells die out.

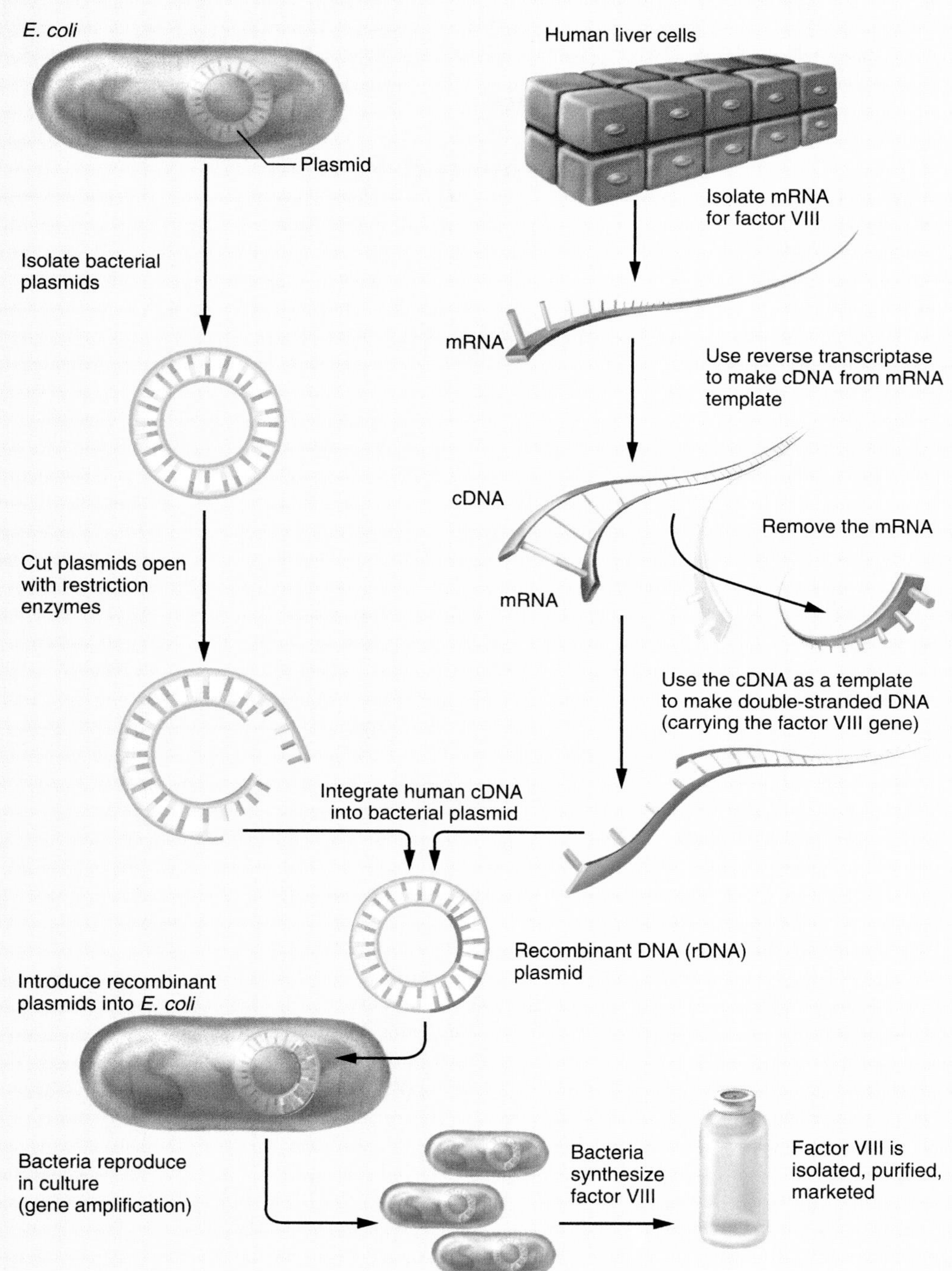

Figure 1 The Recombinant DNA Technique. In this example, a blood-clotting protein called factor VIII is produced for the treatment of patients with hemophilia.

Chapter Review

Online Learning Center

Study resources for this chapter are available at www.mhhe.com/saladin/

Study Outline

The Nucleus and Nucleic Acids (p. 141)

1. Nuclear structure
   a. Nuclear envelope and pores
   b. Nucleoplasm and nucleoli
   c. Chromatin
2. Organization of the chromatin
   a. Nucleosome structure
   b. Supercoiling of DNA
3. DNA structure and function
   a. Nucleotides—monomers of DNA and RNA
   b. Sugar-phosphate backbone
   c. Nitrogenous bases
      - Pyrimidines—single ring
         - Cytosine (C)
         - Thymine (T)
         - Uracil (U)
      - Purines—double ring
         - Adenine (A)
         - Guanine (G)
   d. Double helix
   e. Complementary base pairing
      - A across from T
      - G across from C
   f. A code for protein structure
   g. Genes and the genome
4. RNA structure and function
   a. Smaller and single-stranded
   b. Ribose replaces deoxyribose
   c. Uracil replaces thymine
   d. Interprets code in DNA
   e. Directs protein synthesis

Protein Synthesis and Secretion (p. 146)

1. The genetic code
   a. Four nucleotides representing 20 amino acids
   b. DNA base triplets and RNA codons
   c. Stop and start codons
2. Transcription
   a. Action of RNA polymerase
   b. Pre-mRNA
3. Posttranscriptional modification
4. Movement of mRNA into cytoplasm
5. Translation
   a. Attachment of ribosome to mRNA
   b. Binding of tRNA
   c. Growth of peptide chain
   d. Polyribosomes
6. Chaperones and protein structure
7. Posttranslational modification
   a. Role of leader sequence
   b. Changes occurring in ER cisterna
   c. Formation of transport vesicles
   d. Changes occurring in Golgi complex
8. Packaging and secretion
   a. Golgi vesicles
   b. Secretory vesicles and exocytosis

DNA Replication and the Cell Cycle (p. 153)

1. DNA replication
   a. Unwinding from histones
   b. Uncoiling by DNA helicase
   c. Replication by DNA polymerase
   d. Rewinding around new histones
   e. Semiconservative nature of DNA replication
2. Errors and mutations
   a. Error rates in bacteria
   b. Proofreading and error correction
   c. Effects of mutations
3. The cell cycle
   a. G_1 phase: normal cellular functions
   b. S phase: DNA replication
   c. G_2 phase: preparation for mitosis
   d. M phase: nuclear and cytoplasmic division
   e. G_0 phase: cells that have left the cycle
4. Functions of mitosis
   a. Embryonic development
   b. Tissue growth
   c. Replacement of old and dead cells
   d. Repair of injury
5. Stages of mitosis
   a. Prophase
      - Chromatin condenses into chromosomes
      - Nuclear envelope disintegrates
      - Mitotic spindle begins to form
   b. Metaphase
      - Chromosomes line up on equator
      - Spindle fibers attach to centromeres
   c. Anaphase
      - Centromeres divide
      - Sister chromatids move apart
   d. Telophase
      - Chromosomes decondense
      - Nuclear envelopes form
      - Mitotic spindle breaks down
   e. Cytokinesis: cytoplasmic division
6. Timing of cell division
7. Cancer
   a. Tumors and oncology
   b. Benign versus malignant tumors
   c. Causes of cancer
      - Chemical carcinogens
      - Radiation
      - Viruses
   d. Mutagens
   e. Defenses against cancer
   f. Cancer genes
      - Oncogenes
      - Tumor suppressor genes
   g. Effects of cancer
      - Loss of organ function
      - Compression of tissues and organs
      - Energy diversion

Chromosomes and Heredity (p. 159)

1. The karyotype
   a. Homologous chromosomes
      - Autosomes (22 pairs)
      - Sex chromosomes (X and Y)

b. Germ cells and somatic cells
c. Haploid and diploid cells
d. Maternal and paternal chromosomes
2. Genes and alleles
a. Gene loci
b. Dominant and recessive alleles
c. Homozygous and heterozygous genotypes
d. Phenotypes
e. Heterozygous carriers
3. Multiple alleles and the gene pool
4. Codominance
5. Incomplete dominance
6. Polygenic inheritance
7. Pleiotropy
8. Sex linkage
9. Penetrance and environmental effects
10. Alleles at the population level

Selected Vocabulary

cytogenetics 141
molecular genetics 141
enucleate 141
multinucleate 141
nuclear envelope 141
nuclear pore 141
nucleoplasm 141
chromatin 142
nucleoli 142
nucleosome 143
histone 143
nonhistone 143
nucleotide 143
nitrogenous base 143
cytosine 143
thymine 143
uracil 143
adenine 143
guanine 143
base pair 144
law of complementary base pairing 144
gene 144
genome 144
messenger RNA (mRNA) 146
ribosomal RNA (rRNA) 146
transfer RNA (tRNA) 146
genetic code 146
base triplet 146
codon 146
stop codon 146
start codon 147
transcription 147
RNA polymerase 147
pre-mRNA 148
exon 148
intron 148
posttranscriptional modification 148
genetic translation 148
leader sequence 148
anticodon 148
polyribosome 149
chaperone 150
posttranslational modification 151
transport vesicle 151
Golgi vesicle 151
secretory vesicle 151
DNA helicase 153
DNA polymerase 153
semiconservative replication 153
mutation 153
cell cycle 154
G_1 phase 154
S phase 155
G_2 phase 155
M phase 155
interphase 155
G_0 phase 155
mitosis 155
prophase 156
chromosome 156
sister chromatids 156
centromere 156
metaphase 156
mitotic spindle 156
anaphase 156
kinetochore 156
telophase 156
cytokinesis 156
growth factor 156
contact inhibition 156
neoplasm 156
oncology 156
benign 156
cancer 156
malignant 156
metastasis 158
carcinogen 158
mutagen 158
oncogene 159
proto-oncogene 159
tumor suppressor (TS) gene 159
heredity 159
karyotype 160
homologous chromosomes 160
sex chromosome 160
autosome 160
germ cell 160
somatic cell 160
diploid 160
haploid 161
segregation 161
locus 161
allele 161
dominant 161
recessive 161
homozygous 161
heterozygous 161
genotype 161
phenotype 161
carrier 162
multiple alleles 162
gene pool 162
codominant 162
incomplete dominance 162
polygenic inheritance 163
pleiotropy 163
sex-linked trait 163
penetrance 164

Testing Your Recall

1. Production of more than one phenotypic trait by a single gene is called
   a. pleiotropy.
   b. genetic determinism.
   c. codominance.
   d. penetrance.
   e. genetic recombination.

2. When a ribosome reads a codon on mRNA, it must bind to the __________ of a corresponding tRNA.
   a. start codon
   b. stop codon
   c. intron
   d. exon
   e. anticodon

3. The normal functions of a liver cell—synthesizing proteins, detoxifying wastes, storing glycogen, and so forth—are done during its
   a. anaphase.
   b. telophase.
   c. G_1 phase.
   d. G_2 phase.
   e. synthesis phase.

4. Two genetically identical strands of a chromosome, joined at the centromere, are its
   a. kinetochores.
   b. centrioles.
   c. sister chromatids.
   d. homologous chromatids.
   e. nucleosomes.

5. Which of the following is *not* found in DNA?
   a. thymine
   b. phosphate
   c. cytosine

d. deoxyribose
e. uracil

6. Genetic transcription is performed by
   a. ribosomes.
   b. RNA polymerase.
   c. DNA polymerase.
   d. helicase.
   e. chaperones.
7. A chaperone comes into play in
   a. the folding of a new protein into its tertiary structure.
   b. keeping DNA organized within the nucleus.
   c. escorting sister chromatids to opposite daughter cells during mitosis.
   d. repairing DNA that has been damaged by mutagens.
   e. preventing malignant cells from metastasizing.
8. An allele that is not phenotypically expressed in the presence of an alternative allele of the same gene is said to be
   a. codominant.
   b. lacking penetrance.
   c. heterozygous.
   d. recessive.
   e. subordinate.
9. Semiconservative replication occurs during
   a. transcription.
   b. translation.
   c. posttranslational modification.
   d. the S phase of the cell cycle.
   e. mitosis.
10. Mutagens sometimes cause no harm to cells for all of the following reasons *except*
    a. some mutagens are natural, harmless products of the cell itself.
    b. peroxisomes detoxify some mutagens before they can do any harm.
    c. the body's DNA repair mechanisms detect and correct genetic damage.
    d. change in a codon does not always change the amino acid encoded by it.
    e. some mutations change protein structure in ways that are not critical to normal function.
11. The cytoplasmic division at the end of mitosis is called ___________.
12. The alternative forms in which a single gene can occur are called ___________.
13. The pattern of nitrogenous bases that represents the 20 amino acids of a protein is called the ___________.
14. Several ribosomes attached to one mRNA, which they are all transcribing, form a cluster called a/an ___________.
15. Environmental agents that can cause cells to become malignant are called ___________.
16. Newly synthesized proteins may be escorted to their destination in a cell by other proteins called ___________.
17. At prophase, a cell has ___________ chromosomes, ___________ chromatids, and ___________ molecules of DNA.
18. Tumors that can metastasize to other organs are said to be ___________.
19. Cells are stimulated to divide by chemical signals called ___________ .
20. All chromosomes except the sex chromosomes are called ___________ .

Answers in appendix B

Testing Your Comprehension

1. Reexamine table 4.2 in the previous chapter. In light of what you have learned in the present chapter, explain why nuclear pores are so much larger than pores in the plasma membrane. What is the functional significance of this difference?
2. Suppose the DNA double helix had a backbone of alternating nitrogenous bases and phosphates, with the deoxyribose components facing each other across the middle of the helix. Why couldn't such a molecule function as a genetic code?
3. Given the information in this chapter, present an argument that evolution is not merely possible but inevitable. (Hint: Review the definition of evolution in chapter 1.)
4. What would be the minimum length (approximate number of bases) of an mRNA that coded for a protein 300 amino acids long?
5. Give three examples from this chapter of the complementarity of structure and function.

Answers at the Online Learning Center

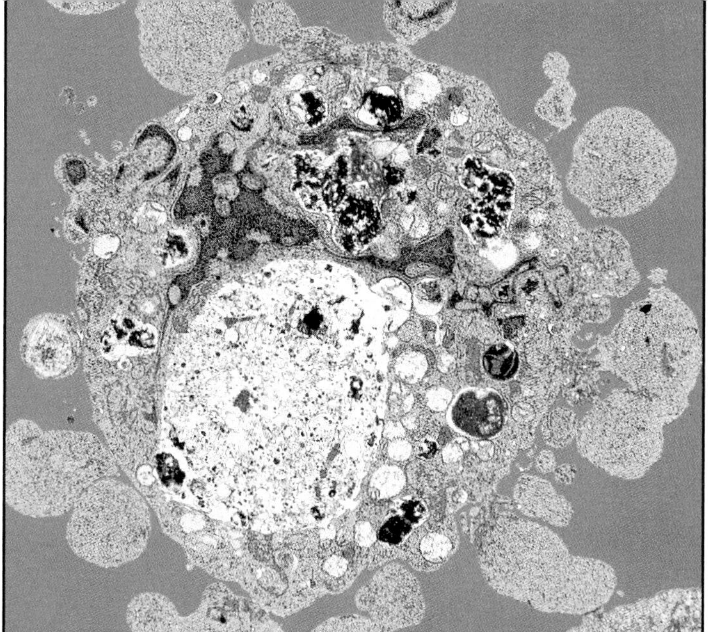

A cell undergoing apoptosis (programmed cell death) (TEM)

CHAPTER 6

Histology

Chapter Outline

Insights

Study with Your ESP CD-ROM

Refer to your ESP Tutorial CD-ROM for helpful animations, activities, and quizzes related to the material covered in this chapter. The following are suggested areas to go to on the CD and are presented according to "Unit/Topic":
- Levels of Organization/Tissues

Brushing Up

To understand this chapter, it is important that you understand or brush up on the following concepts:
- Body cavities and membranes (p. 35)
- Glycoproteins and proteoglycans (p. 83)
- Terminology of cell shapes (p. 104)
- Secretory vesicles and exocytosis (p. 151)

With its 50 trillion cells and thousands of organs, the human body may seem to be a structure of forbidding complexity. Fortunately for our health, longevity, and self-understanding, the biologists of past generations were not discouraged by this complexity, but discovered patterns that made it more understandable. One of these patterns is the fact that these trillions of cells belong to only 200 different types or so, and these cells are organized into tissues that fall into just 4 broad categories—*epithelial, connective, nervous,* and *muscular tissue.* An organ is a structure with discrete structural boundaries that is composed of two or more of these tissue types (usually all four). Organs derive their function not from their cells alone but from how the cells are organized into tissues. Cells are specialized for certain tasks—muscle contraction, defense, enzyme secretion, and so forth. No one cell type has the mechanisms to carry out all of the body's vital functions. Cells work together at certain tasks and form tissues that carry out a particular function, such as nerve signaling or nutrient digestion.

The study of tissues and how they are arranged into organs is called **histology,**[1] or **microscopic anatomy.** That is the subject of this chapter. Here we study the four tissue classes; the variations within each class; how to recognize tissue types microscopically and relate their microscopic anatomy to their function; how tissues are arranged to form an organ; and how tissues change over the life of the individual as they grow, shrink, or change from one tissue type to another. Histology bridges the gap between the *cytology* of the preceding chapters and the *organ system* approach of the chapters that follow.

The Study of Tissues

Objectives

When you have completed this section, you should be able to

- name the four primary classes into which all adult tissues are classified.
- name the three embryonic germ layers and some adult tissues derived from each; and
- visualize the three-dimensional shape of a structure from a two-dimensional tissue section.

The Primary Tissue Classes

Epithelial, connective, nervous, and muscular tissue are called the four **primary tissues** (table 6.1). These tissues differ from each other in the types and functions of their cells, the characteristics of the **matrix (extracellular material)** that surrounds the cells, and the relative amount of space occupied by cells versus matrix. In muscle and epithelium, the cells are so close together that the matrix is scarcely visible, while in connective tissues, the matrix usually occupies much more space than the cells do.

1. *histo* = tissue + *logy* = study of

The matrix is composed of fibrous proteins and, usually, a clear gel variously known as **ground substance, tissue fluid, extracellular fluid (ECF), interstitial**[2] **fluid,** or **tissue gel.** In cartilage and bone, it can be rubbery or stony in consistency. The ground substance contains water, gases, minerals, nutrients, wastes, and other chemicals. In summary, a tissue is composed of cells and matrix, and the matrix is composed of fibers and ground substance.

Embryonic Tissues

Human development begins with a single cell, the fertilized egg, which soon divides to produce scores of identical, smaller cells. The first tissues appear when these cells start to organize themselves into layers—first two, and soon three strata called the **primary germ layers,** which give rise to all of the body's mature tissues. The three layers are called *ectoderm, mesoderm,* and *endoderm.* The **ectoderm**[3] is an outer layer that gives rise to the epidermis and nervous system. The inner layer, the **endoderm,**[4] gives rise to the mucous membranes of the digestive and respiratory tracts and to the digestive glands, among other things. Between these two is the **mesoderm,**[5] a layer of more loosely organized cells. Mesoderm eventually turns to a gelatinous tissue called **mesenchyme,** composed of fine, wispy collagen (protein) fibers and branching cells called *fibroblasts* embedded in a gelatinous ground substance. Mesenchyme closely resembles the connective tissue layer in figure 6.13. It gives rise to muscle, bone, and blood, among other tissues. Most organs are composed of tissues derived from two or more primary germ layers. The rest of this chapter concerns the "mature" tissues that exist from infancy through adulthood.

Interpreting Tissue Sections

To study tissues, histologists use a variety of techniques for preserving, sectioning (slicing), and staining them to show their structural details as clearly as possible. Tissue specimens are preserved in a **fixative**—a chemical such as formalin that prevents decay. After fixation, most tissues are cut into very thin slices called **histological sections,** typically only one or two cells thick, to allow the light of a microscope to pass through and to reduce the confusion of the image that would result from many layers of overlapping cells. These sections are then mounted on slides and artificially colored with histological **stains** to bring out detail. If they were not stained, most tissues would appear very pale gray. With stains that bind to different

2. *inter* = between + *stit* = to stand
3. *ecto* = outer + *derm* = skin
4. *endo* = inner
5. *meso* = middle

Table 6.1 The Four Primary Tissue Classes

Type	Definition	Representative Locations
Epithelial	Tissue composed of layers of closely spaced cells that cover organ surfaces and serve for protection, secretion, and absorption	Epidermis Inner lining of digestive tract Liver and other glands
Connective	Tissue with more matrix than cell volume, often specialized to support, bind together, and protect organs	Tendons and ligaments Cartilage and bone Blood and lymph
Nervous	Tissue containing excitable cells specialized for rapid transmission of coded information to other cells	Brain Spinal cord Nerves
Muscular	Tissue composed of elongated, excitable cells specialized for contraction	Skeletal muscles Heart (cardiac muscle) Walls of viscera (smooth muscle)

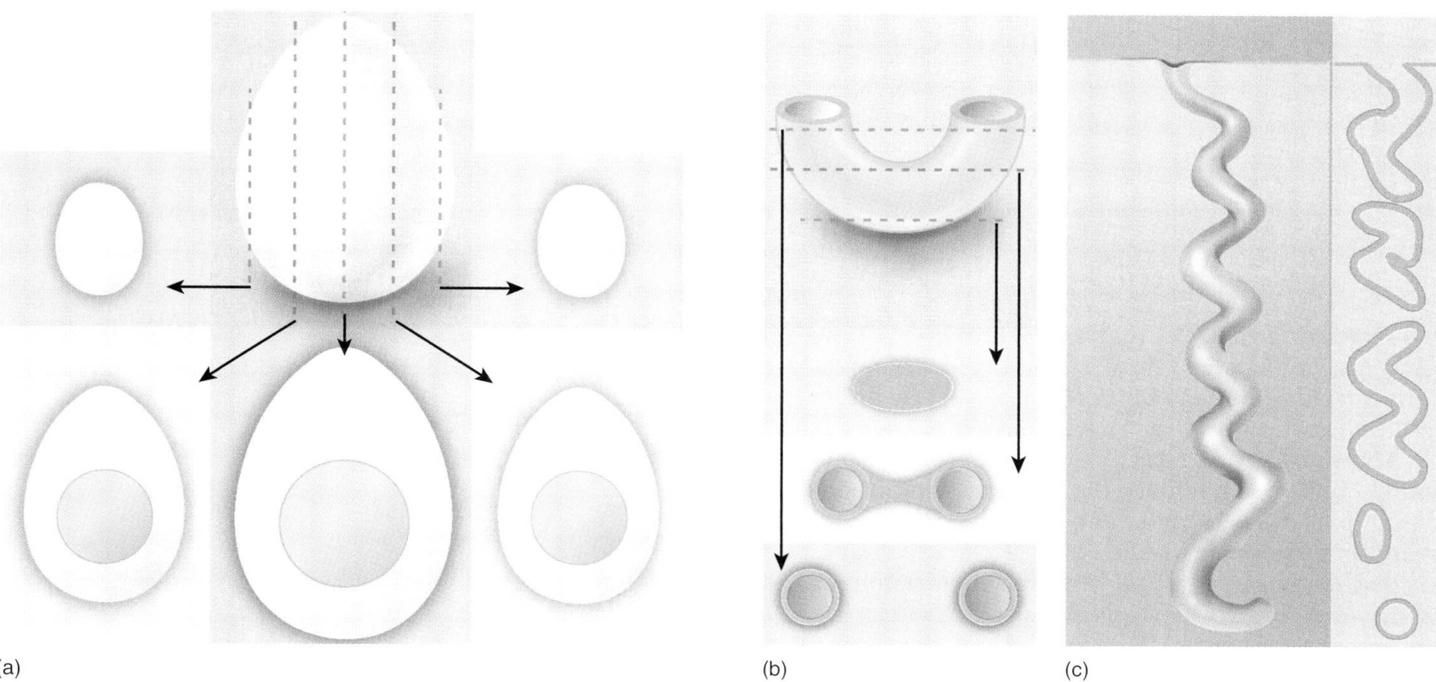

Figure 6.1 Three-Dimensional Interpretation of Two-Dimensional Images. (*a*) A boiled egg. Note that grazing sections (*far left* and *right*) would miss the yolk, just as a tissue section may miss a nucleus or other structure. (*b*) Elbow macaroni, which resembles many curved ducts and tubules. A section far from the bend would give the impression of two separate tubules; a section near the bend would show two interconnected lumina (cavities); and a section still farther down could miss the lumen completely. (*c*) A coiled gland in three dimensions and as it would look in a vertical tissue section.

components of a tissue, however, you may see pink cytoplasm, violet nuclei, and blue, green, or golden brown protein fibers, depending on the stain used.

Sectioning a tissue reduces a three-dimensional structure to a two-dimensional slice. You must keep this in mind and try to translate the microscopic image into a mental image of the whole structure. Like the boiled egg and elbow macaroni in figure 6.1, an object may look quite different when it is cut at various levels, or *planes of section.* A coiled tube, such as a gland of the uterus (fig. 6.1*c*), is often broken up into multiple portions since it meanders in and out of the plane of section. An experienced viewer, however, would recognize that the separated pieces are parts of a single tube winding its way to

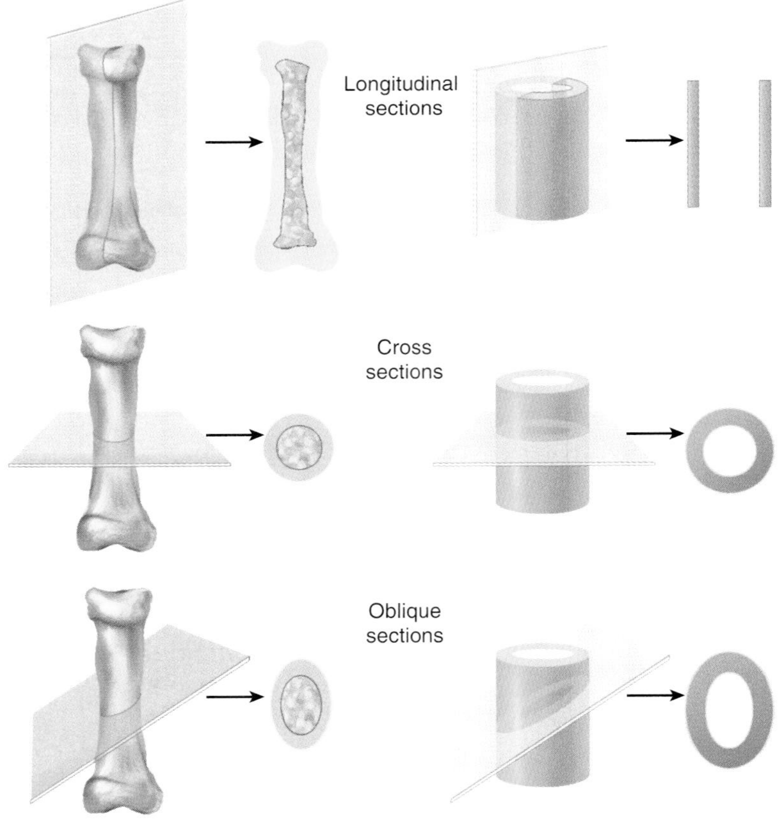

Figure 6.2 Longitudinal, Cross, and Oblique Sections. Note the effect of the plane of section on the two-dimensional appearance of elongated structures such as bones and blood vessels.

the organ surface. Note that a grazing slice through a boiled egg might miss the yolk, just as a tissue section might miss the nucleus of a cell or an egg in the ovary, even though these structures were present.

Many anatomical structures are significantly longer in one direction than another—the humerus and esophagus, for example. A tissue cut in the long direction is called a **longitudinal section (l.s.),** and one cut perpendicular to this is a **cross section** (**c.s.** or **x.s.**), or **transverse section (t.s.).** A section cut at an angle between a longitudinal and cross section is an **oblique section.** Figure 6.2 shows how certain organs look when sectioned on each of these planes.

Before You Go On

Answer the following questions to test your understanding of the preceding section:

1. Classify each of the following into one of the four primary tissue classes: the skin surface, fat, the spinal cord, most heart tissue, bones, tendons, blood, and the inner lining of the stomach.
2. What are tissues composed of in addition to cells?
3. What embryonic germ layer gives rise to nervous tissue? To the liver? To muscle?
4. What is the term for a thin, stained slice of tissue mounted on a microscope slide?

Epithelial Tissue

Objectives

When you have completed this section, you should be able to

- describe the properties that distinguish epithelium from other tissue classes;
- list and classify eight types of epithelium, distinguish them from each other, and state where each type can be found in the body;
- explain how the structural differences between epithelia relate to their functional differences; and
- visually recognize each epithelial type from specimens or photographs.

Epithelial[6] **tissue** covers the surface of the body, lines the body cavities, forms the external and internal linings of many organs, and constitutes most of the bulk of most glands. An epithelium is composed of one or more layers of closely adhering cells. The extracellular material is so thin it is not visible with the light microscope, and epithelia allow no room for blood vessels. They do, however, almost always lie on a layer of loose connective tissue and depend on its blood vessels for nourishment and waste removal.

6. *epi* = upon + *theli* = nipple, female

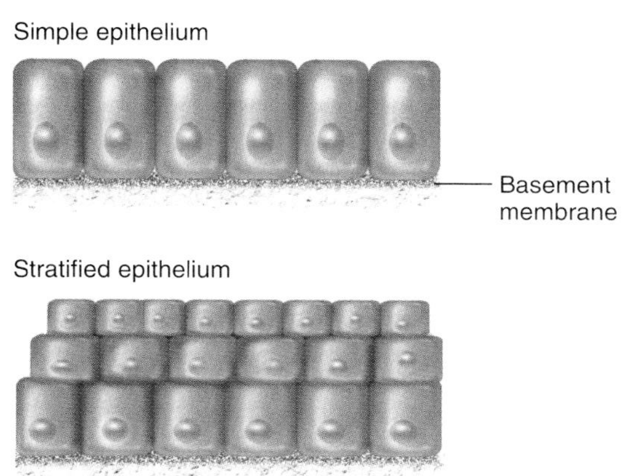

Figure 6.3 Comparison of Simple and Stratified Epithelia.

Between an epithelium and the underlying connective tissue is a layer called the **basement membrane,** usually too thin to be visible with the light microscope. It contains collagen, adhesive glycoproteins called *laminin* and *fibronectin,* and a large protein-carbohydrate complex called *heparan sulfate.* It gradually blends with collagenous and reticular fibers on the connective tissue side. The basement membrane serves to anchor an epithelium to the connective tissue below it. The surface of an epithelial cell that faces the basement membrane is its *basal surface,* and the one that faces away from the basement membrane is the *apical surface.*

Epithelia are classified into two broad categories—*simple* and *stratified epithelia*—with four types in each category. In a simple epithelium, every cell touches the basement membrane, whereas in a stratified epithelium, some cells rest on top of other cells and do not contact the basement membrane (fig. 6.3).

Simple Epithelia

Generally, a **simple epithelium** has only one layer of cells, although this is a somewhat debatable point in the *pseudostratified* type. Three types of simple epithelia are named for the shapes of their cells: **simple squamous**[7] (thin scaly cells), **simple cuboidal** (square or round cells), and **simple columnar** (tall narrow cells) **epithelia.** In the fourth type, **pseudostratified columnar epithelium,** not all cells reach the free surface; the shorter cells are covered over by the taller ones. This epithelium looks stratified in most tissue sections, but careful examination, especially with the electron microscope, shows that every cell reaches the basement membrane. Simple columnar and pseudostratified columnar epithelia often produce protective mucous coatings over the mucous membranes. The mucus is secreted by wineglass-shaped **goblet cells.**

Table 6.2 illustrates and summarizes the structural and functional differences among these four types.

Stratified Epithelia

Stratified epithelia range from 2 to 20 or more layers of cells, with some cells resting directly on others and only the deepest layer resting on the basement membrane. Three of the stratified epithelia are named for the shapes of their surface cells: **stratified squamous, stratified cuboidal,** and **stratified columnar epithelia.** The deeper cells, however, may be of a different shape than the surface cells. The fourth type, **transitional epithelium,** was named when it was thought to represent a transitional stage between stratified squamous and stratified columnar epithelium. This is now known to be untrue, but the name has persisted.

Stratified columnar epithelium is rare—seen only in places where two other epithelial types meet, as in limited regions of the pharynx, larynx, anal canal, and male urethra. We will not consider this type any further. The other three types are illustrated and summarized in table 6.3.

The most widespread epithelium in the body is stratified squamous epithelium, which deserves further discussion. Its deepest layer of cells are cuboidal to columnar and undergo continual mitosis. Their daughter cells push toward the surface and become flatter (more *squamous,* or scalelike) as they migrate farther upward, until they finally die and flake off. Their separation from the surface is called **exfoliation,** or **desquamation** (fig. 6.12); the study of exfoliated cells is called *exfoliate cytology.* You can easily study exfoliated cells by scraping your gums with a toothpick, smearing this material on a slide, and staining it. A similar procedure is used in the **Pap smear** (fig. 6.13), an examination of exfoliated cells from the cervix for signs of uterine cancer.

Stratified squamous epithelia are of two kinds—keratinized and nonkeratinized. A **keratinized** epithelium, found on the skin surface (epidermis), is covered with a layer of compact, dead squamous cells. These cells are packed with the durable protein keratin and coated with water repellent. The skin surface is therefore relatively dry, it retards water loss from the body, and it resists penetration by disease organisms. (Keratin is also the protein of which animal horns are made, hence its name.[8]) The tongue, oral mucosa, esophagus, vagina, and a few other internal membranes are covered with the **nonkeratinized** type, which lacks the surface layer of dead cells. This type provides a surface that is, again, abrasion-resistant, but also moist and slippery. These characteristics are well suited to resist stress produced by the chewing and swallowing of food and by sexual intercourse and childbirth.

7. *squam* = scale

8. *kerat* = horn

Table 6.2 Simple Epithelia

Simple Squamous Epithelium

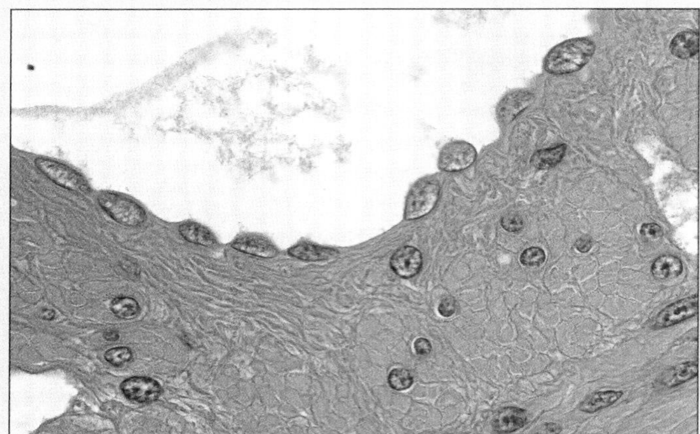
(a)

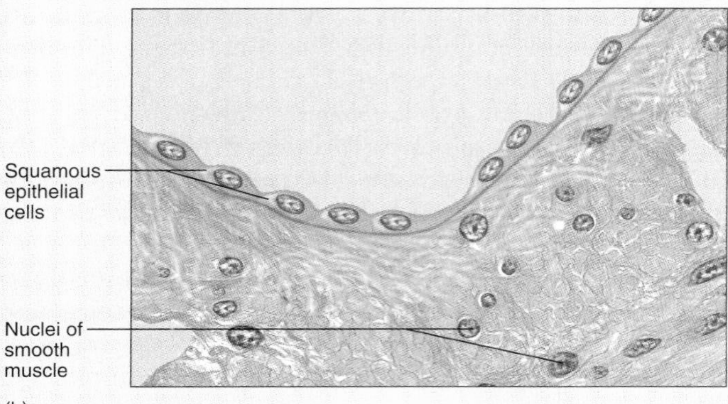

(b)

Figure 6.4 External Surface (serosa) of the Small Intestine.

Microscopic appearance: Single layer of thin cells, shaped like fried eggs with bulge where nucleus is located; nucleus flattened in the plane of the cell, like an egg yolk; cytoplasm may be so thin it is hard to see in tissue sections; in surface view, cells have angular contours and nuclei appear round

Representative locations: Air sacs (alveoli) of lungs; glomerular capsules of kidneys; some kidney tubules; inner lining (endothelium) of heart and blood vessels; serous membranes of stomach, intestines, and some other viscera; surface mesothelium of pleurae, pericardium, peritoneum, and mesenteries

Functions: Allows rapid diffusion or transport of substances through membrane; secretes lubricating serous fluid

Simple Cuboidal Epithelium

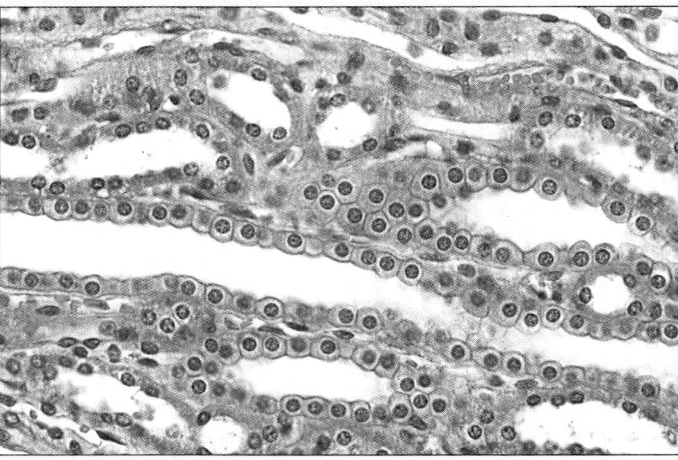
(a)

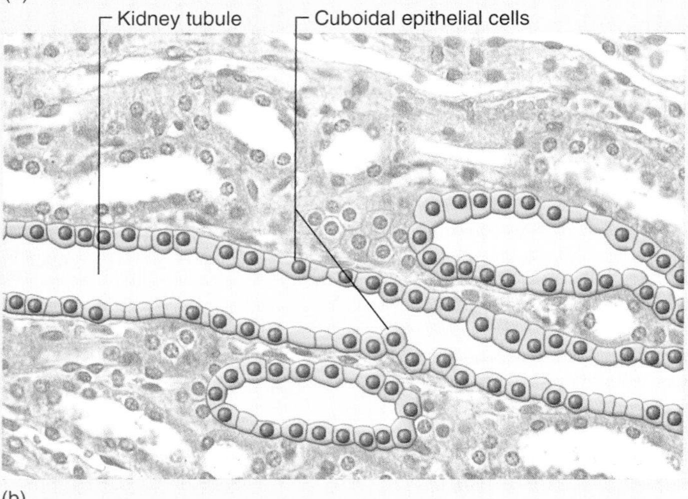

(b)

Figure 6.5 Kidney Tubules.

Microscopic appearance: Single layer of square or round cells; in glands, cells often pyramidal and arranged like segments of an orange around a central space; spherical, centrally placed nuclei; often with a brush border of microvilli in some kidney tubules; ciliated in bronchioles of lung

Representative locations: Liver, thyroid, mammary, salivary, and other glands; most kidney tubules; bronchioles

Functions: Absorption and secretion; production of protective mucous coat; movement of respiratory mucus

Table 6.2 Simple Epithelia (Continued)

Simple Columnar Epithelium

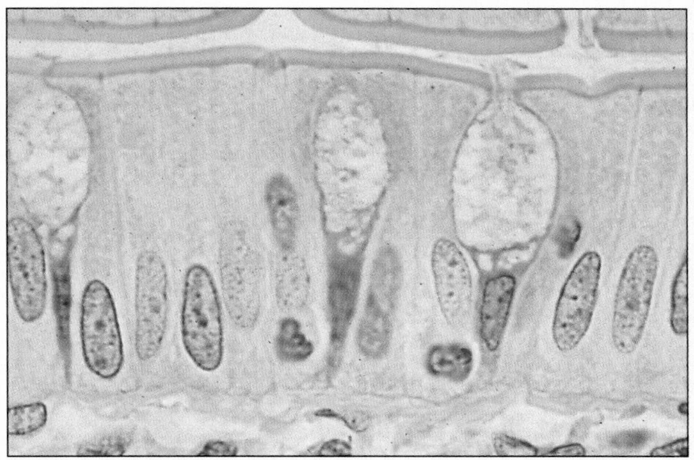

(a)

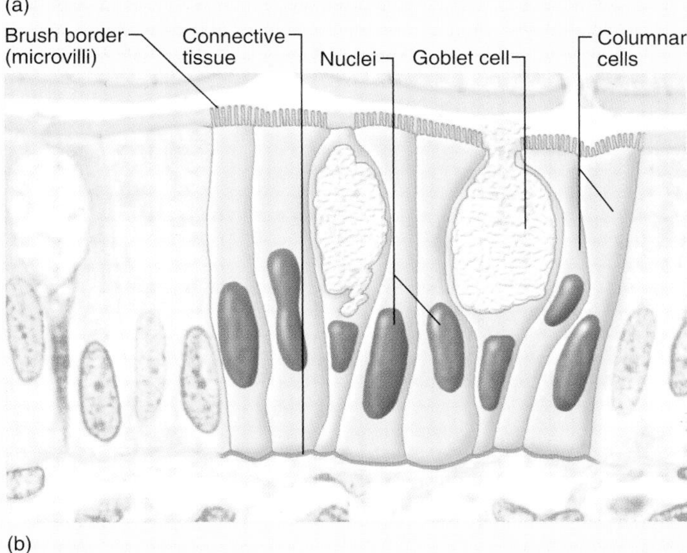

(b)

Figure 6.6 Internal Surface (mucosa) of the Small Intestine.

Microscopic appearance: Single layer of tall, narrow cells; oval or sausage-shaped nuclei, vertically oriented, usually in basal half of cell; apical portion of cell often shows secretory vesicles visible with TEM; often shows a brush border of microvilli; ciliated in some organs; may possess goblet cells

Representative locations: Inner lining of stomach, intestines, gallbladder, uterus, and uterine tubes; some kidney tubules

Functions: Absorption and secretion; movement of egg and embryo in uterine tube; secretion of mucus

Pseudostratified Columnar Epithelium

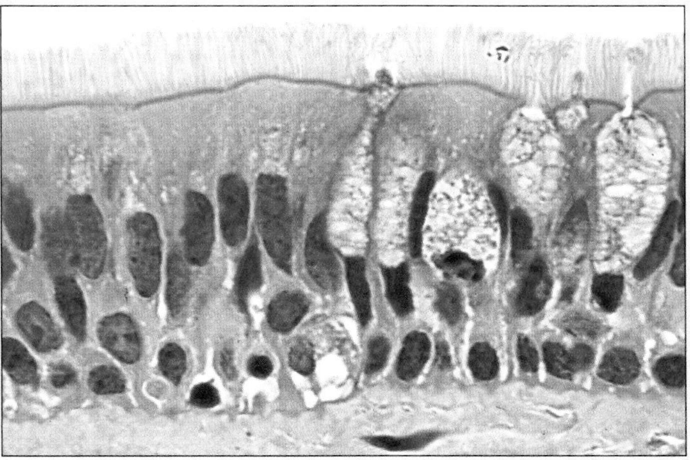

(a)

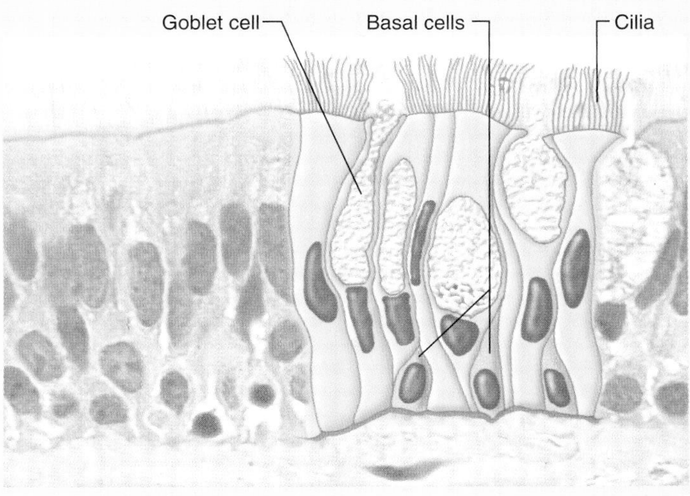

(b)

Figure 6.7 Mucosa of the Trachea.

Microscopic appearance: Looks multilayered; some cells do not reach free surface but all cells reach basement membrane; nuclei at several levels in deeper half of epithelium; often with goblet cells; often ciliated

Representative locations: Respiratory tract from nasal cavity to bronchi; portions of male reproductive tract

Functions: Secretes and propels mucus

Table 6.3 Stratified Epithelia

Stratified Squamous Epithelium—Keratinized

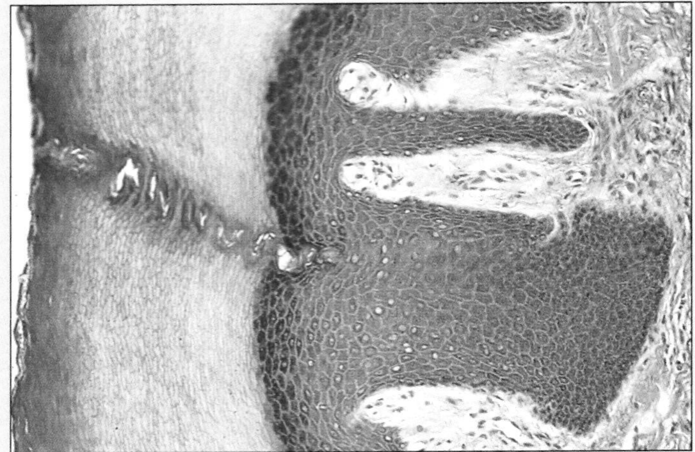

(a)

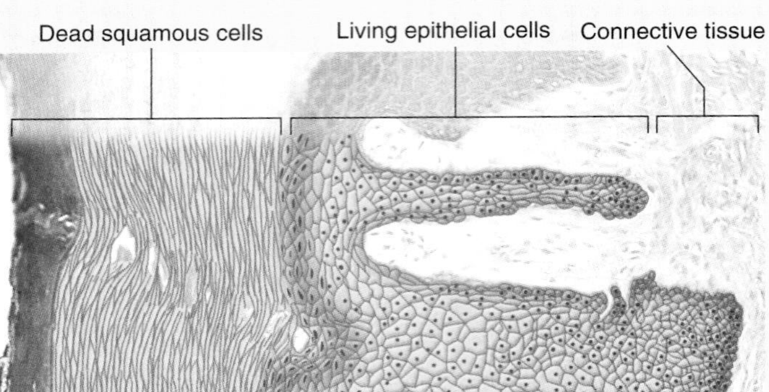

(b)

Figure 6.8 Skin from the Sole of the Foot.

Microscopic appearance: Multiple cell layers with cells becoming increasingly flat and scaly toward surface; surface covered with a layer of compact dead cells without nuclei; basal cells may be cuboidal to columnar

Representative locations: Epidermis; palms and soles are especially heavily keratinized

Functions: Resists abrasion; retards water loss through skin; resists penetration by pathogenic organisms

Stratified Squamous Epithelium—Nonkeratinized

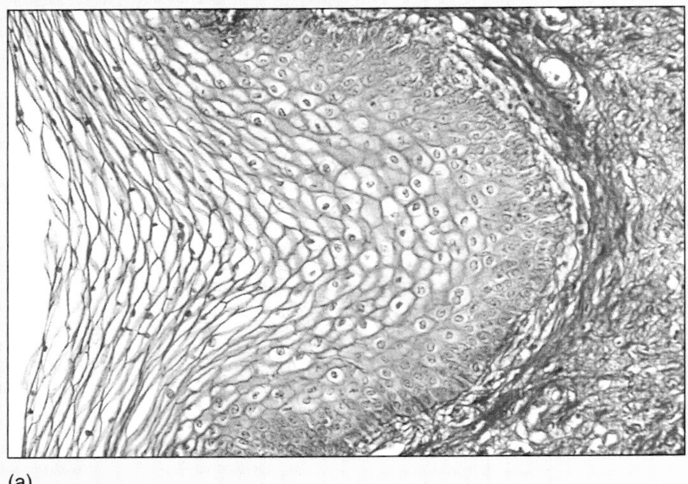

(a)

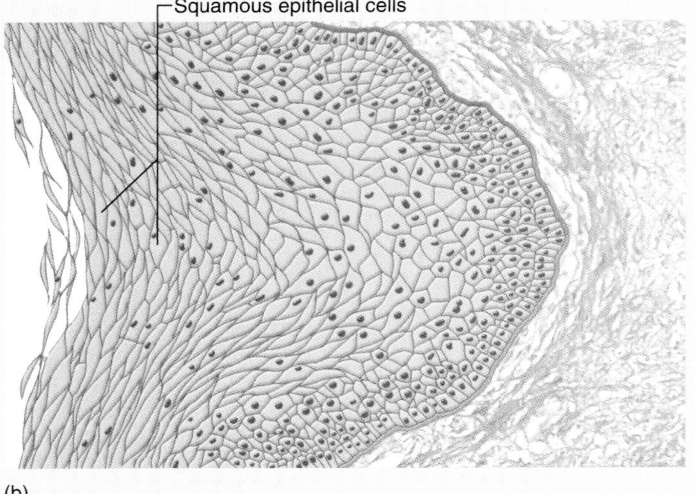

(b)

Figure 6.9 Mucosa of the Vagina.

Microscopic appearance: Same as keratinized epithelium but without the surface layer of dead cells

Representative locations: Tongue, oral mucosa, esophagus, anal canal, vagina

Functions: Resists abrasion and penetration by pathogenic organisms

Table 6.3 Stratified Epithelia (Continued)

Stratified Cuboidal Epithelium

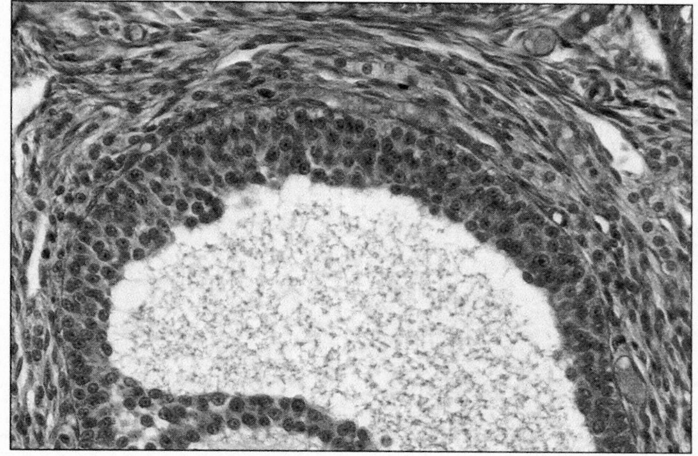

(a)

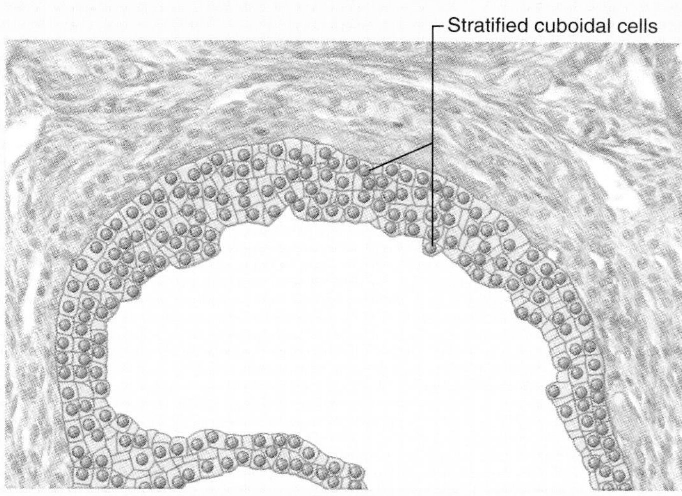

(b)

Figure 6.10 Wall of a Follicle in the Ovary.

Microscopic appearance: Two or more layers of cells; surface cells square or round

Representative locations: Sweat gland ducts; egg-producing vesicles (follicles) of ovaries; sperm-producing ducts (seminiferous tubules) of testis

Functions: Contributes to sweat secretion; secretes ovarian hormones; produces sperm

Transitional Epithelium

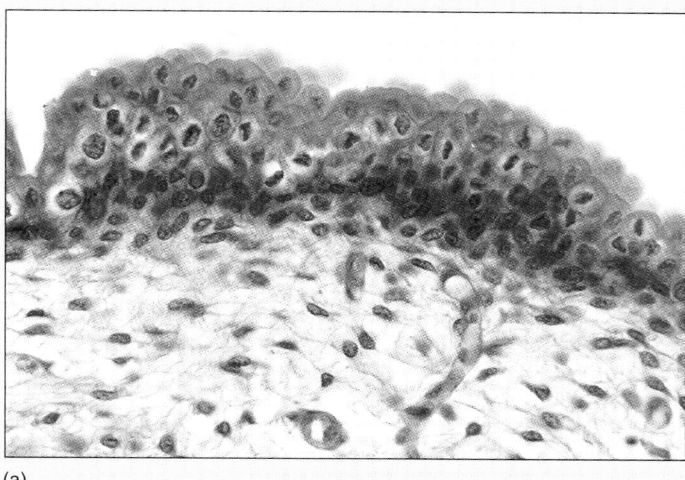

(a)

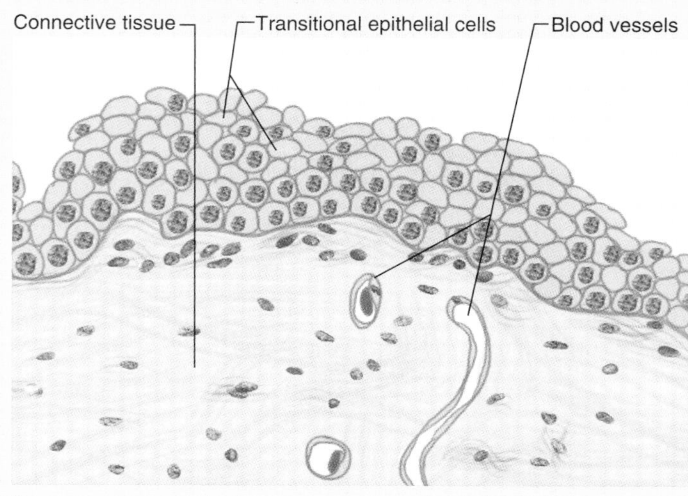

(b)

Figure 6.11 Allantoic Duct of Umbilical Cord.

Microscopic appearance: Somewhat resembles stratified squamous epithelium, but surface cells are rounded, not flattened, and often bulge above surface; typically five or six cells thick when relaxed and two or three cells thick when stretched; cells may be flatter and thinner when epithelium is stretched (as in a distended bladder); some cells have two nuclei

Representative locations: Urinary tract—part of kidney, ureter, bladder, part of urethra; allantoic duct in umbilical cord; external surface of umbilical cord

Functions: Stretches to allow filling of urinary tract

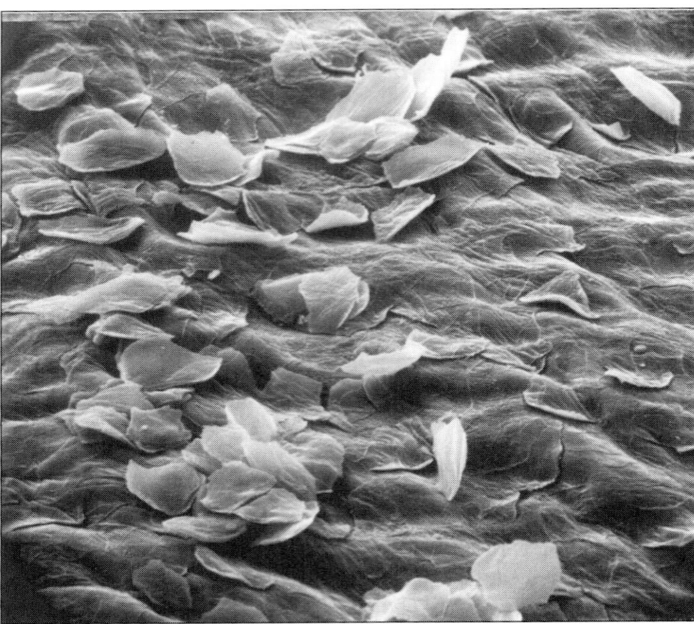

Figure 6.12 **Exfoliation of Squamous Cells from the Mucosal Surface of the Vagina.** From R. G. Kessel and R. H. Kardon, *Scanning Electron Microscopy of Tissues and Organs*, 1979.

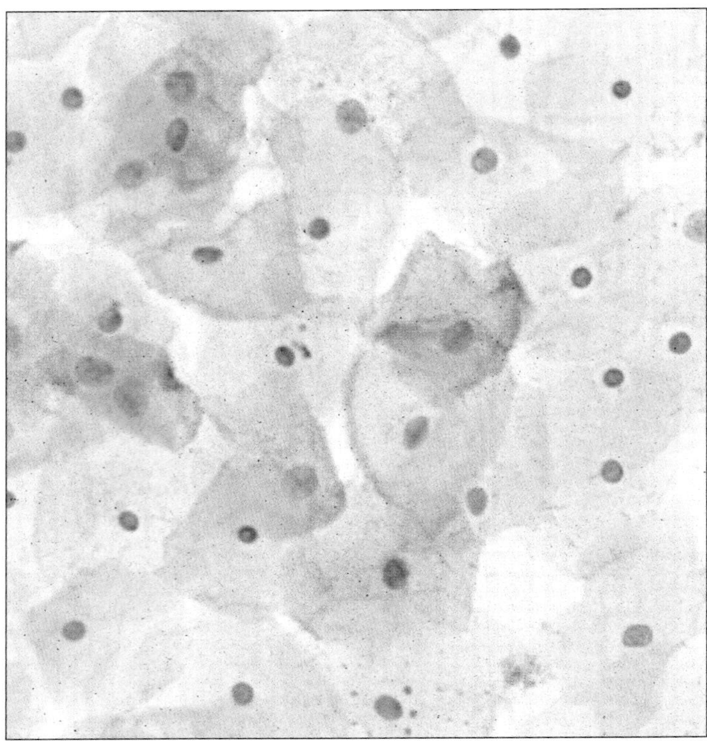

Figure 6.13 **A Normal Pap Smear.** See clinical insight 28.1 for a cancerous Pap smear.

Before You Go On

Answer the following questions to test your understanding of the preceding section:

5. Distinguish between simple and stratified epithelia, and explain why pseudostratified columnar epithelium belongs in the former category.
6. Explain how to distinguish a stratified squamous epithelium from a transitional epithelium.
7. What function do keratinized and nonkeratinized stratified squamous epithelia have in common? What is the structural difference between these two? How is this structural difference related to a functional difference between them?
8. How do the epithelia of the esophagus and stomach differ? How does this relate to their respective functions?

Connective Tissue

Objectives

When you have completed this section, you should be able to

- describe the properties that most connective tissues have in common;
- discuss the types of cells found in connective tissue;
- explain what the matrix of a connective tissue is and describe its components;
- name 10 types of connective tissue, describe their cellular components and matrix, and explain what distinguishes them from each other; and
- visually recognize each connective tissue type from specimens or photographs.

Overview

Connective tissue is the most abundant, widely distributed, and histologically variable of the primary tissues. As the name implies, it often serves to connect organs to each other—for example, the way a tendon connects muscle to bone—or serves in other ways to support, bind, and protect organs. This category includes fibrous tissue, fat, cartilage, bone, blood, and others. It is difficult to generalize about such a broad range of tissues, but in most cases, the volume of extracellular matrix is greater than the volume occupied by cells. Most of the cells are not in direct contact with each other, but are scattered through the matrix.

The functions of connective tissue include the following:

- **Binding of organs.** Tendons bind muscle to bone, ligaments bind one bone to another, fat holds the kidneys and eyes in place, and fibrous tissue binds the skin to underlying muscle.

- **Support.** Bones support the body and cartilage supports the ears, nose, trachea, and bronchi.
- **Physical protection.** The cranium, ribs, and sternum protect delicate organs such as the brain, lungs, and heart; fatty cushions around the kidneys and eyes protect these organs.
- **Immune protection.** Connective tissue cells attack foreign invaders, and connective tissue fiber forms a "battlefield" under the skin and mucous membranes where immune cells can be quickly mobilized against disease agents.
- **Movement.** Bones provide the lever system for body movement, cartilages are involved in movement of the vocal cords, and cartilages on bone surfaces ease joint movements.
- **Storage.** Fat is the body's major energy reserve; bone is a reservoir of calcium and phosphorus that can be drawn upon when needed.
- **Heat production.** Brown fat generates heat in infants and children.
- **Transport.** Blood transports gases, nutrients, wastes, hormones, and blood cells.

The mesenchyme described earlier in this chapter is a form of embryonic connective tissue. The connective tissues present after birth fall into three broad categories: *fibrous connective tissues, supportive connective tissues* (cartilage and bone), and *fluid connective tissue* (blood).

Fibrous Connective Tissue

Fibrous connective tissues are the most diverse type of connective tissue. They are also called *fibroconnective tissue* or *connective tissue proper.* Nearly all connective tissues contain fibers, but the tissues considered here are classified together because the fibers are so conspicuous. Fibers are, of course, just one component of the tissue, which also includes cells and ground substance. Before examining specific types of fibroconnective tissue, let's examine these components.

Components of Fibrous Connective Tissue

Cells. The cells of fibrous connective tissue include the following types:

- **Fibroblasts.**[9] These are large, flat cells that often appear tapered at the ends and show slender, wispy branches. They produce the fibers and ground substance that form the matrix of the tissue. Fibroblasts that have finished this task and become inactive are called *fibrocytes* by some histologists.
- **Macrophages.**[10] These are large phagocytic cells that wander through the connective tissues, where they engulf and destroy bacteria, other foreign particles, and dead or dying cells of our own body, and activate the immune system when they sense foreign matter called *antigens.* They arise from certain white blood cells called *monocytes* or from the same stem cells that produce monocytes.
- **Leukocytes,**[11] or **white blood cells (WBCs).** WBCs travel briefly in the bloodstream, then crawl out through the capillary walls and spend most of their time in the connective tissues. Most of them are *neutrophils,* which wander about in search of bacteria. Our mucous membranes often exhibit dense patches of tiny WBCs called *lymphocytes,* which react against bacteria, toxins, and other foreign agents.
- **Plasma cells.** Certain lymphocytes turn into plasma cells when they detect foreign agents. The plasma cells then synthesize disease-fighting proteins called *antibodies.* Plasma cells are rarely seen except in the walls of the intestines and in inflamed tissue.
- **Mast cells.** These cells, found especially alongside blood vessels, secrete a chemical called *heparin* that inhibits blood clotting, and one called *histamine* that increases blood flow by dilating blood vessels.
- **Adipocytes** (AD-ih-po-sites), or **fat cells.** These appear in small clusters in some fibroconnective tissues. When they dominate an area, the tissue is called *adipose tissue.*

Fibers. Three types of protein fibers are found in fibroconnective tissues:

- **Collagenous** (col-LADJ-eh-nus) **fibers.** These fibers, made of collagen, are tough and flexible and resist stretching. Collagen is about 25% of the body's protein, the most abundant type. It is the base of such animal products as gelatin, leather, and glue.[12] In fresh tissue, collagenous fibers have a glistening white appearance, as seen in tendons and some cuts of meat (fig. 6.14); thus, they are often called *white fibers.* In tissue sections, collagen forms coarse, wavy bundles, often dyed pink, blue, or green by the most common histological stains. Tendons, ligaments, and the deep

9. *fibro* = fiber + *blast* = producing

10. *macro* = big + *phage* = eater
11. *leuko* = white + *cyte* = cell
12. *colla* = glue + *gen* = producing

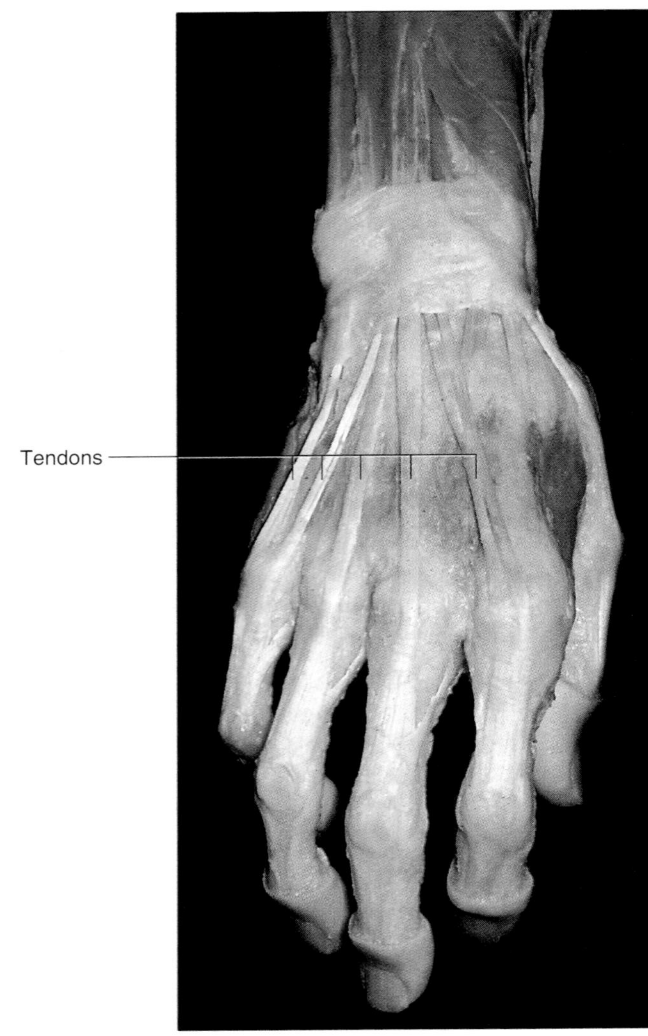

Figure 6.14 Tendons of the Hand. The white glistening appearance results from the collagen of which tendons are composed.

layer of the skin (the dermis) are made mainly of collagen. Less visibly, collagen pervades the matrix of cartilage and bone.

- **Reticular**[13] **fibers.** These are thin collagen fibers coated with glycoprotein. They form a spongelike framework for such organs as the spleen and lymph nodes.
- **Elastic fibers.** These are thinner than collagenous fibers, and they branch and rejoin each other along their course. They are made of a protein called **elastin,** whose coiled structure allows it to stretch and recoil like a rubber band. Elastic fibers account for the ability of the skin, lungs, and arteries to spring back after they are stretched. (Elasticity is not the ability to stretch, but the tendency to recoil when tension is released.) Fresh elastic fibers are yellowish and therefore often called *yellow fibers.*

Ground Substance Amid the cells and fibers in some tissue sections, there appears to be a lot of empty space. In life, this space is occupied by the featureless **ground substance**. Ground substance usually has a gelatinous to rubbery consistency resulting from three classes of large molecules: glycosaminoglycans, proteoglycans, and adhesive glycoproteins.

A **glycosaminoglycan (GAG)** (gly-COSE-ah-MEE-no-GLY-can) is a long polysaccharide composed of unusual disaccharides called *amino sugars* and *uronic acid.* GAGs are negatively charged and thus tend to attract sodium and potassium ions, which in turn causes the GAGs to absorb and hold water. Thus, GAGs play an important role in regulating the water and electrolyte balance of tissues. The most abundant GAG is **chondroitin** (con-DRO-ih-tin) **sulfate.** It is abundant in blood vessels and bones and is responsible for the relative stiffness of cartilage. Some other GAGs that you will read of elsewhere in this book are *heparin* (an anticoagulant) and *hyaluronic* (HY-uh-loo-RON-ic) *acid.* The latter is a gigantic molecule up to 20 μm long, as large as most cells. It is a viscous, slippery substance that forms a very effective lubricant in the joints and constitutes much of the jellylike *vitreous humor* of the eyeball.

A **proteoglycan** is another gigantic molecule. It is shaped somewhat like a test tube brush (fig. 6.15), with the central core composed of protein and the bristlelike outgrowths composed of GAGs. The entire proteoglycan may be attached to hyaluronic acid, thus forming an enormous molecular complex. Proteoglycans form thick colloids similar to those of gravy, pudding, gelatin, and glue. This gel slows the spread of pathogenic organisms through the tissues. Some proteoglycans are embedded in the plasma membranes of cells, attached to the cytoskeleton on the inside and to other extracellular molecules on the outside. Thus, they create a strong structural bond between cells and extracellular macromolecules and help to hold tissues together.

Adhesive glycoproteins are protein-carbohydrate complexes that bind plasma membrane proteins to collagen and proteoglycans outside the cell. They bind all the components of a tissue together and mark pathways that guide migrating embryonic cells to their destinations in a tissue.

Types of Fibrous Connective Tissue

Fibrous connective tissue is divided into two broad categories according to the relative abundance of fiber: *loose* and *dense connective tissue.* In **loose connective tissue,** much of the space is occupied by ground substance, which is dissolved out of the tissue during histological fixation

13. *ret* = network + *icul* = little

Clinical Insight 6.1

Hyaluronidase—A Spreading Factor

The viscosity of hyaluronic acid makes it difficult for microorganisms and other pathogens to spread through the connective tissues. Not surprisingly, many infectious organisms have evolved a counterstrategy—they secrete an enzyme called *hyaluronidase*, or *spreading factor*, that breaks down hyaluronic acid and gives the tissue fluid a more watery consistency in which they can move about more freely.

Hyaluronidase is produced by bacteria such as *Staphylococcus aureus* (the cause of "staph" infections) and by sperm and white blood cells. Bee and snake venoms also contain hyaluronidase, which enables the venoms to spread and disable or kill their victims more quickly. Hyaluronidase is included in some subcutaneous drug injections to promote faster drug absorption and reduce pain.

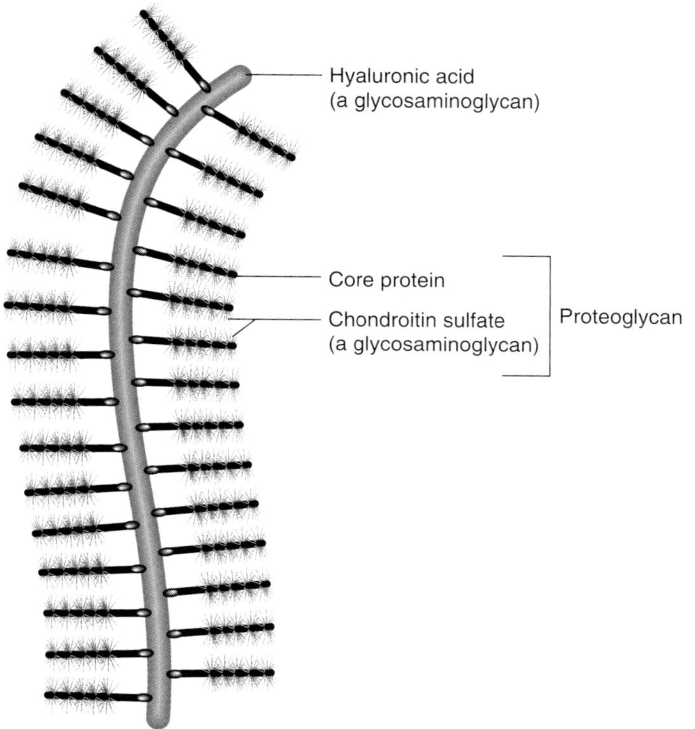

Figure 6.15 Proteoglycan Molecules Linked to a Hyaluronic Acid Core. The resulting hydrophilic complex is larger than many cells.

and leaves empty space in prepared tissue sections. The loose connective tissues we will discuss are *areolar, reticular,* and *adipose tissue.* In **dense connective tissue,** fiber occupies more space than the cells and ground substance, and appears closely packed in tissue sections. The two dense connective tissues we will discuss are *dense regular* and *dense irregular connective tissue.*

Areolar Tissue **Areolar**[14] (AIR-ee-OH-lur) **tissue** exhibits loosely organized fibers, has abundant blood vessels, and shows a lot of seemingly empty space. It possesses all six of the aforementioned cell types. Its fibers run in random directions and are mostly collagenous, but elastic and reticular fibers are also present. Areolar tissue is highly variable in appearance. In many serous membranes, it looks like figure 6.16, but in the skin and mucous membranes, it is more compact (see fig. 6.8) and sometimes difficult to distinguish from dense irregular connective tissue. Some advice on how to tell them apart is given after the discussion of dense irregular connective tissue.

Areolar tissue is found in tissue sections from almost every part of the body. It surrounds blood vessels and nerves and penetrates with them even into the small spaces of muscles, tendons, and other tissues. Nearly every epithelium rests on a layer of areolar tissue, whose blood vessels provide the epithelium with nutrition, waste removal, and a ready supply of infection-fighting WBCs in times of need. Because of the abundance of open, fluid-filled space, WBCs can move about freely in areolar tissue and can easily find and destroy pathogens.

Reticular Tissue **Reticular tissue** is a mesh of reticular fibers and fibroblasts that forms the structural framework (stroma) of such organs and tissues as the lymph nodes, spleen, thymus, and bone marrow. The space amid the fibers is filled with blood cells. If you imagine a kitchen sponge soaked with blood, the sponge fibers would be analogous to the reticular tissue stroma.

Adipose Tissue Adipocytes occur singly or in small clusters in areolar tissue, but when they become the dominant cell type, the tissue is called **adipose tissue,** or **fat.** Adipocytes usually range from 70 to 120 μm in diameter, but they may be five times as large in obese people. The space between adipocytes is occupied by areolar tissue, reticular tissue, and blood capillaries.

Fat is the body's primary energy reservoir. The quantity of stored triglyceride and the number of adipocytes are quite stable in a person, but this doesn't mean stored fat is stagnant. New triglycerides are constantly being synthesized and stored, as others are hydrolyzed and released

14. *areola* = little space

into circulation. Thus, there is a constant turnover of stored triglyceride, with an equilibrium between synthesis and hydrolysis, energy storage and energy use. Adipose tissue also provides thermal insulation, and it contributes to body contours such as the female breasts and hips. Most adipose tissue is of a type called *white fat,* but fetuses, infants, and children also have a heat-generating tissue called *brown fat,* which accounts for up to 6% of an infant's weight. Brown fat gets its color from an unusual abundance of blood vessels and lysosomes. It stores lipid in the form of multiple droplets rather than one large one. Brown fat has numerous mitochondria, but their oxidation pathway is not linked to ATP synthesis. Therefore, when these cells oxidize fats, they release all of the energy as heat. Hibernating animals accumulate brown fat in preparation for winter.

Table 6.4 summarizes the three types of loose fibrous connective tissues.

Think About It

Why would infants and children have more need for brown fat than adults do? (Hint: Smaller bodies have a higher ratio of surface area to volume than larger bodies do.)

Dense Regular Connective Tissue This tissue is named for two properties: (1) the collagen fibers are closely packed and leave relatively little open space, and (2) the fibers are parallel to each other. **Dense regular connective tissue** is found especially in tendons and ligaments. The parallel arrangement of fibers is an adaptation to the fact that tendons and ligaments are pulled in predictable directions. With some minor exceptions such as blood vessels and sensory nerve fibers, the only cells in this tissue are fibroblasts, visible by their slender, violet-staining nuclei squeezed between bundles of collagen. This type of tissue has few blood vessels, so injured tendons and ligaments are slow to heal.

The vocal cords, suspensory ligament of the penis, and some ligaments of the vertebral column are made of a type of dense regular connective tissue called **yellow elastic tissue.** In addition to the densely packed collagen fibers, it exhibits branching elastic fibers and more fibroblasts. The fibroblasts have larger, more conspicuous nuclei than seen in most dense regular connective tissue.

Elastic tissue also takes the form of wavy sheets in the walls of the large and medium arteries. When the heart pumps blood into the arteries, these sheets enable them to expand and relieve some of the pressure on smaller vessels downstream. When the heart relaxes, the arterial wall springs back and keeps the blood pressure from dropping too low between heartbeats. The importance of this elastic tissue becomes especially clear when there is not enough of it—for example, in Marfan syndrome (see clinical insight 6.4, p. 201)—or when it is stiffened by arteriosclerosis (see chapter 19).

Dense Irregular Connective Tissue This tissue also has thick bundles of collagen and relatively little room for cells and ground substance, but the collagen bundles run in random directions. This arrangement enables the tissue to resist unpredictable stresses. **Dense irregular connective tissue** constitutes most of the dermis, where it binds the skin to the underlying muscle and connective tissue. It forms a protective capsule around organs such as the kidneys, testes, and spleen and a tough fibrous sheath around the bones, nerves, and most cartilages.

It is sometimes difficult to judge whether a tissue is areolar or dense irregular. In the dermis, for example, these tissues occur side by side, and the transition from one to the other is not at all obvious (see fig. 7.3). A relatively large amount of clear space suggests areolar tissue, and thicker bundles of collagen with relatively little clear space suggests dense irregular tissue. The dense connective tissues are summarized in table 6.5.

Cartilage

Cartilage (table 6.6) is a supportive connective tissue with a flexible rubbery matrix. It gives shape to the external ear, the tip of the nose, and the larynx (voicebox)—the most easily palpated cartilages in the body. Cells called **chondroblasts**[15] (CON-dro-blasts) secrete the matrix and surround themselves with it until they become trapped in little cavities called **lacunae**[16] (la-CUE-nee). Once enclosed in lacunae, the cells are called **chondrocytes** (CON-dro-sites). Cartilage is free of blood vessels except when transforming into bone; thus nutrition and waste removal depend on solute diffusion through the stiff matrix. Because this is a slow process, chondrocytes have low rates of metabolism and cell division, and injured cartilage heals slowly. The matrix is rich in chondroitin sulfate and contains collagen fibers that range in thickness from invisibly fine to conspicuously coarse. Differences in the fibers provide a basis for classifying cartilage into three types: *hyaline cartilage, elastic cartilage,* and *fibrocartilage.*

Hyaline[17] (HY-uh-lin) **cartilage** is named for its clear, glassy microscopic appearance, which stems from the usually invisible fineness of its collagen fibers. **Elastic cartilage** is named for its conspicuous elastic fibers, and **fibrocartilage** for its coarse, readily visible bundles of collagen. Elastic cartilage and most hyaline cartilage are surrounded by a sheath of dense irregular connective tissue called the

15. *chondro* = cartilage, gristle + *blast* = forming
16. *lacuna* = lake, cavity
17. *hyal* = glass

Table 6.4 Loose Connective Tissues

Areolar Tissue

(a)

(b)

Figure 6.16 Spread of the Mesentery.

Microscopic appearance: Loose arrangement of collagenous and elastic fibers; scattered cells of various types; abundant ground substance; numerous blood vessels

Representative locations: Underlying nearly all epithelia; surrounding blood vessels, nerves, esophagus, and trachea; fascia between muscles; mesenteries; visceral layers of pericardium and pleura

Functions: Loosely binds epithelia to deeper tissues; allows passage of nerves and blood vessels through other tissues; provides an arena for immune defense; blood vessels provide nutrients and waste removal for overlying epithelia

Reticular Tissue

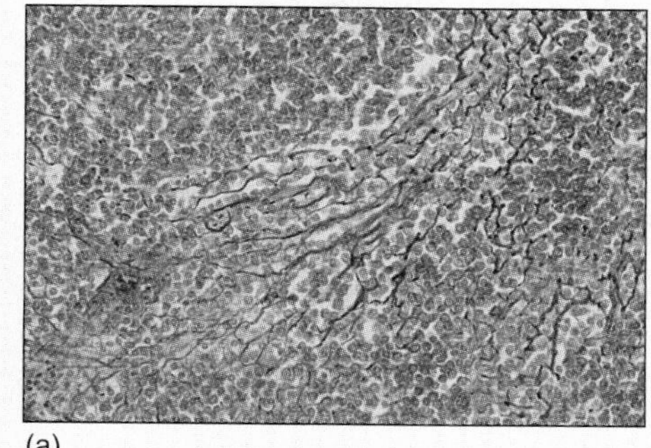

(a)

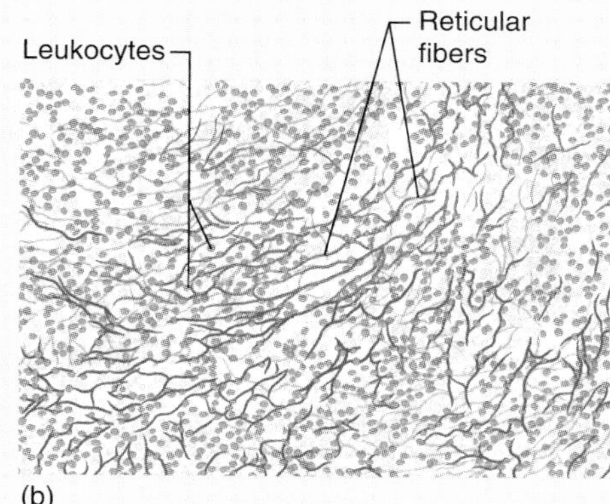

(b)

Figure 6.17 Lymph Node.

Microscopic appearance: Loose network of reticular fibers and cells, infiltrated with numerous lymphocytes and other blood cells

Representative locations: Lymph nodes, spleen, thymus, bone marrow

Functions: Supportive stroma (framework) for lymphatic organs

Adipose Tissue

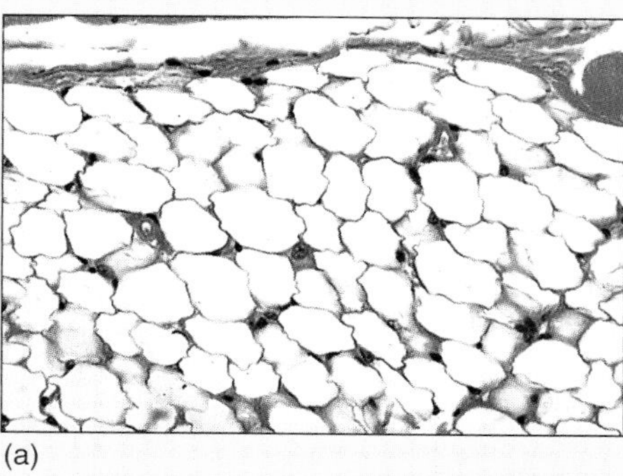

(a)

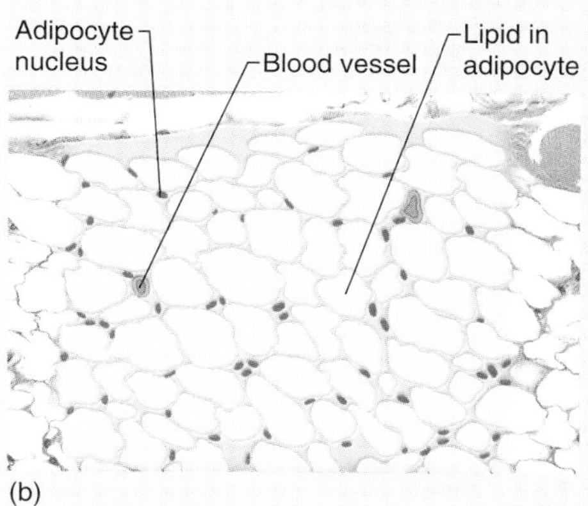

(b)

Figure 6.18 Adipose Tissue.

Microscopic appearance: Dominated by adipocytes—large, empty-looking cells with thin margins; tissue sections often very pale because of scarcity of stained cytoplasm; adipocytes shrunken; nucleus pressed against plasma membrane; blood vessels often present

Representative locations: Subcutaneous fat beneath skin; breast; heart surface; surrounding organs such as kidneys and eyes

Functions: Energy storage; thermal insulation; heat production by brown fat; protective cushion for some organs; filling space, shaping body

Table 6.5 Dense Connective Tissues

Dense Regular Connective Tissue

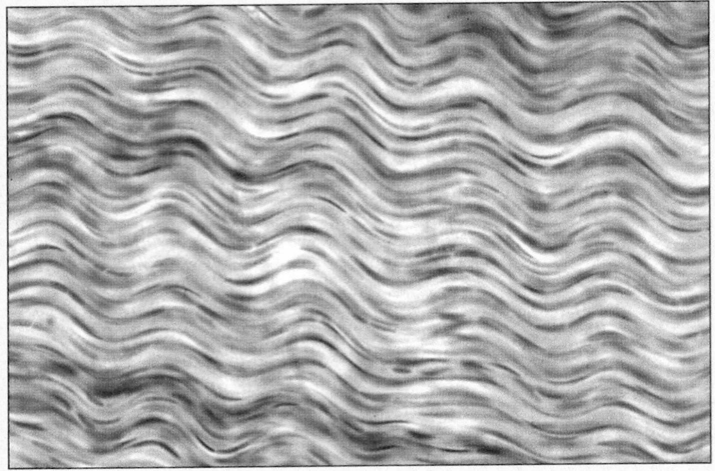

(a)

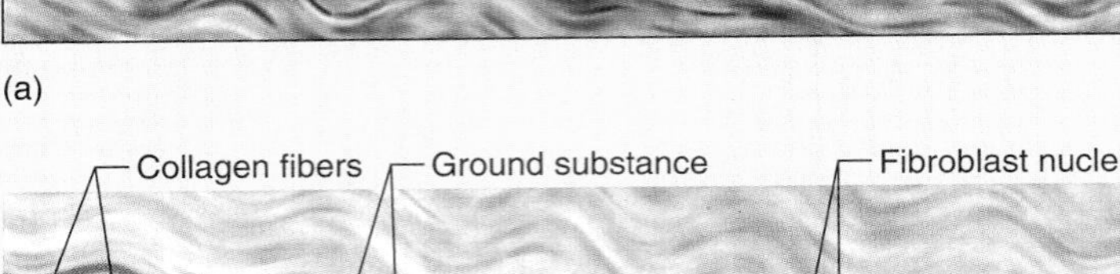

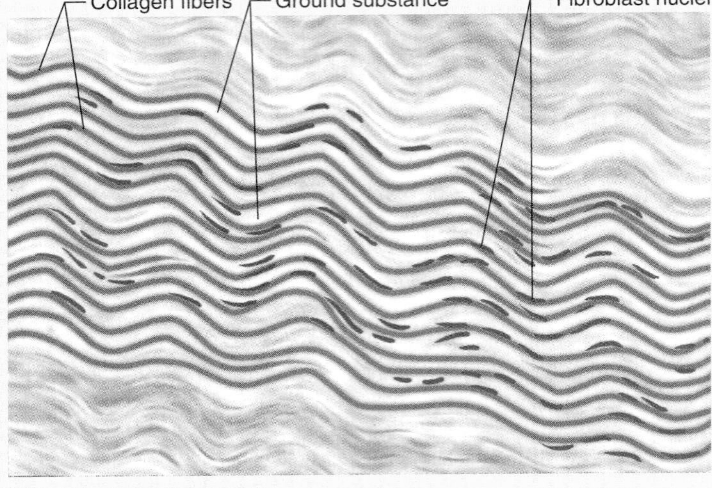

(b)

Figure 6.19 Tendon.

Microscopic appearance: Densely packed, parallel, often wavy collagen fibers; slender fibroblast nuclei compressed between collagen bundles; scanty open space (ground substance); scarcity of blood vessels

Representative locations: Tendons and ligaments

Functions: Ligaments tightly bind bones together; resist stress; tendons attach muscle to bone and transfer muscular tension to bones

Dense Irregular Connective Tissue

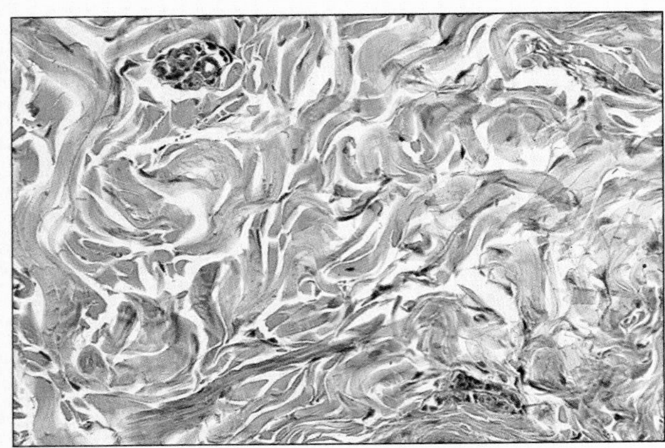

(a)

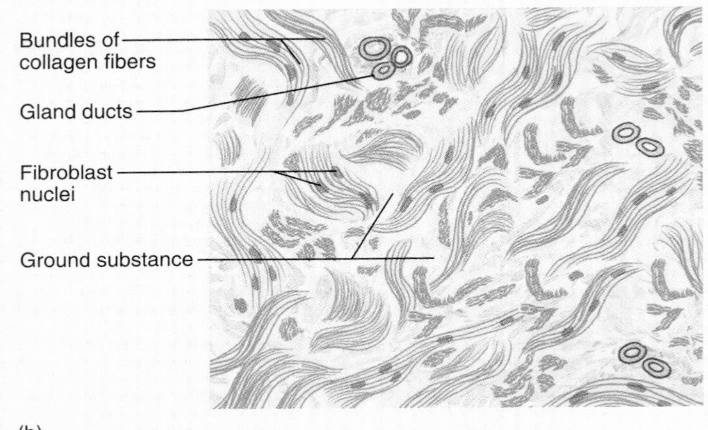

(b)

Figure 6.20 Dermis of the Skin.

Microscopic appearance: Densely packed collagen fibers running in random directions; scanty open space (ground substance); few visible cells; scarcity of blood vessels

Representative locations: Deeper portion of dermis of skin; capsules around viscera such as liver, kidney, spleen; fibrous sheaths around cartilages and bones

Functions: Durable, hard to tear; withstands stresses applied in unpredictable directions

Table 6.6 Types of Cartilage

Hyaline Cartilage

(a)

Perichondrium
Chondrocytes
Matrix

(b)

Figure 6.21 Fetal Skeleton.

Microscopic appearance: Clear, glassy matrix, often stained light blue or pink in tissue sections; fine, dispersed collagen fibers, not usually visible; chondrocytes often in small clusters of three or four cells (*cell nests*), enclosed in lacunae; usually covered by perichondrium

Representative locations: Forms a thin *articular cartilage,* lacking perichondrium, over the ends of bones at movable joints; a *costal cartilage* attaches the end of a rib to the breastbone; forms supportive rings and plates around trachea and bronchi; forms a boxlike enclosure around the larynx; forms much of the fetal skeleton

Functions: Eases joint movements; holds airway open during respiration; moves vocal cords during speech; a precursor of bone in the fetal skeleton and the growth zones of long bones of children

Elastic Cartilage

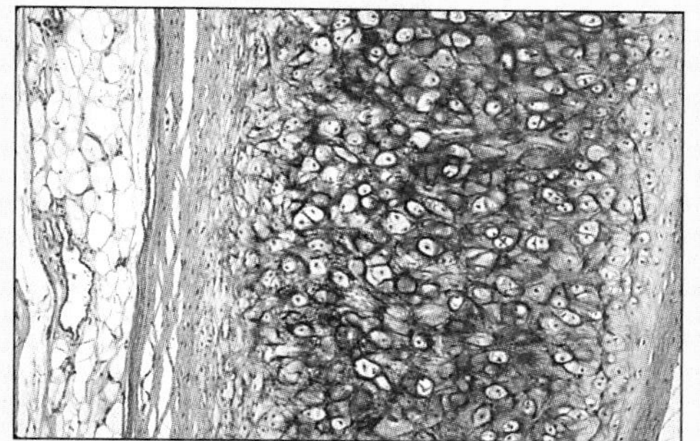

(a)

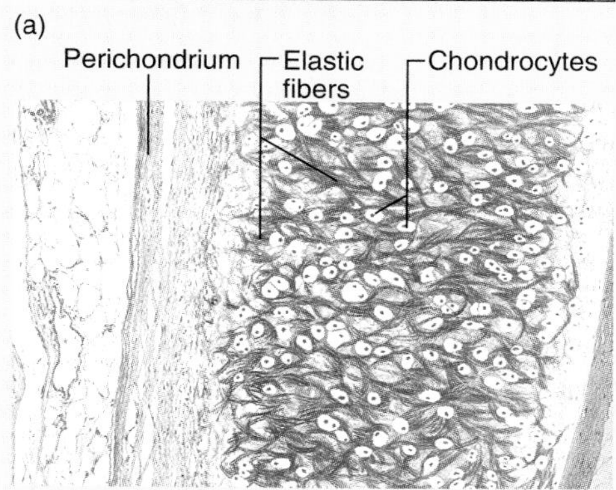

(b)

Figure 6.22 External Ear.

Microscopic appearance: Elastic fibers form weblike mesh amid lacunae; always covered by perichondrium

Representative locations: External ear; epiglottis

Functions: Provides flexible, elastic support

Fibrocartilage

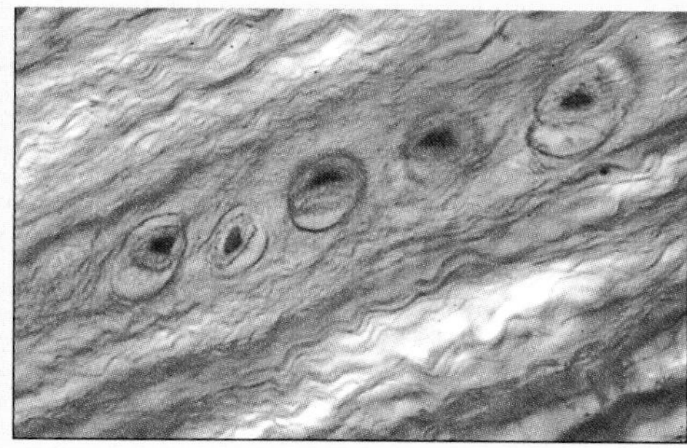

(a)

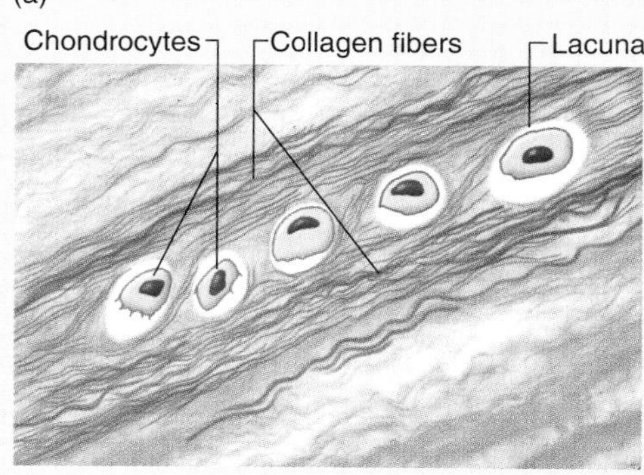

(b)

Figure 6.23 Intervertebral Disc.

Microscopic appearance: Parallel collagen fibers similar to those of tendon; rows of chondrocytes in lacunae between collagen fibers; never has a perichondrium

Representative locations: Pubic symphysis (anterior joint between two halves of pelvic girdle); intervertebral discs that separate bones of spinal column; menisci, or pads of shock-absorbing cartilage, in knee joint; at points where tendons insert on bones near articular hyaline cartilage

Functions: Resists compression and absorbs shock in some joints; often a transitional tissue between dense connective tissue and hyaline cartilage (for example, at some tendon-bone junctions)

perichondrium.[18] A reserve population of chondroblasts between the perichondrium and cartilage contributes to cartilage growth throughout life. There is no perichondrium around fibrocartilage.

You can feel the texture of hyaline cartilage by palpating the tip of your nose, your "Adam's apple" at the front of the larynx (voicebox), and periodic rings of cartilage around the trachea (windpipe) just below the larynx. Hyaline cartilage is easily seen in many grocery items—it is the "gristle" at the ends of pork ribs, on chicken leg and breast bones, and at the joints of pigs' feet, for example. Elastic cartilage gives shape to the external ear. You can get some idea of its springy resilience by folding your ear down and releasing it.

Bone

The term *bone* refers both to organs of the body such as the femur and mandible, composed of multiple tissue types, and to the bone tissue, or **osseous tissue,** that makes up most of the mass of bones. There are two forms of osseous tissue: (1) **Spongy bone** fills the heads of the long bones. Although it is calcified and hard, its delicate slivers and plates give it a spongy appearance. (2) **Compact (dense) bone** is a more dense calcified tissue with no spaces visible to the naked eye. It forms the external surfaces of all bones, so spongy bone, when present, is always covered by compact bone.

The differences between compact and spongy bone are described in chapter 8. Here, we examine only compact bone (table 6.7). Most specimens you study will probably be chips of dead, dried bone ground to microscopic thinness. In such preparations, the cells are absent but spaces reveal their former locations. Most compact bone is arranged in cylinders of tissue that surround **haversian**[19] **(central) canals,** which run longitudinally through the shafts of long bones such as the femur. Blood vessels and nerves travel through the haversian canals in life. The bone matrix is deposited in concentric **lamellae,** onionlike layers around each haversian canal. A haversian canal and its surrounding lamellae are called an **osteon.** Tiny lacunae between the lamellae are occupied in life by mature bone cells, or **osteocytes.**[20] Delicate canals called **canaliculi** radiate from each lacuna to its neighbors and allow the osteocytes to keep in touch with each other. The bone as a whole is covered with a tough fibrous **periosteum** (PERR-ee-OSS-tee-um) similar to the perichondrium of cartilage.

About a third of the dry weight of bone is composed of collagen fibers and chondroitin sulfate; two-thirds consists of minerals (mainly calcium salts) deposited around the collagen fibers.

18. *peri* = around + *chondri* = cartilage
19. Clopton Havers (1650–1702), English anatomist
20. *osteo* = bone + *cyte* = cell

Blood

Blood (table 6.8) is a fluid connective tissue that travels through tubular vessels. Its primary function is to transport cells and dissolved matter from place to place. Blood consists of a ground substance called **plasma** and of cells and cell fragments collectively called **formed elements. Erythrocytes**[21] (eh-RITH-ro-sites), or red blood cells, are the most abundant formed elements. In stained blood films, they look like pink discs with a thin, pale center. They have no nuclei. Erythrocytes transport oxygen and carbon dioxide. **Leukocytes,** or white blood cells, serve various roles in defense against infection and other diseases. They travel from one organ to another in the bloodstream and lymph but spend most of their lives in the connective tissues. Leukocytes are somewhat larger than erythrocytes and have conspicuous nuclei that usually appear violet in stained preparations. There are five kinds, distinguished partly by variations in nuclear shape: neutrophils, eosinophils, basophils, lymphocytes, and monocytes. Their individual characteristics are considered in detail in chapter 18. **Platelets** are small cell fragments scattered amid the blood cells. They are involved in clotting and other mechanisms for minimizing blood loss, and in secreting growth factors that promote blood vessel growth and maintenance.

Before You Go On

Answer the following questions to test your understanding of the preceding section:

9. What features do most or all connective tissues have in common to set this class apart from nervous, muscular, and epithelial tissue?
10. List the cell and fiber types found in fibrous connective tissues and state their functional differences.
11. What substances account for the gelatinous consistency of connective tissue ground substance?
12. What is areolar tissue? How can it be distinguished from any other kind of connective tissue?
13. Discuss the difference between dense regular and dense irregular connective tissue as an example of the relationship between form and function.
14. Describe some similarities, differences, and functional relationships between hyaline cartilage and bone.
15. What are the three basic kinds of formed elements in blood, and what are their respective functions?

21. *erythro* = red + *cyte* = cell

Table 6.7 Bone

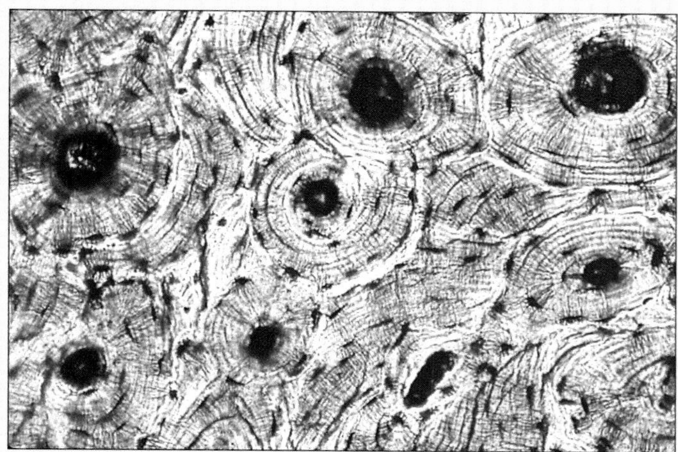

(a)

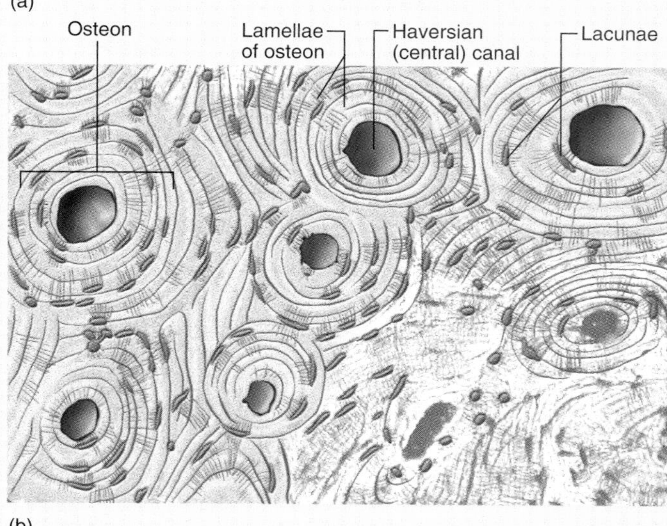

(b)

Figure 6.24 Compact Bone.

Microscopic appearance (compact bone): Calcified matrix arranged in concentric lamellae around haversian canals; osteocytes occupy lacunae between adjacent lamellae; lacunae interconnected by delicate canaliculi

Representative locations: Skeleton

Functions: Physical support of body; leverage for muscle action; protective enclosure of viscera; reservoir of calcium and phosphorus

Table 6.8 Blood

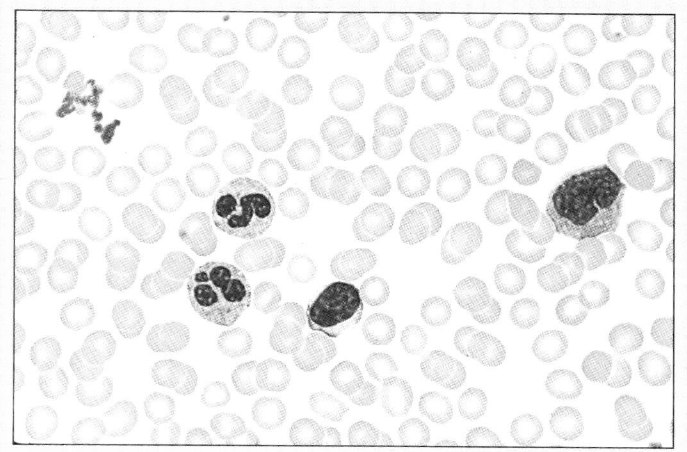

(a)

Platelets Neutrophils Lymphocyte Erythrocytes Monocyte

(b)

Figure 6.25 Blood Smear.

Microscopic appearance: Erythrocytes appear as pale pink discs with light centers and no nuclei; leukocytes are slightly larger, are much fewer, and have variously shaped nuclei, usually stained violet; platelets are cell fragments with no nuclei, about one-quarter the diameter of erythrocytes

Representative locations: Contained in heart and blood vessels

Functions: Transports gases, nutrients, wastes, chemical signals, and heat throughout body; provides defensive leukocytes; contains clotting agents to minimize bleeding; platelets secrete growth factors that promote tissue maintenance and repair

Nervous and Muscular Tissue—Excitable Tissues

Objectives

When you have completed this section, you should be able to

- explain what distinguishes excitable tissues from other tissues;
- name the cell types that compose nervous tissue;
- identify the major parts of a nerve cell;
- visually recognize nervous tissue from specimens or photographs;
- name the three kinds of muscular tissue and describe the differences between them; and
- visually identify any type of muscular tissue from specimens or photographs.

Excitability is a characteristic of all living cells, but it is developed to its highest degree in nervous and muscular tissue, which are therefore described as **excitable tissues.** The basis for their excitation is an electrical charge difference (voltage) called the **membrane potential,** which occurs across the plasma membranes of all cells. Nervous and muscular tissues respond quickly to outside stimuli by means of rapid changes in membrane potential. In nerve cells, these changes result in the rapid transmission of signals to other cells. In muscle cells, they result in contraction, or shortening of the cell.

Nervous Tissue

Nervous tissue (table 6.9) consists of **neurons** (NOOR-ons), or nerve cells, and a much greater number of **neuroglia** (noo-ROG-lee-uh), or **glial** (GLEE-ul) **cells,** which protect and assist the neurons. Neurons are specialized to detect stimuli, respond quickly, and transmit coded information rapidly to other cells. Each neuron has a prominent **soma,** or cell body, that houses the nucleus and most other organelles. This is the cell's center of genetic control and protein synthesis. Somas are usually round, ovoid, or stellate in shape. Extending from the soma, there are usually multiple short, branched processes called **dendrites**[22] that receive signals from other cells and transmit messages to the soma, and a single, much longer **axon,** or **nerve fiber,** that sends outgoing signals to other cells. Some axons are more than a meter long and extend from the brainstem to the foot. Nervous tissue is found in the brain, spinal cord, nerves, and ganglia, which are knotlike swellings in nerves. Local variations in the structure of nervous tissue are described in chapters 13 to 16.

Table 6.9 Nervous Tissue

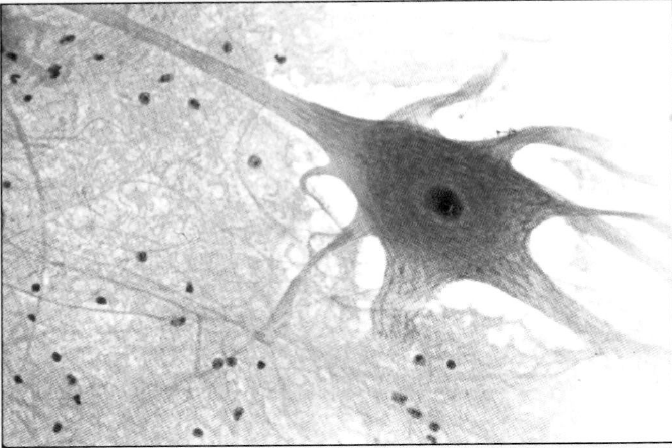

(a)

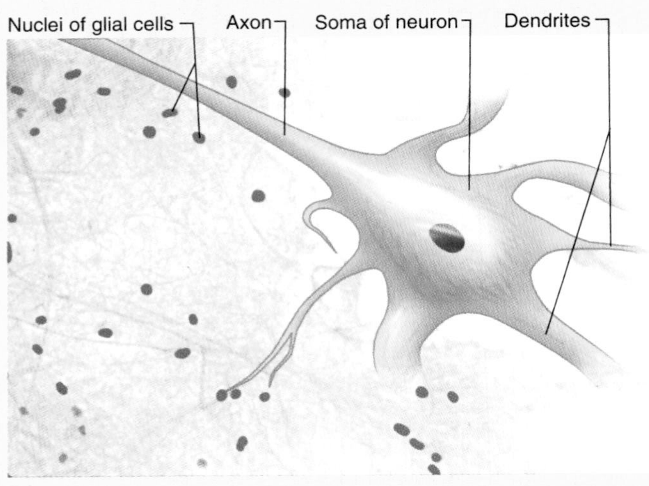

(b)

Figure 6.26 Spinal Cord Smear from a Cow.

Microscopic appearance: Most sections show a few large neurons, usually with rounded or stellate cell bodies (somas) and fibrous processes (axon and dendrites) extending from the somas; neurons are surrounded by a greater number of much smaller glial cells, which lack dendrites and axons

Representative locations: Brain, spinal cord, nerves, ganglia

Function: Internal communication

Muscular Tissue

Muscular tissue responds to stimulation by contracting; thus, its primary job is to exert physical forces on other tissues and organs—for example, when a skeletal muscle pulls on a bone, the heart contracts and expels blood, or the bladder contracts and expels urine. Not only do movements of the body and its limbs depend on muscle, but so do such processes as digestion, waste elimination, breathing, speech, and blood circulation. The muscles are also an important source of body heat. The word *muscle* means

22. *dendr* = tree + *ite* = little

"little mouse," apparently referring to the appearance of rippling muscles under the skin.

There are three histological types of muscle—*skeletal, cardiac,* and *smooth*—which differ in appearance, physiology, and function (table 6.10). **Skeletal muscle** consists of long, cylindrical cells called **muscle fibers.** Most of it is attached to bones, but there are exceptions in the tongue, upper esophagus, some facial muscles, and some **sphincter**[23] (SFINK-tur) muscles (ringlike or cufflike muscles that open and close body passages). Each cell contains multiple nuclei adjacent to the plasma membrane. Skeletal muscle is described as *striated* and *voluntary.* The first term refers to alternating light and dark bands, or **striations** (stry-AY-shuns), created by the overlapping pattern of cytoplasmic protein filaments that cause muscle contraction. The second term, *voluntary,* refers to the fact that we usually have conscious control over skeletal muscle contraction.

Cardiac muscle is essentially limited to the heart, though it extends slightly into the nearby blood vessels. It, too, is striated, but it differs from skeletal muscle in its other features. Its cells are much shorter, so they are commonly called **myocytes**[24] rather than fibers. The myocytes are branched and contain only one nucleus, which is located near the center. A light-staining region of glycogen often surrounds the nucleus. Cardiac myocytes are joined end to end by junctions called **intercalated**[25] (in-TUR-ku-LAY-ted) **discs.** Electrical connections at these junctions enable a wave of excitation to travel rapidly from cell to cell, and mechanical connections keep the myocytes from pulling apart when the heart contracts. The electrical junctions allow all the myocytes of a heart chamber to be stimulated, and contract, almost simultaneously. Intercalated discs appear as dark transverse lines separating each myocyte from the next. They may be only faintly visible, however, unless the tissue has been specially stained for them. Cardiac muscle is considered *involuntary* because it is not usually under conscious control; it contracts even if all nerve connections to it are severed.

Smooth muscle lacks striations and is involuntary. Smooth muscle cells are fusiform (thick in the middle and tapered at the ends) and relatively short. They have only one, centrally placed nucleus. Small amounts of smooth muscle are found in the iris of the eye and in the skin, but most of it, called **visceral muscle,** forms layers in the walls of the digestive, respiratory, and urinary tracts, in the uterus, and in other viscera. In locations such as the esophagus and small intestine, smooth muscle forms adjacent layers, with the cells of one layer encircling the organ and the cells of the other layer running longitudinally. When the circular smooth muscle contracts, it may propel contents such as food through the organ. When the longitudinal layer contracts, it makes the organ shorter and thicker. By regulating the diameter of blood vessels, smooth muscle is very important in controlling blood pressure and flow. Both smooth and skeletal muscle form sphincters that control the emptying of the bladder and rectum.

23. *sphinc* = squeeze, bind tightly
24. *myo* = muscle + *cyte* = cell
25. *inter* = between + *calated* = inserted

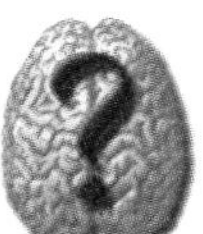

Think About It

How does the meaning of the word *fiber* differ in the following uses: muscle fiber, nerve fiber, and connective tissue fiber?

Before You Go On

Answer the following questions to test your understanding of the preceding section:

16. What do nervous and muscular tissue have in common? What is the primary function of each?
17. What kinds of cells compose nervous tissue, and how can they be distinguished from each other?
18. Name the three kinds of muscular tissue, describe how to distinguish them from each other in microscopic appearance, and state a location and function for each one.

Intercellular Junctions, Glands, and Membranes

Objectives

When you have completed this section, you should be able to

- describe the junctions that hold cells and tissues together;
- describe or define different types of glands;
- describe the typical anatomy of a gland;
- name and compare different modes of glandular secretion;
- describe the way tissues are organized to form the body's membranes; and
- name and describe the major types of membranes in the body.

Intercellular Junctions

Most cells, with the exception of blood and metastatic cancer cells, must be anchored to each other and to the matrix if they are to grow and divide normally. The connections between one cell and another are called **intercellular junctions.** These attachments enable the cells to resist stress and communicate with each other. Without them, cardiac muscle cells would pull apart when they contracted, and every swallow of food would scrape away the lining of your esophagus. The principal types of intercellular junctions are shown in figure 6.30.

Table 6.10 Muscular Tissue

Skeletal Muscle

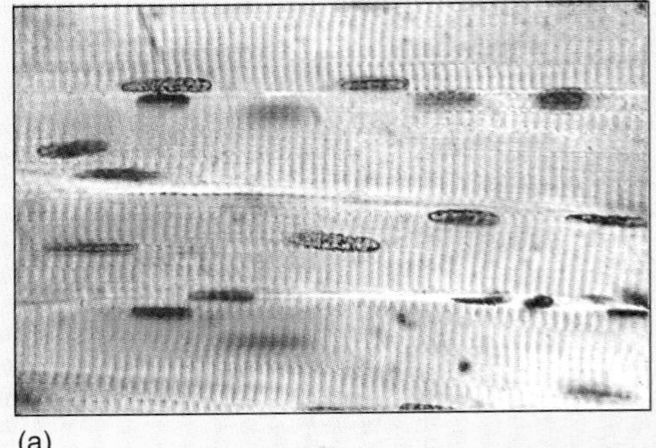

(a)

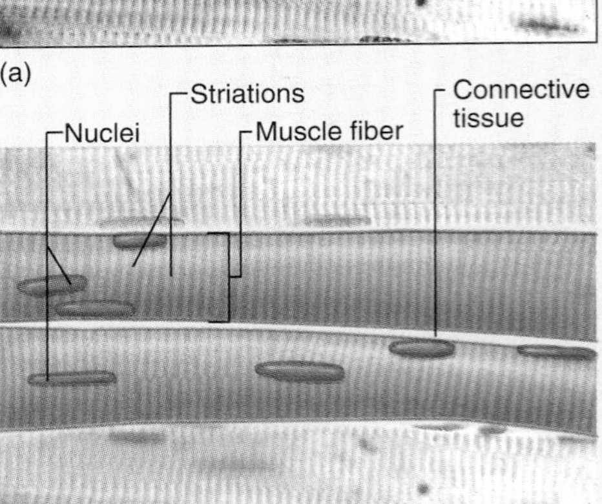

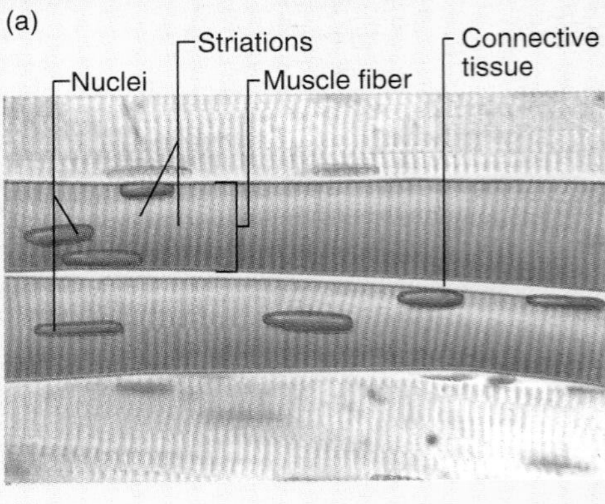

(b)

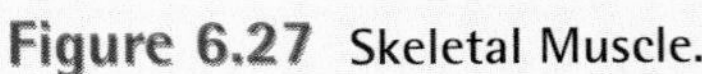

Figure 6.27 Skeletal Muscle.

Microscopic appearance: Long, cylindrical, unbranched cells (fibers), relatively parallel in longitudinal tissue sections; striations; multiple nuclei per cell, near plasma membrane

Representative locations: Skeletal muscles, mostly attached to bones but also including voluntary sphincters of the lips, eyelids, urethra, and anus; diaphragm; some muscles of tongue and esophagus

Functions: Body movements, facial expression, posture, breathing, speech, swallowing, control of urination and defecation, and assistance in childbirth; under voluntary control

Cardiac Muscle

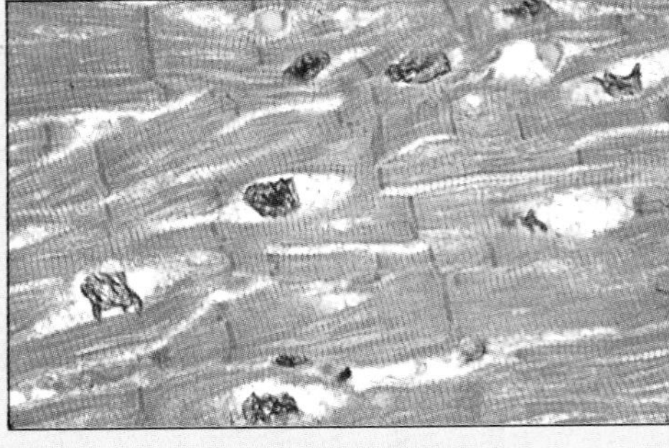

(a)

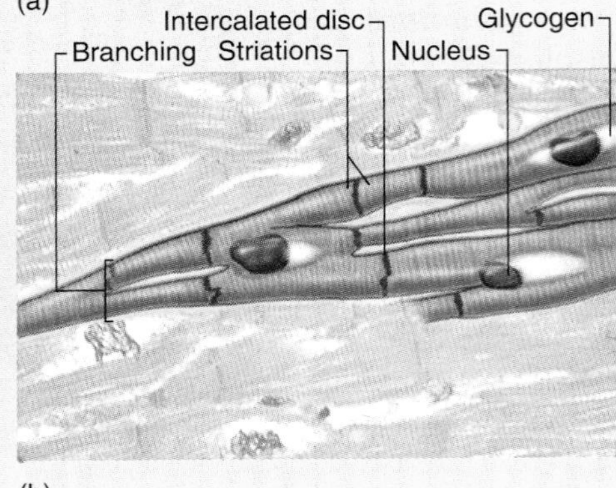

(b)

Figure 6.28 Cardiac Muscle.

Microscopic appearance: Short branched cells (myocytes); less parallel appearance in tissue sections; striations; intercalated discs; one nucleus per cell, centrally located and often surrounded by a light zone

Representative locations: Heart

Functions: Pumping of blood; under involuntary control

Smooth Muscle

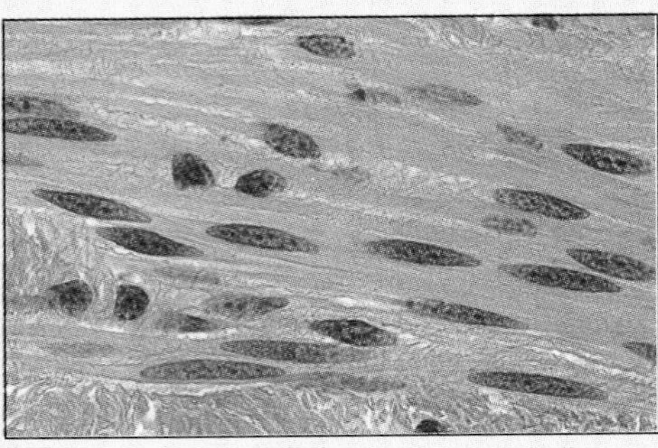

(a)

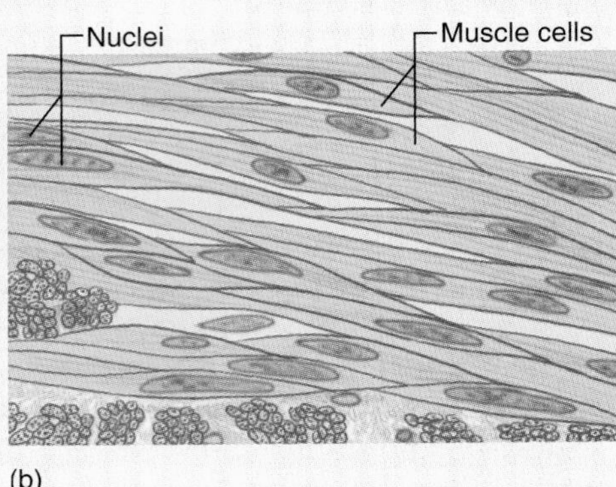

(b)

Figure 6.29 Smooth Muscle, Wall of Small Intestine.

Microscopic appearance: Short fusiform cells overlapping each other; nonstriated; one nucleus per cell, centrally located

Representative locations: Usually found as sheets of tissue in walls of viscera; also in iris and associated with hair follicles; involuntary sphincters of urethra and anus

Functions: Swallowing; contractions of stomach and intestines; expulsion of feces and urine; labor contractions; control of blood pressure and flow; control of respiratory airflow; control of pupillary diameter; erection of hairs; under involuntary control

Tight Junctions

A **tight junction** completely encircles an epithelial cell near its apex and joins it tightly to the neighboring cells. Proteins in the membranes of two adjacent cells form a zipperlike pattern of complementary grooves and ridges. This seals off the intercellular space and makes it difficult for some substances to pass between the cells. In the stomach and intestines, tight junctions prevent digestive juices from seeping between epithelial cells and digesting the underlying connective tissue. They also help to prevent intestinal bacteria from invading the tissues, and they ensure that most digested nutrients pass *through* the epithelial cells and not *between* them.

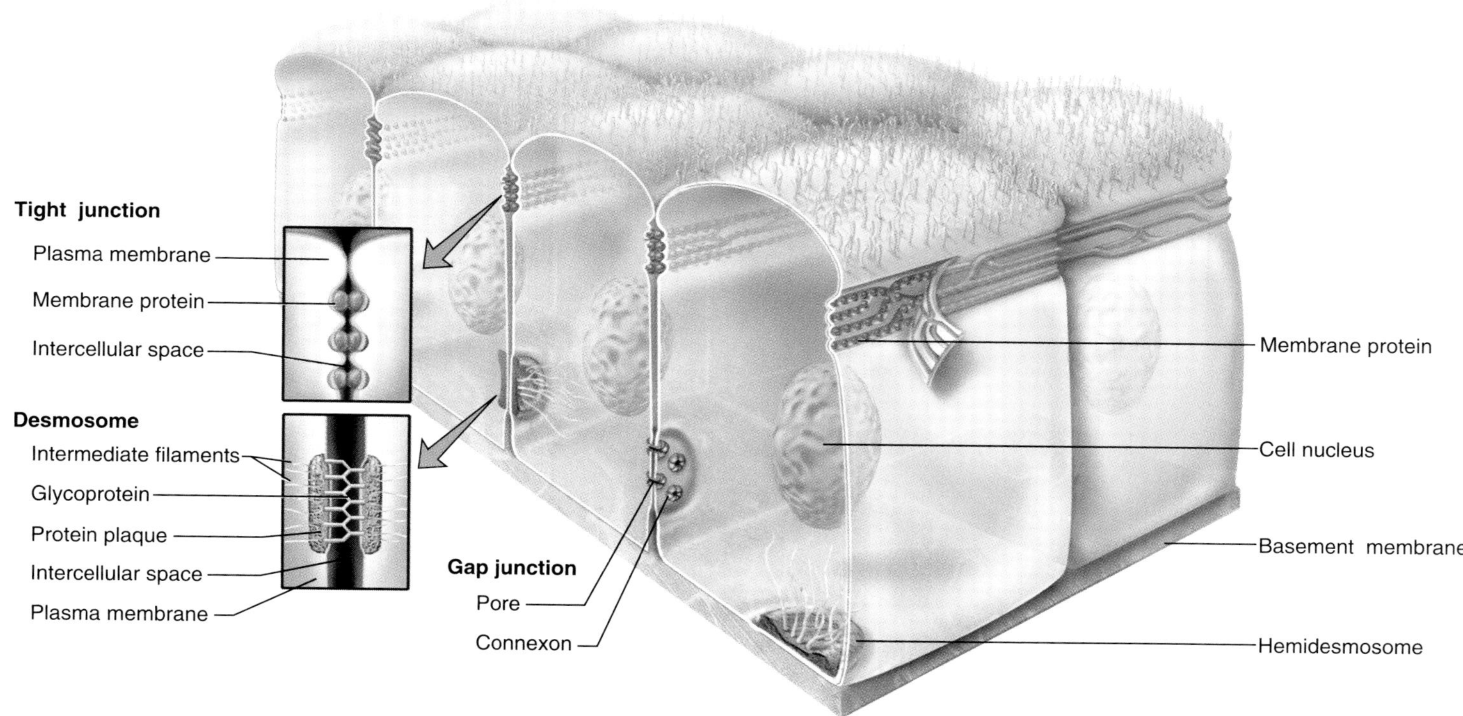

Figure 6.30 Types of Intercellular Junctions.

Desmosomes

A **desmosome**[26] (DEZ-mo-some) is like a "spot weld" between two cells, a patch that holds cells together and enables a tissue to resist mechanical stress, but does not totally encircle a cell. Desmosomes are common in the epidermis, cardiac muscle, and cervix of the uterus. The neighboring cells are separated by a small gap, which is spanned by a fine mesh of glycoprotein filaments. These filaments terminate in a thickened protein plaque at the surface of each cell. On the cytoplasmic side of each plaque, intermediate filaments from the cytoskeleton approach and penetrate the plaque, turn like a J, and return a short distance back into the cytoplasm. Each cell contributes half of the complete desmosome. The basal cells of epithelial tissue have *hemidesmosomes*—half-desmosomes that anchor them to the underlying basement membrane.

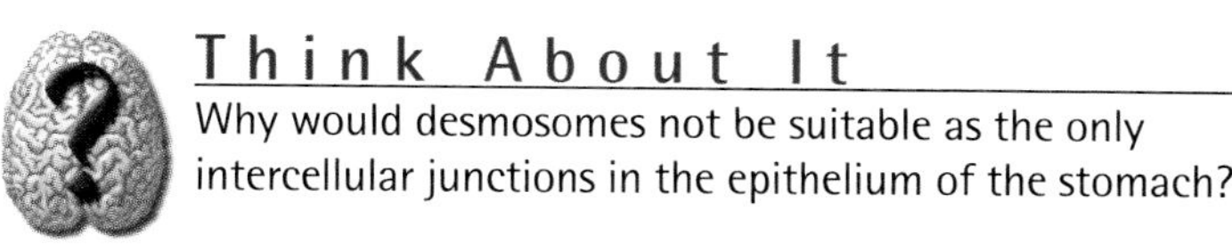

Why would desmosomes not be suitable as the only intercellular junctions in the epithelium of the stomach?

26. *desmo* = band, bond, ligament + *som* = body

Gap (Communicating) Junctions

A **gap junction** is formed by a ringlike *connexon,* which consists of six transmembrane proteins surrounding a water-filled channel. Ions, glucose, amino acids, and other small solutes can pass directly from the cytoplasm of one cell into the next through these channels. In the embryo, nutrients pass from cell to cell through gap junctions until the circulatory system forms and takes over the role of nutrient distribution. Gap junctions are found in the intercalated discs of cardiac muscle and between the cells of some smooth muscle. The flow of ions through these junctions allows electrical excitation to pass directly from cell to cell so that the cells contract in near-unison. Gap junctions are absent from skeletal muscle.

Glands

A **gland** is a cell or organ that secretes substances for use elsewhere in the body or releases them for elimination from the body. The gland product may be something synthesized by the gland cells (such as digestive enzymes) or something removed from the tissues and modified by the gland (such as urine). Glands are composed predominantly of epithelial tissue.

Clinical Insight 6.2

Pemphigus Vulgaris

Some people produce antibodies that turn against their own tissues (*autoantibodies*) and attack the proteins of the desmosomes, especially in the skin. This breaks down the attachments between epidermal cells and produces a skin disease called *pemphigus vulgaris*[27] (PEM-fih-gus vul-GAIR-iss). The effect is widespread blistering of the skin and oral mucosa, loss of tissue fluid, and sometimes death. The condition can be controlled with drugs that suppress the immune system.

27. *pemphigus* = blistering + *vulgaris* = common

Endocrine and Exocrine Glands

Glands are broadly classified as endocrine or exocrine. They originate as invaginations of a surface epithelium. In **exocrine**[28] (EC-so-crin) **glands,** they maintain their contact with the surface by way of a **duct,** an epithelial tube that conveys their secretion to the surface. The secretion may be released to the body surface, as in the case of sweat, mammary, and tear glands, but more often it is released into the cavity (lumen) of another organ such as the mouth or intestine. **Endocrine**[29] (EN-doe-crin) **glands** lose their contact with the surface and have no ducts. They do, however, have a high density of blood capillaries and secrete their products directly into the blood. The secretions of endocrine glands, called *hormones,* function as chemical messengers to stimulate cells elsewhere in the body. Endocrine glands are the subject of chapter 17 and are not further considered here.

The exocrine-endocrine distinction is not always clear. The liver is an exocrine gland that secretes one of its products, bile, through a system of ducts but secretes hormones, albumin, other products directly into the bloodstream. Several glands have both exocrine and endocrine components, such as the pancreas, testis, ovary, and kidney. And nearly all of the viscera have at least some cells that secrete hormones, even though most of these organs are not usually thought of as glands (for example, the heart).

Unicellular glands are exocrine cells found in an epithelium that is predominantly nonsecretory. For example, the respiratory tract, which is lined mainly by ciliated cells, also has a liberal scattering of nonciliated, mucus-secreting goblet cells (see fig. 6.7).

Exocrine Gland Structure

Figure 6.31 shows a generalized multicellular exocrine gland—a structural arrangement found in such organs as the mammary gland, pancreas, and salivary glands. Most glands are enclosed in a fibrous **capsule.** The capsule often gives off extensions called **septa,** or **trabeculae** (trah-BEC-you-lee), that divide the interior of the gland into compartments called **lobes,** which are visible to the naked eye. Finer connective tissue septa may further subdivide each lobe into microscopic **lobules** (LOB-yools). Blood vessels, nerves, and the gland's own ducts generally travel through these septa. The connective tissue framework of the gland, called its **stroma,** supports and organizes the glandular tissue. The cells that perform the tasks of synthesis and secretion are collectively called the **parenchyma** (pa-REN-kih-muh). This is typically simple cuboidal or simple columnar epithelium.

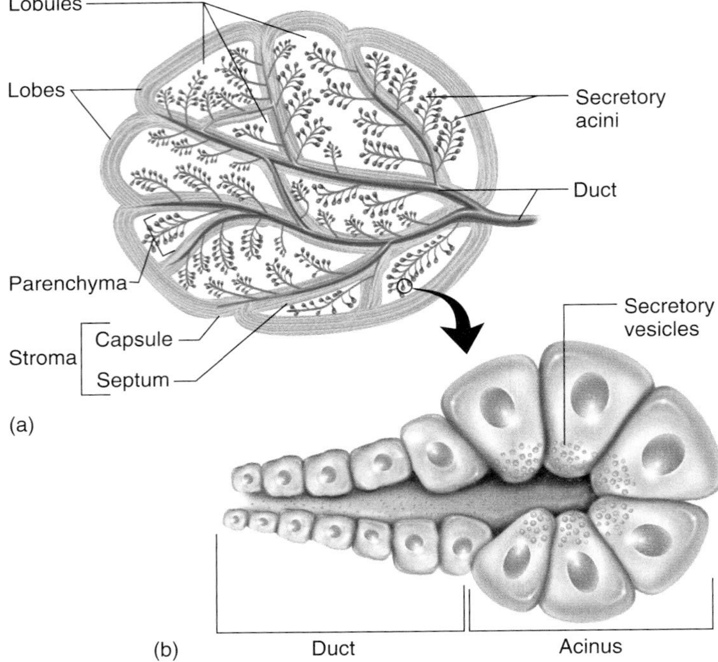

Figure 6.31 General Structure of an Exocrine Gland. (*a*) The gland duct branches repeatedly, following the connective tissue septa, until its finest divisions end on saccular acini of secretory cells. (*b*) Detail of an acinus and the beginning of a duct.

28. *exo* = out + *crin* = to separate, secrete
29. *endo* = in, into

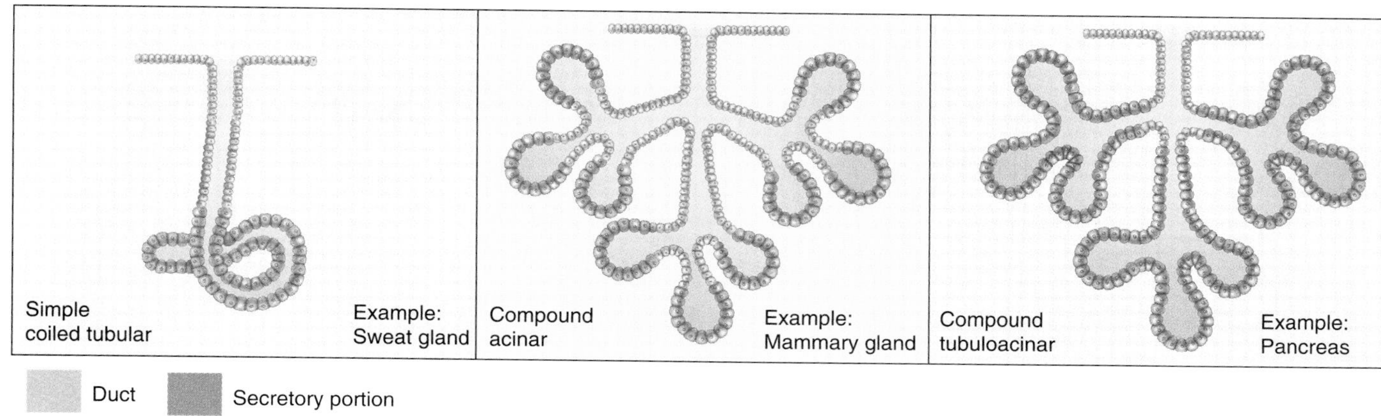

Figure 6.32 **Some Types of Exocrine Glands.** Glands are simple if their ducts do not branch and compound if they do; they are tubular if they have a uniform diameter, acinar if their secretory cells are limited to saccular acini, and tubuloacinar if they have secretory cells in both the acinar and tubular regions.

Exocrine glands are classified as **simple** if they have a single unbranched duct and **compound** if they have a branched duct. If the duct and secretory portion are of uniform diameter, the gland is called **tubular.** If the secretory cells form a dilated sac, the gland is called **acinar** and the sac is an **acinus**[30] (ASS-ih-nus), or **alveolus**[31] (AL-vee-OH-lus). A gland with secretory tubules ending in acini is called a **tubuloacinar gland** (fig. 6.32).

Types of Secretions

Glands are classified not only by their structure but also by the nature of their secretions. **Serous** (SEER-us) **glands** produce relatively thin, watery fluids such as perspiration, milk, tears, and digestive juices. **Mucous glands,** found in the tongue and roof of the mouth among other places, secrete a glycoprotein called *mucin* (MEW-sin). After it is secreted, mucin absorbs water and forms the sticky product *mucus.* (Note that *mucus,* the secretion, is spelled differently from *mucous,* the adjective form of the word.) **Mixed glands,** such as the two pairs of salivary glands in the chin, contain both serous and mucous cells and produce a mixture of the two types of secretions. **Cytogenic**[32] **glands** release whole cells. The only examples of these are the testes and ovaries, which produce sperm and egg cells.

Methods of Secretion

Glands are classified as merocrine or holocrine depending on how they produce their secretions. **Merocrine**[33] (MERR-oh-crin) **glands,** also called **eccrine**[34] (EC-rin) **glands,** have vesicles that release their secretion by exocytosis, as described in chapter 4 (fig. 6.33*a*). These include the tear glands, pancreas, gastric glands, and many others. In **holocrine**[35] **glands,** cells accumulate a product and then the entire cell disintegrates, so the secretion is a mixture of cell fragments and the substance the cell had synthesized prior to its disintegration (fig. 6.33*b*). The oil-producing glands of the scalp are an example. Holocrine secretions tend to be thicker than merocrine secretions.

Some glands, such as the axillary (armpit) sweat glands and mammary glands, are named **apocrine**[36] **glands** from a former belief that the secretion was composed of bits of apical cytoplasm that broke away from the cell surface. Closer study showed this to be untrue; these glands are primarily merocrine in their mode of secretion. These glands are nevertheless different from other merocrine glands in function and histological appearance, and they are still referred to as apocrine glands even though their mode of secretion is not unique.

Membranes

In atlas A, the major cavities of the body were described, as well as some of the membranes that line them and cover their viscera. We now consider some histological aspects of the major body membranes.

The largest membrane of the body is the **cutaneous** (cue-TAY-nee-us) **membrane**—or more simply, the skin (detailed in the next chapter). It consists of a stratified squamous epithelium (epidermis) resting on a layer of

30. *acinus* = berry
31. *alveol* = cavity, pit
32. *cyto* = cell + *genic* = producing
33. *mero* = part + *crin* = to separate, secrete
34. *ec* = *ex* = out
35. *holo* = whole, entire
36. *apo* = from, off, away

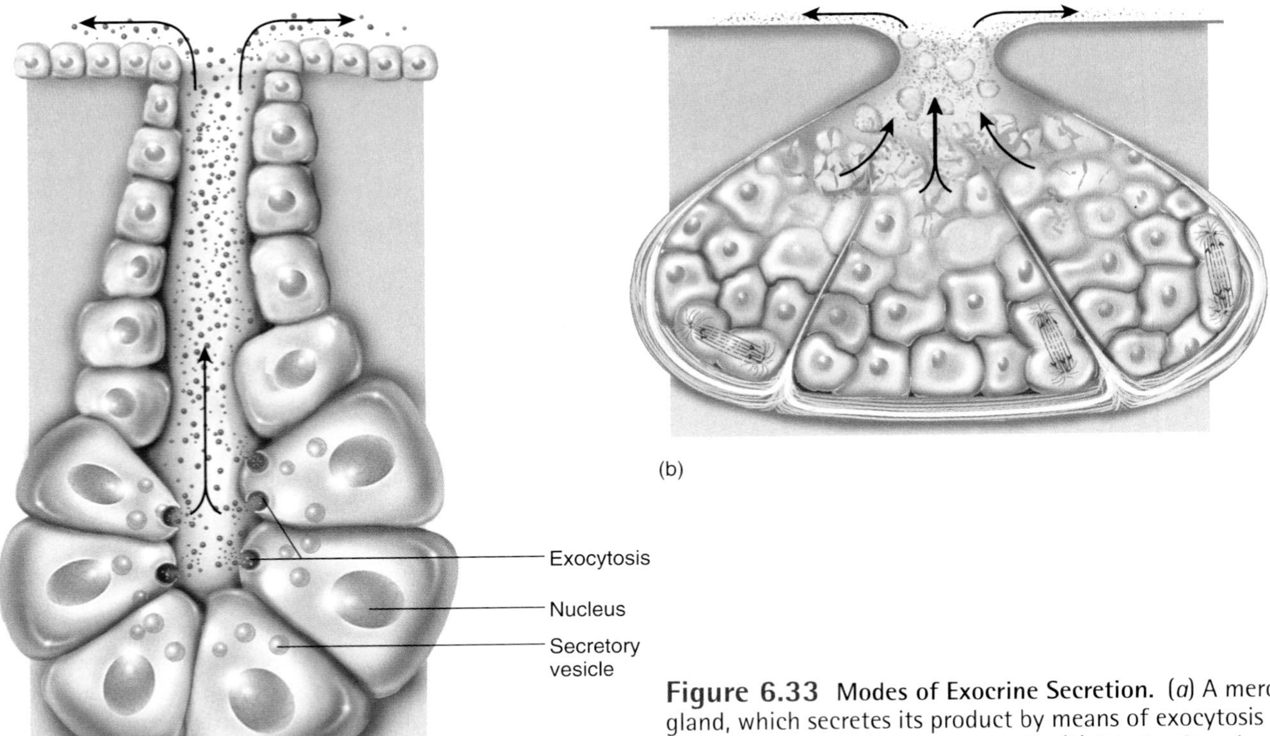

Figure 6.33 **Modes of Exocrine Secretion.** (*a*) A merocrine gland, which secretes its product by means of exocytosis at the apical surfaces of the secretory cells. (*b*) A holocrine gland, whose secretion is composed of disintegrated secretory cells. These cells are replaced by mitosis of the basal cells.

connective tissue (dermis). Unlike the other membranes to be considered, it is relatively dry. It retards dehydration of the body and provides an inhospitable environment for the growth of infectious organisms.

The two principal kinds of internal membranes are mucous and serous membranes. **Mucous membranes** (fig. 6.34), also called **mucosae** (mew-CO-see), line passageways that open to the exterior environment: the digestive, respiratory, urinary, and reproductive tracts. A mucosa consists of two to three layers: (1) an epithelium, (2) an areolar connective tissue layer called the **lamina propria**[37] (LAM-ih-nuh PRO-pree-uh), and, sometimes, (3) a layer of smooth muscle called the **muscularis** (MUSK-you-LAIR-iss) **mucosae.** Mucous membranes have absorptive, secretory, and protective functions. They are often covered with mucus secreted by goblet cells, multicellular mucous glands, or both. The mucus traps bacteria and foreign particles, which keep them from invading the tissues and aids in their removal from the body. The epithelium of a mucous membrane may also include absorptive, ciliated, and other types of cells.

37. *lamina* = layer + *propia* = of one's own

A **serous membrane (serosa)** is composed of a simple squamous epithelium resting on a thin layer of areolar connective tissue. Serous membranes produce watery **serous** (SEER-us) **fluid,** which arises from the blood and derives its name from the fact that it is similar to blood serum in composition. Serous membranes line the insides of some body cavities and form a smooth surface on the outer surfaces of some of the viscera, such as the digestive tract. The pleurae, pericardium, and peritoneum described in atlas A are serous membranes.

The circulatory system is lined with a simple squamous epithelium called **endothelium,** derived from mesoderm. The endothelium rests on a thin layer of areolar tissue, which often rests in turn on an elastic sheet. Collectively, these tissues make up a membrane called the *tunica interna* of the blood vessels and *endocardium* of the heart. The simple squamous epithelium that lines the pleural, pericardial, and peritoneal cavities is called **mesothelium.**

Some joints of the skeletal system are lined by fibrous **synovial** (sih-NO-vee-ul) **membranes,** made only of connective tissue. These membranes span the gap from one bone to the next and secrete slippery *synovial fluid* (rich in hyaluronic acid) into the joint.

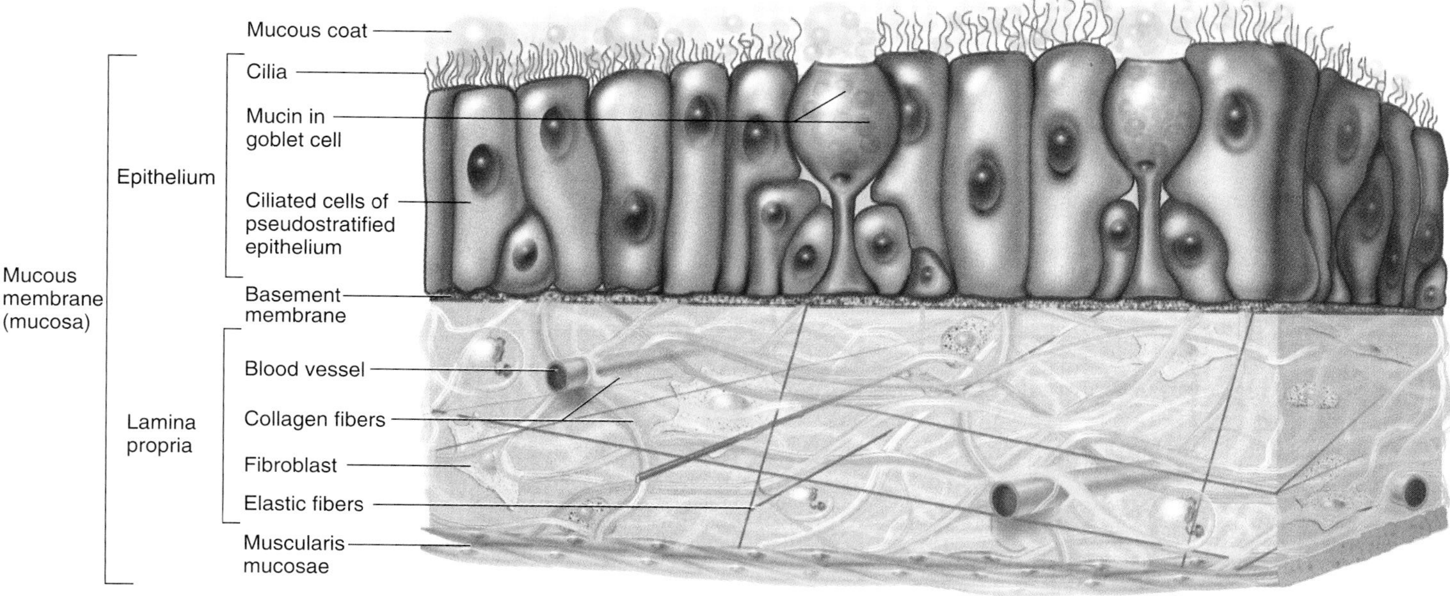

Figure 6.34 Histology of a Mucous Membrane.

Before You Go On

Answer the following questions to test your understanding of the preceding section:

19. Compare the structure of tight junctions and gap junctions. Relate their structural differences to their functional differences.
20. Distinguish between a simple gland and a compound gland, and give an example of each. Distinguish between a tubular gland and an acinar gland, and give an example of each.
21. Contrast the merocrine and holocrine methods of secretion, and name a gland product produced by each method.
22. Describe the differences between a mucous and a serous membrane.
23. Name the layers of a mucous membrane, and state which of the four primary tissue classes composes each layer.

Tissue Growth, Development, Death, and Repair

Objectives

When you have completed this section, you should be able to

- name and describe the ways that a tissue can change from one type to another;
- name and describe the modes of tissue growth;
- name and describe the modes and causes of tissue shrinkage and death; and
- name and describe the ways the body repairs damaged tissues.

Changes in Tissue Type

You have studied the form and function of more than two dozen discrete types of human tissue in this chapter. You should not leave this subject, however, with the impression that once these tissue types are established in the adult, they never change. Tissues are, in fact, capable of changing from one type to another within certain limits. Most obviously, unspecialized tissues of the embryo develop into more diverse and specialized types of mature tissue—mesenchyme to muscle, for example. This development of a more specialized form and function is called **differentiation.**

Epithelia often exhibit **metaplasia,**[38] a change from one type of mature tissue to another. For example, the vagina of a young girl is lined with a simple cuboidal epithelium. At puberty, it changes to a stratified squamous epithelium, better adapted to the future demands of intercourse and childbirth. The nasal cavity is lined with ciliated pseudostratified columnar epithelium. However, if we block one nostril and breathe through the other one for several days, the epithelium in the unblocked passage

38. *meta* = change + *plas* = form, growth

changes to stratified squamous. In smokers, the pseudostratified columnar epithelium of the bronchi may transform into a stratified squamous epithelium.

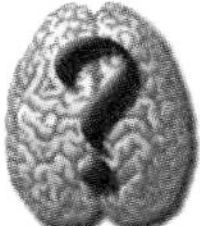

Think About It

What functions of a pseudostratified columnar epithelium could not be served by a stratified squamous epithelium? In light of this, what might be some consequences of bronchial metaplasia in heavy smokers?

Tissue Growth

Tissues grow either because their cells increase in number or because the existing cells grow larger. Most embryonic and childhood growth occurs by **hyperplasia**[39] (HY-pur-PLAY-zhuh), tissue growth through cell multiplication. Exercised muscles grow, however, through **hypertrophy**[40] (hy-PUR-truh-fee), the enlargement of preexisting cells. **Neoplasia**[41] (NEE-oh-PLAY-zhuh) is the development of a tumor (neoplasm)—whether benign or malignant—composed of abnormal, nonfunctional tissue.

Tissue Shrinkage and Death

The shrinkage of a tissue through a loss in cell size or number is called **atrophy**[42] (AT-ruh-fee). Atrophy results from both normal aging (*senile atrophy*) and lack of use of an organ (*disuse atrophy*). Muscles that are not exercised exhibit disuse atrophy as their cells become smaller. This was a serious problem for the first astronauts who participated in prolonged microgravity space flights. Upon return to normal gravity, they were sometimes too weak from muscular atrophy to walk. Space stations and shuttles now include exercise equipment to maintain the crews' muscular condition. Disuse atrophy also occurs when a limb is immobilized, as in a cast.

Necrosis[43] (neh-CRO-sis) is the premature, pathological death of tissue due to trauma, toxins, infection, and so forth. **Gangrene** is any tissue necrosis resulting from an insufficient blood supply. *Gas gangrene* is necrosis of a wound resulting from infection with certain bacteria. **Infarction** is the sudden death of tissue, such as heart muscle (*myocardial infarction*), which occurs when its blood supply is cut off. A *decubitus ulcer* (bed sore) is tissue necrosis that occurs when immobilized persons, such as those confined to a hospital bed or wheelchair, are unable to move, and continual pressure on the skin cuts off blood flow to an area. Cells dying by necrosis usually swell, exhibit *blebbing* (bubbling) of their plasma membranes, and then rupture. The cell contents released into the tissues trigger an inflammation response in which macrophages phagocytize the cellular debris.

Apoptosis[44] (AP-oh-TOE-sis), or **programmed cell death,** is the normal death of cells that have completed their function and best serve the body by dying and getting out of the way. Cells undergoing apoptosis shrink and are quickly phagocytized by macrophages and other cells. The cell contents never escape the cell, so there is no inflammatory response. Although billions of cells die every hour by apoptosis, they are engulfed so quickly that they are almost never seen except within macrophages. For this reason, apoptosis was not discovered until recently.

Apparently every cell has a built-in "suicide program" that enables the body to dispose of it when necessary. In some cases, a receptor protein in the plasma membrane called *Fas* binds to an extracellular suicide signal. *Fas* then activates intracellular enzymes that destroy the cell, including an *endonuclease* that chops up its DNA and a *protease* that destroys cellular proteins. In other cases, cells seem to undergo apoptosis automatically if they stop receiving growth factors from other cells. For example, in embryonic development we produce about twice as many neurons as we need. Those that make connections with target cells survive, while the excess 50% die for lack of *nerve growth factor.* Apoptosis also "dissolves" the webbing between the fingers and toes during embryonic development; it frees the earlobe from the side of the head in people with the genotype for detached earlobes (see chapter 5); and it causes shrinkage of the uterus after pregnancy and of the breasts after lactation ceases.

Tissue Repair

Damaged tissues can be repaired in two ways: *regeneration* or *fibrosis.* **Regeneration** is the replacement of dead or damaged cells by the same type of cells as before. Regeneration restores normal function to the organ. Most skin injuries (cuts, scrapes, and minor burns) heal by regeneration. The liver also regenerates remarkably well. **Fibrosis** is the replacement of damaged tissue with scar tissue, composed mainly of collagen produced by fibroblasts. Scar tissue helps to hold an organ together, but it does not restore normal function. Examples include the healing of severe cuts and burns, the healing of muscle injuries, and scarring of the lungs in tuberculosis.

Figure 6.35 illustrates the following stages in the healing of a cut in the skin, where both regeneration and fibrosis are involved:

1. Severed blood vessels bleed into the cut. Mast cells and cells damaged by the cut release histamine, which dilates blood vessels, increases blood flow to

39. *hyper* = excessive + *plas* = growth
40. *trophy* = nourishment
41. *neo* = new
42. *a* = without
43. *necr* = death + *osis* = process
44. *apo* = away + *ptosis* = falling

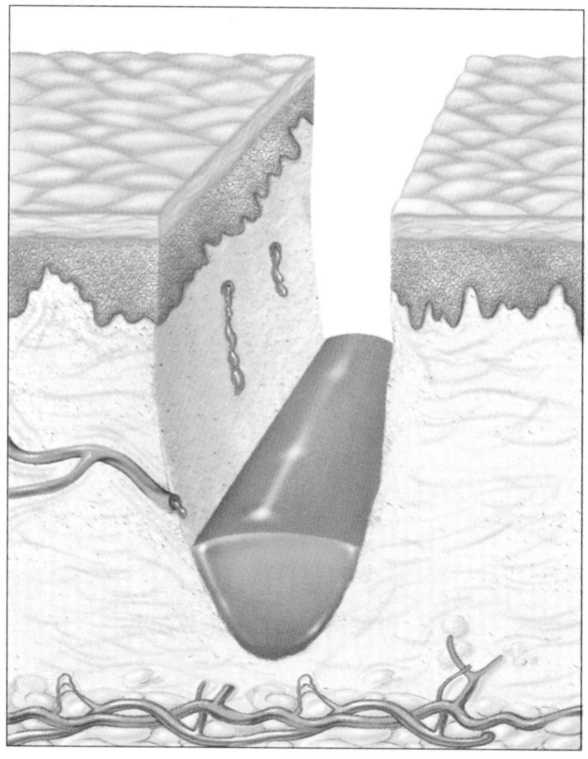

1. Bleeding into the wound

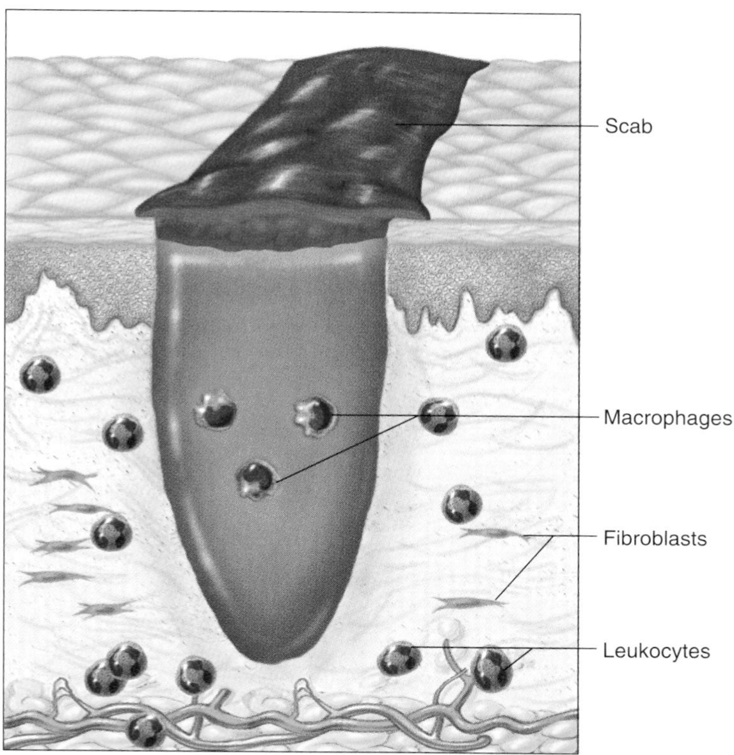

2. Scab formation and macrophage activity

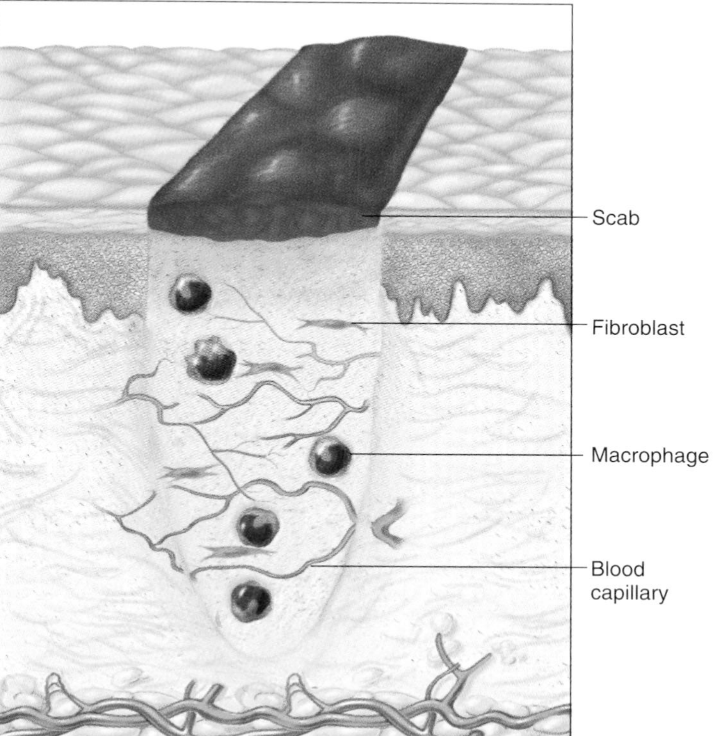

3. Formation of granulation tissue

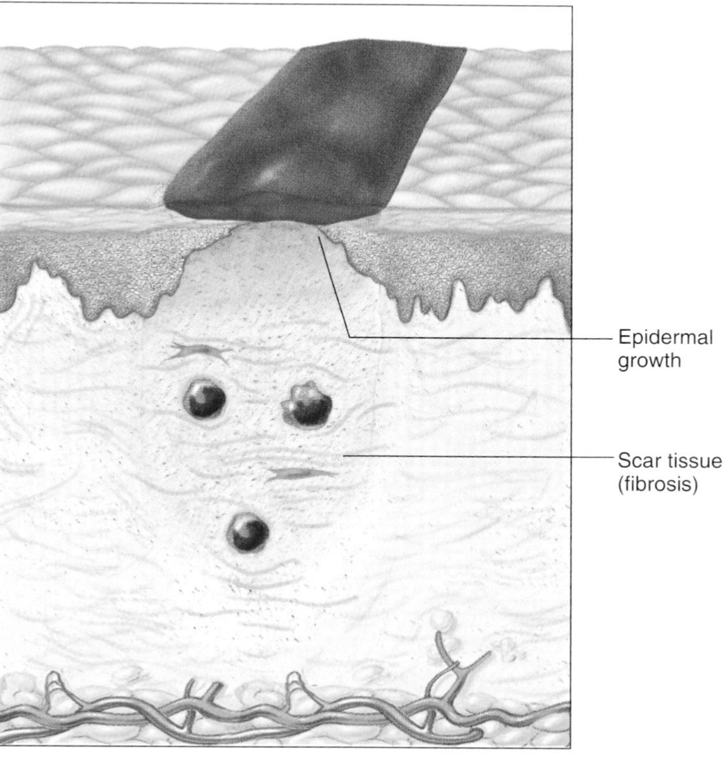

4. Epithelial regeneration and connective tissue fibrosis

Figure 6.35 Stages in the Healing of a Wound to the Skin.

Clinical Insight 6.3

Keloids

In some people, especially dark-skinned adults, healing skin wounds exhibit excessive fibrosis, producing raised, shiny scars called *keloids.* Keloids extend beyond the boundaries of the original wound and tend to return even if they are surgically removed. Keloids may result from the excessive secretion of a fibroblast-stimulating growth factor by macrophages and platelets. They occur most often on the upper trunk and earlobes. Some tribespeople practice *scarification*—scratching or cutting the skin to induce keloid formation as a way of decorating the body.

the area, and makes blood capillaries more permeable. Blood plasma seeps into the wound, carrying antibodies, clotting proteins, and blood cells.

2. A blood clot forms in the tissue, loosely knitting the edges of the cut together and interfering with the spread of pathogens from the site of injury into healthy tissues. The surface of the blood clot dries and hardens in the air, forming a scab that temporarily seals the wound and blocks infection. Beneath the scab, macrophages begin to clean up tissue debris by phagocytizing and digesting it.
3. New blood capillaries sprout from nearby vessels and grow into the wound. The deeper portions of the clot become infiltrated by capillaries and fibroblasts and transform into a soft mass called **granulation tissue.** Macrophages remove the blood clot while fibroblasts deposit new collagen to replace it. This *fibroblastic (reconstructive) phase* of repair begins 3 to 4 days after the injury and lasts up to 2 weeks.
4. Surface epithelial cells around the wound multiply and migrate into the wounded area, beneath the scab. The scab loosens and eventually falls off, and the epithelium grows thicker. Thus, the epithelium *regenerates* while the underlying connective tissue undergoes *fibrosis,* or scarring. Capillaries withdraw from the area as fibrosis progresses. The scar tissue may or may not show through the epithelium, depending on the severity of the wound. The wound may exhibit a depressed area at first, but this is often filled in by continued fibrosis and remodeling from below, until the scar becomes unnoticeable. This *remodeling (maturation) phase* of tissue repair begins several weeks after injury and may last as long as two years.

Before You Go On

Answer the following questions to test your understanding of the preceding section:

24. Distinguish between *differentiation* and *metaplasia.*
25. Tissues can grow through an increase in cell size or cell number. What are the respective terms for these two kinds of growth?
26. Distinguish between *atrophy, necrosis,* and *apoptosis,* and describe a circumstance under which each of these forms of tissue loss may occur.
27. Distinguish between regeneration and fibrosis. Which process restores normal cellular function? What good is the other process if it does not restore function?

Clinical Insight 6.4

Connective Tissue Diseases

Collagen and elastin contribute so greatly to the physical support and resilience of tissues and organs it is not surprising that deficiencies in these connective tissue proteins can have serious consequences.

Marfan[45] *syndrome* is a hereditary defect in elastin fibers, usually resulting from the mutation of a gene on chromosome 15 that codes for *fibrillin,* a glycoprotein that forms a structural scaffold for elastin. Clinical signs of Marfan syndrome include unusually tall stature, long limbs and spidery fingers, abnormal spinal curvature, and a protruding "pigeon breast." Some authorities postulate that Abraham Lincoln's tall, gangly physique and spindly fingers were signs of Marfan syndrome, which perhaps would have ended his life prematurely had he not been assassinated. Some other signs include hyperextensible joints, hernias of the groin, and vision problems resulting from abnormally long eyeballs and deformed lenses. More seriously, however, victims exhibit a weakening of the heart valves and arterial walls. The aorta, where blood pressure is the highest, is sometimes enormously dilated close to the heart and may suddenly rupture. Marfan syndrome is present in about 1 out of 20,000 live births, and it kills most of its victims by their mid-30s.

Several collagen diseases are known, and since collagen is such a widespread protein in the body, the effects are very diverse. *Ehlers-Danlos*[46] *syndrome* is a hereditary defect in collagen synthesis. People with this defect have abnormally long, loose collagen fibers, which show their effects in unusually stretchy skin, loose joints, slow wound healing, and abnormalities in the blood vessels, intestines, and urinary bladder. Infants with this syndrome are often born with dislocated hips. *Osteogenesis imperfecta* results from insufficient collagen deposition in the bones. A collagen deficiency makes the bones exceptionally brittle. Bone fractures are often present at birth; children suffer from frequent spontaneous fractures; the teeth are often deformed; and there may be hearing loss due to deformity of the middle-ear bones.

Not all collagen diseases are hereditary. *Scurvy,* for example, results from a dietary deficiency of vitamin C (ascorbic acid). Ascorbic acid is a cofactor needed for the metabolism of proline and lysine, two amino acids that are especially abundant in collagen. The signs of scurvy include bleeding gums, loose teeth, subcutaneous and intramuscular hemorrhages, and poor wound healing.

45. Antoine Bernard-Jean Marfan (1858–1942), French physician

46. Edward L. Ehlers (1863–1937), Danish dermatologist; Henri A. Danlos (1844–1912), French dermatologist

Chapter Review

Online Learning Center

Study resources for this chapter are available at www.mhhe.com/saladin/

Study Outline

The Study of Tissues (p. 172)

1. The primary tissue classes
   a. Epithelial tissue
   b. Connective tissue
   c. Nervous tissue
   d. Muscular tissue
2. The tissue matrix
   a. Fibrous proteins
   b. Ground substance
3. Embryonic tissues
   a. Primary germ layers
      - Ectoderm
      - Endoderm
      - Mesoderm
   b. Mesenchyme
4. Interpreting tissue sections
   a. Fixatives
   b. Histological sections
   c. Stains
   d. Planes of section

Epithelial Tissue (p. 174)

1. General characteristics
   a. One or more layers of cells
   b. Cover organ surfaces and form glands
   c. Minimal extracellular material
   d. No blood vessels
   e. Basement membrane
   f. Basal and apical cell surfaces

2. Simple epithelia
   a. Simple squamous
   b. Simple cuboidal
   c. Simple columnar
   d. Pseudostratified columnar
3. Stratified epithelia
   a. Stratified squamous
      - Keratinized
      - Nonkeratinized
   b. Stratified cuboidal
   c. Stratified columnar
   d. Transitional

Connective Tissue (p. 180)

1. General characteristics
   a. Abundant extracellular matrix
   b. Cells usually not in direct contact
   c. Functions
      - Binding of organs
      - Support of body
      - Physical protection
      - Immunity
      - Movement
      - Energy and mineral storage
      - Heat production
      - Transport
2. Fibrous connective tissue
   a. Components
      - Cells
        - •• Fibroblasts
        - •• Macrophages
        - •• Leukocytes
        - •• Plasma cells
        - •• Mast cells
        - •• Adipocytes
      - Fibers
        - •• Collagenous
        - •• Reticular
        - •• Elastic
      - Ground substance
        - •• Glycosaminoglycans
        - •• Proteoglycans
        - •• Adhesive glycoproteins
   b. Types
      - Loose connective tissue
        - •• Areolar tissue
        - •• Reticular tissue
        - •• Adipose tissue
      - Dense connective tissue
        - •• Dense regular connective tissue
        - •• Dense irregular connective tissue
3. Cartilage
   a. Components
   b. Types
      - Hyaline cartilage
      - Elastic cartilage
      - Fibrocartilage
4. Bone
   a. Components
   b. Types
      - Spongy bone
      - Compact bone
5. Blood
   a. Plasma
   b. Formed elements
      - Erythrocytes
      - Leukocytes
      - Platelets

Nervous and Muscular Tissue—Excitable Tissues (p. 190)

1. Nervous tissue
   a. Neurons
   b. Neuroglia
2. Muscular tissue
   a. Skeletal muscle
   b. Cardiac muscle
   c. Smooth muscle

Intercellular Junctions, Glands, and Membranes (p. 191)

1. Intercellular junctions
   a. Tight junctions
   b. Desmosomes
   c. Gap junctions
2. Glands
   a. Endocrine and exocrine glands
   b. Exocrine gland structure and classification
   c. Types of secretions
      - Serous
      - Mucous
      - Mixed
      - Cytogenic
   d. Methods of secretion
      - Merocrine
      - Holocrine
      - Apocrine
3. Membranes
   a. Cutaneous membrane
   b. Mucous membranes
   c. Serous membranes
   d. Endothelium
   e. Synovial membranes

Tissue Growth, Development, Death, and Repair (p. 197)

1. Changes in tissue type
   a. Differentiation
   b. Metaplasia
2. Tissue growth
   a. Hyperplasia
   b. Hypertrophy
   c. Neoplasia
3. Tissue shrinkage and death
   a. Atrophy
   b. Necrosis
      - Gangrene
      - Infarction
   c. Apoptosis
4. Tissue repair
   a. Regeneration
   b. Fibrosis

Selected Vocabulary

histology 172
microscopic anatomy 172
primary tissues 172
matrix 172
extracellular material 172
ground substance 172
tissue fluid 172
extracellular fluid 172
interstitial fluid 172
tissue gel 172
primary germ layers 172
ectoderm 172
endoderm 172
mesoderm 172
mesenchyme 172
fixative 172
histological section 172
stain 172
longitudinal section 174
cross section 174
oblique section 174
epithelial tissue 174
basement membrane 175
simple epithelium 175
simple squamous 175
simple cuboidal 175
simple columnar 175
pseudostratified columnar 175
goblet cell 175
stratified epithelium 175
stratified squamous 175
stratified cuboidal 175
stratified columnar 175
transitional 175
exfoliation 175
desquamation 175
Pap smear 175
keratinized 175
nonkeratinized 175
connective tissue 180
fibroblast 181
macrophage 181
leukocyte 181
plasma cell 181
mast cell 181
adipocyte 181
collagenous fibers 181
reticular fibers 182

elastic fibers 182
elastin 182
ground substance 182
glycosaminoglycan 182
chondroitin sulfate 182
proteoglycan 182
adhesive glycoprotein 182
loose connective tissue 183
dense connective tissue 183
areolar tissue 183
reticular tissue 183
adipose tissue 183
dense regular connective tissue 184
yellow elastic tissue 184
dense irregular connective tissue 184
cartilage 184
chondroblast 184
lacuna 184
chondrocyte 184
hyaline cartilage 188
elastic cartilage 188
fibrocartilage 188
perichondrium 188
osseous tissue 188
spongy bone 188
compact bone 188
haversian canal 188
lamellae 188
osteon 188
osteocyte 188
canaliculi 188
periosteum 188
blood 188
plasma 188
formed elements 188
erythrocyte 188
leukocyte 188
platelet 188
excitable tissue 190
membrane potential 190
nervous tissue 190
neuron 190
neuroglia 190
soma 190
dendrite 190
axon 190
muscular tissue 190
skeletal muscle 191
muscle fiber 191
sphincter 191
striation 191
cardiac muscle 191
myocyte 191
intercalated disc 191
smooth muscle 191
visceral muscle 191
intercellular junction 191
tight junction 192
desmosome 193
gap junction 193
gland 193
exocrine gland 194
duct 194
endocrine gland 194
unicellular gland 194
capsule 194
septa 194
lobes 194
lobules 194
stroma 194
parenchyma 194
simple gland 195
compound gland 195
tubular gland 195
acinar gland 195
acinus 195
tubuloacinar gland 195
serous gland 195
mucous gland 195
mixed gland 195
cytogenic gland 195
merocrine gland 195
holocrine gland 195
apocrine gland 195
cutaneous membrane 195
mucous membrane 196
mucosa 196
lamina propria 196
muscularis mucosae 196
serous membrane 196
serous fluid 196
endothelium 196
mesothelium 196
synovial membrane 196
differentiation 197
metaplasia 197
hyperplasia 198
hypertrophy 198
neoplasia 198
atrophy 198
necrosis 198
gangrene 198
infarction 198
apoptosis 198
regeneration 198
fibrosis 198
granulation tissue 200

Testing Your Recall

1. Transitional epithelium is found in
   a. the urinary system.
   b. the respiratory system.
   c. the digestive system.
   d. the reproductive system.
   e. all of the above.
2. The external surface of the stomach is covered by
   a. a mucosa.
   b. a serosa.
   c. the parietal peritoneum.
   d. a lamina propria.
   e. a basement membrane.
3. Which of these is a primary germ layer?
   a. epidermis
   b. mucosa
   c. ectoderm
   d. endothelium
   e. epithelium
4. A seminiferous tubule of the testis is lined with ___________ epithelium.
   a. simple cuboidal
   b. pseudostratified columnar ciliated
   c. stratified squamous
   d. transitional
   e. stratified cuboidal
5. ___________ prevent fluids from seeping between epithelial cells.
   a. Glycosaminoglycans
   b. Hemidesmosomes
   c. Tight junctions
   d. Communicating junctions
   e. Basement membranes
6. A fixative serves to
   a. stop tissue decay.
   b. improve contrast.
   c. repair a damaged tissue.
   d. bind epithelial cells together.
   e. bind cardiac myocytes together.
7. The collagen of areolar tissue is produced by
   a. macrophages.
   b. fibroblasts.
   c. mast cells.
   d. leukocytes.
   e. chondrocytes.
8. Tendons are composed of ___________ connective tissue.
   a. skeletal
   b. areolar
   c. dense irregular
   d. yellow elastic
   e. dense regular
9. The shape of the external ear is due to
   a. skeletal muscle.
   b. elastic cartilage.
   c. fibrocartilage.
   d. articular cartilage.
   e. hyaline cartilage.
10. The most abundant formed element(s) of blood is/are
    a. plasma.
    b. erythrocytes.
    c. platelets.
    d. leukocytes.
    e. proteins.
11. Any form of pathological tissue death is called ___________.
12. The simple squamous epithelium that lines the peritoneal cavity is called ___________.
13. Osteocytes and chondrocytes occupy little cavities called ___________.
14. Muscle cells and axons are often called ___________ because of their shape.
15. Tendons and ligaments are made mainly of the protein ___________.
16. The only type of muscle that lacks gap junctions is ___________.

17. An epithelium rests on a layer called the __________ between its deepest cells and the underlying connective tissue.
18. Fibers and ground substance make up the __________ of a connective tissue.
19. Polysaccharide chains bound to a core protein form giant molecules called __________, an important part of the connective tissue matrix.
20. Any epithelium in which every cell touches the basement membrane is called a/an __________ epithelium.

Answers in appendix B

Testing Your Comprehension

1. A woman in labor is often told to push. In doing so, is she consciously contracting her uterus to expel the baby? Justify your answer based on the muscular composition of the uterus.
2. A major tenet of the cell theory is that all bodily structure and function is based on cells. The structural properties of bone, cartilage, and tendons, however, are due more to their extracellular material than to their cells. Is this an exception to the cell theory? Why or why not?
3. When cartilage is compressed, water is squeezed out of it, and when pressure is taken off, water flows back into the matrix. This being the case, why do you think cartilage at weight-bearing joints such as the knees can degenerate from lack of exercise?
4. The epithelium of the respiratory tract is mostly of the pseudostratified columnar ciliated type, but in the alveoli—the tiny air sacs where oxygen and carbon dioxide are exchanged between the blood and inhaled air—the epithelium is simple squamous. Explain the functional significance of this histological difference. That is, why don't the alveoli have the same kind of epithelium as the rest of the respiratory tract?
5. Which do you think would heal faster, cartilage or bone? Stratified squamous or simple columnar epithelium? Why?

Answers at the Online Learning Center

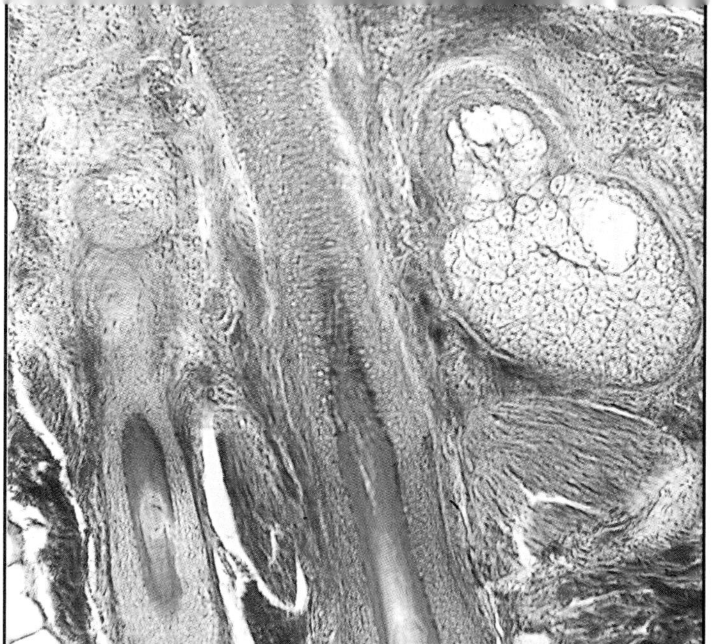

Hair follicles and sebaceous (oil) glands

CHAPTER 7

The Integumentary System

Chapter Outline

Insights

Study with Your ESP CD-Rom

Refer to your ESP Tutorial CD-ROM for helpful animations, activities, and quizzes related to the material covered in this chapter. The following are suggested areas to go to on the CD and are presented according to "Unit/Topic":

- Support and Movement/Integumentary System

Brushing Up

To understand this chapter, it is important that you understand or brush up on the following concepts:

- Stratified squamous epithelium (p. 175)
- Collagenous fibers (p. 181)
- Areolar and dense irregular connective tissue (pp. 183–184)
- Desmosomes (p. 193)
- Merocrine and holocrine glands (p. 195)

The **integumentary system** consists of the skin and its derivatives—hair, nails, and cutaneous glands. We are perhaps more conscious of this system than of any other in the body. Few people venture out of the house each day without first looking in a mirror to see if their skin and hair are presentable. We are well aware of how important this system is to individual recognition, social acceptance, and self-image—and social considerations aside, self-image is important to a person's overall health.

Dermatology is the scientific study and medical treatment of the integumentary system. This is the most easily examined of the organ systems, and its appearance provides important clues not only to its own health but also to deeper disorders such as liver cancer, anemia, and heart failure. It is also the most vulnerable organ system, exposed to radiation, trauma, infection, and injurious chemicals. Consequently, it needs and receives more medical attention than any other organ system.

Structure of the Skin and Subcutaneous Tissue

Objectives

When you have completed this section, you should be able to

- describe the histological structure of the epidermis, dermis, and subcutaneous tissue; and
- discuss the skin's color and markings.

The **skin,** or **integument,** is the body's largest organ (fig. 7.1). In adults, it covers an area of 1.5 to 2.0 m^2 and accounts for about 15% of the body weight. It consists of two layers: (1) a stratified squamous epithelium called the *epidermis* and (2) a deeper connective tissue layer called the *dermis.* Below the skin is another connective tissue layer, the *hypodermis,* which is also discussed in this chapter.

Most of the skin is 1 to 2 mm thick—about half as thick as the cover of this book. It ranges, however, from about 0.5 mm on the eyelids to 6 mm between the shoulder blades. The difference is due mainly to variation in the thickness of the dermis. Skin is classified as *thick* or *thin skin* based on the relative thickness of the epidermis alone, especially the surface layer of dead cells called the *stratum corneum.* **Thick skin** covers the palms, soles, and corresponding surfaces of the fingers and toes. It has an epidermis that is 400 to 600 μm thick, due to a very thick, tough stratum corneum (fig. 7.2*b*). Thick skin has sweat glands but no hair follicles or sebaceous (oil) glands. The rest of the body is covered with **thin skin,** which has an epidermis 75 to 150 μm thick, with a thin stratum corneum (see fig. 7.3). It possesses hair follicles, sebaceous glands, and sweat glands.

The Epidermis

The **epidermis**[1] is a keratinized stratified squamous epithelium, as discussed in the previous chapter. That is, its surface consists of dead cells packed with the tough protein *keratin.* Like other epithelia, the epidermis lacks blood vessels and depends on the diffusion of nutrients from the underlying connective tissue. It has sparse nerve endings for touch and pain, but most sensations of the skin are due to nerve endings in the dermis. The epidermis usually consists of four zones (five in thick skin), described here in order from deep to superficial (fig. 7.2).

Stratum Basale

The **stratum basale** (bah-SAY-lee) consists of a single layer of cuboidal to low columnar cells resting on the basement membrane of the epithelium. There are three types of cells in this layer:

1. **Keratinocytes** (keh-RAT-ih-no-sites) are the majority of cells. They are named for their role in synthesizing the keratin of the skin. Keratinocytes of the stratum basale undergo mitosis and produce new epidermal cells to replace the dead ones that exfoliate (flake off) from the surface.
2. **Melanocytes** synthesize the pigment melanin. They have long branching processes that spread among the basal keratinocytes and continually shed melanin-containing fragments from their tips. The keratinocytes phagocytize these fragments and accumulate melanin granules on the "sunny side" of the nucleus. Like a little parasol, the pigment shields the DNA from ultraviolet radiation. People of all races have about equal numbers of melanocytes. Differences in skin color result from differences in the rate of melanin synthesis and how clumped or spread out the melanin is in the keratinocytes. In light skin, the melanin is less abundant and is relatively clumped near the nucleus, imparting less color to the cells.
3. **Tactile (Merkel**[2]**) cells,** relatively few in number, are receptors for the sense of touch. The tactile cell and its dermal nerve fiber are collectively called a *tactile (Merkel) disc.*

Stratum Spinosum

The **stratum spinosum** (spy-NO-sum) consists of several layers of keratinocytes. The deepest cells undergo mitosis and contribute to the replacement of epidermal cells that exfoliate from the surface. As they are pushed farther upward, however, they cease dividing. Instead, they produce more and more keratin filaments, which cause the cells to flatten.

When skin is histologically fixed, keratinocytes shrink and pull away from each other but remain attached to their neighbors by several desmosomes. These create

1. *epi* = above, upon + *derm* = skin

2. F. S. Merkel (1845–1919), German anatomist

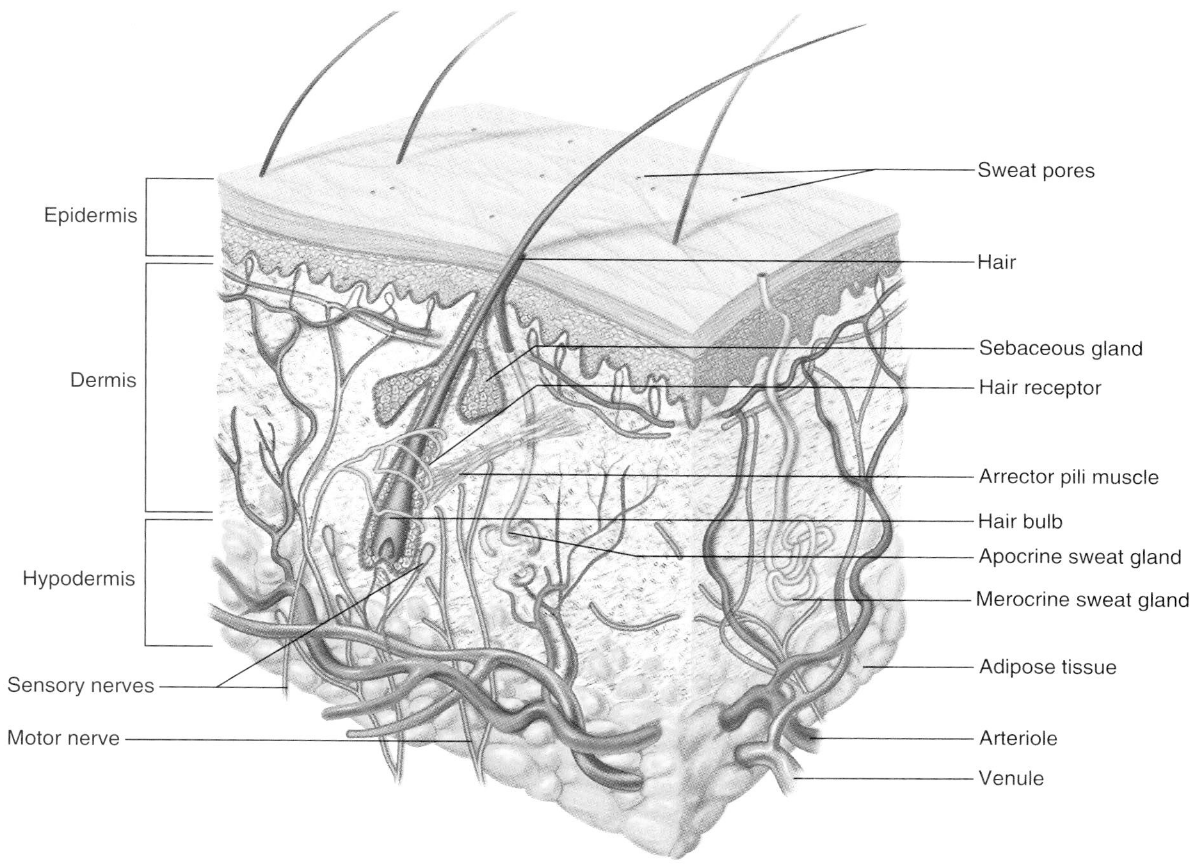

Figure 7.1 Structure of the Skin and Its Derivatives.

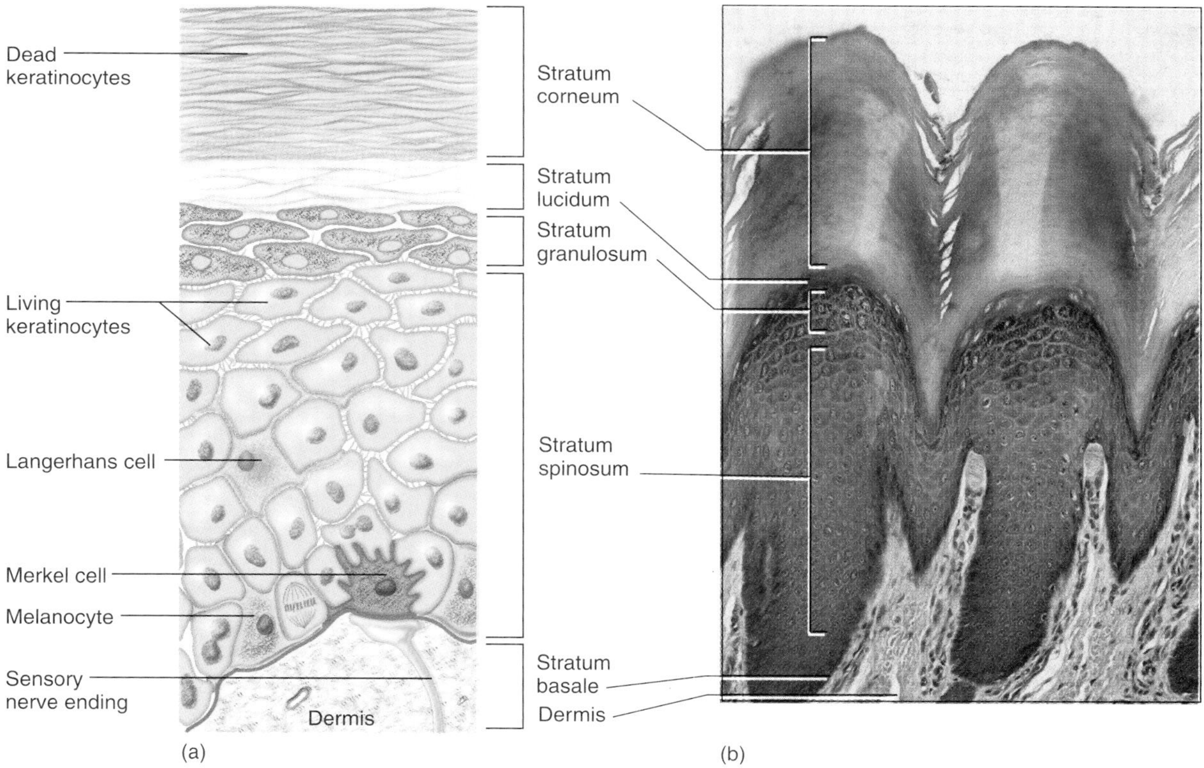

Figure 7.2 The Epidermis. *(a)* Drawing showing epidermal layers and cell types. *(b)* Photograph of thick skin from the fingertip. Shows two of the surface friction ridges responsible for the fingerprints.

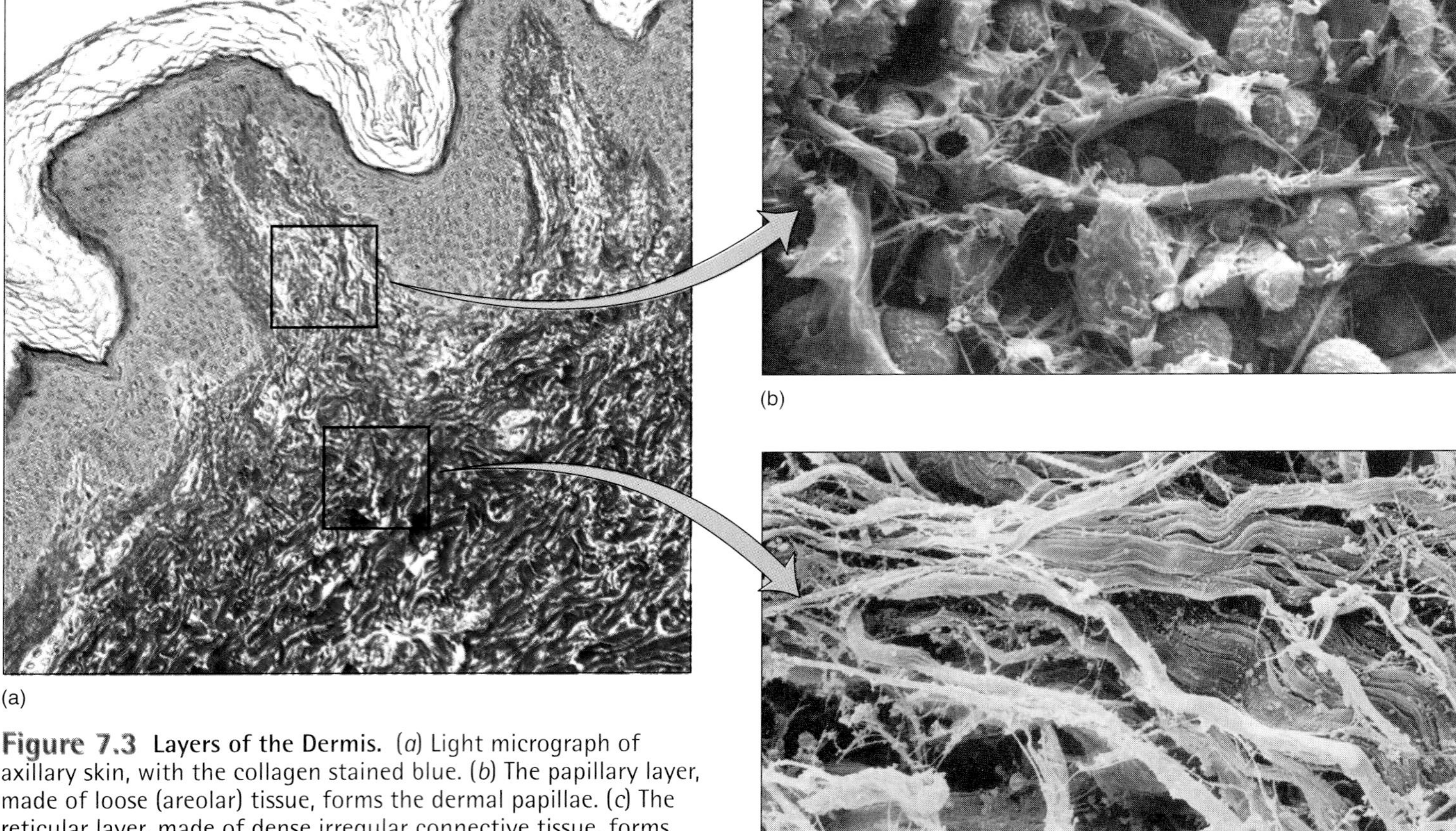

(a) (b) (c)

Figure 7.3 Layers of the Dermis. (*a*) Light micrograph of axillary skin, with the collagen stained blue. (*b*) The papillary layer, made of loose (areolar) tissue, forms the dermal papillae. (*c*) The reticular layer, made of dense irregular connective tissue, forms the deeper four-fifths of the dermis. Credit for *b*: From R. G. Kessel and R. H. Kardon, *Scanning Electron Microscopy of Tissues and Organs,* 1979.

bridgelike extensions where one keratinocyte reaches out to another across the gap—a little like two people holding hands while standing farther apart. These bridges give the cells the spiny appearance for which the stratum spinosum is named.

The stratum spinosum and stratum granulosum also contain **dendritic**[3] (**Langerhans**[4]) **cells.** These are macrophages that arise in the bone marrow but migrate to the stratified squamous epithelia of the epidermis, oral cavity, esophagus, and vagina. The epidermis has as many as 800 dendritic cells per square millimeter. They help to protect the body against pathogens by capturing foreign matter and "presenting" it to the immune system for a response.

Stratum Granulosum

The **stratum granulosum** consists of three to five layers of flat keratinocytes—more in thick skin than in thin skin. These keratinocytes contain coarse, dark-staining *keratohyalin granules* that give this layer its name. These granules consist of a substance that combines with intermediate filaments of the cytoskeleton and converts them to keratin. Keratinocytes in the stratum spinosum and stratum granulosum also produce lipid-filled **membrane-coating vesicles.** Here in the stratum granulosum, these vesicles undergo exocytosis and release a glycolipid that spreads out over the keratinocyte membranes and waterproofs the skin. The glycolipid also constitutes a barrier between the surface cells of the skin and the deeper layers. Cut off from their nutrient supply, cells above the stratum granulosum quickly die.

Stratum Lucidum

The **stratum lucidum**[5] (LOO-sih-dum) is a thin translucent zone superficial to the stratum granulosum, seen only in thick skin. Here, the keratinocytes are densely packed with *eleidin* (ee-LEE-ih-din), an intermediate product in the production of keratin. The cells have no nuclei or other organelles. Because organelles are absent and eleidin does not stain well, this zone has a pale, featureless appearance with indistinct cell boundaries.

3. *dendr* = tree, branch
4. Paul Langerhans (1847–88), German anatomist
5. *lucid* = light, clear

Clinical Insight 7.1

Dead Skin and Dust Mites

In the beams of late afternoon sun that shine aslant through a window, you may see tiny white specks floating through the air. Most of these are flakes of dander; the dust on top of your bookshelves is mostly a film of dead human skin. Composed of protein, this dust in turn supports molds and other microscopic organisms that feed on the skin and each other. One of these organisms is the house dust mite, *Dermatophagoides*[6] (der-MAT-oh-fah-GOY-deez). (What wonders may be found in humble places!)

Dermatophagoides thrives abundantly in pillows, mattresses, and upholstery—warm, humid places that are liberally sprinkled with edible flakes of keratin. No home is without these mites, and it is impossible to entirely exterminate them. What was once regarded as "house dust allergy" is now regarded as an allergy to the inhaled feces of these mites.

Figure 1 *Dermatophagoides*, the House Dust Mite.

6. *dermato* = skin + *phag* = eat

Stratum Corneum

The **stratum corneum** consists of up to 30 layers of dead, scaly, keratinized cells. Those at the surface flake off (*exfoliate* or *desquamate*) as tiny scales called *dander* (see clinical insight 7.1). Dandruff is composed of clumps of dander stuck together by oil from the scalp.

As we surveyed the five layers of epidermis, we followed keratinocytes from their birth to their death. They are produced by mitosis in the stratum basale and lower level of the stratum spinosum. Mitosis occurs here mainly at night, and most skin specimens are taken during the day, so mitotic cells are rarely seen in prepared slides of skin. Over the course of 30 to 40 days, a keratinocyte is pushed farther and farther toward the surface by the dividing cells below, until it finally exfoliates from the surface. This process is slower in old age, but faster in skin that has been injured or stressed. Injured epidermis regenerates more rapidly than any other tissue in the body. Mechanical stress from manual labor or tight shoes accelerates keratinocyte multiplication and results in calluses on the hands or feet.

The Dermis

The **dermis** ranges from 0.6 mm thick in the eyelids to about 3 mm thick in the palms and soles. It is composed mainly of collagen but also contains elastic and reticular fibers, the usual cells of fibrous connective tissue (described in chapter 6), and blood vessels, sweat glands, sebaceous glands, hair follicles, and nail roots (see fig. 7.1). It also contains sensory nerve endings, discussed later in this chapter (p. 213), and muscular tissue. Smooth muscle cells associated with hair follicles make the hairs stand on end in response to cold or fear, cause "goosebumps," and wrinkle the skin in areas such as the scrotum and areola in response to cold or touch. In the face, skeletal muscles attach to dermal collagen fibers and produce such expressions as a smile, a wrinkle of the forehead, and the wink of an eye.

The boundary between the epidermis and dermis is histologically distinct and usually wavy. The upward waves are fingerlike extensions of the dermis called **dermal papillae** and the downward waves are extensions of the epidermis. The dermal and epidermal boundaries thus interlock like corrugated cardboard, an arrangement that resists slippage of the epidermis across the dermis. It forms the ridges of the fingerprints (see fig. 7.2*b*), and in highly sensitive areas such as the lips and genitals, tall dermal papillae allow nerve fibers and blood capillaries to reach close to the skin surface. If you look closely at the skin of your hands, you will see delicate furrows that divide it into tiny rectangular to rhomboidal areas. The dermal papillae produce the raised areas between the furrows.

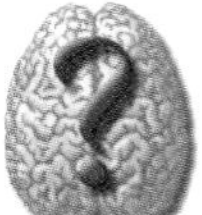

Think About It

Dermal papillae are relatively high and numerous in palmar and plantar skin but low and few in number in skin of the face and abdomen. What do you think is the functional significance of this difference?

There are two zones of dermis called the papillary and reticular layers (fig. 7.3). The **papillary**[7] (PAP-ih-lerr-ee) **layer** is a zone of areolar tissue in and near the dermal papillae that forms about one-fifth of the thickness of the dermis. The loosely organized tissue of the papillary layer allows for mobility of white blood cells and other defenses against organisms introduced through breaks in the epidermis.

The **reticular layer** of the dermis, the deeper four-fifths, consists of dense irregular connective tissue. The boundary between the papillary and reticular layers is often vague. In the reticular layer, the collagen forms thicker bundles with less room for ground substance, and there are often small clusters of adipocytes. Stretching of the skin, as occurs in obesity and pregnancy, can tear the collagen fibers and produce *striae* (STRY-ee), or stretch marks. These occur especially in areas most stretched by weight gain: the thighs, buttocks, abdomen, and breasts.

The Hypodermis

Beneath the skin is a layer called the **hypodermis,**[8] **subcutaneous tissue,** or **superficial fascia**[9] (FASH-ee-uh). The boundary between the dermis and hypodermis is indistinct, but the hypodermis generally has more areolar and adipose tissue. When adipose tissue dominates, the hypodermis is called the **subcutaneous fat** (*panniculus*[10] *adiposus*). The hypodermis binds the skin to the underlying tissues, pads the body, serves as an energy reservoir, and provides thermal insulation. Infants and elderly people are more sensitive to cold than others because they have less fat in this layer. Obesity is due mainly to the accumulation of subcutaneous fat. The hypodermis is about 8% thicker in women than in men and differs in distribution between the sexes (fig. 7.4). Drugs are introduced here by hypodermic injection because the subcutaneous tissue is highly vascular and absorbs them quickly.

The layers of skin and hypodermis are summarized in table 7.1.

7. *papilla* = nipple
8. *hypo* = below + *derm* = skin
9. *fasc* = band
10. *panni* = rag, cloth + *culus* = little

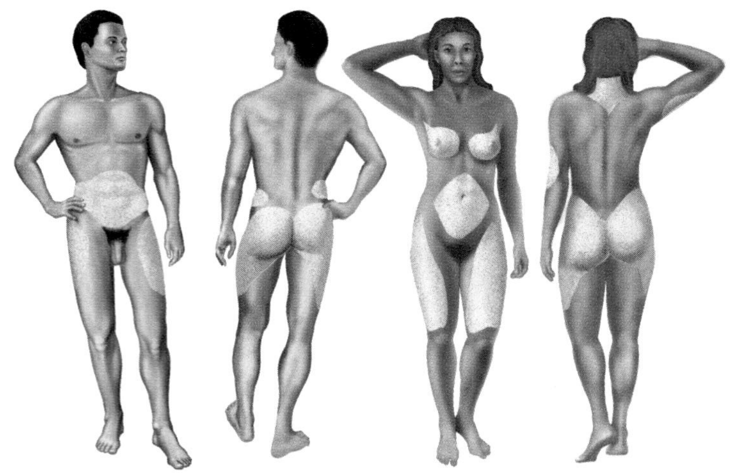

Figure 7.4 Distribution of Subcutaneous Fat in Males and Females.

Skin Color

Three pigments bear most of the responsibility for normal skin colors:

1. **Hemoglobin,** the red pigment carried in the dermal blood vessels, combines with the white color of the dermal collagen fibers to produce the flesh tones typical of Caucasian skin. The skin is redder in places such as the lips, where blood capillaries come closer to the surface and the hemoglobin shows through more vividly.
2. **Melanin**[11] in the stratum basale and stratum spinosum produces a variety of brown, black, tan, yellowish, and reddish hues (fig. 7.5). (The yellowish tint of some Asiatic complexions, once attributed to carotene, is now regarded as being due to melanin.) The amount of melanin in the skin is determined by a combination of heredity and light exposure. Ultraviolet (UV) radiation from the sun stimulates melanin synthesis and darkens the skin. A suntan fades as melanized keratinocytes migrate to the surface and exfoliate. The amount of melanin varies substantially from place to place on the body. It is relatively concentrated in freckles and moles,

11. *melan* = black

Table 7.1 Stratification of the Skin and Hypodermis

Layer	Description
Epidermis	Stratified squamous epithelium.
Stratum corneum	Dead, keratinized cells of the skin surface.
Stratum lucidum	Clear, featureless, narrow zone seen only in thick skin.
Stratum granulosum	Two to five layers of cells with dark-staining keratohyalin granules; scanty in thin skin.
Stratum spinosum	Many layers of keratinocytes, typically shrunken in fixed tissues but attached to each other by desmosomes, which give them a spiny look; progressively flattened the farther they are from the dermis. Dendritic cells occur here but are not visible in routinely stained preparations.
Stratum basale	Single layer of cuboidal to columnar cells resting on basal lamina; site of most mitosis; consists of keratinocytes, melanocytes, and tactile cells, but these are not distinguishable with routine stains. Melanin is conspicuous in keratinocytes of this layer in black to brown skin.
Dermis	Fibrous connective tissue, richly endowed with blood vessels and nerve endings. Sweat glands and hair follicles originate here and in hypodermis.
Papillary layer	Superficial one-fifth of dermis; composed of areolar tissue; often extends upward as dermal papillae.
Reticular layer	Deeper four-fifths of dermis; dense irregular connective tissue.
Hypodermis	Areolar or adipose tissue between skin and muscle.

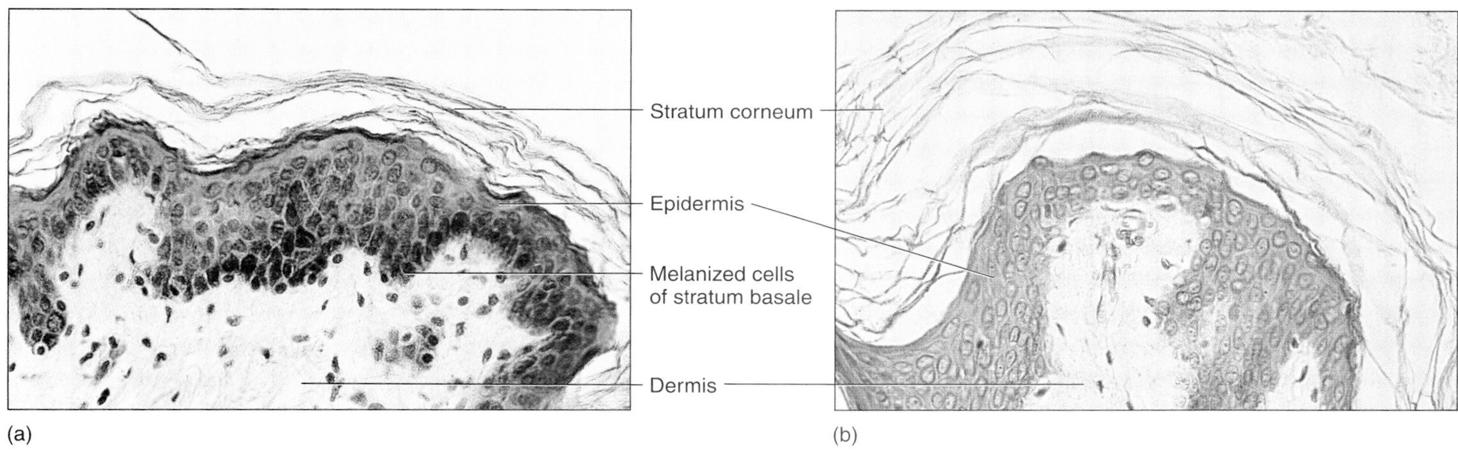

Figure 7.5 Variations in Skin Pigmentation. (*a*) Keratinocytes in and near the stratum basale have heavy deposits of melanin in dark skin. (*b*) Light skin shows little to no visible melanin in the basal keratinocytes.

on the dorsal surfaces of the hands and feet as compared to the palms and soles, in the nipple and surrounding area (areola) of the breast, around the anus, in the scrotum and penis, and on the lateral surface of the female genital folds (labia majora). The contrast between heavily melanized and lightly melanized regions of the skin is more pronounced in some races than in others, but it exists to some extent in nearly everyone.

3. **Carotene**[12] is a yellow pigment acquired from egg yolks and yellow and orange vegetables. Depending on the diet, it can become concentrated to various degrees in the stratum corneum and subcutaneous fat. It is often most conspicuous in skin of the heel and in "corns" or calluses of the feet because this is where the stratum corneum is thickest.

The skin may also exhibit abnormal colors of diagnostic value:

- **Cyanosis**[13] is blueness of the skin resulting from a deficiency of oxygen in the circulating blood. Oxygen deficiency turns the hemoglobin a reddish violet color.

12. *carot* = carrot

13. *cyan* = blue + *osis* = condition

It can result from conditions that prevent the blood from picking up a normal load of oxygen in the lungs, such as airway obstructions in drowning and choking, lung diseases such as emphysema, or respiratory arrest. Cyanosis also occurs in situations such as cold weather and cardiac arrest, when blood flows so slowly through the skin that most of its oxygen is extracted faster than freshly oxygenated blood arrives.

- **Erythema**[14] (ERR-ih-THEE-muh) is abnormal redness of the skin. It occurs in such situations as exercise, hot weather, sunburns, anger, and embarrassment. Erythema is caused by increased blood flow in dilated cutaneous blood vessels or by dermal pooling of red blood cells that have escaped from abnormally permeable capillaries.
- **Jaundice**[15] is a yellowing of the skin and whites of the eyes resulting from high levels of bilirubin in the blood. Bilirubin is produced by the breakdown of free hemoglobin and disposed of by the liver. Jaundice may therefore occur when there is a rapid rate of erythrocyte destruction; when diseases such as cancer, hepatitis, and cirrhosis interfere with liver function; and in premature infants, where the liver is not well enough developed to dispose of bilirubin efficiently.
- **Bronzing** is a golden-brown skin color that results from Addison disease, a deficiency of glucocorticoid hormones from the adrenal cortex (see fig. 17.16, p. 654). John F. Kennedy had Addison disease and bronzing of the skin, which many people mistook for a suntan.
- **Pallor** is a pale or ashen color that occurs when there is so little blood flow through the skin that the white color of the dermal collagen shows through. It can result from emotional stress, low blood pressure, circulatory shock, cold temperatures, or severe anemia.
- **Albinism**[16] is a genetic lack of melanin that results in white hair, pale skin, and pink eyes. Melanin is synthesized from the amino acid tyrosine by the enzyme tyrosinase. People with albinism have inherited a recessive, nonfunctional tyrosinase allele from both parents.
- A **hematoma**,[17] or bruise, is a mass of clotted blood showing through the skin. It is usually due to accidental trauma (blows to the skin), but it may indicate hemophilia, other metabolic or nutritional disorders, or physical abuse.

14. *eryth* = red + *em* = blood
15. *jaun* = yellow
16. *alb* = white + *ism* = state, condition
17. *hemat* = blood + *oma* = mass

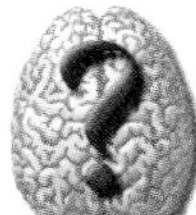

Think About It

An infant brought to a clinic shows abnormally yellow skin. What sign could you look for to help decide whether this was due to jaundice or to a large amount of carotene from strained vegetables in the diet?

Skin Markings

Hemangiomas[18] (he-MAN-jee-OH-mas), or birthmarks, are patches of discolored skin caused by benign tumors of the dermal blood capillaries. *Capillary hemangiomas* (strawberry birthmarks) are bright red to deep purple and are slightly swollen; they usually disappear in childhood. *Cavernous hemangiomas* (port wine stains) are flat, are duller in color, and last for life.

Freckles and moles are tan to black aggregations of melanocytes. **Freckles** are flat melanized patches that vary with heredity and exposure to the sun. A **mole (nevus)** is an elevated patch of melanized skin, often with hair. Moles are harmless and sometimes even regarded as "beauty marks," but they should be watched for changes that may suggest malignancy.

The skin is also marked by many lines, creases, and ridges. **Friction ridges** are the markings on the fingertips that leave oily fingerprints on surfaces we touch. Friction ridges are characteristic of most primates. They help prevent monkeys, for example, from slipping off a branch as they walk across it, and they enable us to manipulate small objects more easily. Friction ridges form during fetal development and remain essentially unchanged for life. Everyone has a unique pattern of friction ridges; not even identical twins have identical fingerprints.

Flexion (palmar) creases are the lines of the palms formed after birth by repeated closing of the hand. Similar **flexion lines** form in other places such as the wrist and elbow (see fig. B.7 in atlas B).

Before You Go On

Answer the following questions to test your understanding of the preceding section:

1. Name the four kinds of epidermal cells and state the function of each.
2. List the five layers of epidermis from deep to superficial. What are the distinctive features of each layer?
3. What are the two layers of the dermis? What type of tissue composes each layer?
4. Name the pigments responsible for normal skin colors, and explain how certain conditions can produce symptomatic discolorations of the skin.

18. *hem* = blood + *angi* = vessels + *oma* = mass

Functions of the Skin

Objective

When you have completed this section, you should be able to

- describe the roles that the skin plays as a barrier and in vitamin D synthesis, absorption of chemicals, sensation, thermoregulation, and social interaction.

The skin is much more than a container for the body. It has a wide variety of important functions that go well beyond appearance, as we shall see here. Although its structural and physiological complexity is commonly underestimated, the skin is indeed a marvel of biological design. It is hard to conceive of any other organ that could serve functions so varied, yet display such strength, flexibility, and remarkable capacity for growth and self-repair.

The Skin As a Barrier

The skin bears the brunt of most physical injuries to the body, but it resists and recovers from trauma better than other organs do. The toughness of keratin and strength of the epidermal desmosomes make the skin a barrier that is not easily breached. Few infectious organisms can penetrate the skin on their own. Those that do either rely on accidental breaks in the skin or have life cycles that involve *vectors*—animals such as mosquitoes, fleas, lice, and ticks, with mouthparts strong enough to puncture the skin.

The epidermal surface is populated by great numbers of bacteria, fungi, and other pathogens poised for any opportunity to get inside. Even vigorous scrubbing in a hot shower does not rid the skin of bacteria. They are, however, discouraged from multiplying on the skin by its relatively dry, unfavorable habitat. The sebum (oil of the skin) contains bactericidal substances, and sweat forms a film called the **acid mantle** (pH 4–6) that is unfavorable to microbial growth. Even when a pathogen breaches the epidermis, it faces an army of dermal macrophages and leukocytes that can quickly migrate to the site of infection and mount a defense.

The skin is also important as a barrier to water. It prevents the body from absorbing excess water when you are swimming or bathing, but even more importantly, it prevents the body from losing excess water. This function becomes especially evident when skin is lost; in burn patients who have lost a large area of skin, fluid replacement therapy becomes one of the most critical needs for survival.

The skin is also a barrier to solar radiation, including ultraviolet (UV) rays. Most UV rays are filtered out by atmospheric ozone, but even the small fraction that reaches us is enough to cause sunburns and skin cancer. Scientists fear that we could see a catastrophic increase in skin cancer because of the loss of atmospheric ozone. Ozone is destroyed by chemicals called chlorofluorocarbons (CFCs), used in air conditioners, refrigerators, spray cans, and other sources. Many countries have agreed to phase out CFC production by the year 2000.

Vitamin D Synthesis

Vitamin D is important in regulating blood calcium and phosphate levels and maintaining the health of the skeleton. The first step in its synthesis occurs in the skin, when UV rays penetrate to the blood capillaries of the dermis. The UV rays convert a cholesterol derivative in the blood, *7-dehydrocholesterol,* to *vitamin D_3.* The further details of vitamin D synthesis and its effect on the bones are explained in the next chapter.

Cutaneous Absorption

Although the skin is impermeable to most chemicals, there are exceptions. The blood receives 1% to 2% of its oxygen by diffusion through the skin, and it gives off some carbon dioxide and volatile organic chemicals. Amino acids and steroids diffusing through the skin are one factor that attracts mosquitoes to people. The fat-soluble vitamins A, D, E, and K can be readily absorbed through the skin, as can many drugs and poisons (see clinical insight 7.2).

Sensory Functions

The skin is our most extensive sense organ. It is equipped with a variety of nerve endings that react to heat, cold, touch, texture, pressure, vibration, and tissue injury. These sensory receptors are especially abundant on the face, palms, fingers, soles, nipples, and genitals. There are relatively few on the back and in skin overlying joints such as the knees and elbows. Some receptors are naked dendrites that penetrate into the epidermis. Most, however, are limited to the dermis and hypodermis, where specialized connective tissues give the nerve cells more selective sensitivity to particular types of stimuli. The individual types of cutaneous receptors are described in chapter 16.

Thermoregulation

The nervous, endocrine, muscular, and integumentary systems are all involved in regulating body temperature. The details of thermoregulation are discussed in chapter 26, but here we briefly consider the role of the skin. The dermis contains naked nerve endings called **thermoreceptors,**

Clinical Insight 7.2

Transdermal Absorption

The ability of the skin to absorb chemicals makes it possible to administer several medicines as ointments or lotions, or by means of adhesive patches that release the medicine steadily through a membrane. For example, inflammation can be treated with a hydrocortisone ointment, nitroglycerine patches are used to relieve heart pain, nicotine patches are used to help overcome tobacco addiction, and other medicated patches are used to control high blood pressure and motion sickness.

Unfortunately, the skin can also be a route for absorption of poisons. These include toxic alkaloids from poison ivy and other plants; metals such as mercury, arsenic, and lead; and solvents such as carbon tetrachloride (dry cleaning fluid), acetone (nail polish remover), paint thinner, and insecticides. Some of these can cause brain damage, liver failure, or kidney failure, which is good reason for using protective gloves when handling such substances.

some of which respond when the skin temperature rises above normal and others when it falls below normal. They transmit signals to a region called the *hypothalamus* in the base of the brain. To cool the body, the hypothalamus sends signals that dilate the arteries of the skin. This increases blood flow close to the body surface and allows more heat to radiate away from the body. If this is not enough, the hypothalamus also stimulates sweating. When sweat evaporates, it carries a lot of heat away from the body. Conversely, when it is necessary to warm the body, the hypothalamus constricts the cutaneous arteries, reducing blood flow and heat loss.

Social Functions

The integumentary system plays an important role in the social relations of all vertebrate animals by enabling them to identify members of their own species and distinguish the sexes. Some mammals recognize each other by the color and distribution of hair. Animals may also accept or reject one another's company and choose mates based on the appearance of the integument, which may indicate an animal's state of health. Thus, it is not surprising that animals allocate a lot of time to grooming.

Humans are no exception. A little reflection will emphasize just how much impact an integumentary condition can have on someone's self-image and emotional state—whether the ravages of adolescent acne, the presence of a birthmark or a disfiguring scar, or just a "bad hair day." The skin is also our most significant means of nonverbal communication. The faces of primates are far more expressive than those of other mammals because complex skeletal muscles insert on dermal collagen fibers and create subtle and varied facial expressions (fig. 7.6). Thus the skin has very important psychological and social functions.

Before You Go On

Answer the following questions to test your understanding of the preceding section:

5. How does the skin inhibit the growth of pathogens on its surface?
6. To what other things besides pathogenic organisms does the skin present a barrier?
7. What step in the synthesis of vitamin D occurs in the skin?
8. Describe two ways in which the skin helps to regulate body temperature.

Hair and Nails

Objectives

When you have completed this section, you should be able to

- distinguish between three types of hair;
- describe the histology of a hair and its follicle;
- explain how a hair grows and discuss the types and causes of baldness;
- discuss some theories of the purposes served by various kinds of hair; and
- describe the structure and function of nails.

The hair, nails, and cutaneous glands are the **accessory organs,** or **appendages,** of the skin. Hair and nails are composed mostly of dead, keratinized cells. While the stratum corneum of the skin is made of pliable **soft keratin,** the hair and nails are composed mostly of **hard keratin.** Hard keratin is more compact than soft keratin and is toughened by a greater number of disulfide bridges between the protein molecules.

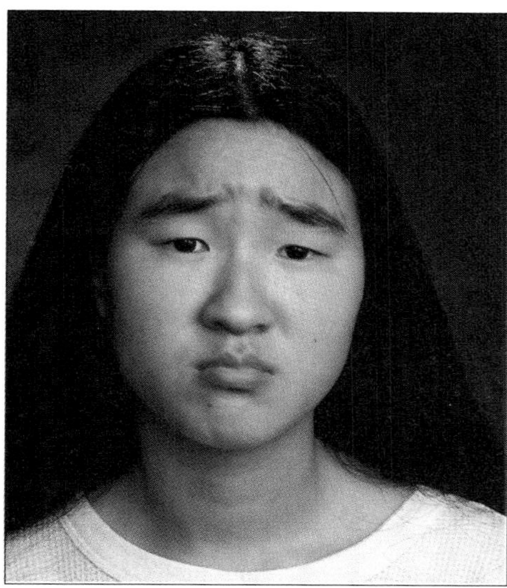

Figure 7.6 Importance of the Skin in Nonverbal Expression. Primates differ from other mammals in having very expressive faces due to facial muscles that insert on the collagen fibers of the dermis and move the skin.

Types and Distribution of Hair

A hair is also known as a **pilus** (PY-lus); in the plural, *pili* (PY-lye). It is a slender filament of keratinized cells that grows from an oblique tube in the skin called a **follicle** (fig. 7.7). Hair is found almost everywhere on the body except the lips, nipples, parts of the genitals, palmar and plantar skin, lateral surfaces of the fingers, toes, and feet, and distal segment of the fingers. The extremities and trunk have about 55 to 70 hairs per square centimeter, and the face has about 10 times as many. The scalp has about 100,000 hair follicles and a man's beard has about 30,000. The number of hairs in a given area does not differ much from one person to another or even between the sexes. Differences in apparent hairiness are due mainly to differences in the texture and pigmentation of the hair.

Not all hair is alike, even on one person. Over the course of our lives, we grow three kinds of hair: lanugo, vellus, and terminal hair. **Lanugo**[19] is fine, downy, unpigmented hair that appears on the fetus in the last 3 months of development. By the time of birth, most of it is replaced by similarly fine, unpigmented hair called **vellus.**[20] Except for the eyebrows, eyelashes, and hair of the scalp, all of the hair of children, two-thirds of the hair of women, and one-tenth of the hair of men is vellus. **Terminal hair** is longer, coarser, and pigmented. It occurs on the scalp, eyebrows, and eyelashes; at puberty it replaces the vellus in the axillary and pubic regions, on the face of males (to form the beard), and to varying degrees on the trunk and extremities.

Structure of the Hair and Follicle

A hair is divisible into three zones along its length: (1) the **bulb,** a swelling at the base where the hair originates in the dermis; (2) the **root,** which is the remainder of the hair within the follicle; and (3) the **shaft,** which is the portion above the skin surface. Except near the bulb, all the tissue is dead. A bit of vascular connective tissue called the **papilla** grows into the bulb and provides the hair with its sole source of nutrition. Although ingested substances can become incorporated into hair (see clinical insight 7.3), nothing you apply externally to your hair can nourish it, notwithstanding advertising pitches to the gullible.

In cross section, a hair reveals three layers: (1) the **medulla,** a core of loosely arranged cells and air spaces; (2) the **cortex,** composed of densely packed keratinized cells; and (3) the **cuticle,** a single layer of scaly cells that overlap each other like roof shingles, with their free edges directed upward (fig. 7.7*c*). Cells lining the follicle are like shingles facing in the opposite direction. They interlock with the scales of the hair cuticle and resist pulling on the hair. When a hair is pulled out, this layer of follicle cells comes with it.

The texture of hair is related to differences in cross-sectional shape (fig. 7.8)—straight hair is round, wavy hair is oval, and kinky hair is relatively flat. The color of black and brown hair is due to melanin produced by melanocytes in the bulb and transferred to cells of the cortex. A blond color is simply due to a scanty amount of

19. *lan* = down, wool
20. *vellus* = fleece

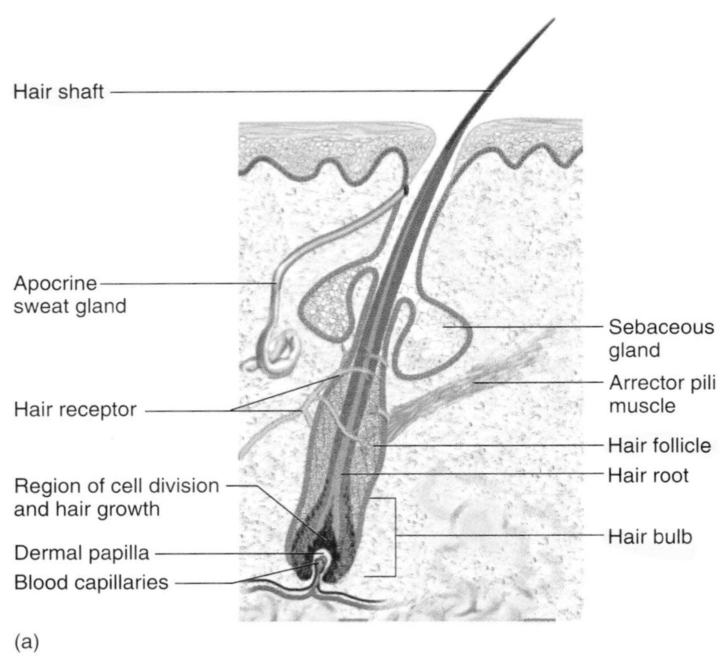

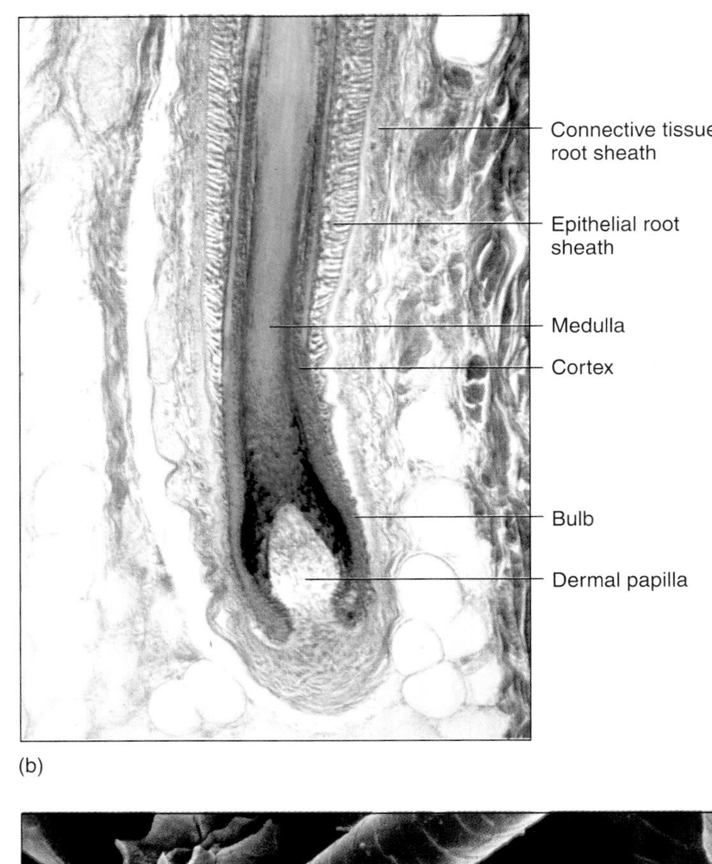

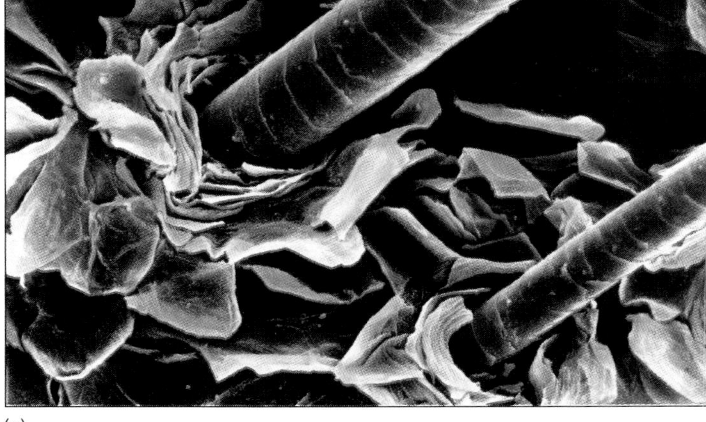

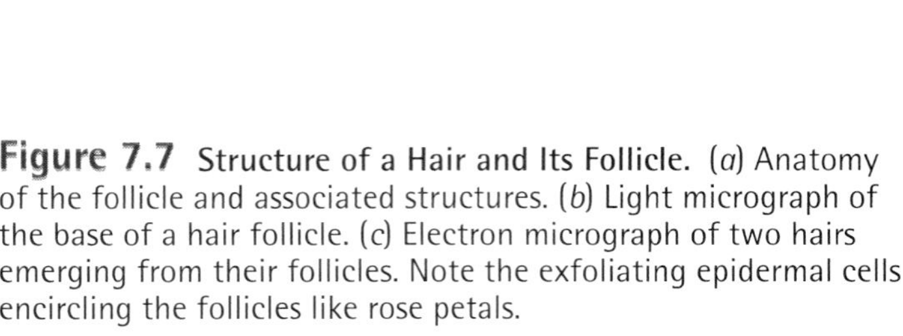

Figure 7.7 **Structure of a Hair and Its Follicle.** (*a*) Anatomy of the follicle and associated structures. (*b*) Light micrograph of the base of a hair follicle. (*c*) Electron micrograph of two hairs emerging from their follicles. Note the exfoliating epidermal cells encircling the follicles like rose petals.

melanin. Red hair is colored by an iron-containing pigment called **trichosiderin.**[21] White hair is due to air in the medulla and lack of pigment in the cortex, and gray hair is a mixture of white and pigmented hairs

The follicle is a diagonal tube that dips deeply into the dermis and sometimes extends as far as the hypodermis. It has two principal layers: an **epithelial root sheath** and a **connective tissue root sheath.** The epithelial root sheath, which is an extension of the epidermis, lies immediately adjacent to the hair root. The connective tissue root sheath, derived from the dermis, surrounds the epithelial sheath.

Associated with the follicle are nerve and muscle fibers. Nerve fibers called **hair receptors** entwine each follicle and respond to hair movements. You can feel their effect by carefully moving a single hair with a pin or by lightly running your finger over the hairs of your arm without touching the skin. Also associated with each hair

21. *tricho* = hair + *sider* = iron

Clinical Insight 7.3

The Science and Pseudoscience of Hair Analysis

Laboratory analysis of hair can provide important clues to metabolic diseases and poisoning. A deficiency of zinc in the hair may indicate malnutrition. A deficiency of both magnesium and calcium is a sign of phenylketonuria (PKU), a hereditary defect in metabolism. A deficiency of calcium along with an excess of sodium may indicate cystic fibrosis.

Several poisons accumulate in the hair. Arsenic, cadmium, lead, and mercury, for example, can be 10 times as concentrated in hair as in blood or urine, and hair analysis can therefore aid in establishing a cause of death. DNA extracted from a single hair can be enough to implicate a suspect in a crime through the technique of *DNA fingerprinting.*

While hair analysis has valuable medical and legal applications, it is also exploited by health-food charlatans who charge high fees for hair analysis and special diets based on supposed deficiencies that the analysis "reveals." In reality, however, hair analysis has not proven very useful for assessing people's general nutritional status.

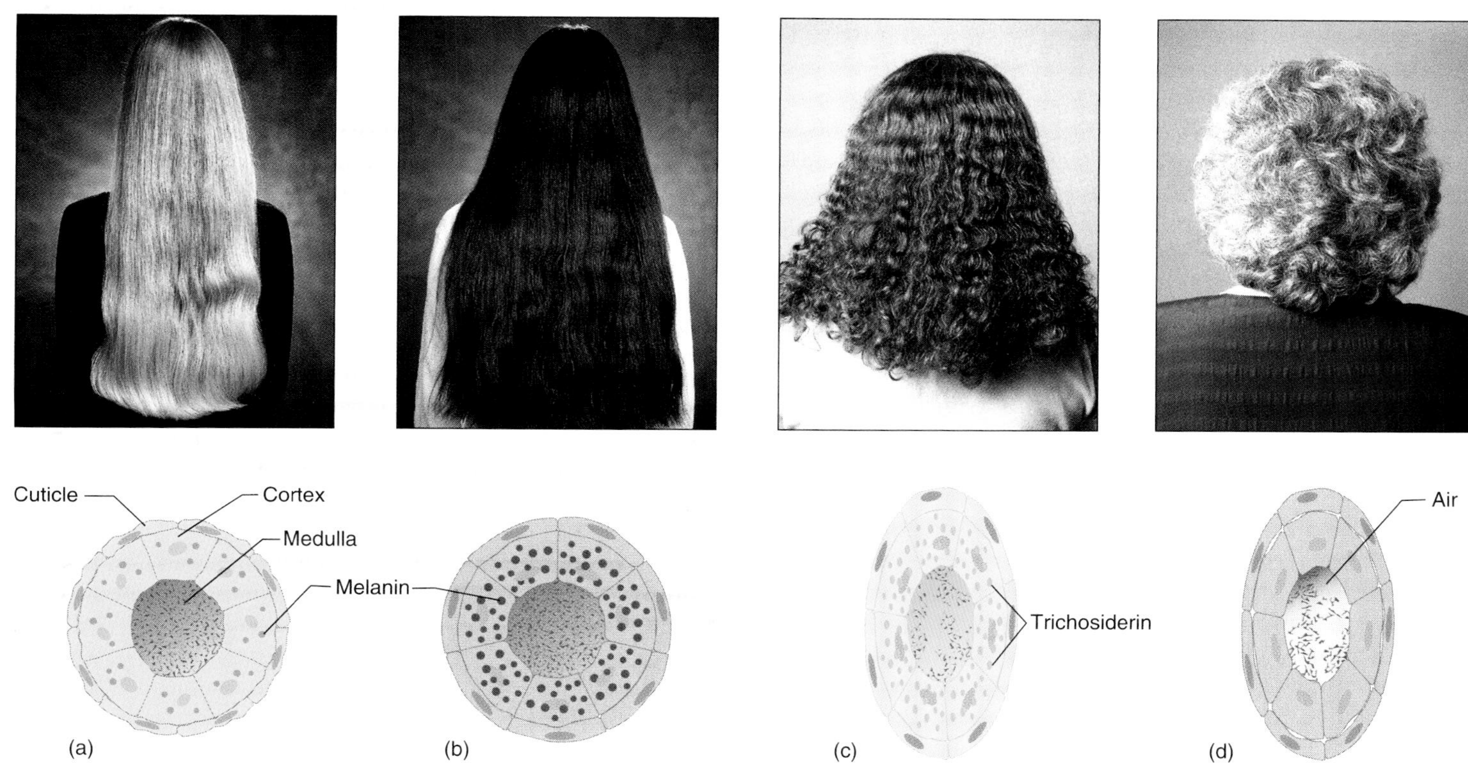

Figure 7.8 Basis of Hair Color and Texture. Straight hair (*a* and *b*) is round in cross section while curly hair (*c* and *d*) is flatter. Blonde hair (*a*) has a sparse amount of melanin in the cortex. Black and brown hair (*b*) have heavier melanin deposits. Red hair (*c*) derives its color from the iron-containing pigment trichosiderin. Gray and white hair (*d*) lack pigment and have air in the medulla.

is an **arrector pili**[22] (ah-REC-tur PY-lye; plural, *arrectores pilorum*), a bundle of smooth muscle fibers extending from dermal collagen fibers to the connective tissue root sheath of the follicle (see figs. 7.1 and 7.7). In response to cold, fear, or other stimuli, the sympathetic nervous system stimulates these muscles and makes the hair stand on end. In other mammals, this traps an insulating layer of warm air next to the skin or makes the animal appear larger and less vulnerable to a potential enemy. In humans, it pulls the follicles into a vertical position and causes "goose bumps" but serves no useful purpose.

22. *arrect* = erect + *pil* = of hair

Growth of Hair

The growth of a hair is due to mitosis of cells in the stratum basale of the epithelial root sheath. Hair cells and follicle cells are pushed together toward the skin surface as the cells below them multiply. The hair cells, like epidermal keratinocytes, become progressively keratinized and die as they are pushed upward, away from the blood supply.

A hair of the scalp grows about 1 mm in 3 days (10–18 cm/yr) and may attain a length of a meter or more. It normally grows for 2 to 4 years and then enters a dormant phase lasting 3 to 4 months. A new hair begins to grow beneath it, pushing the older hair ahead of it until the older hair falls from the follicle. The scalp normally loses 10 to 100 hairs a day. Hairs of the eyebrows and eyelashes grow for only 3 to 4 months; therefore, they never attain the length that scalp hairs do.

Hair grows fastest from adolescence until the 40s. After that, terminal hairs of the scalp are increasingly replaced by wispy vellus, and the hair grows thinner. Thinning of the hair, or baldness, is called **alopecia**[23] (AL-oh-PEE-she-uh). It occurs to some degree in both sexes and may be worsened by disease, poor nutrition, fever, emotional trauma, radiation, or chemotherapy. In the great majority of cases, however, it is simply a matter of aging.

Pattern baldness is the condition in which hair is lost from only some regions of the scalp rather than thinning uniformly across the entire scalp. It results from a combination of genetic and hormonal influences. The relevant gene has two alleles, one for uniform hair growth and a baldness allele for patchy hair growth. The baldness allele is dominant in males and is expressed only in the presence of the high level of testosterone that is characteristic of men. In men who are either heterozygous or homozygous for the baldness allele, testosterone causes the terminal hair of the scalp to be replaced by thinner vellus, beginning on top of the head and later the sides. In women, the baldness allele is recessive. Homozygous dominant and heterozygous women show normal hair distribution; only homozygous recessive women are at risk of pattern baldness. Even then, they exhibit the trait only if their testosterone levels are abnormally high for a woman (for example, because of a tumor of the adrenal gland, a woman's only source of testosterone). Such characteristics in which an allele is dominant in one sex and recessive in the other are called *sex-influenced traits.*

Excessive or undesirable hairiness in areas that are not usually hairy, especially in women and children, is called **hirsutism.**[24] It tends to run in families and usually results from either masculinizing ovarian tumors or hypersecretion of testosterone by the adrenal cortex. It is often associated with menopause.

Functions of Hair

In comparison to other mammals, the relative hairlessness of humans is so unusual that it raises the question, Why do we have any hair at all? What purpose does it serve? There are probably different answers for the different types of hair; furthermore, some of the answers would make little sense if we limited our frame of reference to industrialized societies, where barbers and hairdressers are engaged to alter the natural state of the hair. It is more useful to take a comparative approach to this question and consider the purposes hair serves in other species of mammals.

Most hair of the human trunk and extremities is probably best interpreted as vestigial, with little present function. Body hair undoubtedly served to keep our ancestors warm, but in modern humans it is too scanty for this purpose. Stimulation of the hair receptors, however, alerts us to parasites crawling on the skin, such as lice and fleas.

The scalp is the only place where the hair is thick enough to retain heat. Heat loss from a bald scalp can be extensive and quite uncomfortable for several reasons. First, the brain receives a rich supply of warm blood, and most of the scalp lacks an insulating fat layer. Heat is easily conducted through the bone of the skull and lost to the surrounding air. In addition, without hair there is nothing to break the wind and stop it from carrying away heat. Hair also protects the scalp from sunburn, since the scalp is most directly exposed to the sun's rays. These may be the reasons humans have retained hair on their heads while losing most of it from the rest of the body.

Tufts and patches of hair, sometimes with contrasting colors, are important among mammals in advertising species, age, sex, and individual identity. For the less groomed members of the human species, scalp hair may play a similar role. The indefinitely growing hair of a man's scalp and beard, for example, could provide a striking contrast to a face that is otherwise almost hairless. It creates a badge of recognition instantly visible at a distance.

The beard and pubic and axillary hair signify sexual maturity and aid in the transmission of sexual scents. We will further reflect on this in a later discussion of apocrine sweat glands, whose distribution and function add significant evidence to support this theory.

Stout protective **guard hairs**, or **vibrissae** (vy-BRISS-ee), guard the nostrils and ear canals and prevent foreign particles from entering easily. The eyelashes can shield the eye from windblown debris with a quick blink. In windy or rainy conditions, we can squint so that the eyelashes protect the eye without completely obstructing our vision.

23. *alopecia* = fox mange
24. *hirsut* = shaggy

The eyebrows are often presumed to keep sweat or debris out of the eyes, but this seems a minimal role. It is more plausible that they function mainly to enhance facial expression. Movements of the eyebrows are an important means of nonverbal communication in humans of all cultures. Eyebrow expressiveness is not unique to humans; many species of monkeys and apes use quick flashes of the eyebrows to greet each other, assert their dominance, and break up quarrels.

Nails

Fingernails and toenails are clear, hard derivatives of the stratum corneum. They are composed of very thin, dead, scalelike cells, densely packed together and filled with parallel fibers of hard keratin. Most mammals have claws, whereas flat nails are one of the distinguishing characteristics of primates (see chapter 1). Flat nails allow for more fleshy and sensitive fingertips, while they also serve as strong keratinized "tools" that can be used for digging, grooming, picking apart food, and other manipulations. Fingernails grow at a rate of about 1 mm per week and toenails somewhat more slowly. New cells are added to the nail plate by mitosis in the *nail matrix* at its proximal end.

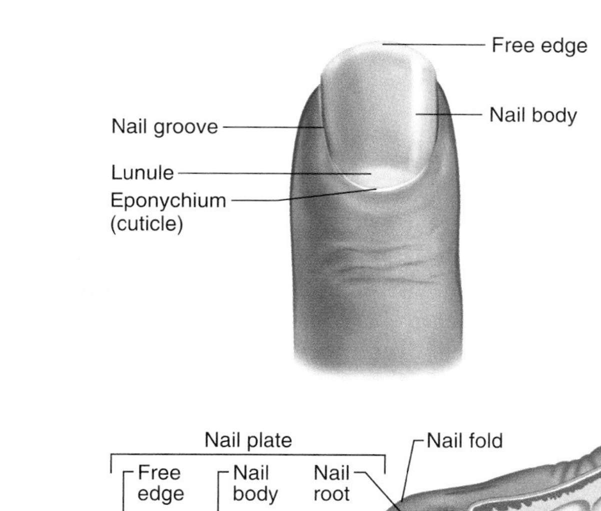

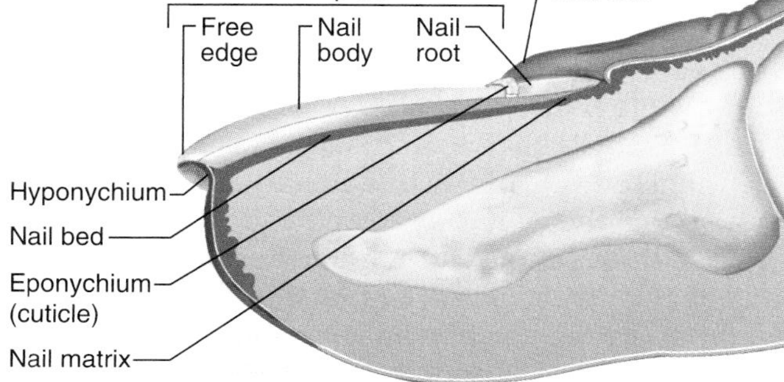

Figure 7.9 Anatomy of a Fingernail.

Contrary to some advertising claims, adding gelatin to the diet has no effect on the growth or hardness of the nails.

The anatomical features of a nail are shown in figure 7.9 and defined in table 7.2. The most important of these are the aforementioned **nail matrix,** a growth zone concealed beneath the skin at the proximal edge of the nail, and the **nail plate,** which is the visible portion covering the fingertip. The nail groove and *hyponychium* (HY-po-NICK-ee-um) accumulate dirt and bacteria and require special attention when scrubbing for duty in an operating room or nursery. The appearance of the nails can be valuable to medical diagnosis. An iron deficiency, for example, may cause the nails to become flat or concave (spoonlike) rather than convex. The nails and fingertips become clubbed in conditions of long-term hypoxemia—deficiency of oxygen in the blood—resulting from congenital heart defects and other causes.

Table 7.2	Anatomical Features of the Nails
Nail Bed	The skin on which the nail plate rests
Nail Plate	The clear, keratinized portion of a nail
Root	The proximal end of a nail, underlying the nail fold
Body	The major portion of the nail plate, overlying the nail bed
Free Edge	The portion of the nail plate that extends beyond the end of the digit
Hyponychium[25]	The space beneath the free edge of the nail plate, where dirt accumulates
Nail Fold	The fold of skin around the margins of the nail plate
Nail Groove	The groove where the nail fold meets the nail plate
Eponychium[26]	Dead epidermis that covers the proximal end of the nail; commonly called the cuticle
Nail Matrix	The growth zone (mitotic tissue) at the proximal end of the nail, corresponding to the stratum basale of the epidermis
Lunule[27]	The region at the base of the nail that appears as a small white crescent because it overlies a thick stratum basale that obscures dermal blood vessels from view

25. *hypo* = under + *onych* = nail
26. *ep* = above + *onych* = nail
27. *lun* = moon + *ule* = little

Before You Go On

Answer the following questions to test your understanding of the preceding section:

9. What is the difference between vellus and terminal hair?
10. Describe the three regions of a hair from its base to its tip, and the three layers of a hair seen in cross section.
11. State the function of the hair papilla, hair receptors, and arrector pili.
12. State a reasonable theory for the different functions of hair of the eyebrows, eyelashes, scalp, nostrils, and axilla.
13. Define or describe the nail body, nail fold, eponychium, hyponychium, and nail matrix.

Cutaneous Glands

Objectives

When you have completed this section, you should be able to

- name two types of sweat glands, and describe the structure and function of each;
- describe the location, structure, and function of sebaceous and ceruminous glands; and
- discuss the distinction between breasts and mammary glands, and explain their respective functions.

The skin has five types of glands: *merocrine sweat glands, apocrine sweat glands, sebaceous glands, ceruminous glands,* and *mammary glands.*

Sweat Glands

Sweat glands, or **sudoriferous**[28] (soo-dor-IF-er-us) **glands,** are of two kinds, described in the previous chapter: merocrine and apocrine. **Merocrine (eccrine) sweat glands,** the most numerous glands of the skin, produce watery perspiration that serves primarily to cool the body (fig. 7.10*a*). There are 3 to 4 million merocrine sweat glands in the adult skin, with a total weight about equal to that of a kidney. They are especially abundant on the palms, soles, and forehead, but they are widely distributed over the rest of the body as well. Each is a simple tubular gland with a twisted coil in the dermis or hypodermis and an undulating or coiled duct leading to a sweat pore on the skin surface. This duct is lined by a stratified cuboidal epithelium in the dermis and by keratinocytes in the epidermis. Amid the secretory cells at the deep end of the gland, there are specialized **myoepithelial**[29] **cells** with properties similar to smooth muscle. They contract in response to stimuli from the sympathetic nervous system and squeeze perspiration up the duct.

Sweat begins as a protein-free filtrate of the blood plasma produced by the deep secretory portion of the gland. Most sodium chloride is reabsorbed from this filtrate as the secretion passes through the duct. Potassium ions, urea, lactic acid, ammonia, and some sodium chloride remain in the sweat. Some drugs are also excreted in the perspiration. On average, sweat is 99% water and has a pH ranging from 4 to 6. Each day, the sweat glands secrete about 500 mL of **insensible perspiration,** which does not produce noticeable wetness of the skin. Under conditions of exercise or heat, however, a person may lose as much as a liter of perspiration each hour. In fact, so much fluid can be lost from the bloodstream by sweating as to cause circulatory shock. Sweating with visible wetness of the skin is called **diaphoresis**[30] (DY-uh-foe-REE-sis).

Apocrine sweat glands occur in the groin, anal region, axilla, and areola, and in mature males, they also occur in the beard area. They are absent from the axillary region of Koreans, however, and are very sparse in the Japanese. Their ducts lead into nearby hair follicles rather than opening directly onto the skin surface (fig. 7.10*b*). They produce their secretion in the same way that merocrine glands do—that is, by exocytosis. The secretory part of an apocrine gland, however, has a much larger lumen than that of a merocrine gland, so these glands have continued to be referred to as apocrine glands to distinguish them functionally and histologically from the merocrine type. Apocrine sweat is thicker and more milky than merocrine sweat because it has more fatty acids in it.

Apocrine sweat glands are scent glands that respond especially to stress and sexual stimulation. They do not develop until puberty, and they apparently correspond to the scent glands that develop in other mammals on attainment of sexual maturity. In women, they enlarge and shrink in phase with the menstrual cycle. Apocrine sweat does not have a disagreeable odor, and indeed it is considered attractive or arousing in some cultures, where it is as much a part of courtship as artificial perfume is to other people. Clothing, however, traps stale sweat long enough for bacteria to degrade the secretion and release free fatty acids with a rancid odor. Disagreeable body odor is called *bromhidrosis.* It occasionally indicates a metabolic disorder, but more often it reflects poor hygiene.

Many mammals have apocrine scent glands associated with specialized tufts of hair. In humans, apocrine glands are found only in the regions covered by the pubic hair, axillary hair, and beard, suggesting that they are similar to other mammalian scent glands in function. The hair serves to retain the aromatic secretion and regulate its rate of evaporation from the skin. Thus, it seems no mere coincidence that women's faces lack both apocrine scent glands and a

28. *sudor* = sweat + *fer* = carry, bear
29. *myo* = muscle
30. *dia* = through + *phoresis* = carrying

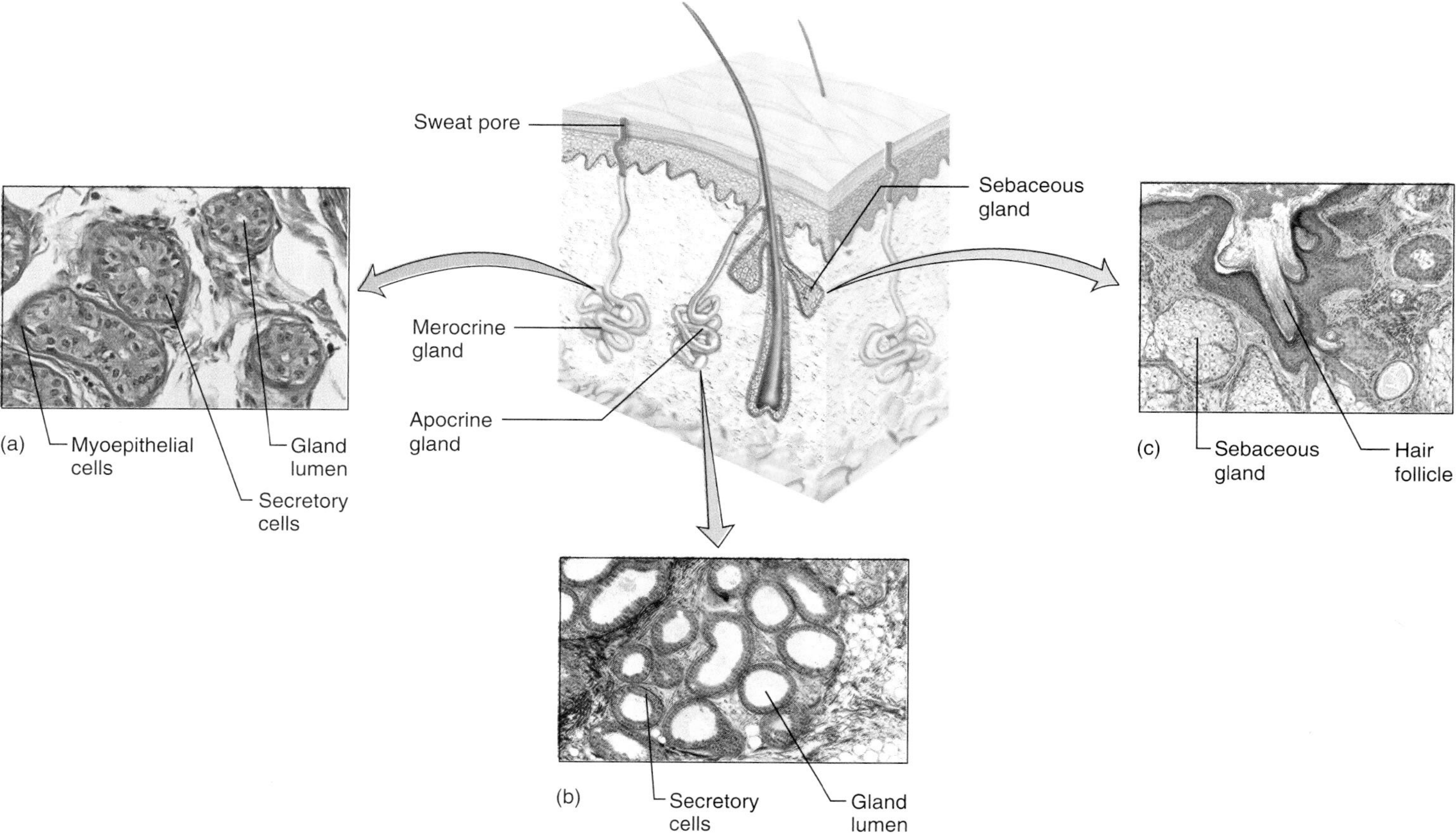

Figure 7.10 Cutaneous Glands. (*a*) Merocrine sweat glands have a narrow lumen and a duct that opens by way of a pore on the skin surface. (*b*) Apocrine sweat glands have a larger lumen and a duct that conveys their aromatic secretion into a hair follicle. (*c*) Sebaceous glands have cells that break down in entirety to form an oily secretion that is released into the hair follicle.

beard. It has also been demonstrated that men's beards grow faster when they live in the presence of women than when they live apart from them.

Sebaceous Glands

Sebaceous[31] (see-BAY-shus) **glands** produce an oily secretion called **sebum** (SEE-bum). They are flask-shaped, with short ducts that usually open into a hair follicle (fig. 7.10*c*), although some of them open directly onto the skin surface. These are holocrine glands with little visible lumen. Their secretion consists of broken-down cells that are replaced by mitosis at the base of the gland. Sebum keeps the skin and hair from becoming dry, brittle, and cracked. The sheen of well-brushed hair is due to sebum distributed by the hairbrush. Ironically, we go to great lengths to wash sebum from the skin, only to replace it with various skin creams and hand lotions made of little more than lanolin, which is sheep sebum.

Ceruminous Glands

Ceruminous (seh-ROO-mih-nus) **glands** are found only in the external ear canal, where their secretion combines with sebum and dead epidermal cells to form earwax, or **cerumen.**[32] They are simple, coiled, tubular glands with ducts leading to the skin surface. Cerumen is waterproof and bitter to the taste. It keeps the eardrum pliable and its bitterness may repel mites, insects, and other pests that would otherwise invade the ear canal.

Breasts and Mammary Glands

The **breasts** (mammae) and **mammary glands** are often mistakenly regarded as one and the same. Breasts, however, are present in both sexes, and even in females they rarely contain more than small traces of mammary gland. The female breast is one of the *secondary sexual characteristics*—anatomical features whose function lies primarily in their

31. *seb* = fat, tallow + *aceous* = possessing

32. *cer* = wax

appeal to the opposite sex. The mammary glands, by contrast, are the milk-producing glands that develop within the female breast only during pregnancy and lactation. Mammary glands are modified apocrine sweat glands that produce a richer secretion and channel it through ducts to a nipple for more efficient conveyance to the offspring. The anatomy and physiology of the mammary gland are discussed in more detail in chapter 28.

In most mammals, two rows of mammary glands form along lines called the *mammary ridges,* or *milk lines.* The most anterior glands on the milk line produce the most milk, so in mammals with litters of young there is an advantage to nursing from those. Primates have dispensed with all but these two most productive glands. A few women, however, develop additional nipples or mammae along the milk line inferior to the primary mammae. In the Middle Ages and Colonial America, this condition, called *polythelia,*[33] was used to incriminate women as supposed "witches."

The glands of the skin are summarized in table 7.3.

33. *poly* = many + *theli* = nipples

Before You Go On

Answer the following questions to test your understanding of the preceding section:

14. How do merocrine and apocrine sweat glands differ in structure and function?
15. What types of hair are associated with apocrine glands? Why?
16. What other type of gland is associated with hair follicles? How does its mode of secretion differ from that of sweat glands?
17. What is the difference between the breast and mammary gland?

Table 7.3 Cutaneous Glands

Gland Type	Definition
Sudoriferous glands	Sweat glands
Merocrine	Sweat glands that function in evaporative cooling; widely distributed over the body surface; open by ducts onto the skin surface
Apocrine	Sweat glands that function as scent glands; found in the regions covered by the pubic, axillary, and male facial hair; open by ducts into hair follicles
Sebaceous glands	Holocrine oil-producing glands associated with hair follicles
Ceruminous glands	Glands of the ear canal that produce cerumen (earwax)
Mammary glands	Milk-producing glands located in the breasts

Diseases of the Skin

Objectives

When you have completed this section, you should be able to

- describe the three most common forms of skin cancer; and
- describe the three classes of burns and the priorities in burn treatment.

Because it is the most exposed of all our organs, skin is not only the most vulnerable to injury and disease but is also the one place where we are most likely to notice anything out of the ordinary. Skin diseases become increasingly common in old age, and the great majority of people over age 70 have complaints about their integumentary system. Aging of the skin is discussed more fully in chapter 29. The healing of cuts and other injuries to the skin occurs by the process described at the end of chapter 6. We focus here on two particularly common and serious disorders, skin cancer and burns. Other skin diseases are briefly summarized in table 7.4, and the previous chapter has clinical insight boxes on two skin disorders, *pemphigus vulgaris* and *keloids.*

Skin Cancer

Skin cancer is induced by the ultraviolet rays of the sun. It occurs most often on the head and neck, where exposure to the sun is greatest; it is most common in fair-skinned people and the elderly, who have had the longest lifetime UV exposure. The ill-advised popularity of suntanning, however, has caused an alarming increase in skin cancer among younger people. Skin cancer is one of the most common cancers, but it is also one of the easiest to treat and has one of the highest survival rates when it is detected and treated early.

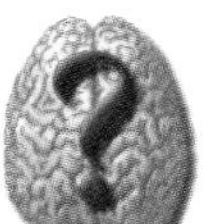

Think About It

Skin cancer is relatively rare in people with dark skin. Other than possible differences in behavior, such as lack of intentional suntanning, why do you think this is so?

There are three types of skin cancer named for the epidermal cells in which they originate—*basal cell carcinoma, squamous cell carcinoma,* and *malignant melanoma.* The three types are also distinguished from each other by the appearance of their **lesions**[34] (zones of tissue injury).

Basal cell carcinoma[35] is the most common type, but it is also the least dangerous because it seldom

34. *lesio* = injure
35. *carcin* = cancer + *oma* = tumor

Table 7.4 Some Skin Diseases

Acne	Inflammation of the sebaceous glands, especially beginning at puberty; follicle becomes blocked with keratinocytes and sebum, and develops into a blackhead (*comedo*) composed of these and bacteria; continued inflammation of follicle results in pus production and appearance of pimples.
Dermatitis	Any inflammation of the skin, typically marked by itching and redness; often *contact dermatitis,* caused by exposure to toxins such as poison ivy.
Eczema (ECK-zeh-mah)	Itchy, red, "weeping" skin lesions cause by an allergy, usually beginning before age five; may progress to thickened, leathery, darkly pigmented patches of skin.
Psoriasis (so-RY-ah-sis)	Recurring, reddened plaques covered with silvery scale; sometimes disfiguring; possibly caused by an autoimmune response (immune system attack on one's own tissues), often running in families.
Ringworm	A fungus infection of the skin (not a worm) that sometimes grows in a circular pattern; common in moist areas such as the axilla, groin, and foot (*athlete's foot*).
Rosacea (ro-ZAY-she-ah)	A red rashlike area, often in the area of the nose and cheeks, marked by fine networks of dilated blood vessels; worsened by hot drinks, alcohol, and spicy food.
Seborrheic (seb-oh-REE-ik) Dermatitis	Recurring patches of scaly white or yellowish inflammation often on the head, face, chest, and back; called *cradle cap* (yellow, crusty scalp lesion) in infants. Cause unknown, but correlated with genetic and climatic factors.
Warts	Benign skin tumors caused by the human papillomavirus (HPV), transmitted by touch; most common in children and on the fingers. *Plantar warts* are located on the soles and *venereal warts* on the genitals and anus.

metastasizes. It arises from cells of the stratum basale and eventually invades the dermis. On the surface, the lesion first appears as a small, shiny bump. As the bump enlarges, it often develops a central depression and a beaded "pearly" edge (fig. 7.11*a*). These lesions usually occur on the face and are treated by surgical removal and radiation therapy.

Squamous cell carcinoma arises from keratinocytes of the stratum spinosum. Lesions usually appear on the scalp, ears, lower lip, or back of the hand. They have a raised, reddened, scaly appearance, later forming a concave ulcer with raised edges (fig. 7.11*b*). The chance of recovery is good with early detection and surgical removal, but if it goes unnoticed or is neglected, this cancer tends to metastasize to the lymph nodes and can be lethal.

Malignant melanoma is the most deadly skin cancer but accounts for only 5% of all cases. It usually arises from the melanocytes of a preexisting mole. It metastasizes quickly and is often fatal if not treated immediately. The risk for malignant melanoma is greatest in people who experienced severe sunburns as children, especially redheads. It is important to distinguish a mole from malignant melanoma. A mole usually has a uniform color and even contour, and it is no larger in diameter than the end of a pencil eraser (about 6 mm). If it becomes malignant, however, it forms a large, flat, spreading lesion with a scalloped border (fig. 7.11*c*). The American Cancer Society suggests an "ABCD rule" for recognizing malignant melanoma: *A* for asymmetry (one side of the lesion looks different from the other); *B* for border irregularity (the contour is not uniform but wavy or scalloped); *C* for color (often a mixture of brown, black, tan, and sometimes red and blue); and *D* for diameter (greater than 6 mm).

Burns

Burns are the leading cause of accidental death. They are usually caused by fires, kitchen spills, or excessively hot bath water, but they also can be caused by sunlight, ionizing radiation, strong acids and bases, or electrical shock. Burn deaths result primarily from fluid loss, infection, and the toxic effects of **eschar** (ESS-car)—the burned, dead tissue.

Burns are classified according to the depth of tissue involvement (fig. 7.12). **First-degree burns** involve only the epidermis and are marked by redness, slight edema, and pain. They heal in a few days and seldom leave scars. Most sunburns are first-degree burns.

Second-degree burns involve the epidermis and part of the dermis but leave at least some of the dermis intact. First- and second-degree burns are therefore also known as **partial-thickness burns.** A second-degree burn may be red, tan, or white and is blistered and very painful. It may take from 2 weeks to several months to heal and may leave scars. The epidermis regenerates by division of epithelial cells in the hair follicles and sweat glands and those around the edges of the lesion. Some sunburns and many scalds are second-degree burns.

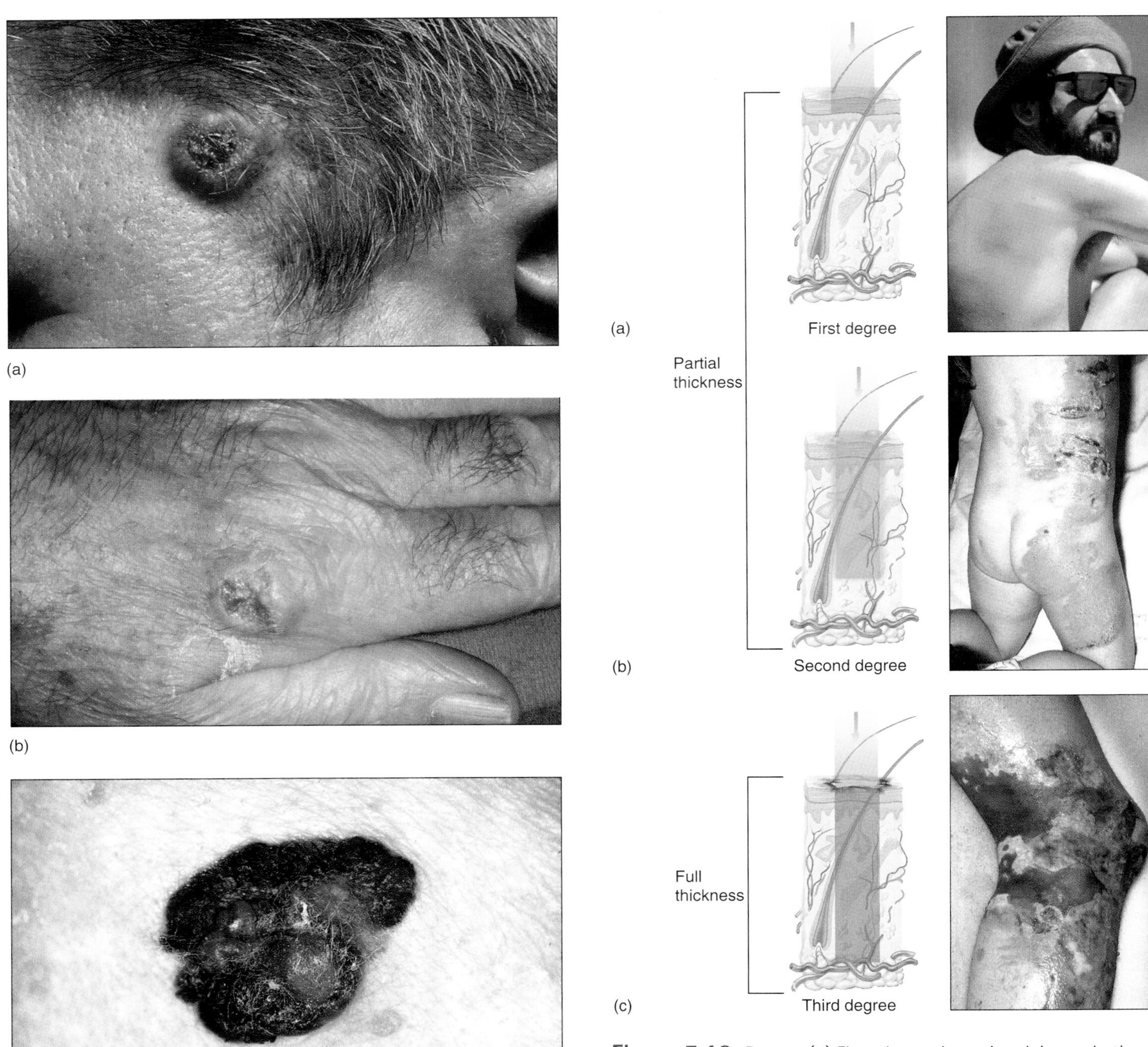

Figure 7.11 Skin Cancer. (*a*) Basal cell carcinoma. (*b*) Squamous cell carcinoma. (*c*) Malignant melanoma.

Figure 7.12 Burns. (*a*) First-degree burn, involving only the epidermis. (*b*) Second-degree burn, involving the epidermis and part of the dermis. (*c*) Third-degree burn, extending through the entire dermis and often involving even deeper tissue.

Clinical Insight 7.4

UVA, UVB, and Sunscreens

Some people distinguish between two forms of ultraviolet radiation and argue, fallaciously, that one is less harmful than the other. UVA has wavelengths ranging from 320 to 400 nm and UVB has wavelengths from 290 to 320 nm. (Visible light starts at about 400 nm, the deepest violet we can see.) UVA and UVB are sometimes called "tanning rays" and "burning rays," respectively. Tanning salons often advertise that the UVA rays they use are safe, but public health authorities know better. UVA can burn as well as tan, and it inhibits the immune system, while both UVA and UVB are now thought to initiate skin cancer. As dermatologists say, there is no such thing as a healthy suntan.

Whether or not sunscreens help to protect against skin cancer remains unproven. As the sale of sunscreen has risen in recent decades, so has the incidence of skin cancer. Indeed, recent studies have shown that people who use sunscreens have a higher incidence of basal cell carcinoma than people who do not, while data relating malignant melanoma to sunscreen use are still contradictory and inconclusive. Some of the chemicals used in sunscreens damage DNA and generate harmful free radicals when exposed to UV—chemicals such as para-aminobenzoic acid (PABA) (now discontinued), zinc oxide, and titanium dioxide. The jury is still out, however, on how sunscreens react on and with the skin; authorities cannot say with certainty, yet, whether sunscreens provide any protection or do any harm. Data are not yet available on older people who have used sunscreen their entire lives. Given the uncertainty about their effectiveness, it is best not to assume that sunscreens protect you from skin cancer and, above all, not to assume that using a sunscreen means that you can safely stay out longer in the sun.

Third-degree burns are also called **full-thickness burns** because the epidermis, dermis, and often some deeper tissue are completely destroyed. Since no dermis remains, the skin can regenerate only from the edges of the wound. Third-degree burns often require skin grafts (see clinical insight 7.5). If a third-degree burn is left to itself to heal, contracture (abnormal connective tissue fibrosis) and severe disfigurement may result.

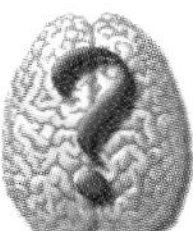

Think About It

A third-degree burn may be surrounded by painful areas of first- and second-degree burns, but the region of the third-degree burn is painless. Explain the reason for this lack of pain.

The two most urgent considerations in treating a burn patient are fluid replacement and infection control. A patient can lose several liters of water, electrolytes, and protein each day from the burned area. As fluid is lost from the tissues, more is transferred from the bloodstream to replace it, and the volume of circulating blood declines. A patient may lose up to 75% of the blood plasma within a few hours, potentially leading to circulatory shock and cardiac arrest—the principal cause of death in burn patients. Intravenous fluid must be given to make up for this loss. A burn patient also requires thousands of extra calories daily to compensate for protein loss and the demands of tissue repair. Supplementary nutrients are given intravenously or through a gastric tube.

Infection is controlled by keeping the patient in an aseptic (germ-free) environment and administering antibiotics. The eschar is sterile for the first 24 hours, but then it quickly becomes infected and may have toxic effects on the digestive, respiratory, and other systems. Its removal, called **debridement**[36] (deh-BREED-ment), is essential to infection control.

36. *de* = un + *bride* = bridle

Before You Go On

Answer the following questions to test your understanding of the preceding section:

18. What types of cells are involved in each type of skin cancer?
19. Which type of skin cancer is most dangerous? What are its early warning signs?
20. What is the difference between a first-, second-, and third-degree burn?
21. What are the two most urgent priorities in treating a burn victim? How are these needs dealt with?

Clinical Insight 7.5

Skin Grafts and Artificial Skin

Third-degree burns leave no dermal tissue to regenerate what was lost, and therefore they generally require skin grafts. The ideal graft is an *autograft*—tissue taken from another location on the same person's body—because it is not rejected by the immune system. An autograft is performed by taking epidermis and part of the dermis from an undamaged area such as the thigh or buttock and grafting it to a burned area. This method is called a *split-skin graft* because part of the dermis is left behind to proliferate and replace the epidermis that was removed—the same way a second-degree burn heals.

An autograft may not be possible, however, if the burns are too extensive. The best treatment option in this case is an *isograft,* which uses skin from an identical twin. Because the donor and recipient are genetically identical, the recipient's immune system is unlikely to reject the graft. Since identical twins are rare, however, the best one can hope for in most cases is donor skin from another close relative.

A *homograft,* or *allograft,* is a graft from an unrelated person. Skin banks provide skin from deceased persons for this purpose. The immune system attempts to reject homografts, but they suffice as temporary coverings for the burned area. They can be replaced by autografts when the patient is well enough for healthy skin to be removed from an undamaged area of the body.

Pig skin is sometimes used on burn patients but presents the same problem of immune rejection. A graft of tissue from a different species is called a *heterograft,* or *xenograft.* This is a special case of a *heterotransplant,* which also includes transplantation of organs such as baboon hearts or livers into humans. Heterografts and heterotransplants are short-term methods of maintaining a patient until a better, long-term solution is possible. The immune reaction can be suppressed by drugs called *immunosuppressants.* This procedure is risky, however, because it lowers a person's resistance to infection, which is already compromised in a burn patient.

Some alternatives to skin grafts are also being used. Burns are sometimes temporarily covered with amnion (the membrane that surrounds a developing fetus) obtained from afterbirths. In addition, tiny keratinocyte patches cultured with growth stimulants have produced sheets of epidermal tissue as large as the entire body surface. These can replace large areas of burned tissue. Dermal fibroblasts also have been successfully cultured and used for autografts. A drawback to these approaches is that the culture process requires 3 or 4 weeks, which is too long a wait for some patients with severe burns.

Various kinds of artificial skin have also been developed as a temporary burn covering. One concept is a sheet with an upper layer of silicone and a lower layer of collagen and chondroitin sulfate. It stimulates growth of connective tissue and blood vessels from the patient's underlying tissue. The artificial skin can be removed after about 3 weeks and replaced with a thin layer of cultured or grafted epidermis. At least two bioengineering companies have developed artificial skins approved in 1997–98 by the U.S. Food and Drug Administration for patient use. The manufacture of one such product begins by culturing fibroblasts on a collagen gel to produce a dermis, then culturing keratinocytes on this dermal substrate to produce an epidermis. Such products are now being used to treat burn patients as well as leg and foot ulcers that result from diabetes mellitus. This is one aspect of the larger field of *tissue engineering,* which biological technology companies hope will lead, within a few decades, even to engineering replacement livers and other organs.

Connective Issues

Interactions Between the INTEGUMENTARY SYSTEM and Other Organ Systems

← indicates ways in which this system affects other systems
→ indicates ways in which other systems affect this one

All Systems

The integumentary system serves all other systems by providing a physical barrier to environmental hazards

Skeletal System

← Role of skin in vitamin D synthesis promotes calcium absorption needed for bone growth and maintenance

→ Supports skin at scalp and other places where bone lies close to surface

Muscular System

← Role of skin in vitamin D synthesis promotes calcium absorption needed for muscle contraction; skin dissipates heat generated by muscles

→ Active muscles generate heat and warm skin; contractions of skeletal muscles pull on skin and produce facial expressions

Nervous System

← Sensory impulses from skin transmitted to nervous system

→ Regulates diameter of cutaneous blood vessels; stimulates perspiration and contraction of arrector pili muscles

Endocrine System

← Vitamin D_3 is synthesized in the dermis and, after further conversion, acts as a hormone that affects bone metabolism

→ Sex hormones cause changes in integumentary features at puberty; some hormone imbalances have pathological effects on skin

Circulatory System

← Dermal vasoconstriction diverts blood to other organs; skin prevents loss of fluid from cardiovascular system; dermal mast cells cause vasodilation and increased blood flow

→ Delivers O_2, nutrients, and hormones to skin and carries away wastes; hemoglobin colors skin

Lymphatic/Immune Systems

← Dendritic cells detect foreign substances

→ Lymphatic system controls fluid balance and prevents edema; immune cells protect skin from infection and promote tissue repair

Respiratory System

← Nasal hairs filter particles that might otherwise be inhaled

→ Provides O_2 and removes CO_2

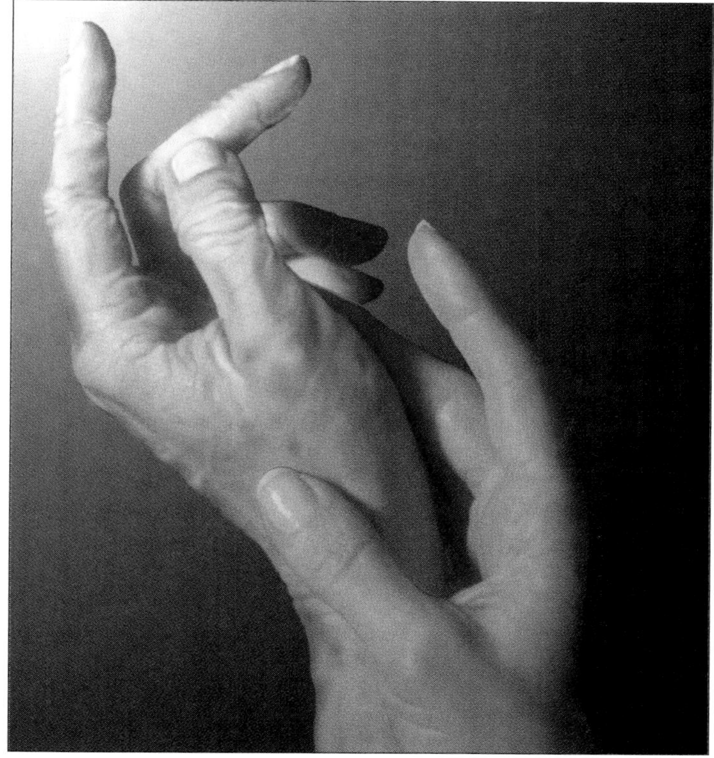

Urinary System

← Skin complements urinary system by excreting salts and some nitrogenous wastes in sweat

→ Disposes of wastes and maintains electrolyte and pH balance

Digestive System

← Role of skin in vitamin D synthesis enables intestinal absorption of calcium

→ Provides nutrients needed for integumentary development and function; nutritional disorders often reflected in appearance of skin, hair, and nails

Reproductive System

← Cutaneous receptors respond to erotic stimuli; mammary glands produce milk to nourish infants; apocrine glands produce scents with subtle sexual functions

→ Gonadal sex hormones promote growth, maturation, and maintenance of skin

Chapter Review

Online Learning Center

Study resources for this chapter are available at www.mhhe.com/saladin/

Study Outline

Structure of the Skin and Subcutaneous Tissue (p. 206)

1. The epidermis
   a. Thin and thick skin
   b. Layers of epidermis
      - Stratum basale
      - Stratum spinosum
      - Stratum granulosum
      - Stratum lucidum
      - Stratum corneum
2. The dermis
   a. Thickness
   b. Composition
   c. Dermal papillae
   d. Layers
      - Papillary layer
      - Reticular layer
3. The hypodermis
4. Skin color
   a. Normal pigments and colors
      - Hemoglobin
      - Melanin
      - Carotene
   b. Abnormal colors
      - Cyanosis
      - Erythema
      - Jaundice
      - Bronzing
      - Pallor
      - Albinism
      - Hematoma
5. Skin markings
   a. Hemangiomas
   b. Freckles and moles
   c. Friction ridges
   d. Flexion creases and lines

Functions of the Skin (p. 213)

1. The skin as a barrier
   a. Toughness of keratin
   b. Dryness
   c. Acid mantle
   d. Importance as water barrier
   e. Importance as UV barrier
2. Vitamin D synthesis
3. Cutaneous absorption
4. Sensory functions
5. Thermoregulation
   a. Cutaneous vasodilation and vasoconstriction
   b. Sweating
6. Social functions
   a. Appearance and social acceptance
   b. Nonverbal communication

Hair and Nails (p. 214)

1. Types and distribution of hair
   a. Soft and hard keratin
   b. Hair distribution
   c. Hair types
      - Lanugo
      - Vellus
      - Terminal hair
2. Hair structure
   a. Bulb, root, and shaft
   b. Medulla, cortex, and cuticle
   c. Texture and shape of hair
   d. Hair colors
3. Follicle structure
   a. Epithelial root sheath
   b. Connective tissue root sheath
   c. Hair receptors
   d. Arrector pili muscle
4. Growth of hair
   a. Mitosis in epithelial root sheath
   b. Rates of growth and hair loss
   c. Alopecia and pattern baldness
   d. Hirsutism
5. Functions of hair
   a. Vestigial nature of body hair
   b. Sensory functions
   c. Heat retention on scalp
   d. Sex and individual recognition
   e. Indication of sexual maturity
   f. Relationship to scent glands
   g. Guard hairs
   h. Facial expression
6. Nails
   a. Significance of flat nails
   b. Growth rate
   c. Anatomy (review table 7.2)
   d. Diagnostic significance

Cutaneous Glands (p. 220)

1. Sweat glands
   a. Merocrine glands
      - Abundance and distribution
      - Structure
      - Cooling function
         - Production of sweat
         - Insensible perspiration
         - Diaphoresis
   b. Apocrine glands
      - Distribution and structure
      - Scent function
      - Association with specialized hair
2. Sebaceous glands
   a. Distribution and structure
   b. Production of sebum
3. Ceruminous glands
   a. Distribution and structure
   b. Functions of cerumen
4. Breasts and mammary glands
   a. Distribution
   b. Function

Diseases of the Skin (p. 222)

1. Miscellaneous skin diseases (table 7.4)
2. Skin cancer
   a. Basal cell carcinoma
   b. Squamous cell carcinoma
   c. Malignant melanoma
3. Burns
   a. Causes of burns
   b. Causes of death
   c. Three degrees of burns
   d. Treatment

Selected Vocabulary

For additional terminology not listed here, see tables 7.2 and 7.4.

integumentary system 206
dermatology 206
skin 206
thick skin 206
thin skin 206
epidermis 206
stratum basale 206
keratinocyte 206
melanocyte 206
tactile cell 206
stratum spinosum 206
dendritic cell 208
stratum granulosum 208
membrane-coating vesicle 208
stratum lucidum 208
stratum corneum 209
dermis 209
dermal papillae 209
papillary layer 210
reticular layer 210
hypodermis 210
subcutaneous tissue 210
superficial fascia 210
subcutaneous fat 210
hemoglobin 210
melanin 210
carotene 211
cyanosis 211
erythema 212
jaundice 212
bronzing 212
pallor 212
albinism 212
hematoma 212
hemangioma 212
freckle 212
mole 212
friction ridge 212
friction crease 212
flexion line 212
acid mantle 213
thermoreceptor 213
accessory organ 214
soft keratin 214
hard keratin 214
pilus 215
follicle 215
lanugo 215
vellus 215
terminal hair 215
hair bulb 215
hair root 215
hair shaft 215
hair papilla 215
hair medulla 215
hair cortex 215
hair cuticle 215
trichosiderin 216
epithelial root sheath 216
connective tissue root sheath 216
hair receptor 216
arrector pili 217
alopecia 218
pattern baldness 218
hirsutism 218
guard hair 218
nail matrix 219
nail plate 219
sudoriferous gland 220
merocrine gland 220
myoepithelial cell 220
insensible perspiration 220
diaphoresis 220
apocrine sweat gland 220
sebaceous gland 221
sebum 221
ceruminous gland 221
cerumen 221
breast 221
mammary gland 221
lesion 222
basal cell carcinoma 222
squamous cell carcinoma 223
malignant melanoma 223
burn 223
eschar 223
first-degree burn 223
second-degree burn 223
partial-thickness burn 223
third-degree burn 225
full-thickness burn 225
debridement 225

Testing Your Recall

1. Cells of the _____ are keratinized and dead.
   a. papillary layer
   b. stratum spinosum
   c. stratum basale
   d. stratum corneum
   e. stratum granulosum

2. Which of the following terms is *least* related to the rest?
   a. panniculus adiposus
   b. superficial fascia
   c. reticular layer
   d. hypodermis
   e. subcutaneous tissue

3. Which of the following skin conditions or appearances would most likely result from liver failure?
   a. pallor
   b. erythema
   c. pemphigus vulgaris
   d. jaundice
   e. melanization

4. All of the following interfere with microbial invasion of the skin *except*
   a. the acid mantle.
   b. melanization.
   c. inflammation.
   d. keratinization.
   e. sebum.

5. The hair on a six-year-old's arms is
   a. vellus.
   b. lanugo.
   c. trichosiderin.
   d. terminal hair.
   e. rosacea.

6. Which of the following terms is *least* related to the rest?
   a. lunule
   b. nail plate
   c. hyponychium
   d. free edge
   e. cortex

7. Which of the following is a scent gland?
   a. eccrine gland
   b. sebaceous gland
   c. apocrine gland
   d. ceruminous gland
   e. merocrine gland

8. _____ are skin cells with a sensory role.
   a. Merkel cells
   b. Dendritic cells
   c. Prickle cells
   d. Melanocytes
   e. Keratinocytes

9. Which of the following glands produce the acid mantle?
   a. merocrine sweat glands
   b. apocrine sweat glands
   c. mammary glands
   d. ceruminous glands
   e. sebaceous glands

10. Which of the following skin cells alert the immune system to pathogens?
    a. fibroblasts
    b. melanocytes
    c. keratinocytes
    d. dendritic cells
    e. Merkel cells

11. __________ is sweating without noticeable wetness of the skin.

12. A muscle that causes a hair to stand on end is called a/an __________.

13. The process of removing burned skin from a patient is called __________.

14. Blueness of the skin due to low oxygen concentration in the blood is called __________.
15. Projections of the dermis toward the epidermis are called __________.
16. Cerumen is more commonly known as __________.
17. The holocrine glands that secrete into a hair follicle are called __________.
18. The space beneath the tip of a fingernail is called the __________.
19. A hair is nourished by blood vessels in a connective tissue projection called the __________.
20. A __________ burn destroys the entire dermis.

Answers in appendix B

Testing Your Comprehension

1. Many organs of the body contain numerous smaller organs, perhaps even thousands. Describe an example of this in the integumentary system.
2. Certain aspects of human form and function are easier to understand when viewed from the perspective of comparative anatomy and evolution. Discuss examples of this in the integumentary system.
3. Explain how the complementarity of form and function is reflected in the fact that the dermis has two histological layers and not just one.
4. Cold weather does not normally interfere with oxygen uptake by the blood, but it can cause cyanosis anyway. Why?
5. Why is it important for the epidermis to be effective, but not *too* effective, in screening out UV radiation?

Answers at the Online Learning Center

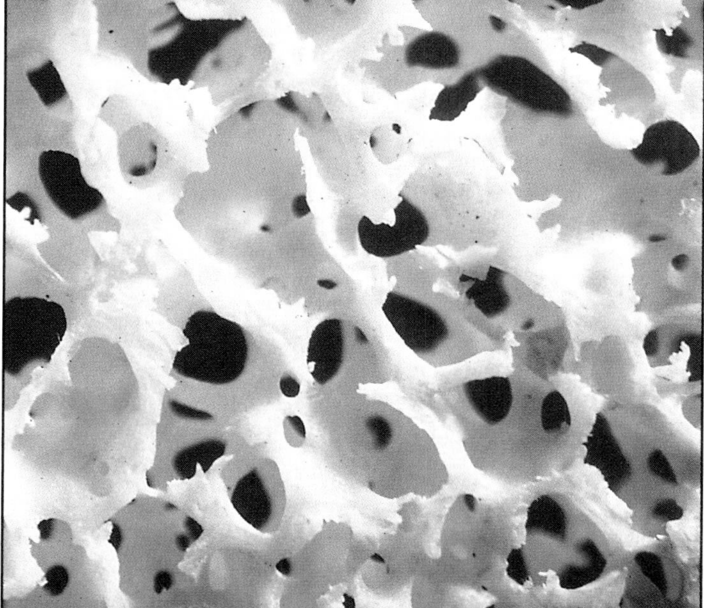

Spongy bone of the human femur

CHAPTER

8

Bone Tissue

Chapter Outline

Insights

Study with Your ESP CD-ROM

Refer to your ESP Tutorial CD-ROM for helpful animations, activities, and quizzes related to the material covered in this chapter. The following are suggested areas to go to on the CD and are presented according to "Unit/Topic":

- Support and Movement/Skeletal System

Brushing Up

To understand this chapter, it is important that you understand or brush up on the following concepts:

- Histology of hyaline cartilage (p. 187)
- Introduction to bone histology (p. 188)

In art and history, nothing has symbolized death more than a skull or skeleton.[1] The dry bones presented for laboratory study suggest that the skeleton is an inert scaffold for the body, like the steel girders of a building. Seeing it in such a sanitized form makes it easy to forget that the living skeleton is made of dynamic tissues, full of cells—that it continually remodels itself and interacts physiologically with all of the other organ systems of the body. The skeleton is permeated with nerves and blood vessels, which attests to its sensitivity and metabolic activity. **Osteology,**[2] the study of bone, is the subject of these next three chapters. In this chapter, we study bone as a tissue—its composition, its functions, how it develops and grows, how its metabolism is regulated, and some of its disorders. This will provide a basis for understanding the skeleton, joints, and muscles in the chapters that follow.

Tissues and Organs of the Skeletal System

Objectives

When you have completed this section, you should be able to

- name the tissues and organs that compose the skeletal system;
- state several functions of the skeletal system;
- describe four types of bones classified by shape;
- describe the general features of a long bone; and
- distinguish between bone as a tissue and as an organ.

The **skeletal system** is composed of bones, cartilages, and ligaments joined tightly to form a strong, flexible framework for the body. Cartilage, the forerunner of most bones in embryonic and childhood development, covers many joint surfaces in the mature skeleton. Ligaments hold bones together at the joints and are discussed in chapter 10. Tendons are structurally similar to ligaments but attach muscle to bone; they are discussed with the muscular system in chapter 11.

1. *skelet* = dried up
2. *osteo* = bone + *logy* = study of

Functions of the Skeleton

The skeleton obviously provides the body with physical support, but it plays many other roles that go unnoticed by most people. These are summarized in table 8.1.

Bones and Osseous Tissue

Bone, or **osseous**[3] **tissue,** is a connective tissue in which the matrix is hardened by the deposition of calcium phosphate and other minerals. The hardening process is called **mineralization** or **calcification.** (Bone is not the hardest substance in the body; that distinction goes to tooth enamel.) Osseous tissue is only one of the tissues that make up a bone. Also present are blood, bone marrow, cartilage, adipose tissue, nervous tissue, and fibrous connective tissue. The word *bone* can denote an organ composed of all these tissues, or it can denote just the osseous tissue.

The Shapes of Bones

Bones are classified into four groups according to their shapes and corresponding functions (fig. 8.1):

1. **Long bones** are conspicuously longer than wide. Like crowbars, they serve as rigid levers that are acted upon by the skeletal muscles to produce body movements. Long bones include the humerus of the

3. *os, osse, oste* = bone

Table 8.1 Functions of the Skeletal System

Function	Examples or Explanation
Support	Bones of the legs, pelvis, and vertebral column hold up the body; the mandible supports the teeth; nearly all bones provide support for muscles; many other soft organs are directly or indirectly supported by nearby bones.
Protection	Bones enclose and protect the brain, spinal cord, lungs, heart, pelvic viscera, and bone marrow.
Movement	Skeletal muscles would serve little purpose if not for the rigid attachment and leverage provided by bones. Leg and arm movements are the most obvious examples of skeletomuscular movement; a less obvious one is that ventilation of the lungs depends on movement of the ribs by skeletal muscles.
Blood formation	Red bone marrow is the major producer of blood cells, including most cells of the immune system.
Electrolyte balance	The skeleton is the body's main mineral reservoir. It stores calcium and phosphate and releases them according to the body's physiological needs.
Acid-base balance	Bone buffers the blood against excessive pH changes by absorbing or releasing alkaline salts.
Detoxification	Bone tissue removes heavy metals and other foreign elements from the blood and thus reduces their effects on nervous and other tissues. It can later release these more slowly for excretion.

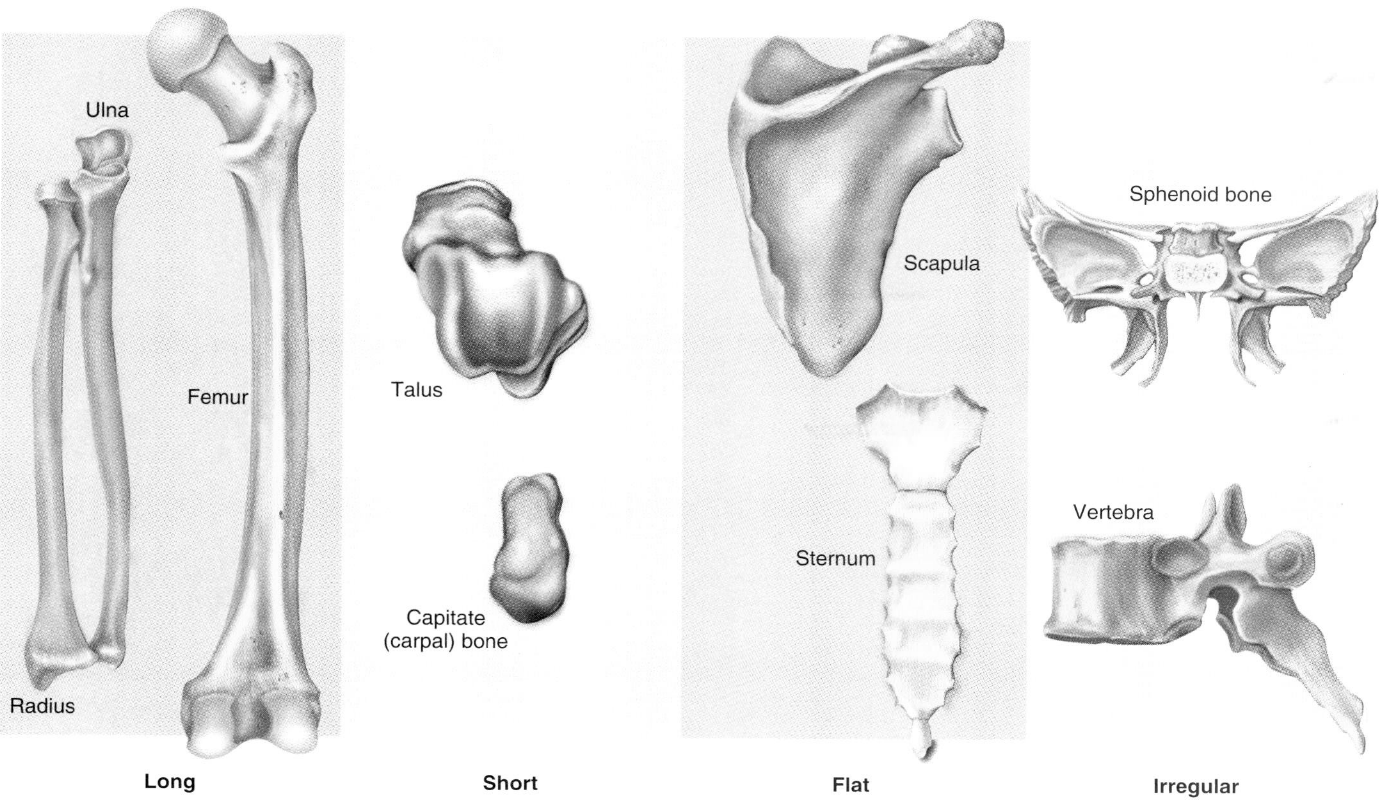

Figure 8.1 Classification of Bones by Shape.

arm, the radius and ulna of the forearm, the metacarpals and phalanges of the hand, the femur of the thigh, the tibia and fibula of the leg, and the metatarsals and phalanges of the feet.

2. **Short bones** are more nearly equal in length and width. They include the carpal (wrist) and tarsal (ankle) bones. They have limited motion and merely glide across one another, enabling the ankles and wrists to bend in multiple directions.
3. **Flat bones** enclose and protect soft organs and provide broad surfaces for muscle attachment. They include most cranial bones and the ribs, sternum (breastbone), scapula (shoulder blade), and os coxae (hipbone).
4. **Irregular bones** have elaborate shapes that do not fit into any of the preceding categories. They include the vertebrae and some of the skull bones, such as the sphenoid and ethmoid.

General Features of Bones

Knowing the terms used to describe a long bone will help you to understand the anatomy of the other types. Figure 8.2*a* shows a longitudinal section through a long bone. You will note immediately that much of it is composed of a cylinder of dense white osseous tissue; this is called **compact (dense) bone.** The cylinder encloses a space called the **medullary** (MED-you-lerr-ee) **cavity,** which contains bone marrow. At the ends of the bone, the central space is occupied by a more loosely organized form of osseous tissue called **spongy (cancellous) bone.** The skeleton is about three-quarters compact bone and one-quarter spongy bone by weight. Spongy bone is found at the ends of the long bones and in the middle of nearly all others. It is always enclosed by more durable compact bone. In flat bones of the skull, two layers of compact bone enclose a middle layer of spongy bone like a sandwich (fig. 8.2*b*). The spongy layer is called the **diploe**[4] (DIP-lo-ee). A moderate blow to the skull can fracture the outer layer of compact bone, but the diploe may absorb the impact and leave the inner layer of compact bone unharmed.

The principal features of a long bone are its shaft, called the **diaphysis**[5] (dy-AF-ih-sis), and an expanded head at each end called the **epiphysis**[6] (eh-PIF-ih-sis).

4. *diplo* = double
5. *dia* = across + *physis* = growth; originally named for a ridge on the shaft of the tibia
6. *epi* = upon, above + *physis* = growth

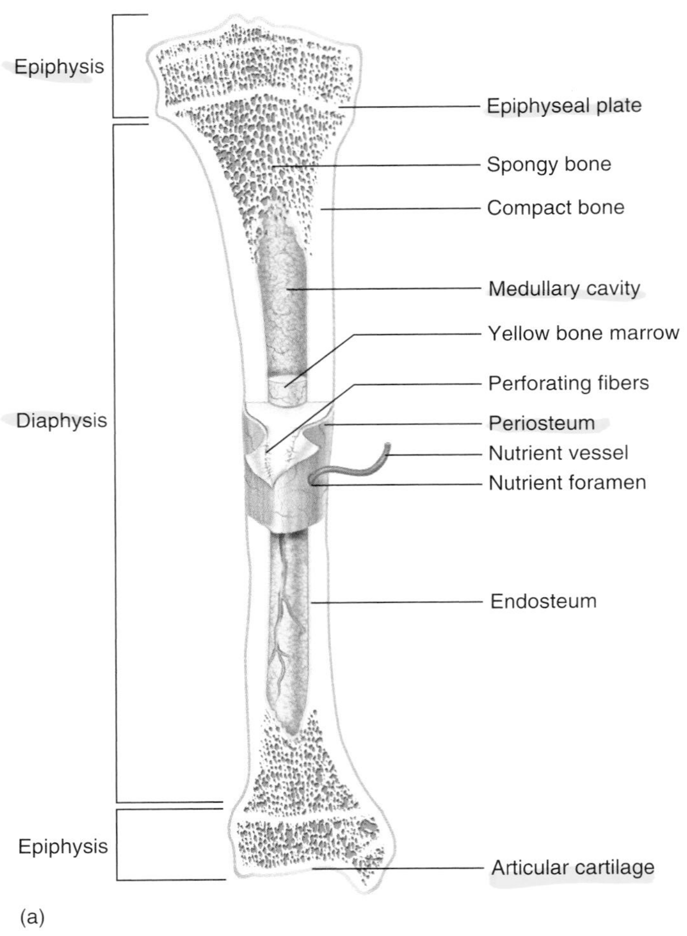

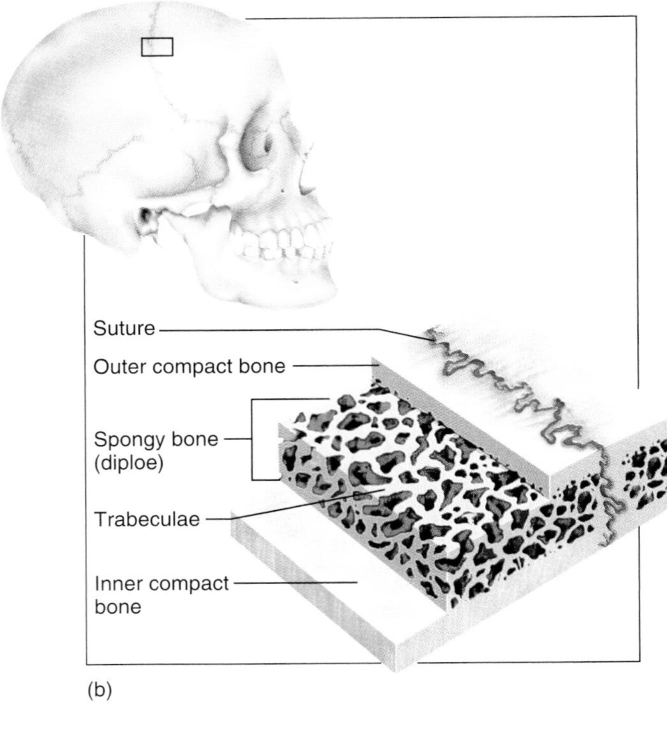

Figure 8.2 Anatomy of Bones. (*a*) A long bone, the tibia. (*b*) Two flat bones, the parietal and frontal bones of the cranium.

The diaphysis provides leverage, while the epiphysis is enlarged to strengthen the joint and provide added surface area for the attachment of tendons and ligaments. The joint surface where one bone meets another is covered with a layer of hyaline cartilage called the **articular cartilage.** Together with a lubricating fluid secreted between the bones, this cartilage enables a joint to move far more easily than it would if one bone rubbed directly against the other. Blood vessels penetrate into the bone through minute holes called **nutrient foramina** (for-AM-ih-nuh); we will trace where they go when we consider the histology of bone.

Externally, a bone is covered with a sheath called the **periosteum.**[7] This has a tough, outer *fibrous layer* of collagen and an inner *osteogenic layer* of bone-forming cells described later in the chapter. Some collagen fibers of the outer layer are continuous with the tendons that bind muscle to bone, and some penetrate into the bone matrix as **perforating (Sharpey**[8]**) fibers.** The periosteum thus provides strong attachment and continuity from muscle to tendon to bone. The osteogenic layer is important to the growth of bone and healing of fractures. There is no periosteum over the articular cartilage. The internal surface of a bone is lined with **endosteum,**[9] a thin layer of reticular connective tissue and cells that deposit and dissolve osseous tissue.

In children and adolescents, an **epiphyseal** (EP-ih-FIZZ-ee-ul) **plate** of hyaline cartilage separates the marrow spaces of the epiphysis and diaphysis (fig 8.2*a*). On X rays, it appears as a transparent line at the end of a long bone (see fig. 8.12). The epiphyseal plate is a zone where the bones elongate by a growth process detailed later in the chapter. In adults, the epiphyseal plate is depleted and the bones no longer grow in length, but an *epiphyseal line* on the bone surface marks where the plate used to be.

7. *peri* = around + *oste* = bone

8. William Sharpey (1802–80), Scottish histologist

9. *endo* = within + *oste* = bone

Before You Go On

Answer the following questions to test your understanding of the preceding section:

1. Name five tissues found in a bone.
2. List three or more functions of the skeletal system other than supporting the body and protecting some of the internal organs.
3. Name the four bone shapes and give an example of each.
4. Explain the difference between compact and spongy bone, and describe their spatial relationship to each other.
5. State the anatomical terms for the shaft, head, growth zone, and fibrous covering of a long bone.

Histology of Osseous Tissue

Objectives

When you have completed this section, you should be able to

- list and describe the cells, fibers, and ground substance of bone tissue;
- state the importance of each constituent of bone tissue;
- compare the histology of the two types of bone tissue; and
- distinguish between the three types of bone marrow.

Cells

Like any other connective tissue, bone consists of cells, fibers, and ground substance. There are four principal types of bone cells (fig. 8.3):

1. **Osteogenic**[10] **cells** occur in the endosteum, the inner layer of the periosteum, and within the haversian canals. They arise from embryonic fibroblasts. Osteogenic cells multiply continually and some of them differentiate into the *osteoblasts* described next.
2. **Osteoblasts**[11] are bone-forming cells that synthesize the organic matter of the bone matrix and help to mineralize the bone. They line up in rows in the endosteum and inner layer of periosteum and resemble a cuboidal epithelium on the bone surface. They are nonmitotic, but their numbers build up rapidly in response to stress or fractures because osteogenic cells multiply and differentiate into osteoblasts.
3. **Osteocytes** are former osteoblasts that have become trapped in the matrix they deposited. They reside in tiny cavities called **lacunae,**[12] which are connected to each other by slender channels called **canaliculi**[13] (CAN-uh-LIC-you-lye). Each osteocyte has delicate cytoplasmic processes that reach into the canaliculi to meet the processes of neighboring osteocytes. The processes of neighboring osteocytes are joined by gap junctions, which allow osteocytes to pass nutrients and chemical signals to each other and to transfer wastes to the nearest blood vessels for disposal. Osteocytes also communicate by gap junctions with the osteoblasts on the bone surface. Osteocytes play no significant role in depositing or resorbing bone. They detect mechanical stress on a bone and communicate this information to osteoblasts at the surface. Osteoblasts then deposit bone where needed—for example, building up bone in response to weight-bearing exercise—and they chemically signal osteoclasts (the cells discussed next) to remove bone elsewhere.
4. **Osteoclasts**[14] are bone-dissolving cells found on the bone surface. They develop from the same bone marrow cells that produce monocytes of the blood. Several of these marrow cells fuse with each other to form an osteoclast; thus osteoclasts are unusually large (up to 150 μm in diameter) and typically have 3 or 4 nuclei, but sometimes up to 50. Osteoclasts often reside in little pits called *resorption bays (Howship*[15] *lacunae)* that they have etched into the bone surface. Their structural details are described later in connection with their bone-dissolving activity.

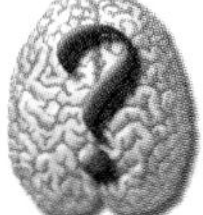

Think About It

What two organelles do you think are especially prominent in osteoblasts? (Hint: Consider the major substances that osteoblasts synthesize.)

Matrix

The matrix of osseous tissue is, by dry weight, about one-third organic and two-thirds inorganic matter. The organic matter includes collagen and various protein-carbohydrate complexes such as glycosaminoglycans, proteoglycans, and glycoproteins. The inorganic matter is about 85% **hydroxyapatite,** a crystallized calcium phosphate salt $[Ca_{10}(PO_4)_6(OH)_2]$, 10% calcium carbonate ($CaCO_3$), and lesser amounts of magnesium, sodium, potassium, fluoride, sulfate, carbonate, and hydroxyl ions. Several foreign elements behave chemically like bone minerals and become

10. *osteo* = bone + *genic* = producing
11. *osteo* = bone + *blast* = form, produce
12. *lac* = lake, hollow + *una* = little
13. *canal* = canal, channel + *icul* = little
14. *osteo* = bone + *clast* = destroy, break down
15. J. Howship (1781–1891), English surgeon

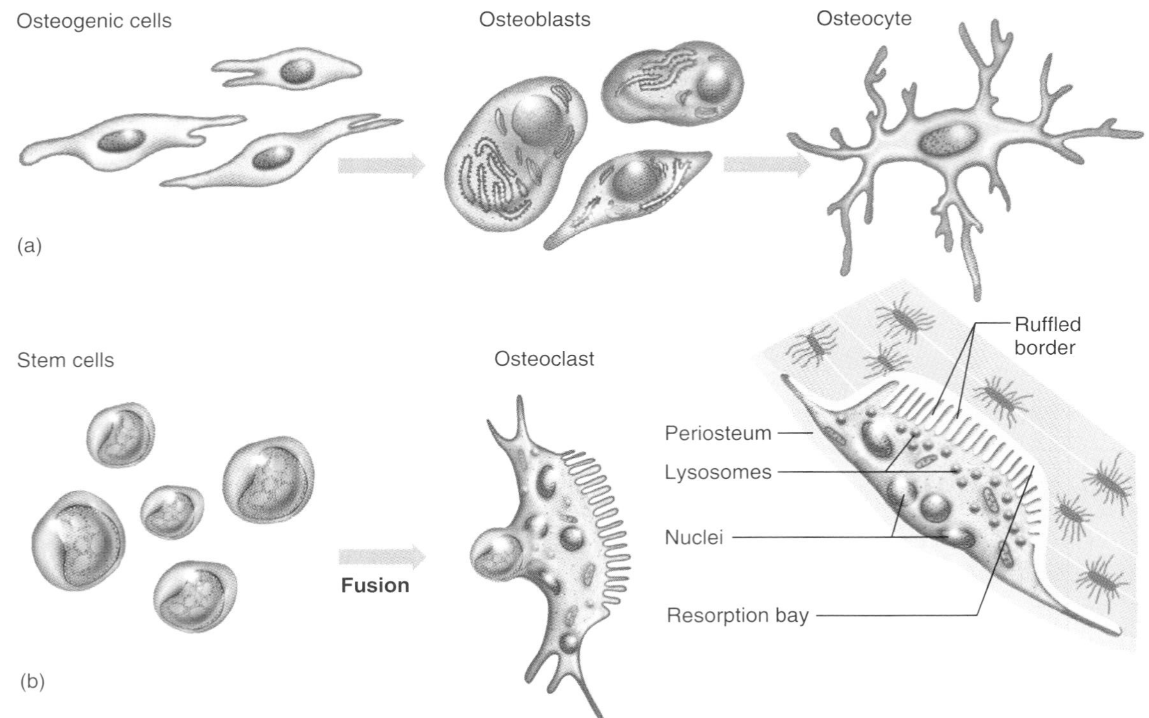

Figure 8.3 **Bone Cells and Their Development.** (*a*) Osteogenic cells give rise to osteoblasts, which deposit matrix around themselves and transform into osteocytes. (*b*) Bone marrow cells fuse to form osteoclasts.

incorporated into osseous tissue as contaminants, sometimes with deadly results (see historical insight 8.1).

The combination of collagen and minerals gives bones a combination of resilience and strength similar to that of reinforced concrete. Like concrete, the minerals resist compression (crumbling or sagging when weight is applied), and like steel reinforcement bars, the collagen resists tension (bending). If the minerals are dissolved by immersing a bone in a weak acid such as vinegar, the bone becomes rubbery (fig. 8.4). If only the collagen is removed, it becomes very brittle. Without collagen, a jogger's bones would shatter under the impact of running. Defective collagen is responsible for *osteogenesis imperfecta* (brittle bone disease) (see clinical insight 6.4, p. 201).

Compact Bone

The histological study of compact bone usually uses slices that have been dried, cut with a saw, and ground to translucent thinness. This procedure destroys the cells and most other organic content but reveals fine details of the inorganic matrix (fig. 8.5). Such sections show onion-like **lamellae**—layers of matrix concentrically arranged around a central **haversian** (ha-VUR-zhun) **canal** and connected with each other by canaliculi. A haversian canal and its lamellae constitute an **osteon** (haversian system)—the basic structural unit of compact bone. In longitudinal views and three-dimensional reconstructions, we can see that an osteon is actually a cylinder of tissue surrounding a haversian canal. Along their length, haversian canals are joined by transverse or diagonal passages.

The skeleton receives about half a liter of blood per minute. Blood vessels, along with nerves, enter the bone tissue through nutrient foramina on the surface. These open into narrow **perforating (Volkmann[16]) canals** that cross the matrix and feed into the haversian canals. The innermost osteocytes around each haversian canal receive nutrients from these blood vessels and pass them along through their gap junctions to neighboring osteocytes. They also receive wastes from their neighbors and convey them to the haversian canal for removal by the bloodstream. Thus, the cytoplasmic processes of the osteocytes maintain a two-way flow of nutrients and wastes between the haversian canal and the outermost cells of the osteon.

Not all of the matrix is organized into osteons. The inner and outer boundaries of dense bone are arranged in *circumferential lamellae* that run parallel to the bone surface. Between osteons, we can find irregular regions called

16. Alfred Volkmann (1800–1877), German physiologist

Historical Insight 8.1

Bone Contamination

When Marie and Pierre Curie and Henri Becquerel received their 1903 Nobel Prize for the discovery of radioactivity (see historical insight 2.3, p. 70), radiation captured the public's imagination. Not for several decades did anyone realize its dangers. For example, watch factories employed women to paint glow-in-the-dark numbers on watch and clock dials with radium paint. The women moistened their paint brushes with their tongues to keep them finely pointed and ingested radium in the process. Their bones readily absorbed the radium, which caused many of the women to develop a form of bone cancer called osteosarcoma.

Even more horrific, in the wisdom of our hindsight, was a deadly health fad in which people drank "tonics" made of radium-enriched water. One famous enthusiast was the millionaire playboy and championship golfer Eben Byers, who drank several bottles of radium tonic each day and praised its virtues as a wonder drug and aphrodisiac. Like the factory women, Byers contracted osteosarcoma. By the time of his death, holes had formed in his skull and doctors had removed his entire upper jaw and most of his mandible in an effort to halt the spreading cancer. Byers's bones and teeth were so radioactive they could expose photographic film in the dark. Brain damage left him unable to speak, but he remained mentally alert to the bitter end. His tragic decline and death in 1932 shocked the world and put an end to the radium tonic fad.

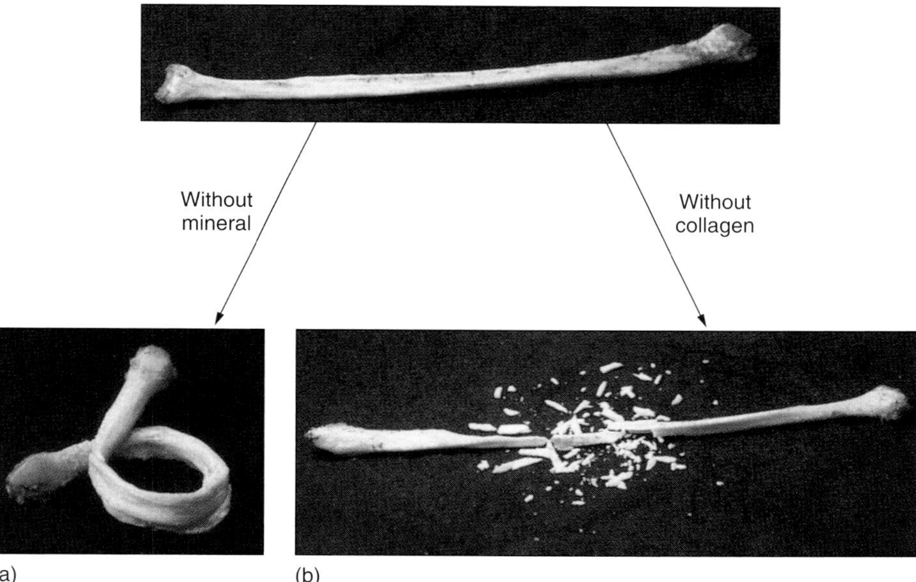

Figure 8.4 Role of the Collagen and Minerals in Bone. (*a*) With the minerals removed, the fibula (a bone of the leg) becomes rubbery. (*b*) With the collagen removed, the fibula becomes very brittle and shatters easily.

interstitial lamellae, the remains of old osteons that broke down as the bone grew and remodeled itself.

Spongy Bone

Spongy bone (fig. 8.5*a*) consists of a lattice of slender rods, plates, and spines called **trabeculae.**[17] Although calcified and hard, spongy bone is named for its sponge-like appearance (see p. 231). It is permeated by spaces filled with bone marrow. The matrix is arranged in lamellae like those of compact bone, but there are few osteons. Haversian canals are not needed here because no osteocyte is very far from the marrow. Spongy bone is well designed to impart strength to a bone while adding a minimum of weight. Its trabeculae are not randomly arranged as they might seem at a glance, but develop along the bone's lines of stress (fig. 8.6).

17. *trabe* = plate + *cul* = little

Scapula
Humerus
Spongy bone
Compact bone

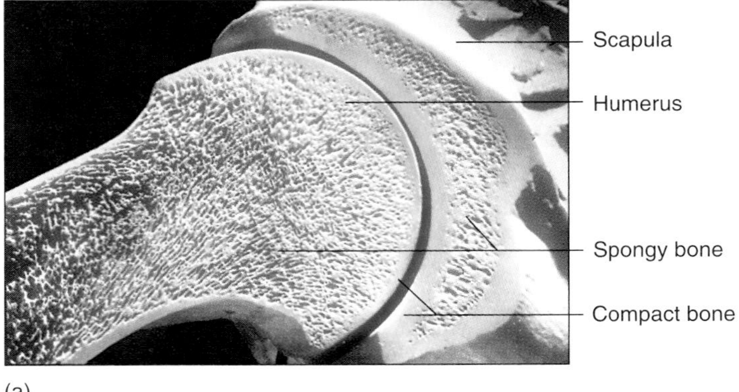

(a)

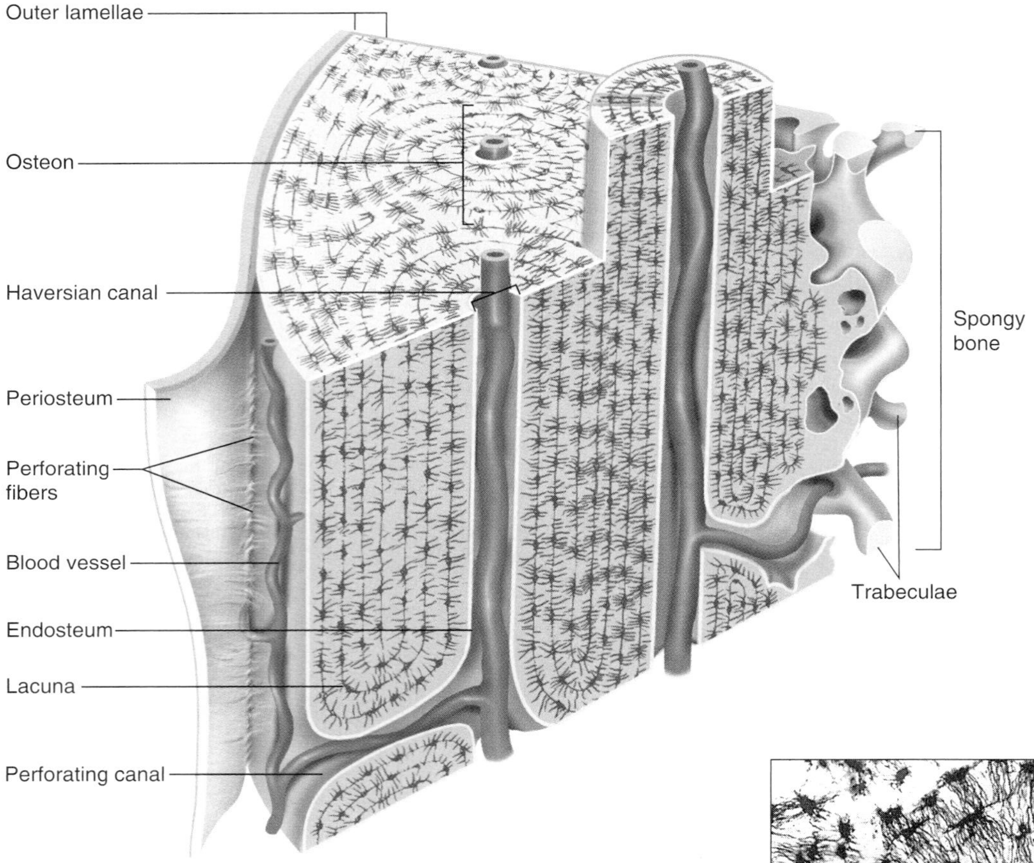

Outer lamellae
Osteon
Haversian canal
Periosteum
Perforating fibers
Blood vessel
Endosteum
Lacuna
Perforating canal
Spongy bone
Trabeculae

(b)

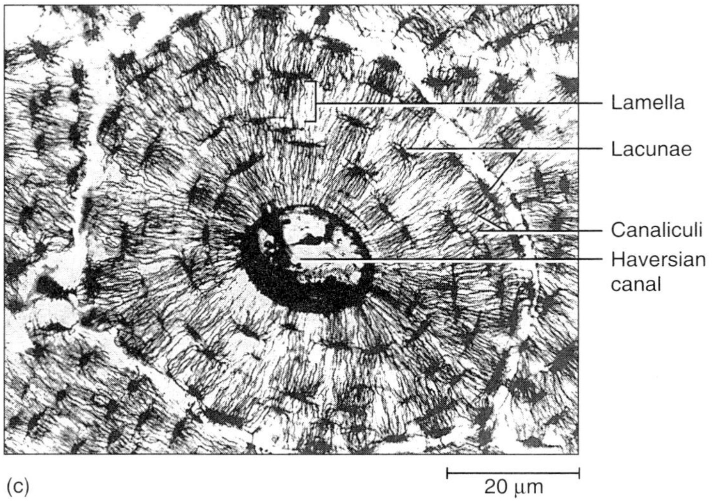

Lamella
Lacunae
Canaliculi
Haversian canal
20 μm

(c)

Figure 8.5 The Histology of Compact Bone. (*a*) Compact and spongy bone in a frontal section of the shoulder joint. (*b*) The three-dimensional structure of compact bone. (*c*) Microscopic appearance of a cross section of compact bone.

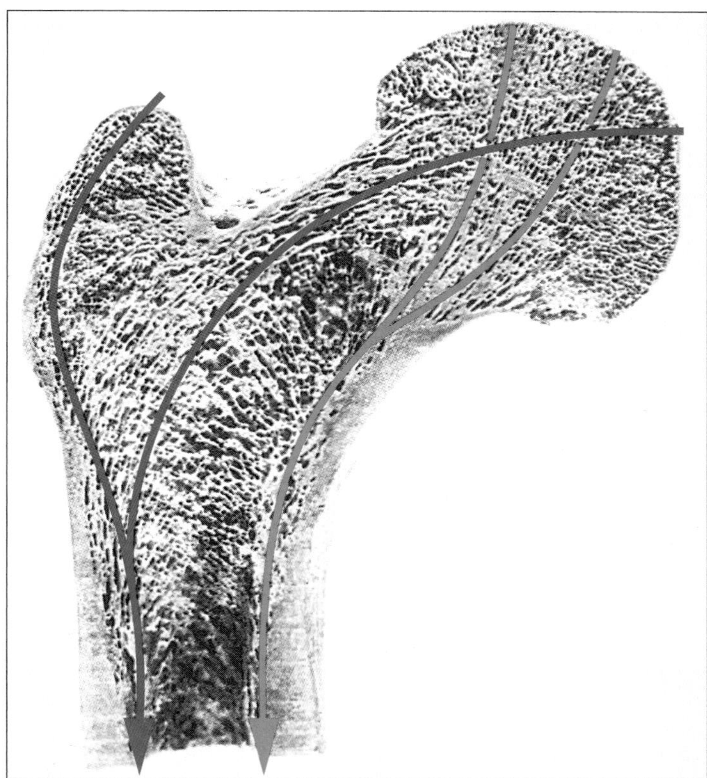

Figure 8.6 Spongy Bone Structure in Relation to Mechanical Stress. In this frontal section of the femur (thigh bone), the trabeculae of spongy bone can be seen oriented along lines of mechanical stress applied by the weight of the body.

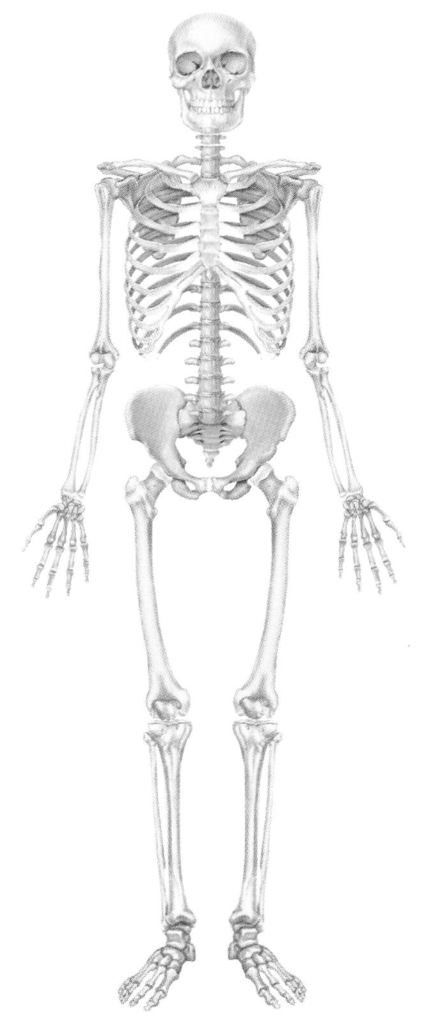

Figure 8.7 Distribution of Red and Yellow Bone Marrow. In an adult, red bone marrow occupies the medullary cavities of the axial skeleton and proximal heads of the humerus and femur. Yellow bone marrow occurs in the long bones of the limbs.

Bone Marrow

Bone marrow is a general term for soft tissue that occupies the medullary cavity of a long bone, the spaces amid the trabeculae of spongy bone, and the larger haversian canals. There are three kinds of marrow—red, yellow, and gelatinous. We can best appreciate their differences by considering how marrow changes over a person's lifetime.

In a child, the medullary cavity of nearly every bone is filled with **red bone marrow (myeloid tissue).** This is a *hemopoietic*[18] (HE-mo-poy-ET-ic) tissue—that is, it produces blood cells. Red bone marrow looks like blood but with a thicker consistency. It consists of a delicate mesh of reticular tissue saturated with immature blood cells and scattered adipocytes.

In young to middle-aged adults, the shafts of the long bones are filled with fatty **yellow bone marrow,** like the fat at the center of a ham bone. Yellow bone marrow no longer produces blood, although in the event of severe or chronic anemia, it can transform back into red marrow. In adults, red marrow is limited to the axial skeleton, pectoral (shoulder) and pelvic (hip) girdles, and proximal heads of the humerus and femur (fig. 8.7). In old age, most of the yellow bone marrow is replaced by a reddish jelly called **gelatinous bone marrow.**

18. *hemo* = blood + *poietic* = forming

Before You Go On

Answer the following questions to test your understanding of the preceding section:

6. Suppose you had unlabeled electron micrographs of the four kinds of bone cells and their neighboring tissues. Name each of the four cells and explain how you could visually distinguish it from the other three.
7. Name three organic components of the bone matrix.
8. What are the mineral crystals of bone called, and what are they made of?
9. Sketch a cross section of an osteon and label its major parts.
10. What are the three kinds of bone marrow? What does *hemopoietic tissue* mean? Which type of bone marrow fits this description?

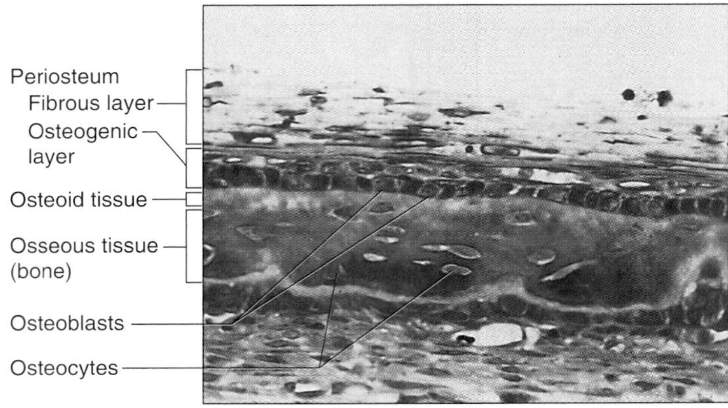

Figure 8.8 Intramembranous Ossification.

Bone Development

Objectives

When you have completed this section, you should be able to

- describe two mechanisms of bone formation; and
- explain how mature bone continues to grow and remodel itself.

The formation of bone is called **ossification** (OSS-ih-fih-CAY-shun), or **osteogenesis.** There are two methods of ossification—*intramembranous* and *endochondral*—which we examine in the following sections.

Intramembranous Ossification

Intramembranous[19] (IN-tra-MEM-bruh-nus) **ossification** produces the flat bones of the skull and most of the clavicle (collarbone). It begins when some of the embryonic connective tissue (mesenchyme) condenses into a sheet of soft tissue with a dense supply of blood capillaries. The cells of this sheet enlarge and differentiate into osteogenic cells, and some of the mesenchyme transforms into a network of soft trabeculae. Osteogenic cells gather on the trabeculae, become osteoblasts, and deposit an organic matrix called **osteoid**[20] **tissue**—soft collagenous tissue similar to bone except for a lack of minerals (fig. 8.8). As the trabeculae grow thicker, calcium phosphate is deposited in the matrix and some osteoblasts become trapped in lacunae. Once trapped, they differentiate into osteocytes. Some of the now-calcified trabeculae form permanent spongy bone. Osteoclasts appear early on these trabeculae, reabsorbing and remodeling bone and creating a marrow space in the middle of the bone. Trabeculae at the surface continue to calcify until the spaces between them are filled in, thereby converting the spongy bone to compact bone. This process gives rise to the typical structure of a flat cranial bone—a sandwichlike arrangement of spongy bone between two surface layers of compact bone. Mesenchyme at the surface of the developing bone remains uncalcified, but becomes increasingly fibrous and eventually gives rise to the periosteum of the bone.

Endochondral Ossification

Endochondral[21] (EN-doe-CON-drul) **ossification** produces most other bones, including the vertebrae, pelvic bones, and bones of the limbs. It is a method in which mesenchyme first transforms into a hyaline cartilage *model* that resembles the shape of the bone to come, and then the cartilage is broken down, reorganized, and calcified to form a bone (fig. 8.9).

The Primary Ossification Center

The first sign of endochondral ossification is the multiplication and swelling of chondrocytes near the center of the model, forming a **primary ossification center.** As the lacunae enlarge, the matrix between them is reduced to thin walls and the model becomes weak at this point. It soon gets reinforcement, however. Some cells of the perichondrium become osteoblasts, which produce a bony collar around the model. This collar acts like a splint to provide temporary support for the model, and it cuts off the diffusion of nutrients to the chondrocytes, hastening their death. Once the collar has formed, the fibrous sheath around it is considered periosteum rather than perichondrium.

Buds of connective tissue grow from this periosteum into the cartilage and penetrate the thin walls between the

19. *intra* = within + *membran* = membrane
20. *oste* = bone + *oid* = like, resembling
21. *endo* = within + *chondr* = cartilage

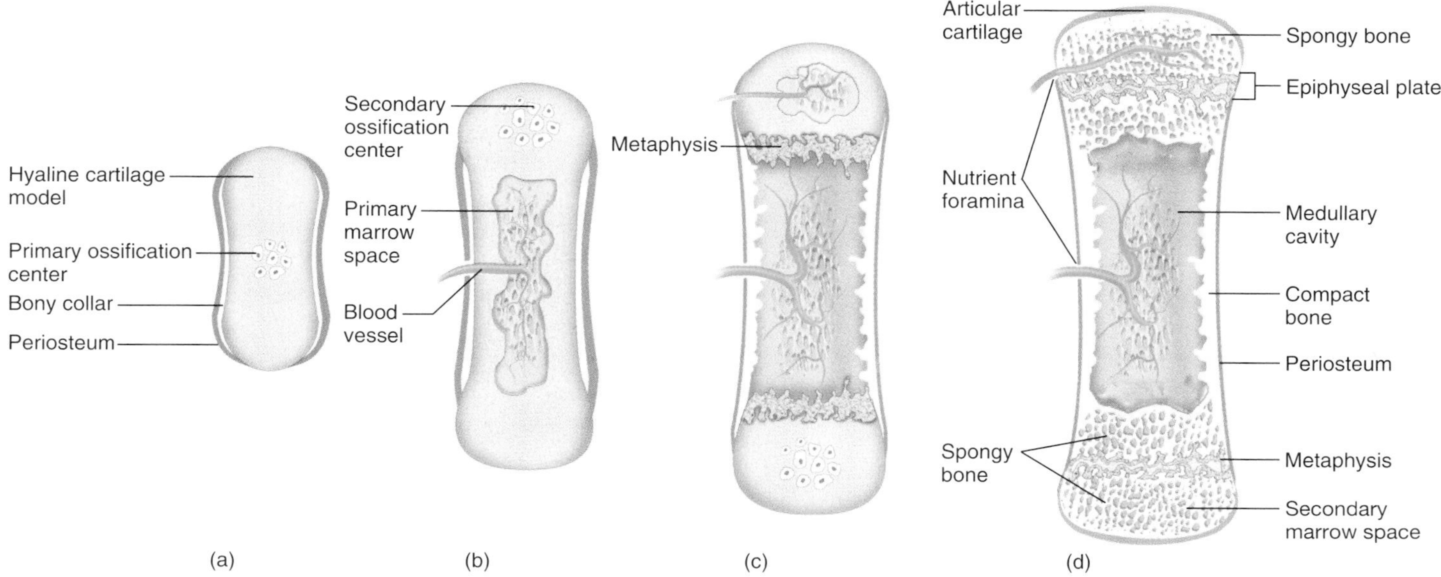

Figure 8.9 **Stages of Endochondral Ossification.** (*a*) Chondrocyte hypertrophy at the center of the cartilage model and formation of a supportive bony collar. (*b*) Invasion of the model by blood vessels and creation of a primary marrow space. (*c*) Typical state of a long bone at the time of birth, with blood vessels growing into the secondary marrow space and well-defined metaphyses at each end of the primary marrow space. (*d*) Appearance of a long bone in childhood. By adulthood, the epiphyseal plates will be depleted and the primary and secondary marrow spaces will be united.

enlarged lacunae. They break down the lacunae and transform the primary ossification center into a cavity called the **primary marrow space.** Osteogenic cells invade the model by way of the connective tissue buds, transform into osteoblasts, and line the marrow space. The osteoblasts deposit osteoid tissue and then calcify it to form a temporary framework of bony trabeculae. As ossification progresses, osteoclasts break down these temporary trabeculae and enlarge the primary marrow space. The ends of the bone are still composed of hyaline cartilage at this stage.

The Metaphysis

Between the cartilaginous head and the shaft of a developing long bone, there is a transitional zone called the **metaphysis** (meh-TAF-ih-sis). It exhibits five histological zones of transformation from cartilage to bone (fig. 8.10):

1. **Zone of reserve cartilage.** In this zone, farthest from the marrow space, the resting cartilage as yet shows no sign of transforming into bone.
2. **Zone of cell proliferation.** A little closer to the marrow space, chondrocytes multiply and become arranged into longitudinal columns of flattened lacunae.
3. **Zone of cell hypertrophy.** Next, the chondrocytes cease to divide and begin to hypertrophy, just as they did in the primary ossification center. The cartilage walls between lacunae become very thin. Cell multiplication in zone 2 and hypertrophy in zone 3 continually push the zone of reserve cartilage toward the ends of the bone and make the bone grow longer.
4. **Zone of calcification.** Minerals are deposited in the matrix between columns of lacunae and calcify the cartilage for temporary support.
5. **Zone of bone deposition.** Within each column, the walls between lacunae break down and the chondrocytes die. This converts each column into a longitudinal channel, which is quickly invaded by marrow and blood vessels from the primary marrow space. Osteoclasts dissolve the calcified cartilage while osteoblasts line up along the walls of these channels and begin depositing concentric layers of bone matrix. The channel therefore grows smaller and smaller as one layer after another is laid down, until only a narrow channel remains in the middle—now a haversian canal. Osteoblasts trapped in their own matrix become osteocytes and stop producing matrix.

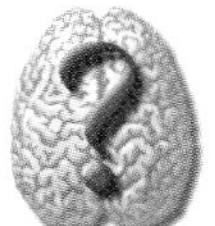

Think About It

In a given osteon, which lamellae are the oldest—those immediately adjacent to the haversian canal or those around the perimeter of the osteon? Explain your answer.

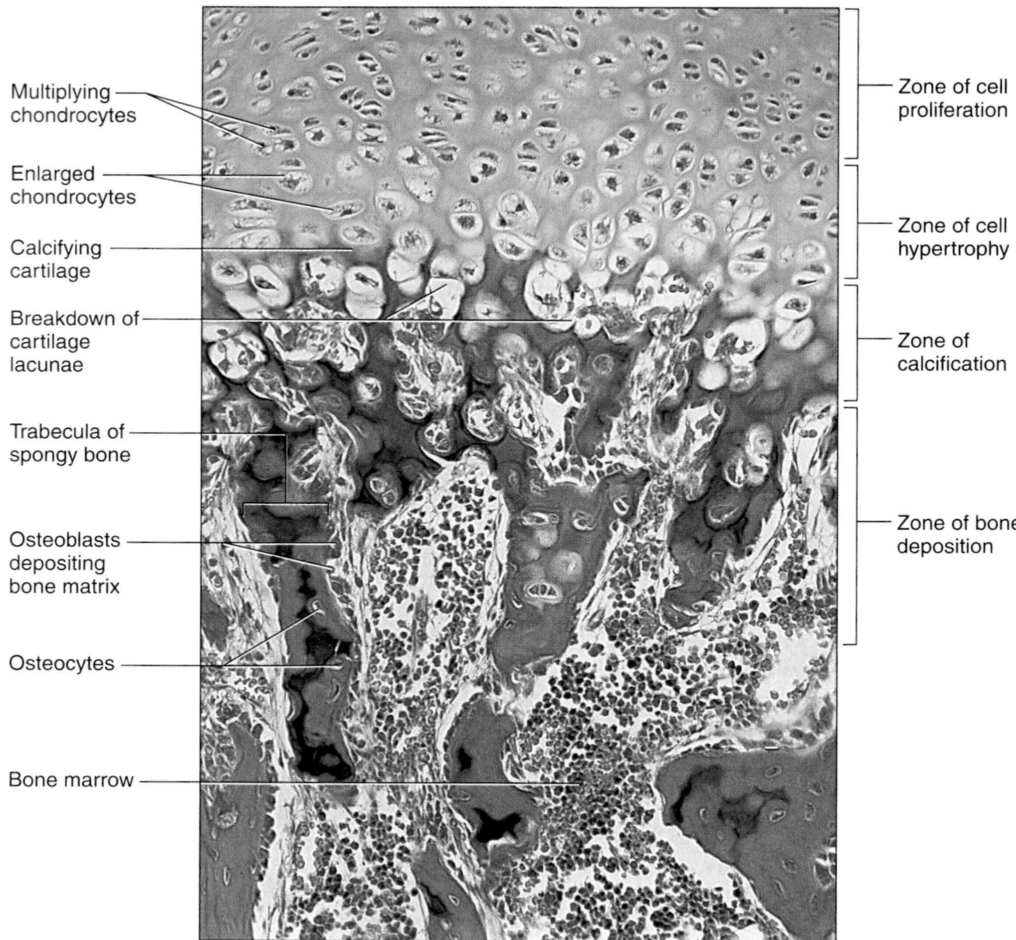

Figure 8.10 Zones of the Metaphysis. This micrograph shows the transition from cartilage to bone in the growth zone of a long bone.

The primary ossification centers of a 12-week-old fetus are shown in figure 8.11. The joints are translucent because they are still composed of cartilage that has not yet ossified. They are still cartilaginous even at the time of birth, which is one reason human newborns cannot walk.

The Secondary Ossification Center

Around the time of birth, **secondary ossification centers** begin to form in the epiphyses (see fig. 8.9*b*). Here, too, chondrocytes enlarge, the walls of matrix between them dissolve, and the chondrocytes die. Vascular buds arise from the perichondrium and grow into the cartilage, bringing osteogenic cells and osteoclasts with them. The cartilage is eroded from the center of the epiphysis outward in all directions. Thin trabeculae of cartilage matrix are calcified to form spongy bone. Hyaline cartilage persists in two places—on the epiphyseal surfaces as the articular cartilages and at the junction of the diaphysis and epiphysis, where it forms the epiphyseal plate (fig. 8.12). Each side of the epiphyseal plate has a metaphysis, where the transformation of cartilage to bone occurs.

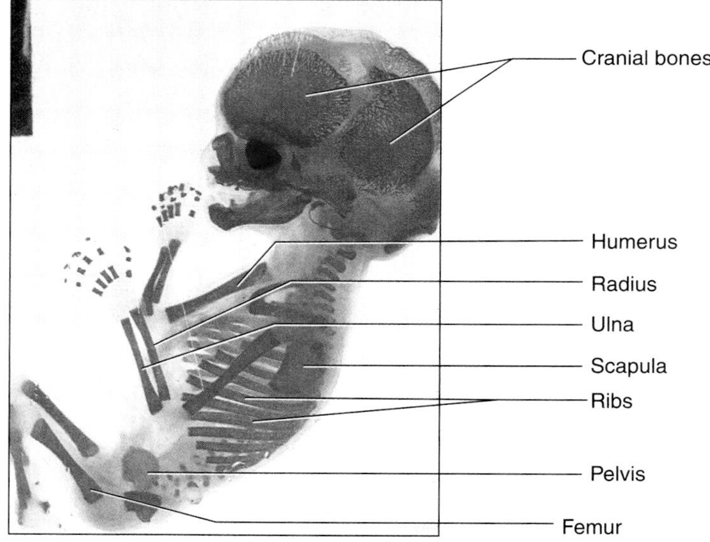

Figure 8.11 The Fetal Skeleton at 12 Weeks. The red-stained regions are calcified at this age, whereas the skull sutures and the elbow, wrist, knee, and ankle joints appear translucent because they are still cartilaginous.

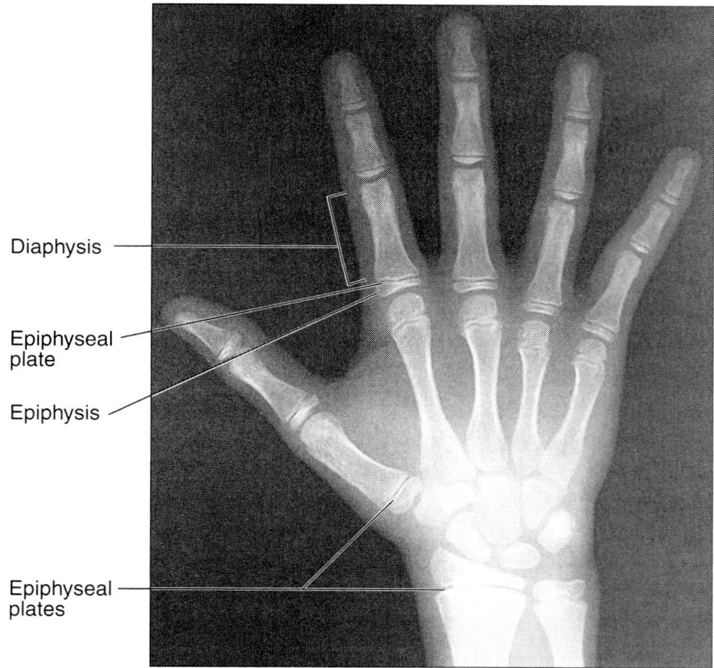

Figure 8.12 **X Ray of a Child's Hand.** The cartilaginous epiphyseal plates are evident at the ends of the long bones. These will disappear, and the epiphyses will fuse with the diaphyses, by adulthood.

Bone Growth and Remodeling

Bones continue to grow and remodel themselves throughout life, changing size and shape to accommodate the changing forces applied to the skeleton. For example, in children the femurs become longer, the curvature of the cranium increases to accommodate a growing brain, and many bones develop surface bumps, spines, and ridges (described in the next chapter) as a child begins to walk and the muscles exert tension on the bones. The prominence of these surface features and the density of bone depend on the amount of stress to which a bone has been subjected. On average, the bones have greater density and mass in athletes and people engaged in heavy manual labor than they do in sedentary people. Anthropologists who study skeletal remains can distinguish between members of different social classes by the degree of bone development—a reflection of the individual's nutritional status and history of manual labor. The skeleton also yields information about sex, race, height, weight, and medical history that can be useful to forensic pathology—study of the body to determine the identity of a person, cause and time of death, and so forth.

Cartilage grows by two mechanisms—**interstitial**[22] **growth** (adding more matrix internally) and **appositional**[23] **growth** (adding more to the surface). Interstitial growth in the epiphyseal plate adds to the length of a bone. A mature bone, however, grows only by the appositional mechanism. Osteocytes have little room as it is and none to spare for the deposition of more matrix. The only way an adult bone can grow, therefore, is by adding more osseous tissue to the surface.

Appositional growth is similar to intramembranous ossification. The osteogenic cells in the inner layer of periosteum differentiate into osteoblasts. These deposit osteoid tissue on the bone surface, calcify it, and become trapped in it as osteocytes. At the bone surface, matrix is laid down in layers parallel to the surface, not in cylindrical osteons like those deeper in the bone. While deposition occurs at the outer surface of a bone, osteoclasts dissolve bone on the inner surface and thus enlarge the marrow cavity as the bone grows. There is a critical balance between bone deposition and removal. If one process outpaces the other, or if both of them occur too rapidly, various bone deformities and other developmental abnormalities occur (see table 8.4, especially *osteitis deformans*).

Before You Go On

Answer the following questions to test your understanding of the preceding section:

11. Describe the stages of intramembranous ossification. Name a bone that is formed in this way.
12. Describe the five zones of a metaphysis and the major distinctions between them.

Physiology of Osseous Tissue

Objectives

When you have completed this section, you should be able to

- explain how minerals are deposited in bone tissue and removed from it;
- discuss the role of the bones in regulating blood calcium and phosphate levels;
- describe how vitamin D is synthesized and how it affects the bones; and
- list other hormones that affect bone physiology and briefly state their effects.

Even after a bone is fully formed, it remains a metabolically active organ with many roles to play. Not only is it involved in its own maintenance, growth, and remodeling, it also exerts a profound influence on the rest of the body by exchanging minerals with the tissue fluid. Disturbances of calcium homeostasis in the skeleton can disrupt the functioning of other organ systems, especially the

22. *inter* = between + *stit* = to place, stand
23. *ap* = *ad* = to, near + *posit* = to place

Clinical Insight 8.2

Achondroplastic Dwarfism

Achondroplastic[24] (ah-con-dro-PLAS-tic) *dwarfism* is a condition in which the long bones of the limbs stop growing in childhood, while the growth of other bones is unaffected. As a result, a person has a short stature but a normal-sized head and trunk. As its name implies, achondroplastic dwarfism results from a failure of cartilage growth—specifically, failure of the chondrocytes in zones 2 and 3 of the metaphysis to multiply and enlarge. This is different from *pituitary dwarfism,* in which a deficiency of growth hormone stunts the growth of all of the bones and a person has short stature but normal proportions throughout the skeletal system (see fig. 17.7*a*).

Achondroplastic dwarfism results from a spontaneous mutation that can arise any time DNA is replicated. Two people of normal height with no family history of dwarfism can therefore have a child with achondroplastic dwarfism. The mutant allele is dominant, so the children of a heterozygous achondroplastic dwarf have at least a 50% chance of exhibiting dwarfism, depending on the genotype of the other parent.

Figure 1 The student on the right, pictured with her roommate of normal height, is an achondroplastic dwarf with a height of about 122 cm (48 in.). Her parents were of normal height. Note the normal proportion of head to trunk but shortening of the limbs.

24. *a* = without + *condro* = cartilage + *plast* = growth

nervous and muscular systems. For reasons explained later, such disturbances can even cause a person to die of suffocation. At this point, we turn our attention to the physiology of mature osseous tissue.

Mineral Deposition

Mineral deposition (mineralization) is a crystallization process in which calcium and phosphate ions, among others, are taken from the blood plasma and deposited in bone tissue. It begins in fetal ossification and continues throughout life. Osteoblasts first produce collagen fibers that wind in a helical pattern along the length of the osteon, spiraling in alternating directions from one lamella to the next in a way that enhances the strength of the bone. These fibers then become encrusted with minerals—especially calcium phosphate—that harden the matrix. Calcium phosphate crystals do not form unless the product of calcium and phosphate concentration in the tissue fluids, represented $[Ca^{2+}] \cdot [PO_4^{3-}]$, reaches a critical value called the **solubility product.** Most tissues have inhibitors to prevent this, so they do not become calcified. Osteoblasts, however, apparently neutralize these inhibitors and thus allow the salts to precipitate in the bone matrix. The first few hydroxyapatite crystals to form act as "seed crystals" that attract more calcium and phosphate from solution. The more hydroxyapatite that forms, the more it attracts additional minerals from the tissue fluid, until the matrix is thoroughly calcified. Crystalization is thus a positive feedback process.

Osseous tissue sometimes forms in the lungs, brain, eyes, muscles, tendons, arteries, and other organs. Such abnormal calcification of tissues is called **ectopic**[25] **ossification.** One example of this is arteriosclerosis, or "hardening of the arteries," which results from calcification of the arterial walls. A calcified mass in an otherwise soft organ such as the lungs is called a **calculus.**[26]

25. *ec* = out of + *top* = place
26. *calc* = stone + *ulus* = little

Mineral Resorption

Mineral resorption is the process of dissolving bone. It releases minerals into the blood and makes them available for other uses. Resorption is carried out by osteoclasts. On the side facing the bone surface, an osteoclast has a *ruffled border* with deep infoldings of the plasma membrane. Hydrogen pumps in the ruffled border secrete hydrogen ions into the extracellular fluid, and chloride ions follow by electrical attraction; thus, the space between the osteoclast and the bone becomes filled with concentrated hydrochloric acid with a pH of about 4. The acid dissolves the bone minerals. The osteoclast also secretes an enzyme called **acid phosphatase** that digests the collagen of the bone matrix. Acid phosphatase is named for its ability to function in this highly acidic environment.

When orthodontic appliances (braces) are used to reposition teeth, a tooth moves because osteoclasts dissolve bone in front of the tooth (where the appliance creates greater pressure of the tooth against the bone) and osteoblasts deposit bone in the low-pressure zone behind it.

Calcium and Phosphate Homeostasis

Calcium and phosphate are used for much more than bone structure. Phosphate groups are a component of DNA, RNA, ATP, phospholipids, and many other compounds. Phosphate ions also help to correct acid-base imbalances in the body fluids (see clinical insight 8.3). Calcium is needed for communication between neurons, muscle contraction, blood clotting, and exocytosis. It is also a second messenger in many cell signaling processes and a cofactor for some enzymes. The skeleton is a reservoir for these minerals. Minerals are deposited in the skeleton when the supply is ample and withdrawn when they are needed for other purposes.

The adult body contains about 1,100 g of calcium, with 99% of it in the bones. Bone has two calcium reserves—a stable pool of calcium, which is incorporated into hydroxyapatite and is not easily exchanged with the blood, and exchangeable calcium, which is 1% or less of the total but is easily released to the tissue fluid. The adult skeleton exchanges about 18% of its calcium with the blood each year.

The calcium concentration in the blood plasma is normally 9.2 to 10.4 mg/dL. This is a rather narrow margin of safety, as we shall soon see. About 45% of it is in the ionized form (Ca^{2+}), which can diffuse through capillary walls and affect neighboring cells. The rest of it is bound to plasma proteins and other solutes. It is not physiologically active, but it serves as a reserve from which free Ca^{2+} can be obtained as needed.

The average adult has 500 to 800 g of phosphorus, of which 85% to 90% is in the bones. The phosphorus concentration in the plasma ranges between 3.5 and 4.0 mg/dL. It occurs in two forms, HPO_4^{2-} and $H_2PO_4^-$ (monohydrogen and dihydrogen phosphate ions, respectively).

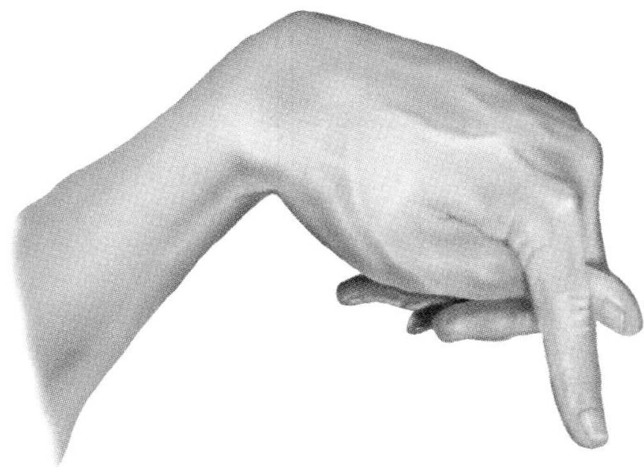

Figure 8.13 **Carpopedal Spasm.** Such muscle tetanus occurring in the hands and feet can be a sign of hypocalcemia.

Changes in phosphate concentration have little immediate effect on the body, but changes in calcium can be serious. A deficiency of blood calcium is called **hypocalcemia**[27] (HY-po-cal-SEE-me-uh). It causes excessive excitability of the nervous system and leads to muscle tremors, spasms, or **tetanus**—inability of the muscle to relax. Tetanus begins to appear as the plasma Ca^{2+} concentration reaches 6 mg/dL. One of the signs of hypocalcemia is a tetanus of the hands and feet called *carpopedal spasm* (fig. 8.13). At 4 mg/dL, muscles of the larynx contract tightly, a condition called *laryngospasm* that can shut off air flow and cause suffocation.

The reason for this hypocalcemic excitability is this: Calcium ions normally bind to negatively charged groups on the cell surface, contributing to the difference between the positively charged outer surface of the membrane and the negatively charged inner surface. In hypocalcemia, fewer of these calcium ions are present, so there is less charge difference between the two sides of the membrane. Sodium channels in the plasma membrane are sensitive to this charge difference, and when the difference is diminished, sodium channels open more easily and stay open longer. This allows sodium ions to enter the cell too freely. As you will see in chapters 12 and 13, an inflow of sodium is the normal process that excites nerve and muscle cells. In hypocalcemia, this excitation is excessive and results in the aforementioned signs.

An excess of blood calcium is called **hypercalcemia.**[28] In hypercalcemia, excessive amounts of calcium bind to the cell surface, increasing the charge difference across the membrane and making sodium channels reluctant to open. Thus, nerve and muscle cells are not as excitable as they should be. At 12 mg/dL and higher, hypercalcemia causes

27. *hypo* = below normal + *calc* = calcium + *emia* = blood condition
28. *hyper* = above normal + *calc* = calcium + *emia* = blood condition

Clinical Insight 8.3

Osseous Tissue and pH Balance

The urinary, respiratory, and skeletal systems cooperate to maintain the body's acid-base balance. If the pH of the blood drops below 7.35, a state of acidosis exists and triggers corrective mechanisms in these three organ systems. The role of the skeleton is to release calcium phosphate. As a base, calcium phosphate helps to prevent the blood pH from dropping lower. Patients with chronic kidney disease may have impaired hydrogen ion excretion in the urine. Their pH stabilizes at a level below normal but is kept from dropping indefinitely by the buffering action of the skeleton. This can have adverse effects on the skeleton, however, and lead to rickets or osteomalacia. Treatment of the acidosis with intravenous bicarbonate restores the pH to normal and prevents damage to the bones.

depression of the nervous system, emotional disturbances, muscle weakness, sluggish reflexes, and sometimes cardiac arrest.

You can see how critical blood calcium level is, but what causes it to deviate from the norm, and how does the body correct such imbalances? Hypercalcemia is rare, but hypocalcemia can result from a wide variety of causes including vitamin D deficiency, diarrhea, thyroid tumors, or underactive parathyroid glands. Pregnancy and lactation put women at risk of hypocalcemia because of the calcium demanded by ossification of the fetal skeleton and synthesis of milk. The leading cause of hypocalcemic tetanus is accidental removal of the parathyroid glands during thyroid surgery. Without hormone replacement therapy, the lack of parathyroid glands can lead to fatal tetanus within 4 days.

Calcium phosphate homeostasis depends on a balance between dietary intake, urinary and fecal losses, and exchanges with the osseous tissue. It is regulated by three hormones: *calcitriol, calcitonin,* and *parathyroid hormone.*

Calcitriol

Calcitriol (CAL-sih-TRY-ol) is an activated form of vitamin D produced mainly by the following process (fig. 8.14):

1. A cholesterol derivative called 7-dehydrocholesterol circulates through dermal capillaries close to the skin surface, where the ultraviolet radiation in sunlight converts it to cholecalciferol (CO-lee-cal-SIF-ur-ol), or vitamin D_3.
2. The liver adds a hydroxyl group to cholecalciferol, converting it to *calcidiol.*
3. The kidneys add another hydroxyl group to calcidiol, converting it to calcitriol.

Calcitriol behaves as a hormone—a blood-borne chemical messenger from one organ to another. It has three effects that influence bone deposition:

1. It stimulates the small intestine to absorb both calcium and phosphate.
2. It reduces the urinary excretion of both calcium and phosphate.
3. It promotes osteoclast activity.

The third effect, by itself, would reduce bone mass, but calcitriol overrides this by the first two mechanisms, which make calcium and phosphate abundantly available for bone deposition.

Vitamin D is called a vitamin only because it is added to the diet, mainly in fortified milk, as a safeguard for people who do not get enough sunlight to synthesize adequate amounts of it themselves. An insufficient amount of sunlight or supplemental vitamin D results in abnormal softness of the bones, a condition called **rickets** in children and **osteomalacia**[29] in adults. In children, this condition typically results in bowed legs as the softened bones are deformed by the weight of the body.

Calcitonin

Calcitonin is secreted by *C cells* (*C* for "calcitonin") of the thyroid gland, a large endocrine gland in the neck (see fig. 17.10). It is secreted when the blood calcium concentration rises too high, and it lowers the concentration by two principal mechanisms (fig. 8.15*a*):

1. **Osteoclast inhibition.** Within 15 minutes after it is secreted, calcitonin reduces osteoclast activity by as much as 70%, so osteoclasts liberate less calcium from the skeleton.
2. **Osteoblast stimulation.** Within an hour, calcitonin increases the number and activity of osteoblasts, which deposit calcium into the skeleton.

29. *osteo* = bone + *malacia* = softening

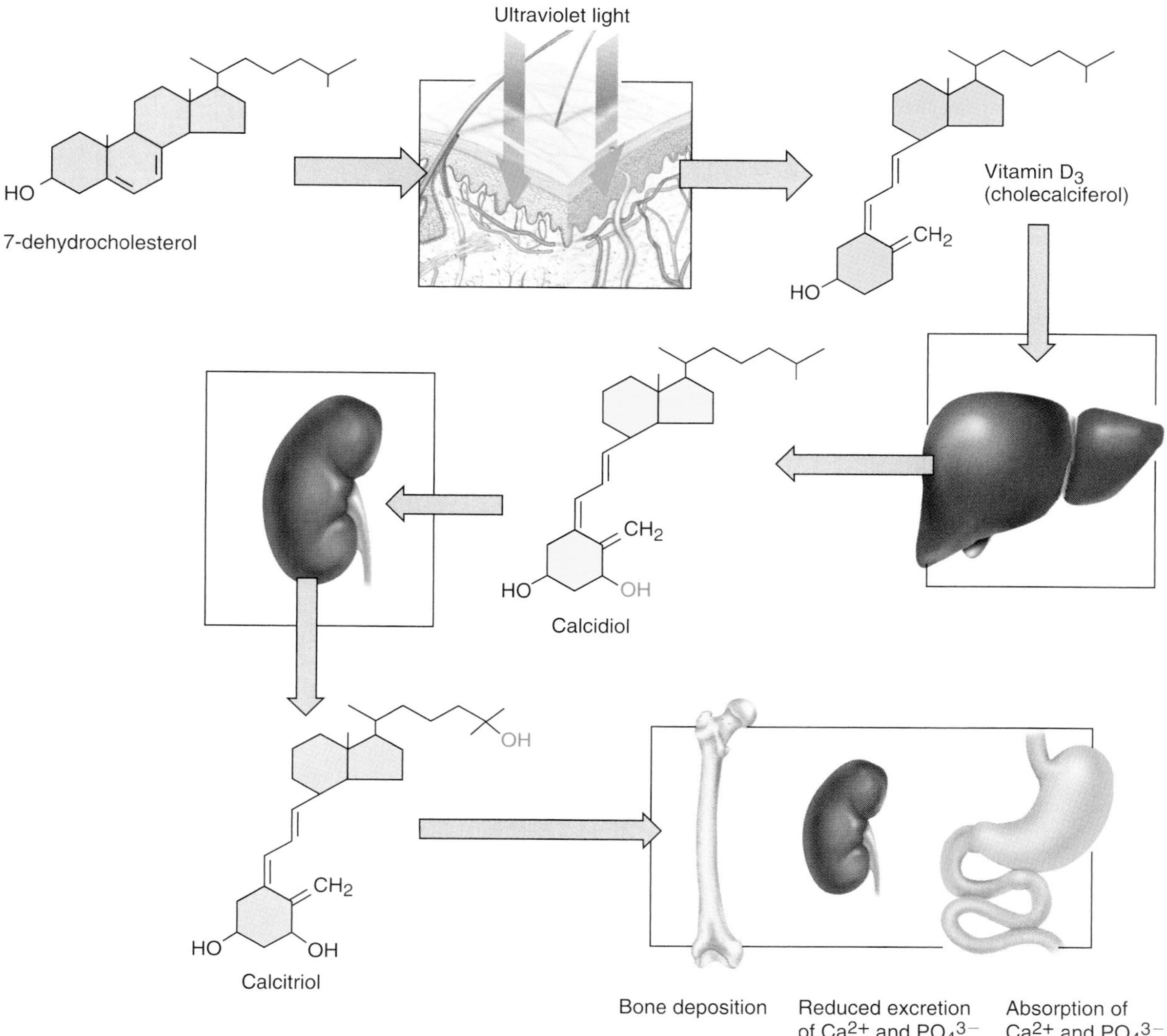

Figure 8.14 Calcitriol Synthesis and Action. Starting at the *upper left*, ultraviolet rays penetrating into the dermis transform 7-dehydrocholesterol in the blood into vitamin D_3. The liver adds an $-OH$ group to vitamin D_3, which converts it to calcidiol; the kidney adds another, which converts this to calcitriol, the most potent form of vitamin D. Calcitriol acts on the bones, kidneys, and small intestine to raise blood calcium and phosphate levels and promote bone deposition.

Calcitonin plays an important role in children but has little effect in most adults. The osteoclasts of children are highly active in skeletal remodeling and release 5 g or more of calcium into the blood each day. By inhibiting this activity, calcitonin can significantly lower the blood calcium level in children. In adults, however, the osteoclasts release only about 0.8 g of calcium per day. Calcitonin cannot change adult blood calcium very much by suppressing this minor contribution. Calcitonin deficiency is not known to cause any adult disease. Calcitonin may, however, prevent bone loss in pregnant and lactating women, and it is useful for reducing bone loss in osteoporosis (see clinical insight 8.4 at the end of the chapter).

Parathyroid Hormone

Parathyroid hormone (PTH) is secreted by the parathyroid glands, which adhere to the posterior surface of the thyroid gland (see fig. 17.12). These glands release PTH when the blood calcium is too low. A mere 1% drop in the blood calcium level doubles the secretion of PTH. PTH raises the blood calcium level by four mechanisms (fig. 8.15*b*):

1. **Osteoblast inhibition.** PTH binds to receptors on the osteoblasts and inhibits their bone-depositing activity, so they remove less Ca^{2+} from the blood.
2. **Osteoclast stimulation.** When stimulated by PTH, the osteoblasts secrete another chemical messenger,

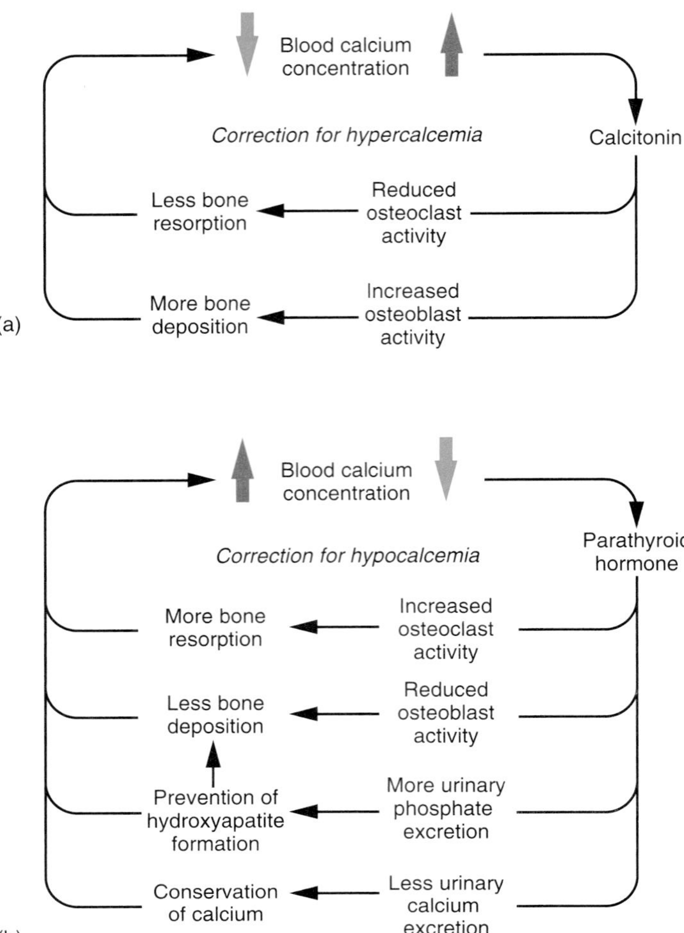

Figure 8.15 Negative Feedback Loops in Calcium Homeostasis. (a) The correction of hypercalcemia by calcitonin. (b) The correction of hypocalcemia by parathyroid hormone.

osteoclast-stimulating factor, that stimulates osteoclasts to multiply rapidly and dissolve osseous tissue.

3. **Urinary excretion.** PTH reduces calcium excretion and increases phosphate excretion in the urine. By holding down the plasma phosphate level, PTH prevents the deposition of calcium phosphate in the bone and ensures that Ca^{2+} remains dissolved in the blood.
4. **Calcitriol synthesis.** PTH stimulates the production of an enzyme in the kidneys that carries out the last step in calcitriol synthesis. As you have already seen, calcitriol raises the blood calcium level.

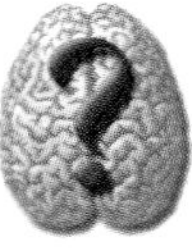

Think About It

Parathyroid hormone does not promote intestinal phosphate absorption. Explain why it would defeat its purpose if it did.

Other Factors Affecting Bone

At least 20 more hormones, growth factors, and vitamins affect osseous tissue in complex ways that are still not well understood (table 8.2). Bone growth is especially rapid in puberty and adolescence, when surges of growth hormone, estrogen, and testosterone promote ossification. These hormones stimulate rapid multiplication of osteogenic cells, matrix deposition by osteoblasts, and multiplication and hypertrophy of the chondrocytes in the metaphyses. Adolescent girls grow faster than boys and attain their full height earlier, not only because they begin puberty earlier but also because estrogen has a stronger effect than testosterone. Since males grow for a longer time, however, they usually grow taller. Growth ceases when the cartilage of the epiphyseal plates is depleted. At this time, the epiphyseal plates are said to "close," the epiphysis and diaphysis of the bone fuse, and the bones can no longer increase in length. Excessive or deficient secretion of these steroids can therefore cause abnormalities ranging from stunted growth to very tall stature (see chapter 17). The use of anabolic steroids by adolescent athletes can cause premature closure of the epiphyseal plates and result in abnormally short adult stature.

Before You Go On

Answer the following questions to test your understanding of the preceding section:

13. Describe the role of collagen and seed crystals in bone mineralization.
14. Why is it important to regulate blood calcium concentration within such a narrow range?
15. What effect does calcitonin have on blood calcium concentration, and how does it produce this effect? Answer the same questions for parathyroid hormone.
16. How is vitamin D synthesized, and what effect does it have on blood calcium concentration?

Bone Disorders

Objectives

When you have completed this section, you should be able to

- name and describe several bone diseases;
- name and describe the types of fractures;
- explain how a fracture is repaired; and
- discuss some clinical treatments for fractures and other skeletal disorders.

Most people probably give little thought to their skeletal system unless they break a bone. This section will describe

Table 8.2 Agents Affecting Calcium and Bone Metabolism

Name	Effect
Hormones	
Parathyroid hormone	Indirectly activates osteoclasts, which resorb bone and raise blood Ca^{2+} concentration; inhibits urinary Ca^{2+} excretion; promotes urinary PO_4^{3-} excretion, calcitriol synthesis, and intestinal Ca^{2+} absorption
Calcitonin	Little effect in adults; promotes mineralization and lowers blood Ca^{2+} concentration in children; may prevent bone loss in pregnant and lactating women
Calcitriol (vitamin D)	Promotes intestinal absorption of Ca^{2+} and PO_4^{3-}; reduces urinary excretion of both; promotes mineralization; stimulates osteoclast activity
Growth hormone	Stimulates bone elongation and cartilage proliferation at epiphyseal plate; increases urinary excretion of Ca^{2+} but also increases intestinal Ca^{2+} absorption, which compensates for the loss
Glucocorticoids	Inhibit osteoclast activity, but if secreted in excess (Cushing disease) can cause osteoporosis by reducing bone deposition (inhibiting cell division and protein synthesis)
Thyroid hormone	Essential to bone growth; enhances effects of growth hormone, but excesses can cause hypercalcemia, increased Ca^{2+} excretion in urine, and osteoporosis
Estrogen	Stimulates osteoblasts and prevents osteoporosis
Testosterone	Stimulates osteoblasts and promotes protein synthesis, thus promoting epiphyseal growth and closure
Insulin	Stimulates bone formation; significant bone loss occurs in untreated diabetes mellitus
Growth Factors	At least 12 hormonelike substances produced in bone itself that stimulate neighboring bone cells, promote collagen synthesis, stimulate epiphyseal growth, and produce many other effects
Vitamins	
Vitamin A	Promotes glycosaminoglycan (chondroitin sulfate) synthesis
Vitamin C (ascorbic acid)	Promotes collagen cross-linking, bone growth, and fracture repair
Vitamin D	Normally functions as a hormone (see calcitriol)

bone fractures, their healing, and their treatment, followed by a summary of other bone diseases and a clinical insight on osteoporosis.

Fractures and Their Repair

There are multiple ways of classifying bone fractures. A **stress fracture** is a break caused by abnormal trauma to a bone, such as fractures incurred in falls, athletics, and military combat. A **pathologic fracture** is a break in a bone weakened by some other disease, such as bone cancer or osteoporosis, usually caused by a stress that would not normally fracture a bone. Fractures are also classified according to the direction of the fracture line, whether or not the skin is broken, and whether a bone is merely cracked or broken into separate pieces (table 8.3; fig. 8.16).

The Healing of Fractures

An uncomplicated fracture heals in about 8 to 12 weeks, but complex fractures take longer and all fractures heal more slowly in older people. The healing process occurs in the following stages (fig. 8.17):

1. **Hematoma formation.** A bone fracture also breaks blood vessels of the bone and periosteum, causing bleeding and the formation of a clot *(fracture hematoma).*
2. **Formation of granulation tissue. Granulation tissue** is a soft, fibrous tissue produced as a hematoma is infiltrated by blood capillaries and fibroblasts. Macrophages, osteoclasts, and osteogenic cells also invade the tissue from the periosteal and medullary sides of the fracture. Osteogenic cells build up in numbers within 48 hours of the injury.
3. **Callus[30] formation.** Fibroblasts deposit collagen in the granulation tissue, while some osteogenic cells become chondroblasts and produce patches of fibrocartilage called **soft callus** tissue. Osteogenic cells also differentiate into osteoblasts that produce a bony collar, the **hard callus,** which they cement to

30. *call* = hard, tough

Table 8.3 Classification of Fractures

Type	Description
Closed	Skin is not broken (formerly called a *simple* fracture)
Open	Skin is broken; bone protrudes through skin or wound extends to fractured bone (formerly called a *compound* fracture)
Complete	Bone is broken into two or more pieces
Incomplete	Partial fracture that extends only partway across bone; pieces remain joined
Greenstick	Bone is bent on one side and has incomplete fracture on opposite side
Hairline	Fine crack in which sections of bone remain aligned; common in skull
Comminuted	Bone is broken into three or more pieces
Displaced	The portions of a fractured bone are out of anatomical alignment
Nondisplaced	The portions of bone are still in correct anatomical alignment
Impacted	One bone fragment is driven into the medullary space or spongy bone of the other
Depressed	Broken portion of bone forms a concavity, as in skull fractures
Linear	Fracture parallel to long axis of bone
Transverse	Fracture perpendicular to long axis of bone
Oblique	Diagonal fracture, between linear and transverse
Spiral	Fracture spirals around axis of long bone, the result of a twisting stress, often produced when an abusive adult roughly picks a child up by the arm
Epiphyseal	Epiphysis is separated from the diaphysis along the epiphyseal plate; seen in juveniles
Avulsion	Body part (such as a finger) is completely severed
Colles[31]	Fracture of the distal end of the radius and ulna; common in osteoporosis
Pott[32]	Fracture at the distal end of the tibia, fibula, or both; a common sports injury

31. Abraham Colles (1773–1843), Irish surgeon
32. Sir Percivall Pott (1713–88), British surgeon

the dead bone around the injury. This collar acts as a temporary splint to join the broken ends or fragments of the bone to each other. It takes about 4 to 6 weeks for a hard callus to form. During this period, it is important that a broken bone be immobilized by traction or a cast to prevent reinjury.

4. **Remodeling.** The hard callus persists for 3 to 4 months as osteoclasts dissolve small fragments of broken bone and osteoblasts bridge the gap between the broken ends with spongy bone. This spongy bone is subsequently remodeled into compact bone. Usually the fracture leaves a slight thickening of the bone visible by X ray, but in some cases healing is so complete that no trace of the fracture can be found.

The Treatment of Fractures

Most fractures are set by **closed reduction,** a procedure in which the bone fragments are manipulated into their normal positions without surgery. **Open reduction** involves the surgical exposure of the bone and the use of plates, screws, or pins to realign the fragments (fig. 8.18*b*). To stabilize the bone during healing, fractures have traditionally been set in plaster casts. Fiberglass casts have become more common, however, because they are lighter, stronger, washable, porous, and cause less itching. Traction is used to treat fractures of the femur in children. It aids in the alignment of the bone fragments by overriding the force of the strong thigh muscles. Traction is rarely used for elderly patients, however, because the risks from long-term confinement to bed outweigh the benefits. Hip fractures are usually pinned and early ambulation (walking) is encouraged because it promotes blood circulation and healing. Fractures that take longer than 2 months to heal may be treated with electrical stimulation, which accelerates repair by suppressing the effects of parathyroid hormone.

Orthopedics[33] is the branch of medicine that deals with the prevention and correction of injuries and disorders

33. *ortho* = straight + *ped* = child, foot

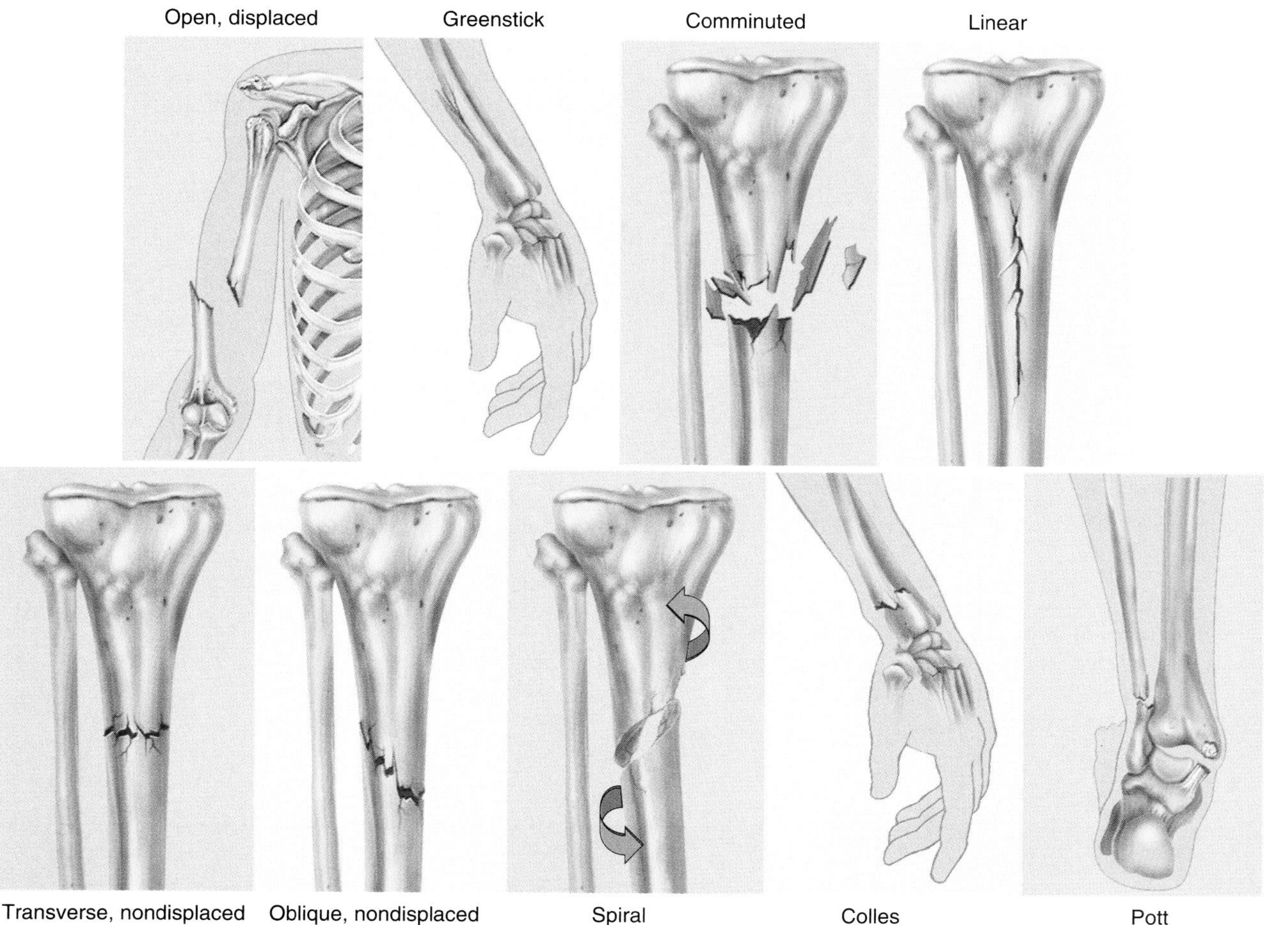

Figure 8.16 Some Types of Bone Fractures.

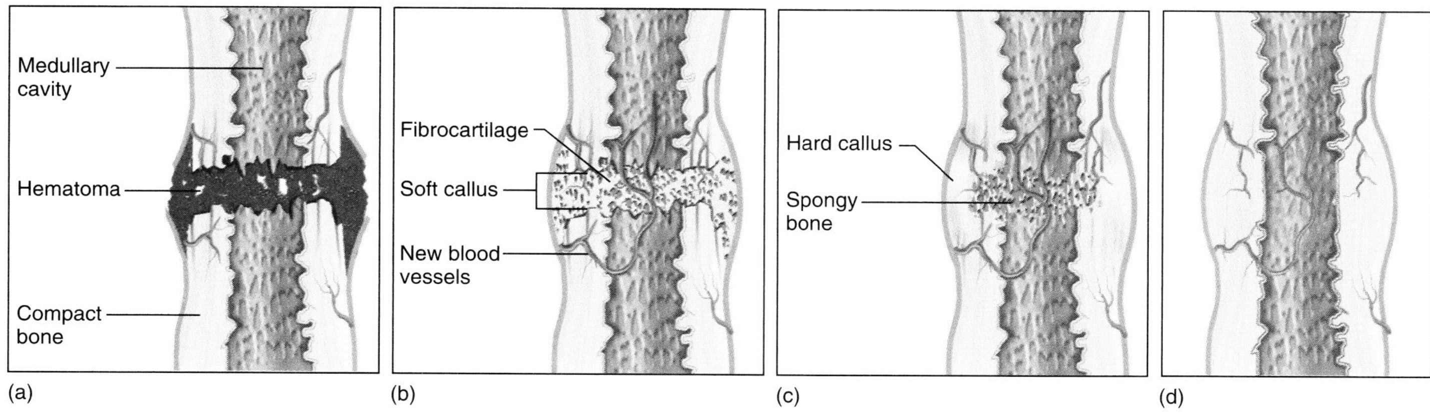

Figure 8.17 **The Healing of a Bone Fracture.** (*a*) Blood vessels are broken at the fracture line; blood clots and forms a fracture hematoma. (*b*) Blood vessels grow into the clot and a soft callus of fibrocartilage forms. (*c*) Mineral deposition hardens the soft callus and converts it to a hard callus of spongy bone. (*d*) Osteoclasts remove excess tissue from the hard callus and the bone eventually resembles its original appearance.

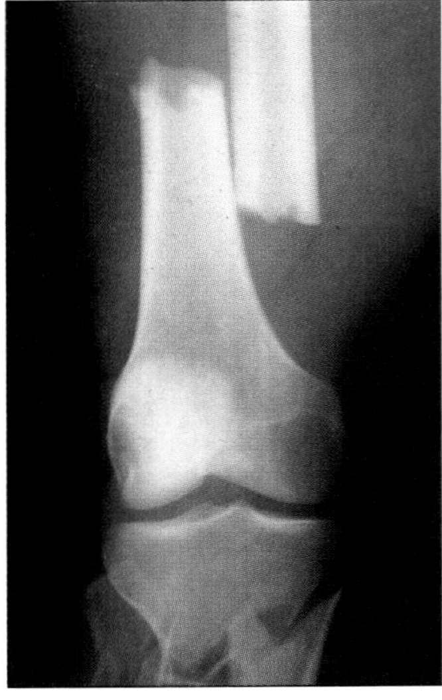

(a)

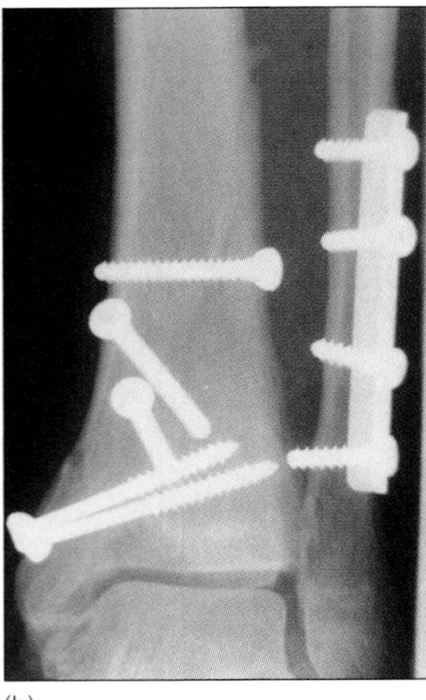

(b)

Figure 8.18 X Rays of Bone Fractures. (*a*) A displaced fracture of the femur. (*b*) An ankle fracture involving both the tibia and fibula. This fracture has been set by open reduction, a process of surgically exposing the bone and realigning the fragments with plates and screws.

of the bones, joints, and muscles. As the word suggests, this field originated as the treatment of skeletal deformities in children, but it is now much more extensive. It includes the design of artificial joints and limbs and the treatment of athletic injuries.

Other Bone Disorders

Several additional bone disorders are summarized in table 8.4. The most common bone disease, **osteoporosis,**[34] receives special consideration in clinical insight 8.4.

Before You Go On

Answer the following questions to test your understanding of the preceding section:

17. Name and describe any five types of bone fractures.
18. Why would osteomyelitis be more likely to occur in an open fracture than in a closed fracture?
19. What is a callus? How does it contribute to fracture repair?

34. *osteo* = bone + *por* = porous + *osis* = condition

Table 8.4	Bone Diseases
Rickets	Defective mineralization of bone in children, usually as a result of insufficient sunlight or vitamin D, sometimes due to a dietary deficiency of calcium or phosphate or to liver or kidney diseases that interfere with calcitriol synthesis. Causes bone softening and deformity, especially in the weight-bearing bones of the lower limbs.
Osteomalacia	Adult form of rickets, most common in poorly nourished women who have had multiple pregnancies. Bones become softened, deformed, and more susceptible to fractures.
Osteoporosis	Loss of bone mass, especially spongy bone, usually as a result of lack of exercise or a deficiency of estrogen after menopause. It results in increasing brittleness and susceptibility to fractures (see clinical insight 8.4 for details).
Osteitis Deformans (Paget[35] disease)	Excessive proliferation of osteoclasts and resorption of excess bone, with osteoblasts attempting to compensate by depositing extra bone. This results in rapid, disorderly bone remodeling and weak, deformed bones. Osteitis deformans usually passes unnoticed, but in some cases it causes pain, disfiguration, and fractures. It is most common in males over the age of 50.
Osteomyelitis[36]	Inflammation of osseous tissue and bone marrow as a result of bacterial infection. This disease was often fatal before the discovery of antibiotics and is still very difficult to treat.
Osteoma[37]	A benign bone tumor, especially in the flat bones of the skull; may grow into the orbits or sinuses.
Osteochondroma[38]	A benign tumor of bone and cartilage; often forms bone spurs at the ends of long bones.
Osteosarcoma[39] (osteogenic sarcoma)	The most common and deadly form of bone cancer. It occurs most often in the tibia, femur, and humerus of males between the ages of 10 and 25. In 10% of cases, it metastasizes to the lungs or other organs; if untreated, death occurs within 1 year.
Chondrosarcoma	A slow-growing cancer of hyaline cartilage, most common in middle age. It requires surgical removal; chemotherapy is ineffective.

35. Sir James Paget (1814–99), English surgeon
36. *osteo* = bone + *myel* = marrow + *itis* = inflammation
37. *oste* = bone + *oma* = tumor
38. *osteo* = bone + *chondr* = cartilage + *oma* = tumor
39. *osteo* = bone + *sarc* = flesh + *oma* = tumor

Clinical Insight 8.4

Osteoporosis

The most common bone disease is osteoporosis (OSS-tee-oh-pore-OH-sis), literally, "porous bones." This is a disease in which the bones lose mass and become increasingly brittle and subject to fractures. It involves loss of both organic matrix and minerals, and it affects spongy bone in particular, since this is the most metabolically active type (fig. 1). The highest incidence of osteoporosis is among elderly white women, where it is closely linked to age and menopause. Osteoporosis also affects men (white males somewhat more than black), although less severely than it does women. It rarely affects black women.

Fractures are the most serious consequence of osteoporosis. They occur especially in the hip, wrist, and vertebral column and under stresses as slight as sitting down too quickly. Hip fractures usually occur at the neck of the femur, while wrist fractures occur at the distal end of the radius and ulna (Colles fracture). Among the elderly, hip fractures lead to fatal complications such as pneumonia in 12% to 20% of cases; they involve a long, costly recovery for half of those who survive. As the weight-bearing bodies of the vertebrae lose spongy bone, they become compressed like marshmallows. Consequently, many people lose height after middle age, and in some women, the spine becomes deformed into a "widow's (dowager's) hump," a condition called *kyphosis* (fig. 2).

Postmenopausal white women are at greatest risk for osteoporosis because women have less bone mass than men to begin with, begin losing it earlier (starting around age 35), and lose it faster than men do. By age 70, the average white woman has lost 30% of her bone mass, and some have lost as much as 50%. Young black women develop more bone mass than whites. Although they, too, lose bone mass after menopause, the loss usually does not reach the threshold for osteoporosis and pathologic fractures. In men, bone loss begins around age 60 and seldom exceeds 25%. Estrogen stimulates osteoblasts and bone deposition, but the ovaries stop producing estrogen after menopause. Ironically, osteoporosis also occurs among young female runners and dancers in spite of their vigorous exercise. Their percentage of body fat is so low that they stop ovulating and the ovaries secrete unusually low levels of estrogen.

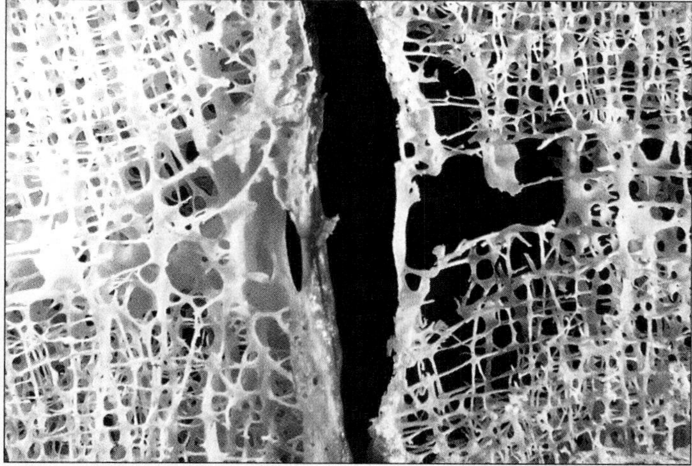

Figure 1 Osteoporosis. The *left* side of the photo is a section through a healthy lumbar vertebra. The *right* side shows a lumbar vertebra in which much of the spongy bone has been lost to osteoporosis.

Figure 2 Woman with osteoporosis. Note the abnormal curvature (kyphosis, widow's hump, or dowager's hump) of the upper thoracic spine. This results from compression fractures of the weakened vertebrae.

Continued

Estrogen replacement therapy cannot reverse osteoporosis, but it can slow its progress. Alternatives to estrogen therapy are becoming available, but each has its own undesirable side effects. Some patients are now treated with a calcitonin nasal spray. Milk and other calcium sources and moderate exercise can also slow the progress of osteoporosis, but only slightly.

As is so often true, an ounce of prevention is worth a pound of cure. The time to minimize the risk for osteoporosis is between the ages of 25 and 40, when the skeleton is building to its maximum mass. The more bone mass a person has going into middle age, the less he or she will be affected by osteoporosis later. Ample exercise and calcium intake (850–1,000 mg/day) are the best preventative measures.

Although osteoporosis has become a major public health problem because of the increasing age of the population, it is not limited to the elderly. *Disuse osteoporosis* can occur at any age as a result of immobilization or inadequate weight-bearing exercise. In the absence of preventive exercise, astronauts on prolonged microgravity missions have experienced disuse osteoporosis. Other risk factors include smoking, low calcium and protein intake, vitamin C deficiency, and diabetes mellitus.

Chapter Review

Online Learning Center

Study resources for this chapter are available at www.mhhe.com/saladin/

Study Outline

Tissues and Organs of the Skeletal System (p. 232)

1. Tissues of the skeletal system
2. Functions of the skeleton (table 8.1)
3. Bones and osseous tissue
4. Shapes of bones
5. General features of bones

Histology of Osseous Tissue (p. 235)

1. Cells
   a. Osteogenic cells
   b. Osteoblasts
   c. Osteocytes
   d. Osteoclasts
2. Matrix
   a. Collagen
   b. Other organic components
   c. Hydroxyapatite
   d. Other minerals
3. Compact bone
   a. Lamellae
   b. Haversian canal
   c. Perforating canals
4. Spongy bone
5. Bone marrow
   a. Red marrow
   b. Yellow marrow
   c. Gelatinous marrow

Bone Development (p. 240)

1. Intramembranous ossification
   a. Condensation of mesenchyme
   b. Formation of osteoid tissue
   c. Calcification
2. Endochondral ossification
   a. Cartilage model
   b. Primary ossification center
   c. Primary marrow space
   d. Metaphysis
      - Zone of reserve cartilage
      - Zone of cell proliferation
      - Zone of cell hypertrophy
      - Zone of calcification
      - Zone of bone deposition
   e. Secondary ossification center
3. Growth and remodeling
   a. Response to stress on bone
   b. Appositional growth

Physiology of Osseous Tissue (p. 243)

1. Mineral deposition
   a. Role of osteoblasts
   b. Solubility product
   c. Ectopic ossification
2. Mineral resorption
   a. Role of hydrochloric acid
   b. Role of acid phosphatase
3. Calcium and phosphate homeostasis
   a. Functions of calcium and phosphate
   b. Calcium and phosphate in the skeleton
   c. Calcium and phosphate in the blood
   d. Calcium imbalances
      - Hypocalcemia
      - Hypercalcemia
   e. Calcium homeostasis
      - Calcitriol
      - Calcitonin
      - Parathyroid hormone
4. Other factors affecting bone (table 8.2)

Bone Disorders (p. 248)

1. Fractures and their repair
   a. Types of fractures
   b. Healing of fractures
      - Fracture hematoma
      - Granulation tissue
      - Callus formation
      - Remodeling
   c. Treatment of fractures
2. Other bone disorders (table 8.4)

Selected Vocabulary

For additional terminology not listed here, see tables 8.3 and 8.4.

osteology 232
skeletal system 232
osseous tissue 232
mineralization 232
calcification 232
long bone 232
short bone 233
flat bone 233
irregular bone 233
compact bone 233
medullary cavity 233
spongy bone 233
diploe 233
diaphysis 233
epiphysis 233
articular cartilage 234
nutrient foramina 234
periosteum 234
perforating fibers 234
endosteum 234
epiphyseal plate 234
osteogenic cell 235
osteoblast 235
osteocyte 235
lacunae 235
canaliculi 235
osteoclast 235
hydroxyapatite 235
lamella 236
haversian canal 236
osteon 236
perforating canal 236
trabeculae 237
red bone marrow 239
yellow bone marrow 239
gelatinous bone marrow 239
intramembranous ossification 240
osteoid tissue 240
endochondral ossification 240
primary ossification center 240
primary marrow space 241
metaphysis 241
zone of reserve cartilage 241
zone of cell proliferation 241
zone of cell hypertrophy 241
zone of calcification 241
zone of bone deposition 241
secondary ossification center 242
interstitial growth 243
appositional growth 243
mineral deposition 244
solubility product 244
ectopic ossification 244
calculus 244
mineral resorption 245
acid phosphatase 245
hypocalcemia 245
tetanus 245
hypercalcemia 245
calcitriol 246
rickets 246
osteomalacia 246
calcitonin 246
parathyroid hormone 247
stress fracture 249
pathologic fracture 249
granulation tissue 249
soft callus 249
hard callus 249
closed reduction 250
open reduction 250
orthopedics 250
osteoporosis 255

Testing Your Recall

1. Which cells have a ruffled border and secrete hydrochloric acid?
   a. C cells
   b. osteocytes
   c. osteogenic cells
   d. osteoblasts
   e. osteoclasts

2. The medullary cavity of an adult bone may contain
   a. myeloid tissue.
   b. hyaline cartilage.
   c. periosteum.
   d. osteocytes.
   e. articular cartilages.

3. The spurt of growth in puberty results from cell proliferation and hypertrophy in
   a. the epiphysis.
   b. the epiphyseal line.
   c. the dense bone.
   d. the epiphyseal plate.
   e. the spongy bone.

4. Osteoclasts are most closely related, by common descent, to
   a. osteoprogenitor cells.
   b. osteogenic cells.
   c. monocytes.
   d. fibroblasts.
   e. osteoblasts.

5. The walls between cartilage lacunae break down in the zone of
   a. cell proliferation.
   b. calcification.
   c. reserve cartilage.
   d. bone deposition.
   e. cell hypertrophy.

6. Which of these is *not* an effect of PTH?
   a. rise in blood phosphate level
   b. reduction of calcium excretion
   c. increased intestinal calcium absorption
   d. increased number of osteoclasts
   e. increased calcitriol synthesis

7. A child jumps to the ground from the top of a playground "jungle gym." His leg bones do not shatter mainly because they contain
   a. an abundance of glycosaminoglycans.
   b. young, resilient osteocytes.
   c. an abundance of calcium phosphate.
   d. collagen fibers.
   e. hydroxyapatite crystals.

8. One long bone meets another at its
   a. diaphysis.
   b. epiphyseal plate.
   c. periosteum.
   d. metaphysis.
   e. epiphysis.

9. Calcitriol is made from
   a. calcitonin.
   b. dehydrocholesterol.
   c. hydroxyapatite.
   d. estrogen.
   e. PTH.

10. One sign of osteoporosis is
    a. osteosarcoma.
    b. osteomalacia.
    c. osteomyelitis.
    d. a Pott fracture.
    e. hypocalcemia.

11. Calcium phosphate crystallizes in bone as a mineral called __________ .

12. Osteocytes contact each other through channels called __________ in the bone matrix.

13. A bone increases in diameter only by __________ growth, the addition of new surface osteons.

14. Seed crystals of hydroxyapatite form only when the levels of calcium and phosphate in the tissue fluid exceed the __________ .

15. A calcium deficiency called __________ can cause death by suffocation.
16. __________ are cells that secrete collagen and stimulate calcium phosphate deposition.
17. The most active form of vitamin D, produced mainly by the kidneys, is __________.
18. The most common bone disease is __________.
19. The transitional region between epiphyseal cartilage and the primary marrow cavity of a young bone is called the __________.
20. A pregnant, poorly nourished woman may suffer a softening of the bones called __________.

Answers in appendix B

Testing Your Comprehension

1. Most osteocytes of an osteon are far removed from blood vessels, but are still able to respond to blood-borne hormones. Explain how this is possible.
2. What positive feedback loop can you recognize in the process of bone deposition?
3. How does the regulation of blood calcium concentration exemplify negative feedback and homeostasis?
4. Describe how the arrangement of trabeculae in spongy bone demonstrates the unity of form and function.
5. Identify two bone diseases you would expect to see if the epidermis were a completely effective barrier to UV radiation and a person took no dietary supplements to compensate for this. Explain your answer.

Answers at the Online Learning Center

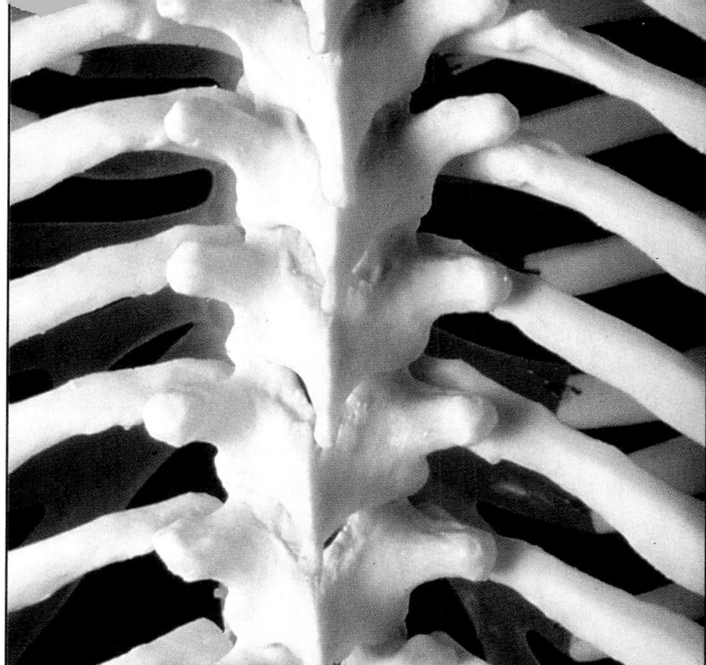

Vertebral column and ribs

CHAPTER 9

The Skeletal System

Chapter Outline

Insights

 Study with Your ESP CD-ROM

Refer to your ESP Tutorial CD-ROM for helpful animations, activities, and quizzes related to the material covered in this chapter. The following are suggested areas to go to on the CD and are presented according to "Unit/Topic":
- Support and Movement/Skeletal System

Brushing Up

To understand this chapter, it is important that you understand or brush up on the following concepts:
- Directional terminology (p. 31)
- Body regions and cavities (pp. 31–37)

Knowledge of skeletal anatomy will be useful as you study later chapters. It provides a foundation for studying the gross anatomy of other organ systems because many organs are named for their relationships to nearby bones. The subclavian artery and vein, for example, are located beneath the clavicles; the temporalis muscle is attached to the temporal bone; the ulnar nerve and radial artery travel beside the ulna and radius of the forearm; and the frontal, parietal, temporal, and occipital lobes of the brain are named for bones of the cranium. An understanding of how the muscles produce body movements also depends on knowledge of skeletal anatomy. Additionally, the positions, shapes, and processes of bones can serve as landmarks for a clinician in determining where to give an injection or record a pulse, what to look for in an X ray, or how to perform physical therapy and other medical procedures.

Overview of the Skeleton

Objectives

When you have completed this section, you should be able to

- state the approximate number of bones in the adult body;
- explain why this number varies with age and from one person to another; and
- define several terms that denote surface features of bones.

Students typically begin by examining an articulated[1] skeleton (dried bones held together by wires and rods to show their spatial relationships to each other) or a disarticulated skeleton (one that is taken apart so that the anatomy of individual bones can be studied in more detail). The skeleton is shown in figure 9.1. Note that it is divided into two regions: the **axial skeleton,** which forms the central supporting axis, and the **appendicular skeleton,** which includes the limbs (*appendages* or *extremities*) and the bones that attach them to the axial skeleton. The axial skeleton includes the skull, vertebral column, and thoracic cage (ribs and sternum). The appendicular skeleton includes the bones of the upper limb and pectoral girdle and the bones of the lower limb and pelvic girdle.

Bones of the Skeletal System

It is often stated that there are 206 bones in the skeleton, but this is only a typical adult count. At birth there are about 270, and even more bones form during childhood. With age, however, the number decreases as separate bones gradually fuse. For example, each half of the adult pelvis is a single bone called the *os coxae* (oss COC-see), which results from the fusion of three childhood bones—the ilium, ischium, and pubis. The fusion of several bones, completed by late adolescence or the early 20s, brings about the average adult number of 206. These bones are listed in table 9.1

This number varies even among adults. One reason is the development of **sesamoid**[2] **bones**—bones that form within some tendons in response to stress. The patella (kneecap) is the largest of these; most of the others are small, rounded bones in such locations as the knuckles. Another reason for adult variation is that some people have extra bones in the skull called **sutural** (SOO-chure-ul), or **wormian,**[3] **bones** (see fig. 9.6).

Surface Features of Bones

The surface of a bone may exhibit a variety of ridges, spines, bumps, depressions, canals, pores, slits, and articular surfaces, often called *surface markings.* It is important to know the names of these features because later descriptions of joints, muscle attachments, and the routes traveled by nerves and blood vessels are based on this terminology. The terms for the most common of these features are listed in table 9.2, and several of them are illustrated in figure 9.2.

The rest of this chapter is divided into four sections: (1) the skull, (2) the vertebral column and thoracic cage, (3) the pectoral girdle and upper limb, and (4) the pelvic girdle and lower limb. At the end of each section, you will find a review and checklist of skeletal features you should know (see tables 9.4, 9.6, 9.7, and 9.9). As you study this chapter, use yourself as a model. You can easily palpate (feel) many of the bones and some of their details through the skin. Rotate your forearm, cross your legs, palpate your skull, and think about what is happening beneath the surface or what you can feel through the skin. You will gain the most from this chapter (and indeed, the entire book) if you are conscious of your own body in relation to what you are studying.

Before You Go On

Answer the following questions to test your understanding of the preceding section:

1. Name the major components of the axial skeleton. Name those of the appendicular skeleton.
2. Explain why an adult does not have as many bones as a child does. Explain why one adult may have more bones than another adult of the same age has.
3. Briefly describe each of the following bone features: a condyle, epicondyle, process, tubercle, fossa, sulcus, and foramen.

1. *artic* = joint

2. *sesam* = sesame seed + *oid* = resembling

3. Ole Worm (1588–1654), Danish physician

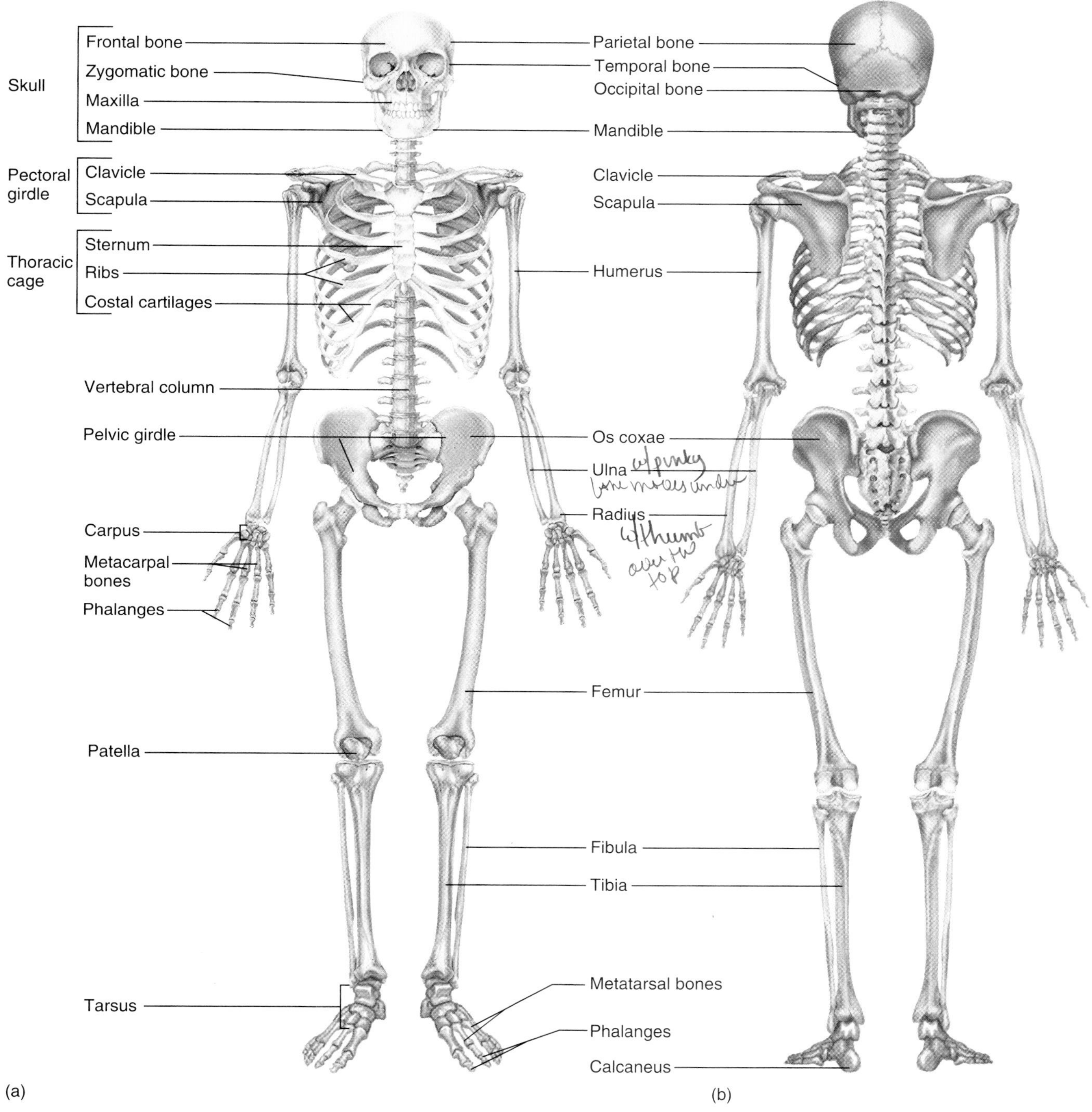

Figure 9.1 The Adult Skeleton. (*a*) Anterior view; (*b*) posterior view. The appendicular skeleton is colored blue, and the rest is axial skeleton.

Table 9.1 Bones of the Adult Skeletal System

Axial Skeleton			
Skull	*Total 22*	*Auditory Ossicles*	*Total 6*
Cranial bones		Malleus (2)	
Frontal bone (1)		Incus (2)	
Parietal bone (2)		Stapes (2)	
Occipital bone (1)		*Hyoid Bone (1)*	*Total 1*
Temporal bone (2)		*Vertebral Column*	*Total 26*
Sphenoid bone (1)		Cervical vertebrae (7)	
Ethmoid bone (1)		Thoracic vertebrae (12)	
Facial bones		Lumbar vertebrae (5)	
Maxilla (2)		Sacrum (1)	
Palatine bone (2)		Coccyx (1)	
Zygomatic bone (2)		*Thoracic Cage*	*Total 25*
Lacrimal bone (2)		Ribs (24)	
Nasal bone (2)		Sternum (1)	
Vomer (1)			
Inferior nasal concha (2)			
Mandible (1)			
Appendicular Skeleton			
Pectoral Girdle	*Total 4*	*Pelvic Girdle*	*Total 2*
Scapula (2)		Os coxa (2)	
Clavicle (2)		*Lower Limb*	*Total 60*
Upper Limb	*Total 60*	Femur (2)	
Humerus (2)		Patella (2)	
Radius (2)		Tibia (2)	
Ulna (2)		Fibula (2)	
Carpals (16)		Tarsals (14)	
Metacarpals (10)		Metatarsals (10)	
Phalanges (28)		Phalanges (28)	
			Grand Total: 206

The Skull

Objectives

When you have completed this section, you should be able to

- name the bones of the skull and their anatomical features; and
- describe the development of the skull from infancy through childhood.

The skull is the most complex part of the skeleton. Figures 9.3 to 9.6 present an overview of its general anatomy. Although it may seem to consist only of the mandible (lower jaw) and "the rest," in reality the skull is composed of 22 bones (and sometimes more). Most of these bones are joined together by immovable joints called **sutures** (SOO-chures), visible as seams on the surface (fig. 9.4). These are important landmarks in the descriptions that follow.

The skull contains several prominent cavities (fig. 9.7). The largest, with an adult volume of 1,300 to 1,350 mL, is the **cranial cavity,** which encloses the brain. Other cavities include the **orbits** (eye sockets), **nasal cavity, buccal** (BUCK-ul) **cavity** (mouth), **middle-** and **inner-ear cavities,** and **paranasal sinuses.** The paranasal sinuses are named for the bones in which they occur (fig. 9.8)—the **frontal, sphenoid, ethmoid,** and **maxillary sinuses.**

Table 9.2 Surface Features (Markings) of Bones

Term	Description and Example
Articulations	
Condyle	A rounded knob (occipital condyles of the skull)
Facet	A smooth, flat, slightly concave or convex articular surface (articular facets of the vertebrae)
Head	The prominent expanded end of a bone, sometimes rounded (head of the femur)
Extensions and Projections	
Process	Any bony prominence (mastoid process of the skull)
Crest	A narrow ridge (iliac crest of the pelvis)
Line	A slightly raised, elongated ridge (nuchal lines of the skull)
Spine	A sharp, slender, or narrow process (spine of the scapula)
Epicondyle	A projection superior to a condyle (medial epicondyle of the femur)
Trochanter	A massive process unique to the femur
Tubercle	A small, rounded process (greater tubercle of the humerus)
Tuberosity	A rough surface (tibial tuberosity)
Depressions	
Alveolus	A pit or socket (tooth socket)
Fossa	A shallow, broad, or elongated basin (mandibular fossa)
Fovea	A small pit (fovea capitis of the femur)
Sulcus	A groove for a tendon, nerve, or blood vessel (intertubercular sulcus of the humerus)
Passages	
Foramen	A hole through a bone, usually round (foramen magnum of the skull)
Canal or meatus	A tubular passage or tunnel through a bone (auditory meatus of the ear)
Fissure	A slit through a bone (orbital fissures behind the eye)
Sinuses	Spaces or cavities within a bone (frontal sinus of the skull)

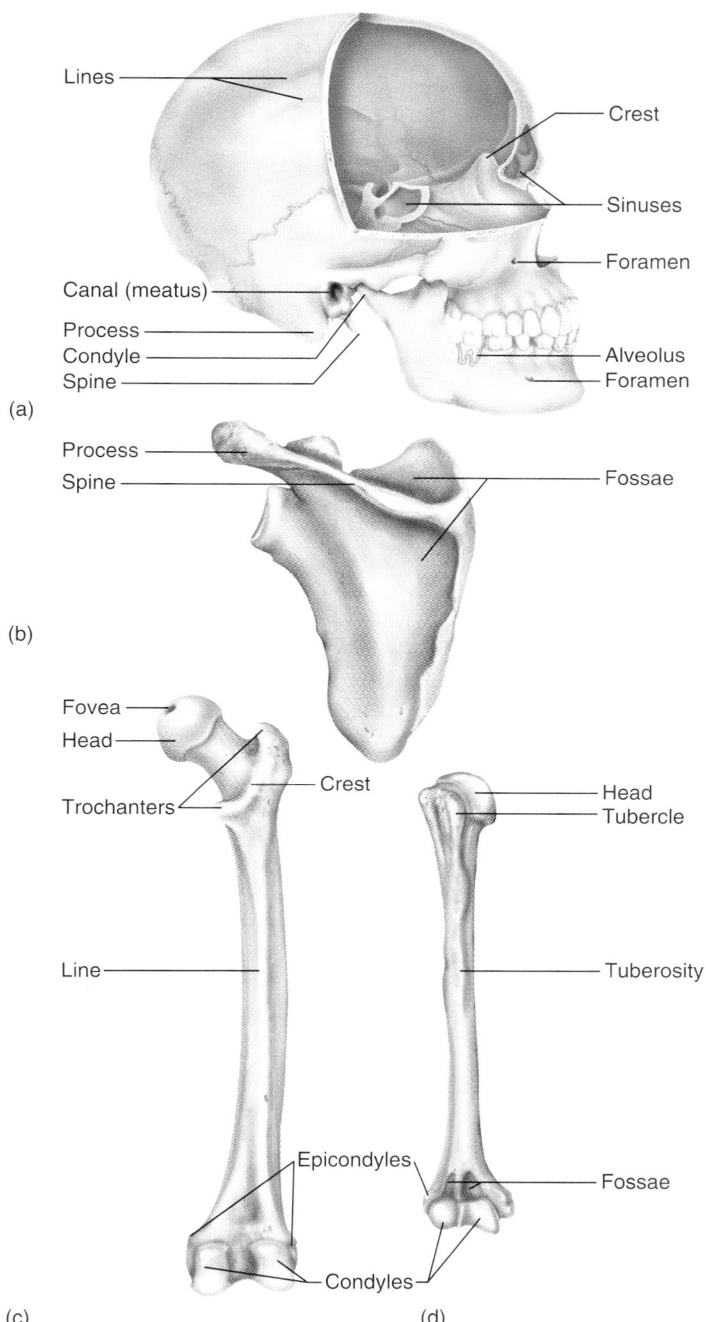

Figure 9.2 Some of the Surface Features of Bones. (*a*) Skull; (*b*) scapula; (*c*) femur; (*d*) humerus.

These cavities are connected with the nasal cavity, lined by a mucous membrane, and filled with air. They lighten the anterior portion of the skull and act as chambers that add resonance to the voice. The latter effect can be sensed in the way your voice changes when you have a cold and mucus obstructs the travel of sound into the sinuses and back.

Bones of the skull have especially conspicuous **foramina**—singular, *foramen* (fo-RAY-men)—holes that allow passage for nerves and blood vessels. The major

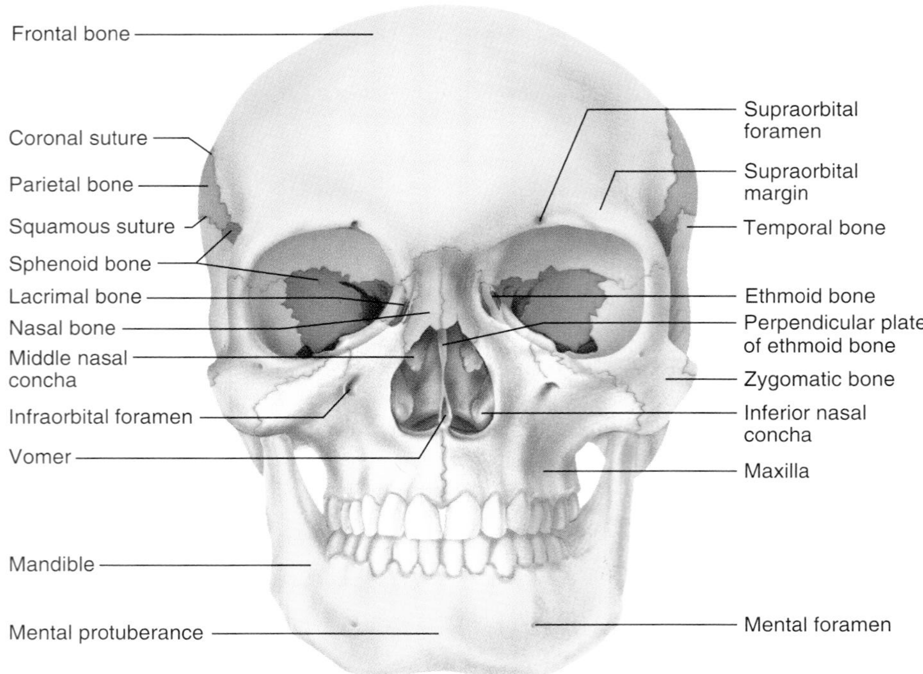

Figure 9.3 The Skull, Anterior View.

foramina are summarized in table 9.3 (p. 266). The details of this table will mean more to you when you study cranial nerves and blood vessels in later chapters.

Cranial Bones

The cranial cavity is enclosed by the **cranium**[4] (braincase), which protects the brain and associated sensory organs. **Cranial bones** are those that directly contact the meninges around the brain. There are eight of these:

1 frontal bone	1 occipital bone
2 parietal bones	1 sphenoid bone
2 temporal bones	1 ethmoid bone

The cranium is a rigid structure with an opening, the **foramen magnum** (literally "large hole"), where the spinal cord enters. An important consideration in treatment of head injuries is swelling of the brain. Since the cranium cannot enlarge, swelling puts pressure on the brain and results in even more tissue damage. Severe swelling may force the brainstem out through the foramen magnum, usually with fatal consequences.

The delicate brain tissue does not come directly into contact with the cranial bones but is separated from them by three membranes called the *meninges* (meh-NIN-jeez) (see chapter 14). The thickest and toughest of these, the *dura mater*[5] (DUE-rah MAH-tur), lies loosely against the inside of the cranium in most places but is firmly attached to it at a few points.

The cranium consists of two major parts—the calvaria and the base. The **calvaria**[6] (skullcap) forms the roof and walls (see fig. 9.6). In study skulls it is often sawed so that part of it can be lifted off for examination of the interior. This reveals the **base** (floor) of the cranial cavity (see fig. 9.5*b*), which is divided into three basins corresponding to the contour of the inferior surface of the brain (fig. 9.9). The relatively shallow **anterior cranial fossa** is shaped like a crescent moon and accommodates the frontal lobes of the brain. The **middle cranial fossa,** which drops abruptly deeper, is shaped like a pair of outstretched bird's wings and accommodates the temporal lobes. The **posterior cranial fossa** is deepest and houses a large posterior division of the brain called the cerebellum.

We now consider the eight cranial bones and their distinguishing features.

Frontal Bone

The **frontal bone** extends from the forehead back to a prominent *coronal suture,* which crosses the crown of

4. *crani* = helmet

5. *dura* = tough, strong + *mater* = mother

6. *calvar* = bald, skull

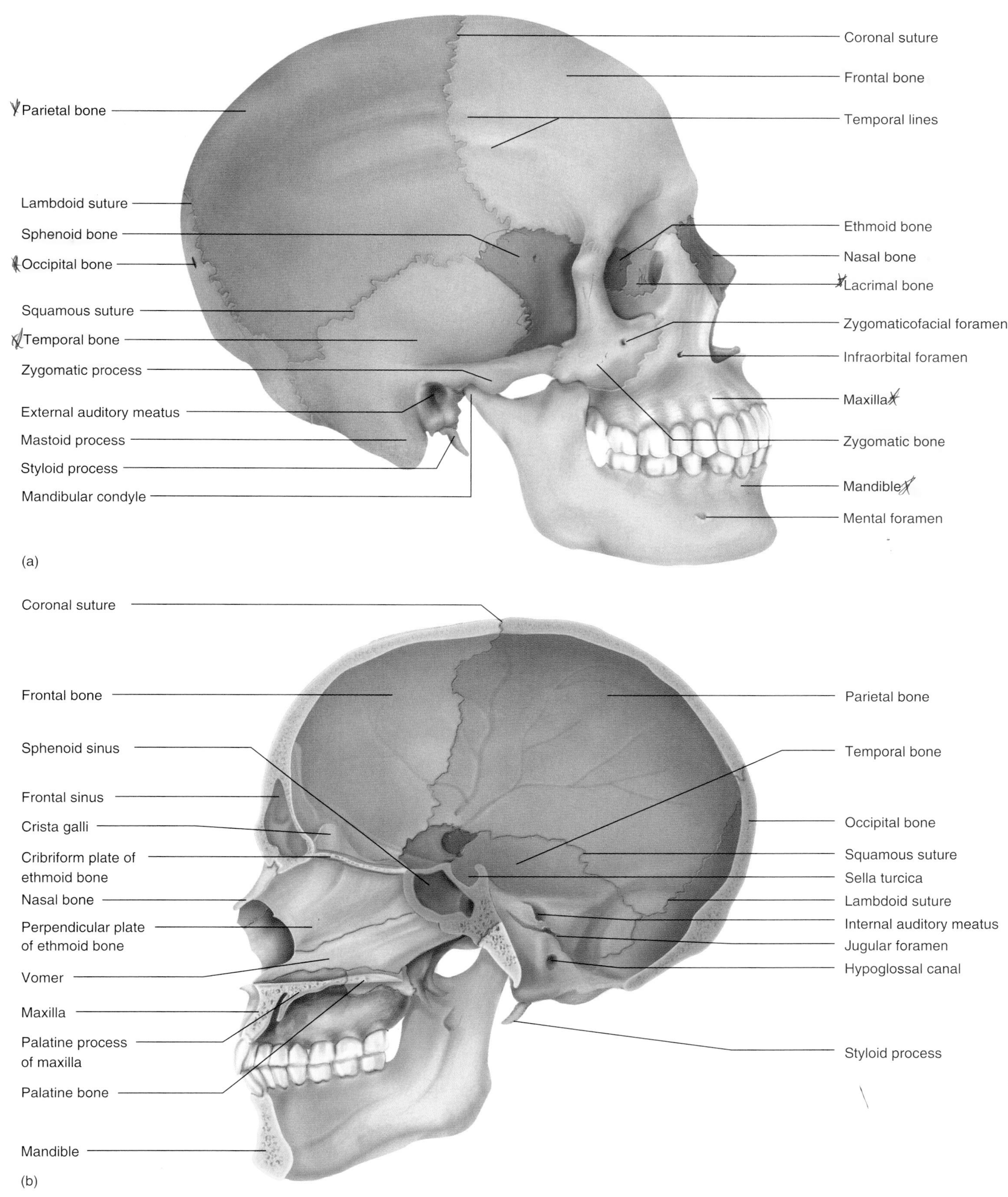

Figure 9.4 **Skull.** (*a*) Right lateral view; (*b*) interior of the right half.

Figure 9.5 Base of the Skull. (*a*) Inferior view; (*b*) internal view of the cranial floor.

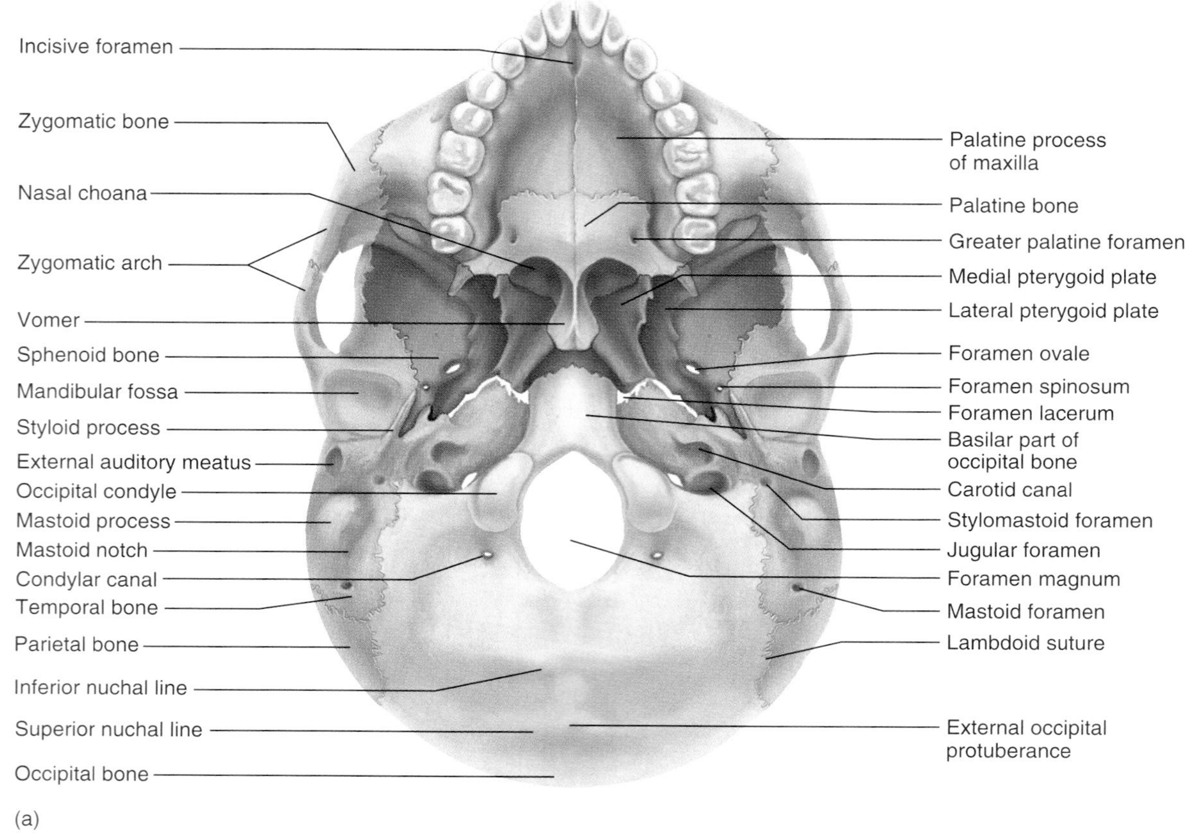

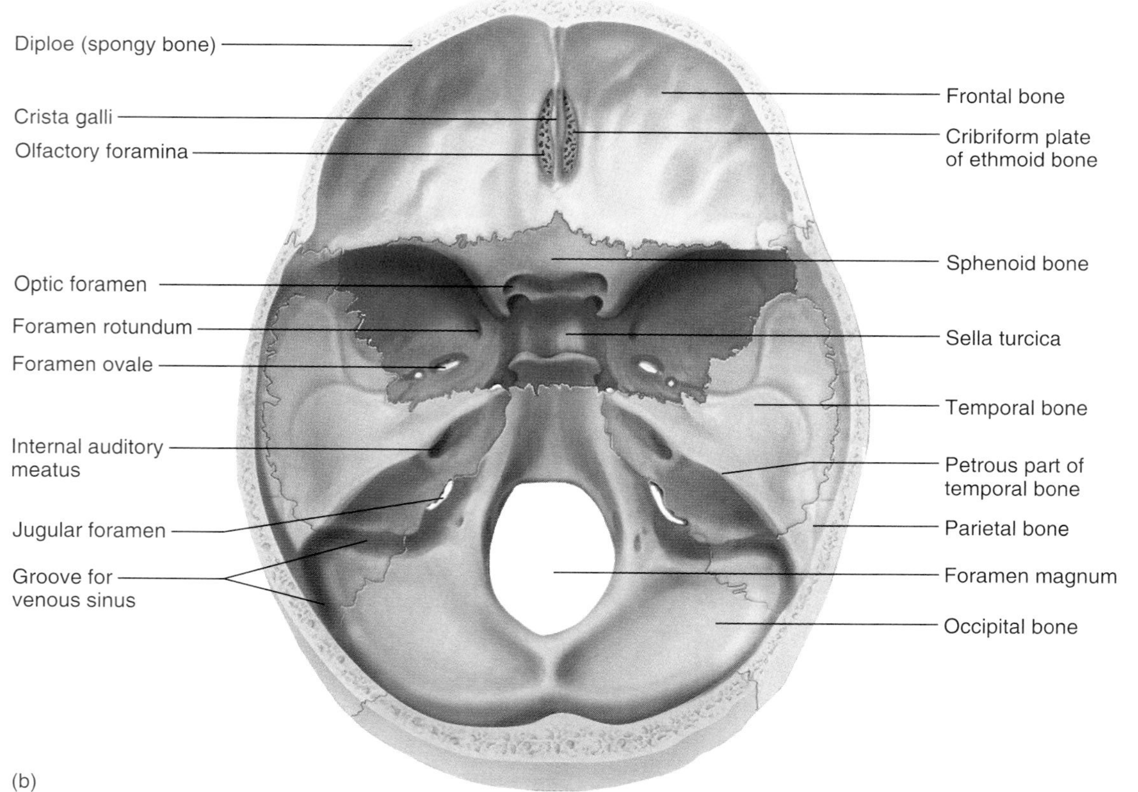

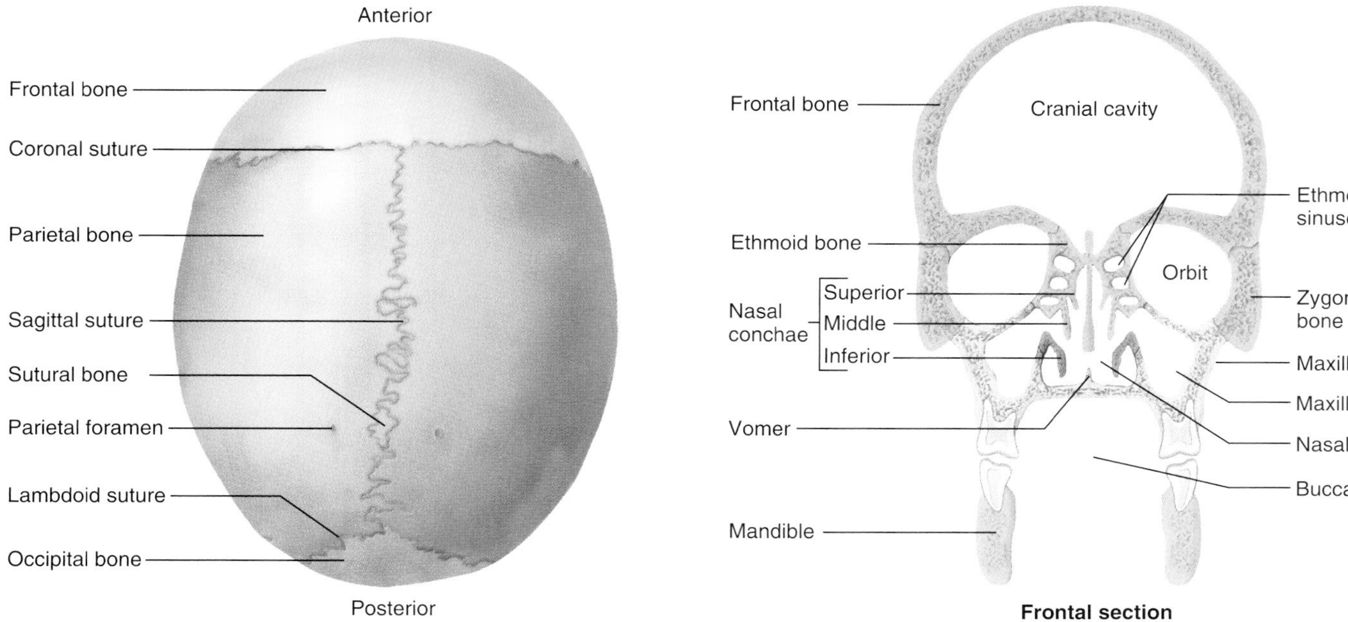

Figure 9.6 The Calvaria (skullcap), Superior View.

Figure 9.7 Major Cavities of the Skull, Frontal Section.

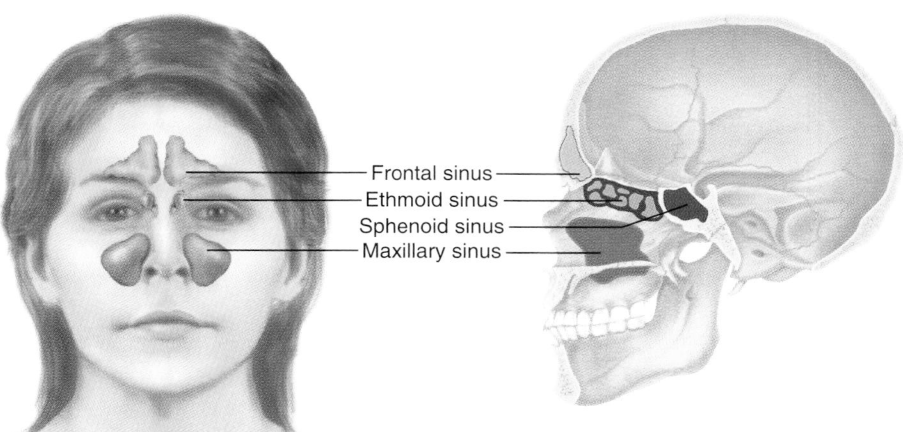

Figure 9.8 The Paranasal Sinuses.

the head from right to left and joins the frontal bone to the parietal bones (see figs. 9.3 and 9.4). It forms the anterior wall and about one-third of the roof of the cranial cavity, and it turns inward to form nearly all of the anterior cranial fossa and the roof of the orbit. Deep to the eyebrows it has a ridge called the **supraorbital margin.** The center of each margin is perforated by a single **supraorbital foramen** (see figs. 9.3 and 9.14), which provides passage for a nerve, artery, and veins. In some people, the edge of this foramen breaks through the margin of the orbit and forms a *supraorbital notch.* The frontal bone also contains the frontal sinus. You may not see this on some study skulls. It is absent from some people, and on some skulls the calvaria is cut too high to show the sinus. Along the cut edge of the calvaria, you can see the diploe—the layer of spongy bone in the middle of the cranial bones (see fig. 9.5*b*).

Parietal Bones

The right and left **parietal** (pa-RY-eh-tul) **bones** form most of the cranial roof and part of its lateral walls (see figs. 9.4 and 9.6). Each is bordered by four sutures that join it to the neighboring bones: (1) a median **sagittal suture** between

Table 9.3 Foramina of the Skull and the Nerves and Blood Vessels Transmitted Through Them

Bones and Their Foramina*	Structures Transmitted
Frontal Bone	
Supraorbital foramen or notch	Supraorbital nerve, artery, and vein; ophthalmic nerve
Parietal Bone	
Parietal foramen	Emissary vein of superior longitudinal sinus
Temporal Bone	
Carotid canal	Internal carotid artery
External auditory meatus	Sound waves to eardrum
Internal auditory meatus	Vestibulocochlear nerve; internal auditory vessels
Stylomastoid foramen	Facial nerve
Mastoid foramen	Meningeal artery; vein from sigmoid sinus
Temporal–Occipital Region	
Jugular foramen	Internal jugular vein; glossopharyngeal, vagus, and accessory nerves
Temporal–Occipital–Sphenoid Region	
Foramen lacerum	No major nerves or vessels; closed by cartilage
Occipital Bone	
Foramen magnum	Spinal cord; accessory nerve; vertebral arteries
Hypoglossal canal	Hypoglossal nerve to muscles of tongue
Condylar canal	Vein from transverse sinus
Sphenoid Bone	
Foramen ovale	Mandibular division of trigeminal nerve; accessory meningeal artery
Foramen rotundum	Maxillary division of trigeminal nerve
Foramen spinosum	Middle meningeal artery; spinosal nerve; part of trigeminal nerve
Optic foramen	Optic nerve; ophthalmic artery
Superior orbital fissure	Oculomotor, trochlear, and abducens nerves; ophthalmic division of trigeminal nerve; ophthalmic veins
Ethmoid Bone	
Olfactory foramina	Olfactory nerves
Maxilla	
Infraorbital foramen	Infraorbital nerve and vessels; maxillary division of trigeminal nerve
Incisive foramen	Nasopalatine nerves
Maxilla–Sphenoid Region	
Inferior orbital fissure	Infraorbital nerve; zygomatic nerve; infraorbital vessels
Lacrimal Bone	
Lacrimal foramen	Tear duct leading to nasal cavity

(continued)

Table 9.3 Foramina of the Skull and the Nerves and Blood Vessels Transmitted Through Them (continued)

Bones and Their Foramina*	Structures Transmitted
Palatine Bone	
Greater palatine foramen	Palatine nerves
Zygomatic Bone	
Zygomaticofacial foramen	Zygomaticofacial nerve
Zygomaticotemporal foramen	Zygomaticotemporal nerve
Mandible	
Mental foramen	Mental nerve and vessels
Mandibular foramen	Inferior alveolar nerves and vessels to the lower teeth

*When two or more bones are listed together (for example, temporal–occipital), it indicates that the foramen passes between them.

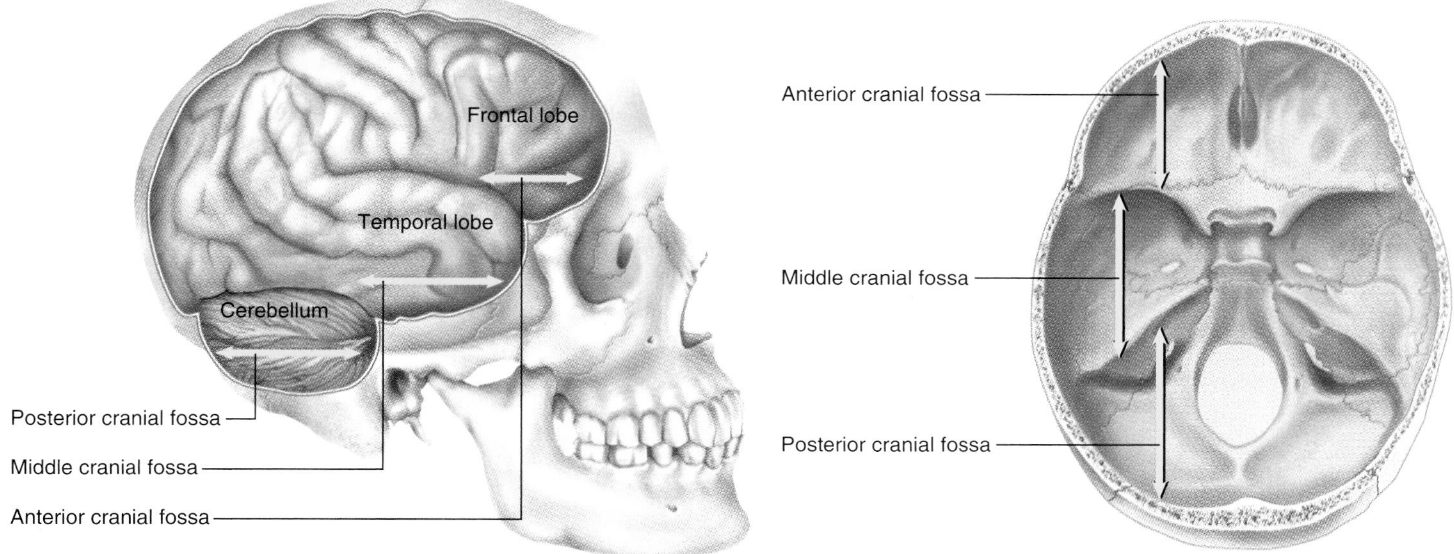

Figure 9.9 **Cranial Fossae.** The three fossae conform to the contours of the base of the brain.

the parietal bones; (2) the **coronal**[7] **suture** at the anterior margin; the **lambdoid**[8] (LAM-doyd) **suture** at the posterior margin; and (4) the **squamous suture** laterally. Small sutural (wormian) bones are often seen along the sagittal and lambdoid sutures, like little islands of bone with the suture lines passing around them. Internally, the parietal and frontal bones have markings that look a bit like aerial photographs of river tributaries (see fig. 9.4*b*). These represent places where the bone has been molded around blood vessels of the dura mater.

Externally, the parietal bones have few features. A **parietal foramen** sometimes occurs near the corner of the lambdoid and sagittal sutures (see fig. 9.6). A pair of slight thickenings, the superior and inferior **temporal lines,** form an arc across the parietal and frontal bones (see fig. 9.4*a*). They mark the attachment of the large, fan-shaped *temporalis* muscle, a chewing muscle that passes between the zygomatic arch and temporal bone and inserts on the mandible.

7. *corona* = crown
8. Shaped like the Greek letter lambda (λ)

Temporal Bones

If you palpate your skull just above and anterior to the ear—that is, the temporal region—you can feel the **temporal bone,** which forms the lower lateral wall and part of the floor of the cranial cavity (fig. 9.10). The temporal bone derives its name from the fact that people often develop their first gray hairs on the temples with the passage of time.[9] The relatively complex shape of the temporal bone is best understood by dividing it into four parts:

1. The **squamous**[10] **part** (which you just palpated) is relatively flat and vertical. It is encircled by the squamous suture. It bears two prominent features: (1) the **zygomatic process,** which extends anteriorly to form part of the **zygomatic arch** (cheekbone), and (2) the **mandibular fossa,** a depression where the mandible articulates with the cranium.
2. The **tympanic**[11] **part** is a small ring of bone that borders the **external auditory meatus** (me-AY-tus), the ear canal. It has a pointed spine on its inferior surface, the **styloid process,** named for its resemblance to the stylus used by ancient Greeks and Romans to write on wax tablets. The styloid process provides attachment for muscles of the tongue, pharynx, and hyoid bone.
3. The **mastoid**[12] **part** lies posterior to the tympanic part. It bears a heavy **mastoid process,** which you can palpate as a prominent lump behind the earlobe. It is filled with small air sinuses that communicate with the middle-ear cavity. These sinuses are subject to infection and inflammation (mastoiditis), which can erode the bone and spread to the brain. Ventrally, there is a groove called the **mastoid notch** medial to the mastoid process (see fig. 9.5*a*). It is the origin of the *digastric muscle,* a muscle that opens the mouth. The notch is perforated by the **stylomastoid foramen** at its anterior end and the **mastoid foramen** at its posterior end.
4. The **petrous**[13] **part** can be seen in the cranial floor, where it resembles a little mountain range separating the middle cranial fossa from the

9. *tempor* = time
10. *squam* = flat + *ous* = characterized by
11. *tympan* = drum (eardrum) + *ic* = pertaining to
12. *mast* = breast + *oid* = resembling
13. *petr* = stone, rock + *ous* = like

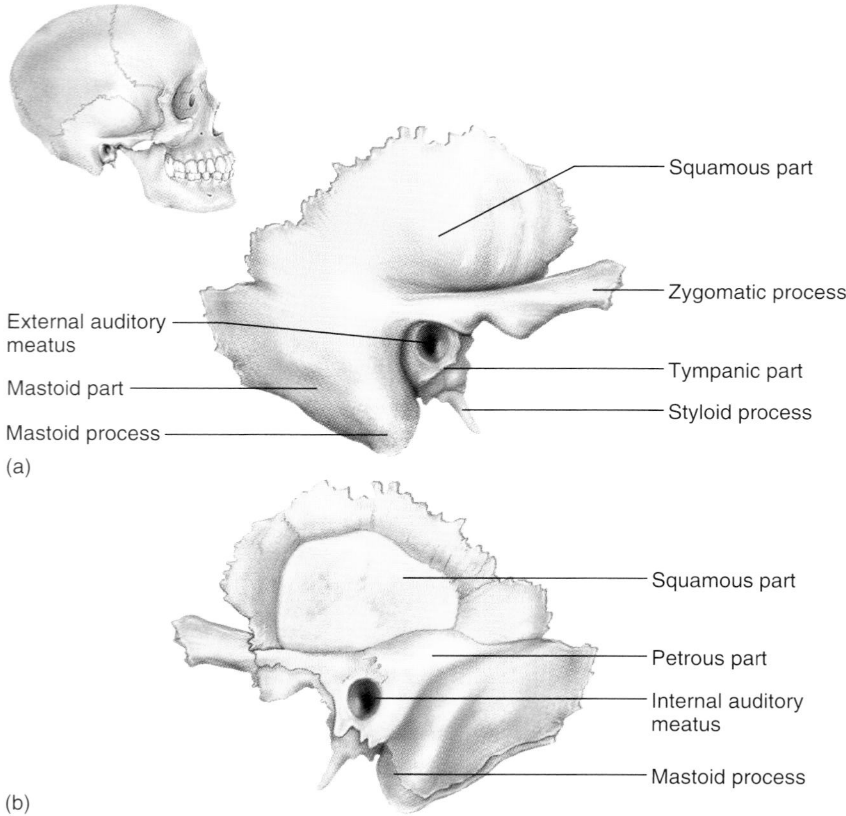

Figure 9.10 The Right Temporal Bone. (*a*) Lateral del view; (*b*) medial view.

posterior fossa (fig. 9.10*b*). It houses the middle- and inner-ear cavities. The **internal auditory meatus,** an opening on its posteromedial surface, allows passage of the vestibulocochlear (vess-TIB-you-lo-COC-lee-ur) nerve, which carries sensations of hearing and balance from the inner ear to the brain. On the ventral surface of the petrous part are two prominent foramina named for the major blood vessels that pass through them (see fig. 9.5*a*): (1) The **carotid canal** is a passage for the internal carotid artery, a major blood supply to the brain. This artery is so close to the inner ear that you can sometimes hear the pulsing of its blood when your ear is resting on a pillow or your heart is beating hard. (2) The **jugular foramen** is a large, irregular opening just medial to the styloid process, between the temporal and occipital bones. Blood from the brain drains through this foramen into the internal jugular vein. Three cranial nerves also pass through this foramen.

Occipital Bone

The **occipital** (oc-SIP-ih-tul) **bone** forms the rear of the skull (*occiput*) and much of its base (see fig. 9.5). Its most conspicuous feature, the foramen magnum, admits the spinal cord to the cranial cavity and provides a point of attachment for the dura mater. The bone continues anterior to the foramen magnum as a thick medial plate, the **basilar part.** On either side of the foramen magnum is a smooth knob called the **occipital condyle** (CON-dile), where the skull rests on the vertebral column. At the anterolateral edge of each condyle is a **hypoglossal**[14] **canal,** named for the hypoglossal nerve that passes through it to supply the muscles of the tongue. A **condylar** (CON-dih-lur) **canal** sometimes occurs posterior to each occipital condyle.

Internally, the occipital bone displays impressions left by large venous sinuses that drain blood from the brain (see fig. 9.5*b*). One of these grooves travels along the midsagittal line. Just before reaching the foramen magnum, it branches into right and left grooves that wrap around the occipital bone like outstretched arms before terminating at the jugular foramina. Here, the venous sinuses drain their blood into the internal jugular veins, which exit the cranial cavity and pass down the neck.

Other features of the occipital bone can be palpated on the back of your head. One is a prominent medial bump called the **external occipital protuberance**—the attachment for the **nuchal**[15] (NEW-kul) **ligament,** which binds the skull to the vertebral column. A ridge, the **superior nuchal line,** can be traced horizontally from the external occipital protuberance toward the mastoid process (see fig. 9.5*a*). It defines the superior limit of the neck and provides attachment for several neck and back muscles to the skull. By pulling down on the occipital bone, some of these muscles help to keep the head erect. The **inferior nuchal line** provides attachment for some of the deep neck muscles. This inconspicuous ridge cannot be palpated on the living body but is visible on an isolated skull.

Sphenoid Bone

The **sphenoid**[16] (SFEE-noyd) **bone** has a complex shape with a thick medial **body** and outstretched **greater** and **lesser wings,** which give the bone as a whole a somewhat ragged, mothlike shape. Most of it is best seen from the superior perspective (fig. 9.11*a*). In this view, the lesser wings form the posterior margin of the anterior cranial fossa and end at a sharp bony crest, where the sphenoid drops abruptly to the greater wings. These form about half of the middle cranial fossa (the temporal bone forming the rest) and are perforated by several foramina to be discussed shortly.

The greater wing forms part of the lateral surface of the cranium just anterior to the temporal bone (see fig. 9.4*a*). The lesser wing forms the posterior wall of the orbit and contains the **optic foramen,** which permits passage of the optic nerve and ophthalmic artery (see fig. 9.14). Superiorly, a pair of bony spines of the lesser wing called the **anterior clinoid processes** appear to guard the optic foramina. A gash in the posterior wall of the orbit, the **superior orbital fissure,** angles upward lateral to the optic foramen. It serves as a passage for nerves that supply some of the muscles that move the eyes.

The body of the sphenoid has a saddlelike prominence named the **sella turcica**[17] (SEL-la TUR-sih-ca). It consists of a deep pit called the *hypophyseal fossa,* which houses the pituitary gland, a raised anterior margin called the *tuberculum sellae* (too-BUR-cu-lum SEL-lee), and a posterior margin called the *dorsum sellae.* In life, a fibrous membrane is stretched over the sella turcica. A stalk penetrates the membrane to connect the pituitary gland to the floor of the brain.

Lateral to the sella turcica, the sphenoid is perforated by several foramina (see fig. 9.5*a*). The **foramen rotundum** and **foramen ovale** (oh-VAY-lee) are passages for two branches of the trigeminal nerve. The **foramen spinosum,** about the diameter of a pencil lead, provides passage for an artery of the meninges. An irregular gash called the **foramen lacerum**[18] (LASS-eh-rum) occurs at the junction

14. *hypo* = below + *gloss* = tongue
15. *nucha* = back of the neck
16. *sphen* = wedge + *oid* = resembling
17. *sella* = saddle + *turcica* = Turkish
18. *lacerum* = torn, lacerated

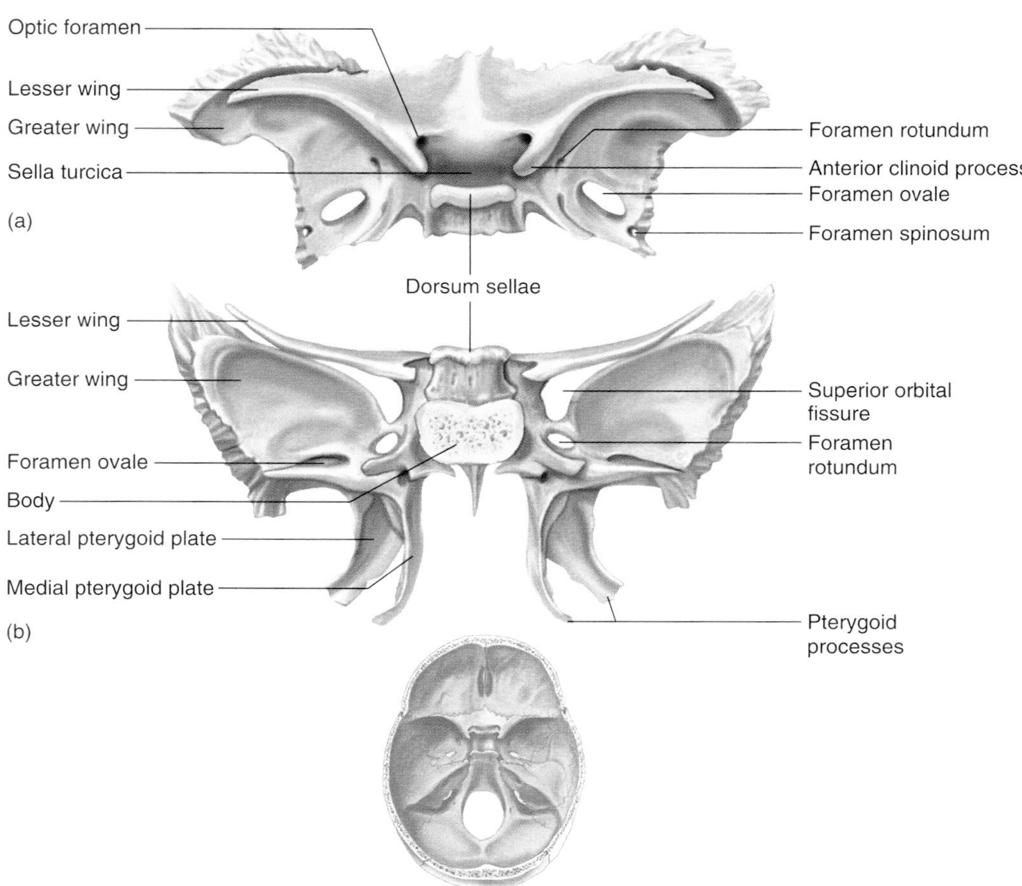

Figure 9.11 The Sphenoid Bone. (*a*) Superior view; (*b*) posterior view.

of the sphenoid, temporal, and occipital bones. It is filled with cartilage in life and transmits no major vessels or nerves.

In an inferior view, the sphenoid can be seen just anterior to the basilar part of the occipital bone. The internal openings of the nasal cavity seen here are called the **nasal choanae**[19] (co-AH-nee), or **internal nares.** Lateral to each choana, the sphenoid bone exhibits a pair of parallel plates—the **medial pterygoid**[20] (TERR-ih-goyd) **plate** and **lateral pterygoid plate** (see fig. 9.5*a*). These provide attachment for some of the jaw muscles. The sphenoid sinus occurs within the body of the sphenoid bone.

Ethmoid Bone

The **ethmoid**[21] (ETH-moyd) **bone** is located between the orbital cavities and forms the roof of the nasal cavity (fig. 9.12). An inferior projection of the ethmoid, called the **perpendicular plate,** forms the superior part of the **nasal**

Cribriform plate
Olfactory foramina
Orbital plate
Ethmoid air cells
Perpendicular plate
Crista galli
Superior nasal concha
Middle nasal concha

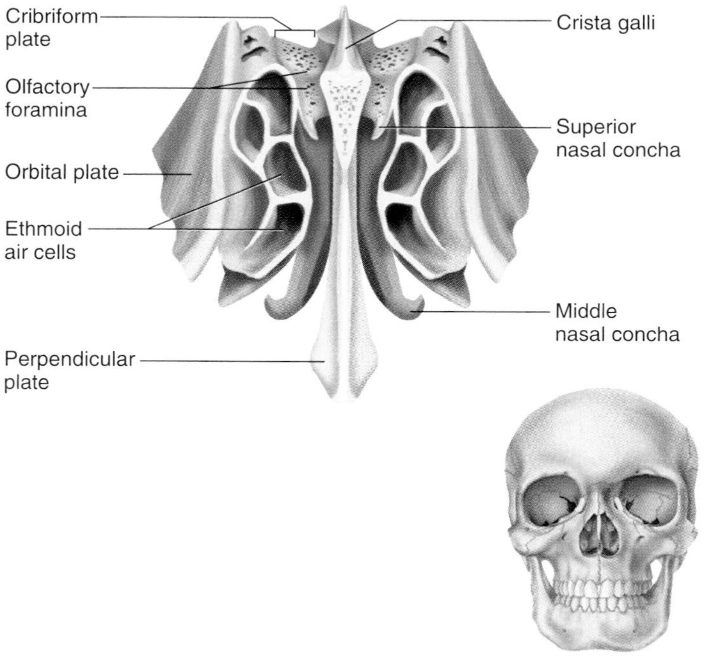

Figure 9.12 The Ethmoid Bone, Anterior View.

19. *choana* = funnel
20. *pterygo* = wing
21. *ethmo* = sieve, strainer + *oid* = resembling

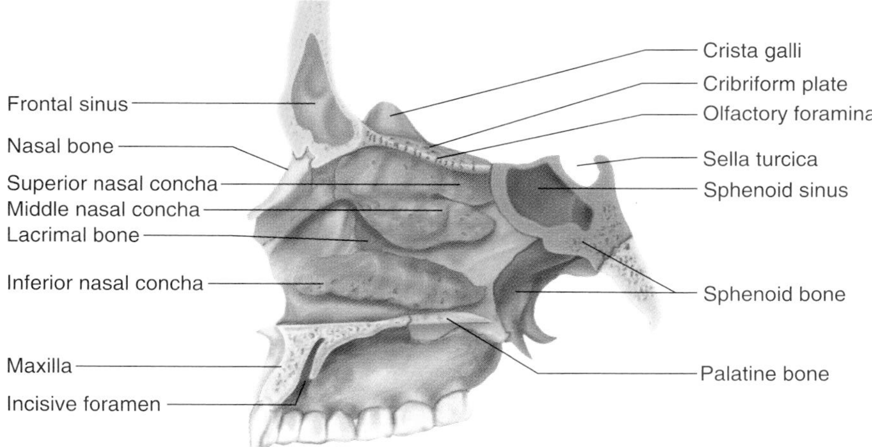

Figure 9.13 The Right Nasal Cavity, Sagittal Section.

septum, which divides the nasal cavity into right and left **nasal fossae** (FOSS-ee). Three curled, scroll-like **nasal conchae**[22] (CON-kee), or **turbinate**[23] **bones,** project into each fossa from the lateral wall (see figs. 9.7 and 9.13). The superior and middle conchae are extensions of the ethmoid bone. The inferior concha—a separate bone—is included in the discussion of facial bones in the next section. The conchae are covered with the mucous membrane of the nasal cavity. The superior concha and the adjacent region of the nasal septum also bear the receptor cells for the sense of smell (olfactory sense). The ethmoid bone also includes a large, delicate mass on each side of the perpendicular plate, honeycombed with chambers called **ethmoid air cells;** collectively, these constitute the ethmoid sinus.

The superior part of the ethmoid, viewed from the interior of the skull, exhibits a midsagittal crest called the **crista galli**[24] (GAL-eye), a point of attachment for the meninges (see figs. 9.4*b* and 9.5*b*). On either side of the crista is a horizontal **cribriform**[25] (CRIB-rih-form) **plate** marked by numerous perforations, the **olfactory foramina.** These foramina allow nerve fibers for the sense of smell to pass from the nasal cavity to the brain (see clinical insight 9.1).

Facial Bones

The **facial bones** are those that have no direct contact with the brain or meninges. They support the teeth, give shape and individuality to the face, form part of the orbital and nasal cavities, and provide attachment for the muscles of facial expression and mastication. There are 14 facial bones:

2 maxillae	2 nasal bones
2 palatine bones	2 inferior nasal conchae
2 zygomatic bones	1 vomer
2 lacrimal bones	1 mandible

Maxillae

The **maxillae** (mac-SILL-ee) are the largest facial bones. They form the upper jaw and meet each other at a midsagittal suture (see figs. 9.3 and 9.4*a*). Small points of maxillary bone called **alveolar processes** grow into the spaces between the bases of the teeth. The root of each tooth is inserted into a deep socket, or **alveolus.** If a tooth is lost or extracted so that chewing no longer puts stress on the maxilla, the alveolar processes are resorbed and the alveolus fills in with new bone, leaving a smooth area on the maxilla. The teeth are discussed in detail in chapter 25.

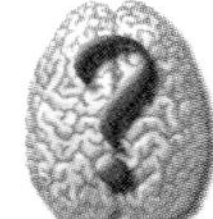

Think About It

Suppose you were studying a skull with some teeth missing. How could you tell whether the teeth had been lost after the person's death or years before it?

Each maxilla extends from the teeth to the inferomedial wall of the orbit. Just below the orbit, it exhibits an **infraorbital foramen,** which provides passage for a blood vessel to the face and a nerve that receives sensations from the nasal region and cheek. This nerve emerges through the foramen rotundum into the cranial cavity. The maxilla forms part of the floor of the orbit, where it exhibits a gash called the **inferior orbital fissure** that angles downward and medially (fig. 9.14). The inferior and

22. *conchae* = conchs (large marine snails)
23. *turbin* = whirling, turning
24. *crista* = crest + *galli* = of a rooster
25. *cribri* = sieve + *form* = in the shape of

Clinical Insight 9.1

Injury to the Ethmoid Bone

The ethmoid bone is very delicate and is easily injured by a sharp upward blow to the nose, such as a person might suffer by striking an automobile dashboard in a collision. The force of a blow can drive bone fragments through the cribriform plate into the meninges or brain tissue. Such injuries are often evidenced by leakage of cerebrospinal fluid into the nasal cavity, and may be followed by the spread of infection from the nasal cavity to the brain. Blows to the head can also shear off the olfactory nerves that pass through the ethmoid bone and cause *anosmia*, an irreversible loss of the sense of smell and a great reduction in the sense of taste (most of which depends on smell). This not only deprives life of some of its pleasures, but can also be dangerous, as when a person fails to smell smoke or gas.

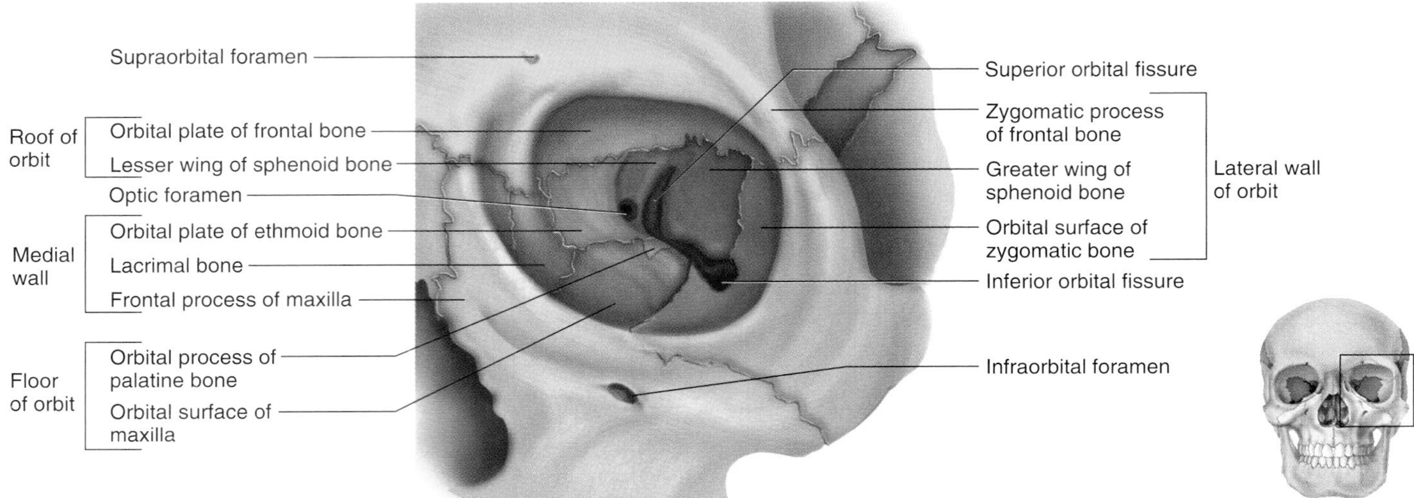

Figure 9.14 The Left Orbit, Anterior View.

superior orbital fissures form a sideways V whose apex lies near the optic foramen. The inferior orbital fissure is a passage for blood vessels and a nerve that supply more of the muscles that control eye movements.

The **palate** forms the roof of the mouth and floor of the nasal cavity. It consists of a bony **hard palate** in front and a fleshy **soft palate** in the rear. Most of the hard palate is formed by horizontal extensions of the maxilla called **palatine** (PAL-uh-tine) **processes** (see fig. 9.5*a*). Near the anterior margin of each palatine process, just behind the incisors, is an **incisive foramen.** The palatine processes normally fuse at about 12 weeks of gestation. Failure to fuse results in a *cleft palate,* often accompanied by a *cleft lip* lateral to the midline. Cleft palate and lip can be surgically corrected with good cosmetic results, but a cleft palate makes it difficult for an infant to generate the suction needed for nursing.

Palatine Bones

The **palatine bones** form the rest of the hard palate, part of the wall of the nasal cavity, and part of the floor of the orbit (see figs. 9.5*a* and 9.13). At the posterolateral corners of the hard palate are two large **greater palatine foramina.**

Zygomatic Bones

The **zygomatic**[26] **bones** form the angles of the cheeks at the inferolateral margins of the orbits and part of the lateral wall of each orbit; they extend about halfway to the ear (see figs. 9.4*a* and 9.5*a*). Each zygomatic bone has an inverted T shape and usually a small **zygomaticofacial** (ZY-go-MAT-ih-co-FAY-shul) **foramen** near the intersection of

26. *zygo* = to join, unite

Evolutionary Insight 9.2

Evolutionary Significance of the Palate

In most vertebrates, the nasal passages open into the oral cavity. Mammals, by contrast, have a palate that separates the nasal cavity from the oral cavity. In order to maintain our high metabolic rate, we must digest our food rapidly; in order to do this, we chew it thoroughly to break it up into small, easily digested particles before swallowing it. The palate allows us to continue breathing during this prolonged chewing.

the stem and crossbar of the T. The prominent zygomatic arch that flares from each side of the skull is formed by the union of the zygomatic and temporal bones.

Lacrimal Bones

The **lacrimal**[27] (LACK-rih-mul) **bones** form part of the medial wall of each orbit (fig. 9.14). A depression called the **lacrimal fossa** houses a membranous *lacrimal sac* in life. Tears from the eye collect in this sac and drain into the nasal cavity.

Nasal Bones

Two small rectangular **nasal bones** form the bridge of the nose (see fig. 9.3) and support cartilages that give shape to the lower portion of the nose. If you palpate the bridge, you can easily feel where the nasal bones end and the cartilages begin. The nasal bones are often fractured by blows to the nose.

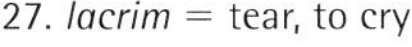

27. *lacrim* = tear, to cry

Inferior Nasal Conchae

There are three conchae in the nasal cavity. The superior and middle conchae, as discussed earlier, are parts of the ethmoid bone. The **inferior nasal concha (inferior turbinate bone)**—the largest of the three—is a separate bone (see fig. 9.13).

Vomer

The **vomer** forms the inferior half of the nasal septum (see figs. 9.3 and 9.4*b*). Its name literally means "plowshare," which refers to its resemblance to the blade of a plow. The superior half of the nasal septum is formed by the perpendicular plate of the ethmoid bone, as mentioned earlier. The vomer and perpendicular plate support a wall of *septal cartilage* that forms most of the anterior part of the nasal septum.

Mandible

The **mandible** (fig. 9.15) is the strongest bone of the skull and the only one that can move. It supports the lower teeth and provides attachment for muscles of mastication and

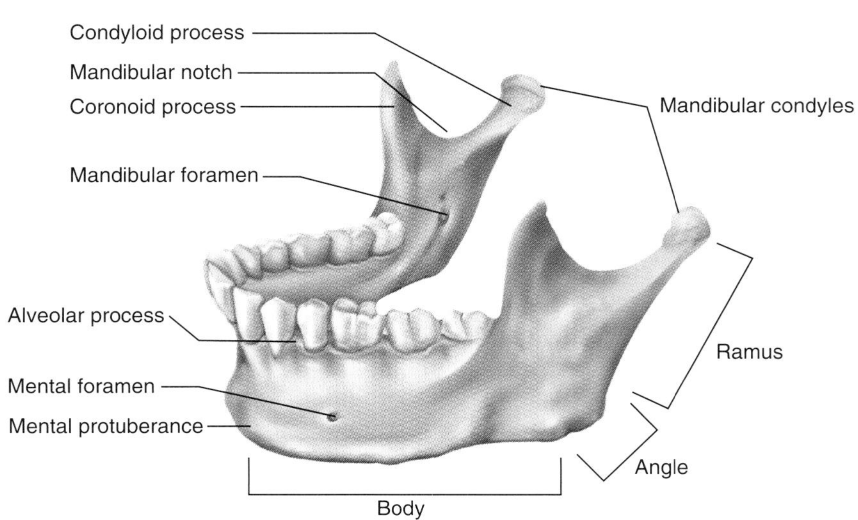

Figure 9.15 The Mandible.

facial expression. The horizontal portion is called the **body;** the vertical-to-oblique posterior portion is the **ramus** (RAY-mus)—plural, *rami* (RAY-my); and these two portions meet at a corner called the **angle.** The mandible develops as separate right and left bones in the fetus, joined by a midsagittal cartilaginous joint called the **mental symphysis** (SIM-fih-sis). This joint ossifies in early childhood, uniting the two halves into a single bone. The point of the chin is the **mental protuberance.** On the anterolateral surface of the body, the **mental foramen** permits the passage of nerves and blood vessels of the chin. The inner surface of the body has a number of shallow depressions and ridges to accommodate muscles and salivary glands. The angle of the mandible has a rough lateral surface for insertion of the *masseter,* a muscle of mastication. Like the maxilla, the mandible has pointed alveolar processes between the teeth.

The ramus is somewhat Y-shaped. Its posterior branch, called the **condyloid** (CON-dih-loyd) **process,** bears the **mandibular condyle**—an oval knob that articulates with the mandibular fossa of the temporal bone. The hinge of the mandible is the **temporomandibular joint (TMJ).** The anterior branch of the ramus, called the **coronoid process,** is the point of insertion for the temporalis muscle, which pulls the mandible upward when you bite. The U-shaped arch between the two processes is called the **mandibular notch.** Just below the notch, on the medial surface of the ramus, is the **mandibular foramen.** The nerve and blood vessels that supply the lower teeth enter this foramen. Dentists inject anesthetic near here to deaden sensation from the lower teeth and near the foramen rotundum to deaden sensation from the upper teeth.

Bones Associated with the Skull

Seven bones are closely associated with the skull but not considered part of it. These are the three auditory ossicles in each middle-ear cavity and the hyoid bone beneath the chin. The **auditory ossicles**[28]—named the malleus (hammer), **incus** (anvil), and **stapes** (STAY-peez) (stirrup)—are discussed in connection with hearing in chapter 16. The **hyoid**[29] **bone** is a slender bone between the chin and larynx (fig. 9.16). It is one of the few bones that does not articulate with any other. The hyoid is suspended from the styloid processes of the skull, somewhat like a hammock, by the small *stylohyoid muscles* and *stylohyoid ligaments.* The medial **body** of the hyoid is flanked on either side by hornlike projections called the **greater** and **lesser cornua**[30] (CORN-you-uh)—singular, *cornu* (COR-new). The hyoid bone serves for attachment of several muscles that control the mandible, tongue, and larynx. Forensic pathologists look for a fractured hyoid as evidence of strangulation.

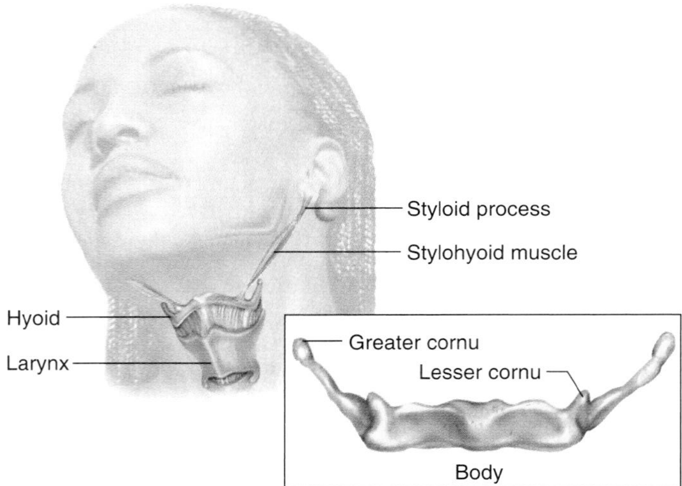

Figure 9.16 The Hyoid Bone.

The Skull in Infancy and Childhood

The head of an infant could not fit through the mother's pelvic outlet at birth were it not for the fact that the bones of its skull are not yet fused. The shifting of the skull bones during birth may cause the infant to appear deformed, but the head soon assumes a more normal shape. Spaces between the unfused cranial bones are called **fontanels,**[31] after the fact that pulsation of the infant's blood can be felt there. The bones are joined at these points only by fibrous membranes, in which intramembranous ossification will be completed later. Four of these sites are especially prominent and regular in location: the **anterior, posterior, sphenoid,** and **mastoid fontanels** (fig. 9.17). Most fontanels ossify by the time the infant is a year old, but the largest one—the anterior fontanel—can still be palpated 18 to 24 months after birth.

The frontal bone and mandible are separate right and left bones at birth, but fuse medially in early childhood. The frontal bones usually fuse by age five or six, but in some children a *metopic*[32] *suture* persists between them. Traces of this suture are evident in some adult skulls.

The face of a newborn is flat and the cranium relatively large. To accommodate the growing brain, the skull grows more rapidly than the rest of the skeleton during childhood. It reaches about half its adult size by 9 months of age, three-quarters by age 2, and nearly final size by 8 or 9 years. The

28. *os* = bone + *icle* = little
29. *hy* = the letter *U* + *oid* = resembling
30. *cornu* = horn
31. *fontan* = fountain + *el* = little
32. *met* = beyond + *op* = the eyes

Clinical Insight 9.3

Cranial Assessment of the Newborn

Obstetric nurses routinely assess the fontanels of newborns by palpation. In a difficult delivery, one cranial bone may override another along a suture line, which calls for close monitoring of the infant. Abnormally wide sutures may indicate hydrocephalus, the accumulation of excessive amounts of cerebrospinal fluid, which causes the cranium to swell. Bulging fontanels suggest abnormally high intracranial pressure, while depressed fontanels indicate dehydration.

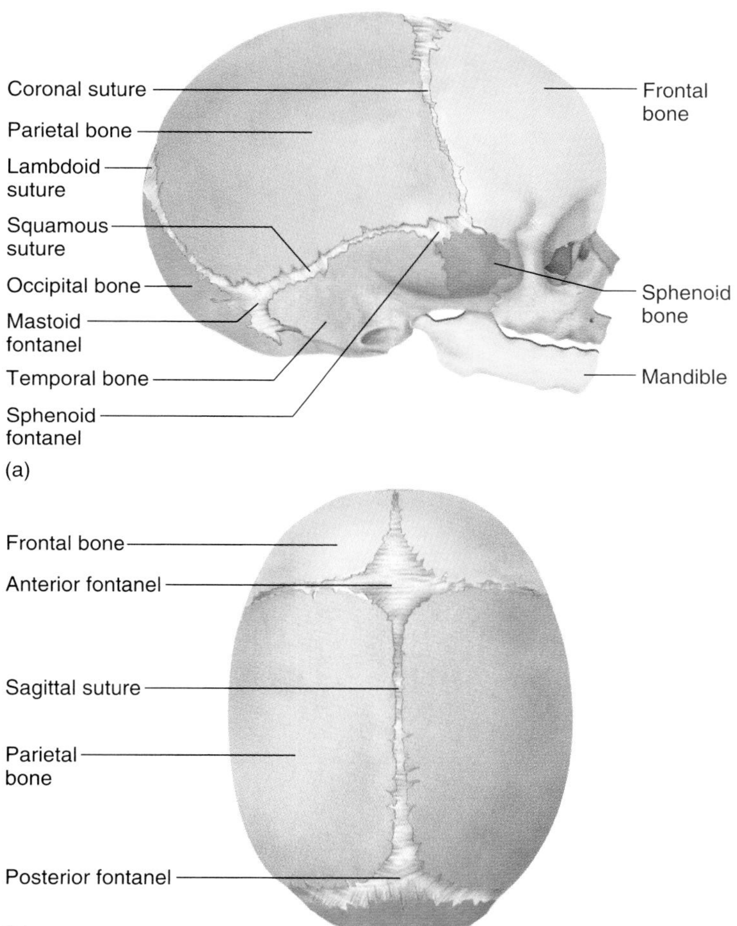

Figure 9.17 The Fetal Skull Near the Time of Birth. (*a*) Right lateral view; (*b*) superior view.

heads of babies and children are therefore much larger in proportion to the trunk than the heads of adults—an attribute thoroughly exploited by cartoonists and advertisers who draw big-headed characters to give them a more endearing or immature appearance. In humans and other animals, the large rounded heads of the young are thought to promote survival by stimulating parental caregiving instincts.

Table 9.4 summarizes the bones of the skull.

Before You Go On

Answer the following questions to test your understanding of the preceding section:

4. Name the paranasal sinuses and state their locations. Name any four other cavities in the skull.
5. Explain the difference between a cranial bone and a facial bone. Give four examples of each.
6. Draw an oval representing a superior view of the calvaria. Draw lines representing the coronal, lambdoid, and sagittal sutures. Label the four bones separated by these sutures.
7. State which bone has each of these features: a squamous part, hypoglossal foramen, greater cornu, greater wing, condyloid process, and cribriform plate.
8. For each of the following bones, name all the other bones with which it articulates: parietal, temporal, zygomatic, and ethmoid.
9. Determine which of the following structures cannot normally be palpated on a living person: the mastoid process, crista galli, superior orbital fissure, palatine processes, zygomatic bone, mental protuberance, and stapes. You may find it useful to palpate some of these on your own skull as you try to answer.

The Vertebral Column and Thoracic Cage

Objectives

When you have completed this section, you should be able to

- describe the general features of the vertebral column and those of a typical vertebra;
- describe the special features of vertebrae in different regions of the vertebral column, and discuss the functional significance of the regional differences; and
- describe the anatomy of the sternum and ribs and how the ribs articulate with the thoracic vertebrae.

Table 9.4 Anatomical Checklist for the Skull and Associated Bones

Cranial Bones

Frontal Bone (figs. 9.4 to 9.7)
Supraorbital margin
Supraorbital foramen or notch
Frontal sinus

Parietal Bones (figs. 9.4a and 9.6)
Temporal lines
Parietal foramen

Temporal Bones (figs. 9.4, 9.5b, and 9.10)
Squamous part
- Zygomatic process
- Mandibular fossa

Tympanic part
- External auditory meatus
- Styloid process

Mastoid part
- Mastoid process
- Mastoid notch
- Mastoid foramen
- Stylomastoid foramen

Petrous part
- Internal auditory meatus
- Carotid canal
- Jugular foramen

Occipital Bone (figs. 9.4, 9.5, and 9.6)
Foramen magnum
Basilar part
Occipital condyles
Hypoglossal canal

Occipital Bone (figs. 9.4, 9.5, and 9.6)—(cont.)
Condylar canal
External occipital protuberance
Superior nuchal line
Inferior nuchal line

Sphenoid Bone (figs. 9.4a, 9.5, and 9.11)
Body
Lesser wing
- Optic foramen
- Anterior clinoid process
- Superior orbital fissure

Greater wing
- Foramen ovale
- Foramen rotundum
- Foramen spinosum
- Foramen lacerum
- Medial and lateral pterygoid plates

Nasal choanae
Sphenoid sinus
Sella turcica
Dorsum sellae

Ethmoid Bone (figs. 9.4, 9.7, and 9.12)
Perpendicular plate
Superior nasal concha (superior turbinate bone)
Middle nasal concha (middle turbinate bone)
Ethmoid sinus (air cells)
Crista galli
Cribriform plate

Facial Bones

Maxilla (figs. 9.3, 9.4, and 9.5a)
Alveoli
Alveolar processes
Infraorbital foramen
Inferior orbital fissure
Palatine processes
Incisive foramen
Maxillary sinus

Palatine Bones (figs. 9.5a and 9.13)
Greater palatine foramen

Zygomatic Bones (figs. 9.4a and 9.5a)
Zygomaticofacial foramen

Lacrimal Bones (figs. 9.3 and 9.14)
Lacrimal fossa

Nasal Bones (figs. 9.3 and 9.13)

Inferior Nasal Concha (fig. 9.13)

Vomer (figs. 9.3 and 9.4b)

Mandible (figs. 9.3 and 9.15)
Body
- Mental symphysis
- Mental protuberance
- Mental foramen

Angle
Ramus
- Condyloid process
- Mandibular condyle
- Coronoid process
- Mandibular notch
- Mandibular foramen

(continued)

Table 9.4 Anatomical Checklist for the Skull and Associated Bones (continued)

Bones Associated with the Skull	
Auditory Ossicles	*Hyoid Bone (fig. 9.16)*
Malleus (hammer)	Body
Incus (anvil)	Greater cornu
Stapes (stirrup)	Lesser cornu

General Features of the Vertebral Column

The **vertebral column** physically supports the skull and trunk, allows for their movement, protects the spinal cord, and absorbs stresses produced by walking, running, and lifting. It also provides attachment for the limbs, thoracic cage, and postural muscles. Although commonly called the backbone, it does not consist of a single bone but a chain of 33 **vertebrae** with **intervertebral discs** of fibrocartilage between most of them. The adult vertebral column averages about 71 cm (28 in.) long, with the 23 intervertebral discs accounting for about one-quarter of the length.

As shown in figure 9.18, the vertebrae are divided into five groups: 7 *cervical* (SUR-vih-cul) *vertebrae* in the neck, 12 *thoracic vertebrae* in the chest, 5 *lumbar vertebrae* in the lower back, 5 *sacral vertebrae* at the base of the spine, and 4 tiny *coccygeal* (coc-SIDJ-ee-ul) *vertebrae.* To help remember the numbers of cervical, thoracic, and lumbar vertebrae—7, 12, and 5—you might think of a typical work day: go to work at 7, have lunch at 12, and go home at 5.

Variations in this arrangement occur in about 1 person in 20. For example, the last lumbar vertebra is sometimes incorporated into the sacrum, producing four lumbar and six sacral vertebrae. In other cases, the first sacral vertebra fails to fuse with the second, producing six lumbar and four sacral vertebrae. The cervical and thoracic vertebrae are more constant in number.

Beyond the age of 3 years, the vertebral column is slightly S-shaped, with four bends called the **cervical, thoracic, lumbar,** and **pelvic curvatures** (fig. 9.19). These are not present in the newborn, whose spine exhibits one continuous C-shaped curve (fig. 9.20) as it does in monkeys, apes, and most other four-legged animals. As an infant begins to crawl and lift its head, the cervical region becomes curved toward the dorsal side, enabling an infant on its belly to look forward. As a toddler begins walking, another curve develops in the same direction in the lumbar region. The resulting S shape makes sustained bipedal walking possible (see evolutionary insight 9.5, p. 301). The thoracic and pelvic curvatures are called *primary curvatures* because they are remnants of the original infantile curvature. The cervical and lumbar curvatures are called *secondary curvatures* because they develop later, in the child's first few years of crawling and walking.

General Structure of a Vertebra

A representative vertebra and intervertebral disc are shown in figure 9.21. The most obvious feature of a vertebra is the **body,** or **centrum**—a mass of spongy bone and red bone marrow covered with a thin layer of compact bone. This is the weight-bearing portion of the vertebra. Its rough superior and inferior surfaces provide firm attachment to the intervertebral discs.

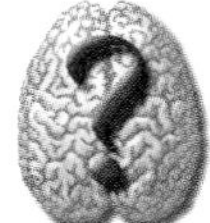

Think About It

The vertebral bodies and intervertebral discs get progressively larger as we look lower and lower on the vertebral column. What is the functional significance of this trend?

Dorsal to the body of each vertebra is a triangular canal called the **vertebral foramen.** The vertebral foramina collectively form the **vertebral canal,** a passage for the spinal cord. The foramen is bordered by a bony **neural arch** composed of two parts on each side: a pillarlike **pedicle**[33] and platelike **lamina.**[34] Extending from the apex of the arch, a projection called the **spinous process** is directed toward the rear and downward. You can see and feel the spinous processes as a row of bumps along the spine. A **transverse process** extends laterally from the point where the pedicle and lamina meet. The spinous and transverse processes provide points of attachment for the spinal muscles.

A pair of **superior articular processes** project upward from one vertebra and meet a similar pair of **inferior articular processes** that project downward from the vertebra just above (fig. 9.22*a*). Each process has a flat articular surface (facet) facing that of the adjacent vertebra. These processes restrict twisting of the vertebral column, which could otherwise severely damage the spinal cord.

33. *ped* = foot + *icle* = little
34. *lamina* = layer, plate

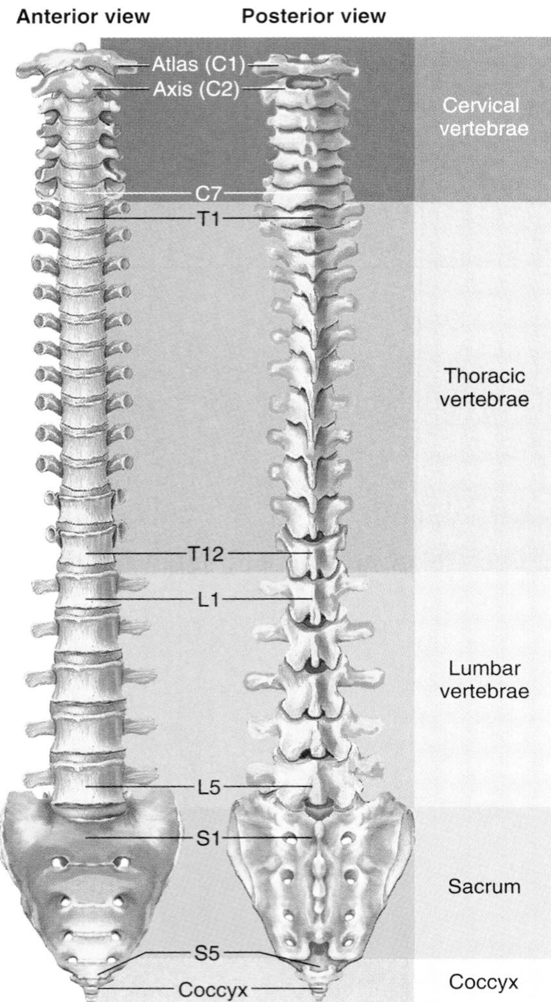

Figure 9.18 The Vertebral Column, Anterior and Posterior views.

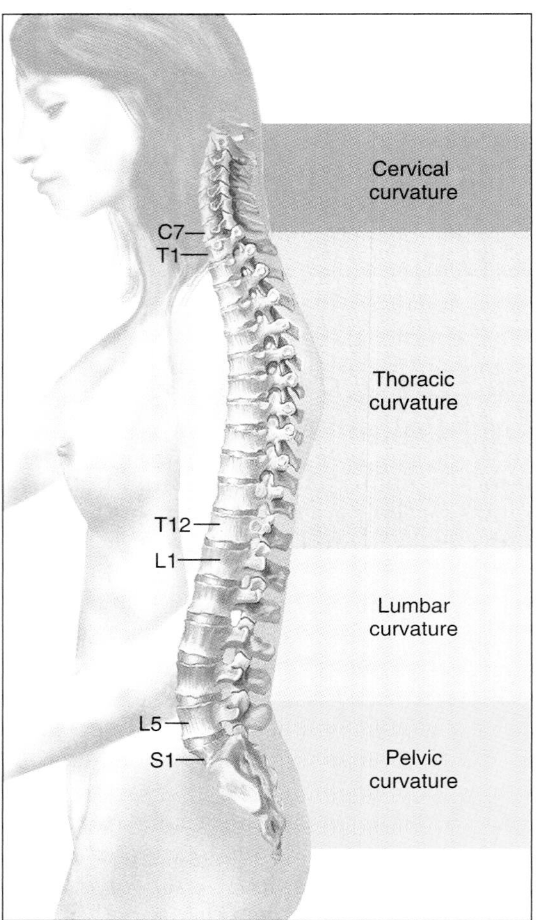

Figure 9.19 Curvatures of the Adult Vertebral Column.

When two vertebrae are joined, they exhibit an opening between their pedicles called the **intervertebral foramen.** This allows passage for spinal nerves that connect with the spinal cord at regular intervals. Each foramen is formed by an **inferior vertebral notch** in the pedicle of the superior vertebra and a **superior vertebral notch** in the pedicle of the one just below it (fig. 9.22*b*).

Intervertebral Discs

An **intervertebral disc** is a pad consisting of an inner gelatinous **nucleus pulposus** surrounded by a ring of fibrocartilage, the **annulus fibrosus** (see fig. 9.21). The discs help to bind adjacent vertebrae together, support the weight of the body, and absorb shock. Under stress—for example, when you lift a heavy weight—the discs bulge laterally. Excessive stress can crack the annulus and cause the nucleus to ooze out. This is called a *herniated disc* ("ruptured" or "slipped" disc in lay terms) and may put painful pressure on the spinal cord or a spinal nerve. To relieve the pressure, a procedure called a *laminectomy* may be performed—each lamina is cut and the laminae and spinous processes are removed. This procedure is also used to expose the spinal cord for anatomical study or surgery.

Regional Characteristics of Vertebrae

We are now prepared to consider how vertebrae differ from one region of the vertebral column to another and from the generalized anatomy just described. Knowing these variations will enable you to identify the region of the spine from which an isolated vertebra was taken. More importantly, these modifications in form reflect functional differences among the vertebrae.

Clinical Insight 9.4

Abnormal Spinal Curvatures

Abnormal spinal curvatures can result from disease, weakness, or paralysis of the trunk muscles, poor posture, or congenital defects in vertebral anatomy. The most common deformity is an abnormal lateral curvature called *scoliosis.* It occurs most often in the thoracic region, particularly among adolescent girls. It sometimes results from a developmental abnormality in which the body and arch fail to develop on one side of a vertebra. If the person's skeletal growth is not yet complete, scoliosis can be corrected with a back brace.

An exaggerated thoracic curvature is called *kyphosis* (hunchback, in lay language). It is usually a result of osteoporosis, but it also occurs in people with osteomalacia or spinal tuberculosis and in adolescent boys who engage heavily in such spine-loading sports as wrestling and weightlifting. An exaggerated lumbar curvature is called *lordosis* (swayback, in lay language). It may have the same causes as kyphosis, or it may result from added abdominal weight in pregnancy or obesity.

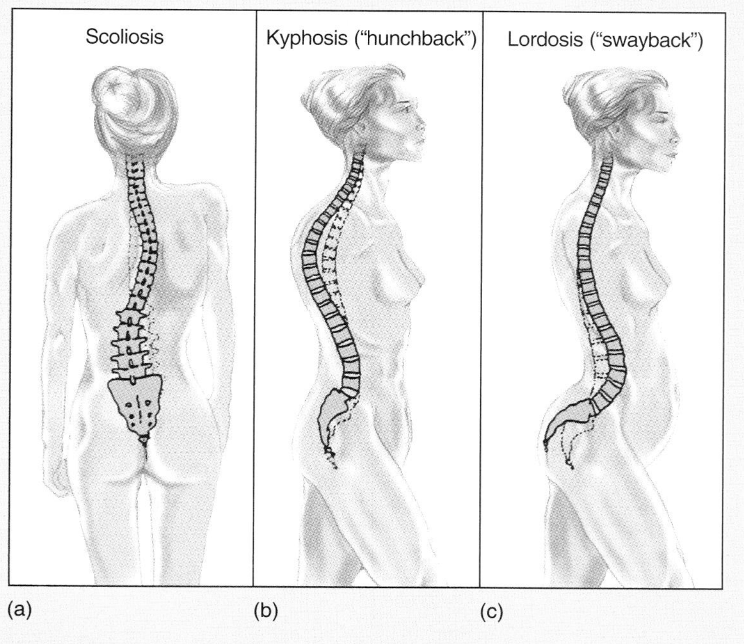

TA 9.1 Abnormal Spinal Curvatures. (*a*) Scoliosis, an abnormal lateral deviation. (*b*) Kyphosis, an exaggerated thoracic curvature. (*c*) Lordosis, an exaggerated lumbar curvature.

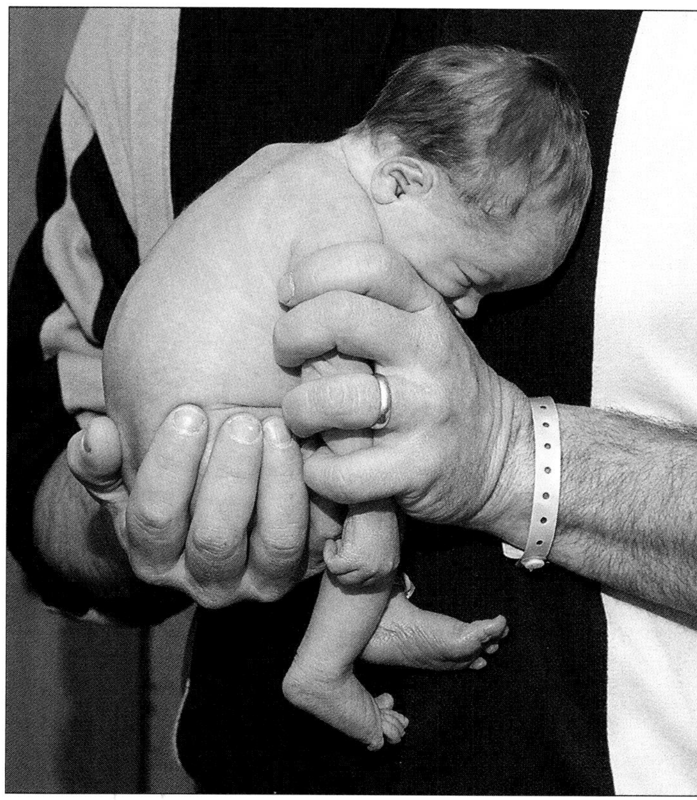

Figure 9.20 Spinal Curvature of the Newborn Infant. At this age, the spine forms a single C-shaped curve.

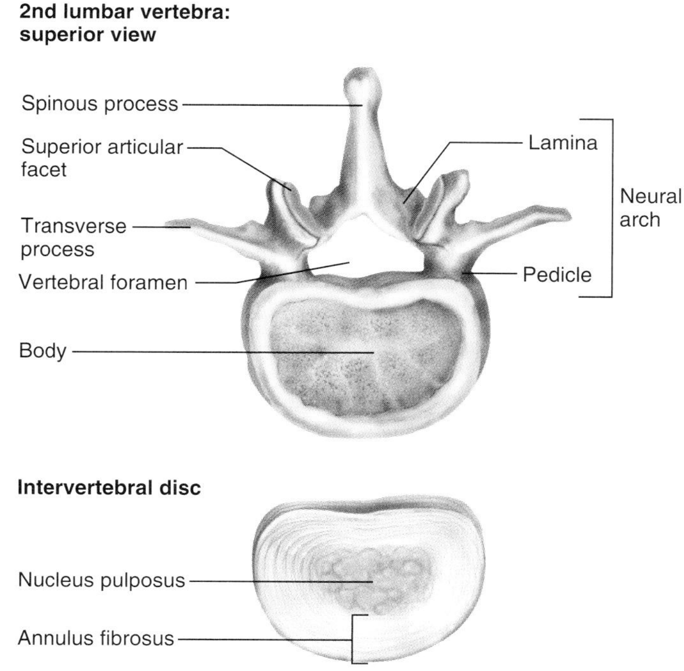

Figure 9.21 A Representative Vertebra and Intervertebral Disc, Superior Views.

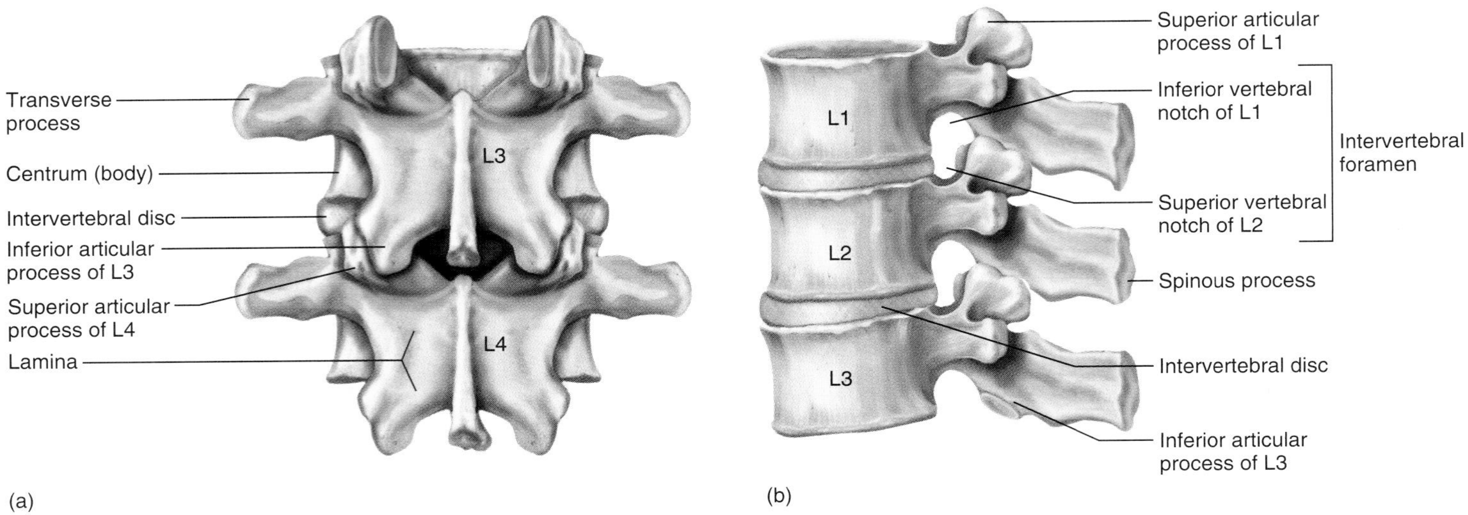

Figure 9.22 Articulated Vertebrae. (*a*) Dorsal view of vertebrae L3 to L4. (*b*) Left lateral view of vertebrae L1 to L3.

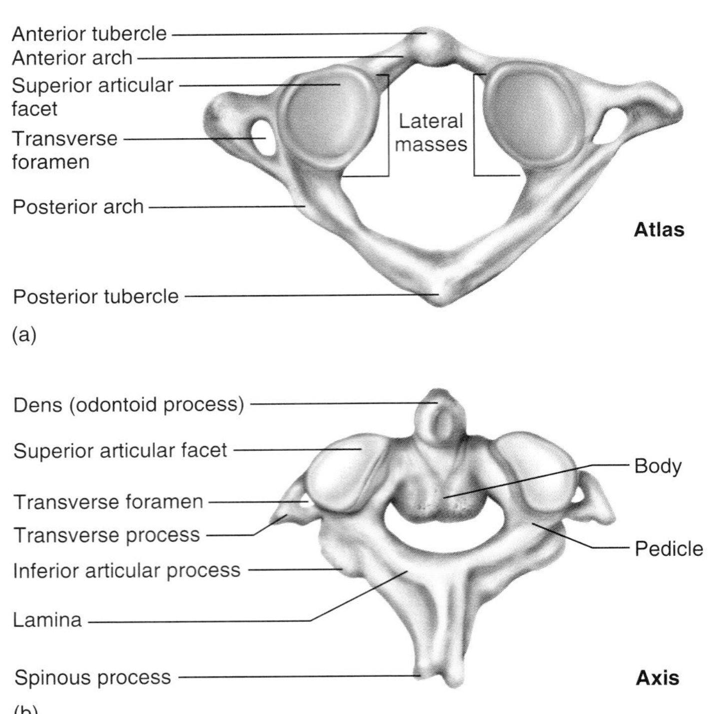

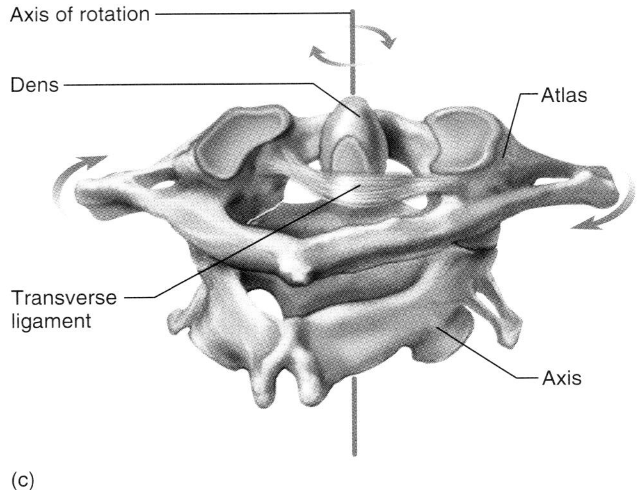

Figure 9.23 The Atlas and Axis, Cervical Vertebrae C1–C2. (*a*) The atlas, superior view. (*b*) The axis, posterosuperior view. (*c*) Articulation of the atlas and axis and rotation of the atlas. This movement turns the head from side to side, as in gesturing "no." Note the transverse ligament holding the dens of the axis in place.

Cervical Vertebrae

The **cervical vertebrae** (C1–C7) are the smallest and lightest. The first two (C1 and C2) have unique structures that allow for head movements (fig. 9.23). Vertebra C1 is called the **atlas** because it supports the head in a manner reminiscent of the Titan of Greek mythology who was condemned by Zeus to carry the world on his shoulders. It scarcely resembles the typical vertebra; it is little more than a delicate ring surrounding a large vertebral foramen. On each side is a **lateral mass** with a deeply concave **superior articular facet** that articulates with the occipital condyle of the skull. A nodding motion of the skull, as in gesturing "yes," causes the occipital condyles to rock back and forth on these facets. The **inferior articular facets,** which are comparatively flat or only slightly concave, articulate with C2. The lateral masses are connected by an **anterior arch** and a **posterior arch,** which bear slight protuberances called the **anterior** and **posterior tubercle,** respectively.

Vertebra C2, the **axis,** allows rotation of the head as in gesturing "no." Its most distinctive feature is a prominent knob called the **dens** (denz), or **odontoid**[35] **process,** on its anterosuperior side. No other vertebra has a dens. It begins to form as an independent ossification center during the first year of life and fuses with the axis by the age of 3 to 6 years. It projects into the vertebral foramen of the atlas, where it is nestled in a facet and held in place by a **transverse ligament** (fig. 9.23*c*). A heavy blow to the top of the head can cause a fatal injury in which the dens is driven through the foramen magnum into the brainstem. The articulation between the atlas and the cranium is called the **atlanto-occipital joint;** the one between the atlas and axis is called the **atlantoaxial joint.**

The axis is the first vertebra that exhibits a spinous process. In vertebrae C2 to C6, the process is forked, or *bifid,*[36] at its tip (fig. 9.24*a*). This fork provides attachment for the nuchal ligament of the back of the neck. All seven cervical vertebrae have a prominent round **transverse foramen** in each transverse process. These foramina provide passage and protection for the *vertebral arteries,* which supply blood to the brain. Transverse foramina occur in no other vertebrae and thus provide an easy means of recognizing a cervical vertebra.

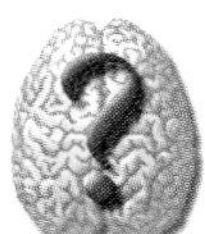

Think About It

How would head movements be affected if vertebrae C1 and C2 had the same structure as C3? What is the functional advantage of the lack of a spinous process in C1?

---

35. *dens* = *odont* = tooth + *oid* = resembling

36. *bifid* = cleft into two parts

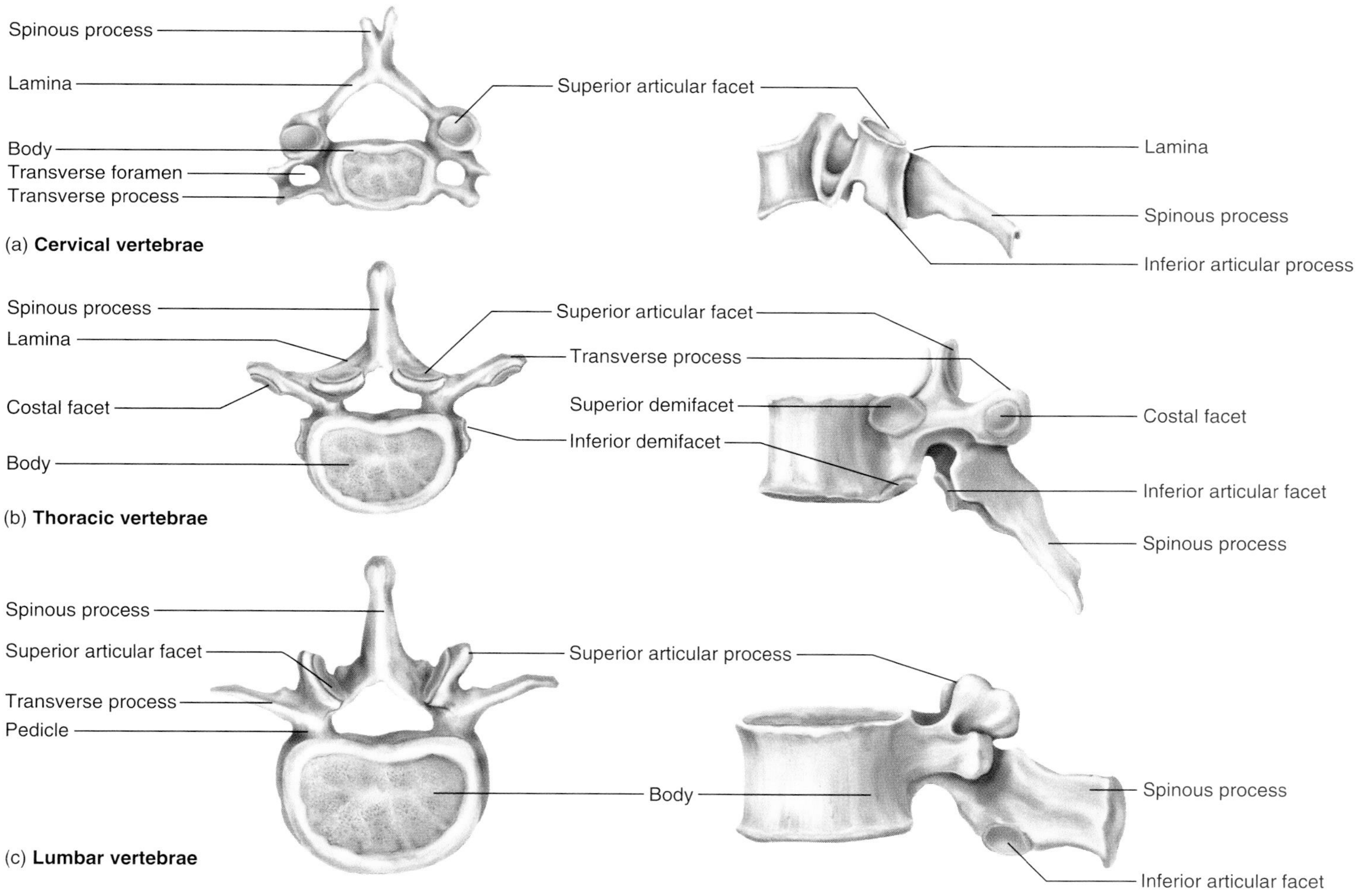

Figure 9.24 Typical Cervical, Thoracic, and Lumbar Vertebrae. The *left-hand* figures are superior views, and the *right-hand* figures are left lateral views.

Cervical vertebrae C3 to C6 are similar to the typical vertebra described earlier, with the addition of the transverse foramina and bifid spinous processes. Vertebra C7 is a little different—its spinous process is not bifid, but it is especially long and forms a prominent bump on the lower back of the neck. This feature is a convenient landmark for counting vertebrae.

Thoracic Vertebrae

There are 12 **thoracic vertebrae** (T1–T12), corresponding to the 12 pairs of ribs attached to them. They lack the transverse foramina and bifid processes that distinguish the cervicals, but possess the following distinctive features of their own (fig. 9.24*b*):

- The spinous processes are relatively pointed and angle sharply downward.
- The body is somewhat heart-shaped, more massive than in the cervical vertebrae but less than in the lumbar vertebrae.
- The body has small, smooth, slightly concave spots called *facets* and *demifacets* (to be described shortly) for attachment of the ribs.
- Vertebrae T1 to T10 have a shallow, cuplike **costal**[37] **facet** at the end of each transverse process. These provide a second point of articulation for ribs 1 to 10. There are no costal facets on T11 and T12 because ribs 11 and 12 attach only to the bodies of the vertebrae.

No other vertebrae have ribs articulating with them. Thoracic vertebrae vary among themselves mainly because of variations in the way the ribs articulate. In most cases, a rib inserts between two vertebrae, so each vertebra contributes one-half of the articular surface—a demifacet. A rib articulates with the **inferior demifacet**[38] (DEM-ee-FASS-et) of the upper vertebra and the **superior demifacet** of the vertebra below that. This terminology may be a little confusing, but note that the demifacets are named for their position on the vertebral body, not for which part of the rib's articulation they provide. Vertebrae T1 and T10 to T12, however, have complete facets on the bodies for ribs 1 and 10 to 12, which articulate on the vertebral body instead of between vertebrae. Vertebrae T11 and T12, as noted, have no costal facets. These variations will be more functionally understandable after you have studied the anatomy of the ribs, so we will return then to the details of these articular surfaces.

Lumbar Vertebrae

There are five **lumbar vertebrae** (L1–L5). Their most distinctive features are a thick, stout body and a blunt, squarish spinous process (fig. 9.24*c*). In addition, their articular processes are oriented differently than on other vertebrae. In thoracic vertebrae, the superior processes face forward and the inferior processes face to the rear. In lumbar vertebrae, the superior processes face medially (like the palms of your hands about to clap), and the inferior processes face laterally, toward the superior processes of the next vertebra. This arrangement makes the lumbar region of the spine especially resistant to twisting. These differences are best observed on an articulated skeleton. Vertebra L1 is an exception to this pattern, as it represents a transition between the thoracic and lumbar pattern. Its superior articular surfaces face forward to meet the inferior processes of T12, while its inferior articular surfaces face laterally like those of the rest of the lumbar vertebrae.

Sacrum

The **sacrum** is a bony plate that forms the dorsal wall of the pelvic cavity (fig. 9.25). It is named for the fact that it was once considered the seat of the soul.[39] In children, there are five separate **sacral vertebrae** (S1–S5). They begin to fuse around age 16 and are fully fused by age 26.

The anterior surface of the sacrum is relatively smooth and concave and has four transverse lines that indicate where the five vertebrae have fused. This surface exhibits four pairs of large **anterior sacral (pelvic) foramina,** which allow for passage of nerves and arteries to the pelvic organs. The dorsal surface of the sacrum is very rough. The spinous processes of the vertebrae fuse into a dorsal ridge called the **median sacral crest.** The transverse processes fuse into a less prominent **lateral sacral crest** on each side of the median crest. Again on the dorsal side of the sacrum, there are four pairs of openings for spinal nerves, the **posterior sacral foramina.** The nerves that emerge here supply the gluteal region and lower limb.

A **sacral canal** runs through the sacrum and ends in an inferior opening called the **sacral hiatus** (hy-AY-tus). This canal contains spinal nerve roots in life. On each side of the sacrum is an ear-shaped region called the **auricular**[40] (aw-RIC-you-lur) **surface.** This articulates with a similarly shaped surface on the os coxae and forms the strong, nearly immovable **sacroiliac** (SAY-cro-ILL-ee-ac) **joint.** At the superior end of the sacrum, lateral to the median crest, are a pair of **superior articular processes** that articulate with

37. *costa* = rib + *al* = pertaining to
38. *demi* = half
39. *sacr* = sacred
40. *auri* = ear + *cul* = little + *ar* = pertaining to

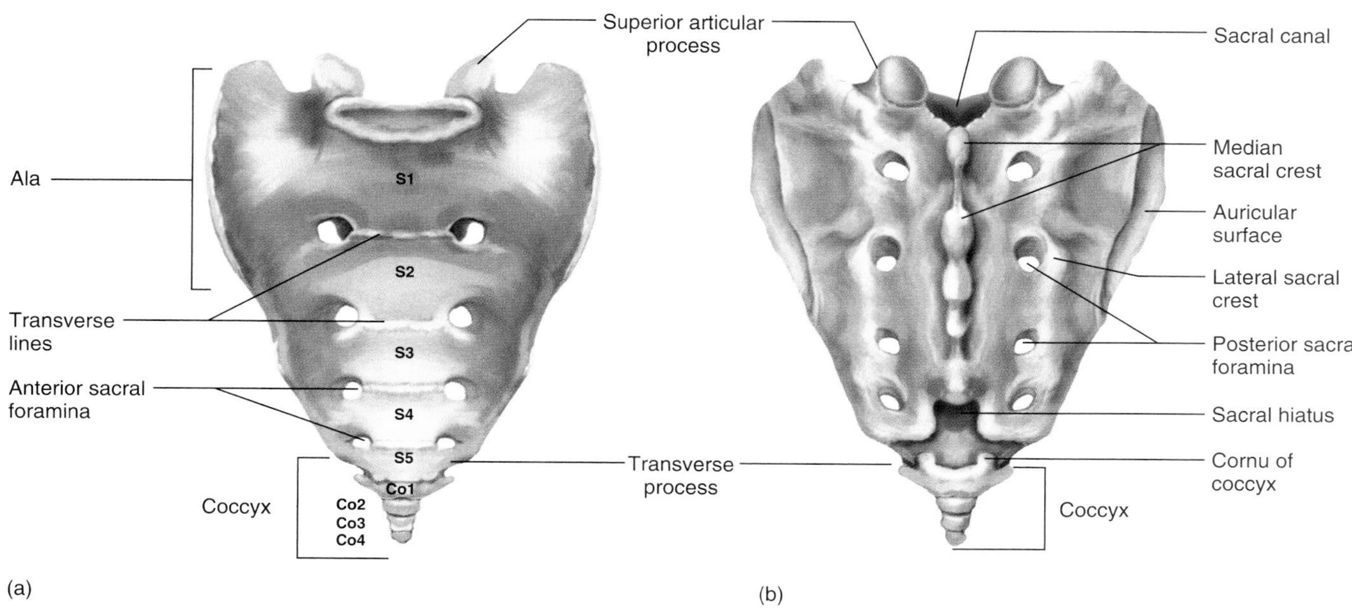

Figure 9.25 **The Sacrum and Coccyx.** (*a*) Anterior surface, which faces the viscera of the pelvic cavity. (*b*) Posterior surface. The processes of this surface can be palpated in the sacral region.

vertebra L5. Lateral to these are a pair of large, rough, winglike extensions called the **alae**[41] (AIL-ee).

Coccyx

The **coccyx**[42] (fig. 9.25) usually consists of four (sometimes five) small vertebrae, Co1 to Co4, which fuse by the age of 20 to 30 into a single triangular bone. Vertebra Co1 has a pair of hornlike projections, the **cornua,** which serve as attachment points for ligaments that bind the coccyx to the sacrum. The coccyx can be fractured by a difficult childbirth or a hard fall to the buttocks. Although it is the vestige of a tail, it is not entirely useless; it provides attachment for muscles of the pelvic floor.

The Thoracic Cage

The **thoracic cage** (fig. 9.26) consists of the thoracic vertebrae, sternum, and ribs. It forms a more or less conical enclosure for the lungs and heart and provides attachment for the pectoral girdle and upper limb. It has a broad base and a somewhat narrower superior apex; it is rhythmically expanded by the respiratory muscles to create a vacuum that draws air into the lungs. The inferior border of the thoracic cage is formed by a downward arc of the ribs called the **costal margin.** The ribs protect not only the thoracic organs but also the spleen, most of the liver, and to some extent the kidneys.

Sternum

The **sternum** (breastbone) is a bony plate anterior to the heart. It is subdivided into three regions: the manubrium, body, and xiphoid process. The **manubrium**[43] (ma-NOO-bree-um) is the broad superior portion. It has a superomedial **sternal notch** (jugular notch), which you can easily palpate between your clavicles (collarbones), and right and left **clavicular notches,** where it articulates with the clavicles. The **body,** or **gladiolus,**[44] is the longest part of the sternum. It joins the manubrium at the **sternal angle,** which can be palpated as a transverse ridge at the point where the sternum projects farthest forward. In some people, however, it is rounded or concave. The second rib attaches here, making the sternal angle a useful landmark for counting ribs in a physical examination. The manubrium and body have scalloped lateral margins where cartilages of the ribs are attached. At the inferior end of the sternum is a small, pointed **xiphoid**[45] (ZIF-oyd) **process** that provides attachment for some of the abdominal muscles.

41. *alae* = wings
42. *coccyx* = cuckoo (named for resemblance to a cuckoo's beak)
43. *manubrium* = handle
44. *gladiolus* = sword
45. *xipho* = sword + *oid* = resembling

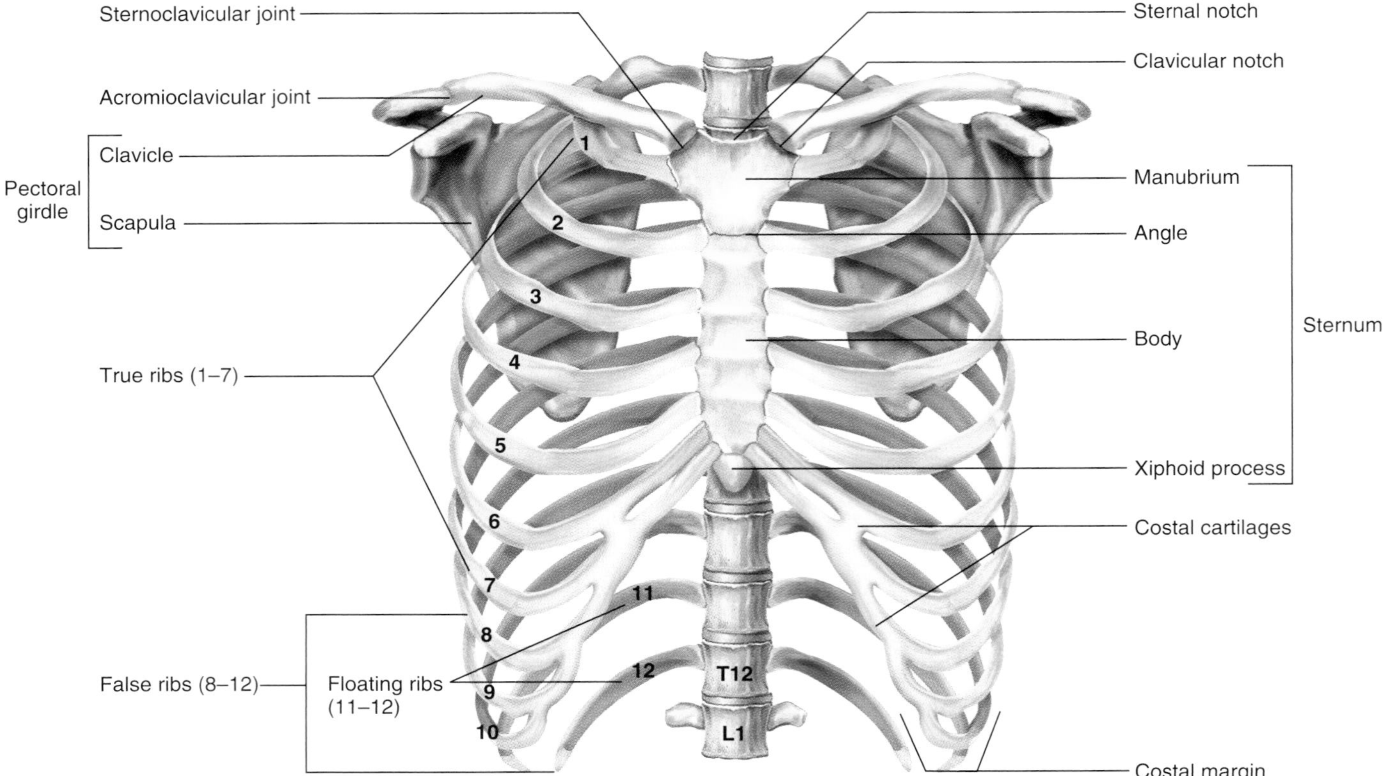

Figure 9.26 The Thoracic Cage and Pectoral Girdle, Anterior View.

Ribs

There are 12 pairs of **ribs.** Each is attached at its posterior (proximal) end to the vertebral column. A strip of hyaline cartilage called the **costal cartilage** extends from the anterior (distal) ends of ribs 1 to 7 to the sternum. Ribs 1 to 7 are thus called **true ribs.** Ribs 8 to 10 attach to the costal cartilage of rib 7, and ribs 11 and 12 do not attach to anything at the distal end but are embedded in thoracic muscle. Ribs 8 to 12 are therefore called **false ribs,** and ribs 11 and 12 are also called **floating ribs** for lack of any connection to the sternum.

Ribs 1 to 10 each have a proximal **head** and **tubercle,** connected by a narrow **neck;** ribs 11 and 12 have a head only (fig. 9.27). Ribs 2 to 9 have beveled heads that come to a point between a **superior articular facet** above and an **inferior articular facet** below. Rib 1, unlike the others, is a flat horizontal plate. Ribs 2 to 10 have a sharp turn called the **angle,** distal to the tubercle, and the remainder consists of a flat blade called the **shaft.** Along the inferior margin of the shaft is a **costal groove** that marks the path of the intercostal blood vessels and nerve.

Variations in rib anatomy relate to the way different ribs articulate with the vertebrae. Once you observe these articulations on an intact skeleton, you will be better able to understand the anatomy of isolated ribs and vertebrae. Vertebra T1 has a complete facet on the body that articulates with rib 1, as well as an inferior demifacet that provides half of the articulation with rib 2. Ribs 2 through 9 all articulate between two vertebrae, so these vertebrae have demifacets. The inferior demifacet of each vertebra articulates with the superior articular facet of the rib; and the superior articular facet of the next vertebra articulates with the inferior articular facet of the same rib. Ribs 10 through 12 each articulate on the body of a single vertebra, so vertebrae T10 to T12 have complete facets on the bodies and no demifacets.

Ribs 1 to 10 each have a second point of attachment to the vertebrae: the tubercle of the rib articulates with the costal facet of the same-numbered vertebra (fig. 9.28). Ribs 11 and 12 articulate only with the vertebral bodies; they do not have tubercles and vertebrae T11 and T12 do not have costal facets.

Table 9.5 summarizes these variations. Table 9.6 provides a checklist that you can use to review your knowledge of the vertebral column and thoracic cage.

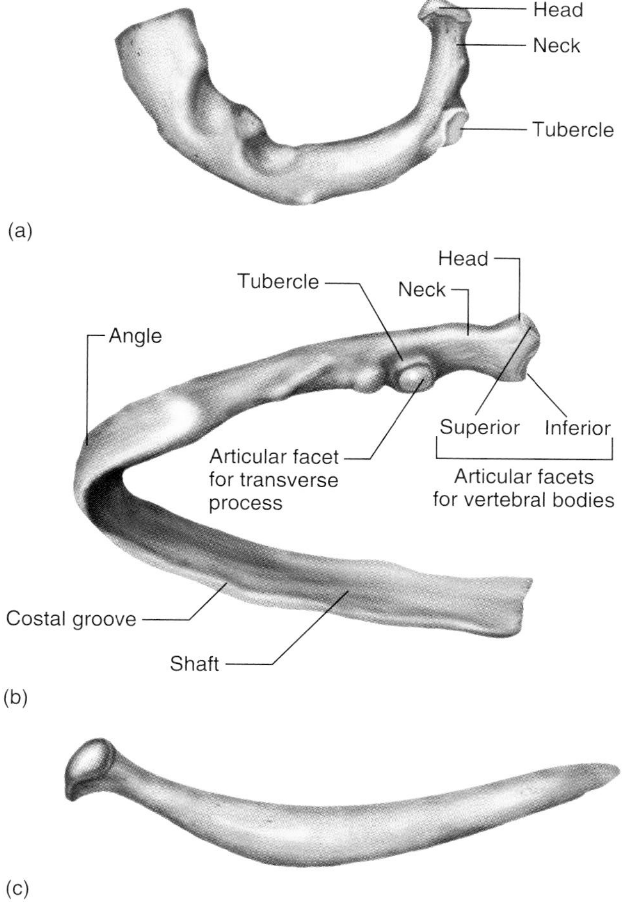

Figure 9.27 Anatomy of the Ribs. (*a*) Rib 1 is an atypical flat plate. (*b*) Typical features of ribs 2 to 10. (*c*) Appearance of the floating ribs, 11 and 12.

Before You Go On

Answer the following questions to test your understanding of the preceding section:

10. Make a table with three columns headed "cervical," "thoracic," and "lumbar." In each column, list the identifying characteristics of each type of vertebra.
11. Describe how rib 5 articulates with the spine. How do ribs 1 and 12 differ from this and from each other in their modes of articulation?
12. Distinguish between true, false, and floating ribs. State which ribs fall into each category.
13. Name the three divisions of the sternum and list the sternal features that can be palpated on a living person.

The Pectoral Girdle and Upper Limb

Objective

When you have completed this section, you should be able to

- identify and describe the features of the clavicle, scapula, humerus, radius, ulna, and bones of the wrist and hand.

Pectoral Girdle

The **pectoral girdle** (shoulder girdle) supports the arm. It consists of two bones on each side of the body: the *clavicle* (collarbone) and the *scapula* (shoulder blade). The medial end of the clavicle articulates with the sternum at the **sternoclavicular joint,** and its lateral end articulates with the scapula at the **acromioclavicular joint** (see fig. 9.26). The scapula also articulates with the humerus at the **humeroscapular joint.** These are loose attachments that result in a shoulder far more flexible than that of most other mammals, but they also make the shoulder joint easy to dislocate.

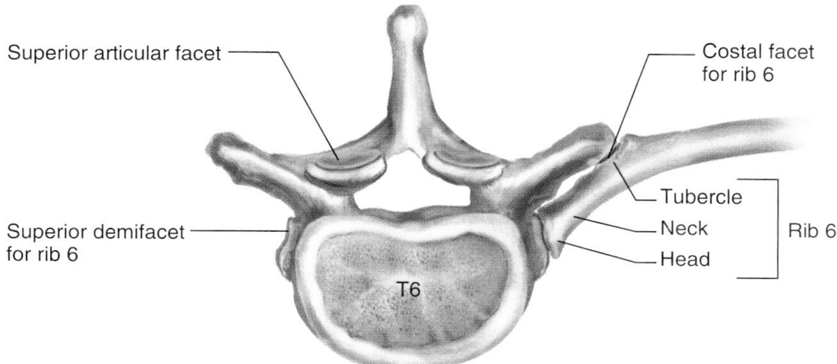

Figure 9.28 Articulation of Thoracic Vertebra T6 with Rib 6. Note that the rib articulates with this vertebra at two points: the vertebral body and transverse process. It also articulates with the body of vertebra T7.

Table 9.5 Articulations of the Ribs

Rib	Type	Costal Cartilage	Articulating Vertebral Bodies	Articulating with a Costal Facet?	Rib Tubercle
1	True	Individual	T1	Yes	Present
2	True	Individual	T1 and T2	Yes	Present
3	True	Individual	T2 and T3	Yes	Present
4	True	Individual	T3 and T4	Yes	Present
5	True	Individual	T4 and T5	Yes	Present
6	True	Individual	T5 and T6	Yes	Present
7	True	Individual	T6 and T7	Yes	Present
8	False	Shared with rib 7	T7 and T8	Yes	Present
9	False	Shared with rib 7	T8 and T9	Yes	Present
10	False	Shared with rib 7	T10	Yes	Present
11	False, floating	None	T11	No	Absent
12	False, floating	None	T12	No	Absent

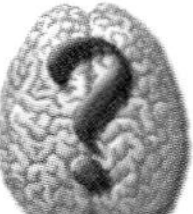

Think About It

How is the unusual flexibility of the human shoulder joint related to the habitat of our primate ancestors?

Clavicle

The **clavicle**[46] (fig. 9.29) is a slightly S-shaped bone, somewhat flattened dorsoventrally and easily seen and palpated on the upper thorax (see fig. B.1*b* in atlas B). The superior surface is relatively smooth, whereas the inferior surface is marked by grooves and ridges for muscle attachment. The medial **sternal end** has a rounded, hammerlike head, and the lateral **acromial end** is markedly flattened. Near the acromial end is a rough tuberosity called the **conoid tubercle**—a ligament attachment that faces toward the rear and slightly downward. The clavicle braces the shoulder and is thickened in people who do heavy manual labor. Without it, the pectoralis major muscles would pull the shoulders forward and medially, as occurs when a clavicle is fractured. Indeed, the clavicle is the most commonly fractured bone in the body because it is so close to the surface and because people often reach out with their arms to break a fall.

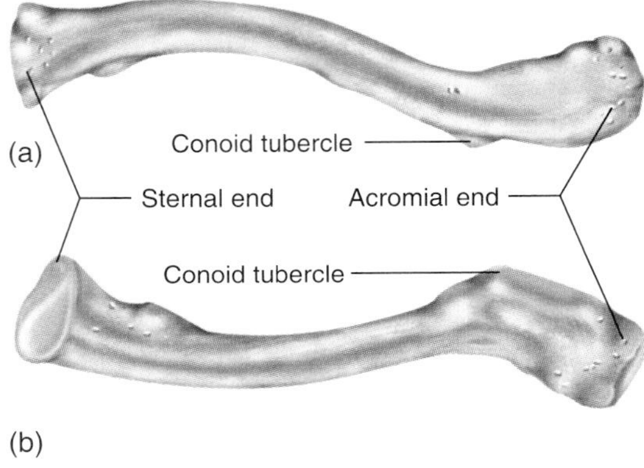

Figure 9.29 The Right Clavicle (collarbone). (*a*) Superior view; (*b*) inferior view.

Scapula

The **scapula** (fig. 9.30) is a triangular plate that dorsally overlies ribs 2 to 7. The three sides of the triangle are called the **superior, medial** (vertebral), and **lateral** (axillary) **borders,** and its three angles are the **superior, inferior,** and **lateral angles.** A conspicuous **suprascapular notch** in the superior border provides passage for a nerve. The broad anterior surface of the scapula, called the **subscapular fossa,** is slightly concave and relatively featureless. The posterior surface has a transverse ridge called the **spine,** a deep indentation superior to the spine called the **supraspinous fossa,** and a broad surface inferior to it called the **infraspinous fossa.**[47] The scapula is held in place by numerous muscles attached to these three fossae.

46. *clav* = hammer, club + *icle* = little

47. *supra* = above; *infra* = below

Table 9.6 Anatomical Checklist for the Vertebral Column and Thoracic Cage

Vertebral Column

Spinal Curvatures (fig. 9.20)
Cervical curvature
Thoracic curvature
Lumbar curvature
Pelvic curvature

General Vertebral Structure (figs. 9.21 and 9.22)
Body (centrum)
Vertebral foramen
Vertebral canal
Neural arch
  Pedicle
  Lamina
Spinous process
Transverse process
Superior articular process
Inferior articular process
Intervertebral foramen
  Inferior vertebral notch
  Superior vertebral notch

Intervertebral Discs (fig. 9.21)
Annulus fibrosus
Nucleus pulposus

Cervical Vertebrae (figs. 9.23 and 9.24a)
Transverse foramina
Bifid spinous process
Atlas
  Anterior arch
  Anterior tubercle
  Posterior arch

Cervical Vertebrae (figs. 9.23 and 9.24a)—(cont.)
  Posterior tubercle
  Lateral mass
  Superior articular facet
  Inferior articular facet
  Transverse ligament
Axis
  Dens (odontoid process)

Thoracic Vertebrae (fig. 9.24b)
Superior demifacet
Inferior demifacet
Costal facet

Lumbar Vertebrae (figs. 9.22 and 9.24c)

Sacral Vertebrae (fig. 9.25)
Sacrum
Anterior sacral foramina
Posterior sacral foramina
Median sacral crest
Lateral sacral crest
Sacral canal
Sacral hiatus
Auricular surface
Superior articular process
Alae

Coccygeal Vertebrae (fig. 9.25)
Coccyx
Cornu

Thoracic Cage

Sternum (fig. 9.26)
Manubrium
  Sternal notch
  Clavicular notch
  Sternal angle
Body (gladiolus)
Xiphoid process

Rib Types (fig. 9.26)
True ribs (ribs 1–7)
False ribs (ribs 8–12)
Floating ribs (ribs 11 and 12)

Rib Features (fig. 9.27)
Head
Superior articular facet
Inferior articular facet
Neck
Tubercle
Angle
Shaft
Costal groove
Costal cartilage

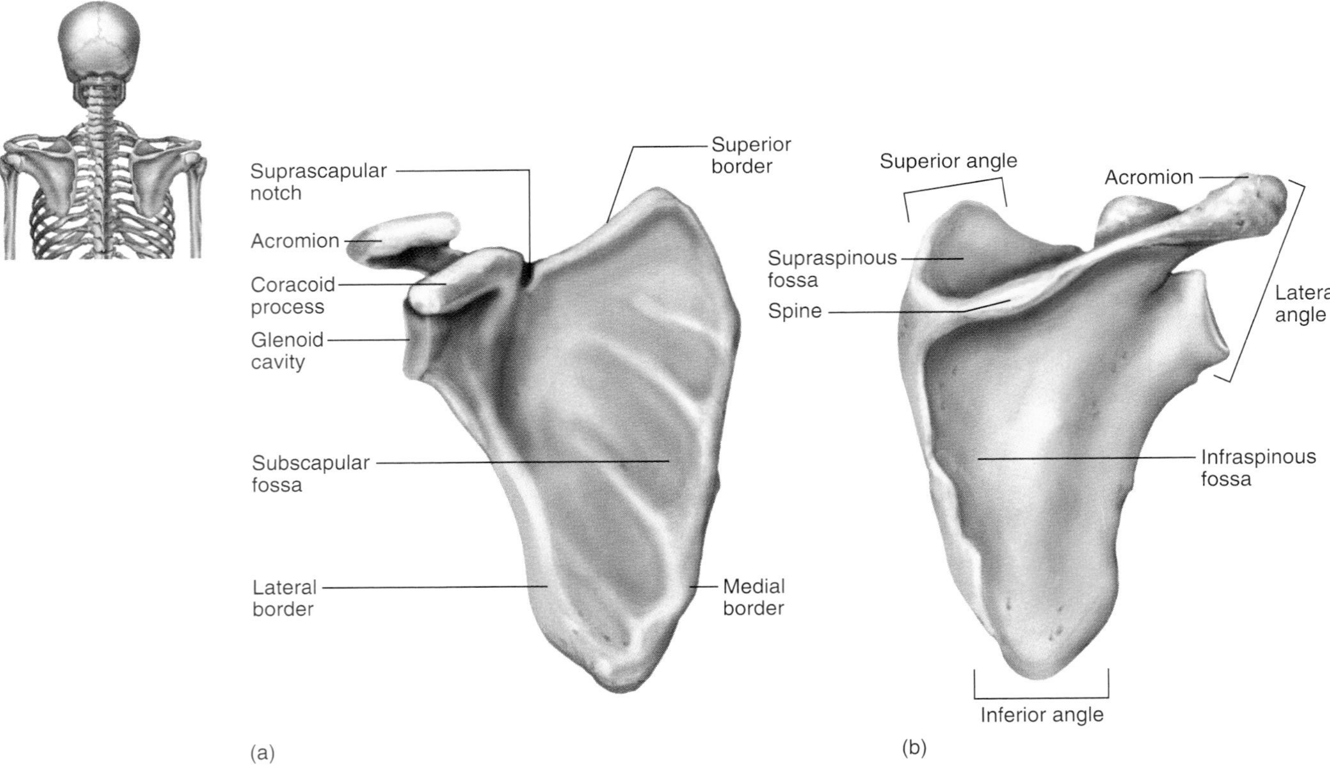

Figure 9.30 The Right Scapula. (*a*) Anterior view; (*b*) posterior view.

The most complex region of the scapula is its lateral angle, which has three main features:

1. The **acromion**[48] (ah-CRO-me-on) is a platelike extension of the scapular spine that forms the apex of the shoulder. It articulates with the clavicle—the sole point of attachment of the arm and scapula to the rest of the skeleton.
2. The **coracoid**[49] (COR-uh-coyd) **process** is shaped like a finger but named for a vague resemblance to a crow's beak; it provides attachment for the biceps brachii and other muscles of the arm.
3. The **glenoid**[50] (GLEN-oyd) **cavity** is a shallow socket that articulates with the head of the humerus.

Upper Limb

The upper limb is divided into four regions containing a total of 30 bones per limb:

1. The **brachium**[51] (BRAY-kee-um), or arm proper, extends from shoulder to elbow. It contains only one bone, the *humerus.*
2. The **antebrachium,**[52] or forearm, extends from elbow to wrist and contains two bones—the *radius* and *ulna.* In anatomical position, these bones are parallel and the radius is lateral to the ulna.
3. The **carpus,**[53] or wrist, contains eight small bones arranged in two rows.
4. The **manus,**[54] or hand, contains 19 bones in two groups—5 *metacarpals* in the palm and 14 *phalanges* in the fingers.

Humerus

The **humerus** has a hemispherical **head** that articulates with the glenoid cavity of the scapula (fig. 9.31). The smooth surface of the head (covered with articular cartilage

48. *acr* = extremity, point + *omi* = shoulder
49. *corac* = crow + *oid* = resembling
50. *glen* = pit, socket
51. *brachi* = arm
52. *ante* = before
53. *carp* = wrist
54. *man* = hand

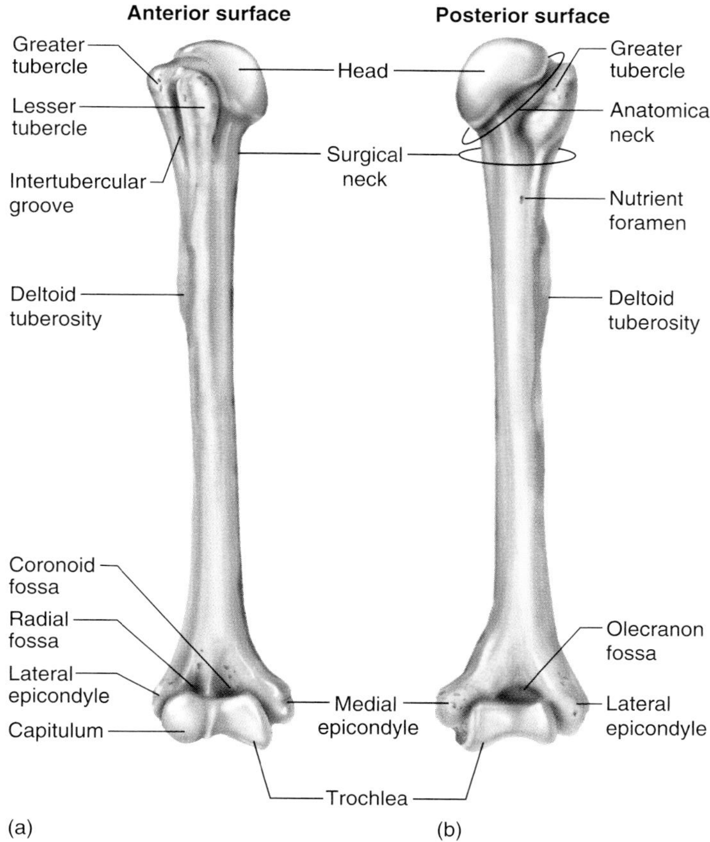

Figure 9.31 The Right Humerus. (*a*) Anterior view; (*b*) posterior view.

in life) is bordered by a groove called the **anatomical neck.** Other prominent features of the proximal end are muscle attachments called the **greater** and **lesser tubercles** and an **intertubercular groove** between them that accommodates a tendon of the biceps muscle. The **surgical neck,** a common fracture site, is a narrowing of the bone just distal to the tubercles, at the transition from the head to the shaft.

The shaft has a rough area called the **deltoid tuberosity** on its lateral surface. This is an insertion for the deltoid muscle of the shoulder. The distal end of the humerus has two smooth condyles. The lateral one, called the **capitulum**[55] (ca-PIT-you-lum), is shaped somewhat like a fat tire and articulates with the radius. The medial one, called the **trochlea**[56] (TROCK-lee-uh), is pulleylike and articulates with the ulna. Immediately proximal to these condyles, the humerus flares out to form two bony processes, the **lateral** and **medial epicondyles.** The medial epicondyle protects the ulnar nerve, which passes close to the surface across the back of the elbow. This epicondyle is popularly known as the "funny bone" because striking the elbow on the edge of a table stimulates the ulnar nerve and produces a sharp tingling sensation.

The distal end of the humerus also shows three deep pits—two anterior and one posterior. The posterior pit, called the **olecranon** (oh-LEC-ruh-non) **fossa,** accommodates the olecranon of the ulna when the arm is extended. On the anterior surface, a medial pit called the **coronoid fossa** accommodates the coronoid process of the ulna when the arm is flexed. The lateral pit is the **radial fossa,** named for the nearby head of the radius.

Radius

The proximal head of the **radius** (fig. 9.32) is a distinctive disc that rotates freely on the humerus when the palm is turned forward and back. It articulates with the capitulum of the humerus and radial notch of the ulna. On the shaft, immediately distal to the head, is a medial rough **tuberosity,** which is the insertion of the biceps muscle. The distal end of the radius has the following features, from lateral to medial:

1. a bony point, the **styloid process,** which can be palpated proximal to the thumb;
2. two shallow depressions (articular facets) that articulate with the scaphoid and lunate bones of the wrist; and
3. the **ulnar notch,** which articulates with the end of the ulna.

Ulna

At the proximal end of the **ulna** (fig. 9.32) is a deep, C-shaped **trochlear notch** that wraps around the trochlea of the humerus. The posterior side of this notch is formed by a prominent **olecranon**—the bony point where you rest your elbow on a table. The anterior side is formed by a less prominent **coronoid process.** Medially, the head of the ulna has a less conspicuous **radial notch,** which accommodates the head of the radius.

At the distal end of the ulna is a medial **styloid process.** The bony lumps you can palpate on each side of your wrist are the styloid processes of the radius and ulna. The radius and ulna are attached along their shafts by a ligament called the **interosseous** (IN-tur-OSS-ee-us) **membrane,** which is attached to an angular ridge called the **interosseous margin** on the medial side of each bone.

Carpal Bones

The **carpal bones,** which form the wrist, are arranged in two rows of four bones each (fig. 9.33). These short bones allow movements of the wrist from side to side and up and

55. *capit* = head + *ulum* = little
56. *troch* = wheel, pulley

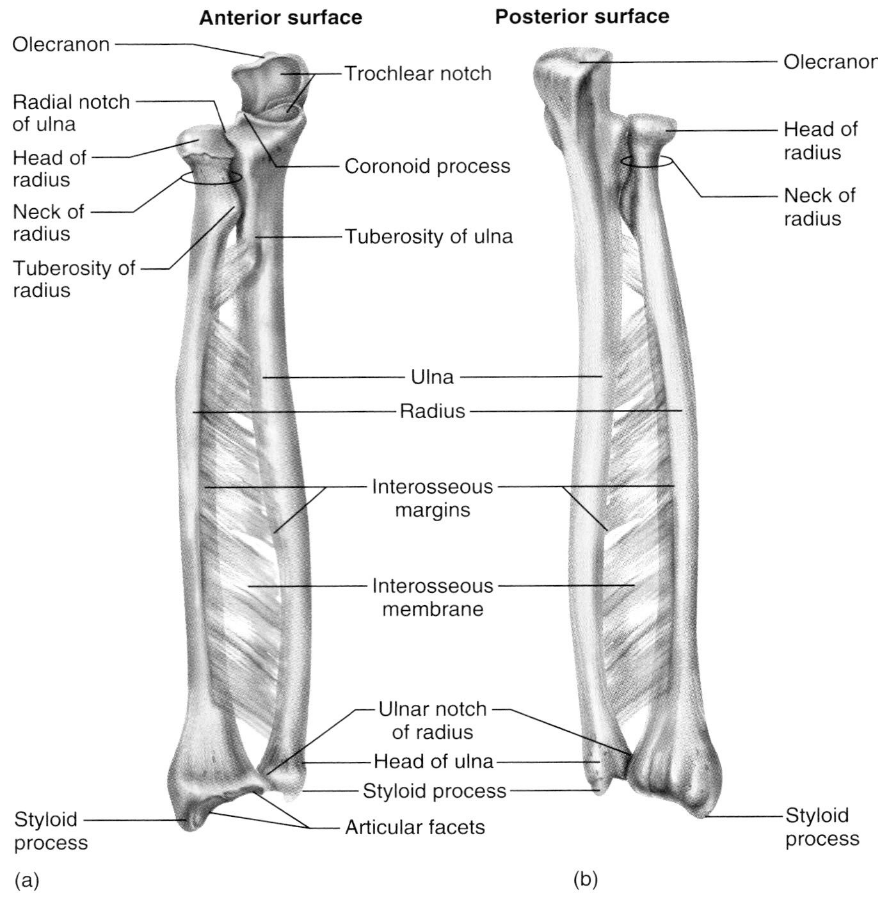

Figure 9.32 The Right Radius and Ulna. (*a*) Anterior view; (*b*) posterior view.

down. The carpal bones of the proximal row, starting at the lateral (thumb) side, are the **scaphoid (navicular), lunate, triquetral** (tri-QUEE-trul), and **pisiform.** Translating from the Latin, these words mean boat-, moon-, triangle-, and pea-shaped, respectively. Unlike the other carpal bones, the pisiform is a sesamoid bone; it develops within the tendon of the *flexor carpi ulnaris muscle.*

The bones of the distal row, again starting on the lateral side, are the **trapezium,**[57] **trapezoid, capitate,**[58] and **hamate.**[59] The hamate can be recognized by a prominent hook, or **hamulus,** on the palmar side.

Metacarpal Bones

Bones of the palm are called **metacarpals.**[60] Metacarpal I is located at the base of the thumb and metacarpal V at the base of the little finger. On a skeleton, the metacarpals look like extensions of the fingers, so that the fingers seem much longer than they really are. The proximal end of a metacarpal bone is called the **base,** the shaft is called the **body,** and the distal end is called the **head.** The heads of the metacarpals form knuckles when you clench your fist.

Phalanges

The bones of the fingers are called **phalanges** (fah-LAN-jeez); in the singular, *phalanx* (FAY-lanks). There are two phalanges in the **pollex** (thumb) and three in each of the other digits. Phalanges are identified by Roman numerals preceded by *proximal, middle,* and *distal.* For example, proximal phalanx I is in the basal segment of the thumb (the first segment beyond the web between the thumb and palm); the left proximal phalanx IV is where people usually wear wedding rings; and distal phalanx V forms the tip of the little finger. The three parts of a phalanx are the same as in a metacarpal: base, body, and head. The ventral surface of a phalanx is slightly concave from end to end and flattened from side to side; the dorsal surface is rounder and slightly convex.

Table 9.7 summarizes the bones of the pectoral girdle and upper limb.

57. *trapez* = table, grinding surface
58. *capit* = head + *ate* = possessing
59. *ham* = hook + *ate* = possessing
60. *meta* = beyond + *carp* = wrist

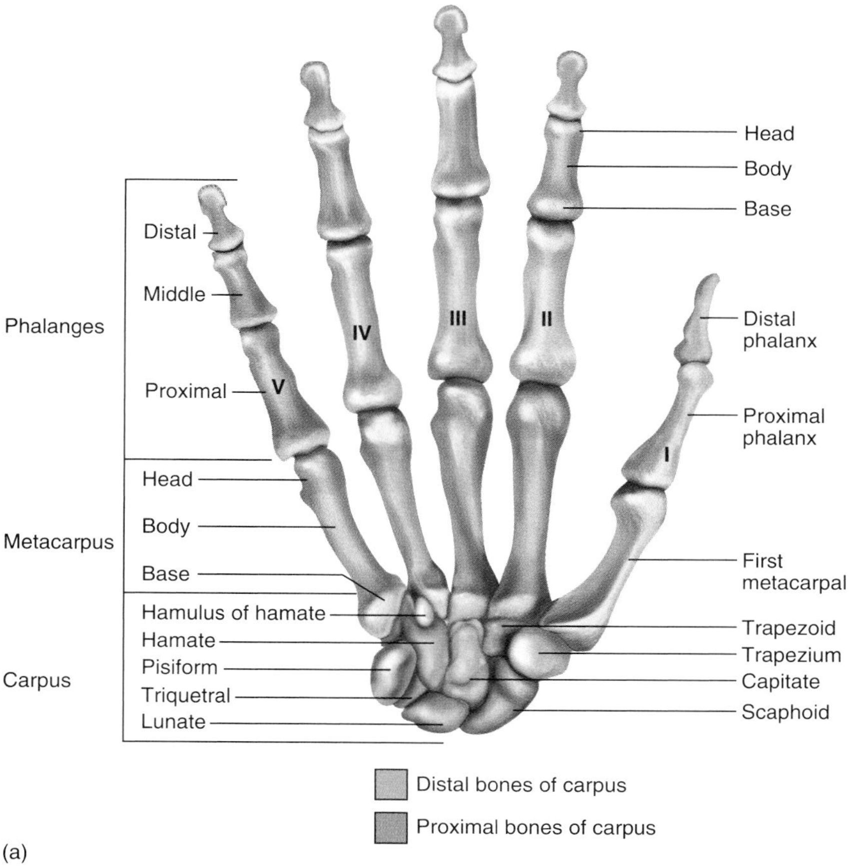

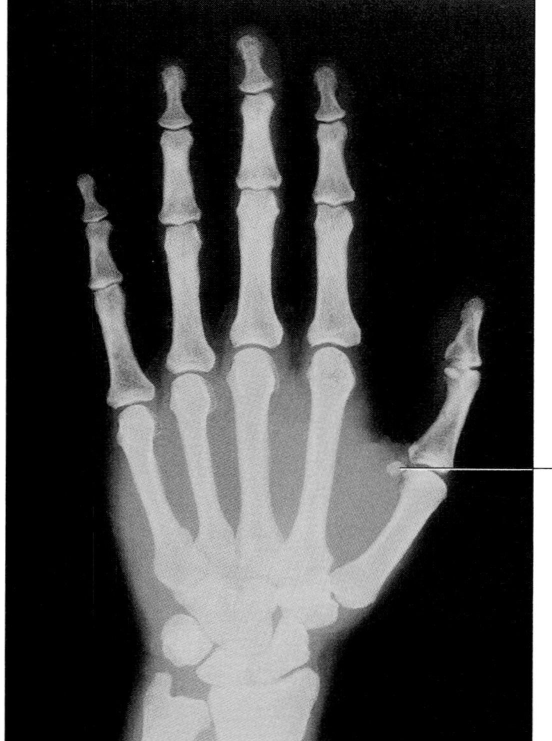

Figure 9.33 The Right Wrist and Hand, Anterior (palmar) View. (*a*) Carpal bones are color-coded to distinguish the proximal and distal rows. (*b*) X ray of an adult hand. Compare to a child's hand, figure 8.12. Identify the unlabeled bones in the X ray by comparing it to the drawing.

Table 9.7 Anatomical Checklist for the Pectoral Girdle and Upper Limb

Pectoral Girdle

Clavicle (fig. 9.29)

Sternal end
Acromial end
Conoid tubercle

Scapula (fig. 9.30)

Borders
- Superior border
- Medial (vertebral) border
- Lateral (axillary) border

Angles
- Superior angle
- Inferior angle
- Lateral angle

Scapula (fig. 9.30)—(cont.)

Suprascapular notch
Spine
Fossae
- Subscapular fossa
- Supraspinous fossa
- Infraspinous fossa

Acromion
Coracoid process
Glenoid cavity
Olecranon

Upper Limb

Humerus (fig. 9.31)

Proximal end
- Head
- Anatomical neck
- Surgical neck
- Greater tubercle
- Lesser tubercle
- Intertubercular groove

Shaft
- Deltoid tuberosity

Distal end
- Capitulum
- Trochlea
- Lateral epicondyle
- Medial epicondyle
- Olecranon fossa
- Coronoid fossa
- Radial fossa

Radius (fig. 9.32)

Head
Tuberosity
Styloid process
Articular facets
Ulnar notch

Ulna (fig. 9.32)

Trochlear notch
Coronoid process

Ulna (fig. 9.32)—(cont.)

Radial notch
Styloid process
Interosseous margin
Interosseous membrane

Carpal Bones (fig. 9.33)

Proximal group
- Scaphoid
- Lunate
- Triquetral
- Pisiform

Distal group
- Trapezium
- Trapezoid
- Capitate
- Hamate
  - Hamulus

Bones of the Hand (fig. 9.33)

Metacarpal bones I–V
- Base
- Body
- Head

Phalanges I–V
- Proximal phalanx
- Middle phalanx
- Distal phalanx

Before You Go On

Answer the following questions to test your understanding of the preceding section.

14. Describe how to distinguish the medial and lateral ends of the clavicle from each other, and how to distinguish its superior and inferior surfaces.
15. Name the three fossae of the scapula and describe the location of each.
16. What three bones meet at the elbow? Identify the fossae, articular surfaces, and processes of this joint and state to which bone each of these features belongs.
17. Name the four bones of the proximal row of the carpus from lateral to medial, and then the four bones of the distal row in the same order.
18. Name the four bones from the tip of the little finger to the base of the hand on that side.

The Pelvic Girdle and Lower Limb

Objectives

When you have completed this section, you should be able to

- identify and describe the features of the pelvic girdle, femur, patella, tibia, fibula, and bones of the foot; and
- compare the anatomy of the male and female pelvis and explain the functional significance of the differences.

Pelvic Girdle

The adult **pelvic**[61] **girdle** is composed of four bones: a right and left **os coxae** (plural, ossa coxae), the sacrum, and the coccyx (fig. 9.34). Another term for the os coxae—arguably the most self-contradictory term in anatomy—is the *innominate*[62] (ih-NOM-ih-nate) *bone,* "the bone with no name." The pelvic girdle supports the trunk on the legs and encloses and protects viscera of the pelvic cavity—mainly the lower colon, urinary bladder, and reproductive organs.

Each os coxae is joined to the vertebral column at one point, the sacroiliac joint, where its **auricular surface** matches the one on the sacrum. On the anterior side of the pelvis is the **pubic symphysis,**[63] the point where the right and left pubic bones are joined by a pad of fibrocartilage (the *interpubic disc*). The symphysis can be palpated immediately above the genitalia.

The pelvic girdle has a bowl-like shape with the broad **greater (false) pelvis** between the flare of the hips and the narrower **lesser (true) pelvis** below. The two are separated by a somewhat round margin called the **pelvic brim.** The opening circumscribed by the brim is called the **pelvic inlet**—an entry into the lesser pelvis through which an infant's head passes during birth. The lower margin of the lesser pelvis is called the **pelvic outlet.**

61. *pelv* = basin, bowl
62. *in* = without + *nomin* = name + *ate* = having
63. *sym* = together + *physis* = growth

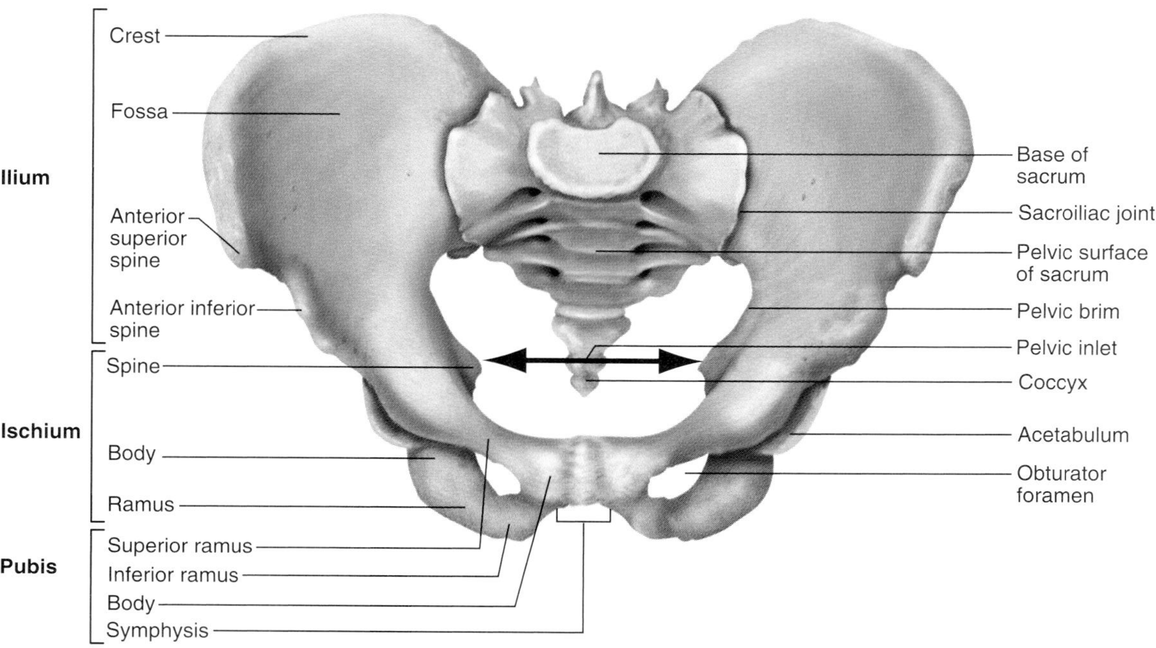

Figure 9.34 **The Pelvic Girdle, Anterosuperior View.** The pelvic girdle consists of the os coxae, sacrum, and coccyx.

The os coxae has three distinctive features that will serve as landmarks for further description. These are the **iliac**[64] **crest** (superior crest of the hip); **acetabulum**[65] (ASS-eh-TAB-you-lum) (the hip socket—named for its resemblance to vinegar cups used in ancient Rome); and **obturator**[66] **foramen** (a large round-to-triangular hole below the acetabulum, closed by a ligament called the *obturator membrane* in life).

The adult os coxae forms by the fusion of three childhood bones called the *ilium* (ILL-ee-um), *ischium* (ISS-kee-um), and *pubis* (PEW-biss), identified by color in figure 9.35. The largest of these is the **ilium,** which extends from the iliac crest to the superior wall of the acetabulum. The iliac crest extends from a point or angle on the anterior side, called the **anterior superior spine,** to a sharp posterior angle, called the **posterior superior spine.** In a lean person, the anterior superior spines form visible anterior protrusions, and the posterior superior spines are sometimes marked by dimples above the buttocks where connective tissue attached to the spines pulls inward on the skin.

64. *ili* = flank, loin + *ac* = pertaining to
65. *acetabulum* = vinegar cup
66. *obtur* = to close, stop up + *ator* = that which

Below the superior spines are the **anterior** and **posterior inferior spines.** Below the posterior inferior spine is a deep **greater sciatic** (sy-AT-ic) **notch,** named for the sciatic nerve that passes through it and continues down the posterior side of the thigh. The ilium is thickened along a line called the **iliac pillar** which extends from the iliac crest to the superior margin of the acetabulum. Both the notch and pillar result from evolution of the pelvis for bipedal locomotion, as explained in evolutionary insight 9.5 (p. 301).

The posterolateral surface of the ilium is relatively rough-textured because it serves for attachment of several muscles of the buttocks and thighs. The anteromedial surface, by contrast, is the smooth, slightly concave **iliac fossa,** covered in life by the broad *iliacus* muscle. Medially, the ilium exhibits an auricular surface that matches the one on the sacrum, so that the two bones form the sacroiliac joint.

The **ischium** forms the inferoposterior portion of the os coxae. Its heavy **body** is marked with a prominent **spine.** Inferior to the spine is a slight indentation, the **lesser sciatic notch,** and then the thick, rough-surfaced **ischial tuberosity,** which supports your body when you are sitting. The tuberosity can be palpated by sitting on your fingers. The **ramus** of the ischium joins the inferior ramus of the pubis anteriorly.

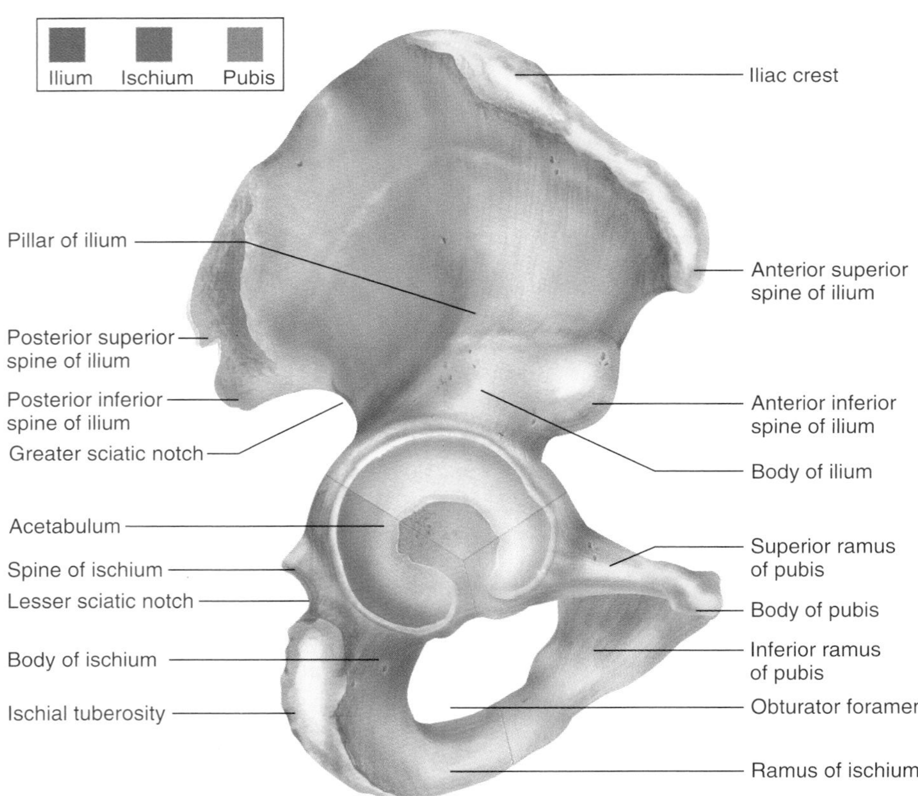

Figure 9.35 **The Right Os Coxae, Lateral View.** The three childhood bones that fuse to form the adult os coxae are identified by color.

Table 9.8 Comparison of the Male and Female Pelves

	Male	Female
General Appearance	More massive; rougher; heavier processes	Less massive; smoother; more delicate processes
Tilt	Upper end of pelvis relatively vertical	Upper end of pelvis tilted forward
Ilium	Deeper; projects farther above sacroiliac joint	Shallower; does not project as far above sacroiliac joint
Sacrum	Narrower and longer	Shorter and wider
Coccyx	Less movable; more vertical	More movable; tilted backward
Width of Greater Pelvis	Anterior superior spines closer together, hips less flared	Anterior superior spines farther apart; hips more flared
Pelvic Inlet	Heart-shaped	Round or oval
Pubic Symphysis	Taller	Shorter
Greater Sciatic Notch	Wider	Narrower
Obturator Foramen	Oval	Triangular
Acetabulum	Faces more laterally	Faces slightly anteriorly
Pubic arch	Usually 90° or less	Usually greater than 100°

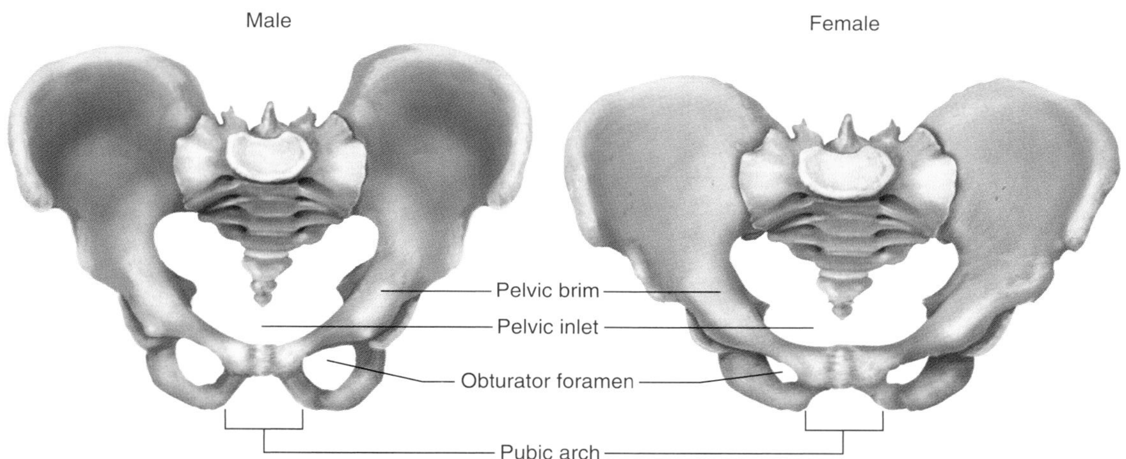

Figure 9.36 Comparison of the Male and Female Pelvic Girdles.

The **pubis** (pubic bone) is the most anterior portion of the os coxae. It has a **superior** and **inferior ramus** and a triangular **body.** The body of one pubis meets the body of the other at the pubic symphysis. The pubis and ischium encircle the obturator foramen.

The female pelvis is adapted to the needs of pregnancy and childbirth. Some of the differences between the male and female pelves are described in table 9.8 and illustrated in figure 9.36.

Lower Limb

The number and arrangement of bones in the lower limb are similar to those of the upper limb. In the lower limb, however, they are adapted for weight-bearing and locomotion and are therefore shaped and articulated differently.

The lower limb is divided into four regions containing a total of 30 bones per limb:

1. The **femoral region,** or thigh, extends from hip to knee and contains the *femur* (the longest bone in the body). The *patella* (kneecap) is a sesamoid bone at the junction of the femoral and crural regions.
2. The **crural** (CROO-rul) **region,** or leg proper, extends from knee to ankle and contains two bones, the medial *tibia* and lateral *fibula.*
3. The **tarsal region (tarsus),** or ankle, is the union of the crural region with the foot. The tarsal bones are treated as part of the foot.
4. The **pedal region (pes),** or foot, is composed of 7 *tarsal bones,* 5 *metatarsals,* and 14 *phalanges* in the toes.

Femur

The **femur** (FEE-mur) (fig. 9.37) has a nearly spherical head that articulates with the acetabulum of the pelvis, forming a quintessential *ball-and-socket joint*. A ligament extends from the acetabulum to a pit, the **fovea capitis**[67] (FOE-vee-uh CAP-ih-tiss), in the head of the femur. Distal to the head is a constricted **neck** and then two massive, rough processes called the **greater** and **lesser trochanters** (tro-CAN-turs), which are insertions for the powerful muscles of the hip. They are connected on the posterior side by a thick oblique ridge of bone, the **intertrochanteric crest,** and on the anterior side by a more delicate **intertrochanteric line.**

The primary feature of the shaft is a posterior ridge called the **linea aspera**[68] (LIN-ee-uh ASS-peh-ruh) at its midpoint. It branches into less conspicuous lateral and medial ridges at its inferior and superior ends.

The distal end of the femur flares into **medial** and **lateral epicondyles,** which serve as sites of muscle and ligament attachment. Distal to these are two smooth round surfaces of the knee joint, the **medial** and **lateral condyles,** separated by a groove called the **intercondylar** (IN-tur-CON-dih-lur) **fossa.** On the anterior side of the femur, a smooth medial depression called the **patellar surface** articulates with the patella.

67. *fovea* = pit + *capitis* = of the head
68. *linea* = line + *asper* = rough

Patella

The **patella,** or kneecap (fig. 9.38), is a roughly triangular sesamoid bone that forms within the tendon of the knee as a child begins to walk. It has a broad superior **base,** a pointed inferior **apex,** and a pair of shallow **articular facets** on its posterior surface where it articulates with the femur. The lateral facet is usually larger than the medial.

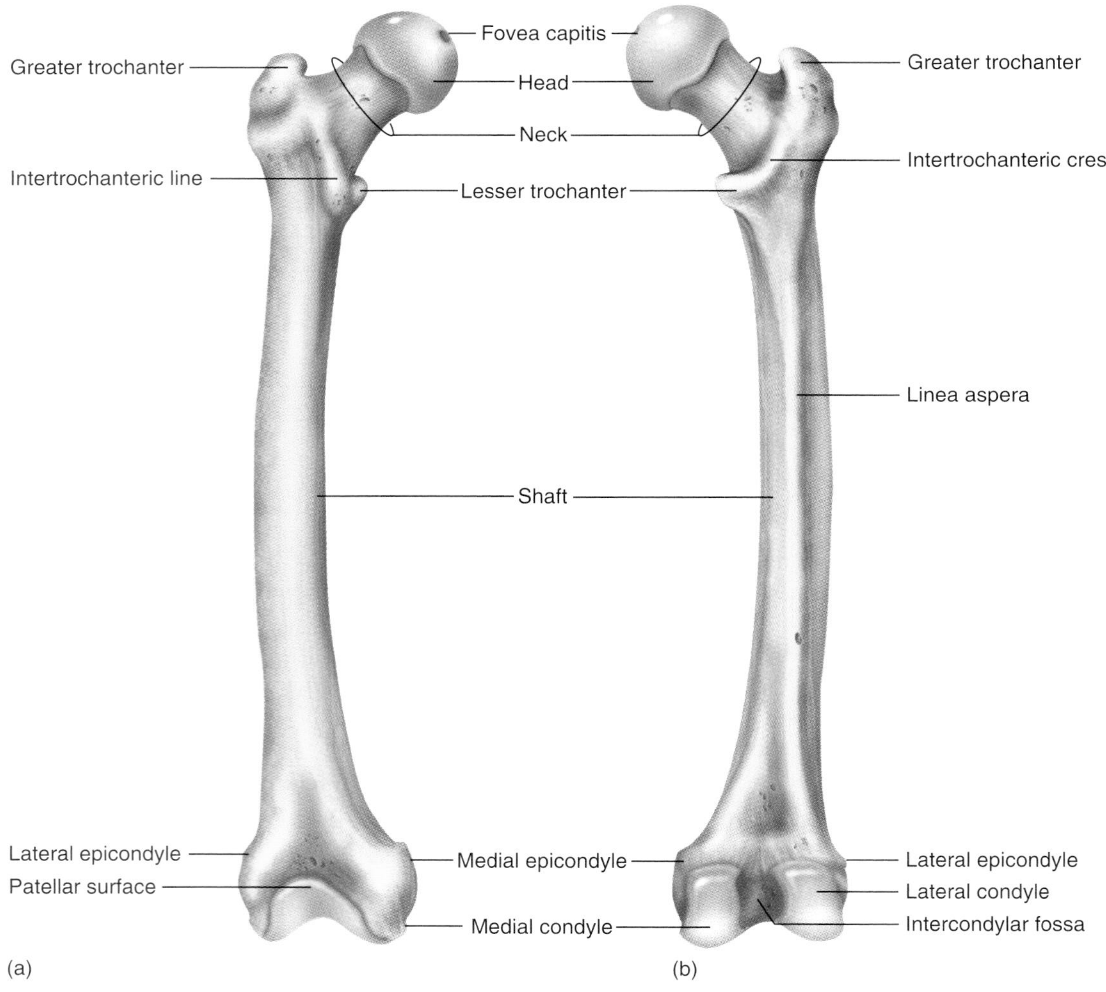

Figure 9.37 The Right Femur. (*a*) Anterior view; (*b*) posterior view.

The *quadriceps femoris tendon* extends from the anterior muscle of the thigh (the *quadriceps femoris*) to the patella, and it continues as the *patellar ligament* from the patella to the tibia.

Tibia

The leg has two bones—a thick, strong tibia (TIB-ee-uh) and a slender, lateral fibula (FIB-you-luh) (fig. 9.38). The **tibia,** on the medial side, is the only weight-bearing bone of the crural region. Its broad superior head has two fairly flat articular surfaces, the **medial** and **lateral condyles,** separated by a ridge called the **intercondylar eminence.** The condyles of the tibia articulate with those of the femur. The rough anterior surface of the tibia, the **tibial tuberosity,** can be palpated just below the patella. This is where the patellar ligament inserts and the thigh muscles exert their pull when they extend the leg. Distal to this, the shaft has a sharply angular **anterior crest,** which can be palpated in the shin region. At the ankle, just above the rim of a standard dress shoe, you can palpate a prominent bony knob on each side. These are the **medial** and **lateral malleoli**[69] (MAL-ee-OH-lie). The medial malleolus is part of the tibia, and the lateral malleolus is the part of the fibula.

Fibula

The **fibula** (fig. 9.38) is a slender lateral strut that helps to stabilize the ankle. It does not bear any of the body's weight; indeed, orthopedic surgeons sometimes remove the fibula and use it to replace damaged or missing bone elsewhere in the body. The fibula is somewhat thicker and broader at its proximal end, the **head,** than at the distal end. The point of the head is called the **apex,** or **styloid process.** The distal expansion is the lateral malleolus.

69. *malle* = hammer + *olus* = little

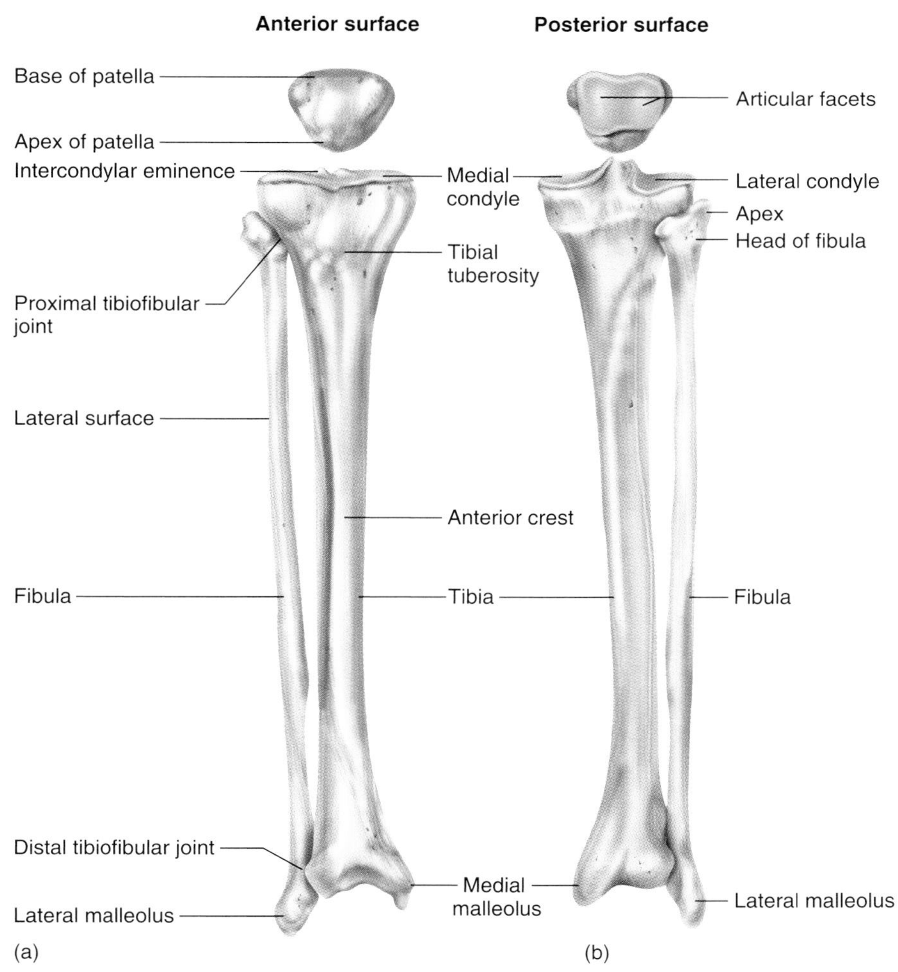

Figure 9.38 The Right Leg and Knee. (*a*) Anterior view; (*b*) posterior view.

The Ankle and Foot

The **tarsal bones** of the ankle are arranged in proximal and distal groups somewhat like the carpal bones of the wrist (fig. 9.39). Because of the load-bearing role of the ankle, however, their shapes and arrangement are conspicuously different from those of the carpal bones, and they are thoroughly integrated into the structure of the foot. The largest tarsal bone is the **calcaneus**[70] (cal-CAY-nee-us), which forms the heel. Its posterior end is the point of attachment for the **calcaneal (Achilles) tendon** from the calf muscles. The second-largest tarsal bone, and the most superior, is the **talus**. It has three articular surfaces: an inferoposterior one that articulates with the calcaneus, a superior **trochlear surface** that articulates with the tibia, and an anterior surface that articulates with a short, wide tarsal bone called the **navicular.** The talus, calcaneus, and navicular are considered the proximal row of tarsal bones. (*Navicular* is also used as a synonym for the scaphoid bone of the wrist.)

The distal group forms a row of four bones. Proceeding from the medial side to the lateral, these are the **first, second,** and **third cuneiforms**[71] (cue-NEE-ih-forms) and the **cuboid.** The cuboid is the largest.

The remaining bones of the foot are similar in arrangement and name to those of the hand. The proximal **metatarsals**[72] are similar to the metacarpals. They are **metatarsals I** to **V** from medial to lateral, metatarsal I being proximal to the great toe. (Note that Roman numeral I represents the *medial* group of bones in the foot but the *lateral* group in the hand. In both cases, however, Roman numeral I refers to the largest digit of the limb.) Metatarsals I to III articulate with the first through third cuneiforms; metatarsals IV and V both articulate with the cuboid.

Bones of the toes, like those of the fingers, are called phalanges. The great toe is the **hallux** and contains only two bones, the proximal and distal phalanx I. The other toes each contain a proximal, middle, and distal phalanx. The metatarsal and phalangeal bones each have a base, body, and head, like the bones of the hand. All of them, especially the phalanges, are slightly concave on the ventral side.

70. *calc* = stone, chalk

71. *cunei* = wedge + *form* = in the shape of

72. *meta* = beyond + *tars* = ankle

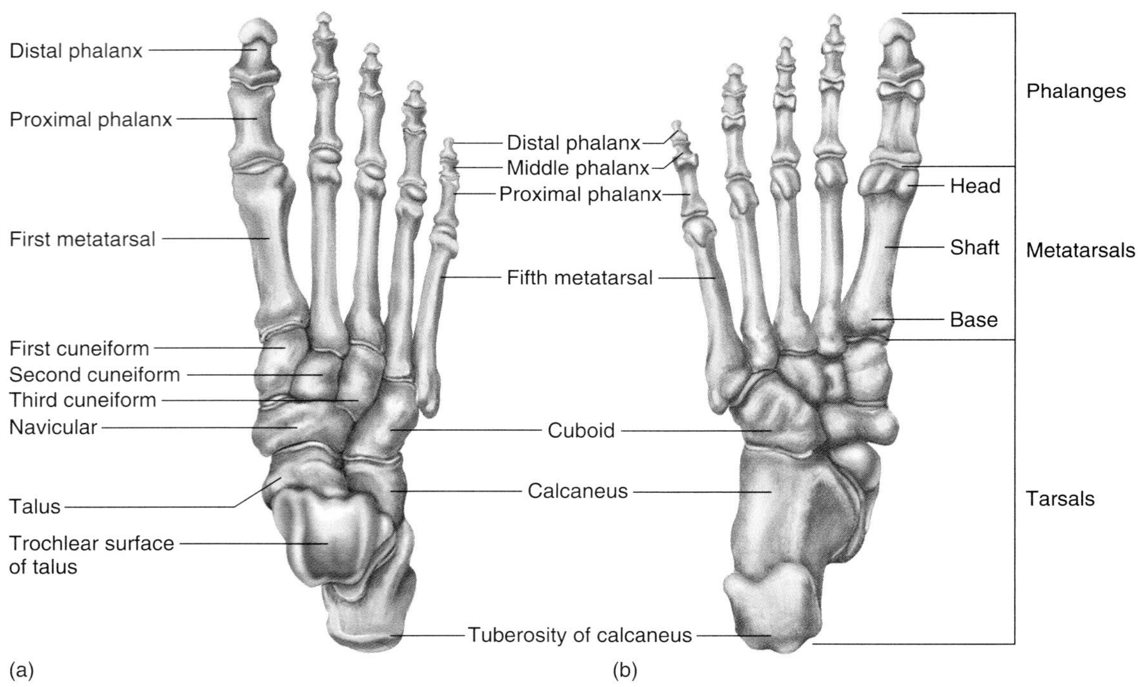

Figure 9.39 **The Right Foot.** (*a*) Superior view; (*b*) inferior view.

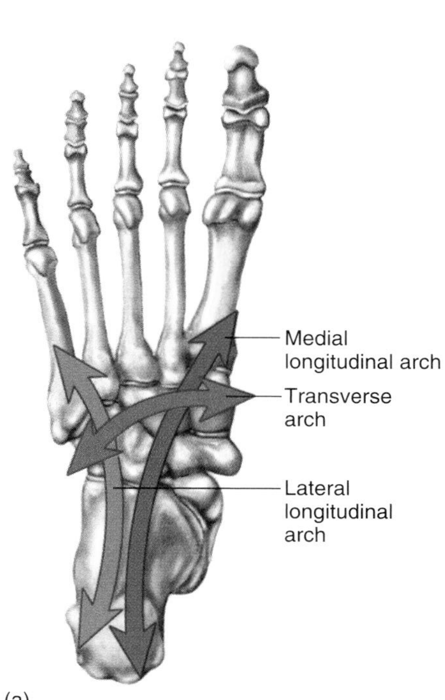

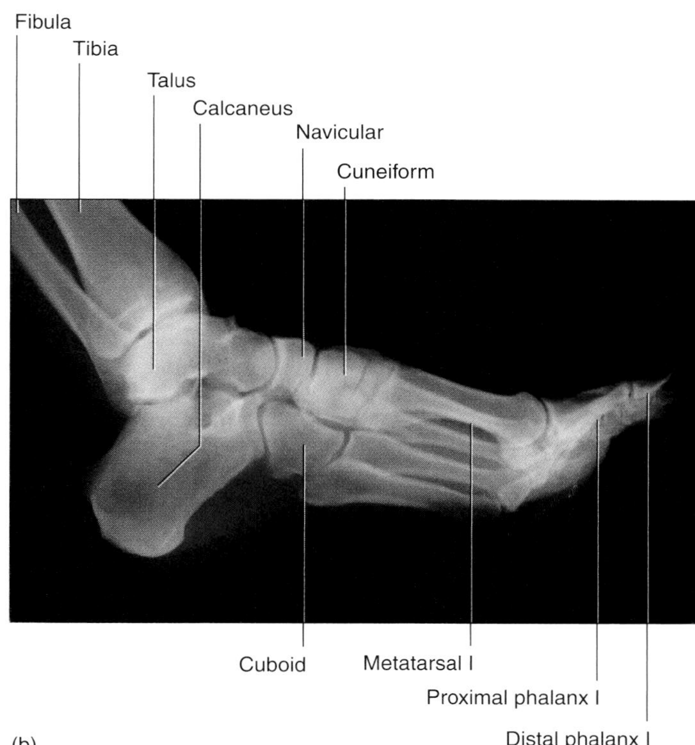

Figure 9.40 Arches of the Foot. (*a*) Inferior view of the right foot. (*b*) X ray of the right foot, lateral view, showing the lateral longitudinal arch.

The sole of the foot normally does not rest flat on the ground; rather, it has three springy arches that absorb the stress of walking (fig. 9.40). The **medial longitudinal arch,** which essentially extends from heel to hallux, is formed from the calcaneus, talus, navicular, cuneiforms, and metatarsals I to III. The **lateral longitudinal arch** extends from heel to little toe and includes the calcaneus, cuboid, and metatarsals IV and V. The **transverse arch** includes the cuboid, cuneiforms, and proximal heads of the metatarsals. These arches are held together by short, strong ligaments. Excessive weight, repetitious stress, or congenital weakness of these ligaments can stretch them, resulting in *pes planis* (commonly called flat feet or fallen arches). This condition makes a person less tolerant of prolonged standing and walking. A comparison of the flat-footed apes with humans underscores the significance of the human foot arches (see evolutionary insight 9.5, p. 301).

Table 9.9 summarizes the pelvic girdle and lower limb.

Before You Go On

Answer the following questions to test your understanding of the preceding section:

19. Name the bones of the adult pelvic girdle. What three bones of a child fuse to form the os coxae of an adult?
20. Name any four structures of the pelvis that you can palpate and describe where to palpate them.
21. What parts of the femur are involved in the hip joint? What parts are involved in the knee joint?
22. Name the prominent knobs on each side of your ankle. What bones contribute to these structures?
23. Name all the bones that articulate with the talus and describe the location of each.

Table 9.9 Anatomical Checklist for the Pelvic Girdle and Lower Limb

Pelvic Girdle

Os Coxae (figs. 9.34 and 9.35)

- Pubic symphysis
- Greater (false) pelvis
- Lesser (true) pelvis
- Pelvic brim
- Pelvic inlet
- Pelvic outlet
- Acetabulum
- Obturator foramen
- Ilium
  - Iliac crest
  - Anterior superior spine
  - Anterior inferior spine
  - Posterior superior spine
  - Posterior inferior spine

Ilium—*(cont.)*

  - Greater sciatic notch
  - Iliac pillar
  - Iliac fossa
  - Auricular surface
- Ischium
  - Body
  - Ischial spine
  - Lesser sciatic notch
  - Ischial tuberosity
  - Ramus
- Pubis
  - Superior ramus
  - Inferior ramus
  - Body

Lower Limb

Femur (fig. 9.37)

- Proximal end
  - Head
  - Fovea capitis
  - Neck
  - Greater trochanter
  - Lesser trochanter
  - Intertrochanteric crest
  - Intertrochanteric line
- Shaft
  - Linea aspera
- Distal end
  - Medial condyle
  - Lateral condyle
  - Intercondylar fossa
  - Medial epicondyle
  - Lateral epicondyle
  - Patellar surface

Patella (fig. 9.38)

- Base
- Apex
- Articular facets

Tibia (fig. 9.38)

- Medial condyle
- Lateral condyle
- Intercondylar eminence
- Tibial tuberosity

Tibia (fig. 9.38)—(cont.)

- Anterior crest
- Medial malleolus

Fibula (fig. 9.38)

- Head
- Apex (styloid process)
- Lateral malleolus

Tarsal Bones (fig. 9.39)

- Proximal group
  - Calcaneus
  - Talus
  - Navicular
- Distal group
  - First cuneiform
  - Second cuneiform
  - Third cuneiform
  - Cuboid

Bones of the Foot (figs. 9.39 and 9.40)

- Metatarsal bones I–V
- Phalanges
  - Proximal phalanx
  - Middle phalanx
  - Distal phalanx
- Arches of the foot
  - Medial longitudinal arch
  - Lateral longitudinal arch
  - Transverse arch

Evolutionary Insight 9.5

Skeletal Adaptations for Bipedalism

Some mammals can stand, hop, or walk briefly on their hind legs, but humans are the only mammals that are habitually bipedal. Footprints preserved in a layer of volcanic ash in Tanzania indicate that hominids walked upright as early as 3.6 million years ago. This bipedal locomotion is possible only because of several adaptations of the human feet, legs, spine, and skull (fig. 1). These features are so distinctive that paleoanthropologists (those who study human fossil remains) can tell with considerable certainty whether a fossil species was able to walk upright.

As important as the hand has been to human evolution, the foot may be an even more significant adaptation. Unlike other mammals, humans support their entire body weight on two feet. While apes are flat-footed, humans have strong, springy foot arches that absorb shock as the body jostles up and down during walking and running. The tarsal bones are tightly articulated with each other, and the calcaneus is strongly developed. The hallux (great toe) is not opposable as it is in most Old World monkeys and apes, but it is highly developed so that it provides the "toe-off" that pushes the body forward in the last phase of the stride. For this reason, loss of the hallux has a more crippling effect than the loss of any other toe.

While the femurs of apes are nearly vertical, in humans they angle medially from the hip to the knee. This places our knees closer together, beneath the body's center of gravity. We lock our knees when standing, allowing us to maintain an erect posture with little muscular effort. Apes cannot do this, and they cannot stand on two legs for very long without tiring—much as you would if you tried to maintain an erect posture with your knees slightly bent.

In apes and other quadrupedal (four-legged) mammals, the abdominal viscera are supported by the muscular wall of the abdomen. In humans, the viscera bear down on the floor of the pelvic cavity, and a bowl-shaped pelvis is necessary to support their weight. This has resulted in a narrower pelvic outlet—a condition quite incompatible with the fact that we, including our infants, are such a large-brained species. The pain of childbirth is unique to humans and, one might say, a price we must pay for having both a large brain and a bipedal stance.

The largest muscle of the buttock, the *gluteus maximus,* serves in apes primarily as an abductor of the thigh—that is, it moves the leg laterally. In humans, however, the ilium has expanded posteriorly, so the gluteus maximus originates behind the hip joint. This changes the function of the muscle—instead of abducting the thigh, it pulls the thigh back in the second half of a stride (pulling back on your right thigh, for example, when your left foot is off the ground and swinging forward). This action accounts for the smooth, efficient stride of a human as compared to the awkward, shuffling gait of a chimpanzee or gorilla when it is walking upright. The posterior growth of the ilium is the reason the greater sciatic notch is so deeply concave.

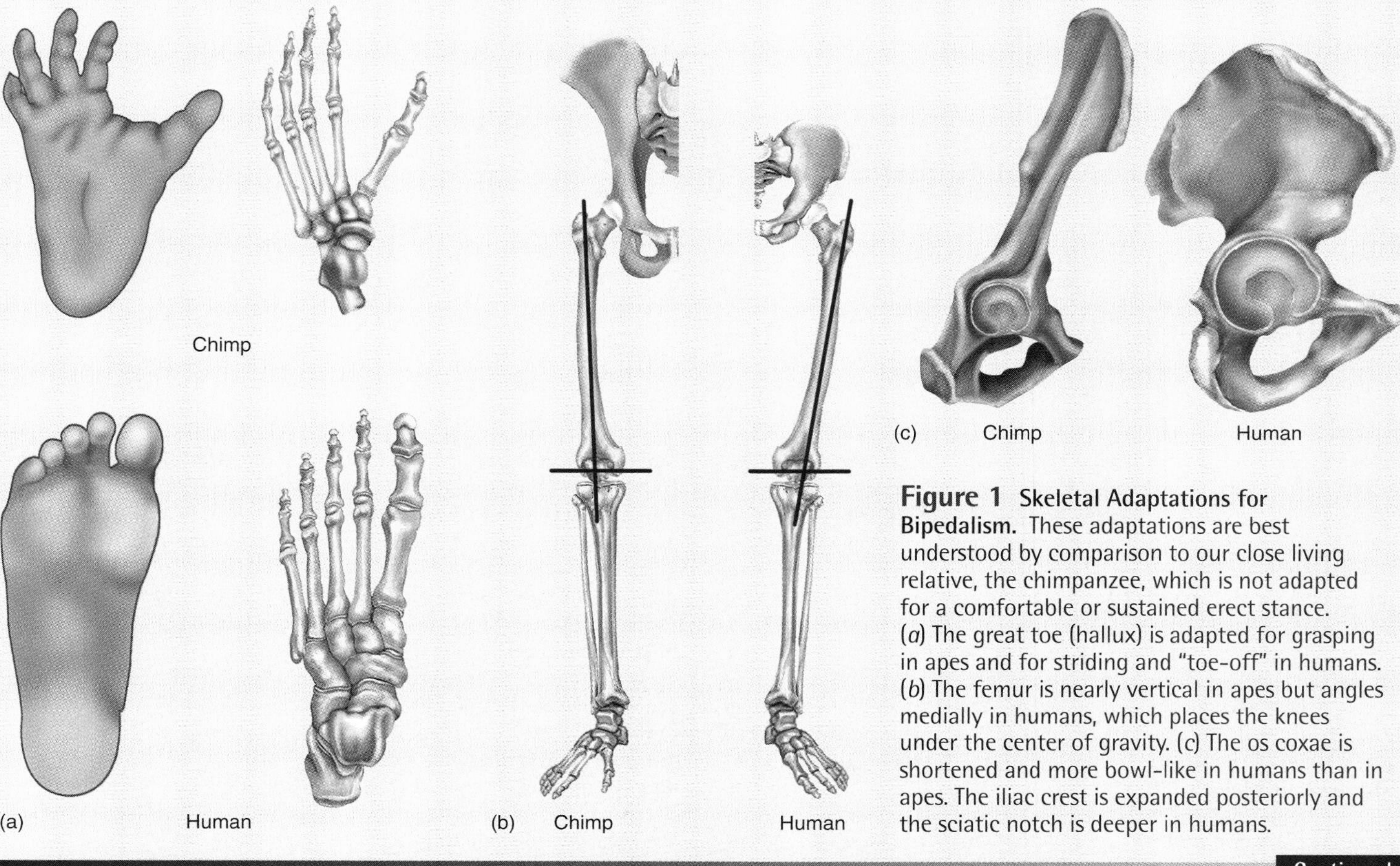

Figure 1 Skeletal Adaptations for Bipedalism. These adaptations are best understood by comparison to our close living relative, the chimpanzee, which is not adapted for a comfortable or sustained erect stance. (*a*) The great toe (hallux) is adapted for grasping in apes and for striding and "toe-off" in humans. (*b*) The femur is nearly vertical in apes but angles medially in humans, which places the knees under the center of gravity. (*c*) The os coxae is shortened and more bowl-like in humans than in apes. The iliac crest is expanded posteriorly and the sciatic notch is deeper in humans.

Continued

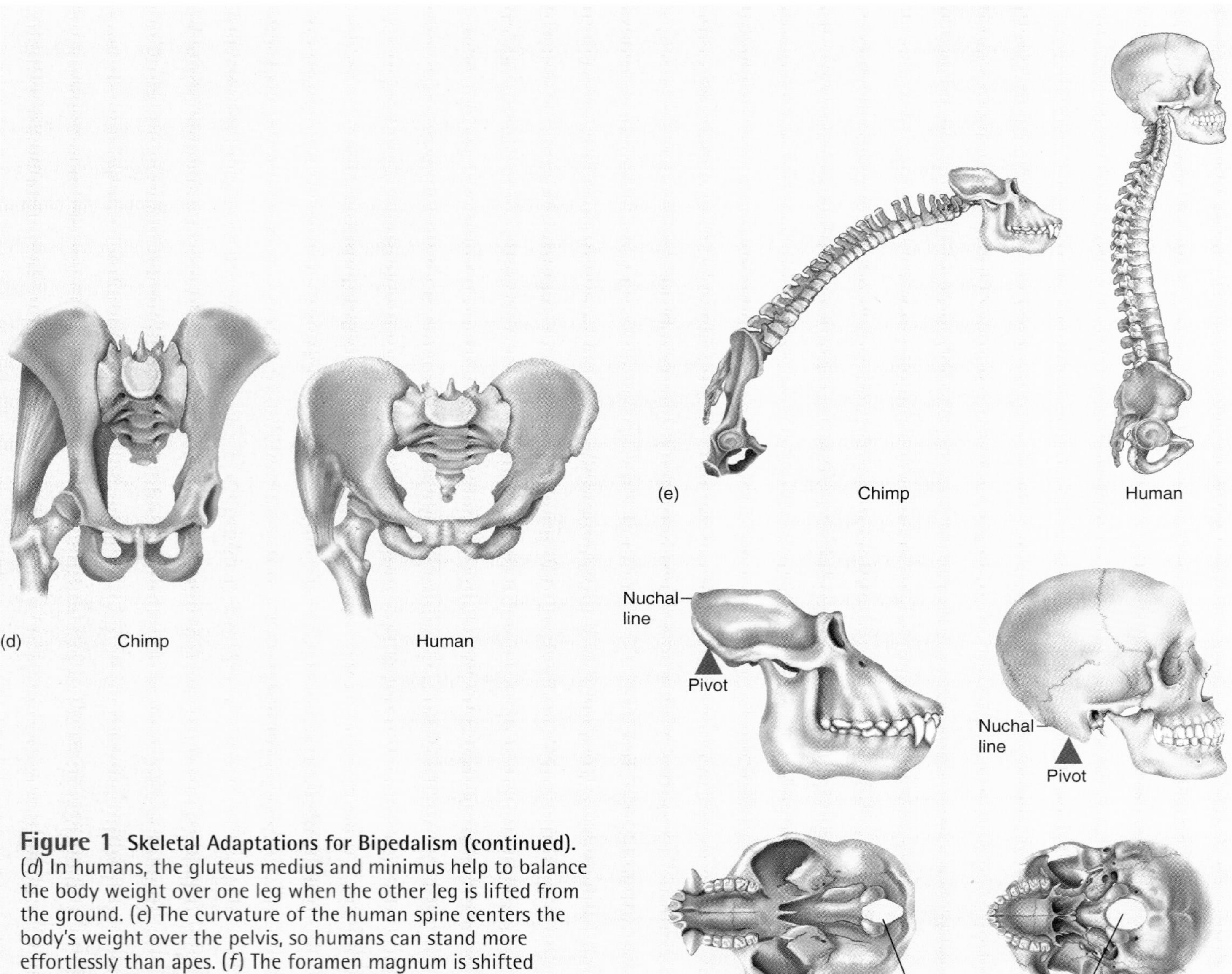

Figure 1 **Skeletal Adaptations for Bipedalism (continued).** (*d*) In humans, the gluteus medius and minimus help to balance the body weight over one leg when the other leg is lifted from the ground. (*e*) The curvature of the human spine centers the body's weight over the pelvis, so humans can stand more effortlessly than apes. (*f*) The foramen magnum is shifted ventrally and the face is flatter in humans; thus the skull is balanced on the vertebral column and the gaze is directed forward when a person is standing.

The iliac pillar is a line of attachment for some of the gluteal muscles. When the left foot is lifted off the ground, these muscles shift the trunk of the body slightly toward the right, and vice versa, which enables us to keep our balance. The presence of an iliac pillar on a fossil primate's pelvis is taken as good evidence that the species was bipedal.

The lumbar curvature of the human spine allows for efficient bipedalism by shifting the body's center of gravity to the rear, above and slightly behind the hip joint. Because of their C-shaped spines, chimpanzees cannot stand as easily. Their center of gravity is anterior to the hip joint when they stand; they must exert a continual muscular effort to keep from falling forward, and fatigue sets in relatively quickly. Humans, by contrast, require little muscular effort to keep their balance. Our australopithecine ancestors probably could travel all day with relatively little fatigue.

The human head is balanced on the vertebral column with the gaze directed forward. The cervical curvature of the spine and remodeling of the skull have made this possible. The foramen magnum has moved to a more inferior location, and the face is much flatter than in an ape, so there is less weight anterior to the occipital condyles. Being balanced on the spine, the head does not require strong muscular attachments to hold it erect. Apes have prominent supraorbital ridges for the attachment of muscles that pull back on the skull. In humans these ridges are much lighter and the muscles of the forehead serve only for facial expression, not to hold the head up.

The forelimbs of apes are longer than the hindlimbs; indeed, some species such as the orangutan and gibbons hold their long forelimbs over their heads when they walk on their hind legs. By contrast, our arms are shorter than our legs and far less muscular than the forelimbs of apes. No longer needed for locomotion, our forelimbs have become better adapted for carrying objects, holding things closer to the eyes, and manipulating them more precisely.

Connective Issues

Interactions Between the SKELETAL SYSTEM and Other Organ Systems

← indicates ways in which this system affects other systems
→ indicates ways in which other systems affect this one

Integumentary System

← Bones lying close to body surfaces shape the skin

→ Initiates synthesis of vitamin D needed for bone deposition

Muscular System

← Bones provide leverage and sites of attachment for muscles; provide calcium needed for muscle contraction

→ Muscles move bones; stress produced by muscles affects patterns of ossification and remodeling, as well as shape of mature bones

Nervous System

← Cranium and vertebral column protect brain and spinal cord; bones provide calcium needed for neural function

→ Sensory receptors provide sensations of body position and pain from bones and joints

Endocrine System

← Bones protect endocrine organs in head, chest, and pelvis

→ Hormones regulate mineral deposition and resorption, bone growth, and skeletal mass and density

Circulatory System

← Myeloid tissue forms blood cells; bone matrix stores calcium needed for cardiac muscle activity

→ Delivers O_2, nutrients, and hormones to bone tissue and carries away wastes; delivers blood cells to marrow

Lymphatic/Immune Systems

← Most types of blood cells produced in myeloid tissue function as part of immune system

→ Maintains balance of interstitial fluid in bones; lymphocytes assist in defense and repair of bones

Respiratory System

← Bones form respiratory passageway through nasal cavity; protect lungs and aid in ventilation

→ Provides O_2 and removes CO_2

Urinary System

← Skeleton physically supports and protects organs of urinary system

→ Kidneys activate vitamin D and regulate calcium and phosphate excretion

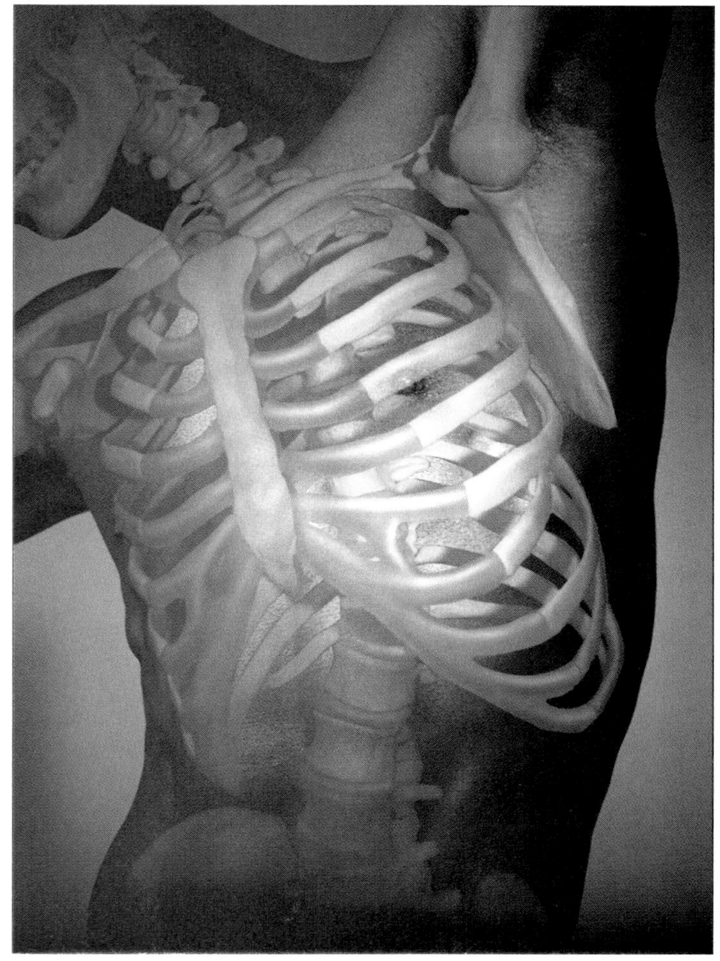

Digestive System

← Skeleton provides bony protection for digestive organs

→ Provides nutrients needed for bone growth and maintenance

Reproductive System

← Skeleton protects some reproductive organs

→ Gonads produce hormones that affect bone growth and closure of epiphyseal plates

Chapter Review

Online Learning Center

Study resources for this chapter are available at www.mhhe.com/saladin/

Study Outline

Overview of the Skeleton (p. 258)

1. Divisions of the skeleton
   a. Axial skeleton
   b. Appendicular skeleton
2. Bones of the skeletal system
   a. Variation in number
   b. Inventory (table 9.1)
3. Surface features of bones (table 9.2)

The Skull (p. 260) (table 9.4)

1. Sutures
2. Cavities
3. Foramina (table 9.3)
4. Cranial bones
   a. Frontal bone
   b. Parietal bones
   c. Temporal bones
   d. Occipital bone
   e. Sphenoid bone
   f. Ethmoid bone
5. Facial bones
   a. Maxillae
   b. Palatine bones
   c. Zygomatic bones
   d. Lacrimal bones
   e. Nasal bones
   f. Inferior nasal conchae
   g. Vomer
   h. Mandible
6. Bones associated with the skull
   a. Auditory ossicles
   b. Hyoid bone
7. The skull in infancy and childhood
   a. Fontanels
   b. Fusion of frontal bone and mandible
   c. Proportions of the skull

The Vertebral Column and Thoracic Cage (p. 275) (table 9.6)

1. Vertebral column
   a. Vertebrae and intervertebral discs
   b. Curvatures of the spine
2. General structure of a vertebra
3. Intervertebral discs
4. Regional characteristics of vertebrae
   a. Cervical vertebrae (C1–C7)
   b. Thoracic vertebrae (T1–T12)
   c. Lumbar vertebrae (L1–L5)
   d. Sacral vertebrae (S1–S5)
   e. Coccygeal vertebrae (Co1–Co4)
5. Thoracic cage
   a. Sternum
   b. Ribs
      • True ribs (1–7)
      • False ribs (8–12)
      • Floating ribs (11–12)

The Pectoral Girdle and Upper Limb (p. 285) (table 9.7)

1. Pectoral girdle
   a. Clavicle
   b. Scapula
2. Upper limb
   a. Brachium and humerus
   b. Antebrachium
      • Radius
      • Ulna
   c. Carpal region
      • Proximal row of carpal bones
         •• Scaphoid
         •• Lunate
         •• Triquetral
         •• Pisiform
      • Distal row of carpal bones
         •• Trapezium
         •• Trapezoid
         •• Capitate
         •• Hamate
   d. Hand
      • Metacarpal bones
      • Phalanges

The Pelvic Girdle and Lower Limb (p. 293) (table 9.9)

1. Pelvic girdle
   a. Os coxae
      • Ilium
      • Ischium
      • Pubis
      • Sexual dimorphism
   b. Sacrum
   c. Coccyx
2. Lower limb
   a. Thigh and femur
   b. Patella
   c. Leg (crural region)
      • Tibia
      • Fibula
   d. Ankle and foot
      • Proximal tarsal bones
         •• Calcaneus
         •• Talus
         •• Navicular
      • Distal tarsal bones
         •• Cuneiforms 1–3
         •• Cuboid
      • Metatarsal bones
      • Phalanges
      • Foot arches

Selected Vocabulary

For additional terminology not listed here, see tables cited in the Study Outline.

axial skeleton 258
appendicular skeleton 258
sesamoid bone 258
sutural bone 258
suture 260
cranial cavity 260
orbit 260
nasal cavity 260
buccal cavity 260
middle-ear cavity 260
inner-ear cavity 260
paranasal sinuses 260
frontal sinus 260
sphenoid sinus 260
ethmoid sinus 260
maxillary sinus 260
foramina 261
cranium 262
cranial bones 262
calvaria 262
base of cranial cavity 262
anterior cranial fossa 262
middle cranial fossa 262
posterior cranial fossa 262
sagittal suture 265
coronal suture 265
lambdoid suture 265
squamous suture 265
zygomatic arch 268
nuchal ligament 269
nasal septum 270
hard palate 272
soft palate 272
temporomandibular joint 274
anterior fontanel 274
posterior fontanel 274
sphenoid fontanel 274
mastoid fontanel 274
atlanto-occipital joint 281
atlantoaxial joint 281
sacroiliac joint 282
costal margin 283
sternoclavicular joint 285
acromioclavicular joint 285
humeroscapular joint 285
brachium 288
antebrachium 288
carpus 288
manus 288
pollex 290
auricular surface 293
femoral region 295
crural region 295
tarsal region 295
pedal region 295
calcaneal tendon 298
trochlear surface 298
hallux 298

Testing Your Recall

1. Which of these is *not* a paranasal sinus?
   a. frontal
   b. temporal
   c. sphenoid
   d. ethmoid
   e. maxillary
2. Which of these is a facial bone?
   a. frontal
   b. ethmoid
   c. occipital
   d. temporal
   e. lacrimal
3. Which of these *cannot* be palpated on a living person?
   a. the crista galli
   b. the mastoid process
   c. the zygomatic arch
   d. the superior nuchal line
   e. the hyoid bone
4. All of the following are groups of vertebrae *except* for ________, which is a spinal curvature.
   a. thoracic
   b. cervical
   c. lumbar
   d. pelvic
   e. sacral
5. Thoracic vertebrae do *not* have
   a. transverse foramina.
   b. costal facets.
   c. demifacets.
   d. transverse processes.
   e. pedicles.
6. The tubercle of a rib articulates with
   a. the sternal notch.
   b. the margin of the gladiolus.
   c. the demifacets of two vertebrae.
   d. the body of a vertebra.
   e. the transverse process of a vertebra.
7. The disc-shaped head of the radius articulates with the ________ of the humerus.
   a. radial tuberosity
   b. trochlea
   c. capitulum
   d. olecranon process
   e. glenoid cavity
8. All of the following are carpal bones, *except* the ________, which is a tarsal bone.
   a. trapezium
   b. cuboid
   c. trapezoid
   d. triquetral
   e. pisiform
9. The bone that supports your body weight when you are sitting down is
   a. the acetabulum.
   b. the pubis.
   c. the ilium.
   d. the coccyx.
   e. the ischium.
10. Which of these is the bone of the heel?
   a. cuboid
   b. calcaneus
   c. navicular
   d. trochlear
   e. talus
11. Gaps between the cranial bones of an infant are called ________.
12. The external auditory meatus is a canal in the ________ bone.
13. Bones of the skull are joined along lines called ________.
14. The ________ bone has greater and lesser wings and protects the pituitary gland.
15. A herniated disc occurs when a ring called the ________ cracks.
16. The transverse ligament of the atlas holds the ________ of the axis in place.
17. The sacroiliac joint is formed where the ________ surface of the sacrum articulates with that of the ilium.
18. The ________ processes of the radius and ulna form bony protuberances on each side of the wrist.
19. The thumb is also known as the ________ and the great toe is also known as the ________.
20. The ________ arch of the foot extends from the heel to the great toe.

Answers in appendix B

Testing Your Comprehension

1. A child was involved in an automobile collision. She was not wearing a safety restraint, and her chin struck the dashboard hard. When the physician looked into her auditory canal, he could see into her throat. What do you infer from this about the nature of her injury?
2. By palpating the hind leg of a cat or dog or by examining a laboratory skeleton, you can see that cats and dogs stand on the heads of their metatarsal bones; the calcaneus does not touch the ground. How is this similar to the stance of a woman wearing high-heeled shoes? How is it different?
3. Contrast the tarsal bones with the carpal bones. Which ones are similar in name, location, or both? Which ones are different?
4. Name some bones you think would be well suited as a source of red bone marrow for a biopsy or transfusion. Why do you suggest these particular bones over others?
5. A teenager is baby-sitting a younger child who is shocked by a frayed electrical cord. The child is not breathing and has no pulse. The baby-sitter has not been trained in cardiopulmonary resuscitation, but attempts to do it the way he has seen it performed on television programs. He performs rhythmic chest compressions on the sternum, but the child dies. An autopsy reveals that the liver has been punctured, resulting in a fatal hemorrhage. From the standpoint of skeletal anatomy, explain how this could have happened.

Answers at the Online Learning Center

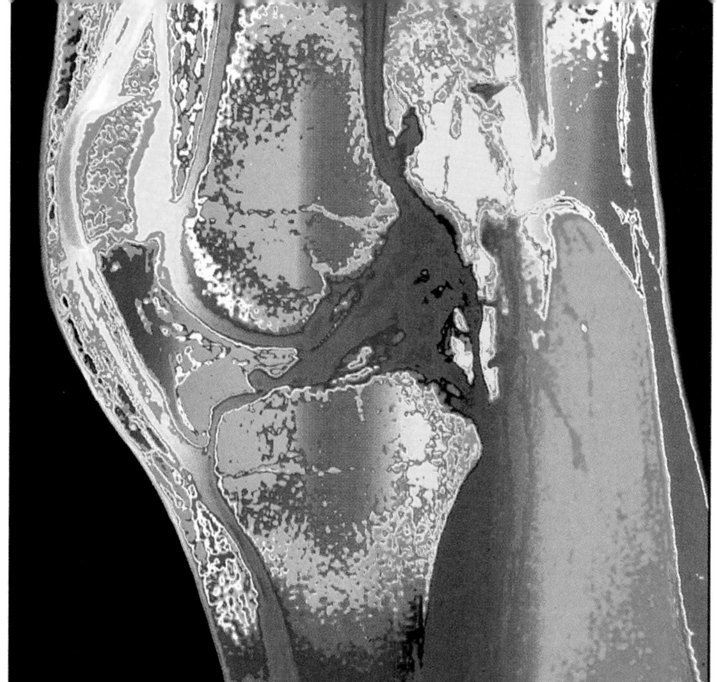

Lateral view of the knee (MRI)

CHAPTER 10

Joints

Chapter Outline

Insights

Study with Your ESP CD-ROM

Refer to your ESP Tutorial CD-ROM for helpful animations, activities, and quizzes related to the material covered in this chapter. The following are suggested areas to go to on the CD and are presented according to "Unit/Topic":

- Support and Movement/Articulations

Brushing Up

To understand this chapter, it is important that you understand or brush up on the following concepts:

- Names of all bones (fig. 9.1, p. 259; table 9.1, p. 260)
- Surface features of bones, especially of their articular surfaces (table 9.2, p. 261)

In order for the skeleton to serve the purposes of protection and movement, the bones must be joined together. A **joint,** or **articulation,** is any point at which two bones meet, regardless of whether they are movable at that point. Your knee, for example, is a very movable joint, whereas the skull sutures described in the preceding chapter are immovable joints. This chapter describes the joints of the skeleton and discusses some basic principles of biomechanics relevant to athletic performance and patient care. Chapter 11, in which the actions of skeletal muscles are described, builds on the discussion of joint anatomy and function presented here.

Figure 10.1 **Joint Flexibility.** This gymnast demonstrates the flexibility, precision, and weight-bearing capacity of many of the body's joints.

Joints and Their Classification

Objectives

When you have completed this section, you should be able to

- explain what joints are, how they are named, and what functions they serve;
- define *arthrology, kinesiology,* and *biomechanics;* and
- name and describe the three major structural classes and three major functional classes of joints.

Arthrology is the science concerned with the anatomy, function, dysfunction, and treatment of joints. The study of musculoskeletal movement is called **kinesiology** (kih-NEE-see-OL-oh-jee). This is a subdiscipline of **biomechanics,** which deals with a broad range of motions and mechanical processes, including the physics of blood circulation, respiration, and hearing.

Joints such as the shoulder, elbow, and knee are remarkable specimens of biological design—self-lubricating, almost frictionless, and able to bear heavy loads and withstand compression while executing smooth and precise movements (fig. 10.1). Yet, it is equally important that other joints be less movable or even immovable. Such joints are better able to support the body and provide protection for delicate organs. The vertebral column, for example, must provide a combination of support and flexibility; thus its joints are only moderately movable. The immovable joints between the cranial bones afford the best possible protection for the brain and sense organs.

The name of a joint is typically derived from the names of the bones involved. For example, the *atlanto-occipital joint* is where the occipital condyles meet the atlas, the *humeroscapular joint* is where the humerus meets the scapula, and the *coxal joint* is where the femur meets the os coxae.

Joints can be classified according to their relative freedom of movement:

- A **diarthrosis**[1] (DY-ar-THRO-sis) is a freely movable joint such as the elbow.
- An **amphiarthrosis**[2] (AM-fee-ar-THRO-sis) is a joint that is slightly movable, such as the intervertebral and intercarpal joints.
- A **synarthrosis**[3] (SIN-ar-THRO-sis) is a joint that is capable of little or no movement, such as a suture of the skull.

Joints are also classified according to the manner in which the adjacent bones are joined. In this system, there are *fibrous, cartilaginous, bony,* and *synovial joints,* defined and described in the sections that follow. These two systems of classification overlap. For example, synovial

1. *dia* = separate, apart + *arthr* = joint + *osis* = condition

2. *amphi* = on all sides

3. *syn* = together

joints may be either diarthroses or amphiarthroses, and amphiarthroses can be any of the three structural types—synovial, fibrous, or cartilaginous (fig. 10.2).

Before You Go On

Answer the following questions to test your understanding of the preceding section:

1. What is the difference between arthrology and kinesiology?
2. Describe the two basic ways of classifying joints. What distinctions are looked for in each system?
3. Explain the distinction between a diarthrosis, amphiarthrosis, and synarthrosis. Give an example of each.

Fibrous, Cartilaginous, and Bony Joints

Objectives

When you have completed this section, you should be able to

- describe the three types of fibrous joints and give an example of each;
- distinguish between the three types of sutures;
- describe the two types of cartilaginous joints and give an example of each; and
- name some joints that become synostoses as they age.

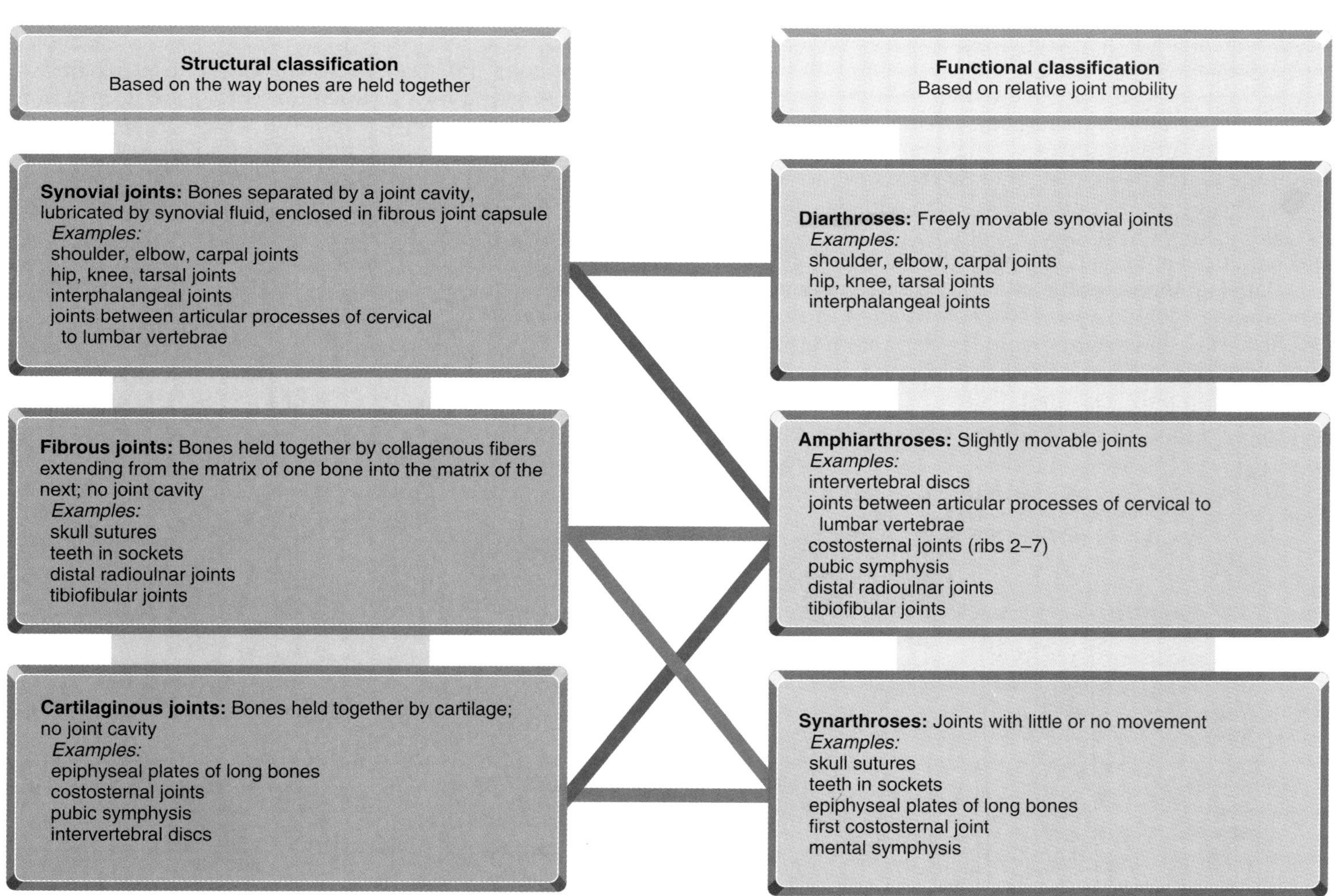

Figure 10.2 Systems of Classifying the Joints. *Left:* A structural classification based on how the bones are joined. *Right:* A functional classification based on relative joint mobility. Connecting lines indicate overlap between the classification systems. For example, synovial joints can be either diarthroses or amphiarthroses, all diarthroses are synovial joints, and amphiarthroses include joints of the synovial, fibrous, and cartilaginous types.

Fibrous Joints

In a **fibrous joint,** collagen fibers emerge from the matrix of one bone and penetrate into the matrix of another, spanning the space between them (fig. 10.3). There are three types of fibrous joints: *sutures, gomphoses,* and *syndesmoses.* In sutures and gomphoses, the collagen fibers are very short and allow for little movement. In syndesmoses, the fibers are longer and the attached bones are more movable.

Sutures

Sutures are immovable fibrous joints that closely bind the bones of the skull to each other; they occur nowhere else. In chapter 9, we did not take much notice of the differences between one suture and another, but some differences may have caught your attention as you studied the diagrams in that chapter or examined laboratory specimens. Sutures can be classified as *serrate, lap,* and *plane sutures.* Readers with some background in woodworking may recognize that the structures and functional properties of these sutures have something in common with basic types of carpentry joints (fig. 10.4).

Serrate sutures appear as wavy lines along which the adjoining bones firmly interlock with each other by their serrated margins. Serrate sutures are analogous to a dovetail wood joint. Examples include the coronal, sagittal, and lambdoid sutures that border the parietal bones.

Lap (squamous) sutures occur where two bones have overlapping beveled edges, like a miter joint in carpentry. On the surface, a lap suture appears as a relatively smooth (nonserrated) line. An example is the squamous suture between the temporal and parietal bones.

Plane (butt) sutures occur where two bones have straight, nonoverlapping edges. The two bones merely border on each other, like two boards glued together in a butt joint. This type of suture is seen between the palatine processes of the maxillae in the roof of the mouth.

Gomphoses

Even though the teeth are not bones, the attachment of a tooth to its socket is classified as a joint called a **gomphosis** (gom-FOE-sis). The term refers to its similarity to a nail hammered into wood.[4] The tooth is held firmly in place by a fibrous **periodontal ligament,** which consists of collagen fibers that extend from the bone matrix of the jaw into the dental tissue (see fig. 10.3*b*). The periodontal ligament allows the tooth to move or "give" a little under the stress of chewing.

Syndesmoses

Syndesmoses[5] (SIN-dez-MO-seez) are joints at which two bones are bound by a ligament only. (Ligaments also bind bones together at synovial joints, but are not the exclusive means of holding those joints together.) Syndesmoses are the most movable of the fibrous joints. The radius and ulna are bound to each other side by side, as are the tibia and fibula, by a syndesmosis in which the ligament forms a broad sheet called an **interosseous membrane** along the shafts of the two bones (see fig. 10.3*c*).

4. *gompho* = nail, bolt
5. *syn* = together + *desm* = band + *osis* = condition

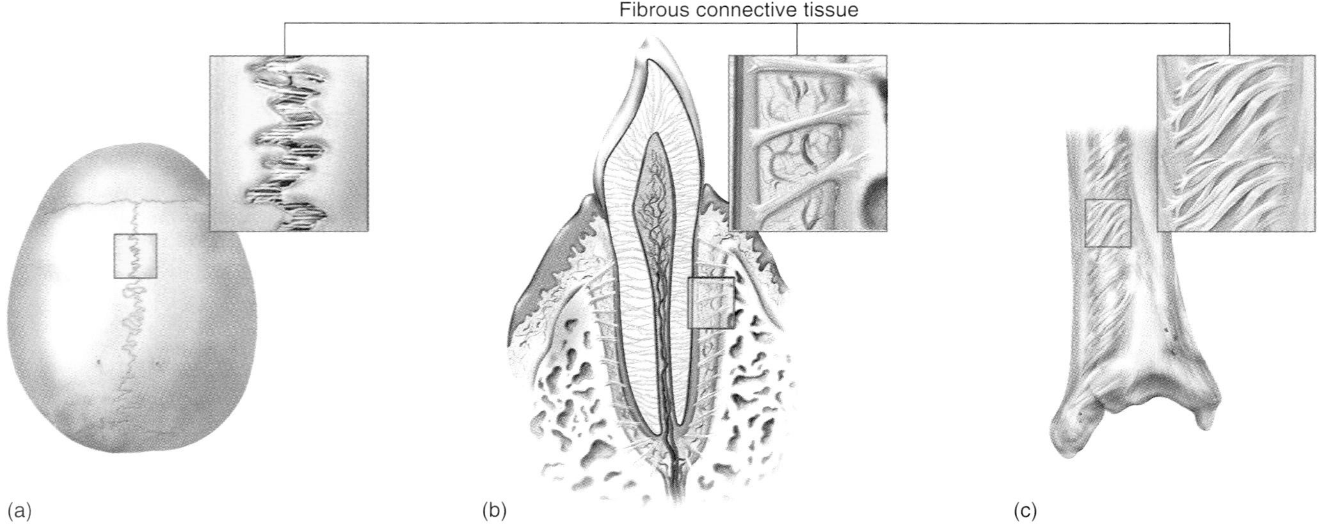

Figure 10.3 Types of Fibrous Joints. (*a*) A suture between the parietal bones; (*b*) a gomphosis between a tooth and the jaw; (*c*) a syndesmosis between the tibia and fibula.

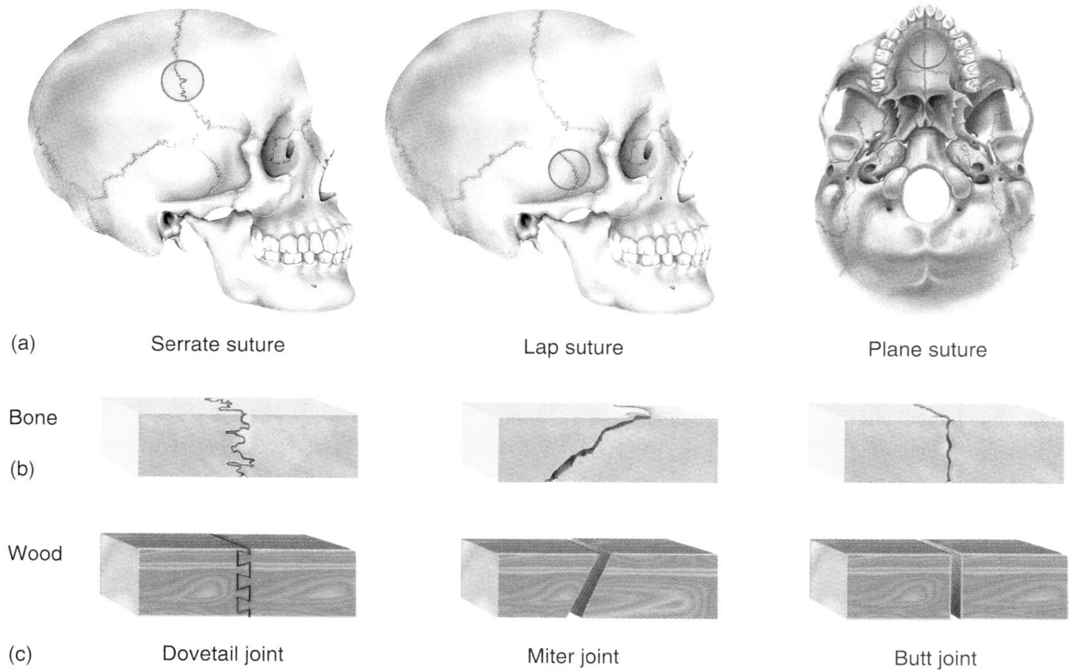

Figure 10.4 Types of Sutures. (*a*) Locations; (*b*) structure of the adjoining bones; (*c*) functional analogies to some common wood joints.

Cartilaginous Joints

In **cartilaginous joints,** the two bones are bound to each other by cartilage. The two types of cartilaginous joints are *synchondroses* and *symphyses,* which involve hyaline cartilage and fibrocartilage, respectively.

Synchondroses

In a **synchondrosis**[6] (SIN-con-DRO-sis), the bones are joined by hyaline cartilage. In children, the hyaline cartilage of the epiphyseal plate forms a synchondrosis that binds the epiphysis and diaphysis of a long bone together. The attachment of a rib to the sternum by a hyaline costal cartilage is also a synchondrosis (fig. 10.5*a*).

Symphyses

In a **symphysis,**[7] two bones are joined by *fibrocartilage* (fig. 10.5*b, c*). One example is the pubic symphysis, in which the right and left pubic bones are joined by the cartilaginous interpubic disc. Another is the joint between the bodies of two vertebrae, united by an intervertebral disc. The surface of each vertebral body is covered with hyaline cartilage. Between the vertebrae, this cartilage becomes infiltrated with collagen bundles to form fibrocartilage. Each intervertebral disc permits only slight movement between adjacent vertebrae, but the collective effect of all 23 discs gives the spine considerable flexibility.

Bony Joints (Synostoses)

A **bony joint,** or **synostosis**[8] (SIN-oss-TOE-sis), is a joint in which two bones, once separate, have become fused by osseous tissue and in most cases are then regarded as a single bone. Some fibrous and cartilaginous joints ossify with age—that is, the gap between adjacent bones becomes filled with osseous tissue until the two bones appear as one. In the skull, for example, both the frontal bone and mandible are represented at birth by separate right and left bones; in early childhood, these bones become fused. In old age, some sutures become obliterated by ossification and adjacent cranial bones fuse seamlessly together. The epiphyses and diaphyses of the long bones are joined by cartilaginous joints in childhood and adolescence, and these become synostoses in early adulthood. The attachment of the first rib to the sternum also becomes a synostosis with age.

6. *syn* = together + *chondr* = cartilage + *osis* = condition
7. *sym* = together + *physis* = growth
8. *ost* = bone

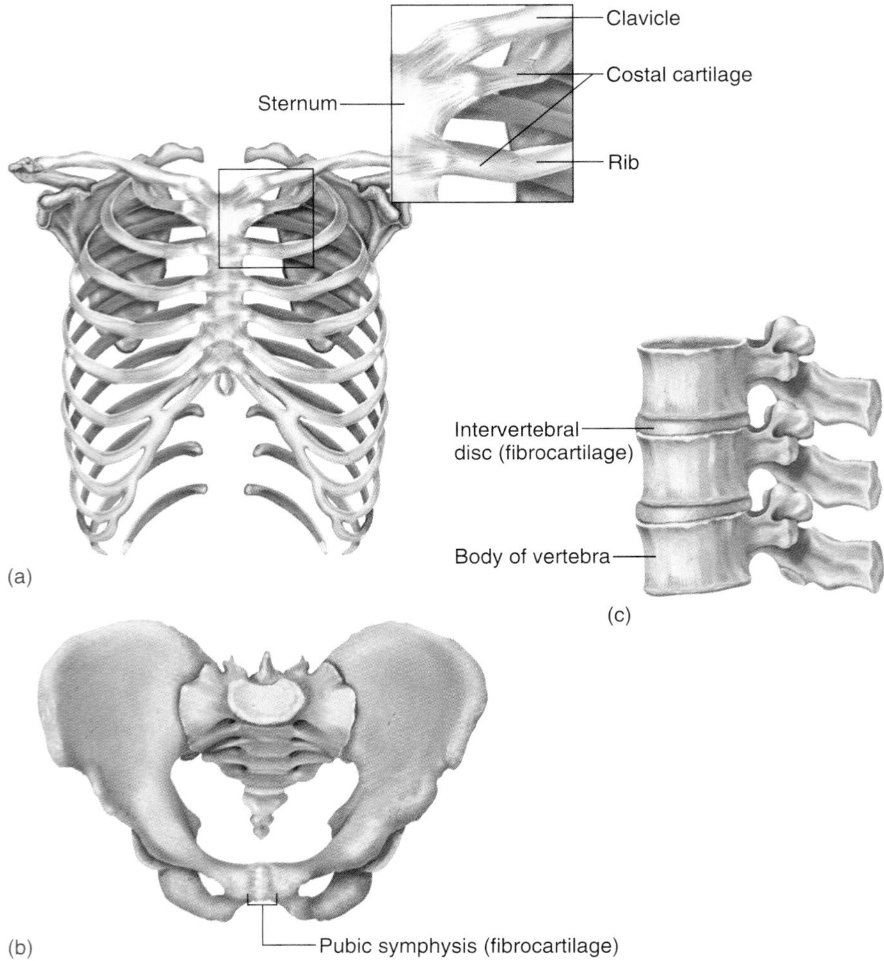

Figure 10.5 Cartilaginous Joints. (*a*) Synchondroses, represented by costal cartilages joining the ribs to the sternum; (*b*) the pubic symphysis; (*c*) intervertebral discs, which join adjacent vertebrae to each other by symphyses.

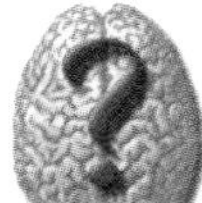

Think About It

The intervertebral joints are symphyses only in the cervical through the lumbar region. How would you classify the intervertebral joints of the sacrum and coccyx in a middle-aged adult?

Before You Go On

Answer the following questions to test your understanding of the preceding section:

4. Define *suture*, *gomphosis*, and *syndesmosis*, and explain what these three joints have in common.
5. Name the three types of sutures and describe how they differ.
6. Name two synchondroses and two symphyses.
7. Give some examples of joints that become synostoses with age.

Synovial Joints

Objectives

When you have completed this section, you should be able to

- describe the anatomy of a synovial joint and its associated structures;
- describe the six types of synovial joints;
- list and demonstrate the types of movements that occur at diarthroses;
- discuss the factors that affect the range of motion of a joint;
- give an anatomical example of a first-, second-, and third-class lever and explain why each is classified as it is; and
- relate the concept of mechanical advantage to the power and speed of joint action.

The rest of this chapter is concerned with synovial joints. A **synovial** (sih-NO-vee-ul) **joint** is one in which two bones are separated by a space that contains a slippery lubricant called **synovial fluid.** Most synovial joints, including the jaw, elbow, hip, and knee joints, are freely movable. These are not only the most common and familiar joints in the body, but they are also the most structurally complex and the most likely to develop uncomfortable and crippling dysfunctions.

General Anatomy

The study of dry bones and models in the laboratory can easily give the impression that the knee, elbow, or hip is a point at which one bone rubs against another. In life, however, the bones do not touch each other; rather, fluid and soft tissues separate them and hold them in proper alignment (fig. 10.6).

The bones of a synovial joint are separated by a **joint (articular) cavity** containing the synovial fluid. Synovial fluid is rich in albumin and hyaluronic acid, which give it a viscous, slippery texture similar to that of raw egg white.[9] It nourishes the articular cartilages and removes their wastes, and it contains phagocytes that clean up tissue debris resulting from wear of the joint cartilages. The adjoining surfaces of the bones are covered with a layer of hyaline **articular cartilage** about 2 mm thick in young, healthy joints. The cartilages and synovial fluid make movements at synovial joints almost friction-free. A fibrous **joint (articular) capsule** encloses the cavity and retains the fluid. It has an outer **fibrous capsule** continuous with the periosteum of the adjoining bones and an inner **synovial membrane** of areolar tissue, which secretes the fluid.

In the jaw, sternoclavicular, and knee joints, cartilage grows inward from the joint capsule and forms a pad called a **meniscus**[10] between the articulating bones (see fig. 10.24). The meniscus absorbs shock and pressure, guides the bones across each other, and reduces the chance of dislocation.

The most important accessory structures of a synovial joint are tendons, ligaments, and bursae. A **tendon** is a strip or sheet of tough, collagenous connective tissue that attaches a muscle to a bone. Tendons are often the most important structures in stabilizing a joint. A **ligament** is a similar tissue that attaches one bone to another. Several ligaments are named and illustrated in our later discussion of individual joints, and tendons are more fully considered in the next chapter along with the gross anatomy of muscles.

A **bursa**[11] is a fibrous sac filled with synovial fluid, located between adjacent muscles or where a tendon passes over a bone (see fig. 10.20). Bursae cushion muscles, help tendons slide more easily over the joints, and sometimes enhance the mechanical effect of a muscle by modifying the direction in which its tendon pulls. Bursae called **tendon sheaths** are elongated cylinders wrapped around a tendon. These are especially numerous in the hand and foot (fig. 10.7).

9. *ovi* = egg

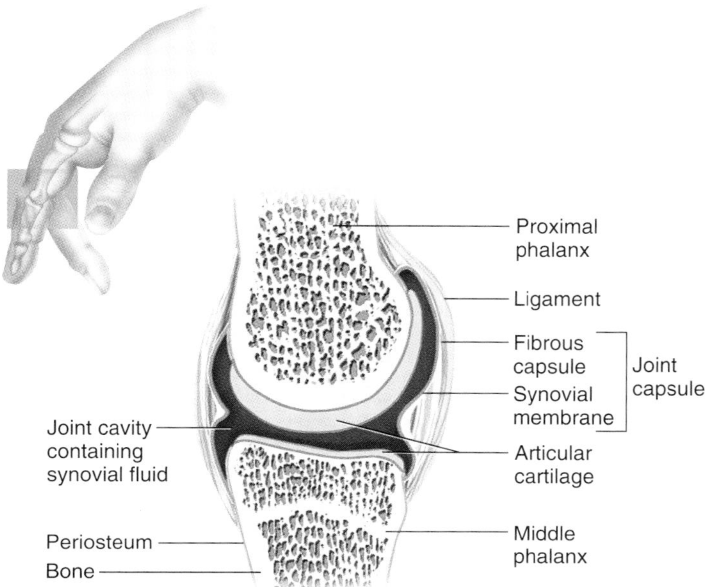

Figure 10.6 Structure of a Simple Synovial Joint.

Types of Synovial Joints

There are six types of synovial joints with distinctive patterns of motion determined by the shapes of the articular surfaces of the bones (fig. 10.8; table 10.1). A bone's movement at a joint can be described with reference to three mutually perpendicular planes in space (*x*, *y*, and *z*). If the bone can move in only one plane, the joint is said to be **monaxial;** if it can move in two planes, the joint is **biaxial;** and if three, it is **multiaxial.**

1. **Ball-and-socket joints.** These occur at the shoulder and hip, where one bone has a smooth hemispherical head that fits within a cuplike depression on the other. The head of the humerus fits into the glenoid cavity of the scapula, and the head of the femur fits into the acetabulum of the os coxae. These are the only multiaxial joints of the skeleton.

10. *men* = moon, crescent + *iscus* = little
11. *burs* = purse

Clinical Insight 10.1

Exercise and Articular Cartilage

When synovial fluid is warmed by exercise, it becomes thinner (less viscous) and more easily absorbed by the articular cartilage. The cartilage then swells and provides a more effective cushion against compression. For this reason, a warm-up period before vigorous exercise helps protect the articular cartilage from undue wear and tear.

Because cartilage is nonvascular, its repetitive compression during exercise is important to its nutrition and waste removal. Each time a cartilage is compressed, fluid and metabolic wastes are squeezed out of it. When weight is taken off the joint, the cartilage absorbs synovial fluid like a sponge, and the fluid carries oxygen and nutrients to the chondrocytes. Lack of exercise causes the articular cartilages to deteriorate more rapidly from lack of nutrition, oxygenation, and waste removal.

Weight-bearing exercise builds bone mass and strengthens the muscles that stabilize many of the joints, thus reducing the risk of joint dislocations. Excessive joint stress, however, can hasten the progression of osteoarthritis by damaging the articular cartilage (see clinical insight 10.4, p. 335). Swimming is a good way of exercising the joints with minimal damage.

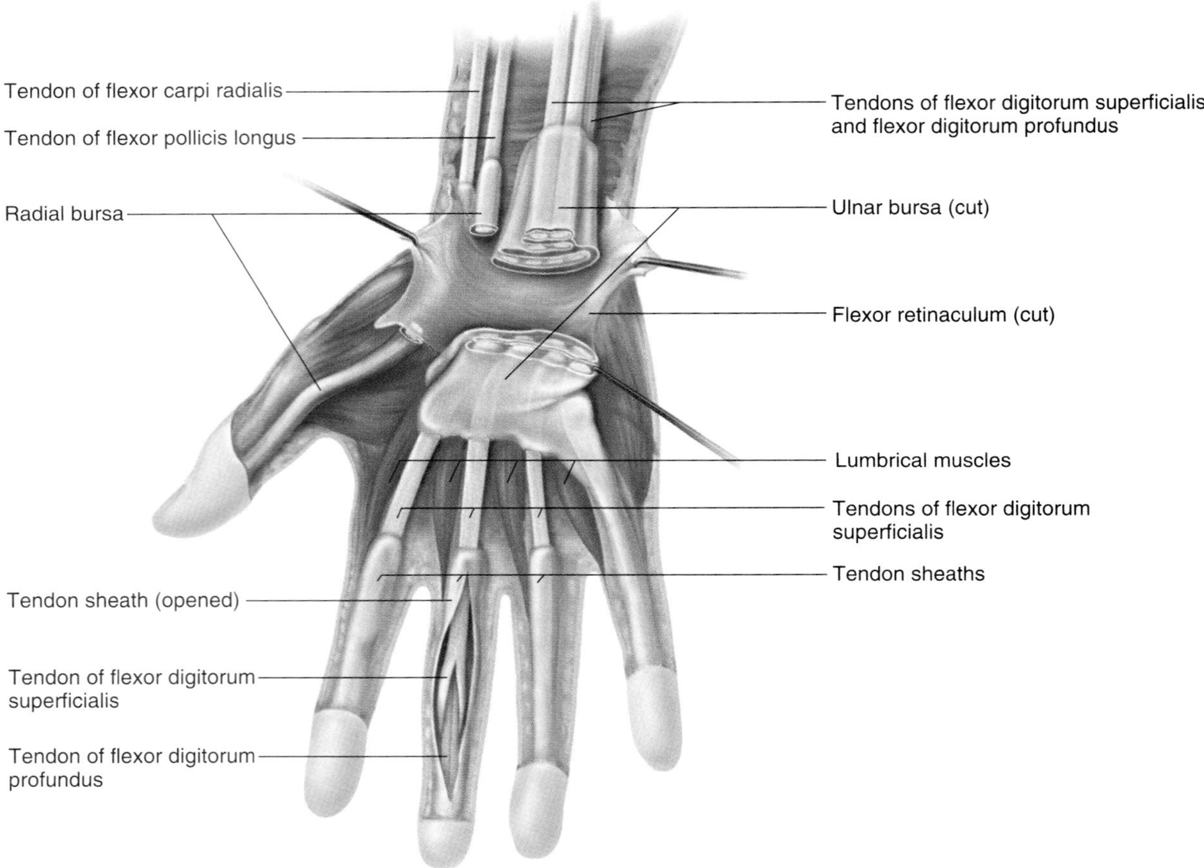

Figure 10.7 Tendon Sheaths and Other Bursae in the Hand and Wrist.

2. **Hinge joints.** At a hinge joint, one bone has a convex surface that fits into a concave depression of the other one. Hinge joints are monaxial—like a door hinge, they can only move in one plane. Examples include the elbow, knee, and interphalangeal (finger and toe) joints.
3. **Saddle joint.** The only saddle joint is the trapeziometacarpal (tra-PEE-zee-oh-MET-uh-CAR-pul) joint at the base of the thumb. Each articular surface—on metacarpal I and the trapezium of the wrist—is shaped like a saddle, concave in one direction and convex in the other. This is a biaxial

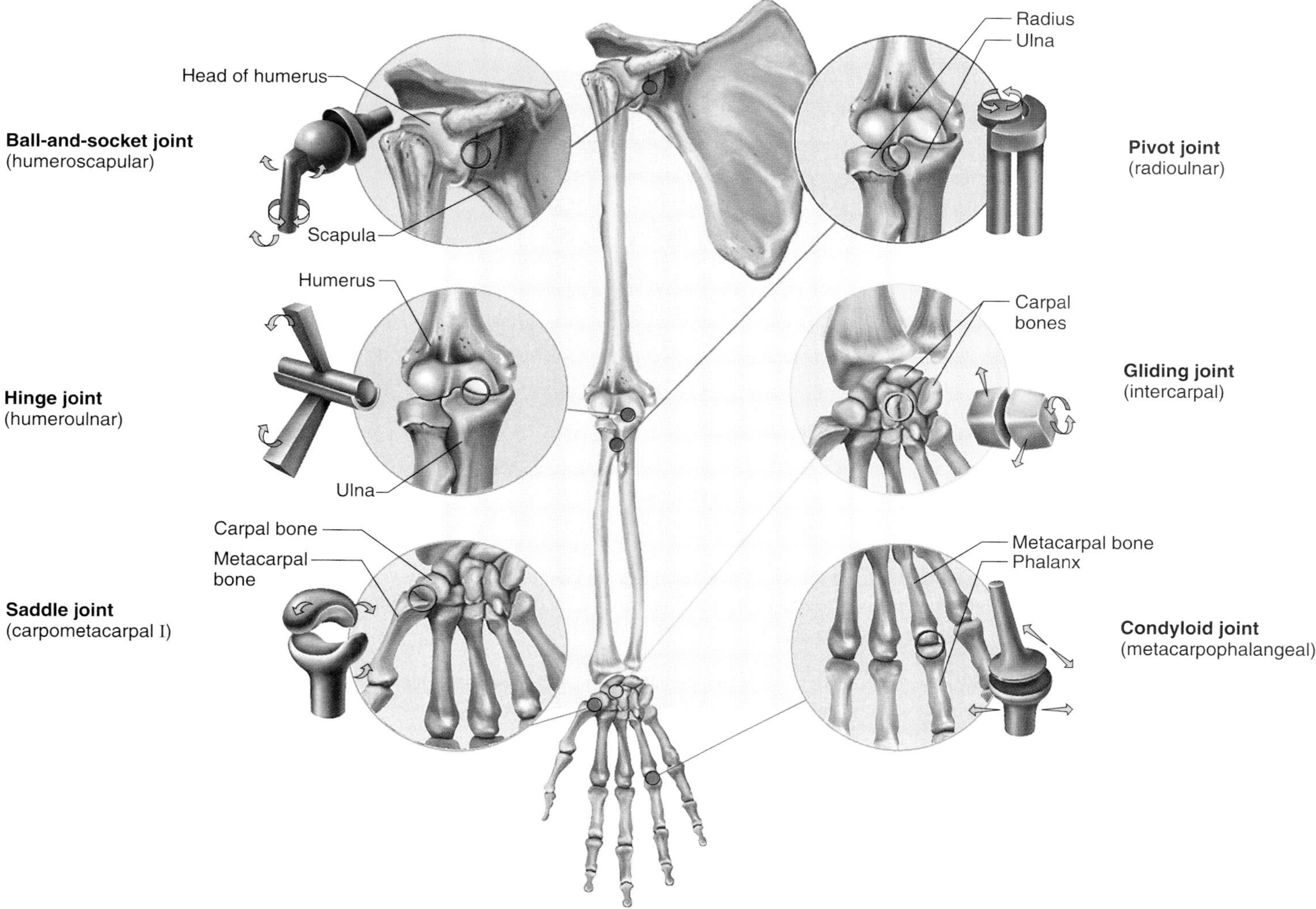

Figure 10.8 The Six Types of Diarthroses. All six have representatives in the forelimb. Mechanical models show the types of motion possible at each joint.

joint. If you compare the range of motion of your thumb with that of your fingers, you can see that a saddle joint is more movable than a condyloid or hinge joint. This is the joint responsible for that hallmark of primate anatomy, the opposable thumb.

4. **Pivot joints.** These are monaxial joints in which one bone has a projection that fits into a ringlike ligament of another, and the first bone rotates on its longitudinal axis relative to the other. One example is the atlantoaxial joint between the first two vertebrae—the dens of the axis projects into the vertebral foramen of the atlas, where it is held against the arch of the atlas by a ligament (see fig. 9.23*c*). This joint pivots when you rotate your head as in gesturing "no." Another example is the proximal radioulnar joint, where the *annular ligament* on the ulna encircles the head of the radius (see fig. 10.21*c, d*) and permits the radius to rotate during pronation and supination of the forearm (motions to be described shortly).
5. **Gliding (plane) joints.** Here, the articular surfaces are flat or only slightly concave and convex. The adjacent bones slide over each other and have rather limited monaxial movement; they are amphiarthroses in contrast to the other five types listed, which are diarthroses. Gliding joints occur between the carpal and tarsal bones, between the articular processes of the vertebrae, and at the sternoclavicular joint. To feel a gliding joint in motion, palpate your sternoclavicular joint as you raise your arm above your head.
6. **Condyloid (ellipsoid) joints.** These joints exhibit an oval convex surface on one bone that fits into a similarly shaped depression on the next. The

Table 10.1 Anatomical Classification of the Joints

Joint	Characteristics and Examples
Fibrous Joints	Adjacent bones bound by collagen fibers extending from the matrix of one into the matrix of the other
Sutures (fig. 10.4)	Immovable fibrous joints between cranial and facial bones
Serrate suture	Bones joined by a wavy line formed by interlocking teeth along the margins. Examples: coronal, sagittal, and lambdoid sutures
Lap suture	Bones beveled to overlap each other; superficial appearance is a smooth line. Example: squamous suture around temporal bone
Plane suture	Bones butted against each other without overlapping or interlocking. Example: palatine suture
Gomphosis (fig. 10.3*b*)	Insertion of a tooth into a socket, held in place by collagen fibers of periodontal ligament
Syndesmosis (fig. 10.3*c*)	Slightly movable joint held together by ligaments or interosseous membranes. Examples: tibiofibular joint and radioulnar joint
Cartilaginous Joints	Adjacent bones bound by cartilage
Synchondrosis (fig. 10.5*a*)	Bones held together by hyaline cartilage. Examples: articulation of ribs with sternum, and epiphyseal plate uniting the epiphysis and diaphysis of a long bone of a child
Symphysis (fig. 10.5*b, c*)	Slightly movable joint held together by fibrocartilage. Examples: intervertebral discs and pubic symphysis
Synostoses	Former fibrous or cartilaginous joints in which adjacent bones have become fused by ossification. Examples: midsagittal line of frontal bone, fusion of epiphysis and diaphysis of an adult long bone, and fusion of ilium, ischium, and pubis to form os coxae
Synovial Joints (figs. 10.6 and 10.8)	Adjacent bones covered with hyaline articular cartilage, separated by lubricating synovial fluid and enclosed in a fibrous joint capsule
Ball-and-socket	Multiaxial diarthrosis in which a smooth hemispherical head of one bone fits into a cuplike depression of another. Examples: shoulder and hip joints
Hinge	Monaxial diarthrosis, able to flex and extend in only one plane. Examples: elbow, knee, and interphalangeal joints
Saddle	Joint in which each bone surface is saddle-shaped (concave on one axis and convex on the perpendicular axis). Unique to the thumb (trapeziometacarpal joint), where it allows opposition (touching of the thumb to the fingertips)
Pivot	Joint in which a projection of one bone fits into a ringlike ligament of another, allowing one bone to rotate on its longitudinal axis. Examples: atlantoaxial joint and proximal radioulnar joint
Gliding	Synovial amphiarthrosis with slightly concave or convex bone surfaces that slide across each other. Examples: intercarpal, intertarsal, and sternoclavicular joints; joints between the articular processes of the vertebrae
Condyloid	Biaxial diarthrosis in which an oval convex surface of one bone articulates with an elliptical depression of another. Examples: radiocarpal and metacarpophalangeal joints

radiocarpal joint of the wrist and the metacarpophalangeal (MET-uh-CAR-po-fuh-LAN-jee-ul) joints at the bases of the fingers are examples. These are considered *biaxial* joints because they can move in two directions, for example up and down and side to side. To demonstrate, hold your hand with your palm facing you. Flex your index finger back and forth as if gesturing to someone, "come here," and then move the finger from side to side toward the thumb and away. This shows the biaxial motion of the condyloid joint.

In table 10.1 the joints are classified by structural criteria. Some joints are difficult to classify, however, because they have elements of more than one type. The jaw joint, for example, has some aspects of condyloid, hinge, and gliding joints for reasons that will be apparent later.

Movements of Diarthroses

In physical therapy, kinesiology, and other medical and scientific fields, specific terms are used to describe the movements of diarthroses. You will need a command of these terms to understand the muscle actions in the next chapter. In the following discussion, many of them are grouped to describe opposite or contrasting movements.

Flexion, Extension, and Hyperextension

Flexion (figs. 10.9 and 10.10*c*) is movement that decreases the angle of a joint, usually in a sagittal plane. Examples are bending the elbow or knee and bending the neck to look down at the floor. Bending at the waist, as if taking a bow, is flexion of the spine. Flexion of the shoulder consists of raising the arm from anatomical position in a sagittal plane, as if to point in front of you or toward the ceiling. Flexion of the hip entails raising the thigh, as in a high-stepping marching stance.

Extension is movement that straightens a joint and generally returns a body part to anatomical position—for example, straightening the elbow or knee, raising the head to look directly forward, straightening the waist, or moving the arm back to a position parallel to the trunk.

Hyperextension is the extension of a joint beyond 180°. For example, if you extend your arm and hand with the palm down, and then raise the back of your hand as if admiring a new ring, you are hyperextending the wrist. If you look up toward the ceiling, you are hyperextending your neck. If you move your arm to a position posterior to the shoulder, you are hyperextending your shoulder.

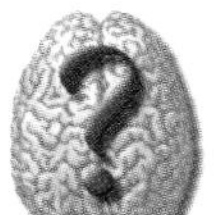

Think About It

Some synovial joints have articular surfaces or ligaments that prevent them from being hyperextended. Try hyperextending some of your synovial joints and list a few for which this is impossible.

Abduction and Adduction

Abduction[12] (ab-DUC-shun) (fig 10.10) is movement of a body part away from the midsagittal line—for example, raising the arm to one side of the body or standing spread-legged. To abduct the fingers is to spread them apart. **Adduction**[13] (ah-DUC-shun) is movement toward the midsagittal line or median axis of the middle digit—

12. *ab* = away + *duc* = to carry, lead
13. *ad* = toward + *duc* = to carry, lead

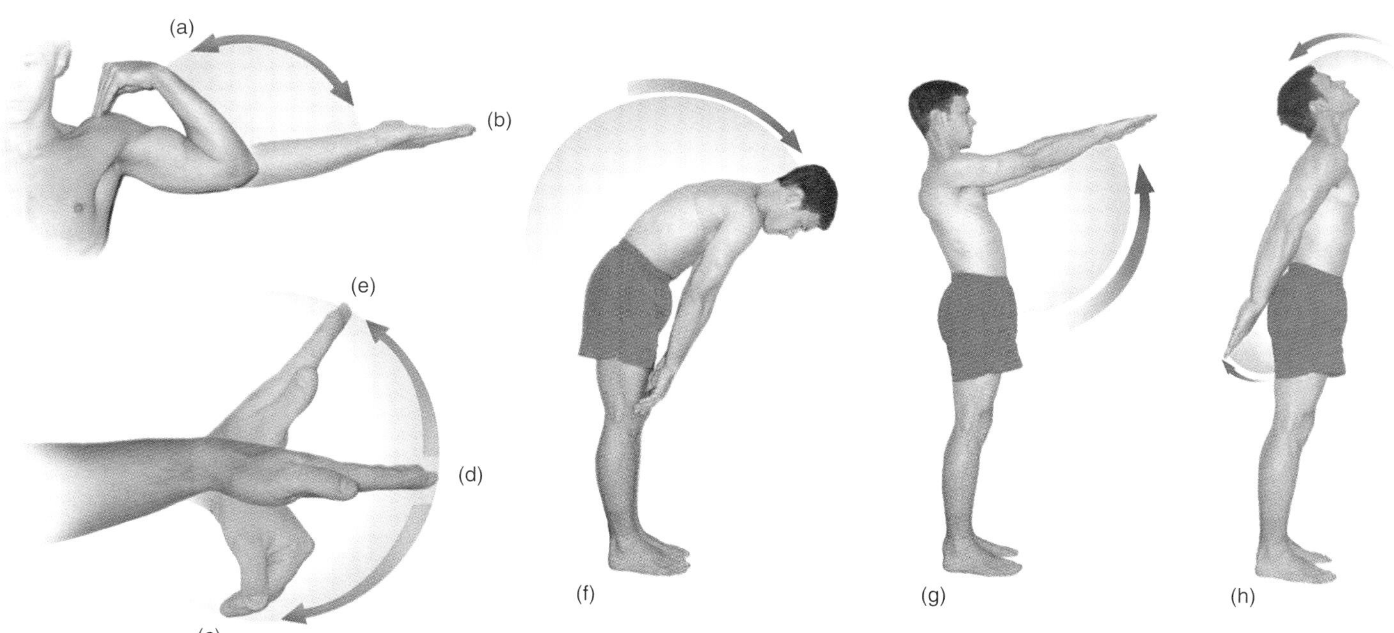

Figure 10.9 Types of Joint Movement. (*a*) Flexion of the elbow; (*b*) extension of the elbow; (*c*) flexion of the wrist; (*d*) extension of the wrist; (*e*) hyperextension of the wrist; (*f*) flexion of the spine; (*g*) extension of the spine and flexion of the shoulder; (*h*) hyperextension of the neck and shoulder.

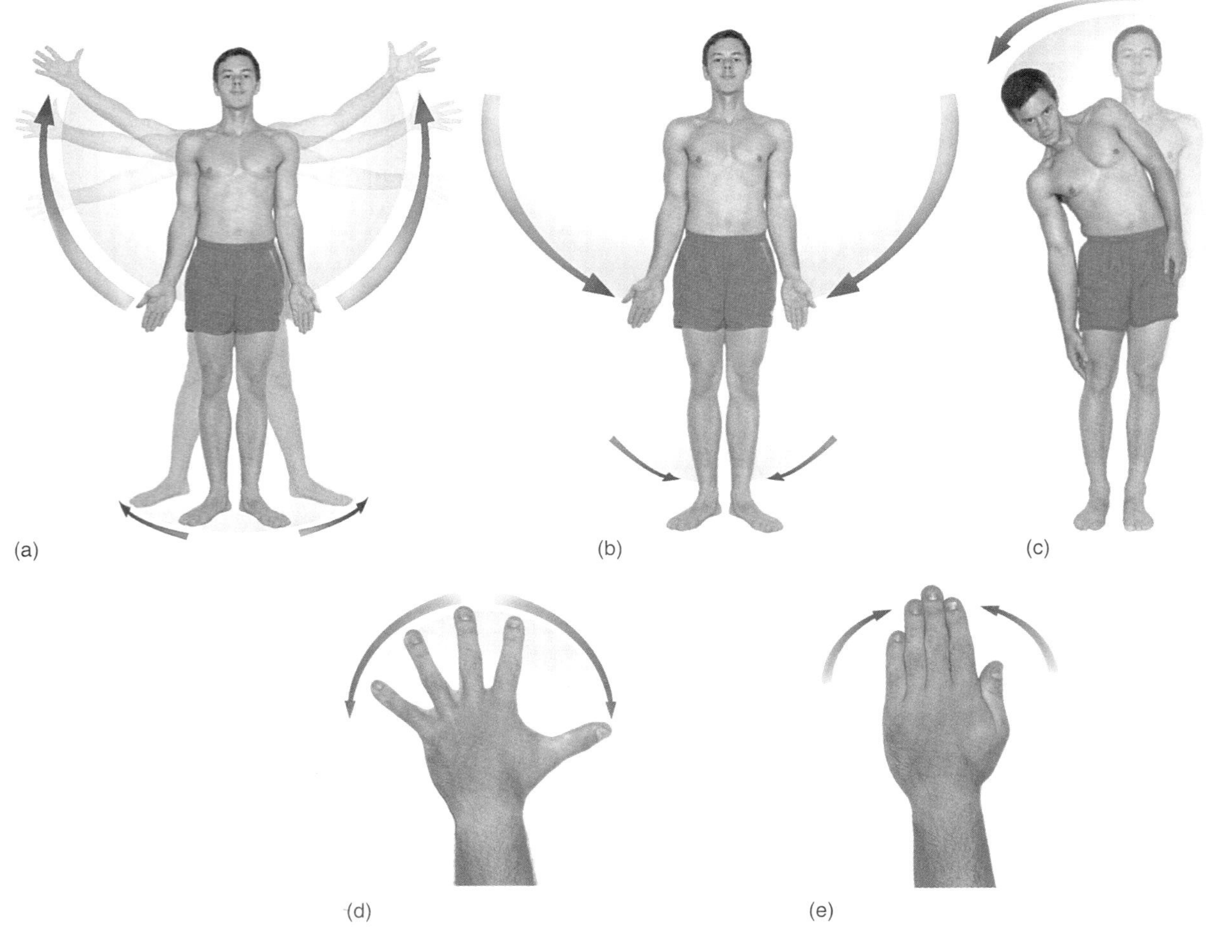

Figure 10.10 Types of Joint Movement, Continued. (*a*) Abduction of the arms and legs; (*b*) adduction of the arms and legs; (*c*) abduction (lateral flexion) of the spine; (*d*) abduction of the fingers; (*e*) adduction of the fingers.

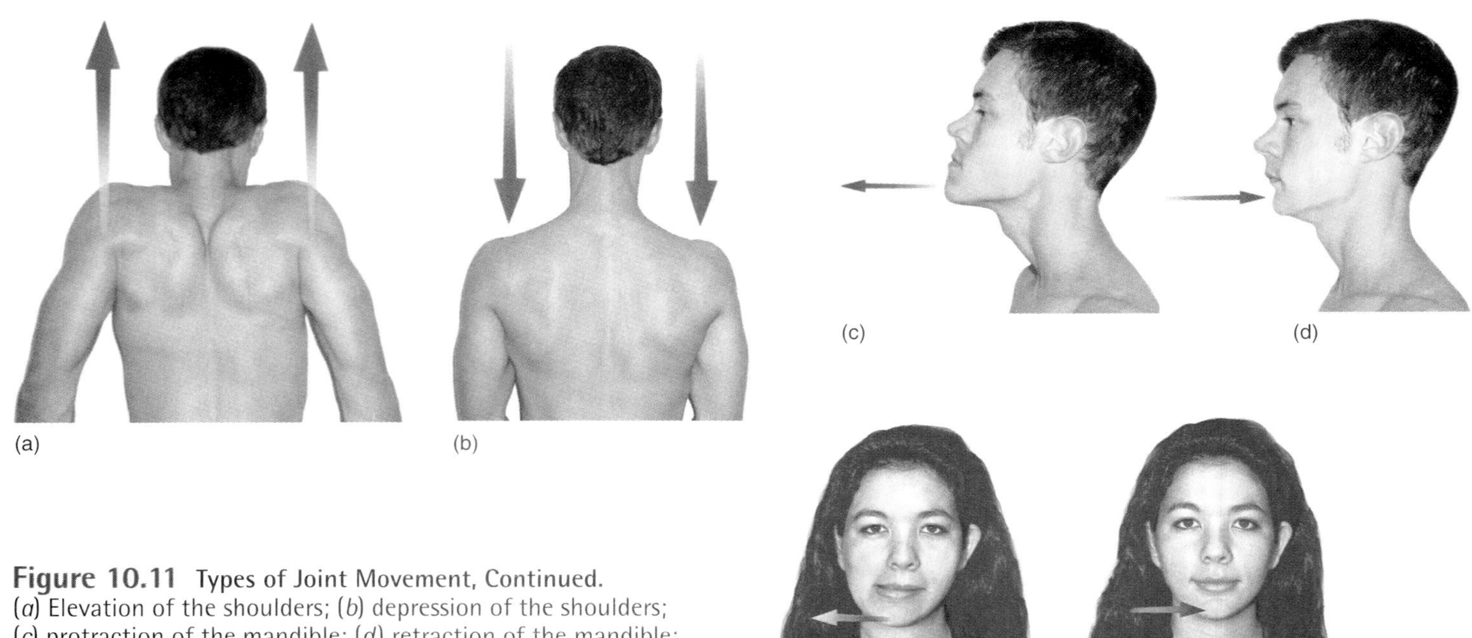

Figure 10.11 Types of Joint Movement, Continued. (*a*) Elevation of the shoulders; (*b*) depression of the shoulders; (*c*) protraction of the mandible; (*d*) retraction of the mandible; (*e*) lateral excursion of the mandible; (*f*) medial excursion of the mandible.

that is, returning the body part to anatomical position. Some movements are open to alternative interpretations. Bending the head to one side or bending sideways at the waist may be regarded as abduction or lateral flexion.

Elevation and Depression

Elevation (fig. 10.11*a*) is movement that raises a bone vertically. The mandible is elevated when biting off a piece of food, and the clavicles are elevated when shrugging the shoulders as if to gesture, "I don't know." The opposite of elevation is **depression**—lowering the mandible to open the mouth or lowering the shoulders, for example (fig. 10.11*b*).

Protraction and Retraction

Protraction[14] is movement of a bone anteriorly (forward) on a horizontal plane, and **retraction**[15] is movement posteriorly (fig. 10.11*c, d*). Jutting the jaw outward, hunching the shoulders forward, or thrusting the pelvis forward are examples of protraction. The clavicles are retracted when standing at military attention. Most people have some degree of overbite and so must protract the mandible to make the incisors meet when taking a bite of fruit, for example. The mandible is then retracted to make the molars meet and grind food between them.

Lateral and Medial Excursion

Biting and chewing food require several movements of the jaw: up and down (elevation-depression), forward and back (protraction-retraction), and side-to-side grinding movements. The last of these are called **lateral excursion** (sideways movement to the right or left) and **medial excursion** (movement back to the midline) (fig. 10.11*e, f*).

Circumduction

Circumduction[16] (fig. 10.12*a*) is movement in which one end of an appendage remains relatively stationary while the other end makes a circular motion. The appendage as a whole thus describes a conical space. For example, if an artist standing at an easel reaches out and draws a circle on the canvas, the shoulder remains stationary while the hand makes a circle. The extremity as a whole thus exhibits circumduction. A baseball player winding up for the pitch circumducts the arm in a more extreme "windmill" fashion. Circumduction is actually a sequence of flexion, abduction, extension, and adduction.

14. *pro* = forward + *trac* = pull, draw
15. *re* = back + *tract* = pull, draw
16. *circum* = around + *duc* = to carry, lead

Rotation

Rotation is a movement in which a bone turns on its longitudinal axis. Figure 10.12 shows the limb movements that occur in **lateral** and **medial rotation** of the femur and humerus. Twisting at the waist is rotation of the trunk. When the head is turned from side to side, the atlas rotates on the axis.

Supination and Pronation

These movements are limited to the forearm. **Supination**[17] (SOO-pih-NAY-shun) (fig. 10.13*a*) is rotation of the forearm so that the palm faces forward or upward; in anatomical position, the forearm is supine. **Pronation**[18] (fig. 10.13*b*) is rotation of the forearm so that the palm faces toward the rear or downward. As an aid to memory, think of it this way: you are *prone* to stand in the most comfortable position, which is with the palm *pronated*. If you were holding a bowl of *soup* in your hand, your forearm would have to be *supinated*. These movements are achieved with muscles discussed in the next chapter. The *supinator* muscle is the most powerful, and supination is the sort of movement you would usually make with the right hand to turn a doorknob clockwise or drive a screw into a piece of wood.

Opposition and Reposition

Opposition[19] is movement of the thumb to approach or touch the fingertips, and **reposition**[20] is its movement back to anatomical position, parallel to the index finger (fig. 10.13*c, d*). Opposition is the movement that enables the hand to grasp objects and is the single most important hand function.

Dorsiflexion and Plantar Flexion

These movements are limited to the foot. **Dorsiflexion** (DOR-sih-FLEC-shun) is a movement in which the toes are raised (as one might do to apply toenail polish) (fig. 10.14*a*). The foot is dorsiflexed in each step you take as your foot comes forward. Dorsiflexion prevents your toes from scraping on the ground and results in a "heel strike" when that foot touches down in front of you. **Plantar** (PLAN-tur) **flexion** is extension of the foot so that the toes point downward, as in standing on tiptoe or pressing the gas pedal of a car (fig. 10.14*c*). This motion also produces the "toe-off" in each step you take, as the heel of the foot behind you lifts off the ground.

17. *supin* = to lay back
18. *pron* = to bend forward
19. *op* = against + *posit* = to place
20. *re* = back + *posit* = to place

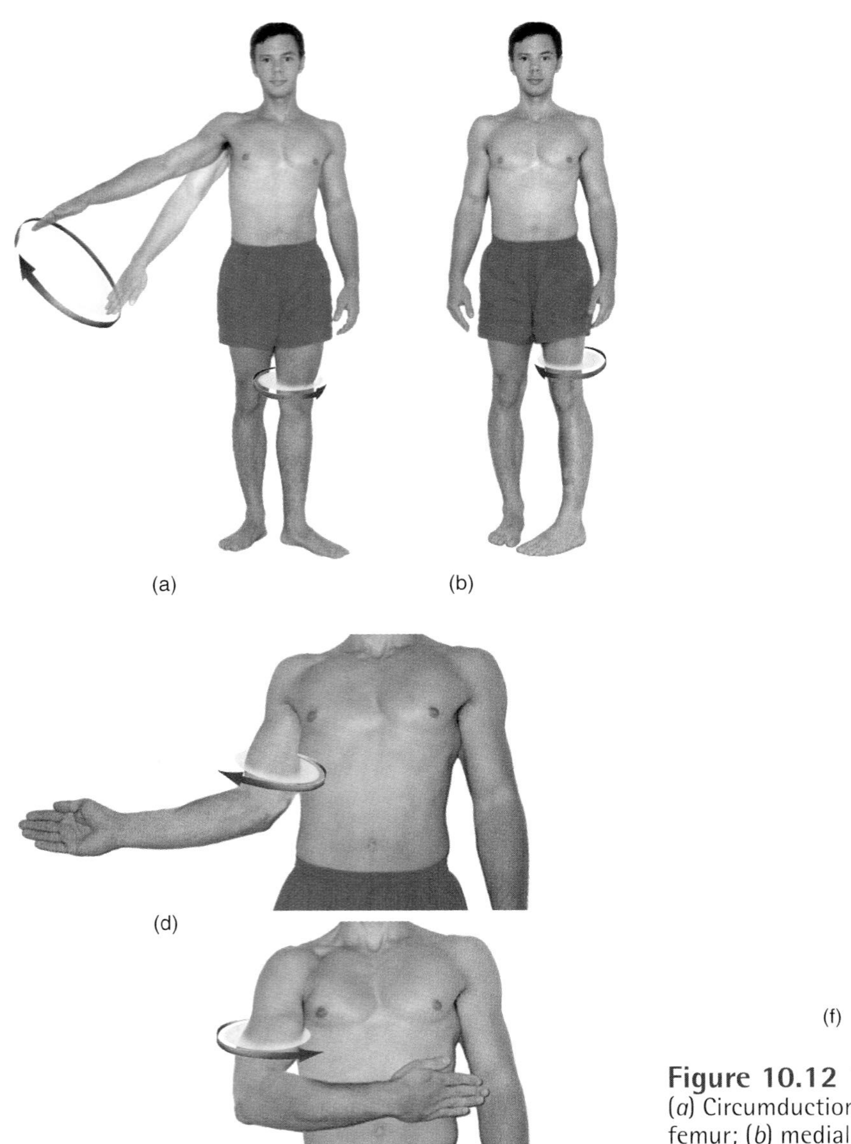

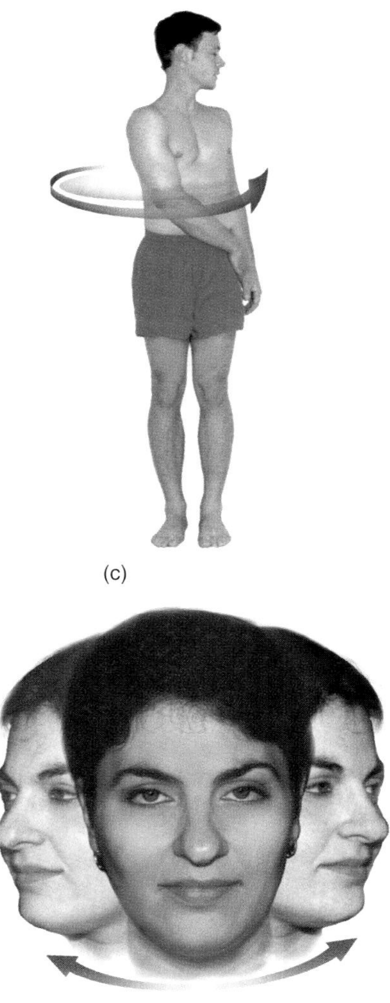

Figure 10.12 Types of Joint Movement, Continued. (*a*) Circumduction of the shoulder and lateral rotation of the femur; (*b*) medial rotation of the femur; (*c*) rotation of the spine; (*d*) lateral rotation of the humerus; (*e*) medial rotation of the humerus; (*f*) rotation of the neck.

Inversion and Eversion

These movements are also unique to the feet (fig. 10.14*d, e*). **Inversion**[21] is a movement in which the soles are turned medially; **eversion**[22] is a turning of the soles to face laterally. Inversion and eversion are common in fast sports such as tennis and football and often result in ankle sprains. These terms also refer to congenital deformities of the feet, which are often corrected by orthopedic shoes or braces.

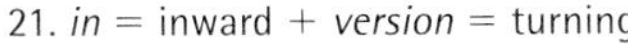

21. *in* = inward + *version* = turning
22. *e* = outward + *version* = turning

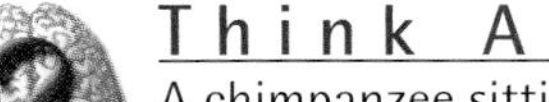

Think About It

A chimpanzee sitting on the ground reaches out and grasps an object between its fingertips. Then it raises its hand to its face and turns the object to examine it. List the movements that would occur at its diarthroses and identify the joint at which each one would occur.

Range of Motion

We can see from the movements just described that the **range of motion (ROM)** of a joint varies greatly from one type to another. ROM obviously affects a person's functional

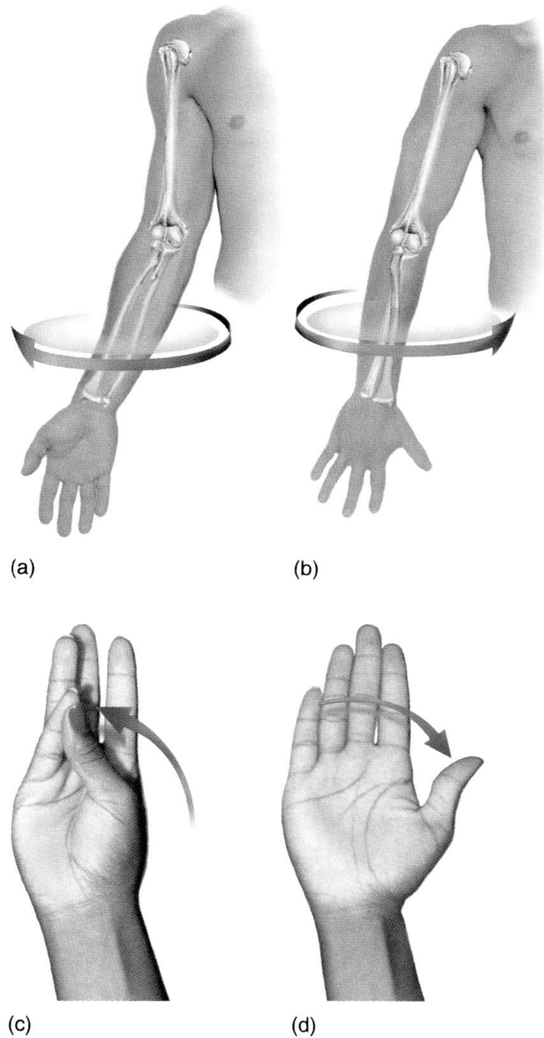

Figure 10.13 Types of Joint Movement Unique to the Upper Limb. (*a*) Supination of the forearm; (*b*) pronation of the forearm; (*c*) opposition of the thumb; (*d*) reposition of the thumb.

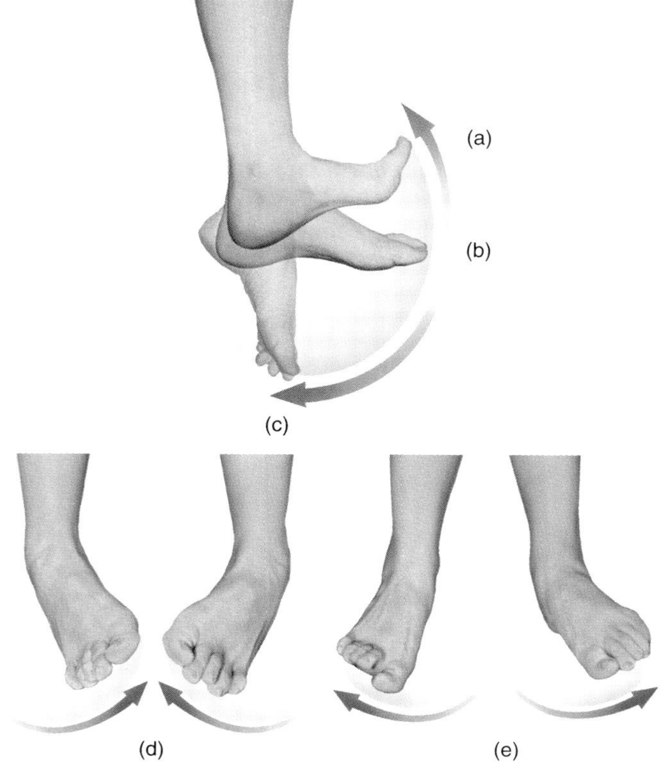

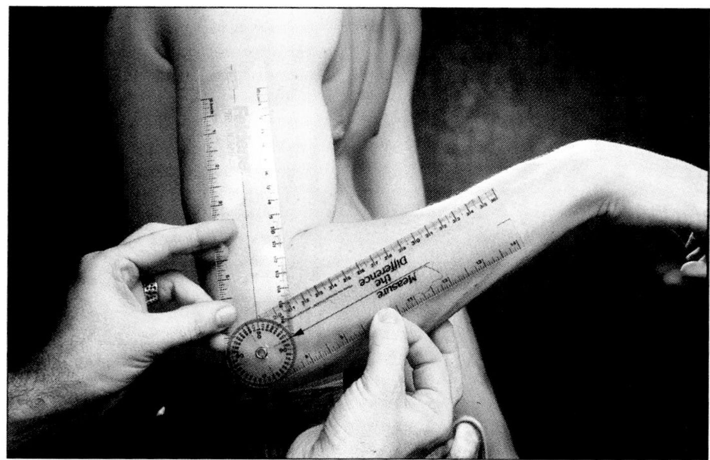

Figure 10.14 Types of Joint Movement Unique to the Foot. (*a*) Dorsiflexion; (*b*) extension; (*c*) plantar flexion; (*d*) inversion; and (*e*) eversion.

independence and quality of life. It is also an important consideration in training for athletics or dance, in clinical diagnosis, and in monitoring the progress of rehabilitation. ROM can be measured with a simple device called a *goniometer*[23] (fig. 10.15).

Several factors affect the ROM and stability of a joint:

- **Structure and action of the muscles.** The two most important factors in stabilizing a joint are tendons and muscle tone (a state of partial contraction of a "resting" muscle). Tendons, ligaments, and muscles have sensory nerve endings called *proprioceptors* (PRO-pree-oh-SEP-turs) that continually monitor joint angle and muscle tension. Upon receiving this information, the spinal cord sends nerve signals back to the muscles to increase or decrease their state of contraction and adjust the position of the joint and tautness of the tendons.

Figure 10.15 Measuring Range of Motion (ROM). The goniometer measures the angle through which a person can move a joint.

23. *gonio* = angle + *meter* = measuring device

- **Structure of the articular surfaces of the bones.** You cannot hyperextend your elbow, for example, because the olecranon of the ulna fits into the olecranon fossa of the humerus and prevents further movement in that direction.
- **Strength and tautness of ligaments, tendons, and the joint capsule.** You cannot hyperextend your knee, for example, because of a *cruciate ligament* that is pulled tight when the knee is extended and prevents further motion. Gymnasts and acrobats increase the ROM of their joints by gradually stretching their ligaments during training. "Double-jointed" people have unusually large ROMs at some joints, not because the joint is actually double or fundamentally different from normal in its anatomy, but because the ligaments are unusually long or slack.

Levers and Biomechanics of the Joints

A **lever** is any elongated, rigid object that rotates around a fixed point called the **fulcrum** (fig. 10.16). Familiar examples include a seesaw and a crowbar. Rotation occurs when an **effort** applied to one point on the lever overcomes a **resistance (load)** located at some other point. The part of a lever from the fulcrum to the point of effort is called the **effort arm,** and the part from the fulcrum to the point of resistance is the **resistance arm.** In the body, a long bone acts as a lever, a joint serves as the fulcrum, and the effort is generated by a muscle attached to the bone.

The function of a lever is to confer an advantage—either to exert more force against a resisting object than the force applied to the lever (for example, in moving a heavy boulder with a crowbar) or to move the resisting object farther or faster than the effort arm is moved (as in swinging a baseball bat). A single lever cannot confer both advantages. There is a trade-off between force on one hand and speed or distance on the other—as one increases, the other decreases.

The **mechanical advantage (MA)** of a lever is the ratio of its output force to its input force. It can be predicted from the length of the effort arm, L_E, divided by the length of the resistance arm, L_R; that is, $MA = L_E/L_R$. If MA is greater than 1.0, the lever will produce more force, but less speed or distance, than the force exerted on it. If MA is less than 1.0, the lever will produce more speed or distance, but less force, than the input. Consider the elbow joint, for example (fig. 10.17*a*). Its resistance arm is longer than its effort arm, so we know from the preceding formula that the mechanical advantage is going to be less than 1.0. The figure shows some representative values for L_E and L_R that yield $MA = 0.15$. The biceps brachii muscle puts more power into the lever than we get out of it, but the hand moves farther and faster than the point where the biceps tendon inserts on the ulna.

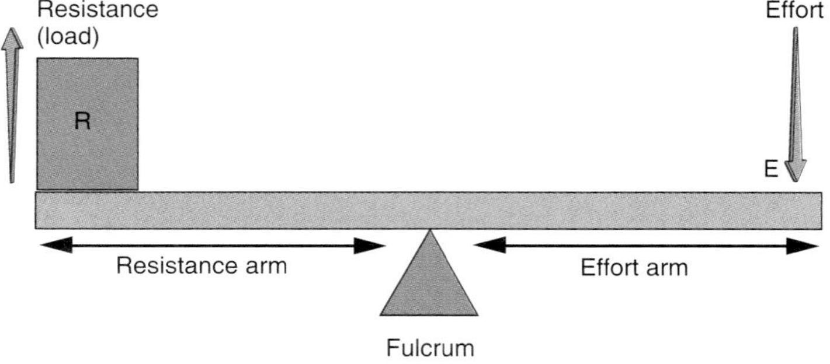

Figure 10.16 **The Basic Components of a Lever.** This example is a first-class lever.

In the next chapter, you will see that several joints have two or more muscles acting on them, seemingly producing the same effect. At first, you might consider this arrangement redundant, but it makes sense if the tendinous insertions of the muscles are at slightly different places and produce different mechanical advantages. A runner taking off from the starting line, for example, uses "low-gear" (high-MA) muscles that do not generate much speed but have the power to overcome the inertia of the body. A runner then "shifts into high gear" by using muscles with different insertions that have a lower mechanical advantage but produce more speed at the feet. This is analogous to the way an automobile transmission works to get a car to move and then cruise at high speed.

Physicists recognize three classes of levers that differ with respect to which component—the fulcrum (F), effort (E), or resistance (R)—is in the middle. A **first-class lever** (EFR) is one with the fulcrum in the middle (fig. 10.18*a*), such as a crowbar. An anatomical example is the atlanto-occipital joint of the neck, where the muscles of the back of the neck pull down on the nuchal lines of the skull and oppose the tendency of the head to tip forward. Loss of muscle tone here can be embarrassing if you nod off in class.

A **second-class lever** (FRE) is one in which the resistance is in the middle (fig. 10.18*b*). Lifting the handles of a wheelbarrow, for example, causes it to pivot on its wheel at the opposite end, and lift a load in the middle. The mandible acts as a second-class lever when the *digastric muscle* pulls down on the chin to open the mouth. The fulcrum is the *temporomandibular (jaw) joint,* the effort is applied to the chin by the digastric muscle, and the resistance is the tension of muscles such as the *temporalis* that is used to bite and to hold the mouth closed. (This arrangement is upside down relative to a wheelbarrow, but the mechanics remain the same.)

In a **third-class lever** (FER), the effort is applied between the fulcrum and resistance (fig. 10.18*c*). A pair of forceps, for example, consists of two third-class levers joined together. Most levers in the human body are third-class levers. The mandible acts as a third-class lever when you close your mouth to bite off a piece of food. Again, the

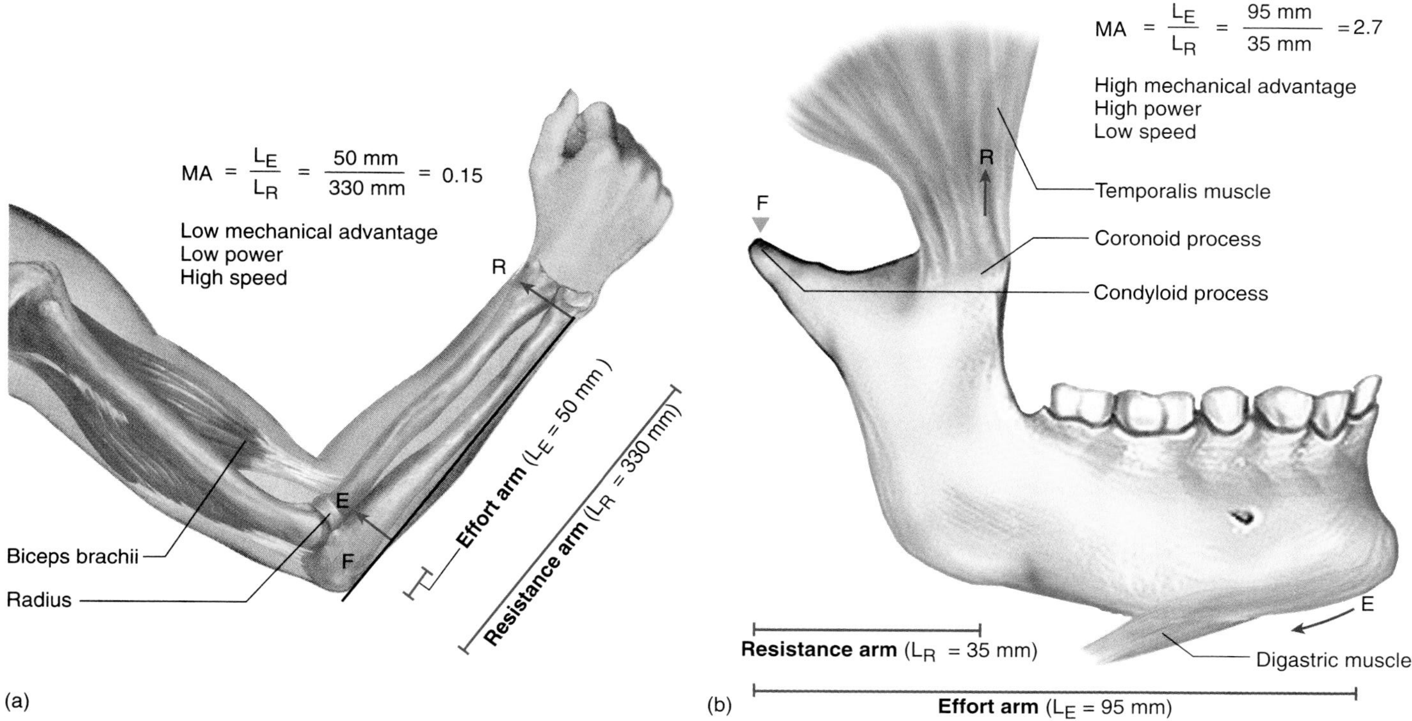

Figure 10.17 **Mechanical Advantage (MA).** MA is calculated from the length of the effort arm divided by the length of the resistance arm. (*a*) The forearm acts as a third-class lever during flexion of the elbow. (*b*) The mandible acts as a second-class lever when the jaw is forcibly opened. The digastric muscle and others provide the effort, while tension in the temporalis muscle and others provide resistance.

temporomandibular joint is the fulcrum, but now the temporalis muscle exerts the effort, while the resistance is supplied by the item of food being bitten. At the elbow, the fulcrum is the joint between the ulna and humerus; the effort is applied by the biceps brachii muscle, and the resistance can be provided by any weight in the hand or the weight of the forearm itself (see fig. 10.17*a*).

Before You Go On

Answer the following questions to test your understanding of the preceding section:

8. What are the two components of a joint capsule? What is the function of each?
9. What types of joints are described as monaxial, biaxial, and multiaxial? Give an example of each and explain the reason for its classification.
10. Name the joints that would be involved if you reached directly overhead and screwed a light bulb into a ceiling fixture. Describe the joint actions that would occur.
11. Where are the effort, fulcrum, and resistance in the act of dorsiflexion? What class of lever is this? Would you expect it to have a mechanical advantage greater or less than 1.0? Why?

Anatomy of Selected Diarthroses

Objectives

When you have completed this section, you should be able to

- identify the major anatomical features of the jaw, shoulder, elbow, hip, knee, and ankle joints; and
- explain how the anatomical differences between these joints are related to differences in function.

We now examine the gross anatomy of certain diarthroses. It is beyond the scope of this book to discuss all of them, but the ones selected here most often require medical attention and many of them have a strong bearing on athletic performance.

The Temporomandibular Joint

The **temporomandibular joint (TMJ)** is the articulation of the condyle of the mandible with the mandibular fossa of the temporal bone (fig. 10.19). You can feel its action by pressing your fingertips against the jaw immediately anterior to the ear while opening and closing

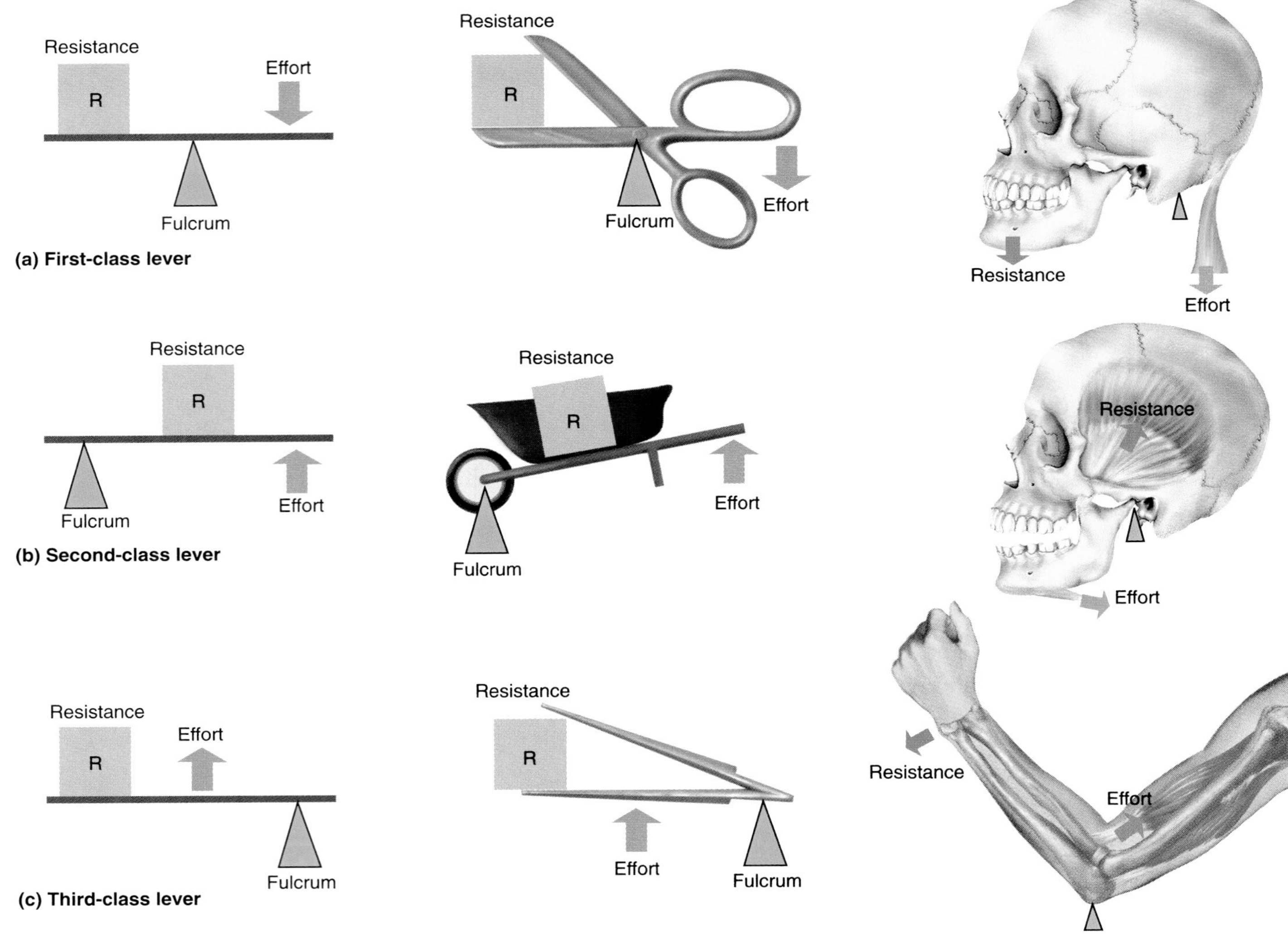

Figure 10.18 The Three Classes of Levers. *Left:* The lever classes defined by the relative positions of the resistance (load), fulcrum, and effort. *Center:* Mechanical examples. *Right:* Anatomical examples. (*a*) Muscles of the back of the neck pull the skull downward to oppose the tendency of the head to drop forward. The fulcrum is the occipital condyles. (*b*) To open the mouth, the digastric muscle pulls down on the chin. It is resisted by the temporalis muscle on the side of the head. The fulcrum is the temporomandibular (jaw) joint. (*c*) In flexing the elbow, the biceps brachii muscle exerts an effort on the radius. Resistance is provided by the weight of the forearm or anything held in the hand. The fulcrum is the elbow joint.

your mouth. This joint combines elements of condyloid, hinge, and gliding joints. It functions in a hingelike fashion when the mandible is elevated and depressed, it glides slightly forward when the jaw is protracted to take a bite, and it glides from side to side to grind food between the molars.

The synovial cavity of the TMJ is divided into superior and inferior chambers by a meniscus called the **articular disc,** which permits lateral and medial excursion of the mandible. Two ligaments support the joint. The **temporomandibular ligament** on the lateral side prevents posterior displacement of the mandible. If the jaw receives a hard blow, this ligament normally prevents the condyloid process from being driven upward and fracturing the base of the skull. The **sphenomandibular ligament** on the medial side of the joint extends from the sphenoid bone to the ramus of the mandible. A *stylomandibular ligament* extends from the styloid process to the angle of the mandible but is not part of the TMJ proper.

A deep yawn or other strenuous depression of the mandible can dislocate the TMJ by making the condyle pop out of the fossa and slip forward. The joint can be relocated by pressing down on the molars while pushing the jaw backward.

Clinical Insight 10.2

TMJ Syndrome

TMJ syndrome has received medical recognition only recently, although it may affect as many as 75 million Americans. It can cause moderate intermittent facial pain, clicking sounds in the jaw, limitation of jaw movement, and in some people, more serious symptoms—severe headaches, vertigo (dizziness), tinnitus (ringing in the ears), and pain radiating from the jaw down the neck, shoulders, and back. It seems to be caused by a combination of psychological tension and malocclusion (misalignment of the teeth). Treatment may involve psychological management, physical therapy, analgesic and anti-inflammatory drugs, and sometimes corrective dental appliances to align the teeth properly.

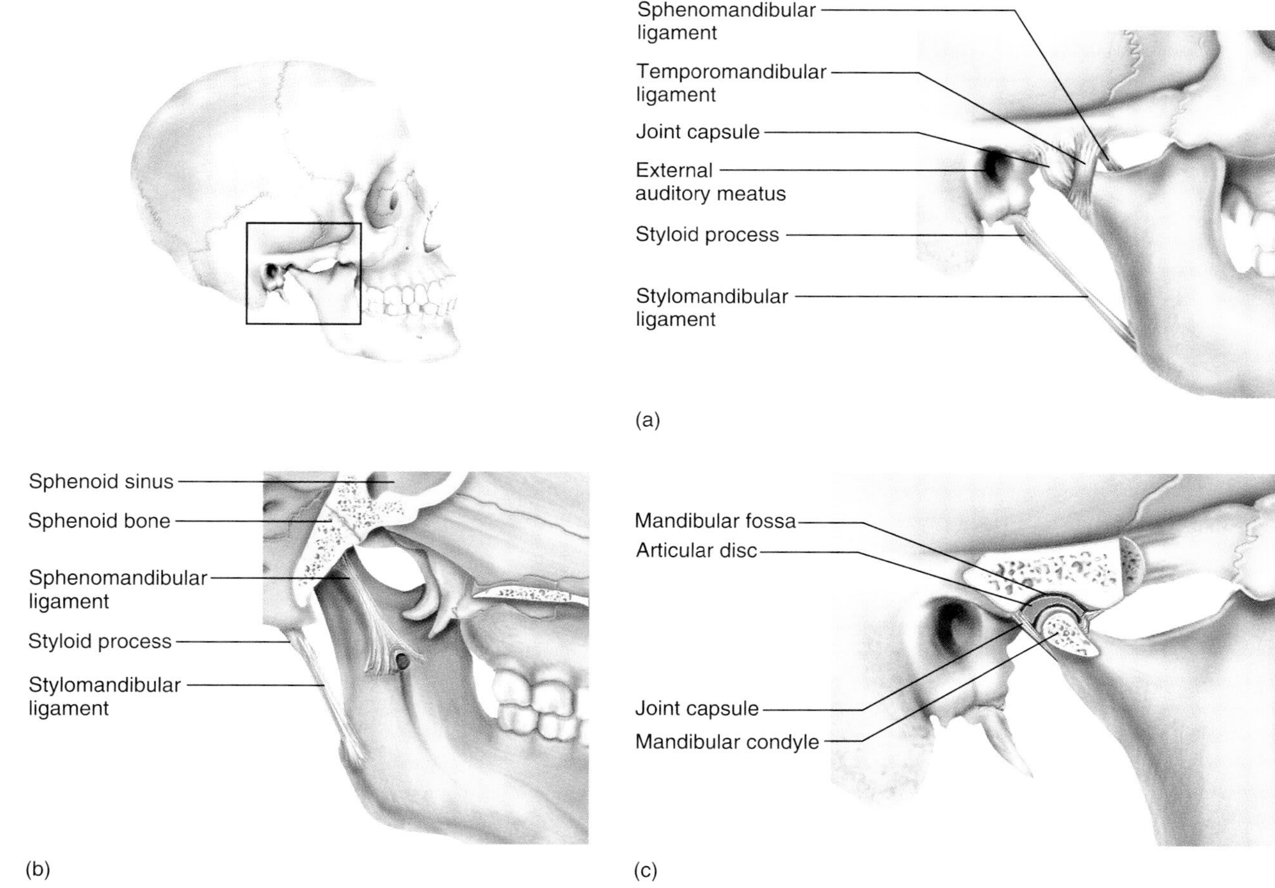

Figure 10.19 **The Temporomandibular Joint (TMJ).** (*a*) Lateral view; (*b*) medial view; (*c*) sagittal section through the joint cavity.

The Humeroscapular Joint

The shoulder joint is called the **humeroscapular (glenohumeral) joint** (fig. 10.20). It is the most freely movable joint of the body but also one of the most commonly injured. The shallowness of the glenoid cavity and looseness of the joint capsule sacrifice joint stability for freedom of movement. The cavity, however, has a ring of fibrocartilage called the **glenoid labrum**[24] around its margin, which makes it somewhat deeper than it appears on a dried skeleton.

Five principal ligaments support this joint. Three of them, called **glenohumeral ligaments,** are relatively weak

24. *labrum* = lip

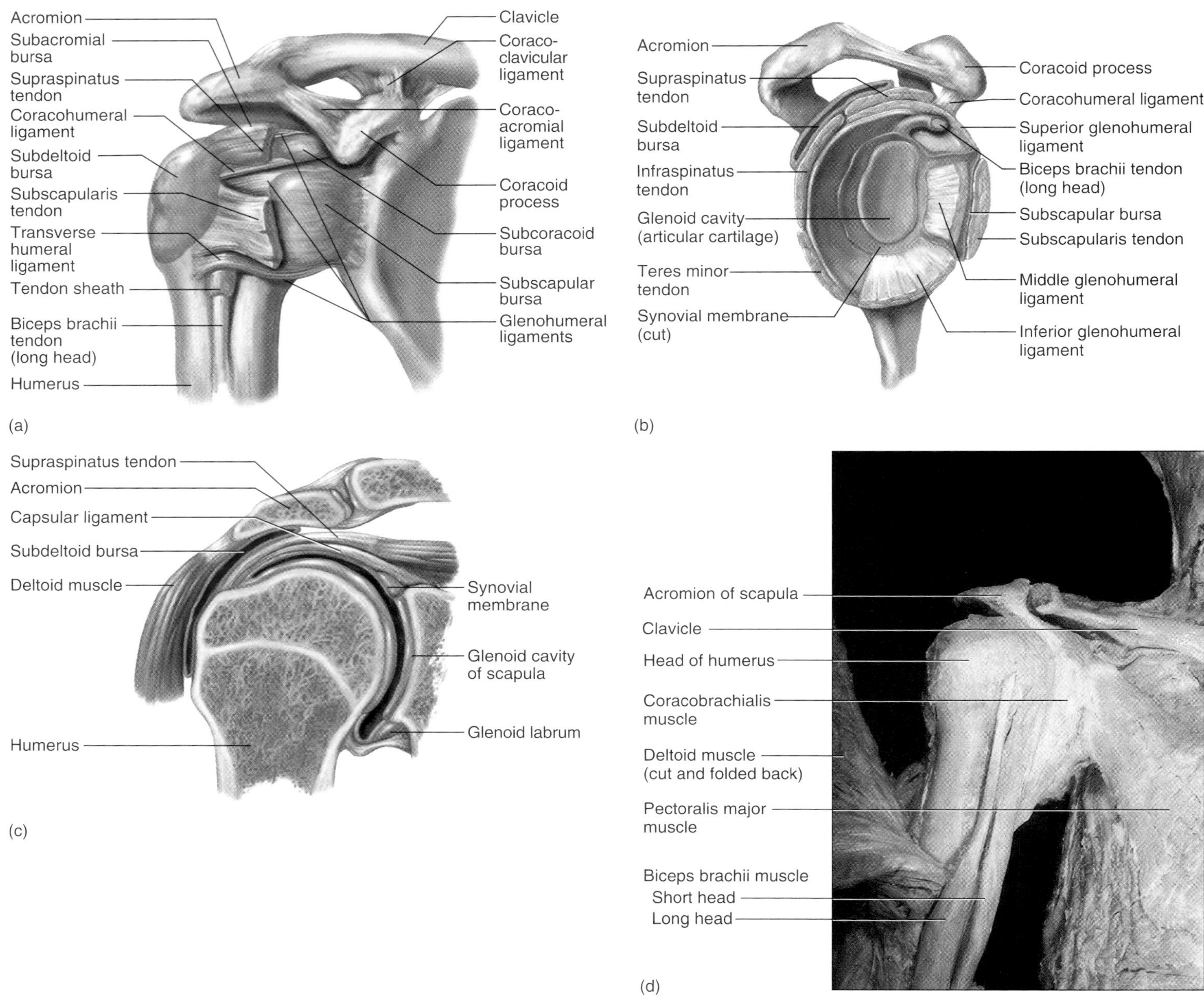

Figure 10.20 The Shoulder (humeroscapular) Joint. (*a*) Anterior view; (*b*) lateral view of the glenoid cavity and labrum with the humerus removed; (*c*) frontal section of the right shoulder joint, anterior view; (*d*) photograph of the joint.

and sometimes absent. The other two are the **coracohumeral ligament,** which extends from the coracoid process of the scapula to the greater tubercle of the humerus, and the **transverse humeral ligament,** which extends from the greater to the lesser tubercle of the humerus, creating a tunnel through which a tendon of the biceps brachii passes.

The biceps tendon is the most important stabilizer of the shoulder. It originates on the margin of the glenoid cavity, passes through the joint capsule, and emerges into the intertubercular groove of the humerus, where it is held by the transverse humeral ligament. Inferior to this groove, it merges into the biceps brachii. Thus, the tendon functions as a taut, adjustable strap that holds the humerus against the glenoid cavity.

In addition to the biceps brachii, four muscles important in stabilizing the humeroscapular joint are the *subscapularis, supraspinatus, infraspinatus,* and *teres minor,* all of which are shown and described more fully in the next chapter. The tendons of these four muscles form the **rotator cuff,** which is fused to the joint capsule on all sides except ventrally. Rotator cuff injuries are common in baseball pitchers, bowlers, and other athletes who often vigorously circumduct the arm.

Shoulder dislocations are very painful and can result in permanent damage. The most common dislocation is downward displacement of the humerus, because (1) the rotator cuff protects the joint in all directions except ventrally, and (2) the joint is protected from above by the coracoid process, acromion, and clavicle. Dislocations most often occur when the arm is abducted and then receives a blow from above—for example, when the outstretched arm is struck by heavy objects falling off a shelf. They also occur in children who are jerked off the ground by one arm or forced to follow by a hard tug on the arm. Children are especially prone to such injury not only because of the inherent stress caused by such abuse but also because a child's shoulder is not fully ossified and the rotator cuff is not strong enough to withstand such stress. Because this joint is so easily dislocated, you should never attempt to move an immobilized person by pulling on his or her arm.

Four bursae are associated with the shoulder joint. Their names describe their locations—the **subdeltoid, subacromial, subcoracoid,** and **subscapular bursae** (fig. 10.20).

The Elbow Joint

The elbow is a hinge joint composed of two articulations—the **humeroulnar joint,** where the trochlea of the humerus joins the trochlear notch of the ulna, and the **humeroradial joint,** where the capitulum of the humerus meets the head of the radius (fig. 10.21). Both are enclosed in a single joint capsule. On the posterior side of the elbow, there is a prominent **olecranon bursa** to ease the movement of tendons over the elbow. Side-to-side motions of the elbow joint are restricted by a pair of ligaments, the **radial (lateral) collateral ligament** and **ulnar (medial) collateral ligament.**

Another joint occurs in the elbow region, the **proximal radioulnar joint,** but it is not involved in the hinge. At this joint, the disclike head of the radius fits into the radial notch of the ulna and is held in place by the **annular ligament,** which encircles the head of the radius and attaches at each end to the ulna. The radial head rotates like a wheel against the ulna as the forearm is rotated.

The Coxal Joint

The **coxal** (hip) **joint** is the point where the head of the femur inserts into the acetabulum of the os coxae (fig. 10.22). Because the coxal joints bear much of the body's weight, they have deep sockets and are much more stable than the shoulder joint. The depth of the socket is somewhat greater than you see on dried bones because a horseshoe-shaped ring of fibrocartilage, the **acetabular labrum,** is attached to its rim. Dislocations of the hip are rare, but some infants suffer congenital dislocations because the acetabulum is not deep enough to hold the head of the femur in place. This condition can be treated by placing the infant in traction until the acetabulum develops enough strength to support the body's weight (fig. 10.23).

Ligaments that support the coxal joint include the **iliofemoral** (ILL-ee-oh-FEM-oh-rul) and **pubofemoral** (PYU-bo-FEM-or-ul) **ligaments** on the anterior side and the **ischiofemoral** (ISS-kee-oh-FEM-or-ul) **ligament** on the posterior side. The name of each ligament refers to the bones to which it attaches—the femur and the ilium, pubis, or ischium. When you stand up, these ligaments become twisted and pull the head of the femur tightly into the acetabulum. The head of the femur has a conspicuous pit called the *fovea capitis.* The **round ligament,** or **ligamentum teres**[25] (TERR-eez), arises here and attaches to the lower margin of the acetabulum. This is a relatively slack ligament, so it is questionable whether it plays a significant role in holding the femur in its socket. It does, however, contain an artery that supplies blood to the head of the femur. A **transverse acetabular ligament** bridges a gap in the inferior margin of the acetabular labrum.

The Knee Joint

The knee joint, or **tibiofemoral joint,** is the largest and most complex diarthrosis of the body (figs. 10.24 and 10.25). It is primarily a hinge joint, but when the knee is flexed it is also capable of slight rotation and lateral gliding. The patella and patellar ligament also form a gliding **patellofemoral joint** with the femur.

The joint capsule encloses only the lateral and posterior aspects of the knee joint, not the anterior. The anterior aspect is covered by the patellar ligament and the **lateral** and **medial patellar retinacula** (not illustrated). These are extensions of the tendon of the *quadriceps femoris* muscle, the large anterior muscle of the thigh. The knee is stabilized mainly by the quadriceps tendon in front and the tendon of the *semimembranosus* muscle on the rear of the thigh. Developing strength in these muscles therefore reduces the risk of knee injury.

The joint cavity contains two cartilages called the **lateral meniscus** and **medial meniscus,** joined by a **transverse ligament.** These menisci absorb the shock of the body weight jostling up and down on the knee joint and prevent the femur from rocking from side to side on the tibia.

---

25. *teres* = round

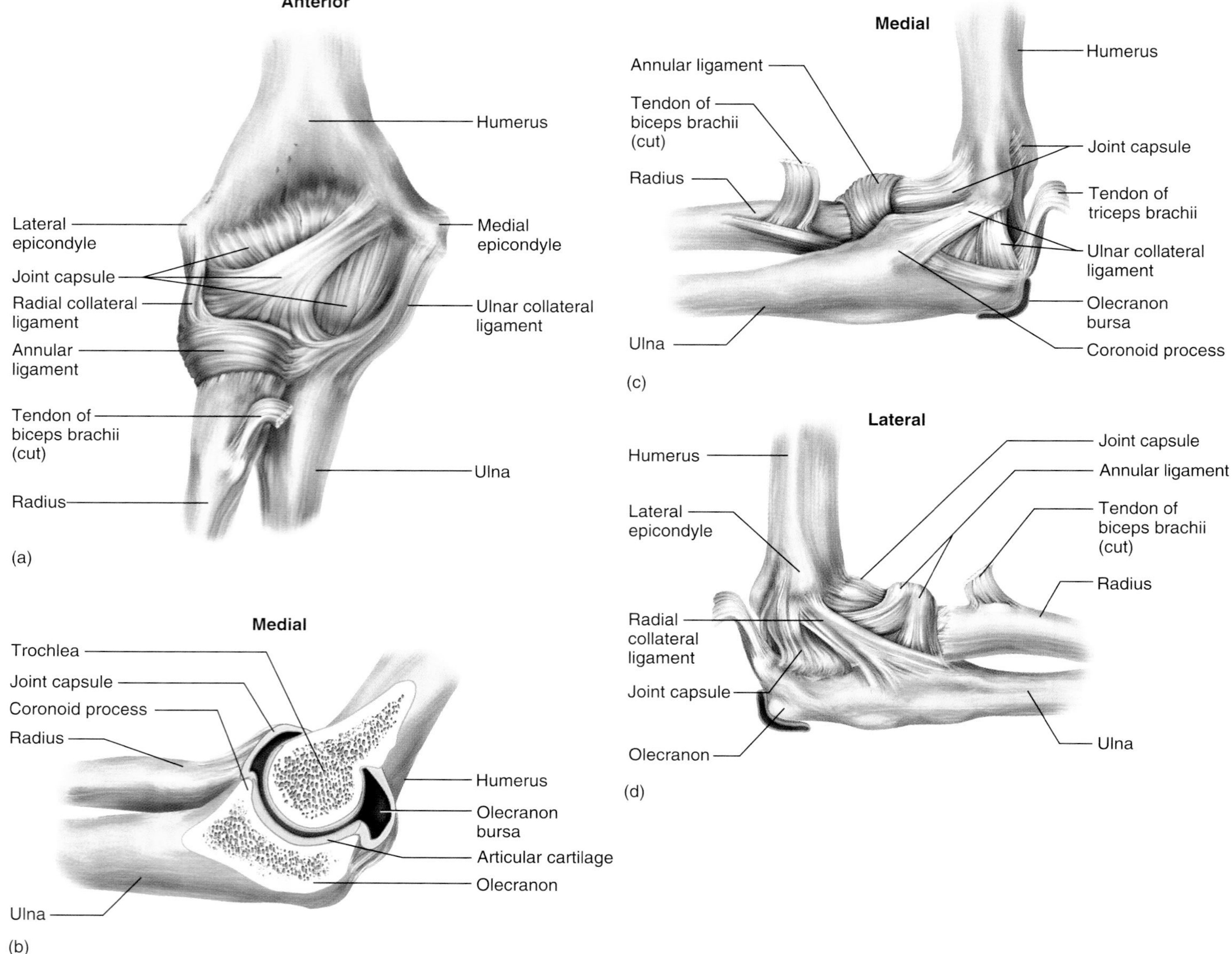

Figure 10.21 The Elbow Joint. (*a*) Anterior view; (*b*) midsagittal section, medial view; (*c*) medial view; (*d*) lateral view.

The posterior "pit" of the knee, the **popliteal** (po-LIT-ee-ul) **region,** is supported by a complex array of **intracapsular ligaments** within the joint capsule and **extracapsular ligaments** external to it. The extracapsular ligaments are the **oblique popliteal ligament** (an extension of the semimembranosus tendon), **arcuate** (AR-cue-et) **popliteal ligament, fibular (lateral) collateral ligament,** and **tibial (medial) collateral ligament.** The two collateral ligaments prevent the knee from rotating when the joint is extended.

There are two intracapsular ligaments deep within the joint cavity. The synovial membrane folds around them, however, so that they are excluded from the fluid-filled synovial cavity. These ligaments cross each other in the form of an X; hence, they are called the **anterior cruciate**[26] (CROO-she-ate) **ligament (ACL)** and **posterior cruciate ligament (PCL).** These are named according to whether they attach to the anterior or posterior side of the tibia, not for their attachments to the femur. When the knee is extended, the ACL is pulled tight and prevents hyperextension. The PCL prevents the femur from sliding off the front of the tibia and prevents the tibia from being displaced backward.

26. *cruci* = cross + *ate* = characterized by

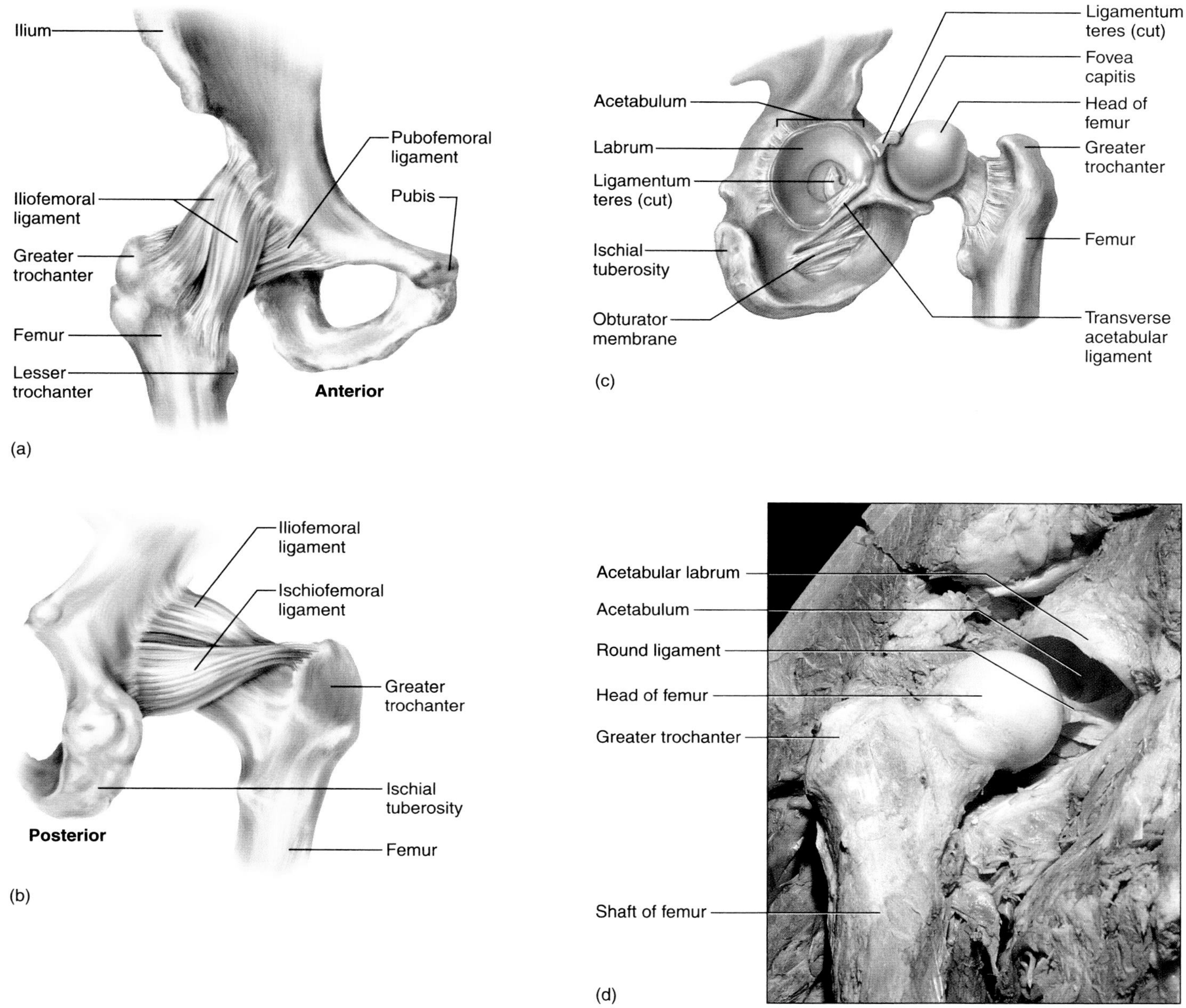

Ilium
Pubofemoral ligament
Pubis
Iliofemoral ligament
Greater trochanter
Femur
Lesser trochanter
Anterior
(a)

Acetabulum
Labrum
Ligamentum teres (cut)
Ischial tuberosity
Obturator membrane
Ligamentum teres (cut)
Fovea capitis
Head of femur
Greater trochanter
Femur
Transverse acetabular ligament
(c)

Iliofemoral ligament
Ischiofemoral ligament
Greater trochanter
Ischial tuberosity
Posterior
Femur
(b)

Acetabular labrum
Acetabulum
Round ligament
Head of femur
Greater trochanter
Shaft of femur
(d)

Figure 10.22 The Coxal (hip) Joint. (*a*) Anterior view; (*b*) posterior view; (*c*) the acetabulum with the femoral head retracted; (*d*) photograph of the right hip with the femoral head retracted, anterior view.

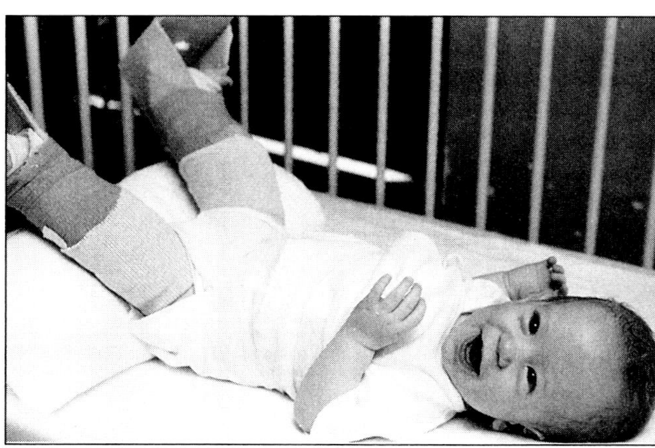

Figure 10.23 Treatment of Congenital Hip Dislocation. Infants are sometimes placed in traction to treat this condition.

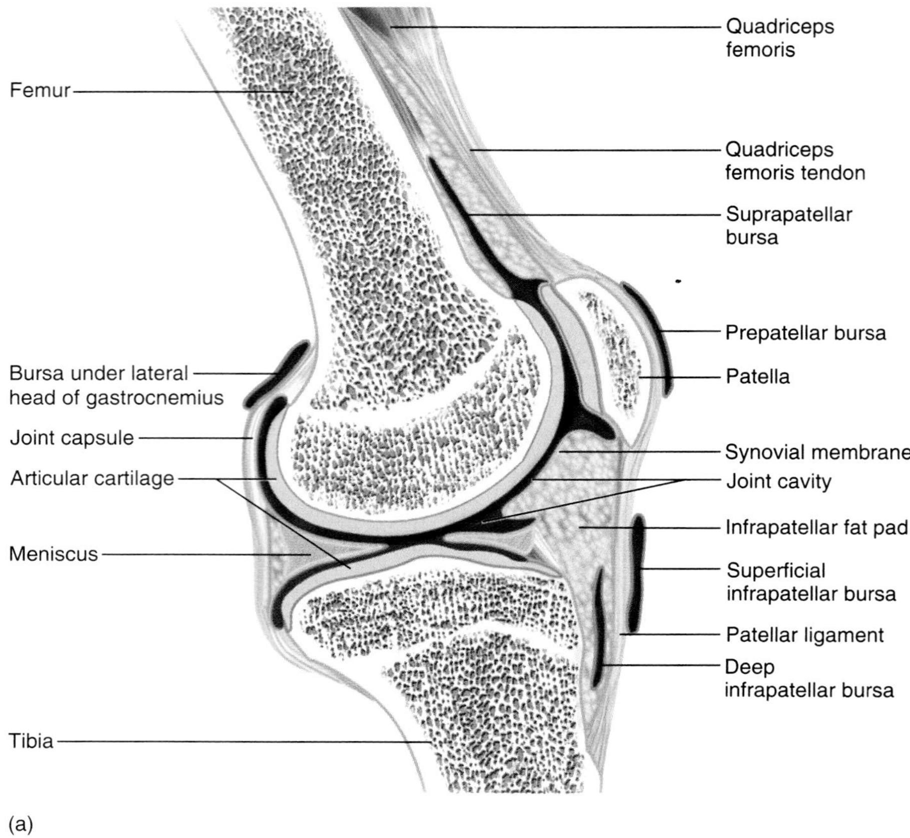

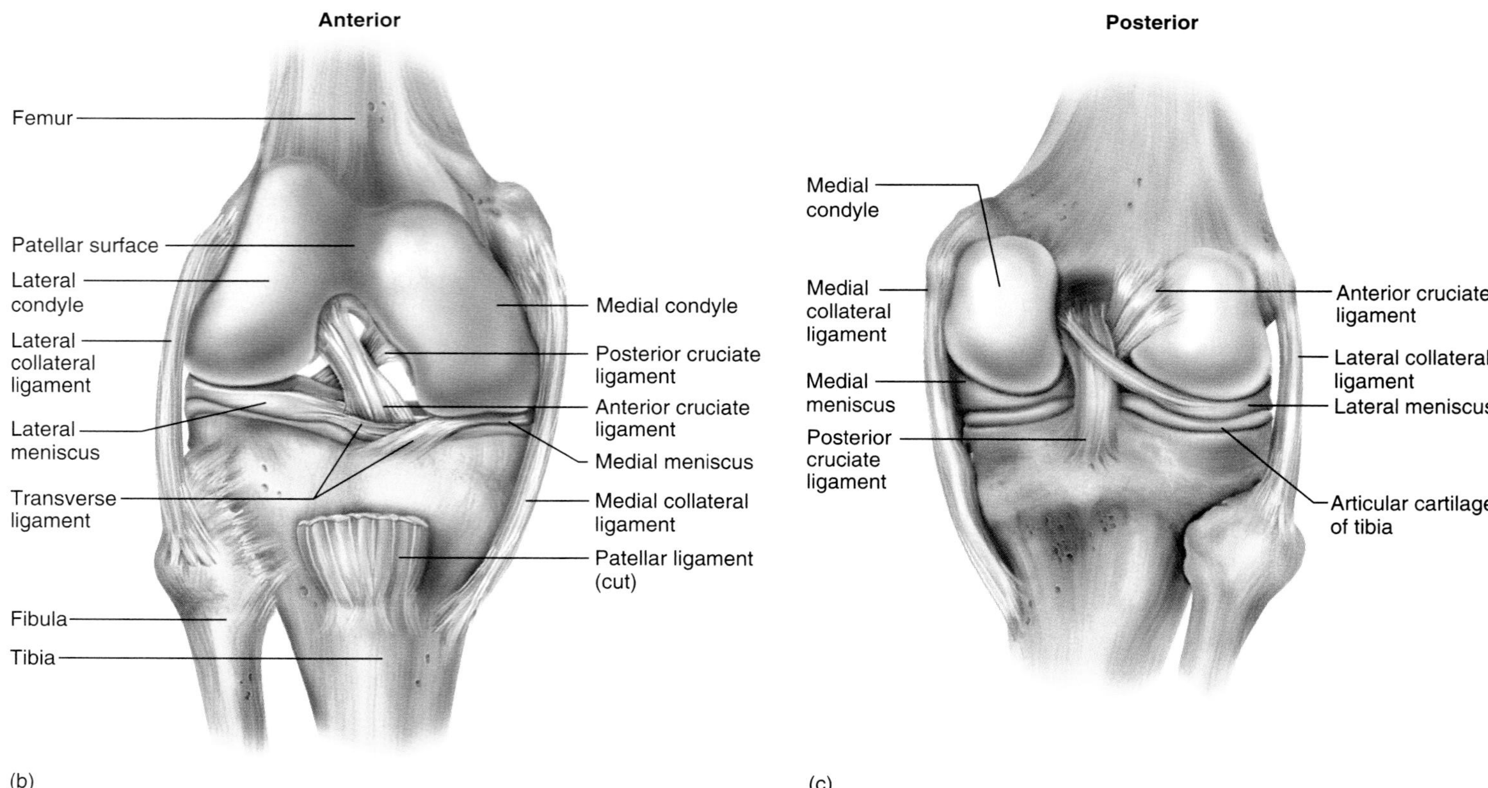

Figure 10.24 **The Knee Joint.** (*a*) Diagram of a midsagittal section; (*b*) anterior view of structures in the joint cavity of the right knee; (*c*) posterior view of the right knee.

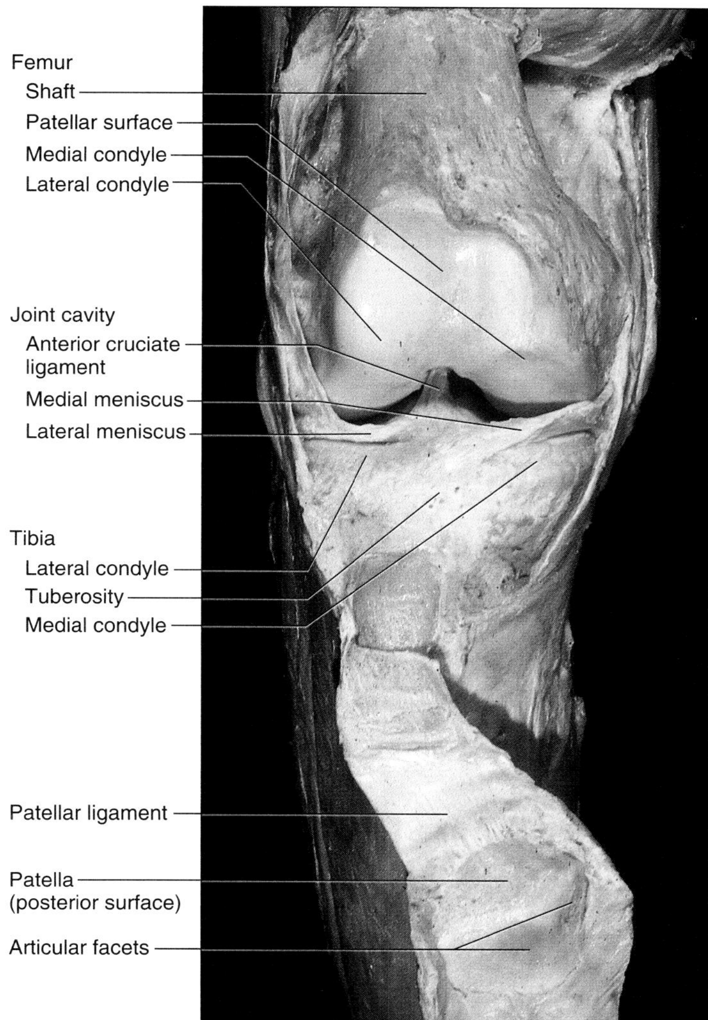

Figure 10.25 **Photograph of the Knee Joint, Anterior View.** The patellar tendon has been cut and folded (reflected) downward to expose the joint cavity and the posterior surface of the patella.

An important aspect of human bipedalism is the ability to "lock" the knees and stand erect without tiring the extensor muscles of the leg. When the knee is extended to the fullest degree allowed by the ACL, the femur rotates medially on the tibia. This action locks the knee, and in this state all the major knee ligaments are twisted and taut. To unlock the knee, the *popliteus* muscle rotates the femur laterally and untwists the ligaments.

The knee joint has at least 13 bursae. Four of these are anterior—the **superficial infrapatellar, suprapatellar, prepatellar,** and **deep infrapatellar.** Located in the popliteal region are the **popliteal bursa** and **semimembranosus bursa** (not illustrated). At least seven more bursae are found on the lateral and medial sides of the knee joint. From figure 10.24*a,* your knowledge of the relevant word elements (*infra-, supra-, pre-*), and the terms *superficial* and *deep,* you should be able to work out the reasoning behind most of these names and develop a system for remembering the locations of these bursae.

The Ankle Joint

The **talocrural** (ankle) **joint** includes two articulations—a medial joint between the tibia and talus and a lateral joint between the fibula and talus, both enclosed in one joint capsule (fig. 10.26). The malleoli of the tibia and fibula overhang the talus on each side like a cap and prevent most side-to-side motion (fig. 10.27). The ankle therefore has a more restricted range of motion than the wrist.

The ligaments of the ankle include (1) **anterior** and **posterior tibiofibular ligaments,** which bind the tibia to the fibula; (2) a multipart **deltoid ligament,** which binds the tibia to the foot on the medial side; and (3) a multipart **lateral collateral ligament,** which binds the fibula to the foot on the lateral side. The **calcaneal (Achilles) tendon** extends from the calf muscles to the calcaneus. It enables plantar flexion and limits dorsiflexion of the joint. Plantar flexion is limited by extensor tendons on the anterior side of the ankle and by the anterior part of the joint capsule.

Sprains (torn ligaments and tendons) occur especially often at the ankle. Excessive eversion can tear the deltoid ligament, and excessive inversion often tears the anterior talofibular ligament and calcaneofibular ligament. These sprains are painful and usually accompanied by immediate swelling. They are best treated by immobilizing the joint and reducing swelling with an ice pack, but in extreme cases they may require a cast or surgery.

The synovial joints described in this section are summarized in table 10.2. A summary of some common joint disorders, including sprains, is presented in table 10.3.

Before You Go On

Answer the following questions to test your understanding of the preceding section:

12. What keeps the mandibular condyle from slipping out of its fossa in a posterior direction?
13. Explain how the biceps tendon braces the shoulder joint.
14. What structure elsewhere in the skeletal system has a structure and function similar to the acetabular labrum of the os coxae?
15. What keeps the femur from slipping backward off the tibia?
16. What keeps the tibia from slipping sideways off the talus?

Clinical Insight 10.3

Knee Injuries and Arthroscopic Surgery

Although the knee can bear a lot of weight, it is highly vulnerable to rotational and horizontal stress, especially when the knee is flexed (as in skiing or running) and receives a blow from behind or from the lateral side. The most common injuries are to a meniscus or the anterior cruciate ligament (ACL). Knee injuries heal slowly because ligaments and tendons have a very scanty blood supply and cartilage has no blood vessels at all.

The diagnosis and surgical treatment of knee injuries has been greatly improved by *arthroscopy,* a procedure in which the interior of a joint is viewed with a pencil-thin instrument, the *arthroscope,* inserted through a small incision. The arthroscope has a light source, a lens, and fiber optics that allow a viewer to see into the cavity, take photographs or videotapes of the joint, and withdraw samples of synovial fluid. Saline is often introduced through one incision to expand the joint and provide a clearer view of its structures. If surgery is required, additional small incisions can be made for the surgical instruments and the procedures can be observed through the arthroscope or on a monitor. Arthroscopic surgery produces much less tissue damage than conventional surgery and enables patients to recover more quickly.

Orthopedic surgeons now often replace a damaged ACL with a graft from the patellar ligament. The surgeon "harvests" a strip from the middle one-third of the patient's patellar ligament, drills a hole into the femur and tibia within the joint cavity, threads the ligament through the holes, and fastens it with screws. The grafted ligament is more taut and "competent" than the damaged ACL. It becomes ingrown with blood vessels and serves as a substrate for the deposition of more collagen, which further strengthens it in time. Following arthroscopic ACL reconstruction, a patient typically must use crutches for 7 to 10 days and undergo supervised physical therapy for 6 to 10 weeks, followed by self-directed exercise therapy. Healing is completed in about 9 months.

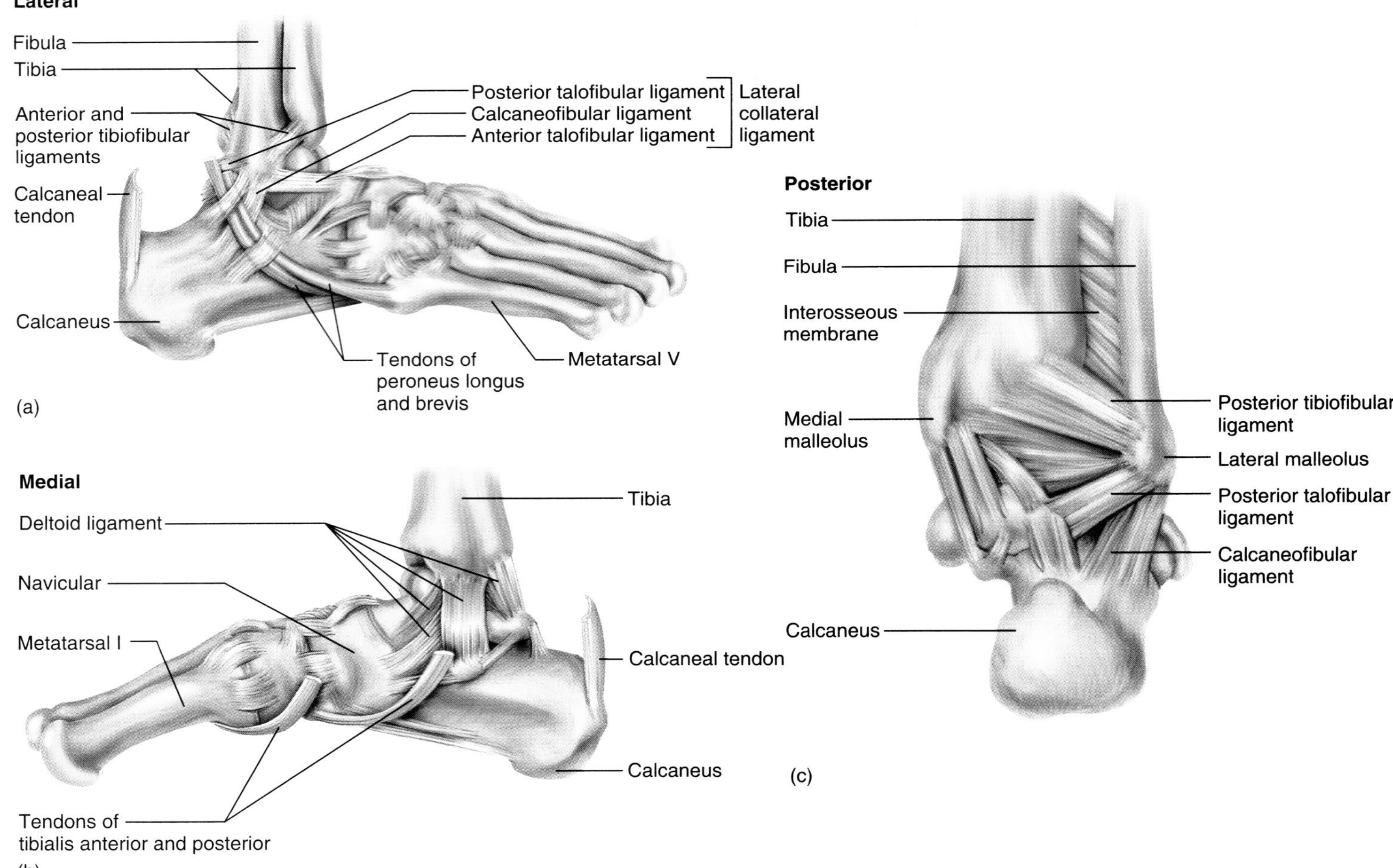

Figure 10.26 The Talocrural Joint and Ligaments of the Right Foot. (*a*) Lateral view; (*b*) medial view; (*c*) posterior view.

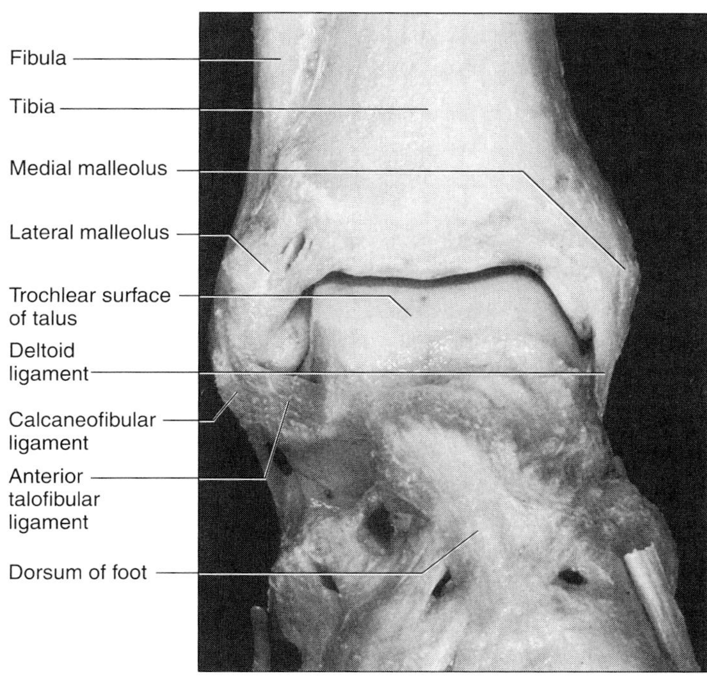

Figure 10.27 Photograph of the Talocrural Joint, Anterior View.

Table 10.2 Review of the Principal Diarthroses

Joint	Major Anatomical Features and Actions
Temporomandibular (fig. 10.19)	*Type:* condyloid, hinge, and gliding *Movements:* elevation, depression, protraction, retraction, lateral and medial excursion *Articulation:* condyle of mandible, mandibular fossa of temporal bone *Ligaments:* temporomandibular, sphenomandibular *Cartilage:* articular disc
Humeroscapular (fig. 10.20)	*Type:* ball-and-socket *Movements:* adduction, abduction, flexion, extension, circumduction, medial and lateral rotation *Articulation:* head of humerus, glenoid fossa of scapula *Ligaments:* coracohumeral, transverse humeral, three glenohumerals *Tendons:* rotator cuff (tendons of subscapularis, supraspinatus, infraspinatus, teres minor), tendon of biceps brachii *Bursae:* subdeltoid, subacromial, subcoracoid, subscapular *Cartilage:* glenoid labrum
Elbow (fig. 10.21)	*Type:* hinge and pivot *Movements:* flexion, extension, pronation, supination, rotation *Articulations:* humeroulnar—trochlea of humerus, trochlear notch of ulna; humeroradial—capitulum of humerus, head of radius; radioulnar—head of radius, radial notch of ulna *Ligaments:* radial collateral, ulnar collateral, annular *Bursa:* olecranon
Coxal (fig. 10.22)	*Type:* ball-and-socket *Movements:* adduction, abduction, flexion, extension, circumduction, medial and lateral rotation *Articulations:* head of femur, acetabulum of os coxae *Ligaments:* iliofemoral, pubofemoral, ischiofemoral, ligamentum teres, transverse acetabular *Cartilage:* acetabular labrum

(continued)

Table 10.2 Review of the Principal Diarthroses (continued)

Joint	Major Anatomical Features and Actions
Knee (fig. 10.24)	*Type:* primarily hinge *Movements:* flexion, extension, slight rotation *Articulations:* tibiofemoral, patellofemoral *Ligaments:* anterior—lateral patellar retinaculum, medial patellar retinaculum; popliteal intracapsular—anterior cruciate, posterior cruciate; popliteal extracapsular—oblique popliteal, arcuate popliteal, lateral collateral, medial collateral *Bursae:* anterior—superficial infrapatellar, suprapatellar, prepatellar, deep infrapatellar; popliteal—popliteal, semimembranosus; medial and lateral—seven other bursae not named in this chapter *Cartilages:* lateral meniscus, medial meniscus (connected by transverse ligament)
Ankle (fig. 10.26)	*Type:* hinge *Movements:* dorsiflexion, plantar flexion, extension, inversion, eversion *Articulations:* tibia-talus, fibula-talus, tibia-fibula *Ligaments:* anterior and posterior tibiofibular, deltoid, lateral collateral *Tendon:* calcaneal (Achilles)

Table 10.3 Some Common Joint Disorders

Strain	*Painful overstretching of a tendon or muscle without serious tissue damage. Often results from inadequate warm-up before exercise.*
Sprain	*Torn ligament or tendon, sometimes with damage to a meniscus or other cartilage.*
Synovitis	*Inflammation of a joint capsule, often as a complication of a sprain.*
Dislocation	*Displacement of a bone from its normal position at a joint, usually accompanied by a sprain of the adjoining connective tissues. Most common at the fingers, thumb, shoulder, and knee.*
Rheumatism	*Broad term for any pain in the supportive and locomotory organs of the body, including bones, ligaments, tendons, and muscles.*
Arthritis	*Broad term embracing more than 100 types of joint rheumatism.*
Osteoarthritis (OA)	*The most common form of arthritis, also known as "wear-and-tear arthritis" because it is apparently a normal consequence of aging. Associated with softening and degeneration of the articular cartilage, exposure of the epiphyseal bone, and development of bony spurs in the joint cavity, causing pain and restricting movement.*
Rheumatoid Arthritis (RA)	*A more severe form of arthritis resulting from an autoimmune attack against the joint tissues (failure of the immune system to recognize the tissues as one's own).*
Gout	*A hereditary disease, most common in men, in which uric acid crystals accumulate in the joints and irritate the articular cartilage and synovial membrane. Causes gouty arthritis, with swelling, pain, tissue degeneration, and sometimes fusion of the joint. Most commonly affects the great toe.*
Bursitis	*Inflammation of a bursa, usually due to overexertion of a joint.*
Tendinitis	*A form of bursitis in which a tendon sheath is inflamed.*

Clinical Insight 10.4

Arthritis and Artificial Joints

Arthritis[27] is a broad term for pain and inflammation of a joint and embraces more than a hundred different diseases of largely obscure or unknown causes. In all of its forms, it is the most common crippling disease in the United States; nearly everyone past middle age develops arthritis to some degree.

The most common form of arthritis is *osteoarthritis (OA),* also called "wear-and-tear arthritis" because it is apparently a normal consequence of years of wear on the joints. As joints age, the articular cartilage softens and degenerates. As the cartilage becomes roughened by wear, joint movement may be accompanied by crunching or crackling sounds called *crepitus.* OA affects especially the fingers, intervertebral joints, hips, and knees. As the articular cartilage wears away, exposed bone tissue often develops spurs that grow into the joint cavity, restrict movement, and cause pain. OA rarely occurs before age 40, but it affects about 85% of people older than 70. It usually does not cripple, but in severe cases it can immobilize the hip.

Rheumatoid arthritis (RA), which is far more severe than osteoarthritis, results from an autoimmune attack against the joint tissues. It begins when the body produces defensive proteins (antibodies) to fight an infection. Failing to recognize the body's own tissues, a misguided antibody known as *rheumatoid factor* also attacks the synovial membranes. Inflammatory cells accumulate in the synovial fluid and produce enzymes that degrade the articular cartilage. The synovial membrane thickens and adheres to the articular cartilage, fluid accumulates in the joint capsule, and the capsule is invaded by fibroconnective tissue. As articular cartilage degenerates, the joint begins to ossify, and sometimes the bones become solidly fused and immobilized, a condition called *ankylosis*[28] (fig. 1). The disease tends to develop symmetrically—if the right wrist or hip develops RA, so does the left.

Rheumatoid arthritis is named for the fact that symptoms tend to flare up and subside (go into remission) periodically.[29] It affects women far more often than men, and because RA typically begins between the ages of 30 and 40, it can cause decades of pain and disability. There is no cure, but joint damage can be slowed with hydrocortisone or other steroids. Because long-term use of steroids weakens the bone, however, aspirin is the treatment of first choice to control the inflammation. Physical therapy is also used to preserve the joint's range of motion and the patient's functional ability.

Arthroplasty,[30] a treatment of last resort, is the replacement of a diseased joint with an artificial device called a *prosthesis.*[31] Joint prostheses were first developed to treat war injuries in World War II and the Korean War. Total hip replacement (THR), first performed in 1963 by English orthopedic surgeon Sir John Charnley, is now the most common orthopedic procedure for the elderly. The first knee replacements were performed in the 1970s. Joint prostheses are now available for finger, shoulder, and elbow joints, as well as for hip and knee joints. Arthroplasty is performed on over 250,000 patients per year in the United States, primarily to relieve pain and restore function in elderly people with OA or RA.

Arthroplasty presents ongoing challenges for biomedical engineering. An effective prosthesis must be strong, nontoxic, and corrosion-resistant. In addition, it must bond firmly to the patient's bones and enable a normal range of motion with a minimum of friction. The heads of long bones are usually replaced with prostheses made of a metal alloy such as cobalt-chrome, titanium alloy, or stainless steel. Joint sockets are made of polyethylene (fig. 2). Prostheses are bonded to the patient's bone with screws or bone cement.

About 80% to 90% of hip replacements and at least 60% of ankle replacements are still functional 2 to 10 years later. The most common form of failure is detachment of the prosthesis from the

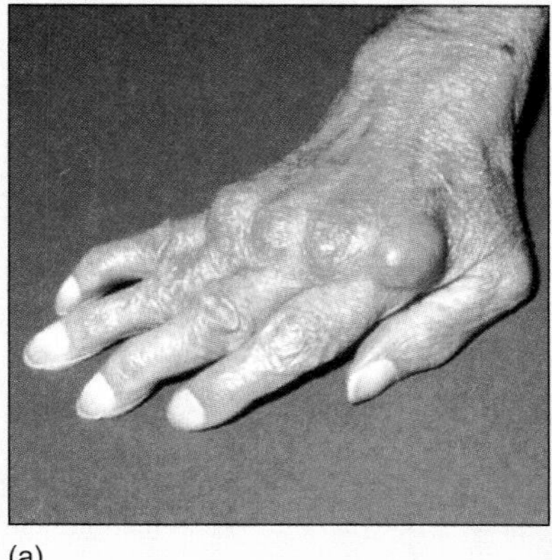
(a)

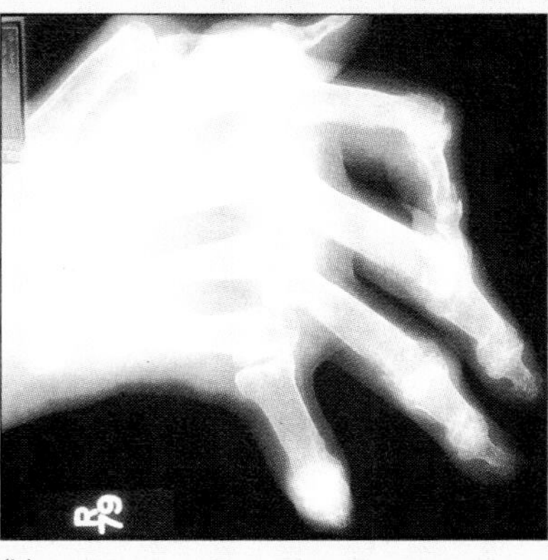
(b)

Figure 1 Rheumatoid arthritis (RA). (*a*) A severe case with ankylosis of the joints. (*b*) X ray of a hand with RA.

Continued

27. *arthr* = joint + *itis* = inflammation
28. *ankyl* = bent, crooked + *osis* = condition
29. *rheumat* = tending to change
30. *arthro* = joint + *plasty* = surgical repair
31. *prosthe* = something added

bone. This problem has been reduced by using *porous-coated prostheses,* which become infiltrated by the patient's own bone and create a firmer bond. A prosthesis is not as strong as a natural joint, however, and is not an option for many young, active patients.

Arthroplasty has been greatly improved by *computer-assisted design and manufacture (CAD/CAM).* A computer scans X rays from the patient and presents several design possibilities for review. Once a design is selected, the computer generates a program to operate the machinery that produces the prosthesis. CAD/CAM has reduced the waiting period for a prosthesis from about 12 weeks to about 2 weeks and has lowered the cost dramatically.

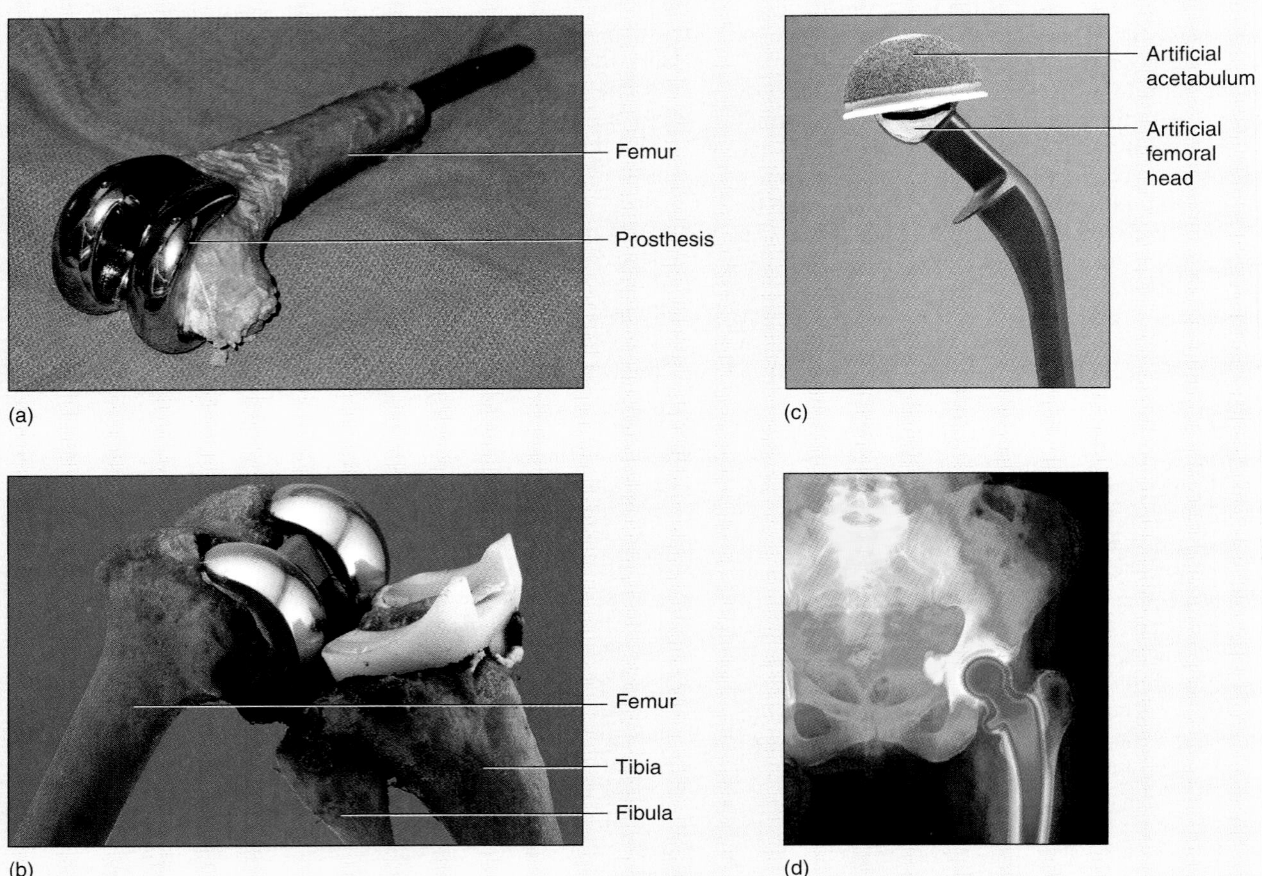

Figure 2 Joint Prostheses. (*a*) An artificial femoral head inserted into the femur. (*b*) An artificial knee joint bonded to a natural femur and tibia. (*c*) A porous-coated hip prosthesis. The caplike portion replaces the acetabulum of the os coxae, and the ball and shaft shown below are bonded to the proximal end of the femur. (*d*) X ray of a patient with a total hip replacement.

Chapter Review

Online Learning Center

Study resources for this chapter are available at www.mhhe.com/saladin/

Study Outline

Joints and Their Classification (p. 308)

1. Sciences of joints
   a. Arthrology
   b. Kinesiology
   c. Biomechanics
2. Classification by freedom of movement
   a. Diarthroses
   b. Amphiarthroses
   c. Synarthroses
3. Classification by manner of union
   a. Fibrous joints
   b. Cartilaginous joints
   c. Bony joints
   d. Synovial joints
4. Overlap between classification schemes

Fibrous, Cartilaginous, and Bony Joints (p. 309)

1. Fibrous joints
   a. Sutures
      - Serrate sutures
      - Lap sutures
      - Plane sutures
   b. Gomphoses
   c. Syndesmoses
2. Cartilaginous joints
   a. Synchondroses
   b. Symphyses
3. Bony joints (synostoses)

Synovial Joints (p. 312)

1. General anatomy
   a. Joint cavity
   b. Synovial fluid
   c. Articular cartilage
   d. Articular capsule
   e. Meniscus
2. Accessory structures
   a. Tendons
   b. Ligaments
   c. Bursae
   d. Tendon sheaths
3. Types of synovial joints
   a. Ball-and-socket joints
   b. Hinge joints
   c. Saddle joint
   d. Pivot joints
   e. Gliding joints
   f. Condyloid joints
4. Movements of diarthroses
   a. Flexion, extension, and hyperextension
   b. Abduction and adduction
   c. Elevation and depression
   d. Protraction and retraction
   e. Lateral and medial excursion
   f. Circumduction
   g. Lateral and medial rotation
   h. Supination and pronation
   i. Opposition and reposition
   j. Dorsiflexion and plantar flexion
   k. Inversion and eversion
5. Range of motion
6. Levers and biomechanics
   a. Fulcrum, effort, and resistance
   b. Effort arm and resistance arm
   c. Mechanical advantage
   d. First- to third-class levers

Anatomy of Selected Diarthroses (p. 323) (table 10.2)

1. Temporomandibular joint
2. Humeroscapular joint
3. Elbow joint
4. Coxal joint
5. Knee joint
6. Ankle joint

Selected Vocabulary

For additional terminology not listed here, see table 10.2.

joint 308
articulation 308
arthrology 308
kinesiology 308
biomechanics 308
diarthrosis 308
amphiarthrosis 308
synarthrosis 308
fibrous joint 310
serrate suture 310
lap suture 310
plane suture 310
gomphosis 310
periodontal ligament 310
syndesmosis 310
interosseous membrane 310
cartilaginous joint 311
synchondrosis 311
symphysis 311
bony joint 311
synostosis 311
synovial joint 313
synovial fluid 313
joint cavity 313
articular cartilage 313
joint capsule 313
fibrous capsule 313
synovial membrane 313
meniscus 313
tendon 313
ligament 313
bursa 313
tendon sheath 313
monaxial 313
biaxial 313
multiaxial 313
ball-and-socket joint 313
hinge joint 314
saddle joint 314
pivot joint 315
gliding joint 315
condyloid joint 315
flexion 317
extension 317
hyperextension 317
abduction 317
adduction 317
elevation 319
depression 319
protraction 319
retraction 319
lateral excursion 319
medial excursion 319
circumduction 319
lateral rotation 319
medial rotation 319
supination 319
pronation 319
opposition 319
reposition 319
dorsiflexion 319
plantar flexion 319
inversion 320
eversion 320
range of motion 320
lever 322
fulcrum 322
effort 322
resistance 322
effort arm 322
resistance arm 322
mechanical advantage 322
first-class lever 322
second-class lever 322
third-class lever 322
tibiofemoral joint 327
patellofemoral joint 327
popliteal region 328
talocrural joint 331

Testing Your Recall

1. Which of the following is unique to the thumb?
   a. gliding joint
   b. hinge joint
   c. saddle joint
   d. condyloid joint
   e. pivot joint
2. Which of the following is the least movable?
   a. diarthrosis
   b. synarthrosis
   c. symphysis
   d. syndesmosis
   e. condyloid joint
3. Which of the following movements are unique to the foot?
   a. dorsiflexion and inversion
   b. elevation and depression
   c. circumduction and rotation
   d. abduction and adduction
   e. opposition and reposition
4. Which of the following joints cannot be circumducted?
   a. carpometacarpal
   b. metacarpophalangeal
   c. humeroscapular
   d. coxal
   e. interphalangeal
5. Which of the following terms denotes a general condition that includes the other four?
   a. gout
   b. arthritis
   c. rheumatism
   d. osteoarthritis
   e. rheumatoid arthritis
6. In the adult, the ischium and pubis are united by
   a. a synchondrosis.
   b. a diarthrosis.
   c. a synostosis.
   d. an amphiarthrosis.
   e. a symphysis.
7. In a second-class lever, the effort
   a. is applied to the end opposite the fulcrum.
   b. is applied to the fulcrum itself.
   c. is applied between the fulcrum and resistance.
   d. always produces an MA less than 1.0.
   e. is applied on one side of the fulcrum to move a resistance on the other side.
8. Which of the following joints has anterior and posterior cruciate ligaments?
   a. the shoulder
   b. the elbow
   c. the hip
   d. the knee
   e. the ankle
9. To bend backward at the waist involves ________ of the vertebral column.
   a. rotation
   b. hyperextension
   c. dorsiflexion
   d. abduction
   e. flexion
10. The rotator cuff includes the tendons of all of the following muscles *except*
    a. the subscapularis.
    b. the supraspinatus.
    c. the infraspinatus.
    d. the biceps brachii.
    e. the teres minor.
11. The lubricant of a diarthrosis is ________.
12. A fluid-filled sac that eases the movement of a tendon over a bone is called a/an ________.
13. A ________ joint allows one bone to swivel on another.
14. ________ is the science of movement.
15. The joint between a tooth and the mandible is called a/an ________.
16. In a ________ suture, the articulating bones have interlocking wavy margins, somewhat like a dovetail joint in carpentry.
17. In kicking a football, what type of action does the knee joint exhibit?
18. The angle through which a joint can move is called its ________.
19. Both the shoulder and hip joints are somewhat deepened and supported by a ring of fibrocartilage called a/an ________.
20. The femur is prevented from slipping sideways off the tibia in part by a pair of cartilages called the lateral and medial ________.

Answers in appendix B

Testing Your Comprehension

1. All second-class levers produce a mechanical advantage greater than 1.0 and all third-class levers produce a mechanical advantage less than 1.0. Explain why.
2. Suppose a lever measures 17 cm from effort to fulcrum and 11 cm from resistance to fulcrum. (a) Calculate its mechanical advantage. (b) Would this lever produce more force, or less, than the force exerted on it? (c) Which of the three classes of levers could *not* have these measurements? Explain.
3. In order of occurrence, list the joint actions (flexion, pronation, etc.) and the joints where they would occur as you (a) sit down at a table, (b) reach out and pick up an apple, (c) take a bite, and (d) chew it. Assume that you start in anatomical position.
4. Suppose you were dissecting a cat or fetal pig with the task of finding examples of each type of synovial joint. Which type of human synovial joint would not be found in either of those animals? For lack of that joint, what human joint actions would those animals be unable to perform?
5. List the six types of synovial joints and for each one, if possible, identify a joint in the upper limb and a joint in the lower limb that falls into each category. Which of these six joints have no examples in the lower limb?

Answers at the Online Learning Center

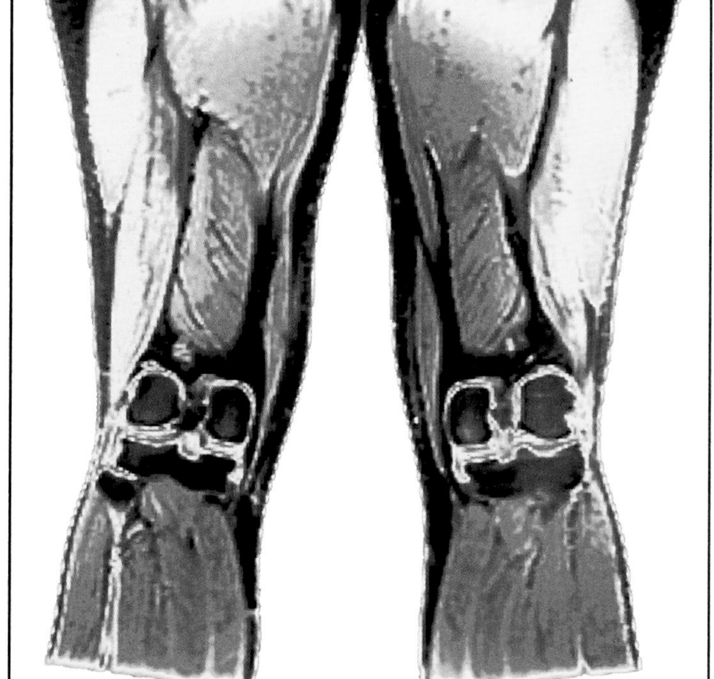
Muscles of the thigh to upper calf (MRI)

CHAPTER 11

The Muscular System

Chapter Outline

Insights

 Study with Your ESP CD-ROM

Refer to your ESP Tutorial CD-ROM for helpful animations, activities, and quizzes related to the material covered in this chapter. The following are suggested areas to go to on the CD and are presented according to "Unit/Topic/":
- Support and Movement/Muscular System

Brushing Up

To understand this chapter, it is important that you understand or brush up on the following concepts:
- Gross anatomy of the skeleton (chapter 9)
- Movements of synovial joints (pp. 316–321)

The **muscular system** consists of about 600 skeletal muscles—striated muscles that are usually attached to bone. (The term does not include smooth or cardiac muscle.) The form and function of the muscular system occupy a place of central importance in several fields of health care and fitness. Physical and occupational therapists must be well acquainted with the muscular system to design and carry out rehabilitation programs. Nurses and other health-care providers often move patients who are physically incapacitated, and to do this safely and effectively requires an understanding of joints and muscles. Even to give intramuscular injections safely requires a knowledge of the muscles and the nerves and blood vessels associated with them. Coaching, movement science, sports medicine, and dance benefit from a knowledge of skeletomuscular anatomy and mechanics.

Myology,[1] the study of muscles, is closely related to what we have covered in the preceding chapters. It relates muscle attachments to the bone structures described in chapter 9 and muscle function to the joint movements described in chapter 10. In this chapter, we consider the gross anatomy of the muscular system and how it relates to joint movements. In the following chapter, we examine the mechanisms of muscle contraction at the cellular and molecular levels.

The Structural and Functional Organization of Muscles

Objectives

When you have completed this section, you should be able to

- List several functions of muscles;
- Describe the connective tissues associated with a skeletal muscle;
- Explain what is meant by the origin, insertion, belly, action, and innervation of a muscle;
- Describe the various shapes of skeletal muscles and relate this to their functions;
- Describe the ways that muscles work in groups to aid, oppose, or moderate each other's actions;
- Distinguish between intrinsic and extrinsic muscles; and
- Translate several Latin words commonly used in the naming of muscles.

The Functions of Muscles

A muscle is an organ specialized to produce movement of a body part. Its cells convert the chemical energy of ATP into the mechanical energy of motion and exert a useful pull on another tissue. More specifically, muscle contraction serves the following overlapping functions:

1. *myo* = muscle + *logy* = study of

- **Movement.** Most obviously, the muscles enable us to move from place to place and to move individual body parts. Muscular contractions also move body contents in the course of respiration, circulation, digestion, defecation, urination, and childbirth.
- **Stability.** Muscles maintain posture by resisting the pull of gravity and preventing unwanted movements. They hold some articulating bones in place by maintaining tension on the tendons.
- **Communication.** Muscles are used for facial expression, other body language, writing, and speech.
- **Control of body openings and passages.** Ringlike *sphincter muscles* around the eyelids, pupils, and mouth control the admission of light, food, and drink into the body; others that encircle the urethral and anal orifices control elimination of waste; and other sphincters control the movement of food, bile, and other materials through the body.
- **Heat production.** The skeletal muscles produce as much as 85% of our body heat, which is vital to the functioning of enzymes and therefore to all of our metabolism.

Some of these functions are shared by skeletal, cardiac, and smooth muscle. The remainder of this chapter, however, is concerned only with skeletal muscles.

Connective Tissues of a Muscle

A skeletal muscle is composed of both muscular tissue and connective tissue (fig. 11.1). A skeletal muscle cell (muscle fiber) is about 10 to 100 μm in diameter and up to 30 cm long. It is surrounded by a sparse layer of areolar connective tissue called the **endomysium**[2] (EN-doe-MIZ-ee-um), which allows room for blood capillaries and nerve fibers to reach each muscle fiber. Muscle fibers are grouped in bundles called **fascicles**[3] (FASS-ih-culs), which are visible to the naked eye as parallel strands. These are the "grain" in a cut of meat; tender meat is easily pulled apart along its fascicles. Each fascicle is separated from neighboring ones by a connective tissue sheath called the **perimysium,**[4] usually somewhat thicker than the endomysium. The muscle as a whole is surrounded by still another connective tissue layer, the **epimysium.**[5] The epimysium grades imperceptibly into connective tissue

2. *endo* = within + *mys* = muscle
3. *fasc* = bundle + *icle* = little
4. *peri* = around
5. *epi* = upon, above

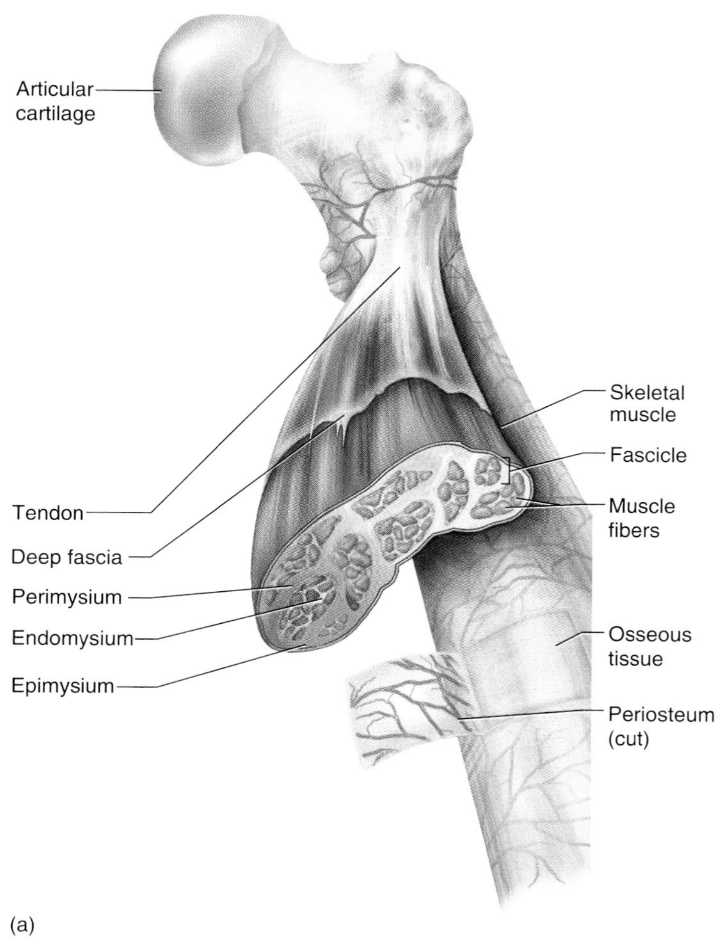

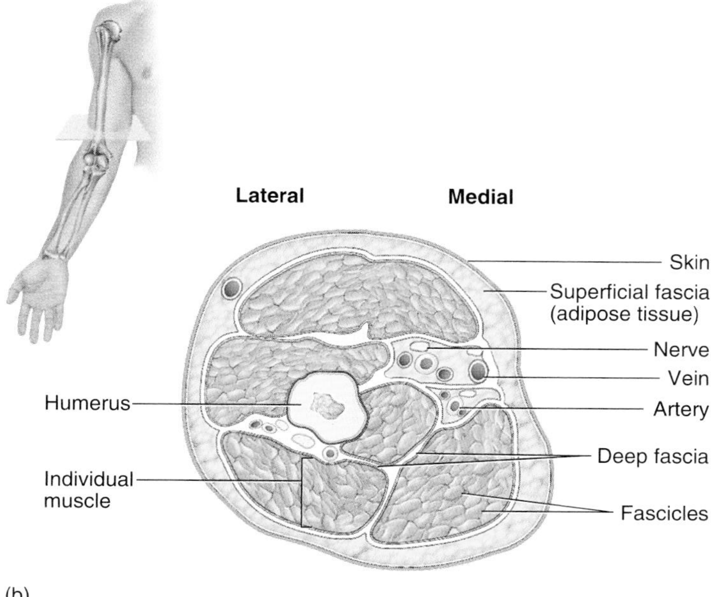

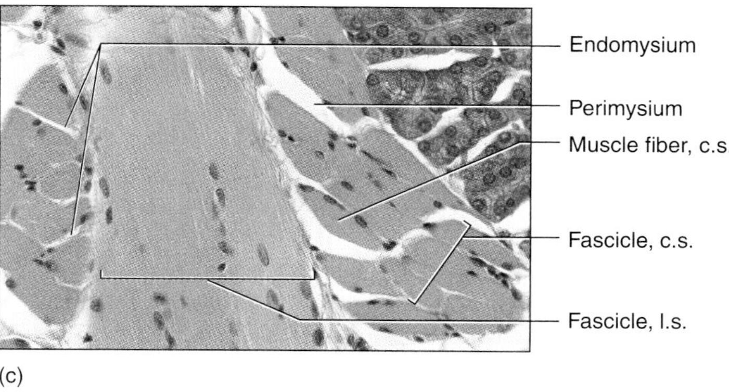

Figure 11.1 The Connective Tissues of a Muscle. (*a*) The muscle-bone attachment. Here there is a continuity of connective tissues from the endomysium around the muscle fibers, to the perimysium, epimysium, deep fascia, and tendon, grading into the periosteum and finally the matrix of the bone. (*b*) A cross section of the arm showing the relationship of neighboring muscles to fascia and bone. (*c*) Muscle fascicles in the tongue. Vertical fascicles passing between the dorsal and ventral surfaces of the tongue are seen alternating with cross-sectioned horizontal fascicles that pass from the tip to the rear of the tongue. A fibrous perimysium can be seen between the fascicles, and endomysium between the muscle fibers within each fascicle.

sheets called fascia (FASH-ee-uh)—**deep fascia** between adjacent muscles and **superficial fascia** (hypodermis) between the muscles and skin. The superficial fascia is very adipose in areas such as the buttocks and abdomen, but the deep fascia is devoid of fat.

There are two ways a muscle can attach to a bone. In a **direct (fleshy) attachment,** collagen fibers of the epimysium are continuous with the periosteum, the fibrous sheath around a bone. The red muscle tissue appears to emerge directly from the bone. The *intercostal muscles* between the ribs show this type of attachment. In an **indirect attachment,** the collagen fibers of the epimysium continue as a strong fibrous **tendon** that merges into the periosteum of a nearby bone (fig. 11.1*a*). The attachment of the *biceps brachii muscle* to the scapula is one of many examples. Some collagen fibers of the periosteum continue into the bone matrix as *perforating fibers* (see chapter 8), so there is a strong structural continuity from endomysium to perimysium to epimysium to tendon to periosteum to bone matrix. Excessive stress is more likely to tear a tendon than to pull it loose from the muscle or bone.

In some cases, the epimysium of one muscle attaches to the fascia or tendon of another or to collagen fibers of the dermis. The ability of a muscle to produce facial expressions depends on the latter type of attachment. Some muscles are connected to a broad sheetlike tendon called an **aponeurosis**[6] (AP-oh-new-RO-sis). This term originally referred to the tendon located beneath the scalp, but now it also refers to similar tendons associated with certain abdominal, lumbar, hand, and foot muscles (see figs. 11.15*a* and 11.16).

In some places, groups of tendons from separate muscles pass under a band of connective tissue called a **retinaculum.**[7] One of these covers each surface of the wrist like a bracelet, for example. The tendons of several forearm muscles pass under them on their way to the hand.

General Anatomy of Skeletal Muscles

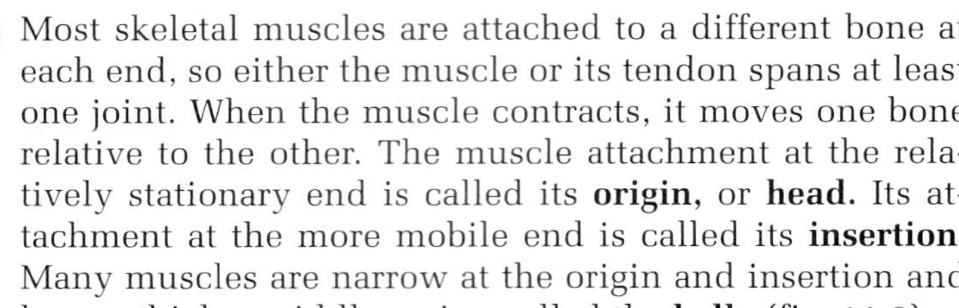

Most skeletal muscles are attached to a different bone at each end, so either the muscle or its tendon spans at least one joint. When the muscle contracts, it moves one bone relative to the other. The muscle attachment at the relatively stationary end is called its **origin,** or **head.** Its attachment at the more mobile end is called its **insertion.** Many muscles are narrow at the origin and insertion and have a thicker middle region called the **belly** (fig. 11.2).

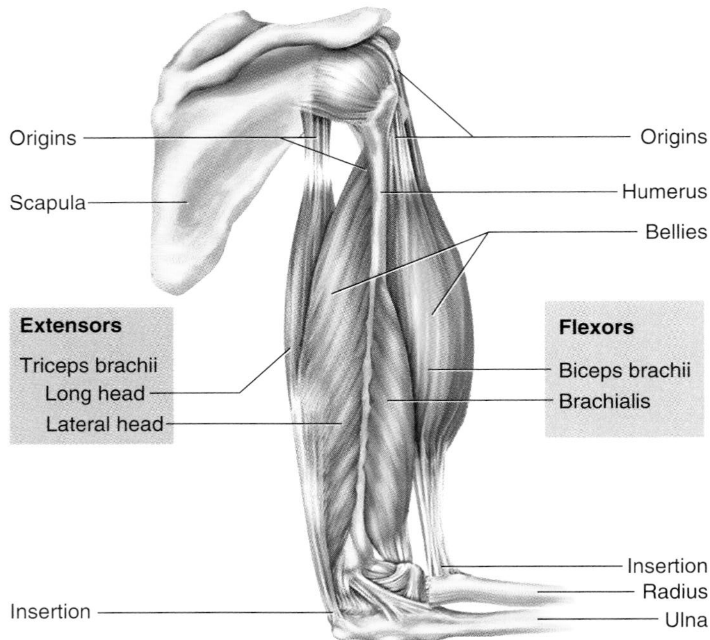

Figure 11.2 **Synergistic and Antagonistic Muscle Pairs.** The biceps brachii and brachialis muscles are synergists in elbow flexion. The triceps brachii is an antagonist of those two muscles and is the prime mover in elbow extension.

The strength of a muscle and the direction in which it pulls are determined partly by the orientation of its fascicles, illustrating the complementarity of form and function. Differences in fascicle orientation are the basis for classifying muscles into five types (fig. 11.3):

1. **Fusiform**[8] **muscles** are thick in the middle and tapered at each end. Their contractions are moderately strong. The *biceps brachii* of the arm and *gastrocnemius* of the calf are examples of this type.
2. **Parallel muscles** are long, straplike muscles of uniform width and parallel fascicles. They can span a great distance and shorten more than other muscle types, but they are weaker than fusiform muscles. Examples include the *rectus abdominis* of the abdomen, *sartorius* of the thigh, and *zygomaticus major* of the face.
3. **Convergent muscles** are fan-shaped—broad at the origin and converging toward a narrower insertion. These muscles are relatively strong because all of their fascicles exert their tension on a relatively small insertion. The *pectoralis major* in the chest is a muscle of this type.
4. **Pennate**[9] **muscles** are feather-shaped. Their fascicles insert obliquely on a tendon that runs the length of the muscle, like the shaft of a feather. There are three types of pennate muscles: *unipennate,* in which all fascicles approach the tendon from one side (for example, the *palmar interosseous muscles* of the hand and *semimembranosus* of the thigh); *bipennate,* in which fascicles approach the tendon from both sides (for example, the *rectus femoris* of the thigh); and *multipennate,* shaped like a bunch of feathers with their quills converging on a single point (for example, the *deltoid* of the shoulder).
5. **Circular muscles (sphincters)** form rings around body openings. These include the *orbicularis oris* of the lips and *orbicularis oculi* of the eyelids.

6. *apo* = upon, above + *neuro* = nerve
7. *retinac* = retainer, bracelet + *cul* = little
8. *fusi* = spindle + *form* = shape
9. *penna* = feather

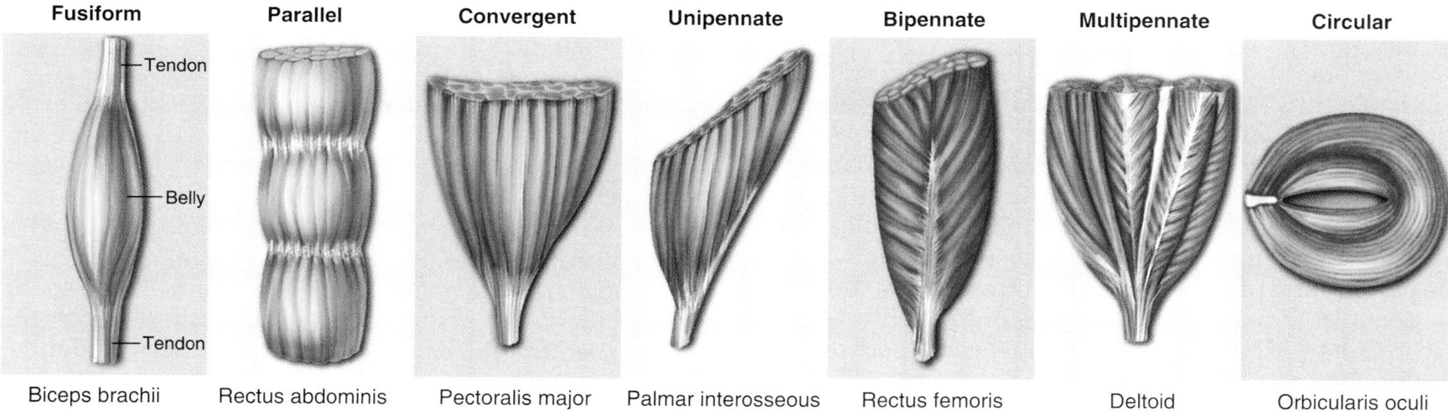

Figure 11.3 Classification of Muscles According to Fascicle Orientation. The fascicles are the "grain" visible in each muscle illustration.

Coordinated Action of Muscle Groups

The movement produced by a muscle is called its **action.** Skeletal muscles seldom act independently; instead, they function in groups whose combined actions produce the coordinated motion of a joint. Muscles can be classified into at least four categories according to their actions, but it must be stressed that a particular muscle can act in a certain way during one joint action and in a different way during other actions of the same joint:

1. The **prime mover (agonist)** is the muscle that produces most of the force during a particular joint action. In flexing the elbow, for example, the prime mover is the biceps brachii.
2. A **synergist**[10] (SIN-ur-jist) is a muscle that aids the prime mover. Several synergists acting on a joint can produce more power than a single larger muscle. The *brachialis,* for example, lies deep to the biceps brachii and works with it as a synergist to flex the elbow. The actions of a prime mover and its synergist are not necessarily identical and redundant. If the prime mover worked alone at a joint, it might cause rotation or other undesirable movements of a bone. A synergist may stabilize a joint and restrict these movements, or modify the direction of a movement, so that the action of the prime mover is more coordinated and specific.
3. An **antagonist**[11] is a muscle that opposes the prime mover. In some cases, it relaxes to give the prime mover almost complete control over an action. More often, however, the antagonist moderates the speed or range of the agonist, thus preventing excessive movement and joint injury. If you extend your arm to reach out and pick up a cup of tea, your *triceps brachii* is the prime mover of elbow extension and your biceps brachii acts as an antagonist to slow the extension and stop it at the appropriate point. If you extend your arm rapidly to throw a dart, the biceps must be quite relaxed. The biceps and triceps brachii represent an **antagonistic pair** of muscles that act on opposite sides of a joint (see fig. 11.2). We need antagonistic pairs at a joint because a muscle can only pull, not push—a single muscle cannot flex *and* extend the elbow, for example. Which member of the pair acts as the agonist depends on the motion under consideration. In flexion of the elbow, the biceps is the agonist and the triceps is the antagonist; when the elbow is extended, their roles are reversed.
4. A **fixator** is a muscle that prevents a bone from moving. To *fix* a bone means to hold it steady, allowing another muscle attached to it to pull on something else. For example, consider again the flexion of the elbow by the biceps brachii. The biceps originates on the scapula and inserts on the radius. The scapula is loosely attached to the axial skeleton, so when the biceps contracts, it seems that it would pull the scapula laterally. There are fixator muscles attached to the scapula, however, that contract at the same time. By holding the scapula firmly in place, they ensure that the force generated by the biceps moves the radius rather than the scapula.

10. *syn* = together + *erg* = work
11. *ant* = against + *agonist* = competitor

Intrinsic and Extrinsic Muscles

In places such as the tongue, larynx, back, hand, and foot, anatomists distinguish between intrinsic and extrinsic muscles. An **intrinsic muscle** is entirely contained within a particular region, having both its origin and insertion there. An **extrinsic muscle** acts upon a designated region but has its origin elsewhere. For example, some movements of the fingers are produced by extrinsic muscles in the forearm, whose long tendons reach to the phalanges; other finger movements are produced by the intrinsic muscles located between the metacarpal bones of the hand.

Muscle Innervation

Innervation means the nerve supply to an organ. Knowing the innervation to each muscle enables clinicians to diagnose nerve and spinal cord injuries from their effects on muscle function and to set realistic goals for rehabilitation. The muscle tables of this chapter identify the innervation of each muscle. This information will be more meaningful after you have studied the peripheral nervous system (see chapter 15), but a brief orientation will be helpful here. Muscles of the head and neck are supplied by *cranial nerves* that arise from the base of the brain and emerge through the skull foramina. Cranial nerves are identified by numerals I to XII, although not all 12 of them innervate skeletal muscles. Muscles elsewhere are supplied by *spinal nerves,* which originate in the spinal cord, emerge through the intervertebral foramina, and branch into a *dorsal* and *ventral ramus.*[12] The spinal nerves are identified by letters and numbers that refer to the vertebrae—for example, T6 for the sixth thoracic nerve and S2 for the second sacral nerve. You will note references to nerve numbers and rami in many of the muscle tables. The term *plexus* in some of the tables refers to weblike networks of spinal nerves adjacent to the vertebral column. All of the nerves named here are illustrated, and most are also discussed, in chapter 15 (see tables 15.2–15.6).

How Muscles Are Named

Most of this chapter is a descriptive inventory of muscles, including their location, action, origin, insertion, and innervation. Learning the names of the muscles is much easier when you have some appreciation of the meanings behind the words.

Early anatomists coined terms in their native languages. This quickly led to a state of confusion and a need for a standard international terminology. In 1895, an international committee of anatomists developed a system of Latin names called the *Nomina Anatomica* (N.A.), which has been updated several times since then. Although this book gives most terms in English rather than Latin, the customary English names for skeletal muscles are, at most, only slight modifications of the Latin names—for example, *anterior scalene muscle* is a derivative of the N.A. term, *musculus scalenus anterior.*

Some muscle names are several words long—for example, the *flexor digiti minimi brevis,* a "short muscle that flexes the little finger." Such names may seem intimidating at first, but they are really more of a help than an obstacle to understanding if you gain a little insight into the most commonly used Latin words. Several of these are interpreted in table 11.1, and others are explained in footnotes throughout the chapter. Familiarity with these terms will help you translate muscle names and remember the location, appearance, and action of the muscles.

A Learning Strategy

In the remainder of this chapter, we consider about 160 muscles. Many of the relatively superficial ones are shown in figure 11.4. The following suggestions may help you develop a rational strategy for learning the muscular system:

- Examine models, cadavers, dissected animals, or a photographic atlas as you read about these muscles. Visual images are often easier to remember than words, and direct observation of a muscle may stick in your memory better than descriptive text or two-dimensional drawings.
- When studying a particular muscle, palpate it on yourself if possible. Contract the muscle to feel it bulge and sense its action. This makes muscle locations and actions less abstract. Atlas B following this chapter shows where you can see and palpate several of these muscles on the living body.
- Locate the origins and insertions of muscles on an articulated skeleton. Some study skeletons are painted and labeled to show these. This helps you visualize the locations of muscles and understand how they produce particular joint actions.
- Study the derivation of each muscle name; the name usually describes the muscle's location, appearance, origin, insertion, or action.
- Say the names aloud to yourself or a study partner. It is harder to remember and spell terms you cannot pronounce, and silent pronunciation is not nearly as effective as speaking and hearing the names. Pronunciation guides are provided in the muscle tables for all but the most obvious cases.

12. *ramus* = branch

Table 11.1 Words Commonly Used to Name Muscles

Criterion	Term and Meaning	Examples of Usage
Size	Major (large)	Pectoralis major
	Maximus (largest)	Gluteus maximus
	Minor (small)	Pectoralis minor
	Minimus (smallest)	Gluteus minimus
	Longus (long)	Abductor pollicis longus
	Brevis (short)	Extensor pollicis brevis
Shape	Rhomboideus (rhomboidal)	Rhomboideus major
	Trapezius (trapezoidal)	Trapezius
	Teres (round, cylindrical)	Pronator teres
	Deltoid (triangular)	Deltoid
Location	Capitis (of the head)	Splenius capitis
	Cervicis (of the neck)	Semispinalis cervicis
	Pectoralis (of the chest)	Pectoralis major
	Thoracis (of the thorax)	Spinalis thoracis
	Intercostal (between the ribs)	External intercostals
	Abdominis (of the abdomen)	Rectus abdominis
	Lumborum (of the lower back)	Quadratus lumborum
	Femoris (of the femur, or thigh)	Quadriceps femoris
	Peroneus (of the fibula)	Peroneus longus
	Brachii (of the arm)	Biceps brachii
	Carpi (of the wrist)	Flexor carpi ulnaris
	Digiti (of a finger or toe, singular)	Extensor digiti minimi
	Digitorum (of the fingers or toes, plural)	Flexor digitorum profundus
	Pollicis (of the thumb)	Opponens pollicis
	Indicis (of the index finger)	Extensor indicis
	Hallucis (of the great toe)	Abductor hallucis
	Superficialis (superficial)	Flexor digitorum superficialis
	Profundus (deep)	Flexor digitorum profundus
Number of heads	Biceps (two heads)	Biceps femoris
	Triceps (three heads)	Triceps brachii
	Quadriceps (four heads)	Quadriceps femoris
Orientation	Rectus (straight)	Rectus abdominis
	Transversus (transverse)	Transversus abdominis
	Oblique (slanted)	External abdominal oblique
Action	Adductor (adducts a body part)	Adductor pollicis
	Abductor (abducts a body part)	Abductor digiti minimi
	Flexor (flexes a joint)	Flexor carpi radialis
	Extensor (extends a joint)	Extensor carpi radialis
	Pronator (pronates forearm)	Pronator teres
	Supinator (supinates forearm)	Supinator
	Levator (elevates a body part)	Levator scapulae
	Depressor (depresses a body part)	Depressor anguli oris

Before You Go On

Answer the following questions to test your understanding of the preceding section:

1. List some functions of the muscular system other than movement of the body.
2. Describe the relationship of endomysium, perimysium, and epimysium to each other. Which of these separates one fascicle from another? Which separates one muscle from another?
3. Distinguish between direct and indirect muscle attachments to bones.
4. Define *origin, insertion, belly, action,* and *innervation.*
5. Describe the five basic muscle shapes (fascicle arrangements).
6. Distinguish among a synergist, antagonist, and fixator. Explain how each of these may affect the action of an agonist.
7. In muscle names, what do the words *brevis, teres, digitorum, pectoralis, triceps,* and *profundus* mean?

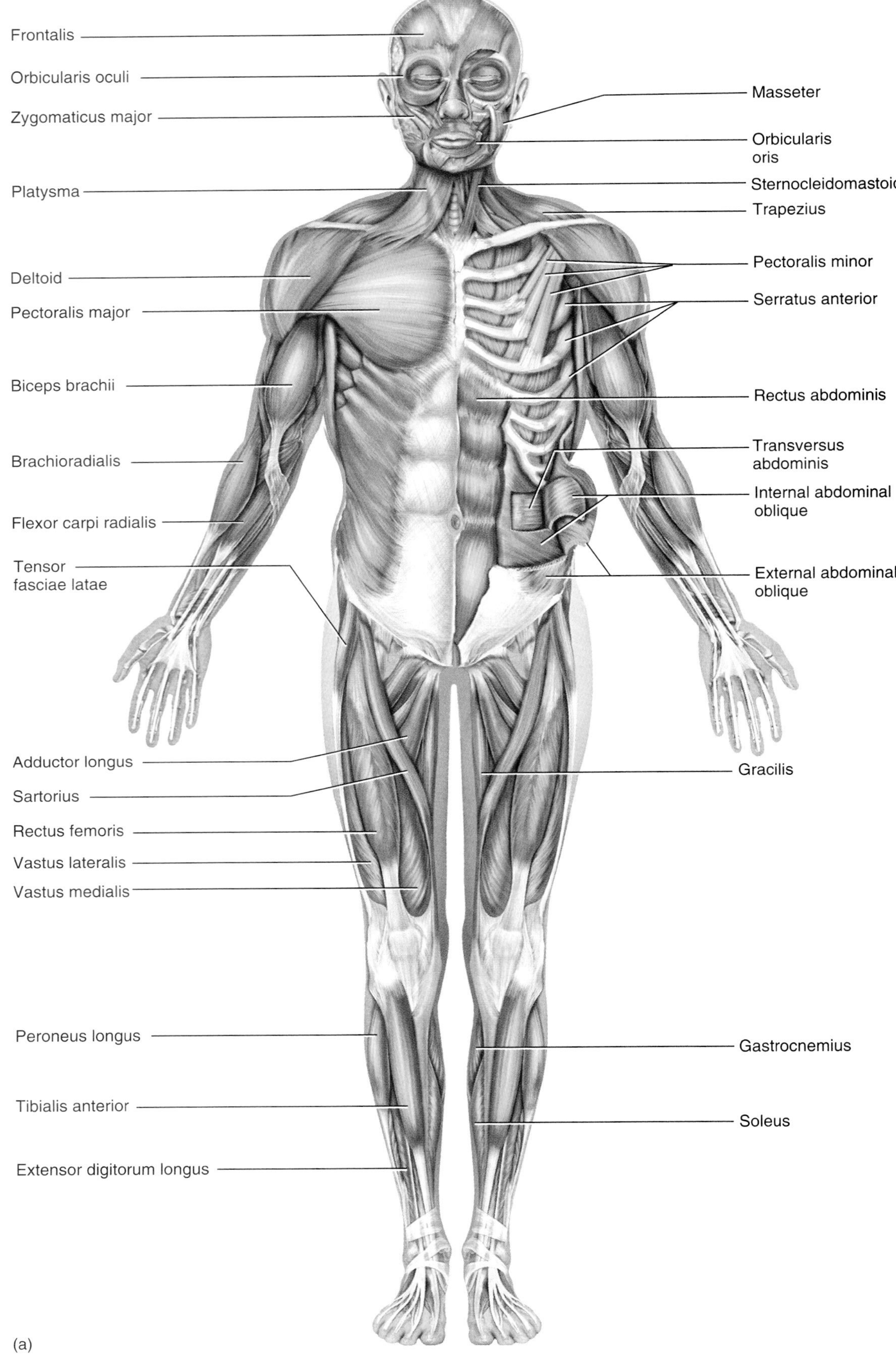

Figure 11.4 The Muscular System. (a) Anterior aspect. *(continued)*

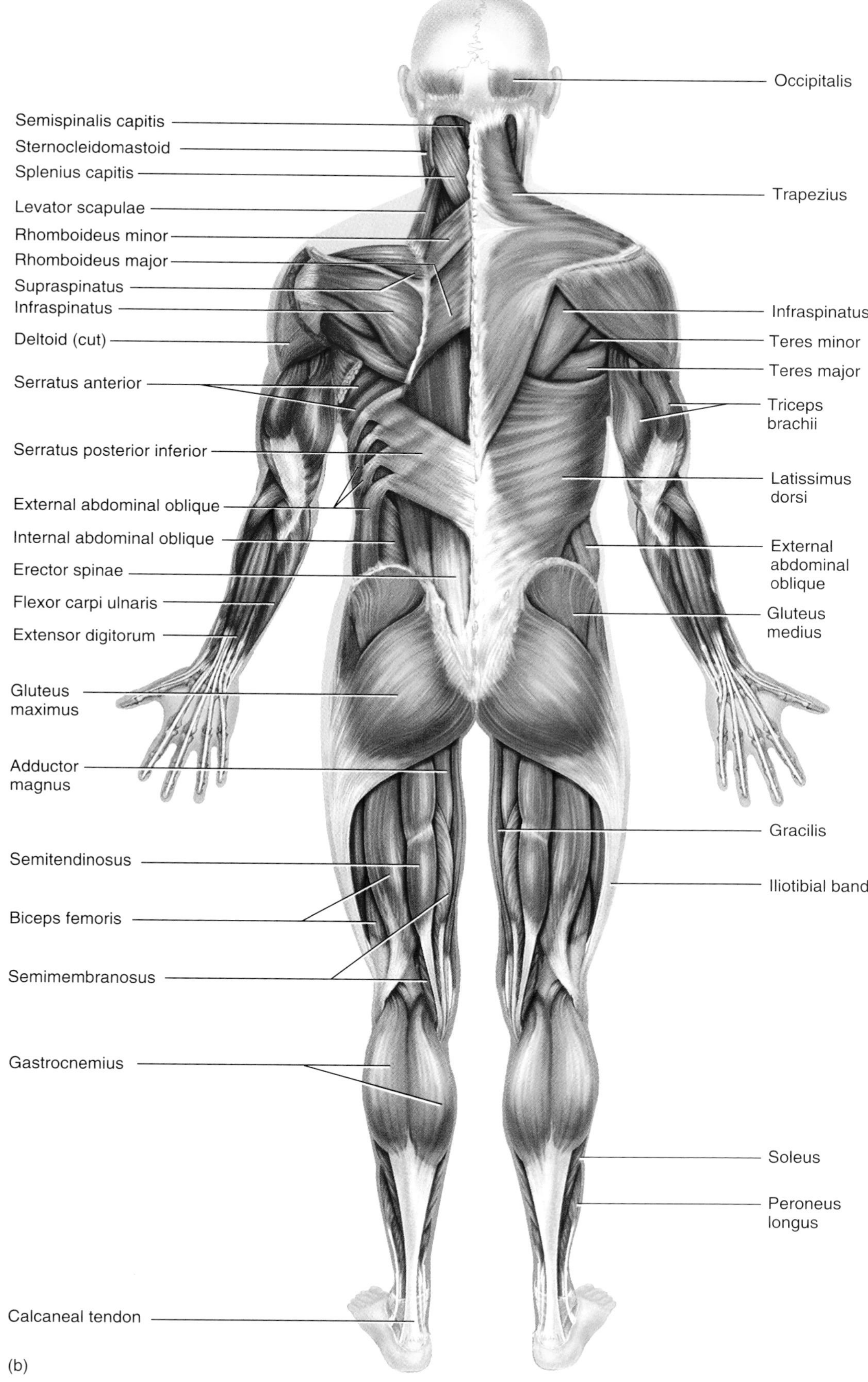

Figure 11.4 **The Muscular System (continued).** (b) Posterior aspect. In each figure, major superficial muscles are shown on the anatomical right, and some of the deeper muscles of the trunk are shown on the left. Muscles not labeled here are shown in more detail in later figures.

Muscles of the Head and Neck

Objectives

When you have completed this section, you should be able to

- Name and locate the muscles that produce facial expressions;
- Name and locate the muscles used for chewing and swallowing;
- Name and locate the neck muscles that move the head; and
- Identify the origin, insertion, action, and innervation of any of these muscles.

Muscles of Facial Expression

One of the most striking contrasts between a human face and that of a rat, horse, or dog, for example, is the variety and subtlety of human facial expression. This is made possible by a complex array of small muscles that insert in the dermis and tense the skin when they contract (fig. 11.5). These muscles produce expressions as diverse as a pleasant smile, a threatening scowl, a puzzled frown, and a flirtatious wink (fig. 11.6). They add subtle shades of meaning to our spoken words and are enormously important in nonverbal communication.

We will briefly "tour" the scalp and face to get a general idea of the locations and actions of these muscles. Table 11.2 gives the details of their origins, insertions, and innervations. All of these muscles but one are innervated by branches of the facial nerve (cranial nerve VII). This nerve is especially vulnerable to damage from lacerations and skull fractures, which can paralyze the innervated muscles and cause parts of the face to sag.

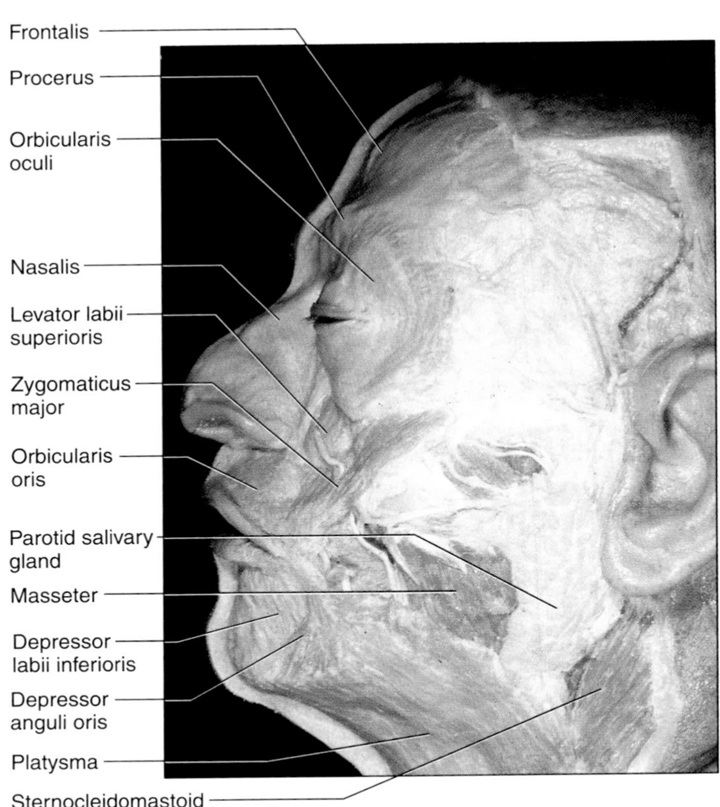

Figure 11.5 Some Muscles of Facial Expression in the Cadaver.

Table 11.2 Muscles of Facial Expression (see fig. 11.7)

O = origin, I = insertion, N = innervation (n. = nerve)

Occipitofrontalis (oc-SIP-ih-toe-frun-TAY-lis)

Occipitalis

Retracts scalp; fixes galea aponeurotica

O: superior nuchal line	I: galea aponeurotica	N: facial n. (VII)

Frontalis

Raises eyebrows and creates wrinkles in forehead when occipitalis is contracted; draws scalp forward when occipitalis is relaxed

O: galea aponeurotica	I: skin of forehead	N: facial n. (VII)

Orbicularis Oculi (or-BIC-you-LERR-iss OC-you-lye)

Closes eye; compresses lacrimal gland to promote flow of tears

O: medial wall of orbit	I: eyelid	N: facial n. (VII)

Levator Palpebrae (leh-VAY-tur pal-PEE-bree) Superioris

Opens eye; raises upper eyelid

O: roof of orbit	I: upper eyelid	N: oculomotor n. (III)

(continued)

Table 11.2 Muscles of Facial Expression (continued)

Corrugator Supercilii (COR-oo-GAY-tur SOO-per-SIL-ee-eye)
Medially depresses eyebrows and draws them closer together; wrinkles skin between eyebrows
O: superciliary ridge | I: skin of eyebrow | N: facial n. (VII)

Procerus (pro-SER-us)
Wrinkles skin between eyebrows; draws skin of forehead down
O: skin on bridge of nose | I: skin of forehead | N: facial n. (VII)

Nasalis (nay-SAY-liss)
One part widens nostrils; another part depresses nasal cartilages and compresses nostrils
O: maxilla and nasal cartilages | I: bridge and alae of nose | N: facial n. (VII)

Orbicularis Oris (or-BIC-you-LERR-iss OR-iss)
Closes lips; protrudes lips as in kissing; aids in speech
O: muscle fibers around mouth | I: mucous membrane of lips | N: facial n. (VII)

Levator Labii Superioris
Elevates upper lip
O: zygomatic bone, maxilla | I: upper lip | N: facial n. (VII)

Levator Anguli (ANG-you-lye) Oris
Elevates corners of mouth, as in smiling and laughing
O: maxilla | I: superior corner of mouth | N: facial n. (VII)

Zygomaticus (ZY-go-MAT-ih-cus) Major, Zygomaticus Minor
Draw corners of mouth laterally and upward, as in smiling and laughing
O: zygomatic bone | I: superolateral corner of mouth | N: facial n. (VII)

Risorius (rih-SOR-ee-us)
Draws corner of mouth laterally, as in grimacing
O: fascia near ear | I: corner of mouth | N: facial n. (VII)

Depressor Anguli Oris, or Triangularis
Depresses corner of mouth, as in frowning
O: mandible | I: inferolateral corner of mouth | N: facial n. (VII)

Depressor Labii Inferioris
Depresses lower lip
O: near mental protuberance | I: lower lip | N: facial n. (VII)

Mentalis (men-TAY-lis)
Pulls skin of chin upward; elevates and protrudes lower lip, as in pouting
O: near mental protuberance | I: skin of chin | N: facial n. (VII)

Buccinator (BUCK-sin-AY-tur)
Compresses cheek; pushes food between teeth; expels air or liquid from mouth; creates suction
O: lateral aspects of maxilla and mandible | I: orbicularis oris | N: facial n. (VII)

Platysma (plah-TIZ-muh)
Depresses mandible, opens and widens mouth, tenses skin of neck
O: fasciae of deltoid and pectoralis major muscles | I: mandible, skin of lower face, muscles at corners of mouth | N: facial n. (VII)

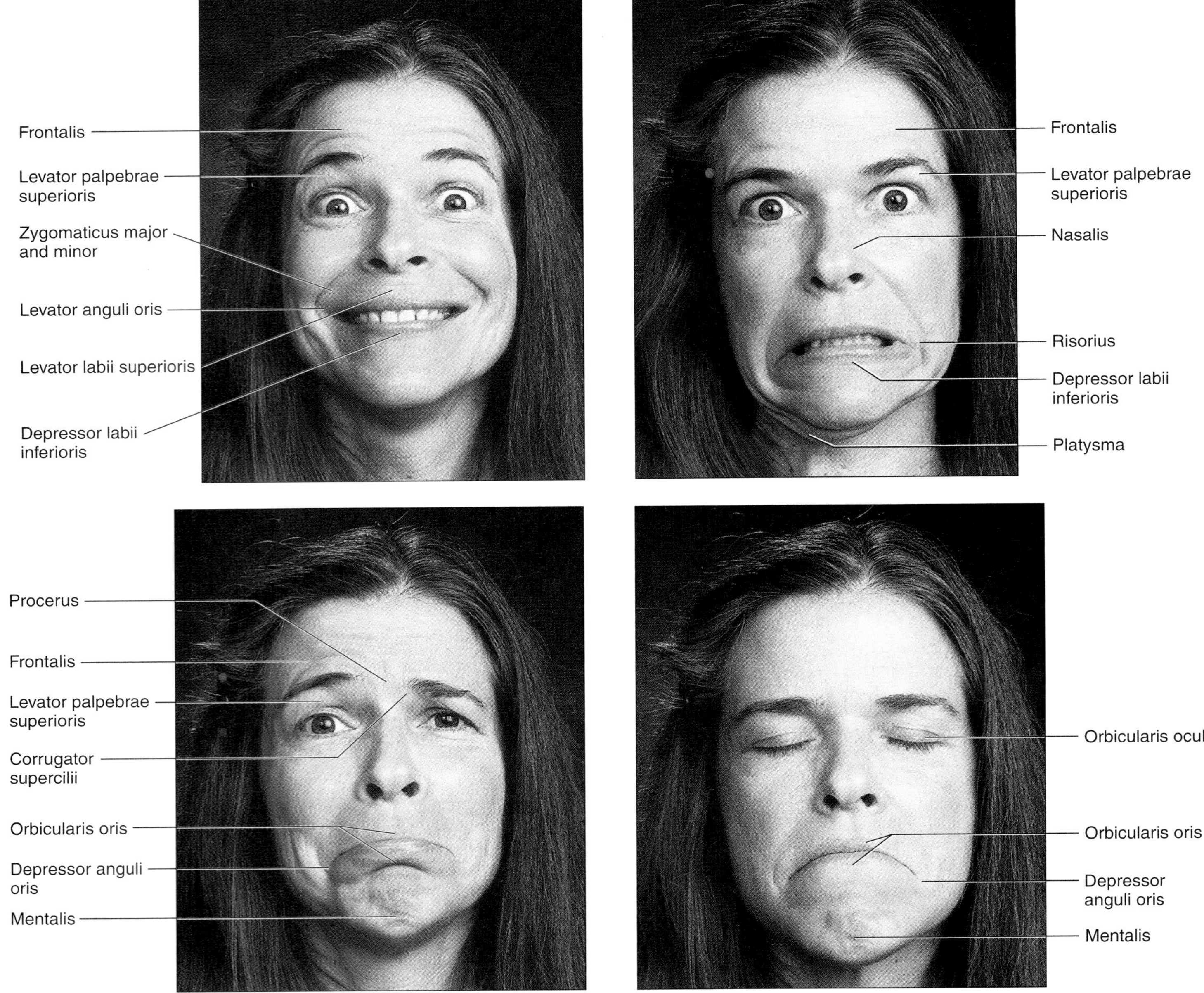

Figure 11.6 **Expressions Produced by Several of the Facial Muscles.** The ordinary actions of these muscles are usually more subtle than these demonstrations.

The **occipitofrontalis** is the muscle of the scalp. It is divided into the **frontalis** of the forehead and **occipitalis** at the rear of the head, connected to each other by a broad aponeurosis, the **galea aponeurotica**[13] (GAY-lee-uh AP-oh-new-ROT-ih-cuh) (fig. 11.7). The occipitofrontalis moves the scalp, forehead skin, and eyebrows.

Each eye is encircled by the **orbicularis oculi**,[14] a sphincter that closes the eye. The **levator palpebrae superioris**[15] opens the eye; it lies deep to the orbicularis oculi in the eyelid and roof of the orbit. Other muscles of the orbital and nasal regions—the **corrugator supercilii**,[16] **procerus**,[17] and **nasalis**[18]—are described in table 11.2. Muscles within the orbit that produce movements of the eye itself are discussed in chapter 16.

The mouth is the most expressive part of the face, so it is not surprising that the muscles here are especially diverse. It is surrounded by the **orbicularis oris**,[19]

13. *galea* = helmet + *apo* = above + *neuro* = nerves, the brain
14. *orb* = circle + *ocul* = eye
15. *levator* = that which raises + *palpebr* = eyelid + *superior* = upper
16. *corrug* = wrinkle + *supercilii* = of the eyebrow
17. *procer* = long, slender
18. *nasalis* = of the nose
19. *oris* = of the mouth

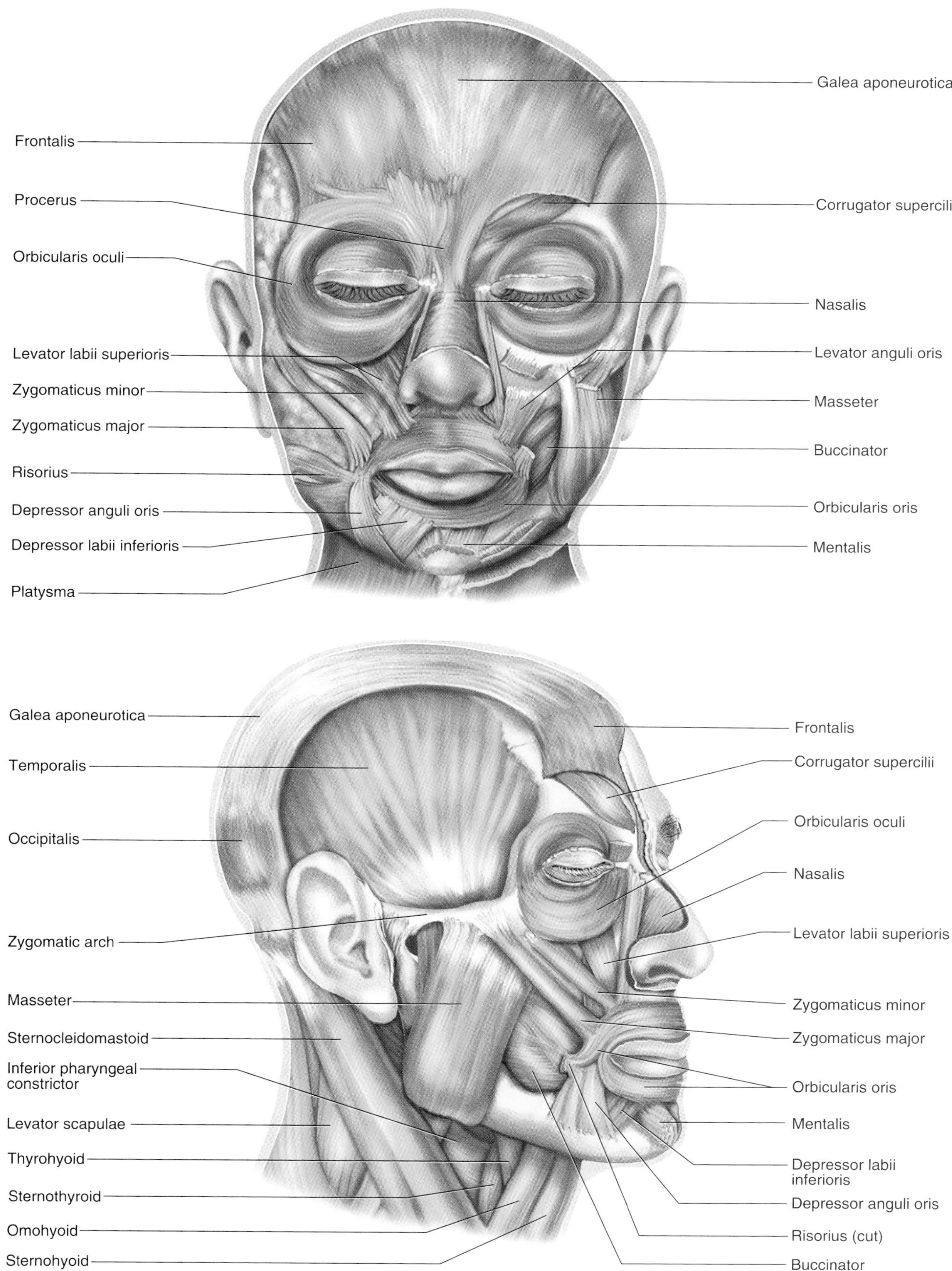

Figure 11.7 The Muscles of Facial Expression.

a sphincter that closes the lips. Several other muscles approach the orbicularis oris from all directions. The first major muscle lateral to the nose is the **levator labii superioris,** a triangular muscle that originates at the middle of the orbicularis oculi. Next is the **zygomaticus**[20] **minor,** which originates near the lateral corner of the eye. Both of these converge on the orbicularis oris. The **zygomaticus major** originates in front of the ear and inserts on the superolateral corner of the mouth. The **levator anguli oris** originates on the maxilla and likewise inserts on the superolateral corner of the mouth. The **risorius**[21] approaches the mouth horizontally and inserts at the junction of the upper and lower lips. From their insertions, you can probably guess that the last five muscles draw the upper lip upward and laterally in such expressions as smiling and laughing.

Along the lower lip are muscles that draw it downward. The most lateral is the **depressor anguli oris,**[22] also known as the *triangularis* because of its shape. Lying deep to it and a little more medially is the **depressor labii inferioris.** Most medially, near the mental protuberance, is a pair of tiny **mentalis**[23] muscles. Unlike the other two, the mentalis muscles do not depress the lip. They originate on the mandible and insert in the dermis of the chin. When they contract, they pull the soft tissues of the chin upward, which wrinkles the chin, pushes the lower lip outward, and creates a pouting expression. People with especially thick mentalis muscles have a groove between them, the *mental cleft,* externally visible as a dimple of the chin.

The muscle of the cheek is the **buccinator,**[24] which has multiple functions in blowing, sucking, and chewing. The name literally means "trumpeter"—if the cheeks are inflated with air, compression of the buccinator muscles blows it out. Sucking is achieved by contracting the buccinators to draw the cheeks inward and then relaxing them. This action is especially important for nursing infants. To appreciate this action, hold your fingertips lightly on your cheeks as you make a kissing noise. You will feel the relaxation of the buccinators at the moment air is sharply drawn in through the pursed lips. The buccinators also aid chewing by pushing and retaining food between the teeth.

The **platysma**[25] is a thin superficial muscle that arises from the shoulder and upper chest and inserts broadly along the mandible and overlying skin. It depresses the mandible, helps to open and widen the mouth, and tenses the skin of the neck (during shaving, for example).

Muscles of Chewing and Swallowing

The following muscles contribute to facial expression and speech but are primarily concerned with manipulation of food, including tongue movements, chewing, and swallowing (table 11.3).

The tongue is a very agile organ. Both intrinsic and extrinsic muscle groups are responsible for its complex movements. The intrinsic muscles consist of variable numbers of vertical muscles that extend from the superior to inferior side of the tongue, transverse muscles that extend from left to right, and longitudinal muscles that extend from root to tip. The extrinsic muscles connect the tongue to other structures in the head and neck. These include the **genioglossus,**[26] **hyoglossus,**[27] **styloglossus,**[28] and **palatoglossus**[29] (fig. 11.8). The tongue and buccinator muscle shift food into position between the molars for chewing (mastication), and the tongue later forces the chewed food into the pharynx for swallowing.

There are four paired muscles of mastication: the temporalis, masseter, and medial and lateral pterygoids. The **temporalis**[30] is a broad, fan-shaped muscle that arises from the temporal lines of the skull, passes behind the zygomatic arch, and inserts on the coronoid process of the mandible (fig. 11.9*a*). The **masseter**[31] is shorter and superficial to the temporalis, arising from the zygomatic arch and inserting on the lateral surface of the angle of the mandible (see fig. 11.7). It is a thick muscle easily palpated on the side of your jaw. The temporalis and masseter elevate the mandible to bite and chew food; they are two of the most powerful muscles in the body. Similar action is provided by the **medial** and **lateral pterygoids.** They arise from the pterygoid processes of the sphenoid bone and insert on the medial surface of the mandible (fig. 11.9*b*). The pterygoids elevate and protract the mandible and produce the lateral excursions used to grind food between the molars.

Several of the actions of chewing and swallowing are aided by eight pairs of *hyoid muscles* associated with the hyoid bone. Four of them, superior to the hyoid, form the **suprahyoid group**—the *digastric, geniohyoid, mylohyoid,* and *stylohyoid.* Those inferior to the hyoid form the **infrahyoid group**—the *thyrohyoid, omohyoid, sternohyoid,* and *sternothyroid.* (See fig. 11.8*a* for the geniohyoid

20. refers to the zygomatic arch
21. *risor* = laughter
22. *depressor* = that which lowers + *angul* = corner, angle
23. *mentalis* = of the chin
24. *bucc* = cheek
25. *platy* = flat
26. *genio* = chin + *gloss* = tongue
27. refers to the hyoid bone
28. refers to the styloid process of the skull
29. *palato* = palate
30. refers to the temporal bone
31. *masset* = chew

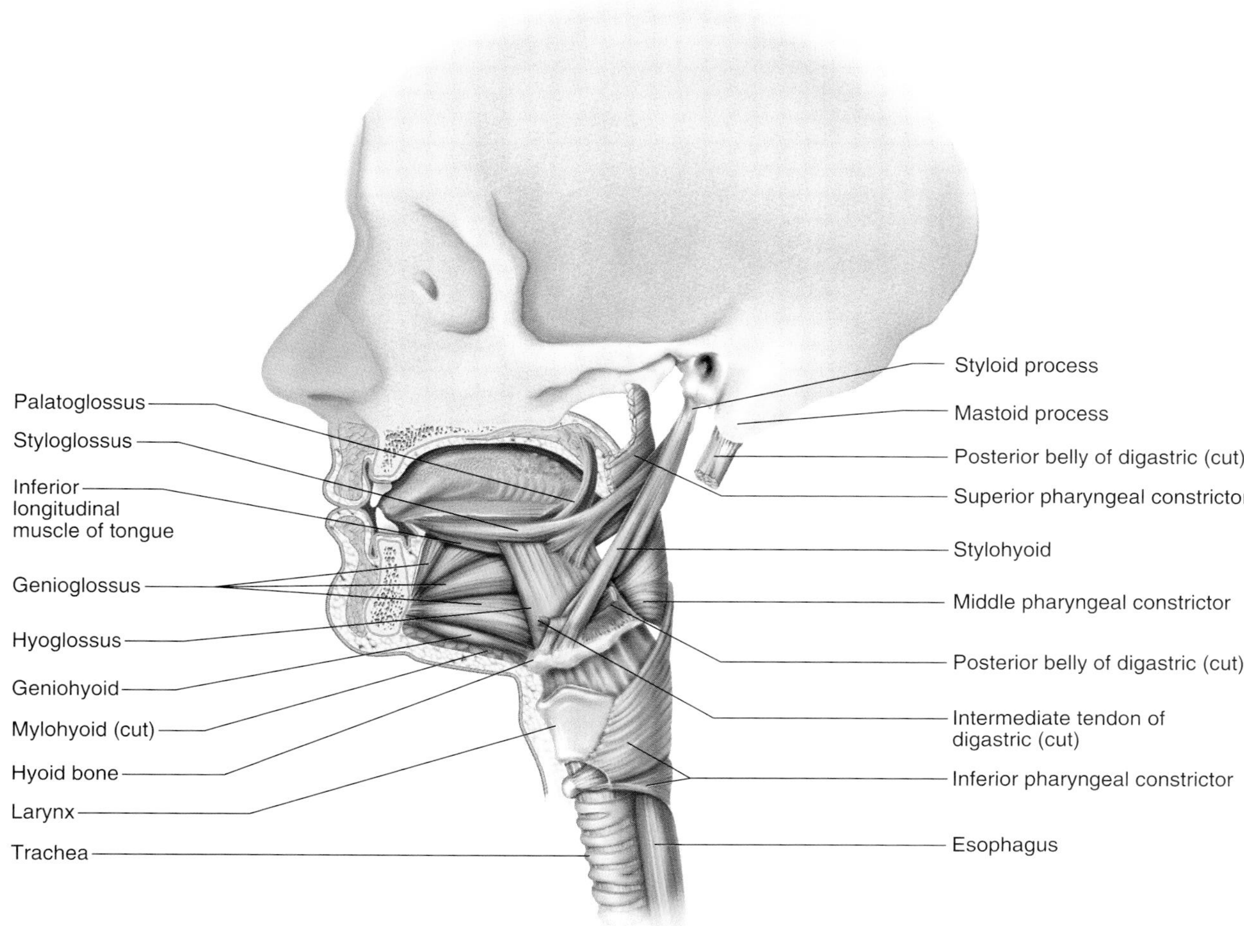

Figure 11.8 Muscles of the Tongue and Pharynx. Left lateral view.

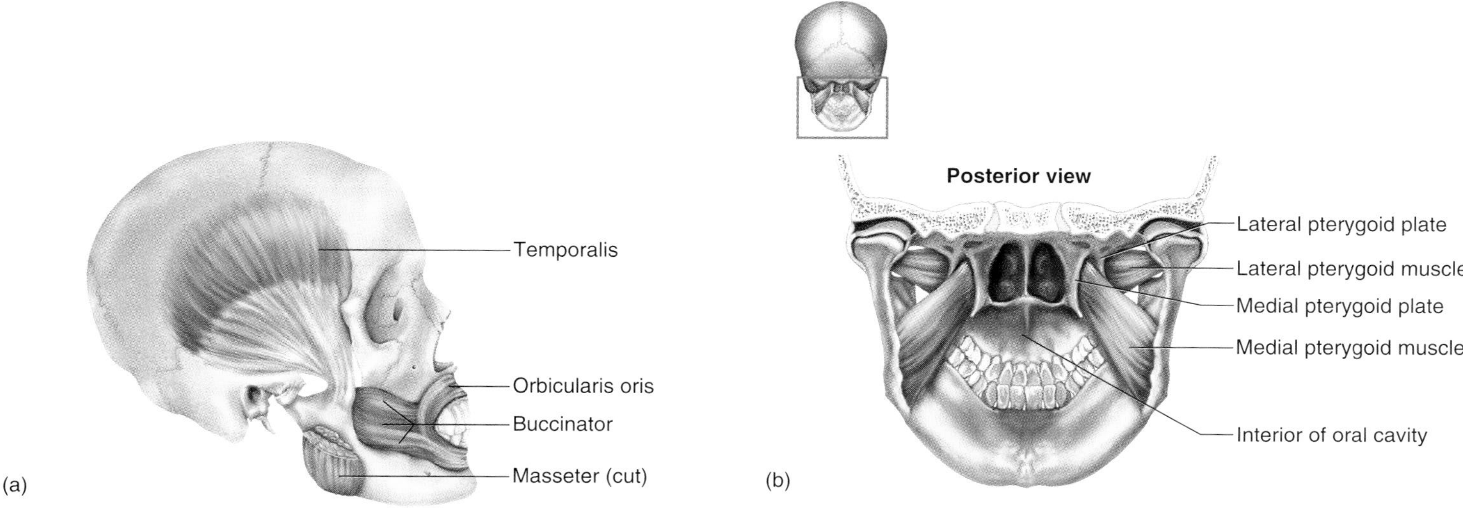

Figure 11.9 Muscles of Chewing. (*a*) Right lateral view. In order to expose the insertion of the temporalis muscle on the mandible, part of the zygomatic arch and masseter muscle are removed. (*b*) View of the pterygoid muscles looking into the oral cavity from behind the skull.

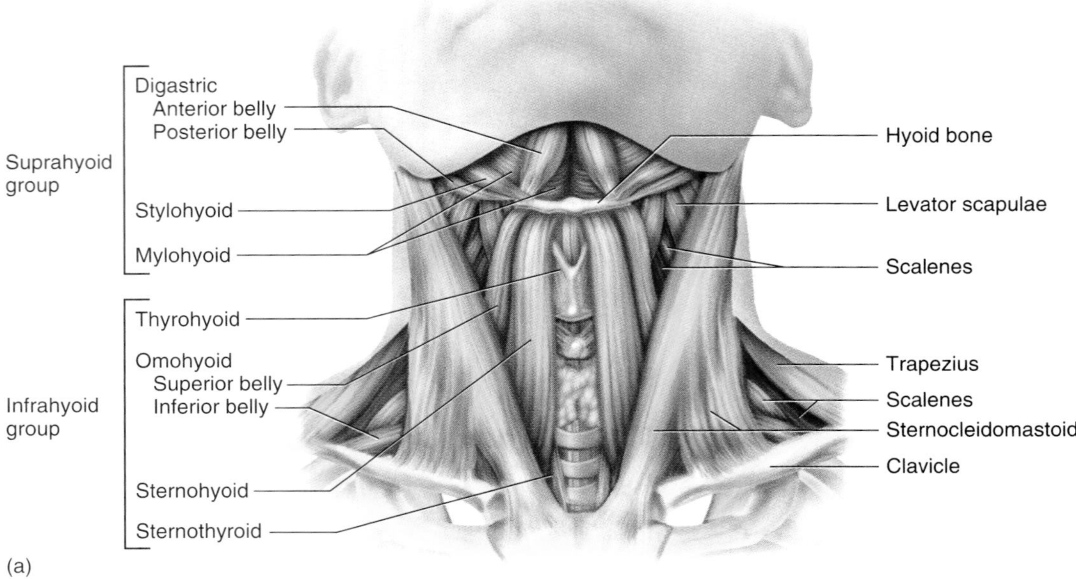

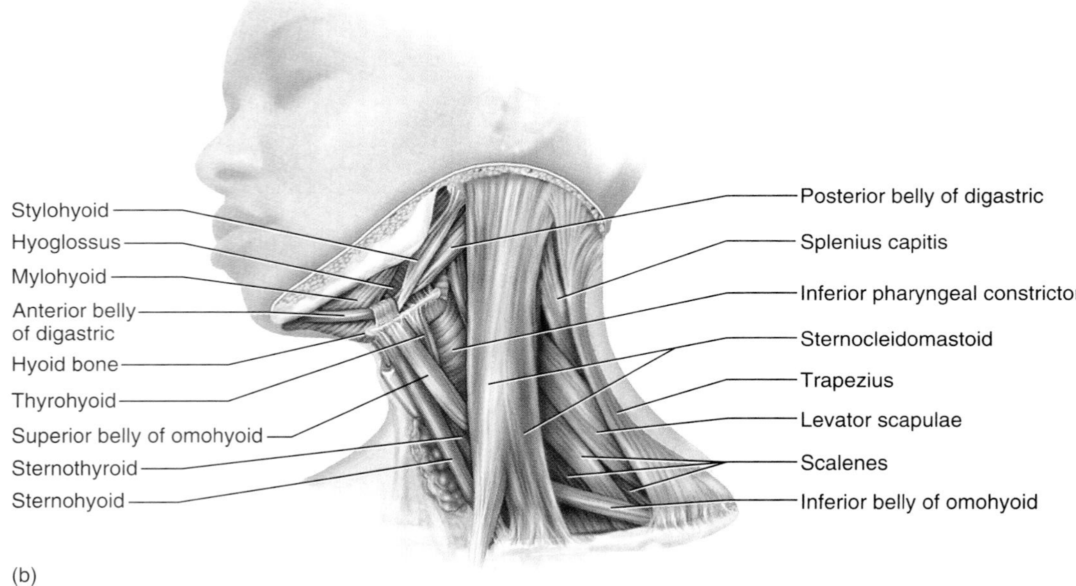

Figure 11.10 **Muscles of the Neck.** (*a*) The hyoid muscles, anterior view. The geniohyoid is deep to the mylohyoid and can be seen in figure 11.8. (*b*) Left lateral view.

and fig. 11.10 for the others.) Most of the hyoid muscles receive their innervation from the *ansa cervicalis,* a loop of nerve at the side of the neck formed by certain fibers of the first through third cervical nerves.

The **digastric**[32] arises from the mastoid process and thickens into a *posterior belly* beneath the margin of the mandible. It then narrows, passes through a connective tissue loop (*fascial sling*) attached to the hyoid bone, widens into an *anterior belly,* and attaches to the mandible near the protuberance. When it contracts, it pulls on the sling and elevates the hyoid bone. When the hyoid is fixed by the infrahyoid muscles, however, the digastric muscle opens the mouth. The mouth normally drops open by itself when the temporalis and masseter muscles are relaxed, but the digastric, platysma, and

32. *di* = two + *gastr* = belly

mylohyoid can open it more widely. The **geniohyoid** protracts the hyoid to widen the pharynx when food is swallowed. The **mylohyoid**[33] muscles fuse at the midline, form the floor of the mouth, and work synergistically with the digastric to forcibly open the mouth. The **stylohyoid,** named for its origin and insertion, elevates the hyoid bone.

When food enters the pharynx, the superior, middle, and inferior **pharyngeal constrictors** contract in that order and force the food downward, into the esophagus. The **thyrohyoid**, named for the hyoid bone and large *thyroid cartilage* of the larynx, helps to prevent choking. It elevates the thyroid cartilage so that the larynx becomes sealed by a flap of tissue, the epiglottis. You can feel this effect by placing your fingers on your "Adam's apple" (a prominence of the thyroid cartilage) and feeling it bob up as you swallow. The **sternothyroid** then pulls the larynx down again. These infrahyoid muscles that act on the larynx are called the extrinsic muscles of the larynx. The larynx also has intrinsic muscles, which are concerned with control of the vocal cords and laryngeal opening (see chapter 22).

33. *mylo* = mill, molar teeth

Table 11.3	**Muscles of Chewing and Swallowing (see figs. 11.8–11.10)**	
O = origin, I = insertion, N = innervation (n. = nerve)		
Extrinsic Muscles of the Tongue		
Genioglossus (JEE-nee-oh-GLOSS-us)		
Depresses and protrudes tongue; creates dorsal groove in tongue that enables infants to grasp nipple and channel milk to pharynx		
O: mental spines of mandible	I: hyoid bone, lateral aspect of tongue	N: hypoglossal n. (XII)
Hyoglossus		
Depresses sides of tongue		
O: hyoid bone	I: lateral aspect of tongue	N: hypoglossal n. (XII)
Styloglossus		
Elevates and retracts tongue		
O: styloid process	I: lateral aspect of tongue	N: hypoglossal n. (XII)
Palatoglossus		
Elevates posterior part of tongue; constricts fauces (entry to pharynx)		
O: soft palate	I: lateral aspect of tongue	N: accessory n. (XI)
Muscles of Mastication		
Temporalis (TEM-po-RAY-liss)		
Elevates mandible for biting and chewing; retracts mandible		
O: temporal lines	I: coronoid process of mandible	N: trigeminal n. (V)

(continued)

Table 11.3 Muscles of Chewing and Swallowing (continued)		
Masseter (ma-SEE-tur) Elevates mandible for biting and chewing; causes some lateral excursion of mandible		
O: zygomatic arch	I: lateral aspect of mandibular ramus and angle	N: trigeminal n. (V)
Medial Pterygoid (TERR-ih-goyd) Elevates mandible; produces lateral excursion		
O: pterygoid process of sphenoid bone	I: medial aspect of mandibular angle	N: trigeminal n. (V)
Lateral Pterygoid (TERR-ih-goyd) Protracts mandible; produces lateral excursion		
O: pterygoid process of sphenoid bone	I: slightly anterior to mandibular condyle	N: trigeminal n. (V)
Muscles of the Pharynx		
Pharyngeal Constrictors (three muscles) Constrict pharynx to force food into esophagus		
O: mandible, medial pterygoid plate, hyoid bone, larynx	I: posterior median raphe (fibrous seam) of pharynx	N: glossopharyngeal n. (IX), vagus n. (X)
Hyoid Muscles—Suprahyoid Group		
Digastric Retracts mandible; elevates and fixes hyoid; depresses mandible when hyoid is fixed		
O: mastoid notch and inner aspect of mandible near protuberance	I: hyoid, via fascial sling	N: trigeminal n. (V), facial n. (VII)
Geniohyoid (JEE-nee-oh-HY-oyd) Elevates and protracts hyoid; dilates pharynx to receive food; opens mouth when hyoid is fixed		
O: inner aspect of mental protuberance	I: hyoid	N: hypoglossal n. (XII)
Mylohyoid Forms floor of mouth; elevates hyoid; opens mouth when hyoid is fixed		
O: inferior margin of mandible	I: hyoid	N: trigeminal n. (V)
Stylohyoid Elevates hyoid		
O: styloid process	I: hyoid	N: facial n. (VII)
Hyoid Muscles—Infrahyoid Group		
Omohyoid Depresses hyoid; fixes hyoid during opening of mouth		
O: superior border of scapula	I: hyoid	N: ansa cervicalis
Sternohyoid Depresses hyoid; fixes hyoid during opening of mouth		
O: manubrium, costal cartilage 1	I: hyoid	N: ansa cervicalis
Thyrohyoid Depresses hyoid; elevates larynx; fixes hyoid during opening of mouth		
O: thyroid cartilage of larynx	I: hyoid	N: hypoglossal n. (XII)
Sternothyroid Depresses larynx; fixes hyoid during opening of mouth		
O: manubrium, costal cartilage 1 or 2	I: thyroid cartilage of larynx	N: ansa cervicalis

Historical Insight 11.1

Discovery of a New Muscle

New discoveries in physiology are an everyday occurrence, but one would think all the muscles of the human body had been discovered long ago. Some have even said that human gross anatomy is a completed science, a "dead discipline." Thus, anatomists were surprised by the 1996 announcement of a new muscle of mastication discovered by U.S. dentists Gary Hack and Gwendolyn Dunn.

Hack and Dunn were studying the muscles of mastication using an unorthodox dissection method in which they entered the head from the front rather than from the side. "There it was," Hack said, "just staring at us"—a muscle, extending from the greater wing of the sphenoid to the medial side of the mandible, that everyone else had either overlooked or dismissed as part of the temporalis or medial pterygoid. Hack and Dunn named it the *sphenomandibularis.*

In chapter 1, we saw that some of history's greatest advances in scientific thinking came from people with the imagination to view things from a different angle than everyone else had done. In the discovery of the sphenomandibularis, we see that even little steps are made this way, and even the "finished" sciences hold surprises for people with imaginative approaches.

Muscles Acting on the Head

Muscles that move the head originate on the vertebral column, thoracic cage, and pectoral girdle and insert on the cranial bones (table 11.4). The principal flexors of the neck are the **sternocleidomastoid**[34] and three **scalenes** on each side (fig. 11.10). The prime mover is the sternocleidomastoid, a thick cordlike muscle that extends from the sternum and clavicle to the mastoid process behind the ear. It is most easily seen and palpated when the head is turned to one side and slightly extended. As it passes obliquely across the neck, the sternocleidomastoid divides it into **anterior** and **posterior triangles.** Other muscles and landmarks subdivide each of these into smaller triangles of surgical importance (fig. 11.11).

When both sternocleidomastoids contract, the neck flexes forward; for example, when you look down at something between your feet. When only the left one contracts, the head tilts down and to the right, and when the right one acts alone, it draws the head down and to the left. To visualize this action, hold the index finger of your left hand on your left mastoid process and the index finger of your right hand on your sternal notch. Now contract the left sternocleidomastoid in a way that brings the two fingertips as close together as possible. You will note that this action causes you to look downward and to the right.

The extensors are located in the back of the neck. Their actions include extension (holding the head erect), hyperextension (as in looking upward toward the sky), abduction (tilting the head to one side), and rotation (as in looking to the left and right). Extension and hyperextension involve equal action of the right and left members of a pair; the other actions require the muscle on one side to contract more strongly than the opposite muscle. Many head movements result from a combination of these actions—for example, looking up over the shoulder involves a combination of rotation and extension.

We will consider only three primary extensors: the trapezius, splenius capitis, and semispinalis capitis (figs. 11.12 and 11.17). The **trapezius** is a vast triangular muscle of the upper back and neck; together, the right and left trapezius muscles form a trapezoid. The origin of the trapezius extends from the occipital protuberance of the skull to thoracic vertebra 12. The trapezius converges to an insertion on the shoulder. The **splenius**[35] **capitis,** which lies just deep to the trapezius on the neck, has oblique fascicles that diverge from the vertebral column toward the ears. It is nicknamed the "bandage muscle" because of the way it tightly binds deeper neck muscles. The **semispinalis capitis** is slightly deeper, and its fascicles travel vertically up the back of the neck to insert on the occipital bone. A complex array of smaller, deeper extensors are synergists of these prime movers; they extend the head, rotate it, or both.

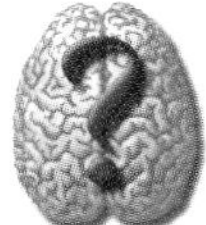

Think About It

Of the muscles you have studied so far, name three that you would consider intrinsic muscles of the head and three that you would classify as extrinsic. Explain your reason for each.

34. *sterno* = sternum + *cleido* = clavicle + *mastoid* = mastoid process of skull

35. *spleni* = bandage

Table 11.4 Muscles Acting on the Head (see figs. 11.10 and 11.17)

O = origin, I = insertion, N = innervation (n. = nerve, nn. = nerves)

Flexors of the Neck

Sternocleidomastoid (STIR-no-CLY-doe-MASS-toyd)

Contraction of either one draws head down and toward the side opposite the contracting muscle; contraction of both draws head forward and down, as in looking between the feet

O: clavicle, manubrium	I: mastoid process	N: accessory n. (XI)

Scalenes (SCAY-leens) (three muscles)

Flex neck laterally; elevate ribs 1 and 2 in inspiration

O: vertebrae C2–C6	I: ribs 1–2	N: C5–C8

Extensors of the Neck

Trapezius (tra-PEE-zee-us)

Abducts and extends neck (see other functions in table 11.9)

O: external occipital protuberance, nuchal ligament, spinous processes of vertebrae C7–T12	I: clavicle, acromion, scapular spine	N: accessory n. (XI), C3–C4

Splenius Capitis (SPLEE-nee-us CAP-ih-tis), Splenius Cervicis (SIR-vih-sis)

Rotate head, extend neck

O: *capitis*—spinous processes of vertebrae C7–T3 or T4; *cervicis*—spinous processes of T3–T6	I: *capitis*—mastoid process, superior nuchal line; *cervicis*—transverse processes of C1–C2 or C3	N: dorsal rami of middle and lower cervical nn.

Semispinalis (SEM-ee-spy-NAY-liss) Capitis

Rotates and extends head (see other parts of semispinalis in table 11.7)

O: transverse processes of vertebrae T1–T6, articular processes of C4–C7	I: occipital bone	N: dorsal rami of cervical nn.

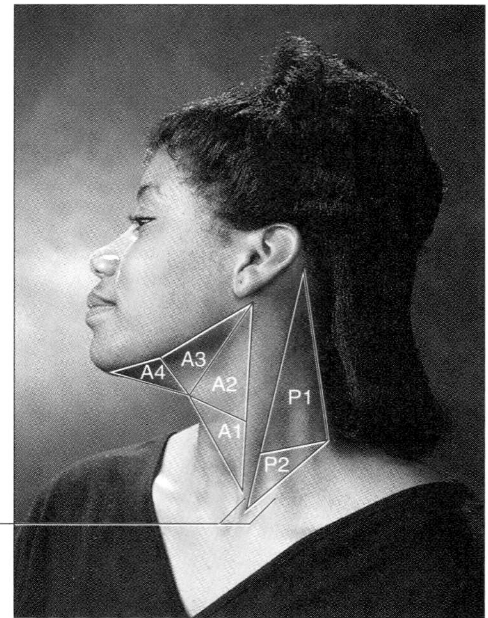

Figure 11.11 **The Triangles of the Neck.** The sternocleidomastoid muscle separates the anterior triangles from the posterior triangles.

Before You Go On

Answer the following questions to test your understanding of the preceding section:

8. Name two muscles that elevate the upper lip and two that depress the lower lip.
9. Name the four paired muscles of mastication and state where they insert on the mandible.
10. Distinguish between the functions of the suprahyoid and infrahyoid muscles.
11. List the prime movers of neck extension and flexion.

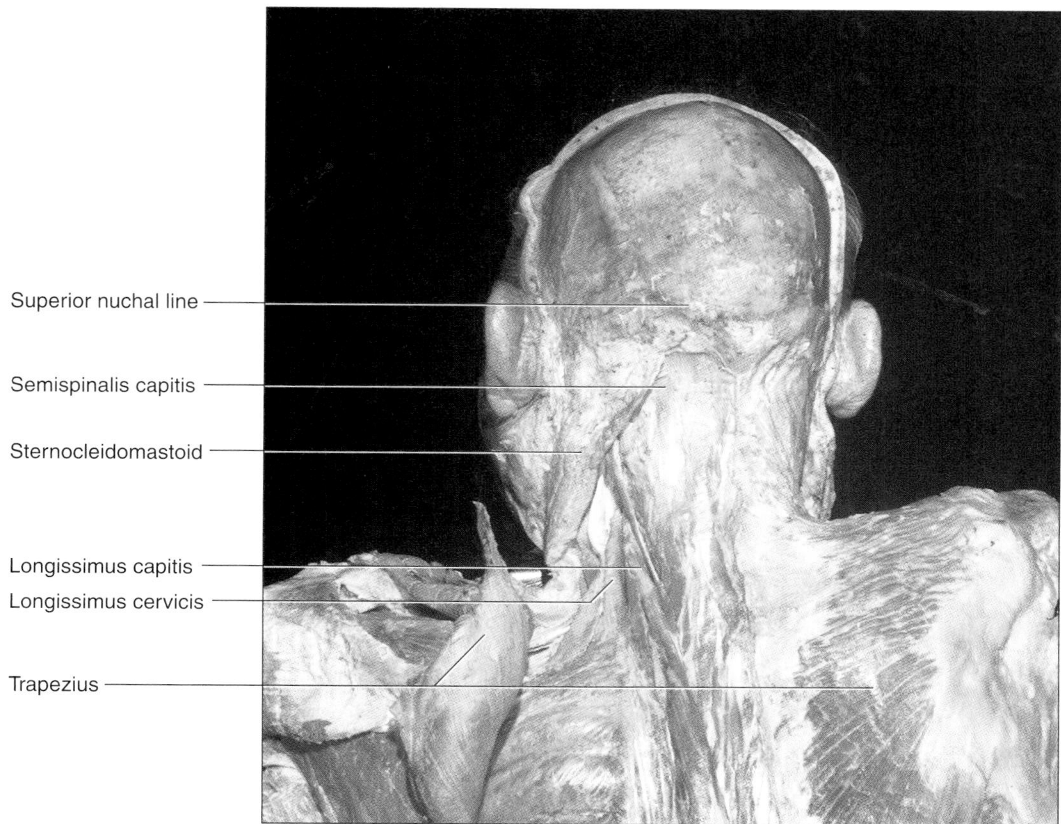

Figure 11.12 Muscles of the Shoulder and Nuchal Regions.

Muscles of the Trunk

Objectives

When you have completed this section, you should be able to

- Name and locate the muscles of respiration and explain how they affect abdominal pressure;
- Name and locate the muscles of the abdominal wall, back, and pelvic floor; and
- Identify the origin, insertion, action, and innervation of any of these muscles.

Muscles of Respiration

We breathe primarily by means of muscles that enclose the thoracic cavity—the *diaphragm,* which forms its floor; 11 pairs of *external intercostal muscles,* which lie superficially between the ribs; and 11 pairs of *internal intercostal muscles* between the ribs deep to the external intercostals (fig. 11.13; table 11.5). The lungs themselves contain no skeletal muscle; they do not play an active part in their own ventilation.

The **diaphragm**[36] is a muscular dome between the abdominal and thoracic cavities. It has openings that allow passage of the esophagus and major blood vessels. Its fascicles converge from the margins toward a fibrous **central tendon.** When the diaphragm contracts, it flattens slightly, increasing the volume of the thoracic cage and creating a partial vacuum that draws air into the lungs. Its contraction also raises pressure in the abdominal cavity below, thus helping to expel the contents of the bladder and rectum and facilitating childbirth—which is why people tend to take a deep breath and hold it during these functions.

The **external intercostals**[37] extend obliquely downward and anteriorly from each rib to the rib below it. When the scalenes fix the first rib, the external intercostals lift the others, pulling them up somewhat like bucket handles. This action pulls the ribs closer together and draws the

36. *dia* = across + *phragm* = partition
37. *inter* = between + *costa* = rib

Table 11.5 Muscles of Respiration (see fig. 11.13)

O = origin, I = insertion, N = innervation (n. = nerve, nn. = nerves)

Diaphragm (DY-uh-fram)

Prime mover of inspiration; compresses abdominal viscera to aid in such processes as defecation, urination, and childbirth

O	I	N
O: xiphoid process, ribs 10–12, costal cartilages 5–9, lumbar vertebrae	I: central tendon	N: phrenic n.

External Intercostals (IN-tur-COSS-tulz)

When scalenes fix rib 1, external intercostals draw ribs 2–12 upward and outward to expand thoracic cavity and inflate lungs

O	I	N
O: inferior margins of ribs 1–11	I: superior margins of ribs 2–12	N: intercostal nn.

Internal Intercostals

When quadratus lumborum and other muscles fix rib 12, internal intercostals draw ribs downward and inward to compress thoracic cavity and force air from lungs; not needed for relaxed expiration

O	I	N
O: inferior margins of ribs 1–11	I: superior margins of ribs 2–12	N: intercostal nn.

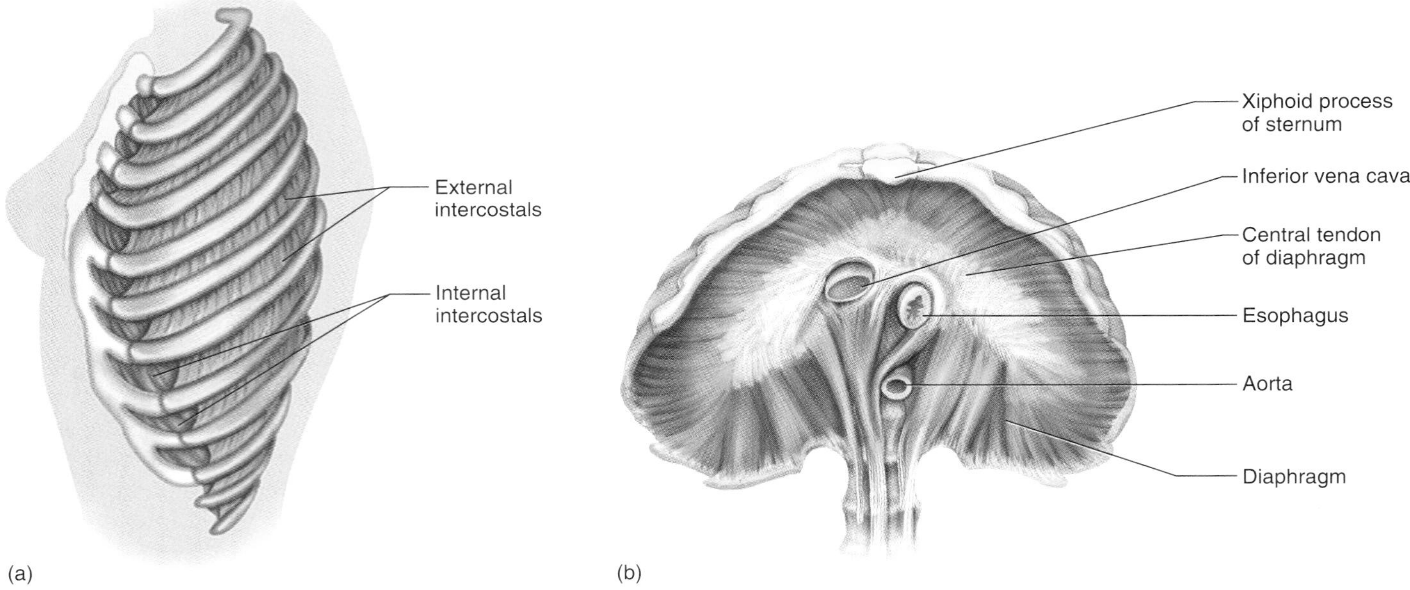

Figure 11.13 Muscles of Respiration. (*a*) The intercostal muscles, viewed from the left. (*b*) The diaphragm, viewed from below.

entire rib cage upward and outward, expanding the thoracic cage and promoting inhalation.

It takes no muscular effort to exhale—when the diaphragm and external intercostals relax, the thoracic cage springs back to its prior size and expels the air. However, forced expiration—exhaling more than the usual amount of air or exhaling quickly as in blowing out a candle—is achieved mainly by the **internal intercostals.** These also extend from one rib to the next, but they lie deep to the external intercostals and have fascicles at right angles to them. The abdominal muscles also aid in forced expiration by pushing the viscera up against the diaphragm.

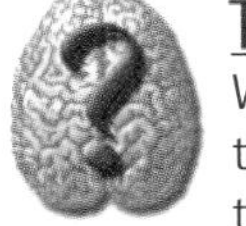

Think About It

What muscles are eaten as "spare ribs"? What is the tough fibrous membrane between the meat and the bone?

Muscles of the Abdomen

The anterior and lateral walls of the abdomen are reinforced by four pairs of sheetlike muscles that support the viscera, stabilize the vertebral column during heavy lifting, and aid in respiration, urination, defecation, vomiting, and childbirth. They are the *rectus abdominis, external abdominal oblique, internal abdominal oblique,* and *transversus abdominis* (table 11.6; figs. 11.14–11.16).

The **rectus**[38] **abdominis** is a medial straplike muscle extending vertically from the pubis to the sternum. It is separated into four segments by fibrous **tendinous intersections** that give the abdomen a segmented appearance in well-muscled individuals. The rectus abdominis is enclosed in a fibrous sleeve called the **rectus sheath,** and the right and left muscles are separated by a vertical fibrous strip called the **linea alba.**[39]

38. *rect* = straight
39. *linea* = line + *alb* = white

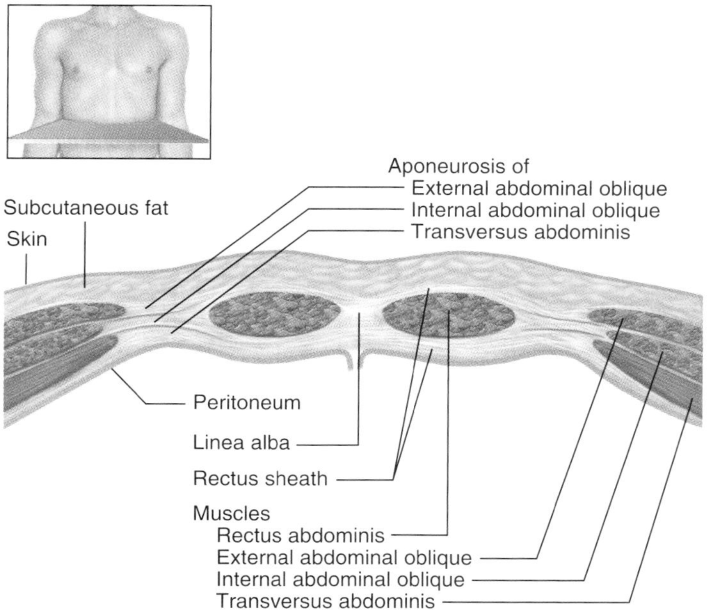

Figure 11.14 Cross Section of the Anterior Abdominal Wall.

Table 11.6 Muscles of the Abdomen (see figs. 11.14 and 11.15)

O = origin, I = insertion, N = innervation (n. = nerve, nn. = nerves)

Rectus Abdominis (ab-DOM-ih-niss)

Supports abdominal viscera; flexes waist as in sit-ups; depresses ribs; stabilizes pelvis during walking; increases intra-abdominal pressure to aid in urination, defecation, and childbirth

O	I	N
O: pubis	I: xiphoid process, costal cartilages 5–7	N: intercostal nn. 7–12

External Abdominal Oblique

Flexes waist as in sit-ups; flexes and rotates vertebral column

O	I	N
O: ribs 5–12	I: xiphoid process, linea alba	N: intercostal nn. 8–12, iliohypogastric n., ilioinguinal n.

Internal Abdominal Oblique

Similar to external oblique

O	I	N
O: inguinal ligament, iliac crest, thoracolumbar fascia	I: xiphoid process, linea alba, pubis, ribs 10–12	N: same as external oblique

Transversus Abdominis

Compresses abdomen, increases intra-abdominal pressure, flexes vertebral column

O	I	N
O: inguinal ligament, iliac crest, thoracolumbar fascia, costal cartilages 7–12	I: xiphoid process, linea alba, pubis, inguinal ligament	N: intercostal nn. 8–12, iliohypogastric n., ilioinguinal n.

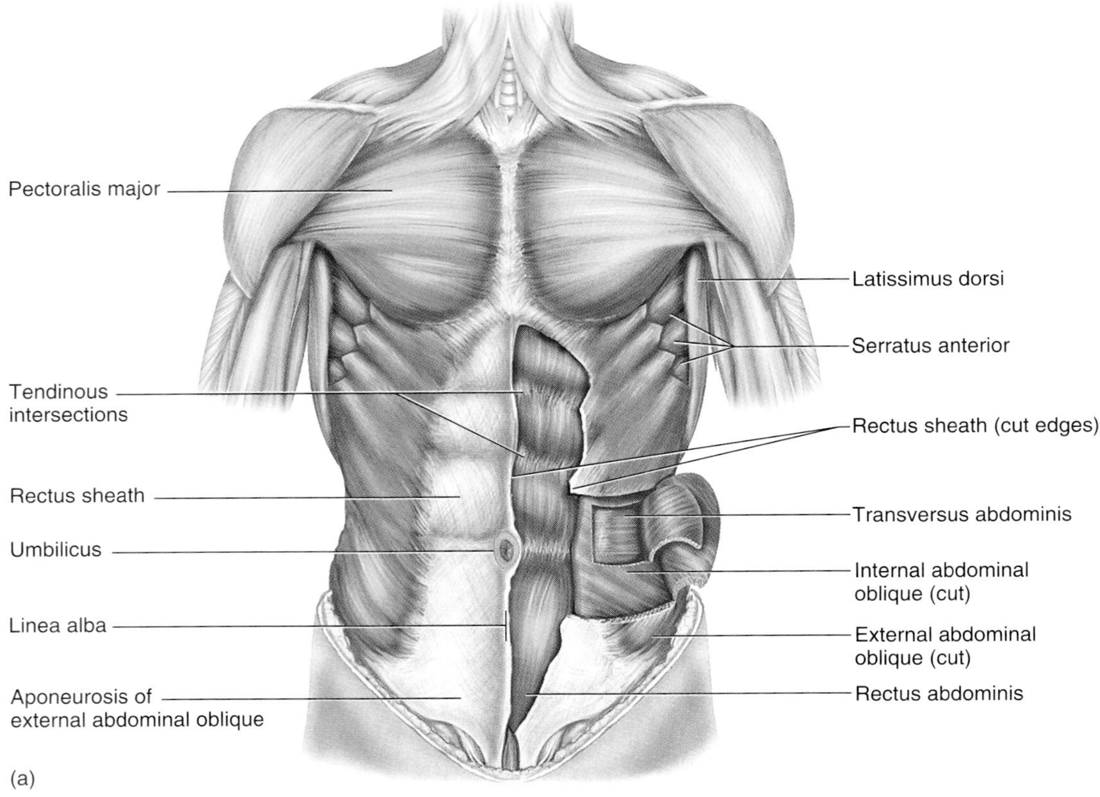

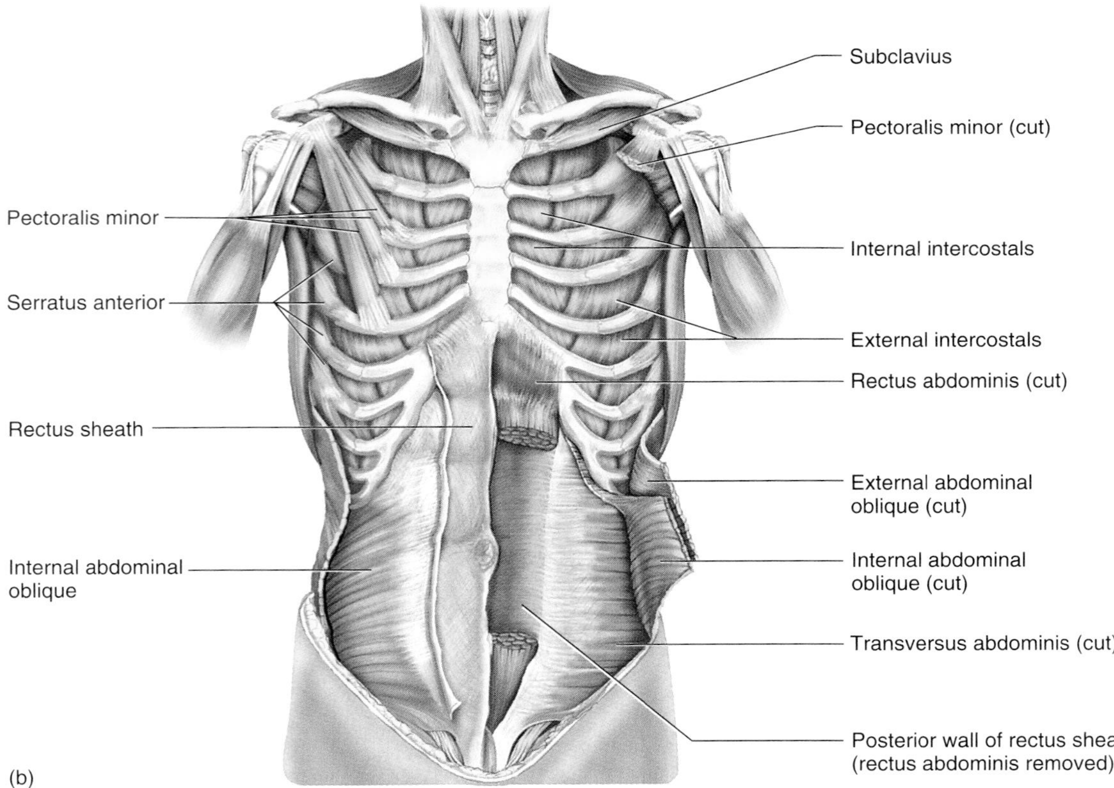

Figure 11.15 **Thoracic and Abdominal Muscles.** (*a*) Superficial muscles. The left rectus sheath is cut away to expose the rectus abdominis muscle. (*b*) Deep muscles. On the anatomical right, the external abdominal oblique has been removed to expose the internal abdominal oblique and the pectoralis major has been removed to expose the pectoralis minor. On the anatomical left, the internal abdominal oblique has been cut to expose the transversus abdominis, and the rectus abdominis has been cut to expose the posterior rectus sheath.

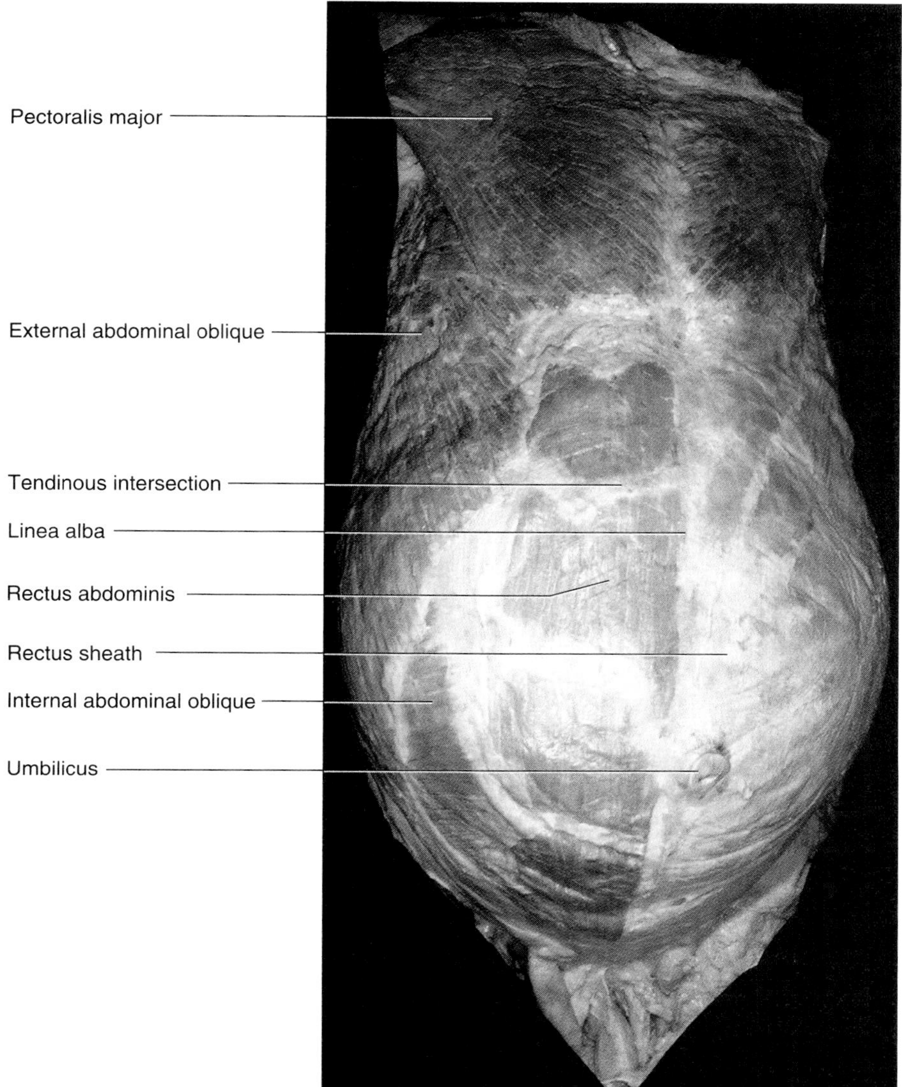

Figure 11.16 Some Thoracic and Abdominal Muscles of the Cadaver.

The **external abdominal oblique** is the most superficial muscle of the lateral abdominal wall. Its fascicles run anteriorly and downward. Deep to it is the **internal abdominal oblique,** whose fascicles run anteriorly and upward. Deepest of all is the **transversus abdominis,** whose fascicles run horizontally across the abdomen. Unlike the thoracic cavity, the abdominal cavity lacks a protective bony enclosure. However, the wall formed by these three muscle layers is strengthened by the way their fascicles run in different directions like layers of plywood.

The tendons of the abdominal muscles are aponeuroses. They continue medially to form the rectus sheath and terminate at the linea alba. At its inferior margin, the aponeurosis of the external abdominal oblique forms a strong, cordlike **inguinal ligament** that extends from the pubis to the anterior superior spine of the ilium.

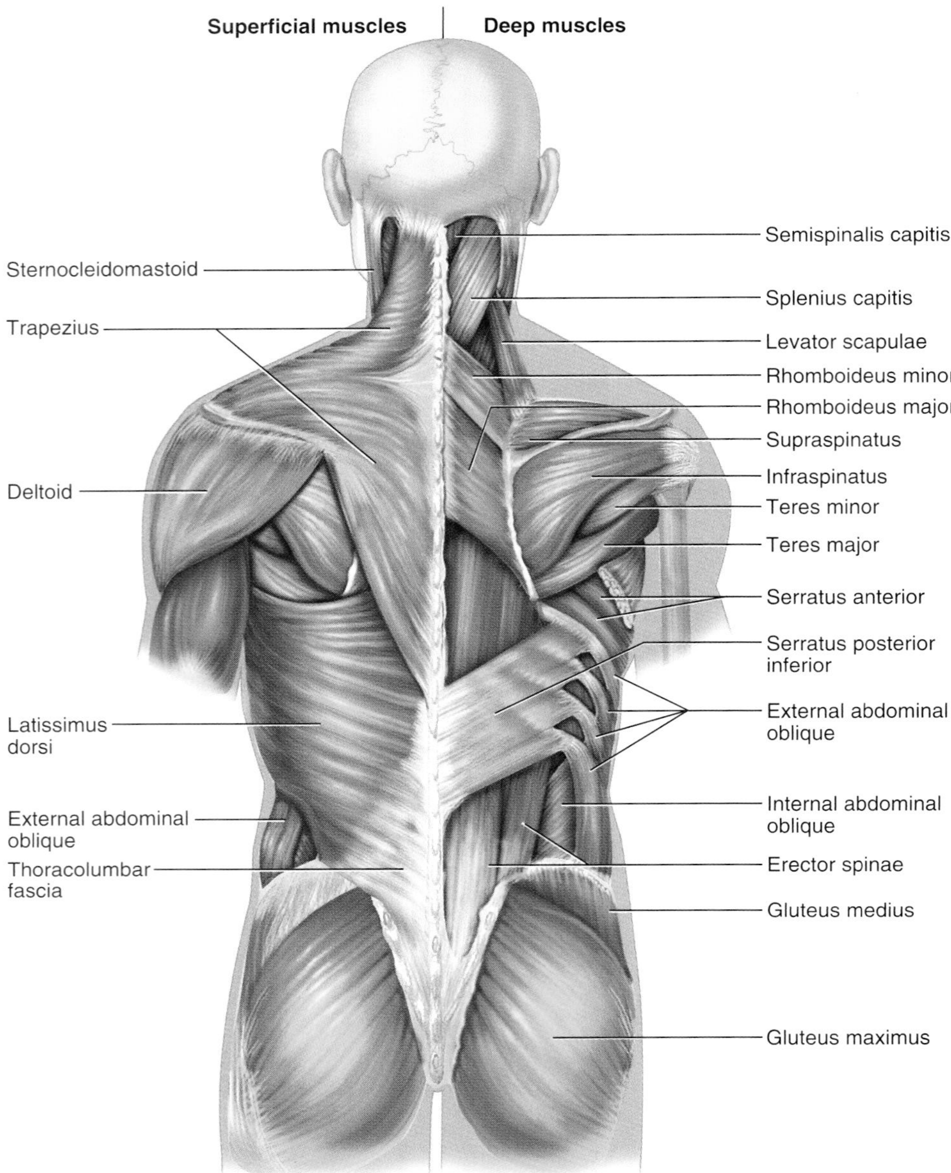

Figure 11.17 **Neck, Back, and Gluteal Muscles.** The most superficial muscles are shown on the left, and the next deeper layer on the right.

Muscles of the Back

We now consider muscles of the back that extend, rotate, and abduct the vertebral column (figs. 11.17–11.19). Back muscles that act on the pectoral girdle and arm are considered later. The muscles associated with the vertebral column moderate your motion when you bend forward and contract to return the trunk to the erect position. They are classified into two groups—a *superficial group,* which extends from the vertebrae to the ribs, and a *deep group,* which connects the vertebrae to each other.

In the superficial group, the prime mover of spinal extension is the **erector spinae.** You use this muscle to maintain your posture and to stand up straight after bending at the waist. It is divided into three "columns"—the **iliocostalis, longissimus,** and **spinalis.** These are complex, multipart

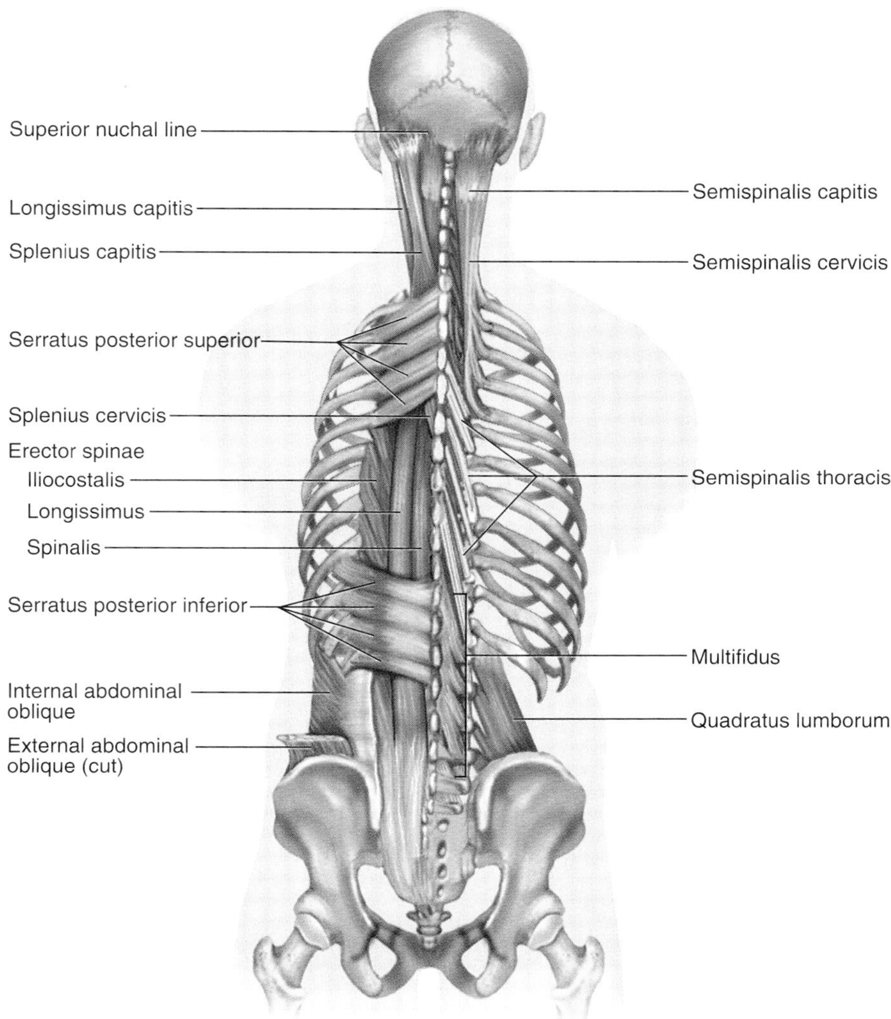

Figure 11.18 **Muscles Acting on the Vertebral Column.** Those on the right are deeper than those on the left.

muscles with cervical, thoracic, and lumbar portions. Some of their portions move the head and have already been discussed, while those that act on cervical and lower parts of the vertebral column are described in table 11.7. Most of the lower back (lumbar) muscles are in the longissimus group. Two **serratus posterior** muscles—one superior and one inferior—overlie the erector spinae and act to move the ribs.

The major deep thoracic muscle is the **semispinalis.** This is divided into three parts, the **semispinalis capitis,** which we have already studied (see table 11.4), the **semispinalis cervicis,**[40] and **semispinalis thoracis,**[41] in that order from superior to inferior. In the lumbar region, the major deep muscle is the **quadratus**[42] **lumborum.** The erector spinae and quadratus lumborum are enclosed in a fibrous sheath called the **thoracolumbar fascia,** which is the origin of some of the abdominal and lumbar muscles. The **multifidus**[43] muscle deep to this connects the vertebrae to each other from the cervical to the lumbar region and acts to extend and rotate the vertebral column.

40. *cervicis* = of the neck
41. *thoracis* = of the thorax
42. *quadrat* = four-sided
43. *multi* = many + *fid* = split, sectioned

Table 11.7 Muscles of the Back (see figs. 11.17 and 11.18)

O = origin, I = insertion, N = innervation (n. = nerve, nn. = nerves)

Origin	Insertion	Innervation
Superficial Group—The Erector Spinae (ee-RECK-tur SPY-nee)		
Iliocostalis Cervicis (ILL-ee-oh-coss-TAH-liss SIR-vih-sis), Iliocostalis Thoracis (tho-RA-sis), and Iliocostalis Lumborum (lum-BORE-um) Extend and laterally flex vertebral column; thoracis and lumborum rotate ribs during forceful inspiration		
O: angles of ribs, sacrum, iliac crest	I: *cervicis*—vertebrae C4–C6; *thoracis*—vertebra C7, angles of ribs 1–6; *lumborum*—angles of ribs 7–12	N: dorsal rami of spinal nn.
Longissimus (lawn-JISS-ih-muss) Cervicis and Longissimus Thoracis Extend and laterally flex vertebral column		
O: *cervicis*—vertebrae T1 to T4 or T5; *thoracis*—sacrum, iliac crest, vertebrae T1–L5	I: *cervicis*—vertebrae C2–C6; *thoracis*— vertebrae T1–T12, ribs 3 or 4 to 12	N: dorsal rami of spinal nn.
Spinalis (spy-NAY-liss) Cervicis, Spinalis Thoracis Extend vertebral column		
O: *cervicis*—nuchal ligament, spinous process of vertebra C7; *thoracis*—spinous processes of T11–L2	I: *cervicis*—spinous process of axis; *thoracis*—spinous processes of upper thoracic vertebrae	N: dorsal rami of spinal nn.
Deep Group		
Semispinalis Cervicis (SEM-ee-spy-NAY-liss SUR-vih-sis), Semispinalis Thoracis (tho-RA-sis) Extend neck; extend and rotate vertebral column		
O: transverse processes of vertebrae T1–T10	I: spinous processes of vertebrae C2–T5	N: dorsal rami of spinal nn.
Quadratus Lumborum (quad-RAY-tus lum-BORE-um) Laterally flexes vertebral column, depresses rib 12		
O: iliac crest, lower lumbar vertebrae, thoracolumbar fascia	I: upper lumbar vertebrae, rib 12	N: ventral rami of L1–L3
Multifidus (mul-TIFF-ih-dus) Extends and rotates vertebral column		
O: sacrum, iliac crest, vertebrae C4–L5	I: laminae and spinous processes of vertebrae above origins	N: dorsal rami of spinal nn.

Clinical Insight 11.2

Heavy Lifting and Back Injuries

When you are fully bent over forward, as in touching your toes, the erector spinae is fully stretched. Because of the *length-tension relationship* explained in the next chapter, muscles that are stretched to such extremes cannot contract very effectively. Standing up from such a position is therefore initiated by the hamstring muscles on the back of the thigh and the gluteus maximus of the buttocks. The erector spinae joins in the action when it is partially contracted.

Standing too suddenly or improperly lifting a heavy weight, however, can strain the erector spinae, cause painful muscle spasms, tear tendons and ligaments of the lower back, and rupture intervertebral discs. The lumbar muscles are adapted for maintaining posture, not for lifting. This is why it is important, in heavy lifting, to kneel and use the powerful extensor muscles of the thighs and buttocks to lift the load.

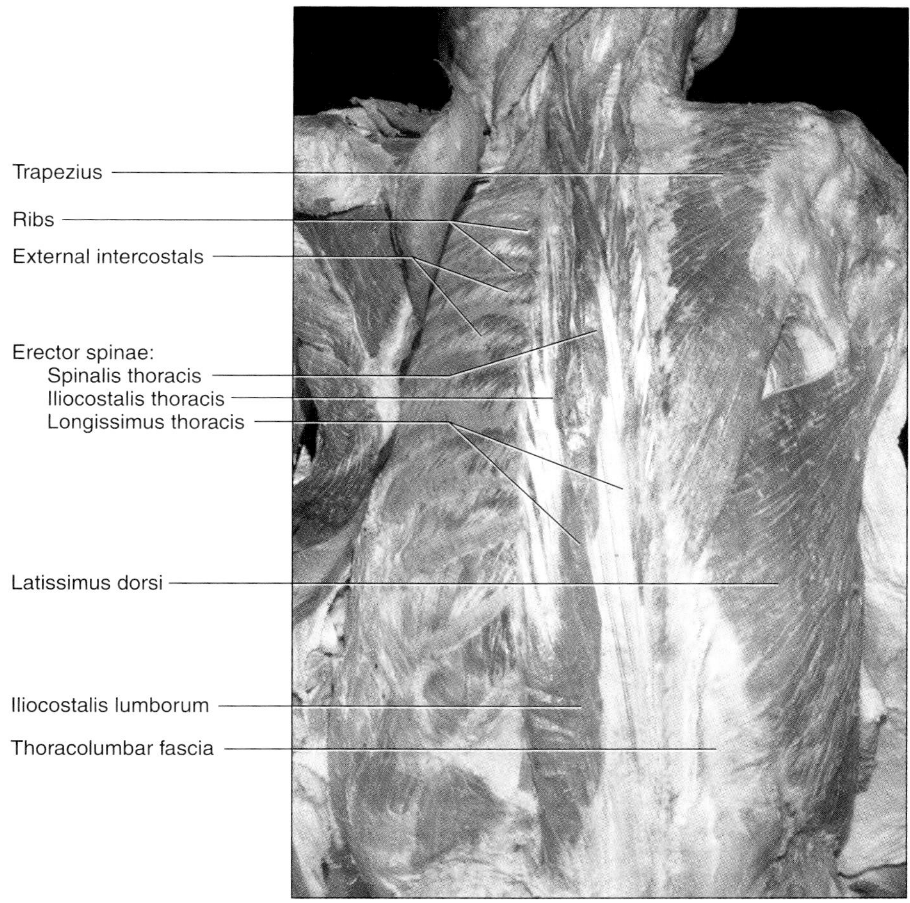

Figure 11.19 Some Deep Back Muscles of the Cadaver.

Muscles of the Pelvic Floor

The floor of the pelvic cavity is formed by three layers of muscles and fasciae that span the pelvic outlet and support the viscera (table 11.8). It is penetrated by the anal canal, urethra, and vagina, which open into a diamond-shaped region between the thighs called the **perineum** (PERR-ih-NEE-um). The perineum is bordered by four bony landmarks—the pubic symphysis anteriorly, the coccyx posteriorly, and the ischial tuberosities laterally. The anterior half of the perineum is the **urogenital triangle** and the posterior half is the **anal triangle** (fig. 11.20*b*). These are especially important landmarks in obstetrics.

The pelvic floor is divided into three layers or "compartments." The one just deep to the skin, called the **superficial perineal space** (fig. 11.20*a, b*), contains three muscles. The **ischiocavernosus** muscles converge like a V from the ischial tuberosities toward the penis or clitoris and assist in erection. In males, the **bulbospongiosus (bulbocavernosus)** forms a sheath around the base (bulb) of the penis; it expels semen during ejaculation. In females, it encloses the vagina like a pair of parentheses and tightens on the penis during intercourse. Voluntary contractions of this muscle in both sexes also help void the last few milliliters of urine. The **superficial transverse perineus** extends from the ischial tuberosities to a strong **central tendon** of the perineum.

In the middle compartment, the urogenital triangle is spanned by a thin triangular sheet called the **urogenital diaphragm.** This is composed of a fibrous membrane and two muscles—the **deep transverse perineus** and the **external urethral sphincter** (fig. 11.20*c, d*). The anal triangle contains the **external anal sphincter.** The deepest compartment, called the **pelvic diaphragm,** is similar in both sexes. It consists of two muscle pairs shown in figure 11.20*e*—the **levator ani** and **coccygeus.**

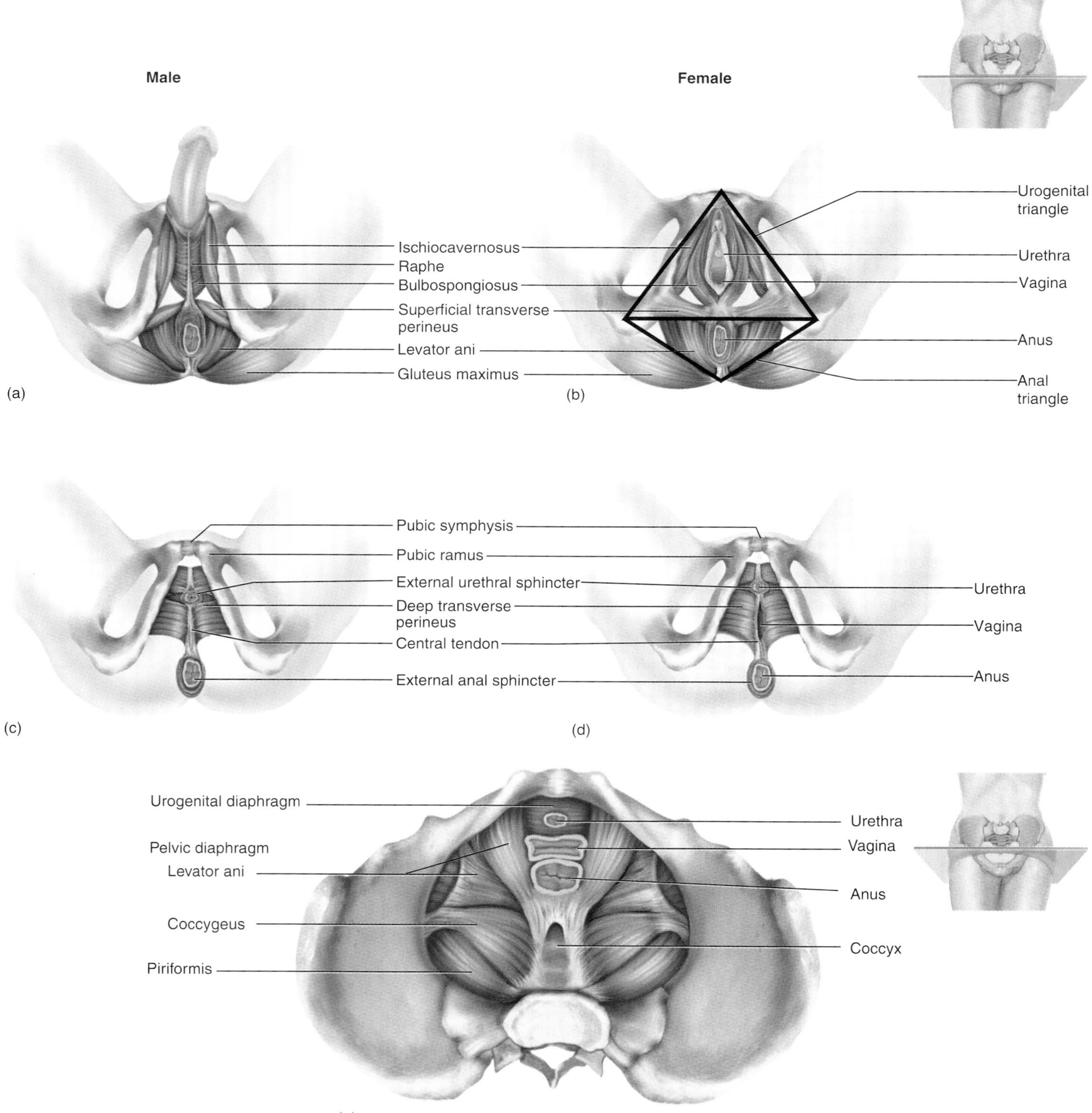

Figure 11.20 Muscles of the Pelvic Floor. (*a, b*) The superficial perineal space, inferior view. Triangles of the perineum are marked in *b*. (*c, d*) The urogenital diaphragm, inferior view; this is the next deeper layer after the muscles in *a* and *b*. (*e*) The pelvic diaphragm, the deepest layer, superior view (seen from within the pelvic cavity).

Table 11.8 Muscles of the Pelvic Floor (see fig. 11.20)

O = origin, I = insertion, N = innervation (n. = nerve)

Superficial Muscles of the Perineum

Ischiocavernosus (ISS-kee-oh-CAV-er-NO-sus)
Aids in erection of penis and clitoris

O: ischial and pubic rami, ischial tuberosity	I: penis, clitoris	N: pudendal n.

Bulbospongiosus (BUL-bo-SPUN-jee-OH-sus)
Male: compresses urethra to expel semen or urine. Female: constricts vaginal orifice.

O: central tendon of perineum, bulb of penis	I: fasciae of perineum, penis or clitoris	N: pudendal n.

Superficial Transverse Perineus (PERR-ih-NEE-us)
Fixes central tendon of perineum, supports pelvic floor

O: ischial ramus	I: central tendon of perineum	N: pudendal n.

Muscles of the Urogenital Diaphragm

Deep Transverse Perineus
Fixes central tendon of perineum; supports pelvic floor; expels last drops of urine in both sexes and semen in male

O: ischial ramus	I: central tendon of perineum	N: pudendal n.

External Urethral Sphincter
Compresses urethra to voluntarily inhibit urination

O: ischial and pubic rami	I: medial raphe of male, vaginal wall of female	N: pudendal n.

Muscle of the Anal Triangle

External Anal Sphincter
Compresses anal canal to voluntarily inhibit defecation

O: anococcygeal raphe	I: central tendon of perineum	N: pudendal n., S4

Muscles of the Pelvic Diaphragm

Levator Ani (leh-VAY-tur AY-nye)
Supports viscera; resists pressure surges in abdominal cavity; elevates anus during defecation; forms vaginal and anorectal sphincters

O: os coxae from pubis to ischial spine	I: coccyx, anal canal, anococcygeal raphe	N: pudendal n., S3–S4

Coccygeus (coc-SIDJ-ee-us)
Draws coccyx anteriorly after defecation or childbirth; supports and elevates pelvic floor; resists abdominal pressure surges

O: ischial spine	I: lower sacrum to upper coccyx	N: S3 or S4

Clinical Insight 11.3

Hernias

A hernia is any condition in which the viscera protrude through a weak point in the muscular wall of the abdominopelvic cavity. The most common type to require treatment is an *inguinal hernia.* In the male fetus, each testis descends from the pelvic cavity into the scrotum by way of a passage called the *inguinal canal* through the muscles of the groin. This canal remains a weak point in the pelvic floor, especially in infants and children. When pressure rises in the abdominal cavity, it can force part of the intestine or bladder into this canal or even into the scrotum. This also sometimes occurs in men who hold their breath while lifting heavy weights. When the diaphragm and abdominal muscles contract, pressure in the abdominal cavity can soar to 1,500 pounds per square inch—more than 100 times the normal pressure and quite sufficient to produce an inguinal hernia, or "rupture." Inguinal hernias rarely occur in women.

Two other sites of hernia are the diaphragm and navel. A *hiatal hernia* is a condition in which part of the stomach protrudes through the diaphragm into the thoracic cavity. This is most common in overweight people over 40. It may cause heartburn due to the regurgitation of stomach acid into the esophagus, but most cases go undetected. In an *umbilical hernia,* abdominal viscera protrude through the navel.

Before You Go On

Answer the following questions to test your understanding of the preceding section:

12. Which muscles are used more often, the external intercostals or internal intercostals? Explain.
13. Explain how pulmonary ventilation affects abdominal pressure and vice versa.
14. Name a major superficial muscle and two major deep muscles of the back.
15. Define *perineum, urogenital triangle,* and *anal triangle.*
16. Name one muscle in the superficial perineal space, one in the urogenital diaphragm, and one in the pelvic diaphragm. State the function of each.

Muscles Acting on the Shoulder and Upper Limb

Objectives

When you have completed this section, you should be able to

- Name and locate the muscles that act on the pectoral girdle, shoulder, elbow, wrist, and hand;
- Relate the actions of these muscles to the joint movements described in chapter 10; and
- Describe the origin, insertion, and innervation of each muscle.

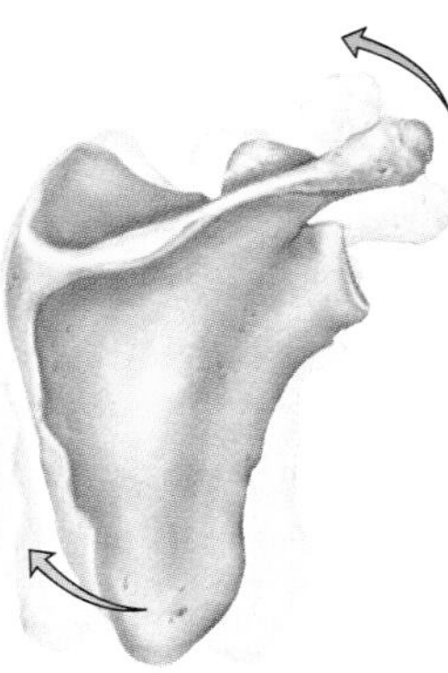

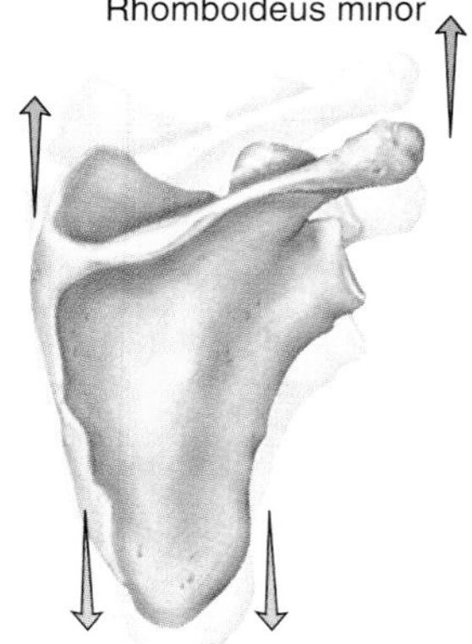

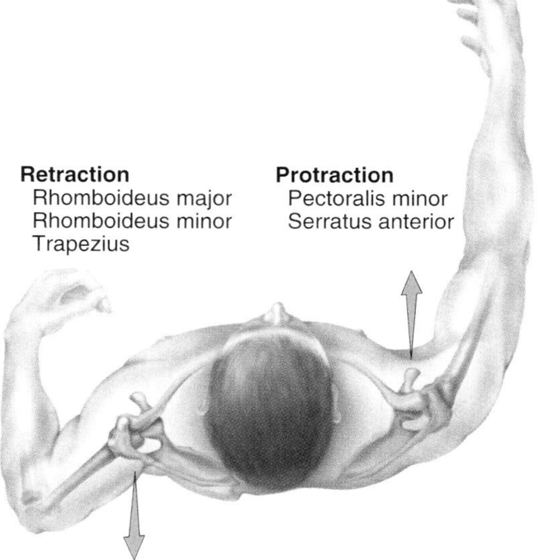

Figure 11.21 Actions of Some Thoracic Muscles on the Scapula. Note that an individual muscle can contribute to multiple actions, depending on which fibers contract and what synergists act with it.

Muscles Acting on the Scapula

The scapula is loosely attached to the thoracic cage and is capable of considerable movement—rotation (as in raising and lowering the apex of the shoulder), elevation and depression (as in shrugging and lowering the shoulders), and protraction and retraction (pulling the shoulders forward or back) (fig. 11.21). The clavicle braces the shoulder and moderates these movements.

The muscles that act on the pectoral girdle originate on the axial skeleton and insert on the clavicle and scapula. They are divided into anterior and posterior groups (table 11.9). The important muscles of the anterior group are the **pectoralis minor** and **serratus anterior** (see fig. 11.15*b*). In the posterior group, we have the large, superficial trapezius, already studied, and three deep muscles, the **levator scapulae, rhomboideus major,** and **rhomboideus minor.** The action of the trapezius depends on whether its superior, middle, or inferior fibers contract and whether it acts alone or with other muscles. The levator scapulae and superior fibers of the trapezius rotate the scapula in opposite directions if either of them acts alone. If both act together, their opposite rotational effects balance each other and they elevate the scapula and shoulder, as when you carry a heavy weight on your shoulder. Depression of the scapula occurs mainly by gravitational pull, but the trapezius and serratus anterior can cause faster, more forcible depression, as in swimming, hammering, and rowing.

Table 11.9 Muscles Acting on the Scapula (see figs. 11.15, 11.17, and 11.21)

O = origin, I = insertion, N = innervation (n. = nerve, nn. = nerves)

Origin	Insertion	Innervation
Anterior Group		
Pectoralis (PECK-toe-RAY-liss) Minor		
Protracts and depresses scapula when ribs are fixed; elevates ribs when scapula is fixed		
O: ribs 3–5	I: coracoid process	N: medial and lateral pectoral nn.
Serratus (serr-AY-tus) Anterior		
Holds scapula against rib cage; elevates ribs; abducts and rotates scapula to tilt glenoid cavity upward; forcefully depresses scapula; abducts and elevates arm; prime mover in forward thrusting, throwing, and pushing ("boxer's muscle")		
O: ribs 1–9	I: medial border of scapula	N: long thoracic n.
Posterior Group		
Trapezius (tra-PEE-zee-us)		
Superior fibers elevate scapula or rotate it to tilt glenoid cavity upward; middle fibers retract scapula; inferior fibers depress scapula. When scapula is fixed, one trapezius acting alone flexes neck laterally and both trapezius muscles working together extend neck		
O: external occipital protuberance, nuchal ligament, spinous processes of C7–T12	I: clavicle, acromion, scapular spine	N: accessory n. (XI), C3–C4
Levator Scapulae (leh-VAY-tur SCAP-you-lee)		
Rotates scapula to tilt glenoid cavity downward; flexes neck when scapula is fixed; elevates scapula when acting with superior fibers of trapezius		
O: transverse processes of vertebrae C1–C4	I: superior angle to medial border of scapula	N: C3–C4, dorsal scapular n.
Rhomboideus (rom-BOY-dee-us) Major, Rhomboideus Minor		
Retract and elevate scapula; rhomboideus major also fixes scapula and rotates it to tilt glenoid cavity downward		
O: spinous processes of vertebrae C7–T1 (r. minor) and T2–T5 (r. major)	I: medial border of scapula	N: dorsal scapular n.

Muscles Acting on the Humerus

Nine muscles cross the humeroscapular (shoulder) joint and insert on the humerus (table 11.10). Two are called *axial muscles* because they originate primarily on the axial skeleton—the **pectoralis major** and **latissimus dorsi**[44] (see figs. 11.15, 11.22, and 11.23). The pectoralis major is the thick, fleshy muscle of the mammary region, and the latissimus dorsi is a broad muscle of the back that extends from the waist to the axilla. These muscles bear the primary responsibility for attachment of the arm to the trunk, and they are the prime movers of the shoulder joint. The pectoralis major flexes the shoulder as in pointing at something in front of you, and the latissimus dorsi

44. *latissimus* = broadest + *dorsi* = of the back

Table 11.10 Muscles Acting on the Humerus (see figs. 11.22–11.24)

O = origin, I = insertion, N = innervation (n. = nerve, nn. = nerves)

Pectoralis (PECK-toe-RAY-liss) Major

Prime mover of shoulder flexion; adducts and medially rotates humerus; depresses pectoral girdle; elevates ribs; aids in climbing, pushing, and throwing

O	I	N
O: clavicle, sternum, costal cartilages 1–6	I: intertubercular groove of humerus	N: medial and lateral pectoral nn.

Latissimus Dorsi (la-TISS-ih-muss DOR-sye)

Adducts and medially rotates humerus; extends shoulder joint; produces strong downward strokes of arm, as in hammering or swimming ("swimmer's muscle"); pulls body upward in climbing

O	I	N
O: vertebrae T7–L5, lower three or four ribs, thoracolumbar fascia, iliac crest, inferior angle of scapula	I: intertubercular groove of humerus	N: thoracodorsal n.

Deltoid

Lateral fibers abduct humerus; anterior fibers flex and medially rotate it; posterior fibers extend and laterally rotate it

O	I	N
O: clavicle, scapular spine, acromion	I: deltoid tuberosity of humerus	N: axillary n.

Teres (TERR-eez) Major

Adducts and medially rotates humerus; extends shoulder joint

O	I	N
O: from inferior angle to lateral border of scapula	I: medial aspect of proximal shaft of humerus	N: subscapular n.

Coracobrachialis (COR-uh-co-BRAY-kee-AL-iss)

Adducts arm; flexes shoulder joint

O	I	N
O: coracoid process	I: medial aspect of shaft of humerus	N: musculocutaneous n.

Rotator Cuff

All rotator cuff muscles hold head of humerus in glenoid cavity and stabilize shoulder joint in addition to performing the functions below.

Infraspinatus (IN-fra-spy-NAY-tus)

Extends and laterally rotates humerus

O	I	N
O: infraspinous fossa of scapula	I: greater tubercle of humerus	N: suprascapular n.

Supraspinatus (SOO-pra-spy-NAY-tus)

Abducts humerus; resists downward displacement when carrying heavy weight

O	I	N
O: supraspinous fossa of scapula	I: lesser tubercle of humerus	N: suprascapular n.

Subscapularis (SUB-SCAP-you-LERR-iss)

Medially rotates humerus

O	I	N
O: subscapular fossa of scapula	I: lesser tubercle of humerus	N: subscapular n.

Teres Minor

Adducts and laterally rotates humerus

O	I	N
O: lateral border of scapula	I: greater tubercle of humerus	N: axillary n.

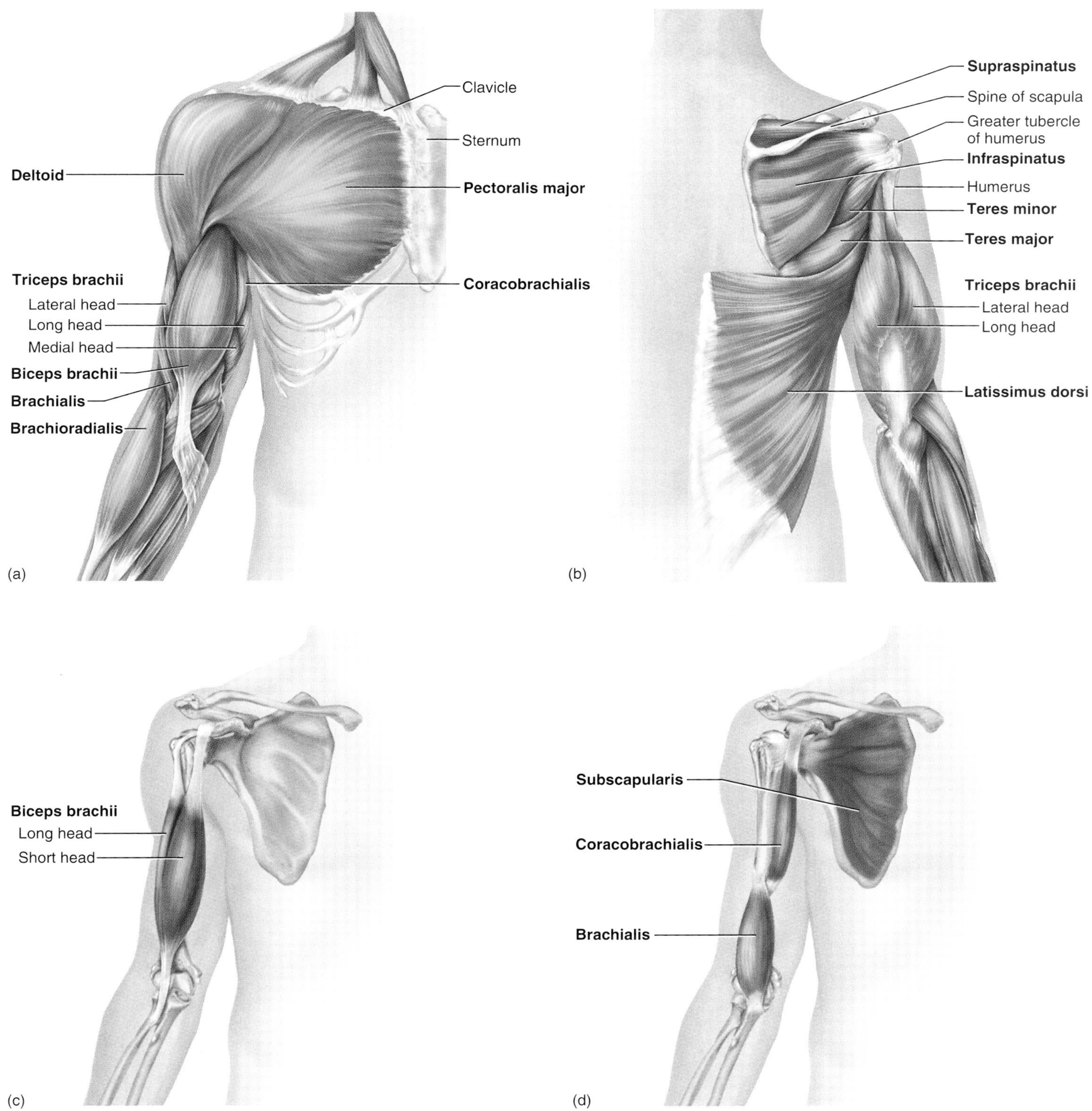

Figure 11.22 Pectoral and Brachial Muscles. (*a*) Anterior view. (*b*) Posterior view. (*c*) The biceps brachii, the superficial flexor of the elbow. (*d*) The brachialis, the deep flexor of the elbow, and the coracobrachialis and subscapularis, which act on the humerus. Figures *c* and *d* are anterior views.

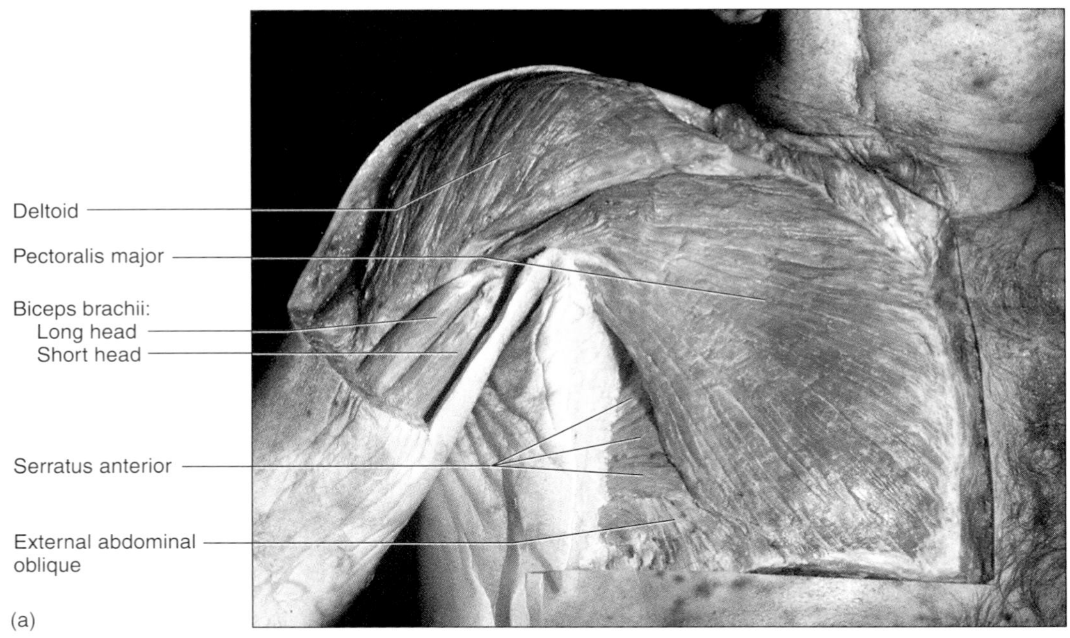

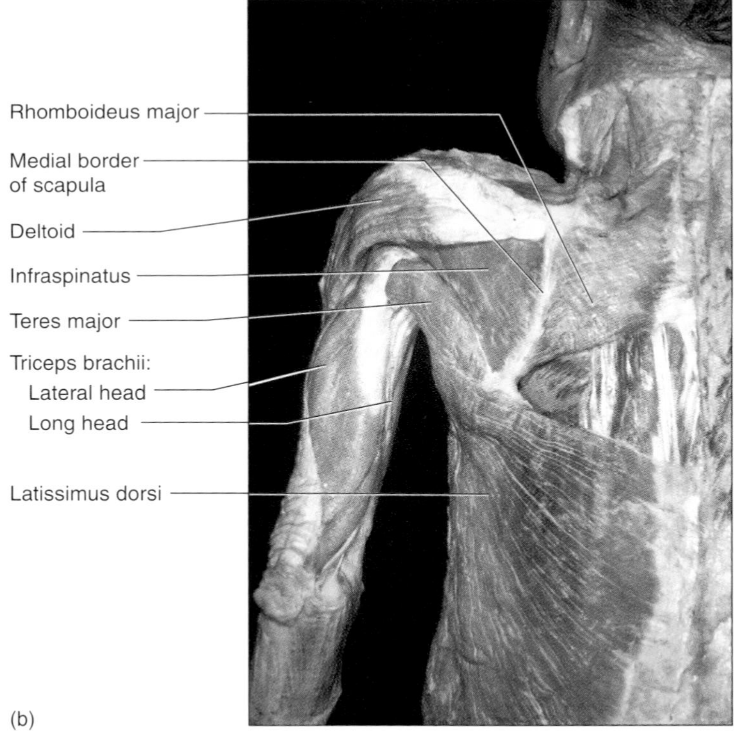

Figure 11.23 Muscles of the Chest and Brachial Region of the Cadaver. (*a*) Anterior view; (*b*) posterior view.

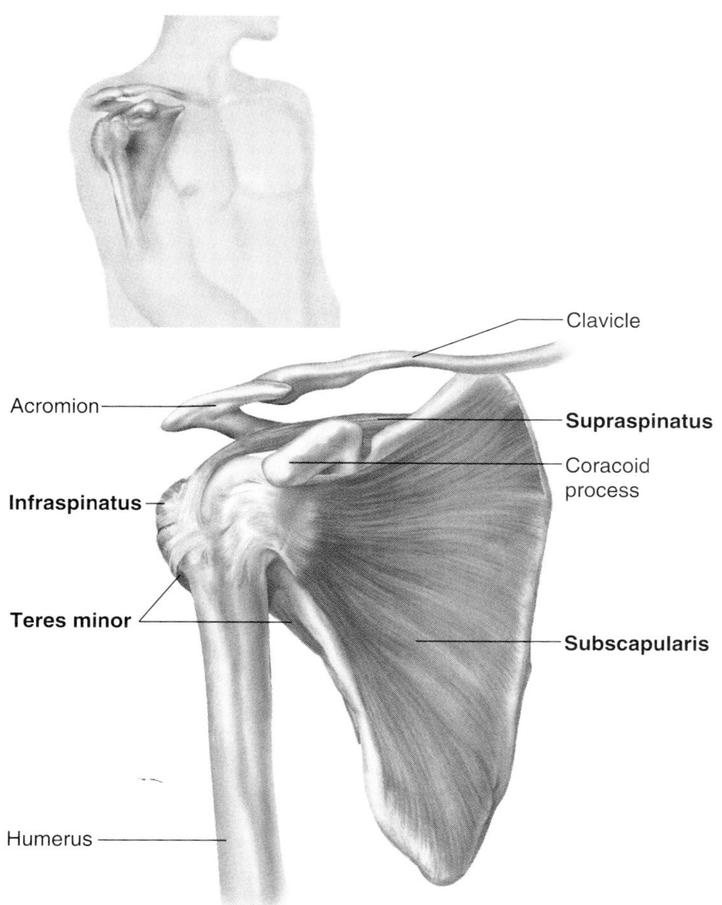

Figure 11.24 **The Rotator Cuff.** Anterolateral view of the right shoulder. The rotator cuff muscles are labeled in boldface.

extends it as in pointing at something behind you—thus, they are antagonists.

The other seven muscles of the shoulder are called *scapular muscles* because they originate on the scapula. Among these, the prime mover is the **deltoid**—the thick muscle that caps the shoulder. It acts like three different muscles. Its anterior fibers flex the shoulder, its posterior fibers extend it, and its lateral fibers abduct it. Abduction by the deltoid is antagonized by the combined action of the pectoralis major and latissimus dorsi. The **teres major** assists extension of the shoulder and the **coracobrachialis** assists flexion and adduction.

Tendons of the other four scapular muscles form the **rotator cuff**—the **infraspinatus, supraspinatus, subscapularis,** and **teres minor** (fig. 11.24). The subscapularis fills most of the subscapular fossa on the anterior surface of the scapula. The other three originate on the posterior surface. The supraspinatus and infraspinatus occupy the corresponding fossae above and below the scapular spine, and the teres minor lies inferior to the infraspinatus. The tendons of these muscles merge with the joint capsule of the shoulder as they pass it en route to the humerus. They insert on the proximal end of the humerus, forming a partial sleeve around it. The rotator cuff reinforces the joint capsule and holds the head of the humerus in the glenoid cavity. These muscles act as synergists in shoulder movements. The rotator cuff, especially the tendon of the supraspinatus, is easily damaged by strenuous circumduction.

Since the humeroscapular joint is capable of such a wide range of movements and is acted upon by so many muscles, its actions are summarized in table 11.11.

Table 11.11 Actions of the Shoulder (Humeroscapular) Joint

Boldface indicates prime movers; others are synergists. Parentheses indicate only a slight effect.

Flexion	Extension
Anterior deltoid **Pectoralis major** Coracobrachialis Biceps brachii	**Posterior deltoid** **Latissimus dorsi** Teres major
Abduction	**Adduction**
Lateral deltoid Supraspinatus	**Pectoralis major** **Latissimus dorsi** Coracobrachialis Triceps brachii Teres major (Teres minor)
Medial Rotation	**Lateral Rotation**
Subscapularis Teres major Latissimus dorsi Deltoid Pectoralis major	**Infraspinatus** **Teres minor** Deltoid

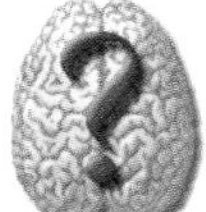

Think About It

Since a muscle can only pull, and not push, antagonistic muscles are needed to produce opposite actions at a joint. Reconcile this fact with the observation that the deltoid muscle both flexes and extends the shoulder.

Muscles Acting on the Forearm

The elbow and forearm are capable of four motions: flexion, extension, pronation, and supination (table 11.12). The principal flexors are on the anterior side of the humerus and include the superficial **biceps brachii**[45] and deeper **brachialis** (see fig. 11.22; table 11.13). In flexion of the elbow, the biceps elevates the radius while the brachialis elevates the ulna. The biceps is named for its two heads, which arise from separate tendons at the scapula. The tendon of the long head is important in holding the humerus in the glenoid cavity and stabilizing the shoulder joint. The two heads converge close to the elbow on a single distal tendon that inserts on the radial tuberosity.

The **brachioradialis** is a synergist in elbow flexion. Its belly lies in the antebrachium (forearm) beside the radius, rather than in the brachium with the other two flexors (see fig. 11.22*a*). It forms the thick, fleshy mass on the lateral side of the forearm just distal to the elbow. Its origin is on the distal end of the humerus, and its insertion is on the distal end of the radius. Since its insertion is so far from the fulcrum, the brachioradialis does not generate as much power as the prime movers; it is effective mainly when the prime movers have partially flexed the elbow.

The prime mover of extension is the **triceps brachii** on the posterior side of the humerus (see figs. 11.2 and

45. *bi* = two + *ceps* = head + *brachi* = arm. Note that *biceps* is singular, there is no such word as *bicep*. The plural form is *bicipites* (by-SIP-ih-teez).

Table 11.12 Actions of the Forearm

Boldface indicates prime movers; others are synergists. Parentheses indicate only a slight effect.

Flexion	Extension
Biceps brachii	**Triceps brachii**
Brachialis	Anconeus
Brachioradialis	
Flexor carpi radialis	
(Pronator teres)	
Pronation	**Supination**
Pronator teres	**Supinator**
Pronator quadratus	Biceps brachii

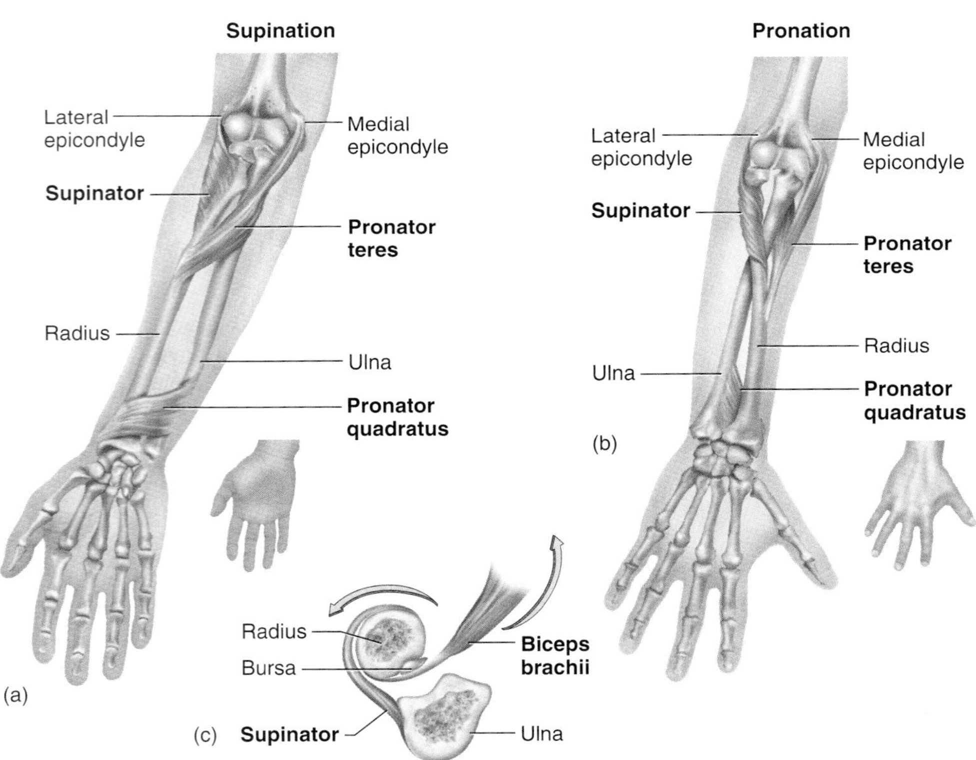

Figure 11.25 Actions of the Rotator Muscles on the Forearm. (*a*) Supination; (*b*) pronation; (*c*) cross section just distal to the elbow, showing how the biceps brachii aids in supination.

Table 11.13 Muscles Acting on the Forearm (see figs. 11.22 and 11.25)

O = origin, I = insertion, N = innervation (n. = nerve, nn. = nerves)

Muscle / Action / Origin	Insertion	Innervation
Muscles with Bellies in the Arm (Brachium)		
Biceps Brachii (BY-seps BRAY-kee-eye)		
Flexes elbow; abducts arm; supinates forearm; holds head of humerus in glenoid cavity		
O: *long head*—supraglenoid tubercle of scapula; *short head*—coracoid process of scapula	I: radial tuberosity	N: musculocutaneous n.
Brachialis (BRAY-kee-AL-iss)		
Flexes elbow		
O: anterior distal shaft of humerus	I: coronoid process of ulna, capsule of elbow joint	N: musculocutaneous n., radial n.
Triceps Brachii (TRI-seps BRAY-kee-eye)		
Extends elbow; long head adducts humerus		
O: *long head*—infraglenoid tubercle of scapula; *lateral head*—proximal posterior shaft of humerus; *medial head*—posterior shaft of humerus	I: olecranon of ulna	N: radial n.
Muscles with Bellies in the Forearm (Antebrachium)		
Brachioradialis (BRAY-kee-oh-RAY-dee-AL-iss)		
Flexes elbow		
O: lateral supracondylar ridge of humerus	I: styloid process of radius	N: radial n.
Anconeus (an-CO-nee-us)		
Extends elbow		
O: lateral epicondyle of humerus	I: olecranon and posterior aspect of ulna	N: radial n.
Pronator Teres (PRO-nay-tur TERR-eez)		
Pronates forearm		
O: medial epicondyle of humerus, coronoid process of ulna	I: lateral midshaft of radius	N: median n.
Pronator Quadratus (PRO-nay-tur quad-RAY-tus)		
Pronates forearm		
O: anterior distal shaft of ulna	I: anterior distal shaft of radius	N: median n.
Supinator (SOO-pih-NAY-tur)		
Supinates forearm		
O: lateral epicondyle of humerus, proximal shaft of ulna	I: proximal shaft of radius	N: radial n.

11.22). The **anconeus**[46] is a weaker synergist of extension that crosses the posterior side of the elbow (see fig. 11.28*d, e*).

Pronation is achieved by two anterior muscles in the forearm—the **pronator teres** near the elbow and **pronator quadratus** near the wrist. Supination is achieved by the biceps brachii and the **supinator** of the posterior forearm (fig. 11.25).

46. *ancon* = elbow

Muscles Acting on the Wrist and Hand

The hand is acted upon by extrinsic muscles in the forearm and intrinsic muscles in the hand itself (table 11.14). The bellies of the extrinsic muscles form the fleshy roundness of the proximal forearm; their tendons extend into the wrist and hand. Their actions are mainly flexion and extension, but the wrist and fingers can be abducted and adducted, and the thumb and fingers can be opposed.

It may seem as if the tendons would stand up like taut bowstrings when these muscles contracted, but this is prevented by the fact that most of them pass under a **flexor retinaculum (transverse carpal ligament)** on the anterior side of the wrist and an **extensor retinaculum (dorsal carpal ligament)** on the posterior side (see fig. 11.29). The **carpal tunnel** is a tight space between the carpal bones and flexor retinaculum (fig. 11.26). The flexor tendons passing through the tunnel are enclosed in tendon sheaths

Table 11.14 Muscles Acting on the Wrist and Hand (see figs. 11.27 and 11.28)

O = origin, I = insertion, N = innervation (n. = nerve, nn. = nerves)

Anterior Compartment—Superficial Layer		
Flexor Carpi Radialis (CAR-pie RAY-dee-AY-liss) Powerful wrist flexor; abducts hand; synergist in elbow flexion		
O: medial epicondyle of humerus	I: base of metacarpals II and III	N: median n.
Flexor Carpi Ulnaris (ul-NAY-riss) Flexes and adducts wrist; stabilizes wrist during extension of fingers		
O: medial epicondyle of humerus	I: pisiform, hamate, metacarpal V	N: ulnar n.
Flexor Digitorum Superficialis (DIDJ-ih-TOE-rum SOO-per-FISH-ee-AY-liss) Flexes fingers II–V at proximal interphalangeal joints; aids in flexion of wrist and metacarpophalangeal joints		
O: medial epicondyle of humerus, radius, coronoid process of ulna	I: four tendons leading to middle phalanges II–V	N: median n.
Palmaris (pall-MERR-iss) Longus Weakly flexes wrist; often absent		
O: medial epicondyle of humerus	I: palmar aponeurosis, flexor retinaculum	N: median n.
Anterior Compartment—Deep Layer		
Flexor Digitorum Profundus Flexes wrist and distal interphalangeal joints		
O: shaft of ulna, interosseous membrane	I: four tendons to distal phalanges II–V	N: median and ulnar nn.
Flexor Pollicis (PAHL-ih-sis) Longus Flexes interphalangeal joint of thumb; weakly flexes wrist		
O: radius, interosseous membrane	I: distal phalanx I	N: median n.
Posterior Compartment—Superficial Layer		
Extensor Carpi Radialis Longus Extends and abducts wrist		
O: lateral epicondyle of humerus	I: base of metacarpal II	N: radial n.
Extensor Carpi Radialis Brevis Extends and abducts wrist; fixes wrist during finger flexion		
O: lateral epicondyle of humerus	I: base of metacarpal III	N: radial n.

(continued)

Table 11.14 Muscles Acting on the Wrist and Hand (continued)		
Posterior Compartment—Superficial Layer (cont.)		
Extensor Carpi Ulnaris Extends and adducts wrist		
O: lateral epicondyle of humerus, posterior shaft of ulna	I: base of metacarpal V	N: radial n.
Extensor Digitorum (DIDJ-ih-TOE-rum) Extends fingers II–V at metacarpophalangeal joints		
O: lateral epicondyle of humerus	I: dorsal aspect of phalanges II–V	N: radial n.
Extensor Digiti Minimi (DIDJ-ih-ty MIN-in-my) Extends metacarpophalangeal joint of little finger; sometimes considered to be a detached portion of extensor digitorum		
O: lateral epicondyle of humerus	I: distal and middle phalanges V	N: radial n.
Posterior Compartment—Deep Layer		
Abductor Pollicis Longus Abducts and extends thumb; abducts wrist		
O: posterior aspect of radius and ulna, interosseous membrane	I: trapezium, base of metacarpal I	N: radial n.
Extensor Indicis (IN-dih-sis) Extends index finger at metacarpophalangeal joint		
O: shaft of ulna, interosseous membrane	I: middle and distal phalanges II	N: radial n.
Extensor Pollicis Longus Extends thumb at metacarpophalangeal joint		
O: shaft of ulna, interosseous membrane	I: distal phalanx I	N: radial n.
Extensor Pollicis Brevis Extends thumb at metacarpophalangeal joint		
O: shaft of radius, interosseous membrane	I: proximal phalanx I	N: radial n.

that enable them to slide back and forth quite easily, although this region is very subject to injury from repetitive motion (see clinical insight 11.4).

Several of these muscles originate on the humerus; therefore, they cross the elbow joint and weakly contribute to flexion and extension of the elbow. This action is relatively negligible, however, and we focus on their action at the wrist and fingers. Although these muscles are numerous and complex, most of their names suggest their actions, and from their actions, their approximate locations in the forearm can generally be deduced.

The deep fascia divide the muscles of the forearm into anterior and posterior compartments and each compartment into superficial and deep layers (fig. 11.27). The muscles are listed and classified this way in table 11.14. Most muscles of the anterior compartment are flexors of the wrist and fingers that arise from a common tendon on the humerus (fig. 11.28). At the distal end, the tendon of the **palmaris longus** passes over the flexor retinaculum while the other tendons pass beneath it. The two prominent tendons that you can palpate at the wrist belong to the palmaris longus on the medial side and the **flexor carpi radialis** on the lateral side. The latter is an important landmark for finding the radial artery, where the pulse is usually taken.

Muscles of the posterior compartment are mostly wrist and finger extensors that share a single proximal tendon arising from the humerus. One of the superficial muscles on this side, the **extensor digitorum,** has four distal tendons that can easily be seen and palpated on the back of the hand when the fingers are strongly hyperextended (fig. 11.28*d*, and see fig. B.7 in the atlas following this

Clinical Insight 11.4

Carpal Tunnel Syndrome

Prolonged, repetitive motions of the wrist and fingers can cause tissues in the carpal tunnel to become inflamed, swollen, or fibrotic. Since the carpal tunnel cannot expand, swelling puts pressure on the median nerve of the wrist, which passes through the carpal tunnel with the flexor tendons. This pressure causes tingling and muscular weakness in the palm and medial side of the hand and pain that may radiate to the arm and shoulder. This condition, called *carpal tunnel syndrome,* is common among keyboard operators, pianists, meat cutters, and others who spend long hours making repetitive wrist motions. Carpal tunnel syndrome is treated with aspirin and other anti-inflammatory drugs, immobilization of the wrist, and sometimes surgical removal of part or all of the flexor retinaculum to relieve pressure on the nerve.

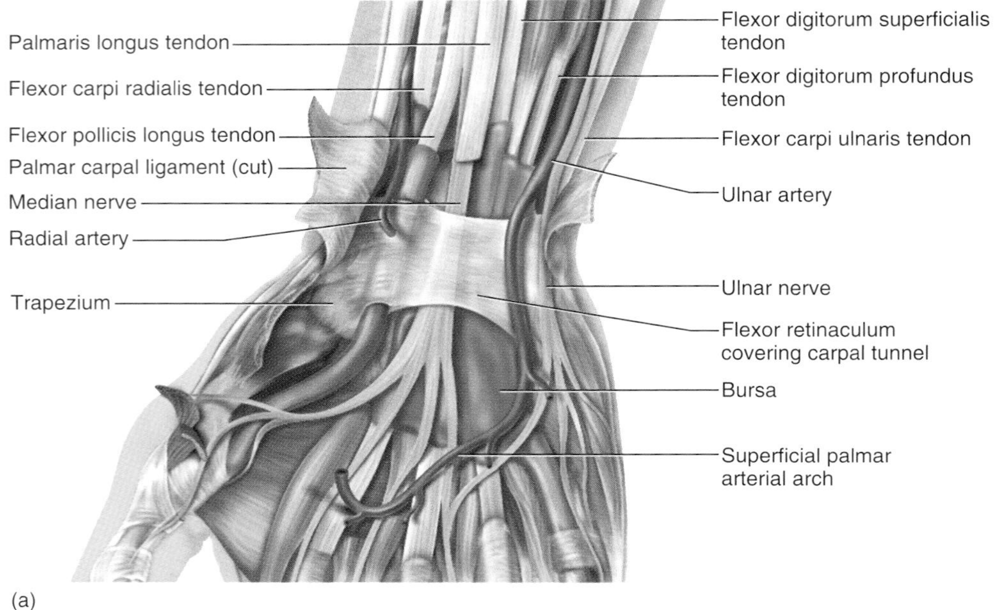

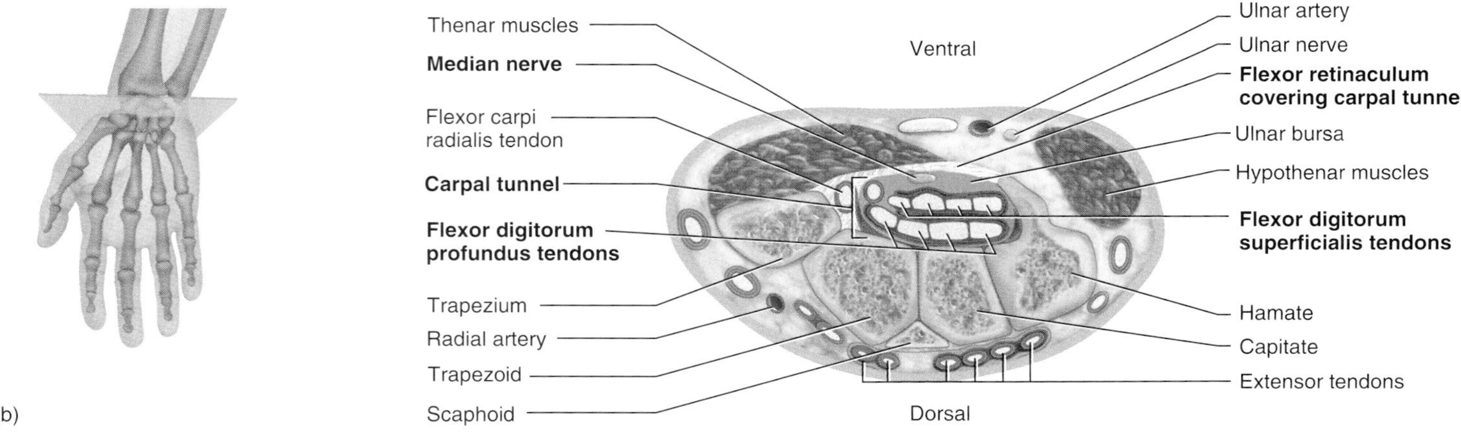

Figure 11.26 The Carpal Tunnel. (*a*) Dissection of the wrist (anterior aspect) showing the tendons, nerve, and bursae that pass under the flexor retinaculum. (*b*) Cross section of the wrist, ventral (anterior side) up. Note how the flexor tendons and median nerve are confined in the tight space between the carpal bones and flexor retinaculum.

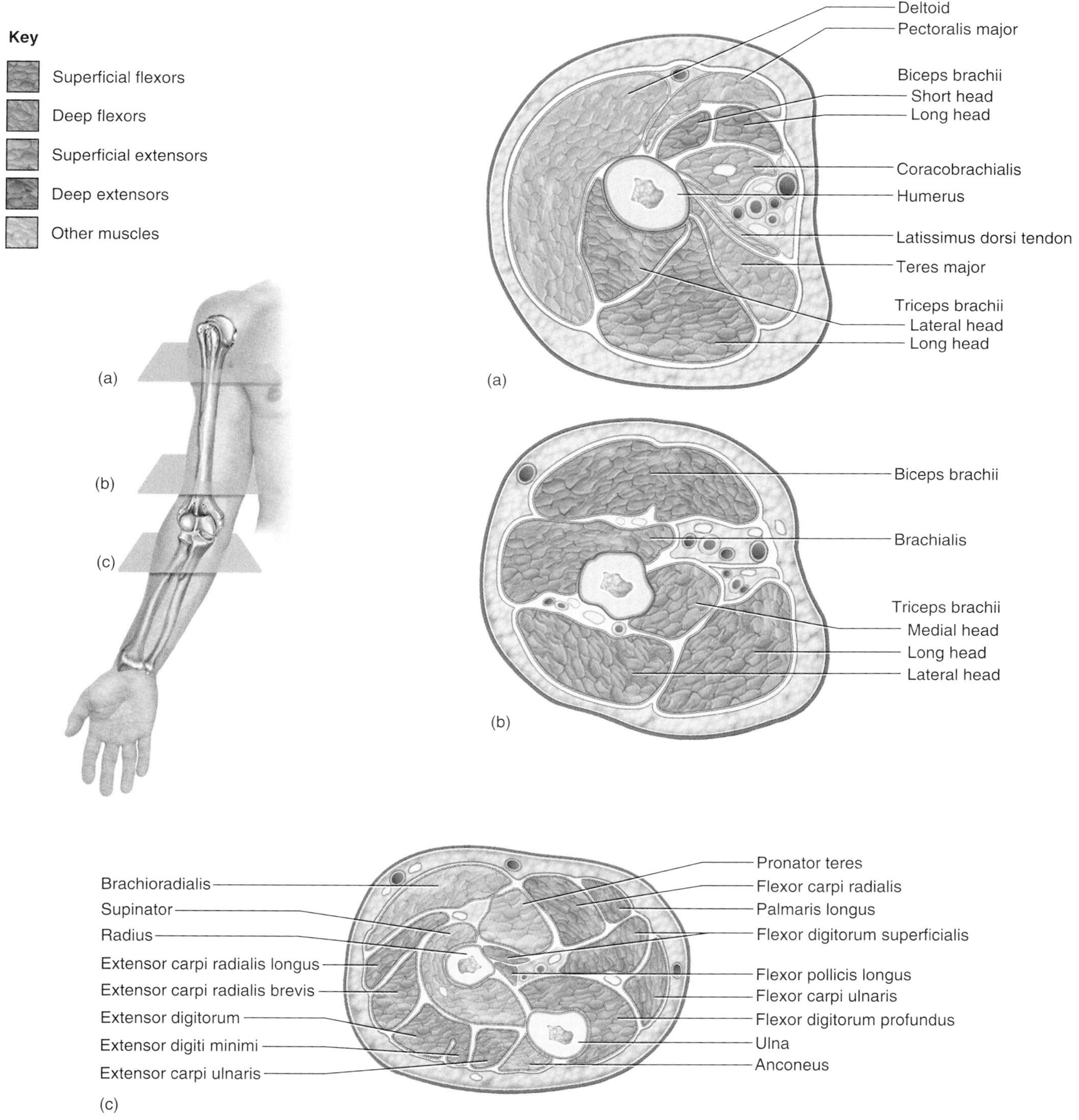

Figure 11.27 Serial Cross Sections Through the Upper Limb. Each section is taken at the correspondingly lettered level in the figure at the left and is pictured with the posterior muscle compartment facing the bottom of the page, as if you were viewing the right arm of a person facing you with the arm extended and the palm up.

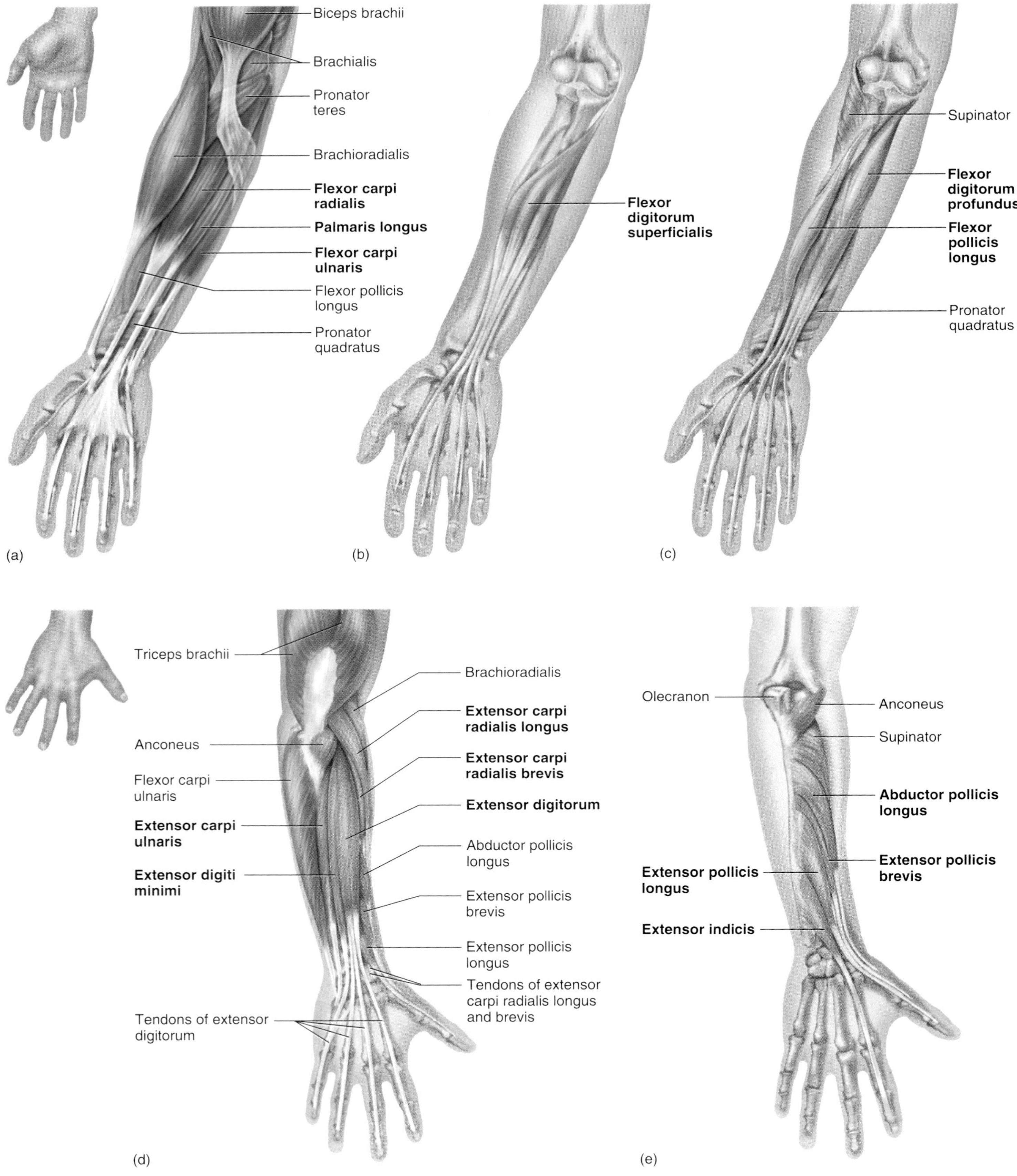

Figure 11.28 Muscles of the Forearm. Figures *a–c* are anterior views and figures *d–e* are posterior. Boldface labels indicate: (*a*) superficial flexors; (*b*) the flexor digitorum superficialis, deep to the muscles in *a* but also classified as a superficial flexor; (*c*) deep flexors; (*d*) superficial extensors; and (*e*) deep extensors.

chapter). By strongly abducting and extending the thumb into a hitchhiker's position, you should also be able to see a deep dorsolateral pit at the base of the thumb, with a taut tendon on each side of it. This depression is called the *anatomical snuffbox* because it was once fashionable to place a pinch of snuff here and inhale it (see fig. B.7). It is bordered laterally by the tendons of the **abductor pollicis longus** and **extensor pollicis brevis** and medially by the tendon of the **extensor pollicis longus.**

Other muscles of the forearm were considered earlier because they act on the radius and ulna rather than on the hand. These are the pronator quadratus, pronator teres, supinator, anconeus, and brachioradialis.

Table 11.15 summarizes the muscles responsible for the major movements of the wrist and hand.

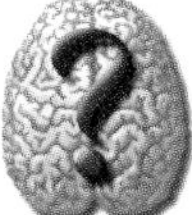

Think About It

Why are the prime movers of finger extension and flexion located in the forearm rather than in the hand, closer to the fingers?

The intrinsic muscles of the hand assist the flexors and extensors of the forearm and make finger movements more precise (fig. 11.29). You will note in table 11.16 that they are divided into three groups. The **thenar group** forms the thick fleshy mass *(thenar eminence)* at the base of the thumb, except for the *adductor pollicis,* which forms the web between the thumb and palm; the **hypothenar group** forms the fleshy mass *(hypothenar eminence)* at the base of the little finger; and the **midpalmar group** occupies the space between these. The midpalmar group consists of 11 muscles divided into three subgroups:

1. **Dorsal interosseous[47] muscles**—four bipennate muscles attached to both sides of the metacarpal bones, serving to abduct (spread) the fingers.
2. **Palmar interosseous muscles**—three unipennate muscles that arise from metacarpals II, IV, and V and adduct the fingers (draw them together).
3. **Lumbrical[48] muscles**—four wormlike muscles that flex the metacarpophalangeal joints (proximal knuckles) but extend the interphalangeal joints (distal knuckles).

47. *inter* = between + *osse* = bone
48. *lumbric* = earthworm

Table 11.15 Actions of the Wrist and Hand

Boldface indicates prime movers; others are synergists. Parentheses indicate only a slight effect.

Wrist Flexion	Wrist Extension
Flexor carpi radialis	**Extensor digitorum**
Flexor carpi ulnaris	Extensor carpi radialis longus
Flexor digitorum superficialis	Extensor carpi radialis brevis
(Palmaris longus)	Extensor carpi ulnaris
(Flexor pollicis longus)	

Wrist Abduction	Wrist Adduction
Flexor carpi radialis	Flexor carpi ulnaris
Extensor carpi radialis longus	Extensor carpi ulnaris
Extensor carpi radialis brevis	
Abductor pollicis longus	

Finger Flexion	Finger Extension	Thumb Opposition
Flexor digitorum superficialis	Extensor pollicis longus	Opponens pollicis
Flexor digitorum profundus	Extensor pollicis brevis	Opponens digiti minimi
Flexor pollicis longus	Extensor digitorum	
	Extensor indicis	

Before You Go On

Answer the following questions to test your understanding of the preceding section:

17. Name a muscle that inserts on the scapula and plays a significant role in each of the following actions: (a) pushing a stalled car, (b) paddling a canoe, (c) squaring the shoulders in military attention, (d) lifting the shoulder to carry a heavy box on it, and (e) lowering the shoulder to lift a suitcase.
18. Describe three contrasting actions of the deltoid muscle.
19. Name the four rotator cuff muscles and identify the scapular surfaces against which they lie.
20. Name the prime movers of elbow flexion and extension.
21. Identify three functions of the biceps brachii.
22. Name three extrinsic muscles and two intrinsic muscles that flex the phalanges.

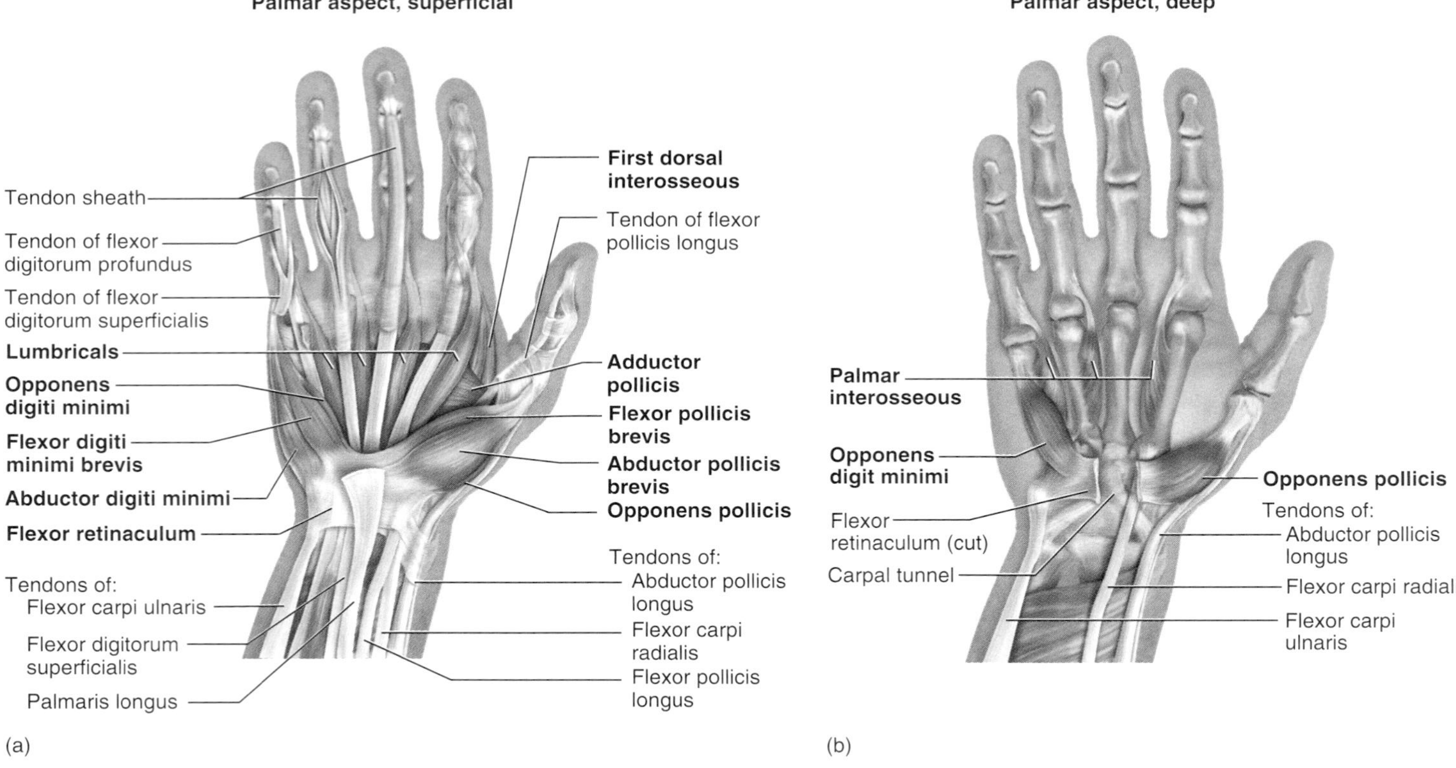

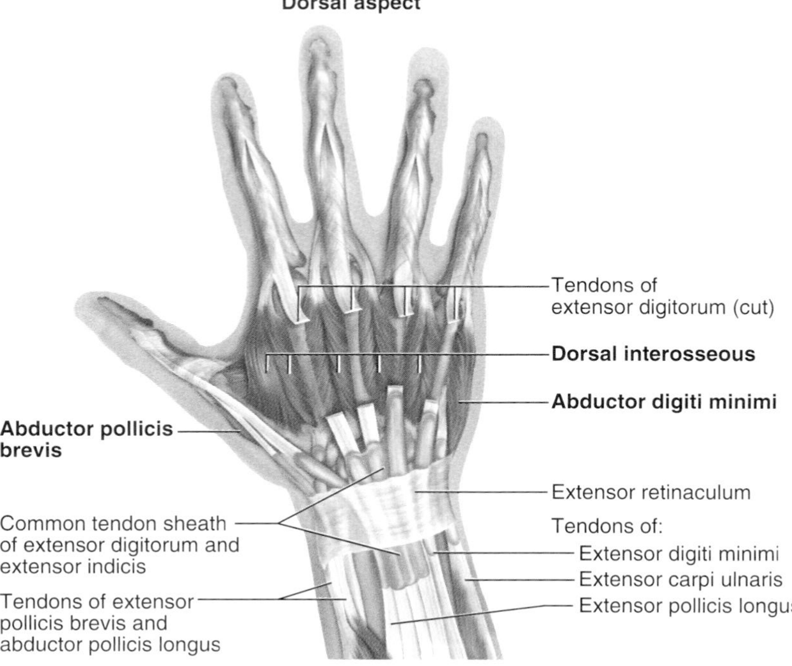

Figure 11.29 **Intrinsic Muscles of the Hand.** Boldface labels indicate: (*a*) superficial muscles, anterior (palmar) view; (*b*) deep muscles, anterior view; (*c*) superficial muscles, posterior (dorsal) view. Figures are oriented to correspond to your own right hand held in front of you palm up (*a*–*b*) and palm down (*c*).

Table 11.16 Intrinsic Muscles of the Hand (see fig. 11.29)

O = origin, I = insertion, N = innervation (n. = nerve, nn. = nerves)

Muscle / Action / Origin	Insertion	Innervation
Thenar Group		
Abductor Pollicis (PAHL-ih-sis) Brevis Abducts thumb O: scaphoid, trapezium, flexor retinaculum	I: lateral aspect of proximal phalanx I	N: median n.
Adductor Pollicis Adducts thumb and opposes it to the fingers O: trapezium, trapezoid, capitate, metacarpals II–IV	I: medial aspect of proximal phalanx I	N: ulnar n.
Flexor Pollicis Brevis Flexes thumb at metacarpophalangeal joint O: trapezium, flexor retinaculum	I: proximal phalanx I	N: median and ulnar nn.
Opponens (op-OH-nens) Pollicis Opposes thumb to fingers O: trapezium, flexor retinaculum	I: metacarpal I	N: median n.
Hypothenar Group		
Abductor Digiti Minimi Abducts little finger O: pisiform, tendon of flexor carpi ulnaris	I: medial aspect of proximal phalanx V	N: ulnar n.
Flexor Digiti Minimi Brevis Flexes little finger at metacarpophalangeal joint O: hamulus of hamate, flexor retinaculum	I: medial aspect of proximal phalanx V	N: ulnar n.
Opponens Digiti Minimi Opposes little finger to thumb; deepens pit of palm O: hamulus of hamate, flexor retinaculum	I: medial aspect of metacarpal V	N: ulnar n.
Midpalmar Group		
Dorsal Interosseous (IN-tur-OSS-ee-us) Muscles (four muscles) Abduct digits II–IV O: two heads on facing sides of adjacent metacarpals	I: proximal phalanges II–IV	N: ulnar n.
Palmar Interosseous Muscles (three muscles) Adduct digits II, IV, and V O: metacarpals II, IV, and V	I: proximal phalanges II, IV, and V	N: ulnar n.
Lumbricals (four muscles) Flex metacarpophalangeal joints; extend interphalangeal joints O: tendons of flexor digitorum profundus	I: proximal phalanges II–V	N: median and ulnar nn.

Clinical Insight 11.5

Intramuscular Injections

Muscles with thick bellies are commonly used for intramuscular (I.M.) drug injections. Since drugs injected into these muscles are absorbed into the bloodstream gradually, it is safe to administer relatively large doses (up to 5 mL) that could be dangerous or even fatal if injected directly into the bloodstream. I.M. injections also cause less tissue irritation than subcutaneous injections.

Knowledge of subsurface anatomy is necessary to avoid damaging nerves or accidentally injecting a drug into a blood vessel. Anatomical knowledge also enables a clinician to position a patient so that the muscle is relaxed, making the injection less painful.

Amounts up to 2 mL are commonly injected into the deltoid muscle about two finger widths below the acromion. Amounts over 2 mL are commonly injected into the gluteus medius, in the superolateral quadrant of the gluteal area, at a safe distance from the sciatic nerve and major gluteal blood vessels. Injections are often given to infants and young children in the vastus lateralis of the thigh, because their deltoid and gluteal muscles are not well developed.

Muscles Acting on the Hip and Lower Limb

Objectives

When you have completed this section, you should be able to

- Name and locate the muscles that act on the hip, knee, ankle, and toe joints;
- Relate the actions of these muscles to the joint movements described in chapter 10; and
- Describe the origin, insertion, and innervation of each muscle.

The largest and strongest muscles in the body are found in the lower limb. Unlike those of the upper limb, they are adapted less for precision than for the strength needed to stand, maintain balance, walk, and run. Several of them cross and act upon two or more joints, such as the hip and knee.

To avoid confusion in this discussion, remember that in the anatomical sense the word *leg* refers only to that part of the limb between the knee and ankle. The term *foot* includes the tarsal region (ankle), metatarsal region, and toes.

Muscles Acting on the Hip and Femur

Most muscles that act on the femur (table 11.17) originate on the os coxae. The two principal anterior muscles are the **iliacus,** which fills most of the broad iliac fossa of the pelvis, and the **psoas major,** a thick, rounded muscle that originates mainly on the lumbar vertebrae. Collectively, they are called the **iliopsoas** (ILL-ee-oh-SO-us) (fig. 11.30). They converge on a single tendon that inserts on the femur and flexes the hip joint—for example, when you bend forward at the waist, swing the leg forward in walking, or raise the thigh in a marching stance.

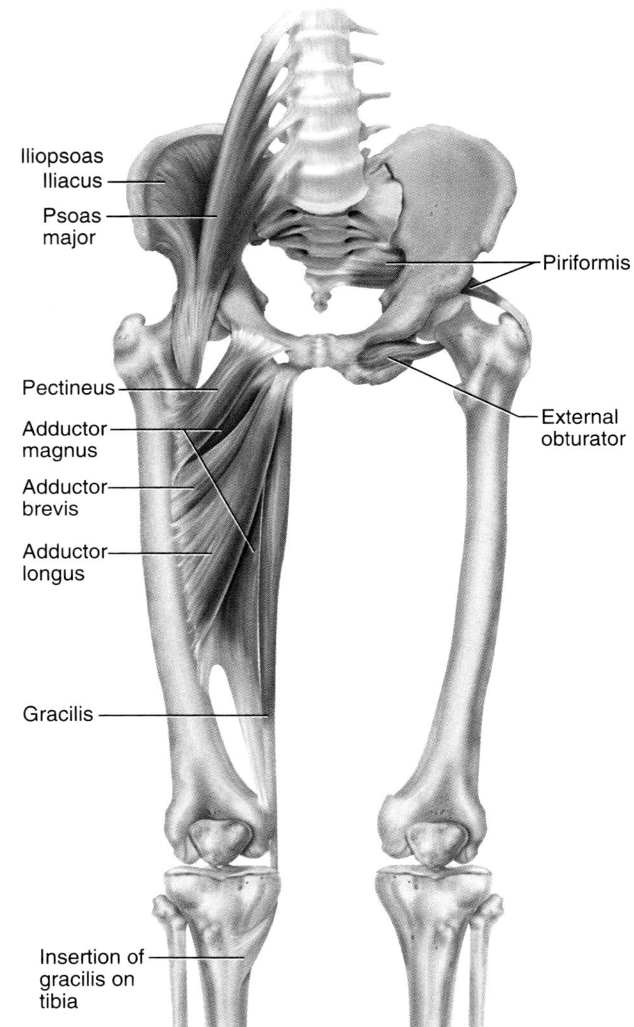

Figure 11.30 Muscles Acting on the Hip and Femur, Anterior View.

Table 11.17 Muscles Acting on the Hip and Femur (see figs. 11.30–11.34)

O = origin, I = insertion, N = innervation (n. = nerve, nn. = nerves)

Muscle / Action	Origin	Insertion	Innervation
Anterior Muscles of the Hip (Iliopsoas)			
Iliacus (ih-LY-uh-cus) Flexes hip joint; medially rotates femur	O: iliac fossa	I: lesser trochanter of femur, capsule of coxal joint	N: femoral n.
Psoas (SO-ass) Major Flexes hip joint; medially rotates femur	O: vertebral bodies T12–L5	I: lesser trochanter of femur	N: lumbar plexus
Lateral and Posterior Muscles of the Hip			
Tensor Fasciae Latae (TEN-sor FASH-ee-ee LAY-tee) Flexes hip joint; abducts and medially rotates femur, tenses fascia lata and braces knee when opposite foot is lifted from ground	O: iliac crest near anterior superior spine	I: lateral condyle of tibia	N: superior gluteal n.
Gluteus Maximus Extends hip joint; abducts and laterally rotates femur; important in the backswing of the stride	O: ilium and sacrum	I: gluteal tuberosity of femur, fascia lata	N: inferior gluteal n.
Gluteus Medius, Gluteus Minimus Abduct and medially rotate femur; maintain balance by shifting body weight during walking	O: ilium	I: greater trochanter of femur	N: superior gluteal n.
Lateral Rotators			
Gemellus (jeh-MEL-us) Superior, Gemellus Inferior Laterally rotate femur	O: body of ischium	I: obturator internus tendon	N: sacral plexus
Obturator (OB-too-RAY-tur) Externus Laterally rotates femur	O: anterior margin of obturator foramen	I: greater trochanter of femur	N: obturator n.
Obturator Internus Abducts and laterally rotates femur	O: posterior margin of obturator foramen	I: greater trochanter of femur	N: sacral plexus
Piriformis (PIR-ih-FOR-miss) Abducts and laterally rotates femur	O: anterolateral aspect of sacroiliac region	I: greater trochanter of femur	N: ventral rami of S1–S2
Quadratus Femoris (quad-RAY-tus FEM-oh-riss) Adducts and laterally rotates femur	O: ischial tuberosity	I: intertrochanteric ridge of femur	N: sacral plexus

(continued)

Table 11.17 Muscles Acting on the Hip and Femur (continued)

Medial (Adductor) Compartment of the Thigh

Adductor Longus,
Adductor Brevis

Adduct and laterally rotate femur; flex hip joint

O: pubis	I: posterior shaft of femur	N: obturator n.

Adductor Magnus

Anterior part adducts and laterally rotates femur and flexes hip joint; posterior part extends hip joint

O: ischium	I: posterior shaft of femur	N: obturator and tibial nn.

Gracilis (GRASS-ih-lis)

Adducts femur; flexes knee; medially rotates tibia

O: pubis	I: medial aspect of proximal tibia	N: obturator n.

Pectineus (pec-TIN-ee-us)

Adducts and laterally rotates femur; flexes hip

O: pubis	I: posterior aspect of proximal femur	N: femoral n.

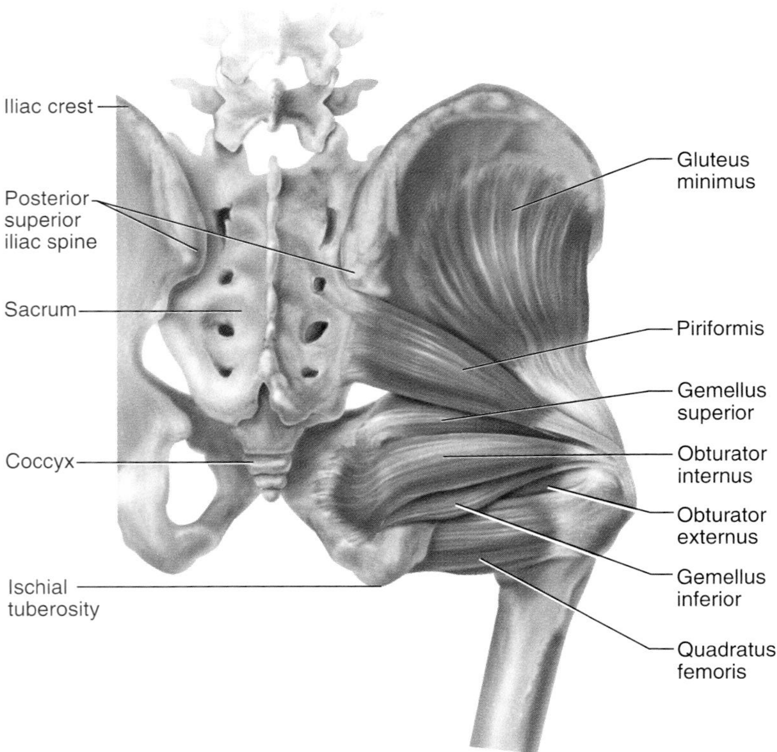

Figure 11.31 Deep Gluteal Muscles. For the superficial gluteal muscles, see figure 11.35.

On the lateral and posterior sides of the hip are the **tensor fasciae latae** and three gluteal muscles—the **gluteus maximus, gluteus medius,** and **gluteus minimus** (figs. 11.31 and 11.34). The gluteus maximus is the largest muscle of this group and forms most of the mass of the buttocks. It is an extensor of the hip joint that produces the backswing of the leg in walking and provides most of the lift when you climb stairs. It generates the most force when the thigh is flexed at a 45° angle to the trunk. This is the advantage in starting a foot race from a crouched position.

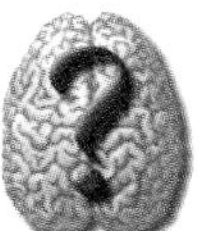

Think About It

What muscles produce the downstroke in pedaling a bicycle? In view of this, what is one reason that racing bicycles are designed to make the rider lean forward?

The deep **lateral rotators** of the pelvic region (table 11.17; fig. 11.31) rotate the femur laterally, as when you cross your legs to rest an ankle on your knee. Thus, they oppose medial rotation by the gluteus medius and minimus. Most of them also abduct or adduct the femur. The abductors are important in walking because, when we lift one foot from the ground, they shift the body weight to the other leg and prevent us from falling over.

The **fascia lata**[49] is a fibrous sheath that encircles the thigh and tightly binds its muscles. On the lateral surface, it combines with the tendons of the gluteus maximus and tensor fasciae latae to form the **iliotibial band,** which extends from the iliac crest to the lateral condyle of the tibia (fig. 11.32). The tensor fasciae latae tautens the iliotibial band and braces the knee, especially when we raise the opposite foot.

The fascia lata divides the thigh muscles into three compartments, each with its own nerve and blood supply: the *anterior (extensor) compartment, medial (adductor) compartment,* and *posterior (flexor) compartment.* Muscles of the anterior compartment function mainly as extensors of the knee, those of the medial compartment as adductors of the femur, and those of the posterior compartment as extensors of the hip and flexors of the knee.

In the medial compartment are five muscles that act on the hip joint—the **adductor longus, adductor brevis, adductor magnus, gracilis,**[50] and **pectineus**[51] (see fig. 11.30). All of them adduct the thigh, but some cross both the hip and knee joints and have additional actions noted in table 11.17.

Muscles Acting on the Knee

The following muscles form most of the mass of the thigh and produce their most obvious actions on the knee joint. Some of them, however, cross both the hip and knee joints and produce actions at both, moving the femur, tibia, and fibula (table 11.18).

The anterior compartment of the thigh contains the large **quadriceps femoris** muscle, the prime mover of knee extension and the most powerful muscle of the body (figs. 11.32 and 11.33). As the name implies, it has four heads—the **rectus femoris, vastus lateralis, vastus medialis,** and **vastus intermedius.** All four converge on a single **quadriceps (patellar) tendon,** which extends to the patella, then continues as the **patellar ligament** and inserts on the tibial tuberosity. (Remember that a tendon usually extends from muscle to bone and a ligament from bone to bone.) The patellar ligament is struck with a rubber mallet to test the knee-jerk reflex. The quadriceps extends the knee when you stand up, take a step, or kick a ball. It is very important in running because, together with the iliopsoas, it flexes the hip in each airborne phase of the leg's cycle of motion. The rectus femoris also flexes the hip in such actions as high kicks or simply in drawing the leg forward during a stride.

Crossing the quadriceps from the lateral side of the hip to the medial side of the knee is the narrow, straplike **sartorius,**[52] the longest muscle of the body. It flexes the hip and knee joints and laterally rotates the thigh, as in crossing the legs. It is colloquially called the "tailor's muscle" after the cross-legged stance of a tailor supporting his work on the raised knee.

The posterior compartment contains the **biceps femoris, semimembranosus,** and **semitendinosus** (fig. 11.34). These muscles are colloquially known as the "hamstrings" because their tendons at the knee of a hog are commonly used to hang a ham for curing. They flex the knee and, aided by the gluteus maximus, they extend the hip during walking and running. The pit at the rear of the knee, called the *popliteal fossa,* is bordered by the biceps tendon on the lateral side and the tendons of the semimembranosus and semitendinosus on the medial side. When wolves attack large prey, they often attempt to sever the hamstring tendons, because this renders the prey helpless. Hamstring injuries are common among sprinters, soccer players, and other athletes who rely on quick acceleration.

49. *fasc* = band + *lata* = broad
50. *gracil* = slender
51. *pectin* = comb

52. *sartor* = tailor

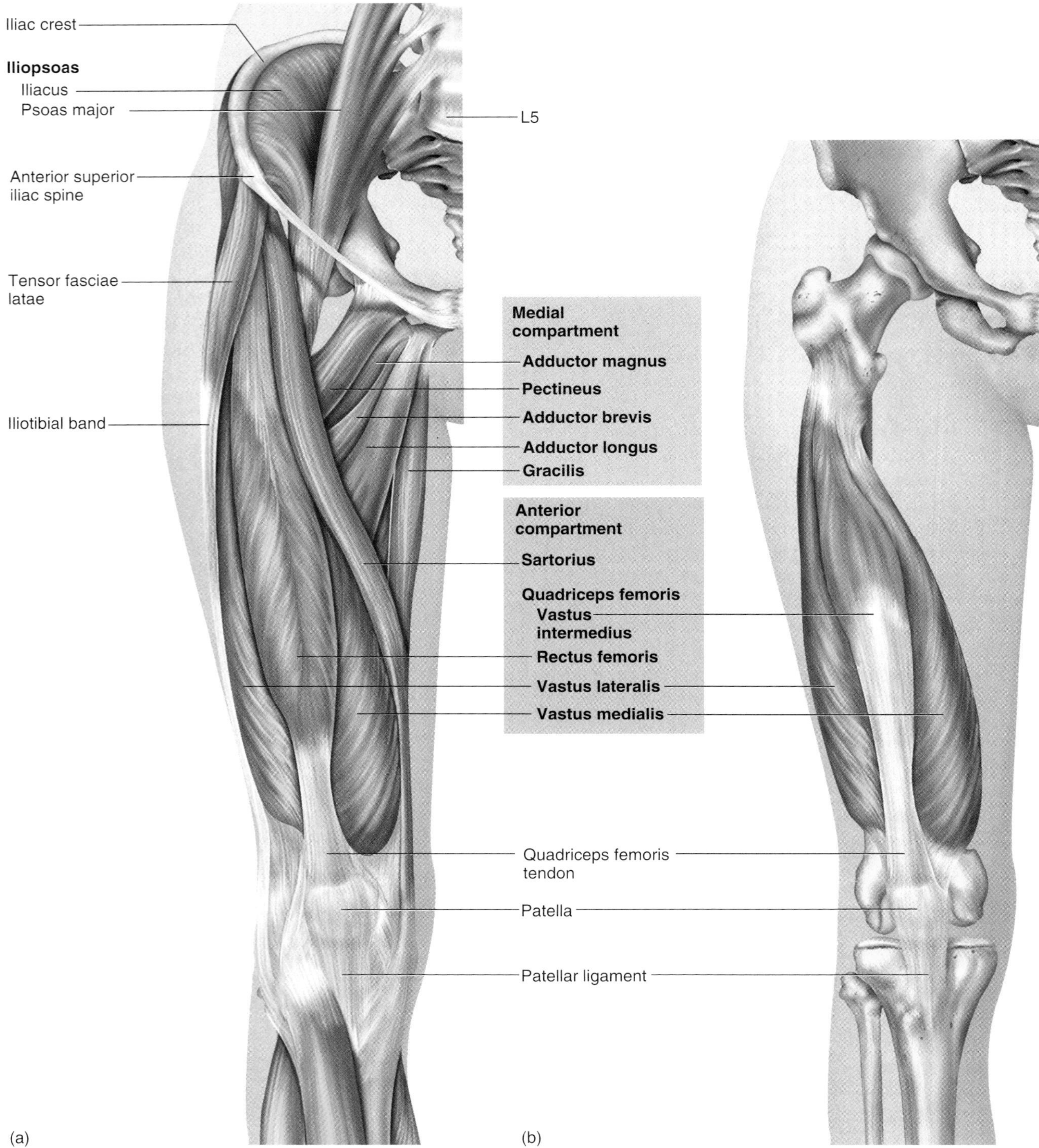

Figure 11.32 Muscles of the Thigh, Anterior View. (*a*) Superficial muscles; (*b*) rectus femoris and other muscles removed to expose the other three heads of the quadriceps femoris.

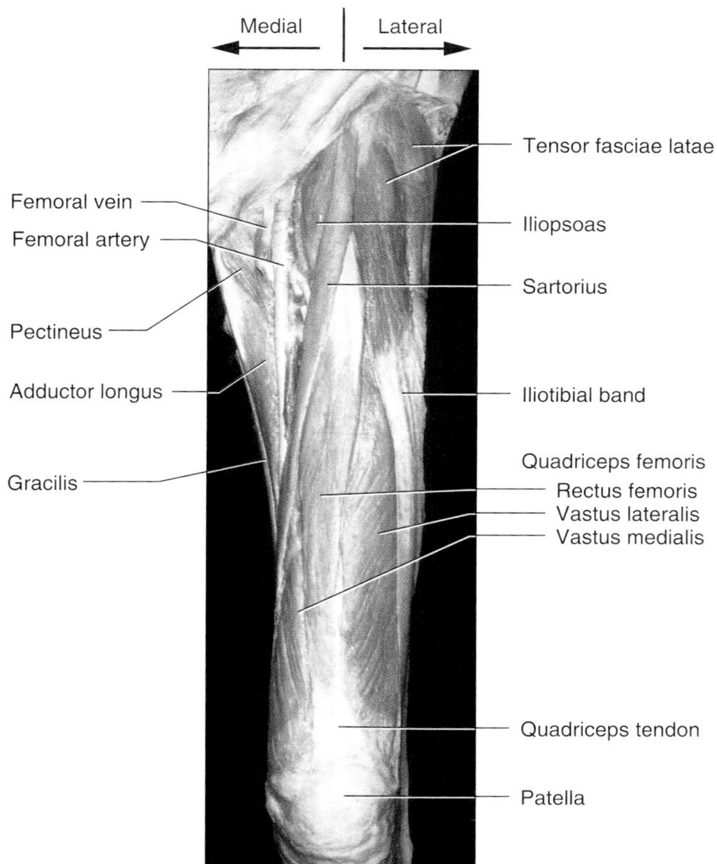

Figure 11.33 Superficial Anterior Muscles of the Thigh of the Cadaver.

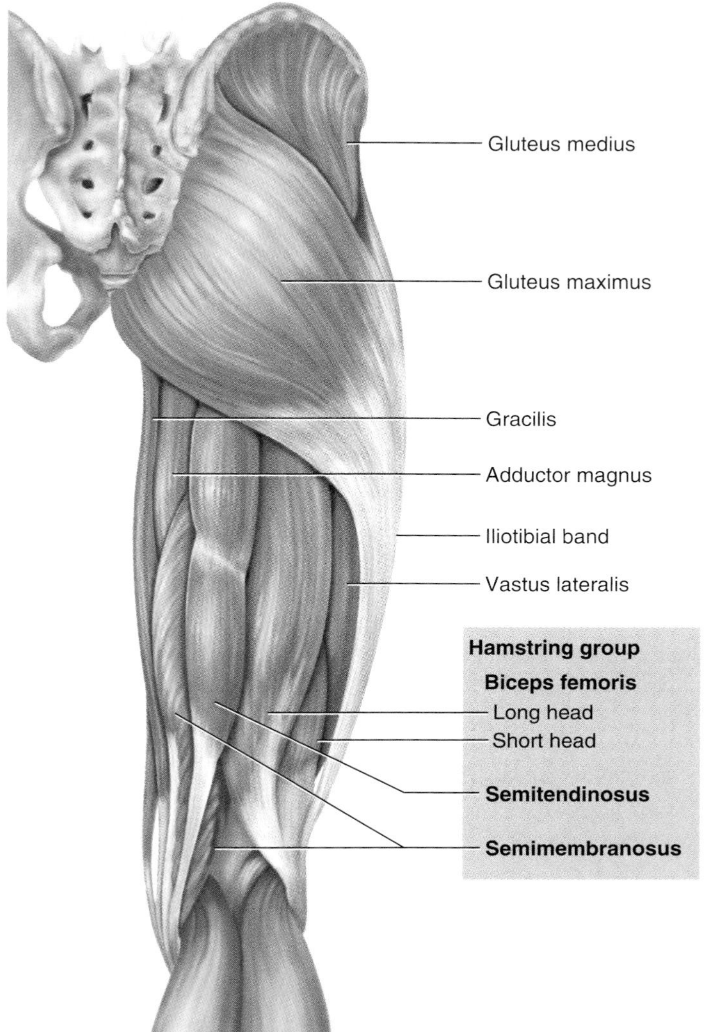

Figure 11.34 Gluteal and Thigh Muscles, Posterior View.

Muscles Acting on the Foot

The fleshy mass of the leg proper (below the knee) is formed by a group of **crural muscles,** which act on the foot (fig. 11.35; table 11.19). These muscles are tightly bound together by deep fasciae, which compress them and aid in the return of blood from the legs. The fasciae separate the crural muscles into anterior, lateral, and posterior compartments, each with its own nerve and blood supply.

Muscles of the anterior compartment dorsiflex the ankle and prevent the toes from scuffing the ground during walking. These are the **extensor digitorum longus** (extensor of toes II–V), **extensor hallucis**[53] **longus** (extensor of the great toe), **peroneus**[54] **(fibularis) tertius,** and **tibialis anterior.** Their tendons are held tightly against the ankle and kept from bowing by two **extensor retinacula** similar to the one at the wrist (fig. 11.36).

The posterior compartment has superficial and deep muscle groups. The three muscles of the superficial group are plantar flexors—the **gastrocnemius,**[55] **soleus,**[56] and **plantaris**[57] (fig. 11.37*a, b*). The first two of these, collectively known as the **triceps surae,**[58] insert on the calcaneus by way of the **calcaneal (Achilles) tendon.** This is the strongest

53. *halluc* = great toe
54. *peroneo* = fibula

55. *gastro* = belly + *cnem* = leg
56. named for its resemblance to a flatfish, the sole
57. *planta* = sole of foot
58. *sura* = calf of leg

Table 11.18 Muscles Acting on the Knee (see figs. 11.32 and 11.34)

O = origin, I = insertion, N = innervation (n. = nerve)

Origin	Insertion	Innervation
Anterior (Extensor) Compartment of Thigh		
Quadriceps Femoris (QUAD-rih-seps FEM-oh-riss)		
Extends knee; rectus femoris also flexes hip		
O: *rectus femoris*—anterior inferior spine of ilium; *vastus lateralis*—posterolateral shaft of femur; *vastus medialis*—linea aspera of femur; *vastus intermedius*—anterior shaft of femur	I: tibial tuberosity	N: femoral n.
Sartorius		
Flexes hip and knee; rotates femur medially; rotates tibia laterally; used in crossing legs		
O: anterior superior spine of ilium	I: medial aspect of tibial tuberosity	N: femoral n.
Posterior (Flexor) Compartment of Thigh (Hamstring Group)		
Biceps Femoris		
Flexes knee; extends hip; laterally rotates leg		
O: *long head*—ischial tuberosity; *short head*—posterior midshaft of femur	I: head of fibula	N: *long head*—tibial n.; *short head*—common peroneal n.
Semimembranosus (SEM-ee-MEM-bran-OH-sis)		
Flexes knee; extends hip; medially rotates tibia; tenses joint capsule of knee		
O: ischial tuberosity	I: medial condyle of tibia, collateral ligament of knee	N: tibial n.
Semitendinosus		
Flexes knee; extends hip; medially rotates tibia		
O: ischial tuberosity	I: near tibial tuberosity	N: tibial n.
Posterior Compartment of Leg		
Popliteus (pop-LIT-ee-us)		
Unlocks knee to allow flexion; flexes knee; medially rotates tibia		
O: lateral condyle of femur	I: posterior proximal tibia	N: tibial n.

Figure 11.35 **Superficial Muscles of the Leg of the Cadaver, Right Lateral View.**

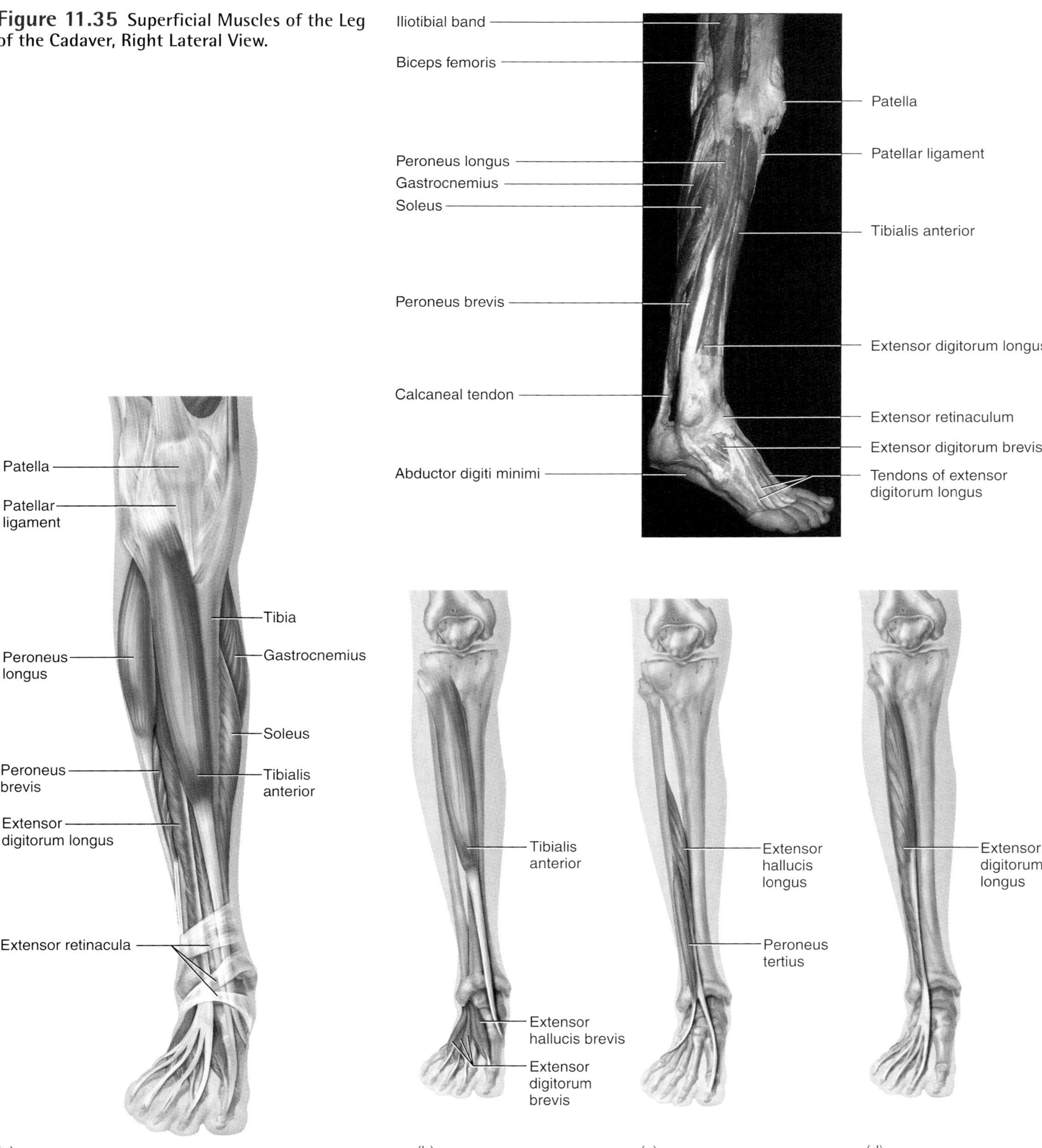

Figure 11.36 **Muscles of the Leg, Anterior View.** (*a*) A view showing some muscles of the anterior, lateral, and posterior compartments. (*b–d*) Individual muscles of the anterior compartment of the leg and dorsal aspect of the foot.

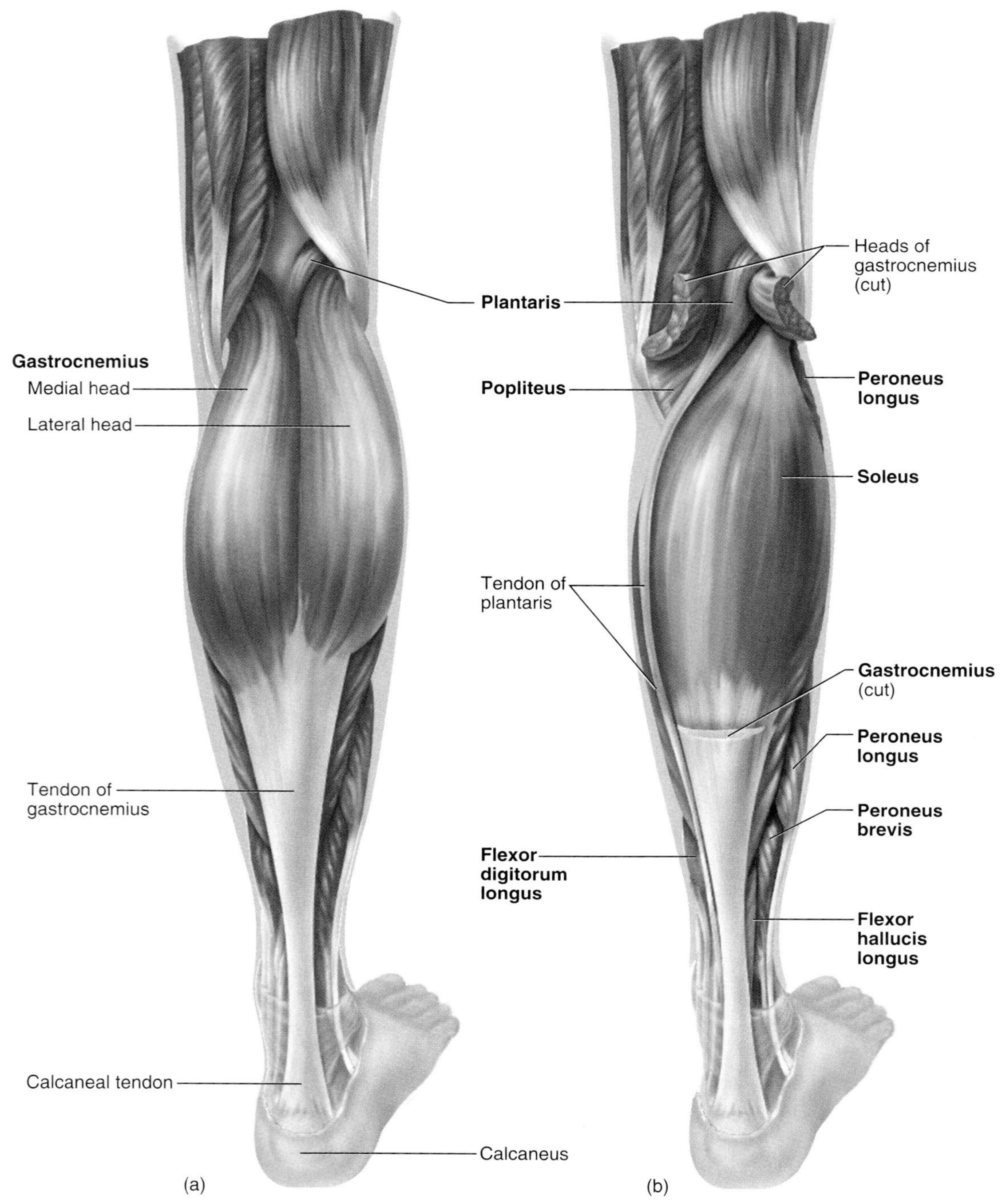

Figure 11.37 Muscles of the Leg, Posterior View. (*a*) The gastrocnemius. (*b*) The soleus, deep to the gastrocnemius and sharing the calcaneal tendon with it. *(continued)*

tendon of the body but is nevertheless a common site of sports injuries resulting from sudden stress. The plantaris inserts medially on the calcaneus by a tendon of its own.

There are four muscles in the deep group (fig. 11.37 *c–f*). The **flexor digitorum longus, flexor hallucis longus,** and **tibialis posterior** are plantar flexors. The **popliteus** unlocks the knee joint so that it can be flexed and functions in flexion and medial rotation at the knee.

The lateral (peroneal) compartment includes the **peroneus (fibularis) brevis** and **peroneus (fibularis) longus** (figs. 11.37*b* and 11.38). They plantar flex and evert the foot. Plantar flexion is important not only in

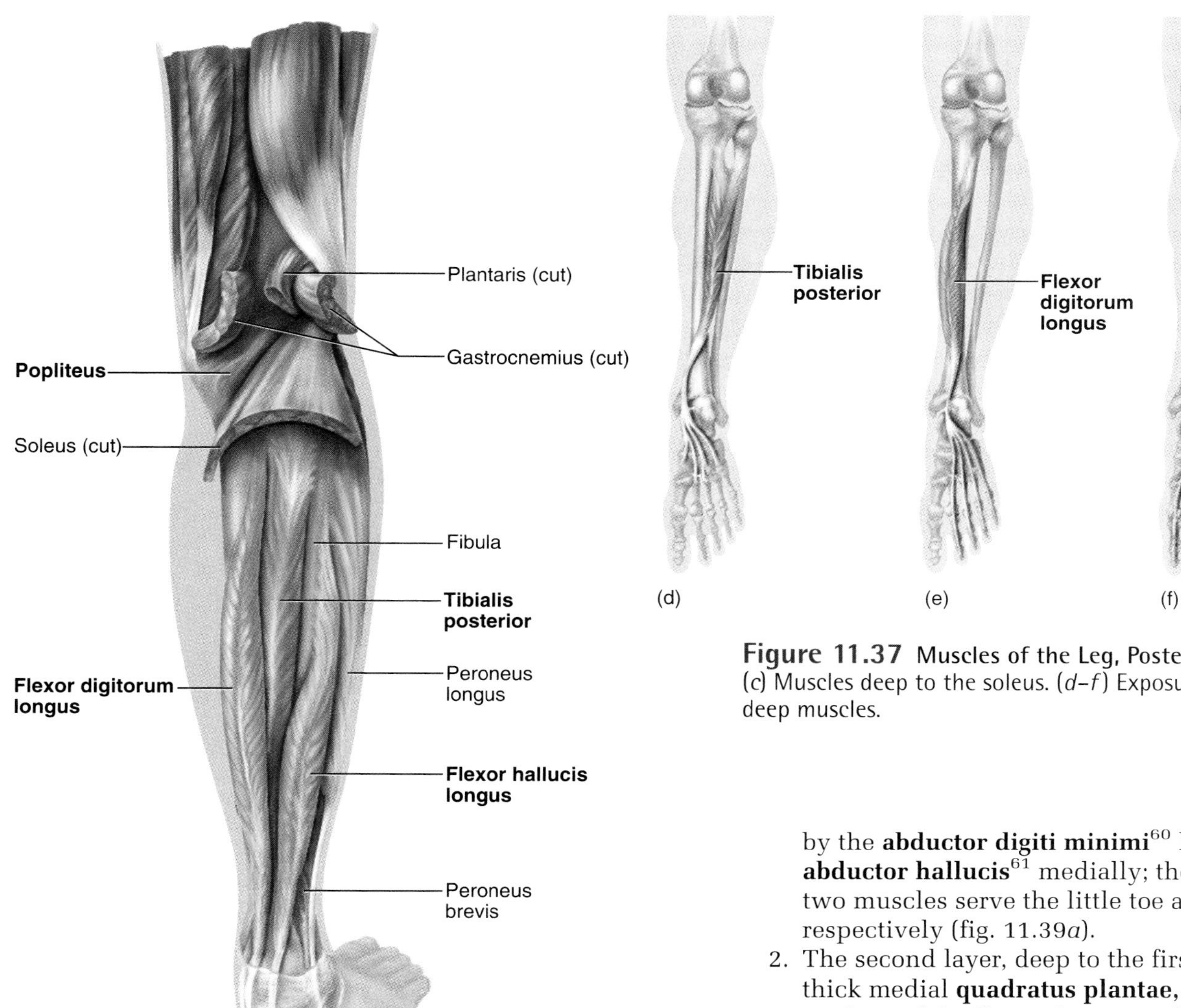

Figure 11.37 Muscles of the Leg, Posterior View (continued). (*c*) Muscles deep to the soleus. (*d–f*) Exposure of some individual deep muscles.

standing on tiptoes but in providing lift and forward thrust each time you take a step.

The intrinsic muscles of the foot support the arches and act on the toes in ways that aid locomotion (table 11.20). Several of them are similar in name and location to the intrinsic muscles of the hand. One of these muscles, the **extensor digitorum brevis,**[59] is on the dorsal side of the foot. The others are ventral or lie between the metatarsals. They are grouped in four layers:

1. The most superficial layer includes the stout **flexor digitorum brevis** medially, with four tendons that supply all the digits except the hallux. It is flanked by the **abductor digiti minimi**[60] laterally and **abductor hallucis**[61] medially; the tendons of these two muscles serve the little toe and great toe, respectively (fig. 11.39*a*).
2. The second layer, deep to the first, consists of the thick medial **quadratus plantae,** which joins the tendons of the flexor digitorum longus, and the four **lumbrical muscles** located between the metatarsals (fig. 11.39*b*).
3. The third layer includes the **adductor hallucis, flexor digiti minimi brevis,** and **flexor hallucis brevis** (fig. 11.39*c*). The adductor hallucis has an oblique head that crosses the foot and inserts at the base of the great toe, and a transverse head that passes across the bases of digits II–V and meets the long head at the base of the hallux.
4. The deepest layer consists of four **dorsal interosseous muscles** and three **plantar interosseous muscles** located between the metatarsals. Each dorsal interosseous muscle is bipennate and originates on two adjacent metatarsals. The plantar interosseous muscles are unipennate and originate on only one metatarsal each (fig. 11.39*d, e*).

59. "short extensor of the digits"

60. "abductor of the little toe"

61. "abductor of the hallux (great toe)"

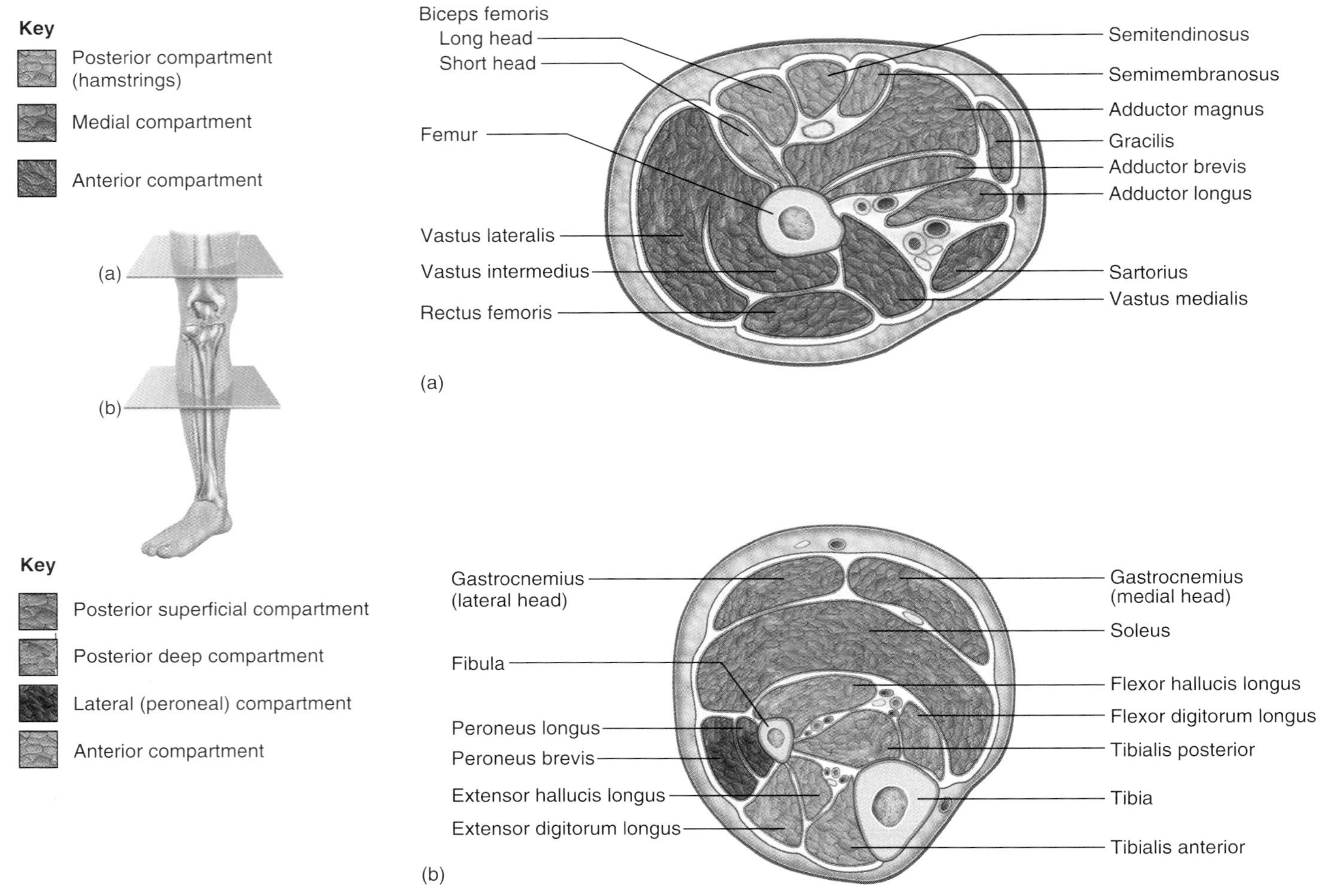

Figure 11.38 Serial Cross Sections Through the Lower Limb. Each section is taken at the correspondingly lettered level in the figure at the left and is pictured with the posterior muscle compartment facing the top of the page.

Table 11.19 Muscles Acting on the Foot (see figs. 11.36 and 11.37)

O = origin, I = insertion, N = innervation (n. = nerve)

Anterior Compartment of Leg

Extensor Digitorum (DIDJ-ih-TOR-um) Longus

Extends toes II–V; dorsiflexes and everts foot

O: lateral condyle of tibia, shaft of fibula, interosseous membrane	I: middle and distal phalanges II–V	N: deep peroneal n.

Extensor Hallucis (hal-OO-sis) Longus

Extends hallux (great toe); dorsiflexes and inverts foot

O: medial aspect of fibula, interosseous membrane	I: distal phalanx I	N: deep peroneal n.

Peroneus Tertius (perr-OH-nee-us TUR-she-us)

Dorsiflexes and everts foot

O: distal shaft of fibula	I: metatarsal V	N: deep peroneal n.

Tibialis (TIB-ee-AY-lis) Anterior

Dorsiflexes and inverts foot

O: lateral tibia, interosseous membrane	I: medial cuneiform, metatarsal I	N: deep peroneal n.

(continued)

Table 11.19 Muscles Acting on the Foot (continued)

Muscle / Action / Origin	Insertion	Nerve
Posterior Compartment of Leg—Superficial Group		
Gastrocnemius (GAS-trock-NEE-me-us) Flexes knee; plantar flexes foot O: medial and lateral epicondyles of femur	I: calcaneus	N: tibial n.
Soleus (SO-lee-us) Plantar flexes foot O: proximal one-third of tibia and fibula	I: calcaneus	N: tibial n.
Plantaris (plan-TERR-is) Flexes knee; plantar flexes foot. Sometimes absent. O: distal femur	I: calcaneus	N: tibial n.
Posterior Compartment of Leg—Deep Group		
Flexor Digitorum Longus Flexes toes II–V; plantar flexes and inverts foot O: midshaft of tibia	I: distal phalanges II–V	N: tibial n.
Flexor Hallucis Longus Flexes hallux (great toe); plantar flexes and inverts foot O: shaft of fibula	I: distal phalanx I	N: tibial n.
Tibialis Posterior Plantar flexes and inverts foot O: proximal half of tibia, fibula, interosseous membrane	I: navicular, cuneiforms, metatarsals II–IV	N: tibial n.
Lateral (Peroneal) Compartment of Leg		
Peroneus Brevis Plantar flexes and everts foot O: shaft of fibula	I: base of metatarsal V	N: superficial peroneal n.
Peroneus Longus Plantar flexes and everts foot O: proximal half of fibula, lateral condyle of tibia	I: medial cuneiform, metatarsal I	N: superficial peroneal n.

Several of the muscles in the first three layers originate on a broad **plantar aponeurosis**, which lies between the plantar skin and muscles. It diverges like a fan from the calcaneus to the bases of all five toes.

Think About It

Not everyone has the same muscles. From the information provided in this chapter, identify two muscles that are lacking in some people.

Before You Go On

Answer the following questions to test your understanding of the preceding section:

23. In the middle of a stride, you have one foot on the ground and you are about to swing the other leg forward. What muscles produce the movements of that leg?
24. Name the muscles that cross both the hip and knee joints and produce actions at both.
25. List the major actions of the muscles of the anterior, medial, and posterior compartments of the thigh.
26. Describe the role of plantar flexion and dorsiflexion in walking. What muscles produce these actions?

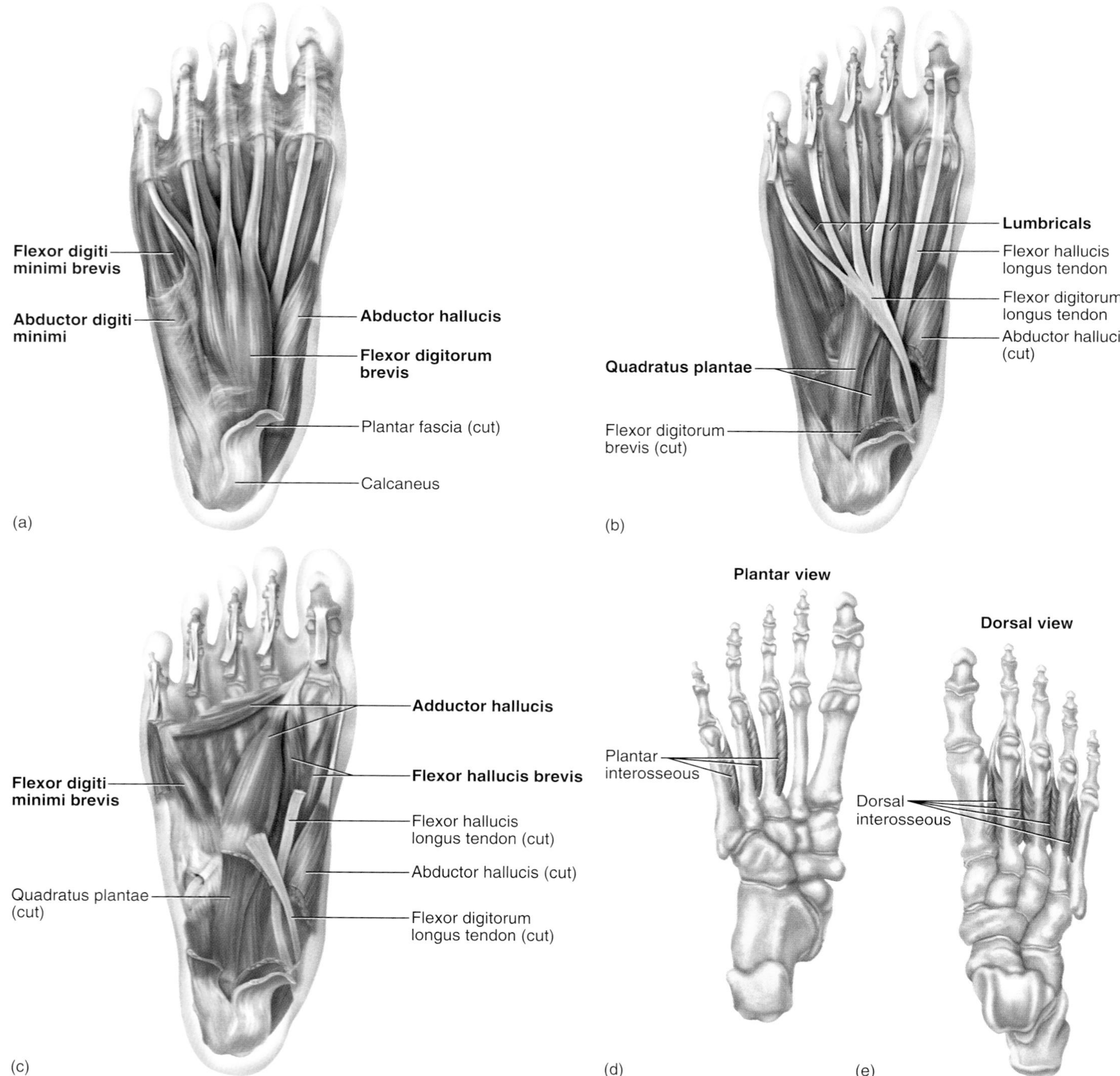

Figure 11.39 Intrinsic Muscles of the Foot. (*a–d*) First through fourth layers, respectively, in ventral (plantar) views; (*e*) fourth layer, dorsal view.

Table 11.20	Intrinsic Muscles of the Foot (see fig. 11.39)	
O = origin, I = insertion, N = innervation (n. = nerve, nn. = nerves)		
Dorsal Aspect of Foot		
Extensor Digitorum (DIDJ-ih-TOE-rum) Brevis Extends toes		
O: dorsal aspect of calcaneus	I: tendons of extensor digitorum longus	N: deep peroneal n.
Ventral Layer 1 (Most Superficial)		
Flexor Digitorum Brevis Flexes toes II–V		
O: calcaneus, plantar aponeurosis	I: middle phalanges II–V	N: medial plantar n.
Abductor Digiti Minimi Abducts and flexes little toe; supports lateral longitudinal arch		
O: calcaneus, plantar aponeurosis	I: proximal phalanx V	N: lateral plantar n.
Abductor Hallucis (hal-OO-sis) Flexes hallux (great toe); supports medial longitudinal arch		
O: calcaneus, plantar aponeurosis	I: proximal phalanx I	N: medial plantar n.
Ventral Layer 2		
Quadratus Plantae (quad-RAY-tus PLAN-tee) Flexes toes		
O: calcaneus, plantar aponeurosis	I: tendons of flexor digitorum	N: lateral plantar n.
Lumbricals (four muscles) Flex metatarsophalangeal joints; extend interphalangeal joints		
O: tendon of flexor digitorum longus	I: extensor tendons to digits II–V	N: lateral and medial plantar nn.
Ventral Layer 3		
Adductor Hallucis Adducts hallux		
O: metatarsals II–IV	I: proximal phalanx I	N: medial plantar n.
Flexor Digiti Minimi Brevis Flexes little toe		
O: metatarsal V, plantar aponeurosis	I: proximal phalanx V	N: lateral plantar n.
Flexor Hallucis Brevis Flexes hallux		
O: cuboid, plantar aponeurosis	I: proximal phalanx I	N: medial plantar n.
Ventral Layer 4 (Deepest)		
Dorsal Interosseous Muscles (four muscles) Abduct toes II–V		
O: each with two heads arising from adjacent metatarsals	I: proximal phalanges II–IV	N: lateral plantar n.
Plantar Interosseous Muscles (three muscles) Adduct toes III–V		
O: medial aspect of metatarsals III–V	I: proximal phalanges III–V	N: lateral plantar n.

Clinical Insight 11.6

Athletic Injuries

Although the muscular system is subject to fewer diseases than most organ systems, it is particularly vulnerable to injuries resulting from sudden and intense stress placed on muscles and tendons. Each year, thousands of athletes from the high school to professional level sustain some type of injury to their muscles, as do the increasing numbers of people who have taken up running and other forms of physical conditioning. Overzealous exertion without proper conditioning and warm-up is frequently the cause. Some of the more common athletic injuries are:

Shinsplints—a general term embracing several kinds of injury with pain in the crural region: tendinitis of the tibialis posterior muscle, inflammation of the tibial periosteum, and anterior compartment syndrome. Shinsplints may result from unaccustomed jogging, walking on snowshoes, or any vigorous activity of the legs after a period of relative inactivity.

Compartment syndrome—a condition in which overuse, contusion, or muscle strain damages blood vessels in a compartment of the arm or leg. Since the fascia enclosing a compartment are tight and cannot stretch, blood or tissue fluid accumulating in the compartment can put pressure on the muscles, nerves, and blood vessels. The lack of blood flow in the compartment can cause destruction of nerves if untreated within 2 to 4 hours and death of muscle tissue if it goes untreated for 6 hours or more. Nerves can regenerate if blood flow is restored, but muscle damage is irreversible. Depending on its severity, compartment syndrome may be treated with immobilization and rest or an incision to drain fluid from the compartment or otherwise relieve the pressure.

Pulled hamstrings—strained hamstring muscles or a partial tear in the tendinous origin, often with a hematoma (blood clot) in the fascia lata. This condition is frequently caused by repetitive kicking (as in football and soccer) or long, hard running.

Tennis elbow—inflammation at the origin of the extensor carpi muscles on the lateral epicondyle of the humerus. It occurs when these muscles are repeatedly tensed during backhand strokes and then strained by sudden impact with the tennis ball. Any activity that requires rotary movements of the forearm and a firm grip of the hand (for example, using a screwdriver) can cause the symptoms of tennis elbow.

Tennis leg—a partial tear in the lateral origin of the gastrocnemius muscle. It results from repeated strains put on the muscle while supporting the body weight on the toes.

Rider's bones—abnormal ossification in the tendons of the adductor muscles of the medial thigh. It results from prolonged abduction of the thighs when riding horses.

Pulled groin—strain in the adductor muscles of the thigh. It is common in gymnasts and dancers who perform splits and high kicks.

Charley horse—any painful tear, stiffness, and blood clotting in a muscle. A charley horse of the quadriceps femoris is often caused by football tackles.

Blocker's arm—ectopic ossification in the lateral margin of the forearm as a result of repeated impact with opposing players.

Pitcher's arm—inflammation at the origin of the flexor carpi resulting from hard wrist flexion in releasing a baseball.

Baseball finger—tears in the extensor tendons of the fingers resulting from the impact of a baseball with the extended fingertip.

Most athletic injuries can be prevented by proper conditioning. A person who suddenly takes up vigorous exercise may not have sufficient muscle and bone mass to withstand the stresses such exercise entails. These must be developed gradually. Stretching exercises keep ligaments and joint capsules supple and therefore reduce injuries. Warm-up exercises promote more efficient and less injurious musculoskeletal function in several ways, discussed in the next chapter. Most of all, moderation is important, as most injuries simply result from overuse of the muscles. "No pain, no gain" is a dangerous misconception.

Muscular injuries can be treated initially with "RICE": **r**est, **i**ce, **c**ompression, and **e**levation. Rest prevents further injury and allows repair processes to occur; ice reduces swelling; compression with an elastic bandage helps to prevent fluid accumulation and swelling; and elevation of an injured limb promotes drainage of blood from the affected area and limits further swelling. If these measures are not enough, anti-inflammatory drugs may be employed, including corticosteroids as well as aspirin and other nonsteroidal agents.

Connective Issues

Interactions Between the MUSCULAR SYSTEM and Other Organ Systems

← Indicates ways in which this system affects other systems
→ Indicates ways in which other systems affect this one

Integumentary System

← Facial muscles pull on skin to provide facial expressions

→ Covers and protects superficial muscles; initiates synthesis of calcitriol, which promotes absorption of calcium needed for muscle contraction; dissipates heat generated by muscles

Skeletal System

← Muscles move and stabilize joints and produce stress that affects ossification, bone remodeling, and shapes of mature bones

→ Provides levers that enable muscles to act; stores calcium needed for muscle contraction

Nervous System

← Muscles give expression to thoughts, emotions, and motor commands that arise in the central nervous system

→ Stimulates muscle contraction; monitors and adjusts muscle tension; adjusts cardiopulmonary functions to meet needs of muscles during exercise

Endocrine System

← Exercise stimulates secretion of stress hormones; skeletal muscles protect some endocrine organs

→ Hormones stimulate growth and development of muscles and regulate levels of glucose and electrolytes important for muscle contraction

Circulatory System

← Muscle contractions help to move blood through veins; exercise stimulates growth of new blood vessels

→ Delivers O_2 and nutrients; carries away wastes and heat generated by muscles; cardiovascular efficiency, RBC count, hemoglobin level, and density of blood capillaries in muscle greatly affect muscular endurance

Lymphatic and Immune Systems

← Muscle contractions promote lymph flow; exercise elevates level of immune cells and antibodies; excess exercise inhibits immune responses

→ Lymphatic system drains fluid from muscles; immune system protects muscles from pathogens and promotes tissue repair

Respiratory System

← Muscle contractions ventilate lungs; muscles of larynx and pharynx regulate air flow; CO_2 generated by exercise stimulates respiratory rate and depth

→ Provides O_2 and removes CO_2; respiratory efficiency greatly affects muscular endurance

Urinary System

← Muscles control voluntary urination; muscles of pelvic floor support bladder

→ Eliminates wastes generated by muscles; regulates levels of electrolytes important for muscle contraction

Digestive System

← Muscles enable chewing and swallowing; control voluntary defecation; abdominal and lumbar muscles protect lower digestive organs

→ Absorbs nutrients needed by muscles; liver regulates blood glucose levels and metabolizes lactic acid generated by anaerobic muscle metabolism

Reproductive System

← Muscles contribute to erection and ejaculation; abdominal and pelvic muscles aid childbirth

→ Gonadal steroids affect muscular growth and development

Chapter Review

Online Learning Center

Study resources for this chapter are available at **www.mhhe.com/saladin/**

Study Outline

The Structural and Functional Organization of Muscles (p. 340)

1. Functions of muscles
   a. Movement
   b. Stability
   c. Communication
   d. Control of body openings and passages
   e. Heat production
2. Connective tissues of a muscle
   a. Endomysium
   b. Perimysium
   c. Epimysium
   d. Deep fascia
3. Muscle-bone attachments
   a. Direct (fleshy)
   b. Indirect (via tendon)
   c. Aponeuroses
   d. Retinacula
4. General anatomy of muscles
   a. Origin, insertion, and belly
   b. Fascicle orientation
5. Coordinated action of muscles
   a. Prime movers
   b. Synergists
   c. Antagonists
   d. Fixators
6. Intrinsic and extrinsic muscles
7. Muscle innervation
8. How muscles are named
   a. By size
   b. By shape
   c. By location
   d. By number of heads
   e. By orientation
   f. By action

Muscles of the Head and Neck (p. 348)

1. Muscles of facial expression (table 11.2)
2. Muscles of chewing and swallowing (table 11.3)
3. Muscles acting on the head (table 11.4)

Muscles of the Trunk (p. 359)

1. Muscles of respiration (table 11.5)
2. Muscles of the abdomen (table 11.6)
3. Muscles of the back (table 11.7)
4. Muscles of the pelvic floor (table 11.8)

Muscles Acting on the Shoulder and Upper Limb (p. 370)

1. Muscles acting on the scapula (table 11.9)
2. Muscles acting on the humerus (table 11.10)
3. Muscles acting on the forearm (table 11.12)
4. Muscles acting on the wrist and hand (table 11.14)

Muscles Acting on the Hip and Lower Limb (p. 386)

1. Muscles acting on the hip and femur (table 11.17)
2. Muscles acting on the knee (table 11.18)
3. Muscles acting on the foot (table 11.19)

Selected Vocabulary

To review the names of individual muscles, see the tables cited in the Study Outline.

muscular system 340
myology 340
endomysium 340
fascicle 340
perimysium 340
epimysium 340
deep fascia 341
superficial fascia 341
direct attachment 341
indirect attachment 341
tendon 341
aponeurosis 342
retinaculum 342
origin 342
head of a muscle 342
insertion 342
belly 342
fusiform muscle 342
parallel muscle 342
convergent muscle 342
pennate muscle 342
circular muscle 342
sphincter 342
action of a muscle 343
prime mover 343
synergist 343
antagonist 343
antagonistic pair 343
fixator 343
intrinsic muscle 344
extrinsic muscle 344
innervation 344
galea aponeurotica 350
suprahyoid group 352
infrahyoid group 352
anterior triangle 357
posterior triangle 357
central tendon of diaphragm 359
tendinous intersections of rectus abdominis 361
rectus sheath 361
linea alba 361
inguinal ligament 363
serratus posterior 365
thoracolumbar fascia 365
perineum 367
urogenital triangle 367
anal triangle 367
superficial perineal space 367
central tendon of perineum 367
urogenital diaphragm 367
pelvic diaphragm 367
rotator cuff 375
flexor retinaculum 378
extensor retinaculum 378
carpal tunnel 378
thenar group 383
hypothenar group 383
midpalmar group 383

Testing Your Recall

1. Which of the following muscles is the prime mover in spitting out a mouthful of liquid?
   a. platysma
   b. buccinator
   c. risorius
   d. masseter
   e. palatoglossus
2. Each muscle fiber has a sleeve of areolar connective tissue around it called
   a. the deep fascia.
   b. the superficial fascia.
   c. the perimysium.
   d. the epimysium.
   e. the endomysium.
3. Which of these is *not* a suprahyoid muscle?
   a. genioglossus
   b. geniohyoid
   c. stylohyoid
   d. mylohyoid
   e. digastric
4. Which of these muscles is an extensor of the neck?
   a. external oblique
   b. sternocleidomastoid
   c. splenius capitis
   d. iliocostalis
   e. latissimus dorsi
5. Which of these muscles of the pelvic floor is the deepest?
   a. superficial transverse perineus
   b. bulbospongiosus
   c. ischiocavernosus
   d. deep transverse perineus
   e. levator ani
6. Which of these actions is *not* performed by the trapezius?
   a. extension of the neck
   b. depression of the scapula
   c. elevation of the scapula
   d. rotation of the scapula
   e. adduction of the humerus
7. Both the hands and feet are acted upon by a muscle or muscles called
   a. the extensor digitorum.
   b. the abductor digiti minimi.
   c. the flexor digitorum profundus.
   d. the abductor hallucis.
   e. the flexor digitorum longus.
8. Which of the following muscles does *not* extend the hip joint?
   a. quadriceps femoris
   b. gluteus maximus
   c. biceps femoris
   d. semitendinosus
   e. semimembranosus
9. Both the gastrocnemius and _____ muscles insert on the heel by way of the calcaneal tendon.
   a. semimembranosus
   b. tibialis posterior
   c. tibialis anterior
   d. soleus
   e. plantaris
10. Which of the following muscles raises the upper lip?
    a. levator palpebrae superioris
    b. orbicularis oris
    c. zygomaticus minor
    d. masseter
    e. mentalis
11. The ________ of a muscle is the point where it attaches to a relatively stationary bone.
12. A bundle of muscle fibers surrounded by perimysium is called a/an ________.
13. The ________ is the muscle primarily responsible for a given movement at a joint.
14. The three large muscles on the posterior side of the thigh are commonly known as the ________ muscles.
15. Connective tissue bands called ________ prevent flexor tendons from rising like bowstrings.
16. The anterior half of the perineum is a region called the ________.
17. The abdominal aponeuroses converge on a midsagittal fibrous band on the abdomen called the ________.
18. A muscle that works with another to produce the same or similar movement is called a/an ________.
19. A muscle somewhat like a feather, with fibers obliquely approaching its tendon from both sides, is called a/an ________ muscle.
20. A circular muscle that closes a body opening is called a/an ________.

Answers in appendix B

Testing Your Comprehension

1. Radical mastectomy, once a common treatment for breast cancer, involved removal of the pectoralis major along with the breast. What functional impairments would result from this? What synergists could a physical therapist train a patient to use to recover some lost function?
2. Removal of cancerous lymph nodes from the neck sometimes requires removal of the sternocleidomastoid on that side. How would this affect a patient's range of head movement?
3. In a disease called tick paralysis, the saliva from a tick bite paralyzes skeletal muscles beginning with the lower limbs and progressing superiorly. What would be the most urgent threat to the life of a tick paralysis patient?
4. Women who habitually wear high heels may suffer painful "high heel syndrome" when they go barefoot or wear flat shoes. What muscle(s) and tendon(s) are involved? Explain.
5. A student moving out of a dormitory kneels down, in correct fashion, to lift a heavy box of books. What prime movers are involved as he straightens his legs to lift the box?

Answers at the Online Learning Center

B

ATLAS

Surface Anatomy

Refer to your ESP Tutorial CD-ROM for helpful animations, activities, and quizzes related to the material covered in this chapter. The following are suggested areas to go to on the CD and are presented according to "Unit/Topic":
- Support and Movement/Muscular System

The Importance of External Anatomy

In the study of human anatomy, it is easy to become so preoccupied with internal structure that we forget the importance of what we can see and feel externally. Yet external anatomy and appearance are major concerns in giving a physical examination and in many aspects of patient care. A knowledge of the body's surface landmarks is essential to one's competence in physical therapy, cardiopulmonary resuscitation, surgery, making X rays and electrocardiograms, giving injections, drawing blood, listening to heart and respiratory sounds, measuring the pulse and blood pressure, and finding pressure points to stop arterial bleeding, among other procedures. A misguided attempt to perform some of these procedures while disregarding or misunderstanding external anatomy can be very harmful and even fatal to a patient.

Having just studied skeletal and muscular anatomy in the preceding chapters, this is an opportune time for you to study the body surface. Much of what we see there reflects the underlying structure of the superficial bones and muscles. A broad photographic overview of surface anatomy is given in atlas A (see fig. A.6). In the following pages, we examine the body literally from head (fig. B.1) to toe (fig. B.10) and study its regions in more detail. To make the most profitable use of this atlas, refer back to the skeletal and muscular anatomy in chapters 9 to 11. Relate drawings of the clavicles in chapter 9 to the photograph in figure B.1, for example. Study the shape of the scapula in chapter 9 and see how much of it you can trace on the photographs in figure B.4. See if you can relate the tendons visible on the hand (fig. B.6) to the muscles of the forearm illustrated in chapter 11.

For learning surface anatomy, there is a resource available to you that is far more valuable than any laboratory model or textbook illustration—your own body. For the best understanding of human structure, compare the art and photographs in this book with your body or with structures visible on a study partner. In addition to bones and muscles, you can palpate a number of superficial arteries, veins, tendons, ligaments, and cartilages, among other structures. By palpating regions such as the shoulder, elbow, or ankle, you can develop a mental image of the subsurface structures better than you can obtain by looking at two-dimensional textbook images. And the more you can study with other people, the more you will appreciate the variations in human structure and be able to apply your knowledge to your future patients or clients, who will not look quite like any textbook diagram or photograph you have ever seen. Through comparisons of art, photography, and the living body, you will get a much deeper understanding of the body than if you were to study this atlas in isolation from the earlier chapters.

At the end of this atlas, you can test your knowledge of externally visible muscle anatomy. The two photographs in figure B.11 have 30 numbered muscles and a list of 26 names, some of which are shown more than once in the photographs and some of which are not shown at all. Identify the muscles to your best ability without looking back at the previous illustrations, and then check your answers in appendix B at the back of the book.

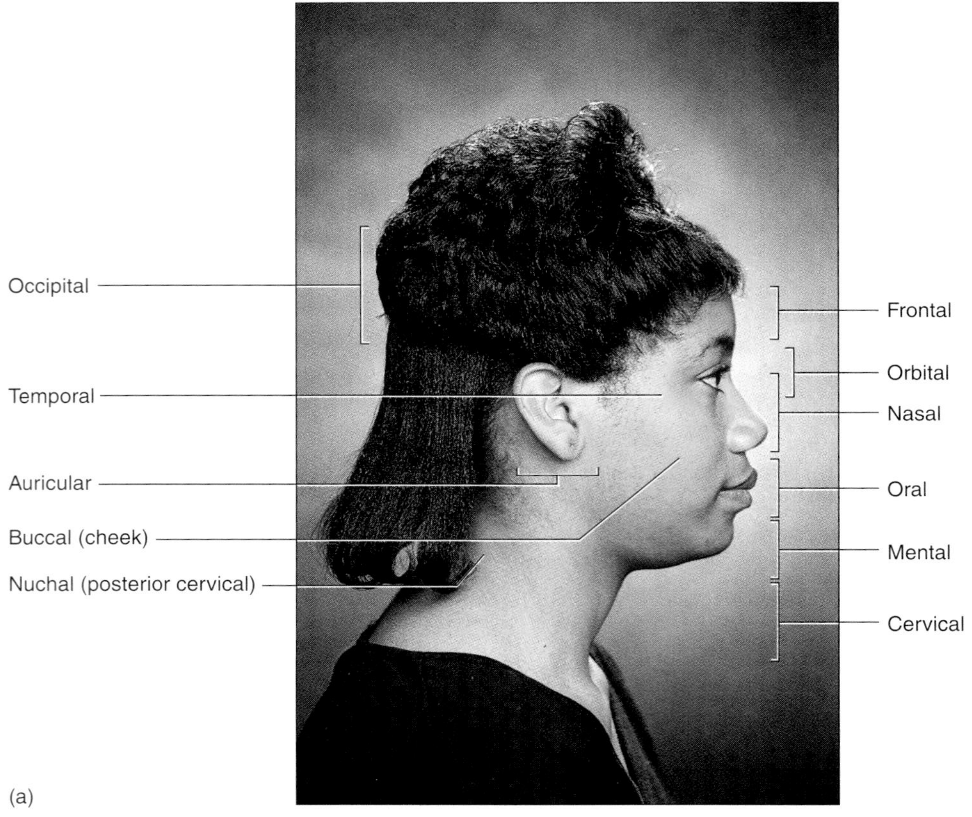

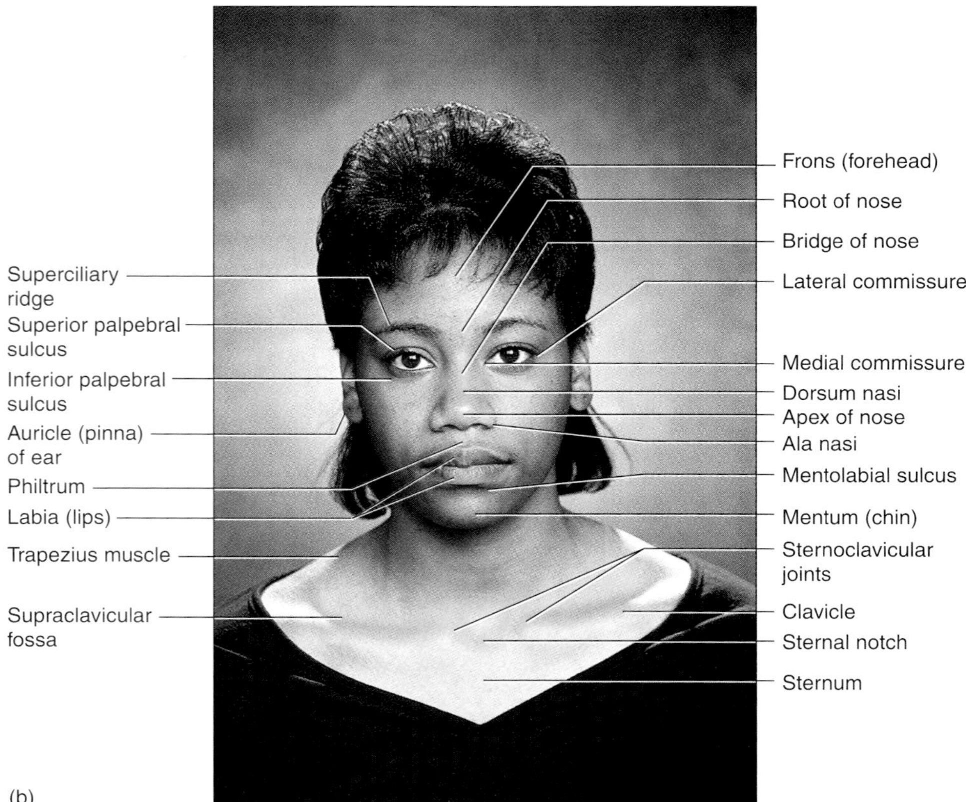

Figure B.1 The Head and Neck. (*a*) Anatomical regions of the head, lateral aspect. (*b*) Features of the facial region and upper thorax.

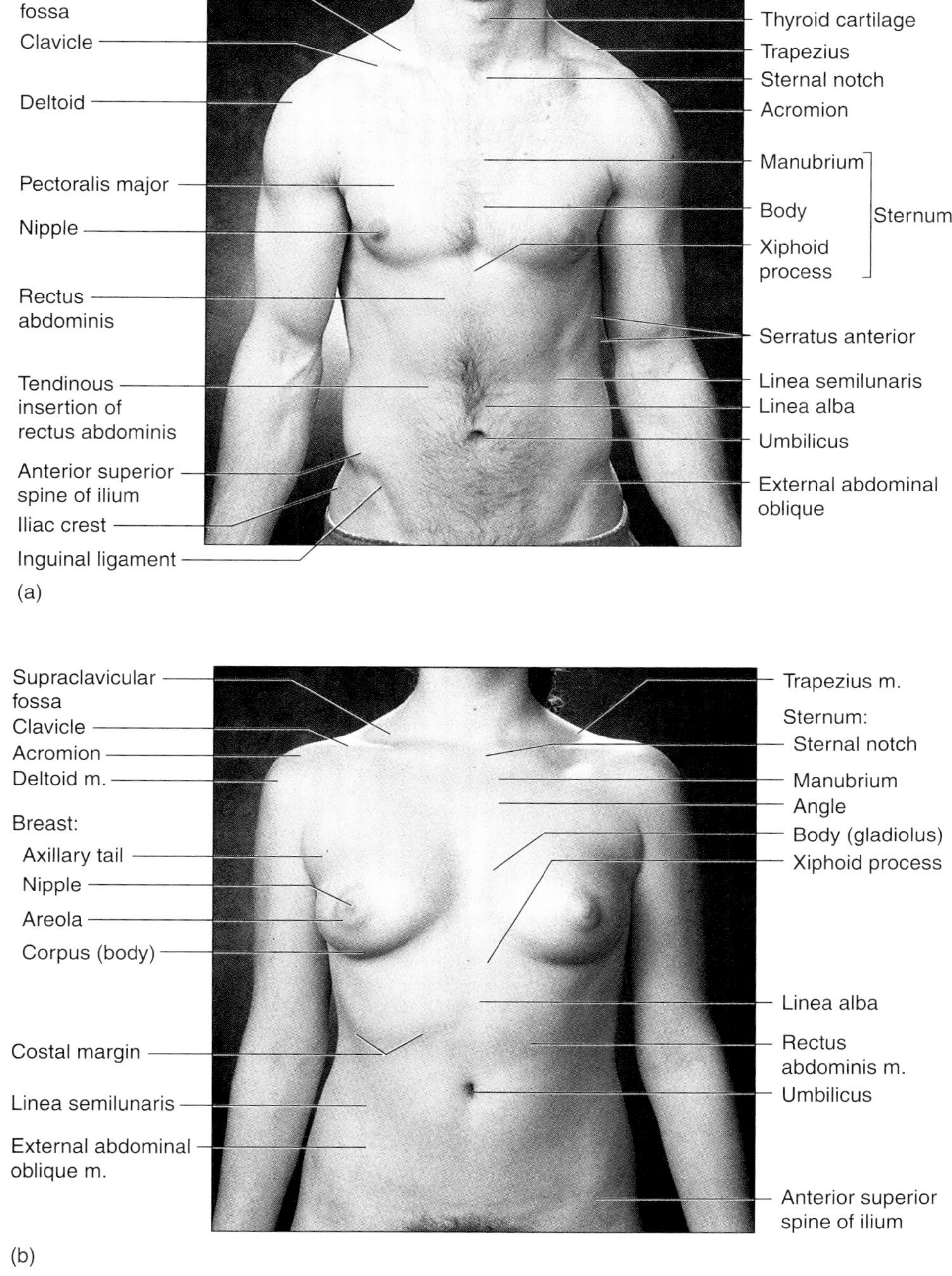

Figure B.2 The Thorax and Abdomen, Anterior (ventral) Aspect. (*a*) Male; (*b*) female. All of the features labeled are common to both sexes, though some are labeled only on the photograph that shows them best. (*m.* = muscle.)

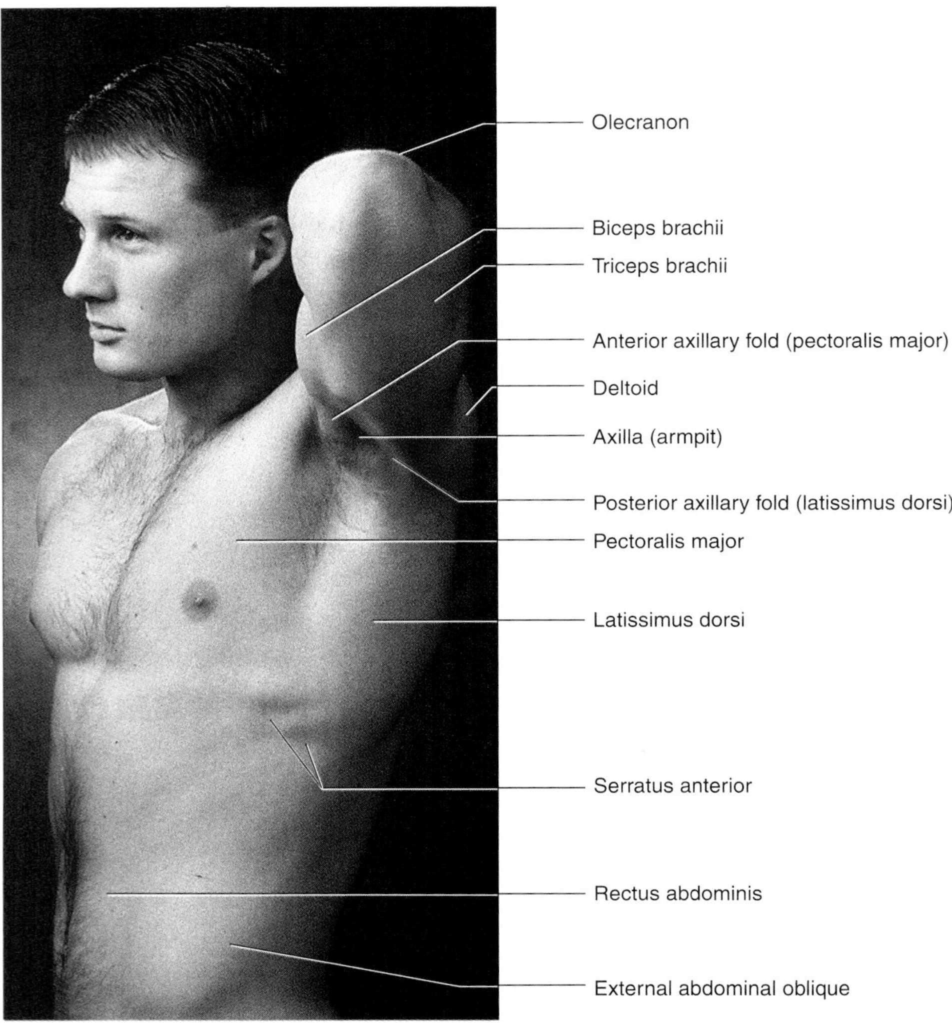

Figure B.3 The Axillary Region.

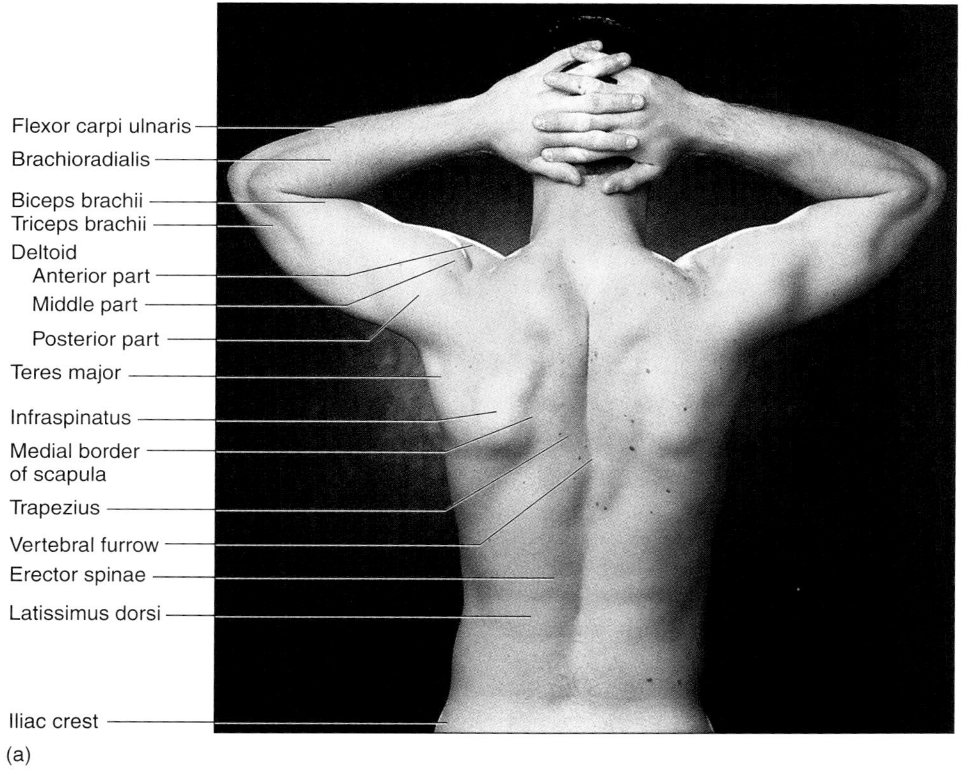

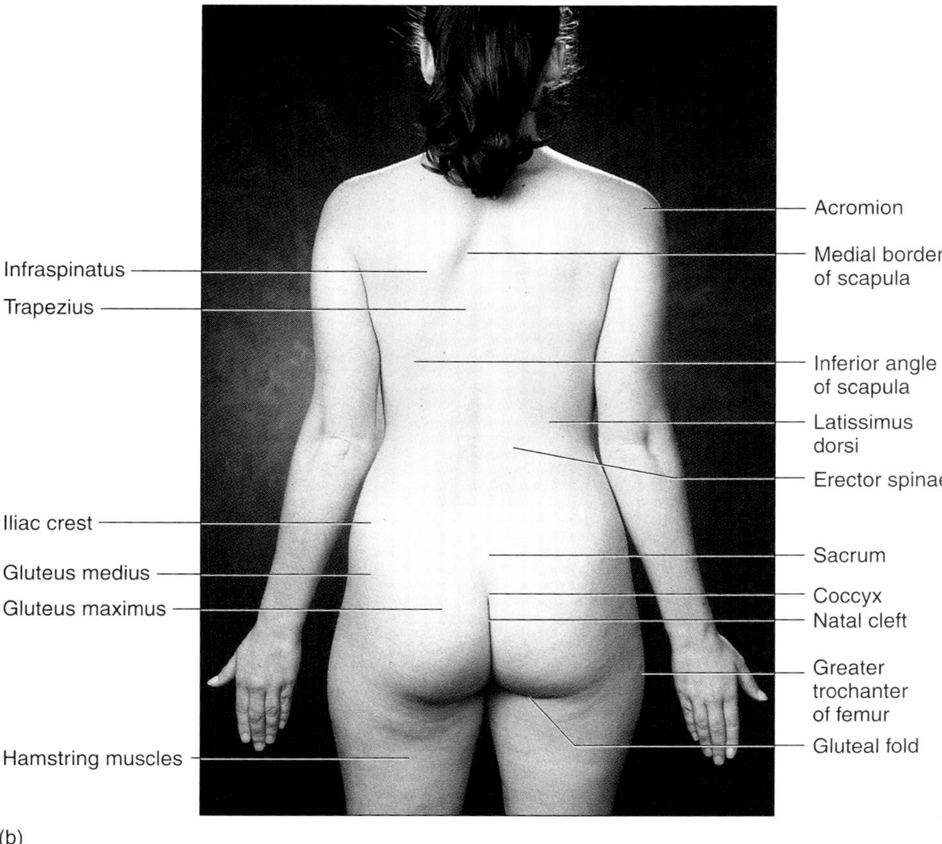

Figure B.4 **The Back and Gluteal Region.** (*a*) Male; (*b*) female. All of the features labeled are common to both sexes, though some are labeled only on the photograph that shows them best.

Trapezius
Acromion
Deltoid
Pectoralis major
Biceps brachii
Triceps brachii
Long head
Lateral head
Brachioradialis
Extensor carpi radialis longus
Lateral epicondyle of humerus
Olecranon
Interphalangeal joints
Metacarpophalangeal joints
Styloid process of ulna
Extensor digitorum

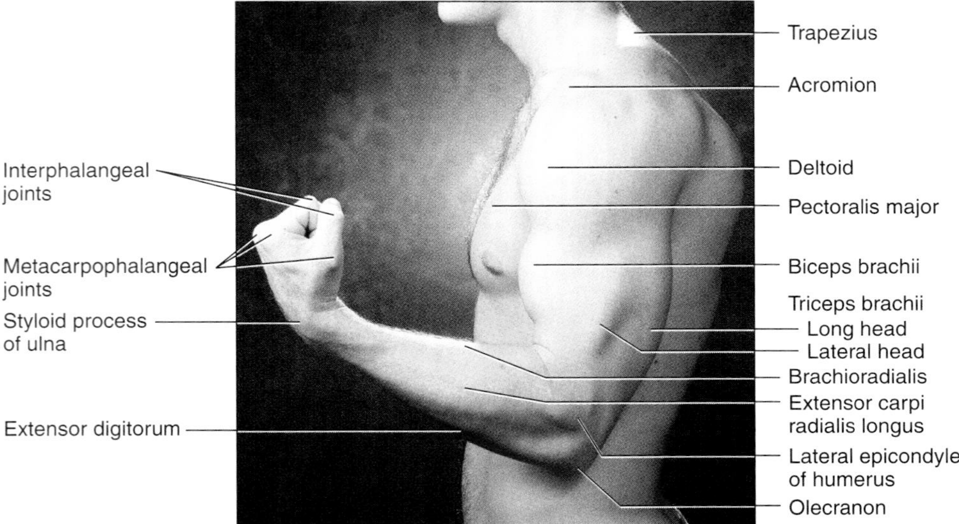

Figure B.5 The Upper Limb, Lateral Aspect.

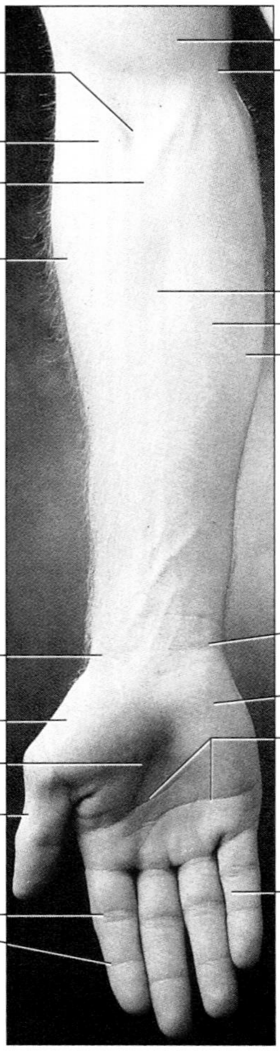

Cubital fossa
Cephalic vein
Median cubital vein
Brachioradialis
Styloid process of radius
Thenar eminence
Palmar surface of hand
Pollex (thumb)
Flexion lines
Biceps brachii
Medial epicondyle of humerus
Flexor carpi radialis
Palmaris longus
Flexor carpi ulnaris
Styloid process of ulna
Hypothenar eminence
Flexion creases
Volar surface of fingers
(a)

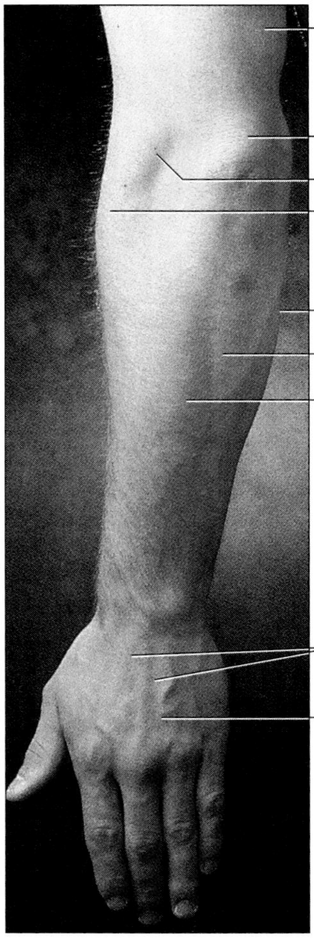

Triceps brachii
Olecranon
Head of radius
Brachioradialis
Flexor carpi ulnaris
Extensor carpi ulnaris
Extensor digitorum
Tendons of extensor digitorum
Dorsum of hand
(b)

Figure B.6 The Right Antebrachium (forearm). (*a*) Ventral (anterior) aspect; (*b*) dorsal (posterior) aspect.

Figure B.7 The Left Wrist and Hand. (*a*) Ventral (anterior) aspect; (*b*) dorsal (posterior) aspect.

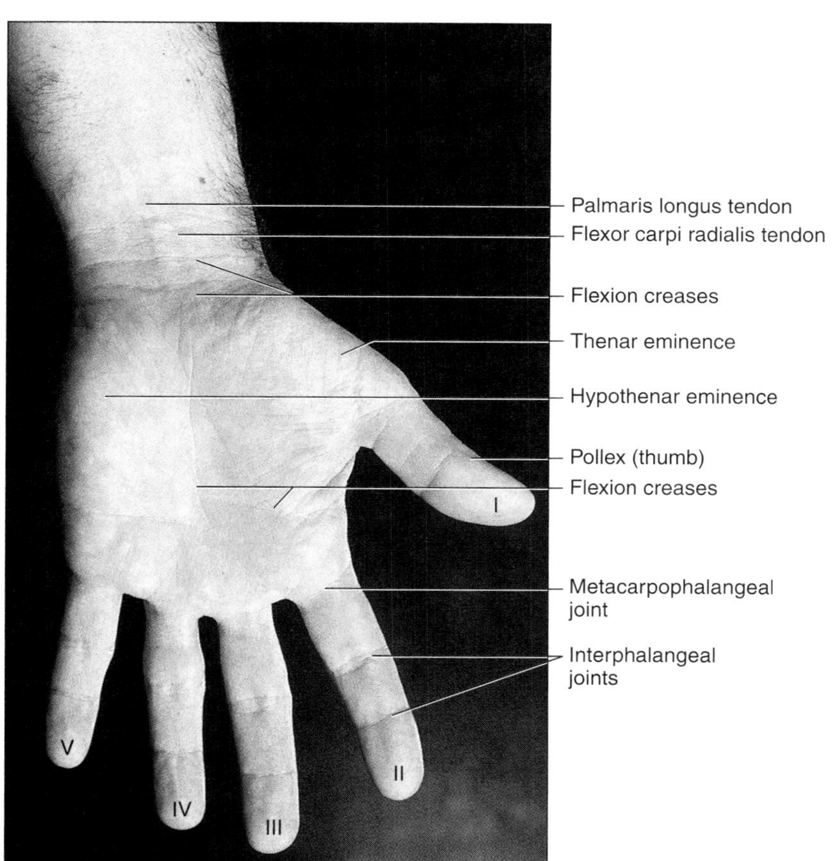

(a)

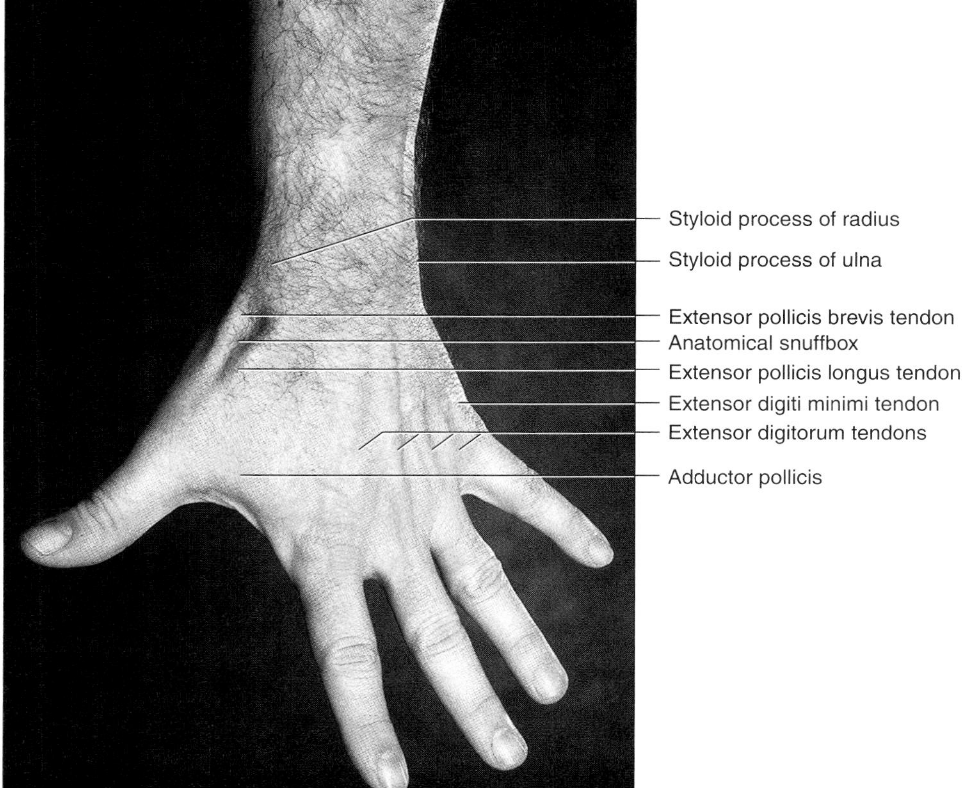

(b)

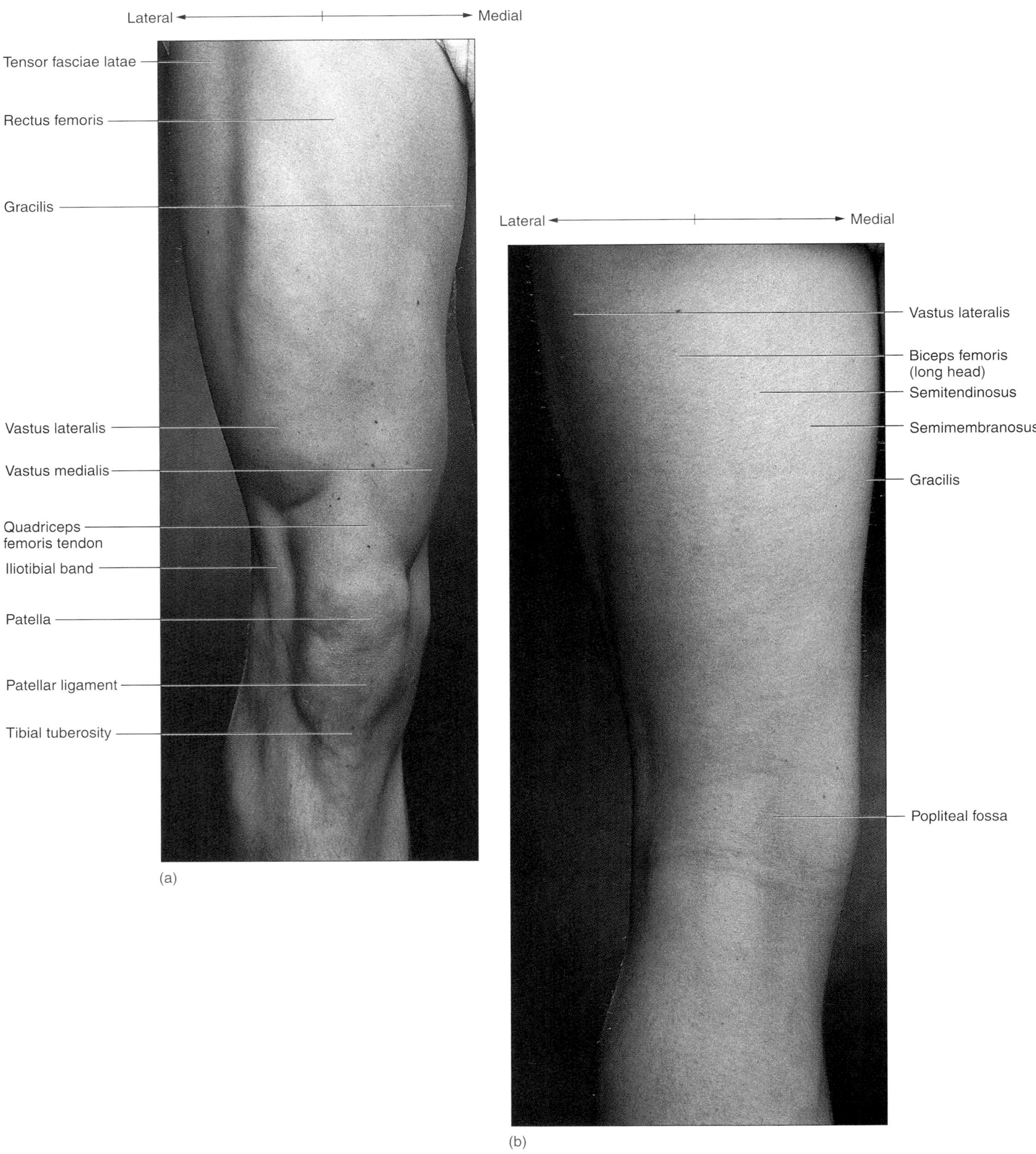

Figure B.8 The Right Thigh and Knee. (*a*) Ventral (anterior) aspect; (*b*) dorsal (posterior) aspect.

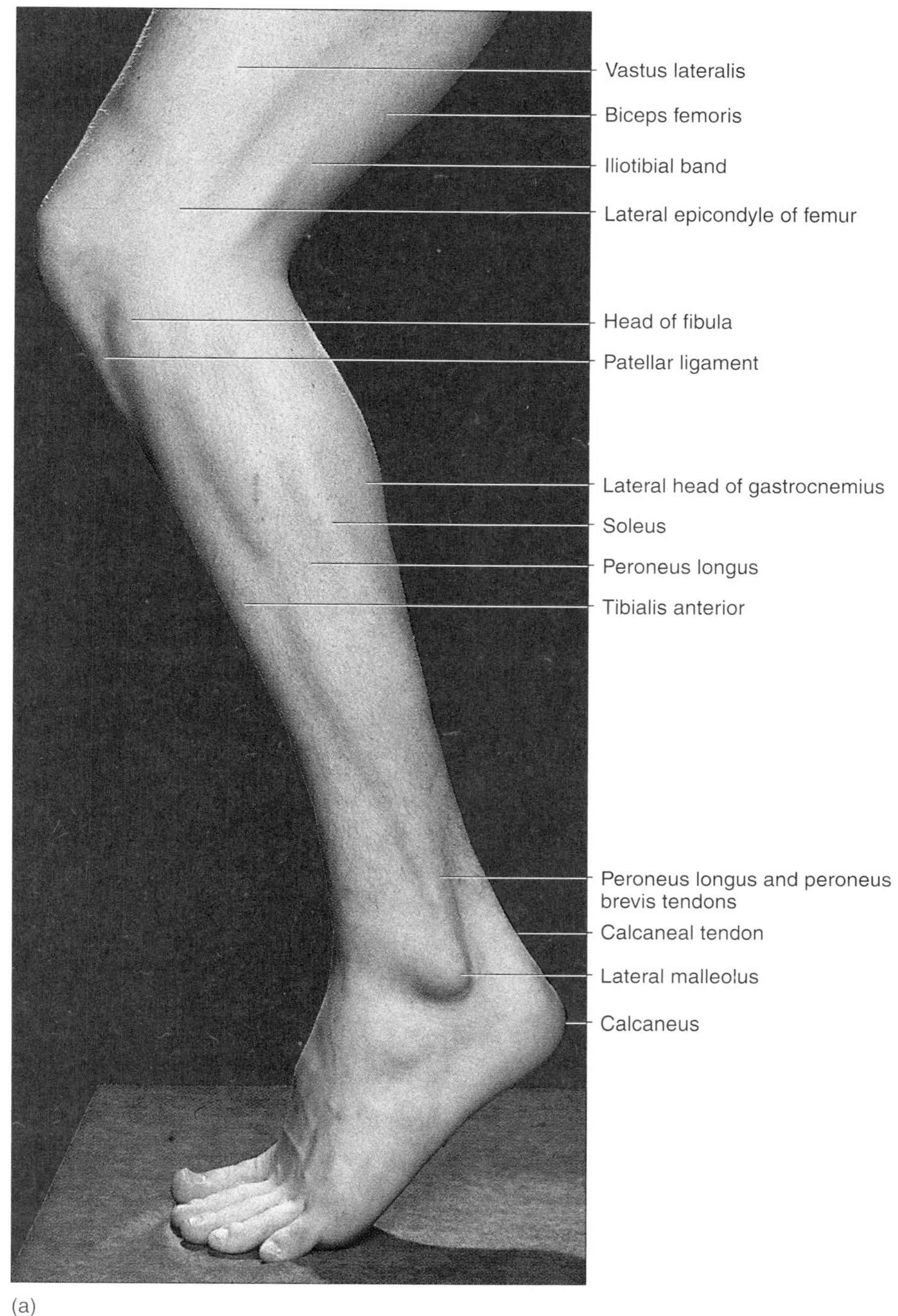

Figure B.9 The Left Leg and Foot. (*a*) Lateral aspect.

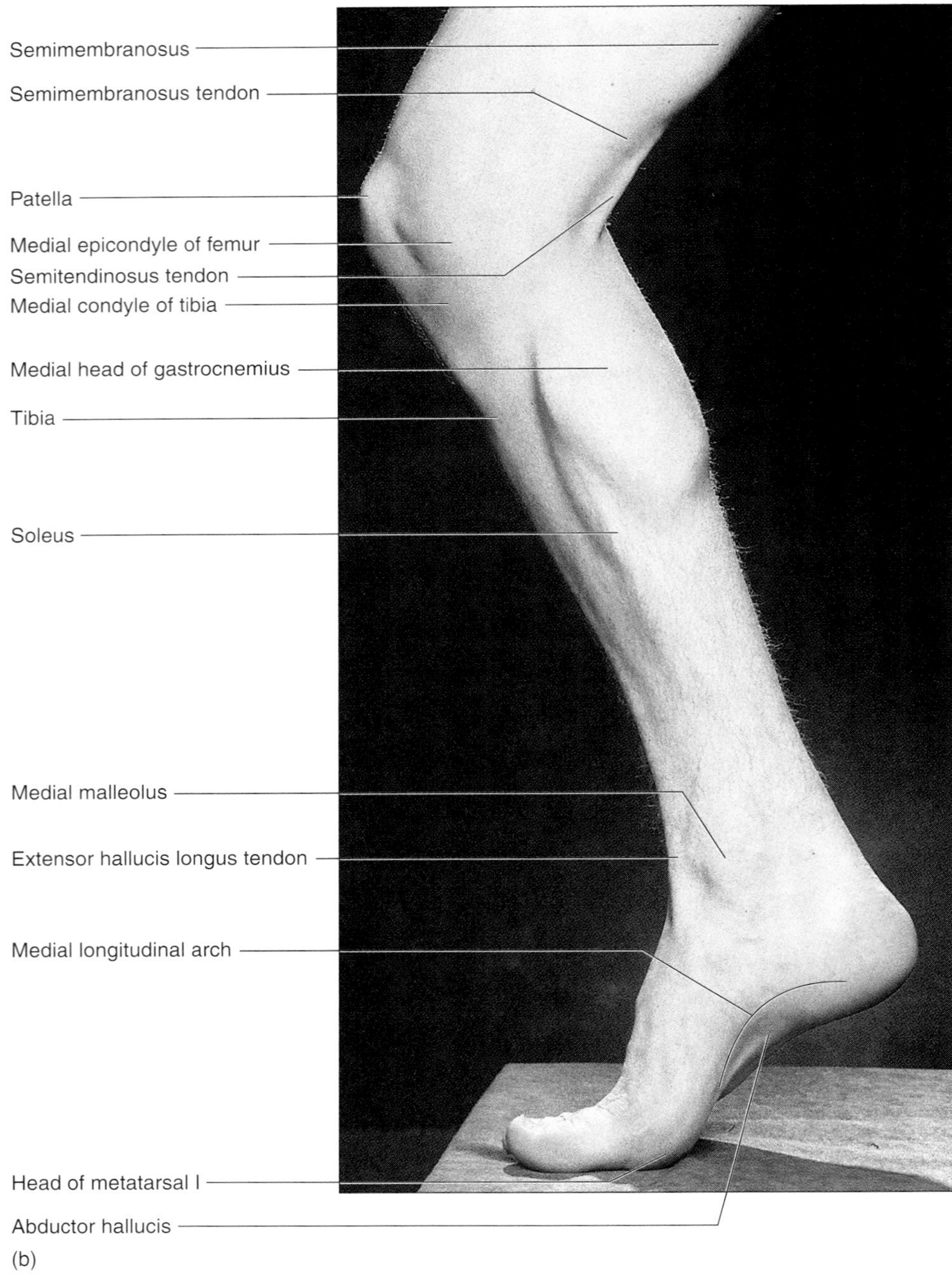

Figure B.9 **The Left Leg and Foot (continued).** (*b*) medial aspect.

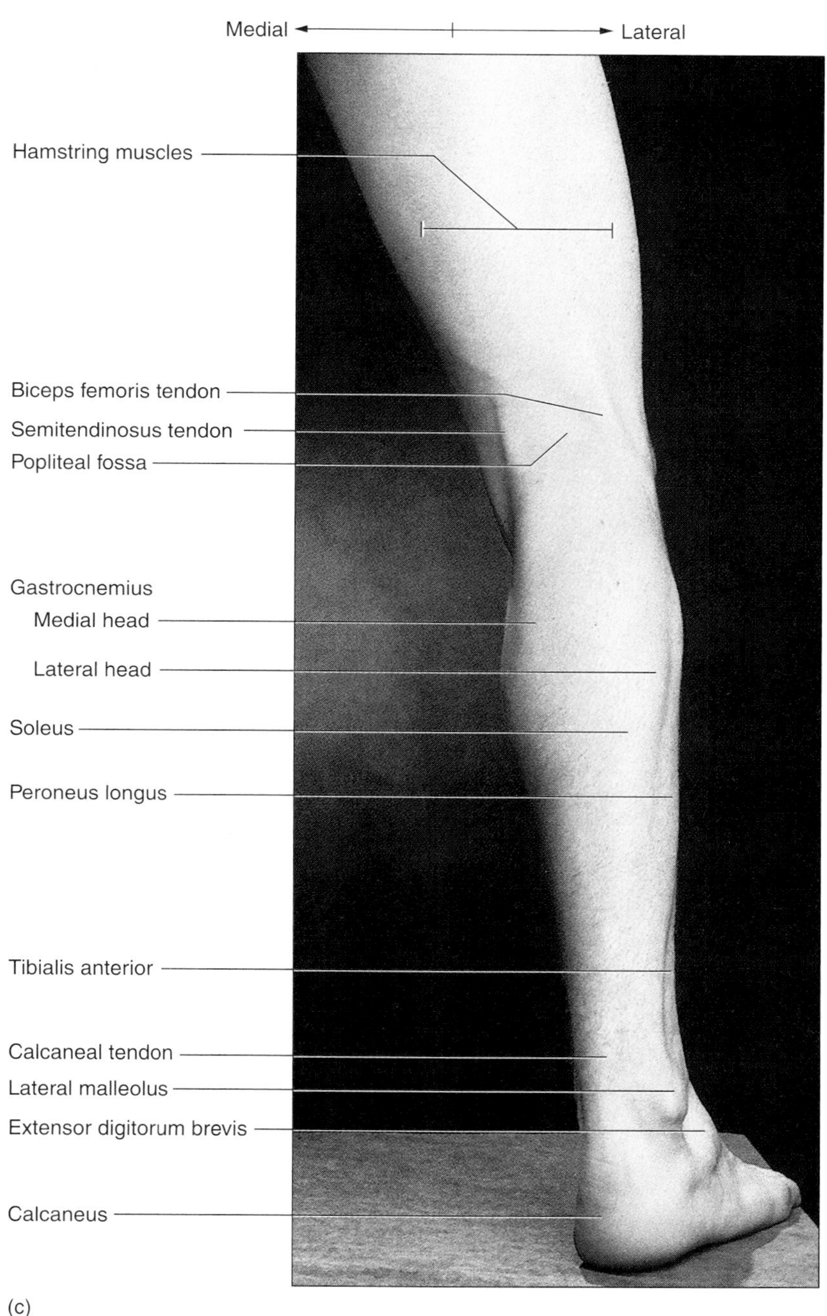

Figure B.9 **The Left Leg and Foot (continued).** (*c*) Dorsal (posterior) aspect.

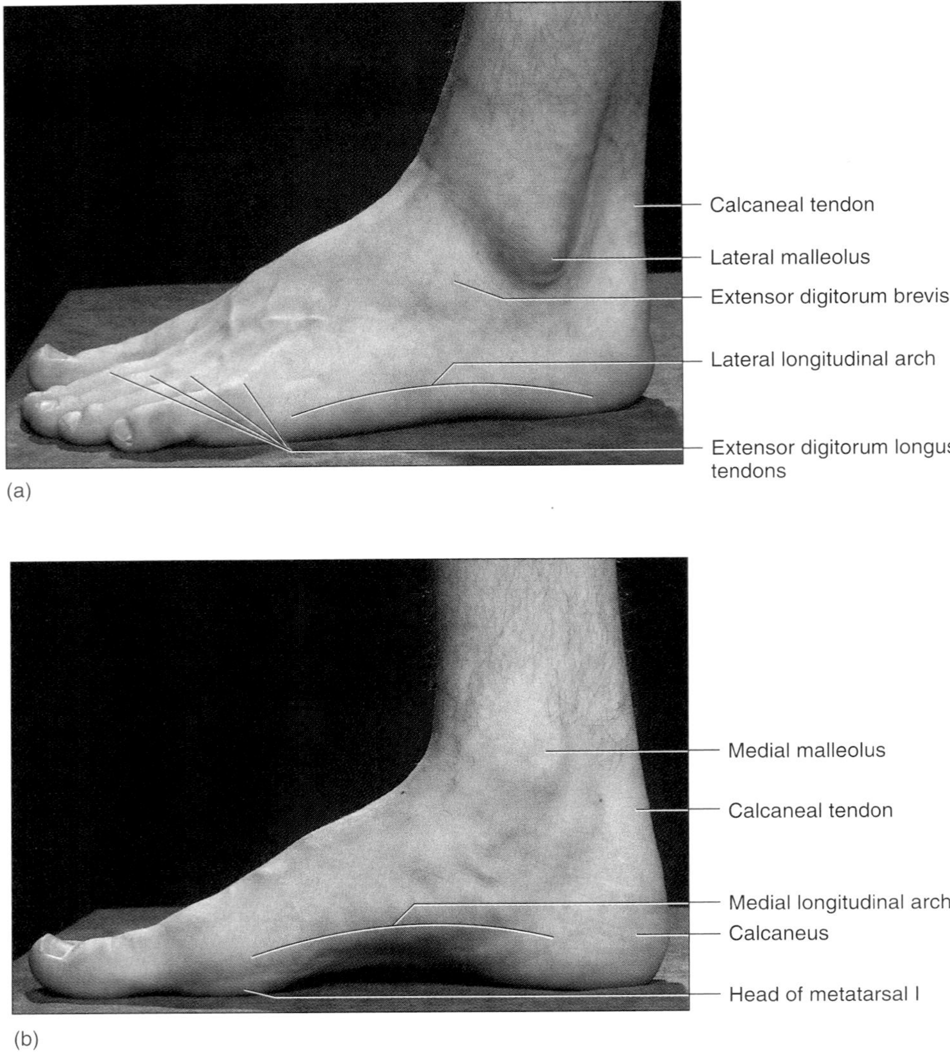

Figure B.10 The Left Foot. (*a*) Lateral aspect; (*b*) medial aspect.

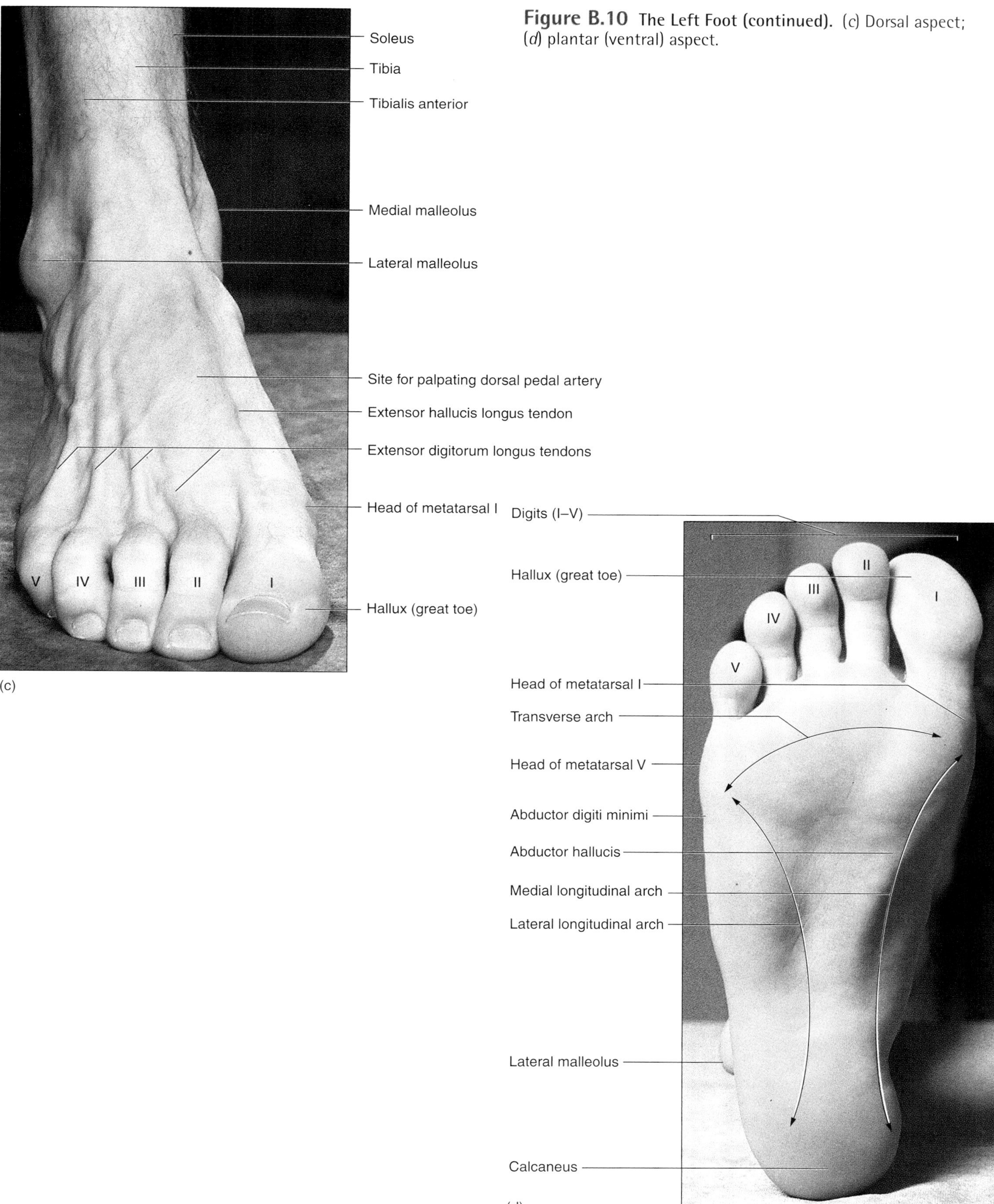

Figure B.10 **The Left Foot (continued).** (*c*) Dorsal aspect; (*d*) plantar (ventral) aspect.

Figure B.11 Muscle Self-Test. To test your knowledge of muscle anatomy, match the 30 labeled muscles on these photographs to the alphabetical list of muscles below. Answer as many as possible without referring back to the previous illustrations. Some of these names are used more than once, since the same muscle may be shown from different perspectives, and some of these names are not used at all. The answers are in appendix B.

a. biceps brachii
b. brachioradialis
c. deltoid
d. erector spinae
e. external oblique
f. flexor carpi ulnaris
g. gastrocnemius
h. gracilis
i. hamstrings
j. infraspinatus
k. latissimus dorsi
l. pectineus
m. pectoralis major
n. rectus abdominis
o. rectus femoris
p. serratus anterior
q. soleus
r. splenius capitis
s. sternocleidomastoid
t. subscapularis
u. teres major
v. tibialis anterior
w. transversus abdominis
x. trapezius
y. triceps brachii
z. vastus lateralis

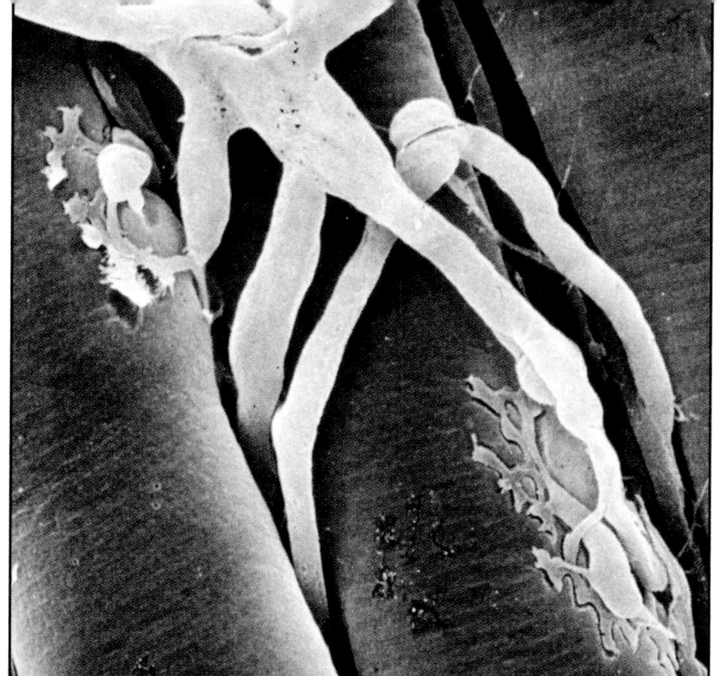
Neuromuscular junctions (SEM)

CHAPTER 12

Muscular Tissue

Chapter Outline

Insights

Study with Your ESP CD-ROM

Refer to your ESP Tutorial CD-ROM for helpful animations, activities, and quizzes related to the material covered in this chapter. The following are suggested areas to go to on the CD and are presented according to "Unit/Topic/":

- Support and Movement/Muscular System

Brushing Up

To understand this chapter, it is important that you understand or brush up on the following concepts:

- Aerobic and anaerobic metabolism (pp. 97–98)
- The functions of membrane proteins, especially receptors and ion gates (pp. 110–111)
- Structure of a neuron (p. 190)
- General histology of the three types of muscle tissue (p. 191)
- Desmosomes and gap junctions (p. 193)
- Connective tissues of a muscle (p. 340)

Muscle tissue is composed of elongated cells that are capable of shortening when stimulated. A muscle cell is essentially a device for converting the chemical energy of ATP into the mechanical energy of contraction. This chapter discusses contraction at the cellular and molecular levels and explains the basis of such aspects of muscle performance as warm-up, strength, endurance, and fatigue. These phenomena have obvious relevance to athletic performance, and they become very important when old age or lack of physical conditioning interferes with a person's ability to carry out everyday motor tasks. The effects of old age on the muscular system are discussed in chapter 29.

The three types of muscle tissue—*skeletal, cardiac,* and *smooth*—were described and compared in chapter 6. The expression "muscular system" refers only to skeletal muscle. This chapter is concerned primarily with the microscopic anatomy and physiology of skeletal muscle. Cardiac and smooth muscle are discussed more briefly to compare their properties and functions with skeletal muscle. Cardiac muscle is discussed more extensively in chapter 19.

Types and Characteristics of Muscle Tissue

Objectives

When you have completed this section, you should be able to

- describe the physiological properties that all muscle types have in common;
- list the defining characteristics of skeletal muscle; and
- describe the elastic functions of the connective tissue components of a muscle.

Universal Characteristics of Muscle

The functions of muscular tissue were listed in the preceding chapter. To carry out those functions, all muscular tissue has the following characteristics:

- **Responsiveness (excitability).** Responsiveness is a property of all living cells, but muscle and nerve cells have developed this property to the highest degree. When stimulated by chemical signals (neurotransmitters), stretch, and other stimuli, muscle cells respond with electrical changes across the plasma membrane.
- **Conductivity.** Stimulation of a muscle fiber produces more than a local effect. The local electrical change triggers a wave of excitation that travels rapidly along the muscle fiber and initiates processes leading to muscle contraction.
- **Contractility.** Muscle fibers are unique in their ability to shorten substantially when stimulated. This enables them to pull on bones and other tissues and create movement of the body and its parts.
- **Extensibility.** In order to contract, a muscle cell must also be extensible—able to stretch again between contractions. Most cells rupture if they are stretched even a little, but skeletal muscle fibers can stretch to as much as three times their contracted length.
- **Elasticity.** When a muscle cell is stretched and the tension is then released, it recoils to its original resting length. Elasticity, commonly misunderstood as the ability to stretch, refers to this tendency of a muscle cell (or other structures) to return to the original length when tension is released.

Skeletal Muscle

Skeletal muscle may be defined as voluntary striated muscle that is usually attached to one or more bones. A typical skeletal muscle cell is about 10 μm in diameter and 100 μm long; some are as thick as 100 μm and as long as 30 cm. Because of their extraordinary length, skeletal muscle cells are usually called *muscle fibers* or *myofibers.* A skeletal muscle fiber exhibits alternating light and dark transverse bands, or **striations,** that reflect the overlapping arrangement of the internal contractile proteins (fig. 12.1). Skeletal muscle is called **voluntary** because it is usually subject to conscious control. The other types of muscle are **involuntary** (not usually under conscious control), and they are never attached to bones.

Series-Elastic Components

The **series-elastic components** of a muscle are a series of connective tissue elements between a muscle fiber and a bone or other attachment. They include the endomysium, perimysium, epimysium, fascia, and tendon described in the previous chapter (see fig. 11.1). Unlike muscle tissue, these tissues are not excitable or contractile, but they are extensible and elastic—they stretch under tension and recoil when released. When you are running, the elastic recoil of

Figure 12.1 Skeletal Muscle Fibers. Note the striations.

the calcaneal tendon may help to lift the heel and produce some of the thrust as your toes push off from the ground.

When a muscle begins to contract, it generates a force called **internal tension** that stretches the series-elastic components. When these are taut, they deliver a force, the **external tension,** to the load to be moved—usually a bone at the muscle's insertion. By analogy, imagine lifting a weight off a table with a rubber band. At first, internal tension stretches the rubber band; then the external tension lifts the weight. Because no energy transfer is 100% efficient, the external tension that a muscle exerts on a bone is always somewhat less than the internal tension. Series-elastic components have the additional function of helping to stretch a muscle to its resting length when contraction ceases.

Before You Go On

Answer the following questions to test your understanding of the preceding section:

1. Define *responsiveness, conductivity, contractility, extensibility,* and *elasticity.* State why each of these properties is necessary for muscle function.
2. How is skeletal muscle different from the other types of muscle?
3. Why would the skeletal muscles perform poorly without their series-elastic components?

Microscopic Anatomy of Skeletal Muscle

Objectives

When you have completed this section, you should be able to

- describe the structural components of a muscle fiber;
- relate the striations of a muscle fiber to the overlapping arrangement of its protein filaments; and
- name the major proteins of a muscle fiber and state the function of each.

The Muscle Fiber

In order to understand muscle function, you must know how the organelles and macromolecules of a muscle fiber are arranged. Muscle fibers have multiple flattened or sausage-shaped nuclei pressed against the inside of the plasma membrane. This unusual condition results from their embryonic development—several unspecialized cells called **myoblasts**[1] fuse to produce each muscle fiber, with each myoblast contributing a nucleus to the mature fiber. Some myoblasts remain as unspecialized **satellite cells** between the muscle fiber and endomysium. When a muscle is injured, satellite cells can multiply and produce new muscle fibers to some degree. Most muscle repair, however, is by fibrosis rather than regeneration of functional muscle.

The plasma membrane, called the **sarcolemma,**[2] has tunnel-like infoldings called **transverse (T) tubules** that penetrate through the fiber and emerge on the other side. The function of a T tubule is to carry an electrical current from the surface of the cell to the interior when the cell is stimulated. The cytoplasm, called **sarcoplasm,** is occupied mainly by long protein bundles called **myofibrils** about 1 μm in diameter (fig. 12.2). Most other organelles of the cell, such as mitochondria and smooth endoplasmic reticulum (ER), are located between adjacent myofibrils. The sarcoplasm also contains an abundance of **glycogen,** which provides stored energy for the muscle to use during exercise, and a red pigment called **myoglobin,** which binds oxygen until it is needed for muscular activity.

The smooth ER of a muscle fiber is called **sarcoplasmic reticulum (SR).** It forms a network around each myofibril, and alongside the T tubules it exhibits dilated sacs called **terminal cisternae.** The SR is a reservoir for calcium ions; it has gated channels in its membrane that can release a flood of calcium into the cytosol, where the calcium activates the muscle contraction process.

Myofilaments

Let's return to the myofibrils just mentioned—the long protein cords that fill most of the muscle cell—and look at their structure at a finer, molecular level. It is here that the key to muscle contraction lies. Each myofibril is a bundle of parallel protein microfilaments called **myofilaments.** There are three kinds of myofilaments:

1. **Thick filaments** (fig. 12.3*a, b*) are about 15 nm in diameter. Each is made of several hundred molecules of a protein called **myosin.** A myosin molecule is shaped like a golf club, with two polypeptides intertwined to form a shaftlike *tail* and a double globular *head* projecting from it at an angle. A thick filament may be likened to a bundle of 200 to 500 such "golf clubs," with their heads directed outward in a spiral array around the bundle. The heads on one half of the thick filament angle to the left, and the heads on the other half angle to the right; in the middle is a *bare zone* with no heads.
2. **Thin filaments** (fig. 12.3*c, d*), 7 nm in diameter, are composed primarily of two intertwined strands of a protein called **fibrous (F) actin.** Each F actin is like a bead necklace—a string of subunits called **globular (G) actin.** Each G actin has an **active site** that can bind to the head of a myosin molecule.

1. *myo* = muscle + *blast* = precursor

2. *sarco* = flesh, muscle + *lemma* = husk

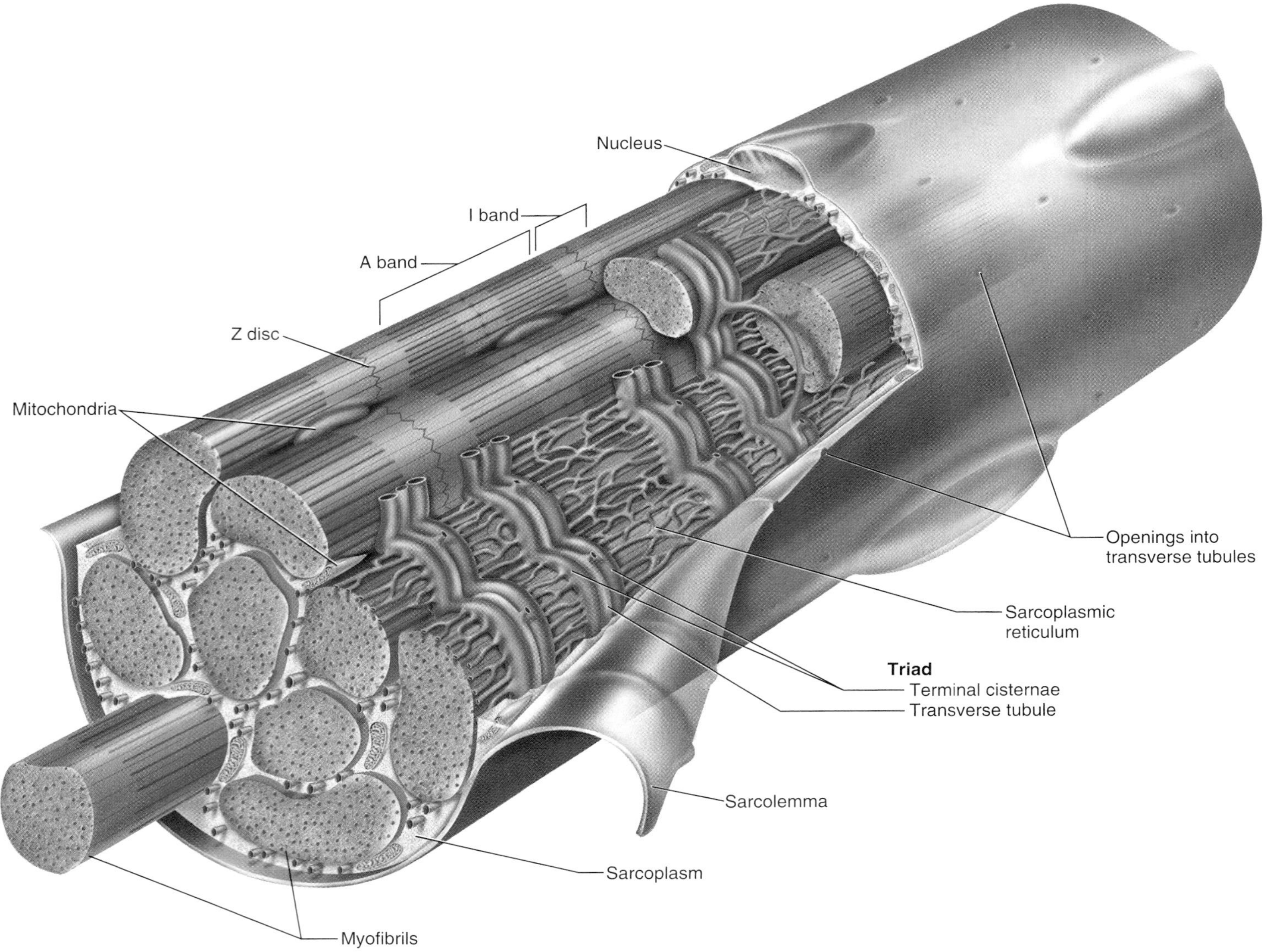

Figure 12.2 **The Structure of a Skeletal Muscle Fiber.** This is a single cell containing 11 myofibrils (9 shown at the *left end* and 2 cut off at midfiber).

A thin filament has a spiral groove along its length that is occupied by 40 to 60 molecules of yet another protein called **tropomyosin.** When a muscle fiber is relaxed, tropomyosin blocks the active sites of actin and prevents myosin from binding to it. Each tropomyosin molecule, in turn, has smaller proteins called **troponin** bound to it at regular intervals. Troponin is a calcium-binding protein.

3. **Elastic filaments** (fig. 12.4*b, c*), 1 nm in diameter, are made of a huge springy protein called **titin**[3] **(connectin).** They run through the core of a thick filament, emerge from the end of it, and connect it to a structure called the *Z disc,* explained shortly. They help to keep thick and thin filaments aligned with each other, resist overstretching of a muscle, and help the cell recoil to resting length after it is stretched.

Myosin and actin are called the **contractile proteins** of muscle because they do the work of shortening the muscle fiber. Tropomyosin and troponin are called the **regulatory proteins** because they act like a switch to determine when it can contract and when it cannot. Several clues as to how they do this may be apparent from what has already been said—calcium ions are released into the sarcoplasm to activate contraction; calcium binds to tro-

3. *tit* = giant + *in* = protein

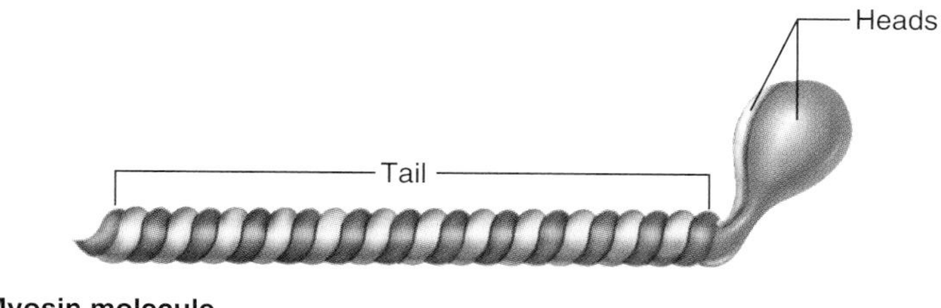

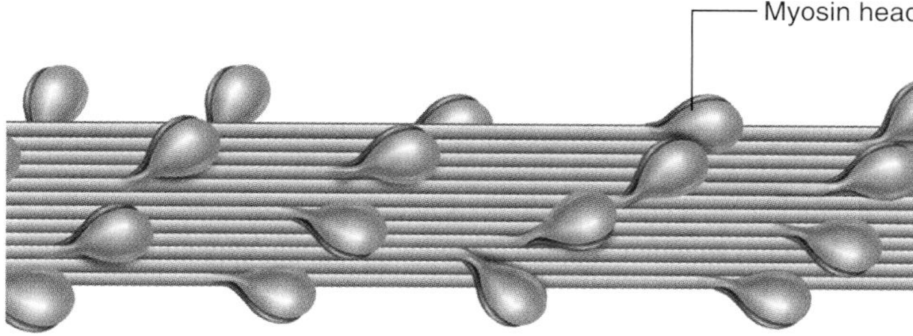

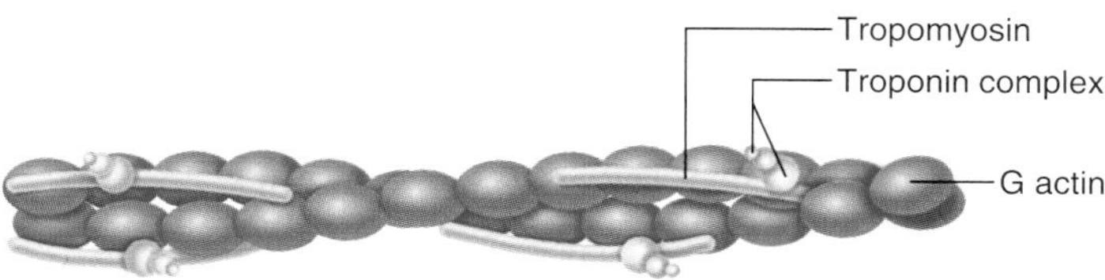

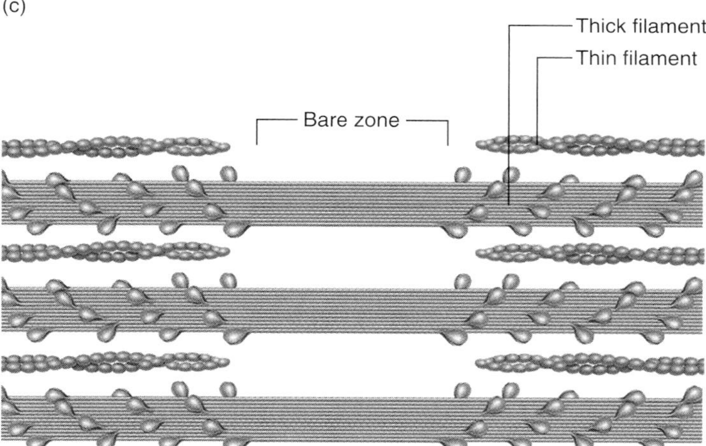

Figure 12.3 Molecular Structure of Thick and Thin Filaments. (*a*) A single myosin molecule consists of two intertwined polypeptides forming a filamentous tail and a double globular head. (*b*) A thick filament consists of 200 to 500 myosin molecules bundled together with the heads projecting outward in a spiral array. (*c*) A thin filament consists of two intertwined chains of G actin molecules, smaller filamentous tropomyosin molecules, and a three-part protein called troponin associated with the tropomyosin. (*d*) A region of overlap between the thick and thin myofilaments.

ponin; troponin is also bound to tropomyosin; and tropomyosin blocks the active sites of actin, so that myosin cannot bind to it when the muscle is not stimulated. Perhaps you are already forming some idea of the contraction mechanism to be explained shortly.

Striations

Myosin and actin are not unique to muscle; these proteins occur in all cells, where they function in cellular motility, mitosis, and transport of intracellular materials. In skeletal and cardiac muscle they are especially abundant, however, and are organized in a precise array that accounts for the striations of these two muscle types (fig. 12.4).

Striated muscle has dark **A bands** alternating with lighter **I bands.** (*A* stands for *anisotropic* and *I* for *isotropic,* which refers to the way these bands affect polarized light. To help remember which band is which, think "d*A*rk" and "l*I*ght.") Each A band consists of thick filaments lying side by side. Part of the A band, where thick and thin filaments overlap, is especially dark. In this region, each thick filament is surrounded by thin filaments. In the middle of the A band, there is a lighter region called the **H band,**[4] into which the thin filaments do not reach.

Each light I band is bisected by a dark narrow **Z disc**[5] (Z line) composed of protein. The Z disc provides anchorage for the thin filaments and elastic filaments. Each segment of a myofibril from one Z disc to the next is called a **sarcomere**[6] (SAR-co-meer), the functional contractile unit of the muscle fiber. A muscle shortens because its individual sarcomeres shorten and pull the Z discs closer to each other, and the Z discs are connected to the sarcolemma by way of the cytoskeleton. As the Z discs are pulled closer together during contraction, they pull on the sarcolemma to achieve overall shortening of the cell.

The terminology of muscle fiber structure is reviewed in table 12.1; this table may be a useful reference as you study the mechanism of contraction.

Before You Go On

Answer the following questions to test your understanding of the preceding section:

4. What special terms are given to the plasma membrane, cytoplasm, and smooth ER of a muscle cell?
5. What is the difference between a myofilament and a myofibril?
6. List five proteins of the myofilaments and describe their physical arrangement.
7. Sketch the overlapping pattern of myofilaments to explain how they account for the A bands, I bands, H bands, and Z discs.

4. H = *helle* = bright
5. Z = *Zwichenscheibe* = "between disc"
6. *sarco* = muscle + *mere* = part, segment

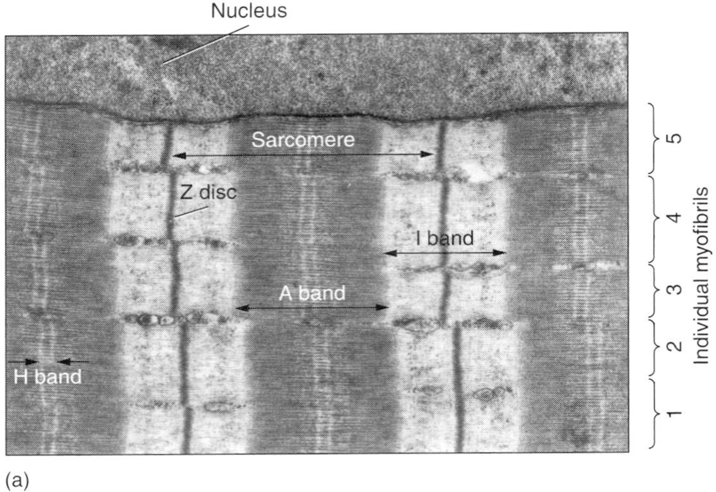

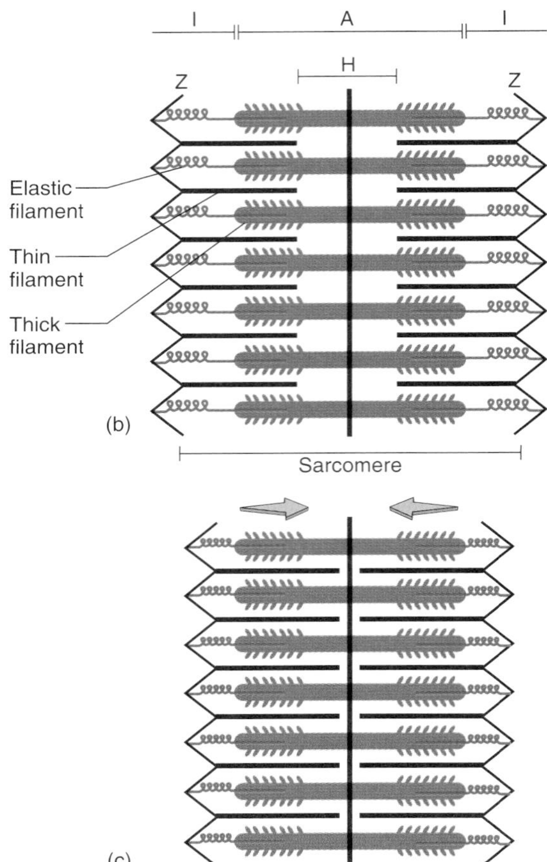

Figure 12.4 **Muscle Striations and Their Molecular Basis.** (*a*) Five myofibrils of a single muscle fiber, showing the striations in the relaxed state. (*b*) The overlapping pattern of thick and thin myofilaments that accounts for the striations seen in figure *a*. (*c*) The pattern of myofilaments in a contracting muscle fiber. Note that all myofilaments are the same length as before, but they overlap to a greater extent. Which band narrows or disappears when muscle contracts?

The Nerve-Muscle Relationship

Objectives

When you have completed this section, you should be able to

- explain what a motor unit is and how it relates to muscle contraction;
- describe the structure of a junction where a nerve fiber meets a muscle fiber; and
- explain why a cell has an electrical charge difference across its plasma membrane and, in general terms, how this relates to muscle contraction.

Skeletal muscle never contracts unless it is stimulated by a nerve (or artificially with electrodes). If its nerve connections are severed or poisoned, a muscle is paralyzed. Thus, muscle contraction cannot be understood without first understanding the relationship between nerve and muscle cells.

Motor Neurons

Skeletal muscles are innervated by *somatic motor neurons.* The cell bodies of these neurons are in the brainstem and spinal cord. Their axons, called **somatic motor fibers,** lead to the skeletal muscles. At its distal end, each somatic motor fiber branches about 200 times, with each branch leading to a different muscle fiber (fig. 12.5). Each muscle fiber is innervated by only one motor neuron.

The Motor Unit

When a nerve signal approaches the end of an axon, it spreads out over all of its terminal branches and stimulates all the muscle fibers supplied by them. Thus, these muscle fibers contract in unison. Since they behave as a single functional unit, one nerve fiber and all the muscle fibers innervated by it are called a **motor unit.** The muscle fibers of a single motor unit are not all clustered together but are dispersed throughout a muscle (fig. 12.6). Thus, when they are stimulated, they cause a weak contraction over a wide area—not just a localized twitch in one small region.

Earlier it was stated that a motor nerve fiber supplies about 200 muscle fibers, but this is just a representative number. Where fine control is needed, we have *small motor units.* In the muscles of eye movement, for example, there are about 23 muscle fibers per nerve fiber. Small motor units are not very strong, but they provide the fine degree of control needed for subtle movements. They also have small neurons that are easily stimulated. Where strength is more important than fine control, we have large motor units. The gastrocnemius muscle of the calf, for example, has about 1,000 muscle fibers per nerve fiber. Large motor units are much stronger, but have larger neurons

Table 12.1 Structural Components of a Muscle Fiber

Term	Definition
General Structure and Contents of the Muscle Fiber	
Sarcolemma	The plasma membrane of a muscle fiber
Sarcoplasm	The cytoplasm of a muscle fiber
Glycogen	An energy-storage polysaccharide abundant in muscle
Myoglobin	An oxygen-storing red pigment of muscle
T tubule	A tunnel-like extension of the sarcolemma extending from one side of the muscle fiber to the other; conveys electrical signals from the cell surface to its interior
Sarcoplasmic reticulum	The smooth ER of a muscle fiber; a Ca^{2+} reservoir
Terminal cisternae	The dilated ends of sarcoplasmic reticulum adjacent to a T tubule
Myofibrils	
Myofibril	A bundle of protein microfilaments (myofilaments)
Myofilament	A threadlike complex of several hundred contractile protein molecules
Thick filament	A myofilament about 11 nm in diameter composed of bundled myosin molecules
Elastic filament	A myofilament about 1 nm in diameter composed of a giant protein, titin, that emerges from the core of a thick filament and links it to a Z disc
Thin filament	A myofilament about 5 to 6 nm in diameter composed of actin, troponin, and tropomyosin
Myosin	A protein with a long shaftlike tail and a globular head; constitutes the thick myofilament
F actin	A fibrous protein made of a long chain of G actin molecules twisted into a helix; main protein of the thin myofilament
G actin	A globular subunit of F actin with an active site for binding a myosin head
Regulatory proteins	Troponin and tropomyosin, proteins that do not directly engage in the sliding filament process of muscle contraction but regulate myosin-actin binding
Tropomyosin	A regulatory protein that lies in the groove of F actin and, in relaxed muscle, blocks the myosin-binding active sites
Troponin	A regulatory protein associated with tropomyosin that acts as a calcium receptor
Titin	A springy protein that forms the elastic filaments and Z discs
Striations and Sarcomeres	
Striations	Alternating light and dark transverse bands across a myofibril
A band	Dark band formed by parallel thick filaments that partly overlap the thin filaments
H band	A lighter region in the middle of an A band that contains thick filaments only; thin filaments do not reach this far into the A band in relaxed muscle
I band	A light band composed of thin filaments only
Z disc	A protein disc to which thin filaments and elastic filaments are anchored at each end of a sarcomere; appears as a narrow dark line in the middle of the I band
Sarcomere	The distance from one Z disc to the next; the contractile unit of a muscle fiber

that are harder to stimulate, and they do not produce such fine control.

One advantage of having multiple motor units in a muscle is that they are able to "work in shifts." Muscle fibers fatigue when subjected to continual stimulation. If all of the fibers in one of your postural muscles fatigued at once, for example, you might collapse. To prevent this, other motor units take over while the fatigued ones rest, and the muscle as a whole can sustain long-term contraction. The role of motor units in muscular strength is discussed later in the chapter.

The Neuromuscular Junction

The functional connection between a nerve fiber and its target cell is called a **synapse** (SIN-aps). When the second cell

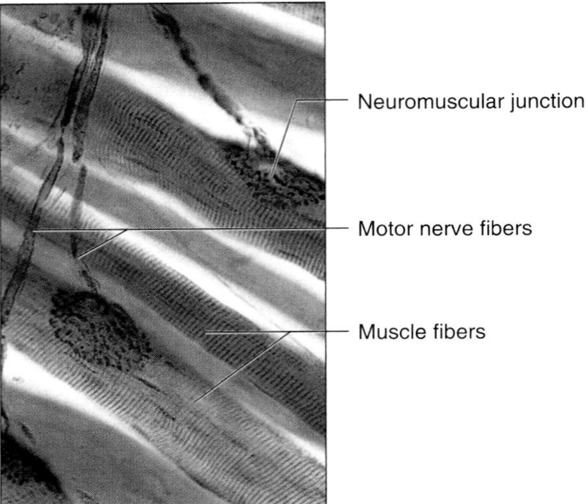

Figure 12.5 Motor Nerve Fibers Innervating a Group of Skeletal Muscle Fibers.

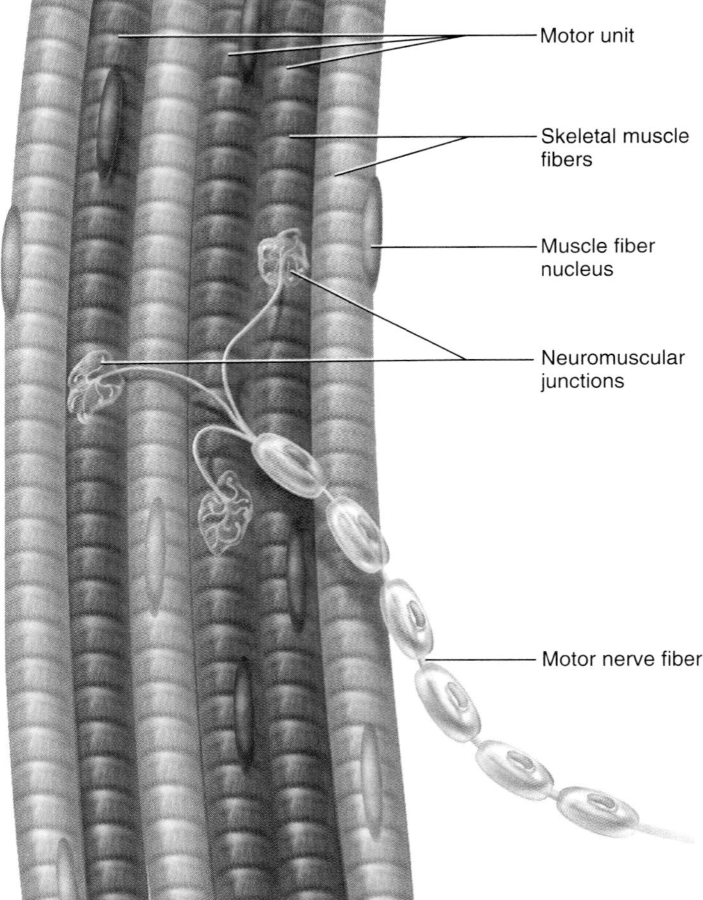

Figure 12.6 **A Motor Unit.** The motor nerve fiber shown here branches to supply those muscle fibers shown in color. The other muscle fibers (gray) belong to other motor units.

is a muscle fiber, the synapse is called a **neuromuscular junction** (fig. 12.7). Each branch of a motor nerve fiber ends in a bulbous swelling called a **synaptic** (sih-NAP-tic) **knob,** which is nestled in a depression on the sarcolemma called the **motor end plate.** The two cells do not actually touch each other but are separated by a tiny gap, the **synaptic cleft,** about 60 to 100 nm wide. A third cell, called a *Schwann cell,* envelops the entire neuromuscular junction and isolates it from the surrounding tissue fluid.

The electrical signal (nerve impulse) traveling down a nerve fiber cannot cross the synaptic cleft like a spark jumping between two electrodes—rather, it causes the nerve fiber to release a neurotransmitter that stimulates the next cell. Although many chemicals function as neurotransmitters, the one released at the neuromuscular junction is **acetylcholine** (ASS-eh-till-CO-leen) **(ACh).** ACh is stored in spherical organelles called **synaptic vesicles.** These are especially concentrated near areas of the plasma membrane called **active zones,** the ACh release sites.

Directly across from the active zones of the nerve cell, the sarcolemma of the muscle cell exhibits infoldings called **junctional folds,** about 1 μm deep. The muscle fiber has about 50 million membrane proteins called **ACh receptors,** which bind the acetylcholine release by the nerve fiber. Most ACh receptors are concentrated in and near these junctional folds. Very few ACh receptors are found anywhere else on a muscle fiber. Junctional folds increase the surface area for receptor sites and ensure a more effective response to ACh. The muscle nuclei beneath the junctional folds are specifically dedicated to the synthesis of ACh receptors and other proteins of the motor end plate. A deficiency of ACh receptors leads to muscle paralysis in the disease *myasthenia gravis* (see clinical insight 12.4, p. 449).

The entire muscle fiber is surrounded by a basement membrane that passes through the synaptic cleft and virtually fills it. Both the sarcolemma and that part of the basement membrane in the cleft contain an enzyme called **acetylcholinesterase** (ASS-eh-till-CO-lin-ESS-ter-ase) **(AChE),** which breaks down ACh, shuts down the stimulation of muscle fibers, and allows a muscle to relax (see clinical insight 12.1).

You must be very familiar with the foregoing terms to understand how a nerve stimulates a muscle fiber and how the fiber contracts. They are summarized in table 12.2 for your later reference.

Electrically Excitable Cells

Muscle fibers and neurons are regarded as *electrically excitable cells* because their plasma membranes exhibit voltage changes in response to stimulation. The study of the electrical activity of cells, called **electrophysiology,** is a key to understanding nervous activity, muscle contraction, the heartbeat, and other physiological phenomena. The details of electrophysiology are presented in the next

Clinical Insight 12.1

Neuromuscular Toxins and Paralysis

Toxins that interfere with synaptic function can paralyze the muscles. Some pesticides, for example, contain *cholinesterase inhibitors* that bind to acetylcholinesterase and prevent it from degrading ACh. This causes *spastic paralysis,* a state of continual contraction of the muscle that poses the danger of suffocation if the laryngeal and respiratory muscles are affected. A person poisoned by a cholinesterase inhibitor must be kept lying down and calm, and sudden noises or other disturbances must be avoided. A minor startle response can escalate to dangerous muscle spasms in a poisoned individual.

Tetanus ("lockjaw") is a form of spastic paralysis caused by a toxin from the bacterium *Clostridium tetani.* In the spinal cord, an inhibitory neurotransmitter called glycine stops motor neurons from producing unwanted muscle contractions. The tetanus toxin blocks glycine release and thus causes overstimulation of the muscles. (At the cost of some confusion, the word *tetanus* also refers to a completely different and normal muscle phenomenon discussed later in this chapter.)

Flaccid paralysis is a state in which the muscles are limp and cannot contract. It can cause respiratory arrest when it affects the thoracic muscles. Flaccid paralysis can be caused by poisons such as curare (cue-RAH-ree) that compete with ACh for receptor sites but do not stimulate the muscle. Curare is extracted from certain plants and used by some South American natives to poison blowgun darts. It has been used to treat muscle spasms in some neurological disorders and to relax abdominal muscles for surgery, but other muscle relaxants have now replaced curare for most purposes.

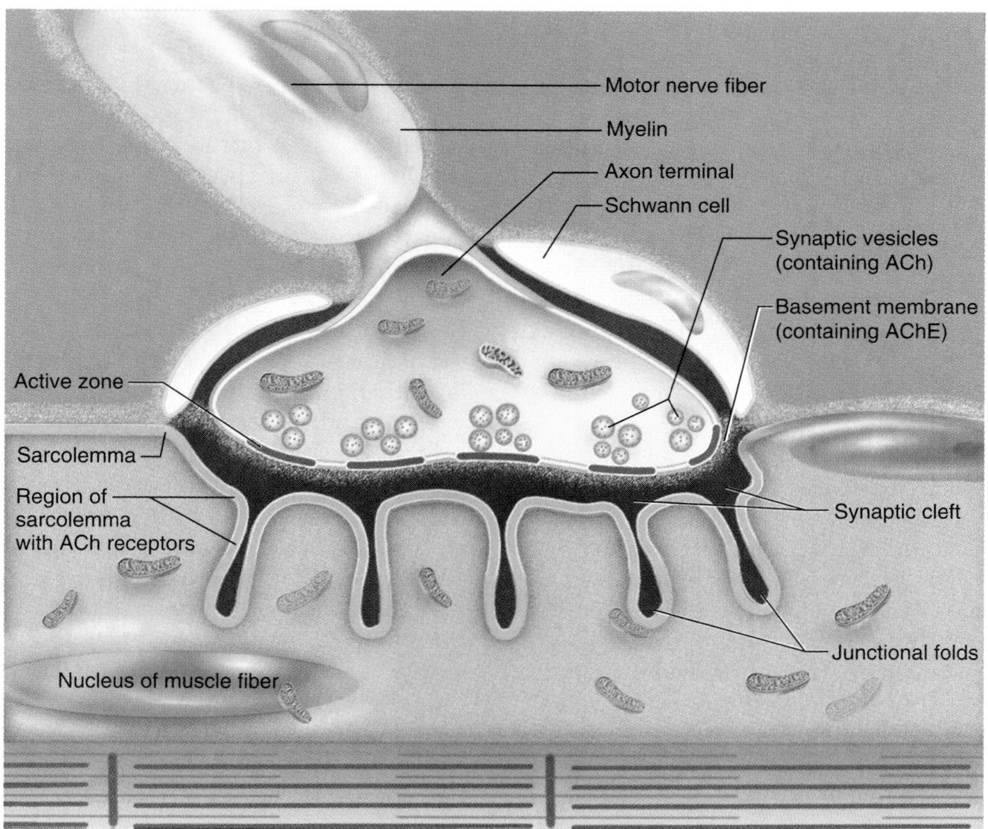

Figure 12.7 A Neuromuscular Junction.

Table 12.2 Components of the Neuromuscular Junction

Term	Definition
Neuromuscular junction	A functional connection between the distal end of a nerve fiber and the middle of a muscle fiber; consists of a synaptic knob and motor end plate
Synaptic knob	The dilated tip of a nerve fiber that contains synaptic vesicles
Motor end plate	A depression in the sarcolemma, near the middle of the muscle fiber, that receives the synaptic knob; contains acetylcholine receptors
Synaptic cleft	A gap of about 60 to 100 nm between the synaptic knob and motor end plate
Synaptic vesicle	A secretory vesicle in the synaptic knob that contains acetylcholine
Active zones	Regions of the plasma membrane of the synaptic knob where synaptic vesicles are clustered; acetylcholine release sites
Junctional folds	Invaginations of the membrane of the motor end plate where ACh receptors are especially concentrated; located across from the active zones
Acetylcholine (ACh)	The neurotransmitter released by a somatic motor fiber that stimulates a skeletal muscle fiber (also used elsewhere in the nervous system)
ACh receptor	An integral protein in the sarcolemma of the motor end plate that binds to ACh
Acetylcholinesterase (AChE)	An enzyme in the sarcolemma and basement membrane of the muscle fiber in the synaptic region; responsible for degrading ACh and stopping the stimulation of the muscle fiber

chapter, but a few fundamental principles must be introduced here so you can understand muscle excitation.

In an unstimulated (resting) cell, there are more anions (negative ions) on the inside of the plasma membrane than on the outside. Thus, the plasma membrane is electrically **polarized,** or charged, like a little battery. In a resting muscle cell, there is an excess of sodium ions (Na^+) in the extracellular fluid (ECF) outside the cell and an excess of potassium ions (K^+) in the intracellular fluid (ICF) within the cell. Also in the ICF, and unable to penetrate the plasma membrane, are anions such as proteins, nucleic acids, and phosphates. These anions make the inside of the plasma membrane negatively charged by comparison to its outer surface.

A difference in electrical charge from one point to another is called an electrical potential, or voltage. The difference is typically 12 volts (V) for a car battery and 1.5 V for a flashlight battery, for example. On a sarcolemma of a muscle cell, the voltage is much smaller, about −90 millivolts (mV), but critically important to life. (The negative sign refers to the relative charge on the intracellular side of the membrane.) This voltage is called the **resting membrane potential (RMP).** It is maintained by the sodium-potassium pump, as explained in chapter 4.

When a nerve or muscle cell is stimulated, dramatic things happen electrically, as we shall soon see in our study of the excitation of muscle. Ion gates in the plasma membrane open and Na^+ rushes down its concentration gradient into the cell. These cations balance or override the negative charges in the ICF, so the inside of the plasma membrane briefly becomes positive. Immediately, Na^+ gates close and K^+ gates open. K^+ rushes out of the cell, partly because it is repelled by the positive sodium charge and partly because it is more concentrated in the ICF than in the ECF, so it diffuses down its concentration gradient when it has the opportunity. The loss of positive potassium ions from the cell turns the inside of the membrane negative again. This quick up-and-down voltage shift, from the negative RMP to a positive value and then back to a negative value again, is called an **action potential.** The RMP is a stable voltage seen in a "waiting" cell, whereas the action potential is a quickly fluctuating voltage seen in an active, stimulated cell.

Action potentials have a way of perpetuating themselves—an action potential at one point on a plasma membrane causes another one to happen immediately in front of it, which triggers another one a little farther along, and so forth. A wave of action potentials spreading along a nerve fiber like this is called a *nerve impulse* or *nerve signal.* Such signals also travel along the sarcolemma of a muscle fiber. We will see shortly how this leads to muscle contraction. Chapter 13 explains the mechanism of action potentials more fully.

Before You Go On

Answer the following questions to test your understanding of the preceding section:

8. What differences would you expect to see between one motor unit where muscular strength is more important than fine control and another motor unit where fine control is more important?

9. Distinguish between acetylcholine, an acetylcholine receptor, and acetylcholinesterase. State where each is found and describe the function it serves.
10. What accounts for the resting membrane potential seen in unstimulated nerve and muscle cells?
11. What is the difference between a resting membrane potential and an action potential?

Behavior of Skeletal Muscle Fibers

Objectives

When you have completed this section, you should be able to

- explain how a nerve fiber stimulates a skeletal muscle fiber;
- explain how stimulation of a muscle fiber activates its contractile mechanism;
- explain the mechanism of muscle contraction;
- explain how a muscle fiber relaxes; and
- explain why the force of a muscle contraction depends on its length prior to stimulation.

The process of muscle contraction and relaxation can be viewed as occurring in four major phases: (1) excitation, (2) excitation-contraction coupling, (3) contraction, and (4) relaxation. Each phase occurs in several smaller steps, which we now examine in detail. The steps are numbered in the following descriptions to correspond to those in figures 12.8 to 12.11.

Excitation

Excitation is the process in which action potentials in the nerve fiber lead to action potentials in the muscle fiber. The steps in excitation are shown in figure 12.8.

1. A nerve signal arrives at the synaptic knob and stimulates voltage-gated calcium channels to open. Calcium ions enter the synaptic knob.
2. Calcium ions stimulate exocytosis of the synaptic vesicles, which release acetylcholine (ACh) into the synaptic cleft. One action potential causes exocytosis of about 60 synaptic vesicles, and each vesicle releases about 10,000 molecules of ACh.
3. ACh diffuses across the synaptic cleft and binds to receptor proteins on the sarcolemma.
4. These receptors are *ligand-gated ion channels.* When ACh (the ligand) binds to them, they change shape and open an ion channel through the middle of the receptor protein. Each channel allows Na^+ to diffuse quickly into the cell and K^+ to diffuse outward. As a result of these ion movements, the sarcolemma reverses polarity—its voltage quickly jumps from the RMP of −90 mV to a peak of +75 mV as Na^+ enters, and then falls back to a level close to the RMP as K^+ diffuses out. This rapid fluctuation in membrane voltage at the motor end plate is called the **end-plate potential (EPP).**
5. Areas of sarcolemma next to the end plate have *voltage-gated ion channels* that open in response to the EPP. Some of the voltage-gated channels are specific for Na^+ and admit it to the cell, while others are specific for K^+ and allow it to leave. These ion movements create an *action potential.* The muscle fiber is now excited.

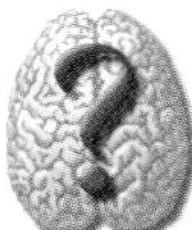

Think About It

An impulse begins at the middle of a 100-mm-long muscle fiber and travels 5 m/sec. How long would it take to reach the ends of the muscle fiber?

Excitation-Contraction Coupling

Excitation-contraction coupling refers to the events that link the action potentials on the sarcolemma to activation of the myofilaments, thereby preparing them to contract. The steps in the coupling process are shown in figure 12.9.

6. A wave of action potentials spreads from the end plate in all directions, like ripples on a pond. When this wave of excitation reaches the T tubules, it continues down them into the sarcoplasm.
7. Action potentials open voltage-regulated ion gates in the T tubules. These are physically linked to calcium channels in the terminal cisternae of the sarcoplasmic reticulum (SR), so gates in the SR open as well and calcium ions diffuse out of the SR, down their concentration gradient and into the cytosol.
8. The calcium ions bind to the troponin of the thin filaments.
9. The troponin-tropomyosin complex changes shape and shifts to a new position. This exposes the active sites on the actin filaments and makes them available for binding to myosin heads.

Contraction

Contraction is the step in which the muscle fiber develops tension and may shorten. (Muscles often "contract," or develop tension, without shortening, as we see later.) How a muscle fiber shortens remained a mystery until sophisticated techniques in electron microscopy enabled cytologists to see the molecular organization of muscle fibers. In 1954, two researchers at the Massachusetts Institute of Technology, Jean Hanson and Hugh Huxley, found evidence for a model now called the **sliding filament theory.** This theory holds that the thin filaments slide over the

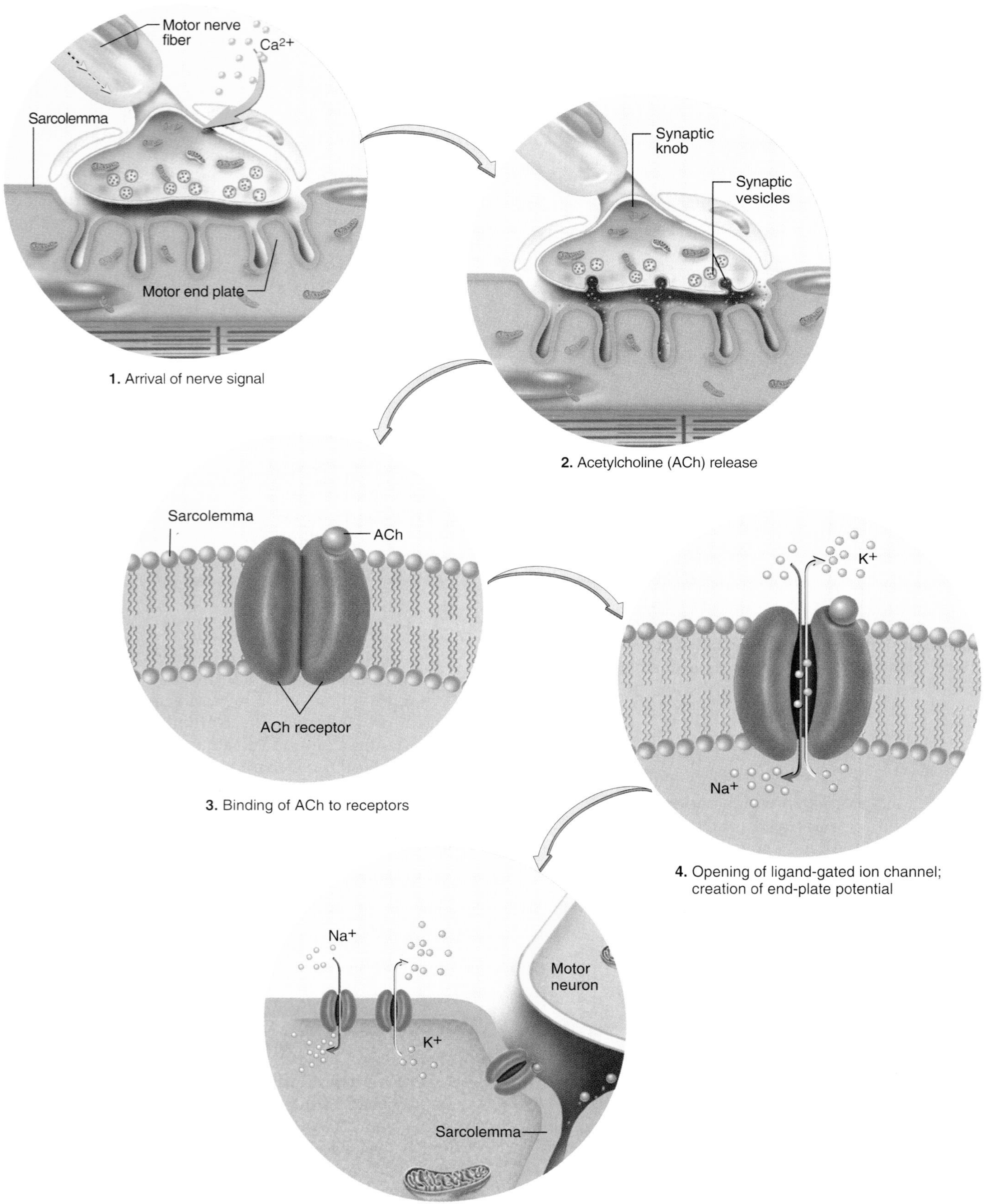

Figure 12.8 **Excitation of a Muscle Fiber.** These events link action potentials in a nerve fiber to the generation of action potentials in the muscle fiber.

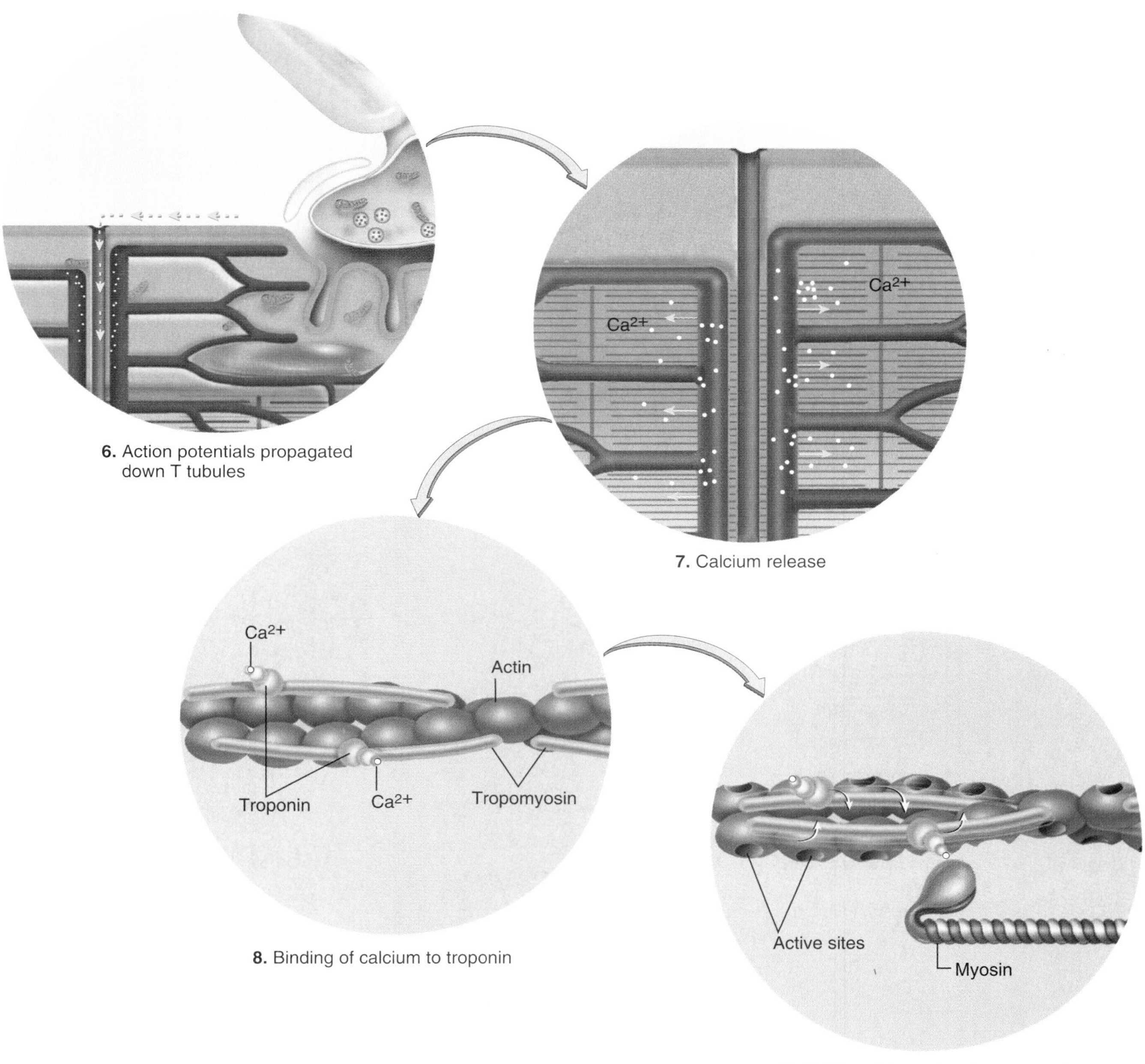

6. Action potentials propagated down T tubules

7. Calcium release

8. Binding of calcium to troponin

9. Shifting of tropomyosin; exposure of active sites on actin

Figure 12.9 **Excitation-Contraction Coupling.** These events link action potentials in the muscle fiber to the release and binding of calcium ions. The numbered steps in this figure begin where the previous figure left off.

thick ones and pull the Z discs behind them, causing the cell as a whole to shorten. The individual steps in this mechanism are shown in figure 12.10.

10. The myosin head must have an ATP molecule bound to it to initiate the contraction process. **Myosin ATPase,** an enzyme in the head, hydrolyzes this ATP. The energy released by this process activates the head, which "cocks" into an extended, high-energy position. The head temporarily keeps the ADP and phosphate group bound to it.
11. The cocked myosin binds to an active site on the thin filament and forms a cross-bridge.
12. Myosin releases the ADP and phosphate and flexes into a bent, low-energy position, tugging the thin filament along with it. This is called the **power stroke.** The head remains bound to actin until it binds a new ATP.
13. Upon binding more ATP, myosin releases the actin. It is now prepared to repeat the whole process—it will hydrolyze the ATP, recock (the **recovery stroke**), form a new cross-bridge, and produce another power stroke.

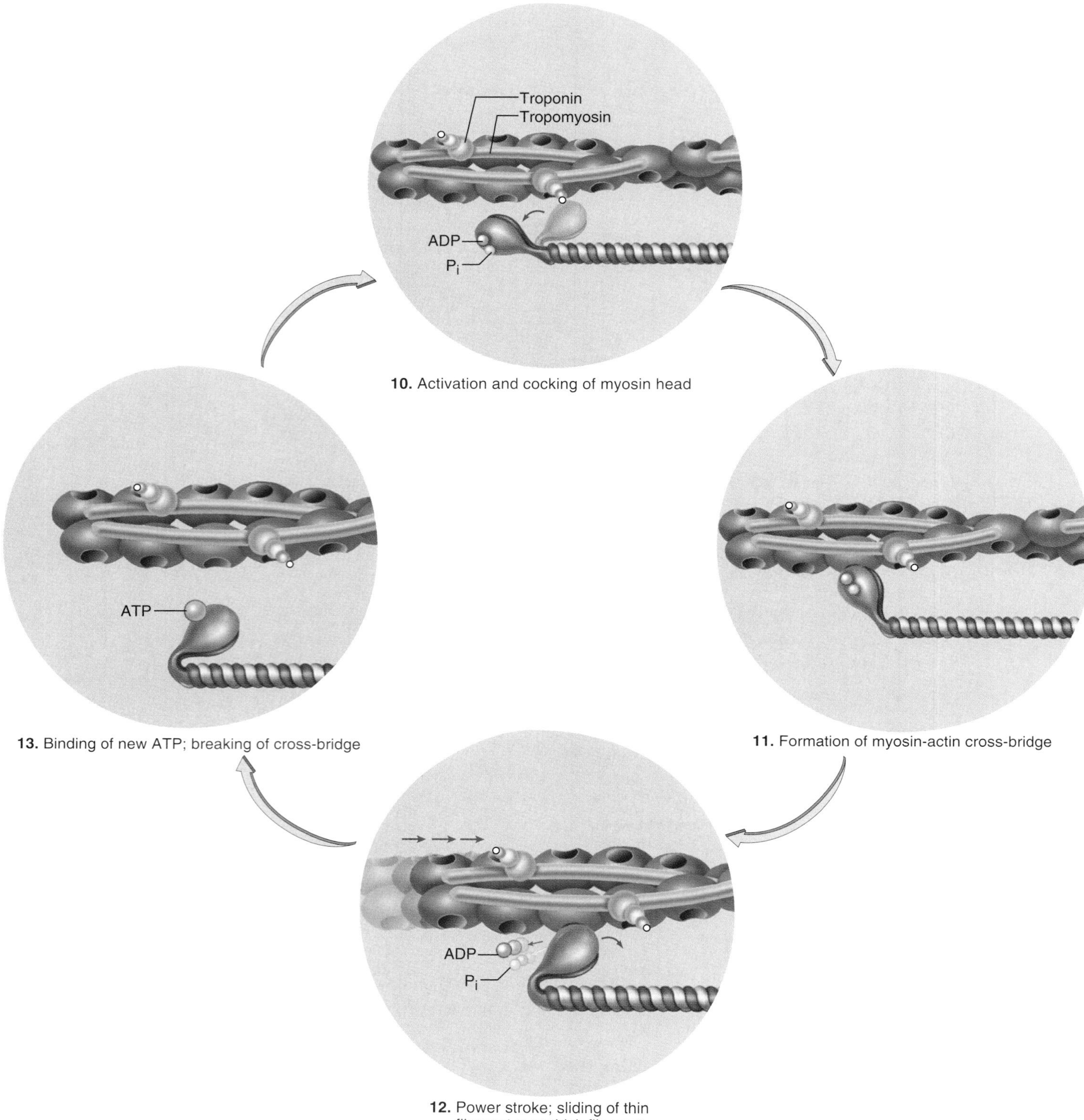

Figure 12.10 The Sliding Filament Mechanism of Contraction. This is a cycle of repetitive events that causes a thin filament to slide over a thick filament and generates tension in the muscle. The numbered steps in this figure begin where the previous figure left off.

It might seem as if releasing the thin filament at step 13 would simply allow it to slide back to its previous position, so that nothing would have been accomplished. Think of the sliding filament mechanism, however, as being similar to the way you would pull in a boat anchor hand over hand. When the myosin head cocks, it is like your hand reaching out to grasp the anchor rope. When it flexes back into the low-energy position, it is like your elbow flexing to pull on the rope and draw the anchor up a little bit. When you let go of the rope with one hand, you hold onto it with the other, alternating hands until the anchor is pulled in. Similarly, when one myosin head releases the actin in preparation for the recovery stroke, there are many other heads on the same thick filament holding onto the thin filament so that it doesn't slide back. At any given moment during contraction, about half of the heads are bound to the thin filament and the other half are extending forward to grasp the filament farther down. That is, the myosin heads of a thick filament do not all "stroke" at once but contract sequentially.

As another analogy, consider a millipede—a little wormlike animal with a few hundred tiny legs. Each leg takes individual jerky steps, but all the legs working together produce smooth, steady movement—just as all the heads of a thick filament collectively produce a smooth, steady pull on the thin filament. Note that even though the muscle fiber contracts, the *myofilaments do not become shorter* any more than a rope becomes shorter as you pull in an anchor. The thin filaments slide over the thick ones, as the name of the theory implies.

A single cycle of power and recovery strokes by all the myosin heads in a muscle fiber would shorten the fiber by about 1%. A fiber, however, may shorten by as much as 40% of its resting length, so obviously the cycle of power and recovery must be repeated many times by each myosin head. Each head carries out about five strokes per second, and each stroke consumes one molecule of ATP.

Relaxation

When its work is done, a muscle fiber relaxes and returns to its resting length. This is achieved by the steps shown in figure 12.11.

14. Nerve signals stop arriving at the neuromuscular junction, so the synaptic knob stops releasing ACh.
15. As ACh dissociates (separates) from its receptor, acetylcholinesterase breaks it down into fragments that cannot stimulate the muscle. The synaptic knob reabsorbs these fragments for recycling. All of this happens continually while the muscle is being stimulated, too; but when nerve signals stop, no new ACh is released to replace that which is broken down. Therefore, stimulation of the muscle fiber by ACh ceases.
16. Active transport pumps in the sarcoplasmic reticulum (SR) begin to pump Ca^{2+} from the cytosol back into the cisternae. Here, the calcium binds to a protein called **calsequestrin** (CAL-see-QUES-trin) and is stored until the fiber is stimulated again. Since active transport requires ATP, you can see that *ATP is needed for muscle relaxation as well as for muscle contraction* (see clinical insight 12.2).
17. As calcium ions dissociate from troponin, they are pumped into the SR and are not replaced.
18. Tropomyosin moves back into the position where it blocks the active sites of the actin filament. Myosin can no longer bind to actin, and the muscle fiber ceases to produce or maintain tension.

A muscle returns to its resting length with the aid of two forces: (1) like a recoiling rubber band, the series-elastic components stretch it; and (2) since muscles often occur in antagonistic pairs, the contraction of an antagonist lengthens the relaxed muscle. Contraction of the triceps brachii, for example, extends the elbow and lengthens the biceps brachii.

The Length-Tension Relationship and Tonus

The amount of tension generated by a muscle, and therefore the force of its contraction, depends on how stretched or contracted it was before it was stimulated. This principle is called the **length-tension relationship.** The reasons for it can be seen in figure 12.12. If a fiber is overly contracted at rest, its thick filaments are rather close to the Z discs. The stimulated muscle may contract a little, but then the thick filaments butt up against the Z discs and can go no farther. The contraction is therefore a weak one. On the other hand, if a muscle fiber is too stretched before it is stimulated, there is relatively little overlap between its thick and thin filaments. When the muscle is stimulated, its myosin heads cannot "get a good grip" on the thin filaments, and again the contraction is weak. (As mentioned in chapter 11, this is one reason you should not bend at the waist to pick up a heavy object. Muscles of the back become overly stretched and cannot contract effectively to straighten your spine against a heavy resistance.)

Between these extremes, there is an optimum resting length at which a muscle produces the greatest force when it contracts. The central nervous system continually monitors and adjusts the length of a resting muscle, maintaining a state of partial contraction called **tonus (muscle tone).** This maintains optimum length and makes the muscles ideally ready for action. The elastic filaments of the sarcomere also help to maintain enough myofilament overlap to ensure an effective contraction when the muscle is called into action.

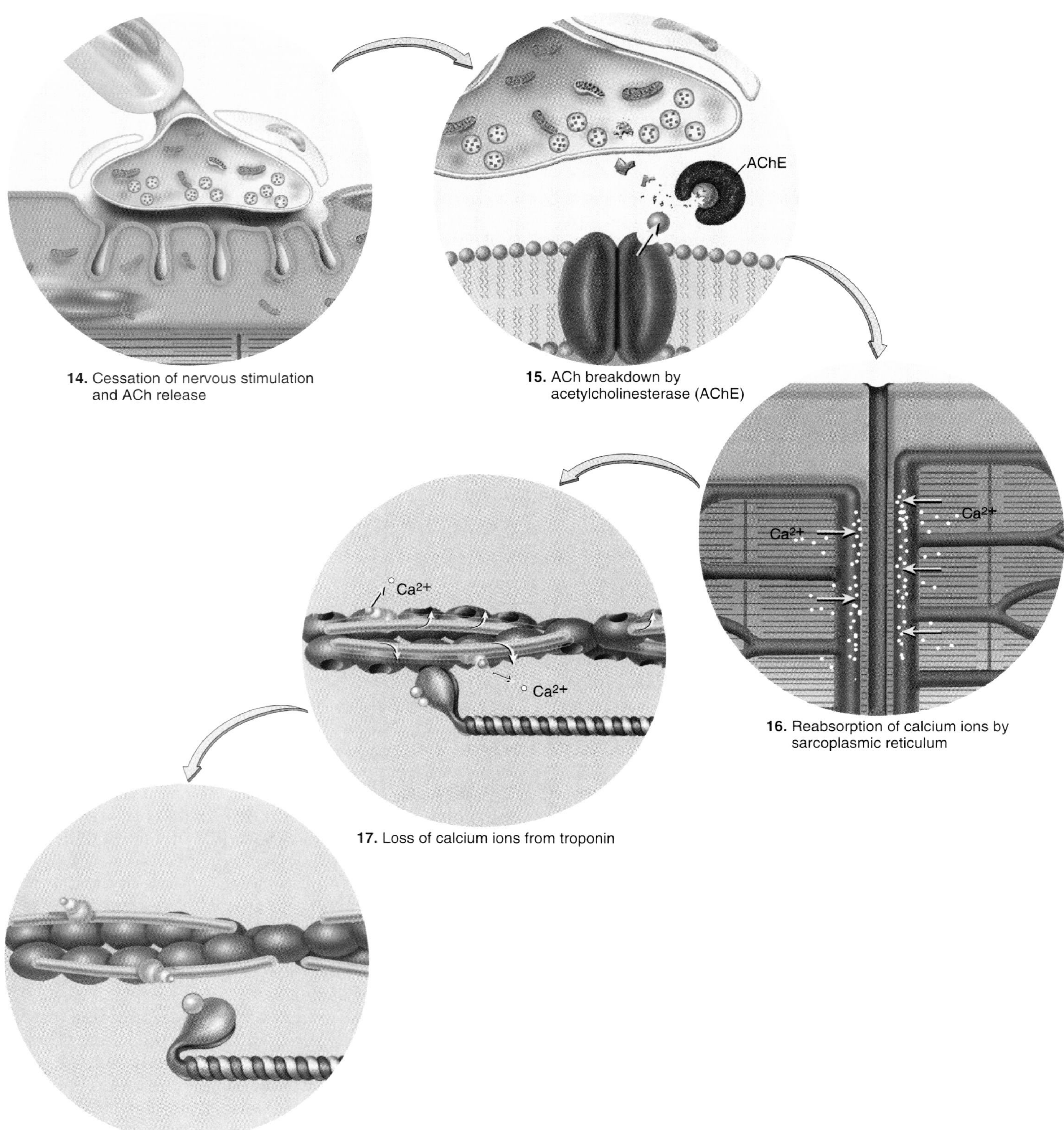

14. Cessation of nervous stimulation and ACh release

15. ACh breakdown by acetylcholinesterase (AChE)

16. Reabsorption of calcium ions by sarcoplasmic reticulum

17. Loss of calcium ions from troponin

18. Return of tropomyosin to position blocking active sites of actin

Figure 12.11 **Relaxation of a Muscle Fiber.** These events lead from the cessation of a nerve signal to the release of thin filaments by myosin. The numbered steps in this figure begin where the previous figure left off.

Clinical Insight 12.2

Rigor Mortis

Rigor mortis[7] is the hardening of the muscles and stiffening of the body that begins 3 to 4 hours after death. It occurs partly because the deteriorating sarcoplasmic reticulum releases calcium ions into the cytosol, and the deteriorating sarcolemma admits more calcium ions from the extracellular fluid. The calcium ions activate myosin-actin cross bridging and muscle contraction. Furthermore, the muscle cannot relax without ATP, and ATP is no longer produced after death. Thus, the fibers remain contracted until the myofilaments begin to decay. Rigor mortis peaks about 12 hours after death and then diminishes over the next 48 to 60 hours.

7. *rigor* = rigidity + *mortis* = of death

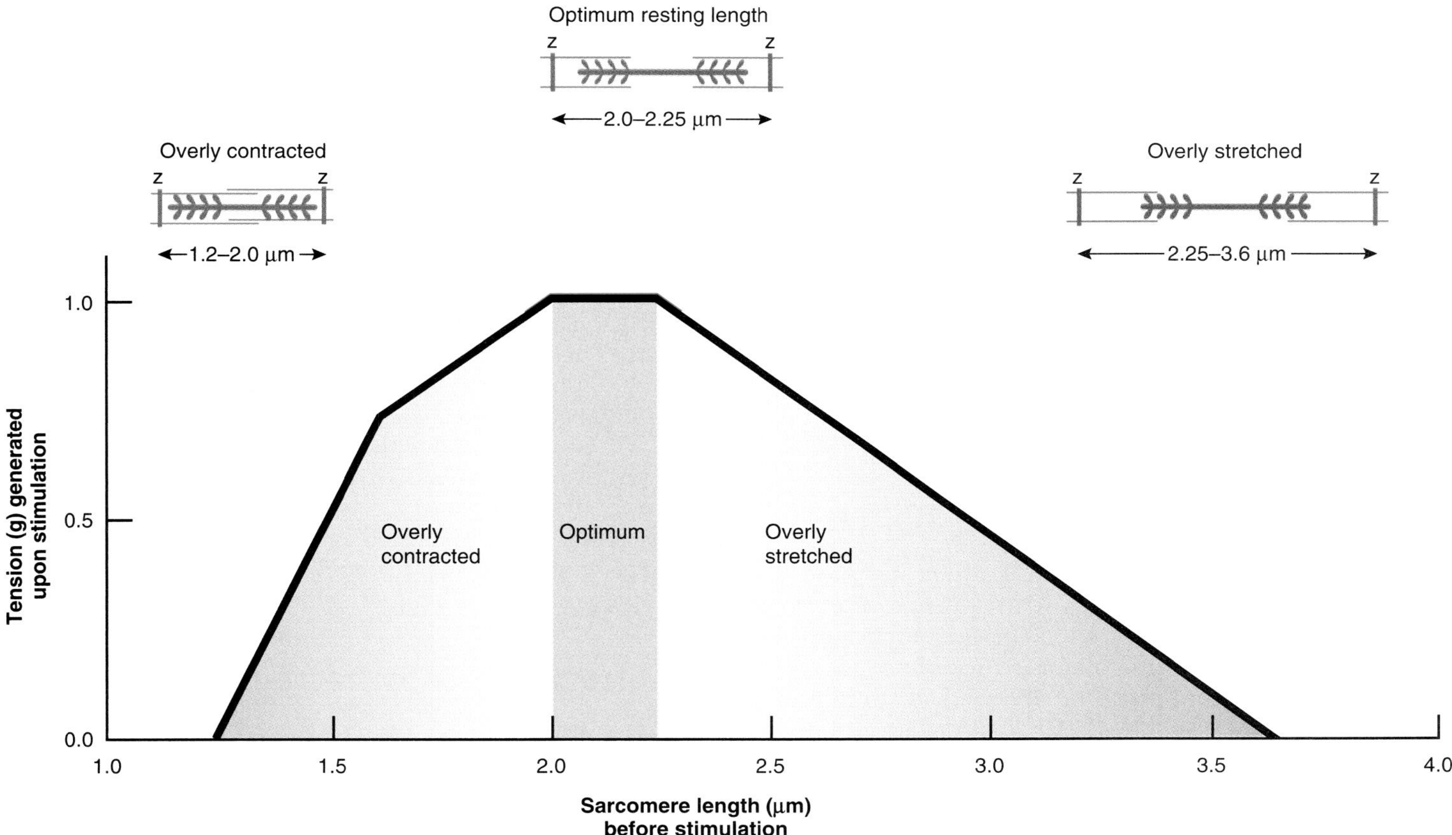

Figure 12.12 The Length-Tension Relationship of Skeletal Muscle. *Center:* In a resting muscle fiber, the sarcomeres are usually 2.0 to 2.25 µm long, the optimum length for producing maximum tension when the muscle is stimulated to contract. Note how this relates to the degree of overlap between the thick and thin filaments. *Left:* If the muscle is overly contracted, the thick filaments butt against the Z discs and the fiber cannot contract very much more when it is stimulated. *Right:* If the muscle is overly stretched, there is so little overlap between the thick and thin myofilaments that few cross-bridges can form between myosin and actin.

Before You Go On

Answer the following questions to test your understanding of the preceding section:

12. What change does ACh cause in an ACh receptor? How does this electrically affect the muscle fiber?
13. How do troponin and tropomyosin regulate the interaction between myosin and actin?
14. Describe the roles played by ATP in the power and recovery strokes of myosin.
15. What steps are necessary for a contracted muscle to return to its resting length?

Behavior of Whole Muscles

Objectives

When you have completed this section, you should be able to

- describe the stages of a muscle twitch;
- describe treppe and explain how it relates to muscle warm-up;
- explain how muscle twitches add up to produce stronger muscle contractions;
- distinguish between isometric and isotonic contraction; and
- distinguish between concentric and eccentric contractions.

Now you know how an individual muscle cell shortens. Our next objective is to move up to the organ grade of construction and consider how this relates to the action of the muscle as a whole.

Threshold, Latent Period, and Twitch

Muscle contraction has often been studied and demonstrated using the gastrocnemius (calf) muscle of a frog, which can easily be isolated from the leg along with its connected sciatic nerve (see historical insight 12.3). This nerve-muscle preparation can be attached to stimulating electrodes and to a recording device that produces a *myogram,* a chart of the timing and strength of the muscle's contraction.

A weak electrical stimulus to the muscle or its nerve causes no contraction. By gradually increasing the voltage and stimulating the muscle again, we can determine the **threshold**—the minimum voltage necessary to produce contraction. At this voltage or higher, a single brief stimulus causes a quick cycle of contraction and relaxation called a **twitch** (fig. 12.13). There is a delay, or **latent period,** of typically 2 milliseconds (msec) or so between the onset of the stimulus and the onset of the twitch. This represents the time required for excitation, excitation-contraction coupling, and tautening of the series-elastic components by the

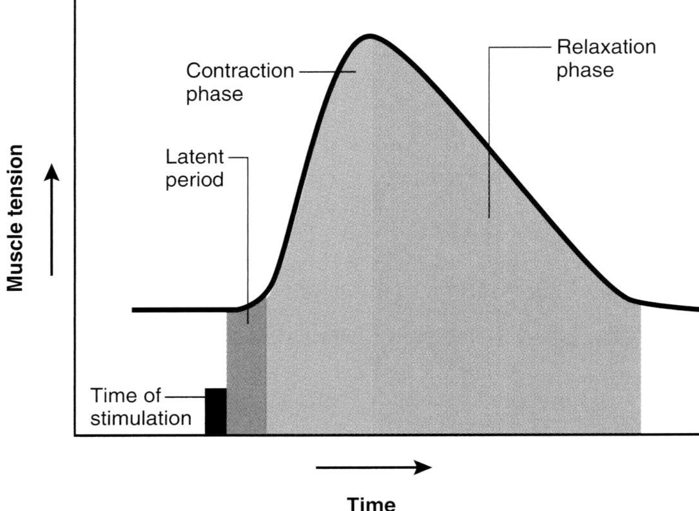

Figure 12.13 A Muscle Twitch.

contracting muscle. The muscle then produces tension and usually moves a resisting object (load) at its distal end. The load could be the sensor of a recording apparatus or, in the body, the bone on which the muscle inserts. The twitch itself lasts from 7 to 100 msec and is divided into two phases: the **contraction phase** in which the muscle develops tension and usually shortens, and the **relaxation phase** in which it loses tension and usually returns to its resting length. For the first 5 msec or so of the contraction phase, there is a **refractory period** in which the muscle will not respond to a new stimulus.

In the body, of course, a muscle is stimulated not by electrodes but by a nerve. In the laboratory, if the electrode is applied to the nerve, it stimulates the nerve to release acetylcholine. If it is applied to the muscle, it opens voltage-regulated gates in the muscle fiber and starts a chain of action potentials, thus bypassing the need for ACh release.

Contraction Strength of Twitches

As long as the voltage of an artificial stimulus delivered directly to a muscle is at threshold or higher, a muscle gives a complete twitch. Increasing the voltage still more does not cause the twitches to become any stronger. There are other factors, however, that can produce stronger twitches. Indeed, muscles *must* be able to contract with variable strength—differently in lifting a glass of champagne than in lifting a heavy barbell, for example.

If we stimulate the nerve rather than the muscle, higher voltages produce stronger muscle contractions because they excite more nerve fibers and therefore more motor units. The more motor units that contract, the more strongly the muscle as a whole contracts (fig. 12.14). The process of bringing more motor units into play is called

Historical Insight 12.3

Galvani, Volta, and Animal Electricity

The invention of modern dry cells can be traced to studies of frog muscle by Italian anatomist Luigi Galvani (1737–98). He suspended isolated frog legs from a copper hook and noticed that they twitched when touched with an iron scalpel. He attributed this to "animal electricity" in the legs. The physicist Alessandro Volta (1745–1827) investigated Galvani's discovery further. He concluded that when two different metals (such as the copper hook and iron scalpel) are separated by an electrolyte solution (a frog's tissue fluids), a chemical reaction occurs that produces an electrical current. This current had stimulated the muscle in the legs of Galvani's frogs and caused the twitch. Based on this principle, Volta invented the first simple voltaic cell, the forerunner of today's dry cells.

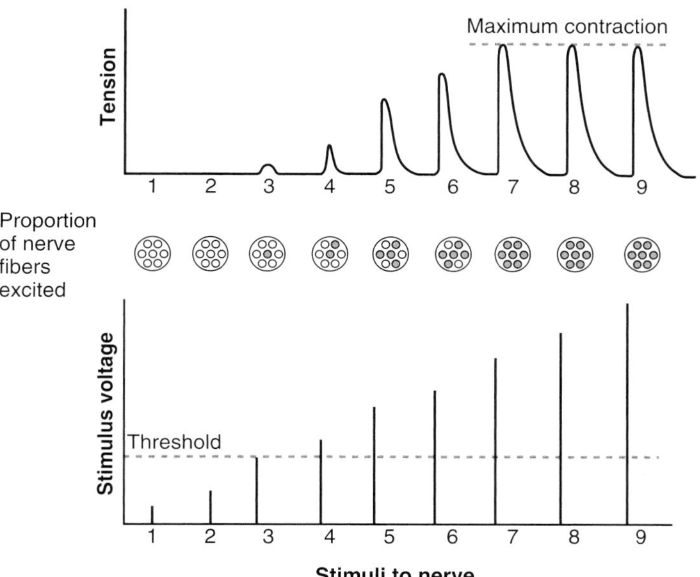

Figure 12.14 The Relationship Between Stimulus Intensity (voltage) and Muscle Tension. Weak stimuli *(1–2)* fail to stimulate any nerve fibers and therefore produce no muscle contraction. When stimuli reach or exceed threshold *(3–7)*, they excite more and more nerve fibers and motor units and produce stronger and stronger contractions. This is multiple motor unit summation (recruitment). Once all of the nerve fibers are stimulated *(7–9)*, further increases in stimulus strength produce no further increase in muscle tension.

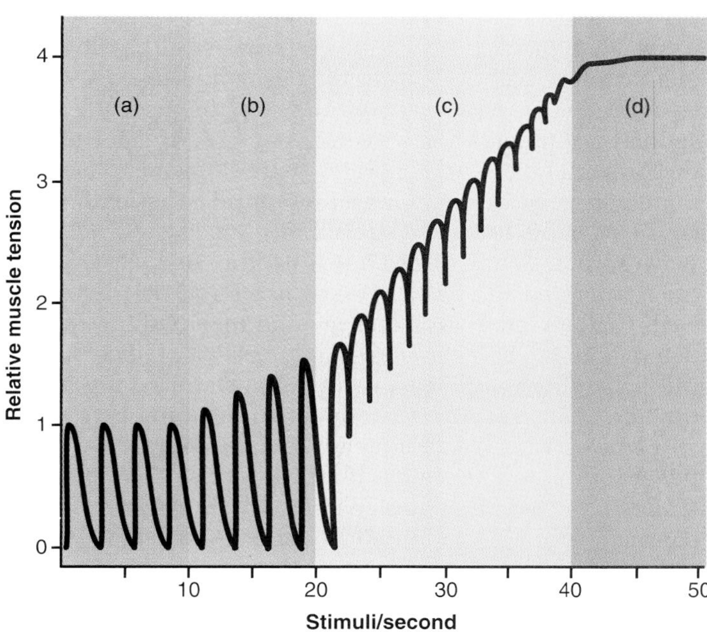

Figure 12.15 The Relationship Between Stimulus Frequency and Muscle Tension. (*a*) At low frequency, the muscle relaxes completely between stimuli and shows twitches of uniform strength. (*b*) Treppe: at a moderate frequency of stimulation, the muscle relaxes fully between contractions, but successive twitches are stronger. (*c*) Wave summation and incomplete tetanus: at still higher stimulus frequency, the muscle does not have time to relax completely between twitches, and the force of each twitch builds on the previous one. (*d*) Complete tetanus: at high stimulus frequency, the muscle does not have time to relax at all between stimuli and exhibits a state of continual contraction with about four times as much tension as a single twitch.

recruitment, or **multiple motor unit (MMU) summation.** It is seen not just in artificial stimulation but is part of the way the nervous system behaves normally to produce variable muscle contractions.

Another way to produce a stronger muscle contraction is to stimulate the muscle at a higher frequency. Even when voltage remains the same, high-frequency stimulation causes stronger contractions than low-frequency stimulation. In figure 12.15*a*, we see that when a muscle is stimulated at a low frequency (up to 10 stimuli/sec in this example), it produces an identical twitch for each stimulus and fully recovers between twitches.

Between 10 and 20 stimuli per second, the muscle still recovers fully between twitches, but each twitch develops more tension than the one before. This pattern of increasing tension with repetitive stimulation is called

treppe[8] (TREP-eh), or the *staircase phenomenon,* after the appearance of the myogram (fig. 12.15*b*). According to one hypothesis, treppe occurs because stimuli arrive so rapidly that the sarcoplasmic reticulum does not have time between stimuli to completely reabsorb all the calcium that it released. Thus, the calcium concentration in the cytosol rises higher and higher with each stimulus and causes subsequent twitches to be stronger. Another hypothesis is that the heat released by each twitch causes muscle enzymes such as myosin ATPase to work more efficiently and produce stronger twitches as the muscle warms up. One purpose of warm-up exercises before athletic competition is to induce treppe, so that the muscle contracts more effectively when the competition begins.

At a still higher stimulus frequency (20–40 stimuli/sec in fig. 12.15*c*), each new stimulus arrives before the previous twitch is over. As long as the new stimulus does not arrive during the refractory period, when the muscle is unresponsive, it triggers a new twitch that "rides piggyback" on the previous one and generates higher tension. This phenomenon goes by two names: **temporal**[9] **summation,** because it results from two stimuli arriving close together, or **wave summation,** because it results from one wave of contraction added to another. Wave is added upon wave, so each twitch reaches a higher level of tension than the one before, and the muscle relaxes only partially between stimuli. This effect produces a state of sustained partial contraction called **incomplete tetanus** (or **incomplete tetany**).

At a still higher frequency, such as 40 to 50 stimuli per second, the muscle has no time to relax at all between stimuli, and the twitches fuse into a smooth, prolonged contraction called **complete tetanus.** A muscle in complete tetanus produces about four times as much tension as a single twitch (fig. 12.15*d*). This type of tetanus should not be confused with the disease of the same name caused by the tetanus toxin, explained in clinical insight 12.1. The tetanus caused by high-frequency stimulation is a normal part of muscle physiology. It results in smooth, sustained contractions rather than jerky, spasmodic ones. Imagine trying to lift a cup of hot tea to your lips if your muscles exhibited short, spasmodic contractions separated by partial relaxations!

Isometric and Isotonic Contraction

In muscle physiology, "contraction" does not always mean the shortening of a muscle—it may mean only that the muscle is producing internal tension while an external resistance causes it to stay the same length or even to become longer. Thus, physiologists speak of different kinds of muscle contraction as *isometric* versus *isotonic,* and *concentric* versus *eccentric* contraction.

Suppose you lift a heavy box of books from a table. When you first contract the muscles of your arms, you can feel the tension building in them even though the box is not yet moving. At this point, your muscles are contracting at a cellular level, but their tension is being absorbed by the series-elastic components and is resisted by the weight of the load; the muscle as a whole is not producing any external movement. This phase is called **isometric**[10] **contraction**—contraction without a change in length (fig. 12.16*a*). **Isotonic**[11] **contraction**—contraction with a change in length but no change in tension—begins when internal tension builds to the point that it overcomes the resistance. The muscle now shortens, moves the load, and maintains essentially the same tension from then on (fig. 12.16*b*). Isometric and isotonic contraction are both phases of normal muscular action (fig. 12.17).

There are two forms of isotonic contraction—concentric and eccentric. In **concentric contraction,** a muscle shortens as it maintains tension—for example, when the biceps brachii contracts and flexes the elbow. In an **eccentric contraction,** a muscle lengthens as it maintains tension. If you set that box of books down again (fig. 12.16*c*), your biceps brachii lengthens as you extend your elbow, but it maintains tension to act as a brake and keep you from simply dropping the box. A weight lifter uses concentric contraction when lifting a barbell and eccentric contraction when lowering it to the floor.

In summary, during isometric contraction, a muscle develops tension without changing length, and in isotonic contraction, it changes length while maintaining constant tension. In concentric contraction, a muscle develops tension while it shortens, and in eccentric contraction, it maintains tension while it is lengthening.

Before You Go On

Answer the following questions to test your understanding of the preceding section:

16. Explain how warm-up is related to treppe and why it improves athletic performance.
17. Explain the role of tetanus in normal muscle action.
18. Describe an everyday activity *not* involving the arms in which your muscles would switch from isometric to isotonic contraction.
19. Describe an everyday activity *not* involving the arms that would involve concentric contraction and one that would involve eccentric contraction.

8. *treppe* = staircase
9. *tempor* = time
10. *iso* = same, uniform + *metr* = length
11. *iso* = same, uniform + *ton* = tension

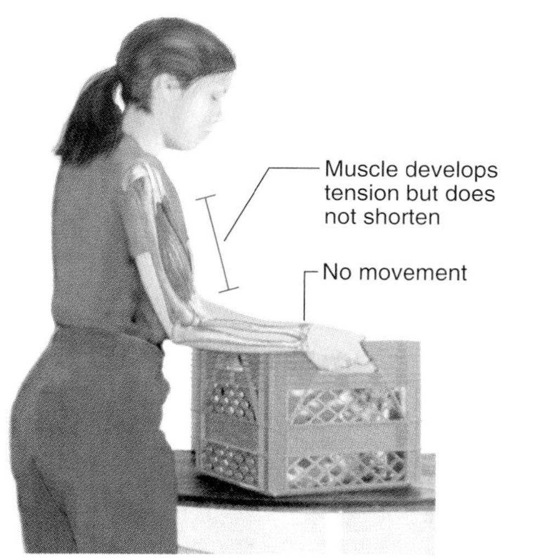

(a) Isometric contraction

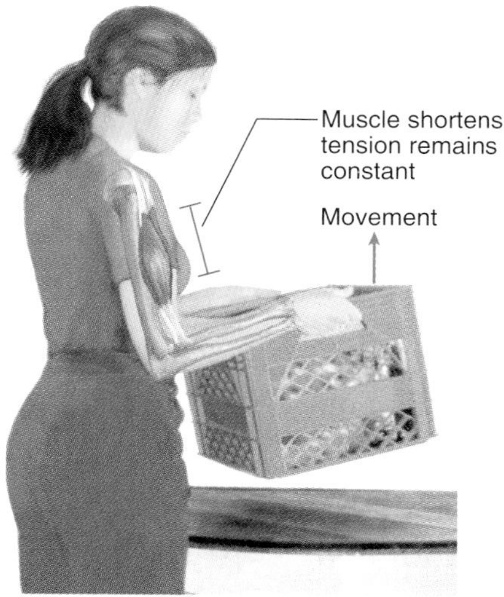

(b) Isotonic, concentric contraction

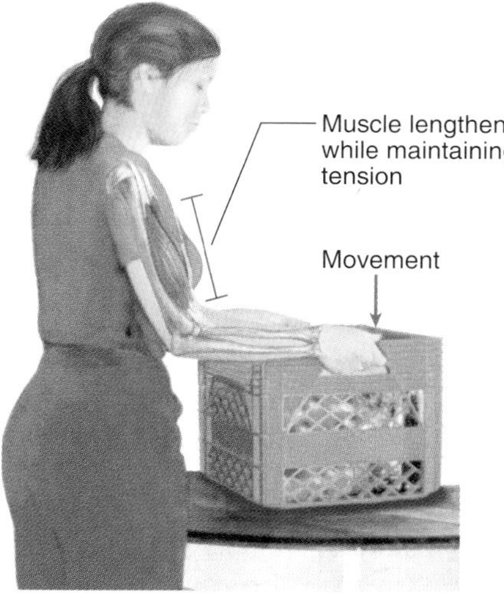

(c) Isotonic, eccentric contraction

Figure 12.16 Isometric and Isotonic Contraction. (*a*) Isometric contraction, in which a muscle develops tension but does not shorten. This occurs at the beginning of any muscle contraction but is prolonged in actions such as lifting heavy weights. (*b*) Isotonic concentric contraction, in which the muscle shortens while maintaining a constant degree of tension. In this phase, the muscle moves a load. (*c*) Isotonic eccentric contraction, in which the muscle maintains tension while it lengthens, allowing a muscle to relax without going suddenly limp.

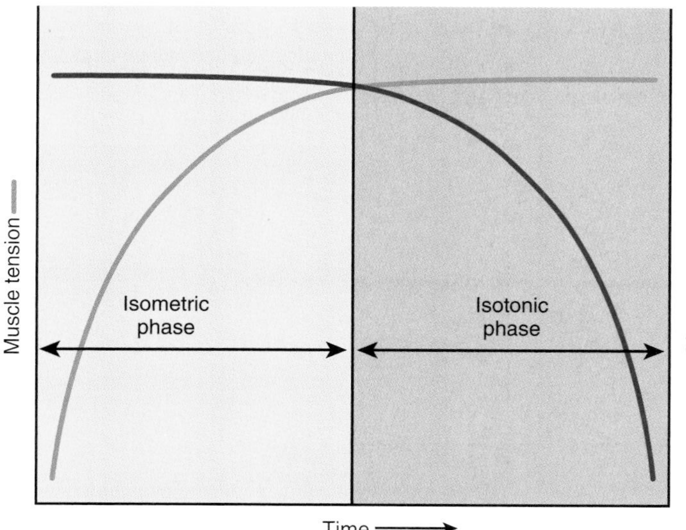

Figure 12.17 Isometric and Isotonic Phases of Muscle Contraction. At the beginning of a contraction (isometric phase), muscle tension rises but the length remains constant (the muscle does not shorten). When tension overcomes the resistance of the load, the tension levels off and the muscle begins to shorten and move the load (isotonic phase). How would you extend this graph in order to show eccentric contraction?

Muscle Metabolism

Objectives

When you have completed this section, you should be able to

- explain how skeletal muscle meets its energy demands during rest and exercise;
- explain the basis of muscle fatigue and soreness;
- define *oxygen debt* and explain why extra oxygen is needed even after an exercise has ended;
- distinguish between slow- and fast-twitch muscle fibers, and explain the functional roles of these two types;
- discuss the factors that affect muscular strength; and
- discuss the effects of resistance and endurance exercises on muscle.

ATP Sources

All muscle contraction depends on ATP; no other energy source can serve in its place. The supply of ATP depends, in turn, on the availability of oxygen and organic energy sources such as glucose and fatty acids. To understand how muscle manages its ATP budget, you must be familiar with the two main pathways of ATP synthesis—*anaerobic fermentation* and *aerobic respiration* (see fig. 3.22, p. 98). Each of these has advantages and disadvantages. Anaerobic fermentation enables a cell to produce ATP in the absence of oxygen, but the ATP yield is very limited and the process produces a toxic end product, lactic acid, which is a major factor in muscle fatigue. By contrast, aerobic respiration produces far more ATP and less toxic end products (carbon dioxide and water), but it requires a continual supply of oxygen. Although aerobic respiration is best known as a pathway for glucose oxidation, it is also used to extract energy from other organic compounds. In a resting muscle, most ATP is generated by the aerobic respiration of fatty acids.

During the course of exercise, different mechanisms of ATP synthesis are used depending on the exercise duration. We will view these mechanisms from the standpoint of immediate, short-term, and long-term energy, but it must be stressed that muscle does not make sudden shifts from one mechanism to another like an automobile transmission shifting gears. Rather, these mechanisms blend and overlap as the exercise continues (fig. 12.18).

Immediate Energy

In a short, intense exercise such as a 100 m dash, the respiratory and cardiovascular systems cannot deliver oxygen to the muscles quickly enough for aerobic respiration to meet the increased ATP demand. The myoglobin in a muscle fiber supplies oxygen for a limited amount of aerobic respiration, but in brief exercises a muscle meets most of its ATP demand by borrowing phosphate (P_i) groups from other molecules and transferring them to ADP. Two enzyme systems control these phosphate transfers (fig. 12.19):

1. **Myokinase** (MY-oh-KY-nase) transfers P_i groups from one ADP to another, converting the latter to ATP.
2. **Creatine kinase** (CREE-uh-tin KY-nase) obtains P_i groups from an energy storage molecule, **creatine phosphate (CP),** and donates them to ADP to make ATP.

ATP and CP, collectively called the **phosphagen system,** provide nearly all the energy used for short bursts of intense activity. Muscle contains about 5 millimoles of ATP and 15 millimoles of CP per kilogram of tissue, which is enough to power about 1 minute of brisk walking or 6 seconds of sprinting or fast swimming. The phosphagen system is especially important in activities requiring brief but maximal effort, such as football, baseball, and weight lifting.

Short-Term Energy

As the phosphagen system is exhausted, the muscles shift to anaerobic fermentation to "buy time" until cardiopulmonary function can catch up with the muscle's oxygen demand. During this period, the muscles obtain glucose from the blood and their own stored glycogen. The pathway from glycogen to lactic acid, called the **glycogen–lactic acid system,** produces enough ATP for 30 to 40 seconds of maximum activity. To play basketball or to run completely around a baseball diamond, for example, depends heavily on this energy-transfer system.

Long-Term Energy

After 40 seconds or so, the respiratory and cardiovascular systems "catch up" and deliver oxygen to the muscles fast enough for aerobic respiration to meet most of the

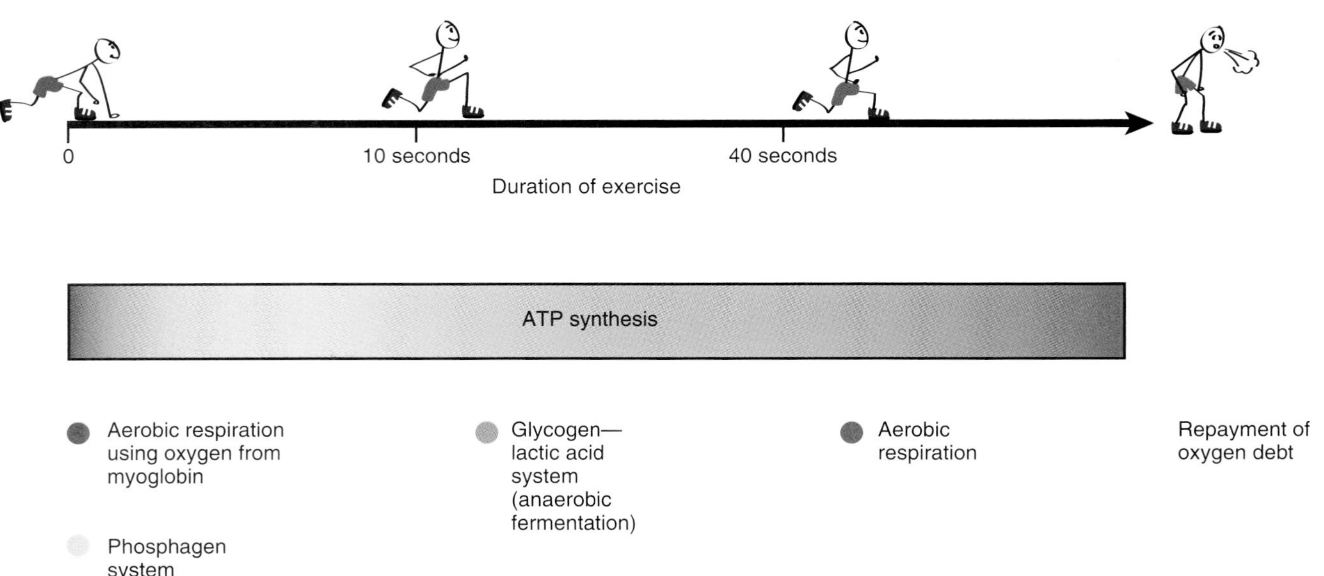

Figure 12.18 Phases of ATP Production During Exercise.

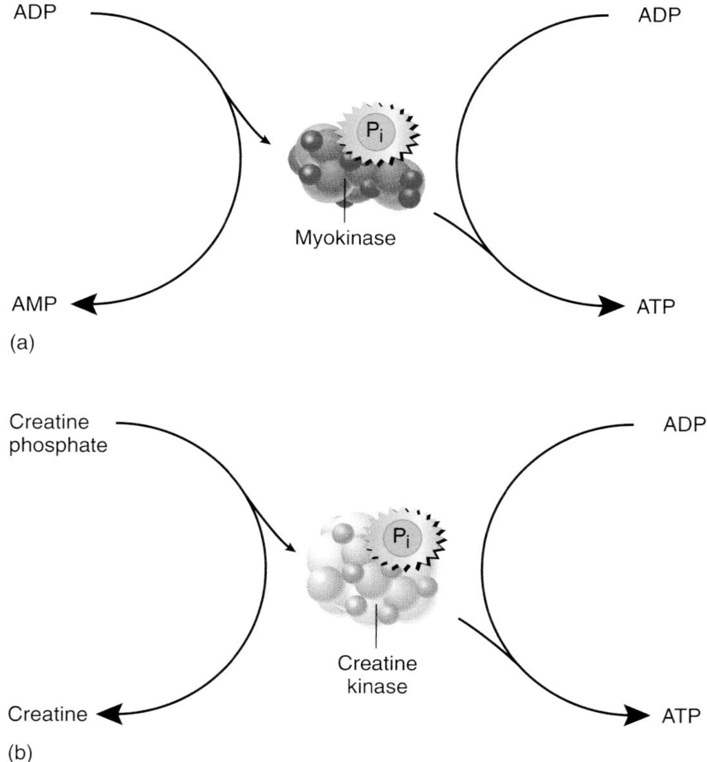

Figure 12.19 **The Phosphagen System.** Two enzymes, myokinase and creatine kinase, generate ATP in the absence of oxygen. Myokinase borrows phosphate groups from ADP, and creatine kinase borrows them from creatine phosphate, to convert an ADP to ATP.

ATP demand. One's rate of oxygen consumption rises for 3 to 4 minutes and then levels off at a *steady state* in which aerobic ATP production keeps pace with the demand. In exercises lasting more than 10 minutes, more than 90% of the ATP is produced aerobically.

Little lactic acid accumulates under steady state conditions, but this does not mean that aerobic exercise can continue indefinitely or that it is limited only by a person's willpower. The depletion of glycogen and blood glucose, together with the loss of fluid and electrolytes through sweating, set limits to endurance and performance even when lactic acid does not.

Fatigue and Endurance

Muscle fatigue is the progressive weakness and loss of contractility that results from prolonged use of the muscles. For example, if you hold this book at arm's length for a minute, you will feel your muscles growing weaker and eventually you will be unable to hold it up. Repeatedly squeezing a rubber ball, pushing a video game button, or trying to take lecture notes from a fast-talking professor produces fatigue in the hand and finger muscles. Fatigue has multiple causes:

- ATP synthesis declines as glycogen is consumed.
- The ATP shortage slows down the sodium-potassium pumps, which are needed to maintain the resting membrane potential and excitability of the muscle fibers.
- Lactic acid lowers the pH of the sarcoplasm, which inhibits the enzymes involved in contraction, ATP synthesis, and other aspects of muscle function.
- Each action potential releases potassium ions from the sarcoplasm to the extracellular fluid. The accumulation of extracellular K^+ lowers the membrane potential and excitability of the muscle fiber.
- Motor nerve fibers use up their acetylcholine, which leaves them less capable of stimulating muscle fibers. This is called *junctional fatigue.*
- The central nervous system, where all motor commands originate, fatigues by processes not yet understood

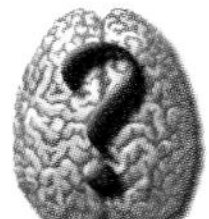

Think About It

Suppose you repeatedly stimulated the sciatic nerve in a frog nerve-muscle preparation until the muscle stopped contracting. What simple test could you do to determine whether this was due to junctional fatigue or to one of the other fatigue mechanisms?

A person's ability to maintain high-intensity exercise for more than 4 to 5 minutes is determined in large part by his or her **maximum oxygen uptake ($\dot{V}O_{2}max$)**—the point at which the rate of oxygen consumption reaches a plateau and does not increase further with an added workload. $\dot{V}O_{2}max$ is proportional to body size; it peaks at around age 20; it is usually greater in males than in females; and it can be twice as great in a trained endurance athlete as in an untrained person (see the later discussion on effects of conditioning).

Physical endurance also depends on the supply of organic nutrients—fatty acids, amino acids, and especially glucose. Many endurance athletes use a dietary strategy called *carbohydrate loading* to "pack" as much as 5 g of glycogen into every 100 g of muscle. This can significantly increase endurance, but an extra 2.7 g of water is also stored with each added gram of glycogen. The resulting "heaviness" and other side effects sometimes outweigh the benefits of carbohydrate loading.

Oxygen Debt

You have probably noticed that you breathe heavily not only during a strenuous exercise but also for several minutes afterwards. This is because your body accrues an oxygen debt that must be "repaid." **Oxygen debt** is the difference between the resting rate of oxygen consumption and the elevated rate following an exercise. The total amount of extra oxygen consumed after a strenuous exercise is typically about 11 L. It is used for the following purposes:

- *Replacing the body's oxygen reserves* that were depleted in the first minute of exercise. These include 0.3 L of oxygen bound to muscle myoglobin, 1.0 L bound to blood hemoglobin, 0.25 L dissolved in the blood plasma and other extracellular fluids, and 0.1 L in the air in the lungs.
- *Replenishing the phosphagen system.* This involves synthesizing ATP and using some of it to donate phosphate groups back to creatine until the resting levels of ATP and CP are restored.
- *Oxidizing lactic acid.* About 80% of the lactic acid produced by muscle enters the bloodstream and is reconverted to pyruvic acid in the kidneys, the cardiac muscle, and especially the liver. Some of this pyruvic acid enters the aerobic (mitochondrial) pathway to make ATP, but the liver converts most of it back to glucose. Glucose is then available to replenish the glycogen stores of the muscle.
- *Serving the elevated metabolic rate.* As long as the body temperature remains elevated by exercise, the total metabolic rate remains high, and this requires extra oxygen.

Slow- and Fast-Twitch Fibers

Not all muscle fibers are alike. There are two primary categories of muscle fibers: *slow-twitch* and *fast-twitch fibers* (table 12.3). All fibers of a single motor unit belong to the same type, but all muscles have a mixture of slow- and fast-twitch motor units in different proportions according to each muscle's function.

Slow-twitch fibers are relatively small and produce twitches up to 100 msec long. They are also known as *slow oxidative fibers, red fibers,* or *type I fibers.* They have more mitochondria, myoglobin, and blood capillaries than fast-twitch fibers do. These features adapt slow-twitch fibers for aerobic respiration—and because aerobic respiration does not generate lactic acid, these fibers are relatively resistant to fatigue. The soleus muscle of the calf and the postural muscles of the back, for example, are composed mainly of slow-twitch, high-endurance fibers.

Fast-twitch fibers are larger than the slow-twitch type and produce twitches as short as 7.5 msec. They are also known as *fast glycolytic fibers, white fibers,* or *type II fibers.* These fibers are specialized to use anaerobic "quick energy" systems—they are rich in enzymes for the phosphagen and glycogen–lactic acid systems. Their sarcoplasmic reticulum releases and reabsorbs calcium ions quickly, which partially accounts for their quick, forceful contractions. Fast-twitch fibers are especially important in sports such as basketball, with stop-and-go activity and frequent changes of pace.

Table 12.3 Classification of Skeletal Muscle Fibers

	Fiber Type	
Properties	**Slow-Twitch**	**Fast-Twitch**
Relative diameter	Smaller	Larger
ATP synthesis	Aerobic	Anaerobic
Fatigue resistance	Good	Poor
ATP hydrolysis	Slow	Fast
Glycolysis	Moderate	Fast
Myoglobin content	Abundant	Low
Glycogen content	Low	Abundant
Mitochondria	Abundant and large	Fewer and smaller
Capillaries	Abundant	Fewer
Color	Red	White, pale
Representative Muscles in Which Fiber Type Is Predominant		
	Soleus	Gastrocnemius
	Erector spinae	Biceps brachii
	Quadratus lumborum	Muscles of eye movement

In the preceding chapter, we noted that the gastrocnemius and soleus muscles of the calf insert on the heel (calcaneus) through a common tendon and have the same action. The muscles may seem redundant until we realize that the soleus is composed mainly of slow-twitch fibers and the gastrocnemius is composed mainly of fast-twitch fibers. The gastrocnemius is well adapted for quick, powerful actions such as jumping, whereas the soleus does most of the work in endurance exercises such as jogging and cross-country skiing.

Some authorities recognize two subcategories of fast-twitch fibers, IIA and IIB. Type IIB is the common type previously described, whereas type IIA (intermediate) fibers are relatively rare except in some endurance athletes. They combine fast-twitch responses with aerobic fatigue-resistant metabolism.

Muscles composed mainly of slow-twitch fibers are relatively dark in color because of their high concentrations of blood capillaries, mitochondria, and myoglobin; they are called *red muscles.* Muscles composed mainly of fast-twitch fibers are called *white muscles* because they are paler.

People with different types and levels of physical activity differ in the proportion of one fiber type to another

Table 12.4 Proportion of Slow- and Fast-Twitch Fibers in the Quadriceps Femoris Muscle of Male Athletes

Sample Population	Slow-Twitch	Fast-Twitch
Marathon runners	82%	18%
Swimmers	74	26
Average males	45	55
Sprinters and jumpers	37	63

(table 12.4). This does not necessarily mean, however, that athletic training can change one fiber type to another. Whether fibers can change type remains an unsettled issue. The preponderance of evidence suggests that people are born with genetically determined differences in the ratios of fiber types, and training produces little if any change. Athletes discover those sports at which they excel and gravitate toward those for which heredity has best equipped them. In effect, some people are "born marathoners" and others are "born sprinters."

Muscular Strength and Conditioning

We have far more muscular strength than we normally use. The gluteus maximus can generate 1,200 kg of tension, and all the muscles of the body can produce a total tension of 22,000 kg (nearly 25 tons). Indeed, the muscles can generate more tension than the bones and tendons can withstand—a fact that accounts for many injuries to the patellar and calcaneal tendons. Muscular strength depends on a variety of anatomical and physiological factors:

- **Muscle size.** The strength of a muscle depends primarily on its size; this is why weight lifting increases the size and strength of a muscle simultaneously. A muscle can exert a tension of about 3 to 4 kg/cm^2 (50 lb/in.^{2}) of cross-sectional area.
- **Fascicle arrangement.** Pennate muscles such as the quadriceps femoris are stronger than parallel muscles such as the sartorius, which in turn are stronger than circular muscles such as the orbicularis oculi.
- **Size of active motor units.** Large motor units produce stronger contractions than small ones.
- **Multiple motor unit summation.** When a stronger muscle contraction is desired, the nervous system activates more motor units. This process is the *recruitment,* or *multiple motor unit (MMU) summation,* described earlier. It can produce extraordinary feats of strength under desperate conditions—rescuing a loved one pinned under an automobile, for example. Getting "psyched up" for athletic competition is also partly a matter of MMU summation.
- **Temporal summation.** Nerve impulses usually arrive at a muscle in a series of closely spaced action potentials. Because of the *temporal summation* described earlier, the greater the frequency of stimulation, the more strongly a muscle contracts.
- **The length-tension relationship.** As noted earlier, a muscle resting at optimum length is prepared to contract more forcefully than a muscle that is excessively contracted or stretched.
- **Fatigue.** Muscles contract more weakly when they are fatigued.

Resistance exercise, such as weight lifting, is the contraction of muscles against a load that resists movement. A few minutes of resistance exercise at a time, a few times each week, is enough to stimulate muscle growth. Growth results primarily from cellular enlargement, not cellular division. The muscle fibers synthesize more myofilaments and the myofibrils grow thicker. Myofibrils split longitudinally when they reach a certain size, so a well-conditioned muscle has more myofibrils than a poorly conditioned one. Muscle fibers themselves are incapable of mitosis, but there is some evidence that as they enlarge, they too may split longitudinally. A small part of muscle growth may therefore result from an increase in the number of fibers, but most results from the enlargement of fibers that have existed since childhood.

Think About It

Is muscle growth mainly the result of hypertrophy or hyperplasia?

Endurance (aerobic) exercise, such as jogging and swimming, improves the fatigue resistance of the muscles. Slow-twitch fibers, especially, produce more mitochondria and glycogen and acquire a greater density of blood capillaries as a result of conditioning. Endurance exercise also improves skeletal strength, increases the red blood cell count and the oxygen transport capacity of the blood, and enhances the function of the cardiovascular, respiratory, and nervous systems. Endurance training does not significantly increase muscular strength, and resistance training does not improve endurance. Optimal performance and skeletomuscular health requires **cross-training,** which incorporates elements of both types. If muscles are not kept sufficiently active, they become *deconditioned*—weaker and more easily fatigued.

Before You Go On

Answer the following questions to test your understanding of the preceding section:

20. From which two molecules can ADP borrow a phosphate group to become ATP? What is the enzyme that catalyzes each transfer?
21. In a long period of intense exercise, why does muscle generate ATP anaerobically at first and then switch to aerobic respiration?
22. List four causes of muscle fatigue.
23. List three causes of oxygen debt.
24. What properties of fast- and slow-twitch fibers adapt them for different physiological purposes?

Cardiac and Smooth Muscle

Objectives

When you have completed this section, you should be able to

- describe the structural and physiological differences between cardiac muscle and skeletal muscle;
- explain why these differences are important to cardiac function;
- describe the structural and physiological differences between smooth muscle and skeletal muscle; and
- relate the unique properties of smooth muscle to its locations and functions.

In this section, we compare cardiac muscle and smooth muscle to skeletal muscle. As you will find, cardiac and smooth muscle have special structural and physiological properties related to their distinctive functions. They also have certain properties in common with each other. The muscle cells of both cardiac and smooth muscle are called **myocytes.** By comparison to the long multinucleate fibers of skeletal muscle, these are relatively short cells with only one nucleus. Cardiac and smooth muscle are *involuntary* muscle tissues, not usually subject to our conscious control.

Cardiac Muscle

Cardiac muscle constitutes most of the heart. Its form and function are discussed extensively in chapter 19 so that you will be able to relate these to the actions of the heart. Here, we only briefly compare it to skeletal and smooth muscle (table 12.5).

Cardiac muscle is striated like skeletal muscle, but its myocytes (*cardiocytes*) are shorter and thicker, they branch like a Y, and each myocyte is linked to several others at its ends (see fig. 19.11). The linkages, called **intercalated** (in-TUR-ku-LAY-ted) **discs,** appear as thick dark lines in stained tissue sections. An intercalated disc has electrical *gap junctions* that allow each myocyte to directly stimulate its neighbors, and mechanical junctions that keep the myocytes from pulling apart when the heart contracts. The sarcoplasmic reticulum is less developed than in skeletal muscle, but the T tubules are larger and admit supplemental Ca^{2+} from the extracellular fluid. Like skeletal muscle fibers, cardiac myocytes are incapable of mitosis. Furthermore, cardiac muscle has no satellite cells, so the repair of damaged cardiac muscle is entirely by fibrosis.

Unlike skeletal muscle, cardiac muscle can contract without the need of nervous stimulation. It contains a built-in **pacemaker** that rhythmically sets off a wave of electrical excitation. This wave travels through the cardiac muscle and triggers the contraction of the heart chambers. Cardiac muscle is said to be **autorhythmic**[12] because of this ability to contract rhythmically and independently. The heart does, however, receive fibers from the *autonomic nervous system* that can either increase or decrease the heart rate and contraction strength. Cardiac muscle does not exhibit quick twitches like skeletal muscle. Rather, it maintains tension for about 200 to 250 msec, enabling the heart to expel blood.

Cardiac muscle uses aerobic respiration almost exclusively. It is very rich in myoglobin and glycogen, and it has especially large mitochondria that fill about 25% of the cell, compared to smaller mitochondria occupying about 2% of a skeletal muscle fiber. Cardiac muscle is very adaptable with respect to the fuel used, but very vulnerable to interruptions in oxygen supply. Because it makes little use of anaerobic fermentation, cardiac muscle is very resistant to fatigue.

Smooth Muscle

Smooth muscle is composed of myocytes with a fusiform shape, about 30 to 200 μm long, 5 to 10 μm wide at the middle, and tapering to a point at each end. There is only one nucleus, located near the middle of the cell. Although thick and thin filaments are both present, they are not aligned with each other and produce no visible striations or sarcomeres; this is the reason for the name *smooth* muscle. Z discs are absent; instead, the thin filaments are attached by way of the cytoskeleton to **dense bodies,** little masses of protein scattered throughout the sarcoplasm and on the inner face of the sarcolemma.

The sarcoplasmic reticulum is scanty, and there are no T tubules. The calcium needed to activate smooth muscle contraction comes mainly from the extracellular fluid (ECF) by way of calcium channels in the sarcolemma. During relaxation, calcium is pumped back out of the cell. Some smooth muscle has no nerve supply, but when nerve fibers are present, they are autonomic (like those of cardiac muscle) and not somatic motor fibers.

12. *auto* = self

Unlike skeletal and cardiac muscle, smooth muscle is capable of mitosis and hyperplasia. Thus, an organ such as the pregnant uterus can grow by adding more myocytes, and injured smooth muscle regenerates well.

Types of Smooth Muscle

There are two functional categories of smooth muscle called *multiunit* and *single-unit* types (fig. 12.20). **Multiunit smooth muscle** occurs in some of the largest arteries and pulmonary air passages, in the arrector pili muscles of the hair follicles, and in the iris of the eye. Its innervation, although autonomic, is otherwise similar to that of skeletal muscle—the terminal branches of a nerve fiber synapse with individual myocytes and form a motor unit. Each motor unit contracts independently of the others, hence the name of this muscle type.

Single-unit smooth muscle is more widespread. It occurs in most blood vessels and in the digestive, respiratory, urinary, and reproductive tracts—thus, it is also called **visceral muscle.** In many of the hollow viscera, it forms two or more layers—typically an inner *circular layer,* in which the myocytes encircle the organ, and an outer *longitudinal layer,* in which the myocytes run lengthwise along the organ (fig. 12.21). The name *single-unit* refers to the fact that the myocytes of this type of muscle are electrically coupled to each other by gap junctions. Thus, they directly stimulate each other and a large number of cells contract as a unit, almost as if they were a single cell.

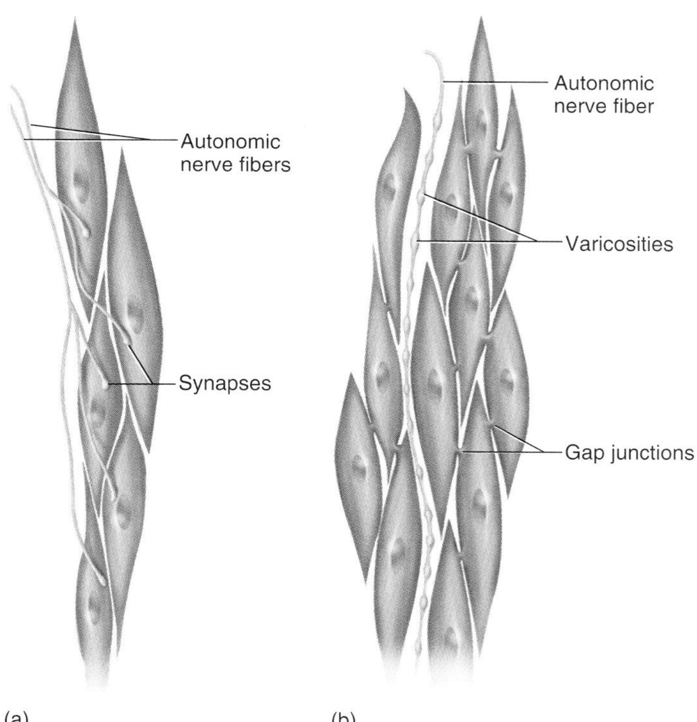

Figure 12.20 Smooth Muscle Innervation. (*a*) Multiunit smooth muscle, in which each myocyte receives its own nerve supply. (*b*) Single-unit smooth muscle, in which a nerve fiber passes through the tissue without synapsing with any specific myocyte.

Stimulation of Smooth Muscle

Like cardiac muscle, smooth muscle is involuntary and capable of contracting without nervous stimulation. Some smooth muscle contracts in response to chemical stimuli such as hormones, carbon dioxide, low pH, and oxygen deficiency and in response to stretch (as in a full stomach or bladder). Some single-unit smooth muscle, especially in the stomach and intestines, has pacemaker cells that spontaneously depolarize and set off waves of contraction throughout an entire layer of muscle. Such smooth muscle is autorhythmic, like cardiac muscle, although with a much slower rhythm.

But like cardiac muscle, smooth muscle is innervated by autonomic nerve fibers that can trigger or modify its contractions. Here too, these nerve fibers can have opposite effects on the smooth muscle; they relax the smooth muscle of arteries while contracting the smooth muscle in the bronchioles of the lungs, for example. In single-unit smooth muscle, each autonomic nerve fiber has up to 20,000 beadlike swellings called **varicosities** along its length (figs. 12.20*b* and 12.22). Each varicosity contains synaptic vesicles and a few mitochondria. Instead of closely approaching any one myocyte, the nerve fiber passes amid several myocytes and stimulates all of them at once when it releases its neurotransmitter. The muscle cells do not have motor end plates or any other specialized area of sarcolemma to bind the neurotransmitter; rather, they have receptor sites scattered throughout the surface.

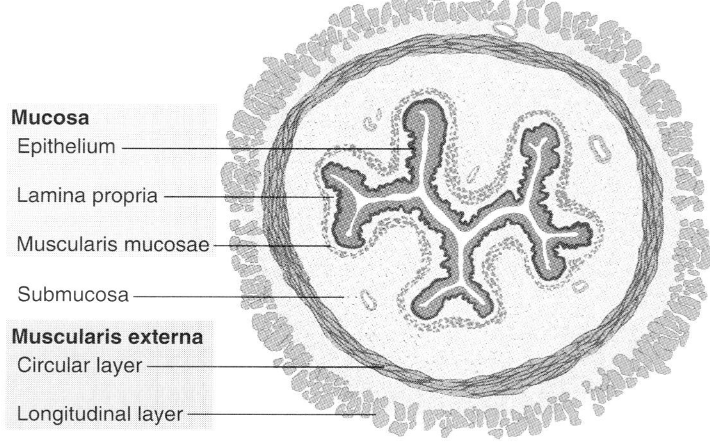

Figure 12.21 Layers of Visceral (single-unit) Smooth Muscle in the Wall of the Esophagus. Many hollow organs have alternating circular and longitudinal layers of smooth muscle.

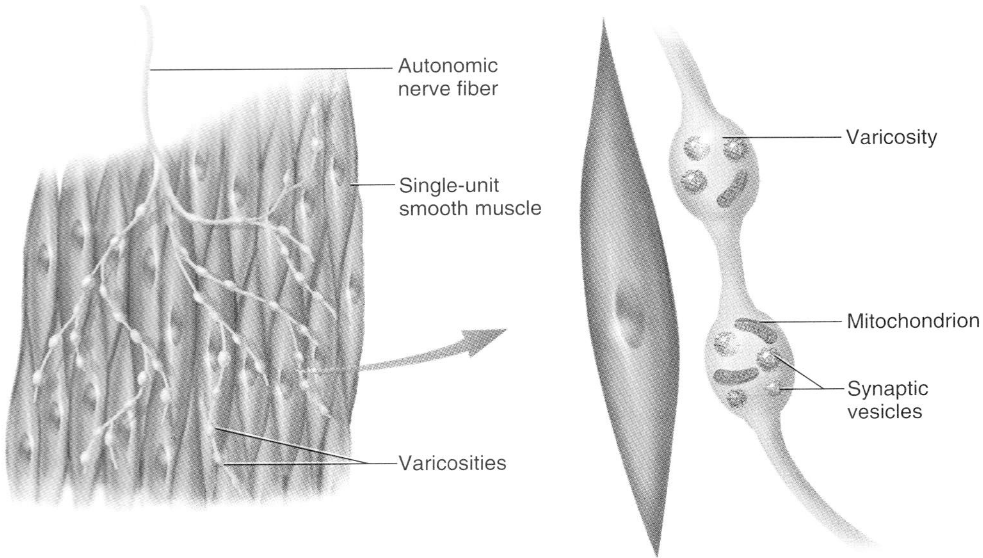

Figure 12.22 Varicosities of an Autonomic Nerve Fiber in Single-Unit Smooth Muscle.

Such nerve-muscle relationships are called **diffuse junctions** because there is no one-to-one relationship between a nerve fiber and a myocyte.

Contraction and Relaxation

Smooth muscle resembles the other muscle types in that contraction is triggered by calcium ions (Ca^{2+}), energized by ATP, and achieved by the sliding of thin filaments over the thick filaments. The mechanism of excitation-contraction coupling, however, is very different. Little of the Ca^{2+} comes from the sarcoplasmic reticulum; most comes from the extracellular fluid and enters the cell through calcium channels in the sarcolemma. Some of these channels are voltage-gated and open in response to changes in membrane voltage; some are ligand-gated and open in response to hormones and neurotransmitters; and some are mechanically gated and open in response to stretching of the cell.

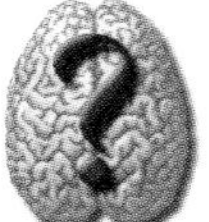

Think About It

How is smooth muscle contraction affected by the drugs called calcium channel blockers? (see p. 112)

Smooth muscle has no troponin. Calcium binds instead to a similar protein called **calmodulin**[13] (cal-MOD-you-lin), associated with the thick filaments. Calmodulin then activates an enzyme called **myosin light-chain kinase,** which transfers a phosphate group from ATP to the head of the myosin. This activates the myosin ATPase and enables it to bind to actin, but in order to execute a power stroke, the myosin must bind and hydrolyze yet another ATP. It then produces power and recovery strokes like those of skeletal muscle.

As thick filaments pull on the thin ones, the thin filaments pull on intermediate filaments, which in turn pull on the dense bodies of the plasma membrane. This shortens the entire cell. When a smooth muscle cell contracts, it twists in a spiral fashion, somewhat like wringing out a wet towel except that the "towel" wrings itself (fig. 12.23).

In skeletal muscle, there is typically a 2 msec latent period between stimulation and the onset of contraction. In smooth muscle, by contrast, the latent period is 50 to 100 msec long. Tension peaks about 500 msec (0.5 sec) after the stimulus and then declines over a period of 1 to 2 seconds. The effect of all this is that compared to skeletal muscle, smooth muscle is very slow to contract and relax. It is slow to contract because its myosin ATPase is a slow enzyme. It is slow to relax because the pumps that remove Ca^{2+} from the cell are also slow. As the Ca^{2+} level falls, myosin is dephosphorylated and is no longer able to hydrolyze ATP and execute power strokes. However, it does not necessarily detach from actin immediately. Its myosin has a *latch-bridge mechanism* that enables it to remain attached to actin for a prolonged time without consuming more ATP.

Smooth muscle often exhibits tetanus and is very resistant to fatigue. It makes most of its ATP aerobically, but its ATP requirement is small and it has relatively few mitochondria. Skeletal muscle requires 10 to 300 times as much ATP as smooth muscle to maintain the same amount of tension. The fatigue-resistance and latch-bridge mechanism of smooth muscle are important in enabling it to maintain a state of continual **smooth muscle tone (tonic**

13. acronym for *cal*cium *modul*ating prote*in*

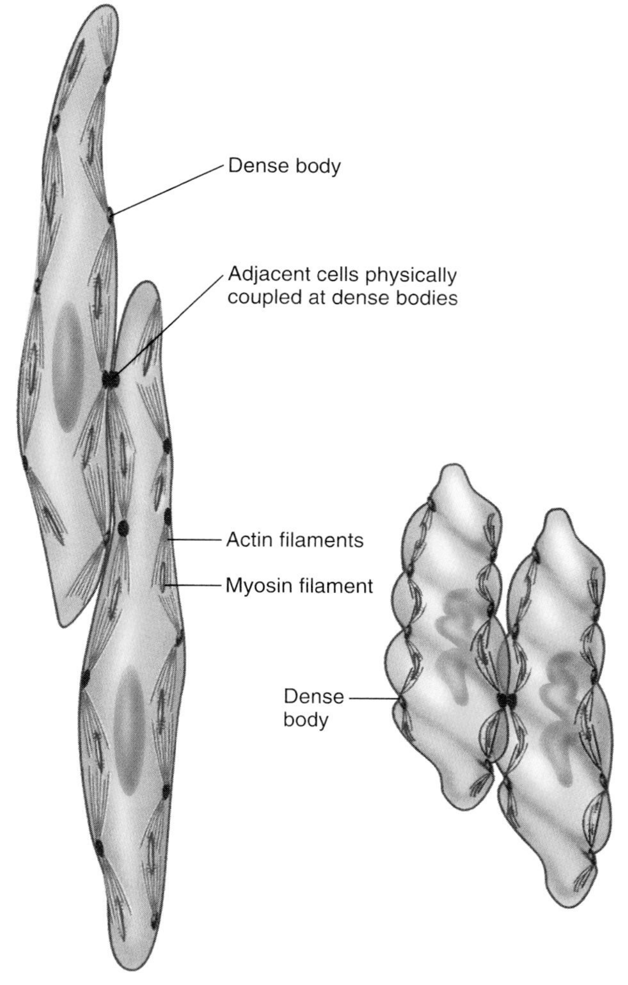

Figure 12.23 Smooth Muscle Contraction. (*a*) Relaxed cells. Actin myofilaments are anchored to dense bodies in the sarcoplasm and on the plasma membrane, rather than to Z discs. (*b*) Contracted cells. Note the twisting effect.

contraction). This tonic contraction keeps the arteries in a state of partial constriction called *vasomotor tone.* A loss of muscle tone in the arteries can cause a dangerous drop in blood pressure. Smooth muscle tone also keeps the intestines partially contracted. The intestines are much longer in a cadaver than they are in a living person because of the loss of muscle tone at death.

Response to Stretch

Stretch alone sometimes causes smooth muscle to contract by opening mechanically gated calcium channels in the sarcolemma. Distension of the esophagus with food or the colon with feces, for example, evokes a wave of contraction called **peristalsis** (PERR-ih-STAL-sis) that propels the contents along the organ.

Smooth muscle exhibits a reaction called the **stress-relaxation** (or **receptive relaxation**) **response.** When stretched, it briefly contracts and resists, but then relaxes. The significance of this response is apparent in the urinary bladder, whose wall consists of three layers of smooth muscle. If the stretched bladder contracted and did not soon relax, it would expel urine almost as soon as it began to fill, thus failing to store the urine until an opportune time.

Remember that skeletal muscle cannot contract very forcefully if it is overstretched. Smooth muscle is not subject to the limitations of this length-tension relationship. It must be able to contract forcefully even when greatly stretched, so that hollow organs such as the stomach and bladder can fill and then expel their contents efficiently. Skeletal muscle must be within 30% of optimum length in order to contract strongly when stimulated. Smooth muscle, by contrast, can be anywhere from half to twice its resting length and still contract powerfully. There are three reasons for this: (1) there are no Z discs, so thick filaments cannot butt against them and stop the contraction; (2) since the thick and thin filaments are not arranged in orderly sarcomeres, stretching of the muscle does not cause a situation where there is too little overlap for cross-bridges to form; and (3) the thick filaments of smooth muscle have myosin heads along their entire length (there is no bare zone), so cross-bridges can form anywhere, not just at the ends. Smooth muscle also exhibits **plasticity**—the ability to adjust its tension to the degree of stretch. Thus, a hollow organ such as the bladder can be greatly stretched yet not become flabby when it is empty.

Table 12.5 compares some properties of skeletal, cardiac, and smooth muscle.

The muscular system suffers fewer diseases than any other organ system. Two disorders of particular importance, muscular dystrophy and myasthenia gravis, are described in clinical insight 12.4; muscle injuries often associated with sports are described in clinical insight 11.6 at the end of the previous chapter; and others are briefly described in table 12.6.

Before You Go On

Answer the following questions to test your understanding of the preceding section:

25. Explain why intercalated discs are important to cardiac muscle function.
26. Explain why it is important for cardiac muscle to have a longer action potential and longer refractory period than skeletal muscle.
27. How do single-unit and multiunit smooth muscle differ in innervation and contractile behavior?
28. How does smooth muscle differ from skeletal muscle with respect to its source of calcium and its calcium receptor?
29. Explain why the stress-relaxation response is an important factor in smooth muscle function.

Table 12.5 Comparison of Skeletal, Cardiac, and Smooth Muscle

Feature	Skeletal Muscle	Cardiac Muscle	Smooth Muscle
Location	Associated with skeletal system	Heart	Walls of viscera and blood vessels, iris of eye, arrector pili of hair follicles
Cell shape	Long cylindrical fibers	Short branched cells	Fusiform cells
Cell length	100 μm–30 cm	50–100 μm	50–200 μm
Cell width	10–100 μm	10–20 μm	2–10 μm
Striations	Present	Present	Absent
Nuclei	Multiple nuclei, adjacent to sarcolemma	Usually one nucleus, near middle of cell	One nucleus, near middle of cell
Connective tissues	Endomysium, perimysium, epimysium	Endomysium only	Endomysium only
Sarcoplasmic reticulum	Abundant	Present	Scanty
T tubules	Present, narrow	Present, wide	Absent
Gap junctions	Absent	Present in intercalated discs	Present in single-unit smooth muscle
Autorhythmicity	Absent	Present	Present in single-unit smooth muscle
Thin filament attachment	Z discs	Z discs	Dense bodies
Regulatory proteins	Tropomyosin, troponin	Tropomyosin, troponin	Calmodulin, light-chain myokinase
Ca^{2+} source	Sarcoplasmic reticulum	Sarcoplasmic reticulum and extracellular fluid	Mainly extracellular fluid
Ca^{2+} receptor	Troponin of thin filament	Troponin of thin filament	Calmodulin of thick filament
Innervation and control	Somatic motor fibers (voluntary)	Autonomic fibers (involuntary)	Autonomic fibers (involuntary)
Nervous stimulation required?	Yes	No	No
Effect of nervous stimulation	Excitatory only	Excitatory or inhibitory	Excitatory or inhibitory
Mode of tissue repair	Limited regeneration, mostly fibrosis	No regeneration; repair by fibrosis only	Relatively good capacity for regeneration

Table 12.6 Some Disorders of the Muscular System

Delayed Onset Muscle Soreness	Pain, stiffness, and tenderness felt from several hours to a day after strenuous exercise. Associated with microtrauma to the muscles, with disrupted Z discs, myofibrils, and plasma membranes; and with elevated levels of myoglobin, creatine kinase, and lactate dehydrogenase in the blood.
Cramps	Painful muscle spasms triggered by heavy exercise, extreme cold, dehydration, electrolyte loss, low blood glucose, or lack of blood flow.
Contracture	Abnormal muscle shortening not caused by nervous stimulation. Can result from failure of the calcium pump to remove Ca^{2+} from the sarcoplasm or from contraction of scar tissue, as in burn patients.
Fibromyalgia	Diffuse, chronic muscular pain and tenderness, often associated with sleep disturbances and fatigue; often misdiagnosed as chronic fatigue syndrome. Can be caused by various infectious diseases, physical or emotional trauma, or medications. Most common in women 30 to 50 years old.
Crush Syndrome	A shocklike state following the massive crushing of muscles; associated with high and potentially fatal fever, cardiac irregularities resulting from K^+ released from the muscle, and kidney failure resulting from blockage of the renal tubules with myoglobin released by the traumatized muscle. Myoglobinuria (myoglobin in the urine) is a common sign.
Disuse Atrophy	Reduction in the size of muscle fibers as a result of nerve damage or muscular inactivity, for example in limbs in a cast and in patients confined to a bed or wheelchair. Muscle strength can be lost at a rate of 3% per day of bed rest.
Myositis	Muscle inflammation and weakness resulting from infection or autoimmune disease.

Clinical Insight 12.4

Muscular Dystrophy and Myasthenia Gravis

Muscular dystrophy[14] is a collective term for several hereditary diseases in which the skeletal muscles degenerate, lose strength, and are gradually replaced by adipose and fibrous tissue. This new connective tissue impedes blood circulation, which in turn accelerates muscle degeneration in a fatal spiral of positive feedback. The most common form of the disease is *Duchenne*[15] *muscular dystrophy (DMD),* caused by a sex-linked recessive allele. Like other sex-linked traits (see chapter 5), DMD is mainly a disease of males. It occurs in about 1 in 3,500 male live births, but is not usually diagnosed until the age of 2 to 10 years. Difficulties begin to appear early on, as a child begins to walk. The child falls frequently and has difficulty standing up again. The disease affects the hips first, then the legs, and progresses to the abdominal and spinal muscles. The muscles shorten as they atrophy, causing postural abnormalities such as scoliosis. DMD is incurable but is treated with exercise to slow the atrophy and with braces to reinforce the weakened hips and correct the posture. Patients are usually confined to a wheelchair by early adolescence and rarely live beyond the age of 20.

The DMD gene was identified in 1987, and genetic screening is now available to inform prospective parents of whether or not they are carriers. The normal allele of this gene makes *dystrophin,* a large protein that links to actin filaments at one end and to membrane glycoproteins on the other. In DMD, dystrophin is absent, the plasma membranes of the muscle fibers become torn, and the muscle fibers die.

A less severe form of muscular dystrophy is *facioscapulohumeral (Landouzy–Dejerine*[16] *muscular dystrophy,* an autosomal dominant trait that begins in adolescence and affects both sexes. It involves the facial and shoulder muscles more than the pelvic muscles and disables some individuals while it barely affects others. A third form, *limb-girdle dystrophy,* is a combination of several diseases of intermediate severity that affect the shoulder, arm, and pelvic muscles.

Myasthenia gravis[17] (MY-ass-THEE-nee-uh GRAV-is) *(MG)* usually occurs in women between the ages of 20 and 40. It is an autoimmune disease in which antibodies attack the neuromuscular junctions and bind ACh receptors together in clusters. The muscle fiber then removes the clusters from the sarcolemma by endocytosis. As a result, the muscle fibers become less and less sensitive to ACh. The effects often appear first in the facial muscle (fig. 1) and commonly include drooping eyelids and double vision (due to weakness of the eye muscles). The initial symptoms are often followed by difficulty in swallowing, weakness of the limbs, and poor physical endurance. Some people with MG die quickly as a result of respiratory failure, but others have normal life spans. One method of assessing the progress of the disease is to use *bungarotoxin,* a protein from cobra venom that binds to ACh receptors. The amount that binds is proportional to the number of receptors that are still functional. The muscle of an MG patient sometimes binds less than one-third as much bungarotoxin as normal muscle does.

Myasthenia gravis is often treated with cholinesterase inhibitors. These drugs retard the breakdown of ACh in the neuromuscular junction and enable it to stimulate the muscle longer. Immunosuppressive agents such as Prednisone and Imuram may be used to suppress the production of the antibodies that destroy ACh receptors. Since certain immune cells are stimulated by hormones from the thymus, removal of the thymus *(thymectomy)* helps to dampen the overactive immune response that causes myasthenia gravis. Also, a technique called *plasmapheresis* may be used to remove harmful antibodies from the blood plasma.

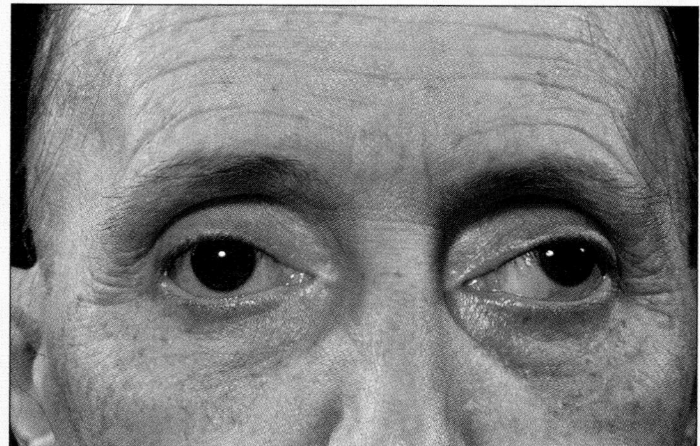

Figure 1 Myasthenia Gravis. This disorder especially affects the muscles of the head. It is characterized by drooping of the eyelids, weakness of the muscles of eye movement, and double vision resulting from the divergence *(strabismus)* of the eyes.

14. *dys* = bad, abnormal + *trophy* = growth
15. Guillaume B. A. Duchenne (1806–75), French physician
16. Louis T. J. Landouzy (1845–1917) and Joseph J. Dejerine (1849–1917), French neurologists
17. *my* = muscle + *asthen* = weakness + *grav* = severe

Chapter Review

Online Learning Center

Study resources for this chapter are available at www.mhhe.com/saladin/

Study Outline

Types and Characteristics of Muscle Tissue (p. 420)

1. Universal characteristics of muscle
   a. Responsiveness
   b. Conductivity
   c. Contractility
   d. Extensibility
   e. Elasticity
2. Properties of skeletal muscle
   a. Voluntary
   b. Striated
3. Series-elastic components

Microscopic Anatomy of Skeletal Muscle (p. 421)

1. The muscle fiber
   a. Multinucleate cells
   b. Sarcolemma and T tubules
   c. Myofibrils
   d. Sarcoplasmic reticulum
   e. Glycogen and myoglobin stores
2. Myofilaments
   a. Thick filaments and myosin
   b. Thin filaments
      - Actin
      - Tropomyosin
      - Troponin
   c. Elastic filaments and titin
3. Striations
   a. A band
   b. I band
   c. H band
   d. Z disc
   e. Sarcomeres

The Nerve-Muscle Relationship (p. 424)

1. Somatic motor neurons
2. The motor unit
3. The neuromuscular junction
   a. Neuron
      - Synaptic knob
      - Synaptic vesicles
      - Acetylcholine (ACh)
   b. Motor end plate
      - Junctional folds
      - ACh receptors
      - Acetylcholinesterase (AChE)
4. Electrically excitable cells
   a. Ion distributions in a resting cell
   b. Resting membrane potentials
   c. Action potentials

Behavior of Skeletal Muscle Fibers (p. 429)

1. Excitation
   a. Arrival of nerve signal
   b. Release of ACh
   c. Binding of ACh to receptors
   d. Generation of end-plate potential
   e. Generation of action potentials
2. Excitation-contraction coupling
   a. Spread of action potentials down T tubules
   b. Ca^{2+} release from sarcoplasmic reticulum
   c. Ca^{2+} binding by troponin
   d. Movement of tropomyosin
3. Contraction (sliding filament theory)
   a. Myosin binds ATP
   b. ATP hydrolysis cocks myosin head
   c. Formation of myosin-actin cross-bridge
   d. Release of ADP and P_i
   e. Power stroke
   f. Myosin binds new ATP
   g. Myosin releases actin
4. Relaxation
   a. Nerve signals cease
   b. AChE breaks down ACh
   c. Sarcoplasmic reticulum reabsorbs Ca^{2+}
   d. Ca^{2+} no longer binds to troponin
   e. Tropomyosin blocks active sites
5. Length-tension relationship
6. Muscle tonus

Behavior of Whole Muscles (p. 436)

1. Response to stimulation
   a. Threshold
   b. Latent period
   c. Twitch
   d. Refractory period
2. Variations in contraction strength
   a. Recruitment
   b. Treppe
   c. Temporal summation
      - Incomplete tetanus
      - Complete tetanus
3. Isometric contraction
4. Isotonic contraction
   a. Concentric
   b. Eccentric

Muscle Metabolism (p. 439)

1. ATP sources
   a. Immediate energy
      - Oxygen from myoglobin
      - ATP from phosphagen system
         - •• Myokinase
         - •• Creatine kinase
   b. Short-term energy
      - Anaerobic fermentation
      - Glycogen–lactic acid system
   c. Long-term energy
      - Aerobic respiration
      - Steady-state O_2 consumption
      - Limitations on endurance
2. Fatigue and endurance
   a. Causes of fatigue
   b. Maximum oxygen uptake
3. Oxygen debt
4. Muscle fiber types
   a. Slow-twitch (red) fibers
      - Twitches up to 100 msec
      - Abundant mitochondria, myoglobin, and capillaries
      - Aerobic respiration
      - Good fatigue resistance
   b. Fast-twitch (white) fibers
      - Twitches as short as 7.5 msec
      - Use phosphagen and glycogen–lactic acid systems
      - Anaerobic
      - Fatigue relatively quickly
   c. Heredity and muscle composition
5. Muscular strength and conditioning
   a. Determinants of strength
   b. Effects of resistance exercise
   c. Effects of endurance exercise

Cardiac and Smooth Muscle (p. 444)

1. Cardiac muscle
   a. Myocyte structure
      - Striated
      - Short, branched cells
      - Intercalated discs
      - Less sarcoplasmic reticulum
      - Larger T tubules
      - Lack of satellite cells
   b. Behavior
      - Internal pacemaker
      - Autorhythmicity
      - Autonomic nerve supply
      - Prolonged contraction
      - Nonmitotic
   c. Metabolism
      - Aerobic
      - Rich in myoglobin and glycogen
      - Large mitochondria
      - Adaptable to various fuels
      - Vulnerable to oxygen deficiency
      - Resistant to fatigue
2. Smooth muscle
   a. Myocytes
      - Fusiform cells
      - One nucleus
      - Nonstriated
      - Dense bodies instead of Z discs
      - Scanty sarcoplasmic reticulum
      - Calcium channels in sarcolemma
      - Autonomic nerve fibers
      - Capable of mitosis
   b. Types
      - Multiunit
      - Single-unit (visceral)
   c. Stimulation
      - Contracts without nervous stimulation
      - Stimulated by chemicals and stretch
      - Sometimes has autorhythmic pacemakers
      - Contraction modified by autonomic nerve fibers
        - •• Varicosities
        - •• Diffuse junctions
   d. Contraction and relaxation
      - ECF as main source of Ca^{2+}
      - Calmodulin as Ca^{2+} receptor
      - Role of myosin light-chain kinase
      - Slow contraction and relaxation
      - Latch-bridge mechanism
      - Smooth muscle tone
   e. Response to stretch
      - Peristalsis
      - Stress-relaxation response
      - Lack of length-tension relationship
      - Plasticity

Selected Vocabulary

For additional terminology not listed here, see tables 12.1 and 12.2.

responsiveness 420
conductivity 420
contractility 420
extensibility 420
elasticity 420
striations 420
voluntary muscle 420
involuntary muscle 420
series-elastic components 420
internal tension 421
external tension 421
myoblast 421
satellite cell 421
active site 421
contractile protein 422
somatic motor fiber 424
motor unit 424
synapse 425
electrophysiology 426
polarized 428
resting membrane potential 428
action potential 428
excitation 429
end-plate potential 429
excitation-contraction coupling 429
sliding filament theory 429
myosin ATPase 431
power stroke 431
recovery stroke 431
calsequestrin 433
length-tension relationship 433
tonus 433
threshold 436
twitch 436
latent period 436
contraction phase 436
relaxation phase 436
refractory period 436
recruitment 437
multiple motor unit summation 437
treppe 438
temporal summation 438
wave summation 438
incomplete tetanus 438
complete tetanus 438
isometric contraction 438
isotonic contraction 438
concentric contraction 438
eccentric contraction 438
myokinase 440
creatine kinase 440
creatine phosphate 440
phosphagen system 440
glycogen–lactic acid system 440
muscle fatigue 441
maximum oxygen uptake 441
oxygen debt 442
slow-twitch fibers 442
fast-twitch fibers 442
resistance exercise 443
endurance exercise 443
cross-training 443
myocytes 444
intercalated discs 444
pacemaker 444
autorhythmic 444
dense bodies 444
multiunit smooth muscle 445
single-unit smooth muscle 445
visceral muscle 445
varicosities 445
diffuse junctions 446
calmodulin 446
myosin light-chain myokinase 446
smooth muscle tone 446
peristalsis 447
stress-relaxation response 447
plasticity 447

Testing Your Recall

1. To make a muscle contract more strongly, the nervous system can activate more motor units. This process is called
   a. recruitment.
   b. summation.
   c. incomplete tetanus.
   d. twitch.
   e. treppe.
2. The ________ is a depression in the sarcolemma that receives a motor nerve ending.
   a. T tubule
   b. terminal cisterna
   c. sarcomere
   d. motor end plate
   e. synapse
3. Before a muscle fiber can contract, ATP must bind to
   a. a Z disc.
   b. the myosin head.
   c. tropomyosin.
   d. troponin.
   e. actin.
4. Before a muscle fiber can contract, Ca^{2+} must bind to
   a. calsequestrin.
   b. the myosin head.
   c. tropomyosin.
   d. troponin.
   e. actin.
5. Skeletal muscle fibers have ________, whereas smooth muscle cells do not.
   a. T tubules
   b. ACh receptors
   c. thick myofilaments
   d. thin myofilaments
   e. dense bodies
6. Smooth muscle cells have ________, whereas skeletal muscle fibers do not.
   a. sarcoplasmic reticulum
   b. tropomyosin
   c. calmodulin
   d. Z discs
   e. myosin ATPase
7. ACh receptors are found mainly in
   a. synaptic vesicles.
   b. terminal cisternae.
   c. thick filaments.
   d. thin filaments.
   e. junctional folds.
8. Single-unit smooth muscle cells can stimulate each other because they have
   a. a latch-bridge.
   b. diffuse junctions.
   c. gap junctions.
   d. tight junctions.
   e. calcium pumps.
9. Warm-up exercises take advantage of ________ to enable muscles to perform at peak strength.
   a. the stress-relaxation response
   b. the length-tension relationship
   c. excitatory junction potentials
   d. oxygen debt
   e. treppe
10. Slow-twitch muscle fibers have all of the following *except*
    a. an abundance of myoglobin.
    b. an abundance of glycogen.
    c. high fatigue resistance.
    d. a red color.
    e. a high capacity to synthesize ATP aerobically.
11. The minimum stimulus intensity that will make a muscle contract is called ___________.
12. A state of prolonged maximum contraction is called ___________.
13. Parts of the sarcoplasmic reticulum called ___________ lie on each side of a T tubule.
14. Thick myofilaments consist mainly of the protein ___________.
15. The neurotransmitter that stimulates skeletal muscle is ___________.
16. Muscle contains an oxygen-binding pigment called ___________.
17. The ___________ of skeletal muscle play the same role as dense bodies in smooth muscle.
18. In autonomic nerve fibers that stimulate single-unit smooth muscle, the neurotransmitter is contained in swellings called ___________.
19. A state of continual partial muscle contraction is called ___________.
20. ___________ is an end product of anaerobic fermentation that causes muscle fatigue.

Answers in appendix B

Testing Your Comprehension

1. Without ATP, relaxed muscle cannot contract and a contracted muscle cannot relax. Explain why.
2. Slight pH variations can cause enzymes to change conformation and can reduce enzyme activity. Explain how this relates to muscle fatigue.
3. Why would skeletal muscle be unsuitable for the wall of the urinary bladder? Explain how this illustrates the complementarity of form and function at a cellular and molecular level.
4. As skeletal muscle contracts, one or more bands of the sarcomere become narrower and disappear, and one or more of them remain the same width. Which bands will change—A, H, or I—and why?
5. Botulism occurs when a bacterium, *Clostridium botulinum*, releases a neurotoxin that prevents motor neurons from releasing ACh. In view of this, what early signs of botulism would you predict? Explain why a person with botulism could die of suffocation.

Answers at the Online Learning Center

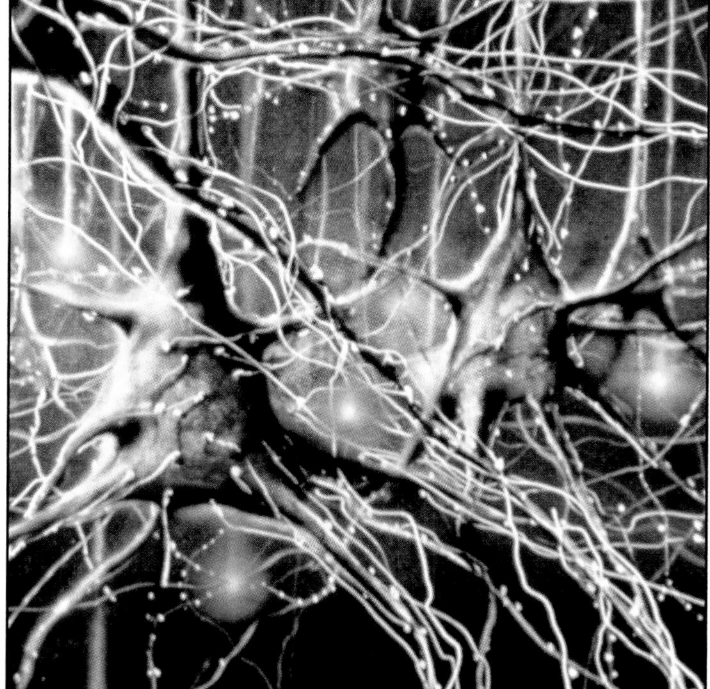

Multiple neurons of the brain "firing" (artist's conception)

CHAPTER 13

Nervous Tissue

Chapter Outline

Insights

Brushing Up

To understand this chapter, it is important that you understand or brush up on the following concepts:
- Cations and anions (p. 54)
- Ligand- and voltage-regulated gates (p. 111)
- Cyclic AMP as a second messenger (p. 112)
- Simple diffusion and active transport (pp. 125, 129)
- The sodium-potassium pump (p. 130)
- Basic structure of nerve cells (p. 190)

Study with Your ESP CD-ROM

Refer to your ESP Tutorial CD-ROM for helpful animations, activities, and quizzes related to the material covered in this chapter. The following are suggested areas to go to on the CD and are presented according to "Unit/Topic":
- Support and Movement/Receptors/Membrane
- Integration and Coordination/Functional Organization

If the body is to maintain homeostasis and function effectively, its trillions of cells must work together in a coordinated fashion. If each cell behaved without regard to what others were doing, the result would be physiological chaos and death. This is prevented by two communication systems—the **nervous system** (fig. 13.1), which is specialized for the rapid transmission of signals from cell to cell, and the **endocrine system,** which is specialized for sending chemical messengers, the hormones, through the blood. The most important aspect of both systems is that they detect changes in an organ, modify its physiology, and modify that of other organs. Thus, these systems functionally coordinate the organs of the body and play a central role in maintaining homeostasis.

The nervous system is the subject of chapters 13 to 16, and the endocrine system is discussed in chapter 17. This chapter is primarily concerned with how individual nerve cells work. The next three chapters are concerned with the organization and function of the nervous system at the organ level. You should reacquaint yourself with the brief introduction to nerve cells in chapter 6 (p. 190) before proceeding, especially with the concepts of the *cell body, dendrites,* and *axon.*

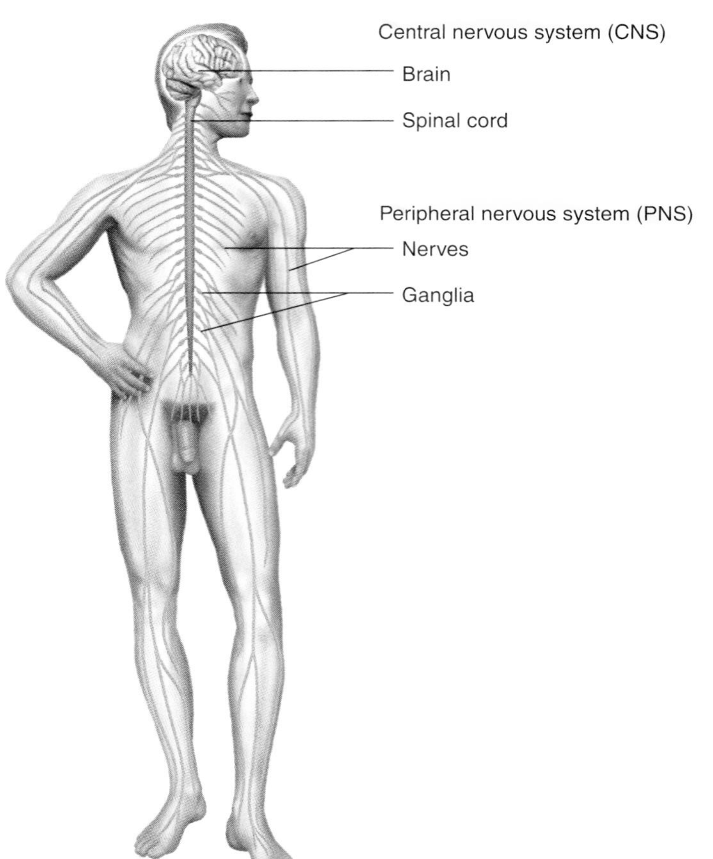

Figure 13.1 The Nervous System.

Overview of the Nervous System

Objectives

When you have completed this section, you should be able to

- describe the major anatomical subdivisions of the nervous system;
- state the general functions of the nervous system and how these relate to the general classes of nerve cells; and
- describe the basic physiological properties of nerve cells that enable them to carry out their functions.

Fundamental Types and Properties of Neurons

The communicative role of the nervous system is carried out by nerve cells, or **neurons.** There are three general classes of neurons corresponding to the three major aspects of nervous system function (fig. 13.2):

1. **Sensory (afferent[1]) neurons.** These neurons are specialized to detect changes in their environment

1. *af* = toward + *fer* = to carry

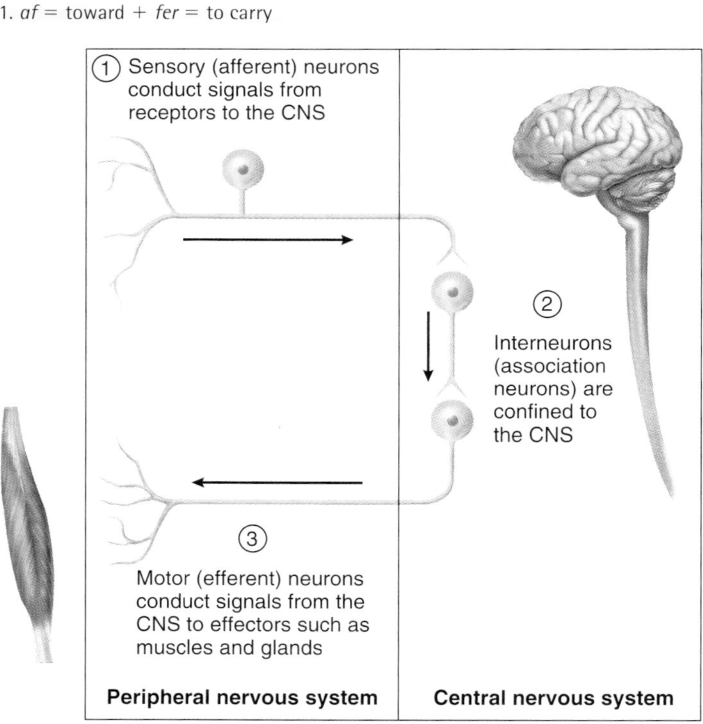

Figure 13.2 **Functional Classes of Neurons.** Sensory (afferent) neurons carry signals to the central nervous system (CNS); interneurons are contained entirely within the CNS and carry signals from one neuron to another; and motor (efferent) neurons carry signals from the CNS to muscles and glands.

called **stimuli**—such as light, heat, pressure, and chemicals—and transmit information about them to the central nervous system, or CNS (the brain and spinal cord). These neurons can begin in almost any organ of the body and end in the CNS; the word *afferent* refers to signal conduction *toward* the CNS. Cells specialized to respond to stimuli are called **receptors** (a word used also for membrane proteins that bind signaling molecules and for sense organs as complex as the eye.) In some cases, such as pain and smell, the receptor cell is a neuron that responds directly to these stimuli. In other cases, such as taste and hearing, the receptor is a separate cell that communicates directly with a sensory neuron.
2. **Interneurons[2] (association neurons).** These neurons lie entirely within the CNS. They receive signals from many other neurons and carry out the *integrative* function of the nervous system—that is, they process, store, and retrieve information and "make decisions" that determine how the body responds to stimuli. About 90% of our neurons are interneurons. The word *interneuron* refers to the fact that they lie *between,* and interconnect, the incoming sensory pathways and the outgoing motor pathways of the CNS.
3. **Motor (efferent[3]) neurons.** These neurons send signals predominantly to muscle and gland cells, which carry out the body's responses to stimuli. They are called *motor* neurons because most of them lead to muscle cells, and *efferent* neurons to signify the signal conduction *away from* the CNS. The cells and organs that carry out the responses to these neurons are called **effectors.**

The communication function of neurons requires them to have three fundamental physiological capabilities:

1. **Excitability (irritability).** Neurons are very excitable and thus highly responsive to stimuli. Although all living cells are excitable, nerve cells have developed this property to the highest degree.
2. **Conductivity.** Neurons respond to stimuli by producing traveling electrical signals that quickly reach other cells at distant locations.
3. **Secretion.** When the electrical signal reaches the end of a nerve fiber, the neuron usually secretes a chemical *neurotransmitter* that "jumps the gap" and stimulates the next cell.

2. *inter* = between
3. *ef* = out of, away + *fer* = to carry

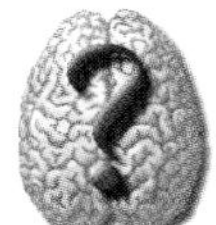

Think About It

What basic physiological properties do a nerve cell and a skeletal muscle fiber have in common? Name a physiological property of each that the other one lacks.

Subdivisions of the Nervous System

The nervous system has two major anatomical subdivisions (see fig. 13.1):

1. The **central nervous system (CNS)** consists of the brain and spinal cord, which are enclosed and protected by the cranium and vertebral column. The CNS is composed of two types of nervous tissue—gray matter and white matter. **Gray matter** forms the surface layer and some deeper masses of the brain and a central H-shaped core of the spinal cord. The cell bodies and synapses of the neurons lie in the gray matter, and all neural integration occurs here. **White matter** lies deep to the gray matter in most of the brain but surrounds the gray matter of the spinal cord. It consists of axons carrying signals from one part of the CNS to another, and it has a bright white color due to lipid sheaths around these axons described later. Details of the central nervous system are discussed in chapter 14.
2. The **peripheral nervous system (PNS)** consists of all the nervous system except the brain and spinal cord. It is composed of nerves and ganglia. A **nerve** is a bundle of nerve fibers wrapped in fibrous connective tissue. Nerves emerge from the CNS through foramina of the skull and vertebral column and carry signals to and from other organs of the body. A **ganglion[4]** (plural, *ganglia*) is a knotlike swelling in a nerve where the cell bodies of neurons are concentrated. The peripheral nervous system is functionally divided into *sensory* and *motor divisions,* and each of these is further divided into *somatic* and *visceral subdivisions.* These distinctions are explained in chapter 15.

The foregoing terminology may give the impression that the body has several nervous systems—central, peripheral, sensory, motor, somatic, and visceral. These are just terms of convenience, however. There is only one nervous system, and these subsystems are interconnected parts of the whole.

4. *gangli* = knot

Before You Go On

Answer the following questions to test your understanding of the preceding section:

1. List the three basic classes of neurons and the basic nervous system functions corresponding to them.
2. What three physiological properties of neurons enable them to perform their communicative function?
3. Distinguish between the central and peripheral nervous systems.

Cells of the Nervous System

Objectives

When you have completed this section, you should be able to

- identify the parts of a neuron;
- explain how neurons transport materials between the cell body and tips of the axon;
- name the cells that aid neuron function and state their functions;
- describe the myelin sheath that is formed around certain nerve fibers and explain its importance; and
- explain how damaged nerve fibers regenerate.

Structure of a Neuron

There are several varieties of neurons, as we shall see, but a good starting point for discussing neuron structure is a neuron of the spinal cord (fig. 13.3). The control center of the neuron is its **soma,**[5] also called the **cell body** or **perikaryon**[6] (PERR-ih-CARE-ee-on). It has a single, centrally located nucleus with a large nucleolus. The cytoplasm contains mitochondria, lysosomes, a Golgi complex, numerous inclusions, and an extensive rough endoplasmic reticulum and cytoskeleton. The cytoskeleton consists of a dense mesh of microtubules and **neurofibrils** (intermediate filaments) that compartmentalize the rough ER into dark-staining regions called *Nissl*[7] *bodies* (fig. 13.3*c, d*). Nissl bodies are unique to neurons and a helpful clue to identifying them in tissue sections with mixed cell types. Mature neurons lack centrioles and apparently undergo no further mitosis after adolescence, but they are unusually long-lived cells, capable of functioning for over a hundred years. Even into old age, however, there are unspecialized *stem cells* in the CNS that can divide and develop into new neurons (see clinical insight 5.3, p. 155).

The major cytoplasmic inclusions in a neuron are glycogen granules, lipid droplets, melanin, and a golden brown pigment called *lipofuscin*[8] (LIP-oh-FEW-sin)—an end product of lysosomal digestion of worn-out organelles and other products. Lipofuscin collects with age and pushes the nucleus to one side of the cell. Lipofuscin granules are also called "wear-and-tear granules" because they are most abundant in old neurons, but they are apparently harmless to the neuron.

The soma of a neuron usually gives rise to a few thick processes that branch into a vast number of **dendrites**[9]—named for their striking resemblance to the bare branches of a tree in winter. The dendrites are the primary site for receiving signals from other neurons. Some neurons have only one dendrite and some have thousands. The more dendrites a neuron has, the more information it can receive from other cells and incorporate into its decision making. As tangled as the dendrites may seem, they provide exquisitely precise pathways for the reception and processing of neural information.

On one side of the soma is a mound called the **axon hillock,** from which the **axon (nerve fiber)** originates. The axon is cylindrical and relatively unbranched for most of its length, although it may give rise to a few branches called *axon collaterals* along the way, and most axons branch extensively at their distal end. An axon is specialized for rapid conduction of nerve signals to points remote from the soma. Its cytoplasm is called the **axoplasm** and its membrane the **axolemma.**[10] A neuron never has more than one axon, and some neurons in the retina and brain have none.

Somas range from 5 to 135 μm in diameter, while axons range from 1 to 20 μm in diameter and from a few millimeters to more than a meter long. Such dimensions are more impressive when we scale them up to the size of familiar objects. If the soma of a spinal motor neuron were the size of a tennis ball, its dendrites would form a huge bushy mass that could fill a 30-seat classroom from floor to ceiling. Its axon would be up to a mile long but a little narrower than a garden hose. This is a worthy point to ponder. The neuron must assemble molecules and organelles in its "tennis ball" soma and deliver them through its "mile-long garden hose" to the end of the axon. How it achieves this remarkable feat is explained shortly.

At the distal end, axons usually have a **terminal arborization**[11]—an extensive complex of fine branches. Each branch ends in a **synaptic knob (terminal button).** As studied in the previous chapter, the synaptic knob is a little swelling that forms a junction **(synapse**[12]**)** with a

5. *soma* = body
6. *peri* = around + *karyo* = nucleus
7. Franz Nissl (1860–1919), German neuropathologist
8. *lipo* = fat, lipid + *fusc* = dusky, brown
9. *dendr* = tree, branch + *ite* = little
10. *lemma* = husk, peel, sheath
11. *arbor* = treelike
12. *syn* = together + *aps* = to touch, join

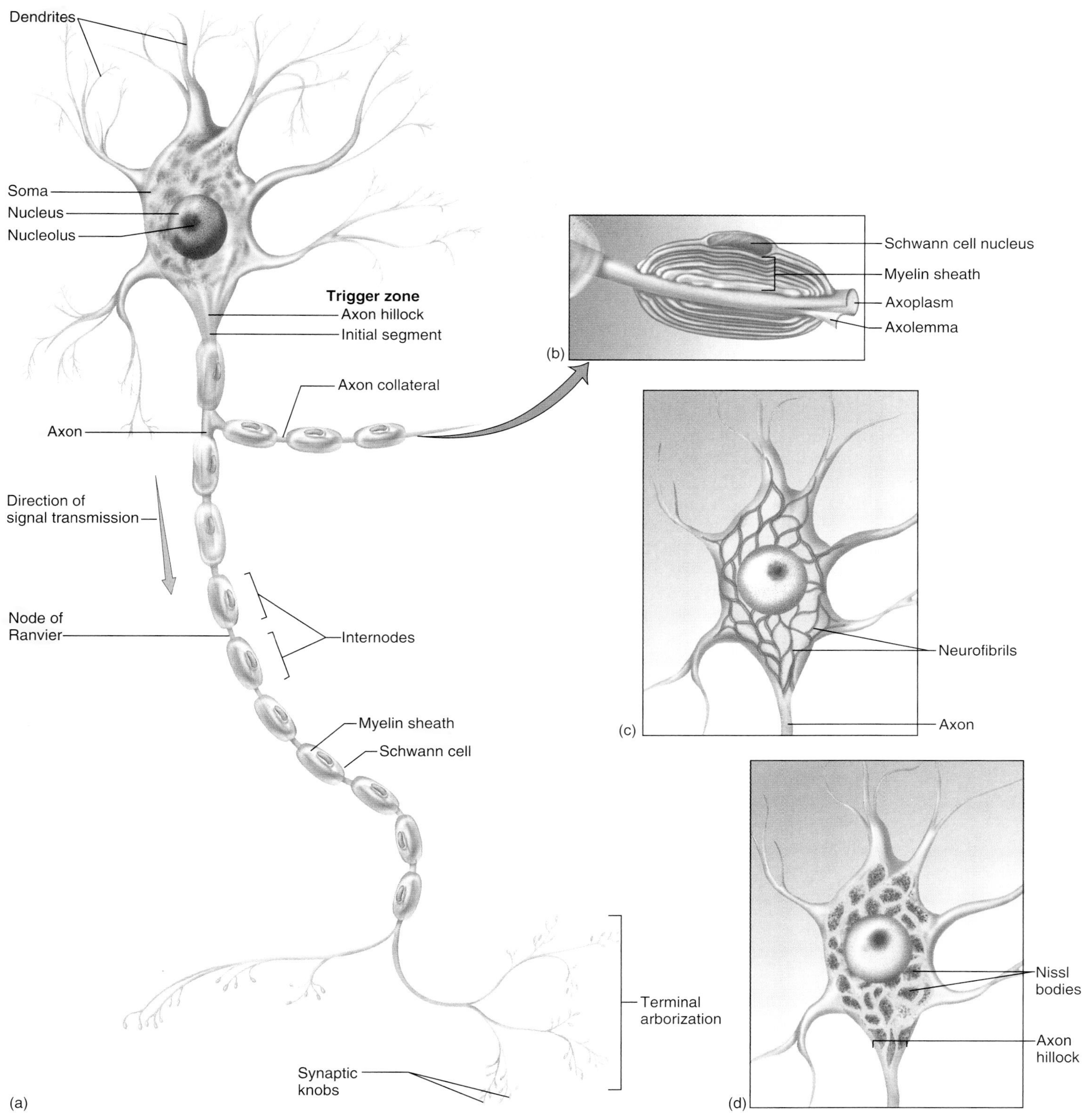

Figure 13.3 **A Representative Neuron.** The Schwann cells and myelin sheath are explained later in this chapter. (*a*) A multipolar neuron such as a spinal motor neuron. (*b*) Detail of myelin sheath. (*c*) Neurofibrils of the soma. (*d*) Nissl bodies, stained masses of rough ER separated by bundles of neurofibrils.

muscle cell, gland cell, or another neuron. It contains **synaptic vesicles** full of neurotransmitter.

Not all neurons fit the preceding description. Neurons are classified structurally according to the number of processes extending from the soma (fig. 13.4):

- **Multipolar neurons** are those, like the preceding, that have one axon and multiple dendrites. This is the most common type of neuron in the body and includes most neurons of the brain and spinal cord.
- **Bipolar neurons** have one axon and one dendrite. Examples include olfactory cells of the nasal cavity, some neurons of the retina, and neurons that carry signals from the inner ear to the brain.
- **Unipolar neurons** have only a single process leading away from the soma. They are represented by the neurons that carry sensory signals from the skin and other organs to the spinal cord. These neurons are also called *pseudounipolar* because they start out as bipolar neurons in the embryo, but their two processes fuse into a single process as the neuron matures. A short distance away from the soma, the process branches like a T, with a *peripheral fiber* carrying signals from the source of sensation and a *central fiber* continuing into the spinal cord. In most other neurons, a dendrite carries signals toward a soma and an axon carries them away. In unipolar neurons, however, there is one long myelinated fiber that bypasses the soma and carries nerve signals directly to the spinal cord. The entire fiber is considered the axon, and the dendrites are the branching receptive endings in the skin or other place of origin.
- **Anaxonic neurons** have multiple dendrites but no axon. They communicate through their dendrites and do not produce action potentials. Some anaxonic neurons are found in the brain and retina. In the retina, they help in visual processes such as the perception of contrast.

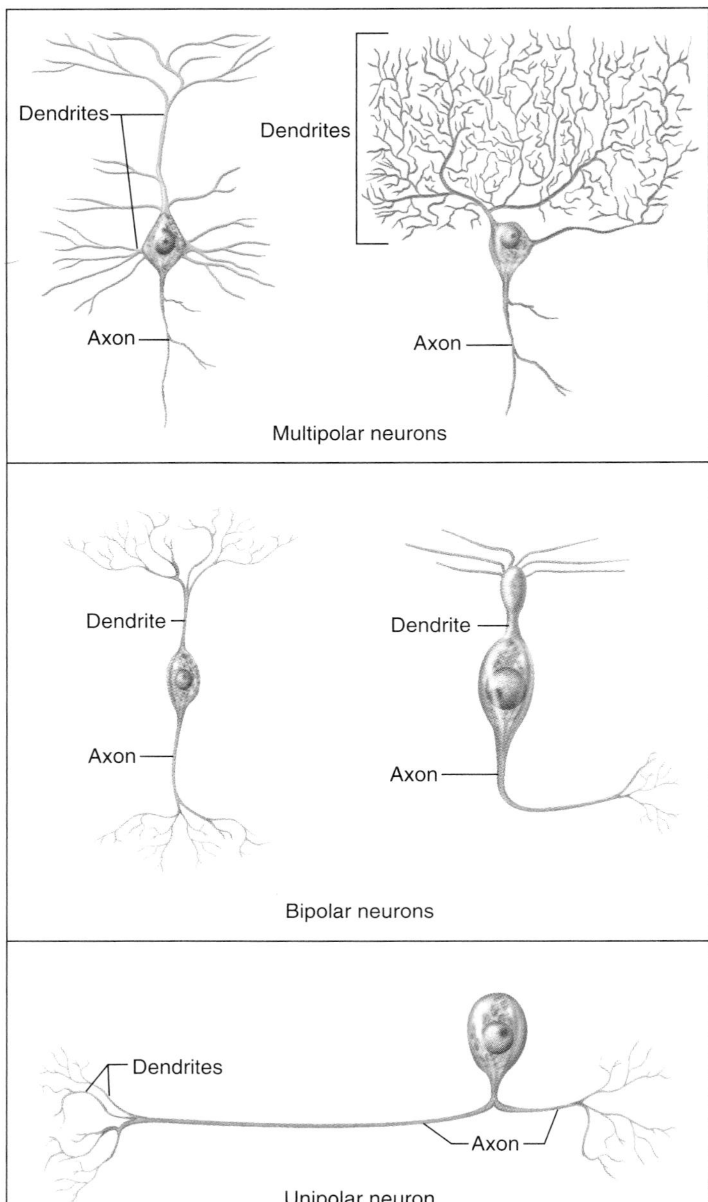

Figure 13.4 Variation in Neuronal Structure. *Top row, left to right:* Two multipolar neurons of the brain—a pyramidal cell and Purkinje cell. *Middle row, left to right:* Two bipolar neurons—a bipolar cell of the retina and an olfactory neuron. *Bottom row:* A unipolar neuron of the type involved in the senses of touch and pain.

Axonal Transport

All of the proteins needed by a neuron must be made in the soma, where the protein-synthesizing organelles such as the nucleus and endoplasmic reticulum are located. Yet many of these proteins are needed in the axon, for example to repair and maintain the axolemma, to furnish ion gates in the membrane, or to act in the synaptic knob as enzymes and signaling molecules. Other substances are transported from the axon terminals back to the soma for disposal or recycling. The two-way passage of proteins, organelles, and other materials along an axon is called **axonal transport.** Movement from the soma down the axon is called **anterograde**[13] **transport** and movement up the axon toward the soma is called **retrograde**[14] **transport.**

13. *antero* = forward + *grad* = to walk, to step
14. *retro* = back + *grad* = to walk, to step

Materials travel along microtubules of the cytoskeleton, which act like railroad tracks to guide them to their destination. But what is the "motor" that drives them along the tracks? Anterograde transport employs a motor protein called *kinesin,* while retrograde transport uses one called *dynein* (the same protein we encountered earlier in cilia and flagella; see chapter 4). These proteins carry materials "on their backs" while they reach out, like the myosin heads of muscle (see chapter 12), to bind repeatedly to the microtubules and crawl along them.

There are two types of axonal transport, fast and slow.

1. **Fast axonal transport** occurs at a rate of 20 to 400 mm/day and may be either anterograde or retrograde:
   - *Fast anterograde transport* moves mitochondria, synaptic vesicles, other organelles, components of the axolemma, enzymes such as acetylcholinesterase, calcium ions, and small molecules such as glucose, amino acids, and nucleotides.
   - *Fast retrograde transport* returns used synaptic vesicles and other materials to the soma and helps to inform the soma of conditions at the axon terminals. Some pathogens exploit retrograde transport to invade neurons, including tetanus toxin and the herpes simplex, rabies, and polio viruses. In such infections, the delay between infection and the onset of symptoms corresponds to the time needed for the pathogens to reach the somas.
2. **Slow axonal transport,** also called *axoplasmic flow,* occurs at a rate of 0.5 to 10.0 mm/day and is always anterograde. It moves enzymes and cytoskeletal components down the axon, renews worn-out axoplasmic components in mature neurons, and supplies new axoplasm for developing or regenerating neurons. Damaged nerve fibers regenerate at a speed governed by slow axonal transport.

Neuroglia

There are about a trillion (10^{12}) neurons in the nervous system—10 times as many neurons in your body as there are stars in our galaxy! Yet the neurons are outnumbered as much as 50 to 1 by attendant cells called **neuroglia** (noo-ROG-lee-uh), or **glial** (GLEE-ul) **cells.** Glial cells protect the neurons and aid their function. The word *glia,* which means "glue," implies one of their roles—they bind neurons together and provide a supportive framework for the nervous tissue. In the fetus, glial cells form a scaffold that guides young migrating neurons to their destinations. Wherever a mature neuron is not in synaptic contact with another cell, it is covered with glial cells. This prevents neurons from contacting each other except at points specialized for signal transmission and thus gives precision to their conduction pathways.

There are six kinds of neuroglia, each with a unique function (table 13.1). Four types occur in the central nervous system (fig. 13.5):

1. **Oligodendrocytes**[15] (OL-ih-go-DEN-dro-sites) somewhat resemble an octopus; they have a bulbous body with as many as 15 armlike processes. Each process reaches out to a nerve fiber and spirals around it like electrical tape wrapped repeatedly around a wire. This spiral wrapping, called the *myelin sheath,* insulates the nerve fiber from the surrounding extracellular fluid. For reasons explained later, it speeds up signal conduction in the nerve fiber. The structure and function of the myelin sheath are discussed in the next section of this chapter.

15. *oligo* = few + *dendro* = branches + *cyte* = cell

Table 13.1 Types of Glial Cells

Type	Location	Functions
Oligodendrocytes	CNS	Form myelin in brain and spinal cord
Protoplasmic astrocytes	CNS	Cover brain surface and nonsynaptic regions of neurons; stimulate formation of blood-brain barrier; remove neurotransmitters and K^+ from extracellular fluid (ECF) of brain and spinal cord; help to regulate composition of ECF
Fibrous astrocytes	CNS	Form supportive framework in CNS; form scar tissue to replace damaged nervous tissue
Ependymal cells	CNS	Line cavities of brain and spinal cord; secrete and circulate cerebrospinal fluid
Microglia	CNS	Phagocytize and destroy microorganisms, foreign matter, and dead nervous tissue
Schwann cells	PNS	Form neurilemma around all PNS nerve fibers and myelin around most of them; aid in regeneration of damaged nerve fibers
Satellite cells	PNS	Surround somas of neurons in the ganglia; function uncertain

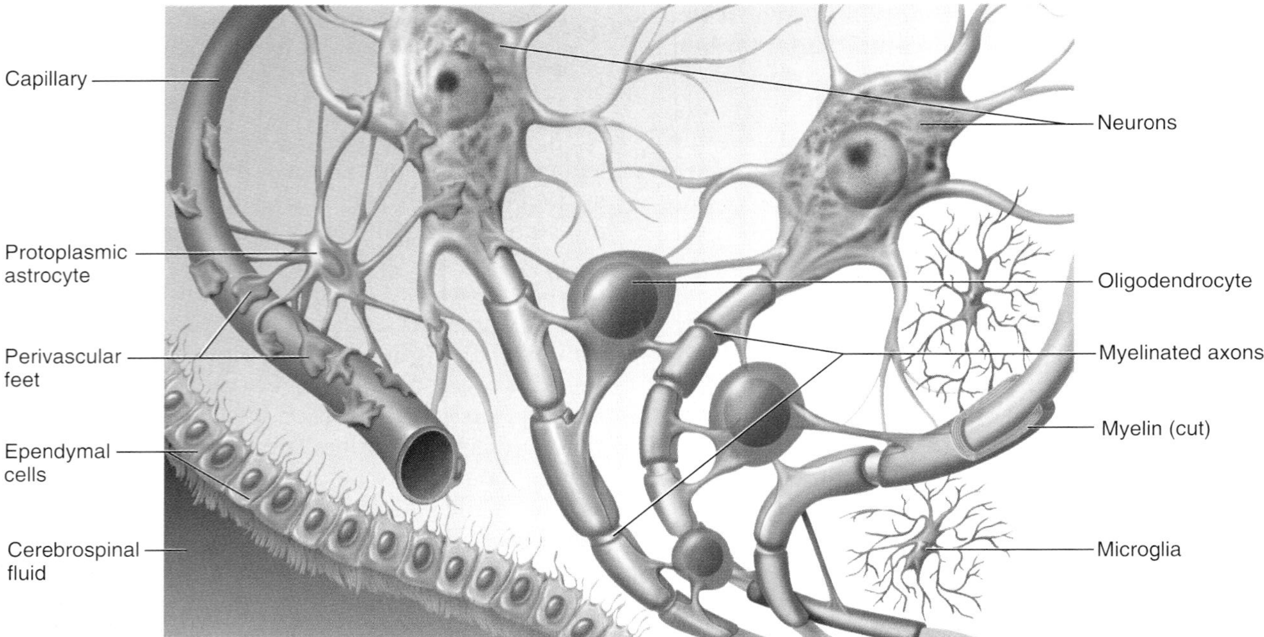

Figure 13.5 Neuroglia of the Central Nervous System.

2. **Astrocytes**[16] are many-branched glial cells with a somewhat starlike shape. They are the most abundant glia in the CNS and constitute over 90% of the tissue in some areas of the brain. There are two types of astrocytes—protoplasmic and fibrous. **Protoplasmic astrocytes** cover the entire brain surface and most nonsynaptic regions of the neurons. They issue numerous extensions, called *vascular processes* or *perivascular feet,* that contact the blood capillaries and stimulate the formation of tight junctions between the capillary endothelial cells. These junctions contribute to a *blood-brain barrier* that strictly controls which substances are able to get from the bloodstream into the brain tissue (see chapter 14). Protoplasmic astrocytes also help to regulate the chemical composition of the tissue fluid. When neurons transmit signals, they release potassium ions and neurotransmitters. Protoplasmic astrocytes absorb these substances and prevent them from reaching excessive levels in the tissue fluid. **Fibrous astrocytes** form a supportive framework for the CNS. When neurons are damaged, fibrous astrocytes form hardened masses of scar tissue and fill space formerly occupied by neurons. This process is called *astrocytosis* or *sclerosis.*[17]
3. **Ependymal**[18] (ep-EN-dih-mul) **cells** resemble a cuboidal epithelium lining the internal cavities of the brain and spinal cord. Unlike epithelial cells, however, they have no basement membrane and they exhibit rootlike processes that penetrate into the underlying nervous tissue. Ependymal cells produce cerebrospinal fluid (CSF), a clear liquid that bathes the CNS and fills its internal cavities. They have patches of cilia on their apical surfaces that help to circulate the CSF. Ependymal cells and CSF are considered in more detail in the next chapter.
4. **Microglia** are small macrophages that develop from white blood cells called monocytes. They wander through the CNS and phagocytize dead nervous tissue, microorganisms, and other foreign matter. They become concentrated in areas damaged by infection, trauma, or stroke. Pathologists look for clusters of microglia in histological sections of the brain as a clue to sites of injury.

The other two types of glial cells occur in the peripheral nervous system:

5. **Schwann**[19] (shwon) **cells** envelop nerve fibers of the PNS. In most cases, a Schwann cell winds repeatedly around a nerve fiber and produces a myelin sheath similar to the one produced by

16. *astro* = star + *cyte* = cell
17. *scler* = hardening + *osis* = process, condition
18. *ependyma* = upper garmet
19. Theodor Schwann (1810–82), German histologist

Clinical Insight 13.1

Glial Cells and Brain Tumors

A tumor consists of a mass of rapidly dividing cells. Mature neurons, however, have little capacity for mitosis and seldom form tumors. Some brain tumors arise from the meninges (protective membranes of the CNS) or arise by metastasis from tumors elsewhere, such as malignant melanoma and colon cancer. Most adult brain tumors, however, are composed of glial cells, which are mitotically active throughout life. Such tumors, called *gliomas,*[20] grow rapidly and are highly malignant. Because of the blood-brain barrier, brain tumors usually do not yield to chemotherapy and must be treated with radiation or surgery.

oligodendrocytes in the CNS. There are some important differences between the CNS and PNS in the way myelin is produced, which we consider shortly. In addition to myelinating peripheral nerve fibers, Schwann cells assist in the regeneration of damaged fibers, which also is discussed later.

6. **Satellite cells** surround the neuron cell bodies in ganglia of the PNS. Little is known of their function.

Myelin

The **myelin** (MY-eh-lin) **sheath** is an insulating layer around a nerve fiber, somewhat like the rubber insulation on a wire. It is formed by oligodendrocytes in the central nervous system and by Schwann cells in the peripheral nervous system. Since it consists of the plasma membranes of these glial cells, its composition is like that of plasma membranes in general. It is about 20% protein and 80% lipid, the latter including phospholipids, glycolipids, and cholesterol. Myelination of the nervous system begins in the fourteenth week of fetal development, yet hardly any myelin exists in the brain at the time of birth. Myelination proceeds rapidly in infancy and isn't completed until late adolescence. Since myelin has such a high lipid content, dietary fat is important to early nervous system development. Well-meaning parents can do their children significant harm by giving them the sort of low-fat diets (skimmed milk, etc.) that may be beneficial to an adult.

In the CNS, each oligodendrocyte reaches out to several nerve fibers in its immediate vicinity. The armlike process of the oligodendrocyte spirals repeatedly around the nerve fiber, laying down many compact layers of its own membrane with almost no cytoplasm between the membranes. These layers constitute the myelin sheath. A nerve fiber is much longer than the reach of a single oligodendrocyte, so it requires many oligodendrocytes to cover one nerve fiber.

In the PNS, a Schwann cell spirals around a single nerve fiber, putting down as many as a hundred layers of membrane (fig. 13.6). External to the myelin sheath is the

20. *glia* = glial cells + *oma* = tumor

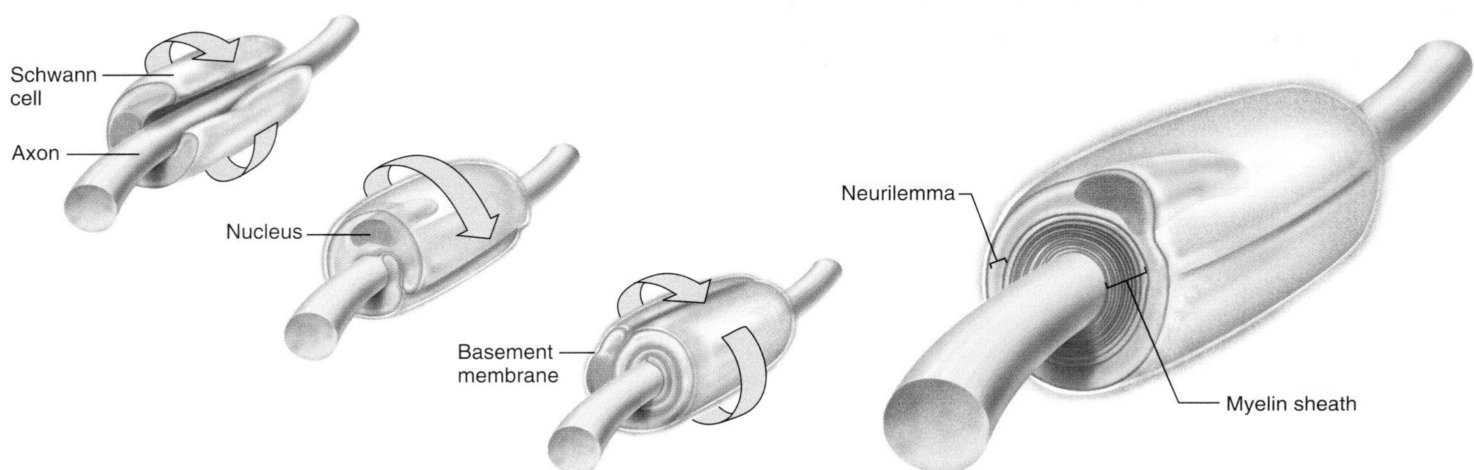

Figure 13.6 Formation of the Myelin Sheath. The Schwann cell coils repeatedly around the axon until it has laid down multiple layers of its membrane, which constitute the myelin sheath.

Clinical Insight 13.2

Diseases of the Myelin Sheath

Multiple sclerosis and Tay-Sachs disease are degenerative disorders of the myelin sheath. In *multiple sclerosis (MS),* the oligodendrocytes and myelin sheaths of the CNS deteriorate and are replaced by hardened scar tissue, especially between the ages of 20 and 40. Nerve conduction is disrupted with effects that depend on what part of the CNS is involved—double vision, blindness, speech defects, neurosis, tremors, and numbness. Patients experience variable cycles of milder and worse symptoms until they eventually become bedridden. Most die from 7 to 32 years after the onset of the disease. The cause of MS remains uncertain; most theories suggest that it results from an immune disorder triggered by a virus in genetically susceptible individuals. There is no cure.

Tay-Sachs[23] *disease* is a hereditary disorder seen mainly in infants of Eastern European Jewish ancestry. It results from the abnormal accumulation of a glycolipid called GM_2 (ganglioside) in the myelin sheath. GM_2 is normally decomposed by a lysosomal enzyme, but this enzyme is lacking from those who inherit the recessive Tay-Sachs gene from both parents. As GM_2 accumulates, it disrupts the conduction of nerve signals and the victim typically suffers blindness, loss of coordination, and dementia. Signs begin to appear before the child is a year old and most victims die by the age of three or four. Asymptomatic adult carriers can be identified by a blood test and advised by genetic counselors on the risk of their children having the disease.

23. Warren Tay (1843–1927), English physician; Bernard Sachs (1858–1944), American neurologist

neurilemma[21] (noor-ih-LEM-ah), the outermost coil of the Schwann cell. Here, the bulging body of the Schwann cell contains its nucleus and most of its cytoplasm. To visualize this, imagine that you wrapped an almost-empty tube of toothpaste tightly around a pencil. The pencil represents the axon, and the spiral layers of toothpaste tube (with the toothpaste squeezed out) represent the myelin. The toothpaste would be forced to one end of the tube, which would form a bulge on the external surface of the wrapping, like the body of the Schwann cell.

External to the neurilemma is a basement membrane and then a thin sleeve of fibrous connective tissue called the *endoneurium.* Nerve fibers of the CNS have no neurilemma or endoneurium.

Since each glial cell (Schwann cell or oligodendrocyte) myelinates only part of an axon, the myelin sheath is segmented. The gaps between the segments of myelin are **nodes of Ranvier**[22] (RON-vee-AY), and the myelin-covered segments from one gap to the next are called **internodes** (see fig. 13.3). The internodes are about 0.2 to 1.0 mm long in the PNS. The short section of nerve fiber between the axon hillock and the first glial cell is called the **initial segment.** Since the axon hillock and initial segment play an important role in initiating a nerve signal, they are collectively called the **trigger zone.**

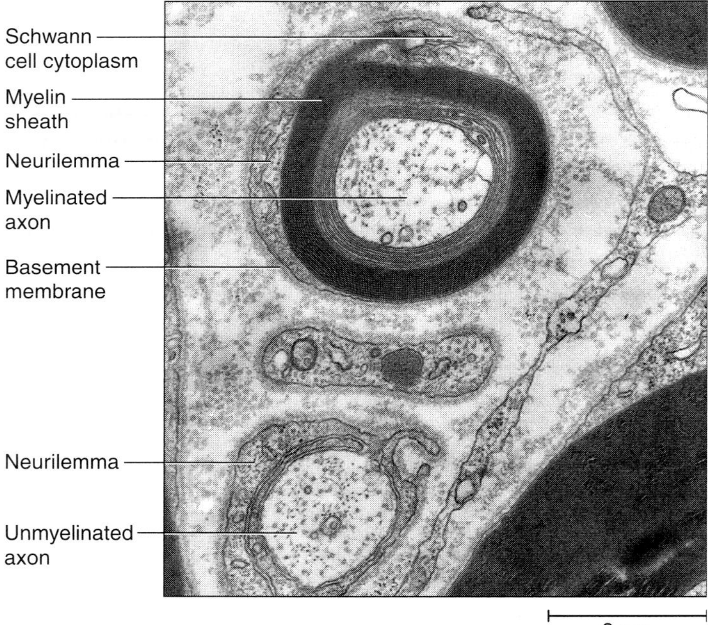

Figure 13.7 Myelinated and Unmyelinated Axons. A Schwann cell produces a neurilemma in both cases, but forms myelin around only the upper axon.

Unmyelinated Nerve Fibers

Not all nerve fibers are myelinated, but even the **unmyelinated** fibers of the PNS are associated with Schwann cells. In such cases, one Schwann cell harbors from 1 to 12 small nerve fibers in grooves in its surface. The Schwann cell's plasma membrane folds once around each nerve fiber and somewhat overlaps itself along the edges (fig. 13.7). This wrapping is the neurilemma. A basement membrane surrounds the entire Schwann cell along with its nerve fibers.

21. *neuri* = nerve + *lemma* = husk, peel, sheat
22. L. A. Ranvier (1835–1922), French histologist and pathologist

Clinical Insight 13.3

Nerve Growth Factor

Nerve growth factor (NGF) is a protein involved in the development of sympathetic nerve fibers and some sensory fibers. It is secreted by effector (gland and muscle) cells, picked up by axon terminals, and carried back to the soma by retrograde axonal transport. NGF guides developing nerve fibers to the appropriate target cells and thus plays a key role in the proper "wiring" of the nervous system to its effectors. The study of NGF and other growth factors is one of the most rapidly developing areas of biomedical research today, especially since recombinant DNA technology has made it possible to produce large quantities of these chemicals for study and clinical use. The use of NGF to treat Alzheimer disease is discussed in clinical insight 13.4 on page 485.

The speed at which a nerve signal travels along a nerve fiber depends on two factors: the diameter of the nerve fiber and the presence or absence of myelin. Signal conduction occurs along the surface of a fiber, not deep within its axoplasm. Larger fibers have more surface area and conduct signals more rapidly than smaller fibers. Myelin further speeds signal conduction for reasons discussed later in this chapter. Nerve signals travel about 0.5 to 2.0 m/sec in small unmyelinated fibers (2–4 μm in diameter) and 3 to 15 m/sec in myelinated fibers of the same size. In large myelinated fibers (up to 20 μm in diameter) they travel as fast as 120 m/sec. One might wonder why all of our nerve fibers are not large, myelinated, and fast, but if this were so, our nervous system would be either impossibly bulky or limited to far fewer fibers. Slow unmyelinated fibers are quite sufficient for processes in which quick responses are not particularly important, such as stimulating the secretion of stomach acid or dilating the pupil of the eye. Fast myelinated fibers are employed where speed is more important, as in motor commands to the skeletal muscles and sensory signals for vision and balance.

Regeneration of Nerve Fibers

Nerve fibers of the PNS are vulnerable to cuts, crushing injuries, and other trauma. A damaged peripheral nerve fiber can regenerate, however, if its soma is intact and at least some neurilemma remains. Within the first few weeks after injury, the severed distal end of an axon and its myelin sheath degenerate and macrophages remove the debris (fig. 13.8). A **regeneration tube,** formed by the neurilemma and endoneurium, is necessary for regeneration. The axon stump puts out several sprouts until one of them finds its way into the tube. This sprout begins to grow rapidly (about 3–5 mm/day), possibly under the influence of chemicals secreted by the tube (see clinical insight 13.3), while the other sprouts are reabsorbed. The regeneration tube guides the growing axon back to its original destination until the neuron reestablishes a connection with the cells that it originally innervated. Skeletal muscle fibers atrophy when their nerve fiber is severed (*denervation atrophy*) but regrow after the connection is reestablished. Damaged neurons in the CNS cannot regenerate. Since the CNS is enclosed in bone, however, it suffers less trauma than the PNS.

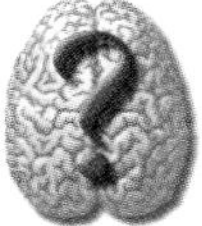

Think About It

What features essential to regeneration, present in the PNS, are lacking from the CNS?

Before You Go On

Answer the following questions to test your understanding of the preceding section:

4. Sketch a multipolar neuron and label its soma, dendrites, axon, terminal arborization, synaptic knobs, myelin sheath, and nodes of Ranvier.
5. How are the materials synthesized in the soma delivered to the axon terminals?
6. How is a glial cell different from a neuron? List the six types of glial cells and the functions of each.
7. How is myelin produced? How does myelin production in the CNS differ from that in the PNS?
8. How can a severed peripheral nerve fiber find its way back to the cells it originally innervated?

Electrophysiology of Neurons

Objectives

When you have completed this section, you should be able to

- explain why a cell has an electrical charge difference (voltage) across its membrane;
- explain how stimulation of a neuron causes a local electrical change in its membrane;
- explain how local electrical changes generate a nerve signal; and
- explain how the nerve signal is transmitted down an axon.

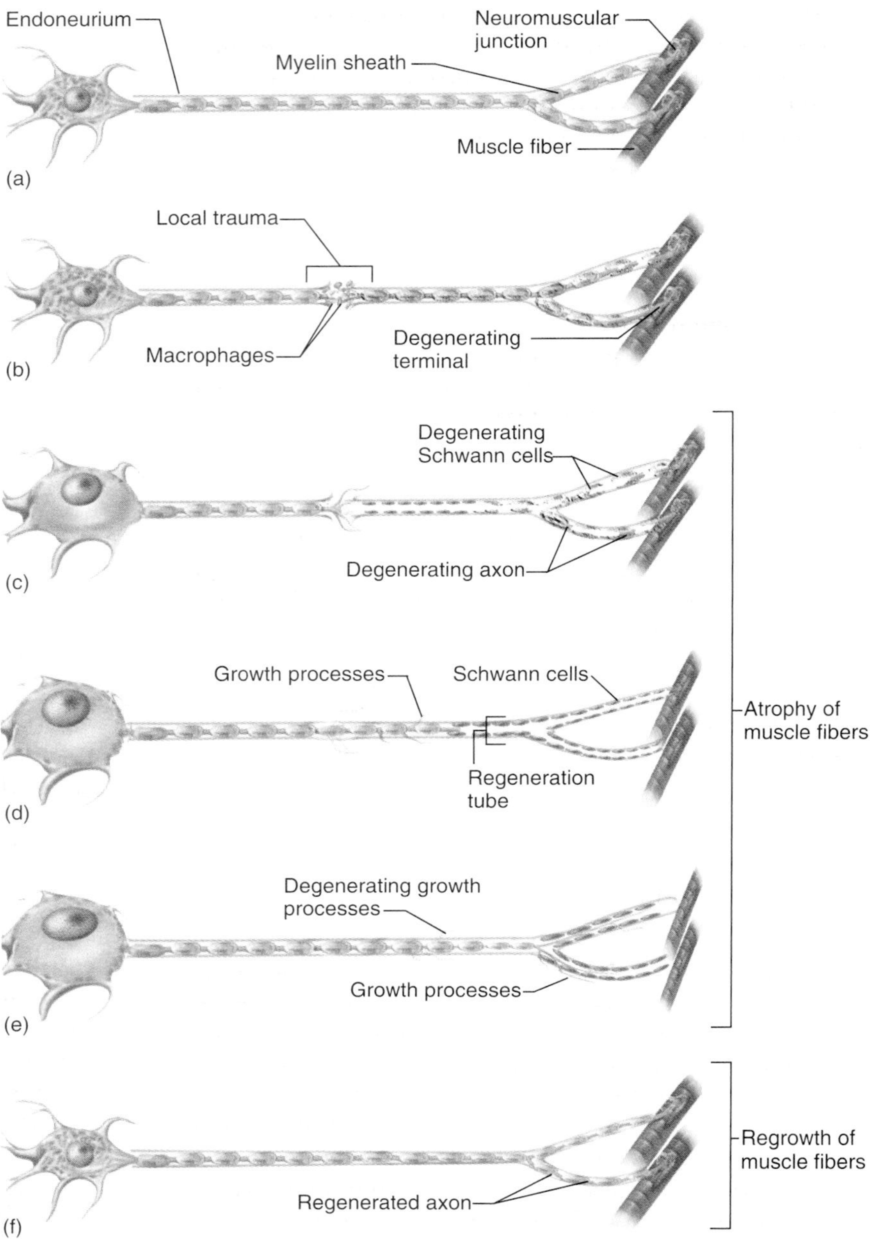

Figure 13.8 **Regeneration of a Damaged Nerve Fiber.** (*a*) Normal nerve fiber. (*b*) Injured fiber with macrophages cleaning up tissue debris at the point of injury, and with early degeneration of the nerve fiber, myelin sheath, and axon terminals distal to that. (*c*) Early regeneration; the soma is swollen, the Nissl bodies have dispersed, and the axon stump has begun producing multiple growth processes, while the severed distal part of the nerve fiber shows further degeneration. In the case of somatic motor neurons, muscle fibers atrophy for lack of innervation. (*d*) Continued regeneration; the Schwann cells, basement membrane, and neurilemma form a regeneration tube, one growth process grows into the tube, and the other growth processes are reabsorbed. (*e*) Continued regeneration; the growth processes have almost reestablished contact with the muscle fibers. (*f*) Regeneration complete; the muscle fibers are reinnervated and have regrown, and the soma of the neuron has returned to its original appearance.

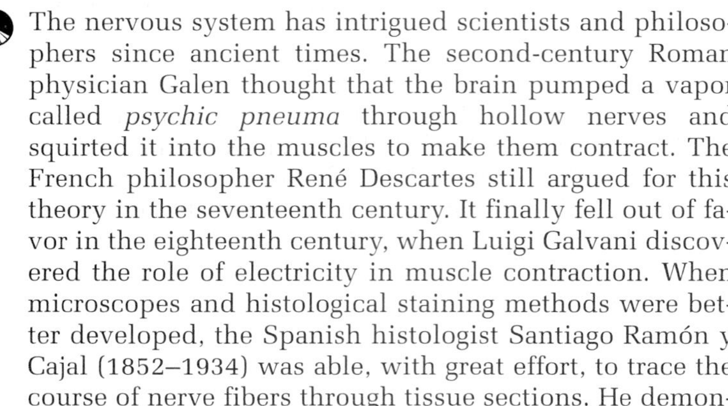

The nervous system has intrigued scientists and philosophers since ancient times. The second-century Roman physician Galen thought that the brain pumped a vapor called *psychic pneuma* through hollow nerves and squirted it into the muscles to make them contract. The French philosopher René Descartes still argued for this theory in the seventeenth century. It finally fell out of favor in the eighteenth century, when Luigi Galvani discovered the role of electricity in muscle contraction. When microscopes and histological staining methods were better developed, the Spanish histologist Santiago Ramón y Cajal (1852–1934) was able, with great effort, to trace the course of nerve fibers through tissue sections. He demonstrated that the nervous pathway was not a continuous "wire" or tube, but a series of separate cells separated by synapses. His theory, now called the *neuron doctrine*, suggested another direction for research—how do neurons communicate? We now know they employ an electrical and chemical code. Our objective in this section is to study how this code is produced and transmitted.

Electrical Potentials and Currents

Neuronal communication, like muscle excitation, is based on electrophysiology—cellular mechanisms for producing electrical potentials and currents. An **electrical potential** is a difference in the concentration of charged particles between one point and another. It is a form of potential energy that, under the right circumstances, can produce a current. An electrical **current** is a flow of charged particles from the one point to the other. A new flashlight battery, for example, typically has a potential, or charge, of 1.5 volts (V). If the two poles of the battery are connected by a wire, electrons flow through the wire from one pole of the battery to the other, creating a current that can be used to light a bulb. As long as the battery has a potential, we say it is **polarized.**

Living cells are also polarized. As we saw in the preceding chapter, the charge difference across the plasma membrane is called the **resting membrane potential (RMP).** It is much less than the potential of a flashlight battery—typically about −70 millivolts (mV) in an unstimulated, "resting" neuron. It is customary to express cellular membrane potentials in terms of the inner (cytoplasmic) side of the plasma membrane. A negative value means there are more negatively charged particles on the inside of the membrane than on the outside.

We do not have free electrons in the body as we do in an electrical circuit. Electrical currents in the body are created, instead, by the flow of ions such as Na^+ and K^+ through openings (channels) in the plasma membrane. As we saw in chapters 4 and 12, gated channels can be opened and closed by various stimuli. This enables cells to turn electrical currents on and off.

The Resting Membrane Potential

The reason that a cell has a resting membrane potential is that electrolytes are unequally distributed between the extracellular fluid (ECF) on the outside of the plasma membrane and the intracellular fluid (ICF) on the inside. The RMP results from the combined effect of three factors: (1) the diffusion of ions down their concentration gradients through the plasma membrane; (2) selective permeability of the plasma membrane, allowing some ions to pass more easily than others; and (3) the electrical attraction of cations and anions to each other.

Potassium ions (K^+) have the greatest influence on the RMP because the plasma membrane is more permeable to K^+ than to any other ions. Imagine a hypothetical cell in which all the K^+ starts out in the ICF, with none in the ECF. Also in the ICF are a number of cytoplasmic anions that cannot escape from the cell because of their size or charge—phosphates, sulfates, small organic acids, proteins, ATP, and RNA. Assuming K^+ can diffuse freely through channels in the plasma membrane, it will diffuse down its concentration gradient and out of the cell, leaving these cytoplasmic anions behind. As a result, the ICF will grow more and more negatively charged. But as the ICF becomes more negative, it will exert a stronger and stronger attraction for the positive potassium ions and attract some of them back into the cell. Eventually an *equilibrium* (balance) will be reached in which K^+ is moving out of the cell (diffusion down its concentration gradient) and into the cell (by electrical attraction) at equal rates. The *net* diffusion of K^+ will then stop. At the point of equilibrium, K^+ is about 40 times as concentrated in the ICF as in the ECF (fig. 13.9).

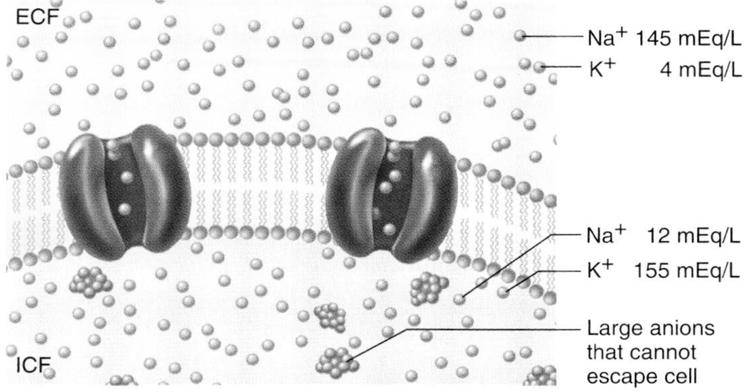

Figure 13.9 **Ionic Basis of the Resting Membrane Potential.** Note that sodium ions are much more concentrated in the extracellular fluid (ECF) than in the intracellular fluid (ICF), while potassium ions are more concentrated in the ICF. Large anions unable to penetrate the plasma membrane give the cytoplasm a negative charge relative to the ECF.

If K^+ were the only ion affecting the RMP, it would give the membrane a potential of about −90 mV. However, sodium ions (Na^+) also enter the picture. Sodium is about 12 times as concentrated in the ECF as in the ICF. The resting plasma membrane is much less permeable to Na^+ than to K^+, but Na^+ does leak into the cell. It diffuses down its concentration gradient into the cell, and it is attracted by the negative charge within the cell. This sodium leak is only a trickle, but it is enough to neutralize some of the negative charge and reduce the voltage across the membrane.

Sodium continually leaks into the cell and potassium leaks out, but the sodium-potassium (Na^+-K^+) pump described in chapter 4 continually compensates for this leakage. It pumps three Na^+ out of the cell for every two K^+ it brings in, consuming one ATP for each exchange cycle. By removing more cations from the cell than it brings in, it contributes about −3 mV to the resting membrane potential. The net effect of all this—K^+ diffusion out of the cell, Na^+ diffusion inward, and the Na^+-K^+ pump—is the resting membrane potential of −70 mV.

The Na^+-K^+ pump accounts for about 70% of the energy (ATP) requirement of the nervous system. Every signal generated by a neuron slightly upsets the distribution of Na^+ and K^+, so the pump must work continually to restore equilibrium. This is why nervous tissue has one of the highest rates of ATP consumption of any tissue in the body and why it demands so much glucose and oxygen.

Although a neuron is said to be resting when it is not producing signals, it is highly active maintaining its RMP and "waiting," as it were, for something to happen.

Local Potentials

We now consider the disturbances in membrane potential that occur when a neuron is stimulated. Typically (but with exceptions), the response of a neuron begins at a dendrite, spreads through the soma, travels down the axon, and ends at the synaptic knobs. We consider the process in that order.

Neurons can be stimulated by chemicals, light, heat, or mechanical distortion of the plasma membrane. We'll take as our example a neuron being chemically stimulated on its dendrite (fig. 13.10). The chemical—perhaps a pain signal from a damaged tissue or odor molecules in a breath of air—binds to receptors on the neuron. These receptors are ligand-regulated sodium gates that open and allow Na^+ to rush into the cell. This inflow of Na^+ neutralizes some of the internal negative charge, so the voltage across the membrane drifts toward zero. Any such case in which membrane voltage shifts to a less negative value is called **depolarization.** The incoming sodium ions diffuse for short distances along the inside of the plasma membrane and produce a current that travels from the point of stimulation toward the cell's trigger zone. Such a local or short-range change in voltage is called a **local potential.**

There are four characteristics that distinguish local potentials from the action potentials we will study shortly (table 13.2). You will appreciate these distinctions more fully after you have studied action potentials.

1. Local potentials are **graded,** meaning that they vary in magnitude (voltage) according to the strength of the stimulus. A more intense or prolonged stimulus opens a greater number of ion gates than a weaker stimulus. Thus, more Na^+ enters the cell and the membrane voltage changes more than it does with a weaker stimulus.
2. Local potentials are **decremental,** meaning they get weaker the farther they spread from the point of stimulation. The decline in signal strength occurs because as Na^+ flows into the cell, spreads out under the plasma membrane, and depolarizes it, K^+ leaks out and reverses the effect of the Na^+ inflow. Therefore, the voltage shift caused by Na^+ diminishes rapidly with distance. This prevents local potentials from having any long-distance effects.
3. Local potentials are **reversible,** meaning that if stimulation ceases, K^+ diffusion out of the cell quickly returns the membrane voltage to its resting potential.

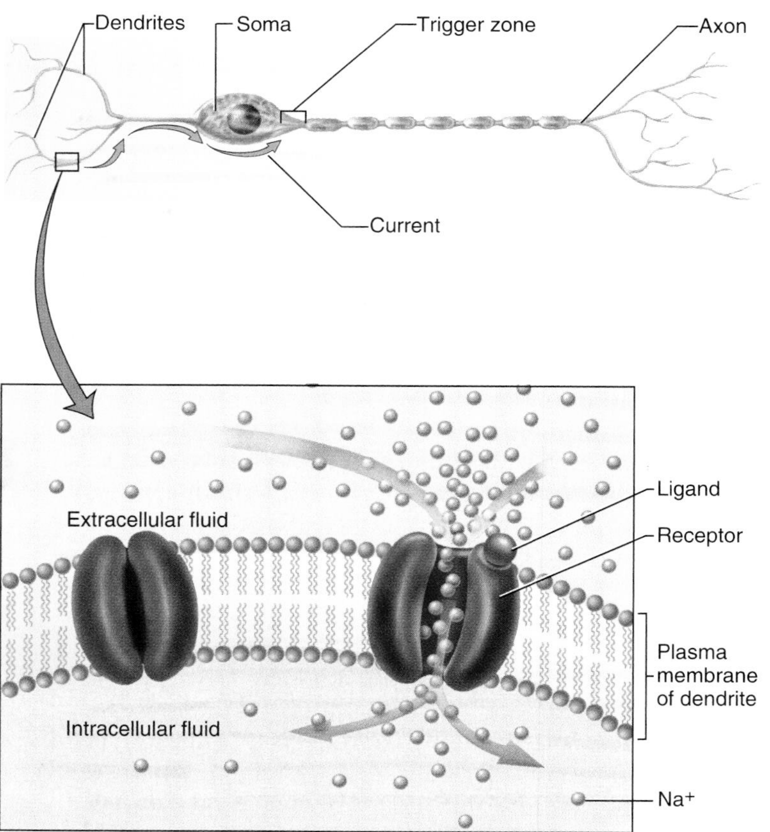

Figure 13.10 **Excitation of a Neuron by a Chemical Stimulus.** When the chemical (ligand) binds to a receptor on the neuron, the receptor acts as a ligand-regulated ion gate through which Na^+ diffuses into the cell. This depolarizes the plasma membrane.

4. Local potentials can be either **excitatory** or **inhibitory.** So far, we have considered only excitatory local potentials, which depolarize a cell and tend to make a neuron more likely to produce an action potential. Acetylcholine usually has this effect. Other neurotransmitters, such as glycine, cause an opposite effect—they **hyperpolarize** a cell, or make the membrane more negative. The neuron is then less sensitive and less likely to produce an action potential. A balance between excitatory and inhibitory potentials is very important to information processing in the nervous system, and we explore this more fully later in the chapter.

Action Potentials

An **action potential** is a more dramatic change produced by voltage-regulated ion gates in the plasma membrane. Action potentials occur only where there is a high enough density of voltage-regulated gates. Most of the soma has only 50 to 75 gates per square micrometer (μm^2) and cannot generate action potentials. The trigger zone, however, has 350 to 500 gates per μm^2. If an excitatory local potential spreads all the way to the trigger zone and is still strong enough when it arrives, it can open these gates and generate an action potential.

The action potential is a rapid up-and-down shift in membrane voltage. Figure 13.11*a* shows an action potential numbered to correspond to the following description. Figure 13.12 correlates these voltage changes with events in the plasma membrane.

1. When sodium ions arrive at the axon hillock, they depolarize the membrane at that point. This appears as a steadily rising local potential.
2. For anything more to happen, this local potential must rise to a critical level called the **threshold** (typically about −55 mV). This is the minimum voltage needed to open voltage-regulated gates.
3. The neuron now "fires," or produces an action potential. At threshold, voltage-regulated Na^+ and K^+ gates open. The Na^+ gates open very quickly and the K^+ gates more slowly, so the initial effect on membrane potential is due to Na^+. Only a few Na^+ gates open initially, but as Na^+ rushes into the cell, it further depolarizes the membrane. This voltage change stimulates still more voltage-regulated Na^+ gates to open and admit even more Na^+ to the cell. Thus, a positive feedback cycle is created that makes the membrane voltage rise rapidly.
4. As the rising membrane potential passes 0 mV, Na^+ gates are *inactivated,* thus closing off the channels that allow Na^+ into the cell. By the time they all close and Na^+ inflow ceases, the voltage peaks at approximately +35 mV. (The peak is as low as 0 mV in some neurons and as high as 50 mV in others.) The membrane is now positive on the inside and negative on the outside—its polarity is reversed compared to the RMP.
5. By the time the voltage peaks, the slow K^+ gates are fully open. Potassium ions rush out of the cell because they are now repelled by the positive intracellular fluid. Their outflow **repolarizes** the membrane—that is, it shifts the voltage back into the negative numbers. The action potential consists of the up-and-down voltage shifts that occur from the time the threshold is reached to the time the voltage returns to the RMP.
6. Potassium gates stay open longer than Na^+ gates do, so the amount of potassium that leaves the cell is greater than the amount of sodium that entered. Therefore, the membrane voltage drops to 1 or 2 mV more negative than the original RMP, producing a negative overshoot called *hyperpolarization.*
7. As you can see, Na^+ and K^+ switch places across the membrane during an action potential. During hyperpolarization, ion diffusion through the membrane and (in the CNS) the removal of extracellular K^+ by the astrocytes gradually restores the original resting membrane potential.

Table 13.2 Comparison of Local Potentials and Action Potentials

Local Potential	Action Potential
Produced by ligand-regulated gates on the dendrites and soma	Produced by voltage-regulated gates on the trigger zone and axon
May be a positive (depolarizing) or negative (hyperpolarizing) voltage change	Always begins with depolarization
Graded; proportional to stimulus strength	All-or-none; either does not occur at all or exhibits same peak voltage regardless of stimulus strength
Reversible; returns to RMP if stimulation ceases before threshold is reached	Irreversible; goes to completion once it begins
Local; has effects for only a short distance from point of origin	Self-propagating; has effects a great distance from point of origin
Decremental; signal grows weaker with distance	Nondecremental; signal maintains same strength regardless of distance

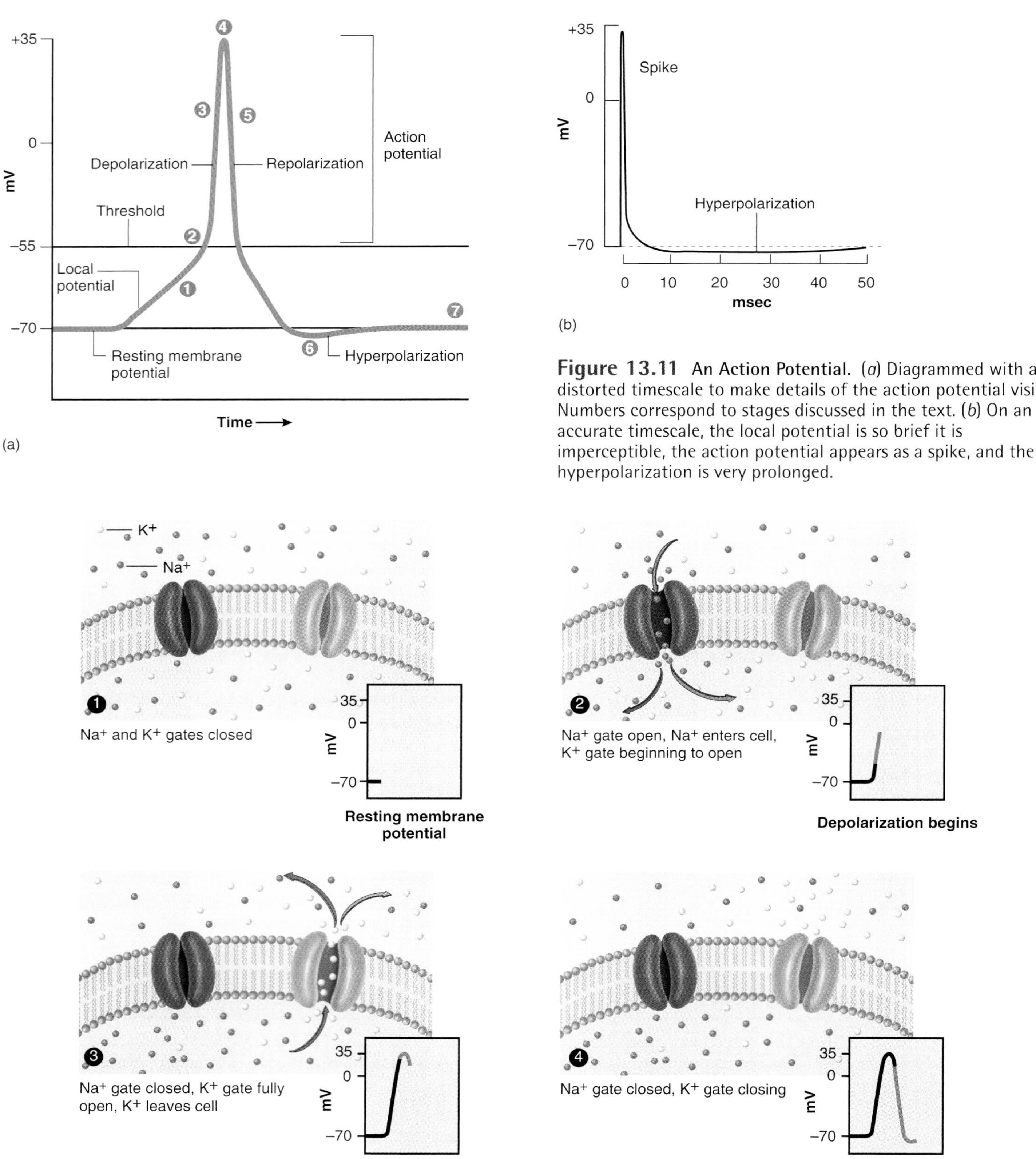

Figure 13.11 An Action Potential. (*a*) Diagrammed with a distorted timescale to make details of the action potential visible. Numbers correspond to stages discussed in the text. (*b*) On an accurate timescale, the local potential is so brief it is imperceptible, the action potential appears as a spike, and the hyperpolarization is very prolonged.

Figure 13.12 Actions of the Sodium and Potassium Gates During an Action Potential.

At the risk of being misleading, figure 13.12 is drawn as if the majority of sodium and potassium ions had traded places. In reality, only about one ion in a million crosses the membrane to produce an action potential, and an action potential affects ion distribution only in a thin layer close to the membrane. If the illustration tried to represent these points accurately, the difference would be so slight you could not see it; indeed the changes in ion concentrations inside and outside the cell are so slight they cannot be measured in the laboratory unless a neuron has been stimulated for a long time. Even after thousands of action potentials, the cytosol still has a higher concentration of K^+ and a lower concentration of Na^+ than the ECF does.

Figure 13.11*a* also is deliberately distorted. In order to demonstrate the different phases of the local potential and action potential, the magnitudes of the local potential and hyperpolarization are exaggerated, the local potential is stretched out to make it seem longer, and the duration of hyperpolarization is shrunken so the graph does not run off the page. When these events are plotted on a more realistic timescale, they look like figure 13.10*b*. The local potential is so brief it is unnoticeable, and hyperpolarization is very long but only slightly more negative than the RMP. An action potential is often called a *spike;* it is easy to see why from this figure.

Earlier we saw that local potentials are graded, decremental, and reversible. Let's see now how action potentials compare on these points.

- Action potentials follow an **all-or-none law.** If a stimulus depolarizes the neuron to threshold, the neuron fires at its maximum voltage (such as +35 mV); if threshold is not reached, the neuron doesn't fire at all. Above threshold, stronger stimuli do not produce stronger action potentials. Thus, action potentials are not graded (proportional to stimulus strength).
- Action potentials are **nondecremental.** For reasons to be examined shortly, they do not get weaker with distance. An action potential at the end of a nerve fiber will be just as strong as an action potential in the trigger zone up to a meter away.
- Action potentials are **irreversible.** If a neuron reaches threshold, the action potential goes to completion; it cannot be stopped once it begins.

In some respects, we can compare the firing of a neuron to the firing of a gun. As the trigger is squeezed, a gun either fires with maximum force or does not fire at all (analogous to the all-or-none law). You cannot fire a fast bullet by squeezing the trigger hard or a slow bullet by squeezing it gently—once the trigger is pulled to its "threshold," the bullet always leaves the muzzle at the same velocity. And, like an action potential, the firing of a gun is irreversible once the threshold is reached. Table 13.2 further contrasts a local potential with an action potential, including some characteristics of action potentials explained in the next section.

The Refractory Period

During an action potential and for a few milliseconds after, it is difficult or impossible to stimulate that region of a neuron to fire again. This period of resistance to restimulation is called the **refractory period.** It is divided into two phases—an *absolute refractory period* in which no stimulus of any strength will trigger a new action potential, and after this, a *relative refractory period* in which it is possible to trigger a new action potential, but only with an unusually strong stimulus (fig. 13.13).

The absolute refractory period lasts from the start of the action potential until the membrane returns to the resting potential—that is, for as long as the Na^+ gates are open and subsequently inactivated. The relative refractory period lasts until hyperpolarization ends. During this period, K^+ gates are still open. A new stimulus tends to admit Na^+ and depolarize the membrane, but K^+ diffuses out through the open gates as Na^+ comes in, and thus opposes the effect of the stimulus. It requires an especially strong stimulus to override the K^+ outflow and depolarize the cell enough to set off a new action potential. By the end of hyperpolarization, K^+ gates are closed and the cell is as responsive as ever.

Note carefully that the refractory period refers only to a small patch of membrane where an action potential has already begun, not to the entire neuron. Other parts of

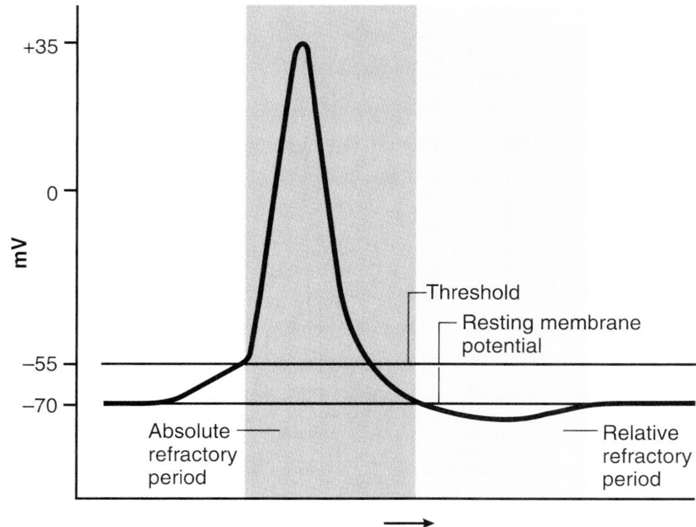

Figure 13.13 The Absolute and Relative Refractory Periods in Relation to the Action Potential.

the neuron can still be stimulated while a small area of it is refractory, and even this area quickly recovers once the nerve signal has passed on.

Signal Conduction in Nerve Fibers

If a neuron is to communicate with another cell, a signal has to travel to the end of the axon. We now see how this is achieved.

Unmyelinated Fibers

Unmyelinated fibers present a relatively simple case of signal conduction, easy to understand based on what we have already covered (fig. 13.14). An unmyelinated fiber has voltage-regulated Na^+ gates along its entire length. When an action potential occurs at the trigger zone, Na^+ enters the axon and diffuses to adjacent regions just beneath the plasma membrane. The resulting depolarization excites voltage-regulated gates immediately distal to the action potential. Sodium and potassium gates open and close just as they did at the trigger zone, and a new action potential is produced. By repetition, this excites the membrane immediately distal to that. This chain reaction continues until the traveling signal reaches the end of the axon.

Note that an action potential itself does not travel along an axon; rather, it stimulates the production of a new action potential in the membrane just ahead of it. Thus, we can distinguish an *action potential* from a *nerve signal.* The nerve signal is a traveling wave of excitation produced by self-propagating action potentials. It is a little like a line of falling dominoes. No one domino travels to the end of the line, but each domino pushes over the next one and there is a transmission of energy from the first domino to the last. Similarly, no one action potential travels to the end of an axon; a nerve signal is a chain reaction of action potentials.

If one action potential can stimulate the production of a new one next to it, you might think that the signal could also start traveling backward and return to the soma. This does not occur, however, because the membrane behind the nerve signal is still in its refractory period and cannot be restimulated. Only the membrane ahead is sensitive to stimulation. The refractory period thus ensures that nerve signals are conducted in the proper direction, from the soma to the synaptic knobs.

A traveling nerve signal is an electrical current, but it is not the same as a current traveling through a wire. A current in a wire travels millions of meters per second and is decremental—it gets weaker with distance. A nerve signal is much slower (not more than 2 m/sec in unmyelinated fibers), but it is *nondecremental.* Even in the longest axons, the last action potential generated in a synaptic knob has the same voltage as the first one generated in the trigger zone. To clarify this concept we can compare the nerve signal to a burning fuse. When a fuse is lit, the heat ignites powder immediately in front of this point, and this event repeats itself in a self-propagating fashion until the end of the fuse is reached. At the end, the fuse burns just as hotly as it did at the beginning. In a fuse, the combustible powder is the source of potential energy that keeps the process going in a nondecremental fashion. In an axon, the potential energy comes from the ion gradient across the plasma membrane. Thus, the signal does not grow weaker with distance; it is self-propagating, like the burning of a fuse.

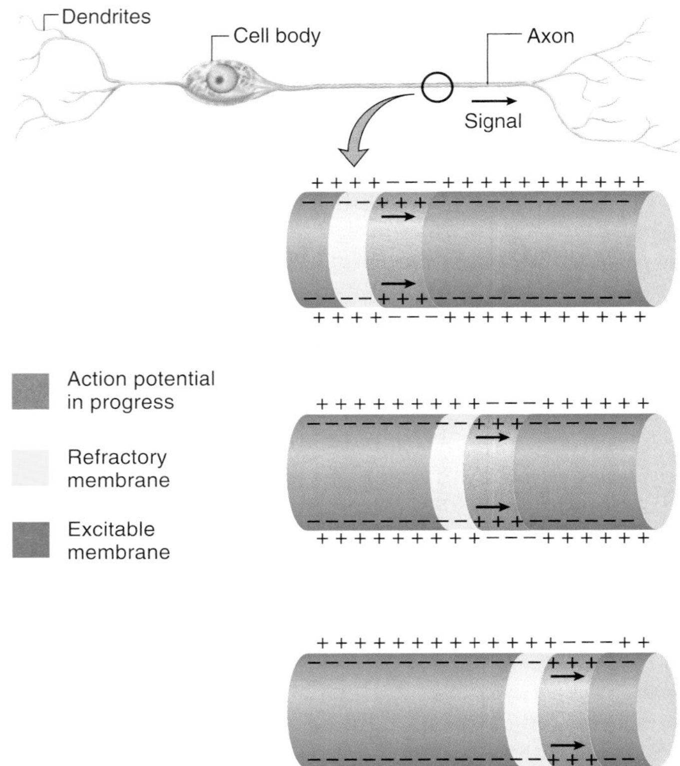

Figure 13.14 Conduction of a Nerve Signal in an Unmyelinated Fiber. Note that the membrane polarity is reversed in the region of the action potential (red). A region of membrane in its refractory period (yellow) trails the action potential and prevents the nerve signals from going backward toward the soma. The other membrane areas (green) are fully polarized and ready to respond.

Myelinated Fibers

Matters are somewhat different in myelinated fibers, because voltage-regulated ion gates are very scarce in the myelin-covered internodes—fewer than 25 per μm^2 in these regions compared with 2,000 to 12,000 per μm^2 at the nodes of Ranvier. There would be little point in having ion gates in the internodes—myelin insulates the fiber from the ECF at these points, and sodium ions from the ECF could not flow into the cell even if more gates were present.

The only way a nerve signal can travel along an internode is for Na^+ that enters at the previous node to diffuse down the fiber under the axolemma (fig. 13.15*a*). This is a very fast process, but the nerve fiber resists their flow (just as a wire resists a current) and the signal becomes weaker the farther it goes. Therefore, this aspect of conduction is decremental. The signal cannot travel much farther than 1 mm before it becomes too weak to open any voltage-regulated gates. But fortunately, there is another node of Ranvier every millimeter or less along the axon, where the axolemma is exposed to ECF and there is an abundance of voltage-regulated gates. When the diffusing ions reach this point, the signal is just strong enough to open these gates and create a new action potential. This action potential has the same strength as the one at the previous node, so each node of Ranvier boosts the signal back to its original strength (+35 mV). This mode of signal conduction is called **saltatory**[24] **conduction**—the propagation of a nerve signal that seems to jump from node to node (fig. 13.15*b*).

In the internodes, saltatory conduction is therefore based on a process that is very fast (diffusion of ions along the fiber) but decremental. In the nodes, conduction is slower but nondecremental. Since most of the axon is

24. from *saltare* = to leap, to dance

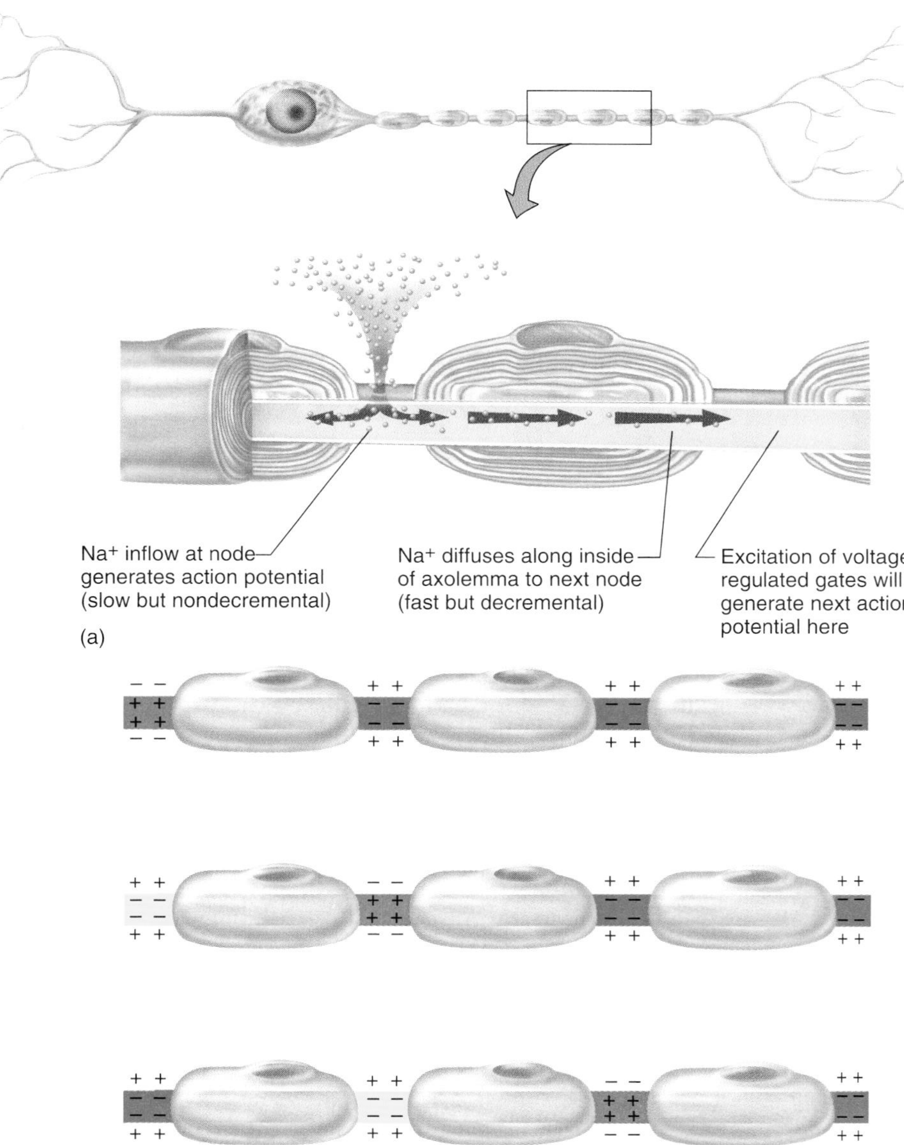

Figure 13.15 Saltatory Conduction of a Nerve Signal in a Myelinated Fiber. (*a*) Ions can be exchanged with the ECF only at the nodes of Ranvier. In the internodes, the nerve signal travels by the rapid diffusion of ions along the inside of the plasma membrane. (*b*) Action potentials (red) occur only at the nodes, and the nerve signal therefore appears to jump from node to node. Yellow areas indicate refractory (recovering) membrane.

covered with myelin, conduction occurs mainly by the fast longitudinal diffusion process. This is why myelinated fibers transmit signals much faster (up to 120 m/sec) than unmyelinated ones (up to 2 m/sec) and why the signal is just as strong at the end of the fiber as it was at the beginning.

Before You Go On

Answer the following questions to test your understanding of the preceding section:

9. What causes K^+ to diffuse out of a resting cell? What attracts it into the cell?
10. What happens to Na^+ when a neuron is stimulated on its dendrite? Why does the movement of Na^+ raise the voltage on the plasma membrane?
11. What does it mean to say a local potential is graded, decremental, and reversible?
12. How does the plasma membrane at the trigger zone differ from that on the soma? How does it resemble the membrane at a node of Ranvier?
13. What makes an action potential rise to +35 mV? What makes it drop again after this peak?
14. List four ways in which an action potential is different from a local potential.
15. Explain why myelinated fibers transmit signals much faster than unmyelinated fibers.

Synapses

Objectives

When you have completed this section, you should be able to

- explain how messages are transmitted from one neuron to another;
- explain how stimulation of a postsynaptic cell is stopped; and
- give examples of neurotransmitters and describe their actions.

A nerve signal soon reaches the ends of an axon and can go no farther. In most cases, it triggers the release of a neurotransmitter that stimulates a new wave of electrical activity in the next cell across the synapse. The most thoroughly studied type of synapse is the neuromuscular junction described in chapter 12, but here we consider synapses between two neurons. The first neuron in the signal path is the **presynaptic neuron,** which releases the neurotransmitter. The second is the **postsynaptic neuron,** which responds to it.

The presynaptic neuron may synapse with a dendrite, the soma, or the axon of a postsynaptic neuron and form an *axodendritic, axosomatic,* or *axoaxonic synapse,* respectively. A neuron can have an enormous number of synapses (fig. 13.16). For example, a spinal motor neuron is covered with about 10,000 synaptic knobs from other neurons—8,000 ending on its dendrites and another 2,000 on the soma. In part of the brain called the cerebellum, one neuron can have as many as 100,000 synapses.

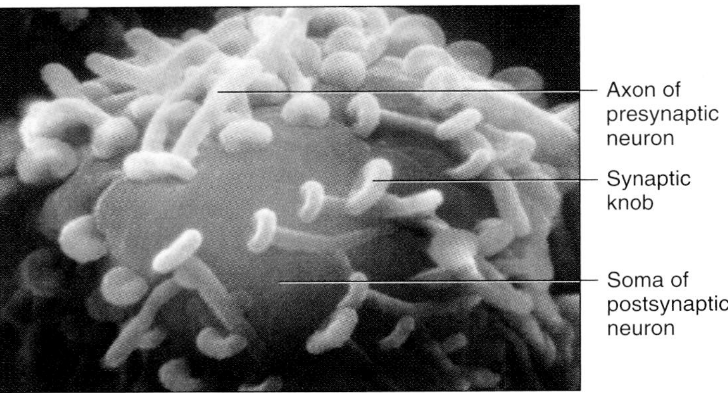

Figure 13.16 Synaptic Knobs on the Soma of a Neuron in a Marine Slug, *Aplysia* (SEM).

The Discovery of Neurotransmitters

As the neuron doctrine became generally accepted, it raised the question of how neurons communicate with each other. In the early twentieth century, biologists assumed that synaptic communication was electrical—a logical hypothesis given that neurons seemed to touch each other and signals were transmitted so quickly from one to the next. Closer histological examination, however, revealed a 20 to 40 nm gap between neurons—the **synaptic cleft**—casting doubt on the possibility of electrical conduction.

In 1921, the German pharmacologist Otto Loewi (1873–1961) conclusively demonstrated that neurons communicate by releasing chemicals. The *vagus nerves* supply the heart, among other organs, and slow it down. Loewi opened two frogs and flooded the hearts with saline to keep them moist. He stimulated the vagus nerve of one frog, and its heart rate dropped as expected. He then removed some of the saline from that heart and squirted it onto the heart of the second frog. The solution alone reduced that frog's heart rate. Evidently it contained something released by the vagus nerve of the first frog. Loewi called it *Vagusstoffe* ("vagus substance") and it was later renamed acetylcholine—the first known neurotransmitter.

Think About It

As described, does the previous experiment conclusively prove that the second frog's heart slowed as a result of something released by the vagus nerves? If you were Loewi, what control experiment would you do to rule out alternative explanations?

Following Loewi's work, the idea of electrical communication between cells fell into disrepute. Now, however, we realize that some neurons, neuroglia, and cardiac and single-unit smooth muscle (see chapter 12) do indeed have **electrical synapses,** where adjacent cells are joined by gap junctions and ions can diffuse directly from one cell into the next. These junctions have the advantage of quick transmission because there is no delay for the release and binding of neurotransmitter. Their disadvantage, however, is that they cannot integrate information and make decisions. The ability to do that is a property of **chemical synapses,** in which neurons communicate by neurotransmitters.

Structure of a Chemical Synapse

The synaptic knob (fig. 13.17) was described in the previous chapter. It contains synaptic vesicles, many of which are "docked" at release sites on the plasma membrane, ready to release their neurotransmitter on demand. A reserve pool of synaptic vesicles are located a little farther away from the membrane, clustered near the release sites and tethered to the cytoskeleton by protein microfilaments.

The postsynaptic neuron does not show such conspicuous specializations. It contains no synaptic vesicles and cannot release neurotransmitters. Its membrane does, however, contain proteins that function as receptors and ligand-regulated ion gates.

Neurotransmitters and Related Messengers

More than 100 confirmed or suspected neurotransmitters have been identified since Loewi discovered acetylcholine. Neurotransmitters fall into three major categories according to chemical composition (fig. 13.18). Some of the best-known ones are listed in table 13.3. Parts of the brain referred to in this table will become familiar to you as you study the next chapter, and you may wish to refer back to this table then to enhance your understanding of brain function.

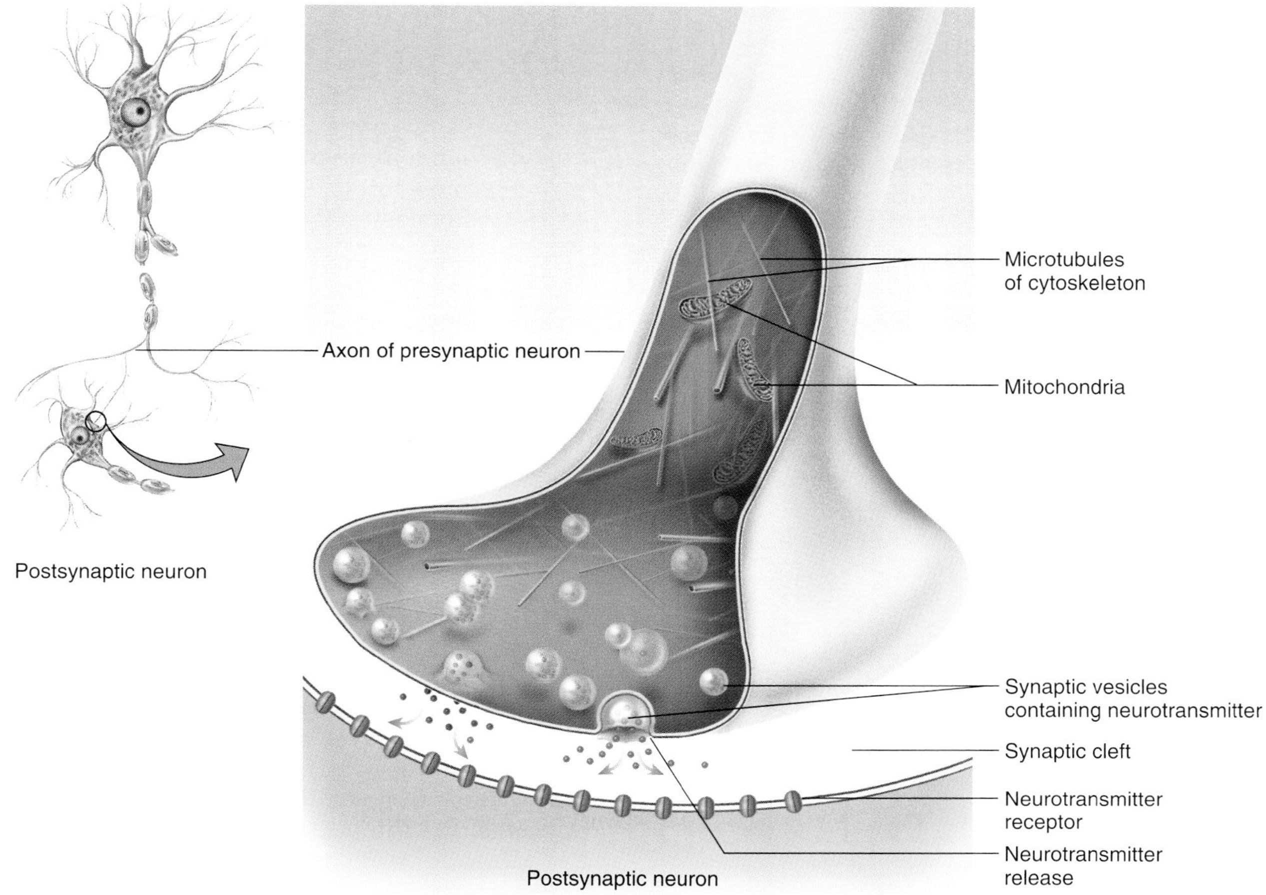

Figure 13.17 Structure of a Chemical Synapse.

1. **Acetylcholine** is in a class by itself. It is formed from acetic acid (acetate) and choline.
2. **Amino acid** neurotransmitters include glycine, glutamate, aspartate, and γ-aminobutyric acid (GABA).
3. **Monoamines (biogenic amines)** are synthesized from amino acids by replacing –COOH with another functional group. They retain the $-NH_2$ (amino group), hence their name. The major monoamines fall into two subclasses defined by the group that replaces –COOH: **catecholamines** (CAT-eh-COAL-uh-meens) (epinephrine, norepinephrine, and dopamine) and **indolamines** (serotonin and histamine).

The more we learn about neurotransmitters, hormones, and other chemical messengers, the harder it is to distinguish them from each other or even to rigorously define "neurotransmitter." Traditionally, neurotransmitters have been conceived as small organic compounds, about the size of amino acids, that function at the synapse as follows: (1) they are synthesized by the presynaptic neuron, (2) they are released in response to stimulation, (3) they bind to specific receptors on the postsynaptic cell, and (4) they alter the physiology of that cell. Neurons have other means of communication, however, that fall outside the scope of this traditional concept.

Neuropeptides are chains of 2 to 40 amino acids that sometimes modify the action of neurotransmitters. It is common, for example, for serotonin to be released together with a neuropeptide that influences its effect on the postsynaptic cell. Some authorities classify neuropeptides as neurotransmitters while others classify them separately. While neurotransmitters are stored in synaptic vesicles about 50 nm in diameter, neuropeptides are stored in larger (100 nm) *secretory granules,* also called *dense-core vesicles* because they exhibit a dense mass of peptide within them.

Some neurons produce both classical neurotransmitters and neuropeptides. They may release them simultaneously, or release one or the other under different conditions—for example, a neurotransmitter under mild stimulation and a neuropeptide under stronger stimulation. Some neuropeptides are produced not only by neurons but also by the digestive tract; thus they are known as *gut-brain peptides.* Some of these cause cravings for specific nutrients such as fat or sugar and may be associated with certain eating disorders.

Think About It

Neuropeptides can be synthesized only in the soma and must be transported to the synaptic knobs. Why is their synthesis limited to the soma?

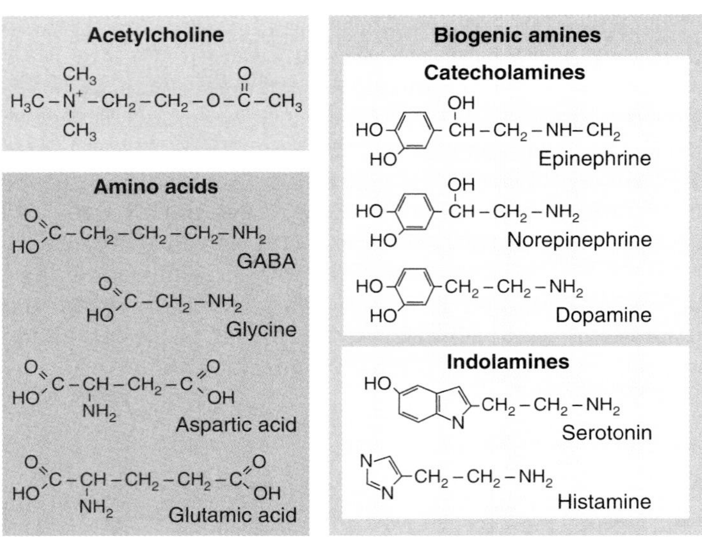

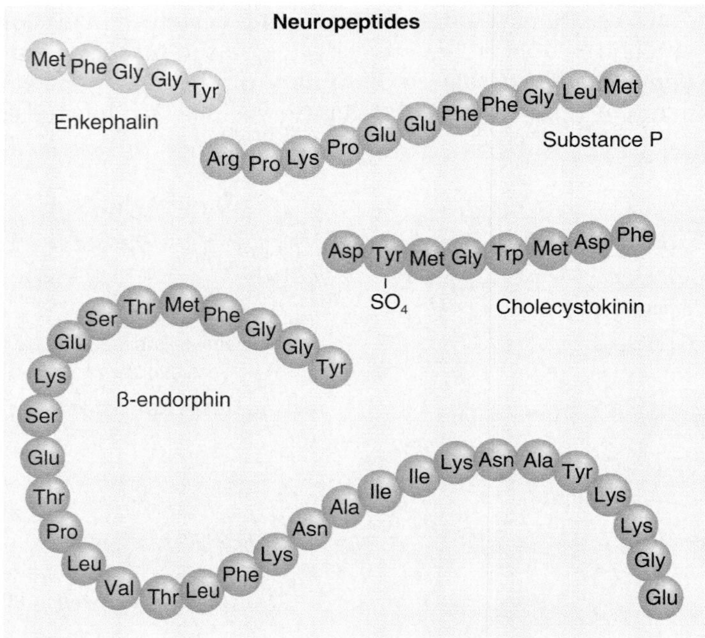

Figure 13.18 Classification of Some Neurotransmitters and Neuropeptides.

Synaptic Transmission

Neurotransmitters are quite diverse in their action. Some are excitatory, some are inhibitory, and some exert either effect depending on what kind of receptor the postsynaptic cell has. Some neurotransmitters open ligand-regulated ion gates while others act through second-messenger systems. Bearing this diversity in mind, let's begin by considering a chemical synapse at which acetylcholine (ACh) opens ligand-regulated gates in the postsynaptic cell and depolarizes (excites) it. A

Table 13.3 Neurotransmitters and Neuropeptides

Name	Locations and Actions
Acetylcholine (ACh)	Neuromuscular junctions, all presynaptic autonomic fibers and postsynaptic parasympathetic fibers, some postsynaptic sympathetic fibers, retina, and many parts of the brain; excites skeletal muscles, inhibits cardiac muscle, and has excitatory or inhibitory effects on smooth muscle and glands depending on location
Excitatory Amino Acids	
Glutamate (glutamic acid)	Cerebral cortex and brainstem; accounts for about 75% of all excitatory synaptic transmission in the brain; involved in learning and memory
Aspartate (aspartic acid)	Spinal cord; effects similar to those of glutamate
Inhibitory Amino Acids	
Glycine	Inhibitory neurons of the brain, spinal cord, and retina; most common inhibitory neurotransmitter in the spinal cord
GABA (γ-aminobutyric acid)	Thalamus, hypothalamus, cerebellum, occipital lobes of cerebrum, and retina; most common inhibitory neurotransmitter in the brain
Monoamines (biogenic amines)	
Catecholamines	
Norepinephrine	Most sympathetic postsynaptic fibers, cerebral cortex, hypothalamus, brainstem, cerebellum, and spinal cord; involved in dreaming, waking, and mood; excites cardiac muscle; can excite or inhibit smooth muscle and glands depending on location
Epinephrine	Hypothalamus, thalamus, spinal cord, and adrenal medulla; effects similar to those of norepinephrine
Dopamine	Hypothalamus, limbic system, cerebral cortex, and retina; highly concentrated in substantia nigra of midbrain; involved in elevation of mood and control of skeletal muscles
Indolamines	
Serotonin	Hypothalamus, limbic system, cerebellum, retina, and spinal cord; also secreted by blood platelets and intestinal cells; involved in sleepiness, alertness, thermoregulation, and mood
Histamine	Hypothalamus; also a potent vasodilator released by mast cells of connective tissue
Neuropeptides	
Substance P	Basal nuclei, midbrain, hypothalamus, cerebral cortex, small intestine, and pain-receptor neurons; mediates pain transmission
Enkephalins	Hypothalamus, limbic system, pituitary, pain pathways of spinal cord, and nerve endings of digestive tract; act as analgesics (pain-relievers) by inhibiting substance P; inhibit intestinal motility; secretion increases sharply in women in labor
β-endorphin	Digestive tract and many parts of the brain; also secreted as a hormone by the pituitary; suppresses pain; reduces perception of fatigue and produces "runner's high" in athletes
Cholecystokinin (CCK)	Cerebral cortex and small intestine; suppresses appetite

synapse where transmission is mediated by ACh is called a **cholinergic**[25] (CO-lin-UR-jic) **synapse.**

The steps in transmission at a cholinergic synapse are as follows (fig. 13.19):

1. The arrival of a nerve signal at the synaptic knob opens voltage-regulated calcium gates.
2. Ca^{2+} enters the synaptic knob and triggers exocytosis of the synaptic vesicles, releasing ACh.
3. Empty vesicles drop back into the cytoplasm to be refilled with ACh, while synaptic vesicles in the reserve pool move forward to the active sites and release their ACh—a bit like a line of Revolutionary War soldiers firing their flintlock rifles and falling back to reload as another line moves to the fore.

25. *cholin* = acetylcholine + *erg* = work, action

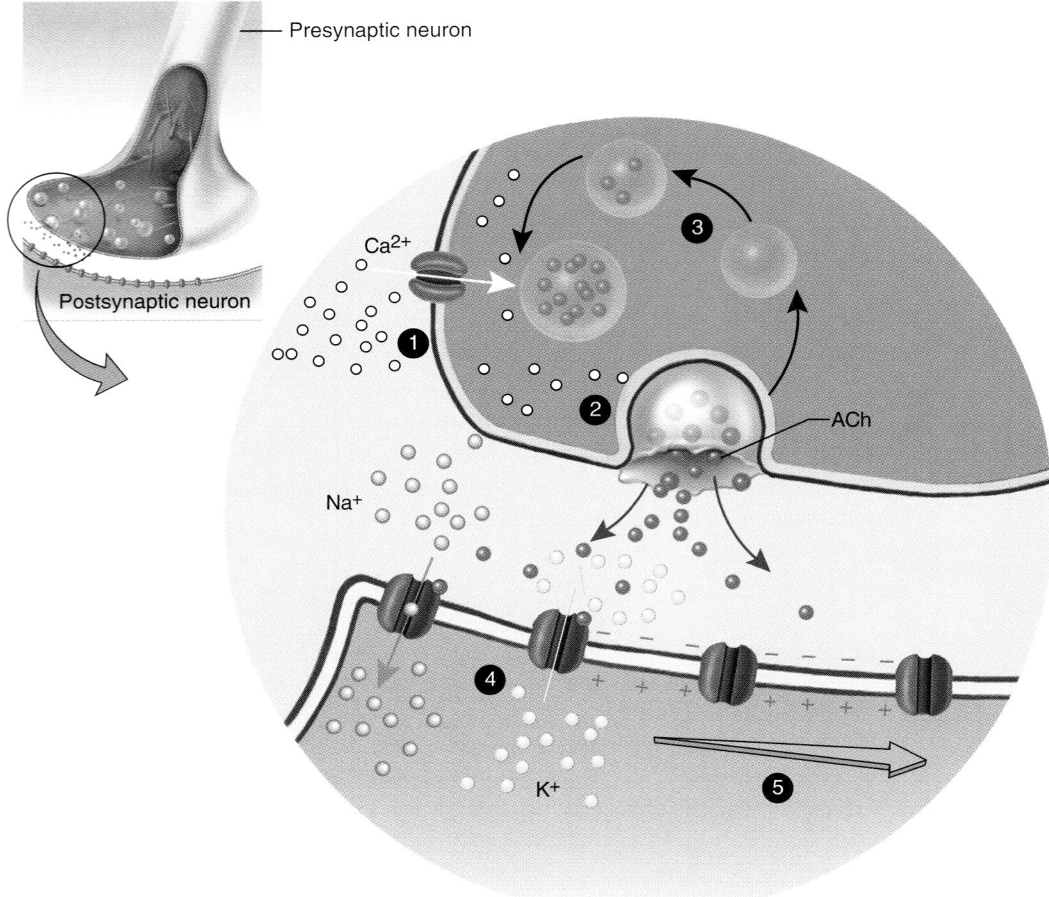

Figure 13.19 Transmission at an Ionotropic Synapse. Here a neurotransmitter directly opens ion gates in the plasma membrane of the postsynaptic neuron. Numbered steps correspond to the description in the text.

4. Meanwhile, ACh diffuses across the synaptic cleft and binds to ligand-regulated gates on the postsynaptic neuron. These gates open, allowing Na^+ to enter the cell and K^+ to leave. Although illustrated separately, Na^+ and K^+ pass in opposite directions through the same gates.
5. As Na^+ enters the cell, it spreads out along the inside of the plasma membrane and depolarizes it, producing a local potential called the **postsynaptic potential.** Like other local potentials, if this is strong and persistent enough (that is, if enough Na^+ makes it to the axon hillock), it opens voltage-regulated ion gates in the trigger zone and causes the postsynaptic neuron to fire.

As complex as the foregoing process may seem, it all happens in 0.5 msec—an interval called **synaptic delay.** This is the time from the arrival of a signal at the axon terminal of a presynaptic cell to the beginning of an action potential in the postsynaptic cell.

The mechanism just described is called an **ionotropic**[26] (eye-ON-oh-TRO-pic) **effect** because the neurotransmitter opens ion gates and causes immediate changes in membrane potential. This is how acetylcholine and the amino acid neurotransmitters stimulate a cell. Monoamines and neuropeptides, by contrast, have **metabotropic** (meh-TAB-oh-TRO-pic) **effects** in which they employ a second messenger such as cyclic AMP (fig. 13.20) to alter the metabolism of the postsynaptic cell. The general features of the cAMP system were described in chapter 4 (see fig. 4.9). In such cases, the receptor does not act as an ion gate but activates the production of cAMP, which then produces various effects on the postsynaptic cell:

- metabolic pathways can be turned on or off;
- internal messengers can enter the nucleus and activate genetic transcription, which leads to the synthesis of new proteins; or

26. *iono* = ions + *trop* = to change, affect

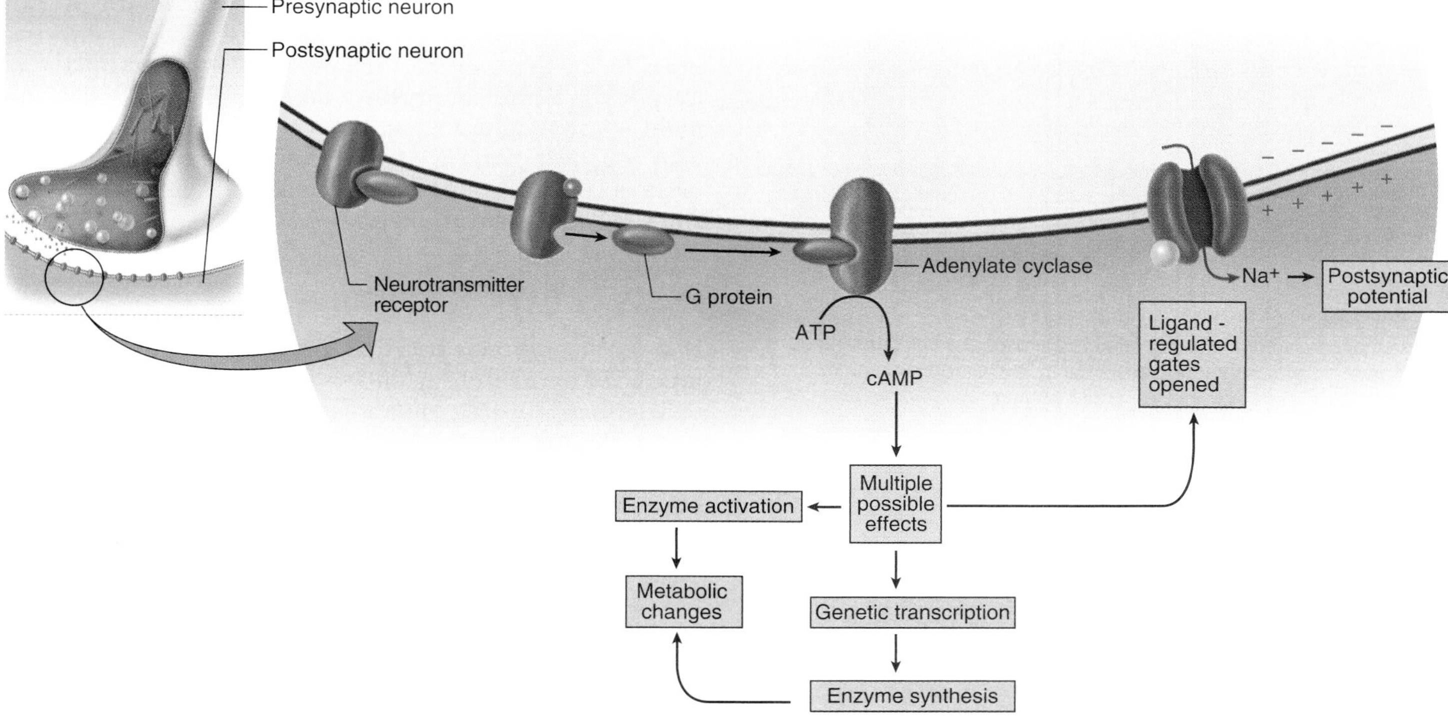

Figure 13.20 Transmission at a Metabotropic Synapse. Here the neurotransmitter receptor is not an ion channel. It activates a second-messenger system with a variety of possible effects in the postsynaptic cell.

- internal messengers can bind to the intracellular side of a membrane protein and open ion gates other than the neurotransmitter receptor.

Many hormones also act through cAMP, underscoring the similarities between the endocrine and nervous systems. Epinephrine is both a metabotropic neurotransmitter and a hormone.

Cessation of the Signal

It is important not only to stimulate a postsynaptic cell but also to turn off the stimulus in due time. Here we examine some ways this is done.

A neurotransmitter molecule binds to its receptor for only 1 msec or so and then dissociates from it. If the presynaptic cell continues to release neurotransmitter, one molecule is quickly replaced by another and the postsynaptic cell is restimulated. This immediately suggests a way of stopping synaptic transmission—stop adding new neurotransmitter and get rid of that which is already there. The first step is achieved simply by the cessation of signals in the presynaptic nerve fiber. The second can be achieved in the following ways:

- **Diffusion.** Neurotransmitter diffuses away from the synapse into the nearby ECF, where astrocytes may absorb it and returned it to the neurons.
- **Reuptake.** The synaptic knob reabsorbs amino acids and monoamines by endocytosis and then breaks them down with an enzyme called **monoamine oxidase (MAO).** Some antidepressant drugs work by inhibiting MAO (see clinical insight 15.5 at the end of chapter 15).
- **Degradation in the synaptic cleft.** The enzyme acetylcholinesterase (AChE), located within the synaptic cleft and on the postsynaptic membrane, breaks acetylcholine down into acetate and choline. These breakdown products have no stimulatory effect on the postsynaptic cell. The synaptic knob reabsorbs the choline and uses it to synthesize more ACh.

Neuromodulators

Neuromodulators are hormones, neuropeptides, and other messengers that modify synaptic transmission. They may stimulate a neuron to raise or lower the number of receptors in the postsynaptic membrane, thus adjusting its sensitivity to a neurotransmitter, or they may alter the rate of neurotransmitter synthesis, release, reuptake, or breakdown. One example—a rather recent and surprising discovery—is nitric oxide (NO), a lightweight gas that is released by postsynaptic neurons in some areas of the brain concerned with learning and memory. NO diffuses into the presynaptic neuron and stimulates it to release more neurotransmitter—like one neuron's way of telling the other, "Give me more." Thus, there is at least some chemical communication that goes backward across a synapse.

Before You Go On

Answer the following questions to test your understanding of the preceding section:

16. Concisely describe five steps that occur in the interval between the arrival of an action potential at the synaptic knob and the beginning of a new action potential in the postsynaptic neuron.
17. Distinguish between ionotropic and metabotropic effects.
18. Describe three mechanisms that cause synaptic transmission to stop.
19. What is the function of neuromodulators? Name two of them.

Neural Integration

Objectives

When you have completed this section, you should be able to

- explain how a neuron "decides" whether or not to produce action potentials;
- explain how the nervous system translates complex information into a simple code;
- explain how neurons work together in groups to process information and produce effective output; and
- describe how memory works at a cellular and molecular level.

Synaptic delay slows the transmission of nerve signals; the more synapses there are in a neural pathway, the longer it takes information to get from its origin to its destination. You might wonder, therefore, why we have synapses—why a nervous pathway is not, indeed, a continuous "wire" as biologists believed before the neuron doctrine was accepted. The presence of synapses is not due to limitations on the length of a neuron—after all, one nerve fiber can reach from your toes to your brainstem, and it is interesting to consider how long some nerve fibers may be in a giraffe or a whale. We also have seen that cells can communicate through gap junctions much more quickly than they can through chemical synapses. So why have chemical synapses at all?

What we value most about our nervous system is its ability to process information, store it, and make decisions—and chemical synapses are the decision-making devices of the system. The more synapses a neuron has, the greater its information-processing capability. At this moment, you are using certain *pyramidal cells* of the cerebral cortex (see fig. 13.4) to read and comprehend this passage. Each pyramidal cell has about 40,000 synaptic contacts with other neurons. The cerebral cortex alone (the main information-processing tissue of your brain) is estimated to have 100 trillion (10^{14}) synapses. To get some impression of this number, imagine trying to count them. Even if you could count two synapses per second, day and night without stopping, and you were immortal, it would take you 1.6 million years. The ability of your neurons to process information, store and recall it, and make decisions is called **neural integration.**

Postsynaptic Potentials

Neural integration is based on the postsynaptic potentials produced by neurotransmitters. Remember that a typical neuron has a resting membrane potential (RMP) of about −70 mV and a threshold of about −55 mV. A neuron has to be depolarized to this threshold in order to produce action potentials. Any voltage change in that direction makes a neuron more likely to fire and is therefore called an **excitatory postsynaptic potential (EPSP)** (fig. 13.21*a*). EPSPs usually result from Na^+ flowing into the cell and canceling some of the negative charge on the inside of the membrane.

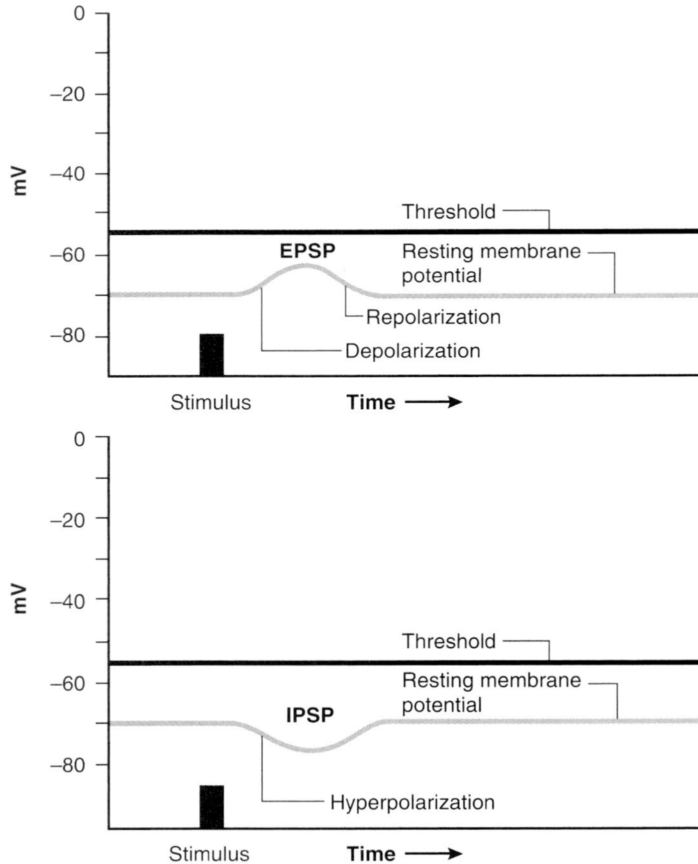

Figure 13.21 **Excitatory and Inhibitory Postsynaptic Potentials.** Note that a single EPSP is not enough to reach threshold and make a neuron fire. The size of the EPSP is greatly exaggerated here for clarity; compare figure 13.23.

In other cases, a neurotransmitter hyperpolarizes the postsynaptic cell and makes it more negative than the RMP. Since this makes the postsynaptic cell less likely to fire, it is called an **inhibitory postsynaptic potential (IPSP)** (fig. 13.21*b*). Some IPSPs are produced by a neurotransmitter opening ligand-regulated chloride gates, causing Cl^- to flow into the cell and make the cytosol more negative. A less common way is to open selective K^+ gates, increasing the diffusion of K^+ out of the cell.

It should be stressed that because of ion leakage through their membranes, all neurons fire at a certain background rate even when they are not being stimulated. EPSPs and IPSPs do not determine whether or not a neuron fires, but only change the rate of firing by stimulating or inhibiting the production of more action potentials.

Glutamate and aspartate are excitatory neurotransmitters that produce EPSPs. Glycine and GABA produce IPSPs and are therefore inhibitory. Acetylcholine (ACh) and norepinephrine are excitatory to some cells and inhibitory to others, depending on the type of receptors present on the target cells. For example, ACh excites skeletal muscle but inhibits cardiac muscle because the two types of muscle have different types of ACh receptors. This is discussed more fully in chapter 15.

Summation, Facilitation, and Inhibition

A neuron may receive input from thousands of presynaptic neurons simultaneously. Some incoming nerve fibers may produce EPSPs while others produce IPSPs. Whether or not the neuron fires depends on whether the *net* input is excitatory or inhibitory. If the EPSPs override the IPSPs, threshold may be reached and the neuron will fire; if the IPSPs prevail, the neuron will not fire. **Summation** is the process of adding up postsynaptic potentials and responding to their net effect. It occurs in the trigger zone.

Suppose, for example, you are working in the kitchen and accidentally touch a cooking pot that is cooling down but still fairly hot. EPSPs in your motor neurons might cause you to jerk your hand back quickly and avoid being burned. Yet a moment later, you might nonchalantly pick up a cup of tea that is even hotter than the pot. Since you are expecting the teacup to be hot, you do not jerk your hand away. You have learned that it will not injure you, so at some level of the nervous system, IPSPs prevail and inhibit the motor response.

It is fundamentally a balance between EPSPs and IPSPs that enables the nervous system to make decisions. A postsynaptic neuron is like a little cellular democracy acting on the "majority vote" of hundreds of presynaptic cells. In the teacup example, some presynaptic neurons are sending messages that signify "hot! danger!" in the form of EPSPs that may activate a hand-withdrawal reflex, while at the same time, others are producing IPSPs that signify "safe" and suppress the withdrawal reflex. Whether the postsynaptic neurons cause you to jerk your hand back depends on whether the EPSPs override the IPSPs or vice versa.

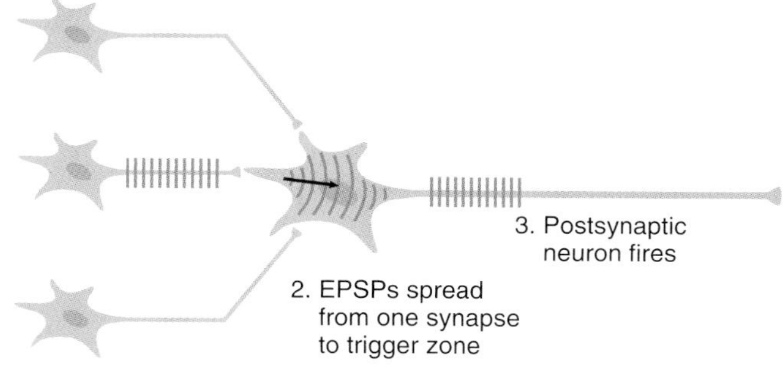

(a) Temporal summation

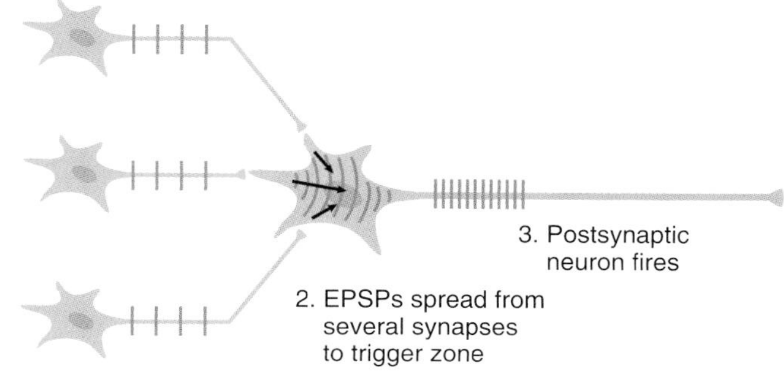

(b) Spatial summation

Figure 13.22 Temporal and Spatial Summation. (*a*) In temporal summation, a single presynaptic neuron stimulates the postsynaptic neuron so intensely that its EPSPs add up to threshold and make it fire. (*b*) In spatial summation, multiple inputs to the postsynaptic cell each produce a moderate amount of stimulation, but collectively they produce enough EPSPs to add up to threshold at the trigger zone and make the cell fire.

One action potential in a synaptic knob does not produce enough activity to make a postsynaptic cell fire. An EPSP may be produced, but it decays before reaching threshold. A typical EPSP has a voltage of 0.5 mV and lasts only 15 to 20 msec. If a neuron has an RMP of −70 mV and a threshold of −55 mV, it will need about 30 EPSPs to reach threshold and fire. There are two ways in which EPSPs can add up to do this, and both may occur simultaneously (fig. 13.22):

1. **Temporal summation.** This occurs when a single synapse generates EPSPs at such short time intervals that each is generated before the previous one decays. This allows the EPSPs to add up over time to a threshold voltage that triggers an action potential (fig. 13.23). Temporal summation can occur if even one presynaptic neuron intensely stimulates the postsynaptic neuron.

2. **Spatial summation.** This occurs when EPSPs from several different synapses add up to threshold at the axon hillock. Any one synapse may admit only a moderate amount of Na^+ into the cell, but several synapses acting together admit enough Na^+ to reach a threshold. The presynaptic neurons cooperate to induce the postsynaptic neuron to fire.

Facilitation is a process in which one neuron enhances the effect of another one. In spatial summation, for example, one neuron acting alone may be unable to induce a postsynaptic neuron to fire. But when they cooperate, their combined "effort" does induce firing in the postsynaptic cell. They each enhance one another's effect, or *facilitate* each other.

Presynaptic inhibition is the opposite of facilitation, a mechanism in which one presynaptic neuron suppresses another one. This mechanism is used to reduce or halt unwanted synaptic transmission. In figure 13.24, we see three neurons which we will call neuron *S* for the stimulator, neuron *I* for the inhibitor, and neuron *R* for the responder. Neuron I forms an axoaxonic synapse with S (that is, I synapses with the axon of S). When presynaptic inhibition is not occurring, neuron S releases its neurotransmitter and triggers a response in R. But when there is a need to block transmission across this pathway, neuron I releases the inhibitory neurotransmitter GABA. GABA prevents the voltage-regulated calcium gates of neuron S from opening. Consequently, neuron S releases less neurotransmitter or none, and fails to stimulate neuron R.

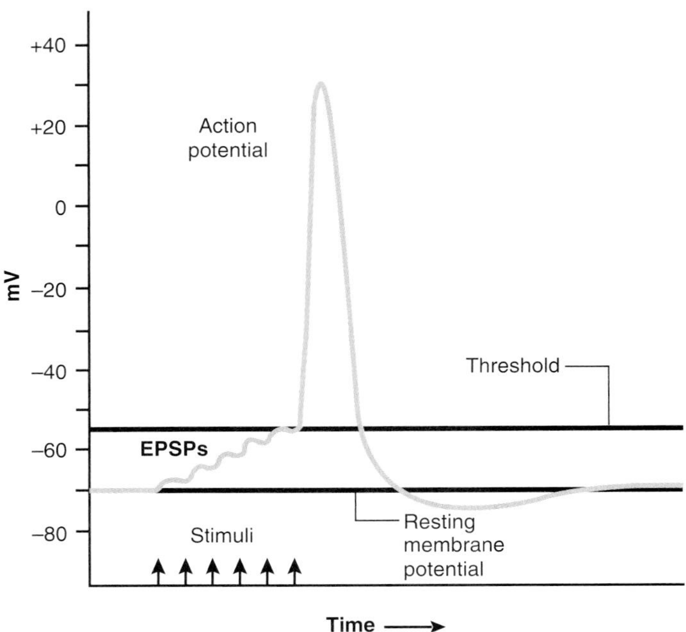

Figure 13.23 Summation of EPSPs. If enough EPSPs arrive at the trigger zone faster than they decay, they can build on each other to bring the neuron to threshold and trigger an action potential.

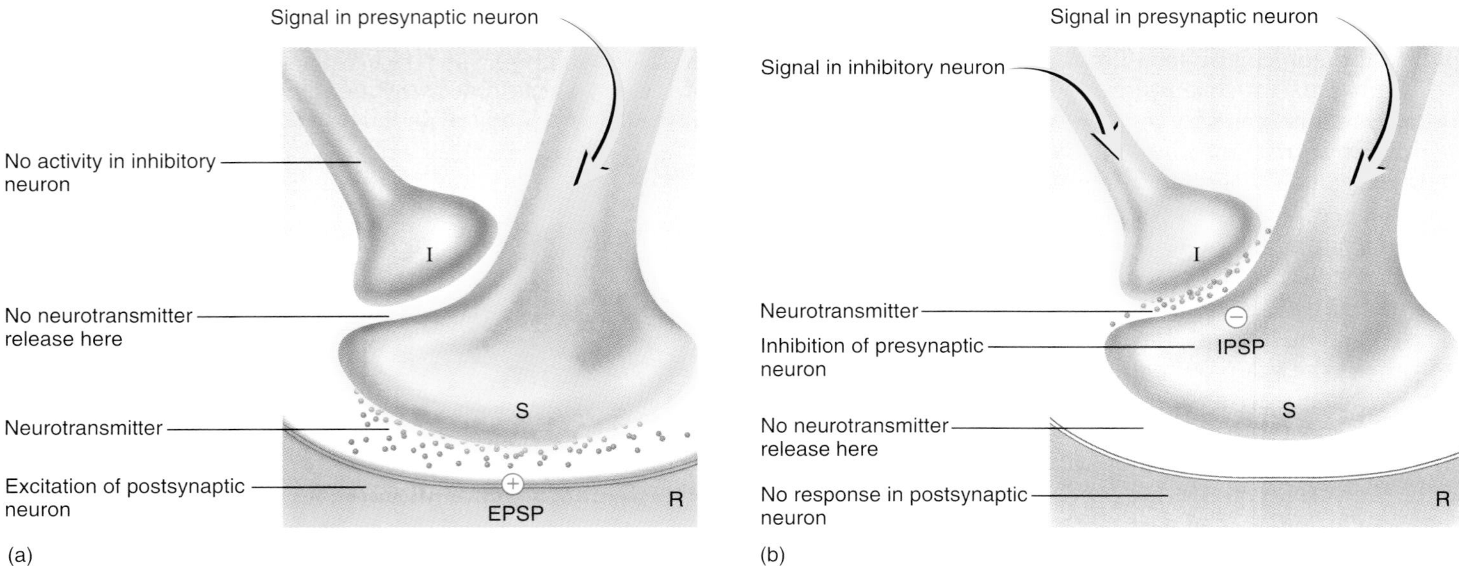

Figure 13.24 Presynaptic Inhibition. (*a*) In the absence of inhibition, neuron *S* releases neurotransmitter and stimulates neuron *R*. (*b*) In presynaptic inhibition, neuron *I* suppresses the release of neurotransmitter by S, and S cannot stimulate R.

Neural Coding

The nervous system must interpret and pass along both quantitative and qualitative information about its environment—whether a light is dim or bright, red or green; whether a taste is mild or intense, salty or sour; whether a sound is loud or soft, high-pitched or low. Considering the complexity of information to be communicated about conditions in and around the body, it is a marvel that it can be done in the form of something as simple as action potentials—particularly since all the action potentials of a given neuron are identical. Yet when we considered the genetic code in chapter 5, we saw that complex messages can indeed be expressed in simple codes. The way in which the nervous system converts information to a meaningful pattern of action potentials is called **neural coding** (or *sensory coding* when it occurs in the sense organs).

Qualitative information is encoded in terms of which neurons are firing. Red light and green light, for example, excite different fibers in the optic nerve; a high-pitched sound and a low-pitched sound excite different fibers in the auditory nerve; a sweet substance and a sour one excite different taste cells. The brain interprets input from different fibers in terms of these stimulus qualities.

Quantitative information—information about the intensity of a stimulus—is encoded in two ways. One depends on the fact that different neurons have different thresholds of excitation. A weak stimulus excites neurons with the lowest thresholds, while a strong stimulus excites less-sensitive high-threshold neurons. Bringing additional neurons into play as the stimulus becomes stronger is called **recruitment.** It enables the nervous system to judge stimulus strength by which neurons, and how many of them, are firing.

Another way of encoding stimulus strength depends on the fact that the more strongly a neuron is stimulated, the more frequently it fires. A weak stimulus may cause a neuron to generate action potentials at a rate of 6 per second and a strong stimulus at 600 per second. Thus, the central nervous system can judge stimulus strength from the firing frequency of afferent neurons (fig. 13.25).

There is a limit to how often a neuron can fire, set by its absolute refractory period. Think of an electronic camera flash by analogy. If you take a photograph and your flash unit takes 15 seconds to recharge, then you cannot take more than four photographs per minute. Similarly, if a nerve fiber takes 1 msec to repolarize after it has fired, then it cannot fire more than 1,000 times per second. Refractory periods may be as short as 0.5 msec, which sets a theoretical limit to firing frequency of 2,000 action potentials per second. The highest frequencies actually observed, however, are between 500 and 1,000 per second.

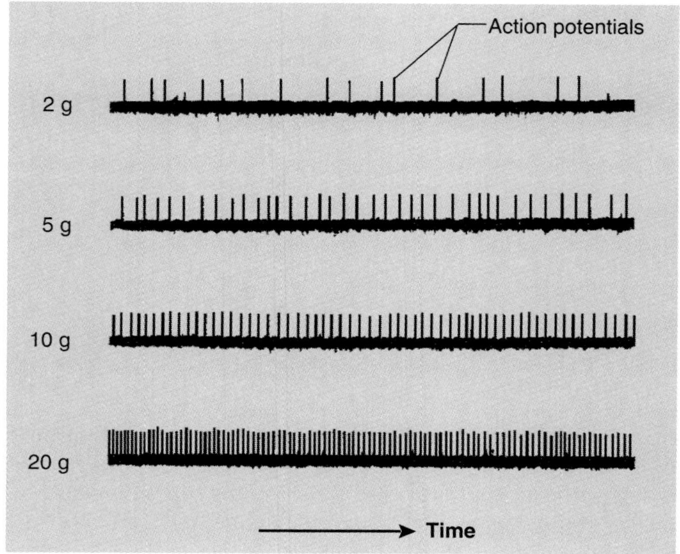

Figure 13.25 An Example of Neural Coding. This figure is based on recordings made from a sensory fiber of the frog sciatic nerve as the gastrocnemius muscle was stretched by suspending weights from it. As the stimulus strength (weight) and stretch increase, the firing frequency of the neuron increases. Firing frequency is a coded message that informs the CNS of stimulus intensity.

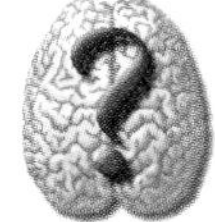

Think About It

How is neuronal recruitment related to the process of multiple motor unit summation described in chapter 12?

Neuronal Pools and Circuits

So far, we have dealt with interactions involving only two or three neurons at a time. Actually, neurons function in larger ensembles called **neuronal pools,** each of which consists of thousands to millions of interneurons concerned with a particular body function—one to control the rhythm of your breathing, one to move your limbs rhythmically as you walk, one to regulate your sense of hunger, and another to interpret smells, for example. At this point, we explore a few ways in which neuronal pools collectively process information.

Information arrives at a neuronal pool through one or more input neurons, which branch repeatedly and synapse with numerous interneurons in the pool. Some input neurons form multiple synapses with a single postsynaptic cell. They can produce EPSPs at all points of contact with that cell and, through spatial summation, make it fire more easily than if they synapsed with it at only one point. Within the **discharge zone** of an input neuron, an input neuron acting alone can make the postsynaptic cells

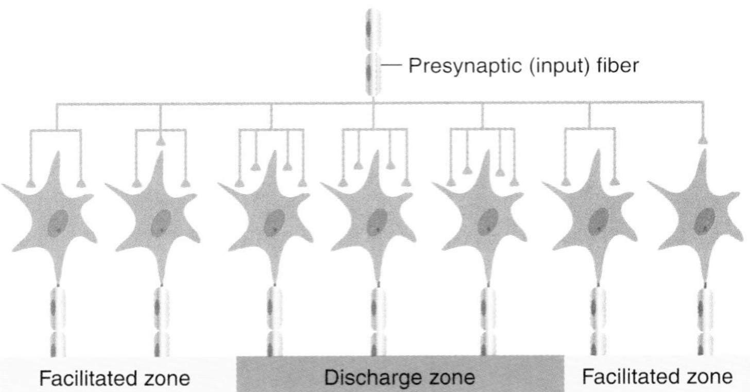

Figure 13.26 **Facilitated and Discharge Zones in a Neuronal Pool.** In a facilitated zone, the input neuron has few synaptic contacts with each output neuron. The input neuron makes it easier for those neurons to respond to stimulation from other sources, but it cannot, by itself, make them fire. In the discharge zone, the input neuron has extensive connections with each output neuron and is capable, by itself, of making the output neurons fire.

fire (fig. 13.26). But in a broader **facilitated zone,** it synapses with still other neurons in the pool, with fewer synapses on each of them. It can stimulate those neurons to fire only with the assistance of other input neurons; that is, it facilitates the other input neurons. Along with other inputs, it "has a vote" on what the postsynaptic cells in the facilitated zone will do, but it cannot alone determine what they do. Such arrangements, repeated thousands of times throughout the central nervous system, give neuronal pools great flexibility in integrating input from several sources and "deciding" on an appropriate output.

The functioning of a radio can be understood from a circuit diagram showing its components and their connections. Similarly, the functions of a neuronal pool are partly determined by its **neuronal circuit**—the pathways among its neurons. Just as a wide variety of electronic devices are constructed from a relatively limited number of circuit types, a wide variety of neuronal functions result from the operation of four principal kinds of neuronal circuits (fig. 13.27):

1. In a **diverging circuit,** one nerve fiber branches and synapses with several postsynaptic cells. Each of those may synapse with several more, so input from just one neuron may produce output through dozens of neurons. Such a circuit allows one motor neuron of the brain, for example, to ultimately cause thousands of muscle fibers to contract.
2. A **converging circuit** is the opposite of a diverging circuit—input from many different nerve fibers is funneled to one neuron or neuronal pool. Such an arrangement allows input from your eyes, inner ears, and stretch receptors in your neck to be channeled to an area of the brain concerned with the sense of balance. Also through neuronal convergence, a respiratory center in your brainstem receives input from other parts of your brain, from receptors for blood chemistry in your arteries, and from stretch receptors in your lungs. The respiratory center can then produce an output that takes all of these factors into account and sets an appropriate pattern of breathing.
3. In a **reverberating circuit,** neurons stimulate each other in a linear sequence such as A → B → C → D, but neuron *C* sends an axon collateral back to *A*. As a result, every time *C* fires it not only stimulates output neuron *D,* but also restimulates *A* and starts the process over. Such a circuit produces a prolonged or repetitive effect that lasts until one or more neurons in the circuit fail to fire, or until an inhibitory signal from another source stops one of them from firing. A reverberating circuit sends repetitious signals to your diaphragm and intercostal muscles, for example, to make you inhale. When the circuit stops firing, you exhale; the next time it fires, you inhale again. Reverberating circuits may also be involved in short-term memory, as discussed in the next section, and they may play a role in the uncontrolled "storms" of neuronal activity that occur in epilepsy.
4. In a **parallel after-discharge circuit,** an input neuron diverges to stimulate several chains of neurons. Each chain has a different number of synapses, but eventually they all reconverge on a single output neuron. Since each pathway differs in total synaptic delay, their signals arrive at the output neuron at different times, and the output neuron may go on firing for some time after input has ceased. Unlike a reverberating circuit, this type has no feedback loop. Once all the neurons in the circuit have fired, the output ceases. Continued firing after the stimulus stops is called *after-discharge.* It explains why you can stare at a lamp, then close your eyes and continue to see an image of it for a while. Such a circuit is also important to withdrawal reflexes, in which a brief pain produces a longer-lasting output to the limb muscles and causes you to draw back your hand or foot from danger.

Memory and Synaptic Plasticity

You may have wondered as you studied this chapter, How am I going to remember all of this? It seems fitting that we end this chapter with the subject of how memory works, for you now have the information necessary to understand its cellular and chemical basis.

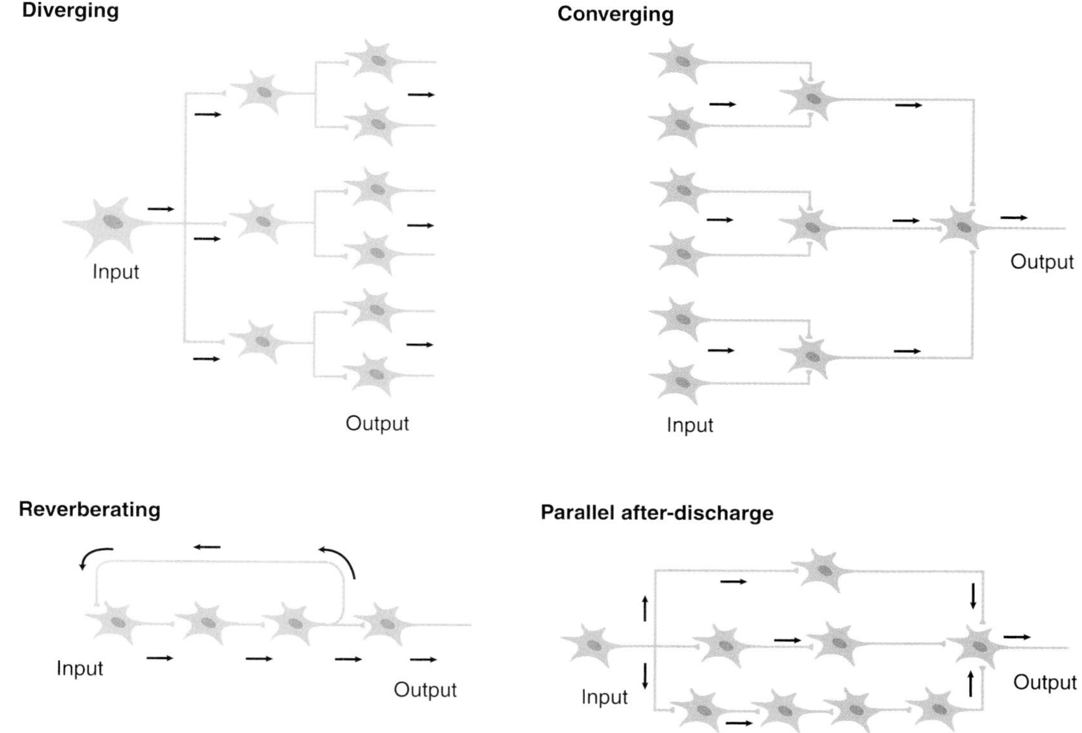

Figure 13.27 **Four Types of Neuronal Circuits.** Arrows indicate the direction of the nerve signal.

The things we learn and remember are not stored in individual "memory cells" in the brain. We do not have a neuron assigned to remember our phone number and another assigned to remember our mother's birthday, for example. Instead, the physical basis of a memory is a *pathway* through the brain called a **memory trace (engram[27]),** in which new synapses have formed or existing synapses have been modified to make transmission easier. In other words, synapses are not fixed for life; in response to experience, they can be added, taken away, or modified to make transmission easier or harder. This ability of synapses to change is called **synaptic plasticity.**

Think about when you learned as a child to tie your shoes. The procedure was very slow, confusing, and laborious at first, but eventually it became so easy you could do it with little thought—like a little motor program playing out in your brain without requiring your conscious attention. It became easier to do because the synapses in a certain pathway were modified to allow signals to travel more easily across them than across "untrained" synapses. The process of making transmission easier is called **synaptic potentiation** (one form of synaptic plasticity).

Neuroscientists still argue about how to classify the various forms of memory, but three kinds often recognized are *immediate memory, short-term memory,* and *long-term memory.* We also know of different modes of synaptic potentiation that last from just a few seconds to a lifetime, and we can correlate these at least tentatively with different forms of memory.

27. *en* = inner + *gram* = mark, trace, record

Immediate Memory

Immediate memory is the ability to hold something in mind for just a few seconds. By remembering what just happened, we get a feeling for the flow of events and a sense of the present. Immediate memory might be based on reverberating circuits. Our impression of what just happened can thus "re-echo" in our minds for a few seconds as we experience the present moment and plan the next one.

Short-Term Memory

Short-term memory (STM) lasts from a few seconds to a few hours and is limited to a few bits of information such as the digits of a telephone number. Information stored in STM may be quickly forgotten if we stop mentally reciting it, we are distracted, or we have to remember something new. **Working memory** is a form of STM that allows us to hold an idea in mind long enough to carry out an action such as calling a telephone number we just looked up, working out the steps of a mathematics problem, or searching for a lost set of keys while remembering where we have already looked. These short-term memory tasks may be carried out by reverberating circuits of neurons.

Somewhat longer-lasting memories, however, probably involve a synaptic effect called **facilitation** (different from the facilitation of one neuron by another that we studied earlier in the chapter). This form of facilitation is induced by *tetanic stimulation,* the rapid arrival of repetitive signals at a synapse. Each signal causes a certain amount of Ca^{2+} to enter the synaptic knob. If signals arrive so close together that all the Ca^{2+} admitted by one action potential cannot be pumped out before the next action potential occurs, then more and more Ca^{2+} will accumulate in the knob. Since Ca^{2+} is what triggers the release of neurotransmitter, each signal will trigger the release of more neurotransmitter than the one before. With more neurotransmitter, the EPSPs in the postsynaptic cell will become stronger and stronger, and that cell will be more likely to fire. Thus, tetanic stimulation facilitates the synapse and makes it easier for the postsynaptic cell to fire.

Memories lasting for a few hours, such as remembering what someone said to you earlier in the day or remembering an upcoming appointment, may involve **posttetanic potentiation.** In this process, the Ca^{2+} level in the synaptic knob stays elevated for so long that another signal, coming along well after the tetanic stimulation has ceased, releases an exceptionally large burst of neurotransmitter. That is, if a synapse has been heavily used in the recent past, a new stimulus can excite the postsynaptic cell more easily. Thus your memory may need only a slight jog to recall something from several hours earlier.

Long-Term Memory

Long-term memory (LTM) lasts up to a lifetime and is less limited than STM in the amount of information it can store. LTM allows you to memorize the lines of a play, the words of a favorite song, or textbook information for an exam. On a still longer timescale, it enables you to remember your name, the route to your home, and your childhood experiences.

There are two forms of LTM—declarative and procedural. **Declarative memory** is the retention of events and facts that you can put into words—numbers, names, dates, and so forth. **Procedural memory** is the retention of motor skills—how to tie your shoes, play a musical instrument, or type on a keyboard. These forms of memory involve different regions of the brain but are probably similar at the cellular level.

Some LTM involves the physical remodeling of synapses or the formation of new ones through the growth and branching of axons and dendrites. In the pyramidal cells of the brain, the dendrites are studded with knoblike *dendritic spines* that increase the area of synaptic contact. Studies on fish and other experimental animals have shown that social and sensory deprivation causes these spines to decline in number, while a richly stimulatory environment causes them to proliferate—an intriguing clue to the importance of a stimulating environment to infant and child development.

LTM can also be grounded in molecular changes called **long-term potentiation.** This involves *NMDA*[28] *receptors,* which occur on the synaptic knobs of the pyramidal cells and bind the neurotransmitter glutamate. NMDA receptors are usually blocked by magnesium ions (Mg^{2+}), but when they bind glutamate *and* are simultaneously subjected to tetanic stimulation, they expel the Mg^{2+} and open to admit Ca^{2+} into the dendrite. When Ca^{2+} enters the dendrite, it acts as a second messenger that leads to a variety of effects:

- The neuron produces an increased number of NMDA receptors, which makes it more sensitive to glutamate.
- It synthesizes proteins concerned with physically remodeling a synapse.
- It releases nitric oxide, which diffuses back to the presynaptic neuron and triggers a chain reaction of events there that ultimately increases glutamate release.

You can see that in all of these ways, long-term potentiation can increase transmission across "experienced" synapses. Remodeling a synapse or increasing the number of neurotransmitter receptors has longer-lasting effects than facilitation or posttetanic potentiation.

Before You Go On

Answer the following questions to test your understanding of the preceding section:

20. Contrast the two types of summation at a synapse.
21. Describe how the nervous system communicates quantitative and qualitative information about stimuli.
22. How does long-term potentiation enhance the transmission of nerve signals along certain pathways?
23. List the four types of neuronal circuits and describe their similarities and differences. Discuss the unity of form and function in these four types—that is, explain why each type would not perform as it does if its neurons were connected differently.

28. N-methyl-D-aspartate, a chemical similar to glutamate

Clinical Insight 13.4

Alzheimer and Parkinson Diseases

Alzheimer and Parkinson diseases are degenerative disorders of the brain associated with neurotransmitter deficiencies.

Alzheimer[29] *disease (AD)* may begin before the age of 50 with symptoms so slight and ambiguous that early diagnosis is difficult. One of its first symptoms is memory loss, especially for recent events. A person with AD may ask the same questions repeatedly, show a reduced attention span, and become disoriented and lost in previously familiar places. Family members often feel helpless and confused as they watch their loved one's personality gradually deteriorate beyond recognition. The AD patient may become moody, confused, paranoid, combative, or hallucinatory—he or she may ask irrational questions such as, Why is the room full of snakes? The patient may eventually lose even the ability to read, write, talk, walk, and eat. Death ensues from pneumonia or other complications of confinement and immobility.

AD affects about 11% of the U.S. population over the age of 65; the incidence rises to 47% by age 85. It accounts for nearly half of all nursing home admissions and is a leading cause of death among the elderly. AD claims about 100,000 lives per year in the United States.

Diagnosis of AD is confirmed on autopsy. There is atrophy of some of the gyri (folds) of the cerebral cortex and the hippocampus, an important center of memory. Nerve cells exhibit *neurofibrillary tangles*—dense masses of broken and twisted cytoskeleton (fig. 1). These were first observed by Alois Alzheimer in 1907 in the brain of a patient who had died of senile dementia. The more severe the disease symptoms, the more neurofibrillary tangles are seen at autopsy. In the intercellular spaces, there are *senile plaques* consisting of aggregations of cells, altered nerve fibers, and a core of *β-amyloid protein*—the breakdown product of a glycoprotein of plasma membranes. β-amyloid protein is rarely seen in elderly people without AD.

AD is marked by the degeneration of cholinergic neurons and a deficiency of ACh. Treatment with ACh precursors is ineffective, but therapy with cholinesterase inhibitors to slow down the degradation of existing ACh has been of some value. AD patients show a deficiency of nerve growth factor (NGF; see clinical insight 13.3) in some regions of the brain. NGF stimulates ACh synthesis; it helps to retard brain degeneration in humans and other animals and improves memory in some AD patients. Intense biomedical research efforts are currently geared toward identifying the cause of AD and developing treatment strategies. Researchers have recently identified three genes on chromosomes 1, 14, and 21 for various forms of early- and late-onset AD.

Parkinson[30] *disease (PD),* also called *paralysis agitans* or *parkinsonism,* is a progressive loss of motor function beginning in a person's 50s or 60s. It is due to degeneration of dopamine-releasing neurons in a portion of the brain called the *substantia nigra* (see chapter 14). A gene has recently been identified for a hereditary form of PD, but most cases are nonhereditary and of little-known cause; some authorities suspect environmental neurotoxins.

Dopamine (DA) is an inhibitory neurotransmitter that normally prevents excessive activity in motor centers of the brain called the *basal nuclei.* Degeneration of the dopamine-releasing neurons leads to an excessive ratio of ACh to DA, which causes the basal nuclei to become hyperactive. As a result, a person with PD suffers involuntary muscle contractions. These take such forms as shaking of the hands (tremor) and compulsive "pill-rolling" motions of the thumb and fingers. In addition, the facial muscles may become rigid and produce a staring, expressionless face with a slightly open mouth. The patient's range of motion diminishes. He or she takes smaller steps and develops a slow, shuffling gait with a forward-bent posture and a tendency to fall forward. Speech becomes slurred and handwriting becomes cramped and eventually illegible. Tasks such as buttoning clothes and preparing food become increasingly laborious.

Patients cannot be expected to recover from PD, but its effects can be alleviated with drugs and physical therapy. Treatment with dopamine is ineffective because it cannot cross the blood-brain barrier, but its precursor, levodopa (L-DOPA), does cross the barrier and has been used to treat PD since the 1960s. L-DOPA affords some relief from symptoms, but it does not slow progression of the disease and it has undesirable side effects on the liver and heart. A newer drug, Deprenyl, is a monoamine oxidase (MAO) inhibitor that retards neuronal degeneration and delays the development of symptoms. Encouraging results also have been obtained by implanting other dopamine-producing tissues into the brains of PD patients—namely, adrenal medulla and fetal brain tissue. Even though the latter tissue has not come from elective abortions, this approach has triggered ethical controversy. A surgical technique called *pallidotomy* has been used since the 1940s to alleviate severe tremors. It involves the destruction of a small portion of cerebral tissue called the *globus pallidus.* Pallidotomy fell out of favor when L-DOPA came into common use, but it has made a comeback now that computer-guided equipment has improved the precision and reduced the risks of surgery.

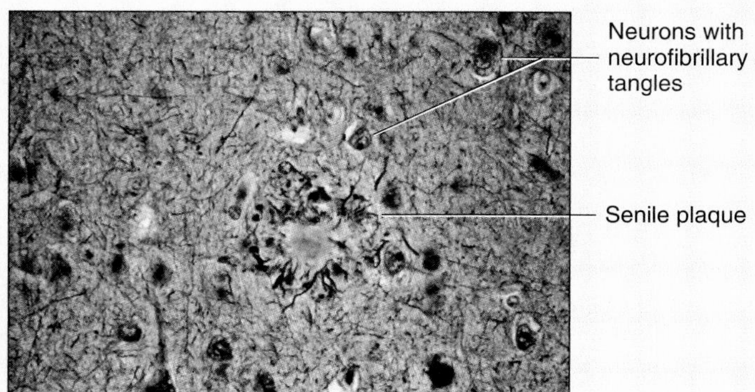

Figure 1 Cerebral Tissue from a Person with Alzheimer Disease. Neurofibrillary tangles appear within the neurons, and a senile plaque is evident in the extracellular matrix.

29. Alois Alzheimer (1864–1915), German neurologist
30. James Parkinson (1755–1824), British physician

Chapter Review

Online Learning Center

Study resources for this chapter are available at www.mhhe.com/saladin/

Study Outline

Overview of the Nervous System (p. 454)

1. Types of neurons
   a. Sensory (afferent) neurons
   b. Interneurons
   c. Motor (efferent) neurons
2. Properties of neurons
   a. Excitability
   b. Conductivity
   c. Secretion
3. Subdivisions of nervous system
   a. Central nervous system (CNS)
   b. Peripheral nervous system (PNS)

Cells of the Nervous System (p. 456)

1. Neuron structure
   a. Soma
   b. Dendrites
   c. Axon (nerve fiber)
   d. Synaptic knobs
2. Shapes of neurons
   a. Multipolar
   b. Bipolar
   c. Unipolar
   d. Anaxonic
3. Axonal transport
   a. Molecular motors
   b. Fast axonal transport
      - Anterograde
      - Retrograde
   c. Slow axonal transport
4. Neuroglia
   a. Oligodendrocytes
   b. Astrocytes
      - Protoplasmic astrocytes
      - Fibrous astrocytes
   c. Ependymal cells
   d. Microglia
   e. Schwann cells
   f. Satellite cells
5. Myelin
   a. Composition
   b. Development
   c. Myelination in the CNS
   d. Myelination in the PNS
   e. Nodes and internodes
   f. Unmyelinated nerve fibers
   g. Myelin and conduction speed
6. Regeneration of nerve fibers
   a. Regeneration tube
   b. Growth of axon
   c. Atrophy and regrowth of muscle fibers

Electrophysiology of Neurons (p. 463)

1. Electrical potentials and currents
2. The resting membrane potential
   a. The role of potassium
   b. The role of cytoplasmic anions
   c. The role of sodium
   d. The role of the Na^+-K^+ pump
3. Local potentials
   a. Effect of stimulus on ion gates
   b. Na^+ inflow
   c. Depolarization of membrane
   d. Characteristics of a local potential
      - Graded
      - Decremental
      - Reversible
      - Excitatory or inhibitory
4. Action potentials
   a. Role of voltage-regulated gates
   b. Events in an action potential
      - Local potential in trigger zone
      - Membrane depolarized to threshold
      - Opening of Na^+ and K^+ gates
      - Na^+ inflow and depolarization
      - K^+ outflow and repolarization
   c. Hyperpolarization
   d. Characteristics of action potentials
      - All-or-none law
      - Irreversibility
      - Self-propagation
      - Nondecremental conduction
5. The refractory period
6. Signal conduction in nerve fibers
   a. Unmyelinated fibers
      - Chain reaction of action potentials
      - Distinction between action potential and nerve signal
      - Unidirectional conduction
      - Nondecremental conduction
   b. Myelinated fibers
      - Distribution of voltage-regulated gates
      - Signal conduction in internodes
      - Signal conduction in nodes of Ranvier
      - Saltatory conduction

Synapses (p. 472)

1. Presynaptic and postsynaptic neurons
2. Discovery of neurotransmitters
3. Electrical synapses
4. Structure of a chemical synapse
   a. Synaptic knob
   b. Synaptic vesicles
   c. Neurotransmitter receptors
5. Neurotransmitters
   a. Acetylcholine (ACh)
   b. Amino acids
   c. Monoamines
6. Neuropeptides
7. Synaptic transmission
   a. At an ionotropic synapse
      - Nerve signal opens Ca^{2+} gates in synaptic knob
      - Ca^{2+} triggers exocytosis of synaptic vesicles
      - Neurotransmitter diffuses across synaptic cleft
      - Neurotransmitter binds to receptors on postsynaptic cell
      - Receptors open and allow Na^+ and K^+ through membrane
      - Ion flow produces postsynaptic potential
   b. At a metabotropic synapse
      - Receptor activates cAMP production

- Effects of cAMP on postsynaptic cell
  - •• Metabolic pathways turned on or off
  - •• Genetic transcription activated
  - •• Ion gates in membrane opened

8. Cessation of signal
   a. Cessation of signals in presynaptic fiber
   b. Removal of neurotransmitter from synapse
      - Diffusion away from synapse
      - Reuptake by presynaptic neuron
      - Degradation in synaptic cleft
9. Neuromodulators

Neural Integration (p. 478)

1. Importance of synapses
2. Postsynaptic potentials
   a. Excitatory (EPSP)
   b. Inhibitory (IPSP)
3. Summation
   a. Net effect of EPSPs and IPSPs
   b. Temporal summation
   c. Spatial summation
4. Facilitation
5. Presynaptic inhibition
6. Neural coding
   a. Encoding qualitative information
   b. Encoding quantitative information
      - Recruitment
      - Firing frequency
7. Neuronal pools and circuits
   a. Discharge and facilitated zones
   b. Neuronal circuits
      - Diverging circuit
      - Converging circuit
      - Reverberating circuit
      - Parallel after-discharge circuit
8. Memory and synaptic plasticity
   a. Memory traces
   b. Synaptic potentiation
   c. Immediate memory
   d. Short-term memory
      - Working memory
      - Facilitation
      - Posttetanic potentiation
   e. Long-term memory
      - Declarative and procedural
      - Mechanisms
        - •• Formation and remodeling of synapses
        - •• Long-term potentiation

Selected Vocabulary

nervous system 454
endocrine system 454
neuron 454
sensory neuron 454
afferent neuron 454
stimulus 455
receptor 455
interneuron 455
association neuron 455
motor neuron 455
efferent neuron 455
effector 455
central nervous system 455
gray matter 455
white matter 455
peripheral nervous system 455
nerve 455
ganglion 455
soma 456
neurofibril 456
dendrite 456
axon hillock 456
axon 456
nerve fiber 456
axoplasm 456
axolemma 456
terminal arborization 456
synaptic knob 456
synapse 456
synaptic vesicle 458
multipolar neuron 458
bipolar neuron 458
unipolar neuron 458
anaxonic neuron 458
axonal transport 458
anterograde transport 458
retrograde transport 458
fast axonal transport 459
slow axonal transport 459
neuroglia 459
oligodendrocyte 459
astrocyte 460
protoplasmic astrocyte 460
fibrous astrocyte 460
ependymal cell 460
microglia 460
Schwann cell 460
satellite cell 461
myelin sheath 461
neurilemma 462
node of Ranvier 462
internode 462
initial segment 462
trigger zone 462
unmyelinated 462
regeneration tube 463
electrical potential 464
current 465
polarized 465
resting membrane potential 465
depolarization 466
local potential 466
hyperpolarize 467
action potential 467
threshold 467
repolarize 467
all-or-none law 469
refractory period 469
saltatory conduction 471
presynaptic neuron 472
postsynaptic neuron 472
synaptic cleft 472
electrical synapse 473
chemical synapse 473
acetylcholine 474
monoamine 474
catecholamine 474
indolamine 474
neuropeptide 474
cholinergic synapse 475
postsynaptic potential 476
synaptic delay 476
ionotropic effect 476
metabotropic effect 476
reuptake 477
monoamine oxidase 477
neuromodulator 477
neural integration 478
excitatory postsynaptic potential 478
inhibitory postsynaptic potential 479
summation 479
temporal summation 479
spatial summation 480
facilitation 480
presynaptic inhibition 480
neural coding 481
recruitment 481
neuronal pool 481
discharge zone 481
facilitated zone 482
neuronal circuit 482
diverging circuit 482
converging circuit 482
reverberating circuit 482
parallel after-discharge circuit 482
memory trace 483
synaptic plasticity 483
synaptic potentiation 483
immediate memory 483
short-term memory 483
working memory 483
facilitation 484
posttetanic potentiation 484
long-term memory 484
declarative memory 484
procedural memory 484
long-term potentiation 484

Testing Your Recall

1. The integrative functions of the nervous system are performed mainly by
   a. afferent neurons.
   b. efferent neurons.
   c. neuroglia.
   d. sensory neurons.
   e. interneurons.
2. The highest density of voltage-regulated ion gates is found on the ________ of a neuron.
   a. dendrites
   b. soma
   c. nodes of Ranvier
   d. internodes
   e. synaptic knobs
3. The soma of a mature neuron lacks
   a. a nucleus.
   b. endoplasmic reticulum.
   c. lipofuscin.
   d. centrioles.
   e. ribosomes.
4. The glial cells that destroy microorganisms in the CNS are
   a. microglia.
   b. satellite cells.
   c. ependymal cells.
   d. oligodendrocytes.
   e. astrocytes.
5. Posttetanic potentiation of a synapse increases the amount of ________ in the synaptic knob.
   a. neurotransmitter
   b. neurotransmitter receptors
   c. calcium
   d. sodium
   e. NMDA
6. An IPSP is ________ of the postsynaptic neuron.
   a. a refractory period
   b. an action potential
   c. a depolarization
   d. a repolarization
   e. a hyperpolarization
7. Saltatory conduction occurs only
   a. at chemical synapses.
   b. in the initial segment of an axon.
   c. in both the initial segment and axon hillock.
   d. in myelinated nerve fibers.
   e. in unmyelinated nerve fibers.
8. Some neurotransmitters can have either excitatory or inhibitory effects depending on the type of
   a. receptors on the postsynaptic neuron.
   b. synaptic vesicles in the axon.
   c. synaptic potentiation that occurs.
   d. postsynaptic potentials on the synaptic knob.
   e. neuromodulator involved.
9. Differences in the volume of a sound are likely to be encoded by differences in ________ in nerve fibers from the inner ear.
   a. neurotransmitters
   b. signal conduction velocity
   c. types of postsynaptic potentials
   d. firing frequency
   e. voltage of the action potentials
10. Motor effects that depend on repetitive output from a neuronal pool are most likely to use
    a. parallel after-discharge circuits.
    b. reverberating circuits.
    c. facilitated circuits.
    d. diverging circuits.
    e. converging circuits.
11. Neurons that convey information to the CNS are called sensory, or ___________ , neurons.
12. To perform their role, neurons must have the properties of excitability, secretion, and ___________ .
13. The ___________ is a period of time in which a neuron is producing an action potential and cannot respond to another stimulus of any strength.
14. Neurons receive incoming signals by way of specialized processes called ___________ .
15. In the central nervous system, cells called ___________ perform one of the same functions that Schwann cells do in the peripheral nervous system.
16. A myelinated nerve fiber can produce action potentials only in specialized regions called ___________ .
17. The trigger zone of a neuron consists of its ___________ and ___________ .
18. Neurotransmitters that work through second-messenger systems are said to have ___________ effects.
19. A presynaptic nerve fiber cannot cause other neurons in its ___________ to fire, but it can make them more sensitive to stimulation from other presynaptic fibers.
20. ___________ are substances released along with a neurotransmitter that modify the neurotransmitter's effect.

Answers in appendix B

Testing Your Comprehension

1. Schizophrenia is sometimes treated with drugs such as chlorpromazine that inhibit dopamine receptors. A side effect is that patients begin to develop muscle tremors, speech impairment, and other disorders similar to Parkinson disease. Explain.
2. Hyperkalemia is an excess of potassium in the extracellular fluid. What effect would this have on the resting membrane potentials of the nervous system and on neuronal excitability?
3. Suppose the Na^+-K^+ pumps of nerve cells were to slow down because of some metabolic disorder. How would this affect the resting membrane potentials of neurons? Would it make neurons more excitable than normal, or make them more difficult to stimulate? Explain.
4. The unity of form and function is an important concept in understanding synapses. Give two structural reasons why nerve signals cannot travel backward across a chemical synapse. What might be the consequences if signals did travel freely in both directions?
5. The local anesthetics tetracaine and procaine (Novocain) prevent voltage-regulated Na^+ gates from opening. Explain why this would block the conduction of pain signals in a sensory nerve.

Answers at the Online Learning Center

14 CHAPTER

The Central Nervous System

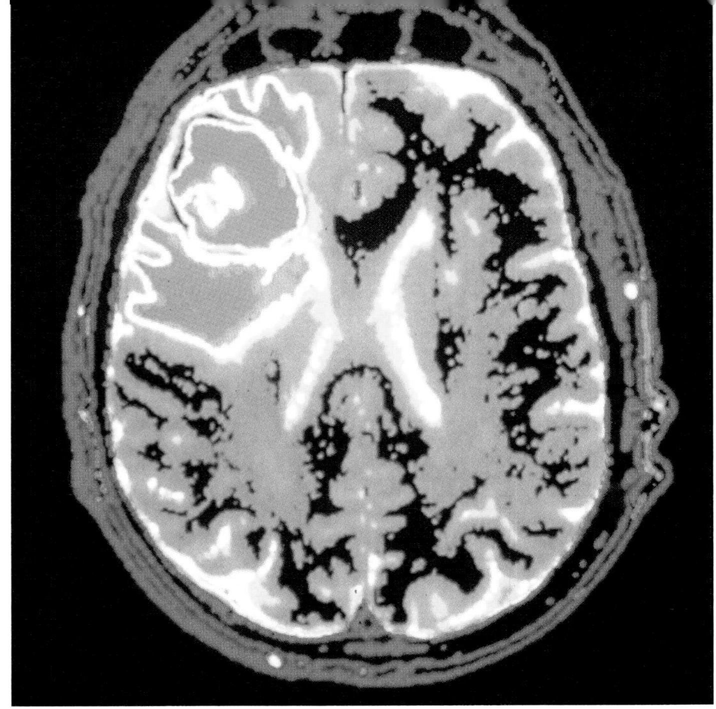

Glioblastoma (red), a fast-growing, highly invasive brain tumor (MRI)

Chapter Outline

Insights

Study with Your ESP CD-ROM

Refer to your ESP Tutorial CD-ROM for helpful animations, activities, and quizzes related to the material covered in this chapter. The following are suggested areas to go to on the CD and are presented according to "Unit/Topic":
- Integration and Coordination/CNS

Brushing Up

To understand this chapter, it is important that you understand or brush up on the following concepts:
- Anatomy of the cranium (pp. 262–271)
- Anatomy of the vertebrae (pp. 277–278)
- Gray matter and white matter (p. 455)
- Glial cells and their functions (pp. 459–461)

The mystique of the brain continues to intrigue modern biologists and psychologists even as it did the philosophers of antiquity. Aristotle dismissed the brain as merely a radiator for cooling the blood, but generations earlier, Hippocrates had expressed a more accurate view of its functions. "Men ought to know," he said, "that from the brain, and from the brain only, arise our pleasures, joy, laughter and jests, as well as our sorrows, pains, griefs and tears. Through it, in particular, we think, see, hear, and distinguish the ugly from the beautiful, the bad from the good, the pleasant from the unpleasant." Brain function is so strongly associated with what it means to be alive and human that the cessation of brain activity is taken as a clinical criterion of death even when other organs of the body are still functioning.

With its hundreds of neuronal pools and trillions of synapses, the brain performs sophisticated tasks beyond our present understanding. Still, all of our mental functions, no matter how complex, are ultimately based on the cellular activities described in the previous chapter. The relationship of the mind or personality to the cellular function of the brain is a question that will provide fertile ground for scientific study and philosophical debate long into the future.

The previous chapter discussed the nervous system at the cellular level. We now move up the structural hierarchy to consider the central nervous system at the organ level. This chapter lays a foundation for understanding some relatively simple reflex functions of the spinal cord and brainstem considered in the next chapter, and it plumbs some of the mysteries of motor control, sensation, emotion, thought, language, personality, memory, dreams, and plans. Your study of this chapter is one brain's attempt to understand itself.

Overview of the Central Nervous System

Objectives

When you have completed this section, you should be able to

- describe the major subdivisions and anatomical landmarks of the brain; and
- describe the embryonic development of the CNS and relate this to adult brain anatomy.

Major Landmarks

Before we consider the form and function of specific regions of the central nervous system (CNS), it will help to get a general overview of its major landmarks, especially of the brain (figs. 14.1 and 14.2). These will provide important points of reference as we progress through a more detailed study of the system.

Two directional terms often used to describe brain anatomy are *rostral* and *caudal*. **Rostral**[1] means "toward the nose" and **caudal**[2] means "toward the tail," descriptions especially appropriate to rats and other mammals on which so much neuroanatomy has been done. In human brain anatomy, one structure is rostral to another if it is closer to the forehead and caudal to another if it is closer to the spinal cord.

The average adult brain weighs about 1,600 g (3.5 lb) in men and 1,450 g in women. Its size is proportional to body size, not intelligence—the Neanderthal people had larger brains than modern humans.

The brain is divided into three major portions—the *cerebrum, cerebellum,* and *brainstem.* The **cerebrum** (SER-eh-brum or seh-REE-brum) constitutes about 83% of its volume and consists of a pair of **cerebral hemispheres.** Each hemisphere is marked by thick folds called **gyri**[3] (JY-rye; singular, *gyrus*) separated by shallow grooves called **sulci**[4] (SUL-sye; singular, *sulcus*) and a few deeper grooves called **fissures.** The **longitudinal fissure** separates the right and left hemispheres from each other. The hemispheres are connected by a thick bundle of nerve fibers called the **corpus callosum**[5]—a prominent landmark for anatomical description (fig. 14.1*d*).

The **cerebellum**[6] (SER-eh-BEL-um) lies inferior to the cerebrum and occupies the posterior cranial fossa. It is also marked by gyri, sulci, and fissures. The cerebellum is the second-largest region of the brain; it constitutes about 10% of its volume but contains over 50% of its neurons.

The **brainstem** is what remains if the cerebrum and cerebellum are removed. In a living person, it is oriented like a vertical stalk with the cerebrum perched on top of it. Postmortem changes give it a more oblique angle in the cadaver and consequently in many medical illustrations. The major components of the brainstem, from rostral to caudal, are the *diencephalon, midbrain, pons,* and *medulla oblongata.* Some authorities, however, regard only the last three of these as the brainstem.The brainstem ends at the foramen magnum of the skull and the CNS continues below this as the spinal cord.

Embryonic Development

To understand the terminology of mature brain anatomy, it is necessary to be aware of the embryonic development of the CNS. The nervous system develops from ectoderm, the outermost germ layer of an embryo. By the third week of development, a dorsal streak called the *neuroectoderm* appears along the length of the embryo and thickens to form a **neural plate** (fig. 14.3). This is destined to give rise to all neurons and glial cells except microglia, which come from mesoderm. As development progresses, the neural plate sinks and forms a **neural groove** with a raised **neural fold**

1. *rostr* = nose
2. *caud* = tail

3. *gyr* = turn, twist
4. *sulc* = furrow, groove
5. *corpus* = body + *call* = thick
6. *cereb* = brain + *ellum* = little

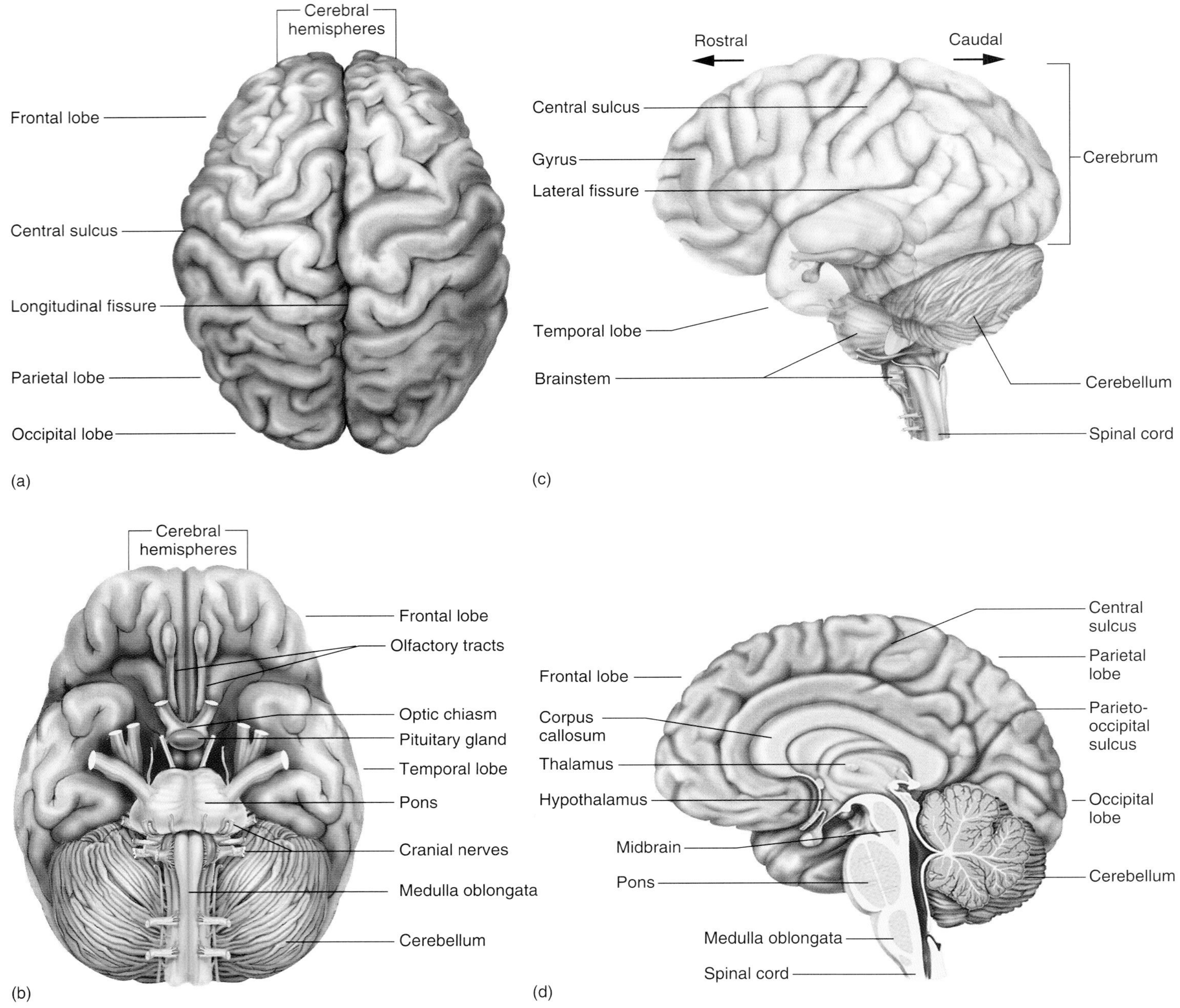

Figure 14.1 Four Views of the Brain. (*a*) Superior; (*b*) inferior (base of the brain); (*c*) left lateral; (*d*) midsagittal section (color indicates brainstem).

along each side. The neural folds then fuse along the midline, somewhat like a closing zipper. By 4 weeks, this process creates a hollow channel called the **neural tube** (see clinical insight 14.1). The neural tube now separates from the overlying ectoderm, sinks a little deeper, and grows lateral processes that later form motor nerve fibers. The lumen of the neural tube develops into fluid-filled spaces called the *central canal* of the spinal cord and *ventricles* of the brain.

As the neural tube develops, some ectodermal cells that originally lay along the margin of the neural groove separate from the rest and form a longitudinal column on each side called the **neural crest.** Some neural crest cells become sensory neurons, while others migrate to other locations and become sympathetic neurons, Schwann cells, and other cell types.

By the fourth week, the neural tube exhibits three anterior dilations, or *primary vesicles,* called the **forebrain**

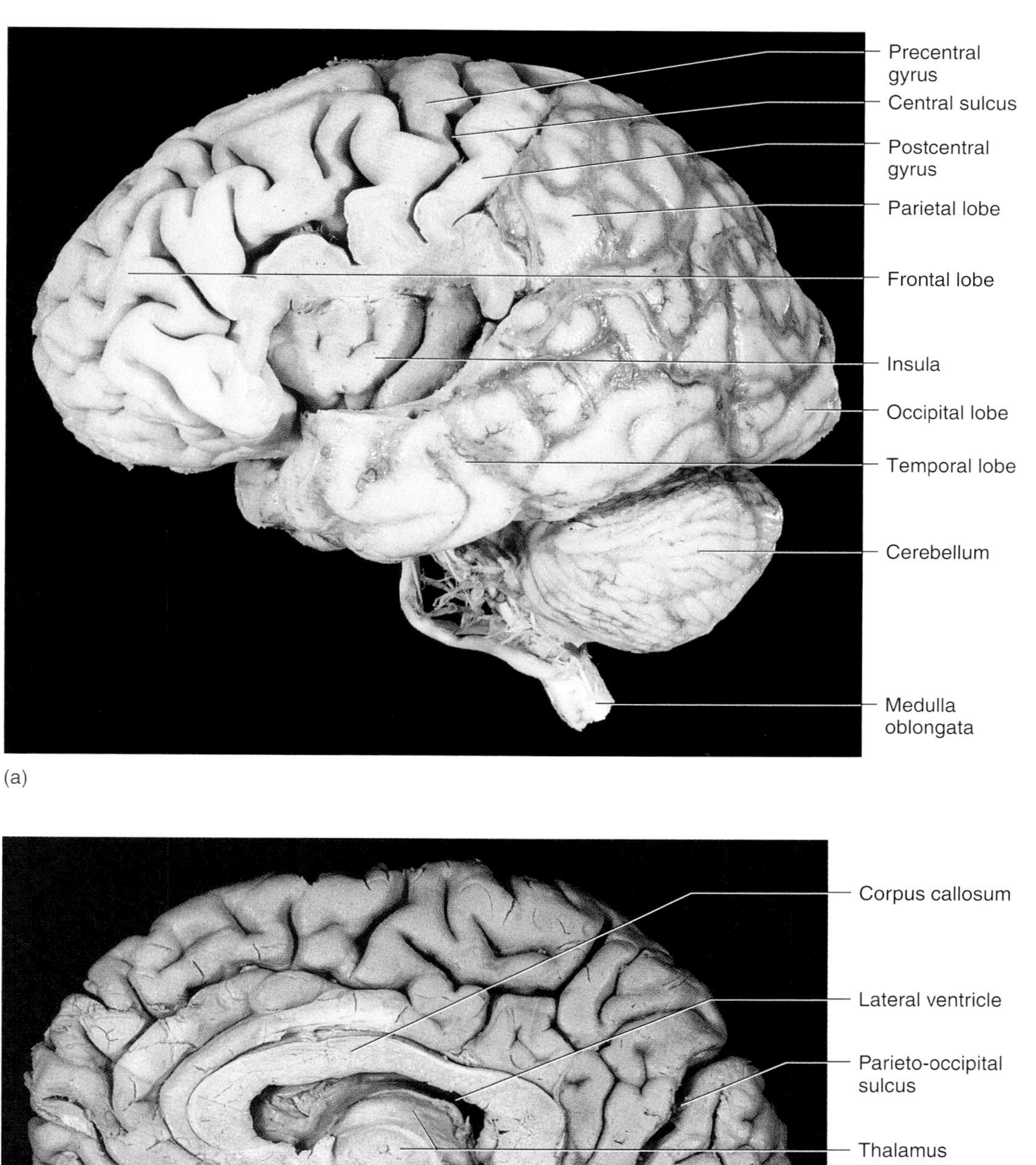

Figure 14.2 Dissections of the Brain. (*a*) Left lateral view with part of the left hemisphere cut away to expose the insula. The pia mater is visible as a delicate translucent membrane over many of the sulci of the occipital lobe. (*b*) Midsagittal section, left lateral view.

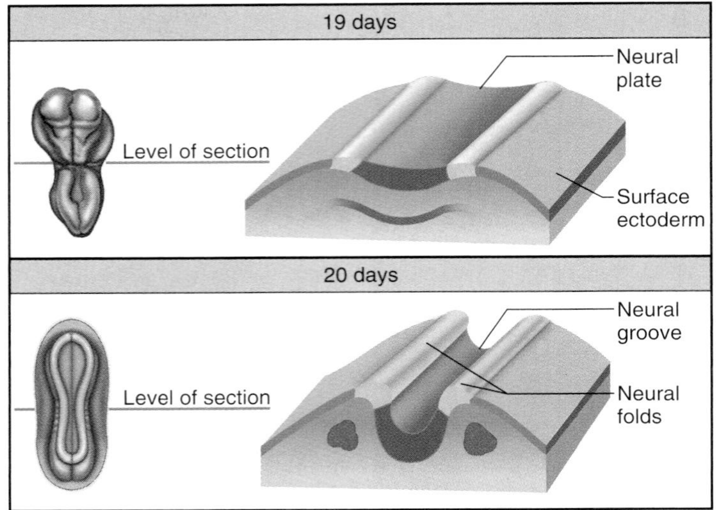

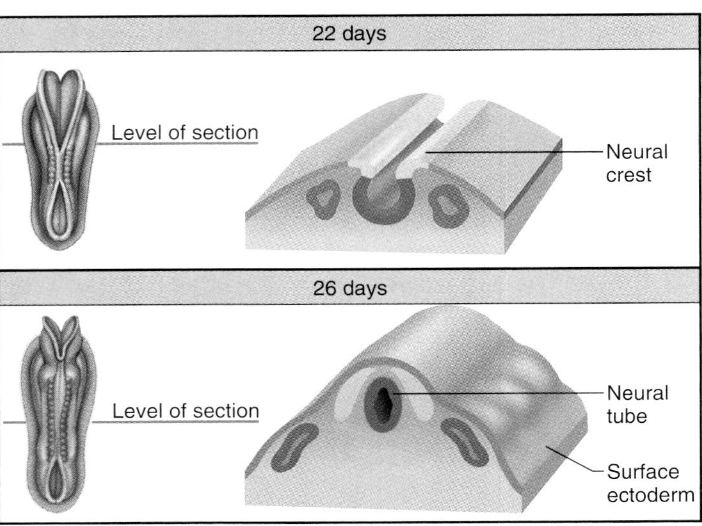

Figure 14.3 Embryonic Development of the CNS to the Neural Tube Stage. The *left-hand* figures represent dorsal views of the embryo and the *right-hand* figures show three-dimensional reconstructions at the indicated levels.

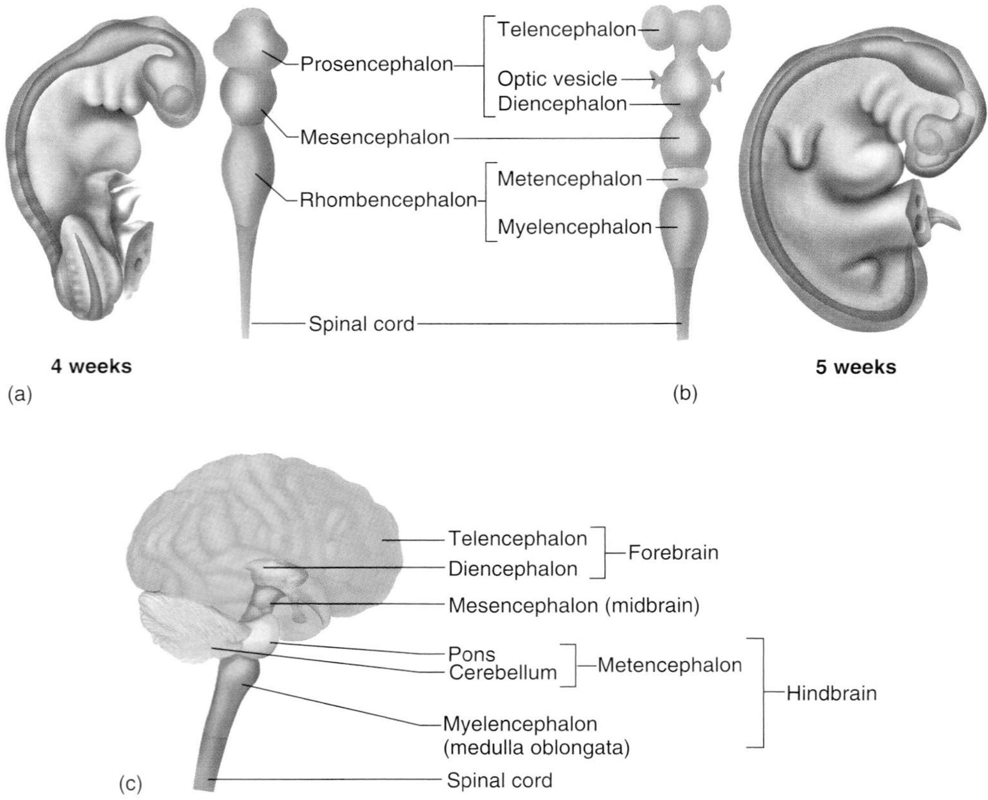

Figure 14.4 Primary and Secondary Vesicles of the Embryonic Brain. (*a*) The primary vesicles at 4 weeks; (*b*) the secondary vesicles at 5 weeks; (*c*) the fully developed brain, color-coded to relate its structures to the secondary embryonic vesicles.

Clinical Insight 14.1

Spina Bifida

Neural tube defects are birth defects that result from abnormal development of the neural tube and vertebral column. About one baby in 1,000 is born with *spina bifida* (SPY-nuh BIF-ih-duh), which occurs when one or more vertebrae fail to form a complete neural arch for enclosure of the spinal cord. It is especially common in the lumbosacral region. One form, *spina bifida occulta,*[7] involves only one to a few vertebrae and causes no functional problems. Its only external sign is a dimple or hairy pigmented spot. *Spina bifida cystica*[8] is more serious. A sac protrudes from the spine and may contain meninges, cerebrospinal fluid, and parts of the spinal cord and nerve roots. In extreme cases, inferior spinal cord function is absent, causing lack of bowel control and paralysis of the lower limbs and urinary bladder. The last of these conditions can lead to chronic urinary infections and renal failure. Pregnant women can significantly reduce the risk of spina bifida by taking supplemental folic acid (a B vitamin) during early pregnancy. Good sources of folic acid include green leafy vegetables, black beans, lentils, and enriched bread and pasta.

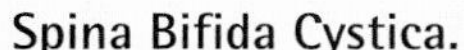
Spina Bifida Cystica.

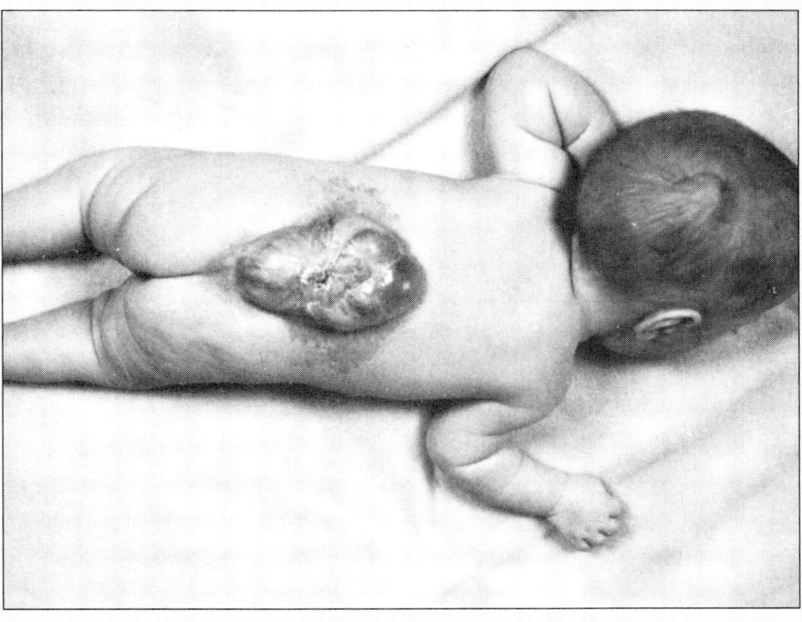

7. *bifid* = divided, forked + *occult* = hidden
8. *cyst* = sac, bladder

(*prosencephalon*[9]) (PROSS-en-SEF-uh-lon), **midbrain** (*mesencephalon*[10]) (MEZ-en-SEF-uh-lon), and **hindbrain** (*rhombencephalon*[11]) (ROM-ben-SEF-uh-lon) (fig. 14.4). By the fifth week, the neural tube undergoes further flexion and subdivides into five *secondary vesicles.* The forebrain divides into two of them, the **telencephalon**[12] (TEL-en-SEF-uh-lon) and **diencephalon**[13] (DY-en-SEF-uh-lon); the midbrain remains undivided and retains the name **mesencephalon;** and the hindbrain divides into two vesicles, the **metencephalon**[14] (MET-en-SEF-uh-lon) and **myelencephalon**[15] (MY-el-en-SEF-uh-lon). The telencephalon has a pair of lateral outgrowths that later become the cerebral hemispheres, and the diencephalon exhibits a pair of small cuplike *optic vesicles* that become the retinas of the eyes. Figure 14.4*c* shows structures of the fully developed brain that arise from each of the secondary vesicles.

9. *pros* = before, in front + *encephal* = brain
10. *mes* = middle
11. *rhomb* = rhombus
12. *tele* = end, remote
13. *di* = through, between
14. *met* = behind, beyond, distal to
15. *myel* = spinal cord

Before You Go On

Answer the following questions to test your understanding of the preceding section:

1. List the three major parts of the brain and describe their locations.
2. Define *gyrus, sulcus,* and *fissure.*
3. Explain how the five secondary brain vesicles arise from the neural tube.

Meninges, Ventricles, Cerebrospinal Fluid, and Blood Supply

Objectives

When you have completed this section, you should be able to

- name the protective membranes that surround the CNS;
- describe the system of fluid-filled chambers within the brain;
- discuss the production, circulation, and function of the cerebrospinal fluid that fills these chambers; and
- explain the significance of the brain barrier system.

Meninges

The **meninges** (meh-NIN-jeez)—singular, *meninx*[16] (MEN-inks)—are three protective fibrous membranes that separate the soft tissue of the brain and spinal cord from the bones of the skull and vertebrae (fig. 14.5). From superficial to deep, they are the **dura mater**[17] (DOO-ruh MAH-tur), **arachnoid**[18] (ah-RACK-noyd) **mater** (or *arachnoid membrane*), and **pia**[19] (PEE-uh) **mater.** The dura mater is a tough membrane similar in thickness and texture to a household rubber glove. In the cranial cavity, it consists of two layers—an outer *periosteal layer,* equivalent to the periosteum of the cranial bone, and an inner *meningeal layer.* In some places, the two layers are separated by **dural sinuses,** which collect blood that has circulated through the brain and empty into the internal jugular veins of the neck. Two major dural sinuses are the **superior sagittal sinus,** found just under the cranium along the midsagittal line, and the **transverse sinus,** which runs horizontally from the rear of the head toward each ear; these sinuses meet like an inverted T at the back of the brain.

In certain places, the meningeal layer of the dura mater folds inward to separate major parts of the brain: the *falx*[20] *cerebri* (falks SER-eh-bry) extends into the longitudinal fissure between the right and left cerebral hemispheres; the *tentorium*[21] (ten-TOE-ree-um) *cerebelli* stretches like a roof over the posterior cranial fossa and separates the cerebellum from the overlying cerebrum; and the *falx cerebelli* partially separates the right and left halves of the cerebellum on the inferior side.

The periosteal layer of the dura is limited to the cranial cavity. The meningeal layer continues into the vertebral canal and forms a **dural sheath** around the spinal cord. The space between the sheath and vertebral bone, called the **epidural space,** is occupied by a protective pad of fat and loose connective tissue (fig. 14.5*b*). During childbirth, some women receive *epidural anesthesia*—introduction of an anesthetic into this space to block pain signals from the pelvic region.

The arachnoid mater adheres to the inner surface of the dura above and has fibrous extensions like a spider web (hence the name) that attach to the pia below. The pia mater is a thin, translucent, richly vascular layer of loose connective tissue. It resembles a sheet of cellophane closely following the contours of the brain and extending into its sulci and fissures. In some places, a *subdural space* separates the dura from the arachnoid and a *subarachnoid space* separates the arachnoid from the pia.

Ventricles and Cerebrospinal Fluid

The brain has four internal chambers called **ventricles** (fig. 14.6). The most rostral ones are the **lateral ventricles,** which form a large arc in each cerebral hemisphere. Through a tiny passage called the **interventricular foramen,** each lateral ventricle is connected to the **third ventricle,** a narrow medial space inferior to the corpus callosum. From here, a canal called the **mesencephalic (cerebral) aqueduct** passes down the core of the midbrain and leads to the **fourth ventricle,** a small chamber between the pons and cerebellum. Caudally, this space narrows and forms a **central canal** that extends through the medulla oblongata into the spinal cord.

These ventricles and canals are lined with ependymal cells, a type of neuroglia that resembles a simple cuboidal epithelium. Each ventricle contains a **choroid** (CO-royd) **plexus,** named for its histological resemblance to a fetal membrane called the chorion. The choroid plexus is a network of blood capillaries anchored to the floor or wall of the ventricle and covered by ependymal cells.

A clear, colorless liquid called **cerebrospinal fluid (CSF)** fills the ventricles and canals of the CNS and bathes its external surface. The brain produces about 500 mL of CSF per day, but the fluid is constantly reabsorbed at the same rate and only 100 to 160 mL are present at one time. About 40% of it is formed in the subarachnoid space external to the brain, 30% by the general ependymal lining of the brain ventricles, and 30% by the choroid plexuses. CSF

16. *menin* = membrane
17. *dura* = tough + *mater* = mother, womb
18. *arachn* = spider, spider web + *oid* = resembling
19. *pia* = tender, soft
20. *falx* = sickle
21. *tentorium* = tent

Clinical Insight 14.2

Meningitis

Meningitis—inflammation of the meninges—is one of the most serious diseases of infancy and childhood. It occurs especially between 3 months and 2 years of age. Meningitis is caused by a variety of bacteria and viruses that invade the CNS by way of the nose and throat, often following respiratory, throat, or ear infections. The pia mater and arachnoid are most likely to be affected, and from here the infection can spread to the adjacent nervous tissue. In bacterial meningitis, the brain swells, the ventricles enlarge, and the brainstem may have hemorrhages. Signs include a high fever, stiff neck, drowsiness, and intense headache and may progress to vomiting, loss of sensory and motor functions, coma, and death.

Meningitis is diagnosed partly by examining the CSF for bacteria and white blood cells. The CSF is obtained by making a *lumbar puncture (spinal tap)* between two lumbar vertebrae and drawing fluid from the subarachnoid space. This site is chosen because it has an abundance of CSF and there is no risk of injury to the spinal cord, which does not extend into the lower lumbar vertebrae.

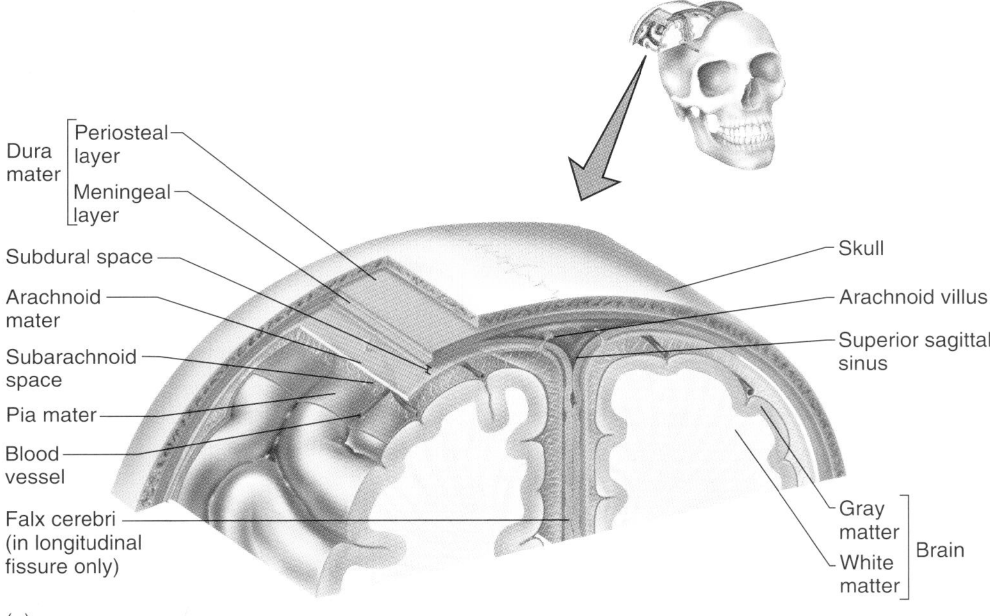

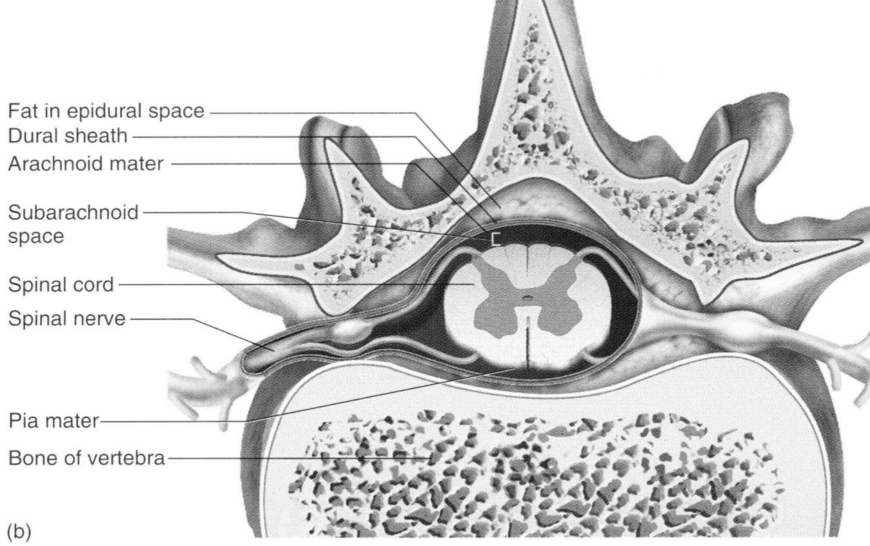

Figure 14.5 The Meninges. (*a*) Frontal section of the head; (*b*) cross section of a vertebra and the spinal cord.

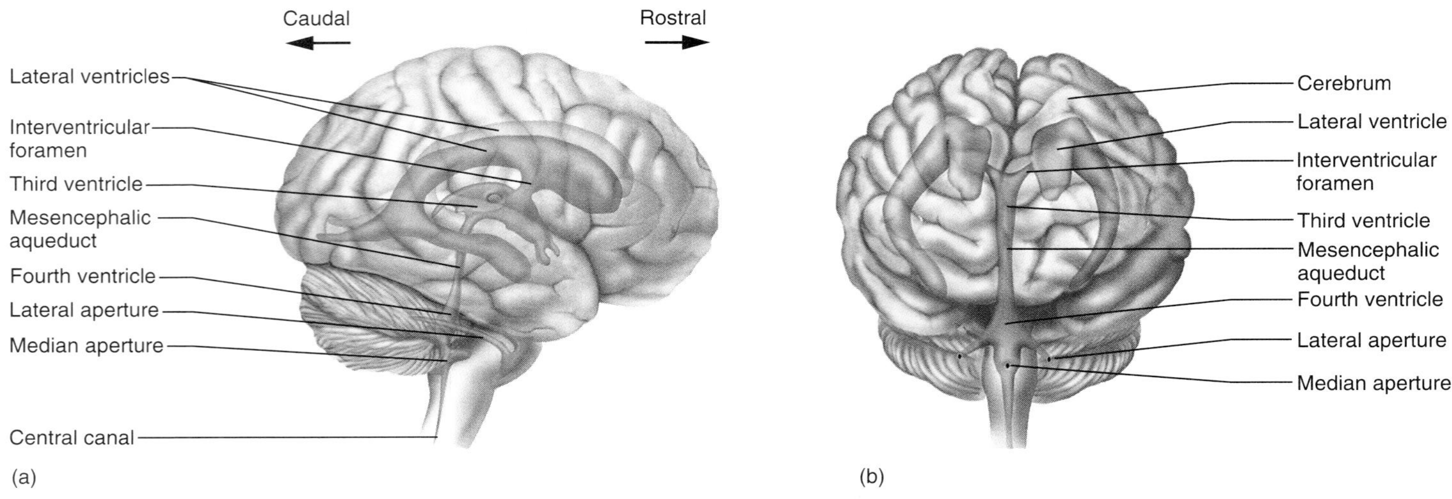

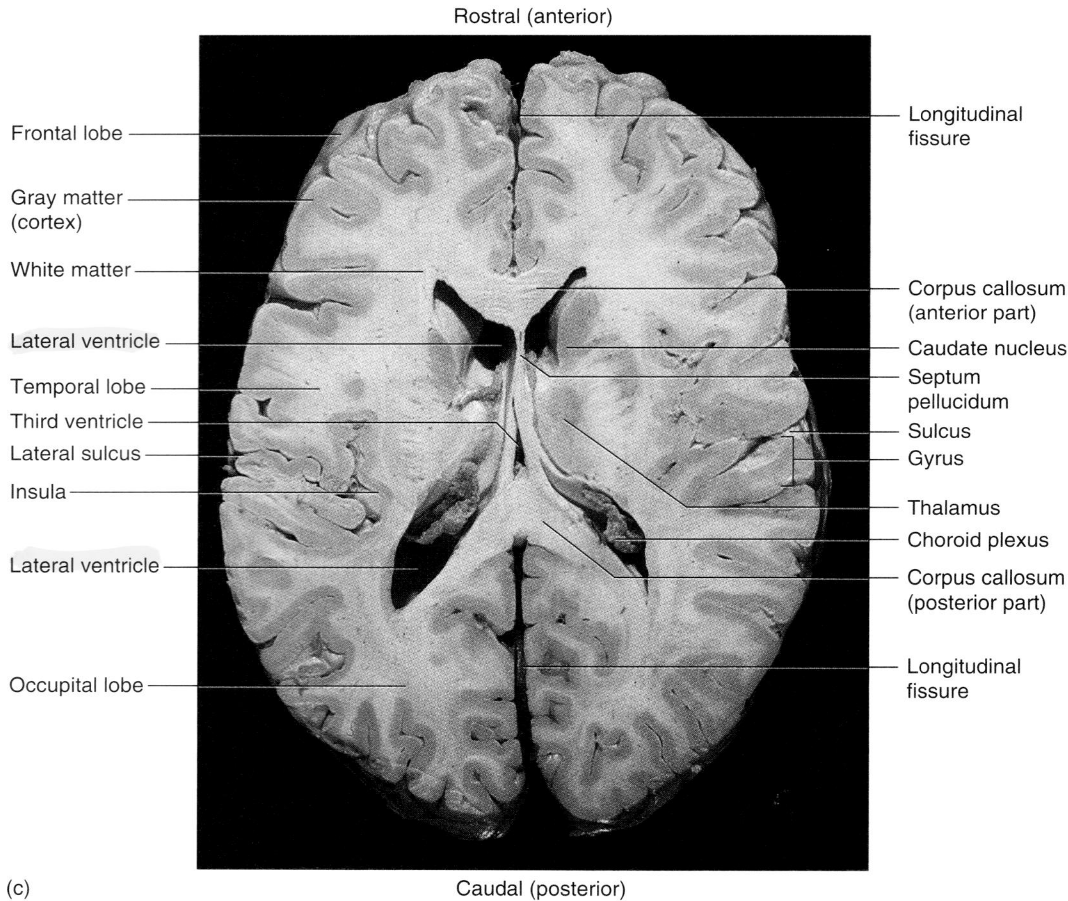

Figure 14.6 Ventricles of the Brain. (*a*) Right lateral aspect; (*b*) anterior aspect; (*c*) superior aspect of a horizontal section of the brain showing the lateral ventricles and other features of the cerebrum.

forms partly by the filtration of blood plasma through the choroid plexuses and other capillaries of the brain. The ependymal cells modify this filtrate, however, so the CSF has more sodium and chloride than the blood plasma, but less potassium, calcium, and glucose and very little protein.

Cerebrospinal fluid serves three purposes:

1. **Buoyancy.** Because the brain and CSF are very similar in density, the brain neither sinks nor floats in the CSF but remains suspended in it—that is, the brain has *neutral buoyancy.* A human brain removed from the body weighs about 1,500 g, but when suspended in CSF its effective weight is only about 50 g. By analogy, consider how much easier it is to lift another person when you are standing in a lake than it is on land. Neutral buoyancy allows the brain to attain considerable size without being impaired by its own weight. If the brain rested heavily on the floor of the cranium, the pressure would kill the nervous tissue.
2. **Protection.** CSF also protects the brain from striking the cranium when the head is jolted. If the jolt is severe, however, the brain still may strike the inside of the cranium or suffer shearing injury from contact with the angular surfaces of the cranial floor. This is one of the common findings in child abuse (shaken child syndrome) and in head injuries (concussions) from auto accidents, boxing, and the like.
3. **Chemical stability.** CSF is secreted into each ventricle of the brain and is ultimately absorbed into the bloodstream. It provides a means of rinsing metabolic wastes from the CNS and homeostatically regulating its chemical environment. Slight changes in its composition can cause malfunctions of the nervous system. For example, a high glycine concentration disrupts temperature and blood pressure control, and a high pH causes dizziness and fainting.

The CSF is not a stationary fluid but continually flows through and around the CNS, driven partly by its own pressure and partly by rhythmic pulsations of the brain produced by each heartbeat. The CSF secreted in the lateral ventricles flows through the interventricular foramina into the third ventricle (fig. 14.7) and then down the mesencephalic aqueduct to the fourth ventricle. The third and fourth ventricles and their choroid plexuses add more CSF along the way. A small amount of CSF fills the central canal of the spinal cord, but ultimately, all of it escapes through three pores in the walls of the fourth ventricle—a *median aperture* and two *lateral apertures.* These lead into the subarachnoid space on the brain surface. From this space, the CSF is absorbed by **arachnoid villi,** cauliflower-like extensions of the arachnoid meninx that protrude through the dura mater into the superior sagittal sinus. CSF penetrates the walls of the arachnoid villi and mixes with the blood in the sinus.

Hydrocephalus[22] is the abnormal accumulation of CSF in the brain, usually resulting from a blockage in its route of flow and reabsorption. Such obstructions occur most often in the interventricular foramen, mesencephalic aqueduct, and the three apertures of the fourth ventricle. The accumulated CSF makes the ventricles expand and compress the nervous tissue, with potentially fatal consequences. In a fetus or infant, it can cause the entire head to enlarge because the cranial bones are not yet fused. Good recovery can be achieved if a tube (shunt) is inserted to drain fluid from the ventricles into a vein of the neck.

Blood Supply and the Brain Barrier System

Although the brain constitutes only 2% of the adult body weight, it receives 15% of the blood (about 750 mL/min) and consumes 20% of the body's oxygen and glucose. Because neurons have such a high demand for ATP, and therefore glucose and oxygen, the constancy of the blood supply is especially critical to the nervous system. A mere 10-second interruption in blood flow can cause loss of consciousness; an interruption of 1 to 2 minutes can significantly impair neural function; and 4 minutes without blood causes irreversible brain damage.

Despite its critical importance to the brain, the blood is also a source of antibodies, macrophages, and other potentially harmful agents. Consequently, there is a **brain barrier system** that strictly regulates what substances can get from the bloodstream into the tissue fluid of the brain. There are two potential points of entry that must be guarded: the blood capillaries throughout the brain tissue and the capillaries of the choroid plexuses.

At the former site, the brain is well protected by the **blood-brain barrier (BBB),** which consists mainly of the tightly joined endothelial cells that form the blood capillaries and partly of the basement membrane surrounding them. In the developing brain, protoplasmic astrocytes reach out and contact the capillaries with their perivascular feet. They stimulate the endothelial cells to form tight junctions, which seal off the capillaries and ensure that anything leaving the blood must pass through the cells and not between them. At the choroid plexuses, the brain is protected by a similar **blood-CSF barrier,** composed of ependymal cells joined by tight junctions. Tight junctions are absent from ependymal cells elsewhere, because it is important to allow exchanges between the brain tissue and CSF. That is, there is no brain-CSF barrier.

The brain barrier system (BBS) is highly permeable to water, glucose, and lipid-soluble substances such as

22. *hydro* = water + *cephal* = head

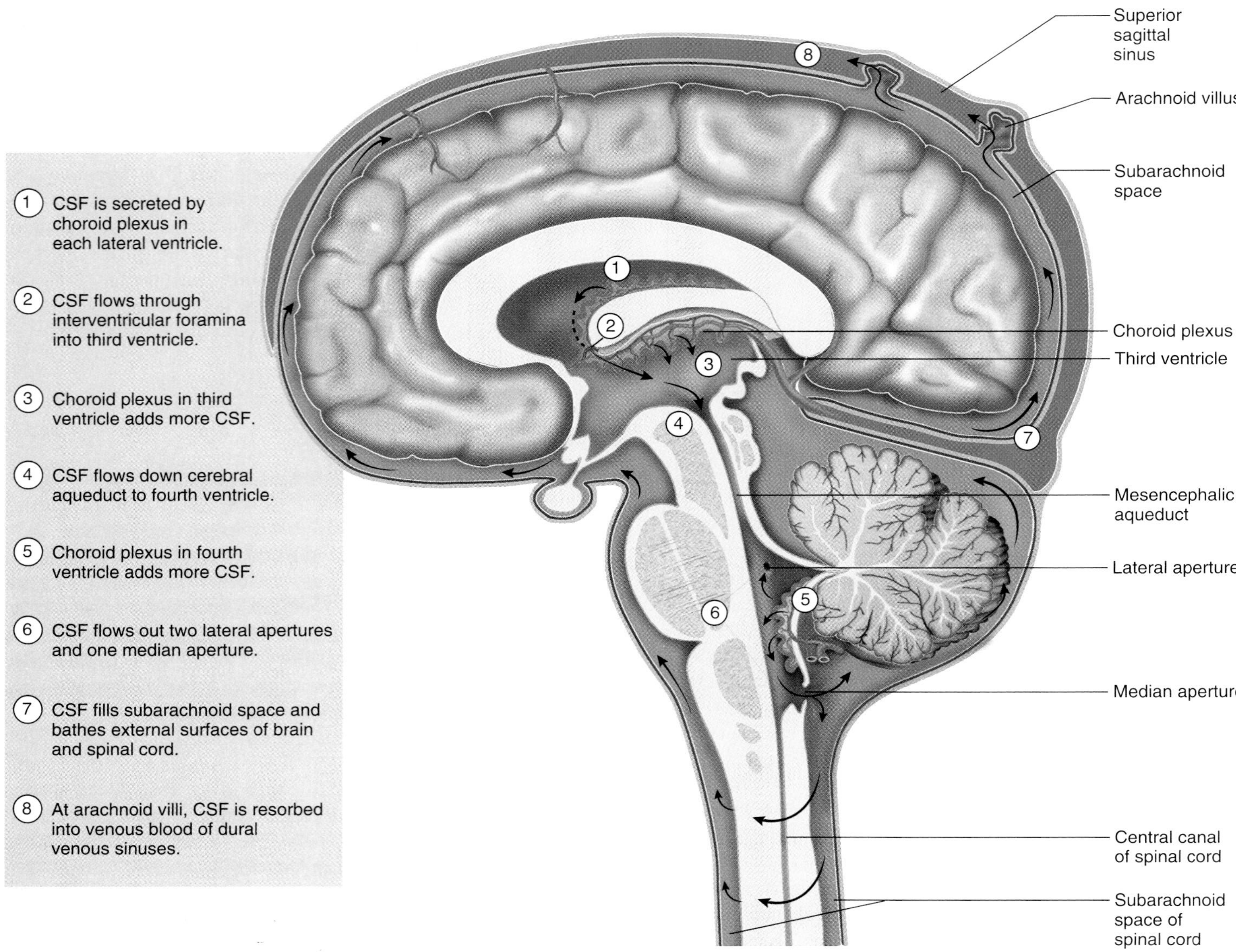

Figure 14.7 The Flow of Cerebrospinal Fluid.

oxygen, carbon dioxide, alcohol, caffeine, nicotine, and anesthetics. It is slightly permeable to sodium, potassium, chloride, and the waste products urea and creatinine. While the BBS is an important protective device, it is an obstacle to the delivery of drugs to treat brain diseases. Some drugs are administered as nasal sprays so they can bypass the BBS by traveling up the olfactory nerve fibers into the brain. Trauma and inflammation sometimes damage the BBS and allow pathogens to enter the brain tissue. Furthermore, there are places called **circumventricular organs (CVOs)** in the third and fourth ventricles where the barrier system is absent and the blood does have direct access to the brain. These enable the brain to monitor and respond to fluctuations in blood glucose, pH, osmolarity, and other variables. Unfortunately, the CVOs also afford a route for the human immunodeficiency virus (HIV, the AIDS virus) to invade the brain.

Before You Go On

Answer the following questions to test your understanding of the preceding section:

4. Name the three meninges from superficial to deep.
5. Describe three functions of the cerebrospinal fluid.
6. Where does the CSF originate and what route does it take through and around the CNS?
7. Name the two components of the brain barrier system and explain the importance of this system.

The Spinal Cord

Objectives

When you have completed this section, you should be able to

- describe the gross and microscopic anatomy of the spinal cord; and
- name the major conduction pathways of the spinal cord and state their functions.

Functions

The spinal cord serves three principal functions:

1. **Conduction.** The white matter consists of bundles of nerve fibers that conduct information up and down the cord. It allows sensory information to reach the brain, motor commands to reach the effectors, and input received at one level of the cord to affect output from another level.
2. **Locomotion.** Walking involves repetitive, coordinated contractions of several muscle groups in the limbs. Motor neurons in the brain initiate walking and determine its speed, distance, and direction, but the simple repetitive muscle contractions that put one foot in front of another, over and over, are coordinated by **central pattern generators,** neuronal pools in the gray matter of the cord. These circuits produce the necessary sequence of outputs to the extensor and flexor muscles to cause alternating movements of the legs.
3. **Reflexes.** Reflexes are involuntary stereotyped responses to stimuli. They involve the brain, spinal cord, and peripheral nerves and are discussed in detail in the next chapter.

Gross Anatomy

The **spinal cord** is a ropelike bundle of nervous tissue that begins at the foramen magnum and passes through the vertebral canal (fig. 14.8) as far as the first lumbar vertebra (L1). It is about as thick as a finger (1.8 cm on average) and about 45 cm long in the adult. Early in fetal development, the spinal cord extends for the full length of the vertebral column. However, the vertebral column grows faster than the spinal cord, so the cord extends only to L3 by the time of birth and to L1 in an adult. Thus, it occupies only the upper two-thirds of the vertebral canal; the lower one-third is described shortly. The cord gives rise to 31 pairs of *spinal nerves* that pass through the intervertebral foramina. Although the spinal cord is not visibly segmented, the part supplied by each pair of spinal nerves is called a *segment.* The cord exhibits longitudinal grooves on its anterior and posterior sides—the *anterior median fissure* and *posterior median sulcus,* respectively.

The spinal cord is divided into **cervical, thoracic, lumbar,** and **sacral regions.** It may seem odd that it has a sacral region when the cord itself ends well above the sacrum. These regions, however, are named for the level of the vertebral column from which the spinal nerves emerge, not for the vertebrae that contain the cord itself.

In the inferior cervical region, a **cervical enlargement** of the cord gives rise to nerves of the upper limbs. In the lumbosacral region, there is a similar **lumbar enlargement** where nerves to the pelvic region and lower limbs arise. Inferior to the lumbar enlargement, the cord tapers to a point called the **medullary cone.** The pia mater continues beyond this point as a fibrous strand, the **terminal filum,** which extends to the coccyx and anchors the lower end of the cord. The canal of vertebrae L2 to S5 contains a bundle of nerve roots that arise from the lumbar enlargement and medullary cone. This bundle, named the **cauda equina**[23] (CAW-duh ee-KWY-nah) for its resemblance to a horse's tail, innervates the pelvic organs and lower limbs.

Cross-Sectional Anatomy

Figure 14.5*b* shows the relationship of the spinal cord to a vertebra and spinal nerve, and figure 14.9 shows the spinal cord in more detail. The cord has a central core of gray matter that looks somewhat butterfly- or H-shaped in cross sections. This is surrounded by white matter. In fresh material, the white matter has a glistening white appearance due to myelin, while the gray matter is somewhat duller.

Gray Matter

The gray matter consists mainly of two **dorsal (posterior) horns,** which extend toward the dorsolateral surfaces of the cord, and two thicker **ventral (anterior) horns,** which extend toward the ventrolateral surfaces. The right and left sides are connected by a **gray commissure.** In the middle of the commissure is the **central canal,** which is collapsed in most areas of the adult spinal cord, but in some places (and in young children) remains open, lined with ependymal cells, and filled with CSF.

As a spinal nerve approaches the cord, it branches into a *dorsal root* and a *ventral root.* The dorsal root carries sensory nerve fibers, which enter the dorsal horn of the cord and sometimes synapse with an interneuron there. Such interneurons are especially numerous in the cervical and lumbar enlargements and are quite evident in histological sections at these levels. The ventral horns contain the large somas of the somatic motor neurons. Axons from these neurons exit by way of the ventral root of the spinal nerve and lead to the skeletal muscles. The spinal nerve roots are described more fully in the next chapter.

23. *cauda* = tail + *equin* = horse

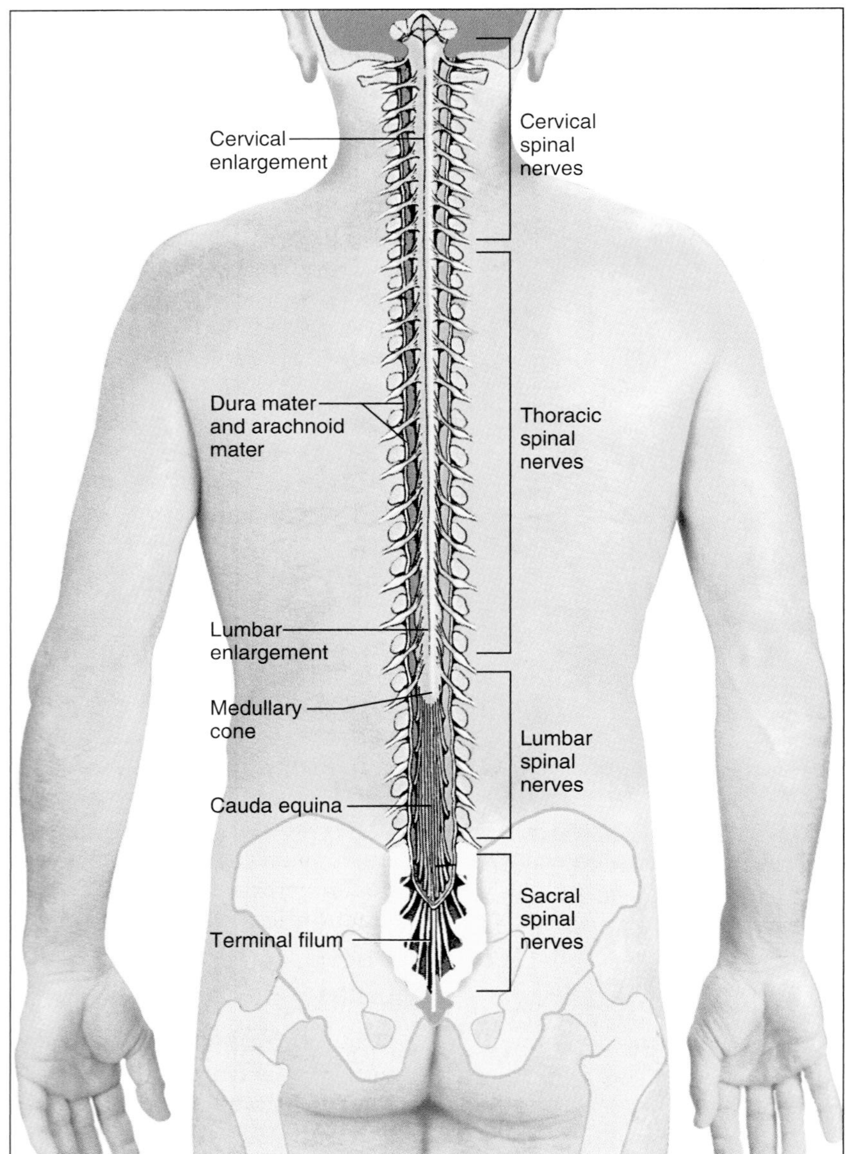

Figure 14.8 The Spinal Cord, Dorsal Aspect.

In the thoracic and lumbar regions, an additional **lateral horn** is visible on each side of the gray matter. It contains neurons of the sympathetic nervous system, which send their axons out of the cord by way of the ventral root along with the somatic efferent fibers.

White Matter

The white matter of the spinal cord consists of bundles of myelinated axons that course up and down the cord and provide avenues of communication between different levels of the CNS. These bundles are arranged in three pairs called **columns** or **funiculi**[24] (few-NIC-you-lie)—a **dorsal (posterior), lateral,** and **ventral (anterior) column** on each side. Each column consists of subdivisions called **tracts** or **fasciculi**[25] (fah-SIC-you-lye).

24. *funicul* = little rope, cord
25. *fascicul* = little bundle

Spinal Tracts

Knowledge of the locations and functions of the spinal tracts is essential in diagnosing and managing spinal cord injuries. **Ascending tracts** carry sensory information up the cord and **descending tracts** conduct motor impulses down. All nerve fibers in a given tract have a similar origin, destination, and function. Many of the fibers exhibit **decussation**[26] (DEE-cuh-SAY-shun)—they cross from one side of the body to the other within the spinal cord or medulla oblongata. As a result, each cerebral hemisphere receives sensory information from and transmits motor commands to the **contralateral**[27] (opposite) side of the body from the neck down. That is, the left hemisphere senses and controls the right side of the body and vice versa. A stroke that damages motor regions of the right cerebral cortex therefore tends to cause paralysis on the left side of the body and vice versa. Most sensation and control in the head and neck and some of that in the lower body is, however, **ipsilateral.**[28] That is, the right brain controls the right side of the body at this level, because signals are carried through nerve fibers that do not decussate.

26. *decuss* = to cross, form an X

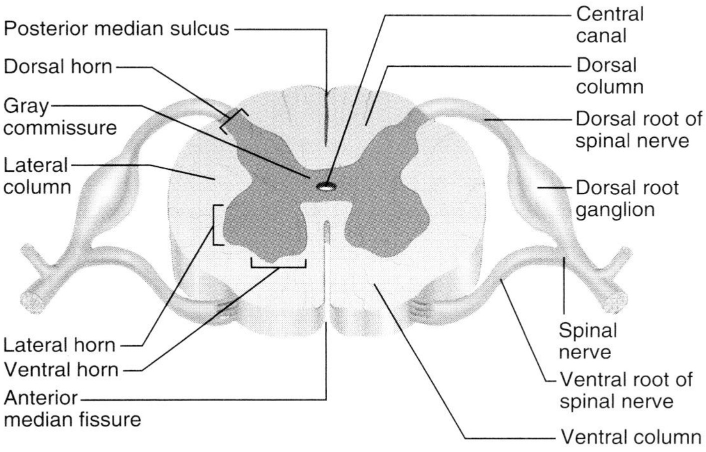

Figure 14.9 Cross Section of the Thoracic Spinal Cord Showing Its Relationship to the Spinal Nerves.

Ascending Tracts

The major spinal cord tracts are summarized in table 14.1 and figure 14.10. Bear in mind that each tract is repeated on the right and left sides of the spinal cord. The ascending tracts are described first, and selected ascending pathways are shown in figure 14.11. The names of most ascending tracts consist of the prefix *spino-* followed by a root denoting the destination of its fibers.

When you become aware of a touch or pain, the sensory signals will have traveled a three-neuron pathway from their origin in the receptors to their destination in the sensory areas of your brain: a **first-order neuron** that

27. *contra* = opposite + *later* = side
28. *ipsi* = same

Table 14.1 Major Spinal Tracts

Tract	Column	Decussation	Functions
Ascending (sensory) Tracts			
Gracile fasciculus	Dorsal	In medulla	Limb and trunk position and movement, deep touch, vibration, below level T6
Cuneate fasciculus	Dorsal	In medulla	Same as gracile fasciculus, from level T6 up
Spinothalamic	Lateral and ventral	In spinal cord	Light touch, tickle, itch, temperature, pain, and pressure
Dorsal spinocerebellar	Lateral	None	Feedback from muscles (proprioception)
Ventral spinocerebellar	Lateral	In spinal cord	Same as dorsal spinocerebellar
Descending (motor) Tracts			
Lateral corticospinal	Lateral	In medulla	Fine control of limbs
Ventral corticospinal	Ventral	None	Fine control of limbs
Tectospinal	Lateral and ventral	In midbrain	Reflexive head-turning in response to visual and auditory stimuli
Lateral reticulospinal	Lateral	None	Balance and posture; regulation of awareness of pain
Medial reticulospinal	Ventral	None	Same as lateral reticulospinal
Vestibulospinal	Ventral	None	Balance and posture

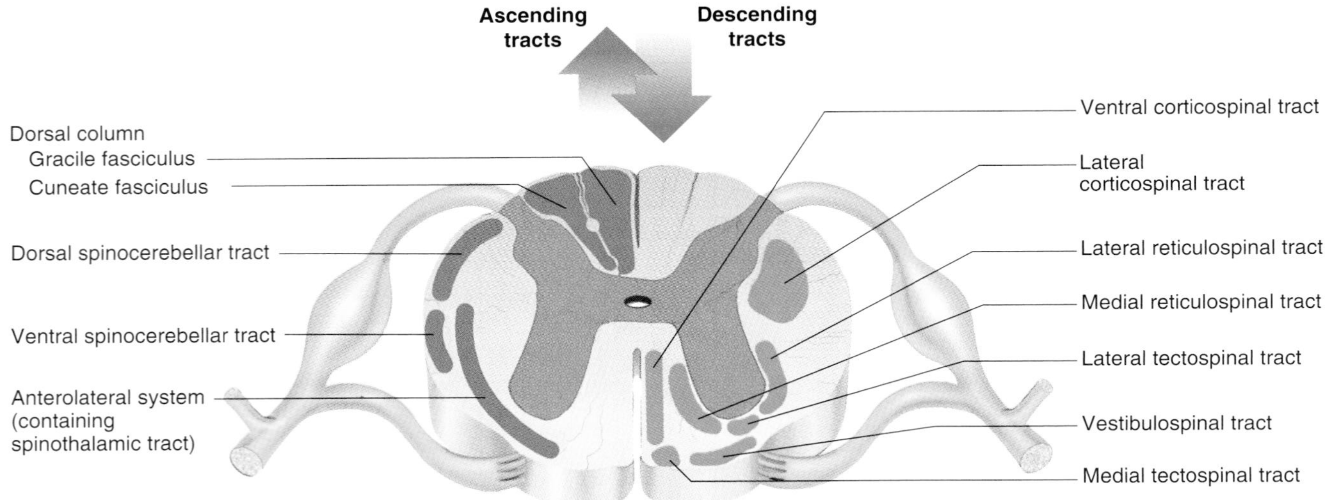

Figure 14.10 **Tracts of the Spinal Cord.** All of the illustrated tracts occur on both sides of the cord, but only the ascending sensory tracts are shown on the *left* (red), and only the descending motor tracts on the *right* (green).

detects a stimulus and transmits a signal to the spinal cord or brainstem, a **second-order neuron** that continues as far as a "gateway" called the *thalamus* at the rostral end of the brainstem, and a **third-order neuron** that carries the signal the rest of the way to the sensory region of the cerebral cortex. The axons of these neurons are called the first- through third-order nerve fibers.

- The **gracile**[29] **fasciculus** (GRAS-el fah-SIC-you-lus) carries signals from the midthoracic and lower parts of the body. Below vertebra T6, it composes the entire dorsal column. At T6, it is joined by the cuneate fasciculus, discussed next. It consists of first-order nerve fibers that travel up the ipsilateral side of the spinal cord and terminate at a nucleus in the medulla oblongata. These fibers carry signals for deep touch, vibration, and especially position and movements of the lower limb and lower trunk. The sense of joint movement and limb position is called *proprioception.*[30]
- The **cuneate**[31] (CUE-nee-ate) **fasciculus** joins the gracile fasciculus at the T6 level. It occupies the lateral portion of the dorsal column and forces the gracile fasciculus medially. It carries the same type of sensory signals, originating from level T6 and up (from the upper limb and chest). Its fibers, too, end in the ipsilateral side of the medulla oblongata. In the medulla, second-order fibers of the gracile and cuneate systems decussate and form the **medial lemniscus**[32] (lem-NIS-cus), a tract of nerve fibers that leads rostrally to the thalamus. Because of this decussation, the signals carried by the gracile and cuneate fasciculi ultimately go to the contralateral cerebral hemisphere.
- The **spinothalamic** (SPY-no-tha-LAM-ic) **tract** and some smaller tracts form the *anterolateral system,* which passes up the anterior and lateral columns of the spinal cord. The spinothalamic tract carries signals for pain, temperature, pressure, light touch, tickle, and itch. In this pathway, first-order neurons end in the dorsal horn of the spinal cord near the point of entry. Second-order neurons decussate to the opposite side of the spinal cord and there form the ascending spinothalamic tract. These fibers lead all the way to the thalamus. Third-order neurons continue from there to the cerebral cortex.
- The **dorsal** and **ventral spinocerebellar** (SPY-no-SERR-eh-BEL-ur) **tracts** travel through the lateral column and carry proprioceptive signals from the limbs and trunk to the cerebellum. The first-order neurons of this system originate in the muscles and tendons and end in the dorsal horn of the spinal cord. Second-order neurons send their fibers up the spinocerebellar tracts and end in the cerebellum. Fibers of the dorsal tract travel up the ipsilateral side of the spinal cord. Those of the ventral tract cross over and travel up the contralateral side but then cross back in the brainstem to enter the ipsilateral cerebellum.

29. *gracil* = thin, slender
30. *proprio* = one's own + *cept* = receive, sense
31. *cune* = wedge
32. *lemniscus* = ribbon

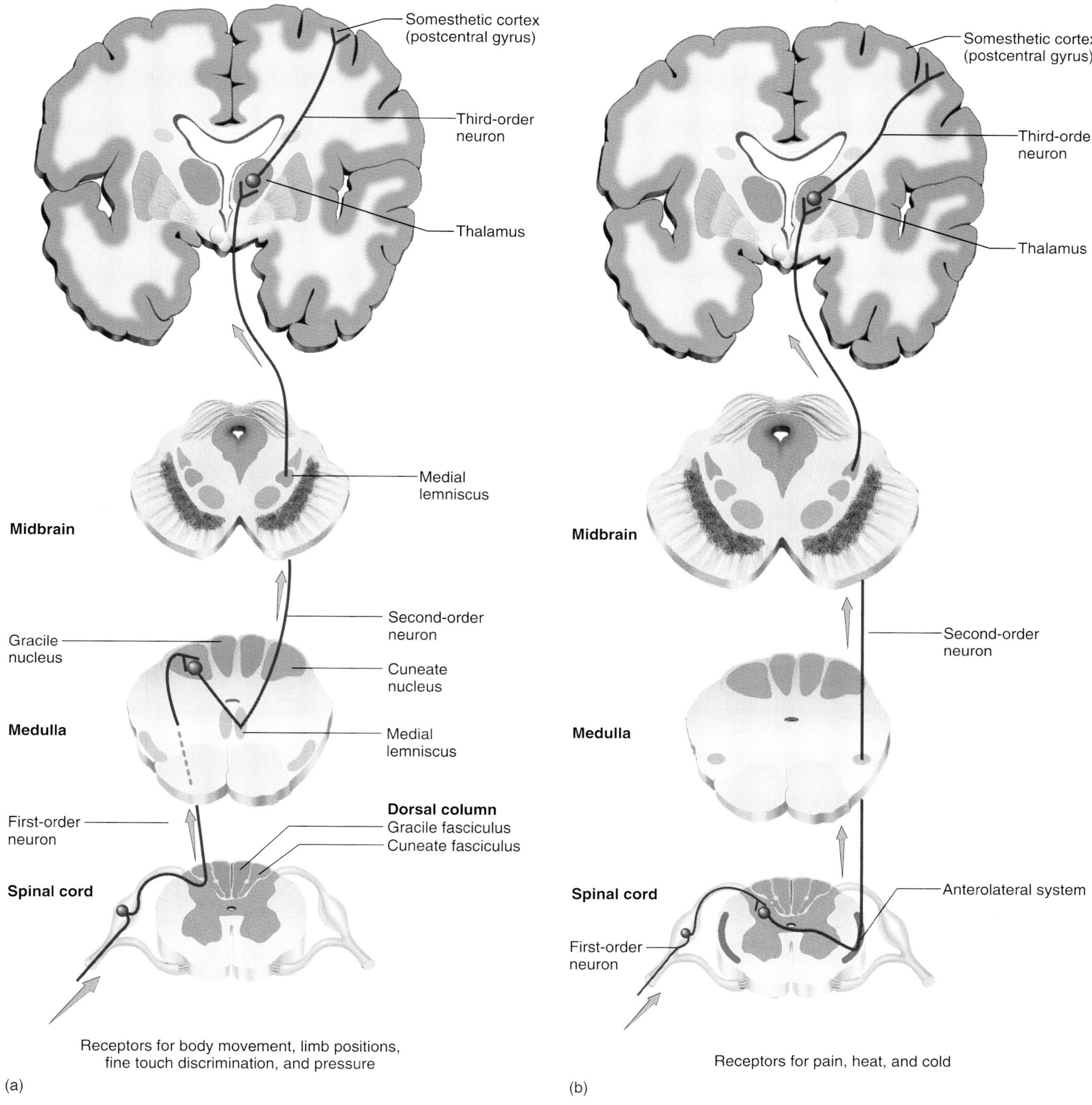

Figure 14.11 **Some Ascending Pathways of the CNS.** The spinal cord, medulla, and midbrain are shown in cross section and the cerebrum and thalamus *(top)* in frontal section. Nerve signals enter the spinal cord at the *bottom* of the figure and carry somatosensory information up to the cerebral cortex. (*a*) The cuneate fasciculus and medial lemniscus; (*b*) the spinothalamic tract.

Both tracts provide the cerebellum with feedback needed to coordinate muscle action, as discussed later in this chapter.

Descending Tracts

The names of most descending tracts consist of a word root denoting the point of origin in the brain followed by *-spinal.* The major descending tracts are described here, and some of their nervous pathways are shown in figure 14.12:

- The **corticospinal** (COR-tih-co-SPY-nul) **tracts** carry motor signals from the cerebral cortex for precise, finely coordinated limb movements. The fibers of this system form pyramid-shaped ridges on the anterior surface of the medulla oblongata; they were once called *pyramidal tracts.* Most corticospinal fibers decussate in the lower medulla and form the **lateral corticospinal tract** on the contralateral side of the spinal cord. A few fibers remain uncrossed and form the **ventral corticospinal tract** on the ipsilateral side. Fibers of the ventral tract decussate lower in the spinal cord, however, so even they control contralateral muscles.
- The **tectospinal** (TEC-toe-SPY-nul) **tract** begins in the midbrain and crosses to the contralateral side of the brainstem. In the lower medulla, it branches into *lateral* and *medial tectospinal tracts* of the upper spinal cord. These are involved in reflex movements of the head, especially in response to visual and auditory stimuli.
- The **lateral** and **medial reticulospinal** (reh-TIC-you-lo-SPY-nul) **tracts** originate in the *reticular formation* of the brainstem. They control muscles of the upper and lower limbs, especially to maintain posture and balance. They also contain *descending analgesic pathways* that reduce the transmission of pain signals to the brain (see chapter 16).
- The **vestibulospinal** (vess-TIB-you-lo-SPY-nul) **tract** begins in a *vestibular nucleus* that receives impulses for balance from the inner ear. The tract passes down the ventral column of the spinal cord and controls limb muscles that maintain balance and posture.

Rubrospinal tracts are prominent in other mammals, where they aid in muscle coordination. Although often pictured in illustrations of human anatomy, they are almost nonexistent in humans and have little functional importance.

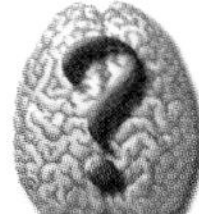

Think About It

You are blindfolded and either a tennis ball or an iron ball is placed in your right hand. What spinal tract(s) would carry the signals that enable you to discriminate between these two objects?

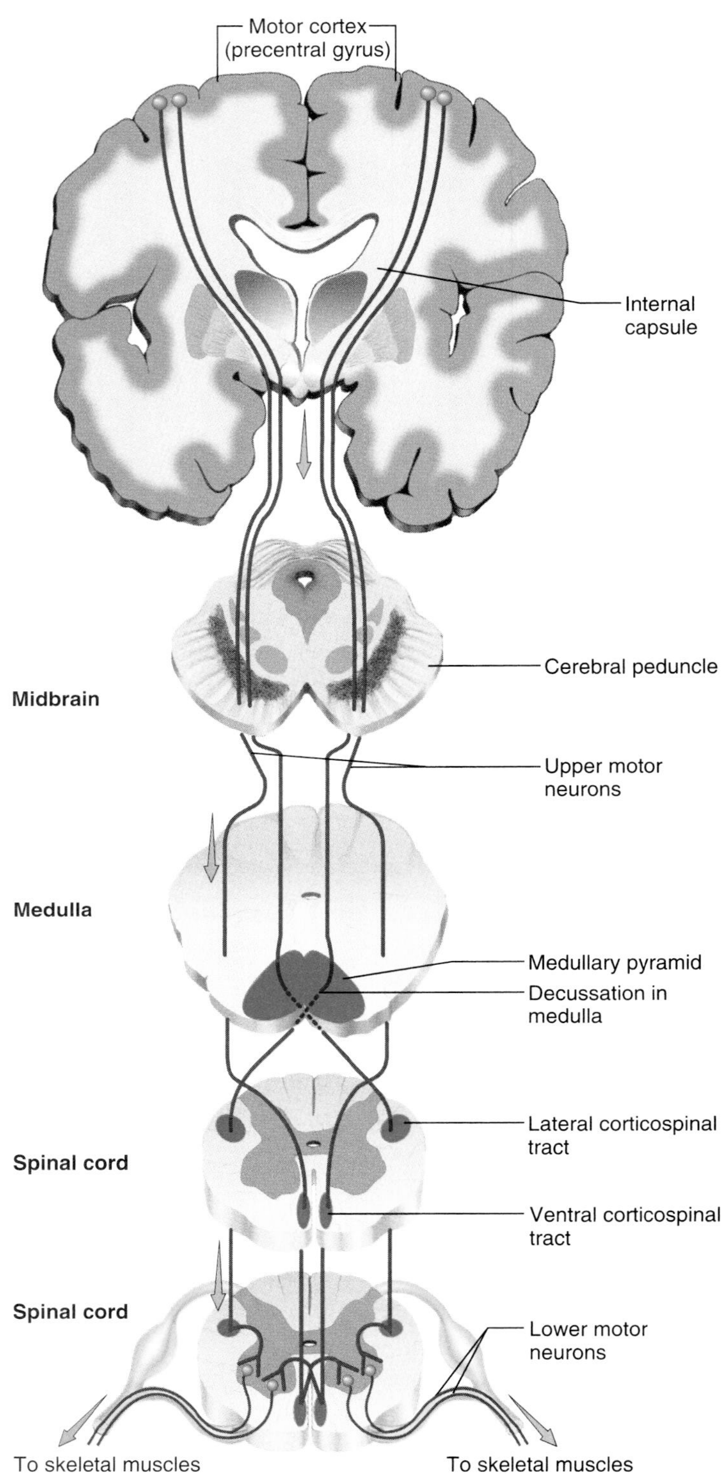

Figure 14.12 **Two Descending Pathways of the CNS, the Corticospinal Tracts.** Nerve signals originate in the cerebral cortex at the *top* of the figure and carry motor commands down the spinal cord. These tracts carry signals for voluntary muscle contraction.

Clinical Insight 14.3

Poliomyelitis and Amyotrophic Lateral Sclerosis

Poliomyelitis[33] and *amyotrophic lateral sclerosis*[34] (ALS) are two diseases that involve destruction of motor neurons. The first is caused by the poliovirus. The cause of ALS is uncertain but also may be viral. In both diseases, the skeletal muscles atrophy from lack of innervation.

The poliovirus destroys motor neurons in the brainstem and ventral horn of the spinal cord. Signs include muscle pain, weakness, and loss of some reflexes, followed by paralysis, muscular atrophy, and sometimes respiratory arrest. The virus spreads by fecal contamination of water. Historically, polio afflicted mainly children, who sometimes contracted the virus in the summer by swimming in contaminated pools. The polio vaccine has nearly eliminated new cases.

ALS is also known as Lou Gehrig disease after the baseball player who contracted it. It is marked not only by the degeneration of motor neurons and atrophy of the muscles, but also sclerosis of the lateral regions of the spinal cord—hence its name. Its early signs include muscular weakness and difficulty in speaking, swallowing, and using the hands. Sensory and intellectual functions remain unaffected, as evidenced by the accomplishments of astrophysicist and best-selling author Stephen Hawking, who was stricken with ALS while he was in college. Despite near-total paralysis, he remains highly productive and communicates with the aid of a speech synthesizer and computer. Tragically, many people are quick to assume that those who have lost most of their ability to communicate their ideas and feelings have no ideas and feelings to communicate. To a victim, this may be more unbearable than the loss of motor function itself.

Stephen Hawking (1942–), Lucasian Professor of Mathematics at Cambridge University.

33. *polio* = gray matter + *myel* = spinal cord + *itis* = inflammation
34. *a* = without + *myo* = muscle + *troph* = nourishment

Before You Go On

Answer the following questions to test your understanding of the preceding section:

8. Name the four major regions and two enlargements of the spinal cord.
9. Describe the distal (inferior) end of the spinal cord and the contents of the vertebral canal from level L2 to S5.
10. Sketch a cross section of the spinal cord showing the dorsal and ventral horns. Where are the gray and white matter? Where are the columns and tracts?
11. Give an anatomical explanation as to why a stroke in the right cerebral hemisphere can paralyze the limbs on the left side of the body.

The Hindbrain and Midbrain

Objectives

When you have completed this section, you should be able to

- list the components of the hindbrain and midbrain and their functions;
- discuss the role of the cerebellum in movement and equilibrium;
- define the term *brainstem* and describe its anatomical relationship to the cerebellum and forebrain; and
- describe the location and functions of the reticular formation.

Our study of the brain is organized around the five secondary vesicles of the embryonic brain and their mature derivatives. We proceed in a caudal to rostral direction, beginning with the hindbrain and its relatively simple functions and progressing to the forebrain, the seat of such complex functions as thought, memory, and emotion.

The Medulla Oblongata

As noted earlier, the embryonic hindbrain differentiates into two subdivisions, the myelencephalon and metencephalon (see fig. 14.4). The myelencephalon develops into just one structure, the **medulla oblongata** (meh-DULL-uh OB-long-GAH-ta). The medulla is about 3 cm long and superficially looks like an extension of the spinal cord, but slightly wider. Significant differences are seen, however, on closer inspection of its gross and microscopic anatomy.

The anterior surface bears a pair of clublike ridges, the **pyramids,** that contain the nerve fibers of the corticospinal (pyramidal) tracts. Resembling two side-by-side baseball bats, the pyramids are wider at the rostral end, taper caudally, and are separated by an *anterior median fissure* continuous with that of the spinal cord (fig. 14.13). Most pyramidal nerve fibers decussate at a visible point near the caudal end. Lateral to each pyramid is an elevated area called the **olive.** It contains a wavy layer of gray matter, the **inferior olivary nucleus,** which is a center that receives information from many levels of the brain and spinal cord and relays it mainly to the cerebellum.

The medulla contains ascending and descending nerve tracts passing through the brainstem; sensory nuclei that receive input from the taste buds, pharynx, and viscera of the thoracic and abdominal cavities; and motor nuclei that control several primitive visceral and somatic functions. Some motor nuclei of the medulla discussed later in the book are:

- the **cardiac center,** which adjusts the rate and force of the heartbeat;
- the **vasomotor center,** which adjusts blood vessel diameter to regulate blood pressure and reroute blood from one part of the body to another; and
- two **respiratory centers,** which control the rate and depth of breathing.

Other nuclei of the medulla are concerned with speech, coughing, sneezing, salivation, swallowing, gagging, vomiting, gastrointestinal secretion, sweating, and movements of the tongue and head. Many of the medulla's sensory and motor functions are mediated through the last four cranial nerves, which begin or end here: the glossopharyngeal (IX), vagus (X), accessory (XI), and hypoglossal (XII).

The Pons and Cerebellum

The embryonic metencephalon develops into two structures, the pons and cerebellum.

The Pons

The **pons**[35] appears as an anterior bulge in the brainstem rostral to the medulla (fig. 14.13). Its white matter includes tracts that conduct signals from the cerebrum to the cerebellum; corticospinal tracts on their way to the medulla; and the medial lemniscus carrying sensory signals upward to the thalamus. Cranial nerve V (trigeminal) arises from the pons, and cranial nerves VI (abducens), VII (facial), and VIII (vestibulocochlear) arise from the junction of the pons and medulla. The pons contains nuclei that relay signals from the cerebrum to the cerebellum, and nuclei concerned with sleep, hearing, equilibrium, taste, eye movements, facial expressions, facial sensation, respiration, swallowing, bladder control, and posture.

The Cerebellum

The **cerebellum** is the largest part of the hindbrain (fig. 14.14). It consists of right and left **cerebellar hemispheres** connected by a narrow bridgelike **vermis.**[36] Three pairs of stalklike **cerebellar peduncles**[37] (peh-DUN-culs) connect the cerebellum to the brainstem: the *inferior peduncles* to the medulla oblongata, the *middle peduncles* to the pons, and the *superior peduncles* to the midbrain. These are composed of the nerve fibers that carry all signals between the cerebellum and the rest of the brain. The cerebellum receives most of its input from the pons, via the middle peduncles.

Each hemisphere exhibits slender, parallel folds called **folia**[38] (gyri) separated by shallow sulci. The cerebellum has a surface cortex of gray matter and a deeper layer of white matter. In a sagittal section, the white matter, called the **arbor vitae,**[39] exhibits a branching, fernlike pattern. Each hemisphere has four **deep nuclei,** masses of gray matter embedded in the white matter. All input to the cerebellum goes to the cortex and all of its output comes from the deep nuclei.

The most distinctive neurons of the cerebellum are the **Purkinje**[40] (pur-KIN-jee) **cells**—unusually large, globose neurons with a tremendous profusion of dendrites (see fig. 13.4, p. 458), arranged in a single row in the cortex. Their axons travel to the deep nuclei, where they synapse on output neurons that issue fibers to the brainstem.

The cerebellum is mostly concerned with muscular coordination and will be discussed in a later section on motor control. Evidence is also emerging that the cerebellum plays a role in judging the passage of time, in some other cognitive (thought) processes, and in emotion.

The Midbrain

The embryonic mesencephalon develops into just one mature brain structure, the **midbrain**—a short segment of the

35. *pons* = bridge

36. *verm* = worm
37. *ped* = foot + *uncle* = little
38. *foli* = leaf
39. "tree of life"
40. Johannes E. von Purkinje (1787–1869), Bohemian anatomist

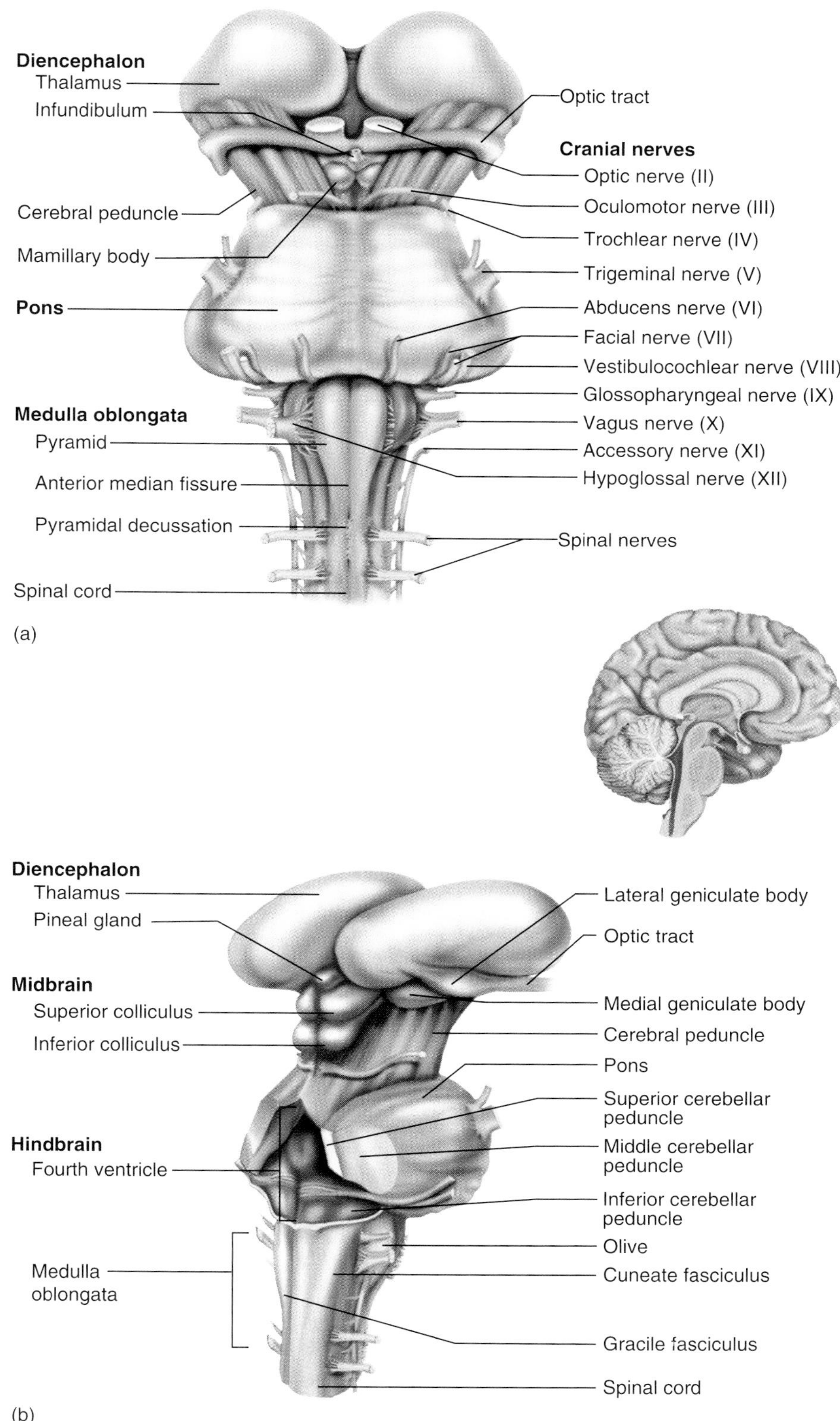

Figure 14.13 **The Brainstem.** (*a*) Anterior aspect; (*b*) right oblique aspect. Some authorities do not include the diencephalon in the brainstem.

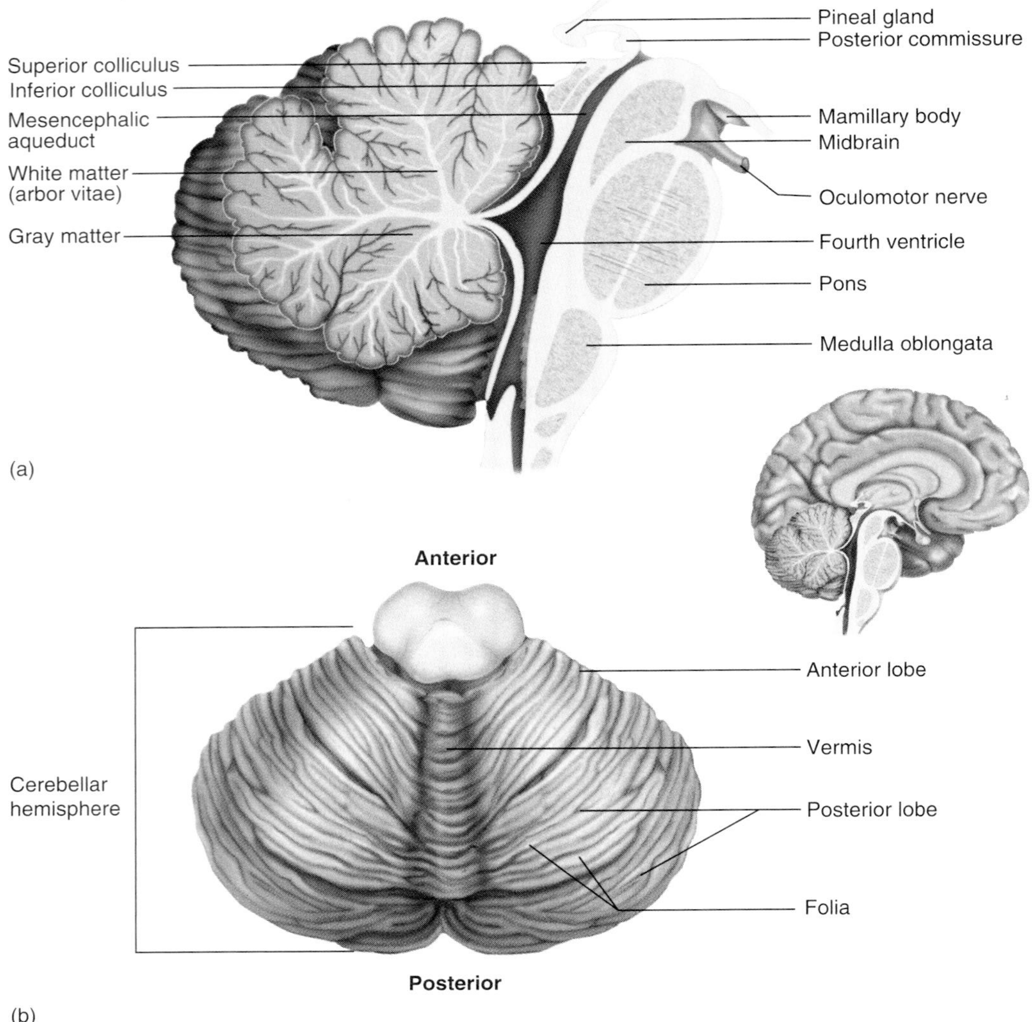

Figure 14.14 The Cerebellum. (*a*) Midsagittal section showing relationship to brainstem; (*b*) superior aspect.

brainstem that connects the hindbrain and forebrain (see figs. 14.1*d*, 14.2*b*, and 14.13). It contains the mesencephalic aqueduct and gives rise to two cranial nerves that control eye movements: the oculomotor (III) and trochlear (IV). Some major regions of the midbrain are (fig. 14.15):

- The **cerebral peduncles,** which help to anchor the cerebrum to the brainstem and contain the corticospinal tracts described earlier.
- The **tegmentum,**[41] the main mass of the midbrain, located between the cerebral peduncles and mesencephalic aqueduct. It contains the **red nucleus,** which has a pink color in life because of its high density of blood vessels. Fibers from the red nucleus form the rubrospinal tract in most mammals, but in humans its connections go mainly to and from the cerebellum, with which it collaborates in fine motor control.
- The **substantia nigra**[42] (sub-STAN-she-uh NY-gruh), a dark gray to black nucleus pigmented with melanin, located between the peduncles and the tegmentum. This is a motor center that relays inhibitory signals to the thalamus and basal nuclei (both of which are discussed later). Degeneration of the neurons in the substantia nigra leads to the muscle tremors of Parkinson disease (see clinical insight 13.4, p. 485).
- The **central (periaqueductal) gray matter,** a large arrowhead-shaped region of gray matter surrounding

41. *tegmen* = cover

42. *substantia* = substance + *nigra* = black

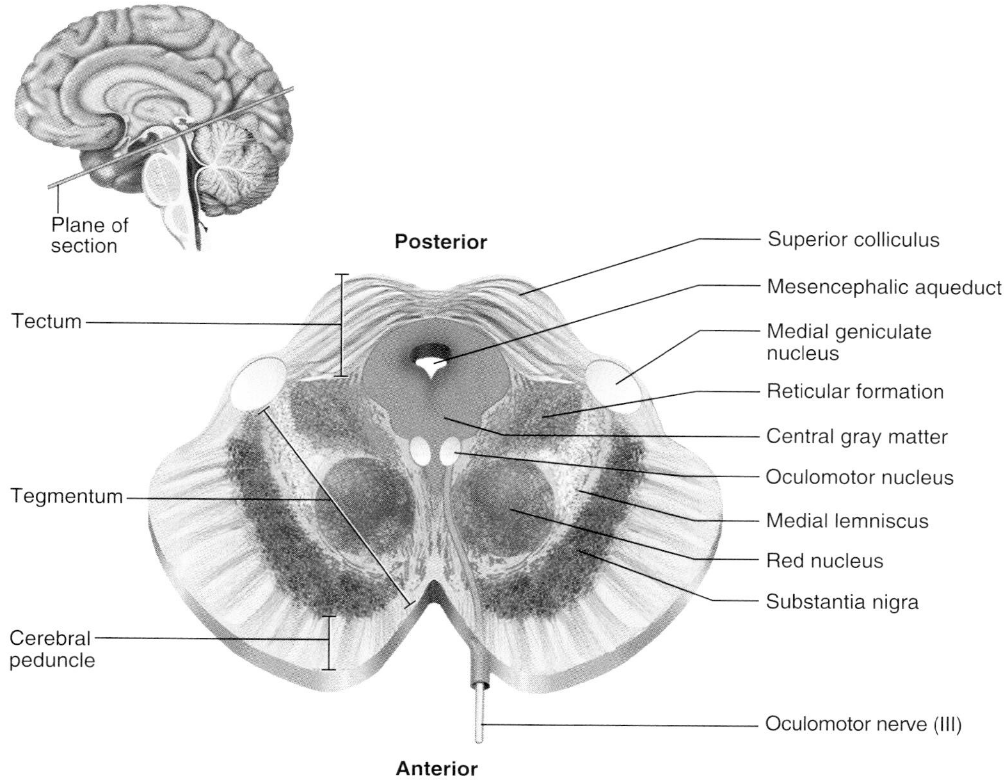

Figure 14.15 Cross Section of the Midbrain.

the mesencephalic aqueduct. It is involved with the reticulospinal tracts in controlling our awareness of pain (see chapter 16).

- The **tectum,**[43] a rooflike region dorsal to the aqueduct. It consists of four nuclei, the **corpora quadrigemina,**[44] which bulge from the midbrain roof. The two superior nuclei, called the **superior colliculi**[45] (col-LIC-you-lye), function in visual attention, visually tracking moving objects, and such reflexes as turning the eyes and head in response to a visual stimulus, for example to look at something that you catch sight of in your peripheral vision. The two **inferior colliculi** receive all afferent signals from the inner ear and relay them to other parts of the brain, especially the thalamus. Among other functions, they mediate the reflexive turning of the head in response to a sound.

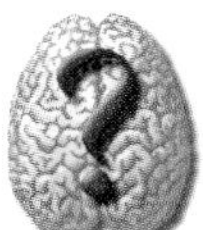

Think About It

Why are the inferior colliculi shown in figure 14.14 but not in figure 14.15? How are these two figures related?

43. *tectum* = roof, cover
44. *corpora* = bodies + *quadrigemina* = quadruplets
45. *colli* = hill + *cul* = little

The Reticular Formation

Running vertically through the core of the midbrain, pons, and medulla is a loosely organized core of gray matter called the **reticular formation,** composed of more than 100 small clusters of neurons mingled with bundles of nerve fibers (fig. 14.16). The functions of the reticular formation include the following:

- **Somatic motor control.** Some motor neurons of the cerebral cortex send their axons to reticular formation nuclei, which then give rise to the reticulospinal tracts of the spinal cord. These tracts modulate (adjust) muscle contraction to maintain tonus, balance, and posture. The reticular formation also relays signals from the eyes and ears to the cerebellum so the cerebellum can integrate visual, auditory, and vestibular (balance and motion) stimuli into its role in motor coordination. Other reticular formation motor nuclei include *gaze centers,* which enable the eyes to track and fixate on objects, and central pattern generators that produce the sequential muscle contractions of breathing and swallowing.
- **Cardiovascular control.** The reticular formation includes the cardiac center and vasomotor center of the medulla oblongata.

- **Pain modulation.** The reticular formation is the origin of the descending analgesic pathways mentioned in the earlier description of the reticulospinal tracts.
- **Sleep and consciousness.** The reticular formation has projections to the cerebral cortex and thalamus that allow it some control over what sensory signals reach the cerebrum. If the thalamus is the "gateway" to the cerebrum (as discussed shortly), then the reticular formation is like a "gatekeeper" that decides what sensory information comes to our conscious attention. It plays a central role in states of consciousness such as alertness and sleep. Injury to the reticular formation can result in irreversible coma. General anesthetics work by blocking signal transmission through the reticular formation.

The reticular formation also is involved in **habituation**—a process in which the brain learns to ignore repetitive, inconsequential stimuli while remaining sensitive to others. In a noisy city, for example, a person can sleep through traffic sounds but wake promptly to the sound of an alarm clock or a crying baby. Reticular formation nuclei that modulate activity of the cerebral cortex used to be called the *reticular activating system*, but this term is falling out of use because of the realization that they do more than activate the cortex. More recently, this network has been called the *extrathalamic cortical modulatory system* because its fibers bypass the thalamus and do not merely activate the cerebral cortex, but modulate it (make it more or less sensitive to stimulation under different circumstances).

Before You Go On

Answer the following questions to test your understanding of the preceding section:

12. Name the visceral functions controlled by nuclei of the medulla.
13. Describe the general functions of the cerebellum.
14. What are some functions of the midbrain nuclei?
15. Describe the reticular formation and list several of its functions.

The Forebrain

Objectives

When you have completed this section, you should be able to

- name the three major components of the diencephalon and describe their locations and functions;
- identify the five lobes of the cerebrum;
- describe the three types of tracts in the cerebral white matter;
- describe the distinctive cell types and histological arrangement of the cerebral cortex; and
- describe the location and functions of the basal nuclei and limbic system.

The forebrain consists of the diencephalon and telencephalon. The diencephalon encloses the third ventricle and is the most rostral part of the brainstem. The telencephalon develops chiefly into the cerebrum, perched on top of the brainstem like a mushroom cap.

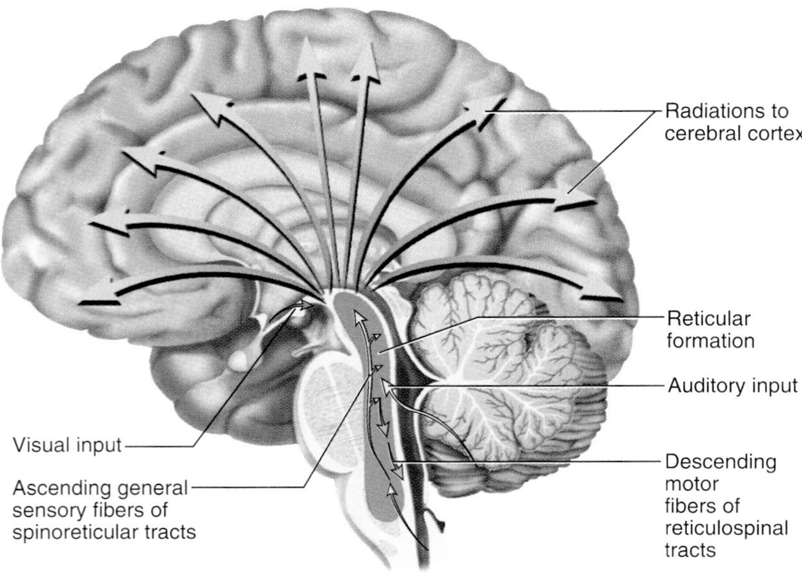

Figure 14.16 The Reticular Formation. The formation consists of over 100 nuclei scattered through the brainstem region indicated in red. *Arrows* represent the breadth of its projections to and from the cerebral cortex and other CNS regions.

The Diencephalon

The embryonic diencephalon has three major derivatives: the *thalamus, hypothalamus,* and *epithalamus.*

The Thalamus

The **thalamus**[46] (fig. 14.17) is about four-fifths of the diencephalon. It consists of an oval mass of gray matter that underlies each cerebral hemisphere, protrudes into the lateral ventricle, and medially protrudes into the third ventricle. In about 70% of people, a narrow *intermediate mass* connects the right and left halves to each other.

The thalamus is the "gateway to the cerebral cortex." Nearly all sensory input and other information going to the cerebrum passes by way of synapses in the thalamus, and the thalamus receives input from several areas of the cerebrum involved in motor control. The thalamus has several nuclei that integrate sensory information originating throughout the body and direct it to the appropriate processing centers in the cerebrum. It is heavily interconnected with the limbic system (discussed later) and thus involved in its emotional and memory functions. It is also involved in arousal, eye movements, taste, smell, hearing, equilibrium, and the somesthetic senses. Its involvement in motor and sensory circuits is further discussed later in this chapter and in chapter 16.

The Hypothalamus

The **hypothalamus** (fig. 14.17) forms part of the walls and floor of the third ventricle. It extends anteriorly to the *optic chiasm* (ky-AZ-um), where the optic nerves meet, and

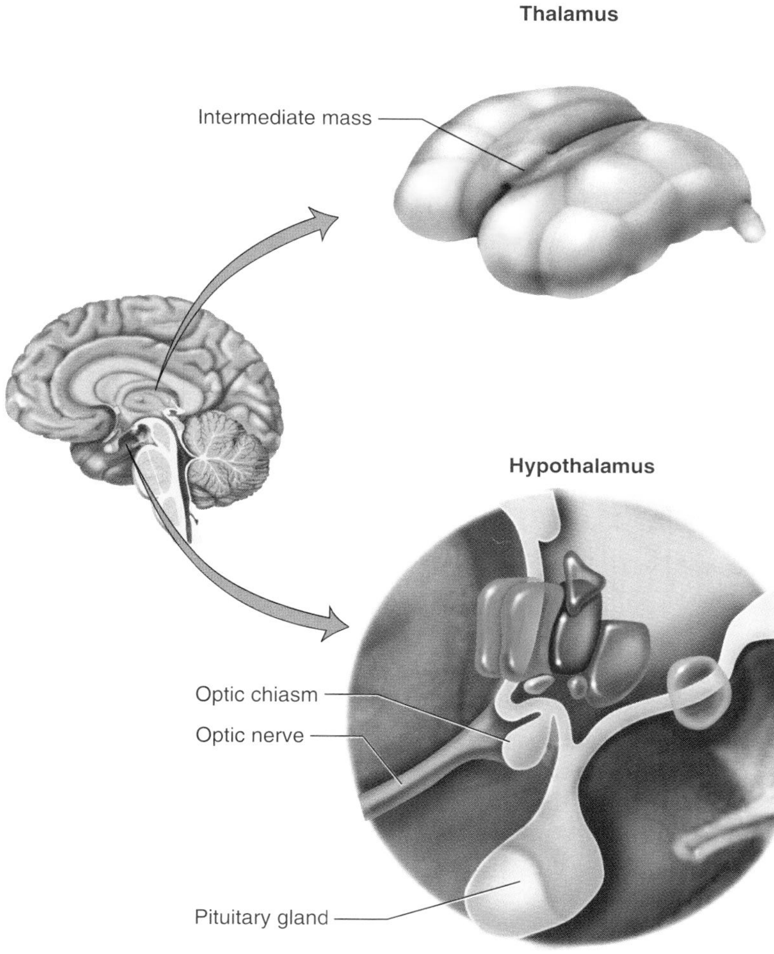

46. *thalamus* = chamber, inner room

Thalamus

Intermediate mass

Lateral group	Somesthetic input to association areas; contributes to limbic system
Medial group	Emotional input to prefrontal cortex; awareness of emotions
Ventral group	Somesthetic input to postcentral gyrus; signals from cerebellum and basal nuclei to primary motor and motor association areas
Anterior group	Part of limbic system
Lateral geniculate nucleus	Visual signals to occipital lobe

Hypothalamus

Optic chiasm

Optic nerve

Pituitary gland

Mamillary body	Relays signals from limbic system to thalamus
Paraventricular nucleus	Produces oxytocin (involved in childbirth, lactation, orgasm); autonomic motor effects; control of posterior pituitary
Anterior nucleus	Thirst center
Ventromedial nucleus	Satiety center (supresses hunger); emotion
Preoptic nucleus	Thermoregulation; control of female reproductive cycle; sexual behavior
Supraoptic nucleus	Produces antidiuretic hormone (involved in water conservation); control of posterior pituitary
Suprachiasmatic nucleus	Biological clock; regulates circardian rhythms and female reproductive cycle

Figure 14.17 **The Diencephalon.** Only some of the nuclei of the thalamus and hypothalamus are shown, and only some of their functions listed. These lists are by no means complete.

posteriorly to a pair of humps called the *mamillary*[47] *bodies.* The mamillary bodies relay signals from the limbic system to the thalamus. The pituitary gland is attached to the hypothalamus by a stalk (*infundibulum*) between the optic chiasm and mamillary bodies.

The hypothalamus is the major control center of the autonomic nervous system and endocrine system and plays an essential role in the homeostatic regulation of nearly all organs of the body. Its nuclei include centers concerned with a wide variety of visceral functions:

- **Hormone secretion.** The hypothalamus secretes hormones that control the anterior pituitary gland. Acting through the pituitary, the hypothalamus regulates growth, metabolism, reproduction, and stress responses. The hypothalamus also produces two hormones that are stored in the posterior pituitary gland, concerned with labor contractions, lactation, and water conservation. These relationships are explored especially in chapter 17.
- **Autonomic effects.** The hypothalamus is a major integrating center for the autonomic nervous system. It sends descending fibers to nuclei lower in the brainstem that influence heart rate, blood pressure, gastrointestinal secretion and motility, and pupillary diameter, among other functions.
- **Thermoregulation.** The *hypothalamic thermostat* is a nucleus that monitors blood temperature. When the temperature becomes too high or too low, the thermostat signals other hypothalamic nuclei—the *heat-losing center* or *heat-producing center,* respectively, which control cutaneous vasodilation and vasoconstriction, sweating, shivering, and piloerection.
- **Food and water intake.** Neurons of the *hunger* and *satiety centers* monitor blood glucose and amino acid levels and produce sensations of hunger and satiety. Hypothalamic neurons called *osmoreceptors* monitor the osmolarity of the blood and stimulate the hypothalamic *thirst center* when we are dehydrated. Dehydration also stimulates the hypothalamus to produce *antidiuretic hormone,* which conserves water by reducing urine output.
- **Sleep and circadian rhythms.** The caudal part of the hypothalamus is part of the reticular formation. It contains nuclei that regulate falling asleep and waking. Superior to the optic chiasm, the hypothalamus contains a *suprachiasmatic nucleus* that controls our 24-hour (circadian) rhythm of activity.
- **Memory.** The mamillary bodies of the hypothalamus lie in the pathway of signals traveling from the hippocampus, an important memory center of the brain, to the thalamus. Thus they are important in memory, and lesions to the mamillary bodies result in memory deficits. (Memory is discussed more fully later in this chapter.)
- **Emotional behavior.** Hypothalamic centers are involved in a variety of emotional responses including anger, aggression, fear, pleasure, and contentment; and in sexual drive, copulation, and orgasm.

The Epithalamus

The **epithalamus** consists mainly of the **pineal gland** (an endocrine gland discussed in chapter 17), the **habenula** (a relay from the limbic system to the midbrain), and a thin roof over the third ventricle.

The Cerebrum

The embryonic telencephalon becomes the cerebrum, the largest and most conspicuous part of the human brain. Your cerebrum enables you to turn these pages, read and comprehend the words, remember ideas, talk about them with your peers, and take an examination. It is the seat of your sensory perception, memory, thought, judgment, and voluntary motor actions. It is the most challenging frontier of neurobiology.

Gross Anatomy

The surface of the cerebrum, including its gray matter (cerebral cortex) and part of the white matter, is folded into gyri that allow a greater amount of cortex to fit in the cranial cavity. These folds give the cerebrum a surface area of about 2,500 cm^2, comparable to 4 1/2 pages of this book. If the cerebrum were smooth-surfaced, it would have only one-third as much area and proportionately less information-processing capability. This extensive folding is one of the greatest differences between the human brain and the relatively smooth-surfaced brains of most other mammals.

Some gyri have consistent and predictable anatomy, while others vary from brain to brain and from the right hemisphere to the left. Certain fissures and unusually deep sulci divide each hemisphere into five anatomically and functionally distinct lobes, listed next. The first four are visible superficially and are named for the cranial bones overlying them (fig. 14.18); the fifth lobe is not visible from the surface.

1. The **frontal lobe** lies immediately behind the frontal bone, superior to the orbits. Its posterior boundary is the **central sulcus.** The frontal lobe is chiefly concerned with voluntary motor functions,

47. *mamill* = nipple

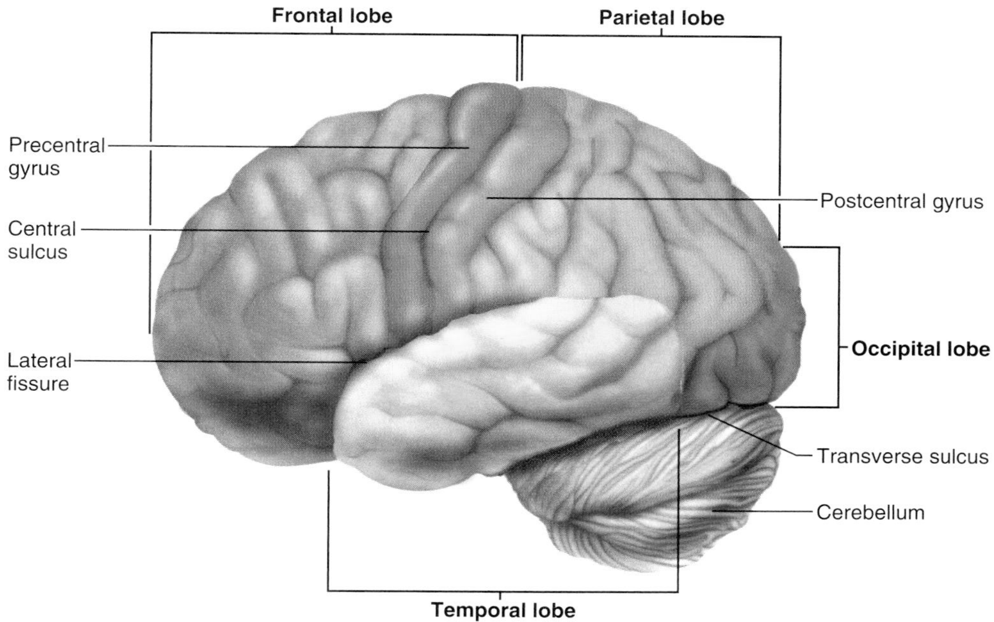

Figure 14.18 Lobes of the Cerebrum. The insula is not visible from the surface (see fig. 14.2*a*).

motivation, foresight, planning, memory, mood, emotion, social judgment, and aggression.

2. The **parietal lobe** forms the uppermost part of the brain and underlies the parietal bone. Its rostral boundary is the central sulcus and its caudal boundary is the relatively indistinct **parieto-occipital sulcus,** visible on the medial surface of each hemisphere. This lobe is concerned with the sensory reception and integration of somesthetic, taste, and some visual information.
3. The **occipital lobe** is at the rear of the head underlying the occipital bone. It is the principal visual center of the brain.
4. The **temporal lobe** is a lateral, horizontal lobe deep to the temporal bone, separated from the parietal lobe above it by a deep **lateral fissure.** It is concerned with hearing, smell, learning, memory, visual recognition, and emotional behavior.
5. The **insula**[48] is a small mass of cortex deep to the lateral sulcus, made visible only by retracting or cutting away some of the overlying cerebrum (see figs. 14.2*a*, 14.6*c*, and 14.21). It is not as accessible to study in living people as other parts of the cortex and is still little-known territory. It apparently plays roles in understanding spoken language, in the sense of taste, and in integrating sensory information from visceral receptors.

The Cerebral White Matter

Most of the volume of the cerebrum is white matter. This is composed of glia and myelinated nerve fibers that transmit signals from one region of the cerebrum to another and between the cerebrum and lower brain centers. These fibers are arranged in three types of tracts (fig. 14.19):

1. **Projection tracts** extend vertically from higher to lower brain or spinal cord centers and carry information between the cerebrum and the rest of the body. The corticospinal tracts, for example, carry motor signals from the cerebrum to the brainstem and spinal cord. Other projection tracts carry signals upward to the cerebral cortex. Superior to the brainstem, such tracts form a dense band called the *internal capsule* between the thalamus and basal nuclei (described shortly), then radiate in a diverging, fanlike array (the *corona radiata*[49]) to specific areas of the cortex.
2. **Commissural tracts** cross from one cerebral hemisphere to the other through bridges called **commissures** (COM-ih-shurs). The great majority of commissural fibers pass through the large C-shaped corpus callosum (see fig. 14.1*d*), which forms the floor of the longitudinal fissure. A few tracts pass through the much smaller **anterior** and **posterior**

48. *insula* = island

49. *corona* = crown + *radiata* = radiating

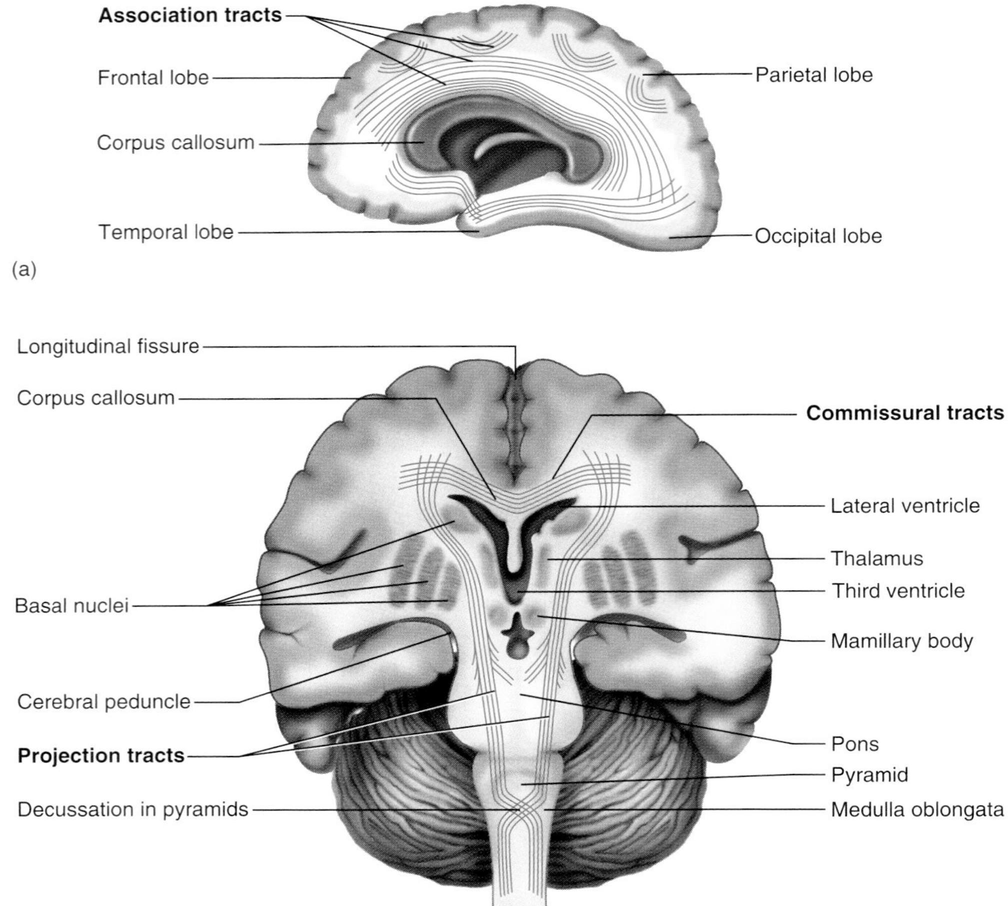

Figure 14.19 Tracts of Cerebral White Matter. (*a*) Left lateral aspect showing association tracts; (*b*) frontal section showing commissural and projection tracts.

commissures. Commissural tracts enable the two sides of the cerebrum to communicate with each other.

3. **Association tracts** connect different regions of the same hemisphere. *Long association fibers* connect different lobes of a cerebral hemisphere to each other, whereas *short association fibers* connect different gyri within a single lobe. Among their roles, association tracts link perceptual and memory centers of the brain; for example, they enable you to smell a rose, name it, and picture what it looks like.

The Cerebral Cortex

Neural integration is carried out in the gray matter of the cerebrum, which is found in three places—the cerebral cortex, basal nuclei, and limbic system. We begin with the **cerebral cortex,**[50] a layer covering the surface of the hemispheres. Even though it is only 2 to 3 mm thick, the cortex constitutes about 40% of the mass of the brain. It is composed of two principal types of neurons (fig. 14.20). **Stellate cells** have spheroidal somas with dendrites projecting for short distances in all directions. They are concerned largely with receiving sensory input and processing information on a local level. **Pyramidal cells** are tall and conical (triangular in tissue sections). Their apex points toward the brain surface and has a thick dendrite with many branches and small, knobby *dendritic spines.* The base gives rise to horizontally oriented dendrites and an axon that passes into the white matter. Pyramidal cells are the output neurons of the cerebrum—they transmit signals to other parts of the CNS. Their axons have collaterals that synapse with other neurons in the cortex or in deeper regions of the brain.

50. *cortex* = bark, rind

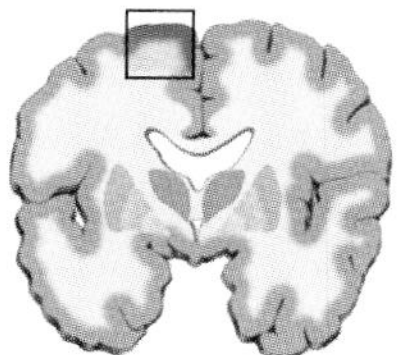

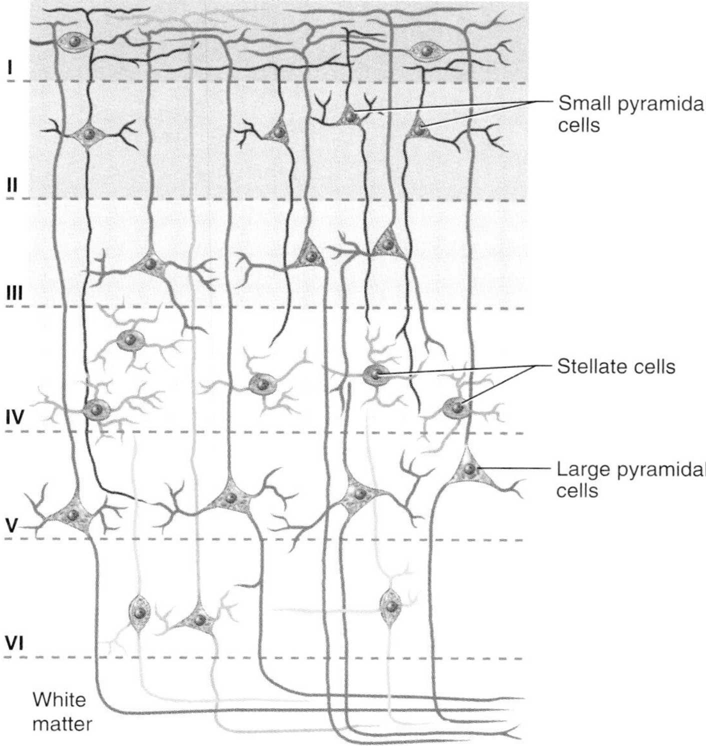

Figure 14.20 Histology of the Neocortex. Neurons are arranged in six layers.

About 90% of the human cerebral cortex is a six-layered tissue called **neocortex**[51] because of its relatively recent evolutionary origin. Although vertebrates have existed for about 600 million years, the neocortex did not develop significantly until about 60 million years ago, when there was a sharp increase in the diversity of mammals. It attained its highest development by far in the primates. The six layers of neocortex, numbered in figure 14.20, vary from one part of the cerebrum to another in relative thickness, cellular composition, synaptic connections, size of the neurons, and destination of their axons. Layer IV is thickest in sensory regions and layer V in motor regions, for example. All axons that leave the cortex and enter the white matter arise from layers III, V, and VI.

51. *neo* = new

Some regions of cerebral cortex have fewer than six layers. The earliest type of cerebral cortex to appear in vertebrate evolution was a one- to five-layered tissue called *paleocortex* (PALE-ee-oh-cor-tex), limited in humans to part of the insula and certain areas of the temporal lobe concerned with smell. The next to evolve was a three-layered *archicortex* (AR-kee-cor-tex), found in the human hippocampus. The neocortex was the last to evolve.

The Basal Nuclei

The **basal nuclei** are masses of cerebral gray matter buried deep in the white matter, lateral to the thalamus (fig. 14.21). They are often called *basal ganglia,* but the word *ganglion* is best restricted to clusters of neurons outside the CNS. Neuroanatomists disagree on how many brain centers to classify as basal nuclei, but agree on at least three: the **caudate**[52] **nucleus, putamen,**[53] and **globus pallidus.**[54] The putamen and globus pallidus are also collectively called the *lentiform*[55] *nucleus,* while the putamen and caudate nucleus are collectively called the *corpus striatum* after their striped appearance. The basal nuclei receive input from the substantia nigra of the midbrain and motor areas of the cerebral cortex and send signals back to both of these locations. They are involved in motor control and are further discussed in a later section on that topic.

The Limbic System

The **limbic**[56] **system,** named for the medial border of the temporal lobe, is a loop of cortical structures surrounding the corpus callosum and thalamus (fig. 14.22). It includes nuclei called the *amygdala* and *hippocampus* on the medial side of the temporal lobe, a tract called the *fornix* leading to the mamillary bodies of the hypothalamus, and a fold of cortex called the *cingulate gyrus* that arches over the corpus callosum. From the earliest investigations of this system in the 1930s, it was thought to be involved in emotion and smell; later, memory was added to the list of functions. Recently, some neuroanatomists have argued that the components of this system have so little in common that there is no point to calling it the limbic system. But whether or not this term is abandoned, it is still agreed that the amygdala is important in emotion and the hippocampus in memory. These functions are explored in later sections of this chapter.

52. *caudate* = tailed, tail-like
53. *putam* = pod, husk
54. *glob* = globe, ball + *pallid* = pale
55. *lenti* = lens + *form* = shape
56. *limbus* = border

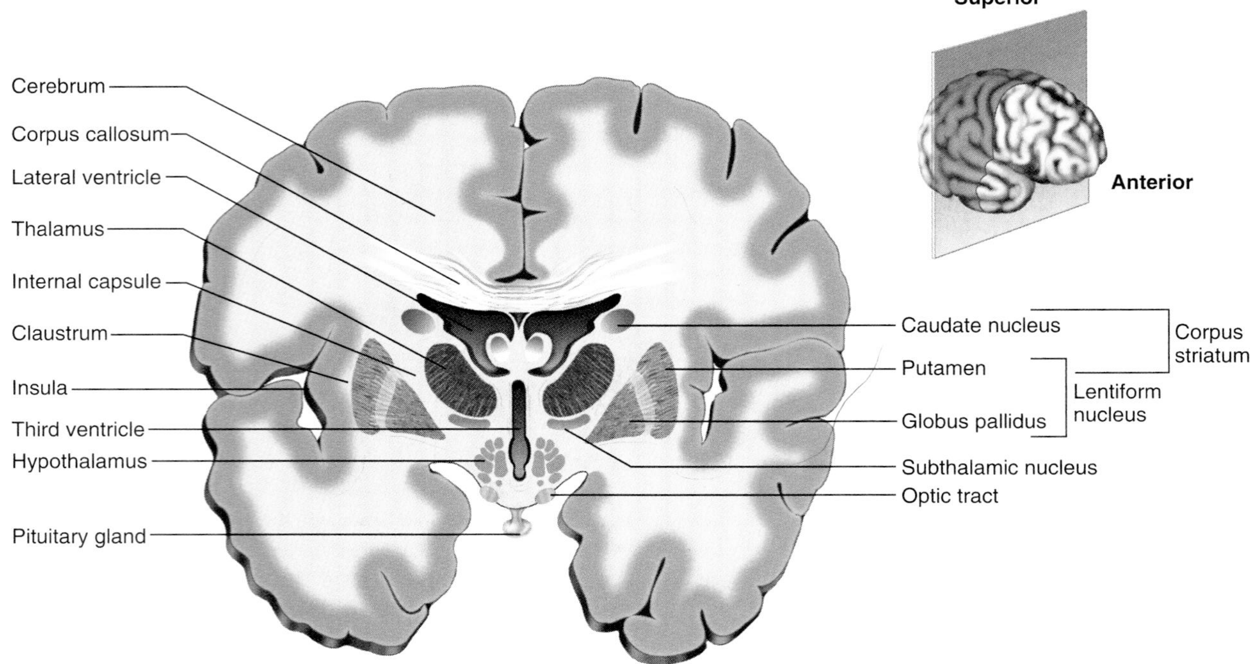

Figure 14.21 The Basal Nuclei (frontal section of the brain).

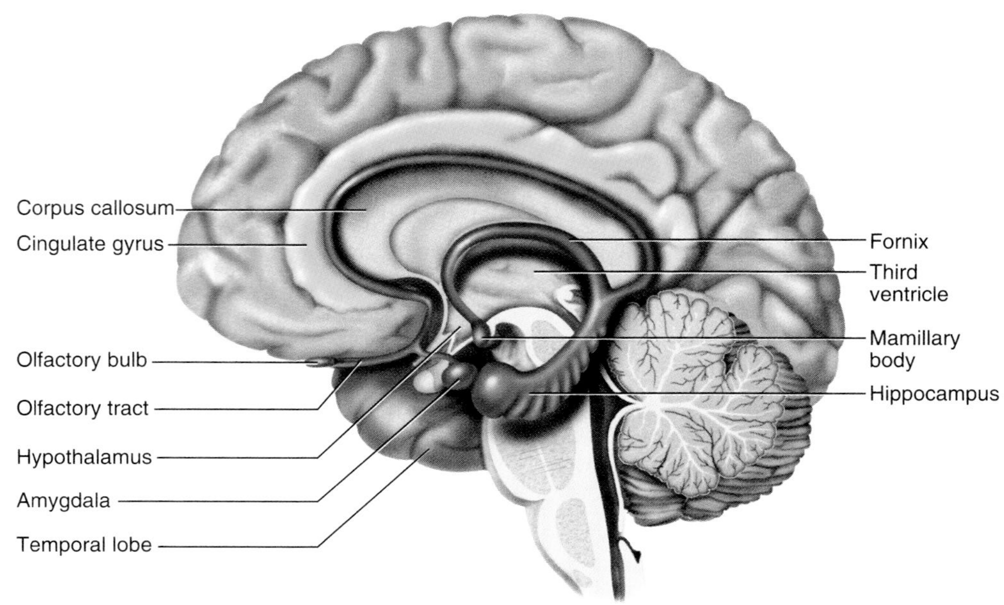

Figure 14.22 The Limbic System.

Before You Go On

Answer the following questions to test your understanding of the preceding section:

16. What are the three major components of the diencephalon? Which ventricle does it enclose?
17. What is the role of the thalamus in sensory function?
18. List at least six functions of the hypothalamus.
19. Name the five lobes of the cerebrum and describe their locations and boundaries.
20. Distinguish between commissural, association, and projection tracts of the cerebrum.
21. Where are the basal nuclei located? What is their general function?
22. Where is the limbic system located? What component of it is involved in emotion? What component is involved in memory?

Higher Brain Functions

Objectives

When you have completed this section, you should be able to

- list the types of brain waves and discuss their relationship to sleep and other mental states;
- explain how the brain controls the skeletal muscles;
- identify the parts of the cerebrum that receive and interpret somatic sensory signals;
- identify the parts of the cerebrum that receive and interpret signals from the special senses;
- describe the locations and functions of the language centers;
- discuss the brain regions involved in memory; and
- discuss the functional relationship between the right and left cerebral hemispheres.

We conclude this chapter with what are often called "higher" or "complex" brain functions, such as sleep, memory, cognition, emotion, sensation, motor control, and language. These are associated especially with the cerebral cortex, but not exclusively; they involve interactions between the cerebral cortex and such areas as the cerebellum, basal nuclei, limbic system, hypothalamus, and reticular formation. These functions present the most difficult challenges for neurobiology, but they are the most intriguing functions of the brain and involve its largest areas.

Brain Waves and Sleep

Brain waves are rhythmic voltage changes resulting predominantly from synchronized postsynaptic potentials in the superficial layers of the cerebral cortex. They can be recorded from electrodes on the scalp. The recording, called an **electroencephalogram**[57] **(EEG),** is useful in studying normal brain functions such as sleep and consciousness and in diagnosing degenerative brain diseases, metabolic abnormalities, brain tumors, sites of trauma, and so forth. States of consciousness ranging from high alert to deep sleep are correlated with changes in the EEG. The complete and persistent absence of brain waves is often used as a clinical and legal criterion of brain death.

There are four types of brain waves, distinguished by differences in amplitude (mV) and frequency (cycles per second, cps; or hertz, Hz) (fig. 14.23):

1. **Alpha (α) waves** have a frequency of 8 to 13 Hz and are recorded especially in the parieto-occipital area. They occur when a person is awake and resting, with the eyes closed and mind wandering. They

57. *electro* = electricity + *encephalo* = brain + *gram* = record

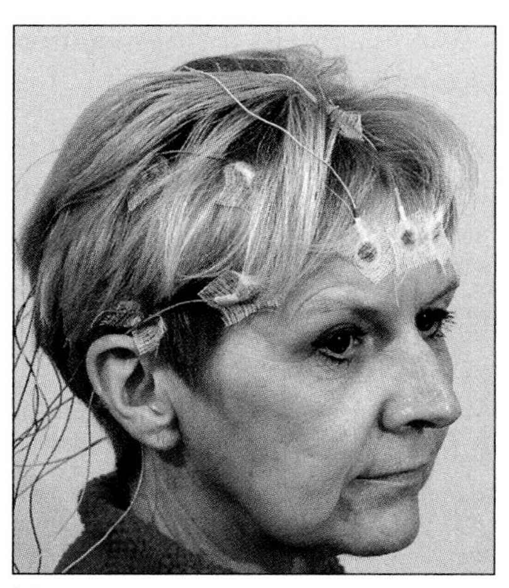
(a)

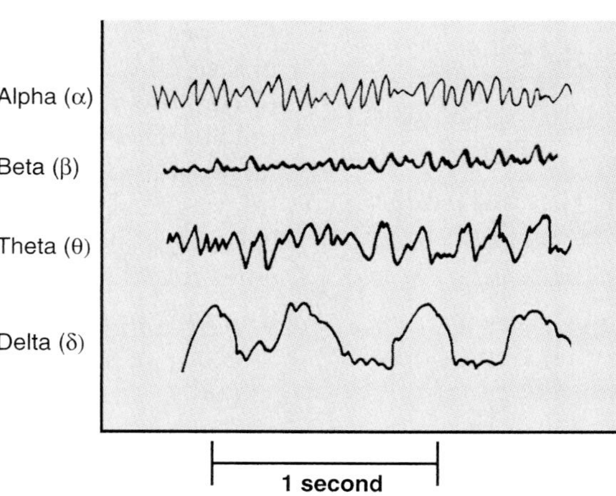

(b)

Figure 14.23 The Electroencephalogram (EEG). (*a*) An EEG is recorded from an array of electrodes on the forehead and scalp. (*b*) Four classes of brain waves are seen in EEGs.

disappear when a person opens the eyes, receives specific sensory stimulation, or engages in a mental task such as performing mathematical calculations. They are also absent during deep sleep.

2. **Beta (β) waves** have a frequency of 14 to 30 Hz and occur in the frontal to parietal region. They are seen during mental activity and sensory stimulation.
3. **Theta (θ) waves** have a frequency of 4 to 7 Hz. They are normal in children and sleeping adults, but in awake adults they suggest emotional stress or brain disorders.
4. **Delta (δ) waves** are high-amplitude "slow waves" with a frequency of less than 3.5 Hz. Infants exhibit delta waves when awake and adults exhibit them in deep sleep. When present in awake adults, they indicate serious brain damage.

Sleep is a temporary state of unconsciousness from which a person can awaken when stimulated. **Coma,** by contrast, is a state of unconsciousness from which no amount of stimulation arouses a person. The cycle of sleep and waking is controlled by nuclei in the hypothalamus and brainstem. In the anterior hypothalamus, just above the optic chiasm, is the **suprachiasmatic** (SOO-pra-KY-az-MAT-ic) **nucleus (SCN),** which acts as a biological clock to regulate our daily **circadian**[58] **rhythm** of sleep and waking. Some nerve fibers lead from the eyes to the SCN and serve to synchronize our body rhythms with the external rhythm of night and day. If the SCN is destroyed, an animal sleeps at random times with no daily rhythm. The reticular formation and hypothalamus have some nuclei that produce wakefulness and alertness and others that bring on sleep. Reticular formation nuclei issue fibers to the thalamus and cerebral cortex and seem to regulate consciousness by modulating communication between those two centers.

In the first 30 to 45 minutes of sleep, our brain waves decline in frequency but increase in amplitude as we pass through four sleep stages (fig. 14.24*a*):

- In **stage 1** we close our eyes and begin to relax; thoughts come and go, and we often have a drifting sensation. We awaken easily if stimulated. The EEG is dominated by alpha waves.
- In **stage 2** we are less easily aroused. The EEG is more irregular, with short bursts of 12- to 14-Hz brain waves called *sleep spindles.*
- In **stage 3,** about 20 minutes after stage 1, sleep deepens, the muscles relax, and the vital signs (body temperature, blood pressure, pulse, and respiratory rate) decline. Theta and delta waves appear in the EEG.
- In **stage 4** the vital signs are at their lowest levels, the muscles are very relaxed, and it is difficult to arouse the sleeper. Stage 4 is also called slow-wave sleep (SWS) because the EEG is dominated by delta waves.

About five times a night, a sleeper backtracks from stage 4 to stage 1 and enters episodes of **rapid eye movement (REM) sleep.** The eyes flicker about under the eyelids as if watching a movie. Vital signs increase and the brain consumes more oxygen than when we are awake. In males, REM sleep is usually accompanied by penile erection. REM sleep is also called *paradoxical sleep* because the EEG resembles that of the waking state, yet the sleeper is harder to arouse than in any other stage. Most dreams occur during REM sleep, although nightmares are usually associated with stages 3 and 4. Another characteristic of REM sleep is **sleep paralysis**—inhibition of the skeletal muscles (except for those of eye movement). Sleep paralysis prevents us from acting out our dreams, and it may have prevented our tree-dwelling ancestors from falling in their sleep. The first period of REM sleep occurs about 90 minutes after we fall asleep. In the second half of the night, REM sleep becomes longer and more frequent (fig. 14.24*b*).

We still know very little about the function of sleep. Non-REM sleep seems to have a restorative effect on the body, and prolonged sleep deprivation is fatal to experimental animals. Yet it is unclear why quiet bed rest is not sufficient—that is, why we must lose consciousness. Unlike total sleep deprivation, selective deprivation of REM sleep (waking a person up whenever he or she begins REM sleep) has no adverse effects. Some have suggested that REM sleep is a period in which the brain "consolidates" and strengthens memories or purges unwanted information from memory, but there is little evidence for either view.

Cognition

Cognition[59] refers to mental processes such as awareness, perception, thinking, knowledge, and memory. Seventy-five percent of our brain tissue consists of **association areas** of the cerebrum—areas concerned with cognition, the integration of information between the points of sensory input and motor output. We know the functions of these areas largely through studies of people with brain lesions—local injuries resulting from such causes as infection, trauma, cancer, and stroke. A few examples of the effects of cerebral lesions reveal some functions of the association areas:

- **Parietal lobe lesions** can cause people to become unaware of objects on one side of the body or even of their own limbs on that side—a condition called *contralateral neglect syndrome.* In typical cases, men shave only one half of the face, women apply makeup to

58. *circa* = approximately + *dia* = a day, 24 hours

59. *cognit* = to know

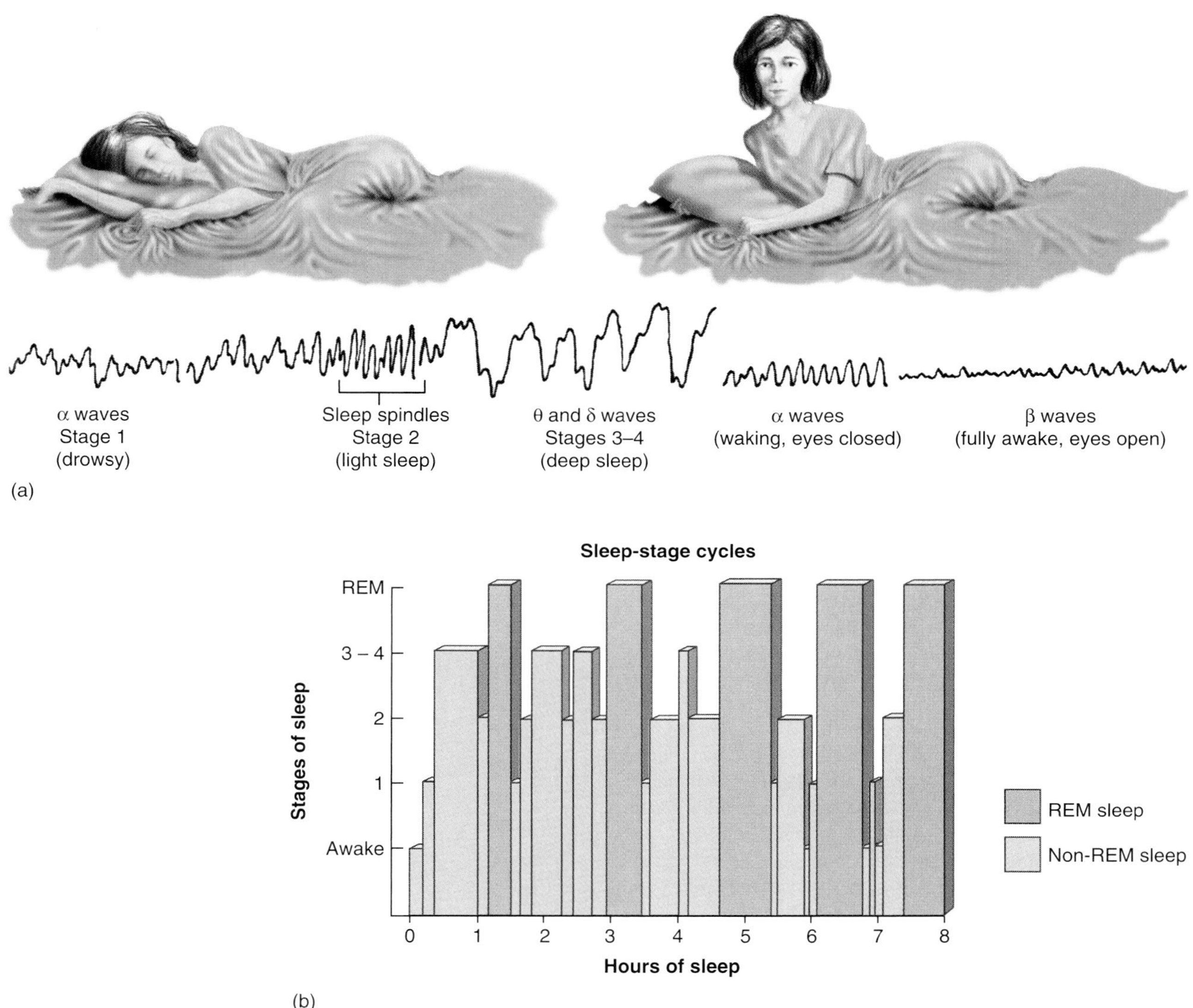

Figure 14.24 **Sleep Stages and Brain Activity.** (*a*) Correlation of the brain waves with the stages of non-REM sleep. (*b*) Stages of sleep over an 8-hour night in an average young adult. Stages 3 to 4 dominate the first half of the night and REM sleep dominates the second half. Most dreaming occurs during REM sleep.

only one side, patients dress only half of the body, and some people deny that one arm or leg belongs to them.

- **Temporal lobe lesions** often result in **agnosia**[60] (ag-NO-zee-ah), the inability to recognize, identify, and name familiar objects. In *prosopagnosia,*[61] a person cannot remember familiar faces, even his or her own reflection in a mirror.
- **Frontal lobe lesions** are especially devastating to the qualities we think of as the personality (see historical insight 14.4). The *frontal association area (prefrontal cortex)* is well developed only in primates, especially humans. It integrates information from sensory and motor regions of the cortex and from other association areas. It gives us a sense of our relationship to the rest of the world enabling us to think about it and to plan and execute appropriate behavior. It is responsible for giving appropriate expression to our emotions.

As a broad generalization, we can conclude that the parietal association cortex is responsible for perceiving and attending to stimuli, the temporal association cortex for identifying them, and the frontal association cortex for planning our responses.

60. *a* = without + *gnos* = knowledge
61. *prosopo* = face, person

Historical Insight 14.4

The Accidental Lobotomy of Phineas Gage

Accidental but nonfatal destruction of parts of the brain has afforded many clues to the function of various regions. One of the most famous incidents occurred in 1848 to Phineas Gage, a laborer on a railroad construction project in Vermont. Gage was packing blasting powder into a hole with a 3 1/2 ft tamping iron when the powder prematurely exploded. The tamping rod was blown out of the hole and passed through Gage's maxilla, orbit, and the frontal lobe of his brain before emerging from his skull near the hairline and landing 50 ft away. Gage went into convulsions, but later sat up and conversed with his crewmates as they drove him to a physician in an oxcart. On arrival, he stepped out on his own and told the physician, "Doctor, here is business enough for you." His doctor, John Harlow, reported that he could insert his index finger all the way into Gage's wound. Yet 2 months later, Gage was walking around town carrying on his normal business.

He was not, however, the Phineas Gage people had known. Before the accident, Gage was a competent, responsible, financially prudent man, well liked by his associates. In his 1868 publication on the incident, Harlow said that following the accident, Gage was "fitful, irreverent, indulging at times in the grossest profanity." He became irresponsible, lost his job, worked for a while as a circus sideshow attraction, and died a vagrant 12 years later.

A 1994 computer analysis of Gage's skull indicated that the brain injury was primarily to the ventromedial region of both frontal lobes. In Gage's time, scientists were reluctant to attribute social behavior and moral judgment to any region of the brain. These functions were strongly tied to issues of religion and ethics and were considered inaccessible to scientific analysis. Based partly on Phineas Gage and other brain injury patients like him, neuroscientists today recognize that planning, moral judgment, and emotional control are among the functions of the prefrontal cortex.

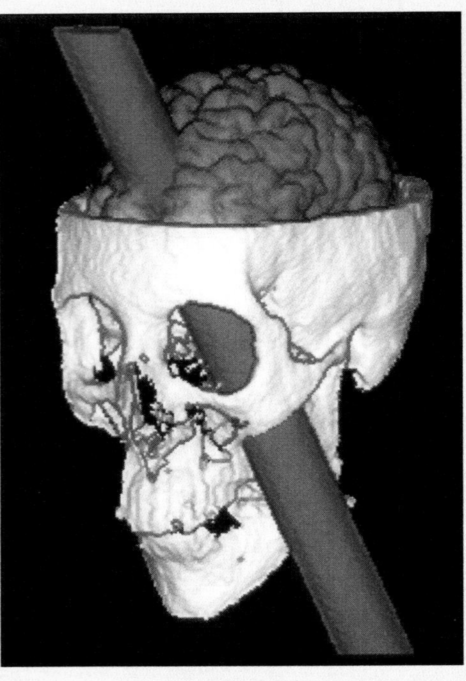

Phineas Gage's 1848 Accident. This is a computer-generated image made in 1994 to show the path taken by the tamping bar through Gage's skull and brain.

Memory

We studied the neuronal and molecular mechanisms of memory in the preceding chapter. Now that you have been introduced to the gross anatomy of the brain, we can consider the anatomical sites of those processes.

Our subject is really a little broader than memory *per se.* Information management by the brain entails learning (acquiring new information), memory proper (information storage and retrieval), and forgetting (eliminating trivial information). Forgetting is as important as remembering. People with a pathological inability to forget trivial information have great difficulty in reading comprehension and other functions that require us to separate what is important from what is not. More often, though, brain-injured people are either unable to store new information **(anterograde amnesia)** or to recall things they knew before the injury **(retrograde amnesia)**. *Amnesia* refers to defects in *declarative* memory (such as the ability to describe past events), not *procedural* memory (such as the ability to tie your shoes) (see definitions in chapter 13).

The **hippocampus**[62] of the limbic system (see fig. 14.22) does not store memories but is essential to **memory consolidation,** the process of organizing diverse sensory and cognitive experiences into a unified long-term memory. Long-term memories are stored elsewhere. Our vocabulary and memory of faces and familiar objects, for example, are stored in the superior temporal lobe, and memories of our plans and social roles are stored in the prefrontal cortex.

Lesions of the hippocampus can cause profound anterograde amnesia. For example, in 1953, a famous patient known as H. M. underwent surgical removal of a large por-

62. *hippocampus* = sea horse, named for its shape

tion of both temporal lobes, including both hippocampi, in an effort to treat his severe epilepsy. The operation had no adverse effect on his intelligence, procedural memory, or declarative memory for things that had happened early in his life, but it left him with an inability to establish new memories. He could hold a conversation with his psychologist, but a few minutes later deny that it had taken place. He worked with the same psychologist for more than 40 years after his operation, yet had no idea of who she was from day to day.

Other parts of the brain involved in memory include the cerebellum, with a role in learning motor skills, and the amygdala, with a role in emotional memory—both examined shortly.

Emotion

The prefrontal cortex is the seat of judgment, intent, and control over the expression of our emotions. However we may feel, it is here that we decide the appropriate way to show our feelings. But the feelings themselves, and our emotional memories, form in deeper regions of the brain—the hypothalamus and **amygdala**[63] (ah-MIG-da-la). Here are the nuclei that stimulate us to recoil in fear from a rattlesnake or yearn for a lost love. (The amygdala also seems to be involved in a broad range of functions including food intake, sexual activity, and drawing our attention to novel stimuli.)

Emotional control centers of the brain have been identified not only by studying people with brain lesions, but also by such techniques as surgical removal, ablation (destruction) of small regions with electrodes, and stimulation with electrodes and chemical implants, especially in experimental animals. Changes in behavior following removal, ablation, or stimulation give clues to the functions that a region performs. However, interpretation of the results is difficult and controversial because of the complex connections between the emotional brain and other regions.

Many important aspects of personality depend on an intact, functional amygdala and hypothalamus. When specific regions of the amygdala or hypothalamus are destroyed or artificially stimulated, humans and other animals exhibit blunted or exaggerated expressions of anger, fear, aggression, self-defense, pleasure, pain, love, sexuality, and parental affection, as well as abnormalities in learning, memory, and motivation.

Much of our behavior is shaped by learned associations between stimuli, our responses to them, and the rewards or punishments that result. Nuclei involved in the senses of reward and punishment have been identified in the hypothalamus of cats, rats, monkeys, and other animals. In a representative experiment, an electrode is implanted in an area of an animal's hypothalamus called the *median forebrain bundle (MFB).* The animal is placed in a chamber with a foot pedal wired to that electrode. When the animal steps on the pedal, it receives a mild electrical stimulus to the MFB. The sensation is apparently strongly rewarding, because the animal soon learns to press the pedal over and over and may spend most of its time doing so—even to the point of neglecting food and water. Rats have been known to bar-press 5,000 to 12,000 times an hour, and monkeys up to 17,000 times an hour, to stimulate their MFBs.

These animals cannot tell us what they are feeling, but electrode implants have also been used to treat people who suffer otherwise incurable schizophrenia, pain, or epilepsy. These patients also bar-press repeatedly, but they do not report feelings of joy or ecstasy. Some are unable to explain why they press the bar, and others report "relief from tension" or "a quiet, relaxed feeling." With electrodes misplaced in other areas of the hypothalamus, subjects report feelings of fear or terror when stimulated.

Sensation

Sensory functions fall into two categories:

1. **somesthetic**[64] **(somatosensory, somatic) sensation,** which comes from receptors widely distributed over the body for such stimuli as touch, pressure, stretch, movement, heat, cold, and pain; and
2. **special senses,** which originate in sense organs of the head and include vision, hearing, equilibrium, taste, and smell.

Somesthetic Sensation

Somesthetic nerve signals travel mainly up the gracile and cuneate fasciculi and spinothalamic tracts of the spinal cord to the brainstem. The thalamus routes these messages to the **postcentral gyrus,** the most anterior gyrus of the parietal lobe, immediately posterior to the central sulcus. The cortex here is the **primary somesthetic cortex (somatosensory area)** (fig. 14.25). Since the somesthetic fibers decussate on their way to the thalamus, the right postcentral gyrus receives signals from the left side of the body and the left gyrus receives signals from the right.

Each gyrus is like an upside-down sensory map of the contralateral side of the body, traditionally diagrammed as a *sensory homunculus*[65] (fig. 14.25*b*). As the diagram shows, receptors in the lower limb project to the superior and medial parts of the gyrus and receptors in the

63. *amygdal* = almond

64. *som* = body + *esthet* = feeling

65. *homunculus* = little man

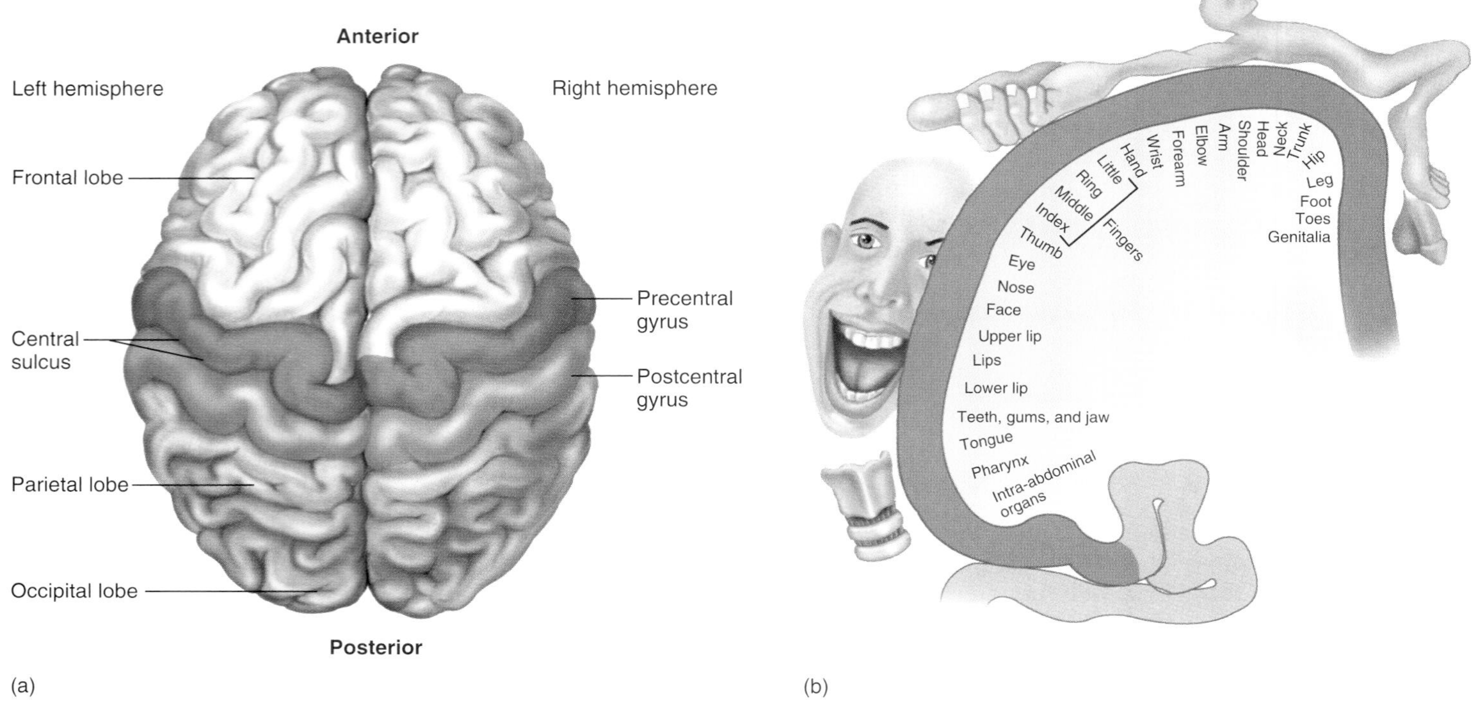

Figure 14.25 The Primary Somesthetic Cortex (postcentral gyrus). (*a*) Location, superior view. (*b*) Sensory homunculus, drawn so that body parts are in proportion to the amount of cortex dedicated to their sensation.

face project to the inferior and lateral parts of the gyrus. The reason for the bizarre, distorted appearance of the homunculus is that the amount of cerebral tissue devoted to a given body region is proportional to how richly innervated and sensitive that part of the body is. Thus, the hands and face are represented by a much larger area of cortex than the trunk. Such a point-for-point correspondence between an area of the body and an area of the CNS is called **somatotopy.**[66]

The Special Senses

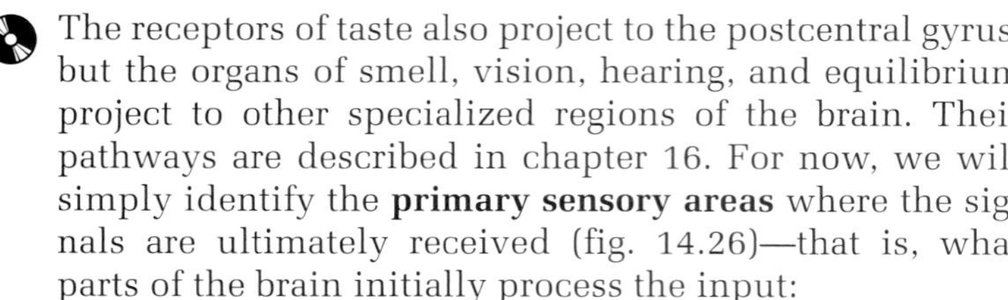

The receptors of taste also project to the postcentral gyrus, but the organs of smell, vision, hearing, and equilibrium project to other specialized regions of the brain. Their pathways are described in chapter 16. For now, we will simply identify the **primary sensory areas** where the signals are ultimately received (fig. 14.26)—that is, what parts of the brain initially process the input:

- **taste** (gustation)—near the inferior lateral end of the postcentral gyrus and part of the insula;
- **smell** (olfaction)—medial surface of the temporal lobe and inferior surface of the frontal lobe;
- **vision**—posterior occipital lobe;
- **hearing**—superior temporal lobe and nearby insula; and
- **equilibrium**—mainly the cerebellum, but also via the thalamus to unknown destinations in the cerebral cortex.

66. *somato* = body + *topy* = place

Sensory Association Areas

Each primary sensory area of the cerebral cortex lies adjacent to an **association area (secondary sensory cortex)** that interprets sensory information and makes it identifiable and useful to us (fig. 14.26). The largest and best-known association areas are:

- The **somesthetic association area** located in the parietal lobe immediately posterior to the postcentral gyrus. It makes us aware of the positions of our limbs, the location of a touch or pain, and the shape, weight, and texture of an object in our hand, for example.
- The **visual association area** is located mainly in the occipital lobe between the somesthetic association area and primary visual cortex. It enables us to identify the things we see. Recognition of faces, however, resides in the inferior temporal lobe.

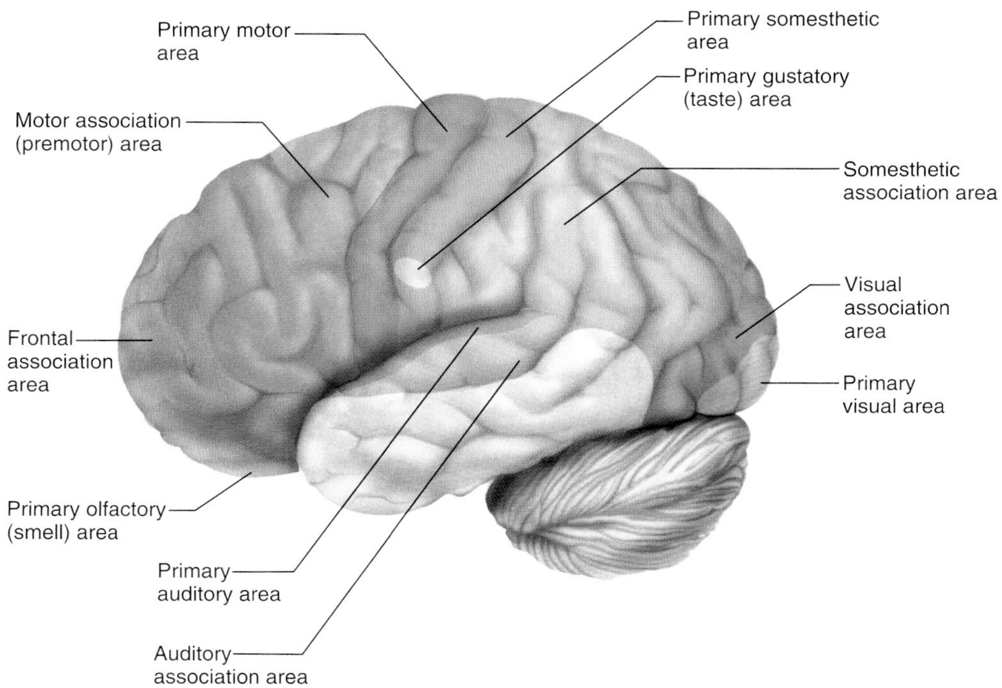

Figure 14.26 Some Functional Regions of the Cerebral Cortex. Left hemisphere.

- The **auditory association area** is located in the superior temporal lobe deep within the lateral sulcus. It enables us to remember the name of a piece of music and to identify a person by his or her voice.

Motor Control

The intention to contract a skeletal muscle begins in the **motor association (premotor) area** of the frontal lobes (fig. 14.26). This is where we plan our behavior, where neurons compile a program for the degree and sequence of muscle contractions required for an action such as dancing, typing, or speaking. The program is then transmitted to neurons of the **precentral gyrus (primary motor area),** which is the most posterior gyrus of the frontal lobe, immediately anterior to the central sulcus. Neurons here send signals to the brainstem and spinal cord that ultimately result in muscle contractions.

The precentral gyrus, like the postcentral one, exhibits somatotopy. The neurons for toe movements, for example, are deep in the longitudinal fissure on the medial side of the gyrus. The summit of the gyrus controls the trunk, shoulder, and arm, and the inferolateral region controls the facial muscles. This map is diagrammed as a *motor homunculus* (fig. 14.27). Like the sensory homunculus, it has a distorted look because the amount of cortex devoted to a given body region is proportional to the number of muscles and motor units in that region, not to the size of the region. Areas of fine control, such as the hands, have more muscles than such areas as the trunk and thigh, more motor units per muscle, and larger areas of motor cortex to control them.

The pyramidal cells of the precentral gyrus are called **upper motor neurons.** Their fibers project caudally, with about 19 million fibers ending in nuclei of the brainstem and 1 million forming the corticospinal tracts, which decussate in the pyramids of the medulla oblongata on their way to the spinal cord. Below the neck, each precentral gyrus controls muscles on the contralateral side of the body. In the brainstem or spinal cord, the fibers from the upper motor neurons synapse with **lower motor neurons** whose axons innervate the skeletal muscles.

Other areas of the brain important to muscle control are the basal nuclei and cerebellum. The basal nuclei receive input from many sensory and motor regions of cerebral cortex—most importantly the prefrontal cortex. Their output goes by way of the thalamus back to the prefrontal cortex, motor association area, and precentral gyrus. The basal nuclei are part of a feedback circuit involved in the planning and execution of movement. They assume responsibility for controlling highly practiced behaviors such as tying your shoes or driving a car—skilled movements that you carry out with little thought. People with lesions of the basal nuclei tend to move slowly and have difficulty initiating movement, such as getting up from a chair. They show changes in muscle tone and posture, the

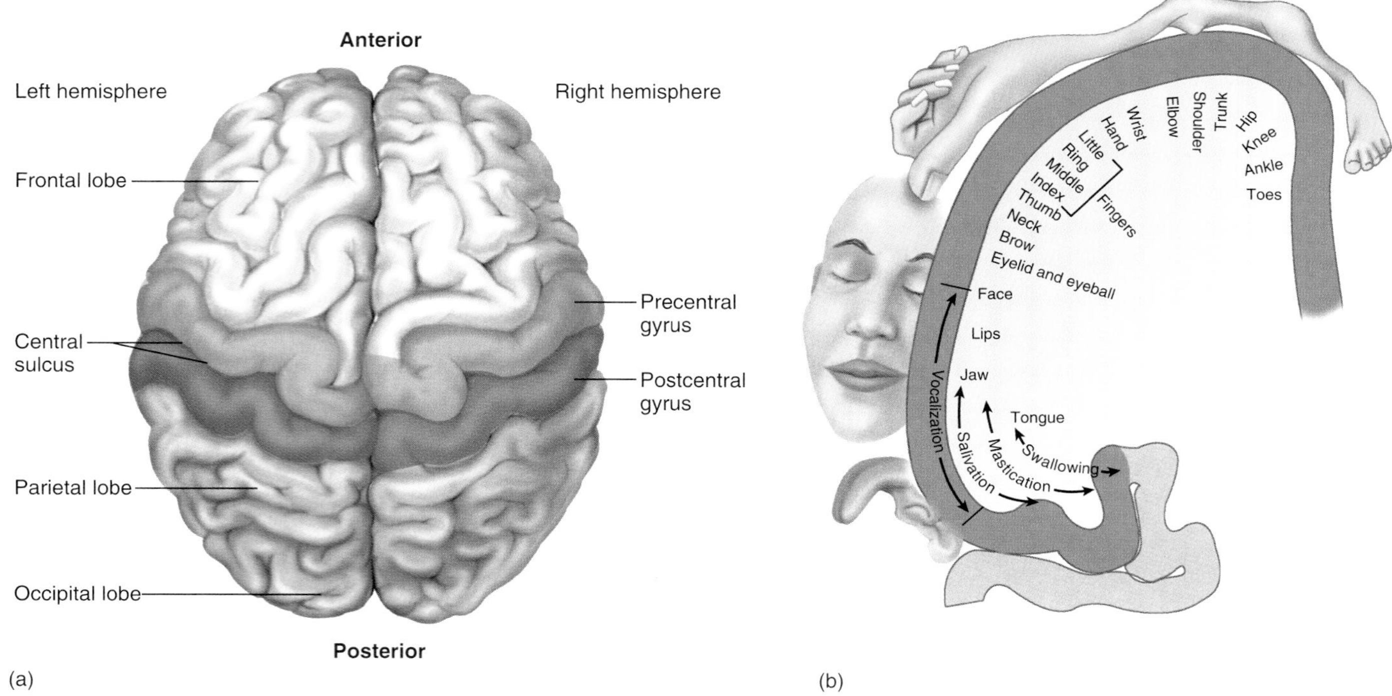

Figure 14.27 The Primary Motor Cortex (precentral gyrus). (*a*) Location, superior view. (*b*) Motor homunculus, drawn so that body parts are in proportion to the amount of primary motor cortex dedicated to their control.

tremors of Parkinson disease, or involuntary movements such as flailing of the limbs (*ballismus*).

The cerebellum smooths muscle contractions, maintains muscle tone and posture, coordinates the motions of different joints with each other (such as the shoulder and elbow joints in pitching a baseball), coordinates eye and body movements, and aids in learning motor skills. It receives signals from the upper motor neurons about intended movements and gets feedback about the actual performance from *proprioceptors* in the muscles and joints, via the spinocerebellar tracts (fig. 14.28). The Purkinje cells of the cerebellum compare what the muscles are doing with what they were told to do and with stored information on learned skills. When there is a discrepancy between the intent and performance, they send signals to the deep cerebellar nuclei, which in turn relay them to the brainstem and cerebral cortex. Motor centers here use information from the cerebellum to correct muscle performance to match the intent. Lesions in the cerebellum can result in a clumsy, awkward gait (*ataxia*) and make some tasks such as climbing a flight of stairs virtually impossible.

Language

Language includes several abilities—reading, writing, speaking, and understanding words—assigned to different regions of cerebral cortex (fig. 14.29). **Wernicke's**[67] (VERR-nick-ez) **area** is responsible for the recognition of spoken and written language. It lies just posterior to the lateral sulcus, usually in the left hemisphere. It is a sensory association area that receives visual and auditory input from the respective regions of primary sensory cortex. The *angular gyrus* just posterior to Wernicke's area processes the words we read into a form that we can speak.

Wernicke's area formulates phrases according to learned rules of grammar and transmits a plan of speech to **Broca's**[68] **area,** located in the inferior prefrontal cortex in the same hemisphere. Broca's area generates a motor program for the muscles of the larynx, tongue, cheeks, and lips to produce speech. This program is then transmitted to the primary motor cortex, which executes it—that is, it issues commands to the lower motor neurons that supply the relevant muscles. PET scans show a rise in the metabolic activity of Broca's area as we prepare to speak.

The emotional aspect of language is controlled by regions in the opposite hemisphere that mirror Wernicke's and Broca's areas. Opposite Broca's area is the *affective language area.* Lesions to this area result in *aprosodia*—

67. Karl Wernicke (1848–1904), German neurologist
68. Pierre Paul Broca (1824–80), French surgeon and anthropologist

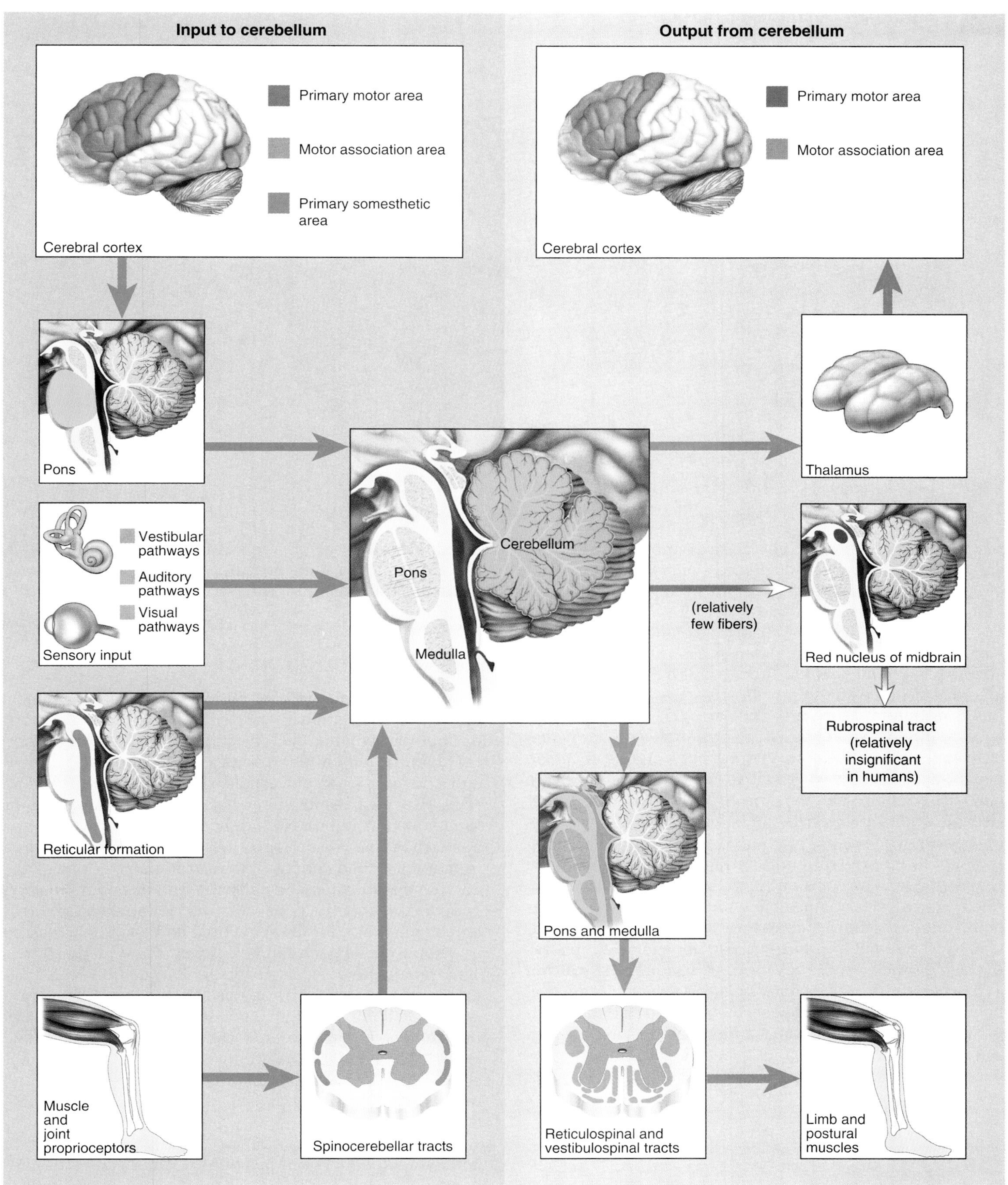

Figure 14.28 **Pathways of Motor Control Involving the Cerebellum.** The cerebellum receives its input from the afferent pathways (red) on the *left* and sends its output through the efferent pathways (green) on the *right*.

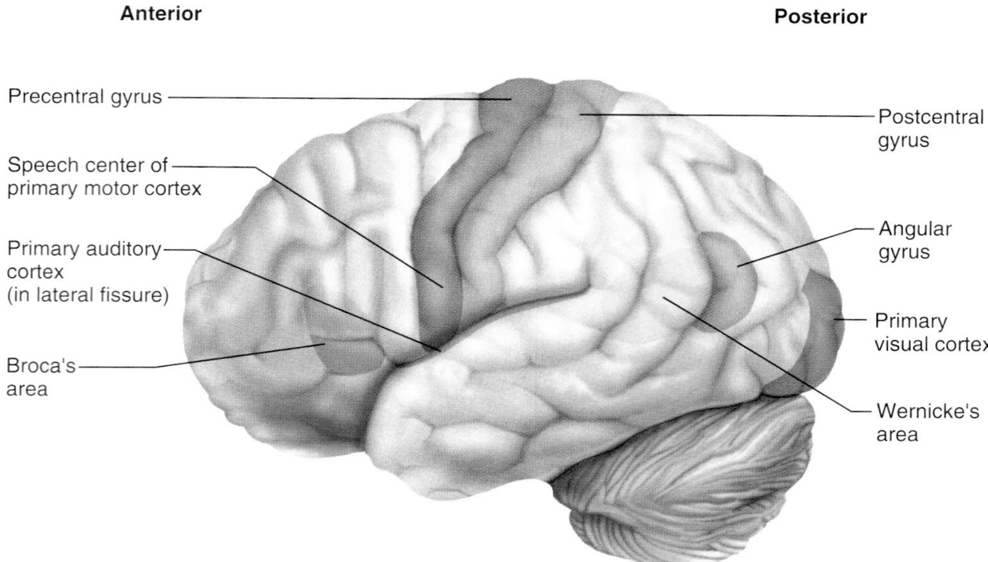

Figure 14.29 Language Centers of the Left Hemisphere.

flat, emotionless speech. The cortex opposite Wernicke's area is concerned with recognizing the emotional content of another person's speech. Lesions here can result in the inability to understand a joke.

Aphasia[69] (ah-FAY-zee-uh) is any language deficit resulting from lesions in the hemisphere (usually the left) containing Wernicke's and Broca's areas. The many forms of aphasia are difficult to classify. *Nonfluent (Broca's) aphasia,* due to a lesion to Broca's area, results in slow speech, difficulty in choosing words, or use of words that only approximate the correct word. For example, a person may say "tssair" when asked to identify a picture of a chair. In extreme cases, the person's entire vocabulary consists of two or three words, sometimes those that were being spoken when a stroke occurred. Patients feel very frustrated with themselves and often maintain a tight-lipped reluctance to talk. A lesion to Wernicke's area may cause *fluent (Wernicke's) aphasia,* in which a person speaks normally and sometimes excessively, but uses jargon and invented words that make little sense (for example, "choss" for chair). Such people also cannot comprehend written and spoken words. In *anomic aphasia,* a person can speak normally and understand speech but cannot identify written words or pictures. Shown a picture of a chair, the person may say, "I know what it is. . . . I have a lot of them," but be unable to name the object.

This represents only a small sample of the complex and puzzling linguistic effects of brain lesions. Other lesions to small areas of cortex can cause impaired mathematical ability, a tendency to write only consonants, or difficulty understanding the second half of each word a person reads.

69. *a* = without + *phas* = speech

Cerebral Lateralization

The two cerebral hemispheres look identical at a glance, but close examination reveals a number of differences. For example, in women the left temporal lobe is longer than the right. In left-handed people, the left frontal, parietal, and occipital lobes are usually wider than those on the right. The two hemispheres also differ in some of their functions (fig. 14.30). Neither hemisphere is "dominant," but each is specialized for certain tasks. This difference in function is called **cerebral lateralization.**

One hemisphere, usually the left, is called the *categorical hemisphere.* It is specialized for spoken and written language and for the sequential and analytical reasoning employed in such fields as science and mathematics. This hemisphere seems to break information into fragments and analyze it in a linear way. The other hemisphere, usually the right, is called the *representational hemisphere.* It perceives information in a more integrated, holistic way. It is the seat of imagination and insight, musical and artistic skill, perception of patterns and spatial relationships, and comparison of sights, sounds, smells, and tastes.

Cerebral lateralization is highly correlated with handedness. Ninety-one percent of Americans are right-handed. The left hemisphere is the categorical one in 96%

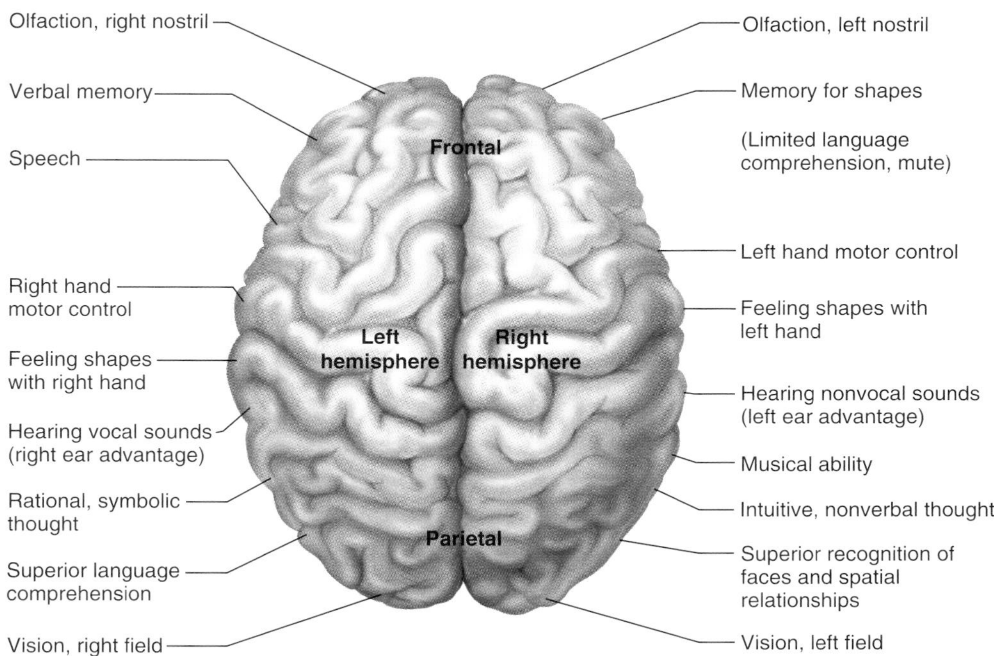

Figure 14.30 Lateralization of Cerebral Functions.

of these people and the right hemisphere in 4%. Among left-handed people, the right hemisphere is categorical in 15%, the left in 70%, and in the remaining 15% neither hemisphere is distinctly specialized.

Lateralization develops with age. In young children, if one cerebral hemisphere is damaged or removed (for example, because of brain cancer), the other hemisphere can often take over its functions. Adult males exhibit more lateralization than females and suffer more functional loss when one hemisphere is damaged. When the left hemisphere is damaged, men are three times as likely as women to become aphasic. The reason for this difference is not yet clear, but it may be related to the corpus callosum. In men, the corpus callosum has a fairly uniform thickness, but in women its posterior portion is thickened by additional commissural fibers, suggesting that women have a more extensive communication between their hemispheres.

Not surprisingly, the brain is the most structurally complex organ of the body. We have barely scratched the surface in this chapter. Table 14.2 summarizes the gross anatomy of the brain and will help you put its structures into context.

Like a machine with a great number of moving parts, the central nervous system is highly subject to malfunctions. Table 14.3 lists a few familiar dysfunctions. Some CNS disorders discussed elsewhere in the book include brain tumors (p. 461), multiple sclerosis and Tay-Sachs disease (p. 462), Alzheimer and Parkinson diseases (p. 485), transient ischemic attack and stroke (p. 769), and earlier in this chapter, spina bifida, meningitis, hydrocephalus, poliomyelitis, amyotrophic lateral sclerosis, contralateral neglect syndrome, agnosia, amnesia, aprosodia, and aphasia.

Before You Go On

Answer the following questions to test your understanding of the preceding section:

23. Suppose you are reading a novel and gradually fall asleep and begin to dream. How would your brain waves change during this sequence of events?
24. Describe the locations and functions of the somesthetic, visual, auditory, and frontal association areas.
25. Describe the somatotopy of the primary motor area and primary sensory area.
26. What are the roles of Wernicke's area, Broca's area, and the precentral gyrus in language?

Table 14.2 Anatomical Checklist for the Brain

Meninges	
Dura mater	Arachnoid mater
Falx cerebri	Arachnoid villi
Falx cerebelli	Pia mater
Tentorium cerebelli	
Ventricle System	
Lateral ventricles	Fourth ventricle
Interventricular foramen	Median and lateral apertures
Third ventricle	Central canal
Mesencephalic (cerebral) aqueduct	Choroid plexuses
Rhombencephalon (hindbrain)	
Myelencephalon	*Metencephalon*
Medulla oblongata	Pons
Medial lemniscus	Cerebellum
Pyramids	Cerebellar hemispheres
Olive	Vermis
Inferior olivary nucleus	Cerebellar peduncles
	Folia
	Arbor vitae
	Deep nuclei
Mesencephalon (midbrain)	
Cerebral peduncles	Tectum
Tegmentum	Corpora quadrigemina
Red nucleus	Superior colliculi
Substantia nigra	Inferior colliculi
Central gray matter	
Reticular Formation	
Prosencephalon (forebrain)	
Diencephalon	
Thalamus	Epithalamus
Hypothalamus	Pineal gland
	Habenula
Telencephalon (cerebrum)	
Cerebral hemispheres	Central sulcus
Major sulci and fissures	Parieto-occipital sulcus
Longitudinal fissure	Lateral sulcus

(continued)

Table 14.2 Anatomical Checklist for the Brain (Continued)

Prosencephalon (forebrain)—*(cont.)*

Telencephalon (cerebrum)

Lobes	
Frontal lobe	Major gyri
Parietal lobe	Precentral gyrus
Occipital lobe	Postcentral gyrus
Temporal lobe	Basal nuclei
Insula	Caudate nucleus
Gray matter (cerebral cortex)	Putamen
White matter	Globus pallidus
Projection tracts	Limbic system
Commissural tracts	Hippocampus
Corpus callosum	Amygdala
Anterior commissure	Fornix
Posterior commissure	Cingulate gyrus
Association tracts	

Table 14.3 A Few Disorders of the Central Nervous System

Cerebral Palsy	Muscular incoordination resulting from damage to the motor areas of the brain during fetal development, birth, or infancy; causes include prenatal rubella infection, drugs, or radiation exposure; oxygen deficiency during birth; and hydrocephalus
Concussion	Damage to the brain typically resulting from a blow, often with loss of consciousness, disturbances of vision or equilibrium, and short-term amnesia
Encephalitis	Inflammation of the brain, accompanied by fever, usually caused by mosquito-borne viruses or herpes simplex virus; causes neuronal degeneration and necrosis; can lead to delirium, seizures, and death
Epilepsy	Disorder causing sudden, massive discharge of neurons (seizures) resulting in motor convulsions, sensory and psychic disturbances, and often impaired consciousness; may result from birth trauma, tumors, infections, drug or alcohol abuse, or congenital brain malformation
Migraine Headache	Recurring headaches often accompanied by nausea, vomiting, dizziness, and aversion to light, often triggered by such factors as weather changes, stress, hunger, red wine, or noise; more common in women and sometimes running in families
Schizophrenia	A thought disorder involving delusions, hallucinations, inappropriate emotional responses to situations, incoherent speech, and withdrawal from society, resulting from hereditary or developmental abnormalities in neuronal networks

Clinical Insight 14.5

Images of the Mind

Enclosed as it is in the cranium, there is no easy way to observe a living brain directly. This has long frustrated neurobiologists, who once had to content themselves with glimpses of brain function afforded by electroencephalograms, patients with brain lesions, and patients who remained awake and conversant during brain surgery and who consented to experimentation while the brain was exposed. New imaging methods, however, are yielding dramatic perspectives on brain function. Two of these—positron emission tomography (PET) and magnetic resonance imaging (MRI)—were explained in clinical insight 1.3 at the end of chapter 1. Both techniques rely on transient increases in blood flow to parts of the brain called into action to perform specific tasks. By monitoring these changes, neuroscientists can identify which parts of the brain are involved in specific tasks.

To produce a PET scan of the brain, the subject is given an injection of radioactively labeled glucose and a PET scan is made in a *control state* before any specific mental task is begun. Then the subject is given a specific task to perform. For example, the subject may be instructed to read the word *car* and speak a verb related to it, such as *drive.* New PET scans are made in the task state while the subject performs this task. Neither control- or *task-state* images are very revealing by themselves, but the computer subtracts the control-state data from the task-state data and presents a color-coded image of the difference. To compensate for chance events and individual variation, the computer also produces an image that is either averaged from several trials with one person or from trials with several different people.

In such averaged images, the busiest areas of the brain seem to "light up" from moment to moment as the task is performed (fig. 1.). This identifies the regions used for various stages of the task, such as reading the word, thinking of a verb to go with it, planning to say *drive,* and actually saying it. Among other things, such experiments demonstrate that Broca's and Wernicke's areas are not involved in simply repeating words; they are active, however, when a subject must evaluate a word and choose an appropriate response—that is, they function in formulating the new word the subject is going to say. PET scans also show that different neuronal pools take over a task as we practice and become more proficient at it.

Functional magnetic resonance imaging (fMRI) depends on a relatively new finding about brain metabolism. Until recently, neurons were thought to employ only aerobic respiration. Now it is known that when they are stimulated to heightened activity, they employ anaerobic fermentation. They receive an increased blood flow but do not increase their oxygen consumption. Thus, the supply of oxygen increases, but the demand for it does not, so blood leaving a region of highly active brain tissue contains more oxygen than the blood leaving less active regions. Since the magnetic properties of hemoglobin depend on how much oxygen is bound to it, changes in brain circulation can be detected by fMRI. fMRI is more precise than PET and pinpoints regions of brain activity with a precision of 1 to 2 mm. It also has the advantages of requiring no injected substances and no exposure to radioisotopes. While it takes about 1 minute to produce a PET scan, fMRI produces images much more quickly, which makes it more useful for determining how the brain responds immediately to sensory input or mental tasks.

PET and fMRI scanning have enhanced our knowledge of neurobiology by identifying shifting patterns of brain activity associated with attention and consciousness, sensory perception, memory, emotion, motor control, reading, speaking, musical judgment, planning a chess strategy, and so forth. In addition to their contribution to basic neuroscience, these techniques show promise as an aid to neurosurgery and psychopharmacology. They also are enhancing our understanding of brain dysfunctions such as depression, schizophrenia, and attention deficit disorders. We are on the threshold of an exciting era in the safe visualization of normal brain function, virtually producing pictures of the mind at work.

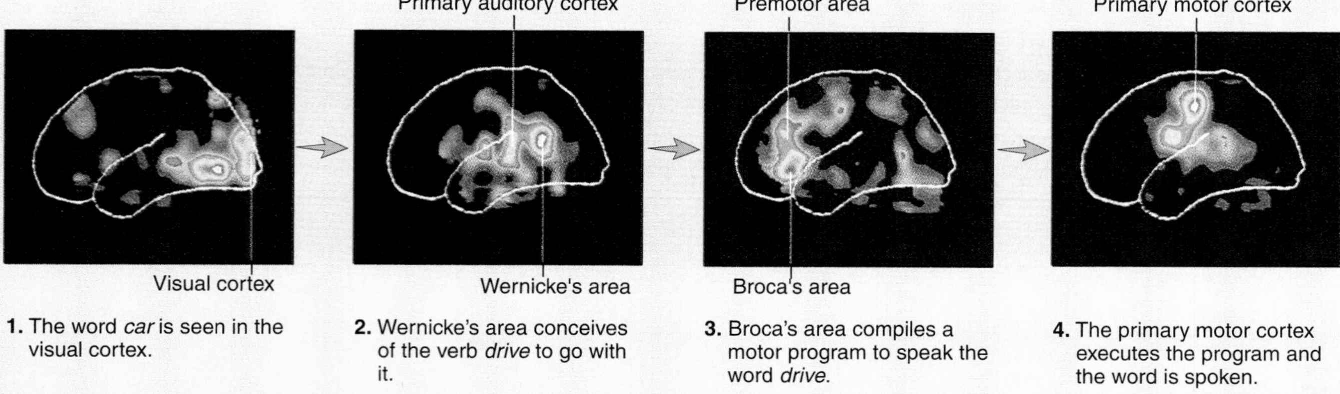

Figure 1 PET Scans of the Brain Made During the Performance of a Language Task. These images show the cortical regions that are active when a person reads words and then speaks them. The most active areas of the cortex show in red and the least active areas show in blue.

Chapter Review

Online Learning Center

Study resources for this chapter are available at www.mhhe.com/saladin/

Study Outline

Overview of Central Nervous System (p. 491)

1. Major landmarks
   a. Terminology: rostral and caudal
   b. Cerebrum
      - Cerebral hemispheres
      - Gyri, sulci, and fissures
      - Corpus callosum
   c. Cerebellum
   d. Brainstem
2. Embryonic development
   a. Neural plate stage
   b. Neural groove stage
   c. Neural tube stage
   d. Neural crest
   e. Brain vesicles
      - Forebrain
         - •• Telencephalon
         - •• Diencephalon
      - Midbrain (mesencephalon)
      - Hindbrain
         - •• Metencephalon
         - •• Myelencephalon

Meninges, Ventricles, Cerebrospinal Fluid, and Blood Supply (p. 496)

1. Meninges
   a. Dura mater
      - Periosteal and meningeal layers
      - Dural sinuses
      - Infoldings of dura
      - Dural sheath of spinal cord
   b. Arachnoid mater
   c. Pia mater
2. Ventricles
   a. Lateral ventricles
   b. Interventricular foramen
   c. Third ventricle
   d. Mesencephalic aqueduct
   e. Fourth ventricle
   f. Choroid plexuses
3. Cerebrospinal fluid
   a. Production
   b. Functions
      - Buoyancy
      - Protection
      - Chemical stability
   c. Pathway of flow
   d. Hydrocephalus
4. Blood supply
5. Brain barrier system
   a. Blood-brain barrier
   b. Blood-CSF barrier
   c. Circumventricular organs

The Spinal Cord (p. 501)

1. Functions
   a. Conduction
   b. Locomotion
   c. Reflexes
2. Gross anatomy
   a. Size and extent
   b. Four regions
   c. Cervical and lumbar enlargements
   d. Inferior end
      - Medullary cone
      - Terminal filum
      - Cauda equina
3. Cross-sectional anatomy
   a. Gray matter
      - Dorsal and ventral horns
      - Gray commissure
      - Lateral horn
   b. White matter
      - Columns (funiculi)
      - Decussation of nerve fibers
      - Ascending tracts
         - •• Gracile and cuneate fasciculi
         - •• Spinothalamic tracts
         - •• Spinocerebellar tracts
      - Descending tracts
         - •• Corticospinal tracts
         - •• Tectospinal tracts
         - •• Reticulospinal tracts
         - •• Vestibulospinal tracts

The Hindbrain and Midbrain (p. 507)

1. Medulla oblongata
   a. Pyramids
   b. Olive
   c. Inferior olivary nucleus
   d. Cardiac center
   e. Vasomotor center
   f. Respiratory centers
   g. Nuclei of cranial nerves IX–XII
2. Pons
   a. Roots of cranial nerves V–VIII
   b. Nuclei for physiological control
   c. Relay from cerebrum to cerebellum
3. Cerebellum
   a. Anatomy
      - Cerebellar hemispheres
      - Peduncles
      - Folia
      - Arbor vitae
      - Deep nuclei
      - Purkinje cells
   b. Functions
      - Motor coordination
      - Cognitive and emotional functions
4. Midbrain
   a. Roots of cranial nerves III–IV
   b. Cerebral peduncles
   c. Tegmentum
   d. Substantia nigra
   e. Central gray matter
   f. Tectum
      - Superior colliculi
      - Inferior colliculi
5. Reticular formation
   a. Somatic motor functions
   b. Cardiovascular functions
   c. Pain modulation
   d. Sleep and consciousness

The Forebrain (p. 512)

1. Diencephalon
   a. Thalamus
      - Anatomy and location
      - Sensory relay function
   b. Hypothalamus
      - Anatomy and location

- Functions
    - Hormone secretion
    - Autonomic effects on viscera
    - Thermoregulation
    - Hunger and thirst
    - Sleep and circadian rhythms
    - Memory pathways
    - Emotional behavior
  - c. Epithalamus
    - Pineal gland
    - Habenula
2. Cerebrum
  - a. Lobes
    - Frontal lobe
    - Parietal lobe
    - Occipital lobe
    - Temporal lobe
    - Insula
  - b. White matter
    - Projection tracts
    - Commissural tracts
    - Association tracts
  - c. Cerebral cortex
    - Types of neurons
      - Stellate cells
      - Pyramidal cells
    - Types of cortex
      - Neocortex
      - Paleocortex
      - Archicortex
  - d. Basal nuclei
    - Caudate nucleus
    - Putamen
    - Globus pallidus
  - e. Limbic system
    - Amygdala
    - Hippocampus

Higher Brain Functions (p. 519)

1. Brain waves and electroencephalograms
  - a. Alpha waves
  - b. Beta waves
  - c. Theta waves
  - d. Delta waves
2. Sleep
  - a. Sleep versus coma
  - b. Control centers
    - Suprachiasmatic nucleus
    - Reticular formation
  - c. Sleep stages
    - Stages 1–4
    - REM sleep
  - d. Sleep functions
3. Cognition
  - a. Evidence from brain lesions
  - b. Functions of association areas
    - Parietal lobe and body awareness
    - Temporal lobe and recognition of objects
    - Frontal lobe and personality
4. Memory
  - a. Importance of forgetting
  - b. Hippocampus and memory consolidation
  - c. Roles of amygdala and cerebellum
5. Emotion
  - a. Role of hypothalamus and amygdala
  - b. Methods of study
6. Sensation
  - a. Somesthetic sensation
    - Decussation of somesthetic pathways
    - Somatotopy and the sensory homunculus
  - b. Special senses
    - Olfaction
    - Gustation
    - Vision
    - Hearing
    - Equilibrium
  - c. Sensory association areas
    - Somesthetic association area
    - Visual association area
    - Auditory association area
7. Motor control
  - a. Motor association area in frontal lobe
  - b. Primary motor area in precentral gyrus
  - c. Somatotopy and the motor homunculus
  - d. Upper and lower motor neurons
  - e. Function of basal nuclei
  - f. Function of cerebellum
8. Language
  - a. Wernicke's area
  - b. Angular gyrus
  - c. Broca's area
  - d. Primary motor cortex
  - e. Affective language area
  - f. Brain lesions and aphasia
9. Cerebral lateralization
  - a. The categorical hemisphere
  - b. The representational hemisphere
  - c. Relation to handedness
  - d. Variations with age and sex

Selected Vocabulary

For additional terminology not listed here, see tables 14.1 and 14.2.

rostral 491
caudal 491
gyri 491
sulci 491
fissures 491
brainstem 491
neural plate 491
neural groove 491
neural fold 491
neural tube 492
neural crest 492
meninges 496
dura mater 496
arachnoid mater 496
pia mater 496
dural sinuses 496
superior sagittal sinus 496
transverse sinus 496
dural sheath 496
epidural space 496
cerebrospinal fluid 496
brain barrier system 499
blood-brain barrier 499
blood-CSF barrier 499
circumventricular organs 500
central pattern generators 501
spinal cord 501
cervical region 501
thoracic region 501
lumbar region 501
sacral region 501
cervical enlargement 501
lumbar enlargement 501
medullary cone 501
terminal filum 501
cauda equina 501
dorsal horn 501
ventral horn 501
gray commissure 501
central canal 501
lateral horn 502
dorsal column 502
lateral column 502
ventral column 502
tract 502
decussation 503
contralateral 503
ipsilateral 503
first-order neuron 503
second-order neuron 504
third-order neuron 504
cardiac center 508
vasomotor center 508
respiratory centers 508
Purkinje cell 508
habituation 512
cerebral cortex 516
stellate cell 516
pyramidal cell 516
neocortex 517
brain waves 519

electroencephalogram 519
alpha waves 520
beta waves 520
theta waves 520
delta waves 520
sleep 520
coma 520
suprachiasmatic nucleus 520
circadian rhythm 520
stage 1–stage 4 sleep 520
REM sleep 520
sleep paralysis 520
cognition 520
association area 520
agnosia 521
anterograde amnesia 522
retrograde amnesia 522
memory consolidation 522
somesthetic sensation 523
special senses 523
primary somesthetic cortex 523
somatotopy 524
primary sensory area 524
somesthetic association area 524
visual association area 524
auditory association area 525
motor association area 525
upper motor neuron 525
lower motor neuron 525
Wernicke's area 526
Broca's area 526
aphasia 526
cerebral lateralization 528

Testing Your Recall

1. Below L2, the vertebral canal is occupied by a bundle of spinal nerve roots called
   a. the filum terminale.
   b. the descending tracts.
   c. the gracile fasciculus.
   d. the conus medullaris.
   e. the cauda equina.

2. Which of these is caudal to the hypothalamus?
   a. the thalamus
   b. the optic chiasm
   c. the mesencephalic aqueduct
   d. the pituitary gland
   e. the corpus callosum

3. If the telencephalon were removed from a 5-week-old embryo, which of the following structures would fail to develop in the fetus?
   a. cerebral hemispheres
   b. the thalamus
   c. the midbrain
   d. the medulla oblongata
   e. the spinal cord

4. The blood-CSF barrier is formed by
   a. blood capillaries.
   b. endothelial cells.
   c. protoplasmic astrocytes.
   d. oligodendrocytes.
   e. ependymal cells.

5. A patient has a gunshot wound that caused a bone fragment to nick the spinal cord. The patient now feels no pain or temperature sensations from that level of the body down. Most likely, the ________ was damaged.
   a. gracile fasciculus
   b. medial lemniscus
   c. tectospinal tract
   d. lateral corticospinal tract
   e. spinothalamic tract

6. The pyramids of the medulla oblongata contain
   a. descending corticospinal fibers.
   b. commissural fibers.
   c. ascending spinocerebellar fibers.
   d. fibers going to and from the cerebellum.
   e. ascending spinothalamic fibers.

7. Which of the following does *not* receive any visual input?
   a. the hypothalamus
   b. the frontal lobe
   c. the thalamus
   d. the occipital lobe
   e. the midbrain

8. While studying in a noisy cafeteria, you get sleepy and doze off for a few minutes. You awaken with a start and realize that all the cafeteria sounds have just "come back." While you were dozing, this auditory input was blocked from reaching your auditory cortex by
   a. the temporal lobe.
   b. the thalamus.
   c. the reticular formation.
   d. the medulla oblongata.
   e. the vestibulocochlear nerve.

9. Because of a brain lesion, a certain patient never feels full, but eats so excessively that she now weighs nearly 600 pounds. The lesion is most likely in her
   a. hypothalamus.
   b. amygdala.
   c. hippocampus.
   d. basal nuclei.
   e. pons.

10. The ________ is most closely associated with the cerebellum in embryonic development and remains its primary source of input fibers throughout life.
    a. telencephalon
    b. thalamus
    c. midbrain
    d. pons
    e. medulla

11. The right and left cerebral hemispheres are connected to each other by a thick C-shaped bundle of fibers called the ___________.

12. The brain has four chambers called ___________ filled with cerebrospinal fluid.

13. On a sagittal plane, the cerebellar white matter exhibits a branching pattern called the ___________.

14. The crossing of nerve fibers between the right and left sides as they ascend or descend the spinal cord and brainstem is called ___________.

15. Cerebrospinal fluid is secreted by a mass of blood capillaries called the ___________ in each ventricle.

16. The primary motor area of the cerebrum is the ___________ gyrus of the frontal lobe.

17. Your personality is determined mainly by which lobe of the cerebrum?

18. Areas of cerebral cortex that identify or interpret sensory information are called ___________.

19. Linear, analytical, and verbal thinking occurs in the ___________ hemisphere of the cerebrum, which is on the left in most people.

20. The motor pattern for speech is generated in an area of cortex called ___________ and then transmitted to the primary motor cortex to be carried out.

Answers in appendix B

Testing Your Comprehension

1. What do you think is the functional relationship between the tectospinal tract of the spinal cord and the corpora quadrigemina of the midbrain? Why is the tract named "tectospinal"?
2. How would a lesion in the cerebellum affect skeletal muscle function differently than a lesion in the basal nuclei?
3. Wallace is the victim of a hunting accident. A bullet grazed his vertebral column and bone fragments severed the left half of his spinal cord at segments T8 through T10. Since the accident, Wallace has had a condition called *dissociated sensory loss,* in which he feels no sensations of deep touch or limb position on the *left* side of his body below the injury and no sensations of pain or heat from the *right* side. Explain what spinal tract(s) the injury has affected and why these sensory losses are on opposite sides of the body.
4. A person can survive destruction of an entire cerebral hemisphere but cannot survive destruction of the hypothalamus, which is a much smaller mass of brain tissue. Explain this difference and describe some ways that destruction of a cerebral hemisphere would affect one's quality of life.
5. Contrast the motor effects of these three CNS injuries: (a) a stroke that destroys tissue of the left precentral gyrus, (b) a diving accident that severs the cervical spinal cord, and (c) a shrapnel wound that shatters the lumbar vertebrae and severs most nerve roots of the cauda equina.

Answers at the Online Learning Center

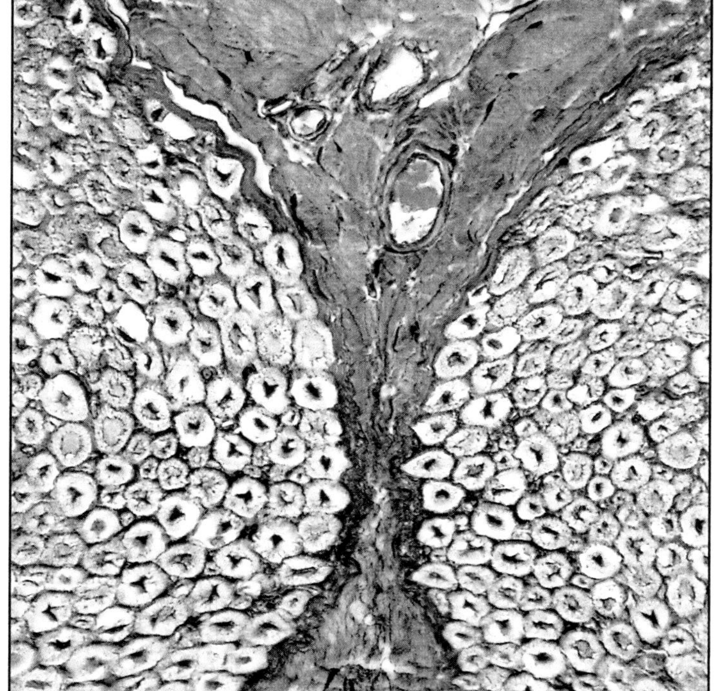

Cross section through two fascicles (bundles) of nerve fibers in a nerve

CHAPTER 15

The Peripheral Nervous System and Reflexes

Chapter Outline

Insights

Study with Your ESP CD-ROM

Refer to your ESP Tutorial CD-ROM for helpful animations, activities, and quizzes related to the material covered in this chapter. The following are suggested areas to go to on the CD and are presented according to "Unit/Topic":
- Integration and Coordination/PNS

Brushing Up

To understand this chapter, it is important that you understand or brush up on the following concepts:
- Function of antagonistic muscles (p. 343)
- Innervation of smooth muscle (p. 445)
- Neurotransmitters and synaptic transmission (pp. 472–477)
- Excitatory and inhibitory postsynaptic potentials (EPSPs and IPSPs) (pp. 478–479)
- Parallel after-discharge circuits (p. 482)

The homeostatic controls exerted by the central nervous system require fast and efficient connections with the rest of the body. These are provided by the peripheral nervous system (PNS), composed of nerves and ganglia. If the entire PNS were suddenly disabled, the most immediate effects would be complete paralysis and loss of sensation from the outside world. But this state of affairs would not last for long, for homeostatic controls throughout the body would break down and death would soon ensue.

In this chapter, we study the peripheral nerves and ganglia, somatic (skeletal muscle) reflexes, the autonomic nervous system, and autonomic (visceral) reflexes.

General Anatomy of the Peripheral Nervous System

Objectives

When you have completed this section, you should be able to

- describe the subdivisions of the peripheral nervous system;
- explain how nerves and nerve fibers are classified; and
- describe the general anatomy of nerves and ganglia.

Subdivisions of the Peripheral Nervous System

In chapter 13, we saw that the nervous system has two major divisions, the *central nervous system (CNS)* and *peripheral nervous system (PNS)*. The CNS consists of the brain and spinal cord and the PNS consists of nerves and ganglia, whose structure we examine shortly. The CNS is entirely encased in bone (the cranium and vertebral column), whereas the PNS lies mostly outside of these bony enclosures.

The PNS conducts sensory signals by way of afferent nerve fibers from receptors to the CNS, and it conducts motor signals by way of efferent nerve fibers from the CNS to the effectors (mostly gland and muscle cells). Thus, it has both sensory and motor subdivisions. Each subdivision can be further divided into somatic and visceral components, as follows:

- The **sensory (afferent) division** carries sensory signals by way of afferent nerve fibers from receptors to the CNS.
    - The **somatic**[1] **sensory division** carries signals from receptors in the skin, muscles, bones, and joints.
    - The **visceral sensory division** carries signals mainly from the viscera of the thoracic and abdominal cavities, such as the heart, lungs, stomach, and urinary bladder.
- The **motor (efferent) division** carries motor signals by way of efferent nerve fibers from the CNS to effectors (mainly gland and muscle cells).
    - The **somatic motor division** carries signals to the skeletal muscles. This output produces muscular contractions that are under voluntary control. It also controls involuntary muscle contractions, such as the *somatic reflexes* described later in this chapter.
    - The **visceral motor division (autonomic**[2] **nervous system)** carries signals to glands, cardiac muscle, and smooth muscle. We usually have no voluntary control over these effectors, and this system operates at an unconscious level. The responses of this system and its effectors are *visceral reflexes*. The autonomic nervous system has two further divisions:
        - The **sympathetic division** tends to arouse the body for action, for example by accelerating the heartbeat and increasing respiratory airflow.
        - The **parasympathetic division** tends to have a calming effect, slowing down the heartbeat, for example, but stimulating digestion.

After describing the peripheral nerves, this chapter focuses largely on the somatic reflexes mediated by the somatic sensory and motor systems, and on the autonomic reflexes mediated by the visceral sensory and motor systems.

Nerve fibers (the axons of nerve cells) of the PNS are classified according to their involvement in motor or sensory, somatic or visceral pathways (table 15.1). **Mixed nerves** contain a mixture of sensory (afferent) fibers and motor (efferent) fibers; thus they transmit signals in two directions. Most nerves are mixed. Purely **sensory nerves,** composed entirely of sensory axons, are less common; they include the optic and olfactory nerves. Nerves with

Table 15.1 The Classification of Nerve Fibers

Class	Description
Afferent fibers	Carry sensory signals from receptors to the CNS
Efferent fibers	Carry motor signals from the CNS to effectors
Somatic fibers	Innervate skin, skeletal muscles, bones, and joints
Visceral fibers	Innervate blood vessels, glands, and viscera
General fibers	Innervate widespread organs such as muscles, skin, glands, viscera, and blood vessels
Special fibers	Innervate more localized organs in the head, including the eyes, ears, olfactory and taste receptors, and muscles of chewing, swallowing, and facial expression

1. *somat* = body + *ic* = pertaining to

2. *auto* = self + *nom* = law, governance

only motor axons are called **motor nerves.** Many nerves often listed as motor are actually mixed because they carry sensory feedback (proprioception) from the muscles they innervate.

Anatomy of Nerves and Ganglia

It is important to make the distinction between a nerve and a nerve fiber. A nerve fiber is the axon of a single neuron, whereas a nerve is an organ composed of multiple nerve fibers bound together by connective tissue (fig. 15.1). If we compare a nerve fiber to a wire carrying an electrical current in one direction only, a nerve would be comparable to an electrical cable composed of thousands of wires carrying currents in both directions. A nerve contains anywhere from a few nerve fibers to more than a million. Nerves usually have a pearly white color; they often look like frayed string as they divide into smaller and smaller branches.

Nerve fibers of the PNS are ensheathed in Schwann cells, which form a neurilemma and often a myelin sheath around the axon (see chapter 13). External to the neurilemma, each fiber is surrounded by a basement membrane and a thin connective tissue sleeve called the **endoneurium.** Nerve fibers have a high metabolic rate and need a good blood supply. The endoneurium ensures that there is enough space between nerve fibers for blood capillaries to reach all of them. In most nerves, the nerve fibers are gathered in bundles called **fascicles,** each wrapped in a fibrous sheath called the **perineurium.** Several fascicles are then bundled together and wrapped in a fibrous **epineurium** to compose the nerve as a whole.

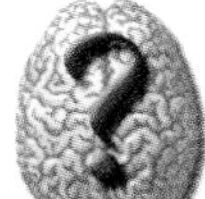

Think About It

How does the structure of a nerve compare to that of a skeletal muscle? (see p. 340) Which of the descriptive terms for nerves have similar counterparts in muscle histology?

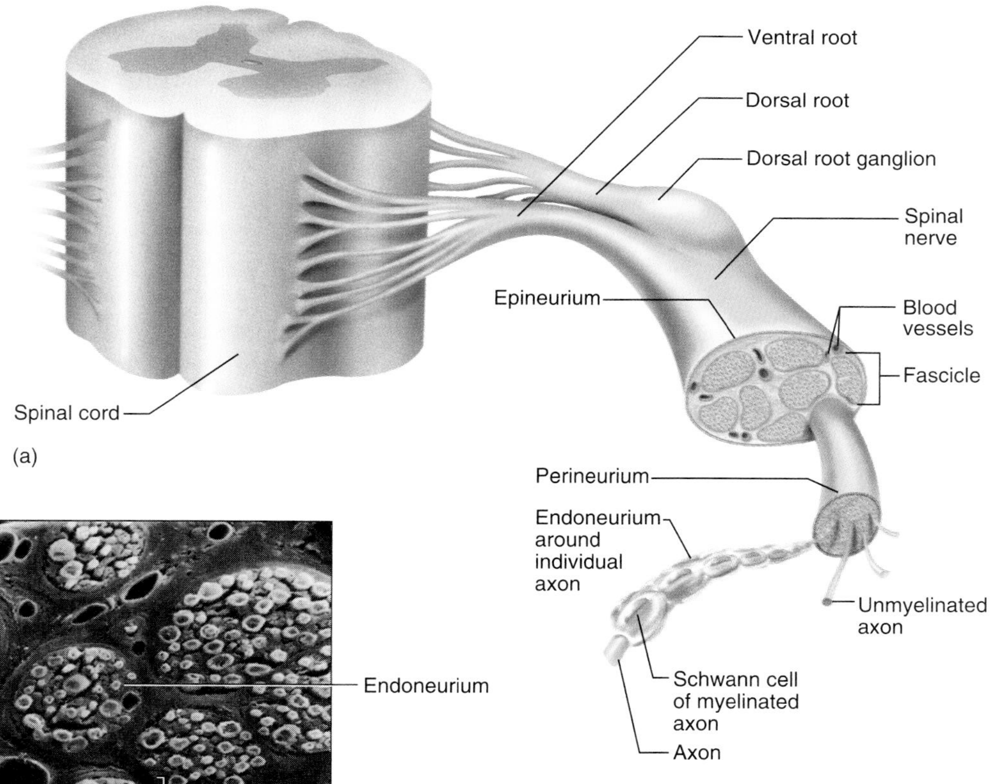

Ventral root
Dorsal root
Dorsal root ganglion
Spinal nerve
Epineurium
Blood vessels
Fascicle
Spinal cord
(a)
Perineurium
Endoneurium around individual axon
Unmyelinated axon
Schwann cell of myelinated axon
Axon

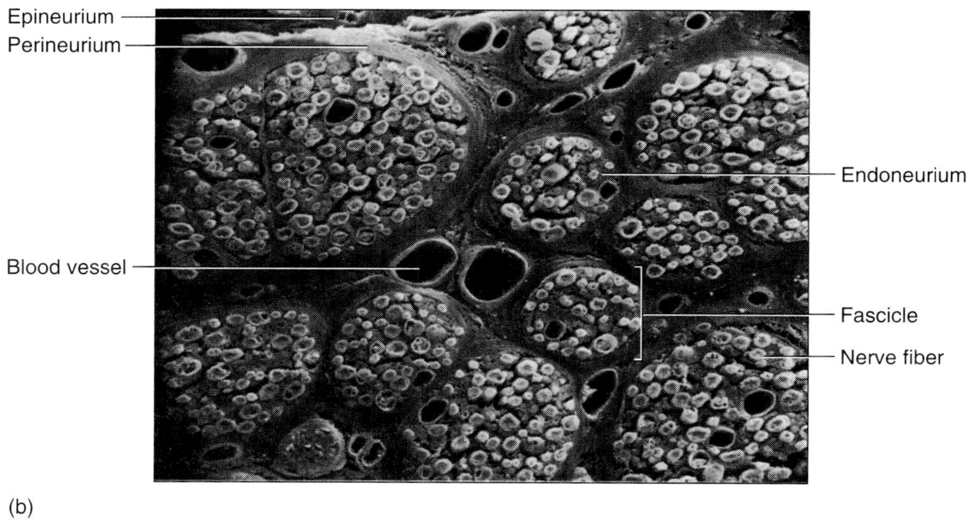

Epineurium
Perineurium
Endoneurium
Blood vessel
Fascicle
Nerve fiber
(b)

Figure 15.1 Anatomy of a Nerve. (*a*) A spinal nerve and its association with the spinal cord. (*b*) Cross section of a nerve (SEM). Credit for *b:* From Richard E. Kessel and Randy H. Kardon, *Tissues and Organs: A Text-Atlas of Scanning Electron Microscopy,* 1979, W. H. Freeman and Company.

A **ganglion**[3] (GANG-lee-un; plural, *ganglia*) is a cluster of neuron cell bodies (somas) in the PNS, enveloped in an epineurium continuous with that of a nerve. It appears as a swelling along the course of a nerve, like a knot in a string. Among the somas, there are bundles of nerve fibers leading into and out of the ganglion. The *dorsal root ganglia* associated with the spinal cord (fig. 15.2) contain the unipolar neurons of the sensory nerve fibers that carry signals to the cord. Each fiber, as we saw in chapter 13, forms a T-shaped connection with its soma in the ganglion. The fiber passes through the ganglion without synapsing. In the autonomic nervous system, however, one nerve fiber (the *preganglionic fiber*) enters the ganglion and in many cases synapses with another neuron. The axon of the second neuron (the *postganglionic fiber*) leaves the ganglion.

3. *gangli* = knot

Before You Go On

Answer the following questions to test your understanding of the preceding section:

1. What structures are supplied by the visceral divisions of the nervous system? What structures are supplied by the somatic divisions?
2. Distinguish between a nerve and a nerve fiber.
3. Describe the relationship of the endoneurium, perineurium, and epineurium to the nerve fibers of the PNS.
4. Distinguish between afferent and efferent nerve fibers, somatic and visceral fibers, and general and special fibers.
5. What structures can be found in a ganglion that are found nowhere else in the PNS?

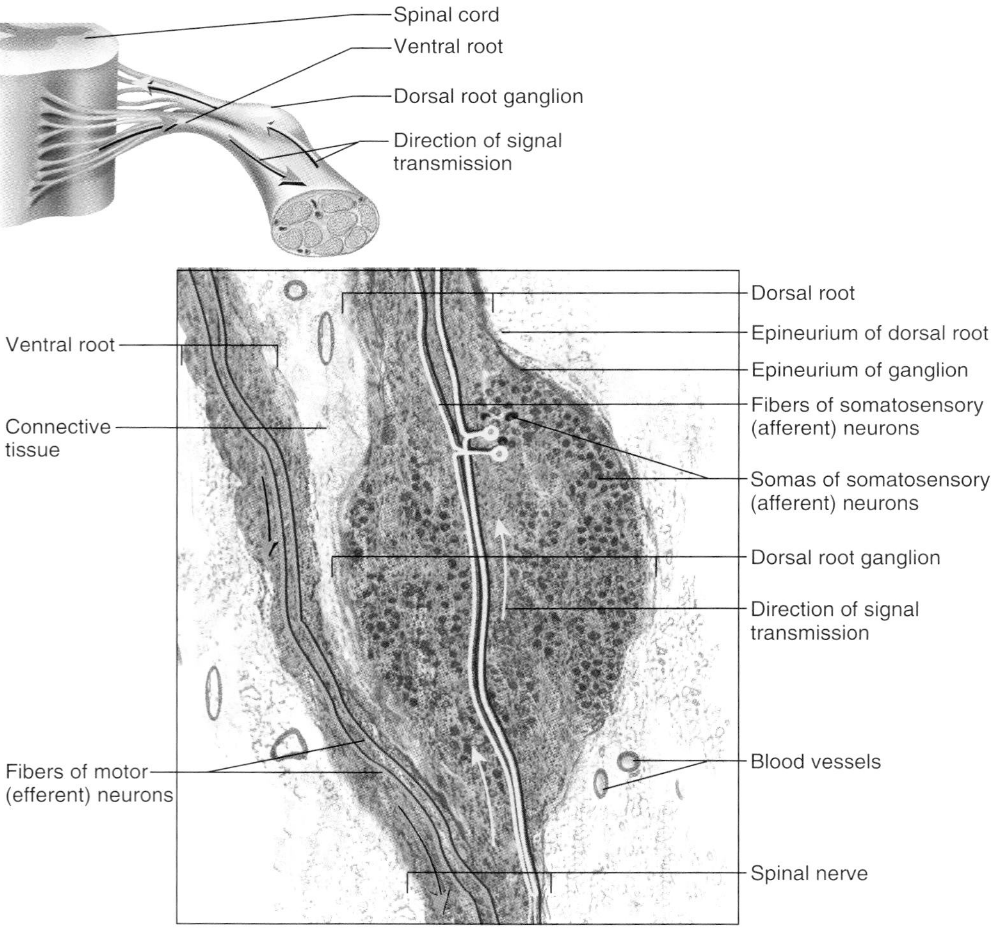

Figure 15.2 Anatomy of a Ganglion. The dorsal root ganglion contains the somas of unipolar sensory neurons conducting signals to the spinal cord. To the left of it is the ventral root of the spinal nerve, which conducts motor signals away from the spinal cord. (The ventral root is not part of the ganglion.)

The Cranial Nerves

Objectives

When you have completed this section, you should be able to

- list the 12 cranial nerves by name and number;
- identify where each cranial nerve originates and terminates; and
- state the functions of each cranial nerve.

Cranial nerves attach to the base of the brain (fig. 15.3), exit the cranium through its foramina, and lead to muscles and sense organs primarily in the head and neck. There are 12 pairs of cranial nerves, numbered I to XII starting with the most rostral. Each nerve also has a descriptive name. The cranial nerves are illustrated and described in table 15.2.

Cranial nerves are traditionally classified as sensory (I, II, and VIII), motor (III, IV, VI, XI, and XII), and mixed (V, VII, IX, and X). In reality, only cranial nerves I and II (for smell and vision) are purely sensory, whereas all the rest contain both afferent and efferent fibers and are therefore mixed nerves. Those traditionally classified as motor not only stimulate muscle contractions but also contain afferent fibers of proprioception, which provide your brain with unconscious feedback for controlling muscle contraction and make you consciously aware of such things as the position of your tongue and orientation of your head. Cranial nerve VIII, concerned with hearing and equilibrium, is traditionally classified as sensory, but it has motor fibers that return signals to the inner ear and "tune" it to sharpen our sense of hearing. The nerves traditionally classified as mixed have sensory functions quite unrelated to their motor functions—for example, the facial nerve (VII) has a sensory role in taste and a motor role in controlling facial expressions.

In order to teach the traditional classification (which is relevant for such purposes as board examinations and comparison to other books), yet remind you that all but two of these nerves are mixed, table 15.2 describes many of the nerves as *predominantly* sensory or motor.

The motor fibers of the cranial nerves begin in nuclei of the brainstem and lead to glands and muscles. The sensory fibers begin in receptors located mainly in the head and neck and lead mainly to the brainstem. Pathways for the special senses are described in chapter 16. Sensory fibers for proprioception begin in the muscles innervated by the motor fibers of the cranial nerves, but they often travel to the brain in a different nerve than the one that provides the motor innervation to the muscle.

In most cases, the cranial nerves carry fibers between the brain and the ipsilateral receptors and effectors. Thus a lesion to one side of the brainstem causes a sensory or motor deficit to the *ipsilateral* side of the head. This contrasts with lesions to the motor and somesthetic cortex of the cerebrum, which, as we saw in the last chapter, cause sensory and motor deficits on the *contralateral* side of the body. The exceptions are the optic nerve, where half the fibers decussate to the opposite side of the brain (see chapter 16), and trochlear nerve, in which all efferent fibers go to a muscle of the contralateral eye.

The Cranial Nerves—An Aid to Memory

Generations of biology and medical students have relied on mnemonic (memory-aiding) phrases and ditties, ranging from the sublimely silly to the unprintably ribald, to help them remember the cranial nerves and other anatomy. An old classic began, "On Old Olympus' towering tops. . .," with the first letter of each word matching the first letter of each cranial nerve (olfactory, optic, oculomotor, etc.). Some cranial nerves have changed names, however, since that passage was devised. One of the author's former students* devised a new and better mnemonic that can remind you of the first two to four letters of most cranial nerves:

Old	olfactory (I)
Opie	optic (II)
occasionally	oculomotor (III)
tries	trochlear (IV)
trigonometry	trigeminal (V)
and	abducens (VI)
feels	facial (VII)
very	vestibulocochlear (VIII)
gloomy,	glossopharyngeal (IX)
vague,	vagus (X)
and	accessory (XI)
hypoactive.	hypoglossal (XII)

* Courtesy of Ms. Marti Haykin

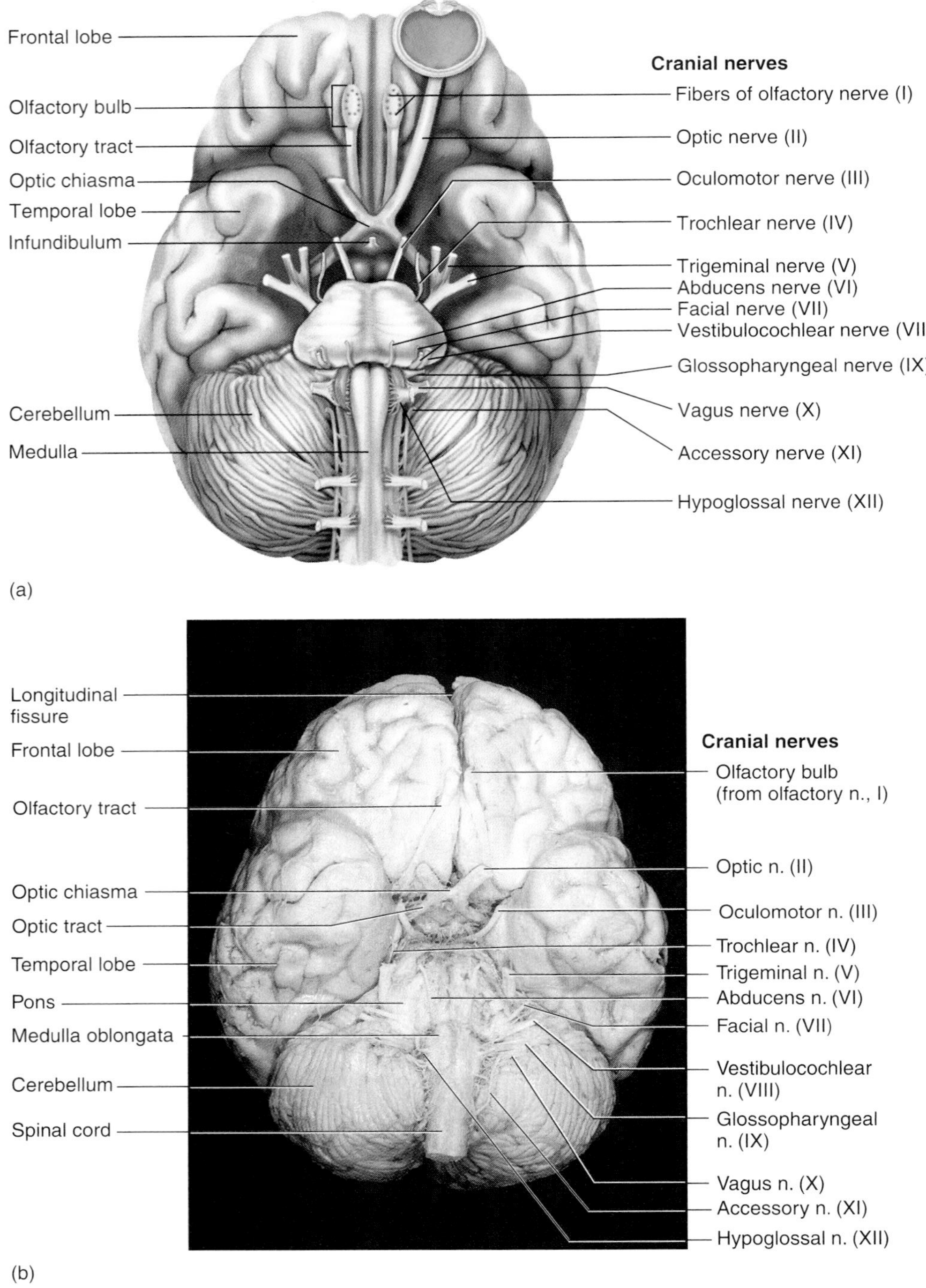

Figure 15.3 **The Cranial Nerves.** (*a*) Base of the brain showing the 12 cranial nerves. (*b*) Photograph of the cranial nerves.

Table 15.2 The Cranial Nerves

Origins of proprioceptive fibers are not tabulated; they are the muscles innervated by the motor fibers. Nerves listed as mixed or sensory are agreed by all authorities to be either mixed or purely sensory nerves. Nerves classified as *predominantly* motor or sensory are traditionally classified that way but contain some fibers of the other type.

I. Olfactory Nerve

Composition: Sensory

Function: Smell

Origin: Olfactory mucosa in nasal cavity

Termination: Olfactory bulbs beneath frontal lobe of brain

Cranial passage: Cribriform plate of ethmoid bone

Effects of damage: Impaired sense of smell

Clinical test: Determine whether subject can smell (not necessarily identify) aromatic substances such as coffee, vanilla, clove oil, or soap

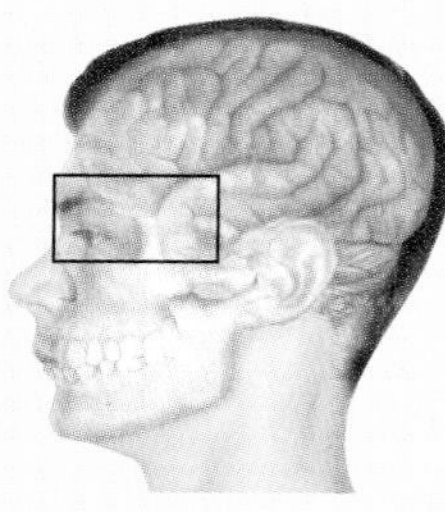

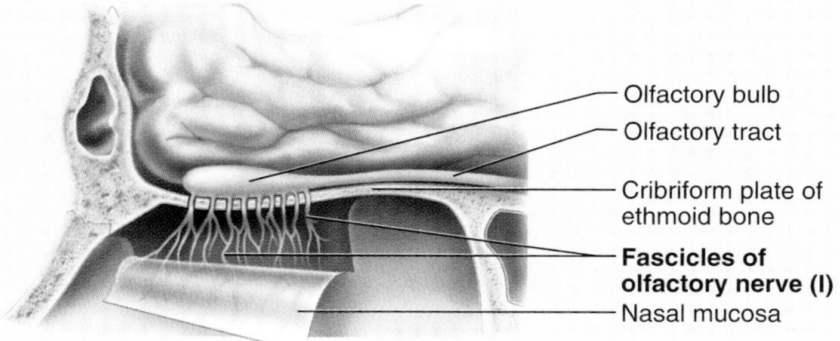

II. Optic Nerve

Composition: Sensory

Function: Vision

Origin: Retina

Termination: Thalamus

Cranial passage: Optic foramen

Effects of damage: Blindness in part or all of the visual field

Clinical test: Inspect retina with ophthalmoscope; test peripheral vision and visual acuity

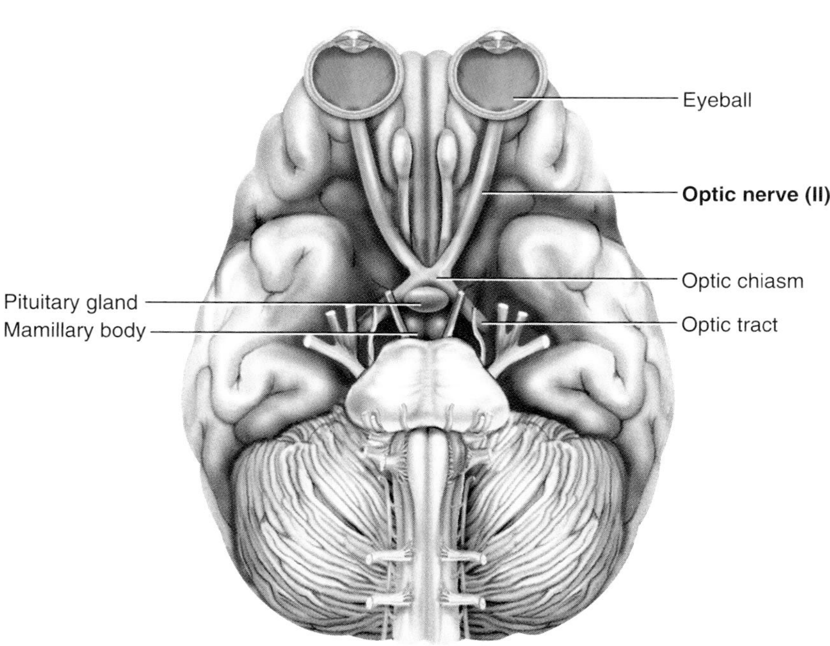

(continued)

Table 15.2 The Cranial Nerves (continued)

III. Oculomotor (OC-you-lo-MO-tur) Nerve

Composition: Predominantly motor

Function: Eye movements, opening of eyelid, constriction of pupil, focusing, proprioception

Origin: Midbrain

Termination: Somatic fibers lead to levator palpebrae superioris, superior, medial, and inferior rectus, and inferior oblique muscles; parasympathetic fibers enter eyeball and lead to constrictor of iris and ciliary muscle of lens

Cranial passage: Superior orbital fissure

Effects of damage: Drooping eyelid, dilated pupil, inability to move eye in certain directions, tendency of eye to rotate laterally at rest, double vision, and difficulty focusing

Clinical test: Look for differences in size and shape of right and left pupil; test pupillary response to light; test ability to track moving objects

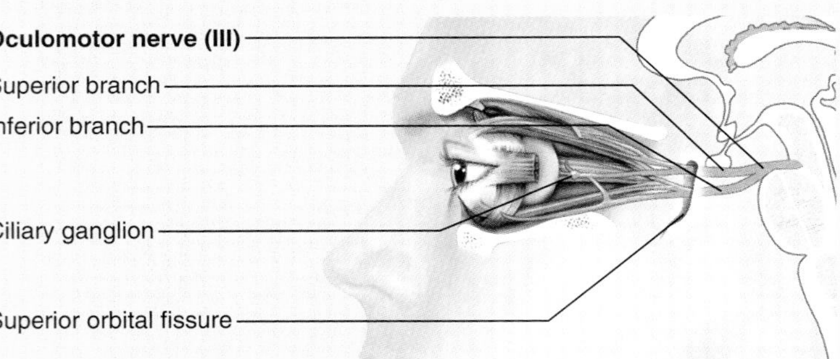

IV. Trochlear (TROCK-lee-ur) Nerve

Composition: Predominantly motor

Function: Eye movements and proprioception

Origin: Midbrain

Termination: Superior oblique muscle of eye

Cranial passage: Superior orbital fissure

Effects of damage: Double vision and inability to rotate eye inferolaterally

Clinical test: Test eye movements with cranial nerve III

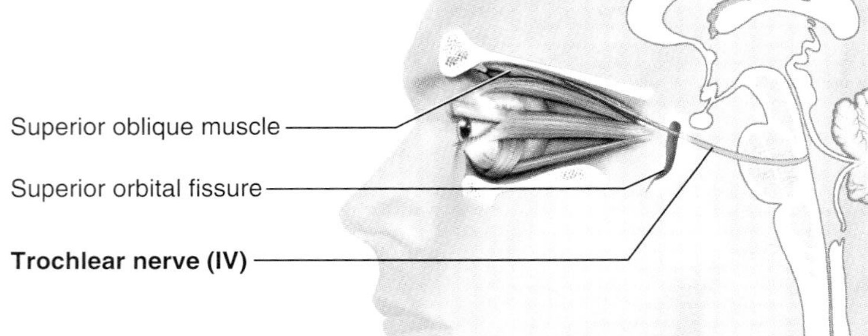

(continued)

Table 15.2 The Cranial Nerves (continued)

V. Trigeminal[4] (tri-JEM-ih-nul) Nerve

Largest of the cranial nerves; consists of three divisions designated V_1 to V_3

V_1, *Ophthalmic Division*

Composition: Sensory

Function: Main sensory nerve of upper face (touch, temperature, pain)

Origin: Superior region of face as illustrated, surface of eyeball, tear gland, superior nasal mucosa, frontal and ethmoid sinuses

Termination: Pons

Cranial passage: Superior orbital fissure

Effects of damage: Loss of sensation

Clinical test: Test corneal reflex—blinking in response to light touch to eyeball

V_2, *Maxillary Division*

Composition: Sensory

Function: Same sensations as V_1 lower on face

Origin: Middle region of face as illustrated, nasal mucosa, maxillary sinus, palate, upper teeth and gums

Termination: Pons

Cranial passage: Foramen rotundum and infraorbital foramen

Effects of damage: Loss of sensation

Clinical test: Test sense of touch, pain, and temperature with light touch, pinpricks, and hot and cold objects

V_3, *Mandibular Division*

Composition: Mixed

Function: Same sensations as V_1–V_2 lower on face; mastication

Sensory origin: Inferior region of face as illustrated, anterior two-thirds of tongue (but not taste buds), lower teeth and gums, floor of mouth, dura mater

Sensory termination: Pons

Motor origin: Pons

Motor termination: Anterior belly of digastric, masseter, temporalis, mylohyoid, pterygoids, and tensor tympani of middle ear

Cranial passage: Foramen ovale

Effects of damage: Loss of sensation; impaired chewing

Clinical test: Assess motor functions by palpating masseter and temporalis muscles while subject clenches teeth; test ability of subject to move mandible from side to side and to open mouth against resistance

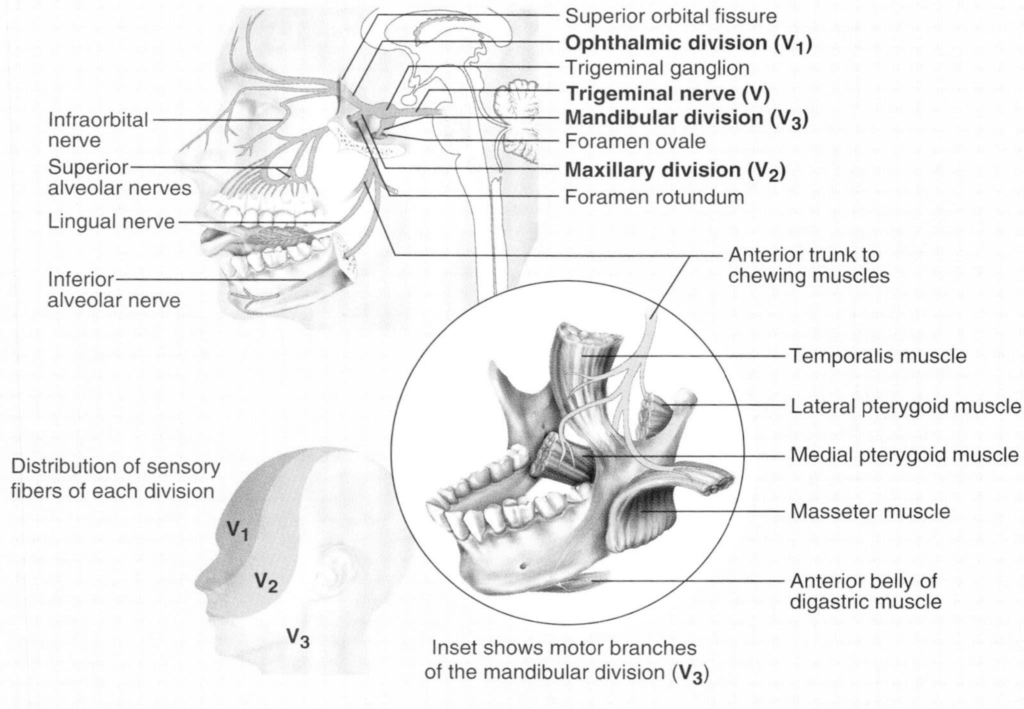

(continued)

4. *tri* = three + *gemin* = twin

Table 15.2 The Cranial Nerves (continued)

VI. Abducens (ab-DOO-senz) Nerve

Composition: Predominantly motor

Function: Eye movements

Origin: Inferior pons

Termination: Lateral rectus muscle of eye

Cranial passage: Superior orbital fissure

Effects of damage: Inability to rotate eye laterally; at rest, eye rotates medially because of action of antagonistic muscles

Clinical test: Test eye movements with cranial nerve III

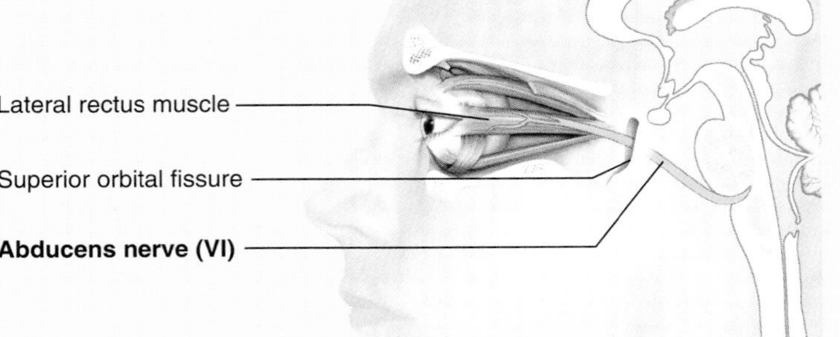

VII. Facial Nerve

Composition: Mixed

Function: Major motor nerve of facial expression; autonomic control of tear glands, nasal and palatine glands, submandibular and sublingual salivary glands; sense of taste

Sensory origin: Taste buds on anterior two-thirds of tongue

Sensory termination: Thalamus

Motor origin: Pons

Motor termination: Divides into *temporal, zygomatic, buccal, mandibular,* and *cervical branches.* Somatic motor fibers end on digastric muscle, stapedius muscle of middle ear, and muscles of facial expression; autonomic fibers end on submandibular and sublingual salivary glands

Cranial passage: Stylomastoid foramen

Effects of damage: Inability to control facial muscles; sagging resulting from loss of muscle tone; distorted sense of taste, especially for sweets

Clinical test: Test anterior two-thirds of tongue with substances such as sugar, salt, vinegar (sour), and quinine (bitter); test response of tear glands to ammonia fumes; test motor functions by asking subject to close eyes, smile, whistle, frown, raise eyebrows, etc.

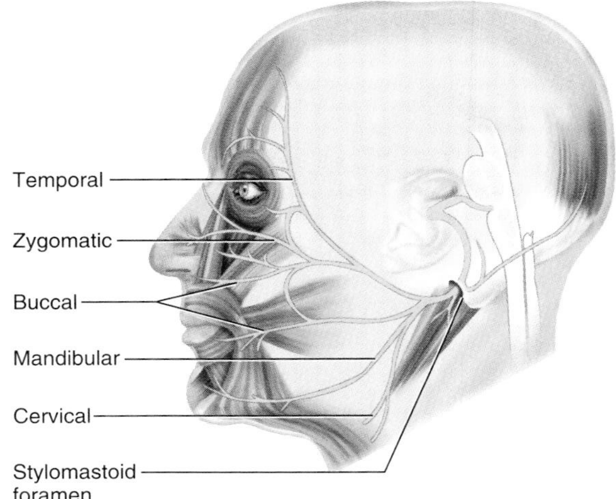

Facial nerve (VII)
Internal acoustic meatus
Geniculate ganglion
Sphenopalatine ganglion
Tear gland
Chorda tympani branch (taste)
Submandibular ganglion
Parasympathetic fibers
Sublingual gland
Submandibular gland
Stylomastoid foramen
Motor branch to muscles of facial expression

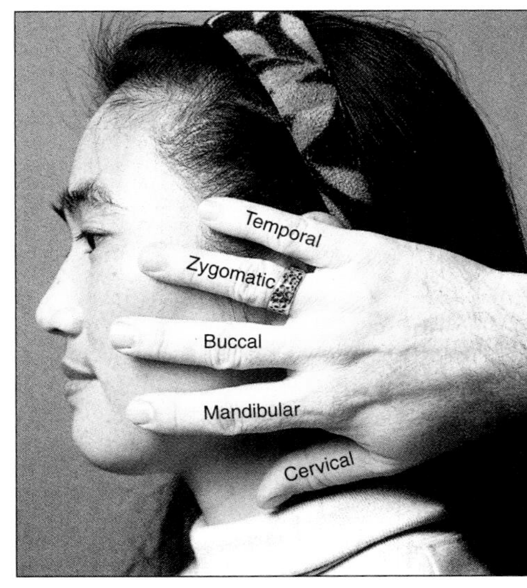

A Way to Remember the Distribution of the Five Major Branches of the Facial Nerve.

(continued)

Table 15.2 The Cranial Nerves (continued)

VIII. Vestibulocochlear (vess-TIB-you-lo-COC-lee-ur) Nerve

Composition: Predominantly sensory

Function: Hearing and equilibrium

Sensory origin: Inner ear

Sensory termination: Fibers for equilibrium end at junction of pons and medulla; fibers for hearing end in medulla

Motor origin: Pons

Motor termination: Outer hair cells of cochlea of inner ear (see chapter 16)

Cranial passage: Internal auditory meatus

Effects of damage: Nerve deafness, dizziness, nausea, loss of balance, and nystagmus (involuntary oscillation of the eyes from side to side)

Clinical test: Test hearing, balance, and ability to walk a straight line

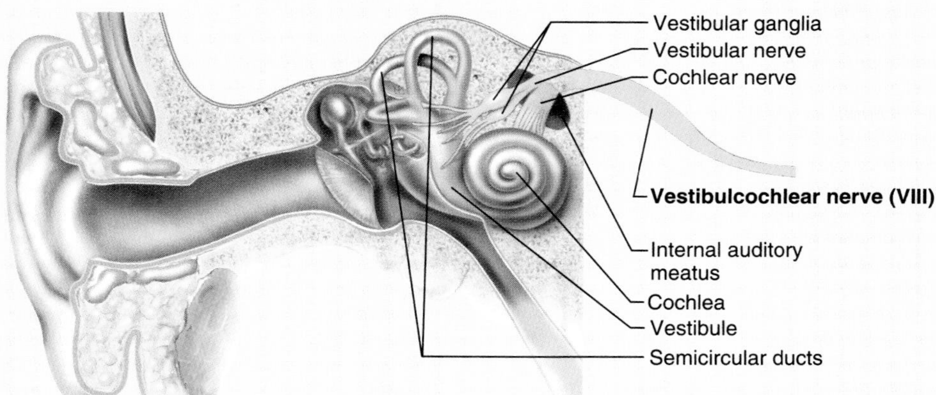

IX. Glossopharyngeal (GLOSS-oh-fah-RIN-jee-ul) Nerve

Composition: Mixed

Function: Swallowing, salivation, gagging; regulation of blood pressure and respiration; touch, pressure, taste, and pain sensations from tongue and pharynx; touch, pain, and temperature sensations from outer ear

Sensory origin: Pharynx, middle and outer ear, posterior one-third of tongue (including taste buds), internal carotid arteries

Sensory termination: Medulla oblongata

Motor origin: Medulla oblongata

Motor termination: Parotid salivary gland, glands of posterior tongue, stylopharyngeal muscle (which dilates the pharynx during swallowing)

Cranial passage: Jugular foramen

Effects of damage: Loss of bitter and sour taste; impaired swallowing

Clinical test: Test gag reflex, swallowing, and coughing; note speech impediments; test posterior one-third of tongue using bitter and sour substances

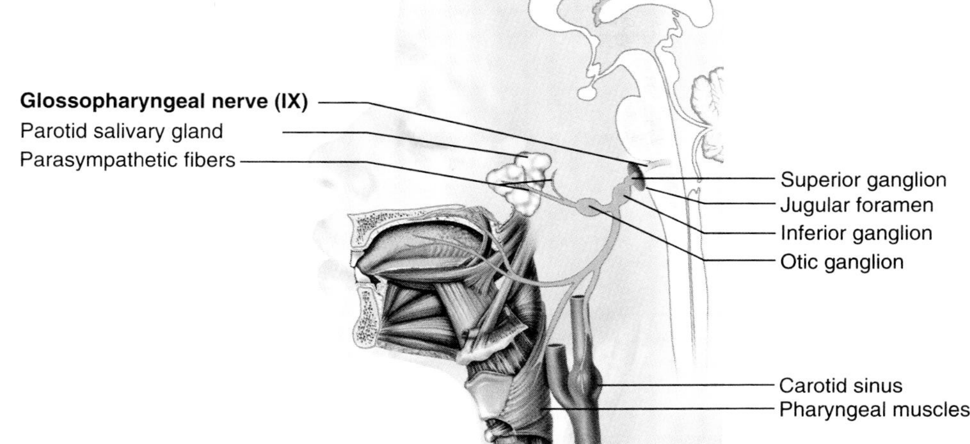

(continued)

Table 15.2 The Cranial Nerves (continued)

X. Vagus[5] (VAY-gus) Nerve

Composition: Mixed

Function: Swallowing; taste; speech; pulmonary, cardiovascular, and gastrointestinal regulation; sensations of hunger, fullness, and intestinal discomfort

Sensory origin: Thoracic and abdominal viscera, root of tongue, epiglottis, pharynx, larynx, outer ear, dura mater

Sensory termination: Medulla oblongata

Motor origin: Medulla oblongata

Motor termination: Tongue, palate, pharynx, larynx, thoracic and abdominal viscera

Cranial passage: Jugular foramen

Effects of damage: Hoarseness or loss of voice; impaired swallowing and gastrointestinal motility; fatal if both vagus nerves are damaged

Clinical test: Test with cranial nerve IX

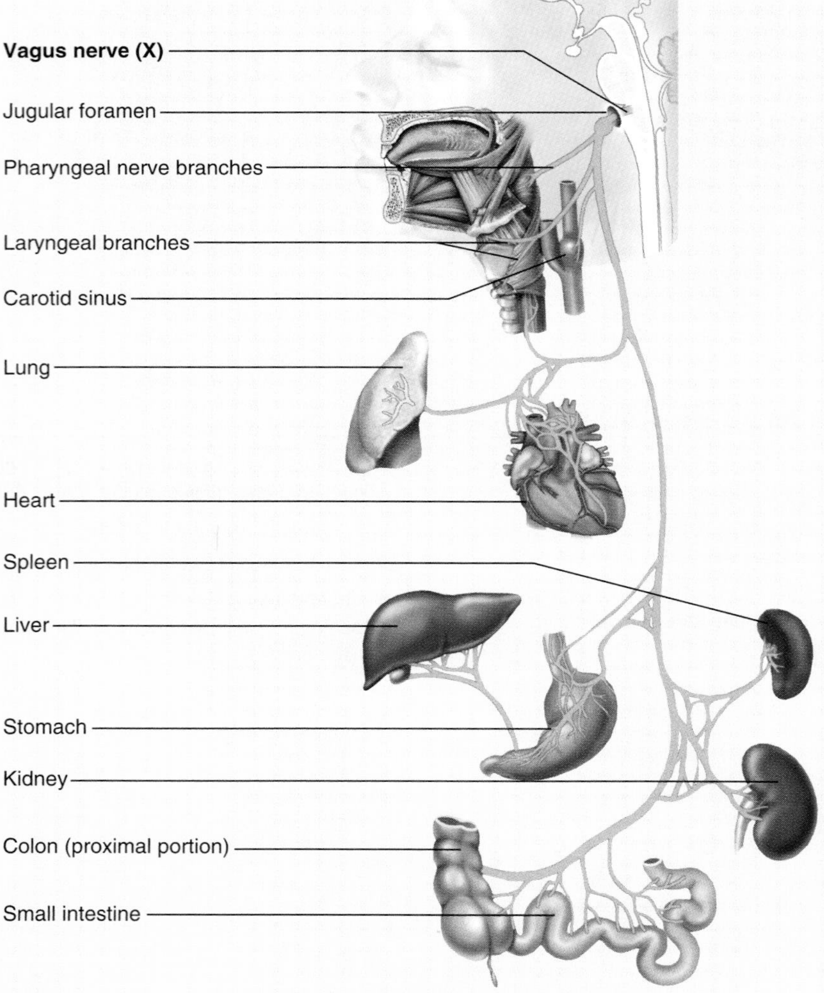

(continued)

5. *vag* = wandering

Table 15.2 The Cranial Nerves (continued)

XI. Accessory (Spinal Accessory) Nerve

Composition: Predominantly motor

Function: Swallowing; head, neck, and shoulder movements

Origin: Medulla oblongata and segments C1 through C5 or C6 of spinal cord

Termination: Palate, pharynx, sternocleidomastoid and trapezius muscles

Cranial passage: Jugular foramen

Effects of damage: Impaired movement of head, neck, and shoulders; difficulty in shrugging shoulders on damaged side; paralysis of sternocleidomastoid, causing head to turn toward injured side

Clinical test: Test ability to rotate head and shrug shoulders against resistance

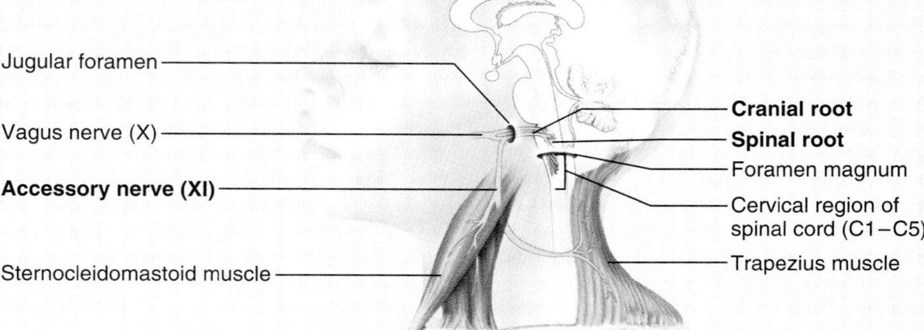

XII. Hypoglossal (HY-po-GLOSS-ul) Nerve

Composition: Predominantly motor

Function: Tongue movements of speech, food manipulation, and swallowing

Origin: Medulla oblongata

Cranial passage: Hypoglossal canal

Termination: Intrinsic and extrinsic muscles of tongue, thyrohyoid and geniohyoid muscles

Effects of damage: Difficulty in speech and swallowing; inability to protrude tongue if both right and left nerves are injured; deviation toward injured side, and atrophy of tongue on that side, if only one nerve is damaged

Clinical test: Note deviations of tongue as subject protrudes and retracts it; test ability to protrude tongue against resistance

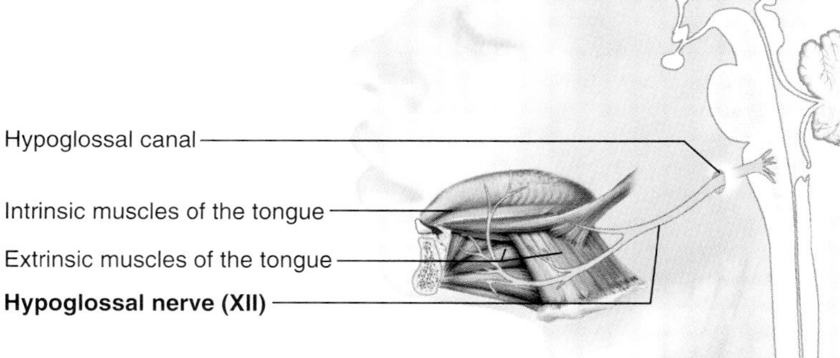

Clinical Insight 15.1

Some Cranial Nerve Disorders

Trigeminal neuralgia[6] *(tic douloureux*[7]*)* is a syndrome characterized by recurring episodes of intense stabbing pain in the trigeminal nerve. The cause is unknown; there is no visible change in the nerve. It usually occurs after the age of 50 and mostly in women. The pain lasts only a few seconds to a minute or two, but it strikes at unpredictable intervals and sometimes up to a hundred times a day. The pain usually occurs in a specific zone of the face, such as around the mouth and nose. It may be triggered by touch, drinking, tooth brushing, or washing the face. Analgesics (pain relievers) give only limited relief. Severe cases are treated by cutting the nerve, but this also deadens most other sensation in that side of the face.

Bell[8] *palsy* is a degenerative disorder of the facial nerve, probably due to a virus. It is characterized by paralysis of the facial muscles on one side with resulting distortion of the facial features, such as sagging of the mouth or lower eyelid. The paralysis may interfere with speech, prevent closure of the eye, and cause excessive tear secretion. There may also be a partial loss of the sense of taste. Bell palsy may appear abruptly, sometimes overnight, and often disappears spontaneously within 3 to 5 weeks.

6. *neur* = nerve + *algia* = pain
7. *douloureux* = painful
8. Sir Charles Bell (1774–1842), Scottish physician

Before You Go On

Answer the following questions to test your understanding of the preceding section:

6. List the purely sensory cranial nerves and state the function of each.
7. What is the only cranial nerve to extend beyond the head-neck region?
8. If the oculomotor, trochlear, or abducens nerve were damaged, the effect would be similar in all three cases. What would that effect be?
9. Which cranial nerve carries sensory signals from the greatest area of the face?
10. Name two cranial nerves involved in the sense of taste and explain where their sensory fibers originate.

The Spinal Nerves

Objectives

When you have completed this section, you should be able to

- describe the attachment of a spinal nerve to the spinal cord;
- trace the branches of a spinal nerve distal to its attachment;
- name the five plexuses of spinal nerves and describe their general anatomy;
- name some major nerves that arise from each plexus; and
- explain the relationship of dermatomes to the spinal nerves.

There are 31 pairs of **spinal nerves:** 8 cervical (C1–C8), 12 thoracic (T1–T12), 5 lumbar (L1–L5), 5 sacral (S1–S5), and 1 coccygeal (Co) (fig. 15.4). The first cervical nerve emerges between the skull and atlas, and the others emerge through intervertebral foramina, including the anterior and posterior foramina of the sacrum.

Proximal Branches

Each spinal nerve has two points of attachment to the spinal cord (fig. 15.5). Dorsally, a branch of the spinal nerve called the **dorsal root** divides into six to eight *nerve rootlets* that enter the spinal cord (fig. 15.6). A little distal to the rootlets is a swelling called the **dorsal root ganglion,** which contains the somas of afferent neurons. Ventrally, another row of six to eight rootlets leave the spinal cord and converge to form the **ventral root.** The dorsal and ventral roots merge, penetrate the dural sac, enter the intervertebral foramen, and there form the spinal nerve proper.

Spinal nerves are mixed nerves, with a two-way traffic of afferent (sensory) and efferent (motor) signals. Afferent signals approach the cord by way of the dorsal root and enter the dorsal horn of the gray matter. Efferent signals begin at the somas of motor neurons in the ventral horn and leave the spinal cord via the ventral root. Some viruses invade the central nervous system by way of these roots (see clinical insight 15.2).

The dorsal and ventral roots are shortest in the cervical region and become longer inferiorly. The roots that arise from segments L2 to Co of the cord form the *cauda*

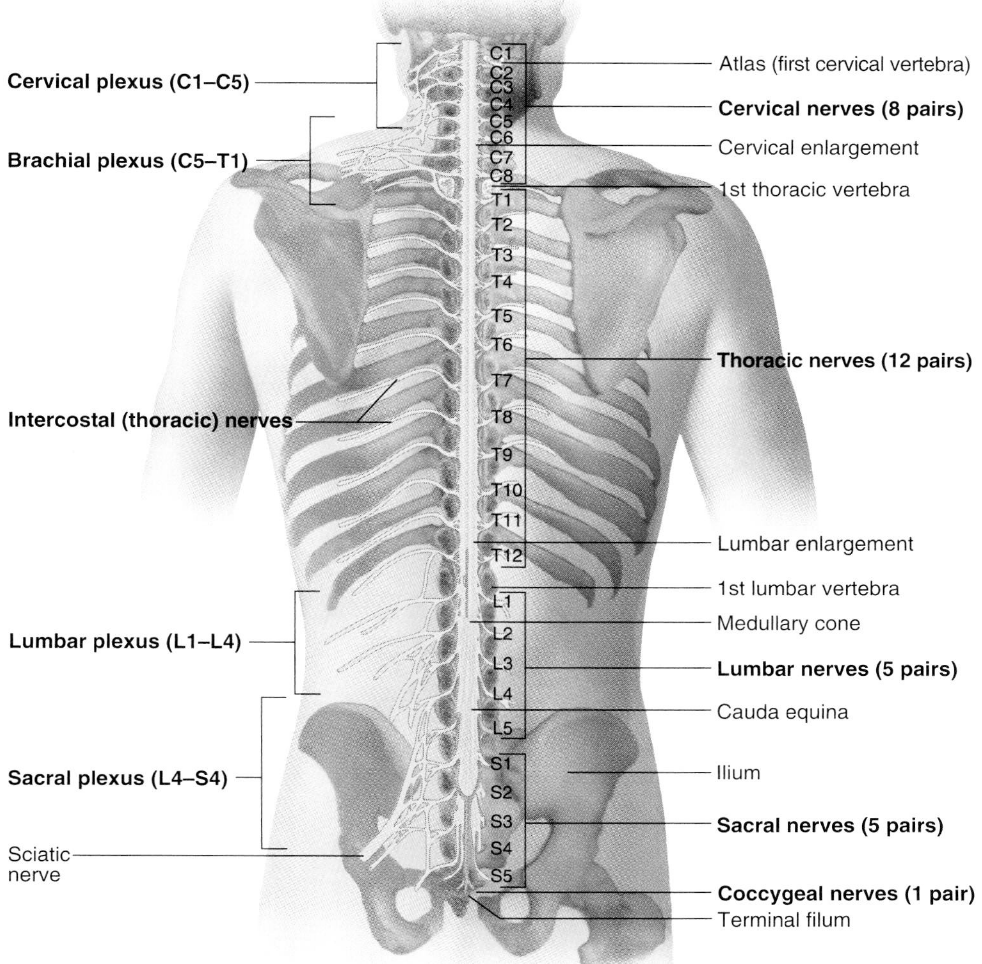

Figure 15.4 The Spinal Nerve Roots and Plexuses, Dorsal View.

equina, which occupies the vertebral canal from level L2 to S5.

Distal Branches

Distal to the vertebrae, the branches of a spinal nerve are more complex (fig. 15.7). Immediately after emerging from the intervertebral foramen, the nerve divides into a **dorsal ramus,**[9] a **ventral ramus,** and a small **meningeal branch.** The meningeal branch (see fig. 15.5) reenters the vertebral canal and innervates the meninges, vertebrae, and spinal ligaments. The dorsal ramus innervates the muscles and joints in that region of the spine and the skin of the back.

9. *ramus* = branch

The ventral ramus innervates the ventral and lateral skin and muscles of the trunk and gives rise to nerves of the limbs.

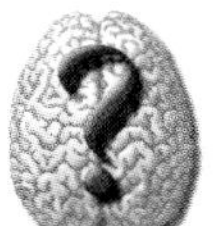

Think About It

Do you think the meningeal branch is sensory, motor, or mixed? Explain your reasoning.

The ventral ramus differs from one region of the trunk to another. In the thoracic region, it forms an **intercostal nerve** that travels along the inferior margin of a rib and innervates the skin and intercostal muscles (thus contributing to breathing), as well as the internal oblique, external oblique, and transversus abdominis. All other ventral rami form the *nerve plexuses* described next.

Clinical Insight 15.2

Shingles

Chickenpox *(varicella),* a common disease of early childhood, is caused by the *varicella-zoster* virus. It produces an itchy rash that usually clears up without complications. The virus, however, remains for life in the dorsal root ganglia. The immune system normally keeps it in check. If the immune system is compromised, however, the virus can travel down the sensory nerves by fast axonal transport and cause *shingles (herpes zoster).* This is characterized by a painful trail of skin discoloration and fluid-filled vesicles along the path of the nerve. These signs usually appear in the chest and waist, often on just one side of the body. Shingles usually occurs after the age of 50. While it can be very painful and may last 6 months or longer, it eventually heals spontaneously and requires no special treatment other than aspirin and steroidal ointment to relieve pain and inflammation.

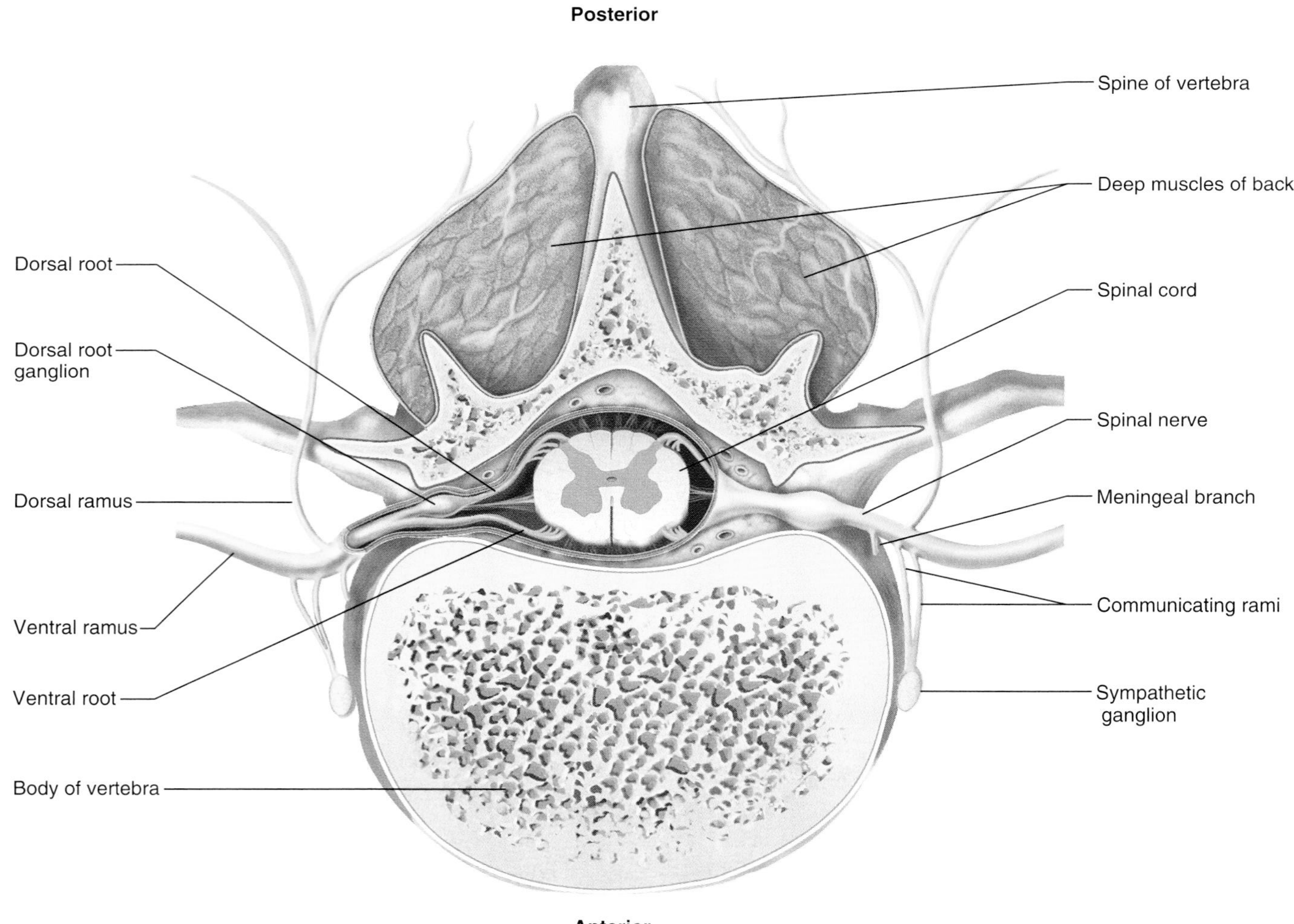

Figure 15.5 Relationship of a Spinal Nerve to the Spinal Cord and Vertebra (cross section).

Clinical Insight 15.3

Spinal Nerve Injuries

The radial and sciatic nerves are especially vulnerable to injury. The radial nerve, which passes through the axilla, may be compressed against the humerus by improperly adjusted crutches, causing *crutch paralysis.* A similar injury often resulted from the now-discredited practice of trying to correct a dislocated shoulder by putting a foot in a person's armpit and pulling on the arm. One consequence of radial nerve injury is *wrist drop*—the fingers, hand, and wrist are chronically flexed because the extensor muscles supplied by the radial nerve are paralyzed.

Because of its position and length, the sciatic nerve of the hip and thigh is the most vulnerable nerve in the body. Trauma to this nerve produces *sciatica,* a sharp pain that travels from the gluteal region along the posterior side of the thigh and leg as far as the ankle. Ninety percent of cases result from a herniated intervertebral disc or osteoarthritis of the lower spine, but sciatica can also be caused by pressure from a pregnant uterus, dislocation of the hip, injections in the wrong area of the buttock, or sitting for a long time on the edge of a hard chair. Men sometimes suffer sciatica from the habit of sitting on a wallet carried in the hip pocket.

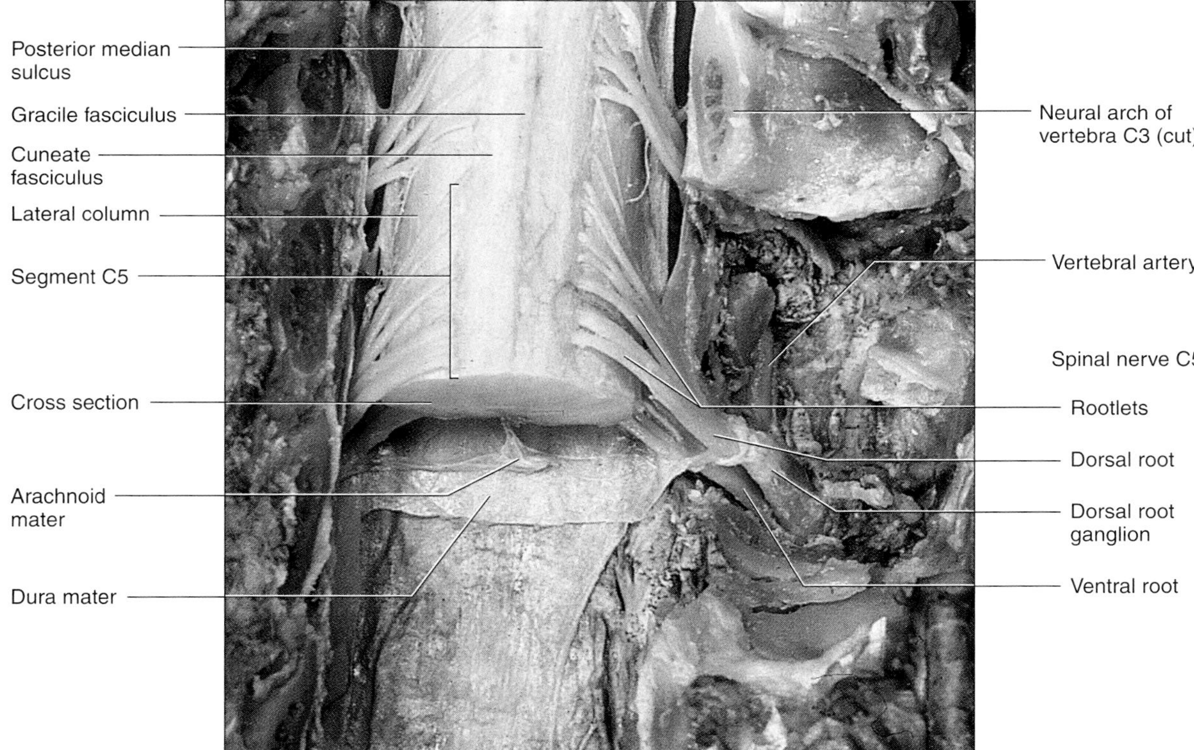

Figure 15.6 The Point of Entry of Two Spinal Nerves into the Spinal Cord. Dorsal view with vertebrae cut away. Note that each dorsal root divides into several rootlets that enter the spinal cord. A segment of the spinal cord is the portion receiving all the rootlets of one spinal nerve.

Nerve Plexuses

Except in the thoracic region, the ventral rami branch and anastomose (merge) repeatedly to form five weblike nerve plexuses: the small **cervical plexus** deep in the neck, the **brachial plexus** near the shoulder, the **lumbar plexus** of the lower back, the **sacral plexus** immediately inferior to this, and finally the tiny **coccygeal plexus** adjacent to the lower sacrum and coccyx. A general view of these plexuses is shown in figure 15.4; they are illustrated and described in tables 15.3 through 15.6.

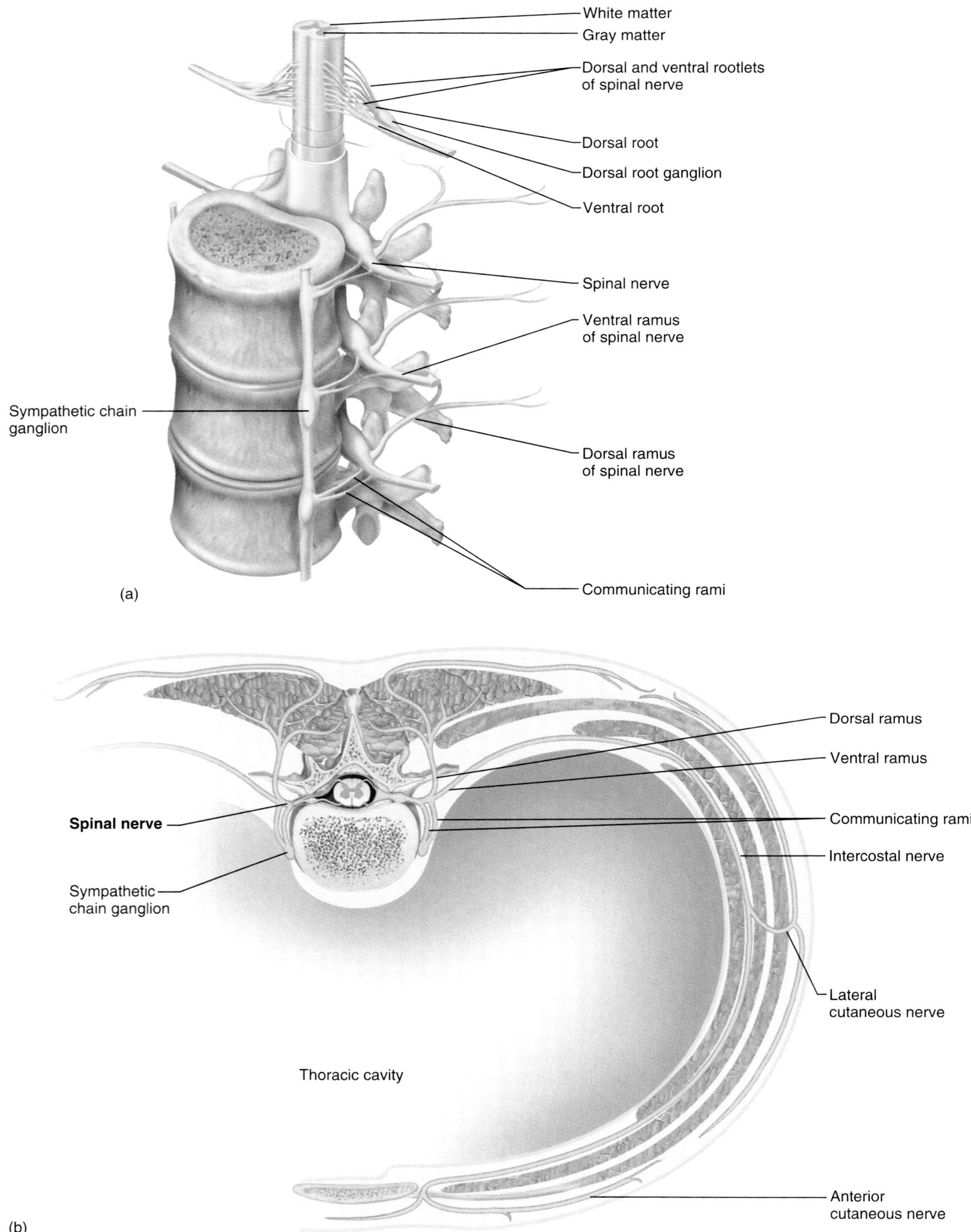

Figure 15.7 Rami of the Spinal Nerves. (*a*) Anterolateral view of the spinal nerves and their subdivisions in relation to the spinal cord and vertebrae. (*b*) Cross section of the body at the thoracic level showing innervation of muscles of the chest and back.

Table 15.3 The Cervical Plexus

The cervical plexus receives fibers from the ventral rami of nerves C1 to C5 and gives rise to the nerves listed, in order from superior to inferior. The most important of these are the *phrenic*[10] *nerves,* which travel down each side of the mediastinum, innervate the diaphragm, and play an essential role in breathing. In addition to the major nerves listed here, there are several motor branches that innervate the geniohyoid, thyrohyoid, scalene, levator scapulae, trapezius, and sternocleidomastoid muscles.

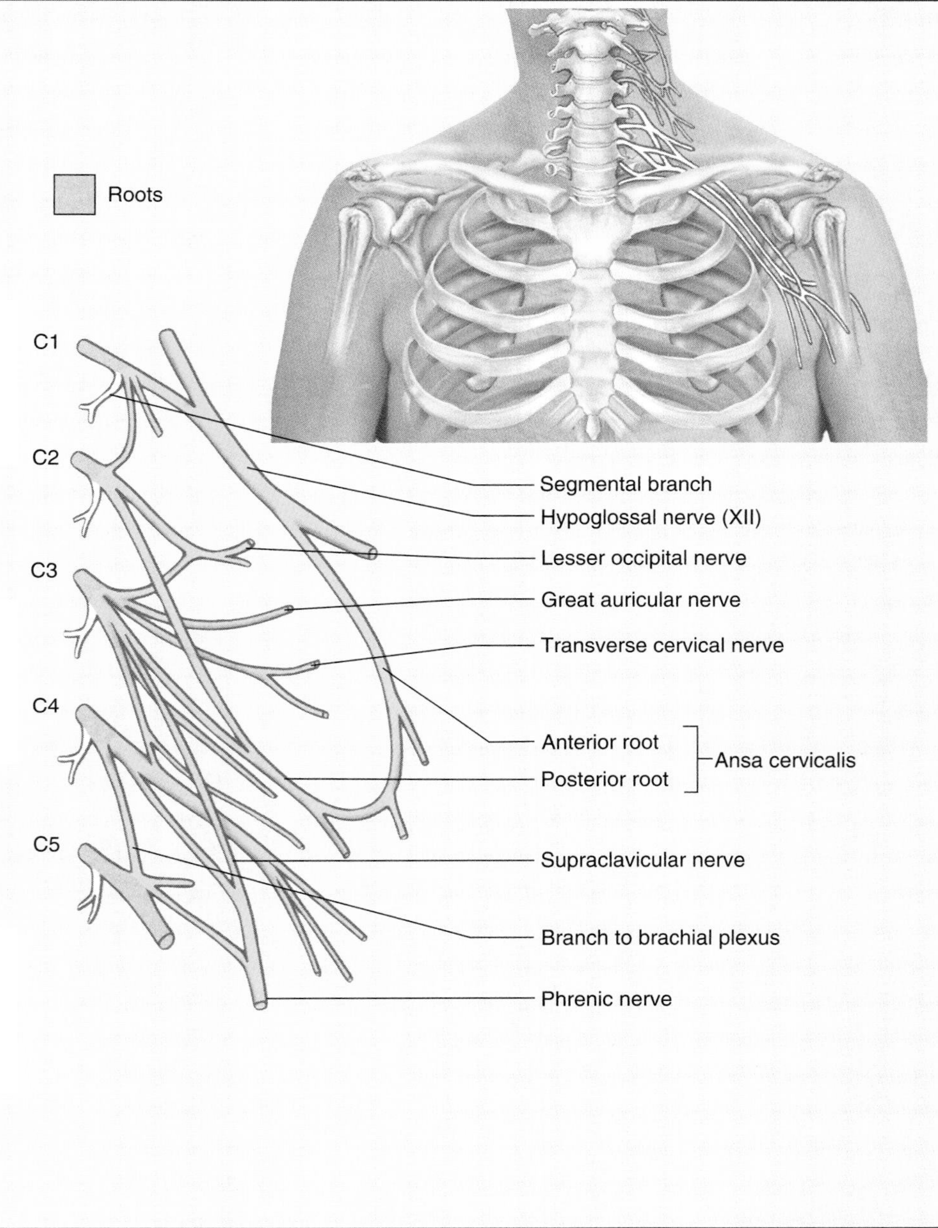

Lesser Occipital Nerve

Composition: Somatosensory

Innervation: Skin of lateral scalp and posterior part of external ear

Great Auricular Nerve

Composition: Somatosensory

Innervation: Skin of and around external ear

Transverse Cervical Nerve

Composition: Somatosensory

Innervation: Skin of anterior and lateral aspect of neck

Ansa Cervicalis

Composition: Motor

Innervation: Omohyoid, sternohyoid, and sternothyroid muscles

Supraclavicular Nerve

Composition: Somatosensory

Innervation: Skin of lower anterior and lateral neck, shoulder, and anterior chest

Phrenic (FREN-ic) Nerve

Composition: Motor

Innervation: Diaphragm

10. *phren* = diaphragm

Table 15.4 The Brachial Plexus

The brachial plexus (fig. 15.8) is formed by the ventral rami of nerves C4 to T2. It passes over the first rib into the axilla and innervates the upper limb and some muscles of the neck and shoulder. It gives rise to nerves for cutaneous sensation, muscle contraction, and proprioception from the joints and muscles. The muscle actions stimulated by these nerves are described in the tables in chapter 11.

The subdivisions of this plexus are called *roots, trunks, divisions,* and *cords* (color-coded in the accompanying figure). The five **roots** are the ventral rami of nerves C5 to T1, which provide most of the fibers to this plexus (C4 and T2 contribute partially). The five roots unite to form the **upper, middle,** and **lower trunks.** Each trunk divides into an **anterior** and **posterior division,** and finally the six divisions merge to form three large fiber bundles—the **posterior, medial,** and **lateral cords.**

Axillary Nerve

Composition: Motor and somatosensory

Origin: Posterior cord of brachial plexus

Sensory innervation: Skin of lateral shoulder and arm; shoulder joint

Motor innervation: Deltoid and teres minor

Radial Nerve

Composition: Motor and somatosensory

Origin: Posterior cord of brachial plexus

Sensory innervation: Skin of posterior aspect of arm, forearm, and wrist; joints of elbow, wrist, and hand

Motor innervation: Muscles of posterior arm and forearm: triceps brachii, supinator, anconeus, brachioradialis, extensor carpi radialis brevis, extensor carpi radialis longus, and extensor carpi ulnaris

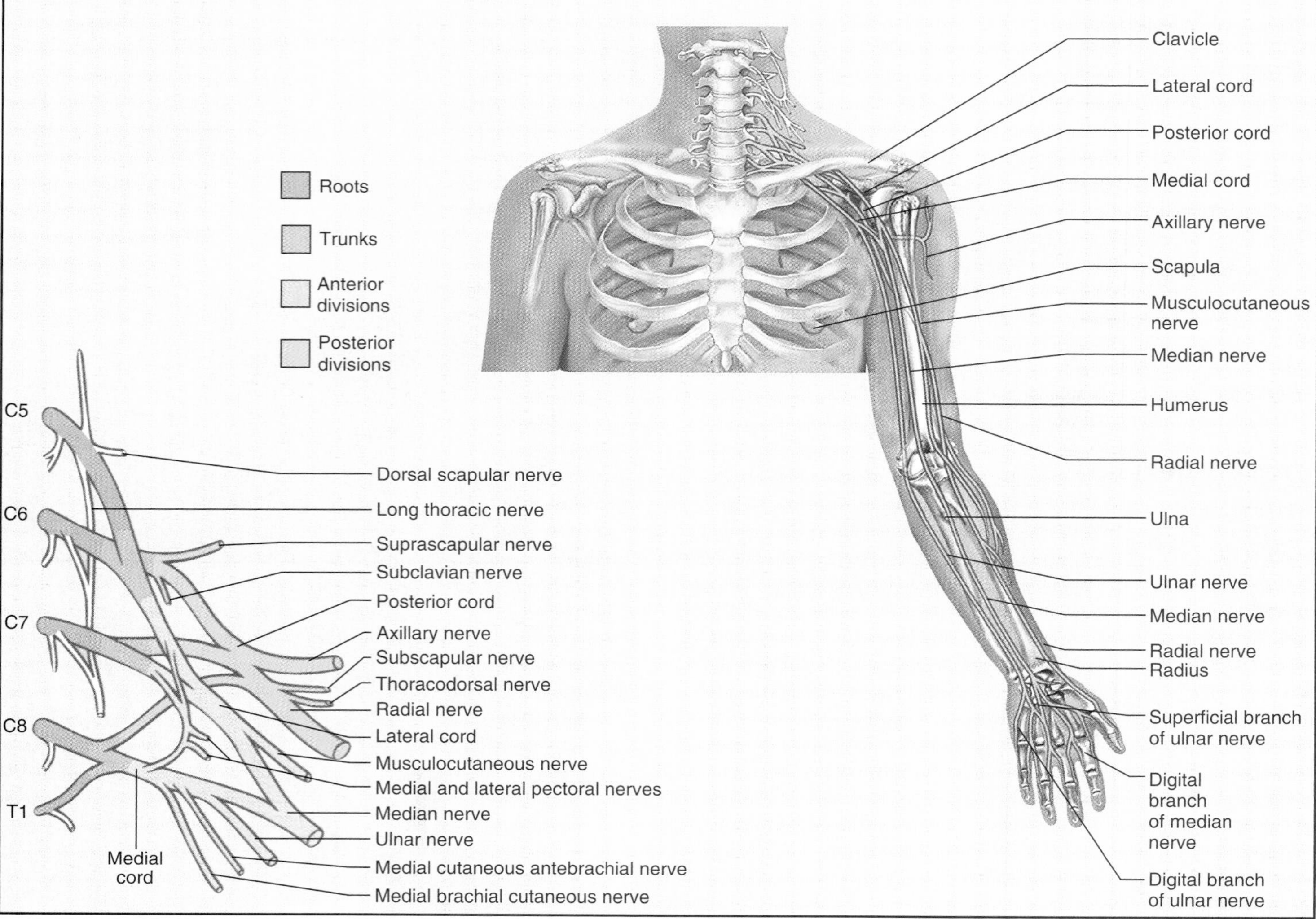

(continued)

Table 15.4 The Brachial Plexus (continued)

Musculocutaneous Nerve

Composition: Motor and somatosensory
Origin: Lateral cord of brachial plexus
Sensory innervation: Skin of lateral aspect of forearm
Motor innervation: Muscles of anterior arm: coracobrachialis, biceps brachii, and brachialis

Median Nerve

Composition: Motor and somatosensory
Origin: Medial cord of brachial plexus
Sensory innervation: Skin of lateral two-thirds of hand, joints of hand
Motor innervation: Flexors of anterior forearm; thenar muscles; first and second lumbricals

Ulnar Nerve

Composition: Motor and somatosensory
Origin: Medial cord of brachial plexus
Sensory innervation: Skin of medial part of hand; joints of hand
Motor innervation: Flexor carpi ulnaris, flexor digitorum profundus, adductor pollicis, hypothenar muscles, interosseous muscles, and third and fourth lumbricals

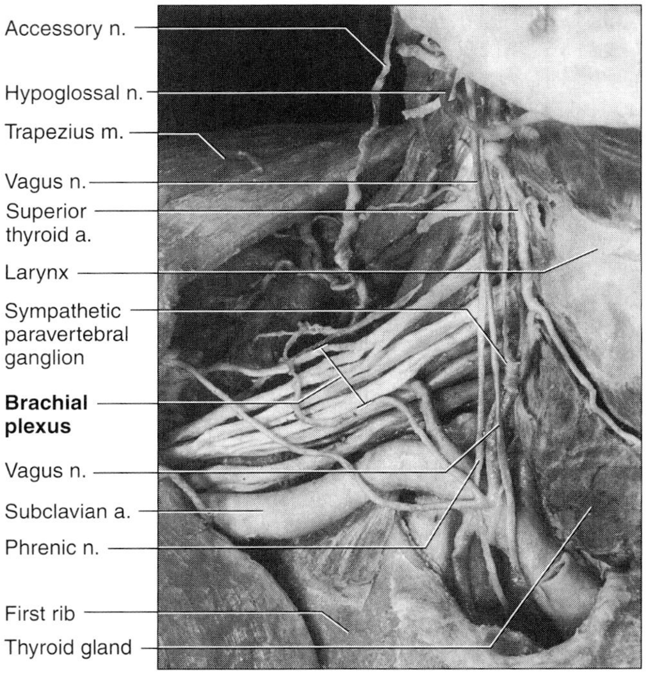

Accessory n.
Hypoglossal n.
Trapezius m.
Vagus n.
Superior thyroid a.
Larynx
Sympathetic paravertebral ganglion
Brachial plexus
Vagus n.
Subclavian a.
Phrenic n.
First rib
Thyroid gland

Figure 15.8 **The Brachial Plexus.** Anterior view of the right shoulder, also showing three of the cranial nerves, the sympathetic trunk, and the phrenic nerve (a branch of the cervical plexus). Most of the other structures resembling nerves in this photograph are blood vessels. (*a.* = artery; *m.* = muscle; *n.* = nerve.)

Table 15.5 The Lumbar Plexus

The lumbar plexus is formed from the ventral rami of nerves L1 to L4 and some fibers from T12. With only five roots and two divisions, it is less complex than the brachial plexus.

Iliohypogastric Nerve

Composition: Motor and somatosensory

Sensory innervation: Skin of anterior abdominal wall

Motor innervation: Internal and external obliques and transversus abdominis

Ilioinguinal Nerve

Composition: Motor and somatosensory

Sensory innervation: Skin of upper medial thigh; male scrotum and root of penis; female labia majora

Motor innervation: Joins iliohypogastric nerve and innervates the same muscles

Genitofemoral Nerve

Composition: Somatosensory

Sensory innervation: Skin of middle anterior thigh; male scrotum and cremaster muscle; female labia majora

Lateral Femoral Cutaneous Nerve

Composition: Somatosensory

Sensory innervation: Skin of lateral aspect of thigh

Femoral Nerve

Composition: Motor and somatosensory

Sensory innervation: Skin of anterior and lateral thigh; medial leg and foot

Motor innervation: Anterior muscles of thigh and extensors of leg; iliacus, psoas major, pectineus, quadriceps femoris, and sartorius

Saphenous (sah-FEE-nus) Nerve

Composition: Somatosensory

Sensory innervation: Skin of medial aspect of leg and foot; knee joint

Obturator Nerve

Composition: Motor and somatosensory

Sensory innervation: Skin of superior medial thigh; hip and knee joints

Motor innervation: Adductor muscles of leg: external obturator, pectineus, adductor longus, adductor brevis, adductor magnus, and gracilis

(continued)

Table 15.5 The Lumbar Plexus (continued)

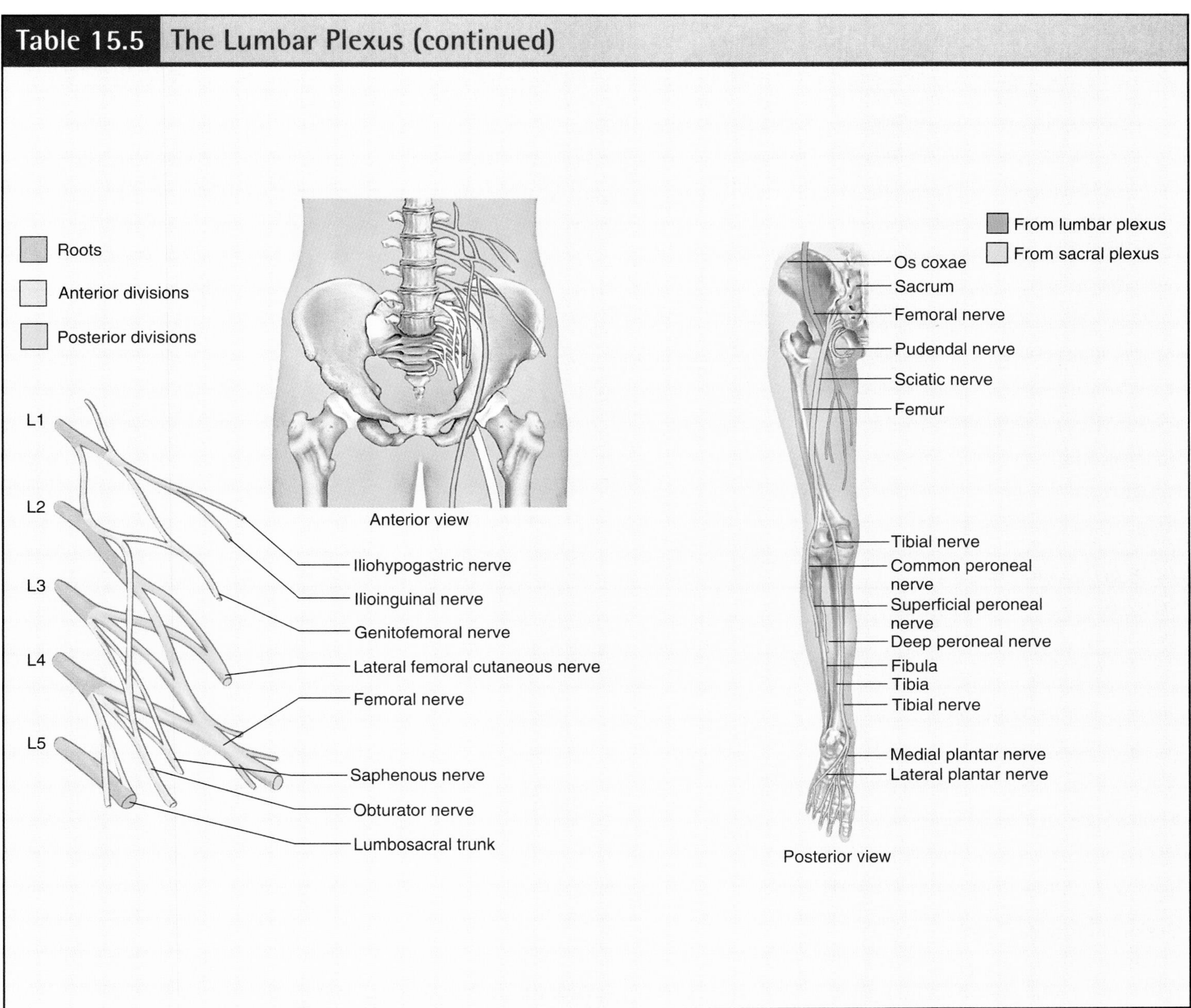

Table 15.6 The Sacral and Coccygeal Plexuses

The sacral plexus is formed from the ventral rami of nerves L4, L5, and S1 to S4. It has six roots and anterior and posterior divisions. Since it is connected to the lumbar plexus by fibers that run through the *lumbosacral trunk,* the two plexuses are sometimes referred to collectively as the *lumbosacral plexus.* The coccygeal plexus is a tiny plexus formed from the ventral rami of S4, S5, and Co.

The *tibial* and *common peroneal nerves* listed in this table travel together through a connective tissue sheath; they are referred to collectively as the **sciatic** (sy-AT-ic) **nerve.** The sciatic nerve passes through the greater sciatic notch of the pelvis, extends for the length of the thigh, and ends at the popliteal fossa. Here, the tibial and common peroneal nerves diverge and follow their separate paths into the leg. The sciatic nerve is a common focus of injury and pain.

Superior Gluteal Nerve

Composition: Motor

Motor innervation: Gluteus minimus, gluteus medius, and tensor fasciae latae

Inferior Gluteal Nerve

Composition: Motor

Motor innervation: Gluteus maximus

Nerve to Piriformis

Composition: Motor

Motor innervation: Piriformis

Nerve to Quadratus Femoris

Composition: Motor and somatosensory

Sensory innervation: Hip joint

Motor innervation: Quadratus femoris and gemellus inferior

Nerve to Internal Obturator

Composition: Motor

Motor innervation: Internal obturator and gemellus superior

Perforating Cutaneous Nerve

Composition: Somatosensory

Sensory innervation: Skin of posterior aspect of buttock

Posterior Cutaneous Nerve

Composition: Somatosensory

Sensory innervation: Skin of lower lateral buttock, anal region, upper posterior thigh, upper calf, scrotum, and labia majora

Tibial Nerve

Composition: Motor and somatosensory

Sensory innervation: Skin of posterior leg and sole of foot; knee and foot joints

Motor innervation: Semitendinosus, semimembranosus, long head of biceps femoris, gastrocnemius, soleus, flexor digitorum longus, flexor hallucis longus, tibialis posterior, popliteus, and intrinsic muscles of foot

(continued)

Table 15.6 The Sacral and Coccygeal Plexuses (continued)

Common Peroneal Nerve

Composition: Motor and somatosensory

Sensory innervation: Skin of anterior distal one-third of leg, dorsum of foot, and toes I and II; knee joint

Motor innervation: Short head of biceps femoris, peroneus tertius, peroneus brevis, peroneus longus, tibialis anterior, extensor hallucis longus, extensor digitorum longus, and extensor digitorum brevis

Pudendal Nerve

Composition: Motor and somatosensory

Sensory innervation: Skin of penis and scrotum of male; clitoris, labia majora and minora, and lower vagina of female

Motor innervation: Muscles of perineum

Coccygeal Nerve

Composition: Motor and somatosensory

Sensory innervation: Skin over coccyx

Motor innervation: Muscles of pelvic floor

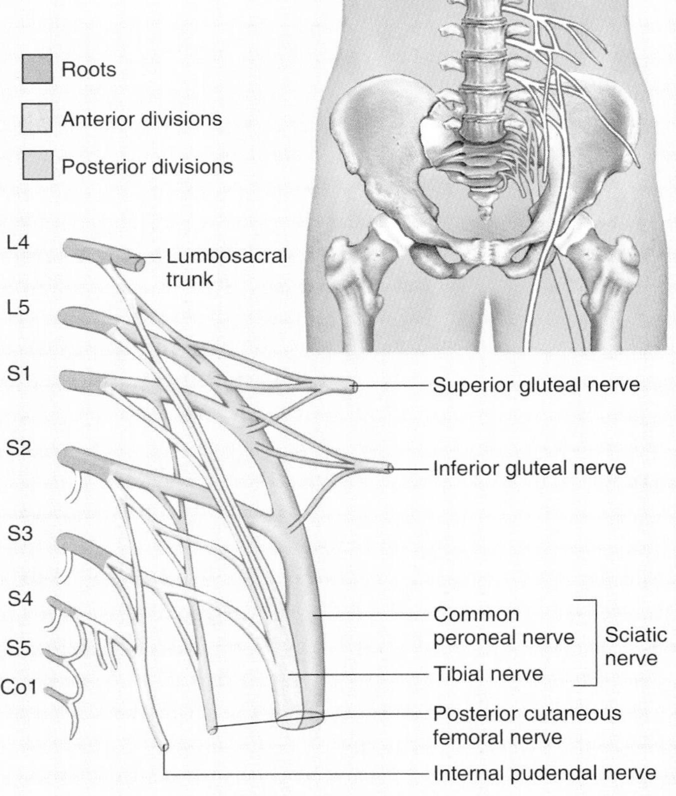

Cutaneous Innervation and Dermatomes

Each spinal nerve except C1 receives sensory input from a specific area of skin called a **dermatome.**[11] A *dermatome map* (fig. 15.9) is a diagram of the cutaneous regions innervated by each spinal nerve. Such a map is oversimplified, however, because the dermatomes overlap at their edges by as much as 50%. Therefore, severance of one sensory nerve root does not entirely deaden sensation from a dermatome. It is necessary to sever or anesthetize three successive spinal nerves to produce a total loss of sensation from one dermatome. Spinal nerve damage is assessed by testing the dermatomes with pinpricks and noting areas in which the patient has no sensation.

11. *derma* = skin + *tome* = segment, part

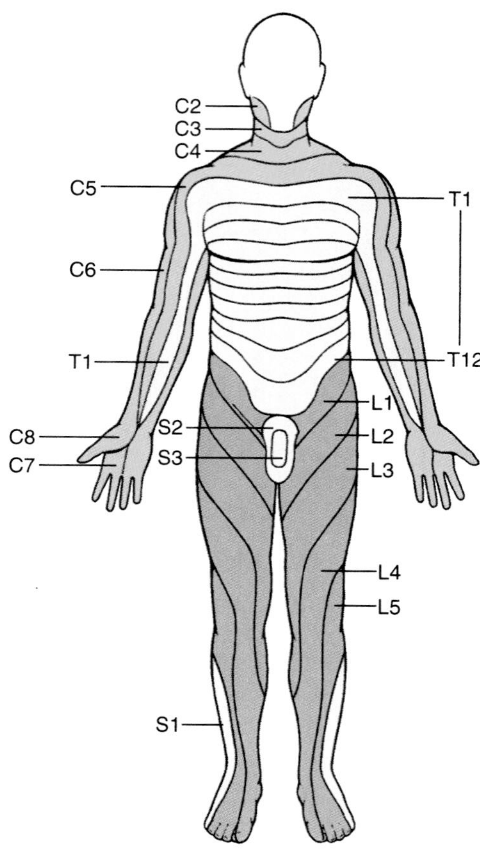

Figure 15.9 A Dermatome Map of the Anterior Aspect of the Body. Each zone of the skin is innervated by sensory branches of the spinal nerves indicated by the labels. Nerve C1 does not innervate the skin.

Before You Go On

Answer the following questions to test your understanding of the preceding section:

11. What is meant by the dorsal and ventral roots of a spinal nerve? Which of these is sensory and which is motor?
12. Where are the somas of the dorsal root located? Where are the somas of the ventral root?
13. List the five plexuses of spinal nerves and state where each one is located.
14. State which plexus gives rise to each of the following nerves: axillary, ilioinguinal, obturator, phrenic, pudendal, radial, and sciatic.

Somatic Reflexes

Objectives

When you have completed this section, you should be able to

- define *reflex* and explain how reflexes differ from other motor actions;
- describe the general components of a typical reflex arc; and
- explain how the basic types of somatic reflexes function.

Most of us have had our reflexes tested with a little rubber hammer; a tap near the knee produces an uncontrollable jerk of the leg, for example. In this section, we discuss what reflexes are and how they are produced by an assembly of receptors, neurons, and effectors. We also survey the different types of neuromuscular reflexes and how they are important to motor coordination.

The Nature of Reflexes

Reflexes are quick, involuntary, stereotyped reactions of glands or muscles to stimulation. This definition sums up four important properties of a reflex:

1. Reflexes *require stimulation*—they are not spontaneous actions but responses to sensory input.
2. Reflexes are *quick*—they generally involve few if any interneurons and minimum synaptic delay.
3. Reflexes are *involuntary*—they occur without intent, often without our awareness, and they are difficult to suppress. Given an adequate stimulus, the response is essentially automatic. You may become conscious of the stimulus that evoked a reflex, and this awareness may enable you to correct or avoid a potentially dangerous situation, but awareness is not a part of the reflex itself. It may come after the reflex action has been completed,

and somatic reflexes can occur even if the spinal cord has been severed so that no stimuli reach the brain.
4. Reflexes are *stereotyped*—they occur in essentially the same way every time; the response is very predictable.

Reflexes include glandular secretion and contractions of all three types of muscle. They also include some learned responses, such as the salivation of dogs in response to a sound they have come to associate with feeding time, first studied by Ivan Pavlov and named *conditioned reflexes.* In this section, however, we are concerned with unlearned skeletal muscle reflexes that are mediated by the brainstem and spinal cord. They result in the involuntary contraction of a muscle—for example, the quick withdrawal of your hand from a hot stove or the lifting of your foot when you step on something sharp. These are **somatic reflexes,** since they involve the somatic nervous system. Later in this chapter, we consider **visceral reflexes.** The somatic reflexes have traditionally been called *spinal reflexes,* although some visceral reflexes also involve the spinal cord, and some somatic reflexes are mediated more by the brain than by the spinal cord.

You are prepared to understand somatic reflexes now that you have studied spinal cord anatomy in the previous chapter and the spinal nerves in this one. A somatic reflex functions by means of a **reflex arc,** in which signals travel along the following pathway:

1. *somatic receptors* in the skin, a muscle, or a tendon;
2. *afferent nerve fibers,* which carry information from these receptors into the dorsal horn of the spinal cord;
3. *interneurons,* which integrate information; these are lacking from some reflex arcs;
4. *efferent nerve fibers,* which carry motor impulses to the skeletal muscles; and
5. *skeletal muscles,* the somatic effectors that carry out the response.

The Muscle Spindle

Many somatic reflexes involve stretch receptors in the muscles called **muscle spindles**. These are among the body's most important **proprioceptors**—sense organs that monitor the position and movements of body parts. Muscle spindles are especially abundant in muscles that require fine control. The hand and foot have 100 or more spindles per gram of muscle, whereas there are relatively few in large muscles with coarse movements and none at all in the middle-ear muscles. Muscle spindles provide the cerebellum with the feedback it needs to regulate the tension in the skeletal muscles.

Muscle spindles are cigar-shaped organs about 4 to 10 mm long scattered throughout the fleshy part of a muscle (fig. 15.10). A spindle contains 3 to 12 modified muscle fibers and a few nerve fibers, all wrapped in a fibrous capsule. The muscle fibers within a spindle are called **intrafusal**[12] **fibers,** while those of the rest of the muscle are called **extrafusal fibers.** Only the two ends of an intrafusal fiber have sarcomeres and are able to contract. The middle portion acts as the stretch receptor. There are two classes of intrafusal fibers: *nuclear chain fibers,* which have a single file of nuclei in the noncontractile region, and *nuclear bag fibers,* which are about twice as long and have nuclei clustered in a thick midregion. Muscle spindles have three types of nerve fibers:

1. **Primary afferent fibers,** which end in *annulospiral endings* that coil around the middle of nuclear chain and nuclear bag fibers. These respond mainly to the onset of muscle stretch.
2. **Secondary afferent fibers,** which have *flower-spray endings,* somewhat resembling the dried head of a wildflower, wrapped primarily around the ends of the nuclear chain fibers. These respond mainly to prolonged stretch.
3. **Gamma (γ) motor neurons,** which originate in the ventral horn of the spinal cord and lead to the contractile ends of the intrafusal fibers. The name distinguishes them from the **alpha (α) motor neurons,** which innervate the extrafusal fibers. Gamma motor neurons adjust the tension in a muscle spindle to variations in the length of the muscle. When a muscle shortens, the γ motor neurons stimulate the ends of the intrafusal fibers to contract slightly. This keeps the intrafusal fibers taut and responsive at all times. Without this feedback, the spindles would become flabby when a skeletal muscle shortened. This feedback is clearly very important, because γ motor neurons constitute about one-third of all the motor fibers in a spinal nerve.

The Stretch Reflex

When a muscle is stretched, it "fights back"—it contracts, maintains increased tonus, and feels stiffer than an unstretched muscle. This response, called the **stretch (myotatic**[13]**) reflex,** helps to maintain equilibrium and posture. For example, if your head starts to tip forward, it stretches muscles such as the semispinalis and splenius capitis of the nuchal region (back of your neck).

12. *intra* = within + *fus* = spindle
13. *myo* = muscle + *tat* (from *tasis*) = stretch

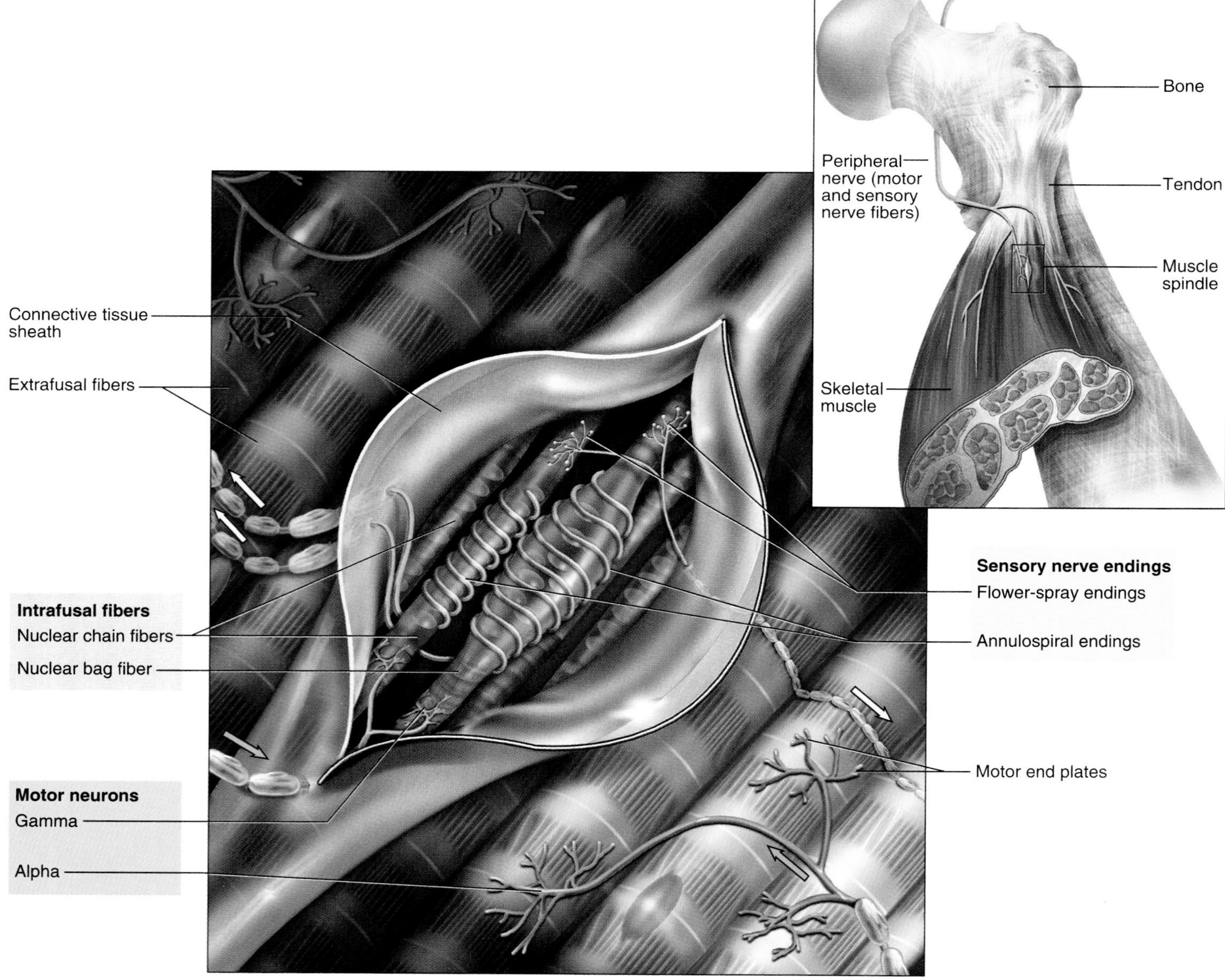

Figure 15.10 A Muscle Spindle and Its Innervation.

This stimulates their muscle spindles, which send afferent signals to the cerebellum by way of the brainstem. The cerebellum integrates this information and relays it to the cerebral cortex, and the cortex sends signals back to the nuchal muscles. The muscles contract and raise your head.

Stretch reflexes often feed back not to a single muscle but to a set of synergists and antagonists. Since the contraction of a muscle on one side of a joint stretches the antagonistic muscle on the other side, the flexion of a joint triggers a stretch reflex in the extensors, and extension stimulates a stretch reflex in the flexors. Consequently, stretch reflexes are valuable in stabilizing joints by balancing the tension of the extensors and flexors. They also dampen (smooth) muscle action. Without stretch reflexes, a person's movements tend to be jerky. Stretch reflexes are especially important in coordinating vigorous and precise movements such as dance.

Although a stretch reflex is mediated primarily by the brain and is not, therefore, strictly a spinal reflex, a weak component of it is spinal and occurs even if the spinal cord is severed below the brain. The spinal component can be more pronounced if a muscle is stretched very suddenly. This occurs in a **tendon reflex**—the reflexive contraction of a muscle when its tendon is tapped, as in the familiar knee-jerk (patellar) reflex. Tapping the patellar ligament with a reflex hammer suddenly stretches the quadriceps femoris muscle of the thigh (fig. 15.11). This stimulates numerous muscle spindles in the quadriceps and sends an intense volley of signals, mainly by way of the primary afferent fibers, to the spinal cord.

In the spinal cord, the primary afferent fibers synapse directly with the α motor neurons that return to the muscle, thus forming **monosynaptic reflex arcs.** That is, there is only one synapse between the afferent and efferent neuron, therefore little synaptic delay and a very prompt response. The α motor neurons excite the quadriceps muscle, making it contract and creating the knee jerk.

There are many other tendon reflexes. A tap on the calcaneal tendon causes plantar flexion of the foot, a tap on the triceps brachii tendon causes extension of the elbow, and a tap on the side of the face causes clenching of the jaw. Testing somatic reflexes is valuable in diagnosing many diseases that cause exaggeration, inhibition, or absence of reflexes, such as neurosyphilis, diabetes mellitus, multiple sclerosis, alcoholism, electrolyte imbalances, and lesions of the nervous system.

Stretch reflexes and other muscle contractions often depend on **reciprocal inhibition,** a reflex phenomenon that prevents muscles from working against each other by inhibiting antagonists. In the knee jerk, for example, the quadriceps femoris would not produce much joint movement if its antagonists, the hamstring muscles, contracted at the same time. But reciprocal inhibition prevents that from happening. Some branches of the sensory fibers from the muscle spindles in the quadriceps stimulate spinal cord interneurons, which, in turn, inhibit the α motor neurons of the hamstring muscles (fig. 15.11). The hamstring muscles therefore remain relaxed and allow the quadriceps to extend the knee.

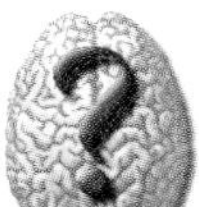

Think About It

In a withdrawal reflex with reciprocal inhibition, where in the nervous pathway do EPSPs do occur? Where do IPSPs occur? Explain your answers. (Review EPSPs and IPSPs in chapter 13 if necessary.)

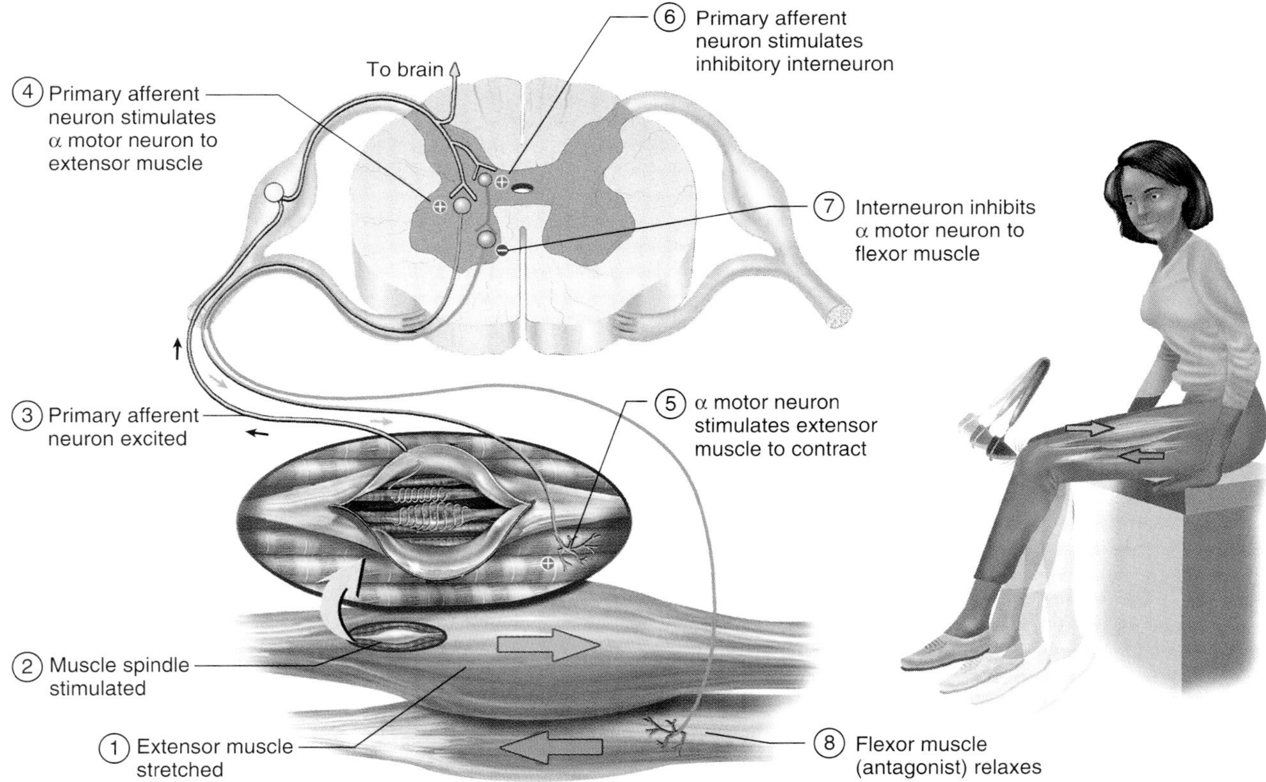

Figure 15.11 The Patellar Tendon Reflex Arc and Reciprocal Inhibition of the Antagonistic Muscle. *Plus signs* indicate excitation of a postsynaptic cell (EPSPs) and *minus signs* indicate inhibition (IPSPs). The tendon reflex is occurring in the quadriceps femoris muscle *(red arrow),* while the hamstring muscles are exhibiting reciprocal inhibition *(blue arrow)* so they do not contract and oppose the quadriceps. Note that all IPSPs occur in the spinal cord. Innervation of the skeletal muscles is excitatory only.

The Flexor (withdrawal) Reflex

A **flexor reflex** is the quick contraction of flexor muscles resulting in the withdrawal of a limb from an injurious stimulus. For example, suppose you are wading in a lake and step on a broken bottle with your right foot (fig. 15.12). Even before you are consciously aware of the pain, you quickly pull your foot away before the glass penetrates any deeper. This action involves contraction of the flexors and relaxation of the extensors in that limb; the latter is another case of reciprocal inhibition.

The protective function of this reflex requires more than a quick jerk like a tendon reflex, so it involves more complex neural pathways. Sustained contraction of the flexors is produced by a parallel after-discharge circuit in the spinal cord (see fig. 13.27, p. 483). This circuit is part of a **polysynaptic reflex arc**—a pathway in which signals travel over many synapses on their way back to the muscle. Some signals follow routes with only a few synapses and return to the flexor muscles quickly. Others follow routes with more synapses, and therefore more delay, so they reach the flexor muscles a little later. Consequently, the flexor muscles receive prolonged output from the spinal cord and not just one sudden stimulus as in a stretch reflex. By the time these efferent signals begin to die out, you will probably be consciously aware of the pain and begin taking voluntary action to prevent further harm.

The Crossed Extensor Reflex

In the preceding situation, if *all* you did was to quickly lift the injured leg from the lake bottom, you would fall over. To prevent this and maintain your balance, other re-

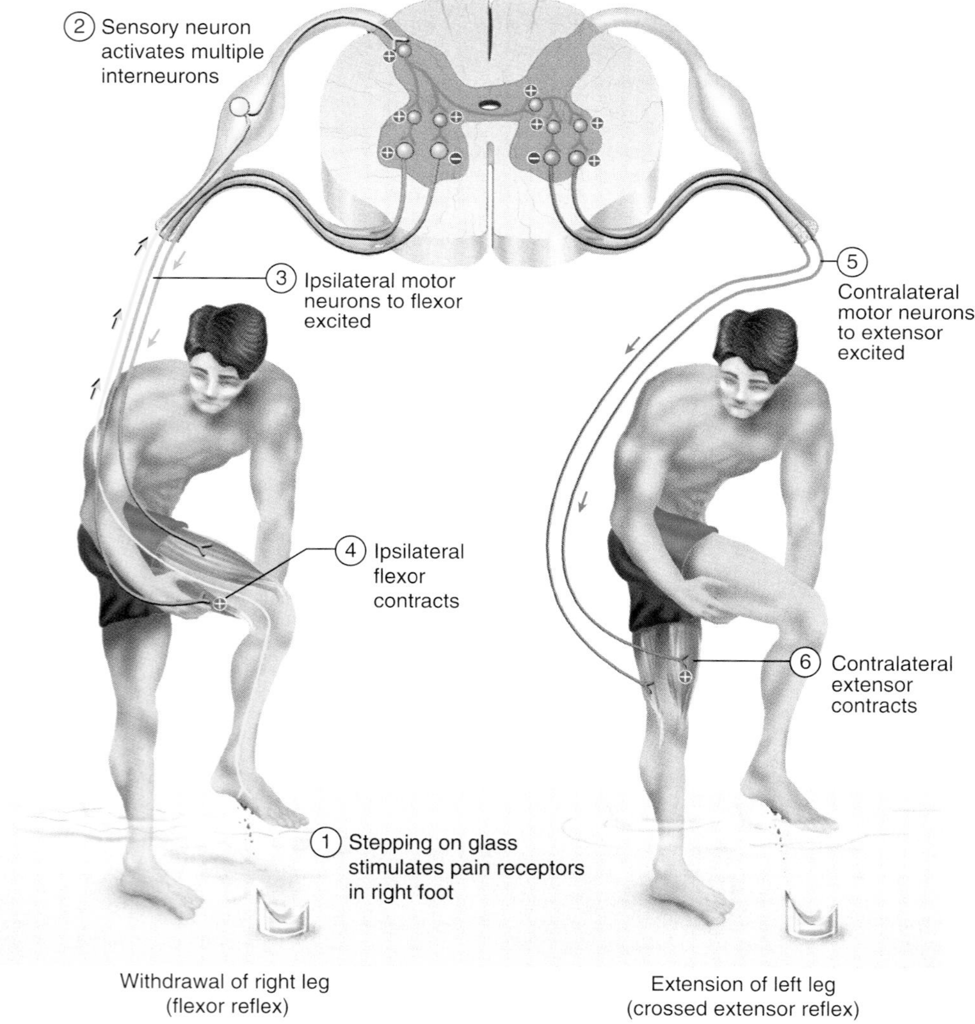

Figure 15.12 The Withdrawal and Crossed Extensor Reflexes. The pain stimulus triggers a withdrawal reflex, which results in contraction of flexor muscles of the injured limb. At the same time, a crossed extensor reflex results in contraction of extensor muscles of the opposite limb. The latter reflex aids in balance when the injured limb is raised. Note that for each limb, while the agonist contracts, the α motor neuron to its antagonist is inhibited, as indicated by the *red minus signs* in the spinal cord.

flexes shift your center of gravity over the leg that is still on the ground. The **crossed extensor reflex** is the contraction of extensor muscles in the limb opposite from the one that is withdrawn (fig. 15.12). It extends that limb and enables you to keep your balance. To produce this reflex, branches of the afferent nerve fibers cross from the stimulated side of the body to the contralateral side of the spinal cord. There, they synapse with interneurons, which, in turn, excite or inhibit α motor neurons to the muscles of the contralateral limb.

In the ipsilateral leg (the side that was hurt), you would contract your flexors and relax your extensors to lift the leg from the ground. On the contralateral side, you would relax your flexors and contract the extensors to stiffen that leg, since it must suddenly support your entire body. At the same time, signals travel up the spinal cord and cause contraction of contralateral muscles of the hip and abdomen to shift your center of gravity over the extended leg. To a large extent, the coordination of all these muscles and maintenance of equilibrium is mediated by the cerebellum and cerebral cortex.

The flexor reflex employs an **ipsilateral reflex arc**—one in which the sensory input and motor output are on the same sides of the spinal cord. The crossed extensor reflex employs a **contralateral reflex arc,** in which the input and output are on opposite sides. An **intersegmental reflex arc** is one in which the input and output occur at different levels (segments) of the spinal cord—for example, when pain to the foot causes contractions of abdominal and hip muscles higher up the body. Note that all of these reflex arcs can function simultaneously to produce a coordinated protective response to pain.

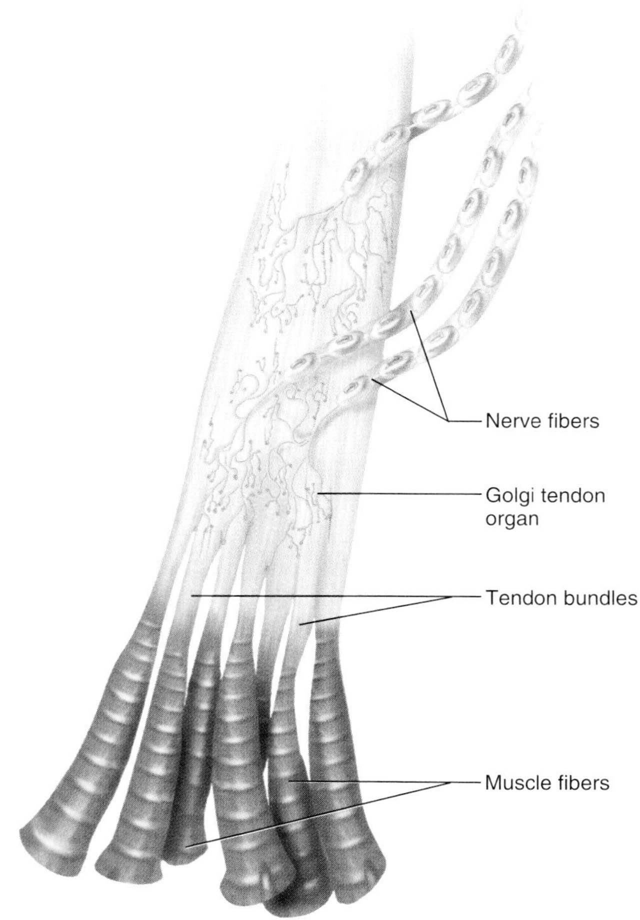

Figure 15.13 A Golgi Tendon Organ.

The Golgi Tendon Reflex

Golgi tendon organs are proprioceptors located in a tendon near its junction with a muscle (fig. 15.13). A tendon organ is about 1 mm long and consists of an encapsulated tangle of knobby nerve endings entwined in the collagen fibers of the tendon. As long as the tendon is slack, its collagen fibers are slightly spread and they put little pressure on the nerve endings woven among them. When muscle contraction pulls on the tendon, the collagen fibers come together like the two sides of a stretched rubber band and squeeze the nerve endings between them. The nerve fiber sends signals to the spinal cord that provide the CNS with feedback on the degree of muscle tension at the joint.

The **Golgi tendon reflex** is a response to excessive tension on the tendon. It inhibits α motor neurons to the muscle so the muscle does not contract as strongly. This serves to moderate muscle contraction before it tears a tendon or pulls it loose from the muscle or bone. Nevertheless, strong muscles and quick movements sometimes damage a tendon before the reflex can occur, causing such athletic injuries as a ruptured calcaneal tendon.

The Golgi tendon reflex also functions when some parts of a muscle contract more than others. It inhibits the fibers connected with overstimulated tendon organs so that their contraction is more comparable to the contraction of the rest of the muscle. This reflex spreads the workload more evenly over the entire muscle, which is beneficial in such actions as maintaining a steady grip on a tool.

Before You Go On

Answer the following questions to test your understanding of the preceding section:

15. Name five structural components of a typical somatic reflex arc. Which of these is absent from a monosynaptic arc?
16. State the function of each of the following in a muscle spindle: intrafusal fiber, annulospiral ending, and γ motor neuron.
17. Explain how nerve fibers in a tendon sense the degree of tension in a muscle.
18. Why must the withdrawal reflex, but not the stretch reflex, involve a polysynaptic reflex arc?
19. Explain why the crossed extensor reflex must accompany a withdrawal reflex of the leg.

The Autonomic Nervous System: Introduction and Anatomy

Objectives

When you have completed this section, you should be able to

- explain how the autonomic and somatic nervous systems differ in form and function;
- explain how the two divisions of the autonomic nervous system differ in general function;
- identify the anatomical components of the sympathetic and parasympathetic divisions; and
- discuss the relationship of the adrenal glands to the sympathetic nervous system.

General Properties

We now turn to visceral reflexes and the **autonomic nervous system (ANS),** which controls them. The ANS acts on all effectors other than skeletal muscles—its target organs are glands, cardiac muscle, and smooth muscle. The ANS is responsible for the homeostasis of all organ systems. Among other roles, it helps to regulate heart rate, blood pressure, body temperature, digestion, and waste elimination.

"Autonomic" literally means "self-governed"—the ANS usually carries out its actions without our conscious intent or awareness. We are unaware, for example, of changes in the diameters of our pupils or blood vessels. Indeed, it is difficult to consciously alter or suppress autonomic responses, and for this reason they form a basis for polygraph ("lie detector") tests. Responses of the somatic nervous system have traditionally been called *voluntary* responses, and those of the autonomic nervous system have been called *involuntary*. This is not as clear-cut a distinction as it once seemed, however. Therapeutic uses of biofeedback, discussed later in this chapter, show that some people can learn to control their autonomic functions. Furthermore, some skeletal muscle responses are quite involuntary, such as the somatic reflexes, and some skeletal muscles cannot be voluntarily controlled at all, such as the middle-ear muscles.

The receptors for autonomic reflex arcs are nerve endings that detect stretch, tissue damage, blood chemicals, and other internal stimuli. For example, high blood pressure stimulates stretch receptors called *baroreceptors*[14] in the carotid arteries and aorta (fig. 15.14). They transmit signals via the glossopharyngeal nerves to the medulla oblongata. The medulla integrates this input with other information and transmits efferent signals back to the heart by way of the vagus nerves. These fibers slow down the heart and reduce blood pressure. A separate autonomic reflex arc accelerates the heart when blood pressure drops below normal.

Nerve fibers of the autonomic reflex arcs travel through the cranial and spinal nerves you have already studied and arise from somas located in both the central and peripheral nervous systems. The ANS is composed of the efferent (motor) neurons of these arcs. It differs from the somatic motor division in several respects. The most conspicuous anatomical difference between them is their efferent pathways. A somatic motor neuron has its soma in the gray matter of the CNS and its axon extends all the way to a skeletal muscle; there are no ganglia or synapses along the way. In the autonomic nervous system, by contrast, *two* neurons span the distance from CNS to effector (fig. 15.15). The first begins with a soma in the CNS and terminates in a ganglion of the PNS. It is therefore called the **preganglionic (presynaptic) neuron.** In the ganglion, it synapses with a **postganglionic (postsynaptic) neuron** whose axon extends the rest of the way to the target cells.

A skeletal muscle cannot contract unless it is stimulated by motor neurons. In the absence of innervation, it exhibits flaccid paralysis. Effectors of the ANS, however, function even in the absence of nervous stimulation—for example, the heart goes on beating even if it is *denervated* (all nerve connections to it are severed). The ANS does not activate its effectors, but *modifies* their activity, for example by accelerating or decelerating the heart. When autonomic nerves are severed from smooth or cardiac muscle, the muscle exhibits not flaccid paralysis but exaggerated responses *(denervation hypersensitivity).* Differences between the somatic and autonomic nervous systems are summarized in table 15.7.

The two divisions of the autonomic nervous system differ in anatomy and function and often have contrasting effects on the same target organs. The *sympathetic division* prepares the body in many ways for physical activity—it makes the heart beat faster and harder, increases pulmonary airflow, elevates blood glucose concentration, increases blood flow to cardiac muscle and the skeletal muscles, and so forth. Walter Cannon, founder of the homeostasis concept (see chapter 1), referred to these sympathetic responses as the "fight or flight" reaction because in animals, it comes into play when the individual must attack, defend itself, or flee from danger. In our own lives it occurs in many situations involving arousal, competition, stress, danger, anger, or fear.

The *parasympathetic division,* by comparison, has a calming effect on many body functions. It is associated with reduced energy expenditure and normal bodily maintenance, including such functions as digestion, defecation, and urination. This can be thought of as the "resting and digesting" state to contrast it with fight or flight.

14. *baro* = pressure

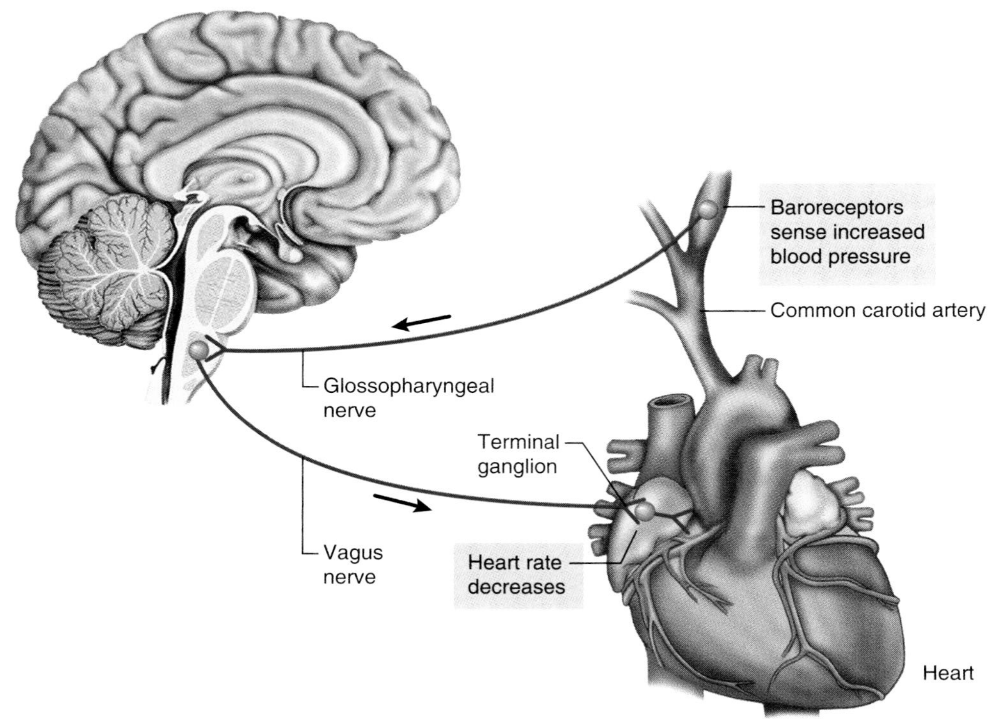

Figure 15.14 An Autonomic Reflex Arc. In this example, a rise in blood pressure is detected by baroreceptors in the carotid artery and the vagus nerve decreases the heart rate, which results in a corrective drop in blood pressure.

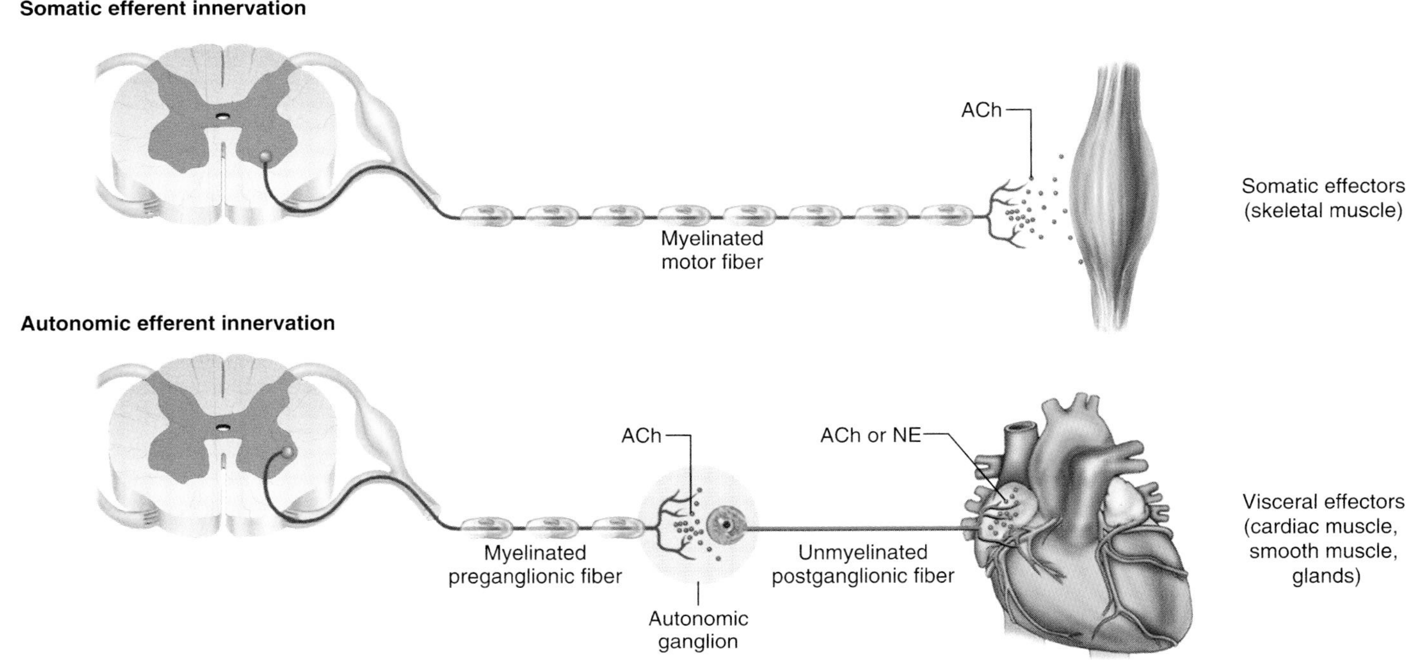

Figure 15.15 Comparison of Somatic and Autonomic Efferent Pathways. The entire distance from CNS to effector is spanned by one neuron in the somatic system and two neurons in the autonomic system. Only acetylcholine (ACh) is employed as a neurotransmitter by the somatic neuron and the autonomic preganglionic neuron, but autonomic postganglionic neurons can employ either ACh or norepinephrine (NE).

Table 15.7 Comparison of the Somatic and Autonomic Nervous Systems

Feature	Somatic	Autonomic
Effectors	Skeletal muscle	Smooth muscle, cardiac muscle, glands
Efferent pathways	One nerve fiber from CNS to effector; no ganglia	Two nerve fibers from CNS to effector; synapse at a ganglion
Neurotransmitters	Acetylcholine (ACh)	ACh and norepinephrine (NE)
Effect on target cells	Always excitatory	Excitatory or inhibitory
Effect of denervation	Flaccid paralysis	Denervation hypersensitivity
Control	Usually voluntary	Usually involuntary

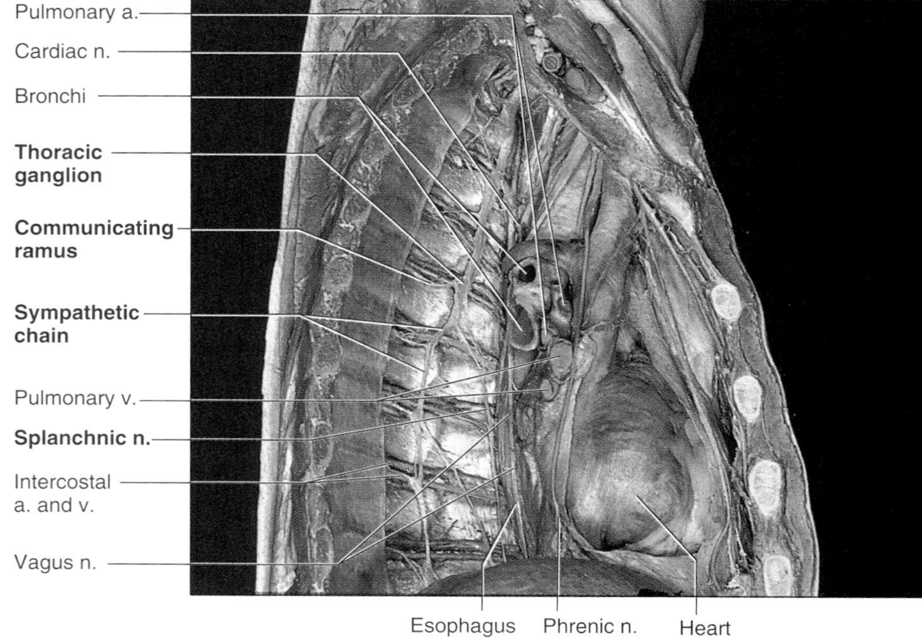

Figure 15.16 The Sympathetic Chain Ganglia. Right lateral view of the thoracic cavity. (*a.* = artery; *n.* = nerve; *v.* = vein.)

Anatomy of the Sympathetic Division

The sympathetic division is also called the *thoracolumbar division* because it arises from the thoracic and lumbar regions of the spinal cord. Its preganglionic somas are in the lateral horns of the gray matter at these levels. Their fibers exit by way of spinal nerves T1 to L2 and lead to the **sympathetic chain** of ganglia **(paravertebral**[15] **ganglia)** along each side of the vertebral column (figs. 15.16 and 15.17). Although these chains receive input from only the thoracolumbar region of the cord, they extend into the cervical and sacral regions as well. Sympathetic ganglia in these regions receive their input from nerve fibers traveling up and down the chains after they leave the spinal cord. The number of ganglia varies from person to person, but usually there are 3 cervical (*superior, middle,* and *inferior*), 11 thoracic, 4 lumbar, 4 sacral, and 1 coccygeal ganglion in each chain.

In the thoracolumbar region, each paravertebral ganglion is connected to a spinal nerve by two branches called *communicating rami* (fig. 15.18). The preganglionic fibers are small myelinated fibers that travel from the spinal nerve to the ganglion by way of the **white communicating ramus,** which gets its color and name from the myelin. The postganglionic fibers leave the ganglion by way of the **gray communicating ramus,** named for its lack of myelin and duller color, and by other routes. They extend the rest of the way to the target organ.

15. *para* = next to + *vertebr* = vertebral column

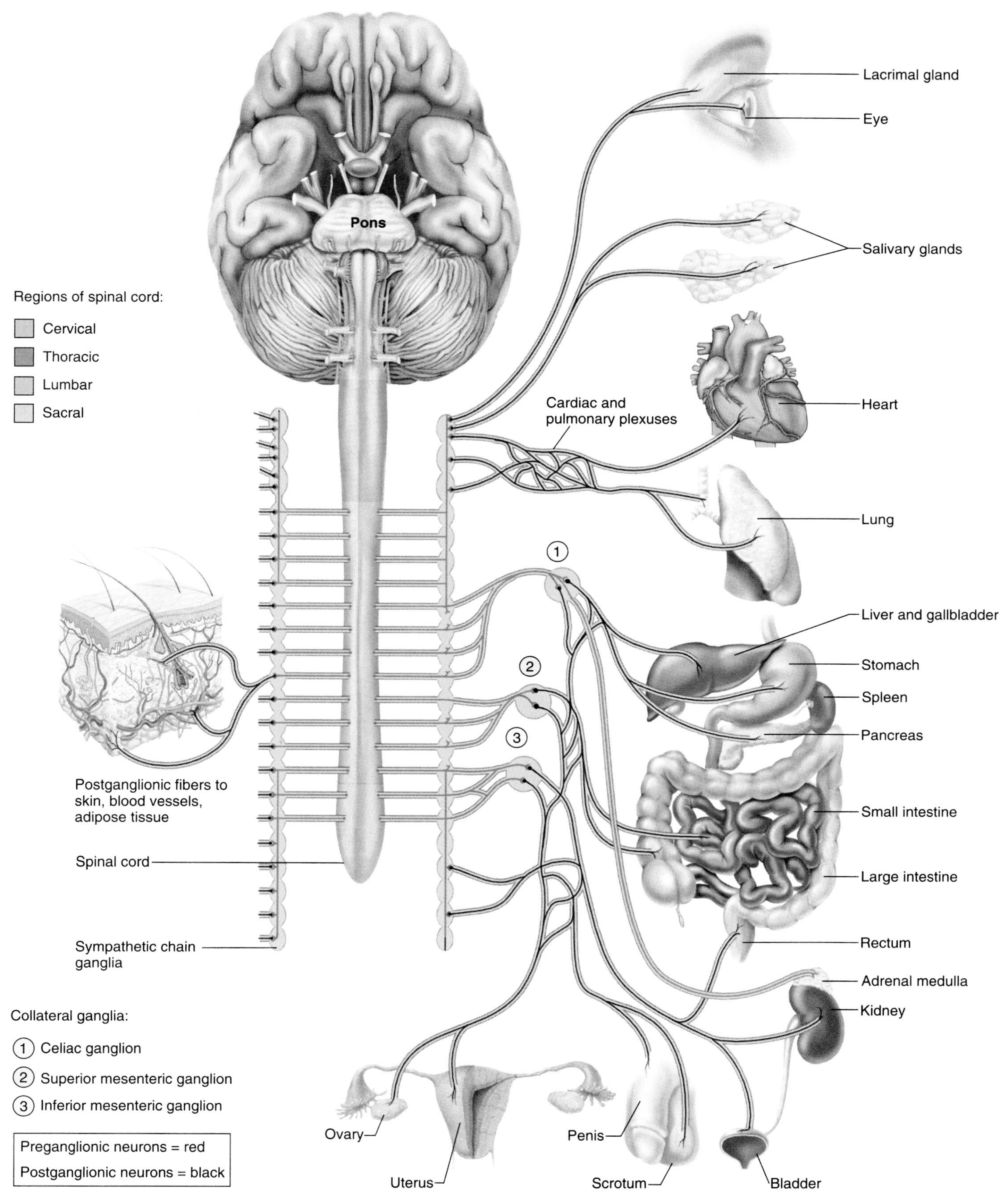

Figure 15.17 Sympathetic Efferent Pathways.

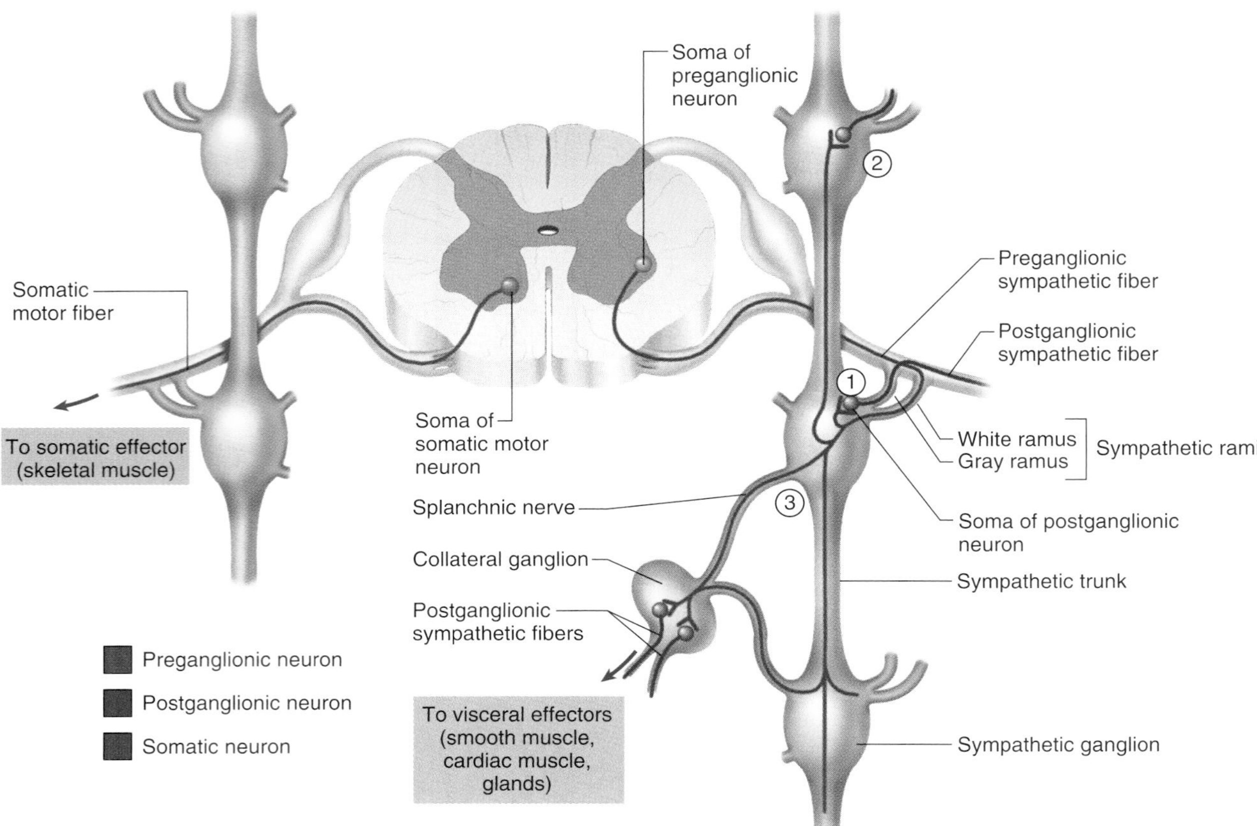

Figure 15.18 **Sympathetic Efferent Pathways** ***(right)*** **Compared to Somatic Efferent Pathways** ***(left).*** Sympathetic efferent fibers can follow any of the three numbered routes: (*1*) the spinal nerve route, (*2*) the sympathetic nerve route, or (*3*) the splanchnic nerve route.

After they enter the sympathetic chain, preganglionic fibers may follow any of these courses:

- Some end in the ganglion that they enter and synapse there with a postganglionic neuron.
- Some travel up or down the chain and synapse in ganglia at other levels. It is these fibers that link the paravertebral ganglia into a chain. They are the only route by which ganglia at the cervical, sacral, and coccygeal levels receive input.
- Some pass through the chain without synapsing and continue as *splanchnic* (SPLANK-nic) *nerves,* to be considered shortly.

Although most preganglionic neurons synapse with postganglionic neurons in the paravertebral ganglia, by no means do the pre- and postganglionics have a simple one-to-one relationship. For one thing, each postganglionic cell may receive synapses from multiple preganglionic cells, thus exhibiting the principle of *neuronal convergence* discussed in chapter 13. Furthermore, each preganglionic fiber branches and synapses with multiple postganglionic fibers, thus showing *neuronal divergence.* Indeed, there are about 17 postganglionic neurons for every preganglionic neuron in the sympathetic division.

This means that when one preganglionic neuron fires, it can excite multiple postganglionic fibers that lead to different target organs. When the sympathetic division is excited, it can therefore affect several organ systems at once—as suggested by the name *sympathetic.*[16] This effect, called **mass activation,** would not be possible without the divergent neural pathways established in the paravertebral ganglia. It demonstrates the complementarity of form and function.

Nerve fibers leave the paravertebral ganglia by three routes: spinal, sympathetic, and splanchnic nerves. These are numbered in figure 15.18 to correspond to the following descriptions:

1. **The spinal nerve route.** Some postganglionic fibers exit by way of the gray ramus, return to the spinal nerve or its subdivisions, and travel the rest of the

16. *sym* = together + *path* = feeling

way to the target organ. This is the route that sympathetic fibers take to most sweat glands, arrector pili muscles, and blood vessels of the skin and skeletal muscles.

2. **The sympathetic nerve route.** Other postganglionic fibers leave by way of **sympathetic nerves** that extend to the heart, lungs, esophagus, and thoracic blood vessels. These nerves form a plexus around each carotid artery and issue fibers from there to effectors in the head—including sweat, salivary, and nasal glands; arrector pili muscles; blood vessels; and dilators of the iris. Some fibers from the superior cervical ganglion form the **cardiac nerves** to the heart.
3. **The splanchnic[17] nerve route.** This route is formed by fibers that pass through the ganglia without synapsing. These fibers originate predominantly from spinal nerves T5 to T12. After passing through the ganglia, they form greater, lesser, and lumbar **splanchnic nerves.** These lead to the **collateral (prevertebral) ganglia,** which contribute to a network called the **abdominal aortic plexus** wrapped around the aorta (fig. 15.19). There are three major collateral ganglia in this plexus—the **celiac** (SEE-lee-ac) **ganglion, superior mesenteric ganglion,** and **inferior mesenteric ganglion—** located at points where arteries of the same names branch off the aorta. The postganglionic fibers accompany these arteries and their branches to the target organs. (The term *solar plexus* is regarded as a collective designation for the celiac and superior mesenteric ganglia by some authorities and as a synonym for only the celiac ganglion by others.) Innervation to and from the three major collateral ganglia is summarized in table 15.8.

17. *splanchn* = viscera

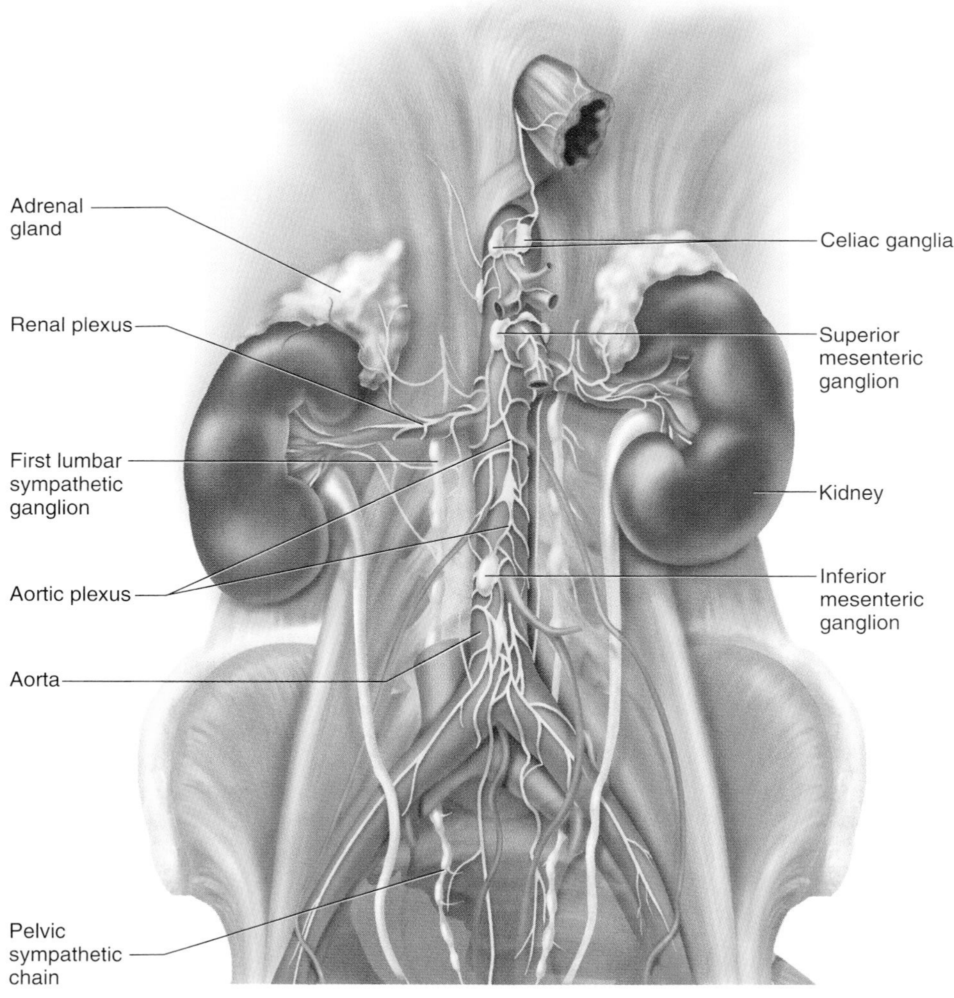

Figure 15.19 Sympathetic Collateral Ganglia and the Abdominal Aortic Plexus.

Table 15.8 Innervation to and from the Collateral Ganglia

Sympathetic Ganglia and Splanchnic Nerve →	Collateral Ganglion →	Postganglionic Target Organs
From thoracic ganglion 5 to 9 or 10 via greater splanchnic nerve	Celiac ganglion	Stomach, spleen, liver, small intestine, and kidneys
From thoracic ganglia 9 and 10 via lesser splanchnic nerve	Celiac and superior mesenteric ganglia	Small intestine and colon
From lumbar ganglia via lumbar splanchnic nerve	Celiac and inferior mesenteric ganglia	Distal colon, rectum, urinary bladder, and reproductive organs

The Adrenal Glands

The paired **adrenal**[18] **glands** rest like hats on the superior pole of each kidney (fig. 15.19). Each adrenal is actually two glands with different functions and embryonic origins. The outer rind, the **adrenal cortex,** secretes steroid hormones discussed in chapter 17. The inner core, the **adrenal medulla,** is a modified sympathetic ganglion. It consists of closely packed round cells that are modified postganglionic neurons. Sympathetic preganglionic fibers penetrate through the cortex and terminate on cells of the medulla.

When stimulated, the adrenal medulla secretes a mixture of hormones into the bloodstream—about 85% epinephrine (adrenalin), 15% norepinephrine (noradrenalin), and a trace of dopamine. These hormones, the *catecholamines,* were briefly considered in chapter 13 because they also function as neurotransmitters. They complement the action of other sympathetic postganglionic neurons, so the adrenal medulla is involved in the mass activation typical of this division. The sympathetic nervous system and adrenal medulla are so closely related in development and function that they are referred to collectively as the *sympathoadrenal system.*

Anatomy of the Parasympathetic Division

The parasympathetic nervous system is also called the *craniosacral division* because it arises from the brain and sacral region of the spinal cord; its fibers travel in certain cranial and sacral nerves. The preganglionic neurons have their somas in the pons, medulla oblongata, and segments S2 to S4 of the spinal cord (fig. 15.20). The parasympathetic ganglia, called **terminal ganglia** (see fig. 15.14), lie in or near the target organs. Thus, the parasympathetic division has long preganglionic fibers, reaching almost all the way to the target cells, and short postganglionic fibers that cover the rest of the distance.

18. *ad* = near + *ren* = kidney

There is a slight degree of neuronal divergence in the parasympathetic division, with a ratio of about two postganglionic fibers to every preganglionic. This divergence is much less than in the sympathetic division, however, and parasympathetic effects do not exhibit mass activation. The parasympathetic division is relatively selective in its stimulation of target organs since the preganglionic fibers reach the organs before any postganglionic divergence occurs.

Parasympathetic fibers leave the brainstem by way of the following four cranial nerves. The first three supply all parasympathetic innervation to the head and the last one supplies viscera of the thoracic and abdominal cavities.

1. **Oculomotor nerve (III).** The oculomotor nerve carries parasympathetic fibers that control the lens and pupil of the eye. The preganglionic fibers enter the orbit and terminate in the *ciliary ganglion.* Postganglionic fibers enter the eyeball and innervate the *ciliary muscle,* which controls the lens, and the *pupillary constrictor,* which narrows the pupil.
2. **Facial nerve (VII).** The facial nerve carries parasympathetic fibers that regulate the secretion of tears, saliva, and nasal secretions. Soon after the facial nerve emerges from the pons, the parasympathetic fibers split away and form two smaller branches. The upper branch ends at the *sphenopalatine ganglion* near the junction of the maxilla and palatine bones. Postganglionic fibers then continue to the tear glands and glands of the nasal cavity, palate, and other areas of the oral cavity. The lower branch crosses the middle-ear cavity and ends at the *submandibular ganglion* near the angle of the mandible. Postganglionic fibers from here supply salivary glands in the floor of the mouth.
3. **Glossopharyngeal nerve (IX).** The glossopharyngeal nerve carries parasympathetic fibers concerned with salivation. The preganglionic fibers leave this nerve soon after its origin and form the *tympanic nerve,* which crosses the eardrum and ends in the

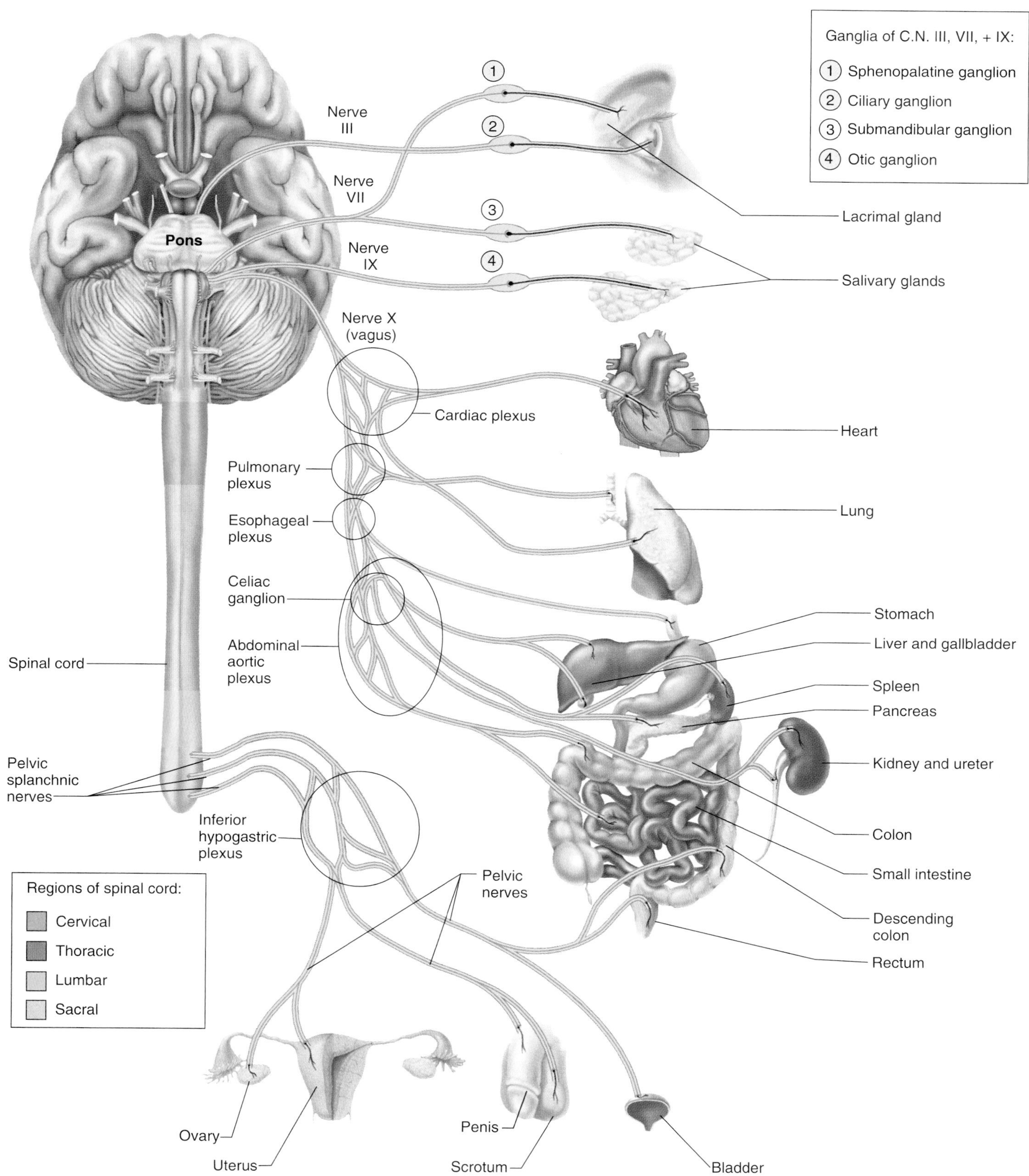

Figure 15.20 Parasympathetic Efferent Pathways.

otic[19] *ganglion* near the foramen ovale. The postganglionic fibers then follow the trigeminal nerve and to the *parotid salivary gland* just in front of the earlobe.

4. **Vagus nerve (X).** The vagus nerve carries about 90% of all parasympathetic preganglionic fibers. It travels down the neck and forms three networks in the mediastinum—the **cardiac plexus,** which supplies fibers to the heart; the **pulmonary plexus,** whose fibers accompany the bronchi and blood vessels into the lungs; and the **esophageal plexus,** whose fibers regulate swallowing.

At the lower end of the esophagus, these plexuses give off the anterior and posterior **vagal trunks,** each of which contains fibers from both the right and left vagus. These penetrate the diaphragm, enter the abdominal cavity, and contribute to the extensive *abdominal aortic plexus* mentioned earlier. As we have seen, sympathetic fibers synapse here. The parasympathetic fibers, however, pass through without synapsing and lead to the liver, pancreas, stomach, small intestine, kidney, ureter, and proximal half of the colon.

The remaining parasympathetic fibers arise from levels S2 to S4 of the spinal cord. They travel a short distance in the ventral rami of the spinal nerves and then form **pelvic splanchnic nerves** that lead to the **inferior hypogastric (pelvic) plexus.** Some parasympathetic fibers synapse here, but most pass through this plexus and travel by way of **pelvic nerves** to the terminal ganglia in their target organs: the distal half of the large intestine, the rectum, urinary bladder, and reproductive organs.

The sympathetic and parasympathetic divisions of the ANS are compared in figure 15.21 and table 15.9.

19. *ot* = ear + *ic* = pertaining to

Before You Go On

Answer the following questions to test your understanding of the preceding section:

20. In what ways is an autonomic reflex arc like a somatic motor reflex arc? In what ways does it differ?
21. Explain why the sympathetic division is also called the thoracolumbar division even though its paravertebral ganglia extend all the way from the cervical to the sacral region.
22. Describe or diagram the structural relationships among the following: preganglionic fiber, postganglionic fiber, ventral ramus, gray ramus, white ramus, and paravertebral ganglion.
23. Explain in anatomical terms why the parasympathetic division affects target organs more selectively than the sympathetic division does.
24. Trace the pathway of a parasympathetic fiber of the vagus nerve from the medulla oblongata to the small intestine.

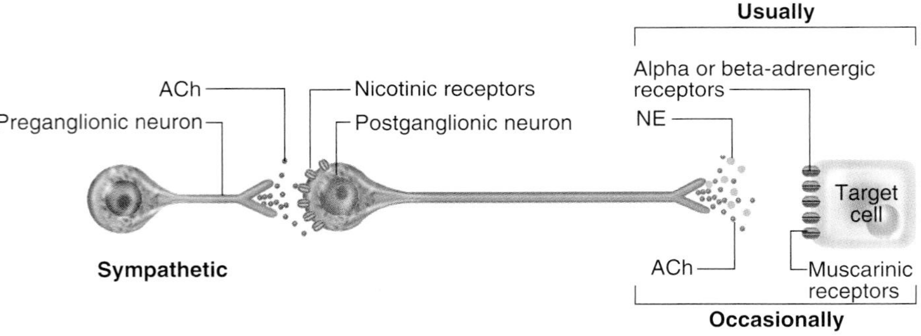

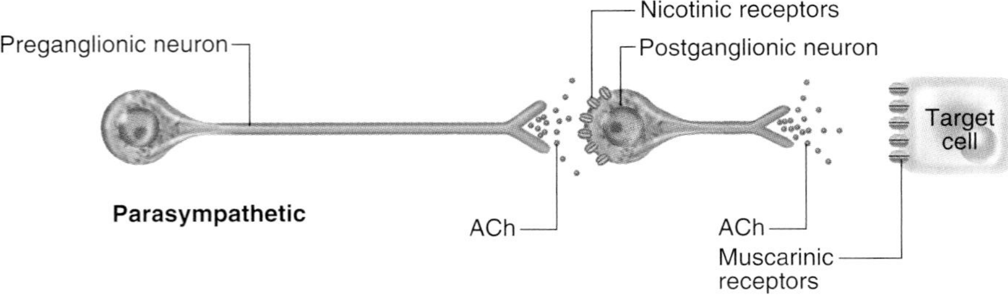

Figure 15.21 Neurotransmitters and Receptors of the Autonomic Nervous System.

Table 15.9 Comparison of the Sympathetic and Parasympathetic Divisions

Feature	Sympathetic	Parasympathetic
Origin in CNS	Thoracolumbar	Craniosacral
Location of ganglia	Paravertebral ganglia adjacent to spinal column and prevertebral ganglia anterior to it	Terminal ganglia near or within target organs
Fiber lengths	Short preganglionic Long postganglionic	Long preganglionic Short postganglionic
Neuronal divergence	Extensive (about 1:17)	Minimal (about 1:2)
Effects of system	Mass activation common	More specific and local effects

The Autonomic Nervous System: Physiology

Objectives

When you have completed this section, you should be able to

- name the neurotransmitters employed at different synapses of the ANS;
- name the receptors for these neurotransmitters and explain how these receptor types relate to autonomic effects;
- explain how the ANS controls many target organs through dual innervation;
- explain how control is exerted in the absence of dual innervation; and
- discuss the ways in which the ANS is governed by various levels of the CNS.

Neurotransmitters and Receptors

The key to understanding the effects of the ANS lies in knowing which neurotransmitters it releases and what kind of receptors occur on the target cells. The ANS has both **cholinergic fibers,** which secrete acetylcholine (ACh), and **adrenergic fibers,** which secrete norepinephrine (NE). Cholinergic fibers include the preganglionic fibers of both divisions, the postganglionic fibers of the parasympathetic division, and a few sympathetic postganglionic fibers (those that innervate sweat glands and some blood vessels in the skeletal muscles). Most sympathetic postganglionic fibers are adrenergic (table 15.10). The receptors for ACh and NE are called **cholinergic** and **adrenergic receptors,** respectively.

Both the sympathetic and parasympathetic divisions have excitatory effects on some effectors and inhibitory effects on others. For example, the parasympathetic division contracts the wall of the urinary bladder but relaxes the internal urinary sphincter—both of which are necessary for the expulsion of urine. It employs ACh for both purposes. Similarly, the sympathetic division constricts most blood vessels but dilates the coronary arteries, and it achieves these contrasting effects with NE. Clearly, the difference is not due to the neurotransmitter. Rather, it is due to the fact that different effector cells have *different kinds of receptors* for it. Knowledge of these receptor types is essential to the field of neuropharmacology (see clinical insight 15.5 at the end of the chapter).

Table 15.10 Locations of Cholinergic and Adrenergic Fibers in the ANS

Division	Preganglionic Fibers	Postganglionic Fibers
Sympathetic	Always cholinergic	Mostly adrenergic; a few cholinergic
Parasympathetic	Always cholinergic	Always cholinergic

Cholinergic Receptors

Acetylcholine binds to two classes of receptors—**nicotinic** (NIC-oh-TIN-ic) and **muscarinic** (MUSS-cuh-RIN-ic) **receptors**—named for the drugs that were used to identify and distinguish them. Nicotine binds only to the former type, while muscarine, a mushroom poison, binds only to the latter. Other drugs also selectively bind to one type or the other—atropine binds only to muscarinic receptors and curare only to nicotinic receptors, for example.

Nicotinic receptors occur on all ANS postganglionic neurons, in the adrenal medulla, and on skeletal muscle fibers. Muscarinic receptors occur on all gland, smooth muscle, and cardiac muscle cells that receive cholinergic innervation. When ACh binds to a nicotinic receptor, it produces an excitatory postsynaptic potential (EPSP). When it binds to a muscarinic receptor, however, the effect can be either excitatory or inhibitory—an EPSP in intestinal smooth muscle or an IPSP in cardiac muscle, for example.

Adrenergic Receptors

There are likewise different classes of adrenergic receptors that account for the effects of norepinephrine (NE) on different target cells. NE receptors fall into two broad classes called **alpha-(α-)adrenergic** and **beta-(β-)adrenergic receptors.** The binding of NE to α-adrenergic receptors is usually excitatory, and its binding to β-adrenergic receptors is usually inhibitory, but there are exceptions to both. For example, NE binds to β receptors in cardiac muscle but has an excitatory effect.

The exceptions result from the existence of subclasses of each receptor type, called α_1 and α_2, β_1 and β_2 receptors. All four types function by means of second messengers. Both β receptors activate the production of cyclic AMP (cAMP), α_2 receptors suppress cAMP production, and α_1 receptors employ calcium ions as the second messenger. Some target cells have both α and β receptors.

The binding of NE to α-adrenergic receptors in blood vessels causes vasoconstriction. The arteries that supply the heart and skeletal muscles, however, have β-adrenergic receptors, which relax the arteries, thereby dilating them and increasing blood flow to these organs. NE also relaxes the bronchioles when it binds to β-adrenergic receptors. These effects are obviously appropriate to the exercise state in which the sympathetic nervous system is most active. The autonomic effects on glandular secretion are often achieved through the adjustment of blood flow to the gland rather than direct stimulation of the gland cells.

The locations and effects of many cholinergic and adrenergic receptor types are summarized in table 15.11. In a number of cases, it is not known yet which subclass of α or β receptor is present in a given organ, which is why several entries in the table have no subscripts.

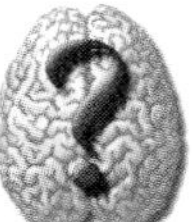

Think About It

It is noted in table 15.11 that the sympathetic nervous system has an α-adrenergic effect on blood platelets and promotes clotting. How can the sympathetic nervous system stimulate platelets, considering that platelets are drifting cell fragments in the bloodstream with no nerve fibers leading to them?

Dual Innervation

Most of the viscera receive nerve fibers from both the sympathetic and parasympathetic divisions and thus are said to have **dual innervation.** In such cases, the two divisions may have either *antagonistic* or *cooperative* effects on the same organ.

Antagonistic effects oppose each other. For example, the sympathetic division speeds up the heart and the parasympathetic division slows it down; the sympathetic division inhibits digestion and the parasympathetic division stimulates it; the sympathetic division dilates the pupil and the parasympathetic division constricts it. In some cases, these effects are exerted through dual innervation of the same effector cells, as in the heart, where nerve fibers of both divisions terminate on the same muscle cells. In other cases, antagonistic effects arise because each division innervates different effector cells with opposite effects on organ function. In the iris of the eye, for example, sympathetic fibers innervate the pupillary dilator and parasympathetic fibers innervate the constrictor (fig. 15.22).

Cooperative effects are seen when the two divisions act on different effectors to produce a unified overall effect. Salivation is a good example. The parasympathetic division stimulates serous cells of the salivary glands to secrete a watery, enzyme-rich secretion, while the sympathetic division stimulates mucous cells of the

Table 15.11 Effects of the Sympathetic and Parasympathetic Nervous Systems

Target	Sympathetic Effect and Receptor Type*	Parasympathetic Effect and Receptor Type*
Adipose Tissue		
	Decreased fat breakdown (α_2) Increased fat breakdown (β_1)	No effect
Eye		
Dilator of pupil	Pupillary dilation (α_1)	No effect
Constrictor of pupil	No effect	Pupillary constriction (M)
Ciliary muscle and lens	Relaxation for far vision (β_1)	Contraction for near vision (M)
Lacrimal (tear) gland	Less secretion (α)	More secretion (M)

(continued)

Table 15.11 Effects of the Sympathetic and Parasympathetic Nervous Systems (continued)

Target	Sympathetic Effect and Receptor Type*	Parasympathetic Effect and Receptor Type*
Integumentary System		
Merocrine sweat glands (cooling)	Secretion (M)	No effect
Apocrine sweat glands (scent)	Secretion (α)	No effect
Arrector pili muscles	Hair erection (α)	No effect
Adrenal Medulla	Hormone secretion (N)	No effect
Circulatory System		
Heart rate and force	Increased (α)	Decreased (M)
Deep coronary arteries	Vasodilation (β_2)	Slight vasodilation (M)
Blood vessels of most viscera	Vasoconstriction (α)	Vasodilation (M)
Blood vessels of skeletal muscles	Vasodilation (β_2)	No effect
Blood vessels of skin	Vasoconstriction (α)	Vasodilation, blushing (M)
Platelets (blood clotting)	Increased clotting (α_2)	No effect
Respiratory System		
Bronchi and bronchioles	Bronchodilation (β_2)	Bronchoconstriction (M)
Mucous glands	Decreased secretion (β_1) Increased secretion (β_2)	No effect
Urinary System		
Kidneys	Reduced urine output (α)	No effect
Bladder	No effect	Contraction (M)
Internal urinary sphincter	Contraction (α)	Relaxation (M)
Digestive System		
Salivary glands	Thick mucous secretion (α_1)	Thin serous secretion (M)
Gastrointestinal peristalsis	Inhibition (α_1)	Stimulation (M)
Gastrointestinal secretion	No effect	Stimulation (M)
Liver	Glycogen breakdown (β_2)	Glycogen synthesis (M)
Pancreatic secretion (exocrine)	Decreased secretion (α)	Increased secretion (M)
Pancreatic insulin secretion	Decreased secretion (α_2) Increased secretion (β_2)	No effect
Reproductive System		
Orgasm, smooth muscle roles	Stimulation (α)	No effect
Penile or clitoral erection	No effect	Stimulation (M)
Glandular secretion	No effect	Stimulation (M)
Uterus in orgasm	Dilation of cervical canal	No effect
Uterus in labor	Contraction (α)	No effect

*Adrenergic receptors are indicated α and β, with subclass if known. Cholinergic receptors are indicated by M for muscarinic and N for nicotinic.

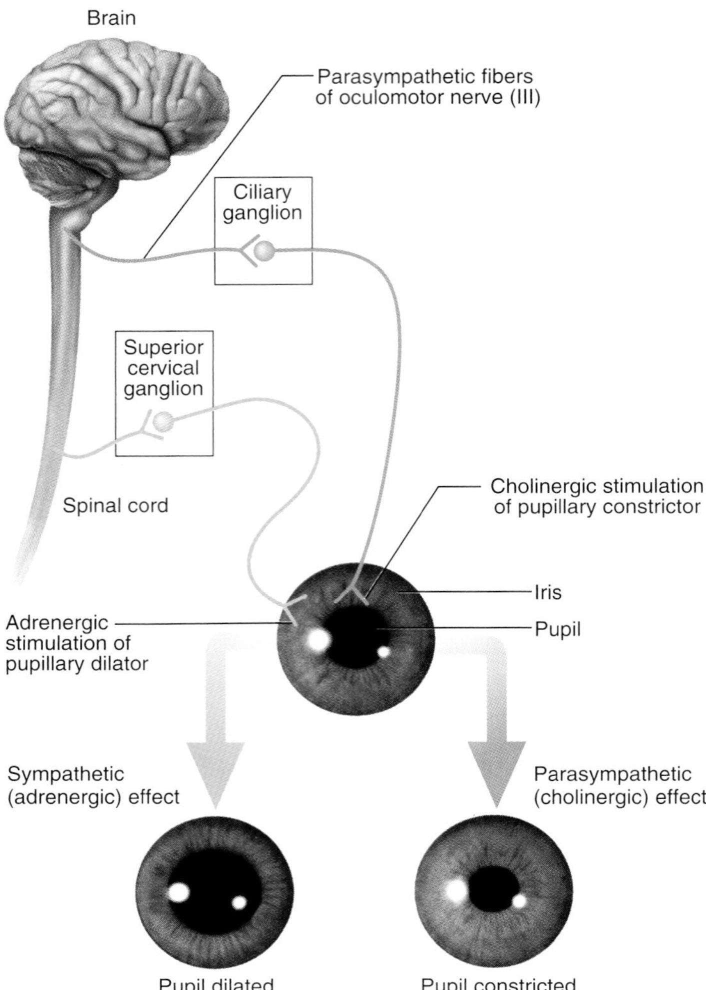

Figure 15.22 Dual Innervation of the Iris. Shows antagonistic effects of the sympathetic and parasympathetic divisions on pupil diameter.

same glands to secrete mucus. The enzymes and mucus are both necessary components of the saliva.

Even when both divisions innervate a single organ, they do not always innervate it equally or exert equal influence. For example, the parasympathetic division forms an extensive plexus in the wall of the digestive tract and exerts much more influence over it than the sympathetic division does. In the ventricles of the heart, by contrast, there is much less parasympathetic than sympathetic innervation.

Control Without Dual Innervation

Dual innervation is not always necessary for the ANS to produce opposite effects on an organ. The adrenal medulla, arrector pili muscles, sweat glands, and many blood vessels receive only sympathetic fibers. The most significant example of control without dual innervation is regulation of blood pressure and routes of blood flow. The sympathetic fibers to a blood vessel have a baseline firing frequency called **sympathetic tone.** This keeps the vessels in a state of partial constriction called **vasomotor tone.** An increase in firing frequency causes vasoconstriction by increasing smooth muscle contraction. A drop in firing frequency causes vasodilation by allowing the smooth muscle to relax. Thus, the sympathetic division alone exerts opposite effects on the vessels.

Sympathetic control of vasomotor tone can shift blood flow from one organ to another according to the changing needs of the body. In times of stress or exercise, the skeletal muscles and heart receive a high circulatory priority, and the sympathetic division dilates the arteries that supply them. Such processes as digestion, nutrient absorption, and urine formation can wait; thus the sympathetic division constricts arteries to the gastrointestinal tract and kidneys. It also reduces blood flow through the skin, which may help to minimize bleeding in the event that the stress-producing situation leads to injury. Furthermore, since there is not enough blood in the body to abundantly supply all the organ systems at once, it is necessary to temporarily divert blood away from some organs in order to supply an adequate amount to the muscular system.

Central Control of Autonomic Function

In spite of its name, the ANS is not an independent nervous system but is regulated by several levels of the CNS (and of course *all* autonomic output originates in the CNS). In this section we briefly consider how four regions of the CNS influence autonomic function:

1. **The cerebral cortex.** Although the word *autonomic* implies that the ANS functions independently of our conscious mind, it is clear that our emotions can have autonomic effects—for example, when anger makes the blood pressure rise, fear makes the heart race, or anxiety inhibits sexual function. Some people can be trained through biofeedback techniques to consciously regulate autonomic functions such as their blood pressure and heart rate (see clinical insight 15.4). The amygdala, a limbic system structure that links emotion and memory, exerts autonomic effects through the hypothalamus and ANS.
2. **The hypothalamus.** The hypothalamus is the most important integrating center for autonomic function. It contains nuclei for numerous visceral functions and is involved in emotions and drives that act through the ANS. The fight or flight responses typical of the sympathetic nervous system originate in the hypothalamus. Output from

Clinical Insight 15.4

Biofeedback

Biofeedback is a technique in which an instrument produces sound or light signals in response to changes in a subject's blood pressure, heart rate, muscle tone, brain waves, or other physiological variables. It gives the subject awareness of changes that he or she would not ordinarily notice. Some people can be trained to control their physiology in order to produce a certain tone or color of light from the apparatus. Eventually they can control it without the aid of the monitor. Biofeedback is not a quick, easy, infallible, or inexpensive cure for all ills, but it has been used successfully to treat hypertension, stress, and migraine headaches.

the hypothalamus travels largely to more caudal regions of the brainstem and from there to the cranial nerves and spinal cord.

3. **The reticular formation.** The reticular formation contains centers for cardiac, vasomotor, and gastrointestinal function. It receives extensive sensory input from the viscera by way of the vagus nerves, and it can integrate and respond directly to this input even without cerebral, cerebellar, or hypothalamic input.
4. **The spinal cord.** Some autonomic responses, such as the defecation and micturition (urination) reflexes, are integrated in the spinal cord without the involvement of the brain. Fortunately, however, the brain is able to inhibit these responses consciously.

Before You Go On

Answer the following questions to test your understanding of the preceding section:

25. Summarize the locations of nicotinic and muscarinic receptors in the ANS.
26. What second messenger is employed by α_1 receptors? What second messenger is involved in the other three classes of adrenergic receptors?
27. Explain what is meant by dual innervation and give some examples.
28. Describe how the sympathetic division can control blood flow in the absence of dual innervation.
29. Cite evidence that the ANS is subject to cerebral influence.

Clinical Insight 15.5

Drugs and the Nervous System

Neuropharmacology is a branch of medicine that deals with the effects of drugs on the nervous system, especially drugs that mimic, enhance, or inhibit the action of neurotransmitters. A few examples will illustrate the clinical relevance of understanding neurotransmitter and receptor functions.

A number of drugs work by stimulating adrenergic and cholinergic neurons or receptors. *Sympathomimetics*[20] are drugs that enhance sympathetic action by stimulating adrenergic receptors or promoting norepinephrine release. For example phenylephrine, found in such cold medicines as Chlor-Trimeton and Dimetapp, aids breathing by stimulating α_1 receptors and dilating the bronchioles. *Sympatholytics*[21] are drugs that suppress sympathetic action by inhibiting norepinephrine release or by binding to adrenergic receptors without stimulating them. Propanolol, for example, is a *beta-blocker.* It reduces hypertension (high blood pressure) partly by blocking β-adrenergic receptors and interfering with the effects of epinephrine and norepinephrine on the heart and blood vessels. (It also reduces the production of *angiotensin II,* a hormone that stimulates vasoconstriction and raises blood pressure.)

Parasympathomimetics enhance parasympathetic effects. Pilocarpine, for example, relieves glaucoma (excessive pressure within the eyeball) by dilating a vessel that drains fluid from the eye. *Parasympatholytics* inhibit ACh release or block its receptors. Atropine, for example, blocks muscarinic receptors and is sometimes used to dilate the pupils for eye examinations and to dry the mucous membranes of the respiratory tract before inhalation anesthesia. It is an extract of the deadly nightshade plant, *Atropa belladonna.* Women of the Middle Ages used nightshade to dilate their pupils, which was regarded as a beauty enhancement.[22]

The drugs we have mentioned so far act on the peripheral nervous system and its effectors. Many others act on the central nervous system. Strychnine, for example, blocks the inhibitory action of glycine on spinal motor neurons. The motor neurons then overstimulate the muscles, causing spastic paralysis and sometimes death by suffocation.

Sigmund Freud predicted that psychiatry would eventually draw upon biology and chemistry to deal with emotional problems once treated only by counseling and psychoanalysis. A branch of neuropharmacology called *psychopharmacology* is fulfilling his prediction. This field dates to the 1950s when chlorpromazine, an antihistamine, was accidentally found to relieve schizophrenia.

The management of clinical depression is one example of how contemporary psychopharmacology has supplemented counseling approaches. Some cases of depression result from deficiencies of the monoamine neurotransmitters. Thus, they yield to drugs that prolong the effects of the monoamines already present at the synapses. One of the earliest discovered antidepressants was imipramine, which blocks the synaptic reuptake of serotonin and norepinephrine. However, it produces undesirable side effects such as dry mouth and irregular cardiac rhythms; it has been largely replaced by Prozac (fluoxetine), which blocks serotonin reuptake and prolongs its mood-elevating effect; thus it is called a *selective serotonin reuptake inhibitor* (SSRI). Prozac is also used to treat fear of rejection, excess sensitivity to criticism, lack of self-esteem, and inability to experience pleasure, all of which were long handled only through counseling, group therapy, and psychoanalysis. After monoamines are taken up from the synapse, they are degraded by monoamine oxidase (MAO). Drugs called *MAO inhibitors* interfere with the breakdown of monoamine neurotransmitters and provide another pharmacological approach to depression.

Our growing understanding of neurochemistry also gives us deeper insight into the action of addictive drugs of abuse such as amphetamines and cocaine. Amphetamines ("speed") chemically resemble norepinephrine and dopamine, two neurotransmitters associated with elevated mood. Dopamine is especially important in sensations of pleasure. Cocaine blocks dopamine reuptake and thus produces a brief rush of good feelings. But when dopamine is not reabsorbed by the neurons, it diffuses out of the synaptic cleft and is degraded elsewhere. Cocaine thus depletes the neurons of dopamine faster than they can synthesize it, so that finally there is no longer an adequate supply to maintain normal mood. The postsynaptic neurons make new dopamine receptors as if "searching" for the neurotransmitter—all of which leads ultimately to anxiety, depression, and the inability to experience pleasure without the drug.

Caffeine exerts its stimulatory effect by competing with adenosine. Adenosine, which you know as a component of DNA, RNA, and ATP, also functions as an inhibitory neurotransmitter in the brain. One theory of sleepiness is that it results when prolonged metabolic activity breaks down so much ATP that the accumulated adenosine has a noticeably inhibitory effect in the brain. Caffeine has enough structural similarity to adenosine (fig. 1) to bind to its receptors, but it does not produce the inhibitory effect. Thus, it prevents adenosine from exerting its effect and a person feels more alert.

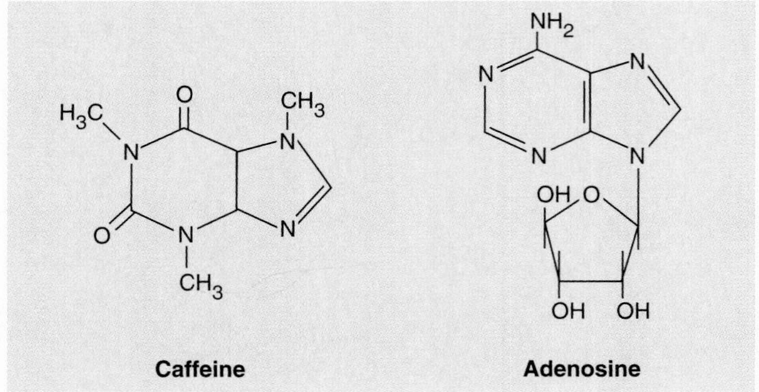

Figure 1 **Caffeine and Adenosine.** Adenosine is an inhibitory neurotransmitter that produces the sense of sleepiness. Caffeine is similar enough in structure to bind to adenosine receptors and block the action of adenosine.

20. *mimet* = imitate, mimic
21. *lyt* = break down, destroy
22. *bella* = beautiful, fine + *donna* = woman

Connective Issues

Interactions Between the NERVOUS SYSTEM and Other Organ Systems

← indicates ways in which this system affects other systems
→ indicates ways in which other systems affect this one

Integumentary System

← Nervous system regulates piloerection and sweating; controls cutaneous blood flow to regulate heat loss

→ Provides sensations of heat, cold, pressure, pain, and vibration; protects peripheral nerves

Skeletal System

← Nervous stimulation generates muscle tension essential for bone development and remodeling

→ Serves as reservoir of Ca^{2+} needed for neural function; protects CNS and some peripheral nerves

Muscular System

← Somatic nervous system activates skeletal muscles and maintains muscle tone

→ Gives expression to thoughts, emotions, and motor commands that arise in the CNS

Endocrine System

← Hypothalamus controls pituitary gland; sympathetic nervous system stimulates adrenal medulla

→ Many hormones affect neuronal growth and metabolism; hormones control electrolyte balance essential for neural function

Circulatory System

← Nervous system regulates heartbeat, blood vessel diameters, blood pressure, and routing of blood; influences blood clotting

→ Delivers O_2 and carries away wastes; transports hormones to and from CNS; CSF produced from and returned to blood

Lymphatic and Immune Systems

← Nerves innervate lymphoid organs and influence development and activity of immune cells; nervous system plays a role in regulating immune response; emotional states influence susceptibility to infection

→ Immune cells provide protection and promote tissue repair

Respiratory System

← Nervous system regulates rate and depth of respiration

→ Provides O_2, removes CO_2, and helps to maintain proper pH for neural function

Urinary System

← Nervous system regulates renal blood flow, thus affecting rate of urine formation; controls emptying of bladder

→ Disposes of wastes and maintains electrolyte and pH balance

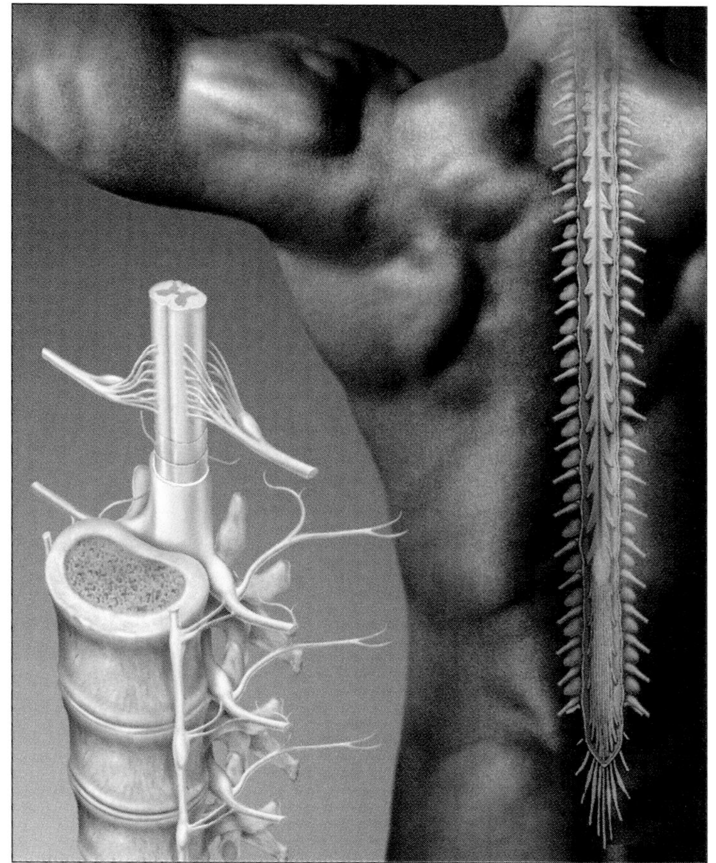

Digestive System

← Nervous system regulates appetite, feeding behavior, digestive secretion and motility, and defecation

→ Provides nutrients; liver provides stable level of blood glucose for neural function during periods of fasting

Reproductive System

← Nervous system regulates sex drive, arousal, and orgasm; secretes or stimulates pituitary release of many hormones involved in menstrual cycle, sperm production, pregnancy, and lactation

→ Sex hormones influence CNS development and sexual behavior; hormones of the menstrual cycle stimulate or inhibit hypothalamus

Chapter Review

Online Learning Center

Study resources for this chapter are available at www.mhhe.com/saladin/

Study Outline

General Anatomy of the Peripheral Nervous System (p. 538)

1. Subdivisions of the PNS
   a. Sensory (afferent) division
      - Somatic sensory division
      - Visceral sensory division
   b. Motor (efferent) division
      - Somatic motor division
      - Visceral motor division (autonomic nervous system)
        - •• Sympathetic division
        - •• Parasympathetic division
2. Nerve fiber classification
   a. Afferent and efferent
   b. Somatic and visceral
   c. General and special
3. Nerve classification
   a. Mixed
   b. Sensory
   c. Motor
4. Anatomy of a nerve
   a. Neurilemma
   b. Myelin
   c. Endoneurium
   d. Perineurium
   e. Epineurium
   f. Fascicles
5. Anatomy of a ganglion

The Cranial Nerves (p. 541)

1. Location and general pathways
2. Cranial nerves I–XII
   a. Numbers and names
   b. Composition (sensory or motor)
   c. Functions
   d. Origin and anatomical course
   e. Effects of damage
   f. Clinical tests

The Spinal Nerves (p. 550)

1. Groups
   a. Cervical nerves (C1–C8)
   b. Thoracic nerves (T1–T12)
   c. Lumbar nerves (L1–L5)
   d. Sacral nerves (S1–S5)
   e. Coccygeal nerve (Co)
2. Proximal branches
   a. Dorsal root
   b. Ventral root
3. Distal branches
   a. Dorsal ramus
   b. Ventral ramus
   c. Meningeal branch
4. Nerve plexuses (tables 15.3–15.6)
   a. Cervical plexus
   b. Brachial plexus
   c. Lumbar plexus
   d. Sacral plexus
   e. Coccygeal plexus
5. Dermatomes

Somatic Reflexes (p. 562)

1. Properties of reflexes
   a. Require stimulation
   b. Quick
   c. Involuntary
   d. Stereotyped
2. Types of reflexes
   a. Conditioned reflexes
   b. Somatic reflexes
   c. Visceral reflexes
3. The somatic reflex arc
4. Muscle spindles
   a. Intrafusal muscle fibers
   b. Nerve fibers
      - Primary afferent fibers
      - Secondary afferent fibers
      - Gamma motor neurons
5. Stretch reflex
   a. Functions
   b. Tendon reflexes as example
   c. Monosynaptic reflex arcs
   d. Reciprocal inhibition
6. Flexor reflex
   a. Withdrawal from injurious stimuli
   b. Involves polysynaptic reflex arc
   c. Involves parallel after-discharge circuit
7. Crossed extensor reflex
   a. Relationship to flexor reflex
   b. Responses of contralateral limb muscles
   c. Reflex pathways
8. Golgi tendon reflex
   a. Golgi tendon organ structure
   b. Response to tension on tendon
   c. Effect on muscle workload

The Autonomic Nervous System: Introduction and Anatomy (p. 568)

1. General properties
   a. Effectors
   b. Homeostatic role
   c. Involuntary action
   d. Receptors
   e. Efferent pathways
      - Preganglionic neuron
      - Postganglionic neuron
   f. Effect of denervation
2. Sympathetic division
   a. Origin of preganglionic neurons
   b. Sympathetic chain (paravertebral) ganglia
      - White communicating ramus
      - Gray communicating ramus
   c. Pathways of preganglionic fibers
      - Enter paravertebral ganglion and synapse there
      - Travel to higher or lower ganglia in chain before synapsing
      - Pass through chain without synapsing, travel in splanchnic nerves
   d. Neuronal convergence and divergence
   e. Routes leading away from sympathetic chain
      - Spinal nerves
      - Sympathetic nerves
      - Splanchnic nerves
      - Collateral ganglia

3. Adrenal glands
   a. Cortex
   b. Medulla
4. Parasympathetic division
   a. Origin of preganglionic neurons
      - Pons and medulla oblongata
      - Spinal cord segments S2–S4
   b. Terminal ganglia in target organs
   c. Pathways of preganglionic fibers
      - Cranial nerves III, VII, and IX
      - Parasympathetic components of vagus nerve (X)
        - •• Cardiac plexus
        - •• Pulmonary plexus
        - •• Exophageal plexus
        - •• Vagal trunks
        - •• Abdominal aortic plexus
      - Components arising from S2–S4
        - •• Pelvic splanchnic nerves
        - •• Inferior hypogastric plexus
        - •• Pelvic nerves

The Autonomic Nervous System: Physiology (p. 577)

1. Neurotransmitters and receptors
   a. Cholinergic fibers and ACh
      - Nicotinic receptors
      - Muscarinic receptors
   b. Adrenergic fibers and NE
      - α_1 and α_2 receptors
      - β_1 and β_2 receptors
2. Dual innervation
   a. Antagonistic effects
   b. Cooperative effects
3. Control without dual innervation
   a. Sympathetic tone
   b. Vasomotor tone
4. Central control of ANS
   a. Influence of cerebrum
   b. Influence of hypothalamus
   c. Influence of reticular formation
   d. Influence of spinal cord

Selected Vocabulary

For additional terms not listed here, see tables 15.1 through 15.6.

sensory division 538
somatic sensory division 538
visceral sensory division 538
motor division 538
somatic motor division 538
visceral motor division 538
sympathetic division 538
parasympathetic division 538
mixed nerve 538
sensory nerve 538
motor nerve 539
endoneurium 539
fascicle 539
perineurium 539
epineurium 539
ganglion 540
cranial nerve 541
spinal nerve 550
dorsal root 550
dorsal root ganglion 550
ventral root 550
dorsal ramus 551
ventral ramus 551
meningeal branch 551
intercostal nerve 551
dermatome 562
reflex 562
somatic reflex 563
visceral reflex 563
reflex arc 563
muscle spindle 563
proprioceptor 563
intrafusal fiber 563
extrafusal fiber 563
primary afferent fiber 563
secondary afferent fiber 563
γ motor neuron 563
α motor neuron 563
stretch reflex 563
tendon reflex 564
monosynaptic reflex arc 565
reciprocal inhibition 565
flexor reflex 566
polysynaptic reflex arc 566
crossed extensor reflex 567
ipsilateral reflex arc 567
contralateral reflex arc 567
intersegmental reflex arc 567
Golgi tendon organ 567
Golgi tendon reflex 567
autonomic nervous system 568
preganglionic neuron 568
postganglionic neuron 568
sympathetic chain 570
white communicating ramus 570
gray communicating ramus 570
mass activation 572
sympathetic nerve 573
cardiac nerve 573
splanchnic nerve 573
collateral ganglion 573
abdominal aortic plexus 573
celiac ganglion 573
superior mesenteric ganglion 573
inferior mesenteric ganglion 573
adrenal gland 574
adrenal cortex 574
adrenal medulla 574
terminal ganglion 574
cardiac plexus 576
pulmonary plexus 576
esophageal plexus 576
vagal trunks 576
pelvic splanchnic nerve 576
inferior hypogastric plexus 576
pelvic nerve 576
cholinergic fiber 577
adrenergic fiber 577
cholinergic receptor 577
adrenergic receptor 577
nicotinic receptor 577
muscarinic receptor 577
α-adrenergic receptor 578
β-adrenergic receptor 578
dual innervation 578
antagonistic effect 578
cooperative effect 578
sympathetic tone 580
vasomotor tone 580

Testing Your Recall

1. Damage to the _______ nerve could result in defects of eye movement.
   a. optic
   b. vagus
   c. trigeminal
   d. facial
   e. abducens
2. The sense of taste is a function of
   a. the glossopharyngeal nerve.
   b. the hypoglossal nerve.
   c. the maxillary nerve.
   d. the abducens nerve.
   e. the trochlear nerve.
3. All of the following *except* the _______ nerve begin or end in the orbit.
   a. optic
   b. oculomotor
   c. trochlear
   d. abducens
   e. accessory
4. The brachial plexus gives rise to all of the following nerves *except*
   a. the axillary nerve.
   b. the radial nerve.
   c. the saphenous nerve.
   d. the median nerve.
   e. the ulnar nerve.
5. Nerve fibers that adjust the tension in a muscle spindle are called
   a. intrafusal fibers.

b. extrafusal fibers.
c. α motor neurons.
d. γ motor neurons.
e. annulospiral fibers.

6. A stretch reflex requires the action of ________ to prevent an antagonistic muscle from interfering with the agonist.
   a. γ motor neurons
   b. a withdrawal reflex
   c. a crossed extensor reflex
   d. reciprocal inhibition
   e. a contralateral reflex

7. The autonomic nervous system innervates all of these *except*
   a. cardiac muscle.
   b. skeletal muscle.
   c. smooth muscle.
   d. salivary glands.
   e. blood vessels.

8. The gray communicating ramus contains
   a. visceral sensory fibers.
   b. parasympathetic motor fibers.
   c. sympathetic preganglionic fibers.
   d. sympathetic postganglionic fibers.
   e. somatic motor fibers.

9. Throughout the autonomic nervous system, the neurotransmitter released by the preganglionic neuron binds to ________ receptors on the postganglionic neuron.
   a. nicotinic
   b. muscarinic
   c. adrenergic
   d. α_1
   e. β_2

10. Which of these does *not* result from sympathetic stimulation?
    a. dilation of the pupil
    b. acceleration of the heart
    c. digestive secretion
    d. enhanced blood clotting
    e. piloerection

11. Outside the CNS, somas are clustered in swellings called __________.

12. Distal to the intervertebral foramen, a spinal nerve branches into a dorsal and ventral __________.

13. The __________ nerves arise from the cervical plexus and innervate the diaphragm.

14. Most parasympathetic preganglionic fibers are found in the __________ nerve.

15. Modified muscle fibers serving primarily to detect stretch are called __________.

16. The sciatic nerve is a composite of two nerves, the __________ and __________.

17. The adrenal medulla consists of modified postganglionic neurons of the __________ nervous system.

18. Nicotinic and muscarinic receptors bind the neurotransmitter __________.

19. Adrenergic receptors classified as α_2, β_1, and β_2 act by changing the level of __________ in the target cell.

20. Sympathetic fibers to blood vessels maintain a background rate of firing called __________, and this causes a state of partial vasoconstriction called __________.

Answers in appendix B

Testing Your Comprehension

1. Stand with your right shoulder, hip, and foot firmly against a wall. Raise your left foot from the floor without losing contact with the wall at any point. What happens? Why? What principle of this chapter does this demonstrate?

2. If the spinal cord were severed at the T2 level, would the diaphragm be paralyzed? Why or why not?

3. You are dicing raw onions while preparing dinner, and your eyes are watering. Describe the afferent and efferent neural pathways involved in this reflex.

4. Which cranial nerves convey pain signals to the brain when (a) sand blows into your eyes, (b) you bite the rear of your tongue, and (c) your stomach hurts from eating too much?

5. When a patient needs a tendon graft, surgeons sometimes use the tendon of the palmaris longus, a relatively dispensable muscle of the forearm. The median nerve lies nearby and looks very similar to this tendon. There have been cases where a surgeon mistakenly removed a section of this nerve instead of the tendon. What effects do you think such a mistake would have on the patient?

Answers at the Online Learning Center

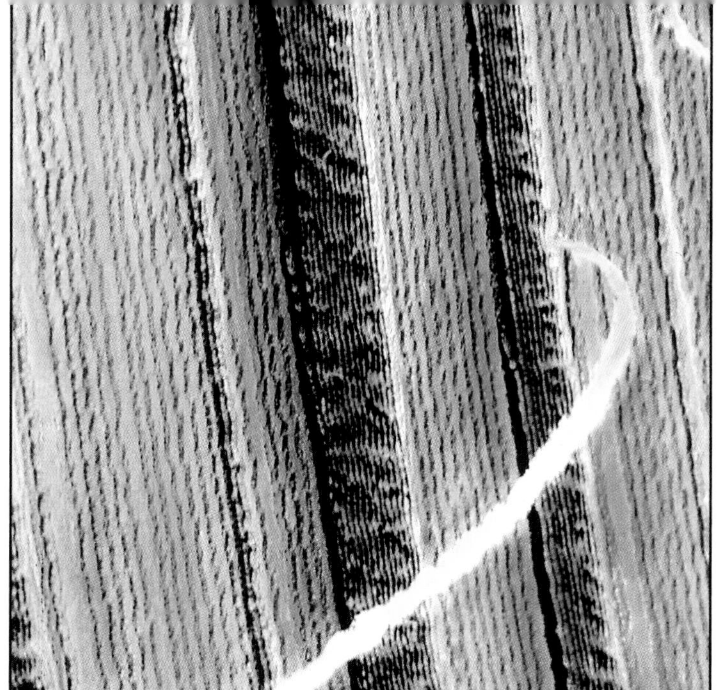

Cells of the lens of a human eye (SEM)

CHAPTER 16

Sense Organs

Chapter Outline

Insights

Brushing Up

To understand this chapter, it is important that you understand or brush up on the following concepts:

- Excitatory and inhibitory postsynaptic potentials (EPSPs and IPSPs) (pp. 478–479)
- Spatial summation (p. 480)
- Neural coding (p. 481)
- Converging circuits of neurons (p. 482)
- Spinal cord tracts (p. 503)
- Decussation (p. 503)

Study with Your ESP CD-ROM

Refer to your ESP Tutorial CD-ROM for helpful animations, activities, and quizzes related to the material covered in this chapter. The following are suggested areas to go to on the CD and are presented according to "Unit/Topic":

- Integration and Coordination/The Senses

Anyone who enjoys music, art, fine food, or a good conversation appreciates the human senses. Yet their importance extends beyond deriving pleasure from the environment. In the 1950s, behavioral scientists at Princeton University studied the methods used by Soviet Communists to extract confessions from political prisoners, including solitary confinement and sensory deprivation. Student volunteers were immobilized in dark soundproof rooms or suspended in dark chambers of water. In a short time, they experienced visual, auditory, and tactile hallucinations, incoherent thought patterns, deterioration of intellectual performance, and sometimes morbid fear or panic. Similar effects have been seen in burn patients who are immobilized and extensively bandaged (including the eyes) and thus suffer prolonged lack of sensory input. Patients connected to life-support equipment and confined under oxygen tents sometimes become delirious. Sensory input is vital to the integrity of personality and intellectual function.

Furthermore, much of the information communicated by the sense organs never comes to our conscious attention—blood pressure, body temperature, and muscle tension, for example. By monitoring such conditions, however, the sense organs initiate somatic and visceral reflexes that are indispensable to homeostasis and to our very survival in a ceaselessly changing and challenging environment.

Properties and Types of Sensory Receptors

Objectives

When you have completed this section, you should be able to

- define *receptor* and *sense organ;*
- list the four kinds of information obtained from sensory receptors, and describe how the nervous system encodes each type; and
- outline three ways of classifying receptors.

A **receptor** is any structure specialized to detect a stimulus. Some receptors are simple nerve endings, whereas others are **sense organs**—nerve endings combined with connective, epithelial, or muscular tissues that enhance or moderate the response to a stimulus. Our eyes and ears are obvious examples of sense organs, but there are also innumerable microscopic sense organs in our skin, muscles, joints, and viscera.

General Properties of Receptors

All sensory receptors are transducers. A *transducer* is any device that converts one form of energy to another—a microphone, light bulb, or gasoline engine, for example. Sensory transducers convert stimulus energy into electrochemical energy—a meaningful pattern of action potentials.

The effect of a stimulus on a receptor is to produce a type of local potential called a **receptor potential**—a graded voltage change across the plasma membrane of a receptor cell caused by a stimulus. The receptor potential may cause a receptor cell (which is not always a neuron) to release a neurotransmitter that stimulates an adjacent neuron. If the receptor itself is a neuron and the voltage reaches threshold, the neuron fires impulses to the central nervous system (CNS). This may result in a **sensation**—a conscious awareness of the stimulus—but much of the sensory information reaching the CNS produces no sensation. We are seldom aware of information concerning muscle tension and blood pH, for example, but it is vitally important to our homeostasis for the CNS to monitor these conditions.

Sensory receptors transmit four kinds of information—*modality, location, intensity,* and *duration:*

1. **Modality** refers to the type of stimulus or the sensation it produces. Vision, hearing, and taste are examples of sensory modalities. The nervous system distinguishes modalities from each other by means of a *labeled line code.* We can think of the brain as having numerous "lines" (nerve fibers) feeding information into it, and each line as being "labeled" to represent a certain modality. All the nerve impulses that arrive at the brain are essentially identical, but impulses arriving on one line have a different meaning than impulses arriving on another. Any impulses from the optic nerve, for example, are interpreted as light. Thus, a blow to the eye may be perceived as a flash of light.
2. **Location** is also encoded by which nerve fibers are firing. **Sensory projection** is the ability of the brain to identify the site of stimulation, including very small and specific areas within a receptor such as the retina. The pathways followed by sensory signals to their ultimate destinations in the CNS are called **projection pathways.**
3. **Intensity** can be encoded in three ways: (a) as stimulus intensity rises, the firing frequencies of sensory nerve fibers rise (see fig. 13.25, p. 481); (b) intense stimuli recruit greater numbers of nerve fibers to fire; and (c) weak stimuli activate only the most sensitive nerve fibers, whereas strong stimuli can activate a less sensitive group of fibers with higher thresholds. Thus, the brain can distinguish intensities based on the number and kind of fibers that are firing and the time intervals between action potentials. These concepts were discussed under *neural coding* in chapter 13.
4. **Duration** is encoded in the way nerve fibers change their firing frequencies over time. **Phasic receptors** generate a burst of action potentials when first stimulated, then quickly adapt and stop transmitting impulses even if the stimulus continues. Some of them fire again when the

stimulus ceases. Lamellated corpuscles, tactile receptors, hair receptors, and smell receptors are rapidly adapting phasic receptors. **Tonic receptors** adapt slowly and generate nerve impulses continually. Proprioceptors are among the most slowly adapting tonic receptors because the brain always must be aware of body position, muscle tension, and joint motions. All receptors, however, exhibit sensory **adaptation**—if the stimulus is prolonged, firing frequency and conscious sensation decline. Adapting to hot bath water is an example.

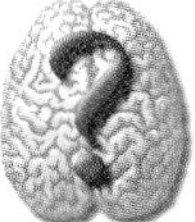

Think About It

Although you may find it difficult to immerse yourself in a tub of hot water or a cold lake, you soon adapt and become more comfortable. In light of this, do you think cold and warm receptors are phasic or tonic? Explain.

Classification of Receptors

Receptors can be classified by several overlapping systems:

1. By stimulus modality:
   - **Chemoreceptors** respond to chemicals, including odors, tastes, and body fluid composition.
   - **Thermoreceptors** respond to heat and cold.
   - **Nociceptors**[1] (NO-sih-SEP-turs) are pain receptors; they respond to tissue damage resulting from trauma (blows, cuts), ischemia (poor blood flow), or excessive stimulation by agents such as heat and chemicals.
   - **Mechanoreceptors** respond to physical deformation of the plasma membrane caused by touch, pressure, stretch, tension, or vibration. They include the organs of hearing and balance and many receptors of the skin, viscera, and joints.
   - **Photoreceptors,** the eyes, respond to light.
2. By the distribution of receptors in the body. As we saw in the preceding chapter, there are two broad classes of receptors:
   - **General (somesthetic) senses,** with receptors that are widely distributed in the skin, muscles, tendons, joint capsules, and viscera. These include the sense of touch, pressure, stretch, heat, cold, and pain, as well as many stimuli that we do not perceive consciously, such as blood pressure and composition.
   - **Special senses,** which are limited to the head and innervated by the cranial nerves. The special senses are vision, hearing, equilibrium, taste, and smell.
3. By the origins of the stimuli:
   - **Interoceptors** detect stimuli in the internal organs and produce feelings of visceral pain, nausea, stretch, and pressure.
   - **Proprioceptors** sense the position and movements of the body or its parts. They occur in muscles, tendons, and joint capsules.
   - **Exteroceptors** sense stimuli external to the body; they include the receptors for vision, hearing, taste, smell, touch, and cutaneous pain.

1. *noci* = pain

Before You Go On

Answer the following questions to test your understanding of the preceding section:

1. What is the difference between a receptor and a nerve ending?
2. Three schemes of receptor classification were presented in this section. In each scheme, how would you classify the receptors for a full bladder? How would you classify taste receptors?
3. What does it mean to say sense organs are transducers? What form of energy do all receptors have as their output?
4. Nociceptors are tonic rather than phasic receptors. Speculate on why this is beneficial to homeostasis.

The General Senses

Objectives

When you have completed this section, you should be able to

- list several types of somesthetic receptors;
- describe the projection pathways for the general senses; and
- explain the mechanisms of pain and the spinal blocking of pain signals.

Receptors for the general senses are relatively simple in structure and physiology. They consist of one or a few sensory nerve fibers and, usually, a sparse amount of connective tissue that enhances the sensitivity or specificity of their responses. These receptors are shown in table 16.1.

Table 16.1 Receptors of the General Senses

Unencapsulated Nerve Endings		Encapsulated Nerve Endings			
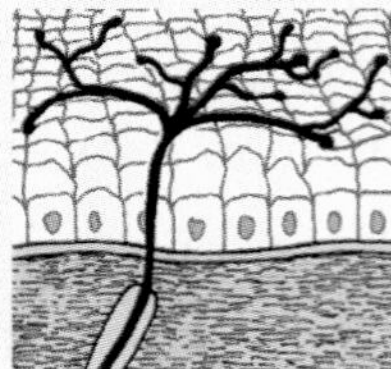	**Free Nerve Endings** Location: Widespread, especially in epithelia and connective tissues Modality: Pain, heat, cold	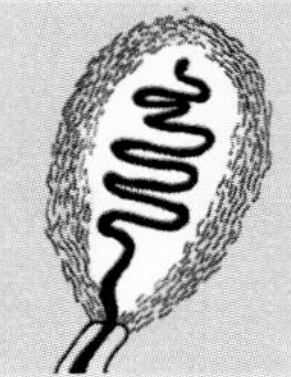	**Tactile Corpuscles** Location: Dermal papillae of fingertips, palms, eyelids, lips, tongue, nipples, and genitals Modality: Light touch, texture	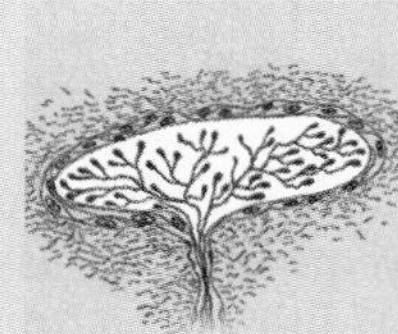	**Ruffini Corpuscles** Location: Dermis, subcutaneous tissue, and joint capsules Modality: Heavy touch, pressure, stretching of skin, joint movements
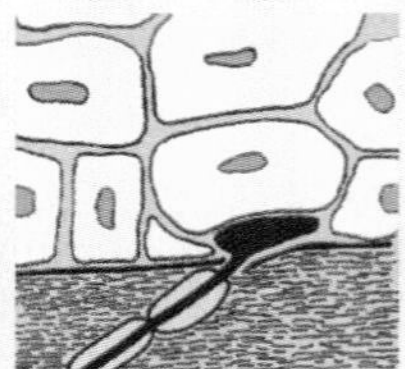	**Tactile Discs** Location: Stratum basale of epidermis Modality: Light touch, texture, edges, shapes	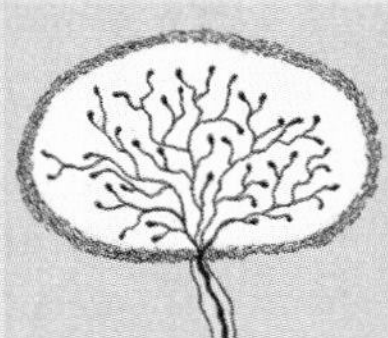	**Krause End Bulbs** Location: Mucous membranes Modality: Similar to tactile corpuscles		**Muscle Spindles** Location: Skeletal muscles near tendon Modality: Muscle stretch (proprioception)
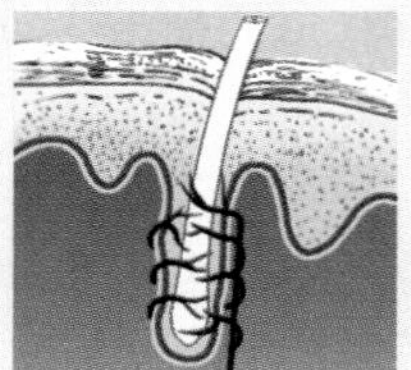	**Hair Receptors** Location: Around hair follicle Modality: Movement of hairs	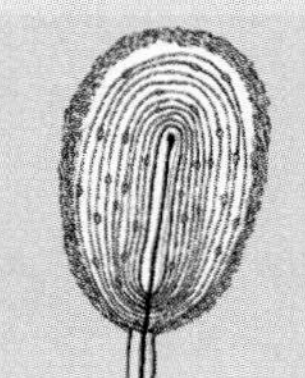	**Lamellated Corpuscles** Location: Dermis, joint capsules, breasts, genitals, and some viscera Modality: Deep pressure, stretch, tickle, vibration		**Golgi Tendon Organs** Location: Tendons Modality: Tension on tendons (proprioception)

Unencapsulated Nerve Endings

Unencapsulated nerve endings are sensory dendrites that are not wrapped in connective tissue. They include free nerve endings, tactile discs, and hair receptors:

- **Free nerve endings** include *warm receptors,* which respond to rising temperature; *cold receptors,* which respond to falling temperature; and *nociceptors* (pain receptors). They are bare dendrites that have no special association with specific accessory cells or tissues. They are most abundant in epithelia and connective tissue.
- **Tactile (Merkel[2]) discs** are tonic receptors for light touch, thought to sense textures, edges, and shapes. They are flattened nerve endings associated with specialized *tactile (Merkel) cells* at the base of the epidermis (see fig. 7.2, p. 207).
- **Hair receptors (peritrichial[3] endings)** monitor the movements of hairs. They consist of a few dendrites entwined around the base of a hair follicle. They respond to any light touch that bends a hair. Because they adapt quickly, we are not constantly annoyed by our clothing bending the body hairs. However, when an ant crawls across our skin, bending one hair after another, we are very aware of it.

Encapsulated Nerve Endings

Encapsulated nerve endings are dendrites wrapped in glial cells or connective tissue. Most of them are mechanoreceptors for touch, pressure, and stretch. The connective tissues around a sensory dendrite enhance the sensitivity or specificity of the receptor. We have already considered some encapsulated nerve endings in chapter 15—muscle spindles and Golgi tendon organs. Others include tactile corpuscles, Krause end bulbs, lamellated corpuscles, and Ruffini corpuscles:

- **Tactile (Meissner[4]) corpuscles** are phasic receptors for light touch and texture. They occur in the dermal papillae of the skin, especially in sensitive hairless areas such as the fingertips, palms, eyelids, lips, nipples, and genitals. They are tall, ovoid- to pear-shaped, and consist of two or three nerve fibers meandering upward through a mass of connective

2. Friedrich S. Merkel (1845–1911), German anatomist and physiologist
3. *peri* = around + *trich* = hair
4. George Meissner (1829–1905), German histologist

tissue. Tactile corpuscles enable you to tell the difference between silk and sandpaper, for example, by light strokes of your fingertips.

- **Krause**[5] **end bulbs** are similar to tactile corpuscles but occur in mucous membranes rather than in the skin.
- **Lamellated (pacinian**[6]**) corpuscles** are phasic receptors for deep pressure, stretch, tickle, and vibration. They consist of numerous concentric layers of Schwann cells surrounding a core of one to several sensory nerve fibers. These receptors occur in the pancreas, some other viscera, and deep in the dermis—especially on the hands, feet, breasts, and genitals.
- **Ruffini**[7] **corpuscles** are tonic receptors for heavy touch, pressure, stretching of the skin, and joint movements. They are flattened, elongated capsules containing a few nerve fibers and are located in the dermis, subcutaneous tissue, ligaments, tendons, and joint capsules.

Somesthetic Projection Pathways

The afferent neuron of a sensory pathway is called the **first-order neuron.** The first-order neurons for touch, pressure, and proprioception have large, fast, myelinated axons. They enter the spinal cord by way of the dorsal root, travel various ascending columns on the same side of the body (see table 14.1, p. 503), and end in the medulla oblongata. Here, they synapse with **second-order neurons** that decussate to the opposite side of the brainstem and continue by way of the medial lemniscus to the thalamus. In the thalamus, they synapse with **third-order neurons** whose axons project to the somesthetic cortex of the postcentral gyrus.

Impulses for heat and cold are carried to the spinal cord by small, unmyelinated, slow fibers. Near the point of entry, they synapse with second-order neurons that decussate at that level and travel up the opposite side of the cord in the spinothalamic tract. In the thalamus, these synapse with third-order neurons that continue to the somesthetic cortex.

Pain

Pain is not merely an effect of overstimulation of our somesthetic receptors. It has its own special receptors (nociceptors) and its own purpose. It makes us conscious of tissue injuries or potentially injurious situations, thus allowing us to "favor" an injured region so that it has a better chance to heal, or enabling us to avoid injury in the first place. A good example of the protective function of pain comes from people with leprosy (Hansen disease). The leprosy bacterium infects the nerves, resulting in a loss of pain and other sensations from affected areas. Victims fail to notice minor injuries such as scrapes and splinter wounds. Their neglect of the wounds leads to serious infections that damage the bone and other deep tissues. About 25% of untreated victims suffer crippling losses of their fingers or toes.

Nociceptors are found in virtually all organs, especially in the skin and mucous membranes, but they are lacking from the brain. In some brain surgery, the patient must remain conscious and able to talk with the surgeon; such patients need only a local scalp anesthetic.

There are two types of nociceptors corresponding to different pain sensations. Myelinated pain fibers conduct at speeds of 12 to 30 m/sec and produce the sensation known as **fast (first) pain**—the initial feeling of sharp, localized, stabbing pain perceived at the time of injury. Unmyelinated pain fibers conduct at speeds of 0.5 to 2.0 m/sec and produce the **slow (second) pain** that follows—a longer-lasting, dull, diffuse feeling.

Pain is also classified by its point of origin. **Somatic pain** arises from the skin, muscles, and joints. *Superficial somatic pain* has a sharp, stabbing, or prickling quality. *Deep somatic pain* is less localized, is longer lasting, and has a more aching, burning, or nauseating quality. **Visceral pain** arises from the viscera as an effect of stretch, chemical irritants, or deficient blood flow (*ischemia*). It is often associated with nausea and may be very intense, but it is poorly localized because the viscera have relatively few nociceptors.

Injured tissues release several chemicals that stimulate nociceptors or sensitize them to other stimuli—bradykinin, serotonin, prostaglandins, histamine, and potassium ions. **Bradykinin** (BRAD-ee-KY-nin), derived from a protein in the blood plasma, is one of the most potent stimuli; it is intensely painful when injected under the skin. Bradykinin not only makes us aware of injuries but also triggers a cascade of reactions that promote healing (see chapter 19).

Projection Pathways for Pain

Pain signals originating in the superficial structures of the head are carried mainly by the trigeminal nerve, whose somas are in the trigeminal ganglion. Below the head, the nociceptors have their somas in the dorsal root ganglia of the spinal nerves. The first-order pain fibers follow the usual route up the spinal nerve, through the dorsal root, into the dorsal horn of the spinal cord. They stimulate the second-order spinal neurons with glutamate and a neuropeptide called **substance P** (think *P* for "pain"). The second-order neurons decussate to the other

5. William J. F. Krause (1833–1910), German anatomist
6. Filippo Pacini (1812–83), Italian anatomist
7. Angelo Ruffini (1864–1929), Italian anatomist

side of the spinal cord and send their fibers mainly up the spinothalamic tract to the thalamus. Third-order neurons from the thalamus then carry the signals to the primary somesthetic cortex of the postcentral gyrus and to the somesthetic association area caudal to that gyrus (fig. 16.1).

Pain signals also travel up the spinoreticular tract to the reticular formation and thus affect arousal—for example, when we are awakened by a muscle cramp or alarmed by a bee sting. From the reticular formation, some third-order nerve fibers travel to the hypothalamus and limbic system, triggering visceral, emotional, and behavioral reactions to pain, and some bypass the thalamus en route to the primary somesthetic cortex.

It is often difficult to precisely locate the source of pain, especially visceral pain. **Referred pain** is a phenomenon in which pain from the viscera is misinterpreted as coming from the skin or other superficial structures. People suffering heart attacks, for example, often perceive cardiac pain as if it were "radiating" along the left shoulder and medial side of the left arm. Spinal cord segments T1 to T5 receive sensory input from the heart as well as the chest and arm. Pain fibers from the heart and skin in this region converge on the same spinal interneurons and follow the same pathway from there to the thalamus and cerebral cortex. The skin has more pain receptors than the heart and suffers injury far more often. Thus the brain "assumes" that signals arriving by this labeled line are most likely coming from the skin, even when the heart is the actual source. Knowledge of the origins of referred pain is essential to the skillful diagnosis of organ dysfunctions (fig. 16.2).

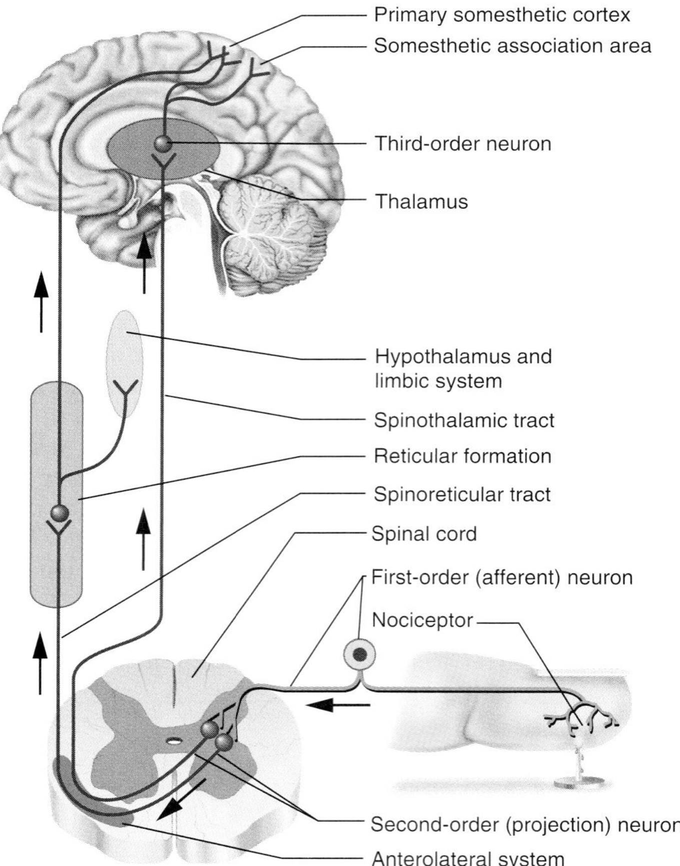

Figure 16.1 Projection Pathways for Pain. A first-order neuron conducts a pain signal to the dorsal horn of the spinal cord, a second-order neuron conducts it to the thalamus, and a third-order neuron conducts it to the cerebral cortex. Signals from the spinothalamic tract pass through the thalamus. Signals from the spinoreticular tract bypass the thalamus on the way to the sensory cortex.

CNS Modulation of Pain

The intensity of pain is not simply a function of the extent of injury and firing of the afferent nociceptors; it can be greatly affected by one's state of mind, such as the joy of childbirth or fear of dental work. The central nervous system (CNS) has **analgesic**[8] (pain-relieving) mechanisms that are just beginning to be understood.

The analgesic effects of opium, morphine, and heroin have been known, used, and abused for many centuries. In 1974, neurophysiologists discovered receptor sites in the brain for these drugs. Since these opiates do not occur naturally in the body, this discovery raised the question, What *normally* binds to these receptors? Physiologists soon found two oligopeptides with 200 times the analgesic potency of morphine and named them **enkephalins.**[9] Larger analgesic neuropeptides, the **endorphins**[10] and **dynorphins,**[11] were discovered later. All three of these peptide families are known collectively as **endogenous opioids** (which means "internally produced opium-like substances").

These opioids are secreted by the CNS, pituitary gland, digestive tract, and other organs in states of stress or exercise. In the CNS, they are found especially in the central gray (periaqueductal) matter of the midbrain (see p. 510) and the dorsal horn of the spinal cord. They are *neuromodulators* (see p. 477) that can block the transmission of pain signals and can produce feelings of pleasure and euphoria. They may be responsible for the "second wind" ("runner's high") experienced by athletes and for the reports of many mortally wounded soldiers that they feel no pain. Their secretion rises sharply in women giving birth. Efforts to employ them in pain therapy have been disappointing, but exercise is an effective part of therapy for chronic pain and may help because it stimulates opioid secretion.

8. *an* = without + *alges* = pain
9. *en* = within + *kephal* = head
10. acronym, from *endo*genous *morphine*like substance
11. *dyn* = pain

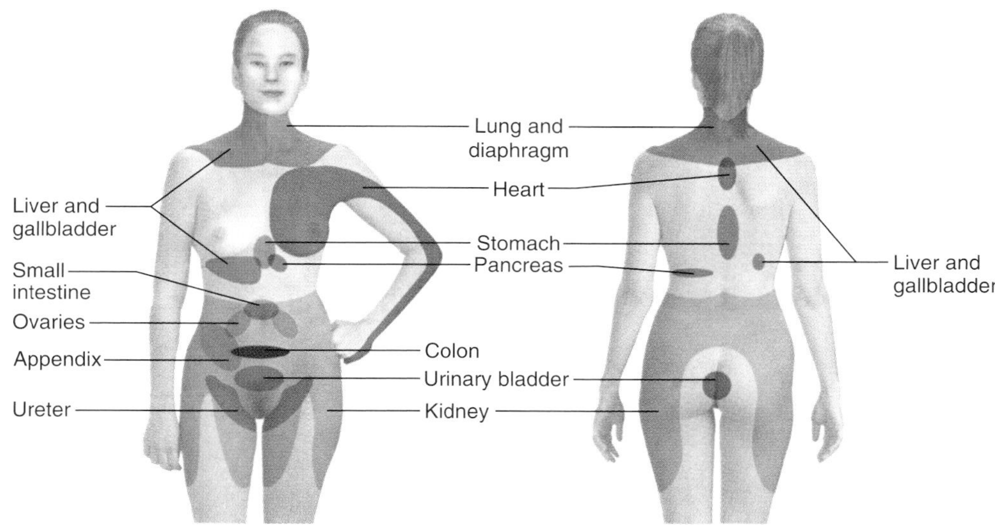

Figure 16.2 **Referred Pain.** Pain from the viscera is often felt in specific areas of the skin.

But how do these opioids block the sensation of pain? For pain to be perceived, signals from the nociceptors must get beyond the dorsal horn of the spinal cord and travel up the spinothalamic tracts to the brain. Through mechanisms called **spinal gating,** pain signals can be stopped at the dorsal horn. Two of these mechanisms are described here (fig. 16.3).

In one mechanism, neurons of the midbrain central gray matter project to certain nuclei of the nearby reticular formation. Neurons from there, in turn, issue **descending analgesic fibers** that travel down the reticulospinal tract and end on interneurons of the dorsal horn at all levels of the spinal cord. The spinal interneurons form axoaxonic synapses with the first-order pain fibers entering the spinal cord. They secrete enkephalins and dynorphins, which inhibit the pain fibers from secreting substance P. Without substance P, the pain signal goes no farther than this; it never ascends the spinal cord to the brain, and we feel no pain.

Pain signals are also modulated by certain dorsal horn interneurons that not only inhibit the second-order neurons of the pain pathway but also receive input from touch receptors. Touch fibers stimulate the inhibitory interneurons, which then block the transmission of signals by the second-order pain fibers. This may explain why rubbing a sore area makes it feel less painful.

Pain control has had a particularly interesting history, some of which is retold in historical insight 16.5 at the end of this chapter.

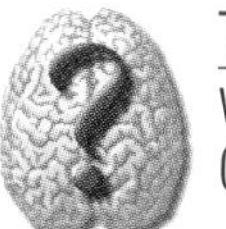

Think About It

Where is the phenomenon of presynaptic inhibition (see p. 480) relevant to the spinal gating of pain?

Before You Go On

Answer the following questions to test your understanding of the preceding section:

5. What stimulus modalities are detected by free nerve endings?
6. Name any four encapsulated nerve endings and identify their stimulus modalities.
7. Where do most second-order somesthetic neurons synapse with third-order neurons?
8. Explain the phenomenon of referred pain in terms of the neural pathways involved.
9. Explain the roles of bradykinin, substance P, and endorphins in the perception of pain.

The Chemical Senses

Objectives

When you have completed this section, you should be able to

- explain how taste and smell receptors are stimulated; and
- describe the receptors and projection pathways for these two senses.

Taste and smell are the chemical senses. In both cases, receptor potentials are created by the action of environmental chemicals on sensory cells.

Taste

Taste **(gustation)** is a sensation that results from the action of chemicals on the **taste buds.** There are about 4,000 of

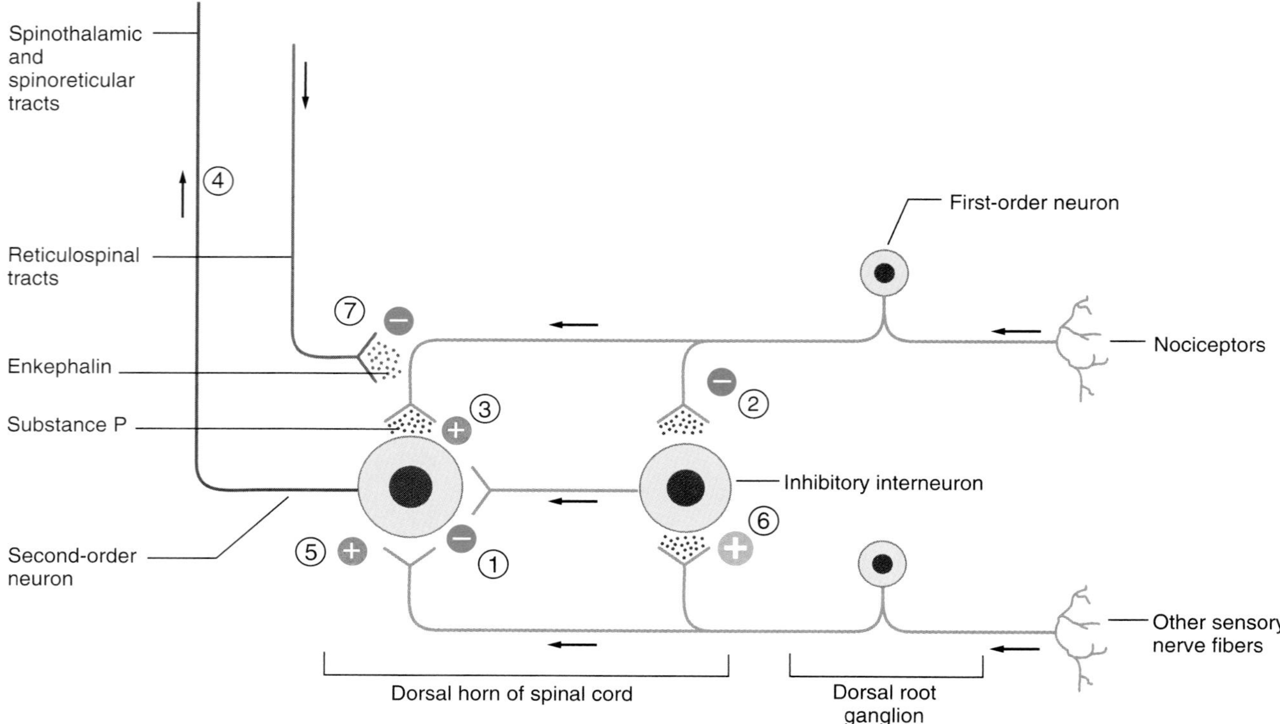

1. In the absence of a pain stimulus, an inhibitory interneuron of the spinal cord prevents transmission of pain signals.
2. When tissue damage stimulates a nociceptor, the nociceptor inhibits the inhibitory interneuron.
3. The nociceptor also releases substance P, which stimulates the second-order neuron.
4. The second-order neuron sends a pain signal to the brain.
5. Some sensory neurons other than nociceptors also stimulate the second-order neuron.
6. These sensory neurons, however, have an even stronger effect on the inhibitory interneuron, thus blocking the transmission of pain signals.
7. Neurons of the reticular formation release enkephalin, which, by presynaptic inhibition, blocks the release of substance P. Thus the brain can reduce the transmission of the pain signal to itself.

Figure 16.3 Spinal Gating of Pain Signals.

these, located on the tongue of course, but also inside the cheeks and on the soft palate, pharynx, and epiglottis.

Anatomy

The tongue, where the sense of taste is best developed, is marked by four types of bumps called **lingual papillae** (fig. 16.4*a*):

1. **Filiform**[12] **papillae** are tiny spikes without taste buds. They are responsible for the rough feel of a cat's tongue and are important to many mammals for grooming the fur. They are the most abundant papillae on the human tongue, but they are small and play no sensory role.
2. **Foliate**[13] **papillae** are also weakly developed in humans. They form parallel ridges on the sides of the tongue about two-thirds of the way back from the tip. Most of their taste buds degenerate by the age of 2 or 3 years.
3. **Fungiform**[14] (FUN-jih-form) **papillae** are shaped somewhat like mushrooms. Each has about three taste buds, located mainly on the apex. These papillae are widely distributed but especially concentrated at the tip and sides of the tongue.
4. **Vallate**[15] **(circumvallate) papillae** are large papillae arranged in a V at the rear of the tongue. Each is surrounded by a deep circular trench. There are

12. *fili* = thread + *form* = shaped
13. *foli* = leaf + *ate* = like
14. *fungi* = mushroom + *form* = shaped
15. *vall* = wall + *ate* = like, possessing

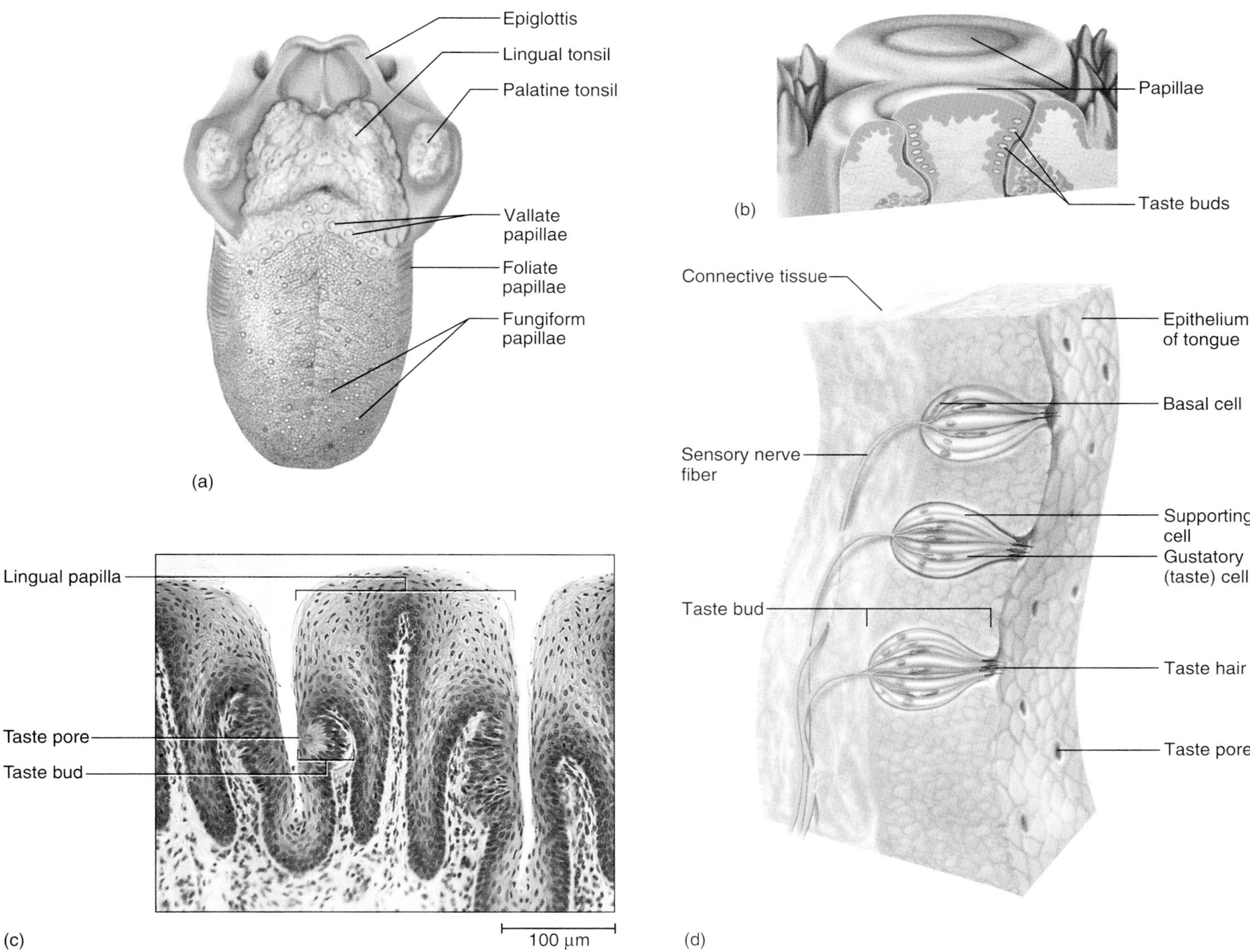

Figure 16.4 **Taste (gustatory) Receptors.** (*a*) Dorsal view of the tongue and locations of its papillae. (*b*) Detail of the vallate papillae. (*c*) Taste buds on the walls of two adjacent foliate papillae. (*d*) Structure of the taste buds.

only 7 to 12 of them, but they contain about half of all our taste buds—around 250 each, located on the wall of the papilla facing the trench (fig. 16.4*b*).

Regardless of location and sensory specialization, all taste buds look alike (fig. 16.4*c, d*). They are lemon-shaped groups of 40 to 60 cells of three kinds—*taste cells, supporting cells,* and *basal cells.* **Taste (gustatory) cells** are more or less banana-shaped and have a tuft of apical microvilli called **taste hairs** that serve as receptor surfaces for taste molecules. The hairs project into a pit called a **taste pore** on the epithelial surface of the tongue. Taste cells are epithelial cells, not neurons, but they synapse with sensory nerve fibers at their base. A taste cell lives 7 to 10 days and is then replaced by mitosis and differentiation of basal cells. Supporting cells have a similar shape but no taste hairs. They lie between the taste cells.

Physiology

To be tasted, molecules must dissolve in the saliva and flood the taste pore. On a dry tongue, sugar or salt has as little taste as a sprinkle of sand. Physiologists currently recognize five primary taste sensations:

1. **Salty,** produced by metal ions such as sodium and potassium. Since these are vital electrolytes, there is obvious value in our ability to taste them.

Electrolyte deficiencies can cause a craving for salty foods; many animals such as deer, elephants, and parrots thus seek out salt deposits when necessary. Pregnancy can lower a woman's electrolyte concentrations and create a craving for salty food.

2. **Sweet,** produced by many organic compounds, especially sugars. Sweetness is associated with carbohydrates and foods of high caloric value. Many flowering plants have evolved sweet nectar and fruits that entice animals to eat them and disperse their pollen and seeds. Thus, our fondness for fruit has coevolved with plant reproductive strategies.
3. **Sour,** usually associated with acids in such foods as citrus fruits.
4. **Bitter,** associated with spoiled foods and with alkaloids such as nicotine, caffeine, quinine, and morphine. Bitter alkaloids are often poisonous, and this sensation usually induces a human or animal to reject a food. While flowering plants make their fruits temptingly sweet, they often load their leaves with bitter, toxic alkaloids—it is of no advantage to a plant to have its leaves eaten!
5. **Umami,** the most recently discovered primary taste, is a "meaty" taste produced by amino acids such as aspartic and glutamic acids. The taste is best known from the salt of glutamic acid, monosodium glutamate (MSG). Pronounced "ooh-mommy," the word is Japanese for "delicious."

The many flavors we perceive are not simply a mixture of these five primary tastes but are also influenced by food texture, aroma, temperature, appearance, and one's state of mind, among other things. Many flavors depend on smell; without its aroma, cinnamon merely has a faintly sweet taste, and coffee and peppermint are bitter. Some flavors such as pepper are due to stimulation of free endings of the trigeminal nerve.

All of the primary tastes can be detected throughout the tongue, but certain regions are more sensitive to one category than to others. The tip of the tongue is most sensitive to sweet tastes, which trigger such responses as licking, salivation, and swallowing. The lateral margins of the tongue are the most sensitive areas for salty and sour tastes. Taste buds in the vallate papillae at the rear of the tongue are especially sensitive to bitter compounds, which tend to trigger rejection responses such as gagging to protect against the ingestion of toxins. The threshold for the bitter taste is the lowest of all—that is, we can taste lower concentrations of alkaloids than of acids, salts, and sugars. The senses of sweet and salty are the least sensitive. It is not yet known whether umami stimulates any particular region of the tongue more than other regions.

Sugars, alkaloids, glutamate, and the umami compounds stimulate taste cells by binding to receptors on the membrane surface, which then activate G proteins and second messenger systems within the cell. Sodium and acids penetrate into the cell and depolarize it directly. By either mechanism, stimulated taste cells then release neurotransmitters that stimulate the sensory dendrites at their base.

Projection Pathways

Taste buds stimulate the facial nerve (VII) in the anterior two-thirds of the tongue, the glossopharyngeal nerve (IX) in the posterior one-third, and the vagus nerve (X) in the palate, pharynx, and epiglottis. All taste fibers project to the *solitary nucleus* in the medulla oblongata. Second-order neurons from this nucleus relay the signals to two destinations: (1) nuclei in the hypothalamus and amygdala that activate autonomic reflexes such as salivation, gagging, and vomiting, and (2) the thalamus, which relays signals to the insula and postcentral gyrus of the cerebrum, where we become conscious of the taste.

Smell

The receptor cells for smell **(olfaction)** form a patch of epithelium, the **olfactory mucosa,** in the roof of the nasal cavity (fig. 16.5). This location places the olfactory cells close to the brain, but it is poorly ventilated; forcible sniffing is often needed to identify an odor or locate its source. Nevertheless, the sense of smell is highly sensitive. We can detect odor concentrations as low as a few parts per trillion. Most people can distinguish 2,000 to 4,000 odors, and some can distinguish up to 10,000. On average, women are more sensitive to odors than men are, and they are measurably more sensitive to some odors near the time of ovulation than during other phases of the menstrual cycle. Olfaction is highly important in the social interactions of other animals and, in more subtle ways, to humans (see evolutionary insight 16.1).

Anatomy

The olfactory mucosa covers about 5 cm^2 of the superior concha and nasal septum. It consists of 10 to 20 million **olfactory cells** as well as epithelial supporting cells and basal cells. The rest of the nasal cavity is lined by a nonsensory *respiratory mucosa.*

Unlike taste cells, which are epithelial, olfactory cells are neurons. They are shaped a little like bowling pins. The widest part, the soma, contains the nucleus. The neck and head of the cell are a modified dendrite with a swollen tip bearing 10 to 20 cilia called **olfactory hairs.** Unlike most cilia, these are immobile, but they have binding sites for odor molecules. They lie in a tangled mass embedded in a thin layer of mucus. The basal end of each cell tapers to become an axon. These axons collect into small fascicles that leave the nasal cavity through pores

Evolutionary Insight 16.1

Human Pheromones

There is an abundance of anecdote, but no clear experimental evidence, that human body odors affect sexual behavior. There is more adequate evidence, however, that a person's sweat and vaginal secretions affect other people's sexual physiology, even when the odors cannot be consciously smelled. Experimental evidence shows that a woman's apocrine sweat can influence the timing of other women's menstrual cycles. This can produce a so-called *dormitory effect* in which women who live together tend to have synchronous menstrual cycles. The presence of women stimulates men's beards to grow faster, and the presence of men seems to influence female ovulation. When a woman is ovulating or close to that time, and therefore fertile, her vaginal secretions contain pheromones called *copulines.* These have been shown to raise men's testosterone levels.

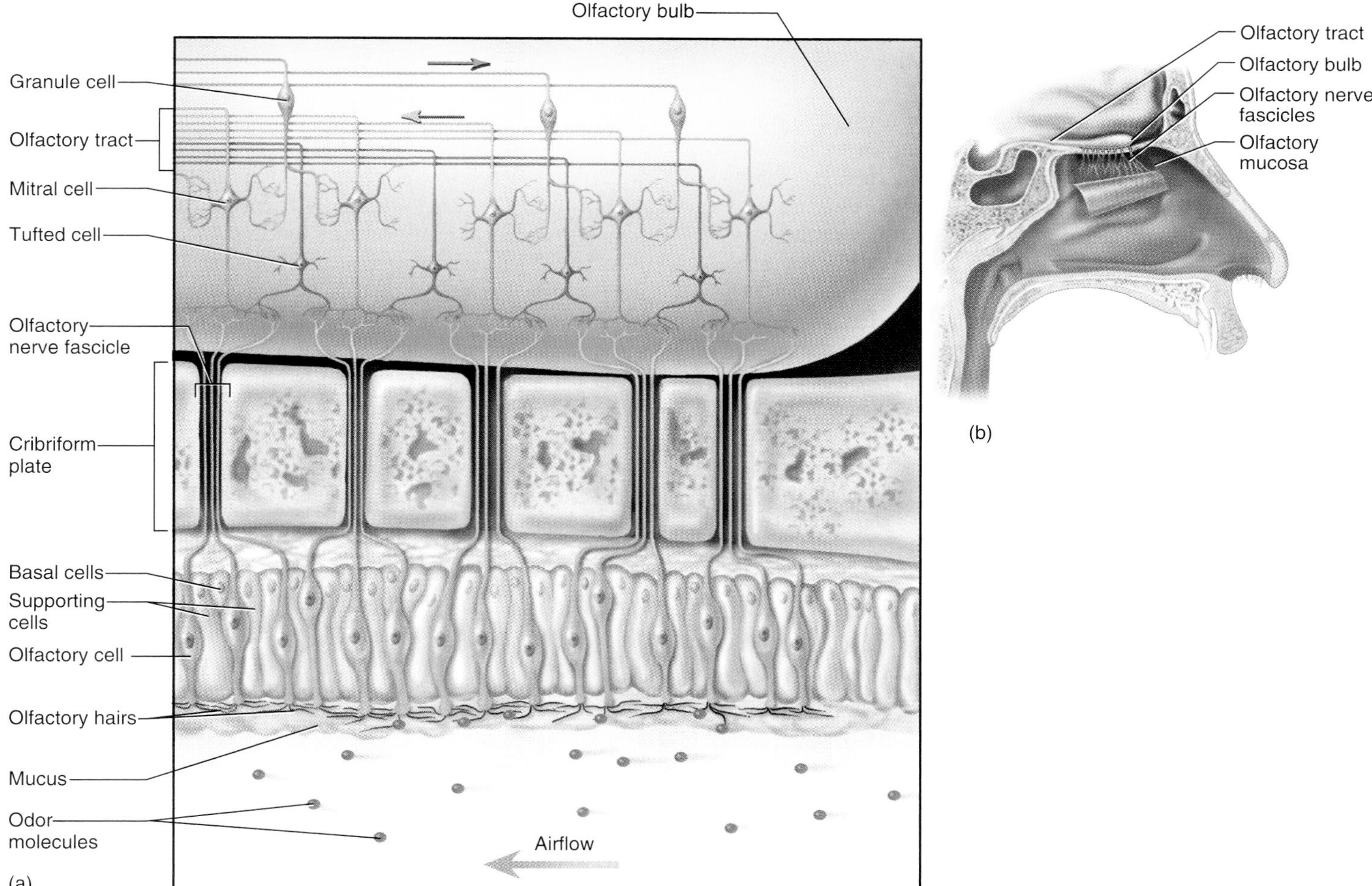

Figure 16.5 Smell (olfactory) Receptors. (*a*) Neural pathways from the olfactory mucosa of the nasal cavity to the olfactory tract of the brain. (*b*) Location of the major structures in relation to the nasal and cranial cavities.

(*olfactory foramina*) in the cribriform plate of the ethmoid bone. Collectively, the fascicles are regarded as cranial nerve I (the olfactory nerve).

Olfactory cells are the only neurons in the body directly exposed to the external environment. Apparently this is hard on them, because they have a life span of only 60 days. Unlike most neurons, however, they are replaceable. The basal cells continually divide and differentiate into new olfactory cells.

Physiology

Efforts to identify a few primary odors comparable to the five primary tastes have been indecisive and controversial. It is difficult even to specify what properties are needed to give a molecule an odor. At a minimum, it must be volatile—able to evaporate and be carried by the inhaled airstream—but the intensity of an odor is not simply proportional to volatility. Water is highly volatile but has no odor, while musk has a pronounced odor but is poorly volatile.

Sensory transduction begins when a molecule binds to a receptor on an olfactory hair. The receptor triggers the production of a second messenger which, in turn, opens ion channels in the membrane. Na^+ enters the cell and creates a receptor potential. In many cases, the second messenger is cAMP, but some odors activate other second-messenger systems.

Some odors stimulate nociceptors of the trigeminal nerve rather than olfactory cells; these include ammonia, menthol, chlorine, and hot peppers. "Smelling salts" are used to revive unconscious persons by strongly stimulating the trigeminal nerve with ammonia fumes.

Olfactory receptors adapt quickly—we may therefore be unaware of our own body odors or have difficulty locating a gas leak in a room. Adaptation does not occur in the receptor cells but is due to synaptic inhibition in the olfactory bulbs of the brain.

Projection Pathways

When olfactory fibers pass through the cribriform plate, they enter a pair of **olfactory bulbs** beneath the frontal lobes of the brain. In the bulbs, they synapse with neurons called *mitral cells* and *tufted cells* (fig. 16.5*a*), whose axons form bundles called the **olfactory tracts.** The tracts follow a complex pathway leading to the medial side of the temporal lobes (fig. 16.6). Olfactory input to the amygdala and hypothalamus can trigger emotional and visceral reactions even without our conscious awareness of an odor. For example, the odor of certain foods, perfume, a hospital, or decaying flesh can evoke strong emotional responses, and some odors can cause us to sneeze, cough, salivate, secrete stomach acid, or vomit. Olfactory signals differ from other sensory inputs in that they reach the cerebral cortex without passing through the thalamus.

The cerebral cortex sends feedback to *granule cells* in the olfactory bulbs. The granule cells, in turn, inhibit the mitral cells. An effect of this is that odors can change in quality and significance under different conditions. Food may smell more appetizing when you are hungry, for example, than it does after you have just eaten.

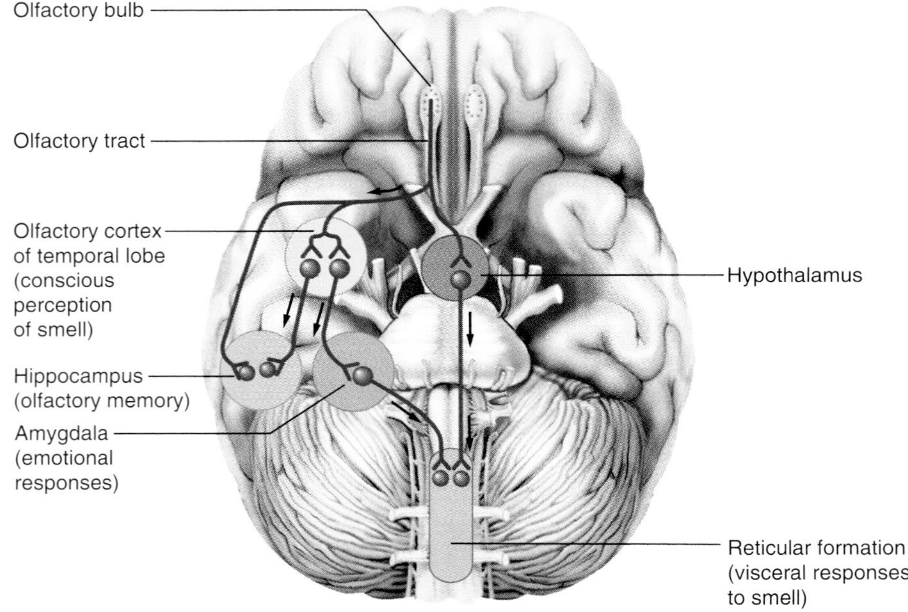

Figure 16.6 Olfactory Projection Pathways in the Brain.

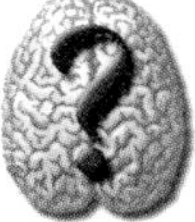

Think About It

Which taste sensations could be lost after damage to (1) the facial nerve or (2) the glossopharyngeal nerve? A fracture of which cranial bone would most likely eliminate the sense of smell?

Before You Go On

Answer the following questions to test your understanding of the preceding section:

10. What is the difference between a lingual papilla and a taste bud? Which is visible to the naked eye?
11. List the primary taste sensations and discuss their adaptive significance (survival value).
12. Which cranial nerves carry gustatory impulses to the brain?
13. What part of an olfactory cell binds odor molecules?
14. What region of the brain receives subconscious olfactory input? What region receives conscious input?

Hearing and Equilibrium

Objectives

When you have completed this section, you should be able to

- identify the properties of sound waves that account for pitch and loudness;
- describe the gross and microscopic anatomy of the ear;
- explain how the ear converts vibrations to nerve signals and discriminates between sounds of different intensity and pitch;
- explain how the vestibular apparatus enables the brain to interpret the body's position and movements; and
- describe the pathways taken by auditory and vestibular signals to the brain.

Hearing is a response to vibrating air molecules and *equilibrium* is the sense of motion and balance. These senses reside in the inner ear, a maze of fluid-filled passages and sensory cells. This section explains how the fluid is set in motion and how the sensory cells convert this motion into an informative pattern of action potentials.

The Nature of Sound

To understand the physiology of hearing, it is necessary to appreciate some basic properties of sound. **Sound** is any audible vibration of molecules. It can be transmitted through water, solids, or air, but not through a vacuum. Our discussion is limited to airborne sound.

Sound is produced by a vibrating object such as a tuning fork, a loudspeaker, or the vocal cords. Consider a loudspeaker producing a pure tone (fig. 16.7). When the speaker cone moves forward, it pushes air molecules closer together and produces a high-density *zone of compression*. When it vibrates back, it leaves a low-density *zone of rarefaction* where air molecules are less closely spaced than at rest. Individual molecules do not travel from the vibrating object to the eardrum. Rather, a molecule collides with one immediately in front of it, that one collides with another, and so on, like a series of billiard balls; finally, some molecules collide with the eardrum and make it vibrate. The sensations we perceive as the pitch and loudness of the sound are related to the physical properties of these vibrations.

Pitch

Pitch is our sense of whether a sound is "high" (treble) or "low" (bass). It is determined by the frequency at which the sound source, eardrum, and other parts of the ear vibrate. One movement of a vibrating object back and forth

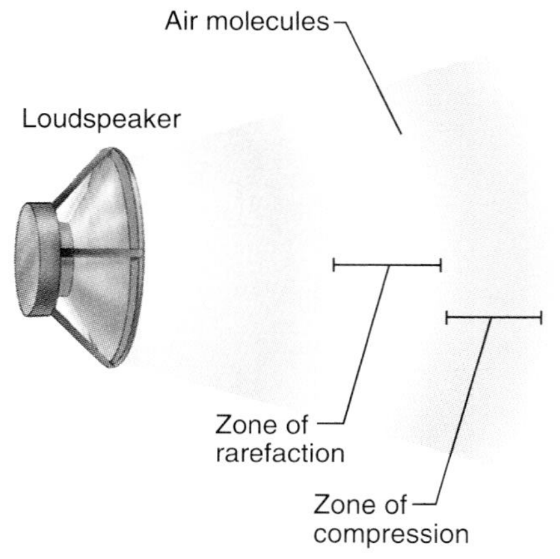

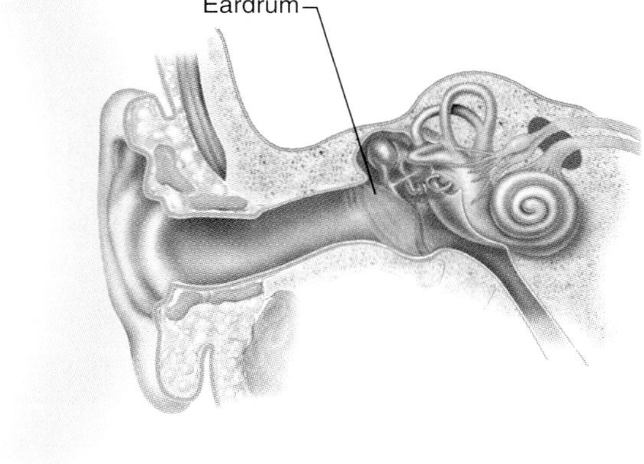

Figure 16.7 The Production and Transmission of Sound Waves. Air molecules are pushed close together in the zones of compression, created by each forward movement of the speaker. In the zones of rarefaction, molecules are more widely spaced.

is called a *cycle,* and the number of cycles per second (cps or hertz, Hz) is called **frequency.** The lowest note on a piano, for example, is 27.5 Hz, middle C is 261 Hz, and the highest note is 4,176 Hz. The most sensitive human ears can hear frequencies from 20 to 20,000 Hz. The *infrasonic* frequencies below 20 Hz are not detected by the ear, but we sense them through vibrations of the skull and skin, and they play a significant role in our appreciation of music. The inaudible vibrations above 20,000 Hz are *ultrasonic.* Human ears are most sensitive to frequencies ranging from 1,500 to 4,000 Hz. In this range, we can hear sounds of relatively low energy (volume), whereas sounds above or below this range must be louder to be audible (fig. 16.8). Normal speech falls within this frequency range. Most of the hearing loss suffered with age is in the range of 250 to 2,050 Hz.

Loudness

Loudness is the perception of sound energy, intensity, or **amplitude** of vibration. In the speaker example, amplitude is a measure of how far forward and back the cone vibrates on each cycle and how much it compresses the air molecules in front of it. Loudness is expressed in decibels (dB), with 0 dB being the threshold of human hearing. Every 10 dB step up the scale represents a sound with 10 times greater intensity. Thus, 10 dB is 10 times threshold, 20 dB is 100 times threshold, 30 dB is 1,000 times threshold, and so forth. Normal conversation has a loudness of about 60 dB. At most frequencies, the threshold of pain is 120 to 140 dB, approximately the intensity of a loud thunderclap. Prolonged exposure to sounds greater than 90 dB can cause permanent loss of hearing.

Anatomy of the Ear

The ear has three regions called the *outer, middle,* and *inner ear.* The first two are concerned only with transmitting sound to the inner ear, which houses the transducer that converts fluid motion to action potentials.

Outer Ear

The **outer (external) ear** is essentially a funnel for conducting air vibrations to the eardrum. It begins with the fleshy **auricle,** or **pinna,** on the side of the head, shaped and supported by elastic cartilage except for the earlobe. It has a predictable arrangement of named whorls and recesses that direct sound into the auditory canal (fig. 16.9).

The **auditory canal (external auditory meatus)** is slightly S-shaped and about 3 cm long in adults (fig. 16.10). It is lined with skin and supported by fibrocartilage at its opening and by the temporal bone for the rest of its length. Ceruminous and sebaceous glands in the canal produce secretions that mix with dead skin cells and form *cerumen* (earwax). Cerumen normally dries and falls from the canal, but sometimes it becomes impacted and interferes with hearing.

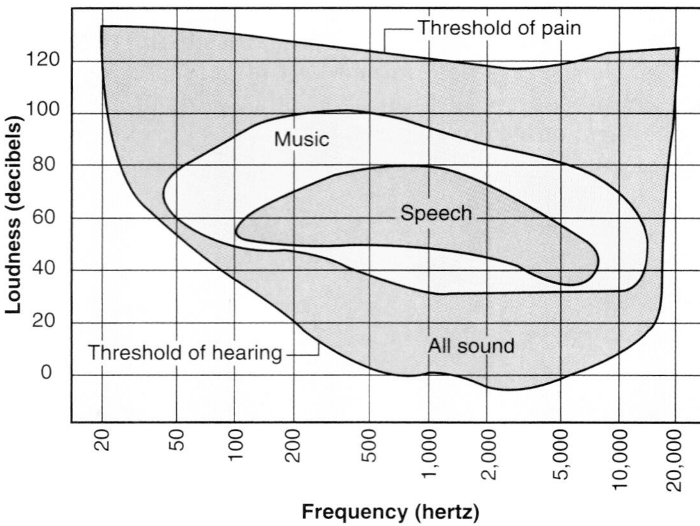

Figure 16.8 The Range of Human Hearing. People with very good hearing can hear sounds from 20 to 20,000 hertz, but sounds must be louder at these extremes than they do in the midrange to be heard. Our ears are most sensitive to frequencies of 1,500 to 5,000 hertz, where we can hear relatively soft sounds. Thus, the threshold of hearing varies with the frequency of the sound. Most sounds above 120 decibels are painful to the ear.

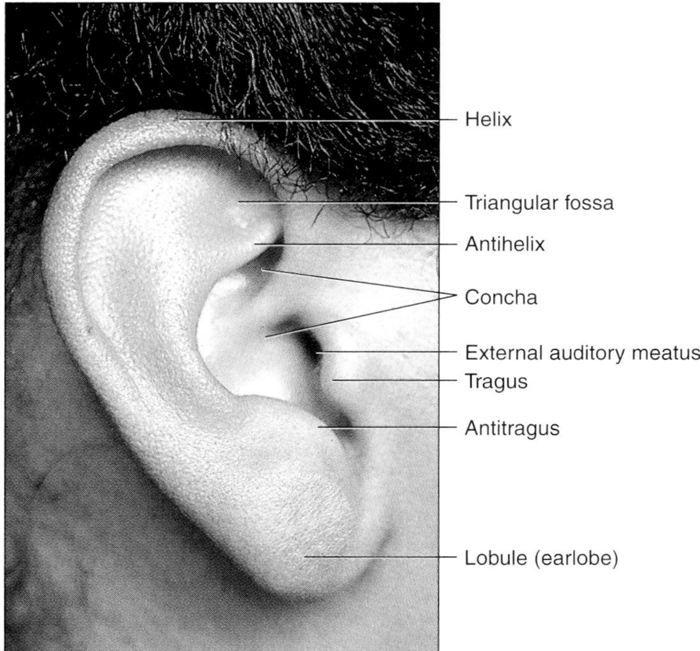

Figure 16.9 Anatomy of the Auricle (pinna) of the Ear.

Clinical Insight 16.2

Middle-Ear Infection

Otitis[16] **media** (middle-ear infection) is common in children because their auditory tubes are relatively short and horizontal. Upper respiratory infections can easily spread from the throat to the tympanic cavity and mastoidal air cells. Fluid accumulates in the cavity and produces pressure, pain, and impaired hearing. If otitis media goes untreated, it may spread from the mastoidal air cells and cause meningitis, a potentially deadly infection (see clinical insight 14.2). Otitis media can also cause fusion of the middle-ear bones and result in hearing loss. It is sometimes necessary to drain fluid from the tympanic cavity by lancing the eardrum and inserting a tiny drainage tube—a procedure called *myringotomy.*[18] The tube, which is eventually sloughed out of the ear, relieves the pressure and permits the infection to heal.

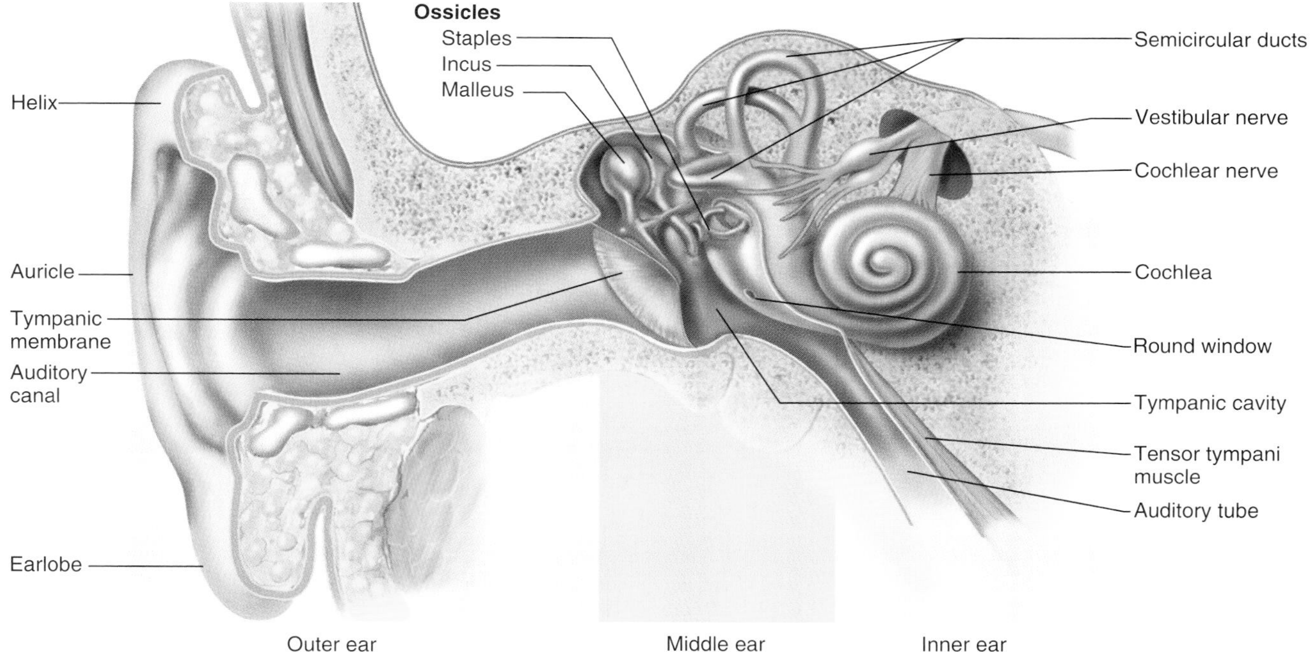

Figure 16.10 Internal Anatomy of the Ear.

Middle Ear

The **middle ear** is located in the **tympanic cavity** of the temporal bone. It begins with the eardrum, or **tympanic**[17] **membrane,** which closes the inner end of the auditory canal and separates it from the middle ear. The membrane is about 1 cm in diameter and slightly concave on its outer surface. It is suspended in a ring-shaped groove in the temporal bone and vibrates freely in response to sound. It is innervated by sensory branches of the vagus and trigeminal nerves and is highly sensitive to pain.

Posteriorly, the tympanic cavity is continuous with the mastoidal air cells in the mastoid process. It is filled with air that enters by way of the **auditory (eustachian**[19]**) tube,** a passageway to the nasopharynx. (Be careful not to confuse *auditory tube* with *auditory canal.*) The auditory tube is normally flattened and closed, but swallowing or yawning opens it and allows air to enter or leave the tympanic cavity. This equalizes air pressure on both sides of the eardrum and allows it to vibrate freely. Excessive pressure on one side or the other dampens the sense of hearing. The auditory tube also allows throat infections to spread to the middle ear.

16. *ot* = ear + *itis* = inflammation
17. *tympan* = drum
18. *myringo* = eardrum + *tomy* = cutting
19. Bartholomeo Eustachio (1520–74), Italian anatomist

The middle ear contains the three smallest bones and two smallest skeletal muscles of the body. The bones, called the **auditory ossicles,**[20] span the 2 to 3 mm distance from the eardrum to the inner ear. Progressing inward, the first is the **malleus,**[21] which has an elongated *handle* attached to the inner surface of the eardrum; a *head,* which is suspended from the wall of the tympanic cavity; and a *short process,* which articulates with the next ossicle. The second bone, the **incus,**[22] articulates in turn with the **stapes**[23] (STAY-peez). The stapes has an arch and *footplate* that give it a shape like a stirrup. The footplate, shaped like the sole of a steam iron, is held by a ringlike ligament in an opening called the **oval window,** where the inner ear begins.

The muscles of the middle ear are the stapedius and tensor tympani. The **stapedius** (stay-PEE-dee-us) arises from the posterior wall of the cavity and inserts on the stapes. The **tensor tympani** (TEN-sor TIM-pan-eye) arises from the wall of the auditory tube, travels alongside it, and inserts on the malleus. The function of these muscles is discussed under the physiology of hearing.

Inner Ear

The **inner ear** is housed in a maze of temporal bone passageways called the **bony labyrinth,** which is lined by a system of fleshy tubes called the **membranous labyrinth** (fig. 16.11). Between the bony and membranous labyrinths is a cushion of fluid called **perilymph** (PER-ih-limf), similar to cerebrospinal fluid. Within the membranous labyrinth is a fluid called **endolymph,** similar to intracellular fluid.

The labyrinths begin with a chamber called the **vestibule,** which contains organs of equilibrium to be discussed later. The organ of hearing is the **cochlea**[24] (COC-lee-uh), a coiled tube that arises from the anterior side of the vestibule. In other vertebrates, the cochlea is straight or slightly curved. In most mammals, however, it assumes the form of a snail-like spiral, which allows a longer cochlea to fit in a compact space. In humans, the spiral is about 9 mm wide at the base and 5 mm high. Its apex points anterolaterally. The cochlea winds for about 2.5 coils around an axis of spongy bone called the **modiolus**[25] (mo-DY-oh-lus). The modiolus is shaped like a screw; its threads form a spiral platform that supports the fleshy tube of the cochlea.

A vertical section cuts through the cochlea about five times (fig. 16.12*a*). A single cross section through it looks like figure 16.12*b*. It is important to realize that the structures seen in cross section actually have the form of spiral strips winding around the modiolus from base to apex.

The cochlea has three fluid-filled chambers separated by membranes. The superior one is called the **scala**[26] **vestibuli** (SCAY-la vess-TIB-you-lye) and the inferior one is the **scala tympani.** These are filled with perilymph and communicate with each other through a narrow channel called the *helicotrema* (HEL-ih-co-tree-muh) at the apex of the cochlea. The scala vestibuli begins near the oval window and spirals to the apex; from there, the scala tympani spirals back down to the base and ends at the **round window** (see fig. 16.10). The round window is covered by a membrane called the *secondary tympanic membrane.*

The middle chamber is a triangular space, the **cochlear duct (scala media).** It is separated from the scala vestibuli above by a thin **vestibular membrane** and from the scala tympani below by a much thicker **basilar membrane.** Unlike those chambers, it is filled with endolymph rather than perilymph. Within the cochlear duct, supported on the basilar membrane, is the **organ of Corti**[27] (COR-tee), a thick epithelium with associated structures (fig. 16.12*c*). It is the transducer that converts vibrations into nerve impulses, so we must pay particular attention to its structural details.

The organ of Corti has an epithelium composed of **hair cells** and **supporting cells.** Hair cells are named for the long, stiff microvilli called **stereocilia**[28] on their apical surfaces. (Stereocilia should not be confused with true cilia. They do not have an axoneme of microtubules as seen in cilia, and they do not move by themselves.) Resting on top of the stereocilia is a gelatinous **tectorial**[29] **membrane.**

The organ of Corti has four rows of hair cells spiraling along its length (fig. 16.13). About 3,500 of these, called **inner hair cells (IHCs),** are arranged in a row by themselves on the medial side of the basilar membrane (facing the modiolus). Each of these has a cluster of 50 to 60 stereocilia, graded from short to tall. Another 20,000 **outer hair cells (OHCs)** are neatly arranged in three rows across from the inner hair cells. Each outer hair cell has about 100 stereocilia arranged in the form of a V, with their tips embedded in the tectorial membrane. All that we hear comes from the IHCs, which supply 90% to 95% of the sensory fibers of the cochlear nerve. The function of the OHCs is to adjust the response of the cochlea to different frequencies and enable the IHCs to work with greater precision. We will see shortly how this is done. Hair cells are not neurons, but they synapse with nerve fibers at their base—the OHCs with both sensory and motor neurons and the IHCs with sensory neurons only.

20. *oss* = bone + *icle* = little
21. *malleus* = hammer
22. *incus* = anvil
23. *stapes* = stirrup
24. *cochlea* = snail
25. *modiolus* = hub
26. *scala* = staircase
27. Alfonso Corti (1822–88), Italian anatomist
28. *stereo* = solid
29. *tect* = roof

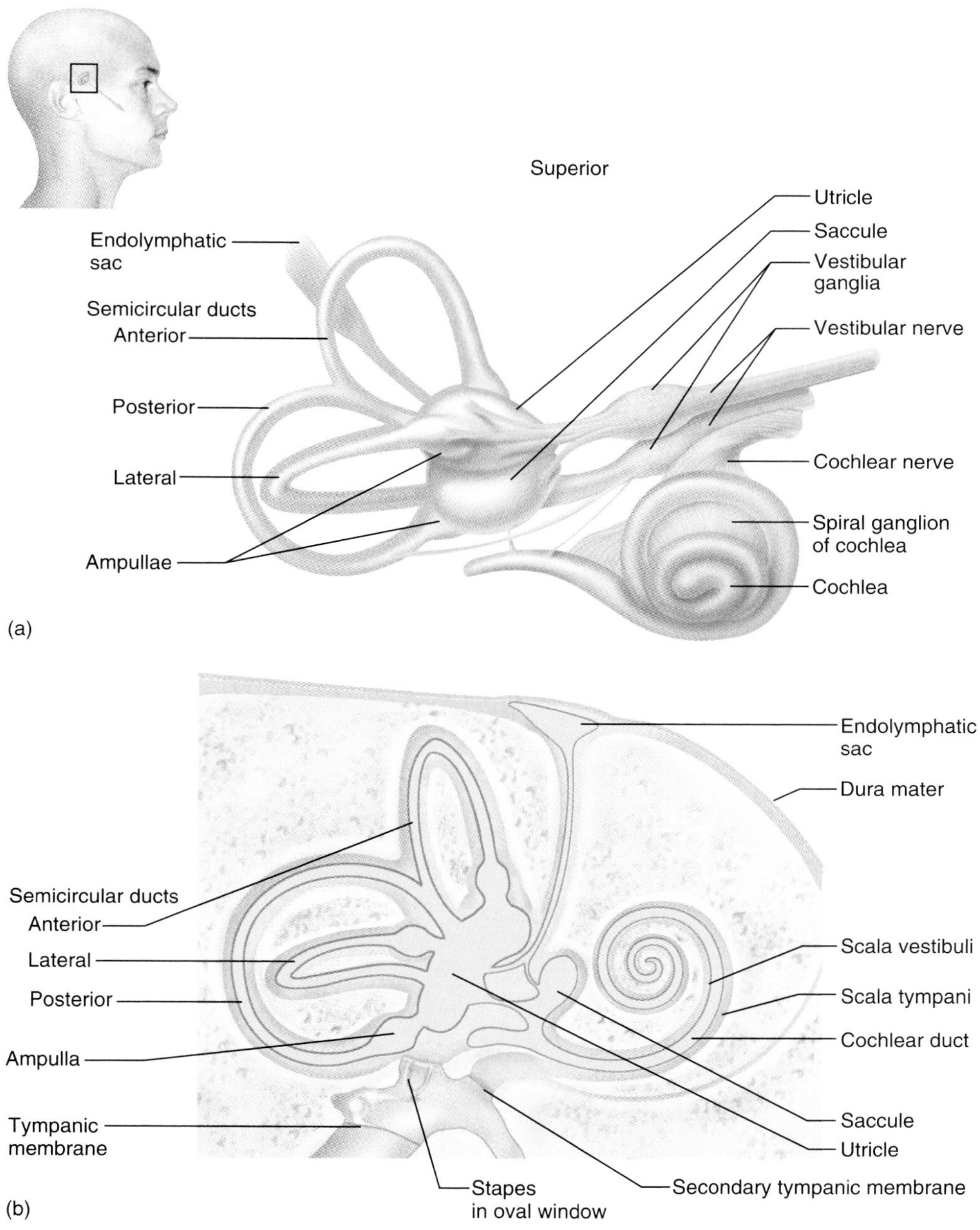

Figure 16.11 Anatomy of the Inner Ear. (*a*) The membranous labyrinth. (*b*) Relationship of the perilymph (blue) and endolymph (yellow) to the labyrinth.

The Physiology of Hearing

We can now examine the way in which sound affects the ear and produces action potentials.

The Middle Ear

The auditory ossicles provide no mechanical advantage, no amplification—vibrations of the stapes against the inner ear normally have the same amplitude as vibrations of the eardrum against the malleus. Why have auditory ossicles, then? There are two answers to this. One is that the eardrum, which moves in air, vibrates quite easily, whereas the stapes footplate must vibrate against the fluid of the inner ear. This fluid puts up a much greater resistance to motion than air does. If airborne sound waves struck the footplate directly, they would not have enough energy to overcome this resistance and move the stapes.

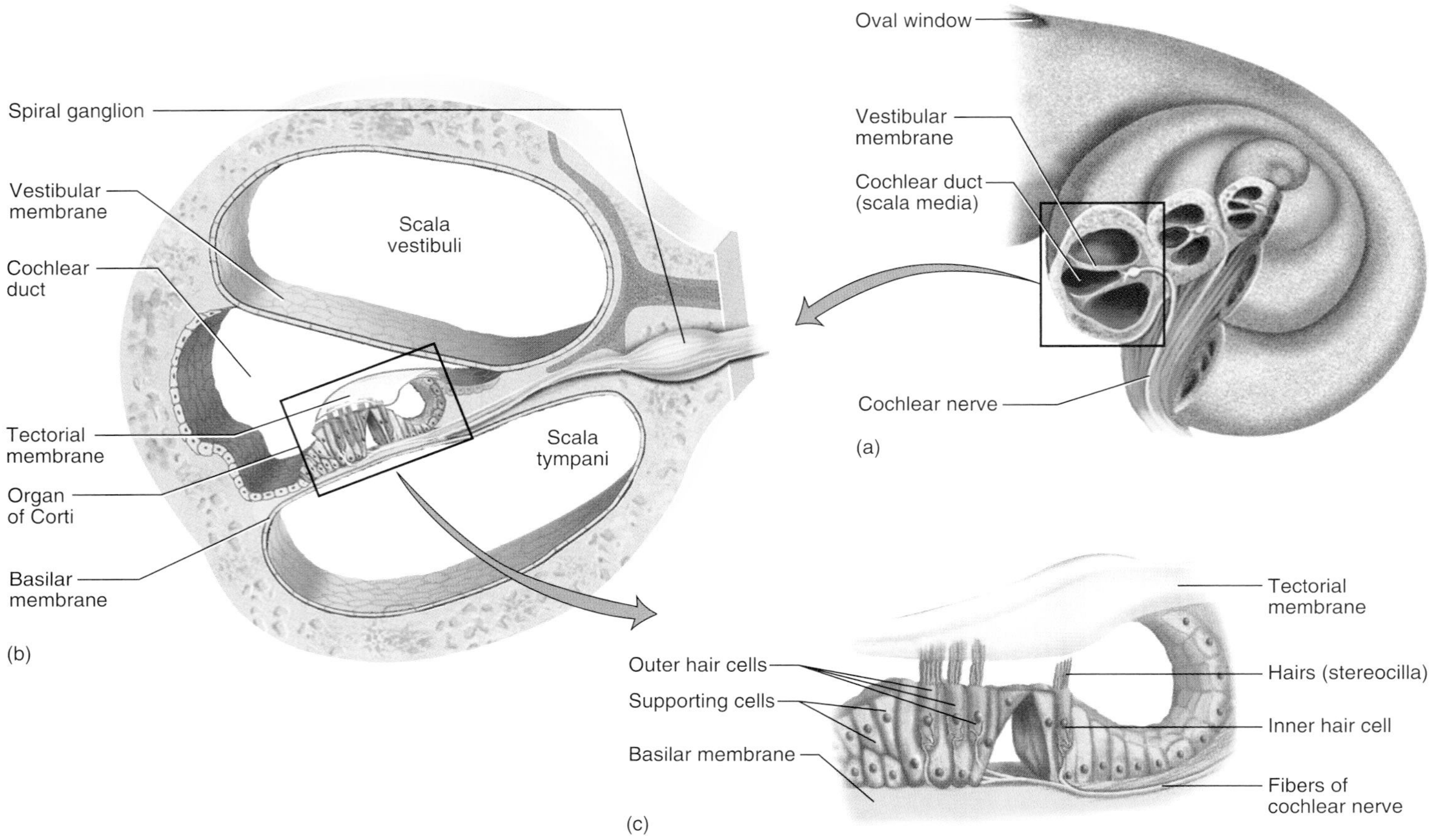

Figure 16.12 **Anatomy of the Cochlea.** (*a*) Vertical section. The apex of the cochlea faces downward and anterolaterally in anatomical position. (*b*) Detail of one section through the cochlea. (*c*) Detail of the organ of Corti.

The eardrum, however, has 18 times the area of the oval window. By concentrating the energy of the vibrating eardrum on an area 1/18 that size, the ossicles create a greater force per unit area at the oval window and overcome the resistance of the endolymph.

The ossicles and their muscles also have a protective function. In response to a loud noise, the tensor tympani pulls the eardrum inward and tenses it, while the stapedius reduces mobility of the stapes. This **tympanic reflex** muffles the transfer of vibrations from the eardrum to the oval window. The reflex probably evolved in part for protection from loud but slowly building noises such as thunder. It has a latency of about 40 msec, which is not quick enough to protect the inner ear from sudden noises such as gunshots. Such noises can irreversibly damage the hair cells of the inner ear by fracturing their stereocilia. It is therefore imperative to wear ear protection when using firearms or working in noisy environments.

The middle-ear muscles also help to coordinate speech with hearing. Without them, the sound of your own speech would be so loud it could damage your inner ear, and it would drown out soft or high-pitched sounds from other sources. Just as you are about to speak, however, the brain signals these muscles to contract. This dampens the sense of hearing in phase with the inflections of your own voice and makes it possible to hear other people while you are speaking.

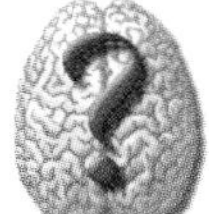

Think About It

What type of muscle fibers—slow-twitch or fast-twitch (see p. 442)—do you think constitute the stapedius and tensor tympani? That is, which type would best suit the purpose of these muscles?

Stimulation of Cochlear Hair Cells

To produce a sensation of sound, vibration of the auditory ossicles leads to vibration of the basilar membrane on which the hair cells rest. A simple mechanical model of the ear (fig. 16.14) makes it easy to see how this happens. The stapes pushes on the perilymph of the scala vestibuli;

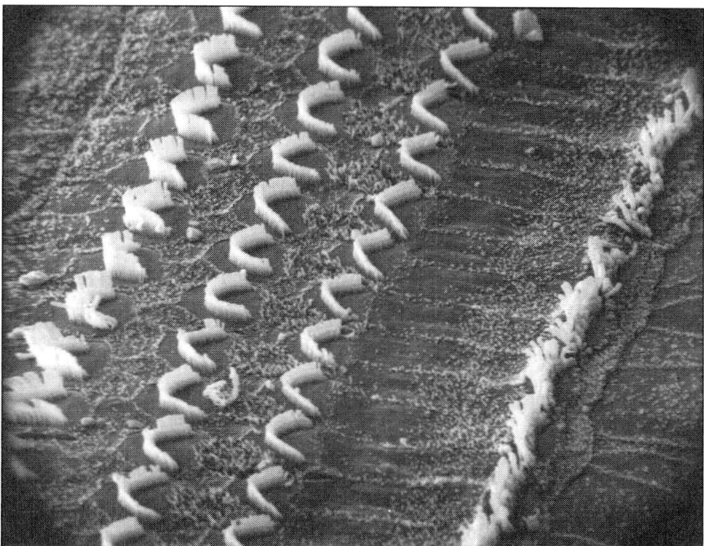

Figure 16.13 Apical Surfaces of the Cochlear Hair Cells. On the *left* are the three rows of outer hair cells, which serve only to tune the cochlea. Each cell has a V-shaped row of stereocilia. On the *right* is the single row of inner hair cells, which generate the signals we hear. From R.G. Kessel and R.H. Kardon, *Tissues and Organs: A Text-Atlas of Scanning Electron Microscopy* (W.H. Freeman & Co., 1979).

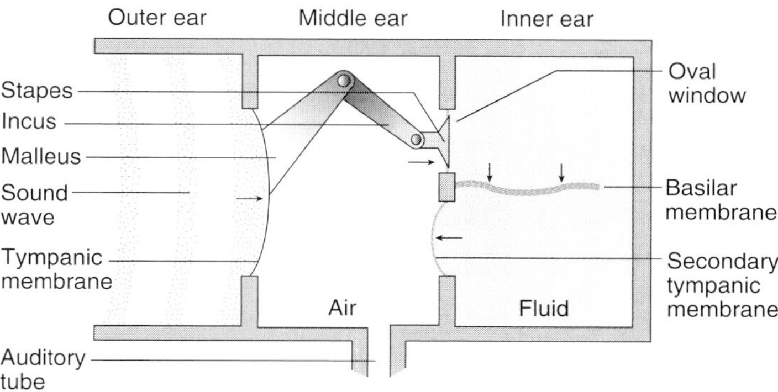

Figure 16.14 Mechanical Model of the Ear. Each inward movement of the tympanic membrane pushes inward on the auditory ossicles of the middle ear and fluid of the inner ear. This pushes down on the basilar membrane, and pressure is relieved by an outward bulge of the secondary tympanic membrane. Thus the basilar membrane vibrates up and down in synchrony with the vibrations of the tympanic membrane.

the perilymph pushes the vestibular membrane down; the vestibular membrane pushes on the endolymph of the cochlear duct; and the endolymph pushes the basilar membrane down. (The vestibular membrane is omitted from the diagram for simplicity; it has no significant effect on the mechanics of the cochlea.) The basilar membrane puts pressure on the perilymph of the scala tympani below it, and the secondary tympanic membrane bulges outward to relieve this pressure. In short, as the stapes goes in-out-in, the secondary tympanic membrane goes out-in-out, and the basilar membrane goes down-up-down. It is not difficult to see how this happens—the only thing hard to imagine is that it can happen as often as 20,000 times per second!

The vestibular membrane separates the perilymph of the scala vestibuli from the endolymph of the cochlear duct. In order for the hair cells to function properly, the tips of their stereocilia must be bathed in endolymph. Endolymph has an exceptionally high K^+ concentration, which creates a strong electrochemical gradient from the tip to the base of a hair cell. This gradient provides the potential energy that ultimately enables a hair cell to work.

The tectorial membrane is especially important in cochlear mechanics. Remember that the stereocilia of the outer hair cells have their tips embedded in it, and those of the inner hair cells come very close to it. The tectorial membrane is anchored to the modiolus (see fig. 16.12*c*), which holds it relatively still as the basilar membrane and hair cells vibrate up and down. Movement of the basilar membrane thus bends the hair cell stereocilia back and forth.

At the tip of each stereocilium of the inner hair cells is a single transmembrane protein that functions as a mechanically gated ion channel. A fine, stretchy protein filament called a **tip link** extends like a spring from the ion channel of one stereocilium to the side of the stereocilium next to it (fig. 16.15). The stereocilia increase in height progressively, so that all but the tallest ones have tip links leading to taller stereocilia beside them. When a taller stereocilium bends away from a shorter one, it pulls on the tip link and opens the ion channel of the shorter stereocilium. The channel is nonselective, but since the predominant ion of the endolymph is K^+, the primary effect of this gating is to allow a quick burst of K^+ to flow into each hair cell. This depolarizes the hair cell while the channel is open, and when the stereocilium bends the other way its channel closes and the cell becomes briefly hyperpolarized. During the moments of depolarization, a hair cell releases a neurotransmitter that stimulates the sensory dendrites synapsing with its base. Each depolarization thus generates action potentials in the cochlear nerve.

Sensory Coding

Our ability to distinguish loudness and pitch depends on the ability of the cochlea to respond differently to vibrations of different amplitude and frequency. Loud sounds produce more vigorous vibrations of the organ of Corti. This excites a greater number of hair cells over a broader area of basilar membrane and triggers a higher frequency of action potentials in the cochlear nerve fibers. If the brain detects intense activity in nerve fibers from a broad region of the organ of Corti, it interprets this as a loud sound.

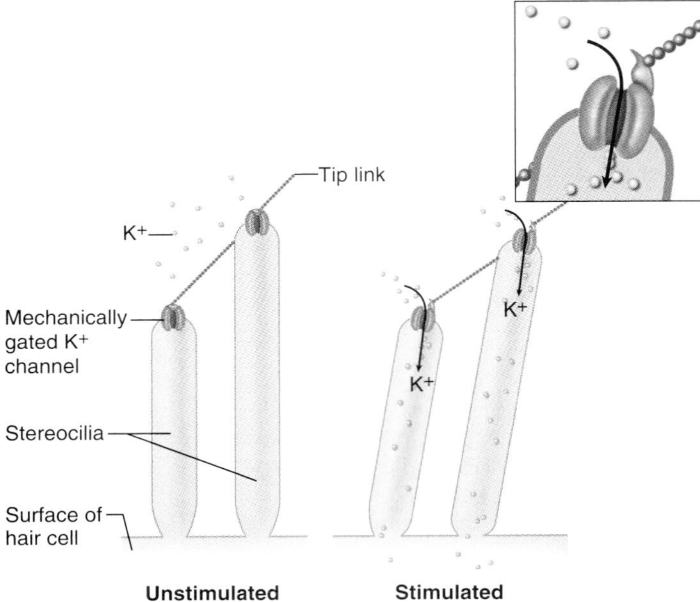

Figure 16.15 Potassium Gates of the Cochlear Hair Cells. Each stereocilium has a gated K^+ channel at its tip. Vibrations of the cochlea cause each stereocilium to bend and, with its tip link, pull open the K^+ channel of the adjacent stereocilium. The inflow of K^+ depolarizes the hair cell.

Frequency discrimination requires a more sophisticated mechanism. The basilar membrane is spanned by short, stiff collagen fibers of various lengths. At its proximal (basal) end, the basilar membrane is attached, narrow, and stiff. At its distal (apical) end, it is unattached, five times wider than at the base, and more flexible. Think of the basilar membrane as analogous to a rope stretched tightly between two posts. If you pluck the rope at one end, a wave of vibration travels down its length and back. This produces a standing wave, with some regions of the rope vibrating more than others. Similarly, a sound causes a standing wave in the basilar membrane. The peak amplitude of this wave is near the distal end in the case of low-frequency sounds and nearer the proximal (attached) end with sounds of higher frequencies. When the brain receives signals mainly from inner hair cells at the distal end, it interprets the sound as low-pitched; when signals come mainly from the proximal end, it interprets the sound as high-pitched (fig. 16.16). Speech, music, and other everyday sounds, of course, are not pure tones—they create complex patterns of vibration in the basilar membrane that must be decoded by the brain.

Cochlear Tuning

Just as we tune a radio to receive a certain frequency, we also tune our cochlea to receive some frequencies better than others. The outer hair cells (OHCs) are supplied with a few sensory fibers (5%–10% of those in the cochlear nerve), but more importantly, they receive motor fibers from the brain.

In response to sound, the OHCs trigger nerve signals to the medulla by way of the sensory neurons, and the pons sends signals immediately to the OHCs by way of the motor neurons. In response, the hair cells contract by about 10% to 15%. Because an OHC is anchored to the basilar membrane below and its stereocilia are embedded in the tectorial membrane above, contraction of an OHC reduces the basilar membrane's freedom to vibrate. This results in some regions of the organ of Corti sending fewer signals to the brain than neighboring regions, so the brain can better distinguish between the more active and less active hair cells and sound frequencies. When OHCs are experimentally incapacitated, the inner hair cells (IHCs) respond much less precisely to differences in pitch.

There is another mechanism of cochlear tuning involving the inner hair cells. The pons sends efferent fibers to the cochlea that synapse with the sensory nerve fibers near the base of the IHCs. The efferent fibers can inhibit the sensory fibers from firing in some areas of the cochlea, and thus enhance the contrast between signals from the more responsive and less responsive regions. Combined with the previously described role of the OHCs, this sharpens the tuning of the cochlea and our ability to discriminate sounds of different pitch.

The Auditory Projection Pathway

A **spiral ganglion**, wound around the modiolus, is composed of the somas of bipolar sensory neurons of the cochlea. Their dendrites originate at the hair cells and their axons form the **cochlear nerve.** The cochlear nerve joins the *vestibular nerve,* discussed later, and the two together become the *vestibulocochlear nerve* (cranial nerve VIII).

The cochlear nerve fibers from each ear lead to *cochlear nuclei* on both sides of the pons. There, they synapse with second-order neurons that ascend to the nearby *superior olivary nucleus* of the pons (fig. 16.17). By way of cranial nerve VIII, the superior olivary nucleus issues the efferent fibers back to the cochlea that are involved in cochlear tuning. By way of cranial nerves V_3 and VII, it issues motor fibers to the tensor tympani and stapedius muscles, respectively. The superior olivary nucleus also functions in **binaural**[30] **hearing**—comparing signals from the right and left ears to identify the direction from which a sound is coming.

Other fibers from the cochlear nucleus ascend to the inferior colliculi of the midbrain. The inferior colliculi help to locate the origin of a sound in space, process fluctuations in pitch that are important for such purposes as understanding another person's speech, and mediate the

30. *bin* = two + *aur* = ears

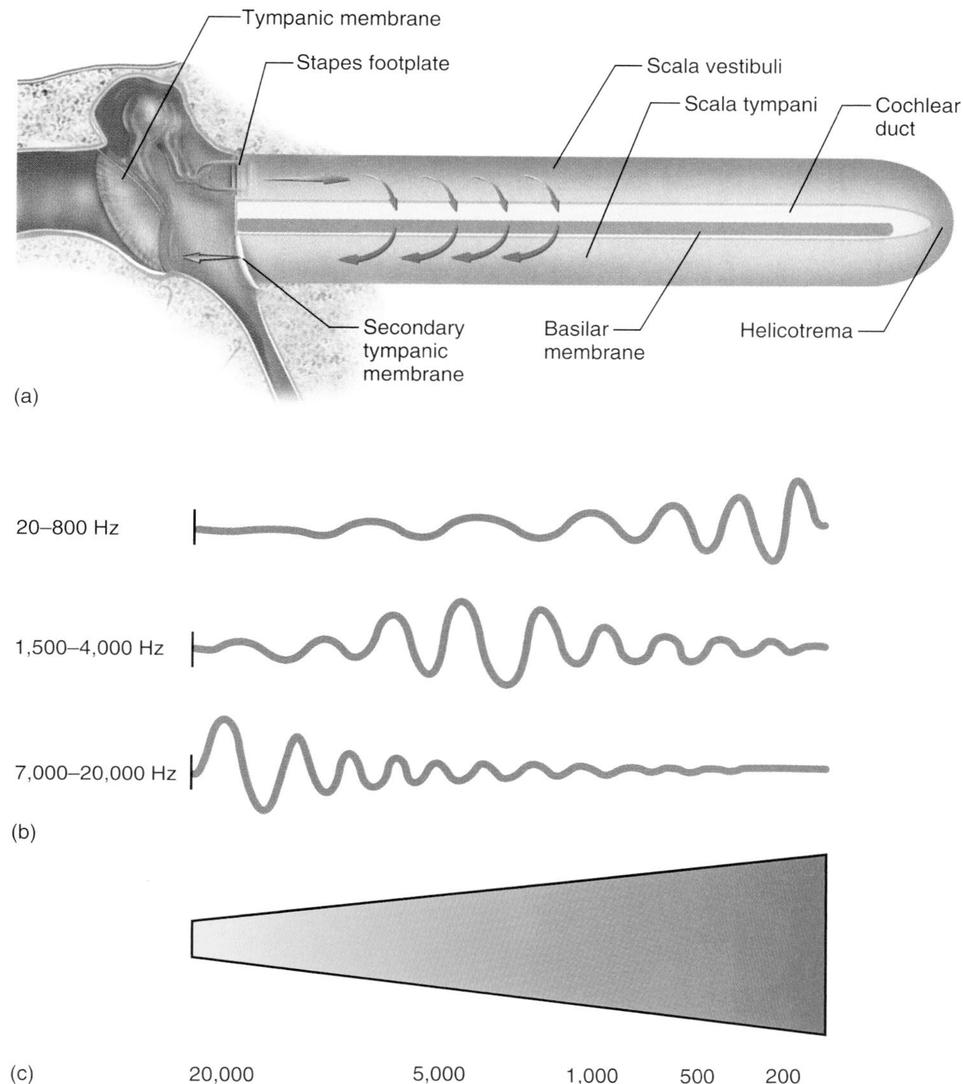

Figure 16.16 Frequency Response of the Basilar Membrane of the Cochlea. (*a*) The cochlea, uncoiled and laid out straight. (*b*) Sounds produce a standing wave of vibration along the basilar membrane. The peak amplitude of the wave varies with the frequency of the sound, as shown here. The amount of vibration is greatly exaggerated in this diagram to clarify the standing wave. (*c*) The taper of the basilar membrane and its correlation with sound frequencies. High frequencies (7,000–20,000 Hz) are best detected by hair cells near the basal (proximal) end at the *left* and low frequencies (20–800 Hz) by hair cells near the apical (distal) end at the *right.*

startle response and rapid head turning that occur in reaction to loud or sudden noises.

Third-order neurons begin in the inferior colliculi and lead to the thalamus. Fourth-order neurons begin here and complete the pathway to the primary auditory cortex, which is in the superior margin of the temporal lobe deep within the lateral sulcus. The temporal lobe is the site of conscious perception of sound, and it completes the information processing essential to binaural hearing. Because of extensive decussation in the auditory pathway, damage to the right or left auditory cortex does not cause a unilateral loss of hearing.

Equilibrium

The original function of the ear in vertebrate evolution was not hearing, it was **equilibrium**—coordination and balance. Only later did vertebrates evolve the cochlea, middle-ear structures, and auditory function of the ear. In humans, the receptors for equilibrium constitute the **vestibular apparatus,** which consists of three **semicircular ducts** and two chambers—an anterior **saccule** (SAC-yule) and a posterior **utricle**[31] (YOU-trih-cul) (see fig. 16.11).

31. *saccule* = little sac; *utricle* = little bag

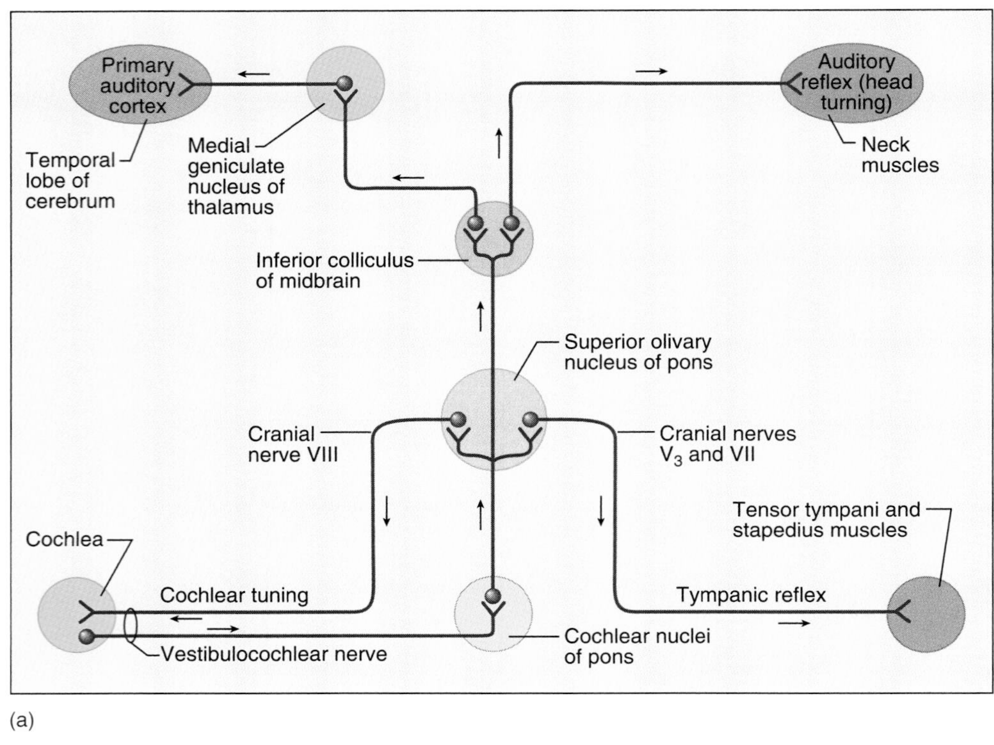

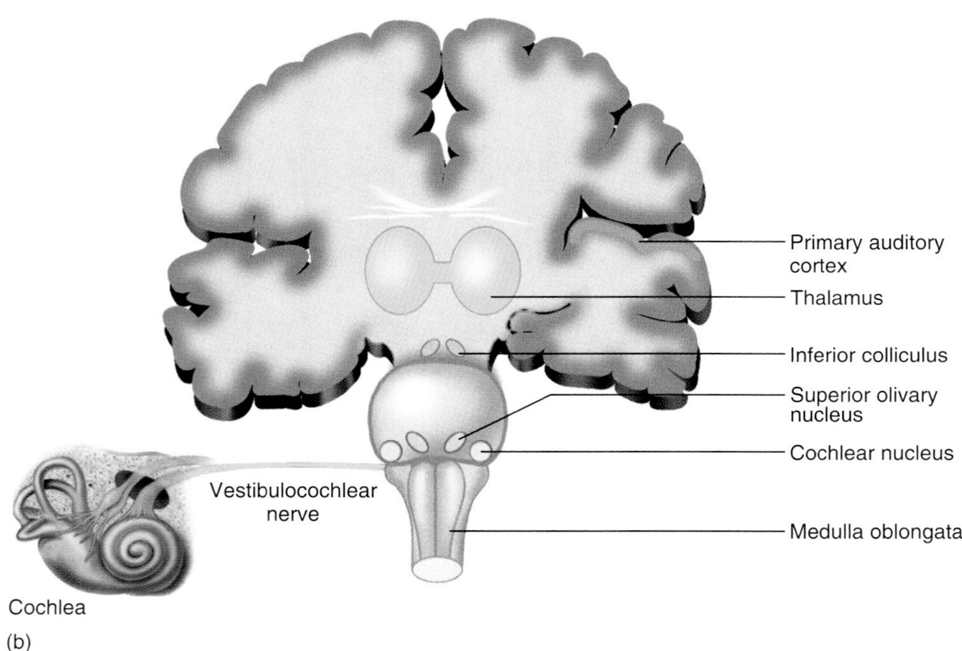

Figure 16.17 Auditory Pathways in the Brain. (*a*) Schematic. (*b*) Brainstem and frontal section of the cerebrum showing the locations of auditory processing centers.

The sense of equilibrium is divided into **static equilibrium,** the perception of the orientation of the head when the body is stationary, and **dynamic equilibrium,** the perception of motion or acceleration. Acceleration is divided into *linear acceleration,* a change in velocity in a straight line, as when riding in a car or elevator, and *angular acceleration,* a change in the rate of rotation. The saccule and utricle are responsible for static equilibrium and the sense of linear acceleration; the semicircular ducts detect only angular acceleration.

Clinical Insight 16.3

Deafness

Deafness means any hearing loss, from mild and temporary to complete and irreversible. *Conduction deafness* results from any condition that interferes with the transmission of vibrations to the inner ear. Such conditions include a damaged eardrum, otitis media, blockage of the auditory canal, and *otosclerosis.*[32] Otosclerosis is the fusion of auditory ossicles to each other or fusion of the stapes to the oval window, which prevents the bones from vibrating freely. *Sensorineural (nerve) deafness* results from the death of hair cells or any of the nervous elements concerned with hearing. It is a common occupational disease of factory and construction workers, musicians, and other people. Deafness leads some people to develop delusions of being talked about, disparaged, or cheated. Beethoven said his deafness drove him nearly to suicide.

The Saccule and Utricle

Each of these chambers has a 2 by 3 mm patch of hair cells and supporting cells called a **macula.**[33] The **macula sacculi** lies nearly vertically on the wall of the saccule, and the **macula utriculi** lies nearly horizontally on the floor of the utricle (fig. 16.18*a*).

Each hair cell of a macula has 40 to 70 stereocilia and one true cilium called a **kinocilium.**[34] The tips of the stereocilia and kinocilium are embedded in a gelatinous **otolithic membrane.** This membrane is weighted with calcium carbonate–protein granules called **otoliths**[35] (fig. 16.18*b*), which add to the density and inertia of the membrane and enhance the sense of gravity and motion.

Figure 16.18*c* shows how the macula utriculi detects tilt of the head. With the head erect, the otolithic membrane bears directly down on the hair cells and stimulation is minimal. When the head is tilted, however, the weight of the membrane bends the stereocilia and stimulates the hair cells. Any orientation of the head causes a combination of stimulation to the utricules and saccules of the two ears. The brain interprets head orientation by comparing these inputs to each other and to other input from the eyes and stretch receptors in the neck.

The inertia of the otolithic membranes is especially important in detecting linear acceleration. Suppose you are sitting in a car at a stoplight and then begin to move. The heavy otolithic membrane of the macula utriculi briefly lags behind the rest of the tissues, bends the stereocilia backward, and stimulates the cells. When you stop at the next light, the macula stops but the otolithic membrane keeps on going for a moment, bending the stereocilia forward. The hair cells convert this pattern of stimulation to nerve signals, and the brain is thus advised of changes in your linear velocity.

If you are standing in an elevator and it begins to move up, the otolithic membrane of the vertical macula sacculi lags behind briefly and pulls down on the hairs. When the elevator stops, the otolithic membrane keeps going for a moment and bends the hairs upward. The macula sacculi thus detects vertical acceleration.

The Semicircular Ducts

Rotational acceleration is detected by the three *semicircular ducts* (fig. 16.19), each housed in an osseous *semicircular canal* of the temporal bone. The **anterior** and **posterior semicircular ducts** are positioned vertically, at right angles to each other. The **lateral semicircular duct** is about 30° from the horizontal plane. The orientation of the ducts causes a different duct to be stimulated by rotation of the head in different planes—turning it from side to side as in gesturing "no," nodding up and down as in gesturing "yes," or tilting it from side to side as in touching your ears to your shoulders.

The semicircular ducts are filled with endolymph. Each duct opens into the utricle and has a dilated sac at one end called an **ampulla.**[36] Within the ampulla is a mound of hair cells and supporting cells called the **crista**[37] **ampullaris.** The hair cells have stereocilia and a kinocilium embedded in the **cupula,**[38] a gelatinous membrane that extends from the crista to the roof of the ampulla. When the head turns the duct rotates, but the endolymph lags behind. It pushes the cupula, bends the stereocilia, and stimulates the hair cells. After 25 to 30 seconds of continual rotation, however, the endolymph catches up with the movement of the duct and stimulation of the hair cells ceases.

32. *oto* = ear + *scler* = hardening + *osis* = process, condition
33. *macula* = spot
34. *kino* = moving
35. *oto* = ear + *lith* = stone
36. *ampulla* = little jar
37. *crista* = crest, ridge
38. *cupula* = little tub

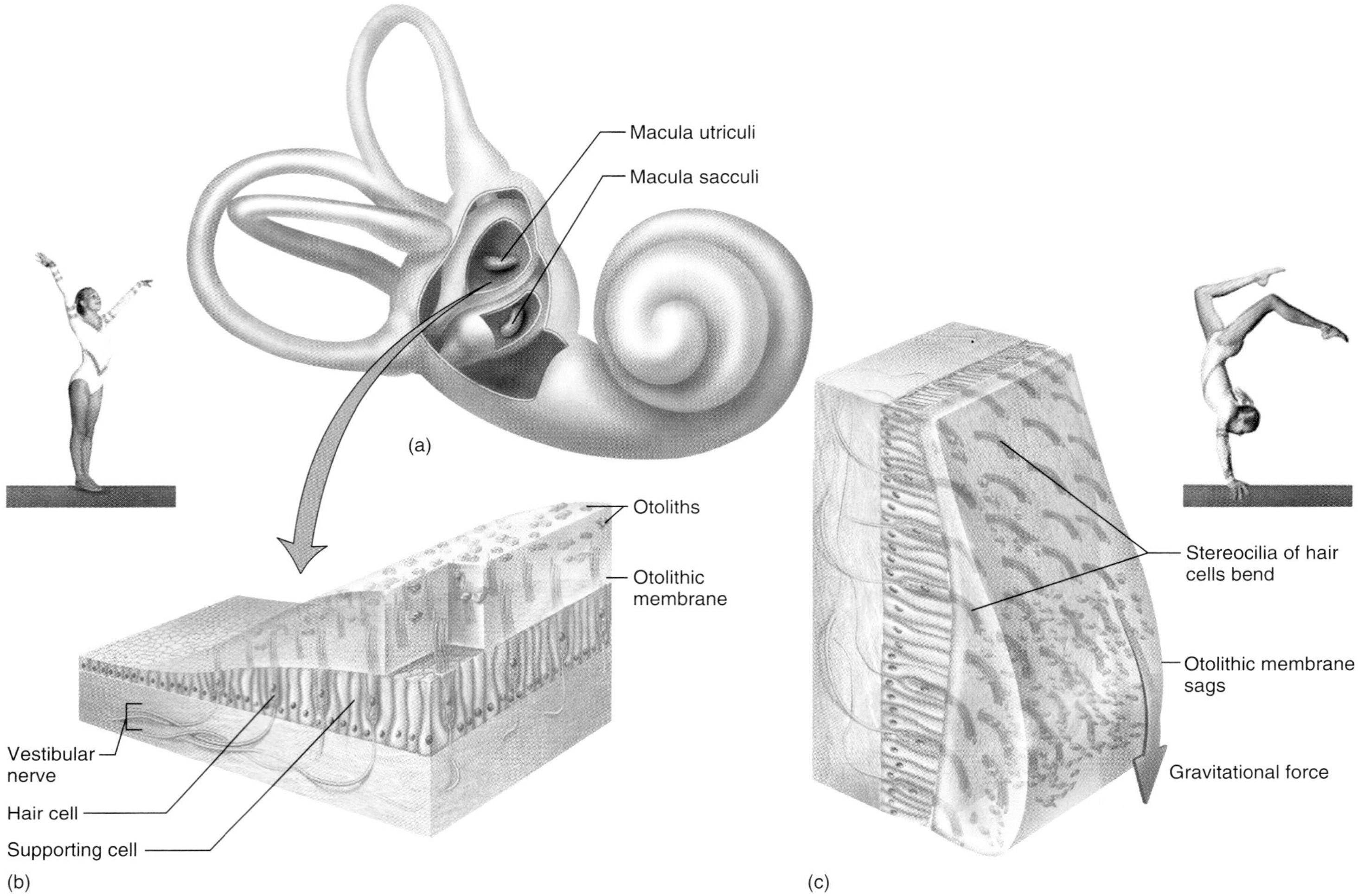

Figure 16.18 **Form and Function of the Saccule and Utricle.** (*a*) Locations of the macula sacculi and macula utriculi. (*b*) Structure of a macula. (*c*) Action of the otolithic membrane on the hair cells when the head is tilted.

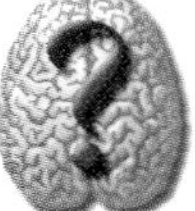

Think About It

The semicircular ducts do not detect motion itself, but only acceleration—a *change in the rate* of motion. Explain.

Equilibrium Projection Pathways

Hair cells of the macula sacculi, macula utriculi, and semicircular ducts synapse at their bases with sensory fibers of the **vestibular nerve.** This nerve joins the cochlear nerve to form the vestibulocochlear nerve (VIII). Most fibers from the vestibular apparatus terminate in the **vestibular nucleus** of the pons, while others project to the cerebellum. Fibers from the vestibular nucleus project caudally to the cervical spinal cord, and from there, fibers lead to the brainstem nuclei for the cranial nerves that control eye movements—the oculomotor (III), trochlear (IV), and abducens (VI) nerves. Other fibers from the cervical spinal cord lead by way of the accessory nerve (XI) to muscles that move the head and neck.

Reflex pathways to the extrinsic eye muscles enable us to fixate visually on a point in space while the head is moving. To observe this effect, hold this book in front of you at a comfortable reading distance and fix your gaze on the middle of the page. Move the book left and right about once per second, and you will be unable to read it. Now hold the book still and shake your head from side to side at the same rate. This time you will be able to read it because the reflex pathway compensates for your head movements and keeps your eyes fixed on the target.

Before we move on to the sense of vision, you may wish to review the anatomy of the ear (table 16.2) and think back on the function of each component.

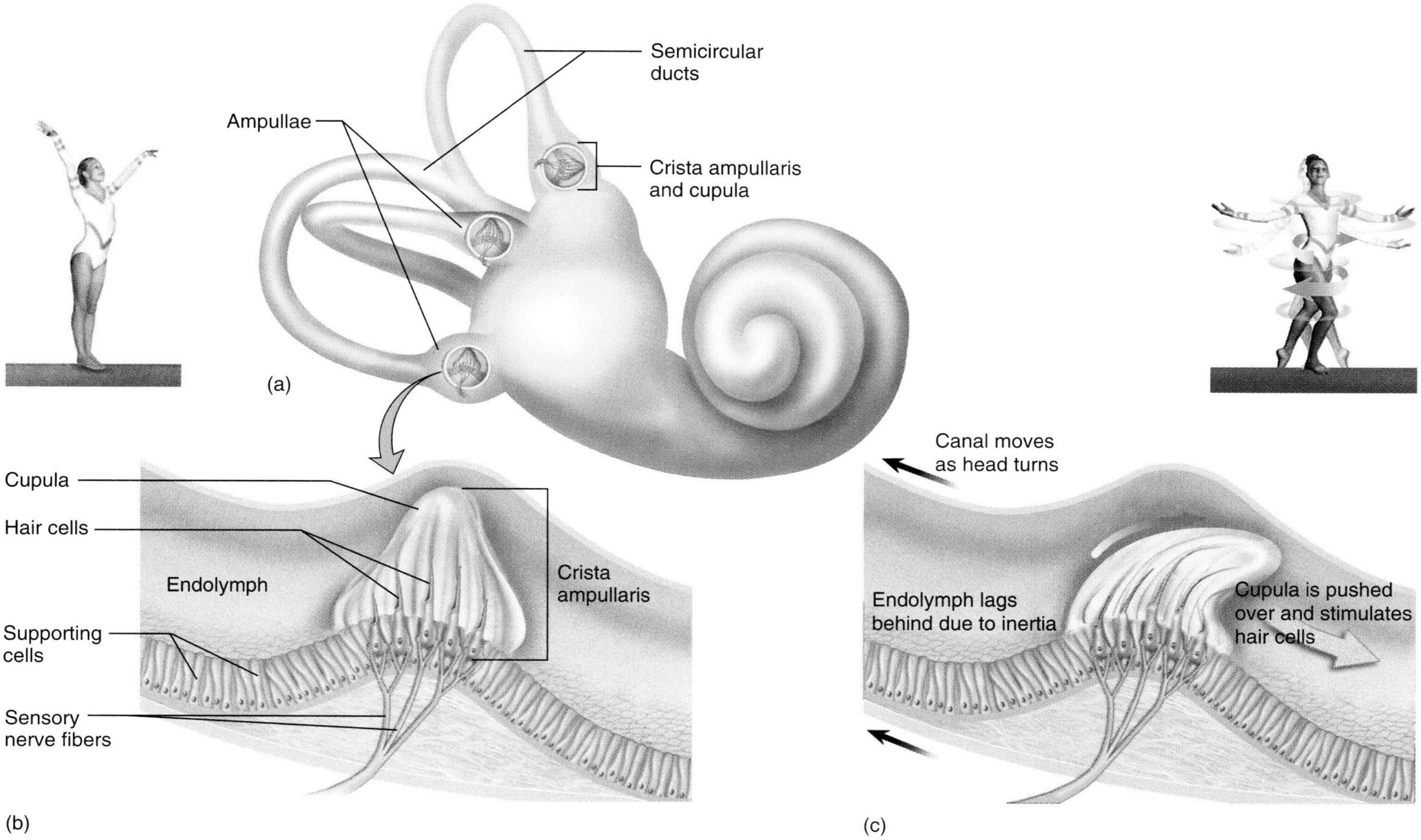

Figure 16.19 Form and Function of the Semicircular Ducts. (*a*) Structure of the semicircular ducts, with each ampulla opened to show the crista ampullaris and cupula. (*b*) Detail of the crista ampullaris. (*c*) Action of the endolymph on the cupula and hair cells when the head is rotated.

Before You Go On

Answer the following questions to test your understanding of the preceding section:

15. What physical properties of sound waves correspond to the sensations of loudness and pitch?
16. What is the benefit of having auditory ossicles and muscles in the middle ear?
17. Explain how vibration of the tympanic membrane ultimately produces fluctuations of membrane voltage in a cochlear hair cell.
18. How does the brain recognize the difference between high C and middle C musical notes? Between a loud sound and a soft one?
19. How does the function of the semicircular ducts differ from the function of the saccule and utricle?
20. How is sensory transduction in the semicircular ducts similar to that in the saccule and utricle?

Vision

Objectives

When you have completed this section, you should be able to

- describe the anatomy of the eye and its accessory structures;
- describe the structure of the retina and its receptor cells;
- explain how the optical system of the eye creates an image on the retina;
- explain how the retina converts this image to nerve impulses;
- explain why different types of receptor cells and neuronal circuits are required for day and night vision; and
- trace the visual projection pathways in the brain.

Table 16.2 Anatomical Review of the Ear

Outer Ear (fig. 16.9)	
Auricle (pinna)	Triangular fossa
Auditory canal (external auditory meatus)	Concha
Ceruminous glands	Tragus
Helix	Antitragus
Antihelix	Lobe
Middle Ear (fig. 16.10)	
Tympanic membrane (eardrum)	*Muscles*
Tympanic cavity	Stapedius
Auditory ossicles	Tensor tympani
Malleus	*Auditory (eustachian) tube*
Incus	
Stapes	
Footplate	
Inner Ear (figs. 16.10 and 16.11)	
Oval window	*Cochlea*
Labyrinths and fluids	*Vestibular apparatus*
Bony labyrinth	*Round window*
Membranous labyrinth	*Secondary tympanic membrane*
Perilymph	
Endolymph	
Cochlea (fig. 16.12)	
Modiolus	Organ of Corti
Scala vestibuli	Supporting cells
Scala tympani	Inner hair cells
Helicotrema	Outer hair cells
Cochlear duct	Stereocilia
Vestibular membrane	Tectorial membrane
	Basilar membrane
	Neural components
	Spiral ganglion
	Cochlear nerve
Vestibular Apparatus (figs. 16.18 and 16.19)	
Vestibule	Semicircular ducts
Saccule	Ampulla
Macula sacculi	Crista ampullaris
Utricle	Cupula
Macula utriculi	Vestibular nerve
Otolithic membranes	
Projection Pathways (fig. 16.17)	
Vestibulocochlear nerve	Inferior and superior colliculi
Cochlear nucleus	Thalamus
Vestibular nucleus	Primary auditory cortex
Superior olivary nucleus	

Vision and Light

Vision (sight) is the perception of objects in the environment by means of the light that they emit or reflect. *Light* is electromagnetic radiation with wavelengths ranging from about 400 to 750 nm. Most solar radiation that reaches the surface of the earth falls within this range; radiation of shorter and longer wavelengths is generally filtered out by ozone, carbon dioxide, and water vapor in the atmosphere. Vision is thus adapted to take advantage of the radiation that is most available to us.

Yet there is further reason for vision to be limited to this range of wavelengths. To produce a physiological response, light must cause a *photochemical reaction*—a change in chemical structure caused by light energy. When an electron absorbs a photon of light, it is boosted to a higher energy level (orbit) around its nucleus and may transfer to another atom. The transfer of an electron from one atom to another is the essence of a chemical reaction. Radiation below 400 nm has so much energy that it ionizes organic molecules and kills cells. It is useful for sterilizing food and instruments, but it has too much energy for the biochemical processes of vision. Radiation above 750 nm has too little energy to activate the visual process. It warms the tissues (heat lamps are based on this principle) but does not usually cause chemical reactions. These other forms of electromagnetic radiation include ultraviolet (UV) rays, X rays, and gamma rays at the high-energy (short-wavelength) end of the spectrum, and infrared (IR) rays, microwaves, and radio waves at the low-energy (long-wavelength) end. Some other animals can see radiation extending in both directions slightly beyond the range visible to humans.

Accessory Structures of the Orbit

Before considering the eye itself, let's survey the accessory structures located in and around the orbit (figs. 16.20 and 16.21). These include the *eyebrows, eyelids, conjunctiva, lacrimal apparatus,* and *extrinsic eye muscles:*

- The **eyebrows** probably serve mainly to enhance facial expressions and nonverbal communication (see p. 219), but they may also protect the eyes from glare and help to keep perspiration from running into the eye.
- The **eyelids,** or **palpebrae** (pal-PEE-bree), block foreign objects from the eye, prevent visual stimuli from disturbing our sleep, and blink periodically to moisten the eye with tears and sweep debris from the surface. The eyelids are separated from each other by the **palpebral fissure** and meet each other at the corners called the **medial** and **lateral commissures (canthi).** The eyelid consists largely of the orbicularis oculi muscle covered with skin (fig. 16.21*a*). It also contains a supportive fibrous **tarsal plate,** which is thickened along the margin of the eyelid. Within the plate are 20 to 25 **tarsal glands** that open along the margin. They secrete an oil that coats the eye and reduces tear evaporation. The **eyelashes** are guard hairs that help to keep debris from the eye. Touching the eyelashes stimulates hair receptors and triggers the blink reflex.

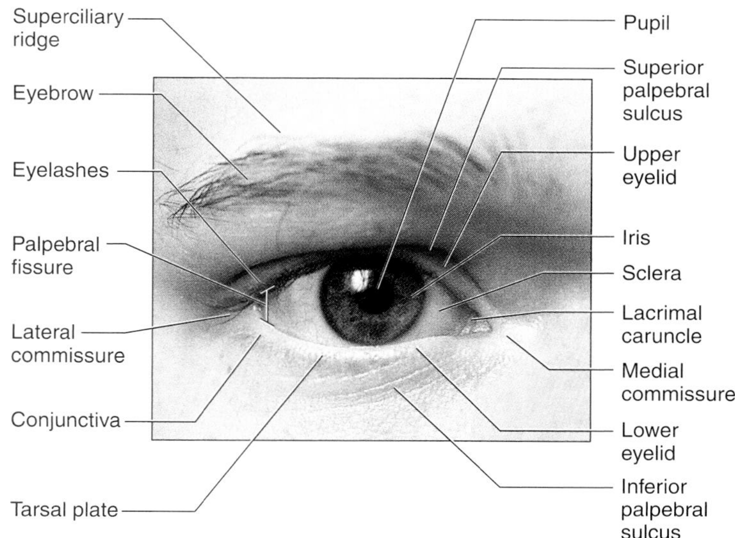

Figure 16.20 External Anatomy of the Orbital Region.

- The **conjunctiva** (CON-junk-TY-vuh) is a transparent mucous membrane that covers the inner surface of the eyelid and anterior surface of the eyeball. Its primary purpose is to secrete a thin mucous film that prevents the eyeball from drying. It is richly innervated and highly sensitive to pain. It is also very vascular, which is especially evident when the vessels are dilated and the eyes are "bloodshot." Because it is vascular and the cornea is not, the conjunctiva heals more readily than the cornea when injured.
- The **lacrimal**[39] **apparatus** (fig. 16.21*b*) consists of the lacrimal (tear) gland and a series of ducts that drain the tears into the nasal cavity. The **lacrimal gland,** about the size and shape of an almond, is nestled in a shallow fossa of the frontal bone in the superolateral corner of the orbit. About 12 short ducts lead from the lacrimal gland to the surface of the conjunctiva. By moistening the conjunctiva, the tears not only wash foreign particles from the eye but also allow for the diffusion of oxygen and carbon dioxide between the air and tissues; such gases cannot diffuse adequately through a dry membrane. The tears also contain a bactericidal enzyme called lysozyme that keeps the eye surface fairly sterile. After washing across the conjunctiva, the tears collect

39. *lacrim* = tear

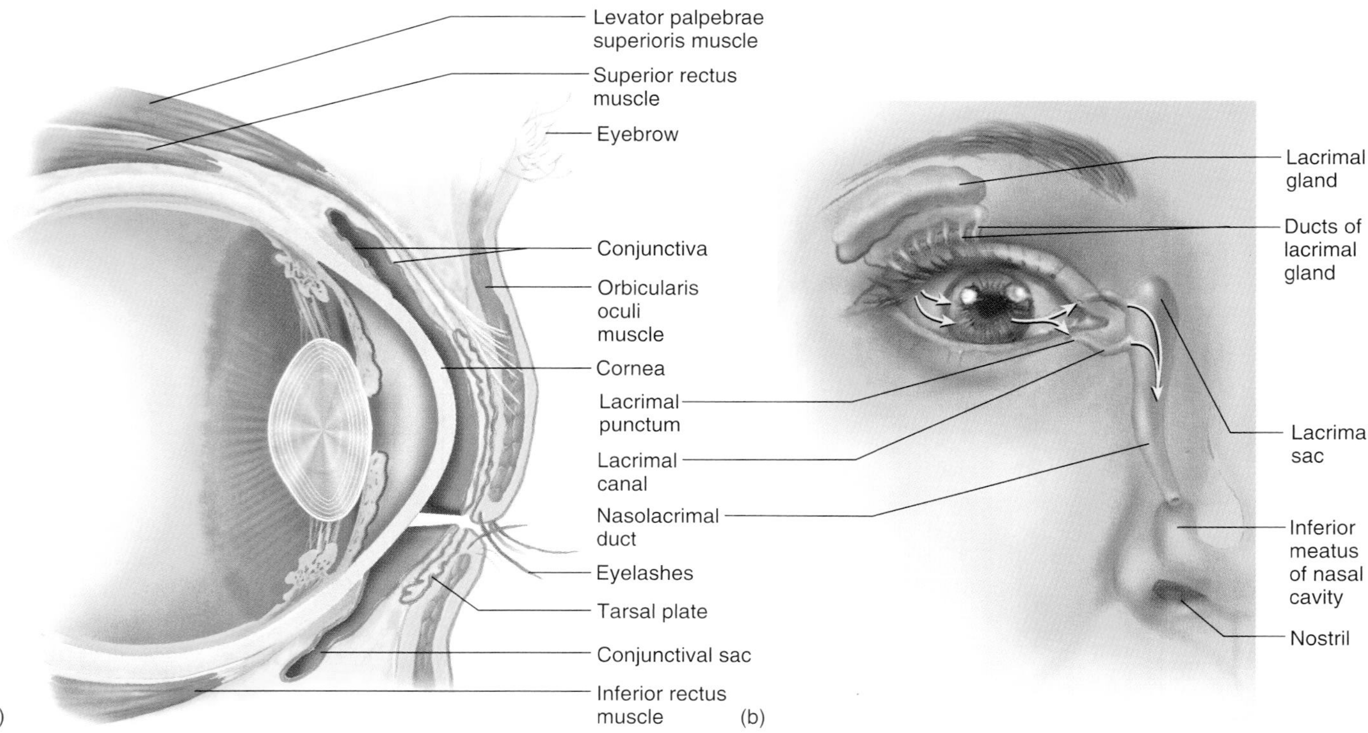

Figure 16.21 Accessory Structures of the Orbit. (*a*) Sagittal section of the eye and orbit. (*b*) The lacrimal apparatus.

at the **lacrimal caruncle**[40] (CAR-un-cul), the pink fleshy mass near the medial commissure of the eye. On the margin of each eyelid near the caruncle is a tiny pore, the **lacrimal punctum.**[41] This is the opening to a short **lacrimal canal,** which leads to the **lacrimal sac** in the medial wall of the orbit. From this sac, a **nasolacrimal duct** carries the tears to the inferior meatus of the nasal cavity—thus an abundance of tears from crying or watery eyes can result in a runny nose. Once the tears enter the nasal cavity, they normally flow back to the throat and we swallow them. When we have a cold, the nasolacrimal ducts become swollen and obstructed, the tears cannot drain, and they may overflow from the brim of the eye.

- The **extrinsic eye muscles** are the six muscles attached to the walls of the orbit and to the external surface of the eyeball. *Extrinsic* means arising externally; it distinguishes these from the *intrinsic* muscles inside the eyeball, to be considered later. The extrinsic muscles move the eye (fig. 16.22). They include four *rectus* ("straight") muscles and two *oblique* muscles. The **superior, inferior, medial,** and **lateral rectus** originate on the posterior wall of the orbit and insert on the anterior region of the eyeball, just beyond the visible "white of the eye." They move the eye up, down, medially, and laterally. The **superior oblique** travels along the medial wall of the orbit. Its tendon passes through a fibrocartilage ring, the **trochlea**[42] (TROCK-lee-uh), and inserts on the superolateral aspect of the eyeball. The **inferior oblique** extends from the medial wall of the orbit to the inferolateral aspect of the eye. To visualize the function of the oblique muscles, suppose you turn your eyes to the right. Your right eye will be abducted (face away from the nose) and the left eye will be adducted (face toward the nose). The primary function of the superior oblique is to depress the abducted eye, and the primary function of the inferior oblique is to elevate the adducted eye. A secondary function of the oblique muscles is to slightly rotate the eye (turn the "twelve o'clock pole" of the eye slightly toward or away from the nose). Most of these muscles are supplied by the oculomotor nerve (cranial nerve III), but the superior oblique is innervated by the trochlear nerve (IV) and the lateral rectus by the abducens (VI).

The eye is surrounded on the sides and back by **orbital fat.** It cushions the eye, gives it freedom of motion, and protects blood vessels and nerves as they pass through the rear of the orbit.

40. *car* = fleshly mass + *uncle* = little
41. *punct* = point
42. *trochlea* = pulley

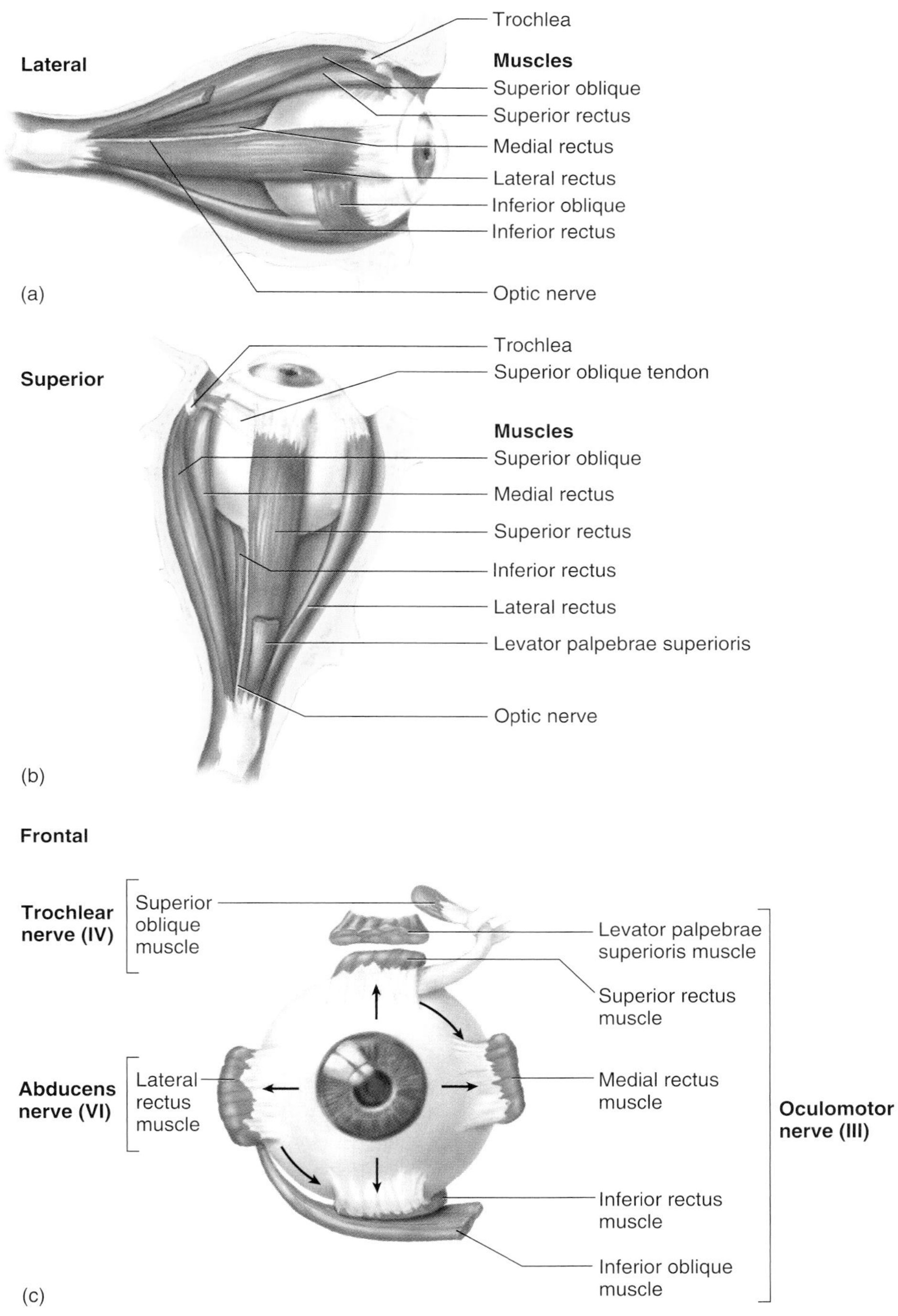

Figure 16.22 Extrinsic Muscles of the Eye. (*a*) Lateral view of the right eye. (*b*) Superior view of the right eye. (*c*) Innervation of the extrinsic muscles; *arrows* indicate the eye movement produced by each muscle.

Anatomy of the Eye

The eyeball itself is a sphere about 24 mm in diameter (fig. 16.23) with three principal components: (1) the *tunics,* three layers that form the wall of the eyeball; (2) the *optical apparatus,* which admits and focuses light; and (3) the *neural apparatus,* which consists of the retina and optic nerve. The retina is not only part of the neural apparatus but also the innermost tunic. The cornea is part of the outermost tunic as well as part of the optical apparatus.

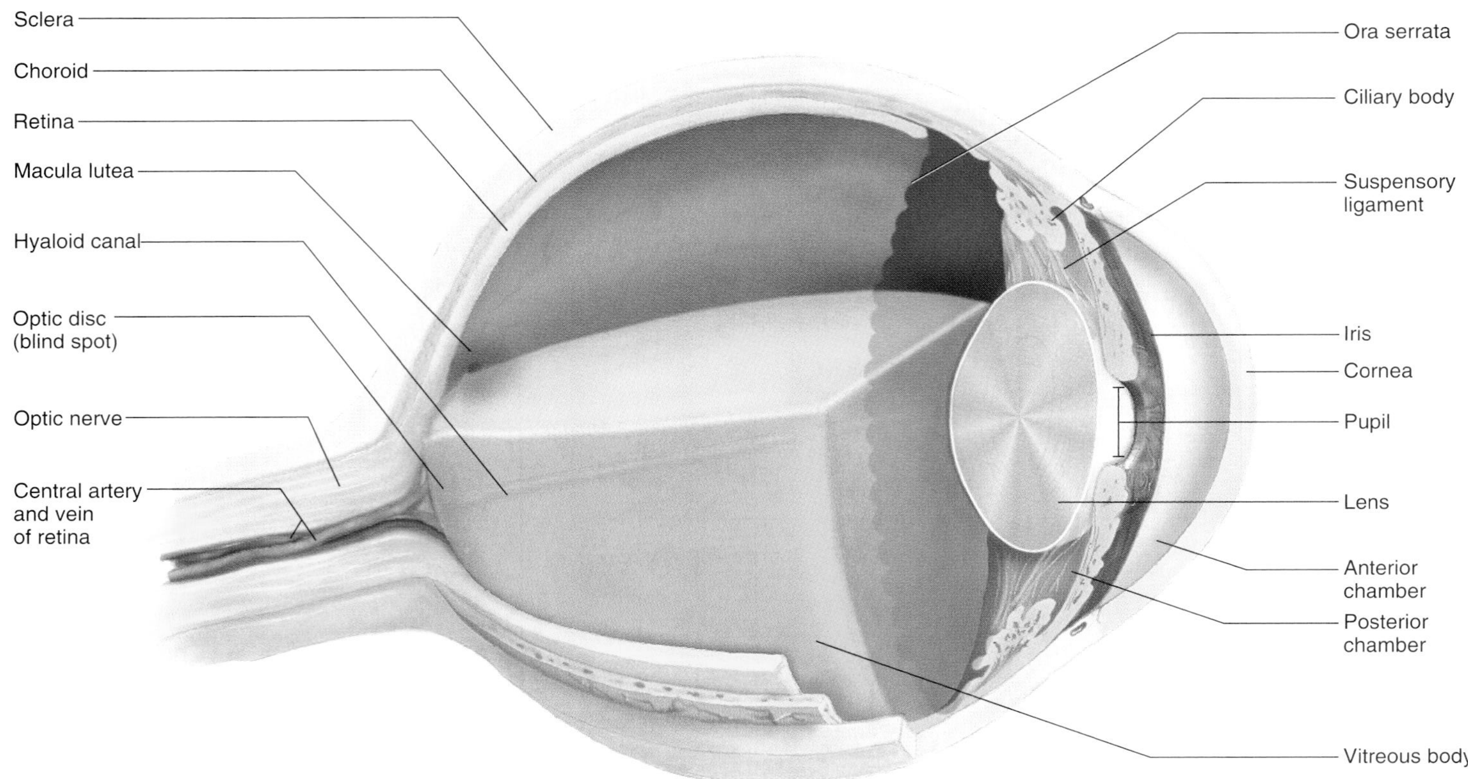

Figure 16.23 **Anatomy of the Eye in Sagittal Section.** The vitreous body has been omitted from the upper half to reveal structures behind it.

The Tunics

The three tunics of the eyeball are as follows:

- The **tunica fibrosa,** the outermost layer. This is divided into two regions: the sclera and cornea. The **sclera**[43] (white of the eye) covers most of the eye surface and consists of dense collagenous connective tissue perforated by blood vessels and nerves. The **cornea** is the transparent region of modified sclera that admits light into the eye.
- The **tunica vasculosa,** the middle layer. This is also called the **uvea**[44] (YOU-vee-uh) because it resembles a peeled grape in fresh dissection. It consists of three regions—the choroid, ciliary body, and iris. The **choroid** (CO-royd) is a highly vascular, deeply pigmented layer of tissue behind the retina. It gets its name from a histological resemblance to the chorion of the pregnant uterus. The **ciliary body,** a thickened extension of the choroid, forms a muscular ring around the lens. It supports the iris and lens and secretes a fluid called the aqueous humor. The **iris** is an adjustable diaphragm that controls the diameter of the **pupil,** its central opening. The iris has two pigmented layers. One is a posterior *pigment epithelium* that blocks stray light from reaching the retina. The other is the *anterior border layer,* which contains pigmented cells called **chromatophores.**[45] High concentrations of melanin in the chromatophores give the iris a black, brown, or hazel color. If the melanin is scanty, light reflects from the posterior pigment epithelium and gives the iris a blue, green, or gray color.
- The **tunica interna** consists of the *retina,* which lines the interior of the eyeball. It is the focus of most of the remainder of this chapter.

The Optical Apparatus

The **optical apparatus** of the eye consists of a series of transparent elements that admit light rays, bend (refract) them, and focus images on the retina. They include the *cornea, aqueous humor, lens,* and *vitreous body.* The cornea has been described already.

43. *scler* = hard, tough
44. *uvea* = grape

45. *chromato* = color + *phore* = bearer

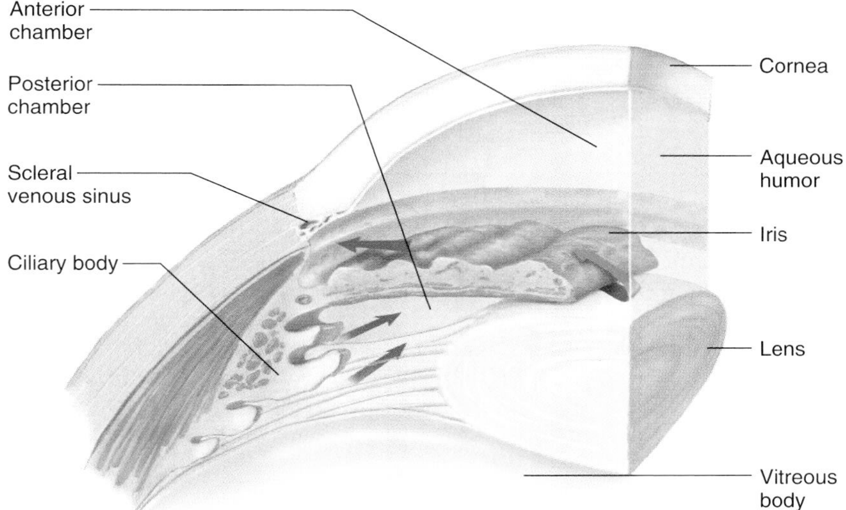

Figure 16.24 Production and Reabsorption of Aqueous Humor.

- The **aqueous humor** is a serous fluid secreted by the ciliary body into the **posterior chamber,** a space between the iris and lens (fig. 16.24). It flows through the pupil into the **anterior chamber** between the cornea and iris. From here, it is reabsorbed by a ringlike blood vessel called the **scleral venous sinus** (*canal of Schlemm*[46]). Normally the rate of reabsorption balances the rate of secretion (see clinical insight 16.4 for an important exception).
- The **lens** is suspended behind the pupil by a ring of fibers called the **suspensory ligament** (figs. 16.23 and 16.25), which attaches it to the ciliary body. Tension on the ligament somewhat flattens the lens so it is about 9.0 mm in diameter and 3.6 mm thick at the middle. When the lens is removed from the eye and not under tension, it relaxes into a more spheroid shape and resembles a plastic bead.
- The **vitreous**[47] **body** (*vitreous humor*) is a transparent jelly that fills the large space behind the lens. An oblique channel through this body, called the *hyaloid canal,* is the remnant of a *hyaloid artery* present in the embryo (see fig. 16.23).

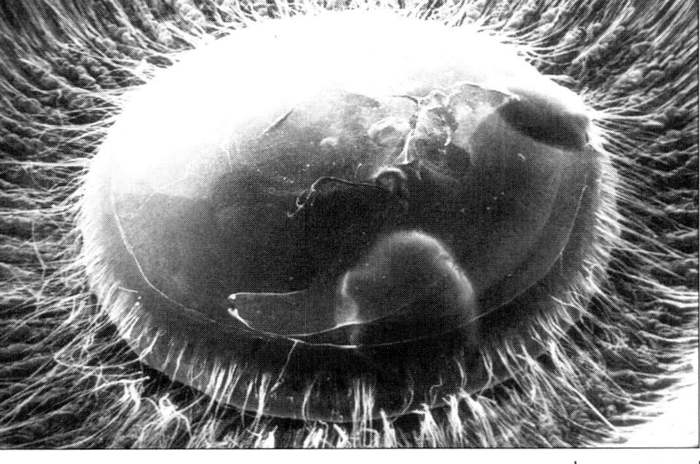

Figure 16.25 **The Lens of the Eye.** This is a posterior view of the lens and the suspensory ligaments that anchor it to the ciliary body.

The Neural Apparatus

The **neural apparatus** includes the retina and optic nerve. The retina forms from a cup-shaped outgrowth of the telencephalon (see chapter 14); it is actually a part of the brain—the only part that can be viewed without dissection. It is a thin transparent membrane attached at only two points—the **optic disc,** where the optic nerve leaves the rear *(fundus)* of the eye, and its scalloped anterior margin, the **ora serrata.** The retina is pressed smoothly against the rear of the eyeball by the pressure of the vitreous body. It can become detached (buckle away from the wall of the eyeball) by blows to the head or because of insufficient pressure from the vitreous body. A *detached retina* may cause blurry areas in the field of vision. It leads to blindness if the retina remains separated for too long from the chorion, on which it depends for oxygen, nutrition, and waste removal.

46. Friedrich S. Schlemm (1795–1858), German anatomist
47. *vitre* = glassy

Clinical Insight 16.4

Cataracts and Glaucoma

The two most common causes of blindness are cataracts and glaucoma. A *cataract* is a clouding of the lens. It occurs as the lens thickens with age, and it is a common complication of diabetes mellitus. It causes the vision to appear milky or as if one were looking from behind a waterfall.[48] Cataracts may also stem from heavy smoking and exposure to the UV radiation of the sun. They can be treated by replacing the natural lens with a plastic one. The implanted lens improves vision almost immediately, but glasses still may be needed for near vision.

Glaucoma is a state of elevated pressure within the eye that occurs when the scleral venous sinus is obstructed and aqueous humor is not reabsorbed as fast as it is secreted. Pressure in the anterior and posterior chambers drives the lens back and puts pressure on the vitreous body. The vitreous body presses the retina against the choroid and compresses the blood vessels that nourish the retina. Without a good blood supply, retinal cells die and the optic nerve may atrophy, producing blindness. Symptoms often go unnoticed until the damage is irreversible. In late stages, they include dimness of vision,[51] reduced visual field, and colored halos around artificial lights. Glaucoma can be halted with drugs or surgery, but lost vision cannot be restored. This disease can be detected at an early stage in the course of regular eye examinations. The field of vision is checked, the optic nerve is examined, and the intraocular pressure is measured with an instrument called a *tonometer.*

The retina is examined with an illuminating and magnifying instrument called an *ophthalmoscope.* Directly posterior to the center of the lens, on the visual axis of the eye, is a patch of cells called the **macula lutea**[49] about 3 mm in diameter (fig. 16.26). In the center of the macula is a tiny pit, the **fovea**[50] **centralis,** which produces the most finely detailed images for reasons that will be apparent later. About 3 mm medial to the macula lutea is the optic disc. Nerve fibers from all regions of the retina converge on this point and exit the eye to form the optic nerve. Blood vessels enter and leave the eye by way of the optic disc. Eye examinations serve for more than evaluating the visual system; they allow for a direct, noninvasive examination of blood vessels for signs of hypertension, diabetes mellitus, atherosclerosis, and other vascular diseases.

The optic disc contains no receptor cells, however, so it produces a **blind spot** in the visual field of each eye. You can detect your blind spot and observe an interesting visual phenomenon with the help of figure 16.27. Close or cover your right eye and hold the page about 30 cm (1 ft) from your face. Fixate on the X with your left eye. Without taking your gaze off the X, move the page slightly forward and back, or right and left, until the red dot disappears. This occurs because the image of the dot is falling on the blind spot of your left eye.

You should notice something else happen at the same time as the dot disappears—a phenomenon called

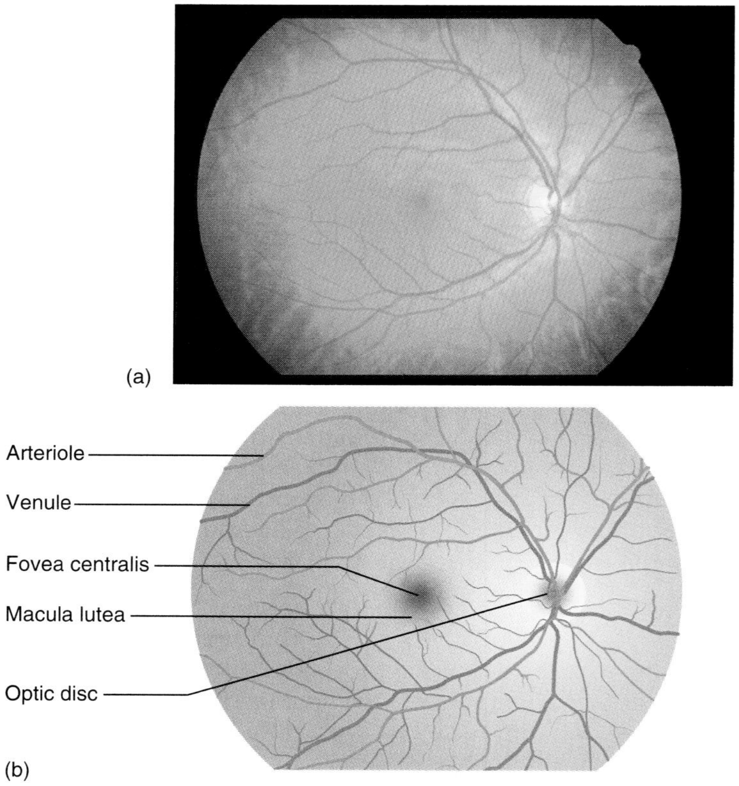

(a)

Arteriole
Venule
Fovea centralis
Macula lutea
Optic disc

(b)

Figure 16.26 The Fundus (rear) of the Eye. (*a*) As seen with an ophthalmoscope; (*b*) anatomical features of the fundus. Note the blood vessels diverging from the optic disc, where they enter the eye with the optic nerve. An eye examination also serves as a partial check on cardiovascular health.

48. *cataract* = waterfall
49. *macula* = spot + *lutea* = yellow
50. *fovea* = pit, depression
51. *glauco* = grayness

visual filling. The green bar seems to fill in the space where the dot used to be. This occurs because the brain uses the image surrounding the blind spot to fill in the area with similar, but imaginary, information. The brain acts as if it is better to assume that the unseen area probably looks like its surroundings than to allow a dark blotch to disturb your vision.

Before we begin to study visual physiology, it would be advisable to review and be sure you thoroughly understand the anatomy of the eye (table 16.3).

Figure 16.27 Demonstration of the Blind Spot and Visual Filling. See text for explanation of how to conduct this demonstration.

Formation of an Image

The visual process begins when light rays enter the eye, become focused on the retina, and produce a tiny inverted image. When fully dilated, the pupil admits five times as much light as it does when fully constricted. Its diameter is controlled by two sets of contractile elements in the iris: (1) The **pupillary constrictor** consists of concentric circles of smooth muscle cells around the pupil. When stimulated by the parasympathetic nervous system, it narrows the pupil and admits less light to the eye. (2) The **pupillary dilator** consists of a spokelike arrangement of modified contractile epithelial cells called *myoepithelial cells.* When stimulated by the sympathetic nervous system, these cells contract, widen the pupil, and admit more light to the eye (see fig. 15.22, p. 580). Pupillary constriction and dilation occur in two situations: when light intensity changes and when we shift our gaze between distant and nearby objects. Pupillary constriction in response to light

Table 16.3 Anatomical Review of the Eye and Accessory Structures

Accessory Organs (figs. 16.20 and 16.21)		
Eyebrows	*Conjunctiva*	*Extrinsic muscles* (fig. 16.22)
Eyelids (palpebrae)	*Lacrimal apparatus (fig. 16.21b)*	Superior oblique
Orbicularis oculi	Lacrimal gland	Trochlea
Tarsal plate	Lacrimal caruncle	Inferior oblique
Tarsal glands	Lacrimal puncta	Superior rectus
Medial and lateral commissures	Lacrimal canal	Inferior rectus
Eyelashes	Lacrimal sac	Lateral rectus
Palpebral fissure	Nasolacrimal duct	Medial rectus
		Orbital fat
Tunics of the Eye (fig. 16.23)		
Tunica fibrosa	*Tunica vasculosa (uvea)*	*Tunica interna (retina)*
Sclera	Choroid	
Cornea	Ciliary body	
	Iris	
Optical Apparatus (figs. 16.23–16.25)		
Cornea	Aqueous humor	Suspensory ligament
Pupil	Scleral venous sinus	Vitreous body
Anterior chamber	Lens	Hyaloid canal
Posterior chamber		
Neural Apparatus (fig. 16.26)		
Retina	Optic disc	Macula lutea
Optic nerve	Ora serrata	Fovea centralis

is called the **photopupillary reflex.** It is also described as a *consensual light reflex* because both pupils constrict even if only one eye is illuminated.

The photopupillary reflex is mediated by a parasympathetic reflex arc. When light intensity rises, signals are transmitted from the eye to the *pretectal region* just rostral to the tectum of the midbrain. Preganglionic parasympathetic fibers originate in the midbrain and travel by way of the oculomotor nerve to the ciliary ganglion in the orbit. From here, postganglionic parasympathetic fibers continue into the eye, where they stimulate the pupillary constrictor. Sympathetic innervation to the pupil originates, like all other sympathetic efferents, in the spinal cord. Preganglionic fibers lead from the thoracic cord to the superior cervical ganglion. From there, postganglionic fibers follow the carotid arteries into the head and lead ultimately to the pupillary dilator.

Refraction

Image formation depends on **refraction,** the bending of light rays. Light travels at a speed of 300,000 km/sec (186,000 mi/sec) in a vacuum, but it slows down slightly in air, water, glass, and other media. The *refractive index* of a medium (symbolized n) is a measure of how much it retards light rays relative to air. The refractive index of air is arbitrarily set at $n = 1.00$. If light traveling through air strikes a medium of higher refractive index at a 90° *angle of incidence,* it slows down but does not change course—the light rays are not bent. If it strikes at any other angle, however, the light is refracted (fig. 16.28*a*). The greater the difference in refractive index between the two media, and the greater the angle of incidence, the stronger the refraction is.

As light enters the eye, it passes from a medium with $n = 1.00$ (air) to one with $n = 1.38$ (the cornea). Light rays striking the very center of the cornea pass straight through, but because of the curvature of the cornea, rays striking off-center are bent toward the center (fig. 16.28*b*). The aqueous humor has a refractive index of 1.33 and does not greatly alter the path of the light. The lens has a refractive index of 1.40. As light passes from air to cornea, the refractive index changes by 0.38; but as it passes from aqueous humor to lens, the refractive index changes by only 0.07. Thus, the cornea refracts light more than the lens does. The lens merely fine-tunes the image, especially as you shift your focus between near and distant objects.

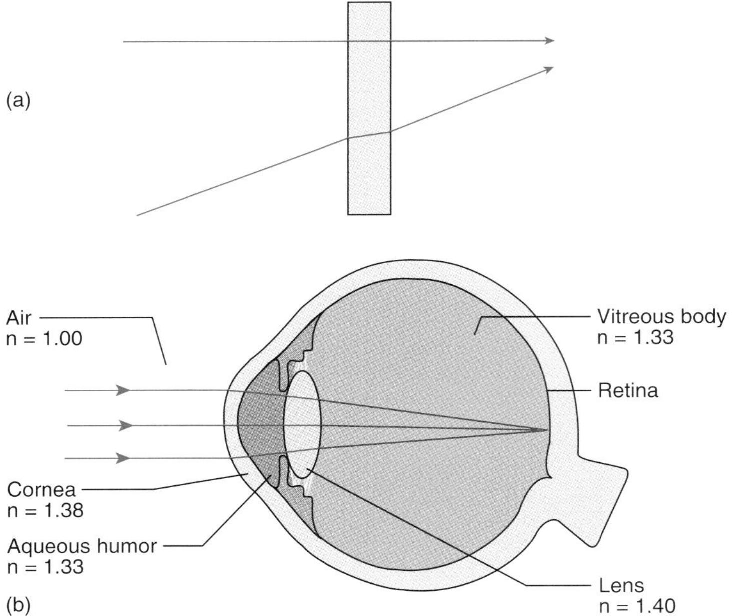

Figure 16.28 Principles of Refraction. (*a*) A refractive medium does not bend light rays that strike it at a 90° angle but does bend rays that enter or leave it at any other angle. (*b*) Refractive indices of the media from air to retina. The greater the difference between the refractive indices of two media, the more strongly light rays are refracted when passing from one to the other. In vision, most refraction (focusing) occurs as light passes from air to cornea.

The Near Response

Emmetropia[52] (EM-eh-TRO-pee-uh) is a state in which the eye is relaxed and focused on an object more than 6 m (20 ft) away, the light rays coming from that object are essentially parallel, and the rays are focused on the retina without effort. (An *emmetropic eye* does not need a corrective lens to focus the image.) If the gaze shifts to something closer, light rays from the source are too divergent to be focused without effort. In other words, the eye is automatically focused on things in the distance unless you make an effort to focus elsewhere. For a wild animal or our prehistoric ancestors, this arrangement would be adaptive because it allows for alertness to predators at a distance.

The **near response** (fig. 16.29), or adjustment to close-range vision, involves three processes to focus an image on the retina:

1. **Convergence of the eyes.** Move your finger gradually closer to a baby's nose and the baby will go cross-eyed. This **convergence** of the eyes orients the visual axis of each eye toward the object in order to focus its image on each fovea. If the eyes cannot converge accurately—for example, when the extrinsic muscles are weaker in one eye than in the other—double vision or *diplopia*[53] results. The images fall on different parts of the two retinas and

52. *em* = in + *metr* = measure + *opia* = vision

53. *dipl* = double + *opia* = vision

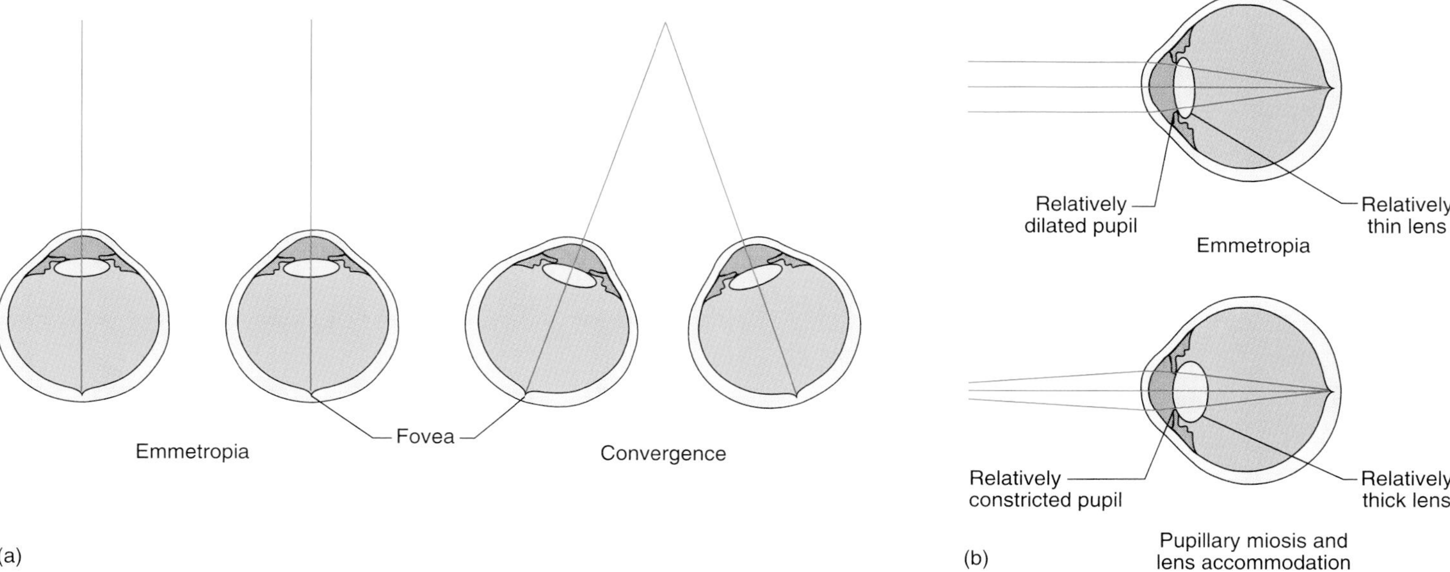

Figure 16.29 **Emmetropia and the Near Response.** (*a*) Superior view of both eyes fixated on an object more than 6 m away *(left)*, and both eyes fixated on an object closer than 6 m *(right)*. (*b*) Lateral view of the eye fixated on a distant object *(top)* and nearby object *(bottom)*.

the brain sees two images. You can simulate this effect by pressing gently on your eyelid as you look at this page; the image of the print will fall on noncorresponding regions of the two eyes and cause you to see double.

2. **Constriction of the pupil.** Lenses cannot refract light rays at their edges as well as they can closer to the center. The image produced by any lens is therefore somewhat blurry around the edges; this *spherical aberration* is quite evident in an inexpensive microscope. It can be minimized by screening out these peripheral light rays, and for this purpose, the pupil constricts as you focus on nearby objects. Like the diaphragm of a camera, the pupil thus has a dual purpose—to adjust the eye to variations in brightness and to reduce spherical aberration.
3. **Accommodation of the lens. Accommodation** is a change in the curvature of the lens that enables you to focus on a nearby object. When you look at something nearby, the ciliary muscle surrounding the lens contracts. This narrows the diameter of the ciliary body, relaxes the fibers of the suspensory ligament, and allows the lens to relax into a more convex shape (fig. 16.30). In emmetropia, the lens is about 3.6 mm thick at the center; in accommodation, it thickens to about 4.5 mm. A more convex lens refracts light more strongly and focuses the divergent light rays onto the retina. The closest an object can be and still come into focus is called the **near point of vision.** It depends on the flexibility of the lens. The lens stiffens with age, so the near point averages about 9 cm at the age of 10 and 83 cm by the age of 60.

Some common defects in image formation are listed in table 16.4.

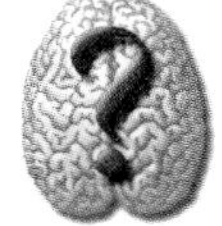

Think About It

Which extrinsic muscles of the eyes are the prime movers in convergence?

Sensory Transduction in the Retina

The conversion of light energy into action potentials occurs in the retina, a transparent layer of nervous tissue pressed against the back of the eye. We begin our exploration of this process with the cellular layout of the retina (fig. 16.31). From there we go to the pigments that absorb light and then to what happens when light is absorbed.

The most posterior layer of the retina is the **pigment epithelium,** a layer of darkly pigmented cuboidal cells whose basal processes interdigitate with receptor cells of the retina. The pigment here is not involved in nerve signal generation; rather, its purpose is to absorb light that is not absorbed first by the receptor cells and to prevent it from degrading the visual image by reflecting back into the eye. It acts like the blackened inside of a camera to reduce stray light.

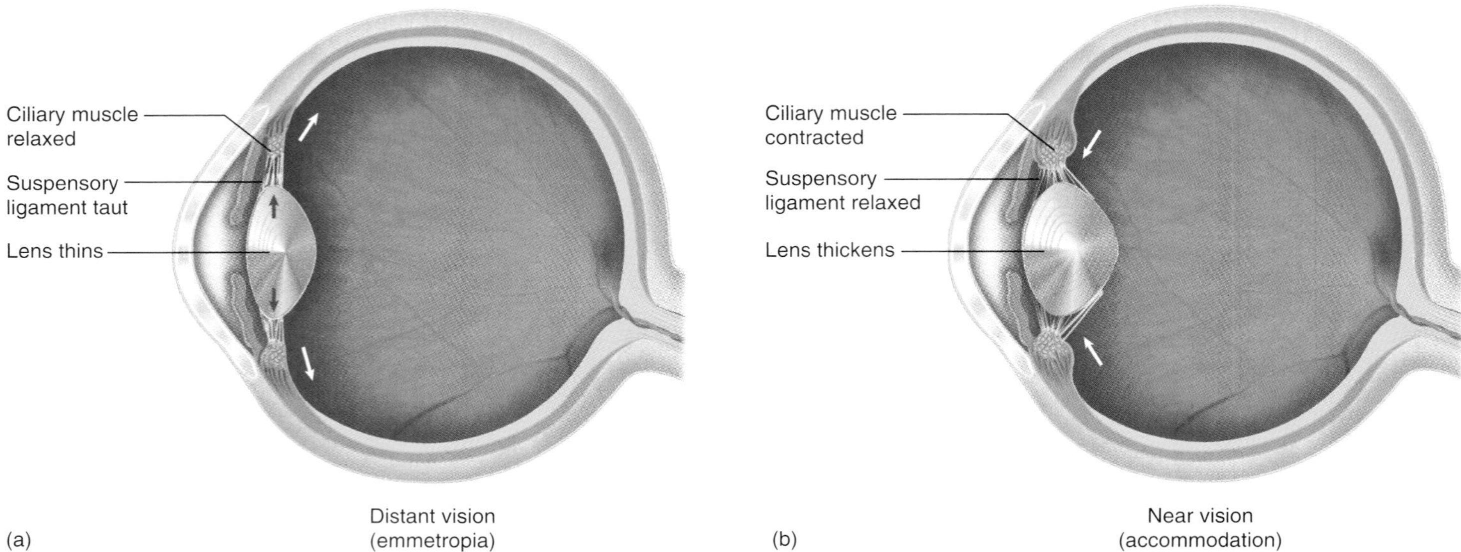

Figure 16.30 **Accommodation of the Lens.** (*a*) In the relaxed (emmetropic) eye, the ciliary muscle is relaxed and dilated. It puts tension on the suspensory ligament and flattens the lens. (*b*) In accommodation, the ciliary muscle contracts and narrows in diameter. This reduces tension on the suspensory ligament and allows the lens to relax into a more convex shape.

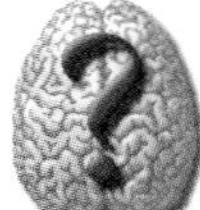

Think About It

The vertebrate eye is often called a *camera eye* for its many resemblances to the mechanisms of a camera. List as many comparisons as you can.

The neural apparatus of the retina consists of three principal cell layers. Progressing from the rear of the eye forward, these are composed of *photoreceptor cells, bipolar cells,* and *ganglion cells:*

1. **Photoreceptor cells.** The cells that absorb light and initiate visual signaling are the **rods** and **cones.** These are not neurons, but are derived from the same stem cells that produce ependymal cells of the brain. Each rod or cone has an **outer segment** that points toward the wall of the eye and an **inner segment** facing the interior (fig. 16.32). The two segments are separated by a narrow constriction containing a cylindrical array of nine pairs of microtubules; the outer segment is actually a highly modified cilium specialized to absorb light. The inner segment contains mitochondria and other organelles. At its base, it gives rise to a cell body, which contains the nucleus, and to processes that synapse with retinal neurons in the next layer.

   In a rod, the outer segment is cylindrical and somewhat resembles a stack of coins in a paper roll—there is a plasma membrane around the outside and a neatly arrayed stack of about 1,000 membranous discs inside. Each disc is densely studded with globular proteins—the visual pigment *rhodopsin,* to be discussed later. The membranes hold these pigment molecules in a position that results in the most efficient light absorption. Rod cells are responsible for **night (scotopic[54]) vision;** they cannot distinguish colors from each other.

   A cone cell is similar except that the outer segment tapers to a point and the discs are not detached from the plasma membrane but are parallel infoldings of it. Cones function in bright light; they are responsible for **day (photopic[55]) vision** as well as color vision.
2. **Bipolar cells.** Rods and cones synapse with the dendrites of **bipolar cells,** the first-order neurons of the visual pathway. They in turn synapse with the ganglion cells described next (see fig. 16.31*b*). There are approximately 130 million rods and 6.5 million cones in one retina, but only 1.2 million nerve fibers in the optic nerve. With a ratio of 114 receptor cells to 1 optic nerve fiber, it is obvious that there must be substantial *neuronal convergence* and information processing in the retina itself before signals are transmitted to the brain proper. Convergence begins with the bipolar cells.
3. **Ganglion cells. Ganglion cells** are the largest neurons of the retina, arranged in a single layer

54. *scot* = dark + *op* = vision
55. *phot* = light + *op* = vision

Table 16.4 Common Defects of Image Formation

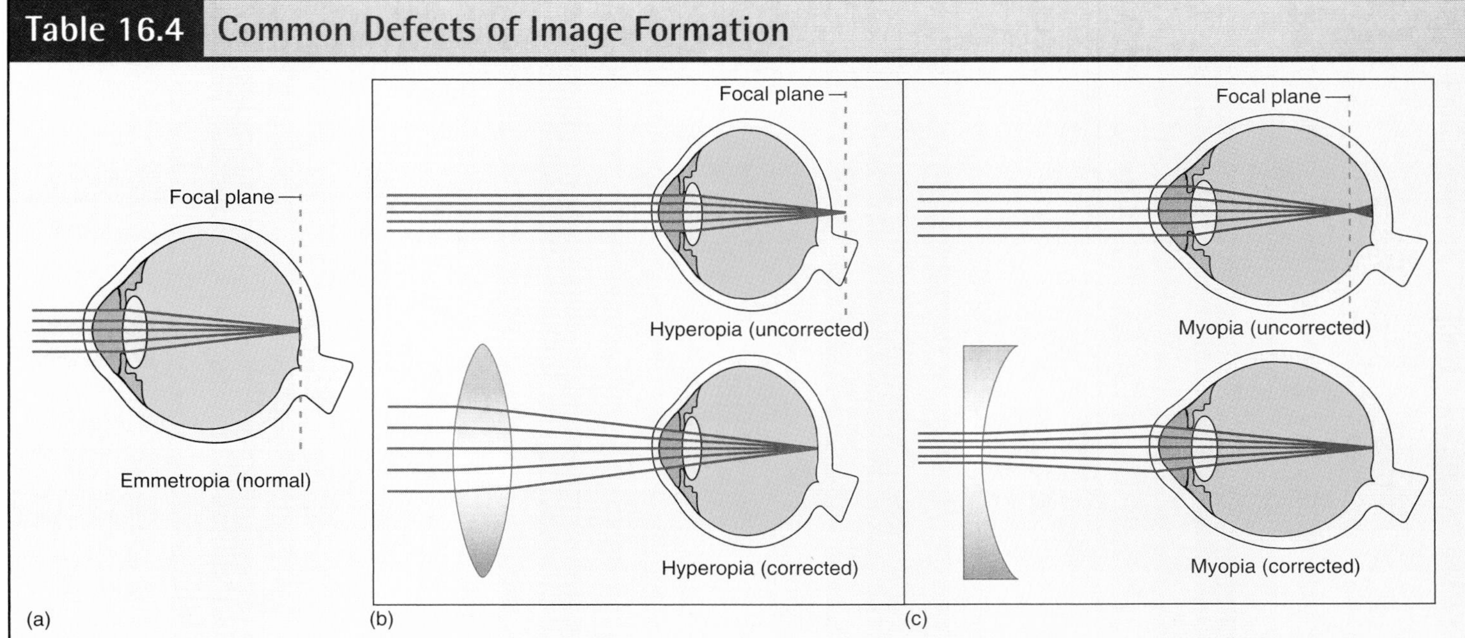

Two Common Visual Defects and the Effects of Corrective Lenses. (*a*) The normal emmetropic eye, with light rays converging on the retina. (*b*) Hyperopia (far-sightedness) and the corrective effect of a convex lens. (*c*) Myopia (near-sightedness) and the corrective effect of a concave lens.

Presbyopia	Reduced ability to accommodate for near vision with age because of declining flexibility of the lens. Results in difficulty in reading and doing close handwork. Corrected with bifocal lenses.
Hyperopia	Farsightedness—a condition in which the eyeball is too short. The retina lies in front of the focal point of the lens, and the light rays have not yet come into focus when they reach the retina (see top of fig. *b*). Causes the greatest difficulty when viewing nearby objects. Corrected with convex lenses, which cause light rays to converge slightly before entering the eye.
Myopia	Nearsightedness—a condition in which the eyeball is too long. Light rays come into focus before they reach the retina and begin to diverge again by the time they fall on it (see top of fig. *c*). Corrected with concave lenses, which cause light rays to diverge slightly before entering the eye.
Astigmatism	Inability to simultaneously focus light rays that enter the eye on different planes. Focusing on vertical lines, such as the edge of a door, may cause horizontal lines, such as a tabletop, to go out of focus. Caused by a deviation in the shape of the cornea so that it is shaped like the back of a spoon rather than like part of a sphere. Corrected with cylindrical lenses, which refract light more in one plane than another.

close to the vitreous body. They are the second-order neurons of the visual pathway. Most ganglion cells receive input from multiple bipolar cells. The ganglion cell axons form the optic nerve.

There are other retinal cells, but they do not form layers of their own. **Horizontal cells** and **amacrine**[56] **cells** form horizontal connections among rod, cone, and bipolar cells. They play diverse roles in enhancing the perception of contrast, the edges of objects, and changes in light intensity. In addition, much of the mass of the retina is composed of astrocytes and other types of glial cells.

56. *a* = without + *macr* = long + *in* = fiber

Visual Pigments

The visual pigment of the rods is called **rhodopsin** (ro-DOP-sin), or *visual purple.* Each molecule consists of two major parts (moieties)—a protein called **opsin** and a vitamin A derivative called **retinal** (rhymes with "pal"), also known as **retinene** (fig. 16.33). Opsin is embedded in the disc membranes of the rod's outer segment. All rod cells contain a single kind of rhodopsin with an absorption peak at a wavelength of 500 nm. The rods are less sensitive to light of other wavelengths.

In cones, the pigment is called **photopsin (iodopsin).** Its retinal moiety is the same as that of rhodopsin, but the opsin moieties have different amino acid sequences that

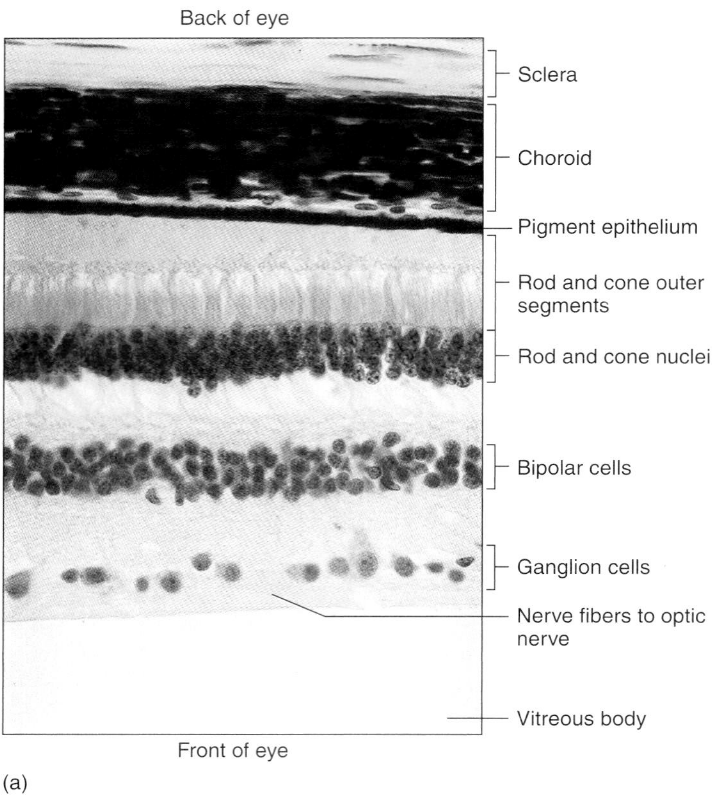

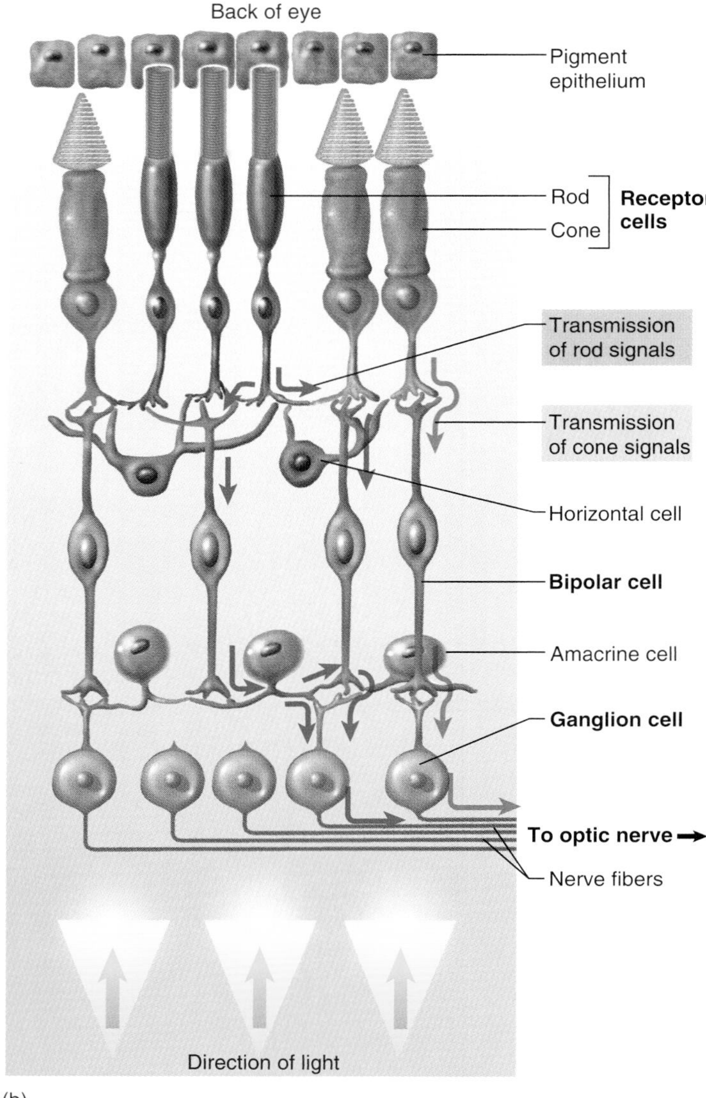

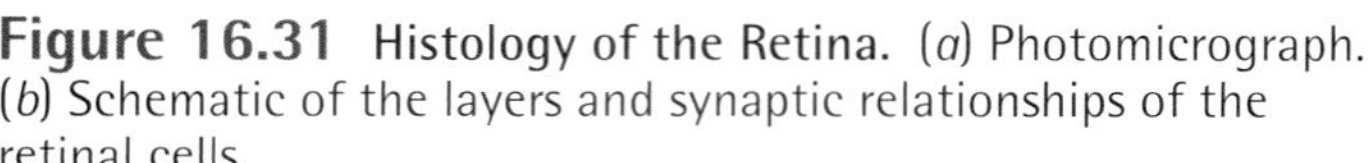
Figure 16.31 Histology of the Retina. (*a*) Photomicrograph. (*b*) Schematic of the layers and synaptic relationships of the retinal cells.

determine which wavelengths of light the pigment absorbs. There are three kinds of cones, which are identical in appearance but optimally absorb different wavelengths of light. These differences, as you will see shortly, enable us to perceive different colors.

The Photochemical Reaction

The events of sensory transduction are probably the same in rods and cones. We describe them here from the standpoint of rhodopsin, the visual pigment of the rods. In the dark, retinal has a bent shape called the 11-*cis* isomer, or ***cis*-retinal.** When it absorbs a photon of light, it is converted to a straight form called the all-*trans* isomer, or ***trans*-retinal** (fig. 16.34). This isomerization leads to the production of a nerve signal by a mechanism explained in the next section.

Rhodopsin can be chemically isolated from animal eyes in a darkroom. When viewed in the light, it briefly exhibits a violet color but quickly fades to colorless. This process, called **bleaching,** occurs in the eye as well as in the test tube. The loss of color occurs as the *trans*-retinal moiety dissociates from the opsin. For a rod to continue to function, it must regenerate rhodopsin at a rate that keeps pace with bleaching. When *trans*-retinal dissociates from rhodopsin, it is transported to the pigment epithelium at the rear of the retina, converted again to *cis*-retinal, transported back to the rod outer segment, and reunited with opsin. It takes about 5 minutes to regenerate 50% of the bleached rhodopsin. Cone cells are less dependent on the pigment epithelium and regenerate half of their pigment in about 90 seconds.

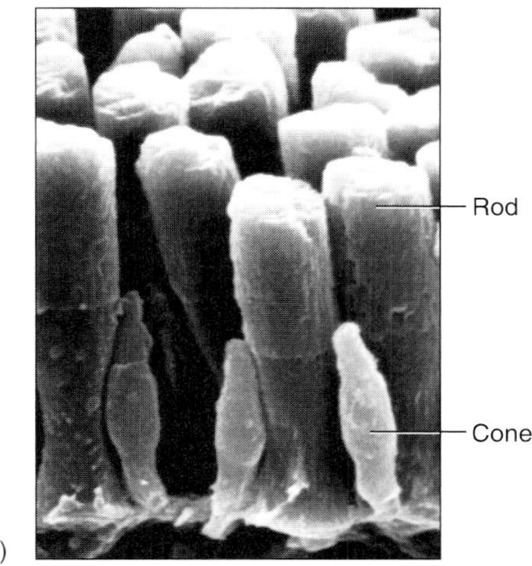

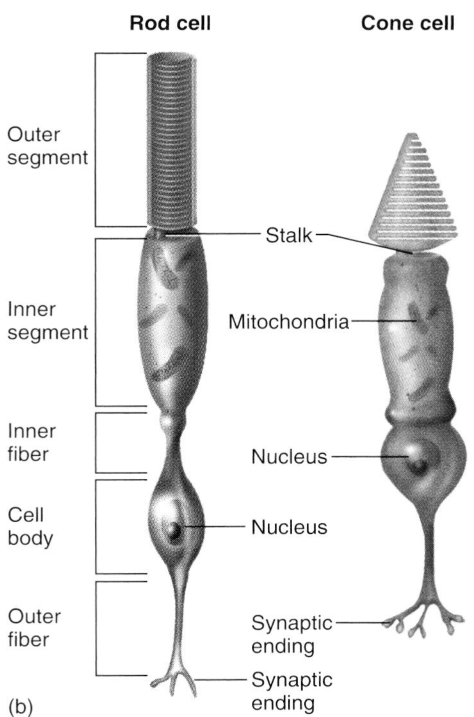

Figure 16.32 **Rod and Cone Cells.** (*a*) Rods and cones of a salamander retina (SEM). (*b*) Structure of rods and cones.

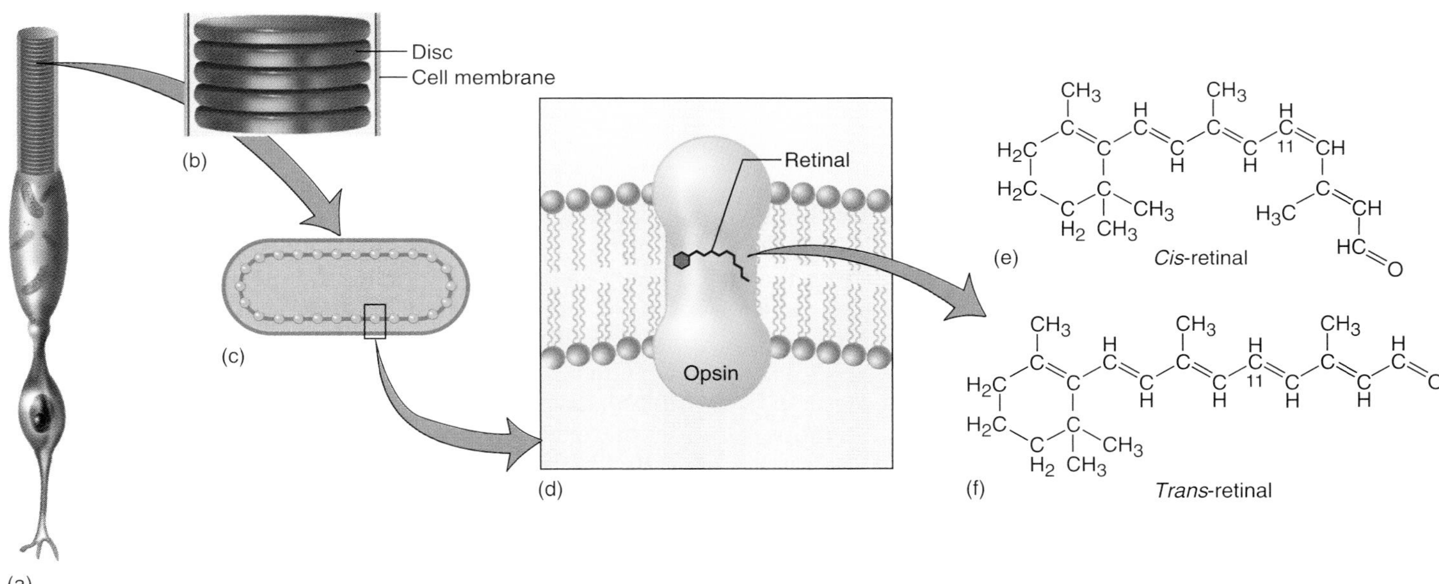

Figure 16.33 **Structure and Location of the Visual Pigments.** (*a*) A rod cell. (*b*) Detail of the rod outer segment. (*c*) One disc of the outer segment showing the membrane studded with pigment molecules. (*d*) A pigment molecule, embedded in the unit membrane of the disc, showing the protein moiety, opsin, and the vitamin A derivative, retinal. (*e*) *Cis*-retinal, the isomer present in the absence of light. (*f*) *Trans*-retinal, the isomer produced when the pigment absorbs a photon of light.

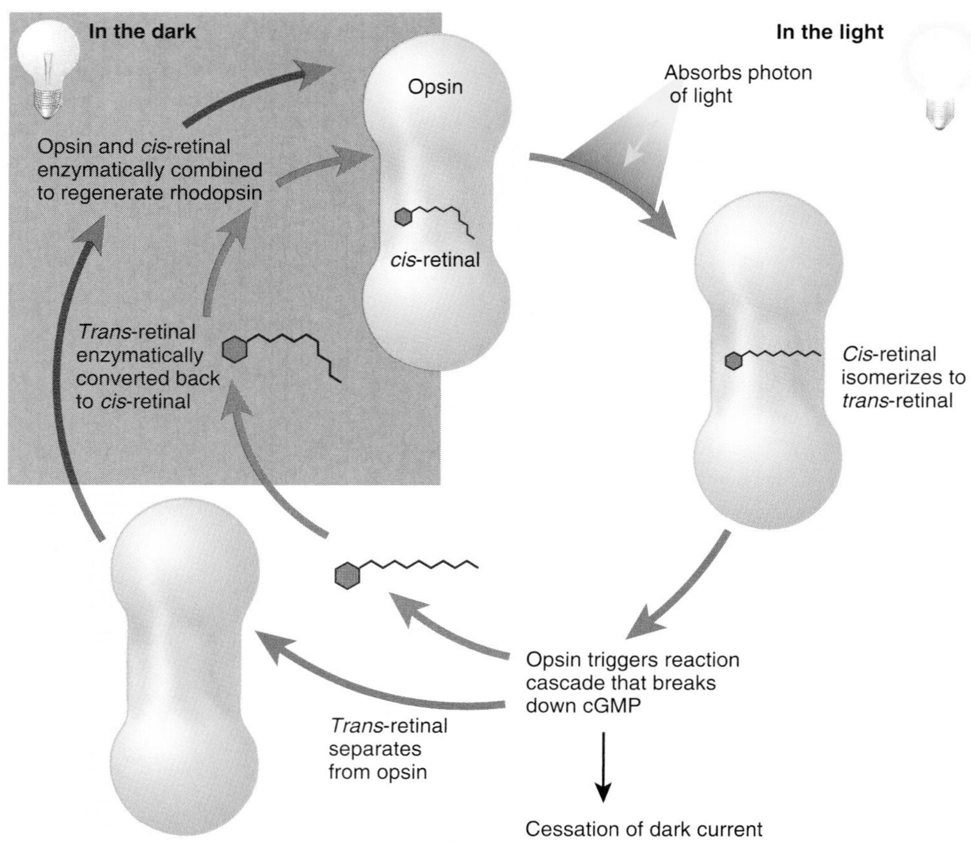

Figure 16.34 The Bleaching and Regeneration of Rhodopsin. The yellow background indicates the bleaching events that occur in the light; the gray background indicates the regenerative events that are independent of light. The latter events occur in light and dark but are able to outpace bleaching only in the dark.

Generating the Optic Nerve Signal

In the absence of light, rod cells do not sit quietly doing nothing. They produce a steady ion flow called the **dark current.** The production of a nerve signal stems from the fact that light *shuts off* this dark current. Here we will see how the current is produced and how light deactivates it (fig. 16.35).

The outer segment of a rod cell has ligand-regulated Na^+ gates. They open when cyclic guanosine monophosphate (cGMP) binds to their intracellular side. Sodium then diffuses into the cell and partially depolarizes it. The expression "dark current" refers to this sodium inflow. The inner segment of the rod cell has nongated K^+ channels which allow K^+ to diffuse out continually. It also has a high density of Na^+-K^+ pumps, which compensate for both the Na^+ and K^+ flows by continually pumping Na^+ out and K^+ back into the cell. As long as these processes continue and the rod is depolarized, the rod releases glutamate from its synaptic end. Its effect on the bipolar cell is considered shortly.

When a rod absorbs light, the dark current ceases. The intact rhodopsin molecule is essentially a dormant enzyme. When it bleaches, it becomes active and triggers a cascade of chemical reactions that ultimately break down several hundred thousand molecules of cGMP. This amplifying effect—the ability of one rhodopsin molecule to affect so many cGMP molecules—makes rod cells extremely sensitive to light. As cGMP is degraded, the Na^+ gates in the outer segment close, the dark current ceases, and glutamate secretion is halted.

There are two kinds of bipolar cells. One type is inhibited (hyperpolarized) by glutamate and thus excited (depolarized) when its secretion drops. This type of cell is excited by rising light intensity. The other type is excited by glutamate and inhibited when its secretion drops, so it is excited by falling light intensity. As your eye scans a scene, it passes areas of greater and lesser brightness. Their images on the retina cause a rapidly changing pattern of bipolar cell responses as the light intensity on a patch of retina rises and falls.

When bipolar cells detect fluctuations in light intensity, they stimulate the ganglion cells. Each ganglion cell receives input from a circular patch of retina called its **receptive field.** The principal function of a ganglion cell is to code for contrast between the center and edge of its receptive field—that is, between an object and its surroundings. Ganglion cells are the only retinal cells that produce

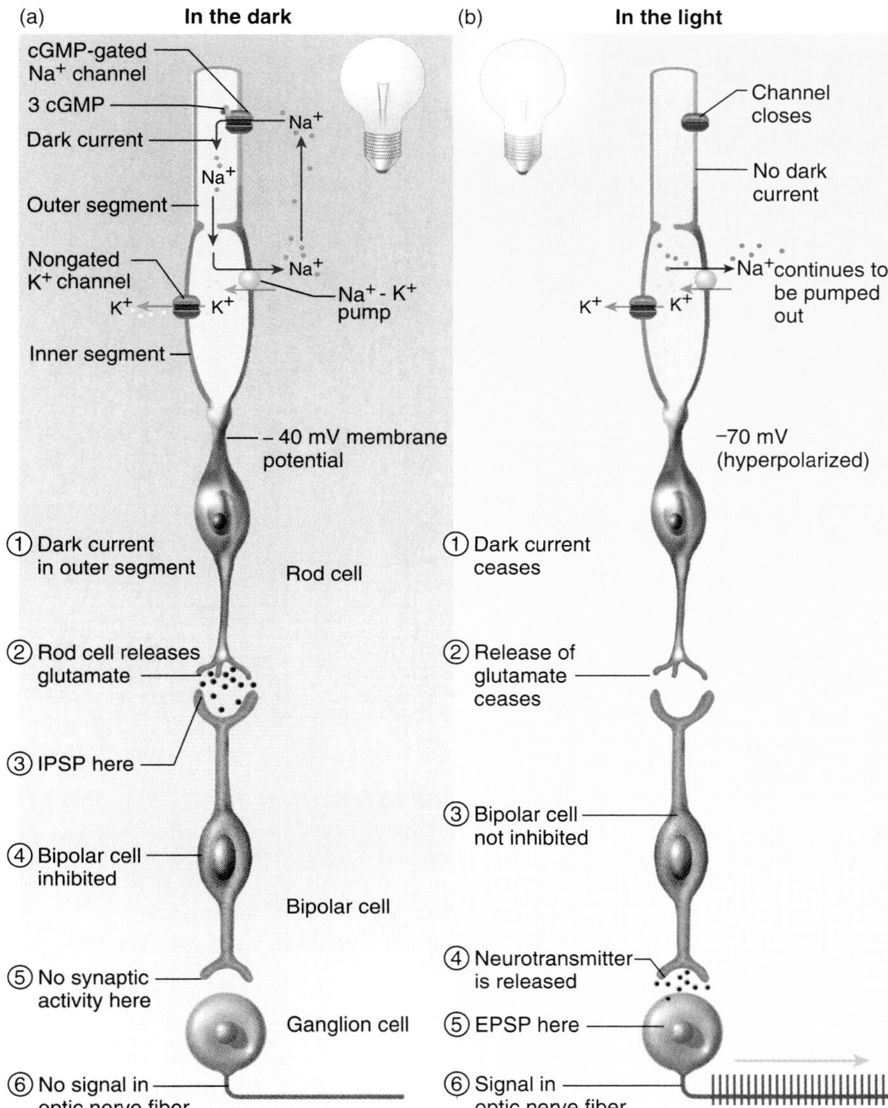

Figure 16.35 Mechanism of Generating Visual Signals. (*a*) In the dark, a dark current in the rod cells stimulates glutamate release, which inhibits the production of nerve signals. (*b*) In the light, cGMP breaks down and shuts off the dark current and glutamate secretion. Bipolar cells then stimulate ganglion cells, which produce an optic nerve signal.

action potentials; all other retinal cells produce only graded local potentials. The ganglion cells respond with rising and falling firing frequencies which, via the optic nerve, provide the brain with a basis for interpreting the image on the retina.

Light and Dark Adaptation

Light adaptation occurs when you go from the dark into bright light. If you wake up in the middle of the night and turn on a lamp, at first you see a harsh glare; you may even experience pain from the overstimulated retinas. Your pupils quickly constrict to reduce the intensity of stimulation, but color vision and visual acuity (the ability to see fine detail) do not become optimal for 5 to 10 minutes—the time needed for pigment bleaching to adjust retinal sensitivity to this light intensity. The rods bleach quickly in bright light, and cones take over. Even in typical indoor light, rod vision is nonfunctional.

On the other hand, suppose you are sitting in a bright room watching television at night, and there is a power failure. Your eyes must undergo **dark adaptation** before you can see well enough to find your way in the dark. Your rod pigment was bleached by the lights in the room while the power was on, but now in the relative absence of light, rhodopsin regenerates faster than it bleaches. In 20 to 30 minutes, the amount of rhodopsin is sufficient for your eyes to have reached essentially the maximum possible sensitivity in the dark. Dilation of the pupils also helps by admitting more light to the eye.

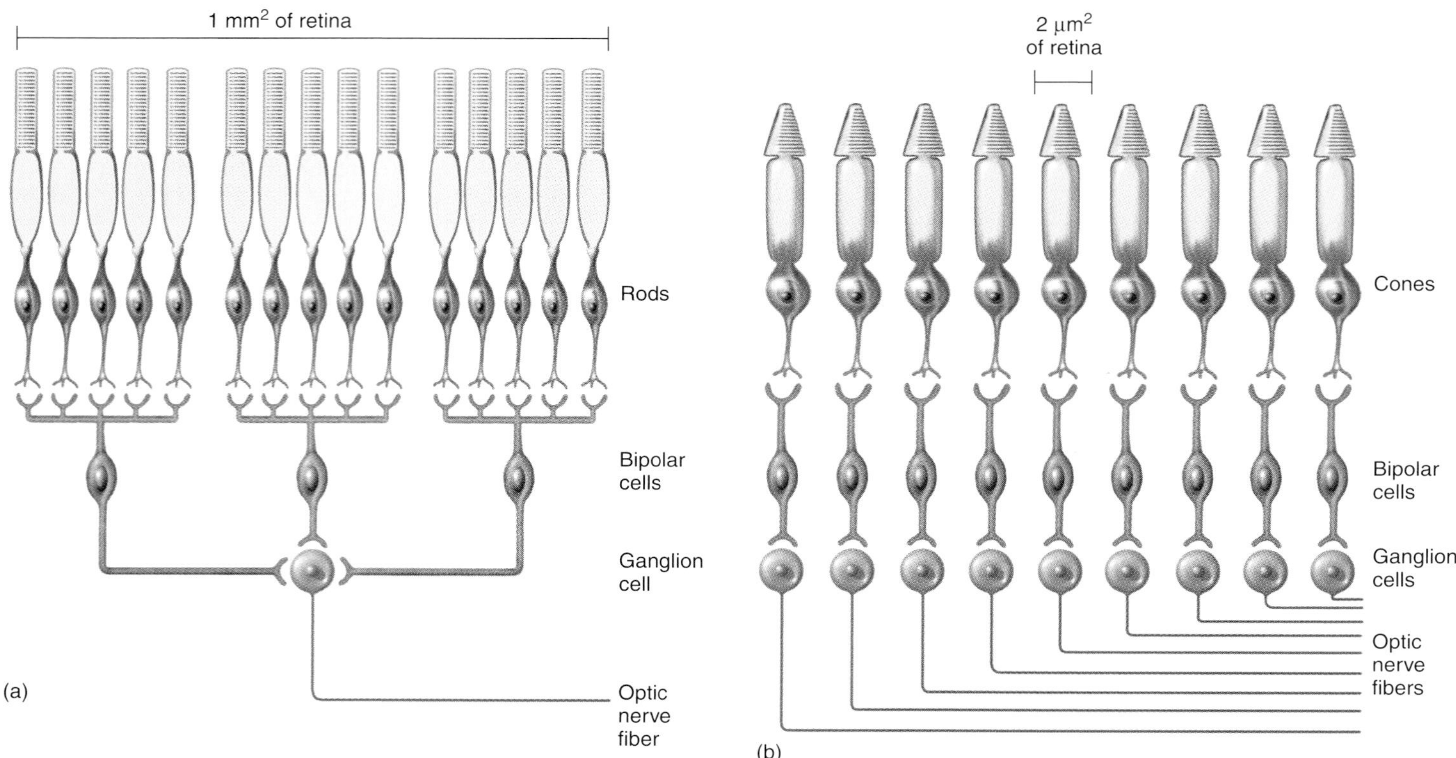

Figure 16.36 **The Duplicity Theory of Vision.** (*a*) In the scotopic (night vision) system, many rod cells converge on each bipolar cell and many bipolar cells converge on each ganglion cell. This allows extensive spatial summation—many rods add up their effects to stimulate a ganglion cell even in dim light. However, it means that each ganglion cell (and its optic nerve fiber) represents a relatively large area of retina and produces a grainy image. (*b*) In the photopic (day vision) system, there is little neuronal convergence. In the fovea, represented here, each cone cell has a "private line" to the brain, so each optic nerve fiber represents a tiny area of retina, and vision is relatively sharp. However, the lack of convergence prevents spatial summation. Photopic vision does not function in dim light because weakly stimulated cone cells cannot collaborate to stimulate a ganglion cell.

The Duplicity Theory

You may wonder why we have two types of photoreceptor cells, the rods and cones. Why can't we simply have one type that would produce detailed color vision, both day and night? The **duplicity theory** of vision holds that a single type of receptor cell cannot produce both high sensitivity and high resolution. It takes one type of cell and neuronal circuit to provide sensitive night vision and a different type of receptor and circuit to provide high-resolution daytime vision.

The high sensitivity of rods in dim light stems partly from the cascade of reactions leading to cGMP breakdown described earlier, but it is also due to the extensive neuronal convergence that occurs between the rods and ganglion cells. Up to 600 rods converge on each bipolar cell, and many bipolar cells converge on each ganglion cell. This allows for a high degree of *spatial summation* in the scotopic system (fig. 16.36*a*). Weak stimulation of many rod cells can produce an additive effect on one bipolar cell, and several bipolar cells can collaborate to excite one ganglion cell. Thus, a ganglion cell can respond in dim light that only weakly stimulates any individual rod. Scotopic (rod) vision is functional even at a light intensity less than starlight reflected from a sheet of white paper. A shortcoming of this system is that it cannot resolve finely detailed images. One ganglion cell receives input from all the rods in about 1 mm^2 of retina—its receptive field. What the brain perceives is therefore a coarse, grainy image similar to an overenlarged newspaper photograph.

Around the edges of the retina, receptor cells are especially large and widely spaced. If you fixate on the middle of this page, you will notice that you cannot read the words near the margins. Visual acuity decreases rapidly as the image falls away from the fovea centralis. Our peripheral vision is a low-resolution system that serves mainly to alert us to motion in the periphery and to stimulate us to look that way to identify what is there.

When you look directly at something, its image falls on the fovea, which is occupied by about 4,000 tiny cones and no rods. The other neurons of the fovea are displaced to one side so they won't interfere with light falling on the cones.

The smallness of these cones is like the smallness of the dots in a high-quality photograph; it is partially responsible for the high-resolution images formed at the fovea. In addition, the cones here show no neuronal convergence. Each cone synapses with only one bipolar cell and each bipolar cell with only one ganglion cell. This gives each foveal cone a "private line to the brain," and each ganglion cell of the fovea reports to the brain on a receptive field of just 2 μm^2 of retinal area (fig. 16.36*b*). Cones distant from the fovea exhibit some neuronal convergence but not nearly as much as rods do. The price of this lack of convergence at the fovea, however, is that cone cells have little spatial summation, and the cone system therefore has less sensitivity to light. The threshold of photopic (cone) vision lies between the intensity of starlight and moonlight reflected from white paper.

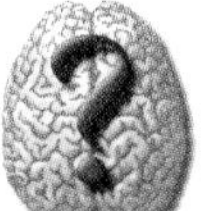

Think About It

If you look directly at a dim star in the night sky, it disappears, and if you look slightly away from it, it reappears. Why?

Color Vision

Most nocturnal vertebrates have only rod cells, but many diurnal animals are endowed with cones and color vision. Color vision is especially well developed in primates for evolutionary reasons discussed in chapter 1. It is based on three kinds of cones named for the absorption peaks of their photopsins: **blue cones,** with peak sensitivity at 420 nm; **green cones,** which peak at 531 nm; and **red cones,** which peak at 558 nm. Red cones do not peak in the red part of the spectrum (558 nm light is perceived as orange-yellow), but they are the only cones that respond at all to red light. Our perception of different colors is based on a mixture of nerve signals representing cones with different absorption peaks. In figure 16.37, note that light at 400 nm excites only the blue cones, but at 500 nm, all three types of cones are stimulated. The red cones respond at 60% of their maximum capacity, green cones at 82% of their maximum, and blue cones at 20%. The brain interprets this mixture of signals as blue-green. The table in figure 16.37 shows how some other color sensations are generated by other response ratios.

Some individuals have a hereditary lack of one photopsin or another and consequently exhibit **color blindness.** The most common form is *red-green color blindness,* which results from a lack of either red or green cones and renders a person incapable of distinguishing these and related shades from each other. For example, a person with normal *trichromatic* color vision sees figure 16.38 as the number 16, whereas a person with red-green color blindness sees no number. Red-green color blindness is a sex-linked recessive trait. It occurs in about 8% of males and 0.5% of females. (See p. 163 to review sex linkage and the reason such traits are more common in males.)

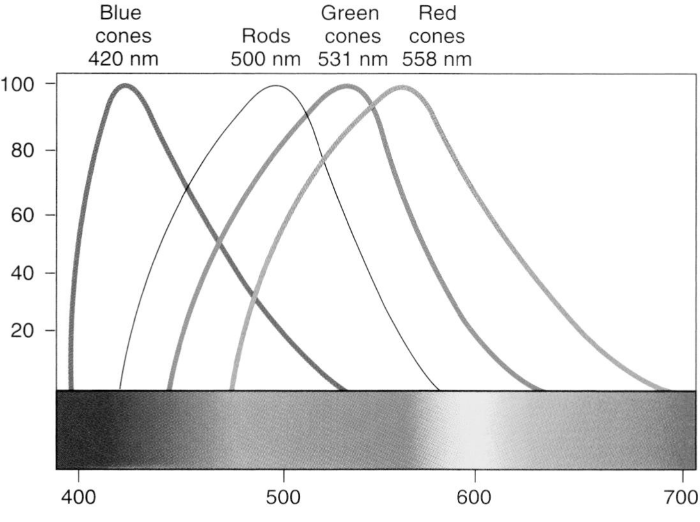

Wavelength (nm)	Percent of maximum cone response (red:green:blue)	Perceived hue
400	0:0:50	Violet
450	0:30:72	Blue
500	60:82:20	Blue-green
550	97:85:0	Green
625	35:3:0	Orange
675	5:0:0	Red

Figure 16.37 Absorption Spectra of the Retinal Cells. In the *middle column* of the table, each number indicates how strongly the respective cone cells respond as a percentage of their maximum capability. At 550 nm, for example, red cones respond at 97% of their maximum, green cones at 85%, and blue cones not at all. The result is a perception of green light.

Stereoscopic Vision

Stereoscopic vision (stereopsis) is depth perception—the ability to judge how far away objects in the visual field are. It depends on having two eyes with overlapping visual fields, which allows each eye to look at the same object from a different angle. Stereoscopic vision contrasts with the *panoramic vision* of mammals such as rodents and horses, where the eyes are on opposite sides of the head. Although stereoscopic vision covers a smaller visual field than panoramic vision and provides less alertness to sneaky predators, it has the advantage of depth perception. The evolutionary basis of depth perception in primates was considered in chapter 1 (p. 11).

When you fixate on something within 30 m (100 ft) away, each eye views it from a slightly different angle and focuses its image on the fovea centralis. The point on which the eyes are focused is called the *fixation point.* Objects farther away than the fixation point cast an image somewhat medial to the foveas, and closer objects cast their images

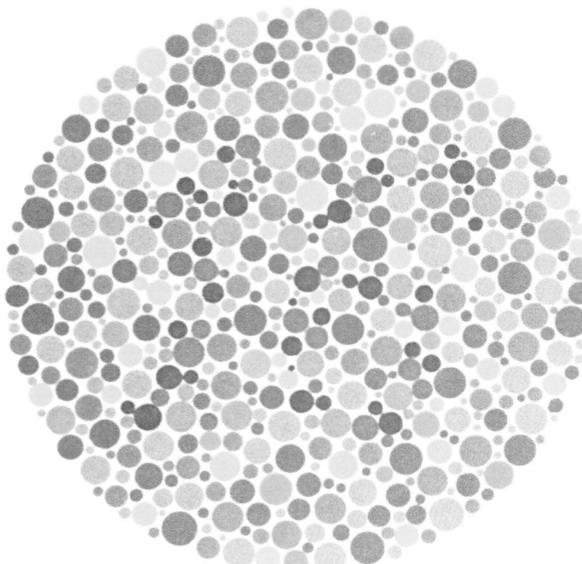

Figure 16.38 **A Test for Red-Green Color Blindness.** Persons with normal vision see the number 16. Persons with red-green color blindness see no discernible number. The above has been reproduced from Ishihara's Tests for Colour Blindness published by KANEHARA Trading Co., Tokyo, Japan, but tests for color blindness cannot be conducted with this material. For accurate testing, the original plates should be used. Copyright © Isshin-Kai Foundation.

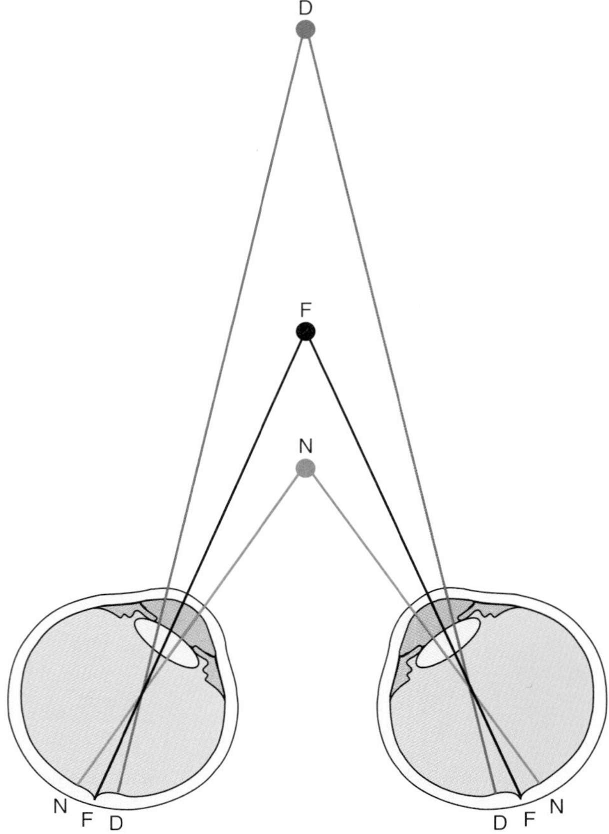

Figure 16.39 **The Retinal Basis of Stereoscopic Vision (depth perception).** When the eyes are fixated on the fixation point (*F*), more distant objects (*D*) are focused on the retinas medial to the fovea and the brain interprets them as being farther away than the fixation point. Nearby objects (*N*) are focused lateral to the fovea and interpreted as being closer.

more laterally (fig. 16.39). The distance of an image from the two foveas provides the brain with information used to judge the position of other points relative to the fixation point.

The Visual Projection Pathway

The first-order neurons in the visual pathway are the bipolar cells of the retina. They synapse with the second-order neurons, the retinal ganglion cells, whose axons are the fibers of the optic nerve. The optic nerves leave each orbit through the optic foramen and then converge on each other to form an X, the **optic chiasm**[57] (ky-AZ-um), immediately inferior to the hypothalamus and anterior to the pituitary. Beyond this, the fibers continue as a pair of **optic tracts** (see p. 543). Within the chiasm, half the fibers of each optic nerve cross over to the opposite side of the brain (fig. 16.40). This is called **hemidecussation,**[58] since only half of the fibers decussate. As a result, objects in the left visual field, whose images fall on the right half of each retina (the medial half of the left eye and lateral half of the right eye), are perceived by the right cerebral hemisphere. Objects in the right visual field are perceived by the left hemisphere. Since the right brain controls motor responses on the left side of the body and vice versa, each side of the brain needs to see what is on the side of the body where it exerts motor control. In animals with panoramic vision, nearly 100% of the optic nerve fibers of the right eye decussate to the left brain and vice versa.

The optic tracts pass laterally around the hypothalamus, and most of their axons end in the **lateral geniculate**[59] (jeh-NIC-you-late) **nucleus** of the thalamus. Third-order neurons arise here and form the **optic radiation** of fibers in the white matter of the cerebrum. These project to the primary visual cortex of the occipital lobe, where the conscious visual sensation occurs. A lesion in the occipital lobe can cause blindness even if the eyes are fully functional.

A few optic nerve fibers take a different route in which they project to the midbrain and terminate in the superior colliculi and pretectal nuclei. The superior colliculi control the visual reflexes of the extrinsic eye muscles, and the pretectal nuclei are involved in the photopupillary and accommodation reflexes.

57. *chiasm* = cross, X
58. *hemi* = half + *decuss* = to cross, form an X
59. *geniculate* = bent like a knee

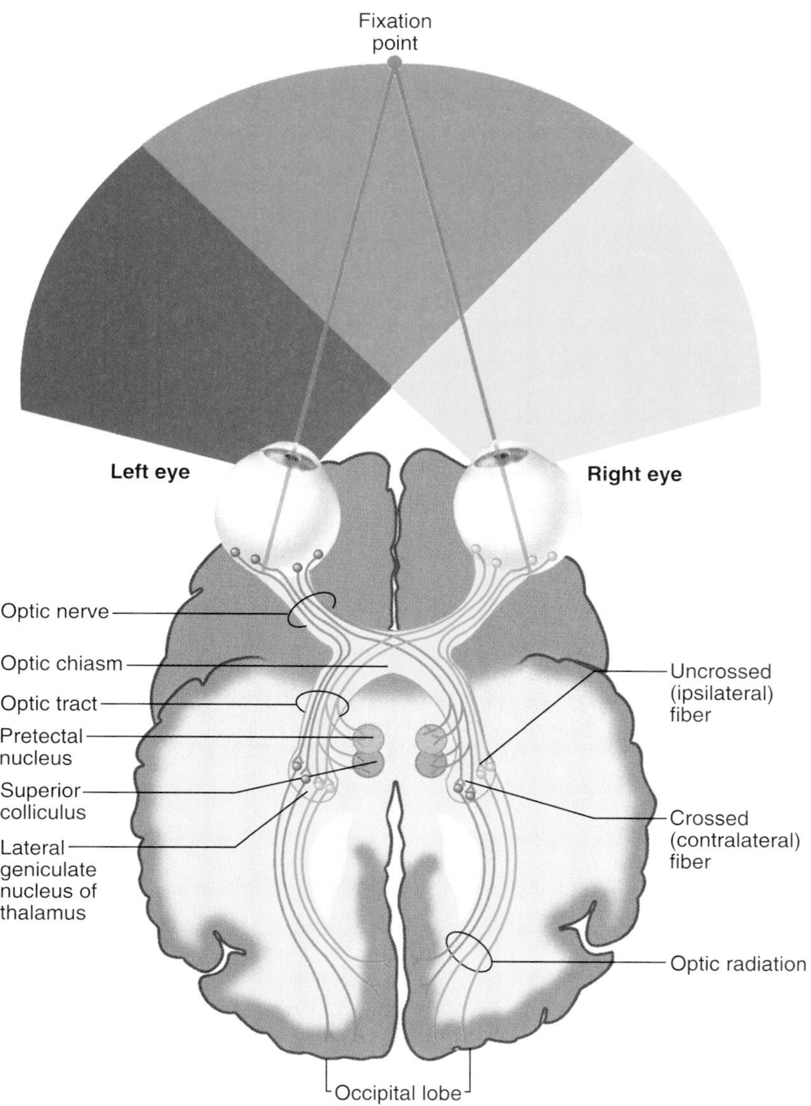

Figure 16.40 **The Visual Projection Pathway.** Diagram of hemidecussation and projection to the primary visual cortex.

Space does not allow us to consider much about the very complex processes of visual information processing in the brain. Some processing, such as contrast, brightness, motion, and stereopsis, begins in the retina. The primary visual cortex in the occipital lobe is connected by association tracts to nearby visual association areas in the posterior part of the parietal lobe and inferior part of the temporal lobe. These association areas process retinal data in ways beyond our present consideration to extract information about the location, motion, color, shape, boundaries, and other qualities of the objects we look at. They also store visual memories and enable the brain to identify what we are seeing—for example, to recognize printed words or name the objects we see. What is yet to be learned about visual processing promises to have important implications for biology, medicine, psychology, and even philosophy.

Before You Go On

Answer the following questions to test your understanding of the preceding section:

21. Why can't we see wavelengths below 350 nm or above 750 nm?
22. Why are light rays bent (refracted) more by the cornea than by the lens?
23. List as many structural and functional differences between rods and cones as you can.
24. Explain how the absorption of a photon of light leads to depolarization of a bipolar retinal cell.
25. Discuss the duplicity theory of vision summarizing the advantage of having two types of retinal photoreceptor cells.

Historical Insight 16.5

Anesthesia—From Ether Frolics to Modern Surgery

Surgery is as old as civilization. People from the Stone Age to the pre-Columbian civilizations of the Americas practiced *trephination*—cutting a hole in the skull to let out "evil spirits" that were thought to cause headaches. The ancient Hindus were expert surgeons for their time, and the Greeks and Romans pioneered military surgery. But until the nineteenth century, surgery was a miserable and dangerous business, done only as a last resort and with little hope of the patient's survival. Surgeons rarely attempted anything more complex than amputations or kidney stone removal. A surgeon had to be somewhat indifferent to the struggles and screams of his patient. Most operations had to be completed in 3 minutes or less, and a strong arm and stomach were more important qualifications for a surgeon than extensive anatomical knowledge.

At least three things were needed before surgery could be effective: better knowledge of anatomy, *asepsis*[60] for the control of infection, and *anesthesia*[61] for the control of pain. Early efforts to control surgical pain were crude and usually ineffective, such as choking a patient into unconsciousness and trying to complete the surgery before he or she awoke. Alcohol and opium were often used as anesthetics, but the dosage was poorly controlled; some patients were underanesthetized and suffered great pain anyway, and others died of overdoses. Often there was no alternative but for a few strong men to hold the struggling patient down as the surgeon worked. Charles Darwin originally intended to become a physician, but left medical school because he was sickened by observing "two very bad operations, one on a child," in the days before anesthesia.

In 1799, Sir Humphrey Davy suggested using nitrous oxide to relieve pain. His student, Michael Faraday, suggested ether. Neither of these ideas caught on for several decades, however. Nitrous oxide ("laughing gas") was a popular amusement in the 1800s, when traveling showmen went from town to town demonstrating its effects on volunteers from the audience. In 1841, at a medicine show in Georgia, some students were impressed with the volunteers' euphoric giggles and antics and asked a young physician in town, Crawford W. Long, if he could make some nitrous oxide for them. Long lacked the equipment to synthesize it, but he recommended they try ether. Ether was commonly used in small oral doses for toothaches and "nervous ailments," but its main claim to popularity was its use as a party drug for so-called ether frolics. Long himself was a bit of a *bon vivant* who put on demonstrations for some of the young ladies, with the disclaimer that he could not be held responsible for whatever he might do under the influence of ether (such as stealing a kiss).

At these parties, Long noted that people sometimes suffered considerable injuries without feeling pain. In 1842, he had a patient who was terrified of pain but needed a tumor removed from his neck. Long excised the tumor without difficulty as his patient sniffed ether from a towel. The operation created a sensation in town, but other physicians ridiculed Long and pronounced anesthesia dangerous. His medical practice declined as people grew afraid of him, but over the next 4 years he performed eight more minor surgeries on patients under ether. Struggling to overcome criticisms that the effects he saw were due merely to hypnotic suggestion or individual variation in sensitivity to pain, Long even compared surgeries done on the same person with and without ether.

Long failed to publish his results quickly enough, and in 1844 he was scooped by a Connecticut dentist, Horace Wells, who had tried nitrous oxide as a dental anesthetic. Another dentist, William Morton of Boston, had tried everything from champagne to opium to kill pain in his patients. He too became interested in ether and gave a public demonstration at Massachusetts General Hospital, where he etherized a patient and removed a tumor. Morton patented a "secret formula" he called Morton's Letheon,[62] which smelled suspiciously of ether, but eventually he went broke trying to monopolize ether anesthesia and he died a pauper. Wells, who had engaged in a bitter feud to establish himself as the inventor of ether anesthesia, committed suicide at the age of 34. Crawford Long went on to a successful career as an Atlanta pharmacist, but to his death he remained disappointed that he had not received credit as the first to perform surgery on etherized patients.

Ether and chloroform became obsolete when safer anesthetics such as cyclopropane, ethylene, and nitrous oxide were developed. These are *general anesthetics* that render a patient unconscious by crossing the blood-brain barrier and blocking nervous transmission through the brainstem. They have little structural similarity, but what they do have in common is a tendency to dissolve in the phospholipids of plasma membranes and make them more fluid. This increases membrane K^+ permeability, hyperpolarizes neurons, and makes them less likely to fire. Diazepam (Valium) deadens pain by activating GABA receptors and causing an inflow of Cl^-, which hyperpolarizes neurons.

Local anesthetics such as procaine (Novocain) and tetracaine selectively deaden specific nerves. They decrease the permeability of membranes to Na^+, thereby reducing their ability to produce action potentials. Spinal anesthesia (spinal block) uses local anesthetics injected into the epidural space for surgery of the abdominopelvic region and lower extremity. It is also an effective and safe method to use in childbirth, since little of the anesthetic enters the mother's bloodstream and the effect on the fetus is minimal.

A sound knowledge of anatomy, control of infection and pain, and development of better tools converged to allow surgeons time to operate more carefully. As a result, surgery became more intellectually challenging and interesting. It attracted a more educated class of practitioner, which put it on the road to becoming the remarkable lifesaving approach that it is today.

60. *a* = without + *sepsis* = infection
61. *an* = without + *esthesia* = feeling, sensation
62. *lethe* = oblivion, forgetfulness

Chapter Review

Online Learning Center

Study resources for this chapter are available at www.mhhe.com/saladin/

Study Outline

Properties and Types of Sensory Receptors (p. 588)

1. Meaning of receptors and sense organs
2. Role as transducers
3. Effects of a stimulus
   a. Receptor potential
   b. Nerve impulses
   c. Neurotransmitter release
   d. Sensation
4. Types of sensory information
   a. Modality
   b. Location
   c. Intensity
   d. Duration
      - Phasic receptors
      - Tonic receptors
5. Types of receptors
   a. Classified by stimulus modality
      - Chemoreceptors
      - Thermoreceptors
      - Nociceptors
      - Mechanoreceptors
      - Photoreceptors
   b. Classified by receptor distribution
      - General senses
      - Special senses
   c. Classified by stimulus origin
      - Interoceptors
      - Proprioceptors
      - Exteroceptors

The General Senses (p. 589)

1. Unencapsulated nerve endings
   a. Free nerve endings
   b. Merkel discs
   c. Hair receptors
2. Encapsulated nerve endings
   a. Tactile corpuscles
   b. Krause end bulbs
   c. Lamellated corpuscles
   d. Ruffini corpuscles
   e. Muscle spindles
   f. Golgi tendon organs
3. Somesthetic projection pathways
   a. First- to third-order neurons
   b. Pathways for touch, pressure, and proprioception
   c. Pathways for heat and cold
4. Pain
   a. Value of nociception
   b. Locations of nociceptors
   c. Types of nociceptors
      - Myelinated fibers and fast pain
      - Unmyelinated fibers and slow pain
   d. Types of pain
      - Somatic pain
      - Visceral pain
   e. Chemical pain stimuli
   f. Pain pathways
      - Trigeminal nerve route
      - Spinal nerve route
        - •• First-order neuron to dorsal horn
        - •• Second-order neuron to contralateral thalamus
        - •• Third-order neuron to somesthetic cortex
        - •• Spinoreticular tract
   g. Referred pain
   h. Modulation of pain by CNS
      - Endogenous opioids
      - Spinal gating of pain

The Chemical Senses (p. 593)

1. Taste (gustation)
   a. Anatomy
      - Lingual papillae and taste buds
      - Structure of taste cells
   b. Physiology
      - Five primary taste sensations
      - Role of other food qualities
      - Regional specialization of tongue
      - Mechanisms of taste cell stimulation
   c. Projection pathways
      - Cranial nerves VII, IX, and X
      - Solitary nucleus of medulla oblongata
      - Nuclei of hypothalamus and amygdala
      - Thalamic route to insula and postcentral gyrus
2. Smell (olfaction)
   a. Anatomy
      - Olfactory mucosa
      - Olfactory neurons
      - Olfactory hairs
      - Olfactory nerve fascicles
   b. Physiology
      - Sensitivity
      - Difficulty of identifying primary odors
      - Properties of odor molecules
      - Stimulation of olfactory cells
      - Stimulation of trigeminal nerve
      - Olfactory adaptation
   c. Projection pathways
      - Olfactory bulbs
      - Olfactory tracts to temporal lobe
   d. Cerebral feedback to olfactory bulbs

Hearing and Equilibrium (p. 599)

1. Nature of sound
   a. Production and transmission of sound waves
   b. Frequency and pitch
   c. Amplitude and loudness
2. Gross anatomy of the ear
   a. Outer ear
      - Auricle
      - Auditory canal
   b. Middle ear
      - Tympanic membrane
      - Auditory tube
      - Auditory ossicles
      - Middle-ear muscles
   c. Inner ear
      - Bony and membranous labyrinths

- Perilymph and endolymph
- Vestibule
- Cochlea

3. Microscopic anatomy of the cochlea
   a. Modiolus
   b. Scala vestibuli and tympani
   c. Cochlear duct
   d. Vestibular membrane
   e. Basilar membrane
   f. Organ of Corti
      - Supporting cells
      - Hair cells
        - •• Inner hair cells (IHCs)
        - •• Outer hair cells (OHCs)
      - Tectorial membrane
4. Physiology of hearing
   a. Auditory ossicles
      - Transfer of vibration to inner ear
      - Protection from loud noise
   b. Stimulation of IHCs
      - Vibration of basilar membrane
      - Gated K^+ channels on stereocilia
   c. Sensory coding
      - Coding for loudness
      - Coding for pitch
   d. Cochlear tuning
      - Vibration damping by OHCs
      - Inhibition of IHCs
5. Auditory projection pathway
   a. Spiral ganglion and cochlear nerve
   b. Cochlear nucleus of medulla oblongata
   c. Superior olivary nucleus
      - Cochlear tuning
      - Control of middle-ear muscles
      - Binaural hearing
   d. Inferior colliculi of midbrain
   e. Thalamic route to temporal lobe
   f. Superior colliculi of midbrain
6. Saccule and utricle
   a. Anatomy
      - Macula sacculi and utriculi
      - Hair cells
      - Stereocilia and kinocilium
      - Otolithic membrane
   b. Detection of linear motion
7. Semicircular ducts
   a. Anatomy
      - Anterior, posterior, and lateral ducts
      - Endolymph
      - Ampulla
      - Crista ampullaris
      - Hair cells
      - Cupula
   b. Detection of angular acceleration
8. Equilibrium projection pathways
   a. Vestibular nerve
   b. Vestibular nucleus of pons
   c. Cerebellum
   d. Cervical spinal cord
   e. Nuclei for cranial nerves II, IV, VI, and XI

Vision (p. 611)

1. Nature of light
   a. Wavelengths available to the eye
   b. Ability to produce photochemical reactions
2. Accessory structures of orbit
   a. Eyebrows
   b. Eyelids
   c. Conjunctiva
   d. Lacrimal apparatus
      - Lacrimal gland
      - Lacrimal caruncle
      - Lacrimal sac
      - Nasolacrimal duct
   e. Extrinsic eye muscles
      - Superior rectus
      - Inferior rectus
      - Medial rectus
      - Lateral rectus
      - Superior oblique
      - Inferior oblique
   f. Orbital fat
3. Gross anatomy of the eye
   a. Tunics
      - Tunica fibrosa (sclera and cornea)
      - Tunica vasculosa (choroid, ciliary body, and iris)
      - Tunica interna (retina)
   b. Optical apparatus
      - Cornea
      - Lens
      - Aqueous humor
      - Vitreous body
   c. Neural apparatus
      - Retina
        - •• Optic disc
        - •• Ora serrata
        - •• Macula lutea
        - •• Fovea centralis
      - Optic nerve
4. Formation of an image
   a. Admission of light by pupil
      - Pupillary constrictor and dilator
      - Photopupillary reflex
   b. Refraction of light rays
   c. Near response
      - Convergence of eyes
      - Constriction of pupil
      - Accommodation of lens
5. Microscopic anatomy of retina
   a. Pigment epithelium
   b. Rod and cone cells
      - Outer segment and discs
      - Inner segment
      - Synaptic connections
   c. Bipolar cells
   d. Ganglion cells and optic nerve fibers
   e. Horizontal and amacrine cells
      - Bipolar cells
      - Ganglion cells
6. Sensory transduction
   a. Visual pigments
      - Rhodopsin
      - Photopsin
   b. The photochemical reaction
      - Bleaching of rhodopsin
      - Regeneration of rhodopsin
   c. The dark current
      - Binding of cGMP to Na^+ channels
      - Inflow of Na^+ (dark current)
      - Depolarization of rod
      - Release of glutamate at synapse
   d. Effect of light
      - cGMP breakdown
      - Closing of Na^+ channels
      - Cessation of dark current
      - Cessation of glutamate release
      - Effect on bipolar cells
        - •• Some excited in absence of glutamate
        - •• Some inhibited by absence of glutamate
      - Bipolar cells stimulate ganglion cells
      - Ganglion cells generate signals in optic nerve
7. Light and dark adaptation
8. Duplicity theory
   a. Rods and scotopic vision
      - High light sensitivity of individual rods
      - Extensive neuronal convergence in retina
      - Rod pathways responsive to dim light
      - Resolution poor in rod pathways
      - Rods and peripheral vision
      - Lack of rods in fovea centralis
   b. Cones and photopic vision
      - High density of cones in fovea centralis
      - Minimal neuronal convergence
      - Cones unresponsive to dim light
      - High resolution in daylight
9. Color vision
   a. Blue, green, and red cones
   b. Color blindness

10. Stereoscopic vision
    a. Necessity of overlapping visual fields
    b. Objects in relation to fixation point
11. Visual projection pathway
    a. Optic nerves and chiasm
    b. Hemidecussation of optic nerve fibers
    c. Optic tracts to lateral geniculate nucleus
    d. Optic radiation to occipital lobe
    e. Superior colliculi and visual reflexes
    f. Pretectal nuclei and photopupillary and accommodation reflexes

Selected Vocabulary

For additional terminology not listed here, see tables 16.1 through 16.3.

receptor 588
sense organ 588
receptor potential 588
sensation 588
modality 588
sensory projection 588
projection pathway 588
phasic receptor 588
tonic receptor 589
adaptation 589
chemoreceptor 589
thermoreceptor 589
nociceptor 589
mechanoreceptor 589
photoreceptor 589
general senses 589
special senses 589
interoceptor 589
proprioceptor 589
exteroceptor 589
first-order neuron 591
second-order neuron 591
third-order neuron 591
fast pain 591
slow pain 591
somatic pain 591
visceral pain 591
bradykinin 591
substance P 591
referred pain 592
analgesic 592
enkephalin 592
endorphin 592
dynorphin 592
endogenous opioid 592
spinal gating 593
descending analgesic fiber 593
gustation 593
taste bud 593
lingual papilla 594
filiform papilla 594
foliate papilla 594
fungiform papilla 594
vallate papilla 594
taste cell 595
taste hair 595
taste pore 595
umami 596
olfaction 596
olfactory mucosa 596
olfactory cell 596
olfactory hair 596
olfactory bulb 598
olfactory tract 598
sound 599
pitch 599
frequency 600
loudness 600
amplitude 600
tympanic reflex 604
tip link 605
binaural hearing 606
static equilibrium 608
dynamic equilibrium 608
kinocilium 609
otolith 609
chromatophore 616
blind spot 618
visual filling 619
pupillary constrictor 619
pupillary dilator 619
photopupillary reflex 620
refraction 620
emmetropia 620
near response 620
convergence 620
accommodation 621
near point of vision 621
pigment epithelium 621
rod 622
cone 622
outer segment 622
inner segment 622
night vision 622
day vision 622
bipolar cell 622
ganglion cell 622
horizontal cell 623
amacrine cell 623
rhodopsin 623
opsin 623
retinal 623
photopsin 623
cis-retinal 624
trans-retinal 624
bleaching 624
dark current 626
receptive field 626
light adaptation 627
dark adaptation 627
duplicity theory 628
blue cones 629
green cones 629
red cones 629
color blindness 629
stereoscopic vision 629
optic chiasm 630
optic tracts 630
hemidecussation 630
lateral geniculate nucleus 630
optic radiation 630

Testing Your Recall

1. Hot and cold stimuli are detected by
    a. free nerve endings.
    b. proprioceptors.
    c. Krause end bulbs.
    d. lamellated corpuscles.
    e. tactile corpuscles.

2. ________ is a neurotransmitter that transmits pain sensations to second-order spinal neurons.
    a. Endorphin
    b. Enkephalin
    c. Substance P
    d. Acetylcholine
    e. Norepinephrine

3. ________ is a neuromodulator that blocks the transmission of pain sensations to second-order spinal neurons.
    a. Endorphin
    b. Enkephalin
    c. Substance P
    d. Acetylcholine
    e. Norepinephrine

4. Taste buds of the vallate papillae are most sensitive to
    a. bitter.
    b. sour.
    c. sweet.
    d. umami.
    e. salty.

5. The higher the frequency of a sound,
    a. the louder it sounds.
    b. the harder it is to hear.
    c. the more it stimulates the distal end of the organ of Corti.
    d. the faster it travels through air.
    e. the higher its pitch.

6. Cochlear hair cells rest on
    a. the tympanic membrane.
    b. the secondary tympanic membrane.

c. the tectorial membrane.
d. the vestibular membrane.
e. the basilar membrane.

7. The acceleration you feel when an elevator begins to rise is sensed by
   a. the anterior semicircular duct.
   b. the organ of Corti.
   c. the crista ampullaris.
   d. the macula sacculi.
   e. the macula utriculi.

8. The color of light is determined by
   a. its velocity.
   b. its amplitude.
   c. its wavelength.
   d. refraction.
   e. how strongly it stimulates the rods.

9. The retina receives its oxygen supply from
   a. the hyaloid artery.
   b. the vitreous body.
   c. the choroid.
   d. the pigment epithelium.
   e. the scleral venous sinus.

10. Which of the following statements about photopic vision is false?
   a. It is mediated by the cones.
   b. It has a low threshold.
   c. It produces fine resolution.
   d. It does not function in starlight.
   e. It does not employ rhodopsin.

11. The most finely detailed vision occurs when an image falls on a pit in the retina called the ___________.

12. The only cells of the retina that generate action potentials are the ___________ cells.

13. The retinal dark current results from the flow of ___________ into the receptor cells.

14. The gelatinous membranes of the macula sacculi and macula utriculi are weighted by calcium carbonate and protein granules called ___________.

15. Three rows of ___________ in the cochlea have V-shaped arrays of stereocilia and tune the frequency sensitivity of the cochlea.

16. The ___________ is a tiny bone that vibrates in the oval window and thereby transfers sound vibrations to the inner ear.

17. The ___________ of the midbrain receive auditory input and trigger the head-turning auditory reflex.

18. The apical stereocilia of a gustatory cell are called ___________.

19. Olfactory neurons synapse with mitral cells and tufted cells in the ___________, which lies inferior to the frontal lobe.

20. In the phenomenon of ___________, pain from the viscera is perceived as coming from an area of the skin.

Answers in appendix B

Testing Your Comprehension

1. The principle of neuronal convergence was explained on page 482. Discuss its relevance to referred pain and scotopic vision.
2. What type of cutaneous receptor enables you to feel an insect crawling through your hair? What type enables you to palpate a patient's pulse? What type enables a blind person to read braille?
3. Contraction of a muscle usually puts more tension on a structure, but contraction of the ciliary muscle puts less tension on the lens. Explain how.
4. Discuss the similarities and differences between presbyopia and hyperopia.
5. Color blindness affects about 10% of males but only 0.7% of females. What does this suggest about its mode of inheritance? (See chapter 5.)

Answers at the Online Learning Center

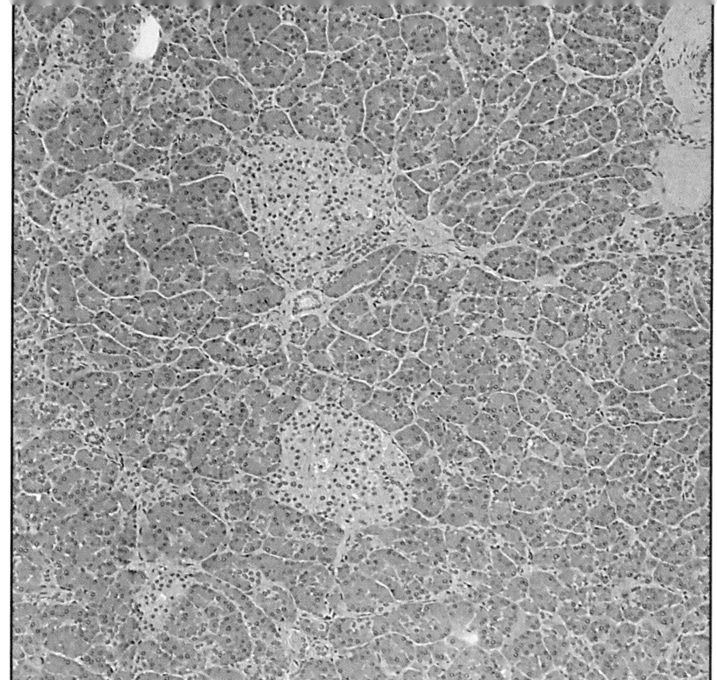

Human pancreas. Light zones in the middle are the insulin-producing islets

CHAPTER 17

The Endocrine System

Chapter Outline

Insights

Brushing Up

To understand this chapter, it is important that you understand or brush up on the following concepts:

- Structure and function of the plasma membrane (pp. 108–112)
- G proteins, cAMP, and other second messengers (p. 112)
- Active transport and the transport maximum (p. 129)
- Monoamines, especially catecholamines (p. 474)
- The hypothalamus (pp. 513–514)

 Study with Your ESP CD-ROM

Refer to your ESP Tutorial CD-ROM for helpful animations, activities, and quizzes related to the material covered in this chapter. The following are suggested areas to go to on the CD and are presented according to "Unit/Topic":

- Integration and Coordination/Endocrine System

We noted in chapter 13 that if the body is to function as an integrated homeostatic whole, it needs mechanisms of internal communication. The major mechanisms are the nervous and endocrine systems, which communicate by means of neurotransmitters and hormones, respectively. This chapter is about **endocrinology**, the study of the endocrine system and the diagnosis and treatment of its dysfunctions.

You probably have at least some prior acquaintance with this system. Perhaps you have heard of the pituitary gland and thyroid gland, secretions such as growth hormone, estrogen, and insulin, and endocrine disorders such as dwarfism, goiter, and diabetes mellitus. Fewer readers, perhaps, are familiar with what hormones are at a chemical level or exactly how they work. Therefore, this chapter starts with the relatively familiar—a survey of the endocrine glands, their hormones, and the dysfunctions that result from hormone deficiencies and excesses. Then we work our way down to the finer and less familiar details at the cellular and molecular level—the chemical identity of hormones, how they are made and transported, and how they produce their effects. This chapter also explains how the body responds to stress through hormonal and other mechanisms and discusses other hormonelike modes of chemical communication.

Overview of the Endocrine System

Objectives

When you have completed this section, you should be able to

- define *hormone* and *endocrine system;*
- list the major organs of the endocrine system;
- recognize the standard abbreviations for many hormones; and
- compare and contrast the nervous and endocrine systems.

The term **hormone**[1] is defined in various ways, but for our purposes it means a chemical messenger that is secreted into the bloodstream and stimulates physiological responses in cells of another tissue or organ. Although a hormone goes everywhere the blood goes, it affects only those cells that have receptors to bind it. We call these the **target cells.** The glands that produce hormones are called **endocrine**[2] **glands** (fig. 17.1), but they are not the full extent of the endocrine system. The **endocrine system** also includes hormone-producing cells in many organs not usually thought of as glands, such as the brain, heart, and small intestine.

In chapter 6, we saw that glands can be classified as exocrine or endocrine. One way in which these differ is that exocrine glands have ducts to carry their secretion to the body surface (as in sweat) or to the cavity of another organ (as in digestive enzymes). Endocrine glands have no ducts but do have dense blood capillary networks. Endocrine cells release their hormones into the surrounding tissue fluid, and then the bloodstream quickly picks up and distributes the hormones. Exocrine secretions have extracellular effects such as the digestion of food, whereas endocrine secretions have intracellular effects—they alter the metabolism of their target cells.

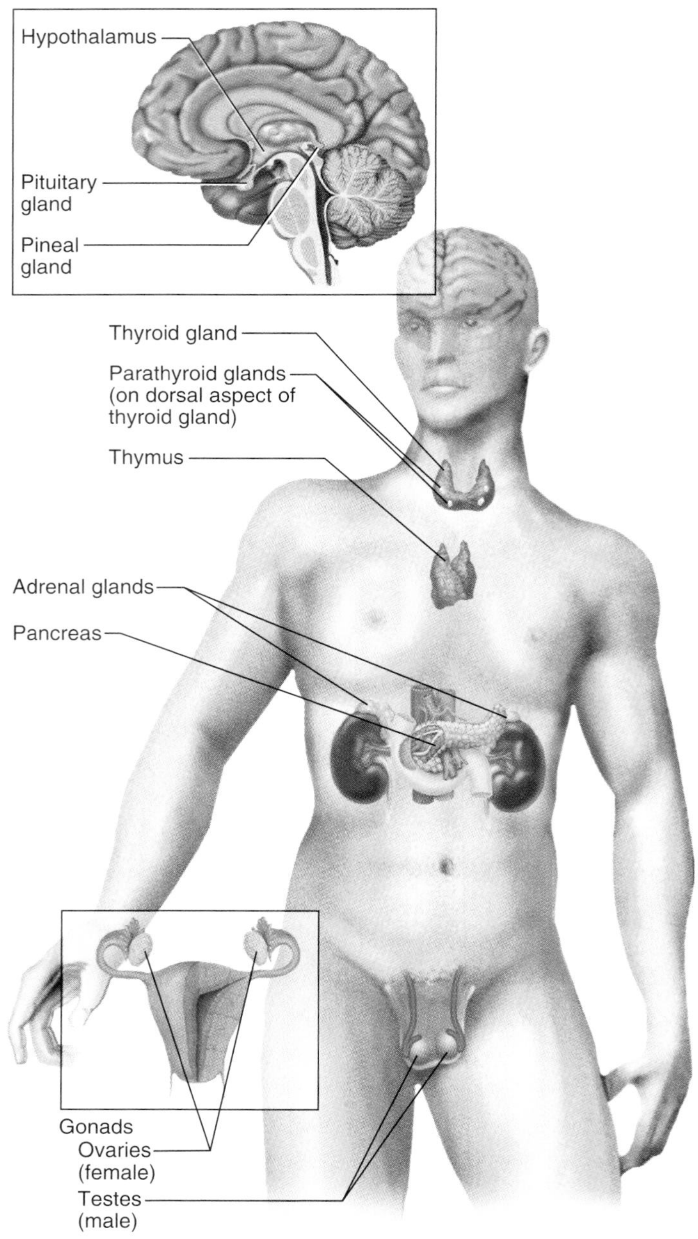

Figure 17.1 Major Organs of the Endocrine System. This system also includes gland cells in many other organs not shown here.

Comparison of the Nervous and Endocrine Systems

Although the nervous and endocrine systems both serve for internal communication, they are not redundant; they complement rather than duplicate each other's

1. *hormone* = to excite, set in motion
2. *endo* = into; *exo* = out of; *crin* = to separate or secrete

function (table 17.1). The systems differ in their means of communication—both electrical and chemical in the nervous system and solely chemical in the endocrine system (fig. 17.2)—yet as we shall see, they have many similarities on this point as well. They differ also in how quickly they start and stop responding to stimuli. The nervous system typically responds in just a few milliseconds, whereas hormone release may follow from several seconds to several days after the stimulus that caused it. Furthermore, when a stimulus ceases, the nervous system stops responding almost immediately, whereas some endocrine effects persist for several days or even weeks. On the other hand, under long-term stimulation, neurons soon adapt and their response declines. The endocrine system shows more persistent responses. For example, thyroid hormone secretion rises in cold weather and remains elevated as long as it remains cold. Another difference between the two systems is that an efferent nerve fiber innervates only one organ and a limited number of cells within that organ; its effects, therefore, are precisely targeted and relatively specific. Hormones, by contrast, circulate throughout the body and some of them, such as growth hormone and thyroid hormone, have very widespread effects.

But these differences should not blind us to the similarities between the two systems. Several chemicals function as both neurotransmitters and hormones, including norepinephrine, cholecystokinin, thyrotropin-releasing hormone, prolactin–inhibiting factor (= dopamine), and antidiuretic hormone (= vasopressin). Some hormones, such as oxytocin and the catecholamines, are secreted by **neuroendocrine cells**—neurons that release their secretions into the extracellular fluid. Some hormones and neurotransmitters produce overlapping effects on the same target cells. For example, norepinephrine and glucagon cause glycogen hydrolysis in the liver. The nervous and endocrine systems continually regulate each other as they coordinate the activities of other organ systems. Neurons often trigger hormone secretion, and hormones often stimulate or inhibit neurons.

Table 17.1 Comparison of the Nervous and Endocrine Systems

Nervous System	Endocrine System
Communicates by means of electrical impulses and neurotransmitters	Communicates by means of hormones carried in blood
Reacts quickly to stimuli, usually within 1 to 10 msec	Reacts more slowly to stimuli, often taking seconds to days
Stops quickly when stimulus stops	May continue responding long after stimulus stops
Adapts relatively quickly to continual stimulation	Adapts relatively slowly; may continue responding for days to weeks
Has relatively local, specific effects on target organs	Sometimes has very general, widespread effects on many organs in body

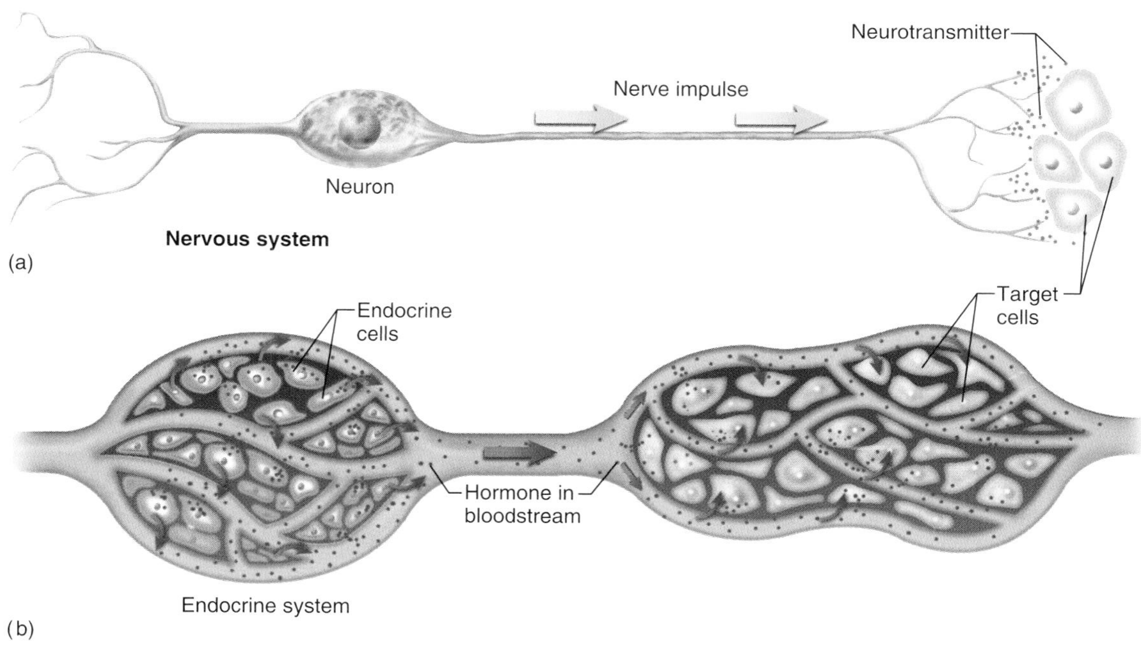

Figure 17.2 Communication by the Nervous and Endocrine Systems. (*a*) A neuron has a long fiber that delivers its neurotransmitter to the immediate vicinity of its target cells. (*b*) Endocrine cells secrete a hormone into the bloodstream. The hormone binds to target cells at places often remote from the gland cells.

Hormone Nomenclature

Many hormones are denoted by standard abbreviations which are used repeatedly in this chapter. These abbreviations are listed alphabetically in table 17.2 so that you can use this as a convenient reference while you work through the chapter. This is by no means a complete list. It does not include hormones that have no abbreviation, such as estrogen and insulin, and it omits hormones that are not discussed much in this chapter. Synonyms used by many authors are indicated in parentheses, but the first name listed is the one that is used in this book.

Table 17.2 Names and Abbreviations for Hormones

Abbreviation	Name	Source
ACTH	Adrenocorticotropic hormone (corticotropin)	Anterior pituitary
ADH	Antidiuretic hormone (vasopressin)	Posterior pituitary
ANF	Atrial natriuretic factor	Heart
CRH	Corticotropin-releasing hormone	Hypothalamus
DHEA	Dehydroepiandrosterone	Adrenal cortex
EPO	Erythropoietin	Kidney
FSH	Follicle-stimulating hormone	Anterior pituitary
GH	Growth hormone (somatotropin)	Anterior pituitary
GHIH	Growth hormone–inhibiting hormone (somatostatin)	Hypothalamus
GHRH	Growth hormone–releasing hormone	Hypothalamus
GnRH	Gonadotropin-releasing hormone	Hypothalamus
IGFs	Insulin-like growth factors (somatomedins)	Liver
LH	Luteinizing hormone	Anterior pituitary
NE	Norepinephrine	Adrenal medulla
OT	Oxytocin	Posterior pituitary
PIF	Prolactin-inhibiting factor (dopamine)	Hypothalamus
PRF	Prolactin-releasing factor	Hypothalamus
PRL	Prolactin (luteotropin)	Anterior pituitary
PTH	Parathyroid hormone (parathormone)	Parathyroids
T_3	Triiodothyronine	Thyroid
T_4	Thyroxine (tetraiodothyronine)	Thyroid
TH	Thyroid hormone (T_3 and T_4)	Thyroid
TRH	Thyrotropin-releasing hormone	Hypothalamus
TSH	Thyroid-stimulating hormone	Anterior pituitary

Before You Go On

Answer the following questions to test your understanding of the preceding section:

1. Define the word *hormone* and distinguish a hormone from a neurotransmitter. Why is this an imperfect distinction?
2. Describe some ways in which endocrine glands differ from exocrine glands.
3. Name some sources of hormones other than purely endocrine glands.
4. List some similarities and differences between the endocrine and nervous systems.

The Hypothalamus and Pituitary Gland

Objectives

When you have completed this section, you should be able to

- list the hormones produced by the hypothalamus and pituitary gland;
- explain how the hypothalamus and pituitary are controlled and coordinated with each other;
- describe the functions of growth hormone; and
- describe the effects of pituitary hypo- and hypersecretion.

There is no "master control center" that regulates the entire endocrine system, but the pituitary gland and a nearby region of the brain, the hypothalamus, have a more wide-ranging influence than any other part of the system. This is an appropriate place to begin a survey of the endocrine glands.

Anatomy

The hypothalamus, shaped like a flattened funnel, forms the floor and walls of the third ventricle of the brain (see fig. 14.17, p. 513). It regulates primitive functions of the body ranging from water balance to sex drive. Many of its functions are carried out by way of the pituitary gland, which is closely associated with it.

The **pituitary gland (hypophysis[3])** is suspended from the hypothalamus by a stalk (*infundibulum*) and housed in the sella turcica of the sphenoid bone. It is usually about 1.3 cm in diameter, but grows about 50% larger in pregnancy. It is actually composed of two structures—the adenohypophysis and neurohypophysis—that arise independently in the embryo and have entirely separate

3. *hypo* = below + *physis* = growth

functions. The adenohypophysis arises from a *hypophyseal pouch* that grows upward from the pharynx of the embryo, while the neurohypophysis arises as a downgrowth of the brain, the *neurohypophyseal bud* (fig. 17.3). They come to lie side by side and are so closely joined to each other that they look like a single gland.

The **adenohypophysis**[4] (AD-eh-no-hy-POFF-ih-sis) constitutes the anterior three-quarters of the pituitary (fig. 17.4*a*). It has two parts: a large **anterior lobe,** also called the *pars distalis* ("distal part") because it is most distal to the pituitary stalk, and the **pars tuberalis,** a small mass of cells adhering to the anterior side of the stalk. In the fetus there is also a **pars intermedia,** a strip of tissue between the anterior lobe and neurohypophysis. During subsequent development, its cells mingle with those of the anterior lobe; in adults, there is no longer a separate pars intermedia.

The anterior pituitary has no nervous connection to the hypothalamus but is connected to it by a complex of blood vessels called the **hypothalamo-hypophyseal portal system** (fig. 17.4*b*). This begins with a network of capillaries in the hypothalamus. These capillaries pick up hormones from the hypothalamus. They lead to *portal venules* (small veins) that travel down the pituitary stalk to a complex of *secondary capillaries* in the anterior pituitary. Here, the hormones from the hypothalamus leave the bloodstream and stimulate cells of the anterior pituitary.

The **neurohypophysis** constitutes the posterior one-quarter of the pituitary. It has three parts: the **posterior lobe** (*pars nervosa*), which constitutes most of it; the **stalk,** which connects it to the hypothalamus; and the **median eminence,** an extension of the hypothalamus. The neurohypophysis is not a true gland but a mass of neuroglia and the axons of hypothalamic neurons. The axons form a bundle called the **hypothalamo-hypophyseal tract,** which runs down through the stalk and ends in the posterior lobe. Hormones are synthesized by the neurons in the hypothalamus, transported down the axons, and stored in the posterior pituitary until a nerve signal triggers their release.

4. *adeno* = gland

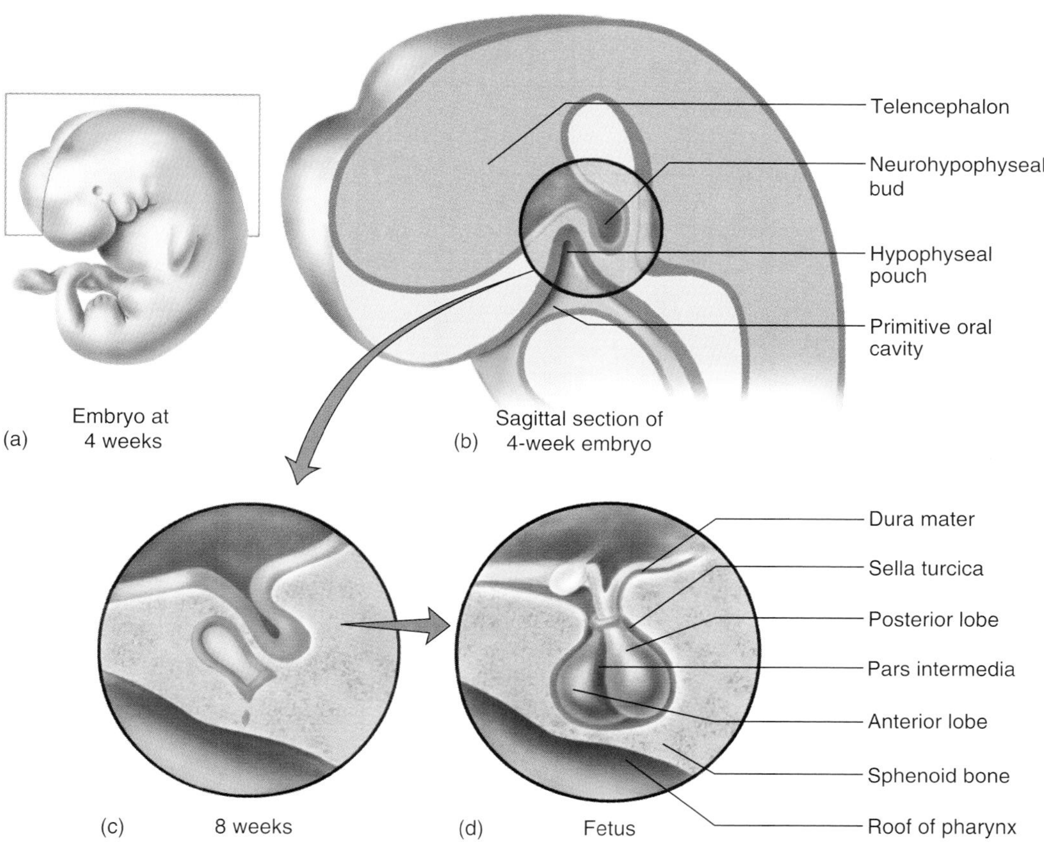

Figure 17.3 **Embryonic Development of the Pituitary Gland.** (*a*) Plane of section seen in *b*. (*b*) Sagittal section of the embryo showing the early beginnings of the adenohypophysis and neurohypophysis. (*c*) Separation of the hypophyseal pouch from the pharynx at about 8 weeks. (*d*) Development nearly completed. The pars intermedia largely disappears by birth.

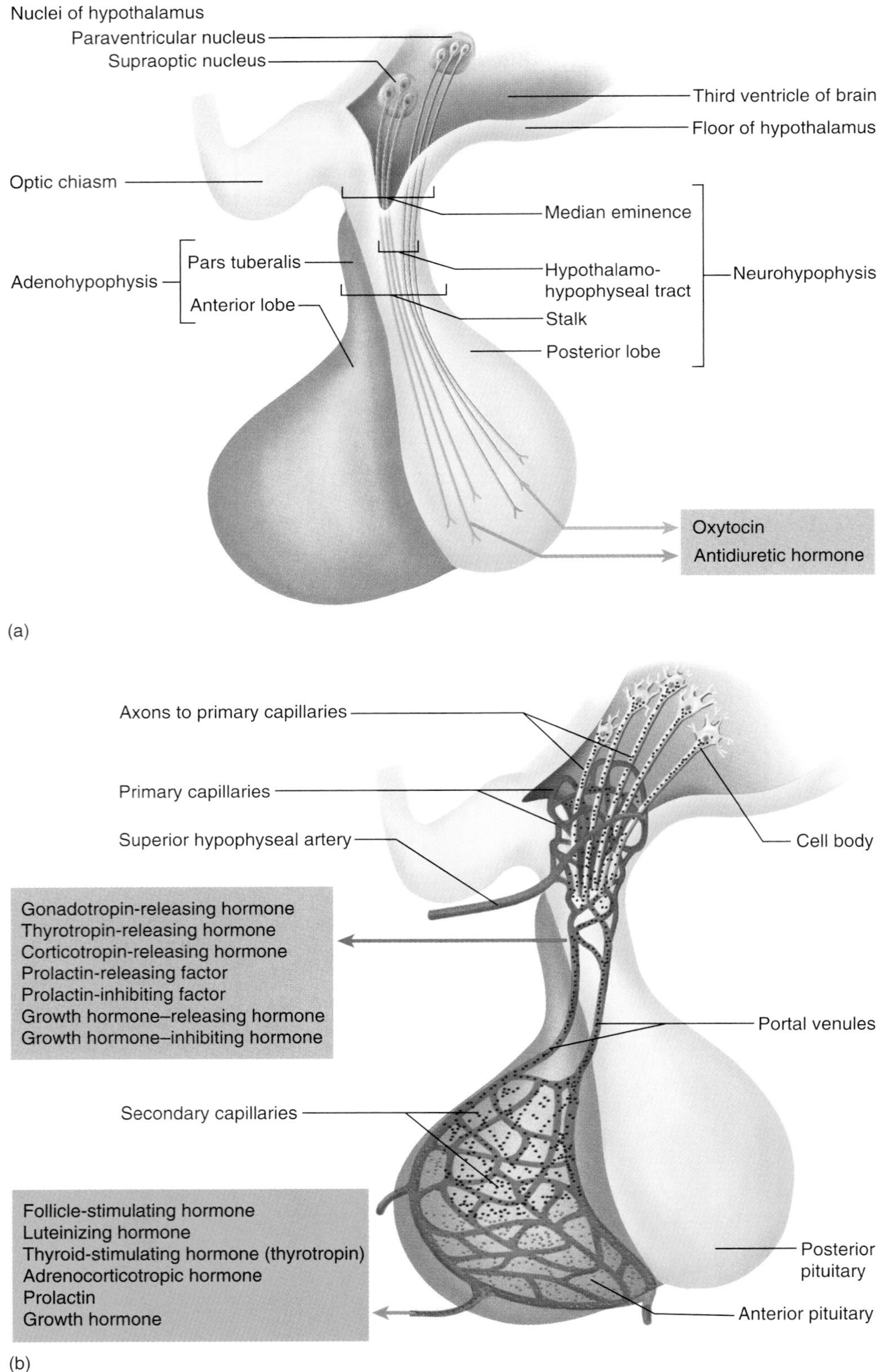

Figure 17.4 Anatomy of the Pituitary Gland. (*a*) Major structures of the pituitary and hormones of the neurohypophysis. Note that these hormones are produced by two nuclei in the hypothalamus and later released from the posterior lobe of the pituitary. (*b*) The hypothalamo-hypophyseal portal system. The hormones in the violet box are secreted by the hypothalamus and travel in the portal system to the anterior pituitary. The hormones in the *red box* are secreted by the anterior pituitary under the control of the hypothalamic releasers and inhibitors.

Hereafter, we can disregard all of the pituitary except the anterior and posterior lobes; they secrete all the pituitary hormones we will consider.

Hypothalamic Hormones

The hypothalamus produces nine hormones important to our discussion. Seven of them, listed in figure 17.4 and table 17.3, travel through the portal system and regulate the activities of the anterior pituitary. Five of these are *releasing hormones* that stimulate the anterior pituitary to secrete its hormones, and two are *inhibiting hormones* that suppress pituitary secretion. Most of these hypothalamic hormones control the release of just one anterior pituitary hormone. Gonadotropin-releasing hormone, however, controls the release of both follicle-stimulating hormone and luteinizing hormone. These are both classified as *gonadotropins* because they stimulate the gonads (ovaries and testes).

The other two hypothalamic hormones are **oxytocin (OT)** and **antidiuretic hormone (ADH).** OT is produced mainly by neurons in the **paraventricular**[5] **nucleus** of the hypothalamus, so-called because it lies in the wall of the third ventricle. ADH is produced mainly by the **supraoptic**[6] **nucleus,** so-called because it lies just above the optic chiasm. (Each nucleus also produces smaller quantities of the other hormone.) OT and ADH are transported down the hypothalamo-hypophyseal tract and stored in the posterior pituitary until needed.

5. *para* = next to + *ventricular* = pertaining to the ventricle
6. *supra* = above

Table 17.3 Hypothalamic Releasing and Inhibiting Hormones That Regulate the Anterior Pituitary

Hormone	Principal Effects
TRH: Thyrotropin-releasing hormone	Promotes TSH and PRL secretion
CRH: Corticotropin-releasing hormone	Promotes ACTH secretion
GnRH: Gonadotropin-releasing hormone	Promotes FSH and LH secretion
PRF: Prolactin-releasing factor	Promotes PRL secretion
PIF: Prolactin-inhibiting factor	Inhibits PRL secretion
GHRH: Growth hormone–releasing hormone	Promotes GH secretion
GHIH: Growth hormone–inhibiting hormone	Inhibits GH and TSH secretion

Pituitary Hormones

The secretions of the three principal regions of the pituitary gland are as follows:

- **The anterior lobe.** The anterior lobe of the pituitary synthesizes and secretes six principal hormones: follicle-stimulating hormone (FSH), luteinizing hormone (LH), thyroid-stimulating hormone (TSH), adrenocorticotropic hormone (ACTH), prolactin (PRL), and growth hormone (GH) (table 17.4). The first four of these are **tropic,** or **trophic,**[7] **hormones**—pituitary hormones whose target organs are other endocrine glands. More specifically, the first two are called **gonadotropins** because their target organs are the gonads.

  The relationship between the pituitary, its tropic hormones, and their target endocrine glands is called an *axis*—a convenient term for referring to the way these endocrine glands influence each other. There are three such axes: the **pituitary-gonadal axis** involving FSH and LH, the **pituitary-thyroid axis** involving TSH, and the **pituitary-adrenal axis** involving ACTH (fig. 17.5).
- **The pars intermedia.** As mentioned earlier, the pars intermedia is absent from the adult human pituitary, but it is present in other animals and the human fetus. In other species, it secretes *melanocyte-stimulating hormone (MSH),* which influences pigmentation of the skin, hair, or feathers. Until recently, it was thought to have a similar effect on the human skin, but evidence now indicates that we produce no circulating MSH. Some cells of the anterior lobe, derived from the pars intermedia, produce a large polypeptide called *pro-opiomelanocortin (POMC).* POMC is not secreted but is processed within the pituitary to yield smaller fragments such as ACTH and endorphins.
- **The posterior lobe.** As we have already seen, the posterior lobe produces no hormones of its own but only stores and releases OT and ADH. Since they are released into the blood by the posterior pituitary, however, these are treated as pituitary hormones for convenience.

Actions of the Pituitary Hormones

Now for a closer look at what all of these pituitary hormones do. Most of these hormones receive their fullest treatment in later chapters on such topics as the urinary and reproductive systems, but growth hormone gets its fullest treatment here.

7. *trop* = to turn, change; *troph* = to feed, nourish

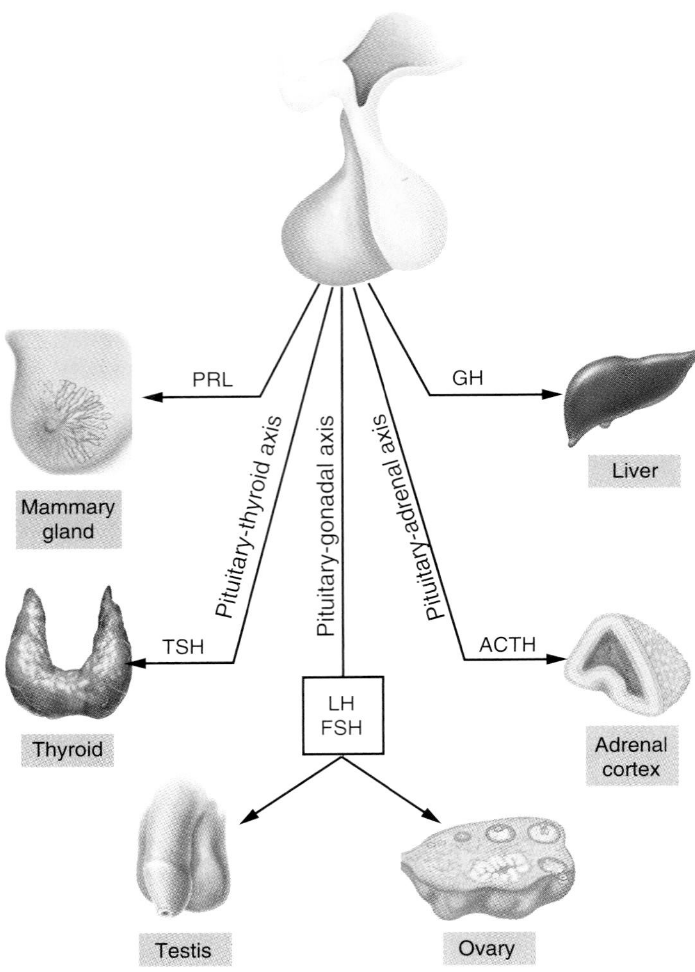

Figure 17.5 Principal Hormones of the Anterior Pituitary Gland and Their Target Organs. The three axes physiologically link pituitary function to the function of other endocrine glands.

Anterior Lobe Hormones

Follicle-Stimulating Hormone (FSH) FSH is one of the two gonadotropins. It is secreted by pituitary cells called *gonadotropes,* and its target organs are the ovaries and testes. In the ovaries, it stimulates the development of eggs and the follicles that contain them. In the testes, it stimulates the production of sperm.

Luteinizing Hormone (LH) LH is another gonadotropin secreted by the gonadotropes. In females, it stimulates *ovulation* (the release of an egg). LH is named for the fact that after ovulation, the remainder of a follicle is called the *corpus luteum* ("yellow body"). LH stimulates the corpus luteum to secrete progesterone, a hormone important to pregnancy. In males, LH is also called *interstitial cell–stimulating hormone (ICSH);* it stimulates *interstitial cells* of the testes to secrete testosterone.

Thyroid-Stimulating Hormone (TSH), or Thyrotropin TSH is secreted by pituitary cells called *thyrotropes.* It stimulates growth of the thyroid gland and the secretion of thyroid hormone, which has widespread effects on the body's metabolism.

Adrenocorticotropic Hormone (ACTH), or Corticotropin ACTH is secreted by pituitary cells called *corticotropes.* It is important in regulating the response to stress. It is named for its effect on the adrenal cortex, which we study later, but it also stimulates insulin secretion by the pancreas. ACTH stimulates the adrenal cortex to secrete *glucocorticoids,* hormones that regulate glucose, fat, and protein metabolism.

Prolactin[8] (PRL) PRL is secreted by *lactotropes (mammotropes),* which increase greatly in size and number during pregnancy. PRL level rises during pregnancy, but it has no effect until after a woman gives birth. Then, it stimulates the mammary glands to synthesize milk and it remains elevated for as long as she nurses. In males, PRL has a gonadotropic effect that makes the testes more sensitive to LH. Thus, it indirectly enhances their secretion of testosterone.

Growth Hormone (GH), or Somatotropin GH is secreted by *somatotropes,* the most numerous cells in the anterior pituitary. The pituitary produces at least a thousand times as much GH as any other hormone. The general effect of GH is to promote mitosis and cellular differentiation and thus to promote widespread tissue growth. However, it does this indirectly by stimulating the liver to produce polypeptides called **somatomedins,**[9] or **insulin-like growth factors (IGFs).** IGFs then stimulate target cells in fat, cartilage, bone, and other tissues. A somatomedin called IGF-I seems to be the most important in bone and cartilage growth.

Hormones have a **half-life,** the time required for half of the hormone to be cleared from the blood. GH is short-lived; it has a half-life of less than 20 minutes. IGFs, by contrast, have half-lives of about 20 hours, so they greatly prolong the effect of GH. The mechanisms of GH-IGF action include:

- **Protein synthesis.** Tissue growth requires protein synthesis, and protein synthesis needs two things: amino acids for "building material" and messenger RNA (mRNA) for "instructions." Within minutes of GH secretion, preexisting mRNA is translated and proteins synthesized; within a few hours, DNA is transcribed and more mRNA is produced. GH enhances amino acid transport into cells, and to ensure that protein synthesis outpaces breakdown, it suppresses protein catabolism.

8. *pro* = favoring + *lact* = milk
9. Acronym for *somato*tropin *med*iating prote*in*

- **Lipid metabolism.** To provide energy for growing tissues, GH stimulates adipocytes to catabolize fat and release free fatty acids (FFAs) and glycerol into the blood. GH has a **protein-sparing effect**—by liberating FFAs and glycerol for energy, it makes it unnecessary for cells to consume their proteins.
- **Carbohydrate metabolism.** GH also has a **glucose-sparing effect.** Its role in mobilizing FFAs reduces the body's dependence on glucose, which is used instead for glycogen synthesis and storage.
- **Electrolyte balance.** GH promotes Na^+, K^+, and Cl^- retention by the kidneys, enhances Ca^{2+} absorption by the small intestine, and makes these electrolytes available to the growing tissues.

The most conspicuous effects of GH are on bone, cartilage, and muscle growth, especially during childhood and adolescence. Through IGF-I, it stimulates bone growth at the epiphyseal plates. It promotes the multiplication of chondrocytes and osteogenic cells and stimulates protein deposition in the cartilage and bone matrix. In adulthood, it stimulates osteoblast activity and the appositional growth of bone; thus, it continues to influence bone thickening and remodeling.

The blood GH level declines gradually with age—averaging about 6 ng/mL (ng = nanogram) in adolescence and one-quarter of that in very old age. The resulting decline in protein synthesis may contribute to aging of the tissues, including wrinkling of the skin and decreasing muscular mass and strength. At age 30, the average adult body is 10% bone, 30% muscle, and 20% fat; at age 75, it averages 8% bone, 15% muscle, and 40% fat.

GH concentration fluctuates greatly over the course of a day. It rises to 20 ng/mL or higher during the first 2 hours of deep sleep and may reach 30 ng/mL in response to vigorous exercise. Smaller peaks occur after high-protein meals, but high-carbohydrate meals tend to suppress GH secretion. Trauma, hypoglycemia (low blood sugar), and other conditions also stimulate GH secretion.

Posterior Lobe Hormones

Antidiuretic[10] Hormone (ADH) ADH acts on the kidneys to increase water retention, reduce urine volume, and help prevent dehydration. We will study this hormone more extensively when we deal with the urinary system. ADH is also called *vasopressin* because it causes vasoconstriction at high concentrations. These concentrations are so unnaturally high for the human body, however, that this effect is of doubtful significance except in pathological states. ADH also functions as a brain neurotransmitter and is usually called vasopressin in the neurobiology literature.

Oxytocin[11] (OT) OT has various reproductive roles. In childbirth, it stimulates smooth muscle cells of the uterus to contract, thus contributing to the labor contractions that expel the infant. In lactating mothers, it stimulates musclelike cells of the mammary glands to squeeze on the glandular acini and force milk to flow down the ducts to the nipple. In both sexes, OT secretion surges during sexual arousal and orgasm. It may play a role in the propulsion of semen through the male reproductive tract, in uterine contractions that help transport sperm up the female reproductive tract, and in feelings of sexual satisfaction and emotional bonding.

Hormones of the pituitary gland are summarized in table 17.4.

Control of Pituitary Secretion

Pituitary hormones are not secreted at a steady rate. GH is secreted mainly at night, LH peaks at the middle of the menstrual cycle, and OT surges during labor and nursing, for example. The timing and amount of pituitary secretion is regulated by the hypothalamus, other brain centers, and feedback from the target organs.

Hypothalamic and Cerebral Control

Both lobes of the pituitary gland are strongly subject to control by the brain. As we have seen, the anterior lobe is regulated by releasing hormones and inhibiting hormones of the hypothalamus. Thus, the brain can monitor conditions within and outside of the body and trigger or suppress the release of anterior lobe hormones appropriately. For example, in cold weather, the hypothalamus triggers the secretion of TSH, which indirectly helps generate body heat; in times of stress, it triggers ACTH secretion, which indirectly mobilizes materials needed for tissue repair; during pregnancy, it triggers PRL secretion so a woman will be prepared to lactate; after a high-protein meal, it triggers the release of GH so we can best use the amino acids for tissue growth.

The posterior lobe of the pituitary is controlled by **neuroendocrine reflexes**—the release of hormones in response to signals from the nervous system. For example, the suckling of an infant stimulates nerve endings in the nipple. Sensory signals are transmitted through the spinal cord and brainstem to the hypothalamus and from there to the posterior pituitary. This triggers the release of oxytocin into the blood, which results in milk ejection.

Antidiuretic hormone (ADH) is also controlled by a neuroendocrine reflex. Dehydration raises the osmolarity of the blood, which is detected by hypothalamic neurons called *osmoreceptors.* The osmoreceptors trigger

10. *anti* = against + *diuret* = to pass through, urinate

11. *oxy* = sharp, quick + *toc* = childbirth

Table 17.4 Pituitary Hormones

Hormone	Target Organ	Principal Effects
Anterior Pituitary		
FSH: Follicle-stimulating hormone	Ovaries, testes	Female: growth of ovarian follicles and secretion of estrogen Male: sperm production
LH: Luteinizing hormone	Ovaries, testes	Female: ovulation, maintenance of corpus luteum Male: testosterone secretion
TSH: Thyroid-stimulating hormone	Thyroid gland	Growth of thyroid, secretion of thyroid hormone
ACTH: Adrenocorticotropic hormone	Adrenal cortex	Growth of adrenal cortex, secretion of glucocorticoids
PRL: Prolactin	Mammary glands, testes	Female: milk synthesis Male: increased LH sensitivity and testosterone secretion
GH: Growth hormone (somatotropin)	Liver	Somatomedin secretion, widespread tissue growth
Posterior Pituitary		
ADH: Antidiuretic hormone	Kidneys	Water retention
OT: Oxytocin	Uterus, mammary glands	Labor contractions, milk release; possibly involved in ejaculation, sperm transport, and sexual affection

ADH release, and ADH promotes water conservation. Excessive blood pressure, by contrast, stimulates stretch receptors in the heart and certain arteries. By another neuroendocrine reflex, this inhibits ADH release, increases urine output, and brings blood volume and pressure back to normal.

Neuroendocrine reflexes can also involve higher brain centers. For example, the milk-ejection reflex can be triggered when a lactating mother simply hears a baby cry. Emotional stress can affect the secretion of gonadotropins, thus disrupting ovulation, the menstrual rhythm, and fertility.

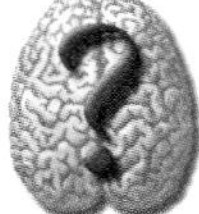

Think About It

Which of the unifying themes at the end of chapter 1 (pp. 21–22) is best exemplified by the neuroendocrine reflexes that govern ADH secretion?

Feedback from Target Organs

The regulation of other endocrine glands by the pituitary is not simply a system of "command from the top down." Those target organs also regulate the pituitary and hypothalamus through various feedback loops.

Most often, this is seen in the form of **negative feedback inhibition**—the pituitary stimulates another endocrine gland to secrete its hormone, and that hormone feeds back to the pituitary and inhibits further secretion. All of the pituitary axes are controlled this way. Figure 17.6 shows negative feedback inhibition in the pituitary-thyroid axis as an example. The figure is numbered to correspond to the following description:

1. The hypothalamus secretes thyrotropin-releasing hormone (TRH).
2. TRH stimulates the anterior pituitary to secrete thyroid-stimulating hormone (TSH). (Plus signs and green arrows indicate stimulatory effects.)
3. TSH stimulates the thyroid gland to secrete the two thyroid hormones, T_3 and T_4.
4. T_3 and T_4 stimulate the metabolism of most cells throughout the body.
5. T_3 and T_4 also *inhibit* the release of TSH by the pituitary. (Minus signs and red arrows indicate inhibitory effects.)
6. To a lesser extent, T_3 and T_4 also *inhibit* the release of TRH by the hypothalamus.

Steps 5 and 6 are negative feedback inhibition of the pituitary and hypothalamus. These steps ensure that when thyroid hormone levels are high, TSH secretion remains low. If thyroid hormone secretion drops, TSH secretion rises and stimulates the thyroid to secrete more hormone. This negative feedback keeps thyroid hormone levels oscillating around a set point in typical homeostatic fashion.

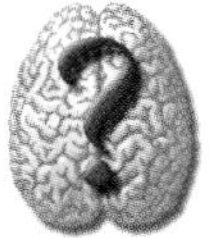

Think About It

If you removed the thyroid gland from a cancer patient, would you expect the level of TSH to rise or fall? Why?

Feedback from a target organ is not always inhibitory. During labor, oxytocin triggers a positive feedback cycle. Uterine stretching sends a nerve signal to the brain that stimulates OT release. OT stimulates uterine contractions,

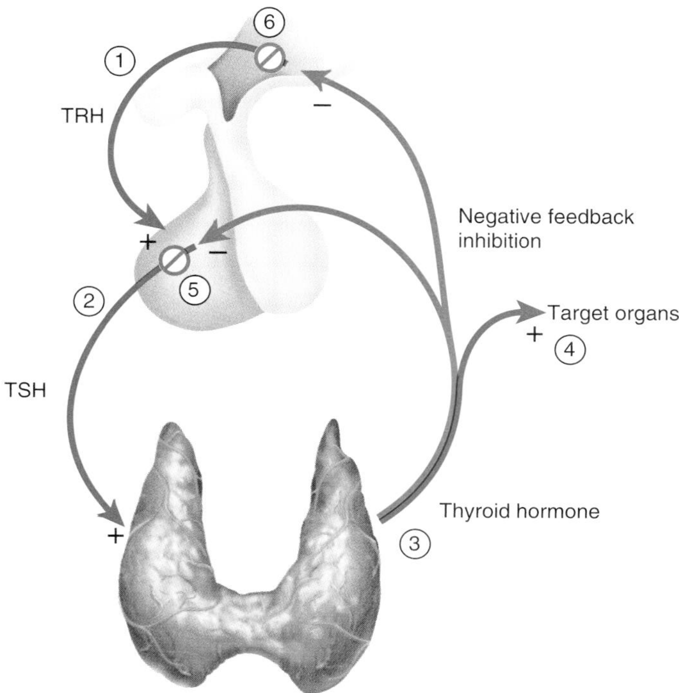

Figure 17.6 **Negative Feedback Inhibition in the Pituitary-Thyroid Axis.** See text for explanation of symbols and numbered steps.

which push the infant downward. This stretches the lower end of the uterus some more, which results in a nerve signal that stimulates still more OT release. This positive feedback cycle continues until the infant is born (see fig. 1.15, p. 21).

Pituitary Disorders

Many diseases result from **hyposecretion** (underactivity) or **hypersecretion** (overactivity) of the endocrine glands. Insufficient pituitary activity, called **hypopituitarism,** can result from prolonged steroid treatment, cerebral hemorrhage, blood clots, or tumors, among other causes. Childhood hypopituitarism results in **pituitary dwarfism** (fig. 17.7*a*) because of the deficiency of growth hormone. If pituitary secretion stops completely (*panhypopituitarism*[12]), the thyroid gland, adrenal cortex, and gonads atrophy. As their activity declines, a broad range of disorders sets in: infertility, loss of pubic and axillary hair, fluid and electrolyte imbalances, and disruption of metabolism. If the pituitary is destroyed, a patient requires lifelong *hormone replacement therapy (HRT)* with thyroxine, cortisone, growth hormone, and sex steroids.

12. *pan* = complete, all

(a)

(b)

Figure 17.7 **Effects of Pituitary Dysfunction.** (*a*) Pituitary dwarfism, the effect of childhood hyposecretion of growth hormone; (*b*) pituitary gigantism, the effect of childhood hypersecretion of growth hormone.

Figure 17.8 **Acromegaly, a Condition Caused by Pituitary Hypersecretion in Adulthood.** These are four photographs of the same person taken at the ages of 9, 16, 33, and 52. Note the characteristic thickening of the face and hands.

Posterior lobe hyposecretion results mainly in **diabetes insipidus** due to a lack of ADH. The term *diabetes* simply means any chronic excessive urine output; not all forms of diabetes are related to insulin and glucose metabolism. Diabetes insipidus usually stems from destruction of the hypothalamus or hypothalamo-hypophyseal tract by a tumor, infection, or basal skull fracture. It causes weakness, intense thirst, and a urine output that may exceed 15 L per day—10 times the normal output. It also creates electrolyte imbalances that place a person at risk for neurological and cardiac disorders. ADH can be taken by injection or nasal spray to treat this disorder.

Pituitary hypersecretion **(hyperpituitarism)** in childhood causes excessive growth, or **gigantism** (fig. 17.7*b*). In adulthood, it causes **acromegaly**—thickening of the bones and soft tissues with especially noticeable effects on the hands, feet, and face (fig. 17.8).

Before You Go On

Answer the following questions to test your understanding of the preceding section:

5. What are two good reasons for considering the pituitary to be two separate glands?
6. Name three anterior lobe hormones that have reproductive functions and three that have nonreproductive roles. What target organs are stimulated by each of these hormones?
7. Briefly contrast hypothalamic control of the anterior pituitary with its control of the posterior pituitary.
8. In what sense does the pituitary "take orders" from the target organs under its command?

Other Endocrine Glands

Objectives

When you have completed this section, you should be able to

- describe the structure and location of the remaining organs of the endocrine system;
- name the hormones these endocrine organs produce and state their functions; and
- describe some common effects of hormone imbalances.

The Pineal Gland

The **pineal**[13] **gland (epiphysis cerebri)** is a pine cone–shaped growth attached to the roof of the third ventricle, beneath the posterior end of the corpus callosum (see fig. 17.1). The philosopher René Descartes (1596–1650) thought it was the seat of the human soul. If so, children must have more soul than adults—a child's pineal gland is about 8 mm long and 5 mm wide, but after age seven it regresses rapidly and is no more than a tiny shrunken mass of fibrous tissue in the adult. Pineal secretion peaks between the ages of 1 and 5 years; by the end of puberty it is 75% lower.

We no longer look for the human soul in the pineal gland, but this little organ remains an intriguing mystery. In animals with seasonal breeding, it regulates the gonads and the annual breeding cycle. It produces **serotonin** by day and converts it to **melatonin** at night. Melatonin may

13. *pineal* = pine cone

Clinical Insight 17.1

Melatonin, SAD, and PMS

There seems to be a relationship between melatonin and mood disorders, including depression and sleep disturbances. Some people experience a mood dysfunction called *seasonal affective disorder (SAD)*, especially in winter when the days are shorter and they get less exposure to sunlight, and in extreme northern and southern latitudes where sunlight may be dim to nonexistent for months at a time. SAD thus affects about 20% of the population in Alaska but only 2.5% in Florida. The symptoms—which include depression, sleepiness, irritability, and carbohydrate craving—can be relieved by 2 or 3 hours of exposure to bright light each day (*phototherapy*). *Premenstrual syndrome (PMS)* is similar to SAD and is also reduced by phototherapy. The melatonin level is elevated in both SAD and PMS and is reduced by phototherapy. However, there is also evidence that casts doubt on any causal link between melatonin and these mood disorders, so for now, "the jury is still out." Many people are taking melatonin for jet lag, and it is quite effective, but it is also risky to use when we know so little, as yet, about its potential effect on reproductive function.

suppress gonadotropin secretion; removal of the pineal from animals causes premature sexual maturation. Some physiologists think that the pineal gland may regulate the timing of puberty in humans, but a clear demonstration of its role has remained elusive. Pineal tumors cause premature onset of puberty in boys, but such tumors also damage the hypothalamus, so we cannot be sure the effect is due specifically to pineal damage.

The Thymus

The **thymus** is located in the mediastinum superior to the heart (fig. 17.9). Like the pineal, it is large in infants and children but undergoes **involution** (shrinkage) after puberty. In elderly people, it is a shriveled vestige of its former self, with most of its parenchyma replaced by fibrous and adipose tissue. The thymus secretes *thymopoietin* and *thymosins*, hormones that regulate the development and later activation of disease-fighting blood cells called T lymphocytes (*T* for *thymus*). This is discussed in detail in chapter 21.

The Thyroid Gland

The **thyroid gland** is the largest endocrine gland; it weighs 20 to 25 g and receives one of the body's highest rates of blood flow per gram of tissue. It is wrapped around the anterior and lateral sides of the trachea, immediately below the larynx. It consists of two large lobes, one on each side of the trachea, connected by a narrow anterior *isthmus* (fig. 17.10*a*).

Histologically, the thyroid is composed mostly of sacs called **thyroid follicles** (fig. 17.10*b*). Each is filled with a protein-rich colloid and lined by a simple cuboidal epithelium of **follicular cells.** These cells secrete two main

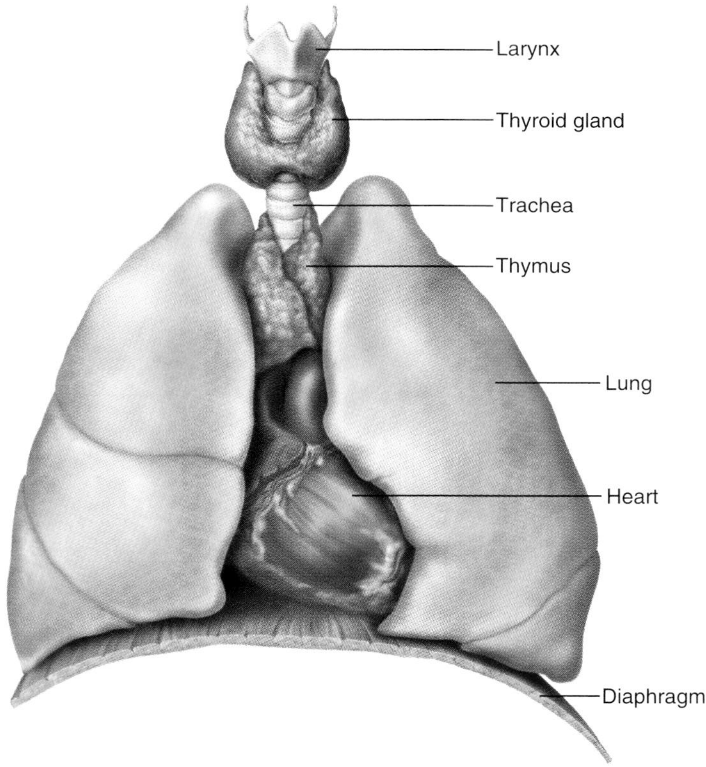

Figure 17.9 Locations of the Thyroid Gland and Thymus.

thyroid hormones—**thyroxine,** also known as $\mathbf{T_4}$ or **tetraiodothyronine** (TET-ra-EYE-oh-doe-THY-ro-neen), and $\mathbf{T_3}$, or **triiodothyronine** (try-EYE-oh-doe-THY-ro-neen). The expression *thyroid hormone* refers to T_4 and T_3 collectively.

The primary effect of thyroid hormone (TH) is to increase the body's metabolic rate. As a result, it raises

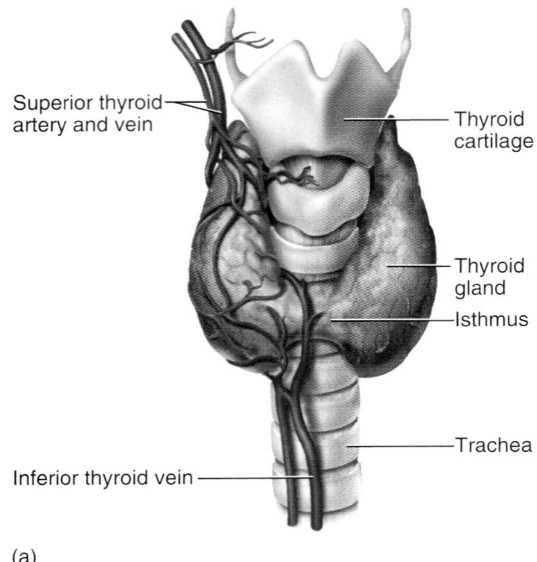

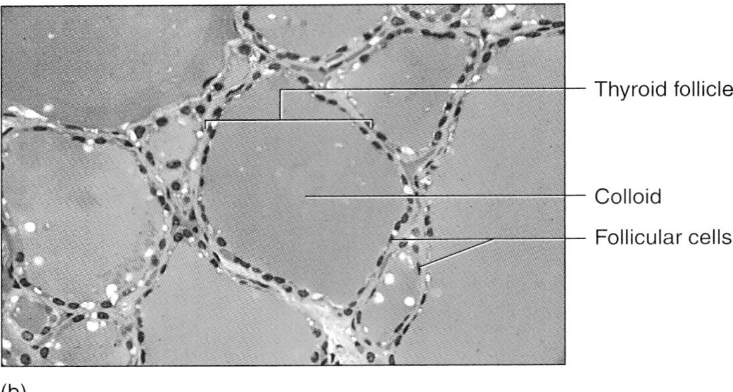

Figure 17.10 The Thyroid Gland. (*a*) Gross anatomy; (*b*) histology.

oxygen consumption and has a **calorigenic**[14] **effect**—it increases the body's heat production. TH secretion rises in cold weather and thus helps to compensate for the increased heat loss we experience. To ensure an adequate blood and oxygen supply to meet this increased metabolic demand, thyroid hormone also raises the heart rate and contraction strength and raises the respiratory rate. It accelerates the breakdown of carbohydrates, fats, and protein for fuel and stimulates the appetite. Thyroid hormone promotes alertness, bone growth and remodeling, the development of the skin, hair, nails, and teeth, and fetal nervous system and skeletal development. It also stimulates the pituitary gland to secrete growth hormone.

Calcitonin is another hormone produced by the thyroid gland. It comes from **C (calcitonin) cells,** also called *parafollicular cells,* found in clusters between the thyroid follicles. Calcitonin is secreted when blood calcium level rises. It antagonizes the action of parathyroid hormone (described shortly) and promotes calcium deposition and bone formation by stimulating osteoblast activity. Calcitonin is important mainly to children. It has relatively little effect in adults for reasons explained earlier (p. 247), although it is sometimes given by injection or nasal spray as a treatment for osteoporosis.

Thyroid Disorders

Thyroid hyposecretion present from birth is called **congenital hypothyroidism** (formerly *cretinism,* now regarded as an insensitive term). Hypothyroidal infants have a normal appearance at birth because they have received an adequate supply of thyroid hormone from their mothers. Removed from this source, however, untreated infants show stunting and other abnormalities of bone development, thickened facial features, low body temperature, lethargy, and irreversible brain damage resulting in mental retardation. Adult hypothyroidism causes **myxedema** (MIX-eh-DEE-muh), a syndrome characterized by low metabolic rate, sluggishness and sleepiness, weight gain, constipation, dry skin and hair, abnormal sensitivity to cold, elevated blood pressure, and swelling of the tissues. Thyroid hormone can be given orally to treat both infant and adult hypothyroidism.

A *goiter* is any pathological enlargement of the thyroid gland. **Endemic goiter** (fig. 17.11) is due to dietary iodine deficiency. There is little iodine in soil or most foods, but seafood and iodized salt are good sources. Without iodine, the gland cannot synthesize TH. Without TH, the pituitary gland receives no feedback and acts as if the thyroid were understimulated. It produces extra TSH, which stimulates hypertrophy of the thyroid gland.

Thyroid hypertrophy also occurs in **toxic goiter (thyrotoxicosis, Graves**[15] **disease),** an autoimmune disease in which abnormal antibodies mimic the effect of TSH on the thyroid. This raises TH secretion (in contrast to endemic goiter), and all of the body's tissues suffer from excessive stimulation. A person with toxic goiter exhibits an elevated metabolic rate and heart rate, nervousness, sleeplessness, weight loss, abnormal sweating and heat sensitivity, and *exophthalmos* (EX-off-THAL-mos), bulging of the eyes because of edema in the orbit.

The Parathyroid Glands

The **parathyroid glands** are partially embedded in the posterior surface of the thyroid (fig. 17.12). There are usually four, each about 3 to 8 mm long and 2 to 5 mm wide.

14. *color* = heat + *genic* = producing

15. Robert James Graves (1796–1853), Irish physician

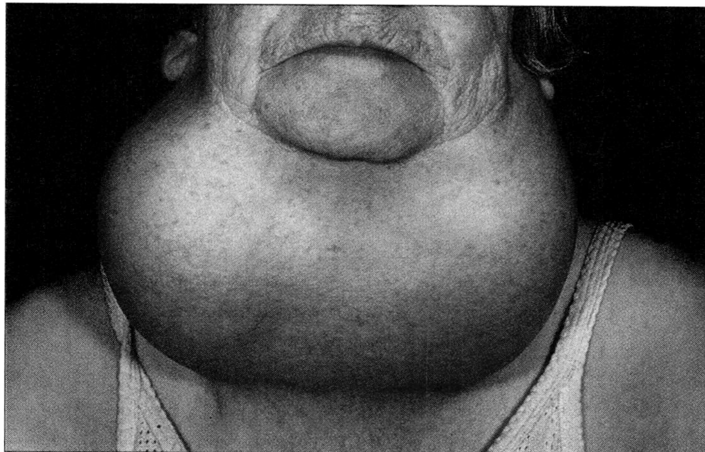

Figure 17.11 Endemic Goiter, a Hypertrophy of the Thyroid Gland Resulting from Iodine Deficiency.

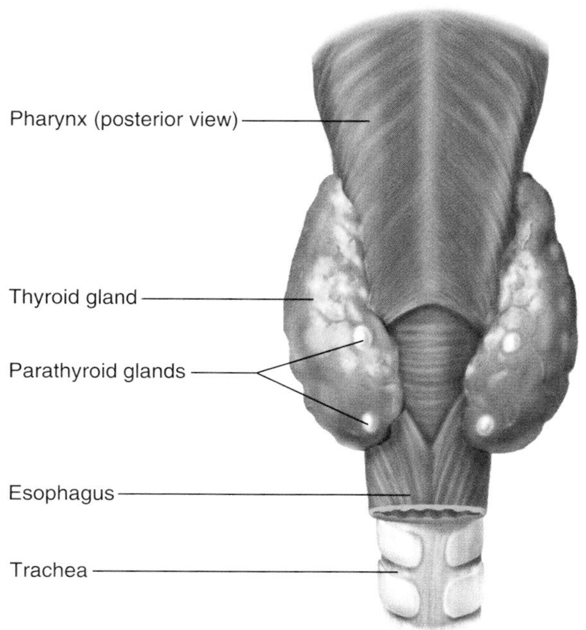

Figure 17.12 The Parathyroid Glands. There are usually four parathyroid glands embedded in the posterior surface of the thyroid gland.

They secrete **parathyroid hormone (PTH)** in response to hypocalcemia. PTH raises blood calcium levels by promoting intestinal calcium absorption, inhibiting urinary calcium excretion, promoting phosphate excretion (so the phosphate does not combine with calcium and deposit into the bones), and indirectly stimulating osteoclasts to resorb bone. PTH and calcium metabolism are discussed in more detail in chapter 8.

Because of their location and small size, the parathyroids are sometimes accidentally removed in thyroid surgery. Without hormone replacement therapy, the resulting **hypoparathyroidism** causes a rapid decline in blood calcium levels and leads to fatal tetany within 3 or 4 days.

Hyperparathyroidism, excess PTH secretion, is usually caused by a parathyroid tumor. It causes the bones to become soft, deformed, and fragile; it raises the blood levels of calcium and phosphate ions; and it promotes the formation of *renal calculi* (kidney stones) composed of calcium phosphate.

The Adrenal Glands

An **adrenal (suprarenal) gland** sits like a cap on the superior pole of each kidney (fig. 17.13). In adults, the adrenal is about 5 cm (2 in.) long, 3 cm (1.2 in.) wide, and weighs about 4 g; it weighs about twice this much at birth. Like the pituitary gland, the adrenal gland is formed by the merger of two fetal glands with different origins and functions. Its inner core, the *adrenal medulla,* is a small portion of the total gland. Surrounding it is a much thicker *adrenal cortex.*

The Adrenal Medulla

The **adrenal medulla** was discussed as part of the sympathetic nervous system in chapter 15 and is reviewed only briefly here. It arises from the neural crest and is not fully formed until the age of three. It is actually a sympathetic ganglion consisting of modified neurons, called *chromaffin cells,* that lack dendrites and axons. These cells are richly innervated by sympathetic preganglionic fibers and respond to stimulation by secreting catecholamines, especially epinephrine and norepinephrine. These hormones mimic the effects of the sympathetic nervous system, but their effects last much longer because the hormones circulate in the blood.

The Adrenal Cortex

The **adrenal cortex** has three layers of glandular tissue (fig. 17.13*b*)—an outer **zona glomerulosa** (glo-MER-you-LO-suh), a thick middle **zona fasciculata** (fah-SIC-you-LAH-ta), and an inner **zona reticularis** (reh-TIC-you-LAR-iss). The cortex synthesizes more than 25 steroid hormones known collectively as the **corticosteroids,** or **corticoids.** These fall into three categories:

1. **Sex steroids.** These are weak **androgens** and smaller amounts of **estrogens** secreted by the zona reticularis. Androgens control many aspects of male

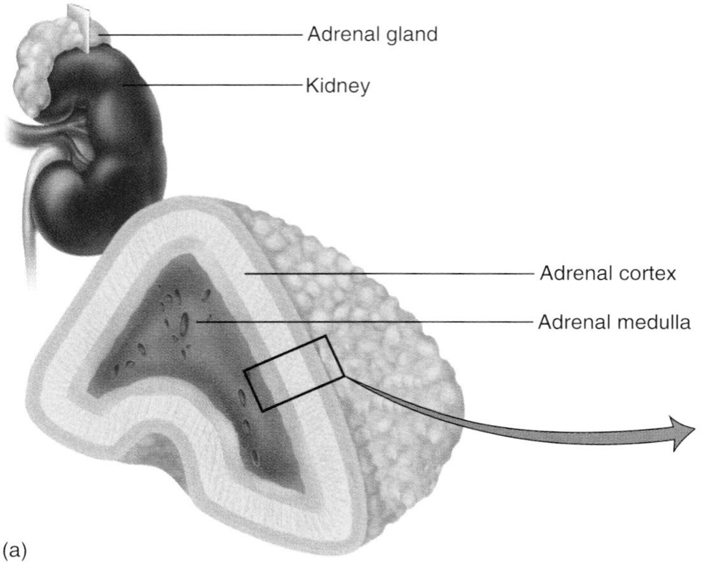

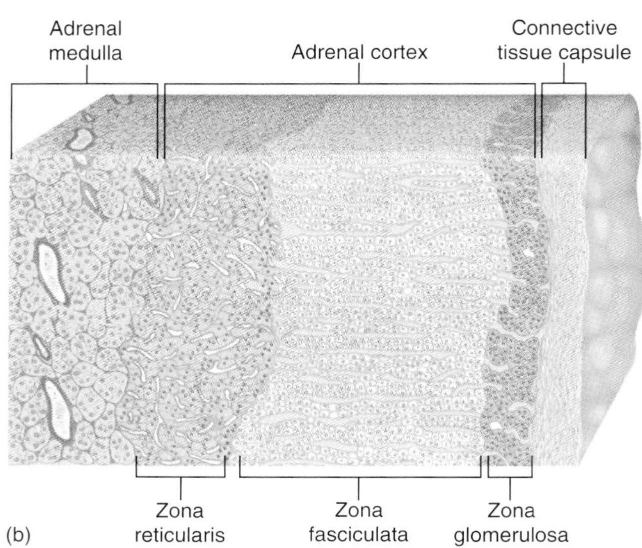

Figure 17.13 The Adrenal Gland. (*a*) Gross anatomy; (*b*) histology.

development and reproductive physiology. The principal androgen secreted by the adrenal gland is **dehydroepiandrosterone (DHEA).** DHEA has weak hormonal effects in itself, but more importantly, the tissues that receive it convert DHEA to the more potent androgen, **testosterone.** This source is relatively unimportant in men because the testes produce so much more testosterone than the adrenals do. In women, however, the adrenal glands meet about 50% of the total androgen requirement. Similarly, adrenal estrogen is of minor importance to women until after menopause, when the ovaries stop functioning and the adrenals are the only remaining estrogen source. Androgens stimulate male fetal development. Later, in both sexes, they stimulate the development of pubic and axillary hair and apocrine scent glands at puberty, and they sustain the libido (sex drive) throughout adult life. Both androgens and estrogens promote adolescent skeletal growth and help to sustain adult bone mass.

2. **Mineralocorticoids.** These steroids, secreted by the zona glomerulosa, control electrolyte balance by acting on the kidneys. The principal mineralocorticoid is **aldosterone,** which acts on the kidneys to promote Na^+ retention and K^+ excretion. It is discussed more fully in chapter 24.
3. **Glucocorticoids.** These steroids are secreted mainly by the zona fasciculata in response to ACTH. The most important glucocorticoid is **cortisol (hydrocortisone); corticosterone** is a less potent relative. Glucocorticoids stimulate fat and protein catabolism, **gluconeogenesis** (the synthesis of glucose from amino acids and fatty acids), and the release of fatty acids and glucose into the blood. This helps the body adapt to stress and to repair damaged tissues. Long-term secretion, however, suppresses the immune system for reasons we will see later in this chapter in the discussion of stress.

Recently it has been discovered that some cells of the adrenal medulla extend into the cortex. When stress activates the sympathetic nervous system, catecholamines from the adrenal medulla stimulate the cortex to secrete corticosterone.

Adrenal Disorders

A tumor of the adrenal medulla called a **pheochromocytoma** (FEE-oh-CRO-mo-sy-TOE-muh) can cause epinephrine and norepinephrine hypersecretion. An excess of these catecholamines mimics overactivity of the sympathetic nervous system, and causes hypertension, elevated metabolic rate, *hyperglycemia* (excess glucose in the blood), *glycosuria* (glucose in the urine), nervousness, indigestion, and sweating. Ultimately, hypersecretion can lead to fatigue, exhaustion, and susceptibility to other diseases.

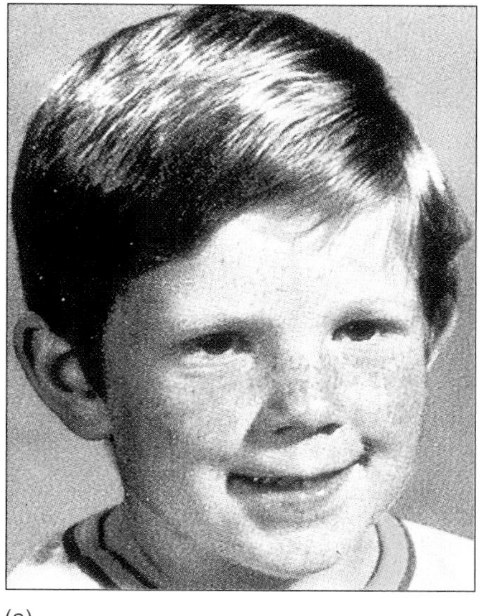
(a)

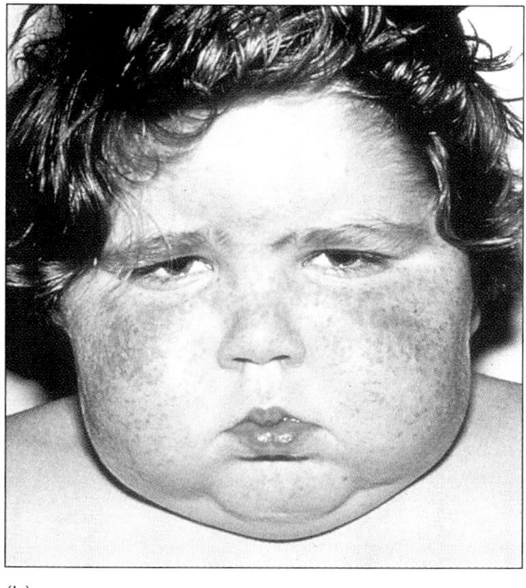
(b)

Figure 17.14. **Cushing Syndrome.** (*a*) Patient before the onset of the syndrome. (*b*) The same boy, only 4 months later, showing the "moon face" characteristic of Cushing syndrome.

Either an adrenal tumor or excess ACTH secretion can cause glucocorticoid hypersecretion and **Cushing[16] syndrome.** This is a disturbance of carbohydrate and protein metabolism with hyperglycemia, hypertension, muscular weakness, and edema. Muscle and bone mass are lost rapidly as protein is catabolized. Some patients exhibit abnormal fat deposition between the shoulders ("buffalo hump") or in the face ("moon face") (fig. 17.14). Long-term corticosteroid therapy has similar effects.

Adrenogenital syndrome (AGS), caused by androgen hypersecretion, commonly accompanies Cushing syndrome. In children, AGS often involves enlargement of the penis or clitoris and the premature onset of puberty. Prenatal AGS can result in newborn girls exhibiting masculinized genitalia and being misidentified as boys (fig. 17.15). In women, AGS produces such masculinizing effects as increased body hair, deepening of the voice, and beard growth.

Addison[17] disease is the hyposecretion of glucocorticoids and mineralocorticoids. Its signs include hypoglycemia, Na^+ and K^+ imbalance, dehydration, hypotension (low blood pressure), weight loss, weakness, and a loss of stress resistance. In the absence of negative feedback inhibition from the adrenal cortex, the pituitary secretes excess ACTH. This stimulates melanin synthesis and causes the skin to darken, often giving it a metallic golden look called *bronzing* (fig. 17.16). President John F. Kennedy had Addison disease and bronzing of the skin, which many mistook for a suntan. Unless treated with corticosteroids, Addison disease can lead to fatal dehydration and electrolyte imbalances.

16. Harvey Cushing (1869–1939), American physician
17. Thomas Addison (1793–1860), English physician

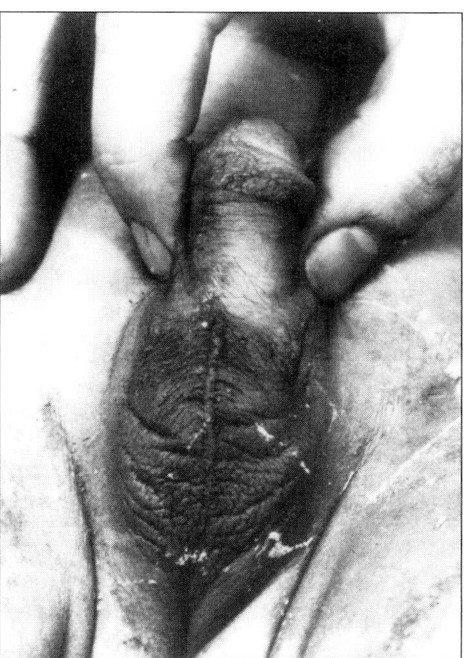

Figure 17.15 **Adrenogenital Syndrome (AGS).** These are the genitals of a female baby with AGS, masculinized by prenatal hypersecretion of adrenal androgens. Note the fusion of the labia majora to resemble a scrotum and enlargement of the clitoris to resemble a penis. Such infants are easily mistaken for boys and raised as such.

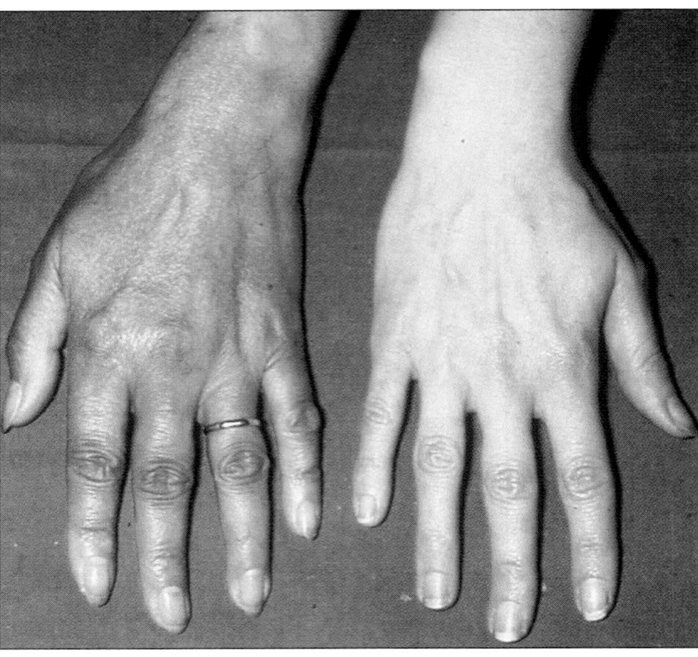

Figure 17.16 Addison Disease. The hand on the *left* exhibits the darkening or "bronzing" of the skin compared to a normal hand on the *right.*

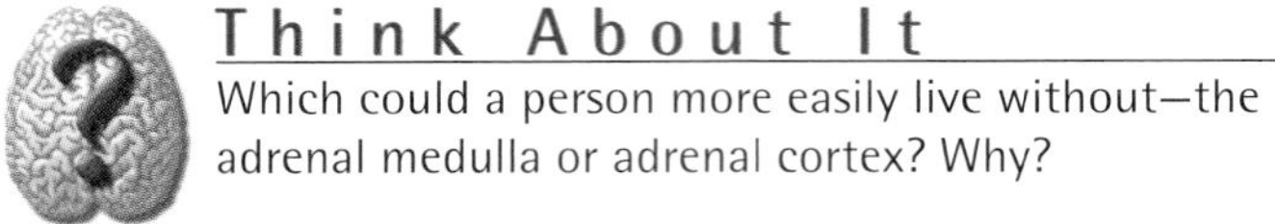

Think About It

Which could a person more easily live without—the adrenal medulla or adrenal cortex? Why?

The Pancreas

The elongated spongy **pancreas** is located retroperitoneally, inferior and dorsal to the stomach (fig. 17.17). It is approximately 15 cm long and 2.5 cm thick. Most of it is an exocrine digestive gland, but scattered among the exocrine acini are endocrine cell clusters called the **pancreatic islets (islets of Langerhans[18]).** The islets secrete at least five hormones or hormonelike messengers, the two most important of which are insulin and glucagon:

- **Insulin** is secreted by the **beta (β) cells** of the islets when we digest a meal and the level of glucose and amino acids in the blood rises. In such "times of plenty," insulin stimulates cells to absorb glucose and amino acids from the blood and especially stimulates muscle and adipose tissue to store glycogen and fat. The liver absorbs glucose even without insulin, but insulin promotes its synthesis of glycogen. The brain, kidneys, and red blood cells also do not depend on insulin for glucose uptake. Essentially, insulin stimulates cells to store excess nutrients for later use, and it suppresses the use of already-stored fuels. The stored nutrients are then available for use between meals and overnight. By stimulating glycogen, fat, and protein synthesis, insulin promotes cell growth and differentiation. Insulin also antagonizes the effects of glucagon, discussed next.
- **Glucagon** is secreted by **alpha (α) cells** when blood glucose levels fall between meals. In the liver, it stimulates gluconeogenesis, **glycogenolysis** (glycogen hydrolysis), and the release of glucose into circulation. In adipose tissue, it stimulates fat catabolism and the release of free fatty acids. Glucagon is also secreted in response to rising amino acid levels in the blood after a high-protein meal. By promoting amino acid absorption, it provides cells with raw material for gluconeogenesis.

Somatostatin (another name for growth hormone–inhibiting hormone, GHIH) is another important pancreatic secretion. It is also produced by the hypothalamus, spinal cord, anterior pituitary, digestive tract, gonads, and other organs. **Delta (δ) cells** of the pancreatic islets secrete somatostatin when blood glucose and amino acids rise after a meal. It acts not as a hormone but as a *paracrine* secretion—a chemical messenger that diffuses to target cells a short distance away in the same organ. Somatostatin modulates the secretion of glucagon and insulin by the neighboring α and β cells.

Any hormone that raises blood glucose concentration is called a *hyperglycemic hormone.* You may have noticed that glucagon is not the only hormone that does so; so do growth hormone, epinephrine, norepinephrine, cortisol, and corticosterone. Insulin is called a *hypoglycemic hormone* because it lowers blood glucose levels.

Diabetes Mellitus

Diabetes mellitus[19] (DM) is a disruption of carbohydrate, fat, and protein metabolism resulting from the hyposecretion or inaction of insulin. It is the most widespread disorder of the endocrine system. Its classic signs are "the three polys": **polyuria[20]** (excessive urine output), **polydipsia[21]** (intense thirst), and **polyphagia[22]** (ravenous hunger). Blood and urine tests to confirm diabetes melli-

18. Paul Langerhans (1847–88), German anatomist

19. *diabet* = to flow through + *melli* = honey
20. *poly* = much, excessive + *uri* = urine
21. *dipsia* = drinking
22. *phagia* = eating

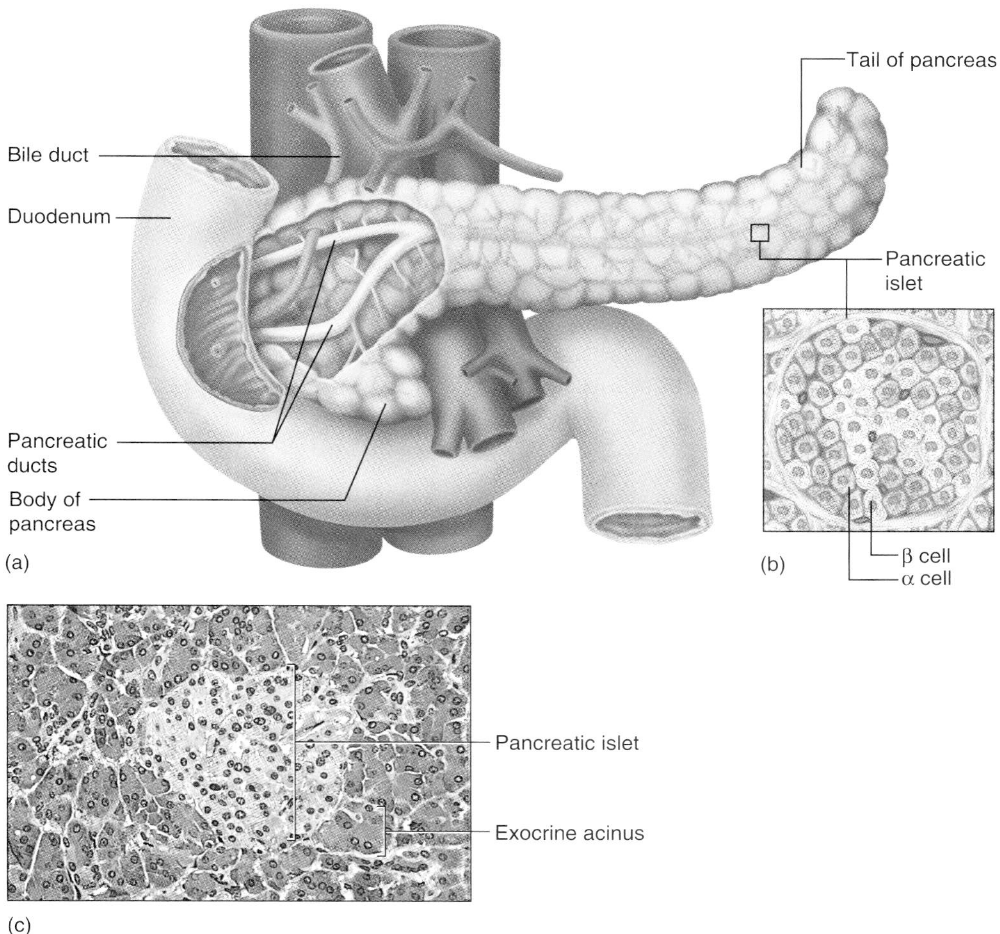

Figure 17.17 The Pancreas. (*a*) Gross anatomy and relationship to the duodenum and other nearby organs; (*b*) the α cells and β cells of a pancreatic islet amid the exocrine acini; *(c)* light micrograph of a pancreatic islet amid the exocrine acini.

tus typically reveal **hyperglycemia**[23] (elevated blood glucose), **glycosuria**[24] (glucose in the urine), and **ketonuria** (ketones in the urine).

A little knowledge of kidney physiology is necessary to understand why glycosuria and polyuria occur. The kidneys filter blood plasma and process the filtrate to produce urine. Normally, the kidney tubules remove all glucose from the filtrate and return it to the blood. Thus there is little or no glucose in the urine of a healthy person. By osmosis, the tubules also reclaim most of the water in the filtrate.

Like any other carrier-mediated transport system, the glucose transporters of the kidney tubules have a transport maximum, T_m (see p. 129). In diabetes mellitus, glucose enters the tubules so rapidly that it exceeds the T_m and the tubules cannot reabsorb it fast enough. The excess passes through into the urine. Glucose and ketones in the tubules also raise the osmolarity of the tubular fluid and cause **osmotic diuresis**—water remains in the tubules with these solutes, so large amounts of water are passed in the urine. This accounts for the polyuria, dehydration, and thirst of diabetes. Diabetics often pass 10 to 15 L of urine per day, compared with 1 or 2 L in a healthy person.

Types and Treatment of Diabetes Mellitus There are two forms of diabetes mellitus—*type I* and *type II.* **Type I,** or **insulin-dependent diabetes mellitus (IDDM),** accounts for about 10% of cases. The cause is still little understood, but at least some cases appear to result from the autoimmune destruction of β cells by misguided antibodies. There are

23. *hyper* = excess + *glyc* = sugar, glucose + *emia* = blood condition
24. *glyco* = glucose, sugar + *uria* = urine condition

several genes implicated in IDDM, but heredity alone is not the cause; IDDM results from the interaction of hereditary and environmental factors. Up to a point, the body can tolerate the loss of β cells. Signs of diabetes begin to appear when 80% to 90% of them are destroyed and insulin falls to critically low levels. However, the signs are not caused by a deficiency of insulin alone. The relative level of glucagon is elevated in IDDM, and it is the abnormally low *ratio* of insulin to glucagon that causes the signs. IDDM is most often diagnosed around age 12 and used to be called *juvenile diabetes,* but it can also appear later in life. IDDM must be treated through meal planning, exercise, the self-monitoring of blood glucose level by the patient, and periodic insulin injections or the continual subcutaneous delivery of insulin by a pump worn on the body.

Most diabetics (90%) have **type II,** or **non-insulin-dependent diabetes mellitus (NIDDM).** About 6.6% of U.S. residents are diagnosed with NIDDM. Type II diabetics usually have normal or even elevated insulin levels; the problem is *insulin resistance*—a failure of target cells to respond to insulin. There may be multiple reasons for resistance; one theory is that target cells have a shortage of insulin receptors. The three major risk factors for NIDDM are heredity, age, and obesity. It tends to run in families; if an identical twin contracts NIDDM it is virtually certain that the other twin will also have it. NIDDM has a gradual onset, with signs usually not appearing until age 40 or beyond; it used to be called *adult-onset DM.* Type II diabetics are usually overweight. In obesity, the adipocytes produce a hormonelike secretion that indirectly interferes with glucose transport into most kinds of cells. NIDDM can often be managed through a weight-loss program of diet and exercise. Some patients are helped with oral medications that improve insulin secretion or target cell sensitivity or help in other ways to lower blood glucose.

Pathology When cells cannot absorb glucose, they rely on fat and protein for energy. Fat and protein breakdown results in emaciation, muscular atrophy, and weakness. Before insulin therapy was introduced in 1922, the victims of type I diabetes wasted away to an astonishing extent (see historical insight 17.4 at the end of the chapter). Diabetes was described in the first century as "a melting down of the flesh and limbs into urine." Adult patients weighed as little as 27 to 34 kg (60–75 lb) and looked like victims of a concentration camp or severe famine. Their breath had a sickeningly sweet ketone smell, like rotten apples. One typical patient was described by historian Michael Bliss as "barely able to lift his head from his pillow, crying most of the time from pain, hunger, and despair." In the terminal stage of the disease, patients became increasingly drowsy, gasped for air, became comatose, and then died within a few hours. Most diabetic children lived less than 1 year after diagnosis—a year of utmost misery at that.

Rapid fat catabolism elevates blood concentrations of free fatty acids and their breakdown products, the ketone bodies (acetoacetic acid, acetone, and β-hydroxybutyric acid). Ketonuria promotes osmotic diuresis and flushes Na^+ and K^+ from the body, thus creating electrolyte deficiencies that can lead to abdominal pain, vomiting, irregular heartbeat, and neurological dysfunction. As acids, ketones lower the pH of the blood and produce a condition called **ketoacidosis.** Ketoacidosis causes *dyspnea,* the gasping for air seen in terminal diabetes, and depresses the nervous system, which produces the diabetic coma.

Diabetes mellitus also leads to long-term degenerative cardiovascular and neurological diseases—signs that were seldom seen before insulin therapy, when patients died too quickly to show the chronic effects. Despite years of debate, it remains unclear exactly how diabetes mellitus causes cardiovascular disease. It appears that chronic hyperglycemia activates a metabolic reaction cascade that leads to cellular damage in small to medium blood vessels and peripheral nerves. Nerve damage (*neuropathy*), the most common complication of diabetes mellitus, can lead to impotence, incontinence, and loss of sensation from affected areas. The last effect makes a patient dangerously unaware of minor injuries, which can thus fester from neglect and contribute (along with lowered resistance to infection) to gangrene and the necessity of amputation. Many diabetics lose their toes, feet, or legs to the disease. The cardiovascular effects include degeneration of the small arteries of the retina and the kidneys, leading to blindness and kidney failure as common complications. People with type I diabetes are much more likely to die of kidney failure than those with type II. Diabetes also promotes *atherosclerosis,* the blockage of blood vessels with fatty deposits, causing poor circulation. In type II diabetes, the most common cause of death is heart failure stemming from atherosclerosis of the coronary arteries. Atherosclerosis also contributes to renal failure and gangrene.

Hyperinsulinism

Hyperinsulinism usually occurs when a diabetic injects too much insulin, but it occasionally results from a tumor of the pancreatic islets. It causes such rapid glucose uptake that a person becomes hypoglycemic, weak, and hungry. Hypoglycemia, in turn, triggers the secretion of corrective hyperglycemic hormones—epinephrine, glucagon, and growth hormone—which have side effects including anxiety, sweating, and elevated heart rate. Uncorrected hyperinsulinism can produce **insulin shock**—a condition in which the brain is so deprived of glucose that a person exhibits disorientation, convulsions, or unconsciousness. It can be fatal if the blood glucose level is not quickly elevated. If the person is conscious, a spoonful of sugar mixed into a glass of fruit juice can be given to restore blood sugar level quickly.

The Gonads

Like the pancreas, the **gonads** are both endocrine and exocrine glands. Their exocrine products are eggs and sperm, and their endocrine products are the gonadal hormones, most of which are steroids.

Each follicle of the ovary contains an egg cell surrounded by a wall of **granulosa cells** (fig. 17.18*a*). The granulosa cells produce an estrogen called **estradiol** in the first half of the menstrual cycle. After ovulation, the corpus luteum secretes **progesterone** for 12 days or so, or for several weeks in the event of pregnancy. The functions of estradiol and progesterone are discussed in chapter 28. In brief, they contribute to the development of the reproductive system and feminine physique, they regulate the menstrual cycle, they sustain pregnancy, and they prepare the mammary glands for lactation. The follicle and corpus luteum also secrete **inhibin,** which suppresses FSH secretion by means of negative feedback inhibition of the anterior pituitary.

The testis consists mainly of microscopic tubules that produce sperm. Nestled between them are clusters of **interstitial cells (cells of Leydig[25])** (fig. 17.18*b*), which produce testosterone and lesser amounts of weaker androgens and estrogen. Testosterone stimulates development of the male reproductive system in the fetus and adolescent, the development of the masculine physique in adolescence, and the sex drive. It sustains sperm production and the sexual instinct throughout adult life.

25. Franz von Leydig (1821–1908), German biologist

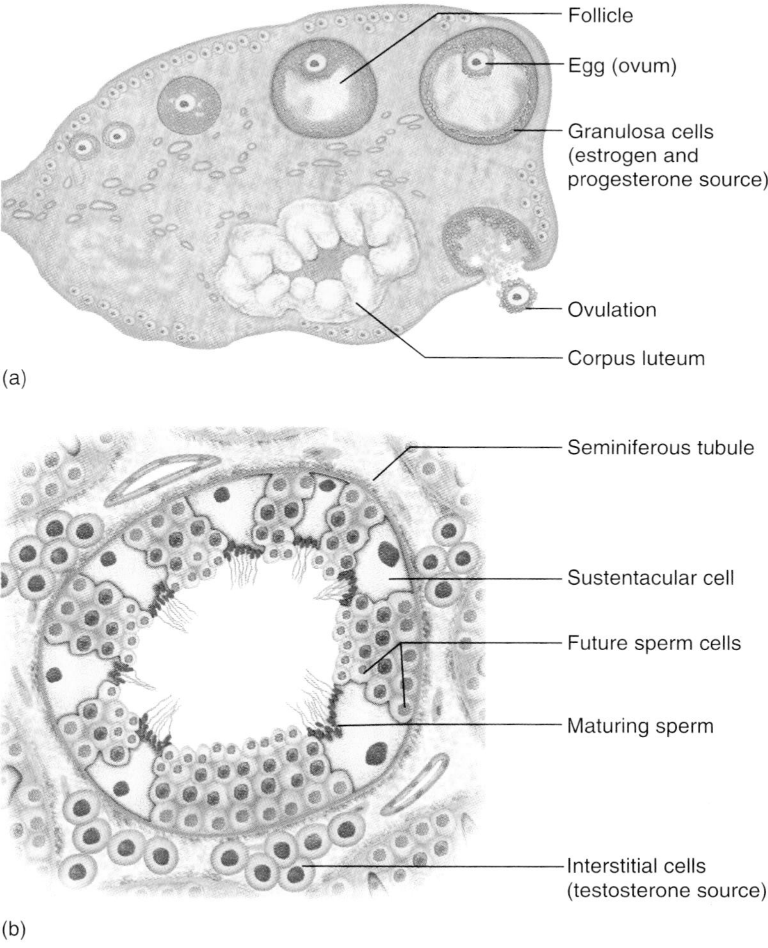

Figure 17.18 **The Gonads.** (*a*) Histology of the ovary; (*b*) histology of the testis. The granulosa cells of the ovary and interstitial cells of the testis are endocrine cells.

Sustentacular (Sertoli[26]) cells of the testis secrete inhibin, which suppresses FSH secretion and thus homeostatically stabilizes the rate of sperm production.

Endocrine Functions of Other Organs

Several other organs have hormone-secreting cells:

- **The heart.** High blood pressure stretches the heart wall and stimulates muscle cells in the atria to secrete **atrial natriuretic[27] factor (ANF).** ANF increases urine output and sodium excretion and interferes with the actions of angiotensin II, described shortly. Together, these effects lower the blood pressure.
- **The kidneys.** The kidneys produce two hormones—calcitriol (a form of vitamin D) and erythropoietin. **Calcitriol** affects the handling of calcium by the kidneys, small intestines, and bones (see chapter 8). **Erythropoietin** (eh-RITH-ro-POY-eh-tin) stimulates the bone marrow to produce red blood cells (see chapter 18).
- **The liver.** The liver secretes somatomedins, which mediate the action of growth hormone; it produces about 15% of the body's erythropoietin; and it secretes a protein called *angiotensinogen.* In the blood, angiotensinogen is converted to angiotensin I by a kidney enzyme (*renin*) and then to angiotensin II by a lung enzyme (*angiotensin-converting enzyme, ACE*). Angiotensin II stimulates vasoconstriction and aldosterone secretion. Together, these effects raise blood pressure (see chapter 20).
- **The stomach and small intestine.** These have various *enteroendocrine cells,*[28] which secrete at least 10 different **enteric hormones.** In general, they coordinate the different regions and glands of the digestive system with each other (see chapter 25).
- **The placenta.** This organ performs many functions in pregnancy, including fetal nutrition and waste removal. But it also secretes estrogen, progesterone, and other hormones that regulate pregnancy and stimulate development of the fetus and the mother's mammary glands (see chapter 28).

You can see that the endocrine system is extensive. It includes numerous discrete glands as well as individual cells in the tissues of other organs. The endocrine organs and tissues other than the hypothalamus and pituitary are reviewed in table 17.5. Table 17.6 summarizes the effects of some hormone imbalances.

26. Enrico Sertoli (1824–1910), Italian histologist
27. *natri* = sodium + *uretic* = pertaining to urine
28. *entero* = intestine

Table 17.5 Hormones from Sources Other than the Hypothalamus and Pituitary

Hormone	Target	Principal Effects
Pineal Gland		
Melatonin and serotonin	Brain	Influence mood; may regulate the timing of puberty
Thymus		
Thymopoietin and thymosins	T lymphocytes	Stimulate T lymphocytes
Thyroid		
Triiodothyronine (T_3) and thyroxine (T_4)	Most tissues	Elevate metabolic rate, O_2 consumption, and heat production; stimulate circulation and respiration; promote nervous system and skeletal development
Calcitonin	Bone	Promotes Ca^{2+} deposition and ossification; reduces blood Ca^{2+} level
Parathyroids		
Parathyroid hormone (PTH)	Bone, small intestine, kidneys	Increases blood Ca^{2+} level by stimulating bone resorption and intestinal Ca^{2+} absorption and reducing urinary Ca^{2+} excretion

(continued)

Table 17.5 Hormones from Sources Other than the Hypothalamus and Pituitary (continued)

Hormone	Target	Principal Effects
Adrenal Medulla		
Epinephrine, norepinephrine, dopamine	Most tissues	Complement effects of sympathetic nervous system
Adrenal Cortex		
Aldosterone	Kidney	Promotes Na^+ retention and K^+ excretion, maintains blood pressure and volume
Cortisol and corticosteroids	Most tissues	Stimulate fat and protein catabolism, gluconeogenesis, stress resistance, and tissue repair; inhibit immune system
Androgen (DHEA) and estrogen	Bone, muscle, integument, many other tissues	Male prenatal development, growth of pubic and axillary hair, bone growth, sex drive
Pancreatic Islets		
Insulin	Most tissues	Stimulates glucose and amino acid uptake; lowers blood glucose level; promotes glycogen, fat, and protein synthesis
Glucagon	Primarily liver	Stimulates gluconeogenesis, glycogen and fat breakdown, release of glucose and fatty acids into circulation
Ovaries		
Estradiol	Many tissues	Stimulates female reproductive development, regulates menstrual cycle and pregnancy, prepares mammary glands for lactation
Progesterone	Uterus, mammary	Regulates menstrual cycle and pregnancy, prepares mammary glands for lactation
Inhibin	Anterior pituitary	Inhibits FSH secretion
Testes		
Testosterone	Many tissues	Stimulates reproductive development, skeletomuscular growth, sperm production, and libido
Inhibin	Anterior pituitary	Inhibits FSH secretion
Heart		
Atrial natriuretic factor	Kidney	Lowers blood volume and pressure by promoting Na^+ and water loss
Kidneys		
Calcitriol	Bone	Promotes bone deposition by increasing calcium and phosphate absorption in small intestine and reducing their urinary loss
Erythropoietin	Red bone marrow	Promotes red blood cell production
Liver		
Angiotensinogen (a prohormone)	Blood vessels	Precursor of angiotensin II, a vasoconstrictor
Erythropoietin	Red bone marrow	Promotes red blood cell production
Insulin-like growth factors (IGFs)	Many tissues	Mediates action of growth hormone
Stomach and Small Intestine		
Enteric hormones	Stomach and intestines	Coordinate digestive motility and secretion
Placenta		
Estrogen, progesterone, and others	Many tissues of mother and fetus	Enhance effects of ovarian hormones on fetal development, maternal reproductive system, and preparation for lactation

Table 17.6 Some Endocrine Disorders Associated with Hormone Hyposecretion and Hypersecretion

Syndrome	Hormonal Cause and Principal Effects
Anterior Pituitary	
Pituitary dwarfism	Cause: Growth hormone hyposecretion Effects: Abnormally short stature but with normal limb-to-torso proportions
Gigantism	Cause: Growth hormone hypersecretion in childhood Effects: Abnormally tall stature
Acromegaly	Cause: Growth hormone hypersecretion in adulthood Effects: Thickening of hands, feet, and face with overgrowth of bone and soft tissues
Posterior Pituitary	
Diabetes insipidus	Cause: Antidiuretic hormone hyposecretion Effects: Copious urine output with dehydration, intense thirst, and electrolyte imbalances
Thyroid Gland	
Congenital hypothyroidism	Cause: Prenatal thyroid hormone hyposecretion Effects: Stunted growth, depressed metabolism, and mental retardation
Myxedema	Cause: Adult thyroid hormone hyposecretion Effects: Depressed metabolism, sluggishness, edema, and weight gain
Endemic goiter	Cause: Thyroid hormone hyposecretion due to dietary iodine deficiency Effects: Thyroid hypertrophy and hypothyroidism
Toxic goiter (Graves disease)	Cause: TSH hypersecretion Effects: Thyroid hypertrophy with hyperthyroidism; elevated metabolism, weight loss, nervousness, hypertension, and exophthalmos
Adrenal Medulla	
Pheochromocytoma	Cause: Tumor of adrenal medulla Effects: Elevated metabolism, nervousness, hyperglycemia, glycosuria, and exhaustion
Adrenal Cortex	
Cushing syndrome	Cause: Glucocorticoid hypersecretion Effects: Deranged carbohydrate and protein metabolism, hypertension, weakness, edema, and fat deposition on face or shoulders
Adrenogenital syndrome	Cause: Androgen hypersecretion Effects: Enlargement of penis or clitoris, premature puberty, and masculinizing effects in females
Addison disease	Cause: Glucocorticoid and mineralocorticoid hyposecretion Effects: Hypoglycemia, hypotension, dehydration, electrolyte imbalances, weakness, and bronzing of skin
Pancreatic Islets	
Diabetes mellitus	Cause: Insulin hyposecretion or inaction Effects: Hyperglycemia, glycosuria, ketonuria, polyuria, insatiable hunger and thirst, ketoacidosis, weight loss, and degenerative nervous and vascular disease
Hyperinsulinism	Cause: Injection of excess insulin or pancreatic tumors Effects: Hypoglycemia, weakness, hunger, disorientation, convulsions, and unconsciousness

Before You Go On

Answer the following questions to test your understanding of the preceding section:

9. Name three endocrine glands that are larger in children than in adults.
10. What is the value of the calorigenic effect of thyroid hormone?
11. Name a glucocorticoid, a mineralocorticoid, and a catecholamine secreted by the adrenal gland.
12. Does the action of glucocorticoids more closely resemble that of glucagon or insulin? Explain.
13. In diabetes mellitus, explain the chain of events that lead to (a) osmotic diuresis, (b) ketoacidosis and coma, and (c) gangrene of the lower extremities.
14. What is the difference between a gonadal hormone and a gonadotropin?

Hormones and Their Actions

Objectives

When you have completed this section, you should be able to

- identify the chemical classes to which various hormones belong;
- describe how hormones are synthesized and transported to their target organs;
- describe how hormones stimulate their target cells;
- explain how target cells regulate their sensitivity to circulating hormones;
- explain how hormones are removed from circulation after they have performed their roles; and
- explain how hormones affect each other when two or more of them stimulate the same target cells.

We now address some questions at the molecular and cellular levels of endocrinology. Exactly what is a hormone? How are hormones produced and transported through the body? How does a hormone work?

Table 17.7 Chemical Classification of Hormones

Steroids	
Aldosterone	Estrogens
Cortisol	Progesterone
Cortisone	Testosterone
Oligopeptides (3–10 amino acids)	
Angiotensin II	Oxytocin
Antidiuretic hormone	Thyrotropin-releasing hormone
Gonadotropin-releasing hormone	
Polypeptides (14–199 amino acids)	
Adrenocorticotropic hormone	Growth hormone–releasing hormone
Atrial natriuretic factor	Insulin
Corticotropin-releasing hormone	Parathyroid hormone
Glucagon	Prolactin
Growth hormone	
Growth hormone–inhibiting hormone	
Glycoproteins (92 amino acids in the α chain, 112–18 amino acids in the β chain)	
Follicle-stimulating hormone	Luteinizing hormone
Human chorionic gonadotropin	Thyroid-stimulating hormone
Inhibin	
Monoamines	
Dopamine	Serotonin
Epinephrine	Thyroxine (T_4)
Melatonin	Triiodothyronine (T_3)
Norepinephrine	

Hormone Chemistry

Most hormones fall into three chemical classes—*steroids, peptides,* and *biogenic amines* (table 17.7).

1. **Steroid hormones** are derived from cholesterol. They include sex steroids produced by the testes and ovaries (such as estrogens, progesterone, and testosterone) and corticosteroids produced by the adrenal gland (such as cortisol, corticosterone, aldosterone, and DHEA). Calcitriol, the calcium-regulating hormone from the thyroid, is not a steroid but is derived from one and has the same hydrophobic character and mode of action as the steroids.
2. **Peptide hormones** are chains of 3 to 200 or more amino acids. The two posterior pituitary hormones, oxytocin and antidiuretic hormone, are very similar oligopeptides; they differ in only two of their nine amino acids (fig. 17.19). All releasing and inhibiting hormones produced by the hypothalamus are polypeptides. Most hormones of the anterior pituitary are polypeptides or glycoproteins—polypeptides conjugated with short carbohydrate chains. All glycoprotein hormones have an identical α chain of 92 amino acids and a variable β chain that distinguishes them from each other.

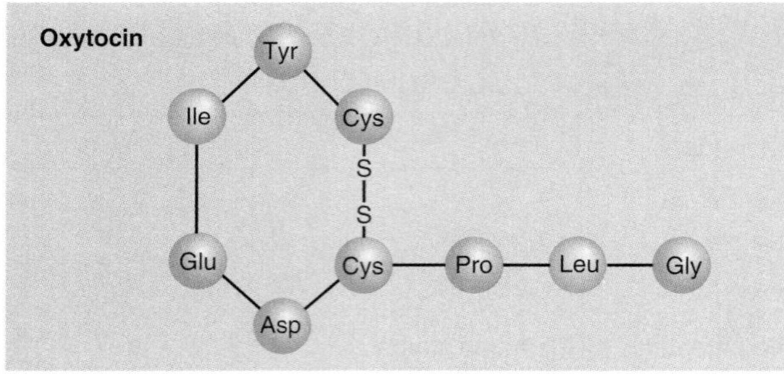

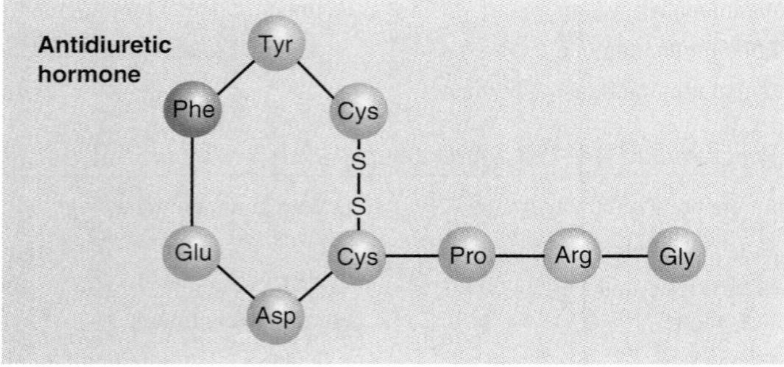

Figure 17.19 Oxytocin and Antidiuretic Hormone. Note the structural similarity of these two hormones of the posterior pituitary.

3. **Monoamines (biogenic amines)** were introduced in chapter 13, since this class also includes several neurotransmitters. All of the hormones in this category (but not all of the monoamine neurotransmitters) are synthesized from the amino acid tyrosine. They retain the amino group, hence the name of this class. These hormones include epinephrine, norepinephrine, dopamine, and the thyroid hormones. The first three of these are also called *catecholamines.*

Hormone Transport

To get from an endocrine cell to a target cell, a hormone must travel in the blood, which is mostly water. Most of the monoamines and peptides are hydrophilic, so mixing with the blood plasma presents no problem for them. Steroids and thyroid hormone, however, are hydrophobic and must bind to hydrophilic **transport proteins** to get to their destination. The transport proteins are albumins and globulins synthesized by the liver. A hormone attached to a transport protein is called **bound hormone,** and one that is not attached is an **unbound (free) hormone.** Only the unbound hormone can leave a blood capillary and get to a target cell (fig. 17.20).

Transport proteins not only enable hydrophobic hormones to travel in the blood, they also prolong their half-life. They protect circulating hormones from being broken down by enzymes in the blood plasma and liver and from being filtered out of the blood by the kidneys. Free hormone may be broken down or removed from the blood in a few minutes, whereas bound hormone may circulate for hours to weeks.

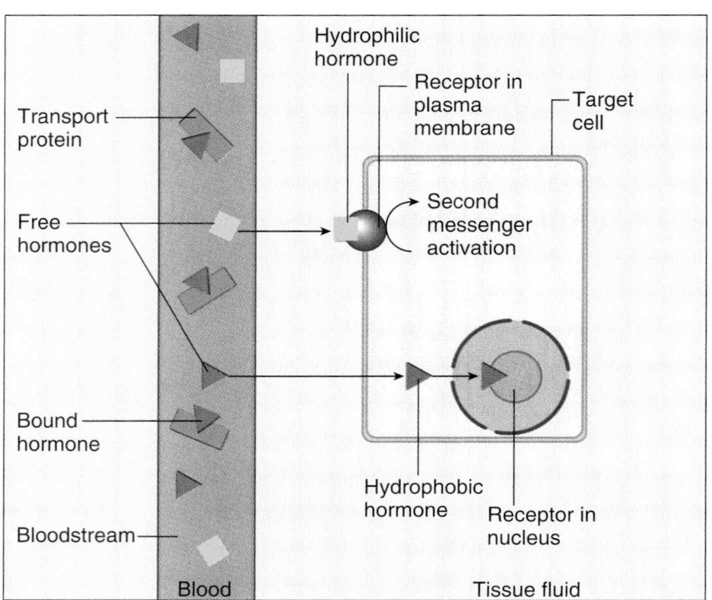

Figure 17.20 Transport and Action of Hormones. Hydrophobic hormones are carried in the blood by transport proteins. They enter the target cell nucleus and bind to receptors associated with the DNA. Hydrophilic hormones usually do not employ carrier proteins. They bind to receptors at the target cell surface and activate the formation of second messengers in the target cell.

Hormone Receptors

Hormones stimulate only those cells that have receptors for them. The receptors are protein or glycoprotein molecules located on the plasma membrane, on mitochondria and other organelles, or in the nucleus. A target cell usually has a few thousand receptors for a given hormone. They act like switches to turn certain metabolic pathways on or off when the hormone binds to them. Receptor defects lie at the heart of several endocrine diseases (see clinical insight 17.2).

Receptor-hormone interactions are similar to the enzyme-substrate interactions described in chapter 3. They exhibit *specificity* and *saturation*—that is, a given receptor will bind only one hormone, and when all the receptors are occupied, the addition of more hormone will not produce any greater effect. But unlike enzymes, hormone receptors do not chemically change their ligands.

Clinical Insight 17.2

Hormone Receptors and Therapy

In treating endocrine disorders, it is essential to understand the role of hormone receptors. For example, type II diabetes mellitus is thought to result from an insulin receptor defect or deficiency (among other possible causes). No amount of insulin replacement can correct this. And while growth hormone is now abundantly available thanks to genetic engineering, it is useless to children with *Laron dwarfism,* who have a hereditary defect in their GH receptors. *Androgen insensitivity syndrome* is due to an androgen receptor defect or deficiency; it causes genetic males to develop feminine genitalia and other features (see clinical insight 27.1). Estrogen stimulates the growth of some malignant tumors with estrogen receptors. For this reason, estrogen replacement therapy should not be used for women with estrogen-dependent cancer.

Peptides and catecholamines (hydrophilic hormones) cannot penetrate into a target cell; they can only bind to surface receptors, which are linked to second-messenger systems on the other side of the plasma membrane (fig. 17.20). Steroids and thyroid hormone (hydrophobic hormones), however, easily penetrate the plasma membrane and enter the cell to reach cytoplasmic and nuclear receptors. Further details of how hormone receptors act are given when we discuss the respective hormone classes.

Steroids

Steroid hormones are synthesized from cholesterol and differ only in the functional groups attached to the four-ringed steroid backbone. Figure 17.21 shows the synthetic pathway for several of the steroid hormones. Notice that while estrogen and progesterone are typically thought of as "female" hormones and testosterone as a "male" hormone, these sex steroids are interrelated in their synthesis and thus have roles in both sexes.

Being hydrophobic, steroids depend on transport proteins to travel to their target cells. Since these transport proteins prolong their half-life, steroids tend to have long-lasting effects. Their hydrophobic (lipophilic) character also enables unbound steroids to penetrate easily through the phospholipids of a plasma membrane and get into their target cells. They enter the nucleus and bind to a receptor associated with the DNA. The receptor has three functional regions that explain its action on the DNA: one that binds the hormone, one that binds to an *acceptor site* on the chromatin, and one that activates DNA transcription at that site. Transcription leads to the synthesis of proteins that alter the metabolism of the target cell. Estrogen, for example, stimulates cells of the uterine lining to synthesize proteins that act as progesterone receptors. Progesterone, which comes later in the menstrual cycle, then binds to these receptors and stimulates the cells to produce enzymes that synthesize glycogen. Glycogen prepares the uterus to nourish an embryo in the event of pregnancy.

Peptides

Peptide hormones are synthesized the same way as any other protein. The gene for the hormone is transcribed as a molecule of mRNA, and ribosomes translate the mRNA and assemble amino acids in the right order to make the hormone. The newly synthesized polypeptide is an inactive **preprohormone.** It has a signal peptide of hydrophobic amino acids that guide it into the cisterna of the rough endoplasmic reticulum (as explained on p. 151). Here, the leader sequence is split off and the remainder of the polypeptide is now a **prohormone.** The prohormone is transferred to the Golgi complex, which may further cut and splice it and then package the hormone for secretion.

Insulin, for example, begins as *preproinsulin.* When the leader sequence is removed, the chain folds back on itself and forms three disulfide bridges. It is now called *proinsulin.* Enzymes in the Golgi complex then remove a large middle segment called the *connecting (C) peptide.* The remainder is now insulin, composed of two polypeptide chains totaling 51 amino acids, connected to each other by two of the three disulfide bridges (fig. 17.22). The C peptide is not wasted; it has recently been found to have some hormonal effects of its own, binding to cellular receptors and reducing some of the pathological effects of diabetes mellitus.

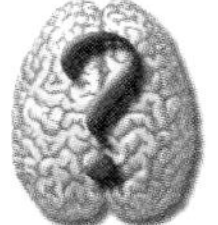

Think About It

During the synthesis of glycoprotein hormones, where in the cell would the carbohydrate be added? (See chapter 5.)

Peptides act through second-messenger systems. The best-known second messenger is cyclic adenosine monophosphate (cAMP). When glucagon binds to liver cells, for example, it activates a G protein, which leads to cAMP production and ultimately the activation of enzymes that hydrolyze glycogen. Figure 17.23 shows the general cAMP mechanism and lists some hormones that employ this method. Growth hormone–inhibiting hormone (GHIH),

Figure 17.21 **The Synthesis of Steroid Hormones from Cholesterol.** The ovaries secrete progesterone and estradiol, the testes secrete testosterone, and the adrenal cortex secretes cortisol and aldosterone.

however, works by *inhibiting* cAMP synthesis. Atrial natriuretic factor (ANF) works through a similar second messenger, cyclic guanosine monophosphate (cGMP). These second messengers do not linger in the cell for long. An enzyme called **phosphodiesterase** breaks down cAMP very quickly after its formation.

Figure 17.24 shows some other second-messenger systems activated by peptide hormones. In oxytocin action, for example, the receptor activates a G protein and the G protein activates a membrane enzyme, *phospholipase.* Phospholipase breaks down certain phospholipids of the plasma membrane into two molecules with second-messenger roles—a **diacylglycerol (diglyceride)** and **inositol triphosphate (IP_3).** Diacylglycerol activates protein kinases just as cAMP does, with similarly diverse effects. IP_3 binds to the smooth endoplasmic reticulum or to ligand-gated channels in the plasma membrane. In either case, it triggers a flood of Ca^{2+} into the cytoplasm. Calcium, the third messenger, can do several things: alter the membrane permeability and membrane potential of the cell; directly activate certain enzymes; or bind to a cytoplasmic receptor called **calmodulin,** which in turn activates protein kinases.

A given hormone doesn't always employ the same second messenger. Antidiuretic hormone (ADH) employs the IP_3-calcium system in smooth muscle but the cAMP system in kidney tubules. Insulin is unusual in comparison to other peptides. Rather than using a second-messenger system, it binds to a plasma membrane enzyme, tyrosine kinase, that directly phosphorylates cytoplasmic proteins.

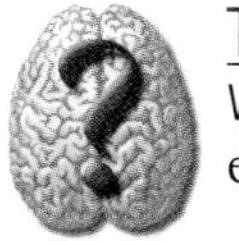

Think About It

Which hormone would have the quickest effect—estradiol or insulin? Why?

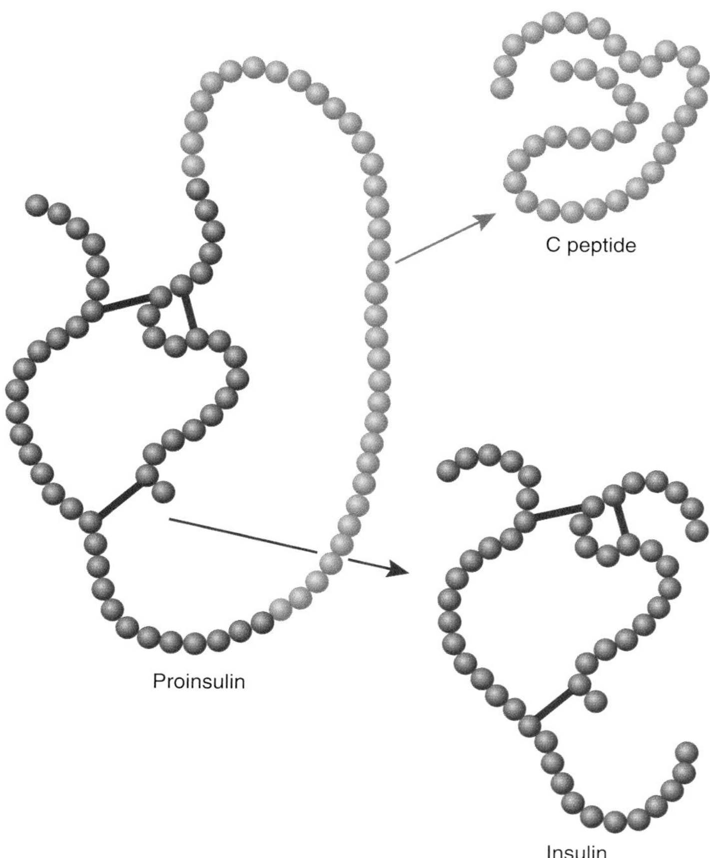

Figure 17.22 The Synthesis of Insulin, a Representative Polypeptide Hormone. Proinsulin has a connecting (C) peptide, 35 amino acids long, that is removed to leave insulin. Insulin has two polypeptide chains, 30 and 21 amino acids long, joined by two disulfide (–S–S–) bridges.

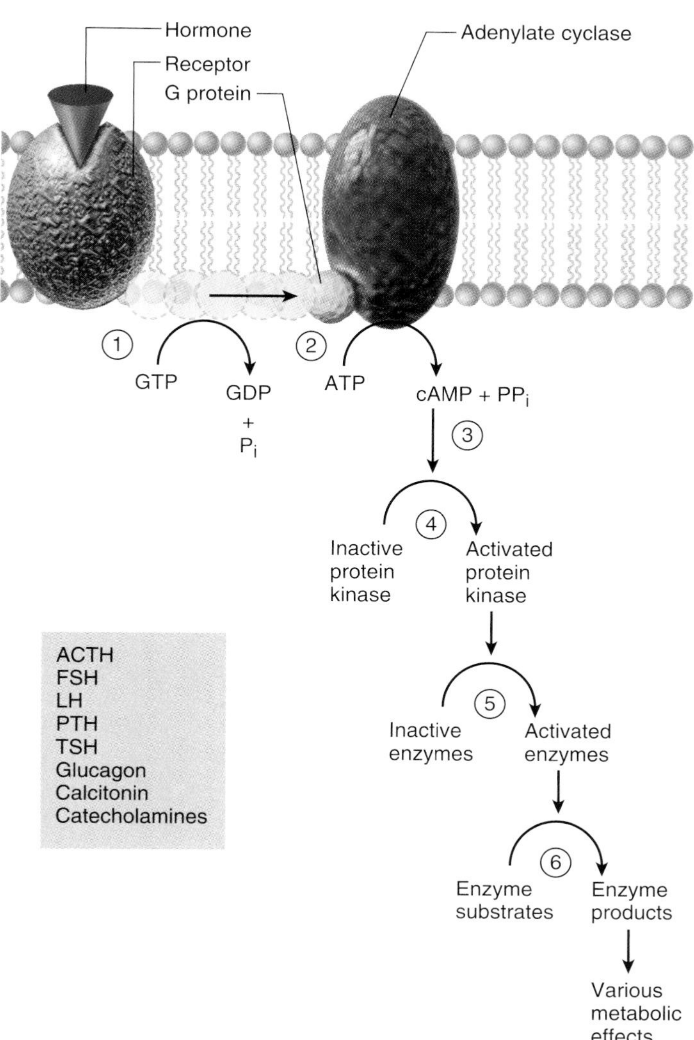

Figure 17.23 Cyclic AMP As a Second Messenger in Hormone Action. (*1*) The binding of a hormone to a membrane receptor activates a G protein. (*2*) The G protein activates adenylate cyclase. (*3*) Adenylate cyclase produces cAMP. (*4*) cAMP activates protein kinases. (*5*) Protein kinases phosphorylate enzymes and other proteins in the cytoplasm. Some enzymes are activated and others are deactivated by this phosphorylation. (*6*) Activated enzymes catalyze metabolic reactions with a wide range of possible effects on the cell, such as synthesis, secretion, and changes in plasma membrane potential. Some hormones acting through cAMP are listed in the box.

Monoamines

The monoamine hormones are synthesized from the amino acid tyrosine. Most of them are hydrophilic and travel without the need of transport proteins. Like peptides, they bind to surface receptors and activate various second messengers.

Thyroid hormone (TH) is the only monoamine made from *two* tyrosine molecules linked together. Remember that there are two slightly different forms of TH: T_3 with three iodine atoms and T_4 (thyroxine) with four. TH is synthesized in the follicles of the thyroid gland by the process shown in figure 17.25, numbered to correspond to the following description:

1. The thyroid follicle cells synthesize a large protein called **thyroglobulin** and secrete it into the lumen of the follicle.

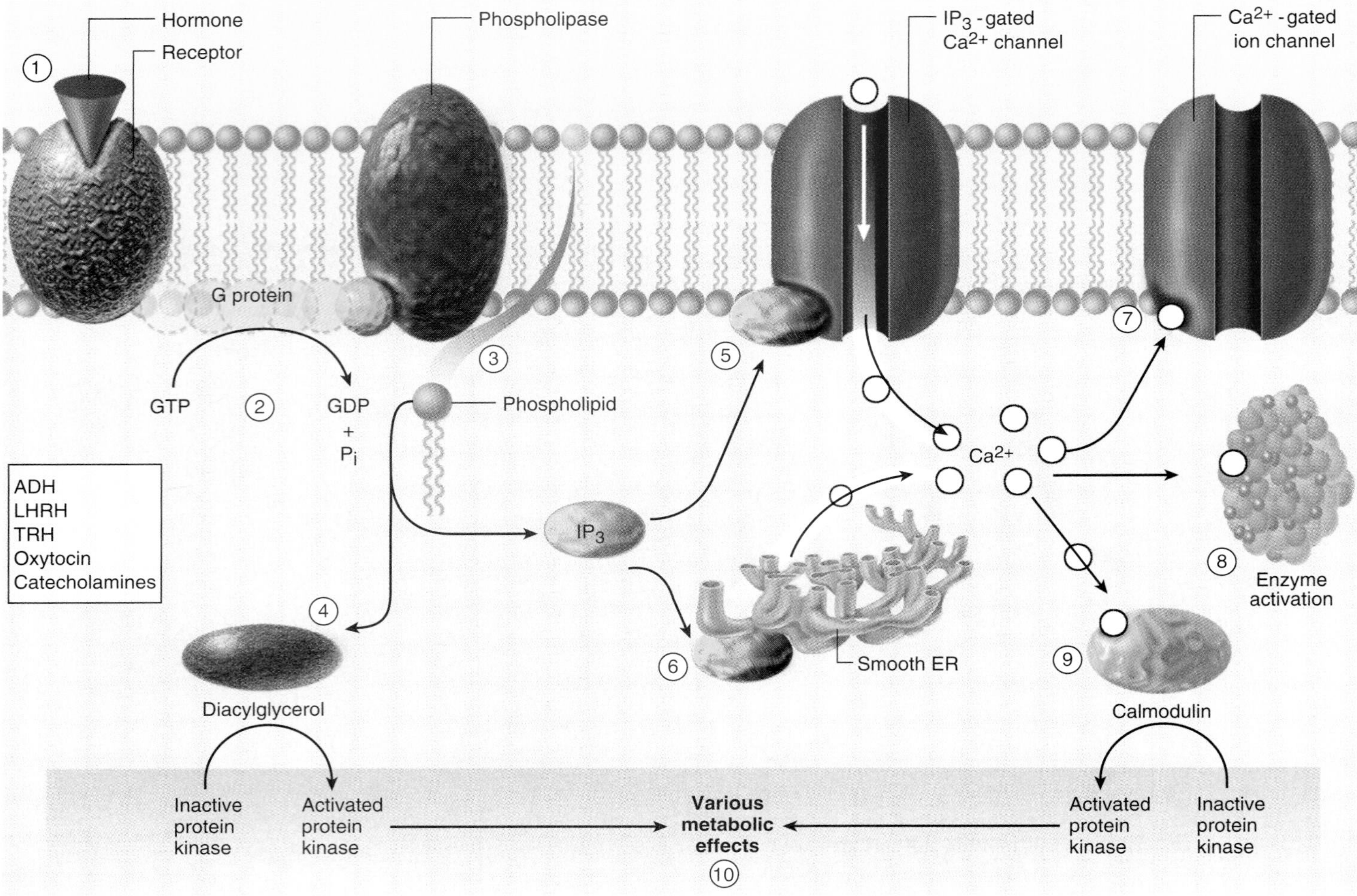

Figure 17.24 **The Role of Diacylglycerol, Inositol Triphosphate (IP_3), and Calcium Ions As Second and Third Messengers in Hormone Action.** (*1*) The binding of a hormone to a membrane receptor activates a G protein. (*2*) The G protein activates a membrane enzyme, phospholipase. (*3*) Phospholipase breaks down membrane phospholipids into diacylglycerol and inositol triphosphate (IP_3). (*4*) Diacylglycerol activates protein kinases. (*5*) Inositol triphosphate binds to gated calcium channels in the plasma membrane, admitting calcium to the cell from the ECF, or (*6*) it triggers calcium release from the smooth ER. Calcium ions can have a variety of effects: (*7*) Ca^{2+} can open other ion channels in the plasma membrane, (*8*) Ca^{2+} can function as a cofactor that activates enzymes, or (*9*) Ca^{2+} can bind to calmodulin, a protein that activates protein kinases. (*10*) Protein kinases exert the same variety of metabolic effects here as they do in the cAMP system.

2. The follicle cells actively transport iodide ions (I^-) from the blood into the cell and then release the I^- into the lumen. Here, the iodine is immediately oxidized to neutral iodine (I).
3. An iodine is added to a tyrosine of the thyroglobulin molecule, converting it to **monoiodotyrosine** (MON-oh-eye-OH-doe-TIE-ro-seen), or **MIT.**
4. Another iodine *may* be added, converting MIT to **diiodotyrosine,** or **DIT.**
5. DIT can combine either with an MIT, as shown in the figure, or with another DIT. The small arrow in the figure shows where the oxygen of DIT bonds with the carbon ring of the MIT. The amino acid chain of the MIT is lost at this stage. At this point, the thyroid hormone triiodothyronine (T_3) is essentially formed, although it is still linked through DIT to the thyroglobulin chain. (If two DITs combine, the hormone produced is tetraiodothyronine, also called thyroxine or T_4. It has another iodine at the point indicated by the red arrow.) Thyroglobulin is stored as a colloid in the lumen of the follicle.

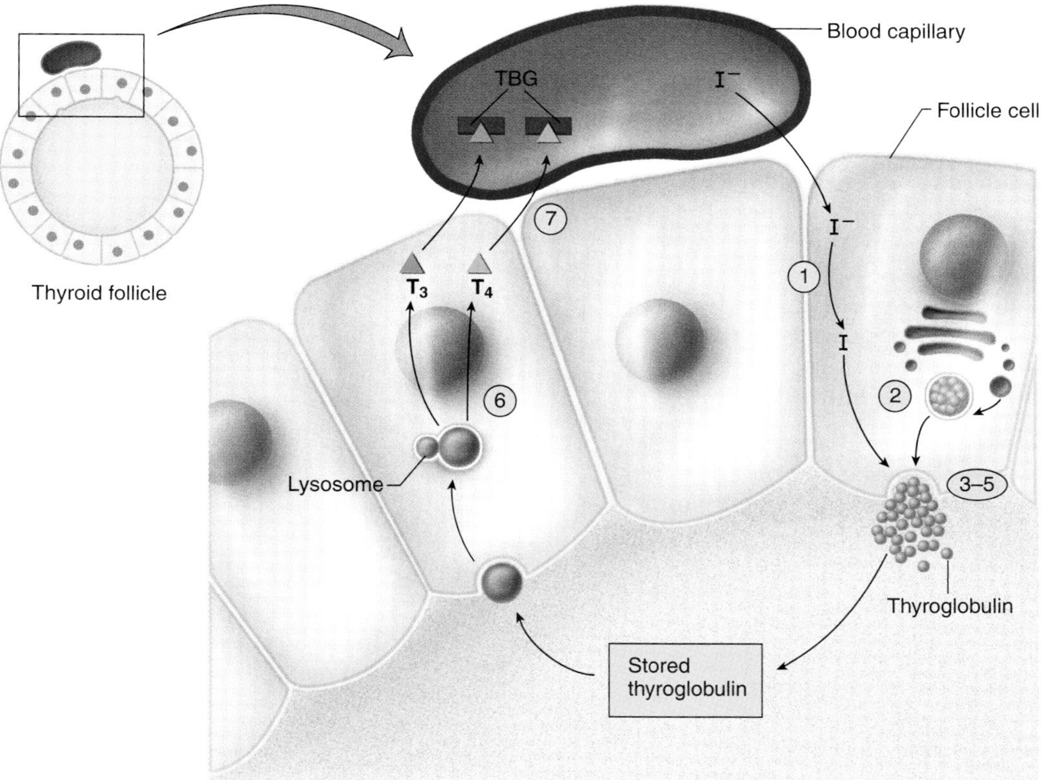

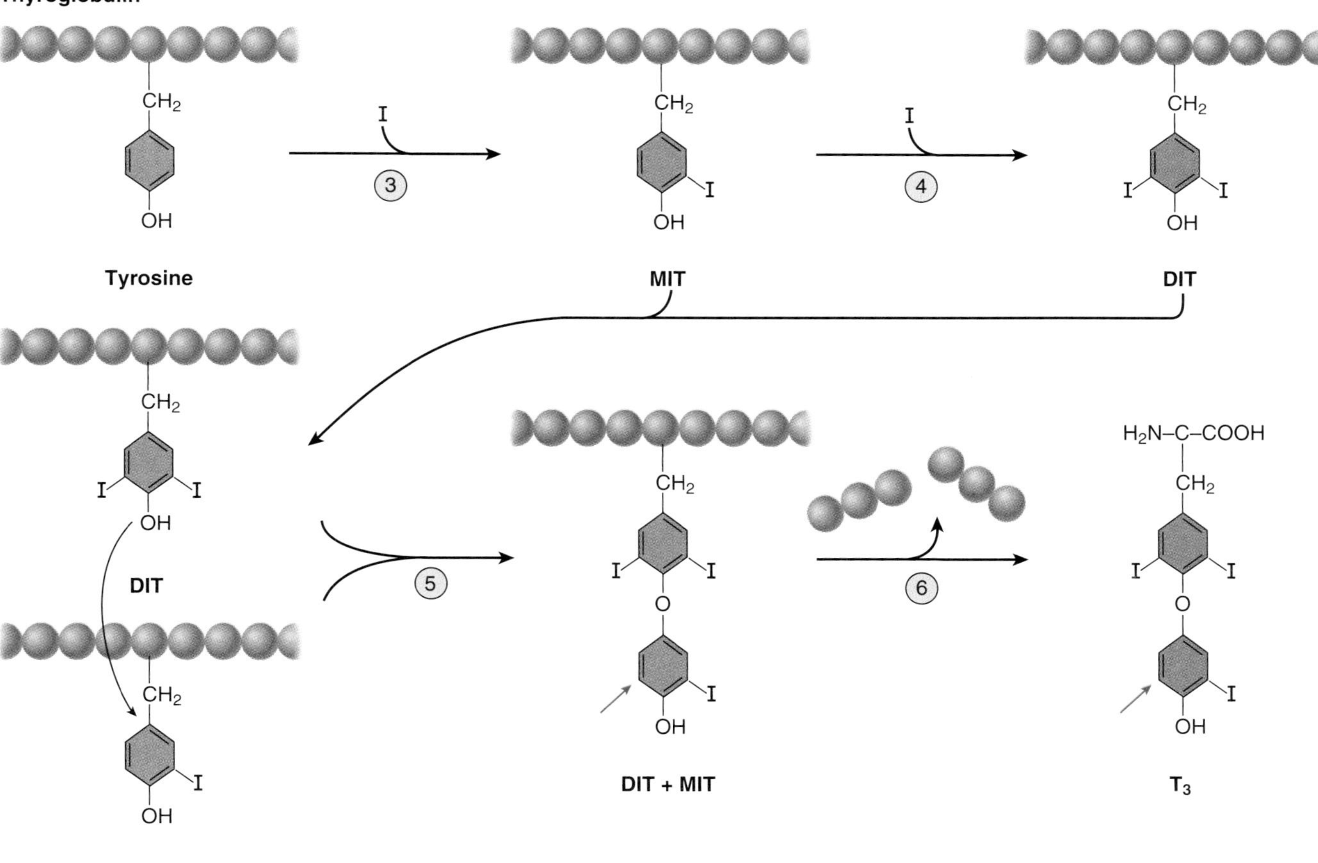

Figure 17.25 Synthesis of Thyroid Hormone (TH). See text for explanation of numbered steps. The upper part of the figure shows where the processes of TH synthesis and secretion occur. The lower part shows the chemistry of steps 3 to 6. T_3 synthesis is used as an example to show the roles played by both MIT and DIT. In T_4 synthesis, two DITs would combine at step 5 and there would be another iodine atom at the point marked by the *red arrows.*

6. When the thyroid gland is stimulated by TSH, the follicle cells take up droplets of colloid by pinocytosis (return to the top of the figure). A lysosome fuses with the pinocytotic vesicle and contributes an enzyme that hydrolyzes the thyroglobulin and liberates the thyroid hormone (step 6 at the bottom).
7. T_3 and T_4 (mostly the latter) are released into the blood. These hormones are hydrophobic, so like the steroids, they must be carried by a transport protein—in this case, **thyroxine-binding globulin (TBG).**

About 99.98% of TH is bound to TBG and 0.02% is unbound. Bound TH serves as a long-lasting blood reservoir, so even if the thyroid is surgically removed (as for cancer surgery), no signs of TH deficiency appear for about 2 weeks. Only unbound TH can enter target cells. Even though most thyroid hormone is T_4, T_4 has little direct metabolic effect on the target cells; the target cells convert most of it to T_3. T_3 binds to receptors on mitochondria and stimulates their rate of oxidative metabolism; it binds to receptors on ribosomes and stimulates the translation of mRNA (increases the rate of protein synthesis); and it enters the nucleus and binds to receptors in the chromatin, where it stimulates DNA transcription (mRNA synthesis) (fig. 17.26). One of the proteins produced under the influence of T_3 is Na^+-K^+ ATPase—the sodium-potassium pump. As we saw in chapter 4 (p. 131), one of the functions of Na^+-K^+ ATPase is to generate heat.

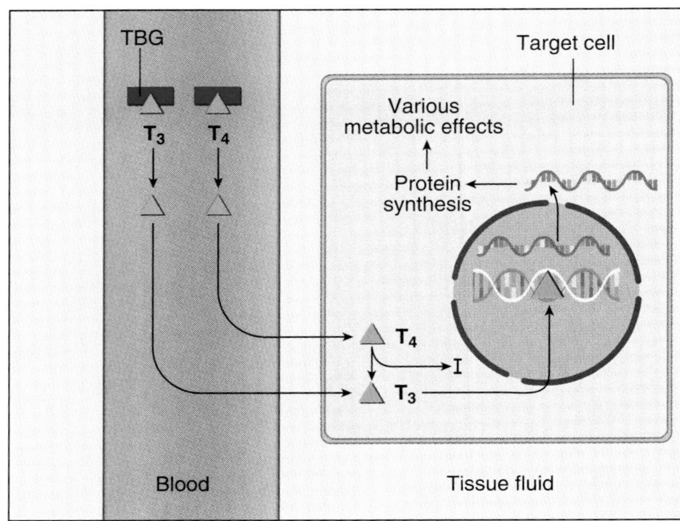

Figure 17.26 **The Action of Thyroid Hormone on a Target Cell.** T_3 and T_4 dissociate from thyroxine-binding globulin (TBG), leave the bloodstream, and enter the target cell cytoplasm. Here, T_4 is converted to T_3 by the removal of one iodine atom. T_3 enters the nucleus and activates genetic transcription.

Enzyme Amplification

One hormone molecule does not trigger the synthesis or activation of just one enzyme molecule. It activates thousands of enzyme molecules through a cascade effect called **enzyme amplification** (fig. 17.27). To put it in a simplistic but illustrative way, suppose one glucagon molecule triggered the formation of 1,000 molecules of cAMP; cAMP activated a protein kinase; each protein kinase activated 1,000 other enzyme molecules; and each of those enzymes produced 1,000 molecules of a reaction product. These are modest numbers as chemical reactions go, and yet even at this low estimate, that one glucagon molecule would have triggered the production of 1 billion molecules of reaction product. Whatever the actual numbers may be, you can see how enzyme amplification enables a very small stimulus to produce a very large effect. Hormones are therefore needed in only very small quantities. Their circulating concentrations are very low compared to other blood substances: on the order of nanograms per deciliter. Blood glucose, for example, is about 100 million times this concentrated. Because of enzyme amplification, target cells do not need a great number of hormone receptors.

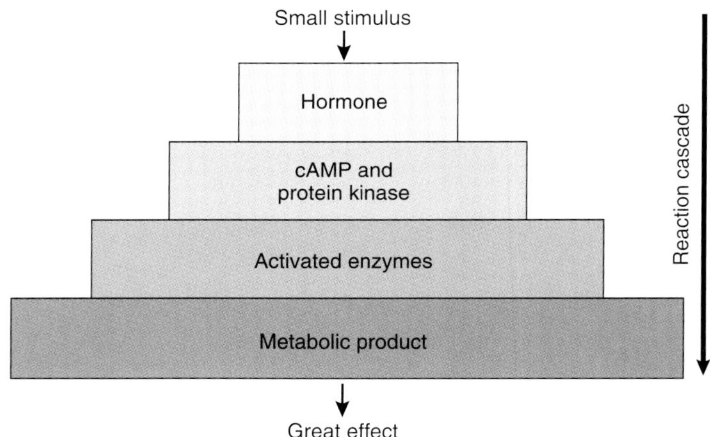

Figure 17.27 **Enzyme Amplification.** A single hormone molecule can trigger the production of many cAMP molecules and activation of many molecules of protein kinase. Every protein kinase molecule can phosphorylate and activate many other enzymes. Each of those enzyme molecules can produce many molecules of a metabolic product. Amplification of the process at each step allows for a very small hormonal stimulus to cause a very large metabolic effect.

Modulation of Hormone Sensitivity

Target cells can modulate (adjust) their sensitivity to a hormone, and one hormone can alter a target cell's sensitivity to another. In **up-regulation,** a cell increases the number of hormone receptors and becomes more sensitive to the hormone (fig. 17.28*a*). In late pregnancy, for example, the uterus produces oxytocin receptors, preparing it for the surge of oxytocin that will occur during childbirth.

Down-regulation is the process in which a cell reduces its receptor population and thus becomes less sensitive to a hormone (fig. 17.28*b*). This sometimes happens in response to long-term exposure to a high hormone concentration. For example, adipocytes down-regulate when exposed to high concentrations of insulin, and cells of the testis down-regulate in response to high concentrations of luteinizing hormone.

Hormone therapy often involves long-term use of abnormally high *pharmacological doses* of hormone, which may have undesirable side effects. Long-term treatment of inflammation with hydrocortisone, for example, has undesirable effects on bone metabolism. Two ways in which these abnormal effects can be produced are: (1) excess hormone may bind to receptor sites for other related hormones and activate effects that mimic those of the other hormones, and (2) a target cell may convert one hormone into another, such as testosterone to estrogen. Thus, long-term high doses of testosterone can, paradoxically, have feminizing effects.

Hormone Interactions

No hormone travels in the bloodstream alone, and no cell is exposed to only one hormone. Rather, there are many hormones in the blood and tissue fluid at once. Cells ignore the majority of them because they have no receptors for them, but most cells are sensitive to more than one. In

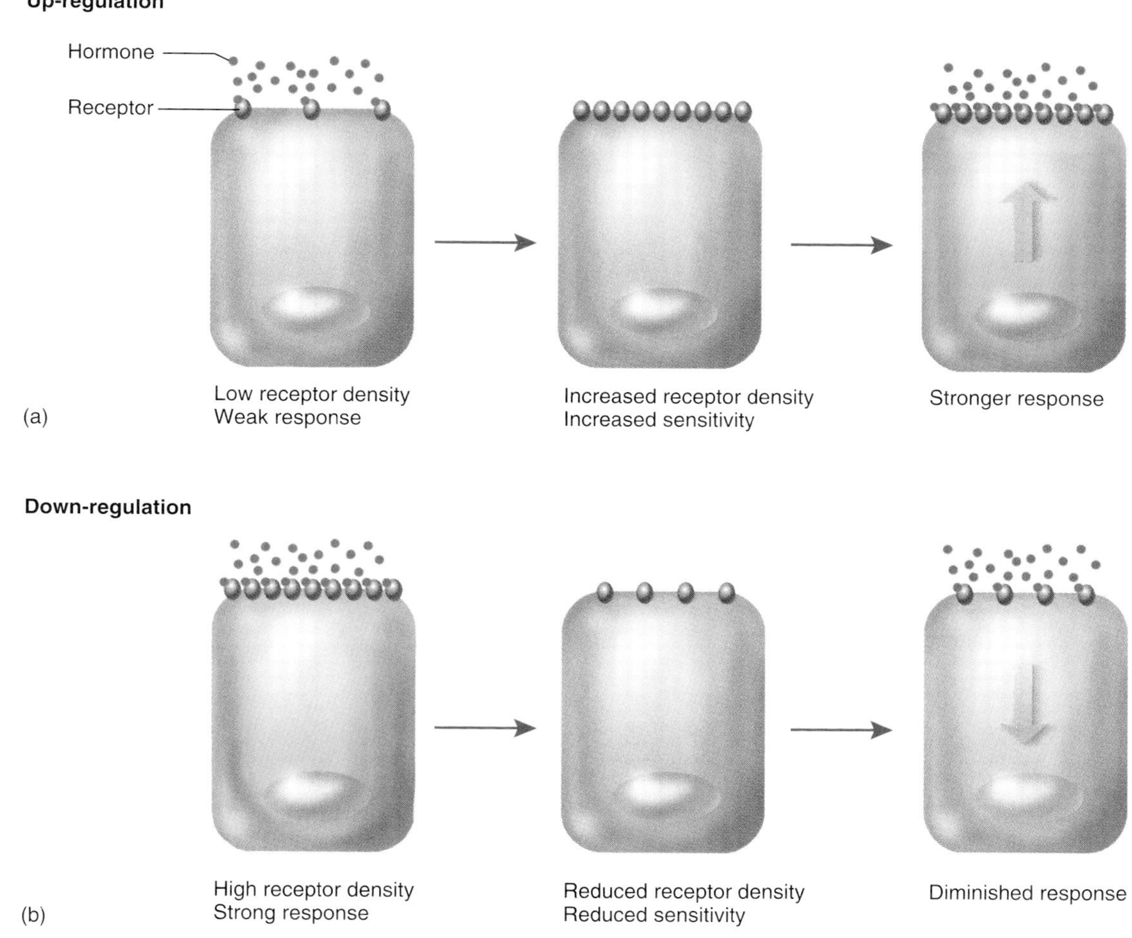

Figure 17.28 Modulation of Target Cell Sensitivity. (*a*) Up-regulation, in which a cell produces more receptors and increases its own sensitivity to a hormone. (*b*) Down-regulation, in which a cell reduces the density of its receptors and lessens its sensitivity to a hormone.

these cases, the hormones may have three kinds of interactive effects:

1. **Synergistic effects,** in which two or more hormones act together to produce an effect that is greater than the sum of their separate effects. Neither FSH nor testosterone alone, for example, can stimulate significant sperm production. When they act together, however, the testes produce some 300,000 sperm per minute.
2. **Permissive effects,** in which one hormone enhances the target organ's response to a second hormone that is secreted later. Estrogen stimulates the up-regulation of progesterone receptors in the uterus. The uterus would respond poorly to progesterone, if at all, had it not been primed by the first hormone. Estrogen thus has a permissive effect on progesterone action.
3. **Antagonistic effects,** in which one hormone opposes the action of another. For example, insulin lowers blood glucose level and glucagon raises it. During pregnancy, estrogen from the placenta inhibits the mammary glands from responding to prolactin; thus milk is not secreted until the placenta is shed at birth.

Before You Go On

Answer the following questions to test your understanding of the preceding section:

15. What are the three chemical classes of hormones? Name at least one hormone in each class.
16. Why do corticosteroids and thyroid hormones require transport proteins to travel in the bloodstream?
17. Explain how MIT, DIT, T_3, and T_4 relate to each other structurally.
18. Where are hormone receptors located in target cells? Name one hormone that employs each receptor location.
19. Explain how one hormone molecule can activate millions of enzyme molecules.

Stress and Adaptation

Objectives

When you have completed this section, you should be able to

- give a physiological definition of stress; and
- discuss how the body adapts to stress through its endocrine system.

Stress affects us all from time to time, and we react to it in ways that are mediated mainly by the endocrine and sympathetic nervous systems. **Stress** is defined as any situation that upsets homeostasis and threatens one's physical or emotional well-being. Physical causes of stress (*stressors*) include injury, surgery, hemorrhage, infection, intense exercise, temperature extremes, pain, and malnutrition. Emotional causes include anger, grief, depression, anxiety, and guilt.

Whatever the cause, the body reacts to stress in a fairly consistent way called the **stress response** or **general adaptation syndrome (GAS).** The response typically involves elevated levels of epinephrine and glucocorticoids, especially cortisol; some physiologists now define stress as any situation that raises the cortisol level. A pioneering researcher on stress physiology, Canadian biochemist Hans Selye, showed in 1936 that the GAS typically occurs in three stages, which he called the *alarm reaction,* the *stage of resistance,* and the *stage of exhaustion.*

The Alarm Reaction

The initial response to stress is an **alarm reaction** mediated mainly by norepinephrine from the sympathetic nervous system and epinephrine from the adrenal medulla. Epinephrine prepares the body to take action such as fighting or escaping danger. It raises the blood pressure and heart rate, increases circulation to the skeletal muscles, and reduces blood flow to organs that can temporarily be assigned a lower priority—the kidneys, digestive tract, and skin. It stimulates glycogenolysis and gluconeogenesis, thus raising the blood glucose level. In order to further ensure an adequate supply of glucose to the brain, epinephrine inhibits insulin secretion and the uptake and use of glucose by the muscles and other organs. Those organs can use other energy sources, whereas nerve cells must have glucose. Aldosterone and angiotensin levels also rise during the alarm reaction. Angiotensin helps to raise the blood pressure, and aldosterone promotes sodium and water conservation, which helps to offset possible losses by sweating and bleeding.

The Stage of Resistance

After a few hours, the body's glycogen reserves are exhausted, and yet the nervous system continues to demand glucose. If a stressful situation is not resolved before the glycogen is gone, the body enters the **stage of resistance,** in which the first priority is to provide alternative fuels for metabolism. This stage is dominated by cortisol. The hypothalamus secretes corticotropin-releasing hormone (CRH), the pituitary responds by secreting adrenocorticotropic hormone (ACTH), and this, in turn, stimulates the adrenal cortex to secrete cortisol and other glucocorticoids. Cortisol promotes the breakdown of fat and protein into glycerol, fatty acids, and amino acids. The liver converts these to glucose. Like epinephrine, cortisol inhibits glucose uptake by most organs. It also inhibits protein synthesis, leaving the free amino acids available for gluconeogenesis. The immune system, which depends heavily on the synthesis of antibodies and other proteins, is depressed by long-term cortisol exposure. Antibody levels drop, the number of circulating leukocytes declines, and lymphoid tissues atrophy. Wounds heal poorly, and a person in chronic stress becomes more susceptible to infections and some forms of cancer. Cortisol stimulates gastric secretion, which may account for the ulcers that occur in chronic stress, but it suppresses the secretion of sex hormones such as estrogen, testosterone, and luteinizing hormone, causing disturbances of fertility and sexual function.

The Stage of Exhaustion

The body's fat reserves can carry it through months of stress, but when fat is depleted, stress overwhelms the body's homeostasis. The **stage of exhaustion** sets in, often marked by rapid decline and death. With its fat stores gone, the body now relies primarily on protein breakdown to meet its energy needs. Thus, there is a progressive wasting away of the muscles and weakening of the body. After prolonged stimulation, the adrenal cortex may stop producing glucocorticoids, making it all the more difficult to maintain glucose homeostasis. Aldosterone sometimes promotes so much water retention that it creates a state of hypertension; and while it conserves sodium, it hastens the elimination of potassium and hydrogen ions. This creates a state of hypokalemia (potassium deficiency in the blood) and alkalosis (excessively high blood pH), resulting in nervous and muscular system dysfunctions. Death frequently results from heart failure, kidney failure, or overwhelming infection.

Before You Go On

Answer the following questions to test your understanding of the preceding section:

20. Define *stress* from the standpoint of endocrinology.
21. Describe the stages of the general adaptation syndrome.
22. List six hormones that show increased secretion in the stress response. Describe how each one contributes to recovery from stress.

Eicosanoids and Other Chemical Messengers

Objectives

When you have completed this section, you should be able to

- explain what eicosanoids are and how they are produced;
- identify some classes and functions of eicosanoids; and
- describe several physiological roles of prostaglandins.

Neurotransmitters and hormones are not the only chemical messengers in the body. Here we briefly consider some others. Unlike neurotransmitters, they are not produced by neurons. Unlike hormones, they do not travel to their target cells by way of the blood. They are commonly called **paracrine**[29] **secretions.** Although we do not use the term *hormone* in such a broad sense in this chapter, some authorities call these messengers *local hormones* because they diffuse short distances and stimulate nearby cells. Histamine is one example. Connective tissue *mast cells* line up along blood vessels and release histamine, which stimulates the smooth muscle of a blood vessel to relax and causes vasodilation. We have already seen some other examples. Somatostatin from the pancreatic δ cells diffuses to the α and β cells in the same islet, where it inhibits their secretion of glucagon and insulin. Catecholamines from the adrenal medulla diffuse to the cortex and stimulate it to secrete corticosterone.

The **eicosanoids**[30] (eye-CO-sah-noyds) are an important family of paracrine secretions. They are 20-carbon compounds derived from a polyunsaturated fatty acid called **arachidonic** (ah-RACK-ih-DON-ic) **acid.** Some peptide hormones and other stimuli liberate arachidonic acid from one of the phospholipids of the plasma membrane, and two enzymes then convert it to various eicosanoids (fig. 17.29).

Lipoxygenase helps to convert arachidonic acid to **leukotrienes,** eicosanoids that mediate allergic and inflammatory reactions (see chapter 21). **Cyclooxygenase** converts arachidonic acid to three other types of eicosanoids:

1. **Prostacyclin** is produced by the walls of the blood vessels, where it inhibits blood clotting and vasoconstriction.
2. **Thromboxanes** are produced by blood platelets. In the event of injury, they override prostacyclin and stimulate vasoconstriction and clotting.

29. *para* = next to + *crin* = secrete
30. *eicosa* (variation of *icosa*) = 20

Clinical Insight 17.3

Anti-Inflammatory Drugs

Cortisol and corticosterone are used as *steroidal anti-inflammatory drugs (SAIDs).* They inhibit inflammation by blocking the release of arachidonic acid from the plasma membrane, thus inhibiting the synthesis of all eicosanoids. Their main disadvantage is that prolonged use causes side effects that mimic Cushing syndrome. Aspirin and ibuprofen (Motrin) are *nonsteroidal anti-inflammatory drugs (NSAIDs)* with more selective effects. They stop the action of cyclooxygenase, thus blocking prostaglandin synthesis without affecting lipoxygenase or the leukotrienes. For similar reasons, aspirin inhibits blood clotting (see chapter 18). One theory of fever is that it results from the action of prostaglandins on the hypothalamus. If true, this would explain the antipyretic (antifever) effect of NSAIDs.

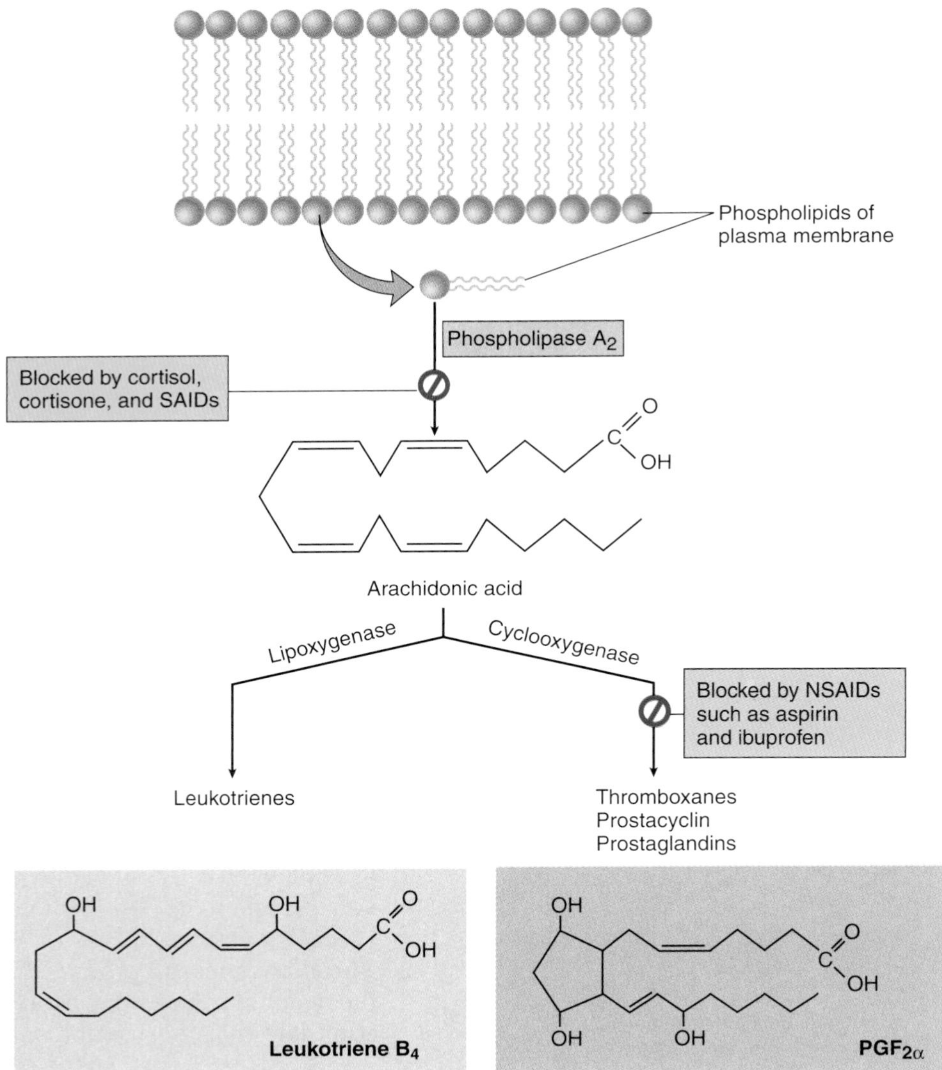

Figure 17.29 Eicosanoid Synthesis and Related Drug Actions. SAIDs are steroidal anti-inflammatory drugs such as hydrocortisone; NSAIDs are nonsteroidal anti-inflammatory drugs such as aspirin and ibuprofen.

Table 17.8 Some of the Roles of Prostagladins

Inflammatory	Promote fever and pain, two cardinal signs of inflammation
Endocrine	Mimic effects of TSH, ACTH, and other hormones; alter sensitivity of anterior pituitary to hypothalamic hormones; work with glucagon, catecholamines, and other hormones in regulation of fat mobilization
Nervous	Function as neuromodulators and alter the release or effects of neurotransmitters in the brain
Reproductive	Promote ovulation and formation of corpus luteum; induce labor contractions
Gastrointestinal	Inhibit gastric secretion
Vascular	Act as vasodilators and vasoconstrictors
Respiratory	Constrict or dilate bronchioles
Renal	Promote blood circulation through the kidney, increase water and electrolyte excretion

Prostacyclin and thromboxanes are further discussed in the next chapter.

3. **Prostaglandins (PGs)** are the most diverse group of eicosanoids. They have a five-sided carbon ring in their backbone. They are named PG for *prostaglandin,* plus a third letter that indicates the type of ring structure (PGE, PGF, etc.) and a subscript that indicates the number of C=C double bonds in the side chain. They were first found in bull semen and the prostate gland, hence their name, but they are now thought to be produced in most organs of the body. The PGEs are usually antagonized by PGFs. For example, the PGE family relaxes smooth muscle in the bladder, intestines, bronchioles, and uterus and stimulates contraction of the smooth muscle of blood vessels. $PGF_{2\alpha}$ has precisely the opposite effects. Some other roles of prostaglandins are described in table 17.8.

Understanding the pathways of eicosanoid synthesis makes it possible to understand the action of several familiar drugs (see clinical insight 17.3). The roles of prostaglandins and other eicosanoids are further explored in later chapters on blood, immunity, and reproduction.

Before You Go On

Answer the following questions to test your understanding of the preceding section:

23. What are eicosanoids and how do they differ from neurotransmitters and hormones?
24. Distinguish between a paracrine and endocrine effect.
25. State four functions of prostaglandins.

Historical Insight 17.4

The Discovery of Insulin

At the start of the twentieth century, physicians felt nearly helpless in the face of diabetes mellitus. They put patients on useless diets—the oatmeal cure, the potato cure, and others—or on starvation diets as low as 750 Calories per day so as not to "stress the system." They were resigned to the fact that their patients were doomed to die, and simple starvation seemed to produce less suffering than the ravenous hunger of diabetes.

After the cause of diabetes was traced to the pancreatic islets in 1901, European researchers tried treating patients and experimental animals with extracts of pancreas, but they became discouraged by the severe side effects of impurities in the extracts and lacked the resources to pursue the problem to completion. By 1913, the scientific community showed signs of giving up on diabetes.

But in 1920, Frederick Banting (1891–1941), a young Canadian physician with a failing medical practice, became intrigued with a possible method for isolating the islets from the pancreas and testing extracts of the islets alone. He returned to his alma mater, the University of Toronto, to present his idea to Professor J. J. R. Macleod (1876–1935), a leading authority on carbohydrate metabolism. Macleod was unimpressed with Banting, finding his knowledge of the diabetes literature and scientific method superficial. Nevertheless, he felt Banting's idea was worth pursuing and thought that with his military surgical training, Banting might be able to make some progress where others had failed. He offered Banting laboratory space for the summer, giving him a marginal chance to test his idea. Banting was uncertain whether to accept, but when his fiancée broke off their engagement and an alternative job offer fell through, he closed his medical office, moved to Toronto, and began work. Little did either man realize that in two years' time, they would share a Nobel Prize and would so thoroughly detest each other they would scarcely be on speaking terms.

A Modest Beginning

Macleod advised Banting on an experimental plan of attack and gave him an assistant, Charles Best (1899–1978). Best had just received his B.A. in physiology and looked forward to an interesting summer job with Banting before starting graduate school. He received no pay for his work, and Banting himself was desperately poor, living in a 7- by 9-foot room and supporting himself on two dollars a week earned by doing tonsillectomies. Over the summer of 1921, Banting and Best removed the pancreases from some dogs to render them diabetic and tied off the pancreatic ducts in other dogs to make most of the pancreas degenerate while leaving the islets intact. Their plan was to treat the diabetic dogs with extracts made from the degenerated pancreases of the others.

It was a difficult undertaking. Their laboratory was tiny, filthy, unbearably hot, and reeked of dog excrement. The pancreatic ducts were very small and difficult to tie, and it was hard to tell if all the pancreas had been removed from the dogs intended to become diabetic. Several dogs died of overanesthesia, infection, and bleeding from Banting's clumsy surgical technique. Banting was also careless in reading his data and interpreting the results, and he had little interest in reading the literature to see what other diabetes researchers were doing. In Banting and Best's first publication, in early 1922, the data in their discussion disagreed with the data in the tables, and both disagreed with the data in their laboratory notebooks. These were not the signs of promising researchers.

In spite of themselves, Banting and Best achieved modest positive results over the summer. Crude pancreatic extracts brought one dog back from its diabetic coma and reduced the hyperglycemia and glycosuria of others. Buoyed by these results, Banting demanded a salary, a better laboratory, and another assistant. Macleod grudgingly obtained salaries for the pair, but Banting began to loathe him for their disagreement over his demands, and he and Macleod grew in mutual contempt as the project progressed.

Success and Bitterness

Biochemist J. B. Collip (1892–1965) was brought into the project that fall in hopes that he could produce purer extracts. More competent in experimental science, Collip was the first to show that pancreatic extracts could eliminate ketosis and restore the liver's ability to store glycogen. He obtained better and better results in diabetic rabbits until, by January 1922, the group felt ready for human trials. Banting was happy to have Collip on the team initially, but grew intensely jealous of him as Collip not only achieved better results than he had, but also developed a closer relationship with Macleod. Banting, who had no qualifications to perform human experiments, feared he would be pushed aside as the project moved to its clinical phase. At one point, the tension between Banting and Collip erupted into a near-fistfight in the laboratory. Banting insisted that the first human trial be done with an extract he and Best prepared, not with Collip's.

The patient was a 14-year-old boy who weighed only 29 kg (65 lb) and was on the verge of death. He was injected on January 11 with the Banting and Best extract, described by one observer as "a thick brown muck." The trial was an embarrassing failure, with only a slight lowering of his blood glucose and a severe reaction to the impurities in the extract. On January 23, the same boy was treated again, but with Collip's extract. This time, his ketonuria and glycosuria were almost completely eliminated and his blood sugar dropped 77%. This was the first successful clinical trial of insulin. Six more patients were treated in February 1922 and quickly became stronger, more alert, and in better spirits. In April, the Toronto group began calling the product insulin, and at a medical conference in May, they gave the first significant public report of their success.

But Banting felt increasingly excluded from the project. He quit coming to the laboratory, stayed drunk much of the time, and daydreamed of leaving diabetes research to work on cancer. He remained only because Best pleaded with him to stay. Banting briefly operated a private diabetes clinic, but fearful of embarrassment over alienating the discoverer of insulin, the university soon lured him back with a salaried appointment and hospital privileges.

Banting had a number of high-profile, successful cases in 1922, such as 14-year-old Elizabeth Hughes, who weighed only 20 kg (45 lb) before treatment. She began treatment in August and showed immediate, dramatic improvement. She was a spirited, optimistic, and articulate girl who kept enthusiastic diaries of being allowed to eat bread, potatoes, and macaroni and cheese for the first time since the onset of her illness. "Oh it is simply too wonderful for words this stuff," she exuberantly wrote to her mother—even though the still-impure extracts caused her considerable pain and swelling. The world quickly beat a path to Toronto begging for insulin. The pharmaceutical firm of Eli Lilly and Company entered into an agreement with the University of Toronto for the mass production of insulin, and by the fall of 1923, over 25,000 patients were being treated at more than 60 Canadian and U.S. clinics.

Recognition and Enmity

Banting's self-confidence was restored. He had become a public hero, and the Canadian Parliament awarded him an endowment generous enough to ensure him a life of comfort. Several distinguished physiologists quickly nominated Banting and Macleod for the 1923 Nobel Prize. When the award was announced, Banting was furious that Macleod would share it. At first, he threatened to refuse it, but when he cooled down, he announced that he would split his share of the prize money with Best. Macleod quickly announced that half of his share would go to Collip.

Banting subsequently made life at the university so unbearable that Macleod left in 1928 to accept a university post in Scotland. Banting stayed on at Toronto. Although now wealthy and surrounded by admiring students, he achieved nothing significant in science for the rest of his career. He was killed in a plane crash in 1941. Best replaced Macleod on the Toronto faculty, led a distinguished career, and developed the anticoagulant heparin. Collip went on to play a lead role in the isolation of PTH, ACTH, and other hormones.

Insulin made an industry giant of Eli Lilly and Company. It became the first protein whose amino acid sequence was determined, for which Frederick Sanger received a Nobel Prize in 1958. Diabetics today no longer depend on a limited supply of insulin extracted from beef and pork pancreas. Human insulin is now in plentiful supply, made by genetically engineered bacteria. Paradoxically, while insulin has dramatically reduced the suffering caused by diabetes mellitus, it has increased the number of people who have the disease—because thanks to insulin, diabetics are now able to live long enough to raise families and pass on the diabetes genes.

Connective Issues

Interactions Between the ENDOCRINE SYSTEM and Other Organ Systems

← indicates ways in which this system affects other systems
→ indicates ways in which other systems affect this one

All Systems

Growth hormone, insulin-like growth factors, insulin, thyroid hormone, and glucocorticoids affect the development and metabolism of most tissues

Integumentary System

- ← Sex hormones affect skin pigmentation, development of body hair and apocrine glands, and subcutaneous fat deposition
- → Protects superficial endocrine glands; skin carries out a step in calcitriol (vitamin D) synthesis

Skeletal System

- ← Parathyroid hormone, calcitonin, calcitriol, growth hormone, thyroid hormone, sex steroids, and other hormones affect bone development
- → Protects some endocrine glands; stores calcium needed for endocrine function

Muscular System

- ← Growth hormone and testosterone stimulate muscular growth; insulin regulates carbohydrate metabolism in muscle; other hormones affect electrolyte balance, which is critical to muscle function
- → Skeletal muscles protect some endocrine glands

Nervous System

- ← Exerts negative feedback inhibition on hypothalamus; several hormones affect nervous system development, mood, and behavior; hormones regulate electrolyte balance, which is critical to neuron function
- → Hypothalamus regulates secretion of anterior pituitary hormones; synthesizes the posterior pituitary hormones; sympathetic nervous system triggers secretion by adrenal medulla

Circulatory System

- ← Angiotensin II, aldosterone, antidiuretic hormone, and atrial natriuretic factor regulate blood volume and pressure; epinephrine, thyroid hormone, and other hormones affect heart rate and contraction force
- → All hormones depend on the bloodstream for their distribution; blood pressure and osmolarity variations trigger secretion of some hormones

Lymphatic/Immune Systems

- ← Thymosin and other hormones activate immune cells; glucocorticoids suppress immunity and inflammation
- → Lymphatic system maintains fluid balance in endocrine glands; lymphocytes protect endocrine glands from infection

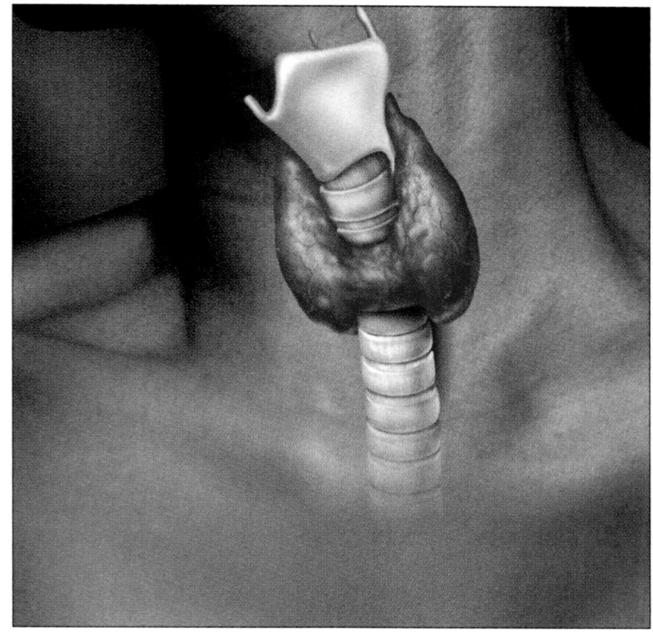

Respiratory System

- ← Epinephrine and norepinephrine increase pulmonary airflow
- → Provides O_2 and removes CO_2; lungs convert angiotensin I to angiotensin II

Urinary System

- ← Antidiuretic hormone regulates water excretion; parathyroid hormone, calcitriol, and aldosterone regulate electrolyte excretion
- → Disposes of wastes and maintains electrolyte and pH balance

Digestive System

- ← Parathyroid hormone affects intestinal calcium absorption; insulin and glucagon modulate nutrient storage and metabolism; several enteric hormones regulate gastrointestinal secretion and motility
- → Provides nutrients; nutrient absorption triggers insulin secretion; gut-brain peptides act on hypothalamus and stimulate specific hungers

Reproductive System

- ← Gonadotropins and sex steroids regulate sexual development, sperm and egg production, sex drive, menstrual cycle, pregnancy, fetal development, and lactation
- → Sex hormones inhibit secretion by hypothalamus and pituitary gland

Chapter Review

Online Learning Center

Study resources for this chapter are available at www.mhhe.com/saladin/

Study Outline

Overview of the Endocrine System (p. 638)

1. Endocrine versus exocrine glands
   a. Presence or absence of ducts
   b. Extracellular or intracellular effects
2. Comparison with nervous system
   a. Means of communication
   b. Speed of response
   c. Persistence of response
   d. Local versus widespread effects
   e. Similarities between the systems
3. Hormone nomenclature

The Hypothalamus and Pituitary Gland (p. 640)

1. Anatomy
   a. Hypothalamus
   b. Pituitary gland
      - Attachment to hypothalamus
         - •• Stalk (infundibulum)
         - •• Hypothalamo-hypophyseal portal system
      - Location and size
      - Embryonic development
      - Adenohypophysis
         - •• Anterior lobe
         - •• Pars tuberalis
         - •• Pars intermedia
      - Neurohypophysis
         - •• Posterior lobe
         - •• Median eminence
         - •• Hypothalamo-hypophyseal tract
2. Hypothalamic hormones
   a. Involved in anterior lobe control
      - Thyrotropin-releasing hormone
      - Corticotropin-releasing hormone
      - Gonadotropin-releasing hormone
      - Prolactin-releasing factor
      - Prolactin-inhibiting factor
      - Growth hormone–releasing hormone
      - Growth hormone–inhibiting hormone
   b. Released by posterior lobe
      - Oxytocin from paraventricular nucleus
      - Antidiuretic hormone from supraoptic nucleus
3. Pituitary hormones
   a. Tropic (trophic) hormones
      - Gonadotropins
         - •• Follicle-stimulating hormone
         - •• Luteinizing hormone
      - Thyroid-stimulating hormone
      - Adrenocorticotropic hormone
   b. Prolactin
   c. Growth hormone
   d. Pituitary axes
   e. Products of the pars intermedia
   f. Hormones of the posterior pituitary
      - Oxytocin
      - Antidiuretic hormone
4. Actions of pituitary hormones
   a. Follicle-stimulating hormone
      - Egg and follicle development
      - Sperm production
   b. Luteinizing hormone
      - Ovulation
      - Maintenance of corpus luteum
      - Testosterone secretion
   c. Thyroid-stimulating hormone
   d. Adrenocorticotropic hormone
   e. Prolactin
      - Milk synthesis in female
      - Testosterone sensitivity in male
   f. Growth hormone
      - Effect on tissue growth
      - Role of somatomedins
      - Mechanisms of action
      - Changes with age
      - Daily fluctuations in secretion
   g. Antidiuretic hormone
      - Water conservation
      - Role as neurotransmitter (vasopressin)
   h. Oxytocin
      - Labor contractions
      - Milk ejection
      - Sexual and emotional roles
5. Control of pituitary secretion
   a. Hypothalamic and cerebral control
   b. Neuroendocrine reflexes
   c. Negative feedback inhibition
   d. Positive feedback control
6. Pituitary disorders
   a. Hypopituitarism
      - Pituitary dwarfism
      - Panhypopituitarism
      - Diabetes insipidus
   b. Hyperpituitarism
      - Gigantism
      - Acromegaly

Other Endocrine Glands (p. 648)

1. Pineal gland
   a. Location and anatomy
   b. Changes with age
   c. Serotonin and melatonin secretion
   d. Roles in puberty and mood
2. Thymus
   a. Location and anatomy
   b. Changes with age
   c. Thymopoietin and thymosins
3. Thyroid gland
   a. Location and anatomy
   b. Thyroid follicles
   c. Thyroid hormones (T_3 and T_4)
      - Metabolic rate and O_2 consumption
      - Calorigenic effect
      - Effects on heartbeat and respiration
      - Other effects
   d. Calcitonin
      - Bone deposition
      - Lowering of blood calcium level

   e. Thyroid disorders
      - Congenital hypothyroidism
      - Myxedema
      - Endemic goiter
      - Toxic goiter
4. Parathyroid glands
   a. Location and anatomy
   b. Parathyroid hormone
      - Bone resorption
      - Elevation of blood calcium level
   c. Parathyroid disorders
      - Hypoparathyroidism
      - Hyperparathyroidism
5. Adrenal glands
   a. Location and anatomy
   b. Adrenal medulla
      - Catecholamine secretion
      - Relationship to sympathetic nervous system
   c. Adrenal cortex
      - Three tissue layers
      - Corticosteroids
         •• Sex steroids
         •• Mineralocorticoids
         •• Glucocorticoids
   d. Adrenal disorders
      - Pheochromocytoma
      - Cushing syndrome
      - Adrenogenital syndrome
      - Addison disease
6. Pancreas
   a. Location and anatomy
   b. Exocrine and endocrine functions
   c. Endocrine secretions
      - Insulin (from β cells)
         •• Secreted after meals
         •• Stimulates glucose and amino acid uptake
         •• Nutrient storage effect
      - Glucagon (from α cells)
         •• Secreted during fasting
         •• Stimulates glycogen, fat, and protein breakdown
         •• Nutrient releasing effect
      - Somatostatin (from δ cells)
         •• Paracrine secretion
         •• Modulation of insulin and glucagon secretion
   d. Diabetes mellitus
      - Signs and symptoms
         •• Polyuria, polydipsia, polyphagia
         •• Hyperglycemia, glycosuria, ketonuria
      - Types and treatment
         •• Type I, insulin-dependent
         •• Type II, non-insulin-dependent
      - Pathology
         •• Weight loss and weakness
         •• Ketonuria and ketoacidosis
         •• Cardiovascular disease
         •• Neuropathy
         •• Coma and death
   e. Hyperinsulinism
7. Gonads
   a. Ovaries
      - Estradiol
      - Progesterone
      - Inhibin
   b. Testes
      - Testosterone
      - Inhibin
8. Endocrine functions of other organs
   a. Heart
   b. Kidneys
   c. Liver
   d. Stomach and small intestine
   e. Placenta

Hormones and Their Actions (p. 661)

1. Hormone chemistry
   a. Steroids
   b. Peptides
   c. Monoamines
2. Hormone transport
   a. Transport proteins
   b. Bound and unbound hormone
   c. Half-life
3. Hormone receptors
   a. Receptor locations
   b. Specificity and saturation
4. Steroids
   a. Synthesized from cholesterol
   b. Transport
   c. Target cell receptors
   d. Effect on protein synthesis
5. Peptides
   a. Synthesis by rough ER and Golgi complex
   b. Role of second messengers
      - cAMP
      - Diacylglycerol
      - Inositol triphosphate
6. Monoamines
   a. Synthesized from tyrosine
   b. Thyroid hormone (TH) synthesis
   c. TH transport and action
7. Enzyme amplification
8. Modulation of hormone sensitivity
   a. Up-regulation
   b. Down-regulation
   c. Effects of long-term therapy
9. Hormone interactions
   a. Synergistic effects
   b. Permissive effects
   c. Antagonistic effects

Stress and Adaptation (p. 670)

1. Stress and stressors
2. Stress response
   a. Alarm reaction
      - Elevated epinephrine and norepinephrine levels
      - Increased heart rate and blood pressure
      - Elevated blood glucose level
      - Sodium and water retention
   b. Stage of resistance
      - Elevated ACTH and cortisol levels
      - Fat and protein breakdown
      - Gluconeogenesis
      - Depressed immune function
      - Susceptibility to infection and ulcers
   c. Stage of exhaustion
      - Protein breakdown and muscle wasting
      - Loss of glucose homeostasis
      - Hypertension and electrolyte imbalances
      - Causes of death

Eicosanoids and Other Messengers (p. 671)

1. Paracrine secretions versus hormones
2. Eicosanoid synthesis
   a. Lipoxygenase and leukotrienes
   b. Cyclooxygenase
      - Prostacyclin
      - Thromboxanes
      - Prostaglandins

Selected Vocabulary

For hormone names, which are not listed here, see tables 17.2 through 17.5.

endocrinology 638
hormone 638
target cell 638
endocrine gland 638
endocrine system 638
neuroendocrine cell 639
pituitary gland 640
adenohypophysis 641
anterior lobe 641
pars tuberalis 641
pars intermedia 641
hypothalamo-hypophyseal portal system 641
neurohypophysis 641
posterior lobe 641
stalk 641
median eminence 641
hypothalamo-hypophyseal tract 641
paraventricular nucleus 643
supraoptic nucleus 643
tropic hormone 643
gonadotropin 643
pituitary-gonadal axis 643
pituitary-thyroid axis 643
pituitary-adrenal axis 643
somatomedin 644
insulin-like growth factor 644
half-life 644
protein-sparing effect 645
glucose-sparing effect 645
neuroendocrine reflex 645
negative feedback inhibition 646
hyposecretion 647
hypersecretion 647
hypopituitarism 647
pituitary dwarfism 647
diabetes insipidus 648
hyperpituitarism 648
gigantism 648
acromegaly 648
pineal gland 648
thymus 649
involution 649
thyroid gland 649
thyroid follicle 649
follicular cell 649
calorigenic effect 650
C cell 650
congenital hypothyroidism 650
myxedema 650
endemic goiter 650
toxic goiter 650
parathyroid glands 650
hypoparathyroidism 651
hyperparathyroidism 651
adrenal gland 651
adrenal medulla 651
adrenal cortex 651
zona glomerulosa 651
zona fasciculata 651
zona reticularis 651
corticosteroid 651
sex steroid 651
androgen 651
mineralocorticoid 652
glucocorticoid 652
gluconeogenesis 652
pheochromocytoma 652
Cushing syndrome 653
adrenogenital syndrome 653
Addison disease 653
pancreas 654
pancreatic islets 654
β cells 654
α cells 654
glycogenolysis 654
somatostatin 654
δ cells 654
diabetes mellitus 654
polyuria 654
polydipsia 654
polyphagia 654
hyperglycemia 655
glycosuria 655
ketonuria 655
osmotic diuresis 655
insulin-dependent diabetes mellitus 655
non-insulin-dependent diabetes mellitus 656
ketoacidosis 656
hyperinsulinism 656
insulin shock 656
gonads 657
granulosa cell 657
interstitial cell 657
sustentacular cell 658
steroid hormone 661
peptide hormone 661
monoamine 662
transport protein 662
bound hormone 662
unbound hormone 662
preprohormone 663
prohormone 663
phosphodiesterase 664
diacylglycerol 664
inositol triphosphate 664
calmodulin 664
thyroglobulin 665
monoiodotyrosine 666
diiodotyrosine 666
thyroxine-binding globulin 668
enzyme amplification 668
up-regulation 669
down-regulation 669
synergistic effect 670
permissive effect 670
antagonistic effect 670
stress 670
stress response 670
general adaptation syndrome 670
alarm reaction 670
stage of resistance 670
stage of exhaustion 671
paracrine secretion 671
eicosanoid 671
arachidonic acid 671
lipoxygenase 671
leukotriene 671
cyclooxygenase 671
prostacyclin 671
thromboxane 671
prostaglandin 673

Testing Your Recall

1. CRH secretion would *not* raise the blood concentration of
   a. ACTH.
   b. thyroxine.
   c. cortisol.
   d. corticosterone.
   e. glucose.
2. Which of the following hormones has the least in common with the others?
   a. adrenocorticotropic hormone
   b. follicle-stimulating hormone
   c. thyrotropin
   d. thyroxine
   e. prolactin
3. Which hormone would no longer be secreted if the hypothalamo-hypophyseal tract were destroyed?
   a. oxytocin
   b. follicle-stimulating hormone
   c. growth hormone
   d. adrenocorticotropic hormone
   e. corticosterone
4. Which of the following is *not* a hormone?
   a. prolactin
   b. prolactin-inhibiting factor
   c. thyroxine-binding globulin
   d. atrial natriuretic factor
   e. cortisol
5. Where are the receptors for insulin located?
   a. in the pancreatic β cells
   b. in the blood plasma
   c. on the target cell membrane
   d. in the target cell cytoplasm
   e. in the target cell nucleus
6. What would be the consequence of defective ADH receptors?
   a. diabetes mellitus
   b. adrenogenital syndrome
   c. dehydration
   d. seasonal affective disorder
   e. none of these

7. Which of these has more exocrine than endocrine tissue?
   a. the pineal gland
   b. the adenohypophysis
   c. the thyroid gland
   d. the pancreas
   e. the adrenal gland
8. Which of these cells stimulate bone deposition?
   a. α cells
   b. β cells
   c. C cells
   d. G cells
   e. T cells
9. Which of these hormones relies on cAMP as a second messenger?
   a. ACTH
   b. progesterone
   c. thyroxine
   d. testosterone
   e. epinephrine
10. Prostaglandins are derived from
    a. pro-opiomelanocortin.
    b. cyclooxygenase.
    c. leukotriene.
    d. lipoxygenase.
    e. arachidonic acid.
11. The __________ develops from the hypophyseal pouch of the embryo.
12. Thyroxine (T_4) is synthesized by combining two amino acids called __________.
13. Growth hormone hypersecretion in adulthood causes a disease called __________.
14. Kidney stones may be caused by hypersecretion of the __________ glands.
15. Adrenal steroids that regulate glucose metabolism are called __________.
16. Sex steroids are secreted by the __________ cells of the ovary and __________ cells of the testis.
17. Target cells can reduce pituitary secretion by a process called __________.
18. Hypothalamic releasing factors are delivered to the anterior pituitary by way of a network of blood vessels called the __________.
19. A hormone is said to have a/an __________ effect when it stimulates the target cell to develop receptors for other hormones to follow.
20. __________ is a process in which a cell increases its number of receptors for a hormone.

Answers in appendix B

Testing Your Comprehension

1. Propose a model of enzyme amplification for the effect of a steroid hormone and construct a diagram similar to figure 17.27 for your model. A review of protein synthesis in chapter 5 may help.
2. Suppose you were browsing in a health-food store and saw a product advertised: "Put an end to heart disease. This herbal medicine will totally rid your body of cholesterol!" Would you buy it? Why or why not? If the product were as effective as claimed, what are some other effects it would produce?
3. A person with toxic goiter tends to sweat profusely. Explain this in terms of homeostasis.
4. How is the action of a peptide hormone similar to the action of a metabotropic neurotransmitter?
5. Review the effects of anabolic steroid abuse (see clinical insight 3.5, p. 99), and explain how some of these effects may relate to the concept of down-regulation explained in this chapter.

Answers at the Online Learning Center

18 CHAPTER

The Circulatory System: Blood

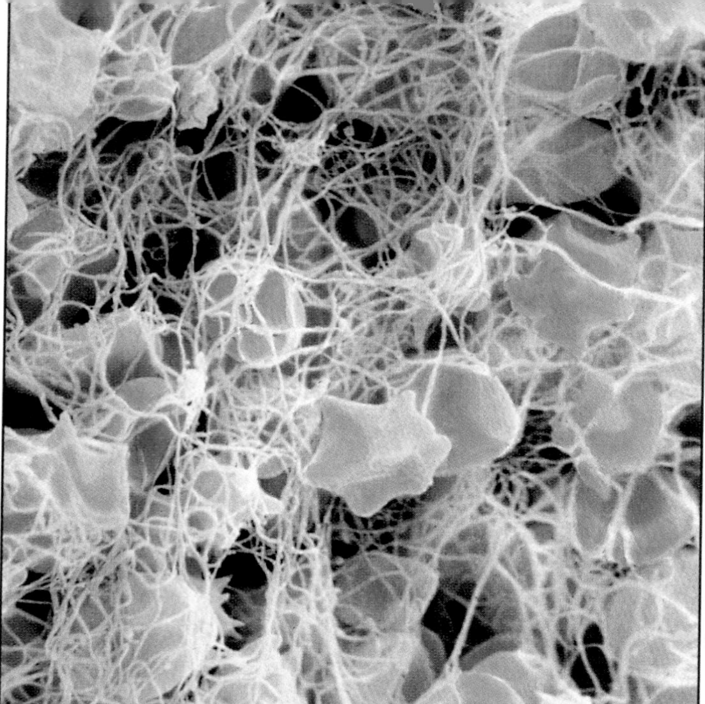

A blood clot showing platelets in a fibrin mesh (SEM)

Chapter Outline

Insights

Study with Your ESP CD-ROM

Refer to your ESP Tutorial CD-ROM for helpful animations, activities, and quizzes related to the material covered in this chapter. The following are suggested areas to go to on the CD and are presented according to "Unit/Topic":
- Transport/Blood

Brushing Up

To understand this chapter, it is important that you understand or brush up on the following concepts:
- Polypeptides and conjugated proteins (pp. 90 and 92)
- Filtration (p. 125)
- Osmosis and osmolarity (pp. 126–127)
- Dominant and recessive alleles (p. 161)
- Sex linkage (p. 163)

Blood has always had a special mystique. From time immemorial, people have seen blood flow from the body and with it, the life of the individual. People thus presumed that blood carried a mysterious "vital force," and Roman gladiators drank it to fortify themselves for battle. Even today, we become especially alarmed when we find ourselves bleeding, and the emotional impact of blood is enough to make many people faint at the sight of it. From ancient Egypt to nineteenth-century America, physicians drained "bad blood" from their patients to treat everything from gout to headaches, from menstrual cramps to mental illness. It was long thought that hereditary traits were transmitted through the blood, and people still use such unfounded expressions as "I have one-quarter Cherokee blood."

Scarcely anything meaningful was known about blood until blood cells were seen with the first microscopes. Even though blood is a uniquely accessible tissue, most of what we know about it dates only to the last 50 years. Recent developments in **hematology**[1]—the study of blood—have empowered us to save and improve the lives of countless people who would otherwise suffer or die.

Functions and Properties of Blood

Objectives

When you have completed this section, you should be able to

- state the various functions of blood;
- list the components of blood;
- explain why the viscosity and osmolarity of blood are important; and
- state what components account for its viscosity and osmolarity.

The blood plays more roles than one might expect (table 18.1); it is involved in respiration, nutrition, waste elimination, thermoregulation, immune defense, water and acid-base balance, and internal communication. Most adults have 4 to 6 L of blood. It is a connective tissue with two main components—the **plasma,** a clear extracellular fluid, and the **formed elements,** which consist of the blood cells and platelets (fig. 18.1).

The formed elements are classified as follows. All of them are cells except for the platelets, which are fragments of certain bone marrow cells. They are called *formed elements* because they are enclosed in a plasma membrane and have a definite shape and visible structure.

Erythrocytes
Platelets
Leukocytes
  Granulocytes
    Neutrophils
    Eosinophils
    Basophils
  Agranulocytes
    Lymphocytes
    Monocytes

Erythrocytes[2] (eh-RITH-ro-sites) are also known as *red blood cells (RBCs)* and leukocytes[3] (LOO-co-sites) are also known as *white blood cells (WBCs).*

The formed elements can be separated from the plasma by placing a sample of blood in a tube and spinning it for a few minutes in a centrifuge (fig. 18.2). RBCs, being more dense than other blood components, become packed into the bottom of the tube and typically constitute about 45% of the total volume. This value is called the *hematocrit.* WBCs and platelets make up a narrow cream-colored zone called the *buffy coat* just above the RBCs. At the top of the tube is the plasma, which has a pale yellow color and accounts for nearly 55% of the total volume.

Table 18.2 lists several properties of blood. Its viscosity and osmolarity warrant special attention. **Viscosity** is the resistance of a fluid to flow due to cohesion between its particles. At a given temperature, mineral oil is more viscous than water, for example, and honey is more viscous than mineral oil. Whole blood is 4.5 to 5.5 times as viscous as water. This is due mainly to the RBCs; plasma alone is 2.0 times as viscous as water, mainly because of

Table 18.1 Functions of the Blood

Functions of the Blood
Transport
Carries oxygen from the lungs to other organs
Carries carbon dioxide to the lungs for removal
Carries nutrients from the digestive system and storage depots to other organs
Carries wastes to the liver and kidneys for detoxification or removal
Carries hormones from endocrine glands to target cells
Carries heat to the skin for removal; helps regulate body temperature
Protection
Plays several roles in inflammation
Leukocytes destroy microorganisms and cancer cells
Antibodies and other proteins neutralize toxins or help to destroy microorganisms
Platelet factors initiate clotting and minimize blood loss
Regulation
Transfers water to and from the tissues; helps stabilize water balance
Buffers acids and bases; helps stabilize pH

1. *hem, hemato* = blood + *logy* = study of

2. *erythro* = red + *cyte* = cell

3. *leuko* = white + *cyte* = cell

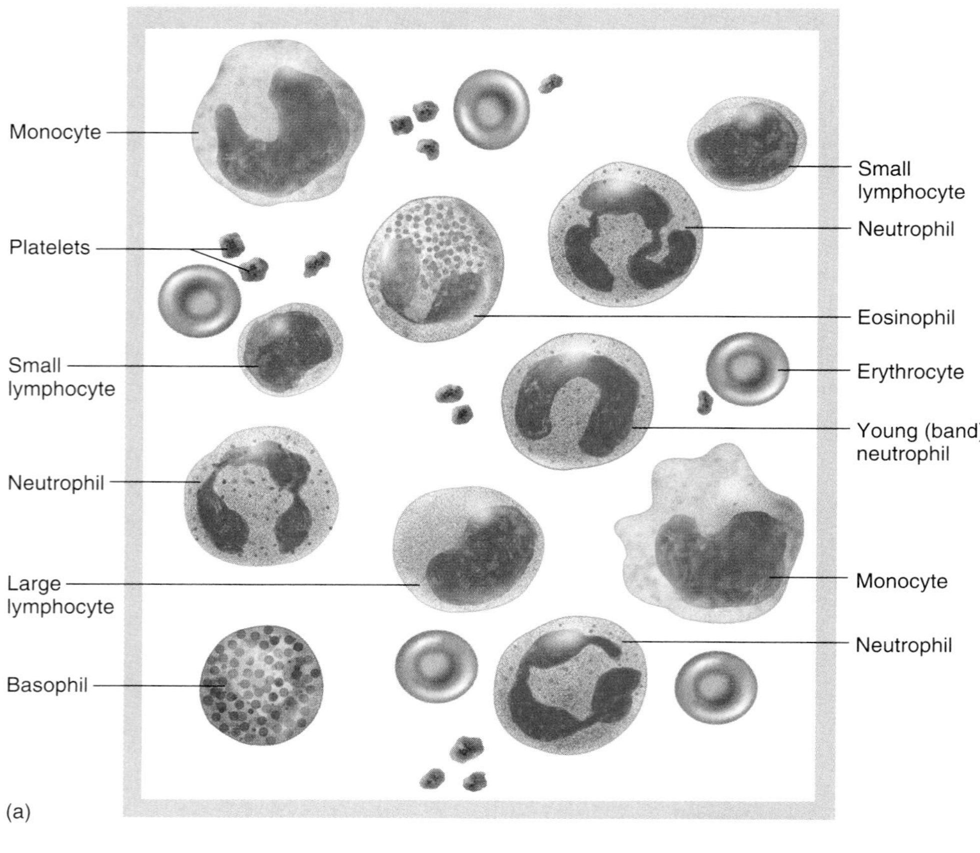

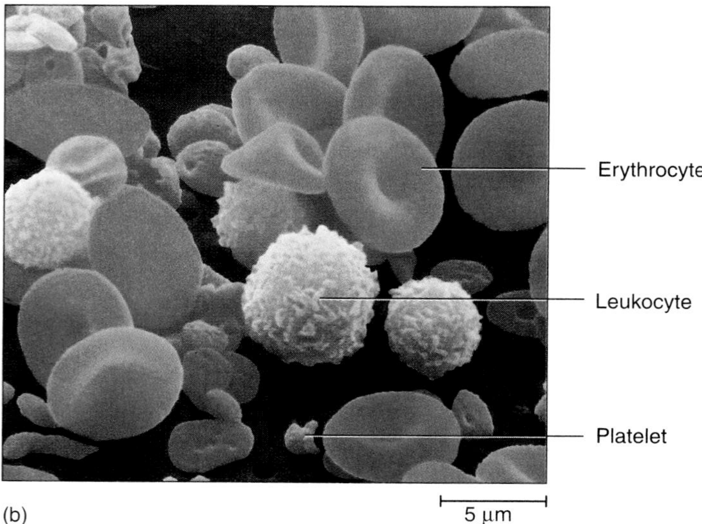

Figure 18.1 The Formed Elements of Blood. (*a*) The structure of red blood cells, white blood cells, and platelets. (*b*) Blood cells and platelets (colorized SEM).

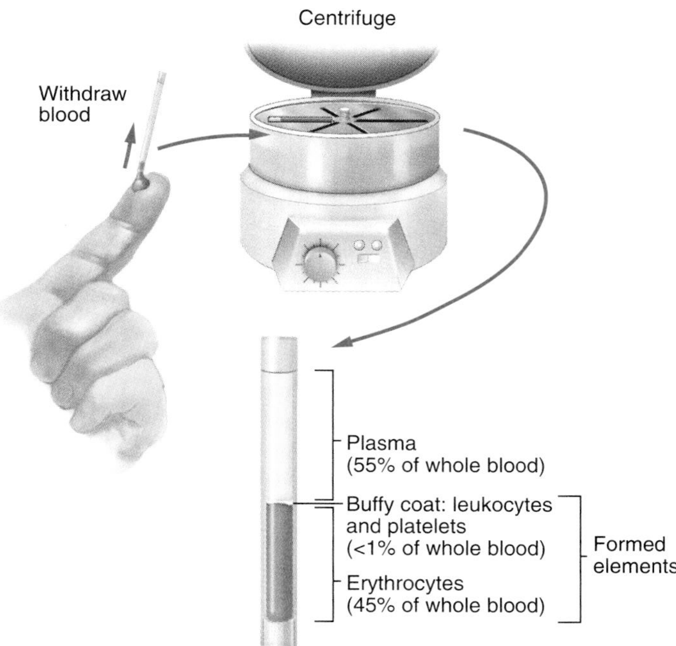

Figure 18.2 **The Hematocrit.** A small sample of blood is taken in a glass tube and spun in a centrifuge to separate the cells from the plasma. The percent volume of red cells (hematocrit) is then measured. In this example, the hematocrit is 45%.

its protein. Viscosity is important in circulatory function because it partially governs the flow of blood through the vessels. An RBC or protein deficiency reduces viscosity and causes blood to flow too easily, whereas an excess causes blood to flow too sluggishly. Either of these conditions puts a strain on the heart that may lead to serious cardiovascular problems if not corrected.

The **osmolarity** of blood (total molarity of its dissolved particles) is another important factor in cardiovascular function. In order to nourish surrounding cells and remove their wastes, substances must pass between the bloodstream and tissue fluid through the capillary walls. This transfer of fluids depends on a balance between the filtration of fluid from the capillary and its reabsorption by osmosis (see fig. 4.22, p. 127). The rate of reabsorption is governed by the relative osmolarity of the blood versus the tissue fluid. If the osmolarity of the blood is too high, the bloodstream absorbs too much fluid, which results in high blood pressure and a potentially dangerous strain on the heart and arteries. If its osmolarity drops too low, too much fluid remains in the tissues. They become edematous (swollen) and the blood pressure may drop to dangerously low levels because of the amount of fluid lost from the bloodstream.

It is therefore important that the blood maintain an optimal osmolarity. The osmolarity of the blood is a product mainly of its sodium ions, protein, and erythrocytes. The contribution of protein to blood osmotic pressure—called the **colloid osmotic pressure (COP)**—is especially important, as we see from the effects of extremely low-protein diets (see clinical insight 18.1).

Table 18.2	General Properties of Blood*
Mean Fraction of Body Weight	8%
Volume in Adult Body	Female: 4–5 L; male: 5–6 L
Volume/Body Weight	80–85 mL/kg
Mean Temperature	38°C (100.4°F)
pH	7.35–7.45
Viscosity (relative to water)	Whole blood: 4.5–5.5; plasma: 2.0
Osmolarity	280–296 mOsm/L
Mean Salinity (mainly NaCl)	0.9%
Hematocrit	Female: 37%–48% male: 45%–52%
Hemoglobin	Female: 12–16 g/dL male: 13–18 g/dL
Mean RBC Count	Female: 4.2–5.4 million/μL male: 4.6–6.2 million/μL
Platelet Count	130,000–360,000/μL
Total WBC Count	5,000–10,000/μL

*Values vary slightly depending on the testing methods used.

Before You Go On

Answer the following questions to test your understanding of the preceding section:

1. From your body weight in kilograms, predict how many kilograms and how many liters of blood you have.
2. What are the two principal components of the blood?
3. What percentage of the blood is composed of erythrocytes? What is the term for this percentage?
4. Why is blood viscosity important? What are the main factors that contribute to blood viscosity?
5. Why is blood osmolarity important? What are the main factors that contribute to blood osmolarity?

Plasma

Objectives

When you have completed this section, you should be able to

- distinguish between plasma and serum;
- list the proteins of blood plasma and state their functions;
- name the nonprotein nitrogenous compounds of blood plasma and explain their significance; and
- list the major nutrients, gases, and electrolytes found in plasma.

Clinical Insight 18.1

Starvation and Plasma Protein Deficiency

Several conditions can lead to hypoproteinemia, a deficiency of plasma protein: extreme starvation or dietary protein deficiency, liver diseases that interfere with protein synthesis, kidney diseases that result in protein loss through the urine, and severe burns that result in protein loss through the body surface. As the protein content of the blood plasma drops, so does its osmolarity. The bloodstream loses more fluid to the tissues than it reabsorbs by osmosis. Thus, the tissues become edematous and a pool of fluid may accumulate in the abdominal cavity—a condition called *ascites* (ah-SY-teez).

Children who suffer severe dietary protein deficiencies often exhibit a condition called *kwashiorkor* (KWASH-ee-OR-cor). The arms and legs are emaciated for lack of muscle, the skin is shiny and tight with edema, and the abdomen is swollen by ascites. *Kwashiorkor* is an African word for a "deposed" or "displaced" child who is no longer breast-fed. Symptoms appear when a child is weaned and placed on a diet consisting mainly of rice or other cereals. Children with kwashiorkor often die of diarrhea and dehydration.

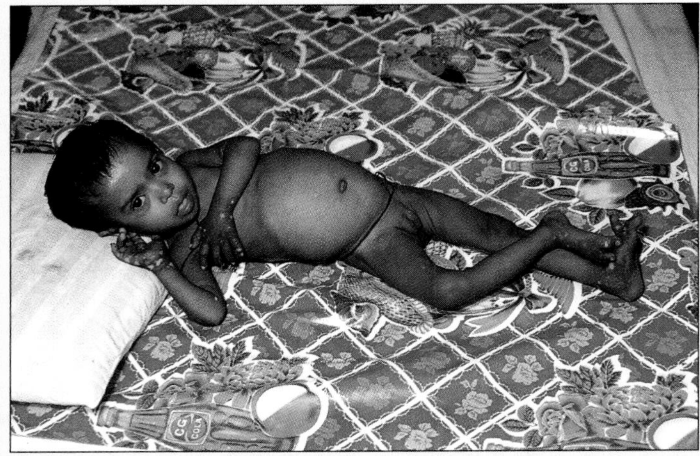

A Child with Kwashiorkor. Note the thin limbs and fluid-distended abdomen.

Blood plasma is a complex mixture of proteins, enzymes, nutrients, wastes, hormones, and gases (table 18.3). If we allow blood to clot and then remove the solids, we are left with a fluid called the blood **serum,** which is essentially identical to plasma except for the absence of clotting proteins. Vaccines are given in blood serum.

Proteins

Protein is the most abundant plasma solute by weight, totaling 6 to 9 g/dL. Plasma proteins play a variety of roles including clotting, defense, and transport. There are three major categories of proteins, the albumins, globulins, and fibrinogen (table 18.4). Many other plasma proteins are indispensable to survival, but they account for less than 1% of the total.

Albumins are the smallest and most abundant plasma proteins. Because of their major contributions to viscosity and osmolarity, pathological changes in albumin concentration strongly influence blood pressure, flow, and fluid balance. **Globulins** are divided into three subclasses; from smallest to largest in molecular weight, they are the alpha (α), beta (β), and gamma (γ) globulins. **Fibrinogen** is a soluble precursor of *fibrin,* a sticky protein that forms the framework of a blood clot. Some other plasma proteins are enzymes involved in the clotting process.

The liver produces as much as 4 g of plasma protein per hour, contributing all of the major proteins except γ globulins. The γ globulins come from *plasma cells*—connective tissue cells that are descended from white blood cells called *B lymphocytes.*

Think About It

What would be the benefit of giving intravenous albumin to a patient who has experienced fluid loss and low blood volume? Relate your answer to the principle of osmosis.

Nonprotein Nitrogenous Substances

Blood plasma contains several important nitrogenous compounds in addition to protein—notably amino acids and nitrogenous wastes. The amino acids come from the digestion of dietary protein or the catabolism of tissue proteins. **Nitrogenous wastes** are toxic end products of catabolism (see table 18.3). The most abundant is *urea,* a product of amino acid catabolism. Nitrogenous wastes are normally cleared from the blood and excreted by the kidneys at a rate that balances their rate of production.

Table 18.3 Composition of Blood Plasma*

Water	92% by weight
Proteins	Total 6–9 g/dL
Albumins	60% of total protein, 3.2–5.5 g/dL
Globulins	36% of total protein, 2.3–3.5 g/dL
Fibrinogen	4% of total protein, 0.2–0.3 g/dL
Nutrients	
Glucose (dextrose)	70–110 mg/dL
Amino acids	33–51 mg/dL
Lactic acid	6–16 mg/dL
Total lipid	450–850 mg/dL
Cholesterol	120–220 mg/dL
Fatty acids	190–420 mg/dL
High-density lipoprotein (HDL)	30–80 mg/dL
Low-density lipoprotein (LDL)	62–185 mg/dL
Neutral fats (triglycerides)	40–150 mg/dL
Phospholipids	6–12 mg/dL
Iron	50–150 μg/dL
Trace elements	Traces
Vitamins	Traces
Electrolytes	
Sodium (Na^+)	135–145 mEq/L
Calcium (Ca^{2+})	9.2–10.4 mEq/L
Potassium (K^+)	3.5–5.0 mEq/L
Magnesium (Mg^{2+})	1.3–2.1 mEq/L
Chloride (Cl^-)	100–106 mEq/L
Bicarbonate (HCO_3^-)	23.1–26.7 mEq/L
Phosphate (HPO_4^{2-})	1.4–2.7 mEq/L
Sulfate (SO_4^{2-})	0.6–1.2 mEq/L
Nitrogenous Wastes	
Urea	8–25 mg/dL
Uric acid	1.5–8.0 mg/dL
Creatinine	0.6–1.5 mg/dL
Creatine	0.2–0.8 mg/dL
Ammonia	0.02–0.09 mg/dL
Bilirubin	0–1.0 mg/dL
Other Components	
Respiratory gases (O_2, CO_2, N_2)	—
Enzymes of diagnostic value	—

*This table is limited to substances of greatest relevance to this and later chapters. Concentrations refer to plasma only, not to whole blood.

Table 18.4 Major Proteins of the Blood Plasma

Proteins	Functions
Albumins (60%)*	Responsible for colloid osmotic pressure; major contributor to blood viscosity; transport lipids, hormones, calcium, and other solutes; buffer blood pH
Globulins (36%)*	
Alpha (α) Globulins	
Haptoglobulin	Transports hemoglobin released by dead erythrocytes
Ceruloplasmin	Transports copper
Prothrombin	Promotes blood clotting
Others	Transport lipids, fat-soluble vitamins, and hormones
Beta (β) Globulins	
Transferrin	Transports iron
Complement proteins	Aid in destruction of toxins and microorganisms
Others	Transport lipids
Gamma (γ) Globulins	Antibodies; combat pathogens
Fibrinogen (4%)*	Becomes fibrin, the major component of blood clots

*Mean percentage of the total plasma protein by weight.

Nutrients

Nutrients absorbed by the digestive tract are transported in the blood plasma. They include glucose, amino acids, fats, cholesterol, phospholipids, vitamins, and minerals.

Gases

Plasma transports some of the oxygen and carbon dioxide carried by the blood. It also contains a substantial amount of dissolved nitrogen, which normally has no physiological role in the body but becomes important under circumstances such as diving and aviation.

Electrolytes

Electrolytes of the blood plasma are listed in table 18.3. Sodium ions constitute about 90% of the plasma cations and account for more of the blood's osmolarity than any other solute. Sodium therefore has a major influence on blood volume and pressure; people with high blood pressure are thus advised to limit their sodium intake. Electrolyte concentrations are carefully regulated by the body and have rather stable concentrations in the plasma.

Before You Go On

Answer the following questions to test your understanding of the preceding section:

6. List the three major classes of plasma proteins. Which one is missing from blood serum?
7. What are the functions of blood albumin?
8. List some organic and inorganic components of plasma other than protein.

Blood Cell Production

Objectives

When you have completed this section, you should be able to

- explain where blood is produced in fetuses, children, and adults;
- describe the stages of blood cell production and state the factors that influence its rate; and
- explain how uncommitted stem cells become committed to forming specific types of blood cells.

A knowledge of **hemopoiesis**[4] (HE-mo-poy-EE-sis), production of the formed elements of blood, provides a foundation for understanding leukemia, anemia, and other blood disorders. The tissues that produce blood are called *hemopoietic tissues.* The earliest of these to develop is the *yolk sac,* a membrane associated with all vertebrate embryos. In most vertebrates, it encloses the yolk of the egg and functions in both hemopoiesis and the transfer of yolk nutrients to the embryo. Even animals that don't lay eggs, however, have a yolk sac that retains its hemopoietic function. (It is also the source of cells that later produce eggs or sperm.) Cell clusters called *blood islands* form in the yolk sac by the third week of human development. They produce primitive *stem cells* that colonize the fetal bone marrow, liver, spleen, and thymus, where they subsequently produce blood cells.

The liver stops producing blood cells around the time of birth. The spleen stops producing RBCs soon after birth, but it continues to produce lymphocytes for life. From infancy onward, all seven formed elements are produced by **myeloid**[5] **hemopoiesis** in the red bone marrow and lymphocytes are additionally produced by **lymphoid hemopoiesis** in widely distributed lymphoid tissues and organs. These sites include the thymus, tonsils, lymph nodes, spleen, and patches of lymphoid tissue in the intestines and elsewhere.

The stages of myeloid hemopoiesis are shown in figure 18.3. The process begins with stem cells called **hemocytoblasts,**[6] which multiply continually to maintain their numbers and which are *pluripotent*—capable of differentiating into multiple cell lines that give rise to all of the formed elements. Differentiation begins when they develop surface receptors for specific stimulatory chemicals—*erythropoietin, thrombopoietin,* and *colony-stimulating factors (CSFs).* At this point, they can no longer produce more hemocytoblasts; they are called *committed cells* because each is destined to continue down one specific developmental pathway. We'll now examine the three principal pathways—*erythropoiesis, leukopoiesis,* and *thrombopoiesis.*

Erythrocyte Production

Erythrocyte production is called **erythropoiesis** (eh-RITH-ro-poy-EE-sis). It normally generates about 2.5 million RBCs per second (20 mL/day). The first committed cell in this pathway is the *proerythroblast,* defined by having receptors for **erythropoietin (EPO),** a hormone secreted mainly by the kidneys. Once EPO receptors are in place, the cell is committed exclusively to producing RBCs. EPO stimulates proerythroblasts to differentiate into erythroblasts, which multiply and synthesize *hemoglobin* (the red oxygen-transport protein). When this task is completed, the nucleus degenerates and is lost from the cell, and the cell is then called a *reticulocyte*—named for a fine network of endoplasmic reticulum (ER) that persists for another day or two. The overall transformation from hemocytoblast to reticulocytes takes 3 to 5 days and involves four major developments—a reduction in cell size, an increase in cell number, the synthesis of hemoglobin, and the loss of the nucleus.

Reticulocytes leave the bone marrow and enter the bloodstream. When the last of the ER disappears, the cell is a mature erythrocyte. About 0.5% to 1.5% of the circulating RBCs are reticulocytes, but this percentage increases under some circumstances. Blood loss, for example, stimulates accelerated erythropoiesis and leads to an increasing number of reticulocytes in circulation—as if the bone marrow were in such a hurry to replenish the lost RBCs that it lets many developing RBCs into circulation a little early.

Erythrocyte Homeostasis

The RBC count is maintained in a classic negative feedback manner (fig. 18.4). If the RBC count should drop (for

4. *hemo* = blood + *poiesis* = formation of
5. *myel* = bone marrow
6. *hemo* = blood + *cyto* = cell + *blast* = precursor

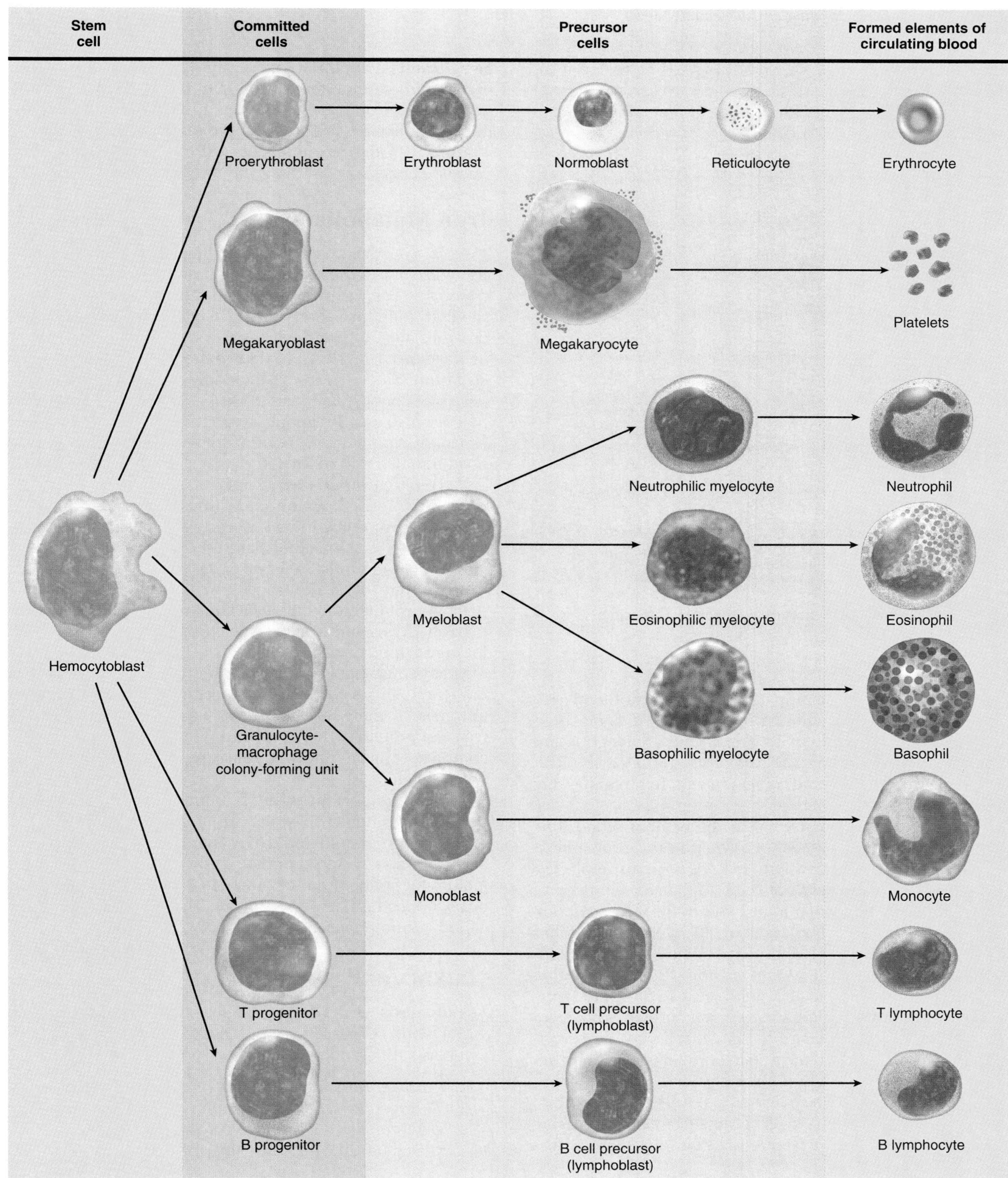

Figure 18.3 **Hemopoiesis.** Stages in the development of all the formed elements of blood.

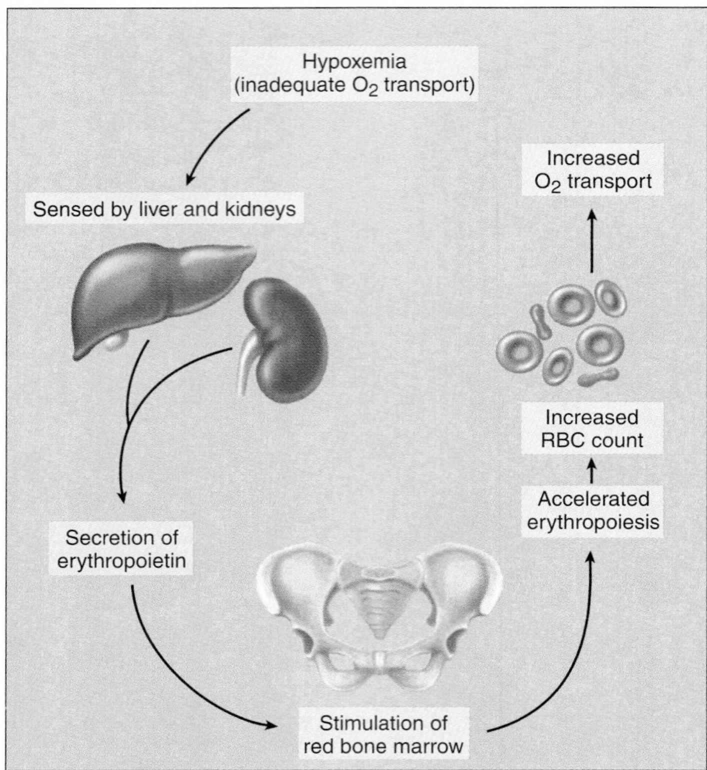

Figure 18.4 The Correction of Hypoxemia Through a Negative Feedback Loop.

example, because of hemorrhaging), then the blood will carry less oxygen—a state of **hypoxemia**[7] (oxygen deficiency in the blood) will exist. The kidneys detect this and increase their EPO output. Three or 4 days later, the RBC count begins to rise and reverses the hypoxemia that started the process.

Hypoxemia has many causes other than blood loss. Another cause is a low level of oxygen in the atmosphere. If you were to move from Miami to Denver, for example, the lower O_2 level at the high altitude of Denver would produce temporary hypoxemia and stimulate EPO secretion and erythropoiesis. The blood of an average adult has about 5 million RBCs/μL, but people who live at high altitudes may have counts of 7 to 8 million RBCs/μL. Another cause of hypoxemia is an abrupt increase in the body's oxygen consumption. If a lethargic person suddenly takes up tennis or aerobics, for example, the muscles consume oxygen more rapidly and create a state of hypoxemia that stimulates erythropoiesis. Endurance-trained athletes commonly have RBC counts as high as 6.5 million RBCs/μL.

Not all hypoxemia can be corrected by increasing erythropoiesis. In emphysema, for example, there is less lung tissue available to oxygenate the blood. Raising the RBC count cannot correct this, but the kidneys and bone marrow have no way of knowing this. The RBC count continues to rise in a futile attempt to restore homeostasis, resulting in a dangerous excess called *polycythemia,* discussed shortly.

Iron Metabolism

Iron is a critical part of the hemoglobin molecule and therefore one of the key nutritional requirements for erythropoiesis. We lose iron daily and must replace it through the diet. Men lose about 0.9 mg/day through the urine, feces, and bleeding, and women of reproductive age lose an average of 1.7 mg/day because of the added factor of menstruation. Since we absorb only a fraction of the iron present in our food, we must consume 5 to 20 mg/day to replace our losses. Pregnant women need 20 to 48 mg/day, especially in the last 3 months, to meet not only their own need but also that of the fetus.

Dietary iron exists in two forms: ferric (Fe^{3+}) and ferrous (Fe^{2+}) ions. Stomach acid converts most Fe^{3+} to Fe^{2+}, the only form that can be absorbed by the small intestine (fig. 18.5). A protein called **gastroferritin,** produced by the stomach, then binds Fe^{2+} and transports it to the small intestine. Here, it is absorbed into the blood, binds to a plasma protein called **transferrin,** and travels to the bone marrow, liver, and other tissues. Bone marrow uses Fe^{2+} for hemoglobin synthesis; muscle cells use it to make their oxygen-storage protein, myoglobin; and nearly all cells use iron to make electron-transport molecules called cytochromes in their mitochondria. The liver binds surplus iron to a protein called **apoferritin,** forming an iron-storage complex called **ferritin.** It releases Fe^{2+} into circulation when needed.

Some other nutritional requirements for erythropoiesis are vitamin B_{12} and folic acid, required for the rapid cell division and DNA synthesis that occurs in erythropoiesis, and vitamin C and copper, which are cofactors for some of the enzymes that synthesize hemoglobin. Copper is transported in the blood by an α globulin called *ceruloplasmin.*[8]

Leukocyte Production

Leukopoiesis (LOO-co-poy-EE-sis) is the production of white blood cells (see fig. 18.3). It begins when some hemocytoblasts differentiate into three types of committed cells:

1. *B progenitors,* destined to become B lymphocytes;
2. *T progenitors,* which become T lymphocytes; and
3. *granulocyte-macrophage colony-forming units,* which become granulocytes and monocytes.

7. *hyp* = below normal + *ox* = oxygen + *emia* = blood condition

8. *cerulo* = blue-green, the color of oxidized copper + *plasm* = blood plasma

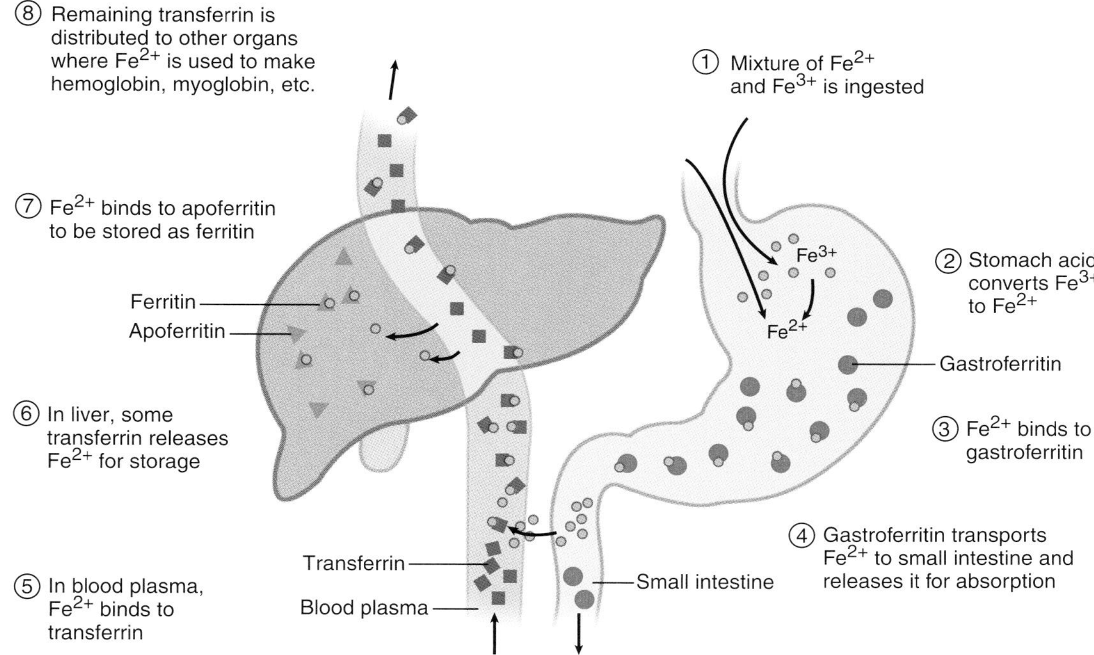

Figure 18.5 The Pathway of Iron Absorption, Transport, and Storage.

These committed cells are distinguished from hemocytoblasts by possessing receptors for colony-stimulating factors (CSFs). Mature lymphocytes and macrophages secrete several types of CSFs in response to infections and other immune challenges. Each CSF stimulates the differentiation of a different WBC type in response to specific needs. Thus, a bacterial infection may trigger the production of neutrophils whereas an allergy triggers the production of eosinophils, each process working through its own CSF.

The red bone marrow stores granulocytes and monocytes until they are needed and contains 10 to 20 times more of these cells than the circulating blood does. Lymphocytes begin developing in the bone marrow but do not stay there. Some types mature there and others migrate to the thymus to complete their development. Mature lymphocytes from both locations then colonize the spleen, lymph nodes, and other lymphoid organs and tissues.

Circulating leukocytes do not stay in the blood for very long. Granulocytes circulate for 4 to 8 hours and then migrate into the tissues, where they live another 4 or 5 days. Monocytes travel in the blood for 10 to 20 hours, then migrate into the tissues and transform into a variety of **macrophages** (MAC-ro-fay-jes). Macrophages can live as long as a few years.

Lymphocytes, responsible for long-term immunity, survive from a few weeks to decades; they leave the bloodstream for the tissues and eventually enter the lymphatic system, which empties them back into the bloodstream. Thus, they are continually recycled from blood to tissue fluid to lymph and finally back to the blood. The biology of leukocytes and macrophages is discussed more extensively in chapter 21.

Platelet Production

The production of platelets is called **thrombopoiesis** because platelets used to be called *thrombocytes.*[9] The latter term is now reserved for nucleated true cells with a blood-clotting function in other animals such as birds and reptiles. Thrombopoiesis begins when a hemocytoblast develops receptors for the hormone *thrombopoietin,* which, like erythropoietin, is produced by the liver and kidneys. With these receptors in place, the hemocytoblast has become a committed cell called a *megakaryoblast.* In response to thrombopoietin, the megakaryoblast replicates its DNA repeatedly without undergoing nuclear or cytoplasmic division. The result is a gigantic cell (up to 100 μm in diameter) called a **megakaryocyte**[10] (meg-ah-CAR-ee-oh-site), with a huge multilobed nucleus and multiple sets of chromosomes (fig. 18.6). Most megakaryocytes live in the bone marrow, but some of them colonize the lungs.

A megakaryocyte exhibits infoldings of the plasma membrane that divide its marginal cytoplasm into little compartments. The cytoplasm breaks up along these lines

9. *thrombo* = clotting + *cyte* = cell
10. *mega* = giant + *karyo* = nucleus + *cyte* = cell

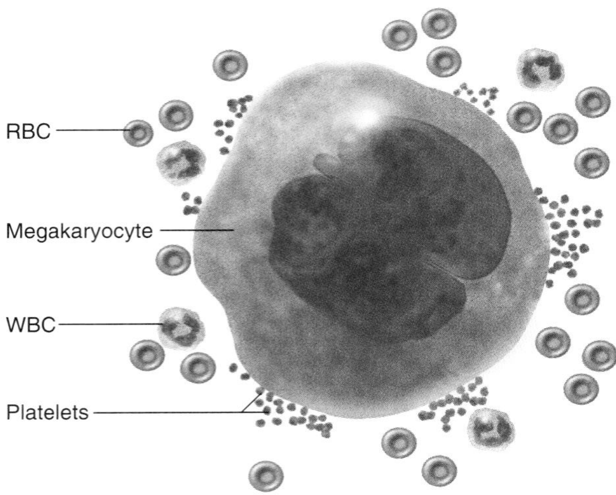

Figure 18.6 **A Megakaryocyte Producing Platelets.** Several red and white blood cells are shown for size comparison.

of weakness into tiny fragments that enter the bloodstream. Some of these are functional platelets, while others are larger particles that break up into platelets as they pass through the lungs. About 25% to 40% of the platelets are stored in the spleen and released as needed. The remainder circulate freely in the blood and live for about 4 days.

Before You Go On

Answer the following questions to test your understanding of the preceding section:

9. List the fetal tissues and organs that produce blood.
10. How do the sites of hemopoiesis differ between children and adults?
11. Distinguish between lymphoid and myeloid hemopoiesis.
12. How is a hemocytoblast different from a committed hemopoietic cell?

Erythrocytes

Objectives

When you have completed this section, you should be able to

- describe the structure of erythrocytes (red blood cells);
- describe the structure and function of hemoglobin;
- describe how the erythrocytes and hemoglobin content of the blood are quantified;
- explain why men and women differ in their erythrocyte count and hemoglobin level;
- describe the life cycle of erythrocytes; and
- describe the types, causes, and effects of anemia and polycythemia.

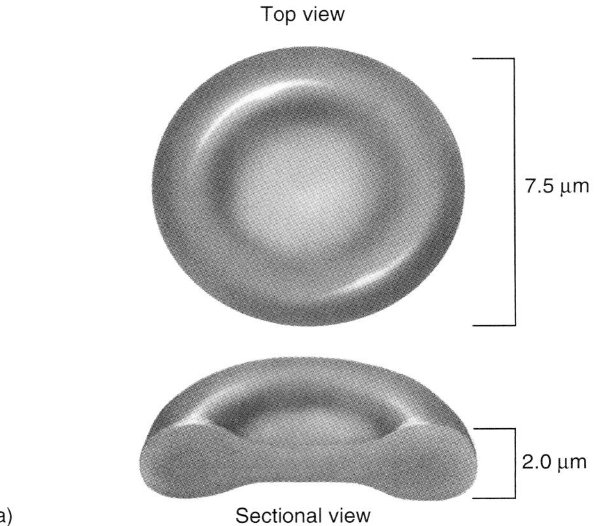

Figure 18.7 **The Structure of Erythrocytes.** (*a*) Dimensions and shape of an erythrocyte. (*b*) Erythrocytes on the tip of a hypodermic needle.

Form and Function

Erythrocytes have two principal functions: (1) to pick up oxygen from the lungs and deliver it to tissues elsewhere and (2) to pick up carbon dioxide from other tissues and unload it in the lungs. An erythrocyte is a disc-shaped cell with a thick rim and a thin sunken center where the nucleus used to be. It is about 7.5 μm in diameter and 2.0 μm thick at the rim (fig. 18.7).

The plasma membrane of a mature RBC has glycoproteins and glycolipids that determine a person's blood type. On its inner surface are two peripheral proteins, *spectrin* and *actin*, that give the membrane resilience and durability. This is especially important when RBCs pass through small

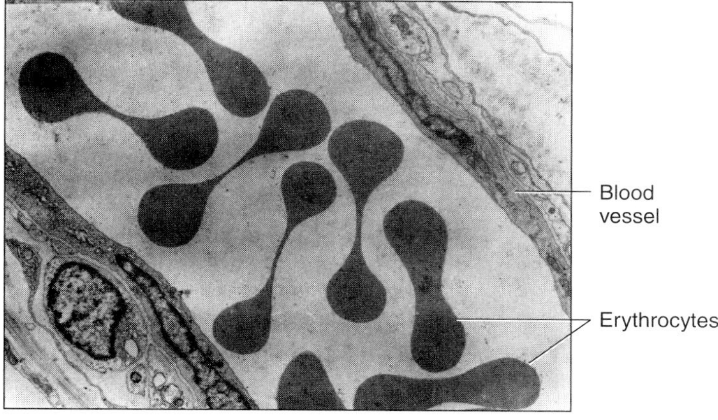

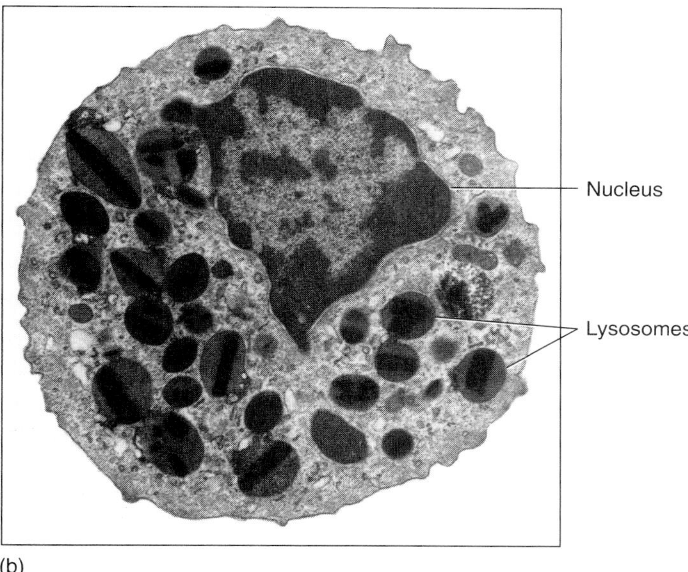

Figure 18.8 Comparison of RBCs and WBCs as Seen with the Transmission Electron Microscope. (*a*) RBCs in a small blood vessel, showing their strongly biconcave shape and lack of organelles. (*b*) An eosinophil, a representative leukocyte, showing a nucleus, many lysosomes, and other organelles.

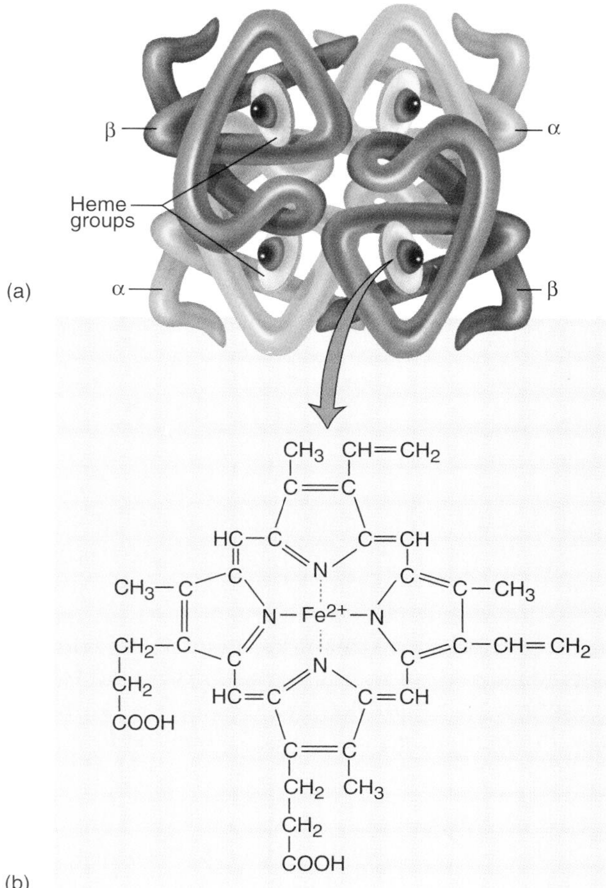

Figure 18.9 The Structure of Hemoglobin. (*a*) The hemoglobin molecule consists of two α proteins and two β proteins, each conjugated to a nonprotein heme group. (*b*) Structure of the heme group. Oxygen binds to Fe^{2+} at the center of the heme.

blood capillaries and sinusoids. Many of these passages are narrower than the diameter of an RBC, forcing the RBCs to stretch, bend, and fold as they squeeze through. When they enter larger vessels, they spring back to their discoid shape.

Most cells, including white blood cells, have an abundance of organelles. RBCs, however, lose nearly all of their organelles during maturation and are almost devoid of internal structure (fig. 18.8). Because they lack mitochondria, RBCs are incapable of aerobic respiration. This prevents them from consuming the oxygen they are meant to transport to other tissues. Erythrocytes are the only cells in the body that carry on anaerobic fermentation indefinitely.

The cytoplasm of an RBC consists mainly of a 33% solution of **hemoglobin (Hb),** the red pigment that gives the RBC its color and name. Hemoglobin carries most of the oxygen and some of the carbon dioxide transported by the blood. The cytoplasm also contains an enzyme, *carbonic anhydrase (CAH),* that catalyzes a reaction between CO_2 and water to produce carbonic acid (H_2CO_3). The role of CAH in gas transport and pH balance is discussed in chapters 22 and 24. The lack of a nucleus makes an RBC unable to repair itself, but it has an overriding advantage: the biconcave shape gives the cell a much greater ratio of surface area to volume, which enables O_2 and CO_2 to diffuse quickly to and from the hemoglobin and CAH.

Hemoglobin

Each erythrocyte contains about 280 million molecules of hemoglobin. Hemoglobin consists of four protein chains called **globins** (fig. 18.9*a*). Two of these, the *alpha* (α) *chains,* are 141 amino acids long, and the other two, the *beta* (β) *chains,* are 146 amino acids long. Each chain is

Evolutionary Insight 18.2

The Packaging of Hemoglobin

The gas-transport pigments of earthworms, snails, and many other animals are dissolved in the plasma rather than contained in blood cells. You might wonder why human hemoglobin must be contained in RBCs. The main reason is osmotic. Remember that the osmolarity of blood depends on the number of particles in solution. A "particle," for this purpose, can be a sodium ion, an albumin molecule, or a whole cell. If all the hemoglobin contained in the RBCs were free in the plasma, it would drastically increase blood osmolarity, since each RBC contains about 280 million molecules of hemoglobin. The circulatory system would become enormously congested with fluid, and circulation would be severely impaired. The blood simply could not contain that much free hemoglobin and support life. On the other hand, if it contained a safe level of free hemoglobin, it could not transport enough oxygen to support the high metabolic demand of the human body. By having our hemoglobin packaged in RBCs, we are able to have much more of it and hence to have more efficient gas transport and more active metabolism.

conjugated with a nonprotein moiety called the **heme** group (fig. 18.9*b*), which binds oxygen to a ferrous ion (Fe^{2+}) at its center. Each heme can carry one molecule of O_2; thus, the hemoglobin molecule as a whole can transport up to 4 O_2. About 20% of the CO_2 in the bloodstream is also transported by hemoglobin but is bound to the globin moiety rather than to the heme. Gas transport by hemoglobin is discussed in detail in chapter 22.

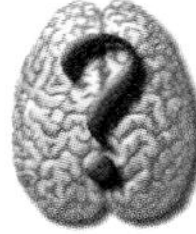

Think About It

Hemoglobin is sometimes considered an enzyme. Based on what you know of hemoglobin and what you have learned about enzymes in chapter 3, do you agree? Why or why not?

Hemoglobin exists in several forms with slight differences in the globin chains. The form we have just described is called *adult hemoglobin (HbA)*. About 2.5% of an adult's hemoglobin, however, is of a form called HbA_2, which has two *delta (δ) chains* in place of the β chains. The fetus produces a form called *fetal hemoglobin (HbF)*, which has two *gamma (γ) chains* in place of the β chains. The δ and γ chains are the same length as the β chains but differ in amino acid sequence. HbF binds oxygen more tightly than HbA does; thus it enables the fetus to extract oxygen from the mother's bloodstream.

Quantities of Erythrocytes and Hemoglobin

The RBC count and hemoglobin concentration are important clinical data because they determine the amount of oxygen the blood can carry. Three of the most common measurements are hematocrit, hemoglobin concentration, and RBC count. The **hematocrit**[11] is the percentage of whole blood volume composed of RBCs (see fig. 18.2). In men, it normally ranges between 42% and 52%; in women, between 37% and 48%. The **hemoglobin concentration** of whole blood is normally 13 to 18 g/dL in men and 12 to 16 g/dL in women. The **RBC count** is normally 4.6 to 6.2 million RBCs/μL in men and 4.2 to 5.4 million/μL in women. This is often expressed as cells per cubic millimeter (mm^3); 1 $\mu L = 1\ mm^3$.

Notice that these values tend to be lower in women than in men. There are three physiological reasons for this: (1) androgens stimulate RBC production, and men have higher androgen levels than women; (2) women of reproductive age have periodic menstrual losses; and (3) the hematocrit is inversely proportional to percent body fat, which is higher in women than in men. In men, the blood also clots faster and the skin has fewer blood vessels than in women. Such differences are not limited to humans. From the evolutionary standpoint, the adaptive value of these differences may lie in the fact that male animals fight more than females and suffer more injuries. The traits described here may serve to minimize or compensate for their blood loss.

Erythrocyte Death and Disposal

Circulating erythrocytes live for about 120 days. The life of an RBC is summarized in figure 18.10. As an RBC ages and its membrane proteins (especially spectrin) deteriorate, the membrane grows increasingly fragile. Without a nucleus or ribosomes, an RBC cannot synthesize new spectrin. Many RBCs die in the spleen, which has been called the "erythrocyte graveyard." The spleen has channels as narrow as 3 μm that severely test the ability of old, fragile RBCs to squeeze through the organ. Old cells become trapped, broken up, and destroyed. An enlarged and tender spleen may indicate diseases in which RBCs are rapidly breaking down.

Table 18.5 outlines the process of disposing of old erythrocytes and hemoglobin. **Hemolysis**[12] (he-MOLL-ih-sis),

11. *hemato* = blood + *crit* = to separate

12. *hemo* = blood + *lysis* = splitting, breakdown

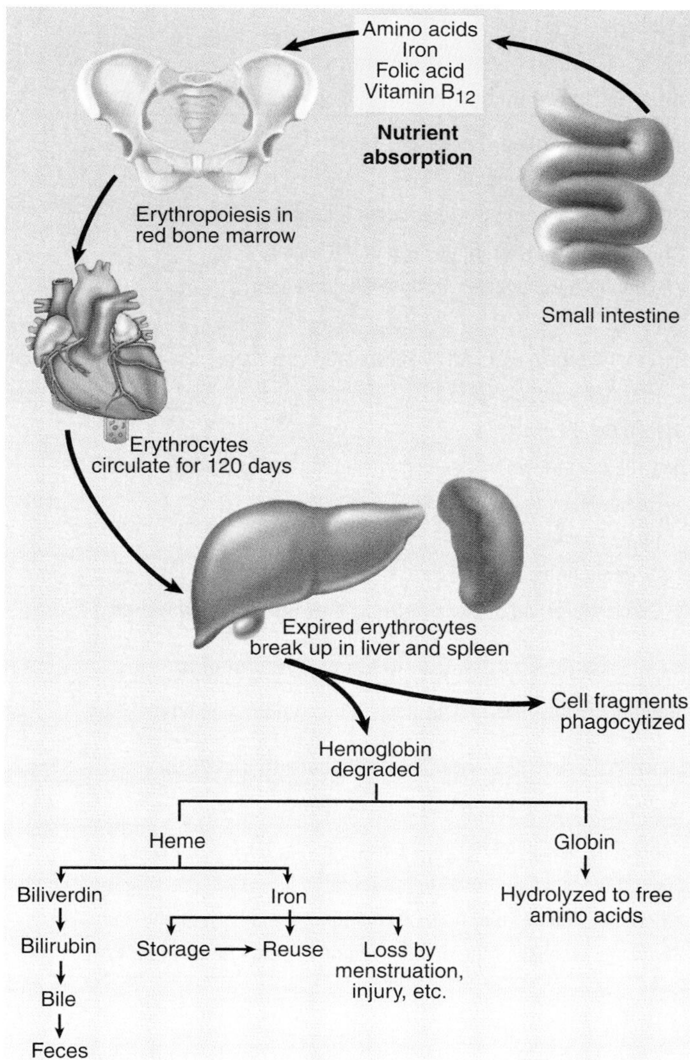

Figure 18.10 The Life and Death of Erythrocytes. Note especially the stages of hemoglobin breakdown and disposal.

the rupture of RBCs, releases hemoglobin and leaves empty plasma membranes. The membrane fragments are easily digested by macrophages in the liver and spleen, but hemoglobin disposal is a bit more complicated. It must be disposed of efficiently, however, or it can block kidney tubules and cause renal failure. Macrophages begin the disposal process by separating the heme from the globin. They hydrolyze the globin into free amino acids, which become part of the body's general pool of amino acids available for protein synthesis or energy-releasing catabolism.

Disposing of the heme is another matter. First, the macrophage removes the iron and releases it into the blood, where it combines with transferrin and is used or stored in the same way as dietary iron. The macrophage converts the rest of the heme into a greenish pigment called **biliverdin**[13] (BIL-ih-VUR-din), then further converts most

Table 18.5 The Fate of Expired Erythrocytes and Hemoglobin

1. RBCs lose elasticity with age
2. RBCs break down while squeezing through blood capillaries and sinusoids
3. Cell fragments are phagocytized by macrophages in the spleen and liver
4. Hemoglobin decomposes into:

   Globin portion—hydrolyzed to amino acids, which can be reused

   Heme portion—further decomposed into:

   Iron
   1. Transported by albumin to bone marrow and liver
   2. Some used in bone marrow to make new hemoglobin
   3. Excess stored in liver as ferritin

   Biliverdin
   1. Converted to bilirubin and bound to albumin
   2. Removed by liver and secreted in bile
   3. Stored and concentrated in gallbladder
   4. Discharged into small intestine
   5. Converted by intestinal bacteria to urobilinogen
   6. Excreted in feces

of this to a yellow-green pigment called **bilirubin.**[14] Bilirubin is released by the macrophages and binds to albumin in the blood plasma. The liver removes bilirubin from the albumin and secretes it into the bile, to which it imparts a dark green color as the bile becomes concentrated in the gallbladder. Biliverdin and bilirubin are collectively known as **bile pigments.** The gallbladder discharges the bile into the small intestine, where bacteria convert bilirubin to *urobilinogen,* responsible for the brown color of the feces. Another hemoglobin breakdown pigment, *urochrome,* produces the yellow color of urine. A high level of bilirubin in the blood causes *jaundice,* a yellowish cast in light-colored skin and the whites of eyes. Jaundice may be a sign of rapid hemolysis or a liver disease or bile duct obstruction that interferes with bilirubin disposal.

Erythrocyte Disorders

Any imbalance between the rates of erythropoiesis and RBC destruction may produce an excess or deficiency of red cells. An RBC excess is called *polycythemia*[15] (POL-ee-sy-THEE-me-uh), and a deficiency of either RBCs or hemoglobin is called *anemia.*[16]

13. *bili* = bile + *verd* = green + *in* = substance
14. *bili* = bile + *rub* = red + *in* = substance
15. *poly* = many + *cyt* = cell + *hem* = blood + *ia* = condition
16. *an* = without + *em* = blood + *ia* = condition

Polycythemia

Primary polycythemia (*polycythemia vera*) is due to cancer of the erythropoietic line of the red bone marrow. It can result in an RBC count as high as 11 million RBCs/μL and a hematocrit as high as 80%. Polycythemia from all other causes, called **secondary polycythemia,** is characterized by RBC counts as high as 6 to 8 million RBCs/μL. It can result from dehydration because water is lost from the bloodstream while erythrocytes remain and become abnormally concentrated. More often, it is caused by smoking, air pollution, emphysema, high altitude, strenuous physical conditioning, or other factors that create a state of hypoxemia and stimulate erythropoietin secretion.

The principal dangers of polycythemia are increased blood volume, pressure, and viscosity. Blood volume can double in primary polycythemia and cause the circulatory system to become tremendously engorged. Blood viscosity may rise to three times normal. Circulation is poor, the capillaries are clogged with viscous blood, and the heart is dangerously strained. Chronic (long-term) polycythemia can lead to embolism, stroke, or heart failure. The deadly consequences of emphysema and some other lung diseases are due in part to polycythemia.

Anemia

The causes of **anemia** fall into three categories: (1) inadequate erythropoiesis or hemoglobin synthesis, (2) **hemorrhagic anemia** from bleeding, and (3) **hemolytic anemia** from RBC destruction. Table 18.6 gives specific examples and causes for each category. We give special attention to the deficiencies of erythropoiesis and some forms of hemolytic anemia.

Anemia often results from kidney failure, because RBC production depends on the hormone erythropoietin (EPO), which is produced mainly by the kidneys. Erythropoiesis also declines with age, simply because the kidneys atrophy with age and produce less and less EPO as we get older. Compounding this problem, elderly people tend to get less exercise and to eat less well, and both of these factors reduce erythropoiesis.

Nutritional anemia results from a dietary deficiency of any of the requirements for erythropoiesis discussed earlier. Its most common form is **iron-deficiency anemia. Pernicious anemia** can result from a deficiency of vitamin B_{12}, but this vitamin is so abundant in meat that a B_{12} deficiency is rare except in strict vegetarians. More often, it occurs when glands of the stomach fail to produce a substance called **intrinsic factor** that the small intestine needs to absorb vitamin B_{12}. This becomes more common in old age because of atrophy of the stomach. Pernicious anemia can also be hereditary. It is treatable with vitamin B_{12} injections; oral B_{12} would be useless because the digestive tract cannot absorb it without intrinsic factor.

Table 18.6 Types and Causes of Anemia

Anemia Due to Inadequate Erythropoiesis
Inadequate nutrition
Iron-deficiency anemia
Folic acid, vitamin B_{12}, or vitamin C deficiency
Pernicious anemia (deficiency of intrinsic factor)
Renal failure (reduced erythropoietin secretion)
Old age
Renal atrophy (reduced erythropoietin secretion)
Nutritional deficiencies
Insufficient exercise
Destruction of myeloid tissue (hypoplastic and aplastic anemia)
Radiation exposure
Viral infection
Autoimmune disease
Some drugs and poisons (arsenic, mustard gas, benzene, etc.)
Hemorrhagic Anemia, Due to Excessive Bleeding
Trauma, hemophilia, menstruation, ulcer, ruptured aneurysm, etc.
Hemolytic Anemia, Due to Erythrocyte Destruction
Mushroom toxins, snake and spider venoms
Some drug reactions (such as penicillin allergy)
Malaria (invasion and destruction of RBCs by certain parasites)
Sickle-cell disease and thalassemia (hereditary hemoglobin defects)
Hemolytic disease of the newborn (mother-fetus Rh mismatch)

Hypoplastic[17] *anemia* is caused by a decline in erythropoiesis, whereas the complete failure or destruction of the myeloid tissue produces *aplastic anemia,* a complete cessation of erythropoiesis. Aplastic anemia leads to grotesque tissue necrosis and blackening of the skin. Most victims die within a year. About half of all cases are of unknown or hereditary cause, especially in adolescents and young adults. Other causes are given in table 18.6.

Anemia has three potential consequences:

1. The tissues suffer **hypoxia** (oxygen deprivation). The individual is lethargic and becomes short of breath upon physical exertion. The skin is pallid because of the deficiency of hemoglobin. Severe anemic hypoxia can cause life-threatening necrosis of brain, heart, and kidney tissues.
2. Blood osmolarity is reduced. More fluid is thus transferred from the bloodstream to the intercellular spaces, resulting in edema.

17. *hypo* = below normal + *plas* = formation + *tic* = pertaining to

3. Blood viscosity is reduced. Because the blood puts up so little resistance to flow, the heart beats faster than normal and cardiac failure may ensue. Blood pressure also drops because of the reduced volume and viscosity.

Sickle-Cell Disease and Thalassemia

Sickle-cell disease and thalassemia are hereditary hemoglobin defects that occur mostly among people of African and Mediterranean descent, respectively. About 1.3% of African Americans have **sickle-cell disease.** This disorder is caused by a recessive allele that modifies the structure of hemoglobin. Sickle-cell hemoglobin (HbS) differs from normal HbA only in the sixth amino acid of the β chain, where HbA has glutamic acid and HbS has valine. People who are homozygous for HbS exhibit sickle-cell disease. People who are heterozygous for it—about 8.3% of African Americans—have *sickle-cell trait* but rarely have severe symptoms. However, if two carriers reproduce, their children each have a 25% chance of being homozygous and having the disease.

Without treatment, a child with sickle-cell disease has little chance of living to age 2, but even with the best available treatment, few victims live to the age of 50. At low oxygen concentrations, HbS turns to a gel and causes the erythrocytes to become elongated and pointed at the ends (fig. 18.11), hence the name of the disease. Sickled erythrocytes are sticky; they **agglutinate**[18] (clump together) and block small blood vessels, causing intense pain in oxygen-starved tissues. Blockage of the circulation can also lead to kidney or heart failure, rheumatism, or paralysis. Hemolysis of the fragile cells causes anemia and hypoxemia, which triggers further sickling in a deadly positive feedback loop. Chronic hypoxemia also causes fatigue, weakness, mental deficiency, and deterioration of the heart and other organs. In a futile effort to counteract the hypoxemia, the hemopoietic tissues become so active that bones of the cranium and elsewhere become enlarged and misshapen. The spleen reverts to a hemopoietic role, while also disposing of dead RBCs, and becomes enlarged and fibrous. Sickle-cell disease is a prime example of *pleiotropy*—the occurrence of multiple phenotypic effects from a change in a single gene (see p. 163).

Why does sickle-cell disease exist? In Africa, where it originated, vast numbers of people have always died of malaria and still do. Malaria is caused by a parasite that invades the RBCs and feeds on hemoglobin. Sickle-cell hemoglobin, HbS, is indigestible to malaria parasites, and people heterozygous for sickle-cell disease are resistant to malaria. The lives saved by this gene outnumber the deaths of homozygous individuals, so the gene persists in the population.

Figure 18.11 Blood of a Person with Sickle-Cell Disease. Note the deformed, pointed erythrocytes.

Thalassemia[19] (thal-ah-SEE-me-ah) refers to a family of hereditary anemias that are typically seen among Greeks, Italians, and other people of the Mediterranean area and their descendants. Thalassemia is characterized by a deficiency or absence of the α or β hemoglobin and by RBC counts that may be less than 2 million cells/μL.

Before You Go On

Answer the following questions to test your understanding of the preceding section:

13. Describe the shape, size, and contents of an erythrocyte, and explain how it acquires its unusual shape.
14. What is the function of hemoglobin? What are its protein and nonprotein moieties called?
15. What happens to each of these moieties when old erythrocytes break up?
16. What is the body's primary mechanism for correcting hypoxemia? How does this illustrate homeostasis?
17. What are the three primary causes or categories of anemia? What are its three primary consequences?

18. *ag* = together + *glutin* = glue

19. *thalass* = the sea; once thought to be caused by "sea air"

Historical Insight 18.3

Charles Drew—Blood Banking Pioneer

Charles Drew was a scientist who lived and died in the arms of bitter irony. After receiving his M.D. from McGill University of Montreal in 1933, Drew became the first black person to pursue the advanced degree of Doctor of Science in Medicine, for which he studied transfusion and blood-banking procedures at Columbia University. He became the director of a new blood bank at Columbia Presbyterian Hospital in 1939 and organized numerous blood banks during World War II.

Drew saved countless lives by convincing physicians to use plasma rather than whole blood for battlefield and other emergency transfusions. Whole blood could be stored for only a week and given only to recipients with compatible blood types. Plasma could be stored longer and was less likely to cause transfusion reactions.

When the U.S. War Department issued a directive forbidding the mixing of Caucasian and Negro blood in military blood banks, Drew denounced the order and resigned his position. He became a professor of surgery at Howard University in Washington, D.C., and later chief of staff at Freedmen's Hospital. He was a mentor for numerous young black physicians and campaigned to get them accepted into the medical community. The American Medical Association, however, firmly refused to admit black members, even Drew himself.

Late one night in 1950, Drew and three colleagues set out to volunteer their medical services to an annual free clinic in Tuskegee, Alabama. Drew fell asleep at the wheel and was critically injured in the resulting accident. Doctors at the nearest hospital administered blood and attempted unsuccessfully to revive him. For all the lives he saved through his pioneering work in blood transfusion, Drew himself bled to death at the age of 45.

Charles Drew (1904–50).

Blood Types

Objectives

When you have completed this section, you should be able to

- explain what determines a person's ABO and Rh blood types and how this relates to transfusion compatibility;
- describe the effect of an incompatibility between mother and fetus in Rh blood type; and
- list some blood groups other than ABO and Rh and explain how they may be useful.

Blood types and transfusion compatibility are a matter of interactions between plasma proteins and erythrocytes. Ancient Greek physicians attempted to transfuse blood from one person to another by squeezing it from a pig's bladder through a porcupine quill into the recipient's vein. While some patients benefited from the procedure, it was fatal to others. The reason some people have compatible blood and some do not remained obscure until 1900, when Karl Landsteiner discovered blood types A, B, AB, and O—a discovery that won him a Nobel Prize in 1930. World War II stimulated great improvements in transfusions, blood banking, and blood substitutes (see historical insight 18.3).

All cells have an inherited combination of proteins, glycoproteins, and glycolipids on their surfaces. The immune system distinguishes our own cells from foreign invaders by means of these *antigens*—molecules that can trigger an immune response. Part of the immune response is the production of γ globulins called *antibodies* to combat the invader. In blood typing, the antigens of RBC surfaces are called **agglutinogens** (ah-glue-TIN-oh-jens) because they are partially responsible for RBC agglutination in mismatched transfusions. The plasma antibodies that react against them are called **agglutinins** (ah-GLUE-tih-nins).

The ABO Group

Blood types A, B, AB, and O form the **ABO blood group** (table 18.7). Your ABO blood type is determined by the

Table 18.7 The ABO Blood Group

	ABO Blood Type			
	Type O	**Type A**	**Type B**	**Type AB**
Possible Genotypes	*ii*	$I^A I^A$, $I^A i$	$I^B I^B$, $I^B i$	$I^A I^B$
RBC Agglutinogen (antigen)	None	A	B	A,B
Plasma Agglutinin (antibody)	Anti-A, anti-B	Anti-B	Anti-A	None
Compatible Donor RBCs	0	0, A	0, B	0, A, B, AB
Incompatible Donor RBCs	A, B, AB	B, AB	A, AB	None
Frequency in U.S. Population				
White	45%	40%	11%	4%
Black	49%	27%	20%	4%
Hispanic	63%	14%	20%	3%
Japanese	31%	38%	22%	9%
Native American	79%	16%	4%	<1%

hereditary presence or absence of agglutinogens A and B on your RBCs. The genetic determination of blood types is explained on page 162. The agglutinogens are glycoproteins and glycolipids—membrane proteins and phospholipids with short carbohydrate chains bonded to them. Figure 18.12 shows how these carbohydrates determine the ABO blood types.

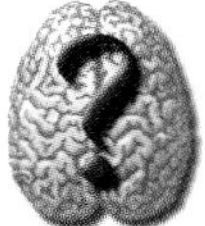

Think About It

Suppose you could develop an enzyme that selectively split N-acetylgalactosamine off the glycolipid of type A blood cells (fig. 18.12). What would be the potential benefit of this product to blood banking and transfusion?

The agglutinins of the ABO group begin to appear in the plasma 2 to 8 months after birth. They reach their maximum concentrations between 8 and 10 years of age and then slowly decline for the rest of one's life. They are produced mainly in response to the bacteria that inhabit our intestines, but they happen to cross-react with RBC antigens and are therefore best known for their significance in transfusions.

AB agglutinins (antibodies) react against any AB agglutinogen (antigen) except those on one's own RBCs. The agglutinin that reacts against antigen A is called α *agglutinin,* or *anti-A;* it is present in the plasma of people with type O or type B blood—that is, anyone who does *not* possess agglutinogen A. The agglutinin that reacts against antigen B is β *agglutinin,* or *anti-B,* and is present in type O and type A individuals—those who do not possess agglutinogen B. Each agglutinin molecule has 10 binding sites where it can attach to either an A or B agglutinogen. An agglutinin can therefore attach to several RBCs at once

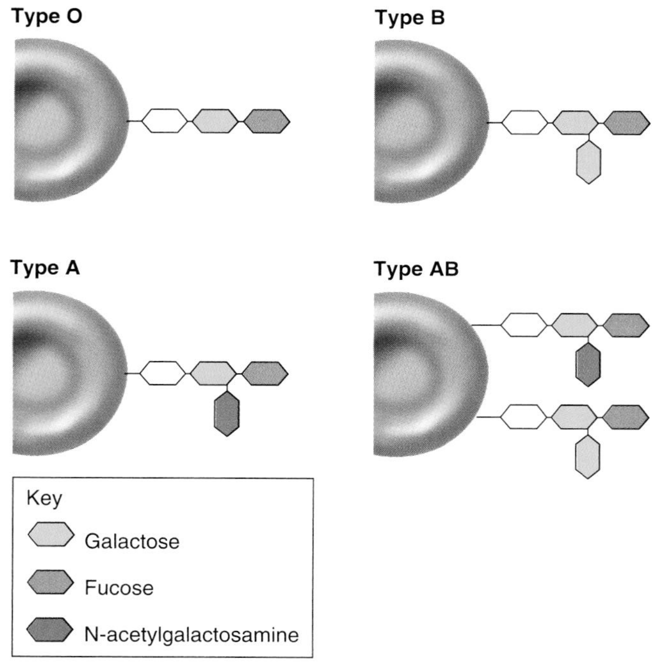

Figure 18.12 Chemical Basis of the ABO Blood Types. The terminal carbohydrates of the antigenic glycolipids are shown. All of them end with galactose and fucose (not to be confused with fructose). In type A, the galactose also has an N-acetylgalactosamine added to it; in type B, it has another galactose; and in type AB, both of these chain types are present.

and bind them together (fig. 18.13). **Agglutination** is the clumping of RBCs bound together by agglutinins.

A person's ABO blood type can be determined by placing one drop of blood in a pool of anti-A serum and another drop in a pool of anti-B serum. Blood type AB

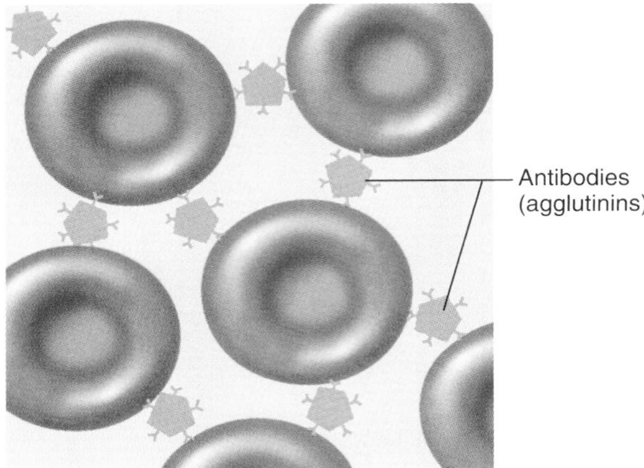

Figure 18.13 Agglutination of RBCs by an Agglutinin. Anti-A and anti-B have 10 binding sites, located at the tips of the *Y*s, and can therefore bind multiple RBCs to each other.

exhibits conspicuous agglutination in both antisera; type A or B agglutinates only in the corresponding antiserum; and type O does not agglutinate in either one (fig. 18.14). The anti-A and anti-B reagents most widely used now, however, are not blood sera but proteins called *lectins* that are extracted from lentils and other plant seeds. These are simpler to obtain and have a longer shelf life than blood serum.

Type O blood is the most common and AB is the rarest in the United States. Percentages differ from one region of the world to another and among ethnic groups because people tend to marry within their locality and ethnic group and perpetuate statistical variations particular to that group.

In giving transfusions, it is imperative that the donor's RBCs not agglutinate as they enter the recipient's bloodstream. For example, if type B blood were transfused into a type A recipient, the recipient's anti-B agglutinins would immediately agglutinate the donor's RBCs (fig. 18.15). A mismatched transfusion causes a **transfusion reaction**—the agglutinated RBCs block small blood vessels, hemolyze, and release their hemoglobin over the next few hours to days. Free hemoglobin can block the kidney tubules and cause death from acute renal failure within a week or so. For this reason, a person with type A (anti-B) blood must never be given a transfusion of type B or AB blood. A person with type B (anti-A) must never receive type A or AB blood. Type O (anti-A and anti-B) individuals cannot safely receive type A, B, or AB blood.

Type AB was once called the *universal recipient* because this blood type lacks both anti-A and anti-B agglutinins; thus, it will not agglutinate donor RBCs of any ABO type. However, this overlooks the fact that the *donor's*

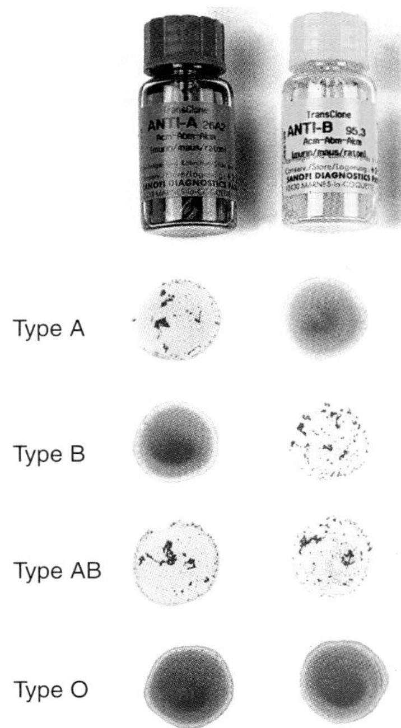

Figure 18.14 ABO Blood Typing. Each row shows the appearance of a drop of blood mixed with anti-A and anti-B reagents. Blood cells become clumped if they possess the agglutinogens for the reagent *(top row left, second row right, third row both),* but otherwise remain uniformly mixed. Thus type A agglutinates only in anti-A; type B agglutinates only in anti-B; type AB agglutinates in both; and type O agglutinates in neither of them.

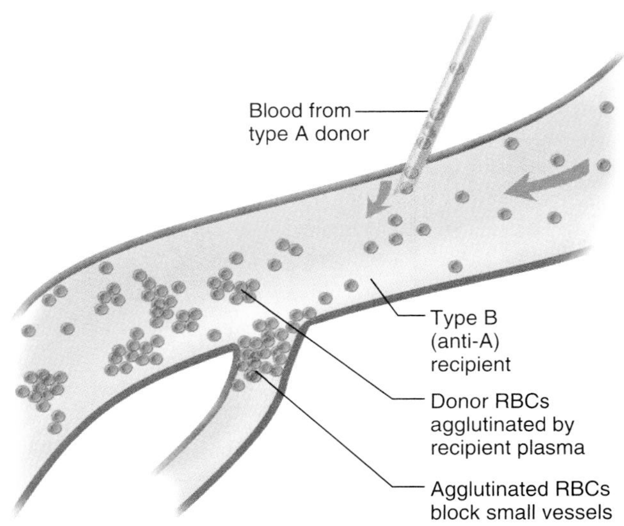

Figure 18.15 Effects of a Mismatched Transfusion. Donor RBCs become agglutinated in the recipient's blood plasma. The agglutinated RBCs lodge in smaller blood vessels downstream from this point and cut off the blood flow to vital tissues.

plasma can agglutinate the *recipient's* RBCs if it contains anti-A, anti-B, or both. For similar reasons, type O was once called the *universal donor.* The plasma of a type O donor, however, can agglutinate the RBCs of a type A, B, or AB recipient. There are procedures for reducing the risk of a transfusion reaction in certain mismatches, such as giving packed RBCs with a minimum of plasma; nevertheless, the concept of universal donors and recipients is obsolete and dangerously misleading.

Contrary to some people's belief, blood type is not changed by transfusion. It is fixed at conception and remains the same for life.

The Rh Group

The **Rh blood group** is named for the rhesus monkey, in which the Rh agglutinogens were discovered in 1940. This group is determined by three genes called C, D, and E, each of which has two alleles: *C, c, D, d, E, e.* Whatever other alleles a person may have, anyone with genotype *DD* or *Dd* has D agglutinogens on his or her RBCs and is classified as *Rh-positive* (Rh^+). In *Rh-negative* (Rh^-) people, the D agglutinogen is lacking. The Rh blood type is tested by using an anti-D reagent. The Rh type is usually combined with the ABO type in a single expression such as O^+ for type O, Rh-positive, or AB^- for type AB, Rh-negative. About 85% of white Americans are Rh^+ and 15% are Rh^-. ABO blood type has no influence on Rh type, or vice versa. If the frequency of type O whites in the United States is 45%, and 85% of these are also Rh^+, then the frequency of O^+ individuals is the product of these separate frequencies: $0.45 \times 0.85 = 0.38$, or 38%. Rh frequencies vary among ethnic groups just as ABO frequencies do. About 99% of Asians are Rh^+, for example.

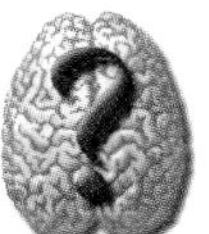

Think About It

Predict what percentage of Japanese Americans have type B^- blood.

In contrast to the ABO group, anti-D agglutinins are not normally present in the blood. They form only in Rh^- individuals who are exposed to Rh^+ blood. If an Rh^- person receives an Rh^+ transfusion, the recipient produces anti-D. Since anti-D does not appear instantaneously, this presents little danger in the first mismatched transfusion. But if that person should later receive another Rh^+ transfusion, his or her anti-D could agglutinate the donor's RBCs.

A related condition sometimes occurs when an Rh^- woman carries an Rh^+ fetus. The first pregnancy is likely to be uneventful because the placenta normally prevents maternal and fetal blood from mixing. However, at the time of birth, or if a miscarriage occurs, placental tearing exposes the mother to Rh^+ fetal blood. She then begins to produce anti-D agglutinins (fig. 18.16). If she becomes pregnant again with an Rh^+ fetus, her anti-D agglutinins may pass through the placenta and agglutinate the fetal erythrocytes. Agglutinated RBCs hemolyze, and the baby is born with a severe anemia called **hemolytic disease of the newborn (HDN),** or **erythroblastosis fetalis.** Not all HDN is due to Rh incompatibility, however. About 2% of cases result from incompatibility of ABO and other blood types. About 1 out of 10 cases of ABO incompatibility between mother and fetus results in HDN.

HDN, like so many other disorders, is easier to prevent than to treat. If an Rh^- woman gives birth to (or miscarries) an Rh^+ child, she can be given an *Rh immune globulin* (sold under trade names such as RhoGAM and Gamulin). The immune globulin binds fetal RBC agglutinogens so they cannot stimulate her immune system to produce anti-D. Some physicians now give immune globulin throughout any pregnancy in which the mother is Rh^- and the father is Rh^+.

If an Rh^- woman has had one or more previous Rh^+ pregnancies, her subsequent Rh^+ children have about a 17% probability of being born with HDN. Infants with HDN are usually severely anemic. As the fetal hemopoietic tissues respond to the need for more RBCs, erythroblasts (immature RBCs) enter the circulation prematurely—hence the name *erythroblastosis fetalis.* Hemolyzed RBCs release hemoglobin, which is converted to bilirubin. High bilirubin levels can cause *kernicterus,* a syndrome of toxic brain damage that may kill the infant or leave it with motor, sensory, and mental deficiencies. HDN can be treated with *phototherapy*—exposing the infant to ultraviolet light, which degrades bilirubin as blood passes through the capillaries of the skin. In more severe cases, an *exchange transfusion* may be given to completely replace the infant's Rh^+ blood with Rh^-. In time, the infant's hemopoietic tissues will replace the donor's RBCs with Rh^+ cells, and by then the mother's agglutinin will have disappeared from the infant's blood.

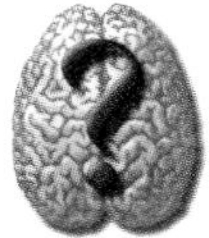

Think About It

A baby with HDN typically has jaundice and an enlarged spleen. Explain these effects.

Other Blood Groups

In addition to the ABO and Rh groups, there are at least 100 other known blood groups with a total of more than 500 antigens, including the MN, Duffy, Kell, Kidd, and Lewis groups. These rarely cause transfusion reactions, but they are useful for such legal purposes as paternity and criminal cases and for research in anthropology and population genetics. The Kell, Kidd, and Duffy groups occasionally cause HDN.

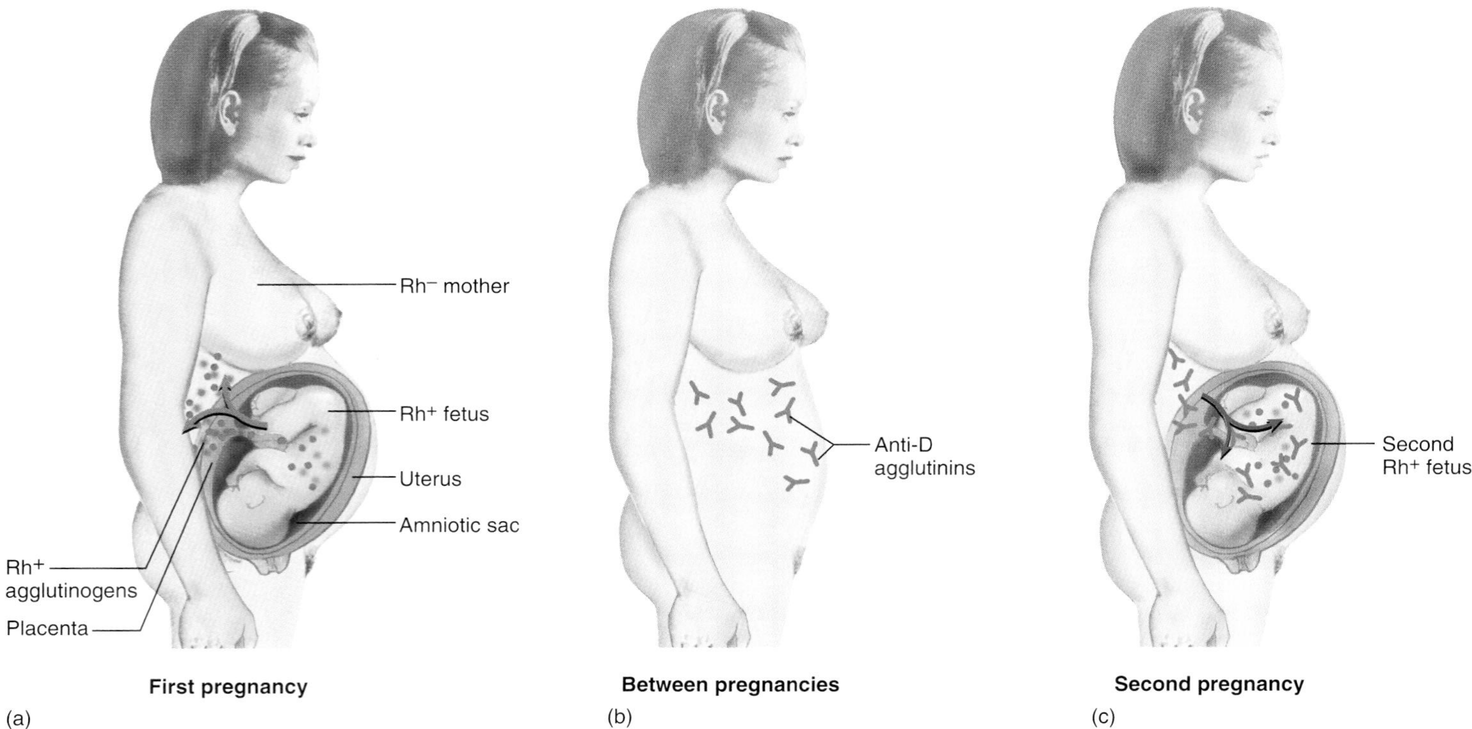

Figure 18.16 **Hemolytic Disease of the Newborn (HDN).** *(a)* When an Rh^- woman is pregnant with an Rh^+ fetus, she is exposed to D (Rh) antigens, especially during childbirth. *(b)* Following that pregnancy, her immune system produces anti-D agglutinins. *(c)* If she later becomes pregnant with another Rh^+ fetus, her anti-D agglutinins can cross the placenta and agglutinate the blood of that fetus, causing that child to be born with HDN.

Before You Go On

Answer the following questions to test your understanding of the preceding section:

18. What are agglutinins and agglutinogens? How do they interact to cause a transfusion reaction?
19. What agglutinins and agglutinogens are present in people with each of the four ABO blood types?
20. Describe the cause, prevention, and treatment of HDN.
21. Why might someone be interested in determining a person's blood type other than ABO/Rh?

Leukocytes

Objectives

When you have completed this section, you should be able to

- state the general function that all leukocytes have in common;
- name and describe the five types of leukocytes; and
- describe the types, causes, and effects of abnormal leukocyte counts.

Leukocytes, or white blood cells (WBCs), play a number of roles in the body's defense against pathogens. Their individual functions are summarized in table 18.8, but they are discussed more extensively in chapter 21. There are five kinds of WBCs. They are easily distinguished from erythrocytes in stained blood films because they contain conspicuous nuclei that stain from light violet to dark purple with the most common blood stains. Three WBC types—the *neutrophils, eosinophils,* and *basophils*—are called **granulocytes** because their cytoplasm contains organelles that appear as colored granules through the microscope. These are missing or relatively scanty in the two types known as **agranulocytes**—the *lymphocytes* and *monocytes.*

Types of Leukocytes

The five leukocyte types are compared in table 18.8. From the photographs and data, take note of their sizes relative to each other and to the size of erythrocytes (which are about 7.5 μm in diameter). Also note how the leukocytes differ from each other in relative abundance—from neutrophils, which constitute about two-thirds of the WBC count, to basophils, which usually account for less than

Table 18.8 The White Blood Cells (Leukocytes)

Neutrophils

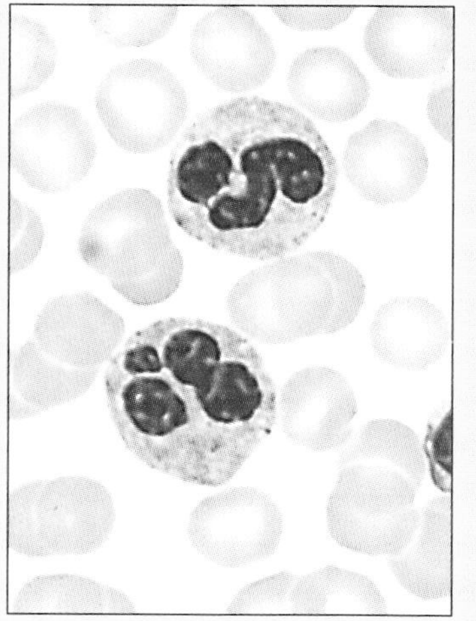

Percent of WBCs	60%–70%
Mean count	4,150 cells/μL
Diameter	9–12 μm

Appearance*
- Nucleus usually with 3–5 lobes in S- or C-shaped array
- Fine reddish to violet granules in cytoplasm

Differential Count
- Increases in bacterial infections

Functions
- Phagocytosis of bacteria
- Release of antimicrobial chemicals

Eosinophils

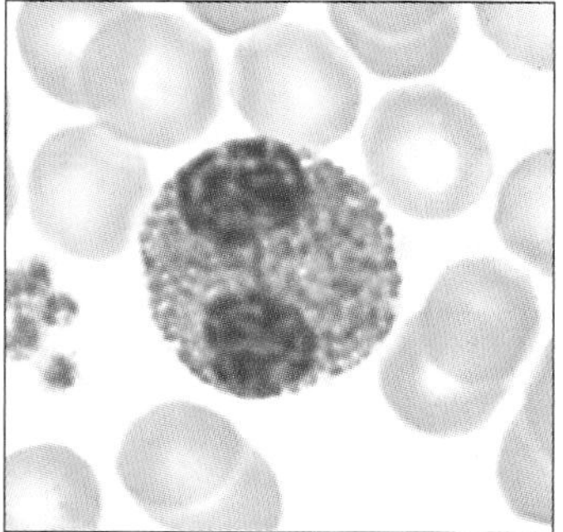

Percent of WBCs	2%–4%
Mean count	165 cells/μL
Diameter	10–14 μm

Appearance
- Nucleus usually has two large lobes connected by thin strand
- Large orange-pink granules in cytoplasm

Differential Count
- Fluctuates greatly from day to night, seasonally, and with phase of menstrual cycle
- Increases in parasitic infections, allergies, collagen diseases, and diseases of spleen and central nervous system

Functions
- Phagocytosis of antigen-antibody complexes, allergens, and inflammatory chemicals
- Release enzymes that weaken or destroy parasites such as worms

Basophils

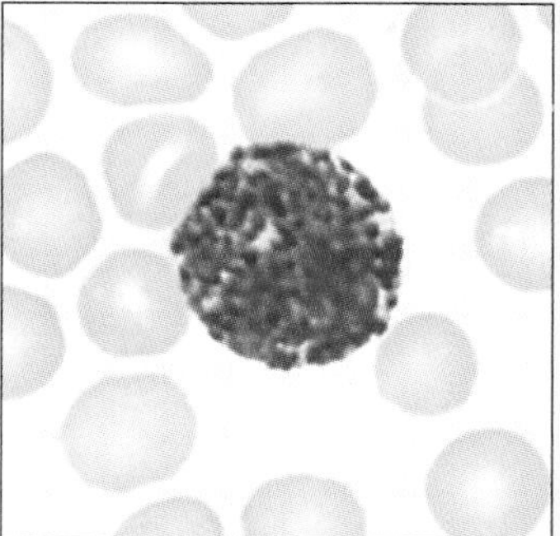

Percent of WBCs	< 0.5%–1%
Mean count	44 cells/μL
Diameter	8–10 μm

Appearance
- Nucleus large and U- to S-shaped, but typically pale and obscured from view
- Coarse, abundant, dark violet granules in cytoplasm

Differential Count
- Relatively stable
- Increases in chicken pox, sinusitis, diabetes mellitus, myxedema, and polycythemia

Functions
- Secrete histamine (a vasodilator), which increases blood flow to a tissue
- Secrete heparin (an anticoagulant), which promotes mobility of other WBCs by preventing clotting

(continued)

Table 18.8 The White Blood Cells (Leukocytes) (continued)

Lymphocytes

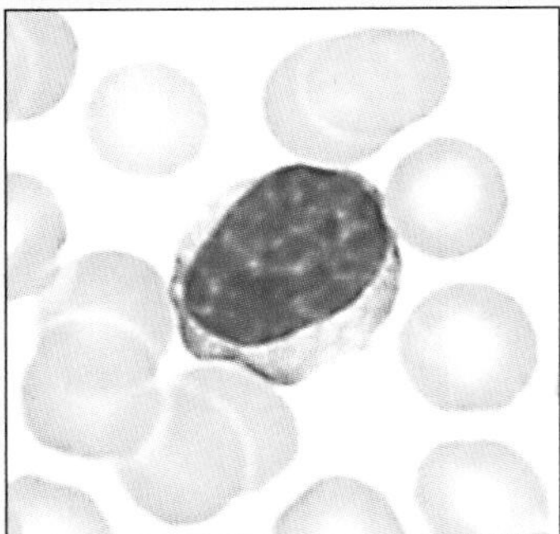

Percent of WBCs	25%–33%
Mean count	2,185 cells/μL
Diameter	
Small class	5–8 μm
Medium class	10–12 μm
Large class	14–17 μm

Appearance

- Nucleus round, ovoid, or slightly dimpled on one side, of uniform dark violet color
- In small lymphocytes, nucleus fills nearly all of the cell and leaves only a scanty rim of clear, light blue cytoplasm
- In larger lymphocytes, cytoplasm is more abundant; large lymphocytes may be hard to differentiate from monocytes

Differential Count

- Increases in diverse infections and immune responses

Functions

- Several functional classes usually indistinguishable by light microscopy
- Destroy cancer cells, cells infected with viruses, and foreign cells
- "Present" antigens to activate other cells of immune system
- Coordinate actions of other immune cells
- Secrete antibodies
- Serve in immune memory

Monocytes

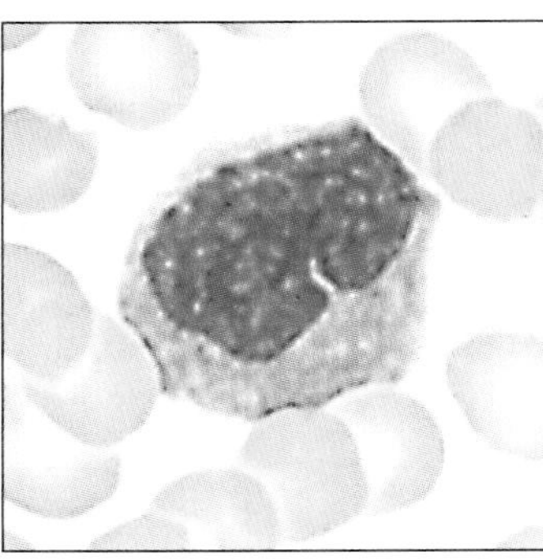

Percent of WBCs	3%–8%
Mean count	456 cells/μL
Diameter	12–15 μm

Appearance

- Nucleus ovoid, kidney-shaped, or horseshoe-shaped; light violet
- Abundant cytoplasm with sparse, fine granules
- Sometimes very large with stellate or polygonal shapes

Differential Count

- Increases in viral infections and inflammation

Functions

- Differentiate into macrophages (large phagocytic cells of the tissues)
- Phagocytize pathogens, dead neutrophils, and debris of dead cells
- "Present" antigens to activate other cells of immune system

*Appearance pertains to blood films dyed with Wright's stain.

1%. Nuclear shape is an important key to identifying leukocytes. The granulocytes are further distinguished from each other by the coarseness, abundance, and staining properties of their cytoplasmic granules.

Granulocytes

Neutrophils have very fine cytoplasmic granules that contain lysozyme, peroxidase, and other antibiotic agents. They are named for the way these granules take up blood stains at pH 7—some stain with acidic dyes and others with basic dyes, and the combined effect gives the cytoplasm a pale lilac color. The nucleus is usually divided into three to five lobes, which are connected by strands of nucleoplasm so delicate that the cell may appear to have multiple nuclei. Young neutrophils often exhibit an undivided nucleus shaped like a band or a knife puncture; they are thus called *band,* or *stab, cells.* Neutrophils are also called *polymorphonuclear* leukocytes (PMNs) because of their variety of nuclear shapes.

Eosinophils (EE-oh-SIN-oh-fills) are easily distinguished by their large rosy to orange-colored granules and prominent, usually bilobed nucleus.

In **basophils,** the nucleus is pale and usually hidden by the coarse, dark violet granules in the cytoplasm. It is sometimes difficult to distinguish a basophil from a lymphocyte, but basophils are conspicuously grainy while the lymphocyte nucleus is more homogeneous, and basophils lack the clear blue rim of cytoplasm usually seen in stained lymphocytes.

Agranulocytes

Lymphocytes are usually comparable to erythrocytes in size, or only slightly larger. They are sometimes classified into three size classes (table 18.8), but there are gradations between these categories. Medium and large lymphocytes are usually seen in fibrous connective tissues and only occasionally in the circulating blood. In small lymphocytes, the nucleus often fills almost the entire cell and leaves only a narrow rim of clear, light blue cytoplasm. Large lymphocytes, however, have ample cytoplasm around the nucleus and are sometimes difficult to distinguish from monocytes. There are several subclasses of lymphocytes with different immune functions (see chapter 21), but they look alike through the light microscope.

Monocytes are the largest of the formed elements, typically about twice the diameter of an erythrocyte but sometimes approaching three times as large. The monocyte nucleus tends to stain a lighter blue than most leukocyte nuclei. The cytoplasm is abundant and relatively clear. In stained blood films monocytes sometimes appear as very large cells with bizarre stellate (star-shaped) or polygonal contours (see fig. 18.1*a*).

Abnormalities of Leukocyte Count

The total WBC count is normally 5,000 to 10,000 WBCs/μL. A count below this range, called **leukopenia**[20] (LOO-co-PEE-nee-uh), is seen in lead, arsenic, and mercury poisoning; radiation sickness; and such infectious diseases as measles, mumps, chicken pox, poliomyelitis, influenza, typhoid fever, and AIDS. It can also be produced by glucocorticoids, anticancer drugs, and immunosuppressant drugs given to organ transplant patients. Since WBCs are protective cells, leukopenia presents an elevated risk of infection and cancer. A count above 10,000 WBCs/μL, called **leukocytosis,**[21] usually indicates infection, allergy, or other diseases but can also occur in response to dehydration or emotional disturbances. More useful than a total WBC count is a *differential WBC count,* which identifies what percentage of the total WBC count consists of each type of leukocyte. A high neutrophil count is a sign of bacterial infection; neutrophils become sharply elevated in appendicitis, for example. A high eosinophil count usually indicates an allergy or a parasitic infection such as hookworms or tapeworms.

Leukemia is a cancer of the hemopoietic tissues that usually produces an extraordinarily high number of circulating leukocytes and their precursors (fig. 18.17*b*). Leukemia is classified as myeloid or lymphoid, acute or chronic. **Myeloid leukemia** is marked by uncontrolled granulocyte production, whereas **lymphoid leukemia** involves uncontrolled lymphocyte or monocyte production. **Acute leukemia** appears suddenly, progresses rapidly, and causes death within a few months if it is not treated. **Chronic leukemia** develops more slowly and may go undetected for many months; if untreated, the typical survival time is about 3 years. Both myeloid and lymphoid leukemia occur in acute and chronic forms. The greatest success in treatment and cure has been with acute lymphoblastic leukemia, the most common type of childhood cancer. Treatment employs chemotherapy and marrow transplants along with the control of side effects such as anemia, hemorrhaging, and infection.

As leukemic cells proliferate, they replace normal bone marrow and a person suffers from a deficiency of normal granulocytes, erythrocytes, and platelets. Although enormous numbers of leukocytes are produced and spill over into the bloodstream, they are immature cells incapable of performing their normal defensive roles. The deficiency of competent WBCs leaves the patient vulnerable to *opportunistic infection*—the establishment of pathogenic organisms that usually cannot get a foothold in people with healthy immune systems. The RBC deficiency

20. *leuko* = white + *penia* = deficiency
21. *leuko* = white + *cyt* = cell + *osis* = condition

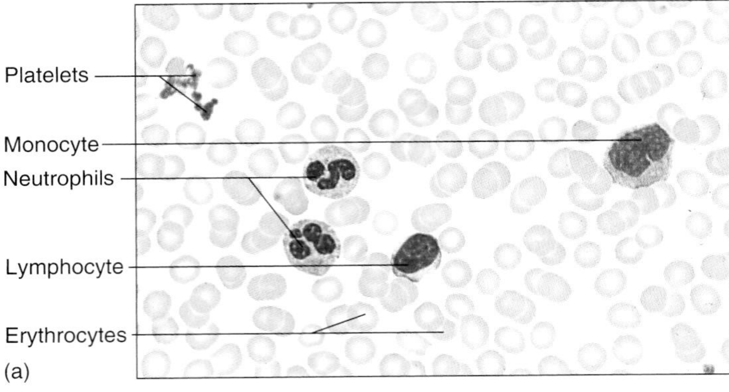

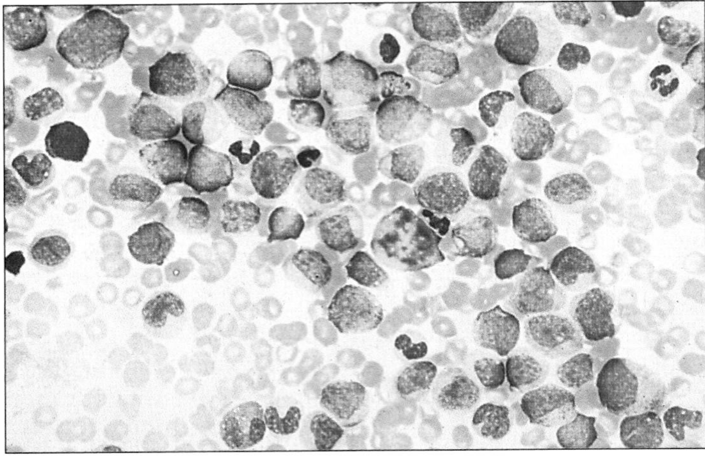

Figure 18.17 **Normal and Leukemic Blood.** (*a*) A normal blood smear; (*b*) blood from a patient with acute monocytic leukemia. Note the abnormally high number of white blood cells, especially monocytes, in *b*.

renders the patient anemic and fatigued, and the platelet deficiency results in hemorrhaging and impaired blood clotting. The immediate cause of death is usually hemorrhage or opportunistic infection. Cancerous hemopoietic tissue often metastasizes from the bone marrow or lymph nodes to other organs of the body, where the cells displace or compete with normal cells. Metastasis to the bone tissue itself is common and leads to bone and joint pain.

Before You Go On

Answer the following questions to test your understanding of the preceding section:

22. What is the overall function of leukocytes?
23. What can cause abnormally high or low leukocyte counts?
24. Define *leukemia*. Distinguish between myeloid and lymphoid leukemia.

Hemostasis—The Control of Bleeding

Objectives

When you have completed this section, you should be able to

- describe the body's mechanisms for controlling bleeding;
- list the functions of platelets;
- describe two reaction pathways that produce blood clots;
- explain what happens to blood clots when they are no longer needed;
- explain what keeps blood from clotting in the absence of injury; and
- describe some disorders of blood clotting.

Circulatory systems developed very early in animal evolution, and with them evolved mechanisms for stopping leaks, which are potentially fatal. **Hemostasis**[22] is the stoppage of bleeding. Although hemostatic mechanisms may not stop a hemorrhage from a large blood vessel, they are quite effective at closing breaks in small ones. Platelets play multiple roles in hemostasis, so we begin with a consideration of their form and function.

Platelets

Platelets (see fig. 18.1) are not cells but small fragments of megakaryocyte cytoplasm. They are 2 to 4 μm in diameter and possess lysosomes, endoplasmic reticulum, a Golgi complex, and Golgi vesicles, or "granules," that contain a variety of factors involved in platelet function. Platelets have pseudopods and are capable of ameboid movement and phagocytosis. In normal blood from a fingerstick, the platelet count ranges from 130,000 to 400,000 platelets/μL (averaging about 250,000/μL). The count can vary greatly, however, under different physiological conditions and in blood from different places in the body. When a blood specimen dries on a slide, platelets clump together; therefore in stained blood films, they often appear in clusters.

Platelets have a broad range of functions, many of which have come to light only in recent years:

- They secrete *procoagulants,* or clotting factors, which promote blood clotting.
- They secrete vasoconstrictors, which cause *vascular spasms* in broken vessels.
- They form temporary *platelet plugs* to stop bleeding.
- They dissolve blood clots that have outlasted their usefulness.

22. *hemo* = blood + *stasis* = stability

- They phagocytize and destroy bacteria.
- They secrete chemicals that attract neutrophils and monocytes to sites of inflammation.
- They secrete growth factors that stimulate mitosis in fibroblasts and smooth muscle and help to maintain the linings of blood vessels.

There are three hemostatic mechanisms—*vascular spasm, platelet plug formation,* and *blood clotting (coagulation)* (fig. 18.18). Platelets play an important role in all three.

Vascular Spasm

The most immediate protection against blood loss is **vascular spasm,** a prompt constriction of the broken vessel. Several things trigger this reaction. An injury stimulates pain receptors, some of which directly innervate nearby blood vessels and cause them to constrict. This effect lasts only a few minutes, but other mechanisms take over by the time it subsides. Injury to the smooth muscle of the blood vessel itself causes a longer-lasting vasoconstriction, and platelets release serotonin, a chemical vasoconstrictor. Thus, the vascular spasm is maintained long enough for the other two hemostatic mechanisms to come into play.

Platelet Plug Formation

Platelets will not adhere to the endothelium (inner lining) of undamaged blood vessels. The endothelium is normally very smooth and coated with **prostacyclin,** a platelet repellent. When a vessel is broken, however, collagen fibers of its wall are exposed to the blood. Upon contact with collagen or other rough surfaces, platelets put out long spiny pseudopods that adhere to the vessel and to other platelets; the pseudopods then contract and draw the walls of the vessel together. The mass of platelets thus formed, called a **platelet plug,** may reduce or stop minor bleeding.

As platelets aggregate, they undergo **degranulation**—the exocytosis of their cytoplasmic granules and release of factors that promote hemostasis. Among these are serotonin, a vasoconstrictor; adenosine diphosphate (ADP), which attracts more platelets to the area and stimulates their degranulation; and **thromboxane A_2,** an eicosanoid that promotes platelet aggregation, degranulation, and vasoconstriction. Thus, a positive feedback cycle is activated that can quickly seal a small break in a blood vessel.

Coagulation

Coagulation (clotting) of the blood is the last but most effective defense against bleeding. It is important for the blood to clot quickly when a vessel has been broken, but equally important for it not to clot in the absence of vessel damage. Because of this delicate balance, coagulation is one of the most complex processes in the body, involving over 30 chemical reactions. It is presented here in a very simplified form.

Perhaps clotting is best understood if we first consider its goal. The objective is to convert the plasma protein fibrinogen into **fibrin,** a sticky protein that adheres to the walls of a vessel. As blood cells and platelets arrive, they become stuck to the fibrin like insects sticking to a

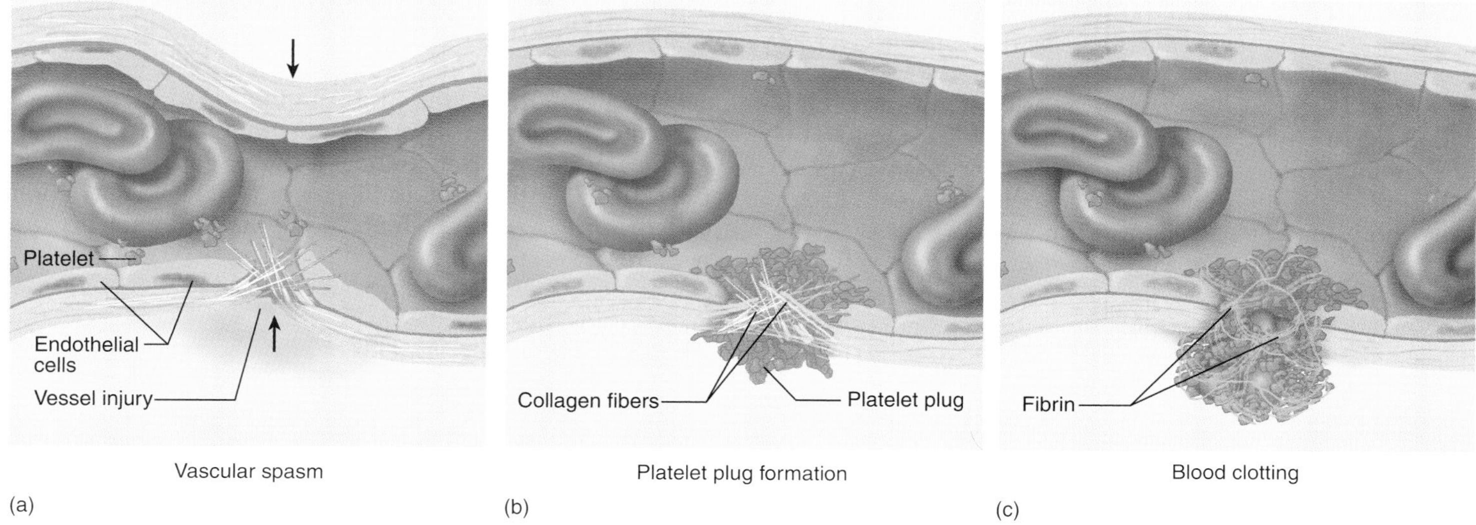

Figure 18.18 Hemostasis. (*a*) Vasoconstriction of a broken vessel reduces bleeding. (*b*) A platelet plug forms as platelets adhere to exposed collagen fibers of the vessel wall. The platelet plug temporarily seals the break. (*c*) A blood clot forms as platelets and erythrocytes become enmeshed in fibrin threads. This forms a longer-lasting seal and gives the vessel a chance to repair itself.

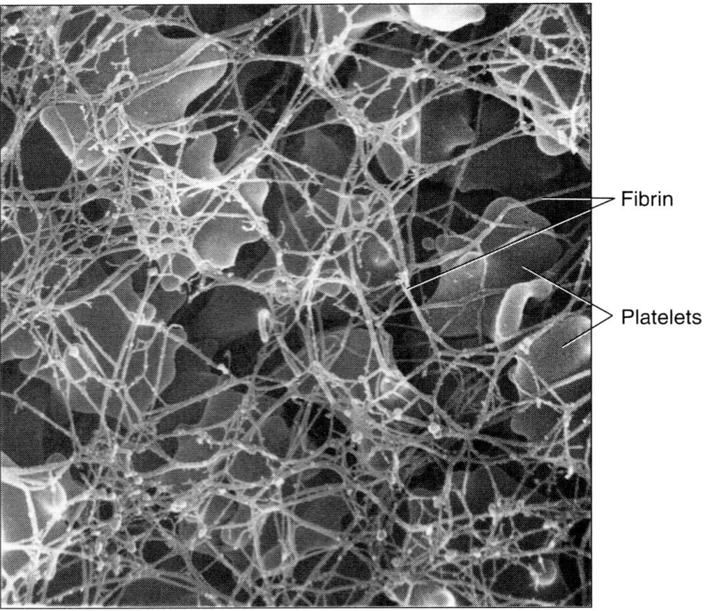

Figure 18.19 A Blood Clot Seen by SEM. Platelets adhere to a mesh of sticky fibrin filaments.

spider web (fig. 18.19). The resulting mass of fibrin, blood cells, and platelets ideally seals the break in the blood vessel. The complexity of clotting lies in how the fibrin is formed.

There are two reaction pathways to coagulation (fig. 18.20). One of them, the **extrinsic mechanism,** is initiated by clotting factors released by the damaged blood vessel and perivascular[23] tissues. The word *extrinsic* refers to the fact that these factors come from sources other than the blood itself. Blood may also clot, however, without these tissue factors—for example, when platelets adhere to a fatty plaque of atherosclerosis or to a test tube. The reaction pathway in this case is called the **intrinsic mechanism** because it uses only clotting factors found in the blood itself. In most cases of bleeding, both the extrinsic and intrinsic mechanisms work simultaneously to contribute to hemostasis.

Clotting factors (table 18.9) are called **procoagulants,** in contrast to the **anticoagulants** discussed later (see clinical insight 18.5, p. 711). Most procoagulants are proteins produced by the liver. They are always present in the plasma in inactive form, but when one factor is activated, it functions as an enzyme that activates the next one in the pathway. That factor activates the next, and so on, in a sequence called a **reaction cascade**—a series of reactions, each of which depends on the product of the preceding one. Many of the clotting factors are identified by Roman numerals, which indicate the order in which they were discovered, not the order of the reactions. Factors IV and VI are not included in table 18.9. These terms were abandoned when it was found that factor IV was calcium and factor VI was activated factor V. The last four procoagulants in the table are called *platelet factors* (PF_1 through PF_4) because they are produced by the platelets.

Initiation of Coagulation

The extrinsic mechanism is diagrammed on the left side of figure 18.20. The damaged blood vessel and perivascular tissues release a lipoprotein mixture called **tissue thromboplastin**[24] (factor III). In the presence of Ca^{2+}, tissue thromboplastin activates factor VII, which, in the presence of factor V, then activates factor X. The extrinsic and intrinsic pathways differ only in how they arrive at active factor X. Therefore, before examining their common pathway from factor X to the end, let's consider how the intrinsic pathway reaches this step.

The intrinsic mechanism is diagrammed on the right side of figure 18.20. Everything needed to initiate it is present in the plasma or platelets. When platelets degranulate, they release factor XII (Hageman factor, named for the patient in whom it was discovered). Through a cascade of reactions, this leads to activated factors XI, IX, and VIII, in that order—each serving as an enzyme that catalyzes the next step—and finally to factor X. This pathway also requires Ca^{2+} and PF_3.

Completion of Coagulation

Once factor X is activated, the remaining events are identical in the intrinsic and extrinsic mechanisms. Factor X combines with factors III and V in the presence of Ca^{2+} and PF_3 to produce *prothrombin activator.* This enzyme acts on a globulin called **prothrombin** (factor II) and converts it to the enzyme **thrombin.** Thrombin then chops up fibrinogen into shorter strands of fibrin. Factor XIII cross-links these fibrin strands to create a dense aggregation called *fibrin polymer,* which forms the structural framework of the blood clot.

Once a clot begins to form, it launches a self-accelerating positive feedback process that seals off the damaged vessel more quickly. Thrombin works with factor V to accelerate the production of prothrombin activator, which in turn produces more thrombin.

The cascade of enzymatic reactions acts as an amplifying mechanism to ensure the rapid clotting of blood (fig. 18.21). Each activated enzyme in the pathway pro-

23. *peri* = around + *vas* = vessel + *cular* = pertaining to

24. *thrombo* = clot + *plast* = forming + *in* = substance

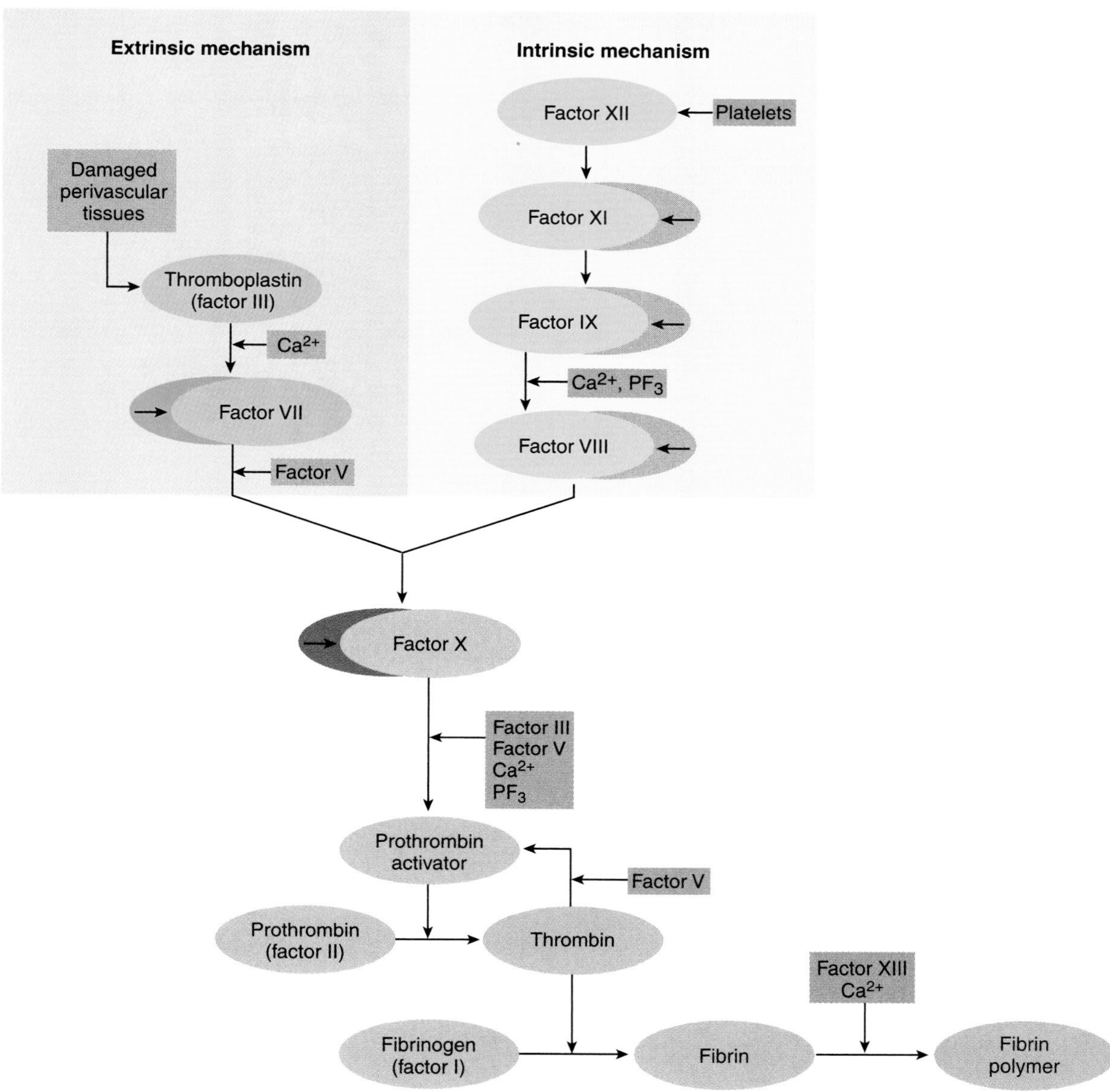

Figure 18.20 **The Pathways of Coagulation.** Most clotting factors act as enzymes that convert the next factor from an inactive form (*shaded ellipse*) to an active form (*lighter ellipse*).

duces a larger number of enzyme molecules at the following step. One activated molecule of factor XII at the start of the intrinsic pathway, for example, causes thousands of fibrin molecules to be produced very quickly. Note the similarity of this process to the *enzyme amplification* that occurs in hormone action (see chapter 17, fig. 17.27).

Notice that the extrinsic mechanism requires fewer steps to activate factor X than the intrinsic mechanism does; it is a "shortcut" to coagulation. It takes 3 to 6 minutes for a clot to form by the intrinsic pathway but only 15 seconds or so by the extrinsic pathway. For this reason, when a small wound bleeds, you can stop the bleeding sooner by massaging the site. This releases thromboplastin from the perivascular tissues and activates or speeds up the extrinsic pathway.

A number of laboratory tests are used to evaluate the efficiency of coagulation. Normally, the bleeding of a fingerstick should stop within 2 to 3 minutes, and a sample of blood in a clean test tube should clot within 15 minutes. Other techniques are available that can separately assess the effectiveness of the intrinsic and extrinsic mechanisms.

Table 18.9 Clotting Factors (Procoagulants)

Number	Name	Origin	Function
I	Fibrinogen	Liver	Precursor of fibrin
II	Prothrombin	Liver	Precursor of thrombin
III	Tissue thromboplastin	Perivascular tissue	Activates factor VII
V	Proaccelerin	Liver	Activates factor VII; combines with factor X to form prothrombin activator
VII	Proconvertin	Liver	Activates factor X in extrinsic pathway
VIII	Antihemophiliac factor A	Liver	Activates factor X in intrinsic pathway
IX	Antihemophiliac factor B	Liver	Activates factor VIII
X	Thrombokinase	Liver	Combines with factor V to form prothrombin activator
XI	Antihemophiliac factor C	Liver	Activates factor IX
XII	Hageman factor	Liver, platelets	Activates factor XI and plasmin; converts prekallikrein to kallikrein
XIII	Fibrin-stabilizing factor	Platelets, plasma	Cross-links fibrin filaments to make fibrin polymer and stabilize clot
PF_1	Platelet factor 1	Platelets	Same role as factor V; also accelerates platelet activation
PF_2	Platelet factor 2	Platelets	Accelerates thrombin formation
PF_3	Platelet factor 3	Platelets	Aids in activation of factor VIII and prothrombin activator
PF_4	Platelet factor 4	Platelets	Binds heparin during clotting to inhibit its anticoagulant effect

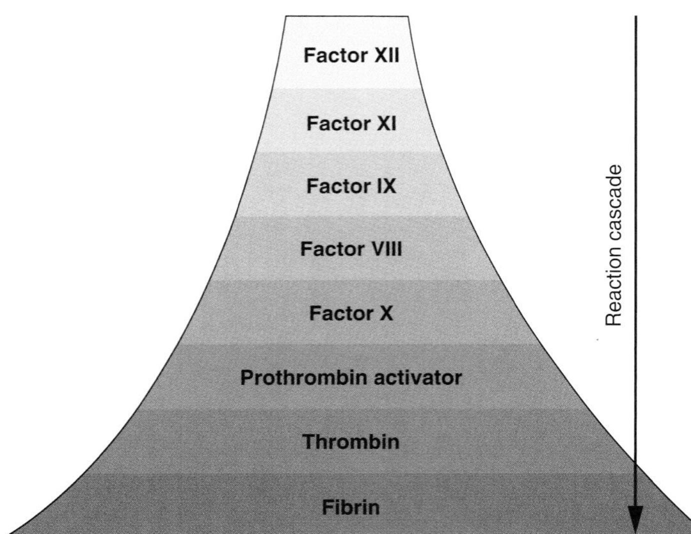

Figure 18.21 Enzyme Amplification in Blood Clotting. Each clotting factor produces many molecules of the next one, so the number of active clotting factors increases rapidly and a large amount of fibrin is quickly formed. The example shown here is for the intrinsic mechanism.

The Fate of Blood Clots

After a clot has formed, spinous pseudopods of the platelets adhere to strands of fibrin and contract. This pulls on the fibrin threads and draws the edges of the broken vessel together, like a drawstring closing a purse. Through this process of **clot retraction,** the clot becomes more compact within about 30 minutes.

Platelets and endothelial cells secrete a mitotic stimulant named *platelet-derived growth factor (PDGF).* PDGF stimulates fibroblasts and smooth muscle cells to multiply and repair the damaged blood vessel. Fibroblasts also invade the clot and produce fibrous connective tissue, which helps to strengthen and seal the vessel while the repairs take place.

Eventually, tissue repair is completed and the clot must be disposed of. **Fibrinolysis,** the dissolution of a clot, is achieved by a small cascade of reactions with a positive feedback component. In addition to promoting clotting, factor XII catalyzes the formation of a plasma enzyme called **kallikrein** (KAL-ih-KREE-in). Kallikrein, in turn, converts the inactive protein *plasminogen* into **plasmin,** a fibrin-dissolving enzyme that breaks up the clot. Thrombin also activates plasmin, and plasmin indirectly promotes the formation of more kallikrein, thus completing a positive feedback loop (fig. 18.22).

Prevention of Inappropriate Coagulation

Precise controls are required to prevent coagulation when it is not needed. These include the following:

- **Platelet repulsion.** As noted earlier, platelets do not adhere to the smooth prostacyclin-coated endothelium of undamaged blood vessels.

Clinical Insight 18.4

Liver Disease, Vitamin K, and Blood Clotting

The liver synthesizes most clotting factors. Liver disease such as hepatitis and cirrhosis can therefore lead to deficiencies of blood clotting. Vitamin K is needed as a cofactor for synthesizing factors II, VII, IX, and X, and it is not absorbed well from the small intestine unless bile is present. Gallstones can therefore lead to a clotting deficiency by blocking the bile duct and interfering with bile secretion and vitamin K absorption. Pregnant women should take supplemental vitamin K to ensure that their blood and the baby's will clot efficiently at birth. A vitamin K injection is frequently given to newborn infants.

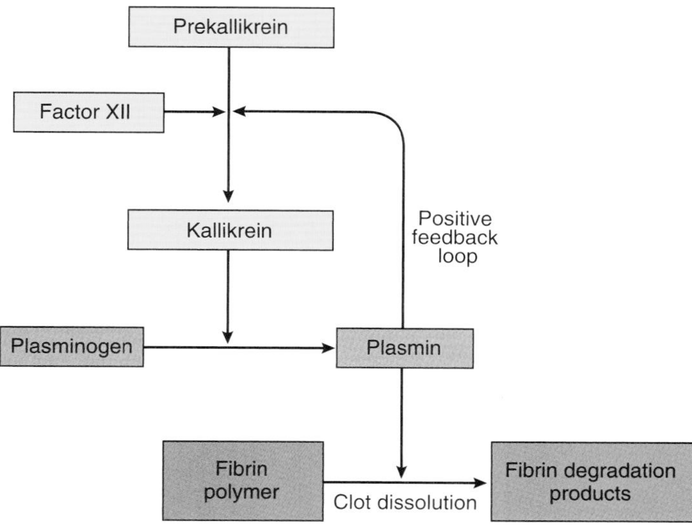

Figure 18.22 The Mechanism for Dissolving Blood Clots. Prekallikrein is converted to kallikrein. Kallikrein is an enzyme that catalyzes the formation of plasmin. Plasmin is an enzyme that dissolves the blood clot.

- **Dilution.** Small amounts of thrombin form spontaneously in the plasma, but at normal rates of blood flow the thrombin is diluted so quickly that a clot has little chance to form. If flow decreases, however, enough thrombin can accumulate to cause clotting. This can happen in circulatory shock, for example, when output from the heart is diminished and circulation slows down.
- **Anticoagulants.** Thrombin formation is suppressed by anticoagulants that are present in the plasma. **Antithrombin,** secreted by the liver, deactivates thrombin before it can act on fibrinogen. **Heparin,** secreted by basophils and mast cells, interferes with the formation of prothrombin activator, blocks the action of thrombin on fibrinogen, and promotes the action of antithrombin. Heparin is given by injection to patients with abnormal clotting tendencies.

Coagulation Disorders

In a process as complex as coagulation, it is not surprising that things can go wrong. Clotting deficiencies can result from causes as diverse as malnutrition, leukemia, and gallstones (see clinical insight 18.4). Destruction of the bone marrow by radiation, drugs, poisons, or leukemia can cause **thrombocytopenia,** a platelet count below 100,000/μL. People with this disorder exhibit small hemorrhagic spots under the skin, and relatively slight blows cause large **hematomas**[25] (bruises or other masses of clotted blood in the tissues).

A deficiency of any clotting factor can shut down the coagulation cascade. This happens in **hemophilia,** a family of hereditary diseases characterized by deficiencies of one factor or another. Because of its sex-linked recessive mechanism of heredity, hemophilia occurs predominantly in males. They can inherit it only from their mothers, however, as happened with the descendants of Queen Victoria. The lack of factor VIII causes *classical hemophilia (hemophilia A),* which accounts for about 83% of cases and afflicts 1 in 5,000 males worldwide. Lack of factor IX causes *hemophilia B,* which accounts for 15% of cases and occurs in about 1 out of 30,000 males. Factors VIII and IX are therefore known as *antihemophilic factors A* and *B.*

Before purified factor VIII became available in the 1960s, more than half of those with hemophilia died before age 5 and only 10% lived to age 21. Physical exertion causes bleeding into the muscles and joints. Intramuscular and joint hematomas can be excruciatingly painful and can eventually immobilize the joints. Hemophilia varies in severity, however. Half of the normal level of clotting factor is enough to prevent the symptoms, and the symptoms are mild even in individuals with as little as 30% of the normal amount. Such cases may go undetected even into adulthood. Bleeding can be relieved for a few days by transfusion of plasma or purified clotting factors. Factor VIII is now produced by transgenic bacteria (see clinical insight 5.5, p. 166).

25. *hemat* = blood + *oma* = mass

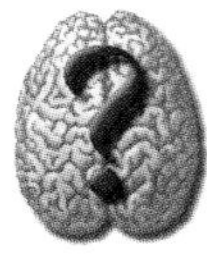

Think About It

Why is it important for people with hemophilia not to use aspirin? (Hint: See p. 673.)

Far more people die from unwanted blood clotting than from clotting failure. Most strokes and heart attacks are due to **thrombosis**—the abnormal clotting of blood in an unbroken vessel. A **thrombus** (clot) may grow large enough to obstruct a small vessel, or a piece of it may break loose and begin to travel in the bloodstream as an **embolus.**[26] An embolus may lodge in a small artery and block blood flow from that point on. If that vessel supplies a vital organ such as the heart, brain, lung, or kidney, *infarction* (tissue death) may result. About 650,000 Americans die annually of *thromboembolism* (traveling blood clots) in the cerebral, coronary, and pulmonary arteries.

Thrombosis is more likely to occur in veins than in arteries because blood flows more slowly in the veins and does not dilute thrombin and fibrin as rapidly. It is especially common in the leg veins of inactive people and patients immobilized in a wheelchair or bed. Most venous blood flows directly to the heart and then to the lungs. Therefore, blood clots arising in the legs or arms commonly lodge in the lungs and cause *pulmonary embolism.* When blood cannot circulate freely through the lungs, it cannot receive oxygen and a person may die of hypoxia.

Disseminated intravascular coagulation (DIC) is the widespread clotting of blood within unbroken vessels. It can be limited to one organ or occur throughout the body. It is most often triggered by *septicemia* (bacteria in the blood), but also occurs if the blood flow slows down or stops (as in cardiac arrest) and in some other situations. DIC is marked by widespread hemorrhaging, congestion of the vessels with clotted blood, and tissue necrosis setting in as organs are deprived of blood flow.

26. *em* = in, within + *bolus* = ball, mass

Before You Go On

Answer the following questions to test your understanding of the preceding section:

25. What are the three basic mechanisms of hemostasis?
26. How do the extrinsic and intrinsic mechanisms of coagulation differ? What do they have in common?
27. In what respect does blood clotting represent a negative feedback loop? What part of it is a positive feedback loop?
28. Describe some of the mechanisms that prevent clotting in undamaged vessels.
29. Describe a common source and effect of pulmonary embolism.

Clinical Insight 18.5

Controlling Coagulation

For many cardiovascular patients, the goal of treatment is to prevent clotting or to dissolve clots that have already formed. Several strategies employ inorganic salts and products of bacteria, plants, and animals with anticoagulant and clot-dissolving effects.

Preventing Clots from Forming

Since calcium is an essential requirement for blood clotting, blood samples can be kept from clotting by adding a few crystals of sodium oxalate, sodium citrate, or EDTA[27]—salts that bind calcium ions and prevent them from participating in the coagulation reactions. Blood-collection equipment such as hematocrit tubes may also be coated with heparin, a natural anticoagulant whose action was explained earlier.

Since vitamin K is required for the synthesis of clotting factors, anything that antagonizes vitamin K usage makes the blood clot less readily. One vitamin K antagonist is *coumarin*[28] (COO-muh-rin), a sweet-smelling extract of tonka beans, sweet clover, and other plants, used in perfume making. Taken orally by patients at risk for thrombosis, coumarin takes up to 2 days to act, but it has longer-lasting effects than heparin. A similar vitamin K antagonist is the pharmaceutical preparation *Warfarin*[29] *(Coumadin)*, which was originally developed as a pesticide—it makes rats bleed to death. Obviously, such anticoagulants must be used in humans with great care.

As explained in the previous chapter, aspirin suppresses the formation of prostaglandins including thromboxane A_2, a factor in platelet aggregation. Low daily doses of aspirin can therefore suppress thrombosis and prevent heart attacks.

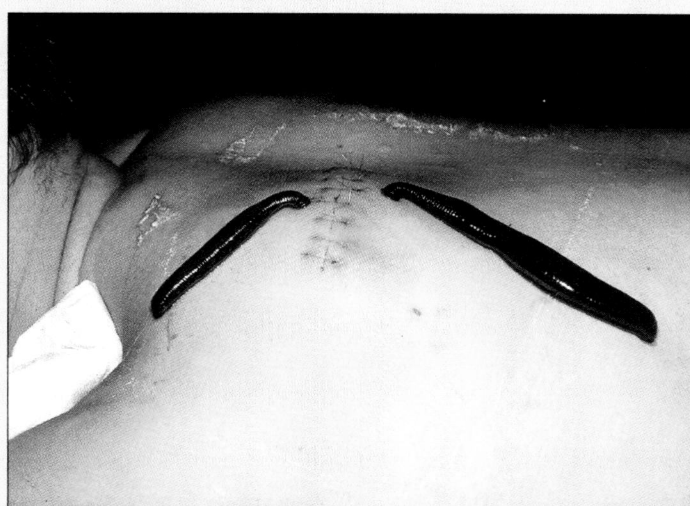

Figure 1 A Modern Use of Leeching. Two medicinal leeches are being used to remove clotted blood from a postsurgical hematoma. These leeches grow up to 20 cm long.

Many parasites feed on the blood of vertebrates and secrete anticoagulants to keep the blood flowing. Among these are segmented worms known as leeches. Leeches secrete a local anesthetic that makes their bites painless; therefore, as early as 1567 B.C.E., physicians used them for bloodletting. This method was less painful and repugnant to their patients than *phlebotomy*[30]—cutting a vein—and indeed, leeching became very popular. In seventeenth-century France it was quite the rage; tremendous numbers of leeches were used to treat headaches, insomnia, whooping cough, obesity, tumors, menstrual cramps, mental illness, and almost anything else doctors or their patients imagined to be caused by "bad blood."

The first known anticoagulant was discovered in the saliva of the medicinal leech, *Hirudo medicinalis,* in 1884. Named *hirudin,* it is a polypeptide of 65 amino acids that prevents clotting by inhibiting thrombin. It causes the blood to flow freely while the leech feeds and for as long as an hour thereafter. While the doctrine of bad blood is now discredited, leeches have lately reentered medical usage (fig. 1). A major problem in reattaching a severed body part such as a finger or ear is that the tiny veins draining these organs are too small to reattach surgically. Since arterial blood flows into the reattached organ and cannot flow out, it pools and clots there. This inhibits the regrowth of veins and the flow of fresh blood through the organ and thus often leads to necrosis. Some vascular surgeons now place leeches on the reattached part. Their anticoagulant keeps the blood flowing freely and allows new veins to grow. After 5 to 7 days, venous drainage is restored and leeching can be stopped.

Anticoagulants also occur in the venom of some snakes. *Arvin,* for example, is obtained from the venom of the Malayan viper. It rapidly breaks down fibrinogen and may have potential as a clinical anticoagulant.

Dissolving Clots That Have Already Formed

When a clot has already formed, it can be treated with clot-dissolving drugs such as *streptokinase,* an enzyme made by certain bacteria (streptococci). Intravenous streptokinase is used to dissolve blood clots in coronary vessels, for example. It is nonspecific, however, and digests almost any protein. *Tissue plasminogen activator (TPA)* works faster, is more specific, and is now made by transgenic bacteria. TPA converts plasminogen into the clot-dissolving enzyme plasmin. Some anticoagulants of animal origin also work by dissolving fibrin. A giant Amazon leech, *Haementeria,* produces one such anticoagulant named *hementin.* This, too, has been successfully produced by genetically engineered bacteria and used to dissolve blood clots in cardiac patients.

27. ethylenediaminetetraacetic acid
28. *coumarú,* tonka bean tree
29. acronym from Wisconsin Alumni Research Foundation
30. *phlebo* = vein + *tomy* = cutting

Chapter Review

Online Learning Center

Study resources for this chapter are available at www.mhhe.com/saladin/

Study Outline

Functions and Properties of Blood (p. 681)

1. Functions (table 18.1)
2. Plasma
3. Formed elements
   a. Erythrocytes
   b. Platelets
   c. Leukocytes
4. Hematocrit
5. Properties of blood
   a. Viscosity
   b. Osmolarity
   c. Other properties (table 18.2)

Plasma (p. 683)

1. Plasma and serum
2. Composition of plasma (table 18.3)
   a. Proteins (table 18.4)
      • Albumins
      • Globulins
      • Fibrinogen
   b. Nonprotein nitrogenous substances
      • Amino acids
      • Nitrogenous wastes
   c. Nutrients
   d. Gases
   e. Electrolytes

Blood Cell Production (p. 686)

1. Hemopoietic tissues
2. Myeloid hemopoiesis
   a. Hemocytoblasts
   b. Committed cells
3. Erythropoiesis
   a. Role of erythropoietin (EPO)
   b. RBC differentiation
   c. Stimulus for erythropoiesis
      • Hypoxemia and EPO
      • Causes of hypoxemia
      • Iron metabolism
      • Other nutritional requirements
4. Leukopoiesis
   a. WBC differentiation
   b. Storage and release of WBCs
   c. WBC circulation, migration, and longevity
5. Thrombopoiesis
   a. Role of thrombopoietin
   b. Megakaryocyte function

Erythrocytes (p. 690)

1. Form and function
   a. Gas transport role
   b. Shape and size
   c. Plasma membrane composition
   d. Lack of organelles
   e. Cytoplasmic proteins
      • Hemoglobin
      • Carbonic anhydrase
2. Hemoglobin
   a. Quantity
   b. Molecular structure
      • Globin moiety
      • Heme moiety
   c. O_2 and CO_2 transport
   d. Varieties of hemoglobin
3. Quantities of RBCs and hemoglobin
   a. Hematocrit
   b. Hemoglobin concentration
   c. RBC count
   d. Differences between sexes
4. Death and disposal
   a. RBC life span and death
   b. Role of spleen in RBC death
   c. Disposal of cell fragments
   d. Disposal of hemoglobin
      • Globin digestion
      • Iron recycling
      • Bile pigments
5. RBC disorders
   a. Polycythemia
      • Primary and secondary
      • Effects
   b. Anemia
      • Causes
         •• Inadequate erythropoiesis
         •• Hemorrhagic anemias
         •• Hemolytic anemias
      • Effects
         •• Tissue hypoxia and necrosis
         •• Low blood osmolarity
         •• Low blood pressure
   c. Sickle-cell disease
      • Abnormal hemoglobin structure
      • Inheritance of trait
      • Pathological effects
      • Evolution
   d. Thalassemia

Blood Types (p. 696)

1. Chemical basis
   a. RBC agglutinogens (antigens)
   b. Plasma agglutinins (antibodies)
2. ABO group
   a. Agglutinogen structures
   b. Agglutinin production
   c. Agglutinin types and structure
   d. Blood typing
   e. Transfusion reactions
3. Rh group
   a. Genetic basis
   b. Rh^+ and Rh^- types
   c. Frequencies of Rh^+ and Rh^-
   d. Anti-D production
   e. Hemolytic disease of the newborn
4. Other blood groups

Leukocytes (p. 700)

1. Types
   a. Granulocytes
      • Neutrophils
      • Eosinophils
      • Basophils
   b. Agranulocytes
      • Lymphocytes
      • Monocytes
2. Relevant comparisons (table 18.8)
   a. Appearance
   b. Size
   c. Abundance
   d. Causes of increased count
   e. Functions

3. WBC disorders
   a. Leukopenia
      - Causes
      - Effects
   b. Leukocytosis
      - Causes
      - Differential count
   c. Leukemia
      - Myeloid and lymphoid
      - Acute and chronic
      - Effects

Hemostasis—The Control of Bleeding (p. 704)

1. Platelets
   a. Structure
   b. Abundance
   c. Functions
      - Secrete clotting factors
      - Secrete growth factors
      - Secrete vasoconstrictors
      - Form platelet plugs
      - Phagocytize bacteria
      - Attract WBCs
      - Dissolve old blood clots
2. Mechanisms of hemostasis
   a. Vascular spasm
      - Role of pain receptors
      - Role of smooth muscle injury
      - Role of serotonin
   b. Platelet plug formation
      - Platelet adhesion to damaged vessel
      - Platelet plug
      - Role of platelet degranulation
   c. Coagulation (blood clotting)
3. Mechanisms of coagulation
   a. Basic goal
      - Fibrin synthesis
      - Clot formation
   b. Procoagulants
   c. Reaction cascades
   d. Extrinsic pathway
      - Initiated by tissue thromboplastin
      - Cascade from factor VII to X
   e. Intrinsic pathway
      - Initiated by factor XII
      - Cascade from factor XI to X
   f. Role of enzyme amplification
   g. Common pathway
      - From factor X to prothrombin activator
      - From prothrombin to thrombin
      - From fibrinogen to fibrin
      - Positive feedback aspects
   h. Coagulation tests
4. Fate of blood clots
   a. Clot retraction
   b. Platelet-derived growth factor
   c. Fibrinolysis
5. Prevention of coagulation
   a. Platelet repulsion
   b. Thrombin dilution
   c. Natural anticoagulants
      - Antithrombin
      - Heparin
6. Coagulation disorders
   a. Thrombocytopenia
   b. Hemophilia
   c. Thrombosis and embolism
   d. Disseminated intravascular coagulation

Selected Vocabulary

hematology 681
plasma 681
formed element 681
viscosity 681
osmolarity 683
colloid osmotic pressure 683
serum 684
albumins 684
globulins 684
fibrinogen 684
nitrogenous waste 684
hemopoiesis 686
myeloid hemopoiesis 686
lymphoid hemopoiesis 686
hemocytoblast 686
erythropoiesis 686
erythropoietin 686
hypoxemia 688
gastroferritin 688
transferrin 688
apoferritin 688
ferritin 688
leukopoiesis 688
macrophage 689
thrombopoiesis 689
megakaryocyte 689
erythrocyte 690
hemoglobin 691
globin 691
heme 692
hematocrit 692
hemoglobin concentration 692
RBC count 692
hemolysis 692
biliverdin 693
bilirubin 693
bile pigment 693
primary polycythemia 694
secondary polycythemia 694
anemia 694
hemorrhagic anemia 694
hemolytic anemia 694
iron-deficiency anemia 694
pernicious anemia 694
intrinsic factor 694
hypoxia 694
sickle-cell disease 695
agglutinate 695
thalassemia 695
agglutinogens 696
agglutinins 696
ABO blood group 696
agglutination 697
transfusion reaction 697
Rh blood group 699
hemolytic disease of the newborn 699
erythroblastosis fetalis 699
leukocyte 700
granulocyte 700
agranulocyte 700
neutrophil 703
eosinophil 703
basophil 703
lymphocyte 703
monocyte 703
leukopenia 703
leukocytosis 703
leukemia 703
myeloid leukemia 703
lymphoid leukemia 703
acute leukemia 703
chronic leukemia 703
hemostasis 704
platelet 704
vascular spasm 705
prostacyclin 705
platelet plug 705
degranulation 705
thromboxane A_2 705
coagulation 705
fibrin 705
extrinsic mechanism 706
intrinsic mechanism 706
procoagulant 706
anticoagulant 706
reaction cascade 706
tissue thromboplastin 706
prothrombin 706
thrombin 706
clot retraction 708
fibrinolysis 708
kallikrein 708
plasmin 708
antithrombin 709
heparin 709
thrombocytopenia 709
hematoma 709
hemophilia 709
thrombosis 710
thrombus 710
embolus 710
disseminated intravascular coagulation 710

Testing Your Recall

1. Antibodies belong to a class of plasma proteins called
   a. albumins.
   b. γ globulins.
   c. α globulins.
   d. procoagulants.
   e. agglutinins.
2. Serum is blood plasma minus its
   a. sodium ions.
   b. calcium ions.
   c. clotting proteins.
   d. globulins.
   e. albumins.
3. Which of the following conditions is most likely to cause hemolytic anemia?
   a. folic acid deficiency
   b. iron deficiency
   c. mushroom poisoning
   d. alcoholism
   e. hypoxemia
4. It is impossible for a type O^+ baby to have a type ______ mother.
   a. AB^-
   b. O^-
   c. O^+
   d. A^+
   e. B^+
5. Which of the following is *not* a component of hemostasis?
   a. platelet plug formation
   b. agglutination
   c. clot retraction
   d. a vascular spasm
   e. degranulation
6. Which of the following contributes most to the viscosity of blood?
   a. albumin
   b. sodium
   c. globulins
   d. erythrocytes
   e. fibrin
7. Which of these is a granulocyte?
   a. a monocyte
   b. a lymphocyte
   c. a macrophage
   d. an eosinophil
   e. an erythrocyte
8. Excess iron is stored in the liver as a complex called
   a. gastroferritin.
   b. transferrin.
   c. ferritin.
   d. hepatoferritin.
   e. erythropoietin.
9. Pernicious anemia is a result of
   a. hypoxemia.
   b. iron deficiency.
   c. malaria.
   d. lack of intrinsic factor.
   e. Rh incompatibility.
10. The first clotting factor that the intrinsic and extrinsic pathways have in common is
    a. thromboplastin.
    b. Hageman factor.
    c. factor X.
    d. prothrombin activator.
    e. factor VIII.
11. Production of all the formed elements of blood is called ________.
12. The percentage of blood volume composed of RBCs is called the ________.
13. The extrinsic pathway of coagulation is activated by ________ from damaged perivascular tissues.
14. The RBC antigens that determine transfusion compatibility are called ________.
15. The hereditary lack of factor VIII causes a disease called ________.
16. The overall cessation of bleeding, involving several mechanisms, is called ________.
17. ________ results from a mutation that changes one amino acid in the hemoglobin molecule.
18. An excessively high RBC count is called ________.
19. Intrinsic factor enables the small intestine to absorb ________.
20. The kidney hormone ________ stimulates RBC production.

Answers in appendix B

Testing Your Comprehension

1. Why would erythropoiesis not correct the hypoxemia resulting from lung cancer?
2. People with chronic kidney disease often have hematocrits of less than half the normal value. Explain why.
3. An elderly white woman is hit by a bus and severely injured. Accident investigators are informed that she lives in an abandoned warehouse, where her few personal effects include several empty wine bottles and an expired driver's license indicating she is 72 years old. At the hospital, she is found to be severely anemic. List all the factors you can think of that may contribute to her anemia.
4. How is coagulation different from agglutination?
5. Although fibrinogen and prothrombin are equally necessary for blood clotting, fibrinogen is about 4% of the plasma protein while prothrombin is present only in small traces. In light of the roles of these clotting factors and your knowledge of enzymes, explain this difference in their abundance.

Answers at the Online Learning Center

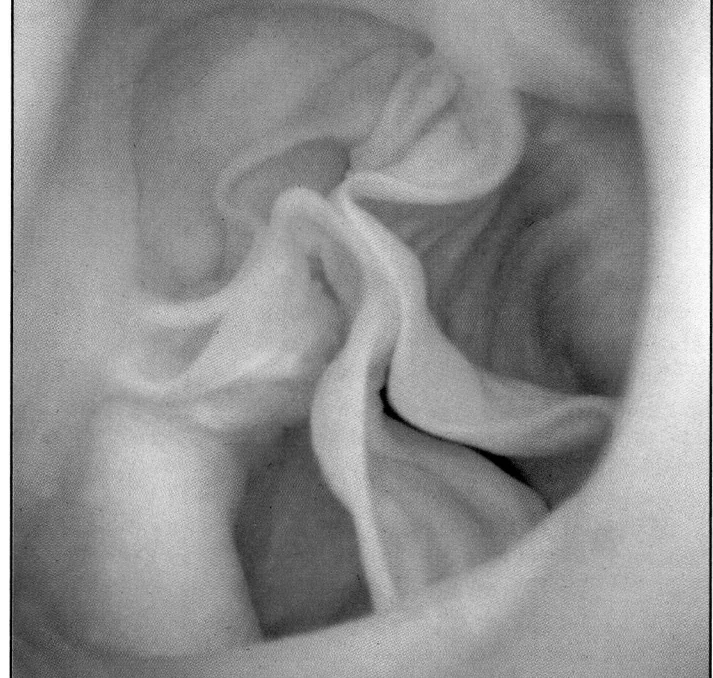

A semilunar valve of the heart (endoscopic photo)

CHAPTER 19

The Circulatory System: The Heart

Chapter Outline

Insights

Study with Your ESP CD-ROM

Refer to your ESP Tutorial CD-ROM for helpful animations, activities, and quizzes related to the material covered in this chapter. The following are suggested areas to go to on the CD and are presented according to "Unit/Topic":
- Transport/Cardiovascular System

Brushing Up

To understand this chapter, it is important that you understand or brush up on the following concepts:
- Desmosomes and gap junctions (p. 193)
- Ultrastructure of striated muscle (pp. 421–23)
- Excitation-contraction coupling in muscle (p. 429)
- Length-tension relationship in muscle fibers (p. 433)
- Action potentials (p. 467)

We are more conscious of our heart than we are of most organs and more wary of its failure. Speculation about the heart is at least as old as written history. Some ancient Chinese, Egyptian, Greek, and Roman scholars correctly surmised that the heart is a pump for filling the vessels with blood. Aristotle's views, however, were a step backward. Perhaps because the heart quickens its pace when we are emotionally aroused, and because grief causes "heartache," he regarded it primarily as the seat of emotion, as well as a source of heat to aid digestion. During the Middle Ages, Western medical schools clung dogmatically to the ideas of Aristotle. Perhaps the only significant advance came from Muslim medicine, when thirteenth-century physician Ibn an-Nafis described the role of the coronary blood vessels in nourishing the heart. The sixteenth-century dissections and anatomical charts of Vesalius, however, greatly improved knowledge of cardiovascular anatomy and set the stage for a more scientific study of the heart and treatment of its disorders—the science we now call **cardiology**.[1]

In the early decades of the twentieth century, little could be recommended for heart disease other than bed rest. Then nitroglycerin was found to improve coronary circulation and relieve the pain resulting from physical exertion, digitalis proved effective for treating abnormal heart rhythms, and diuretics were first used to reduce hypertension. Coronary bypass surgery, replacement of diseased valves, clot-dissolving enzymes, heart transplants, artificial pacemakers, and artificial hearts have made cardiology one of the most dramatic and attention-getting fields of medicine in the last quarter-century.

Gross Anatomy of the Heart

Objectives

When you have completed this section, you should be able to

- describe the relationship of the heart to other thoracic structures;
- identify the chambers and valves of the heart and the features of its wall;
- trace the flow of blood through the heart chambers; and
- describe the blood supply to the heart tissue.

Overview of the Cardiovascular System

The term **circulatory system** refers to the heart, blood vessels, and blood. The term **cardiovascular system,** however, refers only to the passages through which the blood flows—the **heart,** a four-chambered muscular pump; **arteries,** the vessels that carry blood away from the heart; **veins,** the vessels that carry blood back to the heart; and **capillaries,** microscopic blood vessels that connect the smallest arteries to the smallest veins.

The cardiovascular system has two major divisions: a **pulmonary circuit,** which carries blood to the lungs for gas exchange and then returns it to the heart, and a **systemic circuit,** which supplies blood to every organ of the body (fig. 19.1). The right side of the heart serves the pulmonary circuit. It receives blood that has circulated through the body, unloaded its oxygen and nutrients, and picked up a load of carbon dioxide and other wastes. It pumps this oxygen-poor blood into a large artery, the *pulmonary trunk,* which immediately divides into *right* and *left pulmonary arteries.* These transport blood to the lungs, where carbon dioxide is unloaded and oxygen is picked up. The oxygen-rich blood then flows by way of the *pulmonary veins* to the left side of the heart.

The left side of the heart serves the systemic circuit. Oxygenated blood leaves it by way of another large artery, the *aorta.* The aorta takes a sharp U-turn, the *aortic arch,* and passes downward, dorsal to the heart. The aortic arch gives off arteries that supply the head, neck, and upper limbs. The aorta then travels through the thoracic and abdominal cavities and issues smaller arteries to the other organs. After circulating through the body, the now-deoxygenated systemic blood returns to the right side of the heart mainly by way of two large veins, the *superior vena cava* (draining the head, neck, upper limbs, and thoracic organs) and *inferior vena cava* (draining the organs below the diaphragm). The major arteries and veins entering and leaving the heart are called the *great vessels* because of their relatively large diameters.

Size, Shape, and Position of the Heart

The heart is located in the thoracic cavity in the mediastinum, the area between the lungs. About two-thirds of it lies to the left of the midsagittal plane (fig. 19.2). The broad superior portion of the heart, called the **base,** is the point of attachment for the great vessels described previously. Its inferior end, the **apex,** tilts to the left and tapers to a blunt point (figs. 19.3 and 19.4). The adult heart is about 9 cm (3.5 in.) wide at the base, 13 cm (5 in.) from base to apex, and 6 cm (2.5 in.) from anterior to posterior at its thickest point—roughly the size of a fist. It weighs about 300 g (10 oz).

The Pericardium

The heart is enclosed in a double-walled sac called the **pericardium,**[2] which is anchored to the diaphragm below and to the connective tissue of the great vessels above the heart (fig. 19.5). The **parietal pericardium (pericardial sac)** consists of a tough *fibrous layer* of dense irregular connective tissue and a thin, smooth, moist *serous layer.* The serous layer turns inward at the base of the heart and forms the **visceral pericardium** covering the heart surface. Between the parietal and visceral membranes is a space

1. *cardio* = heart + *logy* = study

2. *peri* = around

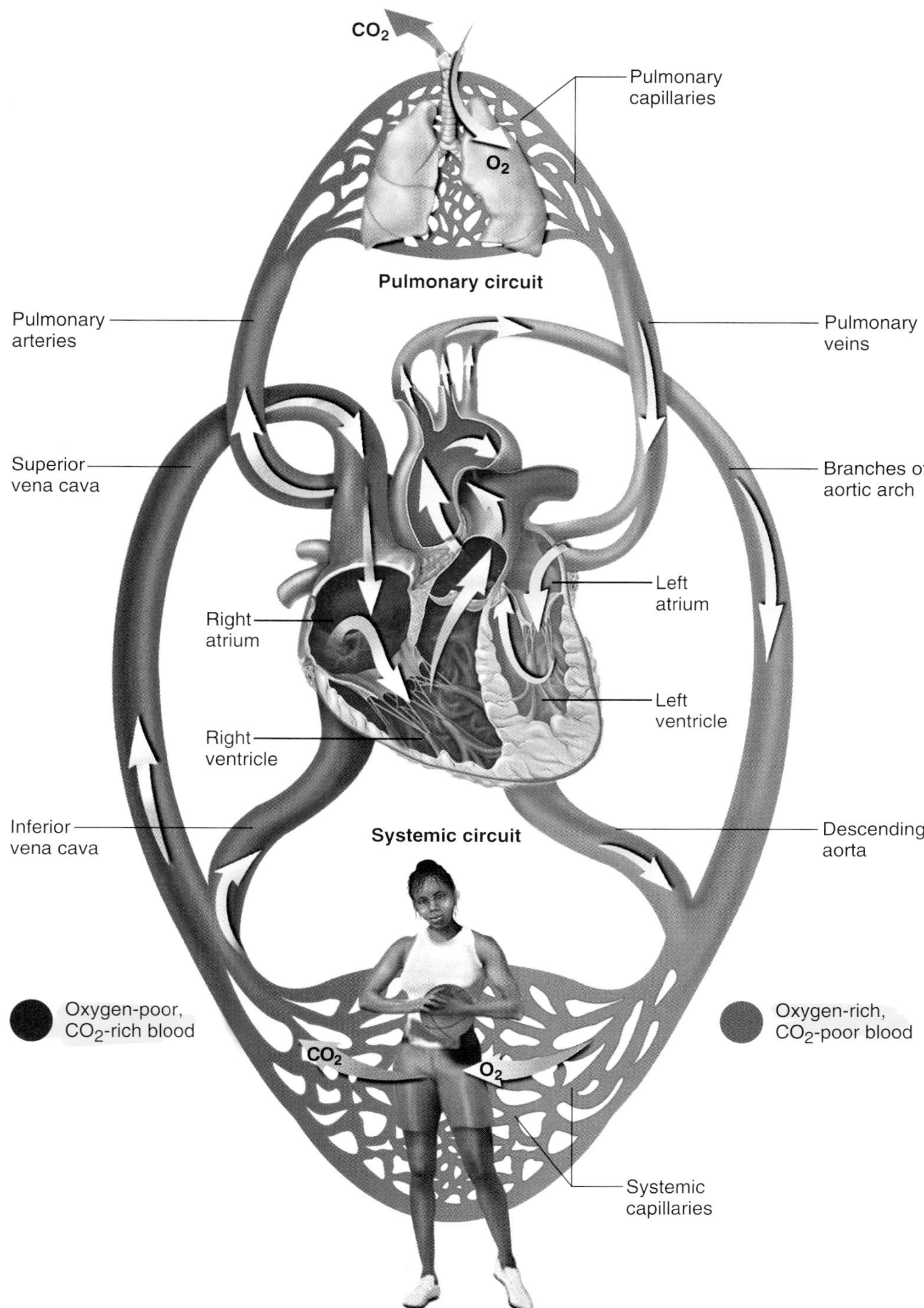

Figure 19.1 General Schematic of the Cardiovascular System.

called the **pericardial cavity.** It contains 5 to 30 mL of **pericardial fluid,** an exudate of the serous pericardium that lubricates the membranes and allows the heart to beat almost without friction. In *pericarditis*—inflammation of the pericardium—the membranes may become dry and produce a painful *friction rub* with each heartbeat. In addition to reducing friction, the pericardium isolates the heart from other thoracic organs, allows the heart room to expand, and resists excessive expansion. (See *cardiac tamponade* in table 20.15 at the end of the next chapter.)

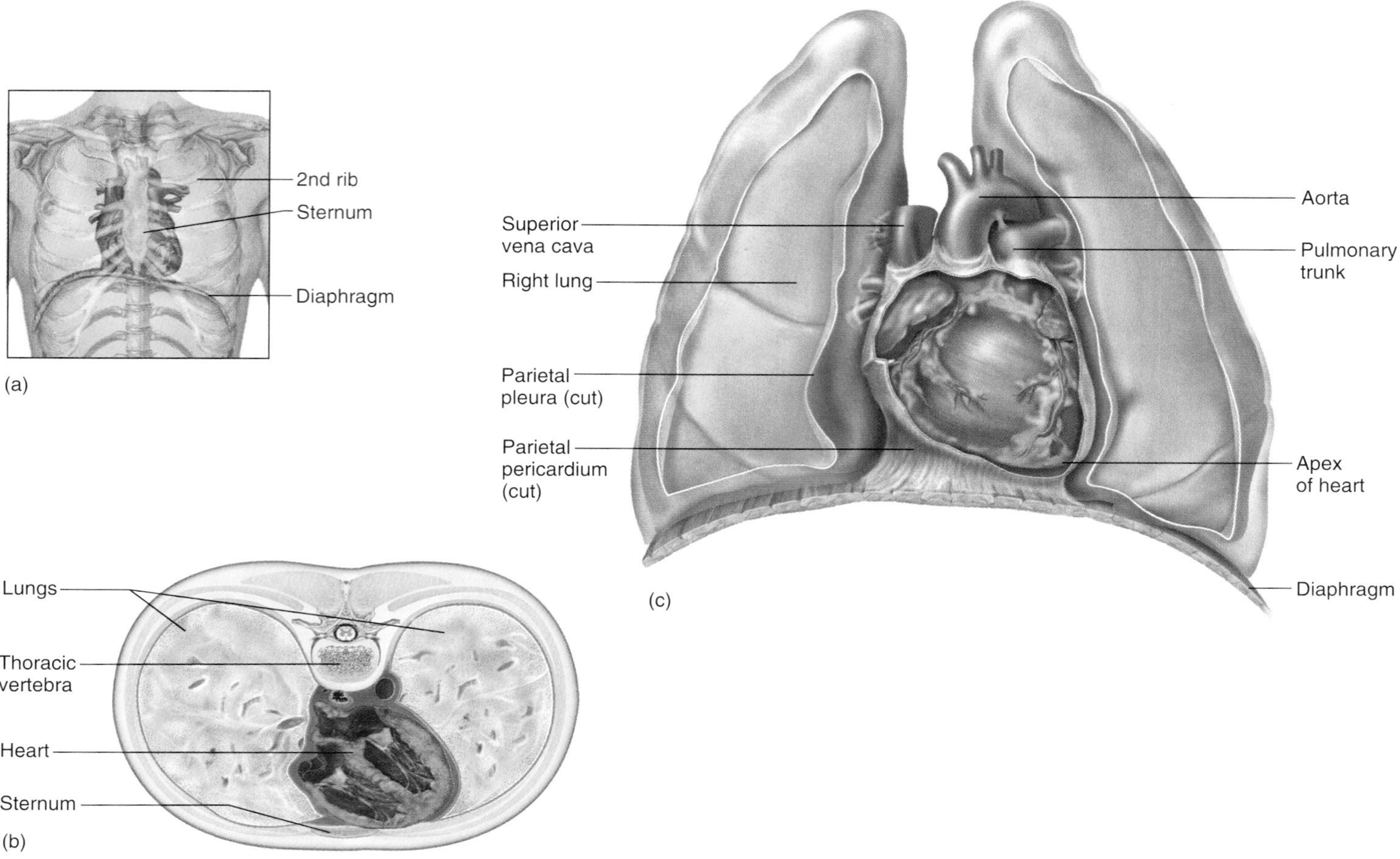

Figure 19.2 Position of the Heart in the Thoracic Cavity. (*a*) Relationship to the thoracic cage; (*b*) cross section of the thorax at the level of the heart; (*c*) frontal section of the thoracic cavity with the lungs slightly retracted and the pericardial sac opened.

The Heart Wall

The heart wall consists of three layers—the epicardium, myocardium, and endocardium (fig. 19.5). The **epicardium**[3] (= visceral pericardium) is a serous membrane composed of a simple squamous epithelium overlying a thin layer of areolar tissue. Over much of the heart, it has thick deposits of fat that fill grooves in the heart surface and protect the coronary blood vessels. In nonfatty areas, the epicardium is thin and translucent, allowing the myocardium to show through.

The **myocardium,**[4] by far the thickest layer, is composed of cardiac muscle and performs the work of the heart. Its muscle cells spiral around the heart and are bound together by a meshwork of collagenous and elastic fibers that make up the **fibrous skeleton.** The fibrous skeleton has at least three functions: to provide structural support for the heart, especially around the valves and the openings of the great vessels; to give the muscle something to pull against; and, as a nonconductor of electricity, to limit the routes by which electrical excitation travels through the heart. This insulation prevents the atria from stimulating the ventricles directly and is important in the timing and coordination of electrical and contractile activity. Elastic recoil of the fibrous skeleton may also aid in refilling the heart with blood after each beat, but physiologists are not in complete agreement about this.

The **endocardium**[5] consists of a simple squamous endothelium overlying a thin areolar tissue layer. It forms the smooth inner lining of the chambers and valves and is continuous with the endothelium of the blood vessels.

The Chambers

The heart has four **chambers** (see fig. 19.4). Blood returning to the heart is received by two superior chambers, the

3. *epi* = upon
4. *myo* = muscle
5. *endo* = internal

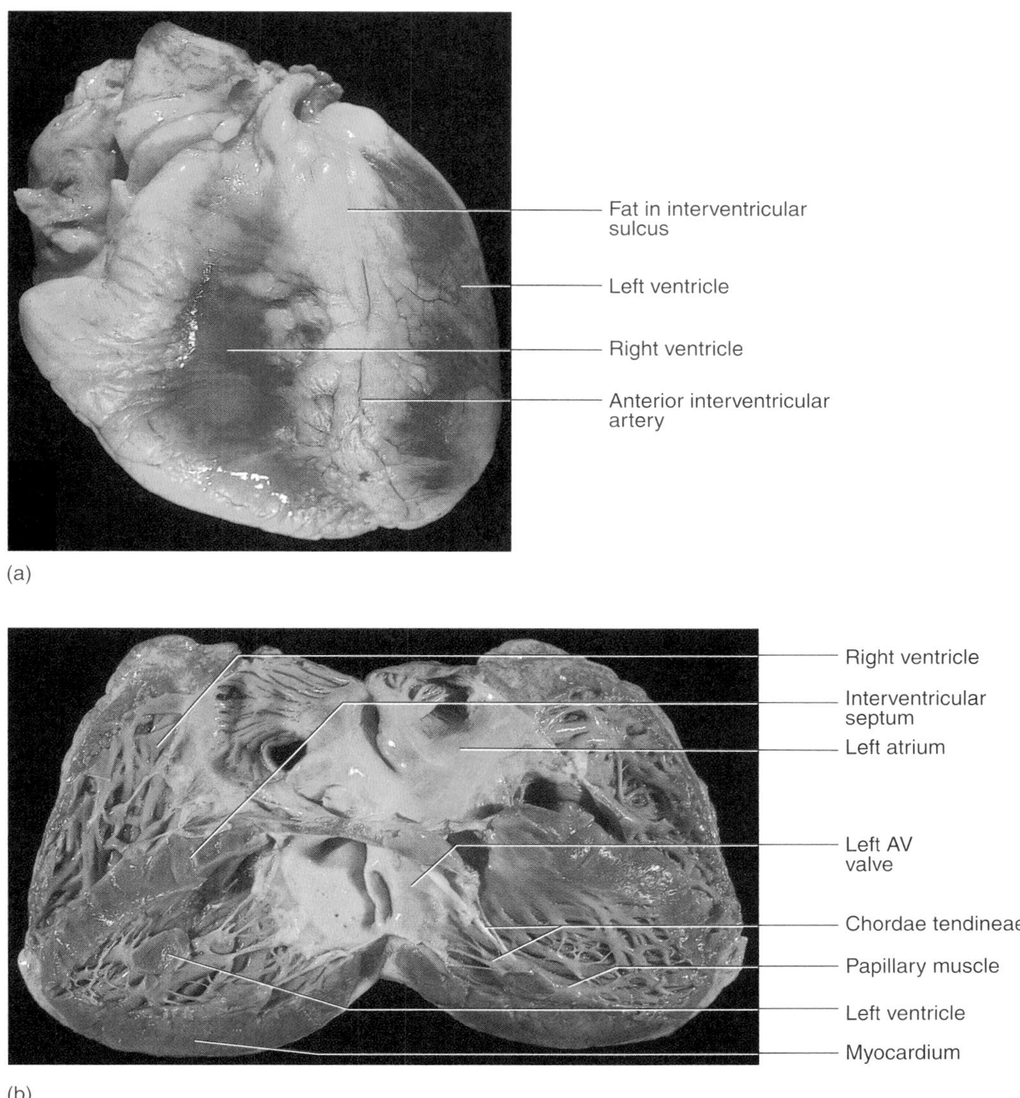

Figure 19.3 **The Human Heart** (*a*) Anterior aspect; (*b*) internal anatomy, with the heart in *a* bisected on the frontal plane and folded open like a book.

right and **left atria** (AY-tree-uh; singular *atrium*[6]). These are mostly posterior in position, so only a small portion of each is visible from the anterior aspect. Each atrium has a small earlike extension called an *auricle*[7] that slightly increases its volume. The two inferior chambers, the **right** and **left ventricles,**[8] are the pumps that eject blood into the arteries. The right ventricle constitutes most of the anterior aspect of the heart, while the left ventricle forms the apex and inferoposterior aspect.

The heart is crisscrossed by sulci (grooves) that mark the boundaries of the four chambers. The sulci are occupied largely by fat and coronary blood vessels. The **atrioventricular (coronary**[9]**) sulcus** encircles the heart near its base and separates the atria from the ventricles. The **anterior** and **posterior interventricular sulci** extend vertically, from the coronary sulcus toward the apex, externally marking the boundary between the right and left ventricles.

The four chambers are best seen in a frontal section (fig. 19.6). The atria exhibit thin flaccid walls corresponding

6. *atrium* = entryway
7. *auricle* = little ear
8. *ventr* = belly, lower part + *icle* = little

9. *coron* = crown + *ary* = pertaining to

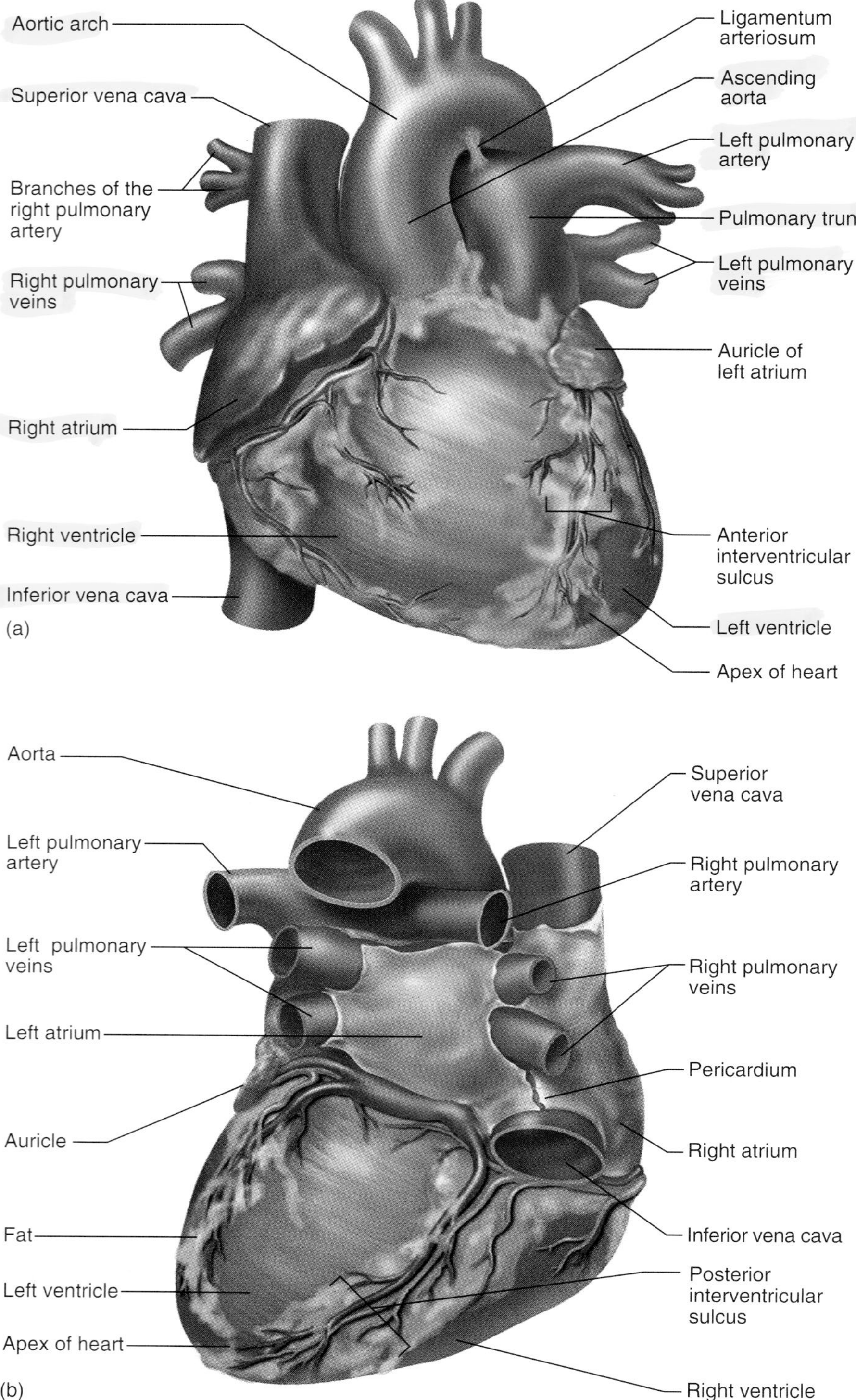

Figure 19.4 **External Anatomy of the Heart.** (*a*) Anterior aspect; (*b*) posterior aspect. The coronary blood vessels on the heart surface are identified in figure 19.10.

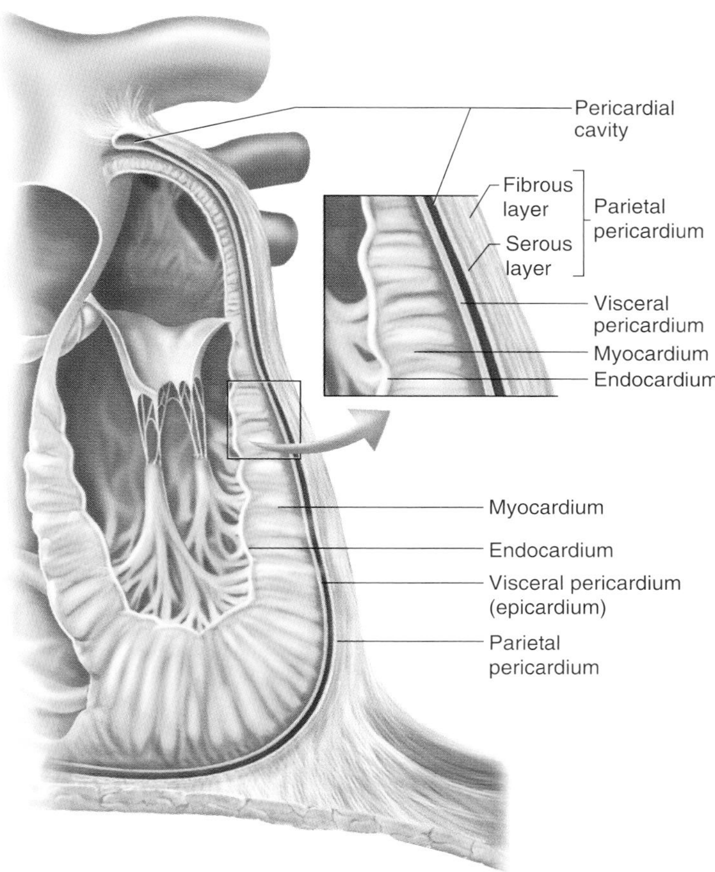

Figure 19.5 **The Pericardium and Heart Wall.** The inset shows the layers of the heart wall in relationship to the pericardium.

to their light workload—all they do is pump blood into the ventricles immediately below. They are separated from each other by a wall, the **interatrial septum.** The right atrium and both auricles exhibit internal ridges of myocardium called the **pectinate**[10] **muscles.** A thicker wall, the **interventricular septum,** separates the right and left ventricles. The right ventricle pumps blood only to the lungs and back, so its wall is only moderately thick and muscular. The left ventricle is two to four times as thick because it bears the greatest workload of all four chambers, pumping blood through the entire body. Both ventricles exhibit internal ridges called **trabeculae carneae**[11] (trah-BEC-you-lee CAR-nee-ee).

The Valves

To pump blood effectively, the heart needs valves that ensure a predominantly one-way flow. There is a valve between each atrium and its ventricle and at the exit from each ventricle into its great artery (figs. 19.6 and 19.7). Each valve consists of two or three fibrous flaps of tissue called **cusps,** covered with endothelium.

The **atrioventricular (AV) valves** regulate the openings between the atria and ventricles. The **right AV (tricuspid) valve** has three cusps and the **left AV (bicuspid) valve** has two. The left AV valve is also known as the **mitral** (MY-trul) **valve** after its resemblance to a miter, the headdress of a church bishop. Stringlike **chordae tendineae** (COR-dee ten-DIN-ee-ee), reminiscent of the shroud lines of a parachute, connect the AV valve cusps to conical **papillary muscles** on the floor of the ventricle.

The **semilunar**[12] **valves** (pulmonary and aortic valves) regulate the flow of blood from the ventricles into the great arteries. The **pulmonary valve** controls the opening from the right ventricle into the pulmonary trunk, and the **aortic valve** controls the opening from the left ventricle into the aorta. Each has three cusps shaped somewhat like shirt pockets (see photograph on p. 715).

The opening and closing of heart valves is the result of pressure gradients between the "upstream" and "downstream" sides of the valve (fig. 19.8). When the ventricles are relaxed, the AV valve cusps hang down limply, both AV valves are open, and blood flows freely from the atria into the ventricles. When the ventricles have filled with blood and begin to contract, their internal pressure rises and blood surges against the AV valves. This pushes their cusps together, seals the openings, and prevents blood from flowing back into the atria. The papillary muscles contract with the rest of the ventricular myocardium and tug on the chordae tendineae, which prevents the valves from bulging excessively (prolapsing) into the atria or turning inside out like windblown umbrellas (see clinical insight 19.1). When rising "upstream" pressure in the ventricles exceeds the "downstream" blood pressure in the great arteries, the ventricular blood forces the semilunar valves open and blood is ejected from the heart. Then as the ventricles relax again and their pressure falls below that in the arteries, arterial blood briefly flows backward and fills the pocketlike cusps of the semilunar valves. The three cusps meet in the middle of the orifice and seal it, thereby preventing blood from reentering the heart.

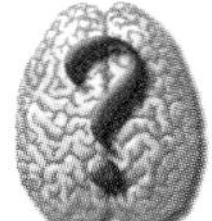

Think About It

How would valvular stenosis (see clinical insight 19.1) affect the amount of blood pumped into the aorta? How might this affect a person's physical stamina? Explain your reasoning.

10. *pectin* = comb + *ate* = like
11. *trabecula* = little beam + *carne* = flesh, meat
12. *semi* = half + *lunar* = like the moon

Clinical Insight 19.1

Valvular Insufficiency

Valvular insufficiency (incompetence) refers to any failure of a value to prevent *reflux (regurgitation)*—the backward flow of blood. *Valvular stenosis*[13] is a form of insufficiency in which the cusps are stiffened and the opening is constricted by scar tissue. It frequently results from rheumatic fever, an autoimmune disease in which antibodies produced to fight a bacterial infection also attack the mitral and aortic valves. As the valves become scarred and constricted, the heart is overworked by the effort to force blood through the openings and may become enlarged. Regurgitation of blood through the incompetent valves creates turbulence that can be heard as a *heart murmur.*

Mitral valve prolapse (MVP) is an insufficiency in which one or both mitral valve cusps bulge into the atrium during ventricular contraction. It is often hereditary and affects about 1 out of 40 people, especially young women. In many cases, it causes no serious dysfunction, but in some people it causes chest pain, fatigue, and shortness of breath. An incompetent valve can eventually lead to heart failure. A defective value can be replaced with an artificial valve or a valve transplanted from a pig heart.

13. *steno* = narrow

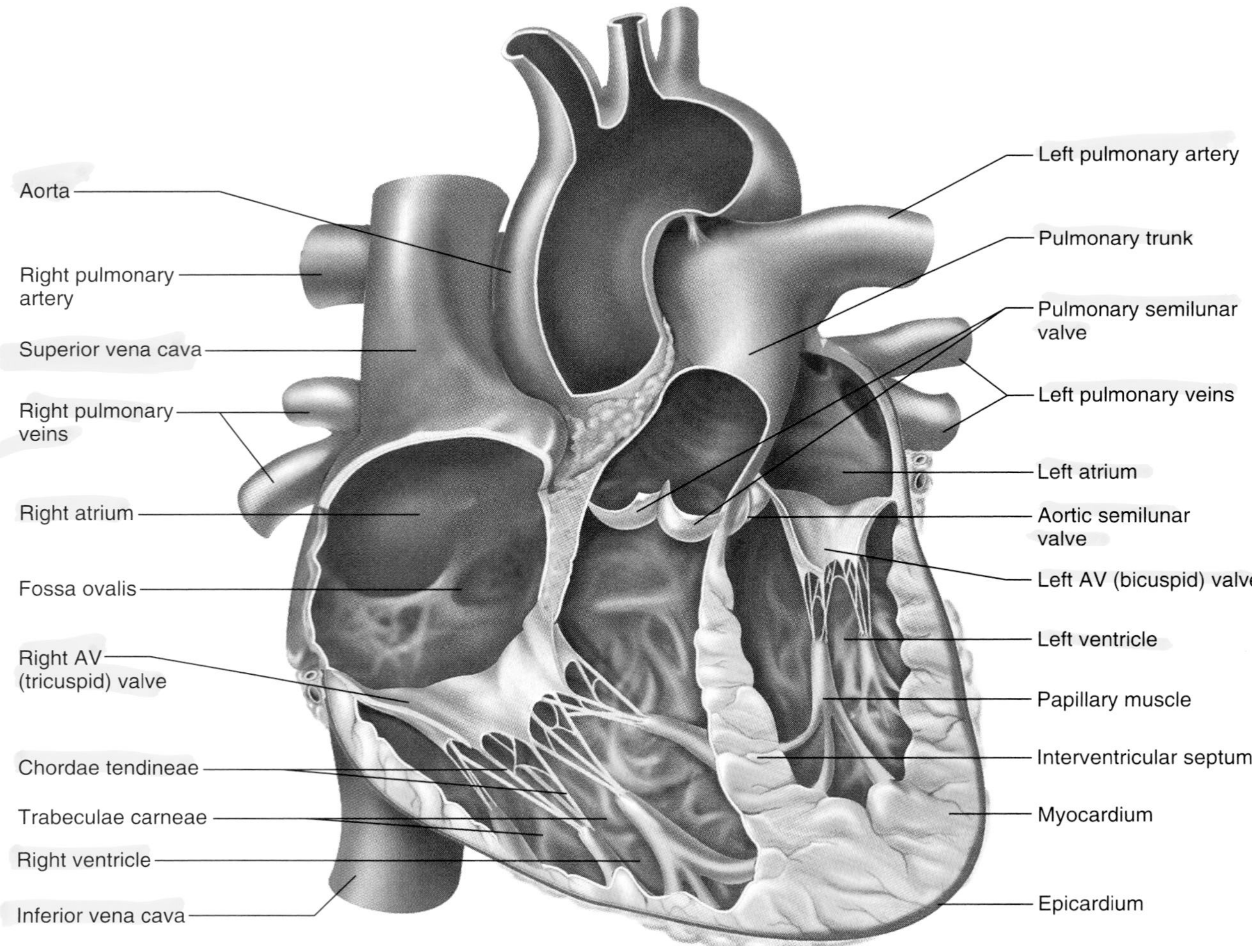

Figure 19.6 Internal Anatomy of the Heart (anterior aspect).

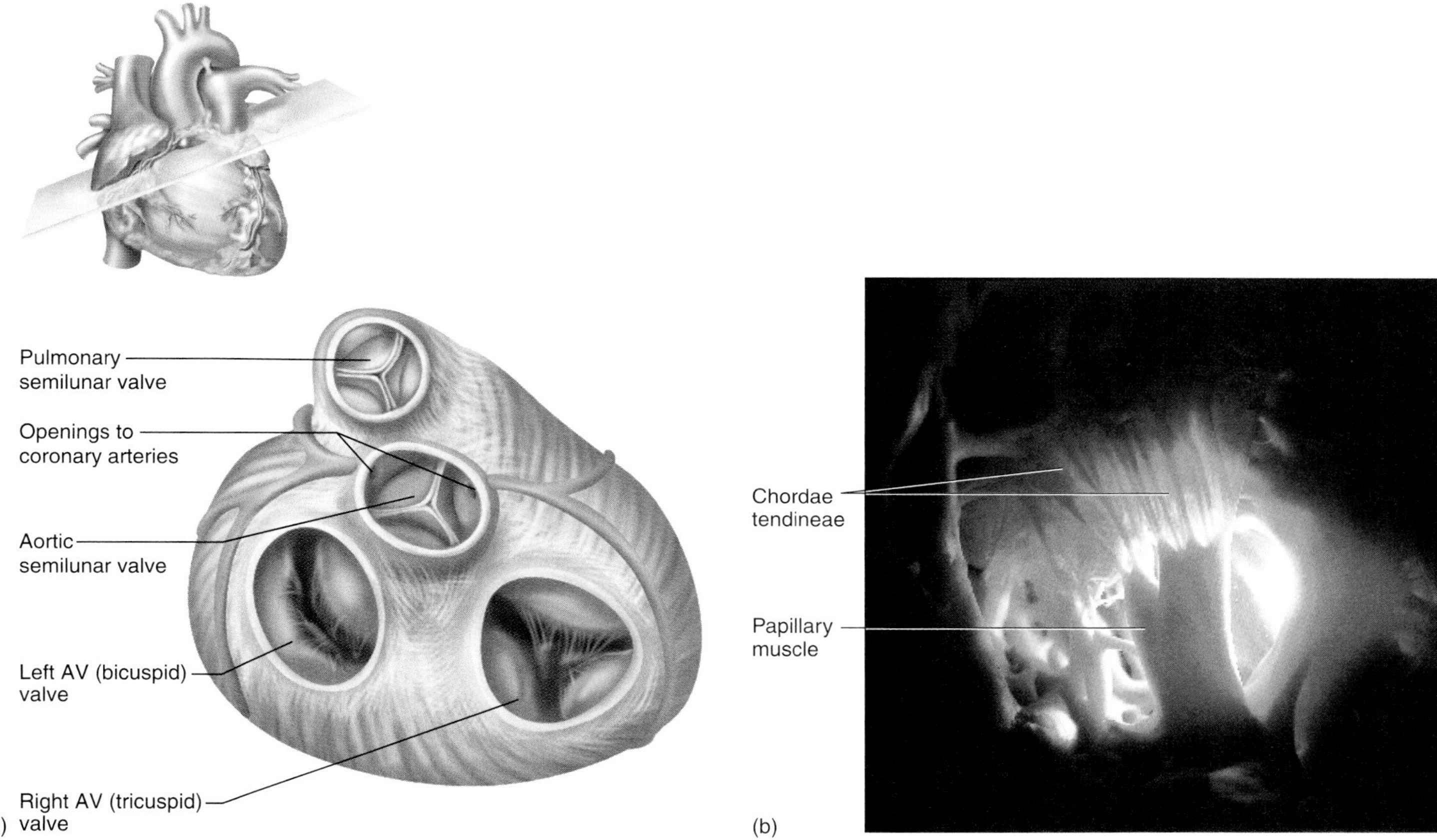

Figure 19.7 The Heart Valves. (*a*) Superior view of the heart with the artia removed; (*b*) papillary muscle and chordae tendineae seen from within the right ventricle. The upper ends of the chordae tendineae are attached to the cusps of the right AV valve.

Blood Flow Through the Heart

Until the sixteenth century, blood was thought to flow directly from the right ventricle into the left through invisible pores in the septum. This of course is not true. Blood on the right and left sides of the heart is kept entirely separate. Figure 19.9 shows the pathway of the blood as it travels from the right atrium through the body and back to the starting point.

The Coronary Circulation

If your heart beats an average of 75 times a minute for 80 years, it will beat more than 3 billion times and pump more than 200 million liters of blood. Understandably, it requires an abundant supply of oxygen and nutrients. Even though the heart is only 0.5% of the body's weight, it uses 5% of the circulating blood to meet its own metabolic needs. The cardiac muscle is not nourished to any great extent by the blood flowing through the heart chambers. Instead, it has its own supply of arteries and capillaries that deliver blood to every cell of the myocardium. The blood vessels of the heart wall constitute the **coronary circulation.** At rest, these vessels supply the myocardium with about 250 mL of blood per minute.

Arterial Supply

Immediately after the aorta leaves the left ventricle, it gives off right and left coronary arteries (fig. 19.10). Each coronary artery begins at an opening deep in the cup formed by a cusp of the aortic valve, like a hole in the bottom of a pocket. The **left coronary artery** passes under the left auricle and divides into two branches:

1. The **anterior interventricular artery** travels down the anterior interventricular sulcus toward the apex. It issues smaller branches to the interventricular septum and anterior walls of both ventricles. Clinically, this vessel is also called the *left anterior descending (LAD) artery.*
2. The **circumflex artery** continues around the left side of the heart in the coronary sulcus. It supplies blood to the left atrium and posterior wall of the left ventricle.

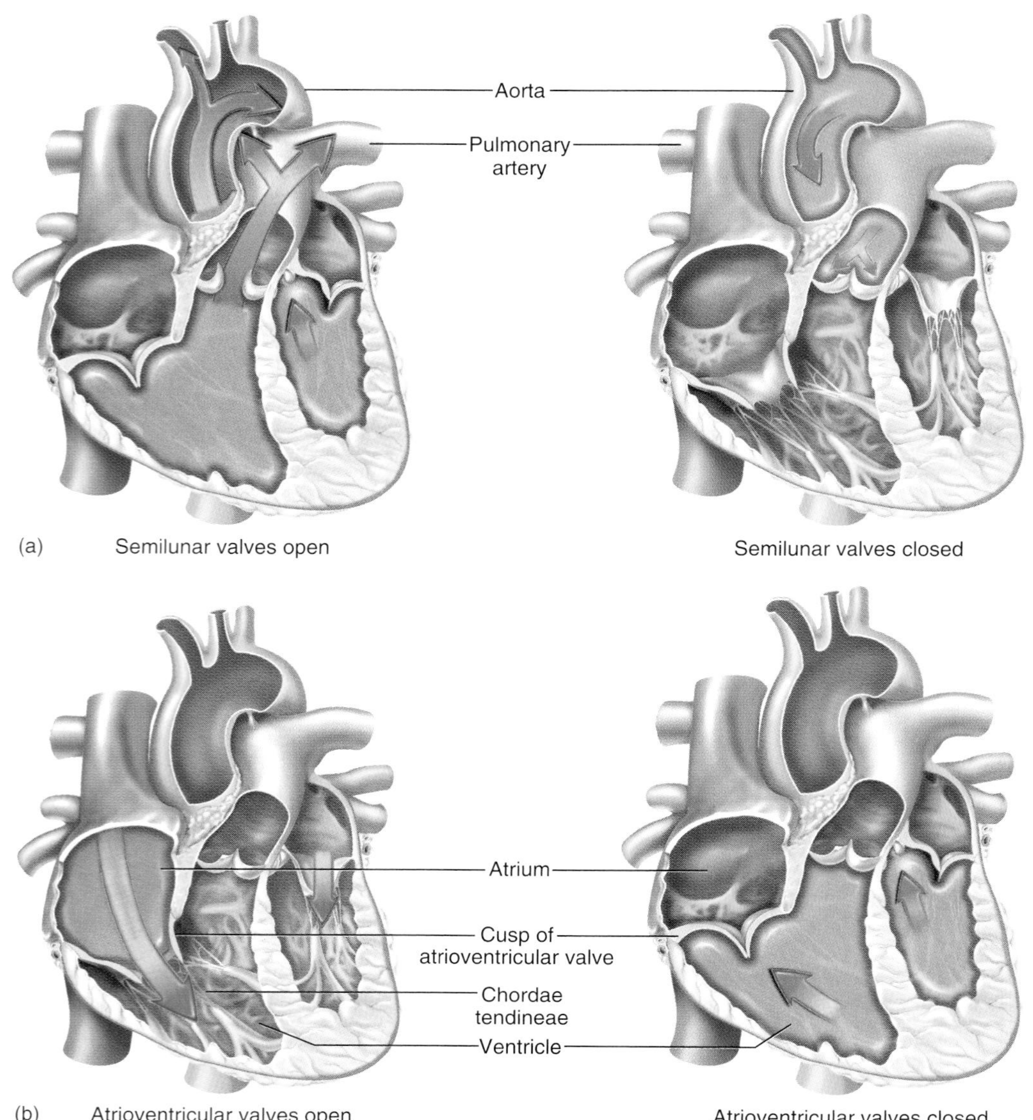

Figure 19.8 Operation of the Heart Valves. (*a*) The semilunar valves. When the pressure in the ventricle is greater than the pressure in the artery, the valve is forced open and blood is ejected. When ventricular pressure is lower than arterial pressure, arterial blood holds the valve closed. (*b*) The atrioventricular valves. When atrial pressure is greater than ventricular pressure, the valve opens and blood flows through. When ventricular pressure rises above atrial pressure, the blood in the ventricle pushes the valve cusps closed.

The **right coronary artery** supplies the right atrium, continues along the coronary sulcus under the right auricle, and then gives off two branches:

1. The **marginal artery** supplies the lateral aspect of the right atrium and ventricle.
2. The **posterior interventricular artery** travels down the corresponding sulcus and supplies the posterior walls of both ventricles.

Think About It

Which ventricle receives the greater coronary blood supply? Why should it receive a greater supply than the other? List the vessels that supply it.

The energy demand of the cardiac muscle is so critical that an interruption of the blood supply to any part of the myocardium can cause necrosis within minutes. A

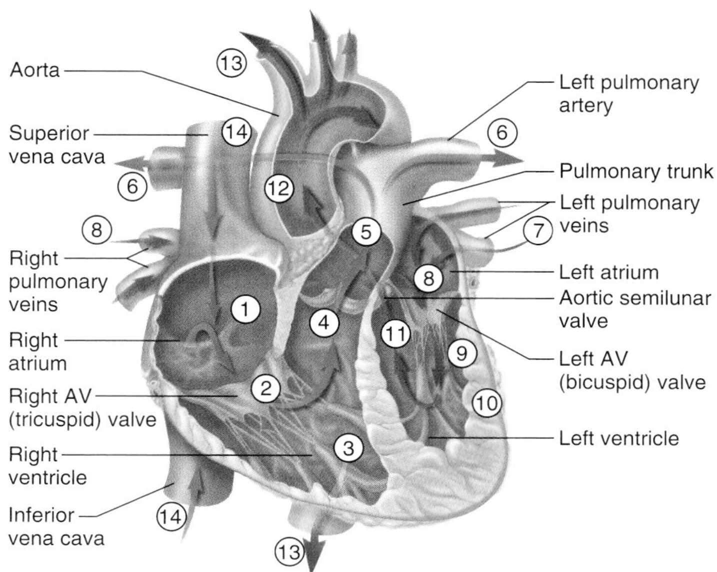

Figure 19.9 The Pathway of Blood from the Right Atrium and Back. *(1)* Right atrium → *(2)* right AV valve → *(3)* right ventricle → *(4)* pulmonary valve → *(5)* pulmonary trunk → *(6)* pulmonary arteries to lungs → *(7)* pulmonary veins returning from lungs → *(8)* left atrium → *(9)* left AV valve → *(10)* left ventricle → *(11)* aortic valve → *(12)* aorta → *(13)* other systemic vessels → *(14)* inferior and superior venae cavae → *(15)* back to the right atrium. The pathway from *5* to *7* is the pulmonary circuit, and the pathway from *12* to *14* is the systemic circuit.

fatty deposit or blood clot in a coronary artery can cause a **myocardial infarction**[14] **(MI),** the sudden death of a patch of tissue deprived of its blood flow (see clinical insight 19.2). The coronary circulation has a defense against such an occurrence—points called *anastomoses* (ah-NASS-tih-MO-seez) where two arteries come together and combine their blood flow to points farther downstream. Thus, if one artery becomes obstructed, blood continues to reach myocardial tissue through the alternative route. The most important anastomosis is the point at which the circumflex artery and right coronary artery meet on the posterior side of the heart; they combine their blood flow into the posterior interventricular artery. Another is the meeting of the anterior and posterior interventricular arteries at the apex of the heart.

Venous Drainage

Venous drainage refers to the route by which blood leaves an organ. After flowing through capillaries of the myocardium, about 20% of the coronary blood empties directly from small veins into the right ventricle. The other 80% returns to the right atrium by the following route (fig. 19.10):

- The **great cardiac vein** collects blood from the anterior aspect of the heart and travels alongside the anterior interventricular artery. It carries blood from the apex of the heart toward the atrioventricular sulcus.

14. *infarct* = to stuff

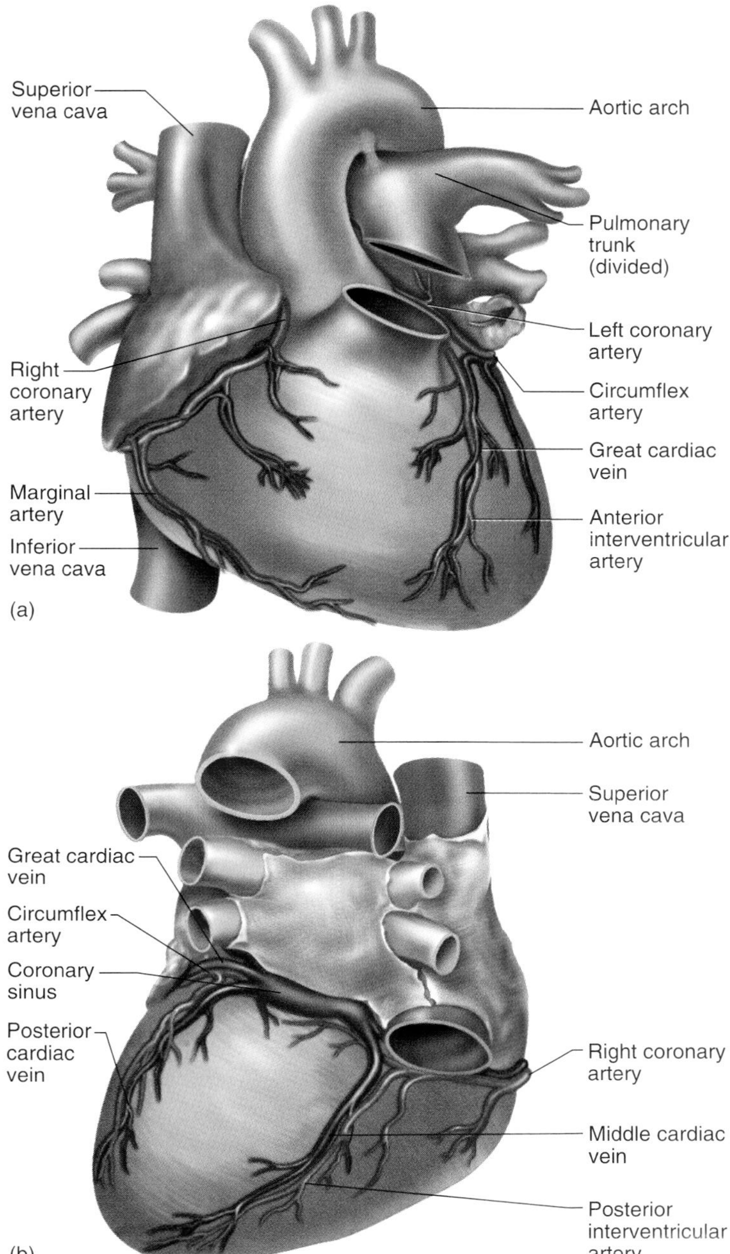

Figure 19.10 The Coronary Blood Vessels. *(a)* Anterior aspect; *(b)* posterior aspect.

Clinical Insight 19.2

Myocardial Infarction and Angina Pectoris

A *myocardial infarction (MI)*—what most people call a heart attack—is the sudden death of a patch of myocardium resulting from *ischemia*[15] (iss-KEE-me-uh), the loss of blood flow. It occurs when a coronary artery becomes obstructed by a blood clot or a fatty deposit (atherosclerosis; see clinical insight 19.5 at the end of this chapter). The myocardium downstream from the obstruction dies from *hypoxia* (oxygen deficiency). This tissue necrosis is felt as a sense of heavy pressure or squeezing pain in the chest, often radiating to the shoulder and arm. Infarctions weaken the heart wall and disrupt electrical conduction pathways, potentially leading to fibrillation and cardiac arrest (discussed later in this chapter). Myocardial infarction is responsible for about half of all deaths in the United States.

A temporary and reversible myocardial ischemia produces a sense of heaviness or pain called *angina pectoris*[16] (an-JY-na PEC-toe-riss). As the myocardium becomes hypoxic, it relies increasingly on anaerobic fermentation. This generates lactic acid, which stimulates pain receptors.

15. *isch* = to hold back + *em* = blood

16. *angina* = to choke, strangle + *pectoris* = of the chest

- The **middle cardiac vein,** found in the posterior sulcus, collects blood from the posterior aspect of the heart. It, too, carries blood from the apex upward.
- The **coronary sinus** collects blood from these and smaller cardiac veins. It passes across the posterior aspect of the heart in the coronary sulcus and empties blood into the right atrium.

Coronary Flow in Relation to the Cardiac Cycle

Most organs receive more arterial blood flow when the ventricles contract than when they relax, but the opposite is true in the coronary arteries. There are two reasons for this. First, contraction of the myocardium compresses the arteries and obstructs blood flow. Second, when the ventricles relax, blood in the aorta surges back toward the heart and fills the semilunar valve cusps. Since the coronary arteries open at the bottom of the pockets created by the cusps, they are filled by this backflow.

Before You Go On

Answer the following questions to test your understanding of the preceding section:

1. Make a two-color sketch of the pericardium; use one color for the fibrous pericardium and another for the serous pericardium and show their relationship to the heart wall.
2. Trace the flow of blood through the heart, naming each chamber and valve in order.
3. Define *pulmonary* and *systemic circuit.*
4. Trace the flow of blood from the left coronary artery to the apex and then to the coronary sinus.

Cardiac Muscle and the Cardiac Conduction System

Objectives

When you have completed this section, you should be able to

- contrast the structure of cardiac and skeletal muscle;
- describe the physiological properties of cardiac muscle and relate its structure to its function;
- explain why the heart does not fatigue; and
- describe the heart's electrical conduction system.

Structure of Cardiac Muscle

Cardiac muscle is striated like skeletal muscle but otherwise differs from it in many structural and physiological ways. Cardiac myocytes (muscle cells), or *cardiocytes,* are relatively short, thick, branched cells, typically 50 to 100 μm long and 10 to 20 μm wide (fig. 19.11). They usually have only one, centrally placed nucleus. The sarcoplasmic reticulum (SR) is less developed than in skeletal muscle. It lacks terminal cisternae, although it does have footlike sacs associated with the T tubules. The T tubules are much larger than in skeletal muscle and admit supplemental calcium ions from the extracellular fluid into the cell during excitation.

The myocytes are joined end to end by thick connections called **intercalated** (in-TUR-kuh-LAY-ted) **discs,** which appear as dark lines (thicker than the striations) in properly stained tissue sections. An intercalated disc is a complex steplike structure with three distinctive features:

1. **Interdigitating folds.** The plasma membrane at the end of the cell is folded somewhat like the inside of an egg carton. The folds of adjoining cells interlock

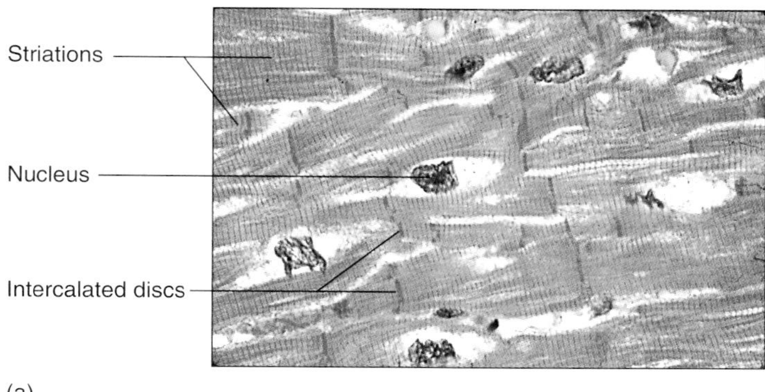

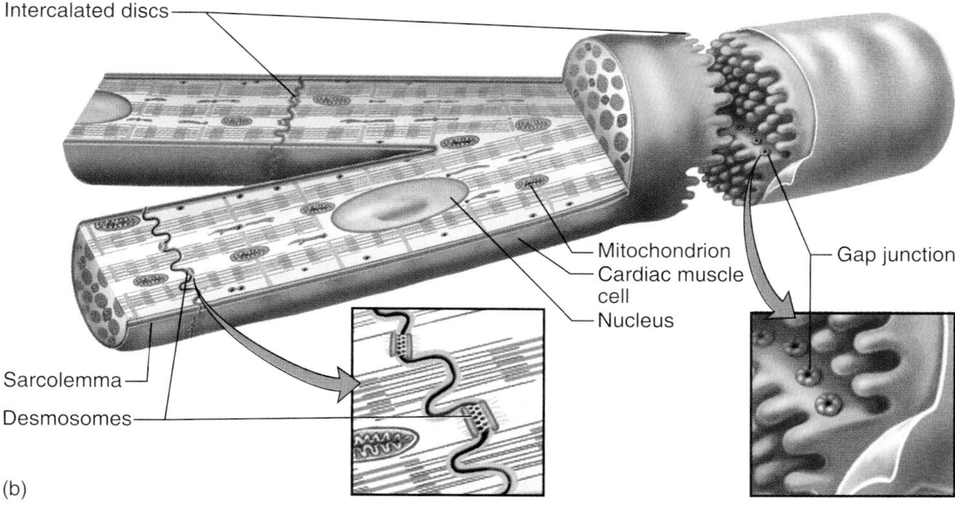

Figure 19.11 Cardiac Muscle. (*a*) Light micrograph. (*b*) Cardiac myocytes and intercalated discs.

with each other and increase the surface area of intercellular contact.

2. **Mechanical junctions.** The cells are tightly joined by two types of mechanical junctions—the fascia adherens and desmosomes. The *fascia adherens*[17] (FASH-ee-ah ad-HEER-enz) is the most extensive. It is a broad band in which the actin of the thin myofilaments is anchored to the plasma membrane and, via transmembrane proteins, one cell is linked to the next. The fascia adherens is interrupted here and there by desmosomes. (Desmosomes and gap junctions, described next, are illustrated in fig. 6.30, p. 193.) These mechanical junctions prevent the myocytes from pulling apart when the heart contracts.
3. **Electrical junctions.** The myocytes are electrically coupled by *gap junctions,* which form channels that allow ions to flow from the cytoplasm of one cell directly into the next. These junctions enable each myocyte to electrically stimulate its neighbors, so the entire myocardium of the atria, and that of the ventricles, each acts almost as if it were a single cell. This unified action is essential for the effective pumping of a heart chamber.

Like skeletal muscle fibers, cardiac myocytes are incapable of mitosis. Furthermore, cardiac muscle lacks satellite cells, which, in skeletal muscle, can divide and replace dead muscle cells to some extent. Thus, the repair of damaged cardiac muscle is entirely by fibrosis (scarring).

Metabolism of Cardiac Muscle

Cardiac muscle depends almost exclusively on aerobic respiration to make ATP. It is very rich in myoglobin (a short-term source of stored oxygen for aerobic respiration) and glycogen (stored energy). It also has especially large mitochondria, which fill about 25% of the myocyte; skeletal muscle fibers, by comparison, have much smaller

17. *fascia* = band + *adherens* = adhering

mitochondria that occupy only 2% of the fiber. Cardiac muscle is relatively adaptable with respect to the organic fuels used. At rest, the heart gets about 60% of its energy from fatty acids, 35% from glucose, and 5% from other fuels such as ketones, lactic acid, and amino acids. Cardiac muscle is more vulnerable to an oxygen deficiency than it is to the lack of any specific fuel. Because it makes little use of anaerobic fermentation or the oxygen debt mechanism, it is not prone to fatigue. You can easily appreciate this fact by clenching and opening your fist once every second for a minute or two. You will soon feel weakness and fatigue in your skeletal muscles and perhaps feel all the more grateful that cardiac muscle can maintain its rhythm, without fatigue, for a lifetime.

The Cardiac Conduction System

Among invertebrates such as clams, crabs, and insects, each heartbeat is triggered by a pacemaker in the nervous system. The vertebrate heartbeat, however, is said to be *myogenic*[18] because the signal originates within the heart itself, from pacemaker cells derived from cardiac muscle. Autonomic nerve fibers to the heart can modify its rhythm, but they do not create it—the heart goes on beating even if all nerve connections to it are severed. Indeed, we can remove a vertebrate heart from the body, keep it in aerated saline, and it will beat for hours. Cut the heart into little pieces, and each piece continues its own rhythmic pulsations.

Cardiac myocytes are said to be **autorhythmic**[19] because they depolarize spontaneously at regular time intervals. Some of them lose the ability to contract and become specialized, instead, for generating action potentials. These constitute the **cardiac conduction system,** which controls the route and timing of electrical conduction to ensure that the four chambers are coordinated with each other. The conduction system consists of the following components (fig. 19.12):

- The **sinoatrial (SA) node,** a patch of modified myocytes in the right atrium, just under the epicardium near the superior vena cava. This is the **pacemaker** that initiates each heartbeat and determines the heart rate. Signals from the SA node spread throughout the atria, as shown by the yellow arrows in figure 19.12.
- The **atrioventricular (AV) node,** located near the right AV valve at the lower end of the interatrial septum. This node acts as an electrical gateway to the ventricles; the fibrous skeleton acts as an insulator to prevent currents from getting to the ventricles by any other route.

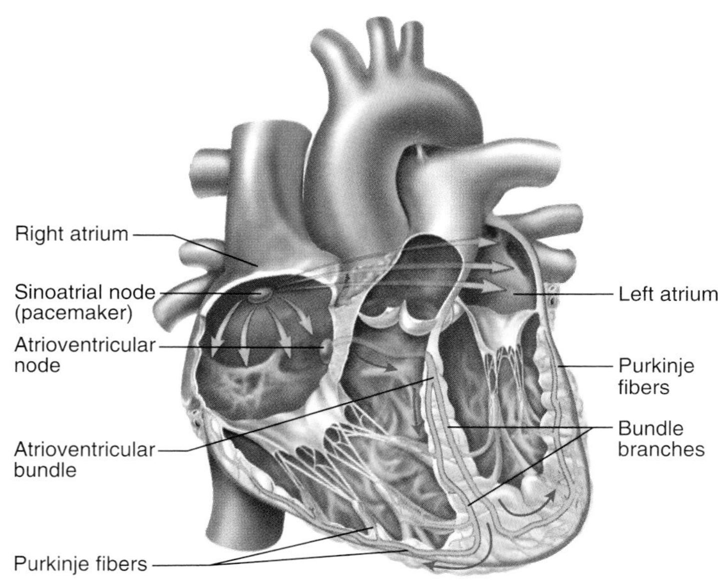

Figure 19.12 The Cardiac Conduction System. Electrical signals travel along the pathway indicated by the *arrows.*

- The **atrioventricular (AV) bundle** *(bundle of His*[20]*),* a pathway by which signals leave the AV node.
- The **right** and **left bundle branches,** divisions of the AV bundle that enter the interventricular septum and descend toward the apex.
- **Purkinje**[21] (pur-KIN-jee) **fibers,** nervelike processes that arise from the bundle branches near the apex of the heart and then turn upward and spread throughout the ventricular myocardium. Purkinje fibers distribute the electrical excitation to the myocytes of the ventricles. They form a more elaborate network in the left ventricle than in the right.

Before You Go On

Answer the following questions to test your understanding of the preceding section:

5. What organelle(s) are less developed in cardiac muscle than in skeletal muscle? What organelle(s) are more developed? What is the functional significance of these differences?
6. Name two types of cell junctions in the intercalated discs and explain their functional importance.
7. Why is the human heart described as myogenic? Where is its pacemaker and what is it called?
8. List the components of the cardiac conduction system in the order traveled by signals from the pacemaker.

18. *myo* = muscle + *genic* = arising from
19. *auto* = self
20. Wilhelm His, Jr. (1863–1934), German physiologist
21. Johannes E. Purkinje (1787–1869), Bohemian physiologist

Clinical Insight 19.3

Cardiac Arrhythmias

Atrial flutter, premature ventricular contractions, and ventricular fibrillation are common cardiac arrhythmias. *Atrial flutter* occurs when ectopic foci in the atria set off extra contractions and the atria beat 200 to 400 times per minute. *Premature ventricular contractions (PVCs)* occur singly or in bursts as a result of early firing of an ectopic focus. PVCs are often due to irritation of the heart by stimulants, emotional stress, or lack of sleep, but they sometimes indicate more serious pathology.

Ventricular fibrillation is a serious arrhythmia caused by electrical signals arriving at different regions of the myocardium at widely different times. A fibrillating ventricle exhibits squirming, uncoordinated contractions; it has been described as looking like a "bag of worms." Since a fibrillating heart does not pump blood, there is no coronary perfusion and the myocardium rapidly dies of ischemia. *Cardiac arrest* is the cessation of cardiac output, with the ventricles either motionless or in fibrillation.

Fibrillation kills quickly if it is not stopped. *Defibrillation* is an emergency procedure in which the heart is given a strong electrical shock with a pair of electrodes. The purpose is to depolarize the entire myocardium and stop the fibrillation, with the hope that the SA node will resume its sinus rhythm. This does not correct the underlying cause of the arrhythmia, but it may sustain a patient's life long enough to allow for other corrective action.

Electrical and Contractile Activity of the Heart

Objectives

When you have completed this section, you should be able to

- explain why the SA node fires spontaneously and rhythmically;
- explain how the SA node excites the myocardium;
- describe the unusual action potentials of cardiac muscle and relate them to the contractile behavior of the heart; and
- interpret a normal electrocardiogram.

In this section, we examine how the electrical events in the heart produce its cycle of contraction and relaxation. Contraction is called **systole** (SIS-toe-lee) and relaxation is **diastole** (dy-ASS-toe-lee). These terms can refer to a specific part of the heart (for example, atrial systole), but if no particular chamber is specified, they usually refer to the more conspicuous and important ventricular action, which ejects blood from the heart.

The Cardiac Rhythm

The normal heartbeat, triggered by the SA node, is called the **sinus rhythm.** At rest, the adult heart rate is usually around 70 to 80 beats per minute (bpm). Left to itself, the SA node would fire more often than this, but the vagus nerves inhibit it and hold it down to this rate at rest.

Stimuli such as hypoxia, electrolyte imbalances, caffeine, nicotine, and other drugs can cause other parts of the conduction system to fire before the SA node does, setting off an extra heartbeat (*extrasystole*). Any region of spontaneous firing other than the SA node is called an **ectopic**[22] **focus.** If the SA node is damaged, an ectopic focus may take over the governance of the heart rhythm. The most common ectopic focus is the AV node, which produces a slower heartbeat of 40 to 50 bpm, called a **nodal rhythm.** If neither the SA nor AV node is functioning, other ectopic foci fire at rates of 20 to 40 bpm. The nodal rhythm is sufficient to sustain life, but a rate of 20 to 40 bpm provides too little flow to the brain to be survivable. This condition calls for an artificial pacemaker.

Any abnormal cardiac rhythm is called **arrhythmia**[23] (see clinical insight 19.3). One cause of arrhythmia is a **heart block**—the failure of any part of the cardiac conduction system to transmit signals, usually as a result of disease and degeneration of conduction system fibers. A *bundle branch block,* for example, is due to damage to one or both bundle branches. Damage to the AV node causes *total heart block,* in which signals from the atria fail to reach the ventricles and the ventricles beat at their own intrinsic rhythm of 20 to 40 bpm.

Physiology of the SA Node

Why does the SA node spontaneously fire 70 or 80 times per minute? Unlike skeletal muscle or neurons, the cells of the SA node do not have a stable resting membrane potential. Their membrane potential starts at about −60 mV and drifts upward, showing a gradual depolarization called the **pacemaker potential** (fig. 19.13). This is thought to result from a slow inflow of Na^+ without a compensating outflow of K^+.

22. *ec* = out of + *top* = place
23. *a* = without + rhythm + *ia* = condition

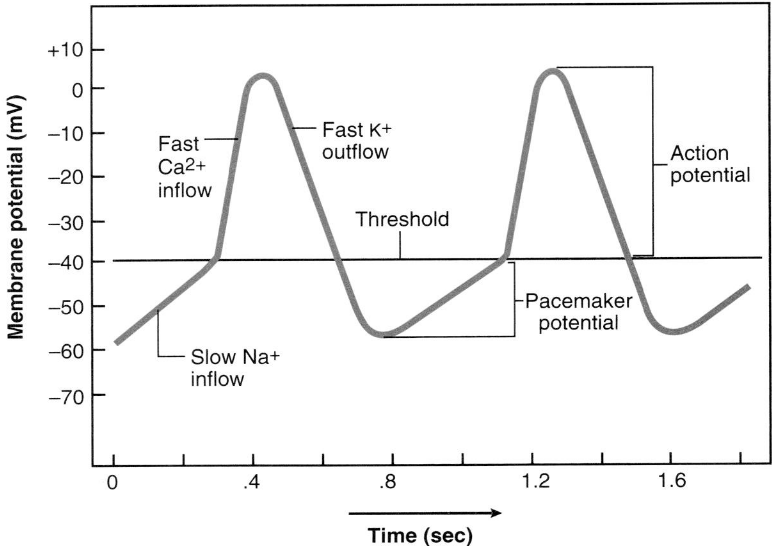

Figure 19.13 Pacemaker Potentials and Action Potentials of the SA Node.

When the pacemaker potential reaches a threshold of −40 mV, voltage-regulated **fast calcium channels** open and Ca^{2+} flows in from the extracellular fluid (ECF). This produces the rising (depolarizing) phase of the action potential, which peaks slightly above 0 mV. At that point, K^+ channels open and potassium ions leave the cell. This makes the cytosol increasingly negative and creates the falling (repolarizing) phase of the action potential. When repolarization is complete, the K^+ channels close and the pacemaker potential starts over, on its way to producing the next heartbeat. Each depolarization of the SA node sets off one heartbeat. When the SA node fires, it excites the other components in the conduction system; thus, the SA node serves as the system's pacemaker. At rest, it fires every 0.8 second or so, creating a heart rate of about 75 bpm.

Impulse Conduction to the Myocardium

Firing of the SA node excites atrial myocytes and stimulates the two atria to contract almost simultaneously. The signal travels at a speed of about 1 m/sec through the atrial myocardium and reaches the AV node in about 50 msec. The AV node has much thinner myocytes that slow the signal down to about 0.05 m/sec. This delays the signal at the AV node for about 100 msec—like highway traffic slowing down at a small town. This delay is essential because it gives the ventricles time to fill with blood before they begin to contract.

The ventricular myocardium has a conduction speed of only 0.3 to 0.5 m/sec. If this were the only route of travel for the excitatory signal, some myocytes would be stimulated much sooner than others. Ventricular contraction would not be synchronized and the pumping effectiveness of the ventricles would be severely compromised. But signals travel through the AV bundle and Purkinje fibers at a speed of 4 m/sec, the fastest in the conduction system. Consequently, the entire ventricular myocardium depolarizes within 200 msec after the SA node fires, causing the ventricles to contract in near unison.

Signals reach the papillary muscles before the rest of the myocardium. Thus, these muscles contract and begin taking up slack in the chordae tendineae an instant before ventricular contraction causes blood to surge against the AV valves. Ventricular systole begins at the apex of the heart, which is first to be stimulated, and progresses upward—pushing the blood upward toward the semilunar valves. Because of the spiral arrangement of ventricular myocytes, the ventricles twist slightly as they contract, like someone wringing out a towel.

Electrical Behavior of the Myocardium

Cardiac myocytes exhibit action potentials that are significantly different from those of neurons and skeletal muscle (fig. 19.14). Unlike the autorhythmic cells of the heart, they have a stable resting potential of −90 mV and depolarize only when stimulated by the cardiac conduction system or by each other. A stimulus opens voltage-regulated sodium gates, causing an Na^+ inflow and depolarizing the cell to its threshold. The threshold voltage rapidly opens additional Na^+ gates and triggers a positive feedback cycle like the one seen in the firing of a neuron (see p. 467). The action potential peaks at nearly +30 mV. The Na^+ gates close quickly, and the rising phase of the action potential is very brief.

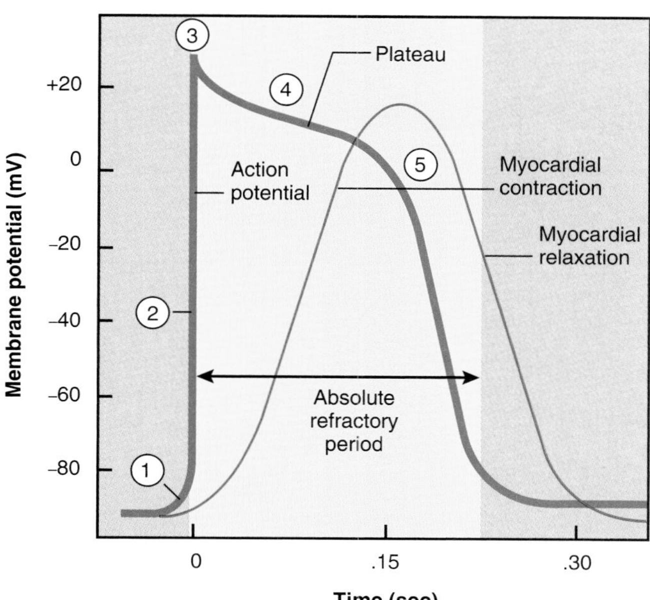

1. Voltage-gated sodium channels open.
2. Na^+ inflow depolarizes the membrane and triggers the opening of still more Na^+ channels, creating a positive feedback cycle and a rapidly rising membrane voltage.
3. Na^+ channels close when the cell depolarizes, and the voltage peaks at nearly +30 mV.
4. Ca^{2+} entering through slow calcium channels prolongs depolarization of membrane, creating a plateau. Plateau falls slightly because of some K^+ leakage, but most K^+ channels remain closed until end of plateau.
5. Ca^{2+} channels close and Ca^{2+} is transported out of cell. K^+ channels open, and rapid K^+ outflow returns membrane to its resting potential.

Figure 19.14 **Action Potential of a Ventricular Myocyte.** The *lighter curve* represents rising and falling muscle tension as the myocardium contracts and relaxes.

As action potentials spread over the plasma membrane, they open voltage-gated **slow calcium channels,** which admit calcium from the extracellular fluid into the cell. This calcium binds to fast calcium channels on the sarcoplasmic reticulum (SR), stimulating them to open and release a greater quantity of Ca^{2+} from the SR into the cytosol. This second wave of Ca^{2+} binds to troponin and triggers contraction of the myocyte, acting in the same way as it does in the excitation-contraction coupling of skeletal muscle (see chapter 12). The SR provides 90% to 98% of the Ca^{2+} needed for myocardial contraction.

In skeletal muscle and neurons, an action potential falls back to the resting potential within 2 msec. In cardiac muscle, however, the depolarization is prolonged for 200 to 250 msec (at a heart rate of 70–80 bpm) producing a long plateau in the action potential—perhaps because the slow Ca^{2+} channels are slow to close or because the SR is slow to remove Ca^{2+} from the cytosol.

As long as the action potential is in its plateau, the myocytes contract. Thus, in figure 19.14, you can see the development of muscle tension (myocardial contraction) following closely behind the depolarization and plateau. Rather than showing a brief twitch like skeletal muscle, cardiac muscle has a more sustained contraction necessary for expulsion of blood from the heart chambers. Both atrial and ventricular myocytes exhibit these plateaus, but they are more pronounced in the ventricles.

At the end of the plateau, Ca^{2+} channels close and K^+ channels open. Potassium diffuses rapidly out of the cell and Ca^{2+} is transported back into the extracellular fluid and SR. Membrane voltage drops rapidly, and muscle tension declines soon afterward.

Cardiac muscle has an *absolute refractory period* of 250 msec, compared with 1 to 2 msec in skeletal muscle. This long refractory period prevents wave summation and tetanus, which would stop the pumping action of the heart.

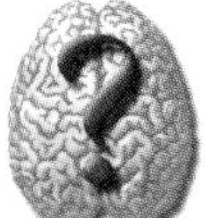

Think About It

With regard to the ions involved, how does the falling (repolarization) phase of a myocardial action potential differ from that of a neuron's action potential? (see p. 467)

The Electrocardiogram

We can detect electrical currents in the heart by means of electrodes (leads) applied to the skin. An instrument called the *electrocardiograph* amplifies these signals and produces a record, usually on a moving paper chart, called an **electrocardiogram**[24] **(ECG** or **EKG**[25]**).** To record an

24. *graph* = recording instrument; *graphy* = recording procedure; *gram* = record of
25. EKG is from the German spelling, Elektrokardiogramm

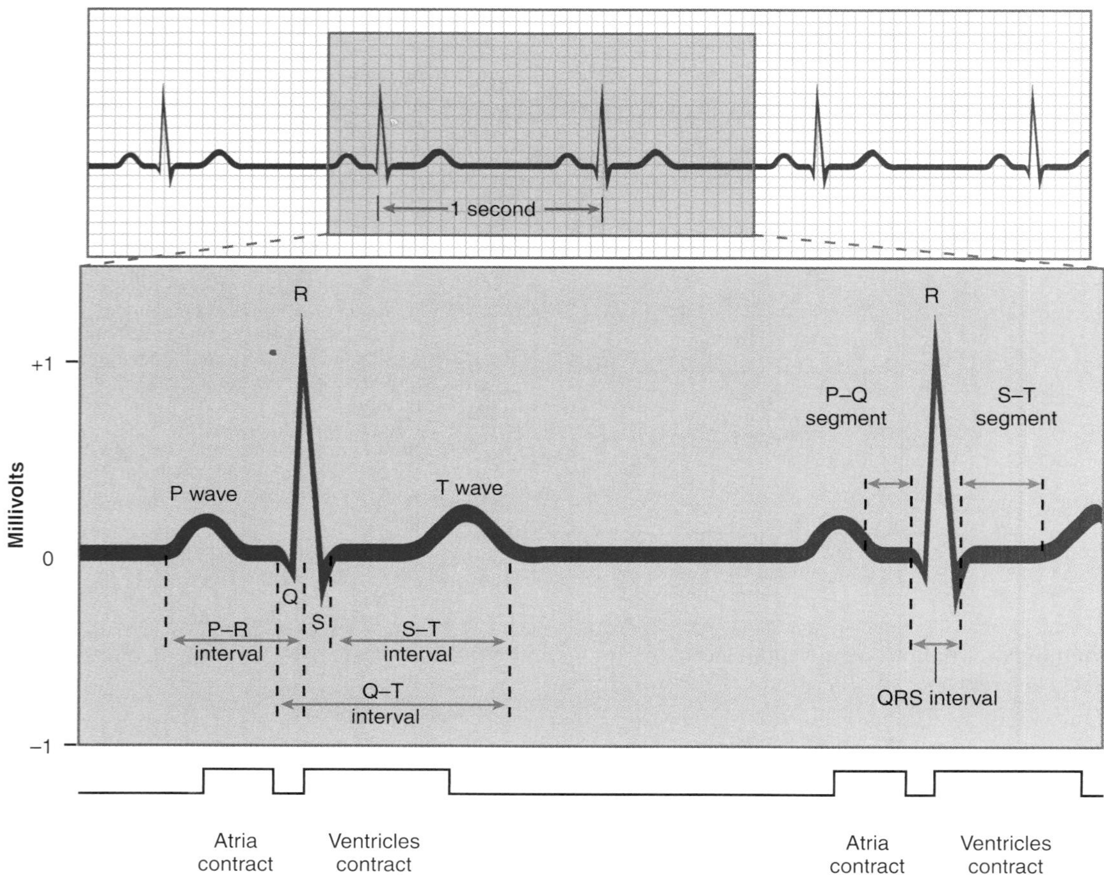

Figure 19.15 The Normal Electrocardiogram.

ECG, electrodes are typically attached to the wrists, ankles, and six locations on the chest. Several simultaneous recordings can be made from electrodes at different distances from the heart; collectively, they provide a comprehensive image of the heart's electrical activity. An ECG is a composite recording of all the action potentials produced by the nodal and myocardial cells—it should not be misconstrued as a tracing of a single action potential.

Figure 19.15 shows a typical ECG. It shows three principal deflections above and below the baseline: the *P wave, QRS complex,* and *T wave.* Figure 19.16 shows how these correspond to regions of the heart undergoing depolarization and repolarization.

The **P wave** is produced when a signal from the SA node spreads through the atria and depolarizes them. Atrial systole begins about 100 msec after the P wave begins, during the *P–Q segment.* This segment is about 160 msec long and represents the time required for impulses to travel from the SA node to the AV node.

The **QRS complex** consists of a small downward deflection (Q), a tall sharp peak (R), and a final downward deflection (S). It marks the firing of the AV node and the onset of ventricular depolarization. Its complex shape is due to the different sizes of the two ventricles and the different times required for them to depolarize. Ventricular systole begins shortly after the QRS complex in the *S–T segment.* Atrial repolarization and diastole also occur during the QRS interval, but atrial repolarization sends a relatively weak signal that is obscured by the electrical activity of the more muscular ventricles. The S–T segment corresponds to the plateau in the myocardial action potential and thus represents the time during which the ventricles contract and eject blood.

The **T wave** is generated by ventricular repolarization immediately before diastole. The ventricles take longer to repolarize than to depolarize; the T wave is therefore smaller and more spread out than the QRS complex, and it has a rounder peak. Even in cases where the T wave is taller than the QRS complex, it can be recognized by its relatively rounded peak.

The ECG affords a wealth of information about the normal electrical activity of the heart. Deviations from

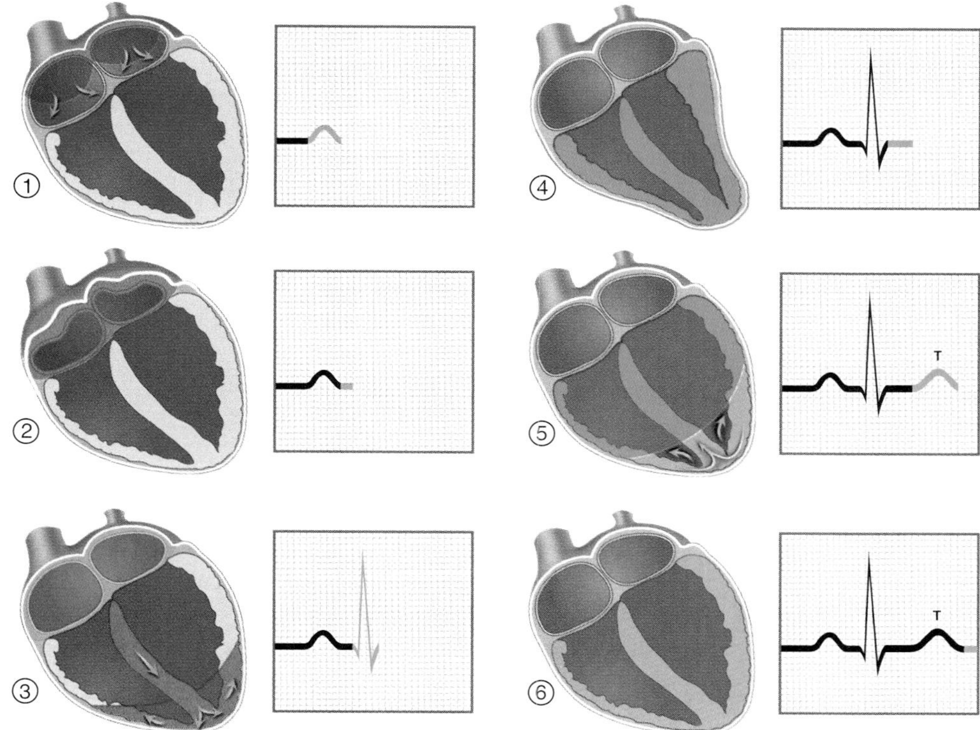

Figure 19.16 **Relationship of the Electrocardiogram (ECG) to Electrical Activity and Contraction of the Myocardium.** Each heart diagram indicates the events occurring at the time of the colored segment of the ECG. *Red* indicates depolarizing or depolarized myocardium, and *green* indicates repolarizing or repolarized myocardium. *Arrows* indicate the direction in which a wave of depolarization or repolarization is traveling.

normal are invaluable for diagnosing abnormalities in the conduction pathways, myocardial infarction, enlargement of the heart, and electrolyte and hormone imbalances. A few examples of abnormal ECGs are given in table 19.1 and figure 19.17.

Before You Go On

Answer the following questions to test your understanding of the preceding section:

9. Define *systole* and *diastole.*
10. How does the pacemaker potential of the SA node differ from the resting membrane potential of a neuron? Why is this important in creating the heart rhythm?
11. How does excitation-contraction coupling in cardiac muscle resemble that of skeletal muscle? How is it different?
12. What produces the plateau in the action potentials of cardiac myocytes? Why is this important to the pumping ability of the heart?
13. Name the waves of the ECG and explain what myocardial events produce each wave.

Blood Flow, Heart Sounds, and the Cardiac Cycle

Objectives

When you have completed this section, you should be able to

- explain how pressure and resistance determine the flow of a fluid;
- explain what causes the sounds of the heartbeat;
- describe in detail one complete cycle of heart contraction and relaxation; and
- relate the events of the cardiac cycle to the volume of blood entering and leaving the heart.

A **cardiac cycle** consists of one complete contraction and relaxation of all four heart chambers. In order of occurrence, its major events are atrial systole, atrial diastole, ventricular systole, ventricular diastole, and a *quiescent period* before the next atrial systole begins. We will examine these events in detail to see how they relate to the entry and expulsion of blood, but first we consider two related issues: (1) some general principles of pressure

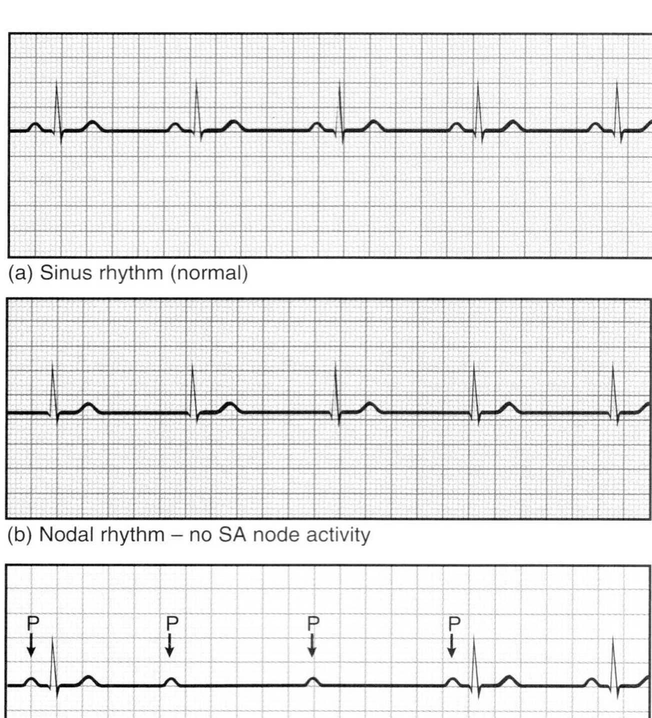

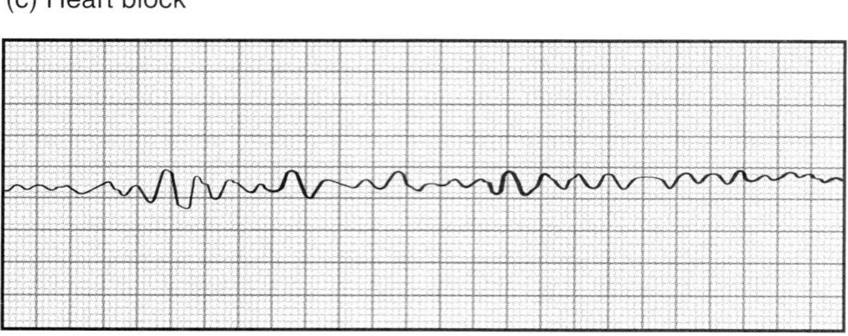

Figure 19.17 Normal and Pathological Electrocardiograms. (*a*) Normal sinus rhythm. (*b*) Nodal rhythm generated by the AV node in the absence of SA node activity; note the lack of P waves. (*c*) Heart block, in which some P waves are not transmitted through the AV node and do not generate QRS complexes. (*d*) Ventricular fibrillation, with grossly irregular waves of depolarization.

changes and how they affect the flow of blood, and (2) the heart sounds produced during the cardiac cycle, which we can then relate to the stages of the cycle.

Principles of Pressure and Flow

A fluid is any liquid or gas—a state of matter that can flow in bulk from one place to another. In this and some forthcoming chapters, we are concerned with factors that govern the flow of fluids such as blood, lymph, air, and urine. Some basic principles of fluid movement (*fluid dynamics*) are therefore important to understand at this time. Fluid dynamics are governed by pressure, which can cause a fluid to flow, and resistance, which opposes flow.

Table 19.1 Examples of the Diagnostic Interpretation of Abnormal Electrocardiograms

Appearance	Suggested Meaning
Enlarged P wave	Atrial hypertrophy, often a result of mitral valve stenosis
Missing or inverted P wave	SA node damage; AV node has taken over pacemaker role
Two or more P waves per cycle	Extrasystole; heart block
Enlarged Q wave	Myocardial infarction
Enlarged R wave	Ventricular hypertrophy
Abnormal T waves	Flattened in hypoxia; elevated in hyperkalemia (K^+ excess)
Abnormally long P–Q segment	Scarring of atrial myocardium, forcing impulses to bypass normal conduction pathways and take slower alternative routes to AV node
Abnormal S–T segment	Elevated above baseline in myocardial infarction; depressed in myocardial hypoxia

Measurement of Pressure

Pressure is often measured by observing how high it can push a column of mercury (Hg) up an evacuated tube called a *manometer*. Mercury is used because it is very dense and enables us to measure pressure with shorter columns than we would need with a less dense liquid such as water. Because pressures are compared to the force generated by a column of mercury, they are expressed in terms of millimeters of mercury (mmHg). Blood pressure is usually measured with a **sphygmomanometer**[26] (SFIG-mo-ma-NOM-eh-tur)—a calibrated tube filled with mercury and attached to an inflatable pressure cuff wrapped around the arm. Blood pressure and the method of measuring it are discussed in greater detail in the next chapter.

Pressure Gradients and Flow

Any change in the volume of a container creates a **pressure gradient,** or difference, between the inside and outside of the container. If there is an opening in the container, fluid flows in or out, "down the gradient," from point A, where pressure is higher, to point B, where pressure is lower. The pressure at point B rises and the pressure at point A falls

26. *sphygmo* = pulse + *mano* = rare, sparse, roomy

until the two are equal. At that time, there is no more pressure gradient and flow stops. Flow also stops, of course, if it is obstructed by the closure of a passage between point A and B—a matter of obvious relevance where the heart valves are concerned.

Suppose you pull back the plunger of a syringe, for example. The volume in the syringe barrel increases and its pressure falls (fig. 19.18). Since pressure outside the syringe is greater than the pressure inside, air flows into it until the pressures inside and outside are equal. If you then push the plunger in, pressure inside rises above the pressure outside, and air flows out—again following a gradient from high pressure to low.

The syringe barrel is analogous to a heart chamber such as the left ventricle. When the ventricle is expanding, its internal pressure falls. If the AV valve is open, blood flows into the ventricle from the atrium above. When the ventricle contracts, its internal pressure rises. When the aortic valve opens, blood is ejected from the ventricle into the aorta.

A pressure difference does not guarantee that a fluid will flow. There is always a positive blood pressure in the aorta, and if it is greater than the pressure in the ventricle, it holds the aortic valve closed and prevents the expulsion of blood. When continuing contraction causes ventricular pressure to rise above aortic pressure, however, the valve is forced open and blood is ejected into the aorta.

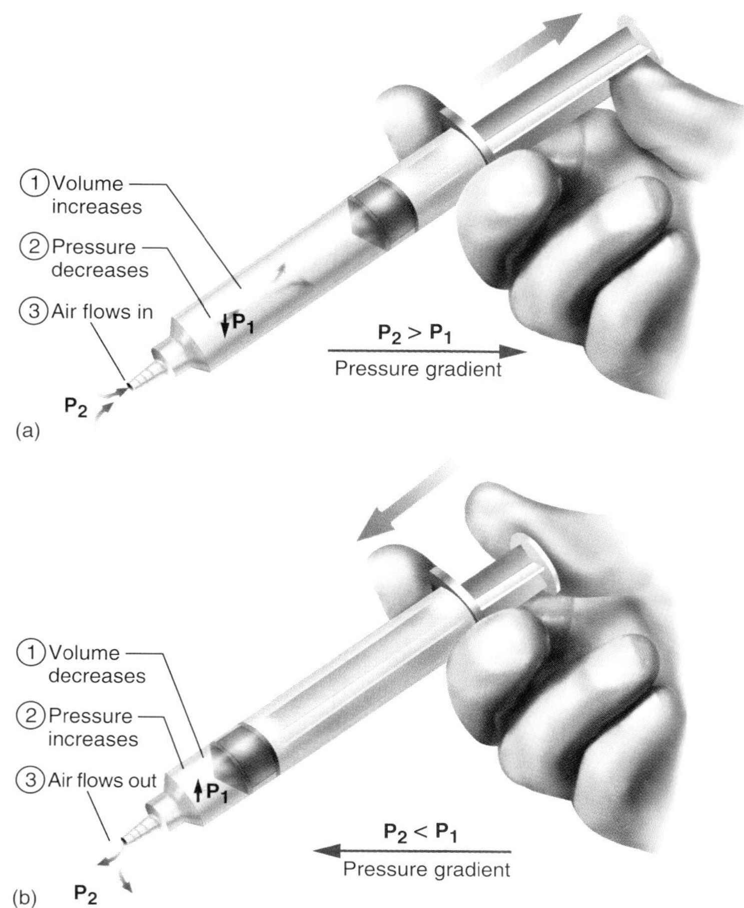

Figure 19.18 **Principles of Volume, Pressure, and Flow Illustrated with a Syringe.** (*a*) As the plunger is pulled back, the volume of the enclosed space increases, its pressure falls, and pressure inside the syringe (P_1) is lower than the pressure outside (P_2). The pressure gradient causes air to flow inward until the pressures are equal. This is analogous to the filling of an expanding heart chamber. (*b*) As the plunger is depressed, the volume of the enclosed space decreases, P_1 rises above P_2, and air flows out until the pressures are equal. This is analogous to the ejection of blood from a contracting heart chamber. In both cases, fluids flow down their pressure gradients.

Heart Sounds

As we follow events through the cardiac cycle, we will note the occurrence of the *heart sounds.* Listening to sounds made by the body is called **auscultation** (AWS-cul-TAY-shun). Each cardiac cycle generates two or three sounds that are audible with a stethoscope. The **first** and **second heart sounds,** symbolized S1 and S2, are often described as a "lubb-dupp"—S1 is louder and longer and S2 a little softer and sharper. In children and adolescents, it is normal to hear a **third heart sound** (S3). This is rarely audible in people older than 30, but when it is, the heartbeat is said to show a *triple rhythm* or *gallop.* If the normal rhythm is roughly simulated by drumming two fingers on a table, a triple rhythm sounds a little like drumming with three fingers. The heart valves themselves operate silently, but S1 and S2 occur in conjunction with the closing of the valves as a result of turbulence in the bloodstream and movements of the heart wall. The cause of each sound is not known with certainty, but the probable factors are discussed in the respective phases of the cardiac cycle.

Phases of the Cardiac Cycle

We now examine the phases of the cardiac cycle, the pressure changes that occur, and how the pressure changes and valves govern the flow of blood. A substantial amount of information about these events is summarized in figure 19.19, which is divided into colored bars numbered to correspond to the phases described here. Closely follow the figure as you study the following text. Where to begin when describing a circular chain of events is somewhat arbitrary. However, in this presentation we begin with the quiescent period, when the heart appears to be preparing for its next beat. Remember that all these events are completed in less than 1 second.

1. The **quiescent period.** In this stage, none of the heart chambers are contracting. Blood is flowing into the atria, and since the AV valves are open,

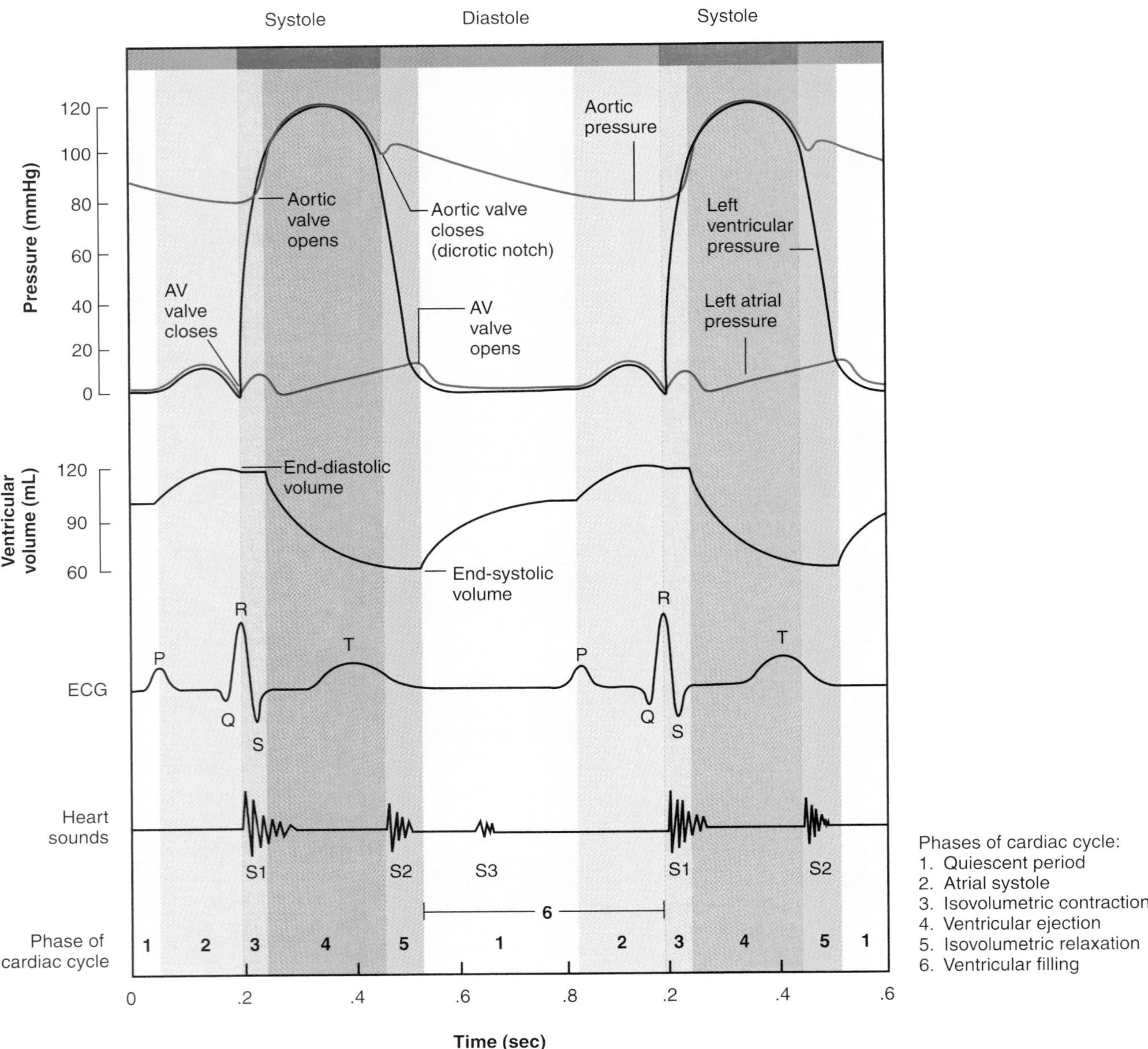

Figure 19.19 Major Events of the Cardiac Cycle. Two complete cycles are shown. The phases are numbered across the bottom to correspond to the text description.

much of it flows directly through into the ventricles. As the ventricles fill, the flaccid cusps of the AV valves begin to float up toward the closed position.

2. **Atrial systole.** The SA node fires and the atria depolarize, producing the P wave of the ECG and stimulating the onset of atrial systole. The right atrium contracts slightly before the left because it is the first to receive the signal from the SA node. Blood pressure in the atria rises and forces additional blood into the ventricles. At the end of this phase, each ventricle contains an **end-diastolic volume (EDV)** of about 130 mL of blood. Only 40 mL (31%) of this is contributed by atrial systole.
3. **Isovolumetric contraction.** The atria repolarize, relax, and remain in diastole for the rest of the cardiac cycle. The ventricles depolarize, generate the QRS complex, and begin to contract. Pressure in the ventricles rises sharply and reverses the pressure gradient between atria and ventricles. The AV valves close as ventricular blood surges back against the cusps. Heart sound S1 occurs at the

beginning of this phase and is produced mainly by the left ventricle; the right ventricle is thought to make little contribution to S1. Causes of the sound are thought to include the tensing of ventricular tissues, acceleration of the ventricular wall, turbulence in the blood as it surges against the closed AV valves, and impact of the heart against the chest wall.

This phase is called *isovolumetric*[27] because even though the ventricles contract, they do not eject blood yet, and there is no change in their volume. This is because pressures in the aorta (80 mmHg) and pulmonary trunk (10 mmHg) are still greater than the pressures in the respective ventricles and thus oppose the opening of the semilunar valves. The myocytes exert force, but they cannot shorten because liquids cannot be compressed and, with all four valves closed, the blood cannot go anywhere.

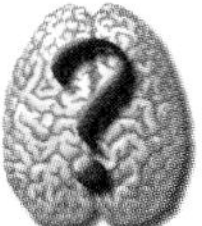

Think About It

Are the ventricular myocytes in isometric or isotonic contraction during the phase just described?

4. **Ventricular ejection.** The ejection of blood begins when ventricular pressure exceeds arterial pressure and forces the semilunar valves open. The pressure peaks at 120 mmHg in the left ventricle and 25 mmHg in the right. Blood spurts out of each ventricle rapidly at first (*rapid ejection*) and then flows out more slowly under less pressure (*reduced ejection*). By analogy, suppose you were to shake up a bottle of soda pop and remove the cap. The soda would spurt out rapidly at high pressure and then more would dribble out at lower pressure, much like the blood leaving the ventricles. Ventricular ejection lasts about 200 to 250 msec, which corresponds to the plateau of the myocardial action potentials but lags somewhat behind it (review the tension curve in fig. 19.14).

   The ventricles do not expel all their blood. In an average resting heart, each ventricle contains an EDV of 130 mL. The amount ejected, about 70 mL, is called the **stroke volume (SV).** The percentage of the EDV ejected, about 54%, is the **ejection fraction.** The blood remaining behind, about 60 mL in this case, is called the **end-systolic volume (ESV).** Note that EDV − SV = ESV. In vigorous exercise, the ejection fraction may be as high as 90%. Ejection fraction is an important measure of cardiac health. A diseased heart may eject much less than 50% of the blood it contains.
5. **Isovolumetric relaxation.** This is early ventricular diastole, when the T wave appears and the ventricles repolarize and begin to expand. There are competing theories as to how they expand. One is that the blood flowing into the ventricles "inflates" them. Another is that contraction of the ventricles deforms the fibrous skeleton, which subsequently springs back like a rubber ball that has been squeezed and released. This elastic recoil and expansion would cause pressure to drop rapidly and suck blood into the ventricles.

   At the beginning of ventricular diastole, blood from the aorta and pulmonary trunk briefly flows backward through the semilunar valves. The backflow, however, quickly fills the cusps and closes them, creating a slight pressure rebound that appears as the *dicrotic notch* of the aortic pressure curve (fig. 19.19). Heart sound S2 occurs as blood rebounds from the closed semilunar valves and the ventricles expand. This phase is called *isovolumetric* because the semilunar valves are closed, the AV valves have not yet opened, and the ventricles are therefore not taking in blood.
6. **Ventricular filling.** Diastolic expansion of the ventricles causes their pressure to drop below that of the atria. As a result, the AV valves open and blood flows rapidly into the ventricles, causing ventricular pressure to rise and atrial pressure to fall. Heart sound S3, if it occurs, is thought to be caused by the transition from expansion of the empty ventricles to their sudden filling with blood. Three phases of ventricular filling occur: (1) The first one-third is *rapid ventricular filling,* when blood enters especially quickly by the same principle as the rapid ejection explained in phase 4 (the soda pop analogy). (2) The second one-third, called *diastasis* (di-ASS-tuh-sis), is marked by slower filling. The P wave occurs at the end of diastasis. (3) During the last one-third, atrial systole completes the filling process. Note, therefore, that phase 6 of the cardiac cycle, as outlined here, includes phases 1 and 2 described earlier. Diastasis occurs in the latter part of phase 1 and atrial systole in phase 2.

In a resting person, atrial systole lasts about 0.1 second; ventricular systole, 0.3 second; and the quiescent period, 0.4 second. Total duration of the cardiac cycle is therefore 0.8 second (800 msec) in a heart beating at 75 bpm.

Overview of Volume Changes

An additional perspective on the cardiac cycle can be gained if we review the volume changes that occur. This

27. *iso* = same

"balance sheet" is from the standpoint of the left ventricle, but for reasons explained shortly, these numbers also must be true of the right. The volumes vary somewhat from one person to another and depend on a person's state of activity.

End-systolic volume (ESV, left from previous heartbeat)	60 mL
Passively added to the ventricle during atrial diastole	+ 30 mL
Added by atrial systole	+ 40 mL
Total: end-diastolic volume (EDV)	130 mL
Stroke volume (SV) ejected by ventricular systole	−70 mL
Leaves: end-systolic volume (ESV)	60 mL

Notice that the ventricle pumps as much blood as it received during diastole—70 mL in this example.

Both ventricles eject the same amount of blood even though pressure in the right ventricle is only about one-fifth the pressure in the left. Blood pressure in the pulmonary trunk is relatively low, so the right ventricle does not need to generate very much pressure to overcome it. It is essential that both ventricles have the same output. If the right ventricle pumped more blood into the lungs than the left side of the heart could handle on return, blood would accumulate in the lungs and cause pulmonary hypertension and edema (fig. 19.20). This would put a person at risk of suffocation as fluid filled the lungs and interfered with gas exchange. Conversely, if the left ventricle pumped out more blood than the right heart could handle on return, pressure would back up in the systemic circuit and cause hypertension and edema there. Over the long term, this could lead to aneurysms (weakened, bulging arteries), stroke, kidney failure, or heart failure (see clinical insight 19.4). To maintain homeostasis, the two ventricles must have equal output.

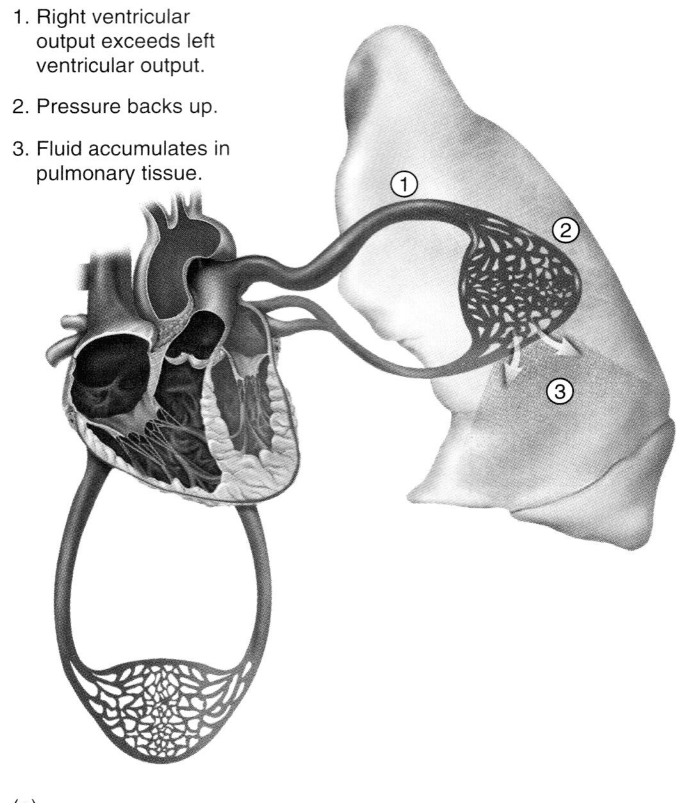

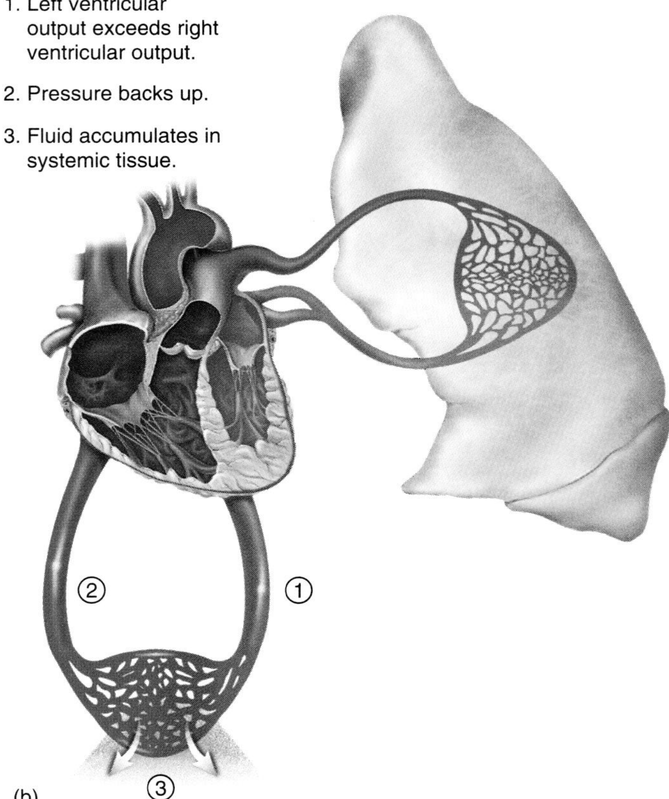

Figure 19.20 The Necessity of Balanced Ventricular Output. (*a*) If the left ventricle pumps less blood than the right, blood pressure backs up into the lungs and causes pulmonary edema. (*b*) If the right ventricle pumps less blood than the left, pressure backs up in the systemic circulation and causes systemic edema. To maintain homeostasis, both ventricles must pump the same average amount of blood.

Clinical Insight 19.4

Congestive Heart Failure

Congestive heart failure (CHF) results from the failure of either ventricle to eject blood effectively. It is usually due to a heart weakened by myocardial infarction, chronic hypertension, valvular insufficiency, or congenital defects in cardiac structure. If the left ventricle fails, blood backs up into the lungs and causes pulmonary edema (fluid in the lungs), shortness of breath, and a sense of suffocation. If the right ventricle fails, blood backs up into the venae cavae and causes systemic, or generalized, edema (formerly called *dropsy*). Systemic edema is marked by enlargement of the liver, ascites (the pooling of fluid in the abdominal cavity), distension of the jugular veins, and swelling of the fingers, ankles, and feet. Failure of one ventricle eventually increases the workload on the other ventricle, which stresses it and leads to its eventual failure as well.

Before You Go On

Answer the following questions to test your understanding of the preceding section:

14. Explain how a pressure gradient across a heart valve determines whether or not a ventricle ejects blood.
15. What factors are thought to cause the first and second heart sounds? When do these sounds occur?
16. What phases of the cardiac cycle are isovolumetric? Explain what this means.

Cardiac Output

Objectives

When you have completed this section, you should be able to

- define *cardiac output* and explain its importance;
- identify the factors that govern cardiac output;
- discuss some of the nervous and chemical factors that alter heart rate, stroke volume, and cardiac output;
- explain how the right and left ventricles achieve balanced output; and
- describe some effects of exercise on cardiac output.

The entire point of all the cardiac physiology we have considered thus far is to eject blood from the heart. The amount ejected by each ventricle in 1 minute is called the **cardiac output (CO).** If HR is heart rate (beats/min) and SV is stroke volume, CO = HR × SV. At typical resting values, CO = 75 beats/min × 70 mL/beat = 5,250 mL/min. Thus, the body's total volume of blood (4–6 L) passes through the heart every minute; or to look at it another way, an RBC leaving the left ventricle will, on average, arrive back at the left ventricle in about 1 minute.

Cardiac output is not constant but varies with the body's state of activity. Vigorous exercise increases CO to as much as 21 L/min in a person in good condition and up to 35 L/min in world-class athletes. The difference between the maximum and resting cardiac output is called **cardiac reserve.** People with severe heart disease may have little or no cardiac reserve and little tolerance of physical exertion.

Given that cardiac output equals HR × SV, you can see that there are only two ways to change it—change the heart rate or change the stroke volume. We will consider factors that influence each of these variables, but bear in mind that heart rate and stroke volume are somewhat interdependent and usually change together.

Heart Rate

Since each beat of the heart produces a surge of pressure in the arteries, the easiest way to measure heart rate is to palpate the **pulse** in a superficial artery, such as the radial artery on the lateral side of the wrist, and count beats per minute. In newborn infants, the resting heart rate is commonly 120 bpm or greater. It declines steadily with age, averaging 72 to 80 bpm in young adult females and 64 to 72 bpm in young adult males. It rises again in the elderly.

Tachycardia[28] is a persistent, resting adult heart rate above 100 bpm. It can be caused by stress, anxiety, drugs, heart disease, or elevated body temperature. As discussed in the next chapter, heart rate also rises to compensate to some extent for a drop in stroke volume. Thus, the heart races when the body has lost a significant quantity of blood or when there is damage to the myocardium.

Bradycardia[29] is a persistent, resting adult heart rate below 60 bpm. It is common during sleep and in endurance-trained athletes. Endurance training enlarges the heart and increases its stroke volume. Thus, it can maintain the same cardiac output with fewer beats. Hypothermia (low body

28. *tachy* = speed, fast + *card* = heart + *ia* = condition
29. *brady* = slow

temperature) also slows the heart rate and may be deliberately induced in preparation for cardiac surgery. Diving mammals such as whales and seals exhibit bradycardia during the dive, as do humans to some extent when the face is immersed in cool water.

Factors that raise the heart rate are called *positive chronotropic*[30] *agents,* and factors that lower it are *negative chronotropic agents.* We next consider some chronotropic effects of the autonomic nervous system, hormones, electrolytes, and blood gases.

Chronotropic Effects of the Autonomic Nervous System

Although the nervous system does not initiate the heartbeat, it does modulate its rhythm and force. The **cardiac center** of the medulla oblongata is an autonomic control center with two neuronal pools, a cardioacceleratory center and cardioinhibitory center. The **cardioacceleratory center** sends signals by way of the thoracic spinal cord and sympathetic **cardiac accelerator nerves** to the SA node, AV node, and myocardium. These nerves secrete norepinephrine, which binds to β-adrenergic receptors in the heart and produces a positive chronotropic effect. Cardiac output peaks when the heart rate is 160 to 180 bpm, although the sympathetic nervous system can get the heart rate up to as much as 230 bpm. This limit is set mainly by the refractory period of the SA node; it cannot fire any more frequently than that. At such a high rate, however, the ventricles beat so rapidly that they have little time to fill between beats; therefore, the stroke volume and cardiac output are less than they are at rest. At a heart rate of 65 bpm, ventricular diastole lasts about 0.62 seconds, but at 200 bpm, it lasts only 0.14 seconds. At that high rate, there is less time available for refilling between beats.

The **cardioinhibitory center** sends signals by way of parasympathetic fibers in the vagus nerves to the SA and AV nodes. The right vagus nerve innervates mainly the SA node, and the left vagus nerve innervates the AV node. The vagus nerves secrete acetylcholine, which binds to muscarinic receptors and opens K^+ channels in the nodal cells. As K^+ leaves the cells, the cells become hyperpolarized and fire less frequently, so the heart slows down.

The vagus nerves maintain a background firing rate called **vagal tone** that inhibits the nodes. If the vagus nerves to the heart are severed, the SA node fires at its own intrinsic, or "natural," frequency of about 100 times per minute. With the vagus nerve intact, however, vagal tone holds the heart rate down to the usual sinus rhythm of 70 to 80 bpm. Maximum vagal stimulation can reduce the heart rate to as low as 20 bpm.

30. *chrono* = time + *trop* = turn, change, influence

The cardiac center receives and integrates input from multiple sources. Sensory and emotional stimuli can act on the cardiac center by way of the cerebral cortex, limbic system, and hypothalamus; therefore, heart rate can climb even as you anticipate taking the first plunge on a roller coaster or competing in an athletic event, and it is influenced by emotions such as love and anger. The cardiac center also receives input from receptors in the muscles, joints, arteries, and brainstem:

- **Proprioceptors** in the muscles and joints (see chapter 16) quickly inform the cardiac center of changes in physical activity. Thus, the heart can increase its output even before the metabolic demands of the muscles rise.
- **Baroreceptors** are pressure sensors found in the aorta and internal carotid arteries (see fig. 15.14, p. 569). They send a continual stream of signals to the cardiac center. If blood pressure drops, the signaling rate drops and the cardiac center sends excitatory signals to the heart. This increases the heart rate and raises the blood pressure. If blood pressure rises too high, the signaling rate from the baroreceptors rises and the cardiac center responds by reducing the heart rate.
- **Chemoreceptors** sensitive to blood pH, carbon dioxide, and oxygen are found in the aortic arch, carotid arteries, and medulla oblongata. They are more important in respiratory control than in cardiovascular control but do have an influence on heart rate. If circulation to the tissues is too slow to remove CO_2 as fast as the tissues produce it, then CO_2 accumulates in the blood and cerebrospinal fluid (CSF) and produces a state of *hypercapnia* (CO_2 excess). Furthermore, CO_2 generates hydrogen ions by reacting with water: $CO_2 + H_2O \rightarrow HCO_3^- + H^+$. The hydrogen ions lower the pH of the blood and CSF and may create a state of acidosis ($pH < 7.35$). Hypercapnia and acidosis stimulate the cardiac center to increase the heart rate, thus improving perfusion of the tissues and restoring homeostasis. The chemoreceptors also respond to extreme *hypoxemia* (oxygen deficiency), such as in suffocation, but the effect is usually to slow down the heart, perhaps so the heart does not compete with the brain for the limited oxygen supply.

Such responses to fluctuations in blood chemistry and blood pressure, called **chemoreflexes** and **baroreflexes,** are good examples of negative feedback loops. They are discussed more fully in the next chapter.

Chronotropic Effects of Chemicals

Epinephrine and norepinephrine are potent cardiac stimulants. They are secreted by the cardiac accelerator nerves and the adrenal medulla in response to arousal, stress, and exercise. These catecholamines act through cAMP as a second messenger. Caffeine and the related stimulants in coffee, tea, and chocolate produce positive chronotropic effects by inhibiting cAMP breakdown. Nicotine also accelerates the heart by stimulating catecholamine secretion. Thyroid hormone increases the number of adrenergic receptors in the cardiac muscle, making the heart more responsive to sympathetic stimulation and thus increasing the heart rate. Hyperthyroidism causes tachycardia, which in the long run can weaken the heart and cause heart failure.

The ion with the greatest chronotropic effects is potassium (K^+). A potassium excess, called *hyperkalemia,*[31] causes increased K^+ diffusion into the myocytes, thereby raising their membrane potential and interfering with their repolarization. The myocardium becomes less excitable, the heart rate becomes slow and irregular, and the heart may arrest in diastole. In *hypokalemia,* a K^+ deficiency, the cells become less excitable because K^+ diffuses out of the cells and leaves them hyperpolarized—their membrane voltage is abnormally low, and it requires more stimulation to bring the cell to threshold. These potassium imbalances are very dangerous and require emergency medical treatment. Chapter 24 further explores the causes and effects of these and other electrolyte imbalances.

Hypercalcemia (a calcium excess) reduces the heart rate and hypocalcemia (a calcium deficiency) increases it. These calcium imbalances are relatively rare, however, and when they do occur, their primary effect is on contraction strength.

Stroke Volume

Stroke volume is governed mainly by three factors called *preload, contractility,* and *afterload.* Increased preload or contractility increases stroke volume, while increased afterload opposes the emptying of the ventricles and reduces stroke volume.

Preload

The amount of tension in the ventricular myocardium immediately before it begins to contract is called the **preload.** To understand how it influences stroke volume, imagine yourself engaged in heavy exercise. As active muscles massage your veins, they drive more blood back to your heart increasing *venous return.* As more blood enters your heart, it stretches the myocardium. Due to the length-tension relationship of striated muscle explained in chapter 12, moderate stretch enables the myocytes to generate more tension when they begin to contract—that is, it increases the preload. If the ventricles contract more forcefully, they expel more blood, thus adjusting your cardiac output to the increase in venous return.

This theory is summarized by the **Frank–Starling law of the heart.**[32] In a concise, symbolic way, it states that $SV \propto EDV$. In other words, the ventricles tend to eject as much blood as they receive. Within limits, the more they are stretched, the harder they contract when stimulated.

While relaxed skeletal muscle is normally at an optimum length for the most forceful contraction, relaxed cardiac muscle is at less than optimum length. Any additional stretch therefore produces a significant increase in contraction force on the next beat. This helps balance the output of the two ventricles. For example, if the right ventricle begins to pump an increased amount of blood, this soon arrives at the left ventricle, stretches it more than before, and causes it to increase its stroke volume to match that of the right.

Contractility

The **contractility** of the myocardium refers to its contraction force *for a given preload.* It does not describe an increase in tension resulting from increased stretch but rather an increase caused by factors that make the myocytes more responsive to stimulation. Factors that increase contractility are called *positive inotropic*[33] *agents,* and those that reduce it are *negative inotropic agents.* (Do not confuse *inotropic* with *ionotropic,* which refers to certain neurotransmitter actions discussed in chapter 13; note the slight difference in spelling.)

Remember that Ca^{2+} is essential to the excitation-contraction coupling of muscle and prolongs the plateau of the myocardial action potential. Calcium therefore has a positive inotropic effect, as do agents that increase its availability to the myofilaments. Epinephrine and norepinephrine act through cAMP to open Ca^{2+} channels. By increasing the supply of Ca^{2+} to the myofilaments, they have a positive inotropic effect. Glucagon acts by stimulating the formation of cAMP; a solution of glucagon and calcium chloride is sometimes used for the emergency treatment of heart attacks. Digitalis, a cardiac stimulant from the foxglove plant, is used to treat congestive heart failure. It acts indirectly by inhibiting the Na^+-K^+ pumps of the myocardium, raising intracellular Na^+ concentration, and increasing the amount of Ca^{2+} in the sarcoplasm. Hypercalcemia causes more than the usual amount of Ca^{2+} to diffuse into the sarcoplasm and thus produces strong, prolonged contractions. In extreme cases, it can cause cardiac arrest in systole. Hypocalcemia can cause a weak, irregular heartbeat and potentially cause diastolic arrest. However,

31. *kal* = potassium (Latin, *kalium*)

32. Otto Frank (1865–1944), German physiologist; Ernest Henry Starling (1866–1927), English physiologist

33. *ino* = fiber

as explained in chapter 8, severe hypocalcemia is likely to kill through skeletal muscle paralysis and suffocation before the cardiac effects are felt.

The vagus nerves have a negative inotropic effect on the atria, but they provide so little innervation to the ventricular myocytes that they have little effect on the ventricles. Hyperkalemia has a negative inotropic effect because it reduces the strength of myocardial action potentials and thus reduces the release of Ca^{2+} into the sarcoplasm. The heart becomes dilated and flaccid. Hypokalemia, however, has little effect on contractility. There are other chronotropic and inotropic agents too numerous to mention here. The ones we have discussed are summarized in table 19.2.

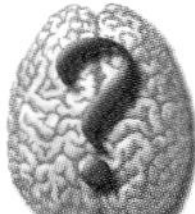

Think About It

Suppose a person has a heart rate of 70 bpm and a stroke volume of 70 mL. A negative inotropic agent then reduces the stroke volume to 50 mL. What would the new heart rate have to be to maintain the same cardiac output?

Afterload

The blood pressure in the arteries just outside the semilunar valves, called the **afterload,** opposes the opening of these valves. An increased afterload therefore reduces stroke volume. Anything that impedes arterial circulation can increase the afterload. For example, in some lung diseases, scar tissue forms in the lungs and restricts pulmonary circulation. This increases the afterload in the pulmonary trunk and opposes emptying of the right ventricle. As the ventricle works harder to overcome this resistance, it gets larger like any other muscle. Stress and hypertrophy of a ventricle can eventually cause it to weaken and fail. Right ventricular failure due to obstructed pulmonary circulation is called *cor pulmonale*[34] (CORE PUL-mo-NAY-lee). It is a common complication of emphysema, chronic bronchitis, and black lung disease (see chapter 22).

Table 19.2 Some Chronotropic and Inotropic Agents

Chronotropic Agents	
Positive	**Negative**
Sympathetic stimulation	Parasympathetic stimulation
Epinephrine and norepinephrine	Acetylcholine
Thyroid hormone	Hyperkalemia
Hypocalcemia	Hypokalemia
Hypercapnia and acidosis	Hypercalcemia
Digitalis	Hypoxia
Inotropic Agents	
Positive	**Negative**
Sympathetic stimulation	(Parasympathetic effect negligible)
Epinephrine and norepinephrine	Hyperkalemia
Hypercalcemia	Hypocalcemia
Digitalis	Myocardial hypoxia
Glucagon	Myocardial hypercapnia

Exercise and Cardiac Output

It is no secret that exercise makes the heart work harder, and it should come as no surprise that this increases cardiac output. The main reason the heart rate increases at the beginning of exercise is that proprioceptors in the muscles and joints transmit signals to the cardiac center signifying that the muscles are active and will quickly need an increased blood flow. As the exercise progresses, muscular activity increases venous return. This increases the preload on the right ventricle and is soon reflected in the left ventricle as more blood flows through the pulmonary circuit and reaches the left heart. As the heart rate and stroke volume rise, cardiac output rises, which compensates for the increased venous return.

A sustained program of exercise causes hypertrophy of the ventricles, which increases their stroke volume. As explained earlier, this allows the heart to beat more slowly and still maintain a normal resting cardiac output. Endurance athletes commonly have low heart rates of 40 to 60 bpm, but because of the higher stroke volume, their resting cardiac output is about the same as that of an untrained person. They have greater cardiac reserve, so they can tolerate more exertion than a sedentary person can.

Before You Go On

Answer the following questions to test your understanding of the preceding section:

17. Define *cardiac output* in words and with a simple formula.
18. Describe the cardiac center and innervation of the heart.
19. Explain what is meant by positive and negative chronotropic and inotropic agents. Give two examples of each.
20. How do preload, contractility, and afterload influence stroke volume and cardiac output?
21. Explain the principle behind the Frank–Starling law of the heart. How does this mechanism normally prevent pulmonary or systemic congestion?

34. *cor* = heart + *pulmo* = lung

Clinical Insight 19.5

Coronary Atherosclerosis

Atherosclerosis[35] is a disorder in which fatty deposits form in an artery, obstruct the lumen, and cause deterioration of the arterial wall. It is especially critical when it occurs in the coronary arteries and threatens to cut off the blood supply to the myocardium. Atherosclerosis is also a leading contributor to stroke and kidney failure.

Cause and Pathogenesis

According to one theory, the stage is set for atherosclerosis when the endothelium of a blood vessel is damaged by hypertension, viral infection, diabetes mellitus, or other causes. Monocytes adhere to the damaged endothelium, penetrate beneath it, and transform into macrophages. Macrophages and smooth muscle cells absorb cholesterol and neutral fats from the blood and acquire a frothy appearance; they are then called *foam cells* and are visible as a *fatty streak* on the vessel wall.

Platelets also adhere to areas of endothelial damage, degranulate, and release platelet-derived growth factor (PDGF); some PDGF also comes from macrophages and endothelial cells. PDGF stimulates mitosis of smooth muscle, leading eventually to a mass of lipid, smooth muscle, and macrophages called an *atheroma (atherosclerotic plaque).* The muscular and elastic tissue of the artery become increasingly replaced with scar tissue. When atheromas become calcified, they are called *complicated plaques.* Advanced atherosclerosis is marked by a state of arterial rigidity called *arteriosclerosis.*

Atherosclerosis is caused primarily by a combination of *low-density lipoproteins (LDLs)* in the blood plasma and defective LDL receptors in the arteries. LDLs are small protein-coated droplets of cholesterol, neutral fat, free fatty acids, and phospholipids (see chapter 26). Most cells have LDL receptors, take up these droplets from the blood by receptor-mediated endocytosis, and stop when they have enough cholesterol. In atherosclerosis, arterial cells have dysfunctional receptors that continue taking up plasma lipids and cause the cells to accumulate excess cholesterol.

As an atheroma grows, more and more of the arterial lumen becomes obstructed (fig. 1). Angina pectoris and other symptoms begin to occur when the lumen of a major coronary artery is reduced by at least 75%. When platelets adhere to lesions of the arterial wall, they release clotting factors, so an atheroma can become a focus for thrombosis. A clot can block what remains of the lumen, or it can break free and become an embolus that travels downstream until it lodges in a smaller artery. Part of an atheroma itself can also break loose and travel as a *fatty embolus.*

Atheromas also contribute to coronary artery spasms. Healthy endothelial cells secrete nitric oxide (NO), which causes the arteries to dilate. Vessels damaged by atherosclerosis release less NO, and the coronary arteries exhibit spasms. With much of the lumen already obstructed by the atheroma and perhaps a thrombus, an arterial spasm can temporarily shut off the remaining flow and precipitate an attack of angina.

Risk and Prevention

Risk factors are personal characteristics or elements of the environment that predispose an individual to a particular disease. Some risk factors for atherosclerosis cannot be avoided—for example, aging,

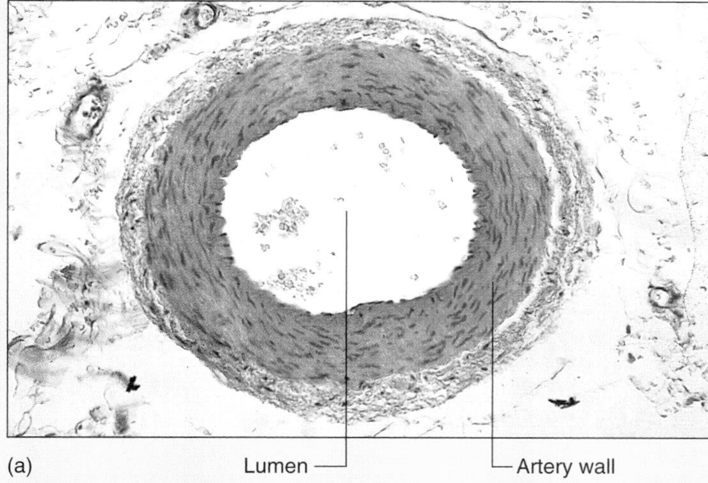

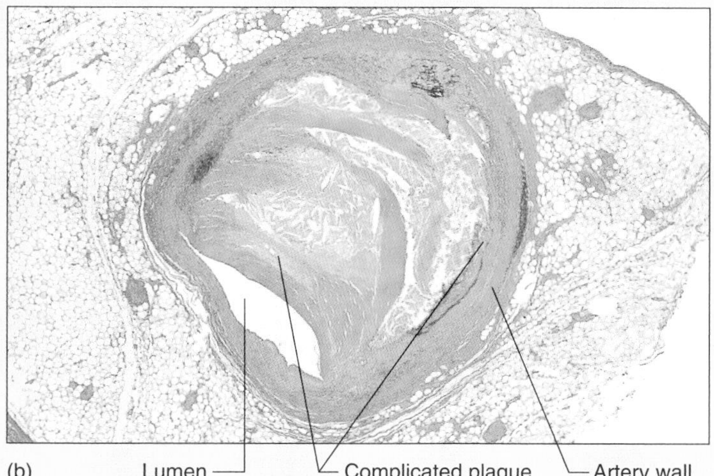

Figure 1 **Atherosclerosis.** (*a*) Cross section of a healthy artery; (*b*) cross section of an artery with advanced atherosclerosis. The lumen is reduced to a small space that can easily be blocked by thrombosis, embolism, or vasoconstriction. Most of the original lumen is obstructed by a complicated plaque composed of calcified scar tissue.

Continued

35. *athero* = fat, fatty + *sclerosis* = hardening

heredity, and being male. One form of hereditary atherosclerosis is *familial hypercholesterolemia* ("elevated blood cholesterol levels running in the family"). Most people have two recessive alleles (*hh*) of the gene for LDL receptors, which leads to the synthesis of normal receptors. One person in 500 is heterozygous *(Hh)* and makes only half the normal number of LDL receptors; one in a million is homozygous dominant *(HH)* and makes no LDL receptors. When the body's cells have few or no LDL receptors, they fail to absorb LDLs from the blood. LDL levels therefore remain high—six times normal in *HH* individuals. Foam cells, however, absorb LDLs even without these receptors, and with excess LDL in the blood plasma, atheromas grow rapidly. Heterozygous individuals usually suffer heart attacks by age 35, and homozygous dominant individuals usually have heart attacks in childhood, sometimes before age 2.

Most risk factors for atherosclerosis, however, are preventable. A sedentary lifestyle promotes LDL formation, whereas exercise promotes the formation of *high-density lipoproteins (HDLs),* which not only don't contribute to coronary disease but also help to lower blood cholesterol. Obesity is a risk factor that can be reduced by exercise. Aggressiveness, anxiety, and emotional stress promote hypertension and atherosclerosis. Smoking is another avoidable risk factor. The incidence of coronary heart disease is proportional to the number of cigarettes smoked per day and the number of years a person has been a smoker. This is reversible; people who quit smoking drop to normal risk levels within 5 years.

Diet, of course, is an overwhelmingly important factor. Eating animal fat reduces the number of LDL receptors and raises plasma LDL levels. Foods high in soluble fiber (such as beans, apples, and oat bran) lower blood cholesterol by an interesting mechanism. The liver normally converts cholesterol to bile acids, which it secretes into the small intestine to aid fat digestion. The bile acids are reabsorbed farther down the intestine and recycled to the liver for reuse. Soluble fiber, however, binds bile acids and carries them out in the feces. To replace them, the liver must synthesize more, thus using more cholesterol and lowering the blood cholesterol.

In the 1970s, scientists found that the Eskimos of Greenland had unusually low rates of coronary atherosclerosis despite the fact that their diet consisted entirely of meat—averaging a pound of whale meat and a pound of fish per day. Japanese and other groups with large amounts of fish in their diets also show low blood cholesterol levels. It is suspected that this is due to *omega-3 polyunsaturated fatty acids (PUFAs)* in fish oil. PUFAs increase the fluidity of plasma membranes and enable cells to remove more lipid from the blood. However, a daily capsule of fish oil does not hold much promise for controlling cholesterol. Doses of PUFAs high enough to reduce blood cholesterol would be prohibitively expensive and have undesirable side effects, including suppression of the immune system. Studies on the effectiveness of PUFAs remain inconclusive.

Treatment Options

The first pioneering approach to treating atherosclerosis, and still a common standby, is *coronary artery bypass surgery.* Sections of the great saphenous vein of the leg or small arteries from the thoracic cavity are used to construct a detour from the aorta to a point on a coronary artery beyond the obstruction.

Balloon angioplasty[36] is a technique in which a thin, flexible catheter is threaded into a coronary artery to the point of obstruction, and then a balloon at its tip is inflated to press the atheroma against the arterial wall, opening up the lumen. Its usefulness is limited to well-localized atheromas. In another method, *laser angioplasty,* an illuminated catheter enables the surgeon to see inside a diseased artery on a monitor and to use a laser to vaporize atheromas and reopen the artery. These methods are cheaper and less risky than bypass surgery. Unfortunately, they are often followed by *restenosis*—atheromas grow back and reobstruct the artery months later. There is also some concern that these procedures may even cause new injuries to the arterial walls, which may be foci for the development of new atheromas.

Clearly, prevention is the least expensive, least risky, and most effective approach to the threat of coronary artery disease.

36. *angio* = vessel + *plasty* = surgical repair

Chapter Review

Online Learning Center

Study resources for this chapter are available at www.mhhe.com/saladin/

Study Outline

Gross Anatomy of the Heart (p. 716)

1. Overview of the cardiovascular system
   a. Circulatory versus cardiovascular system
   b. Pulmonary and systemic circuits
2. Size, shape, and position of the heart
   a. Location in mediastinum
   b. Base and apex
   c. Dimensions
3. The pericardium
   a. Parietal pericardium
      - Fibrous layer
      - Serous layer
   b. Visceral pericardium
   c. Pericardial cavity
   d. Pericardial fluid
4. The heart wall
   a. Epicardium
   b. Myocardium
   c. Fibrous skeleton
   d. Endocardium
5. The chambers
   a. Right and left atria
   b. Right and left ventricles
   c. Atrioventricular sulcus
   d. Interventricular sulci
   e. Interatrial septum
   f. Interventricular septum
6. The valves
   a. Atrioventricular (AV) valves
      - Right AV (tricuspid) valve
      - Left AV (bicuspid, mitral) valve
      - Chordae tendineae
      - Papillary muscles
      - AV valve mechanics
   b. Semilunar valves
      - Pulmonary valve
      - Aortic valve
7. Blood flow through the heart
8. The coronary circulation
   a. Metabolic demands of heart
   b. Arterial supply
      - Left coronary artery
         •• Anterior interventricular artery
         •• Circumflex artery
      - Right coronary artery
         •• Marginal artery
         •• Posterior interventricular artery
      - Myocardial infarction
      - Arterial anastomoses
   c. Venous drainage
      - Direct drainage into right ventricle
      - Great cardiac vein
      - Middle cardiac vein
      - Coronary sinus
   d. Coronary flow in relation to the cardiac cycle
      - Reduced during ventricular contraction
      - Increased during ventricular relaxation

Cardiac Muscle and the Cardiac Conduction System (p. 726)

1. Structure of cardiac muscle
   a. Myocyte shape and size
   b. Myocyte organelles
   c. Intercalated discs
      - Interdigitating folds of membrane
      - Mechanical junctions
         •• Fascia adherens
         •• Desmosomes
      - Electrical (gap) junctions
   d. Repair of cardiac muscle
2. Metabolism of cardiac muscle
   a. Aerobic respiration
   b. Myoglobin and glycogen stores
   c. Large mitochondria
   d. Organic fuels
   e. Fatigue resistance
3. The cardiac conduction system
   a. The myogenic heart
   b. Autorhythmic myocytes
   c. Cardiac conduction system
      - Sinoatrial node (pacemaker)
      - Atrioventricular node
      - Atrioventricular bundle
      - Bundle branches
      - Purkinje fibers

Electrical and Contractile Activity of the Heart (p. 729)

1. The cardiac rhythm
   a. Systole and diastole
   b. Sinus rhythm
   c. Ectopic foci
   d. Nodal rhythm
   e. Arrhythmia and heart block
2. Physiology of the SA node
   a. Generation of the pacemaker potential
   b. Pacemaker potentials and heart rate
3. Impulse conduction to the myocardium
   a. Conduction through the atria
   b. Delay at the AV node
   c. Conduction through the AV bundle and Purkinje fibers
   d. Contraction of the papillary muscles
   e. Contraction of the ventricles
4. Electrical behavior of the myocardium
   a. Rising phase of the action potential
   b. Role of slow and fast calcium channels
   c. Plateau in action potential
   d. Sustained muscle tension
   e. Relaxation
   f. Long refractory period
5. The electrocardiogram (ECG)
   a. Recording method
   b. P wave
      - Atrial depolarization
      - Onset of atrial systole

c. QRS complex
  - Atrial repolarization and diastole
  - Ventricular depolarization
  - Onset of ventricular systole

d. T wave
  - Ventricular repolarization
  - Reason for shape of T wave

e. Diagnostic value of the ECG

Blood Flow, Heart Sounds, and the Cardiac Cycle (p. 734)

1. The cardiac cycle
   a. Atrial systole and diastole
   b. Ventricular systole and diastole
   c. Quiescent period
2. Principles of pressure and flow
   a. Measurement of blood pressure
   b. Pressure gradients and flow
   c. Opposing pressures
3. Heart sounds
   a. First heart sound
   b. Second heart sound
   c. Third heart sound
4. Phases of the cardiac cycle
   a. Quiescent period
   - All heart chambers relaxed
   - AV valves open
   - Blood flowing into ventricles

   b. Atrial systole
   - Firing of SA node
   - P wave appears in ECG
   - Atrial contraction
   - Pumping of blood into ventricles
   - End-diastolic volume

   c. Isovolumetric contraction of ventricles
   - Atria repolarize and relax
   - Ventricles depolarize
   - QRS complex appears in ECG
   - Ventricles contract
   - Rising pressure closes AV valves
   - Heart sound S1 occurs
   - No ejection of blood yet

   d. Ventricular ejection
   - Rising pressure opens semilunar valves
   - Rapid ejection of blood
   - Reduced ejection of blood
   - Stroke volume
   - End-systolic volume

   e. Isovolumetric relaxation of ventricles
   - T waves appears in ECG
   - Ventricles repolarize and relax
   - Semilunar valves close
   - AV valves remain closed
   - Ventricles expand but do not fill
   - Heart sound S2 occurs

   f. Ventricular filling
   - AV valves open
   - Ventricles fill with blood
     - •• Rapid ventricular filling
     - •• Diastasis
     - •• Filling completed by atrial systole
   - Heart sound S3 may occur

   g. Time course of cardiac cycle (average resting values)
   - Atrial systole, 0.1 second
   - Ventricular systole, 0.3 second
   - Quiescent period, 0.4 second
   - Total 0.8 second, heart rate 75 bpm

   h. Overview of volume changes

   i. Balanced ventricular output
   - Pulmonary hypertension if left ventricle fails
   - Systemic hypertension if right ventricle fails
5. Overview of volume changes

Cardiac Output (p. 739)

1. Cardiac output = heart rate × stroke volume
   a. Typically 4–6 L/min
   b. Increases in exercise
   c. Cardiac reserve
2. Heart rate
   a. Measured from pulse
   b. Heart rate and age
   c. Tachycardia and bradycardia
   d. Chronotropic effects
   - Sympathetic nervous system
     - •• Cardioacceleratory center
     - •• Cardiac accelerator nerves
     - •• Heart rate and cardiac output
   - Parasympathetic nervous system
     - •• Cardioinhibitory center
     - •• Vagus nerves and vagal tone
     - •• Heart rate and cardiac output
   - Inputs to cardiac center
     - •• Higher brain centers
     - •• Proprioceptors
     - •• Baroreceptors
     - •• Chemoreceptors
   - Chronotropic chemicals
     - •• Neurotransmitters
     - •• Drugs
     - •• Hormones
     - •• Electrolytes
3. Stroke volume
   a. Preload
   - Presystolic ventricular muscle tension
   - Venous return and myocardial stretching
   - Frank–Starling law of the heart

   b. Contractility
   - Contraction force for a given preload
   - Inotropic agents
     - •• Neurotransmitters
     - •• Drugs
     - •• Hormones
     - •• Electrolytes

   c. Afterload
   - Arterial pressure opposing ventricular ejection
4. Exercise and cardiac output
   a. Effect of proprioceptors
   b Effect of venous return
   c. Effect of ventricular hypertrophy

Selected Vocabulary

cardiology 716
circulatory system 716
cardiovascular system 716
heart 716
artery 716
vein 716
capillary 716
pulmonary circuit 716
systemic circuit 716
base of heart 716
apex of heart 716
pericardium 716
parietal pericardium 716
visceral pericardium 716
pericardial cavity 717
pericardial fluid 717
epicardium 718
myocardium 718
fibrous skeleton 718
endocardium 718
chambers of heart 718
right atrium 719
left atrium 719
right ventricle 719
left ventricle 719
atrioventricular sulcus 719
anterior interventricular sulcus 719
posterior interventricular sulcus 719
interatrial septum 721
pectinate muscle 721
interventricular septum 721
trabeculae carneae 721
valve cusps 721

right atrioventricular valve 721
left atrioventricular valve 721
mitral valve 721
chordae tendineae 721
papillary muscle 721
semilunar valve 721
pulmonary valve 721
aortic valve 721
coronary circulation 723
left coronary artery 723
anterior interventricular artery 723
circumflex artery 723
right coronary artery 723
marginal artery 723
posterior interventricular artery 723
myocardial infarction 725
venous drainage 725
great cardiac vein 725
middle cardiac vein 726
coronary sinus 726
intercalated disc 726
autorhythmic 728
cardiac conduction system 728
sinoatrial node 728
pacemaker 728
atrioventricular node 728
atrioventricular bundle 728
bundle branch 728
Purkinje fiber 728
systole 729
diastole 729
sinus rhythm 729
ectopic focus 729
nodal rhythm 729
arrhythmia 729
heart block 729
pacemaker potential 729
fast calcium channel 730
slow calcium channel 731
electrocardiogram 731
P wave 732
QRS complex 732
T wave 732
cardiac cycle 733
sphygmomanometer 734
pressure gradient 734
auscultation 735
first heart sound 735
second heart sound 735
third heart sound 735
quiescent period 735
atrial systole 736
end-diastolic volume 736
isovolumetric contraction 736
ventricular ejection 737
stroke volume 737
ejection fraction 737
end-systolic volume 737
isovolumetric relaxation 737
ventricular filling 737
cardiac output 739
cardiac reserve 739
pulse 739
tachycardia 739
bradycardia 739
cardiac center 740
cardioacceleratory center 740
cardiac accelerator nerve 740
cardioinhibitory center 740
vagal tone 740
proprioceptor 740
baroreceptor 740
chemoreceptor 740
chemoreflex 740
baroreflex 740
preload 740
Frank–Starling law of the heart 741
contractility 741
afterload 742

Testing Your Recall

1. The cardiac conduction system includes all of the following *except*
   a. the SA node.
   b. the AV node.
   c. the bundle branches.
   d. the chordae tendineae.
   e. the Purkinje fibers.

2. To get from the right atrium to the right ventricle, blood flows through
   a. the pulmonary valve.
   b. the tricuspid valve.
   c. the bicuspid valve.
   d. the aortic valve.
   e. the mitral valve.

3. Assume that one ventricle of a child's heart has an EDV of 90 mL, an ESV of 60 mL, and a cardiac output of 2.55 L/min. What are the child's stroke volume (SV), ejection fraction (EF), and heart rate (HR)?
   a. SV = 60 mL; EF = 33%; HR = 85 bpm
   b. SV = 30 mL; EF = 60%; HR = 75 bpm
   c. SV = 150 mL; EF = 67%; HR = 42 bpm
   d. SV = 30 mL; EF = 33%; HR = 85 bpm
   e. There is not enough information to calculate these.

4. A heart rate of 45 bpm and an absence of P waves suggest
   a. damage to the SA node.
   b. ventricular fibrillation.
   c. cor pulmonale.
   d. extrasystole.
   e. heart block.

5. The fast-rising phase of the action potential of the SA node results from
   a. the opening of slow Ca^{2+} channels.
   b. the closing of K^+ channels.
   c. K^+ outflow.
   d. K^+ inflow.
   e. Ca^{2+} inflow.

6. Cardiac muscle does not exhibit tetanus because it has
   a. fast Ca^{2+} channels.
   b. scanty sarcoplasmic reticulum.
   c. a long absolute refractory period.
   d. electrical synapses.
   e. exclusively aerobic respiration.

7. The atria contract during
   a. the first heart sound.
   b. the second heart sound.
   c. the QRS complex.
   d. the P–Q segment.
   e. the S–T segment.

8. Ventricular pressure peaks during
   a. the first heart sound.
   b. the second heart sound.
   c. the QRS complex.
   d. the P–Q segment.
   e. the S–T segment.

9. The blood contained in a ventricle during isovolumetric relaxation is
   a. the end-systolic volume.
   b. the end-diastolic volume.
   c. the stroke volume.
   d. the ejection fraction.
   e. none of these; the ventricle is empty then.

10. Drugs that increase the heart rate have a ________ effect.
    a. myogenic
    b. negative inotropic
    c. positive inotropic
    d. negative chronotropic
    e. positive chronotropic

11. The contraction of any heart chamber is called ________ and its relaxation is called ________ .

12. The circulatory route from aorta to the venae cavae is the ________ circuit.

13. The circumflex artery travels in a groove called the ________ .

14. The pacemaker potential of the SA node cells results from the slow inflow of ________ .

15. Electrical signals pass quickly from one cardiac myocyte to another through the ________ of the intercalated discs.

16. Repolarization of the ventricles produces the __________ of the electrocardiogram.
17. Closing of the __________ valves produces turbulence in the bloodstream, which contributes to heart sound S2.
18. The procedure for listening to the heart sounds is called cardiac __________ .
19. The end-diastolic volume of blood stretches the ventricles and creates myocardial tension called the __________ .
20. The Frank–Starling law of the heart explains why the __________ of the left ventricle is the same as that of the right ventricle.

Answers in appendix B

Testing Your Comprehension

1. Verapimil is a calcium channel blocker used to treat hypertension. It selectively blocks slow calcium channels. Would you expect it to have a positive or negative inotropic effect? Explain. (See p. 112 to review calcium channel blockers.)
2. To temporarily treat tachycardia and restore the normal resting sinus rhythm, a physician may massage a patient's carotid artery near the angle of the mandible. Propose a mechanism by which this treatment would have the desired effect.
3. Suppose that a patient experienced bleeding into the pericardial cavity because of a ruptured aneurysm. The amount of blood lost was not life-threatening, but about 200 mL of blood accumulated and clotted in the pericardial cavity before the bleeding stopped. Predict how this could affect the heart's end-diastolic volume and stroke volume, and explain your reasoning.
4. In ventricular systole, the left ventricle is the first to begin contracting but the right ventricle is the first to expel blood. Aside from the obvious fact that the pulmonary valve opens before the aortic valve, how can you explain this difference?
5. The action potential of a cardiac myocyte looks very different from the action potential of a neuron. Sketch the two and explain the ionic basis for the differences.

Answers at the Online Learning Center

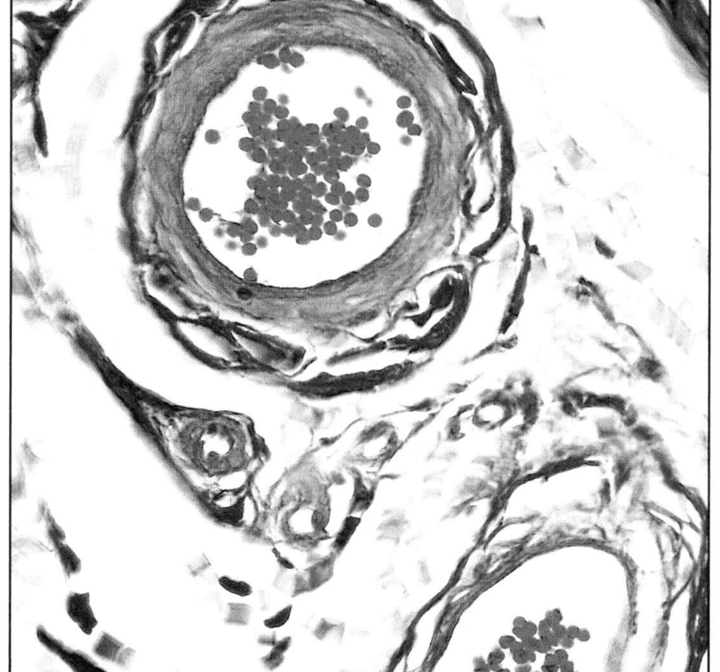

Blood vessels. An arteriole (top), capillaries (middle), and venule (bottom)

CHAPTER 20

The Circulatory System: Blood Vessels and Circulation

Chapter Outline

Insights

Study with Your ESP CD-ROM

Refer to your ESP Tutorial CD-ROM for helpful animations, activities, and quizzes related to the material covered in this chapter. The following are suggested areas to go to on the CD and are presented according to "Unit/Topic":

- Transport/Blood Vessels

Brushing Up

To understand this chapter, it is important that you understand or brush up on the following concepts:

- Set point and dynamic equilibrium in homeostasis (p. 19)
- Diffusion (p. 125)
- Equilibrium between filtration and osmosis (p. 126)
- Transcytosis (p. 132)
- Viscosity and osmolarity of blood (pp. 681, 683)
- Principles of pressure and flow (pp. 734–35)
- Autonomic effects on the heart (p. 740)

The route taken by the blood after it leaves the heart was a point of much confusion until the seventeenth century. Chinese emperor Huang Ti (2697–2597 B.C.E.) correctly believed that it flowed in a complete circuit around the body and back to the heart. But in the second century, Roman physician Claudius Galen (129–c. 199) argued that it flowed back and forth in the veins, like air in the bronchial tubes. He believed that the liver received food from the small intestine and converted it to blood, the heart pumped the blood through the veins to all other organs, and those organs consumed the blood.

Huang Ti was right, but the first experimental demonstration of this did not come until the seventeenth century. English physician William Harvey (1578–1657) studied the filling and emptying of the heart in snakes, tied off the vessels above and below the heart to observe the effects on cardiac filling and output, and measured cardiac output in a variety of living animals. He concluded that (1) the heart pumps more blood in half an hour than there is in the entire body, (2) not enough food is consumed to account for the continual production of so much blood, and (3) since the planets orbit the sun and (as he believed) the human body is modeled after the solar system, it follows that the blood orbits the body. So for a peculiar combination of experimental and superstitious reasons, Harvey argued that the blood returns to the heart rather than being consumed by the peripheral organs. He could not explain how, since the microscope had yet to be developed and he did not know of capillaries—later discovered by Antony van Leeuwenhoek and Marcello Malpighi.

Harvey published his findings in 1628 in a short but elegant book entitled *Exercitio Anatomica de Motu Cordis et Sanguinis in Animalibus (Anatomical Studies on the Motion of the Heart and Blood in Animals).* This landmark in the history of biology and medicine was the first experimental study of animal physiology. But so entrenched were the ideas of Aristotle and Galen in the medical community, and so strange was the idea of doing experiments on living animals, that Harvey's contemporaries rejected his ideas. Indeed, some of them regarded him as a crackpot because his conclusion flew in the face of common sense—if the blood was continually recirculated and not consumed by the tissues, they reasoned, then what purpose could it possibly serve?

Harvey lived to a ripe old age, served as physician to the kings of England, and later did important work in embryology. His case is one of the most interesting in biomedical history, for it shows how empirical science overthrows old theories and spawns better ones, and how common sense and blind allegiance to authority can interfere with acceptance of the truth. But most importantly, Harvey's contributions represent the birth of experimental physiology—the method that generated most of the information in this book.

General Anatomy of the Blood Vessels

Objectives

When you have completed this section, you should be able to

- trace the route usually taken by the blood from the heart and back again;
- describe some variations on this route;
- describe the structure of a blood vessel; and
- describe the different types of arteries, capillaries, and veins.

Circulatory Routes

At its simplest, the usual route of blood flow is heart → arteries → arterioles → capillaries → venules → veins → heart. Blood usually passes through only one network of capillaries from the time it leaves the heart until the time it returns (fig. 20.1*a*). But there are exceptions, notably portal systems and anastomoses. In a **portal system** (fig. 20.1*b*), blood flows through two consecutive capillary networks before returning to the heart. One portal system connects the hypothalamus and anterior pituitary (see chapter 17). Others are found in the kidneys and between the intestines and liver; the latter system is detailed in table 20.13.

An **anastomosis** is a point where two veins or arteries merge with each other. In an **arteriovenous anastomosis (shunt),** blood flows from an artery directly into a vein and bypasses the capillaries (fig. 20.1*c*). Shunts occur in the fingers, palms, toes, and ears, where they reduce heat loss in cold weather by allowing warm blood to bypass these exposed surfaces. Unfortunately, this makes these poorly perfused areas more susceptible to frostbite. In an **arterial**

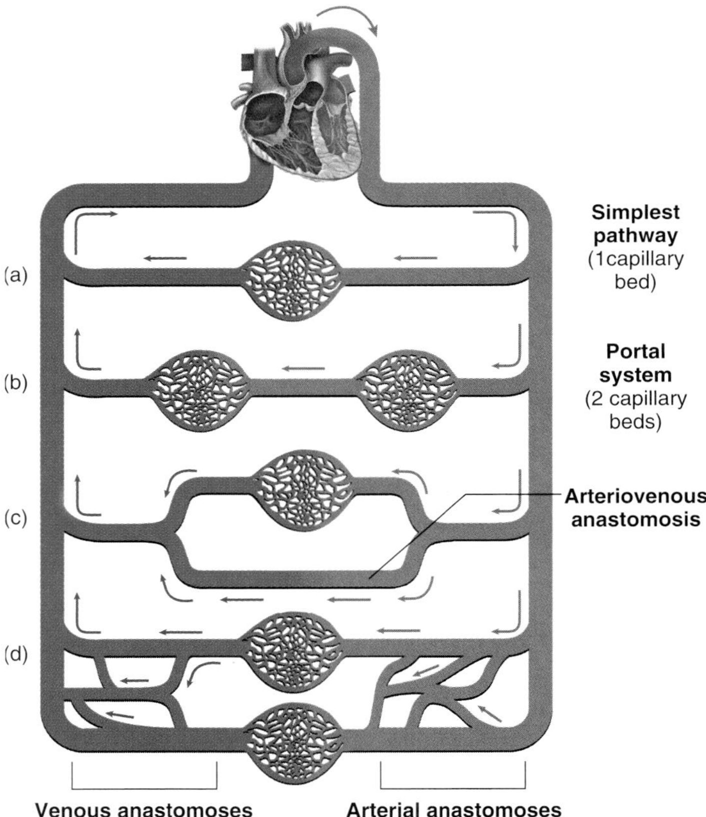

Figure 20.1 Variations in Circulatory Pathways.

anastomosis, two arteries merge and provide *collateral* (alternative) routes of blood supply to a tissue (fig. 20.1*d*). Those of the coronary circulation were mentioned in the previous chapter. They are also common around joints where movement may temporarily obstruct one pathway. **Venous anastomoses** are more common. They provide several alternative routes of drainage from an organ, so blockage of a vein is rarely as life-threatening as blockage of an artery. Several arterial and venous anastomoses are described later in this chapter.

The Vessel Wall

The walls of the arteries and veins have three layers called *tunics* (figs. 20.2 and 20.3):

1. The **tunica externa (tunica adventitia**[1]**)** is the outermost layer. It consists of loose connective tissue that often merges with that of neighboring

1. *advent* = added to

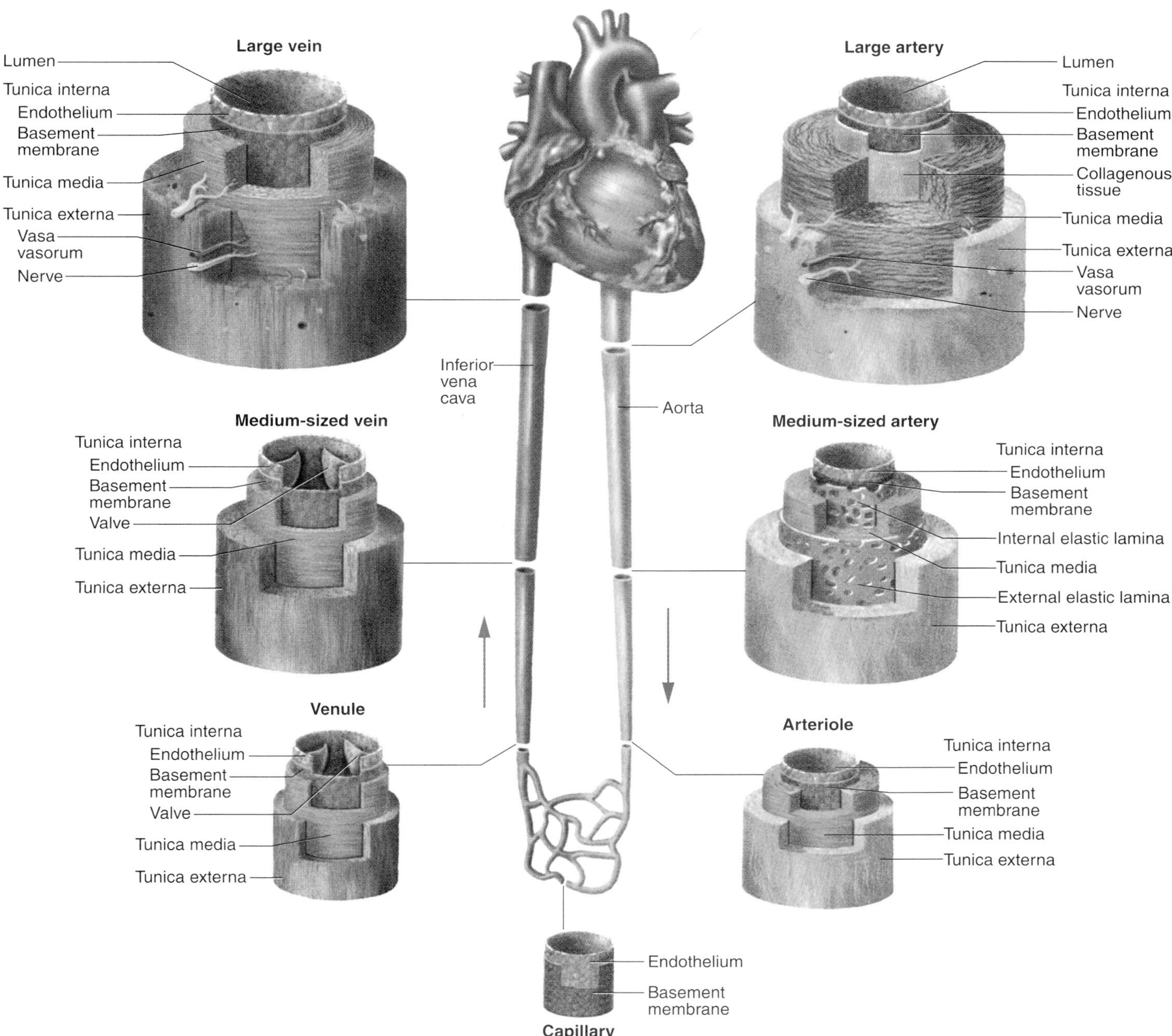

Figure 20.2 The Structure of Arteries and Veins.

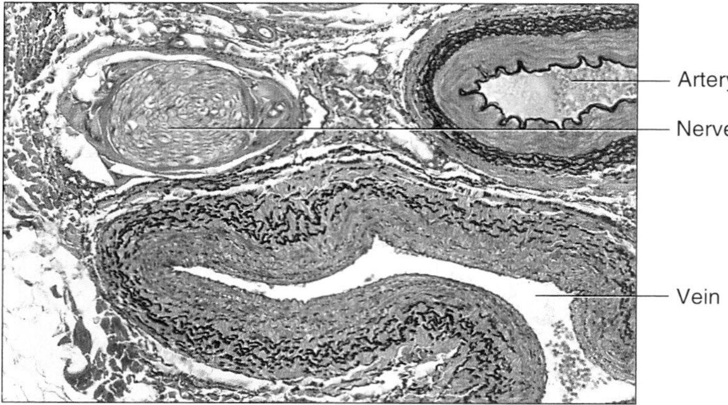

Figure 20.3 A Neurovascular Bundle. A small artery, small vein, and nerve traveling together in a common sheath of connective tissue.

blood vessels, nerves, or other organs. It anchors the vessel and provides passage for small nerves, lymphatic vessels, and smaller blood vessels. Small vessels called the **vasa vasorum**[2] (VAY-za vay-SO-rum) supply blood to at least the outer half of the wall of a larger vessel. Tissues of the inner half of the wall are thought to be nourished by diffusion from blood in the lumen.

2. The **tunica media,** the middle layer, is usually the thickest. It consists of smooth muscle, collagen, and sometimes elastic tissue. The smooth muscle is responsible for the vasoconstriction and vasodilation of blood vessels.
3. The **tunica interna (tunica intima),** the inner layer, is exposed to the blood. It consists of a simple squamous **endothelium** overlying a basement membrane and a sparse layer of fibrous tissue. The endothelium provides a smooth inner lining that normally repels blood cells and platelets. Platelets may adhere to a damaged endothelium, however, and during inflammation, leukocytes adhere loosely to the endothelium by means of *cell-adhesion molecules* produced by the endothelial cells (see next chapter). The endothelium also acts as a selectively permeable barrier to blood solutes, and it secretes vasoconstrictors and vasodilators to be considered later.

Arteries and Metarterioles

Arteries are constructed to withstand the surges of blood pressure generated by ventricular systole. They are more muscular than veins and appear relatively round in tissue sections. They are divided into three categories by size, but of course there is a smooth gradation from one category to the next:

1. **Conducting (elastic) arteries** are the largest. Some examples are the pulmonary arteries, aorta, and common carotid arteries. Their tunica media consists of numerous sheets of elastic tissue, perforated like slices of Swiss cheese, alternating with thin layers of smooth muscle, collagen, and elastic fibers. Conducting arteries expand when the ventricles pump blood into them during systole, and recoil during diastole. This lessens the fluctuations in blood pressure exerted on smaller arteries downstream.
2. **Distributing (muscular) arteries** are smaller branches farther away from the heart that distribute blood to specific organs. You could compare a conducting artery to an interstate highway and distributing arteries to the exit ramps and state highways that serve specific towns. These arteries typically have 25 to 40 layers of smooth muscle cells constituting about three-quarters of the wall thickness. Most arteries to which we give names are in these first two size classes. The brachial, femoral, and splenic arteries are examples of distributing arteries.
3. **Resistance (small) arteries** are usually too variable in number and location to be given names. They exhibit up to 25 layers of smooth muscle cells and relatively little elastic tissue. Their tunica media is thicker in proportion to the lumen than that of larger arteries. The smallest of these arteries, about 40 to 200 μm in diameter and with only one to three layers of smooth muscle, are the **arterioles.** For reasons discussed later, they are the primary points at which the body controls the relative amounts of blood directed to various organs.

Metarterioles[3] are short vessels that link arterioles and capillaries. Instead of a continuous tunica media, they have individual muscle cells spaced a short distance apart, each forming a **precapillary sphincter** that encircles the entrance to a capillary.

Capillaries

Capillaries (fig. 20.4) are the "business end" of the circulatory system. All the rest of the system exists to serve them, because capillaries are almost the only point in the circulatory system where materials are exchanged between the blood and tissue fluid. Capillaries are ideally suited to their role. They consist of endothelium only and

2. *vasa* = vessels + *vasorum* = of the vessels

3. *meta* = beyond, next in a series

have walls as thin as 0.2 to 0.4 μm. They average about 5 μm in diameter at the proximal (arterial) end, widen to about 9 μm in diameter at the distal (venous) end, and often branch along the way. (Recall that an erythrocyte is about 7 μm in diameter.) The number of capillaries has been estimated at a billion and their total surface area at 6,300 m^2. But a more important point is that scarcely any cell in the body is more than 60 to 80 μm away from the nearest capillary. There are a few exceptions. Capillaries are scarce in tendons and ligaments and absent from cartilage, epithelia, and the cornea and lens of the eye.

Capillary Beds

Capillaries are organized in groups called **capillary beds**—usually 10 to 100 capillaries supplied by a single metarteriole (fig. 20.5). The metarteriole continues through the bed as a **thoroughfare channel** leading directly to a venule. Capillaries arise from the proximal end of the metarteriole and lead into its distal end or directly into the venule.

There is a precapillary sphincter at the entrance to each capillary. When the sphincters are open, the capillaries are well perfused with blood and they engage in exchanges with the tissue fluid. When the sphincters are closed, blood bypasses the capillaries, flows through the thoroughfare channel to a venule, and does not engage in significant fluid exchange. There is not enough blood in the body to fill the entire vascular system at once;

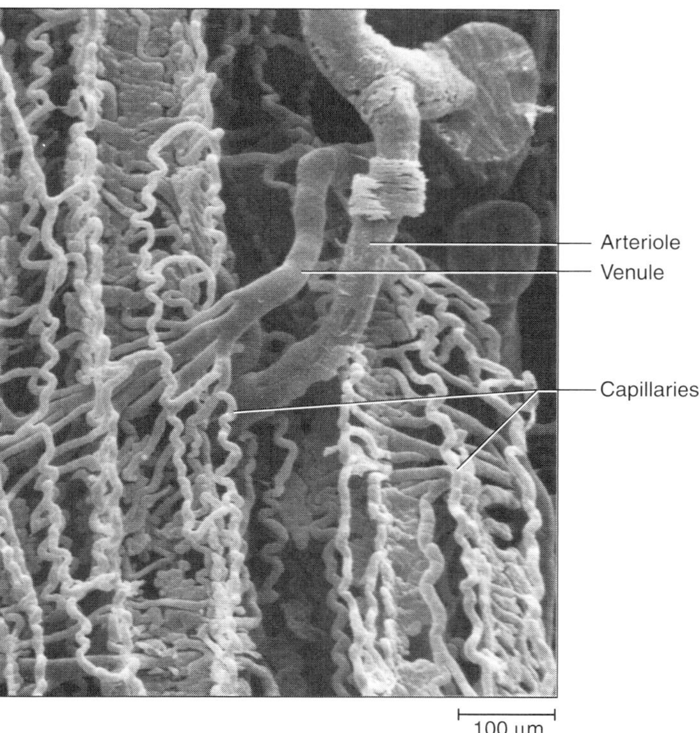

Figure 20.4 Vascular Cast of Blood Vessels in Human Skeletal Muscle. This was prepared by injecting the vessels with a polymer, digesting away all tissue to leave a replica of the vessels, and photographing the cast through the SEM. From R. G. Kessel and R. H. Kardon, *Tissues and Organs: A Text-Atlas of Scanning Electron Microscopy* (W. H. Freeman & Co., 1979).

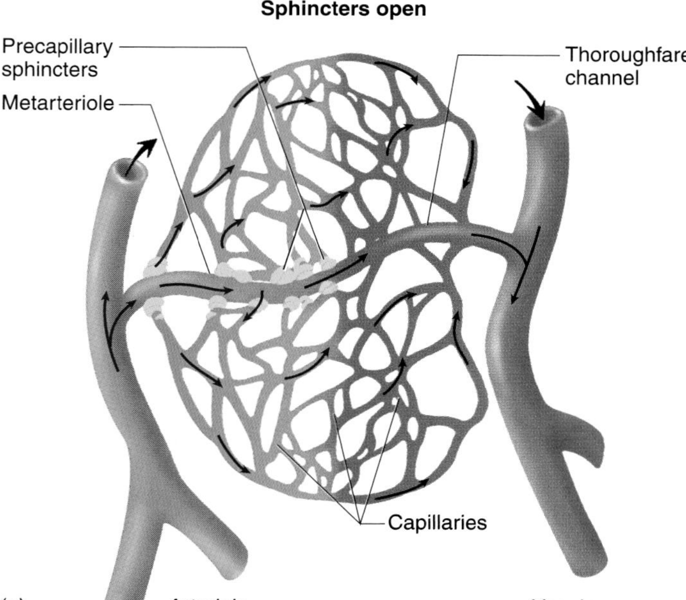

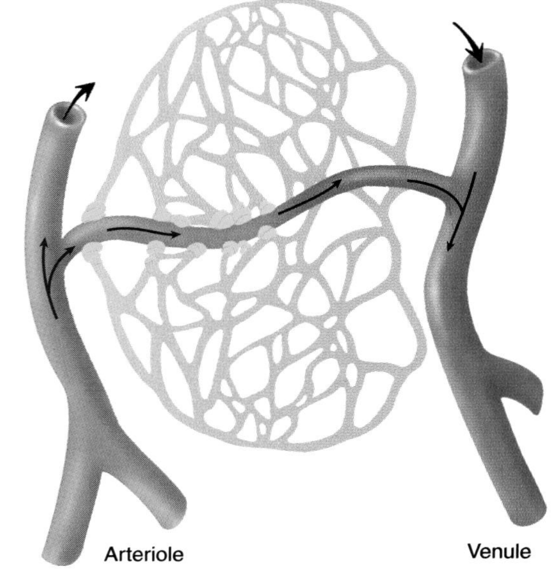

Figure 20.5 Control of Perfusion of a Capillary Bed. (*a*) Precapillary sphincters dilated and capillaries well perfused. (*b*) Precapillary sphincters closed, with blood bypassing the capillaries.

consequently, about three-quarters of the body's capillaries are closed at any given time. The shifting of blood flow from one capillary bed to another is discussed later in the chapter.

Types of Capillaries

Two types of capillaries are distinguished by the ease with which they allow substances to pass through their walls and by structural differences that account for their greater or lesser permeability:

1. **Continuous capillaries** occur in most tissues, such as skeletal muscle. Their endothelial cells, held together by tight junctions, form an uninterrupted tube. The cells usually have narrow **intercellular clefts** about 4 nm wide between them. Small solutes, such as glucose, can pass through these clefts, but plasma proteins, other large molecules, and formed elements are held back. The continuous capillaries of the brain lack intercellular clefts and have more complete tight junctions that form the blood-brain barrier discussed in chapter 14.
2. **Fenestrated capillaries** have endothelial cells that are riddled with holes called **fenestrations**[4] **(filtration pores)** (fig. 20.6). Fenestrations are about 20 to 100 nm in diameter and are usually covered by a thin mucoprotein diaphragm. They allow for the rapid passage of small molecules but still retain proteins and larger particles in the bloodstream. Fenestrated capillaries are important in organs that engage in rapid absorption or filtration—the kidneys, endocrine glands, small intestine, and choroid plexuses of the brain, for example.

Sinusoids are irregular blood-filled spaces in the liver, bone marrow, spleen, and some other organs. They are twisted, tortuous passageways that conform to the shape of the surrounding tissue. Some of them are continuous capillaries with very thin walls; others are fenestrated capillaries with extraordinarily large pores that allow the blood plasma to come into direct contact with the perivascular cells. Even proteins and blood cells can pass through these pores; this is how albumin, clotting factors, and other proteins synthesized by the liver enter the blood and how newly formed blood cells enter the circulation from the bone marrow and lymphatic organs.

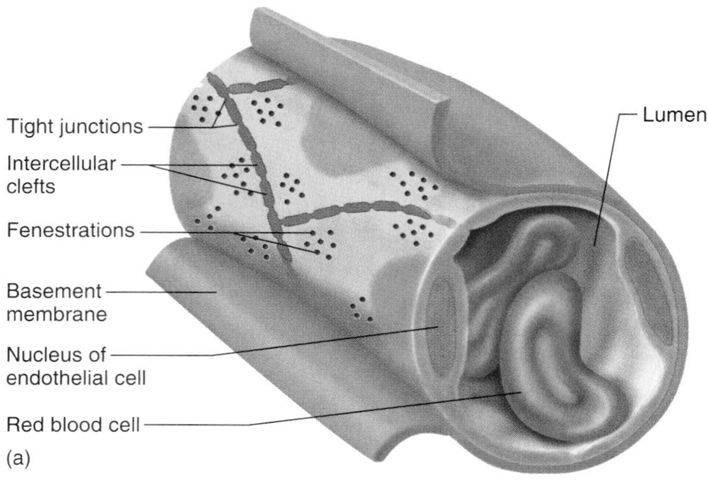

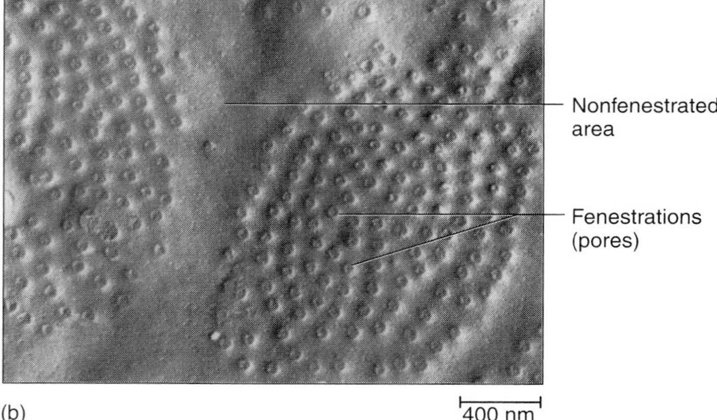

Figure 20.6 A Fenestrated Capillary. (*a*) Structure of the capillary wall. (*b*) Surface view of a fenestrated endothelial cell. The cell has patches of fenestrations separated by nonfenestrated areas.

Veins

After flowing through the capillaries, blood collects in the distal end of the thoroughfare channel and flows into a venule. In the venous circulation, blood flows from smaller vessels into progressively larger ones; hence, instead of giving off *branches* as arteries do, veins receive smaller *tributaries*, just as a river receives water from the many streams that form its tributaries.

Venules range from about 15 to 100 μm in diameter. The proximal part of a venule has only a few fibroblasts around it and is quite porous; therefore venules, like capillaries, can exchange fluid with the surrounding tissues. Farther along, a venule acquires a tunica media of smooth muscle. Even the largest veins, however, have relatively sparse muscular and elastic tissue compared to arteries.

Venous sinuses are veins with especially thin walls, large lumens, and no smooth muscle. Examples include the coronary sinus of the heart and the dural sinuses of the brain.

4. *fenestra* = window

Clinical Insight 20.1

Varicose Veins

In people who stand for long periods, such as dentists and hairdressers, blood tends to pool in the lower limbs and stretch the veins. This is especially true of superficial veins, which are not surrounded by supportive tissue. Stretching pulls the cusps of the venous valves farther apart until the valves become incompetent to prevent the backflow of blood. As the veins become further distended, their walls grow weak and they develop into *varicose veins* with irregular dilations and twisted pathways. Obesity and pregnancy also promote development of varicose veins by putting pressure on large veins of the pelvic region and obstructing drainage from the legs. Varicose veins sometimes develop because of hereditary weakness of the valves. With less drainage of blood, tissues of the leg and foot may become edematous and painful. *Hemorrhoids* are varicose veins of the anal canal.

Because they are farther away from the heart, veins have much lower blood pressure than arteries. In large arteries, it averages 90 to 100 mmHg and surges to 120 mmHg during systole, whereas in veins it averages about 10 mmHg and fluctuates very little with the heartbeat. This has significant implications for the form and function of veins:

- Since they need not withstand high pressure, veins have thinner walls than arteries, with less muscular and elastic tissue. They collapse when empty and look relatively flattened or irregular in histological sections (see fig. 20.3).
- Since their walls are so thin, veins can expand more easily and accommodate more blood than arteries do. About 54% of the blood is found in the systemic veins at rest (fig. 20.7); veins are therefore called *capacitance vessels.*
- The pressure in the veins is not high enough to push blood upward against the pull of gravity to the heart. The upward flow of blood depends in part on the massaging action of skeletal muscles and the presence of one-way **venous valves** that keep the blood from dropping down again when the muscles relax (see fig. 20.2). These valves, similar to the semilunar valves of the heart, occur especially in medium veins of the arms and legs; they are absent from very small and very large veins, veins of the ventral body cavity, and veins of the brain. Varicose veins result in part from the failure of these valves (see clinical insight 20.1).

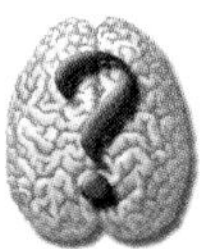

Think About It

From figure 20.7 and your other knowledge, calculate the approximate volume of blood in the *systemic* capillaries. What percentage of the systemic blood is this?

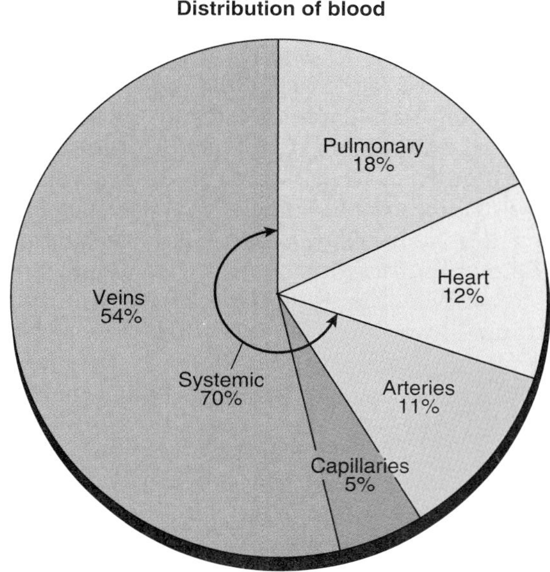

Figure 20.7 Average Distribution of the Blood in a Resting Adult.

Before You Go On

Answer the following questions to test your understanding of the preceding section:

1. Explain how an anastomosis and a portal system differ from the simple artery → capillary → vein scheme of circulation.
2. Name the three tunics of a typical blood vessel and explain how they differ from each other.
3. Describe the route of blood flow through a capillary bed.
4. Contrast the two types of capillaries.
5. Explain why many veins have valves but arteries do not.

Blood Pressure, Resistance, and Flow

Objectives

When you have completed this section, you should be able to

- explain the relationship between blood pressure, resistance, and flow;
- describe how blood pressure is expressed and how pulse pressure and mean arterial pressure are calculated;
- describe three factors that determine resistance to blood flow;
- explain how vasomotion influences blood flow; and
- describe some local, neural, and hormonal influences on vasomotion.

Blood flow is the amount of blood flowing through an organ, tissue, or blood vessel in a given time (such as mL/min). **Perfusion** is the rate of blood flow per given volume or mass of tissue (such as mL/min/g). The principal importance of flow is that it governs the speed of oxygen and nutrient delivery to a tissue and the speed of waste removal. If flow does not keep pace with the metabolic rate of a tissue, the likely result is tissue necrosis and possibly death of the individual. In a resting individual, *total* flow is quite constant and is equal to cardiac output (typically 5.25 L/min). Flow through individual organs, however, varies from minute to minute as blood is redirected from one organ to another. Great variations in regional flow can occur with little or no change in total flow.

Hemodynamics, the physical principles of blood flow, are based mainly on pressure and resistance. The greater the pressure difference (ΔP) between two points, the greater the flow; the greater the resistance (R), the less the flow—in summary, $F \propto \Delta P/R$. Therefore, to understand the flow of blood, we must consider the factors that affect pressure and resistance.

Blood Pressure

Blood pressure (BP) can be measured within a blood vessel or heart chamber by inserting a catheter or needle connected to an external manometer (pressure-measuring device). For routine clinical purposes, however, the measurement of greatest interest is the systemic arterial BP at a point close to the heart. As mentioned in the previous chapter, we customarily measure it with a sphygmomanometer at the brachial artery of the arm. It is easy to encircle the arm with a pressure cuff, and this artery is sufficiently close to the heart to reflect the maximum arterial BP found anywhere in the systemic circuit.

Two pressures are recorded: **systolic pressure** is the peak arterial BP attained during ventricular systole, and **diastolic pressure** is the minimum arterial BP between heartbeats. For a healthy person aged 20 to 30, these pressures are typically about 120 and 75 mmHg, respectively. Arterial BP is written as a ratio of systolic over diastolic pressure: 120/75.

The difference between systolic and diastolic pressure is called **pulse pressure** (not to be confused with pulse *rate*). For the preceding BP, pulse pressure would be $120 - 75 = 45$ mmHg. This is an important measure of the stress exerted on small arteries by the pressure surges generated by the heart. Another measure of stress on the blood vessels is the **mean arterial pressure (MAP)**—the mean pressure you would obtain if you took measurements at several intervals (say every 0.1 sec) throughout the cardiac cycle. Since diastole lasts longer than systole, MAP is not simply the average of systolic and diastolic pressures. The best estimate of MAP is the sum of diastolic pressure and one-third of the pulse pressure. For a blood pressure of 120/75, MAP $\approx 75 + 45/3 = 90$ mmHg, a typical value for vessels at the level of the heart. MAP varies, however, with the influence of gravity. In a standing adult, it is about 62 mmHg in the major arteries of the head and 180 mmHg in major arteries of the ankle.

Hypertension (high BP) is commonly considered to be a chronic resting blood pressure higher than 140/90 (see clinical insight 20.4, p. 796). (*Transient* high BP resulting from emotion or exercise is not hypertension.) Among other effects, it can weaken the small arteries and cause **aneurysms**[5] (AN-you-rizms) (see clinical insight 20.2). **Hypotension** is chronic low resting BP. It may be a consequence of blood loss, dehydration, anemia, or other factors and is normal in people approaching the moment of death.

The ability of the arteries to distend and recoil during the cardiac cycle is important in modulating arterial BP. If the arteries were rigid tubes, pressure would rise much higher in systole and drop to nearly zero in diastole. Blood throughout the circulatory system would flow and stop, flow and stop, thus putting great stress on the small vessels. But when the conducting arteries are healthy, they expand with each systole and absorb some of the force of the ejected blood. Then, when the heart is in diastole, their elastic recoil exerts pressure on the blood and prevents the BP from dropping to zero. The combination of expansion and recoil in the arteries maintains a steady flow of blood downstream, in the capillaries, throughout the cardiac cycle. Thus, the elastic arteries "smooth out" the pressure fluctuations and reduce stress on the smaller arteries.

Nevertheless, blood flow in the arteries is *pulsatile.* Blood in the aorta rushes forward at 120 cm/sec during systole and an average speed of 40 cm/sec over the cardiac cycle. When measured at points farther away from the heart, systolic and diastolic pressures are lower and there

5. *aneurysm* = widening

Clinical Insight 20.2

Aneurysm

An aneurysm is a weak point in a blood vessel or in the heart wall. It forms a thin-walled, bulging sac that pulsates with each beat of the heart and may eventually rupture. In a *dissecting aneurysm,* blood pools between the tunics of a vessel and separates them, usually because of degeneration of the tunica media. The most common sites of aneurysms are the abdominal aorta, the renal arteries, and the arterial circle at the base of the brain. Even without hemorrhaging, aneurysms can cause pain or death by putting pressure on brain tissue, nerves, adjacent veins, pulmonary air passages, or the esophagus. Other consequences include neurological disorders, difficulty in breathing or swallowing, chronic cough, or congestion of the tissues with blood. Aneurysms sometimes result from congenital weakness of the blood vessels and sometimes from trauma or bacterial infections such as syphilis. The most common cause, however, is the combination of atherosclerosis and hypertension.

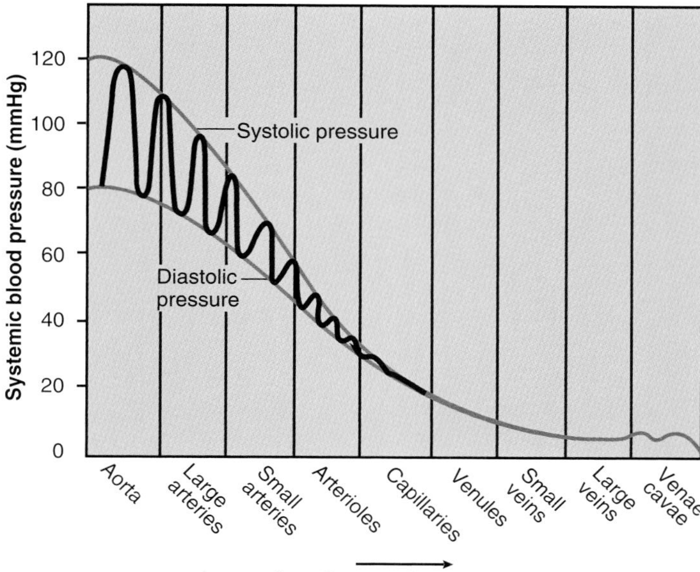

Figure 20.8 Changes in Blood Pressure Related to Distance from the Heart. Because of arterial elasticity and the effect of friction against the vessel wall, all measures of blood pressure decline with distance—systolic pressure, diastolic pressure, pulse pressure, and mean arterial pressure. There is no pulse pressure beyond the arterioles, but there are slight pressure oscillations in the venae cavae caused by the respiratory pump.

Table 20.1 Normal Arterial Blood Pressure at Various Ages*

Age (years)	Male	Female
1	96/66	95/65
5	92/62	92/62
10	103/69	103/70
15	112/75	112/76
20–24	123/76	116/72
25–29	125/78	117/74
30–34	126/79	120/75
40–44	129/81	127/80
50–54	135/83	137/84
60–64	142/85	144/85
70–74	145/82	159/85
80–84	145/82	157/83

*Average for healthy individuals

is less difference between them (fig. 20.8). In capillaries and veins, the blood flows at a steady speed without pulsation because the pressure surges have been damped out by the distance traveled and the elasticity of the arteries. This is why an injured vein exhibits relatively slow, steady bleeding, whereas blood spurts intermittently from a severed artery. In the inferior vena cava near the heart, however, venous flow fluctuates with the respiratory cycle for reasons explained later, and there is some fluctuation in the jugular veins of the neck.

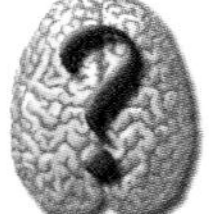

Think About It

Explain how the histological structure of large arteries relates to their ability to stretch during systole and recoil during diastole.

Blood pressure rises with age (table 20.1) as the arteries become less distensible and absorb less systolic force. Atherosclerosis also stiffens the arteries and leads to a rise in BP.

Blood pressure is determined mainly by cardiac output, blood volume, and peripheral resistance. The regulation of cardiac output and blood volume are discussed in chapters 19 and 24, respectively. Here, we turn our attention to peripheral resistance.

Resistance

A moving fluid has no pressure unless it encounters at least some resistance. Thus, pressure and resistance are not independent factors in blood flow—rather, pressure is affected by resistance, and flow is affected by both. **Peripheral resistance** is the resistance that the blood encounters in the vessels as it travels away from the heart. It results from the friction of blood against the walls of the vessels and is proportional to three variables: *blood viscosity, vessel length,* and *vessel radius.*

Blood Viscosity

Blood viscosity ("thickness" of the blood) is due mainly to erythrocytes and albumin. A deficiency of erythrocytes (anemia) or albumin (hypoproteinemia) decreases peripheral resistance and speeds up blood flow. If viscosity increases (as a result of polycythemia or dehydration, for example), resistance increases and flow declines.

Vessel Length

The farther a liquid travels through a tube, the more cumulative friction it encounters; thus, pressure and flow decline with distance. Partly for this reason, if you were to measure MAP in a reclining person, you would obtain a higher value in the arm, for example, than in the ankle. (This would not be true in a standing person because of the influence of gravity, explained earlier.) A strong pulse in the dorsal pedal artery of the foot is a good sign of adequate cardiac output. If perfusion is good at that distance from the heart, it is likely to be good elsewhere in the systemic circulation.

Vessel Radius

In a healthy individual, blood viscosity is quite stable, and of course vessel lengths do not change in the short term. Therefore, the only significant way of controlling peripheral resistance from moment to moment is by adjusting the radius of the blood vessels. A change in vessel radius is called **vasomotion.** This includes **vasoconstriction,** the narrowing of a vessel, and **vasodilation,** the widening of a vessel. Vasoconstriction occurs when the smooth muscle of the tunica media contracts. Vasodilation occurs when this muscle relaxes and allows the blood pressure within the vessel to push its walls outward.

The effect of vessel radius on blood flow is related to the friction of the moving blood against the walls of the vessel. Blood normally exhibits smooth, silent **laminar**[6] **flow.** That is, it flows in "layers"—faster near the center of a vessel, where it encounters less friction, and slower near the walls, where it drags against the vessel. You can observe a similar effect from the vantage point of a riverbank. The current may be very swift in the middle of a river but quite sluggish near shore, where the water encounters more friction against the riverbank and bottom. When a blood vessel dilates, a greater portion of the blood is in the middle of the stream and the average flow may be quite swift. When the vessel constricts, more of the blood is close to the wall and the average flow is slower (fig. 20.9).

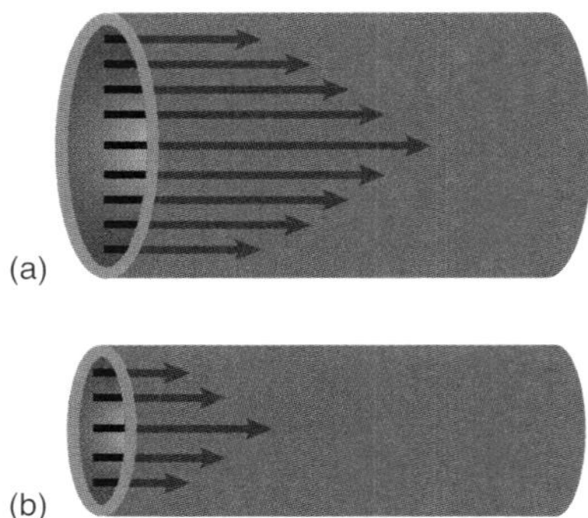

Figure 20.9 Laminar Flow and the Effect of Vessel Radius. Blood flows more slowly near the vessel wall, as indicated by *shorter arrows,* than it does near the center of the vessel. (*a*) When vessel radius is large, the average velocity of flow is high. (*b*) When the radius is less, the average velocity is lower because a larger portion of the blood is slowed down by friction against the vessel wall.

Thus the radius of a vessel markedly affects blood velocity. Indeed, blood flow is proportional not merely to vessel radius but to the *fourth power* of radius—that is, $F \propto r^4$. This makes vessel radius a very potent factor in the control of flow. Arterioles can constrict to as little as one-third of their fully relaxed radius (fig. 20.10). For the sake of simplicity, consider a hypothetical blood vessel with a 1 mm radius when maximally constricted and a 3 mm radius when completely relaxed. At a 1 mm radius, suppose the blood travels 1 mm/sec. By the formula $F \propto r^4$, consider how the velocity would change as radius changed:

$r = 1$ mm $\quad r^4 = 1^4 = 1 \; F = 1$ mm/sec (given)

$r = 2$ mm $\quad r^4 = 2^4 = 16 \; F = 16$ mm/sec

$r = 3$ mm $\quad r^4 = 3^4 = 81 \; F = 81$ mm/sec

These actual numbers do not matter; what matters is that a mere 3-fold increase in radius has produced an 81-fold increase in velocity—a demonstration that vessel radius exerts a very powerful influence over flow; moreover, it is the most adjustable of all variables that govern peripheral resistance.

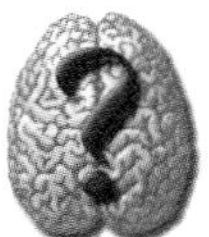

Think About It

Suppose a vessel with a radius of 1 mm had a flow of 5 mm/sec, and then the vessel dilated to a radius of 5 mm. What would be the new flow rate?

6. *lamina* = layer

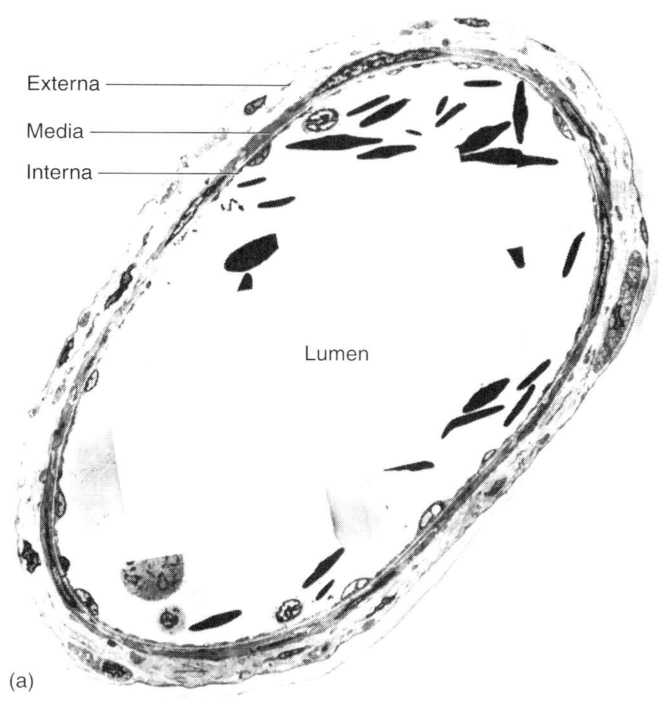

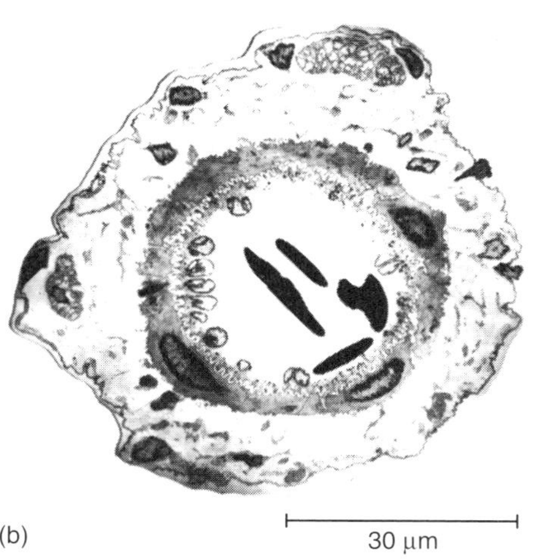

Figure 20.10 The Capacity for Vasoconstriction in an Arteriole. (*a*) A dilated arteriole. (*b*) The same arteriole, at a point just 1 mm from the area photographed in *a*, stimulated by a single drop of epinephrine. The diameter of the dilated region is about three times that of the constricted region.

To integrate this information, consider how the velocity of blood flow differs from one part of the systemic circuit to another (table 20.2). Flow is fastest in the aorta because it is a large vessel close to the pressure source, the left ventricle. From aorta to capillaries, velocity diminishes for three reasons: (1) The blood has traveled a greater distance, so friction has slowed it down. (2) The arterioles and capillaries have smaller radii and therefore

Table 20.2 Blood Velocity in the Systemic Circuit

Vessel	Typical Lumen Diameter	Velocity*
Aorta	2.5 cm	1,200 mm/sec
Arterioles	20–50 μm	15 mm/sec
Capillaries	5–9 μm	0.4 mm/sec
Venules	20 μm	5 mm/sec
Inferior vena cava	3 cm	80 mm/sec

*Peak systolic velocity in the aorta; mean or steady velocity in other vessels

put up more resistance. (3) Even though the radii of individual vessels become smaller as we progress farther from the heart, the number of vessels and their *total* cross-sectional area becomes greater and greater. The aorta has a cross-sectional area of 3 to 5 cm^2, while the total cross-sectional area of all the capillaries is about 4,500 to 6,000 cm^2. Thus, a given volume of aortic blood is distributed over a greater total area in the capillaries, which *collectively* form a wider path in the bloodstream. Just as water slows down when a narrow mountain stream flows into a lake, blood slows down as it enters pathways with a greater total width.

From capillaries to vena cava, velocity rises again. One reason for this is that the veins have larger diameters than the capillaries, so they create less resistance (less friction). Furthermore, since many capillaries converge on one venule, and many venules on a larger vein, a large amount of blood is being forced into a progressively smaller channel—like water flowing from a lake into an outlet stream and thus flowing faster again. Note, however, that blood in the veins never regains the velocity it had in the large arteries. This is because the veins are farther from the heart and the pressure is much lower here.

Regulation of Blood Pressure and Flow

Blood pressure is subject to local, neural, and hormonal controls over vasomotion. We now consider each of these three influences in turn.

Local Control

Autoregulation is the ability of tissues to regulate their own blood supply. According to the *metabolic theory of autoregulation,* if a tissue is inadequately perfused, it becomes hypoxic and its metabolites (waste products) accumulate—CO_2, H^+, K^+, lactic acid, and adenosine, for example. These factors stimulate vasodilation, which increases perfusion. As the bloodstream delivers oxygen and carries away the metabolites, the vessels constrict. Thus, a homeostatic dynamic equilibrium is established that adjusts perfusion to the tissue's metabolic needs.

Blood platelets, endothelial cells, and the perivascular tissues secrete a variety of **vasoactive chemicals**—substances that stimulate vasomotion. Histamine, bradykinin, and prostaglandins stimulate vasodilation under such conditions as trauma, inflammation, and exercise. Endothelial cells secrete prostacyclin and nitric oxide, which are vasodilators, and polypeptides called *endothelins,* which are vasoconstrictors.

If a tissue's blood supply is cut off for a time and then restored, it often exhibits **reactive hyperemia**—an increase above the normal level of flow. This may be due to the accumulation of metabolites during the period of ischemia. Reactive hyperemia can be seen when the skin flushes after a person comes in from the cold. It also occurs in the forearm if a blood pressure cuff is inflated for too long and then loosened.

In the long run, a hypoxic tissue can increase its own perfusion by **angiogenesis**[7]—the growth of new blood vessels. (This term also refers to embryonic development of blood vessels.) Three situations in which this is important are the regrowth of the uterine lining after each menstrual period, the development of a higher density of blood capillaries in the muscles of well-conditioned athletes, and the growth of arterial bypasses around obstructions in the coronary circulation. Several growth factors and inhibitors control angiogenesis, but physiologists are not yet sure how it is regulated. Malignant tumors secrete growth factors that stimulate a dense network of blood vessels to grow into them and provide nourishment to the cancer cells. Oncologists are interested in finding a way to block tumor angiogenesis, which would choke off a tumor's blood supply and perhaps shrink or kill it.

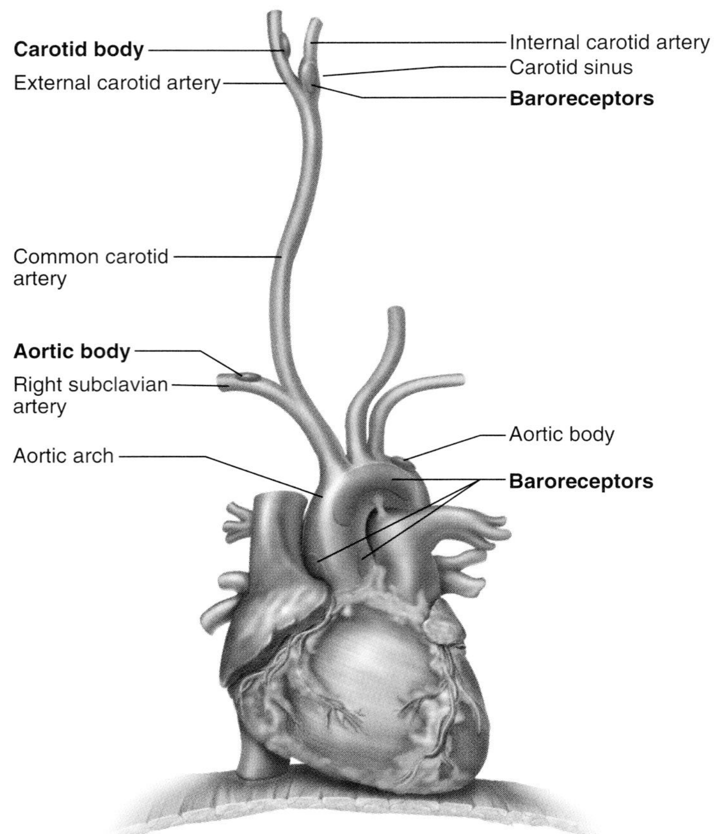

Figure 20.11 Locations of Arterial Baroreceptors and Chemoreceptors. Chemoreceptors are located in the carotid bodies and aortic bodies. Baroreceptors are located in the ascending aorta, aortic arch, and carotid sinus. The structures shown here in the right carotid arteries are repeated in the left carotids.

Neural Control

In addition to local control, the blood vessels are under remote control by hormones and the autonomic nervous system. The **vasomotor center** of the medulla oblongata exerts sympathetic control over blood vessels throughout the body. (Precapillary sphincters have no innervation, however, and respond only to local and hormonal stimuli.) Sympathetic nerve fibers stimulate most blood vessels to constrict, but they cause dilation of the vessels in skeletal and cardiac muscle in order to meet the metabolic demands of exercise. The role of sympathetic tone and vasomotor tone in controlling vessel diameter is explained in chapter 15.

The vasomotor center is an integrating center for three autonomic reflexes—*baroreflexes, chemoreflexes,* and the *medullary ischemic reflex.* A **baroreflex** is an autonomic, negative feedback response to changes in blood pressure. The changes are detected by stretch receptors called **baroreceptors.**[8] These occur in all of the large arteries above the heart but are especially concentrated in the aortic arch, the *aortic sinuses* behind the aortic valve cusps, and the *carotid sinus* at the base of each internal carotid artery near the angle of the mandible (fig. 20.11). They are branched, knobby nerve fibers somewhat resembling Golgi tendon organs (see p. 567). The baroreceptors transmit signals continually to the brainstem. When the blood pressure rises, their signaling rate rises. This input inhibits the sympathetic cardiac and vasomotor neurons and reduces sympathetic tone, and it *excites* the vagal fibers to the heart. Thus, it reduces the heart rate and cardiac output, dilates the arteries and veins, and reduces the blood pressure (fig. 20.12). When BP drops below normal, on the other hand, the opposite reactions occur and BP rises back to normal.

7. *angio* = vessels + *genesis* = production of

8. *baro* = pressure

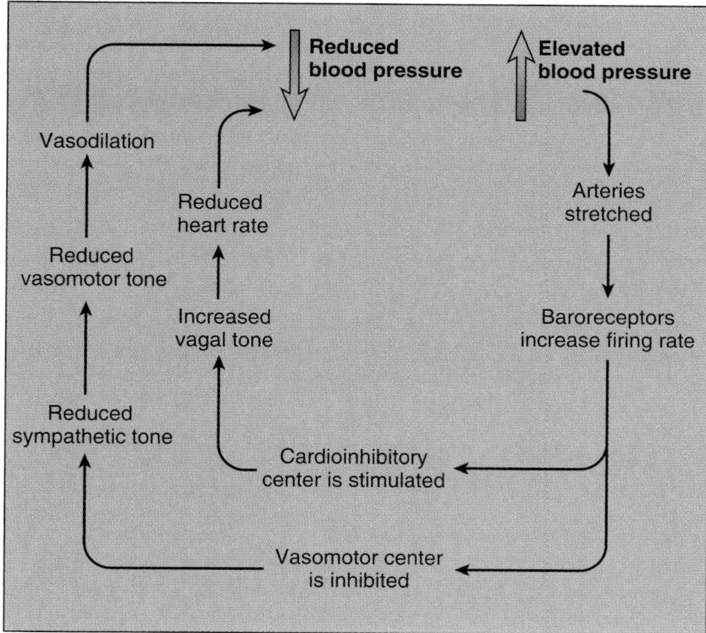

Figure 20.12 Negative Feedback Control of Blood Pressure. The reactions here occur in response to a rise in blood pressure. The opposite responses occur when blood pressure drops below normal.

Baroreflexes are important chiefly in short-term regulation of BP, for example in adapting to changes in posture. Perhaps you have jumped quickly out of bed and felt a little dizzy for a moment. This occurs because gravity draws the blood into the large veins of the abdomen and lower limbs when you stand, which reduces venous return to the heart and cardiac output to the brain. Normally, the baroreceptors respond quickly to this drop in pressure and restore cerebral perfusion. Baroreflexes are not effective in correcting chronic hypertension, however. Apparently they adjust their set point to the higher BP and maintain dynamic equilibrium at this new level.

A **chemoreflex** is an autonomic response to changes in blood chemistry, especially its pH and concentrations of O_2 and CO_2. It is initiated by chemoreceptors within small organs called the **aortic body** and **carotid bodies,** located in the aortic arch, subclavian arteries, and external carotid arteries. The primary role of chemoreflexes is to adjust respiration to changes in blood chemistry, but they have a secondary role in stimulating vasomotion. Hypoxemia, hypercapnia, and acidosis stimulate the chemoreceptors and act through the vasomotor center to cause widespread vasoconstriction. This increases overall BP, thus increasing perfusion of the lungs and the rate of gas exchange. Chemoreceptors also stimulate one's breathing, so increased ventilation of the lungs matches their increased perfusion. Increasing one without the other would be of little use.

The **medullary ischemic** (iss-KEE-mic) **reflex** is an autonomic response to insufficient perfusion of the brainstem. Within seconds after a drop in perfusion of the brain, the cardiac and vasomotor centers of the medulla oblongata send sympathetic signals to the heart and blood vessels that induce (1) an increase in heart rate and contraction force and (2) widespread vasoconstriction. These actions raise the blood pressure and, ideally, restore normal perfusion of the brain. The cardiac and vasomotor centers also receive input from other brain centers. Thus stress, anger, and arousal can also raise the blood pressure. The hypothalamus acts through the vasomotor center to redirect blood flow in response to exercise or changes in body temperature.

Hormonal Control

One of the most important vasoactive hormones is **angiotensin II.** The liver produces its precursor, a prohormone called *angiotensinogen,* and secretes it into the blood. Low blood pressure causes the kidneys to secrete an enzyme, **renin** (REE-nin), that converts angiotensinogen to *angiotensin I.* Then when angiotensin I passes through the lungs, **angiotensin-converting enzyme (ACE)** converts it to angiotensin II. Angiotensin II is a very potent vasoconstrictor that acts to raise the blood pressure. Hypertension is often treated with drugs called *ACE inhibitors* that block the action of this angiotensin-converting enzyme and thus reduce the level of angiotensin II. Other antihypertensive drugs are discussed in clinical insight 20.4 at the end of the chapter.

Epinephrine and norepinephrine bind to α-adrenergic receptors on the smooth muscle of most blood vessels. This stimulates the muscle to contract, thus producing vasoconstriction and raising the blood pressure. In the coronary blood vessels and blood vessels of the skeletal muscles, however, these chemicals bind to β-adrenergic receptors and cause vasodilation, thus increasing blood flow to the myocardium and muscular system during exercise. Antidiuretic hormone (ADH) primarily functions to promote water retention, but at pathologically high concentrations it is also a vasoconstrictor—hence its alternate name, *vasopressin.* Atrial natriuretic factor (ANF), secreted by the heart, is known primarily for increasing urinary sodium excretion, but it also has a generalized vasodilator effect that lowers the blood pressure.

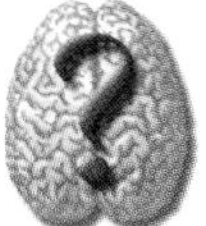

Think About It

Drugs called *renin inhibitors* are currently being developed to treat hypertension. Explain how you think such a drug would produce the desired effect.

Vasomotion and Routing of Blood Flow

If a chemical such as epinephrine causes widespread vasoconstriction, or if it causes vasoconstriction in a large system such as the integumentary or digestive system, it can produce an overall rise in blood pressure. Localized vasoconstriction, however, has a very different effect. If a particular artery constricts, pressure downstream from the constriction drops and pressure upstream from it rises. If blood can travel by either of two routes and one route puts up more resistance than the other, most blood follows the path of least resistance. This mechanism enables the body to redirect blood from one organ to another.

For example, if you are dozing in an armchair after a big meal, vasoconstriction shuts down blood flow to 90% or more of the capillaries in the muscles of your lower limbs. This raises the BP above the limbs, where the aorta gives off a branch, the superior mesenteric artery, supplying the small intestine. High resistance in the circulation of the legs and low resistance in the superior mesenteric artery routes blood to the small intestine, where it is needed to absorb digested nutrients (fig. 20.13*a*).

On the other hand, during vigorous exercise, the arteries in your muscles dilate. To make blood available to the muscles, flow must be reduced elsewhere—notably in the skin, kidneys, and digestive tract (fig. 20.13*b*). Thus, changes in peripheral resistance can shift blood flow from one organ system to another to meet the changing metabolic priorities of the body. Physical exertion increases perfusion of the lungs, myocardium, and skeletal muscles while reducing perfusion of the skin, kidneys and digestive tract (fig. 20.14).

The arterioles are the most significant point of control over peripheral resistance and blood flow because (1) they are on the proximal sides of the capillary beds, so they are best positioned to regulate flow into the capillaries; (2) they greatly outnumber any other class of arteries and thus provide the most numerous control points; and (3) they are more muscular in proportion to their diameters than any other class of blood vessels and are highly capable of vasomotion. Arterioles alone account for about half of the total peripheral resistance of the circulatory system. However, larger arteries and veins are also capable of considerable vasomotion and control of peripheral resistance.

Before You Go On

Answer the following questions to test your understanding of the preceding section:

6. For a healthy 15-year-old girl at rest, what would be typical readings for systolic pressure, diastolic pressure, pulse pressure, and mean arterial pressure?
7. Explain why arterial blood flow is pulsatile and venous flow is not.
8. What three variables affect peripheral resistance to blood flow? Which of these is most able to change from one minute to the next?
9. What are the three primary mechanisms for controlling vessel radius? Briefly explain each.
10. Explain how the baroreflex serves as an example of homeostasis and negative feedback.
11. Explain how the body can shift the flow of blood from one organ system to another.

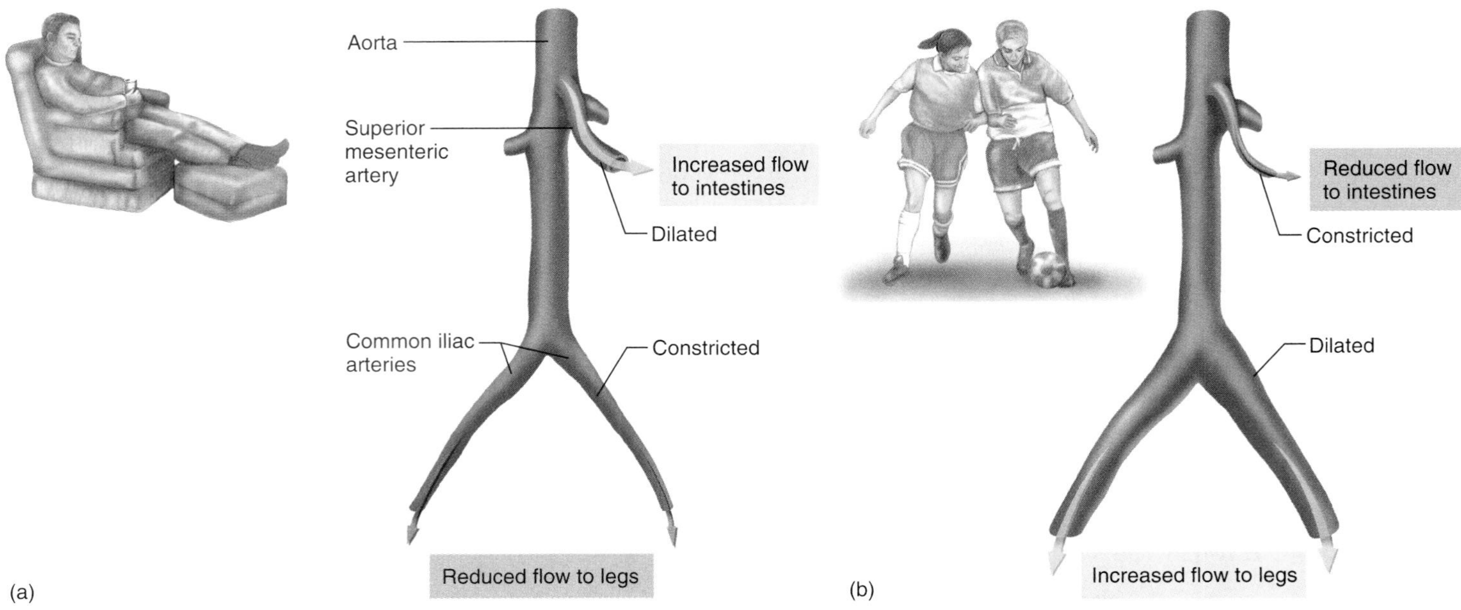

Figure 20.13 Redirection of Blood Flow in Response to Changing Metabolic Needs. (*a*) After a meal, the intestines receive priority and the skeletal muscles receive relatively little flow. (*b*) During exercise, the muscles receive higher priority. Although vasodilation and vasoconstriction are shown here in major arteries for illustration purposes, most control occurs at a microscopic level in the arterioles.

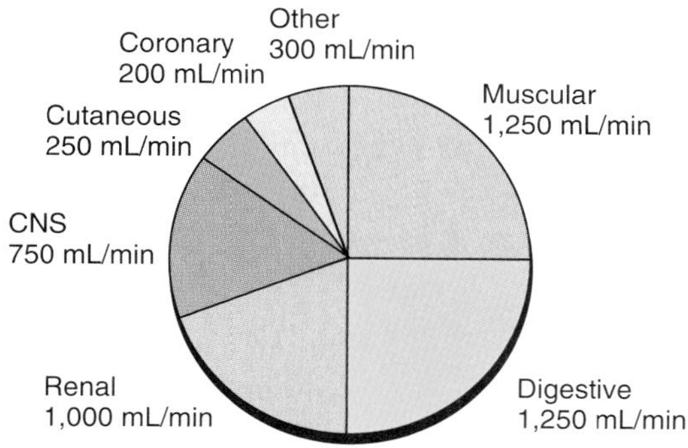

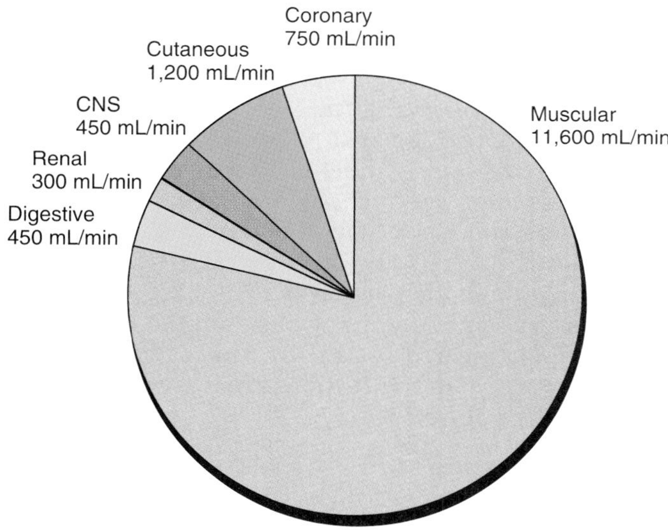

Figure 20.14 **Changes in Systemic Blood Flow During Rest and Exercise.** Areas of the two charts are proportional to total cardiac output under the two conditions.

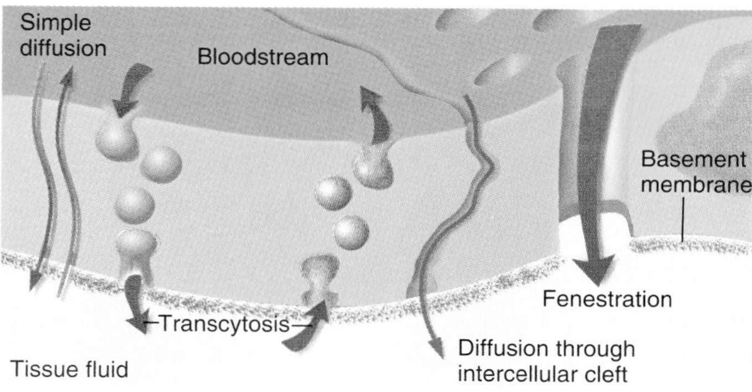

Figure 20.15 **Pathways of Capillary Fluid Exchange.**

Capillary Exchange

Objectives

When you have completed this section, you should be able to

- describe how materials get from the blood to the surrounding tissues;
- describe and calculate the forces that enable capillaries to give off and reabsorb fluid; and
- describe the causes and effects of edema.

Only 250 to 300 mL of blood is in the capillaries at any given time. This is the most important blood in the body, however, for it is only across capillary walls that any appreciable exchanges occur between the blood and surrounding tissues. **Capillary exchange** refers to this two-way movement of fluid.

The mechanisms of capillary exchange are difficult to study quantitatively because it is hard to measure pressure and flow in such small vessels. For this reason, theories of capillary exchange remain in dispute. Few capillaries of the human body are accessible to direct, noninvasive observation, but those of the fingernail bed and eponychium (cuticle) at the base of the nails can be observed with a stereomicroscope and have been the basis for a number of studies. Their BP has been measured at 32 mmHg at the arterial end and 15 mmHg at the venous end, 1 mm away. Capillary BP drops rapidly because of the substantial friction the blood encounters in such narrow vessels. It takes 1 to 2 seconds for an RBC to pass through a nail bed capillary traveling about 0.7 mm/sec.

Substances pass between the blood and tissue fluid by three routes: (1) through the intercellular clefts between endothelial cells, (2) through the fenestrations (pores) of fenestrated capillaries, and (3) through the endothelial cell cytoplasm (fig. 20.15). The mechanisms involved are *diffusion, transcytosis, filtration,* and *reabsorption,* which we examine in that order.

Diffusion

The most important mechanism of exchange is diffusion. Glucose and oxygen, being more concentrated in the systemic blood than in the tissue fluid, diffuse out of the blood. Carbon dioxide and other wastes, being more concentrated in the tissue fluid, diffuse into the blood. (Oxygen and carbon dioxide diffuse in the opposite directions in the pulmonary circuit.) Such diffusion is only possible if the solute can either permeate the plasma membranes of the endothelial cells or find passages large enough to pass through—namely, the fenestrations and intercellular clefts. Such lipid-soluble substances as steroid hormones, O_2, and CO_2 diffuse easily through the plasma membranes. Substances insoluble in lipids, such as glucose and electrolytes, must pass through membrane channels, fenestrations, or intercellular clefts. Large molecules such as proteins are usually held back by the small size of these passages.

Transcytosis

Transcytosis is a process in which endothelial cells pick up droplets of fluid on one side of the plasma membrane by pinocytosis, transport the vesicles across the cell, and discharge the fluid on the other side by exocytosis (see fig. 4.30, p. 134). This probably accounts for only a small fraction of solute exchange across the capillary wall, but fatty acids, albumin, and some hormones such as insulin move across the endothelium by this mechanism.

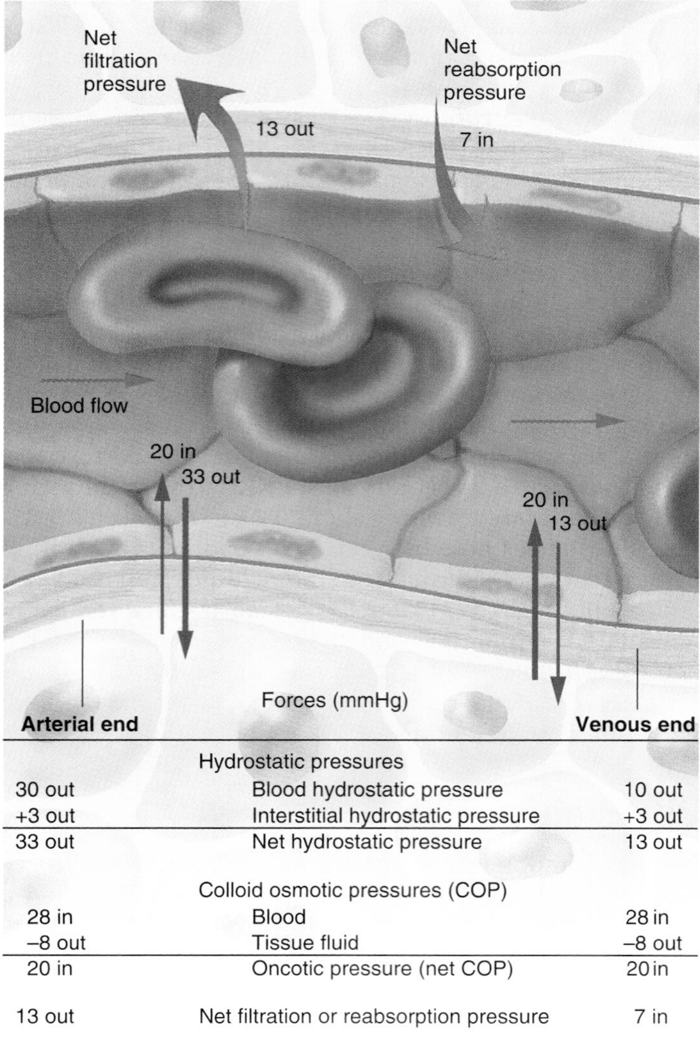

Figure 20.16 The Forces of Capillary Filtration and Reabsorption. Note the shift from net filtration at the arterial (*left*) end to net reabsorption at the venous (*right*) end.

Filtration and Reabsorption

The equilibrium between filtration and osmosis discussed in chapter 4 becomes particularly relevant when we consider capillary fluid exchange. Typically, fluid filters out of the arterial end of a capillary and osmotically reenters it at the venous end (fig. 20.16). This fluid delivers materials to the cells and removes their metabolic wastes.

It may seem odd that a capillary could give off fluid at one point and reabsorb it at another. This comes about as the result of a shifting balance between hydrostatic and osmotic forces. A typical capillary has a blood (hydrostatic) pressure of about 30 mmHg at the arterial end. The hydrostatic pressure of the interstitial space has been difficult to measure and remains a point of controversy, but a typical value accepted by many authorities is −3 mmHg, a slight suction that helps draw fluid out of the capillary. In this case, the positive hydrostatic pressure within the capillary and the negative interstitial pressure work in the same direction, creating a total force of about 33 mmHg, which causes fluid to leave the capillary.

These forces are opposed by **colloid osmotic pressure (COP),** the portion of the blood's osmotic pressure that results from its plasma proteins. The blood has a COP of about 28 mmHg, due mainly to albumin. Tissue fluid has less than one-third the protein concentration of blood plasma and has a COP of about 8 mmHg. The difference between the COP of blood and COP of tissue fluid is called **oncotic pressure:** $28_{in} - 8_{out} = 20_{in}$. Oncotic pressure tends to draw water into the capillary by osmosis, opposing hydrostatic pressure. These opposing forces produce a **net filtration pressure (NFP)** of 13 mmHg out, as follows:

Hydrostatic pressure

Blood pressure	30_{out}
Interstitial pressure	$+ 3_{out}$
Net hydrostatic pressure	33_{out}

Colloid osmotic pressure

Blood COP	28_{in}
Tissue fluid COP	$- 8_{out}$
Oncotic pressure	20_{in}

Net filtration pressure

Net hydrostatic pressure	33_{out}
Oncotic pressure	$- 20_{in}$
Net filtration pressure	13_{out}

The NFP of 13 mmHg causes about 0.5% of the blood plasma to leave the capillaries at the arterial end.

At the venous end, however, capillary blood pressure is lower—about 10 mmHg. All the other pressures are unchanged. Thus, we get:

Hydrostatic pressure

Blood pressure	10_{out}
Interstitial pressure	$+ 3_{out}$
Net hydrostatic pressure	13_{out}

Net reabsorption pressure

Oncotic pressure	20_{in}
Net hydrostatic pressure	$- 13_{out}$
Net reabsorption pressure	7_{in}

The prevailing force is inward at the venous end because osmotic pressure overrides filtration pressure. The **net reabsorption pressure** of 7 mmHg inward causes the capillary to reabsorb fluid at this end.

Now you can see why a capillary gives off fluid at one end and reabsorbs it at the other. The only pressure that changes from the arterial end to the venous end is the capillary blood pressure, and this change is responsible for the shift from filtration to reabsorption. With a reabsorption pressure of 7 mmHg and a net filtration pressure of 13 mmHg, it might appear that far more fluid would leave the capillaries than reenter them. However, since capillaries branch along their length, there are more of them at the venous end than at the arterial end, which partially compensates for the difference between filtration and reabsorption pressures. They also typically have nearly twice the diameter at the venous end that they have at the arterial end, so there is more capillary surface area available to reabsorb fluid than to give it off. Consequently, capillaries reabsorb about 85% of the fluid they filter. The other 15% is absorbed and returned to the blood by way of the lymphatic system, as described in the next chapter.

Of course, water is not the only substance that crosses the capillary wall by filtration and reabsorption. It carries along many of the solutes dissolved in it. This process is called **solvent drag.**

Variations in Capillary Filtration and Reabsorption

The figures used in the preceding discussion serve only as examples; circumstances differ from place to place in the body and from time to time in the same capillaries. Capillaries usually reabsorb most of the fluid they filter, but this is not always the case. The kidneys have capillary networks called *glomeruli* in which there is little or no reabsorption; they are entirely devoted to filtration. Alveolar capillaries of the lungs, by contrast, are almost entirely dedicated to absorption.

Capillary activity also varies from moment to moment. In a resting tissue, most precapillary sphincters are constricted and the capillaries are collapsed. Capillary BP is very low (if there is any flow at all), and reabsorption predominates. When a tissue becomes metabolically active, its capillary flow increases. In active muscles, capillary pressure rises to the point that it overrides reabsorption along the entire length of the capillary. Fluid accumulates in the muscle, and exercising muscles increase in size by as much as 25%. Capillary permeability is also subject to chemical influences. Traumatized tissue releases such chemicals as substance P, bradykinin, and histamine, which increase permeability and filtration.

Edema

Edema is the accumulation of excess fluid in a tissue. It often shows as swelling of the face, fingers, abdomen, or ankles but also affects internal organs, where its effects are hidden from view. Edema occurs when fluid filters into a tissue faster than it is reabsorbed. It has three fundamental causes:

1. **Increased capillary filtration.** This results from increases in capillary BP or permeability. Poor venous return, for example, causes pressure to back up into the capillaries. Congestive heart failure and incompetent heart valves can impede venous return from the lungs and cause pulmonary edema. Systemic edema is a common problem when a person is confined to a bed or wheelchair, with insufficient muscular activity to promote venous return. Kidney failure can lead to edema by causing water retention and hypertension. Histamine causes edema by dilating the arterioles and making the capillaries more permeable. Capillary permeability also increases with age, which puts older people at risk of edema.
2. **Reduced capillary reabsorption.** Capillary reabsorption depends on oncotic pressure, which is proportional to the concentration of blood albumin. A deficiency of blood albumin (hypoproteinemia) produces edema because the capillaries do not osmotically reabsorb fluid as fast as it is given off. Since blood albumin is produced by the liver, liver diseases such as cirrhosis tend to lead to hypoproteinemia and edema. Edema is commonly seen in regions of famine due to dietary protein deficiency. Hypoproteinemia also commonly results from severe burns, radiation sickness, and kidney diseases that allow protein to escape in the urine.
3. **Obstructed lymphatic drainage.** The lymphatic system, described in detail in the next chapter, is a system of one-way vessels that collect fluid from the tissues and return it to the bloodstream. Obstruction of these vessels or the surgical removal of lymph nodes can interfere with fluid drainage and lead to the accumulation of tissue fluid distal to the obstruction.

In severe edema, so much fluid may transfer from the blood vessels to the tissue spaces that blood volume and pressure drop so low as to cause circulatory shock (described later in this chapter). Furthermore, as the tissues become swollen with fluid, oxygen delivery and waste removal are impaired and tissue necrosis may occur. Pulmonary edema presents a threat of suffocation, and cerebral edema can produce headaches, nausea, and sometimes seizures and coma.

Before You Go On

Answer the following questions to test your understanding of the preceding section:

12. List the three mechanisms of capillary exchange and relate each one to the structure of capillary walls.
13. What forces favor capillary filtration? What forces favor reabsorption?
14. How can a capillary shift from a predominantly filtering to a predominantly reabsorbing role?
15. State the three fundamental causes of edema and explain why edema can be dangerous.

Venous Return and Circulatory Shock

Objectives

When you have completed this section, you should be able to

- explain how blood in the veins is returned to the heart;
- discuss the importance of physical activity in venous return;
- discuss several causes of circulatory shock; and
- name and describe the stages of shock.

Hieronymus Fabricius (1537–1619) discovered the valves of the veins and argued that they would allow blood to flow in only one direction, not back and forth as Galen had thought. One of his medical students was William Harvey, who performed simple experiments on the valves that you can easily reproduce. At the top of figure 20.17, from Harvey's book, the experimenter has pressed on a vein at point *H* to block flow from the wrist toward the elbow. With another finger, he has milked the blood out of it up to point *O*, the first valve proximal to *H*. When he tries to force blood downward, it stops at that valve. It can go no farther, and it causes the vein to swell at that point. Blood can flow from right to left through that valve but not from left to right. Figure 20.17*b* is explained in Harvey's own words.

You can easily demonstrate the action of these valves in your own hand. Hold your hand still, below waist level, until veins stand up on the back of it. (Do not apply a tourniquet!) Press on a vein close to your knuckles, and while holding it down, use another finger to milk that vein toward the wrist. It collapses as you force the blood out of it, and if you remove the second finger, it will not refill.

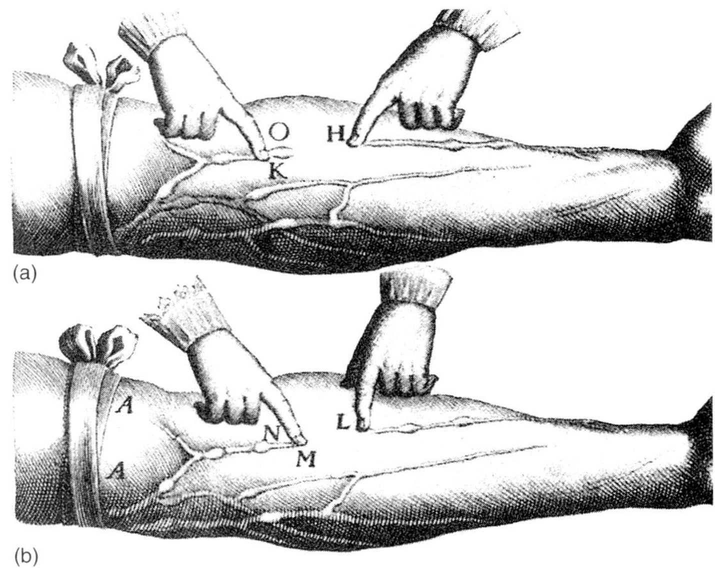

Figure 20.17 An Illustration from William Harvey's *De Motu Cordis (1628).* These experiments demonstrate the existence of one-way valves in veins of the arms. *(a)* See text for explanation. *(b)* In Harvey's words, "With the arm bound as before and the veins swollen, if you will press on a vein a little below a swelling or valve and then squeeze blood upward beyond the valve with another finger, you will see that this part of the vein stays empty, and that no backflow can occur through the valve. But as soon as the finger is removed, the vein is filled from below."

The valves prevent blood from flowing back into it from above. When you remove the first finger, however, the vein fills from below.

Mechanisms of Venous Return

The flow of blood back to the heart, called **venous return,** is achieved by five mechanisms:

1. **The pressure gradient.** Pressure generated by the heart is the most important force in venous flow, even though it is substantially weaker in the veins than in the arteries. Pressure in the venules ranges from 12 to 18 mmHg, and pressure at the point where the venae cavae enter the heart, called **central venous pressure,** averages 4.6 mmHg. Thus, there is a venous pressure gradient (ΔP) of about 7 to 13 mmHg favoring the flow of blood toward the heart. The pressure gradient and venous return increase when blood volume increases. Venous return decreases when the veins constrict (*venoconstriction*) and oppose flow, and it increases when they dilate and offer less resistance. However, it increases if *all* the body's blood vessels constrict, because this reduces the "storage capacity" of the circulatory system and raises blood pressure and flow.

Clinical Insight 20.3

Air Embolism

Injury to the dural sinuses or jugular veins presents less danger from loss of blood than from air sucked into the circulatory system. The presence of air in the bloodstream is called *air embolism.* This is an important concern to neurosurgeons, who sometimes operate with the patient in a sitting position. If a dural sinus is punctured, air can be sucked into the sinus and accumulate in the heart chambers, which blocks cardiac output and causes sudden death. Smaller air bubbles in the systemic circulation can cut off blood flow to the brain, lungs, myocardium, and other vital tissues.

2. **The thoracic (respiratory) pump.** This mechanism aids the flow of venous blood from the abdominal to the thoracic cavity. When you inhale, your thoracic cavity expands and its internal pressure drops, while downward movement of the diaphragm raises the pressure in your abdominal cavity. The *inferior vena cava (IVC),* your largest vein, is a flexible tube passing through both of these cavities. If abdominal pressure on the IVC rises while thoracic pressure on it drops, then blood is squeezed upward toward the heart. It is not forced back into the lower limbs because the venous valves there prevent this. Because of the thoracic pump, central venous pressure fluctuates from 2 mmHg when you inhale to 6 mmHg when you exhale, and blood flows faster when you inhale.
3. **Cardiac suction.** During ventricular systole, the chordae tendineae pull the AV valve cusps downward, slightly expanding the atrial space. This creates a slight suction that draws blood into the atria from the venae cavae and pulmonary veins.
4. **The skeletal muscle pump.** In the limbs, the veins are surrounded and massaged by the muscles. They squeeze the blood out of the compressed part of a vein, and the valves ensure that this blood can go in only one direction—toward the heart (fig. 20.18).
5. **Gravity.** When you are sitting or standing, blood from your head and neck returns to the heart simply by "flowing downhill" by way of the large veins above the heart. Thus the large veins of the neck are normally collapsed or nearly so, and their venous pressure is close to zero. The dural sinuses, however, have more rigid walls and cannot collapse. Their pressure is as low as −10 mmHg, creating a risk of *air embolism* if they are punctured (see clinical insight 20.3).

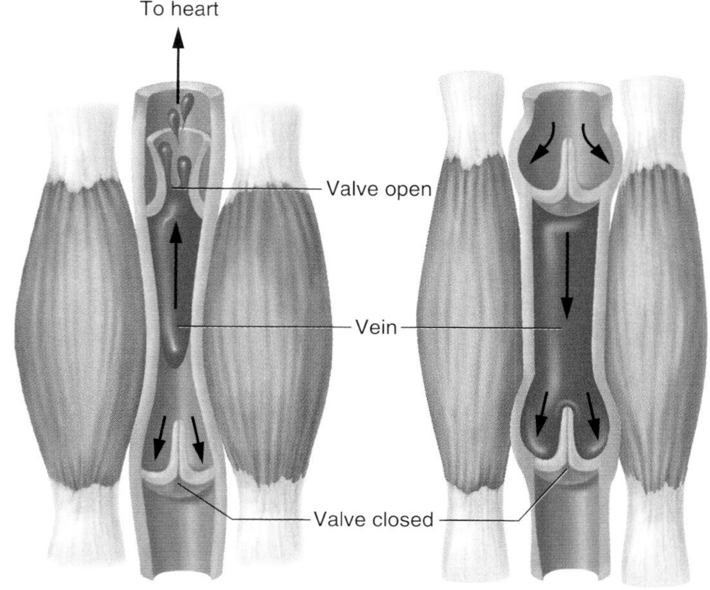

Figure 20.18 The Skeletal Muscle Pump. (*a*) When the muscles contract and compress a vein, blood is squeezed out of it and flows upward toward the heart; valves below the point of compression prevent backflow of the blood. (*b*) When the muscles relax, blood flows back downward under the pull of gravity but can only flow as far as the nearest valve.

Venous Return and Physical Activity

Exercise increases venous return for many reasons. The heart beats faster and harder, increasing cardiac output and blood pressure. Blood vessels of the skeletal muscles, lungs, and heart dilate, increasing flow. The increase in respiratory rate and depth enhances the action of the thoracic pump. Muscle contractions increase venous return by the skeletal muscle pump mechanism. Increased venous return increases cardiac output, which is important in perfusion of the muscles just when they need it most.

Conversely, when a person is still, blood accumulates in the limbs because venous pressure is not high enough to override the weight of the blood and drive it upward. Such accumulation of blood is called **venous pooling.** To demonstrate this effect, hold one hand above your head and the other below your waist for about a minute. Then, quickly bring your two hands together and compare the palms. The hand held above your head usually appears pale because its blood has drained out of it; the hand held below the waist appears redder than normal because of venous pooling in its veins and capillaries. Venous pooling is troublesome to people who must stand for prolonged periods. If enough blood accumulates in the limbs, cardiac output may become so low that the brain is inadequately perfused and a person may experience dizziness or **syncope** (SIN-co-pee) (fainting). This can usually be prevented by periodically tensing the calf and other muscles to keep the skeletal muscle pump active.

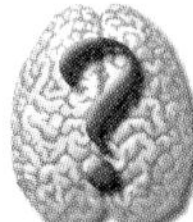

Think About It

Why is venous pooling not a problem when you are sleeping and the skeletal muscle pump is inactive?

Circulatory Shock

Circulatory shock (not to be confused with electrical or spinal shock) is any state in which cardiac output is insufficient to meet the body's metabolic needs. All forms of circulatory shock fall into two categories: (1) **cardiogenic shock,** caused by inadequate pumping by the heart usually as a result of myocardial infarction, and (2) **low venous return (LVR) shock,** in which cardiac output is low because too little blood is returning to the heart.

Types and Causes of LVR Shock

There are three principal forms of LVR shock:

1. **Hypovolemic shock,** the most common form, is produced by a loss of blood volume as a result of hemorrhage, trauma, bleeding ulcers, burns, or dehydration. Dehydration is a major cause of death from heat exposure. In hot weather, the body produces as much as 1.5 L of sweat per hour. Water transfers from the bloodstream to replace lost tissue fluid, and blood volume may drop too low to maintain adequate circulation.
2. **Obstructed venous return shock** occurs when a growing tumor or aneurysm, for example, compresses a nearby vein and impedes its blood flow.
3. **Venous pooling (vascular) shock** occurs when the body has a normal total blood volume, but too much of it accumulates in the limbs. This can result from long periods of standing or sitting or from widespread vasodilation. **Neurogenic shock** occurs when there is a sudden loss of vasomotor tone, which allows the vessels to dilate. This can result from causes as severe as brainstem trauma or as slight as an emotional shock.

Elements of both venous pooling and hypovolemic shock are present in certain cases, such as septic shock and anaphylactic shock, which involve both vasodilation and a loss of fluid through abnormally permeable capillaries. **Septic shock** occurs when bacterial toxins trigger vasodilation and increased capillary permeability. **Anaphylactic shock,** discussed more fully in the next chapter, results from exposure to an antigen to which a person is allergic, such as bee venom. Antigen-antibody complexes trigger the release of histamine, which causes generalized vasodilation and increased capillary permeability.

Responses to Circulatory Shock

In **compensated shock,** several homeostatic mechanisms act to bring about spontaneous recovery. The hypotension resulting from low cardiac output triggers the baroreflex and the production of angiotensin II, both of which counteract shock by stimulating vasoconstriction. Furthermore, if a person faints and falls to a horizontal position, gravity restores blood flow to the brain. Even quicker recovery is achieved if the person's feet are elevated to promote drainage of blood from the legs.

If these mechanisms prove inadequate, **decompensated shock** ensues and several life-threatening positive feedback loops occur. Poor cardiac output results in myocardial ischemia and infarction, which further weakens the heart and reduces output. Slow circulation of the blood can lead to disseminated intravascular coagulation (DIC) (see chapter 18). As the vessels become congested with clotted blood, venous return grows even worse. Ischemia and acidosis of the brainstem depress the vasomotor and cardiac centers, causing loss of vasomotor tone, further vasodilation, and further drop in BP and cardiac output. Before long, damage to the cardiac and brain tissues may be too great to be undone.

Before You Go On

Answer the following questions to test your understanding of the preceding section:

16. Explain how respiration aids venous return.
17. Explain how muscular activity and venous valves aid venous return.
18. Define *circulatory shock.* What are some of the causes of low venous return shock?

Special Circulatory Routes

Objectives

When you have completed this section, you should be able to

- explain how the brain maintains stable perfusion;
- discuss the causes and effects of strokes and transient ischemic attacks;
- explain the mechanisms that increase muscular perfusion during exercise; and
- contrast the blood pressure of the pulmonary circuit with that of the systemic circuit, and explain why the difference is important in pulmonary function.

Certain circulatory pathways have special physiological properties adapted to the functions of their organs. Two of these are described in other chapters: the coronary circulation in chapter 19 and fetal and placental circulation in chapter 29. Here we take a closer look at the circulation to the brain, skeletal muscles, and lungs.

Brain

Total perfusion of the brain fluctuates less than that of any other organ (about 750 mL/min at rest). Such constancy is important because even a few seconds of oxygen deprivation causes loss of consciousness, and 4 or 5 minutes of anoxia is time enough to cause irreversible brain damage. While total cerebral perfusion is fairly stable, blood flow can be shifted from one part of the brain to another in a matter of seconds as different parts engage in motor, sensory, or cognitive functions.

The brain regulates its own blood flow in response to changes in BP and chemistry. The cerebral arteries dilate when the systemic BP drops and constrict when BP rises, thus minimizing fluctuations in cerebral BP. Cerebral blood flow thus remains quite stable even when mean arterial pressure (MAP) fluctuates from 60 to 140 mmHg. An MAP below 60 mmHg produces syncope and an MAP above 160 mmHg causes cerebral edema.

The main chemical stimulus for cerebral autoregulation is pH. Poor cerebral perfusion allows CO_2 to accumulate in the brain tissue. This lowers the pH of the tissue fluid and triggers local vasodilation, which improves perfusion. Extreme hypercapnia (CO_2 excess), however, depresses neural activity. The opposite condition, hypocapnia, raises the pH and stimulates vasoconstriction, thus reducing perfusion and giving CO_2 a chance to rise to a normal level. Hyperventilation (exhaling CO_2 faster than the body produces it) induces hypocapnia, which leads to cerebral vasoconstriction, ischemia, dizziness, and sometimes syncope.

Brief episodes of cerebral ischemia produce **transient ischemic attacks (TIAs),** characterized by temporary dizziness, light-headedness, loss of vision or other senses, weakness, paralysis, headache, or aphasia. A TIA may result from spasms of diseased cerebral arteries. It lasts from just a moment to a few hours and is often an early warning of an impending stroke.

A stroke, or **cerebrovascular accident (CVA),** is the sudden death (infarction) of brain tissue caused by ischemia. Cerebral ischemia can be produced by atherosclerosis, thrombosis, or a ruptured aneurysm. The effects of a CVA range from unnoticeable to fatal, depending on the extent of tissue damage and the function of the affected tissue. Blindness, paralysis, loss of sensation, and loss of speech are common. Recovery depends on the ability of neighboring neurons to take over the lost functions and on the extent of collateral circulation to regions surrounding the cerebral infarction.

Skeletal Muscles

In contrast to the brain, the skeletal muscles receive a highly variable blood flow depending on their state of exertion. At rest, the arterioles are constricted, most of the capillary beds are shut down, and total flow through the muscular system is about 1 L/min. During exercise, the arterioles dilate in response to epinephrine and norepinephrine from the adrenal medulla and sympathetic nerves. Precapillary sphincters, which lack innervation, dilate in response to muscle metabolites such as lactic acid, CO_2, and adenosine. Blood flow increases as much as 100-fold during exercise, which requires that blood be diverted from other organs such as the digestive tract, skin, and kidneys to meet the needs of the working muscles.

Muscular contraction compresses the blood vessels and impedes flow. For this reason, isometric contraction causes fatigue more quickly than intermittent isotonic contraction. If you squeeze a rubber ball as hard as you can without relaxing your grip, you feel the muscles fatigue more quickly than if you intermittently squeeze and relax.

Lungs

After birth, the pulmonary circuit is the only route in which the arteries carry deoxygenated blood and the veins carry oxygenated blood. The pulmonary arteries have thin distensible walls with less elastic tissue than the systemic arteries. Thus, they have a BP of only 25/10. Capillary hydrostatic pressure is about 10 mmHg in the pulmonary circuit as compared with an average of 17 mmHg in systemic capillaries. This lower pressure has two implications for pulmonary circulation: (1) blood flows more slowly through the pulmonary capillaries, and therefore it has more time for gas exchange; and (2) oncotic pressure overrides hydrostatic pressure, so that these capillaries are engaged almost entirely in absorption. This prevents fluid accumulation in the alveolar walls and lumens, which would interfere with gas exchange. In a condition such as mitral valve stenosis, however, BP may back up into the pulmonary circuit, raising the capillary hydrostatic pressure and causing pulmonary edema, congestion, and hypoxemia.

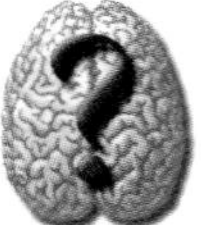

Think About It

What abnormal skin coloration would result from pulmonary edema?

Another unique characteristic of the pulmonary arteries is their response to hypoxia. Systemic arteries dilate in response to local hypoxia and improve tissue perfusion. By contrast, pulmonary arteries constrict. Pulmonary hypoxia indicates that part of the lung is not being ventilated well, perhaps because of mucous congestion of the airway or a degenerative lung disease. Vasoconstriction in poorly ventilated regions of the lung redirects blood flow to better ventilated regions.

Before You Go On

Answer the following questions to test your understanding of the preceding section:

19. In what conspicuous way does perfusion of the brain differ from perfusion of the skeletal muscles?
20. How does a stroke differ from a transient ischemic attack? Which of these bears closer resemblance to a myocardial infarction?
21. How does the low hydrostatic blood pressure in the pulmonary circuit affect the fluid dynamics of the capillaries there?
22. Contrast the vasomotor responses of the lungs versus skeletal muscles to hypoxia.

Anatomy of the Pulmonary Circuit

Objective

When you have completed this section, you should be able to

- trace the route of blood through the pulmonary circuit.

The remainder of this chapter centers on the names and pathways of the principal arteries and veins. The pulmonary circuit is described here, and the systemic arteries and veins are described in the two sections that follow.

The pulmonary circuit (fig. 20.19) begins with the **pulmonary trunk,** a large vessel that ascends diagonally from the right ventricle and branches into the right and left **pulmonary arteries.** Each pulmonary artery enters a medial indentation of the lung called the *hilum* and branches into one **lobar artery** for each lobe of the lung: three on the right and two on the left. These arteries lead ultimately to small basketlike capillary beds that surround the pulmonary alveoli. This is where the blood unloads CO_2 and loads O_2. After leaving the alveolar capillaries, the pulmonary blood flows into venules and veins, ultimately leading to the **pulmonary veins,** which exit the lung at the hilum. The left atrium of the heart receives two pulmonary veins on each side.

The purpose of the pulmonary circuit is to exchange CO_2 for O_2. It does not serve the metabolic needs of the lung tissue itself; there is a separate systemic supply to the lungs for that purpose, the *bronchial arteries,* discussed later.

Before You Go On

Answer the following questions to test your understanding of the preceding section:

23. Trace the flow of an RBC from right ventricle to left atrium and name the vessels along the way.
24. The lungs have two separate arterial supplies. Explain their functions.

Anatomy of the Systemic Arteries

Objectives

When you have completed this section, you should be able to

- identify the principal arteries of the systemic circuit; and
- trace the flow of blood from the heart to any major organ.

The systemic circuit supplies oxygen and nutrients to all the organs and removes their metabolic wastes. Part of it, the coronary circulation, was described in chapter 19. The other systemic arteries (fig. 20.20) are described in tables 20.3 through 20.8. The names of the blood vessels often describe their location by indicating the body region traversed (as in the *axillary* artery or *femoral* artery); an adjacent bone (as in *radial* artery or *temporal* artery); or the organ supplied or drained by the vessel (as in *hepatic* artery or *renal* vein). There is a great deal of anatomical variation in the circulatory system from one person to another. The remainder of this chapter describes the most common pathways.

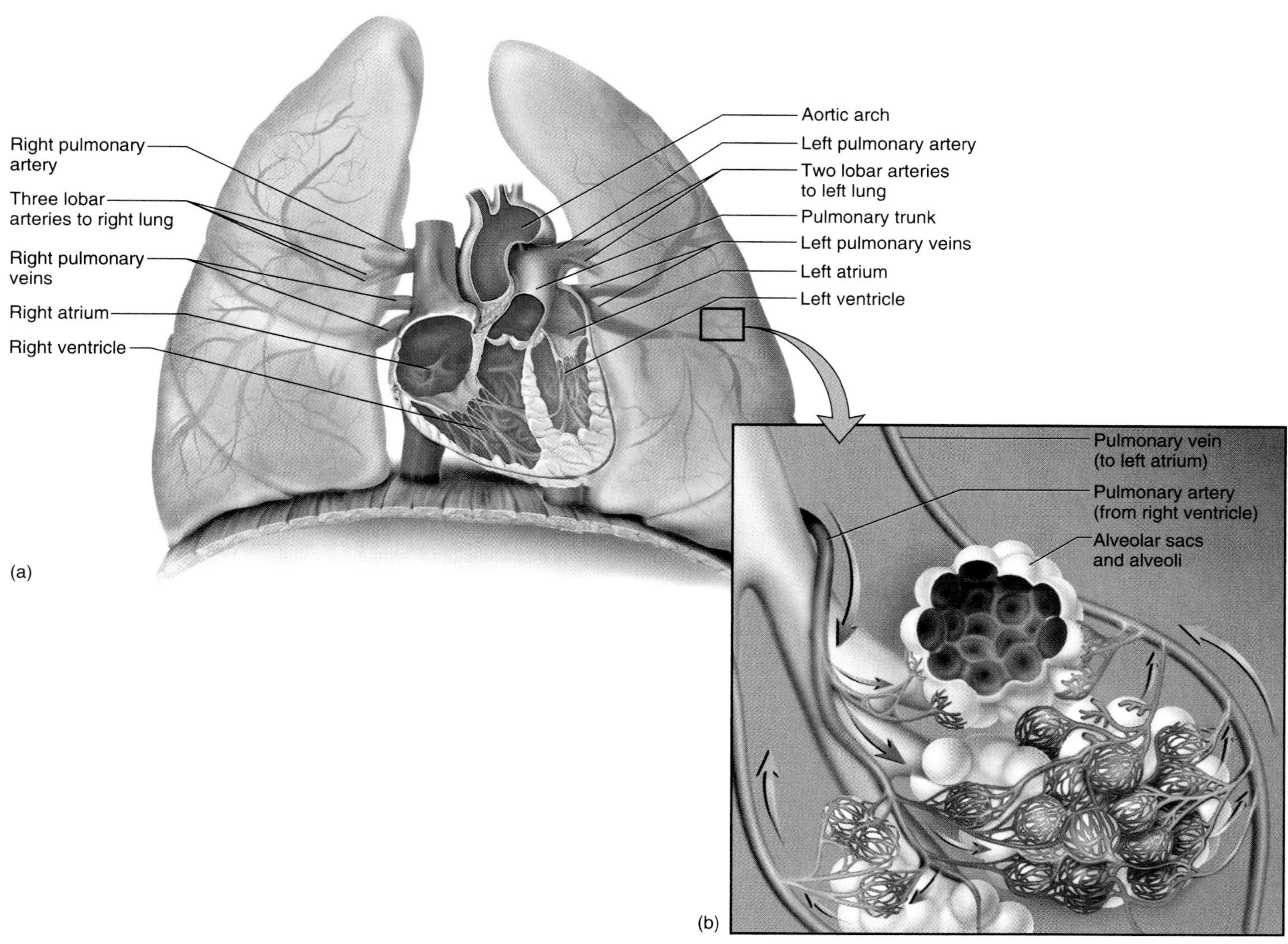

Figure 20.19 The Pulmonary Circulation. (*a*) Gross anatomy. (*b*) Microscopic anatomy of the blood vessels that supply the pulmonary alveoli.

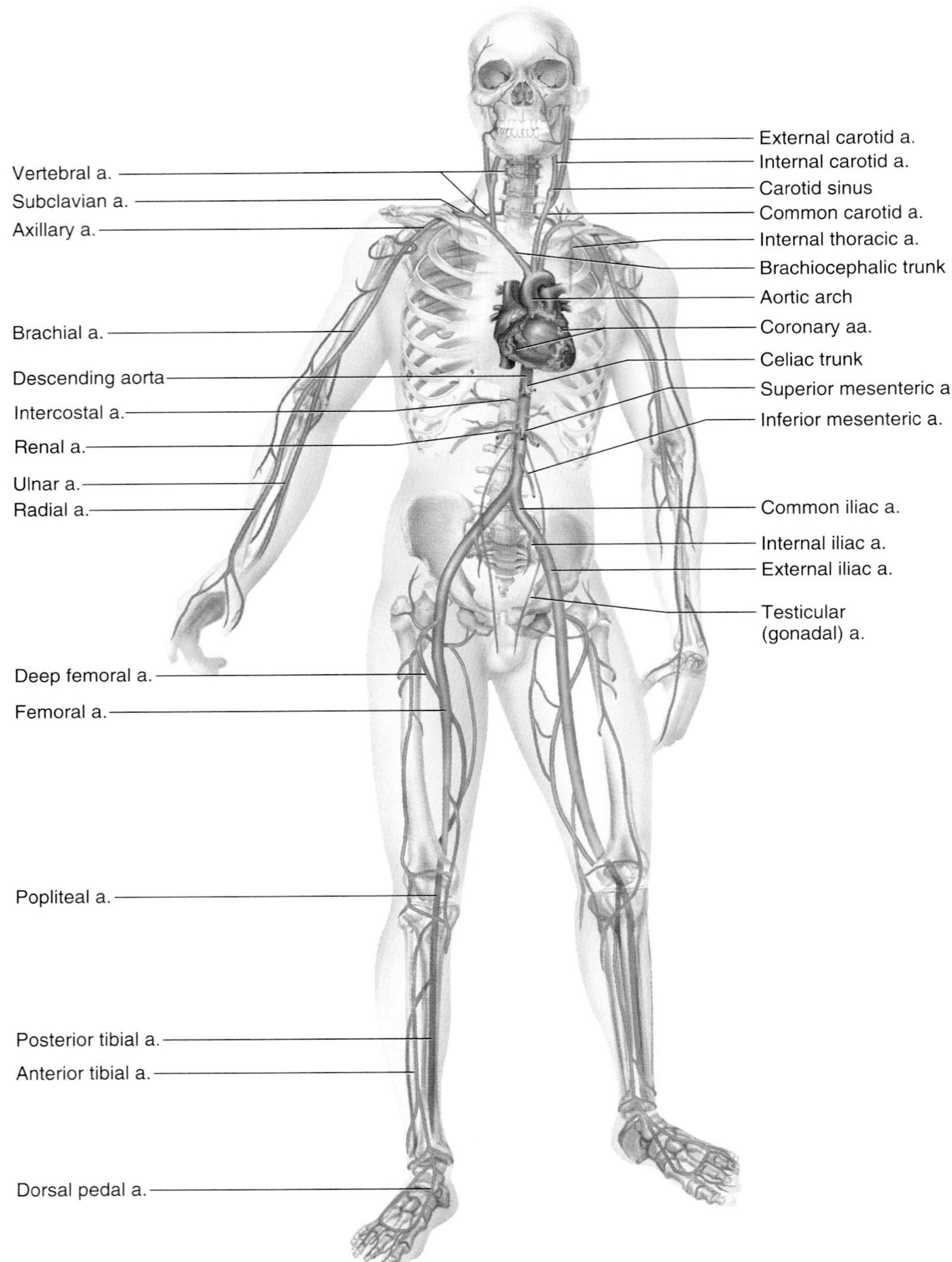

Figure 20.20 The Major Systemic Arteries. (*a.* = artery; *aa.* = arteries)

Table 20.3 The Aorta and Its Major Branches

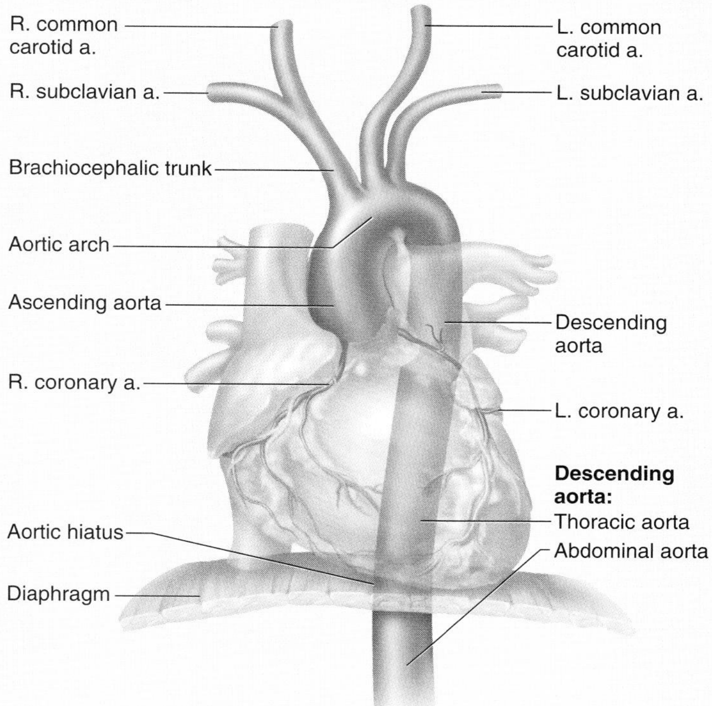

Figure 20.21 Beginning of the Aorta. (R. = right; L. = left; a. = artery)

All systemic arteries arise from the aorta, which has three principal regions (fig. 20.21):

1. The **ascending aorta** rises about 5 cm above the left ventricle. Its only branches are the coronary arteries, which arise behind two cusps of the aortic valve. Opposite each semilunar valve cusp is an **aortic sinus** containing baroreceptors.
2. The **aortic arch** curves to the left like an inverted U superior to the heart. It gives off three major arteries in this order: the **brachiocephalic**[9] (BRAY-kee-oh-seh-FAL-ic) **trunk, left common carotid** (cah-ROT-id) **artery,** and **left subclavian**[10] (sub-CLAY-vee-un) **artery,** which are further traced in tables 20.4 and 20.5.
3. The **descending aorta** passes downward dorsal to the heart, at first to the left of the vertebral column and then anterior to it, through the thoracic and abdominal cavities. It is called the **thoracic aorta** above the diaphragm and the **abdominal aorta** below. It ends in the lower abdominal cavity by forking into the *right* and *left common iliac arteries,* which are further traced in table 20.8.

---

9. *brachio* = arm + *cephal* = head
10. *sub* = below + *clavi* = clavicle, collarbone

Table 20.4 Arterial Supply to the Head and Neck

Origins of the Head-Neck Arteries

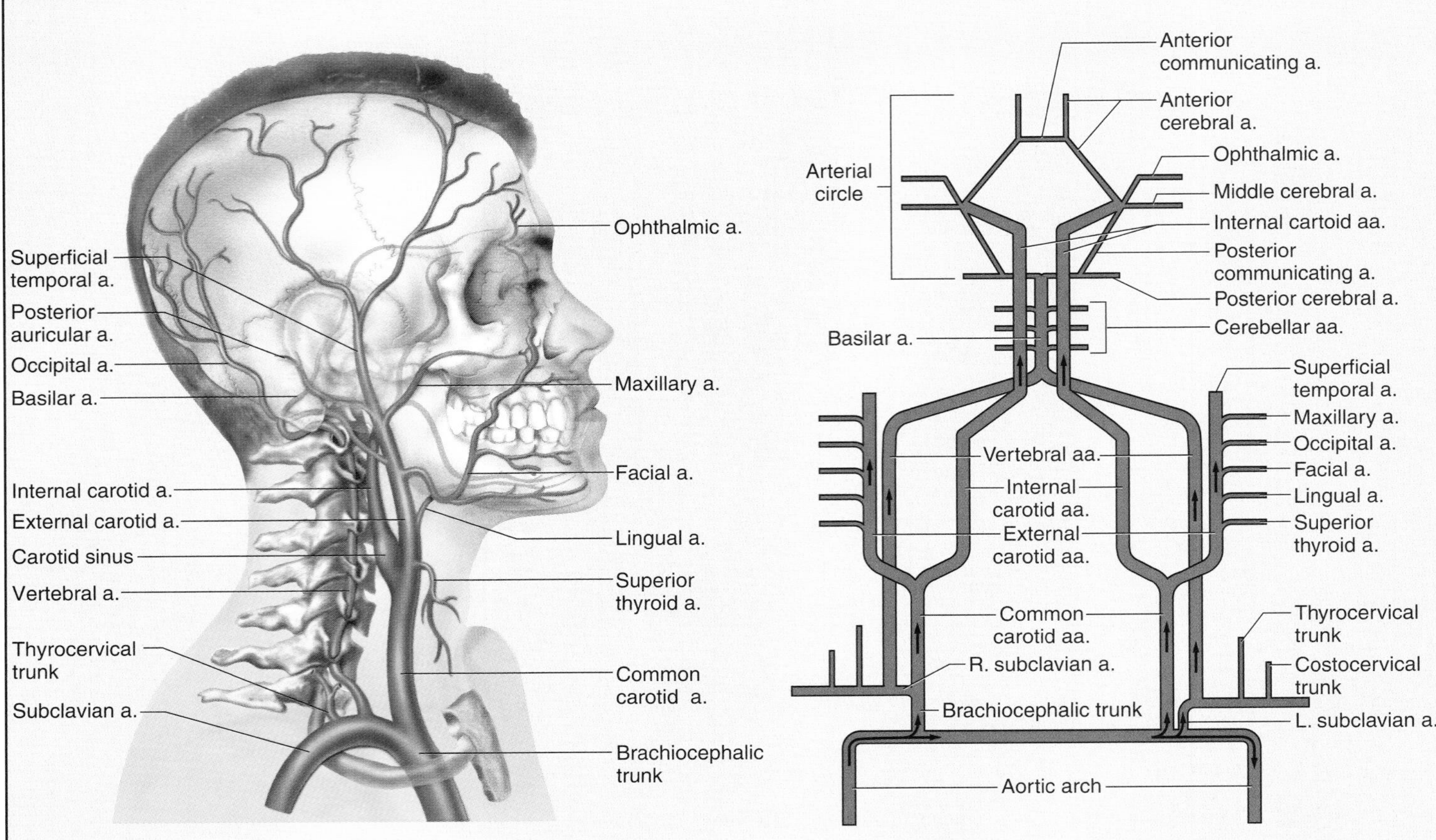

Figure 20.22 Arteries Supplying the Head and Neck.

The head and neck receive blood from four pairs of arteries (fig. 20.22):

1. The **common carotid arteries.** The brachiocephalic trunk divides shortly after leaving the aortic arch and gives rise to the *right subclavian* and *right common carotid arteries.* The *left common carotid* artery arises directly from the aortic arch. The common carotids pass up the anterolateral aspect of the neck, alongside the trachea.
2. The **vertebral arteries** arise from the right and left subclavian arteries. Each travels up the neck through the transverse foramina of the cervical vertebrae and enters the cranial cavity through the foramen magnum.
3. The **thyrocervical**[11] **trunks** are tiny arteries that arise from the subclavian arteries lateral to the vertebral arteries; they supply the thyroid gland and some scapular muscles.
4. The **costocervical**[12] **trunks** (also illustrated in table 20.6) arise from the subclavian arteries a little farther laterally. They perfuse the deep neck muscles and some of the intercostal muscles of the superior rib cage.

Continuation of the Common Carotid Arteries

The common carotid arteries have the most extensive distribution of all the head-neck arteries. Near the laryngeal prominence (Adam's apple), each common carotid branches into an *external carotid artery* and an *internal carotid artery:*

1. The **external carotid artery** ascends the side of the head external to the cranium and supplies most external head structures except the orbits. The external carotid gives rise to the following arteries, in ascending order:
   a. the **superior thyroid artery** to the thyroid gland and larynx,
   b. the **lingual artery** to the tongue,

(continued)

11. *thyro* = thyroid gland + *cerv* = neck
12. *costo* = rib

Table 20.4 Arterial Supply to the Head and Neck (continued)

c. the **facial artery** to the skin and muscles of the face,
d. the **occipital artery** to the posterior scalp,
e. the **maxillary artery** to the teeth, maxilla, buccal cavity, and external ear, and
f. the **superficial temporal artery** to the chewing muscles, nasal cavity, lateral aspect of the face, most of the scalp, and the dura mater surrounding the brain.

2. The **internal carotid artery** passes medial to the angle of the mandible and enters the cranial cavity through the carotid canal of the temporal bone. It supplies the orbits and about 80% of the cerebrum.Compressing the internal carotids near the mandible can therefore cause loss of consciousness.[13] The carotid sinus is located in the internal carotid just above the branch point; the carotid body is nearby. After entering the cranial cavity, each internal carotid artery gives rise to the following branches:
   a. the **ophthalmic artery** to the orbits, nose, and forehead;
   b. the **anterior cerebral artery** to the medial aspect of the cerebral hemisphere (see *arterial circle*); and
   c. the **middle cerebral artery,** which travels in the lateral fissure of the cerebrum and supplies the lateral aspect of the temporal and parietal lobes.

Continuation of the Vertebral Arteries

The vertebral arteries give rise to small branches in the neck that supply the spinal cord and other neck structures, then enter the foramen magnum and merge to form a single **basilar artery** along the anterior aspect of the brainstem. Branches of the basilar artery supply the cerebellum, pons, and inner ear. At the pons-midbrain junction, the basilar artery divides and gives rise to the *arterial circle.*

The Arterial Circle

Blood supply to the brain is so critical that it is furnished by several arterial anastomoses, especially an array of arteries called the **arterial circle (circle of Willis**[14]**)**, which surrounds the pituitary gland and optic chiasm. The arterial circle receives blood from the internal carotid and basilar arteries (fig. 20.23). Only 20% of people have a complete arterial circle. It consists of

1. two **posterior cerebral arteries,**
2. two **posterior communicating arteries,**
3. two **anterior cerebral arteries,** and
4. a single **anterior communicating artery.**

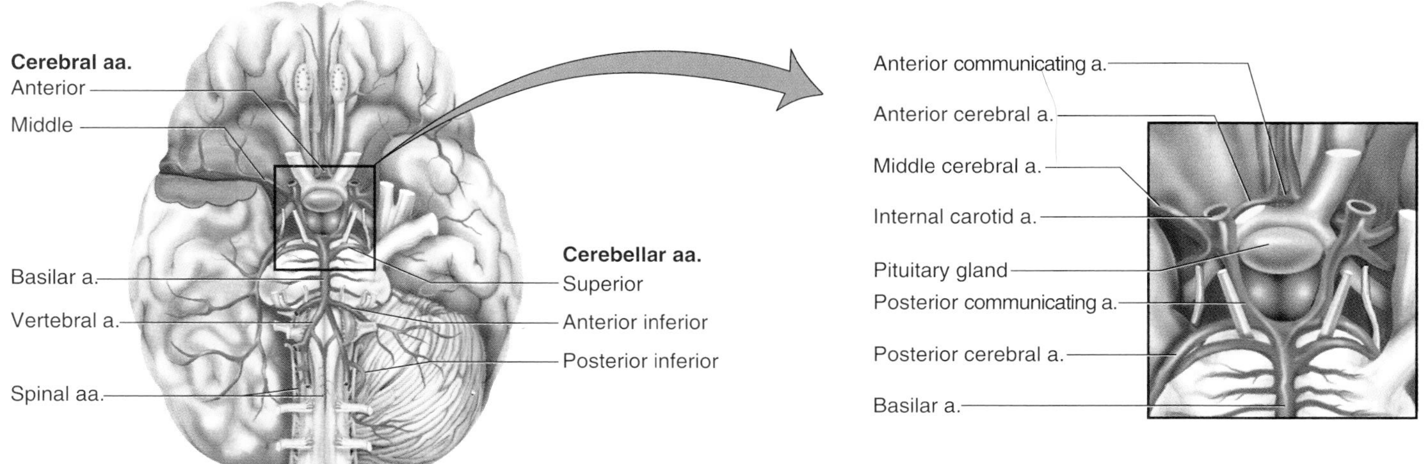

Figure 20.23 The Arterial Circle, Which Supplies the Brain.

13. *carot* = stupor
14. Thomas Willis (1621–75), English anatomist

Table 20.5 Arterial Supply to the Upper Limb

The Shoulder and Arm (brachium)

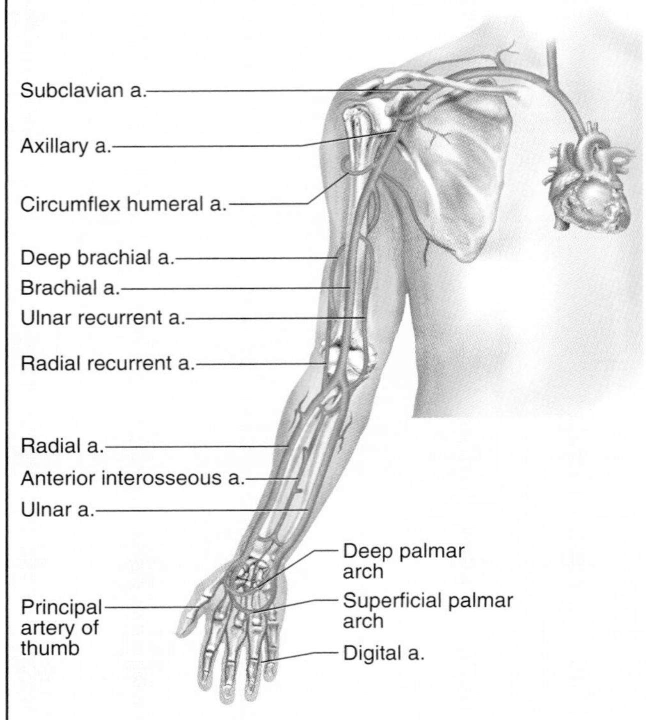

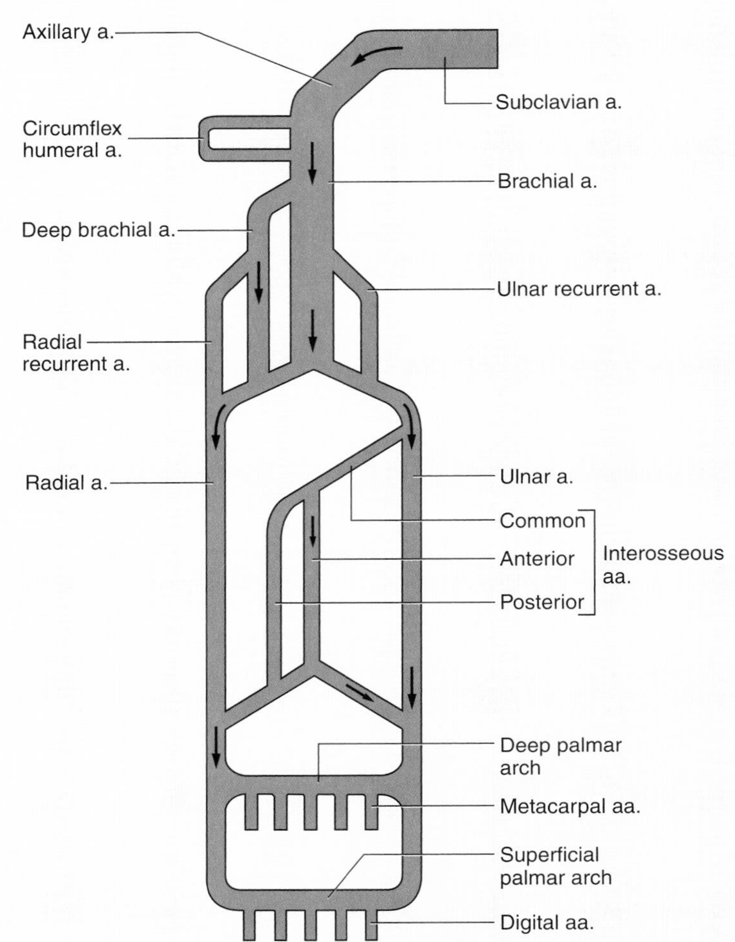

Figure 20.24 Arteries Supplying the Upper Limb.

The origins of the subclavian arteries were described and illustrated in table 20.3. We now trace these further to examine the blood supply to the upper limb (fig. 20.24). This begins with a large artery that changes name from *subclavian* to *axillary* to *brachial* along its course:

1. The **subclavian**[15] **artery** travels between the clavicle and first rib. It gives off several small branches to the thoracic wall and viscera, considered later.
2. The **axillary artery** is the continuation of the subclavian artery through the axillary region. It also gives off small thoracic branches, discussed later, and then ends at the neck of the humerus. Here, it gives off the **circumflex humeral artery,** which encircles the humerus. This loop supplies blood to the shoulder joint and deltoid muscle.
3. The **brachial** (BRAY-kee-ul) **artery** is the continuation of the axillary artery beyond the circumflex. It travels down the medial side of the humerus and ends just distal to the elbow, supplying the anterior flexor muscles of the brachium along the way. It exhibits several anastomoses near the elbow, two of which are noted next. This is the most commonly used artery for routine BP measurements.
4. The **deep brachial artery** arises from the proximal end of the brachial artery and supplies the triceps brachii muscle.
5. The **ulnar recurrent artery** arises about midway along the brachial artery and anastomoses distally with the ulnar artery. It supplies the elbow joint and the triceps brachii.
6. The **radial recurrent artery** leads from the deep brachial artery to the radial artery and supplies the elbow joint and forearm muscles.

(continued)

15. *sub* = below + *clavi* = clavicle

Table 20.5 Arterial Supply to the Upper Limb (continued)

The Forearm (antebrachium)

Just distal to the elbow, the brachial artery divides into the **radial artery** and **ulnar artery,** which travel alongside the radius and ulna, respectively. The most common place to take a pulse is at the radial artery, just proximal to the thumb. Near its origin, the radial artery receives the deep brachial artery. The ulnar artery gives rise, near its origin, to the **anterior** and **posterior interosseous**[16] **arteries,** which travel between the radius and ulna. Structures supplied by these three arteries are as follows:

1. Radial artery: lateral forearm muscles, wrist, thumb, and index finger.
2. Ulnar artery: medial forearm muscles, digits 3 to 5, and medial aspect of index finger.
3. Interosseous arteries: deep flexors and extensors.

The Hand

At the wrist, the radial and ulnar arteries anastomose to form two *palmar arches:*

1. The **deep palmar arch** gives rise to the **metacarpal arteries** of the hand.
2. The **superficial palmar arch** gives rise to the **digital arteries** of the fingers.

16. *inter* = between + osse = bones

Table 20.6 Arterial Supply to the Thorax

The thoracic aorta begins distal to the aortic arch and ends at the **aortic hiatus** (hy-AY-tus), a passage through the diaphragm. Along the way, it sends off numerous small branches to viscera and structures of the body wall (fig. 20.25).

Visceral Branches

These supply the viscera of the thoracic cavity:

1. **Bronchial arteries.** Two of these on the left and one on the right supply the visceral pleura, esophagus, and bronchi of the lungs. They are the systemic blood supply to the lungs mentioned earlier.
2. **Esophageal arteries.** Four or five of these supply the esophagus.
3. **Mediastinal arteries.** Many small mediastinal arteries (not illustrated) supply structures of the posterior mediastinum.

Parietal Branches

The following branches supply chiefly the muscles, bones, and skin of the chest wall; only the first is illustrated:

1. **Posterior intercostal arteries.** Nine pairs of these course around the posterior aspect of the rib cage between the ribs and then anastomose with the anterior intercostal arteries (see following). They supply the skin and subcutaneous tissue, mammary glands, spinal cord and meninges, and the pectoralis, intercostal, and some abdominal muscles.
2. **Subcostal arteries.** A pair of these arise from the aorta, inferior to the twelfth rib, and supply the posterior intercostal tissues, vertebrae, spinal cord, and deep muscles of the back.
3. **Superior phrenic**[17] (FREN-ic) **arteries.** These supply the posterior and superior aspects of the diaphragm.

The thoracic wall is also supplied by the following arteries. The first of these arises from the subclavian artery and the other three from the axillary artery:

1. The **internal thoracic (mammary) artery** supplies the mammary gland and anterior thoracic wall and issues finer branches to the diaphragm and abdominal wall. Near its origin, it gives rise to the **pericardiophrenic artery,** which supplies the pericardium and diaphragm. As the internal thoracic artery descends alongside the sternum, it gives rise to **anterior intercostal arteries** that travel between the ribs and supply the ribs and intercostal muscles.

(continued)

17. *phren* = diaphragm

Table 20.6 Arterial Supply to the Thorax (continued)

2. The **thoracoacromial**[18] (THOR-uh-co-uh-CRO-me-ul) **trunk** supplies the superior shoulder and pectoral regions.
3. The **lateral thoracic artery** supplies the lateral thoracic wall.
4. The **subscapular artery** supplies the scapula, latissimus dorsi, and posterior wall of the thorax.

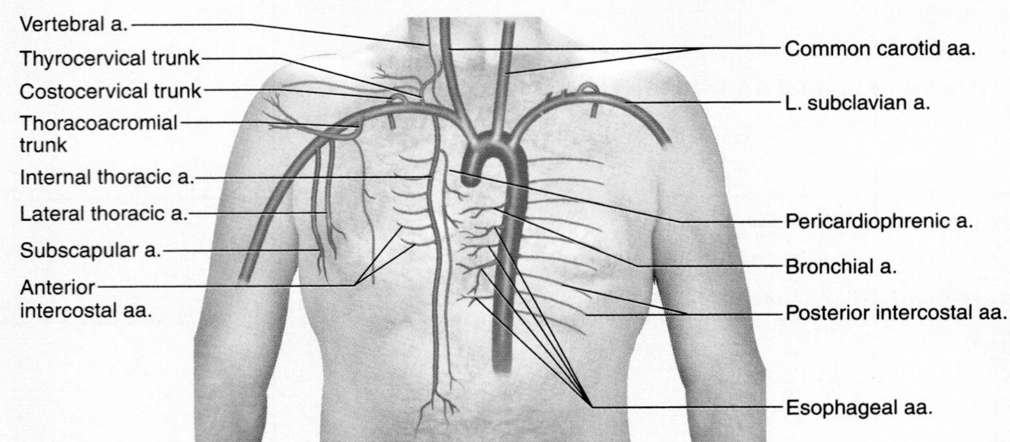

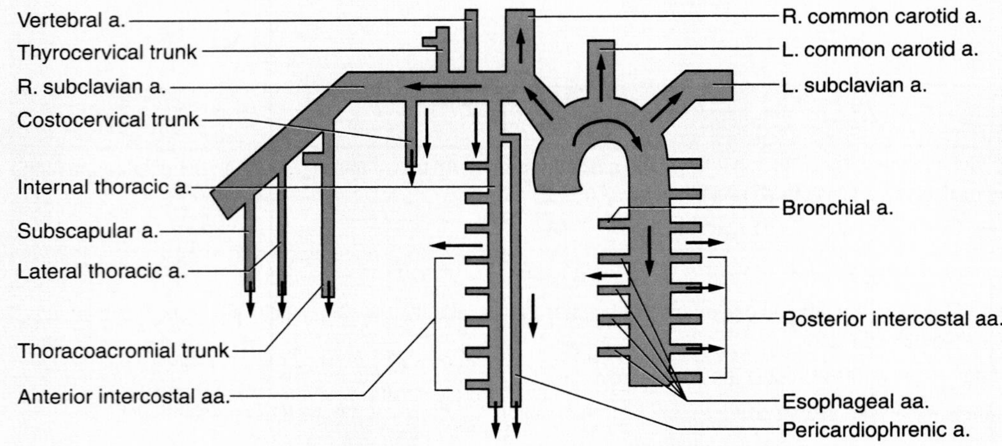

Figure 20.25 Arteries Supplying the Thorax.

18. *thoraco* = chest + *acr* = tip + *om* = shoulder

Table 20.7 Arterial Supply to the Abdomen

Major Branches of Abdominal Aorta

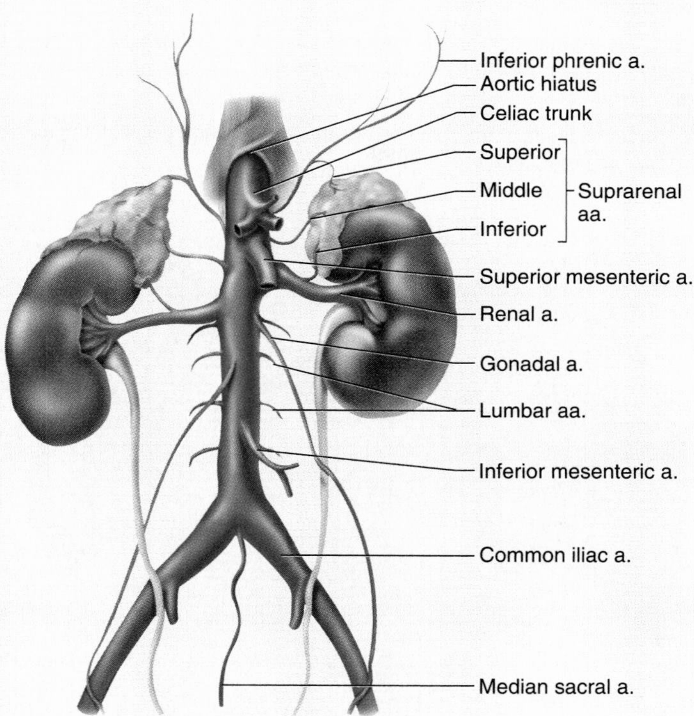

Figure 20.26 The Abdominal Aorta and Its Major Branches.

After passing through the aortic hiatus, the aorta descends through the abdominal cavity. The abdominal aorta is retroperitoneal. It gives off arteries in the order listed here (fig. 20.26). Those indicated in the plural are paired right and left, and those indicated in the singular are single median arteries:

1. The **inferior phrenic arteries** supply the inferior surface of diaphragm and issue a small **superior suprarenal artery** to each adrenal (suprarenal) gland.
2. The **celiac**[19] (SEE-lee-ac) **trunk** issues several branches to the upper abdominal viscera, further traced later in this table.
3. The **superior mesenteric artery** supplies the intestines (see mesenteric circulation later in this table).
4. The **middle suprarenal arteries** arise on either side of the superior mesenteric artery and supply the adrenal glands.
5. The **renal arteries** supply the kidneys and issue a small **inferior suprarenal artery** to each adrenal gland.
6. The **gonadal arteries** are long, narrow, winding arteries that descend from the midabdominal region to the female pelvic cavity or male scrotum. They are called the **ovarian arteries** in females and **testicular arteries** in males. The gonads begin their embryonic development near the kidneys. These arteries acquire their peculiar length and course as the gonads descend to the pelvic cavity during fetal development.
7. The **inferior mesenteric artery** supplies the distal end of the large intestine (see mesenteric circulation).
8. The **lumbar arteries** arise from the lower aorta in four pairs and supply the posterior abdominal wall.
9. The **median sacral artery,** a tiny medial artery at the inferior end of the aorta, supplies the sacrum and coccyx.
10. The **common iliac arteries** arise as the aorta forks at its inferior end. They supply the lower abdominal wall, pelvic viscera (chiefly the urinary and reproductive organs), and lower limbs. They are further traced in table 20.8.

Branches of the Celiac Trunk

The celiac circulation to the upper abdominal viscera is perhaps the most complex route off the abdominal aorta. Because it has numerous anastomoses, the bloodstream does not follow a simple linear path but divides and rejoins itself at several points (fig. 20.27). As you study the following description,

(continued)

19. *celi* = belly, abdomen

Table 20.7 Arterial Supply to the Abdomen (continued)

locate these branches in the figure and identify the points of anastomosis. The short, stubby celiac trunk is a median branch of the aorta. It immediately gives rise to three principal subdivisions—the *common hepatic, left gastric,* and *splenic arteries:*

1. The **common hepatic artery** issues two main branches:
   a. the **gastroduodenal artery,** which supplies the stomach, anastomoses with the right gastroepiploic artery (see following), and then continues as the **inferior pancreaticoduodenal** (PAN-cree-AT-ih-co-dew-ODD-eh-nul) **artery,** which supplies the duodenum and pancreas before anastomosing with the superior mesenteric artery; and
   b. the **proper hepatic artery,** which is the continuation of the common hepatic artery after it gives off the gastroduodenal artery. It enters the inferior surface of the liver and supplies the liver and gallbladder.

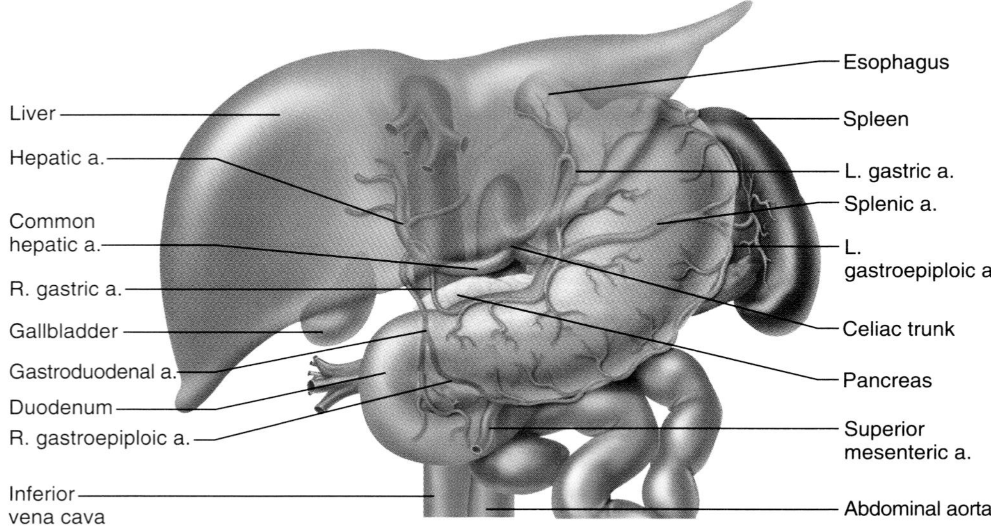

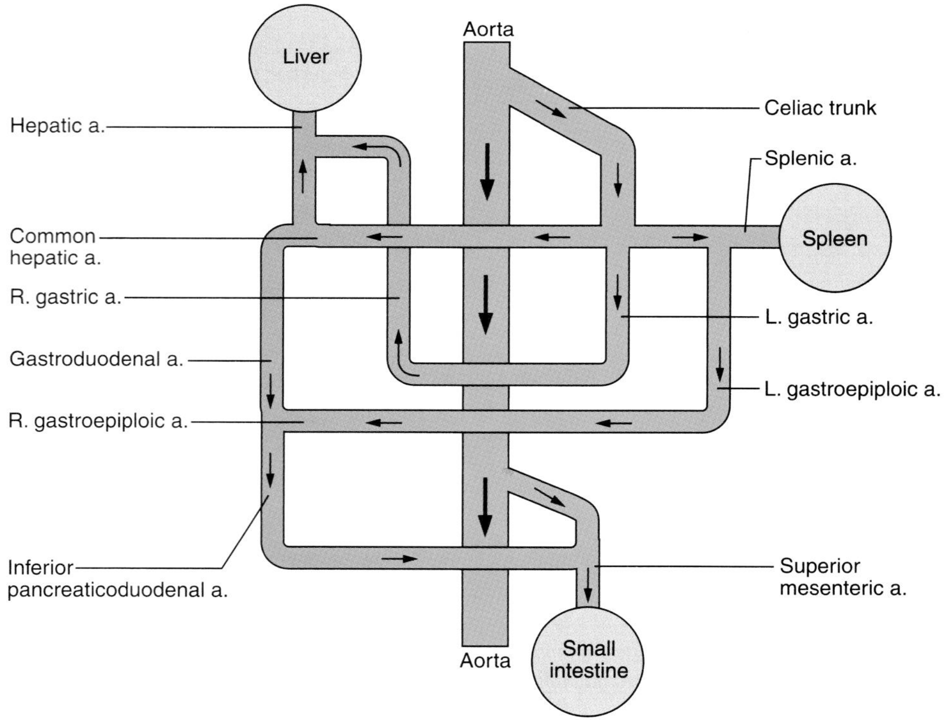

Figure 20.27 Branches of the Celiac Trunk.

(continued)

Table 20.7 Arterial Supply to the Abdomen (continued)

Branches of the Celiac Trunk (continued)

2. The **left gastric artery** supplies the stomach and lower esophagus, arcs around the *lesser curvature* of the stomach, becomes the **right gastric artery** (which supplies the stomach and duodenum), and then anastomoses with the hepatic artery.
3. The **splenic artery** supplies blood to the spleen, but gives off the following branches on its way there:
   a. the **pancreatic arteries** (not illustrated), which supply the pancreas; and
   b. the **left gastroepiploic**[20] (GAS-tro-EP-ih-PLO-ic) **artery,** which arcs around the *greater curvature* of the stomach, becomes the **right gastroepiploic artery,** and then anastomoses with the gastroduodenal artery. Along the way, it supplies blood to the stomach and *greater omentum* (a fatty membrane suspended from the greater curvature).

Mesenteric Circulation

The mesentery (see atlas A, p. 37) contains numerous mesenteric arteries, veins, and lymphatic vessels that perfuse and drain the intestines. The arterial supply issues from the *superior* and *inferior mesenteric arteries* (fig. 20.28); numerous anastomoses between these ensure collateral circulation and adequate perfusion of the intestinal tract even if one route becomes obstructed. The following branches of the **superior mesenteric artery** serve the small intestine and most of the large intestine, among other organs:

1. The **inferior pancreaticoduodenal artery,** already mentioned, is an anastomosis from the gastroduodenal to the superior mesenteric artery; it supplies the pancreas and duodenum.
2. The **intestinal arteries** supply nearly all of the small intestine (jejunum and ileum).
3. The **ileocolic** (ILL-ee-oh-CO-lic) **artery** supplies the ileum of the small intestine and the appendix, cecum, and ascending colon.
4. The **right colic artery** supplies the ascending colon.
5. The **middle colic artery** supplies the transverse colon.

Branches of the *inferior mesenteric artery* serve the distal part of the large intestine:

1. The **left colic artery** supplies the transverse and descending colon.
2. The **sigmoid arteries** supply the descending and sigmoid colon.
3. The **superior rectal artery** supplies the rectum.

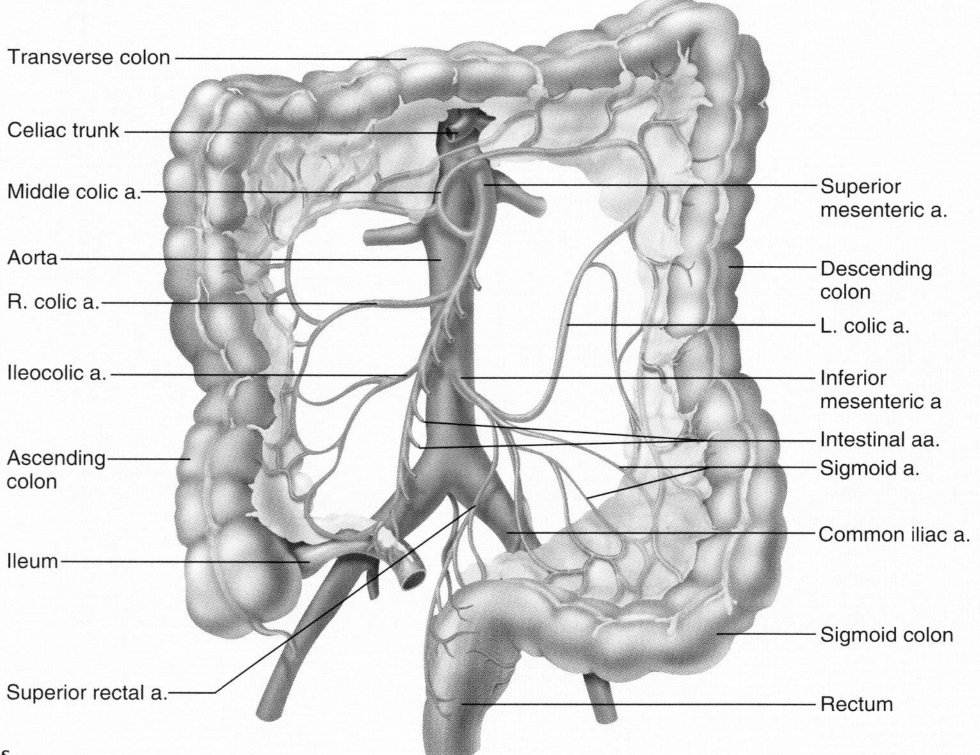

Figure 20.28 The Mesenteric Arteries.

20. *gastro* = stomach + *epi* = upon, above + *ploic* = pertaining to the greater omentum

Table 20.8 Arterial Supply to the Pelvic Region and Lower Limb

The common iliac arteries arise from the aorta at the level of vertebra L4 and continue for about 5 cm. At the level of the sacroiliac joint, each divides into an internal and external iliac artery. The **internal iliac artery** supplies mainly the pelvic wall and viscera, and the **external iliac artery** supplies mainly the lower limb (figs. 20.29 and 29.30).

Branches of the Internal Iliac Artery

1. The **iliolumbar** and **lateral sacral arteries** supply the wall of the pelvic region.
2. The **middle rectal artery** supplies the rectum.
3. The **superior** and **inferior vesical**[21] **arteries** supply the urinary bladder.
4. The **uterine** and **vaginal arteries** supply the uterus and vagina.
5. The **superior** and **inferior gluteal arteries** supply the gluteal muscles.
6. The **obturator artery** supplies the adductor muscles of the medial thigh.
7. The **internal pudendal**[22] (pyu-DEN-dul) **artery** serves the perineum and external genitals; it supplies the blood for vascular engorgement during sexual arousal.

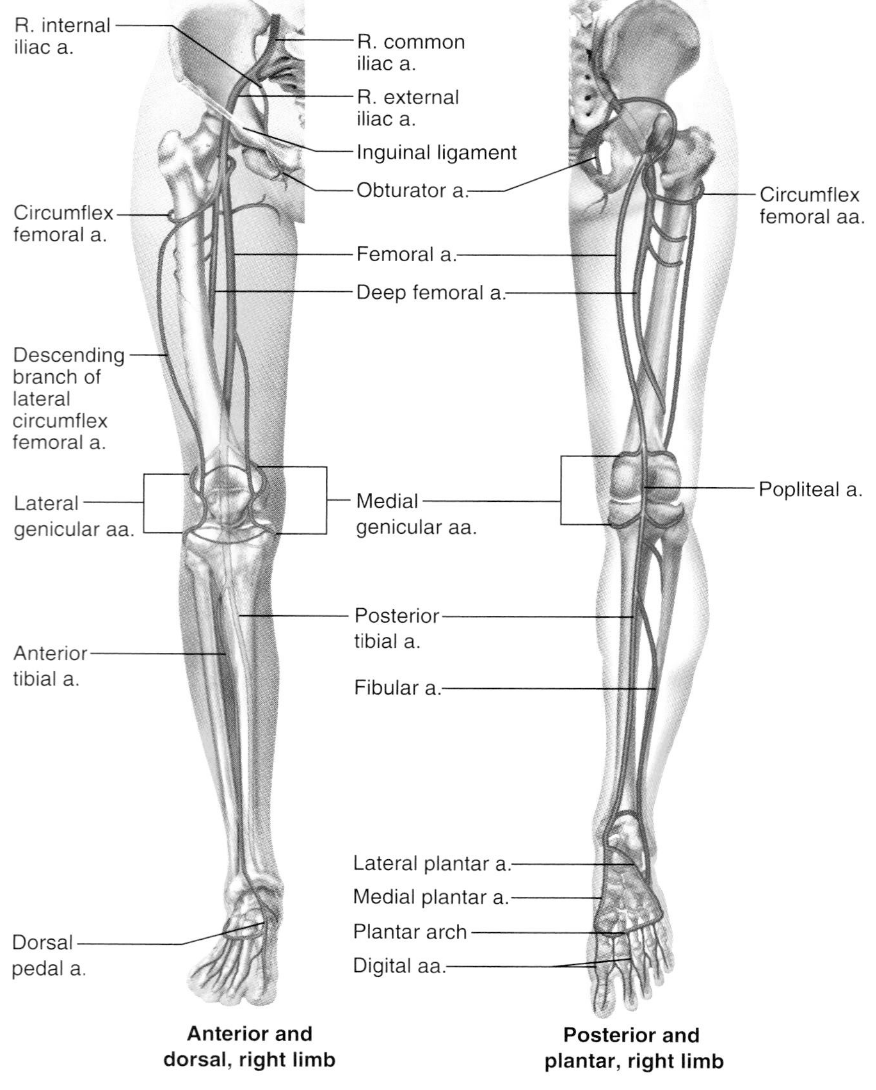

Figure 20.29 Arteries Supplying the Lower Limb.

21. *vesic* = bladder
22. *pudend* = literally "shameful parts"; the external genitals

(continued)

Table 20.8 Arterial Supply to the Pelvic Region and Lower Limb (continued)

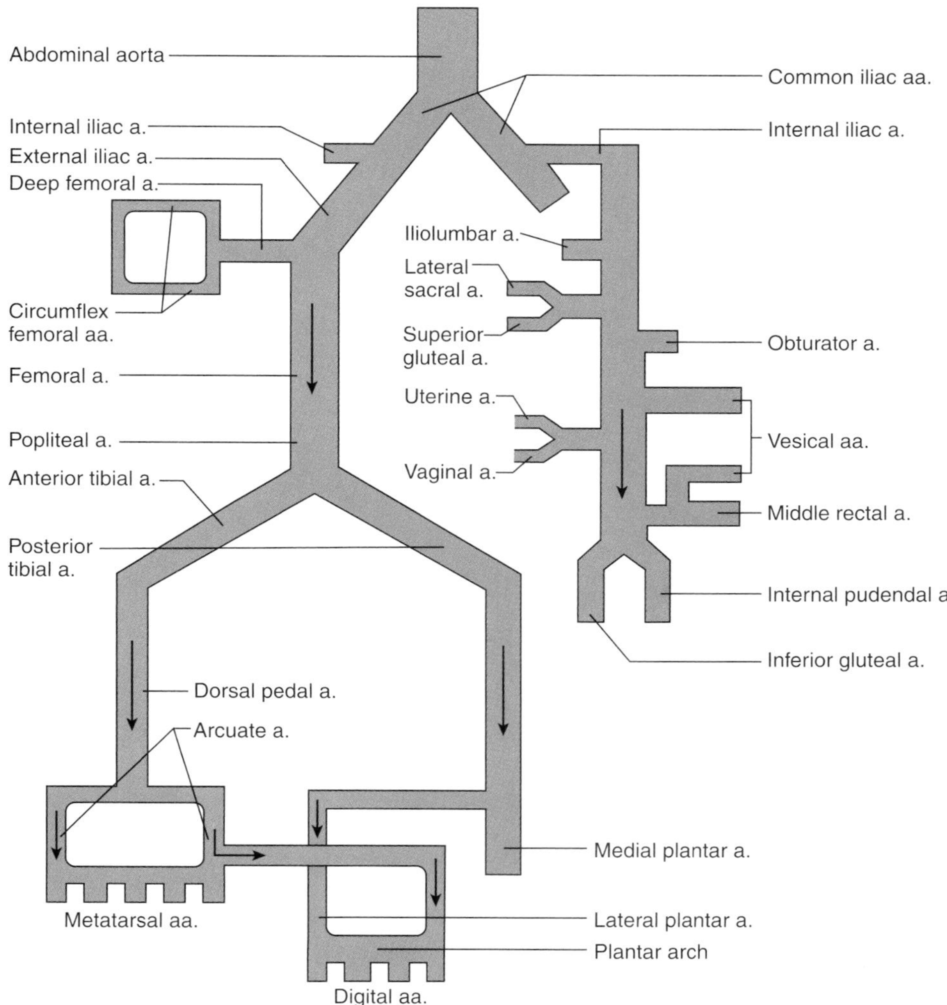

Figure 20.30 Arterial Flowchart of the Lower Limb.

Branches of the External Iliac Artery

The external iliac artery sends branches to the skin and muscles of the abdominal wall and pelvic girdle. It then passes deep to the inguinal ligament and gives rise to branches that serve mainly the lower limbs:

1. The **femoral artery** passes through the femoral triangle of the upper medial thigh, where its pulse can be palpated. It gives off the following branches to supply the thigh region:
   a. The **deep femoral artery,** which supplies the hamstring muscles; and
   b. The **circumflex femoral arteries,** which encircle the neck of the femur and supply the femur and hamstring muscles.
2. The **popliteal artery** is a continuation of the femoral artery in the popliteal fossa at the rear of the knee. It produces anastomoses **(genicular arteries)** that supply the knee and then divides into the anterior and posterior tibial arteries.
3. The **anterior tibial artery** travels lateral to the tibia in the anterior compartment of the leg, where it supplies the extensor muscles. It gives rise to
   a. the **dorsal pedal artery,** which traverses the ankle and dorsum of the foot; and
   b. the **arcuate artery,** a continuation of the dorsal pedal artery that gives off the **metatarsal arteries** of the foot.
4. The **posterior tibial artery** travels through the posteromedial part of the leg and supplies the flexor muscles. It gives rise to
   a. the **fibular (peroneal) artery,** which arises from the proximal end of the posterior tibial artery and supplies the lateral peroneal muscles;
   b. the **lateral** and **medial plantar arteries,** which arise by bifurcation of the posterior tibial artery at the ankle and supply the plantar surface of the foot; and
   c. the **plantar arch,** an anastomosis from the lateral plantar artery to the dorsal pedal artery that gives rise to the **digital arteries** of the toes.

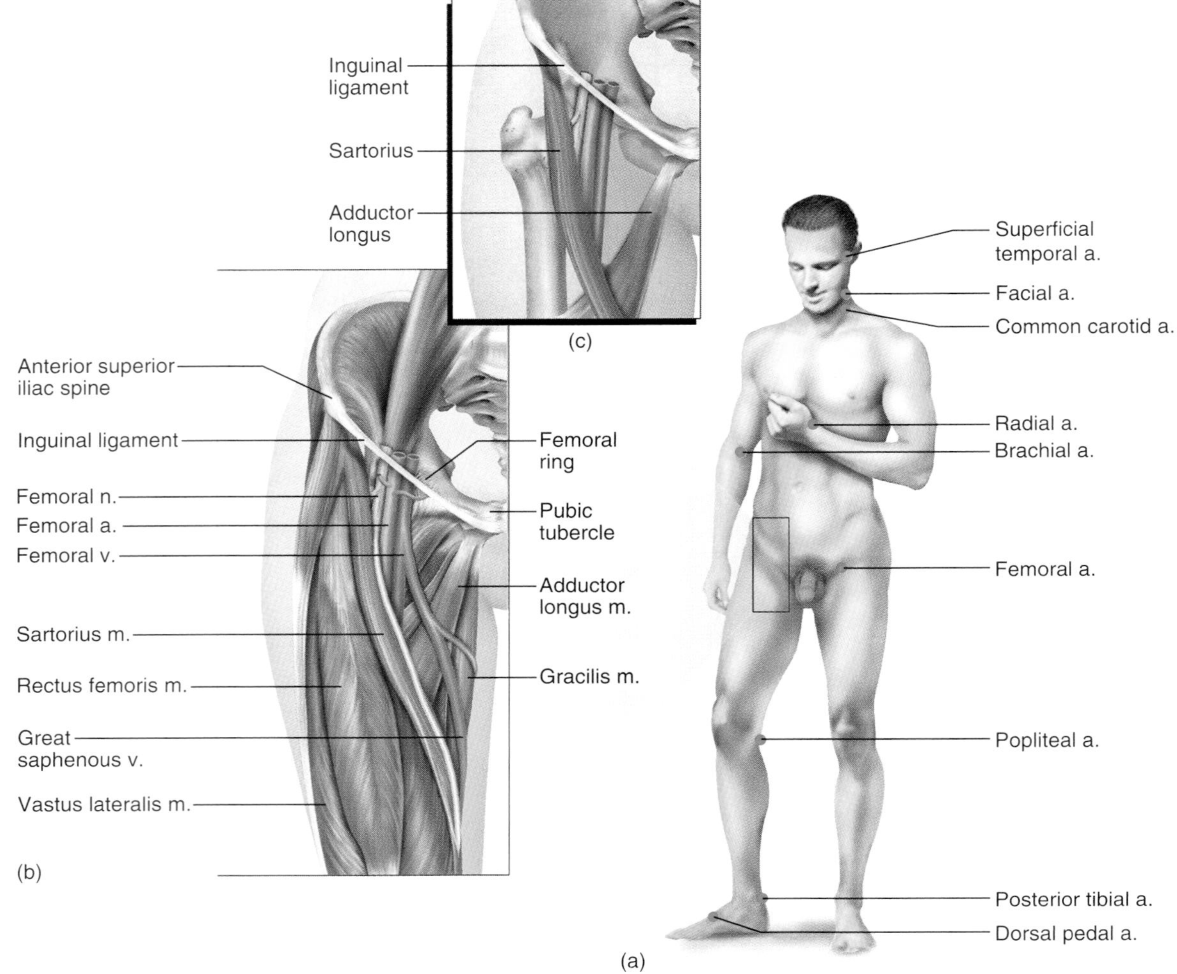

Figure 20.31 Arterial Pressure Points. (*a*) Areas where arteries lie close enough to the surface that a pulse can be palpated or pressure can be applied to reduce arterial bleeding. (*b*) Structures in the femoral triangle. (*c*) Boundaries of the femoral triangle.

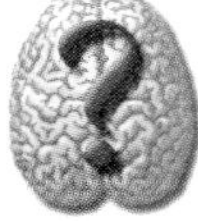

Think About It

There are certain similarities between the arteries of the hand and foot. What arteries of the wrist and hand are most comparable in arrangement and function to the arcuate artery and plantar arch of the foot?

In some places, major arteries come close enough to the body surface to be palpated. These places can be used to take a pulse, and they can serve as emergency **pressure points** where firm pressure can be applied to temporarily reduce arterial bleeding (fig. 20.31*a*). One of these points is the **femoral triangle** of the upper medial thigh (fig. 20.31*b, c*). This is an important landmark for arterial supply, venous drainage, and innervation of the lower limb. Its boundaries are the sartorius muscle laterally, the inguinal ligament superiorly, and the adductor longus muscle medially. The femoral artery, vein, and nerve run close to the surface at this point.

Before You Go On

Answer the following questions to test your understanding of the preceding section:

25. Concisely contrast the destinations of the external and internal carotid arteries.
26. Briefly state the tissues that are supplied with blood by (a) the arterial circle, (b) the celiac trunk, (c) the superior mesenteric artery, and (d) the external iliac artery.
27. Trace the path of an RBC from the left ventricle to the metatarsal arteries. State two places along this path where you can palpate the arterial pulse.

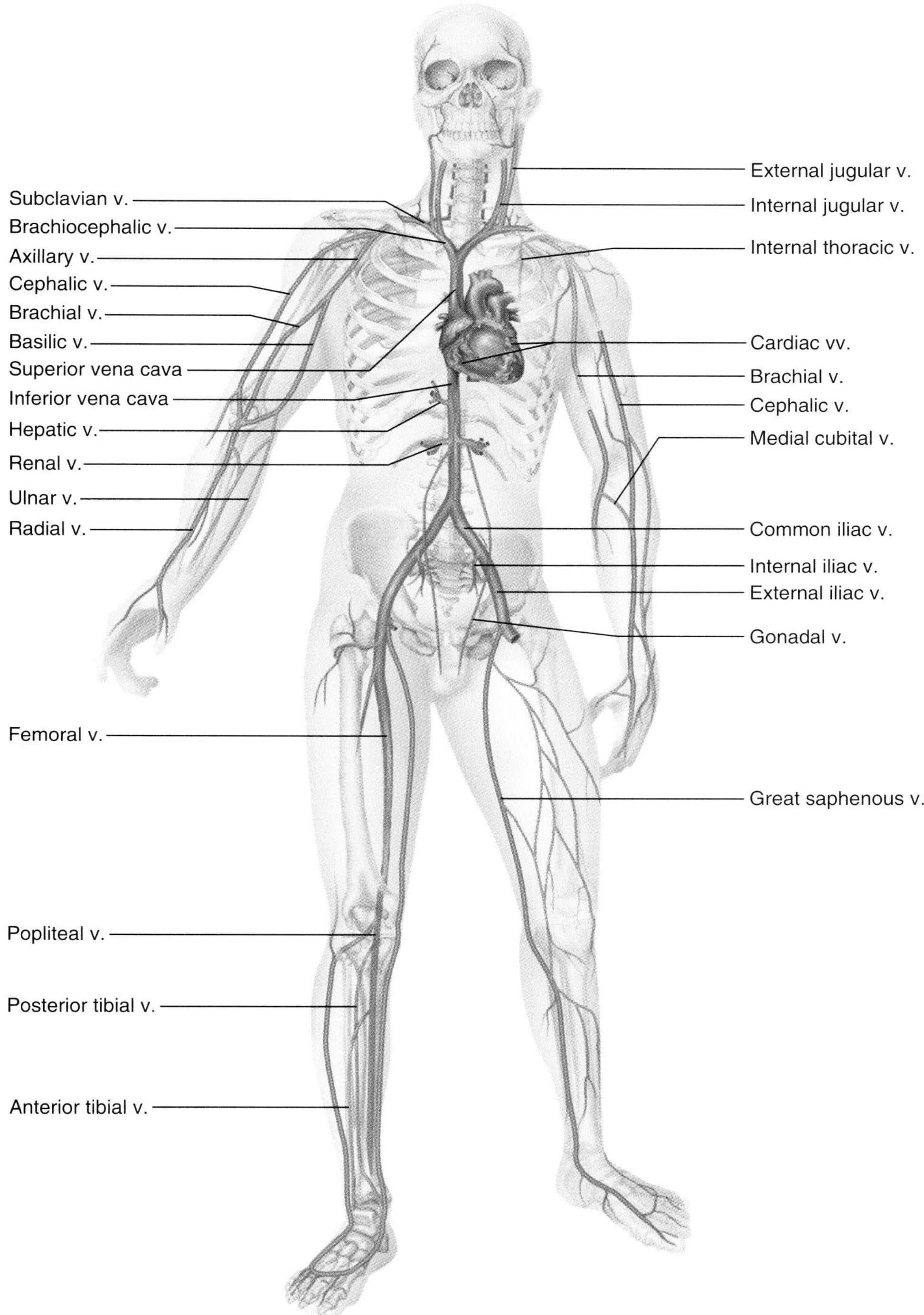

Figure 20.32 **The Major Systemic Veins.** (v. = vein; vv. = veins.)

Anatomy of the Systemic Veins

Objectives

When you have completed this section, you should be able to

- identify the principal veins of the systemic circuit; and
- trace the flow of blood from any major organ to the heart.

The principal veins of the systemic circuit (fig. 20.32) are detailed in tables 20.9 through 20.14. While arteries are usually deep and well protected, veins occur in both deep and superficial groups; you may be able to see quite a few of them in your arms and hands. Deep veins run parallel to the arteries and often have similar names (*femoral artery* and *femoral vein,* for example); this is not true of the superficial veins, however. The deep veins are not described in as much detail as the arteries were, since it can usually be assumed that they drain the same structures as the corresponding arteries supply.

In general, we began the study of arteries with those lying close to the heart and progressed away. In the venous system, by contrast, we begin with those that are remote from the heart and follow the flow of blood as they join each other and approach the heart. Venous pathways have more anastomoses than arterial pathways, so the route of blood flow is often not as clear. Many anastomoses are omitted from the following figures for clarity.

Table 20.9 Venous Drainage of the Head and Neck (continued)

Dural Sinuses

Large thin-walled veins called **dural sinuses** occur within the cranial cavity between layers of dura mater. They receive blood from the brain and face and empty into the internal jugular veins:

1. The **superior** and **inferior sagittal sinuses** are found in the falx cerebri between the cerebral hemispheres; they receive blood that has circulated through the brain.
2. The **cavernous sinuses** occur on each side of the body of the sphenoid bone; they receive blood from the **ophthalmic vein** draining the orbit and the **facial vein** draining the nose and upper lip.
3. The **transverse (lateral) sinuses** encircle the inside of the occipital bone and lead to the jugular foramen on each side. They receive blood from the previously mentioned sinuses and empty into the internal jugular veins.

Major Veins of the Neck

Blood flows down the neck mainly through three veins on each side, all of which empty into the subclavian vein:

1. The **internal jugular**[23] (JUG-you-lur) **vein** courses down the neck, alongside the internal carotid artery, deep to the sternocleidomastoid muscle. It receives most of the blood from the brain, picks up blood from the **facial vein** and **superficial temporal vein** along the way, passes deep to the clavicle, and joins the subclavian vein. (Note that the facial vein empties into both the cavernous sinus and the internal jugular vein.)
2. The **external jugular vein** drains tributaries from the parotid gland, facial muscles, scalp, and other superficial structures. Some of this blood also follows venous anastomoses to the internal jugular vein. The external jugular vein courses down the side of the neck superficial to the sternocleidomastoid muscle and empties into the subclavian vein.
3. The **vertebral vein** travels with the vertebral artery in the transverse foramina of the cervical vertebrae. Although the companion artery leads to the brain, the vertebral vein does not come from there. It drains the cervical vertebrae, spinal cord, and some of the small deep muscles of the neck.

Drainage from Shoulder to Heart

From the shoulder region, blood takes the following path to the heart:

1. The **subclavian vein** drains the arm and travels inferior to the clavicle; receives the external jugular, vertebral, and internal jugular veins in that order; and ends where it receives the internal jugular.
2. The **brachiocephalic vein** is formed by union of the subclavian and internal jugular veins. It continues medially and receives tributaries draining the upper thoracic wall and mammary gland.
3. The **superior vena cava** is formed by the union of the right and left brachiocephalic veins. It travels inferiorly for about 7.5 cm and empties into the right atrium. It drains all structures superior to the diaphragm except the pulmonary circuit and coronary circulation. It also receives considerable drainage from the abdominal cavity by way of the azygos system (see table 20.11).

23. *jugul* = neck, throat

Table 20.9 Venous Drainage of the Head and Neck

Most blood of the head and neck is drained by three pairs of veins—the *internal jugular, external jugular,* and *vertebral veins.* This table traces their origins and drainage and follows them to the formation of the *brachiocephalic veins* and *superior vena cava* (fig. 20.33).

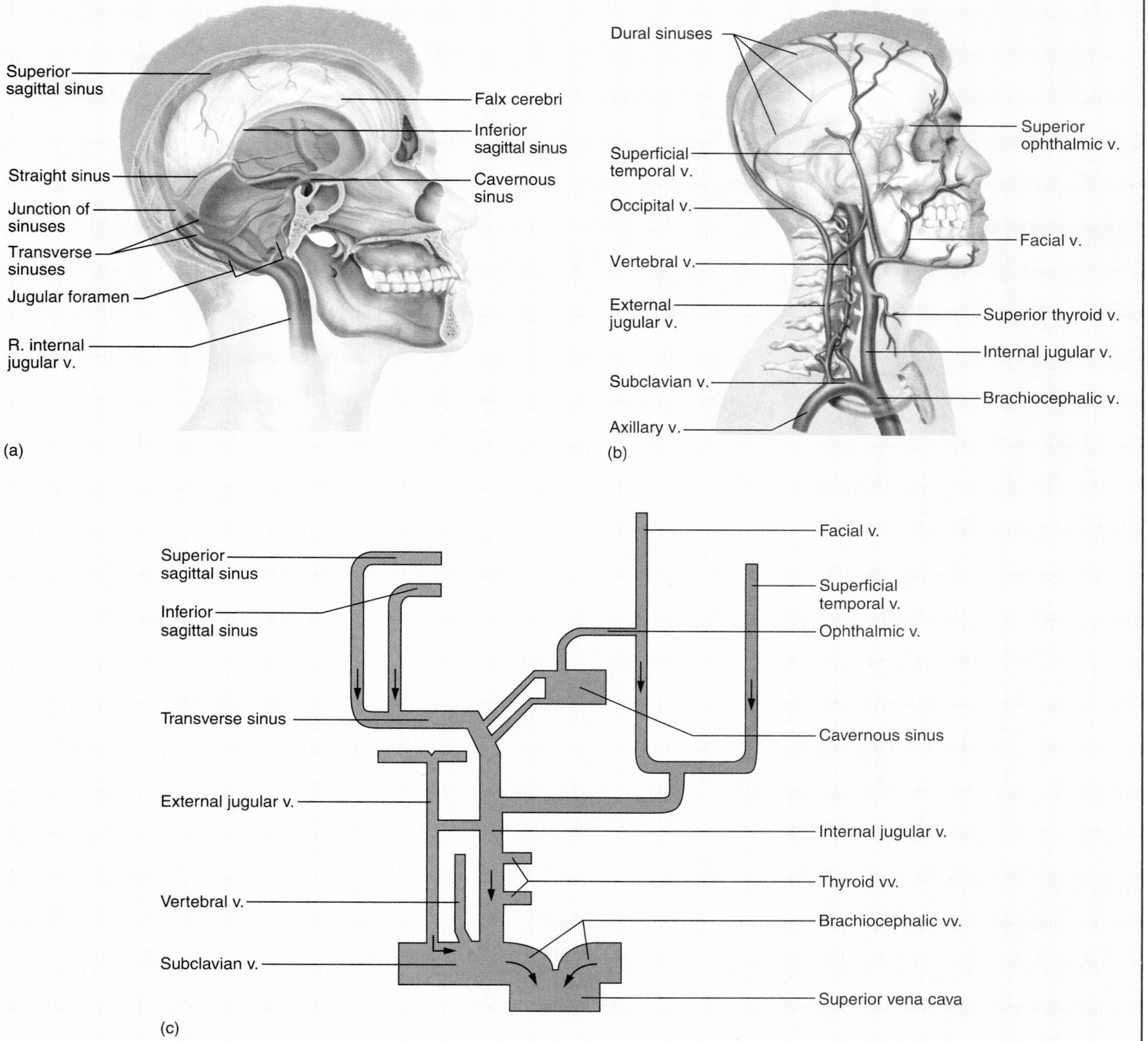

Figure 20.33 Veins Draining the Head and Neck. (*a*) Deep venous drainage. (*b*) Superficial venous drainage. (*c*) Flowchart of venous drainage.

Table 20.10 Venous Drainage of the Upper Limb

Table 20.9 briefly noted the subclavian veins that drain each arm. This table begins distally in the forearm and traces venous drainage to the subclavian vein (fig. 20.34).

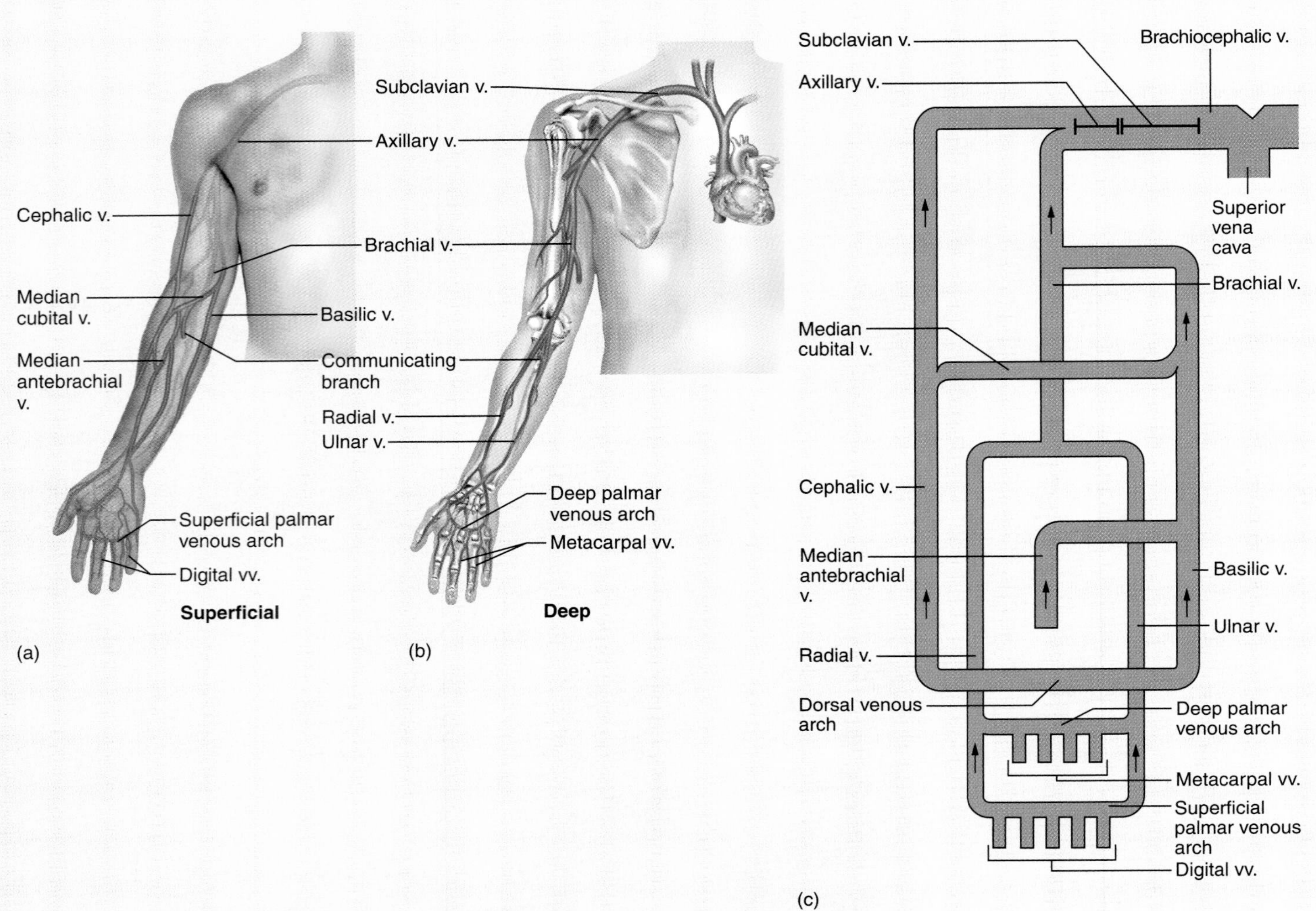

Figure 20.34 Veins Draining the Upper Limb. (*a*) Superficial venous drainage. (*b*) Deep venous drainage. (*c*) Flowchart of venous drainage.

Deep Veins

1. The **digital veins** drain each finger into the **superficial palmar venous arch.**
2. The **metacarpal veins** parallel the metacarpal bones and drain blood from the hand into the **deep palmar venous arch.** Both the superficial and deep palmar venous arches are anastomoses between the next two veins, which are the major deep veins of the forearm.
3. The **radial vein** receives blood from the lateral side of both palmar arches and courses up the forearm alongside the radius.
4. The **ulnar vein** receives blood from the medial side of both palmar arches and courses up the forearm alongside the ulna.
5. The **brachial vein** is formed by the union of the radial and ulnar veins at the elbow; it courses up the brachium.
6. The **axillary vein** is formed at the axilla by the union of the brachial and basilic veins (the basilic vein is described in the next section).
7. The **subclavian vein** is a continuation of the axillary vein into the shoulder inferior to the clavicle. The further course of the subclavian is explained in the previous table.

(continued)

Table 20.10 Venous Drainage of the Upper Limb (continued)

Superficial Veins

These are easily seen through the skin of most people and are larger in diameter than the deep veins:

1. The **dorsal venous arch** (not illustrated) is a plexus of veins visible on the back of the hand; it empties into the major superficial veins of the forearm, the cephalic and basilic.
2. The **cephalic vein** arises from the lateral side of the dorsal venous arch, winds around the radius as it travels up the forearm, continues up the lateral aspect of the brachium to the shoulder, and joins the axillary vein there. Intravenous fluids are often administered through the distal end of this vein.
3. The **basilic**[24] (bah-SIL-ic) **vein** arises from the medial side of the dorsal venous arch, travels up the posterior aspect of the forearm, and continues into the brachium. About midway up the brachium it turns deeper and runs beside the brachial artery. At the axilla it joins the brachial vein, and the union of these two gives rise to the axillary vein.
4. The **median cubital vein** is a short anastomosis between the cephalic and basilic veins that obliquely crosses the cubital fossa (anterior bend of the elbow). It is clearly visible through the skin and is the most common site for drawing blood.
5. The **median antebrachial vein** originates near the base of the thumb, travels up the forearm between the radial and ulnar veins, and terminates at the elbow; it empties into the cephalic vein in some people and into the basilic vein in others.

24. *basilic* = royal, prominent, important

Table 20.11 The Azygos System

The superior vena cava receives extensive drainage from the thoracic and abdominal walls by way of the **azygos** (AZ-ih-goss) **system** (fig. 20.35).

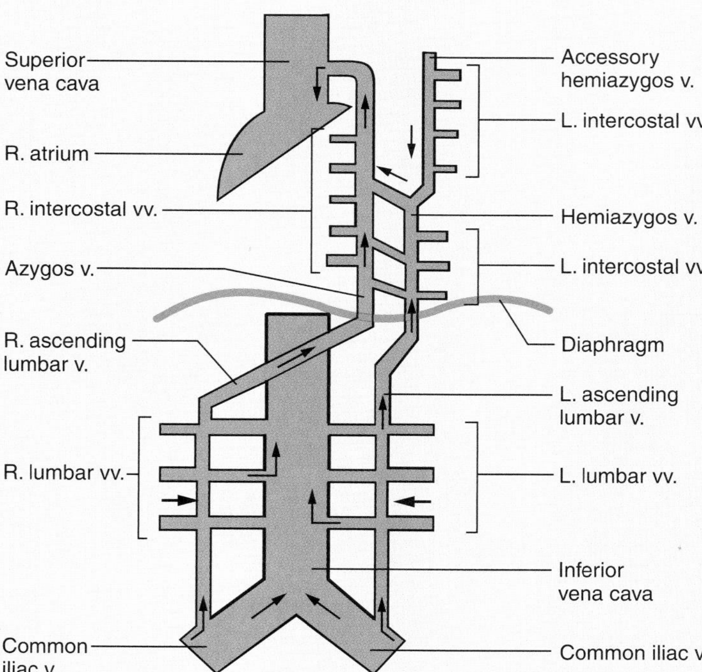

Figure 20.35 Veins of the Azygos System.

Drainage of the Abdominal Wall

A pair of **ascending lumbar veins** receive blood from the common iliac veins below and a series of short horizontal **lumbar veins** that drain the abdominal wall. The ascending lumbar veins anastomose with the inferior vena cava beside them and ascend through the diaphragm into the thoracic cavity.

Table 20.11 The Azygos System (continued)

Drainage of the Thorax
Right side. After penetrating the diaphragm, the right ascending lumbar vein becomes the **azygos**[25] **vein** of the thorax. The azygos receives blood from the right **posterior intercostal veins,** which drain the chest muscles, and from the **esophageal, mediastinal, pericardial,** and **right bronchial veins.** It then empties into the superior vena cava at the level of vertebra T4.
Left side. The left ascending lumbar vein continues into the thorax as the **hemiazygos**[26] **vein.** The hemiazygos drains the ninth through eleventh posterior intercostal veins and some esophageal and mediastinal veins on the left. At midthorax, it crosses over to the right side and empties into the azygos vein.
The **accessory hemiazygos vein** is a superior extension of the hemiazygos. It drains the fourth through eighth posterior intercostal veins and the left bronchial vein. It also crosses to the right side and empties into the azygos vein.

25. unpaired; from *a* = without + *zygo* = union, mate
26. *hemi* = half

Table 20.12 Major Tributaries of the Inferior Vena Cava

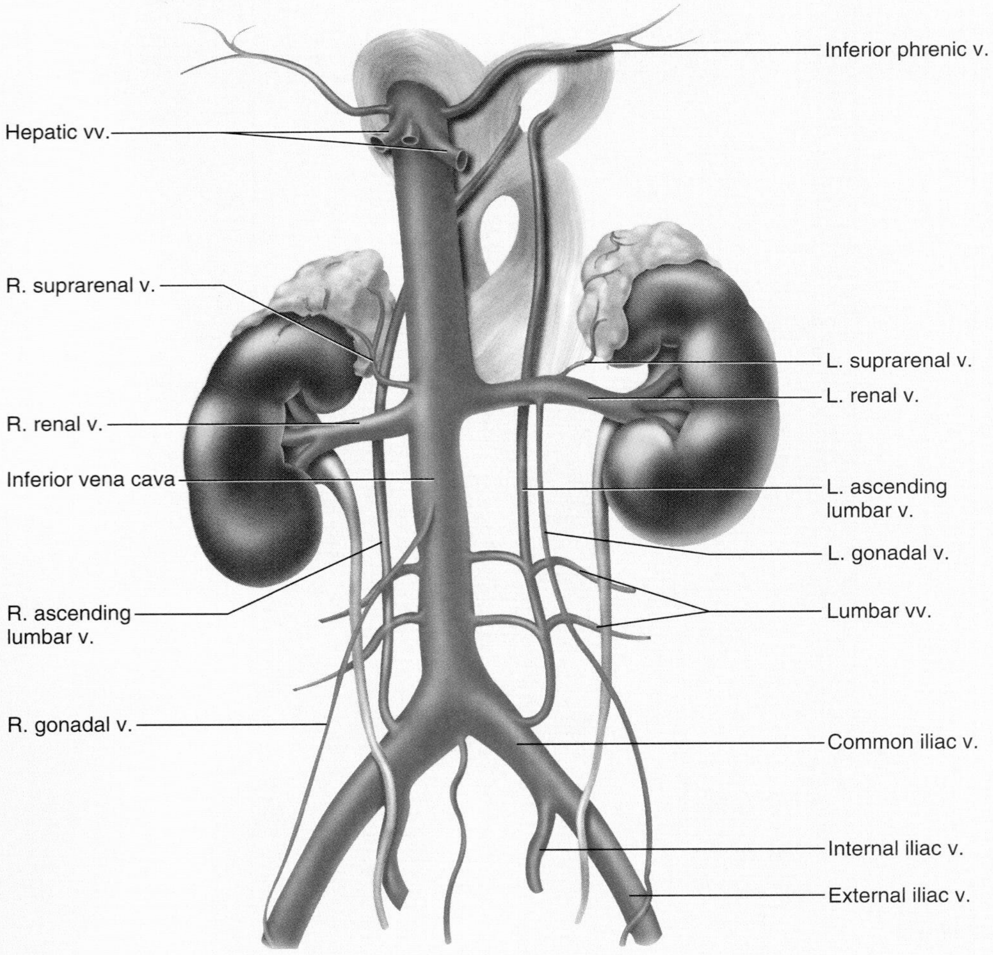

Figure 20.36 The Inferior Vena Cava and Its Tributaries.

Table 20.12 Major Tributaries of the Inferior Vena Cava (continued)

The **inferior vena cava (IVC)** is formed by the union of the right and left common iliac veins at the level of vertebra L5. It is retroperitoneal and lies immediately to the right of the aorta. Its diameter of 3.5 cm is the largest of any vessel in the body. As it ascends the abdominal cavity, the IVC picks up blood from numerous tributaries in the order listed here (fig. 20.36):

1. Some **lumbar veins** empty into the IVC as well as into the ascending lumbar veins described in table 20.11.
2. The **gonadal veins** (**ovarian veins** in the female and **testicular veins** in the male) drain the gonads. The right gonadal vein empties directly into the IVC, whereas the **left gonadal vein** empties into the left renal vein.
3. The **renal veins** drain the kidneys into the IVC. The left renal vein also receives blood from the left gonadal and left suprarenal veins.
4. The **suprarenal veins** drain the adrenal (suprarenal) glands. The right suprarenal empties directly into the IVC, and the left suprarenal empties into the renal vein.
5. The **hepatic veins** drain the liver; they extend a short distance from its superior surface to the IVC.
6. The **inferior phrenic veins** drain the inferior aspect of the diaphragm.

After receiving these inputs, the IVC penetrates the diaphragm and enters the right atrium from below. It does not receive any thoracic drainage.

Table 20.13 The Hepatic Portal System

The **hepatic portal system** connects capillaries of the intestines and other digestive organs to the **hepatic sinusoids** of the liver. The intestinal blood is richly laden with nutrients for a few hours following a meal. The hepatic portal system gives the liver "first claim" to these nutrients before the blood is distributed to the rest of the body. It also allows the blood to be cleansed of bacteria and toxins picked up from the intestines, an important function of the liver. The route from the intestines to the inferior vena cava follows (fig. 20.37):

1. The **inferior mesenteric vein** receives blood from the rectum and distal part of the large intestine. It converges in a fanlike array in the mesentery and empties into the splenic vein.
2. The **superior mesenteric vein** receives blood from the entire small intestine, ascending colon, transverse colon, and stomach. It, too, exhibits a fanlike arrangement in the mesentery and then joins the splenic vein to form the hepatic portal vein.

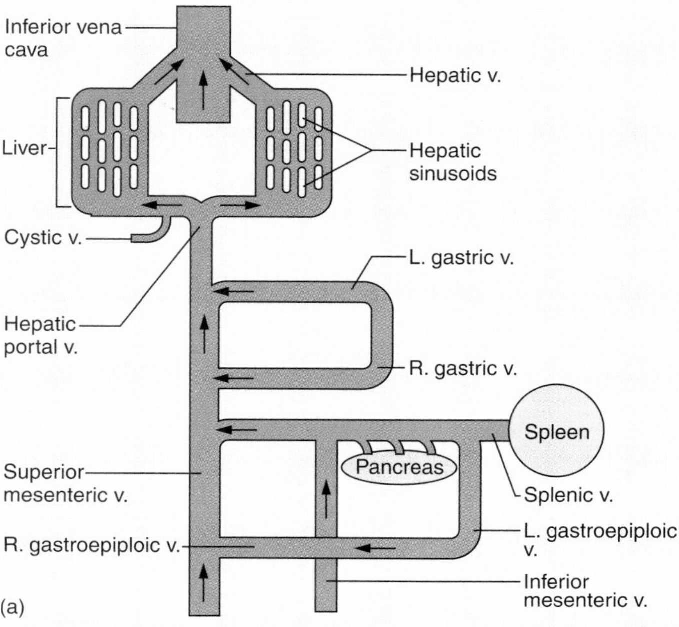

Figure 20.37a Flowchart of the Hepatic Portal System.

(continued)

Table 20.13 The Hepatic Portal System (continued)

3. The **splenic vein** drains the spleen and travels across the abdominal cavity toward the liver. Along the way, it picks up the **pancreatic veins** from the pancreas and the inferior mesenteric vein. It changes name when it joins the superior mesenteric vein, as explained next.
4. The **hepatic portal vein** is formed by convergence of the splenic and superior mesenteric veins. It travels about 8 cm up and to the right and then enters the inferior surface of the liver. Near this point it receives the **cystic vein** from the gallbladder. In the liver, the hepatic portal vein ultimately leads to the innumerable microscopic hepatic sinusoids. Blood from the sinusoids empties into the hepatic veins described earlier. Circulation within the liver is described in more detail in chapter 25.
5. The left and right **gastric veins** form an arch along the lesser curvature of the stomach and empty into the hepatic portal vein.

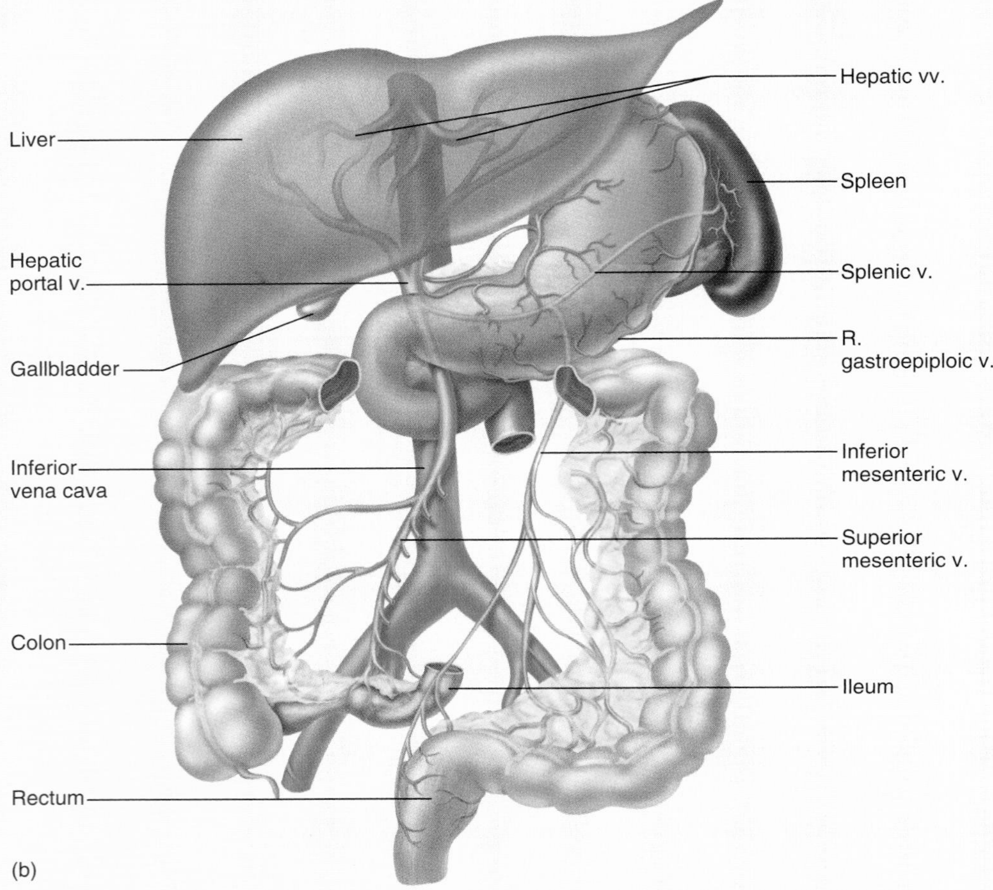

Figure 20.37b Veins of the Hepatic Portal System.

Table 20.14 Venous Drainage of the Lower Limb and Pelvic Organs

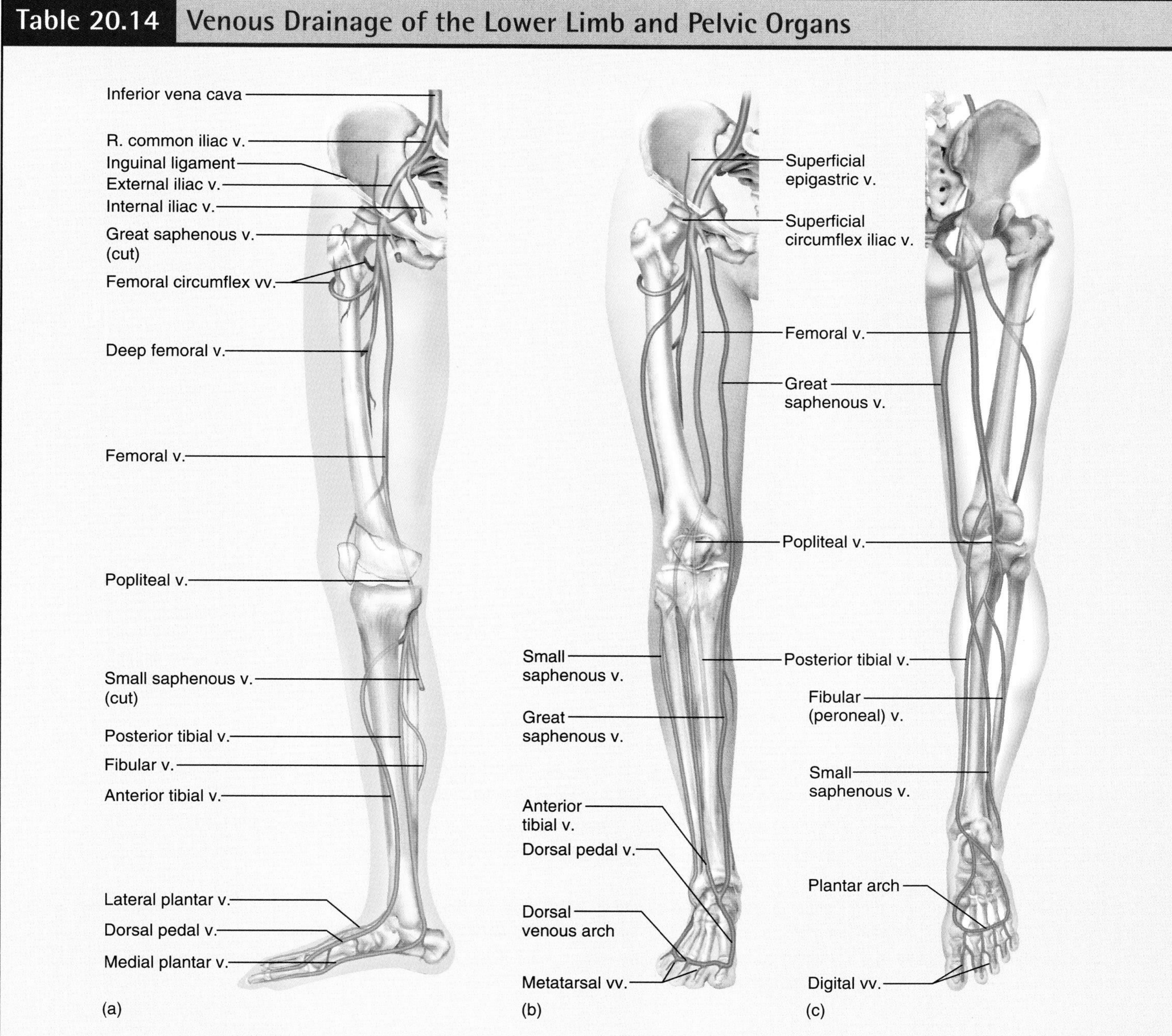

Figure 20.38 Veins Draining the Lower Limb. (*a*) Deep veins, anteromedial view of the right limb. (*b*) Anterior aspect of the right limb and dorsal aspect of the foot. (*c*) Posterior aspect of the right limb and plantar aspect of the foot.

Drainage of the lower limb is described starting at the toes and following the flow of blood to the inferior vena cava (fig. 20.38). As in the upper limb, there are deep and superficial veins with anastomoses between them.

Deep Veins

1. The **plantar arch** drains the plantar aspect of the foot, receives blood from the **digital veins** of the toes, and gives rise to the next vein.
2. The **posterior tibial vein** drains the plantar arch and passes up the leg embedded deep in the calf muscles; it receives drainage along the way from the **fibular (peroneal) vein.**
3. The **dorsal pedal vein** drains the dorsum of the foot.

(continued)

Table 20.14 Venous Drainage of the Lower Limb and Pelvic Organs (continued)

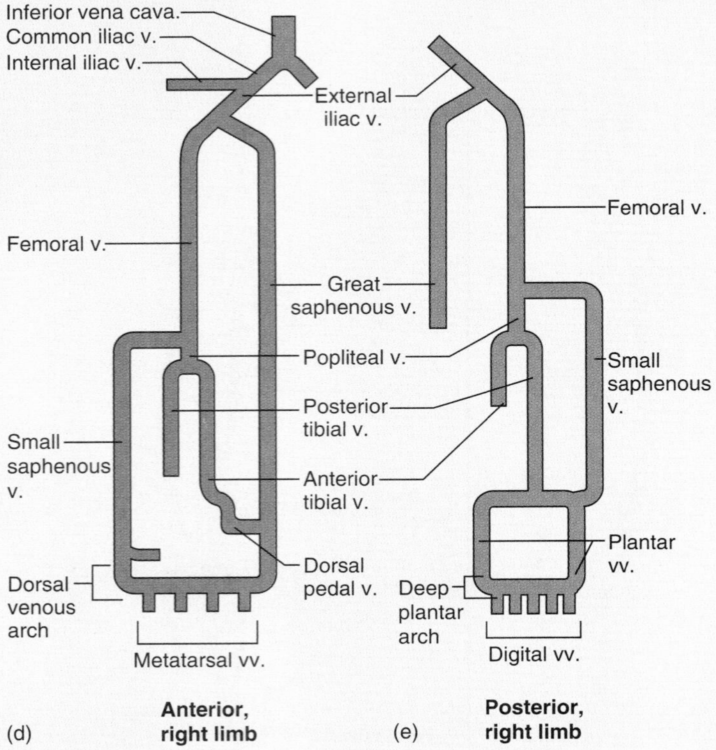

Figure 20.38 Veins Draining the Lower Limb (continued). (*d*) Flowchart of venous drainage of the right limb, anterior aspect. (*e*) Flowchart of the same limb, posterior aspect.

Deep Veins (continued)

4. The **anterior tibial vein** is a continuation of the dorsal pedal vein. It travels up the anterior compartment of the leg between the tibia and fibula.
5. The **popliteal vein** is formed at the back of the knee by the union of the anterior and posterior tibial veins.
6. The **femoral vein** is a continuation of the popliteal vein into the thigh. It receives drainage from the deep thigh muscles and femur.
7. The **external iliac vein,** superior to the inguinal ligament, is formed by the union of the femoral vein and great saphenous vein (one of the superficial veins described next).
8. The **internal iliac vein** follows the course of the internal iliac artery and its distribution. Its tributaries drain the gluteal muscles; the medial aspect of the thigh; the urinary bladder, rectum, prostate, and ductus deferens in the male; and the uterus and vagina in the female.
9. The **common iliac vein** is formed by the union of the external and internal iliac veins; it also receives blood from the ascending lumbar vein. The right and left common iliacs then unite to form the inferior vena cava.

Superficial Veins

1. The **dorsal venous arch** is visible through the skin on the dorsum of the foot. It has numerous anastomoses similar to the dorsal venous arch of the hand.
2. The **great saphenous**[27] (sah-FEE-nus) **vein,** the longest vein in the body, arises from the medial side of the dorsal venous arch. It traverses the medial aspect of the leg and thigh and terminates by emptying into the femoral vein, slightly inferior to the inguinal ligament. It is commonly used as a site for the long-term administration of intravenous fluids; it is a relatively accessible vein in infants and in patients in shock whose veins have collapsed. Portions of this vein are commonly excised and used as grafts in coronary bypass surgery.
3. The **small saphenous vein** arises from the lateral side of the dorsal venous arch, courses up the lateral aspect of the foot and through the calf muscles, and terminates at the knee by emptying into the popliteal vein. It has numerous anastomoses with the great saphenous vein. The great and small saphenous veins are among the most common sites of varicose veins.

27. *saphen* = standing

Table 20.15 Some Circulatory Disorders

Cardiac Tamponade	Compression of the heart by an abnormal accumulation of fluid in the pericardial cavity, interfering with ventricular filling. May result from inflammation of the pericardium (pericarditis).
Dissecting Aneurysm	Splitting of the layers of an arterial wall from each other because of the accumulation of blood between layers. Results from either a tear in the tunica interna or rupture of the vasa vasorum.
Fat Embolism	The presence of fat globules traveling in the bloodstream. Globules originate from bone fractures, fatty degeneration of the liver, and other causes and may block cerebral or pulmonary blood vessels.
Orthostatic Hypotension	A decrease in blood pressure that occurs when one stands, often resulting in blurring of vision, dizziness, and syncope (fainting). Occurs as a result of sluggish or inactive baroreflexes.
Septal Defects	Abnormal openings in the interatrial or interventricular septum of the heart resulting in blood from the high-pressure left atrium or ventricle flowing into the right atrium or ventricle. Results in pulmonary hypertension, difficulty breathing, and fatigue. Often fatal in childhood if uncorrected.
Von Willebrand Disease	Hereditary deficiency of von Willebrand factor (a component of clotting factor VIII) resulting in a failure of platelets to bind to damaged endothelium. Consequences include abnormal gastrointestinal bleeding and excessive menstrual discharge.

Chapters 18 to 20 have touched upon numerous disorders of the circulatory system: hypoproteinemia, hypoxemia, polycythemia, anemia, transfusion reactions, hemolytic disease of the newborn, leukopenia, leukocytosis, leukemia, and clotting disorders in chapter 18; valvular insufficiency, myocardial infarction, angina pectoris, arrhythmia, heart block, congestive heart failure, tachycardia, bradycardia, and atherosclerosis in chapter 19; and varicose veins, hypertension, hypotension, aneurysms, edema, circulatory shock, air embolism, transient ischemic attack, and cerebrovascular accident (stroke) in this chapter. Table 20.15 concludes our study of the circulatory system with brief descriptions of a few more circulatory disorders.

Before You Go On

Answer the following questions to test your understanding of the preceding section:

28. If you were dissecting a cadaver, where would you look for the internal and external jugular veins? What muscle would help you distinguish one from the other?
29. How do the vertebral veins differ from the vertebral arteries in their superior terminations?
30. By what route does blood from the abdominal wall reach the superior vena cava?
31. Trace one possible path of an RBC from the fingertips to the right atrium and name the veins along the way.
32. State two ways in which the great saphenous vein has special clinical significance. Where is this vein located?

Clinical Insight 20.4

Hypertension—The "Silent Killer"

Hypertension, the most common cardiovascular disease, affects about 30% of Americans over age 50 and 50% by age 74. It is a "silent killer" that can wreak its destructive effects for 10 to 20 years before its effects are first noticed. Hypertension is the major cause of heart failure, stroke, and kidney failure. It damages the heart because it increases the afterload, which makes the ventricles work harder to expel blood. The myocardium enlarges up to a point (the *hypertrophic response*), but eventually it becomes excessively stretched and less efficient. Hypertension strains the blood vessels and tears the endothelium, thereby creating lesions that become focal points of atherosclerosis. Atherosclerosis then worsens the hypertension and establishes an insidious positive feedback cycle.

Another positive feedback cycle involves the kidneys. Their arterioles thicken in response to the stress, their lumens become narrower, and renal blood flow declines. When the kidneys detect the resulting drop in blood pressure, they release renin, which leads to the formation of the vasoconstrictor angiotensin II and the release of aldosterone, a hormone that promotes salt retention (described in detail in chapter 24). These effects worsen the hypertension that already existed. If diastolic pressure exceeds 120 mmHg, blood vessels of the eye hemorrhage, blindness ensues, the kidneys and heart deteriorate rapidly, and death usually follows within 2 years.

Primary hypertension, which accounts for 90% of cases, results from such a complex web of behavioral, hereditary, and other factors that it is difficult to sort out any specific underlying cause. It was once considered such a normal part of the "essence" of aging that it continues to be called by another name, *essential hypertension.* That term suggests a fatalistic resignation to hypertension as a fact of life, but this need not be. Many risk factors have been identified, and most of them are controllable.

One of the chief culprits is obesity. Each pound of extra fat requires miles of additional blood vessels to serve it, and all of this added vessel length increases peripheral resistance and blood pressure. Just carrying around extra weight, of course, also increases the workload on the heart. Even a small weight loss can significantly reduce blood pressure. Sedentary behavior is another risk factor. Aerobic exercise helps to reduce hypertension by controlling weight, reducing emotional tension, and stimulating vasodilation.

Dietary factors are also significant contributors to hypertension. Diets high in cholesterol and saturated fat contribute to atherosclerosis. Potassium and magnesium reduce blood pressure; thus, diets deficient in these minerals promote hypertension. The relationship of salt intake to hypertension has been a very controversial subject. The kidneys compensate so effectively for excess salt intake that dietary salt has little effect on the blood pressure of most people. Reduced salt intake may, however, help to control hypertension in older people and in people with reduced renal function.

Nicotine makes a particularly devastating contribution to hypertension because it stimulates the myocardium to beat faster and harder; it stimulates vasoconstriction and thus increases the afterload against which the myocardium must work. Just when the heart needs extra oxygen, nicotine causes coronary vasoconstriction and promotes myocardial ischemia.

Some risk factors cannot be changed at will—race, heredity, and sex. Hypertension runs in some families. A person whose parents or siblings have hypertension is more likely than average to develop it. The incidence of hypertension is about 30% higher, and the incidence of strokes about twice as high, among blacks as among whites. From ages 18 to 54, hypertension is more common in men, but above age 65, it is more common in women. Even people at risk from these factors, however, can minimize their chances of hypertension by changing risky behaviors.

Treatments for primary hypertension include weight loss, diet, and certain drugs. Diuretics lower blood volume and pressure by promoting urination. ACE inhibitors block the formation of the vasoconstrictor angiotensin II. Beta-blockers such as propranolol block the vasoconstrictive action of the sympathetic nervous system. Calcium channel blockers such as verapamil and nifedipine inhibit the inflow of calcium into cardiac and smooth muscle, thus inhibiting their contraction and promoting vasodilation and reduced cardiac workload.

Secondary hypertension, which accounts for about 10% of cases, is high blood pressure that is secondary to (results from) other identifiable disorders. These include kidney disease (which may cause renin hypersecretion), atherosclerosis, hyperthyroidism, Cushing disease, and polycythemia. Secondary hypertension is corrected by treating the underlying disease.

Connective Issues

Interactions Between the CIRCULATORY SYSTEM and Other Organ Systems

← indicates ways in which this system affects other systems
→ indicates ways in which other systems affect this one

All Systems

Circulatory system delivers O_2 and nutrients to all other systems and carries away wastes; carries heat from deeper organs to skin for elimination

Integumentary System

← Dermal blood flow affects sweat production

→ Serves as blood reservoir; helps to regulate blood temperature

Skeletal System

← Provides minerals for bone deposition; delivers erythropoietin to bone marrow and delivers hormones that regulate skeletal growth

→ Skeleton provides protective enclosure for heart and thoracic vessels; serves as reservoir of calcium needed for heart contractions; bone marrow carries out hemopoiesis

Muscular System

← Removes heat generated by exercise

→ Helps to regulate blood temperature; respiratory and limb muscles aid venous return; aerobic exercise enhances circulatory efficiency

Nervous System

← Endothelial cells of blood vessels maintain blood-brain barrier and help to produce CSF

→ Modulates heart rate, strength of contraction, and vasomotion; governs routing of blood flow; monitors blood pressure and composition and activates homeostatic mechanisms to regulate these

Endocrine System

← Transports hormones to their target cells

→ Regulates blood volume and pressure; stimulates hemopoiesis

Lymphatic/Immune System

← Produces tissue fluid, which becomes lymph; provides the WBCs and plasma proteins employed in immunity

→ Lymphatic and circulatory systems jointly regulate fluid balance; lymphatic system returns fluid to bloodstream; spleen acts as RBC and platelet reservoir; lymphatic tissues produce lymphocytes; immune cells protect circulatory system from pathogens

Respiratory System

← Delivers and carries away respiratory gases; low capillary blood pressure keeps alveoli dry

→ Site of exchange for blood gases; helps to regulate blood pH; thoracic pump aids venous return

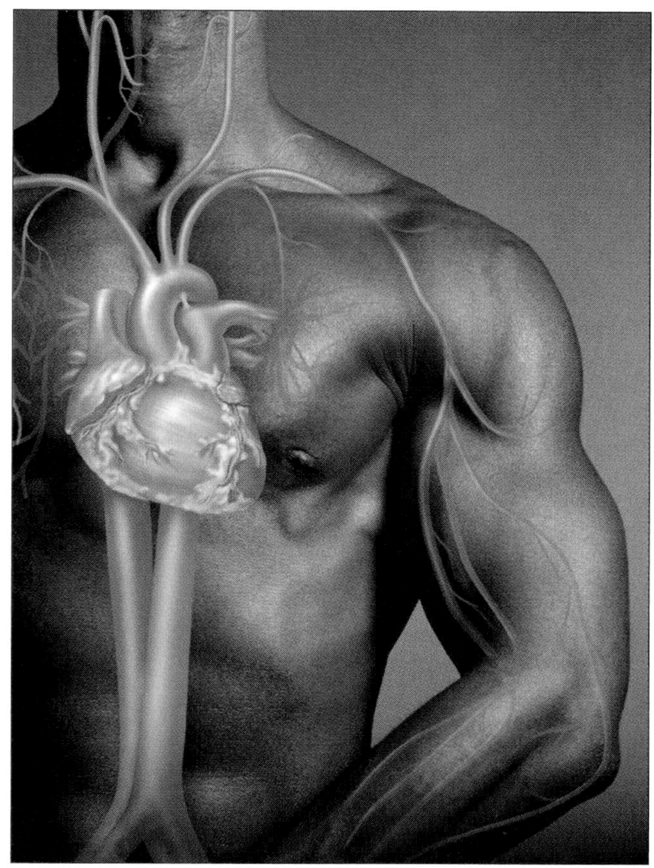

Urinary System

← Blood pressure maintains kidney function

→ Controls blood volume, pressure, and composition; initiates renin-angiotensin-aldosterone mechanism; regulates RBC count by producing erythropoietin

Digestive System

← Carries away absorbed nutrients; helps to reabsorb and recycle bile salts and minerals from intestines

→ Provides nutrients for hemopoiesis; affects blood composition

Reproductive System

← Distributes sex hormones; vasodilation causes erection

→ Estrogens may slow development of atherosclerosis in women; testosterone stimulates erythropoiesis

Chapter Review

Online Learning Center

Study resources for this chapter are available at www.mhhe.com/saladin/

Study Outline

General Anatomy of the Blood Vessels (p. 750)

1. Circulatory routes
   a. Most common route
   b. Portal systems
   c. Anastomoses
2. The vessel wall
   a. Tunica externa
   b. Tunica media
   c. Tunica interna
3. Arteries and metarterioles
   a. Conducting (elastic) arteries
   b. Distributing (muscular) arteries
   c. Resistance (small) arteries
   d. Metarterioles
4. Capillaries
   a. Capillary beds
      - Thoroughfare channels
      - Precapillary sphincters
   b. Types of capillaries
      - Continuous
      - Fenestrated
   c. Sinusoids
5. Veins
   a. Size gradation
   b. Venous sinuses
   c. Low blood pressure
      - Thin walls
      - High capacitance
      - Venous valves

Blood Pressure, Resistance, and Flow (p. 756)

1. General principles
   a. Flow and perfusion
   b. Hemodynamics: principles of flow
   c. Flow, pressure, and resistance
2. Blood pressure
   a. Measurement at brachial artery
   b. Pressures measured
      - Systolic pressure
      - Diastolic pressure
      - Pulse pressure
      - Mean arterial pressure
   c. Abnormalities of blood pressure
      - Hypertension
      - Hypotension
   d. Importance of arterial elasticity
   e. Blood pressure versus distance
   f. Blood pressure and age
3. Resistance
   a. Blood viscosity
   b. Vessel length
   c. Vessel radius
      - Vasomotion
      - Laminar flow
      - Flow in proportion to radius
      - Flow at different points in the circulation
4. Regulation of blood pressure and flow
   a. Local control
      - Autoregulation
         •• Metabolic theory
         •• Vasoactive chemicals
         •• Reactive hyperemia
      - Angiogenesis
   b. Neural control
      - Vasomotor center
      - Baroreflexes
      - Chemoreflexes
      - Medullary ischemic reflex
   c. Hormonal control
      - Angiotensin II
      - Epinephrine and norepinephrine
      - Antidiuretic hormone
      - Atrial natriuretic factor
5. Vasomotion and the routing of blood flow
   a. Effects of exercise
   b. Special importance of arterioles

Capillary Exchange (p. 763)

1. Routes of fluid exchange
   a. Intercellular clefts
   b. Fenestrations
   c. Endothelial cell cytoplasm
2. Mechanisms of fluid exchange
   a. Diffusion
   b. Transcytosis
   c. Filtration and reabsorption
      - Hyrostatic pressure
         •• Of blood
         •• Of interstitial space
         •• Net hydrostatic pressure
      - Colloid osmotic pressure (COP)
         •• COP of blood
         •• COP of tissue fluid
         •• Net COP: oncotic pressure
      - Net filtration pressure
      - Net reabsorption pressure
      - Solvent drag
      - Variations in filtration and reabsorption
3. Edema
   a. Causes
      - Increased capillary filtration
      - Reduced capillary reabsorption
      - Obstructed lymphatic drainage
   b. Consequences
      - Circulatory shock
      - Necrosis
      - Suffocation
      - Cerebral effects

Venous Return and Circulatory Shock (p. 766)

1. Role of venous valves
2. Mechanisms of venous return
   a. Pressure gradient
   b. Thoracic pump
   c. Cardiac suction
   d. Skeletal muscle pump
   e. Gravity
3. Venous return and physical activity
   a. Effects of exercise
   b. Venous pooling
4. Circulatory shock
   a. Cardiogenic shock
   b. Low venous return shock
      - Hypovolemic shock
      - Obstructed venous return shock
      - Venous pooling shock
         •• Neurogenic shock

- Septic shock
- Anaphylactic shock
c. Responses to shock
- Compensated shock
- Decompensated shock

Special Circulatory Routes (p. 789)

1. Brain
   a. Constancy of total flow
   b. Cerebral autoregulation
   c. Effects of CO_2 and pH
   d. Transient ischemic attacks
   e. Cerebrovascular accidents
2. Skeletal muscles
   a. Variability of flow
   b. Effects of exercise
   c. Muscle perfusion and fatigue
3. Lungs
   a. Low pulmonary blood pressure
   - Flow and gas exchange
   - Capillary fluid absorption
   b. Response to hypoxia

Anatomy of the Pulmonary Circuit (p. 770)

1. Pulmonary trunk and pulmonary arteries
2. Lobar arteries
3. Alveolar capillaries
4. Pulmonary veins

Anatomy of the Systemic Arteries (p. 770)

1. The aorta and its major branches (table 20.3)
2. Arterial supply to the head and neck (table 20.4)
3. Arterial supply to the upper limb (table 20.5)
4. Arterial supply to the thorax (table 20.6)
5. Arterial supply to the abdomen (table 20.7)
6. Arterial supply to the pelvic region and lower limb (table 20.8)
7. Arterial pressure points

Anatomy of the Systemic Veins (p. 784)

1. Venous drainage of the head and neck (table 20.9)
2. Venous drainage of the upper limb (table 20.10)
3. The azygos system (table 20.11)
4. Major tributaries of the inferior vena cava (table 20.12)
5. The hepatic portal system (table 20.13)
6. Venous drainage of the lower limb and pelvic organs (table 20.14)

Table of Circulatory Disorders (p. 796)

Selected Vocabulary

For additional terminology not listed here see tables cited in the Study Outline.

portal system 750
anastomosis 750
arteriovenous anastomosis 750
arterial anastomosis 750
venous anastomosis 751
tunica externa 751
vasa vasorum 752
tunica media 752
tunica interna 752
endothelium 752
conducting artery 752
distributing artery 752
resistance artery 752
arteriole 752
metarteriole 752
precapillary sphincter 752
capillary 752
capillary bed 753
thoroughfare channel 753
continuous capillary 754
intercellular cleft 754
fenestrated capillary 754
sinusoid 754
venule 754
venous sinus 754
venous valve 755
blood flow 756
perfusion 756
blood pressure 756
systolic pressure 756
diastolic pressure 756
pulse pressure 756
mean arterial pressure 756
hypertension 756
aneurysm 756
hypotension 756
peripheral resistance 758
vasomotion 758
vasoconstriction 758
vasodilation 758
laminar flow 758
autoregulation 759
vasoactive chemicals 760
reactive hyperemia 760
angiogenesis 760
vasomotor center 760
baroreflex 760
baroreceptor 760
chemoreflex 761
aortic body 761
carotid body 761
medullary ischemic reflex 761
angiotensin II 761
renin 761
angiotensin-converting enzyme 761
capillary exchange 763
colloid osmotic pressure 764
oncotic pressure 764
net filtration pressure 764
net reabsorption pressure 765
solvent drag 765
edema 765
venous return 766
pressure gradient 766
central venous pressure 766
thoracic pump 767
cardiac suction 767
skeletal muscle pump 767
venous pooling 767
syncope 768
circulatory shock 768
cardiogenic shock 768
low venous return shock 768
hypovolemic shock 768
obstructed venous return shock 768
venous pooling shock 768
neurogenic shock 768
septic shock 768
anaphylactic shock 768
compensated shock 768
decompensated shock 768
transient ischemic attack 769
cerebrovascular accident 769
pulmonary trunk 770
pulmonary artery 770
lobar artery 770
pulmonary vein 770
pressure point 784
femoral triangle 784

Testing Your Recall

1. Blood normally flows into a capillary bed from
   a. the distributing arteries.
   b. the conducting arteries.
   c. the metarterioles.
   d. a thoroughfare channel.
   e. the venules.

2. Plasma solutes enter the tissue fluid most easily from
   a. continuous capillaries.
   b. fenestrated capillaries.
   c. arteriovenous anastomoses.
   d. collateral vessels.
   e. venous anastomoses.

3. A blood vessel adapted to withstand a high pulse pressure would be expected to have
   a. an elastic tunica media.
   b. a thick tunica interna.
   c. one-way valves.
   d. a flexible endothelium.
   e. a rigid tunica media.

4. The substance most likely to cause a rapid drop in blood pressure is
   a. epinephrine.
   b. norepinephrine.
   c. angiotensin II.
   d. serotonin.
   e. histamine.

5. A person with a systolic blood pressure of 130 mmHg and a diastolic pressure of 85 mmHg would have a mean arterial pressure of about
   a. 85 mmHg.
   b. 100 mmHg.
   c. 108 mmHg.
   d. 115 mmHg.
   e. 130 mmHg.

6. The velocity of blood flow decreases if
   a. vessel radius increases.
   b. blood pressure increases.
   c. viscosity increases.
   d. afterload increases.
   e. vasomotion decreases.

7. Blood flows faster in a venule than in a capillary because venules
   a. have one-way valves.
   b. exhibit vasomotion.
   c. are closer to the heart.
   d. have higher blood pressures.
   e. have larger diameters.

8. In a case where interstitial hydrostatic pressure is negative, the only force causing capillaries to reabsorb fluid is
   a. colloid osmotic pressure of the blood.
   b. colloid osmotic pressure of the tissue fluid.
   c. capillary hydrostatic pressure.
   d. interstitial hydrostatic pressure.
   e. net filtration pressure.

9. Intestinal blood flows to the liver by way of
   a. the superior mesenteric artery.
   b. the celiac trunk.
   c. the inferior vena cava.
   d. the azygos system.
   e. the hepatic portal system.

10. The brain receives blood from all of the following vessels *except* the ________ artery or vein.
    a. basilar
    b. vertebral
    c. internal carotid
    d. internal jugular
    e. anterior communicating

11. The highest arterial blood pressure attained during ventricular contraction is called ___________ pressure. The lowest attained during ventricular relaxation is called ___________ pressure.

12. The capillaries of skeletal muscles are of the structural type called ___________.

13. ___________ shock occurs as a result of exposure to an antigen to which one is hypersensitive.

14. The role of breathing in venous return is called the ___________.

15. The difference between the colloid osmotic pressure of blood and that of the tissue fluid is called ___________.

16. Movement across the capillary endothelium by the uptake and release of fluid droplets is called ___________.

17. All efferent fibers of the vasomotor center belong to the ___________ division of the autonomic nervous system.

18. The pressure sensors in the major arteries near the head are called ___________.

19. Most of the blood supply to the brain comes from a ring of arterial anastomoses called ___________.

20. The major superficial veins of the arm are the _____ on the medial side and ___________ on the lateral side.

Answers in appendix B

Testing Your Comprehension

1. It is a common lay perception that systolic blood pressure should be 100 plus a person's age. Evaluate the validity of this statement.
2. Calculate the net filtration or reabsorption pressure at a point in a hypothetical capillary assuming a hydrostatic blood pressure of 28 mmHg, an interstitial hydrostatic pressure of −2 mmHg, a blood COP of 25 mmHg, and an interstitial COP of 4 mmHg. Give the magnitude (in mmHg) and direction (in or out) of the net pressure.
3. Aldosterone secreted by the adrenal gland must be delivered to the kidney immediately below. Trace the route that an aldosterone molecule must take from the adrenal gland to the kidney, naming all major blood vessels in the order traveled.
4. People in shock commonly exhibit paleness, cool skin, tachycardia, and a weak pulse. Explain the physiological basis for each of these signs.
5. Discuss why it is advantageous to have baroreceptors in the aortic arch and carotid sinus rather than in some other location such as the common iliac arteries.

Answers at the Online Learning Center

CHAPTER

21

The Lymphatic and Immune Systems

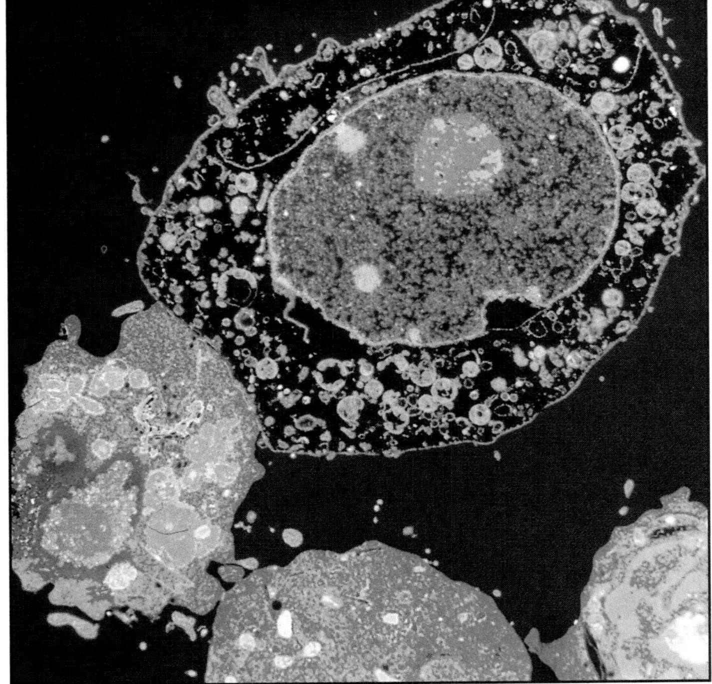

T lymphocytes (green) attacking a cancer cell (with blue nucleus) (TEM)

Chapter Outline

Insights

Study with Your ESP CD-ROM

Refer to your ESP Tutorial CD-ROM for helpful animations, activities, and quizzes related to the material covered in this chapter. The following are suggested areas to go to on the CD and are presented according to "Unit/Topic":

- Transport/Lymphatic System

Brushing Up

To understand this chapter, it is important that you understand or brush up on the following concepts:

- Endocytosis and exocytosis (pp. 131–32)
- General gland structure: capsule, septa, stroma, and parenchyma (p. 194)
- Leukocyte types (pp. 701–702)
- Mechanisms of venous blood flow (p. 766)

Of all the body systems, the lymphatic system is perhaps the least familiar to most people. Yet without it, the circulatory system would quickly shut down from hypovolemic shock and the immune system would be seriously compromised. This chapter discusses the role of the lymphatic system in maintaining fluid balance in the body and protecting it against infection and disease.

The lymphatic system is closely allied with the immune system in providing us with a defense against foreign invaders. As the lymphatic system collects tissue fluid, it picks up pathogens that may be present. On its way back to the bloodstream, the fluid flows through the lymph nodes, where dense populations of lymphocytes and macrophages are strategically positioned to detect such pathogens and mount a quick response. Because the lymphatic system enables the immune system to function efficiently, it is appropriate that we consider these two systems together.

The Lymphatic System

Objectives

When you have completed this section, you should be able to

- list the functions of the lymphatic system;
- explain how lymph is formed and returned to the bloodstream; and
- describe the form and function of the lymph nodes, tonsils, thymus, and spleen.

Fluid continually filters from our blood capillaries into the tissue spaces. The capillaries reabsorb most of it, but by no means all. Each day, the capillaries lose an excess of 2 to 4 L of water and one-quarter to one-half of the plasma protein. The **lymphatic system** absorbs this excess fluid and returns it to the bloodstream by way of a network of lymphatic vessels (fig. 21.1). If not for this fluid recovery, the circulatory system would not have enough blood to function properly. Moreover, even partial interference with lymphatic drainage can lead to severe edema (fig. 21.2). When protein accumulates in the tissue fluid, it increases its colloid osmotic pressure and causes the tissues to osmotically retain extra fluid. In the mucosa of the small intestine, the lymphatic system serves also to absorb dietary lipids that are not absorbed by the blood capillaries (see chapter 25).

The components of the lymphatic system are (1) *lymph,* the recovered fluid on its way back to the bloodstream; (2) *lymphatic vessels,* which transport the lymph; (3) *lymphatic tissue,* consisting of aggregates of lymphocytes and macrophages that populate many organs of the body; and (4) *lymphatic organs,* in which these cells are especially concentrated and which are usually set off from surrounding organs by connective tissue capsules.

Lymph and the Lymphatic Vessels

Lymph is usually a clear, colorless fluid, similar to blood plasma but with much less protein. Its composition varies substantially from place to place. After a meal, for example, lymph draining from the small intestine has a milky appearance because of its high lipid content. Lymph leaving the lymph nodes contains a large number of lymphocytes—indeed, this is the main supply of lymphocytes to the bloodstream. Lymph may also contain bacteria, viruses, cellular debris, or even traveling cancer cells.

Origin of Lymph

Lymph originates in microscopic vessels called **lymphatic capillaries.** These vessels penetrate nearly every tissue of the body but are absent from the central nervous system, cartilage, bone, and bone marrow. They are closely associated with blood capillaries, but unlike them, they are closed at one end (fig. 21.3). A lymphatic capillary consists of a sac of thin endothelial cells that loosely overlap each other like the shingles of a roof. The cells are tethered to surrounding tissue cells by protein filaments that prevent the sac from collapsing. Unlike the endothelial cells of blood capillaries, lymphatic endothelial cells are not joined by tight junctions. The gaps between them are so large that bacteria and other cells can enter along with the fluid. The overlap between endothelial cells creates valve-like flaps that can open and close. When interstitial fluid pressure is high, it pushes the flaps inward (open) and fluid flows into the lymphatic capillary. When pressure is higher in the lymphatic capillary than in the tissue fluid, the flaps are pressed outward (closed).

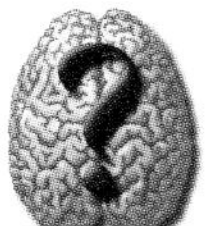

Think About It

Contrast the structure of a lymphatic capillary with that of a continuous blood capillary. Explain how their structural difference is related to their functional difference.

Lymphatic Vessels

Lymphatic vessels form in the embryo by budding from the veins, so it is not surprising that the larger ones have a similar histology. They have three tunics—a *tunica interna,* with an endothelium and valves (fig. 21.4); a *tunica media,* with elastic fibers and smooth muscle; and a thin outer *tunica externa.* Their walls are thinner and their valves are more numerous than those of the veins.

Lymph takes the following route from the tissues back to the bloodstream: lymphatic capillaries → collecting vessels → lymph nodes → more collecting vessels → six lymphatic trunks → two collecting ducts → subclavian veins. Thus, there is a continual recycling of fluid from blood to tissue fluid to lymph and back to the blood (fig. 21.5).

The lymphatic capillaries converge to form **collecting vessels.** These often travel alongside veins and arteries and share a common connective tissue sheath with them.

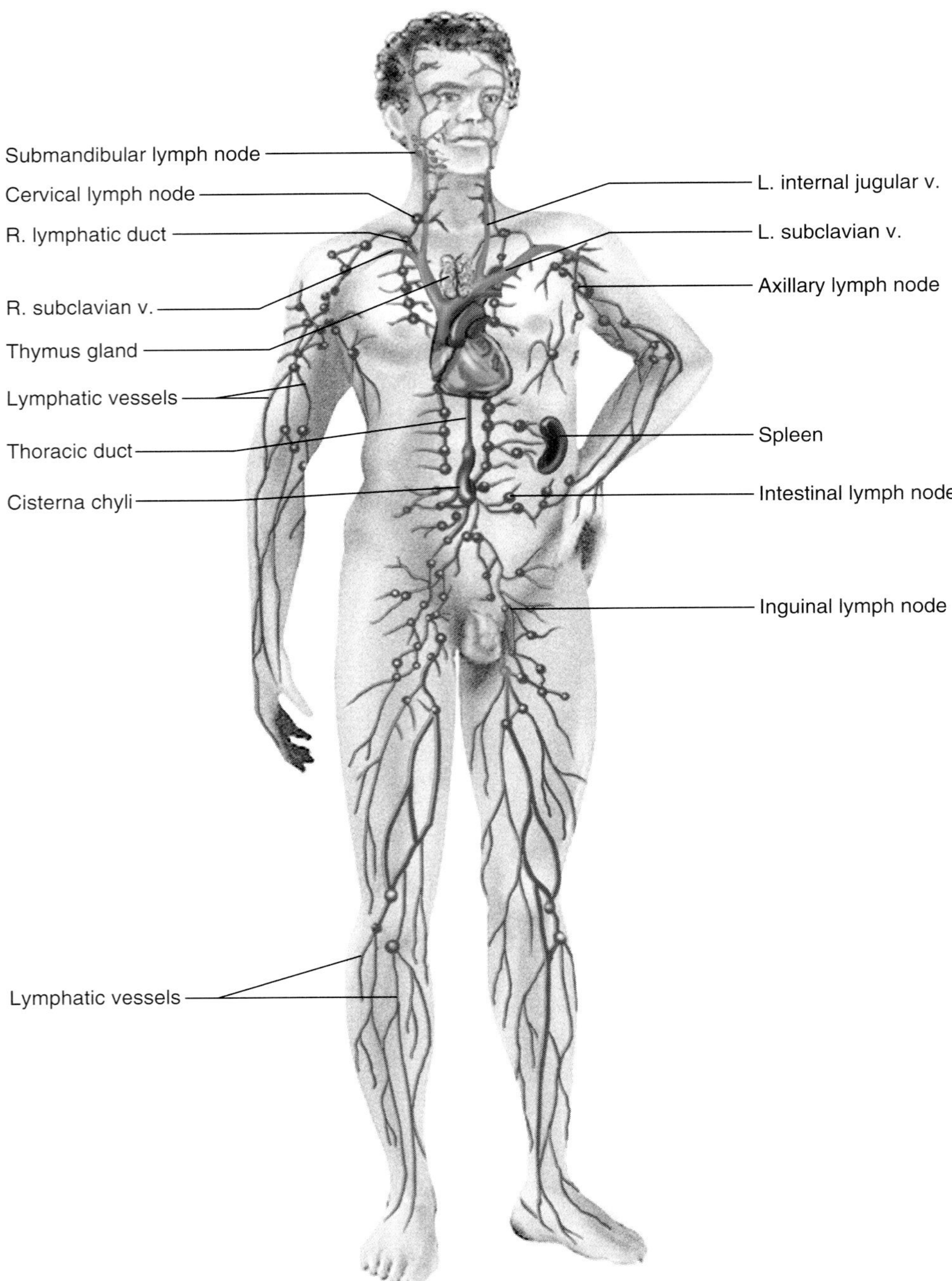

Figure 21.1 The Lymphatic System.

Numerous lymph nodes occur along the course of the collecting vessels, receiving and filtering the lymph, as we discuss shortly. The collecting vessels converge to form larger **lymphatic trunks,** each of which drains a major portion of the body. The principal lymphatic trunks are the *lumbar, intestinal, intercostal, bronchomediastinal, subclavian,* and *jugular trunks.* Their names indicate their locations and parts of the body they drain; the lumbar trunk also drains the lower extremities.

The lymphatic trunks converge to form two **collecting ducts,** the largest of the lymphatic vessels. The **right lymphatic duct** begins in the right thoracic cavity with the union of the right jugular, subclavian, and bronchomediastinal trunks; it receives lymphatic drainage from the right arm and right side of the thorax and head (fig. 21.6*a*). The duct on the left, called the **thoracic duct,** is larger and longer. It begins as a prominent sac in the abdominal cavity called the **cisterna chyli** (sis-TUR-nuh KY-lye) and then

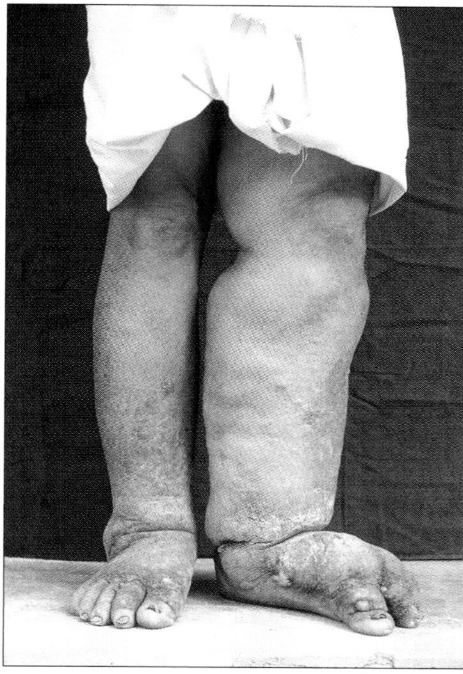

Figure 21.2 Elephantiasis, a Tropical Disease Caused by Lymphatic Obstruction. Mosquito-borne roundworms infect the lymph nodes and block the flow of lymph and recovery of tissue fluid. The resulting chronic edema leads to fibrosis and elephant-like thickening of the skin. The extremities are typically affected as shown here; the scrotum of men and breasts of women are often similarly affected.

passes through the diaphragm and up the mediastinum. It receives lymph from all parts of the body below the diaphragm and from the left arm and left side of the head, neck, and thorax (fig. 21.6*b*). Each collecting duct drains into the subclavian vein on its respective side.

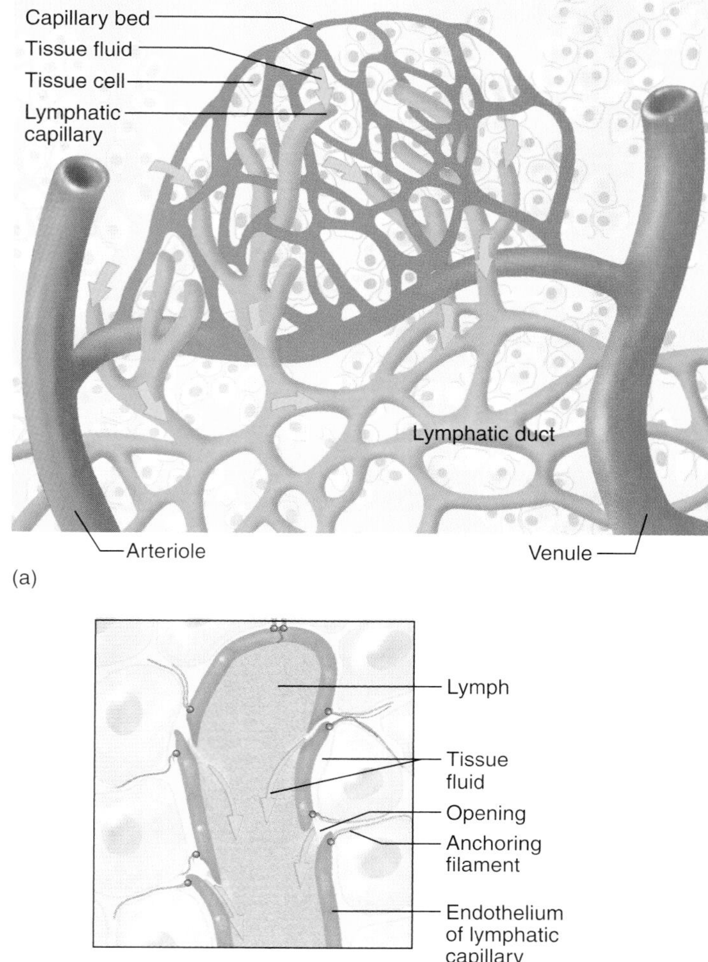

Figure 21.3 Lymphatic Capillaries. (*a*) Relationship of the lymphatic capillaries to a bed of blood capillaries. (*b*) Uptake of tissue fluid by a lymphatic capillary.

Flow of Lymph

Lymph flows under forces similar to those that govern venous return, except that the lymphatic system has no pump like the heart. Like venous blood, lymph flows at relatively low pressure and speed, and the valves in the vessels prevent it from flowing backward. Lymph is moved primarily by rhythmic contractions of the lymphatic vessels themselves. Stretching of the vessels with fluid stimulates them to contract. But the lymphatic vessels, like the veins, are also aided by a skeletal muscle pump that squeezes them and moves the lymph along. Since lymphatic vessels are often wrapped with an artery in a common sheath, arterial pulsation may also contribute to lymph flow. A thoracic (respiratory) pump aids the flow of lymph from the abdominal to the thoracic cavity, just as it does in venous return. Finally, at the point where the collecting ducts join the subclavian veins, the rapidly flowing bloodstream draws the lymph into it. Considering these mechanisms of lymph flow, it should be apparent that physical exercise significantly increases the rate of lymphatic return.

Lymphatic Tissue

The simplest form of lymphatic tissue is **diffuse lymphatic tissue**—a sprinkling of lymphocytes throughout the mucous membranes and connective tissues of many organs. It is particularly prevalent in body passages that are open to the exterior—the respiratory, digestive, urinary, and reproductive tracts. The lymphatic tissue here is called **mucosa-associated lymphatic tissue (MALT)** (formerly *gut-associated lymphatic tissue, GALT*). In some places, the

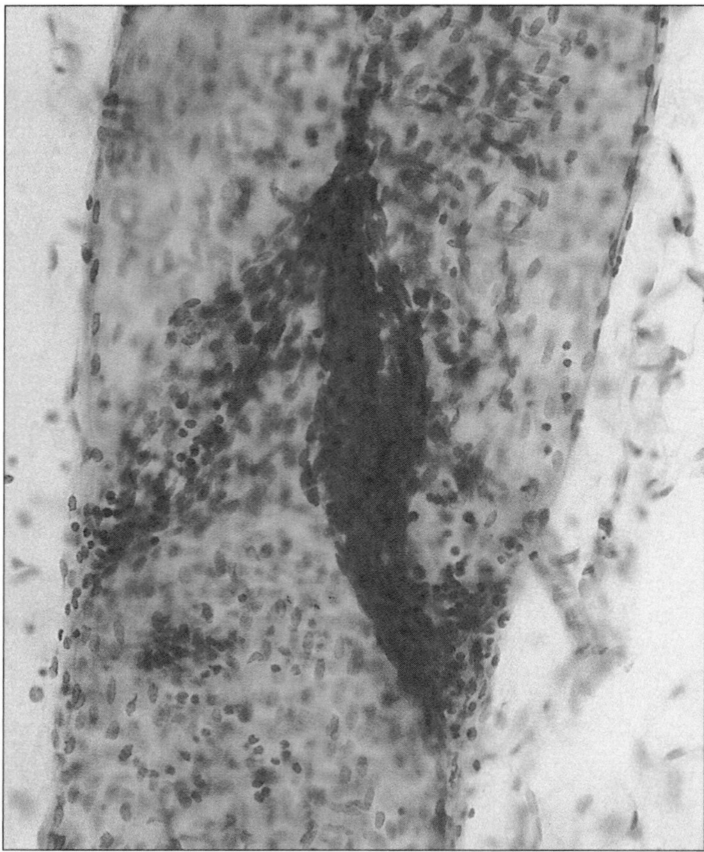

Figure 21.4 Valve in a Lymphatic Vessel.

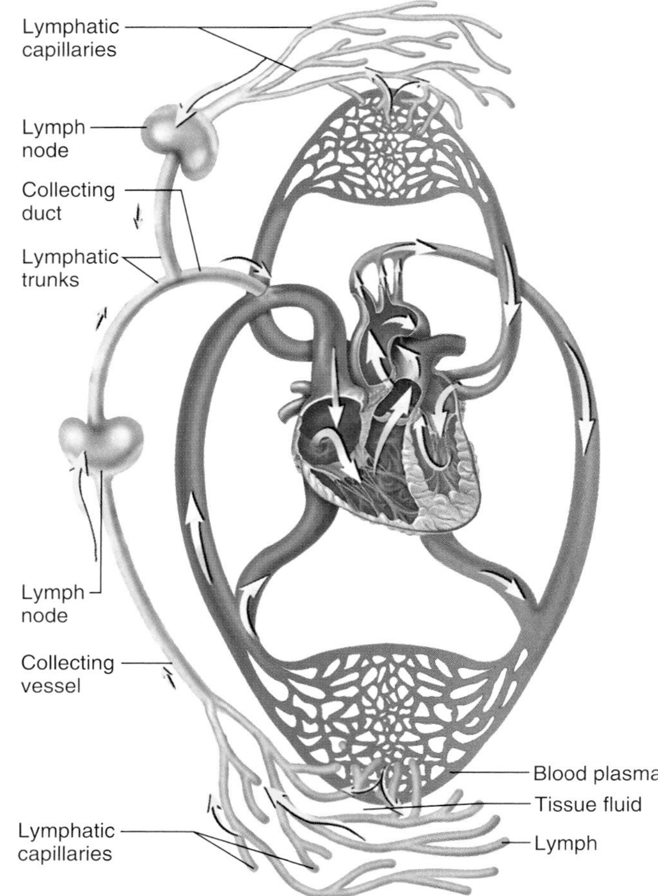

Figure 21.5 The Fluid Cycle from Bloodstream to Tissue Fluid to Lymph and Back to the Bloodstream.

lymphocytes congregate in dense oval masses called **lymphatic nodules (follicles)**, which come and go as pathogens invade the tissues and the immune system answers the challenge. They are, however, a relatively constant feature of the small intestine at its junction with the large intestine, where they form clusters called **Peyer[1] patches.**

Lymphatic Organs

In contrast to the diffuse lymphatic tissue, the lymphatic organs have well-defined anatomical sites and have at least partial connective tissue capsules that separate the lymphatic tissue from neighboring tissues. These organs include the lymph nodes, tonsils, thymus, and spleen.

Lymph Nodes

There are hundreds of **lymph nodes** (fig. 21.7) in the body, mostly embedded in the connective tissues. They are especially concentrated in the following groups (see fig. 21.1):

- *Cervical lymph nodes* behind and below the ear, along the inferior margin of the mandible, and down the side of the neck.
- *Axillary lymph nodes* in the brachium, armpit, the thoracic wall, breast, and superior abdominal wall; also called *mammary lymph nodes,* especially in women.
- *Thoracic lymph nodes* along the trachea and bronchi.
- *Abdominal lymph nodes* along the abdominal aorta and mesenteric arteries.
- *Pelvic lymph nodes* along the iliac arteries and veins.
- *Inguinal lymph nodes* in the groin.

A lymph node is an elongated or bean-shaped structure, usually less than 3 cm long, often with an indentation called the *hilum* on one side. It is enclosed in a fibrous capsule with extensions (trabeculae) that divide the interior of the node into compartments (fig. 21.8). The interior

1. Johann Conrad Peyer (1653–1712), Swiss anatomist

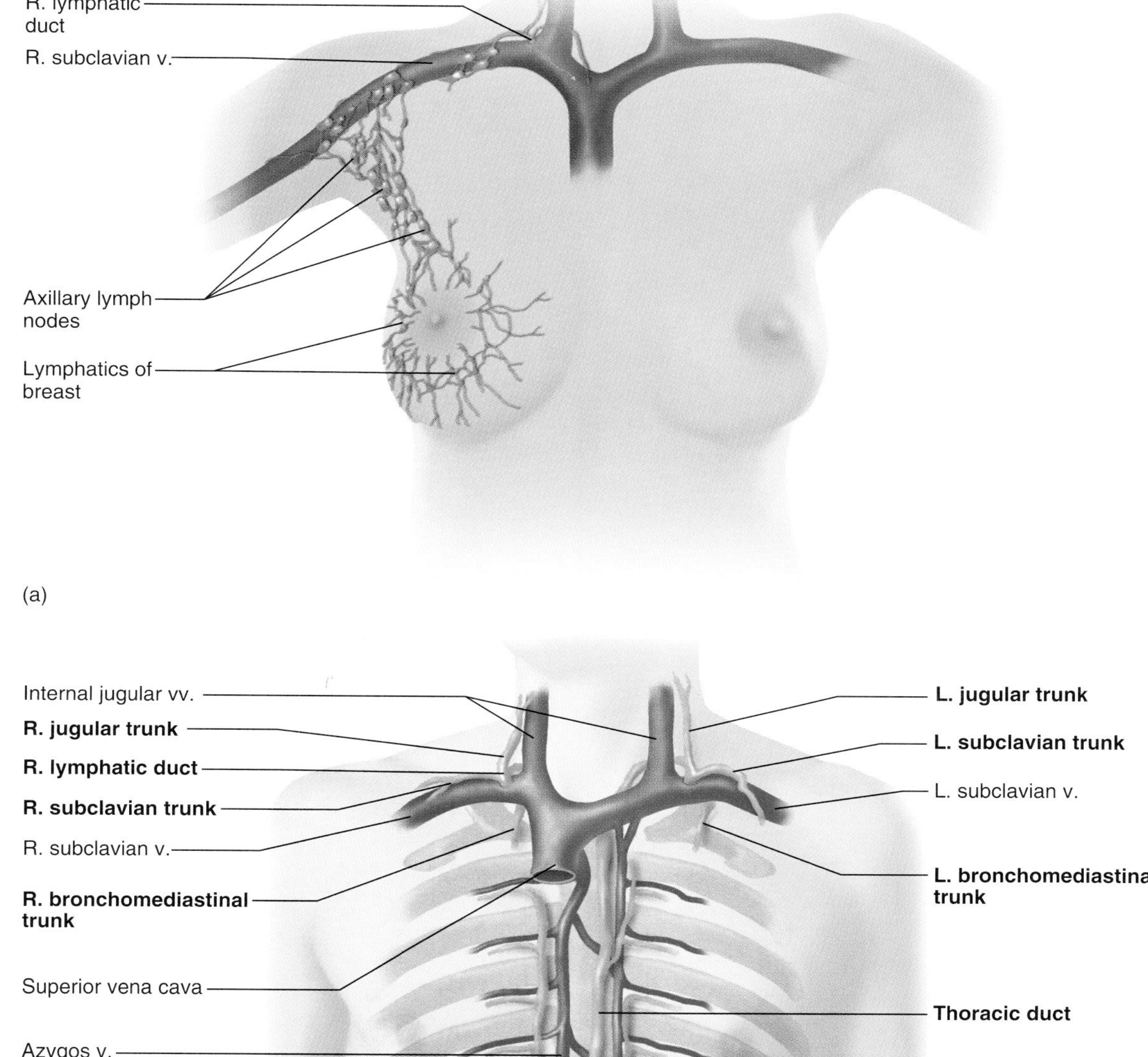

Figure 21.6 **Lymphatic Drainage of the Thoracic Region.** (*a*) Drainage of the right mammary and axillary regions. (*b*) Drainage of the right lymphatic duct and thoracic duct into the subclavian veins.

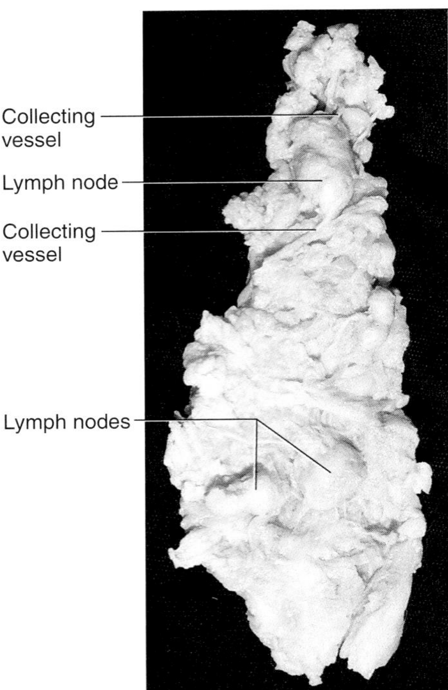

Figure 21.7 Lymph Nodes. Several collecting vessels are especially evident leading to and from the upper lymph node.

consists of a *stroma* of reticular connective tissue (reticular fibers and phagocytic reticular cells) and a *parenchyma* of lymphocytes and macrophages. Between the capsule and parenchyma is a narrow space called the *cortical sinus,* which is relatively clear of lymphocytes.

The parenchyma is divided into an outer *cortex* and an inner *medulla.* The cortex consists mainly of ovoid lymphatic nodules. When the lymph node is fighting a pathogen, these nodules acquire light-staining **germinal centers,** where they rapidly produce B lymphocytes. The other cells of the cortex are mainly T lymphocytes. (The distinctions between B lymphocytes and T lymphocytes are discussed later.) The medulla consists of a loose system of irregular spaces called *medullary sinuses* that converge on the hilum. Macrophages are concentrated in the medulla.

Several **afferent lymphatic vessels** lead into the node along its convex surface. Lymph flows from these vessels into the cortical sinus, filters slowly through the nodules and medullary sinuses, and leaves the node through one to three **efferent lymphatic vessels** that emerge from the hilum. No other lymphatic organs have afferent lymphatic vessels; lymph nodes are the only organs that filter lymph as it flows along its course. With more afferent than efferent vessels, the lymph node is a "bottleneck" that slows down lymph flow and allows time for cleansing the lymph of foreign matter.

The lymph that flows into a node may contain bacteria, antigen molecules, or other undesirable matter. As it filters through the node, reticular cells and other macrophages phagocytize foreign matter, and lymphocytes may respond to antigens by activating an immune attack against them. Lymph flows through several nodes in series before returning to the bloodstream, with each node along the way removing more impurities. Lymph nodes are common sites of metastatic cancer because cancer cells from almost any organ can break loose, enter the lymphatic capillaries, and lodge in the nodes. Cancerous lymph nodes are swollen but relatively firm and usually painless.

Tonsils

The **tonsils** are patches of lymphatic tissue located at the entrance to the pharynx. Each is covered by an epithelium and has deep pits called **tonsillar crypts** lined by lymphatic nodules (fig. 21.9). Ingested and inhaled pathogens get into the crypts, penetrate the epithelium, and encounter the lymphocytes. Many lymphocytes also get through the epithelium and enter the crypts and the saliva.

There are three main sets of tonsils: (1) a pair of **palatine tonsils** at the posterior margin of the oral cavity, (2) a pair of **lingual tonsils** in the root of the tongue, and (3) a single medial **pharyngeal tonsil** on the wall of the pharynx just behind the nasal cavity (see fig. 22.3*b,* p. 846). The palatine tonsils are the largest and most often infected. Their surgical removal, called *tonsillectomy,* used to be one of the most common surgical procedures performed on children, but it is done less often today.

Thymus

The **thymus** is an organ that houses developing lymphocytes and secretes hormones that regulate their later activity; it is a member of both the lymphatic and endocrine systems. It is located between the sternum and aortic arch in the superior mediastinum. The thymus is very large in the fetus and grows slightly during childhood, when it is most active. After age 14, however, it begins to undergo **involution** (shrinkage) so that it is quite small in adults (fig. 21.10). In the elderly, the thymus is replaced almost entirely by fibrous and fatty tissue and is barely distinguishable from the surrounding tissues.

The fibrous capsule of the thymus gives off trabeculae that divide the parenchyma into several angular lobules. Each lobule has a cortex and medulla populated by lymphocytes (fig. 21.11). When residing in the thymus, lymphocytes are also known as *thymocytes. Hassall*[2] *corpuscles,* small

2. Arthur H. Hassall (1817–94), British chemist and physician

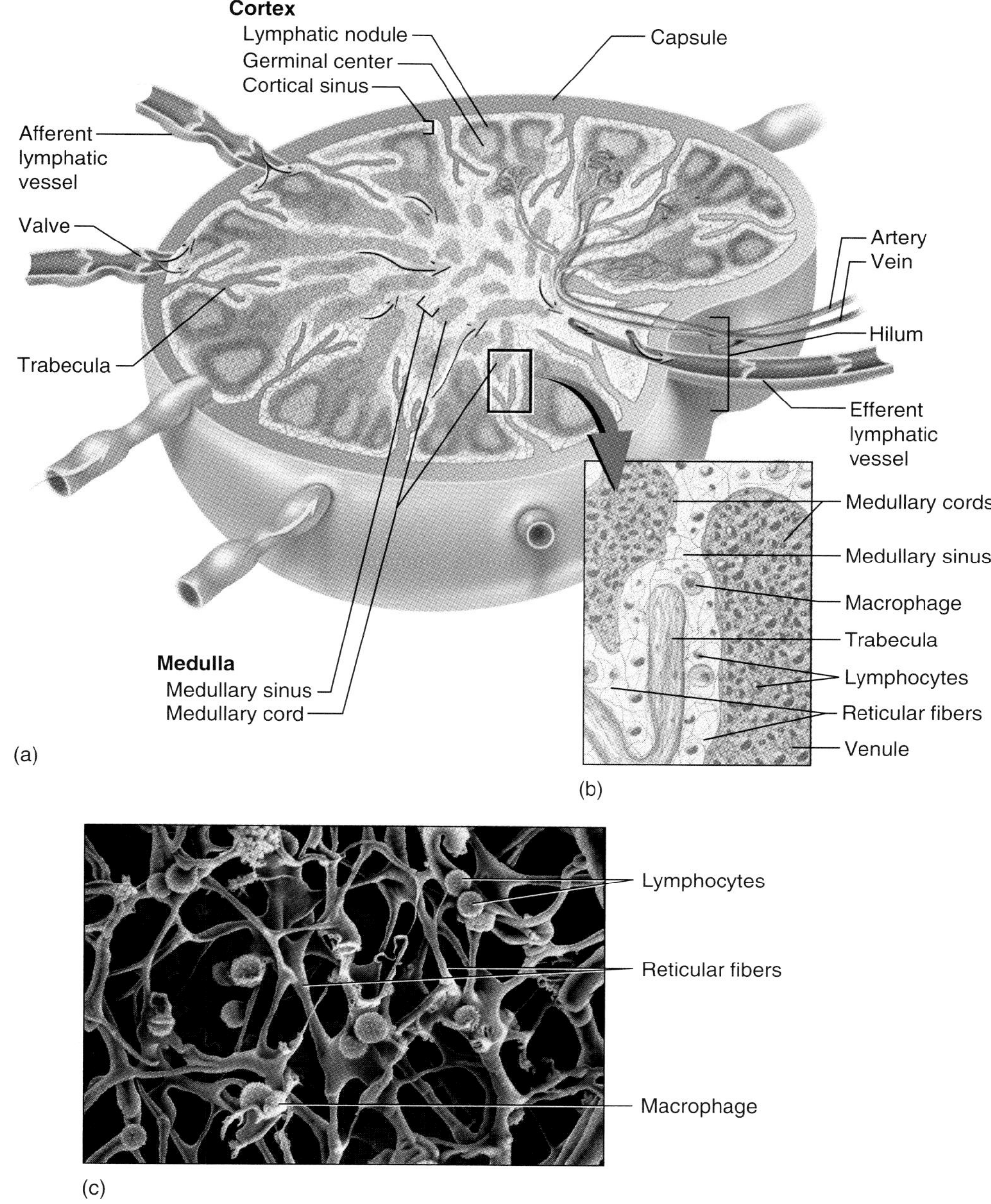

Figure 21.8 Anatomy of a Lymph Node. (*a*) Bisected lymph node showing pathway of lymph flow. (*b*) Detail of the boxed region in *a*. (*c*) Stroma and immune cells in a medullary sinus.

groups of epithelial cells, are seen in the lighter-staining medullary regions; their function is unknown.

The thymus secretes **thymopoietin** and **thymosins,** hormones that stimulate the development and later activity of T lymphocytes and render them capable of fighting pathogens. If the thymus is removed from newborn mammals, they waste away and never develop immunity. Other lymphatic organs also seem to depend on thymosin and develop poorly in such animals.

Spleen

The **spleen** is located in the left hypochondriac region, just inferior to the diaphragm and dorsal to the stomach (fig. 21.12; see also fig. A.16, p. 44). It has a medial hilum penetrated by the splenic artery and vein. Its parenchyma exhibits two types of tissue named for their appearance in fresh specimens (not in stained sections): **red pulp,** which consists of sinuses gorged with concentrated erythrocytes, and **white pulp,** which consists of lymphocytes and macrophages aggregated like sleeves along small branches of the splenic artery.

These two tissue types reflect the multiple functions of the spleen. It produces blood cells in the fetus and may resume this role in adults in the event of extreme anemia.

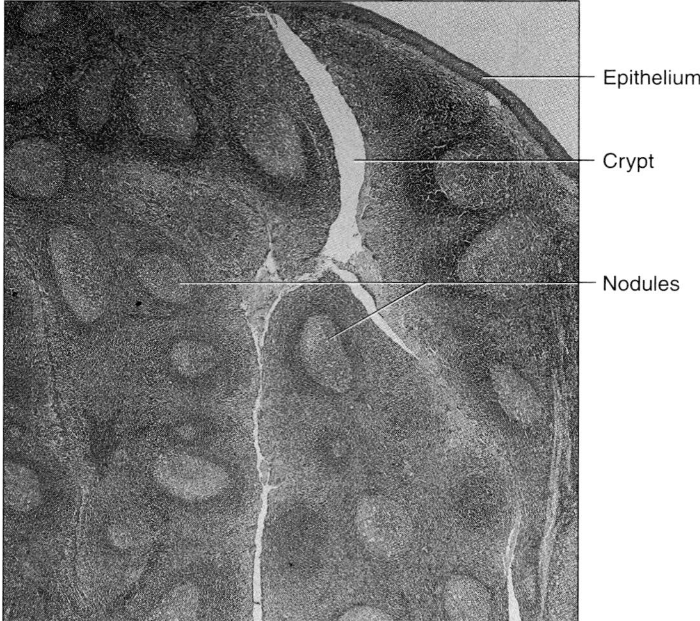

Figure 21.9 Palatine Tonsil. A vertical tissue section showing a branched tonsillar crypt and lymphatic nodules.

The adult spleen is also a blood reservoir. Its capillaries are very permeable; they allow RBCs to leave the bloodstream, accumulate in the sinuses of the red pulp, and reenter the bloodstream later. The spleen is also an "erythrocyte graveyard." Old, fragile RBCs tend to rupture as they squeeze through the capillary walls into the sinuses. Splenic macrophages phagocytize their remains, just as they dispose of blood-borne bacteria and other cellular debris. Thus, the spleen filters the blood somewhat like the lymph nodes do the lymph. Lymphocytes and macrophages of the white pulp are quick to detect foreign antigens in the blood and help to activate immune reactions.

The spleen is highly vascular and vulnerable to trauma and infection. Hemorrhage of the spleen can require immediate *splenectomy* (surgical removal), but the spleen is a relatively dispensable organ whose functions can be taken over by the liver and bone marrow.

Before You Go On

Answer the following questions to test your understanding of the preceding section:

1. List the primary functions of the lymphatic system. What do you think would be the most noticeable effect of clamping the right lymphatic duct closed?
2. How does fluid get into the lymphatic system? What prevents it from draining back out?
3. Predict the relative seriousness of removing the following damaged or diseased organs from a 2-year-old child: (a) a lymph node, (b) the spleen, (c) the thymus, (d) the palatine tonsils.

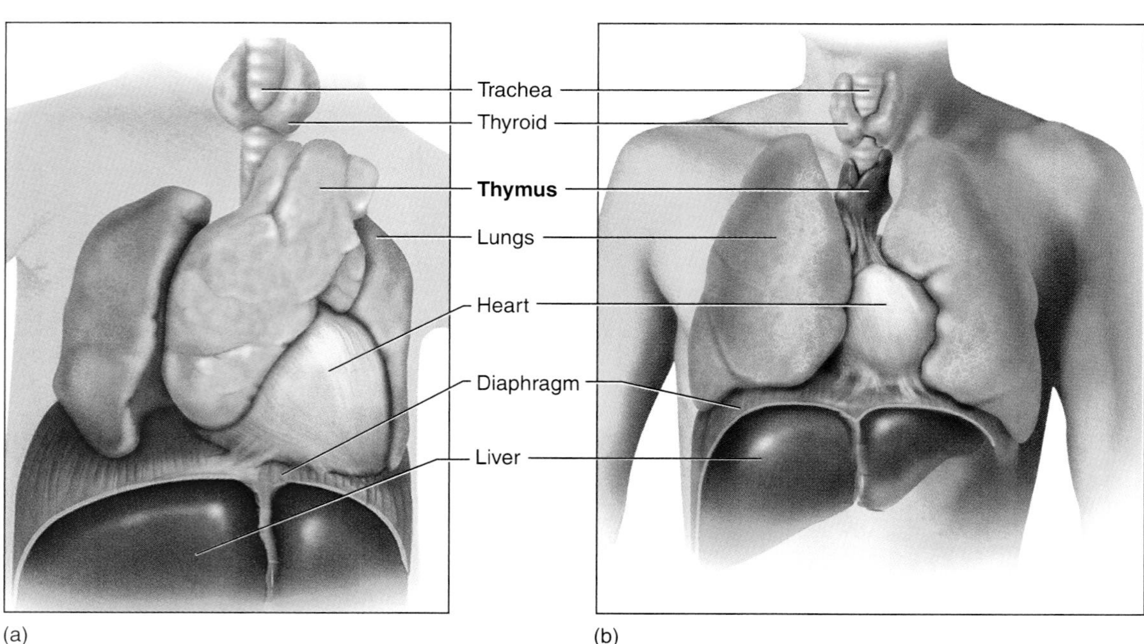

Figure 21.10 The Thymus. (*a*) The thymus is extremely large in a newborn infant. (*b*) The adult thymus is atrophied and often barely noticeable.

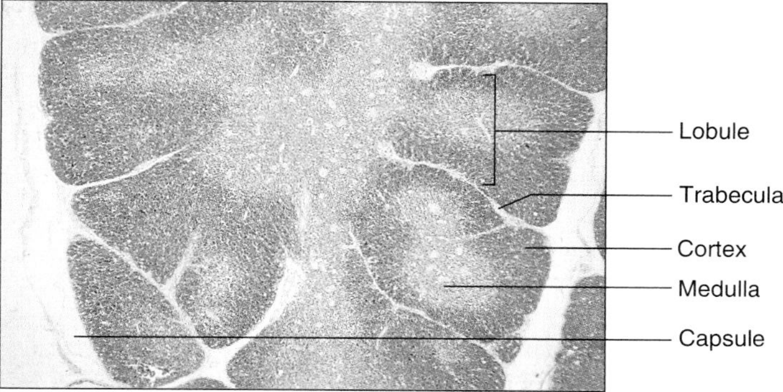

Figure 21.11 Histology of the Thymus.

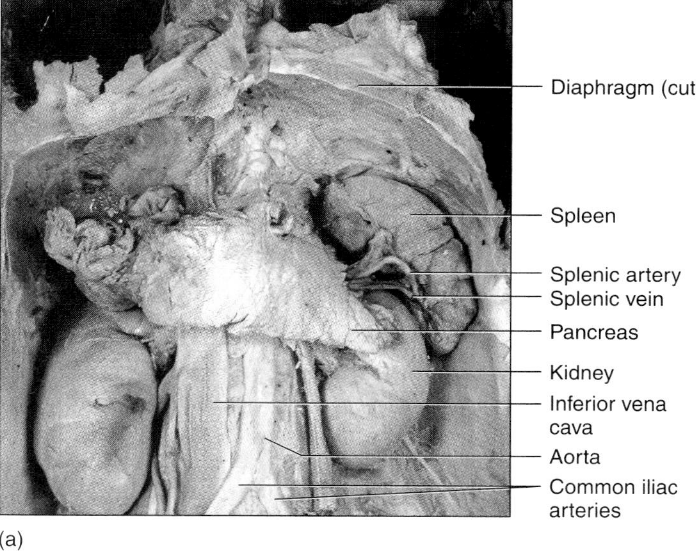

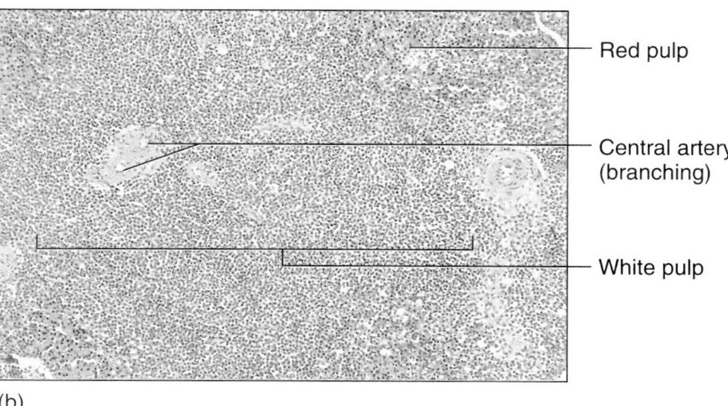

Figure 21.12 The Spleen. (*a*) Position of the spleen in the upper left quadrant of the abdominal cavity. (*b*) Histology.

Nonspecific Resistance

Objectives

When you have completed this section, you should be able to

- identify the body's three lines of defense against pathogens;
- contrast nonspecific resistance with immunity;
- describe the process of inflammation and explain what accounts for its cardinal signs; and
- describe the body's other nonspecific defenses.

The human body harbors a huge number of microorganisms. Indeed, we contain about 10,000 times as many bacterial cells as we do human cells. We are constantly exposed to countless organisms in the food and water we consume, in the air we breathe, and on the objects we touch. Without a means of defense, the human body would be an ideal place for microorganisms to live. Our homeostatic mechanisms would ensure a constant warm temperature, ample water, and a continual supply of nutrients. Upon death, the body is quickly overrun by microbial life, but as long as we are alive and maintaining homeostasis, we resist these hordes of would-be invaders. Our mechanisms for doing so are the subject of the rest of this chapter.

The living body has three lines of defense against pathogens:

1. External barriers, notably the skin and mucous membranes, which are impenetrable to the great majority of pathogens that daily assault us.
2. Inflammation, fever, and other active attacks upon pathogens that break through the first line of defense. These mechanisms are present from birth, are broadly effective against a wide range of pathogens, and work even against pathogens to which the body has never before been exposed.
3. The immune[3] system, which not only defeats a pathogen but leaves the body with a "memory" of that pathogen, enables us to defeat it more quickly in future encounters.

The first two mechanisms are called **nonspecific resistance** because they guard equally against a wide variety of pathogens and their effectiveness does not depend on prior exposure to a given pathogen. Immunity is called a **specific defense** because it results from prior exposure to a pathogen and provides future protection only against that particular pathogen.

In this section we study mechanisms of nonspecific resistance—physical and chemical barriers, phagocytic cells, inflammation, antimicrobial proteins, and fever. These mechanisms are summarized in table 21.4 at the end of the section.

3. *immuno* = free

External Barriers

The skin and mucous membranes make it mechanically difficult for pathogenic organisms to enter the body and spread through its tissues. When the skin has been broken by a scrape or an animal bite, or destroyed by a burn, one of the most urgent treatment concerns is the prevention of infection. This attests to the importance of intact skin as a barrier. The skin surface is composed mainly of keratin, a tough protein that few pathogens can penetrate. Furthermore, the surface of the skin is hostile to microbial reproduction. With exceptions such as the axillary and pubic areas, it is too dry and has too few nutrients to support appreciable microbial growth. The skin is also coated with antimicrobial chemicals such as defensins and lactic acid. *Defensins* are peptides, produced by neutrophils and found also on the skin surface, that kill microbes by causing membrane damage. Lactic acid is a component of perspiration. It coats the skin with a film called the *acid mantle* that inhibits microbial growth.

The digestive, respiratory, urinary, and reproductive tracts are open to the exterior, making them vulnerable to invasion, but they are protected by their mucous membranes. Mucus physically ensnares microbes. Microbes trapped in the respiratory mucus are then moved by cilia to the pharynx, swallowed, and destroyed by stomach acid. Microbes also are flushed from the upper digestive tract by saliva and from the lower urinary tract by urine. Mucus, tears, and saliva also contain **lysozyme,** an enzyme that destroys bacteria by dissolving their cell walls.

Beneath the epithelia of the skin and mucous membranes, there is a layer of areolar tissue. The ground substance of this tissue contains a glycosaminoglycan (see p. 182) called *hyaluronic acid,* which gives it a viscous consistency. It is normally difficult for microbes to migrate through this sticky **tissue gel.** Some organisms overcome this, however, by producing an enzyme called *hyaluronidase,* which breaks it down to a thinner consistency that is more easily penetrated (see clinical insight 6.1, p. 183). Hyaluronidase occurs in some snake venoms and bacterial toxins and is produced by some parasitic protozoans to facilitate their invasion of the connective tissues.

Leukocytes and Macrophages

Organisms that penetrate the skin and mucous membranes are attacked by **phagocytes** (phagocytic cells) that have a voracious appetite for foreign matter. Leukocytes and macrophages play especially important roles in both nonspecific defense and specific immunity.

Leukocytes

The five types of leukocytes are described in chapter 18, and their functions are summarized in table 18.8, pp. 701–702. It is appropriate to now examine their contributions to resistance and immunity:

1. **Neutrophils** are highly mobile cells that spend most of their lives wandering in the connective tissues phagocytizing bacteria. *Neutrophilia,* an elevated neutrophil count, is a common sign of bacterial infection. In addition to their phagocytic activity, neutrophils destroy nearby bacteria by creating a chemical **killing zone.** This process begins with a response called **degranulation,** in which lysosomes migrate to the cell surface and discharge their contents into the tissue fluid. This activates an enzyme in the plasma membrane that triggers a reaction called the **respiratory burst.** The cell rapidly takes up oxygen and reduces it to a highly toxic free radical, the superoxide anion ($O_2\cdot^-$). Superoxide reacts with hydrogen ions to form hydrogen peroxide (H_2O_2). Neutrophils also release an enzyme that uses chloride ion in the tissue fluid to produce hypochlorite (HClO) (the active ingredient in chlorine bleach). Superoxide, hydrogen peroxide, and hypochlorite are highly toxic to bacteria, but they are also deadly to the neutrophils themselves, which die in the course of their attack on the bacteria. These chemicals damage connective tissues as well and contribute to rheumatoid arthritis.
2. **Eosinophils** are less avidly phagocytic than neutrophils, but they do phagocytize antigen-antibody complexes, allergens (allergy-causing antigens), and inflammatory chemicals, as we discuss shortly. They are especially abundant in the mucosae of the respiratory, digestive, and lower urinary tracts. In infections with hookworms, tapeworms, and other parasites too large to phagocytize, eosinophils aggregate near the parasites and release enzymes that weaken or destroy them. Allergies and many parasitic infections are marked by *eosinophilia,* a rise in eosinophil count.
3. **Basophils** are not primarily phagocytic in function, but they aid the mobility and action of other leukocytes by secreting the vasodilator histamine and the anticoagulant heparin. Thus, they increase the blood flow to an infected tissue and inhibit clotting, which could otherwise immobilize the phagocytes.
4. **Lymphocytes** all look more or less alike in blood films, but there are several functional types, most of which are discussed later in this chapter under immunity. **Natural killer (NK) cells,** however, provide a form of nonspecific defense. They are large lymphocytes that attack and lyse *host cells* (cells of your own body) that have either turned

cancerous or become infected with viruses. They do not attack foreign pathogens.

5. **Monocytes** are the circulating precursors of macrophages, discussed next.

Macrophages

The **macrophage system** comprises all of the body's phagocytic cells except leukocytes. It is also known as the *lymphoid-macrophage (L-M)* or *reticuloendothelial (R-E) system*. Some of these phagocytes are wandering cells that actively seek pathogens, while others are fixed in place and can phagocytize only those pathogens that come to them—although they are strategically positioned for this to occur. The phagocytes include the following cell types, the first two of which are introduced in this chapter:

- **macrophages (histiocytes)** of the loose connective tissues;
- **reticular cells** in the stroma of the lymphatic organs;
- **dendritic cells** of the epidermis, oral mucosa, esophagus, and vagina (see chapter 7);
- **microglia** in the central nervous system (see chapter 14);
- **endothelial cells** of the blood vessels (see chapter 20);
- **alveolar macrophages** in the pulmonary alveoli (see chapter 22); and
- **hepatic macrophages** lining the liver sinusoids (see chapter 25).

Inflammation

Inflammation is a defensive response to tissue injury of any kind, including trauma and infection. Words ending in the suffix *-itis* denote inflammation of specific organs and tissues: *arthritis, encephalitis, peritonitis, gingivitis,* and *dermatitis,* for example. Inflammation can occur anywhere in the body, but it is most common and observable in the skin, which is subject to more trauma than any other organ. Examples of cutaneous inflammation include an itchy mosquito bite, a bee sting, sunburn, a poison ivy rash, and the blisters raised by manual labor or tight shoes.

Inflammation is characterized by four **cardinal signs**—redness, swelling, heat, and pain (table 21.1). Some authorities list impaired use as a fifth sign, but this may or may not occur and when it does, it is secondary to the pain that a person feels. The general purposes of inflammation are (1) to limit the spread of pathogens and ultimately destroy them, (2) to remove the debris of damaged tissue, and (3) to initiate tissue repair. The following discussion explains how the cardinal signs are produced and how the purposes of inflammation are achieved. These processes are mediated by the cells listed in table 21.2 and by the **inflammatory chemicals** listed in table 21.3. Inflammatory chemicals are secreted by damaged cells, mast cells, basophils, lymphocytes, macrophages, and platelets; some of them are also components of the blood plasma.

Table 21.1 The Four Cardinal Signs of Inflammation

Sign	Cause
Swelling	Increased capillary permeability and filtration
Redness	Hyperemia (increased blood flow); extravasated RBCs
Heat	Hyperemia
Pain	Inflammatory chemicals such as bradykinin and prostaglandins; fluid pressure on nerves; nerve injury

Table 21.2 Cellular Agents of Inflammation

Agents	Action
Basophils	Secrete histamine, heparin, bradykinin, serotonin, and leukotrienes
Endothelial cells	Produce cell-adhesion molecules to recruit leukocytes; synthesize platelet-derived growth factor to stimulate tissue repair
Fibroblasts	Promote tissue repair by secreting collagen, ground substance, and other tissue components; produce scar tissue
Macrophages	Phagocytize bacteria, tissue debris, dead and dying leukocytes and pathogens; act as antigen-presenting cells, which activate specific immunity
Mast cells	Same actions as basophils
Neutrophils	Phagocytize bacteria; secrete bactericidal oxidizing agents into tissue fluid
Platelets	Secrete clotting factors, which initiate tissue fluid coagulation, and platelet-derived growth factor, which promotes fibroblast activity and tissue repair

Hyperemia, Swelling, Redness, and Heat

Bradykinin, histamine, and *leukotrienes* stimulate vasodilation, which in turn leads to *hyperemia* (increased blood flow) and accounts for the redness and heat of inflamed tissue. Heat increases the local metabolic rate, thus promoting cell multiplication and healing. The increased blood flow dilutes toxins and provides the oxygen, nutrients, and waste removal needed to sustain the elevated metabolic rate.

Table 21.3 Inflammatory Chemicals

Substance	Sources	Effects*
Bradykinin	Plasma and basophils	Pain; vasodilation; increased capillary permeability
Clotting factors	Plasma, platelets, and damaged cells	Coagulation of tissue fluid; isolation of pathogens; formation of temporary framework for tissue repair
Complement	Plasma	Enhanced histamine release; neutrophil chemotaxis; promotion of phagocytosis
Histamine	Basophils and mast cells	Vasodilation; increased capillary permeability
Leukocytosis-promoting factor	Damaged cells	Release of leukocytes from bone marrow
Leukotrienes	Damaged cells, mast cells, and basophils	Vasodilation; increased capillary permeability; neutrophil chemotaxis
Lymphokines	Lymphocytes	Enhanced monocyte and macrophage activity; leukocyte chemotaxis; intercellular signaling among leukocyte types
Platelet-derived growth factor	Platelets and endothelial cells	Stimulation of fibroblast activity; cell division; replacement of damaged cells
Prostaglandins	Damaged cells	Pain; enhanced action of histamine and bradykinin; neutrophil diapedesis

*Some inflammatory chemicals have additional roles in specific immunity, as described in table 21.8.

These chemicals also increase the permeability of blood capillaries, so more fluid seeps from the bloodstream into the inflamed tissue. This produces swelling, but more importantly it allows helpful plasma chemicals and blood cells to get into the tissue. The chemicals include antibodies, complement proteins (described shortly), and fibrinogen, which causes tissue fluid to clot. Clotting sequesters (isolates) bacteria and retards their spread to adjacent tissues, while the fibrin of the clot forms a scaffold for the organization of tissue repair. Swelling reduces venous drainage and increases lymphatic drainage, which favors the removal of bacteria and cellular debris. The unusually permeable capillaries also allow RBCs to escape from the circulation and pool in the tissue. These *extravasated*[4] RBCs contribute to the redness of the tissue, as seen in a sunburn.

Pain and Functional Impairment

Pain arises from direct injury to the nerve endings (nociceptors), their stimulation by inflammatory chemicals, and the pressure caused by tissue swelling. One of the most potent pain stimuli is **bradykinin,** which is secreted by basophils and mast cells and produced from *kininogen,* a protein in the blood plasma. Prostaglandins and some bacterial toxins are also important pain stimuli. Pain is not merely an undesirable side effect of inflammation; it is an important alarm signal that calls attention to tissue injury, causes us to favor an injured body part, or makes it so uncomfortable to use that we keep it still. This gives it the rest it needs for recovery.

4. *extra* = outside of + *vas* = blood vessel

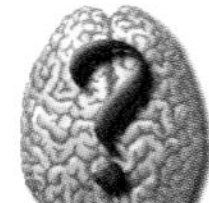

Think About It

Review eicosanoid synthesis (p. 672) and explain why aspirin eases the pain of inflammation.

Leukocyte Deployment

Hyperemia and capillary permeability favor a four-step deployment of leukocytes to the inflamed tissue and attack on invading bacteria (fig. 21.13):

1. **Margination.** Leukocytes adhere to the walls of the blood vessel in inflamed areas, rather than rushing by in the bloodstream. This occurs because chemicals released by damaged tissues stimulate endothelial cells to produce *cell-adhesion molecules (CAMs)* that make their membranes "sticky." The leukocytes adhere loosely to the CAMs and slowly tumble along the endothelium, sometimes coating it so thickly that they obstruct blood flow.
2. **Diapedesis**[5] (DY-uh-peh-DEE-sis). Leukocytes squeeze between the endothelial cells to enter the interstitial space.
3. **Chemotaxis.** Leukocytes are attracted to bradykinin, leukotrienes, and other inflammatory chemicals that guide them to the site of infection or injury.
4. **Phagocytosis.** Neutrophils are the quickest leukocytes to respond to invasion. Hordes of them arrive at the site of damage within an hour, avidly phagocytize bacteria, and secrete the oxidizing agents described earlier.

5. *dia* = through + *pedesis* = stepping

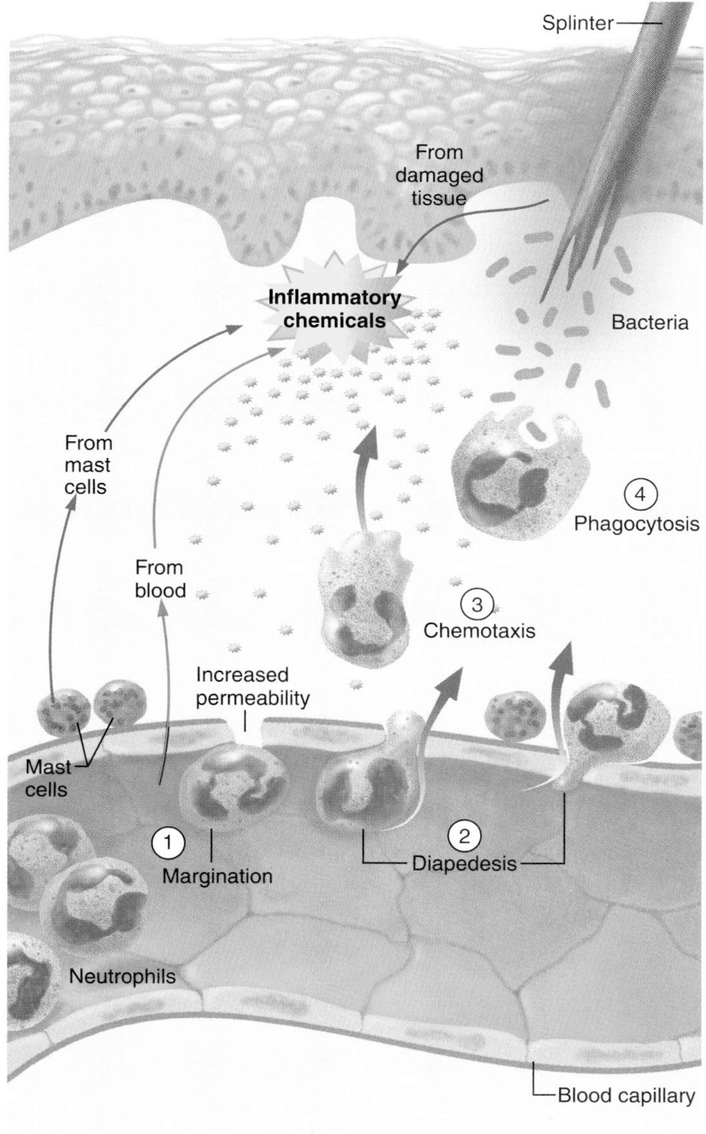

Figure 21.13 Leukocyte Behavior in Inflammation. Chemical messengers are released by basophils, mast cells, blood plasma, and damaged tissue. These inflammatory chemicals stimulate leukocyte margination (adhesion to the capillary wall), diapedesis (crawling through the capillary wall), chemotaxis (movement toward the source of the inflammatory chemicals), and phagocytosis (engulfing bacteria or other pathogens).

Since the attack on pathogens is suicidal to the neutrophils themselves, they require replacements. Damaged tissues release **leukocytosis-promoting factor,** a growth factor that stimulates the rapid production and release of neutrophils by the red bone marrow. Within a few hours, the neutrophil count in the circulating blood may rise to four or five times normal. *Leukocytosis* means any elevated blood leukocyte count.

Basophils release a potent mix of inflammatory chemicals (see table 21.3). Eosinophils are attracted in cases of parasitism and allergy. Monocytes arrive within 8 to 12 hours, differentiate into macrophages, and eventually become the primary agents of cleanup. They phagocytize bacteria, damaged host cells, and dead and dying neutrophils and play a vital role in active immunity as *antigen-presenting cells,* discussed later.

The mixture of tissue fluid, cellular debris, and dead and dying neutrophils and microbes is called **pus.** Pus is usually absorbed as the inflammation subsides, but it sometimes accumulates in a blister and may be released by its rupture.

Tissue Repair

Aside from combating pathogens, it is necessary to repair the damaged tissue. Endothelial cells and platelets secrete **platelet-derived growth factor (PDGF).** PDGF stimulates fibroblasts to multiply and synthesize collagen fibers and matrix, which generally leads to the rebuilding of connective tissues. Some tissues such as skeletal muscle cannot be replaced. In such cases, the fibroblasts form scar tissue. (See the discussion of fibrosis in chapter 6.)

Antimicrobial Proteins

Two groups of proteins, the *interferons* and the *complement system,* provide short-term, nonspecific resistance to viral and bacterial infections.

Interferons

Interferons are polypeptides secreted by cells that have been invaded by viruses. They diffuse to neighboring cells and stimulate them to produce antiviral proteins, which prevent viruses from multiplying within them. Interferons also activate natural killer cells and macrophages, which destroy infected host cells before they release more viruses. Interferons are not specific for a particular virus but provide generalized protection.

Interferons also stimulate the destruction of cancer cells. When discovered in 1957, they seemed to hold promise as a cure for viral diseases and cancer. The body produces such small quantities of interferons, however, that it was difficult to isolate enough for testing, let alone for pharmacological use. In the 1980s, interferons were first produced in quantity by means of transgenic bacteria. Even then, however, they showed little effect against cancer except for one rare form of leukemia and Kaposi's sarcoma, a common cancer of AIDS patients. Interferons have been used with limited success in controlling hepatitis C and some other viral infections, but control of HIV is still in an experimental stage.

Complement System

The **complement system** is a group of 20 or more beta globulins that aid in nonspecific resistance and specific immunity. These proteins are continually present in the blood plasma but must be activated by pathogens to exert their effects. There are two pathways of complement activation (fig. 21.14). In the **classical pathway,** antibodies bind to pathogenic organisms and then bind a complex of three complement proteins called C1, C2, and C4. This step is called **complement fixation.** The **alternate pathway** begins with three complement proteins—factors B, D, and P—binding to the surface polysaccharides of microbes. Both pathways converge on a step where complement C3 is split into two fragments, C3a and C3b. From this point on, the pathway to completion, whether initiated by the classical or alternate pathway, is the same, as shown in the figure. Complement helps destroy pathogens in three ways:

1. **Enhanced inflammation.** Complement C3a stimulates mast cells and basophils to secrete inflammatory chemicals.

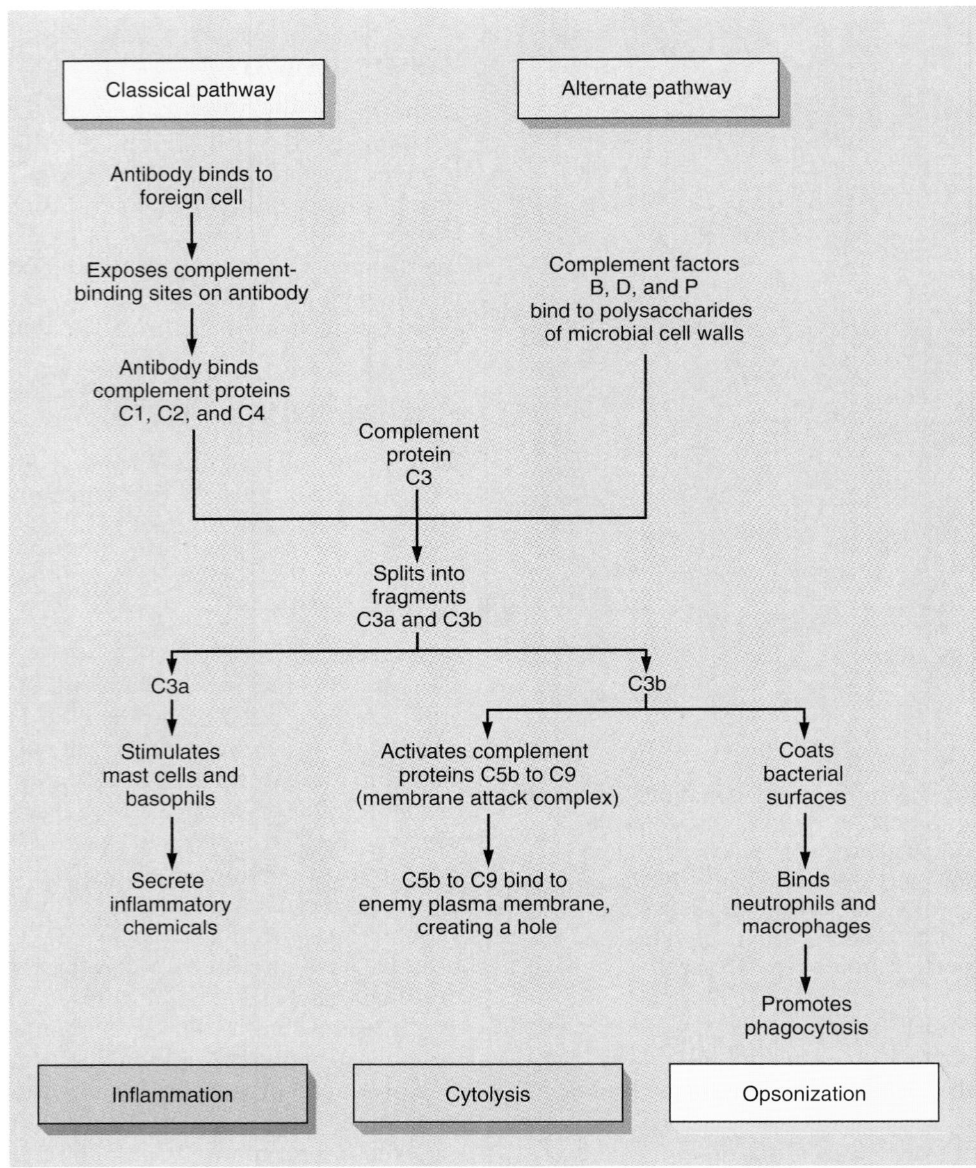

Figure 21.14 **Complement Activation.** Both the classical pathway and the alternate pathway produce complement fragments C3a and C3b. The process is the same from that point to the end. The C3a and C3b fragments promote inflammation, cytolysis, and opsonization.

2. **Cytolysis.** Complement C3b leads to the rupture of target cells. It binds to a cell and triggers the insertion of a group of proteins called the *membrane attack complex (MAC).* The MAC forms a doughnutlike ring in the microbial plasma membrane that allows the cytoplasm to escape (fig. 21.15).
3. **Opsonization.** Complement C3b makes bacteria easier to phagocytize by coating their surfaces and serving as binding sites for macrophages and neutrophils.

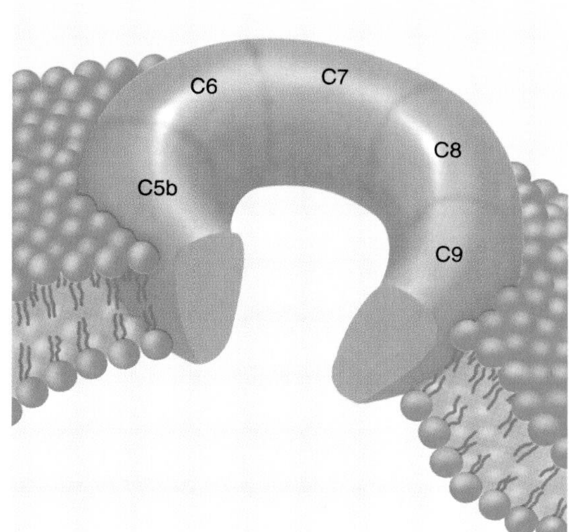

Figure 21.15 The Membrane Attack Complex. Complement proteins C5b to C9 form a doughnutlike ring in the plasma membrane of an enemy cell, thus causing cytolysis.

Fever

Fever *(pyrexia)* can result from trauma, drug reactions, infections, or several other causes. Because of variations in human body temperature, there is no exact criterion for what constitutes a fever—a fever for one person may be a normal temperature for another. Fever was long regarded as an undesirable side effect of illness, and efforts were made to reduce it for the sake of comfort. It is now recognized, however, as an adaptive defense mechanism that, in moderation, does more good than harm. People with colds, for example, recover more quickly and are less infective to others when they allow a fever to run its course rather than using antipyretic (fever-reducing) medications such as aspirin. Fever is beneficial in that (1) it promotes interferon activity, (2) it elevates metabolic rate and accelerates tissue repair, and (3) it inhibits reproduction of bacteria and viruses.

In bacterial infections, macrophages secrete *interleukin-1 (Il-1).* According to one theory, Il-1 is a **pyrogen**[6] (fever-producing agent) that stimulates the anterior hypothalamus to secrete prostaglandin E (PGE). PGE, in turn, stimulates the hypothalamic thermostat to raise the body temperature to a higher set point—say 39°C (102°F) instead of the usual 37°C. Aspirin and ibuprofen reduce fever by inhibiting PGE synthesis.

When the thermostatic set point rises, a person shivers to produce heat and the cutaneous blood vessels constrict to reduce heat loss. In the stage of fever called *onset,* a person exhibits chills, feels cold and clammy to the touch, and has a rising temperature (fig. 21.16).

In the next stage, *stadium,* the body temperature oscillates around the new set point for as long as the pathogen is

6. *pyro* = fire, heat + *gen* = producing

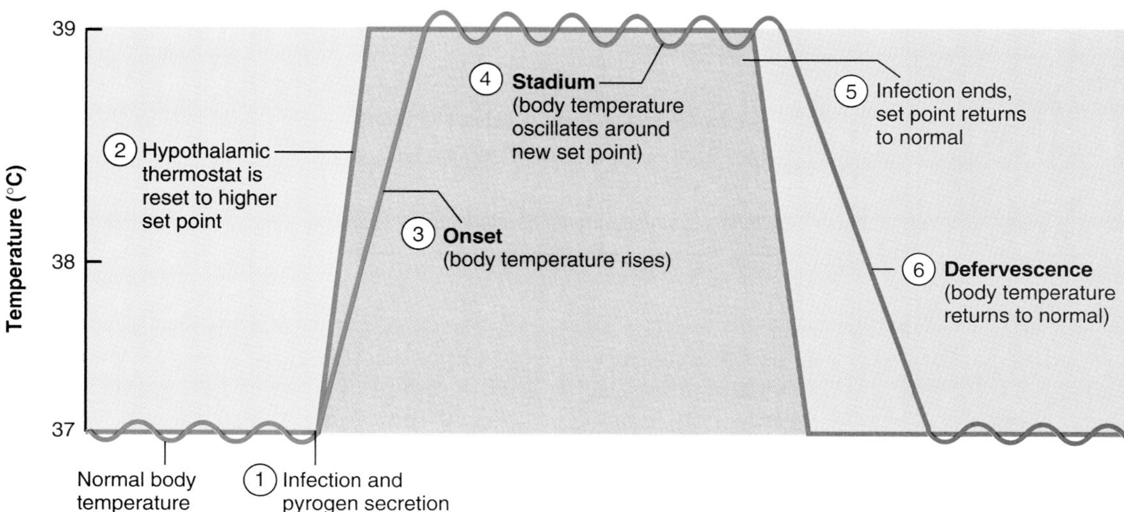

Figure 21.16 The Course of a Fever.

Clinical Insight 21.1

Reye Syndrome

In children under 15, an acute viral infection such as chickenpox or influenza is sometimes followed by a serious disorder called *Reye*[7] *syndrome*. First recognized in 1963, this disease is characterized by swelling of brain neurons and by fatty infiltration of the liver and other viscera. Neurons die from hypoxia and the pressure of the swelling brain, which results in nausea, vomiting, disorientation, seizures, and coma. About 30% of victims die, and the survivors sometimes suffer mental retardation. Reye syndrome is sometimes triggered by the use of aspirin to control fever; parents are now strictly advised never to give aspirin to children with chickenpox or flulike symptoms.

7. R. Douglas Reye (1912–77), Australian pathologist

Table 21.4 Mechanisms of Nonspecific Resistance

First Line of Defense (external barriers)	
Skin	Keratin is impenetrable to most pathogens; surface is poor in moisture and nutrients needed for microbial growth; defensins and lactic acid inhibit microbial growth; tissue gel impedes spread of pathogens
Mucous Membranes	Mucus traps pathogens; saliva and urine flush away pathogens; lysozyme destroys bacteria; tissue gel impedes spread of pathogens
Second Line of Defense (other mechanisms of nonspecific resistance)	
Leukocytes	
Neutrophils	Phagocytize bacteria and secrete strong oxidizing agents that destroy extracellular bacteria
Eosinophils	Phagocytize antigen-antibody complexes, allergens, and inflammatory chemicals; secrete enzymes that destroy pathogens too large to phagocytize
Basophils	Secrete inflammatory chemicals; see table 21.3
Monocytes	Become macrophages; see table 21.2
Natural Killer Cells	Destroy virus-infected or cancerous host cells; other lymphocyte types are involved in specific immunity
Macrophage System	Widespread phagocytic cells of several types; actively phagocytize and destroy pathogens; secrete pyrogens and promote fever; also essential in specific immune responses as antigen-presenting cells
Inflammation	Sequesters pathogens at site of invasion; recruits leukocytes and macrophages to destroy pathogens; removes debris of damaged tissue; initiates tissue repair
Antimicrobial Proteins	
Complement	Lyses pathogenic microbes; enhances action of phagocytes; stimulates release of inflammatory chemicals
Interferons	Produced by virus-infected cells; prevent viruses from reproducing in neighboring cells; activate macrophages and natural killer cells
Fever	Inhibits multiplication of bacteria and viruses; promotes interferon secretion; elevates metabolic rate and accelerates tissue repair

present. The elevated temperature stimulates the liver and spleen to hoard zinc and iron, depriving bacteria of minerals needed for their multiplication. When the infection is defeated, pyrogen secretion ceases and the hypothalamic thermostat is set back to normal. This activates heat-losing mechanisms, especially cutaneous vasodilation and sweating. The skin is warm and flushed during this phase. The phase of falling temperature is called *defervescence* in general, *crisis* if the temperature drops abruptly, or *lysis* if it falls slowly.

Even though most fevers are beneficial, excessively high temperature can be dangerous because it causes cellular dysfunction. Fevers above 40.5°C (105°F) can make a person delirious. Convulsions and coma ensue at higher temperatures, and death or irreversible brain damage commonly results from fevers that range from 44° to 46°C (111° to 115°F).

Table 21.4 summarizes the body's mechanisms of nonspecific resistance.

Before You Go On

Answer the following questions to test your understanding of the preceding section:

4. What are macrophages? Give four examples and state where they are found.
5. List the cardinal signs of inflammation and state the cause of each.
6. How do interferons and the complement system protect against disease?
7. Summarize the benefits of fever and the limits of these benefits.

General Aspects of Specific Immunity

Objectives

When you have completed this section, you should be able to

- define *specific immunity;*
- contrast humoral and cellular immunity;
- describe the chemical constitution of antigens and antibodies and sketch or describe the structure of an antibody molecule; and
- describe the general roles played by lymphocytes, macrophages, and interleukins in the immune response.

The remainder of this chapter is concerned with the immune system and specific immunity, the third line of defense. The immune system is not an organ system but an array of widely distributed cells that recognize foreign substances and act to neutralize or destroy them. Two characteristics distinguish immunity from nonspecific resistance:

1. **Specificity.** Immunity is directed against a particular pathogen. Immunity to one pathogen usually does not confer immunity to others.
2. **Memory.** When reexposed to the same pathogen, the body reacts so quickly that there is no noticeable illness. The reaction time for inflammation and other nonspecific defenses, by contrast, is just as long for later exposures as for the initial one.

These properties of immunity were recognized even in the fifth century B.C.E., when the Greek historian Thucydides remarked that people who recover from a disease often become immune to that one but remain susceptible to others. A person might be immune to measles but still susceptible to polio, for example. In the late 1800s, it was discovered that immunity can be transferred from one animal to another by way of the blood serum. In the mid-1900s, however, it was found that serum does not always confer immunity; sometimes only donor lymphocytes do so. Thus, we now recognize two types of immunity, called humoral and cellular immunity, although these two mechanisms interact extensively and often respond to the same pathogen.

Humoral (antibody-mediated) immunity, named for the fluids, or "humors," of the body, is based on the action of soluble proteins called *antibodies.* Antibodies occur in the body fluids and on the plasma membranes of lymphocytes. Circulating antibodies bind to bacteria, toxins, and extracellular viruses, "tagging" them for destruction by mechanisms described later.

Cellular (cell-mediated) immunity is based on the action of lymphocytes that directly attack diseased or "suspicious" cells, including those of transplanted tissues, cells infected with viruses or parasites, and cancer cells. Lymphocytes can lyse these cells or release chemicals that enhance other defenses such as inflammation.

Specific immunity has the elements of a drama with a large cast of characters—antigens, antibodies, lymphocytes, macrophages, and chemical messengers. These agents are described before we examine the details of humoral and cellular immunity. After you have studied the details, you will find these two types of immunity summarized and compared in table 21.9.

Antigens

An **antigen**[8] **(Ag)** is any molecule that triggers an immune response. Some antigens are free molecules such as venoms and toxins; others are components of plasma membranes and bacterial cell walls. Small universal molecules such as glucose and amino acids are not antigenic; if they were, our immune systems would attack the nutrients and other molecules essential to our very survival. Most antigens have molecular weights over 10,000 amu and are generally complex molecules such as proteins, polysaccharides, glycoproteins, and glycolipids with structures that are unique to each individual. Their uniqueness enables the body to distinguish its own ("self") molecules from those of any other individual or organism ("nonself"). The immune system "learns" to distinguish self from nonself antigens prior to birth; thereafter, it normally attacks only nonself antigens.

A macromolecule typically has several regions called **antigenic determinants (epitopes)** that can trigger separate immune responses. One region of a glycoprotein molecule may stimulate the formation of one antibody, and a different region of the same molecule may stimulate the formation of a different antibody (fig. 21.17).

Some molecules called **haptens**[9] are too small to be antigenic in themselves, but they can stimulate an immune response by binding to a host macromolecule and creating a unique complex that the body recognizes as foreign. After the first exposure, a hapten may stimulate an immune response without needing to bind to another molecule. Cosmetics, detergents, industrial chemicals, poison ivy,

8. acronym from *anti*body *gen*erating
9. from *haptein* = to fasten

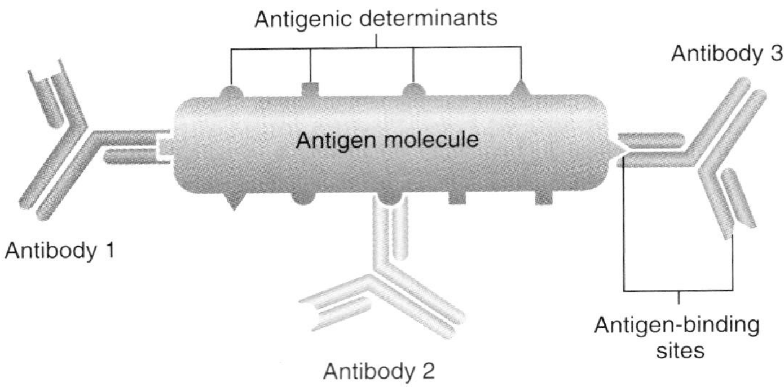

Figure 21.17 **Antigenic Determinants.** An antigen molecule has multiple antigenic determinants that can stimulate formation of a variety of antibodies.

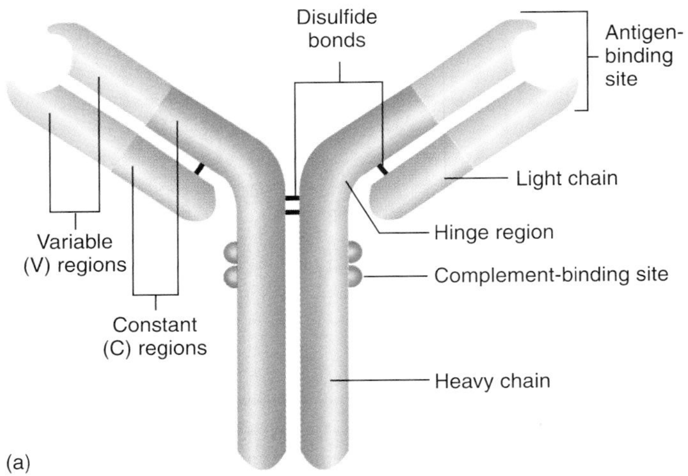

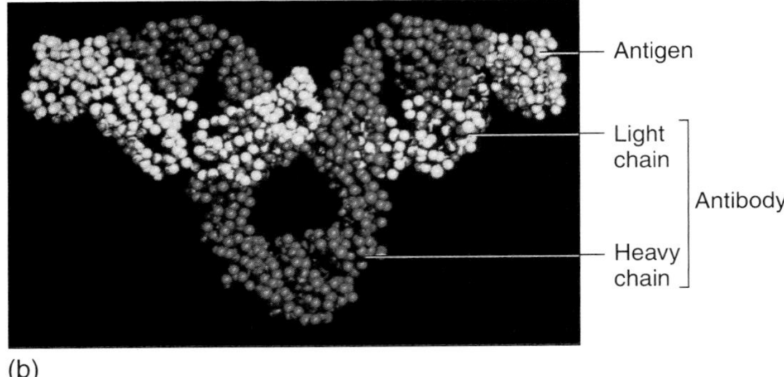

Figure 21.18 **Antibody Structure.** (*a*) A molecule of IgG, a monomer. (*b*) Computer-generated image of IgG bound to an antigen (lysozyme).

and animal dander contain haptens that cause allergies in many people. Penicillin is a hapten that sometimes binds to erythrocytes and triggers an allergic reaction.

Antibodies

An **antibody (Ab),** or **immunoglobulin (Ig),** is a gamma globulin found in the blood plasma, body secretions, and some leukocyte membranes. The basic structural unit of an antibody, an **antibody monomer,** is composed of four polypeptide chains linked by disulfide (–S–S–) bonds (fig. 21.18). The two larger **heavy chains** are about 400 amino acids long and the two **light chains** about half that long. Each heavy chain has a hinge region where the antibody is bent, giving the monomer a T or Y shape.

All four chains have a **variable region** (V region), which gives an antibody its uniqueness. The V regions of a heavy chain and light chain combine to form an **antigen-binding site** on each arm, which attaches to the antigenic determinant of an antigen molecule. The rest of each chain is a **constant region** (C region), which has the same amino acid sequence in all antibodies of a given class (within one person). The C region determines the mechanism of an antibody's action—for example, whether it can bind complement proteins.

There are five classes of antibodies named **IgA, IgD, IgE, IgG,** and **IgM** (table 21.5). These names refer to the structures of their C regions (*alpha, delta, epsilon, gamma,* and *mu*). IgD, IgE, and IgG are monomers. IgA has a monomeric form as well as a dimer composed of two cojoined monomers. IgM is a pentamer composed of five monomers. IgG is particularly important in the immunity of the newborn because it is the only immunoglobulin that crosses the placenta. Thus, it transfers some of the mother's immunity to her fetus. In addition, an infant acquires some maternal IgA through breast milk and colostrum (the secretion an infant feeds on for the first 2 or 3 days, until true milk is secreted).

The immune system is thought to produce as many as 2 million different antibodies. This is an astonishing number considering that normally each protein in the body is encoded by a different gene and we have a total of only 100,000 genes, most of which have functions unrelated to immunity. Obviously there cannot be a different gene for each antibody.

Instead, the genome contains several hundred DNA segments that can be shuffled and combined in various ways to produce antibody genes unique to each antibody-producing cell line. This process is called **somatic recombination** because it forms new combinations of DNA base sequences in somatic (nonreproductive) cells. This explains how we can produce such a tremendous variety of antibodies with a relatively limited number of genes.

Lymphocytes

The major cells of the immune system are lymphocytes and macrophages, which are especially concentrated at strategic places such as the lymphatic organs and mucous membranes. Macrophages are active in nonspecific resistance and were discussed earlier, but they are also crucial in specific immunity. Most lymphocytes can be classified as either **T lymphocytes (T cells)** or **B lymphocytes (B cells).**

Table 21.5 The Five Classes of Antibodies

Class	Structure	Location and Function
IgA	Monomer and dimer forms	Plasma IgA is a monomer found in blood plasma; secretory IgA is a dimer found in mucus, saliva, tears, milk, and intestinal secretions. IgA prevents pathogens from adhering to epithelia and penetrating the underlying tissues.
IgD	Monomer	An integral protein of the B cell membrane; acts as an antigen receptor.
IgE	Monomer	Found mainly in tonsils, skin, and mucous membranes. Stimulates mast cells and basophils to release histamine and other chemical mediators of inflammation and allergy; attracts eosinophils to sites of parasitic infection.
IgG	Monomer	Constitutes 75% to 85% of circulating antibodies in plasma. The predominant antibody secreted in the secondary immune response. Crosses placenta and confers temporary immunity on the fetus. Includes the anti-D antibodies of the Rh blood group. IgG and IgM are the only antibodies able to bind complement.
IgM	Monomer and pentamer forms	Monomer is an antigen receptor of the B cell membrane; pentamer occurs in blood plasma. The predominant antibody secreted in the primary immune response. Very strong agglutinating ability. Includes the anti-A and anti-B agglutinins of the ABO blood group.

T Lymphocytes (T cells)

During fetal development, the bone marrow releases undifferentiated stem cells into the blood. Some of these colonize the thymus for 2 or 3 days; the *T* in T lymphocytes stands for *thymus-dependent.* Hormones of the thymus, called thymosins, stimulate maturing T cells to synthesize about 10,000 to 100,000 identical plasma membrane proteins that serve as antigen receptors. Once these receptors are in place, a T cell is said to be **immunocompetent;** that is, it can now bind to antigens "presented" to it by other cells. An immunocompetent T cell divides rapidly and forms a **clone** of T cells with identical receptors, "programmed" to respond to one particular antigen. All clones that have yet to encounter an antigen are collectively called the **virgin lymphocyte pool.**

Some of the clones in a fetus are capable of responding to self-antigens. In **clonal deletion,** these self-reactive clones are destroyed and leave the body in a state of **self-tolerance** in which the immune system is unresponsive to the person's own tissues. Clonal deletion leaves only the T cells programmed to respond to foreign antigens. These account for 70% to 80% of the lymphocytes in the blood and colonize lymphatic tissues and organs everywhere.

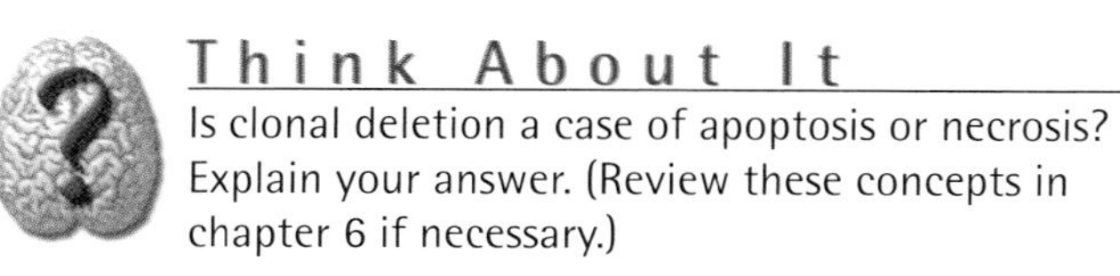

Think About It

Is clonal deletion a case of apoptosis or necrosis? Explain your answer. (Review these concepts in chapter 6 if necessary.)

B Lymphocytes (B cells)

Fetal stem cells that do not colonize the thymus settle temporarily in the liver, the bone marrow, the junction of the small intestine and large intestine, the appendix, and other organs, where they differentiate into B lymphocytes (B cells). B cells are named for a lymphatic organ of chickens, the *bursa of Fabricius,*[10] where they were first discovered. The sites of B cell establishment in humans are sometimes called *bursa equivalents.* Here, the B cells synthesize their antigen receptors (IgD and IgM molecules in the plasma membrane), divide rapidly, and produce immunocompetent clones. Each clone can react to only one antigen—tetanus toxin, for example. If you became infected with the tetanus bacterium, this clone would be called into service. Many clones are never used because the individual is never exposed to the corresponding antigens.

Immunocompetent B cells disperse through the body to colonize the same organs as T cells. They form 20% to 30% of the lymphocytes of the blood, and they are abundant in the lymph nodes, spleen, bone marrow, glands, and mucous membranes.

Antigen-Presenting Cells

In addition to their other roles, B cells and macrophages function as **antigen-presenting cells (APCs).** As you will

10. Hieronymus Fabricius (Girolamo Fabrizzi) (1537–1619), Italian anatomist

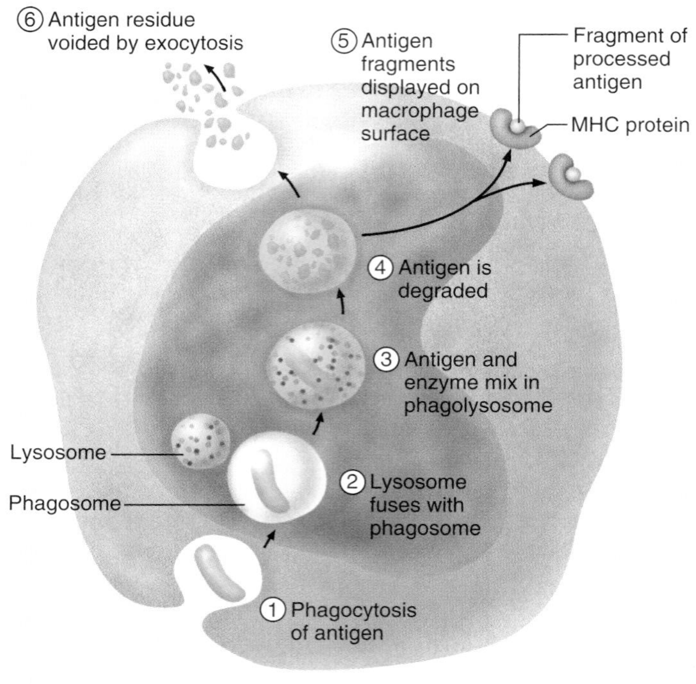

(a)

Macrophages

Pseudopods

Bacteria

(b)

Figure 21.19 The Action of Antigen-Presenting Cells (APCs). (*a*) Stages in the processing and presentation of an antigen by an APC such as a macrophage. (*b*) Macrophages phagocytizing bacteria. Filamentous extensions of the macrophage snare the rod-shaped bacteria and draw them to the cell surface, where they are engulfed.

see later, T cells cannot recognize free antigen molecules but require the help of APCs. APC function hinges on a family of genes on chromosome 6 called the major histocompatibility complex (MHC). These genes code for **MHC proteins,** which are structurally unique to every person except for identical twins. MHC proteins act as "identification tags" that label every cell of your body as belonging to you. The shape of an MHC protein is a little like a hotdog bun—an elongated protein with a central groove.

An antigen-presenting cell internalizes an antigen by endocytosis, digests it into molecular fragments, and "displays" some of these fragments in the groove of its MHC proteins (fig. 21.19). Wandering T cells regularly inspect APCs for displayed antigens (fig. 21.20). If an MHC protein carries a self antigen, the T cells disregard it. If it carries a nonself antigen, however, the T cells initiate an immune attack.

Interleukins

With so many cell types involved in immunity, it is not surprising that they require a special system of chemical messengers to coordinate their activities. **Interleukins**[11] are hormonelike messengers between leukocytes or leukocyte derivatives. Those produced by lymphocytes are called **lymphokines,**[12] and those produced by macrophages are called **monokines** (after monocytes, the macrophage precursors).

Since the terminology of immune cells and chemicals is quite complex, you may find it helpful to refer often to tables 21.7 and 21.8 (p. 831) as you read the following discussions of humoral and cellular immunity.

Before You Go On

Answer the following questions to test your understanding of the preceding section:

8. How does specific immunity differ from nonspecific defense?
9. How does humoral immunity differ from cellular immunity?
10. What structural properties distinguish antigenic molecules from those that are not antigenic?
11. Explain the distinction between the variable region and the antigen-binding site of an immunoglobulin.
12. What is an immunocompetent lymphocyte? What does a lymphocyte have to produce in order to be immunocompetent?
13. Define *T cell* and *B cell.*
14. Define *interleukin, lymphokine,* and *monokine.*

11. *inter* = between + *leuk* = leukocytes

12. *kine* = motion, action

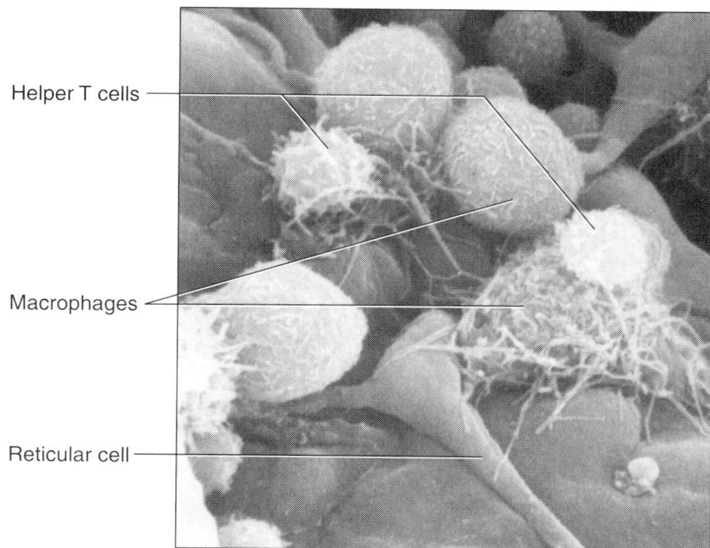

Figure 21.20 **T Cells "Inspecting" Macrophages in a Lymph Node for Antigen Presentation.** From R. G. Kessel and R. H. Kardon, *Tissues and Organs: A Text-Atlas of Scanning Electron Microscopy,* (W. H. Freeman & Co., 1979).

Humoral Immunity

Objectives

When you have completed this section, you should be able to

- explain how B cells recognize and respond to an antigen;
- describe the actions of antibodies; and
- explain how immune memory prevents disease upon reexposure to an antigen.

With the foregoing introduction to the "actors" of the immune system, we can now discuss the "plot"—the humoral and cellular mechanisms of immunity. Each mechanism is like a play in three acts—recognition, attack, and memory. In humoral immunity, the essential features of each stage are as follows:

- **Recognition.** B cells recognize an antigen and divide repeatedly; most of the daughter cells differentiate into *plasma cells,* which synthesize antibodies specific to that antigen.
- **Attack.** Plasma cells release their antibodies, which bind to the antigen, render it harmless, and "tag" it for destruction by other agents.
- **Memory.** Some B cells differentiate into *memory cells,* which provide lasting protection against future exposures to the same pathogen.

You might also think of these as "the three *Rs* of immunity"—*recognize, react,* and *remember.* We now consider some details of these three stages.

Recognition

Antigen recognition and preparation for the immune attack occur in five steps: *capping, endocytosis, display, clonal selection,* and *plasma cell differentiation:*

1. **Capping.** An immunocompetent B cell has thousands of surface receptors (IgD and IgM molecules) for one antigenic determinant. Since an antigen is a large molecule with numerous copies of each antigenic determinant, it binds to several receptors on a B cell and links them together (fig. 21.21*a*). This *capping* process activates the B cell. It helps to explain the action of haptens, which become antigenic only if they bind to host molecules that create a complex large enough to produce a capping reaction. Small molecules cannot reach from receptor to receptor and link them together.
2. **Endocytosis.** The capped receptors of the B cell become drawn into a cluster (fig. 21.21*b*). A depression forms in the membrane at this point, culminating in receptor-mediated endocytosis of the antigen-receptor complex (fig. 21.21*c, d*).
3. **Display.** The internalized antigen is digested into molecular fragments. Some of these are voided from the cell as waste, whereas others are linked to MHC proteins, transported to the plasma membrane, and "presented" or displayed there. For the response to go any further, another type of lymphocyte called a *helper T cell* must get involved. Helper T cells roam the tissues and crawl about on antigen-presenting cells as if "looking for trouble" (see fig. 21.20). This state of continual inspection and "alertness" is called **immunological surveillance.** When a nonself antigen is presented by an MHC protein, the two form a powerful stimulatory complex that induces helper T cells to secrete interleukin-2 and **helper factors.** These lymphokines stimulate the B cell to go on to the next step.
4. **Clonal selection.** Helper factors stimulate a B cell to divide repeatedly into a battalion of identical cells (fig. 21.22). This process is called *clonal selection* because the antigen "selects" only immunocompetent B cells with receptors for it and stimulates those cells to divide into a clone of B cells with the same receptors.
5. **Plasma cell differentiation.** Most cells of a clone differentiate into **plasma cells.** These are larger than B cells and contain an abundance of rough endoplasmic reticulum (fig. 21.23). A plasma cell produces antibodies at the remarkable rate of 2,000 molecules per second over a brief life span of 4 to 5 days. Antibodies are soluble proteins that become distributed throughout the body fluids. The first

Antigenic determinants
Antigen
B cell membrane
Antigen receptors
(a) Antigen binds to several receptor sites on membrane of B cell
(b) Receptor sites are drawn together into a cluster
(c) Antigen is taken into B cell by receptor-mediated endocytosis
(d) Antigen is fully internalized

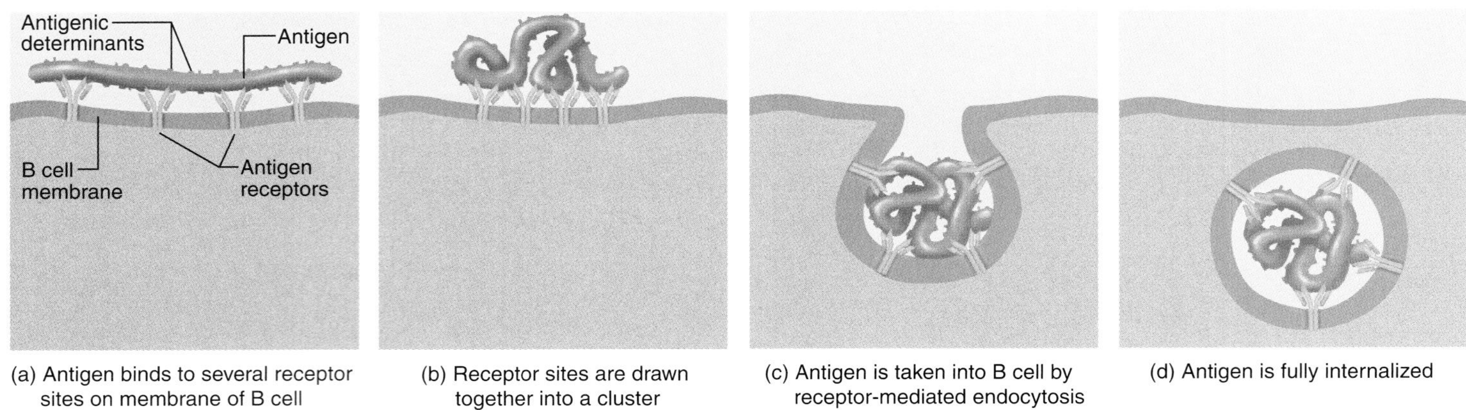

Figure 21.21 Capping and Endocytosis in the Activation of a B Cell by an Antigen.

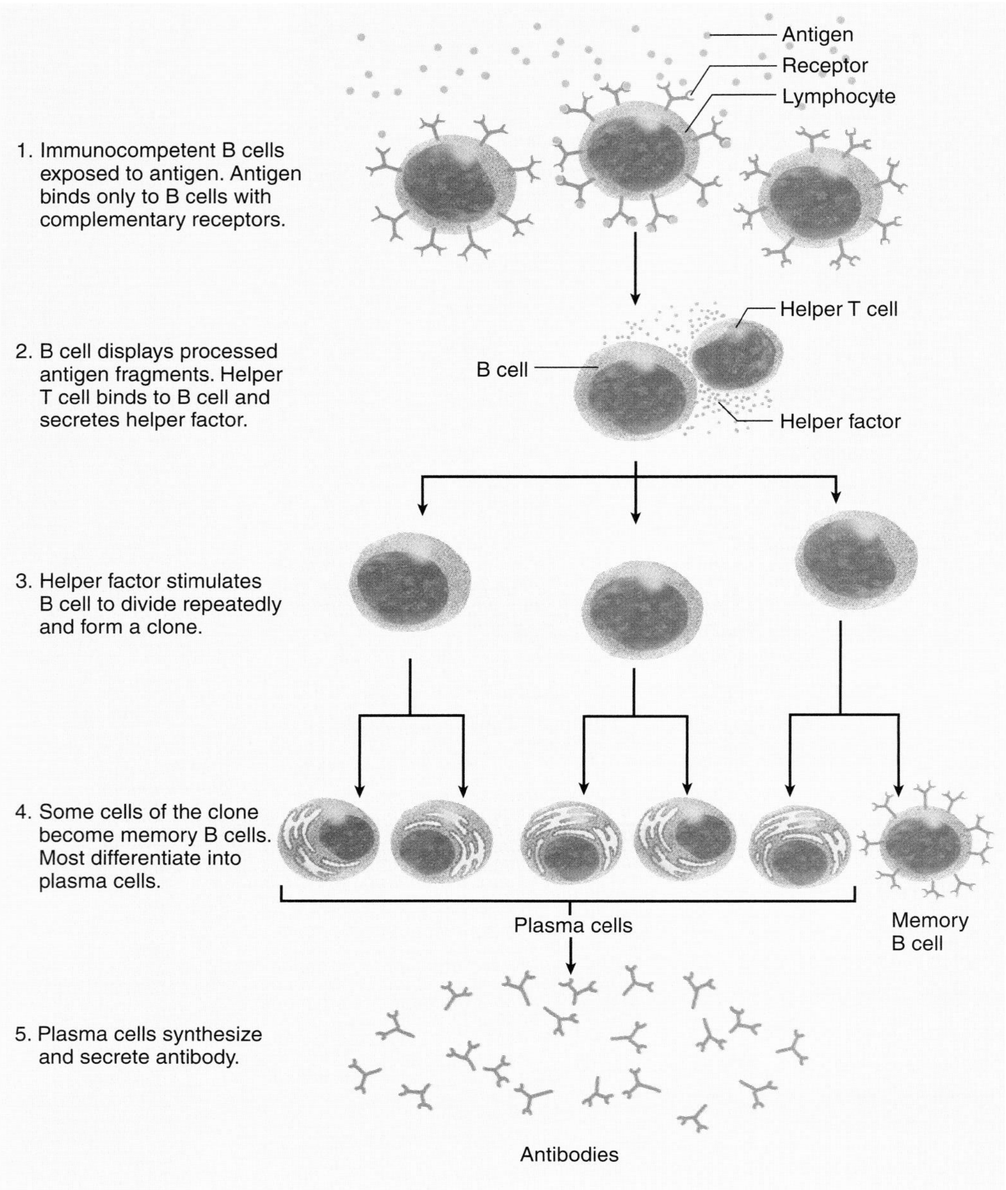

Antigen
Receptor
Lymphocyte
1. Immunocompetent B cells exposed to antigen. Antigen binds only to B cells with complementary receptors.
Helper T cell
B cell
Helper factor
2. B cell displays processed antigen fragments. Helper T cell binds to B cell and secretes helper factor.
3. Helper factor stimulates B cell to divide repeatedly and form a clone.
4. Some cells of the clone become memory B cells. Most differentiate into plasma cells.
Plasma cells
Memory B cell
5. Plasma cells synthesize and secrete antibody.
Antibodies

Figure 21.22 Clonal Selection and Ensuing Events of the Humoral Immune Response.

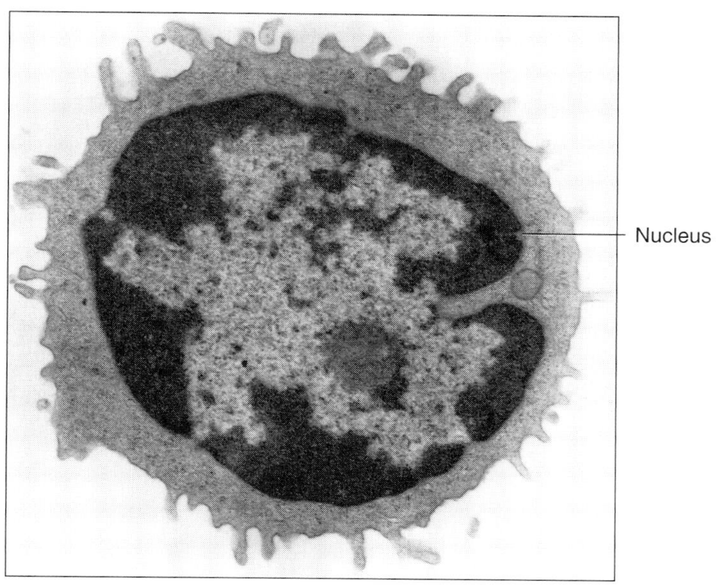

(a)

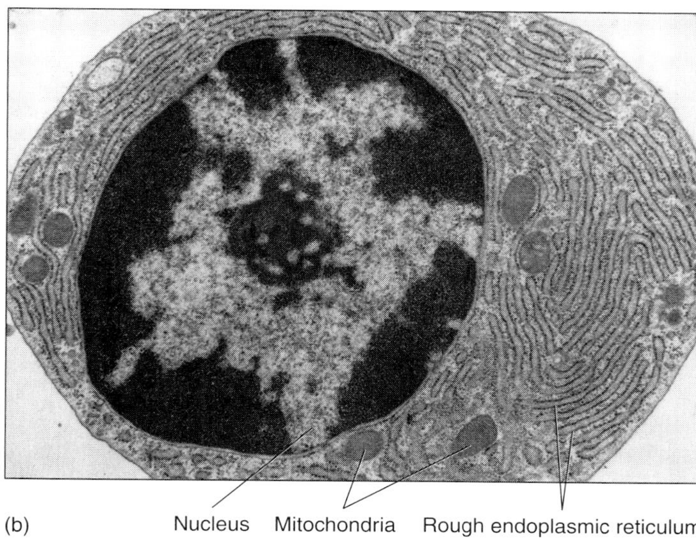

(b)

Figure 21.23 A B Cell and Plasma Cell. (*a*) A B cell has little cytoplasm and scanty organelles. (*b*) A plasma cell, which differentiates from a B cell, has an abundance of rough endoplasmic reticulum, which indicates a high rate of protein (antibody) synthesis.

time you are exposed to a particular antigen, your plasma cells produce mainly IgM. In later exposures to the same antigen, they produce mainly IgG.

Attack

Once released by a plasma cell, antibodies use four mechanisms to render antigens harmless:

1. **Neutralization** is relatively simple. Only certain regions of an antigen are pathogenic—for example, the parts of a toxin molecule and attachment sites of a virus that enable these agents to bind to human cells. Antibodies mask those parts of the antigen.
2. **Complement fixation** is the binding of complement proteins to IgM or IgG. When an antigen binds to one of these antibodies, the antibody changes shape and exposes its complement-binding site (see fig. 21.18*a*). This initiates the binding of complement proteins to an enemy cell surface (see the *classical pathway,* fig. 21.14) and leads to cytolysis, opsonization of bacteria, and enhanced inflammation, as described earlier. Complement fixation is the primary mechanism of defense against such foreign cells as bacteria and mismatched erythrocytes.
3. **Agglutination** was described in chapter 18 in the discussion of ABO and Rh blood types. It is effective not only in mismatched blood transfusions but more importantly as a defense against bacteria. An antibody molecule has 2 to 10 binding sites; thus, it can bind to antigen molecules on two or more enemy cells at once and stick them together (fig. 21.24*a*). This tends to immobilize microbes and antigen molecules and to prevent them from spreading through the tissues.
4. **Precipitation** begins with a similar process in which antibodies link antigen molecules (not whole cells) together. This creates an antigen-antibody complex (fig. 21.24*b*) that is too large to remain in solution. The complex precipitates and an eosinophil may then phagocytize it.

You will note that antibodies do not directly destroy an antigen in any of these mechanisms. They render it harmless by covering its pathogenic sites or agglutinating it, and they mark it for destruction by other agents such as complement, macrophages, eosinophils, or the cell-mediated processes discussed in the next section.

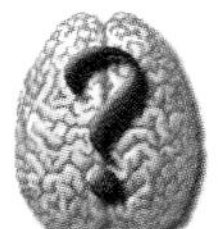

Think About It

Explain why IgM has a stronger power of agglutination than antibodies of any other class.

Memory

When a person is exposed to a particular antigen for the first time, the foregoing reactions are called the **primary response.** It takes 3 to 6 days for the virgin B cells to

Clinical Insight 21.2

Monoclonal Antibodies

Antibodies have long been produced commercially for clinical and research purposes. Commercial production was originally accomplished by injecting animals with purified antigens and later taking blood serum from them. Since each antigen, however, has multiple antigenic determinants capable of stimulating production of different antibodies, this procedure yielded an antibody mixture. The method for making pure antibodies is relatively new.

In the current technology, mice are injected with an antigen and sensitized B cells are later isolated from their spleens. To enable these B cells to survive and multiply, they are fused with certain cancer cells and then grown in thousands of tissue cultures. The hybrid cells multiply rapidly, producing a clone called a *hybridoma* that secretes only one antibody against the antigen to which the original B cell was sensitized. Out of thousands of clones, the one producing the desired antibody is selected and the others are discarded. The selected cell line goes on producing large amounts of pure *monoclonal antibodies.*

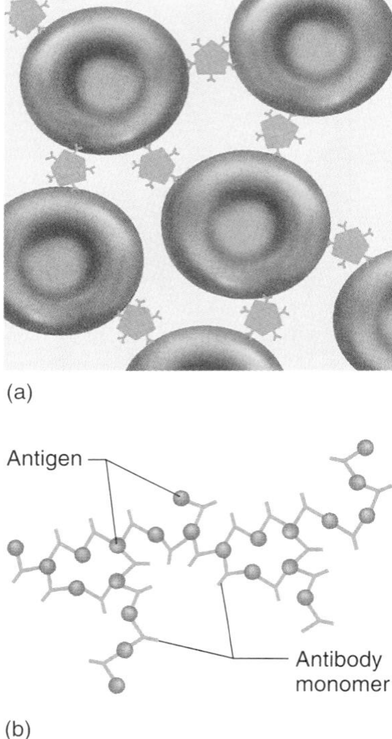

Figure 21.24 Agglutination by Antibodies. (*a*) Agglutination of foreign erythrocytes by IgM. (*b*) An antigen-antibody complex.

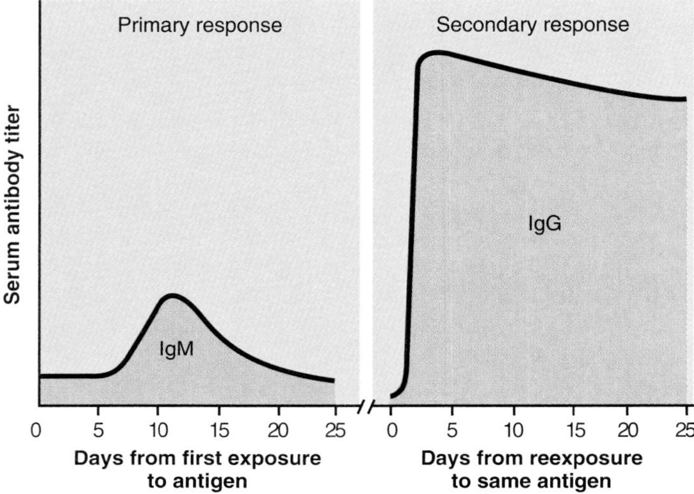

Figure 21.25 The Primary and Secondary (anamnestic) Responses in Humoral Immunity. The individual is exposed to antigen on day 0 in both cases. Note the differences in the speed of response, the height of the antibody titer, and the rate of decline in antibody titer.

proliferate and produce plasma cells. The **antibody titer** (level of antibody in the blood plasma) then rises sharply, peaks in about 10 days, and gradually declines (fig. 21.25). In the primary response, the main antibody is IgM.

During clonal selection, some members of the clone become **memory cells** rather than plasma cells (see fig. 21.22). Memory cells are long-lived and much more numerous than the B cells of a virgin lymphocyte pool. Consequently, if a person is reexposed to the same antigen, memory cells detect it more quickly than the virgin lymphocytes that existed at the time of first exposure. They produce the much quicker **secondary,** or **anamnestic**[13] (an-am-NESS-tic), **response.** Plasma cells form within hours; the antibody titer peaks in about 2 days and remains elevated for weeks to years. The antibody is mainly IgG this time. The secondary response is more intense than the primary response, owing to a quicker reaction time and a much higher antibody titer. It is so rapid that the antigen has little chance to exert a noticeable effect on the body, and no illness results.

13. *ana* = back + *mnes* = remember

Before You Go On

Answer the following questions to test your understanding of the preceding section:

15. What are the three phases of an immune response?
16. What is an MHC protein? How does it relate to B cell and helper T cell function?
17. What is the difference between a B cell and a plasma cell?
18. Describe four ways in which an antibody reacts against an antigen.
19. Why does the secondary immune response prevent a pathogen from causing disease, while the primary immune response does not?

Cellular Immunity

Objectives

When you have completed this section, you should be able to

- list the types of lymphocytes involved in cellular immunity and describe the roles they play;
- describe the role of MHC proteins in the activation of helper and cytotoxic T cells;
- explain how interleukins coordinate the actions of immune system cells;
- contrast the roles played by helper, cytotoxic, and suppressor T cells; and
- compare and contrast the action of memory cells in cellular and humoral immunity.

In cellular (cell-mediated) immunity, lymphocytes directly attack and destroy foreign cells and diseased host cells. This involves four types of T lymphocytes: (1) **cytotoxic (killer) T cells,** which carry out the attack; (2) **helper T cells,** which promote cytotoxic T cell action and other defense mechanisms; (3) **suppressor T cells,** which limit the attack and keep the immune system from running out of control; and (4) **memory T cells,** which are descended from the cytotoxic cells and are responsible for memory in cellular immunity. We can think of the cytotoxic cells as the *effectors* of cellular immunity and the helper and suppressor T cells as *regulatory* lymphocytes that act somewhat like an accelerator and brake on cellular immunity.

Helper T cells are also known as T_H, T_4, or CD4 cells; they have a glycoprotein on their surface called CD4 (CD stands for *cluster designation*—a classification system for many surface molecules). Cytotoxic (T_C) and suppressor (T_S) cells are collectively known as CD8 or T_8 cells after their surface glycoprotein, CD8. The roles of these glycoproteins are discussed later.

Like humoral immunity, cellular immunity occurs in stages of recognition, attack, and memory.

Recognition

The recognition phase has two aspects: (1) antigen presentation, as in humoral immunity, and (2) activation of T cells, so that they can mount the attack.

Antigen Presentation

As in humoral immunity, T cells respond only to antigen fragments displayed by MHC proteins, not to free antigens. Cellular immunity, however, presents some new aspects of antigen recognition: (1) Both helper and cytotoxic T cells are involved in surveillance. (2) T cells examine not only B cells but all cells with MHC proteins. (3) Two kinds of MHC proteins are involved—MHC-I and MHC-II.

MHC-I (class I) proteins occur on nearly every cell of the body; they are not limited to APCs. They display fragments of antigens made within the host cell itself, which may be normal self-antigens, abnormal antigens made by cancer cells, or viral proteins. (When a virus invades a host cell, it causes the cell to manufacture proteins that become part of the structure of new viruses.) When an MHC-I protein is bound to a nonself antigen, it stimulates cytotoxic T cells to attack the displaying cell by a mechanism explained later. The antigen-MHC protein complex is like a tag on the host cell that says, "I'm diseased; kill me." Infected or malignant cells are then destroyed before they can do further harm to the body.

MHC-II (class II) proteins occur only on the surfaces of APCs, including B cells, macrophages, and some T cells. An MHC-II protein displaying an antigen fragment stimulates helper T cells that come into contact with it.

Cytotoxic T cells respond only to MHC-I proteins, and helper T cells respond only to MHC-II—and in both cases, only if the MHC proteins are bound to antigens. While T cells are maturing in the thymus, they undergo a process called **MHC restriction** in which they develop receptors that can bind only to MHC-I or MHC-II proteins. Antigen presentation is similar for cytotoxic and helper T cells, but the presenting cells and MHC proteins differ (table 21.6). It is still unclear how suppressor T cells are activated. Some exhibit MHC restriction and require the aid of APCs, while others bind to free antigens without need of presentation by APCs.

Table 21.6 Comparison of the Responses of Cytotoxic and Helper T Cells

Characteristic	Cytotoxic (killer) T Cells	Helper T Cells
Cells capable of stimulating a response	Nearly any cell	Antigen-presenting cells (macrophages, B cells, some T cells)
MHC protein	MHC-I	MHC-II

T Cell Activation

CD4 and CD8 proteins are cell-adhesion molecules that bind a T cell to a target cell during antigen presentation. In addition to their role in cell adhesion, they are linked to a second-messenger system on the intracellular side of the T cell membrane. The second-messenger system phosphorylates cytoplasmic enzymes and triggers clonal selection: an activated T cell enlarges, multiplies, and forms a clone of identical T cells in much the same manner as the clonal selection of B cells.

Clonal selection requires **costimulation,** a process in which an APC and a T cell communicate with each other by means of interleukins. The binding of a helper T cell to a macrophage stimulates the macrophage to release interleukin-1 (Il-1). Il-1 stimulates the T cell to secrete Il-2 and synthesize Il-2 receptors. Il-2 stimulates more T cells to multiply and secrete still more Il-2—a positive feedback loop that causes a large population of activated T cells to build up rapidly. These T cells then enlarge and divide to form clones of effector and memory cells.

Attack

Helper, cytotoxic, and suppressor T cells play different roles in the attack phase.

Helper T Cells

Most immune responses require the action of helper T cells, which play a central coordinating role in both humoral and cellular immunity (fig. 21.26). When a helper T cell recognizes an MHC-antigen complex, it secretes a variety of lymphokines that attract neutrophils and promote inflammation and other nonspecific responses. One of these lymphokines, **macrophage-activating factor (MAF),** enhances the phagocytic activity of macrophages.

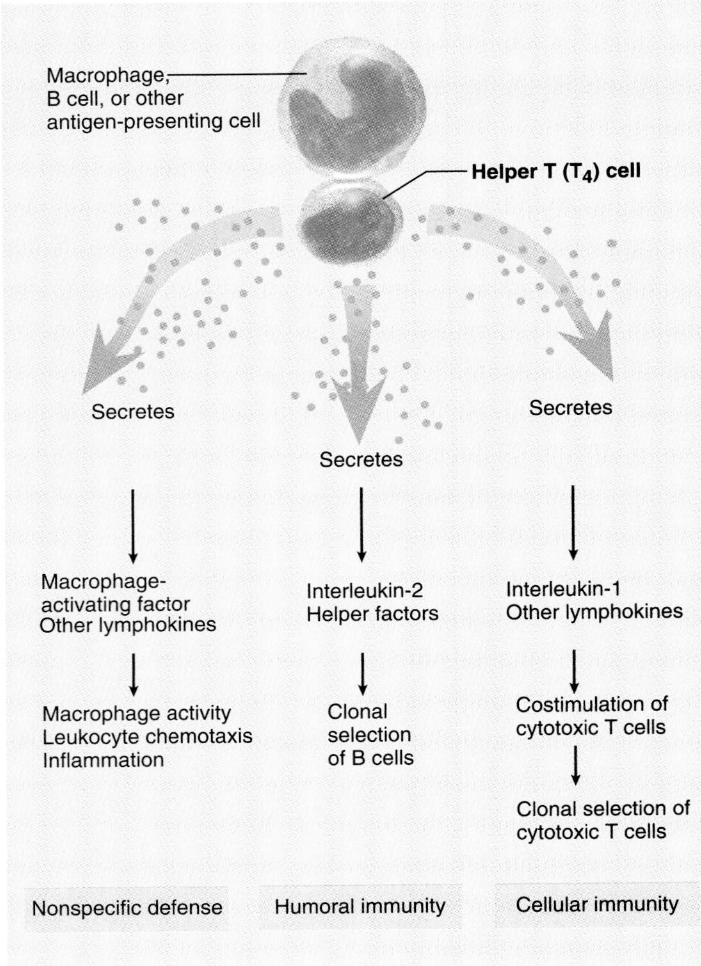

Figure 21.26 The Central Role of Helper T Cells in Defense and Immunity. Note how extensively the body's defenses would be impaired if the helper T cells were destroyed, as happens in AIDS.

Cytotoxic T Cells

Cytotoxic (killer) T cells are the only T lymphocytes that directly attack and kill other cells. They are particularly responsive to cancer cells, cells of transplanted tissues and organs, and host cells that are infected with viruses, intracellular parasites, and bacteria (fig. 21.27). (They are not to be confused with the *natural killer,* or *NK, lymphocytes* of nonspecific defense, although both lymphocyte types destroy their target cells in similar ways.)

When a cytotoxic T cell recognizes a complex of antigen and MHC-I protein on a diseased or foreign cell, it "docks" on that cell and delivers a **lethal hit** that ultimately kills it. To do this, it must be in actual contact with the enemy cell, in contrast to B cells, which attack indirectly. Although the mechanism of the lethal hit is not entirely clear, there is evidence that some cytotoxic T cells degranulate and release a protein, **perforin,** that is inserted into the enemy plasma membrane. The T cell detaches and leaves, in search of other prey, while the perforin gradually polymerizes and creates holes in the enemy cell, similar to the way complement proteins act (fig. 21.28).

Other cytotoxic T cells lack perforin but kill target cells by different means. Some secrete **lymphotoxin,** which destroys the target cell's DNA. Some secrete **tumor necrosis factor (TNF),** which kills cancer cells in 2 to 3 days by unknown mechanisms. Still others secrete interferon, macrophage-activating factor, and **migration-inhibiting factor (MIF).** MIF is a lymphokine that suppresses the migration of macrophages, thus keeping them in the area where their service is most needed.

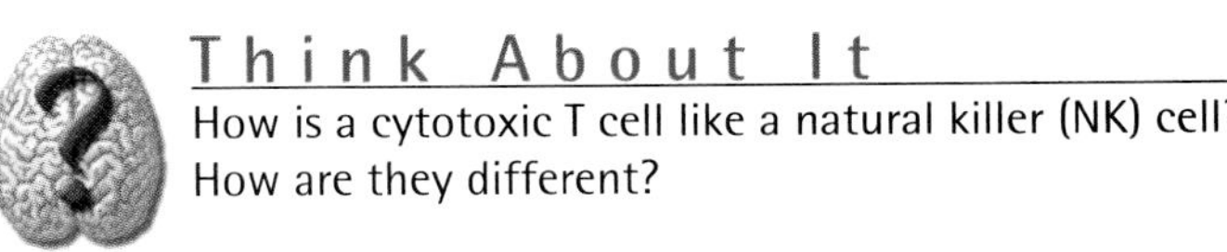

Think About It

How is a cytotoxic T cell like a natural killer (NK) cell? How are they different?

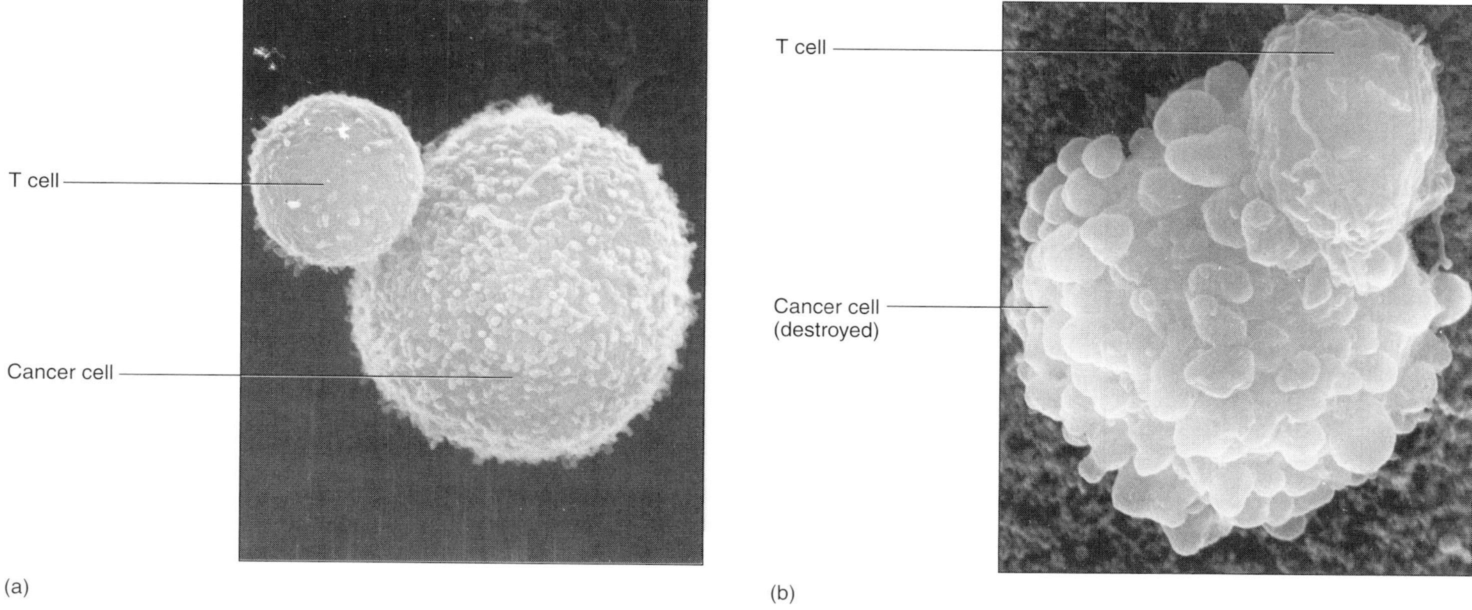

Figure 21.27 **Destruction of a Cancer Cell by a Cytotoxic T Cell.** (*a*) T cell binding to cancer cell. (*b*) Death of the cancer cell due to the lethal hit by the T cell.

Suppressor T Cells

As the pathogen is defeated and disappears from the tissues, suppressor T cells release lymphokines that inhibit T cell and B cell activity. This slows down the immune reaction and keeps it from running out of control. Suppressor T cells may also help prevent autoimmune diseases, as we see shortly.

Memory

As more and more cells are recruited by helper T cells, the immune response exerts an overwhelming force against the pathogen. The primary response, seen on first exposure to a particular pathogen, peaks in about a week and then gradually declines. It is followed by immune memory, which works in much the same way as in humoral immunity although it tends to last longer. Following clonal selection, some T cells become memory cells. Upon reexposure to the same pathogen later in life, memory cells mount a quick attack called the **T cell recall response,** analogous to the secondary response of humoral immunity.

Tables 21.7 and 21.8 summarize many of the cellular and chemical agents involved in humoral and cellular immunity.

Passive and Active Immunity

Immunity can be active or passive, natural or artificial. It is *active immunity* when the body makes its own antibodies or T cells against a pathogen, and *passive immunity* when the body acquires antibodies or T cells produced by another individual. Either type of immunity can occur naturally or, for treatment and prevention purposes, it can be induced artificially. Thus we can recognize four classes of immunity under this scheme:

1. **Natural active immunity.** This is the production of one's own antibodies or T cells as a result of infection or other natural exposure to an antigen.
2. **Artificial active immunity.** This is the production of one's own antibodies or T cells as a result of **vaccination** against diseases such as smallpox, tetanus, polio, measles, or influenza. A **vaccine** consists of either dead or *attenuated* (weakened) pathogens, which can stimulate an immune response but normally cause little or no discomfort or disease. In some cases, periodic "booster shots" are given to restimulate immune memory and maintain a high level of protection. Vaccination has eliminated smallpox worldwide and greatly reduced the incidence of life-threatening childhood diseases, but many people continue to die from influenza and other diseases that could have been prevented by vaccination.
3. **Natural passive immunity.** This is a temporary immunity that results from acquiring antibodies produced by another individual. The only natural way for this to happen is for a fetus to acquire antibodies from the mother through the placenta before birth or for a baby to acquire it through the colostrum or breast milk after birth.

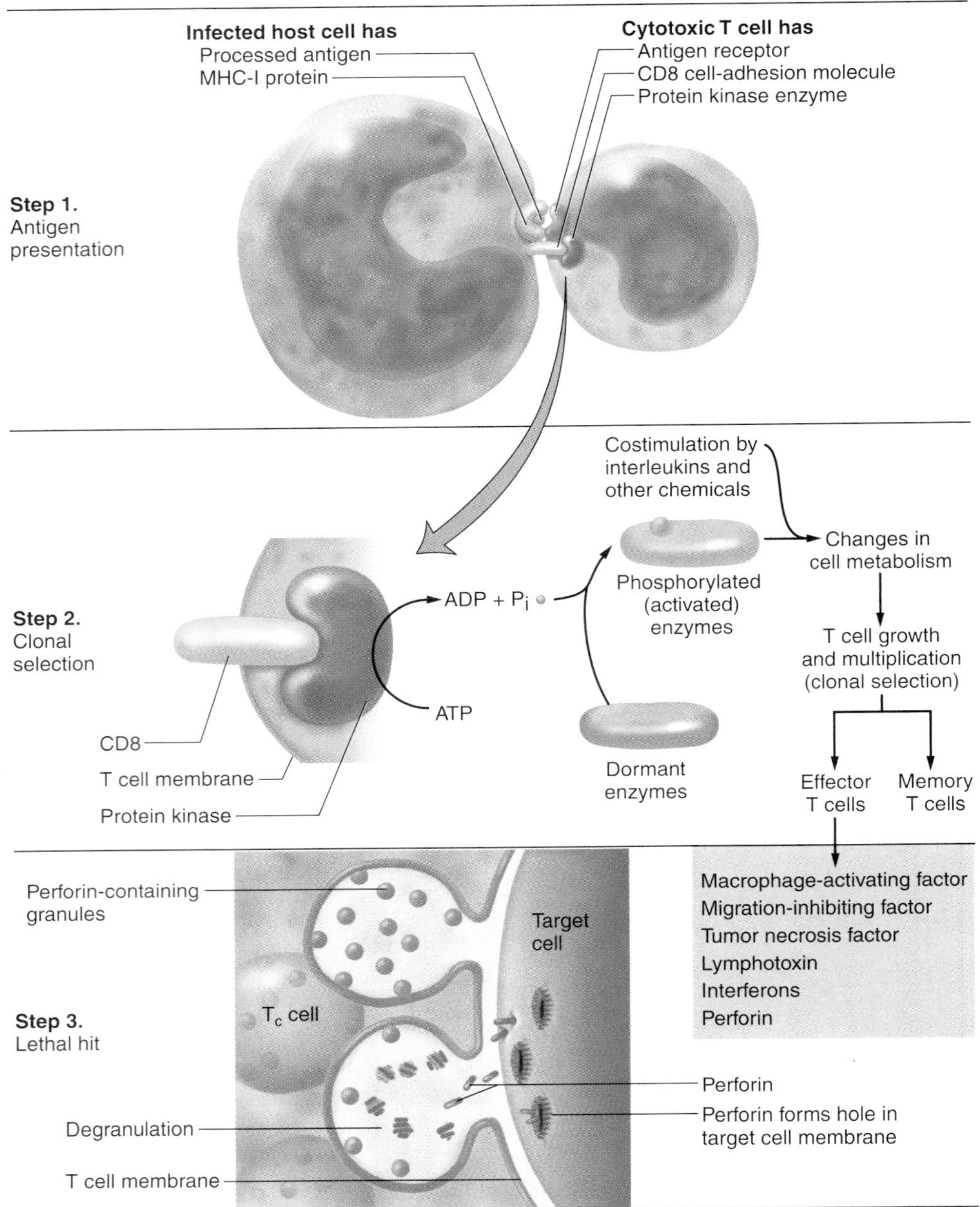

Figure 21.28 Stages in the Activation and Attack of a Cytotoxic T Cell on a Target Cell.

4. **Artificial passive immunity.** This is a temporary immunity that results from the injection of an **immune serum** (gamma globulin) obtained from another individual or from animals (such as horses) that produced antibodies against a certain pathogen. Immune serum is used for emergency treatment of snakebites, botulism, tetanus, rabies, and other diseases.

Only the two forms of active immunity involve immune memory and thus provide future protection. Passive immunity typically lasts only for 2 or 3 weeks, until the injected antibody is degraded.

Humoral Versus Cellular Immunity

Table 21.9 summarizes much of what you have studied in the last two sections by comparing the features of humoral and cellular immunity. Remember that these two processes often occur simultaneously, and in conjunction with inflammation, as a three-pronged attack on the same pathogen.

Table 21.7 Cellular Agents of Immunity

Cell Types	Action
Macrophages	Phagocytize bacteria, viruses, and expended or damaged host cells; act as antigen-presenting cells (APCs), which display antigen fragments to helper T cells; secrete interleukin-1 (see table 21.8)
Lymphocytes	
Natural killer (NK) cells	Attack host cells that have become malignant or infected with viruses; do not depend on specific immune recognition
Virgin lymphocyte pool	A reserve of immunocomponent lymphocytes, capable of responding to an antigen but not yet having encountered one
B cells	Develop in bone marrow; as APCs, recognize and internalize antigens and display antigen fragments on cell surface; give rise to plasma cells
Plasma cells	Develop from B cells that have been activated by helper T cells; synthesize and secrete antibodies
Memory B cells	Activated B cells that do not immediately differentiate into plasma cells; act as a pool of lymphocytes that can execute a quick secondary, humoral immune response upon reexposure to the same antigen that activated them initially
Cytotoxic T cells (killer T cells, T_c cells)	Effector cells of cellular immunity; the only lymphocytes that directly attack foreign cells; also bind to host cells displaying antigen fragments on MHC-I proteins; produce perforin, lymphotoxin, interferon, migration-inhibiting factor, and tumor necrosis factor; produce the lethal hit that destroys target cells
Helper T (T_H) cells	Play a central regulatory role in nonspecific defense and humoral and cellular immunity; cannot recognize free antigens but recognize antigen fragments displayed by antigen-presenting cells with MHC-II proteins; secrete interleukin-2, macrophage-activating factor, and other lymphokines that stimulate other immune cells (see actions of lymphokines in table 21.8)
Suppressor T (T_S) cells	Release lymphokines that inhibit B cell and T cell activity; help "wind down" the immune response as pathogen is defeated
Memory T cells	Activated T lymphocytes that do not immediately differentiate into effector T cells; act as a pool of lymphocytes that can execute a quick T cell recall response upon reexposure to the same antigen that activated them initially
CD4 (T_4) cells	T lymphocytes with CD4 surface glycoproteins; primarily helper T cells
CD8 (T_8) cells	T lymphocytes with CD8 surface glycoproteins, including cytotoxic and suppressor T cells

Table 21.8 Chemical Agents of Immunity

Substance	Source and Action
Pathogenic Agents	
Antigen (Ag)	Molecule capable of triggering an immune response, usually a protein, polysaccharide, or glycolipid
Hapten	Small molecule unable to trigger a primary immune response by itself but able to bind to host molecules and produce a complex that is antigenic
Protective Agents	
Antibody (Ab)	An immunoglobulin (γ globulin) produced by plasma cells in response to an antigen; interferes with antigen's effects by means of complement fixation, neutralization of toxins, agglutination, and precipitation
Complement	Plasma proteins that help to destroy pathogens when activated by antibodies
Interleukins	Hormonelike messengers produced by leukocytes and macrophages to stimulate other leukocytes
Monokines	Interleukins produced by macrophages
Interleukin-1	A monokine that stimulates helper T cells
Lymphokines	Interleukins produced by lymphocytes
Helper factor	A lymphokine produced by helper T cells that stimulates B cells to differentiate into plasma cells and synthesize antibodies
Interleukin-2	A lymphokine produced by helper T cells that stimulates macrophages and other T cells
Macrophage-activating factor (MAF)	A lymphokine produced by cytotoxic and helper T cells that enhances phagocytic activity of macrophages
Migration-inhibiting factor (MAF)	A lymphokine produced by cytotoxic T cells that inhibits macrophages from migrating away from site of immune challenge
Perforin	A protein produced by natural killer cells and cytotoxic T cells that binds to target cells, produces a hole, and causes cytolysis
Tumor necrosis factor (TNF)	A factor secreted by cytotoxic T cells that kills some cancer cells by unknown means
Lymphotoxin	A factor secreted by cytotoxic T cells that destroys the DNA of target cells

Note: Some chemical agents of immunity have additional roles in inflammation, described in table 21.3.

Table 21.9 Some Comparisons Between Humoral and Cellular Immunity

	Humoral Immunity	Cellular Immunity
Pathogens	Bacteria, toxins, mismatched RBCs, extracellular viruses	Transplanted tissues and organs, cancer cells, infected cells
Effector Cells	Plasma cells (develop from B cells)	Cytotoxic T cells
Other Cells Involved in Attack	Helper T cells	Helper T cells, suppressor T cells
Antigen-Presenting Cells	B cells	B cells, macrophages, nearly all cells
MHC Proteins	MHC-II only	MHC-I and MHC-II
Chemical Agents of Attack	Antibodies, complement	Perforins, lymphokines, lymphotoxin, tumor necrosis factor
Mechanisms of Pathogen Destruction	Neutralization, complement fixation, cytolysis, opsonization, agglutination, precipitation, inflammation	Cytolysis (lethal hit), DNA destruction, macrophage activation, inflammation
Memory	Secondary (anamnestic) response	T cell recall response

Before You Go On

Answer the following questions to test your understanding of the preceding section:

20. Name four types of lymphocytes that are involved in cellular immunity. Which of these is also essential to humoral immunity?
21. Explain why cytotoxic T cells are activated by a broader range of host cells than are helper T cells.
22. What positive feedback cycle can you identify in T cell clonal selection?
23. Why would a viral attack on helper T cells be especially devastating to a person's health?
24. Describe one way in which cytotoxic T cells destroy target cells.
25. Contrast active and passive immunity. Give a natural and an artificial example of each.

Immune System Disorders

Objectives

When you have completed this section, you should be able to

- distinguish between the four classes of allergy and give an example of each;
- explain the cause of anaphylaxis and distinguish local anaphylaxis from anaphylactic shock;
- explain the cause of asthma and describe how its symptoms are produced;
- state some reasons immune self-tolerance may fail, and give examples of the resulting diseases; and
- contrast severe combined immunodeficiency disease (SCID) with AIDS.

Because the immune system involves complex cellular interactions controlled by numerous chemical messengers, there are many points at which things can go wrong. The immune response may be too vigorous, too weak, or misdirected against the wrong targets. A few disorders are summarized here to illustrate the consequences.

Hypersensitivity

Hypersensitivity (allergy[14]**)** is an excessive immune reaction against antigens that most people tolerate. Such antigens, called **allergens,** occur in mold, dust, pollen, vaccines, bee and wasp venoms, toxins from poison ivy and other plants, and foods such as milk, nuts, eggs, and shellfish. Drugs such as penicillin, tetracycline, and insulin are allergenic to some people.

Of the four recognized classes of hypersensitivity, type I is described as *acute,* or *immediate,* because the inflammatory response is extremely rapid. Types II and III are described as *subacute* because they exhibit a slower onset (1–3 hours after exposure) and last longer (10–15 hours). Type IV is a cellular immune response whereas the other three are humoral.

- **Type I (acute) hypersensitivity** is the most common form. It begins within seconds of exposure and usually subsides within 30 minutes. **Anaphylaxis**[15] (AN-uh-fih-LAC-sis) is a type I reaction that occurs in sensitized people when an allergen caps IgE molecules on the membranes of mast cells and basophils. The cells degranulate and release inflammatory chemicals such as histamine that cause local edema, mucus hypersecretion, and congestion. Red itchy skin (hives), watery eyes, and a runny nose are typical symptoms. Food and drug allergens may also cause cramps, vomiting, and diarrhea. Local anaphylaxis can be relieved with antihistamines.

14. *allo* = altered + *erg* = action, reaction
15. *ana* = against + *phylax* = protection

Clinical Insight 21.3

Systemic Lupus Erythematosus

Systemic lupus erythematosus (SLE) occurs most often in young women, but it can affect anyone. Its cause is unclear but seems to entail a combination of heredity and other risk factors such as injury, infection, emotional stress, drug reactions, or excessive exposure to sunlight. It involves the formation of autoantibodies against DNA and other nuclear components. Antigen-antibody complexes accumulate in blood vessels and other organs, where they trigger the release of inflammatory chemicals that lead to widespread inflammation of the connective tissues. The disease was named for skin lesions once likened to a wolf bite.[17] Symptoms include fever, fatigue, joint pain, weight loss, hair loss, intolerance of bright light, and sometimes a "butterfly rash" across the nose and cheeks. The skin, liver, spleen, kidneys, lungs, heart, and central nervous system often become inflamed, and death may ensue from renal failure.

17. *lupus* = wolf + *erythema* = redness

- **Asthma**[16] is a local anaphylactic reaction to inhaled allergens. It is the most common chronic illness of children, especially boys. Allergen exposure and even emotional distress can trigger a massive release of histamine, which causes spasmodic constriction of the bronchioles. The consequences include severe coughing and wheezing, *dyspnea* (difficult breathing), and sometimes fatal suffocation. Epinephrine and other β-adrenergic stimulants are used to dilate the airway. Another chemical released by allergen exposure is an enzyme called *tryptase,* which attracts eosinophils. The eosinophils and tryptase itself promote scarring and extensive long-term damage to the lungs. Tryptase inhibitors are being explored as a possible treatment for asthma.
- **Anaphylactic shock** is a severe, widespread response that occurs when an allergen such as bee venom or penicillin is introduced to the bloodstream of an allergic individual. Its effects include bronchiolar constriction, dyspnea, widespread vasodilation, circulatory shock, and sometimes sudden death. Epinephrine relieves the symptoms by dilating the bronchioles and increasing cardiac output.
- **Type II (antibody-dependent cytotoxic) hypersensitivity** occurs when IgG or IgM binds to antigens on certain cells and lyses them. Blood transfusion reactions are an example. IgM agglutinates mismatched erythrocytes, and complement fixation leads to their lysis.
- **Type III (immune complex) hypersensitivity,** also mediated by IgG or IgM, stems from the formation of large amounts of antigen-antibody complex throughout the body. Some complexes become trapped in the basement membrane under the endothelium of blood vessels and trigger intense inflammation. Type III reactions are involved in acute glomerulonephritis, a kidney disease, and in systemic lupus erythematosus (see clinical insight 21.3). Both of these are autoimmune diseases.
- **Type IV (delayed) hypersensitivity** is a reaction that appears from 12 to 72 hours after exposure. It occurs when APCs in the lymph nodes display antigens to helper T cells, and these T cells secrete interferon and other lymphokines that activate cytotoxic T cells and macrophages. The result is a mixture of nonspecific and immune responses. Haptens in cosmetics and poison ivy can stimulate type IV reactions. The skin test for tuberculosis also involves a delayed hypersensitivity reaction.

16. *asthma* = panting

Autoimmune Diseases

Autoimmune diseases are failures of self-tolerance—the immune system fails to distinguish self antigens from foreign antigens and produces **autoantibodies** that attack the body's own tissues. There are at least three reasons why self-tolerance may fail:

1. *Cross-reactivity.* Some antibodies against foreign antigens react to similar self-antigens. In rheumatic fever, for example, a streptococcus infection stimulates production of antibodies that react not only against the bacteria but also against antigens of the heart tissue. It often results in scarring and stenosis (narrowing) of the mitral and aortic valves.
2. *Abnormal exposure of self-antigens to the blood.* Some of our native antigens are normally not exposed to the blood. For example, a blood-testis barrier (BTB) normally isolates sperm cells from the blood. Breakdown of the BTB can cause sterility

when sperm are first formed in adolescence and trigger the production of autoantibodies.

3. *Change in the structure of self-antigens.* Viruses and drugs may change the structure of self-antigens and cause the immune system to perceive them as foreign. One theory of the cause of type I diabetes mellitus is that a viral infection alters the antigens of β cells of the pancreatic islets, which leads to an attack on the β cells by autoantibodies.

Immunodeficiency Diseases

In the foregoing diseases, the immune system reacts too vigorously or directs its attack against the wrong targets. In immunodeficiency diseases, by contrast, the immune system fails to respond vigorously enough.

Severe Combined Immunodeficiency Disease (SCID)

SCID is a group of disorders caused by recessive alleles that result in a scarcity or absence of both T and B cells. Children with SCID are highly vulnerable to opportunistic infections and must live in protective enclosures. Perhaps the most publicized case was David, the "bubble boy" (fig. 21.29). He spent his life in sterile plastic chambers, finally succumbing at age 12 to cancer triggered by a viral infection. Children with SCID are sometimes helped by transplants of bone marrow or fetal thymus, but in some cases the transplanted cells fail to survive and multiply, or transplanted T cells attack the patient's tissues (the *graft-versus-host response*). In David's case, a fatal virus was contracted from his sister through a bone marrow transplant.

Figure 21.29 Severe Combined Immunodeficiency Disease (SCID). David, nicknamed the "bubble boy," lived with SCID from 1971 to 1984. He is seen here at the age of six, when he first received a portable sterile enclosure that enabled him to leave the hospital.

Acquired Immunodeficiency Syndrome (AIDS)

Acquired immunodeficiency diseases are nonhereditary diseases contracted after birth. The most well-known example is **acquired immunodeficiency syndrome (AIDS)**, a group of conditions that involve a severely depressed immune response resulting from infection with the **human immunodeficiency virus (HIV)**.

HIV has a protein coat called the *capsid,* which is surrounded by another protein called P18 and finally by an outer *envelope* of glycoproteins. The virus contains two molecules of RNA and two molecules of an enzyme called *reverse transcriptase* (fig. 21.30*a*). Like other viruses, HIV can only be replicated by a living host cell. It invades helper T (T_H, CD4) cells, macrophages, neutrophils, and brain cells by "tricking" these cells to internalize the viruses by receptor-mediated endocytosis. Within the host cell, its reverse transcriptase uses the viral RNA molecules as templates to synthesize DNA—the opposite of the usual process of genetic transcription. Viruses that carry out this RNA→DNA reverse transcription are called *retroviruses.*[18] The new DNA is inserted into the host cell's DNA, where it may lie dormant for months to years. When activated, however, it causes the host cell to produce new viruses, which emerge from the infected cell (fig. 21.30*b*), invade others, and repeat the process.

By destroying T_H cells, HIV strikes at the central coordinating agent of nonspecific defense, humoral immunity, and cellular immunity. The incubation period—from the time of infection to the time of the first symptoms—can range from a few months to 12 years. Flulike episodes of chills and fever occur as HIV attacks T_H cells. At first, antibodies against HIV are produced and the T_H count re-

18. *retr* = an acronym from *re*verse *tr*anscription

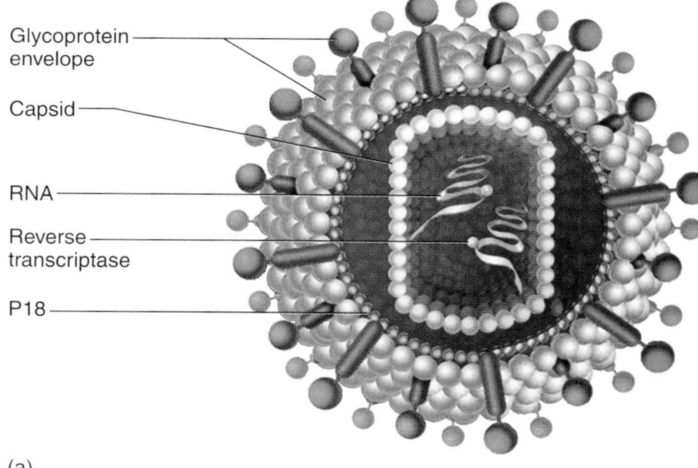

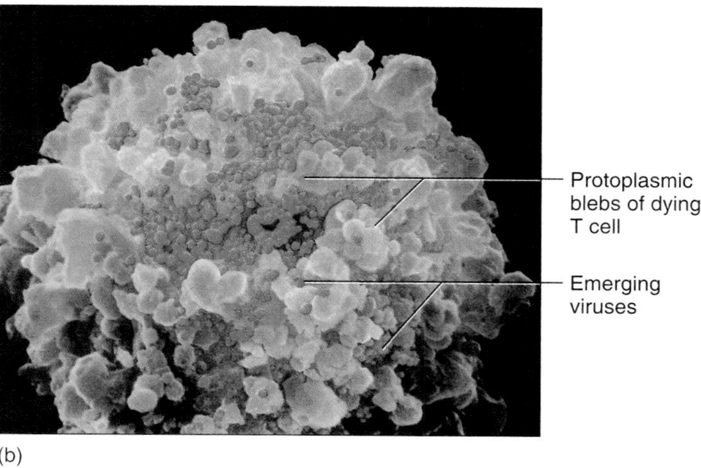

Figure 21.30 The Human Immunodeficiency Virus (HIV). (*a*) Structure of the virus. (*b*) Viruses emerging from a dying helper T cell. Each virus can now invade a new helper T cell and produce a similar number of descendents.

turns nearly to normal. As the virus destroys more and more cells, however, a syndrome called **AIDS-related complex (ARC)** sometimes appears, characterized by night sweats, fatigue, headache, extreme weight loss, and lymphadenitis (inflammation of the lymph nodes).

Normally, the T_H count is 600 to 1,200 cells/μL of blood, but in AIDS, it is less than 200 cells/μL. With such severe depletion of T_H cells, a person succumbs to opportunistic infections with *Toxoplasma* (a protozoan previously known mainly for causing birth defects), *Pneumocystis* (a group of respiratory fungi formerly classified as protozoans), herpes simplex virus, cytomegalovirus (which can cause blindness), or tuberculosis bacteria. White patches may appear on the mucous membranes, a condition called *leukoplakia.* A form of cancer called Kaposi[19] sarcoma, common in AIDS patients, originates in the endothelial cells of the blood vessels and causes bruiselike purple lesions on the skin (fig. 21.31).

Patients with full-blown AIDS show no response to standard skin tests for delayed hypersensitivity. Slurred speech, loss of motor and cognitive functions, and dementia may occur as HIV invades the brain. Death from cancer or infection is inevitable, usually within a few months but sometimes as long as 8 years after diagnosis. Some people, however, have been diagnosed as HIV-positive and yet have survived for 10 years or longer without developing AIDS.

HIV is transmitted through blood, semen, vaginal secretions, and breast milk. It can be transmitted from mother to fetus through the placenta or from mother to infant during childbirth or nursing. HIV occurs in saliva and tears, but is not believed to be transmitted by those fluids. The most common means of transmission are sexual intercourse (vaginal, anal, or oral), contaminated blood products, and drug injections with contaminated needles. Worldwide, about 75% of HIV infections are acquired through heterosexual, predominantly vaginal intercourse. In the United States, homosexual men still account for the majority of cases, but adolescents are the fastest rising group of AIDS patients because of the increasing exchange of unprotected sexual intercourse for drugs. The sharing of needles for drug use remains the chief means of transmission in urban ghettos. Many hemophiliacs became infected with HIV through blood transfusions before preventive measures were implemented in 1984, but all donated blood is now tested for HIV and risk of infection is less than 1%. HIV cannot be contracted by donating blood, but irrational fear has resulted in an alarming drop in blood donors.

AIDS is not known to be transmitted through casual contact—for example, to family members, friends, coworkers, classmates, or medical personnel in charge of AIDS patients. It is not transmitted by kissing. Despite some speculation and fear, it has not been found to be transmitted by mosquitoes or other blood-sucking arthropods.

HIV survives poorly outside the human body. It is destroyed by laundering, dishwashing, exposure to heat (50°C [135°F] for at least 10 minutes), chlorination of swimming pools and hot tubs, and disinfectants such as bleach, Lysol, hydrogen peroxide, rubbing alcohol, and germicidal skin cleansers (Betadine and Hibiclens, for example). A properly used, undamaged latex condom is an effective barrier to HIV, especially if augmented with the spermicide nonoxynol-9. Animal membrane condoms are not effective at blocking HIV transmission because the viruses are smaller than the gaps in the membrane.

19. Moritz Kaposi (1837–1902), Austrian physician

The AIDS epidemic has triggered an effort of unprecedented intensity to find a vaccine or cure. The strategies against HIV include efforts to prevent its binding to the CD4 proteins of T_H cells, disrupting the action of reverse transcriptase, or inhibiting the assembly of new viruses or their release from host cells. HIV is a difficult pathogen to attack. Since it "hides" within host cells, it usually escapes recognition by the immune system. In the brain, it is protected by the blood-brain barrier.

About 1% of HIV's genes mutate every year. This rapid rate of mutation is a barrier to both natural immunity and development of a vaccine. Even when immune cells do become sensitized to HIV, the virus soon mutates and produces new surface antigens that escape recognition. The high mutation rate also would quickly make today's vaccine ineffective against tomorrow's strain of the virus. Another obstacle to treatment and prevention is the lack of animal models for vaccine and drug research and development. Most animals are not susceptible to HIV. The chimpanzee is an exception, but chimpanzees are difficult to maintain, and there are economic barriers and ethical controversies surrounding their use.

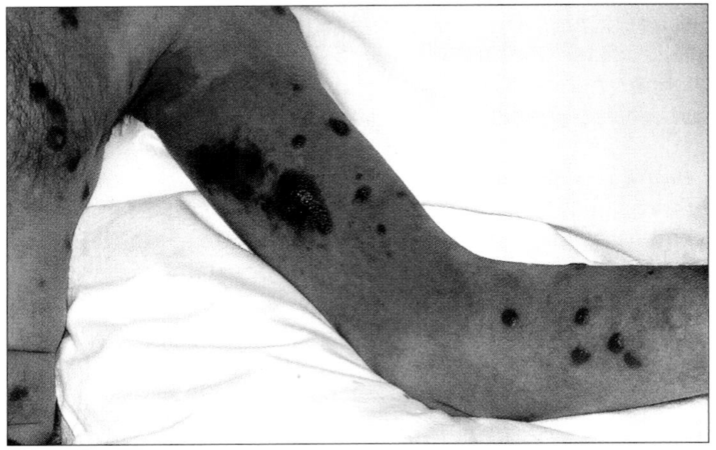

Figure 21.31 Kaposi Sarcoma. Typical lesions on the arm and chest of a patient with AIDS.

Until recently, the only drug approved by the Food and Drug Administration (FDA) was azidothymidine (AZT, or Retrovir), which inhibits reverse transcriptase and prolongs the lives of some HIV-positive individuals. AZT is now recommended for any patient with a CD4 count below 500 cells/μL, but it has undesirable side effects including bone marrow toxicity and anemia. The FDA has approved other drugs, including dideoxyinosine (ddI) and dideoxycytidine (ddC) for patients who do not respond to AZT, but these drugs can also have severe side effects.

Another class of drugs—protease inhibitors—inhibit enzymes (proteases) that HIV needs in order to replicate. In 1995, a "triple cocktail" of two reverse transcriptase inhibitors and a protease inhibitor was proving to be highly effective at inhibiting viral replication, but by 1997, HIV had evolved a resistance to these drugs and this treatment was failing in more than half of all patients. Alpha interferon has shown some success in inhibiting HIV replication and slowing the progress of Kaposi sarcoma.

There remain not only these vexing clinical problems but also a number of unanswered questions about the basic biology of HIV. It remains unknown, for example, why there are such strikingly different patterns of heterosexual versus homosexual transmission in different countries and why some people succumb so rapidly to infection, while others can be HIV-positive for years without developing AIDS. AIDS remains a stubborn problem sure to challenge virologists and epidemiologists for many years to come.

We have surveyed the major classes of immune system disorders and a few particularly notorious immune diseases. A few additional lymphatic and immune system disorders are described in table 21.10.

Table 21.10 Some Disorders of the Lymphatic and Immune Systems

Lymphadenitis[20]	Inflammation of a lymph node in response to infection or other immune challenge, typically marked by swelling and pain.
Hodgkin[21] *Disease*	A lymph node malignancy, with early symptoms including enlarged painful lymph nodes, especially in the neck, and fever of unknown origin; often progresses to neighboring lymph nodes. Radiation and chemotherapy cure about three out of four patients.
Splenomegaly[22]	Enlargement of the spleen, sometimes without underlying disease but often indicating infections, autoimmune diseases, heart failure, cirrhosis, Hodgkin disease, and other cancers. The enlarged spleen may "hoard" erythrocytes, causing anemia, and may become fragile and subject to rupture.
Contact Dermatitis	A form of delayed hypersensitivity that produces skin lesions limited to the site of contact with an allergen or hapten; includes responses to poison ivy, cosmetics, detergents, industrial chemicals, and some topical medicines.
Hives (urticaria[23]*)*	An allergic skin reaction characterized by a "wheal and flare" reaction: white blisters (wheals) surrounded by reddened areas (flares), usually with itching. Caused by local histamine release in response to allergens. Can be triggered by food or drugs, but sometimes by nonimmunological factors such as cold, friction, or emotional stress.

20. *aden* = gland + *itis* = inflammation
21. Thomas Hodgkin (1798–1866), British physician
22. *megaly* = enlargement
23. *urtica* = nettle

Clinical Insight 21.4

Psychoneuroimmunology—The Mind-Body Connection

Psychoneuroimmunology is a relatively new branch of medicine concerned with the relationship between mind and body in health and disease. It is attempting especially to understand how a person's state of mind interacts with the nervous and immune systems to influence health and illness, and how the nervous and immune systems communicate with each other.

As we saw in chapter 17, stress arouses the sympathetic nervous system and increases the secretion of ACTH, cortisol, and epinephrine by the endocrine system. The lymphatic organs receive abundant sympathetic innervation and immune cells have neurotransmitter receptors, so they are strongly influenced by the nervous system. Among hormones, cortisol has especially widespread effects on immunity. It inhibits the release of interleukins and inflammatory chemicals (which is why it is often used to treat inflammation), it reduces antibody synthesis, it promotes atrophy of the thymus, spleen, and lymph nodes, and it lowers the number of circulating lymphocytes, macrophages, and eosinophils. Not surprisingly, therefore, prolonged stress increases a person's vulnerability to infections and some forms of cancer.

But the immune system also has ways of sending messages to the nervous and endocrine systems. Interleukins stimulate the CNS and endocrine glands and affect sleep, thermoregulation, appetite, and hormone secretion in ways that can promote recovery from illness.

Although psychoneuroimmunology has met with some skepticism among physicians, there is less and less room for doubt about the importance of a person's state of mind to immune function. People under stress, such as medical students during examination periods and persons caring for relatives with Alzheimer disease, show more respiratory infections than other people and respond less effectively to hepatitis and flu vaccines. The attitudes, coping abilities, and social support systems of patients significantly influence survival time even in such serious diseases as AIDS and breast cancer. Women with breast cancer die at markedly higher rates if their husbands cope poorly with stress. Attitudes such as optimism, resignation, or despair in the face of disease significantly affect immune function. Religious beliefs can also influence the prospect of recovery. Indeed, ardent believers in voodoo sometimes die just from the belief that someone has cast a spell on them. The stress of hospitalization can interfere with the treatment one gives to a patient, and psychoneuroimmunology has obvious implications for treating patients in ways that minimize this stress and thus promote recovery.

Before You Go On

Answer the following questions to test your understanding of the preceding section:

26. Contrast local anaphylaxis and anaphylactic shock. Include an explanation of why the latter is more serious.
27. Explain why anaphylactic shock occurs immediately upon antigen exposure while the response to poison ivy takes much longer to develop.
28. Give three reasons why the immune system may become activated by the body's own antigens. Name or briefly describe a disease that exemplifies each mechanism.
29. How are SCID and AIDS similar? How are they different?
30. Explain why cancer and opportunistic infections are common in AIDS victims.

Connective Issues

Interactions Between the LYMPHATIC and IMMUNE SYSTEMS and Other Organ Systems

← indicates ways in which these systems affect other systems
→ indicates ways in which other systems affect these systems

Nearly All Systems

Lymphatic system drains excess tissue fluid and removes cellular debris and pathogens. Immune system provides defense against pathogens and immune surveillance against cancer.

Integumentary System

→ Skin provides mechanical and chemical barriers to pathogens; has antigen-presenting cells in epidermis and dermis; and is a common site of inflammation

Skeletal System

→ Lymphocytes and macrophages arise from bone marrow cells; skeleton protects thymus and spleen

Muscular System

→ Skeletal muscle pump moves lymph through lymphatic vessels

Nervous System

→ Neuropeptides and emotional states affect immune function; blood-brain barrier prevents antibodies and immune cells from entering brain tissue

Endocrine System

← Lymph transports some hormones

→ Hormones from thymus stimulate development of lymphatic organs and T cells; stress hormones depress immunity and increase susceptibility to infection and cancer

Circulatory System

← Cardiovascular system would soon fail without return of fluid and protein by lymphatic system; spleen disposes of expired erythrocytes and recycles iron; lymphatic organs prevent accumulation of debris and pathogens in blood

→ Lymphatic vessels develop from embryonic veins; arterial pulsation aids flow of lymph in neighboring lymphatic vessels; leukocytes serve in nonspecific and specific defense; blood transports immune cells, antibodies, complement, interferon, and other immune chemicals; capillary endothelial cells signal areas of tissue injury and stimulate margination and diapedesis of leukocytes; blood clotting restricts spread of pathogens

Respiratory System

← Alveolar macrophages remove debris from lungs

→ Provides immune system with O_2; disposes of CO_2; thoracic pump aids lymph flow; pharynx houses tonsils

Urinary System

← Absorbs fluid and proteins in kidneys, which is essential to enabling kidneys to concentrate the urine and conserve water

→ Eliminates waste and maintains fluid and electrolyte balance important to lymphatic and immune function; urine flushes some pathogens from body; acidic pH of urine protects against urinary tract infection

Digestive System

← Lymph absorbs and transports digested lipids

→ Nourishes lymphatic system and affects lymph composition; stomach acid destroys ingested pathogens

Reproductive System

← Immune system requires that the testes have a blood-testis barrier to prevent autoimmune destruction of sperm

→ Vaginal acidity inhibits growth of pathogens

Chapter Review

Online Learning Center

Study resources for this chapter are available at www.mhhe.com/saladin/

Study Outline

The Lymphatic System (p. 803)

1. Lymph
   a. Source
   b. Importance of absorbing tissue fluid
   c. Composition
   d. Structure and function of lymphatic capillaries
2. Lymphatic vessels
   a. Tunics
   b. Route of lymph flow
      - Collecting vessels
      - Lymphatic trunks
      - Collecting ducts
         - •• Right lymphatic duct
         - •• Thoracic duct
         - •• Cisterna chyli
   c. Mechanisms of lymph flow
      - Contractions of lymphatic vessels
      - Skeletal muscle pump
      - Effect of neighboring arteries
      - Thoracic pump
      - Drainage into subclavian veins
3. Lymphatic tissue
   a. Diffuse lymphatic tissue
   b. Mucosa-associated lymphatic tissue
   c. Lymphatic nodules
   d. Peyer patches
4. Lymphatic organs
   a. Lymph nodes
      - Areas of concentration
      - Structure
         - •• Shape
         - •• Stroma and parenchyma
         - •• Cortex and medulla
         - •• Afferent and efferent vessels
      - Lymph-filtering function
   b. Tonsils
      - Crypts and nodules
      - Locations
         - •• Palatine tonsils
         - •• Lingual tonsils
         - •• Pharyngeal tonsil
   c. Thymus
      - Location
      - Involution
      - Structure
      - Functions
   d. Spleen
      - Location
      - Structure
         - •• Red pulp
         - •• White pulp
      - Functions
         - •• Blood production in fetus
         - •• Blood reservoir
         - •• Erythrocyte disposal
         - •• Immune functions

Nonspecific Resistance (p. 811)

1. Three lines of defense
   a. External barriers
   b. Inflammation, fever, etc.
   c. Immune system
2. Nonspecific and specific defense
3. External barriers
   a. Skin
      - Toughness of keratin
      - Dry and nutrient-poor
      - Defensins
      - Lactic acid (acid mantle)
   b. Mucous membranes
      - Stickiness of mucus
      - Lysozyme
   c. Subepithelial areolar tissue
      - Tissue gel
      - Use of hyaluronidase by pathogens
4. Leukocytes
   a. Neutrophils
      - Phagocytosis of bacteria
      - Killing zone
   b. Eosinophils
      - Phagocytic roles
      - Antiparasitic effects
   c. Basophils
      - Histamine (vasodilator)
      - Heparin (anticoagulant)
   d. Lymphocytes
      - Natural killer cells
      - Other types involved in specific immunity
   e. Monocytes and macrophages
5. Inflammation
   a. Examples ("-itis" diseases)
   b. Cardinal signs (table 21.1)
   c. Inflammatory cells (table 21.2)
   d. Inflammatory chemicals (table 21.3)
   e. Processes of inflammation
      - Vasodilation and hyperemia
      - Redness
      - Heat
      - Capillary permeability and tissue swelling
      - Leukocyte deployment
         - •• Margination
         - •• Diapedesis
         - •• Chemotaxis
         - •• Phagocytosis
      - Actions of leukocytes
      - Pus formation
   f. Tissue repair
6. Antimicrobial proteins
   a. Interferons
      - Antiviral effect
      - Anticancer effect
   b. Complement system
      - Complement proteins
      - Pathways of activation
      - Mechanisms of action
         - •• Enhanced inflammation
         - •• Opsonization
         - •• Cytolysis
7. Fever
   a. Functions of fever
      - Promoting interferon activity
      - Accelerating tissue repair
      - Inhibiting pathogen reproduction
   b. Pyrogens
   c. Stages of fever
      - Onset
      - Stadium
      - Defervescence
   d. Dangers of high fever

General Aspects of Specific Immunity (p. 819)

1. Specificity and memory
2. Humoral and cellular forms
3. Antigens
   a. Molecular characteristics
   b. Antigenic determinants
   c. Haptens
4. Antibodies
   a. Structure
      - Heavy and light chains
      - Variable and constant regions
      - Antigen-binding site
   b. Classes (table 21.5)
   c. Antibody diversity
5. Lymphocytes
   a. T cells
      - Involvement of thymus
      - Development
         - •• Immunocompetence
         - •• Cloning
         - •• Clonal deletion
   b. B cells
      - Sites of development
      - B cell clones
6. Antigen-presenting cells
   a. B cells and macrophages
   b. Major histocompatibility complex
   c. MHC proteins
   d. Antigen display
7. Interleukins
   a. Lymphokines
   b. Monokines

Humoral Immunity (p. 823)

1. Recognition
   a. Capping and B cell activation
   b. Endocytosis
   c. Antigen display
      - Antigen presentation on MHC protein
      - Activation of helper T cell
      - Lymphokine secretion
   d. Clonal selection
   e. Plasma cell differentiation
      - Characteristics of plasma cells
      - Antibody secretion
2. Attack (antibody actions)
   a. Neutralization
   b. Complement fixation
   c. Agglutination
   d. Precipitation
3. Memory
   a. Primary response
   b. Memory cells
   c. Secondary response

Cellular Immunity (p. 827)

1. Lymphocyte types involved
   a. Cytotoxic (killer) T cells
   b. Helper T cells
   c. Suppressor T cells
   d. Memory T cells
2. Recognition
   a. Role of MHC-I proteins
      - Found on nearly all of body's cells
      - Display antigens produced by host cells
      - Stimulate attack by cytotoxic T cells
   b. Role of MHC-II proteins
      - Found only on antigen-presenting cells
      - Stimulate helper T cells
   c. MHC restriction
   d. T cell activation
      - Activation of clonal selection
      - Interleukins and costimulation
3. Attack
   a. Helper T cells
      - Secrete lymphokines
      - Coordinate humoral and cellular immunity
   b. Cytotoxic T cells
      - Directly attack enemy cells
      - Lethal hit mechanism
      - Lymphotoxins
      - Tumor necrosis factor
      - Migration-inhibiting factor
   c. Suppressor T cells
      - Slow down immune response
      - May help prevent autoimmune diseases
4. Memory
   a. Memory T cells
   b. T cell recall response
5. Passive and active immunity
   a. Natural active immunity
   b. Artificial active immunity
   c. Natural passive immunity
   d. Artificial passive immunity

Immune System Disorders (p. 832)

1. Hypersensitivity (allergy)
   a. Allergens
   b. Type I (acute) hypersensitivity
      - Anaphylaxis
      - Asthma
      - Anaphylactic shock
   c. Type II (antibody-dependent cytotoxic) hypersensitivity
      - Involvement of cytolysis
      - Transfusion reactions
   d. Type III (immune complex) hypersensitivity
      - Widespread antigen-antibody complexing
      - Acute glomerulonephritis
      - Systemic lupus erythematosus
   e. Type IV (delayed) hypersensitivity
      - 12- to 72-hour delay
      - Cosmetic and poison ivy allergies
      - Tuberculosis skin test
2. Autoimmune diseases
   a. Failure of self-tolerance
   b. Production of autoantibodies
   c. Reasons for failure of self-tolerance
      - Cross-reactivity
      - Abnormal exposure of self-antigens
      - Changes in antigen structure
3. Immunodeficiency diseases
   a. Severe combined immunodeficiency disease (SCID)
      - Hereditary lack of T and B cells
      - Vulnerability to opportunistic infection
   b. Acquired immunodeficiency syndrome (AIDS)
      - Human immunodeficiency virus (HIV)
         - •• Structure
         - •• Mechanism of infection and replication
         - •• Pathogenic effects
      - Signs and symptoms
      - Modes of HIV transmission
      - Treatment strategies
      - Unanswered questions

Selected Vocabulary

lymphatic system 803
lymph 803
lymphatic capillary 803
collecting vessel 803
lymphatic trunk 804
collecting duct 804
right lymphatic duct 804
thoracic duct 804
cisterna chyli 804
diffuse lymphatic tissue 805
mucosa-associated lymphatic tissue 805
lymphatic nodule 806
Peyer patch 806
lymph node 806
germinal center 808
afferent lymphatic vessel 808
efferent lymphatic vessel 808
tonsil 808
tonsillar crypt 808
palatine tonsil 808
lingual tonsil 808
pharyngeal tonsil 808
thymus 808
involution 808
thymopoietin 809
thymosin 809
spleen 809
red pulp 809
white pulp 810
nonspecific resistance 811
specific defense 811
lysozyme 812
tissue gel 812
phagocyte 812
killing zone 812
degranulation 812
respiratory burst 812
natural killer cell 812
macrophage system 813
macrophage 813
reticular cell 813
dendritic cell 813
microglia 813
endothelial cell 813
alveolar macrophage 813
hepatic macrophage 813
inflammation 813
cardinal sign 813
inflammatory chemical 813
bradykinin 814
margination 814
diapedesis 814
chemotaxis 814
phagocytosis 814
leukocytosis-promoting factor 815
pus 815
platelet-derived growth factor 815
interferon 815
complement system 816
classical pathway 816
complement fixation 816
alternate pathway 816
opsonization 817
cytolysis 817
fever 817
pyrogen 817
humoral immunity 819
cellular immunity 819
antigen 819
antigenic determinant 819
hapten 819
antibody 820
immunoglobulin 820
antibody monomer 820
heavy chain 820
light chain 820
variable region 820
antigen-binding site 820
constant region 820
IgA 820
IgD 820
IgE 820
IgG 820
IgM 820
somatic recombination 820
T lymphocyte 820
B lymphocyte 820
immunocompetent 821
clone 821
virgin lymphocyte pool 821
clonal deletion 821
self-tolerance 821
antigen-presenting cell 821
MHC protein 822
interleukin 822
lymphokine 822
monokine 822
capping 823
endocytosis 823
display 823
immunological surveillance 823
helper factor 823
clonal selection 823
plasma cell 823
neutralization 825
complement fixation 825
agglutination 825
precipitation 825
primary response 825
antibody titer 826
memory cell 826
secondary response 826
cytotoxic T cell 827
helper T cell 827
suppressor T cell 827
memory T cell 827
MHC-I protein 827
MHC-II protein 827
MHC restriction 827
costimulation 828
macrophage-activating factor 828
lethal hit 828
perforin 828
lymphotoxin 828
tumor necrosis factor 828
migration-inhibiting factor 828
T cell recall response 829
natural active immunity 829
artificial active immunity 829
vaccination 829
vaccine 829
natural passive immunity 829
artificial passive immunity 830
immune serum 830
hypersensitivity 832
allergen 832
type I hypersensitivity 832
anaphylaxis 832
asthma 833
anaphylactic shock 833
type II hypersensitivity 833
type III hypersensitivity 833
type IV hypersensitivity 833
autoimmune disease 833
autoantibody 833
acquired immunodeficiency syndrome (AIDS) 834
human immunodeficiency virus (HIV) 834
AIDS-related complex 834

Testing Your Recall

1. The only lymphatic organ with both afferent and efferent lymphatic vessels is
   a. the spleen.
   b. a lymph node.
   c. a tonsil.
   d. a Peyer patch.
   e. the thymus.

2. Which of the following cells are involved in nonspecific defense but not in specific defense?
   a. helper T cells
   b. cytotoxic T cells
   c. natural killer cells
   d. B cells
   e. plasma cells

3. The respiratory burst is used by ______ to kill bacteria.
   a. neutrophils
   b. basophils
   c. mast cells
   d. NK cells
   e. cytotoxic T cells

4. All of the following are macrophages *except*
   a. microglia.
   b. dendritic cells.
   c. reticular cells.
   d. endothelial cells.
   e. mast cells.

5. The cytolytic action of the complement system is most similar to the action of
   a. interleukin-1.
   b. platelet-derived growth factor.
   c. lymphotoxin.
   d. perforin.
   e. IgE.

6. ______ can become antigenic by binding to larger host molecules.
   a. Antigenic determinants
   b. Haptens
   c. Lymphokines
   d. Pyrogens
   e. Cell-adhesion molecules

7. Which of the following correctly states the order of events in humoral immunity? Let *1* = antigen display, *2* = antibody secretion, *3* = capping, *4* = clonal selection, and *5* = endocytosis.
   a. 3–4–1–5–2
   b. 5–3–1–2–4
   c. 3–5–1–4–2
   d. 5–3–1–4–2
   e. 1–2–4–3–5

8. The cardinal signs of inflammation include all of the following *except*
   a. redness.
   b. swelling.
   c. heat.
   d. fever.
   e. pain.

9. A helper T cell can bind only to another cell that has
   a. MHC-II proteins.
   b. an antigenic determinant.
   c. an antigen-binding site.
   d. a complement-binding site.
   e. a CD4 protein.

10. Which of the following results from a lack of immune self-tolerance?
    a. SCID
    b. AIDS
    c. systemic lupus erythematosus
    d. anaphylaxis
    e. asthma

11. Any organism or substance capable of causing disease is called a/an ________ .

12. Mucous membranes contain an antibacterial enzyme called ________ .

13. The pain of inflammation results partly from a polypeptide called ________ .

14. The migration of leukocytes through the capillary wall is called ________ .

15. In the process of ________ , complement proteins coat bacteria and serve as binding sites for phagocytes.

16. Any substance that triggers a fever is called a/an ________ .

17. The hormonelike chemicals produced by lymphocytes to stimulate other leukocytes are called ________ . Those produced by macrophages are called ________ .

18. Part of an antibody called the ________ binds to part of an antigen called the ________ .

19. Self-tolerance results from a process called ________ , in which lymphocytes programmed to react against self-antigens are destroyed.

20. Wandering macrophages of the connective tissues are called ________ .

Answers in appendix B

Testing Your Comprehension

1. Anti-D antibodies of an Rh$^-$ woman sometimes cross the placenta and hemolyze the RBCs of an Rh$^+$ fetus (see p. 699). Yet the anti-B antibodies of a type A mother seldom affect the RBCs of a type B fetus. Explain this difference based on your knowledge of the five immunoglobulin classes.

2. In treating a woman for malignancy in the right breast, the surgeon removes some of her axillary lymph nodes. Following surgery, the patient experiences edema of the right upper limb. Explain why.

3. A girl with a defective heart receives a new heart transplanted from another child who was killed in an accident. The patient is given an antilymphocyte serum containing antibodies against her lymphocytes. The transplanted heart is not rejected, but the patient dies of an overwhelming bacterial infection. Explain why the antilymphocyte serum was given and why the patient was so vulnerable to infection.

4. A burn research center uses mice for studies of skin grafting. To prevent graft rejection, the mice are thymectomized at birth. Even though B cells do not develop in the thymus, these mice show no humoral immune response and are very susceptible to infection. Explain why the removal of the thymus would improve the success of skin grafts and why it would affect humoral immunity.

5. Contrast the structure of a B cell with that of a plasma cell, and explain how their structural difference relates to their functional difference.

Answers at the Online Learning Center

A resin cast of the arteries (red), veins (blue), and alveoli (yellow) of a human lung

CHAPTER 22

The Respiratory System

Chapter Outline

Insights

Study with Your ESP CD-ROM

Refer to your ESP Tutorial CD-ROM for helpful animations, activities, and quizzes related to the material covered in this chapter. The following are suggested areas to go to on the CD and are presented according to "Unit/Topic":

- Absorption and Excretion/Respiration

Brushing Up

To understand this chapter, it is important that you understand or brush up on the following concepts:

- Factors that affect simple diffusion (p. 126)
- The muscles of respiration (pp. 359–60)
- The structure of hemoglobin (pp. 691–92)
- Principles of fluid pressure and flow (p. 734)
- Pulmonary blood circulation (pp. 770–71)

Most metabolic processes of the body depend on ATP, and most ATP production requires oxygen and generates carbon dioxide as a waste product. The respiratory and cardiovascular systems collaborate to provide this oxygen and remove the carbon dioxide. Not only do these two systems have a close spatial relationship in the thoracic cavity, they also have such a close functional relationship that they are often considered jointly under the heading *cardiopulmonary.* A disorder that affects the lungs has direct and pronounced effects on the heart, and vice versa.

Furthermore, as discussed in the next two chapters, the respiratory system works closely with the urinary system to regulate the body's acid-base balance. Changes in the blood pH, in turn, trigger autonomic adjustments of the heart rate and blood pressure. Thus, the cardiovascular, respiratory, and urinary systems have an especially close physiological relationship. It is important that we now address the roles of the respiratory and urinary systems in the homeostatic control of blood gases, pH, blood pressure, and other variables related to the body fluids. This chapter deals with the respiratory system and the next chapter with the urinary system.

Anatomy of the Respiratory System

Objectives

When you have completed this section, you should be able to

- trace the flow of air from the nose to the pulmonary alveoli; and
- relate the function of any portion of the respiratory tract to its gross and microscopic anatomy.

The term **respiration** has three meanings: (1) ventilation of the lungs (breathing), (2) the exchange of gases between air and blood and between blood and tissue fluid, and (3) the use of oxygen in cellular metabolism. In this chapter, we are concerned with the first two processes. Cellular respiration was introduced in chapter 3 and is considered more fully in chapter 26.

The principal organs of the **respiratory system** are the nose, pharynx, larynx, trachea, bronchi, and lungs (fig. 22.1). These organs serve to receive fresh air, exchange gases with the blood, and expel the modified air. Within the lungs, air flows along a dead-end pathway consisting essentially of bronchi → bronchioles → alveoli (with some refinements to be introduced later). Incoming air stops in the alveoli (millions of thin-walled, microscopic air sacs in the lung), exchanges gases with the bloodstream across the alveolar wall, and then flows back out.

The **conducting division** of the respiratory system consists of those passages that serve only for airflow, essentially from the nostrils through the bronchioles. The **respiratory division** consists of the alveoli and other distal gas-exchange regions. The airway from the nose through the larynx is often called the **upper respiratory tract** (that is, the respiratory organs in the head and neck), and the regions from the trachea through the lungs compose the **lower respiratory tract** (the respiratory organs of the thorax).

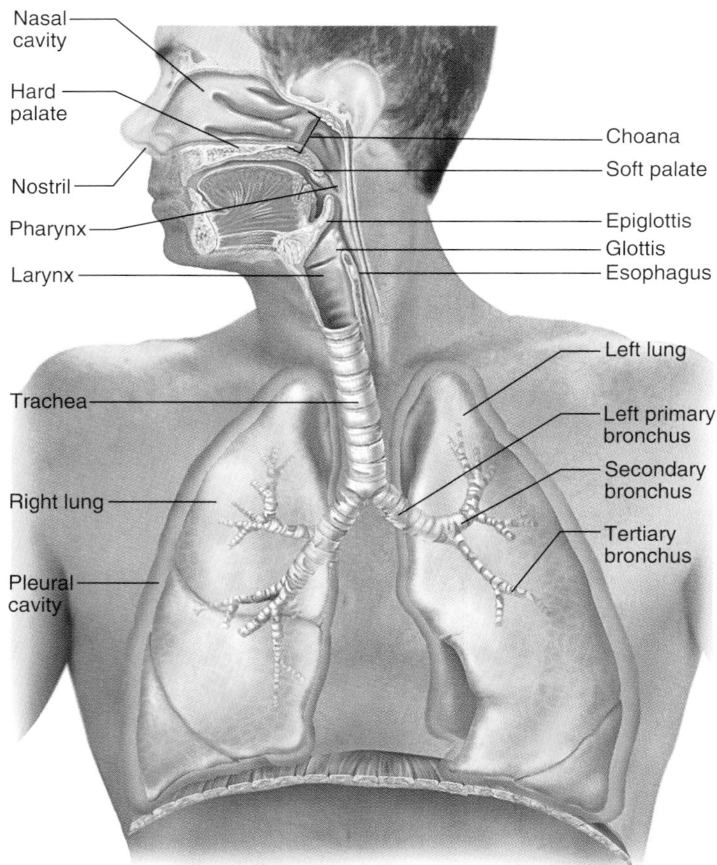

Figure 22.1 The Respiratory System.

The Nose

The **nose** has several functions: it warms, cleanses, and humidifies inhaled air; it detects odors in the airstream; and it serves as a resonating chamber that modifies the voice. The external, protruding part of the nose is supported and shaped by a framework of bone and cartilage. Its superior half is supported by the nasal bones medially and the maxillae laterally. The inferior half is supported by the **lateral** and **alar cartilages** (fig. 22.2). Dense connective tissue shapes the flared portion called the **ala nasi,** which forms the lateral wall of each nostril.

The **nasal cavity** (fig. 22.3) extends from the **anterior (external) nares** (NERR-eez) (singular, *naris*), or **nostrils,** to the **posterior (internal) nares,** or **choanae**[1] (co-AH-nee). The dilated chamber inside the ala nasi is called the **vestibule.** It is lined with stratified squamous epithelium and has stiff **vibrissae** (vy-BRISS-ee), or **guard hairs,** that block the inhalation of large particles.

1. *choana* = funnel

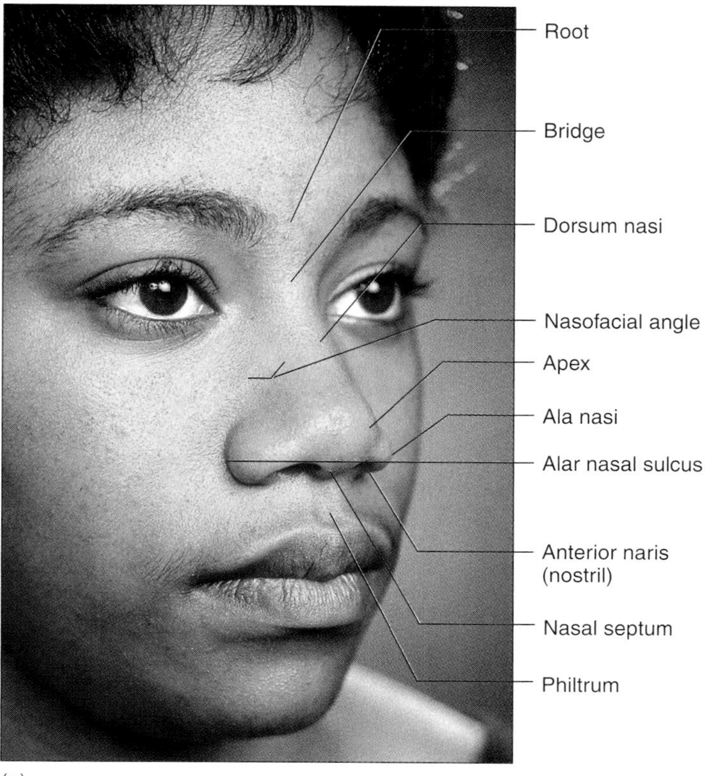

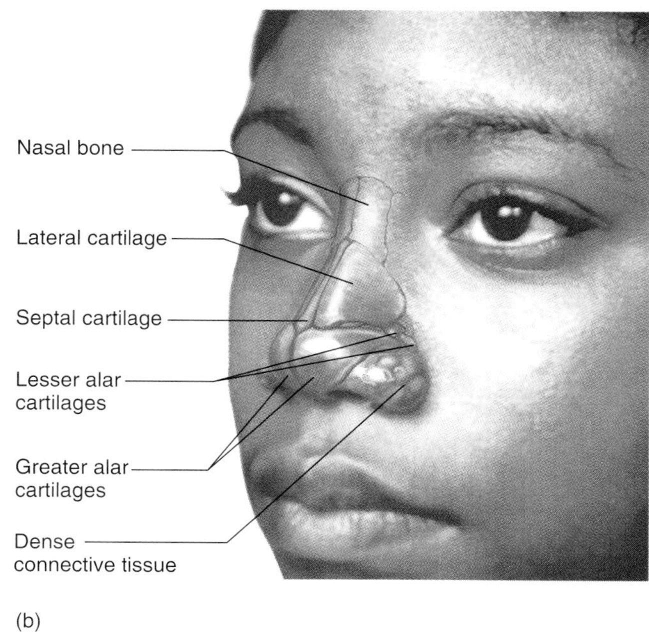

Figure 22.2 **Anatomy of the Nasal Region.** (*a*) External anatomy. (*b*) Connective tissues that shape the nose.

The **nasal septum** divides the nasal cavity into right and left chambers called **nasal fossae** (FOSS-ee). The vomer forms the inferior part of the septum, the perpendicular plate of the ethmoid bone forms its superior part, and the *septal cartilage* forms its anterior part. The ethmoid and sphenoid bones compose the roof of the nasal cavity and the palate forms its floor. The palate separates the nasal cavity from the oral cavity and allows you to breathe while there is food in your mouth.

The lateral wall of the fossa gives rise to three folds of tissue—the **superior, middle,** and **inferior nasal conchae**[2] (CON-kee)—that project toward the septum and occupy most of the fossa. They consist of mucous membranes supported by thin scroll-like **turbinate bones.** Beneath each concha is a narrow air passage called a **meatus** (me-AY-tus). The narrowness of these passages and the turbulence caused by the conchae ensure that most air contacts the mucous membrane on its way through, enabling the nose to cleanse, warm, and humidify it.

The *olfactory mucosa,* concerned with the sense of smell, lines the roof of the nasal fossa and extends over part of the septum and superior concha. The rest of the cavity is lined by ciliated pseudostratified *respiratory mucosa.* The nasal mucosa has an important defensive role. Goblet cells in the epithelium and glands in the lamina propria secrete a layer of mucus that traps inhaled particles. Bacteria are destroyed by lysozyme in the mucus. Additional protection against bacteria is contributed by lymphocytes, which populate the lamina propria in large numbers, and by antibodies (IgE) secreted by plasma cells.

Cilia of the respiratory epithelium continually beat toward the posterior nares and drive debris-laden mucus into the pharynx to be swallowed and digested. Cold air inhibits ciliary motion; thus, in cold weather, some mucus drains from the anterior nares and produces a "runny nose." The four pairs of paranasal sinuses (see chapter 9) and the nasolacrimal ducts of the orbits drain into the nasal cavity and may also contribute to this cold-weather annoyance.

The lamina propria contains large blood vessels that help to warm the air. The inferior concha has an especially extensive venous plexus called the **erectile tissue** (*swell body*). Every 30 to 60 minutes, the erectile tissue on one side becomes engorged with blood and restricts airflow through that fossa. Most air is then directed through the other naris and fossa, allowing the engorged side time to recover from drying. Thus the preponderant flow of air shifts between the right and left nares once or twice each hour. The inferior concha is the most common site of *epistaxis* (nosebleed). Spontaneous epistaxis may be a sign of hypertension.

2. *concha* = seashell

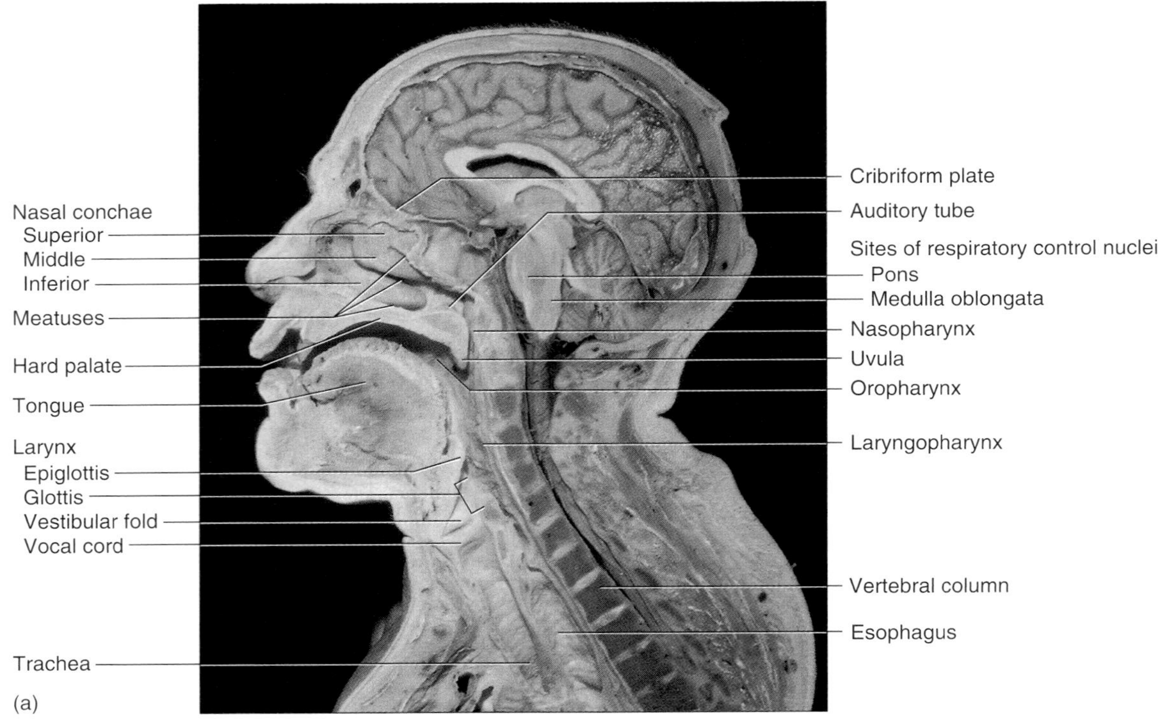

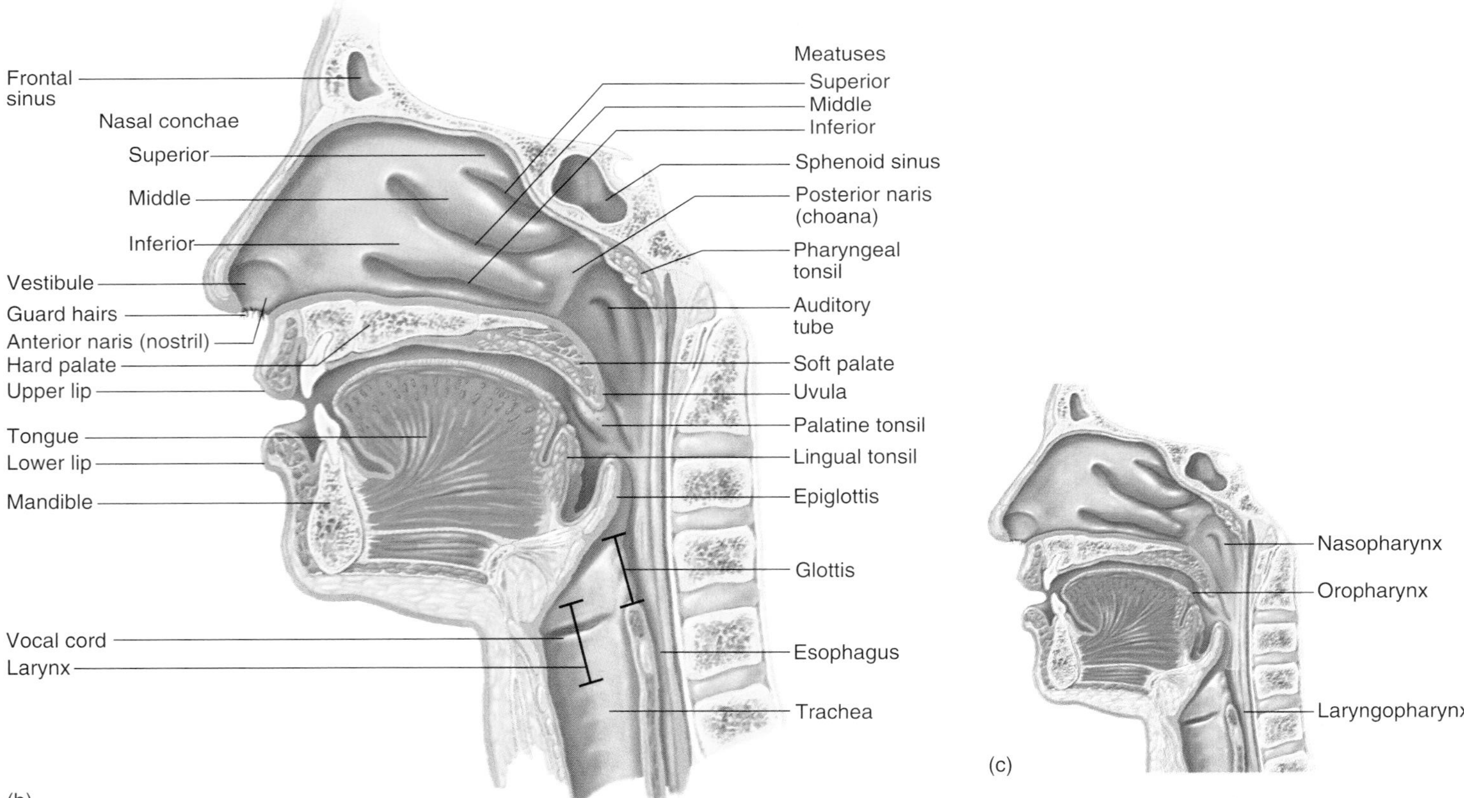

Figure 22.3 Anatomy of the Upper Respiratory Tract. (*a*) Midsagittal section of the head. (*b*) Internal anatomy. (*c*) Regions of the pharynx.

The Pharynx

The **pharynx** (FAIR-inks) is a muscular funnel extending about 13 cm (5 in.) from the choanae to the larynx. It has three regions: the *nasopharynx, oropharynx,* and *laryngopharynx* (fig. 22.3*c*).

The **nasopharynx,** which lies posterior to the choanae and dorsal to the soft palate, receives the auditory (eustachian) tubes from the middle ears and houses the pharyngeal tonsil. Inhaled air turns 90° downward as it passes through the nasopharynx. Dust particles larger than 10 μm generally cannot make the turn because of their inertia. They collide with the posterior wall of the nasopharynx and stick to the mucosa near the tonsil, which is well positioned to respond to airborne pathogens.

The **oropharynx** is a space between the soft palate and root of the tongue that extends inferiorly as far as the hyoid bone. It contains the palatine and lingual tonsils. Its anterior border is formed by the base of the tongue and the *fauces* (FAW-seez), the opening of the oral cavity into the pharynx.

The **laryngopharynx** (la-RING-go-FAIR-inks) begins with the union of the nasopharynx and oropharynx at the level of the hyoid bone. It passes inferiorly and dorsal to the larynx and ends at the level of the *cricoid cartilage* at the inferior end of the larynx (described next). The esophagus begins at that point. The nasopharynx passes only air and is lined by pseudostratified columnar epithelium, whereas the oropharynx and laryngopharynx pass air, food, and drink and are lined by stratified squamous epithelium.

The Larynx

The **larynx** (LAIR-inks), or "voicebox" (figs. 22.4 and 22.5), is a cartilaginous chamber about 4 cm (1.5 in.) long. Its primary function is to keep food and drink out of the airway, but it has evolved the additional role of producing sound.

The superior opening of the larynx, the **glottis,**[3] is guarded by a flap of tissue called the **epiglottis.**[4] During swallowing, *extrinsic muscles* of the larynx pull the larynx upward toward the epiglottis, the tongue pushes the epiglottis downward to meet it, and the epiglottis directs food and drink into the esophagus dorsal to the airway. The *vestibular folds* of the larynx, discussed shortly, play a greater role in keeping food and drink out of the airway, however. People who have had their epiglottis removed because of cancer do not choke any more than when it was present.

In infants, the larynx is relatively high in the throat and the epiglottis touches the soft palate. This creates a more or less continuous airway from the nasal cavity to the larynx and allows an infant to breathe continually while swallowing. The epiglottis deflects milk away from the airstream, like rain running off a tent while it remains dry inside. By age two, the root of the tongue becomes more muscular and forces the larynx to descend to a lower position.

3. *glottis* = back of the tongue
4. *epi* = above, upon

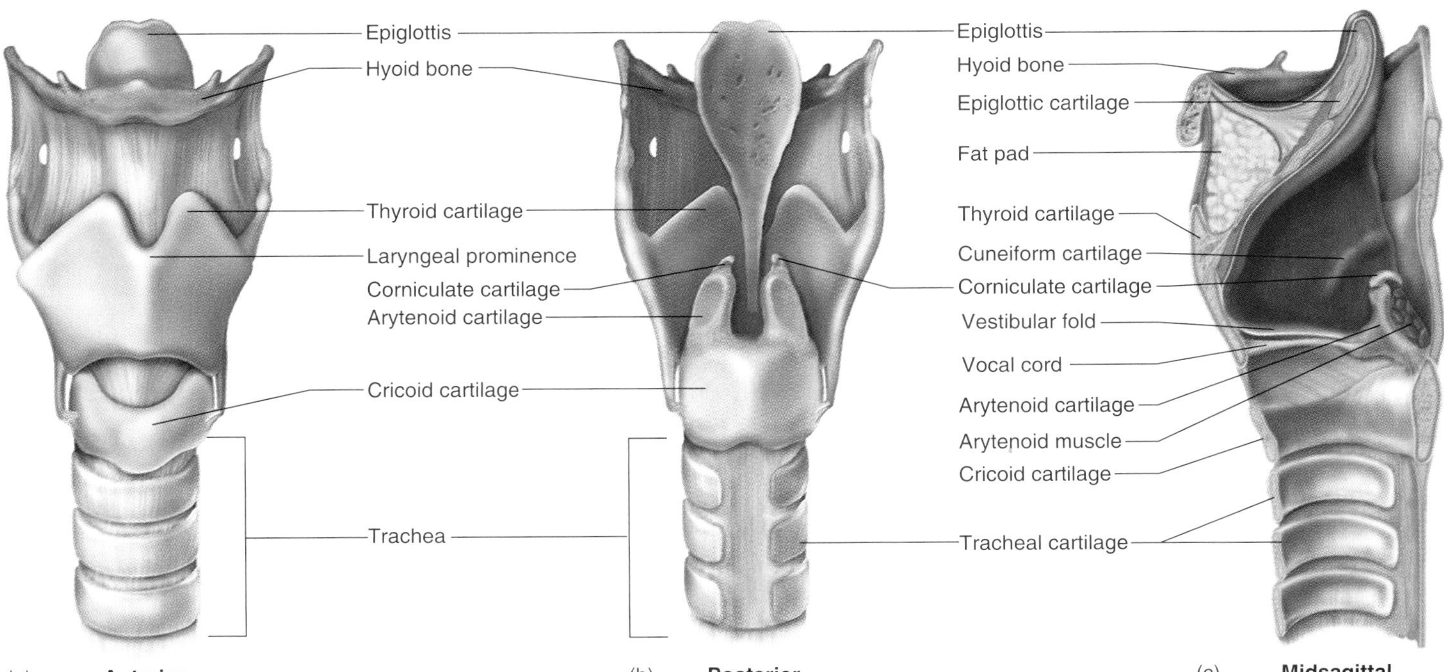

Figure 22.4 Anatomy of the Larynx. (*a*) Anterior aspect. (*b*) Posterior aspect. (*c*) Midsagittal section, anterior aspect facing left.

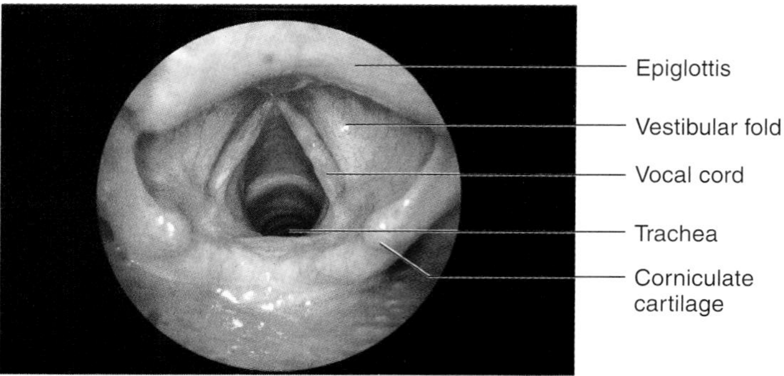

Figure 22.5 Superior View into the Larynx of a Living Person, as Seen with a Laryngoscope.

The framework of the larynx consists of nine cartilages. The first three are relatively large and unpaired. The most superior one, the **epiglottic cartilage,** is a spoon-shaped supportive plate in the epiglottis. The largest, the **thyroid cartilage,** is named for its shieldlike shape. It has an anterior peak, the *laryngeal prominence,* commonly known as the Adam's apple. Testosterone stimulates the growth of this prominence, which is therefore significantly larger in males than in females. Inferior to the thyroid cartilage is a ringlike **cricoid**[5] (CRY-coyd) **cartilage,** which connects the larynx to the trachea.

The remaining cartilages are smaller and occur in three pairs. Posterior to the thyroid cartilage are the two **arytenoid**[6] (AR-ih-TEE-noyd) **cartilages,** and attached to their upper ends are a pair of little horns, the **corniculate**[7] (cor-NICK-you-late) **cartilages.** The arytenoid and corniculate cartilages function in speech, as explained shortly. A pair of **cuneiform**[8] (cue-NEE-ih-form) **cartilages** support the soft tissues between the arytenoids and the epiglottis. The epiglottic cartilage is elastic cartilage; all the others are hyaline.

The walls of the larynx are also quite muscular. The deep **intrinsic muscles** operate the vocal cords, and the superficial **extrinsic muscles** connect the larynx to the hyoid bone and elevate the larynx during swallowing. The extrinsic muscles, also called the *infrahyoid group,* are named and described in chapter 11.

The interior wall of the larynx has two folds on each side that stretch from the thyroid cartilage in front to the arytenoid cartilages in back. The superior pair, called the **vestibular folds** (fig. 22.5), play no role in speech but close the glottis during swallowing. The inferior pair, the **vocal cords** (vocal folds), produce sound when air passes between them. They are covered with stratified squamous epithelium, best suited to endure vibration and contact between the cords.

The intrinsic muscles control the vocal cords by pulling on the corniculate and arytenoid cartilages, causing the cartilages to pivot. Depending on their direction of rotation, the arytenoid cartilages abduct or adduct the vocal cords (fig. 22.6). Air forced between the adducted vocal cords vibrates them, producing a high-pitched sound when the cords are relatively taut and a lower-pitched sound when they are more relaxed. In adult males, the vocal cords are longer and thicker, vibrate more slowly, and produce lower-pitched sounds than in females. Loudness is determined by the force of the air passing between the vocal cords. The crude sounds of the vocal cords are formed into words by actions of the pharynx, oral cavity, tongue, and lips.

The Trachea

The **trachea** (TRAY-kee-uh), or "windpipe," is a rigid tube about 12 cm (4.5 in.) long and 2.5 cm (1 in.) in diameter, lying anterior to the esophagus (fig. 22.7*a*). It is supported by 16 to 20 C-shaped rings of hyaline cartilage, some of which you can palpate between your larynx and sternum. Like the wire spiral in a vacuum cleaner hose, the cartilage rings reinforce the trachea and keep it from collapsing when you inhale (see clinical insight 22.1). The open part of the C faces posteriorly, where it is spanned by a smooth muscle, the **trachealis muscle** (fig. 22.7*c*). The gap in the C allows room for the esophagus to expand as swallowed food passes by. The trachealis muscles can contract or relax to adjust tracheal airflow.

The larynx and trachea are lined mostly by ciliated pseudostratified columnar epithelium (figs. 22.7*b* and 22.8), which functions as a **mucociliary escalator.** That is, the mucus traps inhaled debris and then the ciliary beating drives the mucus up to the pharynx, where it is swallowed.

The Lungs

Each **lung** (fig. 22.9) is a somewhat conical organ with a broad, concave **base** resting on the diaphragm and a blunt peak called the **apex (cupola)** projecting slightly superior to the clavicle. The broad **costal surface** is pressed against the rib cage, and the smaller concave **mediastinal surface** faces medially. The lungs do not fill the entire rib cage. Inferior to the lungs and diaphragm, much of the space within the rib cage is occupied by the liver, spleen, and stomach (see fig. A.15, p. 43).

The lung receives the bronchus, blood vessels, lymphatic vessels, and nerves through its **hilum,** a slit in the mediastinal surface (see fig. 22.26*a*, p. 873). These structures entering the hilum constitute the **root** of the lung. Because the heart tilts to the left, the left lung is a little smaller than the right and has an indentation called the **cardiac**

5. *crico* = ring + *oid* = resembling
6. *aryten* = ladle
7. *corni* = horn + *cul* = little + *ate* = possessing
8. *cune* = wedge + *form* = shape

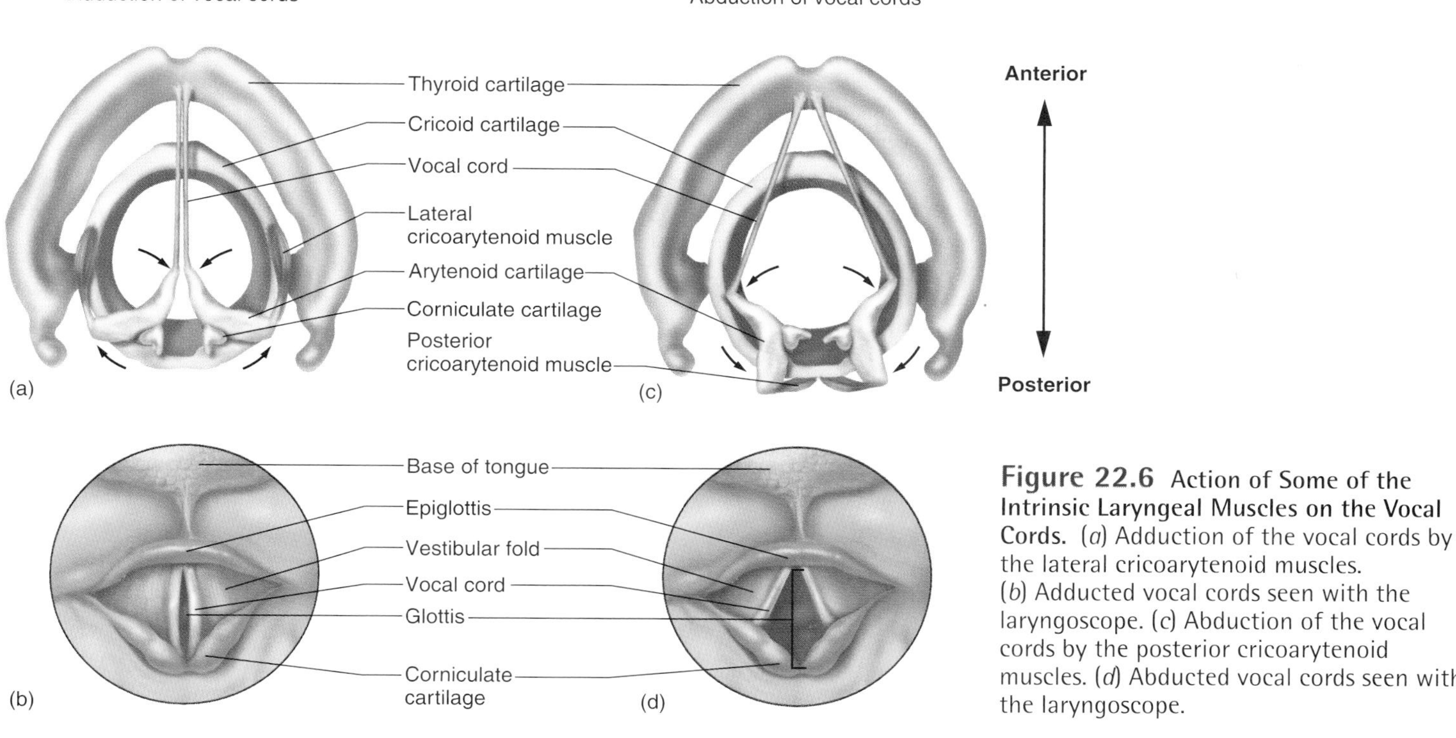

Figure 22.6 Action of Some of the Intrinsic Laryngeal Muscles on the Vocal Cords. (*a*) Adduction of the vocal cords by the lateral cricoarytenoid muscles. (*b*) Adducted vocal cords seen with the laryngoscope. (*c*) Abduction of the vocal cords by the posterior cricoarytenoid muscles. (*d*) Abducted vocal cords seen with the laryngoscope.

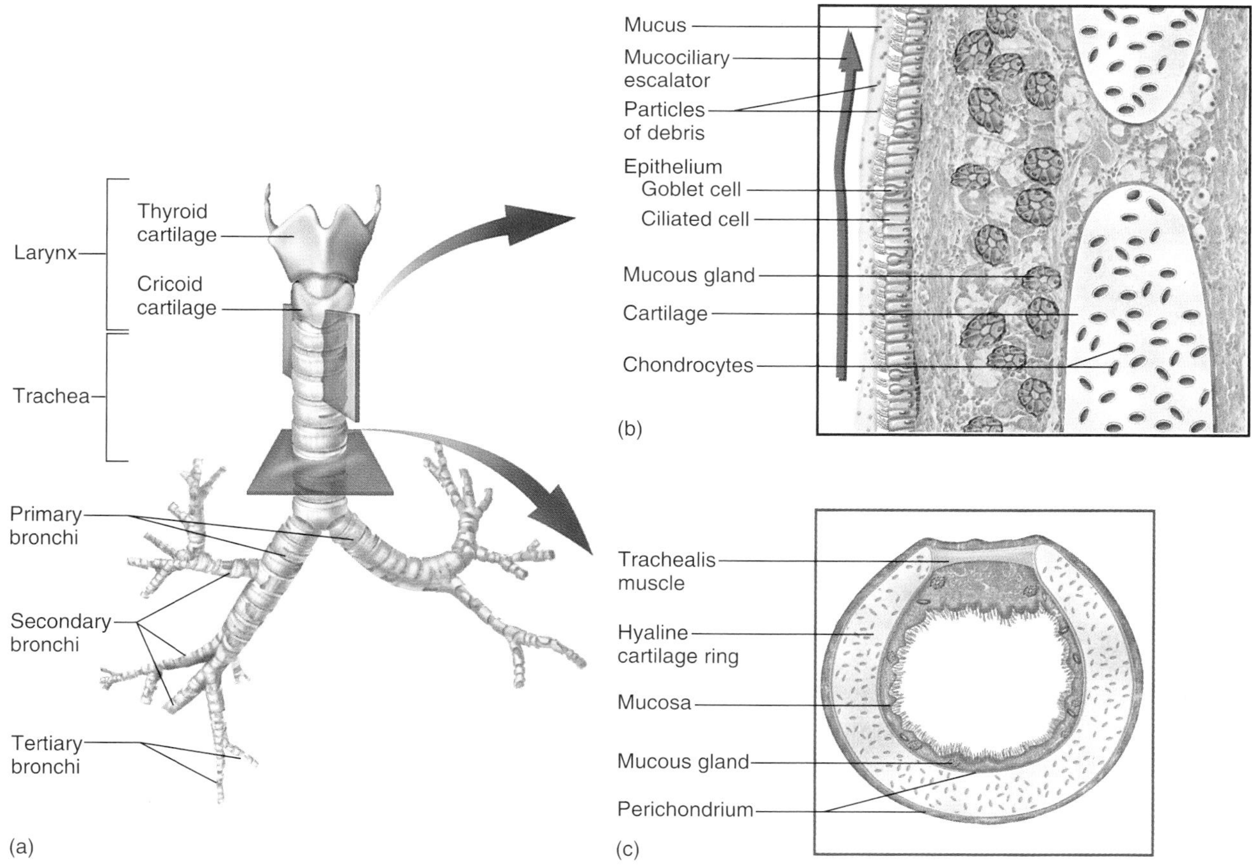

Figure 22.7 Anatomy of the Lower Respiratory Tract. (*a*) Anterior view. (*b*) Longitudinal section of the trachea showing the action of the mucociliary escalator. (*c*) Cross section of the trachea showing the C-shaped tracheal cartilage.

Clinical Insight 22.1

Tracheostomy

If the airway is obstructed with secretions or foreign matter, it may be necessary to make a temporary opening in the trachea inferior to the larynx and insert a tube to allow airflow—a procedure called *tracheostomy.* This prevents asphyxiation, but the inhaled air bypasses the nasal cavity and thus is not humidified. If the opening is left for long, the mucous membranes of the respiratory tract can dry out and become encrusted, interfering with the clearance of mucus from the tract and leading to severe infection. We can understand the functional importance of the nasal cavity especially well when we see the consequences of bypassing it.

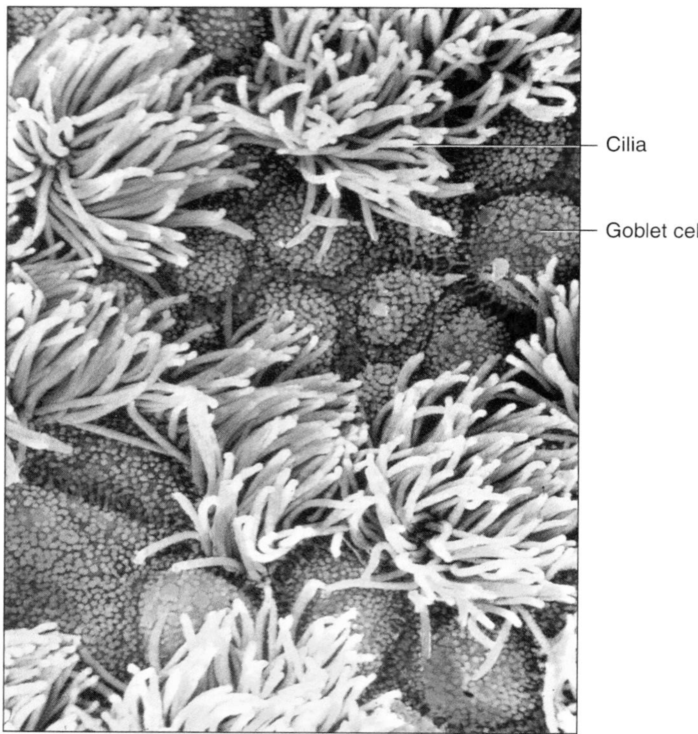

Figure 22.8 The Tracheal Epithelium Showing Ciliated Cells and Nonciliated Goblet Cells. The small bumps on the goblet cells are microvilli. (Colorized SEM micrograph.)

notch to accommodate it. The left lung has a **superior lobe** and an **inferior lobe** with a deep fissure between them; the right lung, by contrast, has three lobes—**superior, middle,** and **inferior**—separated by two fissures.

The Bronchial Tree

The lung has a spongy parenchyma containing the **bronchial tree** (fig 22.10), a highly branched system of air tubes extending from the primary bronchus to about 65,000 *terminal bronchioles.* Two **primary bronchi** (BRONK-eye) arise from the trachea at the level of the angle of the sternum. Each continues for 2 to 3 cm and enters the hilum of its respective lung. The right bronchus is slightly wider and more vertical than the left; consequently, *aspirated* (inhaled) foreign objects lodge in the right bronchus more often than in the left. Like the trachea, the primary bronchi are supported by C-shaped hyaline cartilages. All divisions of the bronchial tree also have a substantial amount of elastic connective tissue, which is important in expelling air from the lungs.

After entering the hilum, the primary bronchus branches into one **secondary (lobar) bronchus** for each pulmonary lobe. Thus, there are two secondary bronchi in the left lung and three in the right.

Each secondary bronchus divides into **tertiary (segmental) bronchi**—10 in the right lung and 8 in the left. The portion of the lung supplied by each tertiary bronchus is called a **bronchopulmonary segment.** Secondary and tertiary bronchi are supported by overlapping plates of cartilage, not rings. Branches of the *pulmonary artery* closely follow the bronchial tree on their way to the alveoli. The bronchial tree itself is nourished by the *bronchial artery,* which arises from the aorta and carries systemic blood.

Bronchioles are continuations of the airway that are 1 mm or less in diameter and lack cartilage. A well-developed layer of smooth muscle in their walls enables them to dilate or constrict, as discussed later. Spasmodic contractions of this muscle at death cause the bronchioles to exhibit a wavy lumen in most histological sections. The portion of the lung ventilated by one bronchiole is called a **pulmonary lobule.**

Each bronchiole divides into 50 to 80 **terminal bronchioles,** the final branches of the conducting division. They measure 0.5 mm or less in diameter and have no mucous glands or goblet cells. They do have cilia, however, so that mucus draining into them from the higher passages can be driven back by the mucociliary escalator, thus preventing congestion of the terminal bronchioles and alveoli.

Each terminal bronchiole gives off two or more smaller **respiratory bronchioles,** which mark the beginning

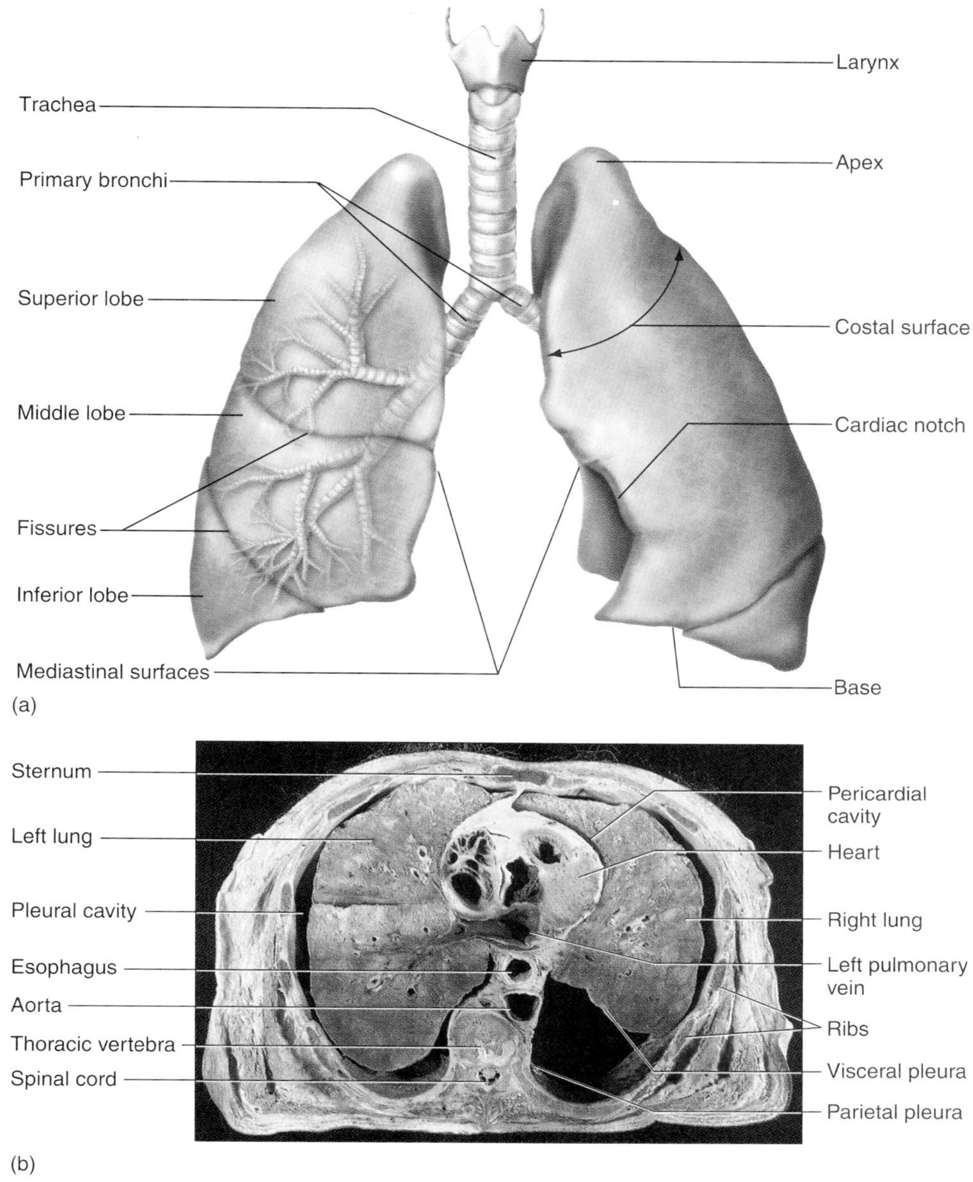

Figure 22.9 Anatomy of the Lungs. (*a*) Anterior view. (*b*) Cross section through the thorax of a cadaver showing the heart, lungs, and pleurae.

of the respiratory division. They have scanty smooth muscle, and the smallest of them are nonciliated. Each respiratory bronchiole divides into 2 to 10 elongated, thin-walled passages called **alveolar ducts** that end in irregularly shaped spaces called **alveolar sacs** (fig. 22.11). Alveoli bud from the walls of the respiratory bronchioles, alveolar ducts, and alveolar sacs. The presence of alveoli defines the respiratory division, where gas exchange with the blood occurs.

The epithelium of the bronchial tree is pseudostratified columnar in the bronchi, simple cuboidal in the bronchioles, and simple squamous in the alveolar ducts, sacs, and alveoli. It is ciliated except in the distal reaches of the respiratory bronchioles and beyond.

Alveoli

The functional importance of human lung structure is best appreciated by comparison to the lungs of a few other animals. In frogs and other amphibians, the lung is a simple sac lined with blood vessels. This is sufficient to meet the oxygen needs of animals with relatively low

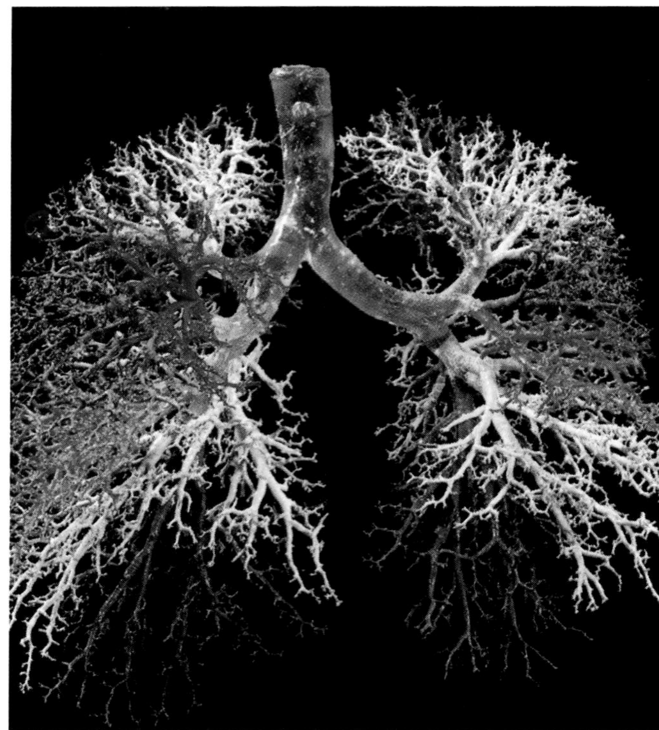

Figure 22.10 **The Bronchial Trees.** Each color identifies a bronchopulmonary segment supplied by a tertiary bronchus.

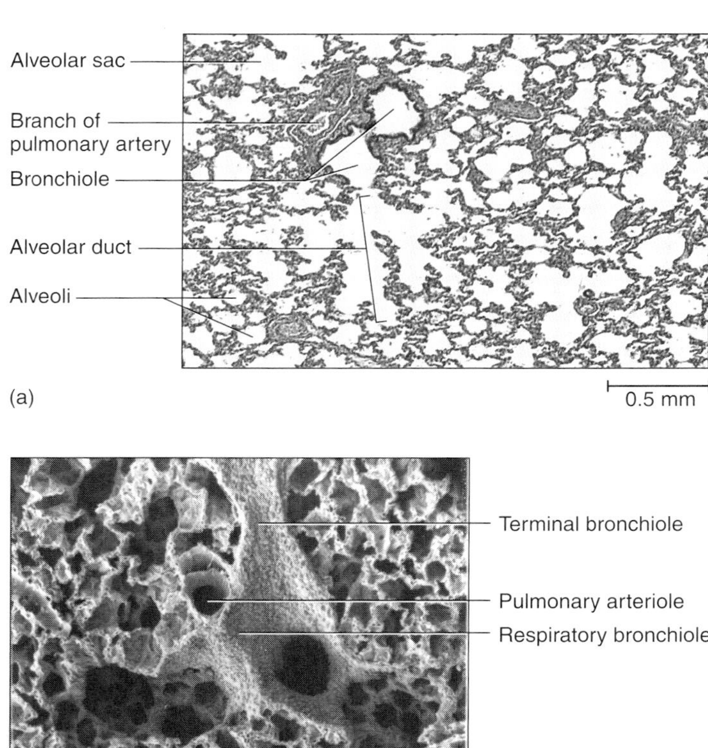

Figure 22.11 **Tissue of the Lung.** (*a*) Light micrograph; (*b*) SEM micrograph. Note the spongy texture of the lung.

metabolic rates. Mammals, with their high metabolic rates, could never have evolved with such a simple lung. Rather than consisting of one large sac, each human lung is a spongy mass composed of 150 million little sacs, the alveoli, which provide about 70 m^2 of surface for gas exchange.

An **alveolus** (AL-vee-OH-lus) (fig. 22.12) is a pouch about 0.2 to 0.5 mm in diameter. Its wall consists predominantly of **squamous (type I) alveolar cells**—thin cells that allow for rapid gas diffusion between the alveolus and bloodstream. About 5% of the alveolar cells are round to cuboidal **great (type II) alveolar cells.** They secrete a detergent-like lipoprotein called **pulmonary surfactant,** which forms a thin film on the insides of the alveoli and bronchioles. Its function is discussed later.

Alveolar macrophages (dust cells) wander the lumens of the alveoli and the connective tissue between them. They are the last line of defense against inhaled matter. Particles measuring 10 μm or larger in diameter are usually strained out by the nasal vibrissae or trapped in the mucus of the upper respiratory tract. Most particles 2 to 10 μm in diameter are trapped in the mucus of the bronchi and bronchioles, where the airflow is relatively slow, and then removed by the mucociliary escalator. Many particles smaller than 2 μm, however, make their way into the alveoli, where they are phagocytized by the macrophages. In lungs that are infected or bleeding, the macrophages also phagocytize bacteria and loose blood cells. Alveolar macrophages greatly outnumber all other cell types in the lung; as many as 50 million perish each day as they ride up the mucociliary escalator to be swallowed.

Each alveolus is surrounded by a basket of blood capillaries supplied by the pulmonary artery. The barrier between the alveolar air and blood, called the **respiratory membrane,** consists only of the squamous type I alveolar cell, the squamous endothelial cell of the capillary, and their fused basement membranes. These have a total thickness of only 0.5 μm.

The pulmonary circulation has very low blood pressure. In alveolar capillaries, the mean blood pressure is 10 mmHg and the oncotic pressure is 25 mmHg. The osmotic uptake of water thus overrides filtration and keeps the alveoli free of fluid. The lungs also have a more extensive lymphatic drainage than any other organ in the body.

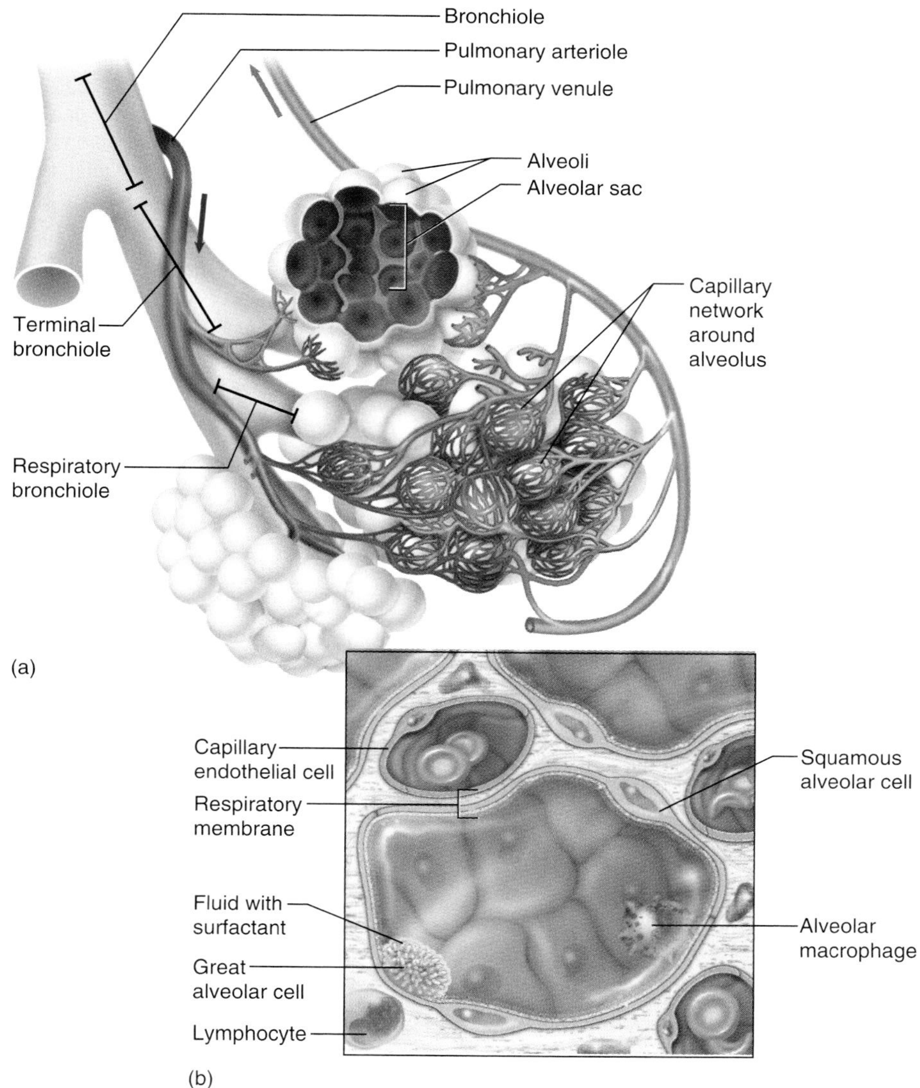

Figure 22.12 **Pulmonary Alveoli.** (*a*) Clusters of alveoli and their blood supply. (*b*) Structure of an alveolus.

The Pleurae

The surface of the lung is covered by a moist serous membrane, the **visceral pleura** (PLOOR-uh), which extends into the fissures. At the hilum, the visceral pleura turns back on itself and forms the **parietal pleura,** which adheres to the mediastinum, superior surface of the diaphragm, and inner surface of the rib cage (see fig. 22.9*b*). An extension of the parietal pleura, the *pulmonary ligament,* extends from the base of each lung to the diaphragm. The space between the parietal and visceral pleurae is called the **pleural cavity.** The two membranes are normally separated only by a film of slippery **pleural fluid;** thus, the pleural cavity is only a *potential space,* meaning there is normally no room between the membranes, but under pathological conditions this space can fill with air or liquid.

The pleurae and pleural fluid have three functions:

1. **Reduction of friction.** Pleural fluid acts as a lubricant that enables the lungs to expand and contract with minimal friction. In some forms of *pleurisy,* the pleurae are dry and inflamed and each breath gives painful testimony to the function that the fluid should be serving.
2. **Creation of pressure gradient.** Pressure in the pleural cavity is lower than atmospheric pressure; as explained later, this assists in inflation of the lungs.
3. **Compartmentalization.** The pleurae, mediastinum, and pericardium compartmentalize the thoracic organs and prevent infections of one organ from spreading easily to neighboring organs.

Think About It

In what ways do the structure and function of the pleurae resemble the structure and function of the pericardium?

Before You Go On

Answer the following questions to test your understanding of the preceding section:

1. A dust particle is inhaled and gets into an alveolus without being trapped along the way. Describe the path it takes, naming all air passages from external naris to alveolus. What would happen to it after arrival in the alveolus?
2. Describe the histology of the epithelium and lamina propria of the nasal cavity and the functions of the cell types present.
3. Describe the roles of the intrinsic muscles, corniculate cartilages, and arytenoid cartilages in speech.
4. Contrast the epithelium of the bronchioles with that of the alveoli and explain how the structural difference is related to their functional differences.

Mechanics of Ventilation

Objectives

When you have completed this section, you should be able to

- explain how pressure gradients cause air to flow into and out of the lungs;
- explain how the respiratory muscles produce these pressure gradients;
- explain the relevance of pulmonary compliance and elasticity to ventilation;
- explain the role of pulmonary surfactant; and
- define various measurements of pulmonary function.

Understanding the ventilation of the lungs, the transport of gases in the blood, and the exchange of gases with the tissues is largely a matter of understanding gas behavior. Several of the gas laws of physics are highly relevant to understanding respiratory function, but since they are named after their discoverers, they are not intuitively easy to remember by name. Table 22.1 lists the gas laws used in this chapter and may be a helpful reference as you progress through respiratory physiology.

A resting adult breathes 10 to 15 times per minute, inhaling about 500 mL of air during **inspiration** and exhaling it again during **expiration.** In this section, we examine the muscular actions and pressure gradients that produce this airflow.

Table 22.1 The Gas Laws of Respiratory Physiology

Boyle's Law[9]	The pressure of a given quantity of gas is inversely proportional to its volume (assuming a constant temperature).
Charles' Law[10]	The volume of a given quantity of gas is directly proportional to its absolute temperature.
Dalton's Law[11]	The total pressure of a gas mixture is equal to the sum of the partial pressures of its individual gases.
Henry's Law[12]	At the air-water interface, the amount of gas that dissolves in water is determined by its solubility in water and its partial pressure in the air (assuming a constant temperature).
Law of Laplace[13]	The force drawing an alveolus in on itself (tending to collapse it) is directly proportional to surface tension and inversely proportional to the radius of the alveolus.

9. Robert Boyle (1627–91), English physicist
10. Jacques A. C. Charles (1746–1823), French physicist
11. John Dalton (1766–1844), British chemist
12. William Henry (1774–1836), British chemist
13. Pierre Simon, Marquis de Laplace (1749–1827), French astronomer and mathematician

Pressure and Flow

Airflow is governed by the same principles of flow, pressure, and resistance as blood flow (see chapter 20). The pressure that drives respiration is **atmospheric (barometric) pressure**—the weight of the air above us. At sea level, a column of air as thick as the atmosphere (60 mi) and 1 in. square weighs 14.7 lb; it is thus said to exert a force of 14.7 pounds per square inch (psi). In standard international (SI) units, this is a column of air 100 km high exerting a force of 1.013×10^6 dynes/cm^2. This pressure, called *1 atmosphere* (1 atm), is enough to force a column of mercury 760 mm up an evacuated tube; therefore, 1 atm = 760 mmHg. This is the average atmospheric pressure at sea level; it fluctuates from day to day and is lower at higher altitudes.

One way to change the pressure of a gas, and thus to change flow, is to change the volume of its container. **Boyle's law** states that *the pressure of a given quantity of gas is inversely proportional to its volume* (assuming a constant temperature). If the lungs contain a quantity of gas and lung volume increases, their **intrapulmonary pressure**—the pressure within the alveoli—falls. If lung volume decreases, intrapulmonary pressure rises. (Compare this to the syringe analogy on p. 735.) To make air flow into the lungs, it is necessary only to lower the intrapulmonary pressure below the atmospheric pressure.

Raising the intrapulmonary pressure above the atmospheric pressure makes air flow out again. These changes are created as skeletal muscles of the thoracic and abdominal walls change the volume of the thoracic cavity.

What matters to flow is the *difference* between atmospheric pressure and intrapulmonary pressure. Since atmospheric pressures vary from one place and time to another, it is more useful for our discussion to refer to relative pressures. A relative pressure of −3 mmHg, for example, means 3 mmHg below atmospheric pressure; a relative pressure of +3 mmHg is 3 mmHg above atmospheric pressure. If the atmospheric pressure was 760 mmHg, these would represent absolute pressures of 757 and 763 mmHg, respectively.

Inspiration

The mechanical process of pulmonary ventilation is achieved by rhythmically changing the pressure in the thoracic cavity so that it drops slightly below atmospheric pressure and draws air in, then rises slightly above atmospheric pressure and expels this air. This is achieved mainly by the dome-shaped diaphragm. When stimulated by the phrenic nerves, the diaphragm flattens somewhat and drops about 1.5 cm in quiet respiration (as much as 7 cm in deep breathing). This enlarges the thoracic cavity and reduces its pressure. In addition, the intercostal nerves stimulate the external intercostal muscles to contract. When the scalenes fix the first pair of ribs, the external intercostals elevate the second through twelfth pairs like bucket handles, making them swing slightly up and out. Forced (deep) inspiration is aided by the pectoralis minor, sternocleidomastoid, and erector spinae muscles.

As the rib cage expands and the diaphragm drops, the parietal pleura clings to them. The **intrapleural pressure** between the parietal and visceral pleurae is about 4 mmHg less than atmospheric pressure at rest (fig. 22.13) and drops to −6 mmHg during inspiration. Because of the surface tension of the pleural fluid, the visceral pleura clings to the parietal pleura, like two sheets of wet paper. Thus, when the parietal pleura is pulled outward, the visceral pleura and lung are pulled along with it, causing the lung to expand. Not all the pressure change in the pleural cavity is transferred to the interior of the lungs, but the intrapulmonary pressure drops to about −3 mmHg. The difference between the intrapleural and intrapulmonary pressures, called **transpulmonary pressure,** is a critical force in inflating the lung. In the case under discussion, this would be $-6 - (-3) = -3$ mmHg. At an atmospheric pressure of 760 mmHg, intrapleural pressure would be 754 mmHg and intrapulmonary pressure would be 757 mmHg. With a gradient of 760 → 757 mmHg from atmospheric to intrapulmonary pressure, air moves into the lungs, and the transpulmonary gradient of 757 → 754 mmHg helps the lungs expand in the enlarging thoracic cavity.

Another force that expands the lungs is warming of the inhaled air. **Charles' law** states that *the volume of a given quantity of gas is directly proportional to its absolute temperature.* On a day when the ambient temperature is 21°C (70°F), inhaled air is heated to 37°C (16°C warmer) by the time it reaches the alveoli. As the inhaled air expands, it helps to inflate the lungs.

When the respiratory muscles stop contracting, the inflowing air quickly achieves an intrapulmonary pressure equal to atmospheric pressure, and flow stops. The dimensions of the thoracic cage increase by only a few millimeters in each direction, but this is enough to increase its total volume by 500 mL. Thus, 500 mL of air flows into the respiratory tract during quiet breathing.

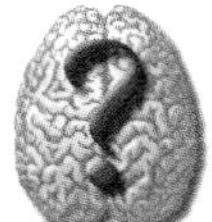

Think About It

When you inhale, does your chest expand because your lungs inflate, or do your lungs inflate because your chest expands? Explain.

Expiration

Inspiration requires a muscular effort and therefore an expenditure of ATP and calories. By contrast, normal expiration during quiet breathing is an energy-saving passive process that requires no muscular contraction. It is achieved by the elasticity of the lungs and thoracic cage—the tendency to return to their original dimensions when released from tension. The bronchial tree has a substantial amount of elastic connective tissue in its walls. The attachments of the ribs to the spine and sternum, and the tendons of the diaphragm and other respiratory muscles, also have a degree of elasticity that causes them to spring back when muscular contraction ceases. As these structures recoil, the thoracic cage diminishes in size. In accordance with Boyle's law, this raises the intrapulmonary pressure; it peaks at about +3 mmHg and expels air from the lungs (fig. 22.13). Diseases that reduce pulmonary elasticity interfere with expiration, as we will see in the discussion of emphysema.

When inspiration ceases, the phrenic nerves continue to stimulate the diaphragm for a little while longer. This produces a slight braking action that prevents the lungs from recoiling too suddenly, so it makes the transition from inspiration to expiration smoother. In relaxed breathing, inspiration usually lasts about 2 seconds and expiration about 3 seconds.

To exhale more completely than usual—say, in blowing out the candles on your birthday cake—you contract your internal intercostal muscles, which depress the ribs. You also contract the abdominal muscles (internal

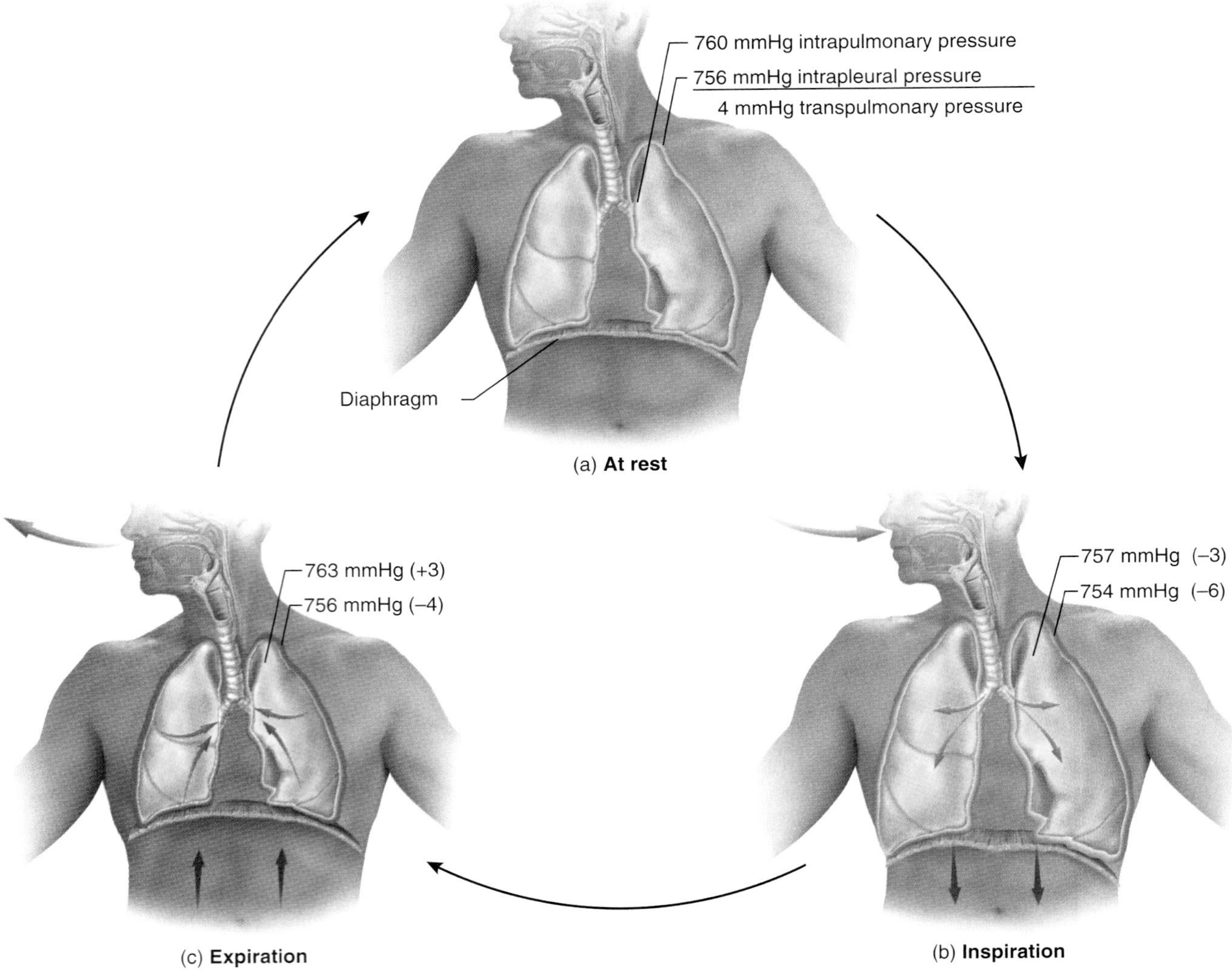

Figure 22.13 The Cycle of Pressure Changes Causing Ventilation of the Lungs. The pressures given here are based on an assumed atmospheric pressure of 760 mmHg (1 atm). Values in parentheses are relative to atmospheric pressure.

and external abdominal obliques, transversus abdominis, and rectus abdominis), which raise the intra-abdominal pressure and force the viscera and diaphragm upward, putting pressure on the thoracic cavity. Intrapulmonary pressure rises as high as 20 to 30 mmHg, causing faster and deeper evacuation of the lungs. Abdominal control of expiration is important in singing and public speaking.

The effect of pulmonary elasticity is evident in a pathological state of pneumothorax and atelectasis. **Pneumothorax** is the presence of air in the pleural cavity. If the thoracic wall is punctured, for example, air is sucked through the wound into the pleural cavity during inspiration and separates the visceral and parietal pleurae. Without the negative intrapleural pressure to keep the lungs inflated, the lungs recoil and collapse. The collapse of a lung or part of a lung is called **atelectasis**[14] (AT-eh-LEC-ta-sis). Atelectasis can also result from airway obstruction—for example, by a lung tumor, aneurysm, swollen lymph node, or aspirated object. Blood absorbs gases from the alveoli distal to the obstruction, and that part of the lung collapses because it cannot be reventilated.

Resistance to Airflow

In discussing blood circulation (p. 756), we noted that flow = pressure/resistance ($F = \Delta P/R$). One factor that affects resistance is **pulmonary compliance**—the distensibility of the lungs, or ease with which they expand. More

14. *atel* = imperfect, incomplete + *ectasis* = extension

exactly, compliance means the change in lung volume relative to a given change in transpulmonary pressure. The lungs normally inflate with ease, but compliance can be reduced by degenerative lung diseases that cause pulmonary fibrosis, such as tuberculosis and black lung disease. In such conditions, the thoracic cage expands normally and transpulmonary pressure falls, but the lungs expand relatively little.

Another factor that governs resistance to airflow is the diameter of the bronchioles. Like arterioles, the large number of bronchioles, their small diameter, and their ability to change diameter make bronchioles the primary means of controlling resistance. Their smooth muscle allows for considerable **bronchoconstriction** and **bronchodilation**—changes in diameter that reduce or increase airflow, respectively. Bronchoconstriction is triggered by airborne irritants, cold air, parasympathetic stimulation, or histamine. Many people have died of extreme bronchoconstriction triggered by asthma and anaphylactic shock. Sympathetic nerves and epinephrine stimulate bronchodilation. Epinephrine inhalants were widely used in the past to halt asthma attacks, but they have been replaced by drugs that produce fewer side effects.

Alveolar Surface Tension

Another factor that resists inspiration and promotes expiration is the surface tension of the water in the alveoli and distal bronchioles. Although the alveoli are relatively dry, they have a thin film of water over the epithelium that is necessary for gas exchange, yet creates a potential problem for pulmonary ventilation. Water molecules are attracted to each other by hydrogen bonds, creating surface tension, as we saw in chapter 3. If you have ever tried to separate two wet microscope slides, you have felt how strong surface tension can be. Such a force draws the walls of the alveoli inward toward the lumen. If it went unchecked, the alveoli would collapse with each expiration and would strongly resist reinflation.

The smaller the diameter of an alveolus, the more crowded the water molecules become, thus increasing the surface tension and the likelihood of alveolar collapse. This is summarized by the **law of Laplace**—*the force drawing the alveolus in on itself is directly proportional to surface tension and inversely proportional to the radius of the alveolus.* If F = force, T = surface tension, and r = radius, this law can be summarized $F = 2T/r$. If surface tension remained the same and r decreased as an alveolus deflated, the force acting on the alveolar wall would increase; the smaller the alveolus became, the more it would tend to collapse.

The solution to this problem takes us back to the great alveolar cells and their surfactant. A *surfactant* is an agent that disrupts the hydrogen bonds of water and reduces surface tension; soaps and detergents are everyday examples. Pulmonary surfactant spreads over the alveolar epithelium and up the alveolar ducts and smallest bronchioles. As these passages contract during expiration, the surfactant molecules are forced closer together; as the local concentration of surfactant increases, it exerts a stronger effect. Therefore, as alveoli shrink during expiration, surface tension decreases to nearly zero; the numerator ($2T$) of the fraction decreases when the denominator (r) does, and thus there is no increase in force (F) favoring alveolar collapse. The importance of this surfactant is especially apparent when it is lacking. Premature infants often have a deficiency of pulmonary surfactant and experience great difficulty breathing (see chapter 29). The resulting *respiratory distress syndrome* is often treated by administering artificial surfactant.

Alveolar Ventilation

Air that actually enters the alveoli becomes available for gas exchange, but not all inhaled air gets that far. About 150 mL of it (typically 1 mL per pound of body weight) fills the conducting division of the airway. Since this air cannot exchange gases with the blood, it is called **dead air,** and the conducting division is called the **anatomic dead space.** In pulmonary diseases, some alveoli may be unable to exchange gases with the blood because they lack blood flow or their pulmonary membrane is thickened by edema. **Physiologic (total) dead space** is the sum of anatomic dead space and any pathological alveolar dead space that may exist. In healthy people, few alveoli are nonfunctional, and the anatomic and physiologic dead spaces are identical.

In a state of relaxation, the bronchioles are constricted by parasympathetic nervous stimulation. This minimizes the dead space so that more of the inhaled air ventilates the alveoli. In a state of arousal, the sympathetic nervous system stimulates bronchodilation, which increases airflow. The increased airflow outweighs the air that is "wasted" by filling the increased dead space.

If a person inhales 500 mL of air and 150 mL of it is dead air, then 350 mL of air ventilates the alveoli. Multiplying this by the respiratory rate gives the **alveolar ventilation rate (AVR)**—for example, 350 mL/breath × 12 breaths/min = 4,200 mL/min. Of all measures of pulmonary ventilation, this one is most directly relevant to the body's ability to get oxygen to the tissues and dispose of carbon dioxide.

Nonrespiratory Air Movements

Breathing serves more purposes than just ventilating the alveoli. It promotes the flow of blood and lymph from abdominal to thoracic vessels, as described in earlier chapters. Variations in pulmonary ventilation also serve the purposes of speaking, expressing emotion (laughing,

Table 22.2 Respiratory Volumes and Capacities for an Average Young Adult Male

Measurement	Typical Value	Definition
Respiratory Volumes		
Tidal volume (TV)	500 mL	Amount of air inhaled or exhaled in one breath during relaxed, quiet breathing
Inspiratory reserve volume (IRV)	3,000 mL	Amount of air in excess of tidal inspiration that can be inhaled with maximum effort
Expiratory reserve volume (ERV)	1,200 mL	Amount of air in excess of tidal expiration that can be exhaled with maximum effort
Residual volume (RV)	1,200 mL	Amount of air remaining in the lungs after maximum expiration; keeps alveoli inflated between breaths and mixes with fresh air on next inspiration
Respiratory Capacities		
Vital capacity (VC)	4,700 mL	Amount of air that can be exhaled with maximum effort after maximum inspiration (ERV + TV + IRV); used to assess strength of thoracic muscles as well as pulmonary function
Inspiratory capacity (IC)	3,500 mL	Maximum amount of air that can be inhaled after a normal tidal expiration (TV + IRV)
Functional residual capacity (FRC)	2,400 mL	Amount of air remaining in the lungs after a normal tidal expiration (RV + ERV)
Total lung capacity (TLC)	5,900 mL	Maximum amount of air the lungs can contain (RV + VC)

crying), yawning, hiccuping, expelling noxious fumes, coughing, sneezing, and expelling abdominal contents. Coughing is triggered by irritants in the lower respiratory tract. To cough, we close the glottis and contract the muscles of expiration, producing high pressure in the lower respiratory tract. We then suddenly open the glottis and release an explosive burst of air at speeds over 900 km/hr (600 mi/hr). This drives mucus and foreign matter toward the pharynx and mouth. Sneezing is triggered by irritants in the nasal cavity. Its mechanism is similar to coughing except that the glottis is continually open, the soft palate and tongue block the flow of air while thoracic pressure builds, and then the uvula (the conical projection of the posterior edge of the soft palate) is depressed to direct part of the airstream through the nose. These actions are coordinated by coughing and sneezing centers in the medulla oblongata.

Taking a deep breath, holding it, and then contracting the abdominal muscles increases the pressure in the abdominal cavity. This process, called the **Valsalva**[15] **maneuver,** is often performed semiconsciously to help expel urine and feces and to aid in childbirth.

Measurements of Ventilation

Ventilation can be measured by having a subject breathe into a device called a **spirometer,**[16] which recaptures the expired breath and records such variables as the rate and depth of breathing, speed of expiration, and rate of oxygen consumption. Four measurements are called **respiratory volumes: tidal volume, inspiratory reserve volume, expiratory reserve volume,** and **residual volume.** Four others, called **respiratory capacities,** are obtained by adding two or more of the respiratory volumes: **vital capacity, inspiratory capacity, functional residual capacity,** and **total lung capacity.** Definitions and representative values for these are given in table 22.2 and figure 22.14.

In general, respiratory volumes and capacities are proportional to body size; consequently, they are generally lower for women than for men. Some values change with age because the lungs become less compliant and the respiratory muscles weaken. Vital capacity declines, residual volume rises, and the lungs are ventilated with less fresh air in a given time. Regular aerobic exercise slows this degeneration by maintaining the strength of the respiratory muscles.

The measurement of respiratory volumes and capacities is important in assessing the severity of a respiratory disease and monitoring improvement or deterioration in a patient's pulmonary function. **Restrictive disorders** of the respiratory system, such as pulmonary fibrosis, stiffen the lungs and thus reduce compliance and vital capacity. **Obstructive disorders** do not reduce respiratory volumes, but they narrow the airway and interfere with airflow; thus, expiration either requires more effort than normal or is less complete than normal. Airflow is measured by having the subject exhale as rapidly as possible into a spirometer and measuring **forced expiratory volume (FEV)**—the percentage of the vital capacity that can be exhaled in a given time interval. A healthy adult should be able to expel 75% to 85% of the vital capacity in 1.0 second (a value called the $FEV_{1.0}$). Significantly lower values may indicate thoracic muscle weakness or obstruction of the airway by mucus, a tumor, or bronchoconstriction (as in asthma).

15. Antonio Maria Valsalva (1666–1723), Italian anatomist
16. *spiro* = breath + *meter* = measuring device

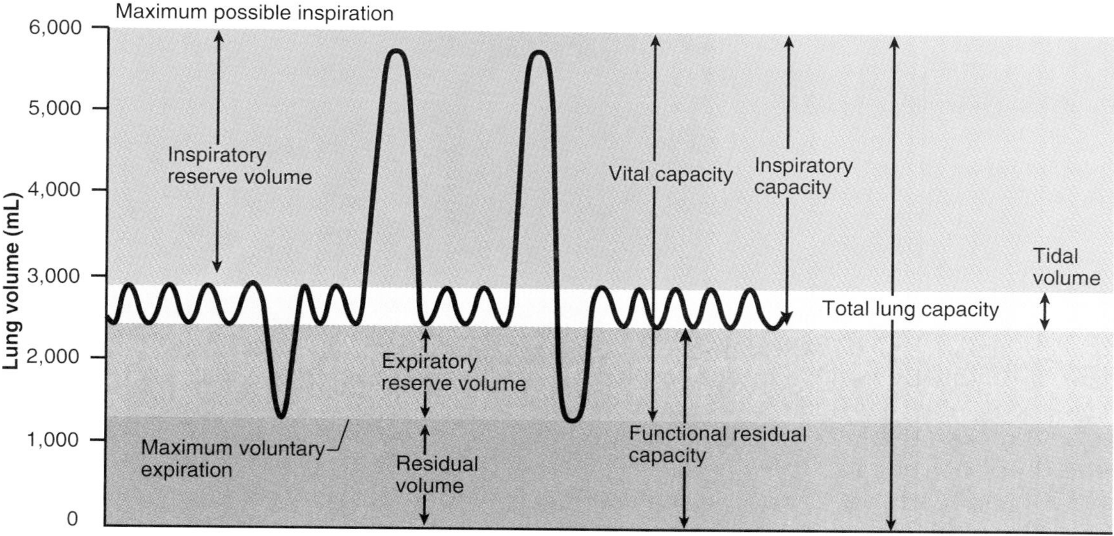

Figure 22.14 **Lung Volumes and Capacities.** The wavy line indicates inspiration when it rises and expiration when it falls. Compare table 22.2.

The amount of air inhaled per minute is called the **minute respiratory volume (MRV).** Its primary significance is that the MRV largely determines the alveolar ventilation rate. MRV can be measured directly with a spirometer or obtained by multiplying tidal volume by respiratory rate. For example, if a person has a tidal volume of 500 mL per breath and a rate of 12 breaths per minute, his or her MRV would be $500 \times 12 = 6{,}000$ mL/min. During heavy exercise, MRV may be as high as 125 to 170 L/min.This is called **maximum voluntary ventilation (MVV),** formerly called *maximum breathing capacity.*

Patterns of Breathing

Some variations in the rhythm of breathing are defined in table 22.3. You should familiarize yourself with these terms before proceeding further in this chapter, as later discussions assume a working knowledge of these terms.

Before You Go On

Answer the following questions to test your understanding of the preceding section:

5. Name the major muscles and nerves involved in inspiration.
6. Relate the action of the respiratory muscles to Boyle's law.
7. Explain the relevance of compliance and elasticity to pulmonary ventilation, and describe some conditions that reduce compliance and elasticity.
8. Explain how pulmonary surfactant and the law of Laplace relate to compliance.
9. Define *vital capacity.* Express it in terms of a formula and define each of the variables.

Table 22.3 Clinical Terminology of Ventilation

Term	Definition
Apnea (AP-nee-uh)	Temporary cessation of breathing (one or more skipped breaths)
Dyspnea[17] *(DISP-nee-uh)*	Labored, gasping breathing; shortness of breath
Eupnea[18] *(yoop-NEE-uh)*	Normal, relaxed, quiet breathing; typically 500 mL/breath, 12 to 15 breaths/min
Hyperpnea (HY-purp-NEE-uh)	Increased rate and depth of breathing in response to exercise, pain, or other conditions
Hyperventilation	Increased pulmonary ventilation in excess of metabolic demand, frequently associated with anxiety; expels CO_2 faster than it is produced, thus lowering the blood CO_2 concentration and raising the blood pH
Hypoventilation	Reduced pulmonary ventilation; leads to an increase in blood CO_2 concentration if ventilation is insufficient to expel CO_2 as fast as it is produced
Kussmaul Respiration	Deep, rapid breathing often induced by acidosis, as in diabetes mellitus
Orthopnea (or-thop-NEE-uh)	Dyspnea that occurs when a person is lying down
Respiratory Arrest	Permanent cessation of breathing (unless there is medical intervention)
Tachypnea (tack-ip-NEE-uh)	Accelerated respiration

17. *dys* = difficult, abnormal, painful
18. *eu* = easy, normal + *pnea* = breathing

Neural Control of Ventilation

Objectives

When you have completed this section, you should be able to

- explain how the brainstem regulates respiration;
- contrast the neural pathways for voluntary and automatic control of the respiratory muscles; and
- describe the stimuli that modify the respiratory rhythm and the pathways that these signals take to the brainstem.

The heartbeat and breathing are the two most conspicuously rhythmic processes in the body. The heart has an internal pacemaker and goes on beating even if all nerves to it are severed. Breathing, by contrast, depends on repetitive stimuli from the brain. There are two reasons for this: (1) Skeletal muscles do not contract without nervous stimulation. (2) Breathing involves the coordinated action of multiple muscles and thus requires a central coordinating mechanism to ensure that they all work together.

This section describes the neural mechanisms that regulate pulmonary ventilation. Neurons in the medulla oblongata and pons provide automatic control of unconscious breathing, whereas neurons in the motor cortex of the cerebrum provide voluntary control.

Control Centers in the Medulla Oblongata

The medulla oblongata contains **inspiratory (I) neurons,** which fire during inspiration, and **expiratory (E) neurons,** which fire during forced expiration (but not during eupnea). Fibers from these neurons travel down the spinal cord and synapse with lower motor neurons in the cervical to thoracic regions. From here, nerve fibers travel in the phrenic and intercostal nerves to the diaphragm and intercostal muscles, respectively. No pacemaker neurons have been found that are analogous to the autorhythmic cells of the heart, and the exact mechanism for setting the rhythm of respiration remains unknown despite intensive research.

The medulla has two respiratory nuclei (fig. 22.15). One of them, called the **inspiratory center,** or **dorsal respiratory group (DRG),** is composed primarily of I neurons, which stimulate the muscles of inspiration. The more frequently they fire, the more motor units are recruited and the more deeply you inhale. If they fire longer than usual, each breath is prolonged and the respiratory rate is slower. When they stop firing, elastic recoil of the lungs and thoracic cage produces passive expiration.

The other nucleus is the **expiratory center,** or **ventral respiratory group (VRG).** It has I neurons in its midregion and E neurons at its anterior and posterior ends. It is not

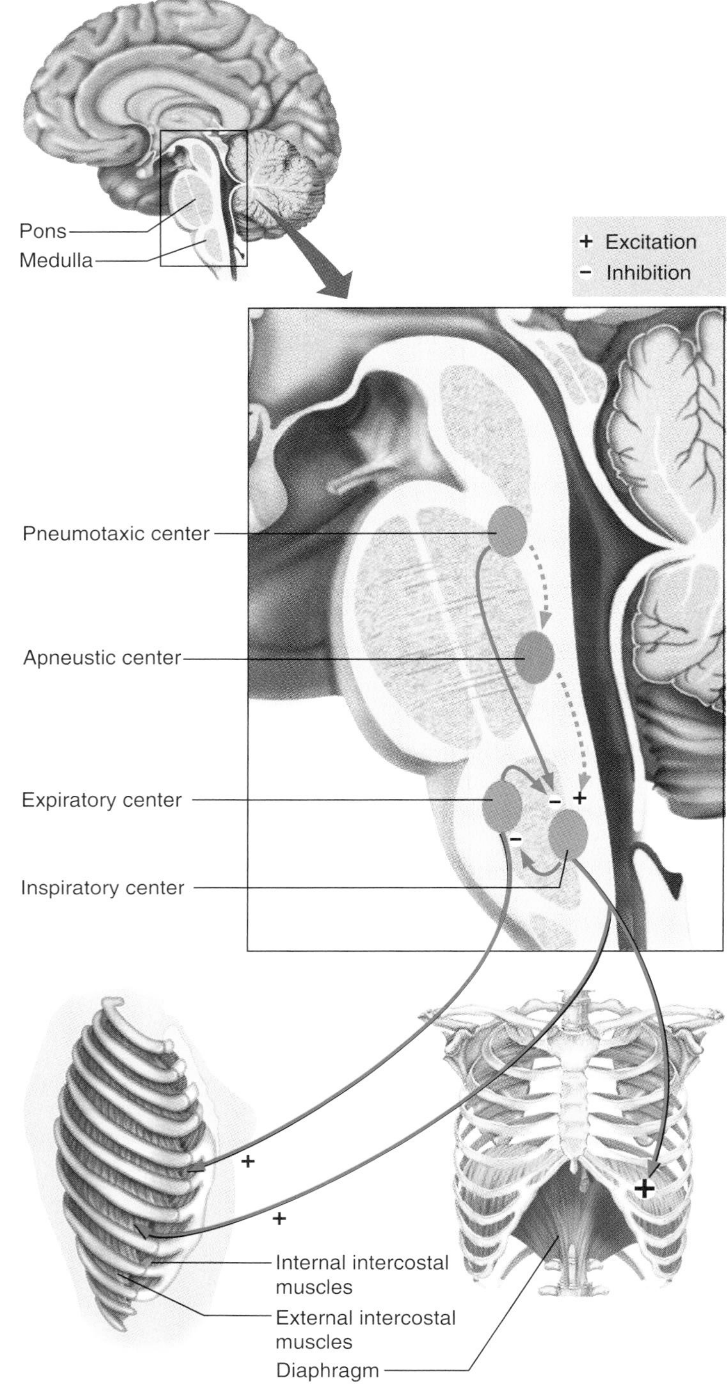

Figure 22.15 Respiratory Control Centers. Functions of the apneustic center are hypothetical and its connections are therefore indicated by *broken lines.* As indicated by the *plus* and *minus signs,* the apneustic center stimulates the inspiratory center, while the pneumotaxic center inhibits it. The inspiratory and expiratory centers inhibit each other.

Clinical Insight 22.2

Ondine's Curse

In German legend, there was a water nymph named Ondine who took a mortal lover. When he was unfaithful to her, the king of the nymphs put a curse on him that took away his automatic physiological functions. Consequently, he had to remember to take each breath, and he could not go to sleep or he would die of suffocation—which, as exhaustion overtook him, was indeed his fate.

Some people suffer a disorder called *Ondine's curse,* in which the automatic respiratory functions are disabled—usually as a result of brainstem damage from poliomyelitis or as an accident of spinal cord surgery. Victims of Ondine's curse must remember to take each breath and cannot go to sleep without the aid of a mechanical ventilator.

involved in eupnea, but its E neurons inhibit the inspiratory center when deeper expiration is needed. Conversely, the inspiratory center inhibits the expiratory center when an unusually deep inspiration is needed.

Control Centers in the Pons

The pons regulates ventilation by means of a *pneumotaxic center* in the upper pons and possibly an *apneustic center* in the lower pons. The **pneumotaxic center** sends a continual stream of inhibitory impulses to the inspiratory center of the medulla. When impulse frequency rises, inspiration lasts as little as 0.5 second and the breathing thus becomes faster and shallower. Conversely, when impulse frequency declines, breathing is slower and deeper, with inspiration lasting as long as 5 seconds.

The role of the **apneustic** (ap-NEW-stic) **center** is still hypothetical. It produces no noticeable effect except when experimenters sever the vagus nerves and the connection from the pneumotaxic center to the apneustic center. Then, signals from the apneustic center to the inspiratory center greatly prolong inspiration and cause exaggerated filling of the lungs interrupted only by occasional, brief expirations. The normal function of the apneustic center remains unclear.

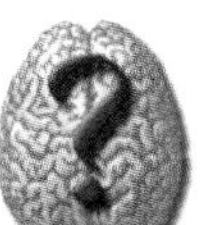

Think About It

Do you think the fibers from the pneumotaxic center produce EPSPs or IPSPs at their synapses in the inspiratory center? Explain.

Afferent Connections to the Brainstem

The brainstem respiratory centers receive input from the limbic system, hypothalamus, chemoreceptors, and lungs themselves. Input from the limbic system and hypothalamus allows pain and emotions to affect respiration—for example, in gasping, crying, and laughing. Anxiety often triggers an uncontrollable bout of hyperventilation. This expels CO_2 from the body faster than it is produced. As blood CO_2 levels drop, the pH rises and causes the cerebral arteries to constrict. The brain thus receives less perfusion, and dizziness and fainting may result. Hyperventilation can be brought under control by having a person rebreathe the expired CO_2 from a paper bag.

Chemoreceptors in the brainstem and arteries monitor blood pH, CO_2, and O_2 levels. They transmit signals to the respiratory centers that enable pulmonary ventilation to be adjusted to keep these variables within homeostatic limits. Chemoreceptors are discussed more extensively later in the chapter.

The vagus nerves transmit sensory signals from the respiratory system to the inspiratory center. Irritants in the airway, such as smoke, dust, noxious fumes, or mucus, stimulate vagal afferent fibers. The medulla then returns signals that result in bronchoconstriction or coughing. Stretch receptors in the bronchial tree and visceral pleura monitor inflation of the lungs. Excessive inflation triggers the **inflation (Hering–Breuer[19]) reflex,** a protective somatic reflex that strongly inhibits the I neurons and stops inspiration. In infants, this may be a normal mechanism of transition from inspiration to expiration, but after infancy it is activated only by extreme stretching of the lungs.

Voluntary Control

Although breathing usually occurs automatically, without our conscious attention, we obviously can hold our breath, take a deep breath, and control ventilation while speaking or singing. This control originates in the motor cortex of the frontal lobe of the cerebrum, which sends impulses down the corticospinal tracts to the respiratory neurons in the spinal cord, bypassing the brainstem respiratory centers.

There are limits to voluntary control. Temperamental children may threaten to hold their breath until they die, but it is impossible to do so. Holding one's breath lowers

19. Heinrich Ewald Hering (1866–1948), German physiologist; Josef Breuer (1842–1925), Austrian physician

the O_2 level and raises the CO_2 level of the blood until a *breaking point* is reached where automatic controls override one's will. This forces a person to resume breathing even if he or she has lost consciousness.

Before You Go On

Answer the following questions to test your understanding of the preceding section:

10. Which of the brainstem respiratory nuclei is (are) indispensable to respiration? What do the other nuclei do?
11. Where do voluntary respiratory commands originate? What pathways do they take to the respiratory muscles?

Gas Exchange and Transport

Objectives

When you have completed this section, you should be able to

- define *partial pressure* and discuss its relationship to a gas mixture such as air;
- contrast the composition of inspired and expired air;
- discuss how partial pressure affects gas transport by the blood;
- describe the mechanisms of transporting O_2 and CO_2;
- describe the factors that govern gas exchange in the lungs and systemic capillaries; and
- explain how gas exchange is adjusted to the metabolic needs of different tissues.

We now consider the stages in which oxygen is obtained from inspired air and delivered to the tissues, while carbon dioxide is removed from the tissues and released into the expired air. First, however, it is necessary to understand the composition of air and the behavior of gases in contact with water.

Composition of Air

Air is a mixture of gases, each of which contributes a share, called its **partial pressure,** to the total atmospheric pressure (table 22.4). Partial pressure is abbreviated P followed by the formula of the gas. The partial pressure of nitrogen is P_{N_2}, for example. Nitrogen constitutes about 78.6% of the atmosphere; thus at 1 atm of pressure, $P_{N_2} = 78.6\% \times 760$ mmHg = 597 mmHg. **Dalton's law** states that *the total pressure of a gas mixture is the sum of the partial pressures of the individual gases.* That is, $P_{N_2} + P_{O_2} + P_{H_2O} + P_{CO_2} = 597.0 + 159.0 + 3.7 + 0.3 = 760.0$ mmHg. These partial pressures are important because they determine the rate of diffusion of a gas and therefore strongly affect the rate of gas exchange between the blood and alveolar air.

Alveolar air can be sampled with an apparatus that collects the last 10 mL of expired air. Its gaseous makeup differs from that of the atmosphere because of three influences: (1) the airway humidifies it, (2) the air exchanges O_2 and CO_2 with the blood, and (3) freshly inspired air mixes with residual air left from the previous respiratory cycle. These factors produce the composition shown in table 22.4.

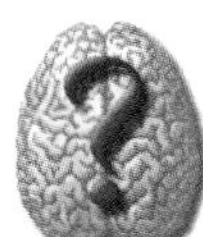

Think About It

Expired air, considered as a whole (not just the last 10 mL), contains about 116 mmHg O_2 and 32 mmHg CO_2. Why do these values differ from the values for alveolar air?

The Air-Water Interface

When air and water are in contact with each other, as in the pulmonary alveolus, gases diffuse down their concentration gradients until the partial pressure of each gas in the air is equal to its partial pressure in the water. If a gas is more abundant in the water than in the air, it diffuses into the air; the smell of chlorine near a swimming pool is evidence of this. If it is more abundant in the air, it diffuses into the water.

Henry's law states that *at the air-water interface, for a given temperature, the amount of gas that dissolves in the water is determined by its solubility in water and its partial pressure in the air* (fig. 22.16). Thus, the greater the P_{O_2} in the alveolar air, the more O_2 the blood picks up. And, since the blood arriving at an alveolus has a higher P_{CO_2} than air, the blood releases CO_2 into the air. At the alveolus, the blood is said to *unload* CO_2 and *load* O_2. Each gas in a mixture behaves independently; the diffusion of one gas does not influence the diffusion of another.

Alveolar Gas Exchange

Alveolar gas exchange is the process of O_2 loading and CO_2 unloading in the lungs. Since both processes depend on erythrocytes (RBCs), their efficiency depends on how

Table 22.4 Composition of Inspired (atmospheric) and Alveolar Air

Gas	Inspired Air*		Alveolar Air	
N_2	78.62%	597.0 mmHg	74.9%	569.0 mmHg
O_2	20.84%	159.0 mmHg	13.6%	104.0 mmHg
H_2O	0.50%	3.7 mmHg	6.2%	47.0 mmHg
CO_2	0.04%	0.3 mmHg	5.3%	40.0 mmHg
Total	100.00%	760.0 mmHg	100.0%	760.0 mmHg

*Typical values for a cool clear day; varies with temperature and humidity. Other gases present in small amounts are disregarded.

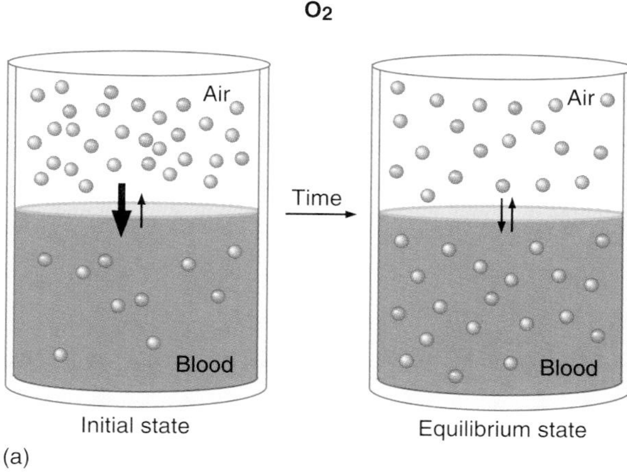

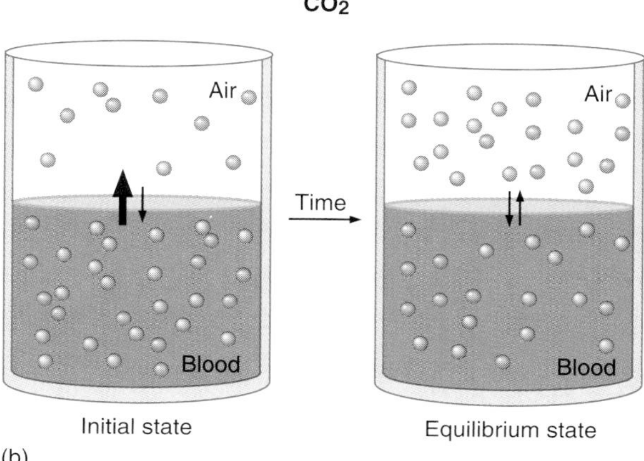

Figure 22.16 Henry's Law and Its Relationship to Alveolar Gas Exchange. (*a*) The P_{O_2} of alveolar air is initially higher than the P_{O_2} of the blood arriving at an alveolus. Oxygen diffuses into the blood until the two are in equilibrium. (*b*) The P_{CO_2} of the arriving blood is initially higher than the P_{CO_2} of alveolar air. Carbon dioxide diffuses into the alveolus until the two are in equilibrium. It takes about 0.25 second for both gases to reach equilibrium.

long an RBC spends in an alveolar capillary compared to how long it takes for O_2 and CO_2 to reach equilibrium concentrations in the capillary blood. It takes only 0.25 second for the gases to equilibrate, while at rest, it takes an RBC about 0.75 second to pass through an alveolar capillary. Even in vigorous exercise when the blood is flowing faster, the transit time for an RBC is at least 0.3 second. Therefore, even at the fastest blood flow, an RBC spends enough time in a capillary to load as much O_2 and unload as much CO_2 as it possibly can.

The following factors especially affect the efficiency of alveolar gas exchange:

- **Concentration gradients of the gases.** The P_{O_2} is about 104 mmHg in the alveolar air and 40 mmHg in the blood arriving at an alveolus. Oxygen therefore diffuses from the air into the blood, where it reaches a P_{O_2} of 104 mmHg. Before the blood leaves the lung, however, this drops to about 95 mmHg because blood in the pulmonary veins receives some oxygen-poor blood from the bronchial veins by way of anastomoses. The P_{CO_2} is about 46 mmHg in the blood arriving at the alveolus and 40 mmHg in the alveolar air. Carbon dioxide therefore diffuses from the blood to the alveoli. These changes are summarized here and at the top of figure 22.17:

Blood Entering Lungs		Blood Leaving Lungs	
P_{O_2}	40 mmHg	P_{O_2}	95 mmHg
P_{CO_2}	46 mmHg	P_{CO_2}	40 mmHg

These gradients differ under special circumstances such as high altitude and *hyperbaric oxygen therapy* (treatment with oxygen at greater than 1 atm of pressure) (fig. 22.18). At high altitudes, the partial pressures of all atmospheric gases are lower. Atmospheric P_{O_2}, for example, is 159 mmHg at sea level and 110 mmHg at 3,000 m (10,000 ft). The O_2 gradient from air to blood is proportionately less, and as we can predict from Henry's law, less O_2 diffuses into the blood. In a hyperbaric oxygen chamber, by contrast, a patient is exposed to 3 to 4 atm of oxygen to treat such conditions as gangrene (to kill anaerobic bacteria) and carbon monoxide poisoning (to displace the carbon monoxide from hemoglobin). The P_{O_2} ranges between 2,300 and 3,000 mmHg. Thus, there is a very steep gradient of P_{O_2} from alveolus to blood and diffusion into the blood is accelerated.

- **Solubility of the gases.** Gases differ in their ability to dissolve in water. Carbon dioxide is about 20 times as soluble as oxygen, and oxygen is about twice as soluble as nitrogen. Even though the concentration gradient of O_2 is much greater than that of CO_2 across the respiratory membrane, equal amounts of the two gases are exchanged because CO_2 is so much more soluble and diffuses more rapidly.

- **Membrane thickness.** The respiratory membrane between the blood and alveolar air is only 0.5 μm thick in most places—much less than the 7 to 8 μm diameter of a single RBC. Thus, it presents little obstacle to diffusion (fig. 22.19*a*). In such heart conditions as left ventricular failure, however, blood pressure backs up into the lungs and promotes capillary filtration into the connective tissues, causing the respiratory membranes to become edematous and thickened (fig. 22.19*b*). The gases have farther to travel between blood and air and cannot equilibrate fast enough to keep pace with blood flow. Under these circumstances, blood leaving the lungs has an unusually high P_{CO_2} and low P_{O_2}.

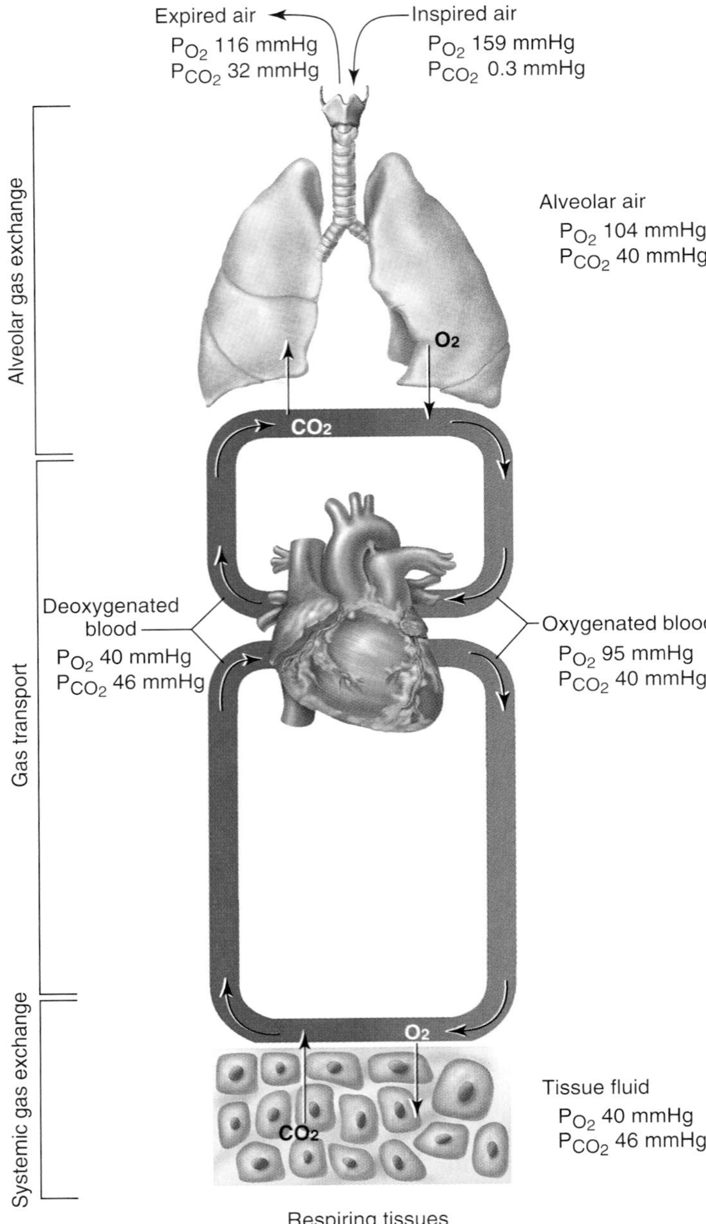

Figure 22.17 **Changes in P_{O_2} and P_{CO_2} Along the Circulatory Route.** Trace the partial pressure of oxygen from inspired air to expired air and explain each change in P_{O_2} along the way. Do the same for P_{CO_2}.

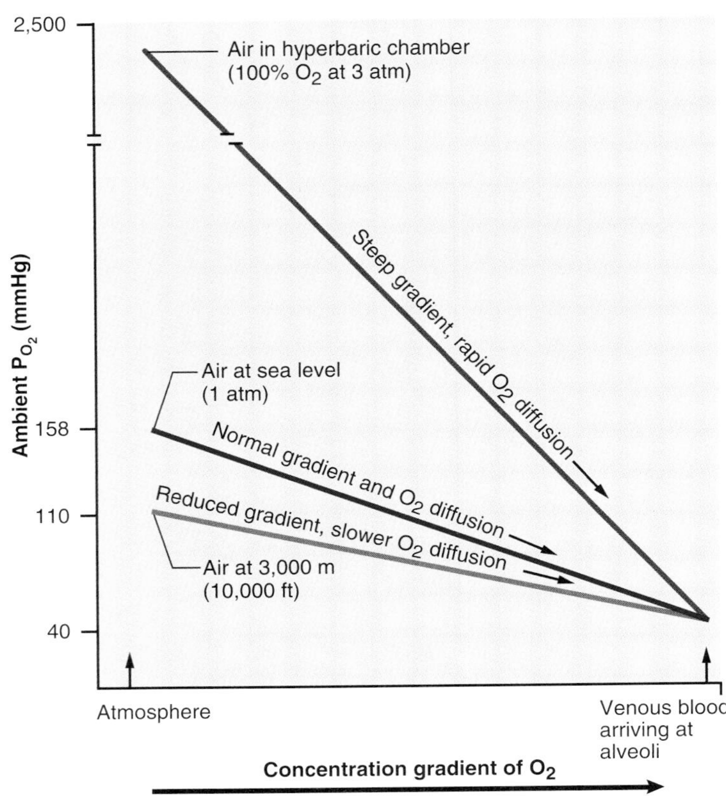

Figure 22.18 **Alveolar Oxygen Loading in Relationship to Concentration Gradient.** The rate of loading depends on the steepness of the gradient from alveolar air to the venous blood arriving at the alveolar capillaries. Compared to the oxygen gradient at sea level (*blue line*), the gradient is less steep at high altitude (*red line*) because the P_{O_2} of the atmosphere is lower. Thus oxygen loading of the pulmonary blood is slower. In a hyperbaric chamber with 100% oxygen, the gradient from air to blood is very steep (*green line*) and oxygen loading is correspondingly rapid. This is an illustration of Henry's law and has important effects in diving, aviation, mountain climbing, and oxygen therapy.

- **Membrane area.** In good health, each lung has about 70 m^2 of respiratory membrane available for gas exchange. Since the alveolar capillaries contain a total of only 100 mL of blood at any one time, this blood is spread very thinly. (Imagine trying to evenly paint an area the size of a tennis court with only 100 mL of paint!) Several pulmonary diseases, however, decrease the alveolar surface area and thus lead to low blood P_{O_2}—for example, emphysema (fig. 22.19*c*), lung cancer, and tuberculosis.
- **Ventilation-perfusion coupling.** Gas exchange not only requires good ventilation of the alveolus but also good perfusion of its capillaries. As a whole, the lungs have a *ventilation-perfusion ratio* of about 0.8—a flow of 4.2 L of air and 5.5 L of blood per minute (at rest). The ratio is somewhat higher in the apex of the lung and lower in the base because more blood is drawn toward the base by gravity. **Ventilation-perfusion coupling** is the ability to match ventilation and perfusion to each other (fig. 22.20). If part of a lung is poorly ventilated because of tissue destruction or airway obstruction, there is little point in directing much blood there. This

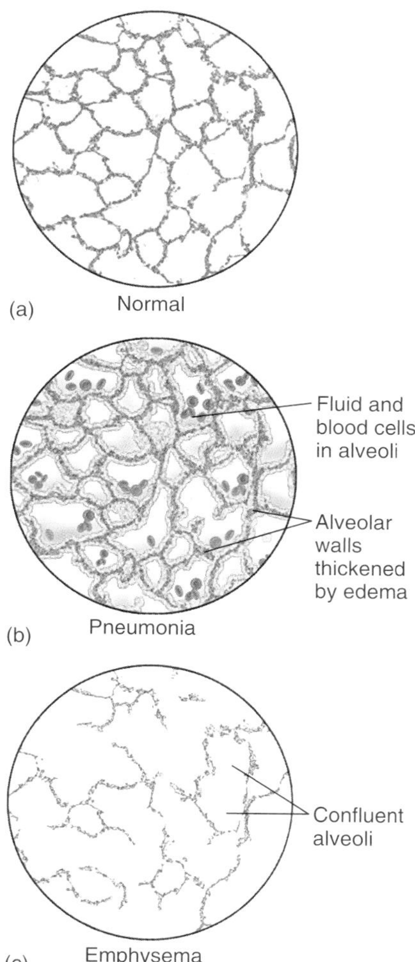

Figure 22.19 Pulmonary Alveoli in Health and Disease. (*a*) In a healthy lung, the alveoli are small and have thin respiratory membranes. (*b*) In pneumonia, the respiratory membranes (alveolar walls) are thick with edema, and the alveoli contain fluid and blood cells. (*c*) In emphysema, alveolar membranes break down and neighboring alveoli join to form larger, fewer alveoli with less total surface area.

blood would leave the lung carrying less oxygen than it should. But poor ventilation causes local constriction of the pulmonary arteries, reducing blood flow to that area and redirecting this blood to better ventilated alveoli. Good ventilation, by contrast, dilates the arteries and increases perfusion so that most blood is directed to regions of the lung where it can pick up the most oxygen. This is opposite from the reactions of systemic arteries, where hypoxia causes vasodilation so that blood flow to a tissue will increase and reverse the hypoxia. Ventilation is also adjustable. Poor ventilation causes local CO_2 accumulation, which stimulates local bronchodilation and improves airflow. Low P_{CO_2} causes local bronchoconstriction.

Gas Transport

Gas transport is the process of carrying gases from the alveoli to the systemic tissues and vice versa. This section explains how the blood loads and transports oxygen and carbon dioxide.

Oxygen

The concentration of oxygen in arterial blood, by volume, is about 20 mL/dL. About 98.5% of this is bound to hemoglobin and 1.5% is dissolved in the blood plasma. Hemoglobin consists of four protein (globin) chains, each with one heme group (see fig. 18.19, p. 691). Each heme group can bind 1 O_2 to the ferrous ion at its center; thus, one hemoglobin molecule can carry up to 4 O_2. If even one molecule of O_2 is bound to hemoglobin, the compound is called **oxyhemoglobin (HbO_2),** whereas hemoglobin with no oxygen bound to it is **deoxyhemoglobin (HHb).** When hemoglobin is 100% saturated, every molecule of it carries 4 O_2; if it is 75% saturated, there is an average of 3 O_2 per hemoglobin molecule; if it is 50% saturated, there is an average of 2 O_2 per hemoglobin; and so forth. The poisonous effect of carbon monoxide stems from its competition for the O_2 binding site (see clinical insight 22.3).

The relationship between hemoglobin saturation and P_{O_2} is shown by an *oxyhemoglobin dissociation curve* (fig. 22.21). As you can see, it is not a simple linear relationship. At low P_{O_2}, the curve rises slowly; then there is a rapid increase in oxygen loading as P_{O_2} rises further; finally, at high P_{O_2}, the curve levels off as the hemoglobin approaches 100% saturation. This reflects the way hemoglobin loads oxygen. When the first heme group binds a molecule of O_2, hemoglobin changes shape in a way that facilitates uptake of the second O_2 by another heme group. This, in turn, promotes the uptake of the third and then the fourth O_2—hence the rapidly rising midportion of the curve.

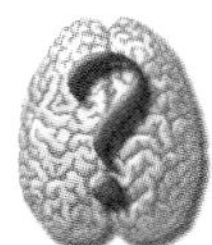

Think About It

Is oxygen loading a positive feedback process or a negative feedback process? Explain.

Carbon Dioxide

Carbon dioxide is transported in three forms—as carbonic acid, carbamino compounds, and dissolved gas:

1. About 90% of the CO_2 is hydrated (reacts with water) to form **carbonic acid,** which then dissociates into bicarbonate and hydrogen ions:

$$CO_2 + H_2O \rightarrow H_2CO_3 \rightarrow HCO_3^- + H^+$$

More will be said about this reaction shortly.

Clinical Insight 22.3

Carbon Monoxide Poisoning

The lethal effect of carbon monoxide (CO) is well known. This colorless, odorless gas occurs in cigarette smoke, engine exhaust, and fumes from furnaces and space heaters. It binds to the ferrous ion of hemoglobin to form *carboxyhemoglobin (HbCO)*. Thus, it competes with oxygen for the same binding site. Not only that, but it binds 210 times as tightly as oxygen. Thus, CO tends to tie up hemoglobin for a long time. Less than 1.5% of the hemoglobin is occupied by carbon monoxide in most nonsmokers, but this figure rises to as much as 3% in residents of heavily polluted cities and 10% in heavy smokers. An atmospheric concentration of 0.1% CO, as in a closed garage, is enough to bind 50% of a person's hemoglobin, and an atmospheric concentration of 0.2% is quickly lethal.

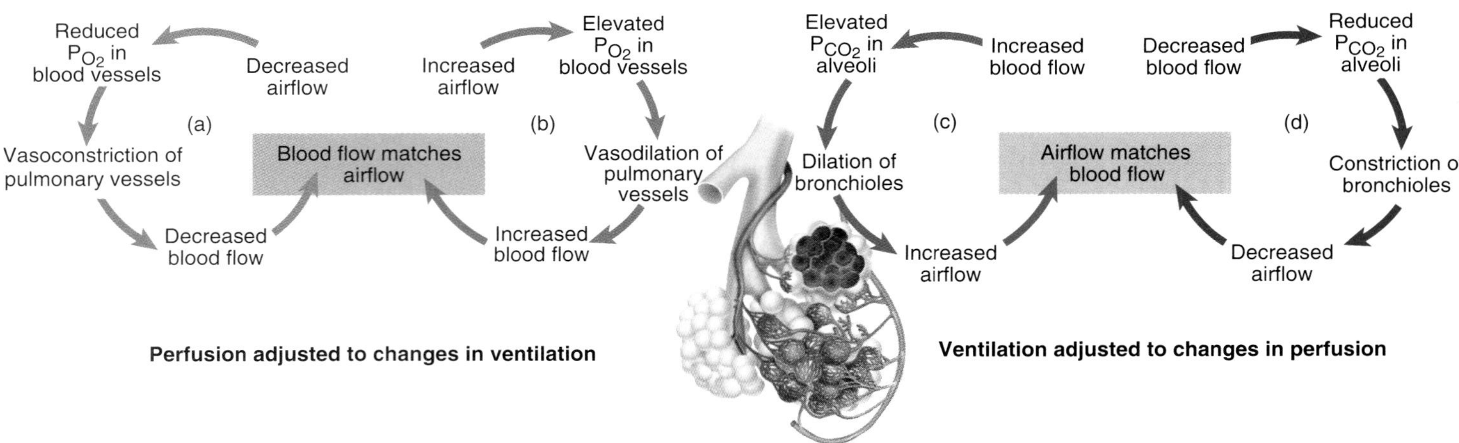

Figure 22.20 **Negative Feedback Loops in the Adjustment of Ventilation-Perfusion Ratio.** (*a*) Effect of reduced ventilation on perfusion. (*b*) Effect of increased ventilation on perfusion. (*c*) Effect of increased perfusion on ventilation. (*d*) Effect of reduced perfusion on ventilation.

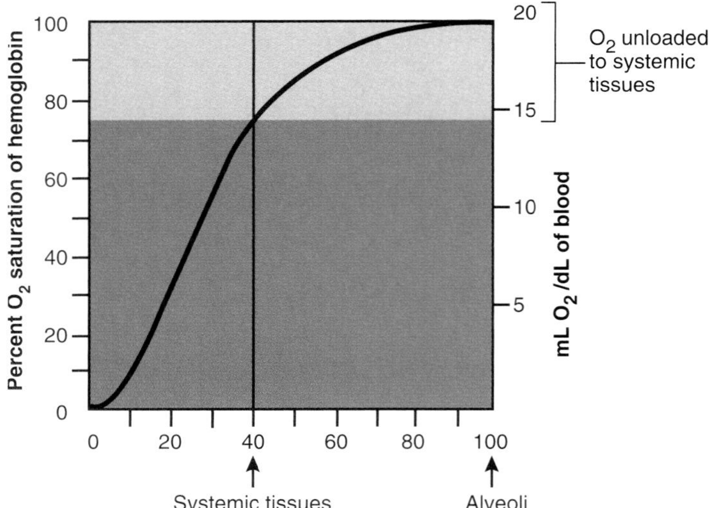

Figure 22.21 **The Oxyhemoglobin Dissociation Curve.** This curve shows the relative amount of hemoglobin that is saturated with oxygen (*y*-axis) as a function of ambient (surrounding) oxygen concentration (*x*-axis). As it passes through the alveolar capillaries where the P_{O_2} is high, hemoglobin becomes saturated with oxygen. As it passes through the systemic capillaries where the P_{O_2} is low, it typically gives up about 22% of its oxygen (*color bar at top of graph*).

2. About 5% binds to the amino groups of plasma proteins and hemoglobin forming **carbamino compounds**—chiefly, **carbaminohemoglobin ($HbCO_2$)**. The reaction with hemoglobin can be symbolized $Hb + CO_2 \rightarrow HbCO_2$. Carbon dioxide does not compete with oxygen because CO_2 and O_2 bind to different sites on the hemoglobin molecule—oxygen to the heme moiety and CO_2 to the polypeptide chains. Hemoglobin can therefore transport both O_2 and CO_2 simultaneously. As we will see, however, each gas somewhat inhibits transport of the other.
3. The remaining 5% of the CO_2 is carried in the blood as dissolved gas, like the CO_2 in soda pop.

The relative amounts of CO_2 exchanged between the blood and alveolar air differ from the percentages just given. About 70% of the *exchanged* CO_2 comes from carbonic acid, 23% from carbamino compounds, and 7% from the dissolved gas. That is, CO_2 has a stronger tendency to remain in the blood in the form of bicarbonate ions than as carbamino compounds or dissolved CO_2, so the last two sources contribute "more than their share" of the CO_2 that diffuses from the blood into the alveoli.

Systemic Gas Exchange

Systemic gas exchange is the unloading of O_2 and loading of CO_2 at the systemic capillaries (see fig. 22.17, *bottom*, and fig. 22.22).

Carbon Dioxide Loading

Aerobic respiration produces a molecule of CO_2 for every molecule of O_2 it consumes. The tissue fluid therefore contains a relatively high P_{CO_2} and there is typically a CO_2 gradient of 46 → 40 mmHg from tissue fluid to blood. Consequently, CO_2 diffuses into the bloodstream, where it is carried in the three forms noted (fig. 22.23). Most of it reacts with water to produce bicarbonate (HCO_3^-) and hydrogen (H^+) ions. This reaction occurs slowly in the blood plasma but much faster in the RBCs, where it is catalyzed by the enzyme *carbonic anhydrase*. In an exchange called the **chloride shift**, most of the HCO_3^- diffuses out of the RBCs in exchange for chloride ions (Cl^-) diffusing in. Most of the H^+ binds to hemoglobin or oxyhemoglobin, which thus buffers the intracellular pH.

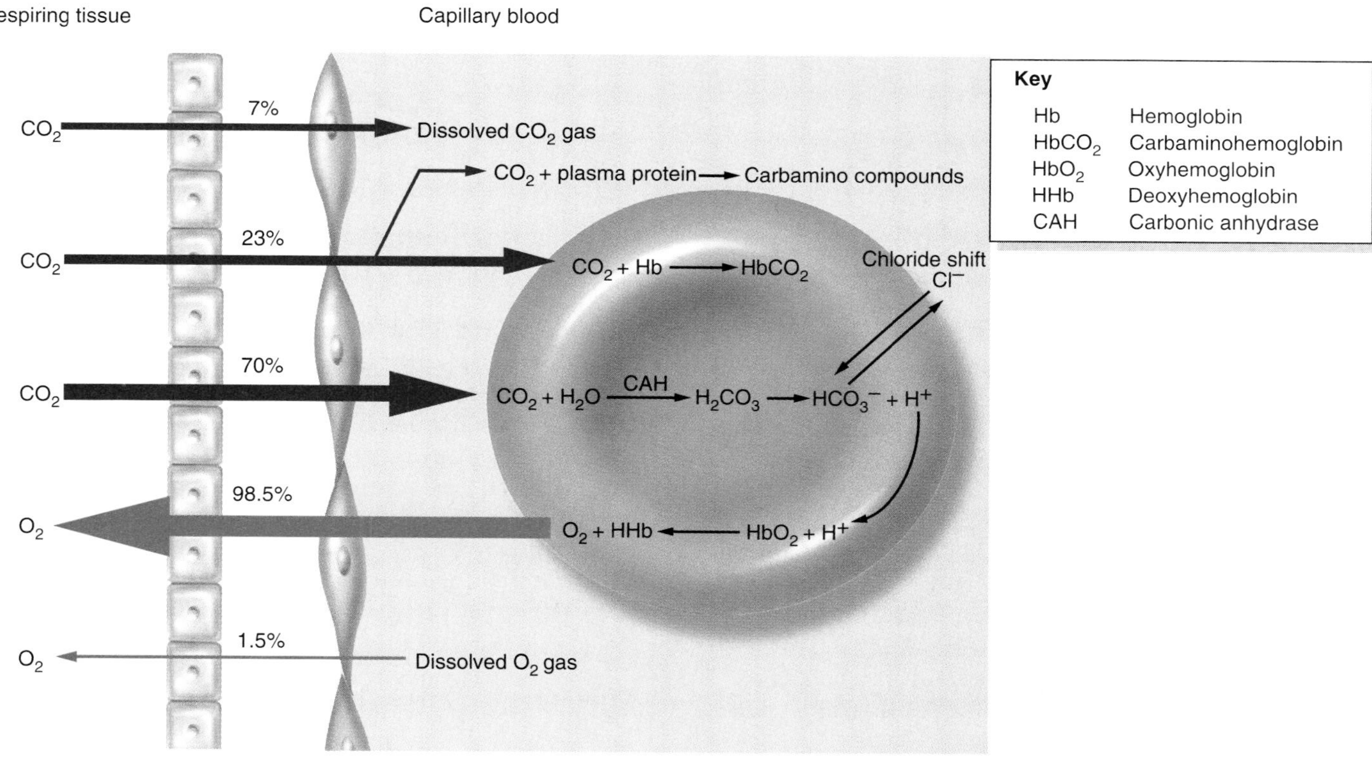

Figure 22.22 Systemic Gas Exchange. *Blue arrows* show the three mechanisms of CO_2 loading and transport; their thickness represents the relative amounts of CO_2 transported in each of the three forms. *Red arrows* show the two mechanisms of O_2 unloading; their thickness indicates the relative amounts unloaded by each mechanism. Note that CO_2 loading releases hydrogen ions in the erythrocyte, and hydrogen ions promote O_2 unloading.

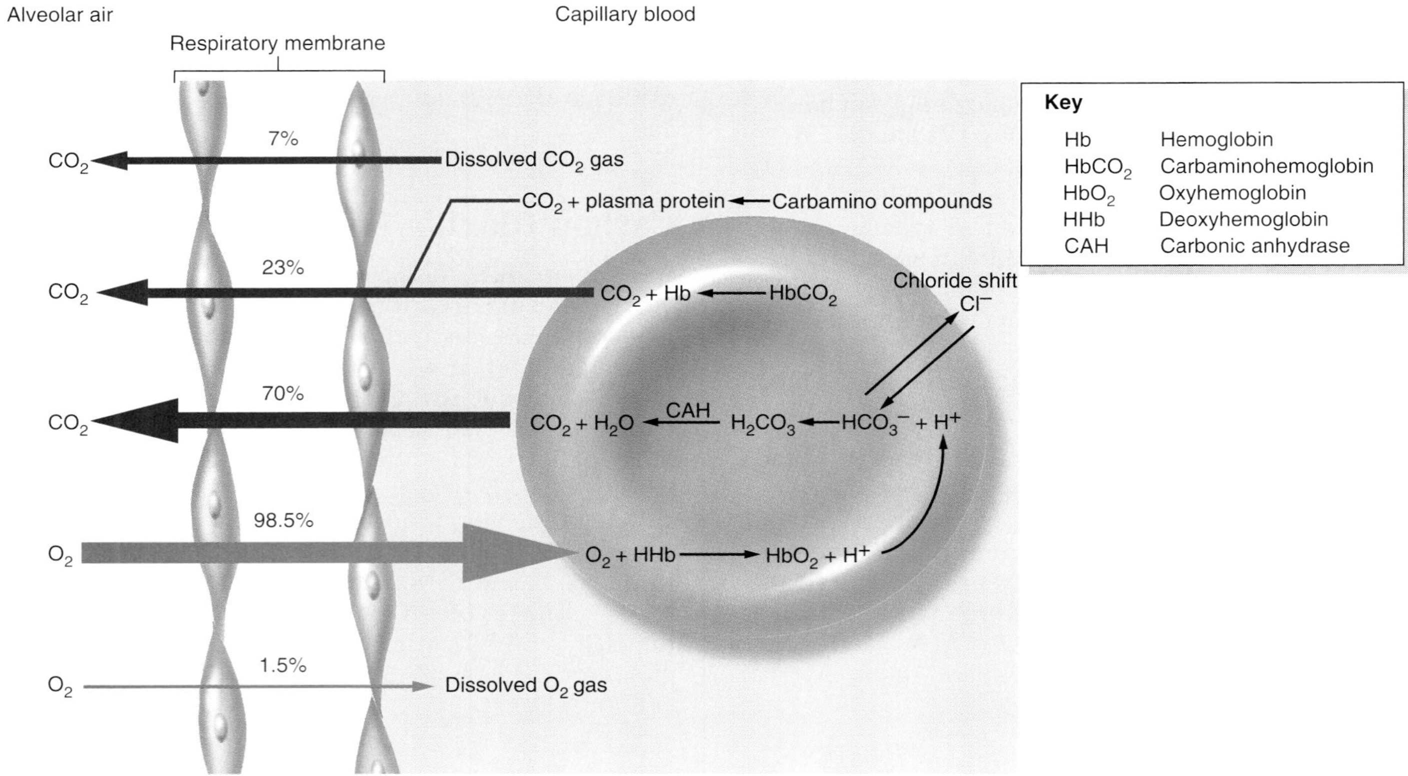

Figure 22.23 Alveolar Gas Exchange. In what fundamental way does this differ from the preceding figure? Following alveolar gas exchange, will the blood contain a higher or lower concentration of bicarbonate ions than it did before?

Oxygen Unloading

When H^+ binds to oxyhemoglobin (HbO_2), it reduces the affinity of hemoglobin for O_2 and tends to make hemoglobin release it. Oxygen consumption by respiring tissues keeps the P_{O_2} of tissue fluid relatively low, and so there is typically a concentration gradient of 95 → 40 mmHg of oxygen from the arterial blood to the tissue fluid. Thus, the liberated oxygen—along with some that was carried as dissolved gas in the plasma—diffuses from the blood into the tissue fluid.

As blood arrives at the systemic capillaries, its oxygen concentration is about 20 mL/dL and the hemoglobin is about 97% saturated. As it leaves the capillaries of a typical resting tissue, its oxygen concentration is about 15.6 mL/dL and the hemoglobin is about 75% saturated. Thus, it has given up 4.4 mL/dL—about 22% of its oxygen load. This fraction is called the **utilization coefficient.** The oxygen remaining in the blood after it passes through the capillary bed provides a **venous reserve** of oxygen, which can sustain life for 4 to 5 minutes even in the event of respiratory arrest. At rest, the circulatory system releases oxygen to the tissues at an overall rate of about 250 mL/min.

Alveolar Gas Exchange Revisited

The processes illustrated in figure 22.22 make it easier to understand alveolar exchange more fully. As shown in figure 22.23, the reactions that occur in the lungs are essentially the reverse of systemic gas exchange. As hemoglobin loads oxygen, its affinity for H^+ declines. Hydrogen ions dissociate from the hemoglobin and bind with bicarbonate (HCO_3^-) ions diffusing from the plasma into the RBCs. Chloride ions diffuse back out of the RBC (a reverse chloride shift). The reaction of H^+ and HCO_3^- reverses the hydration reaction and generates free CO_2. This diffuses into the alveolus to be exhaled—as does the CO_2 released from carbaminohemoglobin and CO_2 gas that was dissolved in the plasma.

Adjustment to the Metabolic Needs of Individual Tissues

Hemoglobin does not unload the same amount of oxygen to all tissues. Some tissues need more and some less, depending on their state of activity. Hemoglobin responds to such variations and unloads more oxygen to the tissues

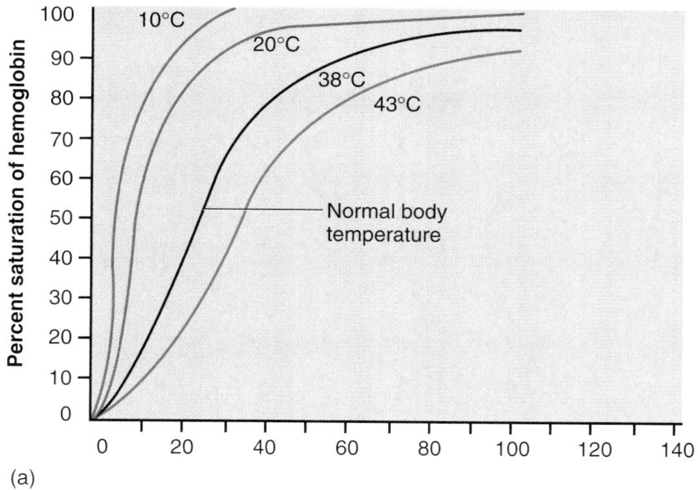

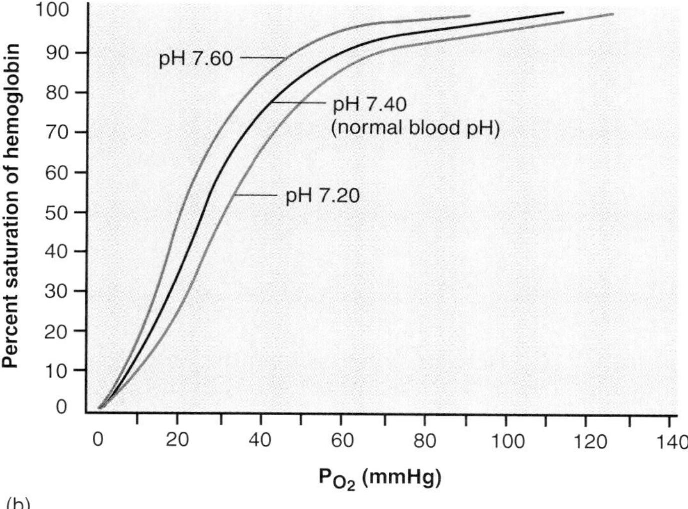

Figure 22.24 Effects of Temperature and pH on Oxyhemoglobin Dissociation. (*a*) For a given P_{O_2}, hemoglobin unloads more oxygen at higher temperatures. (*b*) For a given P_{O_2}, hemoglobin unloads more oxygen at lower pH (the Bohr effect). Both mechanisms cause hemoglobin to release more oxygen to tissues with higher metabolic rates.

that need it most. In exercising skeletal muscles, for example, the utilization coefficient may be as high as 80%. Four factors adjust the rate of oxygen unloading to the metabolic rates of different tissues:

1. **Ambient P_{O_2}.** Since an active tissue consumes oxygen rapidly, the P_{O_2} of its tissue fluid remains low. From the oxyhemoglobin dissociation curve (see fig. 22.21), you can see that at a low P_{O_2}, HbO_2 releases more oxygen.
2. **Temperature.** When temperature rises, the oxyhemoglobin dissociation curve shifts to the right (fig. 22.24*a*); in other words, elevated temperature promotes oxygen unloading. Active tissues are warmer than less active ones and thus extract more oxygen from the blood passing through them.
3. **The Bohr effect.** Active tissues also generate extra CO_2, which raises the H^+ concentration and lowers the pH of the blood. Like elevated temperatures, a drop in pH shifts the oxygen-hemoglobin dissociation curve to the right (fig. 22.24*b*) and promotes oxygen unloading. The increase in HbO_2 dissociation in response to low pH is called the **Bohr[20] effect.** It is less pronounced at the high P_{O_2} present in the lungs, so pH has relatively little effect on pulmonary oxygen loading. In the systemic capillaries, however, P_{O_2} is lower and the Bohr effect is more pronounced.
4. **BPG.** Erythrocytes have no mitochondria and meet their energy needs solely by anaerobic fermentation. One of their metabolic intermediates is **bisphosphoglycerate (BPG)** (formerly called diphosphoglycerate, DPG), which binds to hemoglobin and promotes oxygen unloading. An elevated body temperature (as in fever) stimulates BPG synthesis, as do thyroxine, growth hormone, testosterone, and epinephrine. All of these hormones thus promote oxygen unloading to the tissues.

The rate of CO_2 loading is also adjusted to varying needs of the tissues. A low level of oxyhemoglobin (HbO_2) enables the blood to transport more CO_2, a phenomenon known as the **Haldane effect.**[21] It occurs for two reasons: (1) HbO_2 does not bind CO_2 as well as deoxyhemoglobin (HHb) does. (2) HHb binds more hydrogen ions than HbO_2 does, and by removing H^+ from solution, HHb shifts the $H_2O + CO_2 \rightarrow HCO_3^- + H^+$ reaction to the right. A high metabolic rate keeps oxyhemoglobin levels relatively low and thus allows more CO_2 to be transported by these two mechanisms.

20. Christian Bohr (1855–1911), Danish physiologist
21. John Scott Haldane (1860–1936), British physiologist

Before You Go On

Answer the following questions to test your understanding of the preceding section:

12. Why is the composition of alveolar air different from that of the atmosphere?
13. What four factors affect the efficiency of alveolar gas exchange?
14. Explain how perfusion of a pulmonary lobule changes if it is poorly ventilated. How is the ventilation of a lobule affected by high P_{CO_2}?

15. Describe how oxygen is transported in the blood, and explain why carbon monoxide interferes with this.
16. What are the three ways in which blood transports CO_2?
17. Describe the role of the chloride shift in CO_2 loading.
18. Give two reasons why highly active tissues can extract more oxygen from the blood than less active tissues.

Blood Chemistry and the Respiratory Rhythm

Objectives

When you have completed this section, you should be able to

- explain how blood gases affect the respiratory centers of the brain; and
- explain how the respiratory centers homeostatically control blood gases and pH.

The factors that stimulate a person to breathe, from the most powerful stimulus to the least, are the pH, P_{CO_2}, and P_{O_2} of the body fluids. These conditions are monitored by chemoreceptors in two general locations: **peripheral chemoreceptors** located outside the central nervous system (CNS) and **central chemoreceptors** located in the brainstem. The peripheral chemoreceptors are **aortic bodies** and **carotid bodies** located in the aortic arch and near the branch of the carotid arteries (fig. 22.25). (These are not to be confused with the aortic and carotid *sinuses,* which harbor the baroreceptors that monitor blood pressure.) Although very small, the aortic and carotid bodies are richly supplied with capillaries and receive almost 40 times as much blood per gram of tissue as the brain does. The aortic bodies send signals to the medulla by way of the vagus nerves and the carotid bodies transmit by way of the glossopharyngeal nerves. The central chemoreceptors are paired areas very close to the surface of the medulla oblongata, ventral to the inspiratory center. They primarily monitor the pH of the cerebrospinal fluid (CSF) and the extracellular fluid of the brain tissue.

We now consider how hydrogen ions, carbon dioxide, and oxygen individually affect respiration.

Hydrogen Ions

The most powerful respiratory stimulus is the pH of the CSF; ultimately, pulmonary ventilation is adjusted to maintain the pH of the brain. Hydrogen ions in the blood cannot cross the blood-CSF barrier very easily, but CO_2 crosses easily. In the CSF, CO_2 reacts with water and releases H^+, just as it does in the blood plasma and RBCs. But unlike blood, the CSF contains relatively little protein to buffer this H^+. The H^+ remains free in the CSF and strongly stimulates the central chemoreceptors, which transmit signals to the inspiratory center.

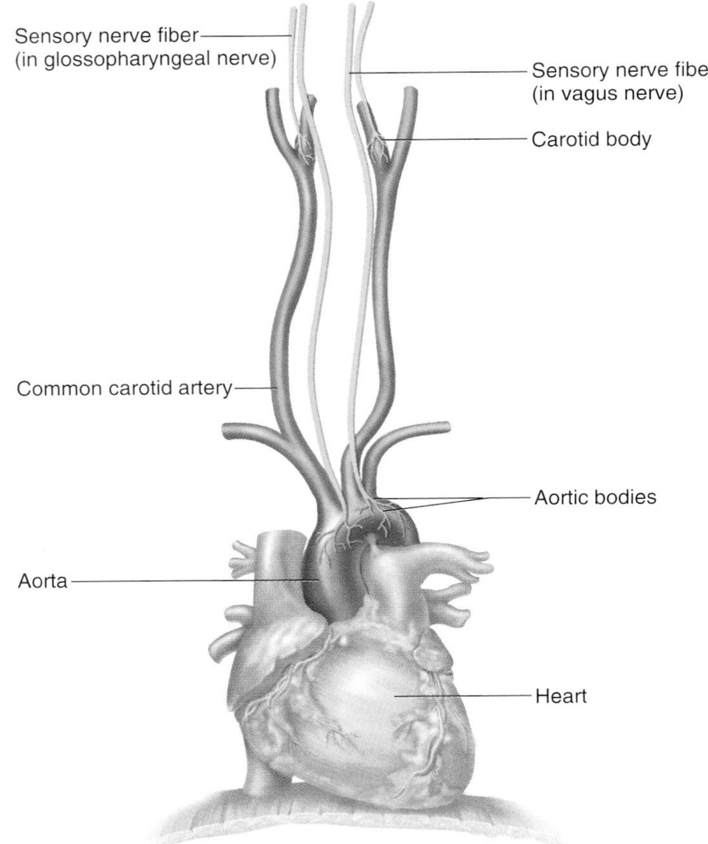

Figure 22.25 Nervous Pathways from the Peripheral Chemoreceptors to the Respiratory Centers of the Medulla Oblongata.

It is important to ask how we know it is the H^+ that does this and not the CO_2 that diffuses into the CSF. Experimentally, it is possible to vary the pH or the P_{CO_2} of the CSF while holding the other variable steady. When pH alone changes, there is a strong effect on respiration; when P_{CO_2} alone changes, there is only a weak effect. Therefore, even though these two variables usually change together, we can see that the chemoreceptors react primarily to the pH.

Normally the blood has a pH of 7.40 ± 0.05. Deviations from this range are called **acidosis** when the pH falls below 7.35 and **alkalosis** when it rises above 7.45. The normal P_{CO_2} of the blood is 40 ± 3 mmHg. The most common cause of acidosis is **hypercapnia,**[22] a $P_{CO_2} > 43$ mmHg; the most common cause of alkalosis is **hypocapnia,** a $P_{CO_2} < 37$ mmHg. Whenever there is a CO_2 imbalance in the blood, CO_2 diffusion across the blood-CSF barrier creates

22. *capn* = smoke

a parallel shift in the pH of the CSF. Therefore even though the brain responds primarily to pH changes, it is CO_2 that usually causes those changes. When these pH imbalances are due to a failure of pulmonary ventilation to match the body's rate of CO_2 production, they are called *respiratory acidosis* and *respiratory alkalosis* (see further discussion in chapter 24).

The corrective homeostatic response to acidosis is hyperventilation, "blowing off" CO_2 faster than the body produces it. This shifts the carbonic acid reaction to the left:

$$CO_2 + H_2O \leftarrow H_2CO_3 \leftarrow HCO_3^- + H^+$$

The CO_2 on the left is expired in the breath. The H^+ on the right is consumed as this reaction proceeds toward the left, and as H^+ concentration declines, the pH rises.

The corrective response to alkalosis is hypoventilation, which allows the body to produce CO_2 faster than it exhales it. Hypoventilation shifts the reaction to the right, raises the H^+ concentration, and lowers the pH to normal:

$$CO_2 + H_2O \rightarrow H_2CO_3 \rightarrow HCO_3^- + H^+$$

Although pH changes usually result from P_{CO_2} changes, they can have other causes. In diabetes mellitus, for example, fat oxidation releases acidic ketone bodies, causing an abnormally low pH called *ketoacidosis* (see chapter 17). Ketoacidosis tends to induce a form of hyperventilation called *Kussmaul*[23] *respiration* (see table 22.3). Hyperventilation cannot reduce the level of ketone bodies in the blood, but by blowing off CO_2, it reduces the concentration of CO_2-generated H^+ and compensates to some degree for the H^+ released by the ketone bodies.

Carbon Dioxide

Although the arterial P_{CO_2} has a strong influence on respiration, we have seen that it is mostly an indirect one, mediated through its effects on the pH of the CSF. Yet the experimental evidence described earlier shows that CO_2 has some effect even when pH remains stable. At the beginning of exercise, the rising blood CO_2 level may directly stimulate the peripheral chemoreceptors and trigger an increase in ventilation more quickly than the central chemoreceptors do.

Oxygen

Oxygen concentration usually has little effect on respiration. Even in eupnea, the hemoglobin is 97% saturated with O_2; therefore, increased ventilation cannot add very much. Only if the arterial P_{O_2} drops below 60 mmHg does it significantly affect ventilation, and such a low P_{O_2} seldom occurs even in prolonged holding of the breath. A moderate drop in P_{O_2} does stimulate the peripheral chemoreceptors, but another effect overrides this: as the level of HbO_2 falls, hemoglobin binds more hydrogen ions (see fig. 22.22). This raises the blood pH, which indirectly inhibits respiration. Only at a $P_{O_2} < 60$ mmHg does the stimulatory effect of hypoxemia override the inhibitory effect of the pH increase. Long-term hypoxemia can significantly stimulate ventilation in situations such as emphysema and pneumonia, which interfere with alveolar gas exchange, and in mountain climbing of at least 2 or 3 days' duration.

In summary, the main chemical stimulus to pulmonary ventilation is the H^+ in the CSF and tissue fluid of the brain. These hydrogen ions arise mainly from CO_2 diffusing into the CSF and brain and generating H^+ through the carbonic acid reaction. Therefore the P_{CO_2} of the arterial blood is an important driving force in respiration, even though its action on the chemoreceptors is indirect. Ventilation is adjusted to maintain arterial pH at about 7.40 and arterial P_{CO_2} at about 40 mmHg. This automatically ensures that the blood is at least 97% saturated with O_2 as well. Under ordinary circumstances, arterial P_{O_2} has relatively little effect on respiration. When it drops below 60 mmHg, however, it excites the peripheral chemoreceptors and stimulates an increase in ventilation. This can be significant at high altitudes and in certain lung diseases.

23. Adolph Kussmaul (1822–1902), German physician

Before You Go On

Answer the following questions to test your understanding of the preceding section:

19. Describe the locations of the chemoreceptors that monitor blood pH and gas concentrations.
20. Define *hypocapnia* and *hypercapnia.* Name the pH imbalances that result from these conditions and explain the relationship between P_{CO_2} and pH.
21. Explain how variations in pulmonary ventilation can correct pH imbalances.

Respiratory Disorders

Objectives

When you have completed this section, you should be able to

- describe the forms and effects of oxygen deficiency and oxygen excess;
- describe the chronic obstructive pulmonary diseases and their consequences; and
- explain how lung cancer begins, progresses, and exerts its lethal effects.

The delicate lungs are exposed to a wide variety of inhaled pathogens and debris; thus, it is not surprising that they are prone to a host of diseases. Several already have been mentioned in this chapter.

Oxygen Imbalances

Hypoxia, discussed in previous chapters, is a deficiency of oxygen in a tissue or the inability to use oxygen. It is not a respiratory disease in itself but is often a consequence of respiratory diseases. Hypoxia is classified according to cause:

- **Hypoxemic hypoxia,** a state of low arterial P_{O_2}, is usually due to inadequate pulmonary gas exchange. Some of its root causes include atmospheric deficiency of oxygen at high altitudes; impaired pulmonary ventilation, as in drowning; aspiration of foreign matter; respiratory arrest; and the degenerative lung diseases discussed shortly. It also occurs in carbon monoxide (CO) poisoning because CO prevents hemoglobin from transporting oxygen.
- **Ischemic hypoxia** results from inadequate circulation of the blood, as in congestive heart failure.
- **Anemic hypoxia** is due to anemia and the resulting inability of the blood to carry adequate oxygen.
- **Histotoxic hypoxia** occurs when a metabolic poison such as cyanide prevents the tissues from using the oxygen delivered to them.

Hypoxia is often marked by **cyanosis,** blueness of the skin. Whatever its cause, the primary effect of hypoxia is the necrosis of oxygen-starved tissues. This is especially critical in organs with the highest metabolic demands, such as the brain, heart, and kidneys.

An oxygen excess is also dangerous. You can safely breathe 100% oxygen at 1 atm for a few hours, but **oxygen toxicity** rapidly develops when pure oxygen is breathed at 2.5 atm or greater. Excess oxygen generates hydrogen peroxide and free radicals that destroy enzymes and damage nervous tissue; thus it can lead to seizures, coma, and death. This is why scuba divers breathe a mixture of oxygen and nitrogen rather than pure compressed oxygen (see clinical insight 22.4, p. 875). Hyperbaric oxygen was formerly used to treat premature infants for respiratory distress syndrome, but it caused retinal deterioration and blinded many infants before the practice was discontinued.

Chronic Obstructive Pulmonary Diseases

Chronic obstructive pulmonary disease (COPD) refers to any disorder in which there is a long-term obstruction of airflow and a substantial reduction in pulmonary ventilation. The major COPDs are *asthma, chronic bronchitis,* and *emphysema.* In asthma, an allergen triggers the release of histamine and other inflammatory chemicals that cause intense bronchoconstriction and sometimes suffocation (see chapter 21). The other COPDs are almost always caused by cigarette smoking but occasionally result from air pollution or occupational exposure to airborne irritants.

Beginning smokers exhibit inflammation and hyperplasia of the bronchial mucosa. In **chronic bronchitis,** the cilia are immobilized and reduced in number, while goblet cells enlarge and produce excess mucus. With extra mucus and fewer cilia to dislodge it, smokers develop a chronic cough that brings up **sputum** (SPEW-tum), a mixture of mucus and cellular debris. Thick, stagnant mucus in the respiratory tract provides a growth medium for bacteria, while cigarette smoke incapacitates the alveolar macrophages and reduces defense mechanisms against respiratory infections. Smokers therefore develop chronic infection and bronchial inflammation, with symptoms that include dyspnea, hypoxia, cyanosis, and attacks of coughing.

In **emphysema**[24] (EM-fih-SEE-muh), alveolar walls break down and the lung exhibits larger but fewer alveoli (see fig. 22.19*c*). Thus, there is much less respiratory membrane available for gas exchange. The lungs become fibrotic and less elastic. The air passages open adequately during inspiration, but they tend to collapse and obstruct the outflow of air. Air therefore becomes trapped in the lungs, and over a period of time a person becomes barrel-chested. The overly stretched thoracic muscles contract weakly, which further contributes to the difficulty of expiration. People with emphysema become exhausted because they expend three to four times the normal amount of energy just to breathe. Even slight physical exertion, such as walking across a room, can cause severe shortness of breath.

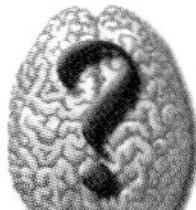

Think About It

Explain how the length-tension relationship of skeletal muscle (see chapter 12) accounts for the weakness of the respiratory muscles in emphysema.

All of the COPDs tend to reduce pulmonary compliance and vital capacity and cause hypoxemia, hypercapnia, and respiratory acidosis. Hypoxemia stimulates the kidneys to secrete erythropoietin, which leads to accelerated erythrocyte production and polycythemia, as discussed in chapter 18. COPD also leads to **cor pulmonale**—hypertrophy and potential failure of the right heart due to obstruction of the pulmonary circulation (see chapter 19).

24. *emphys* = inflamed

Smoking and Lung Cancer

Lung cancer (fig. 22.26) accounts for more deaths than any other form of cancer. The most important cause of lung cancer is cigarette smoking, distantly followed by air pollution. Cigarette smoke contains at least 15 carcinogenic compounds. Lung cancer commonly follows or accompanies COPD.

There are three forms of lung cancer, the most common of which is **squamous-cell carcinoma.** This form begins with the multiplication of basal cells of the bronchial epithelium and transformation of the ciliated pseudostratified columnar epithelium into the stratified squamous type. As the dividing epithelial cells invade the underlying tissues of the bronchial wall, the bronchus develops bleeding lesions. Dense swirled masses of keratin appear in the lung parenchyma and replace functional respiratory tissue. A second form of lung cancer, nearly as common, is **adenocarcinoma,**[25] which originates in mucous glands of the lamina propria. The least common (10%–20% of malignancies) but most dangerous form is **small-cell (oat-cell) carcinoma,** named for clusters of cells that resemble oat grains. This originates in the primary bronchi but invades the mediastinum and metastasizes quickly to other organs.

Over 90% of lung tumors originate in the mucous membranes of the large bronchi. As a tumor invades the bronchial wall and grows around it, it compresses the airway and may cause atelectasis (collapse) of more distal parts of the lung. Growth of the tumor produces a cough, but coughing is such an everyday occurrence among

25. *adeno* = gland

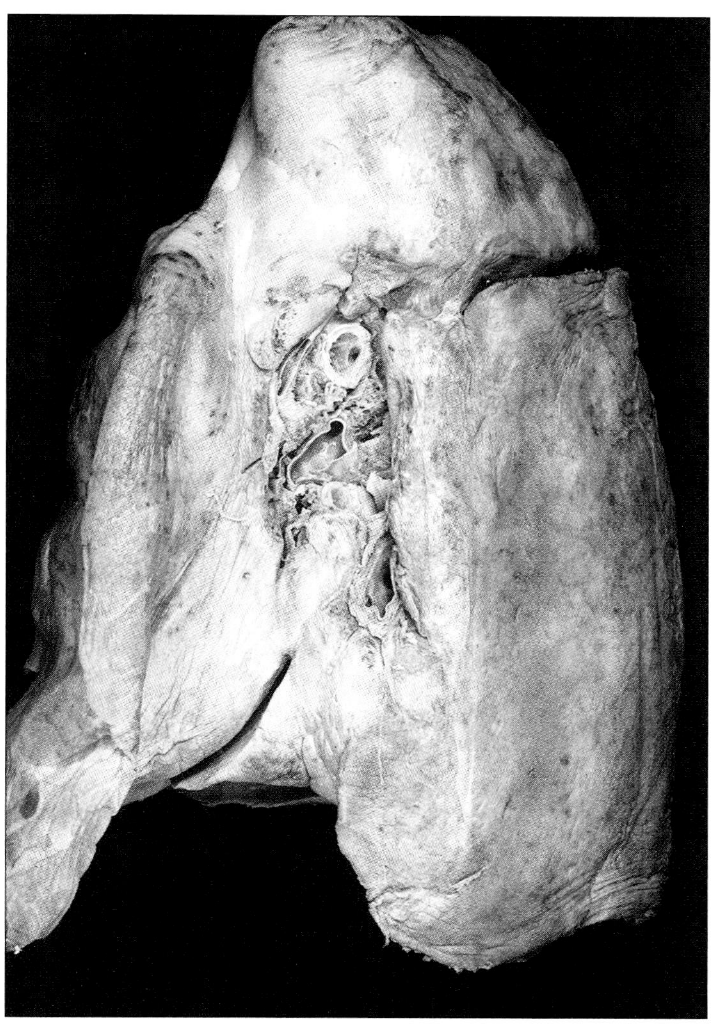

(a)

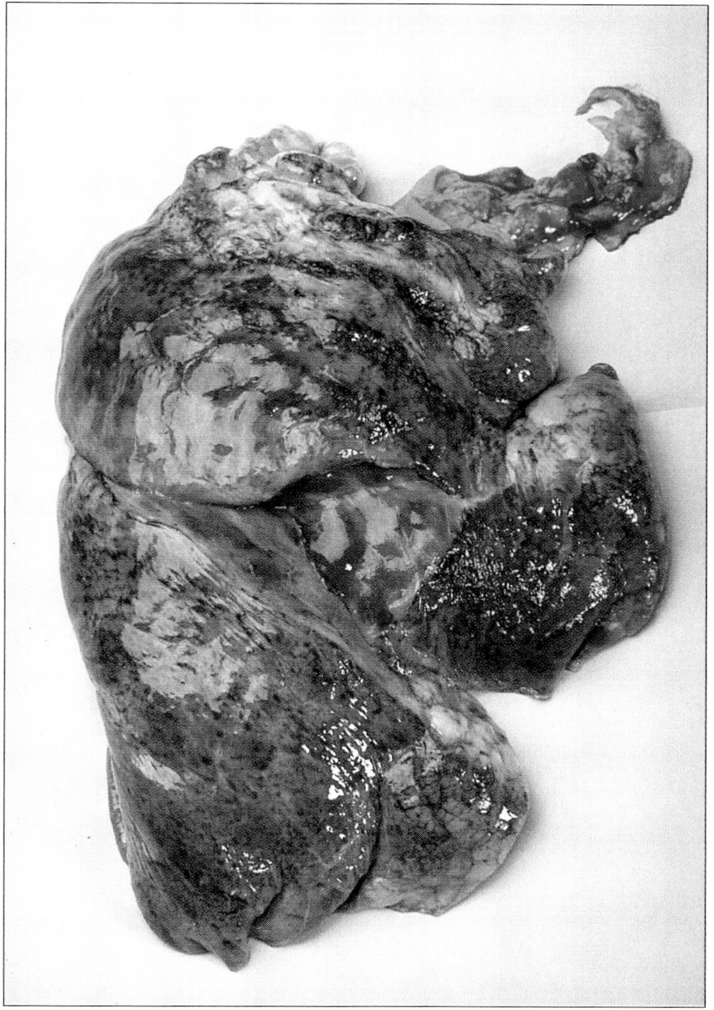

(b)

Figure 22.26 **Effect of Smoking.** (*a*) A healthy adult lung. (*b*) A smoker's lung with carcinoma.

smokers it seldom causes much alarm. Often, the first sign of serious trouble is the coughing up of blood. Lung cancer metastasizes so rapidly that it has usually spread to other organs by the time it is diagnosed. Common sites of metastasis are the pericardium, heart, bones, liver, lymph nodes, and brain. The chance of recovery is poor, with only 7% of patients surviving for 5 years after diagnosis.

Some infectious diseases and other disorders of the respiratory system are briefly described in table 22.5.

Before You Go On

Answer the following questions to test your understanding of the preceding section:

22. Describe the four classes of hypoxia.
23. Name and compare two COPDs and describe some pathological effects that they have in common.
24. In what lung tissue does lung cancer originate? How does it kill?

Table 22.5 Some Disorders of the Respiratory System

Acute Rhinitis	The common cold. Caused by many types of viruses that infect the upper respiratory tract. Symptoms include congestion, increased nasal secretion, sneezing, and dry cough. Transmitted especially by contact of contaminated hands with mucous membranes; not transmitted orally.
Adult Respiratory Distress Syndrome	Acute lung inflammation and alveolar injury stemming from trauma, infection, burns, aspiration of vomit, inhalation of noxious gases, drug overdoses, and other causes. Alveolar injury is accompanied by severe pulmonary edema and hemorrhage, followed by fibrosis that progressively destroys lung tissue. Fatal in about 40% of cases under age 60 and in 60% of cases over age 65.
Pneumonia	A lower respiratory infection caused by any of several viruses, fungi, or protozoans (most often the bacterium *Streptococcus pneumoniae*). Causes filling of alveoli with fluid and dead leukocytes and thickening of the respiratory membrane, which interferes with gas exchange and causes hypoxemia. Especially dangerous to infants, the elderly, and people with compromised immune systems, such as AIDS and leukemia patients.
Sleep Apnea	Cessation of breathing for 10 seconds or longer during sleep; sometimes occurs hundreds of times per night, often accompanied by restlessness and alternating with snoring. Can result from altered function of CNS respiratory centers, airway obstruction, or both. Over time, may lead to daytime drowsiness, hypoxemia, polycythemia, pulmonary hypertension, congestive heart failure, and cardiac arrhythmia. Most common in obese people and in men.
Tuberculosis (TB)	Pulmonary infection with the bacterium *Mycobacterium tuberculosis,* which invades the lungs by way of air, blood, or lymph. Stimulates the lung to form fibrous nodules called tubercles around the bacteria. Progressive fibrosis compromises the elastic recoil and ventilation of the lungs. Especially common among impoverished and homeless people and becoming increasingly common among people with AIDS.

Clinical Insight 22.4

Diving Physiology and Decompression Sickness

Because of the rise in popularity of scuba diving in recent years, many people now know something about the scientific aspects of breathing under high pressure. But diving is by no means a new fascination. As early as the fifth century B.C.E., Aristotle described divers using snorkels and taking containers of air underwater in order to stay down longer. Some Renaissance artists depicted divers many meters deep breathing from tubes to the water surface. In reality, this would be physically impossible. For one thing, such tubes would have so much dead space that fresh air from the surface would not reach the diver. The short snorkels used today are about the maximum length that will work for surface breathing. Another reason snorkels cannot be used at greater depths is that water pressure doubles for every 11 m of depth, and even at 1 m the pressure is so great that a diver cannot expand the chest muscles without help. This is one reason why scuba divers use pressurized air tanks. The tanks create a positive intrapulmonary pressure and enable the diver to inhale with only slight assistance from the thoracic muscles. Scuba tanks also have regulators that adjust the outflow pressure to the diver's depth and the opposing pressure of the surrounding water.

But breathing pressurized (hyperbaric) gas presents its own problems. Divers cannot use pure oxygen because of the problem of oxygen toxicity. Instead, they use compressed air—a mixture of 21% oxygen and 79% nitrogen. On land, nitrogen presents no physiological problems; it dissolves poorly in blood and it is physiologically inert. But under hyperbaric conditions, larger amounts of nitrogen dissolve in the blood. (Which of the gas laws applies here?) Even more dissolves in adipose tissue and the myelin of the brain, since nitrogen is more soluble in lipids. In the brain, it causes *nitrogen narcosis,* or what Jacques Cousteau termed "rapture of the deep." A diver can become dizzy, euphoric, and dangerously disoriented; for every 15 to 20 m of depth, the effect is said to be equivalent to one martini on an empty stomach.

Strong currents, equipment failure, and other hazards sometimes make scuba divers panic, hold their breath, and quickly swim to the surface (a *breath-hold ascent*). Ambient (surrounding) pressure falls rapidly as a diver ascends, and the air in the lungs expands just as rapidly. (Which gas law is demonstrated here?) It is imperative that an ascending diver keep his or her airway open to exhale the expanding gas; otherwise it is likely to cause *pulmonary barotrauma*—ruptured alveoli. Then, when the diver takes a breath of air at the surface, alveolar air goes directly into the bloodstream and causes air embolism. After passing through the heart, the emboli tend to enter the cerebral circulation because the diver is head-up and air bubbles rise in liquid. The resulting cerebral embolism can cause motor and sensory dysfunction, seizures, unconsciousness, and drowning.

Barotrauma can be fatal even at the depths of a backyard swimming pool. In one case, children trapped air in a bucket 1 m underwater and then swam under the bucket to breathe from the air space. Because the bucket was under water, the air in it was compressed. One child filled his lungs under the bucket, did a "mere" 1-m breath-hold ascent, and his alveoli ruptured. He died in the hospital, partly because the case was mistaken for drowning and not treated for what it really was. This would not have happened to a person who inhaled at the surface, did a breath-hold dive, and then resurfaced—nor is barotrauma a problem for those who do breath-hold dives to several meters. (Why? What is the difference?)

Even when not holding the breath, but letting the expanding air escape from the mouth, a diver must ascend slowly and carefully to allow for decompression of the nitrogen that has dissolved in the tissues. *Decompression tables* prescribe safe rates of ascent based on the depth and the length of time a diver has been down. When pressure drops, nitrogen dissolved in the tissues can go either of two places—it can diffuse into the alveoli and be exhaled, or it can form bubbles like the CO_2 in a bottle of soda when the cap is removed. The diver's objective is to ascend slowly, allowing for the former and preventing the latter. If a diver ascends too rapidly, nitrogen "boils" from the tissues—especially in the 3 m just below the surface, where the relative pressure change is greatest. A diver may double over in pain from bubbles in the joints, bones, and muscles—a disease called the "bends," or *decompression sickness (DCS).* Nitrogen bubbles in the pulmonary capillaries cause "chokes"—substernal pain, coughing, and dyspnea. DCS is sometimes accompanied by mood changes, seizures, numbness, and itching. These symptoms usually occur within an hour of surfacing, but they are sometimes delayed for up to 36 hours. DCS is treated by putting the individual in a hyperbaric chamber to be recompressed and then *slowly* decompressed.

DCS is also called *caisson disease.* A caisson is a watertight underwater chamber filled with pressurized air. Caissons are used in underwater construction work on bridges, tunnels, ships' hulls, and so forth. Caisson disease was first reported in the late 1800s among workmen building the foundations of the Brooklyn Bridge.

Connective Issues

Interactions Between the RESPIRATORY SYSTEM and Other Organ Systems

← indicates ways in which this system affects other systems
→ indicates ways in which other systems affect this one

All Systems

The respiratory system serves all other systems by supplying O_2, removing CO_2, and maintaining acid-base balance

Integumentary System

→ Nasal guard hairs reduce inhalation of dust and other foreign matter

Skeletal System

→ Thoracic cage protects lungs; movement of ribs produces pressure changes that ventilate lungs

Muscular System

→ Skeletal muscles ventilate lungs, control position of larynx during swallowing, control vocal cords during speech; exercise strongly stimulates respiration because of the CO_2 generated by active muscles

Nervous System

→ Produces the respiratory rhythm, monitors blood gases and pH, monitors stretching of lungs; phrenic, intercostal, and other nerves control respiratory muscles

Endocrine System

← Lungs produce angiotensin-converting enzyme (ACE), which converts angiotensin I to the hormone angiotensin II

→ Epinephrine and norepinephrine dilate bronchioles and stimulate ventilation

Circulatory System

← Regulates blood pH; thoracic pump aids in venous return; lungs produce blood platelets; production of angiotensin II by lungs is important in control of blood volume and pressure; obstruction of pulmonary circulation leads to right-sided heart failure

→ Blood transports O_2 and CO_2; mitral stenosis or left-sided heart failure can cause pulmonary edema; emboli from peripheral sites often lodge in lungs

Lymphatic/Immune Systems

← Thoracic pump promotes lymph flow

→ Lymphatic drainage from lungs is important in keeping alveoli dry; immune cells protect lungs from infection

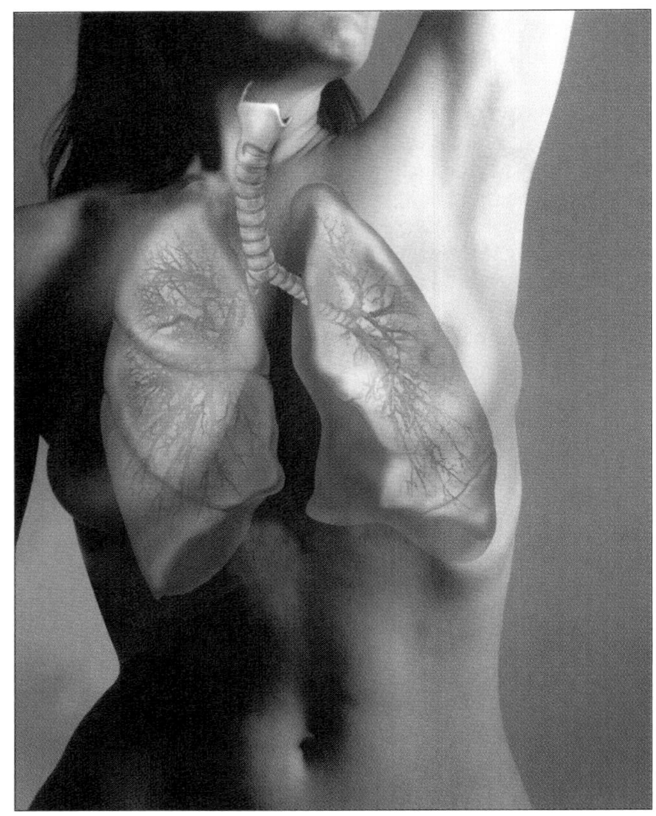

Urinary System

← Valsalva maneuver aids in emptying bladder

→ Disposes of wastes from respiratory organs; collaborates with lungs in controlling blood pH

Digestive System

← Valsalva maneuver aids in defecation

→ Provides nutrients for growth and maintenance of respiratory system

Reproductive System

← Valsalva maneuver aids in childbirth

→ Sexual arousal stimulates respiration

Chapter Review

Online Learning Center

Study resources for this chapter are available at www.mhhe.com/saladin/

Study Outline

Anatomy of the Respiratory System (p. 844)

1. General aspects of respiration
   a. Meanings of *respiration*
   b. Major organs of respiration
   c. Pathway of airflow
   d. Conducting and respiratory divisions
   e. Upper and lower respiratory tract
2. The Nose
   a. Functions
   b. Bony and cartilaginous supports
   c. Nasal cavity
      • Anterior and posterior nares
      • Vestibule
      • Nasal septum and fossae
      • Conchae and meatuses
      • Olfactory and respiratory mucosae
      • Function of cilia
      • Function of erectile tissue
      • Epistaxis
3. The Pharynx
   a. Nasopharynx
   b. Oropharynx
   c. Laryngopharynx
4. The Larynx
   a. Glottis and epiglottis
   b. Role in excluding food and drink
   c. The infant larynx
   d. Supportive cartilages
   e. Intrinsic and extrinsic muscles
   f. Vestibular folds
   g. Vocal cords
5. The Trachea
   a. Size and position
   b. Supportive cartilages
   c. Trachealis muscle
   d. Mucociliary escalator
6. The Lungs
   a. Surface anatomy
      • Base and apex
      • Costal and mediastinal surfaces
      • Hilum
      • Lobes and fissures
   b. Internal anatomy
      • Bronchial tree
         •• Primary bronchi
         •• Secondary bronchi
         •• Tertiary bronchi
         •• Bronchopulmonary segments
         •• Bronchioles and lobules
      • Alveolar ducts and sacs
   c. Alveoli
      • Squamous alveolar cells
      • Great alveolar cells
      • Pulmonary surfactant
      • Alveolar macrophages
   d. The respiratory membrane
   e. Keeping the alveoli dry
7. The Pleurae
   a. Visceral and parietal layers
   b. Pleural cavity and fluid
   c. Functions
      • Reduction of friction
      • Creation of pressure gradient
      • Compartmentalization

Mechanics of Ventilation (p. 854)

1. Gas laws
   a. Boyle's law: pressure and volume
   b. Charles' law: temperature and volume
   c. Dalton's law: partial pressures
   d. Henry's law: gases dissolving in liquids
   e. Law of Laplace: alveolar radius
2. Pressure and flow
   a. Atmospheric pressure
   b. Pressure and lung volume
   c. Pressure gradients
3. Inspiration
   a. Muscles involved
      • Diaphragm
      • External intercostals
      • Others (in deep inspiration)
   b. Pressure changes
      • Drop in intrapleural pressure
      • Drop in intrapulmonary pressure
      • Transpulmonary pressure
   c. Warming of inhaled air
   d. Volume of inhaled air
4. Expiration
   a. Passive expiration
      • Importance of elasticity
      • Relation to Boyle's law
      • Braking action of diaphragm
   b. Forced expiration
      • Role of internal intercostal muscles
      • Role of abdominal muscles
   c. Pneumothorax and atelectasis
5. Resistance to airflow
   a. Importance of pulmonary compliance
   b. Bronchiolar diameter
      • Bronchoconstriction
      • Bronchodilation
6. Alveolar surface tension
   a. Necessity of water film in alveoli
   b. Problem of surface tension
   c. Law of Laplace
   d. Function of pulmonary surfactant
   e. Respiratory distress syndrome
7. Alveolar ventilation
   a. Dead air and dead spaces
   b. Alveolar ventilation rate
8. Nonrespiratory air movements
   a. Functions other than alveolar ventilation
   b. Coughing
   c. Sneezing
   d. The Valsalva maneuver
9. Measurements of ventilation
   a. Use of the spirometer
   b. Respiratory volumes
      • Tidal volume
      • Inspiratory reserve volume
      • Expiratory reserve volume
      • Residual volume
   c. Respiratory capacities
      • Vital capacity
      • Inspiratory capacity

- Functional residual capacity
- Total lung capacity

d. Relationship to sex, age, and exercise
e. Disorders affecting these values
- Restrictive disorders
- Obstructive disorders

f. Forced expiratory volume
g. Minute respiratory volume
h. Maximum voluntary ventilation

10. Patterns of breathing (table 22.3)

Neural Control of Ventilation (p. 860)

1. Neurons involved in ventilation
   a. Inspiratory and expiratory neurons
   b. Elusiveness of the respiratory pacemaker
2. Control centers in the medulla oblongata
   a. Inspiratory center (dorsal respiratory group)
   - Relationship to inspiratory muscles
   - Firing frequency and depth of inspiration
   - Firing duration and respiratory rate

   b. Expiratory center (ventral respiratory group)
   - Relationship to inspiratory center
   - Role in deep breathing
3. Control centers in the pons
   a. Pneumotaxic center
   - Relationship to inspiratory center
   - Effect in rate and depth of respiration

   b. Apneustic center
4. Afferent connections to the brainstem
   a. Input from limbic system and hypothalamus
   - Respiratory effects of emotion
   - Hyperventilation

   b. Input from chemoreceptors
   c. Input from the airway and lungs
   - Response to inhaled irritants
   - Inflation reflex
5. Voluntary control
   a. Neural pathways
   b. Limitations on voluntary control

Gas Exchange and Transport (p. 862)

1. Composition of air
   a. Partial pressures of atmospheric gases
   b. Comparison to expired alveolar air
2. The air-water interface
3. Alveolar gas exchange
   a. O_2 loading and CO_2 unloading
   b. RBC transit time
   c. Factors affecting gas exchange
   - Concentration gradients
   - Gas solubility
   - Membrane thickness
   - Membrane surface area
   - Ventilation-perfusion coupling
4. Gas transport
   a. Oxygen transport
   - Concentration in blood
   - Binding to hemoglobin
   - Oxyhemoglobin dissociation curve

   b. Carbon dioxide transport
   - As carbonic acid
   - As carbamino compounds
   - As dissolved gas
   - Alveolar exchange
5. Systemic gas exchange
   a. CO_2 loading
   - Role of carbonic anhydrase
   - Role of chloride shift

   b. O_2 unloading
   - Effect of H^+ binding to hemoglobin
   - Utilization coefficient
6. Alveolar gas exchange revisited
   a. O_2 loading
   b. CO_2 unloading
7. Adjustment to the metabolic needs of tissues
   a. Factors affecting O_2 unloading
   - Ambient P_{O_2}
   - Temperature
   - Bohr effect
   - Bisphosphoglycerate

   b. Factors affecting CO_2 loading
   - Haldane effect
   - Metabolic rate and HbO_2 level

Blood Chemistry and the Respiratory Rhythm (p. 870)

1. Chemoreceptors
   a. Peripheral
   - Aortic bodies
   - Carotid bodies

   b. Central
2. Effects of hydrogen ions
   a. pH of cerebrospinal fluid
   b. Crossing the blood-brain barrier
   c. Distinguishing pH effect from P_{CO_2} effect
   d. Acidosis
   - Caused by hypercapnia
   - Corrected by hyperventilation

   e. Alkalosis
   - Caused by hypocapnia
   - Corrected by hypoventilation

   f. Other causes of acidosis and alkalosis
3. Carbon dioxide
   a. Indirect effects (through pH)
   b. Direct effects
4. Oxygen
   a. Usually little effect
   b. Effects of hypoxemia

Respiratory Disorders (p. 871)

1. Oxygen imbalances
   a. Hypoxia
   - Hypoxemic hypoxia
   - Ischemic hypoxia
   - Anemic hypoxia
   - Histotoxic hypoxia
   - Hypoxia and cyanosis

   b. Oxygen excess
   - Oxygen toxicity
   - Application to diving
   - Hyperbaric oxygen therapy
2. Chronic obstructive pulmonary diseases
   a. Types
   - Asthma
   - Chronic bronchitis
   - Emphysema

   b. Effects
   - Reduced pulmonary compliance
   - Reduced vital capacity
   - Hypoxemia, hypercapnia, acidosis
   - Polycythemia
   - Cor pulmonale
3. Smoking and lung cancer
   a. Forms of lung cancer
   - Squamous-cell carcinoma
   - Adenocarcinoma
   - Small-cell carcinoma

   b. Progression of disease
   - Origin of tumor
   - Cause of cough
   - Sites of metastasis

Selected Vocabulary

respiration 844
respiratory system 844
conducting division 844
respiratory division 844
upper respiratory tract 844
lower respiratory tract 844
nose 844
lateral cartilage 844
alar cartilage 844
ala nasi 844
nasal cavity 844
anterior nares 844
posterior nares 844
vestibule 844
vibrissa 844
nasal septum 845
nasal fossa 845
nasal concha 845
turbinate bone 845
meatus 845
erectile tissue 845
pharynx 847
nasopharynx 847
oropharynx 847
laryngopharynx 847
larynx 847
glottis 847
epiglottis 847
epiglottic cartilage 848
thyroid cartilage 848
cricoid cartilage 848
arytenoid cartilage 848
corniculate cartilage 848
cuneiform cartilage 848
intrinsic muscle 848
extrinsic muscle 848
vestibular fold 848
vocal cord 848
trachea 848
trachealis muscle 848
mucociliary escalator 848
lung 848
base 848
apex 848
costal surface 848
mediastinal surface 848
hilum 848
cardiac notch 848
superior lobe 850
inferior lobe 850
middle lobe 850
bronchial tree 850
primary bronchus 850
secondary bronchus 850
tertiary bronchus 850
bronchopulmonary segment 850
bronchiole 850
pulmonary lobule 850
terminal bronchiole 850
respiratory bronchiole 850
alveolar duct 851
alveolar sac 851
alveolus 852
squamous alveolar cell 852
great alveolar cell 852
pulmonary surfactant 852
alveolar macrophage 852
respiratory membrane 852
visceral pleura 853
parietal pleura 853
pleural cavity 853
pleural fluid 853
inspiration 854
expiration 854
atmospheric pressure 854
Boyle's law 854
intrapulmonary pressure 854
intrapleural pressure 855
transpulmonary pressure 855
Charles' law 855
pneumothorax 856
atelectasis 856
pulmonary compliance 856
bronchoconstriction 857
bronchodilation 857
law of Laplace 857
dead air 857
anatomic dead space 857
physiologic dead space 857
alveolar ventilation rate 857
Valsalva maneuver 858
spirometer 858
respiratory volume 858
tidal volume 858
inspiratory reserve volume 858
expiratory reserve volume 858
residual volume 858
respiratory capacity 858
vital capacity 858
inspiratory capacity 858
functional residual capacity 858
total lung capacity 858
restrictive disorder 858
obstructive disorder 858
forced expiratory volume 858
minute respiratory volume 859
maximum voluntary ventilation 859
inspiratory neuron 860
expiratory neuron 860
inspiratory center 860
expiratory center 860
pneumotaxic center 861
apneustic center 861
inflation reflex 861
partial pressure 862
Dalton's law 862
Henry's law 862
ventilation-perfusion coupling 864
gas transport 865
oxyhemoglobin 865
deoxyhemoglobin 865
carbonic acid 865
carbamino compound 867
carbaminohemoglobin 867
chloride shift 867
utilization coefficient 868
venous reserve 868
Bohr effect 869
bisphosphyglycerate 869
Haldane effect 869
peripheral chemoreceptor 870
central chemoreceptor 870
aortic body 870
carotid body 870
acidosis 870
alkalosis 870
hypercapnia 870
hypocapnia 870
hypoxemic hypoxia 872
ischemic hypoxia 872
anemic hypoxia 872
histotoxic hypoxia 872
cyanosis 872
oxygen toxicity 872
chronic obstructive pulmonary disease 872
chronic bronchitis 872
sputum 872
emphysema 872
cor pulmonale 872
squamous-cell carcinoma 873
adenocarcinoma 873
small-cell carcinoma 873

Testing Your Recall

1. The nasal cavity is divided by the nasal septum into right and left
   a. nares.
   b. vestibules.
   c. fossae.
   d. choanae.
   e. conchae.
2. The intrinsic laryngeal muscles regulate speech by rotating
   a. the extrinsic laryngeal muscles.
   b. the corniculate cartilages.
   c. the arytenoid cartilages.
   d. the hyoid bone.
   e. the vocal cords.
3. The largest air passages that can engage in gas exchange with the blood are
   a. the respiratory bronchioles.
   b. the terminal bronchioles.
   c. the primary bronchi.
   d. the alveolar ducts.
   e. the alveoli.
4. The law of Laplace describes
   a. the tendency of alveoli to collapse.
   b. the dissolving of gases in liquid.
   c. the effect of pH on O_2 unloading.
   d. the binding of CO_2 to hemoglobin.
   e. the relationship of gas volume to temperature.
5. Which of these values would normally be the highest?
   a. tidal volume
   b. inspiratory reserve volume
   c. expiratory reserve volume
   d. residual volume
   e. vital capacity

6. The ______ protects the lungs from injury by excessive inspiration.
   a. pleura
   b. rib cage
   c. inflation reflex
   d. Haldane effect
   e. Bohr effect
7. According to ______, the warming of air as it is inhaled helps to inflate the lungs.
   a. Boyle's law
   b. Charles' law
   c. Dalton's law
   d. the Bohr effect
   e. the Haldane effect
8. The blood transports most CO_2 in the form of
   a. dissolved gas.
   b. carbaminohemoglobin.
   c. carboxyhemoglobin.
   d. bicarbonate ions.
   e. deoxyhemoglobin.
9. Most of the CO_2 that diffuses from the blood into an alveolus comes from
   a. dissolved gas.
   b. carbaminohemoglobin.
   c. carboxyhemoglobin.
   d. carbonic acid.
   e. expired air.
10. CO_2 affects the pH of the CSF more than it does the pH of the blood because
    a. CSF contains less protein than blood.
    b. CO_2 cannot cross the blood-brain barrier.
    c. all CO_2 crosses the blood-brain barrier, so none remains in the blood.
    d. only the blood contains carbonic anhydrase.
    e. the chloride shift occurs in the blood.
11. The superior opening into the larynx is called the ________.
12. Within each lung, the airway forms a branching complex called the ________.
13. The great alveolar cells secrete a lipoprotein called ________.
14. Intrapulmonary pressure must be greater than ________ pressure for inspiration to occur, but greater than ________ pressure for expiration to occur.
15. ________ disorders reduce the flow of air through the airway.
16. Some inhaled air does not participate in gas exchange because it fills the ________ of the respiratory tract.
17. Inspiration depends on the ease of pulmonary inflation, called ________, whereas expiration depends on ________, which causes pulmonary recoil.
18. Inspiration is caused by the firing of I neurons in the ________ of the medulla oblongata.
19. The matching of airflow to blood flow in any region of the lung is called ________.
20. A blood pH > 7.45 is called ________ and can be caused by a CO_2 deficiency called ________.

Answers in appendix B

Testing Your Comprehension

1. Discuss how the different functions of the conducting division and respiratory division relate to differences in their histology.
2. State whether hyperventilation would raise or lower each of the following—the blood P_{O_2}, P_{CO_2}, and pH—and explain why. Do the same for the effects of emphysema.
3. Some competitive swimmers hyperventilate before an underwater race, thinking they can "load up on extra oxygen" and hold their breaths longer underwater. While they can indeed hold their breaths longer, it is not for the reason they think. Furthermore, some have fainted and drowned because of this practice. What is wrong with this thinking, and what accounts for the loss of consciousness?
4. Consider a man in good health with a 650 mL tidal volume and a respiratory rate of 11 breaths per minute. Report his minute respiratory volume in liters per minute. Assuming his anatomic dead space is 185 mL, calculate his alveolar ventilation rate in liters per minute.
5. An 83-year-old woman is admitted to the hospital, where a critical care nurse attempts to insert a nasoenteric tube ("stomach tube") for feeding. The patient begins to exhibit dyspnea, and a chest X ray reveals air in the right pleural cavity and a collapsed right lung. The patient dies 5 days later from respiratory complications. Name the conditions revealed by the X ray and explain how they could have resulted from the nurse's procedure.

Answers at the Online Learning Center

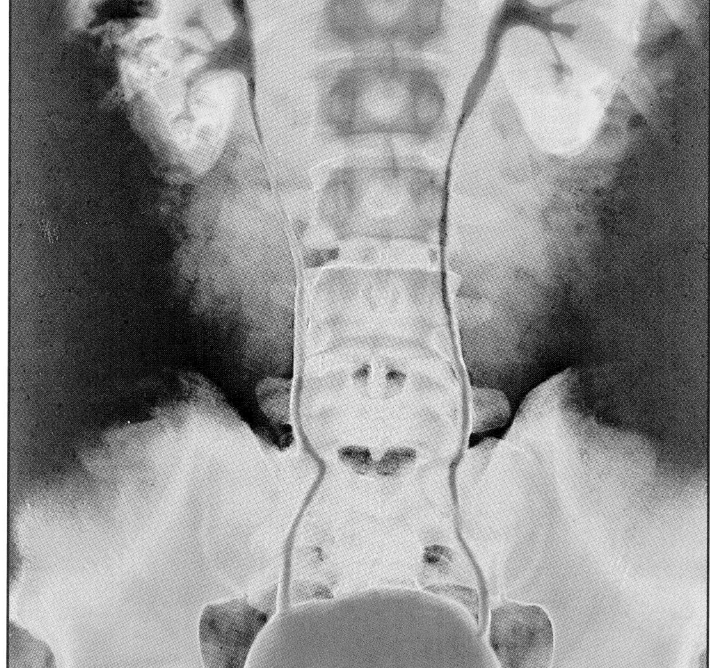

The kidneys (green), ureters, and urinary bladder (red) of a healthy person (colorized X ray)

CHAPTER 23

The Urinary System

Chapter Outline

Insights

Study with Your ESP CD-ROM

Refer to your ESP Tutorial CD-ROM for helpful animations, activities, and quizzes related to the material covered in this chapter. The following are suggested areas to go to on the CD and are presented according to "Unit/Topic/":

- Absorption and Excretion/Urinary

Brushing Up

To understand this chapter, it is important that you understand or brush up on the following concepts:

- Osmosis, tonicity, and osmolarity (pp. 126–27)
- Carrier-mediated transport mechanisms (pp. 128–31)
- Osmotic diuresis (p. 656)
- Blood pressure, resistance, and flow (p. 756)
- Capillary filtration and reabsorption (p. 764)
- Solvent drag (p. 765)

The urinary system is well known for eliminating wastes from the body, but its role in homeostasis goes far beyond that. The kidneys also detoxify poisons, synthesize glucose, and play indispensable roles in controlling electrolyte and acid-base balance, blood pressure, erythrocyte count, and the Po_2 and Pco_2 of the blood. The urinary system thus has a very close physiological relationship with the endocrine, circulatory, and respiratory systems, covered in the preceding chapters.

Anatomically, the urinary system is closely associated with the reproductive system. In many animals the eggs and sperm are emitted through the urinary tract, and the two systems have a shared embryonic development and adult anatomical relationship. This is reflected in humans, where the systems develop together in the embryo and, in the male, the urethra continues to serve as a passage for both urine and sperm. Thus the urinary and reproductive systems are often collectively called the *urogenital (U-G) system,* and *urologists* treat both urinary and reproductive disorders. We examine the anatomical relationship between the urinary and reproductive systems in chapter 27, but the physiological link to the circulatory and respiratory systems is more important to consider at this time.

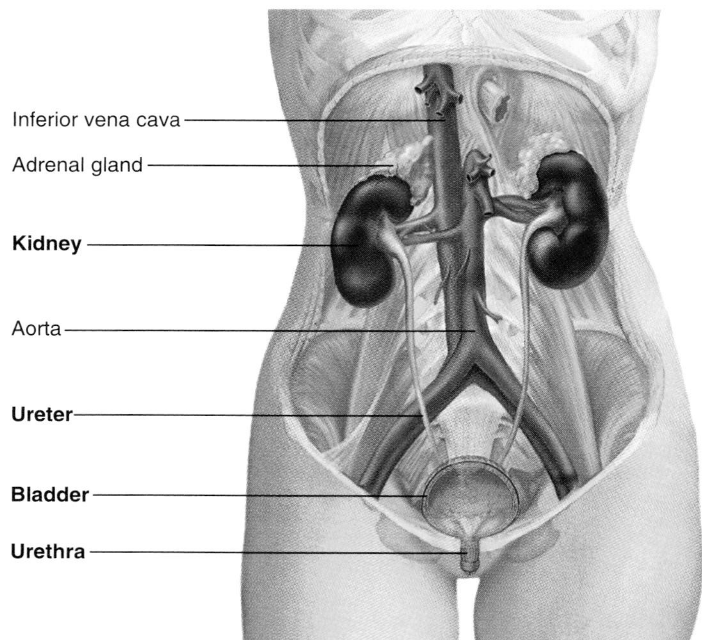

Figure 23.1 **The Urinary System.** Organs of the urinary system are indicated in boldface.

Functions of the Urinary System

Objectives

When you have completed this section, you should be able to

- name and locate the organs of the urinary system;
- list several functions of the kidneys in addition to urine formation;
- name the major nitrogenous wastes and identify their sources; and
- define *excretion* and identify the systems that excrete wastes.

The **urinary system** consists of six organs: two **kidneys,** two **ureters,** the **urinary bladder,** and the **urethra** (fig. 23.1). Most of our focus in this chapter is on the kidneys.

Functions of the Kidneys

The most fundamental role of the kidneys is homeostatic regulation of the volume and composition of the body fluids. The blood plasma, tissue fluid, and intracellular fluid have an osmolarity of about 300 mOsm/L. The urine, however, may be as dilute as 50 mOsm/L when the kidneys are ridding the body of excess water, or as concentrated as 1,200 mOsm/L when the kidneys are conserving water but still need to eliminate other wastes.

All of the following processes are aspects of kidney function:

- They filter blood plasma, separate wastes from the useful chemicals, and eliminate the wastes while returning the rest to the bloodstream.
- They regulate blood volume and pressure by eliminating or conserving water as necessary.
- They regulate the osmolarity of the body fluids by controlling the relative amounts of water and solutes they eliminate.
- They secrete the enzyme *renin,* which activates hormonal mechanisms (angiotensin, aldosterone) that control blood pressure and electrolyte balance.
- They secrete the hormone *erythropoietin,* which controls the red blood cell count and oxygen-carrying capacity of the blood.
- They function with the lungs to regulate the Pco_2 and acid-base balance of the body fluids.
- They contribute to calcium homeostasis through their role in synthesizing calcitriol (vitamin D) (see chapter 8).
- They detoxify free radicals and drugs with the use of peroxisomes.
- In times of starvation, they carry out *gluconeogenesis;* they *deaminate* amino acids (remove the $-NH_2$ group), excrete the amino group as ammonia (NH_3), and synthesize glucose from the rest of the molecule.

Nitrogenous Wastes

A **waste** is any substance that is useless to the body or present in excess of the body's needs. A **metabolic waste,** more specifically, is a waste substance produced by the body. Thus the food residue in feces, for example, is a

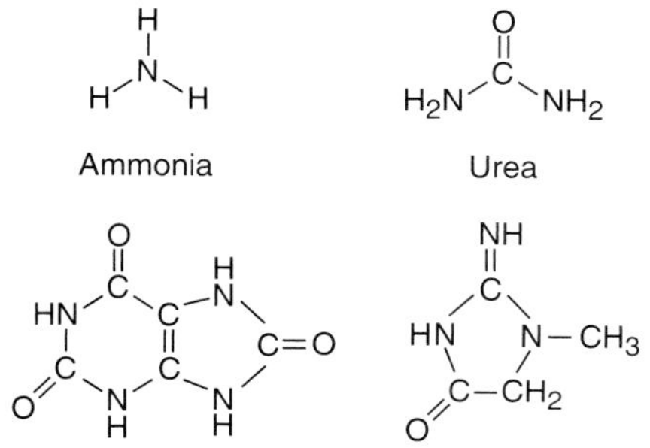

Ammonia

Urea

Uric acid

Creatinine

Figure 23.2 The Major Nitrogenous Wastes.

waste but not a metabolic waste, since it was not produced by the body and, indeed, never entered the body's tissues.

Metabolism produces a great quantity of wastes that are lethal to cells if allowed to accumulate. Some of the most toxic examples are carbon dioxide and small nitrogen-containing organic compounds called **nitrogenous wastes** (fig. 23.2). The respiratory system excretes the former and the urinary system excretes both. About 50% of the nitrogenous waste is **urea,** a by-product of protein catabolism. Proteins are broken down to amino acids, and then the $-NH_2$ group is removed from each amino acid. The $-NH_2$ forms ammonia, which is exceedingly toxic but which the liver quickly converts to urea, a less harmful waste:

$$2\ NH_3 + CO_2 \rightarrow H_2N{-}\overset{\overset{\displaystyle O}{\|}}{C}{-}NH_2 + H_2O$$

Other nitrogenous wastes in the urine include **uric acid** and **creatinine** (cree-AT-ih-neen), produced by the catabolism of nucleic acids and creatinine phosphate, respectively. Although less toxic than ammonia and less abundant than urea, these wastes are far from harmless.

Renal failure can lead to **azotemia**[1] (AZ-oh-TEE-me-uh), the accumulation of nitrogenous wastes in the blood. Azotemia may progress to **uremia** (you-REE-me-uh), a syndrome of diarrhea, vomiting, dyspnea, and cardiac arrhythmia stemming from the toxic effects of these wastes. Convulsions, coma, and death can follow within a few days. Unless a kidney transplant is available, renal failure requires *hemodialysis* to remove nitrogenous wastes from the blood (see clinical insight 23.4, p. 908).

Excretion

Excretion is the process of separating wastes from the body fluids and eliminating them. It is carried out by four organ systems:

1. The respiratory system excretes carbon dioxide, small amounts of other gases, and water.
2. The integumentary system excretes water, inorganic salts, lactic acid, and urea in the sweat.
3. The digestive system not only *eliminates* food residue (which is not a process of excretion) but also actively *excretes* water, salts, carbon dioxide, lipids, bile pigments, cholesterol, and other metabolic wastes.
4. The urinary system excretes a broad variety of metabolic wastes, toxins, drugs, hormones, salts, hydrogen ions, and water.

Before You Go On

Answer the following questions to test your understanding of the preceding section:

1. State four functions of the kidneys other than forming urine.
2. List four nitrogenous wastes and their metabolic sources.
3. Name some wastes eliminated by three systems other than the urinary system.

Anatomy of the Kidney

Objectives

When you have completed this section, you should be able to

- identify the major external and internal features of the kidney;
- trace the flow of blood through the kidney; and
- trace the flow of fluid through the renal tubules.

Gross Anatomy

The kidneys lie against the posterior abdominal wall at the level of vertebrae T12 to L3. The right kidney is slightly lower than the left because of the space occupied by the liver above it. Each kidney weighs about 160 g and measures about 12 cm long, 6 cm wide, and 3 cm thick—about the size of a bar of bath soap. The lateral surface is convex while the medial surface is concave and has a slit, the **hilum,** where it receives the renal nerves, blood vessels, lymphatic vessels, and ureter. The adrenal glands rest on the superior poles of the kidneys. The kidneys, adrenal glands, and ureters are retroperitoneal—they lie between the peritoneum and dorsal body wall (see fig. A.10, p. 37).

The kidney is protected by three layers of connective tissue: (1) a fibrous **renal**[2] **fascia,** immediately deep to the

1. *azot* = nitrogen + *emia* = blood condition

2. *ren* = kidney + *al* = pertaining to

parietal peritoneum, which binds the kidney and associated organs to the abdominal wall; (2) the **adipose capsule,** a layer of fat that cushions the kidney and holds it in place; and (3) the **renal capsule,** a fibrous sac that is anchored at the hilum and encloses the rest of the kidney like a cellophane wrapper, and protects the kidney from trauma and infection.

The renal parenchyma—the glandular tissue that forms the urine—appears C-shaped in frontal section. It encircles a medial space, the **renal sinus,** occupied by blood and lymphatic vessels, nerves, and urine-collecting structures (fig. 23.3). Adipose tissue fills the remaining space and holds these structures in place.

The parenchyma is divided into two zones: an outer **cortex** about 1 cm thick and an inner **medulla** facing the sinus. Extensions of the cortex called **renal columns** project toward the sinus and divide the medulla into 6 to 10 **renal (medullary) pyramids.** Each pyramid is conical, with a broad base facing the cortex and a blunt point called the **papilla** facing the sinus. One pyramid and the overlying cortex constitute one *lobe* of the kidney.

The papilla of each renal pyramid is nestled in a cup called a **minor calyx**[3] (CAY-lix), which collects its urine.

3. *calyx* = cup

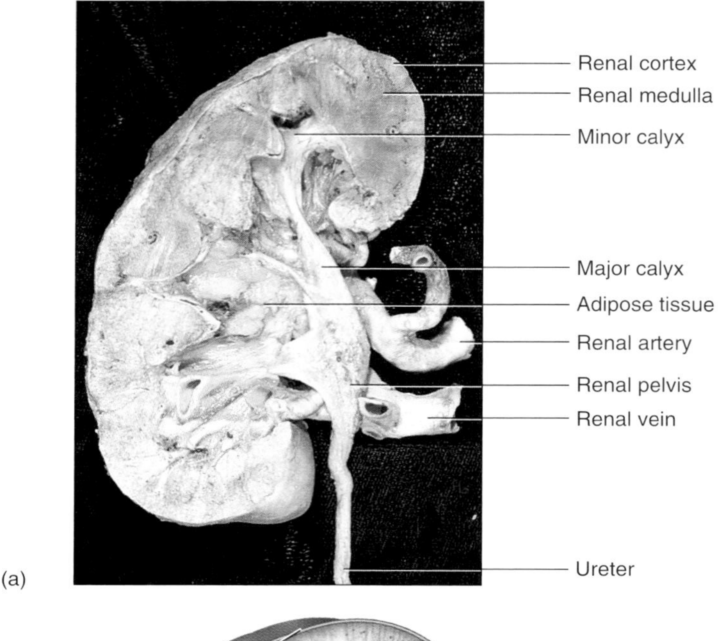

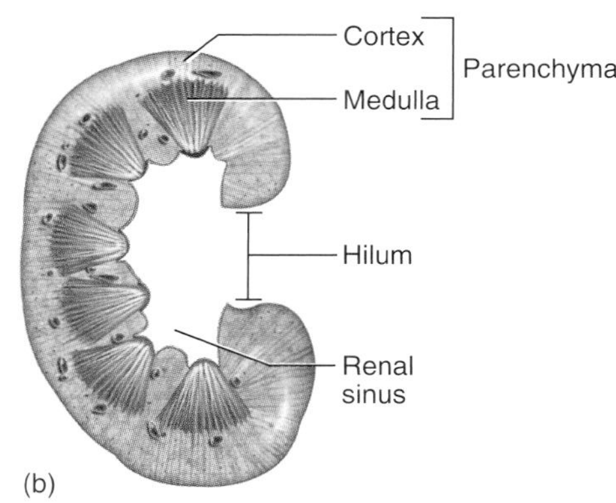

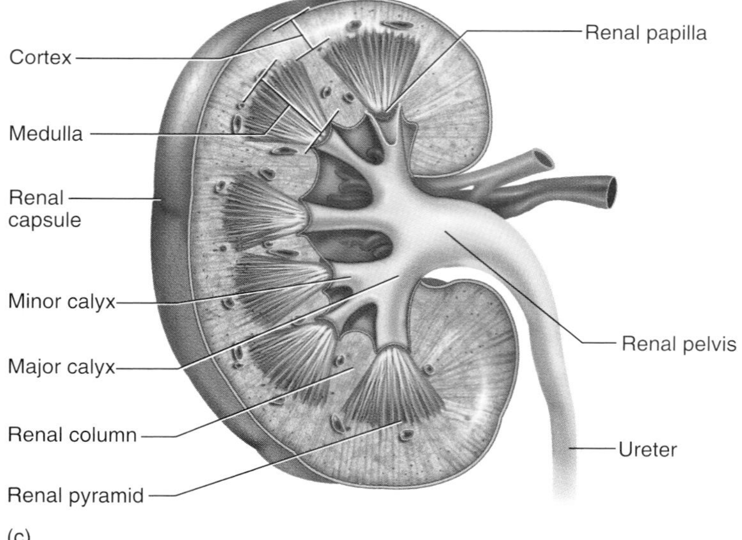

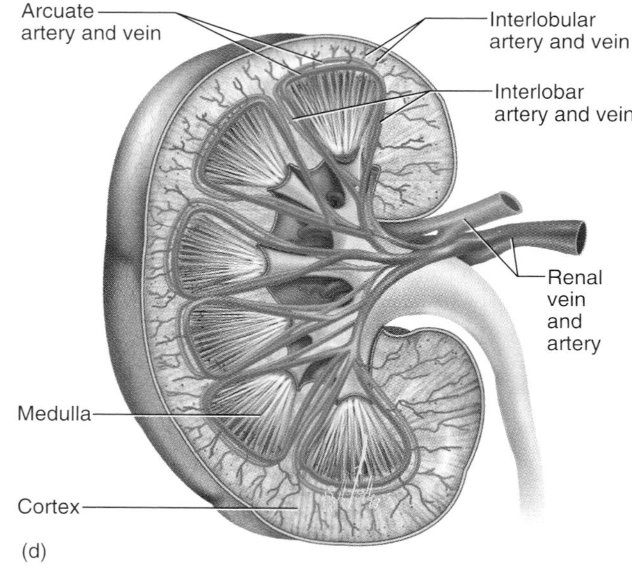

Figure 23.3 Gross Anatomy of the Kidney. (*a*) Photograph of frontal section. (*b*) Ureter, renal pelvis and calices, and blood vessels removed to show the renal sinus. (*c*) Major anatomical features of the kidney. (*d*) The renal blood supply.

Two or three minor calices (CAY-lih-seez) converge to form a **major calyx,** and two or three major calices converge in the sinus to form the funnel-like **renal pelvis.**[4] The ureter is a tubular continuation of the renal pelvis that drains the urine down to the urinary bladder.

The Nephron

Each kidney contains about 1.2 million functional units called **nephrons** (NEF-rons). If you can understand the function of one nephron, you will understand nearly everything necessary about the function of the kidney. A nephron consists of three principal parts: (1) blood vessels, (2) a *renal corpuscle,* and (3) a long *renal tubule* leading to a minor calyx. We examine the structure of these three parts in the following three sections.

Blood Supply

Although the kidneys account for only 0.4% of the body weight, their share of the cardiac output—the *renal fraction*—is about 21% (1.2 L of blood per minute). This attests to their importance in controlling blood volume and composition.

The larger divisions of the renal circulation are shown in figure 23.3*d*. Each kidney is supplied by a **renal artery** (occasionally two or more) arising from the aorta. Just before or after entering the hilum, the renal artery divides and eventually gives rise to a few **interlobar arteries.** One interlobar artery penetrates each renal column and travels between the pyramids to the *corticomedullary junction,* the boundary between the cortex and medulla. Here it branches again to form the **arcuate arteries,** which make a sharp 90° bend and travel along the base of the pyramid. Each arcuate artery gives rise to several **interlobular arteries,** which pass upward into the cortex.

The finer branches of the renal circulation are shown in figure 23.4. As an interlobular artery ascends through the cortex, a series of **afferent arterioles** arise from it like the limbs of a pine tree. Each afferent arteriole supplies blood to one nephron. It ends in a rounded cluster of capillaries called a **glomerulus**[5] (glo-MERR-you-lus), where urine production begins with capillary filtration. The glomerulus is drained by an **efferent arteriole,** which leads next to a plexus of **peritubular capillaries.** These are named for the fact that they form a network amid the renal tubules, described shortly. Blood flows from the peritubular capillaries to, in order, the **interlobular veins, arcuate veins, interlobar veins,** and **renal vein,** which travel parallel to the arteries of the same names. The renal vein leaves the hilum and drains into the inferior vena cava. The principal route of blood flow is thus renal artery → interlobar artery → arcuate artery → interlobular artery → afferent arteriole → glomerulus → efferent arteriole → peritubular capillaries → interlobular vein → arcuate vein → interlobar vein → renal vein → inferior vena cava.

The renal medulla receives only 1% to 2% of the total renal blood flow, supplied by a network of vessels called the **vasa recta.**[6] In nephrons closest to the medulla, the efferent arterioles descend immediately into the medulla and give rise to the vasa recta. The capillaries of the vasa recta lead into venules that ascend and empty into the arcuate and interlobular veins.

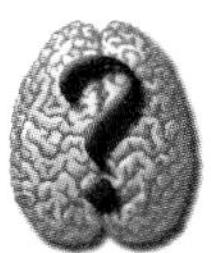

Think About It

Can you identify a portal system in the renal circulation?

The Renal Corpuscle

The **renal corpuscle** (fig. 23.5) consists of a glomerulus enclosed in a two-layered **glomerular (Bowman's**[7]**) capsule.** The parietal layer of the capsule is a simple squamous epithelium, while the visceral layer consists of elaborate cells called **podocytes**[8] wrapped around the capillaries. The podocytes are described in detail later. The fluid that filters from the glomerular capillaries, called the **glomerular filtrate,** collects in the **capsular space** between the parietal and visceral layers and then flows into the renal tubule on one side of the capsule.

The Renal Tubule

The **renal (uriniferous**[9]**) tubule** is a duct that leads away from the glomerular capsule and ends at the tip of a medullary pyramid. It is about 3 cm long and divided into four major regions: the *proximal convoluted tubule, nephron loop, distal convoluted tubule,* and *collecting duct* (see fig. 23.4). The first three of these belong to an individual nephron, whereas many nephrons drain into each collecting duct. Each region has unique physiological properties and roles in the production of urine.

The Proximal Convoluted Tubule The **proximal convoluted tubule (PCT)** arises from the glomerular capsule. It is the longest and most coiled of the four regions and thus dominates histological sections of renal cortex. The PCT has a simple cuboidal epithelium with prominent microvilli (a brush border), which attests to the great deal of

4. *pelvis* = basin
5. *glomer* = ball + *ulus* = little

6. *vasa* = vessels + *recta* = straight
7. Sir William Bowman (1816–92), British physician
8. *podo* = foot + *cyte* = cell
9. *urin* = urine + *fer* = to carry

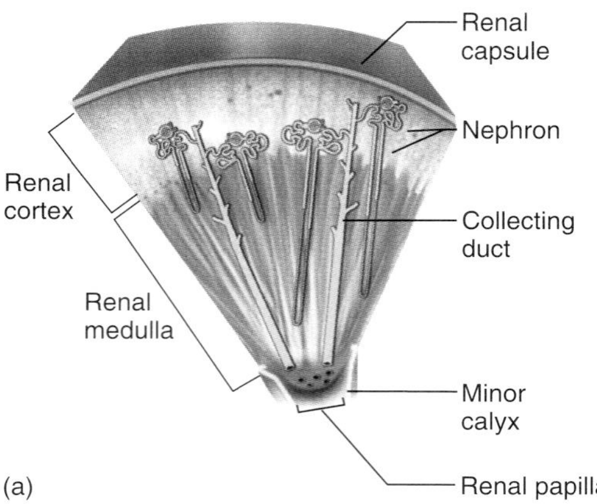

Figure 23.4 Structure of the Nephron. (*a*) Location of the nephrons in one wedge-shaped lobe of the kidney. (*b*) Structure of two nephrons. For clarity, vasa recta are shown only on the *left* and peritubular capillaries only on the *right*. Note that juxtamedullary nephrons are closer to the corticomedullary junction and have longer nephron loops than cortical nephrons. Vasa recta come only from the nephrons closest to the medulla. (*c*) The true proportions of the nephron loops relative to the convoluted tubules.

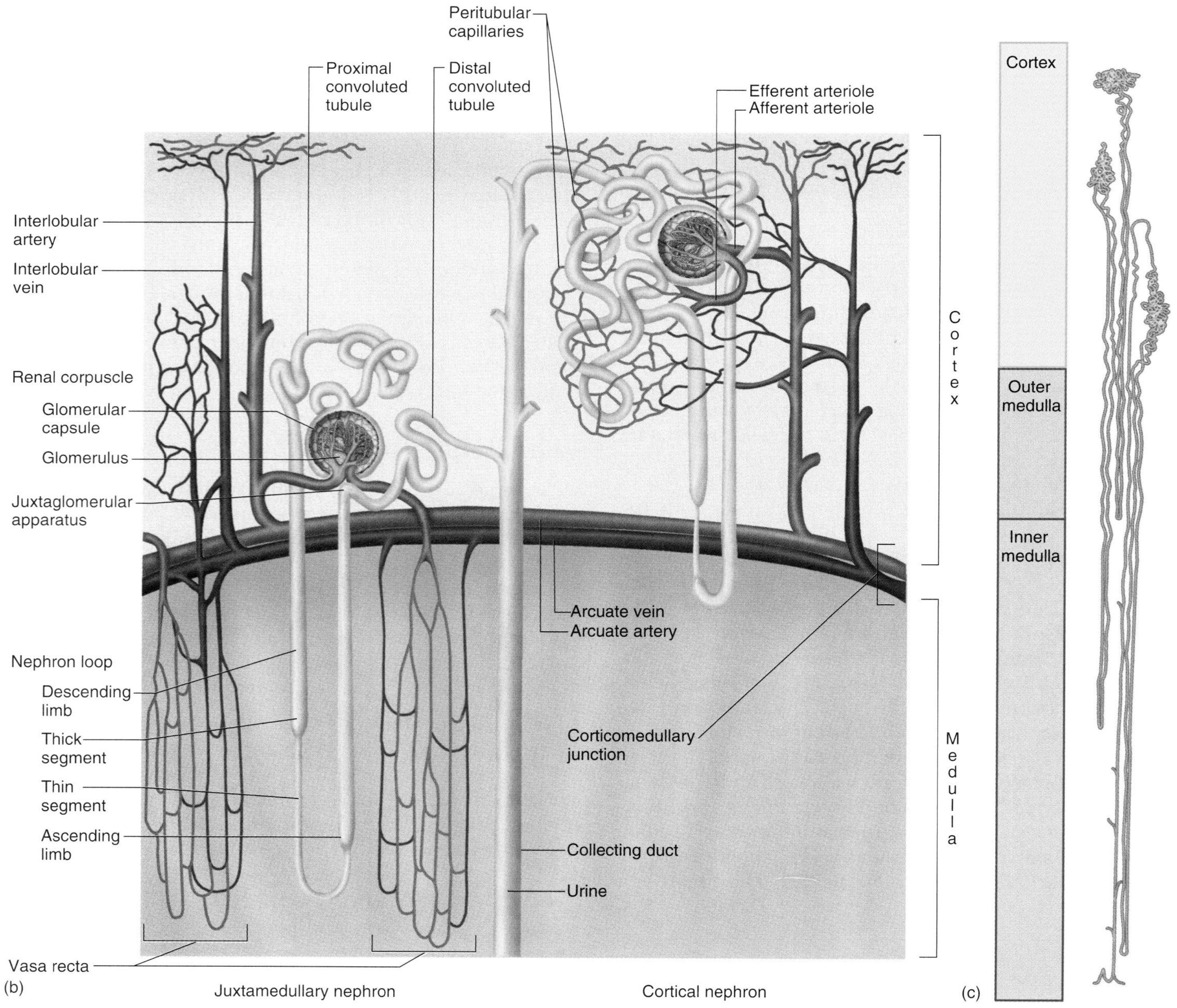

Peritubular capillaries
Proximal convoluted tubule
Distal convoluted tubule
Efferent arteriole
Afferent arteriole
Interlobular artery
Interlobular vein
Renal corpuscle
Glomerular capsule
Glomerulus
Juxtaglomerular apparatus
Cortex
Arcuate vein
Arcuate artery
Nephron loop
Descending limb
Thick segment
Thin segment
Ascending limb
Corticomedullary junction
Medulla
Collecting duct
Urine
Vasa recta
(b)
Juxtamedullary nephron
Cortical nephron

Cortex
Outer medulla
Inner medulla
(c)

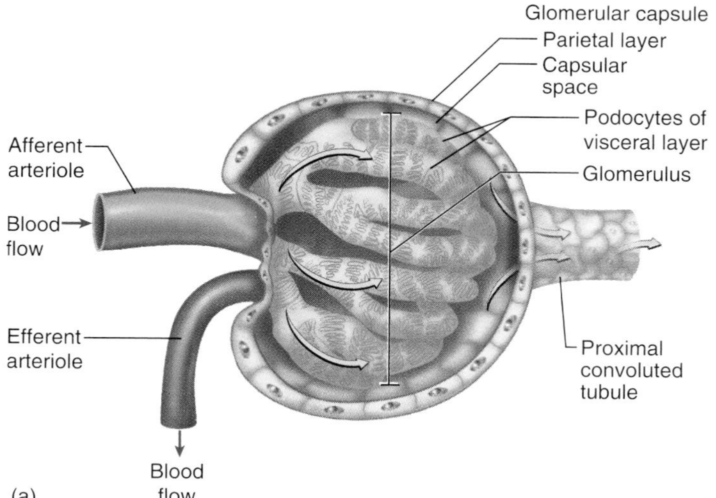

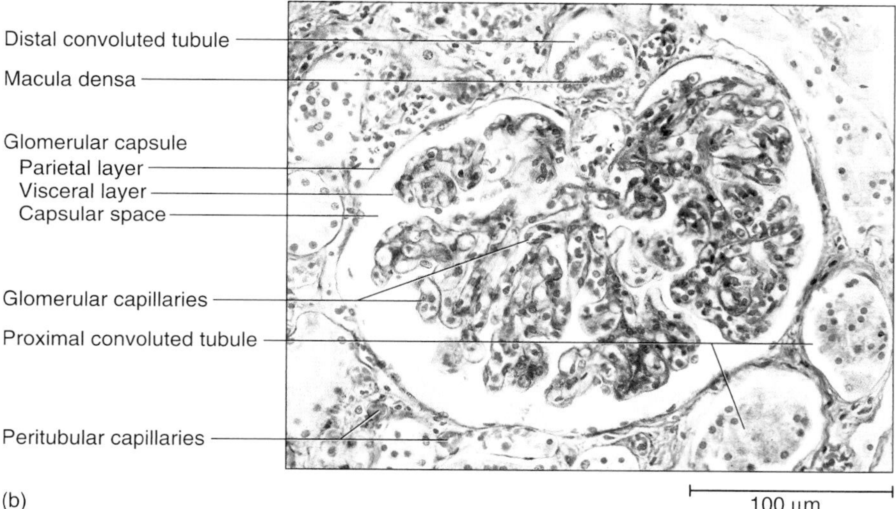

Figure 23.5 **The Renal Corpuscle.** (*a*) Anatomy of the corpuscle; (*b*) light micrograph.

absorption that occurs here. The microvilli give the epithelium a distinctively shaggy look in tissue sections.

The Nephron Loop After coiling extensively near the renal corpuscle, the PCT straightens out and forms a long U-shaped **nephron loop (loop of Henle[10]).** The first portion of the loop, the **descending limb,** passes from the cortex into the medulla. At its lower end it turns 180° and forms an **ascending limb** that returns to the cortex. The nephron loop is divided into thick and thin segments. The **thick segments** have a simple cuboidal epithelium. They form the initial part of the descending limb and part or all of the ascending limb. The cells here are heavily engaged in active transport of salts, so they have very high metabolic activity and are loaded with mitochondria. The **thin segment** has a simple squamous epithelium. It forms the lower part of the descending limb, and in some nephrons, it rounds the bend and continues partway up the ascending limb. The cells here have low metabolic activity but are very permeable to water.

The Distal Convoluted Tubule When the nephron loop returns to the cortex, it coils again and forms the **distal convoluted tubule (DCT).** This is shorter and less convoluted than the PCT, so fewer sections of it are seen in histological sections. It has a cuboidal epithelium with smooth-surfaced cells nearly devoid of microvilli. The DCT is the end of the nephron.

When the DCT returns to the renal corpuscle, and just before it starts coiling, it contacts the afferent and efferent arterioles. The three structures form the **juxtaglomerular** (JUX-tuh-glo-MER-you-lur) **apparatus,** a device that enables a nephron to monitor and stabilize its own performance and to compensate for fluctuations in blood pressure. It will be described in detail when we consider renal autoregulation.

10. Friedrich G.J. Henle (1809–85), German anatomist

The Collecting Duct The DCTs of several nephrons drain into a straight tubule called the **collecting duct,** which passes down into the medulla. Near the papilla, several collecting ducts merge to form a larger **papillary duct;** about 30 of these drain from each papilla into its minor calyx. The collecting and papillary ducts are lined with simple cuboidal epithelium.

The flow of fluid from the point where the glomerular filtrate is formed to the point where urine leaves the body is: glomerular capsule → proximal convoluted tubule → nephron loop → distal convoluted tubule → collecting duct → papillary duct → minor calyx → major calyx → renal pelvis → ureter → urinary bladder → urethra.

Cortical and Juxtamedullary Nephrons

Nephrons just beneath the renal capsule, close to the kidney surface, are called **cortical nephrons.** They have relatively short nephron loops that dip only slightly into the outer medulla before turning back (see fig. 23.4*b, c*) or turn back even before leaving the cortex. Some cortical nephrons have no nephron loops at all. Nephrons close to the medulla are called **juxtamedullary**[11] **nephrons.** They have very long nephron loops that extend to the apex of the renal pyramid. As you will see later, nephron loops are responsible for maintaining a salt gradient in the medulla that helps the body conserve water. This gradient is produced almost solely by the juxtamedullary nephrons. It is also the juxtamedullary nephrons that give rise to the vasa recta, which spring from their efferent arterioles. These arterioles give off few or no peritubular capillaries, although the convoluted tubules of these nephrons are served by peritubular capillaries that originate with other nephrons higher up in the cortex. Only about 15% of the nephrons are juxtamedullary.

Before You Go On

Answer the following questions to test your understanding of the preceding section:

4. Arrange the following in order from the most numerous to the least numerous structures in a kidney: glomeruli, major calices, minor calices, interlobular arteries, interlobar arteries.
5. Trace the path taken by one red blood cell from the renal artery to the renal vein.
6. Consider one molecule of urea in the urine. Trace the route that it took from the point where it left the bloodstream to the point where it left the body.

11. *juxta* = next to

Urine Formation I: Glomerular Filtration

Objectives

When you have completed this section, you should be able to

- describe the glomerular filtration membrane and how it excludes blood cells and proteins from the filtrate;
- explain the forces that promote and oppose glomerular filtration, and calculate net filtration pressure if given the magnitude of these forces; and
- describe how the nervous system, hormones, and the kidney itself regulate glomerular filtration.

The kidney converts blood plasma to urine essentially by four processes: glomerular filtration, tubular reabsorption, tubular secretion, and water conservation (fig. 23.6). As we trace fluid through the nephron, we will refer to it by dif-

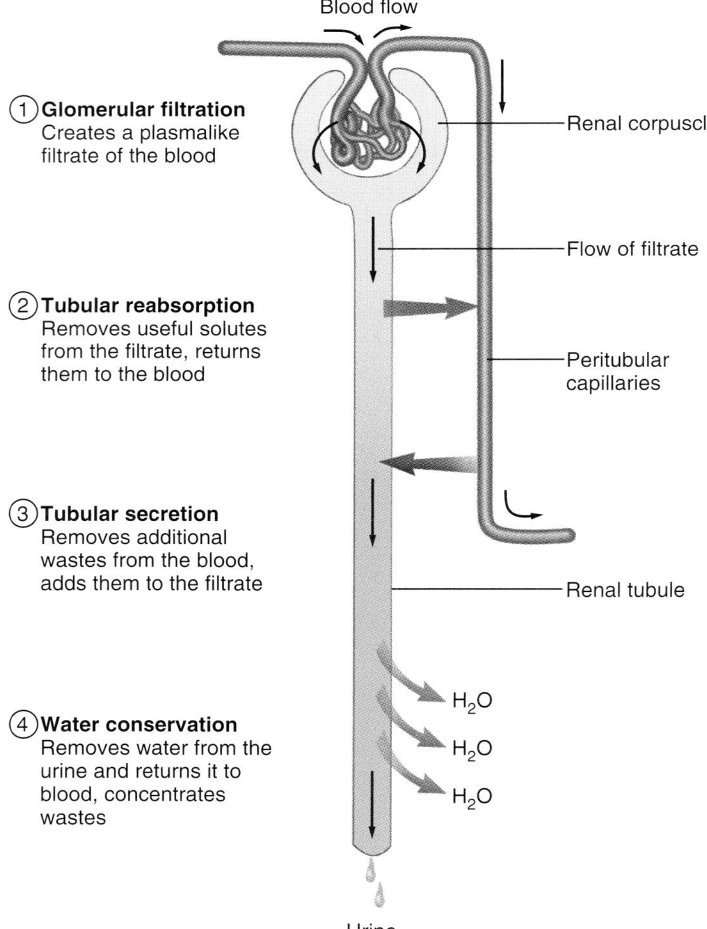

Figure 23.6 Basic Steps in the Formation of Urine.

ferent names that reflect its changing composition: (1) The fluid in the capsular space, called **glomerular filtrate,** is similar to blood plasma except that it has almost no protein. (2) The fluid from the proximal convoluted tubule through the distal convoluted tubule will be called **tubular fluid.** It differs from the glomerular filtrate because of substances removed and added by the tubule cells. (3) The fluid will be called **urine** once it enters the collecting duct.

The Filtration Membrane

Glomerular filtration, discussed in this section, is a special case of the capillary fluid exchange process described in chapter 20. It is a process in which water and some solutes in the blood plasma pass from the capillaries of the glomerulus into the capsular space of the nephron. To do so, fluid passes through three barriers that constitute the **filtration membrane** (fig. 23.7):

1. **The fenestrated endothelium of the capillary.** Endothelial cells of the glomerular capillaries look like honeycombs with large pores about 70 to 90 nm in diameter (see fig. 20.6, p. 754). They are much more permeable than endothelial cells elsewhere, although their pores are small enough to exclude blood cells from the filtrate.
2. **The basement membrane.** This membrane consists of a proteoglycan gel. For large molecules to pass through it is like trying to pass sand through a kitchen sponge. A few particles may penetrate its small spaces, but most are held back. On the basis of size alone, the basement membrane would exclude any molecules larger than 8 nm. Some smaller molecules, however, are also prevented from passing by a negative electrical charge on the proteoglycans. Blood albumin is slightly less than 7 nm in diameter, but it is also negatively charged and thus repelled by the basement membrane. While the blood plasma is 7% protein, the glomerular filtrate is only 0.03% protein. It has traces of albumin and smaller polypeptides, including some hormones.
3. **Filtration slits.** The podocytes of the glomerular capsule are shaped somewhat like octopi, with bulbous cell bodies and several thick arms. Each arm has numerous little extensions called **foot processes (pedicels**[12]**)** that wrap around the capillaries and interdigitate with each other, like wrapping your hands around a pipe and lacing your fingers together. The foot processes have negatively charged **filtration slits** about 30 nm wide between them, which are an additional obstacle to large anions.

12. *pedi* = foot + *cel* = little

Almost any molecule smaller than 3 nm can pass freely through the filtration membrane into the capsular space. This includes water, electrolytes, glucose, amino acids, nitrogenous wastes, and vitamins. Such substances have about the same concentration in the glomerular filtrate as in the blood plasma. Some substances of low molecular weight are retained in the bloodstream because they are bound to plasma proteins that cannot get through the membrane. For example, most calcium, iron, and thyroid hormone in the blood are bound to plasma proteins that retard their filtration by the kidneys. The small fraction that is unbound, however, passes freely through the filtration membrane and appears in the urine.

Kidney infections and trauma commonly damage the filtration membrane and allow albumin or blood cells to filter through. **Proteinuria (albuminuria)** (protein, especially albumin, in the urine) and **hematuria** (blood in the urine) are common signs of kidney disease.

Filtration Pressure

Glomerular filtration follows the same principles that govern filtration in other blood capillaries, but there are significant differences in the magnitude of the forces involved:

- The blood hydrostatic pressure (BHP) is much higher here than elsewhere—about 60 mmHg compared with 10 to 15 mmHg in most other capillaries. This results from the fact that the afferent arteriole is substantially larger than the efferent arteriole, giving the glomerulus a large inlet and small outlet (fig. 23.7*a*).
- The hydrostatic pressure in the capsular space is about 18 mmHg, compared with the slightly negative interstitial pressures elsewhere. This results from the high rate of filtration occurring here and the continual accumulation of fluid in the capsule.
- The colloid osmotic pressure (COP) of the blood is about the same here as anywhere else, 32 mmHg.
- The glomerular filtrate is almost protein-free and has no significant COP. (This can change markedly in kidney diseases that allow protein to filter into the capsular space.)

On balance, then, we have a high outward pressure of 60 mmHg, opposed by two inward pressures of 18 and 32 mmHg (fig. 23.8), giving a net filtration pressure (NFP) of

$$60_{\text{out}} - (18_{\text{in}} + 32_{\text{in}}) = 10 \text{ mmHg}_{\text{out}}$$

In most blood capillaries, the BHP drops low enough at the venous end that osmosis overrides filtration and the capillaries reabsorb fluid. Although BHP also drops along the course of the glomerular capillaries, it remains high

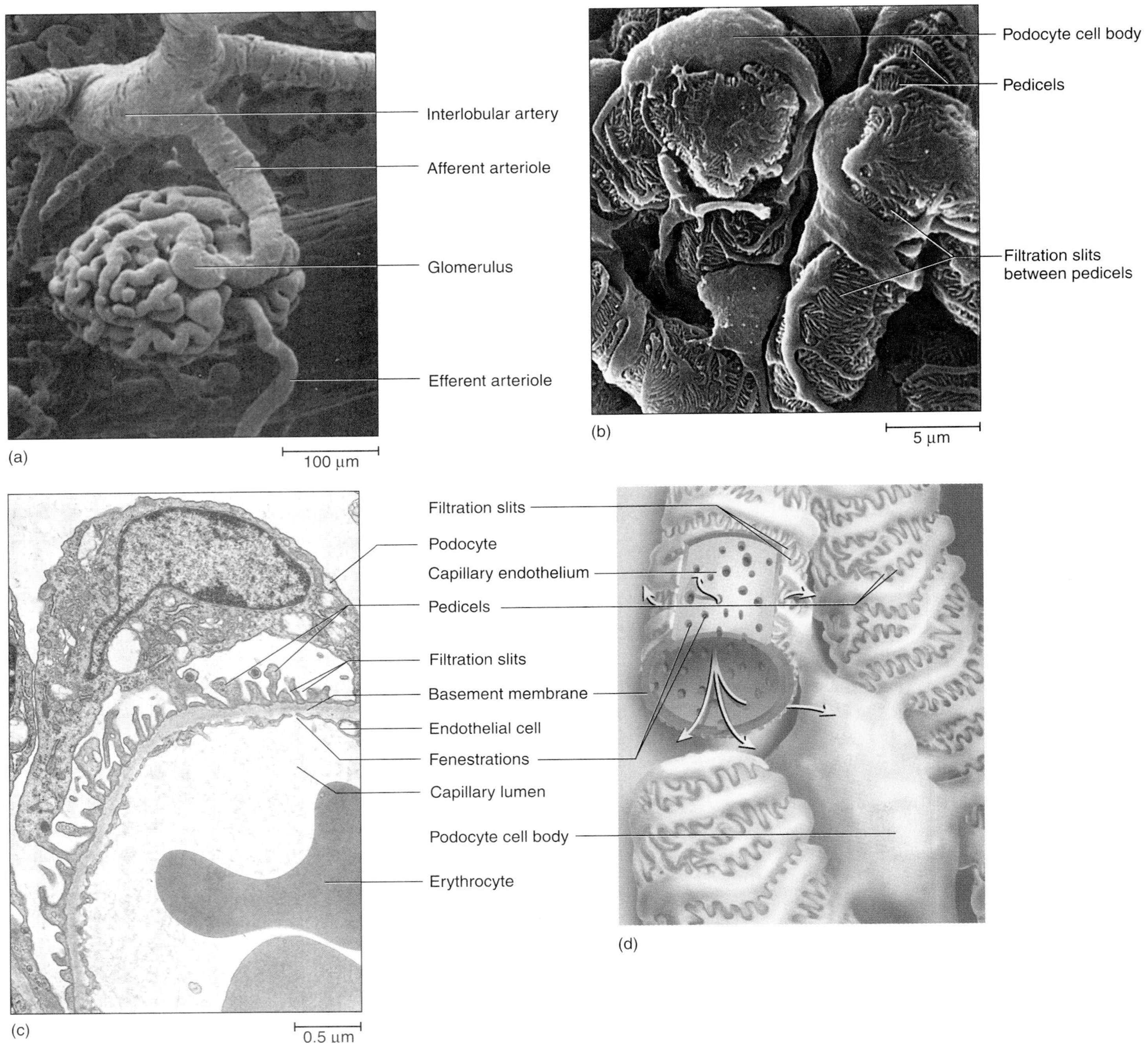

Figure 23.7 Structure of the Glomerulus. (*a*) A resin cast of the glomerulus and nearby arteries (SEM). Note that the efferent arteriole is much narrower than the afferent arteriole, which causes blood pressure in the glomerulus to be unusually high. (*b*) Blood capillaries of the glomerulus closely wrapped in the spidery podocytes that form the visceral layer of the glomerular capsule (SEM). (*c*) A blood capillary and podocyte showing fenestrations and filtration slits (TEM). (*d*) The production of glomerular filtrate by the passage of fluid through the fenestrations and filtration slits. (*a*) From R. G. Kessel and R. H. Kardon, *Tissues and Organs: A Text-Atlas of Scanning Electron Microscopy* (W. H. Freeman, 1979).

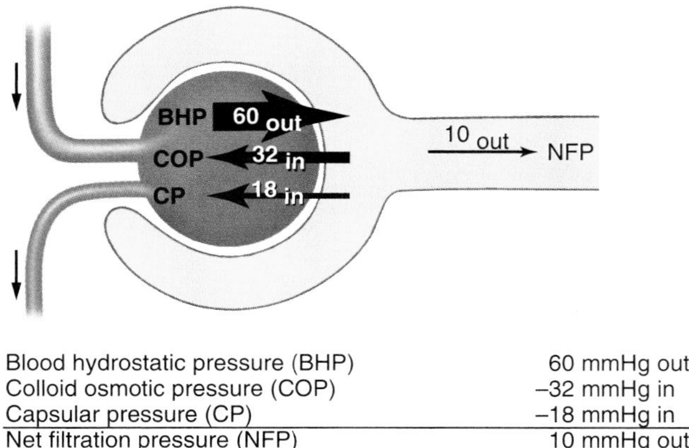

Blood hydrostatic pressure (BHP)	60 mmHg out
Colloid osmotic pressure (COP)	−32 mmHg in
Capsular pressure (CP)	−18 mmHg in
Net filtration pressure (NFP)	10 mmHg out

Figure 23.8 The Forces Involved in Glomerular Filtration.

enough that these capillaries are engaged solely in filtration. They reabsorb little or no fluid.

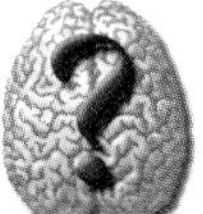

Think About It

According to the preceding equation, how low would capillary BHP have to be for glomerular filtration and urine output to cease?

The high blood pressure in the glomeruli makes the kidneys especially vulnerable to hypertension, which can have devastating effects on renal function. Hypertension ruptures glomerular capillaries and leads to scarring of the kidneys (*nephrosclerosis*). It promotes atherosclerosis of the renal blood vessels just as it does elsewhere in the body and thus diminishes renal blood supply. Over time, hypertension often leads to renal failure.

Glomerular Filtration Rate

Glomerular filtration rate (GFR) is the amount of filtrate formed per minute by the two kidneys combined. For every 1 mmHg of net filtration pressure, the kidneys produce about 12.5 mL of filtrate per minute. This value, called the *filtration coefficient* (K_f), depends on the permeability and surface area of the filtration barrier. K_f is about 10% lower in women than in men. For an average adult male,

$$GFR = NFP \times K_f = 10 \times 12.5 = 125 \text{ mL/min}$$

This is a rate of 180 L/day—an impressive number considering that it is about 60 times the amount of blood plasma in the body and 60 times the amount of filtrate produced by all other capillaries combined. Obviously only a small portion of this is eliminated as urine. An average adult reabsorbs 99% of the filtrate and excretes 1 to 2 L of urine per day.

Regulation of Glomerular Filtration

GFR must be precisely controlled. If it is too high, fluid flows through the renal tubules too rapidly for them to reabsorb the usual amount of water and solutes. Urine output rises and creates a threat of dehydration and electrolyte depletion. If GFR is too low, fluid flows sluggishly through the tubules, they reabsorb wastes that should be eliminated in the urine, and azotemia may occur. The only way to adjust GFR from moment to moment is to change glomerular blood pressure. This is achieved by three homeostatic mechanisms: renal autoregulation, sympathetic control, and a hormonal mechanism involving renin and angiotensin.

Renal Autoregulation

Renal autoregulation is the ability of the kidneys to maintain a relatively stable GFR in spite of changes in arterial blood pressure. If the mean arterial pressure (MAP) rose from 100 to 125 mmHg and there were no renal autoregulation, urine output would increase from the normal 1 to 2 L/day to more than 45 L/day. Because of renal autoregulation, however, urine output increases only a few percent even if MAP rises as high as 160 mmHg.

The nephron has two ways to prevent drastic changes in GFR when blood pressure rises: it can constrict the afferent arteriole and reduce blood flow into the glomerulus, or it can dilate the efferent arteriole and allow the blood to flow out more easily. Conversely, if blood pressure falls, the nephron can compensate by either dilating the afferent arteriole or constricting the efferent arteriole. In actuality, the two arterioles work in unison to fine-tune glomerular blood pressure.

The juxtaglomerular apparatus enables each nephron to respond to blood pressure fluctuations. It has three major components (fig. 23.9):

1. The **juxtaglomerular (JG) cells** are enlarged smooth muscle cells found in the afferent arteriole and to some extent in the efferent arteriole. When stimulated by changes in blood pressure or by the macula densa (discussed next), they dilate or constrict the arterioles accordingly. They also contain granules of renin, which they secrete in response to a drop in blood pressure. This initiates negative feedback mechanisms, described later, that raise blood pressure.
2. The **macula densa**[13] is a patch of slender, closely spaced epithelial cells in the distal convoluted tubule (DCT), directly across from the JG cells. It is thought to monitor the salinity of tubular fluid in the DCT.
3. **Mesangial**[14] **cells** are found in the cleft between the afferent and efferent arterioles and among capillaries

13. *macula* = spot, patch + *densa* = dense
14. *mes* = in the middle + *angi* = vessel

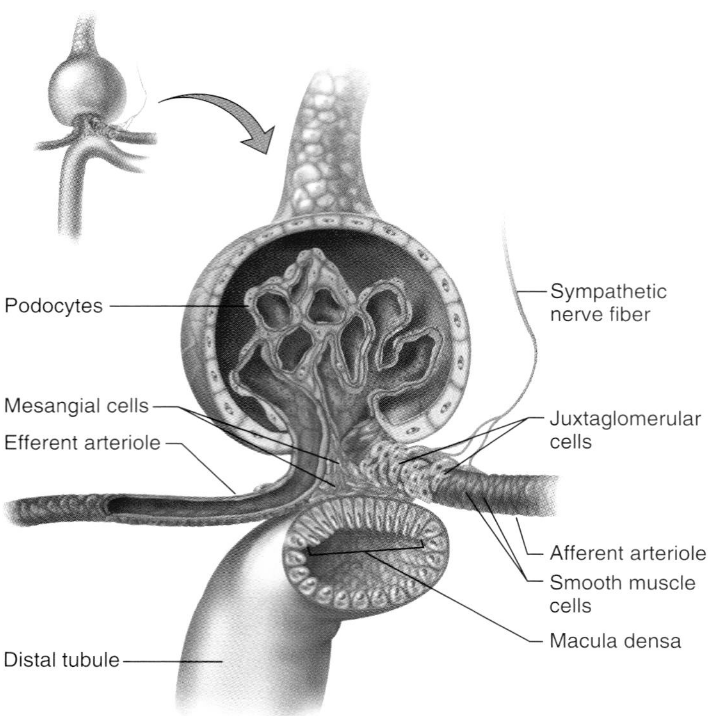

Figure 23.9 The Juxtaglomerular Apparatus.

of the glomerulus. Their role is not yet clearly understood, but they are connected to the macula densa and JG cells by gap junctions and perhaps mediate communication between those cells.

The mode of action of the juxtaglomerular apparatus has not been fully determined. One hypothesis is that a high GFR causes fluid to flow too rapidly through the renal tubule to allow for normal reabsorption. This results in an abnormally high NaCl concentration detected by the macula densa. The macula densa may then release a paracrine secretion that stimulates the afferent arteriole to constrict, stimulates the efferent arteriole to dilate, or both. This would reduce the GFR to a normal level and complete a negative feedback loop (fig. 23.10).

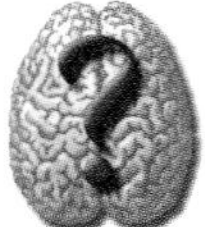

Think About It

Describe or diagram a similar negative feedback loop to show how the macula densa could compensate for a drop in systemic blood pressure.

Two important points must be noted about renal autoregulation. First, it does not completely prevent changes in the GFR. Like any other homeostatic mechanism, it maintains a *dynamic equilibrium;* the GFR fluctuates within narrow limits. Changes in blood pressure do affect the GFR and urine output. When you drink a large amount of water, for

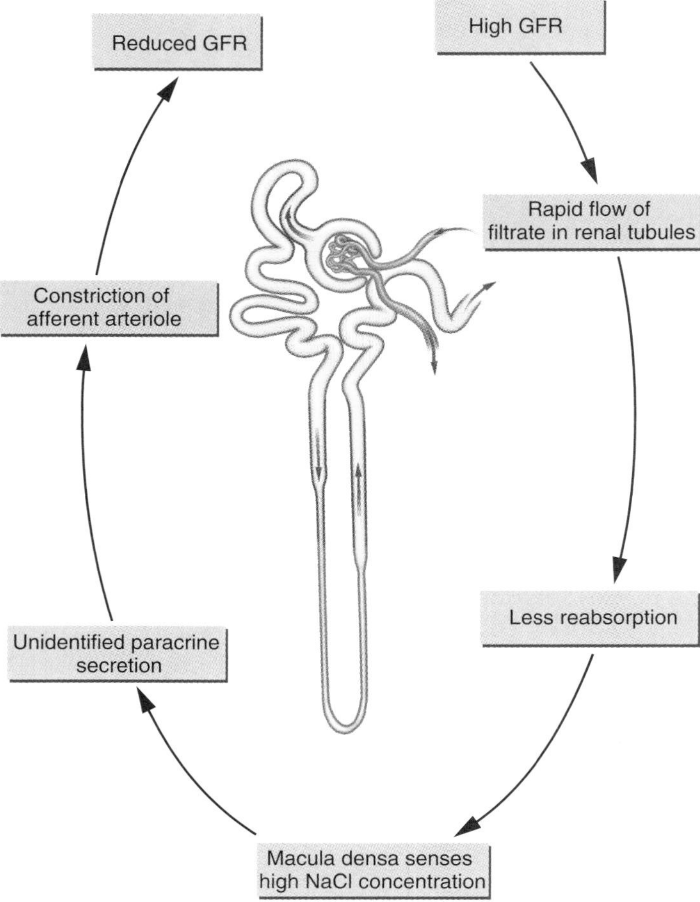

Figure 23.10 Negative Feedback Control of Glomerular Filtration Rate.

example, it increases your blood pressure and GFR causing *water diuresis*[15] (DY-you-REE-sis)—an increased urine output stemming from increased water intake.

Second, renal autoregulation cannot compensate for extreme blood pressure variations. Over a MAP range of 80 to 170 mmHg, the GFR remains quite stable. Below 45 mmHg, however, glomerular filtration and urine output cease. This can happen in hypovolemic shock (p. 768).

Sympathetic Control

Sympathetic nerve fibers richly innervate the renal blood vessels. In strenuous exercise or acute conditions such as circulatory shock, the sympathetic nervous system and adrenal epinephrine stimulate the afferent arterioles to constrict. This reduces GFR and urine production, while redirecting

15. *diuresis* = passing urine

blood from the kidneys to the heart, brain, and skeletal muscles, where it is more urgently needed. Under such conditions, GFR may be as low as a few milliliters per minute.

The Renin-Angiotensin Mechanism

When blood pressure drops, the juxtaglomerular cells secrete the enzyme **renin** (REE-nin), which removes 10 amino acids—a fragment called *angiotensin I*—from the plasma protein *angiotensinogen*. As angiotensin I circulates through the lungs, **angiotensin-converting enzyme (ACE)** removes two more amino acids, converting it to **angiotensin II,** a hormone with multiple effects (fig. 23.11):

- It stimulates widespread vasoconstriction, which raises the MAP throughout the body.
- Although it reduces total renal blood flow, it has an especially strong effect on the efferent arterioles. By constricting these, it prevents the glomerular blood pressure from dropping too low, thus maintaining GFR and waste excretion.
- It increases tubular reabsorption of water by mechanisms explained later and thus reduces water loss.
- It stimulates the adrenal cortex to secrete aldosterone, which in turn promotes sodium and water retention.
- It stimulates the sense of thirst and encourages water intake.

Some of these effects are explained more fully later in this chapter and in the next one.

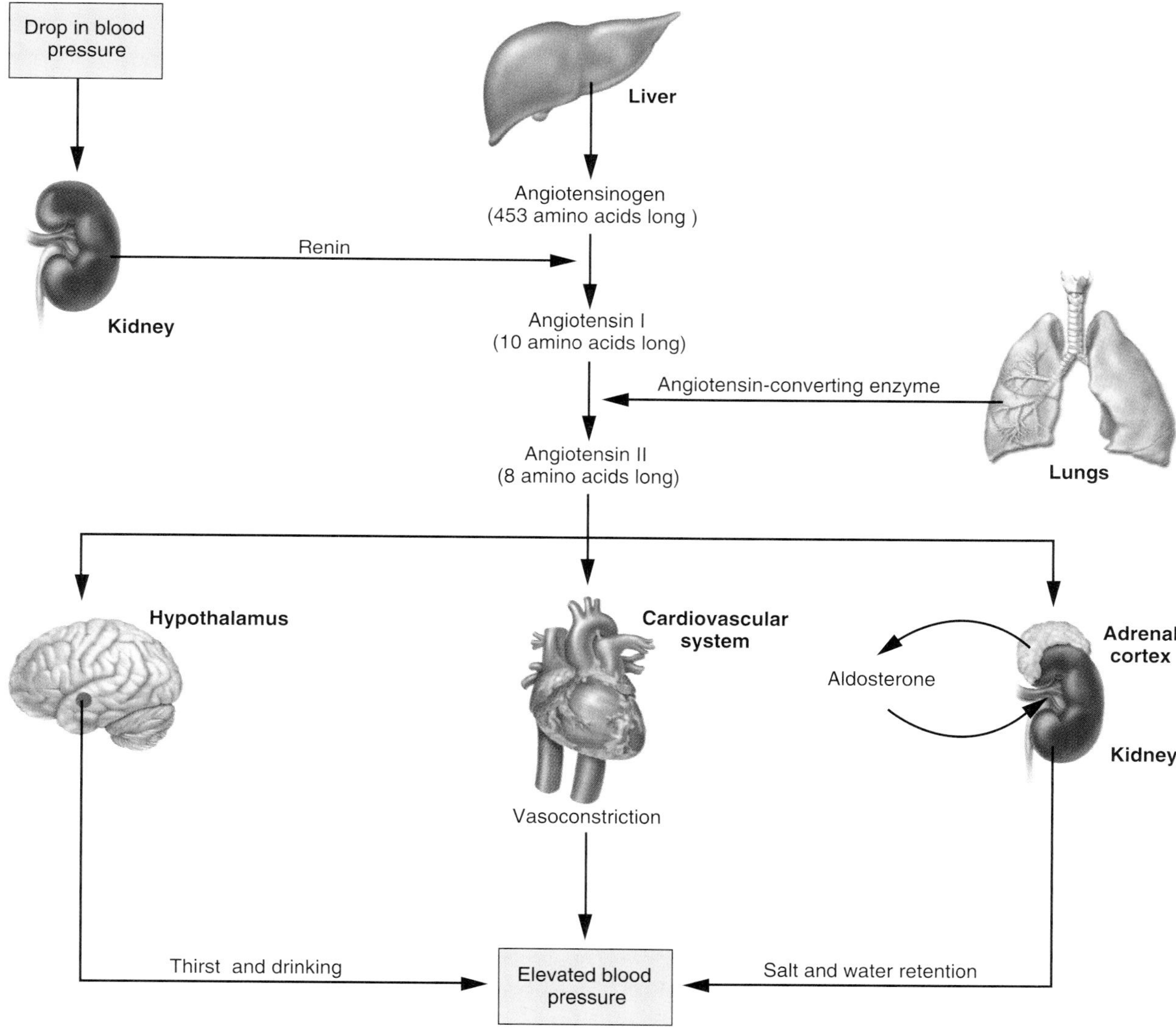

Figure 23.11 The Renin-Angiotensin-Aldosterone Mechanism. This chain of events is activated by a drop in blood pressure and acts to raise it again.

Think About It

What do you predict would be the effect of ACE inhibitors (see p. 761) on the tubular reabsorption of water by the kidneys?

To summarize the events thus far: Glomerular filtration occurs because the high blood pressure of the glomerular capillaries overrides reabsorption. The filtration membrane allows most plasma solutes into the capsular space while retaining formed elements and protein in the bloodstream. Glomerular filtration is maintained at a fairly steady rate of about 125 mL/min in spite of variations in systemic blood pressure. This stability is achieved by renal autoregulation at the juxtaglomerular apparatus, by sympathetic nervous control, and by the actions of epinephrine and the renin-angiotensin system.

Before You Go On

Answer the following questions to test your understanding of the preceding section:

7. Name the four major processes in urine production.
8. Trace the movement of a urea molecule from the blood to the capsular space, and name the barriers it passes through.
9. Calculate the net filtration pressure in a patient whose blood COP is only 10 mmHg because of hypoproteinemia. Assume other relevant variables to be normal.
10. Assume a person is moderately dehydrated and has low blood pressure. Describe the homeostatic mechanisms that would help the kidneys maintain a normal GFR.

Urine Formation II: Tubular Reabsorption and Secretion

Objectives

When you have completed this section, you should be able to

- describe how the renal tubules reabsorb useful solutes from the glomerular filtrate and return them to the blood;
- describe how the tubules secrete solutes from the blood into the tubular fluid; and
- describe how the nephron regulates water excretion.

Conversion of the glomerular filtrate to urine involves the removal and addition of chemicals by tubular reabsorption and secretion, to be described in this section. Here we trace the course of the urine through the nephron, from proximal convoluted tubule to collecting duct, and see how the filtrate is modified by these tubules at each point along the way. Refer to figure 23.6 to put these processes into perspective.

The Proximal Convoluted Tubule

The proximal convoluted tubule (PCT) reabsorbs about 65% of the glomerular filtrate and returns it to the blood in the peritubular capillaries. The importance of this is reflected in the relatively great length and prominent microvilli of the PCT, which increase its absorptive surface area. Its cells also contain abundant large mitochondria that provide ATP for active transport. Your PCTs alone account for about 6% of your resting ATP and calorie consumption.

Tubular Reabsorption

While the glomerular capillaries are engaged entirely in filtration, the peritubular capillaries just downstream from them are engaged mainly in **tubular reabsorption.** One reason for this shift in function is that approximately 20% of the water in the blood is filtered out by the glomerulus while the protein stays behind. Therefore, the blood leaving the glomerulus has an unusually high colloid osmotic pressure (COP). Furthermore, the narrowness of the efferent arteriole reduces the blood hydrostatic pressure (BHP) from about 60 mmHg in the glomerular capillaries to only 8 mmHg in the peritubular capillaries—even lower when this arteriole is constricted by antiogensin II (fig. 23.12). With

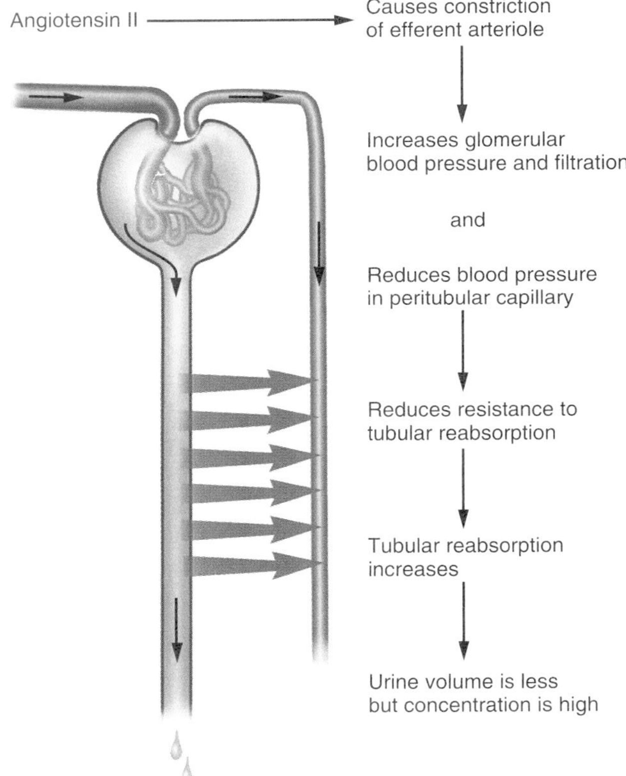

Figure 23.12 The Effect of Angiotensin II on Urine Volume and Concentration.

an elevated COP and reduced BHP, the balance of forces in the peritubular capillaries favors reabsorption. Water is reabsorbed by osmosis and carries other solutes with it.

The PCT reabsorbs a greater variety of chemicals than any other part of the nephron. Some substances are taken up by the tubule epithelial cells, pass through their cytoplasm, and are then transported out, or diffuse out, the base of the cell. This is called the **transcellular**[16] **route** of reabsorption. Other substances pass between the epithelial cells, a path called the **paracellular**[17] **route.** Even though the epithelial cells are held together by tight junctions, these junctions are quite leaky and allow significant amounts of water, minerals, urea, and other matter to pass between the cells. Either way, such materials enter the extracellular fluid (ECF) at the base of the epithelium, and from there they can be taken up by the peritubular capillaries. In the following discussion and figure 23.13, we examine mechanisms for the reabsorption of water and individual solutes.

16. *trans* = across

17. *para* = next to

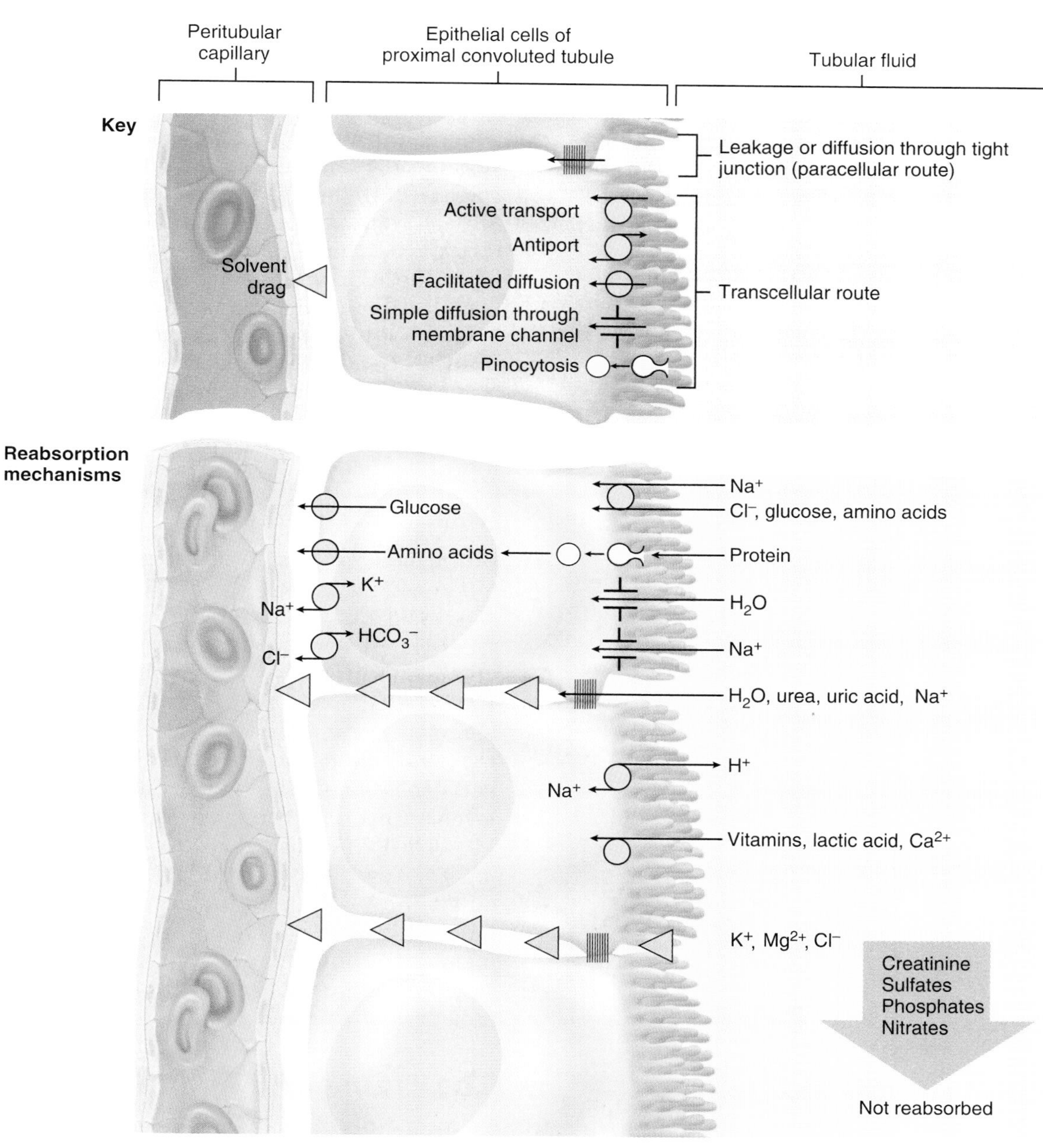

Figure 23.13 Mechanisms of Reabsorption in the Proximal Convoluted Tubule.

Sodium Sodium, the most abundant cation in the glomerular filtrate, is reabsorbed by both transcellular and paracellular routes. It has a concentration of 140 mEq/L in the fluid entering the PCT and only 12 mEq/L in the cytoplasm of the epithelial cells. Thus there is a steep concentration gradient favoring the diffusion of Na^+ into the epithelial cell. Some enters the cell by simple diffusion through membrane channels, and some is reabsorbed by facilitated diffusion using transport proteins in the apical plasma membrane.

This explains how Na^+ enters the epithelial cell, but we also must explain how the cell maintains the low intracellular Na^+ concentration that makes this possible. This is achieved by Na^+-K^+ pumps in the basal and lateral plasma membrane, which continually remove Na^+ from the cytoplasm and pump it into the ECF around the peritubular capillaries. As water enters the peritubular capillaries, it takes sodium with it by solvent drag. Sodium reabsorption is especially important to renal function because it creates a concentration gradient that makes it possible to reabsorb other substances, as we shall see next.

Glucose Some of the apical Na^+ carriers bind simultaneously to a sodium ion and a glucose molecule and transport both of them into the cell. These carries are called **sodium-glucose transport proteins (SGLTs).** Glucose leaves the basal surface of the cell by facilitated diffusion and enters the peritubular capillaries by solvent drag. Normally all glucose in the tubular fluid is reabsorbed and there is none in the urine.

Glucose reabsorption exemplifies certain principles of membrane transport described in chapter 4. The SGLT is an example of a *symport* system, a cotransport carrier that moves two or more solutes across the membrane in the same direction. It is also a case of *secondary active transport,* a system that does not directly consume ATP but depends on other transporters that do. Glucose is more concentrated within the cell than in the tubular fluid, so when it enters the cell it is moving up its concentration gradient. The SGLT itself does not use ATP in doing this, but it depends on the low intracellular Na^+ concentration maintained by the Na^+-K^+ pumps at the base of the cell. Thus the absorption of glucose indirectly depends on ATP.

Amino Acids Amino acids are reabsorbed in the same way as glucose. Some apical Na^+ carriers cotransport amino acids into the cell. The amino acids leave the base of the cell by facilitated diffusion and enter the capillaries by solvent drag.

Water Sodium reabsorption makes the ICF and ECF hypertonic to the tubular fluid. Water follows Na^+ by osmosis through both the paracellular and transcellular routes. Blood is hypertonic to the ECF, so after passing through the tubule epithelium, water enters the peritubular capillaries by osmosis. Because the PCT reabsorbs proportionate amounts of salt and water, the osmolarity of the tubular fluid remains unchanged in the PCT. Elsewhere in the nephron, the amount of water reabsorption is continually modulated by hormones according to the body's state of hydration. In the PCT, however, water is reabsorbed at a constant rate called **obligatory water reabsorption.**

Chloride Chloride is reabsorbed along both the paracellular and transcellular routes. Its reabsorption is favored by two factors: (1) negative chloride ions tend to follow the positive sodium ions by electrical attraction, and (2) water reabsorption raises the Cl^- concentration in the tubular fluid, thereby creating a gradient favorable to Cl^- reabsorption. In the transcellular route, Cl^- is apically absorbed by Na^+-linked cotransport. At the base of the cell, chloride pumps remove Cl^- in exchange for HCO_3^-; this is a *chloride shift* mechanism like the one we saw in red blood cells.

Other Electrolytes Potassium, magnesium, and variable amounts of calcium pass through the paracellular route by solvent drag (being carried along by reabsorbed water). Sulfates, phosphates, and nitrates are not reabsorbed; thus they pass in the urine. Substantial amounts of bicarbonate enter the tubular fluid by glomerular filtration, yet the urine is usually bicarbonate-free. No bicarbonate ions are reabsorbed by the renal tubule, but all bicarbonate in the fluid is usually neutralized by acid in the tubular fluid (see chapter 24 for details).

Protein The glomerulus filters a small amount of protein from the blood. The PCT reclaims it by pinocytosis, hydrolyzes it to amino acids, and releases these to the ECF by facilitated diffusion.

Nitrogenous Wastes Urea diffuses through the tubule epithelium with water. The nephron as a whole reabsorbs 40% to 60% of the urea in the tubular fluid, but since it reabsorbs 99% of the water, urine has a substantially higher urea concentration than blood or glomerular filtrate. When blood enters the kidney, it has about 20 mg of urea per deciliter; when it leaves the kidney, the concentration is down to 10.4 mg/dL. Thus the kidney removes about half of the urea, keeping its concentration down to a safe level but not completely clearing the blood of it.

The PCT reabsorbs nearly all the uric acid entering it, but later parts of the nephron secrete it back into the tubular fluid. Creatinine is not reabsorbed at all. It is too large to diffuse through water channels in the plasma membrane, and there are no transport proteins for it. Therefore, all creatinine filtered by the glomerulus is excreted in the urine.

The Transport Maximum

There is a limit to the amount of solute that the renal tubule can reabsorb because there are a limited number of transport proteins in the plasma membranes. If all the transporters are occupied as solute molecules pass through, some solute will escape reabsorption and appear in the urine. The maximum rate of reabsorption is the *transport maximum* (T_m), which is reached when the transporters are saturated (see chapter 4). Each organic solute reabsorbed by the renal tubule has its own T_m. For glucose, for example, T_m = 320 mg/min. Glucose normally enters the renal tubule at a rate of 125 mg/min, well within the T_m; thus all of it is reabsorbed. But when the plasma concentration of glucose reaches a *threshold* of about 220 mg/dL, more glucose is filtered than the tubule can reabsorb, and we begin to see glucose in the urine, a condition called **glycosuria**[18] (GLY-co-soo-ree-uh). In untreated diabetes mellitus, the plasma glucose concentration may exceed 400 mg/dL, so glycosuria is one of the classic symptoms of this disease.

Tubular Secretion

Tubular secretion is a process in which the renal tubule extracts chemicals from the capillary blood and secretes them into the tubular fluid (see fig. 23.6). Tubular secretion in the distal convoluted tubule is discussed shortly. In the proximal convoluted tubule and nephron loop, it serves two purposes:

1. **Waste removal.** Urea, uric acid, bile salts, ammonia, catecholamines, and a little creatinine are secreted into the tubule. Tubular secretion of uric acid compensates for its reabsorption earlier in the PCT and accounts for all of the uric acid in the urine. Tubular secretion also clears the blood of penicillin, aspirin, and other drugs. One reason that so many drugs must be taken three or four times a day is to keep pace with this rate of clearance and maintain a therapeutically effective drug concentration in the blood.
2. **Acid-base balance.** Tubular secretion of hydrogen and bicarbonate ions serves to regulate the pH of the body fluids. The details are discussed in the next chapter.

The Nephron Loop

The primary function of the nephron loop is to enable the collecting duct to concentrate the urine and conserve water, as discussed later. But in addition, it reabsorbs about 25% of the Na^+, K^+, and Cl^- and 20% of the water in the glomerular filtrate. Cells in the thick segment of the loop have proteins in the apical membranes that simultaneously bind 1 Na^+, 1 K^+, and 2 Cl^- from the tubular fluid and cotransport them into the cytoplasm. These ions leave the basolateral cell surfaces by active transport of Na^+ and diffusion of K^+ and Cl^-. The thick segment is impermeable to water; thus water cannot follow the reabsorbed electrolytes, and tubular fluid becomes very dilute by the time it passes from the nephron loop into the distal convoluted tubule.

18. *glycos* = sugar + *uria* = urine condition

The Distal Convoluted Tubule and Collecting Duct

Fluid arriving in the DCT still contains about 20% of the water and 10% of the salts from the glomerular filtrate. If this were all passed as urine, it would amount to 36 L/day, so obviously a great deal of fluid reabsorption must occur in the DCT and collecting duct. A distinguishing feature of these parts of the renal tubule is that unlike the PCT and nephron loop, they are subject to hormonal control—particularly by aldosterone, atrial natriuretic factor, and antidiuretic hormone.

Aldosterone

Aldosterone, the "salt-retaining hormone," is a steroid secreted by the adrenal cortex. A drop in blood Na^+ concentration or a rise in K^+ concentration directly stimulates aldosterone secretion. A drop in blood pressure does so indirectly—it stimulates the kidney to secrete renin, this leads to the production of angiotensin II, and angiotensin II stimulates aldosterone secretion (see fig. 23.11). The mechanism of aldosterone action on the kidney tubule is detailed in the following chapter, but its general effect is to cause the DCT and cortical portion of the collecting duct to reabsorb more Na^+ (which is followed by Cl^- and water) and to secrete more K^+. Thus the urine volume is reduced, and it contains more K^+ but less NaCl. Salt and water reabsorption helps to maintain blood volume and pressure.

Atrial Natriuretic Factor

Atrial natriuretic factor (ANF) is secreted by the atrial myocardium of the heart in response to high blood pressure. To some degree its effects are opposite those of aldosterone: it inhibits Na^+ and water reabsorption, thus increasing urine and Na^+ output and reducing blood volume and pressure.

Antidiuretic Hormone

Antidiuretic hormone (ADH) is secreted by the posterior lobe of the pituitary gland in response to dehydration and rising blood osmolarity. Its mechanism of action is explained later in more detail. Briefly, it makes the collecting duct more permeable to water, so water in the tubular fluid reenters the tissue fluid and bloodstream rather than being lost in the urine.

In summary, the PCT reabsorbs about 65% of the glomerular filtrate and returns it to the blood of the peritubular capillaries. Much of this reabsorption occurs by osmotic and cotransport mechanisms linked to the active transport of sodium ions. The nephron loop reabsorbs another 25% of the solutes, although its primary role, detailed later, is to aid the function of the collecting duct. The DCT reabsorbs more sodium, chloride, and water, but its rates of reabsorption are subject to control by hormones, especially aldosterone and ANF. These tubules also extract drugs, wastes, and some other solutes from the blood and secrete them into the tubular fluid. The DCT essentially completes the process of determining the chemical composition of the urine. The principal function left to the collecting duct is to conserve body water.

Before You Go On

Answer the following questions to test your understanding of the preceding section:

11. The reabsorption of water, Cl^-, and glucose by the PCT are all linked to the reabsorption of Na^+, but in three very different ways. Contrast these three mechanisms.
12. Explain why a substance appears in the urine if its rate of glomerular filtration exceeds the T_m of the renal tubule.
13. Contrast the effects of aldosterone and ANF on the renal tubule.

Urine Formation III: Water Conservation

Objectives

When you have completed this section, you should be able to

- explain how the collecting duct and antidiuretic hormone regulate the volume and concentration of urine; and
- explain how the kidney maintains an osmotic gradient in the renal medulla that enables the collecting duct to function.

The kidney serves not just to eliminate metabolic waste from the body but to prevent excessive water loss in doing so, and thus to support the body's fluid balance. As the kidney returns water to the body's tissue fluid and bloodstream, the urine remaining in the renal tubule becomes more and more concentrated. In this section, we examine the kidney's mechanisms for conserving water and concentrating the urine.

The Collecting Duct

The collecting duct (CD) begins in the cortex, where it receives tubular fluid from numerous nephrons. As it passes through the medulla, it reabsorbs water and concentrates the urine. When urine enters the upper end of the CD, it is isotonic with blood plasma (300 mOsm/L), but by the time it leaves the lower end, it can be up to four times as concentrated as this. This ability to concentrate wastes and control water loss was crucial to the evolution of terrestrial animals such as ourselves (see evolutionary insight 23.1).

Two facts enable the collecting duct to produce such hypertonic urine: (1) the osmolarity of the extracellular fluid is four times as high deep in the medulla as it is in the cortex, and (2) the medullary portion of the CD is permeable to water but not to NaCl. Therefore, as urine passes down the CD through the increasingly salty medulla, water leaves the tubule by osmosis, NaCl and other wastes remain behind, and the urine becomes more and more concentrated (fig. 23.14).

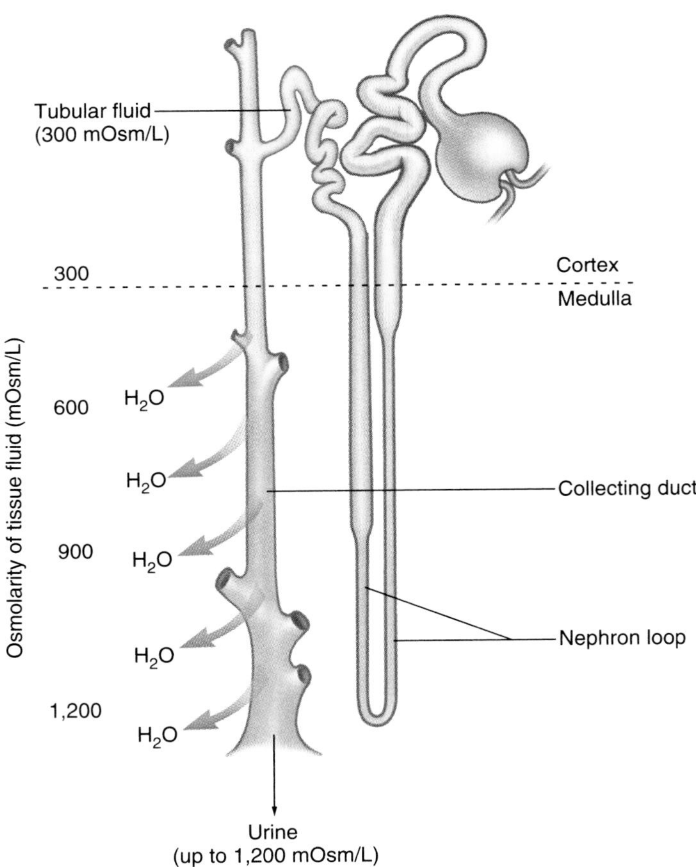

Figure 23.14 Water Reabsorption by the Collecting Duct. Note that the osmolarity of the tissue fluid increases fourfold from 300 mOsm/L in the cortex to 1,200 mOsm/L deep in the medulla. When the collecting duct has open water channels, water leaves the duct by osmosis and urine concentration increases.

Evolutionary Insight 23.1

The Kidney and Life on Dry Land

Physiologists first suspected that the nephron loop plays a role in water conservation because of their studies of a variety of animal species. Animals that must conserve water have longer, more numerous nephron loops than animals with little need to conserve it. Fish and amphibians lack nephron loops and produce urine that is isotonic to their blood plasma. Aquatic mammals such as beavers have short nephron loops and only slightly hypertonic urine.

But the kangaroo rat, a desert rodent, provides an instructive contrast. It lives on seeds and other dry foods and need never drink water. The water produced by its aerobic respiration is enough to meet its needs because its kidneys are extraordinarily efficient at conserving water. They have extremely long nephron loops and produce urine that is 10 to 14 times as concentrated as their blood plasma (compared with about 4 times, at most, in humans) .

Comparative studies thus suggested a hypothesis for the function of the nephron loop and prompted many years of difficult research that led to the discovery of the countercurrent multiplier mechanism for water conservation. This shows how comparative anatomy provides suggestions and insights into function and why physiologists do not study human function in isolation from other species.

Control of Water Loss

Just *how* concentrated the urine becomes depends on the body's state of hydration. For example, if you drink a large volume of water, you will soon produce a large volume of hypotonic urine (water diuresis). Under such conditions, the cortical portion of the CD reabsorbs NaCl but is impermeable to water. Thus salt is removed from the urine, water stays in the CD, and urine concentration may be as low as 50 mOsm/L.

Dehydration, on the other hand, causes your urine to be scanty and more concentrated. One reason for this effect is that when your body has lost a lot of fluid, your blood pressure is lower and your glomerular filtration rate (GFR) drops. When the GFR is low, fluid flows more slowly through the renal tubules and there is more time for tubular reabsorption. Less salt remains in the urine as it enters the collecting duct, so there is less opposition to the osmosis of water out of the duct and into the ECF. More water is reabsorbed and less urine is produced.

In addition, the high blood osmolarity of a dehydrated person stimulates the release of antidiuretic hormone (ADH) from the posterior lobe of the pituitary gland. The mechanism of ADH action on the collecting duct is explained in the next chapter, but its general effect is to increase the water permeability of the epithelial cells of the CD. The cortical portion of the CD, especially, then reabsorbs more water, which is carried away by the peritubular capillaries. Urine output is consequently reduced. By contrast, ADH secretion is inhibited when you are well hydrated. The duct is then less permeable to water, so more water remains in the duct and you produce abundant, dilute urine.

The Countercurrent Multiplier

The ability of the CD to concentrate urine depends on the salinity gradient of the renal medulla. It may seem surprising that the ECF is four times as salty deep in the medulla as it is in the cortex. We would expect the salt to diffuse toward the cortex until it was evenly distributed through the kidney. However, there is a mechanism that overrides this—the nephron loop acts as a **countercurrent multiplier,** which continually recaptures salt and returns it to the deep medullary tissue. It is called a *multiplier* because it multiplies the salinity deep in the medulla and a *countercurrent* mechanism because it is based on fluid flowing in opposite directions in two adjacent tubules—downward in the descending limb and upward in the ascending limb. The countercurrent multiplier works as follows:

1. The descending limb is very permeable to water but not to NaCl. Therefore, as the tubular fluid descends into the increasingly salty medulla, more and more water leaves the descending limb while NaCl remains in the tubule. As the fluid reaches the lower end of the loop, it has a concentration of about 1,200 mOsm/L.
2. The ascending limb, by contrast, is impermeable to water, but the thick segment actively cotransports Na^+, K^+, and Cl^- into the ECF. This keeps the osmolarity of the renal medulla high. Since water cannot follow these ions out of the ascending limb, the tubular fluid becomes more and more dilute as it approaches the cortex. It is about 100 mOsm/L at the top of the loop.

The essence of the *countercurrent* mechanism is that the two limbs of the nephron loop are close enough to influence each other through the positive feedback relationship shown in figure 23.15.

The collecting duct also helps to maintain the osmotic gradient (fig. 23.16). Its lower end is somewhat permeable to urea, which diffuses down its concentration gradient, out of the duct and into the tissue fluid. Some of this urea enters the descending thin segment of the nephron

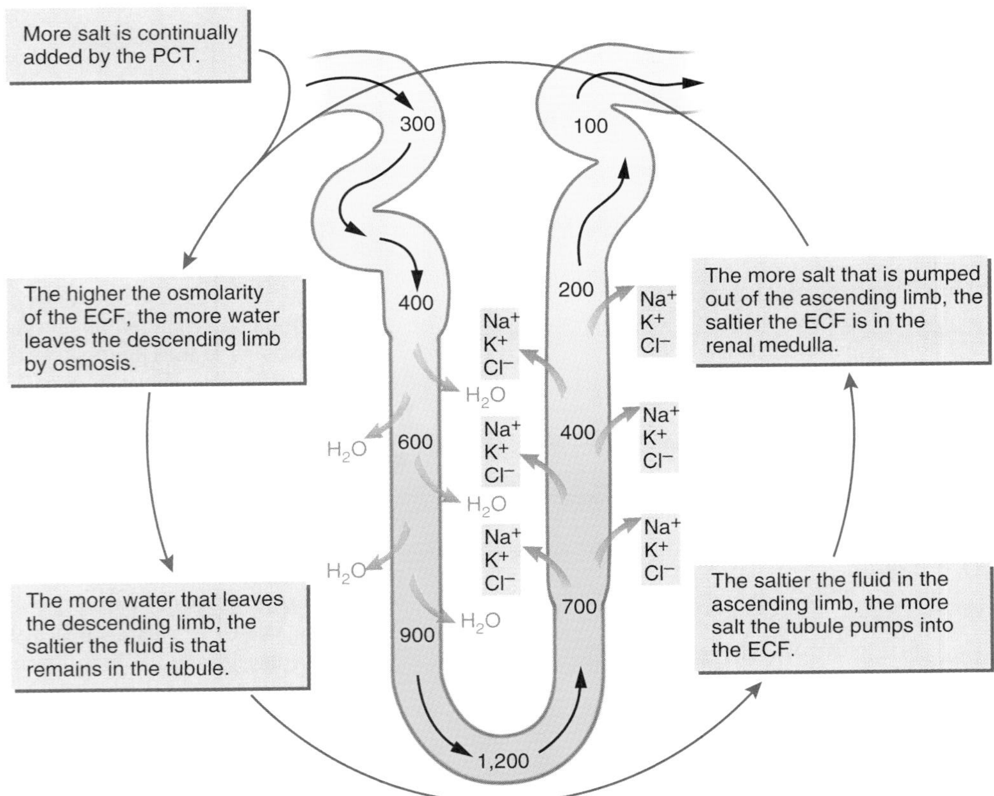

Figure 23.15 The Countercurrent Multiplier of the Nephron Loop.

loop and travels to the distal convoluted tubule. Neither the thick segment of the loop nor the distal tubule is permeable to urea, so urea remains in the tubules and returns to the collecting duct. Combined with new urea being added continually by the glomerular filtrate, urea becomes more and more concentrated in the fluid of the collecting duct, and still more diffuses out into the medulla. Thus there is a continual recycling of urea from the collecting duct to the medulla and back. Urea accounts for about 40% of the high osmolarity deep in the medulla.

The Countercurrent Exchange System

The renal medulla must have a blood supply to meet its metabolic needs, and this creates a potential problem—capillaries of the medulla could carry away the urea and salt that produce the high medullary osmolarity. The vasa recta that supply the medulla, however, form a countercurrent system of their own that prevents this from happening. Blood flows in opposite directions in adjacent parallel capillaries. These capillaries form a **countercurrent exchange** system. Blood in the vasa recta exchanges water for salt as it flows downward into the deep medulla—water diffuses out of the capillaries and salt diffuses in. As the blood flows back toward the cortex, the opposite occurs; it exchanges salt for water. Thus, the vasa recta leave the salt behind and do not subtract from the osmolarity of the medulla. Indeed, they absorb more water on the way out than they unload on the way in; they carry away the water reabsorbed from the urine by the collecting duct.

To summarize what we have studied in this section, the collecting duct can adjust water reabsorption to produce urine as hypotonic as 50 mOsm/L or as hypertonic as 1,200 mOsm/L, depending on the body's need for water conservation or removal. In a state of hydration, ADH is not secreted and the cortical part of the CD reabsorbs salt without reabsorbing water; the water remains to be excreted in the dilute urine. In a state of dehydration, ADH is secreted, the medullary part of the CD reabsorbs water, and the urine is more concentrated. The CD is able to do this because it passes through a salinity gradient in the medulla from 300 mOsm/L near the cortex to 1,200 mOsm/L near the papilla. This gradient is produced by a countercurrent multiplier of the nephron loop, which concentrates NaCl in the lower medulla, and by the diffusion of urea from the col-

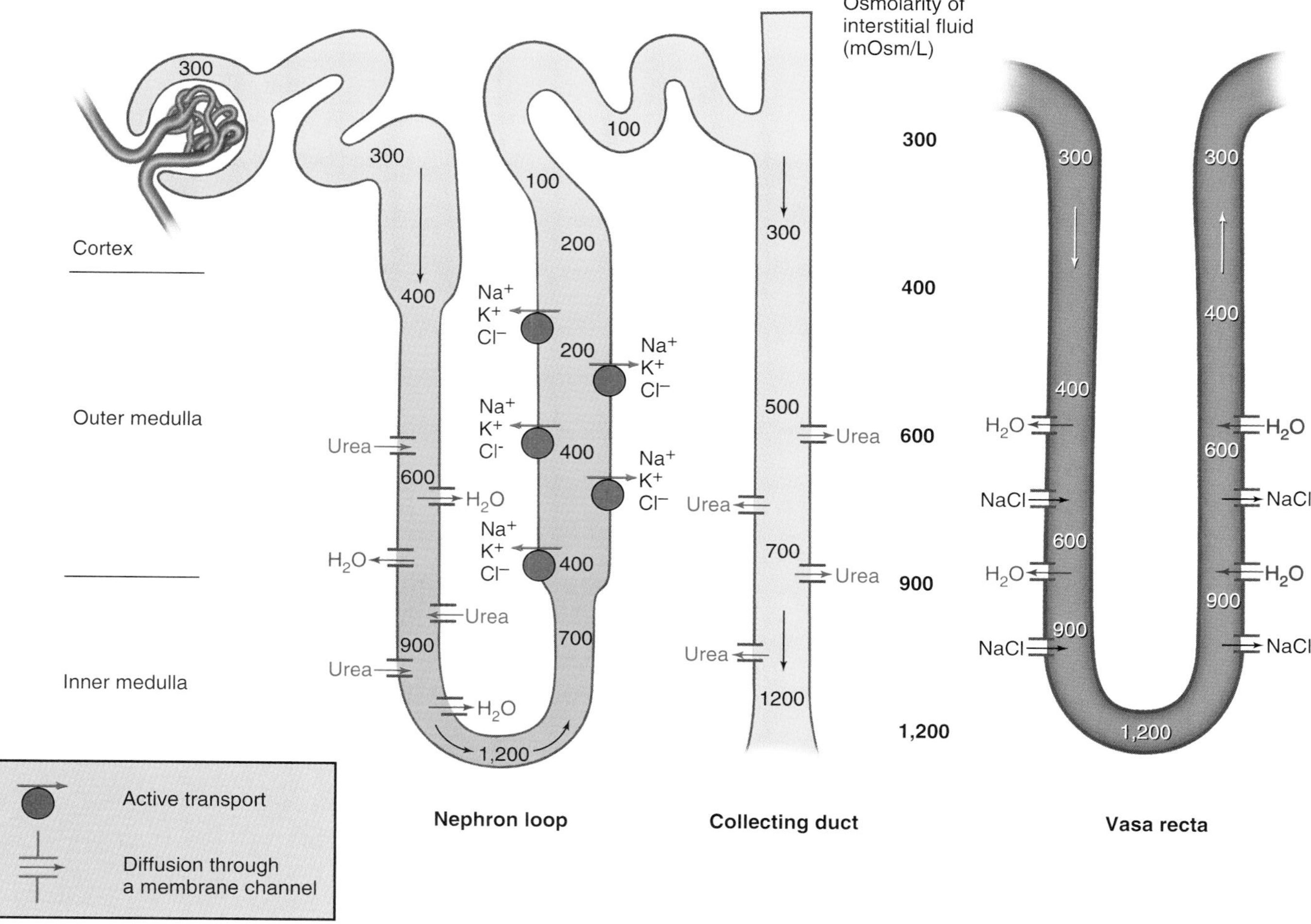

Figure 23.16 The Relationship of the Nephron Loop, Vasa Recta, and Collecting Duct in Maintaining the Osmolarity of the Renal Medulla.

lecting duct into the medulla. The vasa recta are arranged as a countercurrent exchange system that enables them to supply blood to the medulla without subtracting from its salinity gradient. Figure 23.17 summarizes the major solutes reabsorbed and secreted in each part of the renal tubule.

Before You Go On

Answer the following questions to test your understanding of the preceding section:

14. Predict how ADH hypersecretion would affect the sodium concentration of the urine, and explain why.
15. Concisely contrast the role of the countercurrent multiplier with that of the countercurrent exchanger.
16. How would the function of the collecting duct change if the nephron loop did not exist?

Urine and Renal Function Tests

Objectives

When you have completed this section, you should be able to

- describe the composition and properties of urine; and
- carry out some calculations to evaluate renal function.

Medical diagnosis often rests on determining the current and recent physiological state of the tissues. No two fluids are as valuable for this purpose as blood and urine. **Urinalysis,** the examination of the physical and chemical properties of urine, is therefore one of the most routine procedures in medical examinations. The principal characteristics of urine and certain tests used to evaluate renal function are described here.

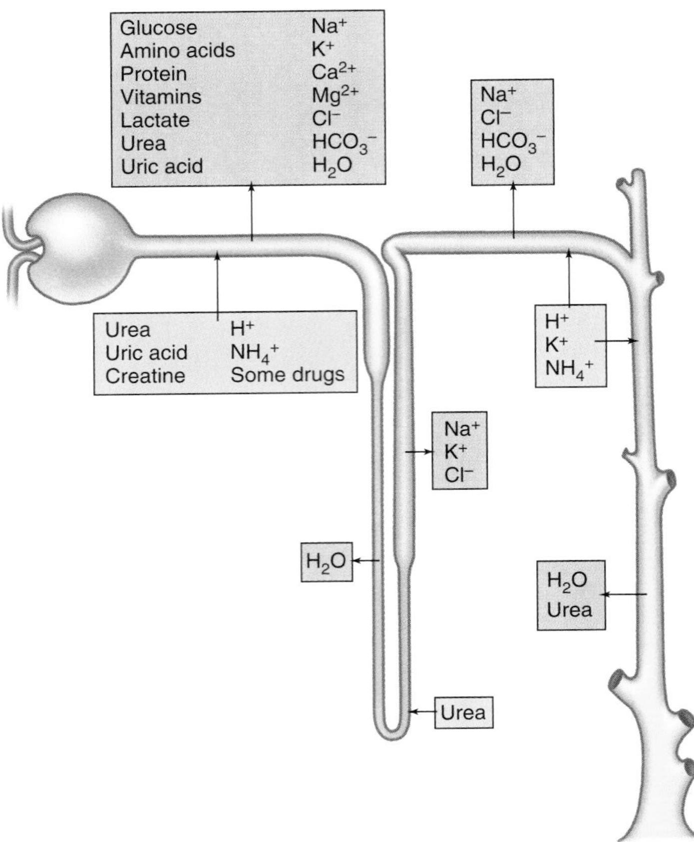

Figure 23.17 Solutes Reabsorbed and Secreted in Different Portions of the Renal Tubule.

Composition and Properties of Urine

The basic composition and properties of urine are as follows:

- **Appearance.** Urine varies from almost colorless to deep amber, depending on the body's state of hydration. The yellow color of urine is due to **urochrome,** a pigment produced by the breakdown of hemoglobin from expired erythrocytes. Pink, green, brown, black, and other colors result from certain foods, vitamins, drugs, and metabolic diseases. Urine is normally clear but turns cloudy upon standing because of bacterial growth. Pus in the urine (**pyuria**) makes it cloudy and suggests kidney infection. Blood in the urine (hematuria) may be due to a urinary tract infection, trauma, or kidney stones. Cloudiness or blood in a urine specimen sometimes, however, simply indicates contamination with semen or menstrual fluid.
- **Odor.** Fresh urine has a distinctive but not repellent odor. As it stands, however, bacteria multiply, degrade urea to ammonia, and produce the pungent odor typical of stale wet diapers. Asparagus and other foods can impart distinctive aromas to the urine. Diabetes mellitus gives it a sweet, "fruity" odor of acetone. A "mousy" odor suggests phenylketonuria (PKU), and a foul odor may indicate urinary tract infection.
- **Specific gravity.** This is a ratio of the density (g/mL) of a substance to the density of distilled water. Distilled water has a specific gravity of 1.000, and urine ranges from 1.001 when it is very dilute to 1.035 when it is very concentrated. Multiplying the last two digits of the specific gravity by a proportionality constant of 2.6 gives an estimate of the grams of solid matter per liter of urine. For example, a specific gravity of 1.025 indicates a solute concentration of $25 \times 2.6 = 65$ g/L.
- **Osmolarity.** Urine can have an osmolarity as low as 50 mOsm/L in a very hydrated person or as high as 1,200 mOsm/L in a dehydrated person. Compared with the osmolarity of blood (300 mOsm/L), then, urine can be either hypotonic or hypertonic under different conditions.
- **pH.** The pH of urine ranges from 4.5 to 8.2 but is usually about 6.0 (mildly acidic). The regulation of urine pH is discussed extensively in chapter 24.
- **Chemical composition.** Urine averages 95% water and 5% solutes by volume (table 23.1). Normally, the most abundant solute is urea, followed by sodium chloride, potassium chloride, and lesser amounts of creatinine, uric acid, phosphates, sulfates, and traces of calcium, magnesium, and sometimes bicarbonate. It is abnormal to find glucose, free hemoglobin, albumin, ketones, or bile pigments in the urine; their presence is an important indicator of disease.

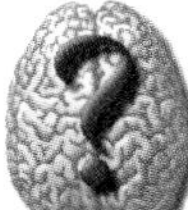

Think About It

In table 23.1, what process accounts for most of the difference between values in the middle column and those on the right?

Urine Volume

An average adult produces 1 to 2 L of urine per day. An output in excess of 2 L/day is called **polyuria**[19] (POL-ee-YOU-ree-uh), or **diuresis** (DIE-you-REE-sis). Fluid intake and some drugs can temporarily increase output to as much as 20 L/day. Chronic diseases such as diabetes (see next) can do so over a long term. **Oliguria**[20] (oll-ih-GUE-ree-uh) is an output of less than 500 mL/day, and **anuria**[21] is an output of 0 to 100 mL/day. Low output can result

19. *poly* = many, much
20. *oligo* = few, a little
21. *an* = without

Table 23.1 Average Composition of Blood Plasma, Glomerular Filtrate, and Urine

Substance	Blood Plasma (total amount)	Glomerular Filtrate (amount per day)	Urine (amount per day)
Water	3 L	180 L	1–2 L
Urea	4.8 g	53 g	25 g
Chloride	10.7 g	639 g	6.3 g
Sodium	9.7 g	580 g	4.6 g
Potassium	0.5 g	30 g	2.0 g
Creatinine	0.03 g	1.6 g	1.6 g
Uric acid	0.15 g	8.5 g	0.8 g
Protein	200 g	2 g	0.1 g
Bicarbonate	4.6 g	275 g	0 g
Glucose	3 g	180 g	0 g

from kidney disease, dehydration, circulatory shock, prostate enlargement, and other causes. If urine output drops to less than 400 mL/day, the body cannot maintain a safe, low concentration of wastes in the blood plasma. The result is azotemia (see *obligatory water loss* in the next chapter).

Diabetes

Diabetes[22] is chronic polyuria resulting from various metabolic disorders. There are at least seven forms of diabetes: *diabetes mellitus type I* and *type II, gestational diabetes, pituitary diabetes, adrenal diabetes, renal diabetes,* and *diabetes insipidus.* In most cases, the polyuria results from a high concentration of glucose in the renal tubule. Glucose opposes the osmotic reabsorption of water, so more water is passed in the urine (*osmotic diuresis*) and a person may become severely dehydrated. In most forms of diabetes, the high glucose concentration in the tubule is a result of hyperglycemia, a high concentration of glucose in the blood: both forms of *diabetes mellitus,* caused by insulin hyposecretion or inactivity (see chapter 17); *pituitary diabetes,* caused by hypersecretion of growth hormone; and *adrenal diabetes,* caused by hypersecretion of cortisol. About 1% to 3% of pregnant women experience *gestational diabetes,* in which pregnancy reduces the mother's insulin sensitivity, resulting in hyperglycemia and glycosuria. In *renal diabetes,* blood glucose level is not elevated, but there is a hereditary deficiency of glucose transporters in the PCT, which causes glucose to remain in the tubular fluid.

All of the preceding forms result in glycosuria. Before chemical tests for urine glucose were developed, physicians diagnosed diabetes mellitus by tasting the patient's urine for sweetness.[23] Tests for glycosuria are now as simple as dipping a chemical test strip into the urine specimen—an advance in medical technology for which urologists are no doubt grateful. In *diabetes insipidus,*[24] the urine contains no glucose and, by the old diagnostic method, does not taste sweet. This disease results from ADH hyposecretion; the collecting duct does not reabsorb as much water as usual, so more water is passed in the urine.

Diuretics

Diuretics are chemicals that increase urine volume. They are used for treating hypertension and congestive heart failure because they reduce the body's fluid volume and blood pressure. Diuretics work by one of two mechanisms—increasing glomerular filtration or reducing tubular reabsorption. For example, caffeine, in the former category, dilates the afferent arteriole and increases GFR. Alcohol, in the latter category, inhibits ADH secretion. Also in the latter category are many osmotic diuretics, which reduce water reabsorption by increasing the osmolarity of the tubular fluid. Many diuretic drugs, such as furosemide (Lasix), produce osmotic diuresis by inhibiting sodium reabsorption.

Renal Function Tests

There are several tests for diagnosing kidney diseases, evaluating their severity, and monitoring their progress. Here we examine two methods used to determine renal clearance and glomerular filtration rate.

22. *diabetes* = passing through

23. *melli* = honey, sweet

24. *insipid* = tasteless

Renal Clearance

Renal clearance is the volume of blood plasma from which a particular waste is completely removed in 1 minute. It represents the net effect of three processes:

Glomerular filtration of the waste
\+ Amount added by tubular secretion
− Amount reclaimed by tubular reabsorption
Renal clearance

In principle, we could determine renal clearance by sampling blood entering and leaving the kidney and comparing their waste concentrations. In practice, it is not practical to draw blood samples from the renal vessels, but clearance can be assessed indirectly by collecting samples of blood and urine, measuring the waste concentration in each, and measuring the rate of urine output.

Suppose the following values were obtained for urea:

U (urea concentration in urine) = 6.0 mg/mL

V (rate of urine output) = 2 mL/min

P (urea concentration in plasma) = 0.2 mg/mL

Renal clearance (C) is

$$C = UV/P$$
$$= \frac{(6.0 \text{ mg/mL})(2 \text{ mL/min})}{0.2 \text{ mg/mL}}$$
$$= 60 \text{ mL/min}$$

This means the equivalent of 60 mL of blood plasma is completely cleared of urea per minute. If this person has a normal GFR of 125 mL/min, then the kidneys have cleared urea from only 60/125 = 48% of the glomerular filtrate. This is a normal rate of urea clearance, however, and is sufficient to maintain safe levels of urea in the blood.

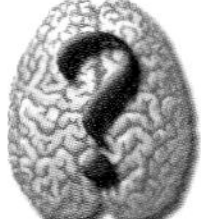

Think About It

What would you expect the value of renal clearance of glucose to be in a healthy individual? Why?

Glomerular Filtration Rate

Assessment of kidney disease often calls for a measurement of GFR. We cannot determine GFR from urea excretion for two reasons: (1) some of the urea in the urine is secreted by the renal tubule, not filtered by the glomerulus, and (2) much of the urea filtered by the glomerulus is reabsorbed by the tubule. To measure GFR requires a substance that is not secreted or reabsorbed at all, so that all of it in the urine gets there by glomerular filtration.

There doesn't appear to be a single urine solute produced by the body that is not secreted or reabsorbed to some degree. However, several plants, including garlic and artichoke, produce a chemical that is useful for GFR measurement: inulin (IN-you-lin), a polymer of fructose. All inulin filtered by the glomerulus remains in the renal tubule and appears in the urine; none is reabsorbed, nor does the tubule secrete it. To measure GFR, inulin is injected into the bloodstream, the rate of urine output is measured, and the concentrations of inulin in blood and urine samples are determined. For this solute, GFR is equal to the renal clearance.

Suppose, for example, that a patient's plasma concentration of inulin is P = 0.5 mg/mL, the urine concentration is U = 30 mg/mL, and urine output is V = 2 mL/min. This person has a normal GFR:

$$GFR = UV/P$$
$$= \frac{(30 \text{ mg/mL})(2 \text{ mL/min})}{0.5 \text{ mg/mL}}$$
$$= 120 \text{ mL/min}$$

A solute that is reabsorbed by the renal tubules will have a renal clearance *less* than the GFR (provided its tubular secretion is less than its rate of reabsorption). This is why the renal clearance of urea is about 60 mL/min. A solute that is secreted by the renal tubules will have a renal clearance *greater* than the GFR (provided its reabsorption does not exceed its secretion). Creatinine, for example, has a renal clearance of 140 mL/min.

Before You Go On

Answer the following questions to test your understanding of the preceding section:

17. Define *oliguria* and *polyuria*. Which of these is characteristic of diabetes?
18. Identify two causes of glycosuria other than diabetes mellitus.
19. How is the diuresis produced by furosemide like the diuresis produced by diabetes mellitus? How are they different?
20. Explain why GFR could not be determined from measurement of the amount of NaCl in the urine.

Urine Storage and Elimination

Objectives

When you have completed this section, you should be able to

- describe the functional anatomy of the ureters, urinary bladder, and male and female urethra; and
- explain how the nervous system and urethral sphincters control the voiding of urine.

Urine is produced continually, but fortunately it does not drain continually from the body. Urination is episodic—occurring when we allow it. This is made possible by an apparatus for storing urine and by neural controls for its timely release.

Clinical Insight 23.2

Kidney Stones

A *renal calculus*[25] (kidney stone) is a hard granule of calcium, phosphate, uric acid, and protein. Renal calculi form in the renal pelvis and are usually small enough to pass unnoticed in the urine flow. Some, however, grow to several centimeters in size and block the renal pelvis or ureter, which can lead to the destruction of nephrons as pressure builds in the kidney. A large, jagged calculus passing down the ureter stimulates strong contractions that can be excruciatingly painful. It can also damage the ureter and cause hematuria. Causes of renal calculi include hypercalcemia, dehydration, pH imbalances, frequent urinary tract infections, or an enlarged prostate gland causing urine retention. Calculi are sometimes treated with stone-dissolving drugs, but often they require surgical removal. A nonsurgical technique called *lithotripsy*[26] uses ultrasonic vibration to pulverize the calculi into fine granules easily passed in the urine.

25. *calc* = calcium, stone + *ul* = little

26. *litho* = stone + *tripsy* = crushing

The Ureters

The renal pelvis funnels urine into the ureter, a retroperitoneal, muscular tube that extends to the urinary bladder. The ureter is about 25 cm long and reaches a maximum diameter of about 1.7 cm near the bladder. The ureters pass dorsal to the bladder and enter it from below, passing obliquely through its muscular wall and opening onto its floor. As pressure builds in the bladder, it compresses the ureters and prevents urine from being forced back to the kidneys.

The ureter has three layers: an adventitia, muscularis, and mucosa. The adventitia is a connective tissue layer that binds it to the surrounding tissues. The muscularis consists of two layers of smooth muscle. When urine enters the ureter and stretches it, the muscularis contracts and initiates a peristaltic wave that "milks" the urine down to the bladder. These contractions occur every few seconds to few minutes, proportional to the rate at which urine enters the ureter. The mucosa has a transitional epithelium continuous with that of the renal pelvis above and urinary bladder below. The lumen of the ureter is very narrow and is easily obstructed or injured by kidney stones (see clinical insight 23.2).

The Urinary Bladder

The urinary bladder (fig. 23.18) is a muscular sac on the floor of the pelvic cavity, inferior to the peritoneum and posterior to the pubic symphysis. It is covered by parietal peritoneum on its flattened superior surface and by a fibrous adventitia elsewhere. Its muscularis, called the **detrusor**[27] (deh-TROO-zur) **muscle,** consists of three layers of smooth muscle. The mucosa has a transitional epithelium, and in the relaxed bladder it has conspicuous wrinkles called **rugae**[28] (ROO-gee). The openings of the two ureters and the urethra mark a smooth-surfaced triangular area called the **trigone**[29] on the bladder floor. This is a common site of bladder infection (see clinical insight 23.3). For photographs of the relationship of the bladder and urethra to other pelvic organs in both sexes, see figures A.23 and A.24 (p. 49).

The bladder is highly distensible. As it fills, it expands superiorly, the rugae flatten, and the wall becomes quite thin. A moderately full bladder contains about 500 mL of urine and extends about 12.5 cm from top to bottom. The maximum capacity is 700 to 800 mL.

27. *de* = down + *trus* = push

The Urethra

The urethra conveys urine out of the body. In the female, it is a tube 3 to 4 cm long bound to the anterior wall of the vagina by connective tissue. Its opening, the **external urethral orifice,** lies between the vaginal orifice and clitoris. The male urethra is about 18 cm long and has three regions: (1) The **prostatic urethra** begins at the urinary bladder and passes for about 2.5 cm through the prostate gland. During orgasm, it receives semen from the reproductive glands. (2) The **membranous urethra** is a short (0.5 cm), thin-walled portion where the urethra passes through the muscular floor of the pelvic cavity. (3) The **penile urethra** is about 15 cm long and passes through the penis to the external urethral orifice. The male urethra assumes an S shape: it passes downward from the bladder, turns anteriorly as it enters the root of the penis, and then turns about 90° downward again as it enters the external, pendant part of the penis. The mucosa has a

28. *ruga* = fold, wrinkle
29. *tri* = three + *gon* = angle

Clinical Insight 23.3

Urinary Tract Infections

Infection of the urinary bladder is called *cystitis*.[30] It is especially common in females because bacteria such as *Escherichia coli* can travel easily from the perineum up the short urethra. Because of this risk, young girls should be taught never to wipe the anus in a forward direction. If cystitis is untreated, bacteria can spread up the ureters and cause *pyelitis*,[31] infection of the renal pelvis. If it reaches the renal cortex and nephrons, it is called *pyelonephritis*. Kidney infections can also result from invasion by blood-borne bacteria. Urine stagnation due to renal calculi or prostate enlargement increases the risk of infection.

30. *cyst* = bladder + *itis* = inflammation

31. *pyel* = pelvis

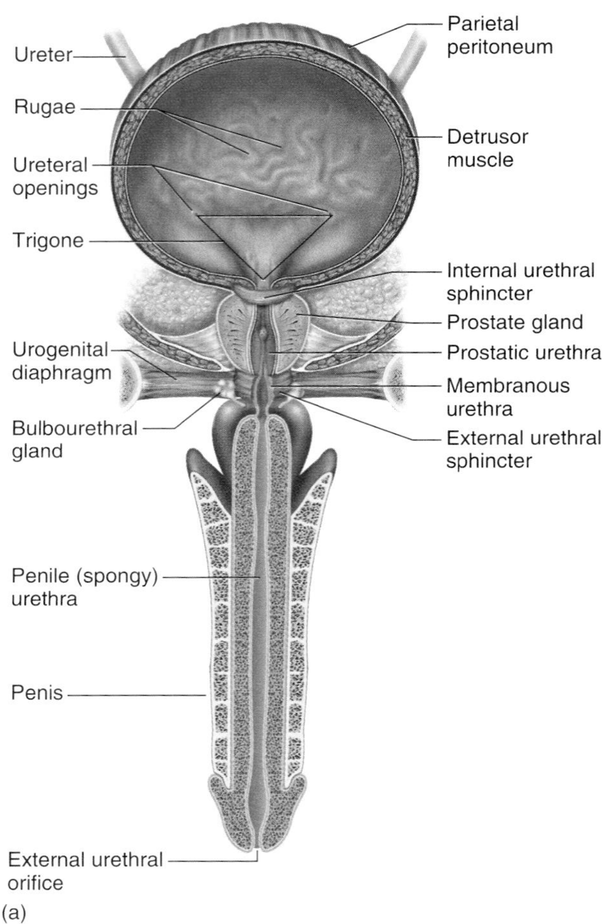

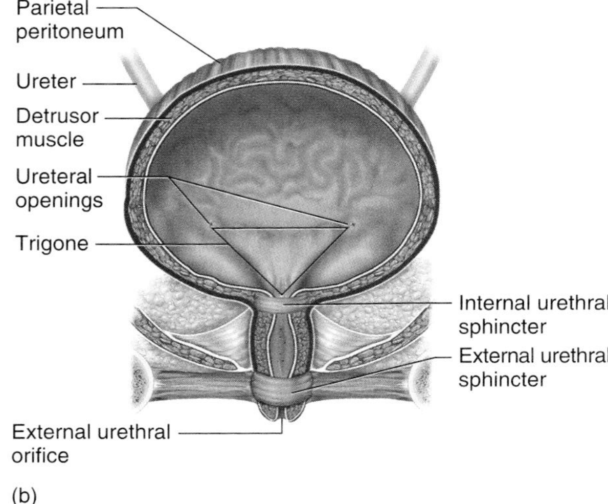

Figure 23.18 Anatomy of the Urinary Bladder and Urethra. (*a*) Male. (*b*) Female.

transitional epithelium near the bladder, a pseudostratified columnar epithelium for most of its length, and finally stratified squamous near the external urethral orifice. There are mucous **urethral glands** in its wall.

In both sexes, the detrusor muscle is thickened near the urethra to form an **internal urethral sphincter,** which compresses the urethra and retains urine in the bladder. Since this sphincter is composed of smooth muscle, it is under involuntary control. Where the urethra passes through the pelvic floor, it is encircled by an **external urethral sphincter** of skeletal muscle, which provides voluntary control over the voiding of urine.

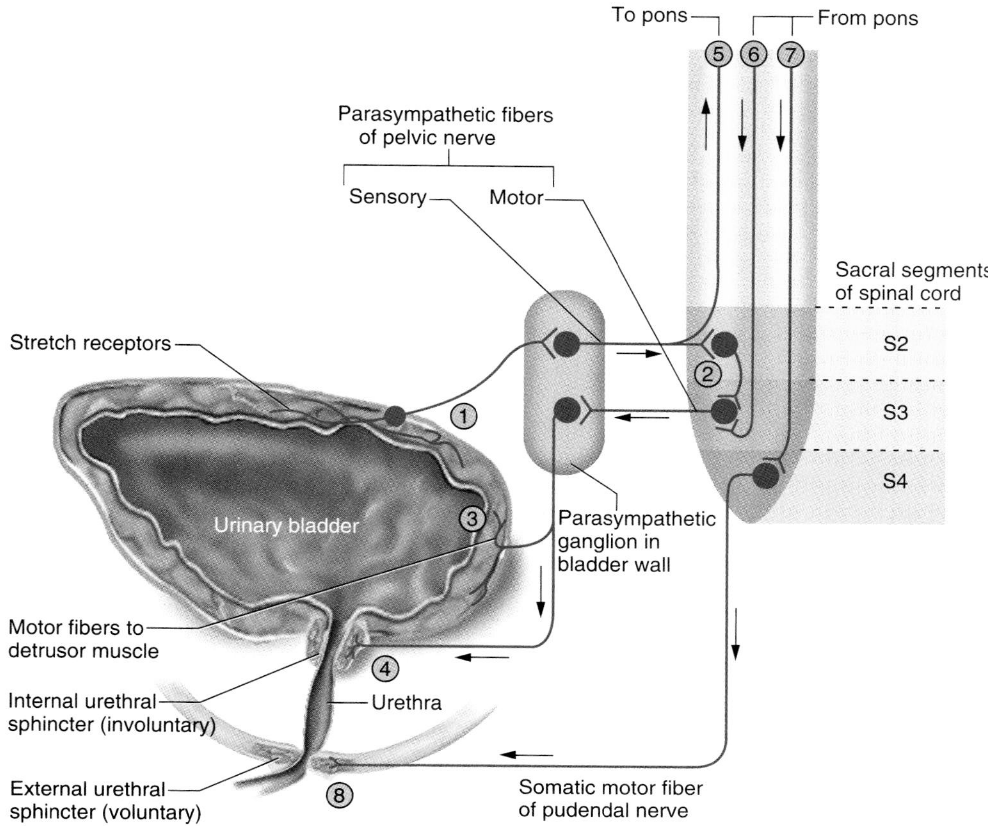

Figure 23.19 Neural Control of Micturition. Circled numbers correspond to text description.

Voiding Urine

Urination, or emptying of the bladder, is also called **micturition**[32] (MIC-too-RISH-un). It is controlled in part by the **micturition reflex** shown in figure 23.19, which is numbered to correspond to the following description:

(1) When the bladder contains about 200 mL of urine, stretch receptors in the wall send afferent nerve impulses to the spinal cord by way of the pelvic nerves. **(2)** By way of a parasympathetic reflex arc through segments S2 to S3 of the cord, signals return to the bladder and stimulate contraction of the detrusor muscle **(3)** and relaxation of the internal urethral sphincter **(4).** This reflex is the predominant mechanism that voids the bladder in infants and young children.

As the brain and spinal cord mature, however, we acquire voluntary control over the external urethral sphincter, and emptying of the bladder is controlled predominantly by a **micturition center** in the pons. This center receives signals from the stretch receptors **(5)** and integrates this information with cortical input concerning the appropriateness of urinating at the moment. It sends back impulses **(6)** that excite the detrusor and relax the internal urethral sphincter. **(7)** At times when it is inappropriate to urinate, a steady train of nerve impulses travel from the brainstem through the pudendal nerve to the external urethral sphincter, thus keeping it contracted. When you wish to urinate, these impulses are inhibited, the external sphincter relaxes **(8),** and contractions of the detrusor muscle expel the urine. The Valsalva maneuver (p. 858) also aids in expulsion of urine by increasing pressure on the bladder. Males voluntarily contract the bulbocavernosus muscle encircling the base of the penis to expel the last few milliliters of urine.

When it is desirable to urinate (for example, before a long trip) but the urge does not yet exist because the bladder is not full enough, the Valsalva maneuver can activate the micturition reflex. Contraction of the abdominal muscles compresses the bladder and may excite the stretch receptors even if there is less than 200 mL of urine in the bladder.

We have considered several disorders associated with the urinary system in this chapter, including uremia, azotemia, diabetes, kidney stones, urinary tract infections, and renal insufficiency. Some additional disorders of this system are briefly described in table 23.2.

32. *mictur* = to urinate

Table 23.2 Some Disorders of the Urinary System

Disorder	Description
Acute Glomerulonephritis	An autoimmune inflammation of the glomeruli, often following a streptococcus infection. Results in destruction of glomeruli leading to hematuria, proteinuria, edema, reduced glomerular filtration, and hypertension. Can progress to chronic glomerulonephritis and renal failure, but most individuals recover from acute glomerulonephritis without lasting effect.
Acute Renal Failure	An abrupt decline in renal function, often due to traumatic damage to the nephrons or a loss of blood flow stemming from hemorrhage or thrombosis.
Chronic Renal Failure	Long-term, progressive, irreversible loss of nephrons; see clinical insight 23.4 for a variety of causes. Requires a kidney transplant or hemodialysis.
Hydronephrosis[33]	Increase in fluid pressure in the renal pelvis and calices owing to obstruction of the ureter by kidney stones, nephroptosis, or other causes. Can progress to complete cessation of glomerular filtration and atrophy of nephrons.
Nephroptosis[34] (NEFF-rop-TOE-sis)	Slippage of the kidney to an abnormally low position *(floating kidney).* Occurs in people with too little body fat to hold the kidney in place and in people who subject the kidneys to prolonged vibration, such as truck drivers, equestrians, and motorcyclists. Can twist or kink the ureter, which causes pain, obstructs urine flow, and potentially leads to hydronephrosis.
Nephrotic Syndrome	Excretion of large amounts of protein in the urine ($\geq$ 3.5 g/day) due to glomerular injury. Can result from trauma, drugs, infections, cancer, diabetes mellitus, lupus erythematosus, and other diseases. Loss of plasma protein leads to edema, ascites, hypotension, and susceptibility to infection (because of immunoglobulin loss).
Urinary Incontinence	Inability to hold the urine; involuntary leakage from the bladder. Can result from incompetence of the urinary sphincters; bladder irritation; pressure on the bladder in pregnancy; an obstructed urinary outlet so that the bladder is constantly full and dribbles urine *(overflow incontinence);* uncontrollable urination due to brief surges in bladder pressure, as in laughing or coughing *(stress incontinence);* and neurological disorders such as spinal cord injuries.

33. *hydro* = water + *nephr* = kidney + *osis* = medical condition
34. *nephro* = kidney + *ptosis* = sagging, falling

Before You Go On

Answer the following questions to test your understanding of the preceding section:

21. Describe the location and function of the detrusor muscle.
22. Compare and contrast the functions of the internal and external urethral sphincters.
23. How would micturition be affected by a spinal cord lesion that prevented voluntary nerve impulses from reaching the sacral part of the cord?

Clinical Insight 23.4

Renal Insufficiency and Hemodialysis

Renal insufficiency is a state in which the kidneys cannot maintain homeostasis due to extensive destruction of their nephrons. Some causes of nephron destruction include the following:

- Chronic or repetitive kidney infections.
- Trauma from such causes as blows to the lower back or continual vibration from machinery.
- Poisoning by heavy metals such as mercury and lead and solvents such as carbon tetrachloride, acetone, and paint thinners. These are absorbed into the blood from inhaled fumes or by skin contact and then filtered by the glomeruli. They kill renal tubule cells.
- Blockage of renal tubules with proteins small enough to be filtered by the glomerulus—for example, myoglobin released by skeletal muscle damage and hemoglobin released by a transfusion reaction.
- Atherosclerosis, which reduces blood flow to the kidney.
- Glomerulonephritis, an autoimmune disease of the glomerular capillaries.

Nephrons can regenerate and restore kidney function after short-term injuries. Even when some of the nephrons are irreversibly destroyed, others hypertrophy and compensate for their lost function. Indeed, a person can survive on as little as one-third of one kidney. When 75% of the nephrons are lost, however, urine output may be as low as 30 mL/hr compared with the normal rate of 50 to 60 mL/hr. This is insufficient to maintain homeostasis and is accompanied by azotemia and acidosis. Uremia develops when there is 90% loss of renal function. Anemia also commonly accompanies renal insufficiency because there are too few nephrons left to produce erythropoietin (EPO), the hormone that stimulates red blood cell formation.

Hemodialysis is a procedure for artificially clearing wastes from the blood when the kidneys are not adequately doing so (fig. 1). Blood is pumped from the radial artery to a *dialysis machine* (artificial kidney) and returned to the patient by way of a vein. In the dialysis machine, the blood flows through a semipermeable cellophane tube surrounded by dialysis fluid. Urea, potassium, and other solutes that are more concentrated in the blood than in the dialysis fluid diffuse through the membrane into the fluid, which is discarded. Glucose, electrolytes, and drugs can be administered by adding them to the dialysis fluid so they will diffuse through the membrane into the blood. People with renal insufficiency also accumulate substantial amounts of body water between dialysis treatments, and dialysis serves also to remove this excess water. Patients are typically given erythropoietin (EPO) to compensate for the lack of EPO from the failing kidneys and maintain the hematocrit.

Hemodialysis patients typically have three sessions per week for 4 to 8 hours per session. In addition to inconvenience, it carries risks of infection and thrombosis. Blood tends to clot when exposed to foreign surfaces, so an anticoagulant such as heparin is added during dialysis. Unfortunately, this inhibits clotting in the patient's body as well, and dialysis patients sometimes suffer internal bleeding.

A procedure called *continuous ambulatory peritoneal dialysis (CAPD)* is more convenient. It can be carried out at home by the patient, who is provided with plastic bags of dialysis fluid. Fluid is introduced into the abdominal cavity through an indwelling catheter. Here, the peritoneum provides over 2 m^2 of blood-rich semipermeable membrane. The fluid is left in the body cavity for 15 to 60 minutes to allow the blood to equilibrate with it; then it is drained, discarded, and replaced with fresh dialysis fluid. The patient is not limited by a stationary dialysis machine and can go about most normal activities. CAPD is less expensive and promotes better morale than conventional hemodialysis, but it is less efficient in removing wastes and it is more often complicated by infection.

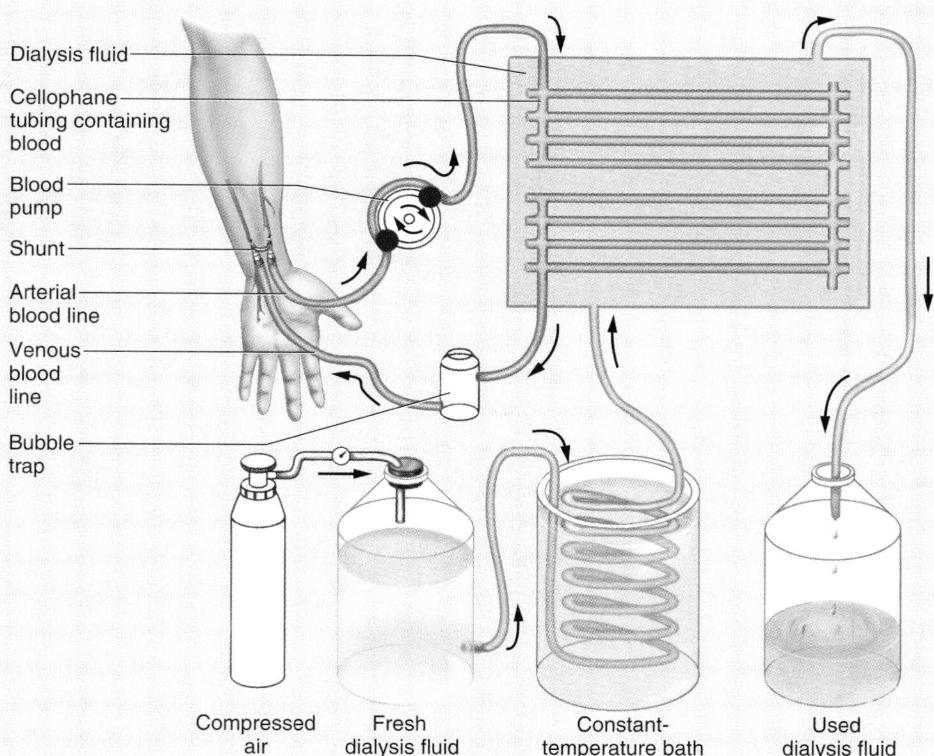

Figure 1 Hemodialysis. Blood is pumped into a dialysis chamber, where it flows through a selectively permeable cellophane membrane surrounded by dialysis fluid. Blood leaving the chamber passes through a bubble trap to remove air before it is returned to the patient's body. Compressed air forces dialysis fluid through a warming bath and then into the dialysis chamber. The fluid picks up excess water and metabolic wastes from the patient's blood and may contain medications that diffuse into the blood.

Connective Issues

Interactions Between the URINARY SYSTEM and Other Organ Systems

← indicates ways in which this system affects other systems
→ indicates ways in which other systems affect this one

All Systems

The urinary system serves all other systems by eliminating metabolic wastes and maintaining fluid, electrolyte, and acid-base balance

Integumentary System

← Renal control of fluid balance essential for sweat secretion

→ Epidermis is normally a barrier to fluid loss; profuse sweating can lead to oliguria; skin and kidneys collaborate in calcitriol synthesis

Skeletal System

← Renal control of calcium and phosphate balance and role in calcitriol synthesis are essential for bone deposition

→ Lower ribs and pelvis protect some urinary system organs

Muscular System

← Renal control of Na^+, K^+, and Ca^{2+} balance important for muscle contraction

→ Some skeletal muscles aid or regulate micturition (external urethral sphincter, male bulbocavernosus muscle, abdominal muscles used in Valsalva maneuver); muscles of pelvic floor support bladder

Nervous System

← Nervous system is very sensitive to fluid, electrolyte, and acid-base imbalances that may result from renal dysfunction

→ Regulates glomerular filtration and micturition

Endocrine System

← Renin secretion by kidneys leads to angiotensin synthesis and aldosterone secretion; kidneys produce erythropoietin

→ Regulates renal function through angiotensin II, aldosterone, atrial natriuretic factor, and antidiuretic hormone

Circulatory System

← Kidneys control blood pressure more than any other organ; erythropoietin from kidneys regulates hematocrit; kidneys regulate plasma composition; cardiac rhythm is very sensitive to electrolyte imbalances that may result from renal dysfunction

→ Perfuses kidneys so wastes can be filtered from blood; blood pressure influences glomerular filtration rate; blood reabsorbs water and solutes from renal tubules

Lymphatic/Immune Systems

← Acidity of urine provides nonspecific defense against infection

→ Return of fluid to bloodstream maintains blood pressure and fluid balance essential for renal function; immune system protects kidneys from infection

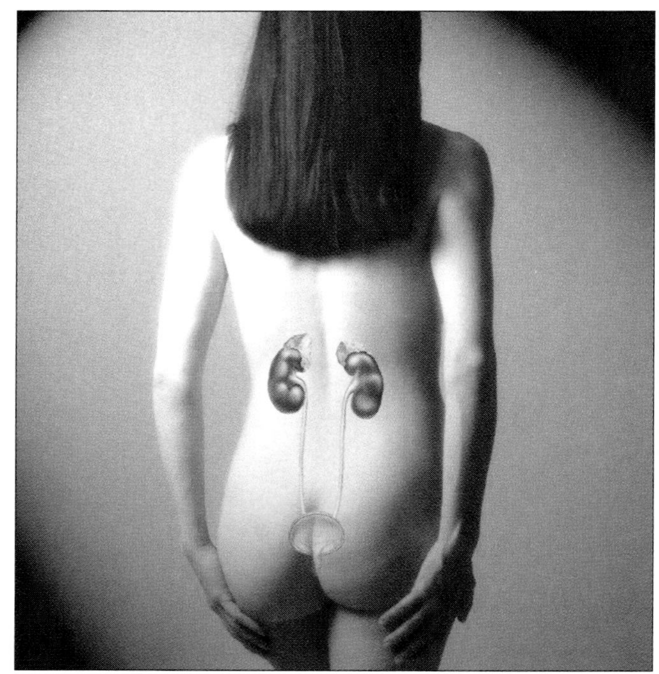

Respiratory System

← Rate of acid secretion by kidneys affects pH of blood and may therefore affect respiratory rhythm

→ Provides O_2 to meet high metabolic demand of kidneys; dysfunctions of pulmonary ventilation may require compensation by kidneys to maintain acid-base balance; inhaled toxic fumes can damage kidneys

Digestive System

← Kidneys excrete toxins absorbed by digestive tract; kidneys excrete hormones and metabolites after liver deactivates them; calcitriol synthesized by kidneys regulates Ca^{2+} absorption by small intestine

→ Liver synthesizes urea, the main nitrogenous waste eliminated by kidneys; urea contributes to osmotic gradient of renal medulla; liver and kidneys collaborate to synthesize calcitriol

Reproductive System

← Urethra serves as common passageway for urine and sperm in males; urinary system of a pregnant woman eliminates metabolic wastes of fetus

→ Enlarged prostate can cause urine retention and kidney damage in males; pregnant uterus compresses bladder and reduces its capacity in females

Chapter Review

Online Learning Center

Study resources for this chapter are available at www.mhhe.com/saladin/

Study Outline

Functions of the Urinary System (p. 882)

1. Organs of the urinary system
2. Functions of the kidneys
3. Nitrogenous wastes
   a. Types
      - Urea
      - Uric acid
      - Creatinine
   b. Azotemia and uremia
4. Excretion
   a. Definition
   b. Organ systems involved

Anatomy of the Kidney (p. 883)

1. Gross anatomy
   a. Position, weight, and size
   b. Shape
   c. Connective tissue coverings
   d. Renal cortex and columns
   e. Renal medulla and pyramids
   f. Renal calices and pelvis
2. The nephron
   a. Blood supply
      - From renal artery to interlobular artery
      - Afferent arteriole
      - Glomerulus
      - Efferent arteriole
      - Peritubular capillaries
      - From interlobular vein to renal vein
      - Vasa recta
   b. Renal corpuscle
      - Glomerular capsule
        - •• Squamous cells of parietal layer
        - •• Podocytes of visceral layer
        - •• Capsular space
      - Glomerulus
   c. Renal tubule
      - Proximal convoluted tubule (PCT)
      - Nephron loop
        - •• Descending and ascending limbs
        - •• Thick and thin segments
      - Distal convoluted tubule (DCT)
      - Juxtaglomerular apparatus
      - Collecting duct (CD)
   d. Cortical and juxtamedullary nephrons

Urine Formation I: Glomerular Filtration (p. 888)

1. The filtration membrane
   a. Components
      - Fenestrated endothelium
      - Basement membrane
      - Filtration slits
   b. Substances passed
   c. Signs of damage
2. Filtration pressure
   a. High blood pressure in glomerulus
   b. Pressure in glomerular capsule
   c. Colloid osmotic pressure of blood
   d. Net filtration pressure (NFP)
3. Glomerular filtration rate (GFR)
   a. Filtration coefficient (K_f)
   b. NFP $\times K_f =$ GFR $\approx$ 125 mL/min
   c. Percent reabsorption
4. Regulation of glomerular filtration
   a. Effects of GFR abnormalities
   b. Renal autoregulation
      - Vasomotion of afferent and efferent arteriole
      - Juxtaglomerular apparatus
        - •• Cellular components
        - •• Mechanism of action
      - Limitations of autoregulation
   c. Sympathetic control
      - Constriction of afferent arteriole
      - Reduction of GFR
   d. Hormonal control
      - Renin secretion
      - Angiotensin II production
      - Angiotensin II effects
        - •• Generalized vasoconstriction
        - •• Constriction of efferent arteriole
        - •• Increased GFR
        - •• Increased tubular reabsorption
        - •• Aldosterone secretion
        - •• Thirst
      - ACE inhibitors

Urine Formation II: Tubular Reabsorption and Secretion (p. 894)

1. The proximal convoluted tubule (PCT)
   a. Relevance of PCT structure
      - Length
      - Microvilli
      - Abundance of mitochondria
   b. Reasons that reabsorption is favored in the PCT
   c. Transcellular and paracellular routes of reabsorption
   d. Role of peritubular capillaries
   e. Reabsorption mechanisms
      - Sodium
      - Glucose
      - Amino acids
      - Water
      - Chloride
      - Other electrolytes
      - Protein
      - Nitrogenous wastes
   f. The transport maximum
   g. Tubular secretion in the PCT
      - Waste removal
      - Acid-base balance
2. The nephron loop
   a. Role in water conservation
   b. Water and electrolyte reabsorption
3. The distal convoluted tubule and collecting duct
   a. Effect of aldosterone
      - Drop in blood pressure causes angiotensin II formation
      - Angiotensin II stimulates adrenal cortex
      - Adrenal cortex secretes aldosterone
      - Aldosterone promotes Na^+ reabsorption

- Na^+ reabsorption promotes water reabsorption
- Water reabsorption reduces urine volume
- Blood pressure drops less rapidly

b. Effect of atrial natriuretic factor (ANF)
- High blood pressure stimulates right atrium
- Atrium secretes ANF
- ANF promotes Na^+ and water excretion
- Blood pressure drops

c. Effect of antidiuretic hormone (ADH)
- Dehydration stimulates hypothalamus
- Hypothalamus stimulates posterior pituitary
- Posterior pituitary releases ADH
- ADH increases water reabsorption
- Urine volume is reduced

Urine Formation III: Water Conservation (p. 898)

1. The collecting duct (CD)
   a. Ability to concentrate urine
   b. Importance in water conservation
2. Control of water loss
   a. Producing hypotonic urine
   - NaCl reabsorbed by cortical CD
   - Water remains in urine

   b. Producing hypertonic urine
   - GFR drops
   - Tubular reabsorption increases
   - Less NaCl remains in CD
   - ADH increases CD's water permeability
   - More water is reabsorbed
   - Urine is more concentrated
3. The countercurrent multiplier
   a. Recaptures NaCl and returns it to renal medulla
   b. Descending limb
   - Reabsorbs water but not salt
   - Concentrates tubular fluid

   c. Ascending limb
   - Reabsorbs Na^+, K^+, and Cl^-
   - Maintains high osmolarity of renal medulla
   - Impermeable to water
   - Tubular fluid becomes hypotonic

   d. Role of collecting duct and urea
4. The countercurrent exchange system
   a. Formed by vasa recta
   - Provide blood to medulla
   - Do not remove NaCl from medulla

   b. Descending capillaries
   - Water diffuses out of blood
   - NaCl diffuses into blood

   c. Ascending capillaries
   - Water diffuses into blood
   - NaCl diffuses out of blood

Urine and Renal Function Tests (p. 901)

1. Composition and properties of urine
   a. Color and clarity
   b. Odor
   c. Specific gravity
   d. Osmolarity
   e. pH
   f. Chemical composition
2. Urine volume
   a. Normal volume
   b. Polyuria
   c. Oliguria and anuria
3. Diabetes
   a. Chronic polyuria of metabolic origin
   b. Types of diabetes
   - With hyperglycemia and glycosuria
     - •• Diabetes mellitus I and II
     - •• Gestational diabetes
     - •• Pituitary diabetes
     - •• Adrenal diabetes
   - With glycosuria but no hyperglycemia
     - •• Renal diabetes
   - With no hyperglycemia or glycosuria
     - •• Diabetes insipidus
4. Diuretics
   a. Effects
   - Increase urine output
   - Reduce blood volume

   b. Uses
   c. Mechanisms of action
   - Increase GFR
   - Reduce tubular reabsorption
5. Renal function tests
   a. Determining renal clearance
   b. Determining GFR

Urine Storage and Elimination (p. 904)

1. The ureters
   a. Length and location
   b. Junction with bladder
   c. Histology of wall
   d. Flow of urine
2. The urinary bladder
   a. Location
   b. Histology of wall
   c. Storage capacity
3. The urethra
   a. Length and location
   b. Regions of male urethra
   c. Urethral sphincters
4. Voiding urine (micturition)
   a. Micturition reflex
   - Stretching of bladder
   - Parasympathetic reflex arc
   - Contraction of detrusor
   - Relaxation of internal urethral sphincter

   b. Voluntary control of micturition
   - Micturition center of pons
   - Stimulation of detrusor
   - Relaxation of external urethral sphincter

   c. Valsalva maneuver

Selected Vocabulary

urinary system 882
kidney 882
ureter 882
urinary bladder 882
urethra 882
waste 882
metabolic waste 882
nitrogenous waste 883
urea 883
uric acid 883
creatinine 883
azotemia 883
uremia 883
excretion 883
hilum 883
renal fascia 883
adipose capsule 884
renal capsule 884
renal sinus 884
cortex 884
medulla 884
renal column 884
renal pyramid 884
papilla 884
minor calyx 884
major calyx 885
renal pelvis 885
nephron 885
renal artery 885
interlobar artery 885
arcuate artery 885
interlobular artery 885
afferent arteriole 885
glomerulus 885
efferent arteriole 885
peritubular capillary 885

interlobular vein 885
arcuate vein 885
interlobar vein 885
renal vein 885
vasa recta 885
renal corpuscle 885
glomerular capsule 885
podocyte 885
glomerular filtrate 885
capsular space 885
renal tubule 885
proximal convoluted tubule 885
nephron loop 887
descending limb 887
ascending limb 887
thick segment 887
thin segment 887
distal convoluted tubule 887
juxtaglomerular apparatus 887
collecting duct 888
papillary duct 888
cortical nephron 888
juxtamedullary nephron 888
glomerular filtrate 889
tubular fluid 889
urine 889
glomerular filtration 889
filtration membrane 889
foot processes 889
filtration slits 889
proteinuria 889
hematuria 889
glomerular filtration rate 891
renal autoregulation 891
juxtaglomerular cell 891
macula densa 891
mesangial cell 891
renin 893
angiotensin-converting enzyme 893
angiotensin II 893
tubular reabsorption 894
transcellular route 894
paracellular route 894
sodium-glucose transport protein 896
obligatory water reabsorption 896
glycosuria 897
tubular secretion 897
countercurrent multiplier 899
countercurrent exchange 900
urinalysis 901
urochrome 902
pyuria 902
polyuria 902
diuresis 902
oliguria 902
anuria 902
diabetes 903
diuretic 903
renal clearance 904
detrusor muscle 905
rugae 905
trigone 905
external urethral orifice 905
prostatic urethra 905
membranous urethra 905
penile urethra 905
urethral gland 906
internal urethral sphincter 906
external urethral sphincter 906
micturition 907
micturition reflex 907
micturition center 907

Testing Your Recall

1. Micturition occurs when the ________ contracts.
   a. detrusor muscle
   b. internal urethral sphincter
   c. external urethral sphincter
   d. muscularis of the ureter
   e. all of the above
2. The compact ball of capillaries in a nephron is called
   a. the nephron loop.
   b. the peritubular plexus.
   c. the renal corpuscle.
   d. the glomerulus.
   e. the vasa recta.
3. Which of these is the most abundant nitrogenous waste in the blood?
   a. uric acid
   b. urea
   c. ammonia
   d. creatinine
   e. albumin
4. Which of these lies closest to the renal cortex?
   a. the parietal peritoneum
   b. the renal fascia
   c. the renal capsule
   d. the adipose capsule
   e. the renal pelvis
5. Most sodium is reabsorbed from the glomerular filtrate by
   a. the vasa recta.
   b. the proximal convoluted tubule.
   c. the distal convoluted tubule.
   d. the nephron loop.
   e. the collecting duct.
6. A glomerulus and glomerular capsule make up one
   a. renal capsule.
   b. renal corpuscle.
   c. kidney lobule.
   d. kidney lobe.
   e. nephron.
7. The kidney has more ________ than any of the other structures listed.
   a. arcuate arteries
   b. minor calices
   c. medullary pyramids
   d. afferent arterioles
   e. collecting ducts
8. The renal clearance of ________ is normally zero.
   a. sodium
   b. potassium
   c. uric acid
   d. urea
   e. amino acids
9. Beavers have relatively little need to conserve water and could therefore be expected to have ______ than humans do.
   a. fewer nephrons
   b. longer nephron loops
   c. shorter nephron loops
   d. longer collecting ducts
   e. longer convoluted tubules
10. Increased ADH secretion should cause the urine to have
    a. a higher specific gravity.
    b. a lighter color.
    c. a higher pH.
    d. a lower urea concentration.
    e. a lower potassium concentration.
11. The __________ reflex is an autonomic reflex activated by pressure in the urinary bladder.
12. __________ is the ability of a nephron to adjust its GFR independently of external nervous or hormonal influences.
13. The two ureters and the urethra form the boundaries of a smooth area called the __________ on the floor of the urinary bladder.
14. The __________ is a group of epithelial cells of the distal convoluted tubule that monitors the flow or composition of the tubular fluid.
15. To enter the capsular space, filtrate must pass between foot process of the __________, cells that form the visceral layer of the glomerular capsule.
16. Glycosuria occurs if the rate of glomerular filtration of glucose exceeds the __________ of the proximal convoluted tubule.

17. ___________ is a hormone that regulates the amount of water reabsorbed by the collecting duct.
18. The ___________ sphincter is under involuntary control and relaxes during the micturition reflex.
19. Very little ___________ is found in the glomerular filtrate because it is negatively charged and is repelled by the basement membrane of the glomerulus.
20. Blood flows through the ___________ arteries just before entering the interlobular arteries.

Answers in appendix B

Testing Your Comprehension

1. How would glomerular filtration rate be affected by kwashiorkor (see p. 684)?
2. A patient produces 55 mL of urine per hour. Urea concentration is 0.25 mg/mL in her blood plasma and 8.6 mg/mL in her urine. (a) What is her rate of renal clearance for urea? (b) About 95% of adults excrete urea at a rate of 12.6 to 28.6 g/day. Is this patient above, within, or below this range? Show how you calculated your answers.
3. A patient with poor renal perfusion is treated with an ACE inhibitor and goes into renal failure. Explain the reason for the renal failure.
4. Suppose the ureters entered the bladder from above instead of from below. Which disorder in table 23.2 would you expect to result from this? Explain why.
5. In what ways do the proximal and distal convoluted tubules differ in structure? How is this related to their functional difference?

Answers at the Online Learning Center

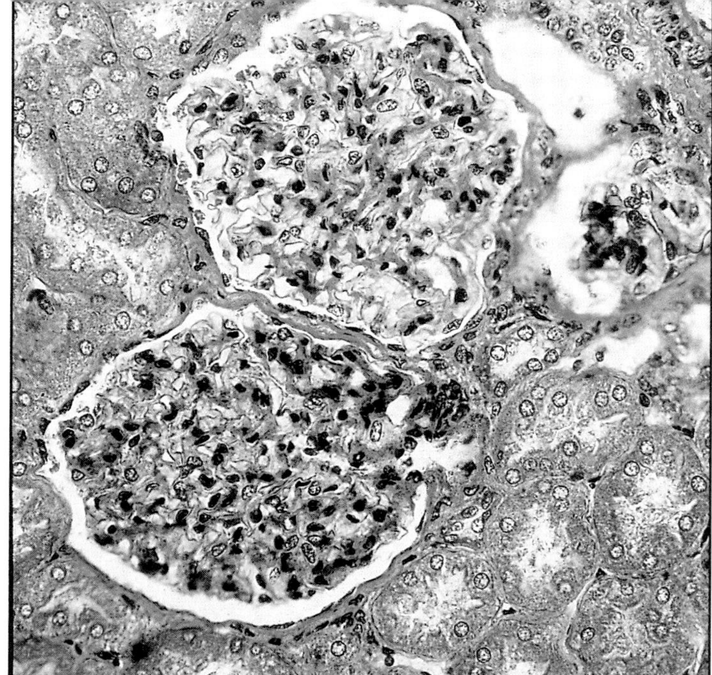

Two glomeruli (green) and renal tubules (violet) of the kidney

CHAPTER 24

Water, Electrolyte, and Acid-Base Balance

Chapter Outline

Insights

Brushing Up

To understand this chapter, it is important that you understand or brush up on the following concepts:

- Electrolytes and milliequivalents/liter (p. 63)
- Acids, bases, and the pH scale (p. 63)
- Osmolarity (p. 127)
- Role of electrolytes in plasma membrane potentials (p. 465)
- Depolarization and hyperpolarization of plasma membranes (fig. 13.21, p. 478)
- The hypothalamus and posterior pituitary (p. 641)
- Influence of CO_2 and pH on pulmonary ventilation (pp. 870–71)
- Structure and physiology of the nephron (pp. 885–88)

Study with Your ESP CD-ROM

Refer to your ESP Tutorial CD-ROM for helpful animations, activities, and quizzes related to the material covered in this chapter. The following are suggested areas to go to on the CD and are presented according to "Unit/Topic":

- Absorption and Excretion/Water & Electrolytes

Cellular function requires a fluid medium with a carefully controlled composition. If the quantity, osmolarity, electrolyte concentration, or pH of this medium is altered, life-threatening disorders of cellular function may result. Consequently, the body has several mechanisms for keeping these variables within narrow limits and maintaining three types of homeostatic balance:

1. *fluid balance,* in which average daily water intake and loss are equal;
2. *electrolyte balance,* in which the amount of electrolytes absorbed by the small intestine balance the amount lost from the body, chiefly through the urine; and
3. *acid-base balance,* in which the body rids itself of acid (hydrogen ions) at a rate that balances its metabolic production, thus maintaining a stable pH.

These balances are maintained by the collective action of the urinary, respiratory, digestive, integumentary, endocrine, nervous, cardiovascular, and lymphatic systems. This chapter describes the homeostatic regulation of water, electrolyte, and acid-base balance and shows the close relationship of these variables to each other.

Water Balance

Objectives

When you have completed this section, you should be able to

- name the major fluid compartments and explain how water moves from one to another;
- list the body's sources of water and routes of water loss;
- describe the mechanisms of regulating water intake and output; and
- describe some conditions in which the body has a deficiency or excess of water or an improper distribution of water among the fluid compartments.

We enter the world in a rather soggy condition, having swallowed, excreted, and floated in amniotic fluid for months. At birth, a baby's weight is as much as 75% water; infants normally lose a little weight in the first day or two as they excrete the excess. Young adult men average 55% to 60% water, while women average slightly less because they have more fat and adipose tissue is nearly free of water. Obese and elderly people are as little as 45% water by weight (see clinical insight 24.1).

Fluid Compartments

The **total body water (TBW)** content of a 70 kg (150 lb) young male is about 40 L. This is distributed among the following **fluid compartments,** which are separated by selectively permeable membranes and differ from each other in chemical composition:

65% *intracellular fluid (ICF)* and
35% *extracellular fluid (ECF),* subdivided into
- 25% *tissue (interstitial) fluid,*
- 8% *blood plasma* and *lymph,* and
- 2% *transcellular fluid,* a catch-all category for cerebrospinal, synovial, peritoneal, pleural, and pericardial fluids; vitreous and aqueous humors of the eye; bile; and fluid in the digestive, urinary, and respiratory tracts.

Fluid is continually exchanged between compartments by way of capillary walls and plasma membranes (fig. 24.1). Water moves by osmosis from the digestive tract to the bloodstream and by capillary filtration from the blood to the tissue fluid. From the tissue fluid, it may be reabsorbed by the capillaries, osmotically absorbed into cells, or taken up by the lymphatic system, which returns it to the bloodstream.

Because water moves so easily through plasma membranes, osmotic gradients between the ICF and ECF never last for very long. If a local imbalance arises, osmosis usually restores the balance within seconds so that intracellular and extracellular osmolarity are equal. If the osmolarity of the tissue fluid rises, water moves out of the cells; if it falls, water moves into the cells.

Osmosis from one fluid compartment to another is determined by the relative concentration of solutes in each compartment. The most abundant solute particles by far are the electrolytes—especially sodium salts in the ECF and potassium salts in the ICF. Electrolytes play the principal role in governing the body's water distribution and total water content; the subjects of water and electrolyte balance are therefore inseparable.

Water Gain and Loss

A person is in a state of **fluid balance** when daily gains and losses are equal. We typically gain and lose about 2,500 mL/day (fig. 24.2). The gains come from two sources: **metabolic water** (about 200 mL/day), which is produced as a by-product of aerobic respiration and dehydration synthesis reactions, and **preformed water,** which is ingested in food (700 mL/day) and drink (1,600 mL/day).

The routes of water loss are more varied:

- 1,500 mL/day is excreted as urine.
- 200 mL/day is eliminated in the feces.
- 300 mL/day is lost in the expired breath. You can easily visualize this by breathing onto a cool surface such as a mirror.
- 100 mL/day of sweat is secreted by a resting adult at an ambient (air) temperature of 20°C (68°F).

Clinical Insight 24.1

Bias in Physiological Values

Many physiological averages refer to the "young adult male." Decades of military recruitment provided a large sample of young, healthy men subject to almost every conceivable anatomical and physiological measurement. This generated a larger pool of data on them than on women or people of other age groups. Clinical problems can occur, however, when female and older patients are treated with therapies based on values typical for young men. Other people may respond very differently to the same treatments, and assumptions based on young males can lead a health-care provider to overlook or misinterpret important clinical signs in other patients.

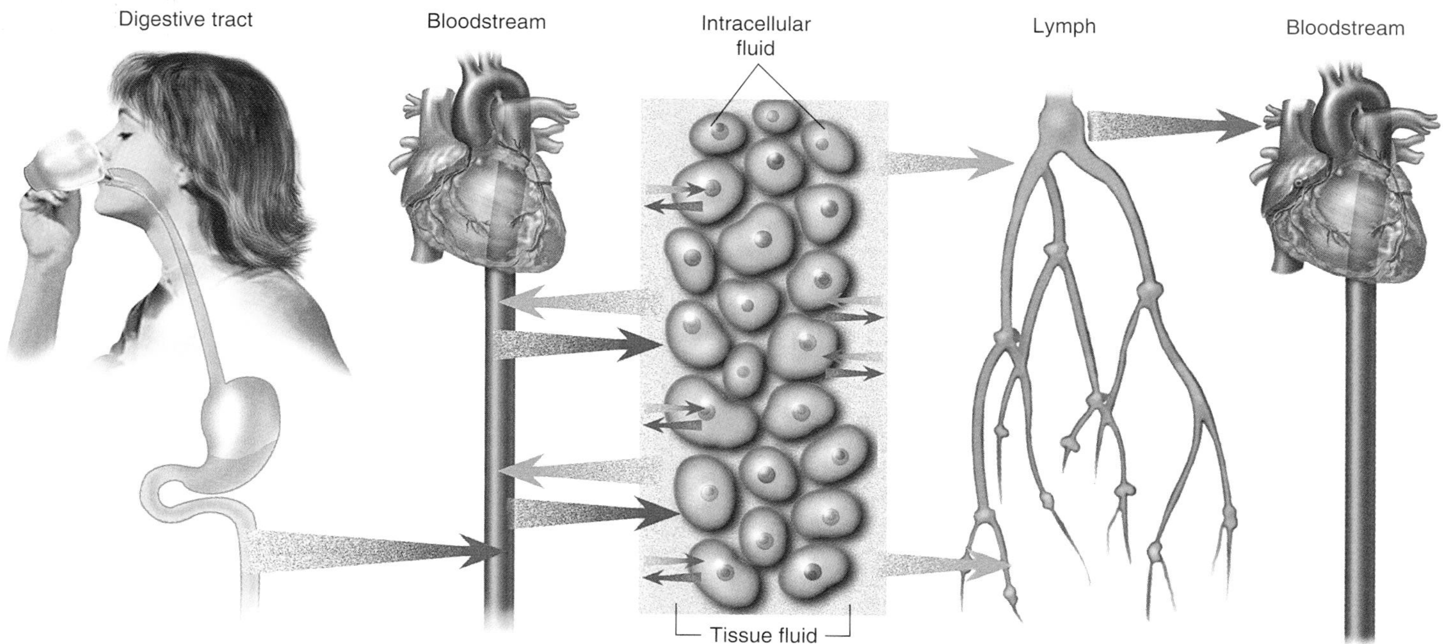

Figure 24.1 **The Movement of Water Between the Major Fluid Compartments.** Ingested water is absorbed by the bloodstream. There is a two-way exchange of water between the blood and tissue fluid and between the tissue and intracellular fluids. Excess tissue fluid is picked up by the lymphatic system, which returns it to the bloodstream.

- 400 mL/day is lost as **cutaneous transpiration,**[1] water that diffuses through the epidermis and evaporates. This is not the same as sweat; it is not a glandular secretion. A simple way to observe it is to cup the palm of your hand for a minute against a cool nonporous surface such as a laboratory benchtop or mirror. When you take your hand away, you will notice the water that transpired through the skin and condensed on that surface.

Water loss varies greatly with physical activity and environmental conditions. Respiratory loss increases in cold weather, for example, because cold air is drier and absorbs more body water from the respiratory tract. Hot, humid weather slightly reduces the respiratory loss but increases perspiration to as much as 1,200 mL/day. Prolonged, heavy work can raise the respiratory loss to 650 mL/day and perspiration to as much as 5 L/day, though it reduces urine output by nearly two-thirds.

Output through the breath and cutaneous transpiration is called **insensible water loss** because we are not usually conscious of it. **Obligatory water loss** is output that is relatively unavoidable: expired air, cutaneous transpiration, sweat, fecal moisture, and the minimum urine output, about 400 mL/day, needed to prevent azotemia. Even dehydrated individuals cannot prevent such losses; thus they become further dehydrated.

1. *trans* = across, through + *spir* = to breathe

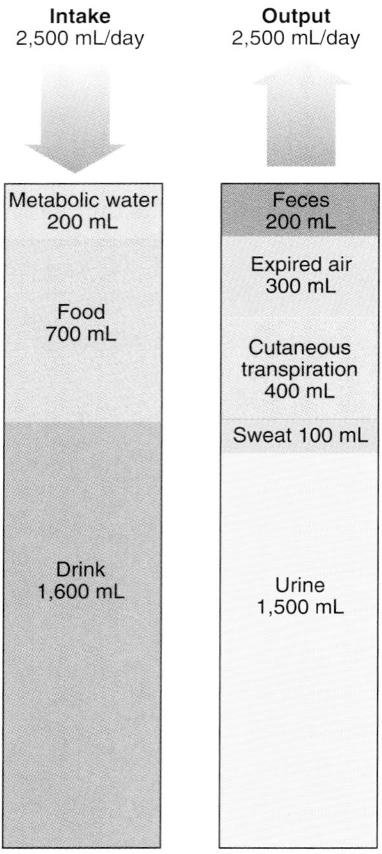

Figure 24.2 Typical Water Intake and Output in a State of Fluid Balance.

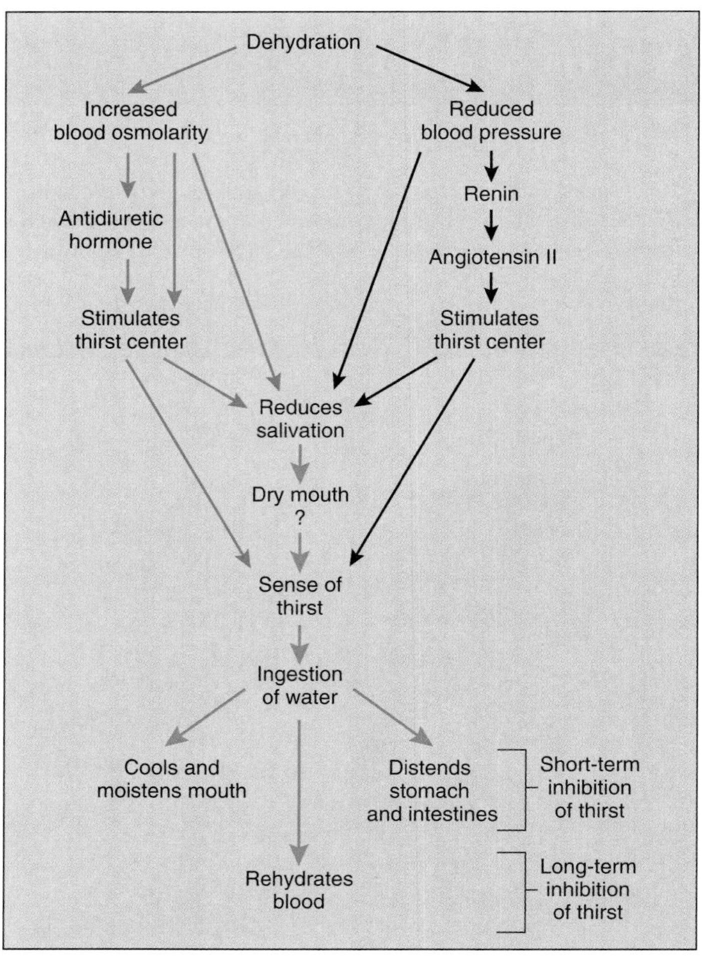

Figure 24.3 Dehydration, Thirst, and Rehydration.

Regulation of Intake

Fluid intake is governed mainly by thirst, which is controlled by the mechanisms shown in figure 24.3. Dehydration reduces blood volume and pressure and raises blood osmolarity. The hypothalamus has a nucleus called the **thirst center** that responds to multiple signs of dehydration: (1) angiotensin II, produced in response to falling blood pressure; (2) antidiuretic hormone, released in response to rising blood osmolarity; and (3) signals from *osmoreceptors,* neurons in the hypothalamus that monitor the osmolarity of the ECF. A 2% to 3% increase in plasma osmolarity makes a person intensely thirsty, as does a 10% to 15% blood loss.

In response to such cues, the thirst center sends sympathetic signals to the salivary glands to inhibit salivation. Salivation is also reduced for another reason. Most of the saliva is produced by capillary filtration, but in dehydration, filtration is reduced by the lower blood pressure and higher osmolarity of the blood. Reduced salivation gives us a dry, sticky-feeling mouth, but it is by no means certain that this is our primary motivation to drink. People who do not secrete saliva and experimental animals that have the salivary ducts tied off do not drink any more than normal individuals except when eating, when they need water to moisten the food.

Long-term satiation of thirst depends on absorbing water from the small intestine and lowering the osmolarity of the blood. Reduced osmolarity stops the osmoreceptor response, promotes capillary filtration, and makes the saliva more abundant and watery. However, these changes require 30 minutes or longer to take effect, and it would be rather impractical if we had to drink that long while waiting to feel satisfied. Water intake would be grossly excessive. Fortunately, there are mechanisms that act more quickly to temporarily quench the thirst and allow time for the change in blood osmolarity to occur.

Experiments with rats and dogs have isolated the stimuli that quench the thirst. One of these is cooling and moistening the mouth; rats drink less if their water is cool than if it is warm, and simply moistening the mouth temporarily satisfies an animal even if the water is drained from its esophagus before it reaches the stomach. Distension of the stomach and small intestine is another inhibitor of thirst. If a dog is allowed to drink while the water is drained from its esophagus but its stomach is inflated with a balloon, its thirst is satisfied for a time. If the water is drained away but the stomach is not inflated, satiation does not last as long. Such fast-acting stimuli as coolness, moisture, and filling of the stomach stop an animal (and presumably a human) from drinking an excessive amount of liquid, but they are effective for only 30 to 45 minutes. If they are not soon followed by absorption of water into the bloodstream, the thirst soon returns. Only a drop in blood osmolarity produces a lasting effect.

Regulation of Output

The only way to control water output significantly is through variations in urine volume. It must be realized, however, that the kidneys cannot completely prevent water loss, nor can they replace lost water or electrolytes. Therefore, they never restore fluid volume or osmolarity, but in dehydration they can support existing fluid levels and slow down the rate of loss until water and electrolytes are ingested.

To understand the effect of the kidneys on water and electrolyte balance, it is also important to bear in mind that if a substance is reabsorbed by the kidneys, it is kept in the body and returned to the ECF, where it will affect fluid volume and composition. If a substance is filtered by the glomerulus or secreted by the renal tubules and not reabsorbed, then it is excreted in the urine and lost from the body fluids.

Changes in urine volume are usually linked to adjustments in sodium reabsorption. As sodium is reabsorbed or excreted, proportionate amounts of water accompany it. The total volume of fluid remaining in the body may change, but its osmolarity remains stable. Controlling water balance by controlling sodium excretion is best understood in the context of electrolyte balance, discussed later in the chapter.

Antidiuretic hormone (ADH), however, provides a means of controlling water output independently of sodium. In true dehydration (defined shortly), blood volume declines and sodium concentration rises. The increased osmolarity of the blood stimulates the hypothalamic osmoreceptors, which stimulate the posterior pituitary to release ADH. In response to ADH, cells of the collecting ducts of the kidneys synthesize membrane proteins called **aquaporins.** When installed in the plasma membrane, these serve as channels that allow water to diffuse out of the duct into the hypertonic tissue fluid of the renal medulla. Thus the kidneys reabsorb more water and produce less urine. Sodium continues to be excreted, so the *ratio* of sodium to water in the urine increases (the urine becomes more concentrated). By helping the kidneys retain water, ADH slows down the decline in blood volume and the rise in its osmolarity. Thus the ADH mechanism forms a negative feedback loop (fig. 24.4).

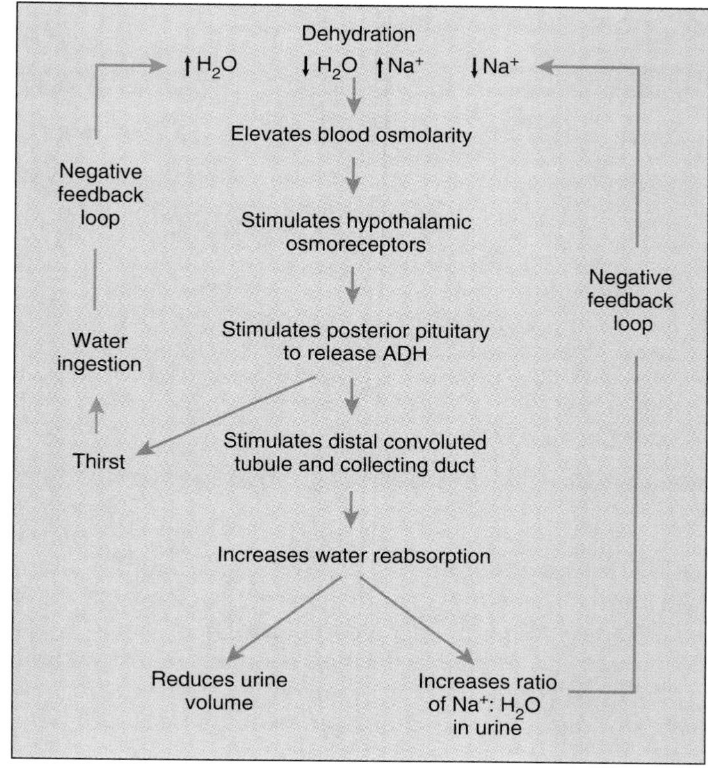

Figure 24.4 **The Secretion and Effects of Antidiuretic Hormone.** Pathways shown in red represent negative feedback.

Conversely, if blood volume and pressure are too high or blood osmolarity is too low, ADH release is inhibited. The renal tubules reabsorb less water, urine output increases, and total body water declines. This is an effective way of compensating for hypertension. Since the lack of ADH increases the ratio of water to sodium in the urine, it raises the sodium concentration and osmolarity of the blood.

Disorders of Water Balance

The body is in a state of fluid imbalance if there is an abnormality of total fluid *volume,* fluid *concentration,* or fluid *distribution* among the compartments.

Fluid Deficiency

Fluid deficiency arises when output exceeds intake over a long period of time. There are two kinds of deficiency, called volume depletion and dehydration, which differ in the relative loss of water and electrolytes and the resulting osmolarity of the ECF. This is an important distinction that

Clinical Insight 24.2

Fluid Balance in Cold Weather

Hot weather and profuse sweating are obvious threats to fluid balance, but so is cold weather. The body conserves heat by constricting the blood vessels of the skin and subcutaneous tissue, thus forcing blood into the deeper circulation. This raises the blood pressure, which inhibits the secretion of antidiuretic hormone and increases the secretion of atrial natriuretic factor. These hormones increase urine output and reduce blood volume. In addition, cold air is relatively dry and increases respiratory water loss. This is why exercise causes the respiratory tract to "burn" more in cold weather than in warm.

These cold-weather respiratory and urinary losses can cause significant hypovolemia. Furthermore, the onset of exercise stimulates vasodilation in the skeletal muscles. In a hypovolemic state, there may not be enough blood to supply them and a person may experience weakness, fatigue, or fainting (hypovolemic shock). In winter sports and other activities such as snow shoveling, it is important to maintain fluid balance. Even if you do not feel thirsty, it is beneficial to take ample amounts of warm liquids such as soup or cider. Coffee, tea, and alcohol, however, have diuretic effects that defeat the purpose of fluid intake.

calls for different strategies of fluid replacement therapy (see clinical insight 24.3 at the end of the chapter).

Volume depletion (hypovolemia[2]) occurs when proportionate amounts of water *and* sodium are lost without replacement. Total body water declines but osmolarity remains normal. Volume depletion occurs in cases of hemorrhage, severe burns, and chronic vomiting or diarrhea. A less common cause is aldosterone hyposecretion (Addison disease), which results in inadequate sodium and water reabsorption.

Dehydration (negative water balance) occurs when the body eliminates significantly more water than sodium, so the ECF osmolarity rises. The simplest cause of dehydration is a lack of drinking water; for example, when stranded in a desert or at sea. It can be a serious problem for elderly and bedridden people who depend on others to provide them with water—especially for those who cannot express their need or whose caretakers are insensitive to it. Diabetes mellitus, ADH hyposecretion (diabetes insipidus), profuse sweating, and overuse of diuretics are additional causes of dehydration. Prolonged exposure to cold weather can dehydrate a person just as much as exposure to hot weather (see clinical insight 24.2).

For three reasons, infants are more vulnerable to dehydration than adults: (1) Their high metabolic rate produces toxic metabolites faster, and they must excrete more water to eliminate them. (2) Their kidneys are not fully mature and cannot concentrate urine as effectively. (3) They have a greater ratio of body surface to volume; consequently, compared to adults, they lose twice as much water per kilogram of body weight by evaporation.

Dehydration affects all fluid compartments. Suppose, for example, that you play a strenuous tennis match on a hot summer day and lose a liter of sweat per hour. Where does this fluid come from? Most of it filters out of the bloodstream through the capillaries of the sweat glands. In principle, 1 L of sweat would amount to about one-third of the blood plasma. However, as the blood loses water its osmolarity rises and water from the tissue fluid enters the bloodstream to balance the loss. This raises the osmolarity of the tissue fluid, so water moves out of the cells to balance that (fig. 24.5). Ultimately, all three fluid compartments (the intracellular fluid, blood, and tissue fluid) lose water. To excrete 1 L of sweat, about 300 mL of water would come from the ECF and 700 mL from the ICF. Immoderate exercise without fluid replacement can lead to even greater loss than 1 L per hour.

The most serious effects of fluid deficiency are circulatory shock due to loss of blood volume and neurological dysfunction due to dehydration of brain cells. Volume depletion by diarrhea is a major cause of infant mortality, especially under unsanitary conditions that lead to intestinal infections such as cholera.

Fluid Excess

Fluid excess is less common than fluid deficiency because the kidneys are highly effective at compensating for excessive intake by excreting more urine (fig. 24.6). Renal failure and other causes, however, can lead to excess fluid retention.

Fluid excesses are of two types called volume excess and hypotonic hydration. In **volume excess,** both sodium and water are retained and the ECF remains isotonic. Volume excess can result from aldosterone hypersecretion or renal failure. In **hypotonic hydration** (also called **water intoxication** or **positive water balance**), more water than sodium is retained or ingested and the ECF becomes hypotonic. This can occur if you lose a large amount of water *and* salt through urine and sweat and you replace it by

2. *hypo* = below normal + *vol* = volume + *emia* = blood condition

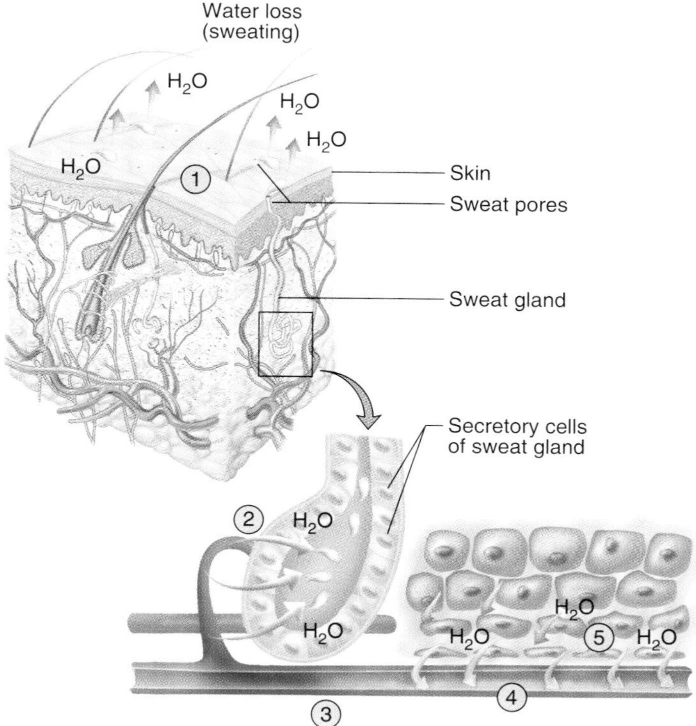

Figure 24.5 Effects of Profuse Sweating on the Fluid Compartments. (*1*) Sweat is released from pores in the skin. (*2*) Sweat is produced by filtration from the blood capillaries. (*3*) As fluid is taken from the bloodstream, blood volume and pressure drop and blood osmolarity rises. (*4*) The blood absorbs tissue fluid to replace its loss. (*5*) Fluid is transferred from the intracellular compartment to the tissue fluid. In severe dehydration, this results in cell shrinkage and malfunction.

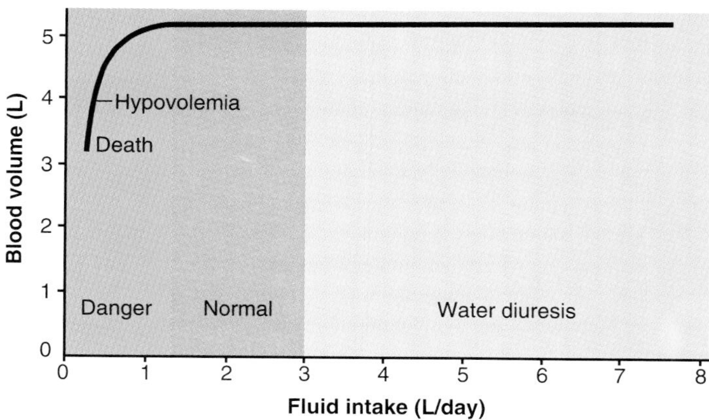

Figure 24.6 The Relationship of Blood Volume to Fluid Intake. The kidneys cannot compensate very well for inadequate fluid intake. Below an intake of about 1 L/day, blood volume drops significantly and there may be a threat of death from hypovolemic shock. The kidneys compensate very well, on the other hand, for abnormally high fluid intake; they eliminate the excess by water diuresis and maintain a stable blood volume.

drinking plain water. Without a proportionate intake of electrolytes, water dilutes the ECF, makes it hypotonic, and causes cellular swelling. ADH hypersecretion can cause hypotonic hydration by stimulating excessive water retention as sodium continues to be excreted. Among the most serious effects of either type of fluid excess are pulmonary and cerebral edema.

Table 24.1 Forms of Fluid Imbalance

	Total Body Water	Osmolarity
Fluid Deficiency		
Volume depletion (hypovolemia)	Reduced	Isotonic (normal)
Dehydration (negative water balance)	Reduced	Hypertonic (elevated)
Fluid Excess		
Volume excess	Elevated	Isotonic (normal)
Hypotonic hydration (positive water balance, water intoxication)	Elevated	Hypotonic (reduced)

Fluid Sequestration

Fluid sequestration[3] (seh-ques-TRAY-shun) is a condition in which excess fluid accumulates in a particular location. Total body water may be normal, but the volume of circulating blood may drop to the point of causing circulatory shock. The most common form of sequestration is *edema,* the abnormal accumulation of fluid in the interstitial spaces, causing swelling of a tissue (discussed in detail in chapter 20). Hemorrhage can be another cause of fluid sequestration; blood that pools and clots in the tissues is lost to circulation. Yet another example is *pleural effusion,* caused by some lung infections, in which several liters of fluid accumulate in the pleural cavity.

The four principal forms of fluid imbalance are summarized and compared in table 24.1.

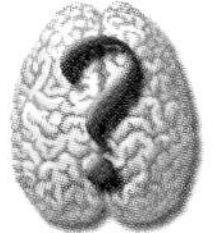

Think About It

Some tumors of the brain, pancreas, and small intestine secrete ADH. What type of water imbalance would this produce? Explain why.

3. *sequestr* = to isolate

Before You Go On

Answer the following questions to test your understanding of the preceding section:

1. List five routes of water loss. Which one accounts for the greatest loss? Which one is most controllable?
2. Explain why even a severely dehydrated person inevitably experiences further fluid loss.
3. Suppose there were no mechanisms to stop the sense of thirst until the blood became sufficiently hydrated. Explain why we would routinely suffer hypotonic hydration.
4. Summarize the effect of ADH on total body water and blood osmolarity.
5. Name and define the four types of fluid imbalance, and give an example of a situation that could produce each type.

Electrolyte Balance

Objectives

When you have completed this section, you should be able to

- describe the physiological roles of sodium, potassium, calcium, chloride, and phosphate;
- describe the hormonal and renal mechanisms that regulate the concentrations of these electrolytes; and
- state the term for an excess or deficiency of each electrolyte and describe the consequences of these imbalances.

Electrolytes are physiologically important for multiple reasons: they are chemically reactive and participate in metabolism, they determine the electrical potential (charge difference) across cell membranes, and they strongly affect the osmolarity of the body fluids and the body's water content and distribution. Strictly speaking, electrolytes are salts such as sodium chloride, not just sodium or chloride ions. In common usage, however, the individual ions are often referred to as electrolytes. The major cations are sodium (Na^+), potassium (K^+), calcium (Ca^{2+}), and hydrogen (H^+), and the major anions are chloride (Cl^-), bicarbonate (HCO_3^-), and phosphate (PO_4^{3-}). Hydrogen and bicarbonate regulation are discussed later under acid-base balance. Here we focus on the other five.

The typical concentrations of these ions and the terms for electrolyte imbalances are listed in table 24.2. Blood plasma is the most accessible fluid for measurements of electrolyte concentration, so excesses and deficiencies are defined with reference to normal plasma concentrations. Concentrations in the tissue fluid differ only slightly from those in the plasma. The prefix *normo-* denotes a normal electrolyte concentration (for example, *normokalemia*), and *hyper-* and *hypo-* denote concentrations that are sufficiently above or below normal to cause physiological disorders.

Sodium

Functions

Sodium is one of the principal ions responsible for the resting membrane potentials of cells, and the inflow of sodium through gated membrane channels is an essential event in the depolarization that underlies nerve and muscle function. Sodium is the principal cation of the ECF; sodium salts account for 90% to 95% of its osmolarity. Sodium is therefore the most significant solute in determining total body water and the distribution of water among fluid compartments. Sodium gradients across the plasma membrane provide the potential energy that is tapped to cotransport other solutes such as glucose, potassium, and calcium. The Na^+-K^+ pump is an important mechanism for generating body heat. Sodium bicarbonate ($NaHCO_3$) plays a major role in buffering the pH of the ECF.

Table 24.2 Electrolyte Concentrations and the Terminology of Electrolyte Imbalances

	Mean Concentration (mEq/L)*			
Electrolyte	**Plasma**	**ICF**	**Deficiency**	**Excess**
Sodium (Na^+)	142	10	Hyponatremia	Hypernatremia[4]
Potassium (K^+)	5	141	Hypokalemia	Hyperkalemia[5]
Calcium (Ca^{2+})	5	<1	Hypocalcemia	Hypercalcemia
Chloride (Cl^-)	103	4	Hypochloremia	Hyperchloremia
Phosphate (PO_4^{3-})	4	75	Hypophosphatemia	Hyperphosphatemia

*Concentrations in mmol/L are the same for Na^+, K^+, and Cl^-, one-half the above values for Ca^{2+}, and one-third the above values for PO_4^{3-}.

4. *natr* = sodium + *emia* = blood condition
5. *kal* = potassium

Homeostasis

An adult needs about 0.5 g of sodium per day, whereas the typical American diet contains 3 to 7 g/day. Thus a dietary sodium deficiency is rare, and the primary concern is adequate excretion of the excess. This is one of the most important roles of the kidneys. There are multiple mechanisms for controlling sodium concentration, tied to its effects on blood pressure and osmolarity and coordinated by three hormones: aldosterone, antidiuretic hormone, and atrial natriuretic factor.

Aldosterone, the "salt-retaining hormone," plays the primary role in adjustment of sodium excretion. Hyponatremia and hyperkalemia directly stimulate the adrenal cortex to secrete aldosterone, and hypotension stimulates its secretion by way of the renin-angiotensin mechanism (fig. 24.7).

Only cells of the distal convoluted tubule and cortical part of the collecting duct have aldosterone receptors. Aldosterone, a steroid, binds to nuclear receptors and activates transcription of a gene for the Na^+-K^+ pump. In 10 to 30 minutes, enough Na^+-K^+ pumps are synthesized and installed in the plasma membrane to produce a noticeable effect—sodium concentration in the urine begins to fall and potassium concentration rises as the tubules reabsorb more Na^+ and secrete more H^+ and K^+. Water and Cl^- passively follow Na^+. Thus the primary effects of aldosterone are that the urine contains less NaCl and more K^+ and has a lower pH. An average adult male excretes 5 g of sodium per day, but the urine can be virtually sodium-free when aldosterone level is high. Although aldosterone strongly influences sodium reabsorption, it has little effect on plasma sodium *concentration* because reabsorbed sodium is accompanied by a proportionate amount of water.

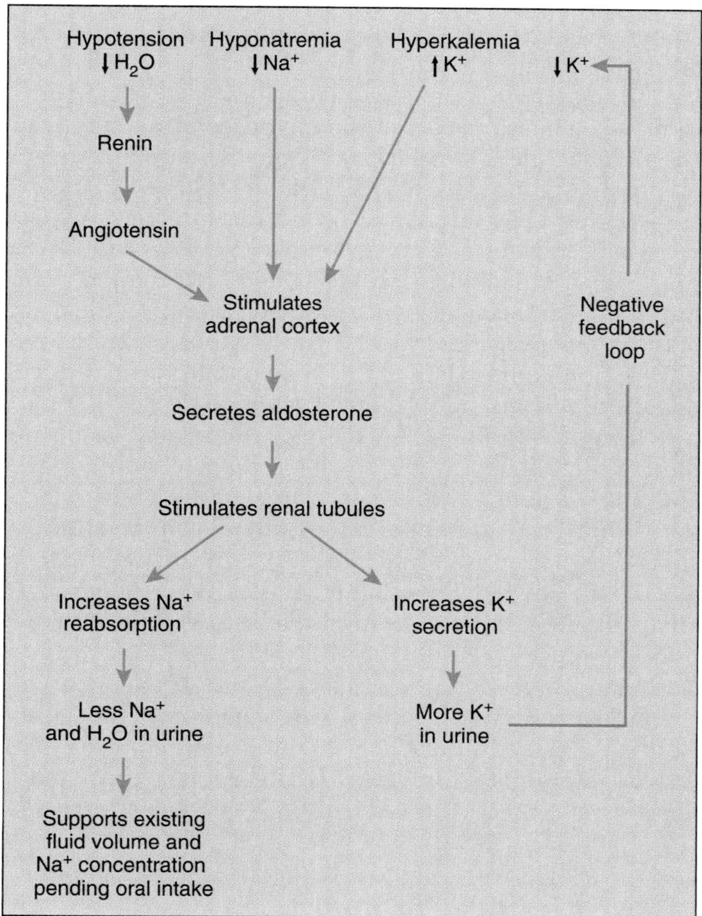

Figure 24.7 The Secretion and Effects of Aldosterone. The pathway shown in red represents negative feedback.

Hypertension inhibits the renin-angiotensin-aldosterone mechanism. The kidneys then reabsorb almost no sodium beyond the proximal convoluted tubule (PCT), and the urine contains up to 30 g of sodium per day.

Aldosterone has only slight effects on urine volume, blood volume, and blood pressure in spite of the tendency of water to follow sodium osmotically. Even in aldosterone hypersecretion, blood volume is rarely more than 5% to 10% above normal. An increase in blood volume increases blood pressure and glomerular filtration rate (GFR). Even though aldosterone increases the tubular reabsorption of sodium and water, this is offset by the rise in GFR and there is only a small drop in urine output.

Antidiuretic hormone modifies water excretion independently of sodium excretion. Thus, unlike aldosterone, it can change sodium *concentration*. A high concentration of sodium in the blood stimulates the posterior lobe of the pituitary gland to release ADH. Thus the kidneys reabsorb more water, which slows down any further increase in blood sodium concentration. ADH alone cannot lower the blood sodium concentration; this requires water ingestion, but remember that ADH also stimulates thirst. A drop in sodium concentration, by contrast, inhibits ADH release. More water is excreted and this raises the concentration of the sodium that remains in the blood.

Atrial natriuretic factor (ANF) is secreted by endocrine cells of the atrial myocardium in response to the stretching produced by hypertension. It inhibits sodium and water reabsorption and the secretion of renin and ADH. The kidneys thus eliminate more sodium and water and lower the blood pressure.

Several other hormones also affect sodium homeostasis. Estrogens mimic the effect of aldosterone and cause women to retain water during pregnancy and certain phases of the menstrual cycle. Progesterone reduces sodium reabsorption and has a diuretic effect. High levels of glucocorticoids promote sodium reabsorption and edema.

In some cases, sodium homeostasis is achieved by regulation of salt intake. A craving for salt occurs in people who are depleted of sodium; for example, by blood loss or Addison disease. Pregnant women sometimes develop a craving for salty foods. Salt craving is not limited to humans; many animals ranging from elephants to butterflies seek out salty soil where they can obtain this vital mineral.

Imbalances

True imbalances in sodium concentration are relatively rare because sodium excess or depletion is almost always accompanied by proportionate changes in water volume. **Hypernatremia** is a plasma sodium concentration in excess of 145 mEq/L. It can result from the administration of intravenous saline (see clinical insight 24.3, p. 934). Its major consequences are water retention, hypertension, and edema. **Hyponatremia** (less than 130 mEq/L) is usually the result of excess body water rather than excess sodium excretion, as in the case mentioned earlier of a person who loses large volumes of sweat or urine and replaces it by drinking plain water. Usually, hyponatremia is quickly corrected by excretion of the excess water, but if uncorrected it produces the symptoms of hypotonic hydration described earlier.

Potassium

Functions

Potassium is the most abundant cation of the ICF and is the greatest determinant of intracellular osmolarity and cell volume. Along with sodium, it produces the resting membrane potentials and action potentials of nerve and muscle cells (fig. 24.8*a*). Potassium is as important as sodium to the Na^+-K^+ pump and its functions of cotransport and thermogenesis (heat production). It is an essential cofactor for protein synthesis and some other metabolic processes.

Homeostasis

Potassium homeostasis is closely linked to that of sodium. Regardless of the body's state of potassium balance, about 90% of the K^+ filtered by the glomerulus is reabsorbed by the PCT and the rest is excreted in the urine. Variations in potassium excretion are controlled later in the nephron by changing the amount of potassium returned to the tubular fluid by the distal convoluted tubule and cortical portion of the collecting duct (CD). When K^+ concentration is high, they secrete more K^+ into the filtrate and the urine may contain more K^+ than the glomerulus filtered from the blood. When blood K^+ level is low, the CD secretes less. The CD has *intercalated cells* in its epithelium that can reabsorb potassium when necessary.

Aldosterone regulates potassium balance along with sodium (see fig. 24.7). A rise in K^+ concentration stimulates the adrenal cortex to secrete aldosterone. Aldosterone stimulates renal secretion of K^+ at the same time that it stimulates reabsorption of sodium. The more sodium there is in the urine, the less potassium, and vice versa.

Imbalances

Potassium imbalances are the most dangerous of all electrolyte imbalances. **Hyperkalemia** ($>$ 5.5 mEq/L) can have completely opposite effects depending on whether K^+ concentration rises quickly or slowly. It can rise quickly when, for example, a crush injury or hemolytic anemia releases large amounts of K^+ from ruptured cells. This can also result from a transfusion with outdated, stored blood because K^+ leaks from erythrocytes into the plasma during storage. A sudden increase in extracellular K^+ tends to make nerve and muscle cells abnormally excitable. Normally, K^+ continually passes in and out of cells at equal rates—leaving by diffusion and reentering by the Na^+-K^+ pump. But in hyperkalemia, there is less concentration difference between the ICF and ECF, so the outward diffusion of K^+ is reduced. More K^+ remains in the cell than normal, and the plasma membrane therefore has a less negative resting potential and is closer to the threshold at which it will set off action potentials (fig. 24.8*b*). This is a very dangerous condition that can quickly produce cardiac arrest. High-potassium solutions are sometimes used by veterinarians to euthanize animals and are used in some states as a lethal injection for capital punishment.

Hyperkalemia can also have a slower onset stemming from such causes as aldosterone hyposecretion, renal failure, or acidosis. (The relationship of acid-base imbalances to potassium imbalances is explained later.) Paradoxically, if the extracellular K^+ concentration rises slowly, nerve and muscle become *less* excitable. *Slow* depolarization of a cell inactivates voltage-gated Na^+ channels, and the channels do not become excitable again until the membrane repolarizes. Inactivated Na^+ channels cannot produce action potentials. For this reason, muscle cramps can be relieved by taking supplemental potassium.

Hypokalemia ($<$ 3.5 mEq/L) rarely results from a dietary deficiency, because most diets contain ample amounts of potassium; it can occur, however, in people with depressed appetites. Hypokalemia more often results from heavy sweating, chronic vomiting or diarrhea, excessive use of laxatives, aldosterone hypersecretion, or alkalosis. As ECF potassium concentration falls, more K^+ moves from the ICF to the ECF. With the loss of these cations from the cytoplasm, cells become hyperpolarized and nerve and muscle cells are less excitable (fig. 24.8*c*). This is reflected in muscle weakness, loss of muscle tone, depressed reflexes, and irregular electrical activity of the heart.

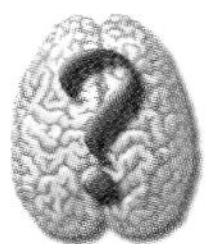

Think About It

Some tumors of the adrenal cortex secrete excess aldosterone and may cause paralysis. Explain this effect and identify the electrolyte and fluid imbalances you would expect to observe in such a case.

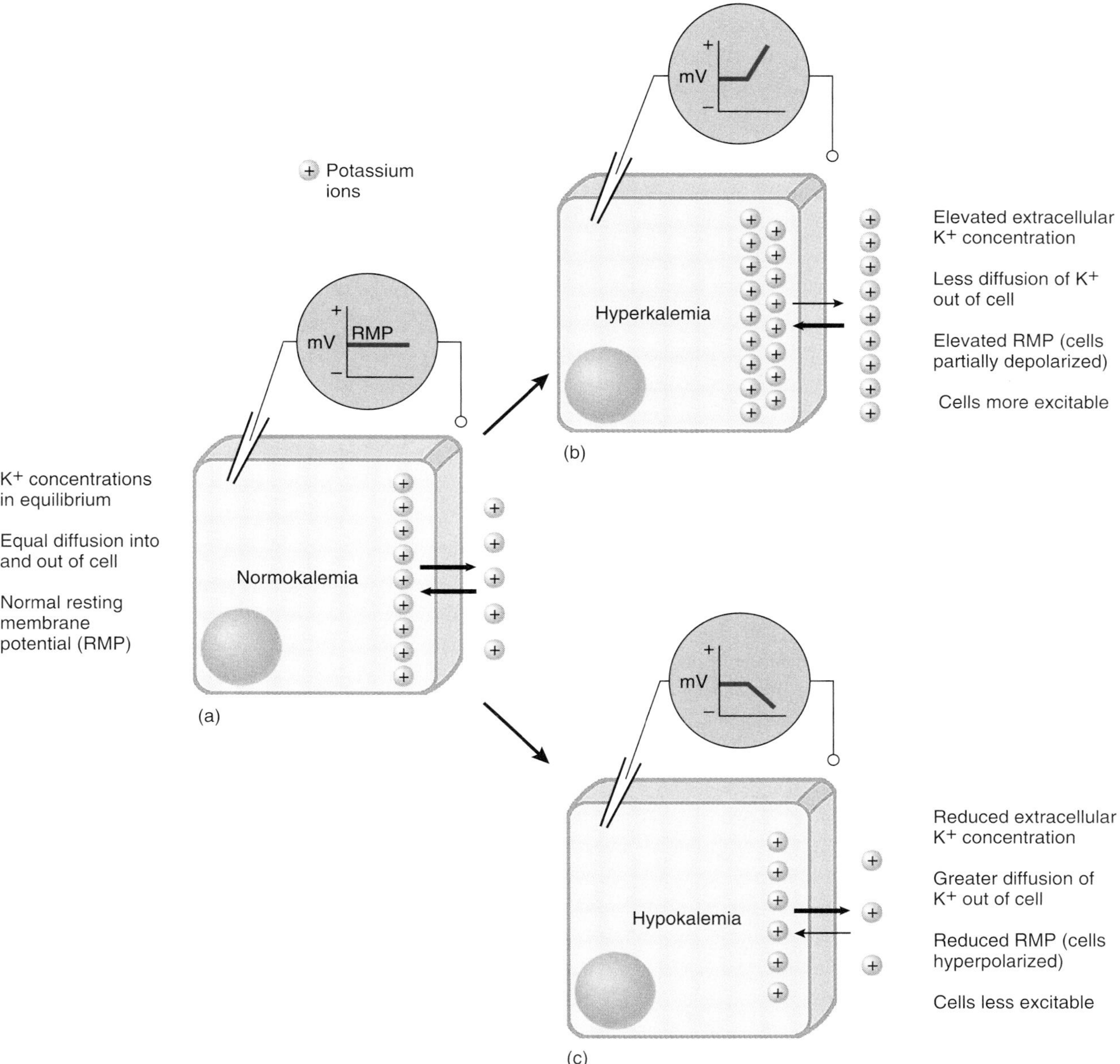

Figure 24.8 Effects of Potassium Imbalances on Membrane Potentials. The diagram at the *top* of each cell represents the voltage measured across the plasma membrane. (*a*) Normokalemia, with a normal resting membrane potential (RMP). (*b*) Hyperkalemia, with an elevated RMP. (*c*) Hypokalemia, with a depressed RMP.

Chloride

Functions

Chloride ions are the most abundant anions of the ECF and thus make a major contribution to its osmolarity. Chloride ions are required for the formation of stomach acid (HCl), and they are involved in the chloride shift that accompanies carbon dioxide loading and unloading by the erythrocytes (see chapter 22). By a similar mechanism explained later, Cl^- plays a major role in the regulation of body pH.

Homeostasis

Cl^- is strongly attracted to Na^+, K^+, and Ca^{2+}. It would require great expenditure of energy to keep it separate from these cations, so Cl^- homeostasis is achieved primarily as an effect of Na^+ homeostasis—as sodium is retained or excreted, Cl^- passively follows.

Imbalances

Hyperchloremia (> 105 mEq/L) is usually the result of dietary excess or administration of intravenous saline. **Hypochloremia** (< 95 mEq/L) is usually a side effect of hyponatremia but sometimes results from hypokalemia. In the latter case, the kidneys retain K^+ by excreting more Na^+, and Na^+ takes Cl^- with it. The primary effects of chloride imbalances are disturbances in acid-base balance, but this works both ways—a pH imbalance arising from some other cause can also produce a chloride imbalance. Chloride balance is therefore discussed further in connection with acid-base balance.

Calcium

Functions

Calcium and phosphate homeostasis were extensively discussed in chapter 8 and are summarized only briefly here. Calcium lends strength to the skeleton, activates the sliding filament mechanism of muscle contraction, serves as a second messenger for some hormones and neurotransmitters, activates exocytosis of neurotransmitters and other cellular secretions, and is an essential factor in blood clotting. Cells maintain a very low intracellular calcium concentration because they require a high concentration of phosphate ions (for reasons discussed shortly). If calcium and phosphate were both very concentrated in a cell, the product of their concentrations, $[Ca^{2+}] \bullet [PO_4^{3-}]$, could reach the critical *solubility product* where calcium phosphate crystals precipitate in the cytoplasm (as described in chapter 8). To maintain a high phosphate concentration but avoid crystallization of calcium phosphate, cells must pump out Ca^{2+} and keep it at a low intracellular concentration or else sequester Ca^{2+} in the smooth ER and release it only when needed. Cells that store Ca^{2+} often have a protein called *calsequestrin,* which binds the stored Ca^{2+} and keeps it chemically unreactive.

Homeostasis

The homeostatic control of Ca^{2+} concentration was discussed extensively in chapter 8. It is regulated chiefly by parathyroid hormone, calcitriol, and in children, by calcitonin. These hormones regulate blood calcium concentration through their effects on bone deposition and resorption, intestinal absorption of calcium, and urinary excretion.

Imbalances

Hypercalcemia (> 5.8 mEq/L) can result from alkalosis, hyperparathyroidism, or hypothyroidism. It reduces the Na^+ permeability of plasma membranes and inhibits the depolarization of nerve and muscle cells. At concentrations ≥ 12 mEq/dL, hypercalcemia causes muscular weakness, depressed reflexes, and cardiac arrhythmia.

Hypocalcemia (< 4.5 mEq/L) can result from vitamin D deficiency, diarrhea, pregnancy, lactation, acidosis, hypoparathyroidism, or hyperthyroidism. It increases the Na^+ permeability of plasma membranes causing the nervous and muscular systems to be overly excitable. Tetanus occurs when calcium concentration drops to 6 mg/dL and may be lethal at 4 mg/dL due to laryngospasm and suffocation.

Phosphates

Functions

The phosphates of the body fluids are an equilibrium mixture of phosphate (PO_4^{3-}), monohydrogen phosphate (HPO_4^{2-}), and dihydrogen phosphate ($H_2PO_4^-$) ions. Phosphates are relatively concentrated in the ICF, where they are generated by the hydrolysis of ATP and other phosphate compounds. They are a component of nucleic acids, phospholipids, ATP, GTP, cAMP, and related compounds. Every process that depends on ATP depends on phosphate ions. Phosphates activate many metabolic pathways by phosphorylating enzymes and substrates such as glucose. They are also important as buffers that help stabilize the pH of the body fluids.

Homeostasis

The average diet provides ample amounts of phosphate ions, which are readily absorbed by the small intestine. Plasma phosphate concentration is usually maintained at about 4 mEq/L, with continual loss of excess phosphate by glomerular filtration. If plasma phosphate concentration drops much below this level, however, the renal tubules reabsorb all filtered phosphate.

Parathyroid hormone increases the excretion of phosphate as part of the mechanism for increasing the concentration of free calcium ions in the ECF. Lowering the ECF phosphate concentration minimizes the formation of calcium phosphate and thus helps support plasma calcium concentration. Rates of phosphate excretion are also strongly affected by the pH of the urine, as discussed shortly.

Imbalances

Phosphate homeostasis is not as critical as that of other electrolytes. The body can tolerate broad variations several times above or below the normal concentration with little immediate effect on physiology.

Before You Go On

Answer the following questions to test your understanding of the preceding section:

6. Which of these do you think would have the most serious effect, and why—a 5 mEq/L increase in the plasma concentration of sodium, potassium, chloride, or calcium?
7. Answer the same question for a 5 mEq/L *decrease.*
8. Explain why ADH is more likely than aldosterone to change the osmolarity of the blood plasma.
9. Explain why aldosterone hyposecretion could cause hypochloremia.
10. Why are more phosphate ions required in the ICF than in the ECF? How does this affect the distribution of calcium ions between these fluid compartments?

Acid-Base Balance

Objectives

When you have completed this section, you should be able to

- define *buffer* and write chemical equations for the bicarbonate, phosphate, and protein buffer systems;
- discuss the relationship between pulmonary ventilation, pH of the extracellular fluids, and the bicarbonate buffer system;
- explain how the kidneys secrete hydrogen ions and how these ions are buffered in the tubular fluid;
- identify some types and causes of acidosis and alkalosis, and describe the effects of these pH imbalances; and
- explain how the respiratory and urinary systems correct acidosis and alkalosis, and compare their effectiveness and limitations.

As we saw in chapter 3, metabolism depends on the functioning of enzymes, and enzymes are very sensitive to pH. Slight deviations from the normal pH can shut down metabolic pathways as well as alter the structure and function of other macromolecules. Consequently, acid-base balance is one of the most important aspects of homeostasis.

The blood and tissue fluid normally have a pH of 7.35 to 7.45. Such a narrow range of variation is remarkable considering that our metabolism constantly produces acid: lactic acid from anaerobic fermentation, phosphoric acids from nucleic acid catabolism, fatty acids and ketones from fat catabolism, and carbonic acid from carbon dioxide. Here we examine mechanisms for resisting these challenges and maintaining acid-base balance.

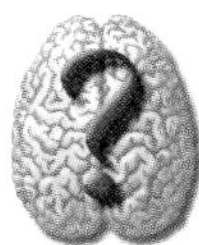

Think About It

In the systemic circulation, arterial blood has a pH of 7.40 and venous blood has a pH of 7.35. What do you think causes this difference?

Acids, Bases, and Buffers

The pH of a solution is determined solely by its hydrogen ions (H^+) (or strictly speaking, hydronium ions, H_3O^+, as explained in chapter 2). An acid is any chemical that releases H^+ in solution. A **strong acid** such as hydrochloric acid (HCl) ionizes freely, gives up most of its hydrogen ions, and can markedly lower the pH of a solution. A **weak acid** such as carbonic acid (H_2CO_3) ionizes only slightly and keeps most hydrogen in a chemically bound form that does not affect pH. A base is any chemical that accepts H^+. A **strong base** such as the hydroxyl ion (OH^-) has a strong tendency to bind H^+ and raise the pH, whereas a **weak base** such as the bicarbonate ion (HCO_3^-) binds less of the available H^+ and has less effect on pH.

A **buffer,** broadly speaking, is any mechanism that resists changes in pH by converting a strong acid or base to a weak one. The body has both physiological and chemical buffers. A **physiological buffer** is a system—namely the respiratory or urinary system—that stabilizes pH by controlling the body's output of acids, bases, or CO_2. Of all buffer systems, the urinary system buffers the greatest quantity of acid or base, but it requires several hours to days to exert an effect. The respiratory system exerts an effect within a few minutes but cannot alter the pH as much as the urinary system can.

A **chemical buffer** is a substance that binds H^+ and removes it from solution as its concentration begins to rise or releases H^+ into solution as its concentration falls. Chemical buffers can restore normal pH within a fraction of a second. They function as mixtures called **buffer systems** composed of a weak acid and a weak base. The three major chemical buffer systems of the body are the bicarbonate, phosphate, and protein systems.

The amount of acid or base that can be neutralized by a chemical buffer system depends on two factors: the concentration of the buffers and the pH of their working environment. Each system has an optimum pH at which it functions best; its effectiveness is greatly reduced if the pH of its environment deviates too far from this. The relevance of these factors will become apparent as you study the following buffer systems.

The Bicarbonate Buffer System

The **bicarbonate buffer system** is a solution of carbonic acid and bicarbonate ions. Carbonic acid (H_2CO_3) forms by the hydration of carbon dioxide and then dissociates into bicarbonate (HCO_3^-) and H^+:

$$CO_2 + H_2O \leftrightarrow H_2CO_3 \leftrightarrow HCO_3^- + H^+$$

This is a reversible reaction. When it proceeds to the right, carbonic acid acts as a weak acid by releasing H^+ and lowering pH. When the reaction proceeds to the left, bicarbonate acts as a weak base by binding H^+, removing the ions from solution, and raising pH.

At a pH of 7.4, the bicarbonate system would not ordinarily have a particularly strong buffering capacity outside of the body. This is too far from its optimum pH of 6.1. If a strong acid were added to a beaker of carbonic acid–bicarbonate solution at pH 7.4, the preceding reaction would shift only slightly to the left. Much surplus H^+ would remain and the pH would be substantially lower. In the body, by contrast, the bicarbonate system works quite well because the lungs and kidneys constantly remove CO_2 and prevent an equilibrium from being reached. This keeps the reaction moving to the left, and more H^+ is neutralized. Conversely, if there is a need to lower the pH, the kidneys excrete HCO_3^-, keep this reaction moving to the right, and elevate the H^+ concentration of the ECF. Thus you can see that the physiological and chemical buffers of the body function together in maintaining acid-base balance.

The Phosphate Buffer System

The **phosphate buffer system** is a solution of HPO_4^{2-} and $H_2PO_4^-$. It works in much the same way as the bicarbonate system. The following reaction can proceed to the right to liberate H^+ and lower pH, or it can proceed to the left to bind H^+ and raise pH:

$$H_2PO_4^- \leftrightarrow HPO_4^{2-} + H^+$$

The optimal pH for this system is 6.8, closer to the actual pH of the ECF. Thus the phosphate buffer system has a stronger buffering effect than an equal amount of bicarbonate buffer. However, phosphates are much less concentrated in the ECF than bicarbonate, so they are less important in buffering the ECF. They are more important in the renal tubules and ICF, where not only are they more concentrated, but the pH is lower and closer to their functional optimum. In the ICF, the constant production of metabolic acids creates pH values ranging from 4.5 to 7.4, probably averaging 7.0. The reason for the low pH in the renal tubules is discussed later.

The Protein Buffer System

Proteins are more concentrated than either bicarbonate or phosphate buffers, especially in the ICF. The **protein buffer system** accounts for about three-quarters of all chemical buffering ability of the body fluids. The buffering ability of proteins is due to certain side groups of their amino acid residues. Some have carboxyl (–COOH) side groups, which release H^+ when pH begins to rise and thus lower pH:

$$-COOH \rightarrow -COO^- + H^+$$

Others have amino (NH_2) side groups, which bind H^+ when pH falls too low, thus raising pH toward normal:

$$-NH_2 + H^+ \rightarrow -NH_3^+$$

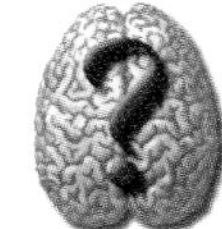

Think About It

What protein do you think is the most important buffer in blood plasma? In erythrocytes?

Respiratory Control of pH

The equation for the bicarbonate buffer system shows that the addition of CO_2 to the body fluids raises H^+ concentration and lowers pH, while the removal of CO_2 has the opposite effects. This is the basis for the strong buffering capacity of the respiratory system. Indeed, this system can neutralize two or three times as much acid as the chemical buffers can.

Carbon dioxide is constantly produced by aerobic metabolism and is normally eliminated by the lungs at an equivalent rate. As explained in chapter 22, rising CO_2 concentration and falling pH stimulate peripheral and central chemoreceptors, which stimulate an increase in pulmonary ventilation. This expels excess CO_2 and thus reduces H^+ concentration. Essentially, the free H^+ becomes part of the water molecules produced by this reaction:

$$HCO_3^- + H^+ \rightarrow H_2CO_3 \rightarrow CO_2 \text{ (expired)} + H_2O$$

Conversely, a drop in H^+ concentration raises pH and reduces pulmonary ventilation. This allows metabolic CO_2 to accumulate in the ECF faster than it is expelled, thus lowering pH to normal.

These are classic negative feedback mechanisms that result in acid-base homeostasis. Respiratory control of pH has some limitations, however, which are discussed later under acid-base imbalances.

Renal Control of pH

The kidneys can neutralize more acid or base than either the respiratory system or the chemical buffers. The essence of this mechanism is that the renal tubules secrete H^+ into the tubular fluid, where most of it binds to bicarbonate, ammonia, and phosphate buffers. Bound and free H^+ are then excreted in the urine. Thus the kidneys are the only organs that actually expel H^+ from the body. The other buffer systems only reduce its concentration by binding it to another chemical.

Figure 24.9 shows the process of H^+ secretion and neutralization. It is numbered to correspond to the following description, and the hydrogen ions are shown in color so you can trace them through the system from blood to urine:

1. Hydrogen ions in the blood are neutralized in two ways: by reacting with bicarbonate ions to produce carbonic acid and with hydroxyl ions to produce water.
2. Carbonic acid dissociates into water and carbon dioxide, which diffuse into the tubule cells.

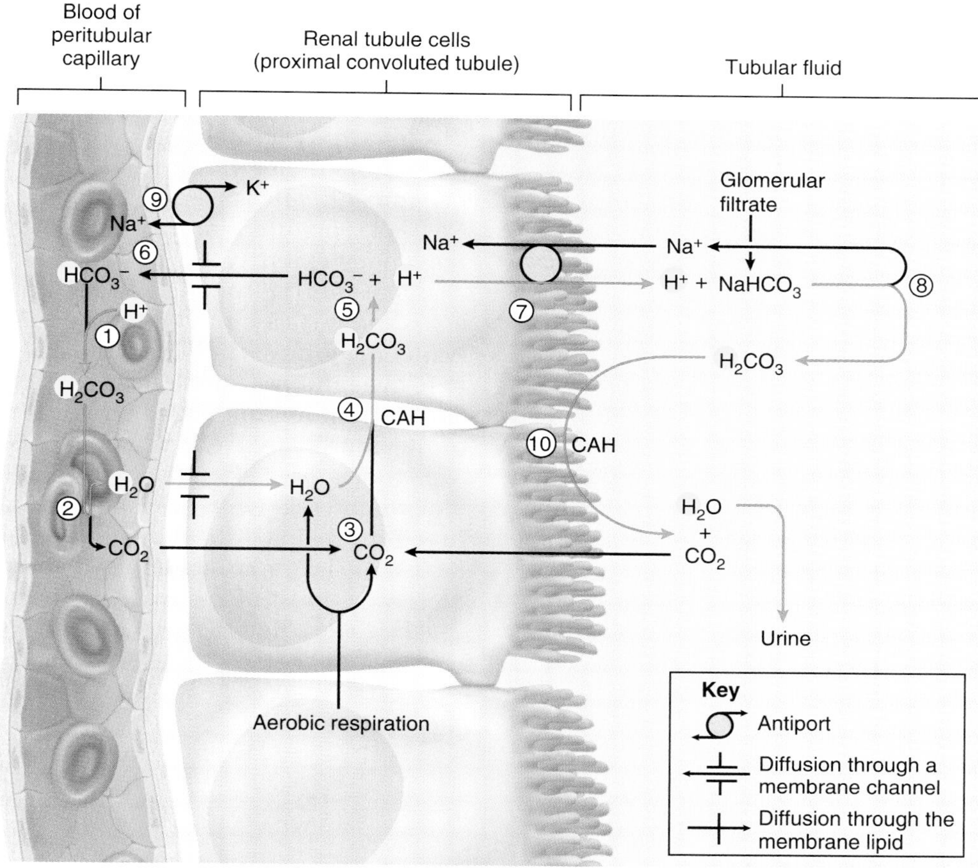

Figure 24.9 **Secretion and Neutralization of Hydrogen Ions in the Kidneys.** *Circled numbers* correspond to the explanation in the text. The *colored arrows* and *hydrogen symbols* allow you to trace hydrogen from H^+ in the blood to H_2O in the urine.

3. The tubule cells obtain CO_2 from three sources: the blood, the tubular fluid, and their own aerobic respiration.
4. Within the tubule cell, carbonic anhydrase (CAH) catalyzes the reaction of CO_2 and H_2O to produce carbonic acid.
5. Carbonic acid dissociates into bicarbonate and hydrogen ions.
6. The bicarbonate ions diffuse back into the bloodstream and may reenter the reaction cycle.
7. An antiport in the tubule cells pumps H^+ into the tubular fluid in exchange for Na^+.
8. Sodium bicarbonate ($NaHCO_3$) in the glomerular filtrate reacts with these hydrogen ions producing free sodium ions and carbonic acid.
9. The sodium ions are pumped into the tubule cells by the antiport at step 7 and then transferred to the blood by a Na^+-K^+ pump in the basal plasma membrane.
10. The carbonic acid in the tubular fluid dissociates into carbon dioxide and water. (The role of CAH is discussed shortly.) The CO_2 is recycled into the tubule cell and the water may be passed in the urine. Thus the hydrogen ions removed from the blood at step 1 are now part of the water molecules excreted in the urine at step 10.

Tubular secretion of H^+ (step 7) continues only as long as there is a sufficient concentration gradient between a high H^+ concentration in the tubule cells and a lower H^+ concentration in the tubular fluid. If the pH of the tubular fluid drops any lower than 4.5, tubular secretion of H^+ (step 7) ceases for lack of a sufficient gradient. Thus, pH 4.5 is the **limiting pH** for tubular secretion of H^+. This has added significance later in our discussion.

In a person with normal acid-base balance, the tubules secrete enough H^+ to neutralize all HCO_3^- in the tubular fluid; thus there is no HCO_3^- in the urine. Bicarbonate ions are filtered by the glomerulus, gradually disappear from the tubular fluid, and appear in the peritubular capillary blood. It *appears* as if HCO_3^- were reabsorbed by the renal tubules, but this is not the case;

indeed, the renal tubules are incapable of HCO_3^- reabsorption. The cells of the proximal convoluted tubule, however, have carbonic anhydrase (CAH) on their brush borders facing the lumen. This breaks down the H_2CO_3 in the tubular fluid to $CO_2 + H_2O$ (step 10). It is the CO_2 that is reabsorbed, not the bicarbonate. For every CO_2 reabsorbed, however, a *new* bicarbonate ion is formed in the tubule cell and released into the blood (steps 5–6). The effect is the same as if the tubule cells had reabsorbed bicarbonate itself.

Note that for every bicarbonate ion that enters the peritubular capillaries, a sodium ion does too. Thus the reabsorption of Na^+ by the renal tubules is part of the process of neutralizing acid. The more acid the kidneys excrete, the less sodium the urine contains.

The tubules secrete somewhat more H^+ than the available bicarbonate can neutralize. The urine therefore contains a slight excess of free H^+, which gives it a pH of about 5 to 6. Yet if all of the excess H^+ secreted by the tubules remained in this free ionic form, the pH of the tubular fluid would drop far below the limiting pH of 4.5, and H^+ secretion would stop. This must be prevented, and there are additional buffers in the tubular fluid to do so.

The glomerular filtrate contains Na_2HPO_4 (dibasic sodium phosphate), which reacts with some of the H^+ (fig. 24.10). A hydrogen ion replaces one of the sodium ions in the buffer, forming NaH_2PO_4 (monobasic sodium phosphate). This is passed in the urine and the displaced Na^+ is transported into the tubule cell and from there to the bloodstream.

In addition, tubular cells catabolize certain amino acids and release ammonia (NH_3) as a product (fig. 24.10). Ammonia diffuses into the tubular fluid, where it acts as a base to neutralize acid. It reacts with H^+ and Cl^- (the most abundant anion in the glomerular filtrate) to form ammonium chloride (NH_4Cl), which is passed in the urine.

Since there is so much chloride in the tubular fluid, you might ask why H^+ is not simply excreted as hydrochloric acid (HCl). Why involve ammonia? The reason is that HCl is a strong acid—it dissociates almost completely, so most of its hydrogen would be in the form of free H^+. The pH of the tubular fluid would drop below the

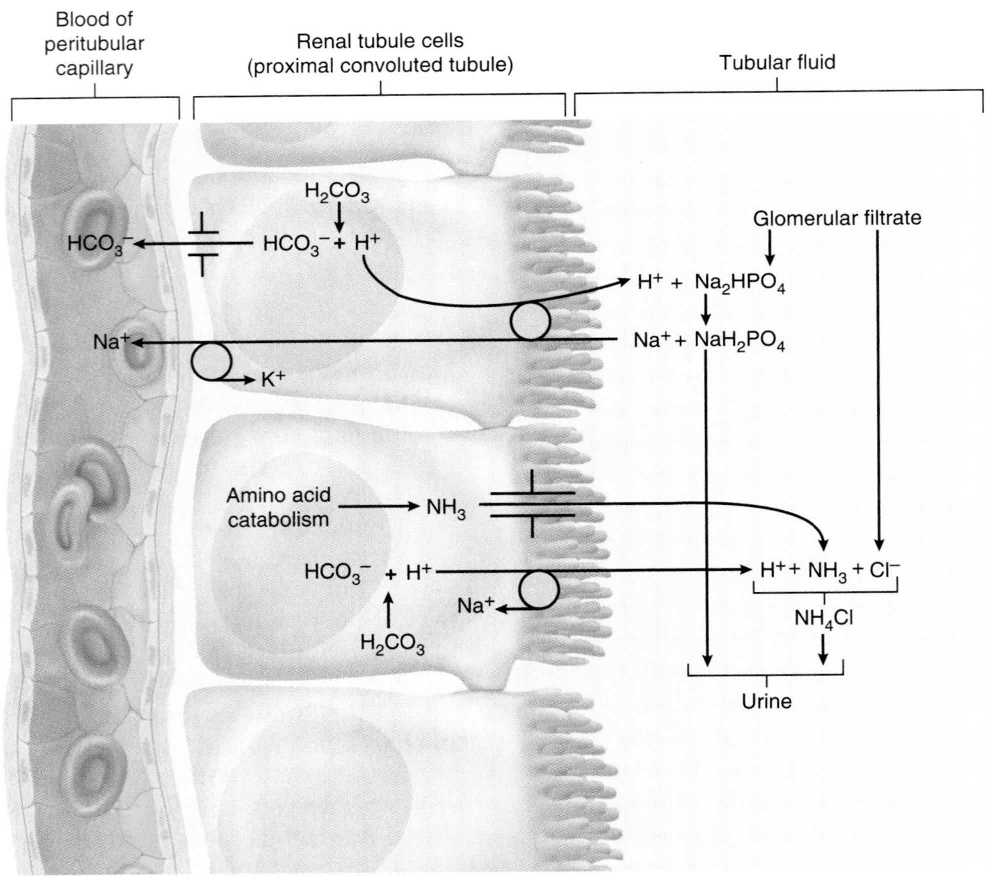

Figure 24.10 Mechanisms of Buffering Acid in the Urine. Reactions in the tubule cells are the same as in figure 24.9 but are simplified in this diagram. The essential differences are the buffering mechanisms shown in the tubular fluid.

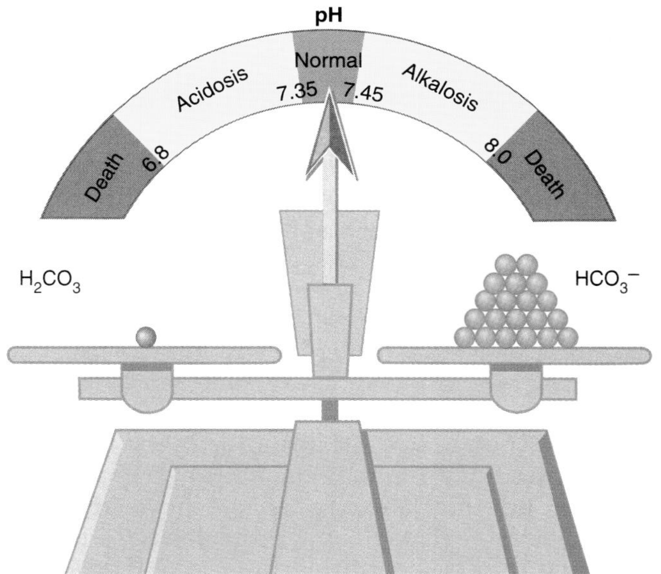

Figure 24.11 The Relationship of Carbonic Acid–Bicarbonate Ratio to pH. At a normal pH of 7.40, there is a 20:1 ratio of bicarbonate ions (HCO_3^-) to carbonic acid (H_2CO_3) in the blood plasma. An excess of HCO_3^- tips the balance toward alkalosis, whereas an excess of H_2CO_3 tips it toward acidosis.

limiting pH and prevent excretion of more acid. Ammonium chloride, by contrast, is a weak acid—most of its hydrogen remains bound to it and does not lower the pH of the tubular fluid.

Disorders of Acid-Base Balance

At pH 7.4, the ECF has a 20:1 ratio of HCO_3^- to H_2CO_3 (fig. 24.11). If the relative amount of H_2CO_3 rises higher than this, it tips the balance to a lower pH. If the pH falls below 7.35, a state of **acidosis** exists. An excess of HCO_3^-, by contrast, tips the balance to a higher pH. A pH above 7.45 is a state of **alkalosis.** Either of these imbalances has potentially fatal effects. A person cannot live more than a few hours if the blood pH is below 7.0 or above 7.7; a pH below 6.8 or above 8.0 is quickly fatal.

In acidosis, H^+ diffuses down its concentration gradient into the cells, and to maintain electrical balance, K^+ diffuses out (fig. 24.12*a*). The H^+ is buffered by intracellular proteins, so this exchange results in a net loss of cations from the cell. This makes the resting membrane potential more negative than usual (hyperpolarized) and makes nerve and muscle cells more difficult to stimulate. This is why acidosis depresses the central nervous system and causes such symptoms as confusion, disorientation, and coma.

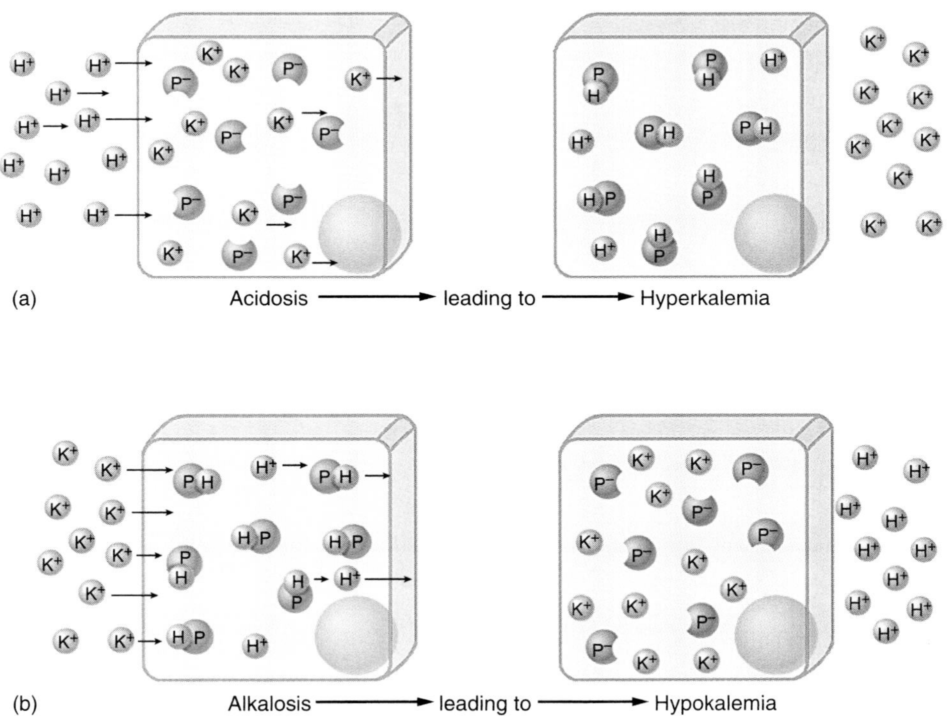

Figure 24.12 The Relationship Between Acid-Base Imbalances and Potassium Imbalances. (*a*) In acidosis, H^+ diffuses into the cells and drives out K^+, elevating the K^+ concentration of the ECF. (*b*) In alkalosis, H^+ diffuses out of the cells and K^+ diffuses in to replace it, lowering the K^+ concentration of the ECF.

In alkalosis, the extracellular H^+ concentration is low. Hydrogen ions diffuse out of the cells and K^+ diffuses in to replace them (fig. 24.12*b*). The net gain in positive intracellular charges shifts the membrane potential closer to firing level and makes the nervous system hyperexcitable. Neurons fire spontaneously and overstimulate skeletal muscles, causing muscle spasms, tetany, convulsions, or respiratory paralysis.

Acid-base imbalances fall into two categories, respiratory and metabolic (table 24.3). **Respiratory acidosis** occurs when the rate of alveolar ventilation fails to keep pace with the body's rate of CO_2 production. Carbon dioxide accumulates in the ECF and lowers its pH. **Respiratory alkalosis** results from excessive ventilation (hyperventilation), in which CO_2 is eliminated faster than it is produced.

Metabolic acidosis can result from increased production of organic acids, such as lactic acid in anaerobic fermentation and ketone bodies in alcoholism and diabetes mellitus. It can also result from the ingestion of acidic drugs such as aspirin or from the loss of base due to chronic diarrhea or overuse of laxatives. Dying persons also typically exhibit acidosis. **Metabolic alkalosis** is rare but can result from overuse of bicarbonates (such as oral antacids and intravenous bicarbonate solutions) or from the loss of stomach acid by chronic vomiting.

Compensation for Acid-Base Imbalances

Uncompensated acidosis or alkalosis is a pH imbalance that the body fails to correct. **Compensated** acidosis or alkalosis is a more temporary condition that the body corrects by means of its physiological and chemical buffers.

The respiratory system compensates for some states of acid-base imbalance by adjusting the partial pressure of carbon dioxide (P_{CO_2}) in the ECF. Hypercapnia (a CO_2 excess) stimulates increased pulmonary ventilation, and hypocapnia (a CO_2 deficiency) reduces it, by the mechanisms discussed in chapter 22. This is very effective in correcting pH imbalances due to abnormal P_{CO_2} but not very effective in correcting other causes of acidosis and alkalosis. In diabetic acidosis, for example, the lungs cannot reduce the concentration of ketone bodies in the blood, although it can somewhat compensate for the H^+ that they release by increasing pulmonary ventilation and exhausting extra CO_2. The respiratory system can adjust a blood pH of 7.0 back to 7.2 or 7.3 but not all the way back to the normal 7.4. Although the respiratory system has a very powerful buffering effect, its ability to stabilize pH is therefore limited.

The kidneys are slower to respond to pH imbalances but better at restoring a fully normal pH. Urine usually has a pH of 5 to 6, but in acidosis it may fall as low as 4.5 because of excess H^+, whereas in alkalosis it may rise as high as 8.2 because of excess HCO_3^-. The kidneys cannot act quickly enough to compensate for short-term pH imbalances, such as the acidosis that might result from an asthmatic attack lasting an hour or two, or the alkalosis resulting from a brief episode of emotional hyperventilation. They are effective, however, at compensating for pH imbalances that last for a few days or longer.

In acidosis, the renal tubules increase the rate of H^+ secretion. The extra H^+ in the tubular fluid must be buffered; otherwise, the fluid pH could exceed the limiting pH and H^+ secretion would stop. Therefore, in acidosis, the renal tubules secrete more ammonia to buffer the added H^+, and the amount of ammonium chloride in the urine may rise to 7 to 10 times normal.

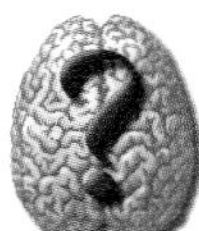

Think About It

Suppose you measured the pH and ammonium chloride concentration of urine from a person with emphysema and urine from a healthy individual. How would you expect the two to differ, and why?

In alkalosis, the bicarbonate concentration and pH of the urine are elevated. This is partly because there is more HCO_3^- in the blood and glomerular filtrate and partly because there is not enough H^+ in the tubular fluid to neutralize all the HCO_3^- in the filtrate.

Table 24.3 Some Causes of Acidosis and Alkalosis

	Acidosis	Alkalosis
Respiratory	Hypoventilation, apnea, or respiratory arrest; asthma, emphysema, chronic bronchitis	Hyperventilation due to emotions or oxygen deficiency (as at high altitudes)
Metabolic	Excess production of organic acids, as in diabetes mellitus and long-term anaerobic fermentation; drugs such as aspirin and laxatives; chronic diarrhea	Rare but can result from chronic vomiting or overuse of bicarbonates (antacids)

Acid-Base Imbalances in Relation to Electrolyte and Water Imbalances

The foregoing discussion once again stresses a point made early in this chapter—we cannot understand or treat imbalances of water, electrolyte, or acid-base balance in isolation from each other, because each of these frequently affects the other two. Table 24.4 itemizes and explains a few of these interactions. This is by no means a complete list of how fluid, electrolytes, and pH affect each other, but it does demonstrate their interdependence. Note that many of these relationships are reciprocal—for example, acidosis can cause hyperkalemia, and conversely, hyperkalemia can cause acidosis.

Before You Go On

Answer the following questions to test your understanding of the preceding section.

11. Write two chemical equations that show how the bicarbonate buffer system compensates for acidosis and alkalosis and two equations that show how the phosphate buffer system compensates for these imbalances.
12. Why are phosphate buffers more effective in the cytoplasm than in the blood plasma?
13. Renal tubules cannot reabsorb HCO_3^-, and yet HCO_3^- concentration in the tubular fluid falls while in the blood plasma it rises. Explain this apparent contradiction.
14. In acidosis, the renal tubules secrete more ammonia. Why?

Table 24.4 Some Relationships Among Fluid, Electrolyte, and Acid-Base Imbalances

Cause	Potential Effect	Reason
Acidosis →	Hyperkalemia	H^+ diffuses into cells and displaces K^+ (see fig. 24.12*a*). As K^+ leaves the ICF, K^+ concentration in the ECF rises.
Hyperkalemia →	Acidosis	Opposite from the above; high K^+ concentration in the ECF causes less K^+ to diffuse out of the cells than normally. H^+ diffuses out to compensate, and this lowers the extracellular pH.
Alkalosis →	Hypokalemia	H^+ diffuses from ICF to ECF. More K^+ remains in the ICF to compensate for the H^+ loss, causing a drop in ECF K^+ concentration (see fig. 24.12*b*).
Hypokalemia →	Alkalosis	Opposite from the above; low K^+ concentration in the ECF causes K^+ to diffuse out of cells. H^+ diffuses in to replace K^+, lowering the H^+ concentration of the ECF and raising its pH.
Acidosis →	Hypochloremia	More Cl^- is excreted as NH_4Cl to buffer the excess acid in the renal tubules, leaving less Cl^- in the ECF.
Alkalosis →	Hyperchloremia	More Cl^- is reabsorbed from the renal tubules, so ingested Cl^- accumulates in the ECF rather than being excreted.
Hyperchloremia →	Acidosis	More H^+ is retained in the blood to balance the excess Cl^-, causing hyperchloremic acidosis.
Hypovolemia →	Alkalosis	More Na^+ is reabsorbed by the kidney. Na^+ reabsorption is coupled to H^+ secretion (see fig. 24.9), so more H^+ is secreted and pH of the ECF rises.
Hypervolemia →	Acidosis	Less Na^+ is reabsorbed, so less H^+ is secreted into the renal tubules. H^+ retained in the ECF causes acidosis.
Acidosis →	Hypocalcemia	Acidosis causes more Ca^{2+} to bind to plasma protein and citrate ions, lowering the concentration of free, ionized Ca^{2+} and causing symptoms of hypocalcemia.
Alkalosis →	Hypercalcemia	Alkalosis causes more Ca^{2+} to dissociate from plasma protein and citrate ions, raising the concentration of free Ca^{2+}.

Clinical Insight 24.3

Fluid Replacement Therapy

One of the most significant problems in the treatment of seriously ill patients is the restoration and maintenance of proper fluid volume, composition, and distribution among the fluid compartments. Fluids may be administered to replenish total body water, restore blood volume and pressure, shift water from one fluid compartment to another, or restore and maintain electrolyte and acid-base balance.

Drinking water is the simplest method of fluid replacement, but it does not replace electrolytes. Heat exhaustion can occur when you lose water and salt in the sweat and replace the fluid by drinking plain water. Broths, juices, and isotonic sports drinks such as Gatorade replace both water and electrolytes. Sports drinks contain a mixture of salts, lactate, and glucose comparable to ECF.

If a patient cannot take fluids by mouth, they must be administered by alternative routes. Some can be given by enema and absorbed through the colon. All routes of fluid administration other than the digestive tract are called *parenteral*[6] routes. The most common of these is the intravenous (I.V.) route, but for various reasons, including inability to find a suitable vein, fluids are sometimes given by subcutaneous (sub-Q), intramuscular (I.M.), or other parenteral routes. Many kinds of sterile solutions are available to meet the fluid replacement needs of different patients.

In cases of extensive blood loss, there may not be time to type and cross-match blood for a transfusion. The more urgent need is to replenish blood volume and pressure. *Normal saline* (isotonic, 0.9% NaCl) is a relatively quick and simple way to raise blood volume while maintaining normal osmolarity, but it has significant shortcomings. It takes three to five times as much saline as whole blood to rebuild normal volume because much of the saline escapes the circulation into the interstitial fluid compartment or is excreted by the kidneys. In addition, normal saline can induce hypernatremia and hyperchloremia, because the body excretes the water but retains much of the NaCl. Hyperchloremia can, in turn, produce acidosis. Normal saline also lacks potassium, magnesium, and calcium. Indeed, it dilutes those electrolytes that are already present and creates a risk of cardiac arrest from hypocalcemia. Saline also dilutes plasma albumin and RBCs, creating still greater risks for patients who have suffered extensive blood loss. Nevertheless, the emergency maintenance of blood volume sometimes takes temporary precedence over these other considerations.

Fluid therapy is also used to correct pH imbalances. Acidosis may be treated with *Ringer's lactate solution,* which includes sodium to rebuild ECF volume, potassium to rebuild ICF volume, lactate to balance the cations, and enough glucose to make the solution isotonic. Alkalosis can be treated with potassium chloride. This must be administered very carefully, because potassium ions can cause painful venous spasms, and even a small potassium excess can cause cardiac arrest. High-potassium solutions should never be given to patients in renal failure or whose renal status is unknown, because in the absence of renal excretion of potassium they can bring on lethal hyperkalemia. Ringer's lactate or potassium chloride also must be administered very cautiously, with close monitoring of blood pH, to avoid causing a pH imbalance opposite the one that was meant to be corrected. Too much Ringer's lactate causes alkalosis and too much KCl causes acidosis.

Plasma volume expanders are hypertonic solutions of chemicals that are retained in the bloodstream and draw interstitial water into it by osmosis. They include albumin, sucrose, mannitol, and dextran. Plasma expanders are also used to combat hypotonic hydration by drawing water out of swollen cells, averting such problems as seizures and coma. A plasma expander can draw several liters of water out of the intracellular compartment within a few minutes.

Patients who cannot eat are often given isotonic 5% dextrose (glucose). A fasting patient loses as much as 70 to 85 g of protein per day from the tissues as protein is broken down to fuel the metabolism. Giving 100 to 150 g of I.V. glucose per day reduces this by half and is said to have a *protein-sparing effect.* More than glucose is needed in some cases—for example, if a patient has not eaten for several days and cannot be fed by nasogastric tube (due to lesions of the digestive tract, for example) or if large amounts of nutrients are needed for tissue repair following severe trauma, burns, or infections. In *total parenteral nutrition (TPN),* or *hyperalimentation,*[7] a patient is provided with complete I.V. nutritional support, including a protein hydrolysate (amino acid mixture), vitamins, electrolytes, 20% to 25% glucose, and on alternate days, a fat emulsion.

The water from parenteral solutions is normally excreted by the kidneys. If the patient has renal insufficiency, however, excretion may not keep pace with intake, and there is a risk of hypotonic hydration. Intravenous fluids are usually given slowly, by *I.V. drip,* to avoid abrupt changes or overcompensation for the patient's condition. In addition to pH, the patient's pulse rate, blood pressure, hematocrit, and plasma electrolyte concentrations are monitored, and the patient is examined periodically for respiratory sounds indicating pulmonary edema.

The delicacy of fluid replacement therapy underscores the close relationships among fluids, electrolytes, and pH. It is dangerous to manipulate any one of these variables without close attention to the others. Parenteral fluid therapy is usually used for persons who are seriously ill. Their homeostatic mechanisms are already compromised and leave less room for error than in a healthy person.

---

6. *para* = beside + *enter* = intestine
7. *hyper* = above normal + *aliment* = nourishment

Chapter Review

Online Learning Center

Study resources for this chapter are available at www.mhhe.com/saladin/

Study Outline

Water Balance (p. 916)

1. Fluid compartments
   a. Total body water
   b. Fluid compartments
      - Intracellular fluid (ICF) (65%)
      - Extracellular fluid (ECF) (35%)
        - •• Tissue (interstitial) fluid (25%)
        - •• Blood plasma and lymph (8%)
        - •• Transcellular fluid (2%)
   c. Exchange between fluid compartments
2. Water gain and loss
   a. Fluid balance
   b. Water gain
      - Metabolic water
      - Preformed (ingested) water
   c. Water loss
      - Avenues of loss
        - •• Urine
        - •• Fecal moisture
        - •• Expired breath
        - •• Sweat
        - •• Cutaneous transpiration
      - Loss in relation to weather
      - Loss in relation to physical activity
      - Insensible water loss
      - Obligatory water loss
3. Regulation of intake
   a. Thirst mechanisms
      - Stimulation of thirst center
        - •• Angiotensin II
        - •• Antidiuretic hormone
        - •• Hypothalamic osmoreceptors
      - Inhibition of salivation
   b. Satiation mechanisms
      - Cooling and moistening of mouth
      - Distension of stomach and intestine
      - Rehydration of blood
4. Regulation of output
   a. By controlling Na^+ reabsorption
   b. By antidiuretic hormone (ADH)
      - ADH secretion
      - Aquaporins
      - Effect on water:sodium ratio
5. Disorders of water balance
   a. Fluid deficiency
      - Volume depletion
      - Dehydration
        - •• Causes
        - •• Special problems for infants
        - •• Effect on fluid compartments
      - Effects of fluid deficiency
   b. Fluid excess
      - Volume excess
      - Hypotonic hydration
      - Effects of fluid excess
   c. Fluid sequestration
      - Edema
      - Hemorrhage
      - Pleural effusion

Electrolyte Balance (p. 922)

1. General perspectives
   a. Functions of electrolytes
   b. Definition of electrolytes
   c. Major electrolyte ions
   d. Normal concentrations
2. Sodium
   a. Functions
      - Membrane potentials
      - Extracellular fluid osmolarity
      - Cotransport
      - Na^+-K^+ pump and heat production
   b. Homeostasis
      - Rarity of Na^+ deficiency
      - Aldosterone
      - Antidiuretic hormone
      - Atrial natriuretic factor
      - Other hormones
      - Variations in intake
   c. Imbalances
      - Hypernatremia
      - Hyponatremia
3. Potassium
   a. Functions
      - Intracellular fluid osmolarity
      - Membrane potentials
      - Cotransport
      - Na^+-K^+ pump and heat production
   b. Homeostasis
      - Renal secretion and reabsorption
      - Aldosterone
   c. Imbalances
      - Hyperkalemia
      - Hypokalemia
4. Chloride
   a. Functions
      - Extracellular fluid osmolarity
      - Part of stomach acid
      - Chloride shift mechanism
      - Regulation of body pH
   b. Homeostasis
   c. Imbalances
      - Hyperchloremia
      - Hypochloremia
5. Calcium
   a. Functions
      - Skeletal mineralization
      - Muscle contraction
      - Second messenger
      - Exocytosis
      - Blood clotting
   b. Necessity of low intracellular Ca^{2+} concentration
   c. Homeostasis
      - Parathyroid hormone
      - Calcitriol (vitamin D)
      - Calcitonin
   d. Imbalances
      - Hypercalcemia
      - Hypocalcemia
6. Phosphates
   a. Forms of phosphate ions

b. Functions
  - Component of ATP, nucleotides, nucleic acids, phospholipids
  - Phosphorylation reactions
  - Buffering of pH

c. Homeostasis
  - Renal control
  - Parathyroid hormone

d. Imbalances (not critical)

Acid-Base Balance (p. 927)

1. Normal pH
   a. Importance of pH
   b. Normal pH range of ECF
   c. Challenges to acid-base balance
2. Acids, bases, and buffers
   a. Strong and weak acids
   b. Strong and weak bases
   c. Buffers
   - Physiological buffers
     - •• Respiratory system
     - •• Urinary system
   - Chemical buffers
     - •• Bicarbonate buffer system
     - •• Phosphate buffer system
     - •• Protein buffer system
3. Respiratory control of pH
   a. Collaboration with bicarbonate system
   b. Adjustment of pulmonary ventilation
4. Renal control of pH
   a. H^+ secretion
   b. Limiting pH
   c. Buffers in tubular fluid
   - Bicarbonate
   - Phosphates
   - Ammonia
5. Disorders of acid-base imbalance
   a. Normal pH range
   b. Types of acid-base imbalance
   - Acidosis
     - •• Respiratory
     - •• Metabolic
   - Alkalosis
     - •• Respiratory
     - •• Metabolic

   c. Effects of acidosis
   - H^+ displaces K^+ from ICF
   - Cells become hyperpolarized
   - Depression of neuromuscular function

   d. Effects of alkalosis
   - K^+ displaces H^+ from ICF
   - Cells become depolarized
   - Excessive neuromuscular excitability
6. Compensation for acid-base imbalances
   a. Respiratory adjustments in P_{CO_2}
   b. Renal adjustments in H^+ secretion
7. pH imbalances in relation to electrolyte and water imbalances

Selected Vocabulary

total body water 916
fluid compartment 916
fluid balance 916
metabolic water 916
preformed water 916
cutaneous transpiration 917
insensible water loss 917
obligatory water loss 917
thirst center 918
aquaporin 919
volume depletion 920
hypovolemia 920
dehydration 920
negative water balance 920
volume excess 920
hypotonic hydration 920
fluid sequestration 921
hypernatremia 924
hyponatremia 924
hyperkalemia 924
hypokalemia 924
hyperchloremia 926
hypochloremia 926
hypercalcemia 926
hypocalcemia 926
strong acid 927
weak acid 927
strong base 927
weak base 927
buffer 927
physiological buffer 927
chemical buffer 927
buffer system 927
bicarbonate buffer system 927
phosphate buffer system 928
protein buffer system 928
limiting pH 929
acidosis 931
alkalosis 931
respiratory acidosis 932
respiratory alkalosis 932
metabolic acidosis 932
metabolic alkalosis 932
uncompensated 932
compensated 932

Testing Your Recall

1. The greatest percentage of the body's water is in
   a. the blood plasma.
   b. the lymph.
   c. the intracellular fluid.
   d. the interstitial fluid.
   e. the extracellular fluid.

2. Hypertension is likely to increase the secretion of
   a. atrial natriuretic factor.
   b. antidiuretic hormone.
   c. bicarbonate ions.
   d. aldosterone.
   e. ammonia.

3. ______ increases water reabsorption without increasing sodium reabsorption.
   a. Antidiuretic hormone
   b. Aldosterone
   c. Atrial natriuretic factor
   d. Parathyroid hormone
   e. Calcitonin

4. Hypotonic hydration can result from
   a. ADH hypersecretion.
   b. ADH hyposecretion.
   c. aldosterone hypersecretion.
   d. aldosterone hyposecretion.
   e. *a* and *d* only.

5. Tetanus is most likely to result from
   a. hypernatremia.
   b. hypokalemia.
   c. hyperkalemia.
   d. hypocalcemia.
   e. *c* and *d* only.

6. The principal determinant of intracellular osmolarity and cellular volume is
   a. protein.
   b. phosphate.
   c. potassium.
   d. sodium.
   e. chloride.

7. Increased excretion of ammonium chloride in the urine most likely indicates
   a. hypercalcemia.
   b. hyponatremia.
   c. hypochloremia.
   d. alkalosis.
   e. acidosis.

8. The most effective buffer in the intracellular fluid is
   a. phosphate.
   b. protein.
   c. bicarbonate.
   d. carbonic acid.
   e. ammonia.
9. Tubular secretion of hydrogen is directly linked to
   a. tubular secretion of potassium.
   b. tubular secretion of sodium.
   c. tubular reabsorption of potassium.
   d. tubular reabsorption of sodium.
   e. tubular secretion of chloride.
10. Hyperchloremia is most likely to result in
   a. alkalosis.
   b. acidosis.
   c. hypernatremia.
   d. hyperkalemia.
   e. hypovolemia.
11. The most abundant cation in the ECF is ________.
12. The most abundant cation in the ICF is ________.
13. Water produced by the body's chemical reactions is called ________.
14. The skin loses water by two processes, sweating and ________.
15. Any abnormal accumulation of fluid in a particular place in the body is called ________.
16. An excessive concentration of potassium ions in the blood is called ________.
17. A deficiency of sodium ions in the blood is called ________.
18. A blood pH of 7.2 caused by inadequate pulmonary ventilation would be classified as ________.
19. Tubular secretion of hydrogen ions would cease if the acidity of the tubular fluid fell below a value called the ________.
20. Long-term satiation of thirst depends on a reduction of the ________ of the blood.

Answers in appendix B

Testing Your Comprehension

1. A duck hunter is admitted to the hospital with a shotgun injury to the abdomen. He has suffered extensive blood loss but is conscious. He complains of being intensely thirsty. Explain the physiological mechanism connecting his injury to his thirst.
2. A woman living at poverty level finds bottled water at the grocery store next to the infant formula. The label on the water states that it is made especially for infants, and she construes this to mean that it can be used as a nutritional supplement. The water is much cheaper than formula, so she gives her baby several ounces of bottled water a day as a substitute for formula. After several days the baby has seizures and is taken to the hospital, where it is found to have edema, acidosis, and a plasma sodium concentration of 116 mEq/L. The baby is treated with anticonvulsants followed by normal saline and recovers. Explain each of the signs.
3. Explain why the respiratory and urinary systems are both necessary for the bicarbonate buffer system to work effectively in the blood plasma.
4. The left column indicates some increases or decreases in blood plasma values. In the right column, replace the question mark with an up or down arrow to indicate the expected effect. Explain each effect.

	Cause	Effect
a.	$\uparrow H_2O$	? Na^+
b.	$\uparrow Na^+$	? Cl^-
c.	$\downarrow K^+$	? H^+
d.	$\uparrow H^+$	? K^+
e.	$\downarrow Ca^{2+}$	? PO_4^{3-}

5. A 4-year-old child is caught up in tribal warfare in Africa. In a refugee camp, the only drinking water is from a sewage-contaminated pond. The child soon develops severe diarrhea and dies 10 days later of cardiac arrest. Explain the possible physiological cause(s) of his death.

Answers at the Online Learning Center

CHAPTER 25

The Digestive System

Surface of the human tongue (SEM)

Chapter Outline

Insights

Brushing Up

To understand this chapter, it is important that you understand or brush up on the following concepts:

- Hydrolysis reactions (p. 80)
- Carbohydrate, fat, and protein structures (pp. 81–92)
- pH and enzyme function (p. 95)
- Brush borders and microvilli (p. 113)
- Membrane transport processes (pp. 123–33)
- The Valsalva maneuver (p. 858)

Study with Your ESP CD-ROM

Refer to your ESP Tutorial CD-ROM for helpful animations, activities, and quizzes related to the material covered in this chapter. The following are suggested areas to go to on the CD and are presented according to "Unit/Topic":

- Absorption and Excretion/Digestion

Most of the nutrients we eat cannot be used in their existing form. They must be broken down into smaller components, such as amino acids and monosaccharides, that are universal to all species. Consider what happens if you eat a piece of beef, for example. The myosin of beef differs very little from that of your own muscles, but the two are not identical, and even if they were, beef myosin could not be absorbed, transported in the blood, and incorporated into your muscles. Like any other dietary protein, it must be broken down into amino acids before it can be used. Since beef and human proteins are made of the same 20 amino acids, those of beef proteins might indeed become part of your own myosin but could equally well end up in your insulin, fibrinogen, collagen, or any other protein. The digestive system is essentially a disassembly line—its primary purpose is to break nutrients down into forms that can be used by the body and to absorb them so they can be distributed to the tissues.

General Anatomy and Digestive Processes

Objectives

When you have completed this section, you should be able to

- list the regions of the digestive tract and the accessory organs of the digestive system;
- list the functions and major physiological processes of the digestive system;
- distinguish between mechanical and chemical digestion; and
- describe the basic chemical process underlying all chemical digestion, and name the major substrates and products of this process.

Subdivisions of the Digestive System

The digestive system has two anatomical subdivisions, the digestive tract and the accessory organs (fig. 25.1). The **digestive tract** is a tube extending from mouth to anus, measuring about 9 m (30 ft) long in the cadaver. It is also known as the *gastrointestinal (GI) tract* or *alimentary*[1] *canal.* It includes the oral cavity, pharynx, esophagus, stomach, small intestine, and large intestine. The **accessory organs** are the teeth, tongue, salivary glands, liver, gallbladder, and pancreas.

The digestive tract is open to the environment at both ends. Most of the material in it has not entered any body tissues and is considered to be external to the body until it is absorbed by epithelial cells of the alimentary canal. In the strict sense, defecated food residue was never in the body.

Relationship to the Peritoneum

The relationship of the stomach and intestines to the peritoneum is shown in figure A.11 (p. 37). Most of the digestive tract is within the peritoneal cavity, but some portions of it are retroperitoneal—notably the duodenum (the first part of the small intestine), most of the pancreas, and parts of the large intestine.

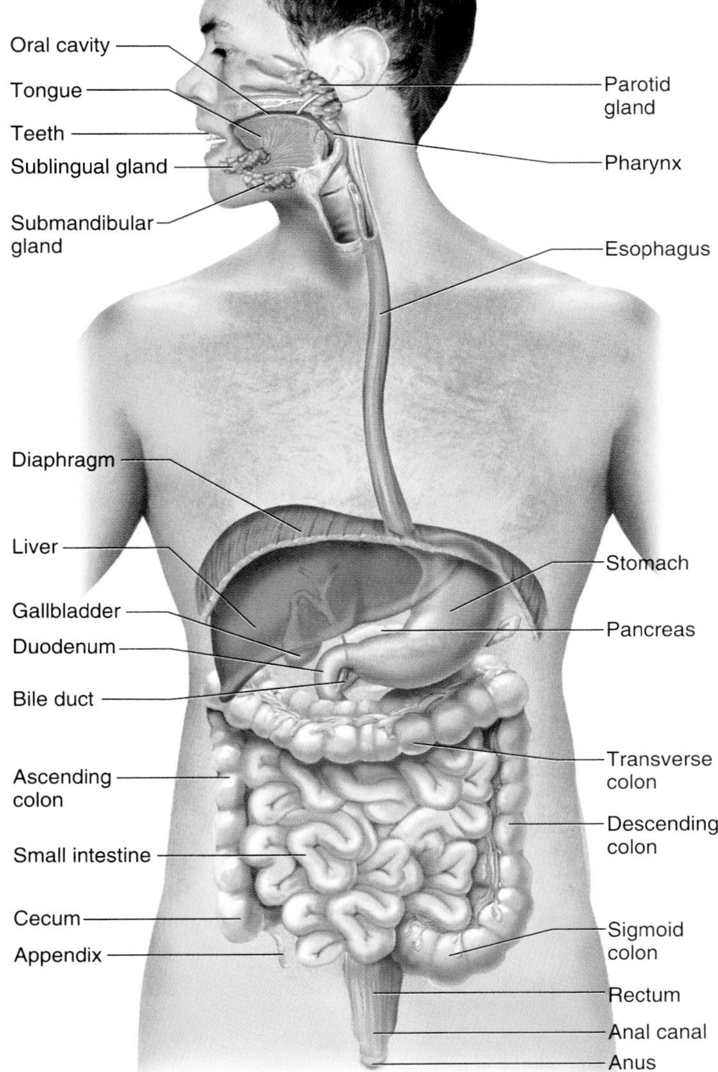

Figure 25.1 The Digestive System.

Along the dorsal midline of the abdominal cavity, the parietal peritoneum turns inward and forms a sheet of tissue, the **dorsal mesentery,** extending to the digestive tract. The membrane then forms the outer covering, or **serosa** (visceral peritoneum), of the stomach and most parts of the intestines. In some places it continues beyond the digestive organs as a sheet of tissue called the **ventral mesentery,** which may hang freely in the abdominal cavity or attach to the ventral abdominal wall or other organs.

Along the right superior margin (*lesser curvature*) of the stomach, the serosae of the anterior and posterior

1. *aliment* = food

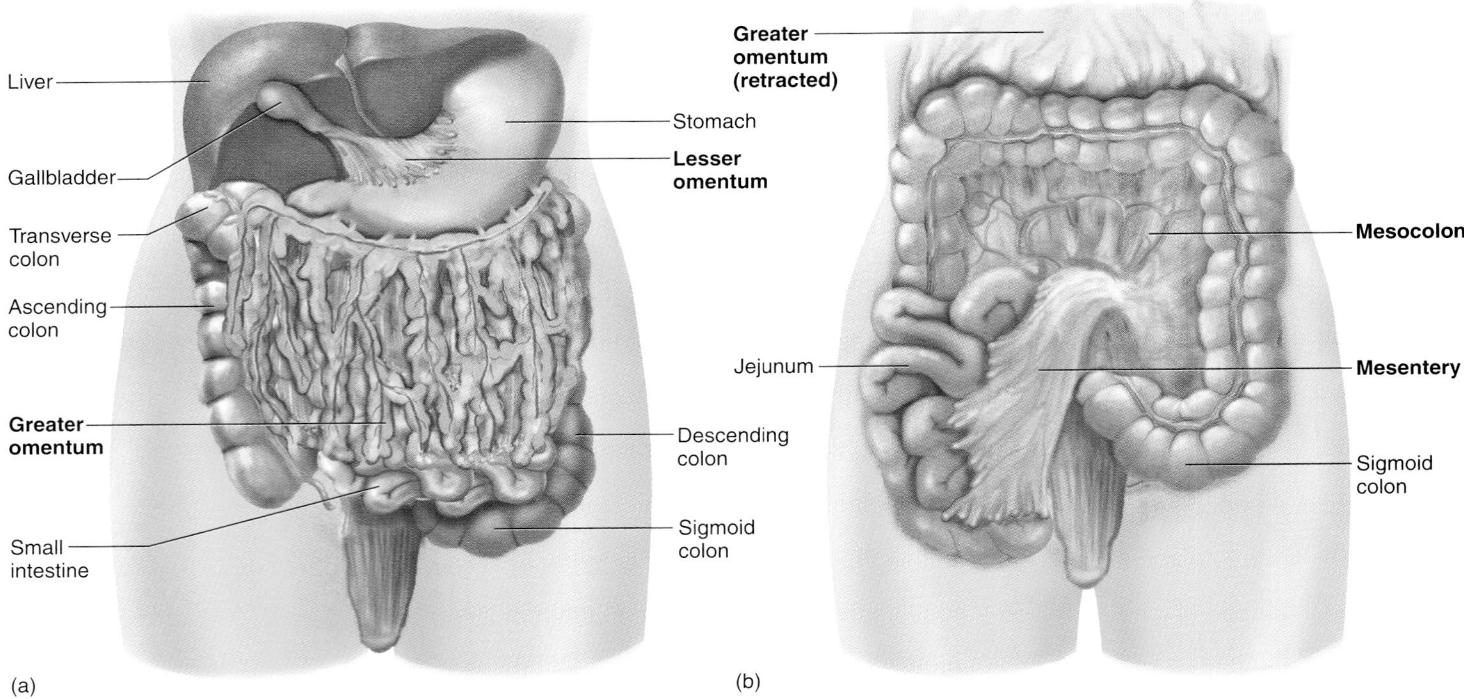

Figure 25.2 Serous Membranes Associated with the Digestive Tract. (*a*) The greater and lesser omenta. (*b*) Greater omentum and small intestine retracted to show the mesocolon and mesentery. These membranes contain the mesenteric arteries and veins.

stomach surfaces meet and continue as a ventral mesentery, the **lesser omentum,** extending from the stomach to the liver (fig. 25.2). Along the left inferior margin (*greater curvature*) of the stomach, the serosae form the **greater omentum,** which hangs loosely over the small intestine like an apron. At its inferior margin, the greater omentum turns back on itself, passes upward, and forms serous membranes around the spleen and transverse colon. Beyond the transverse colon, it continues as a mesentery called the **mesocolon,** which anchors the colon to the posterior abdominal wall.

The omenta have a loosely organized, lacy appearance due partly to many holes or gaps in the membranes and partly to an irregular distribution of fatty tissue. They also contain many lymph nodes, lymphatic vessels, blood vessels, and nerves. The omenta adhere to perforations or inflamed areas of the stomach or intestines, contribute immune cells to the site, and isolate infections that might otherwise give rise to peritonitis.

Digestive Functions and Processes

The digestive system has four functions:

1. **ingestion,** the selective intake of nutrients;
2. **digestion,** the breakdown of large molecules into smaller ones;
3. **absorption,** the uptake of nutrient molecules, first into the epithelial cells of the digestive tract and then into the blood or lymph; and finally
4. **defecation,** the elimination of undigested residue.

These functions are carried out through three principal processes:

1. **motility,** muscular contractions that break up food, propel it through the canal, and mix it with digestive enzymes;
2. **secretion** of enzymes, hormones, and other products that carry out or regulate digestion; and
3. **membrane transport,** mechanisms of absorbing nutrients and transferring them to the blood and lymph.

Stages of Digestion

There are two stages of digestion: mechanical and chemical. **Mechanical digestion** is the physical breakdown of food into smaller particles. It is achieved by the cutting and grinding action of the teeth and the churning contractions of the stomach and small intestine. These actions break food items into smaller particles to expose more of their surface to the action of digestive enzymes.

Chemical digestion is a series of hydrolysis reactions that break dietary macromolecules into their monomers (*residues*): polysaccharides into monosaccharides, proteins into amino acids, fats into glycerol and fatty acids, and nucleic acids into nucleotides. It is carried out by digestive enzymes produced by the salivary glands, stomach, pancreas, and small intestine. Some nutrients are already present in usable form in the ingested food and are absorbed without being digested: vitamins, free amino acids, minerals, cholesterol, and water.

Before You Go On

Answer the following questions to test your understanding of the preceding section:

1. What is the term for the serous membrane that suspends the intestines from the abdominal wall?
2. Which physiological process of the digestive system truly moves a nutrient from the outside to the inside of the body?
3. What one type of reaction is the basis of all chemical digestion?
4. Name some nutrients that are absorbed without being digested.

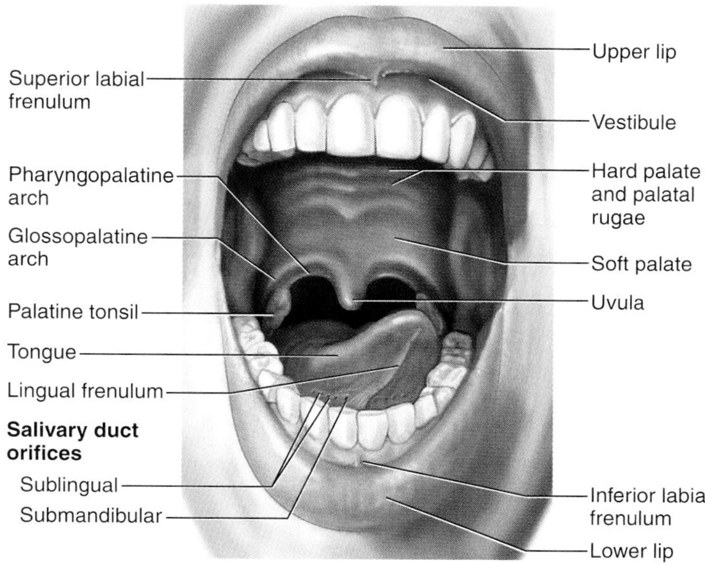

Figure 25.3 The Oral Cavity. For a photographic midsagittal view see figure A.18 (atlas A).

The Mouth Through Esophagus

Objectives

When you have completed this section, you should be able to

- describe the gross anatomy of the digestive tract from the mouth through the esophagus;
- describe the composition and functions of saliva;
- describe the neural control of salivation and swallowing; and
- describe the histological layers of the esophagus and explain the functional significance of each.

The Mouth

The mouth is also known as the **oral,** or **buccal** (BUCK-ul), **cavity.** Its functions include ingestion (food intake), taste and other sensory responses to food, mastication (chewing), chemical digestion (starch is partially digested in the mouth), deglutition (swallowing), speech, and respiration. The mouth is enclosed by the cheeks, lips, palate, and tongue (fig. 25.3). Its anterior opening between the lips is the **oral orifice** and its posterior opening into the throat is the **fauces**[2] (FAW-seez). The oral cavity is lined with nonkeratinized stratified squamous epithelium.

The Cheeks and Lips

The cheeks and lips retain food and push it between the teeth for mastication. They are essential for articulate speech and for sucking and blowing actions, including suckling by infants. Their fleshiness is due mainly to subcutaneous fat, the buccinator muscles of the cheeks, and the orbicularis oris muscle of the lips. Each lip is attached to the gum behind it by a midsagittal fold called the **labial frenulum.**[3] The **vestibule** is the space behind the cheeks and lips, external to the teeth.

Externally the lips are divided into a *cutaneous area* and *red area (vermilion).* Loosely speaking, you could think of these as the mustache area and lipstick area, respectively. The cutaneous area is colored like the rest of the face and has hair and sebaceous glands. The red area is more brightly colored and more sensitive because the skin here has unusually tall dermal papillae, which allow blood capillaries and nerves to come especially close to the surface. The red area has no hair or sebaceous glands.

The Tongue

The tongue, although muscular and bulky, is a remarkably agile and sensitive organ. It manipulates food between the teeth while it avoids being bitten, it can extract food particles from the teeth after a meal, and it is sensitive enough to feel a single stray hair in a bite of food. Its surface is covered with nonkeratinized stratified squamous epithelium and exhibits bumps and projections called **lingual papillae,** the site of the taste buds. The types of papillae and sense of taste are discussed in chapter 16, and the general anatomy of the tongue is shown in figure 16.4 (p. 595).

2. *fauces* = throat

3. *labi* = lip + *frenulum* = little bridle

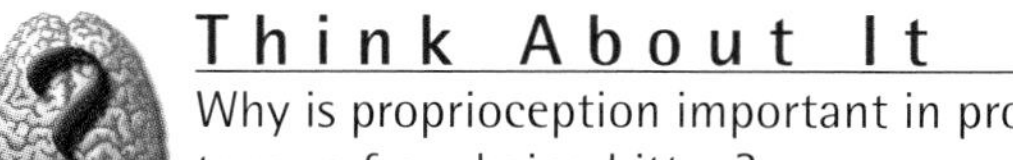

Think About It

Why is proprioception important in protecting the tongue from being bitten?

The anterior two-thirds of the tongue, called the **body,** occupies the oral cavity and the posterior one-third, the **root,** occupies the oropharynx. The boundary between them is marked by a V-shaped row of vallate papillae and, behind them, a groove called the **terminal sulcus.** The body is attached to the floor of the mouth by a midsagittal fold called the **lingual frenulum.** The muscles of the tongue, which compose most of its mass, are described in chapter 11. Amid the muscles are serous and mucous **lingual glands,** which secrete a portion of the saliva. The lingual tonsils are contained in the root.

The Palate

The palate, separating the oral cavity from the nasal cavity, makes it possible to breathe while chewing food. Its anterior portion, the **hard (bony) palate,** is supported by the palatine processes of the maxillae and by the smaller palatine bones. It has transverse *friction ridges (palatal rugae)* that aid the tongue in holding and manipulating food. Posterior to this is the **soft palate,** which has a more spongy texture and is composed mainly of skeletal muscle and glandular tissue, but no bone. It has a conical medial projection, the **uvula,**[4] visible at the rear of the oral cavity.

A pair of muscular arches on each side of the oral cavity begin dorsally near the uvula and follow the wall of the cavity to its floor. The anterior one is the **glossopalatine arch** and the posterior one is the **pharyngopalatine arch.** The latter arch marks the beginning of the pharynx. The palatine tonsils are located on the wall between the arches.

The Teeth

The teeth are collectively called the **dentition.** Adults normally have 16 teeth in the mandible and 16 in the maxilla. On each side of the midline, there are two incisors, a canine, two premolars, and three molars in each jaw (fig. 25.4*a*). The **incisors** are chisel-like cutting teeth used to bite off a piece of food. The **canines** are more pointed and act to puncture and shred it. They serve as weapons in many mammals but became reduced in the course of human evolution until they now project barely above the other teeth. The **premolars** and **molars** have relatively broad surfaces adapted to crushing and grinding.

The meeting of the teeth when the mouth closes is called **occlusion** (ah-CLUE-zhun), and the surfaces where they meet are called the **occlusal** (ah-CLUE-zul) **surfaces.**

4. *uvula* = little grape

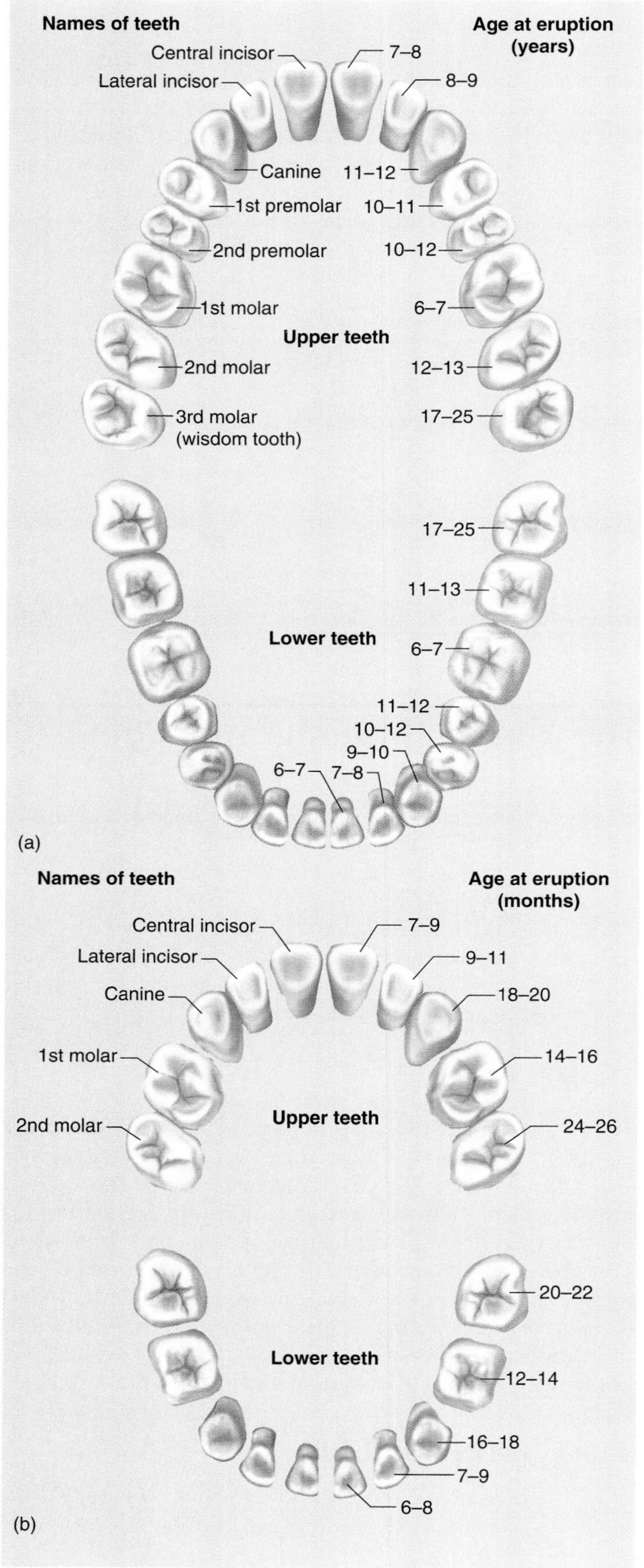

Figure 25.4 **The Dentition and Ages at Which the Teeth Erupt.** (*a*) Permanent teeth; (*b*) deciduous (baby) teeth.

The occlusal surface of a premolar has two rounded bumps called **cusps;** thus the premolars are also known as **bicuspids.** The molars have four to five cusps. Cusps of the upper and lower premolars and molars mesh when the jaws are closed and slide over each other as the jaw makes lateral chewing motions. This grinds and tears food more effectively than if the occlusal surfaces were flat.

Teeth develop beneath the gums and **erupt** (emerge) in predictable order. Twenty **deciduous teeth** (*milk teeth* or *baby teeth*) erupt from the ages of 6 to 30 months, beginning with the incisors (fig. 25.4*b*). Between 6 and 25 years of age, these are replaced by the 32 **permanent teeth.** As a permanent tooth grows below a deciduous tooth (fig. 25.5), the root of the deciduous tooth dissolves and leaves little more than the crown by the time it falls out. The third molars (wisdom teeth) erupt around ages 17 to 25, if at all. Over the course of human evolution, the face became flatter and the jaws shorter, leaving little room for the third molars. Thus, they often remain below the gum and become *impacted*—so crowded against neighboring teeth and bone that they cannot erupt.

Each tooth is embedded in a socket called an **alveolus,** forming a joint called a *gomphosis* between the tooth and bone. The alveolus is lined by a **periodontal** (PERR-ee-oh-DON-tul) **ligament,** a modified periosteum whose collagen fibers penetrate into the bone on one side and into the tooth on the other. This anchors the tooth very firmly in the alveolus. The gum, or **gingiva** (JIN-jih-vuh), covers the alveolar bone. Regions of a tooth are defined by their relationship to the gingiva: the **crown** is the portion visible above the gumline, the **neck** is the portion from the margin of the gum to the beginning of the alveolar bone, and the **root** is the portion inserted into the alveolus (fig. 25.6). The space between the neck and gum is the **gingival sulcus.** The hygiene of this sulcus is especially important to dental health (see clinical insight 25.1).

Most of a tooth consists of hard yellowish tissue called **dentin,** covered with **enamel** in the crown and neck and **cementum** in the root. Dentin and cementum are living connective tissues with cells or cell processes embedded in a calcified matrix. Cells of the cementum (*cementocytes*) are scattered more or less randomly and occupy tiny cavities similar to the lacunae of bone. Cells of the dentin (*odontoblasts*) line the pulp cavity and have slender processes that travel through tiny parallel tunnels in the dentin. Enamel is not a tissue but a noncellular secretion produced before the tooth erupts. Damaged dentin and cementum can regenerate, but damaged enamel cannot—it must be artificially repaired.

Internally, a tooth has a dilated **pulp cavity** in the crown and a narrow **root canal** in the root. These spaces are occupied by **pulp**—a mass of loose connective tissue, blood and lymphatic vessels, and nerves. These nerves and vessels enter the tooth through a pore, the **apical foramen,** at the inferior end of each root canal.

Mastication

Mastication (chewing) breaks food into pieces small enough to be swallowed and exposes more surface to the action of

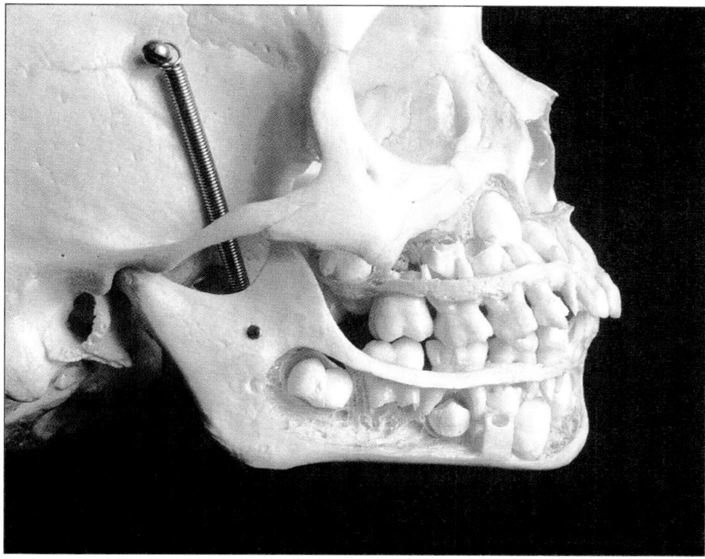

Figure 25.5 Permanent and Deciduous Teeth in a Child's Skull. This dissection shows erupted deciduous teeth and, below them, the permanent teeth waiting to erupt.

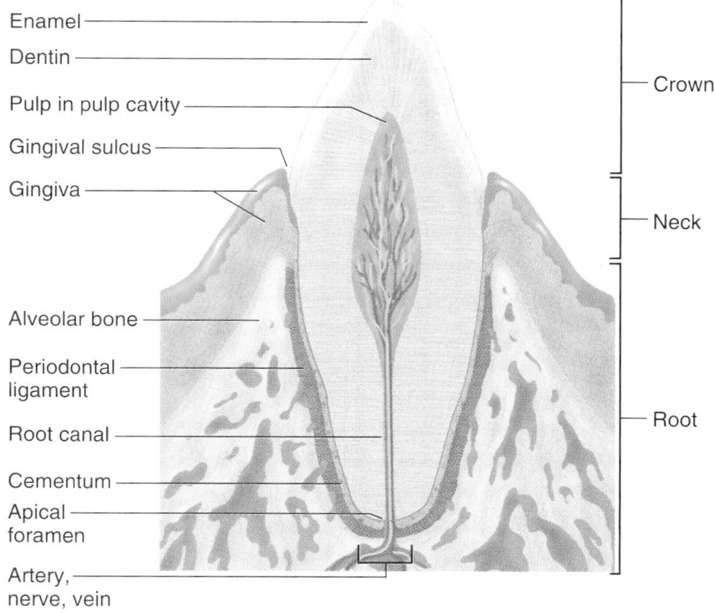

Figure 25.6 Medial Section of a Canine Tooth and Its Alveolus. Shows typical anatomy of a tooth and periodontal tissues.

Clinical Insight 25.1

Tooth and Gum Disease

Food leaves a sticky residue on the teeth called *plaque*, composed mainly of bacteria and sugars. If plaque is not thoroughly removed by brushing and flossing, bacteria accumulate, metabolize the sugars, and release lactic acid and other acids. These acids dissolve the minerals of enamel and dentin, and the bacteria enzymatically digest the collagen and other organic components. The eroded "cavities" of the tooth are known as *dental caries*.[5] If not repaired, caries may fully penetrate the dentin and spread to the pulp cavity. They then require either extraction of the tooth or *root canal therapy*, in which the pulp is removed and replaced with inert material.

When plaque calcifies on the tooth surface, it is called *calculus (tartar)*. Calculus in the gingival sulcus wedges the tooth and gum apart and allows bacterial invasion of the sulcus. This leads to *gingivitis*, or gum inflammation. Nearly everyone has gingivitis at some time. In some cases, bacteria spread from the sulcus into the alveolar bone and begin to dissolve it, producing *periodontal disease*. About 86% of people over age 70 have periodontal disease and many suffer tooth loss as a result. This accounts for 80% to 90% of adult tooth loss.

5. *caries* = rottenness

digestive enzymes. It is the first step in mechanical digestion. Mastication requires little thought because food stimulates receptors that trigger an involuntary chewing reflex. The tongue, buccinator, and orbicularis oris muscles manipulate food and push it between the teeth. The masseter and temporalis muscles produce the up-and-down crushing action of the teeth, and the lateral and medial pterygoid muscles and masseters produce side-to-side grinding action.

Saliva and the Salivary Glands

Saliva moistens the mouth, digests a little starch and fat, cleanses the teeth, inhibits bacterial growth, dissolves molecules so they can stimulate the taste buds, and moistens food and binds particles together to aid in swallowing. It is a hypotonic solution of 97.0% to 99.5% water and the following solutes:

- **salivary amylase,** an enzyme that begins starch digestion in the mouth;
- **lingual lipase,** an enzyme that is activated by stomach acid and digests fat after the food is swallowed;
- **mucus,** which binds and lubricates the food mass and aids in swallowing;
- **lysozyme,** an enzyme that kills bacteria;
- **immunoglobulin A** (IgA), an antibody that inhibits bacterial growth; and
- **electrolytes,** including sodium, potassium, chloride, phosphate, and bicarbonate ions.

Saliva has a pH of 6.8 to 7.0. There are striking differences in pH from one region of the digestive tract to another, with a powerful influence on the activity and deactivation of digestive enzymes. For example, salivary amylase works well at a neutral pH and is deactivated by the low pH of the stomach, whereas lingual lipase does not act in the mouth at all but is activated by the acidity of the stomach. Thus saliva begins to digest starch before the food is swallowed and fat after it is swallowed.

The Salivary Glands

There are two kinds of salivary glands, intrinsic and extrinsic. The **intrinsic salivary glands** are an indefinite number of small glands dispersed amid the other oral tissues. They include *lingual glands* in the tongue, *labial glands* on the inside of the lips, and *buccal glands* on the inside of the cheeks. They secrete relatively small amounts of saliva at a fairly constant rate whether we are eating or not. This saliva contains lingual lipase and lysozyme and serves to moisten the mouth and inhibit bacterial growth.

The **extrinsic salivary glands** are three pairs of larger, more discrete organs located outside of the oral mucosa; they communicate with the oral cavity by way of ducts (fig. 25.7):

1. The **parotid**[6] **gland** is located just beneath the skin anterior to the earlobe. Its duct passes superficially over the masseter, pierces the buccinator, and opens into the mouth opposite the second upper molar tooth. Viral inflammation and swelling of the parotid gland is called *mumps*.
2. The **submandibular gland** is located halfway along the body of the mandible, medial to its margin, just deep to the mylohyoid muscle. Its duct empties into

6. *par* = next to + *ot* = ear

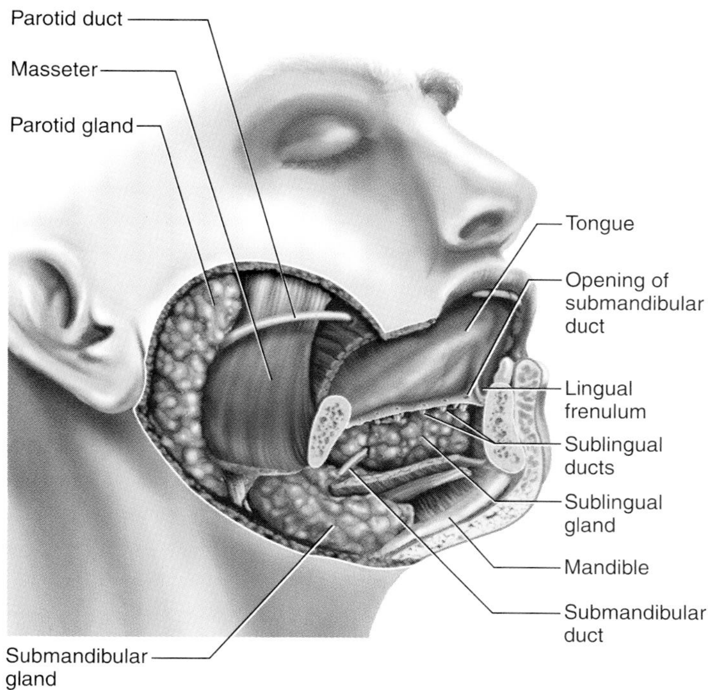

Figure 25.7 The Extrinsic Salivary Glands. Part of the mandible has been removed to expose the sublingual gland medial to it.

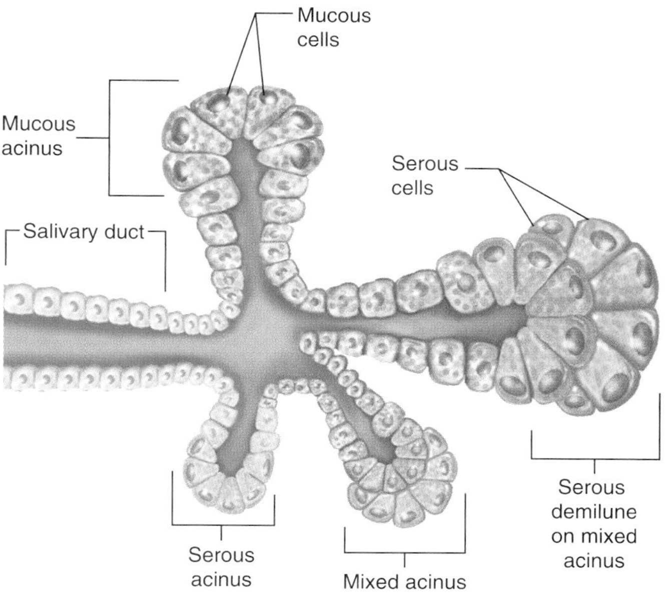

Figure 25.8 Microscopic Anatomy of the Salivary Glands. (*a*) Duct and acini of a generalized salivary gland with a mixture of mucous and serous cells. Serous cells often form crescent-shaped caps called *serous demilunes* over the ends of mucous acini. (*b*) Histology of the sublingual salivary gland.

the mouth at a papilla on the side of the lingual frenulum, near the lower central incisors.

3. The **sublingual gland** is located in the floor of the mouth. It has multiple ducts that empty into the mouth posterior to the papilla of the submandibular duct.

These are all compound tubuloacinar glands with a treelike arrangement of branching ducts ending in acini (see chapter 6). Some acini have only mucous cells, some have only serous cells, and some have a mixture of both (fig. 25.8). Mucous cells secrete salivary mucus, and serous cells secrete a thinner fluid rich in amylase and electrolytes.

Salivation

The extrinsic salivary glands secrete about 1.0 to 1.5 L of saliva per day. Cells of the acini filter water and electrolytes from the blood capillaries and add amylase, mucin, and lysozyme to it. The ducts slightly modify its electrolyte composition.

Food stimulates tactile, pressure, and taste receptors in the mouth, which transmit signals to a group of **salivatory nuclei** in the medulla oblongata and pons. These nuclei also receive input from higher brain centers, so even the odor, sight, or thought of food stimulates salivation. Irritation of the stomach and esophagus by spicy foods, stomach acid, or toxins also stimulates salivation, perhaps serving to dilute and rinse away the irritants.

The salivatory nuclei send signals to the glands by way of autonomic fibers in the facial and glossopharyngeal nerves. In response to such stimuli as the aroma or taste of food, the parasympathetic nervous system stimulates the glands to produce abundant, thin saliva rich in enzymes. Sympathetic stimulation, by contrast, causes the glands to produce less abundant, thicker saliva with more mucus. This is why the mouth may feel sticky or dry under conditions of stress. Dehydration also reduces salivation because it reduces capillary filtration.

Salivary amylase begins to digest starch as the food is chewed, while the mucus in the saliva binds food particles into a soft, slippery, easily swallowed mass called a **bolus.** Without mucus, one must drink a much larger volume of fluid to swallow food.

The Pharynx

The pharynx, described in chapter 22, has a deep layer of longitudinally oriented skeletal muscle and a superficial layer of circular skeletal muscle. The circular muscle is divided into superior, middle, and inferior **pharyngeal constrictors,** which force food downward during swallowing. When food is not being swallowed, the inferior constrictor remains contracted to exclude air from the esophagus.

The Esophagus

The **esophagus** is a straight muscular tube 25 to 30 cm long. It begins at the level of the cricoid cartilage, inferior to the larynx and dorsal to the trachea. After passing downward through the mediastinum, the esophagus penetrates the diaphragm at an opening called the *esophageal hiatus,* continues another 3 to 4 cm, and meets the stomach at an opening called the **cardiac orifice** (named for its proximity to the heart).

The wall of the esophagus is organized into the following tissue layers (fig. 25.9):

- Mucosa
  - Epithelium
  - Lamina propria
  - Muscularis mucosae
- Submucosa
- Muscularis externa
  - Inner circular layer
  - Outer longitudinal layer
- Adventitia (above the diaphragm), or
- Serosa (below the diaphragm)
  - Areolar tissue
  - Mesothelium

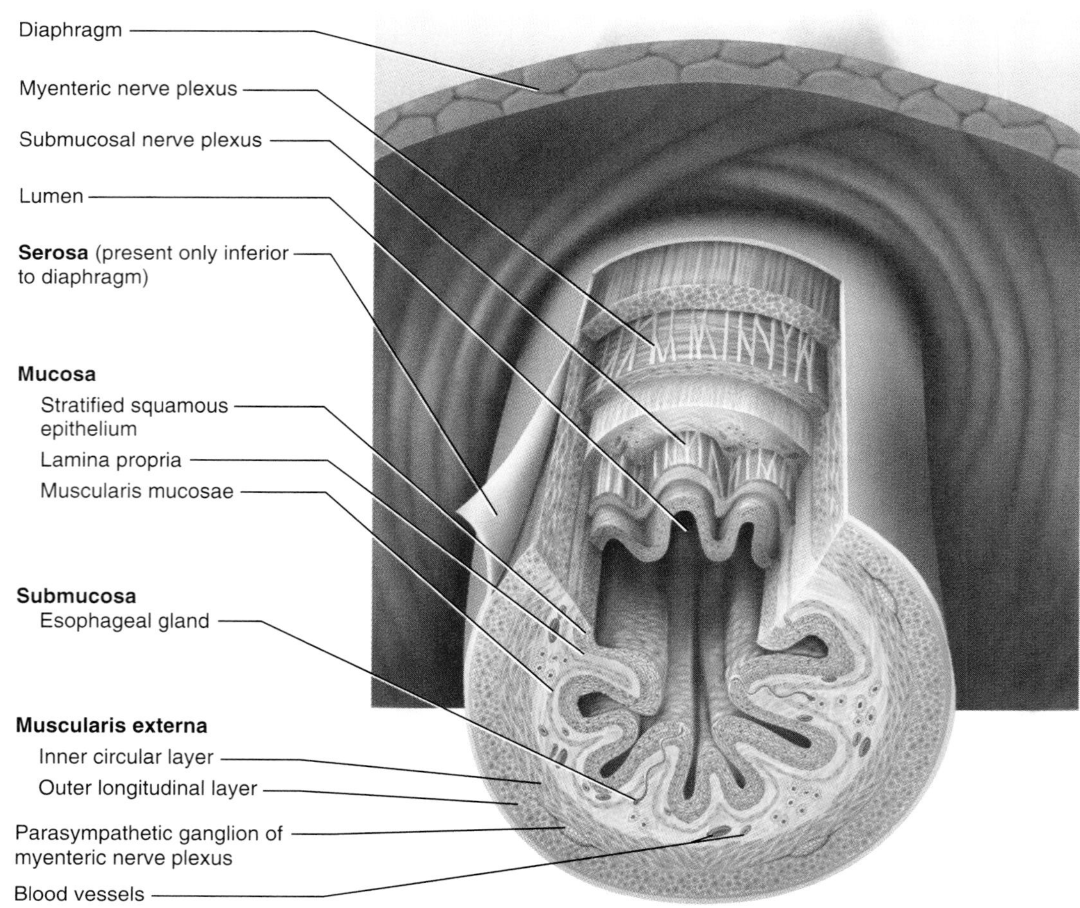

Figure 25.9 Anatomy of the Esophagus Inferior to the Diaphragm.

With some variations to be noted later, these layers are found throughout the rest of the digestive tract. You should familiarize yourself with these terms because they are also used in later discussion of the stomach and intestines.

The **mucosa,** lining the lumen, consists of a nonkeratinized stratified squamous epithelium, a loose connective tissue layer called the **lamina propria,** and a thin layer of smooth muscle called the **muscularis mucosae** (MUSS-cue-LERR-is mew-CO-see).

The **submucosa** is a thicker layer of loose connective tissue containing blood vessels, lymphatic vessels, a nerve plexus, and **esophageal glands** that secrete lubricating mucus into the lumen. When the esophagus is empty, the mucosa and submucosa are deeply folded into longitudinal ridges, giving the lumen a starlike shape in cross section.

The **muscularis externa** consists of two layers of muscle near the outer surface. Fibers of the inner layer encircle the esophagus while those of the outer layer run longitudinally. In the upper one-third of the esophagus, the muscularis externa consists of skeletal muscle, in the middle one-third it is a mixture of skeletal and smooth muscle, and in the lower one-third it is smooth muscle only. This transition in muscle type corresponds to a shift from voluntary to involuntary phases of swallowing as a food bolus passes down the esophagus.

Most of the esophagus is in the mediastinum. Here, it is covered with a connective tissue **adventitia,** which merges into the adventitias of the trachea and thoracic aorta. The short segment below the diaphragm is covered by a serosa composed of a simple squamous mesothelium overlying a thin layer of areolar tissue. Most of the digestive tract from here through the large intestine is covered with such a serosa, but some exceptions will be noted.

The esophagus, stomach, and intestines have a nervous network called the **enteric**[7] **nervous system,** which regulates digestive tract motility, secretion, and blood flow. Two nerve networks make up the enteric nervous system: the **submucosal (Meissner**[8]**) plexus** in the submucosa and the **myenteric (Auerbach**[9]**) plexus** between the two layers of the muscularis externa. These networks control **short reflexes** in which a stimulus to the mucosa triggers changes in the motility or secretion of nearby regions of the GI tract—for example, in the peristaltic contractions of swallowing, discussed shortly. **Long reflexes** act through autonomic nerve fibers that run to and from the central nervous system.

The inferior end of the esophagus is more constricted than the rest, forming the **lower esophageal sphincter.** This is not an anatomical feature—the muscularis externa is no thicker here than it is higher up—but is a physiological constriction that helps close the cardiac orifice. **Gastroesophageal reflux,** the backflow of stomach contents into the esophagus, is normally prevented partly by the tonus of this sphincter but more importantly by constriction of the diaphragm around the lower esophagus. "Heartburn" has nothing to do with the heart, but is the burning sensation produced by acid reflux into the esophagus.

Swallowing

Swallowing, or **deglutition** (DEE-glu-TISH-un), is a complex action involving over 22 muscles in the mouth, pharynx, and esophagus, coordinated by the **swallowing center,** a nucleus in the medulla oblongata and pons. This center communicates with muscles of the pharynx and esophagus by way of the trigeminal, facial, glossopharyngeal, and hypoglossal nerves (cranial nerves V, VII, IX, and XII).

Swallowing occurs in stages called the buccal and pharyngeal-esophageal phases (fig. 25.10). In the *buccal phase,* the tongue collects food, presses it against the palate to form a bolus, and pushes it back into the oropharynx. Here the bolus stimulates tactile receptors and activates the *pharyngeal-esophageal phase.* In that phase, three actions block food and drink from reentering the mouth or entering the nasal cavity or larynx: (1) the root of the tongue blocks the oral cavity, (2) the soft palate rises and blocks the nasopharynx, and (3) the infrahyoid muscles pull the larynx up, the epiglottis covers its opening, and the vestibular folds are adducted to close the airway. The food bolus is driven downward by constriction of the upper, then the middle, and finally the lower pharyngeal constrictors. As the bolus slides off the epiglottis into the esophagus, it stretches the esophagus and triggers **peristalsis,** a wave of muscular contraction that pushes the bolus ahead of it (fig. 25.10*d*).

Peristalsis is moderated partly by a short reflex through the myenteric nerve plexus. The bolus stimulates stretch receptors that feed into the nerve plexus, which transmits signals to the muscularis externa behind and ahead of the bolus. The circular muscle behind the bolus constricts and pushes it downward. Ahead of the bolus, the circular muscle relaxes while the longitudinal muscle contracts. The latter action pulls the wall of the esophagus slightly upward, which makes the esophagus a little shorter and wider and able to receive the descending food.

When we are standing or sitting upright, most food and liquid drop through the esophagus by gravity faster than the peristaltic wave can catch up to it. Peristalsis, however, propels more solid food pieces and ensures that you can swallow regardless of the body's position—even standing on your head! Liquid normally reaches the stomach in 1 to 2 seconds and a food bolus in 4 to 8 seconds. As a bolus reaches the lower end of the esophagus, the lower esophageal sphincter relaxes to let it pass into the stomach.

For a review of the anatomy up to this point, see table 25.1.

7. *enter* = intestine
8. Georg Meissner (1829–1905), German histologist
9. Leopold Auerbach (1828–97), German anatomist

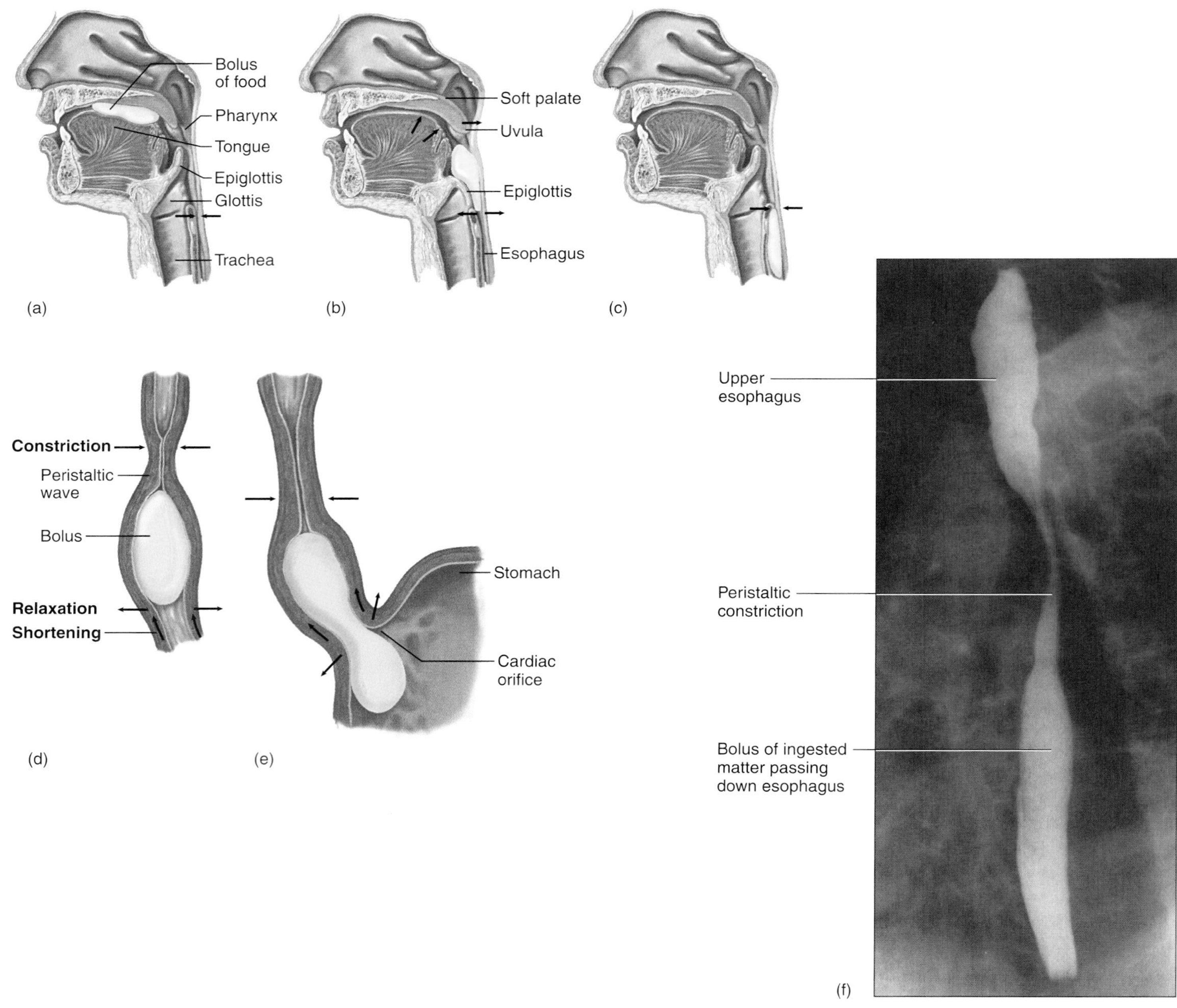

Figure 25.10 Swallowing. (*a*) The tongue compresses food against the palate to form a bolus. (*b*) The bolus passes into the esophagus while the tongue blocks the oral cavity, the soft palate blocks the nasal cavity, and the epiglottis blocks the larynx. (*c*) The superior end of the esophagus constricts as the bolus passes downward. (*d*) A peristaltic wave moves the bolus down the esophagus. The esophagus constricts behind the bolus while it dilates and shortens in front of the bolus. (*e*) The lower esophageal sphincter relaxes to admit the bolus to the stomach. (*f*) X ray of esophagus showing a peristaltic constriction above a bolus of ingested material.

Table 25.1 Anatomical Checklist of the Digestive System from the Mouth Through the Esophagus

Oral (buccal) Cavity

- *Oral orifice*
- *Fauces*
- *Cheeks*
- *Lips*
  - Cutaneous area
  - Red area (vermilion)
  - Labial frenulum
- *Vestibule*
- *Tongue*
  - Body
  - Root
  - Terminal sulcus
  - Lingual frenulum
  - Lingual papillae
  - Taste buds
  - Lingual muscles
  - Lingual tonsils
  - Lingual glands
- *Palate*
  - Hard palate
  - Soft palate
    - Uvula
- *Glossopalatine arch*
- *Pharyngopalatine arch*
- *Dentition (teeth)*
  - Developmental types
    - Deciduous teeth
    - Permanent teeth
  - Functional types
    - Incisors
    - Canines
    - Premolars (bicuspids)
    - Molars
  - Dental tissues
    - Dentin
    - Enamel
    - Cementum
    - Pulp
  - Anatomical features
    - Crown
      - Occlusal surface
      - Cusps
    - Neck
    - Root
    - Pulp cavity
    - Root canal
    - Apical foramen
- *Periodontal tissues*
  - Alveolus
  - Periodontal ligament
  - Gingiva (gum)
    - Gingival sulcus

Salivary Glands

- *Intrinsic*
- *Extrinsic*
  - Parotid gland
  - Submandibular gland
  - Sublingual gland

Pharynx

- *Pharyngeal constrictors*

Esophagus

- *Esophageal glands*
- *Lower esophageal sphincter*
- *Cardiac orifice*

Before You Go On

Answer the following questions to test your understanding of the preceding section:

5. List as many functions of the tongue as you can.
6. Imagine a line from the mandibular bone to the root canal of a tooth. Name the tissues, in order, through which this line would pass.
7. What is the difference in function and location between intrinsic and extrinsic salivary glands? Name the extrinsic salivary glands and describe their locations.
8. Describe the muscularis externa of the esophagus and its action in peristalsis.
9. Describe the mechanisms that prevent food from entering the nasal cavity and larynx during swallowing.

The Stomach

Objectives

When you have completed this section, you should be able to

- describe the gross and microscopic anatomy of the stomach;
- state the function of each type of epithelial cell in the gastric mucosa;
- identify the secretions of the stomach and state their functions;
- explain how the stomach produces hydrochloric acid and pepsin;
- describe the contractile responses of the stomach to food; and
- describe the three phases of gastric function and how gastric activity is activated and inhibited.

The stomach is a muscular sac in the upper left abdominal cavity immediately inferior to the diaphragm. It functions primarily as a food storage organ, with a volume of about 50 mL when empty and 1.0 to 1.5 L after a typical meal. When extremely full, it may hold up to 4 L and extend nearly as far as the pelvis.

Well into the nineteenth century, authorities regarded the stomach as essentially a grinding chamber, fermentation vat, or cooking pot. Some even attributed digestion to a supernatural spirit in the stomach. We now know that it mechanically breaks up food particles, liquefies the food, and begins the chemical digestion of proteins and a small amount of fat. This produces a soupy or pasty mixture of semidigested food called **chyme**[10] (kime). Most digestion occurs after the chyme passes on to the small intestine.

10. *chyme* = juice

Gross Anatomy

The stomach is J-shaped (fig. 25.11), relatively vertical in tall people, and more horizontal in short people. Its medial to superior margin is called the **lesser curvature** and its longer lateral to inferior margin is the **greater curvature.**

The stomach is divided into four regions: (1) The **cardiac region (cardia)** is a small area immediately inside the cardiac orifice. (2) The **fundic region (fundus)** is the dome-shaped portion superior to the esophageal attachment. (3) The **body (corpus)** makes up the greatest part of the stomach inferior to the cardiac orifice. (4) The **pyloric region** is a slightly narrower pouch at the inferior end; it is subdivided into a funnel-like **antrum**[11] and a narrower **pyloric canal.** The latter terminates at the **pylorus,**[12] a narrow passage into the duodenum. The pylorus is surrounded by a thick ring of smooth muscle, the **pyloric (gastroduodenal) sphincter,** which regulates the passage of chyme into the duodenum.

Innervation and Circulation

The stomach receives parasympathetic nerve fibers from the vagus nerves and sympathetic fibers from the celiac ganglia (see p. 573). It is supplied with blood by branches of the celiac trunk (see pp. 779–81). All blood drained from the stomach and intestines enters the hepatic portal circulation and filters through the liver before returning to the heart.

The Stomach Wall

The stomach wall has tissue layers similar to those of the esophagus, with some variations. The mucosa is covered with a simple columnar glandular epithelium (fig. 25.12). The apical regions of its surface cells are filled with mucin, which swells with water and becomes mucus after it is secreted. The mucosa and submucosa are flat and smooth when the stomach is full, but as it empties, these layers form conspicuous longitudinal wrinkles called **rugae** (ROO-gee). The lamina propria is almost entirely occupied by tubular glands, to be described shortly. The muscularis externa has three layers, rather than two—an outer longitudinal, middle circular, and inner oblique layer (see fig. 25.11).

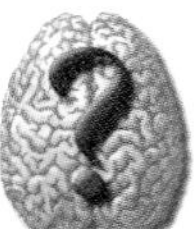

Think About It

Contrast the epithelium of the esophagus with that of the stomach. Why is each epithelial type best suited to the function of its respective organ?

The gastric mucosa is pocked with depressions called **gastric pits** lined with the same columnar epithelium as the surface (fig. 25.12). Cells near the bottom of the gastric pits divide repeatedly and produce new epithelial cells that continually migrate upward toward the lumen and replace old epithelial cells that are sloughed off into the chyme.

Two or three tubular glands open into the bottom of each gastric pit and span the rest of the lamina propria. In the cardiac and pyloric regions they are called **cardiac glands** and **pyloric glands,** respectively. In the rest of the stomach, they are called **gastric glands.** These glands differ in cellular composition, as noted shortly. Collectively, they have the following cell types:

- **Mucous cells,** which secrete mucus, predominate in the cardiac and pyloric glands. In gastric glands, they are called *mucous neck cells* and are concentrated in the narrow *neck* of the gland, where it opens into the gastric pit.
- **Regenerative cells,** found in the base of the pit and neck of the gland, divide rapidly and produce a continual supply of new cells. Newly generated cells migrate upwards to the gastric surface as well as downwards into the glands to replace cells that die.
- **Parietal cells,** found mostly in the upper half of the gland, secrete hydrochloric acid and intrinsic factor. They are found mostly in the gastric glands, but a few occur in the pyloric glands.
- **Chief cells,** so-named because they are the most numerous, secrete rennin and lipase in infancy and pepsinogen throughout life. They dominate the lower half of the gastric glands but are absent from cardiac and pyloric glands.
- **Enteroendocrine cells,** concentrated especially in the lower end of a gland, secrete hormones and paracrine messengers that regulate digestion. There are at least eight different kinds of enteroendocrine cells in the stomach, each of which produces a different chemical messenger.

In general, the cardiac glands secrete mainly mucus; acid and enzyme secretion occur predominantly in the gastric glands; and hormones are secreted throughout the stomach.

Gastric Secretions

The gastric glands produce 2 to 3 L of **gastric juice** per day, composed mainly of water, hydrochloric acid, and pepsin.

Hydrochloric Acid

Gastric juice has a high concentration of hydrochloric acid (HCl) and a pH as low as 0.8. Such concentrated acid could

11. *antrum* = cavity
12. *pylorus* = gatekeeper

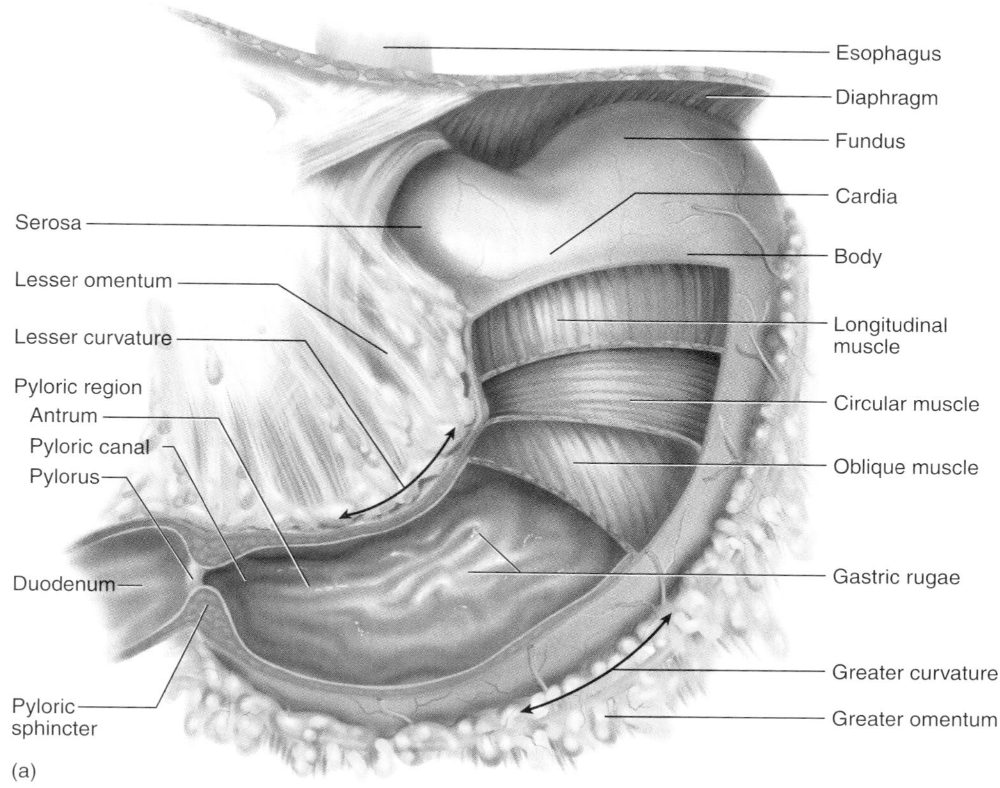

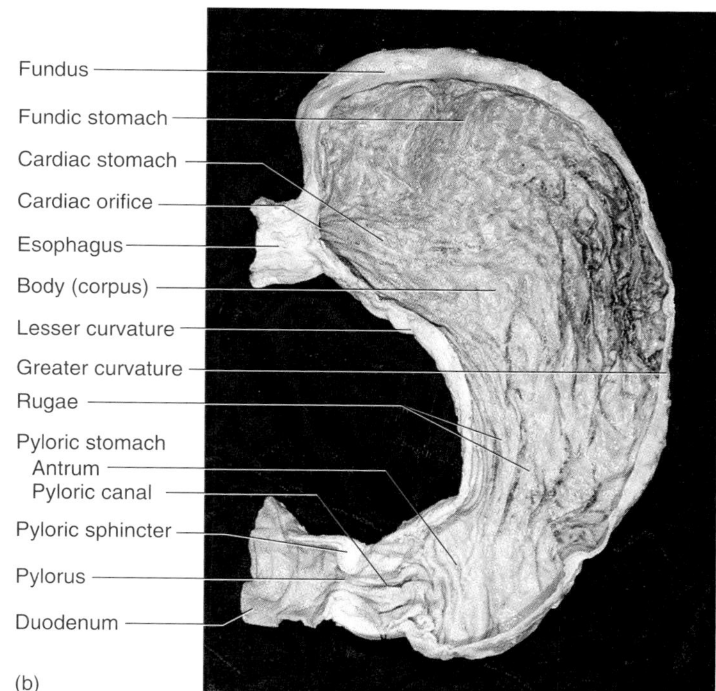

Figure 25.11 **The Stomach.** (*a*) Gross anatomy. (*b*) Photograph of the internal surface.

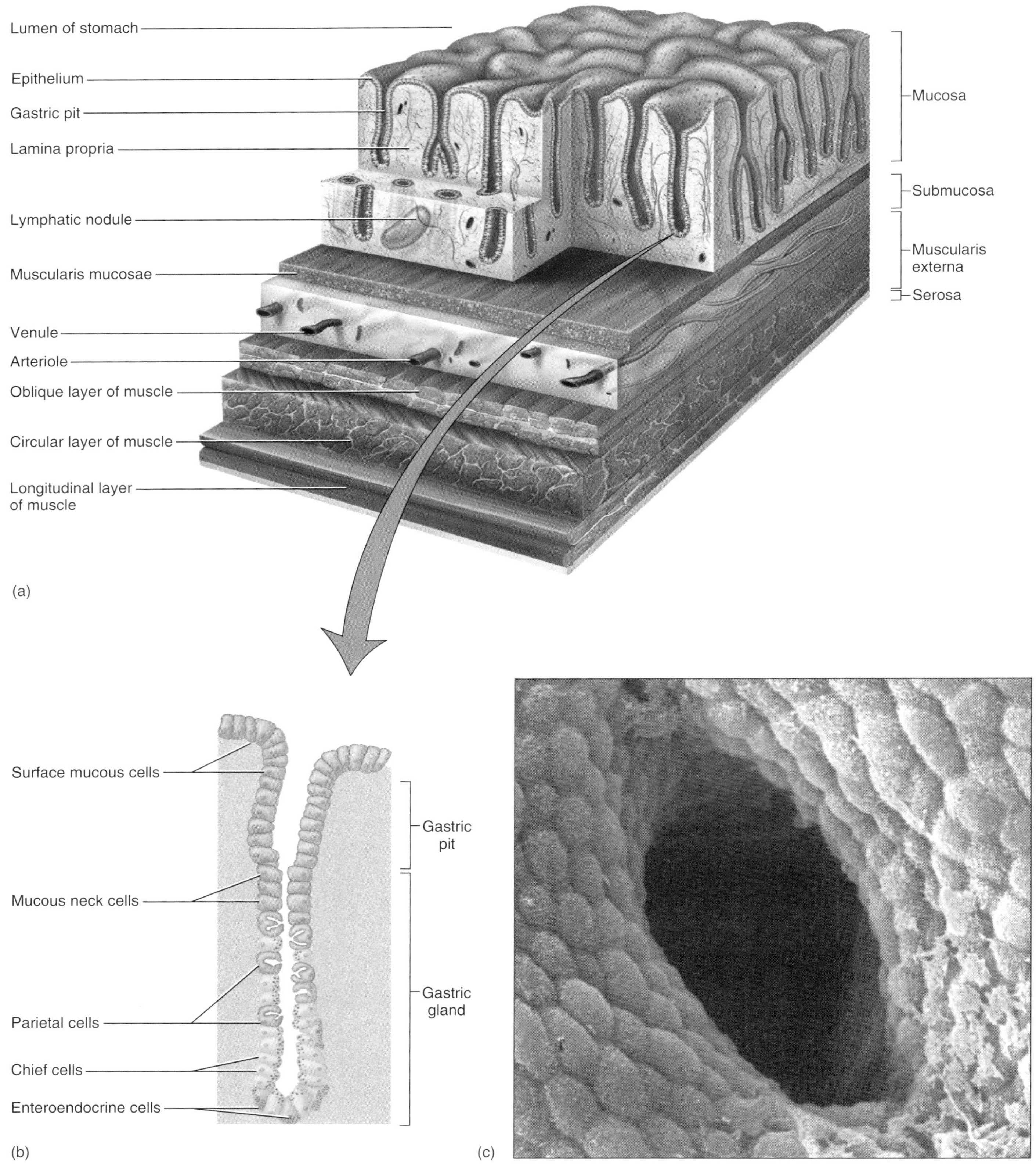

Figure 25.12 **Microscopic Anatomy of the Stomach Wall.** (*a*) A block of tissue showing all layers from the mucosa (*top*) to the serosa (*bottom*). (*b*) Detail of a gastric pit and gastric gland. (*c*) The opening of a gastric pit into the stomach, surrounded by the rounded apical surfaces of the columnar epithelial cells of the mucosa.

cause a serious chemical burn to the skin. How, then, does the stomach produce and tolerate such acidity?

The reactions that produce HCl (fig. 25.13) may seem familiar by now because they have been discussed in previous chapters—most recently in connection with renal excretion of H^+ in chapter 24. Parietal cells contain carbonic anhydrase (CAH), which catalyzes the first step in the following reaction:

$$CO_2 + H_2O \xrightarrow{CAH} H_2CO_3 \rightarrow HCO_3^- + H^+$$

The H^+ produced by this reaction is pumped into the lumen of a gastric gland by an active transport protein similar to the Na^+-K^+ pump, called **H^+-K^+ ATPase.** This is an antiport that uses the energy of ATP to pump H^+ out and K^+ into the cell. HCl secretion does not affect the pH within the parietal cell because H^+ is pumped out as fast as it is generated. The bicarbonate ions (HCO_3^-) generated by the preceding reaction are exchanged for chloride ions (Cl^-) from the blood plasma—the same *chloride shift* process that occurs in the renal tubules and red blood cells—and the Cl^- is pumped into the lumen of the gastric gland to join the H^+.

Thus HCl accumulates in the stomach while bicarbonate ions accumulate in the blood. Because of the bicarbonate, blood leaving the stomach has a higher pH when digestion is occurring than when the stomach is empty. This high-pH blood is called the **alkaline tide.**

Stomach acid has several functions: (1) It activates the enzymes pepsin and lingual lipase, as discussed shortly. (2) It breaks up connective tissues and plant cell walls, helping to liquefy food and form chyme. (3) It converts ingested ferric ions (Fe^{3+}) to ferrous ions (Fe^{2+}), a form of iron that can be absorbed and used for hemoglobin synthesis. (4) It contributes to nonspecific disease resistance by destroying ingested bacteria and other pathogens.

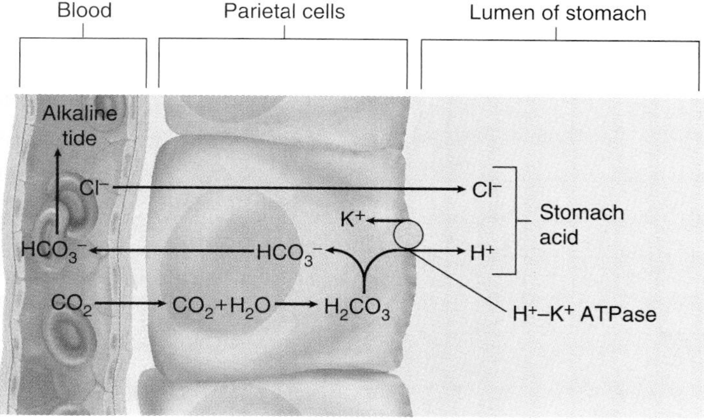

Figure 25.13 Hydrochloric Acid Secretion by Gastric Parietal Cells. The parietal cell takes CO_2 from the blood and combines it with water to make carbonic acid (*bottom line of figure*). Carbonic acid breaks down into bicarbonate ion (HCO_3^-) and hydrogen ion (H^+). The H^+ is pumped into the lumen of the stomach in exchange for K^+ by an H^+-K^+ ATPase. The bicarbonate ion is returned for the blood in exchange for a chloride ion (Cl^-). Cl^- follows H^+ into the lumen, where the two constitute hydrochloric acid.

Intrinsic Factor

Parietal cells also secrete a glycoprotein called **intrinsic factor** that is essential to the absorption of vitamin B_{12} by the small intestine. Without vitamin B_{12}, hemoglobin cannot be synthesized and pernicious anemia develops (see chapter 18). The secretion of intrinsic factor is the only indispensable function of the stomach. Digestion can continue following removal of the stomach (*gastrectomy*), but a person must then take vitamin B_{12} by injection, or vitamin B_{12} and intrinsic factor orally. As we age, the gastric mucosa atrophies, less intrinsic factor is secreted, and the risk of pernicious anemia rises.

Pepsin

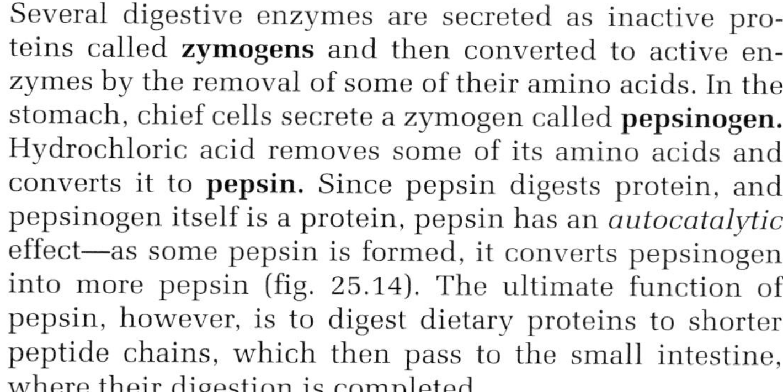

Several digestive enzymes are secreted as inactive proteins called **zymogens** and then converted to active enzymes by the removal of some of their amino acids. In the stomach, chief cells secrete a zymogen called **pepsinogen.** Hydrochloric acid removes some of its amino acids and converts it to **pepsin.** Since pepsin digests protein, and pepsinogen itself is a protein, pepsin has an *autocatalytic* effect—as some pepsin is formed, it converts pepsinogen into more pepsin (fig. 25.14). The ultimate function of pepsin, however, is to digest dietary proteins to shorter peptide chains, which then pass to the small intestine, where their digestion is completed.

Other Enzymes

In infants, the chief cells also secrete **gastric lipase** and **rennin** (not to be confused with renin, the enzyme secreted by the kidneys). Gastric lipase digests some of the *butterfat* of milk, and rennin curdles milk by coagulating its proteins.

Chemical Messengers

The gastric and pyloric glands have various kinds of enteroendocrine cells that collectively produce as many as 20 chemical messengers. Most of these secretions behave as hormones—they travel in the bloodstream and stimulate distant target cells. Some also behave as paracrine secretions, diffusing a short distance away and stimulating other cells in the gastric mucosa. Some of these messengers are also produced by neurons of the enteric nervous system and behave as neurotransmitters.

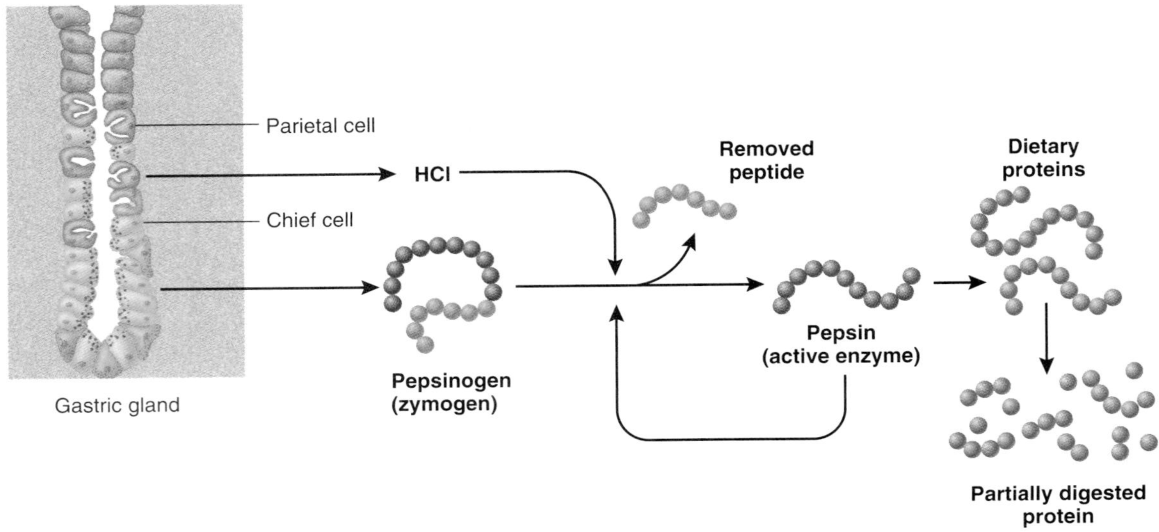

Figure 25.14 The Production and Action of Pepsin. The chief cells secrete pepsinogen and the parietal cells secrete HCl. HCl removes some of the amino acids from pepsinogen and coverts it to pepsin. Pepsin catalyzes the production of more pepsin (autocatalytic effect), as well as partially digesting dietary proteins.

Table 25.2 Major Secretions of the Gastric Glands

Secretory Cells	Secretion	Function
Mucous neck cells	Mucus	Protects mucosa from HCl and enzymes
Parietal cells	Hydrochloric acid	Activates pepsin and lingual lipase; helps liquefy food; reduces dietary iron to usable form (Fe^{2+}); destroys ingested pathogens
	Intrinsic factor	Enables small intestine to absorb vitamin B_{12}
Chief cells	Pepsinogen	Converted to pepsin, which digests protein
	Rennin	Coagulates milk proteins in infant stomach; not secreted in adults
	Gastric lipase	Digests fats in infant stomach; not secreted in adults
Enteroendocrine cells	Gastrin	Stimulates gastric glands to secrete HCl and enzymes; stimulates intestinal motility; relaxes ileocecal valve
	Serotonin	Stimulates gastric motility
	Histamine	Stimulates HCl secretion
	Somatostatin	Inhibits gastric secretion and motility; delays emptying of stomach; inhibits secretion by pancreas; inhibits gallbladder contraction and bile secretion; reduces blood circulation and nutrient absorption in small intestine

Several of the gastric secretions are summarized in table 25.2. Some of the functions listed there are explained later in the chapter.

Gastric Motility

As you begin to swallow, the swallowing center of the medulla oblongata signals the stomach to relax, thus preparing it to receive food. The arriving food stretches the stomach and activates the *receptive-relaxation response* of smooth muscle: the stomach briefly resists stretching but then relaxes and is able to accommodate more food.

Soon, the stomach begins to show a rhythm of peristaltic contractions. These are governed by pacemaker cells in the longitudinal layer of the muscularis externa of the greater curvature. About every 20 seconds, a gentle ripple begins in the fundus and becomes stronger as it progresses toward the pyloric region, where the muscularis externa is thicker. After 30 minutes or so, these contractions become quite strong. They churn the food, mix it with gastric juice, and promote its physical breakup and chemical digestion.

The antrum holds about 30 mL of chyme. As a peristaltic wave passes down the antrum, it squirts about 3 mL of chyme into the duodenum at a time. When the wave reaches the pyloric sphincter, it squeezes the sphincter shut. Chyme that does not get through this time is turned back into the antrum and body of the stomach for further digestion. Allowing only small amounts into the duodenum at a time enables the duodenum to neutralize the stomach acid and digest nutrients little by little. If the duodenum becomes overfilled, it inhibits gastric motility and postpones receiving more chyme; the mechanism for this is discussed shortly. A typical meal is emptied from the stomach in about 4 hours, but it takes less time if the meal is more liquid and as long as 6 hours if the meal is high in fat.

Vomiting

Vomiting is induced by excessive stretching of the stomach, psychological stimuli, and chemical irritants such as alcohol and bacterial toxins. Vomiting is not caused by contraction of the stomach itself, but by the diaphragm and abdominal muscles. The **emetic**[13] **center** in the medulla oblongata stimulates the lower esophageal sphincter to relax while it stimulates the diaphragm and abdominal muscles to contract. These muscles squeeze the stomach like a bagpipe and force its contents up the esophagus. Severe vomiting may expel even the contents of the small intestine.

Digestion and Absorption

Salivary and gastric enzymes partially digest protein and small amounts of starch and fat in the stomach, but most digestion and nearly all nutrient absorption occur after the chyme passes into the small intestine. The stomach does not absorb any significant amount of nutrients but does absorb aspirin and some lipid-soluble drugs. Alcohol is absorbed mainly by the small intestine, so its intoxicating effect depends partly on how rapidly the stomach is emptied.

Protection of the Stomach

You may wonder why the stomach does not digest itself. We can digest tripe (animal stomachs) as readily as any other meat, and when a human dies, the stomach digests holes through itself within hours. The living stomach, however, is protected in three ways from the harsh acidic and enzymatic environment it creates:

1. **Mucous coat.** The thick, highly alkaline mucus resists the action of acid and enzymes.
2. **Epithelial cell replacement.** The stomach's epithelial cells live only 3 to 6 days and are then sloughed off into the chyme and digested with the food. They are replaced just as rapidly, however, by cell division in the gastric pits.
3. **Tight junctions.** The epithelial cells are joined by tight junctions that prevent gastric juice from seeping between them and digesting the connective tissue of the lamina propria or beyond.

The breakdown of these protective mechanisms can result in inflammation and peptic ulcer (see clinical insight 25.2).

Regulation of Gastric Function

Because of the coordinated actions of the endocrine and nervous systems, gastric secretion and motility rise when food is eaten and decline as the stomach empties. Gastric activity is divided into three stages called the cephalic, gastric, and intestinal phases, based on whether the stomach is being controlled by the brain, by itself, or by the small intestine, respectively (fig. 25.15). These phases overlap and all three can occur simultaneously.

The Cephalic Phase

The **cephalic phase** (fig. 25.15*a*) is a stage in which the medulla oblongata sends signals to the stomach by way of the vagus nerves and stimulates gastric secretion and motility. This phase is activated by the sight, smell, taste, or mere thought of food. These sensory and mental inputs converge on the hypothalamus, which relays signals to the medulla. Vagus nerve fibers from the medulla stimulate the enteric nervous system of the stomach, and in turn, the enteric nerves stimulate gastric secretion even before food is swallowed.

The Gastric Phase

The **gastric phase** (fig. 25.15*b*) is a period in which swallowed food and semidigested protein (peptides and amino acids) activate gastric activity. About two-thirds of gastric secretion occurs during this phase. Ingested food stimulates gastric activity in two ways: by stretching the stomach and by raising the pH of its contents. Stretch activates two reflexes: a short **myenteric reflex** mediated through the myenteric nerve plexus, and a long **vagovagal reflex** in which signals travel from the stomach to the brainstem by way of sensory fibers in the vagus nerves and then return to the stomach through parasympathetic motor fibers in the same nerves. (Remember that the parasympathetic nervous system stimulates digestive processes and the sympathetic nervous system inhibits them.)

Gastric secretion is stimulated chiefly by three chemicals: acetylcholine, histamine, and gastrin. Each chemical

13. *emet* = vomiting

Clinical Insight 25.2

Peptic Ulcer

Inflammation of the stomach, called *gastritis,* can lead to a *peptic ulcer* as pepsin and hydrochloric acid erode the stomach wall. Peptic ulcers occur even more commonly in the duodenum and occasionally in the esophagus. If untreated, they can perforate the organ and cause potentially fatal hemorrhaging or peritonitis. Most such fatalities occur in people over age 65.

There is no evidence to support the popular belief that peptic ulcers result from psychological stress. Hypersecretion of acid and pepsin is sometimes involved, but even normal secretion can cause ulceration if the mucosal defense is compromised by other causes. Many or most ulcers involve an acid-resistant bacterium, *Helicobacter pylori,* that invades the mucosa of the stomach and duodenum and opens the way to chemical damage to the tissue. Other risk factors include smoking and the use of aspirin and other nonsteroidal anti-inflammatory drugs (NSAIDs). NSAIDs suppress the synthesis of prostaglandins, which normally stimulate the secretion of protective mucus and acid-neutralizing bicarbonate. Aspirin itself is an acid that directly irritates the gastric mucosa.

Until recently, the most widely prescribed drug in the United States was Cimetidine (Tagamet), which was designed to treat peptic ulcers by reducing acid secretion. Histamine stimulates acid secretion by binding to sites on the parietal cells called *H_2 receptors;* Cimetidine, an *H_2 blocker,* prevents this binding. Lately, however, ulcers have been treated more successfully with antibiotics against *Helicobacter* combined with bismuth suspensions such as Pepto-Bismol. This is a much shorter and less expensive course of treatment and permanently cures about 90% of peptic ulcers, as compared with a cure rate of only 20% to 30% for H_2 blockers.

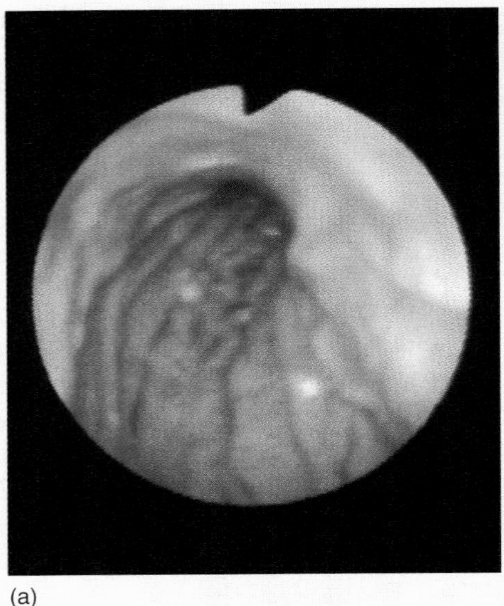

(a)

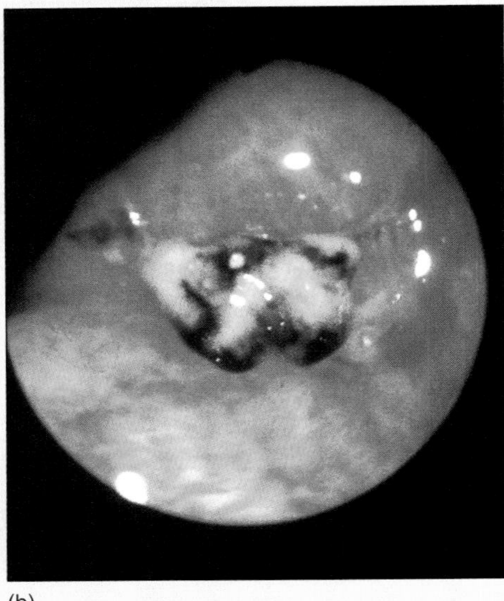

(b)

TA 25.1 Endoscopic Views of the Gastroesophageal Junction. The esophagus can be seen opening into the cardiac stomach. (*a*) A healthy gastric mucosa; the small white spots are reflections of light from the endoscope. (*b*) A bleeding peptic ulcer. A peptic ulcer typically has an oval shape and yellow-white color. Here the yellowish floor of the ulcer is partially obscured by black blood clots, and fresh blood is visible around the margin of the ulcer.

messenger has its own receptors on the acid-secreting parietal cells. Acetylcholine comes from the parasympathetic nerve fibers of the two reflex pathways just described. **Histamine** is a paracrine secretion of enteroendocrine cells in the gastric glands. Like other paracrine messengers, by definition (see chapter 17), it does not get to its destination by way of the blood but by diffusion to its target cells a short distance away. **Gastrin** is a hormone produced by enteroendocrine **G cells** in the pyloric glands. It reaches the gastric glands by way of the bloodstream and stimulates the parietal cells to secrete HCl. The chief cells secrete pepsinogen in response to gastrin and even more to acetylcholine.

As dietary protein is digested, it breaks down into smaller peptides and amino acids, which directly stimulate the G cells to secrete even more gastrin—a positive feedback

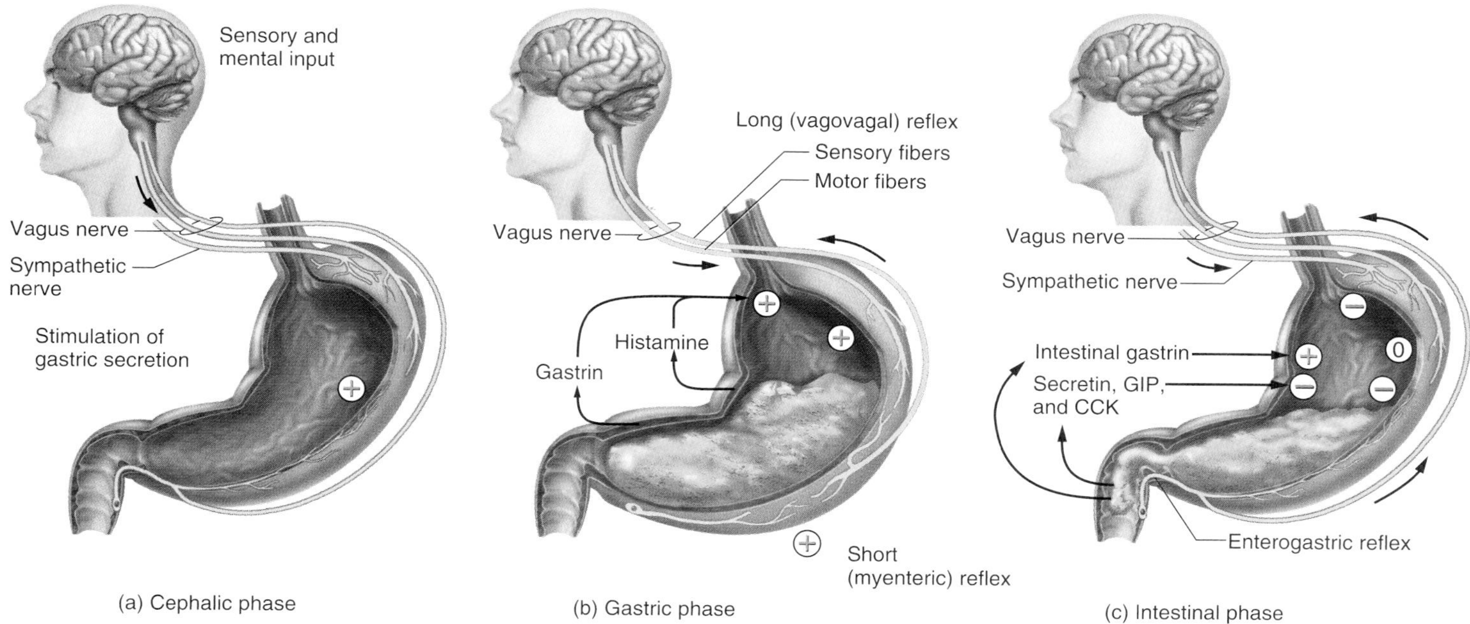

Figure 25.15 **Neural and Hormonal Control of Gastric Secretion.** *Plus signs* indicate stimulation and *minus signs* indicate inhibition. (*a*) The cephalic phase. The vagus nerve stimulates gastric secretion even before food is swallowed. (*b*) The gastric phase. Food in the stomach stretches it and activates a myenteric reflex and a vagovagal reflex. These reflex pathways trigger gastric secretion. Histamine and gastrin also stimulate acid and enzyme secretion. (*c*) The intestinal phase. Intestinal gastrin has a short stimulatory effect on the stomach, but then secretin, GIP, and CCK, as well as the enterogastric reflex, inhibit gastric secretion and motility until the duodenum can process the chyme already in it.

loop that accelerates protein digestion (fig. 25.16). Small peptides also buffer stomach acid so the pH does not fall excessively low. But as digestion continues and these peptides are emptied from the stomach, the pH drops lower and lower. Below a pH of 2, stomach acid inhibits the parietal cells and G cells—a negative feedback loop that winds down the gastric phase as the need for pepsin and HCl declines.

The Intestinal Phase

The **intestinal phase** is a stage in which the duodenum responds to arriving chyme and moderates gastric activity through hormones and nervous reflexes (see fig. 25.15*c*). The duodenum initially enhances gastric secretion, but soon inhibits it. Stretching of the duodenum activates vagovagal reflexes that stimulate the stomach, and peptides and amino acids in the chyme stimulate G cells of the duodenum to secrete more gastrin, which further stimulates the stomach.

Soon, however, the acid and semidigested fats in the duodenum trigger the **enterogastric reflex**—the duodenum sends inhibitory signals to the stomach by way of the enteric nervous system, and sends signals to the medulla that (1) inhibit the vagal nuclei, thus reducing vagal stimulation of the stomach, and (2) stimulate sympathetic neurons, which send inhibitory signals to the stomach. Chyme also stimulates duodenal enterendocrine cells to release **secretin, cholecystokinin** (CO-leh-SIS-toe-KY-nin) **(CCK),** and **gastric inhibitory peptide (GIP).** Secretin and CCK primarily stimulate the pancreas and gallbladder, as discussed later, but all three of these hormones suppress gastric secretion and motility. The effect of all this is that gastrin secretion declines and the pyloric sphincter contracts tightly to limit the admission of more chyme into the duodenum. This gives the duodenum time to work on the chyme it has already received before being loaded with more.

Before You Go On

Answer the following questions to test your understanding of the preceding section:

10. Name four types of epithelial cells of the gastric and pyloric glands and state what each one secretes.
11. Explain how the gastric glands produce hydrochloric acid and how this produces an alkaline tide.
12. What positive feedback cycle can you identify in the formation and action of pepsin?
13. How does food in the duodenum inhibit motility and secretion in the stomach?

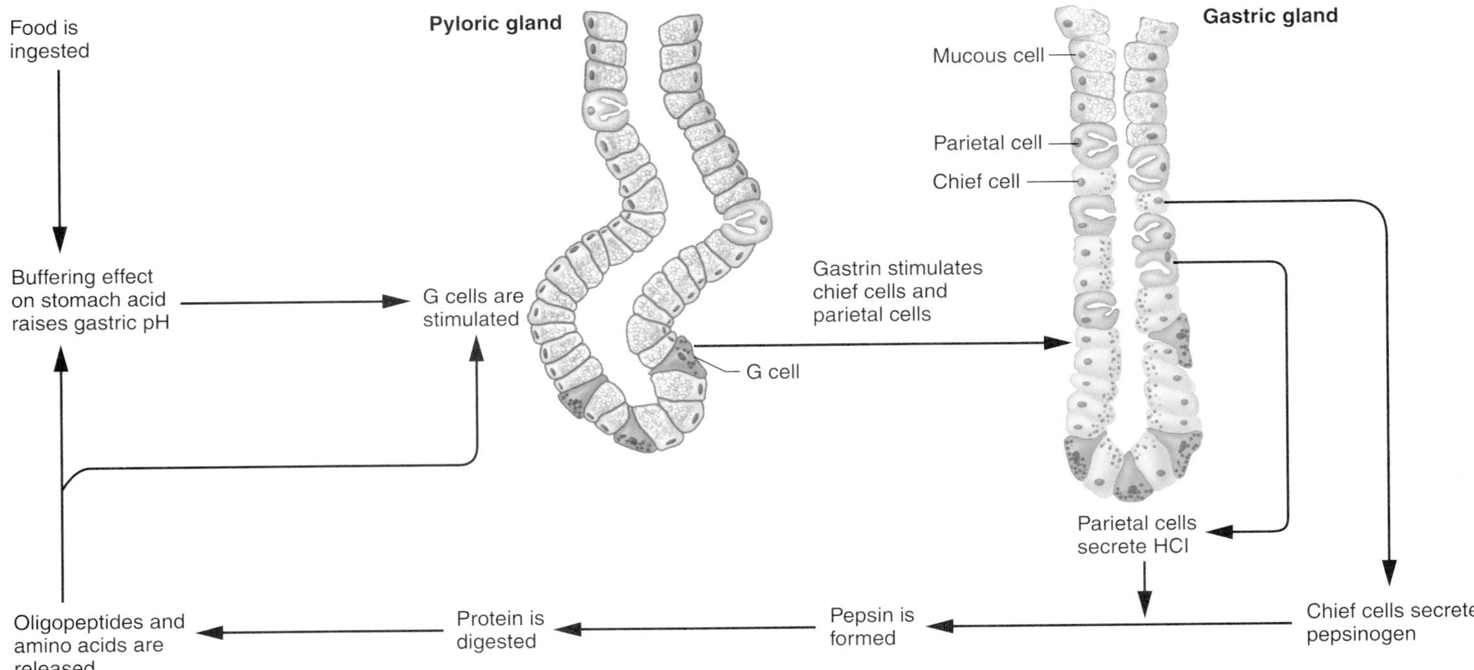

Figure 25.16 **Positive Feedback Control of Gastric Secretion.** This feedback cycle declines and stops as the stomach is emptied and the pH drops.

The Liver, Gallbladder, and Pancreas

Objectives

When you have completed this section, you should be able to

- describe the gross and microscopic anatomy of the liver, gallbladder, bile duct system, and pancreas;
- describe the digestive secretions and functions of the liver, gallbladder, and pancreas; and
- explain how hormones regulate secretions of the liver and pancreas.

The small intestine receives not only chyme from the stomach but also secretions from the liver and pancreas, which enter the digestive tract near the junction of the stomach and small intestine. These secretions are so important to the digestive processes of the small intestine that it is necessary to understand them before continuing with intestinal physiology.

The Liver

The liver (fig. 25.17) is a reddish brown gland located immediately inferior to the diaphragm, filling most of the right hypochondriac and epigastric regions. It is the body's largest gland, weighing about 1.4 kg (3 lb). The liver has a tremendous variety of functions, but only one of them, the secretion of bile, contributes to digestion. Others are discussed in the following chapter, which provides a more thorough physiological basis for understanding liver function.

Gross Anatomy

The liver has four lobes called the right, left, quadrate, and caudate lobes. From an anterior view, we see only a large **right lobe** and smaller **left lobe.** They are separated from each other by the **falciform**[14] **ligament,** a sheet of mesentery that suspends the liver from the diaphragm and anterior abdominal wall. The **round ligament** (*ligamentum teres*), also visible anteriorly, is a fibrous remnant of the umbilical vein, which carries blood from the umbilical cord to the liver of a fetus.

From the inferior view, we also see a squarish **quadrate lobe** next to the gallbladder and a taillike **caudate**[15] **lobe** posterior to that. An irregular opening between these lobes, the **porta hepatis,**[16] is a point of entry for the hepatic portal vein and proper hepatic artery and a point of exit for the bile passages, all of which travel in the lesser

14. *falci* = sickle + *form* = shape
15. *caud* = tail
16. *porta* = gateway, entrance + *hepatis* = of the liver

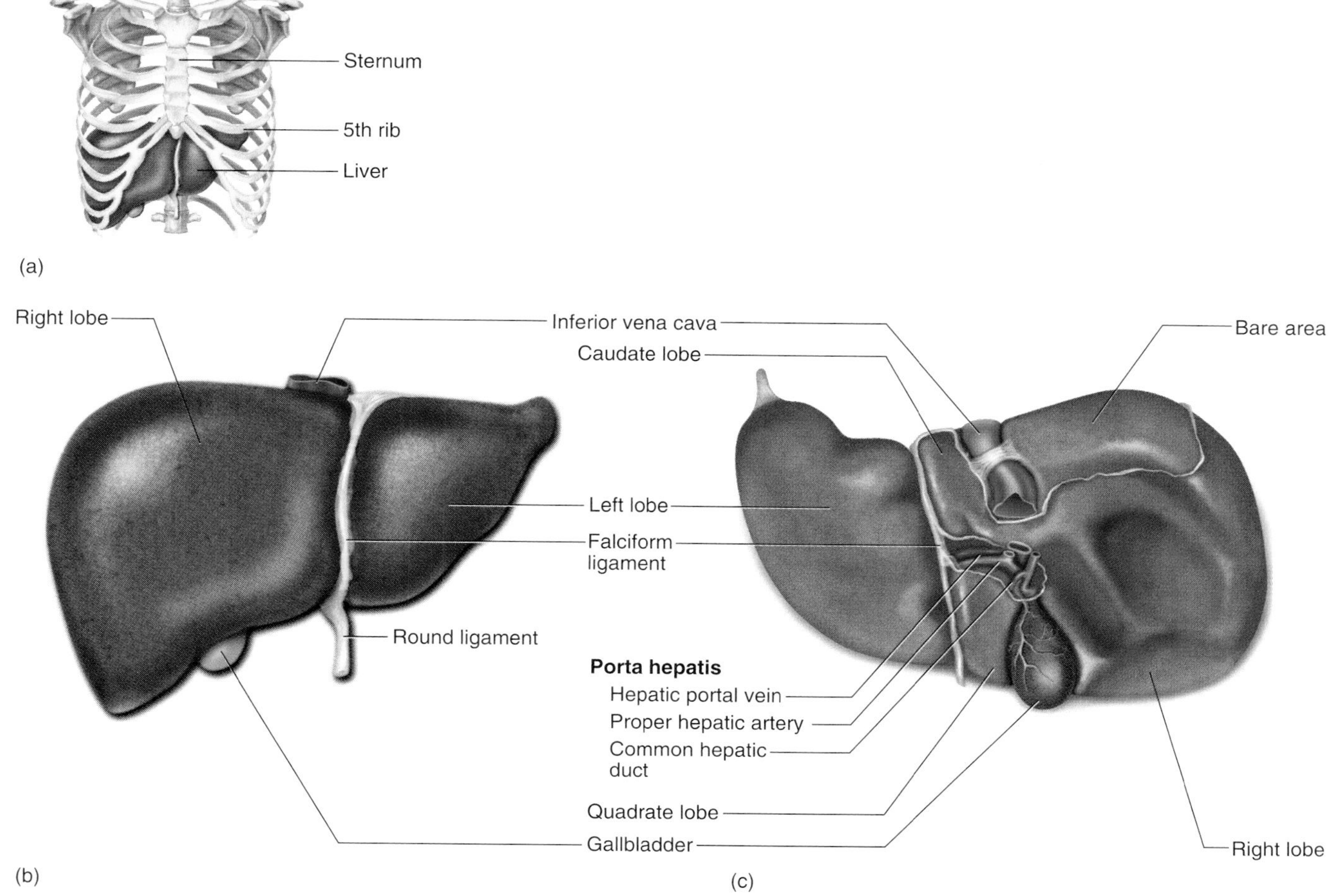

Figure 25.17 Gross Anatomy of the Liver. (*a*) Relationship of the liver to the thoracic cage. (*b*) Anterior aspect. (*c*) Inferior aspect.

omentum. The gallbladder adheres to a depression on the inferior surface of the liver between the right and quadrate lobes. The posterior aspect of the liver has a deep groove (sulcus) that accommodates the inferior vena cava. The superior surface has a *bare area* where it is attached to the diaphragm. The rest of the liver is covered by a serosa.

Microscopic Anatomy

The interior of the liver is filled with innumerable tiny cylinders called **hepatic lobules,** about 2 mm long by 1 mm in diameter. A lobule consists of a **central vein** passing down its core, surrounded by radiating sheets of cuboidal cells called **hepatocytes** (fig. 25.18). Imagine spreading a paperback book wide open until its front and back covers touch. The pages of the book would fan out around the spine somewhat like the plates of hepatocytes fan out from the central vein of a liver lobule.

Each plate of hepatocytes is an epithelium one or two cells thick. The spaces between the plates are blood-filled channels called **hepatic sinusoids.** The sinusoids are lined by a fenestrated endothelium with fenestrations so large that the hepatocytes are in direct contact with the blood. The blood filtering between the hepatocytes comes directly from the intestines. After a meal, as blood filters between the hepatocytes on its way to the central vein, the liver removes glucose, amino acids, iron, vitamins, and other nutrients from it for metabolism or storage. The liver also removes and degrades hormones, toxins, bile pigments, and drugs from the blood. At the same time, the liver secretes albumin, lipoproteins, clotting factors, angiotensinogen, and other proteins into the blood. Between meals, the liver breaks down stored glycogen and releases glucose into the circulation. The sinusoids also contain phagocytic cells called **hepatic macrophages (Kupffer[17] cells),** which remove bacteria and debris from the blood.

The hepatic lobules are separated by a sparse connective tissue stroma. In cross sections, the stroma is especially

17. Karl W. von Kupffer (1829–1902), German anatomist

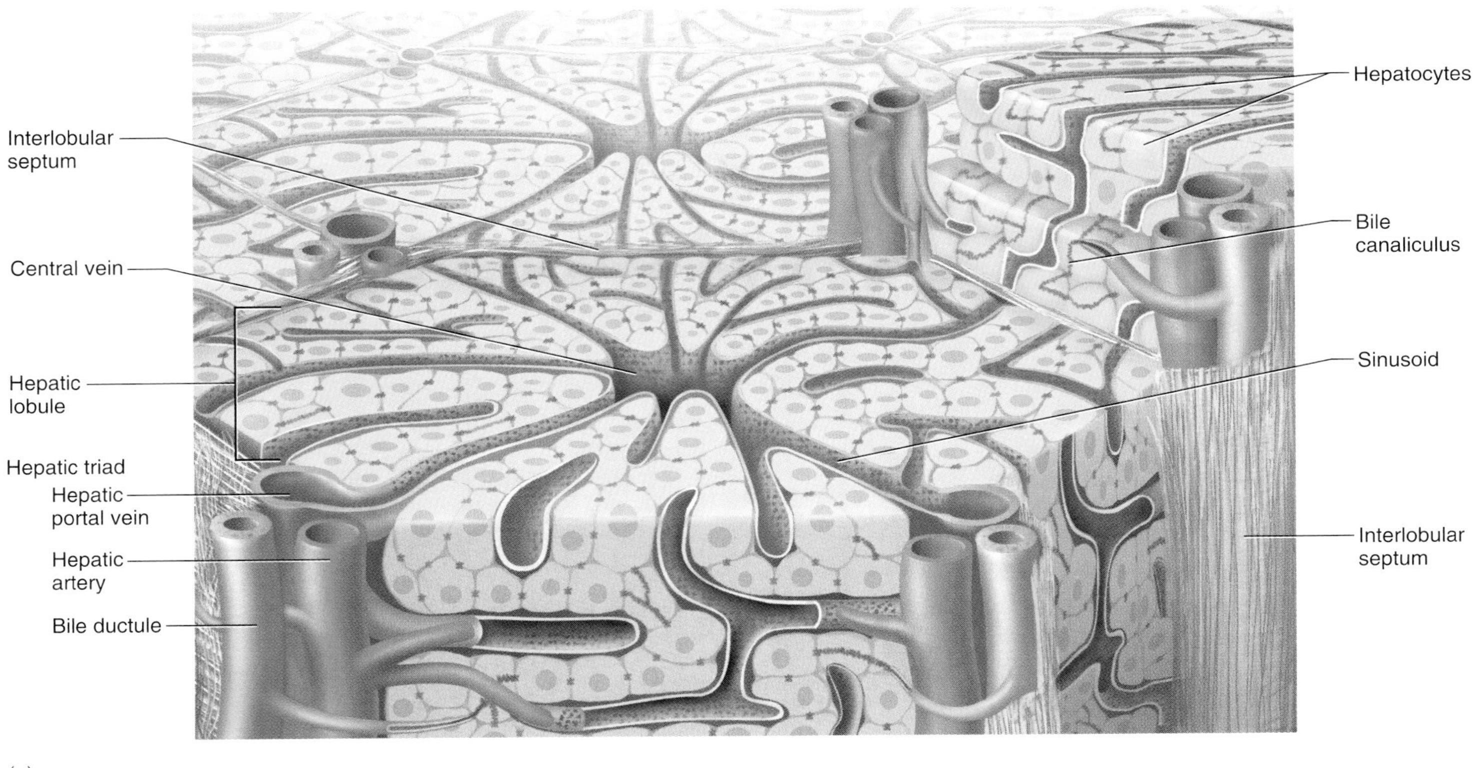

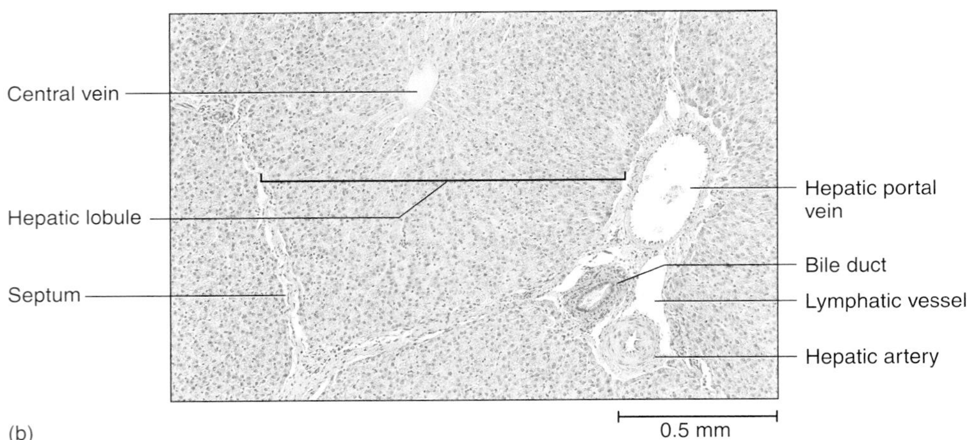

Figure 25.18 Microscopic Anatomy of the Liver. (*a*) The hepatic lobules and their relationship to the blood vessels and bile tributaries. (*b*) Histological section of the liver.

visible in the triangular areas where three or more lobules meet. Here there is often a **hepatic triad** of two blood vessels and a bile ductule. The blood vessels are small branches of the proper hepatic artery and hepatic portal vein. Both of them supply blood to the sinusoids, which therefore receive a mixture of nutrient-laden venous blood from the intestines and freshly oxygenated arterial blood from the celiac trunk. After filtering through the sinusoids, this blood collects in the central vein. From here, it ultimately flows into the right and left hepatic veins, which leave the liver at its superior surface and drain immediately into the inferior vena cava.

The liver secretes bile into narrow channels, the **bile canaliculi,** between sheets of hepatocytes. Bile passes from there into the small **bile ductules** of the triads and ultimately into the **right** and **left hepatic ducts.** The two hepatic ducts converge on the inferior side of the liver to form the **common hepatic duct.** A short distance

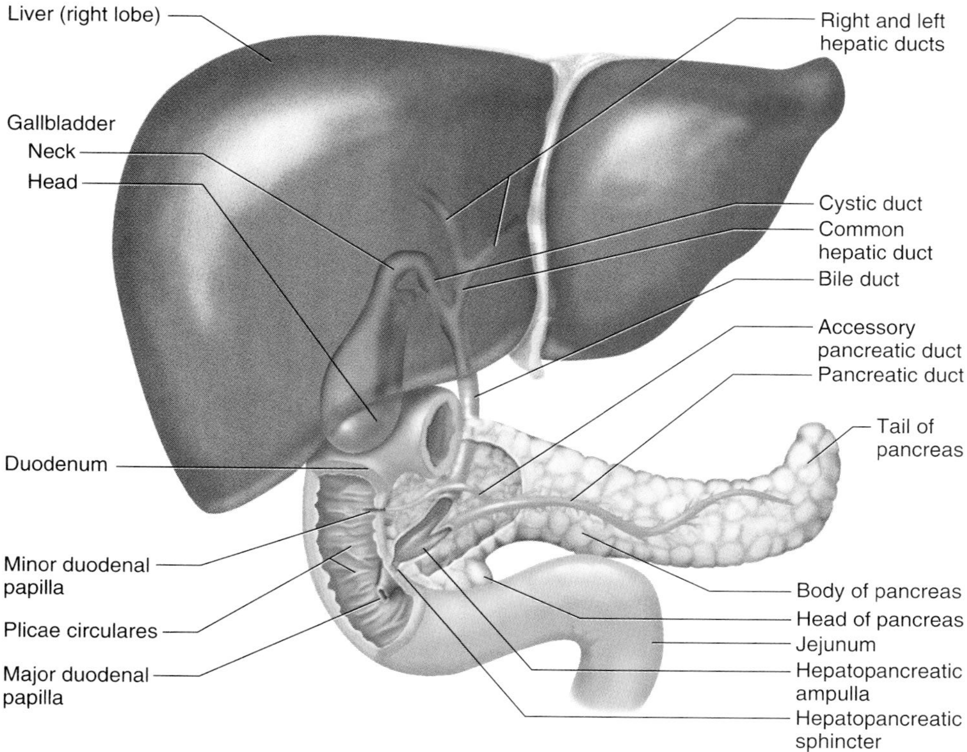

Figure 25.19 Gross Anatomy of the Gallbladder, Pancreas, and Bile Passages.

farther on, this is joined by the **cystic duct** coming from the gallbladder (fig. 25.19). Their union forms the **bile duct,** which descends through the lesser omentum toward the duodenum. Near the duodenum, the bile duct joins the duct of the pancreas and forms an expanded chamber called the **hepatopancreatic ampulla.** The ampulla terminates at a fold of tissue, the **major duodenal papilla,** on the duodenal wall. This papilla contains a muscular **hepatopancreatic sphincter (sphincter of Oddi**[18]**),** which regulates the passage of bile and pancreatic juice into the duodenum. Between meals, this sphincter is closed and prevents the release of bile into the intestine.

The Gallbladder and Bile

The **gallbladder** is a sac on the underside of the liver that serves to store and concentrate bile. It is about 10 cm long and internally lined by a highly folded mucosa with a simple columnar epithelium. Its head (*fundus*) usually projects slightly beyond the inferior margin of the liver. Its neck (*cervix*) leads into the cystic duct, which leads in turn to the bile duct.

18. Ruggero Oddi (1864–1913), Italian physician

The only way bile can get into the gallbladder is by first filling up the bile duct, then overflowing into the gallbladder. Between meals, the gallbladder absorbs water and electrolytes from the bile and concentrates it by a factor of 5 to 20 times. The liver secretes about 500 to 1,000 mL of bile per day.

Bile is a yellow-green fluid containing minerals, cholesterol, neutral fats, phospholipids, bile pigments, and bile acids. The principal pigment is **bilirubin,** derived from the decomposition of hemoglobin. Bacteria of the small intestine metabolize bilirubin to **urobilinogen,** which is responsible for the brown color of feces. In the absence of bile secretion, the feces are grayish white and marked with streaks of undigested fat (*acholic feces*). **Bile acids (bile salts)** are steroids synthesized from cholesterol. Bile acids and lecithin, a phospholipid, aid in fat digestion and absorption, as discussed later. All other components of the bile are wastes destined for excretion in the feces. When these waste products become excessively concentrated, they may form gallstones (see clinical insight 25.3).

About 80% of the bile acids are reabsorbed in the ileum, the last portion of the small intestine, and returned to the liver, where the hepatocytes absorb them and resecrete them. This route of secretion, reabsorption, and resecretion is called the *enterohepatic circulation.* The body's entire supply of bile is reabsorbed and reused two or more

Clinical Insight 25.3

Gallstones

Gallstones (biliary calculi) are hard masses in the gallbladder or bile ducts, usually composed of cholesterol, calcium carbonate, and bilirubin. *Cholelithiasis,* the formation of gallstones, is most common in obese women over the age of 40 and usually results from excess cholesterol. The gallbladder may contain several gallstones, some over 1 cm in diameter. Gallstones cause excruciating pain when they obstruct the bile ducts or when the gallbladder or bile ducts contract. When they block the flow of bile into the duodenum, they cause jaundice (yellowing of the skin due to bile pigment accumulation), poor fat digestion, and impaired absorption of fat-soluble vitamins. Once treated only by surgical removal, gallstones are now often treated with stone-dissolving drugs or by *lithotripsy,* the use of ultrasonic vibration to pulverize them without surgery. The recurrence of gallstones can be prevented by inserting a stent (tube) into the bile duct, which keeps it distended and allows small gallstones to pass.

times during the digestion of an average meal. The 20% of the bile that is not reabsorbed is excreted in the feces. This is the body's only way of eliminating excess cholesterol. The liver synthesizes new bile acids from cholesterol to replace the quantity lost in the feces.

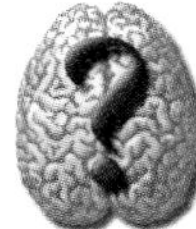

Think About It

Certain drugs designed to reduce blood cholesterol work by blocking the reabsorption of bile acids in the ileum. Explain why they would have this cholesterol-lowering effect.

The Pancreas

The pancreas (fig. 25.19) is a spongy retroperitoneal gland posterior to the greater curvature of the stomach. It has a globose *head* encircled by the duodenum, a midportion called the *body,* and a blunt, tapered *tail* on the left. The pancreas is both an endocrine and exocrine gland. Its endocrine part is the pancreatic islets, which secrete insulin and glucagon (see chapter 17). Most of the pancreas is exocrine tissue, which secretes 1,200 to 1,500 mL of **pancreatic juice** per day. The cells of the secretory acini exhibit a high density of rough ER and *zymogen granules,* which are vesicles filled with secretion (fig. 25.20). These acini open into a system of larger and larger ducts that eventually converge on the main **pancreatic duct.** This duct runs lengthwise through the middle of the gland and joins the common bile duct at the hepatopancreatic ampulla. The hepatopancreatic sphincter thus controls the release of both bile and pancreatic juice into the duodenum. In many people, however, there is a smaller **accessory pancreatic duct** that branches from the main pancreatic duct and opens independently into the duodenum at the **minor duodenal papilla,** proximal to the major papilla. The accessory duct bypasses the sphincter and allows pancreatic juice to be released into the duodenum even when bile is not.

Pancreatic juice is an alkaline mixture of water, enzymes, zymogens, sodium bicarbonate, and other electrolytes. The acini secrete the enzymes and zymogens, whereas the ducts secrete the sodium bicarbonate. The bicarbonate buffers HCl arriving from the stomach.

The pancreatic zymogens are **trypsinogen** (trip-SIN-oh-jen), **chymotrypsinogen** (KY-mo-trip-SIN-o-jen), and **procarboxypeptidase** (PRO-car-BOC-see-PEP-tih-dase). When trypsinogen is secreted into the intestinal lumen, it is converted to **trypsin** by **enterokinase,** an enzyme on the surface of the intestinal epithelial cells (fig. 25.21). Trypsin then converts the other two zymogens into **chymotrypsin** and **carboxypeptidase,** in addition to its primary role of digesting dietary protein.

Other pancreatic enzymes include **pancreatic amylase,** which digests starch; **pancreatic lipase,** which digests fat; and **ribonuclease** and **deoxyribonuclease,** which digest RNA and DNA, respectively. Unlike the zymogens, these enzymes are not altered after secretion. They become active, however, only upon exposure to bile and ions in the intestinal lumen.

The exocrine secretions of the pancreas are summarized in table 25.3. Their specific digestive functions are explained later in more detail.

Regulation of Secretion

Bile and pancreatic juice are secreted in response to parasympathetic (vagal) stimulation and inhibited by sympathetic stimulation, and both are stimulated by the hormones gastrin, cholecystokinin (CCK), and secretin. The duodenum secretes CCK in response to acid and fat arriving from the stomach. CCK triggers three responses: (1) contraction of the gallbladder, which forces bile into the bile duct; (2) secretion of pancreatic enzymes; and (3) relaxation of the hepatopancreatic sphincter, which allows bile and pancreatic juice to be released into the duodenum. Acidic chyme also stimulates the duodenum to secrete **secretin,** the first hormone ever discovered (by

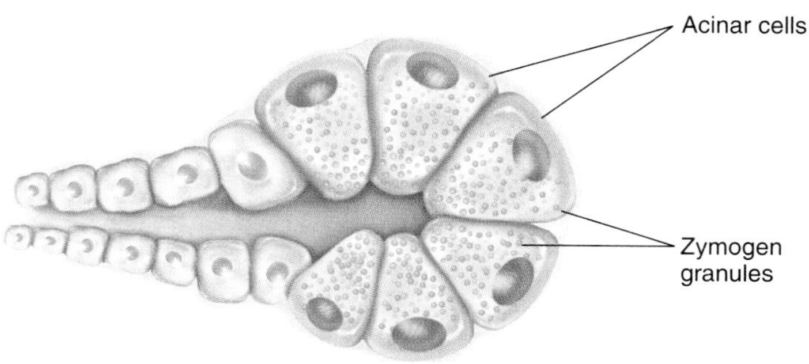

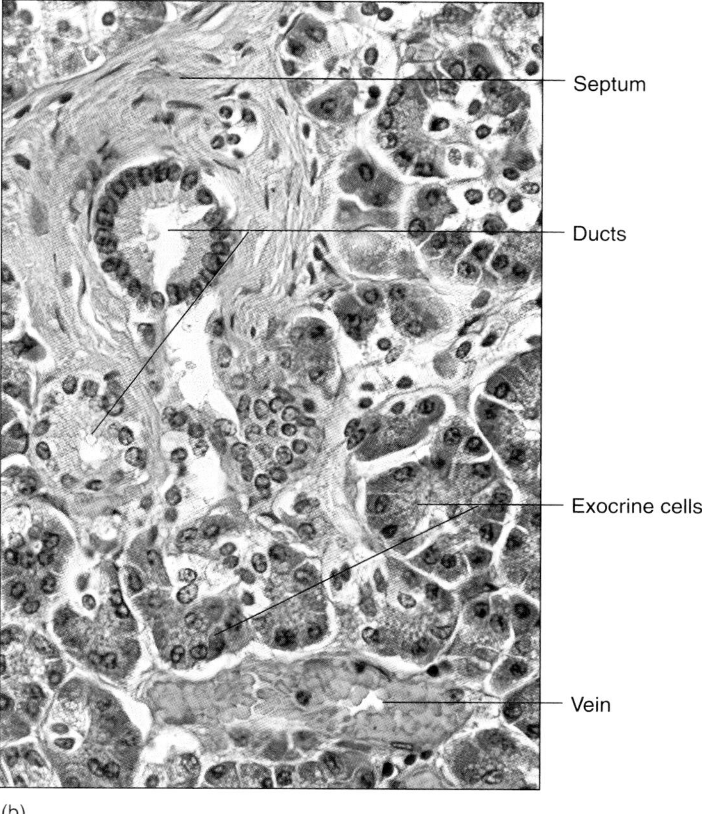

Figure 25.20 Histology of the Pancreas. (*a*) An acinus. (*b*) Histological section of the exocrine tissue and some of the connective tissue stroma.

William Bayliss and Ernest Starling in 1902). Secretin stimulates the hepatic bile ducts and pancreatic ducts to secrete bicarbonate, so the bile and pancreatic juice both help to neutralize stomach acid in the duodenum. Gastrin from the stomach and duodenum stimulates gallbladder contraction and pancreatic enzyme secretion, but only half as strongly as CCK does.

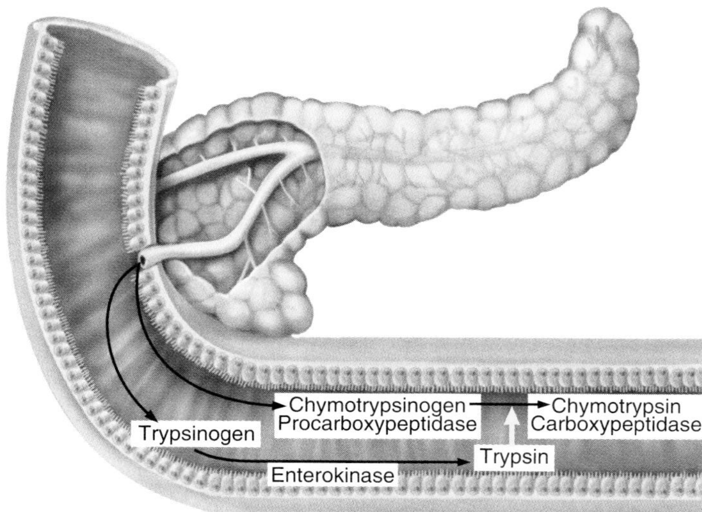

Figure 25.21 The Activation of Pancreatic Enzymes in the Small Intestine. The pancreas secretes trypsinogen, and enterokinase in the intestinal brush border converts it to trypsin. Trypsin not only digests dietary protein but also activates two other pancreatic zymogens, chymotrypsinogen and procarboxypeptidase.

Table 25.3 Exocrine Secretions of the Pancreas

Secretion	Function
Sodium bicarbonate	Neutralizes HCl
Zymogens	Converted to active digestive enzymes after secretion
Trypsinogen	Becomes trypsin, which digests protein
Chymotrypsinogen	Becomes chymotrypsin, which digests protein
Procarboxypeptidase	Becomes carboxypeptidase, which hydrolyzes the terminal amino acid from the carboxyl (–COOH) end of small peptides
Enzymes	
Pancreatic amylase	Digests starch
Pancreatic lipase	Digests fat
Ribonuclease	Digests RNA
Deoxyribonuclease	Digests DNA

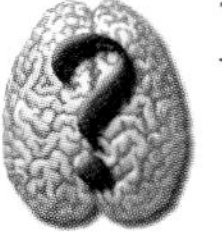

Think About It

Draw a negative feedback loop showing how secretin influences duodenal pH.

Before You Go On

Answer the following questions to test your understanding of the preceding section:

14. What does the liver contribute to digestion?
15. Trace the pathway taken by bile acids from the liver and back. What is this pathway called?
16. Name two hormones, four enzymes, and one buffer secreted by the pancreas, and state the function of each.
17. What stimulates cholecystokinin (CCK) secretion, and how does CCK affect other parts of the digestive system?

The Small Intestine

Objectives

When you have completed this section, you should be able to

- describe the gross and microscopic anatomy of the small intestine;
- state how the mucosa of the small intestine differs from that of the stomach, and explain the functional significance of the differences;
- define *contact digestion* and describe where it occurs; and
- describe the types of movement that occur in the small intestine.

Nearly all chemical digestion and nutrient absorption occur in the small intestine. To perform these roles efficiently, the small intestine must have a large surface area exposed to the chyme. Thus, it is the longest part of the digestive tract—about 6 to 7 m long in a cadaver, but because of muscle tonus, only 2 m long in a living person. The term *small* intestine refers not to its length but to its diameter—about 2.5 cm (1 in.). Further enhancing its surface area, the mucosa of the small intestine is highly folded, with *circular folds* visible to the naked eye, barely visible projections called *villi*, and microscopic *microvilli* forming a brush border on its absorptive cells.

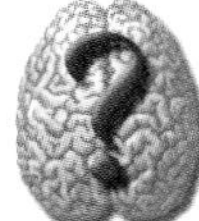

Think About It

The small intestine exhibits some of the same structural adaptations as the proximal convoluted tubule of the kidney, and for the same reasons. Discuss what they have in common, the reasons for it, and how this relates to this book's theme of the unity of form and function.

Gross Anatomy

The small intestine is a coiled mass filling most of the abdominal cavity inferior to the stomach and liver. It is divided into three regions (fig. 25.22): the duodenum, jejunum, and ileum. The **duodenum** (dew-ODD-eh-num, DEW-oh-DEE-num) constitutes the first 25 cm (10 in.). It begins at the pyloric valve, arcs around the head of the pancreas and passes to the left, and ends at a sharp bend called the **duodenojejunal flexure.** Along with the pancreas, it is retroperitoneal. Its name refers to its length, about equal to the width of 12 fingers.[19] The duodenum receives the stomach contents, pancreatic juice, and bile. Stomach acid is neutralized here, fats are physically broken up (emulsified) by the bile acids, pepsin is inactivated by the elevated pH, and pancreatic enzymes take over the job of chemical digestion.

The **jejunum** (jeh-JOO-num) comprises the next 2.5 m (8 ft). Its name refers to the fact that early anatomists typically found it to be empty.[20] The **ileum**[21] forms the last 3.6 m (12 ft). (These lengths are for the cadaver.) The jejunum and ileum are not separated by any conspicuous anatomical landmark, but the jejunum is located largely toward the upper left of the intestinal coils and the ileum to the lower right. The ileum ends at the **ileocecal** (ILL-ee-oh-SEE-cul) **junction,** where it joins the cecum, the first part of the large intestine. The jejunum and ileum lie within the peritoneal cavity and are covered externally with a serosa, which is continuous with the complex, folded mesentery that suspends the small intestine from the dorsal abdominal wall.

19. *duoden* = 12
20. *jejun* = empty, dry
21. from *eilos* = twisted

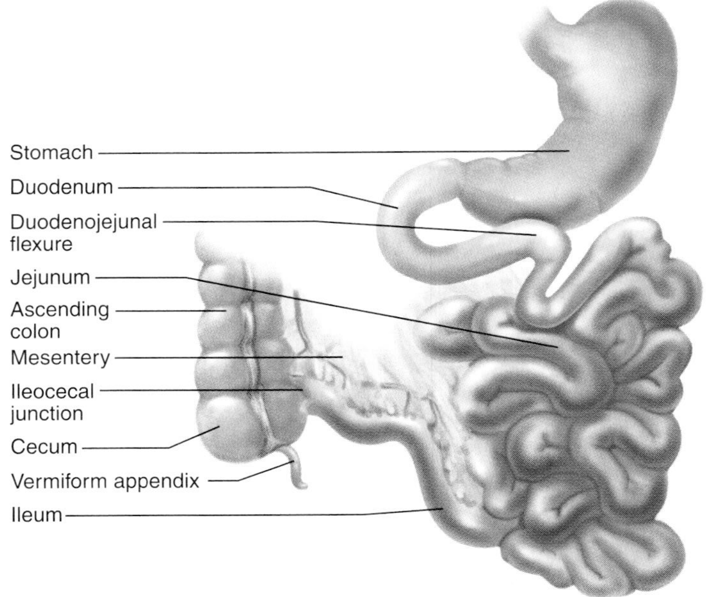

Figure 25.22 Gross Anatomy of the Small Intestine. The intestine is pulled aside to expose the mesentery and ileocecal junction.

Microscopic Anatomy

The largest folds of the intestinal wall are transverse to spiral ridges, up to 10 mm high, called **circular folds** (*plicae circulares*) (see fig. 25.19). These involve only the mucosa and submucosa; they are not visible on the external surface, which is smooth. They occur from the duodenum to the middle of the ileum, where they cause the chyme to flow on a spiral path along the intestine. This slows its progress, causes more contact with the mucosa, and promotes more thorough mixing and nutrient absorption. Circular folds are not found in the distal half of the ileum, but most nutrient absorption is completed by that point.

If the mucosa is examined closely it appears fuzzy, like a terrycloth towel. This is due to projections called **villi** (VIL-eye; singular, *villus*), about 0.5 to 1.0 mm high, with tongue- to fingerlike shapes (fig. 25.23). The villi are largest in the duodenum and become progressively smaller in more distal regions of the small intestine. A villus is covered with two kinds of epithelial cells—columnar **absorptive cells** and mucus-secreting **goblet cells.** Like epithelial cells of the stomach, those of the small intestine are joined by tight junctions that prevent digestive enzymes from seeping between them.

The core of a villus is filled with areolar tissue of the lamina propria. Embedded in this tissue are an arteriole, a capillary network, a venule, and a lymphatic capillary called a **lacteal** (LAC-tee-ul). Most nutrients are absorbed by the blood capillaries, but most fat is absorbed by the lacteal and gives its contents the milky appearance for which the lacteal is named.[22] The core of the villus also has a few fibers of smooth muscle that contract periodically. This enhances mixing of the chyme in the intestinal lumen and milks lymph down the lacteal to the larger lymphatic vessels of the submucosa.

Each absorptive cell of a villus has a fuzzy brush border of microvilli about 1 μm high. The brush border increases the absorptive surface area of the small intestine and contains **brush border enzymes,** integral proteins of the plasma membrane. One of these, enterokinase, activates pancreatic enzymes as explained earlier. Others carry out some of the final stages of enzymatic digestion. They are not released into the lumen; instead, the chyme must contact the brush border for digestion to occur. This process, called **contact digestion,** is one reason that thorough mixing of the chyme is so important.

On the floor of the small intestine, between the bases of the villi, there are numerous pores that open into tubular glands called **intestinal crypts (crypts of Lieberkühn;**[23] LEE-ber-koohn). These crypts, similar to the gastric glands, extend as far as the muscularis mucosae. In the upper half they consist of absorptive and goblet cells like those of the villi. The lower half is dominated by dividing epithelial cells. In its life span of 3 to 6 days, an epithelial cell migrates up the crypt to the tip of the villus, where it is sloughed off and digested. A few **Paneth**[24] **cells** are clustered at the base of each crypt. Their function is uncertain, but they apparently secrete lysozyme as a defense against bacteria.

The duodenum has prominent **duodenal (Brunner**[25]**) glands** in the submucosa. They secrete an abundance of bicarbonate-rich mucus, which neutralizes stomach acid and shields the mucosa from its corrosive effects. Throughout the small intestine, the lamina propria and submucosa have a large population of lymphocytes that intercept pathogens before they can invade the bloodstream. In some places these are aggregated into conspicuous lymphatic nodules. These become more and more numerous closer to the large intestine, where the bacterial population is greatest. In the distal portion of the ileum, the nodules form ever-changing aggregates called **Peyer**[26] **patches** on one side of the intestine.

The muscularis externa consists of a relatively thick inner circular layer and a thinner outer longitudinal layer. Ganglia of the myenteric nerve plexus occur between these layers.

Intestinal Secretion

The intestinal crypts secrete 1 to 2 L of **intestinal juice** per day, especially in response to acid, hypertonic chyme, and distension of the intestine. This fluid has a pH of 7.4 to 7.8. It contains water and mucus but relatively little enzyme. Most enzymes that function in the small intestine are found in the brush border and pancreatic juice.

Intestinal Motility

Contractions of the small intestine serve three functions: (1) to mix chyme with intestinal juice, bile, and pancreatic juice, allowing these fluids to neutralize acid and digest nutrients more effectively; (2) to churn chyme and bring it into contact with the mucosa for contact digestion and nutrient absorption; and (3) to move residue toward the large intestine.

Segmentation is a movement in which ringlike constrictions appear at several places along the intestine and then relax as new constrictions form elsewhere (fig. 25.24*a*). This is the most common movement of the small intestine. Its effect is to knead or churn the contents, which promotes the mixing of food and digestive secretions and enhances contact digestion. Pacemaker cells of the muscularis externa set the rhythm of segmentation, with contractions

22. *lact* = milk
23. Johann N. Lieberkühn (1711–56), German anatomist
24. Josef Paneth (1857–90), Austrian physician
25. Johann C. Brunner (1653–1727), Swiss anatomist
26. Johann K. Peyer (1653–1712), Swiss anatomist

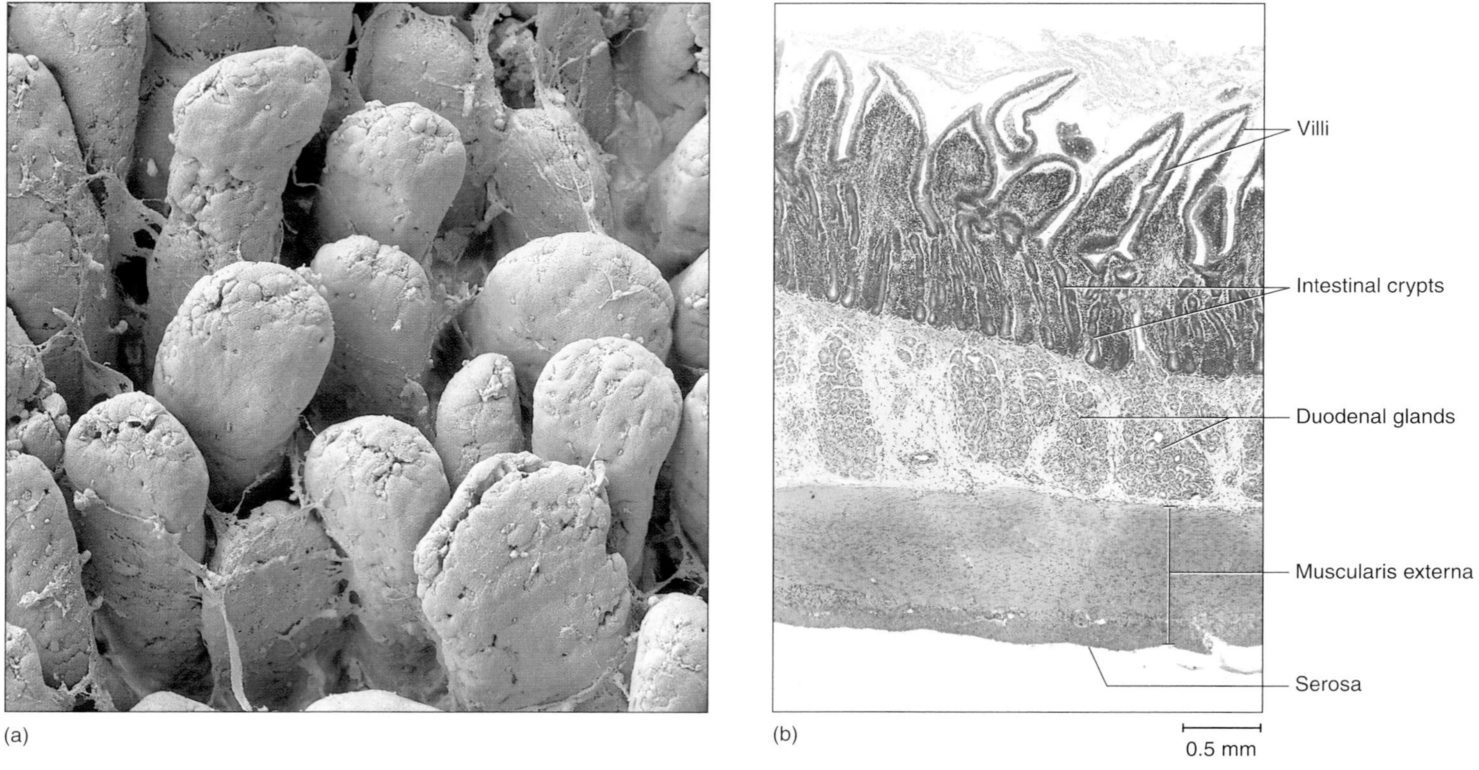

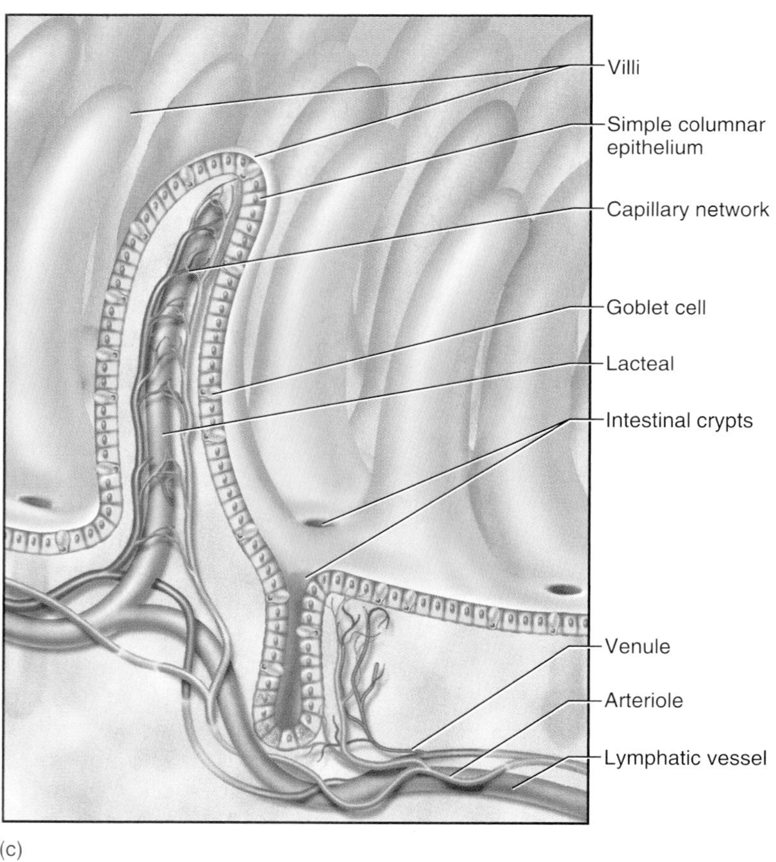

Figure 25.23 **Intestinal Villi.** (*a*) Villi of the small intestine (SEM). (*b*) Histological section of the duodenum showing villi, intestinal crypts, and duodenal glands. (*c*) Structure of a villus.

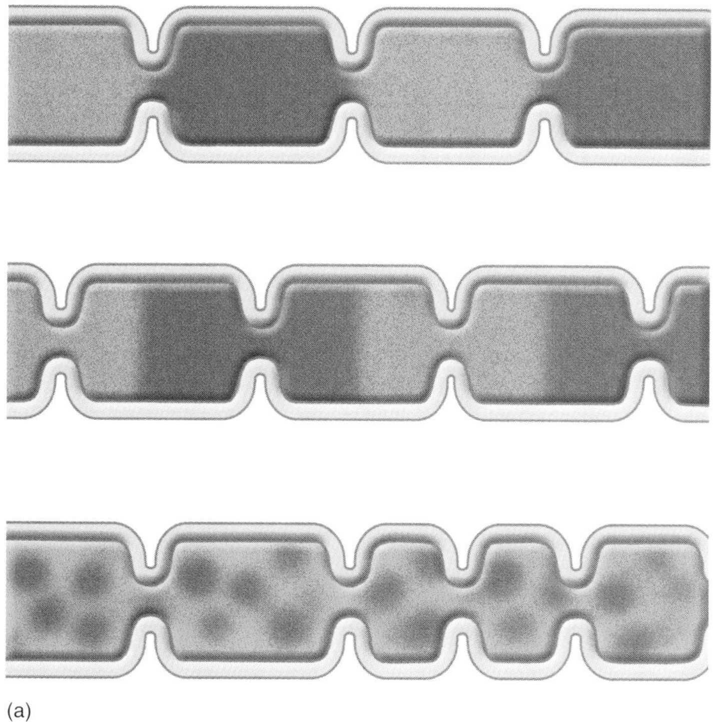

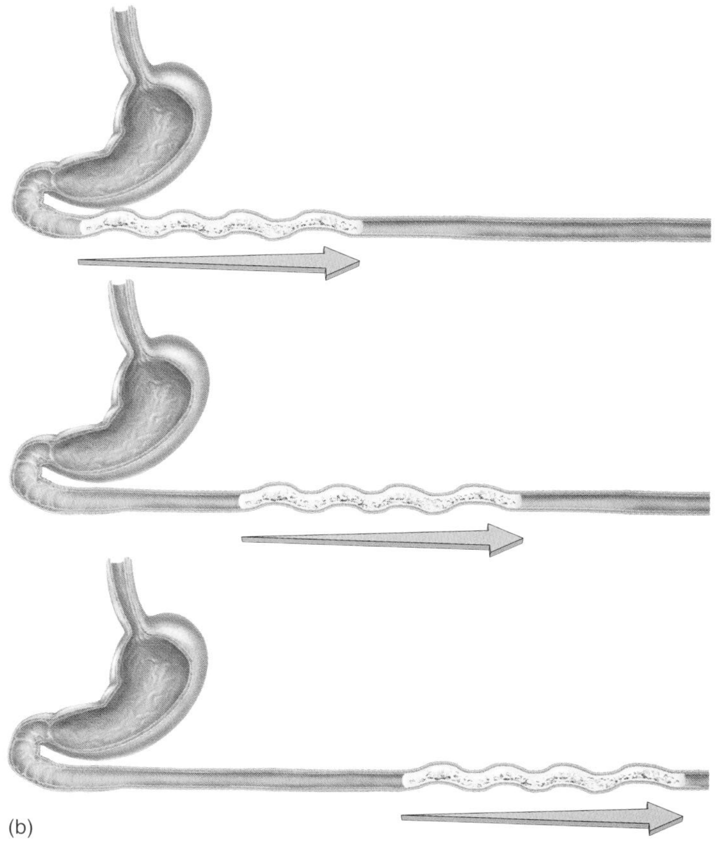

Figure 25.24 **Contractions of the Small Intestine.** (*a*) Segmentation, in which circular constrictions of the intestine cut into the contents, churning and mixing them. (*b*) The migrating motor complex of peristalsis, in which successive waves of peristalsis overlap each other. Each wave travels partway down the intestine and milks the contents toward the colon.

about 12 times per minute in the duodenum and 8 to 9 times per minute in the ileum. Since the contractions are less frequent distally, segmentation causes slow progression of the chyme toward the colon. The intensity (but not frequency) of contractions is modified by nervous and hormonal influences.

When most nutrients have been absorbed and little remains but undigested residue, segmentation declines and peristalsis begins. A peristaltic wave begins in the duodenum, travels 10 to 70 cm, and dies out, only to be followed by another wave that begins a little farther down the tract than the first one did (fig. 25.24*b*). These successive, overlapping waves of contraction are called a **migrating motor complex.** They milk the chyme toward the colon over a period of about 2 hours. A second complex then expels residue and bacteria from the small intestine, thereby helping to limit bacterial colonization. Refilling of the stomach at the next meal suppresses peristalsis and reactivates segmentation.

At the ileocecal junction, the muscularis of the ileum is thickened to form a sphincter, the **ileocecal** (ILL-ee-oh-SEE-cul) **valve,** which protrudes into the cecum like a doughnut (fig. 25.25). This valve is usually closed. Food in

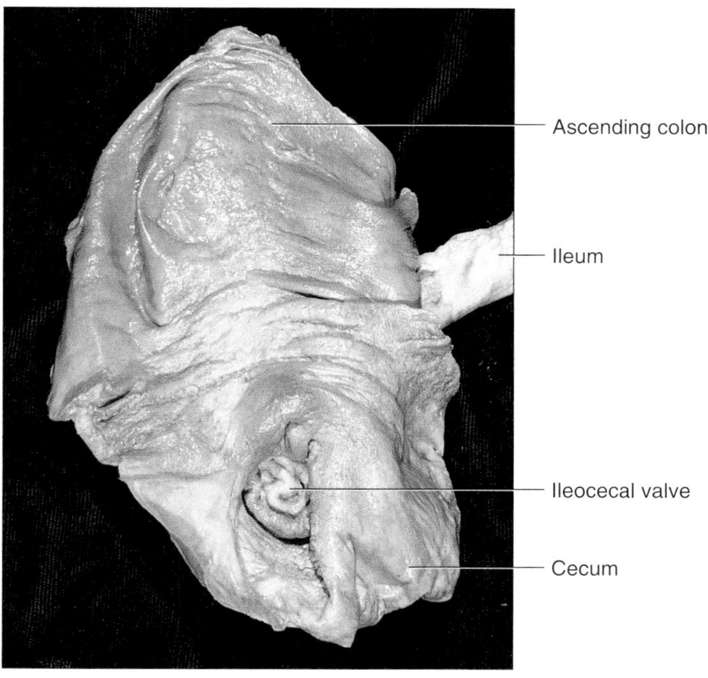

Figure 25.25 **The Cecum, Cut and Folded Open to Show the Ileocecal Valve.**

Table 25.4 Anatomical Checklist of the Digestive System from the Stomach Through the Small Intestine

Stomach

- *General features*
  - Lesser curvature
  - Greater curvature
  - Cardiac region
  - Fundic region
  - Body (corpus)
  - Pyloric region
    - Antrum
    - Pyloric canal
    - Pylorus
    - Pyloric sphincter
- *Mucosa*
  - Rugae
  - Gastric pits
  - Cardiac glands
  - Pyloric glands
  - Gastric glands
    - Mucous neck cells
    - Regenerative cells
    - Parietal cells
    - Chief cells
    - Enteroendocrine cells
- *Submucosa*
- *Muscularis externa*
- *Serosa*

Liver

- *Lobes*
  - Right lobe
  - Left lobe
  - Quadrate lobe
  - Caudate lobe
- *Ligaments*
  - Falciform ligament
  - Round ligament
- *Porta hepatis*
- *Microscopic anatomy*
  - Hepatocytes
  - Hepatic lobules
  - Central vein
  - Hepatic sinusoids
  - Hepatic macrophages
  - Hepatic triads
- *Bile tributaries*
  - Bile canaliculi
  - Bile ductules
  - Hepatic ducts
  - Common hepatic duct
- *Gallbladder and bile duct*
  - Head (fundus)
  - Neck (cervix)
  - Cystic duct
  - Bile duct
  - Hepatopancreatic ampulla
  - Hepatopancreatic sphincter

Pancreas

- *Head*
- *Body*
- *Tail*
- *Exocrine acini*
- *Pancreatic islets*
- *Pancreatic duct*
- *Accessory pancreatic duct*

Small Intestine

- *Duodenum*
  - Major duodenal papilla
  - Minor duodenal papilla
  - Duodenal glands
  - Duodenojejunal flexure
- *Jejunum*
- *Ileum*
  - Peyer patches
  - Ileocecal valve
  - Ileocecal junction
- *Mucosa*
  - Circular folds
  - Villi
    - Goblet cells
    - Absorptive cells
    - Microvilli
    - Lacteal
  - Intestinal crypts
    - Paneth cells

the stomach, however, triggers both the release of gastrin and the **gastroileal reflex,** both of which enhance segmentation in the ileum and relax the valve. As the cecum fills with residue, the pressure pinches the valve shut and prevents the reflux of cecal contents into the ileum.

Table 25.4 summarizes the anatomy of the stomach, accessory glands of the abdomen, and small intestine.

Before You Go On

Answer the following questions to test your understanding of the preceding section:

18. What three structures increase the absorptive surface area of the small intestine?
19. Sketch a villus and label its epithelium, brush border, lamina propria, blood capillaries, and lacteal.
20. Distinguish between segmentation and the migrating motor complex of the small intestine. How do these differ in function?

Chemical Digestion and Absorption

Objectives

When you have completed this section, you should be able to

- describe how each major class of nutrients is chemically digested, name the enzymes involved, and discuss the functional differences among these enzymes; and
- describe how each type of nutrient is absorbed by the small intestine.

Chemical digestion and nutrient absorption are essentially finished by the time food residue leaves the small intestine and enters the cecum. Bacteria in the large intestine provide some nutrients that we consider later. But before going on to the functions of the large intestine, we trace each major class of nutrients—especially carbohydrates, proteins, and fats—from the mouth through the small intestine to see how it is chemically degraded and absorbed.

Clinical Insight 25.4

Lactose Intolerance

Humans are a strange species. Unique among mammals, we go on drinking milk in adulthood, and moreover, we drink the milk of other species! This odd habit is largely limited, however, to Europeans, a few pastoral tribes of Africa, and their descendants. Only they have an ancestral history of milking domestic animals, a practice that goes back about 10,000 years and has coincided with the continued production of lactase into adulthood.

People without lactase have *lactose intolerance.* If they consume milk, lactose passes undigested into the large intestine, increases the osmolarity of the intestinal contents, and causes colonic water retention and diarrhea. In addition, lactose fermentation by intestinal bacteria produces gas, resulting in painful cramps and flatulence.

Lactose intolerance occurs in about 15% of American whites, nearly all people of Asian descent, and about 90% of American blacks, who are predominantly descended from nonpastoral African tribes. People with lactose intolerance can consume products such as yogurt and cheese, in which bacteria have broken down the lactose, and they can digest milk with the aid of lactase drops or tablets.

Carbohydrates

Most of the digestible dietary carbohydrate is starch. Cellulose is indigestible and is not considered here, although its importance as dietary fiber is discussed in the next chapter. The amount of glycogen in the diet is negligible, but it is digested in the same manner as starch.

Carbohydrate Digestion

Starch is digested first to oligosaccharides two to eight glucose residues long, then into the disaccharide maltose, and finally to glucose, which is absorbed by the small intestine. The process begins in the mouth, where salivary amylase hydrolyzes starch into oligosaccharides. Salivary amylase functions best at pH 6.8 to 7.0, typical of the oral cavity. It is quickly denatured upon contact with stomach acid, but it can digest starch for as long as 1 to 2 hours in the stomach as long as it is in the middle of a food mass and escapes contact with the acid. Amylase therefore works longer when the meal is larger, especially in the fundus, where gastric motility is weakest and a food bolus takes longer to break up. As acid, pepsin, and the churning contractions of the stomach break up the bolus, amylase is denatured; it does not function at a pH any lower than 4.5. Being a protein, amylase is then digested by pepsin along with the dietary proteins. About 50% of the dietary starch is digested before it reaches the small intestine.

Starch digestion resumes in the small intestine when the chyme mixes with pancreatic amylase (fig. 25.26). It is entirely converted to oligosaccharides and maltose within 10 minutes. Its digestion is completed as the chyme contacts the brush border of the absorptive cells. Two brush border enzymes, **dextrinase** and **glucoamylase,** hydrolyze oligosaccharides that are three or more residues long. The third, **maltase,** hydrolyzes maltose to glucose.

Maltose is also present in some foods, but the major dietary disaccharides are sucrose (cane sugar) and lactose (milk sugar). They are digested by the brush border enzymes **sucrase** and **lactase,** respectively, and the resulting monosaccharides (glucose and fructose from the former; glucose and galactose from the latter) are immediately absorbed. In most humans, however, lactase is no longer produced after the age of four and lactose is indigestible past that age (see clinical insight 25.4).

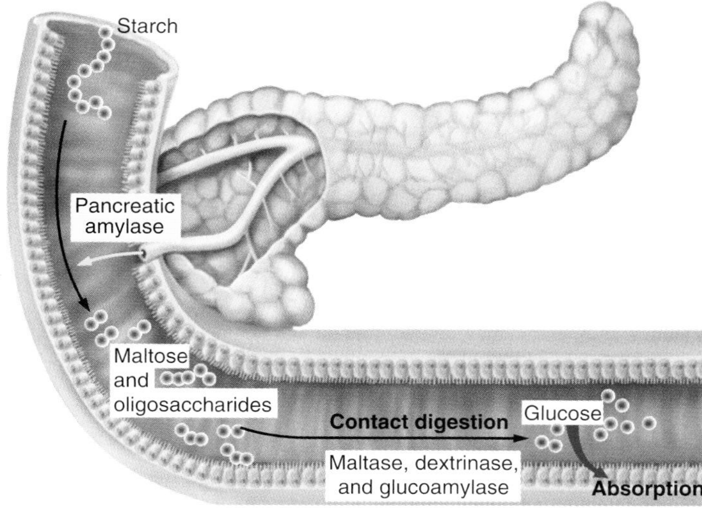

Figure 25.26 **Starch Digestion in the Small Intestine.** Pancreatic amylase digests starch into maltose and small oligosaccharides. Brush border enzymes digest these to glucose, which is absorbed by the epithelial cells.

Carbohydrate Absorption

In the plasma membrane of the absorptive cells, there are transport proteins that absorb monosaccharides as soon as the brush border enzymes release them (fig. 25.27). About 80% of the absorbed sugar is glucose, which is taken up by

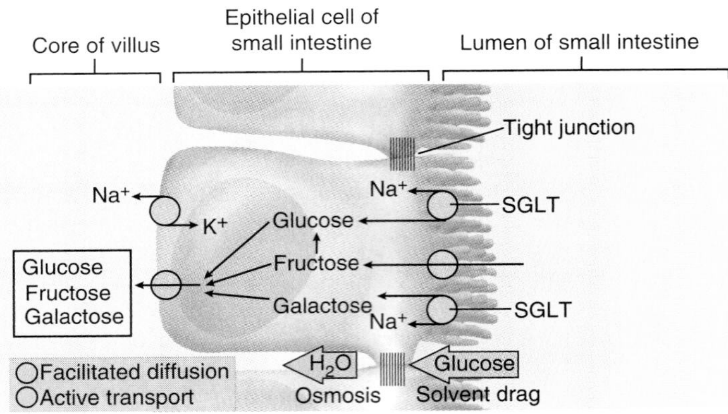

Figure 25.27 Monosaccharide Absorption by the Small Intestine. Glucose and galactose are absorbed by the SGLT cotransporter in the apical membrane of the absorptive cell (*right*). Glucose is also absorbed along with water through the paracellular route (between cells) by solvent drag. Fructose is absorbed by a separate facilitated diffusion carrier. Most fructose is converted to glucose within the epithelial cell. The monosaccharides pass through the basal membrane of the cell by facilitated diffusion (*left*).

a sodium-glucose transport protein (SGLT) like that of the kidney tubules (see p. 896). After a high-carbohydrate meal, however, solvent drag absorbs two to three times as much glucose as the SGLT. Sugars entering the extracellular fluid (ECF) at the base of the intestinal epithelium increase its osmolarity. Water then passes osmotically from the lumen, through the now-leaky tight junctions between the epithelial cells, and into the ECF, carrying glucose and other nutrients with it.

The SGLT also absorbs galactose, whereas fructose is absorbed by facilitated diffusion using a separate carrier that does not depend on Na^+. Inside the epithelial cell, most fructose is converted to glucose. Glucose, galactose, and the small amount of remaining fructose are then transported out the base of the cell by facilitated diffusion and are absorbed by the blood capillaries of the villus. The hepatic portal system delivers them to the liver; chapter 26 follows the fate of these sugars from there.

Proteins

The amino acids absorbed by the small intestine come from three sources: (1) dietary proteins, (2) digestive enzymes digested by each other, and (3) sloughed epithelial cells digested by these enzymes. The endogenous amino acids from the last two sources total about 30 g/day, compared with a recommended dietary intake of 44 to 60 g.

Enzymes that digest proteins are called **proteases (peptidases).** They are absent from the saliva but begin work in the stomach. Here, pepsin hydrolyzes any peptide bond between tyrosine and phenylalanine, thus digesting 10% to 15% of the dietary protein into shorter polypeptides and a small amount of free amino acids (fig. 25.28). Pepsin has an optimal pH of 1.5 to 3.5; thus it is inactivated when it passes into the duodenum and mixes with the alkaline pancreatic juice (pH 8).

In the small intestine, the pancreatic enzymes trypsin and chymotrypsin take over protein digestion by hydrolyzing polypeptides into even shorter oligopeptides. Finally, these are taken apart one amino acid at a time by three more enzymes: (1) **carboxypeptidase** removes amino acids from the –COOH end of the chain; (2) **aminopeptidase** removes them from the $-NH_2$ end; and (3) **dipeptidase** splits dipeptides in the middle and releases the last two free amino acids. All three of these are brush border enzymes, while carboxypeptidase also occurs in the pancreatic juice.

Amino acid absorption is similar to that of monosaccharides. There are several sodium-dependent amino acid cotransporters for different classes of amino acids. Dipeptides and tripeptides can also be absorbed, but they are hydrolyzed within the cytoplasm of the epithelial cells before their amino acids are released to the bloodstream. At the basal surfaces of the cells, amino acids behave like the monosaccharides discussed previously—they leave the cell by facilitated diffusion, enter the capillaries of the villus, and are thus carried away in the hepatic portal circulation.

The absorptive cells of infants can take up intact proteins by pinocytosis and release them to the blood by exocytosis. This allows IgA from breast milk to pass into an infant's bloodstream and confer passive immunity from mother to infant. It has the disadvantage, however, that intact proteins entering the infant's blood are detected as foreign antigens and sometimes trigger food allergies. As the intestine matures, its ability to pinocytose protein declines but never completely ceases.

Lipids

The hydrophobic quality of lipids makes their digestion and absorption more complicated than that of carbohydrates and proteins (fig. 25.29). Fats are digested by enzymes called **lipases.** *Lingual lipase,* secreted by the intrinsic salivary glands of the tongue, is activated by acid in the stomach, where it digests as much as 10% of the ingested fat. In infants, the stomach also secretes *gastric lipase.* Most fat digestion, however, occurs in the small intestine through the action of *pancreatic lipase.*

As chyme enters the duodenum, its fat is in large globules exposed to lipase only at their surfaces. Fat digestion would be rather slow and inefficient if it remained this way, but instead, it is broken up into smaller **emulsification droplets** by certain components of the bile—lecithin (a

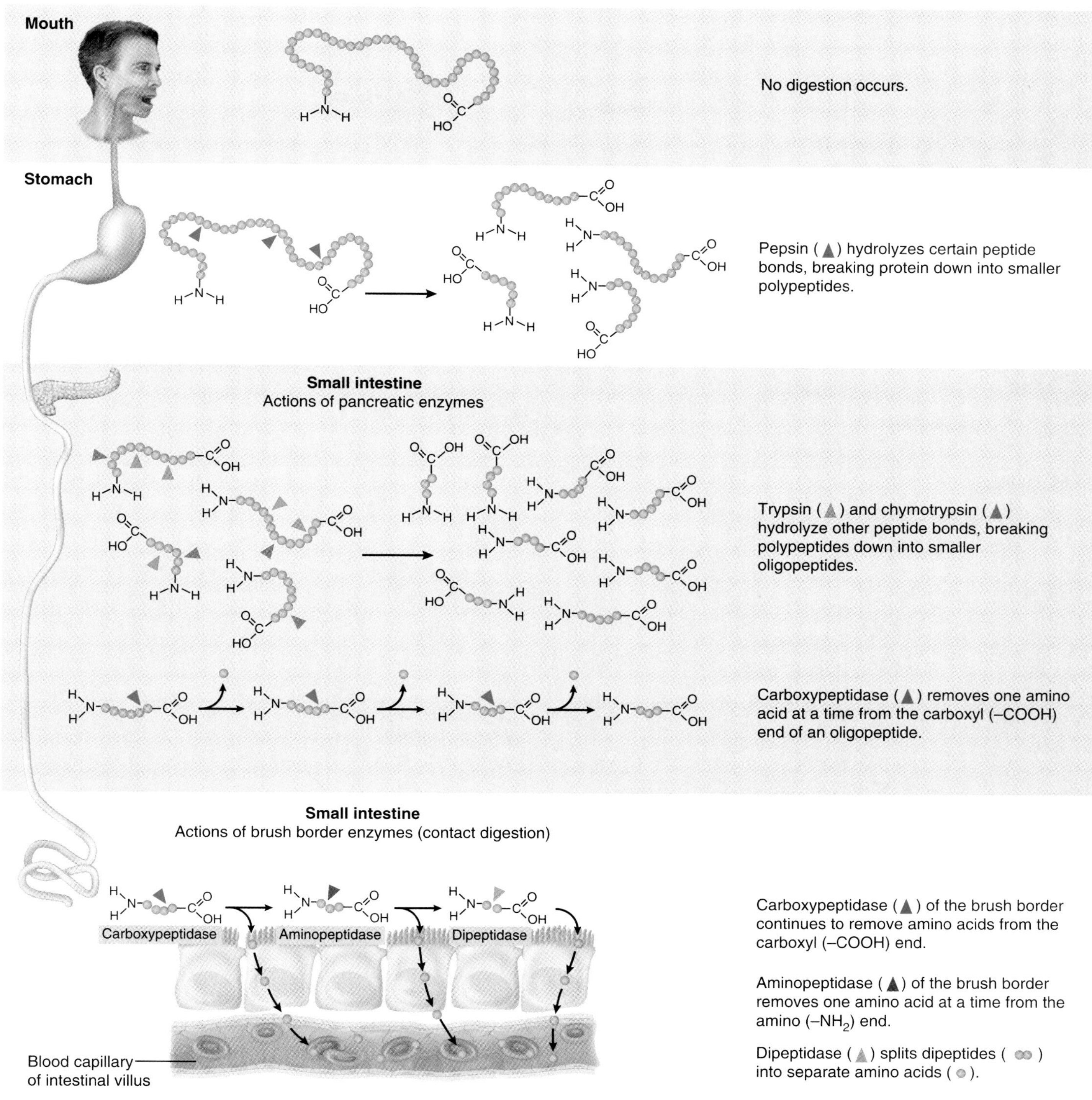

Figure 25.28 Protein Digestion and Absorption.

phospholipid) and bile acids (steroids). These agents have hydrophobic regions attracted to the surface of a fat globule and hydrophilic regions attracted to the surrounding water. The agitation produced by intestinal segmentation contractions breaks the fat up into droplets as small as 1 μm in diameter, and a coating of lecithin and bile acids keeps it broken up, exposing far more of its surface to enzymatic action. There is enough pancreatic lipase in the small intestine after a meal to digest the average total daily fat intake in as little as 1 or 2 minutes.

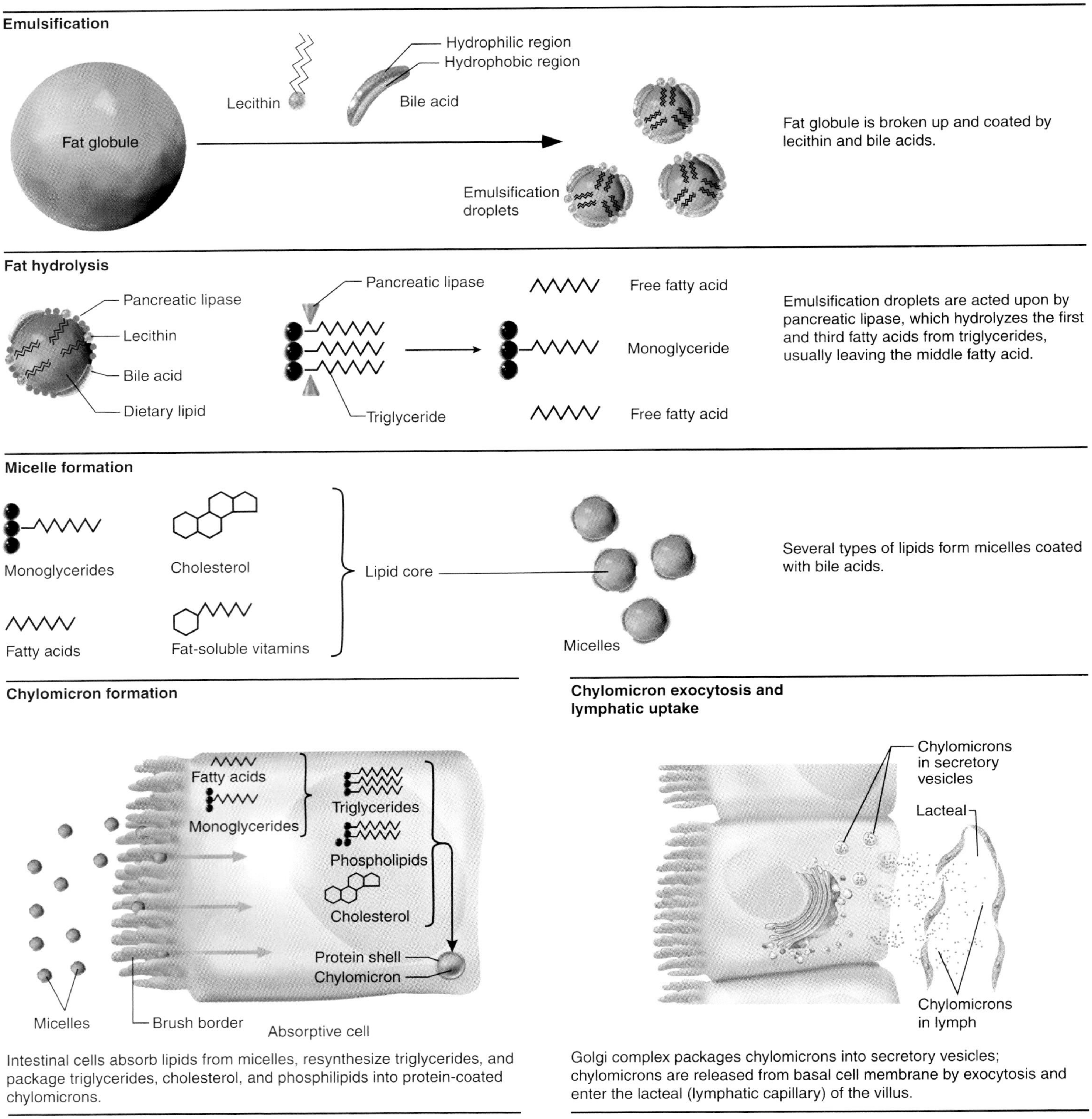

Emulsification
Lecithin
Hydrophilic region
Hydrophobic region
Bile acid
Fat globule
Emulsification droplets
Fat globule is broken up and coated by lecithin and bile acids.
Fat hydrolysis
Pancreatic lipase
Lecithin
Bile acid
Dietary lipid
Pancreatic lipase
Triglyceride
Free fatty acid
Monoglyceride
Free fatty acid
Emulsification droplets are acted upon by pancreatic lipase, which hydrolyzes the first and third fatty acids from triglycerides, usually leaving the middle fatty acid.
Micelle formation
Monoglycerides
Cholesterol
Fatty acids
Fat-soluble vitamins
Lipid core
Micelles
Several types of lipids form micelles coated with bile acids.
Chylomicron formation
Fatty acids
Monoglycerides
Triglycerides
Phospholipids
Cholesterol
Protein shell
Chylomicron
Micelles
Brush border
Absorptive cell
Intestinal cells absorb lipids from micelles, resynthesize triglycerides, and package triglycerides, cholesterol, and phosphilipids into protein-coated chylomicrons.
Chylomicron exocytosis and lymphatic uptake
Chylomicrons in secretory vesicles
Lacteal
Chylomicrons in lymph
Golgi complex packages chylomicrons into secretory vesicles; chylomicrons are released from basal cell membrane by exocytosis and enter the lacteal (lymphatic capillary) of the villus.

Figure 25.29 Fat Digestion and Absorption.

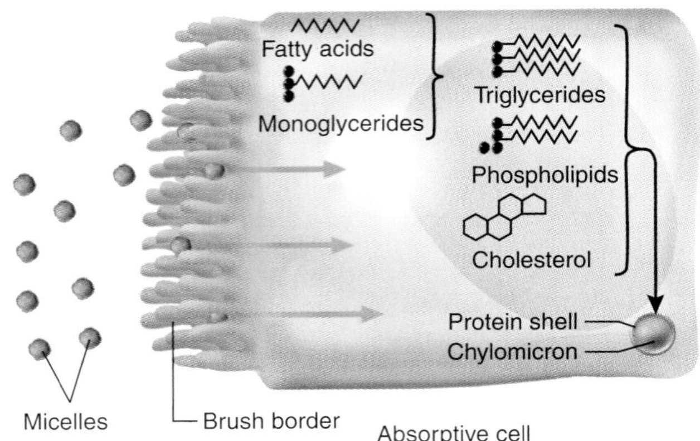

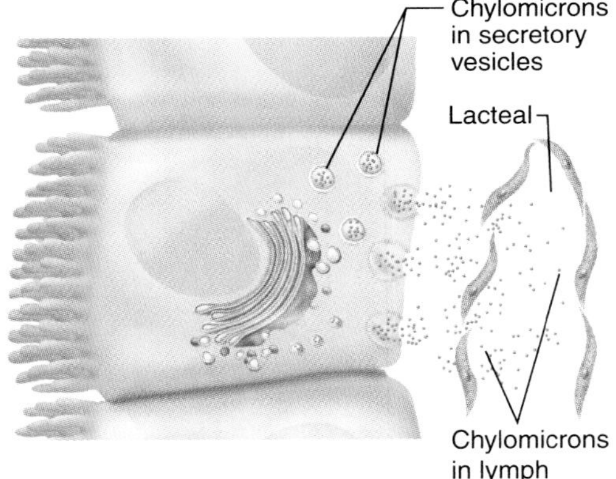

When lipase acts on a triglyceride, it removes the first and third fatty acids from the glycerol backbone and usually leaves the middle fatty acid. The products of lipase action are therefore two free fatty acids (FFAs) and a monoglyceride. Bile acids coat these and other lipids and form droplets called **micelles**[27] (my-SELLS), droplets about 5 nm in diameter that contain about 20 to 30 molecules of bile acids, FFAs, monoglycerides, cholesterol, and fat-soluble vitamins. Micelles pass amid the microvilli of the brush border and release their lipids, which diffuse freely through the plasma membranes (which are themselves mostly lipid) into the absorptive cells. The absorptive cells also have transport proteins that take up cholesterol and fatty acids.

Within the cell, the FFAs and monoglycerides are transported into the smooth endoplasmic reticulum and resynthesized into triglycerides. The Golgi complex combines these with a small amount of cholesterol and coats the complex with a film of phospholipids and protein, forming droplets about 60 to 750 nm in diameter called **chylomicrons**[28] (KY-lo-MY-crons). It packages chylomicrons into secretory vesicles that migrate to the basal surface of the cell and release their contents into the core of the villus. Although some FFAs enter the blood capillaries, chylomicrons are too large to penetrate the endothelium. They are taken up by the more porous lacteal (lymphatic vessel) into the lymph. The white, fatty intestinal lymph (*chyle*) flows through larger and larger lymphatic vessels of the mesenteries, eventually reaching the thoracic duct and entering the bloodstream at the left subclavian vein. The further fate of dietary fat is described in chapter 26.

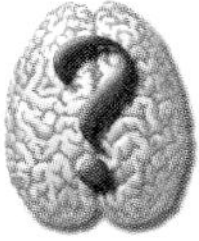

Think About It

Explain why the right lymphatic duct does not empty dietary fat into the bloodstream.

Nucleic Acids

The nucleic acids, DNA and RNA, are present in much smaller quantities than the polymers discussed previously. The **nucleases** (ribonuclease and deoxyribonuclease) of pancreatic juice hydrolyze these to their constituent nucleotides. **Nucleosidases** and **phosphatases** of the brush border then decompose the nucleotides into phosphate ions, ribose (from RNA) or deoxyribose (from DNA), and nitrogenous bases. These products are transported across the intestinal epithelium by membrane carriers and enter the capillary blood of the villus.

27. *mic* = grain, crumb + *elle* = little
28. *chyl* = juice + *micr* = small

Vitamins

Vitamins are absorbed unchanged. The fat-soluble vitamins A, D, E, and K are absorbed with other lipids as just described. Therefore, if they are ingested without fat-containing food, they are not absorbed at all but are passed in the feces and wasted. Water-soluble vitamins (the B complex and vitamin C) are absorbed by simple diffusion, with the exception of vitamin B_{12}. This is an unusually large molecule that can only be absorbed if it binds to intrinsic factor from the stomach. The B_{12}-intrinsic factor complex then binds to receptors on absorptive cells of the distal ileum, where it is taken up by receptor-mediated endocytosis.

Minerals

Minerals (electrolytes) are absorbed along the entire length of the small intestine. Sodium ions are cotransported with sugars and amino acids. Chloride ions are actively transported in the distal ileum by a pump that exchanges them for bicarbonate ions, thus reversing the chloride-bicarbonate exchange that occurs in the stomach. Potassium ions are absorbed by simple diffusion. The K^+ concentration of the chyme rises as water is absorbed from it, creating a gradient favorable to K^+ absorption. In diarrhea, when water absorption is hindered, potassium ions remain in the intestine and pass with the feces; thus chronic diarrhea can lead to hypokalemia.

Iron and calcium are unusual in that they are absorbed in proportion to the body's need, whereas other minerals are absorbed at fairly constant rates regardless of need, leaving it to the kidneys to excrete any excess. Furthermore, iron and calcium absorption is largely limited to the duodenum. Iron enters the absorptive cells by active transport and binds to the cytoplasmic protein apoferritin, forming an iron-ferritin complex. This serves as an intracellular reservoir of iron that is lost when the epithelial cells are sloughed off, unless the iron is needed sooner. In that case intracellular iron is released to the blood and carried to the bone marrow by the plasma protein transferrin (see chapter 18).

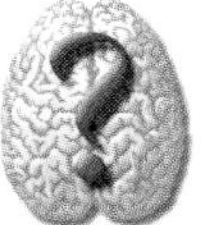

Think About It

Young adult women have four times as many iron transport proteins in the intestinal mucosa as men have. Can you explain this?

The rate of calcium absorption is determined by the availability of the cofactor calcitriol (vitamin D). The synthesis of calcitriol is under the control of parathyroid hormone (PTH) and thus adjusted to the body's state of calcium balance (see chapter 8).

Water

The digestive system is one of several systems involved in water balance. The digestive tract receives about 9 L of water per day—0.7 L in food, 1.6 L in drink, 6.7 L in the gastrointestinal secretions: saliva, gastric juice, bile, pancreatic juice, and intestinal juice. About 8 L of this is absorbed by the small intestine and 0.8 L by the large intestine, leaving 0.2 L voided in the daily fecal output. Water is absorbed by osmosis, following the absorption of salts and organic nutrients that create an osmotic gradient from the intestinal lumen to the ECF.

Diarrhea occurs when the large intestine absorbs too little water. This occurs when the intestine is irritated by bacteria and feces pass through too quickly for adequate reabsorption, or when the feces contain abnormally high concentrations of a solute such as lactose that opposes osmotic absorption of water. *Constipation* occurs when fecal movement is slow, too much water is reabsorbed, and the feces become hardened. This can result from lack of dietary fiber, lack of exercise, emotional upset, or long-term laxative abuse.

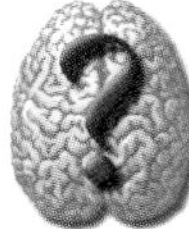

Think About It

Magnesium sulfate (epsom salt) is poorly absorbed by the intestines. In light of this, explain why it has a laxative effect.

Before You Go On

Answer the following questions to test your understanding of the preceding section:

21. What three polymers account for most of the dietary calories? What are the end products of enzymatic digestion of each?
22. What two nutrients are digested by saliva? Why is only one of them digested in the mouth?
23. Name as many enzymes of the intestinal brush border as you can, and identify the substrate or function of each.
24. Explain the distinctions between an emulsification droplet, a micelle, and a chylomicron.
25. What happens to digestive enzymes after they have done their job? What happens to dead epithelial cells that slough off the gastrointestinal mucosa? Explain.

The Large Intestine

Objectives

When you have completed this section, you should be able to

- describe the gross anatomy of the large intestine;
- contrast the mucosa of the colon with that of the small intestine;
- state the physiological significance of intestinal bacteria;
- discuss the types of contractions that occur in the colon; and
- explain the neurological control of defecation.

The large intestine (fig. 25.30) receives about 500 mL of indigestible food residue per day, reduces it to about 150 mL of feces by absorbing water and salts, and eliminates the feces by defecation.

Gross Anatomy

The large intestine measures about 1.5 m (5 ft) long and 6.5 cm (2.5 in.) in diameter in the cadaver. It begins with the **cecum,**[29] a blind pouch in the lower right abdominal quadrant inferior to the ileocecal valve. Attached to the lower end of the cecum is the **vermiform**[30] **appendix,** a blind tube 2 to 7 cm long. The appendix is densely populated with lymphocytes and is a significant source of immune cells.

The **ascending colon** begins at the ileocecal valve and passes up the right side of the abdominal cavity. It makes a 90° turn at the **right colic (hepatic) flexure,** near the right lobe of the liver, and becomes the **transverse colon.** This passes horizontally across the upper abdominal cavity and turns 90° downward at the **left colic (splenic) flexure** near the spleen. Here it becomes the **descending colon,** which passes down the left side of the abdominal cavity. Ascending, transverse, and descending colons thus form a squarish, three-sided frame around the small intestine.

The pelvic cavity is narrower than the abdominal cavity, so at the pelvic inlet the colon turns medially and downward, forming a roughly S-shaped portion called the **sigmoid**[31] **colon.** (Visual examination of this region is performed with an instrument called a *sigmoidoscope.*) In the pelvic cavity, the large intestine straightens and forms the **rectum.**[32] The rectum has three internal transverse folds called **rectal valves** that enable it to retain feces while passing gas.

The final 3 cm of the large intestine is the **anal canal** (fig. 25.30*b*), which passes through the levator ani muscle of the pelvic floor and terminates at the anus. Here, the mucosa forms longitudinal ridges called **anal columns** with depressions between them called **anal sinuses.** As feces pass through the canal, they press against the sinuses and cause them to exude extra mucus and lubricate the canal during defecation. Large **hemorrhoidal veins** form superficial plexuses in the anal columns and around the orifice. Unlike veins in the extremities, they lack valves and are particularly subject to distension and venous pooling.

29. *cec* = blind
30. *vermi* = worm + *form* = shaped
31. *sigm* = sigma or S + *oid* = resembling
32. *rect* = straight

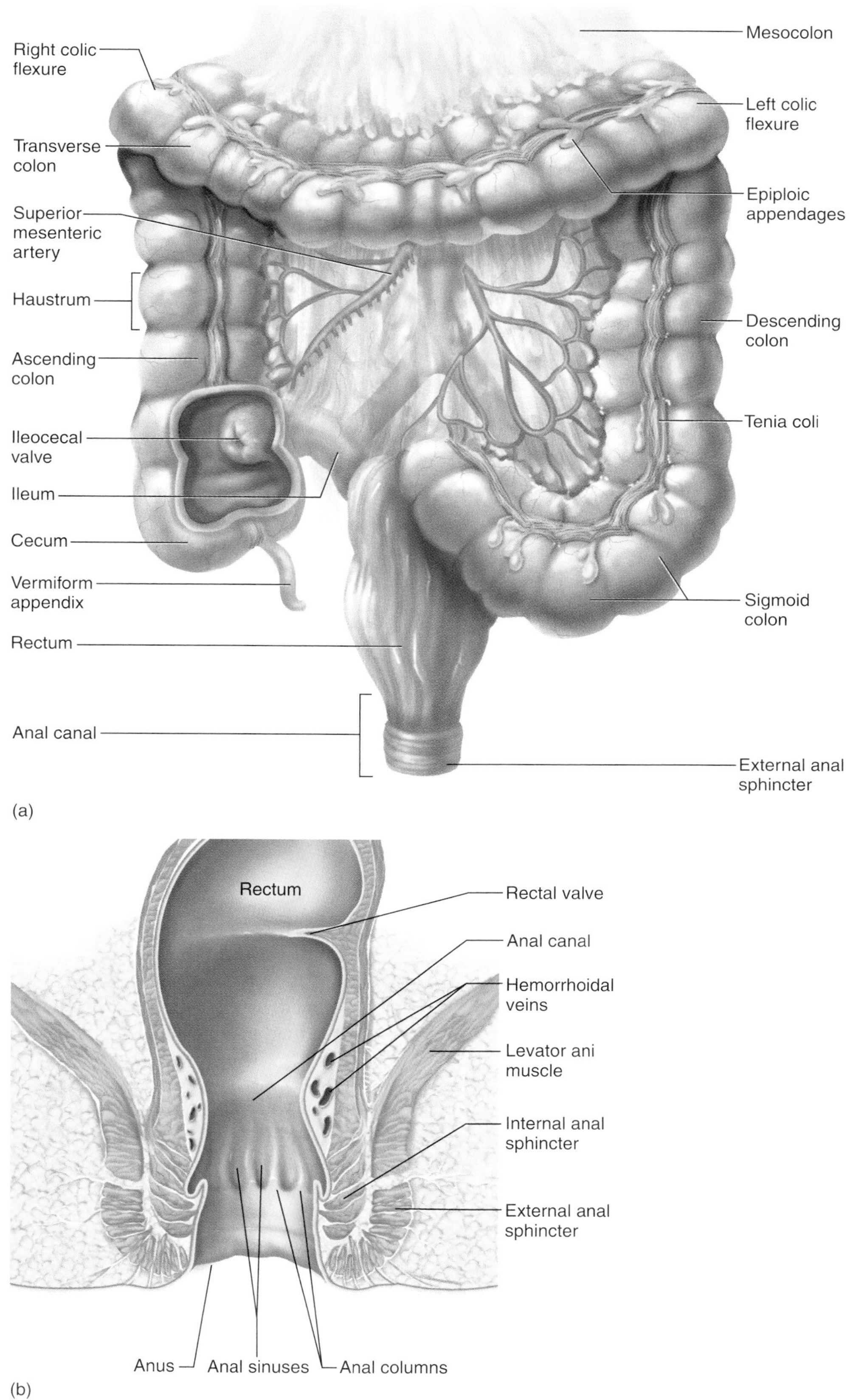

Figure 25.30 **The Large Intestine.** (*a*) Gross anatomy. (*b*) Detail of the anal canal.

Hemorrhoids are permanently distended veins that protrude into the anal canal or form bulges distal to the anus.

The muscularis externa of the colon is unusual in that its longitudinal fibers do not encircle the colon but are divided into three ribbonlike strips called the **teniae coli** (TEE-nee-ee CO-lye). The tonus of the teniae coli contracts the colon lengthwise and causes its wall to form pouches called **haustra**[33] (HAW-stra; singular, *haustrum*). In the rectum and anal canal, however, the longitudinal muscle forms a continuous sheet and haustra are absent. The anus, like the urethra, is regulated by two sphincters—an **internal anal sphincter** composed of smooth muscle of the muscularis externa and an **external anal sphincter** composed of skeletal muscle of the pelvic diaphragm.

The ascending and descending colon are retroperitoneal, whereas the transverse and sigmoid colon are covered with serosa and anchored to the dorsal abdominal wall by the mesocolon. The serosa of these regions often has **epiploic**[34] **appendages,** clublike fatty pouches of peritoneum of unknown function.

Microscopic Anatomy

The mucosa of the large intestine has a simple columnar epithelium in all regions except the anal canal, where it has a nonkeratinized stratified squamous epithelium. The latter provides more resistance to the abrasion caused by the passage of feces. There are no circular folds or villi in the large intestine, but there are intestinal crypts. They are deeper than in the small intestine and have a greater density of goblet cells; mucus is their only significant secretion.

The anatomical features of the digestive tract from the large intestine to the anus are summarized in table 25.5.

Bacterial Flora and Intestinal Gas

The large intestine is densely populated with several species of bacteria collectively called the **bacterial flora.**[35] They ferment cellulose and other undigested carbohydrates and synthesize B vitamins and vitamin K, which are absorbed by the colon. This vitamin K is especially important because the diet alone usually does not provide enough to ensure adequate blood clotting.

The average person expels about 500 mL of **flatus** (gas) per day. Most of this is swallowed air that has worked its way through the digestive tract, but the bacterial flora add to it. Painful cramping can result when undigested nutrients pass into the colon and furnish an abnormal substrate for bacterial action—for example, in lactose intolerance. Flatus is composed of nitrogen (N_2), carbon dioxide (CO_2), hydrogen (H_2), methane (CH_4), hydrogen sulfide (H_2S), and two amines: indole and skatole. Indole, skatole, and H_2S produce the odor of flatus and feces, whereas the others are odorless.

33. *haustr* = to draw
34. *epiploic* = pertaining to an omentum
35. *flora* = flowers, plants

Table 25.5 Anatomical Checklist of the Digestive System from the Large Intestine Through the Anus

Cecum	Anal canal
Vermiform appendix	Internal anal sphincter
Ascending colon	External anal sphincter
Right colic flexure	Anal columns
Transverse colon	Anal sinuses
Left colic flexure	Hemorrhoidal veins
Descending colon	Teniae coli
Sigmoid colon	Haustra
Rectum	Epiploic appendages
Rectal valves	Mesocolon

Absorption and Motility

The large intestine takes about 12 to 24 hours to reduce the residue of a meal to feces. It does not chemically change the residue but reabsorbs water and electrolytes (especially NaCl) from it. The feces consist of about 75% water and 25% solids. The solids are about 30% bacteria, 30% undigested dietary fiber, 10% to 20% fat, and smaller amounts of protein, sloughed epithelial cells, salts, mucus, and other digestive secretions. The fat is not from the diet but from digested epithelial cells and bacteria.

The most common type of colonic motility is a type of segmentation called **haustral contractions,** which occur about every 30 minutes. Distension of a haustrum with feces stimulates it to contract. This churns and mixes the residue, promotes water and salt absorption, and passes the residue distally to another haustrum. Stronger contractions called **mass movements** occur one to three times a day, last about 15 minutes, and occur especially within an hour after breakfast. They are often triggered by the **gastrocolic** and **duodenocolic reflexes,** in which filling of the stomach and duodenum stimulates motility of the colon. Mass movements occur especially in the transverse to sigmoid colon, moving residue for several centimeters with each contraction.

Defecation

Stretching of the rectum stimulates the defecation reflexes, which account for the urge to defecate that is often felt soon after a meal. The predictability of this response is useful in house-training pets and toilet-training children. In the **intrinsic defecation reflex,** stretch signals travel by the myenteric nerve plexus to the muscularis of the descending and sigmoid colons and the rectum. This triggers a peristaltic wave that drives feces downward, and it relaxes the internal anal sphincter. Defecation occurs only if the external anal sphincter is voluntarily relaxed at the same time.

The intrinsic reflex is relatively weak and usually requires the cooperative action of a stronger **parasympathetic defecation reflex** involving the spinal cord. As shown in figure 25.31, this reflex involves three processes: (1) Stretching of the rectum sends signals to the sacral segments of the spinal cord. (2) The spinal cord returns signals by way of the parasympathetic fibers in the pelvic nerves to the rectum, thus intensifying rectal peristalsis. (3) At the same time, parasympathetic fibers relax the internal anal sphincter to allow the passage of feces.

The external anal sphincter, being composed of skeletal muscle, is under voluntary control, enabling us to limit defecation to appropriate circumstances (step 4 in fig. 25.31). As involuntary contractions of the rectum push the feces downward, voluntary contractions of the levator ani pull the anal canal upward and allow the feces to fall away. Defecation is also aided by the Valsalva maneuver, which increases abdominal pressure, compresses the rectum, and squeezes the feces from it. This maneuver can also initiate the defecation reflex by forcing feces from the descending colon into the rectum.

If the defecation urge is suppressed, contractions cease in a few minutes and the rectum relaxes. The defecation reflexes reoccur a few hours later or when another mass movement propels more feces into the rectum.

Throughout this chapter, we have considered such issues of digestive pathophysiology as tooth and gum disease, gastroesophageal reflux, peptic ulcer, gallstones, lactose intolerance, hemorrhoids, diarrhea, and constipation. A few other digestive system disorders are briefly explained in table 25.6, and additional related disorders (especially concerned with appetite, obesity, high blood cholesterol, liver disease, and alcoholism) are discussed in the next chapter.

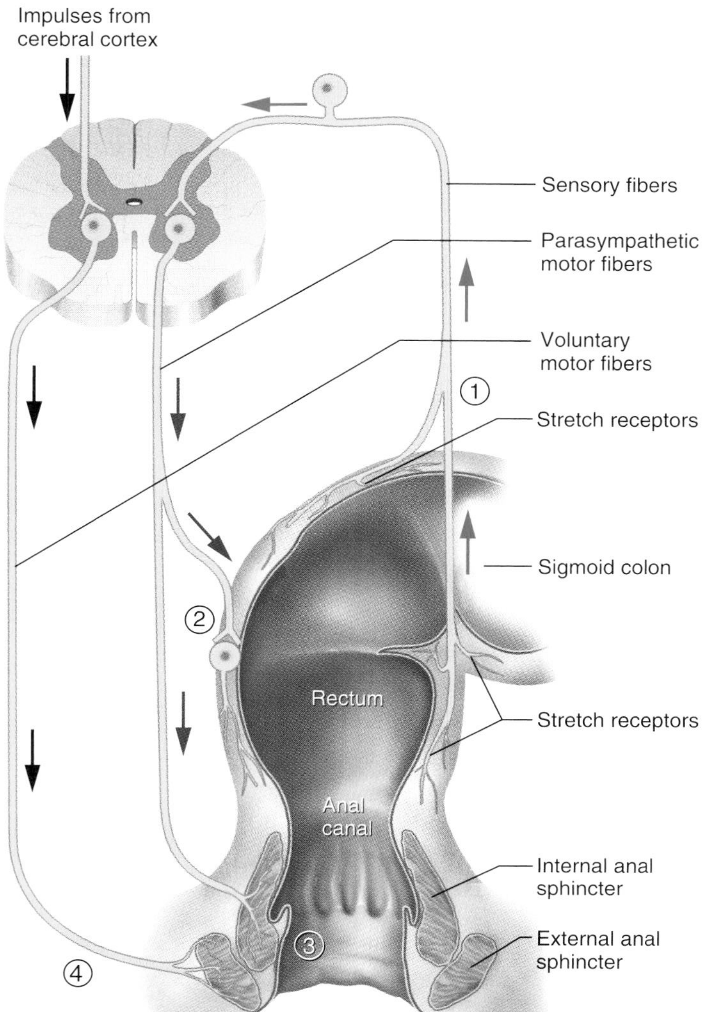

Figure 25.31 Neural Control of Defecation. (*1*) Filling of the rectum with feces stimulates stretch receptors, which transmit impulses to the spinal cord. (*2*) A spinal reflex stimulates contractions of the rectum and relaxation of the internal anal sphincter. (*3*) Defecation normally does not occur unless voluntary impulses relax the external anal sphincter.

Before You Go On

Answer the following questions to test your understanding of the preceding section:

26. How does the mucosa of the large intestine differ from that of the small intestine? How does the muscularis externa differ?
27. Name and briefly describe two types of contractions that occur in the colon and nowhere else in the alimentary canal.
28. Describe the reflexes that cause defecation in an infant. Describe the additional neural controls that function following toilet-training.

Table 25.6	Some Digestive System Diseases
Acute Pancreatitis	Severe pancreatic inflammation perhaps caused by trauma leading to leakage of pancreatic enzymes into parenchyma, where they digest tissue and cause inflammation and hemorrhage.
Appendicitis	Inflammation of the vermiform appendix, with swelling, pain, and sometimes gangrene, perforation, and peritonitis.
Ascites	Accumulation of serous fluid in the peritoneal cavity, often causing extreme distension of the abdomen. Most often caused by cirrhosis of the liver (see chapter 26) and frequently associated with alcoholism. The diseased liver "weeps" fluid into the abdomen. About 25% of people who develop ascites as a consequence of cirrhosis die within one year.
Cancers	Digestive system is subject to cancer especially of the esophagus, stomach, colon, liver, and pancreas, with colon and pancreatic cancer being among the leading causes of cancer death in the United States.
Crohn Disease	Inflammation of small and large intestines, similar to ulcerative colitis. Produces granular lesions and fibrosis of intestine, diarrhea, and lower abdominal pain. Often hereditary.
Diverticulitis	Presence of inflamed herniations (outpocketings, diverticula) of the colon, caused especially by low-fiber diets. Diverticula may rupture, leading to peritonitis.
Dysphagia	Difficulty swallowing. Can result from esophageal obstructions (tumors, constrictions) or impaired peristalsis (due to neuromuscular disorders).
Hiatal Hernia	Protrusion of part of the stomach into the thoracic cavity, where the negative thoracic pressure may cause it to balloon. Often causes gastroesophageal reflux (especially when a person is supine) and esophagitis (inflammation of the esophagus).
Ulcerative Colitis	Chronic inflammation resulting in ulceration of the large intestine, especially the sigmoid colon and rectum. Tends to be hereditary but exact causes are not well known.

Historical Insight 25.5

The Man with a Hole in His Stomach

Perhaps the most famous episode in the history of digestive physiology began in 1822 on Mackinac Island in the strait between Lake Michigan and Lake Huron. Alexis St. Martin, a 19-year-old fur trapper, was standing outside a trading post when he was accidentally hit by a shotgun blast from 3 feet away. An Army doctor stationed at Fort Mackinac, William Beaumont (1785–1853), was summoned to examine St. Martin. As Beaumont later wrote, "a portion of the lung as large as a turkey's egg" protruded through St. Martin's lacerated and burnt flesh. Below that was a portion of the stomach with a puncture in it "large enough to receive my forefinger." Beaumont did his best to pick out bone fragments and dress the wound, though he did not expect St. Martin to survive.

Surprisingly, St. Martin lived. Over a period of months the wound extruded pieces of bone, cartilage, gunshot, and gun wadding. As the wound healed, a fistula (hole) remained in the stomach, so large that Beaumont had to cover it with a compress to prevent food from coming out. A fold of tissue later grew over the fistula, but it was easily opened. A year later, St. Martin was still feeble. Town authorities decided they could no longer support him on public funds and wanted to ship him 2,000 miles to his home in Canada. Beaumont, however, was imbued with a passionate sense of destiny. Very little was known about digestion, and he saw the accident as a unique opportunity to learn. He took St. Martin in at his personal expense and performed 238 experiments on him over several years. Beaumont had never attended medical school and had little idea how scientists work, yet he proved to be an astute experimenter. Under crude frontier conditions and with almost no equipment, he discovered many of the basic facts of gastric physiology discussed in this chapter.

"I can look directly into the cavity of the stomach, observe its motion, and almost see the process of digestion," Beaumont wrote. "I can pour in water with a funnel and put in food with a spoon, and draw them out again with a siphon." He put pieces of meat on a string into the stomach and removed them hourly for examination. He sent vials of gastric juice to the leading chemists of America and Europe, who could do little but report that it contained hydrochloric acid. He proved that digestion required HCl and could even occur outside the stomach, but he found that HCl alone did not digest meat; gastric juice must contain some other digestive ingredient. Theodor Schwann, one of the founders of the cell theory (see chapter 4), identified that ingredient as pepsin. Beaumont also demonstrated that gastric juice is secreted only in response to food; it did not accumulate between meals as previously thought. He disproved the idea that hunger is caused by the walls of the empty stomach rubbing against each other.

For his part, St. Martin felt helpless and humiliated by Beaumont's experiments. His fellow trappers taunted him as "the man with a hole in his stomach," and he longed to return to hunting and trapping in the wilderness. He had a wife and daughter in Canada whom he rarely got to see, and he ran away repeatedly to join them. He was once gone for 4 years before his poverty and physical disability made him yield to Beaumont's financial enticement to come back. Beaumont despised St. Martin for his drunkenness and profanity and was quite insensitive to St. Martin's embarrassment and discomfort over the experiments. Yet St. Martin's temper enabled Beaumont to make the first direct observations of the relationship between emotion and digestion. When St. Martin was particularly distressed, Beaumont noted little digestion occurring—as we now know, the sympathetic nervous system inhibits digestive activity.

Beaumont published a book in 1833 that laid the foundation for modern gastric physiology and dietetics. It was enthusiastically received by the medical community and had no equal until Russian physiologist Ivan Pavlov (1849–1936) performed his celebrated experiments on digestion in animals. Building on the methods pioneered by Beaumont, Pavlov received the 1904 Nobel Prize for Physiology or Medicine.

In 1853, Beaumont slipped on some ice, suffered a blow to the base of his skull, and died a few weeks later. St. Martin continued to tour medical schools and submit to experiments by other physiologists, whose conclusions were often less correct than Beaumont's. Some, for example, attributed chemical digestion to lactic acid instead of hydrochloric acid. St. Martin lived in wretched poverty in a tiny shack with his wife and several children and died 28 years after Beaumont. By then he was senile and believed he had been to Paris, where Beaumont had often promised to take him.

Connective Issues

Interactions Between the DIGESTIVE SYSTEM and Other Organ Systems

◄ indicates ways in which this system affects other systems
► indicates ways in which other systems affect this one

All Systems

◄ The digestive system provides all other systems with nutrients in a form usable for cellular metabolism and building of tissues

Integumentary System

► Skin plays a role in the synthesis of calcitriol, needed for calcium and phosphorus absorption by the small intestine

Skeletal System

◄ Small intestine adjusts calcium absorption in proportion to the needs of the skeletal system

► Provides protective enclosure for some digestive organs, support for the teeth, and movements of mastication

Muscular System

◄ Liver disposes of lactic acid generated by muscles, thus promoting recovery from fatigue

► Essential for chewing, swallowing, and voluntary control of defecation; abdominal muscles protect lower GI organs

Nervous System

► Enteric and autonomic nervous systems regulate GI motility and secretion; somatic nervous system controls chewing, swallowing, and defecation; sense organs involved in food selection and ingestion; hypothalamus contains centers for hunger, thirst, and satiety

Endocrine System

◄ Liver degrades hormones; enteroendocrine cells produce many hormones

► Hormones regulate GI motility, secretion, and processing of absorbed nutrients

Circulatory System

◄ Digestive tract absorbs fluid needed to maintain blood volume; liver degrades heme from dead erythrocytes, secretes clotting factors, albumin, and other plasma proteins, and regulates blood glucose level; intestinal epithelium stores and releases iron as needed for erythropoiesis

► Blood transports hormones that regulate GI activity; absorbs and distributes nutrients; vasomotion alters capillary filtration and salivation

Lymphatic/Immune Systems

◄ GI mucosa is a site of lymphocyte production; acid, lysozyme, and other digestive enzymes provide nonspecific defense against pathogens; infant intestine absorbs maternal IgA to confer passive immunity on infant

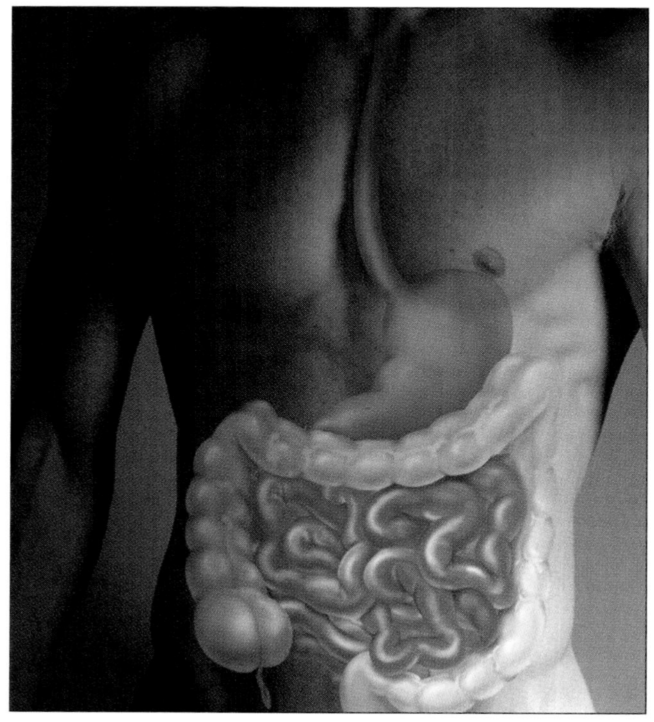

► Lymphatic capillaries (lacteals) absorb digested lipids; immune cells protect GI tract from infection

Respiratory System

◄ Pressure of digestive organs against diaphragm aids in expiration when abdominal muscles contract

► Provides O_2, removes CO_2; Valsalva maneuver aids defecation

Urinary System

◄ Intestines complement kidneys in water and electrolyte reabsorption; liver carries out last step before kidneys in calcitriol synthesis; liver synthesizes urea and kidneys excrete it

► Excretes bile pigments and other products of liver metabolism; completes the synthesis of calcitriol, needed for intestinal absorption of calcium and phosphorus

Reproductive System

◄ Provides nutrients for fetal growth

► Developing fetus crowds digestive organs; may cause constipation and heartburn during pregnancy

Chapter Review

Online Learning Center

Study resources for this chapter are available at www.mhhe.com/saladin/

Study Outline

General Anatomy and Digestive Processes (p. 939)

1. Subdivisions of the digestive system
2. Relationship to the peritoneum
3. Digestive functions and processes
4. Stages of digestion
   a. Mechanical digestion
   b. Chemical digestion

The Mouth Through Esophagus (p. 941)

1. Oral (buccal) cavity
   a. Functions
   b. Cheeks and lips
   c. Tongue
      - Lingual papillae
      - Body and root
      - Lingual muscles
      - Lingual glands
   d. Palate
      - Hard and soft palates
      - Palatine arches
   e. Teeth (dentition)
      - Types
      - Development and eruption
      - Periodontal tissues
      - Tooth structure
2. Mastication
3. Saliva and salivary glands
   a. Composition of saliva
   b. Locations of salivary glands
   c. Salivation
4. Pharynx
5. Esophagus
   a. Gross anatomy
   b. Histology
   c. Enteric nervous system
6. Swallowing (deglutition)
   a. Swallowing center
   b. Buccal phase
   c. Pharyngeal-esophageal phase
   d. Mechanism of peristalsis

The Stomach (p. 949)

1. Functions
2. Gross anatomy
   a. Curvatures
   b. Four regions of stomach
   c. Pyloric sphincter
3. Innervation and circulation
4. Stomach wall
   a. Rugae
   b. Gastric pits
   c. Glands
   d. Cell types
5. Gastric secretions
   a. Hydrochloric acid
   b. Intrinsic factor
   c. Pepsinogen and pepsin
   d. Gastric lipase and rennin
   e. Hormones and paracrines
6. Gastric motility
   a. Receptive-relaxation response
   b. Gastric peristalsis
   c. Entry of chyme into duodenum
7. Vomiting
8. Digestion and absorption
9. Protective mechanisms
   a. Mucous coat
   b. Epithelial replacement
   c. Tight junctions
10. Regulation of gastric function
   a. Cephalic phase
   b. Gastric phase
      - Myenteric reflex
      - Vagovagal reflex
      - Regulatory chemicals
         - •• Acetylcholine
         - •• Histamine
         - •• Gastrin
      - Negative feedback control
   c. Intestinal phase
      - Gastrin
      - Enterogastric reflex
      - Secretin
      - Cholecystokinin
      - Gastric inhibitory peptide

The Liver, Gallbladder, and Pancreas (p. 958)

1. Liver
   a. Gross anatomy
   b. Microscopic anatomy
2. Gallbladder and bile
   a. Gallbladder structure
   b. Composition of bile
   c. Digestive role of bile salts
   d. Enterohepatic circulation
3. Pancreas
   a. Gross anatomy
   b. Endocrine and exocrine parts
   c. Ducts and duodenal papillae
   d. Pancreatic juice
   e. Zymogens and their activation
      - Trypsinogen
      - Chymotrypsinogen
      - Procarboxypeptidase
   f. Other pancreatic enzymes
      - Pancreatic amylase
      - Pancreatic lipase
      - Ribonuclease
      - Deoxyribonuclease
4. Regulation of secretion
   a. Cholecystokinin
   b. Secretin

The Small Intestine (p. 964)

1. Importance of surface area
2. Gross anatomy
   a. Duodenum
   b. Jejunum
   c. Ileum
   d. Ileocecal junction
3. Microscopic anatomy
   a. Circular folds
   b. Villi
      - Absorptive and goblet cells
      - Blood capillaries and lacteal
      - Brush border
   c. Intestinal crypts
      - Absorptive and goblet cells
      - Paneth cells
   d. Duodenal glands
   e. Peyer patches
4. Secretion (intestinal juice)
5. Motility
   a. Functions of intestinal motility
   b. Segmentation
   c. Peristalsis

d. Ileocecal valve
e. Gastroileal reflex

Chemical Digestion and Absorption (p. 968)

1. Carbohydrates
   a. Digestion
      - Action of salivary amylase
      - Action of pancreatic amylase
      - Action of brush border enzymes
         - Dextrinase
         - Glucoamylase
         - Maltase
         - Sucrase
         - Lactase
   b. Absorption
      - Glucose
      - Galactose
      - Fructose
2. Proteins
   a. Sources of amino acids
   b. Digestion
      - Action of pepsin
      - Action of trypsin and chymotrypsin
      - Contact digestion
         - Carboxypeptidase
         - Aminopeptidase
         - Dipeptidase
   c. Absorption
      - Amino acid absorption
      - Protein pinocytosis
3. Lipids
   a. Digestion
      - Action of lingual and gastric lipase
      - Emulsification
      - Action of pancreatic lipase
      - Micelle formation
   b. Absorption
      - Uptake by epithelial cells
      - Formation of chylomicrons
      - Uptake by lacteals
4. Nucleic acids
5. Vitamins
   a. Fat-soluble vitamins
   b. Water-soluble vitamins
   c. Vitamin B_{12} and intrinsic factor
6. Minerals
   a. Na^+, K^+, and Cl^- absorption
   b. Iron absorption and storage
   c. Calcium absorption
7. Water

The Large Intestine (p. 975)

1. Functions
2. Gross anatomy
   a. Cecum and appendix
   b. Ascending to descending
   c. Sigmoid colon, rectum, and anal canal
   d. Taeniae coli and haustra
   e. Anal sphincters
   f. Mesocolon
   g. Epiploic appendages
3. Microscopic anatomy
4. Bacterial flora and intestinal gas
5. Absorption and motility
   a. Absorption
   b. Composition of feces
   c. Motility
      - Haustral contractions
      - Mass movements
   d. Defecation
      - Intrinsic defecation reflex
      - Parasympathetic defecation reflex
      - Voluntary control

Selected Vocabulary

For additional terminology not listed here, see tables 25.1, 25.4, and 25.5.

digestive tract 939
accessory organs 939
dorsal mesentery 939
serosa 939
ventral mesentery 939
lesser omentum 940
greater omentum 940
mesocolon 940
mechanical digestion 940
chemical digestion 940
occlusion 942
eruption of teeth 943
mastication 943
salivary amylase 944
lingual lipase 944
salivatory nuclei 945
bolus 946
mucosa 947
lamina propria 947
muscularis mucosae 947
submucosa 947
esophageal glands 947
muscularis externa 947
adventitia 947
enteric nervous system 947
submucosal plexus 947
myenteric plexus 947
short reflex 947
long reflex 947
lower esophageal sphincter 947
gastroesophageal reflux 947
deglutition 947
swallowing center 947
peristalsis 947
chyme 949
gastric juice 950
H^+-K^+ ATPase 953
alkaline tide 953
intrinsic factor 953
zymogen 953
pepsinogen 953
pepsin 953
gastric lipase 953
rennin 953
emetic center 955
cephalic phase 955
gastric phase 955
myenteric reflex 955
vagovagal reflex 955
histamine 956
gastrin 956
G cell 956
intestinal phase 957
enterogastric reflex 957
secretin 957
cholecystokinin 957
gastric inhibitory peptide 957
bile 961
bilirubin 961
urobilinogen 961
bile acid 961
pancreatic juice 962
trypsinogen 962
chymotrypsinogen 962
procarboxypeptidase 962
trypsin 962
enterokinase 962
chymotrypsin 962
carboxypeptidase 962
pancreatic amylase 962
pancreatic lipase 962
ribonuclease 962
deoxyribonuclease 962
secretin 962
brush border enzymes 965
contact digestion 965
intestinal juice 965
segmentation 965
migrating motor complex 967
gastroileal reflex 967
dextrinase 969
glucoamylase 969
maltase 969
sucrase 969
lactase 969
protease 970
carboxypeptidase 970
aminopeptidase 970
dipeptidase 970
lipase 970
emulsification droplet 970
micelle 973
chylomicron 973
nuclease 973
nucleosidase 973
phosphatase 973
hemorrhoid 976
bacterial flora 976
flatus 976
haustral contraction 976
mass movement 976
gastrocolic reflex 976
duodenocolic reflex 976
intrinsic defecation reflex 977
parasympathetic defecation reflex 977

Testing Your Recall

1. Which of the following enzymes acts in the stomach?
   a. chymotrypsin
   b. lingual lipase
   c. carboxypeptidase
   d. enterokinase
   e. dextrinase
2. Which of the following enzymes does *not* digest any nutrients?
   a. chymotrypsin
   b. lingual lipase
   c. carboxypeptidase
   d. enterokinase
   e. dextrinase
3. Which of the following is *not* an enzyme?
   a. chymotrypsin
   b. enterokinase
   c. secretin
   d. pepsin
   e. nucleosidase
4. The substance in question 3 that is *not* an enzyme is
   a. a zymogen.
   b. a nutrient.
   c. an emulsifier.
   d. a neurotransmitter.
   e. a hormone.
5. The lacteals absorb
   a. chylomicrons.
   b. micelles.
   c. emulsification droplets.
   d. amino acids.
   e. monosaccharides.
6. All of the following contribute to the absorptive surface area of the small intestine *except*
   a. its length.
   b. the brush border.
   c. haustra.
   d. circular folds.
   e. villi.
7. Which of the following is a periodontal tissue?
   a. the gingiva
   b. the enamel
   c. the cementum
   d. the pulp
   e. the dentin
8. The ________ of the stomach most closely resemble the ________ of the small intestine.
   a. gastric pits, intestinal crypts
   b. pyloric glands, intestinal crypts
   c. rugae, Peyer patches
   d. parietal cells, goblet cells
   e. gastric glands, duodenal glands
9. Which of the following cells secrete digestive enzymes?
   a. chief cells
   b. mucous neck cells
   c. parietal cells
   d. goblet cells
   e. enteroendocrine cells
10. What phase of gastric regulation includes inhibition by the enterogastric reflex?
    a. the intestinal phase
    b. the gastric phase
    c. the buccal phase
    d. the cephalic phase
    e. the pharyngo-esophageal phase
11. Cusps are a feature of the ________ surfaces of the molars and premolars.
12. The acidity of the stomach deactivates ________ but activates ________ of the saliva.
13. The ________ salivary gland is named for its proximity to the ear.
14. The submucosal and myenteric nerve plexuses collectively constitute the ________ nervous system.
15. Nervous stimulation of gastrointestinal activity is mediated mainly through the parasympathetic fibers of the ________ nerves.
16. Food in the stomach causes G cells to secrete ________, which in turn stimulates the secretion of HCl and pepsinogen.
17. Hepatic macrophages occur in blood-filled spaces of the liver called ________.
18. The brush border enzyme that finishes the job of starch digestion, producing glucose, is called ________. Its substrate is ________.
19. Fats are transported in the lymph and blood in the form of droplets called ________.
20. Within the absorptive cells of the small intestine, ferritin binds the nutrient ________.

Answers in appendix B

Testing Your Comprehension

1. A physician plans to attend a cocktail party, but he is also on call that night and must avoid intoxication. Therefore he drinks a glass of cream just before leaving for the party. Explain.
2. Which of these do you think would have the most severe effect on digestion: surgical removal of the stomach, gallbladder, or pancreas? Explain.
3. What do carboxypeptidase and aminopeptidase have in common? Identify as many differences between them as you can.
4. What do micelles and chylomicrons have in common? Identify as many differences between them as you can.
5. Explain why most dietary lipids must be absorbed by the lacteals rather than by the blood capillaries of a villus.

Answers at the Online Learning Center

26 CHAPTER

Nutrition and Metabolism

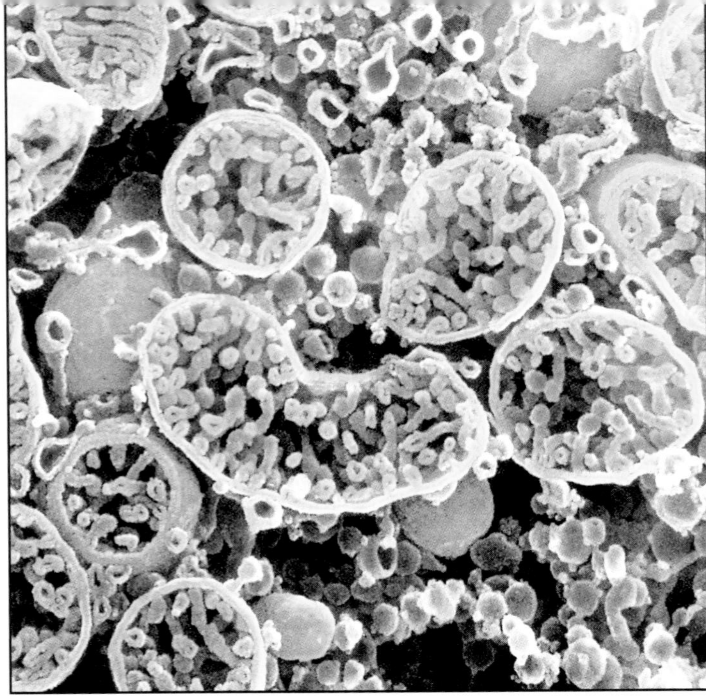

Mitochondria (green) and smooth endoplasmic reticulum in a cell of the ovary (SEM)

Chapter Outline

Insights

Brushing Up

To understand this chapter, it is important that you understand or brush up on the following concepts:

- Free energy (p. 68)
- Metabolism, catabolism, and anabolism (p. 68)
- Oxidation and reduction (pp. 68–69)
- Saturated and unsaturated fats (pp. 85–87)
- Structure and functions of ATP (p. 96)
- Receptor-mediated endocytosis (p. 132)
- Insulin and glucagon (pp. 654–55)

Study with Your ESP CD-ROM

Refer to your ESP Tutorial CD-ROM for helpful animations, activities, and quizzes related to the material covered in this chapter. The following are suggested areas to go to on the CD and are presented according to "Unit/Topic":

- Absorption and Excretion/Nutrition

Nutrition is the starting point and basis for all human form and function. From the time a single-celled, fertilized egg divides in two, nutrition provides the matter needed for cell division, growth, and development. It is the source of fuel that provides the energy for all biological work and of the raw materials for replacement of worn-out biomolecules and cells. The fact that it provides only the *raw* materials means, further, that chemical change—metabolism—lies at the foundation of form and function. In the previous chapter, we saw how the digestive system breaks nutrients down into usable form and absorbs them into the blood and lymph. We now consider these nutrients in more depth, follow their fate after absorption, and explore related issues of metabolism and body heat.

Nutrition

Objectives

When you have completed this section, you should be able to

- describe the mechanisms of hunger and satiety;
- define *nutrient* and list the six major categories of nutrients;
- state the function of each class of macronutrients, the approximate amounts required in the diet, and some major dietary sources of each;
- name the blood lipoproteins, state their functions, and describe how they differ from each other; and
- name the major vitamins and minerals required by the body and the general functions they serve.

Body Weight and Energy Balance

The subject of nutrition quickly brings to mind the subject of body weight and the popular desire to control it. While this is only one aspect of the subject, it is a significant consideration.

Weight is determined by the body's energy balance—if energy intake and output are equal, body weight is stable. We gain weight if intake exceeds output and lose weight if output exceeds intake. Although weight is determined by this ratio, it remains quite stable over many years' time and seems to have a homeostatic set point. This has been experimentally demonstrated in animals. If an animal is force-fed until it becomes quite obese and then allowed to feed at will, it voluntarily reduces its intake and quickly returns to its former weight. Similarly, if an animal is undernourished until it loses much of its weight and then allowed to feed at will, it increases its intake and again returns quickly to its former weight.

In humans the set point varies greatly from person to person and body weight results from a combination of hereditary and environmental influences. From studies of identical twins and other people, it appears that about 30% to 50% of the variation in human weight is due to heredity, and the rest to environmental factors such as eating and exercise habits.

Appetite

The struggle for weight control often seems to be a struggle against the appetite, but despite decades of research, we are still far from a complete understanding of how appetite is regulated. In the 1940s, it was discovered that a region in the lateral area of the hypothalamus seems to trigger the desire for food. When this **feeding center** is destroyed in animals, they exhibit drastic **anorexia**[1] (loss of appetite) and starve to death if not force-fed. It was recently discovered that, at least in rats, this area of the hypothalamus secretes a hormone named *orexin*, which rises during fasting and stimulates intense hunger. The ventromedial hypothalamus has a **satiety center;** damage here causes **hyperphagia**[2] (overeating) and extreme obesity (fig. 26.1).

The satiety center has neurons called **glucostats** that rapidly absorb blood glucose after a meal. According to the *glucostat hypothesis* of hunger, when these neurons absorb glucose, they send inhibitory signals to the feeding center, thus suppressing the appetite. Some weight-loss drugs work by inhibiting the feeding center. As blood glucose concentration drops hours after a meal, inhibitory signals from the satiety center decline, the feeding center becomes active, and the sense of hunger returns. But this is certainly far from the whole story. Hunger and satiety are regulated by a complex interaction of multiple brain centers, hormones, and sensory and motor pathways. Experimental lesions disrupt multiple pathways and may produce abnormal feeding behavior for a variety of reasons.

Other factors influence appetite in ways similar to the control of thirst (see p. 918). Merely chewing and swallowing food briefly satisfies the appetite, even if the food is removed through an esophageal fistula (opening) before reaching the stomach. Inflating the stomach with a balloon inhibits hunger even in an animal that has not actually swallowed any food.

Hunger is stimulated partly by gastric peristalsis. Mild **hunger contractions** begin soon after the stomach is emptied and increase in intensity over a period of hours. They can become quite a painful and powerful incentive to eat, yet they do not affect the amount of food consumed—this remains much the same even when nervous connections to the stomach and intestines are severed to cut off all conscious perception of hunger contractions.

Hormones also play a role in appetite regulation. Amino acids and fatty acids stimulate the duodenum to secrete *cholecystokinin (CCK)*. CCK, a polypeptide, is a well-known appetite suppressant that exerts its effect by stimulating the brain and digestive organs. *Neuropeptide Y,*

1. *an* = without + *orexia* = appetite
2. *hyper* = excessive + *phagia* = eating

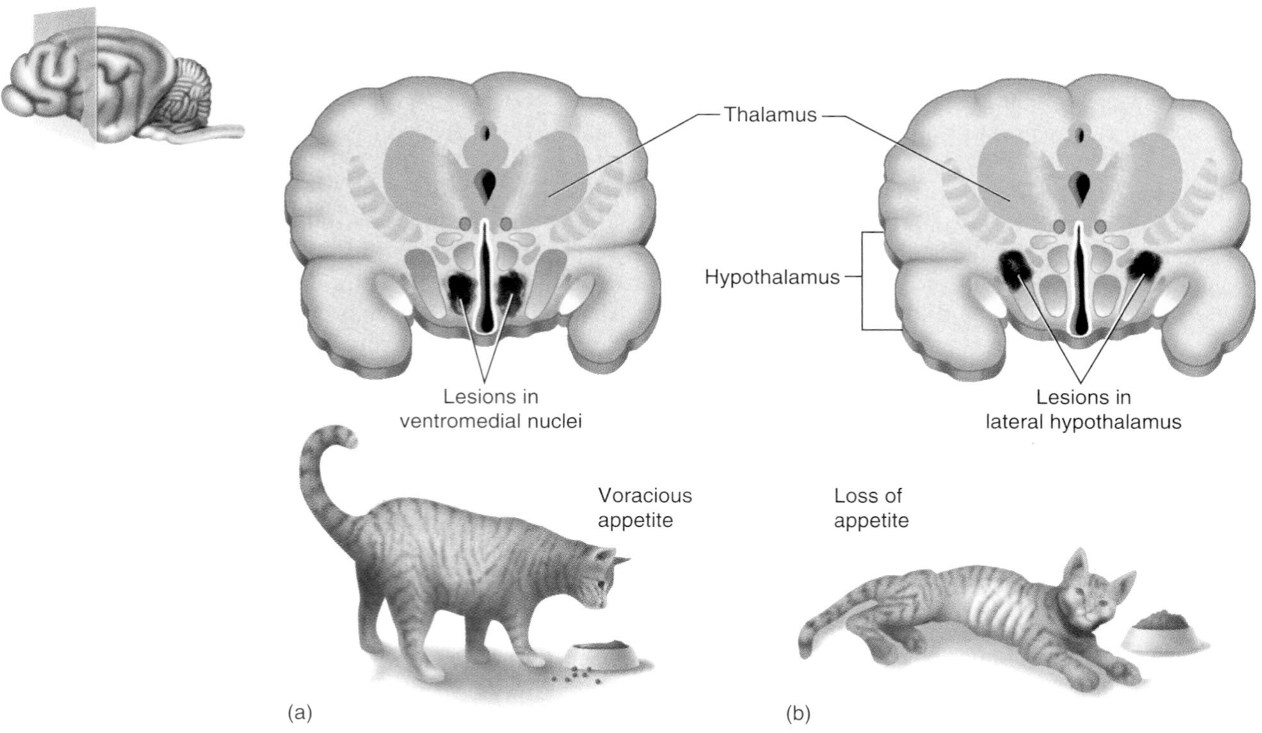

Figure 26.1 Experimental Identification of Appetite-Controlling Regions of the Hypothalamus. (*a*) Destruction of the ventromedial nuclei destroys the sense of satiety and results in extreme overeating. (*b*) Destruction of the lateral hypothalamus destroys the sense of hunger.

produced by the hypothalamus, is a potent appetite stimulant. Its action may be inhibited by a hormone named *leptin* that adipocytes secrete when they have "enough" stored fat (see clinical insight 26.1).

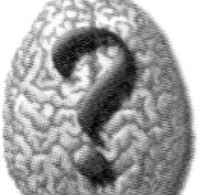

Think About It

Suppose CCK could be economically produced and packaged as tablets to be taken orally. Would this be an effective diet drug? Why or why not?

Appetite is not merely a question of *how much* but also *what kind* of food is consumed. Even animals shift their diets from one kind of food to another, apparently because some foods provide nutrients that others do not. In humans, different neurotransmitters also seem to govern the appetite for different classes of nutrients. For example, norepinephrine stimulates the appetite for carbohydrates, galanin for fatty foods, and endorphins for protein.

Calories

One calorie is the amount of heat that will raise the temperature of 1 g of water 1°C. One thousand calories is called a Calorie (capital *C*) in dietetics and a **kilocalorie** (kcal) in biochemistry. The relevance of calories to physiology is that they are a measure of the capacity to do biological work.

Nearly all dietary calories come from carbohydrates, proteins, and fats. Carbohydrates and proteins yield about 4 kcal/g when they are completely oxidized, and fats yield about 9 kcal/g. Some foods such as sugar and alcohol are said to provide "empty calories" because they provide nothing useful except for calories (see clinical insight 26.4, p. 1012). By suppressing the appetite but failing to provide other nutrients the body requires, they can contribute to malnutrition. In sound nutrition, the body's energy needs are met by more complex foods that simultaneously meet the need for proteins, lipids, vitamins, and other nutrients.

When a chemical is described as **fuel** in this chapter, we mean it is oxidized solely or primarily to extract energy from it. The extracted energy is usually used to make adenosine triphosphate (ATP), which then transfers the energy to other physiological processes (see fig. 3.21).

Nutrients

A **nutrient** is any ingested chemical that is used for growth, repair, or maintenance of the body. Nutrients fall into six major classes: water, carbohydrates, lipids, proteins, minerals, and vitamins (table 26.1). Water, carbohydrates,

Clinical Insight 26.1

Obesity

Obesity is clinically defined as a weight more than 20% above the recommended norm for one's age, sex, and height. The National Institutes of Health estimate that 50% of people living in the United States are overweight and one-third are clinically obese. To judge whether you are one of them, you can calculate your *body mass index (BMI).* If W is your weight in kilograms and H is your height in meters, BMI = W/H^2. (English-metric conversion factors can be found inside the front cover of this book.) A BMI of 20 to 25 kg/m^2 is considered to be optimal for most people. A BMI over 27 kg/m^2 is considered overweight, and above 30 kg/m^2 is considered obese.

Excess weight shortens life expectancy and increases a person's risk of atherosclerosis, hypertension, diabetes mellitus, joint pain and degeneration, kidney stones, and gallstones; cancer of the breast, uterus, and liver in women; and cancer of the colon, rectum, and prostate gland in men. The excess thoracic fat in obese people interferes with breathing and results in increased blood P_{CO_2}, sleepiness, and reduced vitality. Obesity is also a significant obstacle to successful surgery.

A predisposition to obesity is often established by overfeeding in infancy and childhood. During childhood, consumption of excess calories causes the adipocytes to increase in size and number—that is, adipose tissue grows by both hypertrophy and hyperplasia. In adulthood, adipocytes do not divide except in some extreme weight gains; their number remains constant while weight gains and losses result from changes in cell size.

As so many dieters learn, it is very difficult to substantially reduce one's adult weight. Most diets are unsuccessful over the long run as dieters lose and regain the same weight over and over. There are several theories and perhaps equally as many contributing factors for obesity, ranging from psychological to genetic. In the mid-1990s, new genes and hormones were discovered that regulate body weight in mice. Mutations in an obesity *(ob)* gene cause mice to overeat until they weigh as much as three times normal. Another mouse gene, *UCP2,* codes for proteins that cause more dietary calories to be converted to heat instead of ATP synthesis, so animals have to burn more calories to meet their ATP needs and thus store fewer calories as fat. Mutations in genes like these might explain why some people gain more weight than others from the same amount of food. However, mutated *ob* genes and *UCP2* genes have yet to be found in humans.

In both mice and humans, adipocytes produce a protein hormone named *leptin* that seems to inhibit fat storage in adipocytes, possibly by blocking the synthesis of triglycerides from fatty acids. Leptin receptors in the hypothalamus of mice also suggest that this hormone might signal the brain when to reduce the appetite. Leptin injections cause obese mice to slim down markedly. This discovery briefly raised high hopes of weight-controlling injections for humans, but the relationship of leptin to human obesity has proven to be opposite from that seen in mice in some ways. No hormonal treatment for human obesity has emerged yet, but this is a highly active area of biomedical research today, and much more will undoubtedly be known even before this brief summary is published.

Table 26.1 Nutrient Classes and Their Principal Functions

Nutrient	Daily Requirement	Representative Functions
Water	2.5 L	Solvent; coolant; reactant or product in many metabolic reactions (especially hydrolysis and condensation); dilutes and eliminates metabolic wastes; supports blood volume and pressure
Carbohydrates	125–175 g	Fuel; a component of nucleic acids, ATP and other nucleotides, glycoproteins, and glycolipids
Lipids	80–100 g	Fuel; plasma membrane structure; myelin sheaths of nerve fibers; hormones; eicosanoids; bile salts; insulation; protective padding around organs; absorption of fat-soluble vitamins; vitamin D synthesis; some blood-clotting factors
Proteins	44–60 g	Muscle contraction; ciliary and flagellar motility; structure of cellular membranes and extracellular material; enzymes; major component of connective tissues; transport of plasma lipids; some hormones; oxygen binding and transport pigments; blood-clotting factors; blood viscosity and osmolarity; antibodies; immune recognition; neuromodulators; buffers; emergency fuel
Minerals	0.05–3,300 mg	Structure of bones and teeth; component of some structural proteins, hormones, ATP, phospholipids, and other chemicals; cofactors for many enzymes; electrolytes; oxygen transport by hemoglobin and myoglobin; buffers; stomach acid; osmolarity of body fluids
Vitamins	0.002–60 mg	Coenzymes for many metabolic pathways; antioxidants; component of visual pigment; one hormone (vitamin D)

lipids, and proteins are considered **macronutrients** because they must be consumed in relatively large quantities. Minerals and vitamins are called **micronutrients** because only small quantities are required.

Recommended daily allowances (RDAs) of nutrients were first developed in 1943 by the National Research Council and National Academy of Sciences; they have been revised several times since. An RDA is a liberal but safe estimate of the daily intake that would meet the nutritional needs of most healthy people. Consuming less than the RDA of a nutrient does not necessarily mean you will be malnourished, but the probability of malnutrition increases in proportion to the amount of the deficit and how long it lasts.

Many nutrients can be synthesized by the body when they are unavailable from the diet. The body is incapable, however, of synthesizing minerals, most vitamins, eight of the amino acids, and one to three of the fatty acids. These are called **essential nutrients** because it is essential that they be included in the diet.

Carbohydrates

A well-nourished adult has about 440 g of carbohydrate in the body, most of it in three places: about 325 g of muscle glycogen, 90 to 100 g of liver glycogen, and 15 to 20 g of blood glucose.

Sugars function as a structural component of other molecules including nucleic acids, glycoproteins, glycolipids, ATP, and related nucleotides (GTP, cAMP, etc.), and they can be converted to amino acids and fats. Most of the body's carbohydrate, however, serves as fuel—an easily oxidized source of chemical energy. Most cells meet their energy needs from a combination of carbohydrates and fats, but some cells, such as neurons and erythrocytes, depend almost exclusively on carbohydrates. Even a brief period of **hypoglycemia**[3] (deficiency of blood glucose) causes nervous system disturbances felt as weakness or dizziness.

Blood glucose concentration is therefore carefully regulated, mainly through the interplay of insulin and glucagon (see chapter 17 and later in this chapter). Among other effects, these hormones regulate the balance between glycogen and free blood glucose. If blood glucose concentration drops too low, the body draws on its stores of glycogen to meet its energy needs. If glycogen stores are depleted, physical endurance is greatly reduced. Thus it is important to consume enough carbohydrate to ensure that the body maintains adequate stores of glycogen for periods of exercise and fasting (including sleep).

Carbohydrate intake also influences the metabolism of other nutrients. Excess carbohydrate is converted to fat and conversely, fat is oxidized as fuel when glucose and glycogen levels are too low to meet our energy needs. This is why the consumption of starchy and sugary foods has a pronounced effect on body weight. It is unwise, however, to try to "burn off fat" by excessively reducing carbohydrate intake. As shown later in this chapter, the complete and efficient oxidation of fats depends on adequate carbohydrate intake and the presence of certain intermediates of carbohydrate metabolism. If these are lacking, fats are incompletely oxidized to ketone bodies, which may cause metabolic acidosis.

Requirements

Because carbohydrates are rapidly oxidized, they are required in greater amounts than any other nutrient. The RDA is 125 to 175 g. The brain alone consumes about 120 g of glucose per day. Most Americans get about 40% to 50% of their calories from carbohydrates, but highly active people should get up to 60%.

Carbohydrate consumption in the United States has become excessive over the past century due to a combination of fondness for sweets, increased use of sugar in processed foods, and reduced levels of physical activity (see evolutionary insight 26.2). A century ago, Americans consumed an average of 1.8 kg (4 lb) of sugar per year. Now, with sucrose and high-fructose corn syrup so widely used in foods and beverages, the average American ingests 200 to 300 g of carbohydrate per day and the equivalent of 27 kg (60 lb) of table sugar and 21 kg (46 lb) of corn syrup per year. A single nondiet soft drink contains 38 to 43 g (about 8 teaspoons) of sugar per 355 mL (12 oz) serving.

Dietary carbohydrates come in three principal forms discussed in earlier chapters: monosaccharides, disaccharides, and polysaccharides (complex carbohydrates). The only nutritionally significant polysaccharide is starch. Although glycogen is a polysaccharide, only trivial amounts of it are present in cooked meats. Cellulose, another polysaccharide, is not considered a nutrient because it is not digested and never enters the human tissues. Its importance as dietary fiber, however, is discussed shortly.

The three major disaccharides are sucrose, lactose, and maltose. The monosaccharides—glucose, galactose, and fructose—arise mainly from the digestion of starch and disaccharides. The small intestine and liver convert fructose and galactose to glucose, so ultimately all carbohydrate digestion generates glucose. Outside of the hepatic portal system, glucose is the only monosaccharide present in the blood in significant quantity; thus it is known as *blood sugar.* Its concentration is normally maintained at 70 to 110 mg/dL in peripheral venous blood.

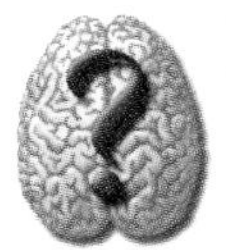

Think About It

Glucose concentration is about 15 to 30 mg/dL higher in arterial blood than in venous blood. Explain why.

3. *hypo* = below normal + *glyc* = sugar + *emia* = blood condition

Evolutionary Insight 26.2

Evolution of the Sweet Tooth

Our craving for sugar doubtlessly originated in our prehistoric ancestors. Not only did they have to work much harder to survive than we do, but high-calorie foods were scarce in their environment and people were at constant risk of starvation. Those who were highly motivated to seek and consume sugary, high-calorie foods passed their "sweet tooth" on to us, their descendents—along with a similarly adaptive appetite for other rare but vital nutrients, namely fat and salt. The tastes that were essential to our ancestors' survival can now be a disadvantage in a culture where salty, fatty, and sugary foods are all too easy to obtain and the food industry is eager to capitalize on these tastes.

Ideally, most carbohydrate intake should be in the form of complex carbohydrates, primarily starch. This is partly because foods that provide starch also usually provide other nutrients. Simple sugars not only provide empty calories but also promote tooth decay. A typical American, however, now obtains only 50% of his or her carbohydrates from starch and the other 50% from sucrose and corn syrup.

Dietary Sources

Nearly all dietary carbohydrates come from plants—particularly grains, legumes, fruits, and root vegetables. Sucrose is refined from sugarcane and sugar beets. Fructose is present in fruits and corn syrup. Maltose is present in some foods such as germinating cereal grains. Lactose is the most abundant solute in cow's milk (about 4.6% lactose by weight).

Fiber

Dietary fiber refers to all fibrous materials of plant and animal origin that resist digestion. Most is plant matter—the carbohydrates cellulose and pectin and such noncarbohydrates as gums and lignin. Although it is not a nutrient, fiber is an essential component of the diet. The recommended daily allowance is about 30 g, but average intake varies greatly from country to country—from 40 to 150 g/day in India and Africa to only 12 g/day in the United States.

Fiber in the intestines absorbs water, swells, softens the stool, and increases its bulk by 40% to 100%. The last effect stretches the colon and stimulates peristalsis, thereby quickening the passage of feces and removal of carcinogenic chemicals from the colon. Fiber itself binds and carries away toxins and other undesirable chemicals, thus reducing the risk of colon cancer.

Pectin is a **water-soluble fiber** found in oats, beans, peas, carrots, brown rice, and fruits. Soluble fiber reduces blood cholesterol and low-density lipoprotein (LDL) levels (see clinical insight 19.5, p. 743). Cellulose, hemicellulose, and lignin, called **water-insoluble fibers,** apparently have no effect on cholesterol or LDLs. While a certain amount of fiber is beneficial to the health, you should not ingest too much. Excess fiber interferes with the absorption of iron, calcium, magnesium, phosphorus, and some trace elements.

Lipids

By weight, the adult body averages about 15% fat in males and 25% fat in females. Fat accounts for most of the body's stored energy. Lesser amounts of phospholipid, cholesterol, and other lipids also play vital structural and physiological roles.

A well-nourished adult meets 80% to 90% of his or her resting energy needs from fat. Fat is superior to carbohydrates for energy storage for two reasons: (1) carbohydrates are hydrophilic, absorb water, and thus expand and occupy more space in the tissues. Fat, however, is hydrophobic, contains almost no water, and is a more compact energy storage substance. (2) Fat is less oxidized than carbohydrate and contains over twice as much energy (9 kcal/g of fat compared with 4 kcal/g of carbohydrate). A man's typical fat reserves contain enough energy for 119 hours of running, whereas his carbohydrate stores would suffice for only 1.6 hours.

Fat has **glucose-sparing** and **protein-sparing effects**—as long as enough fat is available to meet the energy needs of the tissues, protein is not catabolized for fuel and glucose is spared for consumption by cells that cannot use fat, such as neurons.

Vitamins A, D, E, and K are fat-soluble vitamins, which depend on dietary fat for their absorption by the intestine. People who ingest less than 20 g of fat per day are at risk of vitamin deficiency because there is not enough fat in the intestine to transport these vitamins into the body tissues.

Phospholipids and cholesterol are major structural components of plasma membranes and myelin. Cholesterol is also important as a precursor of steroid hormones, bile salts, and vitamin D. Thromboplastin, an essential blood-clotting factor, is a lipoprotein. Two fatty acids—arachidonic acid and linoleic acid—are precursors of prostaglandins and other eicosanoids.

In addition to its metabolic and structural roles, fat has important protective and insulating functions described under adipose tissue in chapter 6.

Requirements

Fat should account for no more than 30% of your daily caloric intake; no more than 10% of your fat intake should be saturated fat; and average cholesterol intake should not exceed 300 mg/day (the amount in one egg yolk). A typical American consumes 30 to 150 g of fat per day, obtains 40% to 50% of his or her calories from fat, and ingests twice as much cholesterol as the recommended limit.

Most fatty acids can be synthesized by the body. **Essential fatty acids** are those we cannot synthesize and therefore must obtain from the diet. These include linoleic acid and possibly linolenic and arachidonic acids; there are differences of opinion about the body's ability to synthesize the last two. We should get 1% to 2% of our daily calories from linoleic acid.

Sources

Saturated fats are predominantly of animal origin. They occur in meat, egg yolks, and dairy products but also in some plant products such as coconut and palm oil (common in nondairy coffee creamers and other products). Processed foods such as hydrogenated oils and vegetable shortening are also high in saturated fat, which is therefore abundant in many baked goods. Unsaturated fats predominate in nuts, seeds, and most vegetable oils. The essential fatty acids are amply provided by the vegetable oils in mayonnaise, salad dressings, and margarine and by whole grains and vegetables.

The richest source of cholesterol is egg yolks, but it is also prevalent in milk products, shellfish (especially shrimp), organ meats such as kidneys, liver, and brains, and other mammalian meat. Cholesterol does not occur in foods of plant origin. It must be noted, however, that saturated fat stimulates cholesterol synthesis. A food may be truthfully advertised as cholesterol-free, but if it is high in saturated fat it can nevertheless raise your blood cholesterol level. Excessive consumption of saturated and unsaturated fats is a risk factor for diabetes mellitus, cardiovascular disease, and breast and colon cancer.

Cholesterol and Serum Lipoproteins

Lipids are an important part of the diet and must be transported to all cells of the body, yet they are hydrophobic and will not dissolve in the aqueous blood plasma. This problem is overcome by complexes called **lipoproteins**—tiny droplets with a core of cholesterol and triglycerides and a coating of proteins and phospholipids. The coating not only enables the lipids to remain suspended in the blood, but also serves as a recognition marker for cells that absorb them.

Lipoproteins are classified into four major categories (and some lesser ones) by their density: **chylomicrons, high-density lipoproteins (HDLs), low-density lipoproteins (LDLs),** and **very low–density lipoproteins (VLDLs).** In table 26.2, notice that the higher the proportion of protein to lipid, the higher the density. (What other trends do you see correlated with their density?) Lipoproteins vary not only in size and density but more importantly in composition and function. Figure 26.2 shows the three primary pathways by which they are made and processed.

Chylomicrons form in the absorptive cells of the small intestine and then pass into the lymphatic system and ultimately the bloodstream (see chapter 25). Endothelial cells of the blood capillaries have a surface enzyme called **lipoprotein lipase** that hydrolyzes triglycerides into glycerol and free fatty acids (FFAs). These products can then pass through the capillary walls into adipocytes, where they are resynthesized into storage triglycerides. Some FFAs, however, remain in the blood plasma bound to albumin. The remainder of a chylomicron after the triglycerides have been extracted, called a *chylomicron remnant,* is removed and degraded by the liver.

VLDLs, produced by the liver, transport lipids to the adipose tissue for storage. When their triglycerides are removed in the adipose tissue, the VLDLs become LDLs and

Table 26.2 Major Classes of Serum Lipoproteins, in Order of Increasing Density

		Approximate Percentage Composition				
Type	**Size (nm)**	**Protein**	**Total Lipid**	**Cholesterol**	**Triglyceride**	**Phospholipid**
Chylomicrons	75–1,200	2	98	5	90	3
VLDLs	30–80	8	92	20	55	17
LDLs	18–25	20	80	53	6	21
HDLs	5–12	50	50	20	5	25

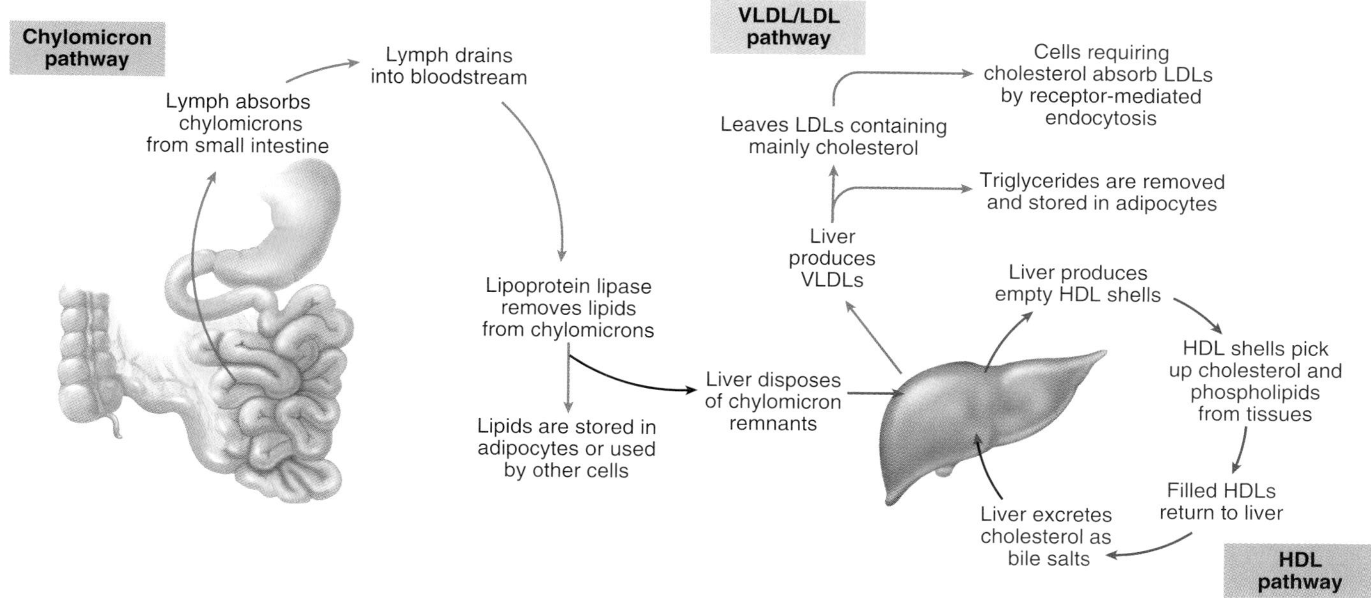

Figure 26.2 Pathways of Lipoprotein Processing.

contain mostly cholesterol. Cells that need cholesterol (usually for membrane structure or steroid hormone synthesis) absorb LDLs by receptor-mediated endocytosis, digest them with lysosomal enzymes, and release the cholesterol for intracellular use.

HDL production begins in the liver, which produces an empty, collapsed protein shell. This shell travels in the blood and picks up cholesterol and phospholipids from other organs. The next time it circulates through the liver, the liver removes the cholesterol and eliminates it in the bile, either as cholesterol or as bile acids. HDLs are therefore a vehicle for removing excess cholesterol from the body.

It is desirable to maintain a total plasma cholesterol concentration of 200 mg/dL or less. From 200 to 239 mg/dL is considered borderline high, and levels over 240 mg/dL are undesirable.

Most of the body's cholesterol is endogenous (internally synthesized) rather than dietary, and to some extent the body compensates for variations in dietary intake. High intake somewhat inhibits hepatic cholesterol synthesis. However, the liver synthesizes a certain amount of cholesterol regardless of intake, and severe restriction of dietary cholesterol does not necessarily result in a proportionate drop in blood cholesterol. Dietary fatty acids also strongly influence cholesterol levels. A moderate reduction of saturated fatty acid intake can lower blood cholesterol by 15% to 20%, while unsaturated fatty acids stimulate cholesterol excretion and lower its concentration in the blood.

Vigorous exercise also lowers blood cholesterol levels. The mechanism is somewhat roundabout: Exercise reduces the sensitivity of the right atrium of the heart to blood pressure, so the heart secretes less atrial natriuretic factor. Consequently, the kidneys excrete less sodium and water, and the blood volume rises. This dilutes the lipoproteins in the blood, and the adipocytes compensate by producing more lipoprotein lipase. Thus, the adipocytes consume more blood triglycerides. This shrinks the VLDL particles, which shed some of their cholesterol in the process, and HDLs pick up this free cholesterol for removal by the liver.

Blood cholesterol is not the only important measure of healthy lipid concentrations, however. A high LDL concentration is a warning sign because, as you can see from the function of LDLs described previously, it signifies a high rate of cholesterol deposition in the arteries. Factors that contribute to high LDL levels include not only saturated fats but also cigarette smoking, coffee, and stress. A high proportion of HDL, on the other hand, is beneficial because it indicates that cholesterol is being removed from the arteries and transported to the liver for disposal. Thus it is desirable to increase your ratio of HDL to LDL. This is best done with a diet low in calories and saturated fats and is promoted by regular aerobic exercise.

Proteins

Protein constitutes about 12% to 15% of the body's mass; 65% of it is in the skeletal muscles. Proteins are responsible for muscle contraction and the motility of cilia and flagella. They are a major structural component of all cellular membranes, with multiple important roles such as

Clinical Insight 26.3

Predigested Protein and Athletic Performance

Some bodybuilders and other "power athletes" take powdered or liquid amino acid mixtures ("predigested protein") in the belief that simple amino acids are absorbed more easily and rapidly or that they somehow contribute more to muscle building. Such beliefs are unfounded. Dietary proteins are rapidly digested and absorbed, and there is no added benefit to taking predigested protein. Amino acid supplements have not been shown to increase muscle mass, strength, or endurance. Moreover, they may be harmful to the health. Concentrated amino acid solutions osmotically retain water in the intestines and cause cramps and diarrhea. The catabolism of excess amino acids produces extra nitrogenous waste, which places undesirable stress on the liver and kidneys.

membrane receptors, pumps, ion channels, and cell-identity markers. Fibrous proteins such as collagen, elastin, and keratin make up much of the structure of bone, cartilage, tendons, ligaments, skin, hair, and nails. Globular proteins include antibodies, hormones, neuromodulators, hemoglobin, myoglobin, and about 2,000 enzymes that control nearly every aspect of cellular metabolism. They also include the albumin and other plasma proteins that maintain blood viscosity and osmolarity and transport lipids and some other plasma solutes. Proteins buffer the pH of body fluids and contribute to the resting membrane potentials of all cells. No other class of biomolecules has such a broad variety of functions.

Requirements

For persons of average weight, the RDA of protein is 44 to 60 g, depending on age and sex. Multiplying your weight in pounds by 0.37 or your weight in kilograms by 0.8 gives an estimate of your RDA of protein in grams. A higher intake is recommended, however, under conditions of stress, infection, injury, and pregnancy. Infants and children require more protein than adults relative to body weight.

Total protein intake is not the only significant measure of dietary adequacy. The nutritional value of a protein depends on whether it supplies the right amino acids in the proportions needed for human proteins. Adults can synthesize 12 of the 20 amino acids from other organic compounds when they are not available from the diet, but there are 8 **essential amino acids** that we cannot synthesize: isoleucine, leucine, lysine, methionine, phenylalanine, threonine, tryptophan, and valine. (Infants also require histidine.) In addition, there are 2 amino acids that can only be synthesized from essential amino acids—cystine from methionine and tyrosine from phenylalanine. The other 10 (9 in infants) are called **inessential amino acids**—not because the body does not require them but because it can synthesize its own when the diet does not supply them.

Cells do not store surplus amino acids for later use. When a protein is to be synthesized, all of the amino acids necessary must be present at once, and if even one is missing, the protein cannot be made. High-quality **complete proteins** are those that provide all of the essential amino acids in the necessary proportions for human tissue growth, maintenance, and nitrogen balance. Lower-quality **incomplete proteins** lack one or more essential amino acids. For example, cereals are low in lysine, and legumes are low in methionine.

Protein quality is also determined by **net protein utilization**—the percent of the amino acids in a protein that the human body uses. We typically use 70% to 90% of animal protein but only 40% to 70% of plant protein. It therefore takes a larger serving of plant protein than animal protein to meet our needs—for example, we need 400 g (about 14 oz) of rice and beans to provide as much usable protein as 115 g (about 4 oz) of hamburger. However, reducing meat intake and increasing plant intake has advantages. Among other considerations, plant foods provide more vitamins, minerals, and fiber; less saturated fat; no cholesterol; and less pesticide. In an increasingly crowded world, it must also be borne in mind that it requires far more land to produce meat than to produce crops.

Dietary Sources

The animal proteins of meat, eggs, and dairy products closely match human proteins in amino acid composition. Thus animal products provide high-quality complete protein, whereas plant proteins are incomplete. Nevertheless, this does not mean that your dietary protein *must* come from meat; indeed, about two-thirds of the world's population receives adequate protein nutrition from diets containing very little meat. We can combine plant foods so that one provides what another lacks—beans and rice, for example, are a complementary combination of legume and cereal. Beans provide the isoleucine and lysine lacking in grains, while rice provides the tryptophan and cysteine lacking in beans.

An exclusively vegetarian diet cannot meet all of your nutritional needs unless supplemented with vitamin B_{12} and essential amino acids. Supplementing a vegetarian diet with milk and one egg a day (an *ovolactovegetar-*

ian diet), however, can provide an adequate intake of complete protein.

Nitrogen Balance

Proteins are our chief dietary source of nitrogen. **Nitrogen balance** is a state in which the rate of nitrogen ingestion equals the rate of excretion (chiefly as nitrogenous wastes). Growing children exhibit a state of **positive nitrogen balance** because they ingest more than they excrete, thus retaining protein for tissue growth. Pregnant women and athletes in resistance training also show positive nitrogen balance. When excretion exceeds ingestion, a person is in a state of **negative nitrogen balance.** This indicates that body proteins are being broken down and used as fuel. Proteins of the muscles and liver are more easily broken down than others; thus negative nitrogen balance tends to be associated with muscle atrophy. Negative nitrogen balance may occur if carbohydrate and fat intake are insufficient to meet the need for energy. Carbohydrates and fats are said to have a protein-sparing effect because they prevent protein catabolism when present in sufficient amounts to meet energy needs.

Nitrogen balance is affected by some hormones. Growth hormones and sex steroids promote protein synthesis and positive nitrogen balance during childhood, adolescence, and pregnancy. Glucocorticoids, on the other hand, promote protein catabolism and negative nitrogen balance in states of stress.

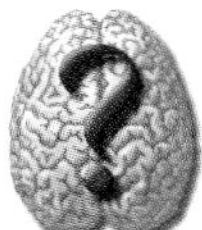

Think About It

Would you expect a person recovering from a long infectious disease to be in a state of positive or negative nitrogen balance? Why?

Minerals and Vitamins

Minerals are inorganic elements and vitamins are small organic compounds. Neither is used as fuel, but both are essential to our ability to use other nutrients. With the exception of a few vitamins, these nutrients cannot be synthesized by the body and must be included in the diet. They are, however, required in relatively small quantities. Mineral RDAs range from 0.05 mg of chromium and selenium to 1,200 mg of calcium and phosphorus. Vitamin RDAs range from about 0.002 mg of vitamin B_{12} to 60 mg of vitamin C. Despite the small quantities involved, minerals and vitamins have very potent effects on physiology. Indeed, excessive amounts are toxic and potentially lethal.

Minerals

Minerals constitute about 4% of the body mass, with three-quarters of this being the calcium and phosphorus in the bones and teeth. Phosphorus is also a key structural component of phospholipids, ATP, cAMP, GTP, and creatine phosphate and is the basis of the phosphate buffer system (see chapter 24). Calcium, iron, magnesium, and manganese function as cofactors for enzymes. Iron is essential to the oxygen-carrying capacity of hemoglobin and myoglobin. Chlorine is a component of stomach acid (HCl). Many mineral salts function as electrolytes and thus govern the function of nerve and muscle cells, osmotically regulate the content and distribution of water in the body, and maintain blood volume.

Table 26.3 summarizes adult mineral requirements and dietary sources. Broadly speaking, the best sources of minerals are vegetables, legumes, milk, eggs, fish, shellfish, and some other meats. Cereals are a relatively poor source.

Sodium chloride has been both a prized commodity and a curse. Animal tissues contain relatively large amounts of salt, and carnivores rarely lack ample salt in their diets. Plants, however, are relatively poor in salt, so herbivores often must supplement their diet by ingesting salt from the soil. As humans developed agriculture and became more dependent on plants, they also became increasingly dependent on supplemental salt. Salt has often been used as a form of payment for goods and services—the word *salary* comes from *sal* (salt). Our fondness for salt and high sensitivity to it undoubtedly stem from its physiological importance and its scarcity in a largely vegetarian diet.

Now, however, this fondness has become a bane. The recommended sodium intake is 1.1 g/day, but a typical American diet contains about 4.5 g/day. This is due not just to the use of table salt but more significantly to the large amounts of salt in processed foods, much of it "disguised" in soy sauce, MSG (monosodium glutamate), baking soda, and baking powder. In some areas of Japan, salt intake averages 27 g/day and the great majority of people die before age 70 of stroke and other complications of hypertension.

Hypertension is a leading cause of death among American blacks, who have twice the risk of hypertension and 10 times the risk of dying from it that American whites have. The reason for this is not excessive salt intake, but rather that people of West African descent have kidneys with an especially strong tendency to retain salt.

Vitamins

Vitamins were originally named with letters in the order of their discovery, but they also have chemically descriptive names such as ascorbic acid (vitamin C) and riboflavin (vitamin B_2). Most vitamins must be obtained from the diet (table 26.4), but the body synthesizes some of them from precursors called *provitamins*—niacin from the amino acid tryptophan, vitamin D from cholesterol, and vitamin A from carotene, which is abundantly present in carrots, squash, and other yellow vegetables and fruits. Vitamin K, pantothenic acid, biotin, and folic acid are produced by the

Table 26.3 Mineral Requirements and Some Dietary Sources

Mineral	RDA (mg)	Some Dietary Sources*
Major Minerals		
Calcium	1,200	Milk, fish, shellfish, greens, tofu, orange juice
Phosphorus	1,200	Red meat, poultry, fish, eggs, milk, legumes, whole grains, nuts
Sodium	1,100	Table salt, processed foods; usually present in excess
Chloride	700	Table salt, some vegetables; usually present in excess
Magnesium	280–350	Milk, greens, whole grains, nuts, legumes, chocolate
Potassium	Unknown	Red meat, poultry, fish, cereals, spinach, squash, bananas, apricots
Sulfur	Unknown	Meats, milk, eggs, legumes; almost any proteins
Trace Minerals		
Zinc	12–15	Red meat, seafood, cereals, wheat germ, legumes, nuts, yeast
Iron	10–15	Red meat, liver, shellfish, eggs, dried fruits, legumes, nuts, molasses
Manganese	2.5–5.0	Greens, fruits, legumes, whole grains, nuts
Copper	1.5–3.0	Red meat, liver, shellfish, legumes, whole grains, nuts, cocoa
Fluoride	1.5–4.0	Fluoridated water and toothpaste, tea, seafood, seaweed
Iodine	0.15	Marine fish, fish oils, shellfish, iodized salt
Molybdenum	0.07–0.25	Beans, whole grains, nuts
Chromium	0.05–0.25	Meats, liver, cheese, eggs, whole grains, yeast, wine
Selenium	0.05–0.07	Red meats, organ meats, fish, shellfish, eggs, cereals
Cobalt	Unknown	Red meat, poultry, fish, liver, milk

*"Red meat" refers to mammalian muscle such as beef and pork. "Organ meat" refers to brain, pancreas, heart, kidney, etc. Liver is specified separately and refers to beef, pork, and chicken livers, which are similar for most nutrients.

bacteria of the large intestine. The feces contain more biotin than food does.

Vitamins are classified as water-soluble or fat-soluble. **Water-soluble vitamins** are absorbed with water from the small intestine, dissolve freely in the body fluids, and are quickly excreted by the kidneys. They cannot be stored in the body and therefore seldom accumulate to excess. The water-soluble vitamins are ascorbic acid (vitamin C) and the B vitamins. Ascorbic acid promotes hemoglobin synthesis, collagen synthesis, and sound connective tissue structure, and it is an antioxidant that scavenges free radicals and possibly reduces the risk of cancer. The B vitamins function as coenzymes or parts of coenzyme molecules; they assist enzymes by transferring electrons from one metabolic reaction to another, making it possible for enzymes to catalyze these reactions. Some of their functions arise later in this chapter as we consider carbohydrate metabolism.

Fat-soluble vitamins are incorporated into lipid micelles in the small intestine and absorbed with dietary lipids. They are more varied in function than water-soluble vitamins. Vitamin A is a component of the visual pigments and promotes proteoglycan synthesis and epithelial maintenance. Vitamin D promotes calcium absorption and bone mineralization. Vitamin K is essential to prothrombin synthesis and blood clotting. Vitamins A and E are antioxidants, like ascorbic acid.

It is common knowledge that various diseases result from vitamin deficiencies, but it is less commonly known that **hypervitaminosis** (vitamin excesses) also causes disease. A *deficiency* of vitamin A, for example, can result in night blindness, dry skin and hair, a dry conjunctiva and cloudy cornea, and increased incidence of urinary, digestive, and respiratory infections. This is the world's most common vitamin deficiency. An *excess* of vitamin A, however, may cause anorexia, nausea and vomiting, headache, pain and fragility of the bones, hair loss, an enlarged liver and spleen, and birth defects. Vitamins B_6, C, D, and E have also been implicated in toxic hypervitaminosis.

Some people take *megavitamins*—doses 10 to 1,000 times the RDA—thinking that they will improve athletic performance. Since vitamins are not burned as fuel, and small amounts fully meet the body's metabolic needs, there is no evidence that vitamin supplements improve performance except when used to correct a dietary deficiency. Indeed, megadoses of fat-soluble vitamins can be extremely harmful.

Before You Go On

Answer the following questions to test your understanding of the preceding section:

1. What regions of the hypothalamus regulate hunger and satiety? Why would it be wrong to say these are the sole controls over appetite?
2. Explain the following statement: Cellulose is an important part of a healthy diet but it is not a nutrient.
3. What class of nutrients provides most of the calories in the diet? What class of nutrients provides the body's major reserves of stored energy?
4. Contrast the functions of VLDLs, LDLs, and HDLs. Explain how this is related to the fact that a high blood HDL level is desirable, but a high VLDL-LDL level is undesirable?
5. Why do some proteins have more nutritional value than others?

Table 26.4 Vitamin Requirements and Some Dietary Sources

Mineral	RDA (mg)	Some Dietary Sources*
Water-Soluble Vitamins		
Ascorbic acid (C)	60	Citrus fruits, strawberries, tomatoes, greens, cabbage, cauliflower, broccoli, Brussels sprouts
B complex		
Thiamine (B_1)	1.5	Red meat, organ meats, liver, eggs, greens, asparagus, legumes, whole grains, seeds, yeast
Riboflavin (B_2)	1.7	Widely distributed, and deficiencies are rare; all types of meat, milk, eggs, greens, whole grains, apricots, legumes, mushrooms, yeast
Pyridoxine (B_6)	2.0	Red meat, organ meats, fish, liver, greens, apricots, legumes, whole grains, seeds
Cobalamin (B_{12})	0.002	Red meat, organ meats, liver, shellfish, eggs, milk; absent from food plants
Niacin (nicotinic acid)	19	Readily synthesized from tryptophan, which is present in any diet with adequate protein; red meat, organ meats, liver, poultry, fish, apricots, legumes, whole grains, mushrooms
Panthothenic acid	4–7	Widely distributed, and deficiencies are rare; red meat, organ meats, liver, eggs, green and yellow vegetables, legumes, whole grains, mushrooms, yeast
Folic acid (folacin)	0.2	Eggs, liver, greens, citrus fruits, legumes, whole grains, seeds
Biotin	0.03–0.10	Red meat, organ meats, liver, eggs, cheese, cabbage, cauliflower, bananas, legumes, nuts
Fat-Soluble Vitamins		
Vitamin A (retinol)	1.0	Fish oils, eggs, cheese, milk, greens, other green and yellow vegetables and fruits, margarine
Vitamin D (calcitriol)	0.01	Formed by exposure of skin to sunlight; fish, fish oils, milk
Vitamin E (α-tocopherol)	10	Fish oils, greens, seeds, wheat germ, vegetable oils, margarine, nuts
Vitamin K (phylloquinone)	0.08	Most of the RDA is met by synthesis by intestinal bacteria; liver, greens, cabbage, cauliflower

*See footnote in table 26.3.

Carbohydrate Metabolism

Objectives

When you have completed this section, you should be able to

- describe the principal reactants and products of each major step of glucose oxidation;
- contrast the functions and products of anaerobic fermentation and aerobic respiration;
- explain where and how cells produce ATP; and
- describe the production, function, and use of glycogen.

Most dietary carbohydrate is burned as fuel within a few hours of absorption. Although three monosaccharides are absorbed from digested food—glucose, galactose, and fructose—the last two are quickly converted to glucose, and all oxidative carbohydrate consumption is essentially a matter of glucose catabolism. The overall reaction for this is

$$C_6H_{12}O_6 + 6\ O_2 \rightarrow 6\ CO_2 + 6\ H_2O$$

The function of this reaction is not to produce carbon dioxide and water but to transfer energy from glucose to ATP.

Along the pathway of glucose oxidation are several links through which other nutrients—especially fats and amino acids—can also be oxidized as fuel. Carbohydrate catabolism therefore provides a central vantage point from which we can view the catabolism of all fuels and the generation of ATP.

Glucose Catabolism

If the preceding reaction were carried out in a single step, it would generate a short, intense burst of heat—like the burning of paper, which has the same chemical equation. Not only would this be useless to the body's metabolism, it would kill the cells. In the body, however, the process is carried out in a series of small steps, each controlled by a separate enzyme. Energy is released in small manageable amounts, and as much as possible is transferred to ATP. The rest is released as heat.

There are three major pathways of glucose catabolism:

1. **glycolysis,** which splits a glucose molecule into two molecules of pyruvic acid;
2. **anaerobic fermentation,** which occurs in the absence of oxygen and reduces pyruvic acid to lactic acid; and
3. **aerobic respiration,** which occurs in the presence of oxygen and oxidizes pyruvic acid to carbon dioxide and water.

You may find it helpful to review figure 3.22 (see p. 98) for a broad overview of these processes and their relationship to ATP production. Figures 26.3 to 26.6 examine these processes in closer detail. The first two figures are labeled with numbers that correspond to reaction steps described shortly.

Coenzymes are vitally important to these reactions. Enzymes remove electrons (as hydrogen atoms) from the intermediate compounds of these pathways, but they do not bind them. Instead, they transfer the hydrogen atoms to coenzymes, and the coenzymes donate them to other compounds later in one of the reaction pathways. Thus the enzymes of glucose catabolism cannot function without their coenzymes.

The two coenzymes of special importance to glucose catabolism are **NAD^+** (nicotinamide adenine dinucleotide) and **FAD** (flavin adenine dinucleotide). Both are derived from B vitamins: NAD^+ from niacin and FAD from riboflavin. Hydrogen atoms are removed from metabolic intermediates in pairs—that is, two protons and two electrons ($2\ H^+$ and $2\ e^-$) at a time—and transferred to a coenzyme. This produces a reduced coenzyme with a higher free energy content than it had before the reaction. Coenzymes thus become the temporary carriers of the energy extracted from glucose metabolites. The reactions for this are

$$FAD + 2\ H \rightarrow FADH_2$$

and

$$NAD^+ + 2\ H \rightarrow NADH + H^+$$

FAD binds two protons and two electrons to become $FADH_2$. NAD^+, however, binds the two electrons but only one of the protons to become NADH. The other proton remains a free hydrogen ion, H^+ (or H_3O^+, but it is represented in this chapter as H^+).

Glycolysis

Upon entering a cell, glucose begins a series of conversions called **glycolysis**[4] (fig. 26.3):

Step 1: phosphorylation. The enzyme *hexokinase* transfers an inorganic phosphate group (P_i) from ATP to glucose, producing glucose 6-phosphate (G6P). This has two effects:

- It keeps the intracellular concentration of glucose very low, thus maintaining a concentration gradient that favors the continued diffusion of more glucose into the cell.
- Phosphorylated compounds cannot pass through the plasma membrane, so this prevents the sugar from leaving the cell. In most cells, step 1 is irreversible

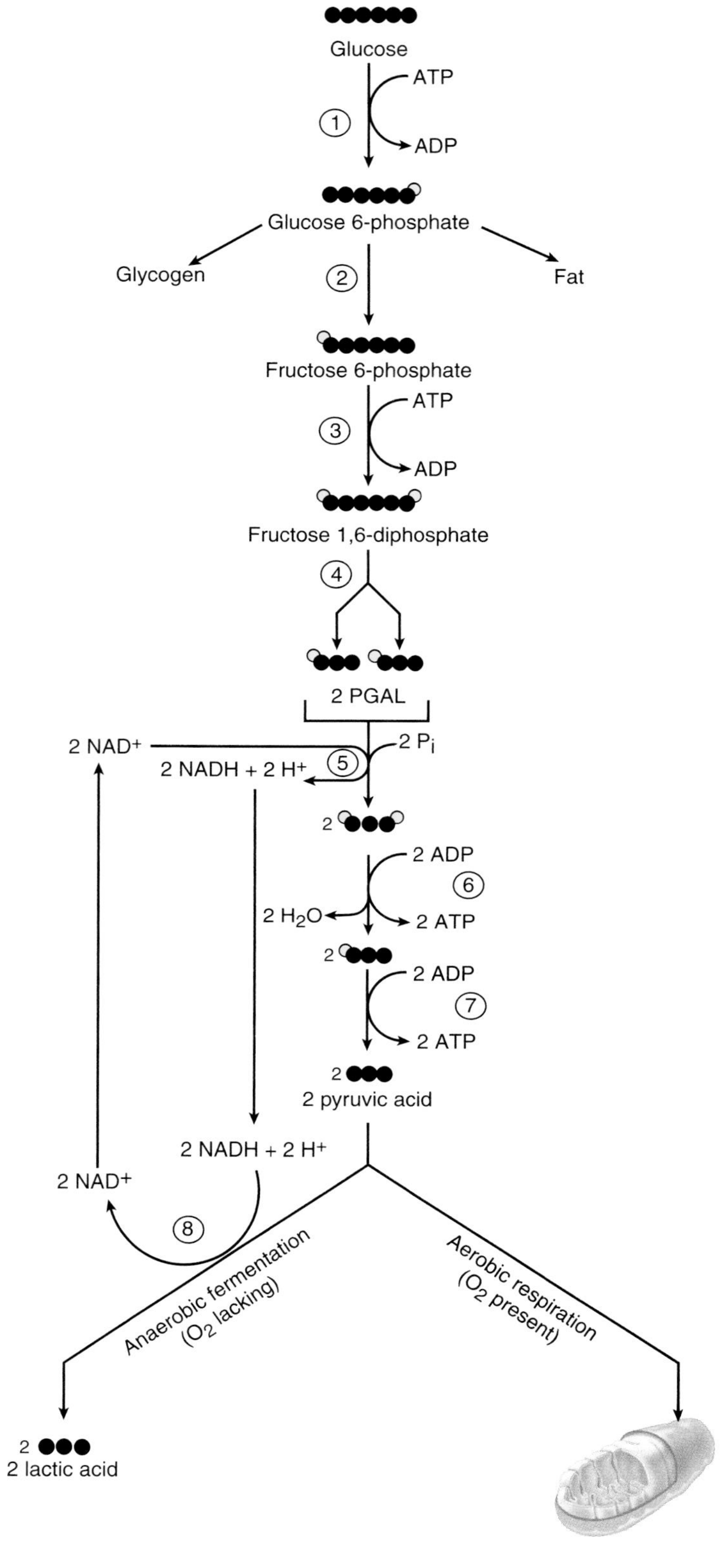

Figure 26.3 Glycolysis and Anaerobic Fermentation. The *black circles* represent carbon atoms in the carbon skeleton of each molecule; the *yellow circles* indicate phosphate groups. Numbered reaction steps are explained in the text.

4. *glyco* = sugar + *lysis* = splitting

because they lack the enzyme to convert G6P back to glucose. The few exceptions are cells that must be able to release free glucose to the blood: absorptive cells of the small intestine, proximal convoluted tubule cells in the kidney, and liver cells.

G6P is a versatile molecule that can be converted to fat or amino acids, polymerized to form glycogen for storage, or further oxidized to extract its energy. For now, we are mainly concerned with its further oxidation (glycolysis), the general effect of which is to split G6P (a six-carbon sugar, C_6) into two three-carbon (C_3) molecules of **pyruvic acid (pyruvate).** Continue tracing these steps in figure 26.3 as you read.

Steps 2 and 3: priming. G6P is rearranged (isomerized) to form fructose 6-phosphate, which is phosphorylated again to form fructose 1,6-diphosphate. This "primes" the process by providing activation energy, somewhat like the heat of a match used to light a fireplace. Two molecules of ATP have already been consumed, but just as a fire gives back more heat than it takes to start it, aerobic respiration eventually gives back far more ATP than it takes to prime glycolysis.

Step 4: cleavage. The "lysis" part of glycolysis occurs when fructose 1,6-diphosphate splits into two three-carbon (C_3) molecules. Through a slight rearrangement of one of them (not shown in the figure), this generates two molecules of **PGAL (phosphoglyceraldehyde,** or **glyceraldehyde 3-phosphate).**

Step 5: oxidation. Each PGAL molecule is then oxidized by removing a pair of hydrogen atoms. The electrons and one proton are picked up by NAD^+ and the other proton is released into the cytosol, yielding $NADH + H^+$. At this step, a phosphate (P_i) group is also added to each of the C_3 fragments. Unlike the earlier steps, this P_i is not supplied by ATP but comes from the cell's pool of free phosphate ions.

Steps 6 and 7: dephosphorylation. In the next two steps, phosphate groups are taken from the glycolysis intermediates and transferred to ADP, phosphorylating it to ATP. This converts the C_3 compound to pyruvic acid. The end products of glycolysis are therefore

$$2 \text{ pyruvic acid} + 2 \text{ NADH} + 2 \text{ H}^+ + 2 \text{ ATP}$$

Note that 4 ATP are actually produced (steps 6–7), but 2 ATP were consumed to initiate glycolysis (steps 1 and 3), so the net gain is 2 ATP per glucose.

Some of the energy originally in the glucose is contained in this ATP, some is in the NADH, and some is lost as heat. Most of the energy, however, remains in the pyruvic acid.

Anaerobic Fermentation

The fate of pyruvic acid depends on whether or not oxygen is available. In an exercising muscle, the demand for ATP may exceed the supply of oxygen, and the only ATP the cells can make under these circumstances is the 2 ATP produced by glycolysis. Cells that lack mitochondria, such as erythrocytes, are also restricted to making ATP by this method.

But glycolysis would quickly come to a halt if the reaction stopped at pyruvic acid. Why? Because it would use up the supply of NAD^+, which is needed to accept electrons at step 5 and keep glycolysis going. NAD^+ must be replenished.

In the absence of oxygen, a cell resorts to a one-step reaction called anaerobic fermentation. (This is often inaccurately called *anaerobic respiration,* but strictly speaking, human cells do not carry out anaerobic respiration; that is a process found only in certain bacteria.) In this pathway (fig. 26.3, **step 8**), NADH donates a pair of electrons to pyruvic acid, thus reducing it to **lactic acid** and regenerating NAD^+.

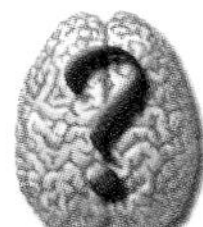

Think About It

Does lactic acid have more free energy than pyruvic acid or less? Explain.

Lactic acid leaves the cells that generate it and travels by way of the bloodstream to the liver. When oxygen becomes available again, the liver oxidizes lactic acid back to pyruvic acid, which can then enter the aerobic pathway described shortly. The oxygen required to do this is part of the *oxygen debt* created by exercising skeletal muscles (see p. 441). The liver can also convert lactic acid back to G6P and can do either of two things with that: (1) polymerize it to form glycogen for storage or (2) remove the phosphate group and release free glucose into the blood.

Although anaerobic fermentation keeps glycolysis running a little longer, it has some drawbacks. One is that it is wasteful, because most of the energy of glucose is still in the lactic acid and has contributed no useful work. The other is that lactic acid is toxic and contributes to muscle fatigue.

Skeletal muscle is relatively tolerant of anaerobic fermentation, and cardiac muscle is less so. The brain employs almost no anaerobic fermentation. During birth, when the infant's blood supply is cut off, almost every organ of the body switches to anaerobic fermentation; thus they do not compete with the brain for the limited supply of oxygen.

Aerobic Respiration

Most ATP is generated in the mitochondria, which require oxygen as the final electron acceptor. In the presence of oxygen, pyruvic acid enters the mitochondria and is oxidized by aerobic respiration. This occurs in two principal steps:

- a group of reactions we will call the **matrix reactions,** because their controlling enzymes are in the fluid of the mitochondrial matrix; and

- reactions we will call the **membrane reactions,** because their controlling enzymes are bound to the membranes of the mitochondrial cristae.

The Matrix Reactions

The matrix reactions are shown in figure 26.4, where the reaction steps are numbered to resume where figure 26.3 ended. Most of the matrix reactions constitute a series called the **citric acid (Krebs[5]) cycle.** Preceding this, however, are three steps that prepare pyruvic acid to enter the cycle and thus link glycolysis to it.

Step 9. Pyruvic acid is *decarboxylated*—CO_2 is removed and pyruvic acid, a C_3 compound, becomes a C_2 compound.

Step 10. NAD^+ removes hydrogen atoms from the C_2 compound (an oxidation reaction) and converts it to an **acetyl group (acetic acid).**

Step 11. The acetyl group binds to coenzyme A, a derivative of pantothenic acid (a B vitamin). The result is **acetyl-coenzyme A (acetyl-CoA).** At this stage the C_2 remnant of the original glucose molecule is ready to enter the citric acid cycle.

Step 12. At the beginning of the citric acid cycle, CoA hands off the acetyl (C_2) group to a C_4 compound, **oxaloacetic acid.** This produces the C_6 compound **citric acid,** for which the cycle is named.

Step 13. Water is removed and the citric acid molecule is reorganized, but it still retains its six carbon atoms.

Step 14. Hydrogen atoms are removed and accepted by NAD^+.

Step 15. Another CO_2 is removed and the substrate becomes a five-carbon chain.

Steps 16 and 17. Steps 14 and 15 are essentially repeated, generating another free CO_2 molecule and leaving a four-carbon chain. No more carbon atoms are removed beyond this point; the substrate remains a series of C_4 compounds from here back to the start of the cycle. The three carbon atoms of pyruvic acid have all been removed as CO_2 at steps 9, 15, and 17. These *decarboxylation reactions* are the source of most of the CO_2 in your breath.

Step 18. Some of the energy in the C_4 substrate goes to phosphorylate guanosine diphosphate (GDP) and convert it to guanosine triphosphate (GTP), a molecule similar to ATP. GTP quickly transfers the P_i group to ADP to make ATP. Coenzyme A participates again in this step but is not shown in the figure.

Step 19. Two hydrogen atoms are removed and accepted by the coenzyme FAD.

Step 20. Water is added.

5. Sir Hans Krebs (1900–1981), German biochemist

Step 21. A final two hydrogen atoms are removed and transferred to NAD^+. This reaction generates oxaloacetic acid, which is available to start the cycle all over again.

It is important to remember that for every glucose molecule that entered glycolysis, all of these matrix reactions occur twice (once for each pyruvic acid). The matrix reactions can be summarized:

$$2 \text{ pyruvate} + 6\ H_2O \rightarrow 6\ CO_2$$
$$+ 2\ ADP + 2\ P_i \rightarrow 2\ ATP$$
$$+ 8\ NAD^+ + 8\ H_2 \rightarrow 8\ NADH + 8\ H^+$$
$$+ 2\ FAD + 2\ H_2 \rightarrow 2\ FADH_2$$

There is nothing left of the organic matter of the glucose; its carbon atoms have all been carried away as CO_2 and exhaled. Although still more of its energy is lost as heat along the way, some is stored in the additional 2 ATP, and most of it, by far, is in the reduced coenzymes—8 NADH and 2 $FADH_2$ molecules generated by the matrix reactions and 2 NADH generated by glycolysis. These must be oxidized to extract the energy from them.

The citric acid cycle not only oxidizes glucose metabolites but is also a pathway and a source of intermediates for the synthesis of fats and nonessential amino acids. The connections between the citric acid cycle and the metabolism of other nutrients are discussed later.

The Membrane Reactions

The membrane reactions have two purposes: (1) to further oxidize NADH and $FADH_2$ and transfer their energy to ATP and (2) to regenerate NAD^+ and FADH and make them available again to earlier reaction steps. The membrane reactions are carried out by a series of compounds called the **mitochondrial electron-transport chain.** Most members of the chain are bound to the inner mitochondrial membrane. They are arranged in a precise order that enables each one to receive a pair of electrons from the member on one side of it (or, in two cases, from NADH and $FADH_2$) and pass these electrons along to the member on the other side—like a row of people passing along a hot potato. By the time the "potato" reaches the last member in the chain it is relatively "cool"—its energy has been used to make ATP.

The members of this transport chain are as follows:

- **Flavin mononucleotide (FMN),** a derivative of riboflavin similar to FAD, bound to a membrane protein. FMN accepts electrons from NADH.
- **Iron-sulfur (Fe-S) centers,** complexes of iron and sulfur atoms bound to membrane proteins.
- **Coenzyme Q (CoQ),** which accepts electrons from $FADH_2$. Unlike the other members, this is a relatively small, mobile molecule that moves about in the membrane.

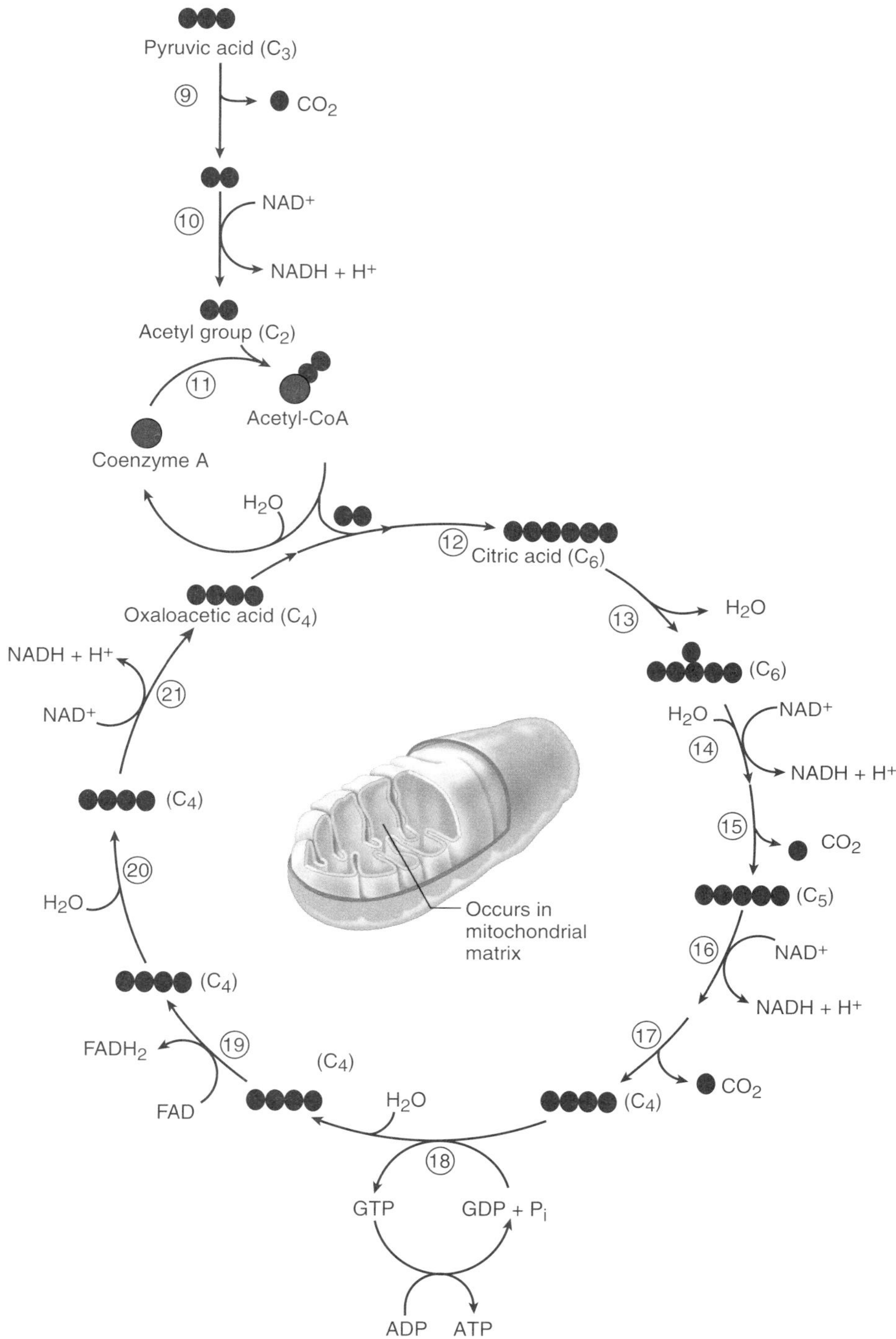

Figure 26.4 **The Mitochondrial Matrix Reactions.** *Black circles* represent carbon atoms in the carbon skeleton of each molecule. Numbered reaction steps are explained in the text.

- **Copper (Cu) ions,** bound to two membrane proteins.
- **Cytochromes,**[6] five enzymes with iron cofactors, so-named because they are brightly colored in pure form. In order of participation in the chain, they are cytochromes b, c_1, c, a, and a_3.

Electron Transport Figure 26.5 shows the order in which electrons are passed along the chain. Hydrogen atoms are split apart as they transfer from coenzymes to the chain. The protons are released to the mitochondrial matrix and the electrons travel in pairs ($2\ e^-$) along the transport chain. Each electron carrier in the chain becomes reduced when it receives an electron pair and oxidized again when it passes the electrons along to the next carrier. Energy is liberated at each transfer.

The final electron acceptor in the chain is oxygen. Each oxygen atom (half of an O_2 molecule) accepts two electrons ($2\ e^-$) from cytochrome a_3 and two protons ($2\ H^+$) from the mitochondrial matrix. The result is a molecule of water:

$$1/2\ O_2 + 2\ e^- + 2\ H^+ \rightarrow H_2O$$

This is the body's primary source of *metabolic water*—water synthesized in the body rather than ingested in food and drink. This reaction also explains why the body requires oxygen. Without it, this reaction stops and, like a traffic jam, stops all the other processes leading to it. As a result, a cell produces too little ATP to sustain life, and death can ensue within a few minutes.

The Chemiosmotic Mechanism Of primary importance is what happens to the energy liberated by the electrons as they pass along the chain. Some of it is unavoidably lost as heat, but some of it drives **proton pumps.** The carriers of the electron-transport chain are clustered in three groups called **respiratory enzyme complexes.** The first complex includes FMN and five or more Fe-S centers; the second complex includes cytochromes b and c_1 and an Fe-S center; and the third complex includes two copper centers and cytochromes a and a_3. Each complex collectively acts as a proton pump that removes H^+ from the mitochondrial matrix and pumps it into the space between the inner and outer mitochondrial membranes (fig. 26.6). Coenzyme Q is a shuttle that transfers electrons from the first pump to the second, and cytochrome c shuttles electrons from the second pump to the third.

These pumps create a very high H^+ concentration (low pH) and positive charge between the membranes compared to a low H^+ concentration and negative charge in the mitochondrial matrix. That is, they create a steep electrochemical gradient across the inner mitochondrial membrane. If the inner membrane were freely permeable to H^+, these ions would have a strong tendency to diffuse down this gradient and back into the matrix.

The inner membrane, however, is permeable to H^+ only through specific channel proteins called **ATP synthase** (separate from the electron-transport system). As H^+

6. *cyto* = cell + *chrom* = color

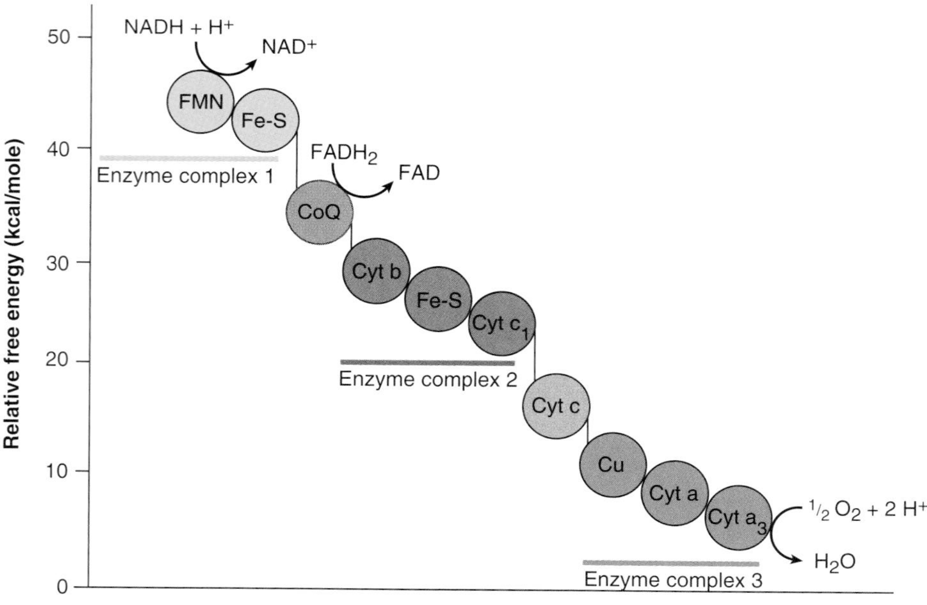

Figure 26.5 The Mitochondrial Electron-Transport Chain. Transport molecules are grouped into three enzyme complexes, each of which acts as a proton pump. Molecules at the *upper left* of the figure have a relatively high free energy content, and molecules at the *lower right* are relatively low in energy.

flows through these channels, it creates an electrical current (which, you may recall, is simply moving charged particles). ATP synthase harnesses the energy of this current to drive ATP synthesis. This process is called the **chemiosmotic[7] mechanism,** which suggests the "push" created by the electrochemical H^+ gradient.

Overview of ATP Production

NADH releases its electron pairs (as hydrogen atoms) to FMN in the first proton pump of the electron-transport system. From there to the end of the chain, this generates enough energy to synthesize 3 ATP molecules per electron pair. $FADH_2$ releases its electron pairs to coenzyme Q, the shuttle between the first and second proton pumps. Therefore, it enters the chain at a point beyond the first pump and does not contribute energy to that pump. Each $FADH_2$ contributes enough energy to synthesize 2 ATP.

With that in mind, we can draw up an energy balance sheet to see how much ATP is produced by the complete aerobic oxidation of glucose to CO_2 and H_2O and where the ATP comes from; see also figure 26.7. This summary refers back to the reaction steps 1 to 21 in figures 26.3 and 26.4. For each glucose molecule, there are

	10 NADH produced at steps 5, 10, 14, 16, and 21
	× 3 ATP per NADH produced by the electron-transport chain
	30 ATP generated by NADH
Plus:	2 $FADH_2$ produced at step 19
	× 2 ATP per $FADH_2$ produced by the electron -transport chain
	= **4 ATP** generated by $FADH_2$
Plus:	**2 ATP** net amount generated by glycolysis (steps 6–7 offset by step 3)
	2 ATP generated by the matrix reactions (step 18)
Total:	**38 ATP** per glucose

This should be viewed as a theoretical maximum. There is some uncertainty about how much H^+ must be pumped between the mitochondrial membranes to generate 1 ATP, and some of the energy from the H^+ current is consumed by pumping ATP from the mitochondrial matrix into

7. *chemi* = chemical + *osmo* = push

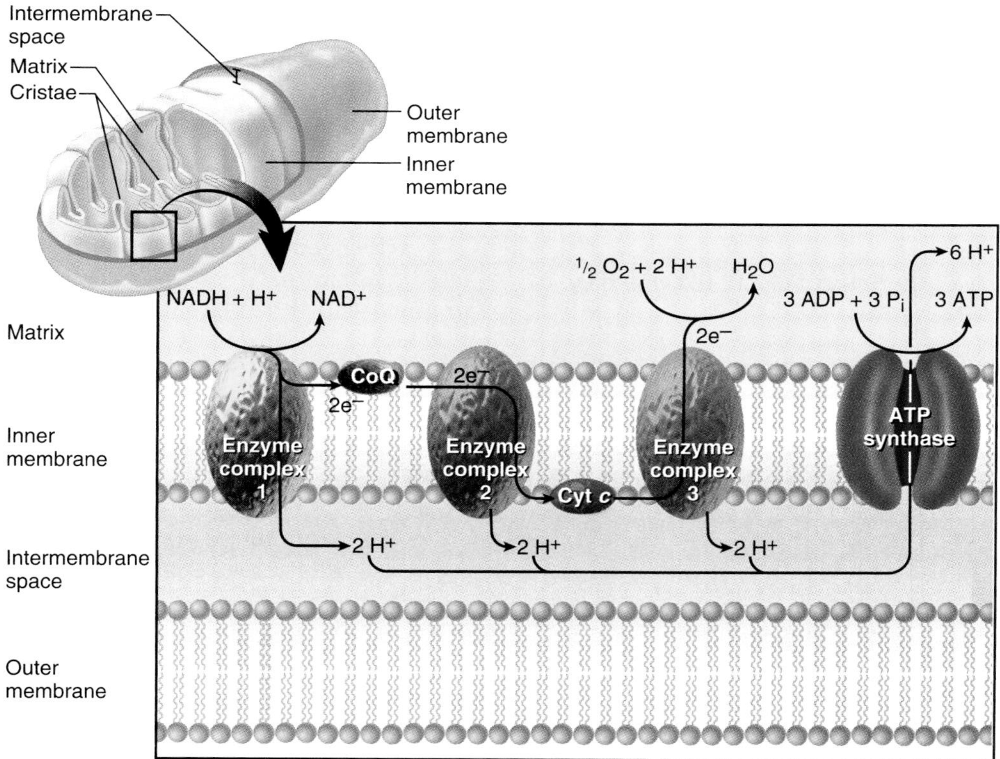

Figure 26.6 The Chemiosmotic Mechanism of ATP Synthesis. Each enzyme complex pumps hydrogen ions into the space between the mitochondrial membranes. These hydrogen ions diffuse back into the matrix by way of ATP synthase, which taps their energy to synthesize ATP. In what ways would you change this figure to show the role of $FADH_2$ in place of NADH?

the cytosol and exchanging it for more raw materials (ADP and P_i) pumped from the cytosol into the mitochondria.

Furthermore, the NADH generated by glycolysis cannot enter the mitochondria and donate its electrons directly to the electron-transport chain. In liver, kidney, and myocardial cells, NADH passes its electrons to *malate,* a "shuttle" molecule that delivers the electrons to the beginning of the electron-transport chain. In this case, each NADH yields enough energy to generate 3 ATP. In skeletal muscle and brain cells, however, the glycolytic NADH transfers its electrons to *glycerol phosphate,* a different shuttle that donates the electrons farther down the electron-transport chain and results in the production of only 2 ATP. Therefore the amount of ATP produced per NADH differs from one cell type to another and is still unknown for others.

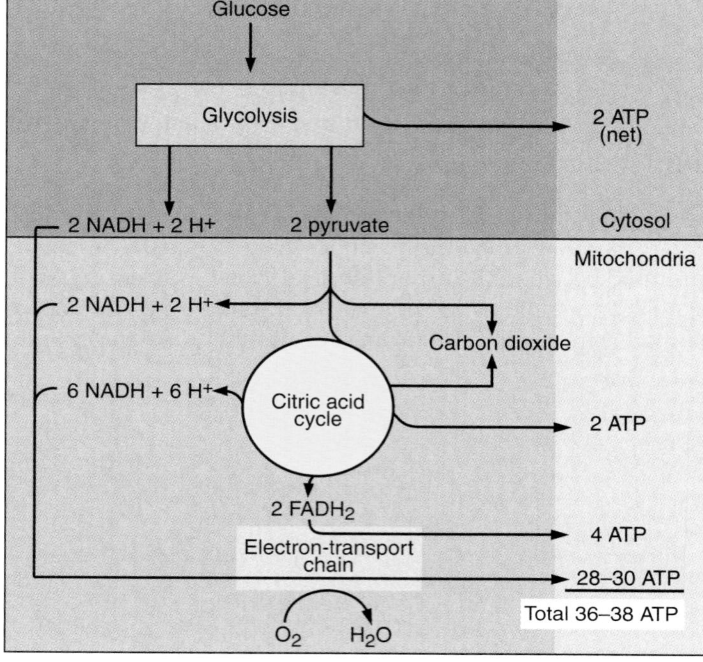

Figure 26.7 Summary of the Sources of ATP Generated by the Complete Oxidation of Glucose.

But if we assume the maximum ATP yield, every mole (180 g) of glucose releases enough energy to synthesize 38 moles of ATP. Glucose has an energy content of 686 kcal/mole and ATP has 7.3 kcal/mole (277.4 kcal in 38 moles). This means that aerobic respiration has an **efficiency** (a ratio of energy output to input) of up to 277.4 kcal/686 kcal = 40%. The other 60% (408.6 kcal) is body heat.

The pathways of glucose catabolism are summarized in table 26.5. The aerobic respiration of glucose can be represented in the summary equation:

$$C_6H_{12}O_6 + 6\ O_2 \longrightarrow 6\ CO_2 + 6\ H_2O$$
$$38\ ADP + P_i \longrightarrow 38\ ATP$$

Glycogen Metabolism

ATP is quickly used after it is synthesized—it is an *energy-transfer* molecule, not an *energy-storage* molecule. Therefore, if the body has an ample amount of ATP and there is still more glucose in the blood, it does not produce and store excess ATP but converts the glucose to other compounds better suited for energy storage—namely glycogen and fat. Fat synthesis is considered later. Here we consider the synthesis and use of glycogen. The average adult body contains about 400 to 450 g of glycogen: nearly one-quarter of it in the liver, three-quarters of it in the skeletal muscles, and small amounts in cardiac muscle and other tissues.

Glycogenesis, the synthesis of glycogen, is stimulated by insulin. Glucose 6-phosphate (G6P) is isomerized to glucose 1-phosphate (G1P). The enzyme *glycogen syn-*

Table 26.5 Pathways of Glucose Catabolism

Stage	Principal Reactants	Principal Products	Purpose
Glycolysis	Glucose, 2 ADP, 2 P_i, 2 NAD^+	2 pyruvic acid, 2 ATP, 2 NADH, 2 H_2O	Reorganizes glucose and splits it in two in preparation for further oxidation by the mitochondria; sole source of ATP in anaerobic conditions
Lactic acid fermentation	2 pyruvic acid, 2 NADH	2 lactic acid, 2 NAD^+	Regenerates NAD^+ so glycolysis can continue to function (and generate ATP) in the absence of oxygen
Matrix reactions	2 pyruvic acid, 8 NAD^+, 2 FAD, 2 ADP, 2 P_i, 8 H_2O	6 CO_2, 8 NADH, 2 $FADH_2$, 2 ATP, 2 H_2O	Remove electrons from pyruvic acid and transfer them to coenzymes NAD^+ and FAD; produce some ATP
Membrane reactions	10 NADH, 2 $FADH_2$, 6 O_2	32–34 ATP, 12 H_2O	Finish oxidation and produce most of the ATP of cellular respiration

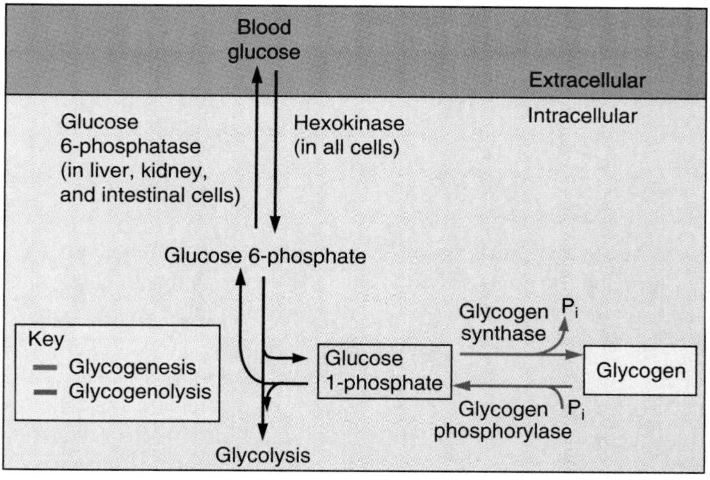

Figure 26.8 Major Pathways of Glucose Storage and Use. In most cells, the glucose 1-phosphate generated by glycogenolysis can only undergo glycolysis. In liver, kidney, and intestinal cells, it can be converted back to free glucose and released into circulation.

Table 26.6 Some Terminology Related to Glucose and Glycogen Metabolism

Anabolic (synthesis) Reactions	
Glycogenesis	The synthesis of glycogen by polymerizing glucose
Gluconeogenesis	The synthesis of glucose from noncarbohydrates such as fats and amino acids
Catabolic (breakdown) Reactions	
Glycolysis	The splitting of glucose into two molecules of pyruvic acid in preparation for anaerobic fermentation or aerobic respiration
Glycogenolysis	The hydrolysis of glycogen to release free glucose or glucose 1-phosphate

thase then cleaves off the phosphate group and attaches the glucose to a growing polysaccharide chain.

Glycogenolysis, the hydrolysis of glycogen, releases glucose between meals when new glucose is not being ingested. The process is stimulated by glucagon and epinephrine. The enzyme *glycogen phosphorylase* begins by phosphorylating a glucose residue and splitting it off the glycogen molecule as G1P. This is isomerized to G6P, which can then enter the pathway of glycolysis.

G6P usually cannot leave the cells that produce it. Liver cells, however, have an enzyme called *glucose 6-phosphatase,* which removes the phosphate group and produces free glucose. This can diffuse out of the cell into the blood, where it is available to any cells in the body. Although muscle cells cannot directly release glucose into the blood, they contribute indirectly to blood glucose concentration because they release pyruvic and lactic acids, which are converted to glucose by the liver.

Gluconeogenesis[8] is the synthesis of glucose from noncarbohydrates such as fats and amino acids. It occurs chiefly in the liver, but after several weeks of fasting, the kidneys also undertake this process and eventually produce just as much glucose as the liver does.

The processes described here are summarized in figure 26.8, and the distinctions among these similar terms are summarized in table 26.6.

Functions of the Liver

We have seen that the liver plays a central role in carbohydrate metabolism. Additional liver functions (table 26.7) were described in previous chapters and will be described later in this chapter. Except for phagocytosis, all of these are performed by the cuboidal hepatocytes described in chapter 25. Such functional diversity is remarkable in light of the uniform structure of these cells. Because of the wide range of functions performed by the liver, degenerative liver diseases such as hepatitis, cirrhosis, and liver cancer are especially life-threatening.

Before You Go On

Answer the following questions to test your understanding of the preceding section:

6. Identify the reaction steps in figures 26.3 and 26.4 at which vitamins are essential to glucose catabolism.
7. In the laboratory, glucose can be oxidized in a single step to CO_2 and H_2O. Why is it done in so many little steps in cells?
8. Explain the origin of the word *glycolysis* and why this is an appropriate name for the function of that reaction pathway.
9. What are two advantages of aerobic respiration over anaerobic fermentation?
10. What important enzyme is found in the inner mitochondrial membrane other than those of the electron-transport chain? Explain how its function depends on the electron-transport chain.
11. Describe how the liver responds to (a) an excess and (b) a deficiency of blood glucose.

8. *gluco* = sugar, glucose + *neo* = new + *genesis* = production of

Table 26.7 Functions of the Liver

Function
Carbohydrate Metabolism
Converts dietary fructose and galactose to glucose. Stabilizes blood glucose concentration by storing excess glucose as glycogen (glycogenesis), releasing glucose from glycogen when needed (glycogenolysis), and synthesizing glucose from fats and amino acids (gluconeogenesis) when glucose demand exceeds glycogen reserves. Receives lactic acid generated by anaerobic fermentation in skeletal muscle and other tissues and converts it back to pyruvic acid or glucose 6-phosphate.
Lipid Metabolism
Degrades chylomicron remnants. Carries out most of the body's lipogenesis (fat synthesis) and synthesizes cholesterol and phospholipids; produces VLDLs to transport lipids to adipose tissue and other tissues for storage or use; and stores fat in its own cells. Carries out most β-oxidation of fatty acids; produces ketone bodies from excess acetyl-CoA. Produces HDL shells, which pick up excess cholesterol from other tissues and return it to the liver; excretes the excess cholesterol in bile.
Protein and Amino Acid Metabolism
Carries out most deamination and transamination of amino acids. Removes $-NH_2$ from glutamic acid and converts the resulting ammonia to urea by means of the ornithine cycle. Synthesizes nonessential amino acids by transamination reactions.
Synthesis of Plasma Proteins
Synthesizes nearly all the proteins of blood plasma, including albumin, α and β globulins, fibrinogen, prothrombin, and several other clotting factors. (Does not synthesize plasma enzymes, peptide hormones, or γ globulins.)
Vitamin and Mineral Metabolism
Converts vitamin D_3 to calcidiol, a step in the synthesis of calcitriol; stores a 3- to 4-month supply of vitamin D. Stores a 10-month supply of vitamin A and enough vitamin B_{12} to last one to several years. Stores iron in ferritin and releases it as needed. Excretes excess calcium by way of the bile.
Digestion
Synthesizes bile acids, which emulsify fat and promote its digestion.
Disposal of Drugs, Toxins, and Hormones
Detoxifies alcohol, antibiotics, and many other drugs. Metabolizes bilirubin from RBC breakdown and excretes it as bile pigments. Deactivates thyroxine and steroid hormones and excretes them or converts them to a form more easily excreted by the kidneys.
Phagocytosis
Macrophages cleanse blood of bacteria and other foreign matter.

Lipid and Protein Metabolism

Objectives

When you have completed this section, you should be able to

- discuss some advantages of lipids over carbohydrates for energy storage and fuel;
- describe the processes of lipid catabolism and anabolism;
- describe the processes of protein catabolism and anabolism; and
- explain the metabolic source of ammonia and how the body disposes of it.

In the foregoing discussion, glycolysis and the mitochondrial reactions were treated from the standpoint of carbohydrate oxidation. These pathways also serve for the oxidation of proteins and lipids as fuel and as a source of metabolic intermediates that can be used for protein and lipid synthesis. Here we examine these related metabolic pathways.

Lipids

Triglycerides are stored primarily in the body's adipocytes, where a given molecule remains for about 2 to 3 weeks. Although the total amount of stored triglyceride remains quite constant, there is a continual turnover as lipids are released, transported in the blood, and either oxidized for energy or redeposited in other adipocytes. Synthesizing fats from other types of molecules is called **lipogenesis**, and breaking down fat for fuel is called **lipolysis** (lih-POL-ih-sis).

Lipogenesis

It is common knowledge that a diet high in sugars causes us to put on fat and gain weight. Lipogenesis employs compounds such as sugars and amino acids to synthesize the triglyceride precursors, glycerol and fatty acids. PGAL, one of the intermediates of glucose oxidation, can be converted to glycerol. As glucose and amino acids enter the citric acid cycle by way of acetyl-CoA, the acetyl-CoA can also be diverted to make fatty acids. The glycerol and fatty acids can then be condensed to form a triglyceride, which can be stored in the adipose tissue or converted to other lipids. These pathways are summarized in figure 26.9.

Lipolysis

Lipolysis, also shown in figure 26.9, begins with the hydrolysis of a triglyceride into glycerol and fatty acids—a process stimulated by epinephrine, norepinephrine, glucocorticoids, thyroid hormone, and growth hormone. The glycerol and fatty acids are further oxidized by separate pathways. Glycerol is easily converted to PGAL and thus

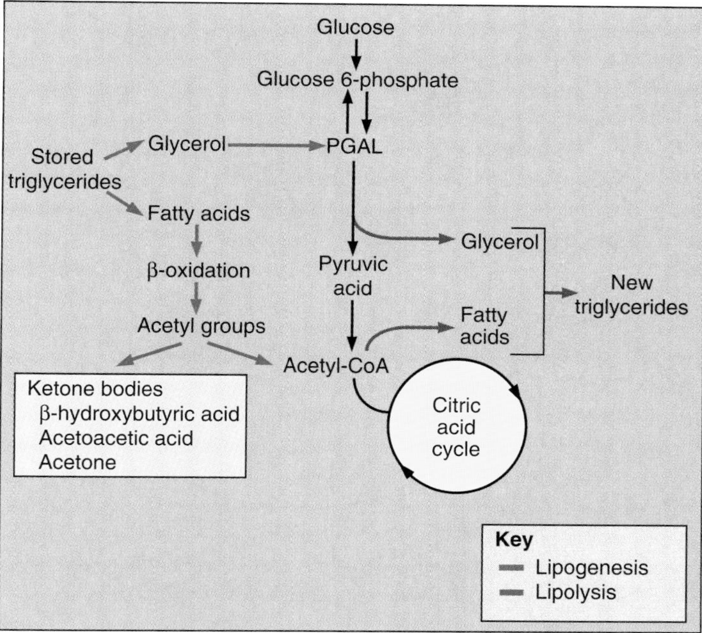

Figure 26.9 Pathways of Lipolysis and Lipogenesis in Relation to Glycolysis and the Citric Acid Cycle.

enters the pathway of glycolysis. It generates only half as much ATP as glucose, however, because it is a C_3 compound compared to glucose (C_6); thus it leads to the production of only half as much pyruvic acid.

The fatty acid component is catabolized in the mitochondrial matrix by a process called **β-oxidation,** which removes two carbon atoms at a time from the fatty acid. The resulting acetyl (C_2) groups are bonded to coenzyme A to make acetyl-CoA—the entry point into the citric acid cycle. A fatty acid of 16 carbon atoms can yield 129 molecules of ATP—obviously a much richer source of energy than a glucose molecule.

Excess acetyl groups can be metabolized by the liver in a process called **ketogenesis.** Two acetyl groups are condensed to form acetoacetic acid, and some of this is further converted to β-hydroxybutyric acid and acetone. These three products are the *ketone bodies.* Some cells convert acetoacetic acid back to acetyl-CoA and thus feed the C_2 fragments into the citric acid cycle to extract their energy. Cardiac muscle and the renal cortex, in fact, use acetoacetic acid as their principal fuel. When the body is rapidly oxidizing fats, however, excess ketone bodies accumulate. This causes the ketoacidosis typical of insulin-dependent diabetes mellitus, in which cells must oxidize fats because they cannot absorb glucose.

Acetyl-CoA cannot go backward up the glycolytic pathway and produce glucose, because this pathway is irreversible past the point of pyruvic acid. While glycerol can be used for gluconeogenesis, fatty acids cannot.

It was mentioned earlier that inadequate carbohydrate intake interferes with the complete fat oxidation. This is because the mitochondrial reactions cannot proceed without oxaloacetic acid as a "pickup molecule" in the citric acid cycle. When carbohydrate intake is deficient, oxaloacetic acid is converted to glucose and becomes unavailable to the citric acid cycle. Under such circumstances, fat oxidation leads to ketogenesis and potentially to ketoacidosis. This is a serious risk of poorly planned diets with excessively restricted carbohydrate ingestion.

Proteins

About 100 g of tissue protein breaks down each day into free amino acids. These combine with the amino acids from the diet to form an **amino acid pool** that cells can draw upon to make new proteins. The fastest rate of tissue protein turnover is in the intestinal mucosa, where epithelial cells are replaced at a very high rate. Dead cells are digested along with the food and thus contribute to the amino acid pool. Of all the amino acid absorbed by the small intestine, about 50% is from the diet, 25% from dead epithelial cells, and 25% from enzymes that have digested each other.

Some amino acids in the pool can be converted to others. Free amino acids also can be converted to glucose and fat or directly used as fuel. Such conversions involve three processes: (1) **deamination,** the removal of an amino group ($-NH_2$); (2) **amination,** the addition of $-NH_2$; or (3) **transamination,** the transfer of $-NH_2$ from one molecule to another. The following discussion shows how these processes are involved in amino acid metabolism.

Use as Fuel

The first step in using amino acids as fuel is to deaminate them. After the $-NH_2$ group is removed, the remainder of the molecule is called a *keto acid.* Depending on which amino acid is involved, the resulting keto acid may be converted to pyruvic acid, acetyl-CoA, or one of the acids of the citric acid cycle (fig. 26.10). It is important to note that some of these reactions are reversible. When there is a deficiency of amino acids in the body, citric acid cycle intermediates can be aminated and converted to amino acids, which are then available for protein synthesis. In gluconeogenesis, keto acids are used to synthesize glucose, essentially through a reversal of the glycolysis reactions.

Transamination, Ammonia, and Urea

When an amino acid is deaminated, its amino group is transferred to a citric acid cycle intermediate, α-ketoglutaric acid, converting it to glutamic acid. Such transamination reactions are the route by which several amino acids enter the citric acid cycle.

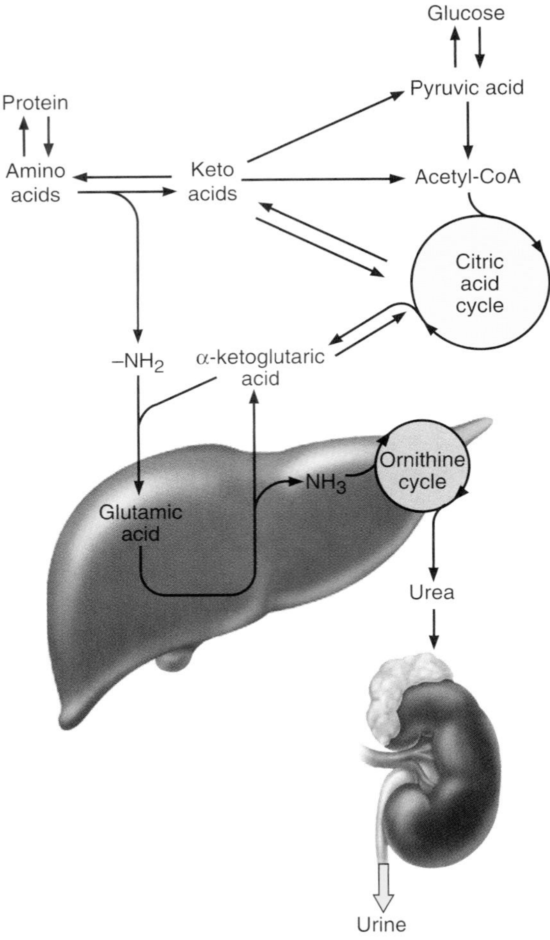

Figure 26.10 Pathways of Amino Acid Metabolism in Relation to Glycolysis and the Citric Acid Cycle.

Glutamic acid can travel from any of the body's cells to the liver. Here its $-NH_2$ group is removed, converting it back to α-ketoglutaric acid. The $-NH_2$ becomes ammonia (NH_3), which is extremely toxic to cells and cannot be allowed to accumulate. The liver quickly converts ammonia to a less toxic form, urea, by a pathway called the **ornithine cycle** (fig. 26.11). Urea is then excreted in the urine as one of the body's nitrogenous wastes. Other nitrogenous wastes and their sources are described in chapter 23 (see p. 882). When a diseased liver cannot carry out the ornithine cycle, NH_3 accumulates in the blood and death from *hepatic coma* may ensue within a few days.

Protein Synthesis

Protein synthesis, described in detail in chapter 5, is a complex process involving DNA, mRNA, tRNA, ribosomes, and often the rough ER. It is stimulated by growth hormone, thyroid hormones, and insulin, and it requires a supply of all the amino acids necessary for a particular protein. The liver can make many of these amino acids from other amino acids or from citric acid cycle intermediates by transamination reactions. The essential amino acids, however, must be obtained from the diet.

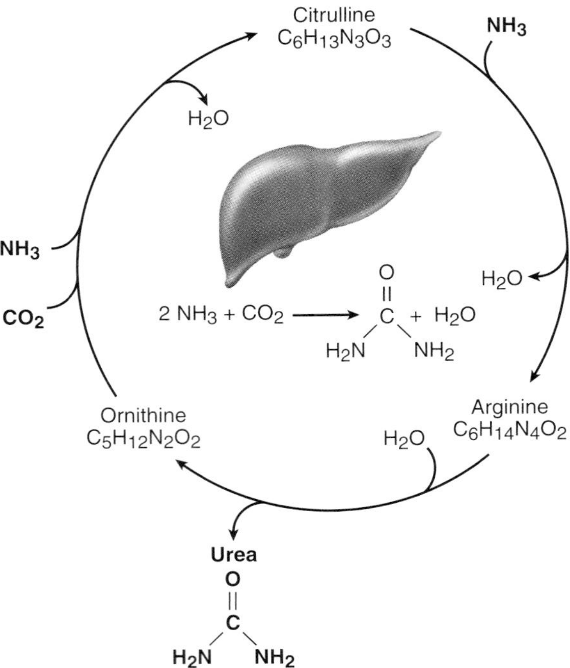

Figure 26.11 Urea Synthesis in the Liver. The ornithine cycle converts ammonia and carbon dioxide to urea.

Before You Go On

Answer the following questions to test your understanding of the preceding section:

12. Which of the processes in table 26.6 is most comparable to lipogenesis? Which is most comparable to lipolysis? Explain.
13. When fats are converted to glucose, only the glycerol component is used in this way, not the fatty acid. Explain why and what happens to the fatty acids.
14. What metabolic process produces ammonia? How does the body dispose of ammonia?

Metabolic States and Metabolic Rate

Objectives

When you have completed this section, you should be able to

- define the absorptive and postabsorptive states;
- explain what happens to carbohydrates, fats, and proteins in each of these states;

Table 26.8 Major Aspects of the Absorptive and Postabsorptive States

	Absorptive	Postabsorptive
Regulatory Hormones	Principally insulin	Principally glucagon
	Also gastrin, secretin, CCK	Also epinephrine, growth hormone
Carbohydrate Metabolism	Blood glucose rising	Blood glucose falling
	Glucose stored by glycogenesis	Glucose released by glycogenolysis
	Gluconeogenesis suppressed	Gluconeogenesis stimulated
Lipid Metabolism	Lipogenesis occurring	Lipolysis occurring
	Lipid uptake from chylomicrons	Fatty acids oxidized for fuel
	Lipid storage in fat and muscle	Glycerol used for gluconeogenesis
Protein Metabolism	Amino acid uptake, protein synthesis	Amino acids oxidized if glycogen and fat stores are inadequate for energy needs
	Excess amino acids burned as fuel	

- describe the hormonal and nervous regulation of each state;
- define *metabolic rate* and *basal metabolic rate;* and
- describe some factors that alter the metabolic rate.

Your metabolism changes from hour to hour depending on how long it has been since your last meal. The **absorptive (fed) state** lasts about 4 hours during and after a meal. This is a time in which nutrients are being absorbed and may be used immediately to meet energy and other needs. The **postabsorptive (fasting) state** prevails in the late morning, late afternoon, and overnight. During this time the stomach and small intestine are empty and the body's energy needs are met from stored fuels. The two states are compared in table 26.8 and explained in the following discussion.

Absorptive State

In the absorptive state, blood glucose is readily available for ATP synthesis. It serves as the primary fuel and spares the body from having to draw on stored fuels. The status of major nutrient classes during this phase is as follows:

- **Carbohydrates.** Absorbed sugars are transported by the hepatic portal system to the liver. Most glucose passes through the liver and becomes available to cells everywhere in the body. Glucose in excess of immediate need, however, is absorbed by the liver and may be converted to glycogen or fat. Most fat synthesized in the liver is released into the circulation; its further fate is comparable to that of dietary fats, discussed next.
- **Fats.** Fats enter the lymph as chylomicrons and initially bypass the liver. As described earlier, lipoprotein lipase removes fats from the chylomicrons for uptake by the tissues, especially adipose and muscular tissue. The liver disposes of the chylomicron remnants. Fats are the primary energy substrate for hepatocytes, adipocytes, and muscle cells.
- **Amino acids.** Amino acids, like sugars, circulate first to the liver. Most pass through and become available to other cells for protein synthesis. Some, however, are removed by the liver and have one of the following fates: (1) to be used for protein synthesis; (2) to be deaminated and used as fuel for ATP synthesis; or (3) to be deaminated and used for fatty acid synthesis.

Regulation of the Absorptive State

The absorptive state is regulated largely by insulin, which is secreted in response to elevated blood glucose and amino acid levels and to the intestinal hormones gastrin, secretin, and cholecystokinin. Insulin regulates the rate of glucose uptake by nearly all cells except neurons, kidney cells, and erythrocytes, which have an independent rate of uptake. On other target cells, insulin has the following effects:

- Within minutes, it increases the cellular uptake of glucose by as much as 20-fold. As cells absorb glucose, the blood glucose concentration falls.
- It stimulates glucose oxidation, glycogenesis, and lipogenesis.
- It inhibits gluconeogenesis, which makes sense since blood glucose concentration is already high and there is no immediate need for more.
- It stimulates the active transport of amino acids into cells and promotes protein synthesis.

Following a high-protein, low-carbohydrate meal, it may seem that the amino acids would stimulate insulin secretion, insulin would accelerate both amino acid and glucose uptake, and since there was relatively little glucose in the ingested food, this would create a risk of hypoglycemia.

In actuality, this is prevented by the fact that a high amino acid level stimulates the secretion of *both* insulin and glucagon. Glucagon, you may recall, is an insulin antagonist (see chapter 17). It supports an adequate level of blood glucose to meet the needs of the brain.

Postabsorptive State

The essence of the postabsorptive state is to homeostatically regulate blood glucose concentration within about 90 to 100 mg/dL. This is especially critical to the brain, which cannot use alternative energy substrates except in cases of prolonged fasting. The postabsorptive status of major nutrients is as follows:

- **Carbohydrates.** Glucose is drawn from the body's glycogen reserves (glycogenolysis) or synthesized from other compounds (gluconeogenesis). The liver usually stores enough glycogen after a meal to support 4 hours of postabsorptive metabolism before significant gluconeogenesis occurs.
- **Fats.** Adipocytes and hepatocytes hydrolyze fats and convert the glycerol to glucose. Free fatty acids (FFAs) cannot be converted to glucose, but they can favorably affect blood glucose concentration. As the liver oxidizes them to ketone bodies, other cells absorb and use these, or use FFAs directly, as their source of energy. By switching from glucose to fatty acid catabolism, they leave glucose for use by the brain (the glucose-sparing effect). After 4 to 5 days of fasting, the brain begins to use ketone bodies as supplemental fuel.
- **Proteins.** If glycogen and fat reserves are depleted, the body begins to use proteins as fuel. Some proteins are more resistant to catabolism than others. Collagen is almost never broken down for fuel, but muscle protein goes quickly. The extreme wasting away seen in cancer and some other chronic diseases, resulting from a loss of appetite (anorexia) as well as altered metabolism, is called **cachexia**[9] (ka-KEX-ee-ah).

Regulation of the Postabsorptive State

Postabsorptive metabolism is more complex than the absorptive state. It is regulated mainly by the sympathetic nervous system and glucagon, but several other hormones are involved. As blood glucose level drops, insulin secretion declines and the pancreatic α cells secrete glucagon. Glucagon promotes glycogenolysis and gluconeogenesis, raising the blood glucose level, and it promotes lipolysis and a rise in FFA levels. Thus it makes both glucose and lipids available for fuel.

The sympathoadrenal system also promotes glycogenolysis and lipolysis, especially under conditions of injury, fear, anger, and other forms of stress. Adipose tissue is richly innervated by the sympathetic nervous system, while adipocytes, hepatocytes, and muscle cells also respond to epinephrine from the adrenal medulla. In circumstances where there is likely to be tissue injury and a need for repair, the sympathoadrenal system therefore mobilizes stored energy reserves and makes them available to meet the demands of tissue repair.

Growth hormone is secreted in response to rapid drops in blood glucose level and in conditions of prolonged fasting. It opposes the effects of insulin and raises blood glucose concentration.

Metabolic Rate

Metabolic rate means the amount of energy liberated in the body per unit of time, expressed in such terms as kcal/hr or kcal/day. Metabolic rate can be measured directly by putting a person in a **calorimeter,** a closed chamber with water-filled walls that absorb the heat given off by the body. The rate of energy release is measured from the temperature change of the water. Metabolic rate can also be measured indirectly with a spirometer, an apparatus described in chapter 22 that can be used to measure the amount of oxygen a person consumes. For every liter of oxygen, approximately 4.82 kcal of energy is released from organic nutrients. This is only an estimate, because the number of kilocalories per liter of oxygen varies slightly with the type of nutrients the person is oxidizing at the time of measurement.

Metabolic rate depends on physical activity, mental state, absorptive or postabsorptive status, thyroid hormone and other hormones, and other factors. The **basal metabolic rate (BMR)** is a baseline or standard of comparison that minimizes the effects of such variables. It is your metabolic rate when you are awake but relaxed, in a room at comfortable temperature, in a postabsorptive state 12 to 14 hours after your last meal. It is not the minimum metabolic rate needed to sustain life, however. When you are asleep, your metabolic rate is slightly lower than your BMR. **Total metabolic rate** is the sum of BMR and energy expenditure for voluntary activities, especially muscular contractions.

The BMR of an average adult is about 2,000 kcal/day for a male and slightly less for a female. Roughly speaking, one must therefore consume at least 2,000 kcal/day to fuel his or her essential metabolic tasks—active transport, muscle tone, brain activity, cardiac and respiratory rhythms, renal function, and other essential processes. Even a relatively sedentary lifestyle requires another 500 kcal/day to support a low level of physical activity, and someone who does hard physical labor may require as much as 5,000 kcal/day.

9. *cac* = bad + *exia* = body condition

Aside from physical activity, some factors that raise the total metabolic rate (TMR) and caloric requirements include pregnancy, anxiety (which stimulates epinephrine release and muscle tension), fever (TMR rises about 14% for each 1°C of body temperature), eating (TMR rises after a meal), and the catecholamine and thyroid hormones. TMR is relatively high in children and declines with age. Therefore, as we reach middle age we often find ourselves gaining weight with no apparent change in food intake.

Some factors that lower TMR include apathy, depression, and prolonged starvation. In weight-loss diets, loss is often rapid at first and then goes more slowly. This is partly because the initial loss is largely water and partly because the TMR drops over time, fewer dietary calories are "burned off," and there is more lipogenesis even with the same caloric intake.

Before You Go On

Answer the following questions to test your understanding of the preceding section:

15. Define the absorptive and postabsorptive states. In which state is the body storing excess fuel? In which state is it drawing from these stored fuel reserves?
16. What hormone primarily regulates the absorptive state, and what are the major effects of this hormone?
17. Explain why triglycerides have a glucose-sparing effect.
18. List a variety of factors and conditions that raise a person's total metabolic rate above basal metabolic rate.

Body Heat and Thermoregulation

Objectives

When you have completed this section, you should be able to

- identify the principal sources of body heat;
- describe some factors that cause variations in body temperature;
- define and contrast the different forms of heat loss;
- describe how the hypothalamus monitors and controls body temperature; and
- describe conditions in which the body temperature is excessively high or low.

Heat generation must be matched by heat loss in order to maintain a stable internal body temperature. **Hypothermia,** an excessively low body temperature, can cause metabolism to slow down to the point that it cannot sustain life, whereas **hyperthermia,** an excessively high temperature, can disrupt the coordination of metabolic pathways and also lead to death. **Thermoregulation,** the balance between heat production and loss, is a critically important aspect of homeostasis. Thermoregulation was introduced in chapter 1 as an example of homeostasis.

Body Temperature

"Normal" body temperature depends on when, where, and in whom it is measured. Body temperature fluctuates about 1°C (1.8°F) in a 24-hour cycle. It tends to be lowest in the early morning and highest in the late afternoon. Temperature also varies from one place in the body to another.

The most important body temperature is the **core temperature**—the temperature of organs in the cranial, thoracic, and abdominal cavities. Rectal temperature is relatively easy to measure and gives an estimate of core temperature: usually 37.2° to 37.6°C (99.0°–99.7°F). It may be as high as 38.5°C (101°F) in active children and some adults.

Shell temperature is the temperature closer to the surface, especially skin and oral temperature. Here, heat is lost from the body and temperatures are slightly lower than rectal temperature. Adult oral temperature is typically 36.6° to 37.0°C (97.9°–98.6°F) but may be as high as 40°C (104°F) during hard exercise.

Heat Production and Loss

Most body heat comes from exergonic (energy-releasing) chemical reactions such as nutrient oxidation and ATP use. A little heat is generated by joint friction, blood flow, and other movements. At rest, most heat is generated by the brain, heart, liver, and endocrine glands; the skeletal muscles contribute about 20% to 30% of the total resting heat. Increased muscle tone or exercise greatly increases heat generation in the muscles, however; in vigorous exercise, they produce 30 to 40 times as much heat as the rest of the body.

The body loses heat in three ways—radiation, conduction, and evaporation:

1. **Radiation.** In essence, heat means molecular motion. All molecular motion produces radiation in the infrared (IR) region of the electromagnetic spectrum. When IR radiation is absorbed by an object, it increases its molecular motion and raises its temperature. Therefore IR radiation removes heat from its source and adds heat to anything that absorbs it. The heat lamps used in bathrooms and restaurants work on this principle. Our bodies continually receive IR from the objects around us and give off IR to our surroundings. Since we are usually warmer than the objects around us, we usually lose more heat this way than we gain.
2. **Conduction.** As the molecules of our tissues vibrate with heat energy, they collide with other molecules and transfer kinetic energy to them. The warmth of your body therefore adds to the

molecular motion and temperature of the clothes you wear, the chair you sit in, and the air around you. Conductive heat loss is aided by **convection,** the motion of a fluid due to uneven heating. Air is a fluid that becomes less dense and therefore rises as it is heated. Thus warm air rises from the body and is replaced by cooler air from below. The same is true of water; for example, when you swim in a lake or take a cool bath. You can easily see convection when you heat water in a clear container. Convection of the air around the human body can be seen by a technique called schlieren photography (fig. 26.12). Convection is not a separate category of heat loss in itself, but increases the rate of conductive heat loss.

3. **Evaporative heat loss.** The cohesion of water molecules hampers their vibratory movement in response to heat input. If the temperature of water is raised sufficiently, however, its molecular motion becomes great enough for molecules to break free and evaporate. Evaporation of water thus carries a substantial amount of heat with it (0.5 kcal/g). This is the significance of perspiration. Sweat wets the skin surface and its evaporation carries heat away. In extreme conditions, the body can lose up to 4 L of sweat per hour, which represents a heat loss of 2,000 kcal/hr. When you are sweaty and stand in front of a fan, you may feel refreshingly cooled or even uncomfortably cold. The rapid movement of heat-laden air away from the body **(forced convection)** accelerates heat removal, so a breeze or a fan enhances heat loss by conduction and evaporation. It has no effect on radiation.

The relative amounts of heat lost by different methods depend on prevailing conditions. A nude body at an air temperature of 21°C (70°F) loses about 60% of its heat by radiation, 18% by conduction, and 22% by evaporation. If air temperature is higher than skin temperature, evaporation becomes the only means of heat loss because radiation and conduction add more heat to the body than they remove. Hot, humid weather hinders even evaporative cooling because there is less of a humidity gradient from skin to air. Such conditions increase the risk of heatstroke (discussed shortly).

Figure 26.12 Heat Loss from the Body Demonstrated by Schlieren Photography. The body loses heat to the surrounding air by conduction, and the rising column of warm air then carries the heat away by convection, replacing it with cooler air from below.

Thermoregulation

Thermoregulation is achieved through several negative feedback loops. The preoptic area of the hypothalamus (anterior to the optic chiasm) has a nucleus called the **hypothalamic thermostat.** It monitors the temperature of the blood and receives signals also from **peripheral thermoreceptors** in the skin. In turn, it sends appropriate signals either to the **heat-losing center,** a nucleus still farther anterior in the hypothalamus, or to the **heat-promoting center,** a more posterior nucleus.

When the heat-losing center senses that the blood temperature is too high, it activates heat-losing mechanisms. The first and simplest of these is cutaneous vasodilation, which increases blood flow close to the body surface and thus promotes heat loss. If this fails to restore normal temperature, the heat-losing center triggers sweating. It also inhibits the heat-promoting center.

When the heat-promoting center senses that the blood temperature is too low, it activates mechanisms to conserve body heat or generate more. By way of the sympathetic nervous system, it causes cutaneous vasoconstriction. Warm blood is then retained deeper in the body and less heat is lost through the skin. The sympathetic nervous system in other mammals also stimulates the arrector pili muscles, which make the hair stand on end. This traps an insulating blanket of still air near the skin. The human sympathetic nervous system attempts to do this as well, but since our body hair is so scanty, the only noticeable effect of this is goosebumps.

If dermal vasoconstriction cannot restore or maintain normal core temperature, the body resorts to **shivering thermogenesis.** If you leave a warm house on a cold day, you may notice that your muscles become tense, sometimes even painfully taut, and you begin to shiver. Shivering involves a spinal reflex that causes tiny alternating contractions in antagonistic muscle pairs. Every muscle contraction releases heat from ATP, and shivering can increase the body's heat production as much as fourfold.

Nonshivering thermogenesis is a more long-term mechanism for generating heat, used especially in the colder seasons of the year. The sympathetic nervous system and thyroid hormone stimulate an increase in metabolic rate, which can rise as much as 30% after several weeks of cold weather. More nutrients are burned as fuel, we consume more calories to "stoke the furnace," and consequently, we have greater appetites in the winter than in the summer. Infants can generate heat by breaking down *brown fat,* a tissue in which lipolysis is not linked to ATP synthesis, so all the energy released from the fat is in the form of heat.

In addition to these physiological mechanisms, and of even greater importance, humans and other animals practice **behavioral thermoregulation**—behaviors that raise or lower the body's heat gains and losses. Just getting out of the sun greatly cuts down heat gain by radiation, for example, while shedding heavy clothing or kicking off a blanket at night greatly helps to cool the body down.

In summary, you can see that thermoregulation is a function of multiple organs: the brain, autonomic nerves, thyroid gland, skin, blood vessels, and skeletal muscles.

Disturbances of Thermoregulation

Chapter 21 described the mechanism of fever and its importance in combating infection. Fever is a normal protective mechanism that should be allowed to run its course if it is not excessively high. A body temperature above 42° to 43°C (108°–110°F), however, can be very dangerous. The high temperature elevates the metabolic rate, and the elevated metabolic rate generates heat faster than the body's heat-losing mechanisms can disperse it (see fig. 1.16). Thus the metabolic rate increases the fever and the fever increases the metabolic rate in a dangerous positive feedback loop. At a core body temperature of 44° to 45°C (111°–113°F), metabolic dysfunction and neurological damage can be fatal.

Exposure to excessive heat causes heat cramps, heat exhaustion, and heatstroke. **Heat cramps** are painful muscle spasms that result from excessive electrolyte loss in the sweat. They occur especially when a person begins to relax after strenuous exertion and heavy sweating. **Heat exhaustion** results from more severe electrolyte loss and is characterized by hypotension, dizziness, vomiting, and sometimes fainting. Prolonged heat waves, especially if accompanied by high humidity, bring on many deaths from **heatstroke** (sunstroke). The body gains heat by radiation and conduction, while the humidity retards evaporative cooling. The body temperature can rise as high as 43°C (110°F), and convulsions, coma, and death may suddenly occur.

Hypothermia can result from exposure to cold weather or immersion in icy water. It, too, entails life-threatening positive feedback loops. If the core temperature falls below 33°C (91°F), the metabolic rate drops so low that heat production cannot keep pace with heat loss, and the temperature falls even more. Death from cardiac fibrillation may occur below 32°C (90°F), but some people survive body temperatures as low as 29°C (84°F) in a state of suspended animation. A body temperature below 24°C (75°F) is usually fatal. It is dangerous to give alcohol to someone in a state of hypothermia; it produces an illusion of warmth but actually accelerates heat loss by dilating cutaneous blood vessels.

Before You Go On

Answer the following questions to test your understanding of the preceding section:

19. What is the primary source of body heat? What are some lesser sources?
20. What mechanisms of heat loss are aided by convection?
21. Describe the major heat-promoting and heat-losing mechanisms of the body.
22. Describe the positive feedback loops that can cause death from hyperthermia and hypothermia.

Clinical Insight 26.4

Alcohol and Alcoholism

Alcohol is not only a popular mind-altering drug but is also regarded in many cultures as a food staple. As a source of empty calories, an addictive drug, and a toxin, it can have a broad spectrum of adverse effects on the body.

Absorption and Metabolism

Alcohol is rapidly absorbed from the digestive tract—about 20% of it in the stomach and 80% in the proximal small intestine. Carbonation, as in beer and sparkling wines, increases its rate of absorption, whereas food reduces its absorption by delaying gastric emptying so the alcohol takes longer to reach its place of maximum absorption. Alcohol is soluble in both water and fat, so it is rapidly distributed to all body tissues and easily crosses the blood-brain barrier to exert its intoxicating effects on the brain.

Alcohol is detoxified by the hepatic enzyme *alcohol dehydrogenase,* which oxidizes it to acetaldehyde. This enters the citric acid cycle and is oxidized to CO_2 and H_2O. The average adult male can clear the blood of about 10 mL of 100% (200 proof) alcohol per hour—the amount in about 30 mL (1 oz) of whiskey or 355 mL (12 oz) of beer. Women have less alcohol dehydrogenase and clear alcohol from the bloodstream less quickly. They are also more vulnerable to alcohol-related illnesses such as cirrhosis of the liver (discussed shortly).

Tolerance to alcohol, the ability to "hold your liquor," results from two factors: behavioral modification, such as giving in less readily to lowered inhibitions, and increased levels of alcohol dehydrogenase in response to routine alcohol consumption. Alcohol dehydrogenase also deactivates other drugs, and drug dosages must be adjusted to compensate for this when treating alcoholics for other diseases.

Physiological Effects

Nervous System

Alcohol is a depressant that inhibits the release of norepinephrine and disrupts the function of GABA receptors. In low doses, it depresses inhibitory synapses and creates sensations of confidence, euphoria, and giddiness. As the dosage rises, however, the breakdown products of ethanol enhance the diffusion of K^+ out of neurons, hyperpolarizing them and making them less responsive to neurotransmitters. Thus, the timing and coordination of communication between neurons is impaired, resulting in such symptoms of intoxication as slurred speech, poor coordination, and slower reaction time. These symptoms begin to become significant at a blood alcohol level of 80 to 100 mg/dL—the legal criterion of intoxication in many states. Above 400 mg/dL, alcohol can so disrupt the electrophysiology of neurons as to induce coma and death.

Liver

The liver's role in metabolizing alcohol makes it especially susceptible to long-term toxic effects. Heavy drinking stresses the liver with a high load of acetaldehyde and acetate; this depletes its oxidizing agents and reduces its ability to catabolize these intermediates as well as fatty acids. Alcoholism often produces a greatly enlarged and fatty liver for multiple reasons: the calories provided by alcohol make it unnecessary to burn fat as fuel, fatty acids are poorly oxidized, and acetaldehyde is converted to new fatty acids. Acetaldehyde also causes inflammation of the liver and pancreas (*hepatitis* and *pancreatitis*), leading to disruption of digestive function. Acetaldehyde and other toxic intermediates destroy hepatocytes faster than they can be regenerated, and the liver exhibits extensive scarring and a lumpy or nodular surface—a state called *cirrhosis.* Many symptoms of alcoholism stem from deterioration of liver functions. Hepatic coma may occur as the liver becomes unable to produce urea, thus allowing ammonia to accumulate in the blood. Jaundice results from the liver's inability to excrete bilirubin.

Circulatory System

Deteriorating liver functions exert several effects on the blood and cardiovascular system. Blood clotting is impaired because the liver cannot synthesize clotting factors adequately. Edema results from inadequate synthesis of blood albumin. Cirrhosis obstructs the hepatic portal blood circulation. Portal hypertension results, and combined with hypoproteinemia, this causes the liver and other organs to "weep" serous fluid into the peritoneal cavity. This leads to *ascites*[10] (ah-SY-teez)—swelling of the abdomen with as much as several liters of serous fluid. The combination of hypertension and impaired clotting often leads to hemorrhaging. *Hematemesis,*[11] the vomiting of blood, may occur as enlarged veins of the esophagus hemorrhage. Alcohol abuse also destroys myocardial tissue, reduces contractility of the heart, and causes cardiac arrhythmia.

Digestive System and Nutrition

Alcohol breaks down the protective mucous barrier of the stomach and the tight junctions between its epithelial cells. Thus it may cause gastritis and bleeding. Alcohol is commonly believed to be a factor in peptic ulcers, but there is little concrete evidence of this. Heavy drinking, especially in combination with smoking, increases the incidence of esophageal cancer. Malnutrition is a typical complication of alcoholism, partly because the empty calories of alcohol suppress the appetite for more nutritious foods. The average American gets about 4.5% of his or her calories from alcohol (more when nondrinkers are excluded), but heavy drinkers may obtain half or more of their calories from alcohol and have less appetite for foods that would meet their other nutritional requirements. In addition, acetaldehyde interferes with vitamin absorption and use. Thiamine deficiency is common in alcoholism, and thiamine is routinely given to alcoholics in treatment.

Addiction

Alcohol is the most widely available addictive drug in America. In many respects it is almost identical to barbiturates in its toxic effects,

10. *asc* = bag + *ites* = like, resembling
11. *hemat* = blood + *emesis* = vomiting

its potential for tolerance and dependence, and the risk of overdose. The difference is that obtaining barbiturates usually requires a prescription, while obtaining alcohol requires, at most, proof of age.

Alcoholism is defined by a combination of criteria, including the pathological changes just described; physiological tolerance of high concentrations; impaired physiological, psychological, and social functionality; and withdrawal symptoms occurring when intake is reduced or stopped. Heavy drinking followed by a period of abstinence—for example, when a patient is admitted to the hospital and cannot get access to alcohol—may trigger *delirium tremens (DT)*, characterized by restlessness, insomnia, confusion, irritability, tremors, incoherent speech, hallucinations, convulsions, and coma. DT has a 5% to 15% mortality rate.

Most alcoholism (type I) sets in after age 25 and is usually associated with stress or peer pressure. These influences lead to increased drinking, which can start a vicious cycle of illness, reduced job performance, family and social problems, arrest, and other stresses leading to still more drinking. A smaller number of alcoholics have type II alcoholism, which is at least partially hereditary. Most people with type II alcoholism are men who become addicted before age 25, especially the sons of other type II alcoholics. There is little evidence for hereditary alcoholism in women. Type II alcoholics show abnormally rapid increases in blood acetaldehyde levels when they drink, and they have unusual brain waves (EEGs) even when not drinking. Male children of alcoholics have a higher than average incidence of becoming alcoholic even when raised by nonalcoholic foster parents. It is by no means inevitable that such people will become alcoholics, but stress or peer pressure can trigger alcoholism more easily in those who are genetically predisposed to it.

Alcoholism is treated primarily through behavior modification—abstinence, peer support, avoidance or correction of the stresses that encourage drinking, and sometimes psychotherapy. The drug disulfiram (Antabuse) is used to support behavior modification programs. It inhibits the breakdown of acetaldehyde and thus heightens its short-term toxic effects. A person who takes Antabuse and drinks alcohol experiences headache, vomiting, tachycardia, chest pain, and hyperventilation and is less likely to look upon alcohol as a means of pleasure or escape.

Chapter Review

Online Learning Center

Study resources for this chapter are available at www.mhhe.com/saladin/

Study Outline

Nutrition (p. 985)

1. Body weight and energy balance
   a. Homeostatic set point
   b. Heredity and environment
2. Appetite
   a. Feeding and satiety centers
   b. Glucostats
   c. Effects of chewing and swallowing
   d. Hunger contractions
   e. Hormones and appetite
      - Cholecystokinin
      - Neuropeptide Y
      - Leptin
3. Calories
4. Nutrients
   a. Macro- and micronutrients
   b. Recommended daily allowances
   c. Essential nutrients
5. Carbohydrates
   a. Functions
   b. Requirements
   c. Types in the diet
   d. Dietary sources
   e. Fiber
      - Types of dietary fiber
      - RDA for fiber
      - Effects of fiber
      - Water-soluble and -insoluble fiber
6. Lipids
   a. Functions
   b. Requirements
   c. Dietary sources
   d. Cholesterol and serum lipoproteins
      - Lipoprotein classes
      - Chylomicron processing
      - VLDL and LDL processing
      - HDL processing
      - Cholesterol and lipoprotein levels
      - Exercise and cholesterol level
7. Proteins
   a. Functions
   b. Requirements
   c. Protein quality
      - Essential and inessential amino acids
      - Complete and incomplete proteins
      - Net protein utilization
   d. Dietary sources
   e. Nitrogen balance
8. Minerals
   a. Functions
   b. Requirements
   c. Dietary sources
   d. Sodium
9. Vitamins
   a. Requirements
   b. Water-soluble vitamins

   - c. Fat-soluble vitamins
   - d. Vitamin excesses

Carbohydrate Metabolism (p. 995)

1. Glucose catabolism
   - a. Major catabolic pathways
   - b. Role of coenzymes
2. Glycolysis
3. Anaerobic fermentation
   - a. Purpose
   - b. Disadvantages
4. Aerobic respiration
   - a. Matrix reactions
   - b. Membrane reactions
     - Electron-transport molecules
     - Electron-transport processes
     - Chemiosmotic mechanism
5. Overview of ATP production
6. Glycogen metabolism
   - a. Glycogenesis
   - b. Glycogenolysis
   - c. Gluconeogenesis
7. Functions of the liver

Lipid and Protein Metabolism (p. 1004)

1. Lipids
   - a. Lipogenesis
   - b. Lipolysis
     - Uses of PGAL
     - β-oxidation of fatty acids
     - Ketogenesis
2. Proteins
   - a. The amino acid pool
   - b. Deamination, amination, and transamination
   - c. Proteins as fuel
   - d. Transamination, ammonia, and urea
   - e. Protein synthesis

Metabolic States and Metabolic Rate (p. 1006)

1. Absorptive state
   - a. Carbohydrate metabolism
   - b. Fat metabolism
   - c. Amino acid metabolism
   - d. Regulation of the absorptive state
     - Role of insulin
     - Role of glucagon
2. Postabsorptive state
   - a. Carbohydrate metabolism
   - b. Fat metabolism
   - c. Protein metabolism
   - d. Regulation of the postabsorptive state
     - Role of glucagon
     - Sympathoadrenal effects
3. Metabolic rate (MR)
   - a. Measurement
   - b. Basal and total MR
   - c. Factors affecting MR

Body Heat and Thermoregulation (p. 1009)

1. Hypothermia and hyperthermia
2. Body temperature
   - a. Core and shell temperature
   - b. Variations in temperature
3. Heat production and loss
   - a. Heat production
   - b. Modes of heat loss
     - Radiation
     - Conduction and convection
     - Evaporative heat loss
4. Thermoregulation
   - a. Hypothalamic thermostat
   - b. Thermoreceptors
   - c. Heat-losing mechanisms
   - d. Heat-promoting mechanisms
     - Shivering thermogenesis
     - Nonshivering thermogenesis
     - Behavioral thermoregulation
5. Disturbances of thermoregulation
   - a. Hyperthermia
     - Heat cramps
     - Heat exhaustion
     - Heatstroke
   - b. Hypothermia

Selected Vocabulary

feeding center 985
anorexia 985
satiety center 985
hyperphagia 985
glucostat 985
hunger contractions 985
kilocalorie 986
fuel 986
nutrient 986
macronutrient 988
micronutrient 988
recommended daily allowance 988
essential nutrient 988
hypoglycemia 988
water-soluble fiber 989
water-insoluble fiber 989
glucose-sparing effect 989
protein-sparing effect 989
essential fatty acid 990
lipoprotein 990
chylomicron 990
high-density lipoprotein 990
low-density lipoprotein 990
very low–density lipoprotein 990
lipoprotein lipase 990
essential amino acid 992
inessential amino acid 992
complete protein 992
incomplete protein 992
net protein utilization 992
nitrogen balance 993
positive nitrogen balance 993
negative nitrogen balance 993
water-soluble vitamin 994
fat-soluble vitamin 994
hypervitaminosis 994
glycolysis 995
anaerobic fermentation 995
aerobic respiration 995
NAD^+ 996
FAD 996
glycolysis 996
pyruvic acid 997
PGAL 997
lactic acid 997
matrix reactions 997
membrane reactions 998
citric acid cycle 998
acetyl group 998
acetyl-coenzyme A 998
oxaloacetic acid 998
citric acid 998
mitochondrial electron-transport chain 998
flavin mononucleotide 998
iron-sulfur center 998
coenzyme Q 998
cytochrome 1000
proton pump 1000
respiratory enzyme complex 1000
ATP synthase 1000
chemiosmotic mechanism 1001
efficiency of respiration 1002
glycogenesis 1002
glycogenolysis 1003
gluconeogenesis 1003
lipogenesis 1004
lipolysis 1004
β-oxidation 1005
ketogenesis 1005
amino acid pool 1005
deamination 1005
amination 1005
transamination 1005
ornithine cycle 1006
absorptive state 1006
postabsorptive state 1006
cachexia 1008
metabolic rate 1008
calorimeter 1008
basal metabolic rate 1008
total metabolic rate 1008
hypothermia 1009
hyperthermia 1009
thermoregulation 1009
core temperature 1009
shell temperature 1009
radiation of heat 1009
conduction of heat 1009
convection 1010
evaporative heat loss 1010
forced convection 1010
hypothalamic thermostat 1010
peripheral thermoreceptor 1010
heat-losing center 1010
heat-promoting center 1010
shivering thermogenesis 1011
nonshivering thermogenesis 1011
behavioral thermoregulation 1011
heat cramps 1011
heat exhaustion 1011
heatstroke 1011

Testing Your Recall

1. ______ are not used as fuel and are required in relatively small quantities.
   a. Micronutrients
   b. Macronutrients
   c. Essential nutrients
   d. Proteins
   e. Lipids
2. The only significant digestible polysaccharide in the diet is
   a. glycogen.
   b. cellulose.
   c. starch.
   d. maltose.
   e. fiber.
3. Which of the following stores the greatest amount of energy for the smallest amount of space in the body?
   a. glucose
   b. triglycerides
   c. glycogen
   d. proteins
   e. vitamins
4. The lipoproteins that remove cholesterol from the tissues are
   a. chylomicrons.
   b. lipoprotein lipases.
   c. VLDLs.
   d. LDLs.
   e. HDLs.
5. Proteins serve all of the following functions *except* to act as
   a. enzymes.
   b. coenzymes.
   c. hormones.
   d. antibodies.
   e. structural support for cells.
6. The primary function of B-complex vitamins is to act as
   a. structural components of cells.
   b. sources of energy.
   c. components of pigments.
   d. antioxidants.
   e. coenzymes.
7. FAD is reduced to $FADH_2$ in
   a. glycolysis.
   b. anaerobic fermentation.
   c. the citric acid cycle.
   d. the electron-transport chain.
   e. β-oxidation of lipids.
8. The primary, direct benefit of anaerobic fermentation is to
   a. regenerate NAD^+.
   b. produce $FADH_2$.
   c. produce lactic acid.
   d. dispose of pyruvic acid.
   e. produce more ATP than glycolysis does.
9. Which of these occurs in the mitochondrial matrix?
   a. glycolysis
   b. chemiosmosis
   c. the cytochrome reactions
   d. the citric acid cycle
   e. anaerobic fermentation
10. When the body emits more infrared energy than it absorbs, it is losing heat by
    a. convection.
    b. forced convection.
    c. conduction.
    d. radiation.
    e. evaporation.
11. A/an ________ protein lacks one or more essential amino acids.
12. In the postabsorptive state, glycogen is hydrolyzed to liberate glucose. This process is called ________.
13. Synthesis of glucose from amino acids or triglycerides is called ________.
14. The major nitrogenous waste resulting from protein catabolism is ________.
15. The organ that synthesizes the nitrogenous waste in question 14 is the ________.
16. The absorptive state is regulated mainly by the hormone ________.
17. The temperature of organs in the body cavities is called ________.
18. The feeding center, satiety center, heat-losing center, and heat-promoting center are nuclei located in part of the brain called the ________.
19. The brightly colored, iron-containing, electron-transfer molecules of the inner mitochondrial membrane are called ________.
20. The flow of H^+ from the intermembrane space to the mitochondrial matrix creates an electrical current used by the enzyme ________ to make ________.

Answers in appendix B

Testing Your Comprehension

1. Cyanide blocks the transfer of electrons from cytochrome a_3 to oxygen. In light of this, explain why it causes sudden death. Also explain whether cyanide poisoning could be treated by giving a patient supplemental oxygen, and justify your answer.
2. Chapter 17 defines and describes some hormone actions that are synergistic and antagonistic. Identify some synergistic and antagonistic hormone interactions in the postabsorptive state of metabolism.
3. Mrs. Jones, a 42-year-old, complains that, "Everything I eat goes to fat. But my husband and my son eat twice as much as I do, and they're both as skinny as can be." How would you explain this to her?
4. A television advertisement proclaims, "Feeling tired? Need more energy? Order your supply of Zippy Megavitamins and feel better fast!" Your friend Cathy is about to send in her order, and you try to talk her out of wasting her money. Summarize the argument you would use.
5. Explain why a patient whose liver has been extensively damaged by hepatitis could show elevated concentrations of thyroid hormone and bilirubin in the blood.

Answers at the Online Learning Center

27 CHAPTER

The Male Reproductive System

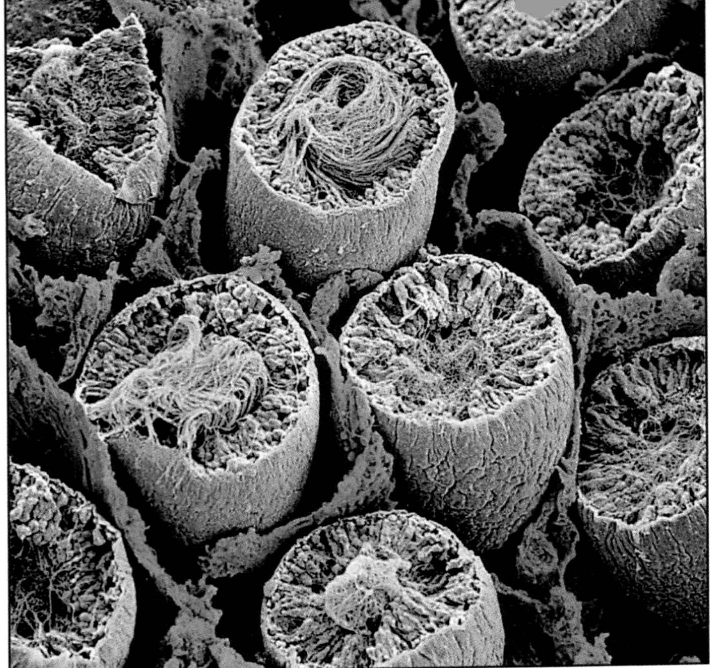

Seminiferous tubules, where sperm are produced (SEM). Sperm tails are seen as hairlike masses in some tubules

Chapter Outline

Insights

Brushing Up

To understand this chapter, it is important that you understand or brush up on the following concepts:

- Stages of mitosis (p.157)
- Chromosome structure (p. 158)
- The karyotype (p. 160)
- Muscles of the pelvic floor (p. 367)
- Hypothalamic releasing factors and pituitary gonadotropins (p. 643)
- Negative feedback inhibition of the pituitary (p. 646)
- Androgens (pp. 652–53)

Study with Your ESP CD-ROM

Refer to your ESP Tutorial CD-ROM for helpful animations, activities, and quizzes related to the material covered in this chapter. The following are suggested areas to go to on the CD and are presented according to "Unit/Topic":

- Reproduction and Development/Reproduction

From all we have learned of the structure and function of the human body, it seems a wonder that it works at all! The fact is, however, that even with modern medicine we cannot keep it working forever. The body suffers various degenerative changes as we age, and eventually we expire. Yet our genes live on in new containers—our offspring. The production of offspring is the subject of these last three chapters. In this chapter, we examine some general aspects of human reproductive biology and then focus on the role of the male in reproduction. The next chapter focuses on the female and the final chapter on the embryonic development of humans and on changes at the other end of the life span—the changes of old age.

Sexual Reproduction

Objectives

When you have completed this section, you should be able to

- define *sexual reproduction;*
- identify the most fundamental biological distinction between male and female; and
- define *primary sex organs, secondary sex organs,* and *secondary sex characteristics.*

The Essence of Sex

Reproduction is one of the fundamental properties of all living things. Many organisms reproduce simply by fission (splitting in two) or by budding off miniature replicas of themselves. These are asexual processes that produce exact genetic copies of the parent. Most organisms, however, reproduce sexually. **Sexual reproduction** does not necessarily entail copulation or even physical contact between the parents—many species merely shed sex cells into the sea and the parents never meet. The essence of sexual reproduction is that each offspring has two parents and a combination of genes from both. Thus the offspring are not genetically identical to their parents and usually not even to each other. Genetic diversity provides the foundation for the survival and evolution of a species and has been such a substantial advantage that it is employed by the great majority of organisms.

The Two Sexes

To reproduce sexually means that the parents must produce **gametes**[1] (sex cells) that can meet and combine their genes in a **zygote**[2] (fertilized egg). The gametes must have two properties for reproduction to be successful: (1) motility, so they can achieve contact, and (2) nutrients for the developing embryo. A single cell cannot perform both of these roles very well, because to contain ample nutrients means to be relatively large and heavy, and this is inconsistent with the need for motility. Therefore, these tasks are usually apportioned to two kinds of gametes. The small motile one—little more than DNA with a propeller—is the **sperm (spermatozoon),** while the large nutrient-laden one is the **egg (ovum).**

It is usually easy to distinguish a human male and female from each other, but this is not so obvious in many other species. By definition, an individual that produces eggs is female and one that produces sperm is male. Even in humans, this criterion is not always that simple—for example, when considering some abnormalities in sexual development to be discussed later. Genetically, any individual with a Y sex chromosome is classified as male and any individual lacking a Y chromosome (usually but not always having two X chromosomes) is classified as female.

In mammals, the female is also the parent that provides a sheltered internal environment for the development and prenatal nutrition of the embryo. For fertilization and development to occur in the female, the male must have a copulatory organ, the penis, for introducing his gametes into the female reproductive tract and the female must have a copulatory organ, the vagina, for receiving the sperm. This is the most obvious difference between the sexes, but appearances can be deceiving (see fig. 17.15, p. 654 and clinical insight 27.1).

Overview of the Reproductive System

The reproductive system consists of primary and secondary sex organs. The **primary sex organs,** or **gonads,**[3] are organs that produce the gametes—**testes (testicles)** of the male and **ovaries** of the female. The **secondary sex organs** are organs other than the gonads that are essential to reproduction. In the male, they constitute a system of ducts, glands, and the penis, concerned with the storage, survival, and conveyance of sperm. In the female, they include the uterine tubes, uterus, and vagina, concerned with uniting the sperm and egg and harboring the developing fetus.

Secondary sex characteristics are features that develop at puberty, distinguish the sexes from each other, and play a role in mate attraction. They typically appear only as an animal approaches sexual maturity (from puberty through adolescence in humans). From the call of a bullfrog to the tail of a peacock, these are well known in the animal kingdom. In human males, secondary sex characteristics include pubic, axillary, and facial hair, relatively coarse and visible hair on the torso and limbs, apocrine glands, a relatively muscular physique, and a relatively low-pitched voice.

1. *gam* = marriage, union
2. *zygo* = yoke, union
3. *gon* = seed

Before You Go On

Answer the following questions to test your understanding of the preceding section:

1. Define *gonad* and *gamete*. Explain the relationship between the terms.
2. Define *male, female, sperm,* and *egg.*

Sex Determination and Development

Objectives

When you have completed this section, you should be able to

- explain the role of the sex chromosomes in determining sex;
- explain how the Y chromosome determines the response of the fetal gonad to prenatal hormones;
- identify which of the male and female external genitalia are homologous to each other; and
- describe the descent of the testes and explain why it is important.

Role of the Sex Chromosomes

Most of our cells have 23 pairs of chromosomes: 22 pairs of *autosomes* and 1 pair of *sex chromosomes* (see fig. 5.16, p. 160). A sex chromosome can either be a large X chromosome or a small Y chromosome. Every egg contains an X chromosome, but half of the sperm carry an X while the other half carry a Y. If an egg is fertilized with an X-bearing sperm, it produces an XX zygote that is destined to become a female. If it is fertilized with a Y-bearing sperm, it produces an XY zygote destined to become a male. Thus the sex of a child is determined at conception (fertilization), and not by the mother's egg but by the sperm that fertilizes it (fig. 27.1).

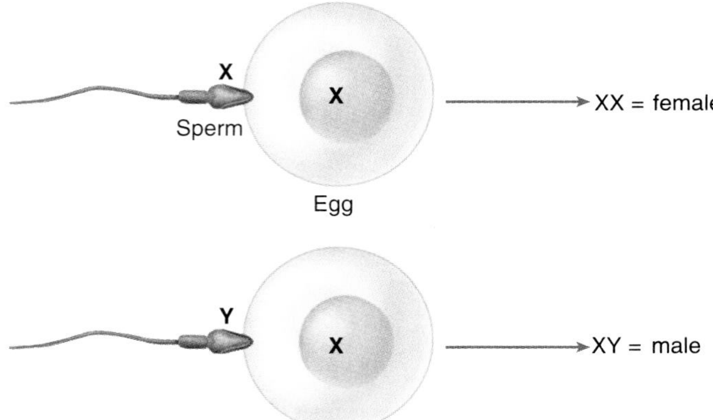

Figure 27.1 Chromosomal Sex Determination. All eggs carry the X chromosome. The sex of a child is determined by whether the egg is fertilized by an X-bearing sperm or a Y-bearing sperm.

Hormones and Sex Differentiation

Sex determination does not end with fertilization, however. It requires an interaction between genetics and the hormones produced by the mother and fetus. Just as we have seen with other hormones (see chapter 17), those involved here require specific receptors on their target cells to exert an effect.

Up to a point, a fetus is sexually undifferentiated, or "noncommittal" as to which sex it will become. Its gonads begin to develop at 5 to 6 weeks as *gonadal ridges,* each lying alongside a primitive kidney, the *mesonephros,* which later degenerates. Adjacent to each gonadal ridge are two ducts, the **mesonephric**[4] (MEZ-oh-NEF-ric) **(wolffian**[5]**) duct,** which originally serves the mesonephros, and the **paramesonephric**[6] **(müllerian**[7]**) duct.** In males, the mesonephric ducts develop into the reproductive tract and the paramesonephric ducts degenerate. In females, the opposite occurs (fig. 27.2).

But why? The Y chromosome has a gene called **SRY** (sex-determining region of the Y) that codes for a protein called **testis-determining factor (TDF).** TDF then interacts with genes on some of the other chromosomes, including a gene on the X chromosome for androgen receptors, and the activity of those genes initiates the development of male anatomy. By 8 to 9 weeks, the male has an identifiable testis that begins to secrete testosterone. Each testis stimulates the mesonephric duct on its own side to develop into the system of male reproductive ducts. By this time, the testis also secretes a hormone called **müllerian-inhibiting factor (MIF)** that causes atrophy of the paramesonephric (müllerian) duct on that side. Even an adult male, however, retains a tiny Y-shaped vestige of the paramesonephric ducts, like a vestigial uterus and uterine tubes, in the area of the prostatic urethra. It is named the *uterus masculinus.*

It may seem as if androgens should induce the formation of a male reproductive tract and estrogens induce a female reproductive tract. However, estrogen levels are always high during pregnancy, so if this mechanism were the case, they would have a feminizing effect on all fetuses. Thus the development of a female results from the absence of androgens, not the presence of estrogens (see clinical insight 27.1).

4. *meso* = middle + *nephr* = kidney; named for a temporary embryonic kidney, the mesonephros
5. Kaspar F. Wolff (1733–94), German anatomist
6. *para* = next to
7. Johannes P. Müller (1801–58), German physician

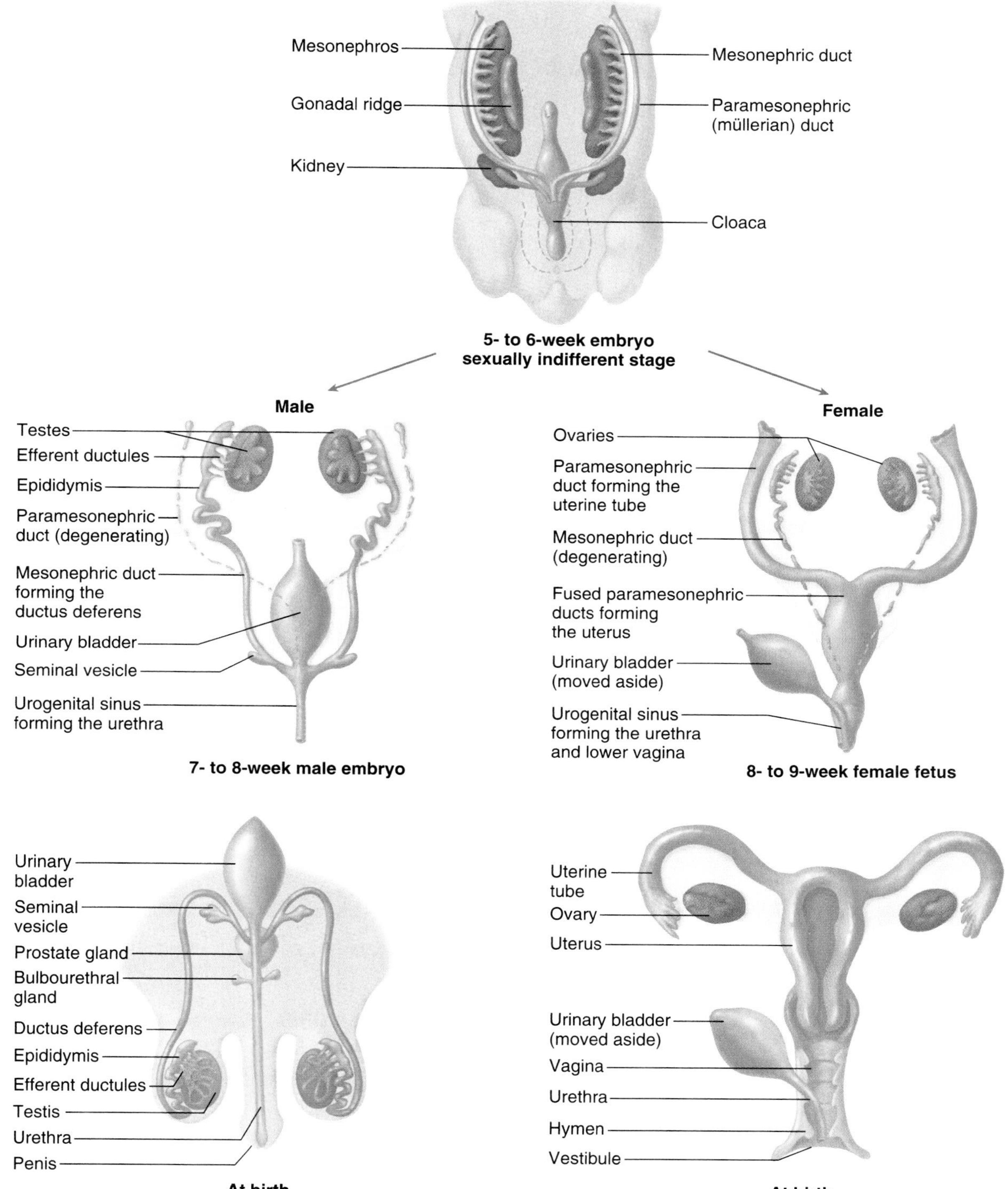

Figure 27.2 Embryonic Development of the Male and Female Reproductive Tracts. Note that the male tract develops from the mesonephric duct and the female tract from the paramesonephric duct, while the other duct in each sex degenerates.

Clinical Insight 27.1

Androgen-Insensitivity Syndrome

Occasionally, a girl shows all the usual changes of puberty except that she fails to menstruate. Physical examination shows the presence of testes in the abdomen and a karyotype reveals that she has the XY chromosomes of a male. The testes produce normal male levels of testosterone, but the target cells lack receptors for it. This is called *androgen-insensitivity syndrome (AIS)*, or *testicular feminization.*

The external genitals develop female anatomy as if no testosterone were present. At puberty, breasts and other feminine secondary sex characteristics develop because the testes secrete small amounts of estrogen and there is no overriding influence of testosterone. However, there is no uterus or menstruation. If the abdominal testes are not removed, the person has a high risk of testicular cancer.

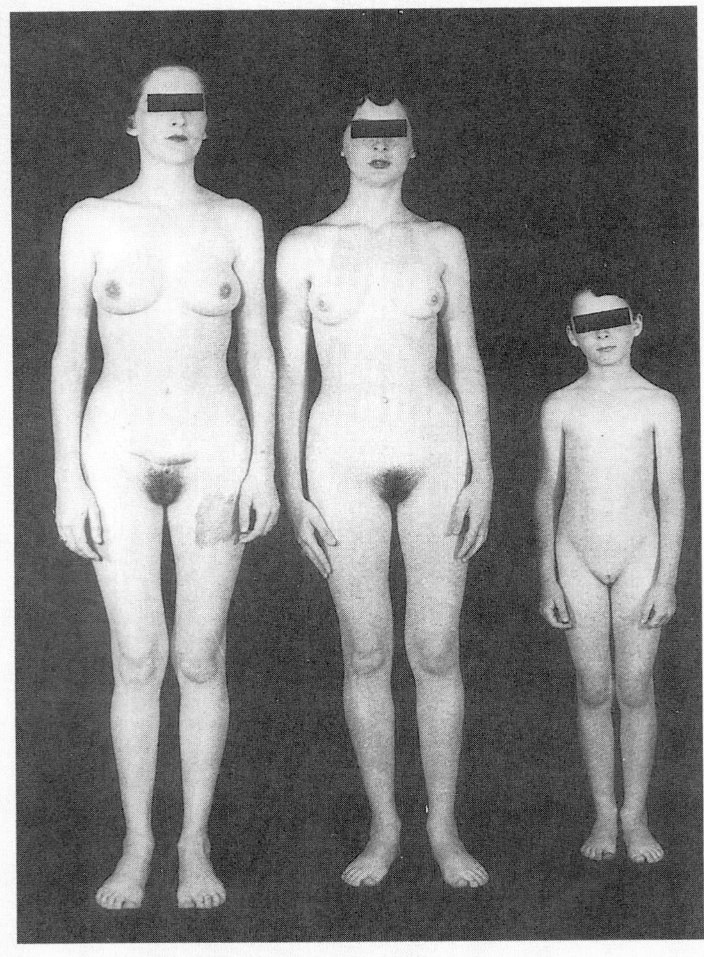

Androgen-Insensitivity Syndrome. These siblings are genetically male (XY). Testes are present and secrete testosterone, but the target cells lack receptors for it, so testosterone cannot exert its masculinizing effects. The external genitalia and secondary sex characteristics are feminine, but there are no ovaries, uterus, or vagina.

Development of the External Genitalia

You perhaps regard the external genitals as the most definitive characteristics of a male or female, yet there is more similarity between the sexes than most people realize. In the fetus, they begin developing from identical structures in both sexes. By 8 weeks, the fetus has the following (fig. 27.3):

- a **phallus,**[8] a small shaft of tissue with a swollen *glans* (head);
- **urogenital folds,** a pair of medial tissue folds slightly posterior to the phallus; and
- **labioscrotal folds,** a larger pair of tissue folds lateral to the urogenital folds.

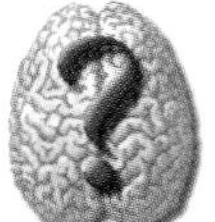

Think About It

In what way is androgen-insensitivity syndrome similar to non-insulin-dependent diabetes mellitus and Laron dwarfism? (see pp. 656 and 663.)

By the end of week 9, the fetus begins to show sexual differentiation, and either male or female genitalia are distinctly formed by the end of week 12. In the female, the three structures just listed become the clitoris, labia minora, and labia majora, respectively; all of these are more fully described in the next chapter. In the male, the phallus

8. *phallo* = penis

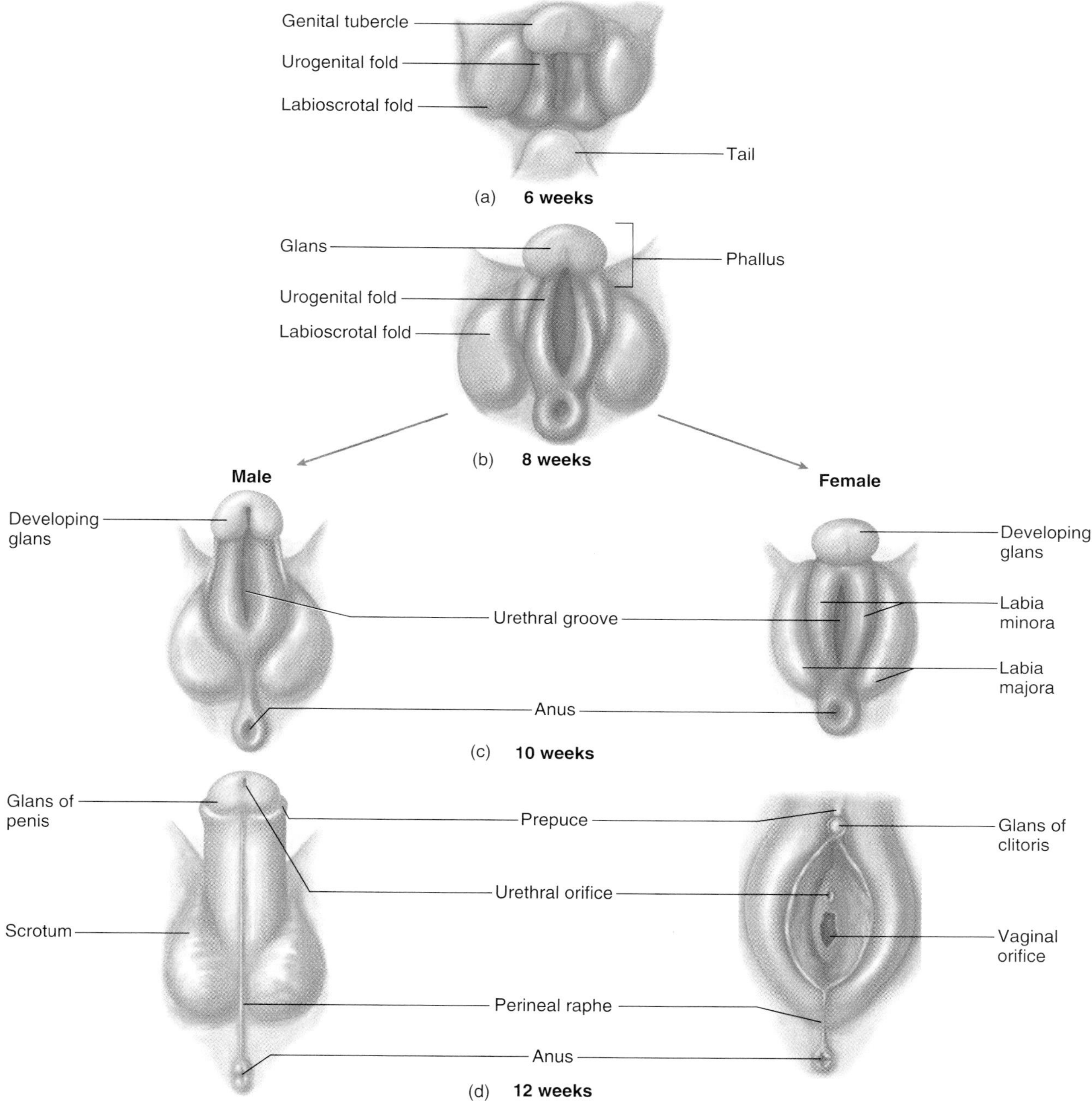

Figure 27.3 Development of the External Genitalia. (*a*) By 6 weeks, the embryo has three primordial structures—the phallus, urogenital folds, and labioscrotal folds—that will become the male or female genitalia. (*b*) At 8 weeks these structures have grown, but the sexes are still indistinguishable. (*c*) Slight sexual differentiation is noticeable at 10 weeks. (*d*) The sexes are fully distinguishable by 12 weeks. Matching colors identify homologous structures of the male and female.

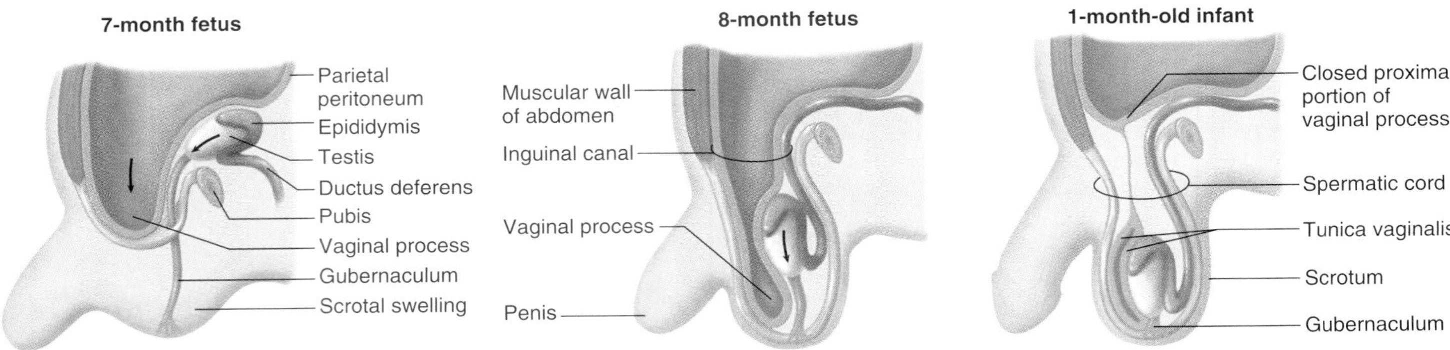

Figure 27.4 **Descent of the Testis.** Note that the testis and spermatic ducts are retroperitoneal. An extension of the peritoneum called the *vaginal process* follows the testis through the inguinal canal and becomes the tunica vaginalis.

elongates to form the penis, the urogenital folds fuse to enclose the urethra within the penis, and the labioscrotal folds fuse to form the scrotum, a sac that will later contain the testes.

Male and female organs that develop from the same embryonic structure are said to be **homologous.** Thus the penis is homologous to the clitoris and the scrotum is homologous to the labia majora. This becomes strikingly evident in some abnormalities of sexual development. In the presence of excess androgen, the clitoris may become greatly enlarged and resemble a small penis. In other cases, the ovaries descend into the labia majora as if they were testes descending into a scrotum. Such abnormalities sometimes result in mistaken identification of the sex of an infant at birth.

Descent of the Testes

The testes begin development near the kidney. How do they end up in the scrotum, and why? In the embryo, a cord of muscle and connective tissue called the **gubernaculum**[9] (GOO-bur-NACK-you-lum) extends from the gonad to the floor of the abdominopelvic cavity. It shortens as the fetus grows and guides the testis through a passageway in the groin called the **inguinal canal.** This **descent of the testes** (fig. 27.4) begins in weeks 6 to 10; by 28 weeks, the testes enter the scrotum. As they descend, they are accompanied by ever-elongating testicular arteries and veins and by lymphatic vessels, nerves, sperm ducts, and extensions of the internal abdominal oblique muscle.

In the scrotum, the testes are kept at a temperature of about 35°C—2°C cooler than in the pelvic cavity. This is essential for sperm production. About 3% of boys are born with undescended testes, or *cryptorchidism* (see table 27.2).

9. *gubern* = rudder, to steer, guide

Before You Go On

Answer the following questions to test your understanding of the preceding section:

3. What are mesonephric and paramesonephric ducts? What factors determine which one develops and which one regresses in the fetus?
4. What male structures develop from the phallus and labioscrotal folds?
5. Define the *gubernaculum* and describe its function.

Male Reproductive Anatomy

Objectives

When you have completed this section, you should be able to

- describe the anatomy of the testes, scrotum, and penis;
- describe the pathway taken by a sperm cell from its formation to its ejaculation; and
- describe the accessory glands and other secondary sex organs of the male, and state their functions.

The external genitalia occupy the *perineum* (PERR-ih-NEE-um), the diamond-shaped region marked by the pubic symphysis, ischial tuberosities, and coccyx (fig. 27.5). The internal reproductive anatomy of the male is shown in figure 27.6.

Testes

Each testis is oval, about 4 cm long and 2.5 cm in diameter (fig. 27.7). Its anterior and lateral surfaces are covered by the **tunica vaginalis,**[10] a saclike extension of the peri-

10. *tunica* = coat + *vagina* = sheath

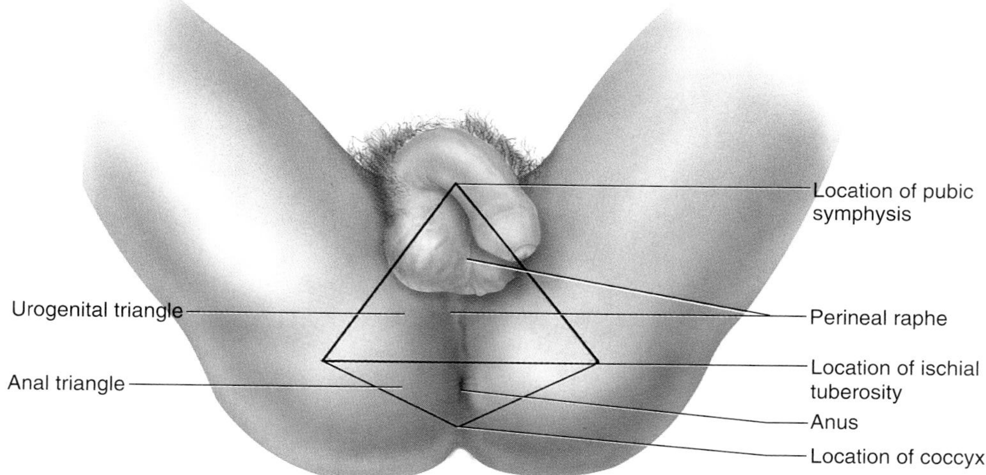

Figure 27.5 The Male Perineum, Inferior Aspect.

toneum that follows the testis as it descends into the scrotum. The testis itself has a white fibrous capsule called the **tunica albuginea**[11] (TOO-nih-ca AL-byu-JIN-ee-uh). Connective tissue septa divide the organ into 250 to 300 wedge-shaped lobules. Each lobule contains one to three **seminiferous**[12] (SEM-ih-NIF-er-us) **tubules**—slender ducts up to 70 cm long in which the sperm are produced. Between the seminiferous tubules are clusters of **interstitial (Leydig**[13]**) cells,** the source of testosterone.

A seminiferous tubule has a narrow lumen lined by a thick **germinal epithelium** (fig. 27.8). The epithelium consists of several layers of **germ cells** in the process of becoming sperm and a much smaller number of tall **sustentacular**[14] **(Sertoli**[15]**) cells,** which protect the germ cells and promote their development. The germ cells depend on the sustentacular cells for nutrients, waste removal, growth factors, and other needs. The sustentacular cells also secrete a hormone, *inhibin,* that regulates the rate of sperm production, as we will see later.

A sustentacular cell is shaped a little like a tree trunk whose roots spread out over the basement membrane, forming the boundary of the tubule, and whose thick trunk reaches to the tubule lumen. Tight junctions between adjacent sustentacular cells form a **blood-testis barrier (BTB),** which prevents proteins and other large molecules in the blood and intercellular fluid from getting to the germ cells. This is important because the germ cells, being genetically different from other cells of the body, would otherwise be attacked by the immune system. Some cases of sterility occur when the BTB fails to form adequately in adolescence and the immune system produces autoantibodies against the germ cells.

11. *alb* = white
12. *semin* = seed, sperm + *fer* = to carry
13. Franz von Leydig (1821–1908), German anatomist
14. *sustentacul* = support
15. Enrico Sertoli (1842–1910), Italian histologist

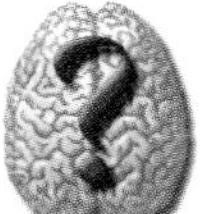

Think About It

Would you expect to find blood capillaries in the walls of the seminiferous tubules? Why or why not?

The seminiferous tubules lead into a network called the **rete**[16] (REE-tee) **testis,** embedded in the capsule on the posterior side. Sperm partially mature in the rete. They are moved along by the flow of fluid secreted by the sustentacular cells and possibly by the cilia seen on some rete cells. Sperm do not swim while they are in the male reproductive tract.

Each testis is supplied by a **testicular artery** that arises from the abdominal aorta just below the renal artery. This is a very long, slender artery that winds its way down the posterior abdominal wall before passing through the inguinal canal into the scrotum. Its blood pressure is very low, and indeed this is one of the few arteries to have no pulse. Consequently, blood flow to the testes is quite meager and the testes receive a poor oxygen supply. In response to this, the sperm develop unusually large mitochondria, which may precondition them for survival in the hypoxic environment of the female reproductive tract.

Blood leaves the testis by way of a **testicular vein.** The right testicular vein drains into the inferior vena cava and the left one drains into the left renal vein. Lymphatic vessels also drain each testis and lead to the inguinal lymph nodes. **Testicular nerves** lead to the gonads from

16. *rete* = network

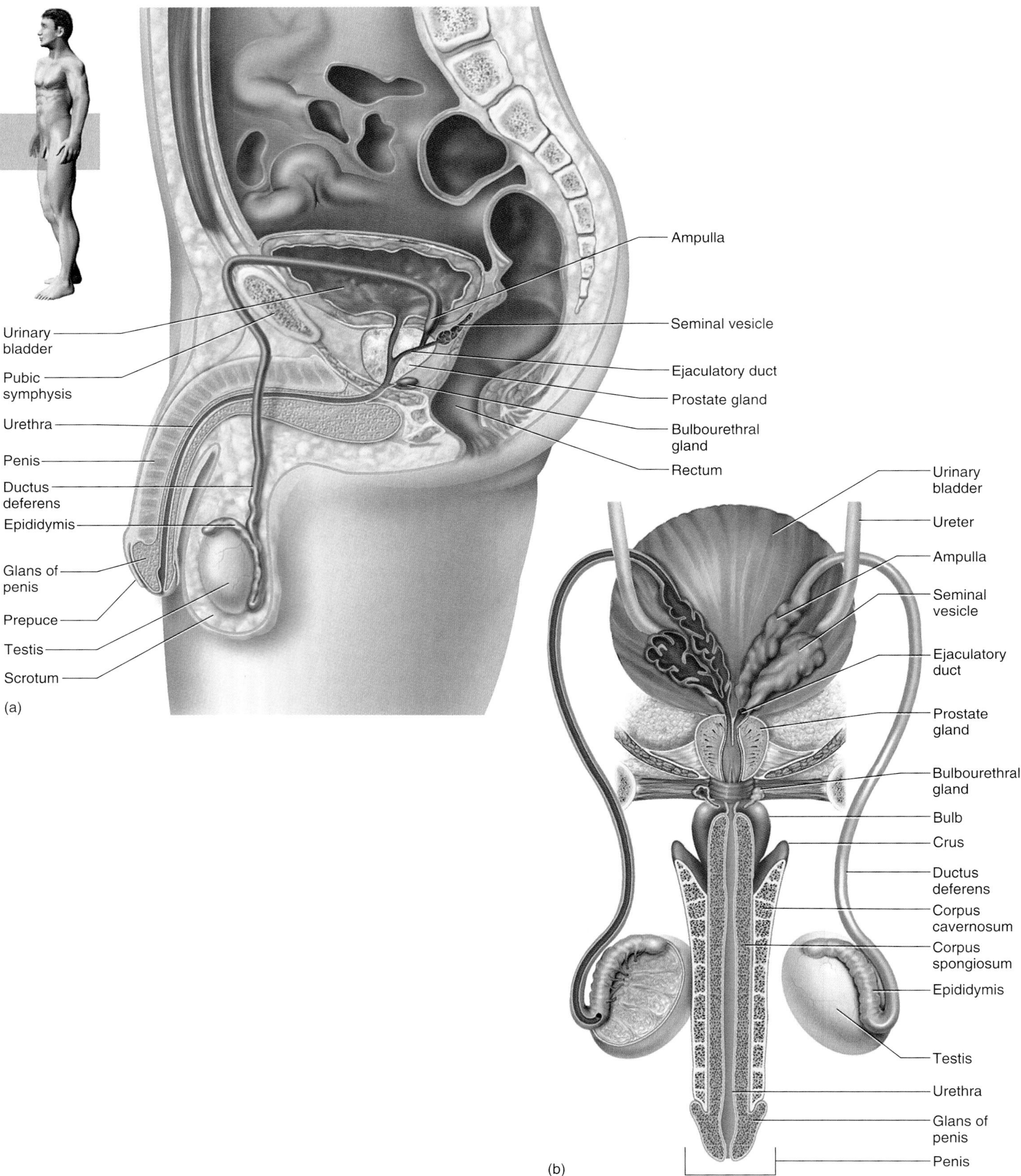

Figure 27.6 **The Male Reproductive System.** (*a*) Midsagittal section of the pelvic cavity viewed from the left. (*b*) Posterior view of the reproductive organs.

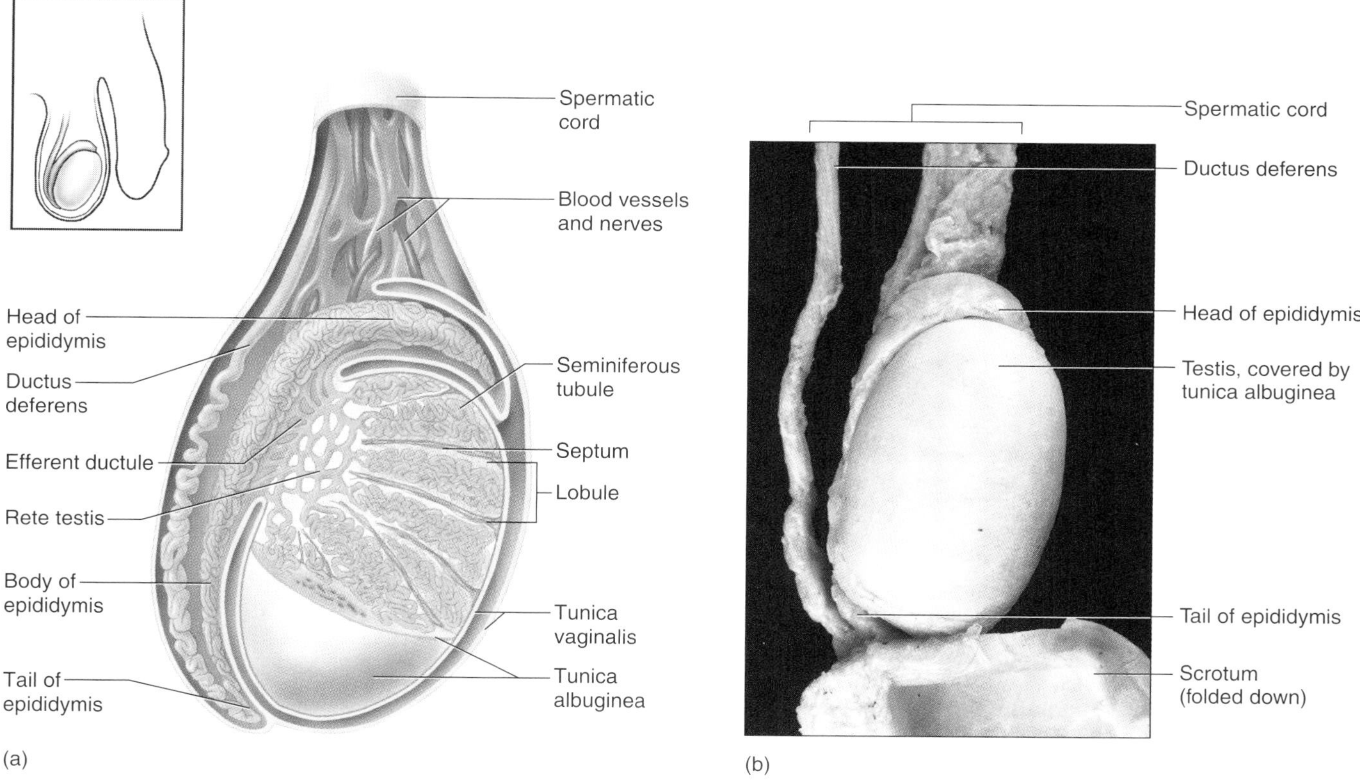

Figure 27.7 **The Testis and Associated Structures.** (*a*) Anatomy of the testis, epididymis, and spermatic cord. (*b*) The testis and associated structures dissected free of the scrotum, shown life size.

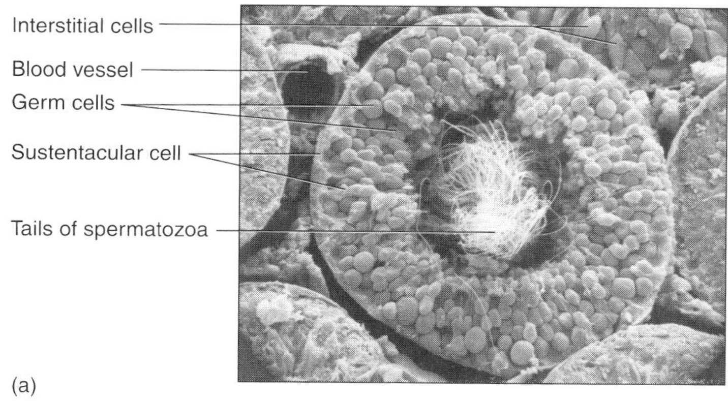

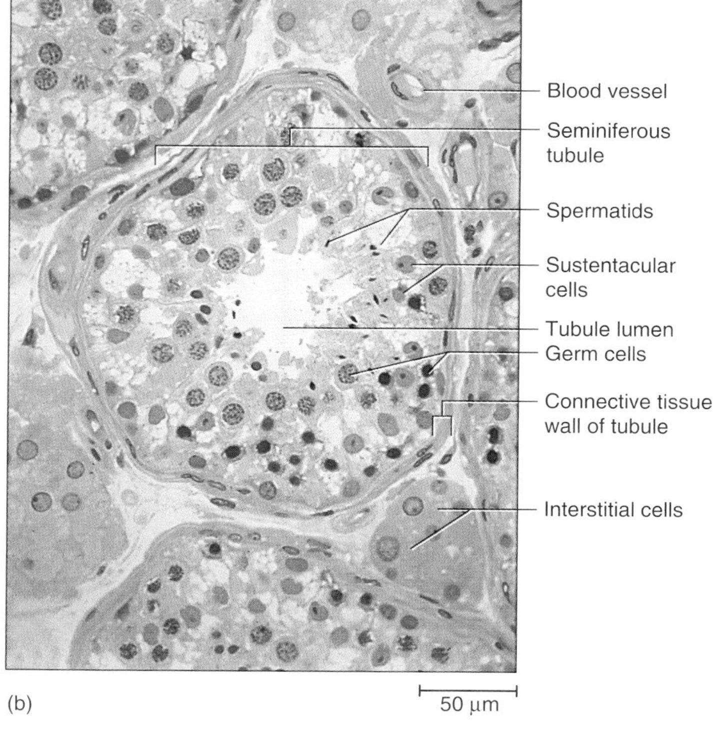

Figure 27.8 **Histology of the Testis.** (*a*) Scanning electron micrograph; (*b*) light micrograph. Figure *b* is from a region of the tubule that did not have mature sperm at the time.
(*a*) From R. G. Kessel and R. H. Kardon, *Tissues and Organs: A Text-Atlas of Scanning Electron Microscopy* (W. H. Freeman, 1979).

spinal cord segment T10. They are mixed sensory and motor nerves containing predominantly sympathetic but also some parasympathetic fibers.

Scrotum

The testes are contained in a pendulous pouch, the **scrotum**[17] (fig. 27.9). The left testis is usually suspended lower than the right so the two are not compressed against each other between the thighs. The skin of the scrotum has sebaceous glands, sparse hair, rich sensory innervation, and somewhat darker pigmentation than skin elsewhere. The scrotum is divided into right and left compartments by an internal **median septum,** which protects each testis from infections of the other one. The location of the septum is externally marked by a seam called the **perineal raphe**[18] (RAY-fee), which also extends anteriorly along the ventral side of the penis and posteriorly as far as the margin of the anus (see fig. 27.5).

17. *scrotum* = bag
18. *raphe* = seam

A structure called the **spermatic cord** passes up the back of the scrotum, then anterior to the pubis, and through the *inguinal ring* into the inguinal canal. The canal is a passage about 4 cm long through the abdominal muscle to the pelvic cavity. The spermatic cord can be palpated through the skin on the back of the scrotum and anterior to the pubic bone. It contains a sperm duct called the *ductus deferens* and other structures that descended with the testis as it passed through the inguinal canal—the aforementioned blood and lymphatic vessels, testicular nerves, and so forth.

The original reason that a scrotum evolved has long been a subject of debate among reproductive biologists and still has no universally accepted answer. Now that human testes reside in the scrotum, however, they have adapted to this cooler environment and cannot produce sperm at the core body temperature of 37°C. The scrotum has three mechanisms for regulating the temperature of the testes:

1. The **cremaster**[19] **muscle** consists of strips of the internal abdominal oblique muscle that enmesh the

19. *cremaster* = suspender

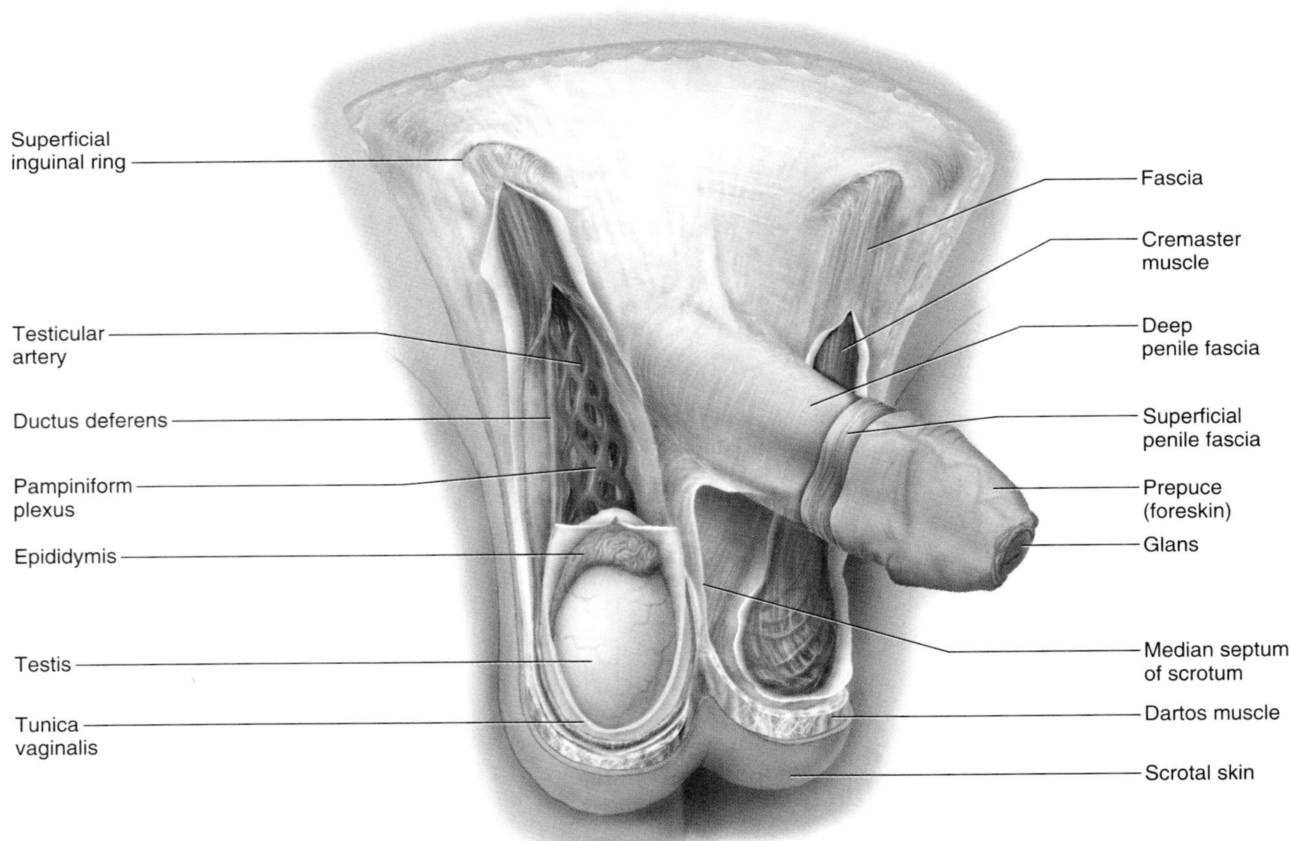

Figure 27.9 Anatomy of the Male Inguinal Region and External Genitalia.

spermatic cord. When it is cold, the cremaster contracts and draws the testes closer to the body to keep them warm. When it is warm, the cremaster relaxes and the testes are suspended farther from the body.

2. The **dartos**[20] **muscle (tunica dartos)** is a subcutaneous layer of smooth muscle. It, too, contracts when it is cold, and the scrotum becomes taut and wrinkled. The tautness of the scrotum helps to hold the testes snugly against the warm body and it reduces the surface area of the scrotum, thus reducing heat loss.
3. The **pampiniform**[21] **plexus** is an extensive network of veins from the testis that surround the testicular artery in the spermatic cord. As they pass through the inguinal canal, these veins converge to form the testicular vein, which emerges from the inguinal canal into the pelvic cavity. Without the pampiniform plexus, warm arterial blood would heat the testis and inhibit spermatogenesis. The pampiniform plexus, however, prevents this by acting as a *countercurrent heat exchanger* (fig. 27.10). Imagine that a house had uninsulated hot and cold water pipes running close to each other. Much of the heat from the hot water pipe would be absorbed by the cold water pipe next to it—especially if the water flowed in opposite directions so the cold water carried the heat away instead of running alongside the hot water. In the spermatic cord, such a mechanism removes heat from the descending arterial blood, so by the time it reaches the testis this blood is 1.5° to 2.5°C cooler than the core body temperature.

20. *dartos* = skinned
21. *pampin* = tendril + *form* = shape

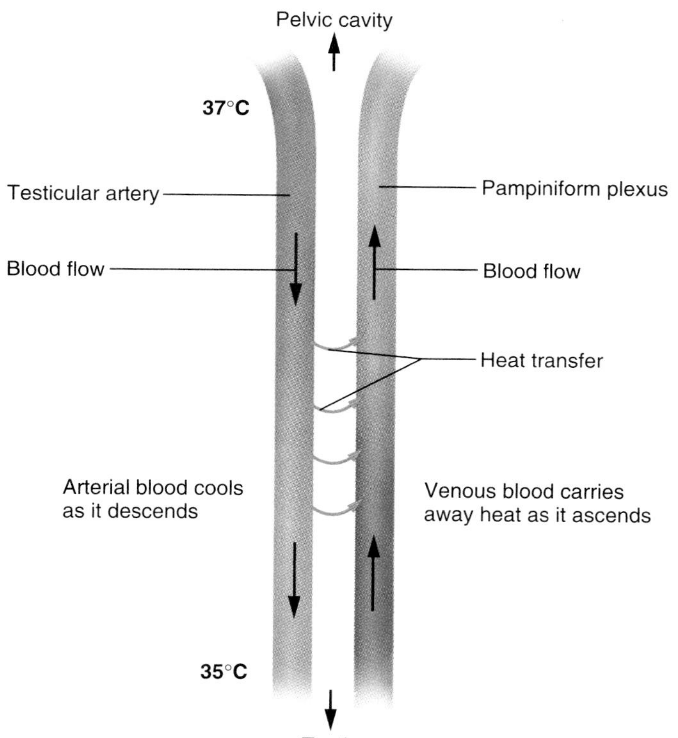

Figure 27.10 The Countercurrent Heat Exchange Mechanism of the Pampiniform Plexus. Warm blood flowing down the testicular artery loses some of its heat to the cooler blood flowing in the opposite direction through the pampiniform plexus of veins (represented as a single vessel for simplicity). Arterial blood is about 2°C cooler by the time it reaches the testis.

Spermatic Ducts

After leaving the testis, the sperm travel through a series of *spermatic ducts* to reach the urethra. These include the following:

- **Efferent ductules.** About 12 small efferent ductules arise from the posterior side of each testis and carry sperm to the epididymis. They have clusters of ciliated cells that help drive the sperm along.
- **Duct of the epididymis.** The **epididymis**[22] (EP-ih-DID-ih-miss; plural, *epididymides*) is a site of sperm maturation and storage. It adheres to the posterior side of the testis (see fig. 27.7), measures about 7.5 cm long, and consists of a clublike *head* at the superior pole of the testis, a long middle *body*, and a slender *tail* at its inferior end. It contains a single coiled duct, about 6 m (18 ft) long, embedded in connective tissue. This duct reabsorbs about 90% of the fluid secreted by the testis. Sperm are physiologically immature (incapable of fertilizing an egg) when they leave the testis but mature as they travel through the head and body of the epididymis. In 20 days or so, they reach the tail. They are stored here and in the adjacent portion of the ductus deferens. Stored sperm remain fertile for 40 to 60 days, but if they become too old without being ejaculated, they disintegrate and the epididymis reabsorbs them.
- **Ductus (vas) deferens.** The duct of the epididymis straightens out at the tail, turns 180°, and becomes the ductus deferens. This is a muscular tube about 45 cm long and 2.5 mm in diameter. It passes upward from the scrotum, travels through the inguinal canal, and enters the pelvic cavity (see fig. 27.7). There, it turns medially and approaches the urinary bladder. After passing between the bladder and ureter, the duct turns downward behind the bladder and widens into a

22. *epi* = upon + *didym* = twins, testes

Clinical Insight 27.2

Prostate Diseases

The prostate gland weighs about 20 g by age 20, remains at that weight until age 45 or so, and then begins to grow slowly again. By age 70, over 90% of men show some degree of *benign prostatic hyperplasia*—noncancerous enlargement of the gland. The major complication of this is that it compresses the urethra, obstructs the flow of urine, and may promote bladder and kidney infections.

Prostate cancer is the second most common cancer in men (after lung cancer), affecting about 9% of men over the age of 50. Prostate tumors tend to form near the periphery of the gland, where they do not obstruct urine flow; therefore, they often go unnoticed until they cause pain. Prostate cancer often metastasizes to nearby lymph nodes and then to the lungs and other organs. It is more common among American blacks than whites and very uncommon among Japanese. It is diagnosed by digital rectal examination and by detecting *prostate specific antigen (PSA)* and *acid phosphatase* (a prostatic enzyme) in the blood. Up to 80% of men with prostate cancer survive when it is detected and treated early, but only 10% to 50% survive if it spreads beyond the prostatic capsule.

terminal **ampulla.** The ductus deferens ends by uniting with the duct of the seminal vesicle, a gland considered later. The duct has a very narrow lumen and a thick wall of smooth muscle well innervated by sympathetic nerve fibers. During orgasm, the sympathetic nervous system stimulates peristaltic contractions that drive sperm from the epididymis to the ejaculatory duct.

- **Ejaculatory duct.** Where the ductus deferens and duct of the seminal vesicle meet, they form a short (2 cm) ejaculatory duct, which passes through the prostate gland and empties into the urethra. The ejaculatory duct is the last of the spermatic ducts.

The male urethra is shared by the reproductive and urinary systems. It is about 20 cm long and consists of three regions: the *prostatic, membranous,* and *penile urethra* (see fig. 23.18, p. 906). Although it serves both urinary and reproductive roles, it cannot pass urine and semen simultaneously for reasons discussed later.

Accessory Glands

There are three sets of *accessory glands* in the male reproductive system: the seminal vesicles, prostate gland, and bulbourethral glands:

1. The **seminal vesicles** are a pair of glands posterior to the urinary bladder; one is associated with each ductus deferens. A seminal vesicle is about 5 cm long, or approximately the dimensions of your little finger. It has a connective tissue capsule and underlying layer of smooth muscle. The secretory portion is a very convoluted duct with numerous branches that form a complex labyrinth. The duct empties into the ampulla of the ductus deferens. The secretion of the seminal vesicles constitutes about 60% of the semen; its composition and functions are discussed later.
2. The **prostate**[23] (PROSS-tate) **gland** surrounds the urethra and ejaculatory duct immediately inferior to the urinary bladder. It measures about 2 by 4 by 3 cm and is an aggregate of 30 to 50 compound tubuloacinar glands enclosed in a single fibrous capsule. These glands empty through about 20 pores in the urethral wall. The stroma of the prostate consists of connective tissue and smooth muscle, like that of the seminal vesicles. The thin, milky secretion of the prostate contributes about 30% of the semen. Its functions, too, are considered later. The position of the prostate immediately anterior to the rectum allows it to be palpated through the rectal wall to check for lumps suggestive of prostate cancer. This procedure is known as *digital rectal examination (DRE)* (see clinical insight 27.2).
3. The **Bulbourethral (Cowper**[24]**) glands** are named for their position near a dilated bulb at the inner end of the penis and their short (2.5 cm) ducts leading into the penile urethra. They are brownish, spherical glands about 1 cm in diameter. During sexual arousal, they produce a clear slippery fluid that lubricates the head of the penis in preparation for intercourse. Perhaps more importantly, though, it neutralizes the acidity of residual urine in the urethra, which would be harmful to the sperm.

Penis

The **penis**[25] serves to deposit semen in the vagina. Half of it is an internal **root** and half is the externally visible **shaft**

23. *pro* = before + *stat* = to stand; commonly misspelled and mispronounced "prostrate"
24. William Cowper (1666–1709), British anatomist
25. *penis* = tail

and **glans**[26] (see figs. 27.6 and 27.11). The external portion is about 8 to 10 cm (3–4 in.) long and 3 cm in diameter when flaccid (nonerect); the typical dimensions of an erect penis are 13 to 18 cm (5–7 in.) long and 4 cm in diameter. The glans is the expanded head at the distal end of the penis with the external urinary meatus at its tip.

The skin is very loosely attached to the shaft, allowing for expansion during erection. It continues over the glans as the **prepuce,** or foreskin, which is often removed by circumcision. A ventral fold of tissue called the *frenulum* attaches the skin to the glans. The skin of the glans itself is thinner and firmly attached to the underlying erectile tissue. The glans and facing surface of the prepuce have sebaceous glands that produce a waxy secretion called **smegma.**[27]

The penis consists mainly of three cylindrical bodies called **erectile tissues,** which fill with blood during sexual arousal and account for its enlargement and erection. A single erectile body, the **corpus spongiosum,** passes along the ventral side of the penis and encloses the penile urethra. It expands at the distal end to fill the entire glans. The dorsal side of the penis, proximal to the glans, has a **corpus cavernosum** (plural, *corpora cavernosa*) on each side. Each is ensheathed in a fibrous **tunica albuginea,** and they are separated from each other by a **median septum.** (Note that the testes also have a tunica albuginea and the scrotum also has a median septum.)

All three cylinders of erectile tissue are spongy in appearance and contain numerous tiny blood sinuses called **lacunae.** The partitions between lacunae, called **trabeculae,** are composed of connective tissue and smooth **trabecular muscle.** In the flaccid penis, tonus of the trabecular muscle collapses the lacunae, which appear as tiny slits in the tissue.

At the body surface, the penis turns 90° dorsally and continues inward as the root. The corpus spongiosum terminates internally as a dilated **bulb,** which is ensheathed in the bulbospongiosus muscle and attached to the lower surface of the perineal membrane within the urogenital triangle (see pp. 368–69). The corpora cavernosa diverge like the arms of a Y. Each arm, called a **crus** (cruss; plural, *crura*), attaches the penis to the pubic arch (ischiopubic ramus) and perineal membrane on its respective side. Each crus is enveloped by an ischiocavernosus muscle. The innervation and blood supply to the penis are discussed later in connection with the mechanism of erection.

26. *glans* = acorn
27. *smegma* = unguent, ointment, soap

Before You Go On

Answer the following questions to test your understanding of the preceding section:

6. State the names and locations of two muscles that help regulate the temperature of the testes.
7. Name three types of cells in the testis, and describe their locations and functions.
8. Name all the ducts that the sperm follow, in order, from the time they leave the testis to the time of ejaculation.
9. Describe the locations and functions of the seminal vesicles, prostate, and bulbourethral glands.
10. Name the erectile tissues of the penis, and describe their locations relative to each other.

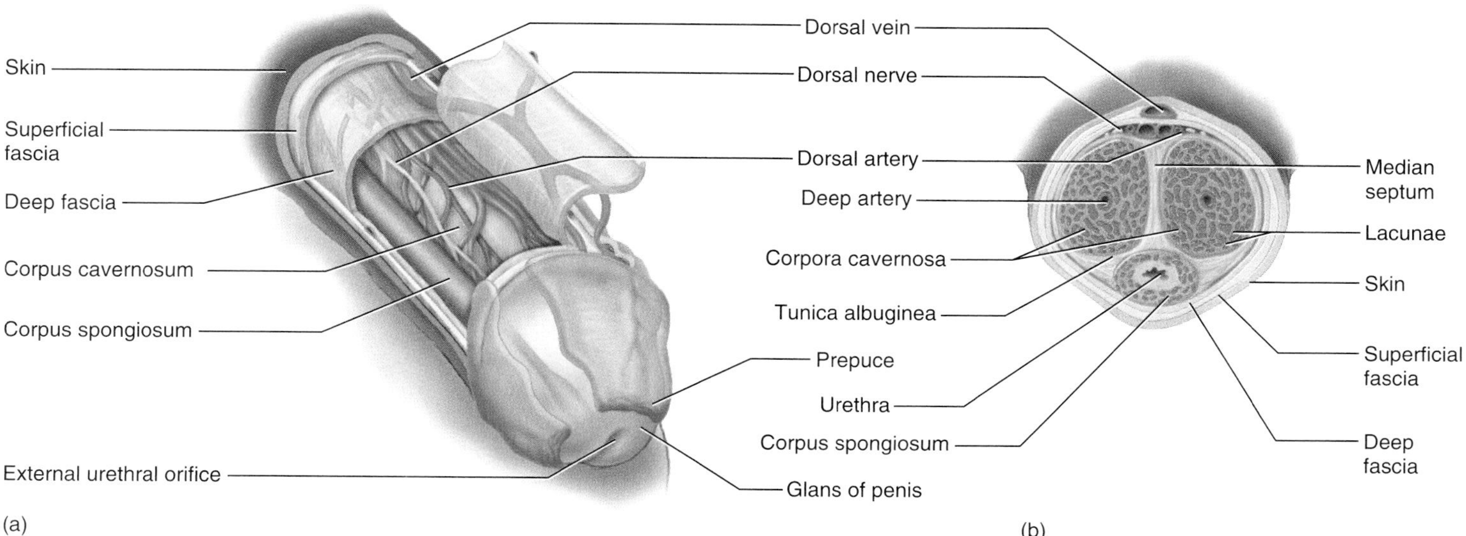

Figure 27.11 Anatomy of the Penis. (*a*) Dissection in lateral view. (*b*) Cross section.

Puberty and Climacteric

Objectives

When you have completed this section, you should be able to

- describe the hormonal control of puberty;
- describe the resulting changes in the male body; and
- define and describe male climacteric and the effect of aging on male reproductive function.

Unlike any other organ system, the reproductive system remains dormant for several years after birth. Around age 10 to 12 in most boys and 8 to 10 in most girls, however, a surge of pituitary gonadotropins "awakens" the reproductive system and begins preparing it for adult reproductive function. This is the onset of puberty.

Definitions of *puberty* and *adolescence* vary. This book uses **puberty**[28] to mean the period from the onset of gonadotropin secretion and reproductive development until a person attains the ability to reproduce sexually. The end of puberty is marked by the first menstrual period in girls and the first ejaculation of viable sperm in boys. In North America, this is typically attained around age 12 in girls and age 13 in boys. **Adolescence**[29] is used here to mean the period including puberty but not ending until a person attains full adult height.

Endocrine Control of Puberty

The testes secrete substantial amounts of testosterone in the first trimester (3 months) of fetal development. Even in the first few months of infancy, testosterone levels are about as high as they are in midpuberty, but then the testes become dormant for the rest of infancy and childhood. From puberty through adulthood, reproductive function is regulated by hormonal links between the hypothalamus, pituitary, and gonads—the **brain-testicular axis** (fig. 27.12).

As the hypothalamus matures, it begins producing **gonadotropin-releasing hormone (GnRH),** which travels by way of the hypothalamo-hypophyseal portal system to the anterior lobe of the pituitary. Here it stimulates cells called *gonadotropes* to secrete **follicle-stimulating hormone (FSH)** and **luteinizing hormone (LH),** two *gonadotropins* that stimulate different cells in the testis. Enlargement of the testes is the first sign of puberty.

LH stimulates the interstitial cells of the testis to secrete androgens, mainly testosterone. In the male, LH is sometimes called *interstitial cell–stimulating hormone (ICSH).* FSH stimulates the sustentacular cells to secrete **androgen-binding protein (ABP).** ABP is thought to raise testosterone levels in the seminiferous tubules and epididymis, but this remains unproven. Without FSH, however, testosterone has no effect on the testis. Germ cells have no androgen receptors and do not respond to it. Nevertheless, testosterone has the following effects:

- It stimulates spermatogenesis in the presence of ABP. If testosterone secretion ceases, the sperm count and semen volume decline rapidly and a male becomes sterile.

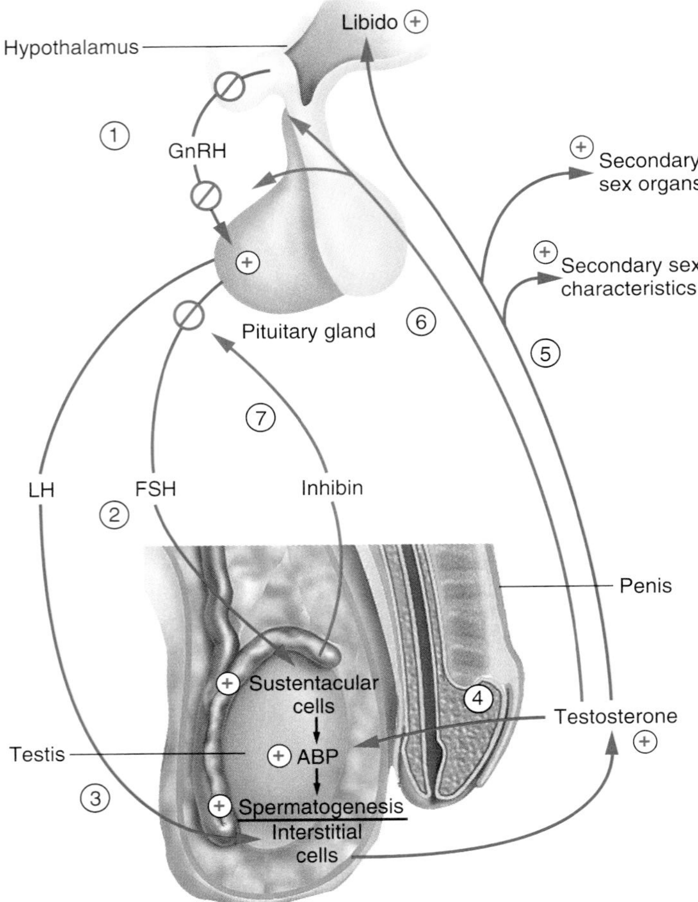

Figure 27.12 Hormonal Relationships in the Brain-Testicular Axis. *Green arrows* and *plus signs* indicate stimulatory effects; *red arrows* and *bars* indicate inhibitory effects. (*1*) GnRH from the hypothalamus stimulates the anterior lobe of the pituitary to secrete FSH and LH. (*2*) FSH stimulates sustentacular cells to secrete androgen-binding protein (ABP). (*3*) LH stimulates interstitial cells to secrete testosterone. (*4*) In the presence of ABP, testosterone stimulates spermatogenesis. (*5*) Testosterone also stimulates development of secondary sex organs and secondary sex characteristics and stimulates the libido. (*6*) Testosterone has a negative feedback effect on the hypothalamus and pituitary, which reduces GnRH secretion and pituitary sensitivity to GnRH. (*7*) To reduce sperm production without reducing testosterone secretion, the sustentacular cells secrete inhibin, which selectively inhibits FSH secretion.

28. *puber* = grown up
29. *adolesc* = to grow up

- It inhibits GnRH secretion by the hypothalamus and reduces the GnRH sensitivity of the pituitary. Consequently, FSH and LH secretion are held in check. Over the course of puberty, however, the pituitary becomes less sensitive to this negative feedback and begins to secrete increasing amounts of the gonadotropins.
- It stimulates development of the secondary sex characteristics and other somatic changes characteristic of puberty. Many of the physiological processes of puberty go unnoticed at first. The first visible sign of puberty is usually enlargement of the testes and scrotum around age 13. The penis continues to grow for about two more years. One of the androgens, dihydrotestosterone (DHT), stimulates development of the pubic, axillary, and facial hair. The skin becomes darker and thicker and secretes more sebum, which often leads to acne. The skin of acne patients contains 2 to 20 times as much DHT as normal. The apocrine sweat (scent) glands of the perineal, axillary, and beard areas develop in conjunction with the hair in those regions.
- It causes the ducts and accessory glands of the reproductive system to enlarge. Erections occur frequently and ejaculation may occur during sleep (nocturnal emissions, or "wet dreams").
- Testosterone stimulates growth hormone secretion, which results in a burst of growth, an increase in muscle mass, a higher basal metabolic rate, and a larger larynx. The last of these effects deepens the voice and makes the thyroid cartilage more prominent at the front of the neck.
- It stimulates erythropoiesis, which gives men a higher hematocrit and RBC count than those of women.
- Testosterone also stimulates the brain and awakens the **libido** (sex drive)—although, perhaps surprisingly, the neurons convert it to estrogen (a supposedly "female" sex hormone), which is what directly affects the behavior.

Throughout adulthood, testosterone sustains the male reproductive tract, spermatogenesis, and libido. At times, it is necessary to reduce FSH secretion and sperm production without reducing LH and testosterone secretion. To achieve this, the sustentacular cells secrete a hormone called **inhibin** that selectively suppresses FSH output from the pituitary. When the sperm count drops below 20 million sperm/mL, however, inhibin secretion drops and FSH secretion rises.

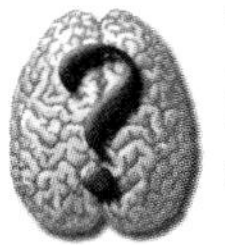

Think About It

If a male animal is castrated, would you expect FSH and LH levels to rise, fall, or be unaffected. Why?

Aging and Sexual Function

Testosterone secretion peaks at about 7 mg/day at age 20 and then declines steadily to as little as one-fifth of this level by age 80. There is a corresponding decline in the number and secretory activity of the interstitial cells (the source of testosterone) and sustentacular cells (the source of inhibin). As testosterone and inhibin levels decline, so does feedback inhibition of the pituitary. Consequently, FSH and LH levels rise significantly after age 50 and produce changes called **male climacteric.** Most men pass through climacteric with little or no effect, but in a few cases there are mood changes, hot flashes, and illusions of suffocation—symptoms similar to those of menopause in women. Despite references to "male menopause," the term *menopause* refers to the cessation of menstruation and therefore makes little sense in the context of male physiology.

About 20% of men in their 60s and 50% of men in their 80s experience *impotence (erectile dysfunction),* the frequent inability to maintain a sufficient erection for intercourse (see table 27.2). Over 90% of impotent men, however, remain able to ejaculate. Impotence and declining sexual activity can have a major impact on older people's perception of the quality of life.

Before You Go On

Answer the following questions to test your understanding of the preceding section:

11. State the source, target organ, and effect of GnRH.
12. Identify the target cells and effects of FSH and LH.
13. Explain how testicular hormones affect the secretion of FSH and LH.
14. Describe the major effects of androgens on the body.
15. Define *male climacteric.*

Spermatogenesis, Spermatozoa, and Semen

Objectives

When you have completed this section, you should be able to

- describe the stages of meiosis and contrast meiosis with mitosis;
- describe the sequence of cell types in spermatogenesis, and relate these to the stages of meiosis;
- describe the role of the sustentacular cell in spermatogenesis;
- draw or describe a sperm cell; and
- describe the composition of semen and functions of its components.

The most significant event of puberty is the onset of **spermatogenesis.** This is a process in which germ cells undergo two divisions called **meiosis** and the four daughter cells differentiate into spermatozoa. It occurs in the seminiferous tubules.

Meiosis

In the eukaryotes (all living organisms except bacteria), there are two forms of cell division—mitosis and meiosis. Mitosis, described in chapter 5, is the basis for division of the single-celled fertilized egg, growth of an embryo, and all postnatal growth and tissue repair. It is essentially the splitting of a cell with a distribution of chromatids that results in two genetically identical daughter cells. It consists of four stages—prophase, metaphase, anaphase, and telophase.

You may find it beneficial to review figure 5.14 (p. 157) because of the important similarities and differences between mitosis and meiosis. There are three important differences:

1. In mitosis, each double-stranded chromosome divides into two single-stranded ones, but each daughter cell still has 46 chromosomes (23 pairs). Meiosis, by contrast, reduces the chromosome number by half. The parent cell is **diploid (2*n*),** meaning it has 46 chromosomes in 23 homologous pairs (see fig. 5.16, p. 160), whereas the daughter cells are **haploid (*n*),** with 23 unpaired chromosomes.
2. In mitosis, the chromosomes do not change their genetic makeup. In an early stage of meiosis, however, the chromosomes of each homologous pair join and exchange portions of their DNA. This creates new combinations of genes, so the chromosomes we pass on to our offspring are not the same chromosomes that we inherit from our parents.
3. In mitosis, each parent cell produces only two daughter cells. In meiosis, it produces four. In the male, four sperm therefore develop from each original germ cell. The situation is somewhat different in the female, as we will see in the next chapter.

Why use such a relatively complicated process for gametogenesis? Why not use mitosis, as we do for all other cell replication in the body? The answer is basically that sexual reproduction is, by definition, biparental. If we are going to combine gametes from two parents to make a child, there must be a mechanism for keeping the chromosome number constant from generation to generation. Mitosis would produce eggs and sperm with 46 chromosomes each. If these gametes combined, the zygote and the next generation would have 92 chromosomes per cell, the generation after that would have 184, and so forth. To prevent the chromosome number from doubling in every generation, the number is reduced by half during gametogenesis. Meiosis is sometimes called *reduction division* for this reason.[30]

The stages of meiosis are fundamentally the same in both sexes. Briefly, it consists of two cell divisions in succession and occurs in the following phases: prophase I, metaphase I, anaphase I, telophase I, interkinesis, prophase II, metaphase II, anaphase II, and telophase II. These events are detailed in figure 27.13, but let us note some of its unique and important aspects.

In prophase I, each pair of homologous chromosomes line up side by side and form a **tetrad** (*tetra* denoting the four chromatids). One chromosome of each tetrad is from the individual's father (the paternal chromosome) and the other is from the mother (the maternal chromosome). The paternal and maternal chromosomes exchange segments of DNA in a process called **crossing-over.** This creates new combinations of genes and thus contributes to genetic variety in the offspring.

After crossing-over, the chromosomes line up at the midline of the cell in metaphase I, they separate at anaphase I, and the cell divides in two at telophase I. This looks superficially like mitosis, but there is an important difference: the centromeres do not divide and the chromatids do not separate from each other at anaphase I; rather, each homologous chromosome parts company with its twin. Therefore, at the conclusion of meiosis I, each chromosome is still double-stranded, but each daughter cell has only 23 chromosomes—it has become haploid.

Meiosis II is more like mitosis—the chromosomes line up on the cell equator at metaphase II, the centromeres divide, and each chromosome separates into two chromatids. These chromatids are drawn to opposite poles of the cell at anaphase II. At the end of meiosis II, there are four haploid cells, each containing 23 single-stranded chromosomes. Fertilization combines 23 chromosomes from the father with 23 chromosomes from the mother and reestablishes the diploid number of 46 in the zygote.

Spermatogenesis

Now let's consider how meiosis relates to sperm production (fig. 27.14). The first cells destined to become sperm are **primordial germ cells.** Like the first blood cells, these form in the yolk sac, a membrane associated with the developing embryo. In the fifth to sixth week of development, they migrate into the embryo itself and colonize the gonadal ridges. Here they differentiate into **spermatogonia,** which lie along the periphery of the seminiferous tubule, outside the blood-testis barrier (BTB).

Spermatogonia multiply by mitosis producing two types of daughter cells called type A and type B spermatogonia. Type A cells remain outside the BTB and continue

30. *meio* = less, fewer

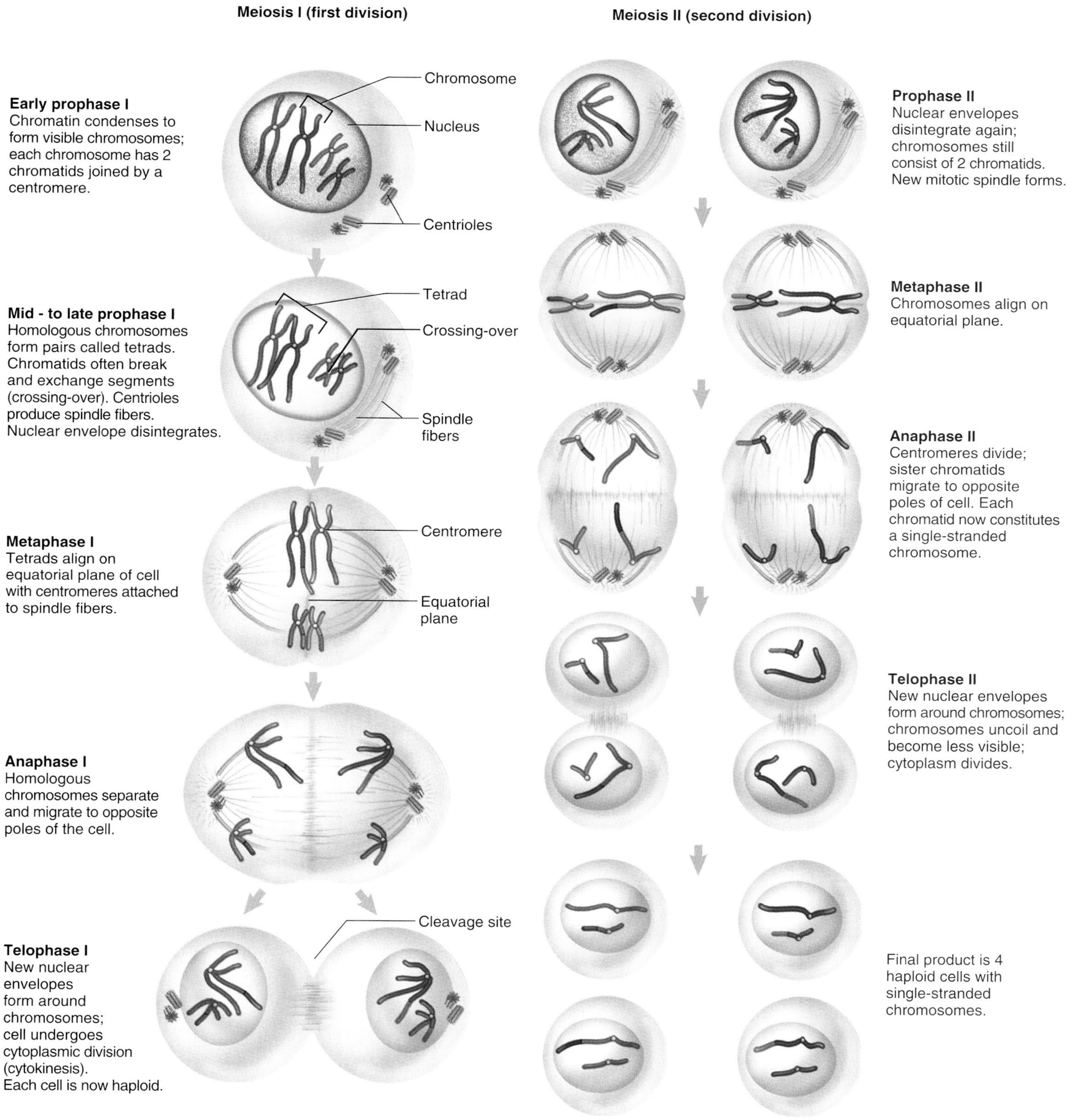

Meiosis I (first division)

Early prophase I
Chromatin condenses to form visible chromosomes; each chromosome has 2 chromatids joined by a centromere.

Chromosome
Nucleus
Centrioles

Mid - to late prophase I
Homologous chromosomes form pairs called tetrads. Chromatids often break and exchange segments (crossing-over). Centrioles produce spindle fibers. Nuclear envelope disintegrates.

Tetrad
Crossing-over
Spindle fibers

Metaphase I
Tetrads align on equatorial plane of cell with centromeres attached to spindle fibers.

Centromere
Equatorial plane

Anaphase I
Homologous chromosomes separate and migrate to opposite poles of the cell.

Telophase I
New nuclear envelopes form around chromosomes; cell undergoes cytoplasmic division (cytokinesis). Each cell is now haploid.

Cleavage site

Meiosis II (second division)

Prophase II
Nuclear envelopes disintegrate again; chromosomes still consist of 2 chromatids. New mitotic spindle forms.

Metaphase II
Chromosomes align on equatorial plane.

Anaphase II
Centromeres divide; sister chromatids migrate to opposite poles of cell. Each chromatid now constitutes a single-stranded chromosome.

Telophase II
New nuclear envelopes form around chromosomes; chromosomes uncoil and become less visible; cytoplasm divides.

Final product is 4 haploid cells with single-stranded chromosomes.

Figure 27.13 Meiosis. For simplicity, the cell is shown with only two pairs of homologous chromosomes. Human cells begin meiosis with 23 pairs.

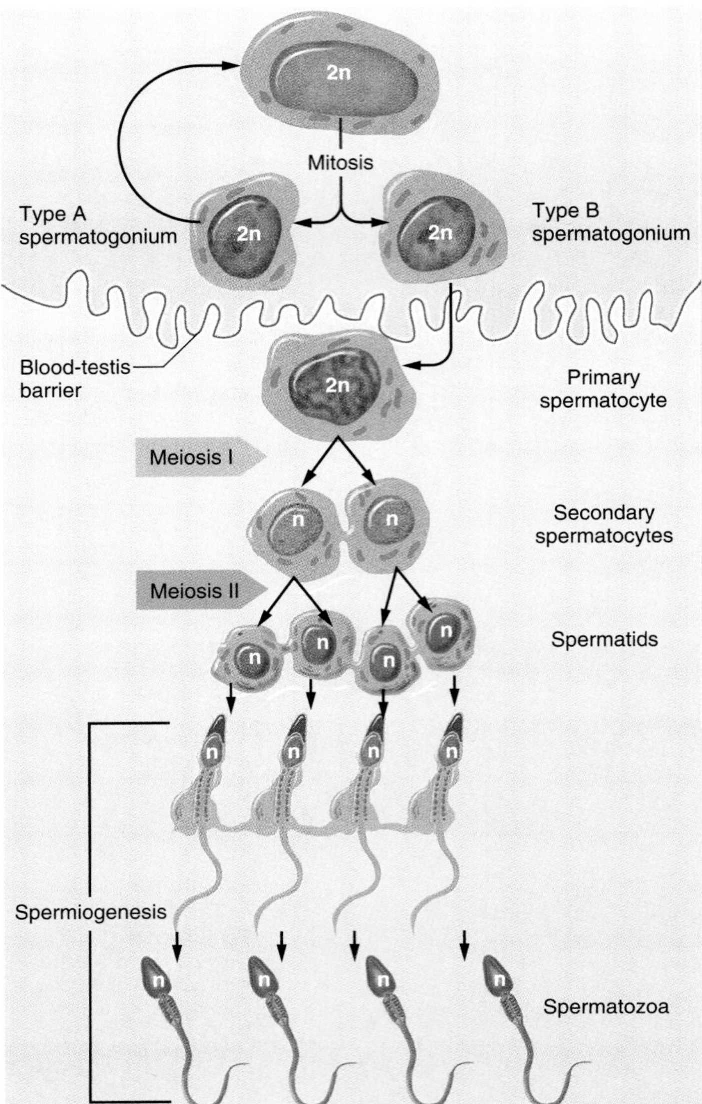

Figure 27.14 Spermatogenesis. 2*n* indicates diploid cells and *n* indicates haploid cells. Note that the daughter cells from secondary spermatocytes through spermatids remain connected by slender cytoplasmic processes until spermiogenesis is complete and individual spermatozoa are released.

to multiply from puberty until death. Thus men never exhaust their supply of gametes and normally remain fertile throughout old age.

Type B spermatogonia migrate closer to the tubule lumen and differentiate into slightly larger cells called **primary spermatocytes.** These cells must pass through the BTB and move toward the lumen of the tubule. Ahead of the primary spermatocyte, the tight junction between two sustentacular cells is dismantled, while a new tight junction forms on the other side, like closing the door behind the spermatocyte (fig. 27.15). The spermatocyte moves forward toward the lumen and is now separated from blood-borne agents, such as antibodies, held back by the tight junction.

Now safely isolated from the blood, the primary spermatocyte undergoes meiosis I, which gives rise to two equal-sized, haploid **secondary spermatocytes.** Each of these undergoes meiosis II, dividing into two **spermatids**—or a total of four for each spermatogonium. Each stage is a little closer to the lumen of the tubule than the earlier stages. All stages on the lumenal side of the BTB are bound to the sustentacular cells by tight junctions and gap junctions and are closely enveloped in tendrils of the sustentacular cells. Throughout these meiotic divisions, the daughter cells do not completely separate, but remain connected to each other by narrow cytoplasmic bridges.

The rest of spermatogenesis is called **spermiogenesis** (fig. 27.16). It involves no further cell division, but a gradual transformation of each spermatid into a spermatozoon. A spermatid sprouts a flagellum (tail) and discards most of its cytoplasm, becoming as small and lightweight as possible. Eventually, the sperm cell is released and is washed down the tubule by fluid from the sustentacular cells. It takes about 74 days for a spermatogonium to become a mature spermatozoon. A young man produces about 300,000 sperm per minute, or 400 million per day.

The Spermatozoon

The spermatozoon has two parts: a pear-shaped head and a long tail (fig. 27.17). The **head,** about 4 to 5 μm long and 3 μm wide at its broadest part, contains three structures: a nucleus, acrosome, and flagellar basal body. The most important of these is the nucleus, which fills most of the head and contains a haploid set of condensed, genetically inactive chromosomes. The **acrosome**[31] is a lysosome in the form of a thin cap covering the apical half of the nucleus. It contains enzymes that are later used to penetrate the egg if the sperm is successful. The basal body of the tail flagellum is nestled in an indentation at the basal end of the nucleus.

The **tail** is divided into three regions called the midpiece, principal piece, and endpiece. The **midpiece,** a cylinder about 5 to 9 μm long and half as wide as the head, is the thickest part. It contains numerous large mitochondria that spiral tightly around the axoneme of the flagellum. They produce the ATP needed for the beating of the tail when the sperm migrates up the female reproductive tract. The **principal piece,** 40 to 45 μm long, constitutes most of the tail and consists of the axoneme surrounded by a sheath of fibers. The **endpiece,** 4 to 5 μm long, consists of the axoneme only and is the narrowest part of the sperm.

31. *acro* = tip, peak + *some* = body

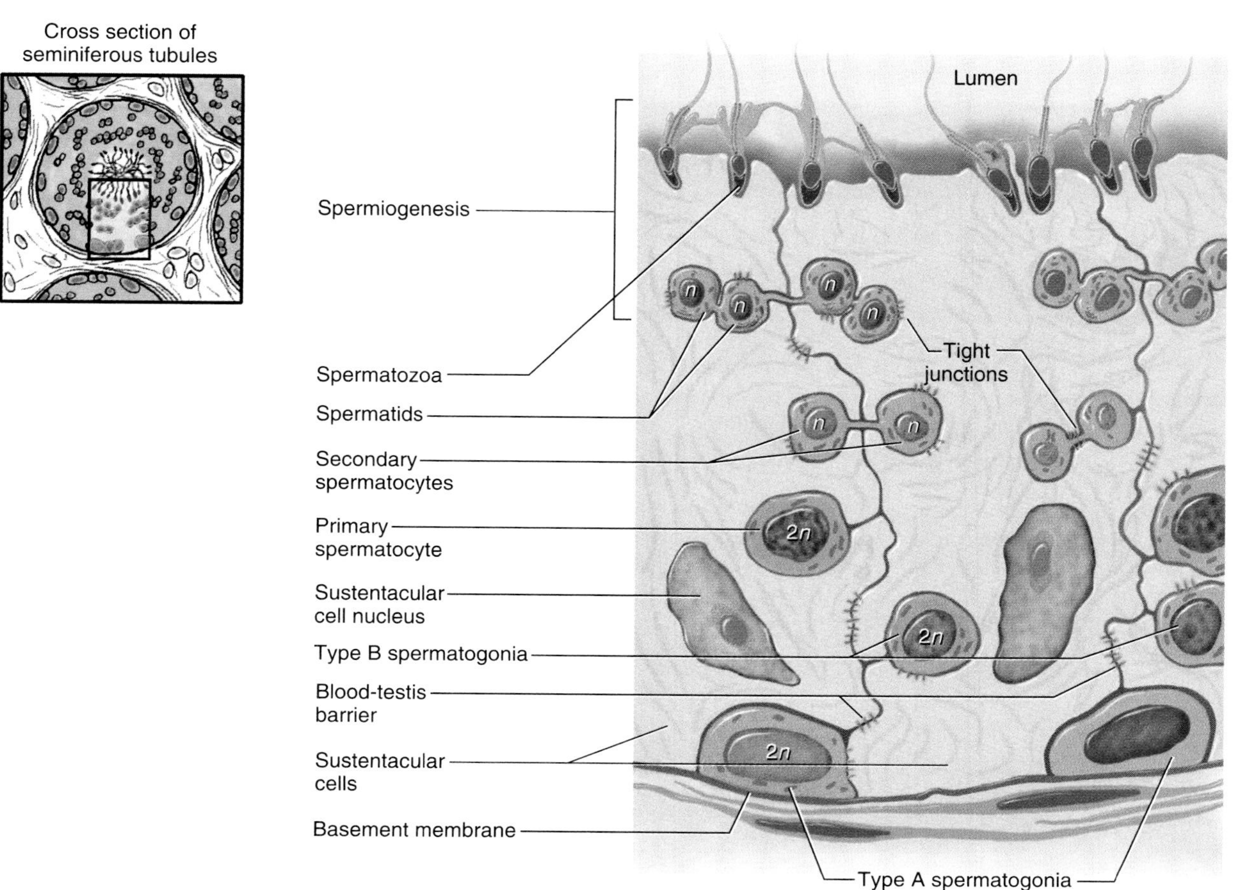

Figure 27.15 Spermatogenesis in Relation to the Sustentacular Cells. Before the germ cells can become haploid, they must be passed through the blood-testis barrier and be isolated from blood-borne elements of the immune system.

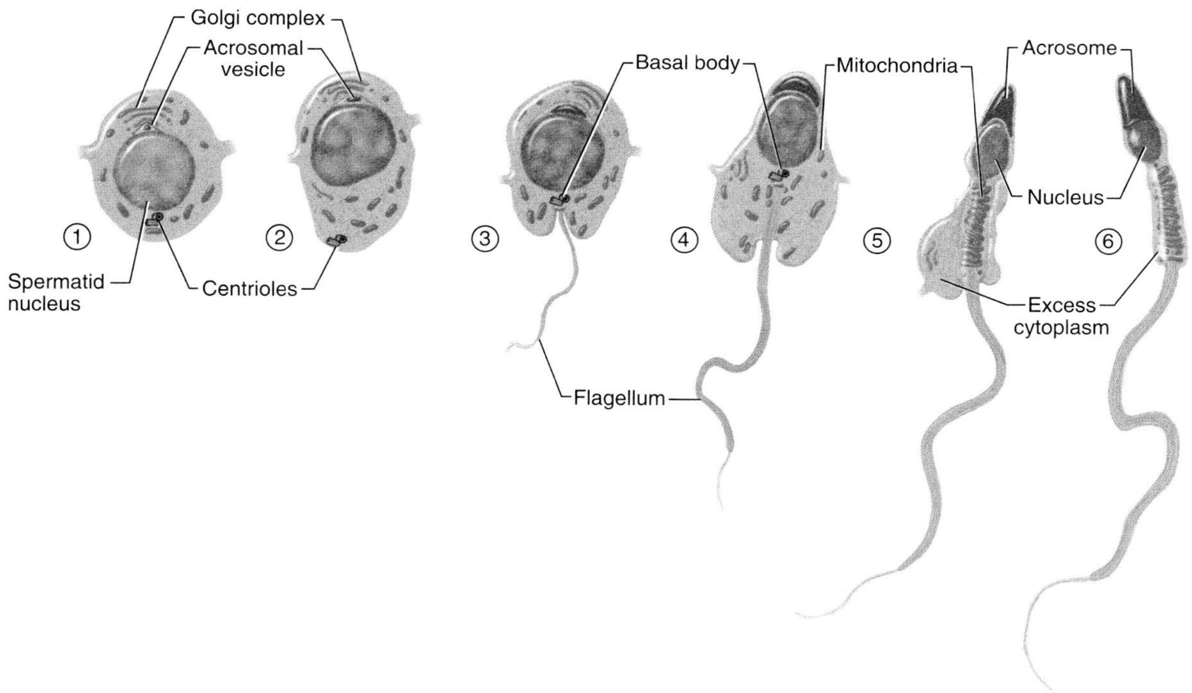

Figure 27.16 Spermiogenesis. In this process, the spermatids discard excess cytoplasm, grow tails, and become spermatozoa.

Clinical Insight 27.3

Reproductive Effects of Pollution

In recent decades, wildlife biologists have noticed increasing numbers of male birds, fish, and alligators with a variety of abnormalities in reproductive development. These deformities have been attributed to chemical pollutants called *endocrine disruptors* or *estrogen mimics.* Evidence is mounting that humans, too, are showing declining fertility and increasing anatomical abnormalities due to pollutants in water, meat, vegetables, and even breast milk and the uterine environment.

Over the last 50 years, there has been an alarming increase in the incidence of cryptorchidism (undescended testes) and hypospadias (a condition in which the urethra opens on the ventral side of the penis instead of at the tip). The rate of testicular cancer has more than tripled in that time. Data on 15,000 men from several countries show a sharp drop in average sperm count—from 113 million/mL in 1940 to only 66 million/mL in 1990. Total sperm production has decreased even more, because the average volume of semen per ejaculate has dropped 19% over this period.

The pollutants implicated in this trend include a wide array of common herbicides, insecticides, industrial chemicals, and breakdown products of materials ranging from plastics to dishwashing detergents. Some authorities think these chemicals act by mimicking estrogens or blocking the action of testosterone by binding to its receptors. Other scientists, however, question the data and feel the issue may be overstated. While the debate continues, the U.S. Environmental Protection Agency is screening thousands of industrial chemicals for endocrine effects.

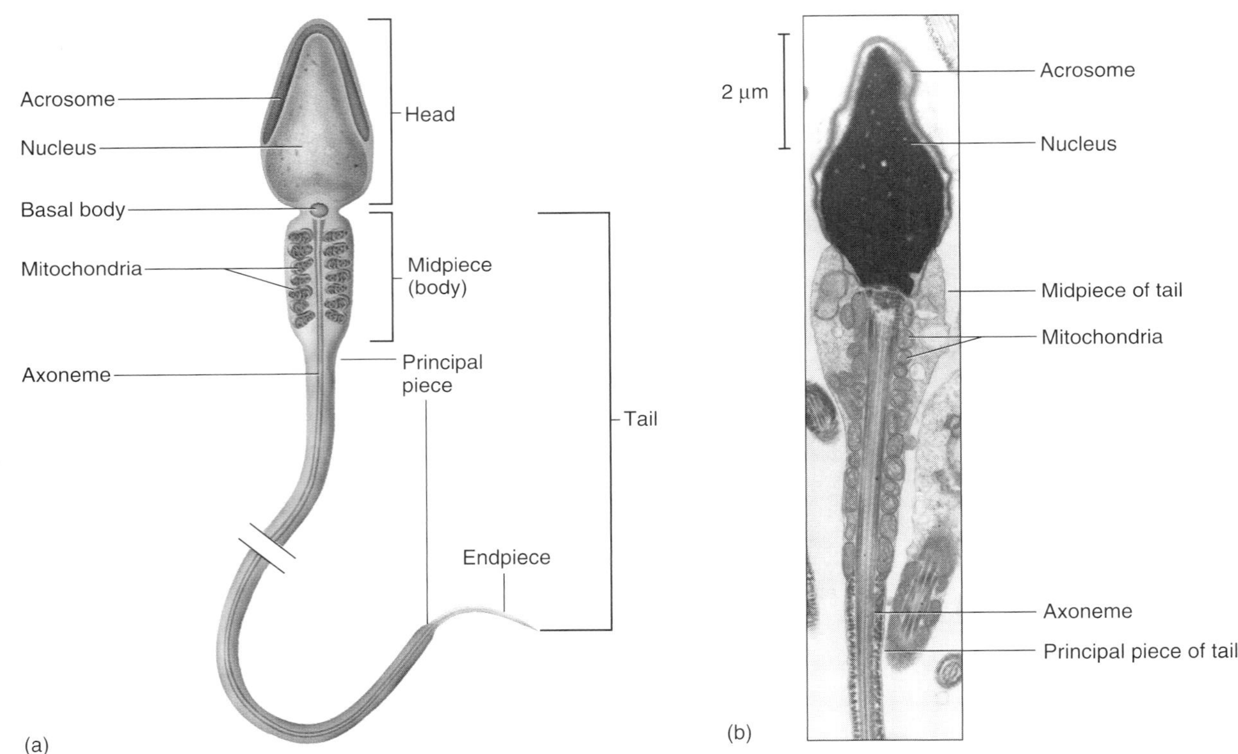

Figure 27.17 The Mature Spermatozoon. (*a*) Structure. (*b*) Head and part of the tail of a spermatozoon (TEM).

Semen

The fluid expelled during orgasm is called **semen,**[32] or **seminal fluid.** A typical ejaculation discharges 2 to 5 mL of semen, composed of about 60% seminal vesicle fluid, 30% prostatic fluid, 10% sperm and spermatic duct secretions, and a trace of bulbourethral fluid. The semen usually has a **sperm count** of 50 to 120 million sperm/mL. A sperm count any lower than 20 to 25 million sperm/mL is usually associated with **infertility (sterility),** the inability to fertilize an egg (see table 27.2). The major constituents of semen are listed in table 27.1.

32. *semen* = seed

Table 27.1 Composition of Semen

Spermatozoa	Serve to digest a path through cervical mucus and to fertilize egg.
Fructose	Provides energy for sperm motility. Secreted by seminal vesicles.
Fibrinogen	Causes semen to clot when acted upon by prostatic clotting enzymes. Secreted by seminal vesicles.
Clotting Enzymes	Convert fibrinogen to fibrin after ejaculation, causing semen to clot and adhere to vagina and cervix.
Fibrinolysin	Dissolves fibrin and liquefies semen about 15 to 30 minutes after ejaculation, thus liberating sperm from the clot.
Prostaglandins	Stimulate peristaltic contractions of female reproductive tract; may help draw sperm into uterus or spread it through the uterus. Reduce viscosity of cervical mucus, making it easier for sperm to travel up cervical canal into uterus. Secreted by prostate and seminal vesicles.
Spermine	A base that gives the semen a pH of 7.2 to 7.6. Helps to neutralize vaginal acidity (pH 4–5) and thus activate sperm (sperm motility is poor below pH 6) and protect them from vaginal acid.

Before You Go On

Answer the following questions to test your understanding of the preceding section:

16. State how many chromosomes a cell has, and how many chromatids each chromosome has, at the conclusion of meiosis I and meiosis II.
17. Name the stages of spermatogenesis from spermatogonium to spermatozoon.
18. Describe the three major parts of a spermatozoon and state what organelles or cytoskeletal components are contained in each.
19. List the major contributions of the seminal vesicles and prostate gland to the semen, and state the functions of these components.

Male Sexual Response

Objectives

When you have completed this section, you should be able to

- describe the blood and nerve supply to the penis; and
- explain how these govern erection and ejaculation.

The physiology of sexual intercourse was unexplored territory before the 1950s because of repressive attitudes toward the subject. British psychologist Havelock Ellis (1859–1939) suffered severe professional sanctions merely for surveying people on their sexual behavior. In the 1950s, William Masters and Virginia Johnson daringly launched the first physiological studies of sexual response in the laboratory. In 1966, they published *Human Sexual Response,* which detailed measurements and observations on more than 10,000 sexual acts by nearly 700 volunteer men and women. Masters and Johnson then turned their attention to disorders of sexual function and pioneered modern therapy for sexual dysfunctions.

Sexual intercourse is also known as **coitus, coition,**[33] or **copulation.**[34] Masters and Johnson divided intercourse into four recognizable phases, which they called excitement, plateau, orgasm, and resolution. The following discussion is organized around this model, although other authorities have modified it or proposed alternatives.

Anatomical Foundations

To understand male sexual function, we must give closer attention to the blood circulation and nerve supply to the penis.

Each internal iliac artery gives rise to an **internal pudendal (penile) artery,** which enters the root of the penis and divides in two. One branch, the **dorsal artery,** travels dorsally along the penis not far beneath the skin (see fig. 27.11). The other branch, the **deep artery,** travels through the core of the corpus cavernosum and gives off smaller **helicine**[35] **arteries,** which penetrate the trabeculae and empty into the lacunae. When the deep artery dilates, the lacunae fill with blood and the penis becomes erect. When the penis is flaccid, most of its blood supply comes from the dorsal arteries.

The penis is richly innervated by sensory and motor nerve fibers. The glans has an abundance of tactile, pressure, and temperature receptors, especially on its proximal margin and frenulum. Sensory fibers of the shaft, scrotum, perineum, and elsewhere are also highly important to erotic stimulation. They lead by way of a prominent **dorsal nerve** of the penis to the **internal pudendal nerve,** then to the sacral plexus, and finally to the sacral region of the spinal cord (fig. 27.18).

Both autonomic and somatic motor fibers carry impulses from integrating centers in the spinal cord to the penis and other pelvic organs. Sympathetic fibers arise

33. *coit* = to come together
34. *copul* = link, bond
35. *helic* = coil, helix

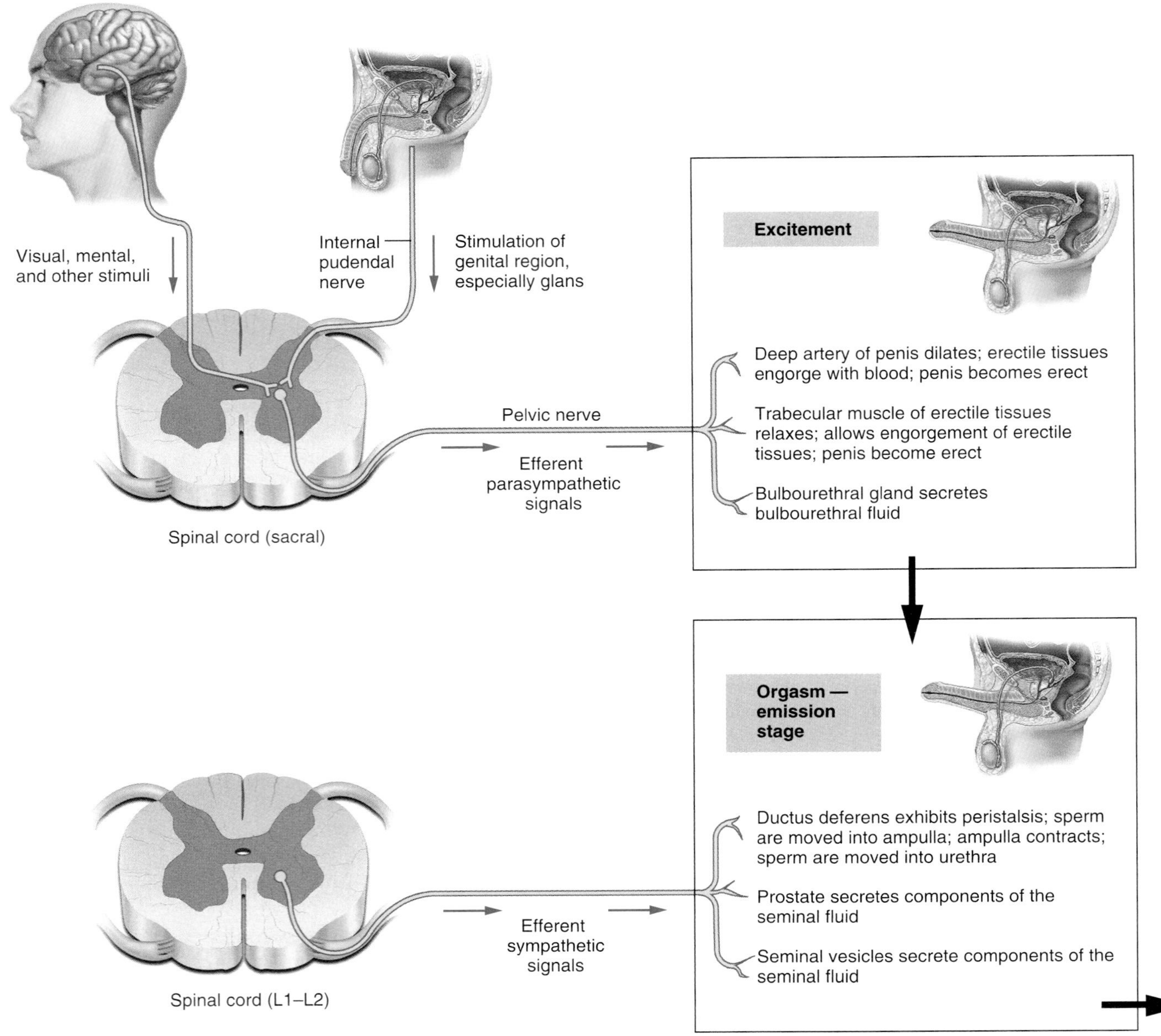

Figure 27.18 Neural Control of Male Sexual Response *(continued).*

from levels T12 to L2, pass through the hypogastric and pelvic nerve plexuses, and innervate the penile arteries, trabecular muscle, spermatic ducts, and accessory glands. They dilate the penile arteries and can induce erection even when the sacral region of the spinal cord is damaged. They also initiate erection in response to input to the special senses and to sexual thoughts.

Parasympathetic fibers extend from segments S2 to S4 of the spinal cord through the pudendal nerves to the arteries of the penis. They are involved in an autonomic reflex arc that causes erection in response to direct stimulation of the penis and other perineal organs.

Excitement and Plateau

The **excitement phase** is characterized by **vasocongestion** (swelling of the genitals with blood), **myotonia** (muscle tension), and increases in heart rate, blood pressure, and pulmonary ventilation. The bulbourethral glands secrete their fluid during this phase. The excitement phase can be initiated by a broad spectrum of erotic stimuli—sights, sounds, aromas, touch—and even by dreams or thoughts. Conversely, emotions can inhibit sexual response and make it difficult to function when a person is anxious, stressed, or preoccupied with other thoughts.

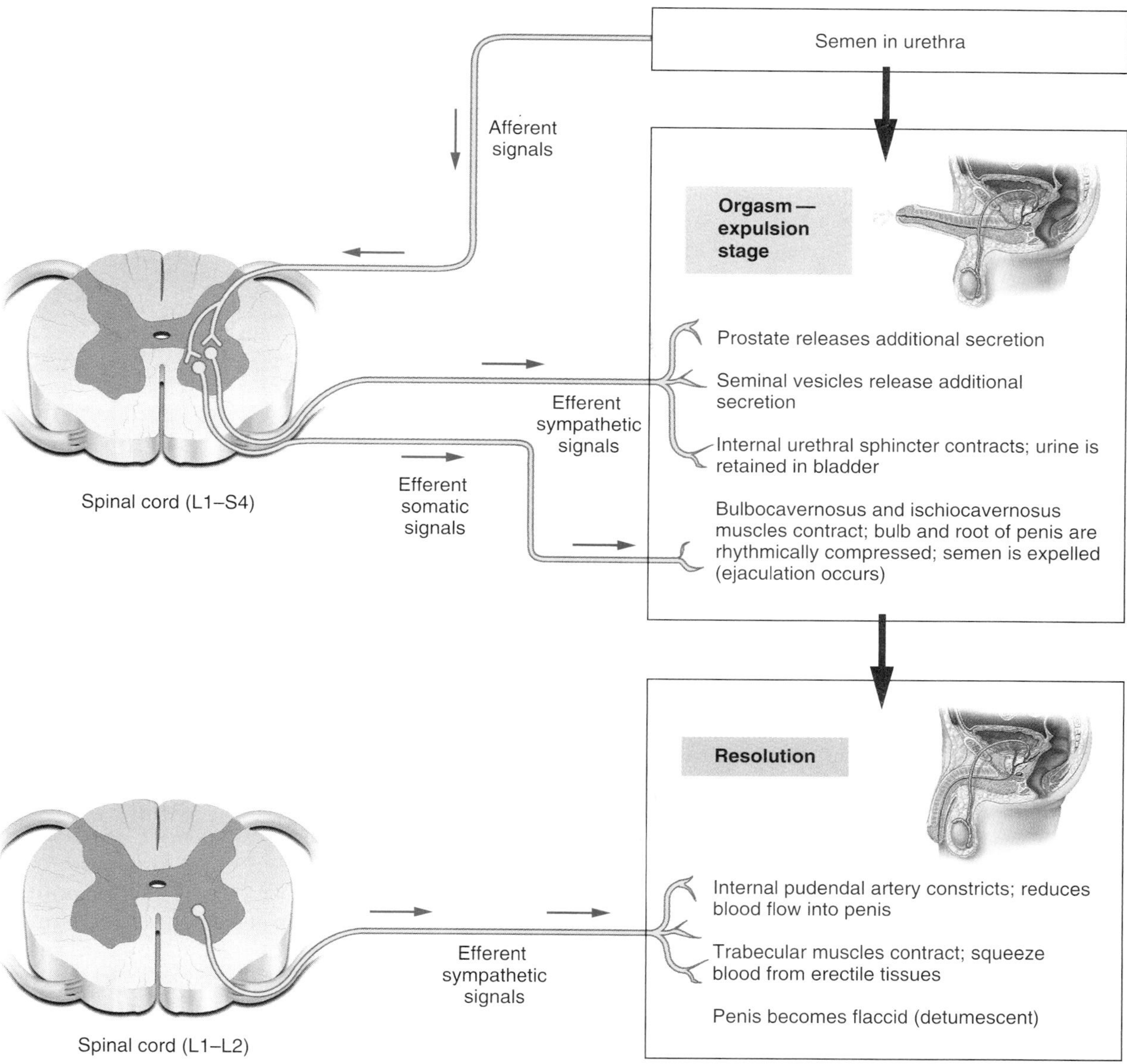

Figure 27.18 (Continued) Neural Control of Male Sexual Response.

The most obvious manifestation of male sexual excitement is **erection** of the penis, which makes entry of the vagina possible. Erection is an autonomic reflex mediated predominantly by parasympathetic nerve fibers that travel alongside the deep and helicine arteries of the penis. These fibers trigger the secretion of nitric oxide (NO), which leads to the relaxation of the deep arteries and lacunae (see clinical insight 27.4). Whether this is enough to cause erection, or whether it is also necessary to block the outflow of blood from the penis, is still being debated. According to one hypothesis, as lacunae near the deep arteries fill with blood, they expand and compress lacunae closer to the periphery of the erectile tissue. This is where blood leaves the erectile tissues, so the compression of the peripheral lacunae helps retain blood in the penis. Their compression is aided by the fact that each corpus cavernosum is wrapped in a tunica albuginea, which fits over the erectile tissue like a tight fibrous sleeve and contributes to its tension and firmness.

As the corpora cavernosa expand, the penis becomes enlarged, rigid, and elevated to an angle conducive to entry of the vagina. Once **intromission** (entry) is achieved, the tactile and pressure sensations produced by vaginal massaging of the penis further accentuate the erection reflex.

The corpus spongiosum has neither a central artery nor a tunica albuginea. It swells and becomes more visible as a cordlike ridge along the ventral surface of the penis, but it does not become nearly as engorged and hardened

Clinical Insight 27.4

Viagra—A Treatment for Erectile Dysfunction

Viagra, a medication for erectile dysfunction, has received more attention than most newly developed drugs. Its development was one of the offshoots of the discovery of the cell-signaling function of nitric oxide (NO). When sexual stimulation triggers NO secretion, NO activates an enzyme called guanylate cyclase. This enzyme produces cyclic guanosine monophosphate (cGMP). cGMP then relaxes the smooth muscle of the deep arteries and lacunae of the corpora cavernosa, increasing blood flow into these erectile tissues. Eventually, cGMP is broken down by an enzyme called phosphodiesterase type 5 (PDE5), and the erection subsides. Viagra helps to prolong erection by inhibiting PDE5, thus slowing the rate of cGMP breakdown.

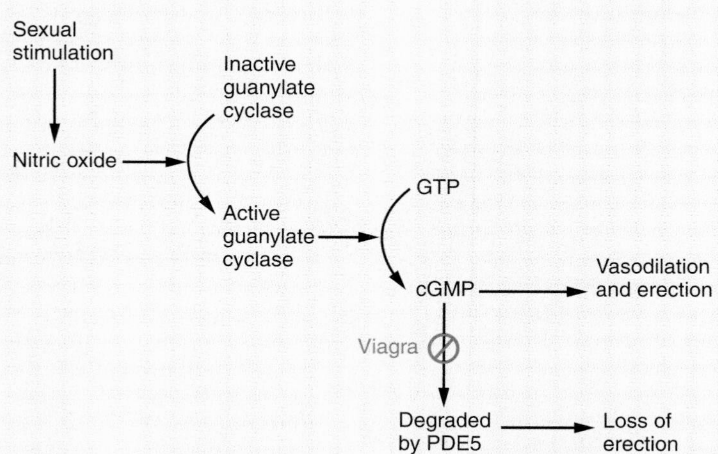

The Mechanism by Which Viagra Prolongs Erection.

as the corpora cavernosa. Vasocongestion is not limited to the penis; the testes also become as much as 50% larger during excitement.

In the **plateau phase,** variables such as respiratory rate, heart rate, and blood pressure are sustained at a high level, or rise slightly, for a few seconds to a few minutes before orgasm. This phase may be marked by increased vasocongestion and myotonia.

Think About It

Why is it important that the corpus spongiosum not become as engorged and rigid as the corpora cavernosa?

Orgasm and Ejaculation

The **orgasm,**[36] or **climax,** is a short but intense reaction that lasts 3 to 15 seconds and usually is marked by the discharge of semen. The heart rate increases to as high as 180 beats per minute, blood pressure rises proportionately, and the respiratory rate becomes as high as 40 breaths per minute. From the standpoint of producing offspring, the most significant aspect of male orgasm is the **ejaculation**[37] of semen into the vagina.

Ejaculation occurs in two stages called emission and expulsion. In **emission,** the sympathetic nervous system stimulates peristalsis in the smooth muscle of the ductus deferens, which propels sperm from the tail of the epididymis, along the ductus, and into the ampulla. Contractions of the ampulla propel the sperm into the prostatic urethra, and contractions of smooth muscle in the prostate gland force prostatic fluid into the urethra. Sperm first begin moving when they mix with these glandular secretions in the urethra. During storage, they remain immobile. Secretions of the seminal vesicles join the semen soon after the prostatic secretion. The contractions and seminal flow of this phase create an urgent sensation that ejaculation is inevitable.

Semen in the urethra activates somatic and sympathetic reflexes that produce **expulsion** of the semen. Sensory impulses travel to the spinal cord via the internal pudendal nerve. Efferent signals come from a reflex center in the upper lumbar region of the spinal cord by way of sympathetic fibers to the prostate and seminal vesicles, causing further glandular secretion. The sympathetic reflex also constricts the internal urinary sphincter so urine cannot enter the urethra.

Somatic motor impulses leave the third and fourth sacral segments of the cord and travel to the bulbospongiosus, ischiocavernosus, and levator ani muscles. The bulbospongiosus, which envelops the root of the corpus spongiosum, undergoes five or six strong, spasmodic contractions that compress the urethra and forcibly expel the semen. Most sperm are ejected in the first one or two jets of semen, mixed primarily with prostatic fluid. The seminal vesicle secretion follows and flushes most remaining sperm from the ejaculatory ducts and urethra. Some sperm may

36. *orgasm* = swelling
37. *e* = *ex* = out + *jacul* = to throw

seep from the penis prior to ejaculation, and pregnancy can therefore result from genital contact even without orgasm.

Orgasm is accompanied by an intense feeling of release from pent-up tension. Ejaculation and orgasm are not the same. Although they usually occur together, it is possible to have all of the sensations of orgasm without ejaculating, and ejaculation occasionally occurs with little or no sensation of orgasm.

Resolution

Immediately following orgasm comes the **resolution** phase. Discharge of the sympathetic nervous system constricts the internal pudendal artery and reduces the flow of blood into the penis. It also causes contraction of the trabecular muscles, which squeeze blood from the lacunae of the erectile tissues. The penis may remain semierect long enough to continue intercourse, which may be important to the female's attainment of climax, but gradually the penis undergoes **detumescence**—it becomes soft and flaccid again. The resolution phase is also a time in which cardiovascular and respiratory functions return to normal. Many people break out in a "cold sweat" during the resolution phase. In men, resolution is followed by a **refractory period,** lasting anywhere from 10 minutes to a few hours, in which it is usually impossible to attain another erection and orgasm.

Men and women have many similarities and a few significant differences in sexual response. The response cycle of women is described in the next chapter.

Two of the most common concerns related to sex are sexually transmitted diseases (STDs) and contraception. Understanding most contraceptive methods requires a prior understanding of female anatomy and physiology, so contraceptive techniques for both sexes are discussed at the end of the next chapter, while STDs are discussed in this chapter (clinical insight 27.5). This is not to imply, of course, that STDs are only a male concern and contraception only a female concern. Reproductive disorders specific to males and females are briefly summarized in tables 27.2 and 28.5, respectively (but see clinical insight 27.2 for prostatic hyperplasia and prostate cancer).

Before You Go On

Answer the following questions to test your understanding of the preceding section:

20. Explain how penile blood circulation changes during sexual arousal and why the penis becomes enlarged and stiffened.
21. State the roles of the sympathetic and parasympathetic nervous systems in male sexual response.

Table 27.2 Some Male Reproductive Disorders

Disorder	Description
Breast Cancer	Accounts for 0.2% of male cancers in the United States; usually seen after age 60 but sometimes in children and adolescents. About 175 females get breast cancer for every male who does so. Usually felt as a lump near the nipple, often with crusting and discharge from nipple. Often quite advanced by the time of diagnosis, with poor prospects for recovery, because of denial and delay in seeking treatment.
Cryptorchidism[38] *(crip-TOR-ki-dizm)*	Failure of one or both testes to descend completely into the scrotum. Leads to infertility if not corrected, because undescended testes are in too warm an environment for spermatogenesis. In most cases, the testes descend spontaneously in the first year of infancy; otherwise, the condition can be corrected with hormone injections or surgery.
Erectile Dysfunction (impotence)	Inability to maintain an erection adequate for vaginal entry in half or more of one's attempts. Can stem from aging and declining testosterone level as well as cardiovascular and neurological diseases, diabetes mellitus, medications, fear of failure, depression, and other causes.
Hypospadias[39] *(HY-po-SPAY-dee-us)*	A congenital defect in which the urethra opens on the ventral side or base of the penis rather than at the tip; usually corrected surgically at about 1 year of age.
Infertility	Inability to fertilize an egg because of a low sperm count (lower than 20 to 25 million/mL), poor sperm motility, or a high percentage of deformed sperm (two heads, defective tails, etc.). May result from malnutrition, gonorrhea and other infections, toxins, or testosterone deficiency.
Penile Cancer	Accounts for 1% of male cancers in the United States; most common in black males aged 50 to 70 and of low income. Most often seen in men with nonretractable foreskins (phimosis) combined with poor penile hygiene; least common in men circumcised at birth.
Testicular Cancer	The most common solid tumor in men 15 to 34 years old, especially white males of middle to upper economic classes. Typically begins as a painless lump or enlargement of the testis. Highly curable if detected early. Men should routinely palpate the testes for normal size and smooth texture.
Varicocele (VAIR-ih-co-seal)	Abnormal dilation of veins of the spermatic cord, so that they resemble a "bag of worms." Occurs in 10% of males in the United States. Caused by absence or incompetence of venous valves. Reduces testicular blood flow and often causes infertility.

38. *crypt* = hidden + *orchid* = testes
39. *hypo* = below + *spad* = to draw off (the urine)

Clinical Insight 27.5

Sexually Transmitted Diseases

Sexually transmitted diseases (STDs) have been well known since the writings of Hippocrates and Galen. They have been called by a number of euphemisms aimed at downplaying their mode of transmission—for example, "social diseases" and "venereal diseases" (after Venus, the goddess of love).

Here we discuss three bacterial STDs—chlamydia, gonorrhea, and syphilis—and three viral STDs—genital herpes, genital warts, and hepatitis. AIDS, another important viral STD, is discussed in chapter 21.

All of these STDs have certain points in common. They have an *incubation period* in which the pathogen multiplies in the body and begins to produce disease but symptoms have not appeared yet, and they have a *communicable period* in which an infected person can transmit the disease to others. The communicable period can begin before symptoms are noticed, it can persist after symptoms have disappeared, or it can exist in *symptomless carriers*—people who show no evidence of the disease. STDs usually do not stimulate an immune response, because the pathogens live inside the host cells where the immune system cannot sense their presence. Thus, there are no vaccines against them. STDs often cause fetal deformity, stillbirth, and neonatal death. The focus here is on adults, whereas STDs of the newborn are discussed in chapter 29.

Chlamydia (cla-MID-ee-uh) is currently the most common bacterial STD in the United States, with 3 to 5 million cases per year. It is caused by *Chlamydia trachomatis.* After an incubation period of 1 to 3 weeks, there appears a scanty, watery discharge from the urethra and pain in the testes or in the rectal or abdominal region. *Nongonococcal urethritis (NGU)* is an STD caused by agents other than the gonorrhea bacterium (see next)—for example, *Chlamydia, Mycoplasma hominis,* and *Ureaplasma urealyticum.* NGU is characterized by inflammation of the urethra, causing pain or discomfort on urination.

Gonorrhea (GON-oh-REE-uh), nicknamed the "clap" or "drip," is caused by the bacterium *Neisseria gonorrhoeae.* It acquired the name *gonorrhea* ("flow of seed") because Galen thought the pus discharged from the penis was semen. In addition to penile or vaginal discharge, gonorrhea causes abdominal discomfort, genital pain, painful urination (dysuria), and abnormal uterine bleeding. It may cause the oviducts to become scarred and obstructed, thus resulting in infertility. About 20% of infected women, however, are asymptomatic. Gonorrhea is treated with antibiotics. Gonorrhea and chlamydia frequently occur together and require treatment with one antibiotic for chlamydia and a different one for gonorrhea.

Pelvic inflammatory disease (PID) is a bacterial infection of the female pelvic organs, usually with *Chlamydia* or *Neisseria.* It often results in sterility and may require surgical removal of infected oviducts or other organs. The incidence of PID in the United States has increased from 17,800 cases in 1970 to about a million cases per year currently. PID has rendered tens of thousands of American women sterile.

Syphilis (SIFF-ih-liss), one of the most potentially devastating STDs, is named for a shepherd boy in a sixteenth-century poem by the physician Fracastoro. It is caused by a corkscrew-shaped bacterium named *Treponema pallidum.* After an incubation period of 2 to 6 weeks, an ulcer called a *chancre* (SHAN-kur) appears at the site of infection—usually on the penis of a male but sometimes out of sight in the vagina of a female. A chancre is a small, hard lesion with no discharge. It disappears in 4 to 6 weeks, ending the first stage of syphilis and often creating an illusion of recovery. In the second stage, however, the disease reappears, with a pink rash over the body, other skin eruptions, fever, joint pain, and hair loss. These symptoms disappear in 3 to 12 weeks. Symptoms then come and go for up to 5 years, but the infection is detectable by a blood test and the infected person is contagious even when symptoms are not occurring. The disease may progress to a third stage, *tertiary syphilis,* or *neurosyphilis,* in which there is damage to the blood vessels and heart valves, thickening of the meninges, and lesions of the brain that can cause paralysis and dementia. Syphilis is treated with antibiotics.

Genital herpes is the most common STD in the United States, with 20 to 40 million infected people and 500,000 new cases per year. It is usually caused by the herpes simplex virus type 2 (HSV-2). A close relative, HSV-1, causes cold sores (fever blisters) of the mouth and occasionally causes genital herpes, probably transmitted through oral-genital sex. After an incubation period of 4 to 10 days, the virus causes red blisters on the penis of the male; on the labia, vagina, or cervix of the female; and sometimes on the thighs and buttocks of either sex. Over a period of 2 to 10 days, these blisters rupture, seep fluid, and begin to form scabs. The initial infection may be painless or may cause intense pain, urethritis, and watery discharge from the penis or vagina. The lesions heal in 2 to 3 weeks and leave no scars.

During this time, however, the viruses travel by way of sensory nerve fibers to the dorsal root ganglia, where they become dormant. Later, they can migrate along the nerves and cause small epithelial lesions at various places on the body. The movement from place to place is the basis of the name *herpes.*[40] These secondary lesions are generally smaller and heal more quickly than the primary lesions. Most patients have five to seven recurrences, ranging from several times a year to several years apart. An infected person is contagious to a sexual partner when the lesions are present and sometimes even when they are not. Although mostly a painful nuisance, HSV has been implicated as a risk factor in cervical cancer. The drug Acyclovir reduces the shedding of viruses and the spread of lesions but does not necessarily prevent recurrences.

Genital warts (condylomas) are one of the most rapidly increasing STDs today. There are about a million new cases per year, especially among young adults with multiple sex partners. Genital warts are caused by 60 or more viruses collectively called *human papillomaviruses (HPVs).* In the male, lesions usually appear on the penis, perineum, or anus, and in the female they are usually on the cervix, vaginal wall, perineum, or anus. Lesions are sometimes small and virtually invisible. A few of the 60 varieties of HPV have been implicated in cervical cancer and cancer of the penis, vagina, and anus. HPV is

40. *herp* = to creep

found in about 90% of all cervical cancers tested. About 90% of cases of genital warts, however, involve forms of HPV that have not been linked to cancer. There is still considerable difference of opinion on how to treat genital warts; they are sometimes treated with cryosurgery (freezing and excision), laser surgery, or interferon.

Hepatitis B and *C* are inflammatory liver diseases caused by the hepatitis B and C viruses (HBV, HCV). Hepatitis B and C are usually transmitted through blood and other body fluids but are becoming increasingly common as sexually transmitted diseases. They have a pattern of transmission quite similar to that of AIDS. In most cases, HBV produces no symptoms and confers permanent immunity. It can, however, cause liver cancer and cirrhosis. Infection can be prevented by vaccination. About 80% to 90% of cases of HCV, by contrast, become chronic infections and lead to liver disease and often liver failure. While an effective vaccine is available against HBV, there is no immediate prospect of a vaccine against HCV because, like HIV, HCV has a rapid mutation rate and quickly escapes the body's immune defenses. Far more people already have HCV than AIDS, and HCV threatens to become a major epidemic of the twenty-first century. HCV is already the leading reason for liver transplants in the United States.

Chapter Review

Online Learning Center

Study resources for this chapter are available at www.mhhe.com/saladin/

Study Outline

Sexual Reproduction (p. 1017)

1. Meaning of sexual reproduction
2. Advantage of genetic diversity
3. The two sexes
   a. Functions of the gametes
   b. Definitions of male and female
4. Overview of the reproductive system
   a. Primary sex organs
   b. Secondary sex organs
   c. Secondary sex characteristics

Sex Determination and Development (p. 1018)

1. Sex chromosomes
   a. Autosomes and sex chromosomes
   b. X and Y chromosomes
   c. Chromosomes and sex determination
2. Hormones and sex differentiation
   a. Embryonic anatomy
      - Gonadal ridges
      - Mesonephric ducts
      - Paramesonephric ducts
   b. SRY gene and testis-determining factor
   c. Influence of HCG
3. Development of external genitals
   a. Phallus
   b. Urogenital folds
   c. Labioscrotal folds
4. Descent of testes

Male Reproductive Anatomy (p. 1022)

1. Testes
   a. Tunica vaginalis
   b. Tunica albuginea
   c. Seminiferous tubules
      - Germinal epithelium
      - Germ cells
      - Sustentacular cells
      - Blood-testis barrier
   d. Interstitial cells
   e. Rete testis
   f. Blood vessels and nerves
2. Scrotum
   a. Median septum
   b. Spermatic cord
   c. Testicular thermoregulation
      - Cremaster muscle
      - Dartos muscle
      - Pampiniform plexus
3. Spermatic ducts
   a. Efferent ductules
   b. Epididymis
   c. Ductus deferens
   d. Ejaculatory duct
4. Urethra
5. Accessory glands
   a. Seminal vesicles
   b. Prostate gland
   c. Bulbourethral glands
6. Penis
   a. Bulb, root, shaft, and glans
   b. Erectile tissues
      - Corpus spongiosum
      - Corpora cavernosa
      - Lacunae and trabeculae

Puberty and Climacteric (p. 1030)

1. Puberty
   a. GnRH secretion
   b. FSH and androgen-binding protein
   c. LH and androgen secretion
   d. Effects of androgens
      - Spermatogenesis
      - Regulation of FSH and LH secretion
      - Growth of reproductive organs
      - Secondary sex characteristics
      - Secretion of growth hormone
      - Erythropoiesis
      - Libido
   e. Inhibin
2. Aging and sexual function
   a. Decline in testosterone secretion
   b. Rise in FSH and LH secretion
   c. Male climacteric
   d. Impotence

Spermatogenesis, Spermatozoa, and Semen (p. 1031)

1. Meiosis
   a. Major effects
      - Production of four haploid cells
      - Crossing-over
   b. Necessity of meiosis
   c. Stages
      - Outcome of meiosis I
      - Outcome of meiosis II
2. Spermatogenesis
   a. Primordial germ cells
   b. Stages of spermatogenesis
      - Spermatogonia
      - Primary spermatocytes
      - Secondary spermatocytes
      - Spermatids
   c. Spermiogenesis
   d. Importance of blood-testis barrier
3. The spermatozoon
   a. Head
      - Nucleus
      - Acrosome
      - Basal body
   b. Tail
      - Midpiece
      - Principal piece
      - Endpiece
4. Semen
   a. Spermatozoa and sperm count
   b. Fructose
   c. Clotting and anticoagulant factors
   d. Prostaglandins
   e. Spermine

Male Sexual Response (p. 1037)

1. Anatomical foundations
   a. Arteries of the penis
   b. Nerves of the penis
2. Excitement and plateau
   a. Vasocongestion
   b. Myotonia
   c. Bulbourethral secretion
   d. Erection
3. Orgasm and ejaculation
   a. Cardiovascular and respiratory responses
   b. Ejaculation
      - Emission
      - Expulsion
4. Resolution
   a. Detumescence
   b. Cardiovascular and respiratory responses
   c. Refractory period

Selected Vocabulary

sexual reproduction 1017
gamete 1017
zygote 1017
sperm 1017
egg 1017
primary sex organ 1017
gonad 1017
testes 1017
ovaries 1017
secondary sex organ 1017
secondary sex characteristic 1017
mesonephric duct 1018
paramesonephric duct 1018
SRY gene 1018
testis-determining factor 1018
müllerian-inhibiting factor 1018
phallus 1020
urogenital fold 1020
labioscrotal fold 1020
homologous organs 1022
gubernaculum 1022
inguinal canal 1022
descent of the testes 1022
tunica vaginalis 1022
tunica albuginea of testis 1023
seminiferous tubule 1023
interstitial cell 1023
germinal epithelium 1023
germ cell 1023
sustentacular cell 1023
blood-testis barrier 1023
rete testis 1023
testicular artery 1023
testicular vein 1023
testicular nerve 1023
scrotum 1026
median septum 1026
perineal raphe 1026
spermatic cord 1026
cremaster muscle 1026
dartos muscle 1027
pampiniform plexus 1027
efferent ductules 1027
epididymis 1027
ductus deferens 1027
ampulla 1028
ejaculatory duct 1028
seminal vesicles 1028
prostate gland 1028
bulbourethral glands 1028
penis 1028
root of penis 1028
shaft of penis 1028
glans of penis 1029
prepuce 1029
smegma 1029
erectile tissue 1029
corpus spongiosum 1029
corpus cavernosum 1029
tunica albuginea of penis 1029
median septum of penis 1029
lacuna 1029
trabecula 1029
trabecular muscle 1029
bulb of penis 1029
crus of penis 1029
puberty 1030
adolescence 1030
brain-testicular axis 1030
gonadotropin-releasing hormone 1030
follicle-stimulating hormone 1030
luteinizing hormone 1030
androgen-binding protein 1030
libido 1031
inhibin 1031
male climacteric 1031
spermatogenesis 1031
meiosis 1031
diploid 1032
haploid 1032
tetrad 1032
crossing-over 1032
primordial germ cell 1032
spermatogonium 1032
primary spermatocyte 1034
secondary spermatocyte 1034
spermatid 1034
spermiogenesis 1034
head of sperm 1034
acrosome 1034
tail of sperm 1034
midpiece 1034
principal piece 1034
endpiece 1034
semen 1036
sperm count 1036
infertility 1036
coitus 1037
internal pudendal artery 1037
dorsal artery 1037
deep artery 1037
helicine artery 1037
dorsal nerve 1037
internal pudendal nerve 1037
excitement phase 1038
vasocongestion 1038
myotonia 1038
erection 1039
intromission 1039
plateau phase 1040
orgasm 1040
ejaculation 1040
emission 1040
expulsion 1040
resolution 1041
detumescence 1041
refractory period 1041

Testing Your Recall

1. The ductus deferens develops from the ______ of the embryo.
   a. mesonephric duct
   b. paramesonephric duct
   c. phallus
   d. labioscrotal folds
   e. urogenital folds

2. Descent of the testes is achieved by contraction of a cord called
   a. the gubernaculum.
   b. the spermatic cord.
   c. the ductus deferens.
   d. the pampiniform plexus.
   e. the rete testis.

3. The expulsion of semen occurs when the bulbospongiosus muscle is stimulated by
   a. somatic efferent neurons.
   b. somatic afferent neurons.
   c. sympathetic efferent neurons.
   d. parasympathetic efferent neurons.
   e. prostaglandins.

4. Prior to ejaculation, sperm are stored primarily in
   a. the seminiferous tubules.
   b. the rete testis.
   c. the epididymis.
   d. the seminal vesicles.
   e. the ejaculatory ducts.

5. The penis is attached to the pubic arch by crura of
   a. the corpora cavernosa.
   b. the corpus spongiosum.
   c. the perineal membrane.
   d. the bulbocavernosus.
   e. the ischiocavernosus.

6. The first hormone secreted at the onset of puberty is
   a. follicle-stimulating hormone.
   b. interstitial cell–stimulating hormone.
   c. human chorionic gonadotropin.
   d. gonadotropin-releasing hormone.
   e. testosterone.

7. When it is necessary to reduce the rate of sperm production without reducing the rate of testosterone secretion, the sustentacular cells secrete
   a. dihydrotestosterone.
   b. androgen-binding protein.
   c. LH.
   d. FSH.
   e. inhibin.

8. Four spermatozoa arise from each
   a. primordial germ cell.
   b. type A spermatogonium.
   c. type B spermatogonium.
   d. secondary spermatocyte.
   e. spermatid.

9. The point in meiosis at which sister chromatids separate from each other is
   a. prophase I.
   b. metaphase I.
   c. anaphase I.
   d. anaphase II.
   e. telophase II.

10. Blood is forced out of the penile lacunae by contraction of the ______ muscles.
    a. bulbocavernosus
    b. ischiocavernosus
    c. cremaster
    d. trabecular
    e. dartos

11. Under the influence of androgens, the embryonic ________ duct develops into the male reproductive tract.

12. Spermatozoa obtain energy for locomotion from ________ in the semen.

13. The ________, a network of veins in the spermatic cord, helps keep the testes cooler than the core body temperature.

14. All germ cells beginning with the ________ are genetically different from the rest of the body cells and therefore must be protected by the blood-testis barrier.

15. The corpora cavernosa as well as the testes have a fibrous capsule called the ________.

16. Over half of the semen consists of secretions from a pair of glands called the ________.

17. The blood-testis barrier is formed by tight junctions between the ________ cells.

18. The earliest haploid stage of spermatogenesis is the ________.

19. Erection of the penis occurs when nitric oxide causes the ________ arteries to dilate.

20. A sperm penetrates the egg by means of enzymes in its ________.

Answers in appendix B

Testing Your Comprehension

1. Explain why testosterone may be considered both an endocrine and a paracrine secretion of the testes. (Review paracrines in chapter 17 if necessary.)

2. A young man is in a motorcycle accident that severs his spinal cord at the neck and leaves him paralyzed from the neck down. When informed of the situation, his wife asks the physician if her husband will be able to have erections and father any children. What should the doctor tell her? Explain your answer.

3. Considering the temperature in the scrotum, would you expect hemoglobin to unload more oxygen to the testes, or less, than it unloads in the warmer internal organs? Why? (Hint: See figure 22.24.) How would you expect this fact to influence sperm development?

4. Why is it possible for spermatogonia to be outside the blood-testis barrier, yet necessary for primary spermatocytes and later stages to be within the barrier, isolated from the blood?

5. A 68-year-old man taking medication for hypertension complains to his physician that it has made him impotent. Explain why this could be an effect of antihypertension drugs.

Answers at the Online Learning Center

CHAPTER 28

The Female Reproductive System

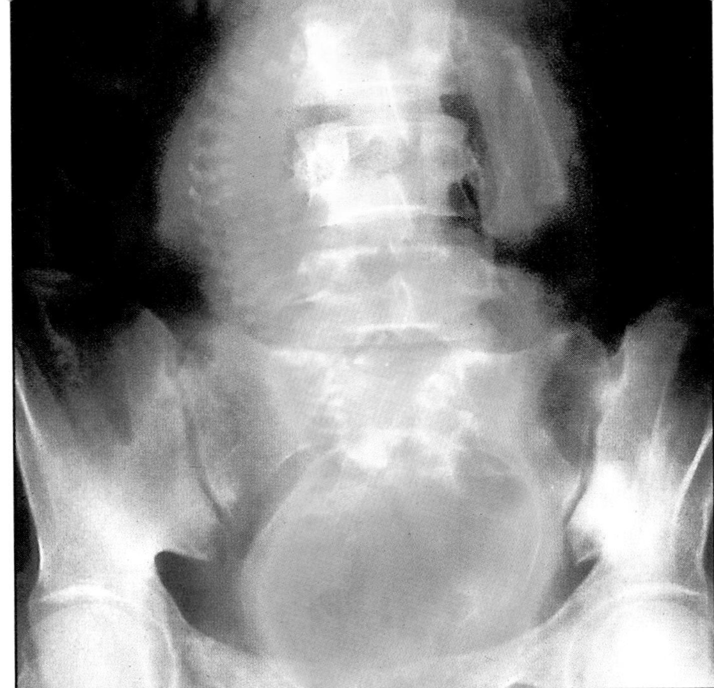

Full-term fetus, head-down in the female abdominal cavity (colorized X ray)

Chapter Outline

Insights

Study with Your ESP CD-ROM

Refer to your ESP Tutorial CD-ROM for helpful animations, activities, and quizzes related to the material covered in this chapter. The following are suggested areas to go to on the CD and are presented according to "Unit/Topic":
- Reproduction and Development/Reproduction

Brushing Up

To understand this chapter, it is important that you understand or brush up on the following concepts:
- Negative feedback inhibition of the pituitary (p. 646)
- Synergistic, permissive, and antagonistic hormone interactions (p. 670)
- Fetal development of the reproductive system (pp. 1018–22)
- Meiosis (pp. 1032–33)

The female reproductive system is more complex than the male's because it serves more purposes. Whereas the male needs only to produce and deliver gametes, the female must do this as well as provide nutrition and safe harbor for fetal development, then give birth and nourish the infant. Furthermore, female reproductive physiology is cyclic and female hormones are secreted in a more complex sequence compared to the relatively steady, simultaneous secretion of regulatory hormones in the male.

This chapter discusses the anatomy of the female reproductive system, the production of gametes and how it relates to the ovarian and menstrual cycles, the female sexual response, and the physiology of pregnancy, birth, and lactation. Embryonic and fetal development are treated in the next chapter.

Reproductive Anatomy

Objectives

When you have completed this section, you should be able to

- describe the structure of the ovary;
- trace the female reproductive tract and describe the gross anatomy and histology of each organ;
- identify the ligaments that support the female reproductive organs;
- describe the blood supply to the female reproductive tract;
- identify the external genitals of the female; and
- describe the structure of the nonlactating breast.

Sex Differentiation

The female reproductive system (fig. 28.1) is conspicuously different from that of the male, but as we saw earlier the two sexes are indistinguishable for the first 8 to 10 weeks of development (see fig. 27.3, p. 1021). The female reproductive tract develops from the paramesonephric duct not because of the positive action of any hormone, but because of the absence of testosterone and müllerian-inhibiting factor (MIF). Without testosterone, the mesonephric duct degenerates while the phallus becomes a clitoris, the urogenital folds develop into labia minora, and the labioscrotal folds develop into labia majora. Without MIF, the paramesonephric duct develops into the uterine tubes, uterus, and vagina. This developmental pattern can be disrupted, however, by abnormal hormonal exposure before birth, as happens in adrenogenital syndrome (see p. 654).

The Ovary

The female gonads are the **ovaries,**[1] which produce egg cells (ova) and sex hormones. The ovary is an almond-shaped organ nestled in the *ovarian fossa,* a depression of the posterior pelvic wall. The ovary measures about 3 cm long, 1.5 cm wide, and 1 cm thick. Its capsule, like that of the testis, is called the **tunica albuginea.** The interior of the ovary is indistinctly divided into an outer **cortex,** where the germ cells develop, and a central **medulla** occupied by the major arteries and veins (fig. 28.2).

The ovary lacks ducts comparable to the seminiferous tubules. Instead, each egg develops in its own fluid-filled, bubblelike **follicle** and is released by **ovulation,** the bursting of the follicle. Figure 28.2 shows several types of follicles that coincide with different stages of egg maturation, as discussed later.

The ovary is held in place by several connective tissue ligaments (fig. 28.3). Its medial pole is attached to the uterus by the *ovarian ligament,* and its lateral pole is attached to the pelvic wall by the *suspensory ligament.* A sheet of peritoneum called the *broad ligament* flanks the uterus on each side and encloses the uterine tube in its superior margin. The anterior margin of the ovary is anchored to the broad ligament by a peritoneal fold called the *mesovarium.*[2]

The ovary is supplied with an **ovarian artery, ovarian veins,** and **ovarian nerves,** which travel through the suspensory ligament. It receives an additional blood supply from the ovarian branches of the uterine arteries.

Secondary Sex Organs (Genitalia)

The internal genitalia are the uterine tubes, uterus, and vagina, which constitute a duct system from the vicinity of the ovary to the outside of the body. The external genitalia include principally the clitoris, labia minora, and labia majora. These occupy the perineum, which is defined by the same skeletal landmarks as in the male (see fig. 11.20*b*, p. 369). Beneath the skin of the perineum are several accessory glands that provide most of the lubrication for intercourse.

The Uterine Tubes

The **uterine tube,** also called the **oviduct** or **fallopian tube,**[3] is a canal about 10 cm long from the ovary to the uterus. At the distal (ovarian) end, it flares into a trumpet-shaped **infundibulum**[4] with feathery projections called **fimbriae**[5] (FIM-bree-ee); the middle part of the tube is the **ampulla;** and near the uterus it forms a narrower **isthmus.** The uterine tube is enclosed in the *mesosalpinx*[6] (MEZ-oh-SAL-pinks), which is the superior margin of the broad ligament.

1. *ov* = egg + *ary* = place for

2. *mes* = middle + *ovari* = ovary
3. Gabriele Fallopio (1528–62), Italian anatomist and physician
4. *infundibulum* = funnel
5. *fimbria* = fringe
6. *meso* = mesentery + *salpin* = trumpet

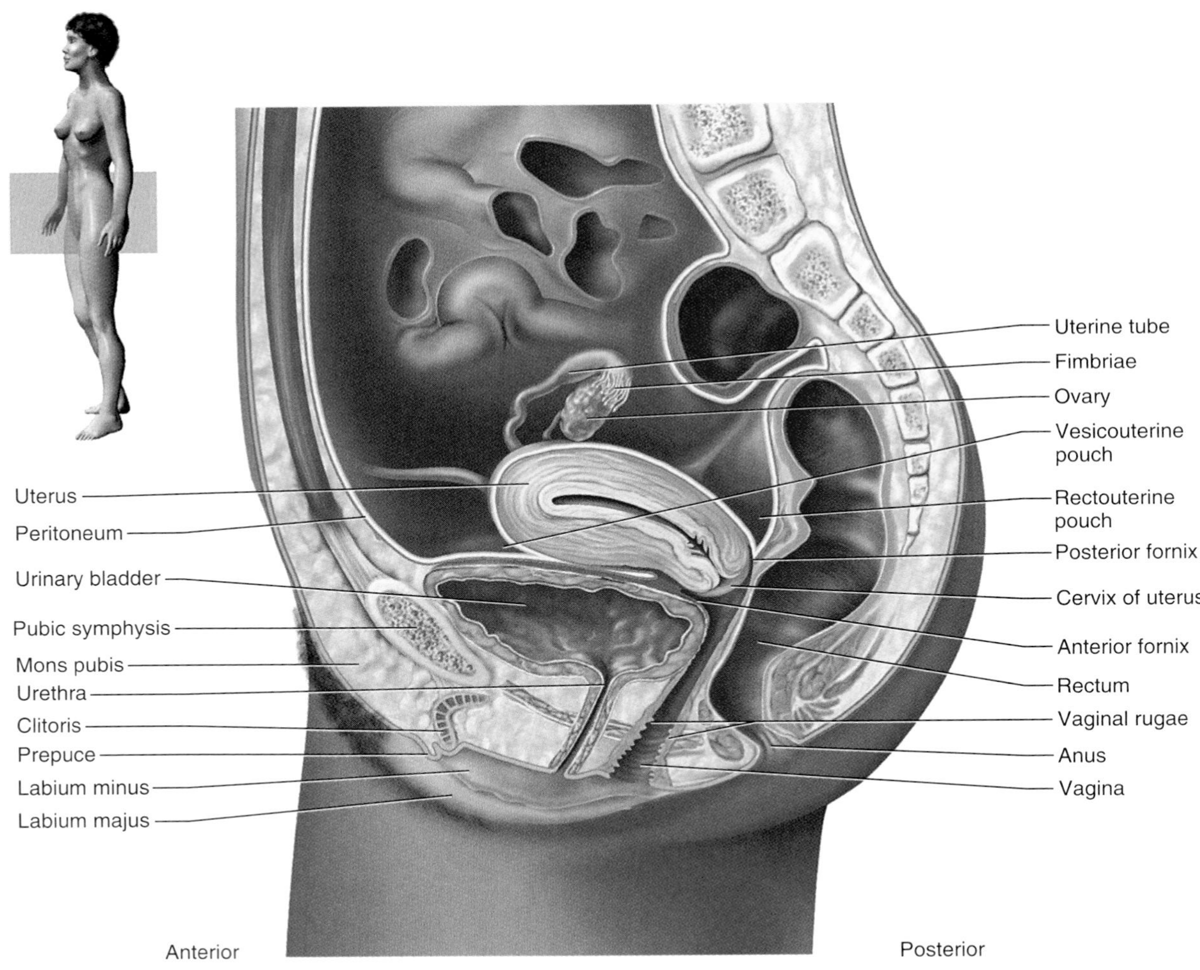

Figure 28.1 The Female Reproductive System.

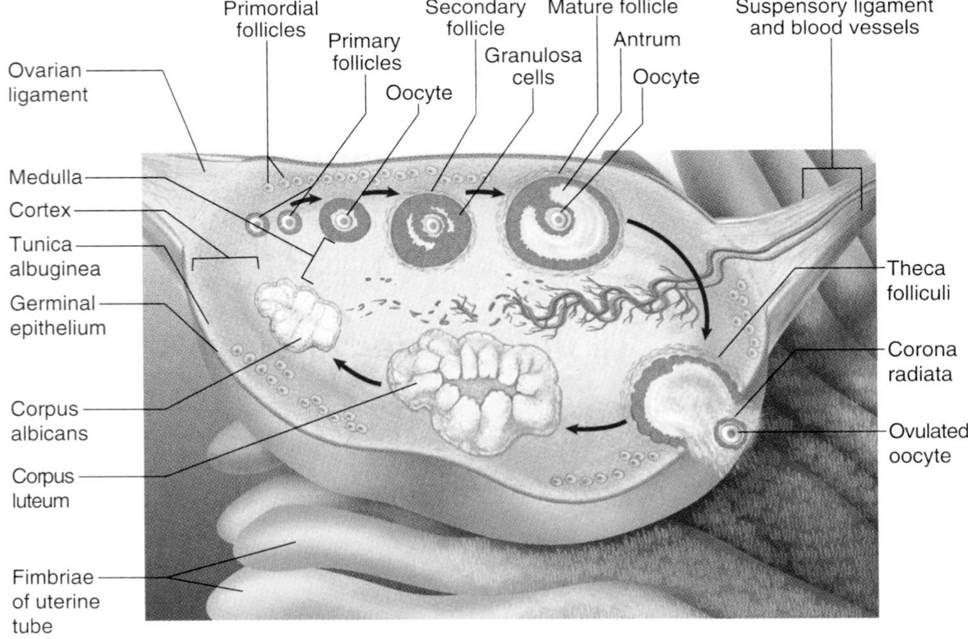

Figure 28.2 Structure of the Ovary and the Developmental Sequence of the Ovarian Follicles.

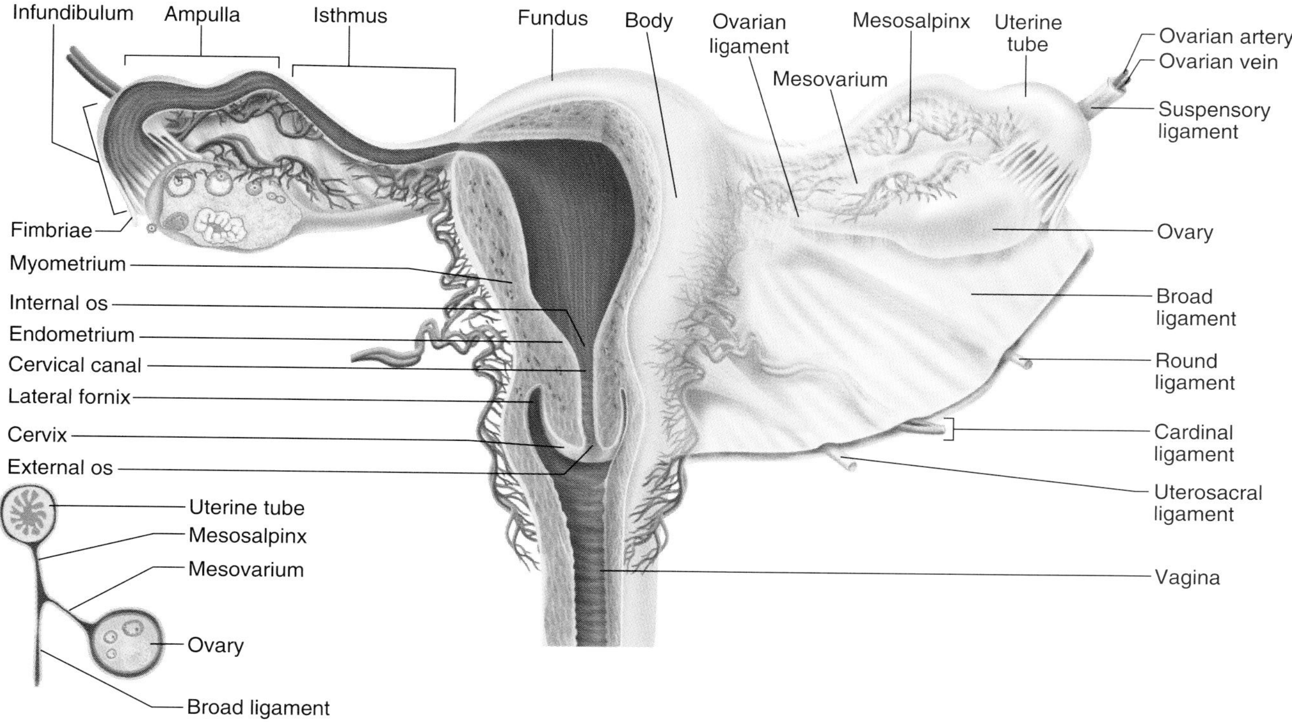

Figure 28.3 The Female Reproductive Tract and Supportive Ligaments, Posterior View.

The wall of the uterine tube is well endowed with smooth muscle. Its mucosa is extremely folded and convoluted and has an epithelium of ciliated cells and a smaller number of secretory cells (fig. 28.4). The cilia beat toward the uterus and, with the help of muscular contractions of the tube, convey the egg in that direction.

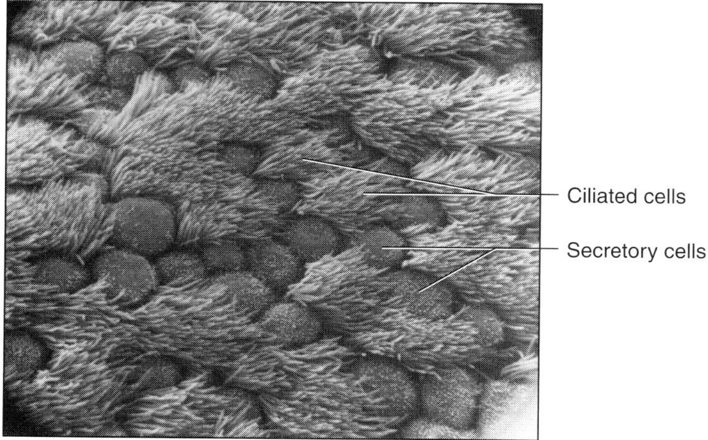

Figure 28.4 Epithelium of the Uterine Tube. Note the mixture of domelike secretory cells with microvilli and ciliated cells, which propel the egg to the uterus.
From R. G. Kessel and R. H. Kardon, *Tissues and Organs: A Text-Atlas of Scanning Electron Microscopy* (W. H. Freeman, 1979).

The Uterus

The **uterus**[7] (fig. 28.5) is a thick muscular chamber that opens onto the roof of the vagina and tilts forward over the urinary bladder. Its function is to harbor the embryo, provide a source of nutrition, and expel the fetus at the end of its development. It is somewhat pear-shaped, with a broad superior curvature called the **fundus,** a midportion called the **body (corpus),** and a cylindrical inferior end called the **cervix** (see fig. 28.3). The uterus measures about 7 cm from cervix to fundus, 4 cm wide at its broadest point, and 2.5 cm thick, but it is somewhat larger in women who have been pregnant.

The lumen of the uterus is roughly triangular, with its two upper corners opening into the uterine tubes. It communicates with the vagina by way of a narrow passage through the cervix called the **cervical canal.** The superior opening of this canal into the body of the uterus is the *internal os*[8] (oss) and its opening into the vagina is the *external os.* The canal contains **cervical glands** that secrete mucus, thought to prevent the spread of microorganisms

7. *uterus* = womb

8. *os* = mouth

Clinical Insight 28.1

Pap Smears and Cervical Cancer

Cervical cancer is common among women from ages 30 to 50, especially those who smoke, who began sexual activity at an early age, and who have histories of frequent sexually transmitted diseases or cervical inflammation. It begins in the epithelial cells of the lower cervix, develops slowly, and remains a local, easily removed lesion for several years. If the cancerous cells spread to the subepithelial connective tissue, however, the cancer is said to be *invasive* and is much more dangerous, potentially requiring *hysterectomy*[9] (removal of the uterus).

The best protection against cervical cancer is early detection by means of a *Pap*[10]*smear*—a procedure in which loose cells are scraped from the cervix and vagina and microscopically examined. The findings are rated on a five-point scale:

Class I—no abnormal cells seen

Class II—atypical cells suggestive of inflammation, infection, or irritation

Class III—nonmalignant but mildly abnormal cell growth *(dysplasia)*

Class IV—cells typical of localized cancer

Class V—cells typical of invasive cancer

An average woman is typically advised to have annual Pap smears for 3 years and may then have them less often at the discretion of her physician. Women with any of the risk factors listed may be advised to have more frequent examinations.

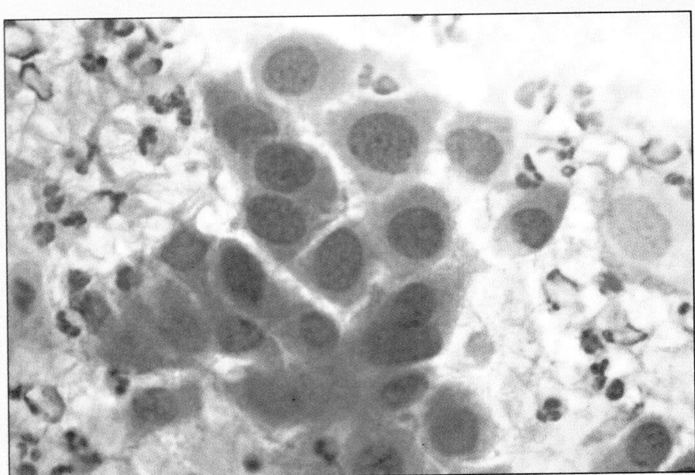

TA28.1 A Cancerous (class V) Pap Smear. Compare this with the normal (class I) Pap smear in figure 6.13 (p. 180).

9. *hyster* = uterus + *ectomy* = *cutting out*
10. George N. Papanicolaou (1883–1962), Greek-American physician and cytologist

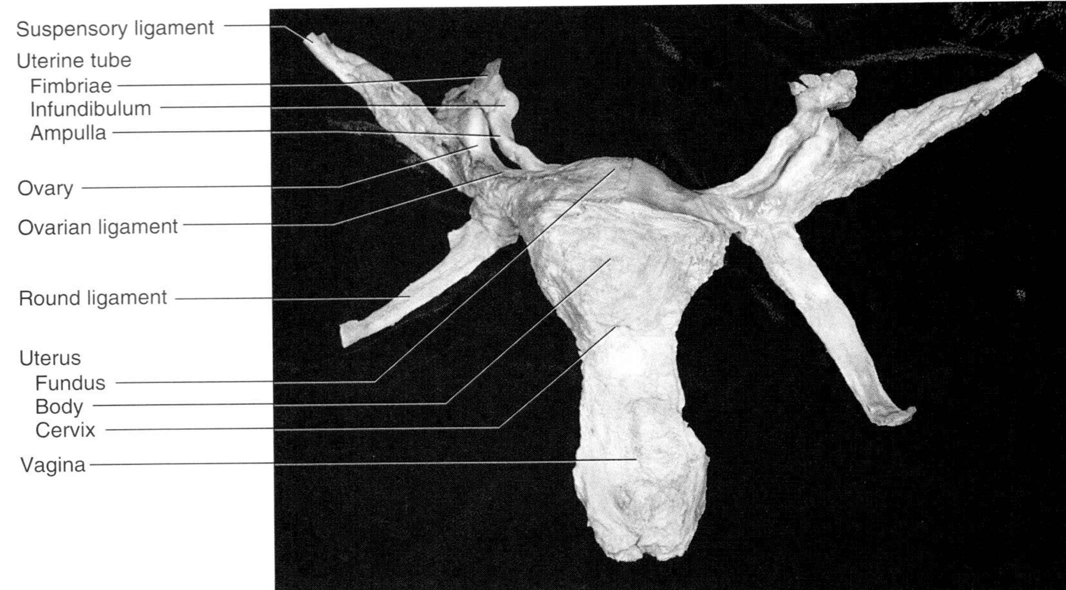

Figure 28.5 The Female Reproductive Tract.

from the vagina into the uterus. Near the time of ovulation, the mucus becomes thinner than usual and allows easier passage for sperm.

Uterine Wall The uterine wall consists of an external serosa called the *perimetrium,* a middle muscular layer called the *myometrium,* and an inner mucosa called the *endometrium.* The **myometrium**[11] constitutes most of the wall; it is about 1.25 cm thick in the nonpregnant uterus. It is composed of bundles of smooth muscle running in all directions, but it is less muscular and more fibrous near the cervix; the cervix itself is almost entirely collagenous. The smooth muscle cells of the myometrium are about 40 μm long immediately after menstruation, but they are twice this long at the middle of the menstrual cycle and 10 times as long in pregnancy. The function of the myometrium is to produce the labor contractions that help to expel the fetus.

The **endometrium**[12] is the mucosa. It has a simple columnar epithelium, compound tubular glands, and a stroma populated by leukocytes, macrophages, and other cells (fig. 28.6). The superficial half to two-thirds of it, called the **stratum functionalis,** is shed in each menstrual period. The deeper layer, called the **stratum basalis,** stays behind and regenerates a new functionalis in the next cycle. When pregnancy occurs, the endometrium is the site of attachment of the embryo and forms the maternal part of the *placenta* from which the fetus is nourished.

11. *myo* = muscle + *metr* = uterus
12. *endo* = inside + *metr* = uterus

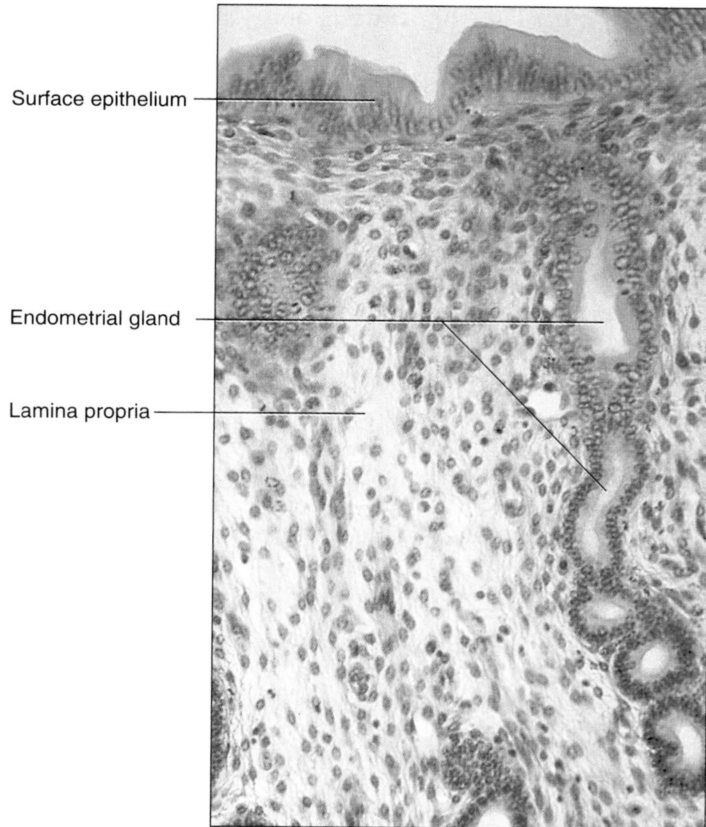

Figure 28.6 Histology of the Endometrium.

Blood Supply The uterine blood supply is particularly important to the menstrual cycle and pregnancy. A **uterine artery** arises from each internal iliac artery and travels through the broad ligament to the uterus (fig. 28.7). It gives off several branches that penetrate into the myometrium and lead to **arcuate arteries.** Each arcuate artery travels in a circle around the uterus and anastomoses with the arcuate artery on the other side. Along its course, it gives rise to smaller arteries that penetrate the rest of the way through the myometrium, into the endometrium, and produce the **spiral arteries.** The spiral arteries wind tortuously between the endometrial glands toward the surface of the mucosa. They rhythmically constrict and dilate, making the mucosa alternately blanch and flush with blood.

Ligaments The uterus is supported mainly by the muscular floor of the pelvic outlet but also by folds of peritoneum that form supportive ligaments around the organ, as they do for the ovary and uterine tube (see fig. 28.3). The broad ligament has two parts: the mesosalpinx mentioned earlier and the *mesometrium* each side of the uterus. The cervix and superior part of the vagina are supported by the *cardinal (lateral cervical) ligaments* extending to the pelvic wall. A pair of *uterosacral ligaments* attach the posterior side of the uterus to the sacrum, and a pair of *round ligaments* attach the anterior surface of the uterus to the abdominal wall. The round ligaments continue through the inguinal canals and terminate in the labia majora, much like the gubernaculum of the male terminating in the scrotum.

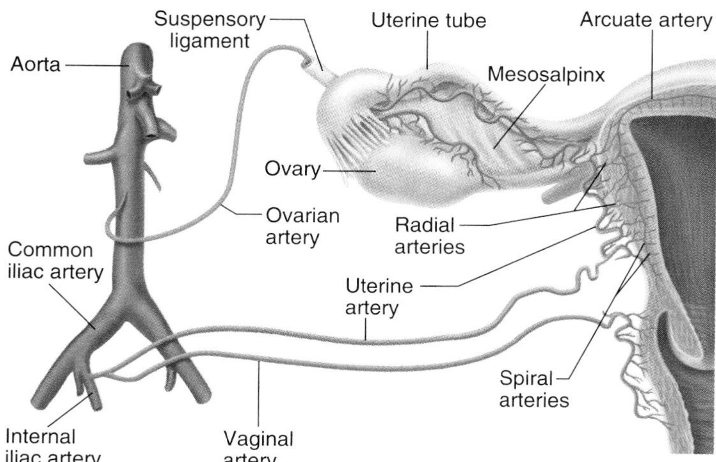

Figure 28.7 Blood Supply to the Female Reproductive Tract.

As the peritoneum folds around the various pelvic organs, it creates several dead-end recesses and pouches. Two major ones are the *vesicouterine*[13] *pouch,* which forms the space between the uterus and urinary bladder, and *rectouterine pouch* between the uterus and rectum (see fig. 28.1).

Vagina

The **vagina,**[14] or birth canal, is a tube about 8 to 10 cm long that allows for the discharge of menstrual fluid, receipt of the penis and semen, and birth of a baby. The vaginal wall is thin but very distensible. It consists of an outer adventitia, a middle muscularis, and an inner mucosa. The vagina tilts posteriorly between the urethra and rectum; the urethra is embedded in its anterior wall. The vagina has no glands, but it is lubricated by the *transudation* ("vaginal sweating") of serous fluid through its walls and by mucus from the cervical glands above it. The vagina extends slightly beyond the cervix and forms blind-ended spaces called *fornices*[15] (FOR-nih-sees; singular, *fornix*) (see figs. 28.1 and 28.3).

At its lower end, the vaginal mucosa folds inward and forms a membrane, the **hymen,** which stretches across the orifice. It has one or more openings to allow menstrual fluid to pass through, but it usually must be ruptured to allow for intercourse. A little bleeding often accompanies the first act of intercourse; however, the hymen is commonly ruptured before then by tampons, medical examinations, or strenuous exercise. The lower end of the vagina also has transverse friction ridges, or **vaginal rugae,** which stimulate the penis and help induce ejaculation.

The vaginal epithelium is simple cuboidal in childhood, but the estrogens of puberty stimulate it to transform into a stratified squamous epithelium. This is an example of *metaplasia,* the transformation of one tissue type to another. The epithelial cells are rich in glycogen. Bacteria ferment this to lactic acid, which produces a low vaginal pH (about 3.5–4.0) that inhibits the growth of pathogens. Recall from the previous chapter that this acidity is neutralized by the semen so it does not harm the sperm. The mucosa also has antigen-presenting cells (APCs) called **dendritic cells,** which are thought to be a route by which HIV from infected semen invades the female body.

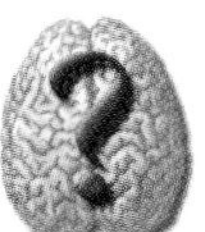

Think About It

Why do you think the vaginal epithelium changes type at puberty? Of all types of epithelium it might become, why stratified squamous?

13. *vesico* = bladder
14. *vagina* = sheath
15. *fornix* = arch, vault

The Vulva (Pudendum)

The external genitalia of the female are collectively called the **vulva**[16] **(pudendum**[17]**);** this includes the mons pubis, labia majora and minora, clitoris, vaginal and urethral orifices, and accessory glands and erectile tissues. It occupies most of the perineum (fig. 28.8*a*).

The **mons**[18] **pubis** consists mainly of a mound of adipose tissue overlying the pubic symphysis. The **labia majora**[19] (singular, *labium majus*) are a pair of thick folds of skin and adipose tissue inferior to the mons; the slit between them is the *pudendal cleft.* Pubic hair grows on the mons pubis and lateral surfaces of the labia majora at puberty, but the medial surfaces of the labia remain hairless. Medial to the labia majora are the much thinner, entirely hairless **labia minora**[20] (singular, *labium minus*). The area enclosed by them, called the **vestibule,** contains the urinary and vaginal orifices. At the anterior margin of the vestibule, the labia minora meet and form a hoodlike **prepuce** over the clitoris.

The **clitoris** is structured much like a miniature penis but has no urinary role. Its function is entirely sensory, serving as the primary center of erotic stimulation. Unlike the penis, it is almost entirely internal, it has no corpus spongiosum, and it does not enclose the urethra. Essentially, it is a pair of corpora cavernosa enclosed in connective tissue. Its **glans** protrudes slightly from the prepuce. The **body (corpus)** passes internally, inferior to the pubic symphysis (see fig. 28.1). At its internal end, the corpora cavernosa diverge like a Y as a pair of **crura,** which, like those of the penis, attach the clitoris to each side of the pubic arch. Like the penis, the clitoris is supplied by the internal pudendal arteries, also called the **clitoral arteries** in the female.

Just deep to the labia majora, a pair of subcutaneous erectile tissues called the **vestibular bulbs** bracket the vagina like parentheses. They become congested with blood during sexual excitement and cause the vagina to tighten somewhat around the penis, enhancing sexual stimulation.

On each side of the vagina is a pea-sized **greater vestibular (Bartholin**[21]**) gland** with a short duct opening into the vestibule or lower vagina (fig. 28.8*b*). These glands are homologous to the bulbourethral glands of the male. They keep the vulva moist, and during sexual excitement they provide most of the lubrication for intercourse. The vestibule is also lubricated by a number of

16. *vulva* = covering
17. *pudend* = shameful
18. *mons* = mountain
19. *labi* = lip + *major* = larger, greater
20. *minor* = smaller, lesser
21. Caspar Bartholin (1655–1738), Danish anatomist

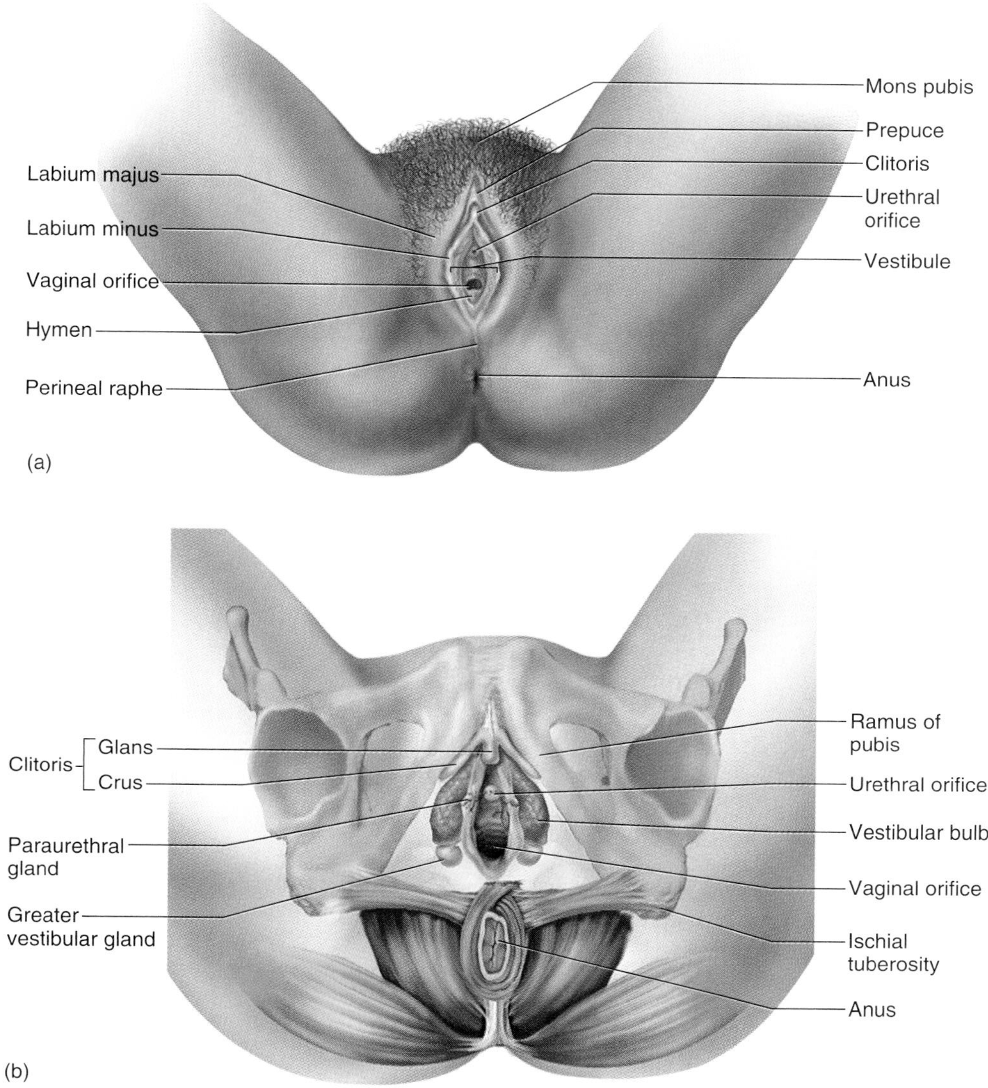

Figure 28.8 **The Female Perineum.** (*a*) Surface anatomy; (*b*) subcutaneous structures.

lesser vestibular glands. A pair of mucous **paraurethral (Skene[22]) glands,** homologous to the male prostate, open into the vestibule near the external urethral orifice.

Secondary Sex Characteristics

The secondary sex characteristics of the female, like those of the male, are features other than the genitalia that develop at puberty, serve to distinguish the sexes from each other, and serve the purpose of sexual attraction. These include the distribution of body fat, flare of the pelvis, the breasts, and the resulting feminine physique; the relatively fine body hair; and a voice pitched higher than the male's.

The Breasts

The breast (fig. 28.9) is a mound of tissue overlying the pectoralis major. It has two principal regions: the conical to pendulous **body,** with the nipple at its apex, and an extension toward the armpit called the **axillary tail.** Lymphatics of the axillary tail are especially important as a route of breast cancer metastasis.

22. Alexander J. C. Skene (1838–1900), American gynecologist

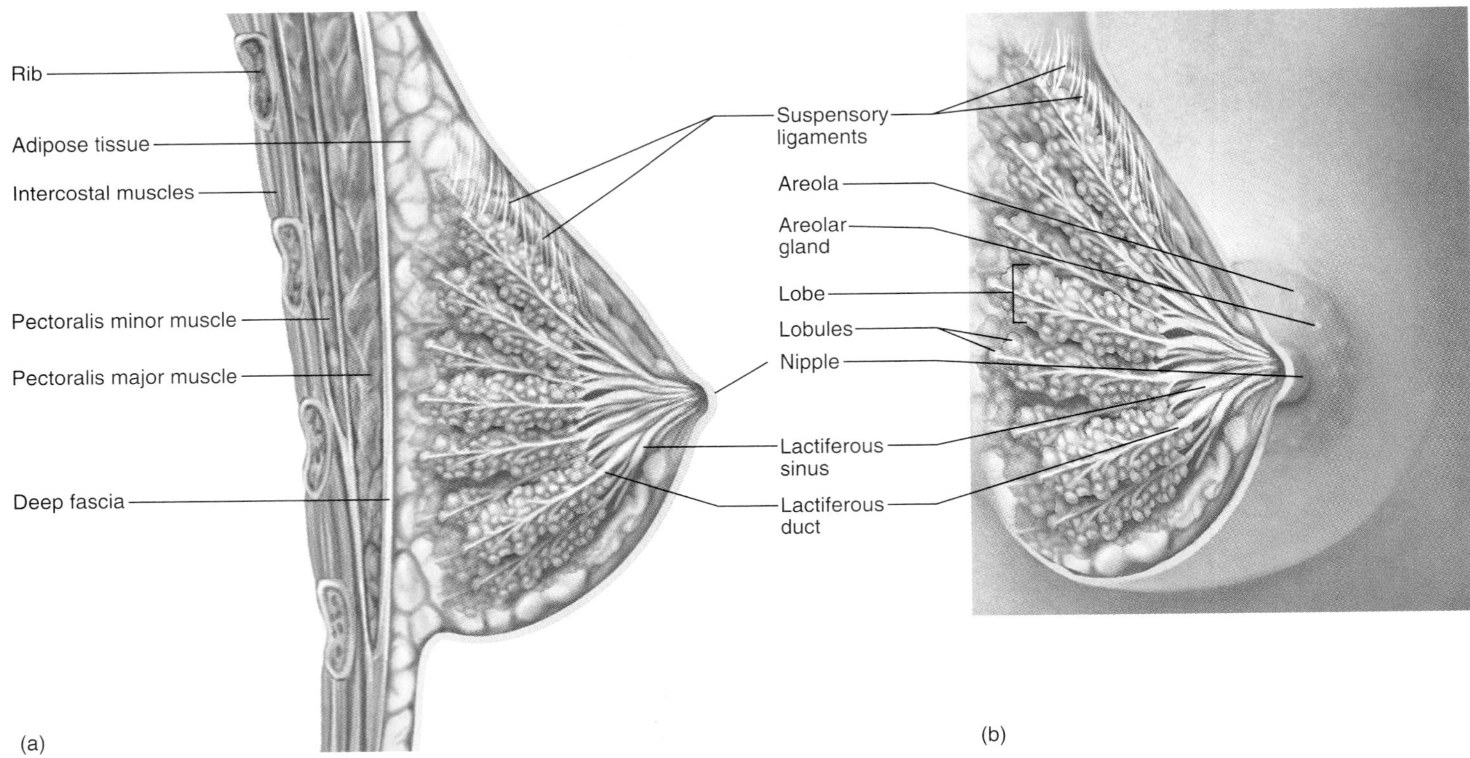

Figure 28.9 Anatomy of the Lactating Breast. (*a*) Sagittal section of breast. (*b*) Surface anatomy with cutaway view of the lobes of mammary gland, anterior view.

The nipple is surrounded by a circular colored zone, the **areola.** Dermal blood capillaries and nerves come closer to the surface here than in the surrounding skin and make the areola more sensitive and more reddish in color. In pregnancy, the areola and nipple often become melanized and make the nipple more visible to the indistinct vision of a nursing infant. Sensory nerve fibers of the areola are important in triggering a *milk ejection reflex* when an infant nurses. The areola has sparse hairs and **areolar glands,** visible as small bumps on the surface. These glands are intermediate between sweat glands and mammary glands in their degree of development. When a woman is nursing, the areola is protected from chapping and cracking by secretions of the areolar glands and sebacous glands of the areola. The dermis of the areola has smooth muscle fibers that contract in response to cold, touch, and sexual arousal, wrinkling the skin and erecting the nipple.

Internally, the nonlactating breast consists mostly of adipose and collagenous tissue (fig. 28.10). Breast size is determined by the amount of adipose tissue and has no relationship to the amount of milk the mammary gland can produce. **Suspensory ligaments** attach the breast to the dermis of the overlying skin and to the fascia of the pectoralis major. The nonlactating breast contains very little glandular tissue, but it does have a system of ducts branching through its connective tissue stroma and converging on the

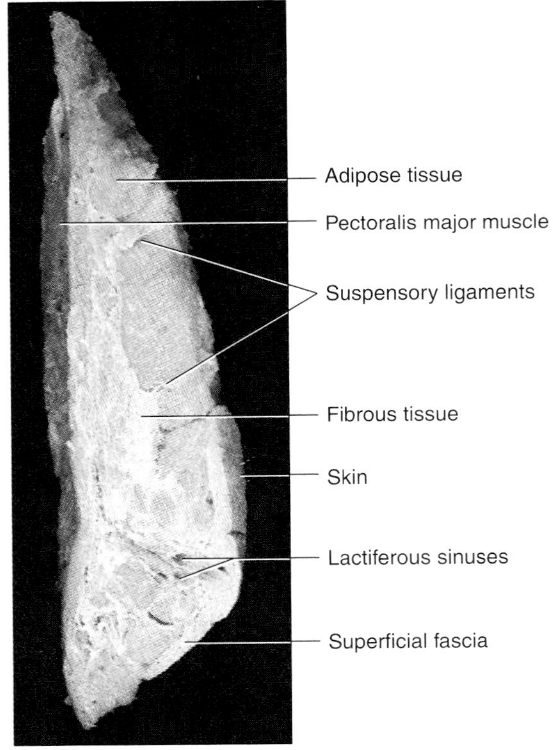

Figure 28.10 Sagittal Section of the Breast of a Cadaver.

nipple. When the mammary gland develops during pregnancy, it exhibits 15 to 20 lobes arranged radially around the nipple, separated from each other by fibrous stroma. Each lobe is drained by a **lactiferous**[23] **duct,** which dilates to form a **lactiferous sinus** opening onto the nipple. Lactation and the associated changes in breast structure are described at the end of this chapter.

Breast Cancer

Breast cancer (fig. 28.11) occurs in one out of every eight or nine American women and is one of the leading causes of female mortality. Breast tumors begin with cells of the mammary ducts and may metastasize to other organs by way of the mammary and axillary lymphatics. Symptoms of breast cancer include a palpable lump (the tumor), puckering of the skin, changes in skin texture, and drainage from the nipple.

Two breast cancer genes were discovered in the 1990s, named BRCA1 and BRCA2, but most breast cancer is nonhereditary. Some breast tumors are stimulated by estrogen. Consequently, breast cancer is more common among women who begin menstruating early in life and who reach menopause relatively late—that is, women who have a long period of fertility and estrogen exposure. Other risk factors include aging, exposure to ionizing radiation and carcinogenic chemicals, excessive alcohol and fat intake, and smoking. Over 70% of cases, however, lack any identifiable risk factors.

23. *lact* = milk + *fer* = to carry

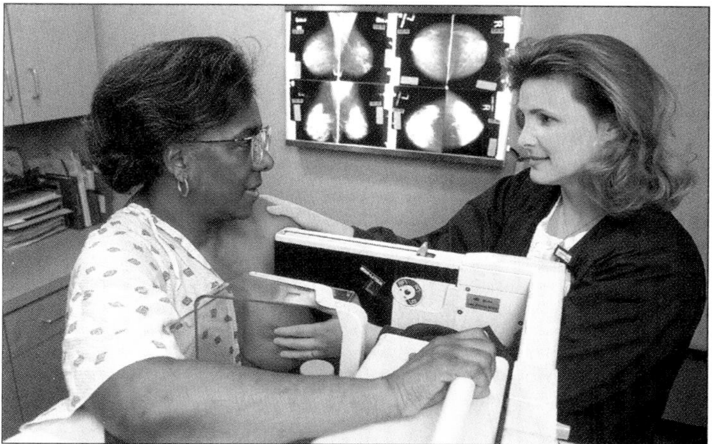
(a)

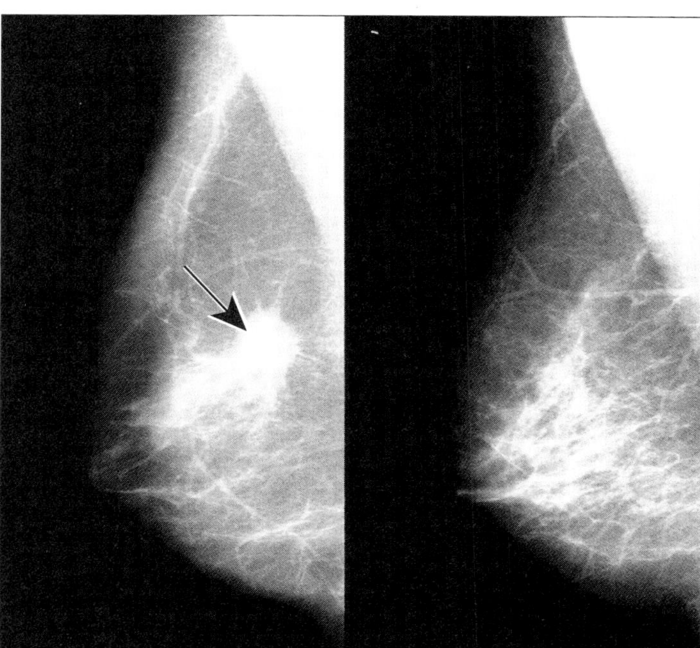
(b)

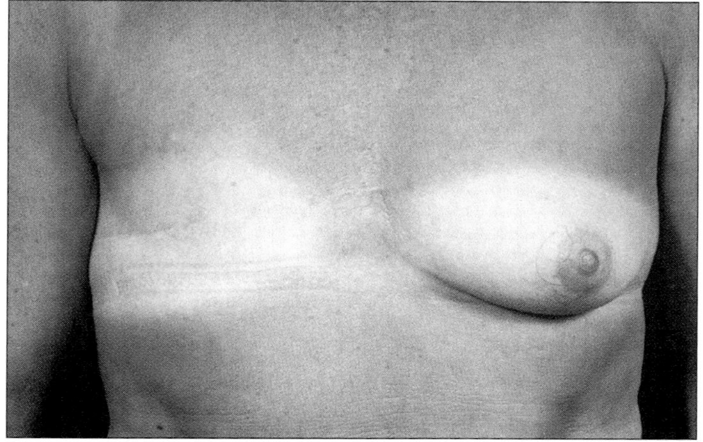
(c)

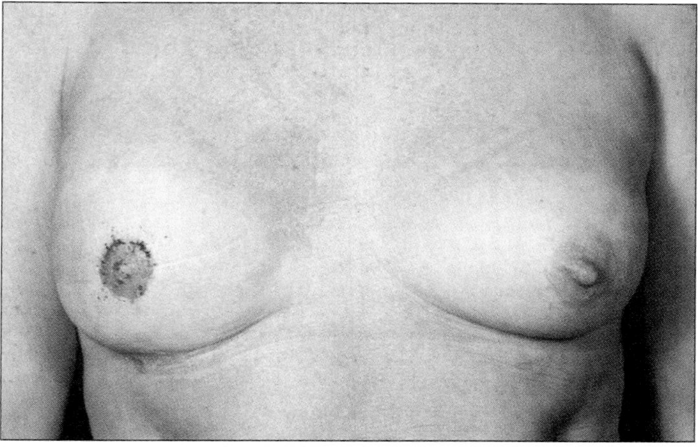
(d)

Figure 28.11 Breast Cancer Screening and Treatment. (*a*) Nurse assisting a patient in mammography. (*b*) Mammogram of a breast with a tumor visible at the arrow (*left*), compared to the appearance of normal connective tissue of the breast (*right*). (*c*) Patient following mastectomy of the right breast. (*d*) The same patient following surgical breast reconstruction.

The majority of tumors are discovered during breast self-examination (BSE), which should be a monthly routine for all women. *Mammograms* (breast X rays), however, can detect tumors too small to be noticed by BSE. Although opinions vary, a schedule commonly recommended is to have a baseline mammogram in the late 30s and then have one every 2 years from ages 40 to 49 and every year beginning at age 50.

Treatment of breast cancer is usually by *lumpectomy* (removal of the tumor only) or *simple mastectomy* (removal of the breast tissue only or breast tissue and some axillary lymph nodes). *Radical mastectomy,* rarely done since the 1970s, involves the removal of not only the breast but also the underlying muscle, fascia, and lymph nodes. Although very disfiguring, it proved to be no more effective than simple mastectomy or lumpectomy. Surgery is generally followed by radiation or chemotherapy, and estrogen-sensitive tumors may also be treated with an estrogen blocker such as tamoxifen. A natural-looking breast can often be reconstructed from skin, fat, and muscle from other parts of the body.

Before You Go On

Answer the following questions to test your understanding of the preceding section:

1. How do the site of female gamete production and mode of release from the gonad differ from those in the male?
2. How is the structure of the uterine tube mucosa related to its function?
3. Contrast the function of the endometrium with that of the myometrium.
4. Describe the similarities and differences between the clitoris and penis.

Puberty and Menopause

Objectives

When you have completed this section, you should be able to

- name the hormones that regulate female reproductive function, and state their roles;
- describe the principal signs of puberty;
- describe the hormonal changes of female climacteric and their effects; and
- define and describe menopause, and distinguish menopause from climacteric.

Puberty and menopause are physiological transitions at the beginning and end of a female's reproductive life.

Puberty

Puberty begins at ages 9 to 10 for most girls in the United States and Europe but significantly later in many countries. However, a 1997 study of more than 17,000 girls in the United States showed 3% of black girls and 1% of white girls beginning puberty by age 3, and 27% and 7%, respectively, by age 7. Puberty is triggered by the same hypothalamic and pituitary hormones in girls as it is in boys. Rising levels of gonadotropin-releasing hormone (GnRH) stimulate the anterior lobe of the pituitary to secrete follicle-stimulating hormone (FSH) and luteinizing hormone (LH). FSH, especially, stimulates development of the ovarian follicles, which, in turn, secrete estrogens, progesterone, inhibin, and a small amount of androgen. These hormone levels rise gradually from ages 8 to 12 and then more sharply in the early teens. The **estrogens**[24] are feminizing hormones with widespread effects on the body. They include *estradiol* (the most abundant), *estriol,* and *estrone.* Most of the visible changes at puberty result from estradiol and androgens.

The earliest noticeable sign of puberty is **thelarche**[25] (thee-LAR-kee), the development of the breasts. This is soon followed by **pubarche** (pyu-BAR-kee), the growth of pubic and axillary hair, sebaceous glands, and apocrine glands. Thelarche is induced by estrogens and pubarche is induced by androgens from the ovaries and adrenal cortex. Androgens are also responsible for the libido of both sexes. Women secrete about 0.5 mg of androgens per day, compared with 6 to 8 mg/day in men.

Next comes **menarche**[26] (men-AR-kee), the first menstrual period. In Europe and America, the average age at menarche declined from age 16.5 in 1860 to age 12 in 1997, probably due mostly to improved nutrition. Menarche cannot occur until a girl has reached at least 17% body fat, so it is delayed until the mid- to late teens in some athletes and dancers. Adult menstruation generally ceases if a woman drops below 22% body fat. This is about the minimum needed to sustain pregnancy and lactation; thus the body reacts as if to prevent a futile pregnancy when it is too lean. Menarche does not necessarily signify fertility. For the first year, a girl's menstrual periods are typically *anovulatory* (no egg is ovulated). Most girls begin ovulating about a year after they begin menstruating.

Estradiol stimulates many other changes of puberty. It causes the vaginal metaplasia described earlier. It stimulates growth of the ovaries and secondary sex organs. It stimulates growth hormone secretion and causes a rapid increase in height and widening of the pelvis. Estradiol is

24. *estro* = desire, frenzy + *gen* = to produce
25. *thel* = breast, nipple + *arche* = beginning
26. *men* = monthly

Evolutionary Insight 28.2

The Evolution of Menopause

There has been considerable speculation about why women do not remain fertile to the end of their lives, as men do. Some theorists argue that menopause served a biological purpose for our prehistoric foremothers. Human offspring take a long time to rear. Beyond a certain point, the frailties of age make it unlikely that a woman could rear another infant to maturity or even survive the stress of pregnancy. She might do better in the long run to become infertile and finish rearing her last child instead of having another one. In this view, menopause was biologically advantageous for our ancestors—in other words, an evolutionary adaptation.

Others argue against this hypothesis on the grounds that Ice Age skeletons indicate that early hominids rarely lived past age 40. If this is true, menopause setting in at 45 to 55 years of age could have served little purpose. In this view, Ice Age women may indeed have been fertile to the end of their lives; menopause now may be just an artifact of modern nutrition and medicine, which have made it possible for us to live much longer than our ancestors did.

largely responsible for the feminine physique because it stimulates fat deposition in the mons pubis, labia majora, hips, thighs, buttocks, and breasts. It makes a girl's skin thicken, but the skin remains thinner, softer, and warmer than in males of corresponding age.

Progesterone[27] acts primarily on the uterus, preparing it for possible pregnancy in the second half of each menstrual cycle and playing roles in pregnancy discussed later. Estrogens and progesterone also suppress FSH and LH secretion through negative feedback inhibition of the anterior pituitary. **Inhibin** selectively suppresses FSH secretion.

Thus we see many hormonal similarities in males and females from puberty onward. The sexes differ less in the hormones that are present than in the relative amounts of those hormones—high levels of androgens and low levels of estrogens in males and the opposite in females. Another difference is that these hormones are secreted more or less continually and simultaneously in males, whereas in females secretion is distinctly cyclic and the hormones are secreted in sequence rather than simultaneously. This will be very apparent as you read about the ovarian and menstrual cycles.

Climacteric and Menopause

Women, like men, go through a midlife change in hormone secretion called the **climacteric.** In women, it is accompanied by **menopause,** the cessation of menstruation (see evolutionary insight 28.2).

With age, the ovaries have fewer remaining follicles and those that remain are less responsive to gonadotropins. Consequently, they secrete less estrogen and progesterone.

27. *pro* = favoring + *gest* = pregnancy + *sterone* = steroid hormone

Without these steroids, the uterus, vagina, and breasts atrophy. Intercourse may become uncomfortable, and vaginal infections more common, as the vagina becomes thinner, less distensible, and drier. The skin becomes thinner, cholesterol levels rise (increasing the risk of cardiovascular disease), and bone mass declines (increasing the risk of osteoporosis). Blood vessels constrict and dilate in response to shifting hormone balances, and the sudden dilation of cutaneous arteries may cause **hot flashes**—a spreading sense of heat from the abdomen to the thorax, neck, and face. Hot flashes may occur several times a day, sometimes accompanied by headaches resulting from the sudden vasodilation of arteries in the head. In some people, the changing hormonal profile also causes mood changes. Many physicians prescribe hormone replacement therapy (HRT)—low doses of estrogen and progesterone taken orally or by a skin patch—to relieve some of these symptoms. The risks and benefits of HRT are still being debated.

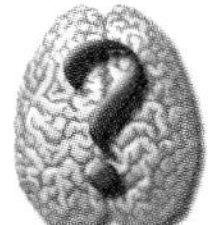

Think About It

FSH and LH secretion *rise* at climacteric and these hormones attain high concentrations in the blood. Explain this using the preceding information and what you know about the pituitary-gonadal axis.

Menopause is the cessation of menstrual cycles, usually occurring between the ages of 45 and 55. The average age has increased steadily in the last century and is now about 52. It is difficult to precisely establish the time of menopause because the menstrual periods can stop for several months and then begin again. Menopause is generally considered to have occurred when there has been no menstruation for a year or more.

Before You Go On

Answer the following questions to test your understanding of the preceding section:

5. Describe the similarities and differences between male and female puberty.
6. Describe the major changes that occur in female climacteric and the principal cause of these changes.
7. What is the difference between climacteric and menopause?

Oogenesis and the Sexual Cycle

Objectives

When you have completed this section, you should be able to

- describe the process of egg production (oogenesis);
- describe how the ovarian follicles change in relation to oogenesis;
- describe the hormonal events that regulate the ovarian cycle;
- describe how the uterus changes during the menstrual cycle; and
- construct a chart of the phases of the monthly sexual cycle showing the hormonal, ovarian, and uterine events of each phase.

The reproductive lives of female humans are conspicuously cyclic. They include the **reproductive cycle,** which encompasses the sequence of events from fertilization to giving birth, and the **sexual cycle,** which encompasses the events that recur every month when pregnancy does not intervene. The sexual cycle, in turn, consists of two interrelated cycles controlled by shifting patterns of hormone secretion: the **ovarian cycle,** consisting of events in the ovaries, and the **menstrual cycle,** consisting of parallel changes in the uterus. In this section, we examine the process of egg production, or oogenesis, and the monthly events of the ovarian and menstrual cycles.

Oogenesis

Egg production is called **oogenesis**[28] (OH-oh-JEN-eh-sis) (fig. 28.12). Like spermatogenesis, it produces a haploid gamete by means of meiosis. There are, however, numerous differences between oogenesis and spermatogenesis. The most obvious, perhaps, is that spermatogenesis goes on continually, while oogenesis is a distinctly cyclic event that normally produces only one egg per month. Oogenesis is accompanied by cyclic changes in hormone secretion and in the histological structure of the ovaries and uterus; the uterine changes result in the monthly menstrual flow.

28. *oo* = egg + *genesis* = production

The female germ cells arise, like those of the male, from the yolk sac of the embryo. They colonize the gonadal ridges in the first 5 to 6 weeks of development and then differentiate into **oogonia** (OH-oh-GO-nee-uh). Oogonia multiply until the fifth month of development, reach 6 to 7 million in number, and then go into a state of arrested development until shortly before birth. At that time, some of them transform into **primary oocytes** and go as far as early meiosis I. Any stage from the primary oocyte to the time of fertilization can be called an egg, or **ovum.**

Most primary oocytes undergo a process of degeneration called **atresia** (ah-TREE-zhee-uh) before a girl is born. Only 2 million remain at the time of birth, and most of those undergo atresia during childhood. By puberty, only 400,000 oocytes remain. This is the female's lifetime supply of gametes, but it is more than ample; even if she ovulated every 28 days from the ages of 14 to 50, she would ovulate only 480 times.

Beginning in adolescence, FSH stimulates the primary oocytes to complete meiosis I, which yields two haploid daughter cells of unequal size and different destinies. In oogenesis it is important to produce an egg with as much cytoplasm as possible, because if it is fertilized it must divide repeatedly and produce numerous daughter cells. Splitting each oocyte into four equal but small parts would run counter to this purpose. Rather, meiosis I produces a large daughter cell called the **secondary oocyte** and a much smaller one called the **first polar body.** The polar body sometimes undergoes meiosis II but ultimately disintegrates. It is merely a means of discarding the extra haploid set of chromosomes.

The secondary oocyte proceeds as far as metaphase II and then arrests until after ovulation. If it is not fertilized, it dies and never finishes meiosis. If it is fertilized, it completes meiosis II and produces a **second polar body,** which disposes of one chromatid from each chromosome. The chromosomes of the large remaining egg unite with those of the sperm. Further development of the fertilized egg is discussed in the next chapter.

Each developing germ cell is contained in a follicle, surrounded by one or more layers of follicular cells. The development and types of follicles are described shortly.

The Sexual Cycle

The sexual cycle averages 28 days in length, which is the basis for the timetable described in the following pages. It commonly varies from 20 to 45 days, however, so be aware that the timetable given in this discussion may differ from person to person and from month to month. As you study this cycle, bear in mind that hormones of the hypothalamus and anterior pituitary gland regulate the ovaries, and hormones from the ovaries, in turn, regulate the uterus. That is, the basic hierarchy of control can be represented: hypothalamus → pituitary → ovaries → uterus. As you

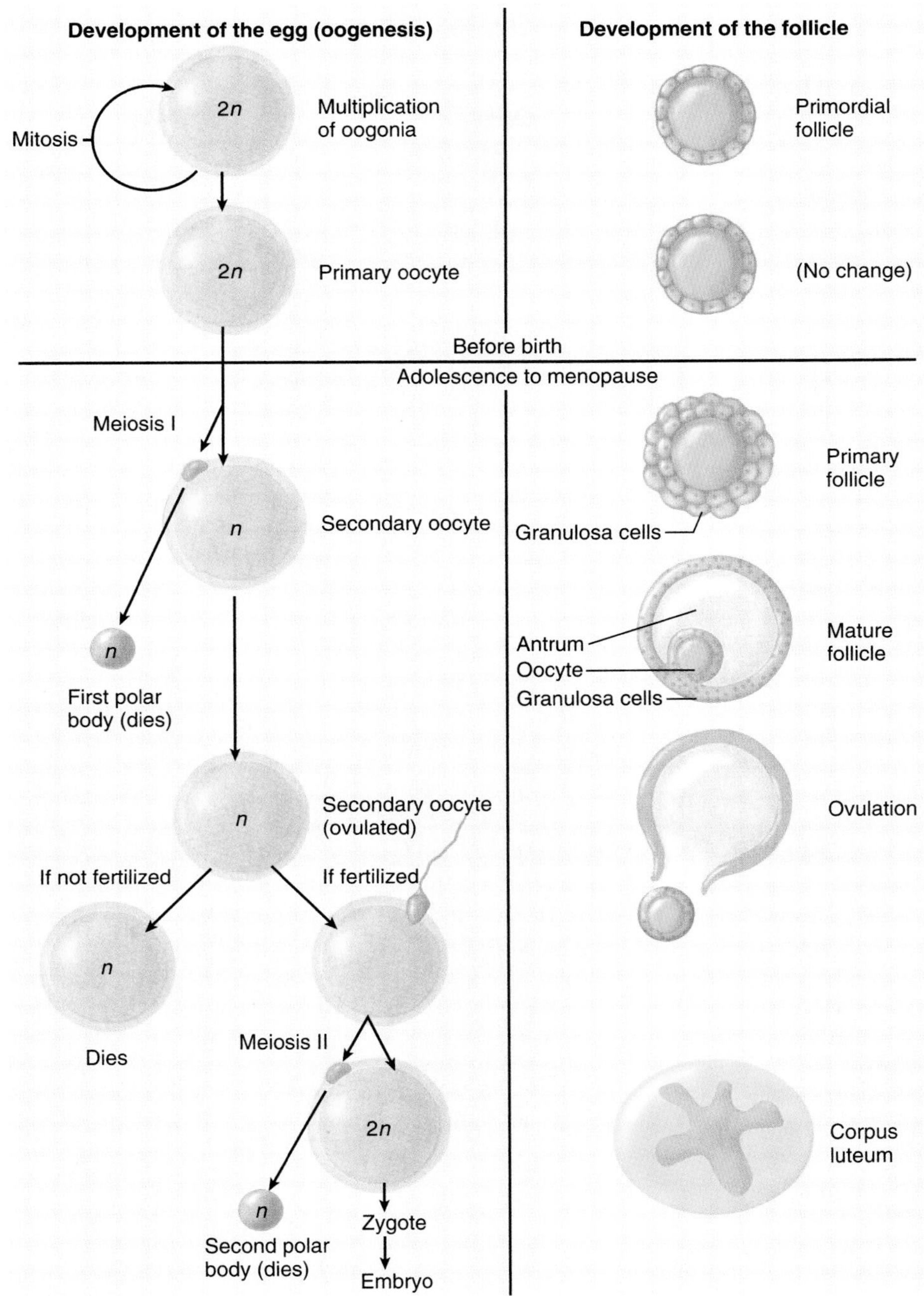

Figure 28.12 Oogenesis *(left)* and Corresponding Development of the Follicle *(right)*.

will see, however, there is also feedback control from the ovaries to the hypothalamus and pituitary.

We begin with a brief preview of the sexual cycle as a whole. The cycle begins with a 2-week *follicular phase.* Menstruation occurs during the first 3 to 5 days, and then the uterus replaces the lost endometrial tissue by mitosis. The ovarian follicles grow during this phase, and one of them ovulates around day 14. After ovulation, the remainder of the follicle becomes a body called the corpus luteum. Over the next 2 weeks, called the *postovulatory phase,* the corpus luteum stimulates endometrial secretion, and the endometrium thickens still more. If pregnancy does not occur, it breaks down again in the last 2 days. As loose tissue and blood accumulate, menstruation begins and the cycle starts over. This cycle is summarized in figure 28.13 and table 28.1.

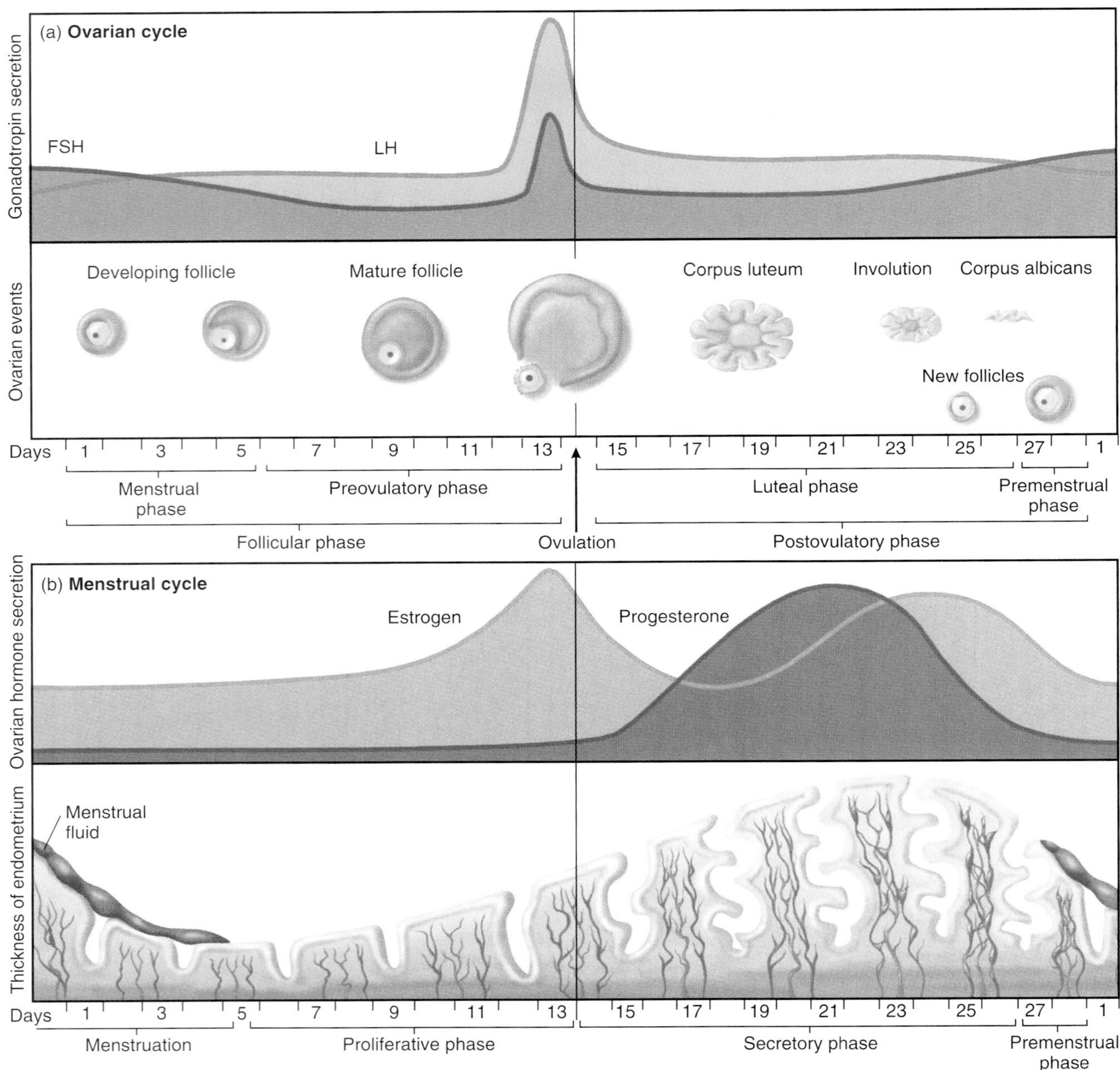

Figure 28.13 The Female Sexual Cycle. (*a*) The ovarian cycle (events in the ovary); (*b*) the menstrual cycle (events in the uterus).

The Ovarian Cycle

Now we can examine the ovarian cycle in more detail. We will see, step by step, what happens in the ovaries and in their relationship to the hypothalamus and pituitary gland.

The **follicular phase** of the cycle extends from the beginning of menstruation (day 1) to ovulation (day 14). This is the most variable part of the cycle and it is seldom possible to predict the date of ovulation reliably. The follicular phase is divided into the *menstrual phase* and *preovulatory phase.*

Menstrual Phase The **menstrual phase** is the period of time when a woman is discharging menstrual fluid (days 1–5 of the typical cycle). As menstruation is occurring, important developments are taking place in the ovaries:

1. Beginning around day 25 of the previous cycle (3 days before menstruation begins), FSH secretion rises and stimulates 20 to 25 primary oocytes to begin meiosis I.
2. Meanwhile, the follicles around these oocytes develop. The primary oocyte is enclosed in a

Table 28.1 Phases of the Female Sexual Cycle

Days	Phase	Major Events
1–14	**Follicular Phase**	
1–5	Menstrual phase (menses)	Menstruation occurring; FSH level high; primordial follicles developing into primary and then secondary follicles
6–13	Preovulatory (proliferative) phase	Rapid growth of one follicle and atresia of the lagging follicles; drop in FSH level; endometrial regeneration and growth by cell proliferation; development of mature follicle; completion of meiosis I, producing secondary oocyte, which arrests at metaphase II; sharp rise in LH and estrogen levels
14	Ovulation	Rupture of follicle and release of oocyte
15–28	**Postovulatory Phase**	
15–26	Luteal (secretory) phase	Formation of corpus luteum; secretion of progesterone; mucus and glycogen secretion by endometrium, causing endometrial thickening; later involution of corpus luteum and falling progesterone level
27–28	Premenstrual (ischemic) phase	Endometrial ischemia and necrosis; sloughing of necrotic tissue from uterine wall, mixing with blood and forming menstrual fluid

primordial follicle, which has a single layer of squamous follicular cells around the oocyte (see fig. 28.12). These cells enlarge and become cuboidal; the follicle is then known as a **primary follicle.** The cuboidal cells multiply and stratify, the follicle as a whole enlarges, and connective tissue condenses around it to form the **theca**[29] **folliculi** (THEE-ca fol-IC-you-lye). Its outer layer, the *theca externa,* becomes a fibrous capsule. Its inner layer, the *theca interna,* secretes androgen, which the granulosa cells convert to estrogen.

3. In the first few days of the follicular phase (during menstruation), the follicular cells begin to secrete an estrogen-rich **follicular fluid,** which accumulates in little pools amid the cells. These pools soon merge and become a fluid-filled cavity, the **antrum.** The follicle is now called a **secondary (antral) follicle** (fig. 28.14) and the follicular cells lining it are called **granulosa cells.** A mound of granulosa cells covers the oocyte and secures it to the follicle wall. The granulosa cells secrete a clear layer of gel called the **zona pellucida**[30] between themselves and the oocyte. The innermost layer of cells in contact with the zona pellucida is the **corona radiata.**[31] This is the state of development when menstruation ceases around day 5.

Preovulatory Phase The **preovulatory phase** spans days 6 to 14 of the average cycle—from the end of menstruation to the day of ovulation. Its major developments are as follows:

4. The growing follicle secretes increasing amounts of estrogen, which has two seemingly contradictory effects: it reduces FSH secretion by the pituitary, and it makes the follicle more sensitive to FSH. The latter effect comes about as estrogen stimulates the granulosa cells of its own follicle to produce an increasing number of FSH receptors. FSH in turn stimulates this follicle to produce still more estrogen, completing a positive feedback loop. Overall, the most advanced follicle reduces the FSH supply to other follicles while at the same time it makes itself more sensitive to the FSH that remains.
5. The less developed, less sensitive follicles undergo atresia, while the most developed follicle attains a diameter of up to 2.5 cm. This follicle, called a **mature (graafian**[32]**) follicle,** protrudes from the surface of the ovary like a blister.
6. As the follicle matures, the primary oocyte completes meiosis I and becomes a secondary oocyte. This cell begins meiosis II but stops at metaphase II. It is now ready for ovulation.
7. FSH and estrogen also stimulate the maturing follicle to produce LH receptors, which are important to the next phase of the cycle.

Ovulation Ovulation, the release of an oocyte, typically occurs on day 14, the midpoint of the average cycle. It takes only 2 or 3 minutes. The events immediately leading up to and including ovulation are as follows:

8. In the last day or two of the preovulatory phase, the estrogen level is very high. This estrogen stimulates the anterior pituitary to secrete LH and the

29. *theca* = box, case
30. *zona* = zone + *pellucid* = clear, transparent
31. *corona* = crown + *radiata* = radiating
32. Reijnier de Graaf (1641–73), Dutch physiologist and histologist

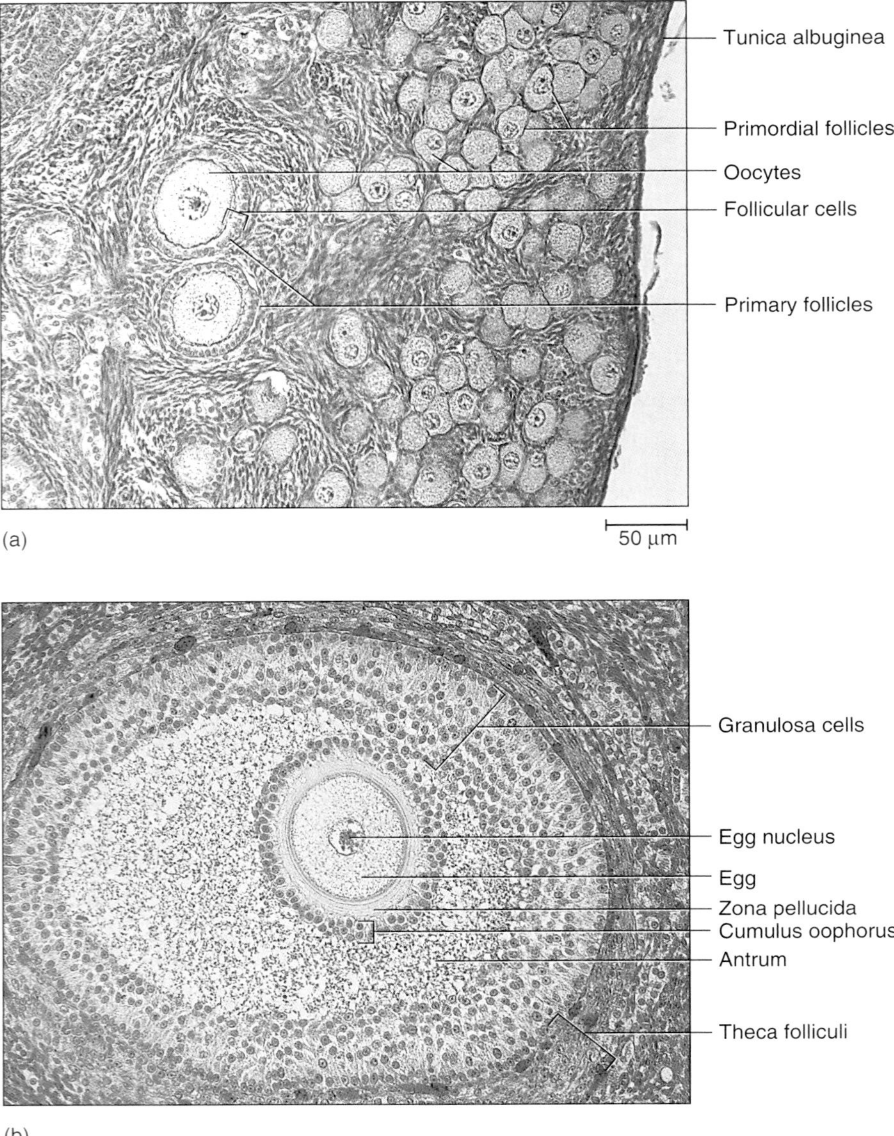

Figure 28.14 Ovarian Follicles. (*a*) Primordial follicles consist of an oocyte (egg cell) surrounded by a thin simple squamous epithelium of follicular cells. The primary follicle has a simple cuboidal epithelium of follicular cells, which will soon become stratified cuboidal. (*b*) A mature (graafian) follicle.

hypothalamus to secrete GnRH. GnRH further induces a surge in FSH and LH secretion by the pituitary (fig. 28.15). The FSH level therefore rises in the last day or two before ovulation, but the LH level rises even more markedly (see fig. 28.13).

9. The uterine tube becomes edematous, its fimbriae envelop and caress the ovary in time with the woman's heartbeat, and its cilia create a gentle current in the nearby peritoneal fluid—all in preparation for receiving the oocyte.
10. LH increases blood flow in the follicle. More serous fluid filters from the capillaries into the antrum and causes the follicle to swell rapidly. Meanwhile, LH also stimulates the theca interna to secrete *collagenase,* an enzyme that weakens the ovarian wall over the swelling follicle.
11. A nipplelike **stigma** appears on the ovarian surface over the follicle. Follicular fluid seeps from the stigma for 1 or 2 minutes, and then the follicle ruptures. The remaining follicular fluid oozes out, carrying the oocyte and the surrounding cells of the corona radiata (fig. 28.16).
12. The oocyte and its attendant cells are normally swept up by the ciliary current and taken into the uterine tube, although many oocytes fall into the pelvic cavity and die.

Clinical Insight 28.3

Signs of Ovulation

If a couple is attempting to conceive a child or to avoid pregnancy, it is important to be able to tell when ovulation occurs. The signs are subtle but detectable. For one, the cervical mucus becomes thinner and more stretchy. Also, the resting body temperature *(basal temperature)* rises 0.2° to 0.3°C (0.4°–0.6°F). This is best measured first thing in the morning, before rising from bed; the change can be detected if basal temperatures are recorded for several days before ovulation in order to see the difference. The LH surge that occurs about 24 hours before ovulation can be detected with a home testing kit. Finally, some women experience twinges of ovarian pain known by the German name, *mittelschmerz,*[33] which last from a few hours to a day or so at the time of ovulation. The most likely time to become pregnant is within 24 hours after the cervical mucus changes consistency and the basal temperature rises.

33. *mittel* = in the middle + *schmerz* = pain

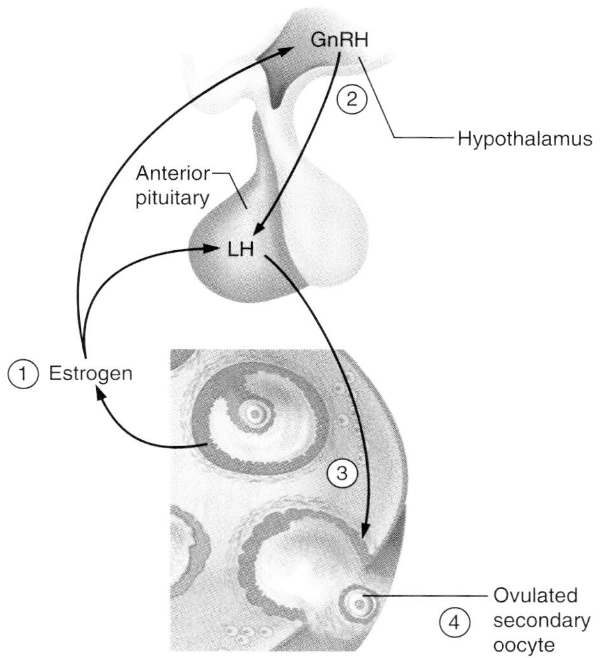

Figure 28.15 Control of Ovulation by Hormones of the Pituitary-Ovarian Axis. (*1*) The maturing follicle secretes high levels of estrogen, which stimulates the hypothalamus and anterior pituitary. (*2*) The hypothalamus secretes GnRH. (*3*) In response to estrogen and GnRH, the anterior pituitary secretes LH. (*4*) LH triggers ovulation.

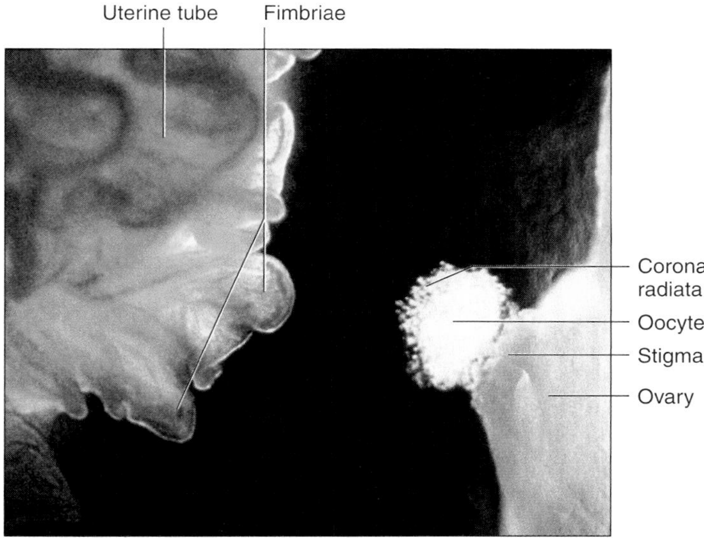

Figure 28.16 Ovulation of a Human Follicle, Viewed by Endoscopy.

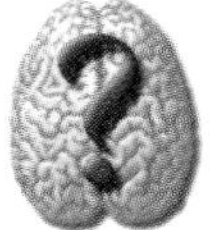

Think About It

In chapter 17, review the concepts of *up-regulation* and the *permissive effect* in hormone interactions; explain their relevance to the ovarian cycle.

Postovulatory Phase The **postovulatory phase** extends from days 15 to 28, from ovulation to the beginning of menstruation. This phase of the sexual cycle is the most predictable in length. Its first 12 days are called the **luteal phase** and its last 2 days are the **premenstrual phase.** The major developments in this phase, assuming pregnancy does not occur, are as follows:

13. When the follicle expels the oocyte, it collapses and bleeds into the antrum. As the clotted blood is slowly absorbed, granulosa and theca interna cells multiply and fill the antrum, and a dense bed of blood capillaries grows amid them. The ovulated follicle has now become a structure called the **corpus luteum,**[34] named for a yellow lipid that accumulates in the theca interna cells. These cells are now called **lutein cells.**

34. *corpus* = body + *lute* = yellow

14. The anterior pituitary continues to secrete LH, which regulates the further growth and activity of the corpus luteum. For this reason, LH is also called *luteotropic hormone.*
15. The lutein cells produce mainly androgen, which the granulosa cells convert to progesterone and a smaller amount of estrogen. Progesterone stimulates developments in the uterus that we survey shortly.
16. The lutein cells also secrete inhibin, which suppresses FSH secretion and prevents new follicles from developing.
17. If pregnancy occurs, the corpus luteum remains active for about 3 months, but in the absence of pregnancy, it begins to atrophy about 10 days after ovulation. At first, its own increasing secretion of progesterone inhibits the secretion of GnRH and LH. As the LH level drops, the corpus luteum exhibits atrophy, or *involution.* This, in turn, results in a drop in the level of progesterone, which brings on menstruation as we shall soon see. Involution is complete by day 26, leaving an inactive scar called the **corpus albicans.**[35]
18. Involution of the corpus luteum also ends its negative feedback inhibition of the hypothalamus. The hypothalamus therefore begins to secrete GnRH anew, the anterior pituitary secretes FSH in response, and a new crop of follicles begins to develop.

The ovarian cycle has now come full circle to the point where we began. We now go on to see how the foregoing events are correlated with changes in the uterus—that is, how the ovarian cycle regulates the menstrual cycle.

The Menstrual Cycle

The menstrual cycle consists of a buildup of the endometrium through most of the sexual cycle, followed by its breakdown and vaginal discharge. The menstrual cycle is divided into a *menstrual phase, proliferative phase, secretory phase,* and *premenstrual phase,* in that order. The menstrual phase averages 5 days long, and the first day of noticeable vaginal discharge is defined as day 1 of the sexual cycle. The reason for menstruation is best understood after you become acquainted with the buildup of endometrial tissue that precedes it, so we begin our survey of the cycle with the proliferative phase.

Proliferative Phase The **proliferative phase** is a time of rebuilding of endometrial tissue lost at the last menstruation. At the end of menstruation, around day 5, the endometrium is about 0.5 mm thick and consists only of the stratum basalis. The stratum functionalis is rebuilt by mitosis from day 6 to day 14. The principal processes in this phase are:

1. Estrogen from the ovaries stimulates mitosis in the stratum basalis as well as the prolific regrowth of blood vessels. By day 14, the endometrium is about 2 to 3 mm thick.
2. Estrogen also stimulates the endometrium to produce progesterone receptors, thereby preparing it for the progesterone-dominated secretory phase.

Secretory Phase The **secretory phase** is a period of further endometrial thickening, but this results from secretion and fluid accumulation rather than mitosis. It extends from day 15 (after ovulation) to day 26 of a typical cycle. The principal processes in this phase are:

3. After ovulation, the corpus luteum secretes mainly progesterone. Progesterone stimulates the endometrial glands and cells of the stroma to accumulate glycogen. The glands grow wider, longer, and more coiled and secrete a glycogen-rich fluid into the lumen. The lamina propria swells with tissue fluid.
4. By the end of the secretory phase, the endometrium is about 5 to 6 mm thick—a soft, wet, nutrient-rich bed available for embryonic development in the event of pregnancy.

Premenstrual Phase The premenstrual phase is a period of endometrial degeneration occurring in the last two days or so of the menstrual cycle.

5. As we have seen, in the absence of pregnancy, the corpus luteum atrophies and the progesterone level falls sharply. In the absence of progesterone, the spiral arteries of the endometrium exhibit spasmodic contractions, that cause endometrial ischemia (interrupted blood flow). The premenstrual phase is therefore also called the **ischemic** (iss-KEE-mic) **phase.**
6. Ischemia leads to tissue necrosis. As the endometrial glands, stroma, and blood vessels degenerate, pools of blood accumulate in the stratum functionalis.
7. Necrotic endometrium falls away from the uterine wall, mixes with blood in the lumen, and forms the **menstrual fluid.**

Menstrual Phase The menstrual phase (*menses*) is the period in which blood, serous fluid, and degenerated endometrial tissue are discharged from the vagina. It commences when enough menstrual fluid accumulates in the uterus. The first day of external discharge marks day 1 of a new cycle. The average woman discharges about 40 mL of blood and 35 mL of serous fluid over a 5-day period. Menstrual fluid contains fibrinolysin, so it does not clot. The vaginal discharge of clotted blood may indicate uterine pathology rather than normal menstruation.

35. *corpus* = body + *alb* = white

In summary, the ovaries go through a follicular phase characterized by growing follicles; then ovulation; and then a postovulatory (mostly luteal) phase dominated by the corpus luteum. The uterus, in the meantime, goes through a menstrual phase in which it discharges its stratum functionalis; then a proliferative phase in which it replaces that tissue by mitosis; then a secretory phase in which the endometrium thickens not by further mitosis but by the accumulation of secretions; and finally, a premenstrual (ischemic) phase in which the stratum functionalis breaks down again. The first half of the cycle is governed largely by follicle-stimulating hormone (FSH) from the pituitary gland and estrogen from the ovaries. Ovulation is triggered by luteinizing hormone (LH) from the pituitary, and the second half of the cycle is governed mainly by LH and progesterone, the latter secreted by the ovaries.

Before You Go On

Answer the following questions to test your understanding of the preceding section:

8. Name the sequence of cell types in oogenesis and identify the ways oogenesis differs from spermatogenesis.
9. Distinguish between a primordial, primary, and secondary follicle. Describe the major structures of a mature follicle.
10. Describe what happens in the ovary during the follicular and postovulatory phases.
11. Describe what happens in the uterus during the menstrual, proliferative, secretory, and premenstrual phases.
12. Describe the effects of FSH and LH on the ovary.
13. Describe the effects of estrogen and progesterone on the uterus, hypothalamus, and anterior pituitary.

Female Sexual Response

Objectives

When you have completed this section, you should be able to

- describe the female sexual response at each phase of intercourse; and
- compare and contrast the female and male responses.

Female sexual response, the physiological changes that occur during intercourse, may be viewed in terms of the four phases identified by Masters and Johnson and discussed in the previous chapter—excitement, plateau, orgasm, and resolution. The neurological and vascular controls of female sexual response are essentially the same as in the male and need not be repeated here. The emphasis here is on ways the female response differs from that of the male.

Excitement and Plateau

Excitement is marked by myotonia, vasocongestion, and increased heart rate, blood pressure, and respiratory rate. Although vasocongestion works by the same mechanism in both sexes, its effects are quite different. In females, the labia minora become congested and often protrude beyond the labia majora. The labia majora become reddened and enlarged and then flatten and spread away from the vaginal orifice.

The vaginal wall becomes purple due to hyperemia, and serous fluid called the **vaginal transudate** seeps through the wall into the canal. Along with secretions of the greater vestibular glands, this moistens the vestibule and provides lubrication. The inner end of the vagina dilates and becomes cavernous, while the lower one-third of it constricts to form a narrow passage called the **orgasmic platform.** Thus it tightens on the penis, and combined with the vaginal rugae (friction ridges), this enhances stimulation and helps induce orgasm in both partners (fig. 28.17).

The uterus, which normally tilts forward over the urinary bladder, stands more erect during excitement and the cervix withdraws from the vagina. In plateau, the uterus is nearly vertical and extends into the false pelvis. This is called the **tenting effect.**

Although the vagina is the female copulatory organ, the clitoris is most comparable to the penis in structure, physiology, and importance as the primary focus of erotic stimulation. It has a high concentration of sensory nerve endings, which, by contrast, are almost absent from the vagina. Recall that the penis and clitoris are homologous structures, both arising from the phallus of the fetus. Both have a pair of corpora cavernosa with central (deep) arteries, and both become engorged by the same mechanism. The glans and shaft of the clitoris swell to two or three times their unstimulated size, but since the clitoris cannot swing upward away from the body like the penis, it tends to withdraw beneath the prepuce. Thrusting of the penis in the vagina tugs on the labia minora and, by extension, pulls on the prepuce and stimulates the clitoris. The clitoris may also be stimulated by pressure between the pubic symphyses of the partners.

The breasts also become congested and swollen during the excitement phase, and the nipples become erect. Stimulation of the breasts also enhances sexual arousal.

Orgasm

Late in plateau, many women experience involuntary pelvic thrusting, followed by 1 to 2 seconds of "suspension" or "stillness" preceding orgasm. Orgasm is commonly described as an intense sensation spreading from the clitoris through the pelvis, sometimes with pelvic throbbing and a spreading sense of warmth. The orgasmic platform gives three to five strong contractions about

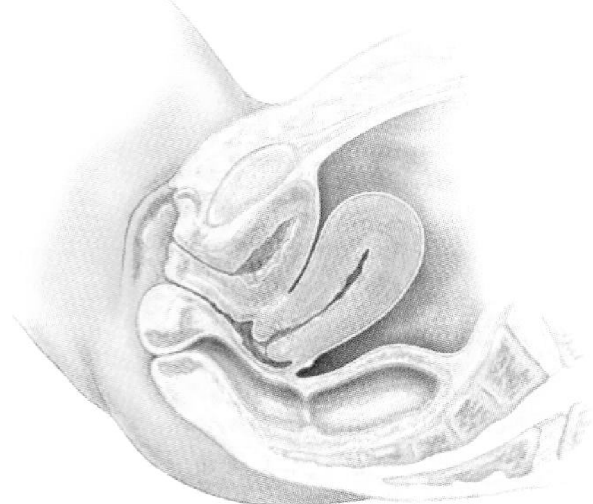

1. Unstimulated
Uterus tilts forward over urinary bladder; vagina relatively narrow; labia minora retracted

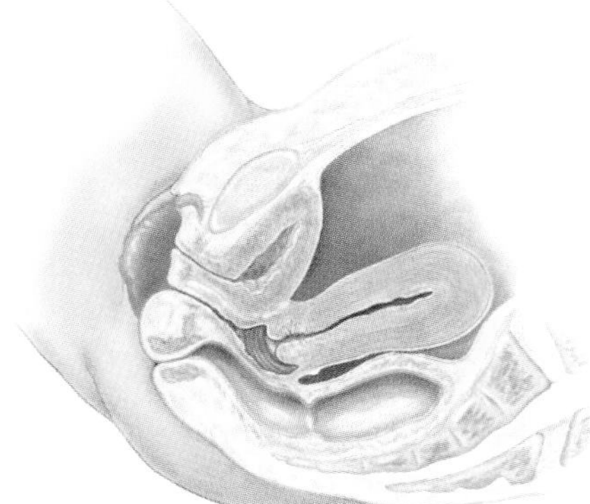

2. Excitement
Uterus stands more vertically; inner end of vagina dilates; labia minora vasocongested, may extend beyond labia majora; labia minora and vaginal mucosa reddened or violet due to hyperemia; vaginal transudate moistens vagina and vestibule

3. Plateau
Uterus is tented (vertical) and cervix is withdrawn from vagina; orgasmic platform (lower one-third) of vagina constricts penis; clitoris is engorged and its glans is withdrawn beneath prepuce; labia bright red or violet

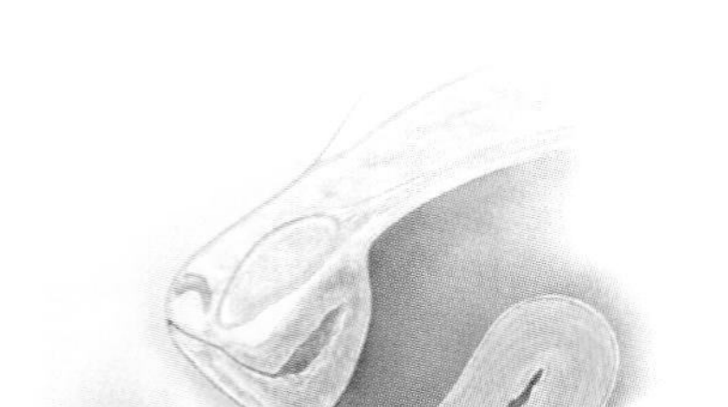

4. Orgasm
Orgasmic platform contracts rhythmically; uterus exhibits peristaltic contractions; anal and urinary sphincters constrict

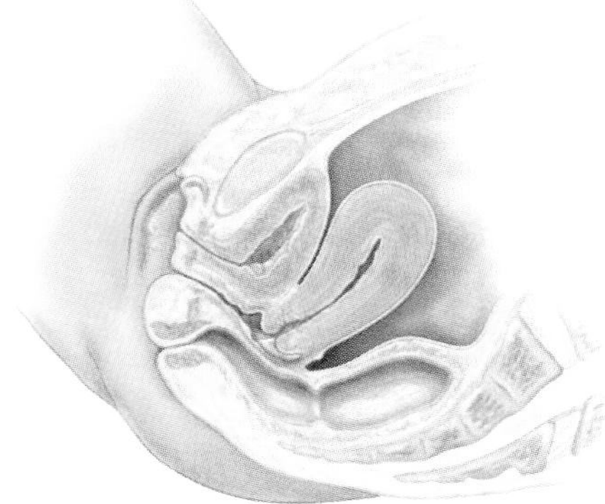

5. Resolution
Uterus returns to original position; cervix may dip into pool of semen; orgasmic platform relaxes; inner end of vagina constricts and returns to original dimensions

Figure 28.17 Stages of Female Sexual Response.

0.8 seconds apart, while the cervix plunges spasmodically into the vagina and into the pool of semen, should this be present. The uterus exhibits peristaltic waves of contraction; it is still debated whether or not this helps to draw semen from the vagina. The anal and urinary sphincters constrict, and the paraurethral glands, which are homologous to the prostate, sometimes expel fluid similar to prostatic fluid. Tachycardia and hyperventilation occur; the breasts enlarge still more and the areolae often become engorged; and in many women a reddish, rashlike flush appears on the lower abdomen, chest, neck, and face.

Resolution

During resolution the uterus drops forward to its resting position. The orgasmic platform quickly relaxes, while the inner end of the vagina returns more slowly to its normal dimensions. The flush disappears quickly and the areolae and nipples undergo rapid detumescence, but it may take 5 to 10 minutes for the breasts to return to their normal size. In many women (and men) there is a postorgasmic outbreak of perspiration. Unlike men, women do not have a refractory period and may quickly experience additional orgasms.

Before You Go On

Answer the following questions to test your understanding of the preceding section:

14. What are the female sources of lubrication in coitus?
15. What female tissues and organs become vasocongested?
16. Describe the actions of the uterus throughout the sexual response cycle.

Pregnancy and Childbirth

Objectives

When you have completed this section, you should be able to

- list the major hormones that regulate pregnancy, and explain their roles;
- describe a woman's adaptations to pregnancy;
- identify the physical and chemical stimuli that increase uterine contractility in late pregnancy;
- describe the mechanisms that produce labor contractions;
- name and describe the three stages of labor; and
- describe the physiological changes that occur in the weeks following childbirth.

This section treats pregnancy from the maternal standpoint—that is, adjustments of the woman's body to pregnancy and the mechanism of childbirth. Development of the fetus is described in the next chapter.

Gestation (pregnancy) lasts an average of 266 days from conception to childbirth, but the gestational calendar is usually measured from the first day of the woman's last menstrual period (LMP). Thus the birth is predicted to occur 280 days (about 40 weeks) from LMP. The duration of pregnancy, called its *term,* is commonly described with reference to 3-month intervals called **trimesters.**

Prenatal Development

A few fundamental facts of fetal development must be introduced here as a foundation for understanding maternal physiology. Fertilization must occur in the distal half of the uterine tube if it is to occur at all, since an unfertilized egg does not live long enough to reach the uterus alive. If fertilized, the egg divides five or six times before it reaches the uterus. All the products of conception—the embryo or fetus as well as the placenta and membranes associated with it—are collectively called the *conceptus.* The developing individual is called a *blastocyst* for most of the first 2 weeks, an *embryo* from 3 to 8 weeks, and a *fetus* from week 9 to birth. The exact meanings and distinctions among these terms are detailed in the next chapter. For the first 6 weeks after birth, the infant is called a *neonate.*[36]

Soon after it arrives in the uterus, the conceptus becomes a *blastocyst,* which consists of an inner cell mass destined to develop into the embryo and an outer cell mass, the *trophoblast,* which plays various supporting roles. About 4 days after arrival, the conceptus attaches to the endometrium by means of rootlike growths of the trophoblast. This process is called *implantation.* It is soon followed by development of the *placenta,* a disclike organ composed of a combination of maternal (uterine) and trophoblastic tissues. This has several functions, including fetal nutrition, waste disposal, and secretion of hormones that regulate pregnancy, mammary development, and fetal development. The placenta is connected to the fetus by way of the *umbilical cord.* The fetus is enclosed in a transparent sac called the *amnion* and floats in the *amniotic fluid* filling the sac. External to the amnion is a membrane called the *chorion* (CO-ree-on), which is the source of the placental hormones. In the seventh month of gestation, the fetus normally turns into a head-down *vertex position.* Consequently, most babies are born head first, the head acting as a wedge that widens the mother's cervix, vagina, and vulva during birth.

Hormones of Pregnancy

The hormones with the strongest influences on pregnancy are estrogens, progesterone, human chorionic gonadotropin, and human chorionic somatomammotropin. These are secreted primarily by the placenta, but the corpus luteum is an important source of hormones in the first 7 to 12 weeks. If the corpus luteum is removed before the seventh week, abortion almost always occurs. From weeks 7 to 17, the corpus luteum degenerates and the placenta takes over its endocrine functions.

36. *neo* = new + *nat* = born, birth

Human Chorionic Gonadotropin

Human chorionic gonadotropin (HCG) is secreted by the trophoblast. Its presence in the urine is the basis of pregnancy tests and can be detected with home testing kits as early as 8 or 9 days after conception. HCG secretion peaks around 10 to 12 weeks and then falls to a relatively low level for the rest of gestation (fig. 28.18). Like LH, it stimulates growth of the corpus luteum, which doubles in size and secretes increasing amounts of progesterone and estrogen. Without HCG, the corpus luteum would atrophy and the uterus would expel the conceptus.

Estrogens

Estrogen secretion increases to about 30 times normal by the end of gestation. The corpus luteum is an important source of estrogen for the first 12 weeks; after that, it comes mainly from the placenta. The adrenal glands of the mother and fetus secrete androgens, which the placenta converts to estrogens. The most abundant estrogen of pregnancy is estriol, but its effects are relatively weak; estradiol is less abundant but accounts for most of the estrogenic effects in pregnancy.

Estrogen stimulates tissue growth in the fetus and mother. It causes the mother's uterus and external genitalia to enlarge, the mammary ducts to grow, and the breasts to increase to nearly twice their former size. It makes the pubic symphysis more elastic and the sacroiliac joints more limber, so the pelvis widens during pregnancy and the pelvic outlet expands during childbirth.

Progesterone

The placenta secretes a great deal of progesterone, and early in the pregnancy, so does the corpus luteum. Progesterone and estrogen suppress pituitary secretion of FSH and LH, thereby preventing more follicles from developing during pregnancy. (This is the basis for contraceptive pills and implants; see clinical insight 28.4, p. 1077.) Progesterone also suppresses uterine contractions so the conceptus is not prematurely expelled. It prevents menstruation and promotes the proliferation of *decidual cells* of the endometrium, on which the blastocyst feeds. Once estrogen has stimulated growth of the mammary ducts, progesterone stimulates development of the secretory acini—another step toward lactation.

Human Chorionic Somatomammotropin

The amount of **human chorionic somatomammotropin (HCS)** secreted in pregnancy is several times that of all the other hormones combined, yet its function is the least understood. The placenta begins secreting HCS around the fifth week and HCS output increases steadily from then until term, in direct proportion to the size of the placenta.

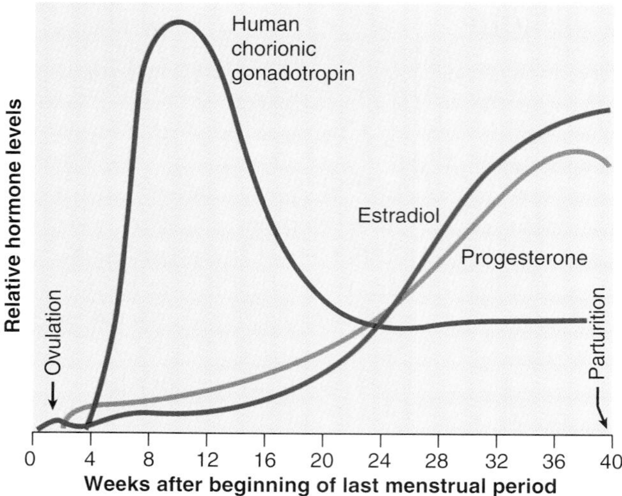

Figure 28.18 Hormone Levels Over the Course of Pregnancy.

HCS is sometimes called *human placental lactogen* because, in other mammals, it causes mammary development and lactation; however, it does not induce lactation in humans. Its effects seem similar to those of growth hormone, but weaker. It also seems to reduce the mother's insulin sensitivity and glucose usage such that the mother consumes less glucose and leaves more of it for use by the fetus. HCS promotes the release of free fatty acids from the mother's adipose tissue, providing an alternative energy substrate for her cells to use in lieu of glucose.

Other Hormones

Pregnancy affects many other aspects of endocrine function. A woman's pituitary gland grows about 50% larger during pregnancy and produces markedly elevated levels of thyrotropin, prolactin, and ACTH. The thyroid gland also becomes about 50% larger under the influence of HCG, pituitary thyrotropin, and *human chorionic thyrotropin* from the placenta. Elevated thyroid hormone secretion increases the metabolic rate of the mother and fetus. The parathyroid glands enlarge and stimulate osteoclast activity, liberating calcium from the mother's bones for fetal use. ACTH stimulates glucocorticoid secretion, which may serve primarily to mobilize amino acids for fetal protein synthesis. Aldosterone secretion rises and promotes fluid retention, contributing to the mother's increased blood volume. The corpus luteum and placenta secrete *relaxin,* which relaxes the pubic symphysis in other animals but does not seem to have this effect in humans. It is thought by some to soften the cervix in preparation for childbirth.

The hormones of pregnancy are summarized in table 28.2.

Table 28.2 The Hormones of Pregnancy

Hormone	Effects
Human chorionic gonadotropin (HCG)	Prevents involution of corpus luteum and stimulates its growth and secretory activity; basis of pregnancy tests
Estrogens	Stimulate maternal and fetal tissue growth, including enlargement of uterus and maternal genitalia; stimulate development of mammary ducts; soften pubic symphysis and sacroiliac joints, facilitating pelvic expansion in pregnancy and childbirth; suppress FSH and LH secretion
Progesterone	Suppresses premature uterine contractions; prevents menstruation; stimulates proliferation of decidual cells, which nourish embryo; stimulates development of mammary acini; suppresses FSH and LH secretion
Human chorionic somatomammotropin (HCS)	Weak growth-stimulating effects similar to growth hormone; glucose-sparing effect on mother, making glucose more available to fetus; mobilization of fatty acids as maternal fuel
Pituitary thyrotropin	Stimulates thyroid activity and metabolic rate
Human chorionic thyrotropin	Same effect as pituitary thyrotropin
Parathyroid hormone	Stimulates osteoclasts and mobilizes maternal calcium
Adrenocorticotropic hormone	Stimulates glucocorticoid secretion; thought to mobilize amino acids for fetal protein synthesis
Aldosterone	Causes fluid retention, contributing to increased maternal blood volume
Relaxin	Uncertain function; may soften the cervix to facilitate childbirth

Adjustments to Pregnancy

Pregnancy places a considerable stress on a woman's body and requires adjustments in nearly all the organ systems. A few of the major adjustments and effects of pregnancy are described here.

Digestive System, Nutrition, and Metabolism

For many women, one of the first signs of pregnancy is morning sickness—nausea, especially after rising from bed, in the first few months of gestation. The cause of morning sickness is unknown. One hypothesis is that it stems from the reduced intestinal motility caused by the steroids of pregnancy. Another is that it is an evolutionary adaptation to protect the fetus from toxins. The fetus is most vulnerable to toxins at the same time that morning sickness peaks. Women with morning sickness tend to prefer bland foods and to avoid spicy and pungent foods, which are highest in toxic compounds. In some women, the nausea progresses to vomiting.

Constipation and heartburn are common in pregnancy. The former is another result of reduced intestinal motility. The latter is due to the enlarging uterus pressing upward on the stomach, causing the reflux of gastric contents into the esophagus.

The basal metabolic rate rises about 15% in the second half of gestation. Pregnant women often feel overheated because of this and the effort of carrying the extra weight. The appetite may be strongly stimulated, but a pregnant woman needs only 300 extra Calories per day even in the last trimester. With poor prenatal care and little self-control, however, some women greatly overeat and gain as much as 34 kg (75 lb) of weight compared with a healthy average of 11 kg (24 lb). Maternal nutrition should emphasize the quality of food eaten, not quantity.

During the last trimester, the fetus needs more nutrients than the mother's digestive tract can absorb. In preparation for this, the placenta stores nutrients early in gestation and releases them in the final trimester. The demand is especially high for protein, iron, calcium, and phosphates. A pregnant woman needs an extra 600 mg of iron for her own hemopoiesis and 375 mg for the fetus. She is likely to become anemic if she does not ingest enough iron during late pregnancy. Supplemental vitamin K is often given late in pregnancy to promote prothrombin synthesis in the fetus. This reduces the risk of neonatal hemorrhage, especially in the brain, caused by the stresses of birth. Supplemental folic acid reduces the risk of neurological disorders in the fetus, such as spina bifida (see clinical insight 14.1, p. 495) and anencephaly (failure of the cerebrum, cerebellum, and calvaria to develop). A vitamin D supplement helps to ensure adequate calcium absorption to meet fetal demands.

Circulatory System

By full term, the placenta requires about 625 mL of blood per minute from the mother. The mother's blood volume rises about 30% during pregnancy because of fluid retention and hemopoiesis; she eventually has about 1 to 2 L of extra blood. Cardiac output rises about 30% to 40% above normal by 27 weeks, but for unknown reasons, it falls almost to normal in the last 8 weeks. As the pregnant uterus puts pressure on the large pelvic blood vessels, it interferes with venous return from the legs and pelvic region. This can result in hemorrhoids and varicose veins.

Respiratory System

Minute ventilation increases about 50% during pregnancy for two reasons: (1) Oxygen demands are about 20% higher by late pregnancy in order to supply the fetus and support the woman's increased metabolic rate. (2) Progesterone increases the sensitivity of her respiratory chemoreceptors to carbon dioxide, and ventilation is adjusted to keep her arterial P_{CO_2} lower than normal. While there is a demand for increased ventilation, the expanding uterus pushes the abdominal viscera up against the diaphragm and interferes with breathing. Consequently, the respiratory rate increases to compensate for the lack of depth. Pressure on the diaphragm may be great enough to cause breathing difficulty (dyspnea) by late pregnancy. In the last month, however, the pelvis usually expands enough for the fetus to drop lower in the abdominopelvic cavity, taking some pressure off the diaphragm and allowing the woman to breathe more easily.

Urinary System

Aldosterone and the steroids of pregnancy promote water and salt retention by the kidneys. Nevertheless, the glomerular filtration rate increases by 50% and urine output is slightly elevated. This enables a woman to dispose of both her own and the fetus's metabolic wastes. As the pregnant uterus compresses the bladder and reduces its capacity, urination becomes more frequent and some women experience uncontrollable leakage of urine, or *incontinence.*

Integumentary System

The skin must grow to accommodate expansion of the abdomen and breasts and the added fat deposition in the hips and thighs. Stretching of the dermis often tears the connective tissue and causes *striae,* or *stretch marks.* These appear reddish at first but fade to a silvery white color after pregnancy. Melanocyte activity increases in some areas and causes darkening of the areolae and linea alba. The latter often becomes a dark line, the **linea nigra**[37] (LIN-ee-uh NY-gruh), from the umbilical to the pubic region. Some women also acquire a temporary blotchy darkening of the skin over the nose and cheeks called the "mask of pregnancy," or **chloasma**[38] (clo-AZ-muh), which usually disappears when the pregnancy is over.

37. *linea* = line + *nigra* = black
38. *chloasma* = to be green

Uterine Growth and Weight Gain

The uterus weighs about 50 g when a woman is not pregnant and about 900 g by the end of pregnancy. Its growth is monitored by palpating the fundus, which eventually reaches almost to the xiphoid process (fig. 28.19). Table 28.3 shows how the weight gained in pregnancy is distributed.

Childbirth

The process of giving birth is called **parturition** (PAR-too-RISH-un). The ancients thought that the fetus kicks against the uterus and pushes itself out head first. The fetus, however, is a rather passive player in its own birth; its expulsion is achieved only by the contractions of the mother's uterine and abdominal muscles. Yet there is evidence that the fetus may play some role in its birth by chemically stimulating labor contractions and perhaps even sending chemical messages that signify when it is ready to be born.

Uterine Contractility

Over the course of gestation, the uterus exhibits relatively weak **Braxton Hicks**[39] **contractions.** These become stronger in late pregnancy and often send women rushing to the hospital with "false labor." At term, however, these contractions transform suddenly into the more powerful **labor contractions.**

Progesterone and estrogen balance may be one factor in this pattern of increasing contractility. Both hormone levels increase over the course of gestation. Progesterone inhibits uterine contractions, but its secretion levels off or declines slightly after 6 months, while estrogen secretion continues to rise (see fig. 28.18). Estrogen stimulates uterine contractions and may be a factor in the irritability of the uterus in late pregnancy.

Also, as the pregnancy nears full term, the posterior pituitary releases more oxytocin (OT) and the uterus produces more OT receptors. Oxytocin promotes labor in two ways: (1) it directly stimulates muscle of the myometrium, and (2) it stimulates the fetal membranes to secrete prostaglandins, which are synergists of OT in producing labor contractions. Labor is prolonged if OT or prostaglandins are lacking, and it may be induced or accelerated by giving a vaginal prostaglandin suppository or an intravenous OT "drip." The conceptus itself may produce chemical stimuli promoting its own birth. Fetal cortisol secretion rises in late pregnancy and may enhance estrogen secretion by the placenta. The fetal pituitary gland also produces oxytocin, which does not enter the maternal circulation but may stimulate the fetal membranes to secrete prostaglandins.

39. John Braxton Hicks (1823–97), British gynecologist

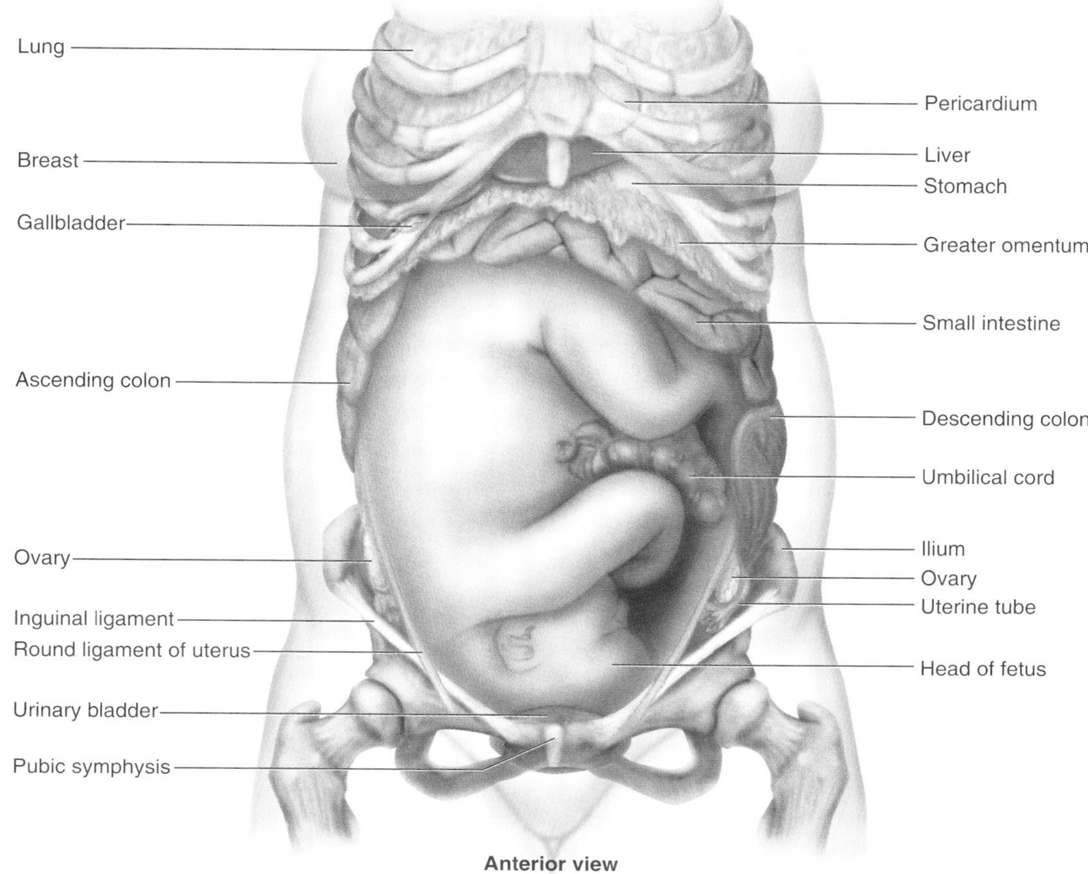

Figure 28.19 The Full-Term Fetus in Vertex Position.

Table 28.3 Distribution of Weight Gain in Pregnancy

Fetus	3 kg (7 lb)
Placenta, fetal membranes, and amniotic fluid	1.8 kg (4 lb)
Blood and tissue fluid	2.7 kg (6 lb)
Fat	1.4 kg (3 lb)
Uterus	0.9 kg (2 lb)
Breasts	0.9 kg (2 lb)
Total	**11 kg (24 lb)**

Uterine stretching is also thought to play a role in initiating labor. Stretching any smooth muscle increases its contractility, and movements of the fetus produce the sort of intermittent stretch that is especially stimulatory to the myometrium. Twins are born an average of 19 days earlier than single infants, probably because of the greater stretching of the uterus. When the fetus is in the *vertex position* (head down), its head pushes against the cervix, which is especially sensitive to stretch.

Labor Contractions

Labor contractions begin about 30 minutes apart. As labor progresses, they become more intense and eventually occur every 1 to 3 minutes. It is important that they be intermittent rather than one long, continual contraction. Each contraction sharply reduces maternal blood flow to the placenta, so the uterus must periodically relax to restore flow and oxygen delivery to the fetus. Contractions are strongest in the fundus and body of the uterus and weaker near the cervix, thus pushing the fetus downward.

According to the **positive feedback theory of labor,** labor contractions are induced by stretching of the cervix. This triggers a reflex contraction of the uterine body that pushes the fetus downward and stretches the cervix still more. Thus there is a self-amplifying cycle of stretch and contraction. In addition, cervical stretching induces a neuroendocrine reflex through the spinal cord, hypothalamus, and posterior pituitary. The posterior pituitary releases oxytocin, which is carried in the blood and stimulates the uterine muscle both directly and through the action of prostaglandins. This, too, is a positive feedback cycle: cervical stretching → oxytocin secretion → uterine contraction → cervical stretching (see fig. 1.15, p. 21).

As labor progresses, a woman feels a growing urge to "bear down." A reflex arc extends from the uterus to the spinal cord and back to the skeletal muscles of the abdomen. Contraction of these muscles—partly reflexive and partly voluntary—aids in expelling the fetus.

The pain of labor is due at first mainly to ischemia of the myometrium—muscle hurts when deprived of blood, and each labor contraction temporarily restricts uterine circulation. As the fetus enters the vaginal canal, the pain becomes stronger because of increased stretching of the cervix, vagina, and perineum and sometimes the tearing of vaginal tissue. At this stage the obstetrician may perform an *episiotomy*—an incision in the vulva to widen the vaginal canal and prevent random tearing. The pain of human childbirth, compared to the relative ease with which other mammals give birth, is an evolutionary product of two factors: the unusually large brain and head of the human infant, and the narrowing of the pelvic outlet, which adapted hominids to bipedal locomotion (see p. 301).

Stages of Labor

Labor occurs in three stages. The duration of each stage tends to be longer in a **primipara** (a woman giving birth for the first time) than in a **multipara** (a woman who has previously given birth).

Dilation (First) Stage

This is the longest stage, lasting 8 to 24 hours in a primipara but as little as a few minutes in a multipara. It is marked by the **dilation** (widening) of the cervical canal and by **effacement** (thinning) of the cervix (fig. 28.20*a, b*). The cervix reaches a maximum diameter of about 10 cm (the diameter of the baby's head). During dilation, the fetal membranes usually rupture and the amniotic fluid is discharged (the "breaking of the waters").

Expulsion (Second) Stage

This stage typically lasts about 30 minutes in a primipara and as little as 1 minute in a multipara. It begins when the

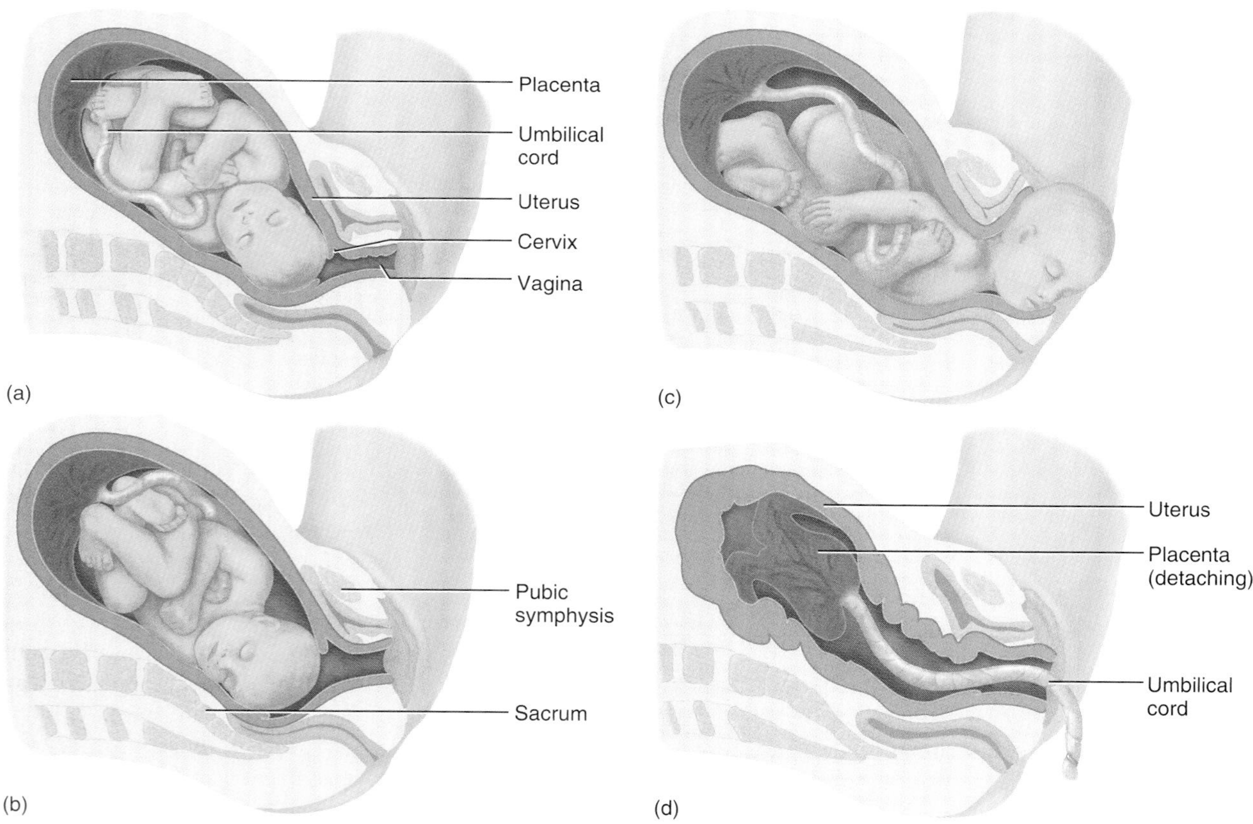

Figure 28.20 **The Stages of Childbirth.** (*a*) Early dilation, (*b*) late dilation, (*c*) expulsion, and (*d*) the placental stage.

baby's head enters the vagina and lasts until the baby is entirely expelled (fig. 28.20*c*). The baby is said to be **crowning** when the top of its head is visible, stretching the vulva (fig. 28.21*a*). Delivery of the head is the most difficult part, with the rest of the body following much more easily. An episiotomy may be performed during this stage. Women in labor employ the *Valsalva maneuver*—holding the breath while contracting the abdominal muscles (see p. 858)— to help expel the infant. An attendant often uses a suction bulb to remove mucus from the baby's mouth and nose even before it is fully delivered. When the baby is fully expelled, the umbilical cord is clamped in two places and cut between the clamps.

Placental (Third) Stage

The uterus continues to contract after expulsion of the baby. The placenta, however, is a nonmuscular organ that cannot contract, so it buckles away from the uterine wall (see fig. 28.20*d*). About 350 mL of blood is typically lost at this stage, but contractions of the myometrium compress the blood vessels and prevent more extensive bleeding. The placenta, amnion, and other fetal membranes are expelled by uterine contractions, which may be aided by a gentle pull on the umbilical cord. The membranes (*afterbirth*) must be carefully inspected to be sure everything has been expelled (fig. 28.21*c*). If any of these structures remain in the uterus, they can cause postpartum hemorrhaging. The umbilical blood vessels are counted because an abnormal number in the cord may indicate cardiovascular abnormalities in the infant.

Puerperium

The first 6 weeks **postpartum** (after birth) are called the **puerperium**[40] (PYU-er-PEER-ee-um), a period in which the mother's anatomy and physiology stabilize and the reproductive organs return nearly to the pregravid state (their condition prior to pregnancy). The shrinkage of the uterus during this period is called **involution.** In a lactating woman, it loses about 50% of its weight in the first week and is nearly at its pregravid weight in 4 weeks. Involution is achieved through the **autolysis** (self-digestion) of uterine cells by their own lysosomal enzymes. For about 10 days, this produces a vaginal discharge called **lochia,** which is bloody at first and then turns clear and serous. Breast-feeding promotes involution because (1) it suppresses estrogen secretion, which would otherwise cause the uterus to remain more flaccid; and (2) it stimulates oxytocin secretion, which causes the myometrium to contract and firm up the uterus sooner. It is important for the puerperium to be undisturbed, as emotional upset can suppress lactation in some women.

40. *puer* = child + *per*, from *par* = birth

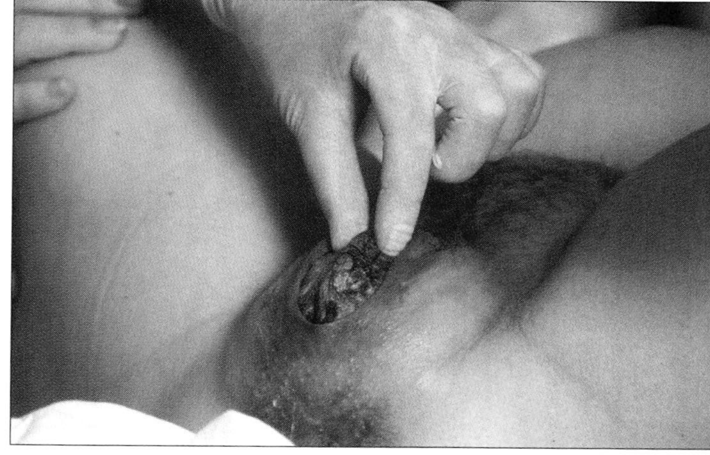

(a)

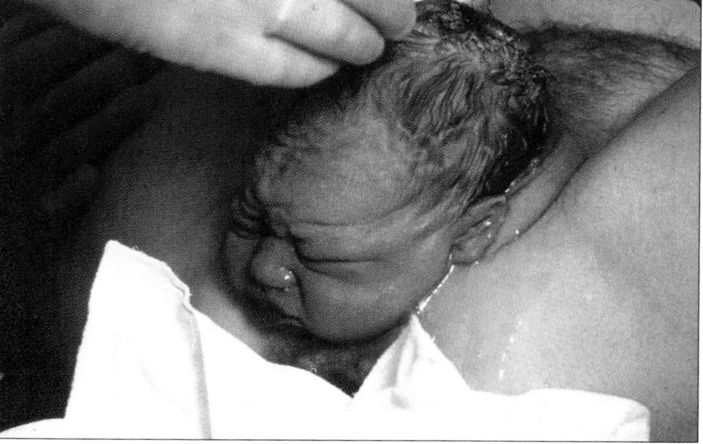

(b)

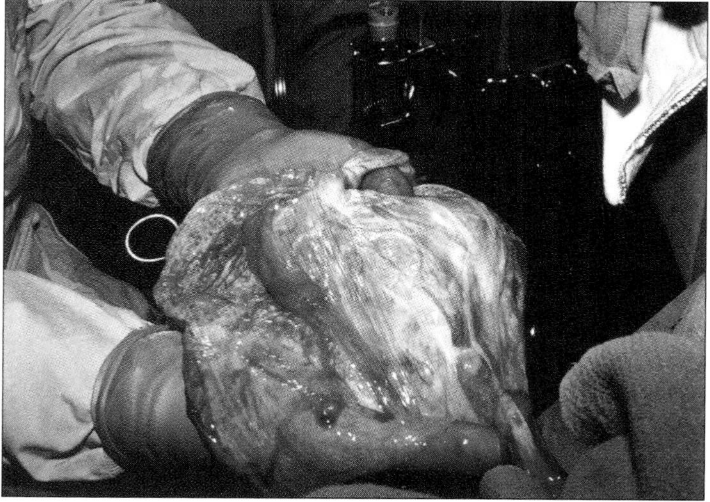

(c)

Figure 28.21 Childbirth. (*a*) Crowning. The baby's head, between the attendant's fingers, has begun to dilate the vulva. (*b*) Emergence of the head. (*c*) The placental and fetal membranes (afterbirth).

Before You Go On

Answer the following questions to test your understanding of the preceding section:

17. List the roles of HCG, estrogen, progesterone, and HCS in pregnancy.
18. What is the role of the corpus luteum in pregnancy? What eventually takes over this role?
19. List and briefly explain the special nutritional requirements of pregnancy.
20. How much weight does the average woman gain in pregnancy? What contributes to this weight gain other than the fetus?
21. Describe the positive feedback theory of labor.
22. What major events define the three stages of labor?

Lactation

Objectives

When you have completed this section, you should be able to

- describe development of the breasts in pregnancy;
- describe the shifting hormonal balance that regulates the onset and continuation of lactation;
- describe the mechanism of milk ejection;
- contrast colostrum with breast milk; and
- discuss the benefits of breast-feeding.

La Leche League, an organization that promotes breast-feeding, reports that worldwide, the average woman breast-feeds her children until the age of 4.2 years. **Lactation,** the synthesis and ejection of milk from the mammary glands, can continue indefinitely as long as the breast is stimulated by a nursing infant or a mechanical device (breast pump).

Development of the Mammary Glands in Pregnancy

The high estrogen level in pregnancy causes the ducts of the mammary glands to grow and branch extensively. Growth hormone, insulin, glucocorticoids, and prolactin also contribute to this development. Once the ducts are complete, progesterone stimulates the budding and development of acini at the ends of the ducts. The fully developed mammary gland is of the compound tubuloacinar type. The acini are organized into grapelike clusters (lobules) within each lobe of the breast (see fig. 28.9).

Colostrum and Milk Synthesis

In late pregnancy, the mammary acini and ducts are distended with a secretion called **colostrum.** This is similar to breast milk in protein and lactose content but contains about one-third less fat. It is the infant's only natural source of nutrition for the first 1 to 3 days postpartum. Colostrum has a thin watery consistency and a cloudy yellowish color. The amount of colostrum secreted per day is at most 1% of the amount of milk secreted later, but since infants are born with excess body water and ample fat, high calorie and fluid intake are not required at first. A major benefit of colostrum is that it contains immunoglobulins, especially IgA. IgA resists digestion and may protect the infant from gastroenteritis. It is also thought to be pinocytosed by the small intestine and to confer wider, systemic immunity to the neonate.

Milk synthesis is promoted by prolactin, a hormone of the anterior pituitary gland. In the nonpregnant state, prolactin secretion is suppressed by dopamine (= prolactin-inhibiting factor) from the hypothalamus. Prolactin secretion begins 5 weeks into the pregnancy, and by full term it is 10 to 20 times its normal level. Even so, prolactin has little effect on the mammary glands until after birth. While the steroids of pregnancy prepare the mammary glands for lactation, they antagonize the effect of prolactin and suppress milk synthesis. When the placenta is discharged at birth, the steroid levels abruptly drop and allow prolactin to have a stronger effect. Milk is synthesized in increasing quantity over the following week. Milk synthesis also requires the action of growth hormone, cortisol, insulin, and parathyroid hormone to mobilize the necessary amino acids, fatty acids, glucose, and calcium.

At the time of birth, baseline prolactin secretion drops to the nonpregnant level. Every time the infant nurses, however, it jumps to 10 to 20 times this level for the next hour and stimulates the synthesis of milk for the next feeding (fig. 28.22). These prolactin surges are accompanied by smaller increases in estrogen and progesterone secretion. If the mother does not nurse or these hormone surges are absent (due to pituitary damage, for

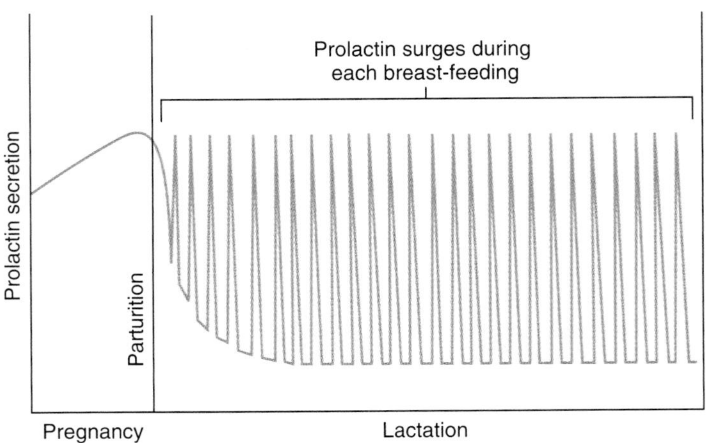

Figure 28.22 Prolactin Secretion in the Lactating Female.

example), the mammary glands stop producing milk in about a week. Even if she does nurse, the rate of milk production declines after 7 to 9 months, although a woman can continue to lactate for years.

Only 5% to 10% of women become pregnant again while nursing an infant full time. Apparently, either prolactin or nerve signals from the breast inhibit GnRH secretion, which, in turn, results in reduced gonadotropin secretion and ovarian cycling. This mechanism may have evolved as a natural means of spacing births, but breast-feeding is not a reliable means of contraception. Even in women who breast-feed, the ovarian cycle sometimes resumes several months postpartum. In those who do not breast-feed, the cycles resume in a few weeks, but for the first 6 months they are usually anovulatory (no egg is ovulated).

Milk Ejection

Milk is continually secreted into the mammary acini, but it does not easily flow into the ducts. Its flow into the ducts, called **milk ejection** (let-down), is controlled by a neuroendocrine reflex. The infant's suckling stimulates nerve endings of the nipple and areola, which in turn signal the hypothalamus and posterior pituitary to release oxytocin. Oxytocin stimulates **myoepithelial cells,** which form a basketlike mesh around each gland acinus (fig. 28.23). These cells are of epithelial origin, but are packed with actin and contract like smooth muscle to squeeze milk from the acinus into the duct. The infant does not get any milk for the first 30 to 60 seconds of suckling, but milk soon fills the ducts and lactiferous sinuses and is then easily sucked out.

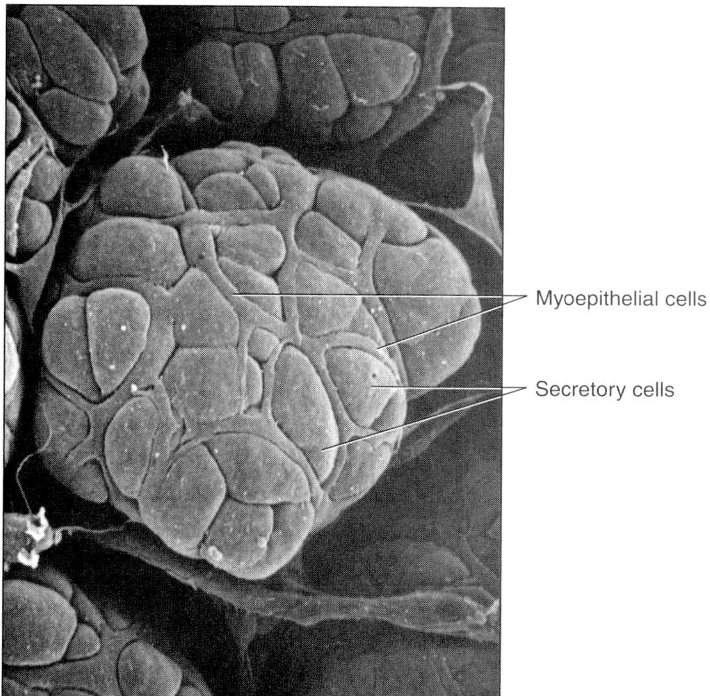

Figure 28.23 Acinus of a Mammary Gland. Myoepithelial cells can be seen forming a mesh around the secretory cells. The myoepithelial cells respond to oxytocin by contracting and forcing milk from the acinus into the duct.

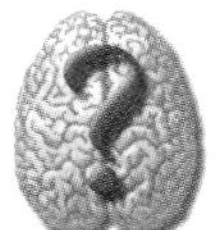

Think About It

When a woman is nursing her baby at one breast, would you expect only that breast, or both breasts, to eject milk? Explain why.

Breast Milk

Table 28.4 compares the composition of colostrum, human milk, and cow's milk. Breast milk changes composition over the first 2 weeks, varies from one time of day to another, and changes even during the course of a single feeding. For example, at the end of a feeding there is less lactose and protein in the milk, but six times as much fat, as there is at the beginning.

Cow's milk is not a good substitute for human milk. It has one-third less lactose but three to five times as much protein and minerals. The excess protein forms a harder curd in the infant's stomach, so cow's milk is not digested and absorbed as efficiently as mother's milk. It also increases the infant's nitrogenous waste excretion, which increases the incidence and severity of diaper rash, particularly as bacteria in the diaper break urea down to ammonia.

Colostrum and milk have a laxative effect that helps to clear the neonatal intestine of *meconium,* a greenish black, sticky fecal matter composed of bile, epithelial cells, and other wastes that accumulated during fetal development. By clearing bile and bilirubin from

Table 28.4 A Comparison of Colostrum, Human Milk, and Cow's Milk*

	Human Colostrum	Human Milk	Cow's Milk
Total Protein (g/L)	22.9	10.6	30.9
Lactalbumin (g/L)	—	3.7	25.0
Casein (g/L)	—	3.6	2.3
Immunoglobulins (g/L)	19.4	0.09	0.8
Fat (g/L)	29.5	45.4	38.0
Lactose (g/L)	57	71	47
Calcium (mg/L)	481	344	1370
Phosphorus (mg/L)	157	141	910

*Colostrum data are for the first day postpartum, and human milk data are for "mature milk" at ≈15 days postpartum.

the system, breast-feeding also reduces the incidence and degree of jaundice in neonates. Breast milk promotes colonization of the neonatal intestine with beneficial bacteria and continues to supply antibodies that lend protection against infection by pathogenic bacteria. Breast-feeding also tends to promote a closer bond between mother and infant.

A woman nursing one baby eventually produces about 1.5 L of milk per day; women with twins produce more. Lactation places a great metabolic demand on the mother. It is equivalent to losing 50 g of fat, 100 g of lactose (made from her blood glucose), and 2 to 3 g of calcium phosphate per day. A woman is at greater risk of bone loss when breast-feeding than when she is pregnant, because much of the infant's skeleton is still cartilage at birth and becomes mineralized at her expense in the first year postpartum. If a nursing mother does not have enough calcium and vitamin D in her own diet, lactation stimulates parathyroid hormone secretion and osteoclast activity, taking calcium from her bones to supply her baby.

To conclude this chapter, table 28.5 briefly describes some of the common disorders of pregnancy. Other reproductive disorders are discussed elsewhere: cervical cancer in clinical insight 28.1, breast cancer on page 1055, and sexually transmitted diseases at the end of the previous chapter.

Before You Go On

Answer the following questions to test your understanding of the preceding section:

23. Why is little or no milk secreted while a woman is pregnant?
24. How does a lactating breast differ from a nonlactating breast in structure? What stimulates these differences to develop during pregnancy?
25. What is colostrum and what is its significance?
26. How does suckling stimulate milk ejection?
27. Why is breast milk superior to cow's milk for an infant?

Table 28.5 Some Disorders of Pregnancy

Abruptio Placentae[41]	Premature separation of the placenta from the uterine wall, often associated with pre-eclampsia or cocaine use. May require birth by cesarian section.
Ectopic[42] *Pregnancy*	Implantation of the conceptus anywhere other than the uterus; usually starts in the uterine tube *(tubal pregnancy)* and may progress to *abdominal pregnancy* if the tube ruptures. See clinical insight 29.2 in the next chapter for further details.
Gestational Diabetes	A form of diabetes mellitus that develops in about 1% to 3% of pregnant women, characterized by insulin insensitivity, hyperglycemia, glycosuria, and a risk of excessive fetal size and birth trauma. Glucose metabolism often returns to normal after delivery of the infant, but 40% to 60% of women with gestational diabetes develop diabetes mellitus within 15 years after the pregnancy.
Hyperemesis Gravidarum[43]	Prolonged vomiting, dehydration, alkalosis, and weight loss in early pregnancy, often requiring hospitalization to stabilize fluid, electrolyte, and acid-base balance; sometimes associated with liver damage.
Placenta Previa[44]	Blockage of the cervical canal by the placenta, preventing birth of the infant before the placenta separates from the uterus. Requires birth by cesarian section.
Pre-eclampsia[45] *(toxemia of pregnancy)*	Rapid onset of hypertension and edema, swelling especially of the face and hands; proteinuria and reduced glomerular filtration rate; increased blood clotting; sometimes with headaches, visual disturbances, and small cerebral infarctions. Seen in about 4% of pregnancies, especially in the third trimester of women pregnant for the first time. Can progress to *eclampsia,* with seizures and widespread vascular spasms sometimes fatal to the mother, fetus, or both. Eclampsia usually occurs shortly before or after parturition.
Spontaneous Abortion	Occurs in 10% to 15% of pregnancies, usually because of fetal deformities or chromosomal abnormalities incompatible with survival, but may also result from maternal abnormalities, infectious disease, and drug abuse.

41. *ab* = away + *rupt* = to tear + *placentae* = of the placenta
42. *ec* = out of + *top* = place
43. *hyper* = excessive + *emesis* = vomiting + *gravida* = pregnant woman
44. *pre* = before + *via* = the way (obstructing the way)
45. *ec* = forth + *lampsia* = shining

Clinical Insight 28.4

Methods of Contraception

The term *contraception* is used here to mean any procedure or device intended to prevent pregnancy (the presence of an implanted conceptus in the uterus). This essay describes the most common methods of contraception, some issues involved in choosing among them, and the relative reliability of the various methods. Several of those options are shown in figure 1.

Behavioral Methods

Abstinence (refraining from intercourse) is, obviously, a completely reliable method if used consistently. The *rhythm method (periodic abstinence)* is based on avoiding intercourse near the time of expected ovulation. Among typical users, it has a 25% failure rate, partly due to lack of restraint and partly because it is difficult to predict the exact date of ovulation. Intercourse must be avoided for at least 48 hours before ovulation so there will be no surviving sperm in the reproductive tract when the egg is ovulated, and for at least 24 hours after ovulation so there will be no fertile egg present when sperm are introduced. The rhythm method is valuable, however, for couples who are trying to conceive a child by having intercourse at the time of apparent ovulation.

Withdrawal (coitus interruptus) requires the male to withdraw the penis before ejaculation. This often fails because of lack of willpower, because some sperm are present in the preejaculatory fluid, and because sperm ejaculated anywhere in the female genital region can potentially get into the reproductive tract.

Barrier and Spermicidal Methods

Barrier methods are designed to prevent sperm from getting into or beyond the vagina. They are most effective when used with chemical *spermicides*, available as over-the-counter foams, creams, and jellies.

The *male condom* is a sheath of latex, rubber, or animal membrane (lamb intestine) that is unrolled over the erect penis and collects the semen. It is inexpensive, convenient, and very reliable when used carefully. About 25% of American couples who use contraceptives use only condoms, which rank second to birth-control pills in popularity.

The *female condom* is less used. It is a polyurethane sheath with a flexible ring at each end. The inner ring fits over the cervix and the outer ring covers the external genitalia. Male and female condoms are the only contraceptives that also protect against disease transmission. Animal membrane condoms, however, are porous to HIV and hepatitis B viruses and do not afford dependable protection from disease.

The *diaphragm* is a latex or rubber dome that is placed over the cervix to block sperm transmission. It requires a physical examination and prescription to ensure proper fit, but is otherwise comparable to the condom in convenience and reliability, provided it is used with a spermicide. Without a spermicide, it is not very effective.

Hormonal Methods

Birth-control pills are composed of estrogen and progesterone. They mimic the negative feedback effect of ovarian hormones on FSH secretion, thus preventing follicle development and ovulation. They are effective for most women with minimal complications, but they can increase the risk of heart attack or stroke in smokers and in women with a history of diabetes, hypertension, or clotting disorders. Birth-control pills are the most widely used contraceptives in the United States. Efforts to develop a birth-control pill for men have so far been unsuccessful.

Norplant is a porous silicone tube of synthetic progesterone. Six tubes the size of matchsticks are inserted under a woman's skin, usually on the upper arm. By slowly releasing progesterone, these tubes suppress ovulation and prevent pregnancy for up to 5 years. Even if ovulation should occur, Norplant interferes with egg transport by altering the cilia in the uterine tube and it makes the endometrium unreceptive to implantation.

Depo-Provera is a synthetic progesterone administered by injection two to four times per year. It provides highly reliable, long-term contraception without the need of daily pills or implants, although in some women it causes headaches, nausea, or weight gain, and fertility may not return immediately when its use is discontinued.

Surgical Sterilization

People who are confident that they do not want more children (or any) often elect to be surgically sterilized. This entails the cutting and tying or clamping of the genital ducts, thus blocking the passage of sperm or eggs. Surgical sterilization has the advantage of convenience, since it requires no further attention. Its initial cost is higher, however, and for people who later change their minds, surgical reversal is much more expensive than the original procedure and is often unsuccessful. *Vasectomy* is the severing of the ductus (vas) deferens, done through a small incision in the back of the scrotum. In *tubal ligation*,[46] the uterine tubes are cut. This can be done through small abdominal incisions to admit a cutting instrument and laparoscope (viewing device).

Preventing Implantation

The *intrauterine device (IUD)* is a springy plastic device inserted through the cervical canal into the uterus and left in place for 1 to 4 years. It prevents the implantation of a blastocyst in the uterine wall. Some IUDs were removed from the market because they caused serious complications such as uterine perforation, and many other models were removed for fear of legal liability. Only two T-shaped models are currently available in the United States.

Mifepristone (RU486), the "morning after pill," can be taken after intercourse has occurred around the expected time of ovulation and any time within 10 days of when the next projected menstrual period would ordinarily occur. By blocking the action of progesterone, it induces menstruation and the discharge of any conceptus that might have formed. Mifepristone has been widely used in France without complications, but its availability in the United States remains stalled by political controversy.

Continued

46. *ligat* = to tie

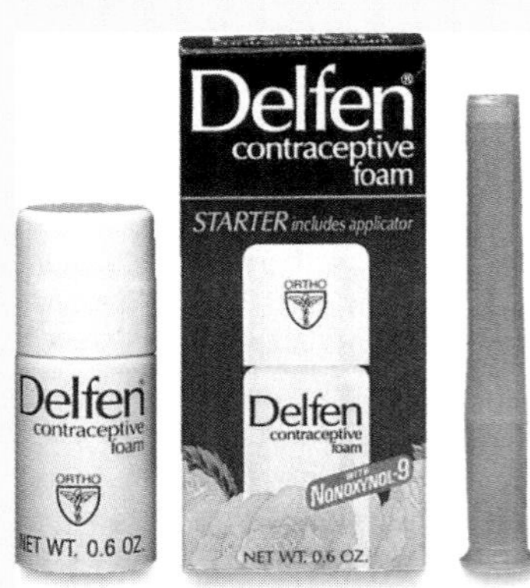

Contraceptive foam with vaginal applicator

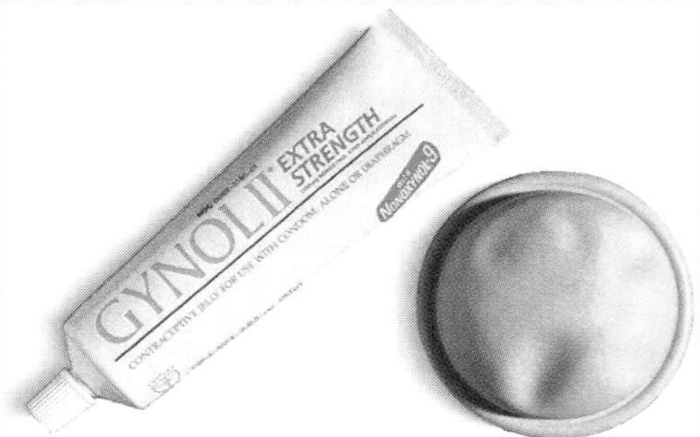

Diaphragm with contraceptive jelly

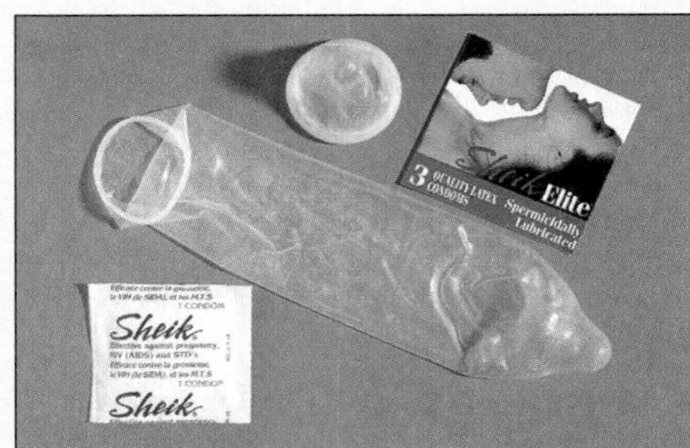

Male condom

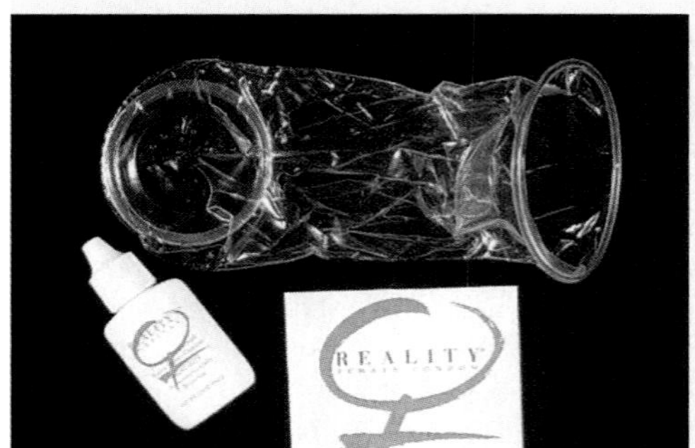

Female condom

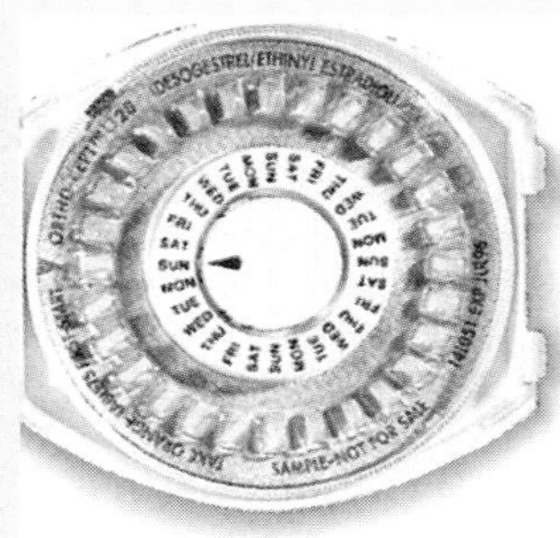

Birth control pills

Norplant

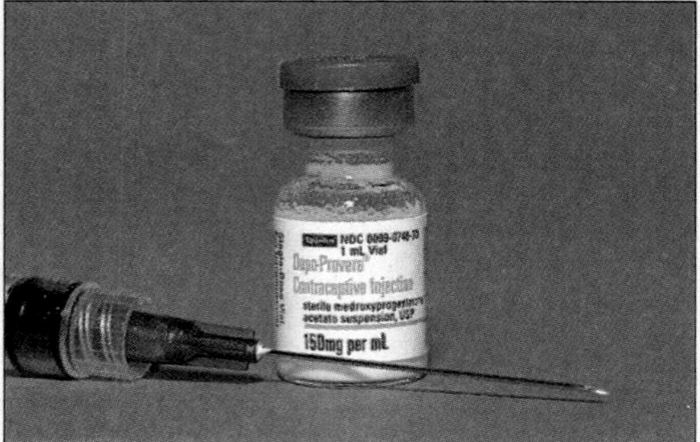

Depo-Provera

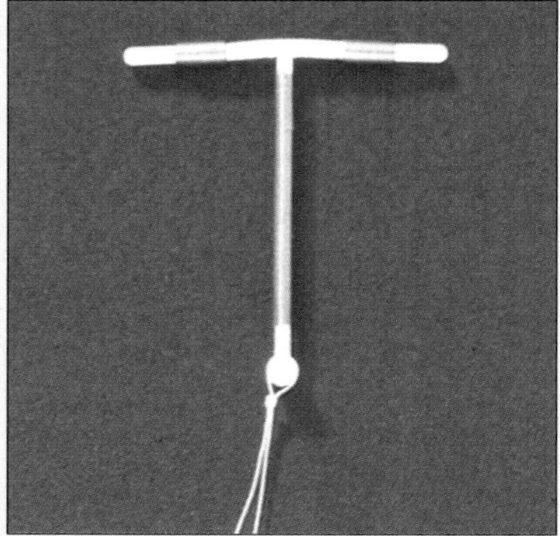
Intrauterine device (IUD)

Figure E28.1 Contraceptive Devices.

Continued

Issues in Choosing a Contraceptive

Many issues enter into the appropriate choice of a contraceptive, including personal preference, pattern of sexual activity, medical history, religious views, convenience, initial and ongoing costs, and disease prevention. For most people, however, the two primary issues are safety and reliability.

The following table shows the expected rates of failure for several types of contraception as reported in the 2000 *Physician's Desk Reference.* Each column shows the number of sexually active women who typically become pregnant within 1 year while they or their partners are using the indicated contraceptives. The lowest rate is for those who use the method correctly and consistently, while the higher rate is based on random surveys of users and takes human error (lapses and incorrect usage) into account.

We have not considered all the currently available methods of contraception or all the issues important to the choice of a contraceptive. No one contraceptive method can be recommended as best for all people. Further information necessary to a sound choice and proper use of contraceptives should be sought from a health department, college health service, physician, or other such sources.

Failure Rates of Contraceptive Methods

	Rate of Failure (pregnancies per 100 users)	
Method	**Lowest**	**Typical**
No protection	85	85
Rhythm method	1–9	25
Withdrawal	4	19
Spermicide alone	6	26
Condom (male or female)	3–5	14–21
Diaphragm with spermicide	6	20
Birth-control pill	0.1–0.5	5
Norplant	0.05	0.05
Depo-Provera	0.3	0.3
Vasectomy	0.10	0.15
Tubal ligation	0.5	0.5
Intrauterine device	0.1–2.0	0.1–1.5

Chapter Review

Online Learning Center

Study resources for this chapter are available at www.mhhe.com/saladin/

Study Outline

Reproductive Anatomy (p. 1047)

1. Sex differentiation
2. Ovary
   a. Tunica albuginea
   b. Cortex and medulla
   c. Follicles
   d. Supportive ligaments
   e. Ovarian artery, veins, and nerves
3. Secondary sex organs
   a. Uterine tubes
      - Infundibulum and fimbriae
      - Ampulla and isthmus
      - Smooth muscle and mucosa
   b. Uterus
      - Regions of organ
         - •• Fundus
         - •• Body
         - •• Cervix
      - Uterine wall
         - •• Myometrium
         - •• Endometrium
      - Blood supply
      - Supportive ligaments
      - Peritoneal pouches
   c. Vagina
      - Vaginal wall
      - Vaginal transudate
      - Fornices
      - Hymen
      - Rugae
      - Epithelium
   d. Vulva
      - Mons pubis
      - Labia majora and minora
      - Vestibule
      - Prepuce
      - Clitoris
   e. Accessory glands
      - Greater vestibular glands
      - Lesser vestibular glands
      - Paraurethral glands
   f. Vestibular bulbs
4. Secondary sex characteristics
   a. Body fat distribution
   b. Fine body hair
   c. Voice
   d. Breasts
      - Body and axillary tail
      - Areola and nipple
      - Areolar glands
      - Suspensory ligaments
      - Mammary gland
   e. Breast cancer
      - Symptoms

(continued)

Connective Issues

Interactions Between the REPRODUCTIVE SYSTEM and Other Organ Systems

← indicates ways in which this system affects other systems
→ indicates ways in which other systems affect this one

Integumentary System

← At puberty, androgens stimulate development of body hair and apocrine glands and increased sebaceous secretion; estrogens stimulate fat deposition and breast development in females; pregnancy may cause pigmentation changes and stretch marks

→ Sensory stimulation of skin important to sexual arousal; mammary glands nourish infant

Skeletal System

← Androgens and estrogens stimulate adolescent skeletal growth; estrogens maintain bone mass in females

→ Encloses and protects pelvic organs; provides minerals for fetal growth and lactation; narrow pelvic outlet makes childbirth difficult

Muscular System

← Gonadal steroids stimulate muscle growth

→ Muscles of pelvic floor support reproductive organs, aid erection of penis and clitoris, and function in orgasm; abdominal muscles aid in childbirth; cremaster muscle helps maintain temperature of testes

Nervous System

← Androgens stimulate libido; hormones from gonads and placenta exert negative feedback control on hypothalamus

→ Hypothalamus initiates gonadotropin function and lactation

Endocrine System

← Gonads and placenta are part of endocrine system and exert feedback control on anterior pituitary

→ Hormones regulate puberty, gametogenesis, libido, pregnancy, lactation, and climacteric

Circulatory System

← Androgens stimulate erythropoiesis; estrogens may inhibit development of atherosclerosis in females; pregnancy increases blood volume and cardiac output and sometimes causes varicose veins

→ Changes in blood flow produce vasocongestion and erection in sexual arousal; blood distributes sex hormones, transports nutrients to fetus, and removes fetal wastes; pampiniform plexus prevents overheating of testes

Lymphatic/Immune Systems

← Blood-testis barrier isolates sperm and protects them from immune system

→ Testes, scrotum, and breasts have extensive lymphatic drainage; immune cells protect reproductive organs; IgA in colostrum and breast milk confers passive immunity on neonate

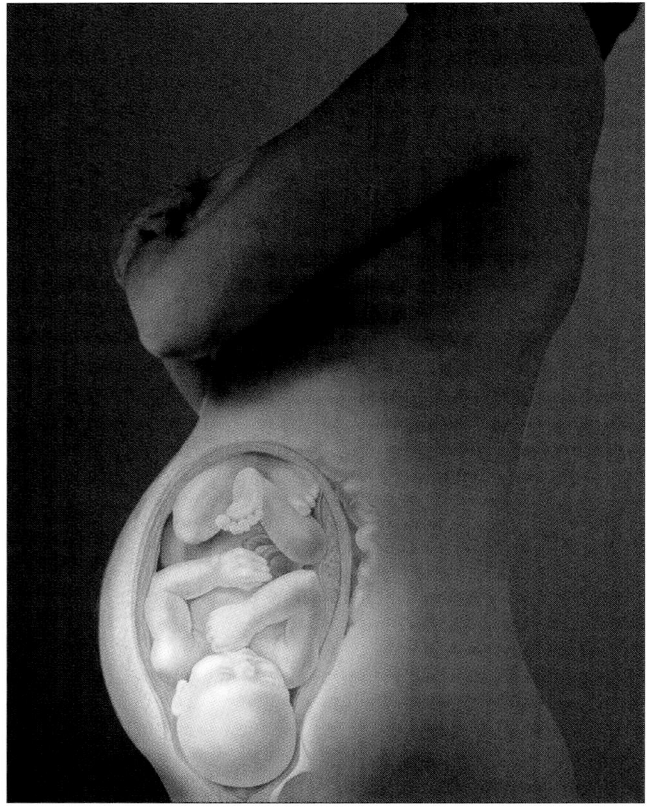

Respiratory System

← Sexual arousal increases pulmonary ventilation; pregnancy reduces depth of inspiration but increases respiratory rate

→ Provides O_2, removes CO_2; Valsalva maneuver aids childbirth

Urinary System

← Sexual arousal constricts internal urinary sphincter; prostatic hyperplasia may impede urine flow; pregnancy crowds urinary bladder and often causes incontinence

→ Disposes of wastes, maintains electrolyte and pH balance of mother and fetus; urethra serves as passageway for semen

Digestive System

← Fetus crowds digestive organs, thus contributing to heartburn and constipation

→ Provides nutrients for gametogenesis and fetal development

- Breast cancer genes
- Risk factors
- Preventive care
- Treatment

Puberty and Menopause (p. 1056)

1. Puberty
   a. Hormones and their effects
      - Gonadotropin-releasing hormone
      - Follicle-stimulating hormone
      - Luteinizing hormone
      - Estrogens
      - Progesterone
      - Inhibin
   b. Visible changes in body
      - Thelarche
      - Pubarche
      - Menarche
      - Growth of sex organs
      - Bone growth
      - Fat deposition
   c. Estrogens
2. Climacteric
   a. Declining ovarian function
   b. Atrophy of genitalia
   c. Loss of bone mass
   d. Hot flashes
   e. Mood changes
3. Menopause

Oogenesis and the Sexual Cycle (p. 1058)

1. Female physiological cycles
   a. Reproductive cycle
   b. Sexual cycle
      - Ovarian cycle
      - Menstrual cycle
2. Oogenesis
   a. Oogonium
   b. Primary oocytes
   c. Prenatal and childhood atresia
   d. Secondary oocytes
   e. Ovulation and fertilization
   f. Polar bodies
3. Sexual cycle
   a. Typical timetable
   b. Hormonal control
   c. Ovarian cycle
      - Follicular phase (days 1–14)
        - •• Menstrual phase (days 1–5)
        - •• Preovulatory phase (days 6–14)
      - Ovulation (day 14)
      - Postovulatory phase (days 15–28)
        - •• Luteal phase (days 15–26)
        - •• Premenstrual phase (days 27–28)
   d. Menstrual cycle
      - Menstrual phase (days 1–5)
      - Proliferative phase (days 6–14)
      - Secretory phase (days 15–26)
      - Premenstrual phase (days 27–28)

Female Sexual Response (p. 1065)

1. Excitement and plateau
   a. Myotonia
   b. Vasocongestion
   c. Vaginal transudation and dilation
   d. Tenting effect and orgasmic platform
   e. Function of clitoris
   f. Congestion of breasts
2. Orgasm
   a. Vaginal contractions
   b. Contractions of uterus and cervix
3. Resolution
   a. Uterine and vaginal relaxation
   b. Lack of refractory period

Pregnancy and Childbirth (p. 1067)

1. Timetable of gestation
2. Overview of prenatal development
3. Hormones of pregnancy
   a. Human chorionic gonadotropin
   b. Estrogens
   c. Progesterone
   d. Human chorionic somatomammotropin
   e. Others
4. Adjustments to pregnancy
   a. Digestive system, nutrition, and metabolism
      - Morning sickness
      - Constipation and heartburn
      - Metabolic rate and appetite
      - Weight gain
      - Nutritional needs
   b. Circulatory system
      - Increased blood volume
      - Cardiac output
      - Circulatory obstruction by uterus
   c. Respiratory system
      - Increased pulmonary ventilation
      - Pressure of uterus on diaphragm
   d. Urinary system
      - Increased glomerular filtration rate
      - Incontinence
   e. Integumentary system
      - Growth of skin
      - Striae
      - Changes in pigmentation
   f. Uterine growth and weight gain
5. Childbirth (parturition)
   a. Braxton Hicks and labor contractions
   b. Role of estrogen, oxytocin, and prostaglandins
   c. Role of uterine stretching
   d. Labor contractions
      - Oxytocin and positive feedback theory
      - Role of voluntary abdominal muscles
      - Cause of labor pains
   e. Stages of labor
      - First stage: dilation and effacement
      - Second stage: expulsion of baby
      - Third stage: expulsion of placenta and membranes
6. Puerperium
   a. Involution of uterus
   b. Effect of breast-feeding

Lactation (p. 1074)

1. Mammary gland development
2. Colostrum and milk synthesis
   a. Composition of colostrum
   b. Immune benefits of colostrum
   c. Role of prolactin and other hormones
   d. Effect of not breast-feeding
   e. Contraceptive effect of breast-feeding
   f. Resumption of menstrual cycles
3. Milk ejection
   a. Role of oxytocin
   b. Function of myoepithelial cells
4. Breast milk
   a. Composition
   b. Changes during breast-feeding
   c. Comparison to cow's milk
   d. Benefits of breast-feeding
   e. Metabolic demands on mother

Selected Vocabulary

ovary 1047
tunica albuginea 1047
cortex 1047
medulla 1047
follicle 1047
ovulation 1047
ovarian artery 1047
ovarian vein 1047
ovarian nerve 1047
uterine tube 1047
infundibulum 1047
fimbriae 1047
ampulla 1047
isthmus 1047
uterus 1049
fundus of uterus 1049
body of uterus 1049
cervix of uterus 1049
cervical canal 1049
cervical gland 1049
myometrium 1051
endometrium 1051
stratum functionalis 1051
stratum basalis 1051
uterine artery 1051
arcuate artery 1051
spiral artery 1051
vagina 1052
hymen 1052
vaginal rugae 1052
dendritic cell 1052
vulva 1052
mons pubis 1052
labia majora 1052
labia minora 1052
vestibule 1052
prepuce 1052
clitoris 1052
glans 1052
body of clitoris 1052
crura 1052
clitoral artery 1052
vestibular bulb 1052
greater vestibular gland 1052
lesser vestibular gland 1053
paraurethral gland 1053
body of breast 1053
axillary tail 1053
areola 1054
areolar glands 1054
suspensory ligament 1054
lactiferous duct 1055
lactiferous sinus 1055
estrogens 1056
thelarche 1056
pubarche 1056
menarche 1056
progesterone 1057
inhibin 1057
climacteric 1057
menopause 1057
hot flashes 1057
reproductive cycle 1058
sexual cycle 1058
ovarian cycle 1058
menstrual cycle 1058
oogenesis 1058
oogonium 1058
primary oocyte 1058
ovum 1058
atresia 1058
secondary oocyte 1058
polar body 1058
follicular phase 1060
menstrual phase 1060
primordial follicle 1061
primary follicle 1061
theca folliculi 1061
follicular fluid 1061
antrum 1061
secondary follicle 1061
granulosa cells 1061
zona pellucida 1061
corona radiata 1061
preovulatory phase 1061
mature (graafian) follicle 1061
stigma 1062
postovulatory phase 1063
luteal phase 1063
premenstrual phase 1063
corpus luteum 1063
lutein cells 1063
corpus albicans 1064
proliferative phase 1064
secretory phase 1064
ischemic phase 1064
menstrual fluid 1064
vaginal transudate 1065
orgasmic platform 1065
tenting effect 1065
gestation 1067
trimester 1067
human chorionic somatomammotropin 1068
linea nigra 1070
chloasma 1070
parturition 1070
Braxton Hicks contractions 1070
labor contractions 1070
positive feedback theory of labor 1071
primipara 1072
multipara 1072
dilation stage of labor 1072
dilation 1072
effacement 1072
expulsion stage of labor 1072
crowning 1073
placental stage of labor 1073
postpartum 1073
puerperium 1073
involution 1073
autolysis 1073
lochia 1073
lactation 1074
colostrum 1074
milk ejection 1075
myoepithelial cells 1075

Testing Your Recall

1. Of the following organs, the one(s) most comparable to the penis is (are)
   a. the clitoris.
   b. the vagina.
   c. the vestibular bulbs.
   d. the labia minora.
   e. the prepuce.
2. The ovaries secrete all of the following *except*
   a. estrogens.
   b. progesterone.
   c. androgens.
   d. follicle-stimulating hormone.
   e. inhibin.
3. The first haploid stage in oogenesis is
   a. the oogonium.
   b. the primary oocyte.
   c. the secondary oocyte.
   d. the second polar body.
   e. the zygote.
4. FSH secretion begins to rise and new follicles begin to grow during
   a. the premenstrual phase.
   b. the menstrual phase.
   c. the preovulatory phase.
   d. the luteal phase.
   e. the plateau phase.
5. The hormone that most directly influences the secretory phase of the menstrual cycle is
   a. HCG.
   b. FSH.
   c. LH.
   d. estrogen.
   e. progesterone.
6. The ischemic phase of the uterus results from
   a. rising progesterone levels.
   b. falling progesterone levels.
   c. stimulation by oxytocin.
   d. stimulation by prostaglandins.
   e. stimulation by estrogens.
7. Before secreting milk, the mammary glands secrete
   a. prolactin.
   b. colostrum.
   c. lochia.
   d. meconium.
   e. chloasma.

8. Few women become pregnant while nursing because ______ inhibits GnRH secretion.
   a. FSH
   b. prolactin
   c. prostaglandins
   d. oxytocin
   e. HCG
9. Smooth muscle cells of the myometrium and myoepithelial cells of the mammary glands are the target cells for
   a. prostaglandins.
   b. LH.
   c. oxytocin.
   d. progesterone.
   e. FSH.
10. Which of these is *not* true of the luteal phase of the sexual cycle?
   a. Progesterone level is high.
   b. The endometrium stores glycogen.
   c. Ovulation occurs.
   d. Fertilization may occur.
   e. The endometrial glands enlarge.
11. Each egg cell develops in its own fluid-filled space called a/an ________.
12. The mucosa of the uterus is called ________.
13. A girl's first menstrual period is called ________.
14. A yellowish structure called the ________ secretes progesterone during the secretory phase of the sexual cycle.
15. The first layer of cells around a mature secondary oocyte is the ________.
16. A secondary follicle differs from a primary follicle in having a cavity called the ________.
17. Menopause occurs during a midlife period of changing hormone secretion called ________.
18. All the products of fertilization, including the embryo or fetus, the placenta, and the embryonic membranes, are collectively called the ________.
19. The funnellike distal end of the uterine tube is called the ________ and has feathery processes called ________.
20. Postpartum uterine involution produces a vaginal discharge called ________.

Answers in appendix B

Testing Your Comprehension

1. Would you expect puberty to create a state of positive or negative nitrogen balance? Explain. (See chapter 26 for nitrogen balance.)
2. Aspirin and ibuprofen can inhibit the onset of labor and are sometimes used to prevent premature birth. Review your knowledge of these drugs and the mechanism of labor, and explain this effect.
3. At 6 months postpartum, a nursing mother is in an automobile accident that fractures her skull and severs the hypothalamo-hypophyseal portal vessels. How would you expect this to affect her milk production? How would you expect it to affect her future ovarian cycles? Explain the difference.
4. If the ovaries are removed in the first 6 weeks of pregnancy, the embryo will be aborted. If they are removed later in pregnancy, the pregnancy can go to a normal full term. Explain the difference.
5. A breast-feeding woman leaves her baby at home and goes shopping. There, she hears another woman's baby crying and notices her blouse becoming wet with a little exuded milk. Explain the physiological link between hearing that sound and the ejection of milk.

Answers at the Online Learning Center

29 CHAPTER

Human Development

Boy meets girl: the union of sperm and egg (SEM)

Chapter Outline

Insights

Study with Your ESP CD-ROM

Refer to your ESP Tutorial CD-ROM for helpful animations, activities, and quizzes related to the material covered in this chapter. The following are suggested areas to go to on the CD and are presented according to "Unit/Topic":
- Reproduction and Development/Growth and Development

Brushing Up

To understand this chapter, it is important that you understand or brush up on the following concepts:
- Structure of the spermatozoon (p. 1034)
- Anatomy and histology of the uterus (pp. 1049–51)
- Stages of oogenesis (p. 1058)
- Structure of a mature ovarian follicle (p. 1062)

Perhaps the most dramatic, seemingly miraculous aspect of human life is the transformation of a one-celled fertilized egg into an independent, fully developed individual. From the beginning of recorded thought, people have pondered how a baby forms in the mother's body and how two parents can produce another human being who, although unique, possesses characteristics of each. Aristotle dissected the embryos of various birds, established their sequence of organ development, and speculated that the hereditary traits of a child resulted from the mixing of the male's semen with the female's menstrual blood. The human egg was first observed and the modern science of **developmental biology** was born in the nineteenth century. In its broad sense, developmental biology embraces the scientific study of changes in form and function from fertilized egg through old age—which is the scope of this chapter.

Fertilization and Preembryonic Development

Objectives

When you have completed this section, you should be able to

- describe the process of sperm migration and fertilization;
- explain how an egg prevents fertilization by more than one sperm;
- describe the major events that transform a fertilized egg into an embryo; and
- describe the implantation of the preembryo in the uterine wall.

Sperm Migration

If it is to survive, an egg must be fertilized within 12 to 24 hours of ovulation, yet it takes about 72 hours for an egg to reach the uterus. In order to fertilize it before it dies, sperm must therefore encounter the egg somewhere in the distal one-third of the uterine tube. The vast majority of sperm never make it that far. Many are destroyed by vaginal acid or drain out of the vagina. Others fail to penetrate the mucus of the cervical canal, and those that do are often destroyed by leukocytes in the uterus. Half of the remainder are likely to go up the wrong uterine tube. Finally, 2,000 to 3,000 spermatozoa reach the vicinity of the egg—not many of the 300 million that were ejaculated.

Spermatozoa migrate mainly by means of the snakelike lashing of their tails as they crawl along the female mucosa, but they are assisted by certain aspects of female physiology. Strands of mucus guide them through the cervical canal. Although female orgasm is not required for fertilization, orgasm does involve uterine contractions that may suck semen from the vagina and spread it throughout the uterus, like hand lotion pressed between your palms. Within close range of the egg, sperm may be chemically attracted to it; this has been demonstrated for some animals but remains unproven for humans.

Capacitation

Spermatozoa can reach the distal uterine tube within 5 to 10 minutes of ejaculation, but they cannot fertilize an egg for about 10 hours. While migrating, they undergo a process of **capacitation** that makes it possible to penetrate an egg. Prior to ejaculation, the membrane of the sperm head contains a substantial amount of cholesterol, which toughens it and prevents premature release of the acrosomal enzymes. This avoids wastage of sperm and enzymatic damage to the spermatic ducts. After ejaculation, however, fluids of the female reproductive tract wash away cholesterol and other inhibitory factors in the semen. The membrane of the sperm head becomes more fragile and more permeable to calcium ions, which diffuse into the sperm and cause more powerful lashing of the tail.

Most sperm are fertile for a maximum of 48 hours after ejaculation, so there is little chance of fertilizing an egg if intercourse occurs more than 48 hours before ovulation. Fertilization also is unlikely if intercourse takes place more than 14 hours after ovulation because the egg would no longer be viable by the time the sperm became capacitated. For those wishing to conceive a child, the "window of opportunity" is therefore 48 hours before ovulation to 14 hours after. Those wishing to avoid pregnancy, however, should allow a wider margin of safety for variations in sperm and egg longevity, capacitation time, and time of ovulation.

Fertilization

When the sperm encounters an egg, it undergoes an **acrosomal reaction**—exocytosis of the acrosome, releasing the enzymes needed to penetrate into the egg. But the first sperm to reach an egg is not the one to fertilize it. Sperm must first penetrate the granulosa cells and zona pellucida that surround it (fig. 29.1). It may require hundreds of sperm to clear a path for the one that penetrates the egg proper.

Two of the acrosomal enzymes are **hyaluronidase,** which digests the hyaluronic acid that binds granulosa cells together, and **acrosin,** a protease similar to the trypsin of pancreatic juice. When a path has been cleared, a sperm binds to the zona pellucida and releases its enzymes, digesting a pathway through the zona until it contacts the egg itself. The membranes of the two gametes then fuse, and the sperm nucleus enters the egg cytoplasm.

Fertilization combines the haploid (n) set of sperm chromosomes with the haploid set of egg chromosomes and produces a diploid ($2n$) set. Fertilization by two or more sperm, called **polyspermy,** would produce a triploid ($3n$) or larger set of chromosomes and the egg would die. The egg has two mechanisms for preventing this: a fast block and slow block to polyspermy. In the **fast block,** binding of the sperm to the egg opens Na^+ channels in the egg membrane.

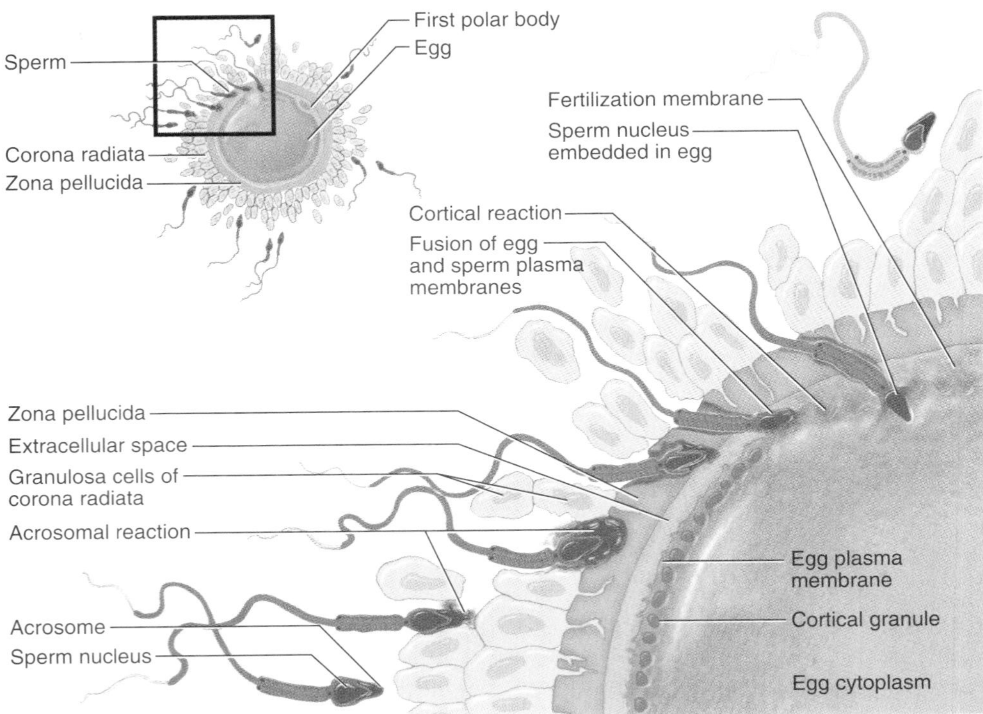

Figure 29.1 Fertilization and the Slow Block to Polyspermy.

The rapid inflow of Na^+ depolarizes the membrane and inhibits the binding of any more sperm to it. The **slow block** involves secretory vesicles called **cortical granules** just beneath the membrane. Sperm penetration triggers an inflow of Ca^{2+}; this, in turn, triggers a **cortical reaction** in which the cortical granules release their secretion beneath the zona pellucida. The secretion swells with water, pushes any remaining sperm away from the egg, and creates an impenetrable **fertilization membrane.**

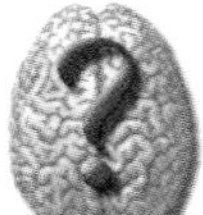

Think About It

What similarity can you see between the slow block to polyspermy and the release of acetylcholine from synaptic vesicles of a neuron? (Compare pp. 474–76.)

Meiosis II

A secondary oocyte begins meiosis II before ovulation and completes it only if fertilized. Through the formation of a second polar body, the fertilized egg discards one chromatid from each chromosome. The sperm and egg nuclei then swell and become **pronuclei.** A mitotic spindle forms between them, each pronucleus ruptures, and the chromosomes of the two gametes mix into a single diploid set (fig. 29.2). The fertilized egg, now called a **zygote,** is ready for its first mitotic division.

The Preembryonic Stage

The **preembryonic stage** consists of the first 2 weeks of development and culminates in the existence of an embryo. It involves three major processes: cleavage, implantation, and embryogenesis.

Cleavage

Cleavage refers to mitotic divisions that occur in the first 3 days after fertilization. The first cleavage occurs about 30 hours after fertilization and produces the first two daughter cells, or **blastomeres.**[1] These divide simultaneously at shorter and shorter time intervals, doubling the number of blastomeres each time. By the time the conceptus arrives in the uterus, about 72 hours after ovulation, it consists of 16 or more cells and somewhat resembles a mulberry—hence it is called a **morula.**[2] The morula is no larger than the zygote; cleavage merely produces smaller and smaller blastomeres. This increases the ratio of cell surface area to volume, which favors rapid nutrient uptake and waste removal, and it produces a larger number of cells from which to form different types of embryonic tissue.

1. *blast* = bud, precursor + *mer* = segment, part
2. *mor* = mulberry + *ula* = little

Clinical Insight 29.1

Twins

There are two ways in which a woman can produce twins (and, by extension, other multiple births). *Dizygotic (DZ) twins* are produced when two eggs are ovulated and fertilized by separate sperm. They are no more or less genetically similar than any other siblings and may be of different sexes. Multiple ovulation can also result in triplets, quadruplets, or even greater numbers of offspring.

Monozygotic (MZ) twins are produced when a single egg is fertilized and the cell mass (embryoblast) later divides into two. MZ twins are genetically identical, or nearly so, and are therefore of the same sex and nearly identical appearance. Identical triplets and quadruplets occasionally result from the splitting of a single embryoblast. Reproductive biologists are beginning to question whether MZ siblings are truly genetically identical. They have suggested that one blastomere may undergo a mutation in the course of DNA replication, and the splitting of the embryoblast may represent an attempt of each cell mass to reject the other one as foreign.

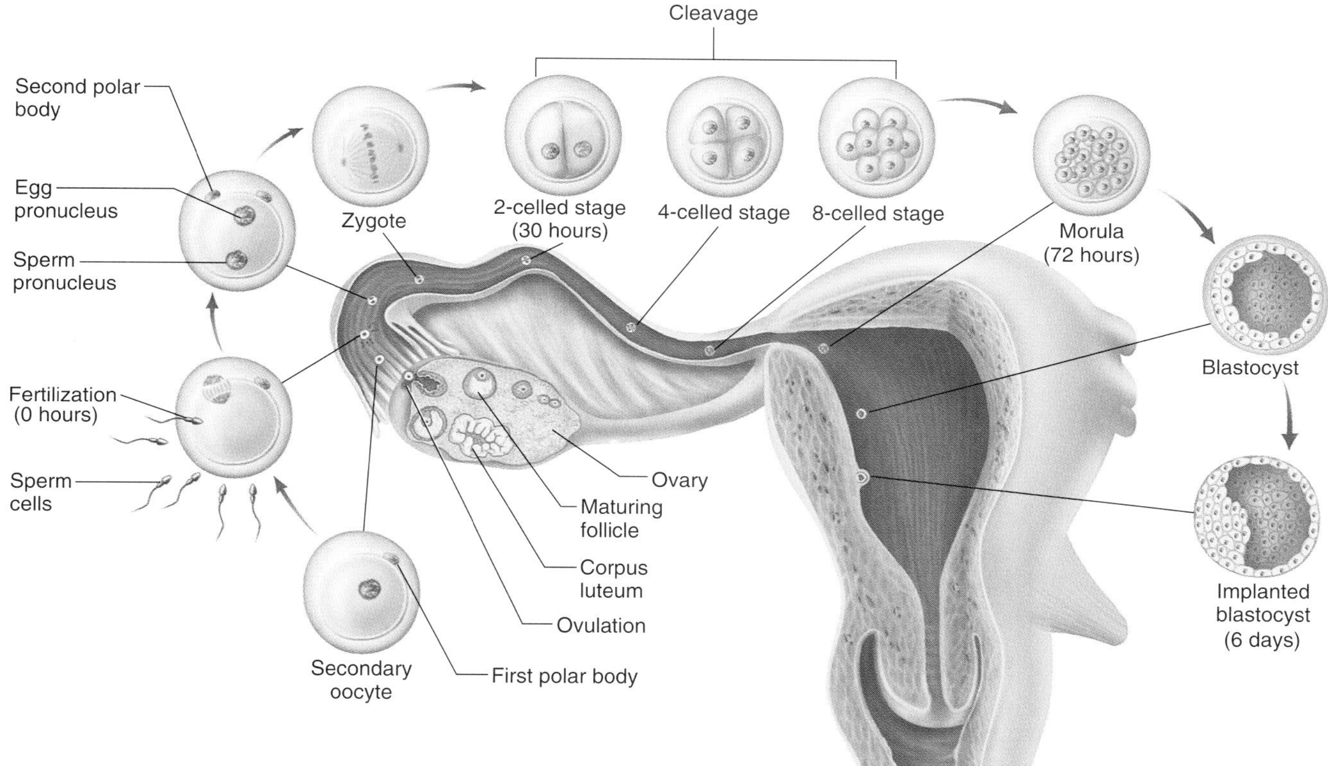

Figure 29.2 Events from Ovulation to Implantation. These processes occur during the first week after ovulation, approximately at the locations indicated in the female reproductive tract.

The morula lies free in the uterine cavity for 4 to 5 days and divides into 100 cells or so. During this time, it is nourished by nutrients that were stored in the egg cytoplasm during oogenesis and by an endometrial secretion called **uterine milk,** which accumulates in a cavity in the morula. Meanwhile, the zona pellucida disintegrates and releases the conceptus, which is now at a stage called the **blastocyst**—a hollow sphere with an outer layer of squamous cells called the **trophoblast**[3] and an inner cell mass called the **embryoblast.** The trophoblast is destined to form part of the placenta and play an important role in nourishment of the embryo, whereas the embryoblast is destined to become the embryo itself.

3. *troph* = food, nourishment

Implantation

About 6 days after ovulation, the blastocyst attaches to the endometrium, usually on the posterior wall of the fundus or body of the uterus (see clinical insight 29.2 for exceptions). This is the beginning of **implantation,** a process in which the conceptus becomes buried in the endometrium (fig. 29.3). The trophoblast cells adjacent to the embryoblast secrete enzymes that stimulate thickening of the endometrium and they separate into two layers. The deep layer is called the **cytotrophoblast** because it retains individual cells divided by membranes. In the superficial layer, however, the plasma membranes break down and the cells fuse into a multinucleate mass called the **syncytiotrophoblast**[4] (sin-SISH-ee-oh-TRO-foe-blast). (A syncytium is any body of protoplasm containing multiple nuclei.)

The syncytiotrophoblast grows into the endometrium like little roots and digests endometrial cells along the way (fig. 29.3*b*). The endometrium reacts to this injury by growing over the trophoblast and eventually enclosing it. Implantation takes about a week and is completed about the time the next menstrual period would have occurred if the woman had not become pregnant.

The trophoblast also secretes human chorionic gonadotropin (HCG). HCG stimulates the corpus luteum to secrete estrogen and progesterone, and progesterone suppresses menstruation. The level of HCG in the mother's blood rises until the end of the second month. During this time, the trophoblast develops into a membrane called the *chorion,* which takes over the role of the corpus luteum and makes HCG unnecessary. The ovaries then become inactive for the rest of the pregnancy, but estrogen and progesterone levels rise dramatically as they are secreted by the ever-growing chorion (see fig. 28.18, p. 1068).

Embryogenesis

During implantation, the embryoblast undergoes **embryogenesis**—arrangement of the blastomeres into the three primary germ layers: ectoderm, mesoderm, and endoderm. Embryogenesis begins with the formation of a narrow space called the **amniotic cavity** between the embryoblast and cytotrophoblast (fig. 29.4). The embryoblast flattens into an **embryonic disc** composed of ectoderm and endoderm. As the disc elongates, a raised groove called the **primitive streak** forms along the midline of the ectoderm. Cells on the surface migrate medially toward this groove, down into it, and then laterally between the ectoderm and endoderm. This forms the middle germ layer, the mesoderm (fig. 29.5). The ectoderm and endoderm are epithelia composed of tightly joined cells, but the mesoderm is a more loosely organized, gelatinous connective tissue called **mesenchyme.** Its cells are more mobile and wispy than those of the ectoderm and endoderm. At the conclusion of embryogenesis, the individual is an **embryo,** about 2 mm long and 2 weeks old.

Blastocyst cavity
Trophoblast
Embryoblast
Endometrial epithelium
Endometrial gland
Endometrial capillary

(a)

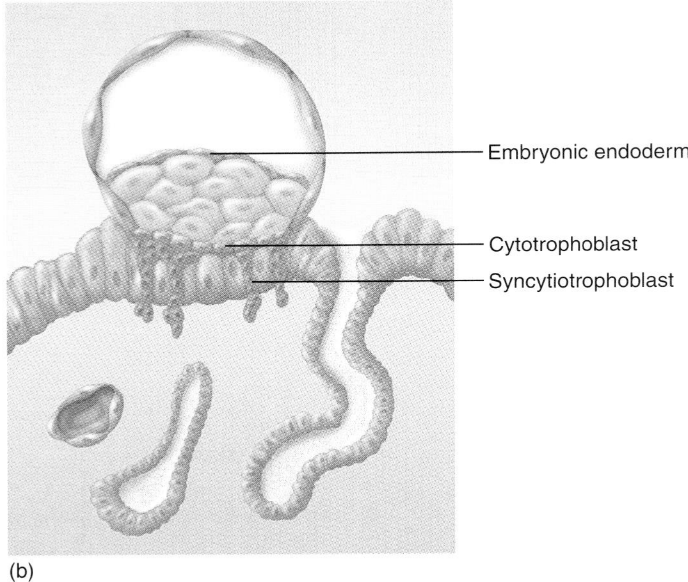

Figure 29.3 Implantation. (*a*) Structure of the blastocyst about 6 to 7 days after ovulation. (*b*) The progress of implantation about 1 day later; the syncytiotrophoblast has begun growing rootlets, which penetrate the endometrium.

4. *syn* = together + *cyt* = cell

Clinical Insight 29.2

Ectopic Pregnancy

In about 1 out of 300 pregnancies, the blastocyst implants somewhere other than the uterus and produces an *ectopic*[5] *pregnancy.* Most cases begin as *tubal pregnancies,* in which the blastocyst implants in the uterine tube. This usually occurs because it has encountered an obstruction such as a constriction resulting from earlier pelvic inflammatory disease, tubal surgery, previous ectopic pregnancies, or repeated abortions. The uterine tube cannot expand enough to accommodate the growing conceptus for long; if the situation is not detected and treated early, the tube usually ruptures within 12 weeks. This can be fatal to the mother. If she survives, the conceptus may reimplant in the abdominopelvic cavity and produce an *abdominal pregnancy.* About 1 pregnancy in 7,000 is abdominal. The conceptus can grow anywhere it finds an adequate blood supply—for example, on the broad ligament or the outside of the uterus, colon, or bladder. This is a serious threat to the mother's life and usually requires abortion, but about 9% of abdominal pregnancies result in live birth by cesarian section.

5. *ec* = outside + *top* = place

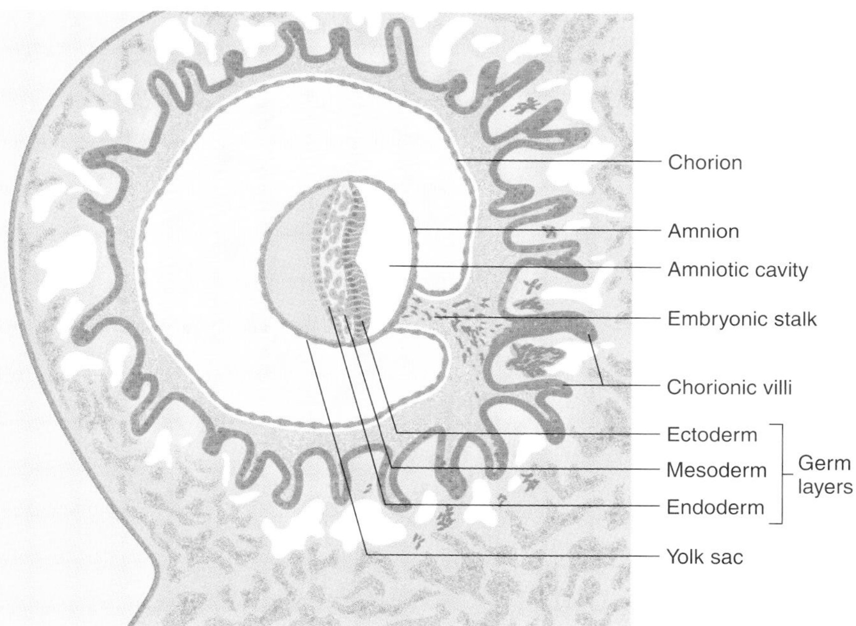

Figure 29.4 The Implanted Conceptus at 2 Weeks. The primary germ layers and extraembryonic membranes have formed, and the conceptus is now an embryo.

Before You Go On

Answer the following questions to test your understanding of the preceding section:

1. How soon can a sperm reach an egg after ejaculation? How soon can it fertilize an egg? What accounts for the difference?
2. Describe two ways a fertilized egg prevents the entry of excess sperm.
3. In the blastocyst, what are the cells called that eventually give rise to the embryo? What are the cells that carry out implantation?
4. What major characteristic distinguishes an embryo from a preembryo?

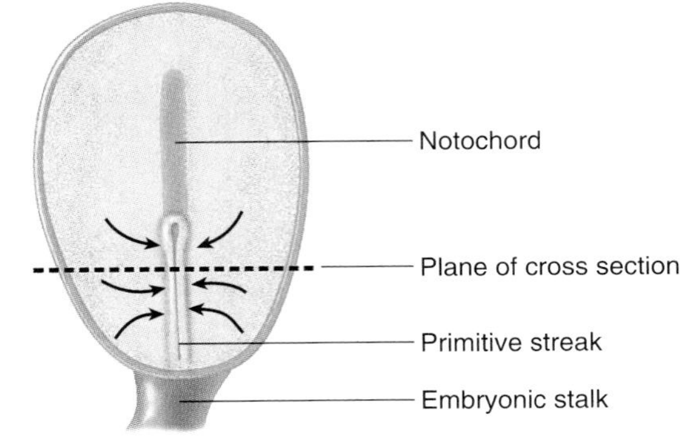

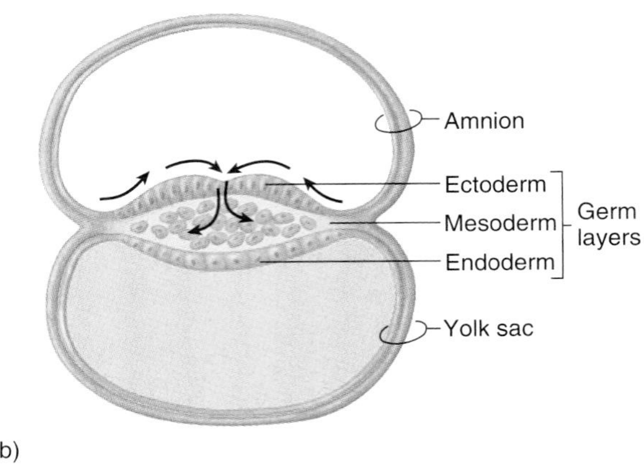

Figure 29.5 Formation of the Primary Germ Layers. (*a*) Dorsal view of the embryonic disc at 16 days showing the migration of ectodermal cells into the primitive streak. (*b*) Cross section of the embryonic disc at the level indicated in *a*, showing ectodermal cells entering the primitive streak and forming the mesodermal layer.

Embryonic and Fetal Development

Objectives

When you have completed this section, you should be able to

- describe the formation and functions of the placenta;
- explain how the conceptus is nourished before the placenta takes over this function;
- describe the embryonic membranes and their functions;
- identify the major tissues derived from the primary germ layers;
- describe the major events of fetal development; and
- describe the fetal circulatory system.

Two weeks after conception, all three germ layers are present and the embryonic stage of development begins. Over the next 6 weeks, the conceptus forms a set of membranes external to the embryo, the embryo begins receiving its nutrition primarily from the placenta, and the germ layers differentiate into organs and organ systems. Although these organs are still far from functional, it is their presence at 8 weeks that marks the transition from the embryonic stage to the fetal stage.

Prenatal Nutrition

One of the important changes to occur in the embryonic phase is the mode of nutrition. In **trophoblastic (deciduous) nutrition,** the conceptus is nourished by digesting cells of the endometrium. Progesterone stimulates the development of **decidual**[6] **cells,** which are rich in glycogen, proteins, and lipids. The trophoblast digests these cells and the embryoblast imbibes the resulting nutritious fluid. This is gradually followed by **placental nutrition,** in which the conceptus is nourished by the mother's bloodstream. Nutrients diffuse from her blood through the **placenta,**[7] a vascular organ that develops on the uterine wall. Figure 29.6 shows the time course of the transition from trophoblastic to placental nutrition. Trophoblastic nutrition is the only means of nutrition for the first week after implantation. The placenta begins to develop about 11 days after conception, but trophoblastic nutrition remains dominant for 8 weeks, called the **trophoblastic phase** of the pregnancy. After that, the placenta provides most nutrition; the trophoblastic mode declines and, by 12 weeks, it ends. The **placental phase** of the pregnancy lasts from the beginning of week 9 until birth.

Placentation, the formation of the placenta, extends from 11 days through 12 weeks after conception. Most development occurs in the embryonic stage. It begins when extensions of the syncytiotrophoblast, called **chorionic villi,** penetrate more and more deeply into the endometrium, like the roots of a tree penetrating into the nourishing "soil" of the uterus (fig. 29.7). As they digest their way through uterine blood vessels, the villi become surrounded by pools of free blood that eventually merge to form the **placental sinus.** This maternal blood stimulates increasingly rapid growth of the chorionic villi. Mesenchyme grows into the villi and develops into blood vessels.

The fully developed placenta is a disc of tissue about 20 cm in diameter and 3 cm thick. At term, it weighs about one-sixth as much as the baby. The surface facing the fetus is smooth and gives rise to the **umbilical cord** (fig. 29.8), which contains three blood vessels that connect the fetus to the placenta. The surface attached to the uterine wall is rougher. It consists of the chorionic villi contributed by the fetus and a region of the mother's endometrium called the *decidua basalis.* The chorionic villi are extensively branched treelike structures surrounded by the maternal blood in the placental sinus.

---

6. *decid* = falling off
7. *placenta* = flat cake

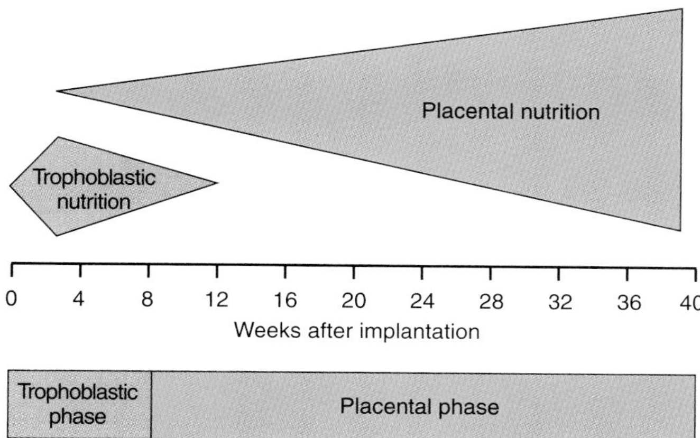

Figure 29.6 The Timetable of Trophoblastic and Placental Nutrition. Trophoblastic nutrition peaks at 2 weeks and ends by 12 weeks. Placental nutrition begins at 2 weeks and becomes increasingly important until birth, 39 weeks after implantation. The transition from the trophoblastic phase to the placental phase of pregnancy, named for the mode that supplies most nutrition to the conceptus, is at 8 weeks.

The fetal heart pumps blood to the placenta by way of two umbilical arteries. Fetal blood flows into the capillaries of the villi and then back to the fetus by way of a single umbilical vein. The villi are *filled with* fetal blood and *surrounded by* maternal blood; the two bloodstreams do not mix unless there is damage to the placental barrier. The barrier, however—the membrane covering the villi—is only 3.5 μm thick, about half the diameter of a single red blood cell. Early in development, the chorionic villi have thicker membranes that are not very permeable to nutrients and wastes, and their total surface area is relatively small. As the villi grow and branch, their surface area increases and the membranes become thinner and more permeable. Thus there is a dramatic increase in **placental conductivity,** the rate at which substances diffuse through the membrane. Oxygen and nutrients pass from the maternal blood to the fetal blood, while fetal wastes pass the other way to be eliminated by the mother.

Gases, electrolytes, fatty acids, and steroids pass through the membrane by simple diffusion, glucose by facilitated diffusion, amino acids by active transport, and insulin by receptor-mediated endocytosis. Unfortunately, the placenta is also permeable to nicotine, alcohol, and most other drugs in the maternal bloodstream. Nutrition, excretion, and other functions of the placenta are summarized in table 29.1.

Embryonic Membranes

The placenta is not the only accessory organ of the conceptus. There are also four membranes—the *amnion, yolk sac, allantois,* and *chorion* (fig. 29.9). To understand these, it helps to realize that all mammals evolved from egg-laying reptiles. Within the shelled, self-contained egg of a reptile, the embryo rests atop a yolk, which is enclosed in the yolk sac; it floats in a little sea of liquid contained in the amnion; it stores its toxic wastes in the allantois; and to breathe, it has a chorion permeable to gases (the "skin" that you may have noticed just beneath the shell of a boiled egg). All of these membranes persist in mammals, including humans, but are modified in their functions.

The **amnion** is a transparent sac that develops from cells of the embryonic disc. It grows to completely enclose the embryo and is penetrated only by the umbilical cord. The amnion becomes filled with **amniotic fluid,** which protects the embryo from trauma and temperature fluctuations, allows the freedom of movement important to muscle development, enables the embryo to develop symmetrically, and prevents adhesion of, for example, an arm to the trunk. At first, the amniotic fluid forms by filtration of the mother's blood plasma, but beginning at 8 to 9 weeks, the fetus urinates into the amniotic cavity about once an hour and contributes substantially to the fluid volume. The volume increases only slowly, however, because the fetus swallows amniotic fluid at a nearly equal rate. At term, the amnion contains 700 to 1,000 mL of fluid.

The **yolk sac** arises partly from cells of the embryonic disc opposite the amnion. It is a small sac suspended from the ventral side of the embryo. It contributes to the formation of the digestive tract and produces the first blood cells and future egg or sperm cells.

The **allantois** (ah-LON-toe-iss) is an outpocketing of the posterior end of the yolk sac. It forms the foundation for the umbilical cord and becomes part of the urinary bladder. It can be seen in proximal cross sections of the cord.

The **chorion** is the outermost membrane, enclosing all the rest of the membranes and the embryo. Initially it has villi around its entire surface, but as the pregnancy advances, the villi of the placenta grow and branch while the rest of them degenerate. The chorion forms the fetal portion of the placenta; its functions are essentially those listed in table 29.1.

Organogenesis

Organogenesis is the formation of organs and organ systems from the primary germ layers. This is the primary developmental process to occur in the embryo itself. The major structures that arise from the primary germ layers are listed in table 29.2.

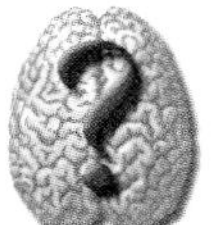

Think About It

List the four primary tissue types of the adult body (see chapter 6) and identify which of the three primary germ layers of the embryo predominantly gives rise to each.

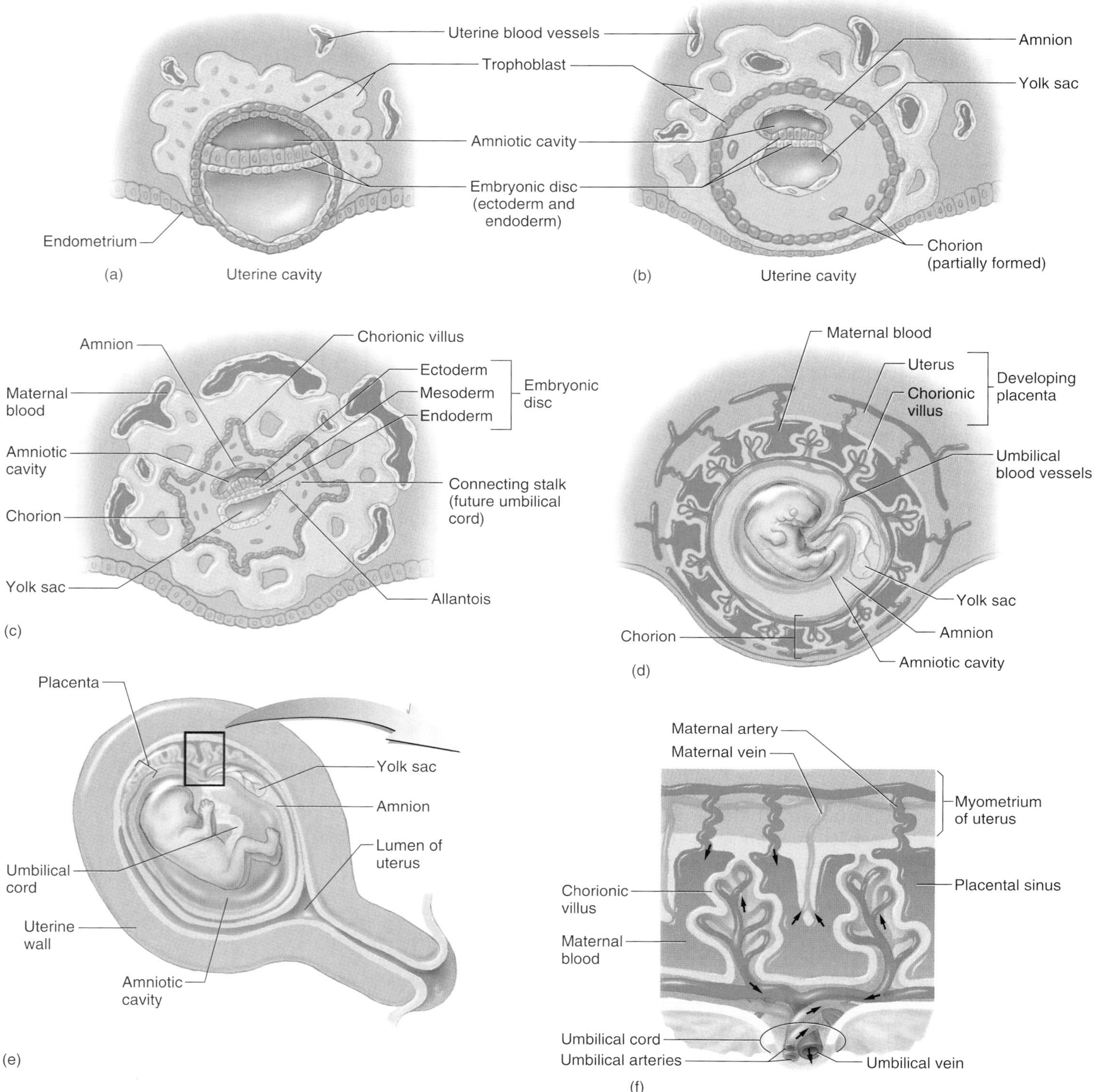

Figure 29.7 **Development of the Placenta and Embryonic Membranes.** (*a*) Implantation nearly complete; an amniotic cavity has formed within the embryonic disc, lined by cells that will become the amnion. (*b*) Conceptus about 9 days after fertilization; embryonic disc cells have now formed the yolk sac. (*c*) Conceptus at 16 days; the allantois is beginning to form, and the chorion is forming from the trophoblast and embryonic mesoderm. (*d*) Embryo at 4.5 weeks, enclosed in the amnion and chorion. (*e*) Embryo at 13.5 weeks; placentation is complete. (*f*) A portion of the mature placenta and umbilical cord.

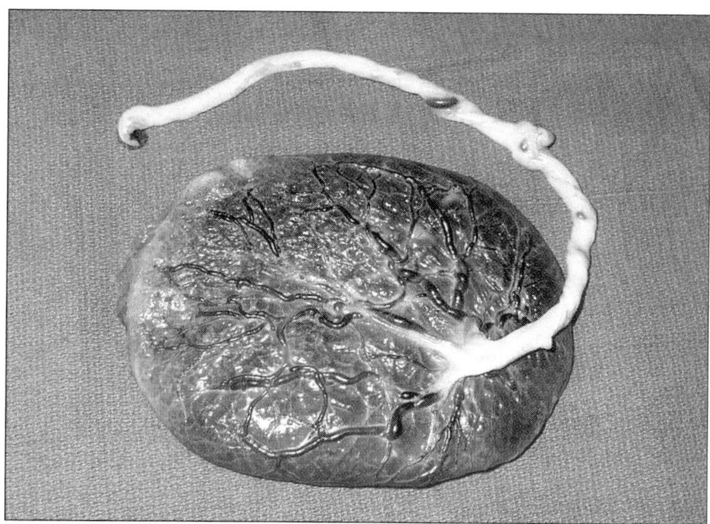

Figure 29.8 The Placenta and Umbilical Cord Viewed from the Fetal Side.

Table 29.1 Functions of the Placenta

Nutritional Roles	Transports nutrients such as glucose, amino acids, fatty acids, minerals, and vitamins from the maternal blood to the fetal blood; stores nutrients such as carbohydrates, protein, iron, and calcium in early pregnancy and releases them to the fetus later, when fetal demand is greater than the mother can absorb from the diet
Excretory Roles	Transports nitrogenous wastes such as ammonia, urea, uric acid, and creatinine from the fetal blood to the maternal blood
Respiratory Roles	Transports O_2 from mother to fetus and CO_2 from fetus to mother
Endocrine Roles	Secretes estrogens, progesterone, relaxin, human chorionic gonadotropin, and human chorionic somatomammotropin; allows other hormones synthesized by the conceptus to pass into the mother's blood and maternal hormones to pass into the fetal blood
Immune Role	Transports maternal antibodies (especially IgG) into fetal blood to confer passive immunity on fetus

At the end of 8 weeks, all of the organ systems are present, the individual is about 3 cm long, and it is now considered a **fetus** (fig. 29.9). Its bones have just begun to ossify and the skeletal muscles exhibit spontaneous contractions, although these are too weak to be felt by the mother. The heart, which has been beating since the fourth week, now

Table 29.2 Derivatives of the Three Primary Germ Layers

Layer	Major Derivatives
Ectoderm	Epidermis; hair follicles and arrector pili muscles; cutaneous glands; nervous system; adrenal medulla; pineal and pituitary glands; lens, cornea, and intrinsic muscles of the eye; internal and external ear; salivary glands; epithelia of the nasal cavity, oral cavity, and anal canal
Mesoderm	Skeleton; skeletal, cardiac, and most smooth muscle; cartilage; adrenal cortex; middle ear; dermis; blood; blood and lymphatic vessels; bone marrow; lymphoid tissue; epithelium of kidneys, ureters, gonads, and genital ducts; mesothelium of ventral body cavity
Endoderm	Most of the mucosal epithelium of the digestive and respiratory tracts; mucosal epithelium of urinary bladder and parts of urethra; epithelial components of accessory reproductive and digestive glands (except salivary glands); thyroid and parathyroid glands; thymus

circulates blood. The heart and liver are very large and form a prominent ventral bulge; the head is nearly half the total body length.

The stages of prenatal development are summarized in table 29.3.

Fetal Development

The fetus is the final stage of prenatal development, extending from the end of the eighth week until birth. The organs that formed during the embryonic stage now undergo growth and cellular differentiation and acquire the functional capability to support life outside the mother.

The circulatory system shows the most conspicuous anatomical changes from a prenatal state, dependent on the placenta, to the independent neonatal (newborn) state. The first trace of its development is the appearance of small spaces in the mesoderm before the third week. These become lined with endothelium and merge with each other to form the future blood vessels, lymphatic vessels, and heart. Two side-by-side endothelial tubes fuse to form a heart tube, which folds into the four-chambered heart.

The unique aspects of fetal circulation (fig. 29.10) are the umbilical-placental circuit and the presence of three circulatory shortcuts called *shunts.* The internal iliac arteries give rise to a pair of **umbilical arteries,** which pass on either side of the bladder into the umbilical cord. The blood in these arteries is low in oxygen

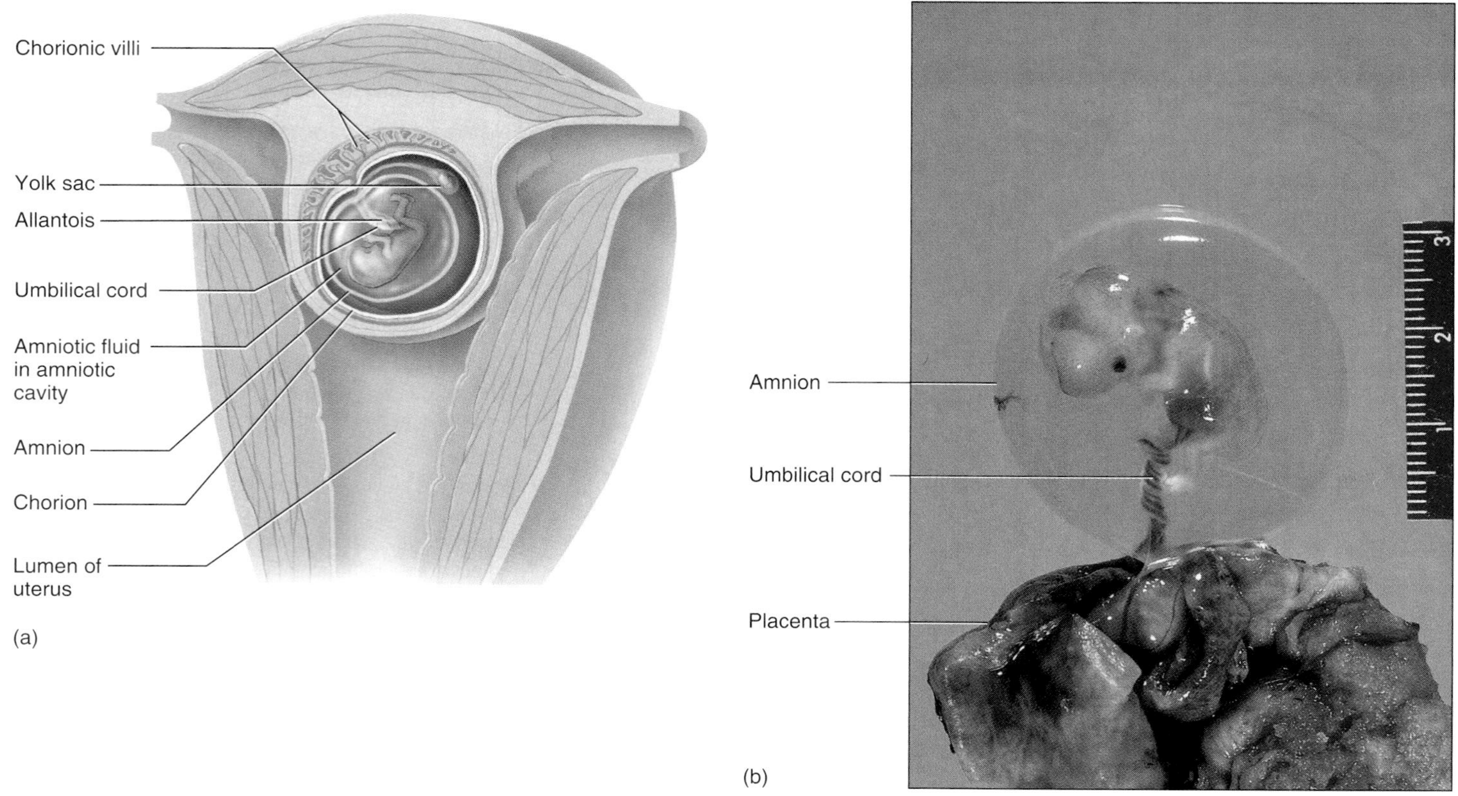

Figure 29.9 **The Embryonic Membranes.** (*a*) Diagram of a frontal section of the uterus. (*b*) Photograph of a human fetus at 8 weeks of gestation. The chorion has been stripped back to reveal the transparent amnion and the fetus. The scale bar is 3 cm.

Table 29.3 The Stages of Prenatal Development

Stage	Age*	Major Developments and Defining Characteristics
Zygote	24–30 hours	A single diploid cell formed by the union of egg and sperm
Cleavage	30–72 hours	Mitotic division of the zygote into smaller, identical blastomeres
Morula	3–4 days	A hollow, spherical stage consisting of 16 or more blastomeres
Blastocyst	4–14 days	A fluid-filled, spherical stage with an outer mass of trophoblast cells and inner mass of embryoblast cells; becomes implanted in the endometrium; inner cell mass forms an embryonic disc and differentiates into the three primary germ layers
Embryo	2–8 weeks	A stage in which the primary germ layers differentiate into organs and organ systems; ends when all organ systems are present
Fetus	8–40 weeks	A stage in which organs grow and mature at a cellular level to the point of being capable of supporting life independently of the mother

*From the time of ovulation

and high in carbon dioxide and other fetal wastes. It discharges these wastes in the placenta, loads oxygen and nutrients, and returns to the fetus by way of a single **umbilical vein,** which leads toward the liver. Some of this blood filters through the liver to nourish it, but most of it bypasses the liver by way of a shunt called the **ductus venosus,** which leads directly to the inferior vena cava. The immature liver is not capable of performing many of its postpartum functions; many of these are performed by the placenta.

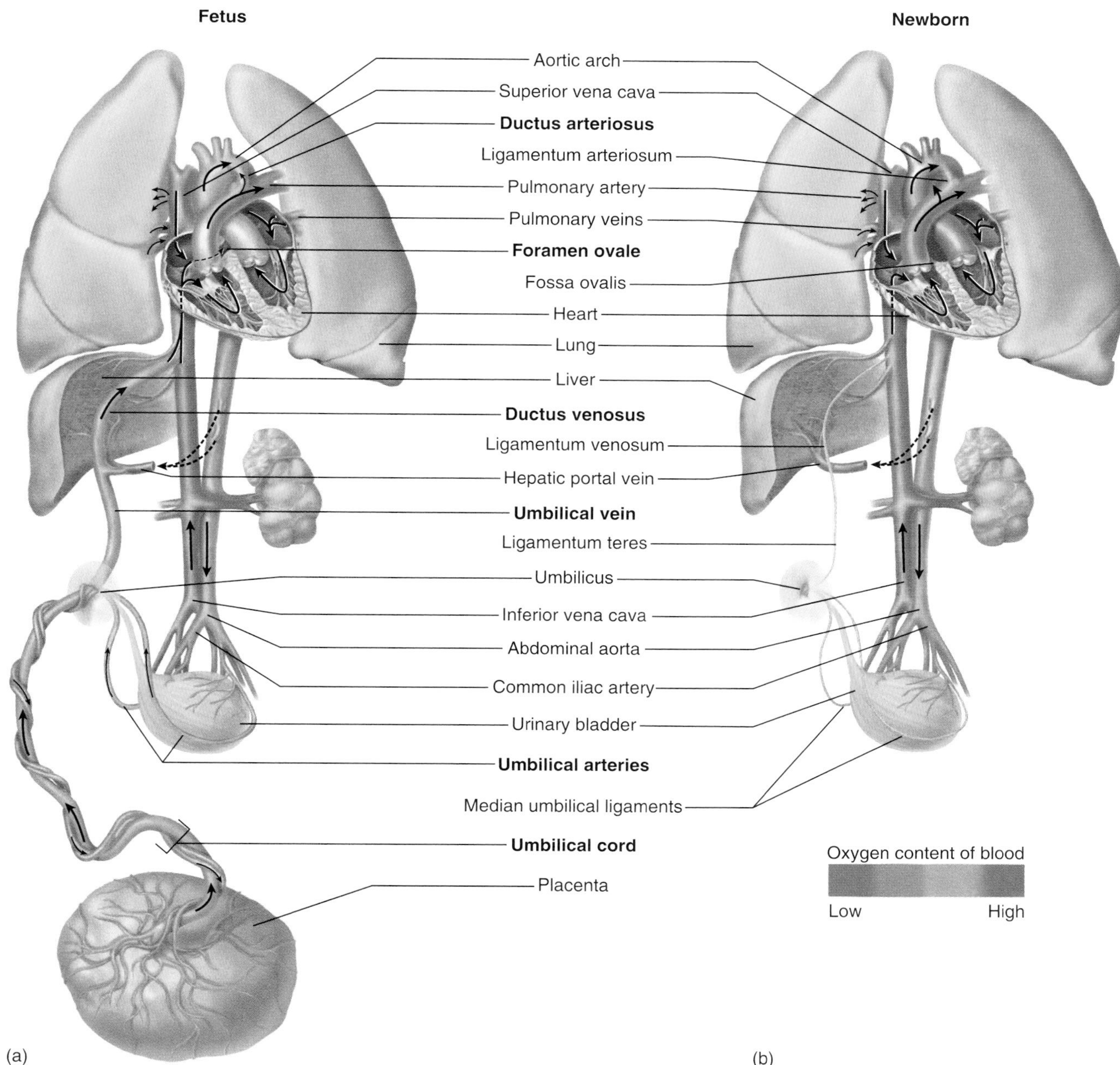

Figure 29.10 Changes in the Circulatory System from Fetus *(a)* to Newborn *(b)*.

In the inferior vena cava, placental blood mixes with venous blood from the fetus's body and flows to the right atrium of the heart. While the right heart normally pumps all of its blood into the lungs after birth, there is little need for this in the fetus because the lungs are not yet functional. Most of the fetal blood therefore bypasses the pulmonary circuit. Some passes directly from the right atrium to the left through the **foramen ovale,** a hole in the interatrial septum. Some goes into the right ventricle and is pumped into the pulmonary trunk, but most of this is shunted directly into the aorta by way of a short passage called the **ductus arteriosus.** This occurs because the collapsed state of the fetal lungs causes resistance and pressure in the pulmonary circuit to be high, so blood in the pulmonary trunk flows through the ductus into the aorta, where the blood pressure is lower. The lungs receive only a trickle of blood, sufficient to meet their metabolic needs during development. Blood leaving the left ventricle enters the general systemic circulation, and some of this returns to the placenta.

Table 29.4 Major Events of Prenatal Development, with Emphasis on the Fetal Stage

Calendar Month	Length* and Weight at End of Month	Developmental Events
1	0.6 cm	Spinal column and central nervous system begin to form; appendages represented by small **limb buds;** heart begins beating around day 22; no visible eyes, nose, or ears
2	3 cm; 1 g	Eyes form, eyelids fused shut; nose flat, nostrils evident but plugged with mucus; head nearly as large as the rest of the body; brain waves detectable; ossification begins; limb buds form paddlelike hands and feet with ridges called **digital rays,** which then separate into distinct fingers and toes; blood cells and major blood vessels form; genitals present but sexes not yet distinguishable
3	9 cm; 30 g	Eyes well developed, eyelids still fused; nose develops bridge; external ears present; limbs well formed, digits exhibit nails; fetus swallows amniotic fluid and produces urine; fetus moves but too weakly for mother to feel it; liver is prominent and produces bile; palate is fusing; sexes can be distinguished visually
4	14 cm; 100 g	Face looks more distinctly human; body larger in proportion to head; skin is bright pink, scalp has hair; joints forming; lips exhibit sucking movements; kidneys well formed; digestive glands forming and **meconium**[8] (fetal feces) accumulating in intestinal lumen; heartbeat can be heard with a stethoscope
5	19 cm; 200–450 g	Body covered with fine hair called **lanugo;**[9] skin has cheeselike sebaceous secretion called **vernix caseosa,**[10] which protects it from amniotic fluid; skin bright pink; brown fat forms and will be used for postpartum heat production; fetus is now bent forward into "fetal position" due to crowding; **quickening** occurs—mother can feel fetal movements
6	27–35 cm; 550–800 g	Eyes open, eyelashes form; skin wrinkled, pink, and translucent; lungs begin producing surfactant; rapid weight gain
7	32–42 cm; 1,100–1,350 g	Skin wrinkled and red; fetus turns into upside-down **vertex position;** bone marrow is now the sole site of hemopoiesis; testes descend into scrotum; can usually survive if born after 25 weeks
8	41–45 cm; 2,000–2,300 g	Subcutaneous fat deposition gives fetus a more plump, babyish appearance, with lighter, less wrinkled skin; twins usually born at this stage
9	50 cm; 3,200–3,400 g	More subcutaneous fat deposited; lanugo is shed; nails extend to or beyond fingertips

*Crown-to-rump (CR) length, measured from the top of the head to the bottom of the buttocks

8. *mecon* = poppy juice, opium
9. *lan* = down, wool
10. *vernix* = varnish + *caseo* = cheese

Other major aspects of embryonic and fetal development are listed in table 29.4 and depicted in figures 29.11 and 29.12.

Before You Go On

Answer the following questions to test your understanding of the preceding section:

5. Distinguish between trophoblastic and placental nutrition.
6. Identify the two sources of blood to the placenta. Where do these two bloodstreams come closest to each other? What keeps them separated?
7. State the functions of the placenta, amnion, chorion, yolk sac, and allantois.
8. What developmental characteristic distinguishes a fetus from an embryo? At what gestational age is this attained?
9. Identify the three circulatory shunts of the fetus. Why does the blood take these "shortcuts" before birth?

The Neonate

Objectives

When you have completed this section, you should be able to

- describe how and why the circulatory system changes at birth;
- explain why the first breaths of air are relatively difficult for a neonate;
- describe the major physiological problems of a premature infant; and
- discuss some common causes of birth defects.

Development is by no means complete at birth. For example, the liver and kidneys still are not fully functional, most joints are not yet ossified, and myelination of the nervous system is not completed until adolescence. Indeed, humans are born in a very immature state compared to other mammals.

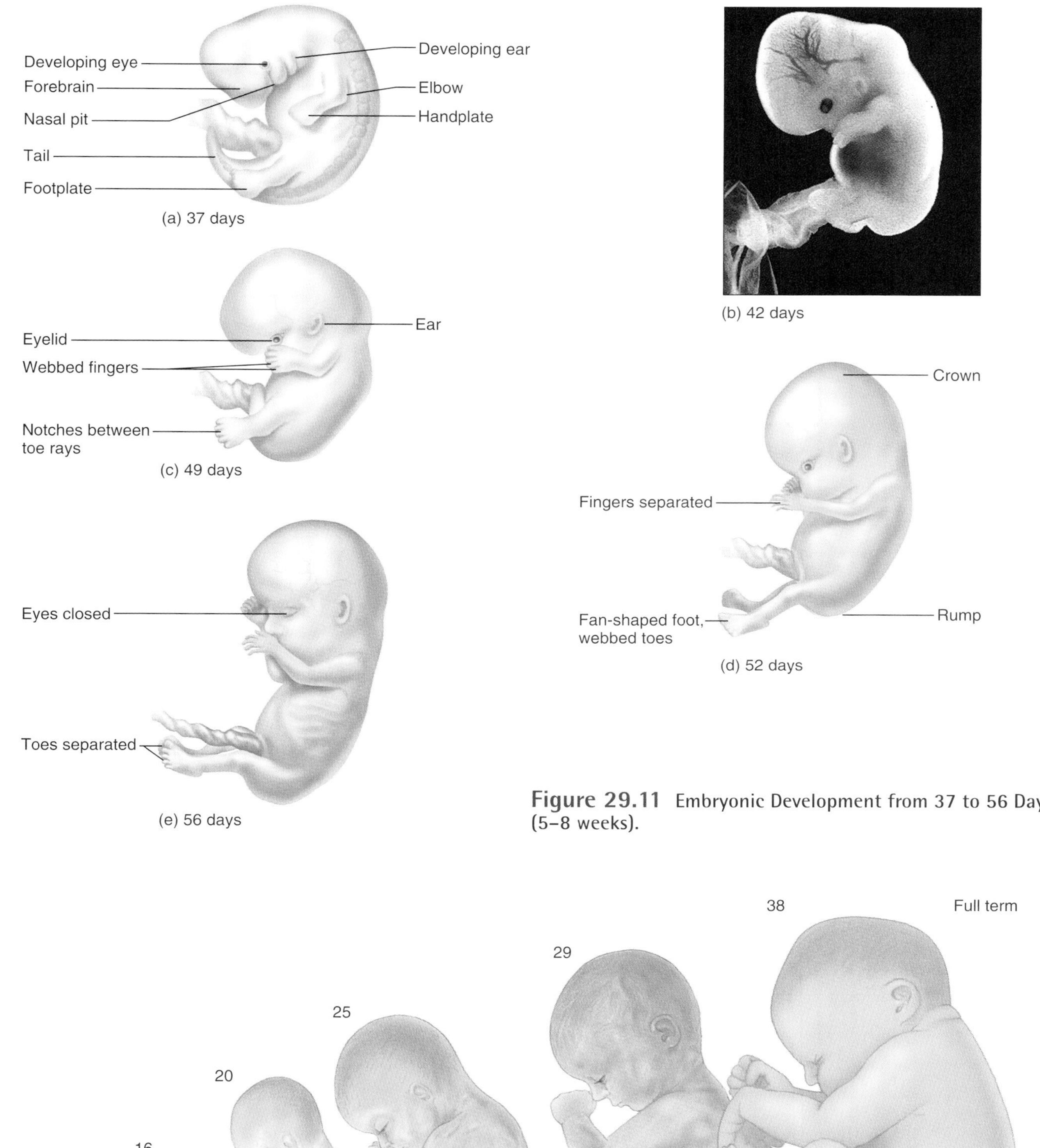

Figure 29.11 Embryonic Development from 37 to 56 Days (5–8 weeks).

Figure 29.12 Fetal Development from 9 to 38 Weeks (full term).

Clinical Insight 29.3

Neonatal Assessment

A newborn infant is immediately evaluated for general appearance, vital signs (temperature, pulse, and respiratory rate), weight, length, and head circumference and other dimensions, and it is screened for congenital disorders such as phenylketonuria (PKU). At 1 minute and 5 minutes after birth, the heart rate, respiratory effort, muscle tone, reflexes, and skin color are noted and given a score of 0 (poor), 1, or 2 (excellent). The total (0–10), called the *Apgar*[11] *score*, is a good predictor of infant survival. Infants with low Apgar scores may have neurological damage and need immediate attention if they are to survive. A low score at 1 minute suggests asphyxiation and may demand assisted ventilation. A low score at 5 minutes indicates a high probability of death.

11. Virginia Apgar (1909–74), American anesthesiologist

The Transitional Period

The period immediately following birth is a crisis in which the neonate suddenly must adapt to life outside the mother's body. The first 6 to 8 hours are a **transitional period** in which the heart and respiratory rates increase and the body temperature falls. Physical activity then declines and the baby sleeps for about 3 hours. In its second period of activity, the baby often gags on mucus and debris in the pharynx. The baby then sleeps again, becomes more stable, and begins a cycle of waking every 3 to 4 hours to feed. The first 6 weeks of life constitute the **neonatal period.**

Circulatory Adaptations

After the umbilical cord is clamped and cut, the umbilical arteries and veins collapse and become fibrotic. The proximal part of each umbilical artery becomes the *superior vesical artery,* which remains to supply the bladder. Other obliterated vessels become fibrous cords or ligaments: the distal parts of the umbilical arteries become the *medial umbilical ligaments* of the abdominal wall; the umbilical vein becomes the *round ligament (ligamentum teres)* of the liver; and the ductus venosus (the former shunt around the liver) becomes the *ligamentum venosum* on the inferior surface of the liver (see fig. 29.10*b*).

When the lungs expand with air, blood pressure in the pulmonary circuit drops rapidly and pressure in the right heart falls below that in the left. Blood flows briefly from the left atrium to the right through the foramen ovale and pushes two flaps of tissue into place to close this shunt. In most people these flaps fuse and permanently seal the foramen during the first year, leaving a depression, the *fossa ovalis,* in the interatrial septum. In about 25% of people, however, the foramen ovale remains unsealed and the flaps are held in place only by the relatively high blood pressure in the left atrium. Pressure changes in the pulmonary trunk and aorta also cause the ductus arteriosus to collapse. It closes permanently around 3 months of age and leaves a permanent cord, the *ligamentum arteriosum,* between the two vessels.

Respiratory Adaptations

It is an old and untrue cliché that a neonate must be spanked to stimulate it to breathe. During birth, CO_2 accumulates in the baby's blood and strongly stimulates the respiratory chemoreceptors. Unless the infant is depressed by oversedation of the mother, it normally begins breathing spontaneously. It requires a great effort, however, to take the first few breaths and inflate the collapsed alveoli. For the first 2 weeks, a baby takes about 45 breaths per minute but subsequently stabilizes at about 12 breaths per minute.

Immunological Adaptations

Cellular immunity begins to appear early in fetal development, but the immune responses of the neonate are still weak. Fortunately, an infant is born with a near-adult level of IgG acquired from the mother through the placenta. This maternal IgG breaks down rapidly after birth, declining to about half the initial level in the first month and to essentially none by 10 months. Nevertheless, maternal IgG levels remain high enough for 6 months to protect the infant from measles, diphtheria, polio, and most other infectious diseases (but not whooping cough). By 6 months, the infant's own IgG reaches about half the typical adult level. The lowest total (maternal + infant) level of IgG exists around 5 to 6 months of age, and respiratory infections are especially common at that age. A breast-fed neonate also acquires protection from gastroenteritis from the IgA present in colostrum.

Other Adaptations

Thermoregulation and fluid balance are also critical aspects of neonatal physiology. An infant has a larger ratio of surface area to volume than an adult does, so it loses heat more easily. One of its defenses against hypothermia is brown fat, a special adipose tissue deposited from weeks 17 to 20 of fetal development. The mitochondria of brown fat release all the energy of pyruvic acid as heat rather than using it to make ATP; thus, this is a heat-generating tissue. As a baby grows, its metabolic rate increases and it accumulates even more subcutaneous fat, thus producing and retaining more heat. Nevertheless, body temperature is more variable in infants and children than in adults.

The kidneys are not fully developed at birth and cannot concentrate the urine as much as a mature kidney can. Consequently, infants have a relatively high rate of water loss and require more fluid intake, relative to body weight, than adults do.

Premature Infants

Neonates weighing under 2.5 kg (5.5 lb) are generally considered **premature.** They have multiple difficulties in respiration, thermoregulation, excretion, digestion, and liver function.

The respiratory system is adequately developed by 7 months' gestation to support independent life. Infants born before this have a deficiency of pulmonary surfactant, causing **respiratory distress syndrome (RDS),** also called *hyaline membrane disease.* The alveoli collapse each time the infant exhales and a great effort is needed to reinflate them. The infant becomes very fatigued by the high energy demand of breathing. RDS may be treated by ventilating the lungs with oxygen-enriched air at a positive pressure to keep the lungs inflated between breaths and by administering surfactant as an inhalant. Nevertheless, RDS remains the most common cause of neonatal death.

A premature infant has an incompletely developed hypothalamus and therefore cannot thermoregulate effectively. Body temperature must be controlled by placing the infant in a warmer. If the infant is more than 8 weeks premature, its digestive tract is too poorly developed for a normal diet of breast milk. The baby requires a low-fat formula because it cannot absorb fats very well, and it is given calcium and vitamin D supplements to promote ossification.

The liver is also poorly developed, and bearing in mind its very diverse functions (see table 26.7, p. 1004), you can probably understand why this would have several serious consequences. The liver synthesizes inadequate amounts of albumin, so the baby suffers hypoproteinemia. This upsets the balance between capillary filtration and reabsorption and leads to edema. The infant bleeds easily because of a deficiency of the clotting factors synthesized by the liver. This is true to some degree even in full-term infants, however, because the baby's intestines are not yet colonized by the bacteria that synthesize vitamin K, which is essential for the synthesis of clotting factors. Jaundice is common in neonates, especially premature babies, because the liver cannot dispose of bile pigments efficiently.

Congenital Anomalies

The most anxious moment for many new parents is awaiting reassurance that the child has no visible birth defects. Any abnormality present at birth is called a **congenital anomaly.**[12] Here we consider some anomalies that result from infectious diseases, teratogens, mutagens, and genetic defects.

Infectious Diseases

Infectious diseases are largely beyond the scope of this book, but it must be noted at least briefly that several microorganisms can cross the placenta and cause serious congenital anomalies, stillbirth, or neonatal death. Common viral infections of the fetus and newborn include herpes simplex, rubella, cytomegalovirus, and human immunodeficiency virus (HIV). Congenital bacterial infections include gonorrhea and syphilis. *Toxoplasma,* a protozoan contracted from meat, unpasteurized milk, and housecats, is another common cause of fetal deformity. Some of these pathogens have relatively mild effects on adults, but because of its immature immune system, the fetus is vulnerable to devastating effects such as blindness, hydrocephalus, cerebral palsy, seizures, and profound physical and mental retardation. These diseases are treated in greater detail in microbiology textbooks.

Teratogens

Teratogens[13] are viruses, chemicals, and other agents that cause anatomical deformities in the fetus. Perhaps the most notorious teratogenic drug is thalidomide, a sedative first marketed in 1957. Thalidomide was taken by women in early pregnancy, often before they knew they were pregnant, and caused over 5,000 babies to be born with unformed arms or legs (fig. 29.13) and often with defects of the ears, heart, and intestines. It was taken off the market in 1961 but has recently been reintroduced for limited purposes. Many teratogens produce less obvious effects, including physical or mental retardation, hyperirritability, inattention, strokes, seizures, respiratory arrest, crib death, and cancer. A general lesson to be learned from the thalidomide tragedy and other cases is that pregnant women should avoid all sedatives, barbiturates, and opiates. Even the acne medicine Acutane has caused severe birth defects.

Alcohol causes more birth defects than any other drug. Even one drink a day has noticeable effects on fetal

---

12. *con* = with + *gen* = born + *a* = without + *nom* = rule, regularity
13. *terato* = monster + *gen* = producing

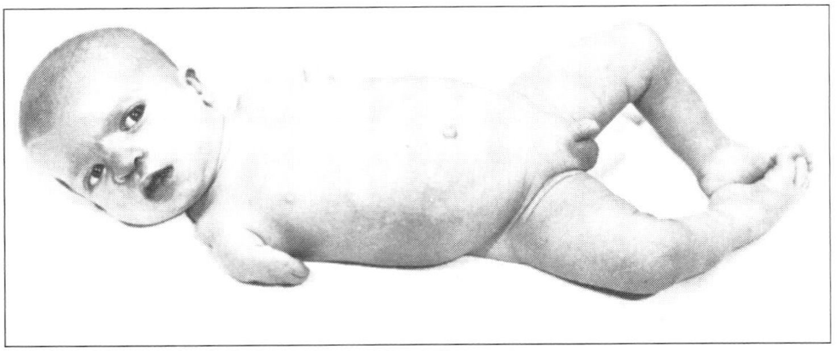

Figure 29.13 Infant with Undeveloped Arms as a Result of Thalidomide, a Sedative Taken by the Mother in Early Pregnancy.

and postpartum development, some of which are not noticed until a child begins school. Alcohol abuse during pregnancy can cause **fetal alcohol syndrome (FAS),** characterized by a small head, malformed facial features, cardiac and central nervous system defects, stunted growth, and behavioral signs such as hyperactivity, nervousness, and a poor attention span. Cigarette smoking also contributes to fetal and infant mortality, ectopic pregnancy, anencephaly (failure of the cerebrum to develop), cleft lip and palate, and cardiac abnormalities. Diagnostic X rays should be avoided during pregnancy because radiation can have teratogenic effects.

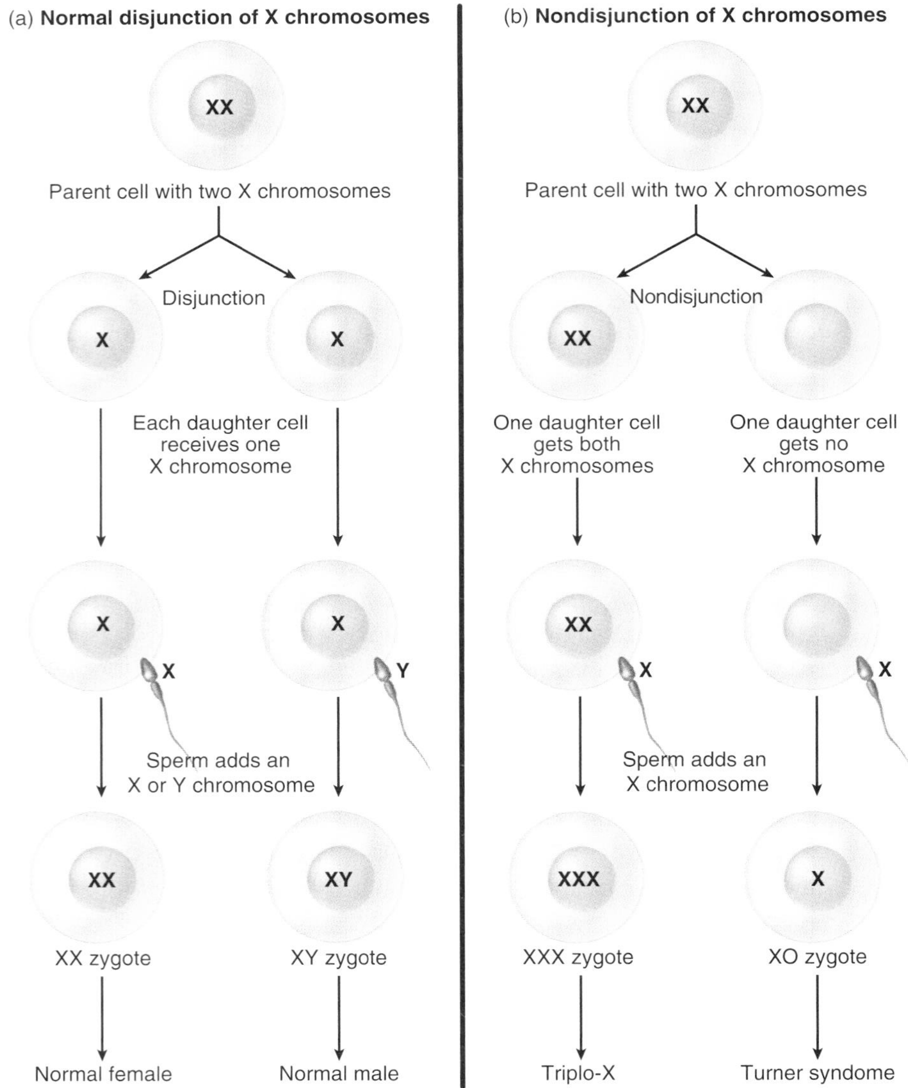

Figure 29.14 Disjunction and Nondisjunction. (*a*) The outcome of normal disjunction and fertilization by X- or Y-bearing spermatozoa. (*b*) Two of the possible outcomes of nondisjunction followed by fertilization with an X-bearing spermatozoon. If the spermatozoon had carried a Y chromosome, it would have produced either an XXY individual with Klinefelter syndrome or a YO zygote that would die with little further development.

Mutagens and Genetic Anomalies

A **mutagen** is any agent that alters DNA or chromosome structure. Ionizing radiation and some chemicals have mutagenic, teratogenic, and carcinogenic effects, with extremely diverse results. Prenatal exposure to mutagens may result, for example, in stillbirths or in increased risk of childhood cancer.

Some of the most common genetic disorders result not from mutagens, however, but from the failure of homologous chromosomes to separate during meiosis. Recall that homologous chromosomes pair up during prophase I and normally separate from each other at anaphase I (see p. 1033). This separation, called *disjunction,* produces daughter cells with 23 chromosomes each.

In **nondisjunction,** a pair of chromosomes fails to separate. Both chromosomes then go to the same daughter cell, which receives 24 chromosomes while the other daughter cell receives 22. **Aneuploidy**[14] (AN-you-PLOY-dee), the presence of an extra chromosome or lack of one, accounts for about 50% of spontaneous abortions. It can be detected prior to birth by *amniocentesis,* the examination of cells in a sample of amniotic fluid, or by *chorionic villus sampling,* the removal and examination of cells from the chorion.

Figure 29.14 compares normal disjunction of the X chromosomes with some effects of nondisjunction. In nondisjunction, an egg cell may receive both X chromosomes. If it is fertilized by an X-bearing sperm, the result is an XXX zygote and a set of anomalies called the **triplo-X syndrome.** Triplo-X females are sometimes infertile and sometimes have mild intellectual impairments. If an XX egg is fertilized by a Y-bearing sperm, the result is an XXY combination and **Klinefelter**[15] **syndrome.** People with Klinefelter syndrome are sterile males, usually of average intelligence, but with undeveloped testes, sparse body hair, unusually long arms and legs, and enlarged breasts (gynecomastia) (fig. 29.15*a*). This syndrome often goes undetected until puberty, when failure to develop the secondary sex characteristics may prompt genetic testing.

The other possible outcome of X chromosome nondisjunction is that an egg cell may receive no X chromosome (both X chromosomes are discarded in the first polar body). If fertilized by a Y-bearing sperm, it dies for lack of the indispensable genes on the X chromosome. If it is fertilized by an X-bearing sperm, however, the result is **Turner**[16] **syndrome,** with an XO combination (*O* represents the absence of one sex chromosome). Only 3% of those with Turner syndrome survive to birth. Girls with Turner syndrome show no serious impairments as children but tend to have a webbed neck and widely spaced nipples. At puberty, the secondary sex characteristics fail to develop (fig. 29.15*b*). The ovaries are nearly absent, the girl remains sterile, and she usually has a short stature.

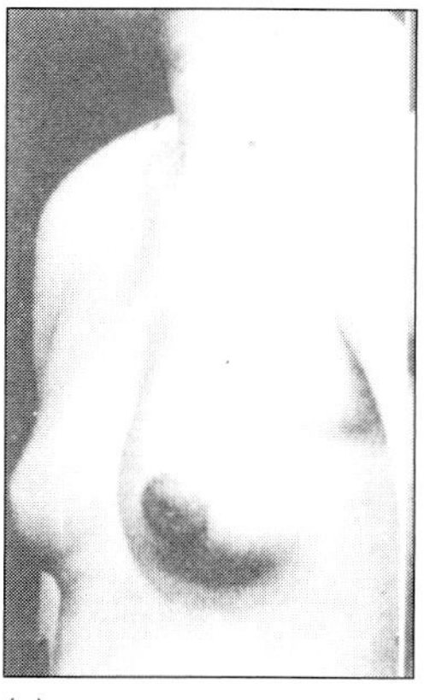

(a)

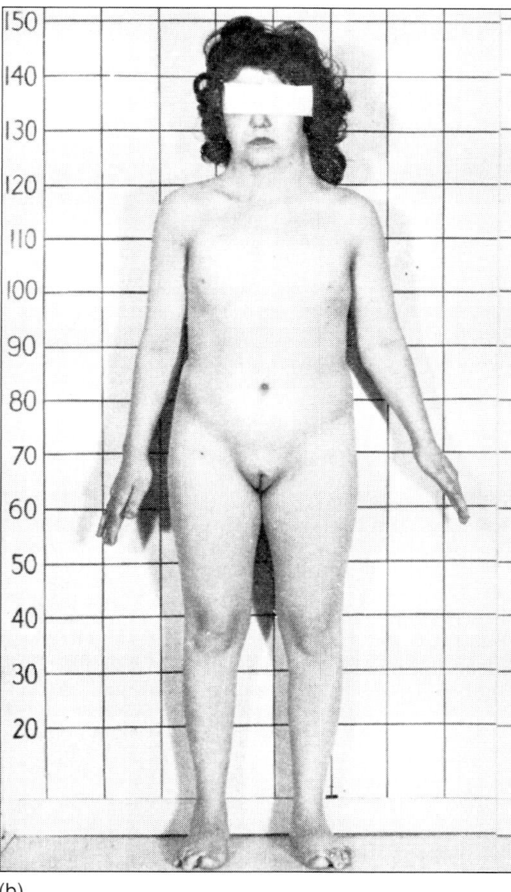

(b)

Figure 29.15 Syndromes Resulting from Nondisjunction of X Chromosomes. (*a*) Gynecomastia in a male with Klinefelter syndrome. (*b*) A 22-year-old woman with Turner syndrome. Note her short stature (about 145 cm, or 4 ft 9 in.), lack of sexual development, webbed neck, and widely spaced nipples.

14. *an* = not, without + *eu* = true, normal + *ploid,* from *diplo* = double, paired
15. Harry F. Klinefelter, Jr. (1912–), American physician
16. Henry H. Turner (1892–1970), American endocrinologist

The other 22 pairs of chromosomes (the autosomes) are also subject to nondisjunction. The most common autosomal anomaly is **Down[17] syndrome (trisomy-21),** which results from three copies of chromosome 21. The signs include retarded physical development; short stature; a relatively flat face with a flat nasal bridge; low-set ears; *epicanthal[18] folds* at the medial corner of the eyes; an enlarged, protruding tongue; stubby fingers; and a short broad hand with only one palmar crease (fig. 29.16). People with Down syndrome tend to have outgoing, affectionate personalities. Mental retardation is common and sometimes severe, but it is not inevitable. Nearly 50% of victims die

17. John Langdon H. Down (1828–96), British physician

18. *epi* = upon + *canth* = commissure of the eyelids

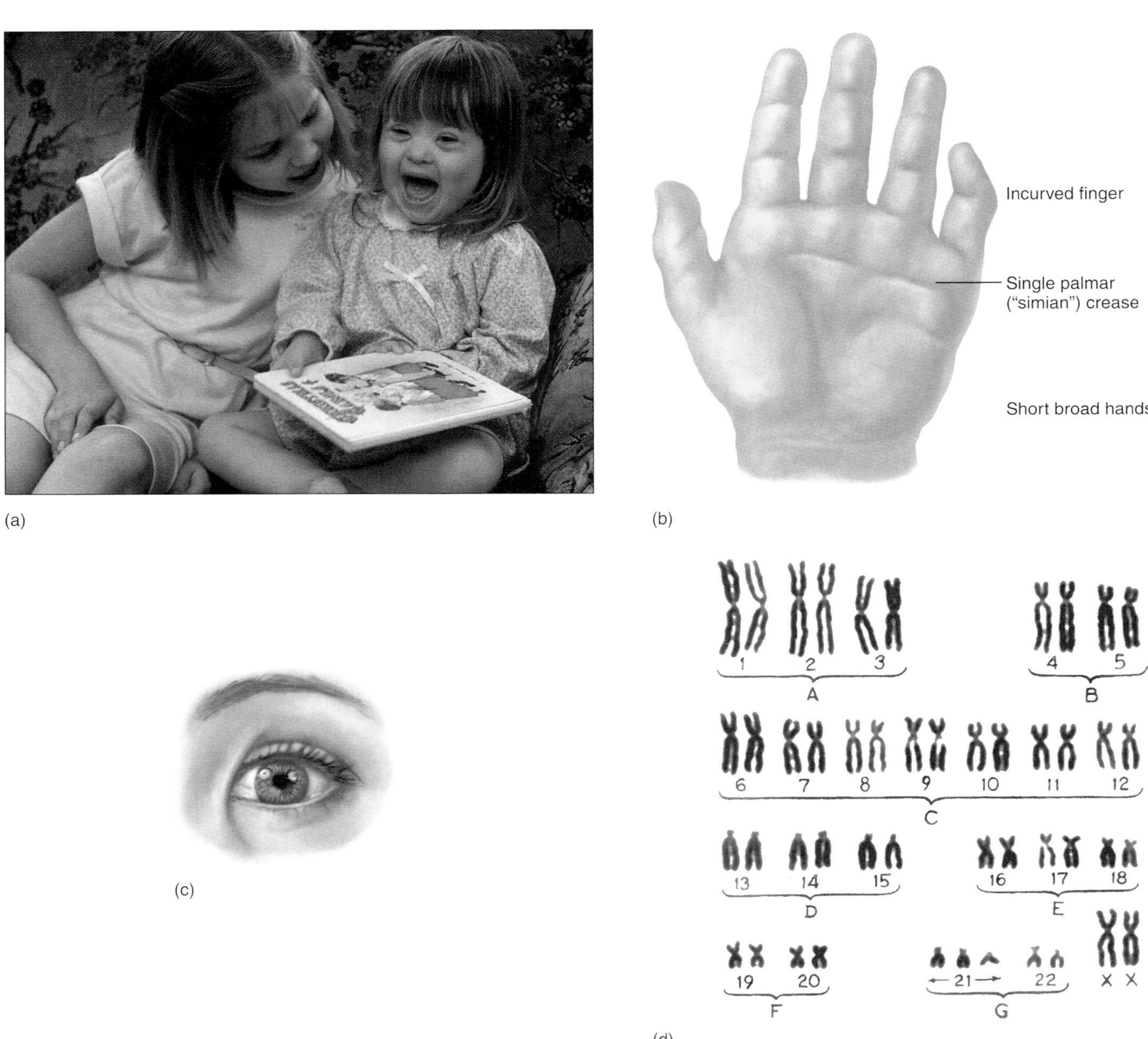

Figure 29.16 Down Syndrome. (*a*) A child with Down syndrome (*right*) plays with her big sister. (*b*) Characteristics of the hand seen in Down syndrome. (*c*) The epicanthal fold over the medial commissure (canthus) of the left eye. (*d*) The karyotype of Down syndrome showing the trisomy of chromosome 21.

within the first year of life due to immune deficiency or abnormalities of the heart or kidneys. Down syndrome occurs in about 1 out of 700 to 800 live births in the United States and increases in proportion to the age of the mother. The chance of having a child with Down syndrome is about 1 in 3,000 for a woman under 30, 1 in 365 by age 35, and 1 in 9 by age 48.

Before You Go On

Answer the following questions to test your understanding of the preceding section:

10. How does inflation of the lungs at birth affect the route of blood flow through the heart?
11. Why is respiratory distress syndrome common in premature infants?
12. Define nondisjunction and explain how it causes aneuploidy. Name two syndromes resulting from aneuploidy.

Aging and Senescence

Objectives

When you have completed this section, you should be able to

- define *senescence* and distinguish it from aging;
- describe some major changes that occur with aging in each organ system;
- summarize some current theories of senescence; and
- be able to explain how exercise and other factors can slow the rate of senescence.

Like Ponce de León searching for the legendary fountain of youth in the Florida wilderness, people yearn for a way to preserve their youthful appearance and function. Our real concern, however, is not aging but senescence. The term **aging** is used in various ways but is taken here to mean all changes that occur in the body with the passage of time—including the growth, development, and increasing functional efficiency that occur from childhood to adulthood, as well as the degenerative changes that occur later in life. **Senescence** is the degeneration that occurs in an organ system after the age of peak functional efficiency. It includes a gradual loss of reserve capacities, reduced ability to repair damage and compensate for stress, and increased susceptibility to disease.

Senescence is not just a personal concern but an important issue for health-care providers. One in nine Americans is 65 or older. As the average age of the population rises, health-care professionals will find themselves increasingly occupied by the prevention and treatment of the diseases of age. The leading causes of death change markedly with age. Accidents, homicide, suicide, and AIDS figure prominently in the deaths of people 18 to 34 years old, whereas the major causes of death after age 55 are clearly related to senescence of the organ systems: heart disease, cancer, stroke, diabetes, and lung disease. The causes of senescence, however, remain as much a scientific mystery today as cancer was 50 years ago and heredity was 100 years ago.

As we survey the senescence of the organ systems, you should notice many points relevant not only to caring for an aging population but also to your personal health and fitness practices that can lessen the effects of senescence and improve the quality of life in your later years. In addition, the study of senescence calls renewed attention to the multiple interactions among organ systems. As you will see, the senescence of one organ system typically contributes to the senescence of others. Your study of this topic will bring together many concepts introduced in earlier chapters of the book. You may find the glossary helpful in refreshing your memory of concepts revisited in the following discussion.

Such terms as *middle age, old age,* and *elderly* have no precise, universal definitions, but for present purposes we consider ages 40 to 59 to be middle age and ages 60 and over to be old age. Unless otherwise stated, when describing the functional decline in the organ systems, we are comparing average, healthy people in their 60s with people in their 20s.

Senescence of the Organ Systems

Organ systems do not all degenerate at the same rate. For example, from ages 30 to 80, the speed of nerve conduction declines only 10% to 15%, but the number of functional glomeruli in the kidneys declines about 60%. Some physiological functions show only moderate changes at rest but more pronounced differences when tested under exercise conditions. The organ systems also vary widely in the age at which senescence becomes noticeable. There are traces of atherosclerosis, for example, even in infants, and visual and auditory sensitivity begin to decline soon after puberty. By contrast, the female reproductive system does not show significant senescence until menopause and then its decline is relatively abrupt. Aside from these unusual examples, most physiological measures of performance peak between the late teens and age 30 and then decline at a rate influenced by the level of use of the organs.

Integumentary System

Two-thirds of people aged 50 and over, and nearly all people over age 70, have medical concerns or complaints about their skin. Senescence of the integumentary system often becomes most noticeable in the late 40s. The hair turns grayer and thinner as melanocytes die out, mitosis slows down, and dead hairs are not replaced. The atrophy of sebaceous glands leaves the skin and hair drier. As epidermal mitosis declines and collagen is lost from the dermis, the skin becomes almost paper-thin and translucent. It becomes looser because of a loss of elastic fibers and flattening of the dermal papillae, which normally form a stress-resistant corrugated boundary between the dermis and epidermis. If you pinch a fold of skin on the back of a child's hand, it quickly springs back when you let go; do the same on an older person and the skinfold remains longer. Because of its loss of elasticity, aged skin sags to various degrees and may hang loosely from the arm and other places.

Aged skin has fewer blood vessels than younger skin, and those that remain are more fragile. The skin can become reddened as broken vessels leak into the connective tissue. Many older people exhibit **rosacea**—patchy networks of tiny, dilated blood vessels visible especially on the nose and cheeks. Because of the fragility of the dermal blood vessels, aged skin bruises more easily. Injuries to the skin are more common and severe in old age, partly because the cutaneous nerve endings decline by two-thirds from age 20 to 80, leaving one less aware of touch, pressure, and injurious stimuli. Injured skin heals slowly in old age because of poorer circulation and a relative scarcity of immune cells and fibroblasts. Antigen-presenting dendritic cells decline by as much as 40% in the aged epidermis, leaving the skin more susceptible to recurring infections.

Thermoregulation is a serious problem in old age because of the atrophy of cutaneous blood vessels, sweat glands, and subcutaneous fat. Older people are more vulnerable to hypothermia in cold weather and heatstroke in hot weather. Heat waves and cold spells take an especially heavy toll among the elderly poor, who suffer from a combination of reduced homeostasis and inadequate housing.

These are all "normal" changes in the skin, or **intrinsic aging**—changes that occur more or less inevitably with the passage of time. In addition, there is **photoaging**—degenerative changes in proportion to a person's lifetime exposure to ultraviolet radiation. UV radiation accounts for more than 90% of the integumentary changes that people find medically troubling or cosmetically disagreeable: skin cancer; yellowing and mottling of the skin; age spots, which resemble enlarged freckles on the back of the hand and other sun-exposed areas; and wrinkling, which affects the face, hands, and arms more than areas of the body that receive less exposure. A lifetime of outdoor activity can give the skin a leathery, deeply wrinkled, "outdoorsy" appearance (fig. 29.17), but beneath this rugged exterior is a less happy histological appearance. The sun-damaged skin shows many malignant and premalignant cells, extensive damage to the dermal blood vessels, and dense masses of coarse, frayed elastic fibers underlying the surface wrinkles and creases.

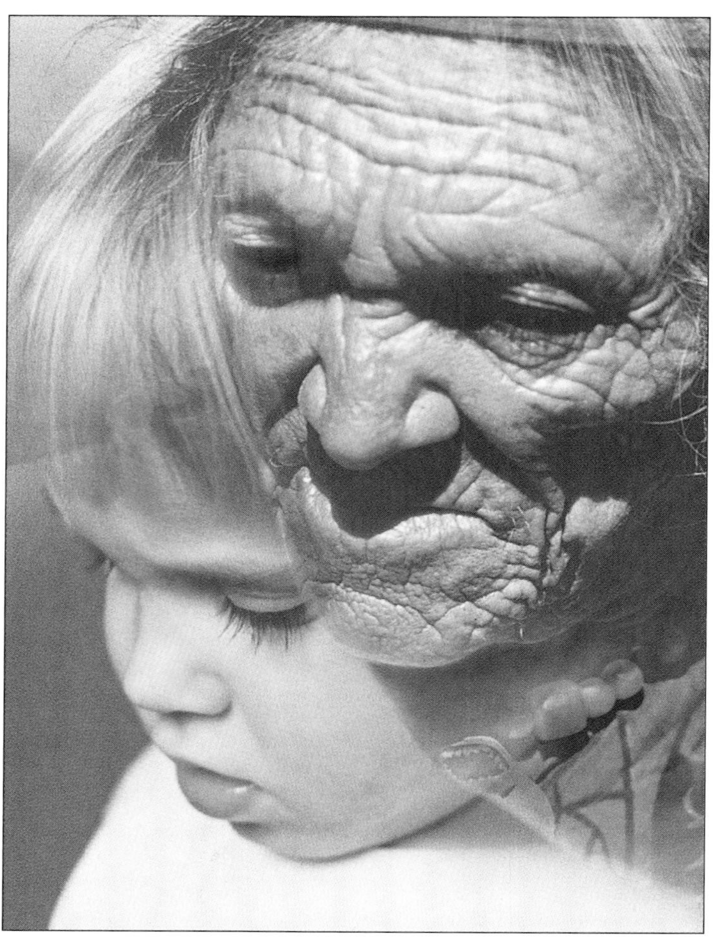

Figure 29.17 Aging of the Skin. The deep creases seen in the elderly woman resulted mainly from photoaging.
Photograph of child from *The 1974 Science Year.* 1973 Field Enterprises Educational Corporation. By permission of World Book, Inc.

Senescence of the skin has far-reaching effects on other organ systems. Cutaneous vitamin D production declines as much as 75% in old age. This is all the more significant because the elderly spend less time outdoors, and because of increasing lactose intolerance, they often avoid dairy products, the only dietary source of vitamin D. Consequently, the elderly are at high risk of insufficient calcium absorption, which, in turn, contributes to bone loss, muscle weakness, and impaired glandular secretion and synaptic transmission.

Skeletal System

After age 30, osteoblasts become less active than osteoclasts. This imbalance results in **osteopenia,** the loss of bone; when the loss is severe enough to compromise a person's physical activity and health, it is called *osteoporosis* (see p. 253). After age 40, women lose about 8% of their bone mass per decade and men about 3%. Bone loss from the jaws is a contributing factor in tooth loss.

Not only does bone density decline with age, but the bones become more brittle as the cells synthesize less protein. Fractures occur more easily and heal more slowly. A fracture may impose a long period of immobility, which makes a person more vulnerable to pneumonia and other infectious diseases.

People notice more stiffness and pain in the synovial joints as they age, and degenerative joint diseases affect the lifestyle of 85% of people over age 75. Synovial fluid is less abundant and the articular cartilage is thinner or absent. Exposed bone surfaces abrade each other and cause friction, pain, and reduced mobility. *Osteoarthritis* is the most common joint disease of older people and one of the most common causes of physical disability (p. 335). Even breathing becomes more difficult and tiring in old age because expansion of the thorax is restricted by calcification of the sternocostal joints. Degeneration of the intervertebral discs causes back pain and stiffness, but herniated discs are less common in old age than in youth because the discs become more fibrous and stronger, with less nucleus pulposus.

Muscular System

One of the most noticeable changes we experience with age is the replacement of lean body mass (muscle) with fat. The change is dramatically exemplified by CT scans of the thigh. In a young well-conditioned male, muscle accounts for 90% of the cross-sectional area of the midthigh, whereas in a frail 90-year-old woman, it is only 30%. Muscular strength and mass peak in the 20s; by the age of 80, most people have only half as much strength and endurance. A large percentage of people over age 75 cannot lift a 4.5 kg (10 lb) weight with their arms; such simple tasks as carrying a sack of groceries into the house may become impossible. The loss of strength is a major contributor to falls, fractures, and dependence on others for the routine activities of daily living. Fast-twitch fibers exhibit the earliest and most severe atrophy, thus increasing reaction time and reducing coordination.

There are multiple reasons for the loss of strength. Aged muscle fibers have fewer myofibrils, so they are smaller and weaker. The sarcomeres are increasingly disorganized, and muscle mitochondria are smaller and have reduced quantities of oxidative enzymes. Aged muscle has less ATP, creatine phosphate, glycogen, and myoglobin; consequently, it fatigues quickly. Muscles also exhibit more fat and fibrosis with age, which limits their movement and blood circulation. With reduced circulation, muscle injuries heal more slowly and with more scar tissue.

But the weakness and easy fatigue of aged muscle also stems from the senescence of other organ systems. There are fewer motor neurons in the spinal cord, and some muscle shrinkage may represent denervation atrophy. The remaining neurons produce less acetylcholine and show less efficient synaptic transmission, which makes the muscles slower to respond to stimulation. As muscle atrophies, motor units have fewer muscle fibers per motor neuron, and more motor units must be recruited to perform a given task. Tasks that used to be easy, such as buttoning the clothes or eating a meal, take more time and effort. The sympathetic nervous system is also less efficient in old age; consequently, blood flow to the muscles does not respond efficiently to exercise and this contributes to their rapid fatigue.

Nervous System

The nervous system reaches its peak development around age 30. From age 35 on, it is estimated that 100,000 brain cells die each day. The average brain weighs 56% less at age 75 than at age 30. The cerebral gyri are narrower, the sulci are wider, the cortex is thinner, and there is more space between the brain and meninges. The remaining cortical neurons have fewer synapses, and for multiple reasons, synaptic transmission is less efficient: the neurons produce less neurotransmitter, they have fewer receptors, and the neuroglia around the synapses is more leaky and allows neurotransmitter to diffuse away. The degeneration of myelin sheaths with age also slows down signal conduction.

Neurons exhibit less rough ER and Golgi complex with age, which indicates that their metabolism is slowing down. Old neurons accumulate lipofuscin pigment and show more neurofibrillary tangles—dense mats of cytoskeletal elements in their cytoplasm. In the extracellular material, plaques of fibrillar protein (amyloid) appear, especially in people with Down syndrome and Alzheimer disease (AD). AD is the most common nervous disability of old age (p. 485).

Not all functions of the central nervous system are equally affected by senescence. Motor coordination, intellectual function, and short-term memory decline more than language skills and long-term memory. Elderly people are often better at remembering things in the distant past than remembering recent events.

The sympathetic nervous system loses adrenergic receptors with age and becomes less sensitive to norepinephrine. This contributes to a decline in homeostatic control of such variables as body temperature and blood pressure. Many elderly people experience *orthostatic hypotension*—a drop in blood pressure when they stand, which sometimes results in dizziness, loss of balance, or fainting.

Sense Organs

Some sensory functions decline shortly after adolescence. Presbyopia (loss of flexibility in the lenses) makes it more difficult for the eyes to focus on nearby objects. Visual acuity declines and often requires corrective lenses by middle age. Cataracts (cloudiness of the lenses) are more common in old age. Night vision is impaired as more and more light is needed to stimulate the retina. This has several causes: there are fewer receptor cells in the retina, the vitreous body becomes less transparent, and the pupil becomes narrower as the pupillary dilators atrophy. Dark adaptation takes longer as the enzymatic reactions of the photoreceptor cells become slower. Changes in the structure of the iris, ciliary body, or lens can block the reabsorption of aqueous humor, thereby increasing the risk of glaucoma. Having to give up reading and driving can be among the most difficult changes of lifestyle in old age.

Auditory sensitivity peaks in adolescence and declines afterward. The tympanic membrane and the joints between the auditory ossicles become stiffer, so vibrations are transferred less effectively to the inner ear, creating a degree of conduction deafness. Nerve deafness occurs as the number of cochlear hair cells and auditory nerve fibers declines. The greatest hearing loss occurs at high frequencies and in the frequency range of most conversation. The death of receptor cells in the semicircular ducts, utricle, and saccule, and of nerve fibers in the vestibular nerve and neurons in the cerebellum, results in poor balance and dizziness—another factor in falls and bone fractures.

The senses of taste and smell are blunted as taste buds, olfactory cells, and second-order neurons in the olfactory bulbs decline in number. Food may lose its appeal, and thus declining sensory function can be a factor in malnutrition.

Endocrine System

The endocrine system degenerates less than any other organ system. The reproductive hormones drop sharply and growth hormone and thyroid hormone secretion decline steadily after adolescence, but other hormones continue to be secreted at fairly stable levels even into old age. Target cell sensitivity declines, however, so some hormones have less effect. For example, the pituitary gland is less sensitive to negative feedback inhibition by adrenal glucocorticoids; consequently, the response to stress is more prolonged than usual. Diabetes mellitus is more common in old age, largely because target cells have fewer insulin receptors. In part, this is an effect of the greater percentage of body fat in the elderly. The more fat at any age, the less sensitive other cells are to insulin. Body fat increases as the muscles atrophy, and muscle is one of the body's most significant glucose-buffering tissues. Because of the blunted insulin response, glucose levels remain elevated longer than normal after a meal.

Circulatory System

Cardiovascular disease is a leading cause of death in old age. Senescence has multiple effects on the blood, heart, arteries, and veins. Anemia may result from nutritional deficiencies, inadequate exercise, disease, and other causes. The factors that cause anemia in older people are so complicated it is almost impossible to control them enough to determine whether aging alone causes it. Evidence suggests that there is no change in the baseline rate of erythropoiesis in old age. Hemoglobin concentration, cell counts, and other variables are about the same among healthy people in their 70s as in the 30s. However, older people do not adapt well to stress on the hemopoietic system, perhaps because of the senescence of other organ systems. As the gastric mucosa atrophies, for example, it produces less of the intrinsic factor needed for vitamin B12 absorption. This increases the risk of pernicious anemia. As the kidneys age and the number of nephrons declines, less erythropoietin is secreted. There may also be a limit to how many times the hemopoietic stem cells can divide and continue giving rise to new blood cells. Whatever its cause, anemia limits the amount of oxygen that can be transported and thus contributes to the atrophy of tissues everywhere in the body.

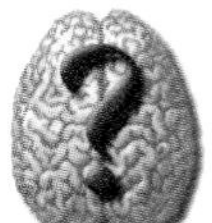

Think About It

Draw a positive feedback loop showing how anemia and senescence of the kidneys could affect each other.

Everyone exhibits coronary atherosclerosis with age. Consequently, myocardial cells die, angina pectoris and myocardial infarction become more common, the heart wall becomes thinner and weaker, and stroke volume, cardiac output, and cardiac reserve decline. Like other connective tissues, the cardiac skeleton becomes less elastic. This limits the distension of the heart and reduces the force of cardiac systole. Degenerative changes in the nodes and conduction pathways of the heart lead to a higher incidence of cardiac arrhythmia and heart block. Physical endurance is compromised by the drop in cardiac output.

As atherosclerosis stiffens the arteries, they cannot expand as effectively to accommodate the pressure surges of cardiac systole. Blood pressure therefore rises steadily with age (see table 20.1, p. 757). Atherosclerosis also narrows the arteries and reduces the perfusion of most organs. The effects of reduced circulation on the skin, skeletal muscles, and brain have already been noted. The combination of atherosclerosis and hypertension also weakens the arteries and increases the risk of aneurysm and stroke.

Atherosclerotic plaques trigger thrombosis, especially in the lower extremities, where flow is relatively slow and the blood clots more easily. About 25% of people over age 50 experience venous blockage by thrombosis—especially people who do not exercise regularly.

Degenerative changes in the veins are most evident in the extremities. The valves become weaker and less able to stop the backflow of blood. Blood pools in the legs and feet, raises capillary blood pressure, and causes edema. Chronic stretching of the vessels often produces varicose veins and hemorrhoids. Support hose can reduce edema by compressing the tissues and forcing tissue fluid to return to the bloodstream, but physical activity is even more important in promoting venous return.

Immune System

The amounts of lymphatic tissue and red bone marrow decline with age; consequently there are fewer hemopoietic stem cells, disease-fighting leukocytes, and antigen-presenting cells (APCs). Also, the lymphocytes produced in these tissues often fail to mature and become immunocompetent. Both humoral and cellular immunity depend on APCs and helper T cells, and therefore both types of immune response are blunted. As a result, an older person has less protection against cancer and infectious diseases. It becomes especially important in old age to be vaccinated against influenza and other acute seasonal infections.

Respiratory System

Pulmonary ventilation declines steadily after the 20s and is one of several factors in the gradual loss of stamina. The costal cartilages and joints of the thoracic cage become less flexible, the lungs have less elastic tissue, and the lungs have fewer alveoli. Vital capacity, minute respiratory volume, and forced expiratory volume fall. The elderly are also less capable of clearing the lungs of irritants and pathogens and are therefore increasingly vulnerable to respiratory infections. Pneumonia causes more deaths than any other infectious disease and is often contracted in hospitals and nursing homes.

The chronic obstructive pulmonary diseases (COPDs)—emphysema and chronic bronchitis—are more common in old age since they represent the cumulative effects of a lifetime of degenerative change. They are among the leading causes of death in old age. Pulmonary obstruction also contributes to cardiovascular disease, hypoxemia, and hypoxic degeneration in all the organ systems. Respiratory health is therefore a major concern in aging.

Urinary System

The kidneys exhibit a striking degree of atrophy with age. From ages 25 to 85, the number of nephrons declines 30% to 40% and up to a third of the remaining glomeruli become atherosclerotic, bloodless, and nonfunctional. The kidneys of a 90-year-old are 20% to 40% smaller than those of a 30-year-old and receive only half as much blood. The glomerular filtration rate is proportionately lower and the kidneys are less efficient at clearing wastes from the blood. Although baseline renal function is adequate even in old age, there is little reserve capacity; thus other diseases can lead to surprisingly rapid renal failure. Drug doses often need to be reduced in old age because the kidneys cannot clear drugs from the blood as rapidly; this is a contributing factor in overmedication among the aged.

Water balance becomes more precarious in old age because the kidneys are less responsive to antidiuretic hormone and because the sense of thirst is sharply reduced. Even when given free access to water, elderly people may not drink enough to maintain normal blood osmolarity. Dehydration is therefore common. It is often said that aged kidneys are deficient in maintaining electrolyte balance, but the evidence for this remains questionable.

Voiding and bladder control become a problem for both men and women. About 80% of men over the age of 80 are affected by benign prostatic hyperplasia. The enlarged prostate compresses the urethra and interferes with emptying of the bladder. Urine retention causes pressure to back up in the kidneys, aggravating the failure of the nephrons. Older women are subject to incontinence (leakage of urine), especially if their history of pregnancy and childbearing has weakened the pelvic muscles and urethral sphincters. Senescence of the sympathetic nervous system and nervous disorders such as stroke and Alzheimer disease can also cause incontinence.

Digestive System and Nutrition

Less saliva is secreted in old age, making food less flavorful, swallowing more difficult, and the teeth more prone to caries. Nearly half of people over age 65 wear dentures because they have lost their teeth to caries and periodontitis. The stratified squamous epithelium of the oral cavity and esophagus is thinner and more vulnerable to abrasion.

The gastric mucosa atrophies and secretes less acid and intrinsic factor. Acid deficiency reduces the absorption of calcium, iron, zinc, and folic acid. Heartburn becomes more common as the weakening lower esophageal sphincter fails to prevent reflux into the esophagus. The most common digestive complaint of older people is constipation, which results from the reduced muscle tone and weaker peristalsis of the colon. This seems to stem from a combination of factors: atrophy of the muscularis externa, reduced sensitivity to neurotransmitters, less fiber and water in the diet, and less exercise. The liver, gallbladder, and pancreas show only slightly reduced function. Any drop in liver function, however, makes it harder to detoxify drugs and can contribute to overmedication.

Older people tend to reduce their food intake because of lower energy demand and appetite, because declining sensory functions make food less appealing, and because reduced mobility makes it more troublesome to shop and prepare meals. However, they need fewer calories than younger people because they have lower basal metabolic rates and tend to be less physically active. Protein, vitamin, and mineral requirements remain essentially unchanged, although vitamin and mineral supplements may be needed to compensate for reduced food intake and intestinal absorption. Malnutrition is common among older people and is an important factor in anemia and reduced immunity.

Reproductive System

In men, the senescent changes in the reproductive system are relatively gradual; they include declining testosterone secretion, sperm count, and libido. By age 65, sperm count is about one-third of what it was in a man's 20s. Men remain fertile (capable of fathering a child) well into old age, but impotence (inability to maintain an erection) can occur because of atherosclerosis, hypertension, medication, and psychological reasons.

In women, the changes are more pronounced and develop more rapidly, over the course of menopause. The ovarian follicles are used up, gametogenesis ceases, and the ovaries stop producing sex steroids. This may result in vaginal dryness, genital atrophy, and reduced libido and make sex less enjoyable. With the loss of ovarian steroids, a postmenopausal woman has an elevated risk of osteoporosis and atherosclerosis.

Exercise and Senescence

Other than the mere passage of time, senescence results from obesity and insufficient exercise more than from any other causes. Conversely, good nutrition and exercise are the best ways to slow its progress.

There is no clear evidence that exercise will prolong your life, but there is little doubt that it improves the quality of life in old age. It maintains endurance, strength, and joint mobility while it reduces the incidence and severity of hypertension, osteoporosis, obesity, and diabetes mellitus. This is especially true if you begin a program of regular physical exercise early in life and continue it throughout old age. If you stop exercising regularly after middle age, the body rapidly becomes deconditioned, although appreciable reconditioning can be achieved even when an exercise program is begun late in life. A person in his or her 90s can increase muscle strength two- or threefold in 6 months with as little as 40 minutes of isometric exercise a week. The improvement results from a combination of muscle hypertrophy and neural efficiency.

Resistance exercises may be the most effective way of reducing accidental injuries such as bone fractures, whereas endurance exercises reduce body fat and increase cardiac output and maximum oxygen uptake. A general guideline for ideal endurance training is to have three to five periods of aerobic exercise per week, each 20 to 60 minutes long and vigorous enough to reach 60% to 90% of your maximum heart rate. The maximum is best determined by a stress test but averages about 220 beats per minute minus your age in years.

An exercise program should ideally be preceded by a complete physical examination and stress test. Warm-up and cool-down periods are especially important in avoiding soft tissue injuries. Because of their lower capacity for thermoregulation, older people must be careful not to overdo exercise, especially in hot weather. At the outset of a new exercise program, it is best to "start low and go slow."

Theories of Senescence

Why do our organs wear out? Why must we die? There still is no general theory on this. The question actually comes down to two issues: What are the mechanisms that cause the organs to deteriorate with age? and Why hasn't natural selection eliminated these and produced bodies capable of longer life?

Mechanisms of Senescence

Numerous theories have been proposed and discarded to explain why organ function degenerates with age. Some authorities maintain that senescence is an intrinsic process governed by inevitable or even preprogrammed changes in cellular function. Others attribute senescence to extrinsic (environmental) factors that progressively damage our cells over the course of a lifetime.

There is good evidence of a hereditary component to longevity. Unusually long and short lives tend to run in families. Monozygotic (identical) twins are more likely than dizygotic twins to die at a similar age. One striking genetic defect called *progeria*[19] is characterized by greatly accelerated senescence (fig. 29.18). Symptoms begin to appear by age two. The child's growth rate declines, the muscles and skin become flaccid, most victims lose their hair, and most die in early adolescence from advanced atherosclerosis. There is some controversy over the relevance or similarity of progeria to normal senescence, but it does demonstrate that many of the changes associated with old age can be brought on by a genetic anomaly.

Knowing that senescence is partially hereditary, however, does not answer the question about why tissues degenerate. Quite likely, no one theory explains all forms of senescence, but let's briefly examine some of them.

Programmed Cell Death Theory Some developmental biologists believe that atrophy of the organs is an effect of programmed cell death (apoptosis). After a certain number of cell divisions or at a certain age, cells may activate a "suicide program" that destroys their own DNA.

The Hayflick Phenomenon Normal organ function usually depends on a rate of cell renewal that keeps pace with cell death. There may be a limit, however, to how many times

19. *pro* = before + *ger* = old age

Figure 29.18 **Progeria.** This is a genetic disorder in which senescence appears to be greatly accelerated. The individuals here, from *left* to *right*, are 15, 12, and 26 years old. Few people with progeria live as long as the woman on the *right*.

cells can divide. In 1950, Leonard Hayflick discovered that cultured fibroblasts from human fetuses divided about 50 times and then degenerated and died. If cells in the body have similar limitations on the number of divisions, it could mean an end of cell renewal and the onset of organ atrophy and functional decline.

Telomere Theory All chromosomes have a "cap" at each end called a **telomere,**[20] like the plastic tip of a shoelace. In humans, it consists of a noncoding nucleotide sequence CCCTAA repeated 1,000 times or more. One of its functions may be to stabilize the chromosome and prevent it from unraveling or sticking to other chromosomes. Also, during DNA replication, DNA polymerase cannot reproduce the very ends of the DNA molecule. If there were functional genes at the end of the molecule, they would not get duplicated. The telomere may therefore provide a bit of "disposable" DNA at the end, so that DNA polymerase doesn't fail to replicate genes that would otherwise be there. Every time DNA is replicated, 50 to 100 bases are lost from the telomere. In old age, the telomere may be exhausted and the polymerase may then indeed fail to replicate some of the terminal genes of DNA. Old chromosomes may therefore be more vulnerable to damage, replication errors, or both, causing old cells to be increasingly dysfunctional. This might be a factor in the Hayflick limit on the number of times cells can divide.

20. *telo* = end + *mer* = piece

Cross-Linking Theory About one-fourth of the body's protein is collagen. With age, collagen molecules become cross-linked by more and more disulfide bridges, thus making the fibers less soluble and more stiff. This is thought to be a factor in several of the most noticeable changes of the aging body, including stiffening of the joints, lenses, and arteries. Similar cross-linking of DNA and enzyme molecules could progressively impair their functions as well.

Other Protein Abnormalities Not only collagen but also many other proteins exhibit increasingly abnormal structure in older tissues and cells. The changes are not in amino acid sequence—therefore not attributable to DNA mutations—but lie in the way the proteins are folded and other moieties such as carbohydrates are attached to them. This is another reason that cells accumulate more dysfunctional proteins as they age.

Free Radical Theory Free radicals have very destructive effects on macromolecules (see chapter 2). We have a number of antioxidants to protect us from their effects, but it is believed that some of these antioxidants become less abundant with age and are eventually overwhelmed by free radicals, or that some of the molecules damaged by free radicals are long-lived and accumulate in cells. Free radical damage may therefore be a contributing factor in some of the other mechanisms of senescence discussed here.

Autoimmune Theory Some of the altered macromolecules described previously may be recognized as foreign antigens and stimulate lymphocytes to mount an immune response against the body's own tissues. Autoimmune diseases such as rheumatoid arthritis do, in fact, become more common in old age.

Evolution and Senescence

If certain genes contribute to senescence, it raises an evolutionary question—Why doesn't natural selection eliminate them? In an attempt to answer this, biologists once postulated that senescence and death were for the good of the species—a way for older, worn-out individuals to make way for younger, healthier ones. We can see the importance of death for the human population by imagining that science had put an end to senescence and people died at a rate of 1 per 1,000 per year regardless of age (the rate at which American 18-year-olds now die). If so, the median age of the population would be 163, and 13% of us would live to be 2,000 years old. The implications for world population and competition for resources would be staggering. Thus it is easy to understand why death was once interpreted as a self-sacrificing phenomenon for the good of the species.

But this hypothesis has several weaknesses. One of them is the fact that natural selection works exclusively through the effects of genes on the reproductive rates of individuals. A species evolves only because some members reproduce more than others. A gene that does not affect reproductive rate can be neither eliminated nor favored by natural selection. Genes for disorders such as Alzheimer disease have little or no effect until a person is past reproductive age. Our prehistoric and even fairly recent ancestors usually died of accidents, predation, starvation, weather, and infectious diseases at an early age. Few people lived long enough to be affected by atherosclerosis, colon cancer, or Alzheimer disease. Natural selection would have been "blind" to such death-dealing genes, which would escape the selection process and remain with us today.

Death

Life expectancy, the average length of life in a given population, has steadily increased in industrialized countries. People born in the United States at the beginning of the twentieth century had a life expectancy of only 45 to 50 years; nearly half of them died of infectious disease. The average boy born today can expect to live 72 years and the average girl 79 years. This is due mostly to victories over infant and child mortality, not to advances at the other end of the life span. **Life span,** the maximum age attainable by humans, has not increased for many centuries and there seems to be little prospect that it ever will. There is no verifiable record of anyone living past the age of 122 years.

There is no definable instant of biological death. Some organs function for an hour or more after the heart stops beating. During this time, even if a person is declared legally dead, living organs may be removed for transplant. For legal purposes, death was once defined as the loss of a spontaneous heartbeat and respiration. Now that cardiopulmonary functions can be artificially maintained for years, this criterion is less distinct. Clinical death is now widely defined in terms of **brain death**—a lack of cerebral activity indicated by a flat electroencephalogram for 30 minutes to 24 hours (depending on state laws), accompanied by a lack of reflexes or lack of spontaneous respiration and heartbeat.

Death usually results from the failure of a particular organ, which then has a cascading effect on other organs. Kidney failure, for example, leads to the accumulation of toxic wastes in the blood, which in turn leads to loss of consciousness, brain function, respiration, and heartbeat.

Ninety-nine percent of us will die before age 100, and there is little chance that this outlook will change within our lifetimes. We cannot presently foresee any "cure for old age" or significant extension of the human life span. The real issue is to maintain the best possible quality of life, and when the time comes to die, to do so in comfort and dignity.

Before You Go On

Answer the following questions to test your understanding of the preceding section:

13. Define *aging* and *senescence*.
14. List some tissues or organs in which changes in collagenous and elastic connective tissues lead to senescence.
15. Many older people have difficulty with mobility and simple self-maintenance tasks such as dressing and cooking. Name some organ systems whose senescence is most relevant to these limitations.
16. Explain why both endurance and resistance exercises are important in old age.
17. Summarize any five cellular mechanisms that may be responsible for senescence.

Clinical Insight 29.4

Reproductive Technology—Making Babies in the Laboratory

Fertile heterosexual couples who have frequent intercourse and use no contraception have an 85% chance of conceiving within 1 year. About one in six American couples, however, are *infertile*—unable to conceive. Infertility can sometimes be corrected by hormone therapy or surgery, but when this fails, parenthood may still be possible through other reproductive technologies discussed here.

Artificial Insemination

If only the male is infertile, the oldest and simplest solution is *artificial insemination (AI),* in which a physician introduces donor semen into or near the cervix. This was first done in the 1890s, when the donor was often the physician himself or a medical student who donated semen for payment. In 1953, a technique was developed for storing semen in glass ampules frozen in liquid nitrogen; the first commercial sperm banks opened in 1970. Most women use sperm from anonymous donors but are able to select from a catalog that specifies the donors' physical and intellectual traits. A man with a low sperm count can donate semen at intervals over a course of several weeks and have it pooled, concentrated, and used to artificially inseminate his partner. Men planning vasectomies sometimes donate sperm for storage as insurance against the death of a child, divorce and remarriage, or a change in family planning. Some cases of infertility are due to sperm destruction by the woman's immune system. This can sometimes be resolved by *sperm washing*—a technique in which the sperm are collected, washed to remove antigenic proteins from their surfaces, and then introduced by AI.

Oocyte Donation

The counterpart to sperm donation is *oocyte donation,* in which fresh oocytes are obtained from one woman, fertilized, and transplanted to the uterus of another. A woman may choose this procedure for a variety of reasons: being past menopause, having had her ovaries removed, or having a hereditary disorder she does not want to pass to her children, for example. The donated oocytes are sometimes provided by a relative or may be left over from another woman's in vitro fertilization (see next). The first baby conceived by oocyte donation was born in 1984. This procedure has a success rate of 20% to 50%.

In Vitro Fertilization

In some women the uterus is normal but the uterine tubes are scarred by pelvic inflammatory disease or other causes. *In vitro fertilization (IVF)* is an option in some of these cases. The woman is given gonadotropins to induce the "superovulation" of multiple eggs. The physician views the ovary with a laparoscope and removes eggs by suction. These are placed in a solution that mimics the chemical environment of her reproductive tract, and sperm are added to the dish. The term *in vitro fertilization* refers to the fact that fertilization occurs in laboratory glassware; children conceived by IVF are often misleadingly called "test-tube babies." In some cases, fertilization is assisted by piercing the zona pellucida before the sperm are added *(zona drilling)* or by injecting sperm directly into the egg through a micropipet. By the day after fertilization, some of the preembryos reach the 8- to 16-celled stage. Several of these are transferred to the mother's uterus through the cervix and her blood HCG level is monitored to determine whether implantation has occurred. Excess IVF preembryos may be donated to other infertile couples or frozen and used in later attempts. In cases where a woman has lost her ovaries to disease, the oocytes may be provided by another donor, often a relative.

IVF costs up to $10,000 per attempt and succeeds only 14% of the time. A couple can easily spend $100,000 before IVF is successful, and then some attempts are "too successful." Multiple preembryos are usually introduced to the uterus as insurance against the low probability that any one of them will implant and survive. Sometimes, however, this results in multiple births—in rare cases, up to seven babies (septuplets). One advantage of IVF is that when the preembryo reaches the eight-celled stage, one or two cells can be removed and tested for genetic defects before the preembryo is introduced to the uterus.

IVF has been used in animal breeding since the 1950s, but the first child conceived this way was Joy Louise Brown (fig. 1), born in England in 1978. It is now estimated that worldwide, about one IVF child is born every day.

Surrogate Mothers

IVF is an option only for women who have a functional uterus. A woman who has had a hysterectomy or is otherwise unable to become pregnant or maintain a pregnancy may contract with a *surrogate mother* who provides a "uterus for hire." Some surrogates are both genetic and gestational mothers, and others gestational only. In the former case, the surrogate is artificially inseminated by a man's sperm and agrees to give the baby to the man and his partner at birth. In the latter case, oocytes are collected from one woman's ovaries, fertilized in vitro, and the preembryos are placed in the surrogate's uterus. This is typical of cases in which a woman has functional ovaries but no functional uterus. A surrogate typically receives a fee of about $10,000 plus medical and legal costs. Several hundred babies have been produced this way in the United States. In at least one case, a woman carried the child of her infertile daughter, thus giving birth to her own granddaughter.

Gamete Intrafallopian Transfer

The low success rate of IVF led to a search for more reliable and cost-effective techniques. *Gamete intrafallopian transfer (GIFT)* was developed in the mid-1980s on the conjecture that pregnancy would be more successful if the oocyte were fertilized and began cleavage in a more natural environment. Eggs are obtained from a woman after a weeklong course of ovulation-inducing drug treatment. The most active sperm cells are isolated from the semen, and the eggs and sperm are introduced into her uterine tube proximal to any existing obstruction. GIFT is about half as expensive as IVF and succeeds about 40% of the time. In a modification called *zygote intrafallopian transfer (ZIFT),* fertilization occurs in vitro and the preembryo is introduced into the uterine tube. Traveling down the uterine tube seems to improve the chance of implantation when the conceptus reaches the uterus.

Continued

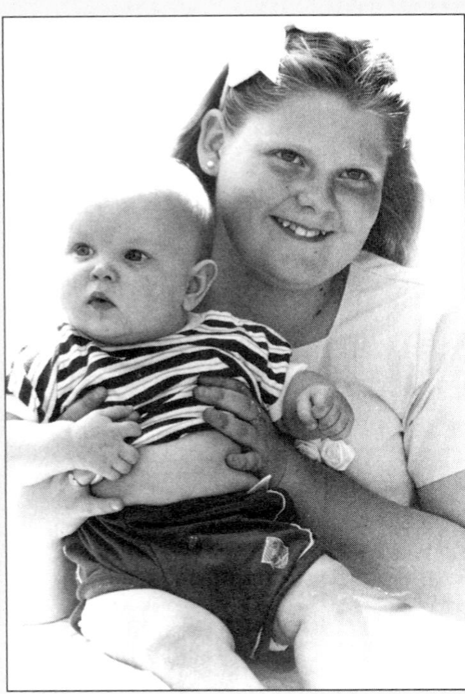

Figure 1 New Beginnings Through Reproductive Technology. Joy Louise Brown, shown here at age 10, was the first child ever conceived by in vitro fertilization (IVF). She is holding Andrew Macheta, another IVF baby, at a 10-year anniversary celebration at the clinic near London where both were conceived.

Embryo Adoption

Embryo adoption is used when a woman has malfunctioning ovaries but a normal uterus. A man's sperm are used to artificially inseminate another woman. A few days later, the preembryo is flushed from the donor's uterus before it implants and is transferred to the uterus of the woman who wishes to have a child.

Ethical and Legal Issues

Like many other advances in medicine, reproductive technology has created its own ethical and legal dilemmas, some of which are especially confounding. Perhaps the most common problem is the surrogate mother who changes her mind. Surrogates enter into a contract to surrender the baby to a couple at birth, but after carrying a baby for 9 months and giving birth, they sometime feel differently. This raises questions about the definition of motherhood, especially if she is the gestational but not the genetic mother.

The converse problem is illustrated by a case in which the child had hydrocephalus and neither the contracting couple nor the surrogate mother wanted it. In this case, genetic testing showed that the child actually had been fathered by the surrogate's husband, not the man who had contracted for her service. The surrogate then accepted the baby as her own. Nevertheless, the case raised the question of whether the birth of a genetically defective child constituted fulfillment of the contract and obligated the contracting couple to accept the child, or whether such a contract implies that the surrogate mother must produce a healthy child.

In another case, a wealthy couple was killed in an accident and left frozen preembryos in an IVF clinic. A lawsuit was filed on behalf of the preembryos on the grounds that they were heirs to the couple's estate and should be carried to birth by a surrogate mother so they could inherit it. The court ruled against the suit and the preembryos were allowed to die. In still another widely publicized case, a man sued his wife for custody of their frozen preembryos as part of a divorce settlement.

IVF also creates a question of what to do with the excess preembryos. Some people view their disposal as a form of abortion, even if the preembryo is only a mass of 8 to 16 undifferentiated cells. On the other hand, there are those who see such excess preembryos as a research opportunity to obtain information that could not be obtained in any other way. In 1996, an IVF clinic in England was allowed to destroy 3,300 unclaimed preembryos, but only after heated public controversy.

It is common for scientific advances to require new advances in law and ethics. The parallel development of these disciplines is necessary if we are to benefit from the developments of science and ensure that knowledge is applied in an ethical and humane manner.

Chapter Review

Online Learning Center

Study resources for this chapter are available at www.mhhe.com/saladin/

Study Outline

Fertilization and Preembryonic Development (p. 1085)

1. Sperm migration
2. Sperm capacitation
3. Fertilization
4. Meiosis II
5. Preembryonic stage
   a. Cleavage
      - Mitotic production of blastomeres
      - Morula stage
      - Blastocyst stage
        - •• Trophoblast
        - •• Embryoblast
   b. Implantation
      - Rootlets of syncytiotrophoblast
      - Secretion of HCG
   c. Embryogenesis
      - Amniotic cavity formation
      - Embryonic disc formation
      - Primary germ layer formation
        - •• Ectoderm
        - •• Mesoderm
        - •• Endoderm

Embryonic and Fetal Development (p. 1090)

1. Prenatal nutrition
   a. Trophoblastic nutrition
   b. Placental nutrition
2. The placenta
   a. Chorionic villi
   b. Placental sinus
   c. Umbilical cord
   d. Placental conductivity
   e. Functions of placenta
3. Embryonic membranes
   a. Amnion and amniotic fluid
   b. Yolk sac
   c. Allantois
   d. Chorion
4. Organogenesis
   a. Derivatives of the ectoderm
   b. Derivatives of the mesoderm
   c. Derivatives of the endoderm
5. Fetal development
   a. Developmental timetable
   b. Fetal circulation
      - Umbilical-placental circuit
      - Circulatory shunts
        - •• Ductus venosus
        - •• Foramen ovale
        - •• Ductus arteriosus

The Neonate (p. 1096)

1. The transitional period
2. Circulatory adaptations
   a. Closure of umbilical vessels
   b. Closure of ductus venosus
   c. Closure of foramen ovale
   d. Closure of ductus arteriosus
3. Respiratory adaptations
   a. Onset of breathing
   b. Difficulty inflating lungs
4. Immunological adaptations
5. Other adaptations
   a. Thermoregulation
   b. Fluid balance
6. Premature infants
   a. Respiratory distress syndrome
   b. Thermoregulation
   c. Digestion
   d. Liver functions
7. Congenital anomalies
   a. Infectious diseases
   b. Teratogens
      - Thalidomide incident
      - Fetal alcohol syndrome
      - Effects of smoking
      - Effects of X rays
   c. Mutagens
   d. Nondisjunction and aneuploidy
      - Nondisjunction of sex chromosomes
        - •• Triplo-X syndrome (XXX)
        - •• Klinefelter syndrome (XXY)
        - •• Turner syndrome (XO)
      - Nondisjunction of autosomes
        - •• Down syndrome (trisomy-21)

Aging and Senescence (p. 1103)

1. Senescence of the organ systems
   a. Integumentary system
      - Intrinsic aging
      - Photoaging
   b. Skeletal system
      - Loss of bone mass
      - Joint diseases
   c. Muscular system
      - Muscular atrophy
      - Effects of nervous system senescence
   d. Nervous system
      - Cerebral atrophy
      - Declining synaptic function
      - Declining autonomic control
   e. Sense organs
      - Vision
      - Hearing
      - Taste and smell
   f. Endocrine system
      - Reproductive hormones
      - Stress hormones
      - Diabetes mellitus
   g. Circulatory system
      - Blood values and anemia
      - Atherosclerosis
      - Cardiac function
      - Varicose veins
      - Thrombosis
   h. Immune system
   i. Respiratory system
      - Declining pulmonary ventilation
      - Chronic obstructive pulmonary diseases
   j. Urinary system
      - Renal atrophy
      - Fluid balance
      - Bladder control
   k. Digestive system
      - Dental health
      - Gastric atrophy
      - Intestinal motility
      - Nutrition

1. Reproductive system
   - Male
   - Female
2. Exercise and senescence
3. Theories of senescence
   a. Intrinsic and environmental factors
   b. Heredity and longevity
   c. Programmed cell death theory
   d. Hayflick phenomenon
   e. Telomere theory
   f. Cross-linking theory
   g. Other protein abnormalities
   h. Free radical theory
   i. Autoimmune theory
4. Evolution and senescence
5. Death
   a. Life expectancy and life span
   b. Defining death
   c. Cascade of organ failures

Selected Vocabulary

developmental biology 1085
sperm capacitation 1085
acrosomal reaction 1085
hyaluronidase 1085
acrosin 1085
polyspermy 1085
fast block 1085
slow block 1086
cortical granules 1086
cortical reaction 1086
fertilization membrane 1086
pronuclei 1086
zygote 1086
preembryonic stage 1086
cleavage 1086
blastomere 1086
morula 1086
uterine milk 1087
blastocyst 1087
trophoblast 1087
embryoblast 1087
implantation 1088
cytotrophoblast 1088
syncytiotrophoblast 1088
embryogenesis 1088
amniotic cavity 1088
embryonic disc 1088
primitive streak 1088
mesenchyme 1088
embryo 1088
trophoblastic nutrition 1090
decidual cells 1090
placental nutrition 1090
placenta 1090
trophoblastic phase 1090
placental phase 1090
placentation 1090
chorionic villi 1090
placental sinus 1090
umbilical cord 1090
placental conductivity 1091
amnion 1091
amniotic fluid 1091
yolk sac 1091
allantois 1091
chorion 1091
organogenesis 1091
fetus 1093
umbilical arteries 1093
umbilical vein 1094
ductus venosus 1094
foramen ovale 1095
ductus arteriosus 1095
limb buds 1096
digital rays 1096
meconium 1096
lanugo 1096
vernix caseosa 1096
quickening 1096
vertex position 1096
transitional period 1098
neonatal period 1098
premature 1099
respiratory distress syndrome 1099
congenital anomaly 1099
teratogen 1099
fetal alcohol syndrome 1100
mutagen 1101
nondisjunction 1101
aneuploidy 1101
triplo-X syndrome 1101
Klinefelter syndrome 1101
Turner syndrome 1101
Down syndrome 1102
aging 1103
senescence 1103
rosacea 1104
intrinsic aging 1104
photoaging 1104
osteopenia 1104
telomere 1109
life expectancy 1110
life span 1110
brain death 1110

Testing Your Recall

1. When a conceptus arrives in the uterus, it is at what stage of development?
   a. zygote
   b. morula
   c. blastomere
   d. blastocyst
   e. embryo
2. The entry of a sperm nucleus into an egg must be preceded by
   a. the cortical reaction.
   b. the acrosomal reaction.
   c. the fast block.
   d. implantation.
   e. cleavage.
3. The stage of the conceptus that implants in the uterine wall is
   a. a blastomere.
   b. a morula.
   c. a blastocyst.
   d. an embryo.
   e. a zygote.
4. Chorionic villi develop from
   a. the zona pellucida.
   b. the endometrium.
   c. the syncytiotrophoblast.
   d. the embryoblast.
   e. the corona radiata.
5. Which of these results from aneuploidy?
   a. Turner syndrome
   b. fetal alcohol syndrome
   c. nondisjunction
   d. progeria
   e. rubella
6. Fetal urine accumulates in the ______ and contributes to the fluid there.
   a. placental sinus
   b. yolk sac
   c. allantois
   d. chorion
   e. amnion
7. One theory of senescence is that it results from a lifetime of damage by
   a. teratogens.
   b. aneuploidy.
   c. free radicals.
   d. cytomegalovirus.
   e. nondisjunction.
8. Photoaging is a major factor in the senescence of
   a. the integumentary system.
   b. the eyes.
   c. the nervous system.
   d. the skeletal system.
   e. the cardiovascular system.

9. Which of these is *not* a common effect of senescence?
   a. reduced synthesis of vitamin D
   b. atrophy of the kidneys
   c. atrophy of the cerebral gyri
   d. increased herniation of intervertebral discs
   e. reduced respiratory vital capacity
10. For the first 8 weeks of gestation, a conceptus is nourished mainly by
    a. the placenta.
    b. amniotic fluid.
    c. colostrum.
    d. decidual cells.
    e. yolk cytoplasm.
11. Viruses and chemicals that cause congenital anatomical deformities are called ________ .
12. Aneuploidy is caused by ________, the failure of two homologous chromosomes to separate in meiosis.
13. The maximum age attainable by the human species is called the ________ .
14. The average age attained by humans in a given population is called the ________ .
15. Fetal blood flows through growths called ________, which project into the placental sinus.
16. The enzymes with which a sperm penetrates an egg are contained in an organelle called the ________ .
17. Stiffening of the arteries, joints, and lenses in old age may be a result of cross-linking between ________ molecules.
18. An enlarged tongue, epicanthal folds of the eyes, and mental retardation are characteristic of a genetic anomaly called ________ .
19. The fossa ovalis is a remnant of a fetal shunt called the ________ .
20. A developing individual is first classified as a/an ________ when the three primary germ layers have formed.

Answers in appendix B

Testing Your Comprehension

1. Suppose a woman had a mutation resulting in a tough zona pellucida that did not disintegrate after the egg was fertilized. How would this affect her fertility? Why?
2. Suppose a drug were developed that could slow down the rate of collagen cross-linking with age. What diseases of old age could be made less severe with such a drug?
3. Some health-food stores market the enzyme superoxide dismutase (SOD) as an oral antioxidant to retard senescence. Explain why it would be a waste of your money to buy it.
4. In some children the ductus arteriosus fails to close after birth—a condition that eventually requires surgery. Predict how this condition would affect (a) pulmonary blood pressure, (b) systemic diastolic pressure, and (c) the right ventricle of the heart.
5. Only one sperm is needed to fertilize an egg, yet a man who ejaculates fewer than 10 million sperm is usually infertile. Explain this apparent contradiction. Supposing 10 million sperm were ejaculated, predict how many would come within close range of the egg. How likely is it that any one of these sperm would fertilize it?

Answers at the Online Learning Center

Appendix A

Periodic Table of the Elements

Nineteenth-century chemists discovered that when they arranged the known elements by atomic weight, certain properties reappeared periodically. In 1869, Russian chemist Dmitri Mendeleev published the first modern periodic table of the elements, leaving gaps for those that had not yet been discovered. He accurately predicted properties of the missing elements, which helped other chemists discover and isolate them.

Each row in the table is a *period* and each column is a *group (family)*. Each period has one electron shell more than the period above it, and as we progress from left to right within a period, each element has one more proton and electron than the one before. The dark steplike line from boron (5) to astatine (85) separates the metals to the left of it (except hydrogen) from the nonmetals to the right. Each period begins with a soft, light, highly reactive *alkali metal,* with one valence electron, in family IA. Progressing from left to right, the metallic properties of the elements become less and less pronounced. Elements in family VIIA are highly reactive gases called *halogens,* with seven valence electrons. Elements in family VIIIA, called *noble (inert) gases,* have a full valence shell of eight electrons, which makes them chemically unreactive. Ninety-one of the elements occur naturally on earth. Physicists have created elements up to atomic number 118 in the laboratory, but the International Union of Pure and Applied Chemistry has established formal names only through element 109 to date.

The 24 elements with normal roles in human physiology are color-coded according to their relative abundance in the body (see chapter 2). Others, however, may be present as contaminants with very destructive effects (such as arsenic, lead, and radiation poisoning).

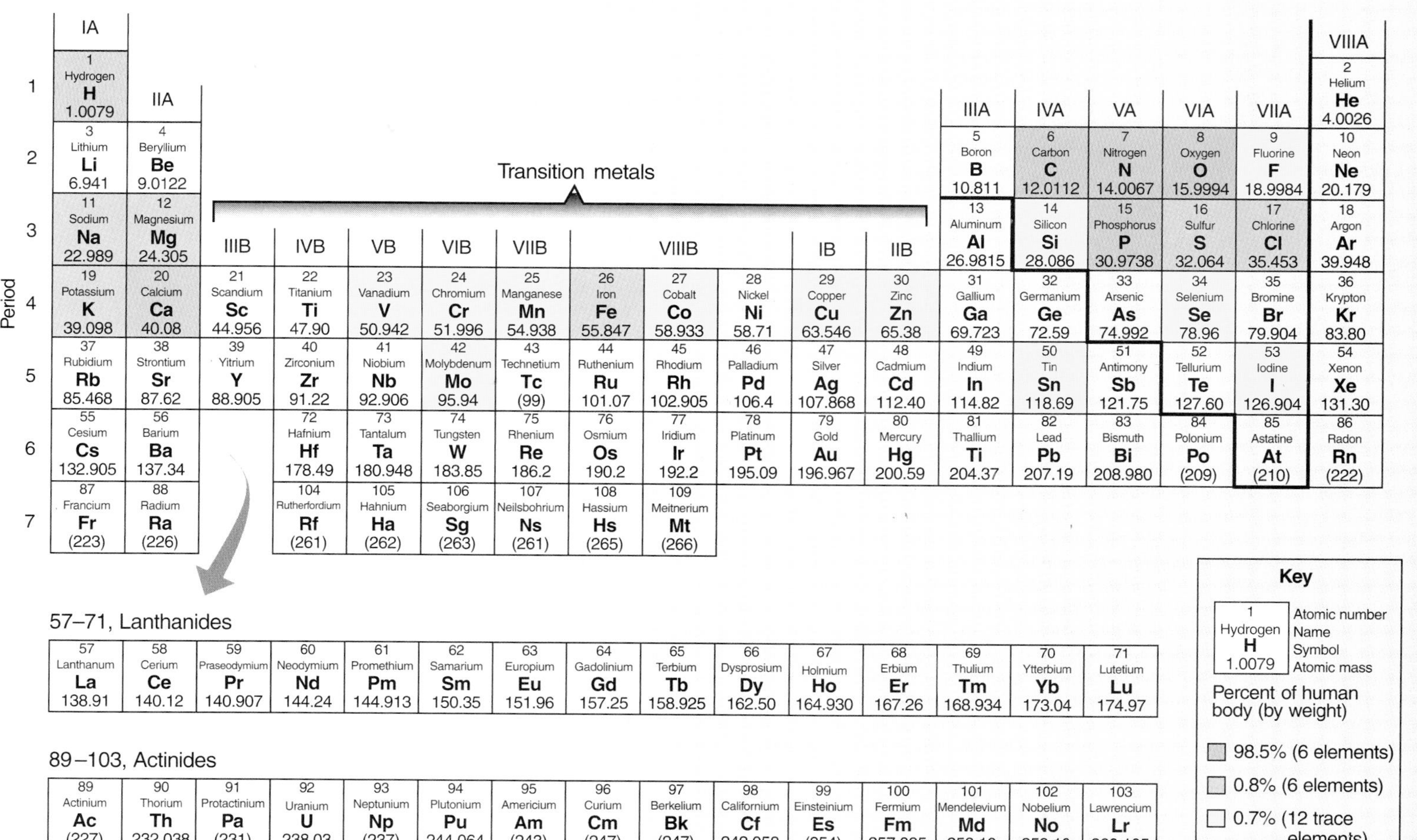

Appendix B

Answers to Testing Your Recall Questions

Answers to the objective chapter review questions are given here. Answers to the Think About It and Testing Your Comprehension questions are in the *Instructor's Manual*. Answers to the Testing Your Comprehension questions are also available to students at the Online Learning Center at www.mhhe.com/saladin/.

Chapter 1

1. a
2. e
3. d
4. a
5. c
6. c
7. a
8. c
9. e
10. b
11. dissection
12. Robert Hooke
13. deduction
14. psychosomatic
15. homeostasis
16. set point
17. negative feedback
18. organ
19. stereoscopic
20. prehensile, opposable

Atlas A

1. d
2. c
3. e
4. d
5. d
6. a
7. a
8. d
9. b
10. e
11. supine
12. parietal
13. mediastinum
14. nuchal
15. hand, foot
16. meninges
17. retroperitoneal
18. medial
19. inferior
20. cubital, popliteal

Chapter 2

1. b
2. c
3. a
4. c
5. a
6. a
7. e
8. d
9. a
10. e
11. valence
12. cation
13. polar covalent
14. isomers
15. free radicals
16. catalyst
17. emulsion
18. mole
19. 10^{-9} M
20. anabolism

Chapter 3

1. a
2. e
3. e
4. a
5. c
6. b
7. c
8. e
9. b
10. d
11. hydration spheres
12. condensation
13. functional groups
14. -ose, -ase
15. monosaccharides, amino acids
16. hydrogen
17. peptide
18. phospholipids
19. cAMP
20. anaerobic fermentation

Chapter 4

1. e
2. b
3. d
4. b
5. e
6. e
7. a
8. c
9. d
10. b
11. micrometers (μm)
12. second messenger
13. voltage-gated
14. hydrostatic pressure
15. hypertonic
16. exocytosis
17. mitochondria, nucleus
18. peroxisomes, smooth ER
19. ligand-gated channel
20. cisterna

Chapter 5

1. a
2. e
3. c
4. c
5. e
6. b
7. a
8. d
9. d
10. a
11. cytokinesis
12. alleles
13. genetic code
14. polyribosome
15. carcinogens
16. chaperones
17. 46, 92, 92
18. malignant (cancerous)
19. cyclins
20. autosomes

Chapter 6

1. a
2. b
3. c
4. e
5. c
6. a
7. b
8. e
9. b
10. b
11. necrosis
12. mesothelium
13. lacunae
14. fibers
15. collagen
16. skeletal
17. basement membrane
18. matrix (extracellular material)
19. proteoglycans
20. simple

Chapter 7

1. d
2. c
3. d
4. b
5. a
6. e
7. c
8. a
9. a
10. d
11. insensible perspiration
12. arrector pili
13. debridement
14. cyanosis
15. dermal papillae

16. earwax
17. sebaceous glands
18. hyponychium
19. papilla
20. third-degree (full-thickness)

Chapter 8

1. e
2. a
3. d
4. c
5. d
6. a
7. d
8. e
9. b
10. d
11. hydroxyapatite
12. canaliculi
13. appositional
14. solubility product
15. hypocalcemia
16. osteoblasts
17. calcitriol
18. osteoporosis
19. metaphysis
20. osteomalacia

Chapter 9

1. b
2. e
3. a
4. d
5. a
6. e
7. c
8. b
9. e
10. b
11. fontanels
12. temporal
13. sutures
14. sphenoid
15. annulus fibrosus
16. dens (odontoid process)
17. auricular
18. styloid
19. pollex, hallux
20. medial longitudinal

Chapter 10

1. c
2. b
3. a
4. e
5. c
6. c
7. a
8. d
9. b
10. d
11. synovial fluid
12. bursa
13. pivot
14. kinesiology
15. gomphosis
16. serrate
17. extension
18. range of motion
19. labrum
20. menisci

Chapter 11

1. b
2. e
3. a
4. c
5. e
6. e
7. b
8. a
9. d
10. c
11. origin
12. fascicle
13. prime mover (agonist)
14. hamstring
15. retinacula
16. urogenital triangle
17. linea alba
18. synergist
19. bipennate
20. sphincter

Atlas B

1. f
2. b
3. k
4. p
5. h
6. z
7. o
8. x
9. c
10. a
11. y
12. m
13. n
14. e
15. g
16. v
17. f
18. c
19. y
20. x
21. k
22. d
23. s
24. b
25. a
26. u
27. j
28. i
29. g
30. q

Chapter 12

1. a
2. d
3. b
4. d
5. a
6. c
7. e
8. c
9. e
10. b
11. threshold
12. tetanus
13. terminal cisternae
14. myosin
15. acetylcholine
16. myoglobin
17. Z discs
18. varicosities
19. tonus
20. lactic acid

Chapter 13

1. e
2. c
3. d
4. a
5. c
6. e
7. d
8. a
9. d
10. b
11. afferent
12. conductivity
13. absolute refractory period
14. dendrites
15. oligodendrocytes
16. nodes of Ranvier
17. axon hillock, initial segment
18. metabotropic
19. facilitated zone
20. neuromodulators

Chapter 14

1. e
2. c
3. a
4. e
5. e
6. a
7. b
8. c
9. a
10. d
11. corpus callosum
12. ventricles
13. arbor vitae
14. decussation
15. choroid plexus
16. precentral
17. frontal lobe
18. association areas
19. categorical
20. Broca's area

Chapter 15

1. e
2. a
3. e
4. c
5. d
6. d
7. b
8. d
9. a
10. c
11. ganglia
12. ramus
13. phrenic
14. vagus
15. intrafusal fibers
16. tibial, common peroneal
17. sympathetic
18. acetylcholine
19. cAMP
20. sympathetic tone, vasomotor tone

Chapter 16

1. a
2. c
3. b
4. a
5. e
6. e
7. d
8. c
9. c
10. b
11. fovea centralis
12. ganglion
13. Na^+
14. otoliths
15. hair cells
16. stapes
17. inferior colliculi
18. taste hairs
19. olfactory bulb
20. referred pain

Chapter 17

1. b
2. d
3. a
4. c
5. c
6. c
7. d
8. c
9. e
10. e
11. adenohypophysis
12. tyrosine
13. acromegaly
14. parathyroid
15. glucocorticoids
16. granulosa, interstitial
17. negative feedback inhibition
18. hypothalamo-hypophyseal portal system
19. permissive
20. up-regulation

Chapter 18

1. b
2. c
3. c
4. a
5. b
6. d
7. d
8. c
9. d
10. c
11. hemopoiesis
12. hematocrit
13. thromboplastin
14. agglutinogens
15. hemophilia
16. hemostasis
17. sickle-cell disease
18. polycythemia
19. vitamin B_{12}
20. erythropoietin

Chapter 19

1. d
2. b
3. d
4. a
5. e
6. c
7. d
8. e
9. a
10. e
11. systole, diastole
12. systemic
13. atrioventricular (coronary) sulcus
14. Na^+
15. gap junctions
16. T wave
17. semilunar
18. auscultation
19. preload
20. cardiac output

Chapter 20

1. c
2. b
3. a
4. e
5. b
6. c
7. e
8. a
9. e
10. d
11. systolic, diastolic
12. continuous capillaries
13. anaphylactic
14. thoracic pump
15. oncotic pressure
16. transcytosis
17. sympathetic
18. baroreceptors
19. arterial circle
20. basilic, cephalic

Chapter 21

1. b
2. c
3. a
4. e
5. d
6. b
7. c
8. d
9. a
10. c
11. pathogen
12. lysozyme
13. bradykinin
14. diapedesis
15. opsonization
16. pyrogen
17. lymphokines, monokines
18. antigen-binding site, antigenic determinant
19. clonal deletion
20. autoimmune

Chapter 22

1. c
2. c
3. a
4. a
5. e
6. c
7. b
8. d
9. d
10. a
11. glottis
12. bronchial tree
13. pulmonary surfactant
14. intrapleural, atmospheric
15. obstructive
16. anatomic dead space
17. compliance, elasticity
18. inspiratory center
19. ventilation-perfusion coupling
20. alkalosis, hypocapnia

Chapter 23

1. a
2. d
3. b
4. c
5. b
6. b
7. d
8. e
9. c
10. a
11. micturition
12. renal autoregulation
13. trigone
14. macula densa
15. podocytes
16. transport maximum
17. antidiuretic hormone
18. internal urethral
19. protein
20. arcuate

Chapter 24

1. c
2. a
3. a
4. a
5. d
6. c
7. e
8. b
9. d
10. b
11. Na^+
12. K^+
13. metabolic water
14. cutaneous transpiration
15. fluid sequestration
16. hyperkalemia
17. hyponatremia
18. respiratory acidosis
19. limiting pH
20. osmolarity

Chapter 25

1. b
2. d
3. c
4. e
5. a
6. c
7. a
8. a
9. a
10. a
11. occlusal
12. amylase, lipase
13. parotid
14. enteric
15. vagus
16. gastrin
17. sinusoids
18. maltase, maltose
19. chylomicrons
20. iron

Chapter 26

1. a
2. c
3. b
4. e
5. b
6. e
7. c
8. a
9. d
10. d
11. incomplete
12. glycogenolysis
13. gluconeogenesis
14. urea
15. liver
16. insulin
17. core temperature
18. hypothalamus
19. cytochromes
20. ATP synthase, ATP

Chapter 27

1. a
2. a
3. a
4. c
5. a
6. d
7. e
8. c
9. d
10. d
11. mesonephric

12. fructose
13. pampiniform plexus
14. secondary spermatocytes
15. tunica albuginea
16. seminal vesicles
17. sustentacular
18. secondary spermatocyte
19. deep
20. acrosome

Chapter 28

1. a
2. d
3. c
4. a
5. e
6. b
7. b
8. b
9. c
10. c
11. follicle
12. endometrium
13. menarche
14. corpus luteum
15. corona radiata
16. antrum
17. climacteric
18. conceptus
19. infundibulum, fimbriae
20. lochia

Chapter 29

1. b
2. b
3. c
4. c
5. a
6. e
7. c
8. a
9. d
10. d
11. teratogens
12. nondisjunction
13. life span
14. life expectancy
15. chorionic villi
16. acrosome
17. collagen
18. Down syndrome
19. foramen ovale
20. embryo

Appendix C

Understanding Biomedical Vocabulary

One of the greatest challenges confronted by students of anatomy and physiology is the scientific vocabulary. Biomedical science has long had an especially large vocabulary of technical terms, which continues to grow to match the fast pace of scientific discovery. *Stedman's Medical Dictionary,* for example, added 12,000 new terms from 1990 to 1995.

The purpose of this appendix is to share with you some basic skills that can help you feel more comfortable with the technical language of medicine. People who find scientific terms confusing and difficult to pronounce, spell, and remember usually feel more confident once they realize the logic of how terms are composed. The editors of *Stedman's* point out that about 90% of scientific terms are composed of various combinations of just 1,200 Greek and Latin word elements. A term such as *hyponatremia* is less forbidding once we recognize that it is composed of three common word elements: *hypo-* (below normal), *natr-* (sodium), and *-emia* (blood condition). Thus, hyponatremia is a deficiency of sodium in the blood. Those three word elements appear over and over in many other terms in this book: *hypothermia, natriuretic, anemia,* and so on. Once you learn the meanings of *hypo-, natri-,* and *-emia,* you already have the tools at least to partially understand hundreds of other biomedical terms.

This appendix explains how scientific words are composed. Inside the back cover, you will find a lexicon of the 400 word elements most commonly footnoted in this book. The more familiar you are with these principles of etymology (the study of word origins) and this list of elements, the more easily you can pronounce, spell, define, and remember the terms you encounter in this and other biomedical textbooks. This knowledge can also enhance your ability to communicate with your peers and supervisors in the health professions.

Basic Word Elements

Scientific terms are typically composed of one or more of the following elements:

- At least one **root,** or **stem,** that bears the core meaning of the word. In *cardiology,* for example, the root is *card-* (heart). Many words have two or more roots. In *cytochrome,* the roots are *cyt-* (cell) and *chrom-* (color).
- **Combining vowels** that are often inserted between roots to join them and make the word easier to pronounce. In *cytochrome,* for example, the letter *o* is a combining vowel. Although *o* is the most common combining vowel, all vowels of the alphabet are used in this way, such as *a* in *ligament, e* in *vitreous,* the first *i* in *spermicidal, u* in *ovulation,* and *y* in *tachycardia.* Some words, however, have no combining vowel. *Periosteum,* for example, does not need one because the vowels that belong to the roots already join them together in an easy-to-pronounce form. A combination of a root and combining vowel is called a **combining form:** for example, *ost* (bone) + *e* (a combining vowel) make the combining form *oste-* (also interpreted "bone"), as in *osteology.*
- A **prefix** may be present to modify the core meaning of the word. For example, *gastric* (pertaining to the stomach or to the belly of a muscle) takes on a wide variety of new meanings when prefixes are added to it: *epigastric* (above the stomach), *hypogastric* (below the stomach), *endogastric* (within the stomach), and *digastric* (a muscle with two bellies).
- A **suffix** may be added to the end of a word to modify its core meaning. For example, *microscope, microscopy, microscopic,* and *microscopist* have different meanings because of their suffixes alone. Often two or more suffixes, or a root and suffix, occur so commonly together that they are treated jointly as a **compound suffix;** for example, *log* (study) + *y* (process) form the compound suffix *-logy* (the study of).

To summarize these basic principles, consider the word *gastroenterology,* a branch of medicine dealing with the stomach and small intestine. It breaks down into:

gastr/o/enter/o/log/y

gastro = a combining form meaning "stomach"

entero = a combining form meaning "small intestine"

logy = a compound suffix meaning "the study of"

Obscure Word Origins

Before you get *too* excited about your new insight, you should be warned that the literal translation of a word doesn't always provide great insight into its modern meaning. The history of language is full of twists and turns that are fascinating in their own right and say much about the history of the whole of human culture, but they can create confusion for students. For example, the *amnion* is a transparent sac that forms around the developing fetus. The word is derived from *amnos,* from the Greek for "lamb." From this origin, *amnos* came to mean a bowl for catching the blood of sacrificial lambs, and from there the word found its way into biomedical usage for the membrane that emerges (quite bloody) as part of the afterbirth. The *acetabulum,* the socket of the hip joint, literally means "vinegar cup." Apparently the hip socket reminded early anatomists of the little cups used to serve vinegar as a condiment on dining tables in ancient Rome. The word *testicles* literally means "little witnesses"; the history of medical vocabulary has several amusing conjectures as to why this word was chosen to name the male gonads.

Many anatomical terms and medical conditions are named in honor of authorities in the field, such as *Alzheimer disease* and the *fallopian tubes.* Terms named after people are called **eponyms** (from *epo* = after, related to + *nym* = name). These, too, afford little clue as to what a structure or condition is. There is an increasing trend in medical education to avoid eponyms in favor of more descriptive terms (such as *uterine tubes* instead of *fallopian tubes*). This book avoids eponyms when possible.

Another case where an attempt at literal translation may lead you to a dead end is in **acronyms**—words composed of the first letter, or first few letters, of a series of words. For example, the word *calmodulin* is cobbled together from a few letters of each of three words: *calcium modulating protein.*

Words from Other Languages

Although Latin and Greek words form the basis for most biomedical terms, it is important not to overlook the contributions of other cultures. Many of the most important medical advances in the Middle Ages were made in the Muslim world, and such modern terms as *alcohol, alkaline, chemistry,* and *sugar* are Arabic in origin. Some medical terms are from modern foreign languages, such as *treppe* from German and *tic douloureux* from French. Some are derived from the languages of the aborigines of various continents, such as *curare* from the language of the Carib Indians and *kwashiorkor* from the Bantu language of East Africa.

Singular and Plural Noun Terminals

A point of confusion for many beginning students is how to recognize the plural forms of medical terms. Few people would fail to recognize that *ovaries* is the plural of *ovary,* but the connection is harder to make in some other cases: for example, the plural of *cortex* is *cortices* (COR-ti-sees), the plural of *corpus* is *corpora,* and the plural of *epididymis* is *epididymides* (EP-ih-DID-ih-MID-ees). The following reference table will help you make the connection between common singular and plural noun terminals.

Singular and Plural Forms of Some Noun Terminals

Singular Ending	Plural Ending	Examples
-a	-ae	axilla, axillae
-ax	-aces	thorax, thoraces
-en	-ina	lumen, lumina
-ex	-ices	cortex, cortices
-is	-es	diagnosis, diagnoses
-is	-ides	epididymis, epididymides
-ix	-ices	appendix, appendices
-ma	-mata	carcinoma, carcinomata
-on	-a	ganglion, ganglia
-um	-a	septum, septa
-us	-era	viscus, viscera
-us	-i	villus, villi
-us	-ora	corpus, corpora
-x	-ges	phalanx, phalanges
-y	-ies	ovary, ovaries
-yx	-ices	calyx, calices

Glossary

The 1,000 terms defined here are not necessarily the most important ones in the book, but they are terms that are reintroduced most often and, for lack of space, are not redefined each time they arise. The index indicates where you can find definitions or explanations of additional terms. Terms are defined only in the sense that they are used in this book. Some have broader meanings, even within biology and medicine, that are beyond its scope. Terms that are commonly abbreviated, such as *ATP* and *PET scan,* are defined under the full spelling. See the list of abbreviations inside the front cover for complete spellings. The glossary gives pronunciation guides for many terms. Syllables in capital letters should be accented, and common letter sequences should be pronounced as shown in the following list:

ah	as in father
al	as in pal
ay	as in day
bry	as in bribe
byu	as in bureau
c	as in calculus
cue	as in ridiculous
cuh	as in cousin
cul	as in bicycle
cus	as in custard
dew	as in dual
eez	as in ease
eh	as in feather
err	as in merry
fal	as in fallacy
few	as in fuse
ih	as in fit
iss	as in sister
lerr	as in lair
lur	as in learn
ma	as in man
mah	as in mama
me	as in meat
merr	as in merry
mew	as in music
muh	as in mother
na	as in corona
nerr	as in nary
new	as in news
nuh	as in nothing
odj	as in dodger
oe	as in go
oh	as in home
ol	as in alcohol
oll	as in doll
ose	as in gross
oss	as in floss
perr	as in pair
pew	as in pewter
ruh	as in rugby
serr	as in serration
sterr	as in stereo
sy	as in siren
terr	as in terrain
tirr	as in tyranny
thee	as in theme
uh	as in mother
ul	as in bicycle
verr	as in very
y	as in why
zh	as in measure
zy	as in enzyme

A

abdominal cavity The body cavity between the diaphragm and pelvic brim. fig. A.8

abduction (ab-DUC-shun) Movement of a body part away from the midsagittal plane, as in raising an arm away from the side of the body. fig. 10.10

absorption **1.** Process in which a chemical passes through a membrane or tissue surface and becomes incorporated into a body fluid or tissue. **2.** Any process in which one substance passes into another and becomes a part of it. *Compare* adsorption.

acetate A two-carbon carboxylic acid; the ionized form of acetic acid (CH_3COO^-). Serves as the monomer of fatty acids and the intermediate of aerobic metabolism that enters the citric acid cycle.

acetylcholine (ACh) (ASS-eh-till-CO-leen) A neurotransmitter released by somatic motor fibers, parasympathetic fibers, and some other neurons, composed of choline and an acetyl group. fig. 13.18

acetylcholinesterase (AChE) (ASS-eh-till-CO-lin-ESS-ter-ase) An enzyme that hydrolyzes acetylcholine, thus halting signal transmission at a cholinergic synapse.

acid A proton (H^+) donor; a chemical that releases protons into solution.

acidosis An acid-base imbalance in which the blood pH is lower than 7.35.

acinus (ASS-ih-nus) A sac of secretory cells at the inner end of a gland duct. fig. 6.31

acquired immunodeficiency syndrome (AIDS) A group of conditions that indicate severe immunosuppression related to infection with the human immunodeficiency virus (HIV); typically characterized by a very low T_4 lymphocyte count and high susceptibility to certain forms of cancer and opportunistic infections.

actin A filamentous intracellular protein that provides cytoskeletal support and interacts with other proteins, especially myosin, to cause cellular movement; important in muscle contraction, ciliary and flagellar beating, and membrane actions such as phagocytosis, ameboid movement, and cytokinesis.

action The movement produced by the contraction of a particular muscle.

action potential A rapid voltage change in which a plasma membrane briefly reverses electrical polarity; has a self-propagating effect that produces a traveling wave of excitation in nerve and muscle cells.

active site The region of a protein that binds to a ligand, such as the substrate-binding site of an enzyme or the hormone-binding site of a receptor.

active transport The movement of a solute through a cellular membrane, against its concentration gradient, involving a carrier protein that expends ATP.

acute Pertaining to a disease with abrupt onset, intense symptoms, and short duration. *Compare* chronic.

adaptation **1.** An evolutionary process leading to the establishment of species characteristics that favor survival and reproduction. **2.** Any characteristic of anatomy, physiology, or behavior that promotes survival and reproduction. **3.** A sensory process in which a receptor adjusts its sensitivity or response to the prevailing level of stimulation, such as dark adaptation of the eye.

adduction (ah-DUC-shun) Movement of a body part toward the midsagittal plane, such as bringing the feet together from a spread-legged position. fig. 10.10

adenine (AD-eh-neen) A double-ringed nitrogenous base (purine) found in such molecules as DNA, RNA, and ATP; one of the four bases of the genetic code; complementary to thymine in the double helix of DNA. fig. 5.3

adenosine triphosphate (ATP) (ah-DEN-oh-seen tri-FOSS-fate) A molecule composed of adenine, ribose, and three phosphate groups that functions as a universal energy-transfer molecule; yields adenosine diphosphate (ADP) and an inorganic phosphate group (P_i) upon hydrolysis.

adenylate cyclase (ah-DEN-ih-late SY-clase) An enzyme of the plasma membrane that removes two phosphate molecules from ATP and makes cyclic adenosine monophosphate (cAMP); important in the activation of the cAMP second-messenger system.

adipocyte (AD-ih-po-site) A fat cell.

adipose tissue A connective tissue composed predominantly of adipocytes; fat.

adrenal gland (ah-DREE-nul) An endocrine gland on the superior pole of each kidney. fig. 17.13

adrenergic (AD-reh-NUR-jic) Pertaining to epinephrine (adrenalin) or norepinephrine (noradrenalin), as in adrenergic neurons that secrete one of these chemicals or adrenergic effects on a target organ.

adrenocorticotropic hormone (ACTH) (ah-DREE-no-COR-tih-co-TRO-pic) A hormone secreted by the anterior pituitary gland that stimulates the adrenal cortex.

adsorption The binding of one substance to the surface of another without becoming a part of the latter. *Compare* absorption.

adventitia (AD-ven-TISH-uh) Loose connective tissue forming the outermost sheath around organs such as a blood vessel or the esophagus.

aerobic exercise (air-OH-bic) Exercise in which oxygen is used to produce ATP; endurance exercise.

aerobic respiration Oxidation of organic compounds in a reaction series that requires oxygen and produces ATP.

afferent (AFF-uh-rent) Carrying toward, as in *afferent neurons*, which carry signals toward the central nervous system, and *afferent arterioles*, which carry blood toward a tissue. *Compare* efferent.

afterload The force exerted by arterial blood pressure that opposes the openings of the aortic and pulmonary valves of the heart.

agglutination (ah-GLUE-tih-NAY-shun) Clumping of cells or molecules by antibodies. fig. 18.13

aging Any changes in the body that occur with the passage of time, including growth, development, and senescence.

agonist *See* prime mover.

agranulocyte Either of the two leukocyte types (lymphocytes and monocytes) that lack prominent cytoplasmic granules.

albumin (al-BYU-min) A class of small proteins constituting about 60% of the protein fraction of the blood plasma; plays roles in blood viscosity, colloid osmotic pressure, and solute transport.

aldosterone (AL-doe-steh-RONE, al-DOSS-teh-rone) A steroid hormone secreted by the adrenal cortex that acts on the kidneys to promote sodium retention and potassium excretion.

alkalosis An acid-base imbalance in which the blood pH is higher than 7.45.

allele (ah-LEEL) Any of the alternative forms that one gene can take, such as dominant and recessive alleles.

all-or-none law The statement that a neuron either produces an action potential of maximum strength if it is depolarized to or above threshold, or produces no action potential at all if the stimulus is not strong enough to reach threshold; there are no action potentials of intermediate strength.

alveolus (AL-vee-OH-lus) **1.** A microscopic air sac of the lung. **2.** A gland acinus. **3.** A tooth socket. **4.** Any small anatomical space.

Alzheimer disease (AD) (ALTS-hy-mur) A degenerative disease of the senescent brain, typically beginning with memory lapses and progressing to severe losses of mental and motor functions and ultimately death.

ameboid movement (ah-ME-boyd) Movement of a cell by means of pseudopods, in a manner similar to that of an ameba; seen in leukocytes and some macrophages.

amino acids Small organic molecules with an amino group and a carboxyl group; the monomers of which proteins are composed.

amino group A functional group with the formula $-NH_2$, found in amino acids and some other organic molecules.

amitotic (AY-my-TOT-ic) Incapable of mitosis, as in mature skeletal muscle fibers and osteocytes.

ampulla (am-PULL-uh) A wide or saclike portion of a tubular organ such as a semicircular duct or uterine tube.

anabolism (ah-NAB-oh-lizm) Any metabolic reactions that consume energy and construct more complex molecules with higher free energy from less complex molecules with lower free energy; for example, the synthesis of proteins from amino acids. *Compare* catabolism.

anaerobic fermentation (AN-err-OH-bic) A reduction reaction independent of oxygen that converts pyruvic acid to lactic acid and enables glycolysis to continue under anaerobic conditions.

anaphylactic shock A severe systemic form of anaphylaxis involving bronchiolar constriction, impaired breathing, vasodilation, and a rapid drop in blood pressure with a threat of circulatory failure.

anaphylaxis (AN-uh-fih-LAC-sis) A form of immediate hypersensitivity in which an antigen triggers the release of inflammatory chemicals, causing edema, congestion, hives, and other, usually local, symptoms.

anastomosis (ah-NASS-tih-MO-sis) An anatomical convergence, the opposite of a branch; a point where two blood vessels merge and combine their bloodstreams or where two nerves or ducts converge. fig. 20.1

anatomical position A reference posture that allows for standardized anatomical terminology. A subject in anatomical position is standing with the feet flat on the floor, arms down to the sides, and the palms and eyes directed forward. fig. A.1

anatomy **1.** Structure of the body. **2.** The study of structure.

androgen (AN-dro-jen) Testosterone or a related steroid hormone. Stimulates somatic changes at puberty in both sexes, adult libido in both sexes, development of male anatomy in the fetus and adolescent, and spermatogenesis.

anemia (ah-NEE-me-uh) A deficiency of erythrocytes or hemoglobin.

aneurysm (AN-you-rizm) A weak, bulging point in the wall of a heart chamber or blood vessel that presents a threat of hemorrhage.

angiogenesis (AN-jee-oh-GEN-eh-sis) The growth of new blood vessels.

angiotensin II (AN-jee-oh-TEN-sin) A hormone produced from angiotensinogen (a plasma protein) by the kidneys and lungs; raises blood pressure by stimulating vasoconstriction and stimulating the adrenal cortex to secrete aldosterone.

anion (AN-eye-on) An ion with more electrons than protons and consequently a net negative charge.

antagonist **1.** A muscle that opposes the agonist at a joint. **2.** Any agent, such as a hormone or drug, that opposes another.

antebrachium (AN-teh-BRAY-kee-um) The region from elbow to wrist; the forearm.

anterior Pertaining to the front (facial-abdominal aspect) of the body; ventral.

antibody A protein of the gamma globulin class that reacts with an antigen; found in the blood plasma, in other body fluids, and on the surfaces of certain leukocytes and their derivatives.

anticoagulant (AN-tee-co-AG-you-lent) A chemical agent that opposes blood clotting.

antidiuretic hormone (ADH) (AN-tee-DYE-you-RET-ic) A hormone released by the posterior lobe of the pituitary gland in response to low blood pressure; promotes water retention by the kidneys. Also known as *vasopressin.*

antigen (AN-tih-jen) Any large molecule capable of binding to an antibody and triggering an immune response.

antigen-presenting cell (APC) A cell that phagocytizes an antigen and displays fragments of it on its surface for recognition by other cells of the immune system; chiefly macrophages and B lymphocytes.

antioxidant A chemical that binds and neutralizes free radicals, minimizing their oxidative damage to a cell; for example, selenium and vitamin E.

antiport A cotransport protein that moves two or more solutes in opposite directions through a cellular membrane; for example, the Na^+-K^+ pump.

aorta A large artery that extends from the left ventricle to the lower abdominal cavity and gives rise to all other arteries of the systemic circulation. fig. 20.21

apical surface The uppermost surface of an epithelial cell, usually exposed to the lumen of an organ. fig. 4.4

apocrine Pertaining to certain sweat glands with large lumens and relatively thick, aromatic secretions and to similar glands such as the mammary gland; formerly thought to form secretions by pinching off bits of apical cytoplasm.

apoptosis (AP-oh-TOE-sis) Programmed cell death; the normal death of cells that have completed their function. *Compare* necrosis.

appendicular (AP-en-DIC-you-lur) Pertaining to the extremities and their supporting skeletal girdles. fig. 9.1

aqueous humor A serous fluid between the cornea and lens of the eye.

arcuate (AR-cue-et) Making a sharp L- or U-shaped bend (arc), as in the *arcuate arteries* of the kidneys and uterus.

areolar tissue (AIR-ee-OH-lur) A fibrous connective tissue with loosely organized, widely spaced fibers and cells and an abundance of fluid-filled space; found under nearly every epithelium, among other places. fig. 6.16

arrector pili (ah-REC-tur PY-lye) (plural, *arrectores pilorum;* ah-rec-TORE-eez py-LOR-um) A bundle of smooth muscle cells associated with a hair follicle, responsible for erection of the hair. fig. 7.7

arrhythmia (ah-RITH-me-uh) An irregularity in the cardiac rhythm.

arteriole (ar-TEER-ee-ole) A small artery that empties into a metarteriole or capillary.

artery Any blood vessel that conducts blood away from the heart.

articular cartilage A thin layer of hyaline cartilage covering the articular surface of a bone at a synovial joint serving to reduce friction and ease joint movement. fig. 10.6

articulation A skeletal joint; any point at which two bones meet; may or may not be movable.

ascorbic acid Vitamin C; a dietary antioxidant.

aspect A particular view of the body or one of its structures, or a part that faces in a particular direction, such as the anterior aspect.

atheroma (ATH-ur-OH-muh) A fatty deposit (plaque) in a blood vessel consisting of lipid, smooth muscle, and macrophages; characteristic of atherosclerosis. fig. 1, p. 743

atherosclerosis (ATH-ur-oh-skleh-ROE-sis) A degenerative disease of the blood vessels characterized by the presence of atheromas and often leading to calcification of the vessel wall.

atrial natriuretic factor (ANF) (AY-tree-ul NAY-tree-you-RET-ic) A hormone secreted by the heart that lowers blood pressure by promoting sodium excretion and antagonizing aldosterone.

atrioventricular (AV) node (AY-tree-oh-ven-TRIC-you-lur) A group of autorhythmic cells in the interatrial septum of the heart that relays excitation from the atria to the ventricles.

atrioventricular (AV) valves The bicuspid (right) and tricuspid (left) valves between the atria and ventricles of the heart.

atrophy (AT-ro-fee) Shrinkage of a tissue due to age, disuse, or disease.

auditory ossicles Three small middle-ear bones that transfer vibrations from the tympanic membrane to the inner ear; the malleus, incus, and stapes.

autoantibody An antibody that fails to distinguish the body's own molecules from foreign molecules and thus attacks host tissues, causing autoimmune diseases.

autoimmune disease Any disease in which antibodies fail to distinguish between foreign and self-antigens and attack the body's own tissues; for example, systemic lupus erythematosus and rheumatic fever.

autolysis (aw-TOLL-ih-sis) Digestion of cells by their own internal enzymes.

autonomic nervous system (ANS) (AW-toe-NOM-ic) A division of the nervous system that innervates glands, smooth muscle, and cardiac muscle; consists of sympathetic and parasympathetic divisions and functions largely without voluntary control. *Compare* somatic nervous system.

autoregulation The ability of a tissue to adjust its own blood supply through vasomotion or angiogenesis.

autorhythmic (AW-toe-RITH-mic) Pertaining to cells that spontaneously produce action potentials at regular time intervals, chiefly cardiac and smooth muscle cells.

autosome (AW-toe-some) Any chromosome except the sex chromosomes. Genes on the autosomes are inherited without regard to the sex of the individual.

axial (AC-see-ul) Pertaining to the head, neck, and trunk; the part of the body excluding the appendicular portion. fig. 9.1

axillary (ACK-sih-LERR-ee) Pertaining to the armpit.

axolemma The plasma membrane of a nerve fiber.

axon A process of a neuron that transmits action potentials; also called a *nerve fiber.* There is only one axon to a neuron, and it is usually much longer and much less branched than the dendrites. fig. 13.3

axoneme (AC-so-neem) The core of microtubules, usually in a "9 + 2" array, at the center of a cilium or flagellum. fig. 4.11

B

baroreceptors (BAR-oh-re-SEP-turz) Pressure sensors located in the heart, aortic arch, and carotid sinuses that trigger autonomic reflexes in response to fluctuations in blood pressure.

basal metabolic rate (BMR) The rate of energy consumption of a person who is awake, relaxed, at a comfortable temperature, and has not eaten for 12 to 14 hours; usually expressed as kilocalories per square meter of body surface per hour. *Compare* metabolic rate.

basal nuclei Masses of deep cerebral gray matter that play a role in the coordination of posture and movement. fig. 14.21

base **1.** A chemical that binds protons from solution; a proton acceptor. **2.** Any of the purines or pyrimidines of a nucleic acid (adenine, thymine, guanine, cytosine, or uracil), serving in part to code for protein structure. **3.** The broadest part of a tapered organ such as the uterus or the inferior aspect of an organ such as the brain.

basement membrane A thin layer of glycoproteins, collagen, and glycosaminoglycans beneath the deepest cells of an epithelium, serving to bind the epithelium to the underlying tissue. fig. 6.34

base triplet A sequence of three DNA nucleotides that codes indirectly (through mRNA) for one amino acid of a protein.

basophil (BAY-so-fill) A granulocyte with coarse cytoplasmic granules that produces heparin, histamine, and other chemicals involved in inflammation. table 18.8

belly The thick part of a skeletal muscle between its origin and insertion. fig. 11.2

bicarbonate buffer system An equilibrium mixture of carbonic acid, bicarbonate ions, and hydrogen ions ($H_2CO_3 \leftrightarrow HCO_3^- + H^+$) that stabilizes the pH of the body fluids.

bicarbonate ion An anion, HCO_3^-, that functions as a base in the buffering of body fluids.

bile A secretion produced by the liver, concentrated and stored in the gallbladder, and released into the small intestine; consists mainly of wastes such as excess cholesterol, salts, and bile pigments but also contains lecithin and bile acids, which aid in fat digestion.

bile pigments Strongly colored organic compounds produced by the breakdown of hemoglobin, including biliverdin and bilirubin.

bilirubin (BIL-ih-ROO-bin) A yellow to orange bile pigment produced by the breakdown of hemoglobin and excreted in the bile; causes jaundice and neurotoxic effects if present in excessive concentration.

biogenic amines A class of chemical messengers with neurotransmitter and hormonal functions, synthesized from amino acids and retaining an amino group; also called *monoamines.* Examples include epinephrine and thyroxine.

bipedalism The habit of walking on two legs; a defining characteristic of the family Hominidae that underlies many skeletal and other characteristics of humans.

blood-brain barrier (BBB) A barrier between the bloodstream and nervous tissue of the brain that is impermeable to many blood solutes and thus prevents them from affecting the brain tissue; formed by the tight junctions between capillary endothelial cells, the basement membrane of the endothelium, and the perivascular feet of astrocytes.

B lymphocyte A lymphocyte that functions as an antigen-presenting cell and, in humoral immunity, differentiates into an antibody-producing plasma cell; also called a *B cell.*

body **1.** The entire organism. **2.** Part of a cell, such as a neuron, containing the nucleus and most other organelles. **3.** The largest or principal part of an organ such as the stomach or uterus; also called the *corpus.*

bolus A mass of matter, especially food or feces traveling through the digestive tract.

bone **1.** A calcified connective tissue; also called *osseous tissue.* **2.** An organ of the skeleton composed of osseous tissue, fibrous connective tissue, marrow, cartilage, and other tissues.

Bowman's capsule *See* glomerular capsule.

brachial (BRAY-kee-ul) Pertaining to the arm proper, the region from shoulder to elbow.

bradykinin (BRAD-ee-KY-nin) An oligopeptide produced in inflammation that stimulates vasodilation, increases capillary permeability, and stimulates pain receptors.

brainstem The stalklike lower portion of the brain, composed of all of the brain except the cerebrum and cerebellum. (Many authorities also exclude the diencephalon and regard only the medulla oblongata, pons, and midbrain as the brainstem.) fig. 14.13

bronchiole (BRONK-ee-ole) A pulmonary air passage that is usually 1 mm or less in diameter and lacks cartilage but has relatively abundant smooth muscle, elastic tissue, and a simple cuboidal, usually ciliated epithelium.

bronchus (BRONK-us) A relatively large pulmonary air passage with supportive cartilage in the wall; any passage beginning with the primary bronchus at the fork in the trachea and ending with tertiary bronchi, from which air continues into the bronchioles.

brush border A fringe of microvilli on the apical surface of an epithelial cell, serving to enhance surface area and promote absorption. fig. 4.10

buffer **1.** A mixture of chemicals that resists changes in pH when acid or base is added to the solution. **2.** A physiological system that contributes to acid-base balance, specifically the respiratory and urinary systems.

bulk transport The movement of particles or fluid droplets through the plasma membrane by the process of endocytosis or exocytosis.

bursa A sac filled with synovial fluid at a diarthrosis, serving to facilitate muscle or joint action. fig. 10.20

C

calcaneal tendon (cal-CAY-nee-ul) A thick tendon at the heel that attaches the triceps surae muscles to the calcaneus; also called the *Achilles tendon.* fig. 11.37

calcification The hardening of a tissue due to the deposition of calcium salts.

calcitonin (CAL-sih-TOE-nin) A hormone secreted by C cells of the thyroid gland that promotes calcium deposition in the skeleton and lowers blood calcium concentration.

calmodulin An intracellular protein that binds calcium ions and mediates many of the second-messenger effects of calcium.

calorie The amount of thermal energy that will raise the temperature of 1 g of water by 1°C. Also called a *small calorie.*

Calorie *See* kilocalorie.

calorigenic (ca-LOR-ih-JEN-ic) Heat-producing, as in the calorigenic effect of thyroid hormone.

calsequestrin A protein found in smooth endoplasmic reticulum that reversibly binds and stores calcium ions, rendering calcium chemically unreactive until needed for such processes as muscle contraction.

calvaria (cal-VERR-ee-uh) The rounded bony dome that forms the roof of the cranium; skullcap.

calyx (CAY-lix) (plural, *calices*) A cuplike structure, as in the kidneys. fig. 23.3

canaliculus (CAN-uh-LIC-you-lus) A microscopic canal, as in osseous tissue. fig. 8.5

capillary (CAP-ih-LERR-ee) The narrowest type of vessel in the cardiovascular and lymphatic systems; engages in fluid exchanges with surrounding tissues.

capillary exchange The process of fluid transfer between the bloodstream and tissue fluid.

capsule The fibrous covering of a structure such as the spleen or a diarthrosis.

carbohydrate A hydrophilic organic compound composed of carbon and a 2:1

ratio of hydrogen to oxygen; includes sugars, starches, glycogen, and cellulose.

carbonic anhydrase An enzyme found in erythrocytes and kidney tubule cells that catalyzes the decomposition of carbonic acid into carbon dioxide and water or the reverse reaction ($H_2CO_3 \leftrightarrow CO_2 + H_2O$).

carboxyl group (car-BOC-sil) An organic functional group with the formula –COOH, found in many organic acids such as amino acids and fatty acids.

carcinogen (car-SIN-oh-jen) An agent capable of causing cancer, including certain chemicals, viruses, and ionizing radiation.

cardiac center A nucleus in the medulla oblongata that regulates autonomic reflexes for controlling the rate and strength of the heartbeat.

cardiac cycle One complete cycle of cardiac systole and diastole.

cardiac muscle Striated involuntary muscle of the heart.

cardiac output (CO) The amount of blood pumped by each ventricle of the heart in 1 minute.

cardiac reserve The difference between maximum and resting cardiac output; determines a person's tolerance for exercise.

cardiovascular system An organ system consisting of the heart and blood vessels, serving for the transport of blood. *Compare* circulatory system.

carotid body (ca-ROT-id) A small cellular mass immediately superior to the branch in the common carotid artery, containing sensory cells that detect changes in blood pH and carbon dioxide and oxygen content. fig. 20.11

carotid sinus A dilation of the common carotid artery at the point where it branches into the internal and external carotids; contains baroreceptors, which monitor changes in blood pressure.

carpal Pertaining to the wrist (carpus).

carrier **1.** A protein in a cellular membrane that performs carrier-mediated transport. **2.** A person who is heterozygous for a recessive allele and does not exhibit the associated phenotype, but may transmit this allele to his or her children; for example, a carrier for sickle-cell disease.

carrier-mediated transport A process of transporting materials through a cellular membrane that involves reversible binding to a membrane protein.

cartilage A connective tissue with a rubbery matrix, cells (chondrocytes) contained in lacunae, and no blood vessels; covers the articular surfaces of many bones and supports organs such as the ear and larynx.

catabolism (ca-TAB-oh-lizm) Any metabolic reactions that release energy and break relatively complex molecules with high free energy into less complex molecules with lower free energy; for example, digestion and glycolysis. *Compare* anabolism.

catalyst (CAT-uh-list) Any chemical that lowers the activation energy of a chemical reaction and thus makes the reaction proceed more rapidly; a role served in cells by enzymes.

catecholamine (CAT-eh-COAL-uh-meen) A subclass of biogenic amines that includes epinephrine, norepinephrine, and dopamine. fig. 13.18

cation (CAT-eye-on) An ion with more protons than electrons and consequently a net positive charge.

caudal (CAW-dul) **1.** Pertaining to a tail or narrow tail-like part of an organ. **2.** Pertaining to the inferior part of the trunk of the body, where the tail of other animals arises. *Compare* cranial.

celiac (SEE-lee-ac) Pertaining to the abdomen.

cell The smallest subdivision of a tissue considered to be alive; consists of a plasma membrane enclosing cytoplasm and, in most cases, a nucleus.

cellular membrane Any unit membrane enclosing a cell or organelle. *See also* unit membrane.

central Located relatively close to the medial axis of the body, as in the central nervous system; opposite of peripheral.

central nervous system (CNS) The brain and spinal cord.

centriole (SEN-tree-ole) An organelle composed of a short cylinder of nine triplets of microtubules, usually paired with another centriole perpendicular to it; origin of the mitotic spindle; identical to the basal body of a cilium or flagellum. fig. 4.18

cephalic (seh-FAL-ic) Pertaining to the head.

cerebellum (SERR-eh-BEL-um) A large portion of the brain posterior to the brainstem and inferior to the cerebrum, responsible for equilibrium, motor coordination, and memory of learned motor skills. fig. 14.14

cerebrospinal fluid (CSF) (SERR-eh-bro-SPY-nul, seh-REE-bro-SPY-nul) A liquid that fills the ventricles of the brain, the central canal of the spinal cord, and the space between the CNS and dura mater.

cerebrovascular accident (CVA) (SERR-eh-bro-VASS-cue-lur, seh-REE-bro-VASS-cue-lur) The loss of blood flow to any part of the brain due to obstruction or hemorrhage of an artery, leading to the necrosis of nervous tissue; also called *stroke* or *apoplexy.*

cerebrum (SERR-eh-brum, seh-REE-brum) The largest and most superior part of the brain, divided into two convoluted cerebral hemispheres separated by a deep longitudinal fissure.

cervical (SUR-vih-cul) Pertaining to the neck or any cervix.

cervix (SUR-vix) **1.** The neck. **2.** A narrow or necklike part of an organ such as the uterus and gallbladder. fig. 28.3

channel protein A protein in the plasma membrane that has a pore through it for the passage of materials between the cytoplasm and extracellular fluid. fig. 4.6

chemical bond A force that attracts one atom to another, such as their opposite charges or the sharing of electrons.

chemical digestion Hydrolysis reactions that occur in the digestive tract and covert dietary polymers into monomers that can be absorbed by the small intestine.

chemical synapse A meeting of a nerve fiber and another cell with which the neuron communicates by releasing neurotransmitters. fig. 13.17

chemoreceptor An organ or cell specialized to detect chemicals, as in the carotid bodies and taste buds.

chemotaxis (KEM-oh-TAC-sis) The movement of a cell along a chemical concentration gradient, especially the attraction of neutrophils to chemicals released by pathogens or inflamed tissues.

chief cells The majority type of cell in an organ or tissue such as the parathyroid glands or gastric glands.

choanae (co-AH-nee) Openings of the nasal cavity into the pharynx; also called *posterior nares.* fig. 22.3

cholecystokinin (CCK) (CO-leh-SIS-toe-KY-nin) A polypeptide employed as a hormone and neurotransmitter, secreted by some brain neurons and cells of the digestive tract. fig. 13.18

cholesterol (co-LESS-tur-ol) A steroid that functions as part of the plasma membrane and as a precursor for all other steroids in the body.

cholinergic (CO-lin-UR-jic) Pertaining to acetylcholine (ACh), as in cholinergic nerve fibers that secrete ACh, cholinergic receptors that bind it, or cholinergic effects on a target organ.

chondrocyte (CON-dro-site) A cartilage cell; a former chondroblast that has become enclosed in a lacuna in the cartilage matrix. fig. 8.10

chorion (CO-ree-on) A fetal membrane external to the amnion; forms part of the placenta and has diverse functions including fetal nutrition, waste removal, and hormone secretion. fig. 29.9

chromatid (CRO-muh-tid) One of two genetically identical rodlike bodies of a metaphase chromosome, joined to its sister chromatid at the centromere. fig. 5.15

chromatin (CRO-muh-tin) Filamentous nuclear material composed of DNA and associated proteins.

chromosome A condensed form of chromatin visible in a cell undergoing mitosis or meiosis. fig. 5.15

chronic **1.** Long-lasting. **2.** Pertaining to a disease that progresses slowly and has a long duration. *Compare* acute.

chronic bronchitis A chronic obstructive pulmonary disease characterized by damaged and immobilized respiratory cilia, excessive mucus secretion, infection of the lower respiratory tract, and bronchial inflammation; caused especially by cigarette smoking. *See also* chronic obstructive pulmonary disease.

chronic obstructive pulmonary disease (COPD) A group of lung diseases (asthma, chronic bronchitis, and emphysema) that result in long-term obstruction of airflow and substantially reduced pulmonary ventilation; one of the leading causes of death in old age.

chylomicron (KY-lo-MY-cron) A protein-coated lipid droplet formed in the small intestine and found in the lymph and blood after a meal; a means of lipid transport in the bloodstream and lymph.

chyme (kime) A slurry of partially digested food in the stomach and small intestine.

cilium (SIL-ee-um) A hairlike process, with an axoneme, projecting from the apical surface of an epithelial cell; usually motile and serving to propel matter across the surface of an epithelium but sometimes serving sensory roles. fig. 4.12

circulatory shock A state of cardiac output inadequate to meet the metabolic needs of the body.

circulatory system An organ system consisting of the heart, blood vessels, and blood. *Compare* cardiovascular system.

circumduction A joint movement in which one end of an appendage remains relatively stationary and the other end is moved in a circle. fig. 10.12

cirrhosis (sih-RO-sis) A degenerative liver disease characterized by replacement of functional parenchyma with fibrous and adipose tissue; causes include alcohol, other poisons, and viral and bacterial inflammation.

cisterna (sis-TUR-nuh) A fluid-filled space or sac, such as the cisterna chyli of the lymphatic system and a cisterna of the endoplasmic reticulum or Golgi complex. fig. 4.14

citric acid cycle A cyclic reaction series involving several carboxylic acids in the mitochondrial matrix; oxidizes acetyl groups to carbon dioxide while reducing NAD^+ to NADH and FADH to $FADH_2$, making these reduced coenzymes available for ATP synthesis. Also called the *Krebs cycle* or *tricarboxylic acid (TCA) cycle.* fig. 26.4

climacteric A period in the lives of men and women, usually in the early 50s, marked by changes in the level of reproductive hormones, a variety of somatic and psychological effects, and in women, cessation of ovulation and menstruation (menopause).

clone A population of cells that are mitotically descended from the same parent cell and are identical to each other genetically or in other respects.

coagulation (co-AG-you-LAY-shun) The clotting of blood, lymph, tissue fluid, or semen.

codominant (co-DOM-ih-nent) A condition in which neither of two alleles is dominant over the other, and both are phenotypically expressed when both are present in an individual; for example, blood type alleles I^A and I^B produce blood type AB when inherited together.

codon A series of three nucleotides in mRNA that codes for one amino acid in a protein or, as a *stop codon,* signals the end of a gene.

coenzyme (co-EN-zime) A small organic molecule, usually derived from a vitamin, that is needed to make an enzyme catalytically active; acts by accepting electrons from an enzymatic reaction and transferring them to a different reaction chain.

cofactor A nonprotein such as a metal ion or coenzyme needed for an enzyme to function.

cohesion The clinging of identical molecules such as water to each other.

collagen (COLL-uh-jen) The most abundant protein in the body, forming the fibers of many connective tissues in places such as the dermis, tendons, and bones.

colloid An aqueous mixture of particles that are too large to pass through most selectively permeable membranes but small enough to remain evenly dispersed through the solvent by the thermal motion of solvent particles; for example, the proteins in blood plasma.

colloid osmotic pressure (COP) A portion of the osmotic pressure of a body fluid that is due to its protein. *Compare* oncotic pressure.

colostrum (co-LOS-trum) A watery, low-fat secretion of the mammary gland that nourishes and immunizes an infant for the first 2 to 3 days postpartum, until true milk is secreted.

commissure (COM-ih-shur) **1.** A bundle of nerve fibers that crosses from one side of the brain or spinal cord to the other. **2.** A corner or angle at which the eyelids, lips, or genital labia meet; in the eye, also called the *canthus.* fig. 16.20

complement **1.** To complete or enhance the structure or function of something else, as in the coordinated action of two different hormones. **2.** A system of plasma proteins involved in nonspecific defense against pathogens.

computerized tomography (CT) A method of medical imaging that uses X rays and a computer to create an image of a thin section of the body; also called a *CT scan.*

concentration gradient A difference in chemical concentration from one point to another, as on two sides of a plasma membrane.

conception The fertilization of an egg, producing a zygote.

conceptus All products of conception, ranging from a fertilized egg to the full-term fetus with its embryonic membranes, placenta, and umbilical cord. *Compare* embryo, fetus.

condyle (CON-dile) A rounded knob on a bone serving to produce smooth motion at a joint. fig. 9.2

conformation The three-dimensional structure of a protein that results from interaction among its amino acid side groups, its interactions with water, and the formation of disulfide bonds.

congenital Present at birth; for example, an anatomical defect, a syphilis infection, or a hereditary disease.

conjugated A state in which one organic compound is bound to another compound of a different class, such as a protein conjugated with a carbohydrate to form a glycoprotein.

connective tissue A tissue usually composed of more extracellular than cellular volume and usually with a substantial amount of extracellular fiber; forms supportive frameworks and capsules for organs, binds structures together, holds them in place, stores energy (as in adipose tissue), or transports materials (as in blood).

contractility 1. The ability to shorten. 2. The amount of force that a contracting muscle fiber generates for a given stimulus; may be increased by epinephrine, for example, while stimulus strength remains constant.

contralateral On opposite sides of the body, as in reflex arcs where the stimulus comes from one side of the body and a response is given by muscles on the other side. *Compare* ipsilateral.

convergent Coming together, as in a convergent muscle and a converging neuronal circuit.

cooperative effects Effects in which two hormones, or both divisions of the autonomic nervous system, work together to produce a single overall result.

cornified Having a heavy deposit of keratin, as in the stratum corneum of the epidermis.

corona A halo- or crownlike structure, as in the corona radiata or the coronal suture of the skull.

coronal plane *See* frontal plane.

corona radiata 1. An array of nerve tracts in the brain that arise mainly from the thalamus and fan out to different regions of the cerebral cortex. 2. The first layer of cuboidal cells immediately external to the zona pellucida around an egg cell.

coronary circulation A system of blood vessels that serve the wall of the heart. fig. 19.10

corpus Body or mass; the main part of an organ, as opposed to such regions as a head, tail, or cervix.

corpus callosum (COR-pus ca-LO-sum) A prominent C-shaped band of nerve tracts that connect the right and left cerebral hemispheres to each other, visible in a midsagittal section of the brain superior to the third ventricle. fig. 14.1

corpus luteum (LOO-tee-um) A yellowish cellular mass that forms in the ovary from a follicle that has ovulated; secretes progesterone, hormonally regulates the second half of the menstrual cycle, and is essential to sustaining the first 7 weeks of pregnancy.

cortex (plural, *cortices*) The outer layer of some organs such as the adrenal gland, cerebrum, lymph node, and ovary; usually covers or encloses tissue called the medulla.

corticosteroid (COR-tih-co-STERR-oyd) Any steroid hormone secreted by the adrenal cortex, such as aldosterone, cortisol, and sex steroids.

costal (COSS-tul) Pertaining to the ribs.

costal cartilage A bladelike plate of hyaline cartilage that attaches the distal end of a rib to the sternum.

cotransport A form of carrier-mediated transport in which a membrane protein transports two solutes simultaneously or within the same cycle of action; for example the sodium-dependent glucose transporter and the Na^+-K^+ pump.

countercurrent A situation in which two fluids flow side by side in opposite directions, as in the countercurrent multiplier of the kidney and the countercurrent heat exchanger of the scrotum.

cranial (CRAY-nee-ul) 1. Pertaining to the cranium. 2. In a position relatively close to the head or a direction toward the head. *Compare* caudal.

cranial nerve Any of 12 pairs of nerves connected to the base of the brain and passing through foramina of the cranium.

creatine phosphate (CP) (CREE-uh-teen FOSS-fate) An energy-storage molecule in muscle tissue that donates a phosphate group to ADP and thus regenerates ATP in periods of hypoxia.

crista A crestlike structure, such as the crista galli of the ethmoid bone or the crista of a mitochondrion.

cross section A cut perpendicular to the long axis of the body or an organ.

crural (CROO-rul) Pertaining to the leg proper or to the crus of a organ. *See* crus.

crus (cruss) (plural, *crura*) 1. The leg proper; the region from the knee to the ankle. 2. A leglike extension of an organ such as the penis and clitoris. fig. 28.7

cuboidal (cue-BOY-dul) A cellular shape that is roughly like a cube or in which the height and width are about equal.

cuneiform (cue-NEE-ih-form) Wedge-shaped, as in the cuneiform cartilages of the larynx and cuneiform bone of the wrist.

current A moving stream of charged particles such as ions or electrons.

cusp 1. One of the flaps of a valve of the heart, veins, and lymphatic vessels. 2. A conical projection on the occlusal surface of a premolar or molar tooth.

cutaneous (cue-TAY-nee-us) Pertaining to the skin.

cyanosis (SY-uh-NO-sis) A bluish color of the skin due to ischemia or hypoxemia.

cyclic adenosine monophosphate (cAMP) A cyclic molecule produced from ATP by the action of adenylate cyclase; serves as a second messenger in many hormone and neurotransmitter actions.

cyclooxygenase An enzyme that converts arachidonic acid to prostacyclin, prostaglandins, and thromboxanes.

cytochromes Enzymes on the mitochondrial cristae that transfer electrons in the final reaction chain of aerobic respiration.

cytokinesis (SY-toe-kih-NEE-sis) Division of the cytoplasm of a cell into two cells following nuclear division.

cytology The study of cell structure and function.

cytolysis (sy-TOL-ih-sis) The rupture and destruction of a cell by such agents as complement proteins and hypotonic solutions.

cytoplasm The contents of a cell between its plasma membrane and its nuclear envelope, consisting of cytosol, organelles, inclusions, and the cytoskeleton.

cytosine A single-ringed nitrogenous base (pyrimidine) found in DNA; one of the four bases of the genetic code; complementary to guanine in the double helix of DNA. fig. 5.3

cytoskeleton A system of protein microfilaments, intermediate filaments, and microtubules in a cell, serving in physical support, cellular movement, and the routing of molecules and organelles to their destinations within the cell. fig. 4.19

cytosol A clear, featureless, gelatinous colloid in which the organelles and other internal structures of a cell are embedded.

cytotoxic T cell A T lymphocyte that directly attacks and destroys infected body cells, cancerous cells, and the cells of transplanted tissues.

D

daughter cells Cells that arise from a parent cell by mitosis or meiosis.

deamination (dee-AM-ih-NAY-shun) Removal of an amino group from an organic molecule; a step in the catabolism of amino acids.

decomposition reaction A chemical reaction in which a larger molecule is broken down into smaller ones. *Compare* synthesis reaction.

decussation (DEE-cuh-SAY-shun) The crossing of nerve fibers from the right side of the central nervous system to the left or vice versa, especially in the spinal cord, medulla oblongata, and optic chiasm.

deep Relatively far from the body surface; opposite of *superficial*. For example, the bones are deep to the skeletal muscles.

degranulation Exocytosis and disappearance of cytoplasmic granules, especially in platelets and granulocytes.

dehydration synthesis A reaction in which two chemical monomers are joined together with water produced as a by-product; also called a *condensation reaction. Compare* hydrolysis.

denaturation A change in the three-dimensional conformation of a protein that destroys its enzymatic or other functional properties, usually caused by extremes of temperature or pH.

dendrites Processes of a neuron that receive information from other cells or from environmental stimuli and conduct signals to the soma. Dendrites are usually shorter, more branched, and more numerous than the axon and are incapable of producing action potentials. fig. 13.3

dendritic cell An antigen-presenting cell of the epidermis and mucous membranes. fig. 7.2

denervation atrophy The shrinkage of skeletal muscle that occurs when its motor neuron dies or is severed from the muscle.

dense connective tissue A connective tissue with a high density of fiber, relatively little ground substance, and scanty cells; seen in tendons and the dermis, for example.

deoxyribonucleic acid (DNA) (dee-OCK-see-RY-bo-new-CLAY-ic) A very large nucleotide polymer that carries the genes of a cell; composed of a double helix of intertwined chains of deoxyribose and phosphate, with complementary pairs of nitrogenous bases facing each other between the helices. fig. 5.4

depolarization A shift in the electrical potential across a plasma membrane toward 0 mV, associated with excitation of a nerve or muscle cell. *Compare* hyperpolarization.

dermal papillae Bumps or ridges of dermis that extend upward to interdigitate with the epidermis and create a wavy boundary that resists stress and slippage of the epidermis.

dermis The deeper of the two layers of the skin, underlying the epidermis and composed of fibrous connective tissue.

desmosome (DEZ-mo-some) A patchlike intercellular junction that mechanically links two cells together. fig. 6.30

dextrose An isomer of glucose; the only form of glucose with a normal role in physiology.

diabetes (DY-uh-BEE-teez) Any disease characterized by chronic polyuria of metabolic origin; diabetes mellitus unless otherwise specified.

diabetes insipidus (in-SIP-ih-dus) A form of diabetes that results from hyposecretion of antidiuretic hormone; unlike other forms, it is not characterized by hyperglycemia or glycosuria.

diabetes mellitus (DM) (mel-EYE-tus) A form of diabetes that results from hyposecretion of insulin or from a deficient target cell response to it; symptoms include hyperglycemia and glycosuria.

dialysis (dy-AL-ih-sis) **1.** The separation of some solute particles from others by diffusion through a selectively permeable membrane. **2.** Hemodialysis, the process of separating wastes from the bloodstream and sometimes adding other substances to it (such as drugs and nutrients) by circulating the blood through a machine with a selectively permeable membrane, used to treat cases of renal or hepatic insufficiency.

diapedesis (DY-uh-peh-DEE-sis) Migration of formed elements of the blood through a capillary wall into the interstitial space. fig. 21.13

diaphysis (dy-AFF-ih-sis) The shaft of a long bone. fig. 8.2

diarthrosis (DY-ar-THRO-sis) A freely movable synovial joint such as the knuckle, elbow, shoulder, or knee.

diastole (dy-ASS-tuh-lee) A period in which a heart chamber relaxes and fills with blood; especially ventricular relaxation.

diencephalon (DY-en-SEFF-uh-lon) A portion of the brain between the midbrain and corpus callosum; composed of the thalamus, epithalamus, and hypothalamus. fig. 14.17

differentiation Development of a relatively unspecialized cell into one with a more specific structure and function.

diffusion The movement of molecules resulting from their spontaneous thermal vibration; *net diffusion* is the movement of more molecules in one direction than in another.

dilation (dy-LAY-shun) Widening of an organ or passageway such as a blood vessel or the pupil of the eye.

diploid (2n) Pertaining to a cell or organism with chromosomes in homologous pairs.

disaccharide (dy-SAC-uh-ride) A carbohydrate composed of two simple sugars (monosaccharides) joined by a glycosidic bond; for example, lactose, sucrose, and maltose. fig. 3.5

disseminated intravascular coagulation (DIC) Widespread clotting of the blood within unbroken blood vessels, leading to widespread hemorrhaging, congestion of the vessels with clotted blood, and ischemia and necrosis of organs.

distal Relatively distant from a point of origin or attachment; for example, the wrist is distal to the elbow. *Compare* proximal.

disulfide bond A covalent bond between the sulfur atoms of two cysteine residues, serving to link one polypeptide chain to another or to hold a single chain in its three-dimensional conformation.

diuretic (DY-you-RET-ic) A chemical that increases urine output.

dizygotic (DZ) twins Two individuals who developed simultaneously in one uterus but originated from separate fertilized eggs and therefore are not genetically identical.

dominant **1.** Pertaining to a genetic allele that is phenotypically expressed in the presence of any other allele. **2.** Pertaining to a trait that results from a dominant allele.

dopamine (DOE-puh-meen) An inhibitory catecholamine neurotransmitter of the central nervous system, especially of the basal nuclei, where it acts to suppress unwanted motor activity. fig 13.18

dorsal Toward the back (spinal) side of the body.

dorsal root A branch of a spinal nerve that enters the spinal cord on its dorsal side, composed of sensory fibers. fig. 15.1

dorsiflexion (DOR-sih-FLEC-shun) A movement of the ankle that reduces the joint angle and raises the toes. fig. 10.14

Down syndrome *See* trisomy-21.

duodenum (DEW-oh-DEE-num, dew-ODD-eh-num) The first portion of the small intestine extending for about 25 cm from the pyloric valve of the stomach to a sharp bend called the duodenojejunal flexure; receives chyme from the stomach and secretions from the liver and pancreas. fig. 25.22

dynamic equilibrium **1.** A state of continual change that is controlled within narrow limits, as in homeostasis and chemical equilibrium. **2.** The sense of motion or acceleration of the body.

dynein (DINE-een) A motor protein involved in the beating of cilia and flagella and in the movement of molecules and organelles within cells, as in retrograde transport in a nerve fiber.

E

ectoderm The outermost of the three primary germ layers of an embryo; gives rise to the nervous system and epidermis.

ectopic (ec-TOP-ic) In an abnormal location; for example, ectopic pregnancy and ectopic pacemakers of the heart.

edema (eh-DEE-muh) Abnormal accumulation of tissue fluid resulting in swelling of the tissue.

effector A molecule, cell, or organ that carries out a response to a stimulus.

efferent (EFF-ur-unt) Carrying away or out, such as a blood vessel that carries blood away from a tissue or a nerve fiber that conducts signals away from the central nervous system. *Compare* afferent.

eicosanoids (eye-CO-sah-noyds) Twenty-carbon derivatives of arachidonic acid that function as intercellular messengers; includes prostaglandins, prostacyclin, leukotrienes, and thromboxanes.

elastic fiber A connective tissue fiber, composed of the protein elastin, that stretches under tension and returns to its original length when released; responsible for the resilience of organs such as the skin and lungs.

elasticity The tendency of a stretched structure to return to its original dimensions when tension is released.

electrical synapse A gap junction that enables one cell to stimulate another directly, without the intermediary action of a neurotransmitter; such synapses connect the cells of cardiac muscle and single-unit smooth muscle.

electrochemical gradient A difference in ion concentration from one point to another (especially across a plasma membrane) resulting in a gradient of both chemical concentration and electrical charge.

electrolyte A salt that ionizes in water and produces a solution that conducts electricity; loosely speaking, any ion that results from the dissociation of such salts, such as sodium, potassium, calcium, chloride, and bicarbonate ions.

elevation A joint movement that raises a body part, as in hunching the shoulders or closing the mouth.

embolism (EM-bo-lizm) The obstruction of a blood vessel by an embolus.

embolus (EM-bo-lus) Any abnormal traveling object in the bloodstream, such as agglutinated bacteria or blood cells, a blood clot, or an air bubble.

embryo A developing individual from the end of the second week of gestation when the three primary germ layers have formed, through the end of the eighth week when all of the organ systems are present. *Compare* conceptus, fetus.

emphysema (EM-fih-SEE-muh) A degenerative lung disease characterized by a breakdown of alveoli and diminishing surface area available for gas exchange; occurs with aging of the lungs but is greatly accelerated by smoking or air pollution.

emulsion A suspension of one liquid in another, such as oil in water or fat in the lymph.

endocrine gland (EN-doe-crin) A ductless gland that secretes hormones into the bloodstream; for example, the thyroid and adrenal glands. Compare *exocrine gland.*

endocytosis (EN-doe-sy-TOE-sis) Any process in which a cell forms vesicles from its plasma membrane and takes in large particles, molecules, or droplets of extracellular fluid; for example, phagocytosis and pinocytosis.

endoderm The innermost of the three primary germ layers of an embryo; gives rise to the mucosae of the digestive and respiratory tracts and to their associated glands.

endogenous (en-DODJ-eh-nus) Originating internally, such as the endogenous cholesterol synthesized in the body in contrast to the exogenous cholesterol coming from the diet. *Compare* exogenous.

endometrium (EN-doe-MEE-tree-um) The mucosa of the uterus; the site of implantation and source of menstrual discharge.

endoplasmic reticulum (ER) (EN-doe-PLAZ-mic reh-TIC-you-lum) An extensive system of interconnected cytoplasmic tubules or channels; classified as rough ER or smooth ER depending on the presence or absence of ribosomes on its membrane. fig. 4.14

endothelium (EN-doe-THEEL-ee-um) A simple squamous epithelium that lines the lumens of the blood vessels, heart, and lymphatic vessels.

endurance exercise A form of physical exercise, such as running or swimming, that promotes cardiopulmonary efficiency and fatigue resistance more than muscular strength. *Compare* resistance exercise.

enteric (en-TERR-ic) Pertaining to the small intestine, as in enteric hormones.

enzyme A protein that functions as a catalyst.

enzyme amplification A series of chemical reactions in which the product of one step is an enzyme that produces an even greater number of product molecules at the next step, resulting in a rapidly increasing amount of reaction product. Seen in hormone action and blood clotting, for example.

eosinophil (EE-oh-SIN-oh-fill) A granulocyte with a large, often bilobed nucleus and coarse cytoplasmic granules that stain with eosin; phagocytizes antigen-antibody complexes, allergens, and inflammatory chemicals and secretes enzymes that combat parasitic infections. table 18.8

epidermis A stratified squamous epithelium that constitutes the superficial layer of the skin overlying the dermis. fig. 7.2

epinephrine (EP-ih-NEFF-rin) A catecholamine that functions as a neurotransmitter in the sympathetic nervous system and as a hormone secreted by the adrenal medulla; also called *adrenalin.* fig. 13.18

epiphyseal plate (EP-ih-FIZZ-ee-ul) A plate of hyaline cartilage between the epiphysis and diaphysis of a long bone in a child or adolescent, serving as a growth zone for bone elongation. fig. 8.12

epiphysis (eh-PIF-ih-sis) **1.** The head of a long bone. **2.** The pineal gland (epiphysis cerebri).

epithelium A type of tissue consisting of one or more layers of closely adhering cells with little intercellular material and no blood vessels; forms the coverings and linings of many organs and the parenchyma of the glands.

erectile tissue A tissue that functions by swelling with blood, as in the penis and clitoris and inferior concha of the nasal cavity.

erythema (ERR-ih-THEE-muh) Abnormal redness of the skin due to such causes as burns, inflammation, and vasodilation.

erythrocyte (eh-RITH-ro-site) A red blood cell.

erythropoiesis (eh-RITH-ro-poy-EE-sis) The production of erythrocytes.

erythropoietin (eh-RITH-ro-POY-eh-tin) A hormone that is secreted by the kidneys in response to hypoxemia and stimulates erythropoiesis.

estrogens (ESS-tro-jenz) A family of steroid hormones known especially for producing female secondary sex characteristics and regulating various aspects of the menstrual cycle and pregnancy; major forms are estradiol, estriol, and estrone.

evolution A change in the relative frequencies of alleles in a population over a period of time; the mechanism that produces adaptations in human form and function. *See also* adaptation.

excitability The ability of a cell to respond to a stimulus, especially the ability of nerve and muscle cells to produce membrane voltage changes in response to stimuli; irritability.

excitation-contraction coupling Events that link the synaptic stimulation of a muscle cell to the onset of contraction.

excitatory postsynaptic potential (EPSP) A partial depolarization of a postsynaptic neuron in response to a neurotransmitter, making it more likely to reach threshold and produce an action potential.

excretion The process of eliminating metabolic waste products from a cell or from the body. *Compare* secretion.

exocrine gland (EC-so-crin) A gland that secretes its products into another organ or onto the body surface, usually by way of a duct; for example, salivary and gastric glands. *Compare* endocrine gland.

exocytosis (EC-so-sy-TOE-sis) A process in which a vesicle in the cytoplasm of a cell fuses with the plasma membrane and releases its contents from the cell; used in the elimination of cellular wastes and in the release of gland products and neurotransmitters.

exogenous (ec-SODJ-eh-nus) Originating externally, such as exogenous (dietary) cholesterol; extrinsic. *Compare* endogenous.

expiration 1. Exhaling. 2. Dying.

extension Movement of a joint that increases the angle between articulating bones (straightens the joint). *Compare* flexion. fig. 10.9

extracellular fluid (ECF) Any body fluid that is not contained in the cells; for example, blood, lymph, and tissue fluid.

extrinsic (ec-STRIN-sic) 1. Originating externally, such as extrinsic blood-clotting factors; exogenous. 2. Not fully contained within an organ but acting on it, such as the extrinsic muscles of the hand and eye. *Compare* intrinsic.

exude (ec-SUDE) To seep out, such as fluid filtering from blood capillaries.

F

facilitated diffusion The process of transporting a chemical through a cellular membrane, down its concentration gradient, with the aid of a membrane carrier that does not consume ATP; enables substances to diffuse through the membrane that would do so poorly, or not at all, without a carrier.

facilitation Making a process more likely to occur, such as the firing of a neuron, or making it occur more easily or rapidly, as in facilitated diffusion.

fallopian tube (fah-LO-pee-un) *See* uterine tube.

fascia (FASH-ee-uh) A layer of connective tissue between the muscles (deep fascia) or separating the muscles from the skin (superficial fascia). fig. 11.1

fascicle (FASS-ih-cul) A bundle of muscle or nerve fibers ensheathed in connective tissue; multiple fascicles bound together constitute a muscle or nerve as a whole. fig. 11.1

fat 1. A triglyceride molecule. 2. Adipose tissue.

fatty acid An organic molecule composed of a chain of an even number of carbon atoms with a carboxyl group at one end and a methyl group at the other; one of the structural subunits of triglycerides and phospholipids.

fenestrated (FEN-eh-stray-ted) Perforated with holes or slits, as in fenestrated blood capillaries and the elastic sheets of large arteries. fig. 20.6

fetus In human development, an individual from the beginning of the ninth week when all of the organ systems are present, through the time of birth. *Compare* conceptus, embryo.

fibrin (FY-brin) A sticky fibrous protein formed from fibrinogen in blood, tissue fluid, lymph, and semen; forms the matrix of a blood clot.

fibroblast A connective tissue cell that produces collagen fibers and ground substance; the only type of cell in tendons and ligaments.

fibrosis Replacement of damaged tissue with fibrous scar tissue rather than by the original tissue type; scarring. *Compare* regeneration.

fibrous connective tissue Any connective tissue with a preponderance of fiber, such as areolar, reticular, dense regular, and dense irregular connective tissues.

filtrate A fluid formed by filtration, as at the renal glomerulus and other capillaries.

filtration A process in which hydrostatic pressure forces a fluid through a selectively permeable membrane (especially a capillary wall).

fire To produce an action potential, as in nerve and muscle cells.

fix 1. To hold a structure in place, for example, by fixator muscles that prevent unwanted joint movements. 2. To preserve a tissue by means of a fixative.

fixative A chemical that preserves tissues from decay, such as formalin.

flagellum (fla-JEL-um) A long, motile, usually single hairlike extension of a cell; the tail of a sperm cell is the only functional flagellum in humans. fig. 27.17

flexion A joint movement that, in most cases, decreases the angle between two bones. *Compare* extension. fig. 10.9

fluid balance An equilibrium between fluid intake and output or between the amounts of fluid contained in the body's different fluid compartments.

fluid compartments Any of the major categories of fluid in the body, separated by selectively permeable membranes and differing from each other in chemical composition. Primary examples are the intracellular fluid, the tissue fluid, and the blood and lymph.

fluid-mosaic model The current theory of the structure of a plasma membrane, depicting it as a bilayer of phospholipids with embedded proteins, many of which are able to move about in the lipid film. fig. 4.6

follicle (FOLL-ih-cul) 1. A small space, such as a hair follicle, thyroid follicle, or ovarian follicle. 2. An aggregation of lymphocytes in a lymphatic organ or mucous membrane.

follicle-stimulating hormone (FSH) A hormone secreted by the anterior pituitary gland that stimulates development of the ovarian follicles and egg cells.

foramen (fo-RAY-men) A hole through a bone or other organ, in many cases providing passage for blood vessels and nerves.

formed element An erythrocyte, leukocyte, or platelet; any cellular component of blood or lymph as opposed to the extracellular fluid component.

fossa (FOSS-uh) A depression in an organ or tissue, such as the fossa ovalis of the heart or a cranial fossa of the skull.

fovea (FOE-vee-uh) A small pit, such as the fovea capitis of the femur or fovea centralis of the retina.

free energy The potential energy in a chemical that is available to do work.

free radical An uncharged particle derived from an atom or molecule, having an unpaired electron that makes it highly reactive and destructive to cells; produced by intrinsic processes such as aerobic respiration and by extrinsic agents such as chemicals and ionizing radiation.

frontal plane An anatomical plane that passes through the body or an organ from right to left and superior to inferior; also called a *coronal plane.* fig. A.3

functional group A group of atoms, such as a carboxyl or amino group, that determines the functional characteristics of an organic molecule.

fundus The base, the broadest part, or the part farthest from the opening of certain viscera such as the stomach and uterus.

fusiform (FEW-zih-form) Spindle-shaped; elongated, thick in the middle, and tapered at both ends, such as the shape of a smooth muscle cell or a muscle spindle.

G

gamete (GAM-eet) An egg or sperm cell.

gametogenesis (GAM-eh-toe-JEN-eh-sis) The production of eggs or sperm.

gamma-(γ-)aminobutyric acid (GABA) (ah-MEE-no-byu-TIRR-ic) An inhibitory neurotransmitter of the central nervous system in the biogenic amine class. fig. 13.18

gamma (γ) globulins (GLOB-you-lins) A class of relatively large proteins found in the blood plasma and on the surfaces of immune cells, functioning as antibodies. *See also* globulin.

ganglion (GANG-glee-un) A cluster of nerve cell bodies in the peripheral nervous system, often resembling a knot in a string.

gangrene Tissue necrosis resulting from ischemia.

gap junction A junction between two cells consisting of a pore surrounded by a ring of proteins in the plasma membrane of each cell, allowing solutes to diffuse from the cytoplasm of one cell to the next; functions include cell-to-cell nutrient transfer in the developing embryo and electrical communication between cells of cardiac and smooth muscle. *See also* electrical synapse. fig. 6.30

gastric Pertaining to the stomach.

gate A protein channel in a unit membrane that can open or close in response to chemical, electrical, or mechanical stimuli, thus controlling when substances are allowed to pass through the membrane.

gene A segment of DNA that codes for the synthesis of one protein.

gene locus The site on a chromosome where a given gene is located.

generator potential A graded, reversible rise in the local voltage across the plasma membrane of a nerve or muscle cell in response to a stimulus; triggers an action potential if it reaches threshold.

genetic engineering Any of several techniques that alter the genetic constitution of a cell or organism, including recombinant DNA technology and gene substitution therapy.

genome (JEE-nome) All the genes of one individual, estimated at 100,000 genes in humans.

genotype (JEE-no-type) The pair of alleles possessed by an individual at one gene locus on a pair of homologous chromosomes; strongly influences the individual's phenotype for a given trait.

germ cell A gamete or any precursor cell destined to become a gamete.

germ layer Any of first three tissue layers of an embryo: ectoderm, mesoderm, or endoderm.

gestation (jess-TAY-shun) Pregnancy.

gland Any organ specialized to produce a secretion; in some cases a single cell, such as a goblet cell.

glaucoma (glaw-CO-muh) A visual disease in which an excessive amount of aqueous humor accumulates and creates pressure that is transmitted through the lens and vitreous body to the retina; pressure on the blood vessels of the choroid causes ischemia, retinal necrosis, and blindness.

globulin (GLOB-you-lin) A globular protein such as an enzyme, antibody, or albumin; especially a family of proteins in the blood plasma that includes albumin, antibodies, fibrinogen, and prothrombin.

glomerular capsule (glo-MERR-you-lur) A double-walled capsule around each glomerulus of the kidney; receives glomerular filtrate and empties into the proximal convoluted tubule. Also called *Bowman's capsule*. fig. 23.5

glomerulus A spheroid mass of blood capillaries in the kidney that filters plasma and produces glomerular filtrate, which is further processed to form the urine. fig. 23.5

glucagon (GLUE-ca-gon) A hormone secreted by α cells of the pancreatic islets in response to hypoglycemia; promotes glycogenolysis and other effects that raise blood glucose concentration.

glucocorticoid (GLUE-co-COR-tih-coyd) Any hormone of the adrenal cortex that affects carbohydrate, fat, and protein metabolism; chiefly cortisol and corticosterone.

gluconeogenesis (GLUE-co-NEE-oh-JEN-eh-sis) The synthesis of glucose from noncarbohydrates such as fats and amino acids.

glucose A monosaccharide ($C_6H_{12}O_6$) also known as blood sugar; glycogen, starch, cellulose, and maltose are made entirely of glucose, and glucose constitutes half of a sucrose or lactose molecule. The isomer involved in human physiology is also called *dextrose*.

glucose-sparing effect An effect of fats or other energy substrates in which they are used as fuel by most cells, so that those cells do not consume glucose; this makes more glucose available to cells such as neurons that cannot use alternative energy substrates.

glycerol (GLISS-er-ol) A viscous three-carbon alcohol that forms the structural backbone of triglyceride and phospholipid molecules; also called *glycerin*.

glycocalyx (GLY-co-CAY-licks) A layer of carbohydrate molecules covalently bonded to the phospholipid and protein molecules of a plasma membrane; forms a surface coat on all human cells.

glycogen (GLY-co-jen) A glucose polymer synthesized by liver, muscle, uterine, and vaginal cells that serves as an energy-storage polysaccharide.

glycogenesis (GLY-co-JEN-eh-sis) The synthesis of glycogen.

glycogenolysis (GLY-co-jeh-NOLL-ih-sis) The hydrolysis of glycogen, releasing glucose.

glycolipid (GLY-co-LIP-id) A phospholipid molecule with a carbohydrate covalently bonded to it, found in the plasma membranes of cells.

glycolysis (gly-COLL-ih-sis) A series of anaerobic oxidation reactions that break a glucose molecule into two molecules of pyruvic acid and produce a small amount of ATP.

glycoprotein (GLY-co-PRO-teen) A protein molecule with a smaller carbohydrate covalently bonded to it; found in mucus and the glycocalyx of cells, for example.

glycosaminoglycan (GAG) (GLY-cose-am-ih-no-GLY-can) A polysaccharide composed of modified sugars with amino groups; the major component of a proteoglycan. GAGs are largely responsible for the viscous consistency of tissue gel and the stiffness of cartilage.

glycosuria (GLY-co-SOOR-ee-uh) The presence of glucose in the urine, typically indicative of a kidney disease, diabetes mellitus, or other endocrine disorder.

goblet cell A mucus-secreting gland cell, shaped somewhat like a wineglass, found in the epithelia of many mucous membranes. fig. 6.34

Golgi complex (GOAL-jee) An organelle composed of several parallel cisternae, somewhat like a stack of saucers, that modifies and packages newly synthesized proteins and synthesizes carbohydrates. fig. 4.15

Golgi vesicle A membrane-bounded vesicle pinched from the Golgi complex, containing its chemical product; may be retained in the cell as a lysosome or become a secretory vesicle that releases the product by exocytosis.

gonad The ovary or testis.

gonadotropin (go-NAD-oh-TRO-pin) A pituitary hormone that stimulates the gonads; specifically FSH and LH.

G protein A protein of the plasma membrane that is activated by a membrane receptor and, in turn, opens an ion channel or activates an intracellular physiological response; important in linking ligand-receptor binding to second-messenger systems.

graded potential A variable change in voltage across a plasma membrane, as opposed to the all-or-none quality of an action potential.

gradient A difference or change in any variable, such as pressure or chemical concentration, from one point in space to another; provides a basis for molecular movements such as gas exchange, osmosis, and facilitated diffusion, and for bulk movements such as blood flow and airflow.

granulocyte (GRAN-you-lo-site) Any of three types of leukocytes (neutrophils, eosinophils, or basophils) with prominent cytoplasmic granules.

granulosa cells Cells that form a stratified cuboidal epithelium lining an ovarian follicle; source of steroid sex hormones. fig. 28.13

gray matter A zone or layer of tissue in the central nervous system where the neuron cell bodies, dendrites, and synapses are found; forms the core of the spinal cord, nuclei of the brainstem, basal nuclei of the cerebrum, cerebral cortex, and cerebellar cortex. fig. 14.6

gross anatomy Bodily structure that can be observed without magnification.

growth factor A chemical messenger that stimulates mitosis and differentiation of target cells that have receptors for it; important in such processes as fetal development, tissue maintenance and repair, and hemopoiesis; sometimes a contributing factor in cancer.

growth hormone (GH) A hormone of the anterior pituitary gland with multiple effects on many tissues, generally promoting tissue growth.

guanine A double-ringed nitrogenous base (purine) found in DNA and RNA; one of the four bases of the genetic code; complementary to cytosine in the double helix of DNA. fig. 5.3

gyrus (JY-rus) A wrinkle or fold in the cortex of the cerebrum or cerebellum.

H

hair cells Sensory cells of the cochlea, semicircular ducts, utricle, and saccule, with a fringe of surface microvilli that respond to the relative motion of a gelatinous membrane at their tips; responsible for the senses of hearing and equilibrium.

hair follicle An oblique epidermal pit that contains a hair and extends into the dermis or hypodermis.

half-life ($T_{1/2}$) **1.** The time required for one-half of a quantity of a radioactive element to decay to a stable isotope *(physical half-life)* or to be cleared from the body through a combination of radioactive decay and physiological excretion (*biological half-life*). **2.** The time required for one-half of a quantity of hormone to be cleared from the bloodstream.

haploid (n) In humans, having 23 unpaired chromosomes instead of the usual 46 chromosomes in homologous pairs; in any organism or cell, having half the normal diploid number of chromosomes for that species.

helper T cell A type of lymphocyte that performs a central coordinating role in humoral and cellular immunity; target of the human immunodeficiency virus (HIV).

hematocrit (he-MAT-oh-crit) The percentage of blood volume that is composed of erythrocytes.

hematoma (HE-muh-TOE-muh) A mass of clotted blood in the tissues; forms a bruise when visible through the skin.

heme (heem) The nonprotein, iron-containing prosthetic group of hemoglobin or myoglobin; oxygen binds to its ferrous ion. fig. 18.9

hemocytoblast (HE-mo-SY-toe-blast) An undifferentiated stem cell of the bone marrow that can give rise to any of the formed elements of the blood. fig. 18.3

hemoglobin (HE-mo-GLO-bin) The red gas-transport pigment of an erythrocyte.

hemolysis (he-MOLL-ih-sis) The rupturing of erythrocytes from such causes as a hypotonic medium, parasitic infection, or a complement reaction.

hemopoiesis (HE-mo-poy-EE-sis) Production of any of the formed elements of blood.

heparin (HEP-uh-rin) A polysaccharide secreted by basophils and mast cells that inhibits blood clotting.

hepatic (heh-PAT-ic) Pertaining to the liver.

hepatic portal system A network of blood vessels that connect capillaries of the intestines to capillaries (sinusoids) of the liver, thus delivering newly absorbed nutrients directly to the liver.

hepatitis (HEP-uh-TY-tiss) Inflammation of the liver.

heterozygous (HET-er-oh-ZY-gus) Having nonidentical alleles at the same gene locus of two homologous chromosomes.

hiatus (hy-AY-tus) An opening or gap, such as the esophageal hiatus through the diaphragm.

high-density lipoprotein (HDL) A lipoprotein of the blood plasma that is about 50% lipid and 50% protein; functions to transport phospholipids and cholesterol from other organs to the liver for disposal. A high proportion of HDL to low-density lipoprotein (LDL) is desirable for cardiovascular health.

hilum (HY-lum) A point on the surface of an organ where blood vessels, lymphatic vessels, or nerves enter and leave, usually marked by a depression and slit; the midpoint of the concave surface of any organ that is roughly bean-shaped, such as the lymph nodes, kidneys, and lungs. Also called the *hilus.* fig. 23.3

histamine (HISS-ta-meen) An amino acid derivative secreted by basophils, mast cells, and some neurons; functions as a paracrine secretion and neurotransmitter to stimulate effects such as gastric secretion, bronchoconstriction, and vasodilation. fig. 13.18

histological section A thin slice of tissue, usually mounted on a slide and artificially stained to make its microscopic structure more visible.

histology **1.** The microscopic structure of tissues and organs. **2.** The study of such structure.

homeostasis (HO-me-oh-STAY-sis) The tendency of a living body to maintain relatively stable internal conditions in spite of greater changes in its external environment.

homologous (ho-MOLL-uh-gus) **1.** Having the same embryonic or evolutionary origin but not necessarily the same function, such as the scrotum and labia majora. **2.** Pertaining to two chromosomes with identical structures and gene loci but not necessarily identical alleles; each member of the pair is inherited from a different parent.

homozygous (HO-mo-ZY-gus) Having identical alleles at the same gene locus of two homologous chromosomes.

hormone A chemical messenger that is secreted into the blood by an endocrine gland or isolated gland cell and triggers a physiological response in distant cells with receptors for it.

host cell Any cell belonging to the human body, as opposed to foreign cells introduced to it by such causes as infections and tissue transplants.

human Any species of primate classified in the family Hominidae, characterized by bipedal locomotion, relatively large brains, and usually articulate speech; currently represented only by *Homo sapiens* but including extinct species of *Homo* and *Australopithecus.*

human chorionic gonadotropin (HCG) (COR-ee-ON-ic) A hormone of pregnancy secreted by the chorion that stimulates continued growth of the corpus luteum and secretion of its hormones. HCG in urine is the basis for pregnancy testing.

human immunodeficiency virus (HIV) A virus that infects human helper T cells and other cells, suppresses immunity, and causes AIDS.

hyaline cartilage (HY-uh-lin) A form of cartilage with a relatively clear matrix and fine collagen fibers but no conspicuous elastic fibers or coarse collagen bundles as in other types of cartilage.

hyaluronic acid (HY-uh-loo-RON-ic) A glycosaminoglycan that is particularly abundant in connective tissues, where it becomes hydrated and forms the tissue gel.

hydrogen bond A weak attraction between a slightly positive hydrogen atom on one molecule and a slightly negative oxygen or nitrogen atom on another molecule, or between such atoms on different parts of the same molecule; responsible for the cohesion of water and the coiling of protein and DNA molecules, for example.

hydrolysis (hy-DROL-ih-sis) A chemical reaction that breaks a covalent bond in a molecule by adding an –OH group to one side of the bond and –H to the other side, thus consuming a water molecule. *Compare* dehydration synthesis.

hydrophilic (HY-dro-FILL-ic) Pertaining to molecules that attract water or dissolve in it because of their polar nature.

hydrophobic (HY-dro-FOE-bic) Pertaining to molecules that do not attract water or dissolve in it because of their nonpolar nature; such molecules tend to dissolve in lipids and other nonpolar solvents.

hydrostatic pressure The physical force generated by a liquid such as blood or tissue fluid, as opposed to osmotic and atmospheric pressures.

hydroxyl group (hy-DROCK-sil) A functional group with the formula –OH found on many organic molecules such as carbohydrates and alcohols.

hypercalcemia (HY-pur-cal-SEE-me-uh) An excess of calcium ions in the blood.

hypercapnia (HY-pur-CAP-nee-uh) An excess of carbon dioxide in the blood.

hyperextension A joint movement that increases the angle between two bones beyond 180°. fig. 10.9

hyperglycemia (HY-pur-gly-SEE-me-uh) An excess of glucose in the blood.

hyperkalemia (HY-pur-ka-LEE-me-uh) An excess of potassium ions in the blood.

hypernatremia (HY-pur-na-TREE-me-uh) An excess of sodium ions in the blood.

hyperplasia (HY-pur-PLAY-zhuh) The growth of a tissue through cellular multiplication, not cellular enlargement. *Compare* hypertrophy.

hyperpolarization A shift in the electrical potential across a plasma membrane to a value more negative than the resting membrane potential, tending to inhibit a nerve or muscle cell. *Compare* depolarization.

hypersecretion Excessive secretion of a hormone or other gland product; can lead to endocrine disorders such as Cushing syndrome or gigantism, for example.

hypertension Excessively high blood pressure; criteria vary but it is often considered to be a condition in which systolic pressure exceeds 150 mmHg or diastolic pressure exceeds 90 mmHg.

hyperthermia Excessively high core body temperature, as in heatstroke or fever.

hypertonic Having a higher osmotic pressure than human cells or some other reference solution and tending to cause osmotic shrinkage of cells.

hypertrophy (hy-PUR-tro-fee) The growth of a tissue through cellular enlargement, not cellular multiplication; for example, the growth of muscle under the influence of exercise. *Compare* hyperplasia.

hypocalcemia (HY-po-cal-SEE-me-uh) A deficiency of calcium ions in the blood.

hypocapnia (HY-po-CAP-nee-uh) A deficiency of carbon dioxide in the blood.

hypodermis (HY-po-DUR-miss) A layer of connective tissue deep to the skin; also called *superficial fascia, subcutaneous tissue,* or when it is predominantly adipose, *subcutaneous fat.*

hypoglycemia (HY-po-gly-SEE-me-uh) A deficiency of glucose in the blood.

hypokalemia (HY-po-ka-LEE-me-uh) A deficiency of potassium ions in the blood.

hyponatremia (HY-po-na-TREE-me-uh) A deficiency of sodium ions in the blood.

hyposecretion Inadequate secretion of a hormone or other gland product; can lead to endocrine disorders such as diabetes mellitus or pituitary dwarfism, for example.

hypothalamic thermostat (HY-po-thuh-LAM-ic) A nucleus in the hypothalamus that monitors body temperature and sends afferent signals to hypothalamic heat-promoting or heat-losing centers to maintain thermal homeostasis.

hypothalamus (HY-po-THAL-uh-mus) The inferior portion of the diencephalon of the brain, forming the walls and floor of the third ventricle and giving rise to the posterior pituitary gland; controls many fundamental physiological functions such as appetite, thirst, and body temperature and exert many of its effects through the endocrine and autonomic nervous systems. fig. 14.17

hypothermia (HY-po-THUR-me-uh) A state of abnormally low core body temperature.

hypothesis An informed conjecture that is capable of being tested and potentially falsified by experimentation or data collection.

hypotonic Having a lower osmotic pressure than human cells or some other reference solution and tending to cause osmotic swelling and lysis of cells.

hypovolemic shock (HY-po-vo-LEE-mic) Insufficient cardiac output resulting from a drop in blood volume. *See also* shock.

hypoxemia (HY-pock-SEE-me-uh) A deficiency of oxygen in the bloodstream.

hypoxia (hy-POCK-see-uh) A deficiency of oxygen in any tissue.

I

immune system A population of cells, including leukocytes and macrophages, that occur in most organs of the body and protect against foreign organisms, some foreign chemicals, and cancerous or other aberrant host cells.

immunity The ability to ward off a specific infection or disease, usually as a result of prior exposure and the body's production of antibodies or lymphocytes against a pathogen. *Compare* resistance.

immunoglobulin (IM-you-no-GLOB-you-lin) *See* antibody.

implantation The attachment of a conceptus to the endometrium of the uterus.

inclusion Any visible object in the cytoplasm of a cell other than an organelle or cytoskeletal element; usually a foreign body or a stored cell product, such as a virus, dust particle, lipid droplet, glycogen granule, or pigment.

infarction (in-FARK-shun) **1.** The sudden death of tissue from a lack of blood perfusion; also called an *infarct.* **2.** An area of necrotic tissue produced by this process.

inferior Lower than another structure or point of reference from the perspective of anatomical position; for example, the stomach is inferior to the diaphragm.

inflammation (IN-fluh-MAY-shun) A complex of tissue responses to trauma or infection serving to ward off a pathogen and tissue repair; recognized by the cardinal signs of redness, heat, swelling, and pain.

infundibulum (IN-fun-DIB-you-lum) Any funnel-shaped passage or structure, such as the distal portion of the uterine tube and the stalk that attaches the pituitary gland to the hypothalamus.

inguinal (IN-gwih-nul) Pertaining to the groin.

inhibin A hormone produced by the testes and ovaries that inhibits the secretion of FSH.

inhibitory postsynaptic potential (IPSP) Hyperpolarization of a postsynaptic neuron in response to a neurotransmitter, making it less likely to reach threshold and fire.

innervation (IN-ur-VAY-shun) The nerve supply to an organ.

insertion The point at which a muscle attaches to another tissue (usually a bone) and produces movement, opposite from its stationary origin; the origin and insertion of a given muscle sometimes depend on what muscle action is being considered. *Compare* origin.

in situ (in SY-too) In the normal anatomical location (Latin, *in place*).

inspiration Inhaling.

insulin (IN-suh-lin) A hormone produced by β cells of the pancreatic islets in response to a rise in blood glucose concentration; accelerates glucose uptake and metabolism by most cells of the body, thus lowering blood glucose concentration.

integral (transmembrane) protein A protein that extends through a plasma membrane and contacts both the extracellular and intracellular fluid. fig. 4.7

integration A process in which a neuron receives input from multiple sources and their combined effects determine its output; the cellular basis of information processing by the nervous system.

integumentary system (in-TEG-you-MEN-tah-ree) An organ system consisting of the skin, cutaneous glands, hair, and nails.

interatrial septum (IN-tur-AY-tree-ul) The wall between the atria of the heart.

intercalated disc (in-TUR-kuh-LAY-ted) A complex of fascia adherens, gap junctions, and desmosomes that join two cardiac muscle cells end to end, microscopically visible as a dark line which helps to histologically distinguish this muscle type; functions as a mechanical and electrical link between cells. fig. 12.24

intercellular Between cells.

intercostal (IN-tur-COSS-tul) Between the ribs, as in the intercostal muscles, arteries, veins, and nerves.

interdigitate (IN-tur-DIDJ-ih-tate) To fit together like the fingers of two folded hands; for example, at the dermal-epidermal boundary, intercalated discs of the heart, and pedicels of the podocytes in the kidney. fig. 23.7

interleukin (IN-tur-LOO-kin) A hormonelike chemical messenger from one leukocyte to another, serving as a means of communication and coordination during immune responses.

interneuron (IN-tur-NEW-ron) A neuron that is contained entirely in the central nervous system and, in the path of signal conduction, lies anywhere between an afferent pathway and an efferent pathway.

interosseous membrane (IN-tur-OSS-ee-us) A fibrous membrane that connects the radius to the ulna and the tibia to the fibula along most of the shaft of each bone. fig. 9.32

interphase That part of the cell cycle between one mitotic phase and the next, from the end of cytokinesis to the beginning of the next prophase.

interstitial (IN-tur-STISH-ul) **1.** Pertaining to the extracellular spaces in a tissue. **2.** Located between other structures, as in the interstitial cells of the testis.

interstitial fluid Fluid in the interstitial spaces of a tissue, also called *tissue fluid.*

intervertebral disc A cartilaginous pad between the bodies of two adjacent vertebrae.

intracellular Within a cell.

intracellular fluid (ICF) The fluid contained in the cells; one of the major fluid compartments.

intravenous (I.V.) **1.** Present or occurring within a vein, such as an intravenous blood clot. **2.** Introduced directly into a vein, such as an intravenous injection or I.V. drip.

intrinsic (in-TRIN-sic) **1.** Arising from within, such as intrinsic blood-clotting factors; endogenous. **2.** Fully contained within an organ, such as the intrinsic muscles of the hand and eye. *Compare* extrinsic.

intrinsic factor A secretion of the gastric glands required for the intestinal absorption of vitamin B_{12}. Hyposecretion of intrinsic factor leads to pernicious anemia.

in vitro (in VEE-tro) In a laboratory container; removed from the body and observed in isolation (Latin, *in glass*).

in vivo (in VEE-vo) In the living state; in the body (Latin, *in life*).

involuntary Not under conscious control, including tissues such as smooth and cardiac muscle and events such as reflexes.

involution (IN-vo-LOO-shun) Shrinkage of a tissue or organ by autolysis, such as involution of the thymus after childhood and of the uterus after pregnancy.

ion A chemical particle with unequal numbers of electrons or protons and consequently a net negative or positive charge; it may have a single atomic nucleus as in a sodium ion or a few atoms as in a bicarbonate ion, or it may be a large molecule such as a protein.

ionic bond The force that binds a cation to an anion.

ionizing radiation High-energy electromagnetic rays that eject electrons from atoms or molecules and convert them to ions, frequently causing cellular damage; for example, X rays and gamma rays.

ionotropic effect (eye-ON-oh-TRO-pic) Effect in which a neurotransmitter or hormone directly causes a gated channel to open in a plasma membrane, thus triggering ion flow and voltage changes across the target cell membrane.

ipsilateral (IP-sih-LAT-ur-ul) On the same side of the body, as in reflex arcs in which a muscular response occurs on the same side of the body as the stimulus. *Compare* contralateral.

ischemia (iss-KEE-me-uh) Insufficient blood flow to a tissue, typically resulting in metabolite accumulation and possibly tissue death.

isometric contraction A muscle contraction in which internal tension rises but the muscle does not shorten.

isotonic Having the same osmotic pressure as human cells or some other reference solution.

isotonic contraction A muscle contraction in which the muscle shortens and moves a load while its internal tension remains constant.

J

jaundice (JAWN-diss) A yellowish color of the skin, corneas, mucous membranes, and body fluids due to an excessive concentration of bilirubin; usually indicative of a liver disease, obstructed bile secretion, or hemolytic disease.

K

ketone (KEE-tone) Any organic compound with a carbonyl (C=O) group covalently bonded to two other carbons.

ketone bodies Certain ketones (acetone, acetoacetic acid, and β-hydroxybutyric acid) produced by the incomplete oxidation of fats, especially when fats are being rapidly catabolized. *See also* ketosis.

ketonuria (KEE-toe-NEW-ree-uh) The abnormal presence of ketones in the urine as an effect of ketosis.

ketosis (kee-TOE-sis) An abnormally high concentration of ketone bodies in the blood, occurring in pregnancy, starvation, diabetes mellitus, and other conditions; tends to cause acidosis and to depress the nervous system.

kilocalorie The amount of heat energy needed to raise the temperature of 1 kg of water by 1°C; 1,000 calories. Also called a *Calorie* or *large calorie.*

L

labium (LAY-bee-um) A lip, such as those of the mouth and the labia majora and minora of the vulva.

lactation The secretion of milk.

lactic acid A small organic acid produced as an end product of the anaerobic fermentation of pyruvic acid; a contributing factor in muscle fatigue.

lacuna (la-CUE-nuh) A small cavity or depression in a tissue such as bone, cartilage, and the erectile tissues.

lamella (la-MELL-uh) A little plate, such as the lamellae of bone. fig. 8.5

lamina (LAM-ih-nuh) A thin layer, such as the lamina of a vertebra or the lamina propria of a mucous membrane. fig. 9.21

lamina propria (PRO-pree-uh) A thin layer of areolar tissue immediately deep to the epithelium of a mucous membrane. fig. 6.34

larynx (LAIR-inks) A cartilaginous chamber in the neck containing the vocal cords; the voicebox.

latent period The interval between a stimulus and response, especially in the action of nerve and muscle cells.

lateral Away from the midline of an organ or midsagittal plane of the body; toward the side. *Compare* medial.

law A verbal or mathematical description of a predictable natural phenomenon or of the relationships between variables; for example, Boyle's law and the second law of thermodynamics.

law of mass action A law that states that the speed and direction of a reversible chemical reaction is determined by the relative quantities of the reactants. A reversible reaction $A + B \leftrightarrow C + D$ proceeds left to right if the quantity of $A + B$ is greater than the quantity of $C + D$ and right to left if the latter is greater. This principle governs such reactions as the binding and dissociation of oxygen and hemoglobin.

leader sequence A sequence of bases in mRNA that is not translated to protein but serves as a binding site for a ribosome.

length-tension relationship A law that relates the tension generated by muscle contraction to the length of the muscle fiber prior to stimulation; it shows that the greatest tension is generated when the fiber exhibits an intermediate degree of stretch before stimulation.

lesion A circumscribed zone of tissue injury, such as a skin abrasion or myocardial infarction.

leukocyte (LOO-co-site) A white blood cell.

leukotrienes (LOO-co-TRY-eens) Eicosanoids that promote allergic and inflammatory responses such as vasodilation and neutrophil chemotaxis; secreted by basophils, mast cells, and damaged tissues.

libido (lih-BEE-do) Sex drive.

ligament A cord or band of tough collagenous tissue binding one organ to another, especially one bone to another, and serving to hold organs in place; for example, the cruciate ligaments of the knee, broad ligament of the uterus, and falciform ligament of the liver.

ligand (LIG-and, LY-gand) A chemical that binds reversibly to a receptor site on a protein, such as a neurotransmitter that binds to a membrane receptor or a substrate that binds to an enzyme.

ligand-regulated gate A channel protein in a plasma membrane that opens or closes when a ligand binds to it, enabling the ligand to determine when substances can enter or leave the cell.

light microscope (LM) A microscope that produces images with visible light.

linea (LIN-ee-uh) An anatomical line, such as the linea alba.

lingual (LING-gwul) Pertaining to the tongue, as in lingual papillae.

lipase (LY-pace) An enzyme that hydrolyzes a triglyceride into fatty acids and glycerol.

lipid A hydrophobic organic compound composed mainly of carbon and a high ratio of hydrogen to oxygen; includes fatty acids, fats, phospholipids, steroids, and prostaglandins.

lipoprotein (LIP-oh-PRO-teen) A protein-coated lipid droplet in the blood plasma or lymph, serving as a means of lipid transport; for example, chylomicrons and the high- and low-density lipoproteins.

load **1.** To pick up a gas for transport in the bloodstream. **2.** The resistance acted upon by a muscle.

lobe **1.** A structural subdivision of an organ such as a gland, a lung, or the brain, bounded by a visible landmark such as a fissure or septum. **2.** The inferior, noncartilaginous, often pendant part of the ear pinna.

lobule (LOB-yool) A small subdivision of an organ or of a lobe of an organ, especially of a gland.

locus *See* gene locus.

long bone A bone such as the femur or humerus that is markedly longer than it is wide and that generally serves as a lever.

longitudinal Oriented along the longest dimension of the body or of an organ.

loose connective tissue *See* areolar tissue.

low-density lipoprotein (LDL) A blood-borne droplet of about 20% protein and 80% lipid (mainly cholesterol) that transports cholesterol from the liver to other tissues.

lower limb The appendage that arises from the hip, consisting of the thigh from hip to knee; the crural region from knee to ankle; the ankle; and the foot. Loosely called the leg, although that term properly refers only to the crural region.

lumbar Pertaining to the lower back and sides, between the thoracic cage and pelvis.

lumen (LOO-men) The internal space of a hollow organ such as a blood vessel or the esophagus, or a space surrounded by cells as in a gland acinus.

luteinizing hormone (LH) (LOO-tee-in-eye-zing) A hormone of the anterior pituitary gland that stimulates ovulation in females and testosterone secretion in males; in males, also called interstitial cell–stimulating hormone.

lymph The fluid contained in lymphatic vessels and lymph nodes, produced by the absorption of tissue fluid.

lymphatic system (lim-FAT-ic) An organ system consisting of lymphatic vessels, lymph nodes, the tonsils, spleen, and thymus; functions include tissue fluid recovery and immunity.

lymph node A small organ found along the course of a lymphatic vessel that filters the lymph and contains lymphocytes and macrophages, which respond to antigens in the lymph. fig. 21.8

lymphocytes (LIM-foe-sites) A class of relatively small agranulocytes with numerous types and roles in nonspecific defense, humoral immunity, and cellular immunity. table 18.8

lymphokine Any interleukin secreted by a lymphocyte.

lysosome (LY-so-some) A membrane-bounded organelle containing a mixture of enzymes with a variety of intracellular and extracellular roles in digesting foreign matter, pathogens, and expired organelles.

lysozyme (LY-so-zime) An enzyme found in tears, milk, saliva, mucus, and other body fluids that destroys bacteria by digesting their cell walls. Also called *muramidase.*

M

macromolecule Any molecule of large size and high molecular weight, such as a protein, nucleic acid, polysaccharide, or triglyceride.

macrophage (MAC-ro-faje) Any cell of the body, other than a leukocyte, that is specialized for phagocytosis; usually derived from blood monocytes and often functioning as antigen-presenting cells.

macula (MAC-you-luh) A patch or spot, such as the *macula lutea* of the retina.

malignant (muh-LIG-nent) Pertaining to a cell or tumor that is cancerous; capable of metastasis.

maltose A disaccharide composed of two glucose monomers.

mammary gland The milk-secreting gland that develops within the breast in pregnancy and lactation; only minimally developed in the breast of a nonpregnant or nonlactating woman.

mast cell A connective tissue cell, similar to a basophil, that secretes histamine, heparin, and other chemicals involved in inflammation; often concentrated along the course of blood capillaries. fig. 6.16

matrix **1.** The extracellular material of a tissue. **2.** The fluid within a mitochondrion containing enzymes of the citric acid cycle. **3.** The substance or framework within which other structures are embedded, such as the fibrous matrix of a blood clot. **4.** A patch of thickened epidermis from which a nail root develops.

mechanoreceptor A sensory nerve ending or organ specialized to detect mechanical stimuli such as touch, pressure, and vibration.

medial Toward the midline of an organ or midsagittal plane of the body. *Compare* lateral.

mediastinum (ME-dee-ass-TY-num) The thick medial partition of the thoracic cavity that separates one pleural cavity from the other and contains the heart, great blood vessels, and thymus. fig. A.8

medulla (meh-DULE-uh, meh-DULL-uh) Tissue deep to the cortex of certain two-layered organs such as the adrenal glands, lymph nodes, hairs, and kidneys.

medulla oblongata (OB-long-GAH-ta) The most inferior part of the brainstem, immediately superior to the foramen magnum of the skull, connecting the spinal cord to the rest of the brain. fig. 14.1

meiosis (my-OH-sis) A form of cell division in which a diploid cell divides twice and produces four haploid daughter cells; occurs only in gametogenesis.

melanocyte A cell of the stratum basale of the epidermis that synthesizes melanin and transfers it to the keratinocytes.

meninges (meh-NIN-jeez) (singular, *meninx*) Three fibrous membranes between the central nervous system and surrounding bone: the dura mater, arachnoid mater, and pia mater. fig. 14.5

menopause Cessation of the menstrual cycles, occurring during female climacteric.

merocrine (MERR-oh-crin) Pertaining to gland cells that release their product by exocytosis; also called *eccrine.*

mesenchyme (MEZ-en-kime) A gelatinous embryonic connective tissue derived from the mesoderm; differentiates into all permanent connective tissues and most muscle.

mesentery (MESS-en-tare-ee) A serous membrane that binds the intestines together and suspends them from the abdominal wall; the visceral continuation of the peritoneum. fig. 25.2

mesoderm (MEZ-oh-durm) The middle layer of the three primary germ layers of an embryo; gives rise to muscle and connective tissue.

mesothelium (MEZ-oh-THEE-lee-um) A simple squamous epithelium that covers the serous membranes.

metabolic pathway A series of linked chemical reactions, most of which are catalyzed by a separate enzyme; glycolysis, for example.

metabolic rate The overall rate of the body's metabolic reactions at any given time, which determines the rates of nutrient and oxygen consumption; often measured from the rate of oxygen consumption or heat production. *Compare* basal metabolic rate.

metabolic waste A product of metabolism that is not useful to the body but is potentially toxic and must be excreted.

metabolism (meh-TAB-oh-lizm) The sum of all chemical reactions in the body.

metabolite (meh-TAB-oh-lite) Any chemical produced by metabolism.

metabotropic effect (meh-TAB-oh-TRO-pic) An effect in which a neurotransmitter or hormone changes the metabolism of a target cell but does not directly stimulate the opening of ion gates in the plasma membrane.

metaplasia Transformation of one mature tissue type into another; for example, a change from pseudostratified to stratified squamous epithelium in an overventilated nasal cavity.

metastasis (meh-TASS-tuh-sis) The spread of cancer cells from the original tumor to a new location, where they seed the development of a new tumor.

microtubule An intracellular cylinder composed of the protein tubulin, forming centrioles, the axonemes of cilia and flagella, and part of the cytoskeleton.

microvillus An outgrowth of the plasma membrane that increases the surface area of a cell and functions in absorption and some sensory processes; distinguished from cilia and flagella by its smaller size and lack of an axoneme.

midsagittal plane (MID-SADJ-ih-tul) The plane that divides the body into equal right and left halves. *Compare* sagittal plane. fig. A.3

milliequivalent One one-thousandth of an equivalent, which is the amount of an electrolyte that would neutralize 1 mole of H^+ or OH^-. Electrolyte concentrations are commonly expressed in milliequivalents per liter (mEq/L).

mineralocorticoid (MIN-ur-uh-lo-COR-tih-coyd) A steroid hormone, chiefly aldosterone, that is secreted by the adrenal cortex and acts to regulate electrolyte balance.

mitochondrion (MY-toe-CON-dree-un) An organelle specialized to synthesize ATP, enclosed in a double unit membrane with infoldings of the inner membrane called cristae.

mitosis (my-TOE-sis) A form of cell division in which a cell divides once and produces two genetically identical daughter cells; sometimes used to refer only to the division of the genetic material or nucleus and not to include cytokinesis, the subsequent division of the cytoplasm.

moiety (MOY-eh-tee) A chemically distinct subunit of a macromolecule, such as the heme and globin moieties of hemoglobin or the lipid and carbohydrate moieties of a glycolipid.

molarity A measure of chemical concentration expressed as moles of solute per liter of solution.

mole The mass of a chemical equal to its molecular weight in grams, containing 6.023×10^{23} molecules.

monocyte An agranulocyte specialized to migrate into the tissues and transform into a macrophage. table 18.8

monokine Any interleukin secreted by a monocyte or macrophage.

monomer (MON-oh-mur) **1.** One of the identical or similar subunits of a larger molecule in the dimer to polymer range; for example, the glucose monomers of starch, the amino acids of a protein, or the nucleotides of DNA. **2.** One subunit of an antibody molecule, composed of four polypeptides.

monosaccharide (MON-oh-SAC-uh-ride) A simple sugar, or sugar monomer; chiefly glucose, fructose, and galactose.

monozygotic (MZ) twins Two individuals who developed from the same fertilized egg and are therefore genetically identical.

motor end plate A depression in a muscle fiber where it has synaptic contact with a nerve fiber and has a high density of neurotransmitter receptors. fig. 12.5

motor neuron A neuron that transmits signals from the central nervous system to any effector (muscle or gland cell); its axon is an efferent nerve fiber.

motor protein Any protein that produces movements of a cell or its components owing to its ability to undergo quick repetitive changes in conformation and to bind reversibly to other molecules; for example, myosin, dynein, and kinesin.

motor unit One motor neuron and all the skeletal muscle fibers innervated by it.

mucosa (mew-CO-suh) A tissue layer that forms the inner lining of an anatomical tract that is open to the exterior, such as the respiratory, digestive, urinary, and reproductive tracts. Composed of epithelium, connective tissue (lamina propria), and often smooth muscle (muscularis mucosae). fig. 6.34

mucous membrane A mucosa.

mucus A viscous, slimy or sticky secretion produced by mucous cells and mucous membranes and consisting of a hydrated glycoprotein, mucin; serves to bind particles together, such as bits of masticated food, and to protect the mucous membranes from infection and abrasion.

muscle fiber One muscle cell, especially of skeletal muscle.

muscle tone A state of continual, partial contraction of resting skeletal or smooth muscle.

muscularis externa The external muscular wall of certain viscera such as the esophagus and small intestine. fig. 25.9

muscularis mucosae (MUSS-cue-LERR-iss mew-CO-see) A layer of smooth muscle immediately deep to the lamina propria of a mucosa. fig. 6.34

muscular system An organ system composed of the skeletal muscles, specialized mainly for maintaining postural support and producing movements of the bones.

muscular tissue A tissue composed of elongated, electrically excitable cells specialized for contraction; the three types are skeletal, cardiac, and smooth muscle.

mutagen (MEW-tuh-jen) Any agent that causes a mutation, including viruses, chemicals, and ionizing radiation.

mutation Any change in the structure of a chromosome or a DNA molecule, often resulting in a change of organismal structure or function.

myelin (MY-eh-lin) A lipid sheath around a nerve fiber, formed from closely spaced spiral layers of the plasma membrane of a Schwann cell or oligodendrocyte. fig. 13.7

myocardium (MY-oh-CAR-dee-um) The middle, muscular layer of the heart.

myocyte A muscle cell, especially a cell of cardiac or smooth muscle.

myoepithelial cell An epithelial cell that has become specialized to contract like a muscle cell; important in dilation of the pupil and ejection of secretions from gland acini.

myofibril (MY-oh-FY-bril) A bundle of myofilaments forming an internal subdivision of a cardiac or skeletal muscle cell. fig. 12.2

myofilament A protein microfilament responsible for the contraction of a muscle cell, composed mainly of myosin or actin. fig. 12.3

myoglobin (MY-oh-GLO-bin) A red oxygen-storage pigment of muscle; supplements hemoglobin in providing oxygen for aerobic muscle metabolism.

myosin A motor protein that constitutes the thick myofilaments of muscle and has globular, mobile heads of ATPase that bind to actin molecules.

N

necrosis (neh-CRO-sis) Pathological tissue death due to such causes as infection, trauma, or hypoxia. *Compare* apoptosis.

negative feedback A self-corrective mechanism that underlies most homeostasis, in which a bodily change is detected and responses are activated that reverse the change and restore stability and normal body function.

negative feedback inhibition A mechanism for limiting the secretion of a pituitary tropic hormone. The tropic hormone stimulates another endocrine gland to secrete its own hormone, and that hormone inhibits further release of the tropic hormone.

neonate (NEE-oh-nate) An infant up to 6 weeks old.

neoplasia (NEE-oh-PLAY-zee-uh) Abnormal growth of new tissue, such as a tumor, with no useful function.

nephron One of approximately 1 million blood-filtering, urine-producing units in each kidney; consists of a glomerulus, glomerular capsule, proximal convoluted tubule, nephron loop, distal convoluted tubule, and the blood vessels supplying them. fig. 23.4

nerve A cordlike organ of the peripheral nervous system composed of multiple nerve fibers ensheathed in connective tissue.

nerve fiber The axon of a single neuron.

nerve impulse A wave of self-propagating action potentials traveling along a nerve fiber.

nervous system An organ system composed of the brain, spinal cord, nerves, and ganglia, specialized for rapid communication of information.

nervous tissue A tissue composed of neurons and neuroglia.

net filtration pressure A net force favoring filtration of fluid from a capillary when all the hydrostatic and osmotic pressures of the blood and tissue fluids are taken into account.

neural tube A dorsal hollow tube in the embryo that develops into the central nervous system. fig. 14.3

neuroglia (noo-ROG-lee-uh) All cells of nervous tissue except neurons; cells that perform various supportive and protective roles for the neurons.

neuromuscular junction A synapse between a nerve fiber and a muscle fiber. fig. 12.7

neuron (NOOR-on) A nerve cell; an electrically excitable cell specialized for producing and transmitting action potentials and secreting chemicals that stimulate adjacent cells.

neuronal pool (noor-OH-nul) A group of interconnected neurons of the central nervous system that perform a single collective function; for example, the vasomotor center of the brainstem and speech centers of the cerebral cortex.

neuropeptide A peptide secreted by a neuron, often serving to modify the action of a neurotransmitter; for example, endorphins, enkephalin, and cholecystokinin. fig. 13.18

neurotransmitter A chemical released at the distal end of an axon that stimulates an adjacent cell; for example, acetylcholine, norepinephrine, or serotonin.

neutral fat A triglyceride.

neutrophil (NOO-tro-fill) A granulocyte, usually with a multilobed nucleus, that serves especially to destroy bacteria by means of phagocytosis, intracellular digestion, and secretion of bactericidal chemicals. table 18.8

nitrogenous base (ny-TRODJ-eh-nus) An organic molecule with a single or double carbon-nitrogen ring that forms one of the building blocks of ATP, other nucleotides, and nucleic acids; the basis of the genetic code. fig. 5.3

nitrogenous waste Any nitrogen-containing substance produced as a metabolic waste and excreted in the urine; chiefly ammonia, urea, uric acid, and creatinine.

nociceptor (NO-sih-SEP-tur) A nerve ending specialized to detect tissue damage and produce a sensation of pain; pain receptor.

norepinephrine (nor-EP-ih-NEF-rin) A catecholamine that functions as a neurotransmitter and adrenal hormone, especially in the sympathetic nervous system. fig. 13.18

nuclear envelope (NEW-clee-ur) A pair of unit membranes enclosing the nucleus of a cell, with prominent pores allowing traffic of molecules between the nucleoplasm and cytoplasm. fig. 5.1

nucleic acid (new-CLAY-ic) An acidic polymer of nucleotides found or produced in the nucleus, functioning in heredity and protein synthesis; of two types, DNA and RNA.

nucleotide (NEW-clee-oh-tide) An organic molecule composed of a nitrogenous base, a monosaccharide, and a phosphate group; the monomer of a nucleic acid.

nucleus (NEW-clee-us) **1.** A cellular organelle containing DNA and surrounded by a double unit membrane. **2.** A mass of neurons (gray matter) surrounded by white matter of the brain, including the basal nuclei and brainstem nuclei. **3.** The positively charged core of an atom, consisting of protons and neutrons. **4.** A central structure, such as the nucleus pulposus of an intervertebral disc.

nucleus pulposus The gelatinous center of an intervertebral disc.

O

olfaction (ole-FAC-shun) The sense of smell.

oncotic pressure (ong-COT-ic) The difference between the colloid osmotic pressure of the blood and that of the tissue fluid, usually favoring fluid absorption by the blood capillaries. *Compare* colloid osmotic pressure.

oocyte (OH-oh-site) In the development of an egg cell, any haploid stage between meiosis I and fertilization.

oogenesis (OH-oh-JEN-eh-sis) The production of a fertilizable egg cell through a series of mitotic and meiotic cell divisions; female gametogenesis.

ophthalmic (off-THAL-mic) Pertaining to the eye or vision; optic.

opposition A movement of the thumb in which it touches any fingertip of the same hand.

optic Pertaining to the eye or vision.

orbit The eye socket of the skull.

organ Any anatomical structure that is composed of at least two different tissue types, has recognizable structural boundaries, and has a discrete function different from the structures around it. Many organs are microscopic and many organs contain smaller organs, such as the skin containing numerous microscopic sense organs.

organelle Any structure within a cell that carries out one of its metabolic roles, such as mitochondria, centrioles, endoplasmic reticulum, and the nucleus; an intracellular structure other than the cytoskeleton and inclusions.

organic Pertaining to compounds of carbon.

origin The relatively stationary attachment of a skeletal muscle. *Compare* insertion.

osmolality (OZ-mo-LAL-ih-tee) The molar concentration of dissolved particles in 1 kg of water.

osmolarity (OZ-mo-LERR-ih-tee) The molar concentration of dissolved particles in 1 L of solution.

osmoreceptor (OZ-mo-re-SEP-tur) A neuron of the hypothalamus that responds to changes in the osmolarity of the extracellular fluid.

osmosis (oz-MO-sis) The net diffusion of water through a selectively permeable membrane.

osmotic diuresis (oz-MOT-ic DY-you-REE-sis) Increased urine output due to an increase in the concentration of osmotically active particles in the tubular fluid.

osmotic pressure The amount of pressure that would have to be applied to one side of a selectively permeable membrane to stop osmosis; proportional to the concentration of nonpermeating solutes on that side and therefore serving as an indicator of solute concentration.

osseous (OSS-ee-us) Pertaining to bone.

ossification (OSS-ih-fih-CAY-shun) Bone formation.

osteoarthritis (OA) A chronic degenerative joint disease characterized by loss of articular cartilage, growth of bone spurs, and impaired movement; occurs to various degrees in almost all people with age.

osteoblasts Bone-forming cells that arise from osteogenic cells, deposit bone matrix, and eventually become osteocytes.

osteoclasts Macrophages of the bone surface that dissolve the matrix and return minerals to the extracellular fluid.

osteocyte A mature bone cell formed when an osteoblast becomes surrounded by its own matrix and entrapped in a lacuna.

osteon A structural unit of compact bone consisting of a haversian canal surrounded by concentric cylindrical lamellae of matrix. fig. 8.5

osteoporosis (OSS-tee-oh-pore-OH-sis) A degenerative bone disease characterized by a loss of bone mass, increasing susceptibility to spontaneous fractures, and sometimes deformity of the vertebral column; causes include aging, estrogen hyposecretion, and insufficient resistance exercise.

ovary The female gonad; produces eggs, estrogen, and progesterone.

ovulation (OV-you-LAY-shun) The release of a mature oocyte by the bursting of an ovarian follicle.

ovum Any stage of the female gamete from the conclusion of meiosis I until fertilization; a primary or secondary oocyte; an egg.

oxidation A chemical reaction in which one or more electrons are removed from a molecule, lowering its free energy content; opposite of reduction and always linked to a reduction reaction.

oxytocin (OT) (OCK-see-TOE-sin) A hormone released by the posterior pituitary gland that stimulates labor contractions and milk release.

P

pancreas (PAN-cree-us) A gland of the upper abdominal cavity, near the stomach, that secretes digestive enzymes and sodium bicarbonate into the duodenum and secretes hormones into the blood.

pancreatic islets (PAN-cree-AT-ic EYE-lets) Small clusters of endocrine cells in the pancreas that secrete insulin, glucagon, somatostatin, and other intercellular messengers; also called *islets of Langerhans.* fig. 17.17

papilla (pa-PILL-uh) A conical or nipplelike structure, such as a lingual papilla of the tongue or the papilla of a hair bulb.

papillary (PAP-ih-lerr-ee) **1.** Pertaining to or shaped like a nipple, such as the papillary muscles of the heart. **2.** Having papillae, such as the papillary layer of the dermis.

paracrine (PERR-uh-crin) **1.** A chemical messenger similar to a hormone whose effects are restricted to the immediate vicinity of the cells that secrete it; sometimes called a local hormone. **2.** Pertaining to such a secretion, as opposed to *endocrine.*

parasympathetic nervous system (PERR-uh-SIM-pa-THET-ic) A division of the autonomic nervous system that issues efferent fibers through the cranial and sacral nerves and exerts cholinergic effects on its target organs.

parathyroid glands (PERR-uh-THY-royd) Small endocrine glands, usually four in number, adhering to the posterior side of the thyroid gland. fig. 17.12

parathyroid hormone (PTH) A hormone secreted by the parathyroid glands that raises blood calcium concentration by stimulating bone resorption by osteoclasts, promoting intestinal absorption of calcium, and inhibiting urinary excretion of calcium.

parenchyma (pa-REN-kih-muh) The tissue that performs the main physiological functions of an organ, especially a gland, as opposed to the tissues (stroma) that mainly provide structural support.

parietal (pa-RY-eh-tul) **1.** Pertaining to a wall, as in the parietal cells of the gastric glands and parietal bone of the skull. **2.** The outer or more superficial layer of a two-layered membrane such as the pleura, pericardium, or glomerular capsule. *Compare* visceral. fig. A.9

pathogen Any disease-causing organism or chemical.

pedicle (PED-ih-cul) A small footlike process, as in the vertebrae and the renal podocytes; also called a *pedicel.*

pelvis A basinlike structure such as the pelvic girdle of the skeleton or the urine-collecting space near the hilum of the kidney. fig. 23.3

peptide Any chain of two or more amino acids. *See also* polypeptide, protein.

peptide bond A group of four covalently bonded atoms (a –C=O group bonded to an –NH group) that links two amino acids in a protein or other peptide. fig. 3.13

perfusion The amount of blood supplied to a given mass of tissue in a given period of time.

perichondrium (PERR-ih-CON-dree-um) A layer of fibrous connective tissue covering the surface of hyaline or elastic cartilage.

perineum (PERR-ih-NEE-um) The region between the thighs bordered by the coccyx, pubic symphysis, and ischial tuberosities; contains the orifices of the urinary, reproductive, and digestive systems. figs. 27.5, 28.7

periosteum (PERR-ee-OSS-tee-um) A layer of fibrous connective tissue covering the surface of a bone. fig. 8.2

peripheral (peh-RIF-eh-rul) Away from the center of the body or of an organ, as in peripheral vision and peripheral blood vessels.

peripheral nervous system (PNS) A subdivision of the nervous system composed of all nerves and ganglia; all of the nervous system except the central nervous system.

peristalsis (PERR-ih-STAL-sis) A wave of constriction traveling along a tubular organ such as the esophagus or ureter, serving to propel its contents.

peritoneum (PERR-ih-toe-NEE-um) A serous membrane that lines the peritoneal cavity of the abdomen and covers the mesenteries and viscera.

perivascular (PERR-ih-VASS-cue-lur) Pertaining to the region surrounding a blood vessel.

pernicious anemia A deficiency of hemoglobin synthesis resulting from inadequate vitamin B_{12} ingestion or absorption.

phagocytosis (FAG-oh-sy-TOE-sis) A form of endocytosis in which a cell surrounds a foreign particle with pseudopods and engulfs it, enclosing it in a cytoplasmic vesicle called a phagosome.

pharynx (FAIR-inks) A muscular passage in the throat at which the respiratory and digestive tracts cross.

physiology **1.** The functional processes of the body. **2.** The study of such function.

pineal gland (PIN-ee-ul) A small conical endocrine gland arising from the roof of the third ventricle of the brain; produces melatonin and serotonin and may be involved in timing the onset of puberty. fig. 14.17

pinocytosis (PIN-oh-sy-TOE-sis) A form of endocytosis in which the plasma membrane sinks inward and imbibes droplets of extracellular fluid.

pituitary gland (pih-TOO-ih-terr-ee) An endocrine gland suspended from the hypothalamus and housed in the sella turcica of the sphenoid bone; secretes numerous hormones, most of which regulate the activities of other glands. fig. 17.4

placenta (pla-SEN-tuh) A thick discoid organ on the wall of the pregnant uterus, composed of a combination of maternal and fetal tissues, serving multiple functions in pregnancy including gas, nutrient, and waste exchange between mother and fetus. fig. 29.8

plantar (PLAN-tur) Pertaining to the sole of the foot.

plaque A small scale or plate of matter, such as dental plaque, the fatty plaques of atherosclerosis, and the amyloid plaques of Alzheimer disease.

plasma The noncellular portion of the blood.

plasma membrane The unit membrane that encloses a cell and controls the traffic of molecules in and out of the cell. fig. 4.6

platelet A formed element of the blood derived from the peripheral cytoplasm of a megakaryocyte, known especially for its role in stopping bleeding but also serves in dissolving blood clots, stimulating inflammation, promoting tissue growth, and destroying bacteria.

pleura (PLOOR-uh) A double-walled serous membrane that encloses each lung.

plexus A network of blood vessels, lymphatic vessels, or nerves, such as a choroid plexus of the brain or brachial plexus of nerves.

polymer A molecule that consists of a long chain of identical or similar subunits, such as protein, DNA, or starch.

polypeptide Any chain of more than 10 or 15 amino acids.

polysaccharide (POL-ee-SAC-uh-ride) A polymer of simple sugars; for example, glycogen, starch, and cellulose.

polyuria (POL-ee-YOU-ree-uh) Excessive output of urine.

popliteal (po-LIT-ee-ul) Pertaining to the posterior aspect of the knee.

positron emission tomography (PET) A method of producing a computerized image of the physiological state of a tissue using injected radioisotopes that emit positrons.

posterior Near or pertaining to the back or spinal side of the body; dorsal.

postganglionic (POST-gang-glee-ON-ic) Pertaining to a neuron that transmits signals from a ganglion to a more distal target organ.

postsynaptic (POST-sih-NAP-tic) Pertaining to a neuron or other cell that receives signals from the presynaptic neuron at a synapse. fig. 13.17

potential A difference in electrical charge from one point to another, especially on opposite sides of a plasma membrane; usually measured in millivolts.

potential space An anatomical space that is usually obliterated by contact between two membranes but opens up if air, fluid, or other matter comes between the membranes. Examples include the pleural cavity and the lumen of the uterus.

preganglionic (PRE-gang-glee-ON-ic) Pertaining to a neuron that transmits signals from the central nervous system to a ganglion.

presynaptic (PRE-sih-NAP-tic) Pertaining to a neuron that transmits signals toward a synapse. fig. 13.17

prime mover The muscle primarily responsible for a given joint action; agonist.

programmed cell death *See* apoptosis.

prolactin (PRL) A pituitary hormone that promotes milk synthesis.

pronation (pro-NAY-shun) A rotational movement of the forearm that turns the palm downward or posteriorly. fig. 10.13

proprioception (PRO-pree-oh-SEP-shun) The nonvisual perception, usually subconscious, of the position and movements of the body, resulting from input from proprioceptors and the vestibular apparatus of the inner ear.

proprioceptor (PRO-pree-oh-SEP-tur) A sensory receptor of the muscles, tendons, and joint capsules that detects muscle contractions and joint movements.

prostaglandins (PROSS-ta-GLAN-dinz) A family of eicosanoids with a five-sided carbon ring in the middle of a hydrocarbon chain, playing a variety of roles in inflammation, neurotransmission, vasomotion, reproduction, and metabolism. fig. 3.10

prostate gland (PROSS-tate) A male reproductive gland that encircles the urethra immediately inferior to the bladder and contributes to the semen. fig. 27.6

protein A large polypeptide; while criteria for a protein are somewhat subjective and variable, polypeptides over 100 amino acids long are generally classified as proteins.

proximal Relatively near a point of origin or attachment; for example, the shoulder is proximal to the elbow. *Compare* distal.

pseudopod (SOO-doe-pod) A temporary cytoplasmic extension of a cell used for locomotion (ameboid movement) and phagocytosis.

pulmonary Pertaining to the lungs.

pulmonary circuit A route of blood flow that supplies blood to the pulmonary alveoli for gas exchange and then returns it to the heart; all blood vessels between the right ventricle and the left atrium of the heart.

pyrogen (PY-ro-jen) A fever-producing agent.

pyruvic acid The three-carbon end product of glycolysis; occurs at the branch point between glycolysis, anaerobic fermentation, and aerobic respiration and is thus an important metabolic intermediate linking these pathways to each other.

R

ramus (RAY-mus) An anatomical branch, as in a nerve or in the pubis.

receptor 1. A cell or organ specialized to detect a stimulus, such as a taste cell or the eye. 2. A protein molecule that binds and responds to a chemical such as a hormone, neurotransmitter, or odor molecule.

receptor-mediated endocytosis A process in which certain molecules in the extracellular fluid bind to receptors in the plasma membrane, these receptors gather together, the membrane sinks inward at that point, and the molecules become incorporated into vesicles in the cytoplasm.

receptor potential A variable change in membrane voltage produced by a stimulus acting on a receptor cell; generates an action potential if it reaches threshold.

recombinant DNA (rDNA) A molecule composed of the DNA of two different species spliced together, such as a combination of bacterial and human DNA used to produce transgenic bacteria that synthesize human proteins.

reduction 1. A chemical reaction in which one or more electrons are added to a molecule, raising its free energy content; opposite of oxidation and always linked to an oxidation reaction. 2. Treatment of a fracture by restoring the broken parts of a bone to their proper alignment.

reflex A stereotyped, automatic, involuntary response to a stimulus; includes somatic reflexes, in which the effectors are skeletal muscles, and visceral (autonomic) reflexes, in which the effectors are usually visceral muscle, cardiac muscle, or glands.

reflex arc A simple neural pathway that mediates a reflex; involves a receptor, an afferent nerve fiber, sometimes one or more interneurons, an efferent nerve fiber, and an effector.

reflux A backward flow, such as the movement of stomach contents back into the esophagus.

refractory period 1. A period of time after a nerve or muscle cell has responded to a stimulus in which it cannot be reexcited by a threshold stimulus. 2. A period of time after male orgasm when it is not possible to reattain erection or ejaculation.

regeneration Replacement of damaged tissue with new tissue of the original type. *Compare* fibrosis.

renal (REE-nul) Pertaining to the kidney.

renin (REE-nin) An enzyme secreted by the kidneys in response to hypotension; converts the plasma protein angiotensinogen to angiotensin I, leading indirectly to a rise in blood pressure.

repolarization Reattainment of the resting membrane potential after a nerve or muscle cell has depolarized.

reproductive system An organ system specialized for the production of offspring.

resistance 1. A nonspecific ability to ward off infection or disease regardless of whether the body has been previously exposed to it. *Compare* immunity. 2. A force that opposes the flow of a fluid such as air or blood. 3. A force, or load, that opposes the action of a muscle or lever.

resistance exercise A physical exercise such as weight lifting that promotes muscle strength more than it promotes cardiopulmonary efficiency, endurance, or fatigue resistance. *Compare* endurance exercise.

respiratory system An organ system specialized for the intake of air and exchange of gases with the blood, consisting of the lungs and the air passages from the nose to the bronchi.

resting membrane potential (RMP) A stable voltage across the plasma membrane of an unstimulated cell.

reticular cell (reh-TIC-you-lur) A delicate, branching macrophage found in the reticular connective tissue of the lymphatic organs.

reticular fiber A fine, branching collagen fiber coated with glycoprotein, found in the stroma of lymphatic organs and some other tissues and organs.

reticular tissue A connective tissue composed of reticular cells and reticular fibers, found in bone marrow, lymphatic organs, and in lesser amounts elsewhere.

ribonucleic acid (RY-bo-new-CLAY-ic) Any of three types of nucleotide polymers smaller than DNA that play various roles in protein synthesis. Composed of ribose, phosphate, adenine, uracil, cytosine, and guanine forming a single nucleotide chain.

ribosome A granule found free in the cytoplasm or attached to the rough endoplasmic reticulum, composed of ribosomal RNA and enzymes; specialized to read the nucleotide sequence of messenger RNA and assemble a corresponding sequence of amino acids to make a protein.

risk factor Any environmental factor or characteristic of an individual that increases one's chance of developing a particular disease; includes such intrinsic factors as age, sex, and race and such extrinsic factors as diet, smoking, and occupation.

ruga (ROO-ga) **1.** An internal fold or wrinkle in the mucosa of a hollow organ such as the stomach and urinary bladder; typically present when the organ is empty and relaxed but not when the organ is full and stretched. **2.** Tissue ridges in such locations as the hard palate and vagina. fig. 25.11

S

saccule (SAC-yule) A saclike receptor in the inner ear with a vertical patch of hair cells, the macula sacculi; senses the orientation of the head and responds to vertical acceleration, as when riding in an elevator or standing up. fig. 16.11

sagittal plane (SADJ-ih-tul) Any plane that extends from ventral to dorsal and cephalic to caudal and divides the body into right and left portions. *Compare* midsagittal plane.

sarcomere (SAR-co-meer) In skeletal and cardiac muscle, the portion of a myofibril from one Z disc to the next, constituting one contractile unit. fig. 12.4

sarcoplasmic reticulum (SR) The smooth endoplasmic reticulum of a muscle cell, serving as a calcium reservoir. fig. 12.2

scanning electron microscope (SEM) A microscope that uses an electron beam in place of light to form high-resolution, three-dimensional images of the surfaces of objects; capable of much higher magnifications than a light microscope. fig. 1c, p. 136

sclerosis (scleh-RO-sis) Hardening or stiffening of a tissue, as in multiple sclerosis of the central nervous system or atherosclerosis of the blood vessels.

sebum (SEE-bum) An oily secretion of the sebaceous glands that keeps the skin and hair pliable.

secondary active transport A mechanism in which solutes are moved through a plasma membrane by a carrier that does not itself use ATP but depends on a concentration gradient established by an active transport pump elsewhere in the cell.

secondary sex characteristic Any feature that develops at puberty, distinguishes the sexes from each other, and promotes attraction between the sexes; examples include the distribution of subcutaneous fat, pitch of the voice, female breasts, male facial hair, and apocrine scent glands.

secondary sex organ An organ other than the ovaries and testes that is essential to reproduction, such as the external genitalia, internal genital ducts, and accessory reproductive glands.

second messenger A chemical that is produced within a cell (such as cAMP) or that enters a cell (such as calcium ions) in response to the binding of a messenger to a membrane receptor, and that triggers a metabolic reaction in the cell.

secretion **1.** A chemical released by a cell to serve a physiological function, such as a hormone or digestive enzyme. **2.** The process of releasing such a chemical, often by exocytosis. *Compare* excretion.

section *See* histological section.

selectively permeable membrane A membrane that allows some substances to pass through while excluding others; for example, the plasma membrane and dialysis membranes.

semen (SEE-men) The fluid ejaculated by a male, including spermatozoa and the secretions of the prostate and seminal vesicles.

semicircular ducts Three ring-shaped, fluid-filled tubes of the inner ear that detect angular accelerations of the head; each is enclosed in a bony passage called the semicircular canal. fig. 16.11

semilunar valve A valve that consists of crescent-shaped cusps, including the aortic and pulmonary valves of the heart and valves of the veins and lymphatic vessels. fig. 19.6

semipermeable membrane *See* selectively permeable membrane.

senescence (seh-NESS-ense) Degenerative changes that occur with age.

sensation Conscious perception of a stimulus; pain, taste, and color, for example, are not stimuli but sensations resulting from stimuli.

sensory nerve fiber An axon that conducts information from a receptor to the central nervous system; an afferent nerve fiber.

serosa (seer-OH-sa) *See* serous membrane.

serous fluid (SEER-us) A watery, low-protein fluid similar to blood serum, formed as a filtrate of the blood or tissue fluid or as a secretion of serous gland cells; moistens the serous membranes.

serous membrane A membrane such as the peritoneum, pleura, or pericardium that lines a body cavity or covers the external surfaces of the viscera; composed of a simple squamous mesothelium and a thin layer of areolar connective tissue.

serum **1.** The fluid that remains after blood has clotted and the solids have been removed; essentially the same as blood plasma except for a lack of fibrinogen. Used as a vehicle for vaccines. **2.** Serous fluid.

sex chromosomes The X and Y chromosomes, which determine the sex of an individual.

shock **1.** Circulatory shock, a state of cardiac output that is insufficient to meet the body's physiological needs, with consequences ranging from fainting to death. **2.** Insulin shock, a state of severe hypoglycemia caused by administration of insulin. **3.** Spinal shock, a state of depressed or lost reflex activity inferior to a point of spinal cord injury. **4.** Electrical shock, the effect of a current of electricity passing through the body, often causing muscular spasm and cardiac arrhythmia or arrest.

sinus **1.** An air-filled space in the cranium. **2.** A modified, relatively dilated vein that lacks smooth muscle and is incapable of vasomotion, such as the dural sinuses of the cerebral circulation and coronary sinus of the heart. **3.** A small fluid-filled space in an organ such as the spleen and lymph nodes. **4.** Pertaining to the sinoatrial node of the heart, as in *sinus rhythm.*

skeletal muscle Striated voluntary muscle, almost all of which is attached to the bones.

skeletal system An organ system consisting of the bones, ligaments, bone marrow, periosteum, articular cartilages, and other tissues associated with the bones.

smooth muscle Nonstriated involuntary muscle found in the walls of the blood vessels, many of the viscera, and other places.

somatic **1.** Pertaining to the body as a whole. **2.** Pertaining to the skin, bones, and skeletal muscles as opposed to the viscera. **3.** Pertaining to cells other than germ cells.

somatic nervous system A division of the nervous system that includes efferent fibers mainly from the skin, muscles, and skeleton and afferent fibers to the skeletal muscles. *Compare* autonomic nervous system.

somesthetic **1.** Pertaining to widely distributed *general senses* in the skin, muscles, tendons, joint capsules, and viscera, as opposed to the *special senses* found in the head only; also called *somatosensory.* **2.** Pertaining to the cerebral cortex of the postcentral gyrus, which receives input from such receptors.

sperm **1.** The fluid ejaculated by the male; semen. Contains spermatozoa and glandular secretions. **2.** A spermatozoon.

spermatogenesis (SPUR-ma-toe-JEN-eh-sis) The production of sperm cells through a series of mitotic and meiotic cell divisions; male gametogenesis.

spermatozoon (SPUR-ma-toe-ZOE-on) A sperm cell.

sphincter (SFINK-tur) A ring of muscle that opens or closes an opening or passageway; found, for example, in the eyelids, around the urinary orifice, and at the beginning of a blood capillary.

spinal column *See* vertebral column.

spinal cord The nerve cord that passes through the vertebral column and constitutes all of the central nervous system except the brain.

spinal nerve Any of the 31 pairs of nerves that arise from the spinal cord and pass through the intervertebral foramina.

spindle **1.** An elongated structure that is thick in the middle and tapered at the ends (fusiform). **2.** A football-shaped complex of microtubules that guide the movement of chromosomes in mitosis and meiosis. fig. 5.14 **3.** A stretch receptor in the skeletal muscles. fig. 15.10

spine **1.** The vertebral column. **2.** A pointed process or sharp ridge on a bone, such as the styloid process of the cranium and spine of the scapula.

splanchnic (SPLANK-nic) Pertaining to the digestive tract.

stem cell Any undifferentiated cell that can divide and differentiate into more functionally specific cell types such as blood cells and germ cells.

stenosis (steh-NO-sis) The narrowing of a passageway such as a heart valve or uterine tube; a permanent, pathological constriction as opposed to physiological constriction of a passageway.

stereocilium An unusually long, sometimes branched microvillus lacking the axoneme and motility of a true cilium; serves such roles as absorption in the epididymis and sensory transduction in the inner ear.

steroid (STERR-oyd, STEER-oyd) A lipid molecule that consists of four interconnected carbon rings; cholesterol and several of its derivatives.

stimulus A chemical or physical agent in a cell's surroundings that is capable of creating a physiological response in the cell; especially agents detected by sensory cells, such as chemicals, light, and pressure.

stress **1.** A mechanical force applied to any part of the body; important in stimulating bone growth, for example. **2.** A condition in which any environmental influence disturbs the homeostatic equilibrium of the body and stimulates a physiological response, especially involving the increased secretion of hormones of the pituitary-adrenal axis.

stroke *See* cerebrovascular accident.

stroke volume The volume of blood ejected by one ventricle of the heart in one contraction.

stroma The connective tissue framework of a gland, lymphatic organ, or certain other viscera, as opposed to the tissue (parenchyma) that performs the physiological functions of the organ.

subcutaneous (SUB-cue-TAY-nee-us) Beneath the skin.

substrate **1.** A chemical that is acted upon and changed by an enzyme. **2.** A chemical used as a source of energy, such as glucose and fatty acids.

substrate specificity The ability of an enzyme to bind only one substrate or a limited range of related substrates.

sulcus (SUL-cuss) A groove in the surface of an organ, as in the cerebrum or heart.

summation **1.** A phenomenon in which multiple stimuli combine their effects on a cell to produce a response; seen especially in nerve and muscle cells. **2.** A phenomenon in which multiple muscle twitches occur so closely together that a muscle fiber cannot fully relax between twitches but develops more tension than a single twitch produces. fig. 12.15

superficial Relatively close to the surface; opposite of deep. For example, the ribs are superficial to the lungs.

superior Higher than another structure or point of reference from the perspective of anatomical position; for example, the lungs are superior to the diaphragm.

supination (SOO-pih-NAY-shun) A rotational movement of the forearm that turns the palm so that it faces upward or forward. fig. 10.13

surfactant (sur-FAC-tent) A chemical that reduces the surface tension of water and enables it to penetrate other substances more effectively. Examples include pulmonary surfactant and bile acids.

sympathetic nervous system A division of the autonomic nervous system that issues efferent fibers through the thoracic and lumbar nerves and usually exerts adrenergic effects on its target organs; includes a chain of paravertebral ganglia adjacent to the vertebral column, and the adrenal medulla.

symphysis (SIM-fih-sis) A joint in which two bones are held together by fibrocartilage; for example, between bodies of the vertebrae and between the right and left pubic bones.

symport A cotransport protein that moves two solutes simultaneously through a plasma membrane in the same direction, such as the sodium-glucose transport protein.

synapse (SIN-aps) **1.** A junction at the end of an axon where it stimulates another cell. **2.** A gap junction between two cardiac or smooth muscle cells at which one cell electrically stimulates the other; called an *electrical synapse.*

synaptic cleft (sih-NAP-tic) A narrow space between the synaptic knob of an axon and the adjacent cell, across which a neurotransmitter diffuses. fig. 13.17

synaptic knobs The swollen tips of the distal branches of an axon; the site of synaptic vesicles and neurotransmitter release. fig. 13.3

synaptic vesicle A spheroid organelle in a synaptic knob containing neurotransmitter.

synergist (SIN-ur-jist) A muscle that works with the agonist to contribute to the same overall action at a joint.

synergistic An effect in which two agents working together (such as two hormones) exert an effect that is greater than the sum of their separate effects. For example, neither follicle-stimulating hormone nor testosterone alone stimulates significant sperm production, but the two of them together stimulate production of vast numbers of sperm cells.

synovial fluid (sih-NO-vee-ul) A lubricating fluid similar to egg white in consistency, found in the synovial joint cavities and bursae.

synovial joint A point where two bones are separated by a narrow, encapsulated space filled with lubricating synovial fluid; most such joints are relatively mobile.

synthesis reaction A chemical reaction in which smaller molecules combine to form a larger one. *Compare* decomposition reaction.

systemic (sis-TEM-ic) Widespread or pertaining to the body as a whole, as in the systemic circulation.

systemic circuit All blood vessels that convey blood from the left ventricle to all organs of the body and back to the right atrium of the heart; all of the cardiovascular system except the heart and pulmonary circuit.

systole (SIS-toe-lee) The contraction of any heart chamber; ventricular contraction unless otherwise specified.

systolic (sis-TOLL-ic) **pressure** The peak arterial blood pressure measured during ventricular systole.

T

target cell A cell acted upon by a nerve fiber, hormone, or other chemical messenger.

tarsal Pertaining to the ankle (tarsus).

T cell A type of lymphocyte involved in nonspecific defense, humoral immunity, and cellular immunity; occurs in several forms including helper, cytotoxic, and suppressor T cells and natural killer cells.

tendon A collagenous band or cord associated with a muscle, usually attaching it to a bone and transferring muscular tension to it.

testis The male gonad; produces spermatozoa and testosterone.

tetanus **1.** A state of sustained muscle contraction produced by temporal summation as a normal part of contraction; also called *tetany.* **2.** Spastic muscle paralysis produced by the toxin of the bacterium *Clostridium tetani.*

tetraiodothyronine (TET-ra-EYE-oh-doe-THY-ro-neen) *See* thyroxine.

thalamus (THAL-uh-muss) The largest part of the diencephalon, located immediately inferior to the corpus callosum and bulging into each lateral ventricle; a point of synaptic relay of nearly all signals passing from lower levels of the CNS to the cerebrum. fig. 14.17

theory An explanatory statement, or set of statements, that concisely summarizes the state of knowledge on a phenomenon and provides direction for further study; for example, the fluid mosaic theory of the plasma membrane and the sliding filament theory of muscle contraction.

thermogenesis The production of heat, for example, by shivering or by the action of thyroid hormones.

thermoreceptor A neuron specialized to respond to heat or cold, found in the skin and hypothalamus, for example.

thermoregulation Homeostatic regulation of the body temperature within a narrow range by adjustments of heat-promoting and heat-losing mechanisms.

thorax A region of the trunk between the neck and the diaphragm; the chest.

threshold **1.** The minimum voltage to which the plasma membrane of a nerve or muscle cell must be depolarized before it produces an action potential. **2.** The minimum combination of stimulus intensity and duration needed to generate an afferent signal from a sensory receptor.

thrombosis (throm-BO-sis) The formation or presence of a thrombus.

thrombus A clot that forms in a blood vessel or heart chamber; may break free and travel in the bloodstream as a thromboembolus.

thymine A single-ringed nitrogenous base (pyrimidine) found in DNA, complementary to adenine in the double helix of DNA. fig. 5.3

thymus A lymphatic organ in the mediastinum superior to the heart; the site where T lymphocytes differentiate and become immunocompetent. fig. 21.10

thyroid gland An endocrine gland in the neck, partially encircling the trachea immediately inferior to the larynx. fig. 17.10

thyroid hormone Either of two similar hormones, thyroxine and triiodothyronine, synthesized from iodine and tyrosine.

thyroid-stimulating hormone (TSH) A hormone of the anterior pituitary gland that stimulates the thyroid gland.

thyroxine (T_4) (thy-ROCK-seen) The thyroid hormone secreted in greatest quantity, with four iodine atoms; also called *tetraiodothyronine.* fig. 17.25

tight junction A zipperlike junction between epithelial cells that limits the passage of substances between them. fig. 6.30

tissue An aggregation of cells and extracellular materials, usually forming part of an organ and performing some discrete function for it; the four primary classes are epithelial, connective, muscular, and nervous tissue.

tissue gel The viscous colloid that forms the ground substance of many tissues; gets its consistency from hyaluronic acid or other glycosaminoglycans.

trabecula (tra-BEC-you-la) A thin plate or layer of tissue, such as the calcified trabeculae of spongy bone or the fibrous trabeculae that subdivide a gland. fig. 8.2

trachea (TRAY-kee-uh) A cartilage-supported tube from the inferior end of the larynx to the origin of the primary bronchi; conveys air to and from the lungs; the "windpipe."

transcription The process of enzymatically reading the nucleotide sequence of a gene and synthesizing a pre-mRNA molecule with a complementary sequence.

transducer Any device that converts one form of energy to another, such as a sense organ, which converts a stimulus into an encoded pattern of action potentials.

transgenic bacteria Genetically engineered bacteria that contain genes from humans or other species and produce proteins of that species; used commercially to produce clotting factors, interferon, insulin, and other products.

translation The process of enzymatically reading an mRNA molecule and synthesizing the protein encoded in its nucleotide sequence.

transmission electron microscope (TEM) A microscope that uses an electron beam in place of light to form high-resolution, two-dimensional images of ultrathin slices of cells or tissues; capable of extremely high magnification. fig. 1*b*, p. 136

triglyceride (try-GLISS-ur-ide) A lipid composed of three fatty acids joined to a glycerol; also called a *triacylglycerol* or *neutral fat.* fig. 3.8

triiodothyronine (T_3) (try-EYE-oh-doe-THY-ro-neen) A thyroid hormone with three iodine atoms, secreted in much lesser quantities than thyroxine. fig. 17.25

trisomy-21 The presence of three copies of chromosome 21 instead of the usual two; causes variable degrees of mental retardation, a shortened life expectancy, and structural anomalies of the face and hands.

tropic hormone (TROPE-ic) A hormone of the anterior pituitary gland that stimulates secretion by another endocrine gland. The four tropic hormones are FSH, LH, TSH, and ACTH.

trunk **1.** That part of the body excluding the head, neck, and appendages. **2.** A major blood vessel, lymphatic vessel, or nerve that gives rise to smaller branches; for example, the pulmonary trunk and spinal nerve trunks.

T tubule A tubular extension of the plasma membrane of a muscle cell that conducts action potentials into the sarcoplasm and excites the sarcoplasmic reticulum. fig. 12.2

tunic (TOO-nic) A layer that encircles or encloses an organ, such as the tunics of a blood vessel or eyeball.

tympanic membrane The eardrum.

U

ultraviolet radiation Invisible, ionizing, electromagnetic radiation with shorter wavelength and higher energy than violet light; causes skin cancer and photoaging of the skin but is required in moderate amounts for the synthesis of vitamin D.

umbilical (um-BIL-ih-cul) **1.** Pertaining to the cord that connects a fetus to the placenta. **2.** Pertaining to the navel (umbilicus).

unit membrane Any cellular membrane composed of a bilayer of phospholipids and embedded proteins. A single unit membrane forms the plasma membrane and encloses many organelles of a cell, whereas double unit membranes enclose the nucleus and mitochondria.

unmyelinated (un-MY-eh-lih-nay-ted) Lacking a myelin sheath. fig. 13.7

upper limb The appendage that arises from the shoulder, consisting of the brachium from shoulder to elbow, the antebrachium from elbow to wrist, the wrist, and the hand; loosely called the *arm*, but that term properly refers only to the brachium.

uracil A single-ringed nitrogenous base (pyrimidine) found in RNA; one of the four bases of the genetic code; occupies the place in RNA that thymine does in DNA. fig. 5.3

urea (you-REE-uh) A nitrogenous waste produced from two ammonia molecules and carbon dioxide; the most abundant nitrogenous waste in the blood and urine. fig. 23.2

urinary system An organ system specialized to filter the blood plasma, excrete waste products from it, and regulate the body's water, acid-base, and electrolyte balance.

uterine tube A duct that extends from the ovary to the uterus and conveys an egg or conceptus to the uterus; also called *fallopian tube* or *oviduct.*

utricle (YOU-trih-cul) A saclike receptor in the inner ear with a horizontal patch of hair cells, the macula utriculi; senses the orientation of the head and responds to horizontal acceleration, as when riding in a car that starts and stops. fig. 16.11

V

varicose vein A vein that has become permanently distended and convoluted due to a loss of competence of the venous valves; especially common in the lower extremity, esophagus, and anal canal (where they are called hemorrhoids).

vas (vass) (plural, *vasa*) A vessel or duct.

vascular Pertaining to blood vessels.

vasoconstriction (VAY-zo-con-STRIC-shun) The narrowing of a blood vessel due to muscular constriction of its tunica media.

vasodilation (VAY-zo-dy-LAY-shun) The widening of a blood vessel due to relaxation of the muscle of its tunica media and the outward pressure of the blood exerted against the wall.

vasomotion (VAY-zo-MO-shun) Collective term for vasoconstriction and vasodilation.

vasomotor center A nucleus in the medulla oblongata that transmits efferent signals to the blood vessels and regulates vasomotion.

vein Any blood vessel that carries blood toward either atrium of the heart.

ventral Pertaining to the front of the body, the regions of the chest and abdomen; anterior.

ventral (anterior) root The branch of a spinal nerve that emerges from the anterior side of the spinal cord and carries efferent (motor) nerve fibers.

ventricle (VEN-trih-cul) A fluid-filled chamber of the brain or heart.

venule (VEN-yool) The smallest type of vein, receiving drainage from capillaries.

vertebra (VUR-teh-bra) One of the bones of the vertebral column.

vertebral column (VUR-teh-brul) A dorsal series of usually 33 vertebrae; encloses the spinal cord, supports the skull and thoracic cage, and provides attachment for the limbs and postural muscles. Also called *spine* or *spinal column.*

vesicle (VESS-ih-cul) A fluid-filled tissue sac or an organelle such as a synaptic or secretory vesicle.

viscera (VISS-er-uh) (singular, *viscus*) The organs contained in the dorsal and ventral body cavities, such as the brain, heart, lungs, stomach, intestines, and kidneys.

visceral (VISS-er-ul) **1.** Pertaining to the viscera. **2.** The inner or deeper layer of a two-layered membrane such as the pleura, pericardium, or glomerular capsule. *Compare* parietal. fig. A.9

visceral muscle Single-unit smooth muscle found in the walls of blood vessels and the digestive, respiratory, urinary, and reproductive tracts.

viscosity The resistance of a fluid to flow; the thickness or stickiness of a fluid.

vitamins A small organic nutrient that is absorbed undigested and serves a purpose other than being oxidized for energy; often serve as coenzymes. Most vitamins cannot be synthesized by the body and are therefore a dietary necessity.

vitreous body (VIT-ree-us) A transparent, gelatinous mass that fills the space between the lens and retina of the eye.

voluntary muscle Muscle that is usually under conscious control; skeletal muscle.

vulva The female external genitalia; the mons, labia majora, and all superficial structures between the labia majora.

W

white matter White myelinated nervous tissue deep to the cortex of the cerebrum and cerebellum and superficial to the gray matter of the spinal cord.

X

X chromosome The larger of the two sex chromosomes; males have one X chromosome and females have two in each somatic cell.

xiphoid process (ZIFF-oyd, ZYE-foyd) A small pointed cartilaginous or bony process at the inferior end of the sternum.

X ray **1.** A high-energy, penetrating electromagnetic ray with wavelengths in the range of 0.1 to 10 nm; used in diagnosis and therapy. **2.** A photograph made with X rays; radiograph.

Y

Y chromosome Smaller of the two sex chromosomes, found only in males and having little if any genetic function except development of the testis.

yolk sac An embryonic membrane that encloses the yolk in vertebrates that lay eggs and serves in humans as the origin of the first blood and germ cells.

Z

zygomatic arch An arch of bone anterior to the ear, formed by the zygomatic processes of the temporal, frontal, and zygomatic bones; origin of the masseter muscle.

zygote A single-celled, fertilized egg.

Credits

Chapter 1

Chapter Opener 1: © Dr. Yorgos Nikas/SPL/Photo Researchers, Inc., **1.1:** © National Library of Medicine/Peter Arnold, Inc., **1.2:** © National Library of Medicine/Photo Researchers, Inc., **1.3:** © SPL/Photo Researcher, Inc., **1.4a-b:** © Kathy Talaro/Visuals Unlimited, **1.5a:** Courtesy of the Armed Forces Institute of Pathology, **1.5b, 1.6:** © Corbis-Bettmann, **1.8:** © Tim Davis/ Photo Researchers, Inc., **1.11:** © Dr. Landrum Shettles, **Box 1.3,1b:** © Custom Medical Stock, **Box 1.3, 2b:** © Scott Camazine/Sue Trainor/Photo Researchers, Inc., **Box 1.3,1a:** © U.H.B. Trust/Tony Stone Images, **Box 1.3,2a:** © Alexander Tsiaras/Photo Researchers, Inc., **Box 1.3,3:** © CNR/ Phototake, **Box 1.3,4:** © Tony Stone Images, **Box 1.3,5:** © Monte S. Buschbaum, Mt. Sinai School of Medicine, New York.

Atlas A

A1, A3, A6a-d: © The McGraw-Hill Companies, Inc./Joe DeGrandis, photographer, **A18:** © The McGraw-Hill Companies, Inc./Rebecca Gray, photographer, **A19:** © The McGraw-Hill Companies, Inc., **A20.a, A21, A22.a:** © The McGraw-Hill Companies, Inc./ Rebecca Gray, photographer, **A23:** © The McGraw-Hill Companies, Inc./ Dennis Strete, photographer, **A24:** © The McGraw-Hill Companies, Inc./ Rebecca Gray, photographer.

Chapter 2

Chapter Opener 2: © Michael Seigel/Phototake NYC, **2.7b:** © The McGraw-Hill Companies, Inc./Dennis Strete, photographer, **2.11 a-d:** © Ken Saladin, pg. 70 © American Institute of Physics/Emilio Segre Visual Archives, W.F. Meggers Collection, **Box 2.4,1a:** © Science VU/Visuals Unlimited, **Box 2.4,1c:** © SIU/Peter Arnold, Inc.

Chapter 3

Chapter Opener 3: © Hermann Eisenbeiss/Photo Researchers, Inc., **Box 3.5:** © Sylvan Legrand/Photo Researchers, Inc.

Chapter 4

Chapter Opener 4: © P.M. Motta, G. Macchiarelli, S.A. Nottola/SPL/Photo Researchers, Inc., **4.3:** © K.G. Murti/Visuals Unlimited, **4.5a:** From *Cell Ultrastructure* by William A. Jenson and Roderick B. Park. © 1967 by Wadsworth Publishing Co., Inc. Reprinted by permission of the publisher., **4.5a-b:** From *Cell Ultrastructure* by William A. Jenson and Roderick B. Park. © 1967 by Wadsworth Publishing Co., Inc. Reprinted by permission of the publisher., **4.6a:** © Don Fawcett/Photo Researchers, Inc., **4.10:** Courtesy of Dr. S. Ito, Harvard Medical School, **4.11a-b:** © Biophoto Associates/ Photo Researchers, Inc., **4.12, 4.14a-b:** © Don Fawcett/ Photo Researchers, Inc., **4.15a:** © Visuals Unlimited, **4.16a-b:** © D.W. Fawcett/ Visuals Unlimited, **4.17a:** Courtesy of Dr. Keith Porter, **4.18a:** From: Manley McGill, D.P. Highfield, T.M. Monahan, and B.R. Brinkley, “Effects of Nucleic Acid Specific Dyes on Centrioles of Mammalian Cells”, published in the *Journal of Ultrastructure Research 57,* 43-53 (1976), **pg. 48, fig. 6** © Academic Press, **4.19a1:** © K.G. Murti/ Visuals Unlimited, **4.19a2:** © Biology Media/ Photo Researchers, Inc., **4.19b:** © K.G. Murti/Visuals Unlimited, **4.23a-c:** © David M. Phillips/Visuals Unlimited, **4.29b-1:** © Company of Biologists, Ltd, **4.29b-2:** © Company of Biologists, Ltd, **4.29b-3:** © Company of Biologists, Ltd, **4.29b-4:** © Company of Biologists, Ltd, **4.30:** © Don Fawcett/ Visuals Unlimited, **4.31c:** Courtesy of Dr. Birgit Satir, Albert Einstein College of Medicine, **Box 4.5,1a:** © Ed Reschke, **Box 4.5,1b:** © David M. Phillips/Photo Researchers, Inc., **Box 4.5,1c:** © David M. Phillips/Visuals Unlimited.

Chapter 5

Chapter Opener 5: © Gopal Murti/SPL/Photo Researchers, Inc., **5.1a:** © Richard Chao, **5.1b:** © E.G. Pollock, **5.2a:** © P. Motta & T. Naguro/SPL/Photo Researchers, Inc., **ST 5.1a:** From “The Double Helix” by James D. Watson, 1968, Aentheneam Press, NY. Courtesy of Cold Springs Harbor Laboratory, **ST 5.1b:** Courtesy of King’s College, London, **ST 5.1c:** © Corbis-Bettman, **5.9:** © E.V. Kiseleva, February 5 Letter, 257:251-253, 1989, **5.14a-d:** © Ed Reschke, **5.16:** © CNRI/SPL/Photo Researchers, Inc., **5.17a-b:** © The McGraw-Hill Companies, Inc./Joe DeGrandis, photographer.

Chapter 6

Chapter Opener 6: © Gopal Murti/Phototake NYC, **6.4a, 6.5a:** © The McGraw-Hill Companies, Inc./Dennis Strete, photographer, **6.6a:** © Ed Reschke, **6.7a:** © The McGraw-Hill Companies, Inc./Dennis Strete, photographer, **6.8a, 6.9a:** © Ed Reschke, **6.10a, 6.11a:** © The McGraw-Hill Companies, Inc./Dennis Strete, photographer, **6.12:** From: *Tissues and Organs: A Text-Atlas of Scanning Electron Microscopy,* © Richard G. Kessel and Randy H. Kardon, published by W.H. Freeman and Company 1979., **6.13:** © SPL/Photo Researchers, Inc., **6.14:** © The McGraw-Hill Companies, Inc./Rebecca Gray, photographer, **6.16a, 6.17a, 6.18a, 6.19a, 6.20a, 6.21a, 6.22a:** © The McGraw-Hill Companies, Inc./Dennis Strete, photographer, **6.23a:** © Ed Reschke/Peter Arnold, Inc., **6.24a:** © The McGraw-Hill Companies, Inc./Dennis Strete, photographer, **6.25a, 6.26a, 6.27a, 6.28a:** © Ed Reschke, **6.29a:** © The McGraw-Hill Companies, Inc./Dennis Strete, photographer.

Chapter 7

Chapter Opener 7: © Ed Reschke/Peter Arnold, **7.2b, 7.3a:** © The McGraw-Hill Companies, Inc./Dennis Strete, photographer, **7.3b:** From R.G. Kessel and R.H. Kardon, *Scanning Electron Microscopy of Tissues and Organs,* 1979, **7.3c:** © SPL/Custom Medical Stock Photos, pg. 209 © Meckes/Ottawa/Photo Researchers, Inc., **7.5a-b:** © The McGraw-Hill Companies, Inc./Dennis Strete, photographer, **7.6-1:** © Tom McHugh/ Photo Researchers, Inc., **7.6-2:** © The McGraw-Hill Companies, Inc./Joe DeGrandis, photographer, **7.7b:** © CBS/Phototake, **7.7c:** © P.M. Motta/SPL/ Custom Medical Stock Photos, **7.8a-d:** © The McGraw-Hill Companies, Inc./Joe DeGrandis, photographer, **7.10a-c:** © The McGraw-Hill Companies, Inc./Dennis Strete, photographer, **7.11a:** © NMSB/ Custom Medical Stock Photos, **7.11b:** © Biophoto Associates/ Photo Researchers, Inc., **7.11c:** © James Stevensons/ SPL/ Photo Researchers, Inc., **7.12a:** © SPL/Custom Medical Stock Photos, **7.12b-c:** © John Radcliffe/ Photo Researchers, Inc., pg. 227 © Images copyright 1997 PhotoDisc, Inc.

Chapter 8

Chapter Opener 8: © Cabisco/ Visuals Unlimited, **8.4-1:** Courtesy of Trent Stephens, **8.4a-b:** Courtesy of Trent Stephens, **8.5a, 8.5c:** © D.W. Fawcett/ Visuals Unlimited, **8.6:** © Robert Calentine/ Visuals Unlimited, **8.8:** © Ken Saladin, **8.10:** © Ed Reschke, **8.11:** © Biophoto Associates/ Photo Researchers, Inc., **8.12:** Courtesy of Utah Valley Regional Medical Center, Department of Radiology, **Box 8.2:** © The McGraw-Hill Companies, Inc./Joe DeGrandis, photographer, **8.18a:** © Doug Sizemore/ Visuals Unlimited, **8.18b:** © SIU/Visuals Unlimited, **Box 8.4,1:** © Michael Klein/ Peter Arnold, Inc., **Box 8.4,2:** © Yoav Levy/ Phototake.

Chapter 9

Chapter Opener 9: © SIU/ Visuals Unlimited, **9.20:** © The McGraw-Hill Companies, Inc./Bob Coyle, photographer, **9.33b:** © NHS Trust/Tony Stone Images, **9.40b:** © Walter Reiter/Phototake, pg. 303 © Images copyright 1997 PhotoDisc, Inc.

Chapter 10

Chapter Opener 10: © Simon Fraser/SPL/Photo Researchers, Inc., **10.1:** © Gerard Vandystadt/Photo Researchers, Inc., **10.15:** © The McGraw-Hill Companies, Inc./Joe DeGrandis, photographer, **10.20d,10.22d:** © The McGraw-Hill Companies, Inc./ Rebecca Gray, photographer, **10.23:** © Richard Anderson, **10.25b:** © The McGraw-Hill Companies, Inc./Rebecca Gray, photographer, **10.27:** From Vidic/Saurez, *Photographic Atlas of the Human Body.* Copyright © 1984 Mosby-Year Book, Inc., St. Louis, MO., **Box 10.4,1a:** © Richard Anderson, **Box 10.4,1b:** © Richard Anderson, **Box 10.4,2a:** © SIU/Visuals Unlimited, **Box 10.4,2b:** © Ron Mensching/Phototake, **Box 10.4,2c:** © SIU/Peter Arnold, Inc., **Box 10.4,2d:** © Mehau Kulyk/SPL/Photo Researchers, Inc.

Chapter 11

Chapter Opener 11: © Scott Camazine/Photo Researchers, Inc., **11.1c:** © The McGraw-Hill Companies, Inc./Dennis Strete, photographer, **11.5:** © The McGraw-Hill Companies, Inc./Rebecca Gray, photographer, **11.6a-d, 11.11:** © The McGraw-Hill Companies, Inc./Joe DeGrandis, photographer, **11.12, 11.16, 11.19, 11.23a-b, 11.33, 11.35:** © The McGraw-Hill Companies, Inc./Rebecca Gray, photographer.

Atlas B

B1a, B1b, B2a, B2b, B3, B4a, B4b, B5, B6a, B6b, B7a, B7b, B8a, B8b, B9a, B9b, B9c, B10a, B10b, B10c, B10d, B11a, B11b: © The McGraw-Hill Companies, Inc./Joe DeGrandis, photographer, pg. 401 © Images copyright 1997 PhotoDisc, Inc.

Chapter 12

Chapter Opener 12: © Don W. Fawcett/Photo Researchers, Inc., **12.1:** © Ed Reschke, **12.4a:** © Don W. Fawcett/ Visuals Unlimited, **12.5:** © Victor B. Eichler, **Box 12.4,1:** © NMSB/Custom Medical Stock Photos.

Chapter 13

Chapter Opener 13: © John Allison/Peter Arnold, Inc., **13.7:** © The McGraw-Hill Companies, Inc./Dr. Dennis Emery, Dept. of Zoology and Genetics, Iowa State University, **13.16:** © Omikron/ Science Source/Photo Researchers, Inc., **Box 13.4,1:** © Martin R. Rotker/Photo Researchers, Inc.

Chapter 14

Chapter Opener 14: © CNRI/ Phototake NYC, **14.2a:** © The McGraw-Hill Companies, Inc./Rebecca Gray, photographer, **14.2b:** © The McGraw-Hill Companies, Inc./Dennis Strete, photographer, **Box 14.1:** Alfred I. DuPont Institute, **Box 14.3:** © Corbis-Bettmann, **14.6c:** © The McGraw-Hill Companies, Inc./ Rebecca Gray, photographer, **14.23a:** © The McGraw-Hill Companies, Inc./Bob Coyle, photographer, **Box 14.4:** Courtesy Hanna Damasio, from: Damasio H., Grabowski, T., Frank R., Galaburda A.M., Damasio A.R.: *The return of Phineas Gage; Clues about the brain from a famous patient.* As seen in Science, 264:1102-1105, 1994. Department of Neurology and Image Analysis Facility, University of Iowa, **Box 14.5,1:** © Marcus Raichle/Peter Arnold, Inc.

Chapter 15

Chapter Opener 15: © Ed Reschke, **15.1b:** From *Tissues and Organs: A Text-Atlas of Scanning Electron Microscopy,* © Richard G. Kessel and Randy H. Kardon, published by W.H. Freeman and Company 1979., **15.3b:** © The McGraw-Hill Companies, Inc./Rebecca Gray, photographer, **Box 15.9:** © The McGraw-Hill Companies, Inc./Joe DeGrandis, photographer, **15.6:** © From *A Stereoscopic Atlas of Anatomy* by David L. Bassett. Courtesy of Dr. Robert A. Chase, M.D., **15.8:** Vidic/ Saurez: *Photographic Atlas of the Human Body* © 1984 Mosby-Year Book, Inc., St. Louis, **15.16:** © From *A Stereoscopic Atlas of Anatomy* by David L. Bassett. Courtesy of Dr. Robert A. Chase, M.D., **CI 15:** © Images copyright 1997 PhotoDisc, Inc.

Chapter 16

Chapter Opener 16: © P. Motta/SPL/Photo Researchers, Inc., **16.4c:** © Ed Reschke, **16.9:** © The McGraw-Hill Companies, Inc./Joe DeGrandis, photographer, **16.13:** From: *Tissues and Organs: A Text-Atlas of Scanning Electron Microscopy,* © Richard G. Kessel and Randy H. Kardon, published by W.H. Freeman and Company 1979., **16.20:** © The McGraw-Hill Companies, Inc./Joe DeGrandis, photographer, **16.25:** © Ralph C. Eagle/MD/Photo Researchers, Inc., **16.26a:** © Lisa Klancher, **16.31a:** © The McGraw-Hill Companies, Inc./Dennis Strete, photographer, **16.32a:** Courtesy of Beckman Vision Center at UCSF School of Medicine/D. Copeahagen, S. Miltman, and M. Maglio.

Chapter 17

Chapter Opener 17: © Ed Reschke/Peter Arnold, Inc., **17.7a:** © Daniel Margulies,

17.7b: © Bettina/Cirone/Photo Researchers, Inc., **17.8a-d:** From Clinical Pathological Conference "Acromegaly, Diabetes, Hypermetabolism, Protein Use and Heart Failure" in *American Journal of Medicine,* 20:133, 1956. Copyright © 1956 by Excerpta Media, Inc., **17.10b:** © Ed Reschke, **17.11b:** © CNR/ Phototake, **17.14a:** From: Atlas of Pediatric Physical Diagnosis, 2/e, by Zitelli & Davis, 1992. Mosby-Wolfe Europe Limited, London. UK., **17.14b:** From: Atlas of Pediatric Physical Diagnosis, 2/e, by Zitelli & Davis, 1992. Mosby-Wolfe Europe Limited, London. UK., **17.15:** © John Money, Ph.D., John Hopkins School of Medicine, **17.16:** © NMSB/Custom Medical Stock, **17.17c:** © Ed Reschke, **CI 17:** © Images copyright 1997 PhotoDisc, Inc.

Chapter 18
Chapter Opener 18: © P. Motta/ SPL/ Photo Researchers, Inc., **18.1b:** © National Cancer Institute/SPL/Photo Researchers, Inc., **Box 18.1:** © John Paul Kay/ Peter Arnold, Inc., **18.7b:** © Richard J. Poole/Polaroid International Photomicrography Competition, **18.8a-b:** © Don W. Fawcett/Visuals Unlimited, **18.11:** © Walter Reinhart/ Phototake, **Box 18.3:** Courtesy Schomburg Center for Research in Black Culture, The New York Public Library, **18.14:** © Claude Revy/ Phototake, **TA 18.3,** page 701: © Ed Reschke, **TA 18.4,** page 701: © Ed Reschke, **TA 18.5,** page 701: © Ed Reschke, **TA 18.6,** page 702: © Ed Reschke, **TA 18.7,** page 703: © Ed Reschke, **18.17a:** © Ed Reschke, **18.17b:** © SIU/ Photo Researchers, Inc., **18.19:** © David M. Phillips/ Photo Researchers, Inc., **Box 18.5,1:** © SPL/ Photo Researchers, Inc.

Chapter 19
Chapter Opener 19: © Manfred Kage/Peter Arnold, Inc., **19.3a-b, 19.7b:** © The McGraw-Hill Companies, Inc., **19.11a, Box 19.5,1a, 19.5,1b:** © Ed Reschke.

Chapter 20
Chapter Opener 20: © Ed Reschke, **20.3:** The McGraw-Hill Companies, Inc./Dennis Strete, photographer, **20.4:** From: *Tissues and Organs: A Text-Atlas of Scanning Electron Microscopy,* © Richard G. Kessel and Randy H. Kardon, published by W.H. Freeman and Company 1979., **20.6b:** Courtesy of S. McNutt, **20.10a:** P.C. Phelps and J.H. Luft, 1969 Am. Anat. 125: 3999, **20.10b:** P.C. Phelps and J.H. Luft, 1969 Am. Anat. 125: 3999, pg. 797 © Images copyright 1997 PhotoDisc, Inc.

Chapter 21
Chapter Opener 21: © Volker Steger/Peter Arnold, Inc., **21.2:** © SPL/Custom Medical Stock Photo, **21.4:** © The McGraw-Hill Companies, Inc./Dennis Strete, photographer, **21.7:** © The McGraw-Hill Companies, Inc./Rebecca Gray, photographer, **21.8c:** © Francis Leroy, SPL/Photo Researchers, Inc., **21.9:** © The McGraw-Hill Companies, Inc./Dennis Strete, photographer, **21.11:** The McGraw-Hill Companies, Inc./ Dennis Strete, photographer, **21.12a:** © The McGraw-Hill Companies, Inc./ Dennis Strete, photographer, **21.12b:** © The McGraw-Hill Companies, Inc./Dennis Strete, photographer, **21.18b:** © Len Lessin/Peter Arnold, Inc., **21.19b:** © Manfred Kage/Peter Arnold, Inc., **21.20:** From: *Tissues and Organs: A Text-Atlas of Scanning Electron Microscopy,* © Richard G. Kessel and Randy H. Kardon, published by W.H. Freeman and Company 1979., **21.23a:** © Don W. Fawcett/Visuals Unlimited, **21.23b:** © Don W. Fawcett/Visuals Unlimited, **21.27a:** Dr. Andrejs Liepins, **21.27b:** Dr. Andrejs Liepins, **21.29:** © Corbis-Bettmann, **21.30b:** © NIBSC/SPL/Photo Researchers, Inc., **21.31:** © SPL/Photo Researchers, Inc., pg. 838 © Images copyright 1997 PhotoDisc, Inc.

Chapter 22
Chapter Opener 22: © Manfred Kage/Peter Arnold, Inc., **22.2a:** © The McGraw-Hill Companies, Inc./Joe DeGrandis, photographer, **22.3a:** © The McGraw-Hill Companies, Inc./Rebecca Gray, photographer, **22.5:** © CNR/ Phototake, **22.8:** © SPL/ Custom Medical Stock, **22.9b:** © The McGraw-Hill Companies, Inc./Karl Rubin, photographer, **22.10:** John Watney Photo Library, **22.11a:** © Carolina Biological Supply/Phototake, **22.11b:** © C.Y. Shih-R. Kessel/Visuals Unlimited, **22.26a:** © The McGraw-Hill Companies, Inc./Dennis Strete, photographer, pg. 876 © Images copyright 1997 PhotoDisc, Inc.

Chapter 23
Chapter Opener 23: © CNRI/ SPL/Photo Researchers, Inc., **23.3a:** © The McGraw-Hill Companies, Inc./Rebecca Gray, photographer, **23.5b:** © Manfred Kage/Peter Arnold, Inc., **23.7a:** From *Tissues and Organs: A Text-Atlas of Scanning Electron Microscopy,* © Richard G. Kessel and Randy H. Kardon, published by W.H. Freeman and Company 1979., **23.7b:** © Don Fawcett/Photo Researchers, Inc., **23.7c:** © B.F. King, Univ. of California School of Medicine/Biological Photo Service, pg. 910 © Digital Stock/Corbis-Body Beautiful Vol. 76.

Chapter 24
Chapter Opener 24: © Manfred Kage/ Peter Arnold, Inc.

Chapter 25
Chapter Opener 25: © Omikron/Photo Researchers, Inc., **25.5:** © The McGraw-Hill Companies, Inc./Rebecca Gray, photographer, **25.8b:** © The McGraw-Hill Companies, Inc./Dennis Strete, photographer, **25.10f:** © The McGraw-Hill Companies, Inc./Jim Shaffer, photographer, **25.11b:** © The McGraw-Hill Companies, Inc./Rebecca Gray, photographer, **25.12c:** © Robert Caughem/Visuals Unlimited, **Box 25.2a, Box 25.2b:** © CNR/SPL/ Photo Researchers, Inc., **25.18b, 25.20b, 25.23b:** © The McGraw-Hill Companies, Inc. /Dennis Strete, photographer, **25.23a:** © Meekes/Ottawa/ Photo Researchers, Inc., **25.25:** © The McGraw-Hill Companies, Inc./Rebecca Gray, photographer, pg. 980 © Images copyright 1997 PhotoDisc, Inc.

Chapter 26
Chapter Opener 26: © P.M. Motta/S. Makabe & T. Nagur/ SPL/Photo Researchers, Inc., **26.12:** © R.P. Clark & M. Goff/Photo Researchers, Inc.

Chapter 27
Chapter Opener 27: © Secchi-Lecaque/CNRI/SPL/ Photo Researchers, Inc., **Box 27.1:** Courtesy Mihaly Bartalos, from M. Bartalos and T.A. Baramki, 1967 Medical Cytogenetics, Williams & Wilkins., **27.7b:** © The McGraw-Hill Companies, Inc./ Dennis Strete, photographer, **27.8a:** From *Tissues and Organs: A Text-Atlas of Scanning Electron Microscopy,* © Richard G. Kessel and Randy H. Kardon, published by W. H. Freeman and Company 1979., **27:8b:** © Ed Reschke, **27.17b:** © Visuals Unlimited

Chapter 28
Chapter Opener 28: © Mehau Kulyk/SPL/Photo Researchers,

Inc., **28.4:** From *Tissues and Organs: A Text-Atlas of Scanning Electron Microscopy,* © Richard G. Kessel and Randy H. Kardon, published by W.H. Freeman and Company 1979., **28.5:** © The McGraw-Hill Companies, Inc./Rebecca Gray, photographer, **Box 28.1:** © Martin Rotker/Phototake NYC, **28.6:** © Ed Reschke, **28.10:** © The McGraw-Hill Companies, Inc./Dennis Strete, photographer, **28.11a:** © Charles Gupton/Tony Stone Images, **28.11b:** © SBNH/Tony Stone Images, **28.11c-d:** © Biophoto Associates/Photo Researchers, Inc., **28.14a:** © Ed Reschke, **28.14b:** © Manfred Kage/Peter Arnold, Inc., **28.16:** © Dr. Landrum Shettles, **28.21a-b:** © D. Van Rossum/ Photo Researchers, Inc., **28.21c:** © Bruce Curtis/Visuals Unlimited, **28.23:** Courtesy Toshikazu Nagato, from Cell and Tissue Research, 209:1-10, **3,** pg. 6, Springer-Verlag, 1980., **Box 28.4a-e:** © The McGraw Hill Companies, Inc./ Bob Coyle, photographer, **Box 28.4f:** © Hank Morgan/Photo Researchers, Inc., **Box 28.4g:** © The McGraw Hill Companies, Inc./ Bob Coyle, photographer, **Box 28.4h:** © The McGraw Hill Companies, Inc., Vincent Ho, photographer., pg. 1080 © Images copyright 1997 PhotoDisc, Inc.

Chapter 29

Chapter Opener 29: © Francis Leroy/SPL/Photo Researchers, Inc., **29.8:** © SIU/Visuals Unlimited, **29.9b:** © Martin Rotker/Phototake, Inc., **29.11b:** © Neil Harding/Tony Stone Images, **29.13:** Courtesy Dr. Almuth Lenz, **29.15a:** Courtesy Mihalay Bartalos, from Bartalos 1976: Medical Cytogenetics, fg 9.2 pg 133, Waverly, a division of Williams & Wilkins, **29.15b:** Courtesy Mihalay Bartalos, from Bartalos 1976: Medical Cytogenetics, fg 10.2 pg 154, Waverly, a division of Williams & Wilkins, **29.16a:** © Steve King/Photo Researchers, Inc., **29.16d:** © Biophoto Associates/Photo Researchers, Inc., **29.17a:** Photograph of child from The 1974 Science Year. © 1973 Field Enterprises Educational Corporation. By permission of World Book, Inc., **29.17b:** © Black Star Publishing, **29.18:** © Corbis/ Bettman, **Box 29.4:** © AP/Wide World Photo.

Index

Note: Page numbers followed by *f* designate figures; page numbers followed by *t* designate tables. Specific arteries, bones, bursae, foramina, fossae, joints, ligaments, muscles, nerves, plexuses, and veins are listed under *Arteries, Bones,* and so forth.

A

C

D

F

G

I

J

K

L

M

N

O

Q

R

S

T

U

X

Y

Z